Lehninger
Principles of Biochemistry

SIXTH EDITION

David L. Nelson

Professor of Biochemistry
University of Wisconsin–Madison

Michael M. Cox

Professor of Biochemistry
University of Wisconsin–Madison

W. H. FREEMAN AND COMPANY · New York

Publisher:	SUSAN WINSLOW
Senior Acquisitions Editor:	LAUREN SCHULTZ
Senior Developmental Editor:	SUSAN MORAN
Developmental Editor:	MATTHEW TONTONOZ
Associate Director of Marketing:	DEBBIE CLARE
Marketing Director:	JOHN BRITCH
Marketing Assistant	LINDSAY NEFF
Media Editor:	ALLISON MICHAEL
Managing Editor:	PHILIP McCAFFREY
Project Editor:	JANE O'NEILL
Photo Editor:	TED SZCZEPANSKI
Photo Researcher:	ELYSE RIEDER
Art Director:	DIANA BLUME
Illustration Coordinator:	JANICE DONNOLA
Illustrations:	H. ADAM STEINBERG and DRAGONFLY MEDIA GROUP
Molecular Graphics:	H. ADAM STEINBERG
Production Manager:	SUSAN WEIN
Composition:	APTARA, INC.
Printing and binding:	QUAD/GRAPHICS VERSAILLES

North American Edition
Cover image: The network of interactions in an animal mitochondrion. Each dot represents a compound, and each line, an enzyme that interconverts the two compounds. The major nodes include ADP, ATP, NAD^+, and NADH. The image was constructed with Cytoscape software by Anthony Smith in the laboratory of Alan Robinson, Medical Research Council Mitochondrial Biology Unit, Cambridge, UK, using data from MitoMiner (Smith, A.C., Blackshaw, J.A., & Robinson, A.J. (2012) MitoMiner: a data warehouse for mitochondrial proteomics data. *Nucleic Acids Res.* 40, D1160–D1167). *Background*: Transmission electron micrograph of interscapular brown adipose cell from a bat. (Don W. Fawcett/Science Source/Photo Researchers)

International Edition
Cover design: Dirk Kaufman
Cover image: Nastco/iStockphoto.com

Library of Congress Control Number: 2012948755

North American Edition
ISBN-13: 978-1-4292-3414-6
ISBN-10: 1-4292-3414-8

International Edition
ISBN-13: 978-1-4641-0962-1
ISBN-10: 1-4641-0962-1

W. H. Freeman and Company
41 Madison Avenue
New York, NY 10010
www.whfreeman.com

Macmillan Higher Education
Houndmills, Basingstoke
RG21 6XS, England
www.macmillanhighered.com/international

To Our Teachers

Paul R. Burton

Albert Finholt

William P. Jencks

Eugene P. Kennedy

Homer Knoss

Arthur Kornberg

I. Robert Lehman

Earl K. Nelson

Wesley A. Pearson

David E. Sheppard

Harold B. White

About the Authors

David L. Nelson, born in Fairmont, Minnesota, received his BS in Chemistry and Biology from St. Olaf College in 1964 and earned his PhD in Biochemistry at Stanford Medical School under Arthur Kornberg. He was a postdoctoral fellow at the Harvard Medical School with Eugene P. Kennedy, who was one of Albert Lehninger's first graduate students. Nelson joined the faculty of the University of Wisconsin–Madison in 1971 and became a full professor of biochemistry in 1982. He was for eight years the Director of the Center for Biology Education at the University of Wisconsin–Madison.

Nelson's research has focused on the signal transductions that regulate ciliary motion and exocytosis in the protozoan *Paramecium*. The enzymes of signal transductions, including a variety of protein kinases, are primary targets of study. His research group has used enzyme purification, immunological techniques, electron microscopy, genetics, molecular biology, and electrophysiology to study these processes.

Dave Nelson has a distinguished record as a lecturer and research supervisor. For 40 years he has taught an intensive survey of biochemistry for advanced biochemistry undergraduates in the life sciences. He has also taught a survey of biochemistry for nursing students, and graduate courses on membrane structure and function and on molecular neurobiology. He has sponsored numerous PhD, MS, and undergraduate honors theses and has received awards for his outstanding teaching, including the Dreyfus Teacher–Scholar Award, the Atwood Distinguished Professorship, and the Unterkofler Excellence in Teaching Award from the University of Wisconsin System. In 1991–1992 he was a visiting professor of chemistry and biology at Spelman College. His second love is history, and in his dotage he has begun to teach the history of biochemistry to undergraduates and to collect antique scientific instruments for use in a laboratory course he teaches.

Michael M. Cox was born in Wilmington, Delaware. In his first biochemistry course, Lehninger's *Biochemistry* was a major influence in refocusing his fascination with biology and inspiring him to pursue a career in biochemistry. After graduating from the University of Delaware in 1974, Cox went to Brandeis University to do his doctoral work with William P. Jencks, and then to Stanford in 1979 for postdoctoral study with I. Robert Lehman. He moved to the University of Wisconsin–Madison in 1983 and became a full professor of biochemistry in 1992.

Cox's doctoral research was on general acid and base catalysis as a model for enzyme-catalyzed reactions. At Stanford, he began work on the enzymes involved in genetic recombination. The work focused

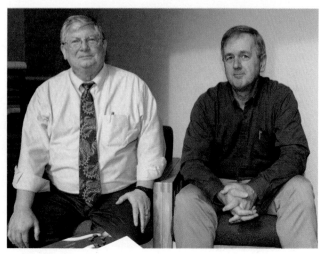

David L. Nelson and Michael M. Cox

particularly on the RecA protein, designing purification and assay methods that are still in use, and illuminating the process of DNA branch migration. Exploration of the enzymes of genetic recombination has remained the central theme of his research.

Mike Cox has coordinated a large and active research team at Wisconsin, investigating the enzymology, topology, and energetics of genetic recombination. A primary focus has been the mechanism of RecA protein–mediated DNA strand exchange, the role of ATP in the RecA system, and the regulation of recombinational DNA repair. Part of the research program now focuses on organisms that exhibit an especially robust capacity for DNA repair, such as *Deinococcus radiodurans*, and the applications of those repair systems to biotechnology.

For almost 30 years he has taught (with Dave Nelson) the survey of biochemistry to undergraduates and has lectured in graduate courses on DNA structure and topology, protein-DNA interactions, and the biochemistry of recombination. More recent projects have been the organization of a new course on professional responsibility for first-year graduate students and the establishment of a systematic program to draw talented biochemistry undergraduates into the laboratory at an early stage of their collegiate career. He has received awards for both his teaching and his research, including the Dreyfus Teacher–Scholar Award, the 1989 Eli Lilly Award in Biological Chemistry, and the 2009 Regents Teaching Excellence Award from the University of Wisconsin. He is also highly active in national efforts to provide new guidelines for undergraduate biochemistry education. His hobbies include turning 18 acres of Wisconsin farmland into an arboretum, wine collecting, and assisting in the design of laboratory buildings.

A Note on the Nature of Science

In this twenty-first century, a typical science education often leaves the philosophical underpinnings of science unstated, or relies on oversimplified definitions. As you contemplate a career in science, it may be useful to consider once again the terms **science**, **scientist**, and **scientific method**.

Science is both a way of thinking about the natural world and the sum of the information and theory that result from such thinking. The power and success of science flow directly from its reliance on ideas that can be tested: information on natural phenomena that can be observed, measured, and reproduced and theories that have predictive value. The progress of science rests on a foundational assumption that is often unstated but crucial to the enterprise: that the laws governing forces and phenomena existing in the universe are not subject to change. The Nobel laureate Jacques Monod referred to this underlying assumption as the "postulate of objectivity." The natural world can therefore be understood by applying a process of inquiry—the scientific method. Science could not succeed in a universe that played tricks on us. Other than the postulate of objectivity, science makes no inviolate assumptions about the natural world. A useful scientific idea is one that (1) has been or can be reproducibly substantiated and (2) can be used to accurately predict new phenomena.

Scientific ideas take many forms. The terms that scientists use to describe these forms have meanings quite different from those applied by nonscientists. A *hypothesis* is an idea or assumption that provides a reasonable and testable explanation for one or more observations, but it may lack extensive experimental substantiation. A *scientific theory* is much more than a hunch. It is an idea that has been substantiated to some extent and provides an explanation for a body of experimental observations. A theory can be tested and built upon and is thus a basis for further advance and innovation. When a scientific theory has been repeatedly tested and validated on many fronts, it can be accepted as a fact.

In one important sense, what constitutes science or a scientific idea is defined by whether or not it is published in the scientific literature after peer review by other working scientists. About 16,000 peer-reviewed scientific journals worldwide publish some 1.4 million articles each year, a continuing rich harvest of information that is the birthright of every human being.

Scientists are individuals who rigorously apply the scientific method to understand the natural world. Merely having an advanced degree in a scientific discipline does not make one a scientist, nor does the lack of such a degree prevent one from making important scientific contributions. A scientist must be willing to challenge any idea when new findings demand it. The ideas that a scientist accepts must be based on measurable, reproducible observations, and the scientist must report these observations with complete honesty.

The **scientific method** is actually a collection of paths, all of which may lead to scientific discovery. In the *hypothesis and experiment* path, a scientist poses a hypothesis, then subjects it to experimental test. Many of the processes that biochemists work with every day were discovered in this manner. The DNA structure elucidated by James Watson and Francis Crick led to the hypothesis that base pairing is the basis for information transfer in polynucleotide synthesis. This hypothesis helped inspire the discovery of DNA and RNA polymerases.

Watson and Crick produced their DNA structure through a process of *model building and calculation*. No actual experiments were involved, although the model building and calculations used data collected by other scientists. Many adventurous scientists have applied the process of *exploration and observation* as a path to discovery. Historical voyages of discovery (Charles Darwin's 1831 voyage on H.M.S. *Beagle* among them) helped to map the planet, catalog its living occupants, and change the way we view the world. Modern scientists follow a similar path when they explore the ocean depths or launch probes to other planets. An analog of hypothesis and experiment is *hypothesis and deduction*. Crick reasoned that there must be an adaptor molecule that facilitated translation of the information in messenger RNA into protein. This adaptor hypothesis led to the discovery of transfer RNA by Mahlon Hoagland and Paul Zamecnik.

Not all paths to discovery involve planning. *Serendipity* often plays a role. The discovery of penicillin by Alexander Fleming in 1928 and of RNA catalysts by Thomas Cech in the early 1980s were both chance discoveries, albeit by scientists well prepared to exploit them. *Inspiration* can also lead to important advances. The polymerase chain reaction (PCR), now a central part of biotechnology, was developed by Kary Mullis after a flash of inspiration during a road trip in northern California in 1983.

These many paths to scientific discovery can seem quite different, but they have some important things in common. They are focused on the natural world. They rely on *reproducible observation* and/or *experiment*. All of the ideas, insights, and experimental facts that arise from these endeavors can be tested and reproduced by scientists anywhere in the world. All can be used by other scientists to build new hypotheses and make new discoveries. All lead to information that is properly included in the realm of science. Understanding our universe requires hard work. At the same time, no human endeavor is more exciting and potentially rewarding than trying, and occasionally succeeding, to understand some part of the natural world.

Preface

As we complete our work on this sixth edition of *Lehninger Principles of Biochemistry*, we are again struck by the remarkable changes in the field of biochemistry that have occurred between editions. The sheer volume of new information from high-throughput DNA sequencing, x-ray crystallography, and the manipulation of genes and gene expression, to cite only three examples, challenges both the seasoned researcher and the first-time biochemistry student. Our goal here is to strike a balance: to include new and exciting research findings without making the book overwhelming for students. The primary criterion for inclusion is that the new finding helps to illustrate an important *principle of biochemistry*.

The image on our cover, a map of the known metabolic transformations in a mitochondrion, illustrates the richness of factual material now available about biochemical transformations. We can no longer treat metabolic "pathways" as though they occurred in isolation; a single metabolite may be simultaneously part of many pathways in a three-dimensional network of metabolic transformations. Biochemical research focuses more and more upon the interactions among these pathways, the regulation of their interactions at the level of gene and protein, and the effects of regulation upon the activities of a whole cell or organism.

This edition of *LPOB* reflects these realities. Much of the new material that we have added reflects our increasingly sophisticated understanding of regulatory mechanisms, including those involved in altering the synthesis of enzymes and their degradation, those responsible for the control and timing of DNA synthesis and the cell cycle, and those that integrate the metabolism of carbohydrates, fats, and proteins over time in response to changes in the environment and in different cell types.

Even as we strive to incorporate the latest major advances, certain hallmarks of the book remain unchanged. We continue to emphasize the relevance of biochemistry to the molecular mechanisms of disease, highlighting the special role that biochemistry plays in advancing human health and welfare. A special theme is the metabolic basis of diabetes and the factors that predispose to the disease. This theme is interwoven through many chapters and serves to integrate the discussion of metabolism. We also underscore the importance of evolution to biochemistry. Evolutionary theory is the bedrock upon which all biological sciences rest, and we have not wasted opportunities to highlight its important role in our discipline.

To a significant degree, research progress in biochemistry runs in parallel with the development of better tools and techniques. We have therefore highlighted some of these crucial developments. Chapter 9, DNA-Based Information Technologies, in particular, has been significantly revised to include the latest advances in genomics and next-generation sequencing.

Finally, we have devoted considerable attention to making the text and the art even more useful to students learning biochemistry for the first time. To those familiar with the book, some of these changes will be obvious as soon as you crack the cover.

With every revision of this textbook, we have striven to maintain the qualities that made the original Lehninger text a classic—clear writing, careful explanations of difficult concepts, and insightful communication to students of the ways in which biochemistry is understood and practiced today. The authors have written together for almost 25 years and taught introductory biochemistry together for nearly 30. Our thousands of students at the University of Wisconsin–Madison over those years have been an endless source of ideas about how to present biochemistry more clearly; they have enlightened and inspired us. We hope that this sixth edition of *Lehninger* will in turn enlighten and inspire current students of biochemistry everywhere, and perhaps lead some of them to love biochemistry as we do.

New Art

The most obvious change to the book is the completely revamped art program. Our goal throughout has been to improve pedagogy, drawing on modern graphic resources to make our subject as clear as humanly possible. Many figures illustrate new topics, and much of the art has been reconceived and modernized in style. Defining features of the new art program include:

▶ **Smarter renditions of classic figures** are easier to interpret and learn from;

Chaperonins in protein folding

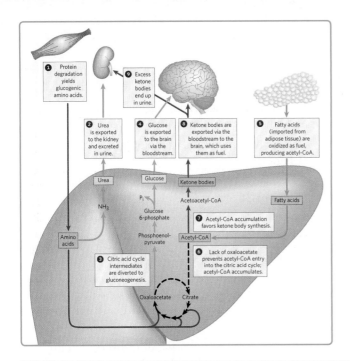

Fuel metabolism in the liver during prolonged fasting or in uncontrolled diabetes mellitus

▶ **Figures that pair molecular models with schematic cartoons**, generated specifically for this book, use shapes and color schemes that are internally consistent;

▶ Figures with **numbered, annotated steps** help explain complex processes; in many cases, we have moved descriptive text out of the legends and into the figure itself;

▶ **Summary figures** help the student to keep the big picture in mind while learning the specifics.

Updated Genomics

Modern genomic techniques have transformed our understanding of biochemistry. In this edition, we have dramatically updated our coverage of genomic methods and their applications. Chapter 9, DNA-Based Information Technologies, has been completely revised to incorporate the latest genomic methods. Many other chapters have been updated to reflect advances gained from these methods. Among the new genomic methods discussed in this edition are:

▶ Next-generation DNA sequencing, including the Illumina and 454 sequencing methods and platforms (Chapter 9)

▶ Applications of genomics, including the use of haplotypes to trace human migrations and phylogenetics to locate human genes associated with inherited diseases (Chapter 9)

▶ Forensic genotyping and the use of personalized genomics in medicine (Chapter 9)

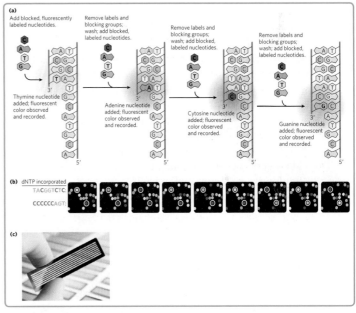

Next-generation reversible terminator sequencing

New Science

Every chapter has been thoroughly revised and updated to include both the most important advances in biochemistry and information needed in a modern biochemistry text. Among the new and updated topics in this edition are:

▶ Prebiotic evolution, black smokers, and the RNA world (Chapter 1)

▶ Intrinsically disordered proteins (Chapter 4)

▶ Transition-state analogs and irreversible inhibition (Chapter 6)

▶ Blood coagulation pathways in the context of enzymatic regulation (Chapter 6)

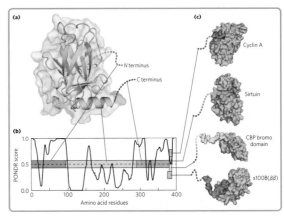

Binding of the intrinsically disordered carboxyl terminus of p53 to its binding partners

- Asymmetric lipid distribution in bilayers (Chapter 11)
- Role of BAR superfamily proteins in membrane curvature (Chapter 11)
- Scaffold proteins (AKAPS and others) and their regulatory roles (Chapter 12)
- Reactive oxygen species as byproducts and as signals (Chapter 19)
- Structure and function of the oxygen-evolving metal cluster in PSII (Chapter 19)
- Formation and transport of lipoproteins in mammals, including the roles of SREBP SCAP, and Insig in cholesterol regulation (Chapter 21)
- Integration of carbohydrate and lipid metabolism by PPARs, SREBPs, mTORC1, and LXR (Chapters 21, 23)
- Creatine phosphate and the role of creatine kinase in moving ATP to the cytosol (Chapter 23)
- Microbial symbionts in the gut and their influence on energy metabolism and adipogenesis (Chapter 23)
- Nucleosomes: their modification and positioning and higher-order chromatin structure (Chapter 24)
- DNA polymerases and homologous recombination (Chapter 25)
- Loading of eukaryotic RNA polymerase II (Chapter 26)
- Mutation-resistant nature of the genetic code (Chapter 27)
- Regulation of eukaryotic gene expression by miRNAs (Chapters 26 and 28).
- DNA looping, combinatorial control, chromatin remodeling, and positive regulation in eukaryotes (Chapter 28)
- Regulation of the initiation of transcription in eukaryotes (Chapter 28)
- Steroid-binding nuclear receptors (Chapter 28)

New Biochemical Methods

An appreciation of biochemistry often requires an understanding of how biochemical information is obtained. Some of the new methods or updates described in this edition are:

- Modern Sanger protein sequencing and mass spectrometry (Chapter 3)
- Mass spectrometry applied to proteomics, glycomics, lipidomics, and metabolomics (Chapters 3, 7, 10)
- Oligosaccharide microarrays to explore protein-oligosaccharide interactions and the "carbohydrate code" (Chapter 7)
- Modern genomic methods (Chapter 9)
- Genetic engineering of photosynthetic organisms (Chapter 20)
- Use of positron emission tomography (PET) to visualize tumors and brown adipose tissue (Chapter 23)
- Development of bacterial strains with altered genetic codes for site-specific insertion of novel amino acids into proteins (Chapter 27)

New Medical Applications

This icon is used throughout the book to denote material of special medical interest. As teachers, our goal is for students to learn biochemistry and to understand its relevance to a healthier life and a healthier planet. Many sections explore what we know about the molecular mechanisms of disease. A few of the new or revised medical applications in this edition are:

- Box 4-6, Death by Misfolding: The Prion Diseases
- Paganini and Ehlers-Danlos syndrome (Chapter 4)
- HIV protease inhibitors and how basic enzymatic principles influenced their design (Chapter 6)
- Blood coagulation cascade and hemophilia (Chapter 6)
- Curing African sleeping sickness with an enzymatic suicide inhibitor (Chapter 6)
- How researchers locate human genes involved in inherited diseases (Chapter 9)
- Multidrug resistance transporters and their importance in clinical medicine (Chapter 11)
- Multistep progression to colorectal cancer (Chapter 12)
- Cholesterol metabolism, cardiovascular disease, and mechanism of plaque formation in atherosclerosis (Chapter 21)
- P-450 and drug interactions (Chapter 21)
- HMG-CoA reductase (Chapter 21) and Box 21–3, The Lipid Hypothesis and the Development of Statins
- Box 24–1, Curing Disease by Inhibiting Topoisomerases, describing the use of topoisomerase inhibitors in the treatment of bacterial infections and cancer, including material on ciprofloxacin (the antibiotic effective for anthrax)
- Stem cells (Chapter 28)

Special Theme: Understanding Metabolism through Obesity and Diabetes

Obesity and its medical consequences—cardiovascular disease and diabetes—are fast becoming epidemic in the industrialized world, and we include new material on the biochemical connections between obesity and health throughout this edition. Our focus on diabetes provides an integrating theme throughout the chapters on metabolism and its control, and this will, we hope, inspire some students to find solutions for this disease. Some of the sections and boxes that highlight the interplay of metabolism, obesity, and diabetes are:

▶ Untreated Diabetes Produces Life-Threatening Acidosis (Chapter 2)

▶ Box 7–1, Blood Glucose Measurements in the Diagnosis and Treatment of Diabetes, introduces hemoglobin glycation and AGEs and their role in the pathology of advanced diabetes

▶ Glucose Uptake Is Deficient in Type 1 Diabetes Mellitus (Chapter 14)

▶ Ketone Bodies Are Overproduced in Diabetes and during Starvation (Chapter 17)

▶ Some Mutations in Mitochondrial Genomes Cause Disease (Chapter 19)

▶ Diabetes Can Result from Defects in the Mitochondria of Pancreatic β Cells (Chapter 19)

▶ Adipose Tissue Generates Glycerol 3-phosphate by Glyceroneogenesis (Chapter 21)

▶ Diabetes Mellitus Arises from Defects in Insulin Production or Action (Chapter 23)

▶ Section 23.4, Obesity and the Regulation of Body Mass, includes a new discussion of the roles of TORC1 in regulating cell growth

▶ Section 23.5, Obesity, the Metabolic Syndrome, and Type 2 Diabetes, discusses the role of ectopic lipids and inflammation in the development of insulin resistance and the management of type 2 diabetes with exercise, diet, and medication

Overloading adipocytes with triacylglycerols triggers inflammation in fat tissue, ectopic lipid deposition, and insulin resistance.

Special Theme: Evolution

Every time a biochemist studies a developmental pathway in nematodes, identifies key parts of an enzyme active site by determining what parts are conserved between species, or searches for the gene underlying a human genetic disease, he or she is relying on evolutionary theory. Funding agencies support the work in nematodes knowing that the insights will be relevant to humans. The conservation of functional residues in an enzyme active site telegraphs the shared history of every organism on the planet. More often than not, the search for a disease gene is a sophisticated exercise in phylogenetics. Evolution is thus a foundational concept to our discipline. Some of the many sections and boxes that deal with evolution include:

▶ Section 1.5, Evolutionary Foundations, discusses how life may have evolved and recounts some of the early milestones in the evolution of eukaryotic cells

▶ Genome Sequencing Informs Us about Our Humanity (Chapter 9)

▶ Genome Comparisons Help Locate Genes Involved in Disease (Chapter 9)

▶ Genome Sequences Inform Us about Our Past and Provide Opportunities for the Future (Chapter 9)

▶ Box 9–3, Getting to Know the Neanderthals

▶ ABC Transporters Use ATP to Drive the Active Transport of a Wide Variety of Substrates (Chapter 11)

▶ Signaling Systems of Plants Have Some of the Same Components Used by Microbes and Mammals (Chapter 12)

▶ The β-Oxidation Enzymes of Different Organelles Have Diverged during Evolution (Chapter 17)

▶ Section 19.10, The Evolution of Oxygenic Photosynthesis

▶ Mitochondria and Chloroplasts Evolved from Endosymbiotic Bacteria (Chapter 19)

▶ Photosystems I and II Evolved from Bacterial Photosystems (Chapter 19)

▶ RNA Synthesis Offers Important Clues to Biochemical Evolution (Chapter 26)

▶ Box 27–1, Exceptions That Prove the Rule: Natural Variations in the Genetic Code

▶ Box 27–2, From an RNA World to a Protein World

▶ Box 28-1, Of Fins, Wings, Beaks, and Things

Lehninger Teaching Hallmarks

Students encountering biochemistry for the first time often have difficulty with two key aspects of the course: approaching quantitative problems and drawing on what they learned in organic chemistry to help them understand biochemistry. Those same students must also learn a complex language, with conventions that are often unstated. To help students cope with these challenges, we provide the following study aids:

Focus on Chemical Logic

▸ Section 13.2, **Chemical Logic and Common Biochemical Reactions**, discusses the common biochemical reaction types that underlie all metabolic reactions, helping students to connect organic chemistry with biochemistry.

▸ **NEW chemical logic figures** highlight the conservation of mechanism and illustrate patterns that make learning pathways easier. Chemical logic figures are provided for each of the central metabolic pathways, including glycolysis (Fig. 14–3), citric acid cycle (Fig. 16–7), and fatty acid oxidation (Fig. 17–9).

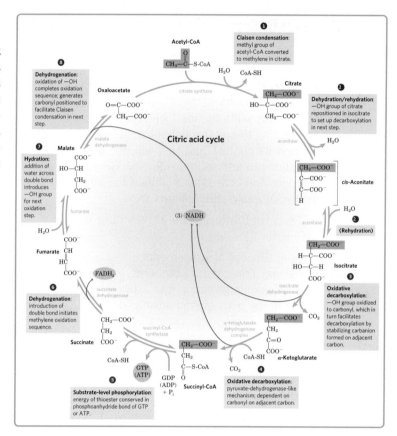

Reactions of the citric acid cycle

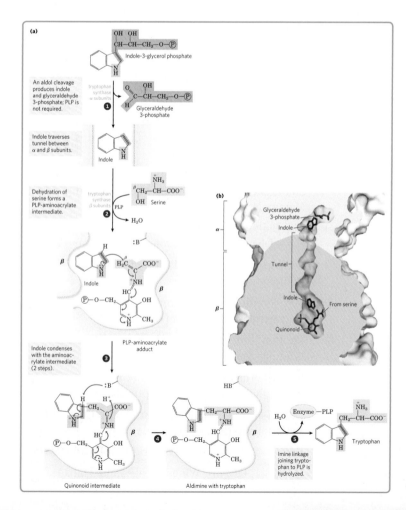

▸ **Mechanism figures** feature step-by-step descriptions to help students understand the reaction process. These figures use a consistent set of conventions introduced and explained in detail with the first enzyme mechanism encountered (chymotrypsin, pp. 216–217).

Tryptophan synthase reaction

Problem-Solving Tools

▶ **In-text Worked Examples** help students improve their quantitative problem-solving skills, taking them through some of the most difficult equations. New worked examples appear in Chapters 1, 2, and 19.

▶ **More than 600 end-of-chapter problems (about 25 of them new)** give students further opportunity to practice what they have learned.

▶ **Data Analysis Problems** (one at the end of each chapter), contributed by Brian White of the University of Massachusetts–Boston, encourage students to synthesize what they have learned and apply their knowledge to the interpretation of data from the literature.

Key Conventions

Many of the conventions that are so necessary for understanding each biochemical topic and the biochemical literature are broken out of the text and highlighted. These **Key Conventions** include clear statements of many assumptions and conventions that students are often expected to assimilate without being told (for example, peptide sequences are written from amino- to carboxyl-terminal end, left to right; nucleotide sequences are written from $5'$ to $3'$ end, left to right).

Media and Supplements

A full package of media resources and supplements provides instructors and students with innovative tools to support a variety of teaching and learning approaches. All these resources are fully integrated with the style and goals of the sixth-edition textbook.

NEW BiochemPortal (courses.bfwpub.com/lehninger6e)

This comprehensive and robust online teaching and learning tool incorporates the e-Book, all instructor and student resources, instructor assignment and gradebook functionality, and a new LearningCurve quizzing tool.

▶ BiochemPortal includes the **e-Book**, with the full contents of the text, highlighting and note-taking tools, and links to important media assets (listed below).

▶ In addition to all **instructor resources** (listed below), BiochemPortal provides instructors with the **ability to assign** any resource, as well as e-Book readings, discussion board posts, and their own materials. A gradebook tracks all student scores and can be easily exported to Excel or a campus Course Management System.

▶ **New** BiochemPortal also includes **LearningCurve**, a self-paced adaptive quizzing tool. With questions tailored to students' target difficulty level and an engaging scoring system, LearningCurve encourages students to incorporate content from the text into their study routine and provides them with a study plan on completion.

▶ Students can access any of the **student resources** provided with the text (see below) through links in the e-Book or the handy Resources tab.

e-Book (ebooks.bfwpub.com/lehninger6e)

This online version of the textbook combines the contents of the printed book with electronic study tools and a full complement of student media specifically created to support the text. The e-Book also provides useful material for instructors.

▶ **e-Book study tools** include instant navigation to any section or page of the book, bookmarks, highlighting, note-taking, instant search for any term, pop-up key-term definitions, and a spoken glossary.

▶ The text-specific **student media**, fully integrated throughout the e-Book, include animated enzyme mechanisms, animated biochemical techniques, problem-solving videos, molecular structure tutorials in Jmol, Protein Data Bank IDs in Jmol, and Living Graphs (each described under "Student Resources" below).

▶ **Instructor features** include the ability to add notes or files to any page and to share these notes with students. Notes may include text, Web links, animations, or photos. Instructors can also assign the entire text or a custom version of the e-Book.

WORKED EXAMPLE 19–2 **Stoichiometry of ATP Production: Effect of c Ring Size**

(a) If *bovine* mitochondria have 8 c subunits per c ring, what is the predicted ratio of ATP formed per NADH oxidized? (b) What is the predicted value for *yeast* mitochondria, with 10 c subunits? (c) What are the comparable values for electrons entering the respiratory chain from $FADH_2$?

Solution: (a) The question asks us to determine how many ATP are produced per NADH. This is another way of asking us to calculate the P/O ratio, or x in Equation 19–11. If the c ring has 8 c subunits, then one full rotation will transfer 8 protons to the matrix and produce 3 ATP molecules. But this synthesis also requires the transport of 3 P_i into the matrix, at a cost of 1 proton each, adding 3 more protons to the total number required. This brings the total cost to (11 protons)/(3 ATP) = 3.7 protons/ATP. The consensus value for the number of protons pumped out per pair of electrons transferred from NADH is 10 (see Fig. 19–19). So, oxidizing 1 NADH produces (10 protons)/(3.7 protons/ATP) = 2.7 ATP.

(b) If the c ring has 10 c subunits, then one full rotation will transfer 10 protons to the matrix and produce 3 ATP molecules. Adding in the 3 protons to transport the 3 P_i into the matrix brings the total cost to (13 protons)/(3 ATP) = 4.3 protons/ATP. Oxidizing 1 NADH produces (10 protons)/(4.3 protons/ATP) = 2.3 ATP.

(c) When electrons enter the respiratory chain from $FADH_2$ (at ubiquinone), only 6 protons are available to drive ATP synthesis. This changes the calculation for bovine mitochondria to (6 protons)/(3.7 protons/ ATP) = 1.6 ATP per pair of electrons from $FADH_2$. For yeast mitochondria, the calculation is (6 protons)/(4.3 protons/ATP) = 1.4 ATP per pair of electrons from $FADH_2$.

These calculated values of x or the P/O ratio define a range that includes the experimental values of 2.5 ATP/NADH and 1.5 ATP/$FADH_2$, and we therefore use these values throughout this book.

Instructor Resources

Instructors are provided with a comprehensive set of teaching tools, each developed to support the text, lecture presentations, and individual teaching styles. All instructor media are available for download on the **book website** (www.whfreeman.com/lehninger6e) and on the **Instructor Resource DVD** (ISBN 1-4641-0969-9).

▶ **New clicker questions** provide instructors with dynamic multiple-choice questions to be used with iClicker or other classroom response systems. The clicker questions have been written specifically to foster active learning in the classroom and better inform instructors on student misunderstandings.

▶ **Fully optimized JPEG files** of every figure, photo, and table in the text feature enhanced color, higher resolution, and enlarged fonts. The files have been reviewed by course instructors and tested in a large lecture hall to ensure maximum clarity and visibility. The JPEGs are also offered in separate files and in **PowerPoint** format for each chapter.

▶ **Animated Enzyme Mechanisms** and **Animated Biochemical Techniques** are available in Flash files and preloaded into PowerPoint, in both PC and Macintosh formats, for lecture presentation.

▶ A list of **Protein Data Bank IDs** for the structures in the text are arranged by figure number. A new feature in this edition is an index to all structures in the Jmol interactive Web browser applet.

▶ **Living Graphs**, illustrating key equations from the textbook, show the graphic results of changing parameters.

▶ A comprehensive **Test Bank** in PDF and editable Word formats includes 150 multiple-choice and short-answer problems per chapter, rated by level of difficulty.

Student Resources

Students are provided with media designed to enhance their understanding of biochemical principles and improve their problem-solving ability. All student media, along with the **PDB Structures** and **Living Graphs**, are also in the e-Book, and many are available on the book website (www.whfreeman.com/lehninger6e). Icons in the text indicate the availability of relevant animation, Living Graph, or Molecular Structure Tutorial.

▶ **Problem-Solving Videos**, created by Scott Ensign of Utah State University, provide 24/7 online problem-solving help to students. Through a two-part approach, each 10-minute video covers a key textbook problem representing a topic that students traditionally struggle to master. Dr. Ensign first describes a proven problem-solving strategy and then applies the strategy to the problem at hand in clear, concise steps. Students can easily pause, rewind, and review any steps as they wish until they firmly grasp not just the solution but also the reasoning behind it. Working through the problems in this way is designed to make students better and more confident at applying key strategies as they solve other textbook and exam problems.

▶ Student versions of the **Animated Enzyme Mechanisms** and **Animated Biochemical Techniques** help students understand key mechanisms and techniques at their own pace.

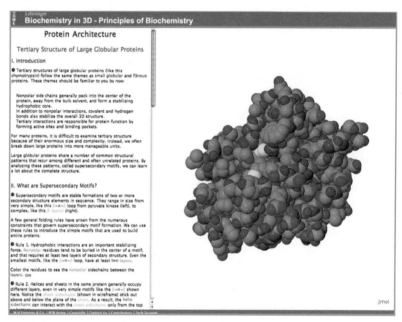

Protein Architecture Molecular Structure Tutorial

▶ **Molecular Structure Tutorials**, using the Jmol-Web browser applet, allow students to explore in more depth the molecular structures included in the textbook, including:

Protein Architecture

Bacteriorhodopsin

Lac Repressor

Nucleotides

MHC Molecules

Trimeric G Proteins

Oxygen-Binding Proteins

Restriction Endonucleases

Hammerhead Ribozyme

The Absolute, Ultimate Guide to Lehninger Principles of Biochemistry, Sixth Edition, Study Guide and Solutions Manual, by Marcy Osgood (University of New Mexico School of Medicine) and Karen Ocorr (Sanford-Burnham Medical Research Institute); ISBN 1429294760

The Absolute, Ultimate Guide combines an innovative study guide with a reliable solutions manual (providing extended solutions to end-of-chapter problems) in one convenient volume. Thoroughly class-tested, the study guide includes for each chapter:

- ► **Major Concepts:** a road map through the chapter
- ► **What to Review:** questions that recap key points from previous chapters
- ► **Discussion Questions:** provided for each section; designed for individual review, study groups, or classroom discussion
- ► **A Self-Test:** "Do you know the terms?"; crossword puzzles; multiple-choice, fact-driven questions; and questions that ask students to apply their new knowledge in new directions—plus answers!

Acknowledgments

This book is a team effort, and producing it would be impossible without the outstanding people at W. H. Freeman and Company who supported us at every step along the way. Susan Moran (Senior Developmental Editor), Susan Winslow (Publisher), and Lauren Schultz (Senior Acquisitions Editor) helped develop the revision plan for this edition, made many helpful suggestions, encouraged us, and tried valiantly (if not always successfully) to keep us on schedule. Matthew Tontonoz (Developmental Editor) provided extremely helpful feedback on many chapters. Our outstanding Project Editor, Jane O'Neill, somehow kept the book moving through production in spite of our missed deadlines and last-minute changes, and did so with her usual grace and skill. We thank Art Director Diana Blume for her artistry in designing both the text and cover for the book. We appreciate the work of present and past copyeditors, including Karen Taschek, Liz Geller, and Linda Strange. Although Linda did not copyedit this edition, her lasting contributions from the first through the fifth editions are still clearly evident in the text. We thank Photo Research Manager Ted Szczepanski and Photo Researcher Elyse Rieder for their help in locating images and Courtney Lyons for help in orchestrating reviews and providing administrative assistance at many turns. We also thank Allison Michael, Media Editor, for assembling the ever more important media components to accompany the text. Our gratitude also goes to Debbie Clare, Associate Director of Marketing, for her creativity and good humor in coordinating the sales and marketing effort. A very special thanks goes to Kate Parker, who oversaw this project for the past three editions, and contributed much to its success, before moving on to other things; we will miss her insight, humor, and excellent taste in restaurants.

In Madison, Brook Soltvedt is (and has been for all the editions we have worked on) our first-line editor and critic. She is the first to see manuscript chapters, aids in manuscript and art development, ensures internal consistency in content and nomenclature, and keeps us on task with more-or-less gentle prodding. As she did for the fourth and fifth editions, Shelley Lusetti of New Mexico State University read every word of the text in proofs, caught numerous mistakes, and made many suggestions that improved the book.

The new art in this edition, including the new molecular graphics, was done by Adam Steinberg, here in Madison, who often made valuable suggestions that led to better and clearer illustrations. We feel very fortunate to have such gifted partners as Brook, Shelley, and Adam on our team.

We are also deeply indebted to Brian White of the University of Massachusetts–Boston, who wrote the data analysis problems at the end of each chapter.

Many colleagues played a special role through their interest in the project and their timely input. Prominent among these are Jeffrey D. Esko of the University of California, San Diego; and Jack Kirsch and his students at the University of California, Berkeley. Charles G. Hoogstraten of Michigan State University made many incisive and helpful comments on the manuscript and figures. We also thank Jeffrey A. Cohlberg of California State University at Long Beach for his critical comments. Many others helped us shape this sixth edition with their comments, suggestions, and criticisms. To all of them, we are deeply grateful:

Richard Amasino, *University of Wisconsin–Madison*
Laurens Anderson, *University of Wisconsin–Madison*
Alan Attie, *University of Wisconsin–Madison*
Kenneth Balazovich, *University of Michigan*
James Blankenship, *Cornell University*
Tracey Boncher, *Ferris State College of Pharmacy*
Brian Bothner, *Montana State University*
Mary Bryk, *Texas A&M University*
Sharada Buddha, *Saint Xavier University*
Jeff DeJong, *University of Texas, Dallas*
Keith Dunker, *Indiana University*
Kelly Elkins, *Metropolitan State College of Denver*
Gerald Feigenson, *Cornell University*
Brent Feske, *Armstrong Atlantic State University*

Marcello Forconi, *College of Charleston*
Wilson Francisco, *Arizona State University*
Greta Giles, *Georgia Gwinnett College*
Glenda Gillaspy, *Virginia Tech University*
Margaret Glasner, *Texas A&M University*
James Gober, *University of California, Los Angeles*
Burt Goldberg, *New York University*
Julie Gosse, *University of Maine*
Lesley Greene, *Old Dominion University*
Eric Hegg, *Michigan State University*
Justin Hines, *Lafayette College*
Peter Hinkle, *Cornell University*
Pui Ho, *Colorado State University*
David Hurley, *Gatton College of Pharmacy, ETSU*
Joseph Jez, *Washington University in St. Louis*
Kelly Johanson, *Xavier University of Louisiana*
Douglas Julin, *University of Maryland*
Mark Kearley, *Florida State University*
Dmitry Kolpashchikov, *University of Central Florida*
Min-Hao Kuo, *Michigan State University*
Nicole LaRonde-LeBlanc, *University of Maryland*
Scott Lefler, *Arizona State University*
Andy LiWang, *University of California, Merced*
Thomas Marsh, *University of St. Thomas*
Michele McGuirl, *The University of Montana*
Michael Mendenhall, *University of Kentucky*
David Merkler, *University of South Florida*
Debra Moriarity, *University of Alabama: Huntsville*
Hunter Moseley, *University of Louisville*
Allen Nicholson, *Temple University*
James Ntambi, *University of Wisconsin–Madison*
Neil Osheroff, *Vanderbilt University School of Medicine*
Donald Ourth, *University of Memphis*
Terry Platt, *University of Rochester*
Wendy Pogozelski, *State University of New York, Geneseo*
Joseph Provost, *Minnesota State University, Moorhead*
Gregory Raner, *University of North Carolina, Greensboro*
Lisa Rezende, *University of Arizona*
Douglas Root, *University of North Texas*
Johannes Rudolph, *University of Colorado*

Phillip Ryals, *University of West Florida*
Kevin Siebenlist, *Marquette University*
Kerry Smith, *Clemson University*
Julian Snow, *University of the Sciences*
Alejandra Stenger, *University of Illinois, Urbana–Champaign*
Amy Stockert, *Ohio Northern University*
Jon Stoltzfus, *Michigan State University*
Toni Vidal-Puig, *University of Cambridge*
Chuan Xiao, *University of Texas, El Paso*
Michael Yaffe, *Massachusetts Institute of Technology*
Laura Zapanta, *University of Pittsburgh*

We lack the space here to acknowledge all the other individuals whose special efforts went into this book. We offer instead our sincere thanks—and the finished book that they helped guide to completion. We, of course, assume full responsibility for errors of fact or emphasis.

We want especially to thank our students at the University of Wisconsin–Madison for their numerous comments and suggestions. If something in the book does not work, they are never shy about letting us know it. We are grateful to the students and staff of our research groups, who helped us balance the competing demands on our time; to our colleagues in the Department of Biochemistry at the University of Wisconsin–Madison, who helped us with advice and criticism; and to the many students and teachers who have written to suggest ways of improving the book. We hope our readers will continue to provide input for future editions.

Finally, we express our deepest appreciation to our wives, Brook and Beth, and our families, who showed extraordinary patience with, and support for, our book writing.

David L. Nelson
Michael M. Cox
Madison, Wisconsin
June 2012

Contents in Brief

Contents

19 Oxidative Phosphorylation and Photophosphorylation 731

The Foundations of Biochemistry

About fourteen billion years ago, the universe arose as a cataclysmic explosion of hot, energy-rich subatomic particles. Within seconds, the simplest elements (hydrogen and helium) were formed. As the universe expanded and cooled, material condensed under the influence of gravity to form stars. Some stars became enormous and then exploded as supernovae, releasing the energy needed to fuse simpler atomic nuclei into the more complex elements. Atoms and molecules formed swirling masses of dust particles, and their accumulation led eventually to the formation of rocks, planetoids, and planets. Thus were produced, over billions of years, Earth itself and the chemical elements found on Earth today. About four billion years ago, life arose—simple microorganisms with the ability to extract energy from chemical compounds and, later, from sunlight, which they used to make a vast array of more complex **biomolecules** from the simple elements and compounds on the Earth's surface. We and all other living organisms are made of stardust.

Biochemistry asks how the remarkable properties of living organisms arise from the thousands of different biomolecules. When these molecules are isolated and examined individually, they conform to all the physical and chemical laws that describe the behavior of inanimate matter—as do all the processes occurring in living organisms. The study of biochemistry shows how the collections of inanimate molecules that constitute living organisms interact to maintain and perpetuate life animated solely by the physical and chemical laws that govern the nonliving universe.

Yet organisms possess extraordinary attributes, properties that distinguish them from other collections of matter. What are these distinguishing features of living organisms?

A high degree of chemical complexity and microscopic organization. Thousands of different molecules make up a cell's intricate internal structures **(Fig. 1–1a)**. These include very long polymers, each with its characteristic sequence of subunits, its unique three-dimensional structure, and its highly specific selection of binding partners in the cell.

Systems for extracting, transforming, and using energy from the environment (Fig. 1–1b), enabling organisms to build and maintain their intricate structures and to do mechanical, chemical, osmotic, and electrical work. This counteracts the tendency of all matter to decay toward a more disordered state, to come to equilibrium with its surroundings.

Defined functions for each of an organism's components and regulated interactions among them. This is true not only of macroscopic structures, such as leaves and stems or hearts and lungs, but also of microscopic intracellular structures and individual chemical compounds. The interplay among the chemical components of a living organism is dynamic; changes in one component cause coordinating or compensating changes in another, with the whole ensemble displaying a character beyond that of its individual parts. The collection of molecules carries out a program, the end result of which is reproduction of the program and self-perpetuation of that collection of molecules—in short, life.

Mechanisms for sensing and responding to alterations in their surroundings. Organisms constantly adjust to these changes by adapting their internal chemistry or their location in the environment.

A capacity for precise self-replication and self-assembly (Fig. 1–1c). A single bacterial cell placed in a sterile nutrient medium can give rise to

FIGURE 1–2 Diverse living organisms share common chemical features. Birds, beasts, plants, and soil microorganisms share with humans the same basic structural units (cells) and the same kinds of macromolecules (DNA, RNA, proteins) made up of the same kinds of monomeric subunits (nucleotides, amino acids). They utilize the same pathways for synthesis of cellular components, share the same genetic code, and derive from the same evolutionary ancestors. Shown here is a detail from "The Garden of Eden," by Jan van Kessel the Younger (1626–1679).

FIGURE 1–1 Some characteristics of living matter. (a) Microscopic complexity and organization are apparent in this colorized image of a thin section of a secretory cell from the pancreas, viewed with the electron microscope. **(b)** A prairie falcon acquires nutrients and energy by consuming a smaller bird. **(c)** Biological reproduction occurs with near-perfect fidelity.

a billion identical "daughter" cells in 24 hours. Each cell contains thousands of different molecules, some extremely complex; yet each bacterium is a faithful copy of the original, its construction directed entirely from information contained in the genetic material of the original cell. On a larger scale, the progeny of a vertebrate animal share a striking resemblance to their parents, also the result of their inheritance of parental genes.

A capacity to change over time by gradual evolution. Organisms change their inherited life strategies, in very small steps, to survive in new circumstances. The result of eons of evolution is an enormous diversity of life forms, superficially very different **(Fig. 1–2)** but fundamentally related through their shared ancestry. This fundamental unity of living organisms is reflected at the molecular level in the similarity of gene sequences and protein structures.

Despite these common properties, and the fundamental unity of life they reveal, it is difficult to make generalizations about living organisms. Earth has an enormous diversity of organisms. The range of habitats, from hot springs to Arctic tundra, from animal intestines

to college dormitories, is matched by a correspondingly wide range of specific biochemical adaptations, achieved within a common chemical framework. For the sake of clarity, in this book we sometimes risk certain generalizations, which, though not perfect, remain useful; we also frequently point out the exceptions to these generalizations, which can prove illuminating.

Biochemistry describes in molecular terms the structures, mechanisms, and chemical processes shared by all organisms and provides organizing principles that underlie life in all its diverse forms, principles we refer to collectively as *the molecular logic of life*. Although biochemistry provides important insights and practical applications in medicine, agriculture, nutrition, and industry, its ultimate concern is with the wonder of life itself.

In this introductory chapter we give an overview of the cellular, chemical, physical, and genetic backgrounds to biochemistry and the overarching principle of evolution—how life emerged and evolved into the diversity of organisms we see today. As you read through the book, you may find it helpful to refer back to this chapter at intervals to refresh your memory of this background material.

1.1 Cellular Foundations

The unity and diversity of organisms become apparent even at the cellular level. The smallest organisms consist of single cells and are microscopic. Larger, multicellular organisms contain many different types of cells, which vary in size, shape, and specialized function. Despite these obvious differences, all cells of the simplest and most complex organisms share certain fundamental properties, which can be seen at the biochemical level.

Cells Are the Structural and Functional Units of All Living Organisms

Cells of all kinds share certain structural features **(Fig. 1–3)**. The **plasma membrane** defines the periphery of the cell, separating its contents from the surroundings. It is composed of lipid and protein molecules that form a thin, tough, pliable, hydrophobic barrier around the cell. The membrane is a barrier to the free passage of inorganic ions and most other charged or polar compounds. Transport proteins in the plasma membrane allow the passage of certain ions and molecules; receptor proteins transmit signals into the cell; and membrane enzymes participate in some reaction pathways. Because the individual lipids and proteins of the plasma membrane are not covalently linked, the entire structure is remarkably flexible, allowing changes in the shape and size of the cell. As a cell grows, newly made lipid and protein molecules are inserted into its plasma membrane; cell division produces two cells, each with its own membrane. This growth and cell division (fission) occurs without loss of membrane integrity.

The internal volume enclosed by the plasma membrane, the **cytoplasm** (Fig. 1–3), is composed of an aqueous solution, the **cytosol**, and a variety of suspended particles with specific functions. These particulate components (membranous organelles such as mitochondria and chloroplasts; supramolecular structures such as **ribosomes** and **proteasomes**, the sites of protein synthesis and degradation) sediment when cytoplasm is centrifuged at 150,000 g (g is the gravitational force of Earth). What remains as the supernatant fluid is the cytosol, a highly concentrated solution containing enzymes and the RNA molecules that encode them; the components (amino acids and nucleotides) from which these macromolecules are assembled; hundreds of small organic molecules called **metabolites**, intermediates in biosynthetic and degradative pathways; **coenzymes**, compounds essential to many enzyme-catalyzed reactions; and inorganic ions.

All cells have, for at least some part of their life, either a **nucleoid** or a **nucleus**, in which the **genome**—the complete set of genes, composed of DNA—is replicated and stored, with its associated proteins. The nucleoid, in bacteria and archaea, is not separated from the cytoplasm by a membrane; the nucleus, in **eukaryotes**, is enclosed within a double membrane, the nuclear envelope. Cells with nuclear envelopes make up the large domain Eukarya (Greek *eu*, "true," and *karyon*, "nucleus"). Microorganisms without nuclear membranes, formerly grouped together as **prokaryotes** (Greek *pro*, "before"), are now recognized as comprising two very distinct groups: the domains Bacteria and Archaea, described below.

Cellular Dimensions Are Limited by Diffusion

Most cells are microscopic, invisible to the unaided eye. Animal and plant cells are typically 5 to 100 μm in diameter, and many unicellular microorganisms are only 1 to 2 μm long (see the inside back cover for information on units and their abbreviations). What limits the dimensions of a cell? The lower limit is probably set by the minimum number of each type of biomolecule required by the cell. The smallest cells, certain bacteria known as mycoplasmas, are 300 nm in diameter and have a volume of about 10^{-14} mL. A single bacterial ribosome is about 20 nm in its longest dimension, so a few ribosomes take up a substantial fraction of the volume in a mycoplasmal cell.

The upper limit of cell size is probably set by the rate of diffusion of solute molecules in aqueous systems. For example, a bacterial cell that depends on oxygen-consuming reactions for energy extraction must obtain molecular oxygen by diffusion from the surrounding medium through its plasma membrane. The cell is so small, and the ratio of its surface area to its volume is so large, that every part of its cytoplasm is easily reached by O_2 diffusing into the cell. With increasing cell size, however, surface-to-volume ratio decreases, until metabolism consumes O_2 faster than diffusion can supply it. Metabolism that requires O_2 thus becomes impossible as cell size increases beyond a certain point, placing a theoretical upper limit on the size of cells. Oxygen is only one of many low molecular weight species that must diffuse from outside the cell to various regions of its interior, and the same surface-to-volume argument applies to each of them as well.

There Are Three Distinct Domains of Life

All living organisms fall into one of three large groups (domains) that define three branches of the evolutionary tree of life originating from a common progenitor

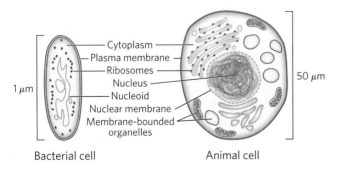

Cytoplasm
Plasma membrane
Ribosomes
Nucleus
Nucleoid
Nuclear membrane
Membrane-bounded organelles

1 μm

50 μm

Bacterial cell

Animal cell

FIGURE 1–3 The universal features of living cells. All cells have a nucleus or nucleoid containing their DNA, a plasma membrane, and cytoplasm. The cytosol is defined as that portion of the cytoplasm that remains in the supernatant after gentle breakage of the plasma membrane and centrifugation of the resulting extract at 150,000 g for 1 hour. Eukaryotic cells contain a variety of membrane-bounded organelles (mitochondria, chloroplasts) and large particles (ribosomes, for example), which are sedimented by this centrifugation and can be recovered from the pellet.

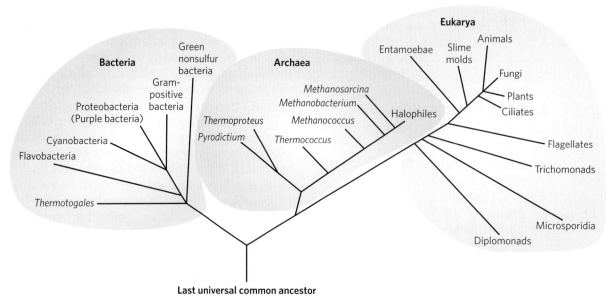

FIGURE 1–4 Phylogeny of the three domains of life. Phylogenetic relationships are often illustrated by a "family tree" of this type. The basis for this tree is the similarity in nucleotide sequences of the ribosomal RNAs of each group; the more similar the sequence, the closer the location of the branches, with the distance between branches representing the degree of difference between two sequences. Phylogenetic trees can also be constructed from similarities across species of the amino acid sequences of a single protein. For example, sequences of the protein GroEL (a bacterial protein that assists in protein folding) were compared to generate the tree in Figure 3-35. The tree in Figure 3-36 is a "consensus" tree, which uses several comparisons such as these to derive the best estimates of evolutionary relatedness among a group of organisms.

(Fig. 1–4). Two large groups of single-celled microorganisms can be distinguished on genetic and biochemical grounds: **Bacteria** and **Archaea**. Bacteria inhabit soils, surface waters, and the tissues of other living or decaying organisms. Many of the Archaea, recognized as a distinct domain by Carl Woese in the 1980s, inhabit extreme environments—salt lakes, hot springs, highly acidic bogs, and the ocean depths. The available evidence suggests that the Archaea and Bacteria diverged early in evolution. All eukaryotic organisms, which make up the third domain, **Eukarya**, evolved from the same branch that gave rise to the Archaea; eukaryotes are therefore more closely related to archaea than to bacteria.

Within the domains of Archaea and Bacteria are subgroups distinguished by their habitats. In **aerobic** habitats with a plentiful supply of oxygen, some resident organisms derive energy from the transfer of electrons from fuel molecules to oxygen within the cell. Other environments are **anaerobic**, virtually devoid of oxygen, and microorganisms adapted to these environments obtain energy by transferring electrons to nitrate (forming N_2), sulfate (forming H_2S), or CO_2 (forming CH_4). Many organisms that have evolved in anaerobic environments are *obligate* anaerobes: they die when exposed to oxygen. Others are *facultative* anaerobes, able to live with or without oxygen.

Organisms Differ Widely in Their Sources of Energy and Biosynthetic Precursors

We can classify organisms according to how they obtain the energy and carbon they need for synthesizing cellular material (as summarized in **Fig. 1–5**). There are two broad categories based on energy sources: **phototrophs** (Greek *trophē*, "nourishment") trap and use sunlight, and **chemotrophs** derive their energy from oxidation of a chemical fuel. Some chemotrophs oxidize inorganic fuels— HS^- to S^0 (elemental sulfur), S^0 to SO_4^-, NO_2^- to NO_3^-, or Fe^{2+} to Fe^{3+}, for example. Phototrophs and chemotrophs may be further divided into those that can synthesize all of their biomolecules directly from CO_2 (**autotrophs**) and those that require some preformed organic nutrients made by other organisms (**heterotrophs**). We can describe an organism's mode of nutrition by combining these terms. For example, cyanobacteria are photoautotrophs; humans are chemoheterotrophs. Even finer distinctions can be made, and many organisms can obtain energy and carbon from more than one source under different environmental or developmental conditions.

Bacterial and Archaeal Cells Share Common Features but Differ in Important Ways

The best-studied bacterium, *Escherichia coli*, is a usually harmless inhabitant of the human intestinal tract. The *E. coli* cell **(Fig. 1–6a)** is an ovoid about 2 μm long and a little less than 1 μm in diameter, but other bacteria may be spherical or rod-shaped. It has a protective outer membrane and an inner plasma membrane that encloses the cytoplasm and the nucleoid. Between the inner and outer membranes is a thin but strong layer of a high molecular weight polymer (peptidoglycan) that gives the cell its shape and rigidity. The plasma membrane and the layers outside it constitute

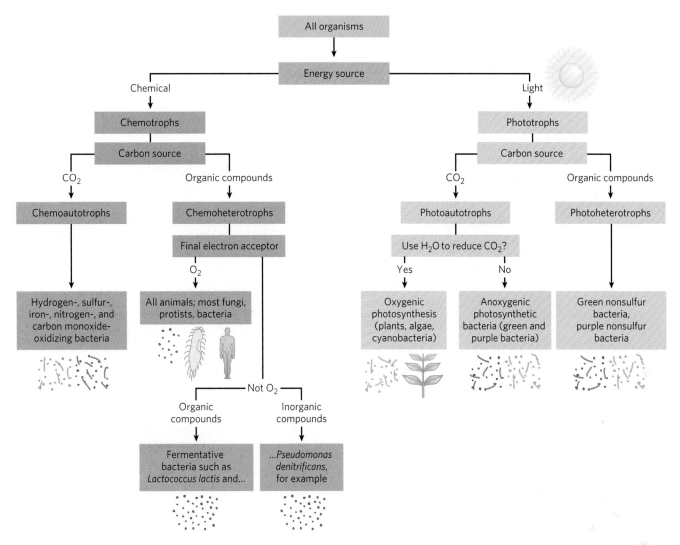

FIGURE 1–5 All organisms can be classified according to their source of energy (sunlight or oxidizable chemical compounds) and their source of carbon for the synthesis of cellular material.

the **cell envelope**. The plasma membranes of bacteria consist of a thin bilayer of lipid molecules penetrated by proteins. Archaeal plasma membranes have a similar architecture, but the lipids can be strikingly different from those of bacteria (see Fig. 10–12). Bacteria and archaea have group-specific specializations of their cell envelopes (Fig. 1–6b–d). Some bacteria, called gram-positive because they are colored by Gram's stain (introduced by Hans Peter Gram in 1882), have a thick layer of peptidoglycan outside their plasma membrane but lack an outer membrane. Gram-negative bacteria have an outer membrane composed of a lipid bilayer into which are inserted complex lipopolysaccharides and proteins called porins that provide transmembrane channels for low molecular weight compounds and ions to diffuse across this outer membrane. The structures outside the plasma membrane of archaea differ from organism to organism, but they, too, have a layer of peptidoglycan or protein that confers rigidity on their cell envelopes.

The cytoplasm of *E. coli* contains about 15,000 ribosomes, various numbers (10 to thousands) of copies of each of 1,000 or so different enzymes, perhaps 1,000 organic compounds of molecular weight less than 1,000 (metabolites and cofactors), and a variety of inorganic ions. The nucleoid contains a single, circular molecule of DNA, and the cytoplasm (like that of most bacteria) contains one or more smaller, circular segments of DNA called **plasmids**. In nature, some plasmids confer resistance to toxins and antibiotics in the environment. In the laboratory, these DNA segments are especially amenable to experimental manipulation and are powerful tools for genetic engineering (see Chapter 9).

Other species of bacteria, as well as archaea, contain a similar collection of biomolecules, but each species has physical and metabolic specializations related to its environmental niche and nutritional sources. Cyanobacteria, for example, have internal membranes specialized to trap energy from light (see Fig. 19–67). Many archaea live in extreme environments and have biochemical

(a) Ribosomes Bacterial and archaeal ribosomes are smaller than eukaryotic ribosomes, but serve the same function—protein synthesis from an RNA message.

(b) Gram-positive bacteria

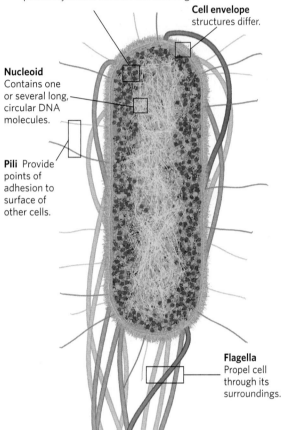

Nucleoid
Contains one or several long, circular DNA molecules.

Pili Provide points of adhesion to surface of other cells.

Cell envelope structures differ.

Flagella Propel cell through its surroundings.

Polysaccharide

Solid layer

Glycoprotein

Peptidoglycan

Plasma membrane

Cytoplasm   Lipoprotein

(c) Gram-negative bacteria (shown at left)

LPS

Outer membrane

Lipoprotein   Porin

Peptidoglycan

Periplasm

Plasma membrane

Cytoplasm   Lipoprotein

(d) *Methanothermus*, an extremely heat-tolerant archaeon

Solid layer

Glycoprotein

Pseudo-peptidoglycan

Plasma membrane

Cytoplasm

FIGURE 1–6 Some common structural features of bacterial and archaeal cells. (a) This correct-scale drawing of *E. coli* serves to illustrate some common features. **(b)** The cell envelope of gram-positive bacteria is a single membrane with a thick, rigid layer of peptidoglycan on its outside surface. A variety of polysaccharides and other complex polymers are interwoven with the peptidoglycan, and surrounding the whole is a porous "solid layer" composed of glycoproteins. **(c)** *E. coli* is gram-negative and has a double membrane. Its outer membrane has a lipopolysaccharide (LPS) on the outer surface and phospholipids on the inner surface. This outer membrane is studded with protein channels (porins) that allow small molecules, but not proteins, to diffuse through. The inner (plasma) membrane, made of phospholipids and proteins, is impermeant to both large and small molecules. Between the inner and outer membranes, in the periplasm, is a thin layer of peptidoglycan, which gives the cell shape and rigidity, but does not retain Gram's stain. **(d)** Archaeal membranes vary in structure and composition, but all have a single membrane surrounded by an outer layer that includes either a peptidoglycanlike structure, a porous protein shell (solid layer), or both.

adaptations to survive in extremes of temperature, pressure, or salt concentration. Differences in ribosomal structure gave the first hints that Bacteria and Archaea constituted separate domains. Most bacteria (including *E. coli*) exist as individual cells, but often associate in biofilms or mats, in which large numbers of cells adhere to each other and to some solid substrate beneath or at an aqueous surface. Cells of some bacterial species (the myxobacteria, for example) show simple social behavior, forming many-celled aggregates in response to signals between neighboring cells.

Eukaryotic Cells Have a Variety of Membranous Organelles, Which Can Be Isolated for Study

Typical eukaryotic cells **(Fig. 1–7)** are much larger than bacteria—commonly 5 to 100 μm in diameter, with cell volumes a thousand to a million times larger than those of bacteria. The distinguishing characteristics of eukaryotes are the nucleus and a variety of membrane-enclosed organelles with specific functions. These organelles include **mitochondria**, the site of most of the energy-extracting reactions of the cell; the **endoplasmic reticulum** and **Golgi complexes**, which play central roles in the synthesis and processing of lipids and membrane proteins; **peroxisomes**, in which very long-chain fatty acids are oxidized; and **lysosomes**, filled with digestive enzymes to degrade unneeded cellular debris. In addition to these, plant cells also contain **vacuoles** (which store large quantities of organic acids) and **chloroplasts** (in which sunlight drives the synthesis of ATP in the process of photosynthesis) (Fig. 1–7). Also present in the cytoplasm of many cells are granules or droplets containing stored nutrients such as starch and fat.

(a) Animal cell

Ribosomes are protein-synthesizing machines.

Peroxisome oxidizes fatty acids.

Cytoskeleton supports cell, aids in movement of organelles.

Lysosome degrades intracellular debris.

Transport vesicle shuttles lipids and proteins between ER, Golgi, and plasma membrane.

Golgi complex processes, packages, and targets proteins to other organelles or for export.

Smooth endoplasmic reticulum (SER) is site of lipid synthesis and drug metabolism.

Nuclear envelope segregates chromatin (DNA + protein) from cytoplasm.

Nucleolus is site of ribosomal RNA synthesis.

Nucleus contains the genes (chromatin).

Rough endoplasmic reticulum (RER) is site of much protein synthesis.

Plasma membrane separates cell from environment, regulates movement of materials into and out of cell.

Nuclear envelope

Ribosomes

Cytoskeleton

Mitochondrion oxidizes fuels to produce ATP.

Chloroplast harvests sunlight, produces ATP and carbohydrates.

Golgi complex

Starch granule temporarily stores carbohydrate products of photosynthesis.

Thylakoids are site of light-driven ATP synthesis.

Cell wall provides shape and rigidity; protects cell from osmotic swelling.

Cell wall of adjacent cell

Vacuole degrades and recycles macromolecules, stores metabolites.

Plasmodesma provides path between two plant cells.

Glyoxysome contains enzymes of the glyoxylate cycle.

(b) Plant cell

FIGURE 1–7 Eukaryotic cell structure. Schematic illustrations of two major types of eukaryotic cell: **(a)** a representative animal cell and **(b)** a representative plant cell. Plant cells are usually 10 to 100 μm in diameter—larger than animal cells, which typically range from 5 to 30 μm. Structures labeled in red are unique to animal cells; those labeled in green are unique to plant cells. Eukaryotic microorganisms (such as protists and fungi) have structures similar to those in plant and animal cells, but many also contain specialized organelles not illustrated here.

In a major advance in biochemistry, Albert Claude, Christian de Duve, and George Palade developed methods for separating organelles from the cytosol and from each other—an essential step in investigating their structures and functions. In a typical cell fractionation (**Fig. 1–8**), cells or tissues in solution are gently disrupted by physical shear. This treatment ruptures the plasma membrane but leaves most of the organelles intact. The homogenate is then centrifuged; organelles such as nuclei, mitochondria, and lysosomes differ in size and therefore sediment at different rates.

These methods were used to establish, for example, that lysosomes contain degradative enzymes, mitochondria contain oxidative enzymes, and chloroplasts contain photosynthetic pigments. The isolation of an organelle enriched in a certain enzyme is often the first step in the purification of that enzyme.

Differential centrifugation

Tissue homogenization

Low-speed centrifugation (1,000 *g*, 10 min)

Supernatant subjected to medium-speed centrifugation (20,000 *g*, 20 min)

Supernatant subjected to high-speed centrifugation (80,000 *g*, 1 h)

Supernatant subjected to very high-speed centrifugation (150,000 *g*, 3 h)

Tissue homogenate

Pellet contains whole cells, nuclei, cytoskeletons, plasma membranes

Pellet contains mitochondria, lysosomes, peroxisomes

Pellet contains microsomes (fragments of ER), small vesicles

Supernatant contains soluble proteins

Pellet contains ribosomes, large macromolecules

FIGURE 1–8 Subcellular fractionation of tissue. A tissue such as liver is first mechanically homogenized to break cells and disperse their contents in an aqueous buffer. The sucrose medium has an osmotic pressure similar to that in organelles, thus balancing diffusion of water into and out of the organelles, which would swell and burst in a solution of lower osmolarity (see Fig. 2–13). The large and small particles in the suspension can be separated by centrifugation at different speeds. Larger particles sediment more rapidly than small particles, and soluble material does not sediment. By careful choice of the conditions of centrifugation, subcellular fractions can be separated for biochemical characterization.

The Cytoplasm Is Organized by the Cytoskeleton and Is Highly Dynamic

Fluorescence microscopy reveals several types of protein filaments crisscrossing the eukaryotic cell, forming an interlocking three-dimensional meshwork, the **cytoskeleton**. There are three general types of cytoplasmic filaments—actin filaments, microtubules, and intermediate filaments (**Fig. 1–9**)—differing in width (from about 6 to 22 nm), composition, and specific function. All types provide structure and organization to the cytoplasm and shape to the cell. Actin filaments and microtubules also help to produce the motion of organelles or of the whole cell.

Each type of cytoskeletal component is composed of simple protein subunits that associate noncovalently to form filaments of uniform thickness. These filaments are not permanent structures; they undergo constant disassembly into their protein subunits and reassembly into filaments. Their locations in cells are not rigidly fixed but may change dramatically with mitosis, cytokinesis, amoeboid motion, or changes in cell shape. The assembly, disassembly, and location of all types of filaments are regulated by other proteins, which serve to link or bundle the filaments or to move cytoplasmic organelles along the filaments. (Bacteria contain actinlike proteins that serve similar roles in those cells.)

The picture that emerges from this brief survey of eukaryotic cell structure is of a cell with a meshwork of structural fibers and a complex system of membrane-enclosed compartments (Fig. 1–7). The filaments disassemble and then reassemble elsewhere. Membranous vesicles bud from one organelle and fuse with another. Organelles move through the cytoplasm along protein filaments, their motion powered by energy-dependent motor proteins. The **endomembrane system** segregates specific metabolic processes and provides surfaces on which certain enzyme-catalyzed reactions occur. **Exocytosis** and **endocytosis**, mechanisms of transport (out of and into cells, respectively) that involve membrane fusion and fission, provide paths between the cytoplasm and surrounding medium, allowing for secretion of substances produced in the cell and uptake of extracellular materials.

(a)

(b)

FIGURE 1-9 **The three types of cytoskeletal filaments: actin filaments, microtubules, and intermediate filaments.** Cellular structures can be labeled with an antibody (that recognizes a characteristic protein) covalently attached to a fluorescent compound. The stained structures are visible when the cell is viewed with a fluorescence microscope. **(a)** Endothelial cells from the bovine pulmonary artery. Bundles of actin filaments called "stress fibers" are stained red; microtubules, radiating from the cell center, are stained green; and chromosomes (in the nucleus) are stained blue. **(b)** A newt lung cell undergoing mitosis. Microtubules (green), attached to structures called kinetochores (yellow) on the condensed chromosomes (blue), pull the chromosomes to opposite poles, or centrosomes (magenta), of the cell. Intermediate filaments, made of keratin (red), maintain the structure of the cell.

Although complex, this organization of the cytoplasm is far from random. The motion and positioning of organelles and cytoskeletal elements are under tight regulation, and at certain stages in its life, a eukaryotic cell undergoes dramatic, finely orchestrated reorganizations, such as the events of mitosis. The interactions between the cytoskeleton and organelles are noncovalent, reversible, and subject to regulation in response to various intracellular and extracellular signals.

Cells Build Supramolecular Structures

Macromolecules and their monomeric subunits differ greatly in size **(Fig. 1-10)**. An alanine molecule is less than 0.5 nm long. A molecule of hemoglobin, the oxygen-carrying protein of erythrocytes (red blood cells), consists of nearly 600 amino acid subunits in four long chains, folded into globular shapes and associated in a structure 5.5 nm in diameter. In turn, proteins are much smaller than ribosomes (about 20 nm in diameter), which are in turn much smaller than organelles such as mitochondria, typically 1,000 nm in diameter. It is a long jump from simple biomolecules to cellular structures that can be seen with the light microscope. **Figure 1-11** illustrates the structural hierarchy in cellular organization.

The monomeric subunits of proteins, nucleic acids, and polysaccharides are joined by covalent bonds. In supramolecular complexes, however, macromolecules are held together by noncovalent interactions—much weaker, individually, than covalent bonds. Among these noncovalent interactions are hydrogen bonds (between polar groups), ionic interactions (between charged groups), hydrophobic interactions (among nonpolar groups in aqueous solution), and van der Waals interactions (London forces)—all of which have energies much smaller than those of covalent bonds. These noncovalent interactions are described in Chapter 2. The large numbers of weak interactions between macromolecules in supramolecular complexes stabilize these assemblies, producing their unique structures.

In Vitro Studies May Overlook Important Interactions among Molecules

One approach to understanding a biological process is to study purified molecules in vitro ("in glass"—in the test tube), without interference from other molecules present in the intact cell—that is, in vivo ("in the living"). Although this approach has been remarkably revealing, we must keep in mind that the inside of a cell is quite different from the inside of a test tube. The "interfering" components eliminated by purification may be critical to the biological function or regulation of the molecule purified. For example, in vitro studies of pure enzymes are commonly done at very low enzyme concentrations in thoroughly stirred aqueous solutions. In the cell, an enzyme is dissolved or suspended in the gel-like cytosol with thousands of other proteins, some of which bind to that enzyme and influence its activity. Some enzymes are components of multienzyme complexes in which reactants are channeled from one enzyme to another, never entering the bulk solvent. When all of the known macromolecules in a cell are represented at their known dimensions and concentrations **(Fig. 1–12)**, it is clear that the cytosol is very crowded and that diffusion of macromolecules within the cytosol must be slowed by collisions with other large structures. In short, a given molecule may behave quite differently in the cell and in vitro. A central challenge of biochemistry is to understand the influences of cellular organization and macromolecular associations on the function of individual enzymes and other biomolecules—to understand function in vivo as well as in vitro.

SUMMARY 1.1 Cellular Foundations

▶ All cells are bounded by a plasma membrane; have a cytosol containing metabolites, coenzymes, inorganic ions, and enzymes; and have a set of genes contained within a nucleoid (bacteria and archaea) or nucleus (eukaryotes).

▶ All organisms require a source of energy to perform cellular work. Phototrophs obtain energy from sunlight; chemotrophs oxidize chemical fuels, passing electrons to good electron acceptors: inorganic compounds, organic compounds, or molecular oxygen.

▶ Bacterial and archaeal cells contain cytosol, a nucleoid, and plasmids, all contained within a cell envelope. Eukaryotic cells have a nucleus and are multicompartmented, with certain processes

FIGURE 1–10 The organic compounds from which most cellular materials are constructed: the ABCs of biochemistry. Shown here are **(a)** six of the 20 amino acids from which all proteins are built (the side chains are shaded light red); **(b)** the five nitrogenous bases, two five-carbon sugars, and phosphate ion from which all nucleic acids are built; **(c)** five components of membrane lipids; and **(d)** D-glucose, the simple sugar from which most carbohydrates are derived. Note that phosphate is a component of both nucleic acids and membrane lipids.

(a) Some of the amino acids of proteins

(b) The components of nucleic acids

(c) Some components of lipids

(d) The parent sugar

FIGURE 1–11 **Structural hierarchy in the molecular organization of cells.** The organelles and other relatively large components of cells are composed of supramolecular complexes, which in turn are composed of smaller macromolecules and even smaller molecular subunits. For example, the nucleus of this plant cell contains chromatin, a supramolecular complex that consists of DNA and basic proteins (histones). DNA is made up of simple monomeric subunits (nucleotides), as are proteins (amino acids).

segregated in specific organelles; organelles can be separated and studied in isolation.

▶ Cytoskeletal proteins assemble into long filaments that give cells shape and rigidity and serve as rails along which cellular organelles move throughout the cell.

▶ Supramolecular complexes held together by noncovalent interactions are part of a hierarchy of structures, some visible with the light microscope. When individual molecules are removed from these complexes to be studied in vitro, interactions important in the living cell may be lost.

1.2 Chemical Foundations

Biochemistry aims to explain biological form and function in chemical terms. By the late eighteenth century, chemists had concluded that the composition of living matter is strikingly different from that of the inanimate world. Antoine-Laurent Lavoisier (1743–1794) noted the relative chemical simplicity of the "mineral world" and contrasted it with the complexity of the "plant and animal worlds"; the latter, he knew, were composed of compounds rich in the elements carbon, oxygen, nitrogen, and phosphorus.

During the first half of the twentieth century, parallel biochemical investigations of glucose breakdown in yeast and in animal muscle cells revealed remarkable chemical similarities in these two apparently very different cell types; the breakdown of glucose in yeast and muscle cells involved the same 10 chemical intermediates, and the same 10 enzymes. Subsequent studies of many other biochemical processes in many different organisms have confirmed the generality of this observation, neatly summarized in 1954 by Jacques Monod: "What is true of *E. coli* is true of the elephant." The current understanding that all organisms share a common evolutionary origin is based in part on this observed

FIGURE 1–12 **The crowded cell.** This drawing by David Goodsell is an accurate representation of the relative sizes and numbers of macromolecules in one small region of an *E. coli* cell. This concentrated cytosol, crowded with proteins and nucleic acids, is very different from the typical extract of cells used in biochemical studies, in which the cytosol has been diluted manyfold and the interactions between diffusing macromolecules have been strongly altered.

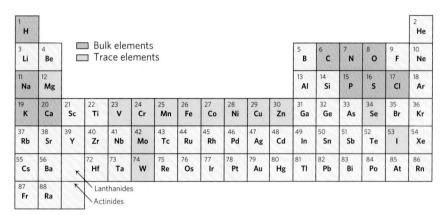

FIGURE 1-13 Elements essential to animal life and health. Bulk elements (shaded light red) are structural components of cells and tissues and are required in the diet in gram quantities daily. For trace elements (shaded yellow), the requirements are much smaller: for humans, a few milligrams per day of Fe, Cu, and Zn, even less of the others. The elemental requirements for plants and microorganisms are similar to those shown here; the ways in which they acquire these elements vary.

universality of chemical intermediates and transformations, often termed "biochemical unity."

Fewer than 30 of the more than 90 naturally occurring chemical elements are essential to organisms. Most of the elements in living matter have relatively low atomic numbers; only three have atomic numbers above that of selenium, 34 **(Fig. 1–13)**. The four most abundant elements in living organisms, in terms of percentage of total number of atoms, are hydrogen, oxygen, nitrogen, and carbon, which together make up more than 99% of the mass of most cells. They are the lightest elements capable of efficiently forming one, two, three, and four bonds, respectively; in general, the lightest elements form the strongest bonds. The trace elements (Fig. 1–13) represent a miniscule fraction of the weight of the human body, but all are essential to life, usually because they are essential to the function of specific proteins, including many enzymes. The oxygen-transporting capacity of the hemoglobin molecule, for example, is absolutely dependent on four iron ions that make up only 0.3% of its mass.

Biomolecules Are Compounds of Carbon with a Variety of Functional Groups

The chemistry of living organisms is organized around carbon, which accounts for more than half the dry weight of cells. Carbon can form single bonds with hydrogen atoms, and both single and double bonds with oxygen and nitrogen atoms **(Fig. 1–14)**. Of greatest significance in biology is the ability of carbon atoms to form very sta-

ble single bonds with up to four other carbon atoms. Two carbon atoms also can share two (or three) electron pairs, thus forming double (or triple) bonds.

The four single bonds that can be formed by a carbon atom project from the nucleus to the four apices of a tetrahedron **(Fig. 1–15)**, with an angle of about 109.5° between any two bonds and an average bond length of 0.154 nm. There is free rotation around each single bond, unless very large or highly charged groups are attached to both carbon atoms, in which case rotation may be restricted. A double bond is shorter (about 0.134 nm) and rigid, and allows only limited rotation about its axis.

Covalently linked carbon atoms in biomolecules can form linear chains, branched chains, and cyclic structures. It seems likely that the bonding versatility of carbon, with itself and with other elements, was a major factor in the selection of carbon compounds for the molecular machinery of cells during the origin and evolution of living organisms. No other chemical element can form molecules of such widely different sizes, shapes, and composition.

Most biomolecules can be regarded as derivatives of hydrocarbons, with hydrogen atoms replaced by a variety of functional groups that confer specific chemical properties on the molecule, forming various families of organic compounds. Typical of these are alcohols, which have one or more hydroxyl groups; amines, with amino groups; aldehydes and ketones, with carbonyl groups; and carboxylic acids, with carboxyl groups **(Fig. 1–16)**. Many biomolecules are polyfunctional, containing two or more types of functional groups

$$\cdot\overset{\cdot}{\underset{\cdot}{C}}\cdot + \cdot H \longrightarrow \cdot\overset{\cdot}{\underset{\cdot}{C}}{:}H \qquad -\overset{|}{\underset{|}{C}}-H$$

$$\cdot\overset{\cdot}{\underset{\cdot}{C}}\cdot + \cdot\overset{\cdot\cdot}{\underset{\cdot\cdot}{O}}{:} \longrightarrow \cdot\overset{\cdot}{\underset{\cdot}{C}}{:}\overset{\cdot\cdot}{\underset{\cdot\cdot}{O}}\cdot \qquad -\overset{|}{\underset{|}{C}}-O-$$

$$\cdot\overset{\cdot}{\underset{\cdot}{C}}\cdot + \cdot\overset{\cdot\cdot}{\underset{\cdot}{O}}{:} \longrightarrow \cdot\overset{\cdot}{\underset{\cdot}{C}}{::}\overset{\cdot\cdot}{\underset{\cdot\cdot}{O}}{:} \qquad \overset{}{\underset{}{>}}C=O$$

$$\cdot\overset{\cdot}{\underset{\cdot}{C}}\cdot + \cdot\overset{\cdot}{\underset{\cdot}{N}}{:} \longrightarrow \cdot\overset{\cdot}{\underset{\cdot}{C}}{:}\overset{\cdot}{\underset{\cdot}{N}}{:} \qquad -\overset{|}{\underset{|}{C}}-N\overset{}{\underset{}{<}}$$

$$\cdot\overset{\cdot}{\underset{\cdot}{C}}\cdot + \cdot\overset{\cdot}{\underset{\cdot}{N}}{:} \longrightarrow \cdot\overset{\cdot}{\underset{\cdot}{C}}{::}\overset{\cdot}{\underset{\cdot}{N}}\cdot \qquad \overset{}{\underset{}{>}}C=N-$$

$$\cdot\overset{\cdot}{\underset{\cdot}{C}}\cdot + \cdot\overset{\cdot}{\underset{\cdot}{C}}\cdot \longrightarrow \cdot\overset{\cdot}{\underset{\cdot}{C}}{:}\overset{\cdot}{\underset{\cdot}{C}}\cdot \qquad -\overset{|}{\underset{|}{C}}-\overset{|}{\underset{|}{C}}-$$

$$\cdot\overset{\cdot}{\underset{\cdot}{C}}\cdot + \cdot\overset{\cdot}{\underset{\cdot}{C}}\cdot \longrightarrow \cdot\overset{\cdot}{\underset{\cdot}{C}}{::}\overset{\cdot}{\underset{\cdot}{C}}\cdot \qquad \overset{}{\underset{}{>}}C=C\overset{}{\underset{}{<}}$$

$$\cdot\overset{\cdot}{\underset{\cdot}{C}}\cdot + \cdot\overset{\cdot}{\underset{\cdot}{C}}\cdot \longrightarrow \cdot C{:::}C\cdot \qquad -C\equiv C-$$

FIGURE 1-14 Versatility of carbon bonding. Carbon can form covalent single, double, and triple bonds (all bonds in red), particularly with other carbon atoms. Triple bonds are rare in biomolecules.

FIGURE 1–15 Geometry of carbon bonding. (a) Carbon atoms have a characteristic tetrahedral arrangement of their four single bonds. **(b)** Carbon–carbon single bonds have freedom of rotation, as shown for the compound ethane (CH_3—CH_3). **(c)** Double bonds are shorter and do not allow free rotation. The two doubly bonded carbons and the atoms designated A, B, X, and Y all lie in the same rigid plane.

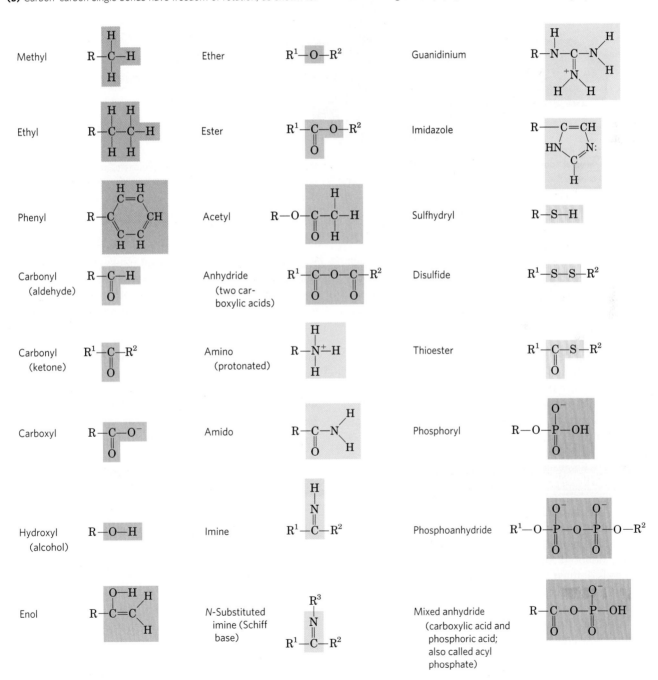

FIGURE 1–16 Some common functional groups of biomolecules. Functional groups are screened with a color typically used to represent the element that characterizes the group: gray for C, red for O, blue for N, yellow for S, and orange for P. In this figure and throughout the book, we use R to represent "any substituent." It may be as simple as a hydrogen atom, but typically it is a carbon-containing group. When two or more substituents are shown in a molecule, we designate them R^1, R^2, and so forth.

FIGURE 1–17 Several common functional groups in a single biomolecule. Acetyl-coenzyme A (often abbreviated as acetyl-CoA) is a carrier of acetyl groups in some enzymatic reactions. Its functional groups are screened in the structural formula. As we will see in Chapter 2, several of these functional groups can exist in protonated or unprotonated forms, depending on the pH. In the space-filling model, N is blue, C is black, P is orange, O is red, and H is white. The yellow atom at the left is the sulfur of the critical thioester bond between the acetyl moiety and coenzyme A.

(Fig. 1–17), each with its own chemical characteristics and reactions. The chemical "personality" of a compound is determined by the chemistry of its functional groups and their disposition in three-dimensional space.

Cells Contain a Universal Set of Small Molecules

Dissolved in the aqueous phase (cytosol) of all cells is a collection of perhaps a thousand different small organic molecules (M_r ~100 to ~500), with intracellular concentrations ranging from nanomolar to millimolar (see Fig. 15–4). (See Box 1–1 for an explanation of the various ways of referring to molecular weight.) These are the central metabolites in the major pathways occurring in nearly every cell—the metabolites and pathways that have been conserved throughout the course of evolution. This collection of molecules includes the common amino acids, nucleotides, sugars and their phosphorylated derivatives, and mono-, di-, and tricarboxylic acids. The molecules may be polar or charged, and are water-soluble. They are trapped in the cell because the plasma membrane is impermeable to them, although

---

BOX 1–1   Molecular Weight, Molecular Mass, and Their Correct Units

There are two common (and equivalent) ways to describe molecular mass; both are used in this text. The first is *molecular weight*, or *relative molecular mass*, denoted M_r. The molecular weight of a substance is defined as the ratio of the mass of a molecule of that substance to one-twelfth the mass of carbon-12 (^{12}C). Since M_r is a ratio, it is dimensionless—it has no associated units. The second is *molecular mass*, denoted m. This is simply the mass of one molecule, or the molar mass divided by Avogadro's number. The molecular mass, m, is expressed in daltons (abbreviated Da). One dalton is equivalent to one-twelfth the mass of carbon-12; a kilodalton (kDa) is 1,000 daltons; a megadalton (MDa) is 1 million daltons.

Consider, for example, a molecule with a mass 1,000 times that of water. We can say of this molecule either $M_r = 18,000$ or $m = 18,000$ daltons. We can also describe it as an "18 kDa molecule." However, the expression $M_r = 18,000$ daltons is incorrect.

Another convenient unit for describing the mass of a single atom or molecule is the atomic mass unit (formerly amu, now commonly denoted u). One atomic mass unit (1 u) is defined as one-twelfth the mass of an atom of carbon-12. Since the experimentally measured mass of an atom of carbon-12 is 1.9926×10^{-23} g, $1\ u = 1.6606 \times 10^{-24}$ g. The atomic mass unit is convenient for describing the mass of a peak observed by mass spectrometry (see Chapter 3, p. 100).

specific membrane transporters can catalyze the movement of some molecules into and out of the cell or between compartments in eukaryotic cells. The universal occurrence of the same set of compounds in living cells reflects the evolutionary conservation of metabolic pathways that developed in the earliest cells.

There are other small biomolecules, specific to certain types of cells or organisms. For example, vascular plants contain, in addition to the universal set, small molecules called **secondary metabolites**, which play roles specific to plant life. These metabolites include compounds that give plants their characteristic scents and colors, and compounds such as morphine, quinine, nicotine, and caffeine that are valued for their physiological effects on humans but used for other purposes by plants.

The entire collection of small molecules in a given cell under a specific set of conditions has been called the **metabolome**, in parallel with the term "genome." **Metabolomics** is the systematic characterization of the metabolome under very specific conditions (such as following administration of a drug or a biological signal such as insulin).

Macromolecules Are the Major Constituents of Cells

Many biological molecules are **macromolecules**, polymers with molecular weights above ~5,000 that are assembled from relatively simple precursors. Shorter polymers are called **oligomers** (Greek *oligos*, "few"). Proteins, nucleic acids, and polysaccharides are macromolecules composed of monomers with molecular weights of 500 or less. Synthesis of macromolecules is a major energy-consuming activity of cells. Macromolecules themselves may be further assembled into supramolecular complexes, forming functional units such as ribosomes. Table 1–1 shows the major classes of biomolecules in an *E. coli* cell.

Proteins, long polymers of amino acids, constitute the largest fraction (besides water) of a cell. Some proteins have catalytic activity and function as enzymes; others serve as structural elements, signal receptors, or transporters that carry specific substances into or out of cells. Proteins are perhaps the most versatile of all biomolecules; a catalog of their many functions would be very long. The sum of all the proteins functioning in a given cell is the cell's **proteome**, and **proteomics** is the systematic characterization of this protein complement under a specific set of conditions. The **nucleic acids**, DNA and RNA, are polymers of nucleotides. They store and transmit genetic information, and some RNA molecules have structural and catalytic roles in supramolecular complexes. The **genome** is the entire sequence of a cell's DNA (or in the case of RNA viruses, its RNA), and **genomics** is the characterization of the comparative structure, function, evolution, and mapping of genomes. The **polysaccharides**, polymers of simple sugars such as glucose, have three major functions: as energy-rich fuel stores, as rigid structural components of cell walls (in plants and bacteria), and as extracellular recognition elements that bind to proteins on other cells. Shorter polymers of sugars (oligosaccharides) attached to proteins or lipids at the cell surface serve as specific cellular signals. A cell's **glycome** is all its carbohydrate-containing molecules. The **lipids**, water-insoluble hydrocarbon derivatives, serve as structural components of membranes, energy-rich fuel stores, pigments, and intracellular signals. The lipid-containing molecules in a cell constitute its **lipidome**. With the application of sensitive methods with great resolving power (mass spectrometry, for example), it is possible to distinguish and quantify hundreds or thousands of these components, and therefore to quantify their variations in response to changing conditions, signals, or drugs. Systems biology is an approach that tries to integrate the information from genomics, proteomics, glycomics, and lipidomics to give a molecular picture of all the activities of a cell under a given set of conditions, and the changes that occur when the system is perturbed by external signals or circumstances or by mutations.

Proteins, polynucleotides, and polysaccharides have large numbers of monomeric subunits and thus high molecular weights—in the range of 5,000 to more than 1 million for proteins, up to several *billion* for nucleic acids, and in the millions for polysaccharides such as starch. Individual lipid molecules are much smaller (M_r 750 to 1,500) and are not classified as macromolecules. But they can associate noncovalently into very large structures. Cellular membranes are built of enormous noncovalent aggregates of lipid and protein molecules.

Given their characteristic information-rich subunit sequences, proteins and nucleic acids are often referred to as **informational macromolecules**. Some oligosaccharides, as noted above, also serve as informational molecules.

TABLE 1–1	Molecular Components of an *E. coli* Cell	
	Percentage of total weight of cell	Approximate number of different molecular species
Water	70	1
Proteins	15	3,000
Nucleic acids		
DNA	1	1–4
RNA	6	>3,000
Polysaccharides	3	10
Lipids	2	20
Monomeric subunits and intermediates	2	500
Inorganic ions	1	20

Three-Dimensional Structure Is Described by Configuration and Conformation

The covalent bonds and functional groups of a biomolecule are, of course, central to its function, but so also is the arrangement of the molecule's constituent atoms in three-dimensional space—its stereochemistry. Carbon-containing compounds commonly exist as **stereoisomers**, molecules with the same chemical bonds and same chemical formula but different **configuration**, the fixed spatial arrangement of atoms. Interactions between biomolecules are invariably **stereospecific**, requiring specific configurations in the interacting molecules.

Figure 1–18 shows three ways to illustrate the stereochemistry, or configuration, of simple molecules. The perspective diagram specifies stereochemistry unambiguously, but bond angles and center-to-center bond lengths are better represented with ball-and-stick models. In space-filling models, the radius of each "atom" is proportional to its van der Waals radius, and the contours of the model define the space occupied by the molecule (the volume of space from which atoms of other molecules are excluded).

Configuration is conferred by the presence of either (1) double bonds, around which there is little or no freedom of rotation, or (2) chiral centers, around which

(a)   **(b)**   **(c)**

FIGURE 1–18 Representations of molecules. Three ways to represent the structure of the amino acid alanine (shown here in the ionic form found at neutral pH). **(a)** Structural formula in perspective form: a solid wedge (—) represents a bond in which the atom at the wide end projects out of the plane of the paper, toward the reader; a dashed wedge (····) represents a bond extending behind the plane of the paper. **(b)** Ball-and-stick model, showing bond angles and relative bond lengths. **(c)** Space-filling model, in which each atom is shown with its correct relative van der Waals radius.

substituent groups are arranged in a specific orientation. The identifying characteristic of stereoisomers is that they cannot be interconverted without temporarily breaking one or more covalent bonds. **Figure 1–19a** shows the configurations of maleic acid and its isomer, fumaric acid. These compounds are **geometric**

FIGURE 1–19 Configurations of geometric isomers. (a) Isomers such as maleic acid (maleate at pH 7) and fumaric acid (fumarate) cannot be interconverted without breaking covalent bonds, which requires the input of much more energy than the average kinetic energy of molecules at physiological temperatures. **(b)** In the vertebrate retina, the initial event in light detection is the absorption of visible light by 11-*cis*-retinal. The energy of the absorbed light (about 250 kJ/mol) converts 11-*cis*-retinal to all-*trans*-retinal, triggering electrical changes in the retinal cell that lead to a nerve impulse. (Note that the hydrogen atoms are omitted from the ball-and-stick models of the retinals.)

Maleic acid (cis)   Fumaric acid (trans)

(a)

11-*cis*-Retinal   All-*trans*-Retinal

(b)

FIGURE 1–20 Molecular asymmetry: chiral and achiral molecules. (a) When a carbon atom has four different substituent groups (A, B, X, Y), they can be arranged in two ways that represent nonsuperposable mirror images of each other (enantiomers). This asymmetric carbon atom is called a chiral atom or chiral center. **(b)** When a tetrahedral carbon has only three dissimilar groups (i.e., the same group occurs twice), only one configuration is possible and the molecule is symmetric, or achiral. In this case the molecule is superposable on its mirror image: the molecule on the left can be rotated counterclockwise (when looking down the vertical bond from A to C) to create the molecule in the mirror.

isomers, or **cis-trans isomers**; they differ in the arrangement of their substituent groups with respect to the nonrotating double bond (Latin *cis*, "on this side"— groups on the same side of the double bond; *trans*, "across"—groups on opposite sides). Maleic acid (maleate at the neutral pH of cytoplasm) is the cis isomer and fumaric acid (fumarate) the trans isomer; each is a well-defined compound that can be separated from the other, and each has its own unique chemical properties. A binding site (on an enzyme, for example) that is complementary to one of these molecules would not be complementary to the other, which explains why the two compounds have distinct biological roles despite their similar chemical makeup.

In the second type of stereoisomer, four different substituents bonded to a tetrahedral carbon atom may be arranged in two different ways in space—that is, have two configurations **(Fig. 1–20)**—yielding two stereoisomers that have similar or identical chemical properties but differ in certain physical and biological properties. A carbon atom with four different substituents is said to be asymmetric, and asymmetric carbons are called **chiral centers** (Greek *chiros*, "hand"; some stereoisomers are related structurally as the right hand is to the left). A molecule with only one chiral carbon can have two stereoisomers; when two or more (n) chiral carbons are present, there can be 2^n stereoisomers. Stereoisomers that are mirror images of each other are called **enantiomers** (Fig. 1–20). Pairs of stereoisomers that are not mirror images of each other are called **diastereomers (Fig. 1–21)**.

As Louis Pasteur first observed in 1843 (Box 1–2), enantiomers have nearly identical chemical reactivities but differ in a characteristic physical property: their interaction with plane-polarized light. In separate solutions, two enantiomers rotate the plane of plane-polarized light in opposite directions, but an equimolar solution of the two enantiomers (a **racemic mixture**) shows no optical rotation. Compounds without chiral centers do not rotate the plane of plane-polarized light.

FIGURE 1–21 Enantiomers and diastereomers. There are four different stereoisomers of 2,3-disubstituted butane ($n = 2$ asymmetric carbons, hence $2^n = 4$ stereoisomers). Each is shown in a box as a perspective formula and a ball-and-stick model, which has been rotated to allow the reader to view all the groups. Two pairs of stereoisomers are mirror images of each other, or enantiomers. All other possible pairs are not mirror images and so are diastereomers.

BOX 1–2 Louis Pasteur and Optical Activity: *In Vino, Veritas*

Louis Pasteur encountered the phenomenon of **optical activity** in 1843, during his investigation of the crystalline sediment that accumulated in wine casks (a form of tartaric acid called paratartaric acid—also called racemic acid, from Latin *racemus*, "bunch of grapes"). He used fine forceps to separate two types of crystals identical in shape but mirror images of each other. Both types proved to have all the chemical properties of tartaric acid, but in solution one type rotated plane-polarized light to the left (levorotatory), the other to the right (dextrorotatory). Pasteur later described the experiment and its interpretation:

Louis Pasteur, 1822-1895

> In isomeric bodies, the elements and the proportions in which they are combined are the same, only the arrangement of the atoms is different . . . We know, on the one hand, that the molecular arrangements of the two tartaric acids are asymmetric, and, on the other hand, that these arrangements are absolutely identical, excepting that they exhibit asymmetry in opposite directions. Are the atoms of the dextro acid grouped in the form of a right-handed spiral, or are they placed at the apex of an irregular tetrahedron, or are they disposed according to this or that asymmetric arrangement? We do not know.*

Now we do know. X-ray crystallographic studies in 1951 confirmed that the levorotatory and dextrorotatory forms of tartaric acid are mirror images of each other at the molecular level and established the absolute configuration of each (Fig. 1). The same approach has been used to demonstrate that although the amino acid alanine has two stereoisomeric forms (designated D and L), alanine in proteins exists exclusively in one form (the L isomer; see Chapter 3).

$$HOOC^1\underset{H\quad OH}{\overset{2}{C}}-\underset{H\quad OH}{\overset{3}{C}}\ ^4COOH$$

(2R,3R)-Tartaric acid
(dextrorotatory)

$$HOOC^1\underset{HO\quad H}{\overset{2}{C}}-\underset{H\quad OH}{\overset{3}{C}}\ ^4COOH$$

(2S,3S)-Tartaric acid
(levorotatory)

FIGURE 1 Pasteur separated crystals of two stereoisomers of tartaric acid and showed that solutions of the separated forms rotated plane-polarized light to the same extent but in opposite directions. These dextrorotatory and levorotatory forms were later shown to be the (R,R) and (S,S) isomers represented here. The RS system of nomenclature is explained in the text.

*From Pasteur's lecture to the Société Chimique de Paris in 1883, quoted in DuBos, R. (1976) *Louis Pasteur: Free Lance of Science*, p. 95, Charles Scribner's Sons, New York.

KEY CONVENTION: Given the importance of stereochemistry in reactions between biomolecules (see below), biochemists must name and represent the structure of each biomolecule so that its stereochemistry is unambiguous. For compounds with more than one chiral center, the most useful system of nomenclature is the RS system. In this system, each group attached to a chiral carbon is assigned a *priority*. The priorities of some common substituents are

—OCH$_3$ > —OH > —NH$_2$ > —COOH >
—CHO > —CH$_2$OH > —CH$_3$ > —H

For naming in the RS system, the chiral atom is viewed with the group of lowest priority (4 in the following diagram) pointing away from the viewer. If the priority of the other three groups (1 to 3) decreases in clockwise order, the configuration is (*R*) (Latin *rectus*, "right"); if counterclockwise, the configuration is (*S*) (Latin *sinister*, "left"). In this way each chiral carbon is designated either (*R*) or (*S*), and the inclusion of these designations in the name of the compound provides an unambiguous description of the stereochemistry at each chiral center.

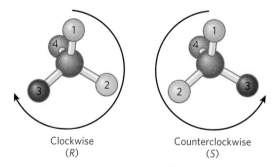

Clockwise
(R)

Counterclockwise
(S)

Another naming system for stereoisomers, the D and L system, is described in Chapter 3. A molecule with a single chiral center (the two isomers of glyceraldehyde, for example) can be named unambiguously by either system.

$$\underset{\text{L-Glyceraldehyde}}{HO-\overset{\overset{\displaystyle CHO}{|}}{\underset{\underset{\displaystyle CH_2OH}{|}}{C}}-H}$$

≡

$$\underset{\text{(S)-Glyceraldehyde}}{\overset{CHO_{(2)}}{\underset{CH_2OH_{(3)}}{H_{(4)}}}-OH_{(1)}}$$

■

Distinct from configuration is molecular **conformation**, the spatial arrangement of substituent groups that, without breaking any bonds, are free to assume different positions in space because of the freedom of rotation about single bonds. In the simple hydrocarbon ethane, for example, there is nearly complete freedom of rotation around the C—C bond. Many different, interconvertible conformations of ethane are possible, depending on the degree of rotation (**Fig. 1–22**). Two conformations are of special interest: the staggered, which is more stable than all others and thus predominates, and the eclipsed, which is least stable. We cannot isolate either of these conformational forms, because they are freely interconvertible. However, when one or more of the hydrogen atoms on each carbon is replaced by a functional group that is either very large or electrically charged, freedom of rotation around the C—C bond is hindered. This limits the number of stable conformations of the ethane derivative.

Interactions between Biomolecules Are Stereospecific

When biomolecules interact, the "fit" between them must be stereochemically correct. The three-dimensional structure of biomolecules large and small—the combination of configuration and conformation—is of the utmost importance in their biological interactions: reactant with its enzyme, hormone with its receptor on a cell surface, antigen with its specific antibody, for example (**Fig. 1–23**). The study of biomolecular stereochemistry, with precise physical methods, is an important part of modern research on cell structure and biochemical function.

In living organisms, chiral molecules are usually present in only one of their chiral forms. For example, the

FIGURE 1–23 Complementary fit between a macromolecule and a small molecule. A glucose molecule fits into a pocket on the surface of the enzyme hexokinase (PDB ID 3B8A), and is held in this orientation by several noncovalent interactions between the protein and the sugar. This representation of the hexokinase molecule is produced with software that can calculate the shape of the outer surface of a macromolecule, defined either by the van der Waals radii of all the atoms in the molecule or by the "solvent exclusion volume," the volume a water molecule cannot penetrate.

amino acids in proteins occur only as their L isomers; glucose occurs only as its D isomer. (The conventions for naming stereoisomers of the amino acids are described in Chapter 3; those for sugars, in Chapter 7. The RS system, described above, is the most useful for some biomolecules.) In contrast, when a compound with an asymmetric carbon atom is chemically synthesized in the laboratory, the reaction usually produces all possible chiral forms: a mixture of the D and L forms, for example. Living cells produce only one chiral form of a biomolecule because the enzymes that synthesize that molecule are also chiral.

Stereospecificity, the ability to distinguish between stereoisomers, is a property of enzymes and other proteins and a characteristic feature of the molecular logic of living cells. If the binding site on a protein is complementary to one isomer of a chiral compound, it will not be complementary to the other isomer, for the same reason that a left glove does not fit a right hand. Some striking examples of the ability of biological systems to distinguish stereoisomers are shown in **Figure 1–24**.

The common classes of chemical reactions encountered in biochemistry are described in Chapter 13, as an introduction to the reactions of metabolism.

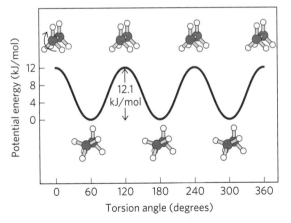

FIGURE 1–22 Conformations. Many conformations of ethane are possible because of freedom of rotation around the C—C bond. In the ball-and-stick model, when the front carbon atom (as viewed by the reader) with its three attached hydrogens is rotated relative to the rear carbon atom, the potential energy of the molecule rises to a maximum in the fully eclipsed conformation (torsion angle 0°, 120°, etc.), then falls to a minimum in the fully staggered conformation (torsion angle 60°, 180°, etc.). Because the energy differences are small enough to allow rapid interconversion of the two forms (millions of times per second), the eclipsed and staggered forms cannot be separately isolated.

SUMMARY 1.2 Chemical Foundations

▶ Because of its bonding versatility, carbon can produce a broad array of carbon–carbon skeletons with a variety of functional groups; these groups give biomolecules their biological and chemical personalities.

▶ A nearly universal set of about a thousand small molecules is found in living cells; the interconversions of these molecules in the central

(a)

(b)

(c)

FIGURE 1–24 Stereoisomers have different effects in humans. (a) Two stereoisomers of carvone: (R)-carvone (isolated from spearmint oil) has the characteristic fragrance of spearmint; (S)-carvone (from caraway seed oil) smells like caraway. **(b)** Aspartame, the artificial sweetener sold under the trade name NutraSweet, is easily distinguishable by taste receptors from its bitter-tasting stereoisomer, although the two differ only in the configuration at one of the two chiral carbon atoms. **(c)** The antidepressant medication citalopram (trade name Celexa), a selective serotonin reuptake inhibitor, is a racemic mixture of these two steroisomers, but only (S)-citalopram has the therapeutic effect. A stereochemically pure preparation of (S)-citalopram (escitalopram oxalate) is sold under the trade name Lexapro. As you might predict, the effective dose of Lexapro is one-half the effective dose of Celexa.

metabolic pathways have been conserved in evolution.

▶ Proteins and nucleic acids are linear polymers of simple monomeric subunits; their sequences contain the information that gives each molecule its three-dimensional structure and its biological functions.

▶ Molecular configuration can be changed only by breaking covalent bonds. For a carbon atom with four different substituents (a chiral carbon), the substituent groups can be arranged in two different ways, generating stereoisomers with distinct properties. Only one stereoisomer is biologically active. Molecular conformation is the position of atoms in space that can be changed by rotation about single bonds, without breaking covalent bonds.

▶ Interactions between biological molecules are almost invariably stereospecific: they require a close fit between complementary structures in the interacting molecules.

1.3 Physical Foundations

Living cells and organisms must perform work to stay alive and to reproduce themselves. The synthetic reactions that occur within cells, like the synthetic processes in any factory, require the input of energy. Energy input is also needed in the motion of a bacterium or an Olympic sprinter, in the flashing of a firefly or the electrical discharge of an eel. And the storage and expression of information require energy, without which structures rich in information inevitably become disordered and meaningless.

In the course of evolution, cells have developed highly efficient mechanisms for coupling the energy obtained from sunlight or chemical fuels to the many energy-requiring processes they must carry out. One goal of biochemistry is to understand, in quantitative and chemical terms, the means by which energy is extracted, stored, and channeled into useful work in living cells. We can consider cellular energy conversions—like all other energy conversions—in the context of the laws of thermodynamics.

Living Organisms Exist in a Dynamic Steady State, Never at Equilibrium with Their Surroundings

The molecules and ions contained within a living organism differ in kind and in concentration from those in the organism's surroundings. A paramecium in a pond, a shark in the ocean, a bacterium in the soil, an apple tree in an orchard—all are different in composition from their surroundings and, once they have reached maturity, maintain a more or less constant composition in the face of constantly changing surroundings.

Although the characteristic composition of an organism changes little through time, the population of molecules within the organism is far from static. Small molecules, macromolecules, and supramolecular complexes are continuously synthesized and broken down in chemical reactions that involve a constant flux of mass and energy through the system. The hemoglobin molecules carrying oxygen from your lungs to your brain at this moment were synthesized within the past month; by next month they will have been degraded and entirely replaced by new hemoglobin molecules. The glucose you ingested with your most recent meal is now circulating in your bloodstream; before the day is over these particular glucose molecules will have been converted into something else—carbon dioxide or fat, perhaps—and will have been replaced with a fresh supply of glucose, so that your blood glucose concentration is more or less constant over the whole day. The amounts of hemoglobin and glucose in the blood remain nearly constant because the rate of synthesis or intake of each just balances the rate of its breakdown, consumption, or conversion into some other product. The constancy of concentration is the result of a *dynamic steady state*, a steady state that is far from equilibrium. Maintaining this steady state requires the constant investment of energy; when a cell can no longer obtain energy, it dies and begins to decay toward equilibrium with its surroundings. We consider below exactly what is meant by "steady state" and "equilibrium."

Organisms Transform Energy and Matter from Their Surroundings

For chemical reactions occurring in solution, we can define a **system** as all the constituent reactants and products, the solvent that contains them, and the immediate atmosphere—in short, everything within a defined region of space. The system and its surroundings together constitute the **universe**. If the system exchanges neither matter nor energy with its surroundings, it is said to be **isolated**. If the system exchanges energy but not matter with its surroundings, it is a **closed** system; if it exchanges both energy and matter with its surroundings, it is an **open** system.

A living organism is an open system; it exchanges both matter and energy with its surroundings. Organisms obtain energy from their surroundings in two ways: (1) they take up chemical fuels (such as glucose) from the environment and extract energy by oxidizing them (see Box 1–3, Case 2); or (2) they absorb energy from sunlight.

The first law of thermodynamics describes the principle of the conservation of energy: *in any physical or chemical change, the total amount of energy in the universe remains constant, although the form of the energy may change.* This means that while energy is "used" by a system, it is not "used up"; rather it is converted from one form into another—from potential energy in chemical bonds, say, into kinetic energy of heat and motion. Cells are consummate transducers of energy, capable of interconverting chemical, electromagnetic, mechanical, and osmotic energy with great efficiency **(Fig. 1–25)**.

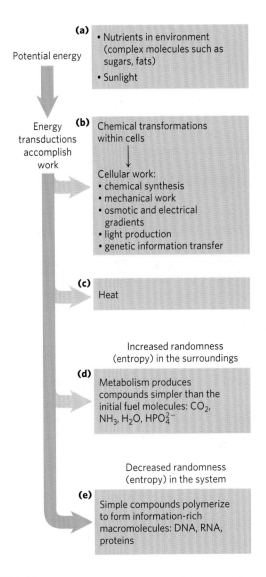

FIGURE 1–25 Some energy transformations in living organisms. As metabolic energy is spent to do cellular work, the randomness of the system plus surroundings (expressed quantitatively as entropy) increases as the potential energy of complex nutrient molecules decreases. **(a)** Living organisms extract energy from their surroundings; **(b)** convert some of it into useful forms of energy to produce work; **(c)** return some energy to the surroundings as heat; and **(d)** release end-product molecules that are less well organized than the starting fuel, increasing the entropy of the universe. One effect of all these transformations is **(e)** increased order (decreased randomness) in the system in the form of complex macromolecules. We return to a quantitative treatment of entropy in Chapter 13.

BOX 1–3   Entropy: Things Fall Apart

The term "entropy," which literally means "a change within," was first used in 1851 by Rudolf Clausius, one of the formulators of the second law of thermodynamics. It refers to the randomness or disorder of the components of a chemical system. Entropy is a central concept in biochemistry; life requires continual maintenance of order in the face of nature's tendency to increase randomness. A rigorous quantitative definition of entropy involves statistical and probability considerations. However, its nature can be illustrated qualitatively by three simple examples, each demonstrating one aspect of entropy. The key descriptors of entropy are *randomness* and *disorder*, manifested in different ways.

Case 1: The Teakettle and the Randomization of Heat

We know that steam generated from boiling water can do useful work. But suppose we turn off the burner under a teakettle full of water at 100 °C (the "system") in the kitchen (the "surroundings") and allow the teakettle to cool. As it cools, no work is done, but heat passes from the teakettle to the surroundings, raising the temperature of the surroundings (the kitchen) by an infinitesimally small amount until complete equilibrium is attained. At this point all parts of the teakettle and the kitchen are at precisely the same temperature. The free energy that was once concentrated in the teakettle of hot water at 100 °C, *potentially* capable of doing work, has disappeared. Its equivalent in heat energy is still present in the teakettle + kitchen (i.e., the "universe") but has become completely randomized throughout. This

energy is no longer available to do work because there is no temperature differential within the kitchen. Moreover, the increase in entropy of the kitchen (the surroundings) is irreversible. We know from everyday experience that heat never spontaneously passes back from the kitchen into the teakettle to raise the temperature of the water to 100 °C again.

Case 2: The Oxidation of Glucose

Entropy is a state not only of energy but of matter. Aerobic (heterotrophic) organisms extract free energy from glucose obtained from their surroundings by oxidizing the glucose with O_2, also obtained from the surroundings. The end products of this oxidative metabolism, CO_2 and H_2O, are returned to the surroundings. In this process the surroundings undergo an increase in entropy, whereas the organism itself remains in a steady state and undergoes no change in its internal order. Although some entropy arises from the dissipation of heat, entropy also arises from another kind of disorder, illustrated by the equation for the oxidation of glucose:

$$C_6H_{12}O_6 + 6O_2 \longrightarrow 6CO_2 + 6H_2O$$

We can represent this schematically as

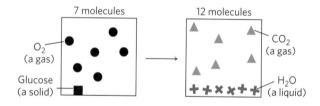

The Flow of Electrons Provides Energy for Organisms

Nearly all living organisms derive their energy, directly or indirectly, from the radiant energy of sunlight. In the photoautotrophs, light-driven splitting of water during photosynthesis releases its electrons for the reduction of CO_2 and the release of O_2 into the atmosphere:

$$6CO_2 + 6H_2O \xrightarrow{\text{light}} C_6H_{12}O_6 + 6O_2$$
(light-driven reduction of CO_2)

Nonphotosynthetic organisms (chemotrophs) obtain the energy they need by oxidizing the energy-rich products of photosynthesis stored in plants, then passing the electrons thus acquired to atmospheric O_2 to form water, CO_2, and other end products, which are recycled in the environment:

$$C_6H_{12}O_6 + 6O_2 \longrightarrow 6CO_2 + 6H_2O + \text{energy}$$
(energy-yielding oxidation of glucose)

Thus autotrophs and heterotrophs participate in global cycles of O_2 and CO_2, driven ultimately by sunlight, mak-

ing these two large groups of organisms interdependent. Virtually all energy transductions in cells can be traced to this flow of electrons from one molecule to another, in a "downhill" flow from higher to lower electrochemical potential; as such, this is formally analogous to the flow of electrons in a battery-driven electric circuit. All these reactions involved in electron flow are **oxidation-reduction reactions**: one reactant is oxidized (loses electrons) as another is reduced (gains electrons).

Creating and Maintaining Order Requires Work and Energy

As we've noted, DNA, RNA, and proteins are informational macromolecules; the precise sequence of their monomeric subunits contains information, just as the letters in this sentence do. In addition to using chemical energy to form the covalent bonds between these subunits, the cell must invest energy to order the subunits in their correct sequence. It is extremely improbable that amino acids in a mixture would spontaneously condense into a single type of protein, with a unique sequence. This would represent increased order in a population of molecules; but

The atoms contained in 1 molecule of glucose plus 6 molecules of oxygen, a total of 7 molecules, are more randomly dispersed by the oxidation reaction and are now present in a total of 12 molecules ($6CO_2 + 6H_2O$).

Whenever a chemical reaction results in an increase in the number of molecules—or when a solid substance is converted into liquid or gaseous products, which allow more freedom of molecular movement than solids—molecular disorder, and thus entropy, increases.

Case 3: Information and Entropy

The following short passage from *Julius Caesar*, Act IV, Scene 3, is spoken by Brutus, when he realizes that he must face Mark Antony's army. It is an information-rich nonrandom arrangement of 125 letters of the English alphabet:

> There is a tide in the affairs of men,
> Which, taken at the flood, leads on to fortune;
> Omitted, all the voyage of their life
> Is bound in shallows and in miseries.

In addition to what this passage says overtly, it has many hidden meanings. It not only reflects a complex sequence of events in the play, it also echoes the play's ideas on conflict, ambition, and the demands of leadership. Permeated with Shakespeare's understanding of human nature, it is very rich in information.

However, if the 125 letters making up this quotation were allowed to fall into a completely random, chaotic pattern, as shown in the following box, they would have no meaning whatsoever.

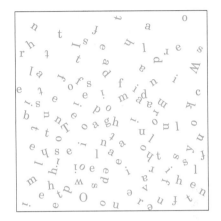

In this form the 125 letters contain little or no information, but they are very rich in entropy. Such considerations have led to the conclusion that information is a form of energy; information has been called "negative entropy." In fact, the branch of mathematics called information theory, which is basic to the programming logic of computers, is closely related to thermodynamic theory. Living organisms are highly ordered, nonrandom structures, immensely rich in information and thus entropy-poor.

according to the second law of thermodynamics, the tendency in nature is toward ever-greater disorder in the universe: *the total entropy of the universe is continually increasing*. To bring about the synthesis of macromolecules from their monomeric units, free energy must be supplied to the system (in this case, the cell).

KEY CONVENTION: The randomness or disorder of the components of a chemical system is expressed as **entropy**, **S** (Box 1–3). Any change in randomness of the system is expressed as entropy change, ΔS, which by convention has a positive value when randomness increases. J. Willard Gibbs, who developed the theory of energy changes during chemical reactions, showed that the **free-energy content**, **G**, of any closed system can be defined in terms of three quantities: **enthalpy**, **H**, reflecting the number and kinds of bonds;

J. Willard Gibbs, 1839–1903

entropy, S; and the absolute temperature, T (in Kelvin). The definition of free energy is $G = H - TS$. When a chemical reaction occurs at constant temperature, the **free-energy change**, ΔG, is determined by the enthalpy change, ΔH, reflecting the kinds and numbers of chemical bonds and noncovalent interactions broken and formed, and the entropy change, ΔS, describing the change in the system's randomness:

$$\Delta G = \Delta H - T\Delta S$$

where, by definition, ΔH is negative for a reaction that releases heat, and ΔS is positive for a reaction that increases the system's randomness. ■

A process tends to occur spontaneously only if ΔG is negative (if free energy is *released* in the process). Yet cell function depends largely on molecules, such as proteins and nucleic acids, for which the free energy of formation is positive: the molecules are less stable and more highly ordered than a mixture of their monomeric components. To carry out these thermodynamically unfavorable, energy-requiring (**endergonic**) reactions, cells couple them to other reactions that liberate free

(P)—(P)—(P)—Adenosine (Adenosine *tri*phosphate, ATP)

Inorganic phosphate (P$_i$)

Inorganic pyrophosphate (PP$_i$)

(P)—(P)—OH + (P)—(P)—Adenosine (Adenosine *di*phosphate, ADP)

(P)—Adenosine (Adenosine *mono*phosphate, AMP)

FIGURE 1-26 Adenosine triphosphate (ATP) provides energy. Here, each (P) represents a phosphoryl group. The removal of the terminal phosphoryl group (shaded light red) of ATP, by breakage of a phosphoanhydride bond to generate adenosine diphosphate (ADP) and inorganic phosphate ion (HPO$_4^{2-}$), is highly exergonic, and this reaction is coupled to many endergonic reactions in the cell (as in the example in Fig. 1-27b). ATP also provides energy for many cellular processes by undergoing cleavage that releases the two terminal phosphates as inorganic pyrophosphate (H$_2$P$_2$O$_7^{2-}$), often abbreviated PP$_i$.

energy (**exergonic** reactions), so that the overall process is exergonic: the *sum* of the free-energy changes is negative.

The usual source of free energy in coupled biological reactions is the energy released by breakage of phosphoanhydride bonds such as those in adenosine triphosphate (ATP; **Fig. 1–26**) and guanosine triphosphate (GTP). Here, each (P) represents a phosphoryl group:

Amino acids → protein ΔG_1 is positive (endergonic)
ATP → AMP + (P)—(P) ΔG_2 is negative (exergonic)
  [or ATP → ADP + (P)]

When these reactions are coupled, the sum of ΔG_1 and ΔG_2 is negative—the overall process is exergonic. By this coupling strategy, cells are able to synthesize and maintain the information-rich polymers essential to life.

Energy Coupling Links Reactions in Biology

The central issue in *bioenergetics* (the study of energy transformations in living systems) is the means by which energy from fuel metabolism or light capture is coupled to a cell's energy-requiring reactions. In think-

ing about energy coupling, it is useful to consider a simple mechanical example, as shown in **Figure 1–27a**. An object at the top of an inclined plane has a certain amount of potential energy as a result of its elevation. It tends to slide down the plane, losing its potential energy of position as it approaches the ground. When an appropriate string-and-pulley device couples the falling object to another, smaller object, the spontaneous downward

(a) Mechanical example

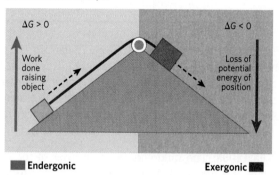

■ Endergonic   Exergonic ■

FIGURE 1-27 Energy coupling in mechanical and chemical processes. **(a)** The downward motion of an object releases potential energy that can do mechanical work. The potential energy made available by spontaneous downward motion, an exergonic process (red), can be coupled to the endergonic upward movement of another object (blue). **(b)** In reaction 1, the formation of glucose 6-phosphate from glucose and inorganic phosphate (P$_i$) yields a product of higher energy than the two reactants. For this endergonic reaction, ΔG is positive. In reaction 2, the exergonic breakdown of adenosine triphosphate (ATP) has a large, negative free-energy change (ΔG_2). The third reaction is the sum of reactions 1 and 2, and the free-energy change, ΔG_3, is the arithmetic sum of ΔG_1 and ΔG_2. Because ΔG_3 is negative, the overall reaction is exergonic and proceeds spontaneously.

(b) Chemical example

Reaction coordinate

motion of the larger can lift the smaller, accomplishing a certain amount of work. The amount of energy available to do work is the **free-energy change, ΔG**; this is always somewhat less than the theoretical amount of energy released, because some energy is dissipated as the heat of friction. The greater the elevation of the larger object, the greater the energy released (ΔG) as the object slides downward and the greater the amount of work that can be accomplished. The larger object can lift the smaller only because, at the outset, the larger object was *far from its equilibrium position*: it had at some earlier point been elevated above the ground, in a process that itself required the input of energy.

How does this apply in chemical reactions? In closed systems, chemical reactions proceed spontaneously until **equilibrium** is reached. When a system is at equilibrium, the rate of product formation exactly equals the rate at which product is converted to reactant. Thus there is no net change in the concentration of reactants and products. The energy change as the system moves from its initial state to equilibrium, with no changes in temperature or pressure, is given by the free-energy change, ΔG. The magnitude of ΔG depends on the particular chemical reaction *and on how far from equilibrium the system is initially*. Each compound involved in a chemical reaction contains a certain amount of potential energy, related to the kind and number of its bonds. In reactions that occur spontaneously, the products have less free energy than the reactants, thus the reaction releases free energy, which is then available to do work. Such reactions are exergonic; the decline in free energy from reactants to products is expressed as a negative value. Endergonic reactions require an input of energy, and their ΔG values are positive. As in mechanical processes, only part of the energy released in exergonic chemical reactions can be used to accomplish work. In living systems some energy is dissipated as heat or lost to increasing entropy.

K_{eq} and $\Delta G°$ Are Measures of a Reaction's Tendency to Proceed Spontaneously

The tendency of a chemical reaction to go to completion can be expressed as an equilibrium constant. For the reaction in which a moles of A react with b moles of B to give c moles of C and d moles of D,

$$aA + bB \longrightarrow cC + dD$$

the equilibrium constant, K_{eq}, is given by

$$K_{eq} = \frac{[C]_{eq}^{c}[D]_{eq}^{d}}{[A]_{eq}^{a}[B]_{eq}^{b}}$$

where $[A]_{eq}$ is the concentration of A, $[B]_{eq}$ the concentration of B, and so on, *when the system has reached equilibrium*. A large value of K_{eq} means the reaction tends to proceed until the reactants are almost completely converted into the products.

WORKED EXAMPLE 1–1   Are ATP and ADP at Equilibrium in Cells?

The equilibrium constant, K_{eq}, for the following reaction is 2×10^5 M:

$$ATP \longrightarrow ADP + HPO_4^{2-}$$

If the measured cellular concentrations are [ATP] = 5 mM, [ADP] = 0.5 mM, and [P$_i$] = 5 mM, is this reaction at equilibrium in living cells?

Solution: The definition of the equilibrium constant for this reaction is:

$$K_{eq} = [ADP][P_i]/[ATP]$$

From the measured cellular concentrations given above, we can calculate the mass-action ratio, Q:

$$Q = [ADP][P_i]/[ATP] = [0.5 \text{ mM}][5 \text{ mM}]/[5 \text{ mM}]$$
$$= 0.5 \text{ mM} = 5 \times 10^{-4} \text{ M}$$

This value is *far* from the equilibrium constant for the reaction (2×10^5 M), so the reaction is *very far* from equilibrium in cells. [ATP] is far higher, and [ADP] is far lower, than is expected at equilibrium. How can a cell hold its [ATP]/[ADP] ratio so far from equilibrium? It does so by continuously extracting energy (from nutrients such as glucose) and using it to make ATP from ADP and P$_i$.

WORKED EXAMPLE 1–2   Is the Hexokinase Reaction at Equilibrium in Cells?

For the reaction catalyzed by the enzyme hexokinase:

$$\text{Glucose} + ATP \longrightarrow \text{glucose 6-phosphate} + ADP$$

the equilibrium constant, K_{eq}, is 7.8×10^2. In living *E. coli* cells, [ATP] = 5 mM; [ADP] = 0.5 mM; [glucose] = 2 mM; and [glucose 6-phosphate] = 1 mM. Is the reaction at equilibrium in *E. coli*?

Solution: At equilibrium,

$$K_{eq} = 7.8 \times 10^2 = [ADP][\text{glucose 6-phosphate}]/[ATP][\text{glucose}]$$

In living cells, [ADP][glucose 6-phosphate]/[ATP][glucose] = [0.5 mM][1 mM]/[5 mM][2 mM] = 0.05. The reaction is therefore *far* from equilibrium: the cellular concentrations of products (glucose 6-phosphate and ADP) are much lower than expected at equilibrium, and those of the reactants are much higher. The reaction therefore tends strongly to go to the right.

Gibbs showed that ΔG (the actual free-energy change) for any chemical reaction is a function of the **standard free-energy change, $\Delta G°$**—a constant that is characteristic of each specific reaction—and a term that expresses the initial concentrations of reactants and products:

$$\Delta G = \Delta G^\circ + RT \ln \frac{[C]_i^c [D]_i^d}{[A]_i^a [B]_i^b} \qquad (1\text{–}1)$$

where $[A]_i$ is the initial concentration of A, and so forth; R is the gas constant; and T is the absolute temperature.

ΔG is a measure of the distance of a system from its equilibrium position. When a reaction has reached equilibrium, no driving force remains and it can do no work: $\Delta G = 0$. For this special case, $[A]_i = [A]_{eq}$, and so on, for all reactants and products, and

$$\frac{[C]_i^c [D]_i^d}{[A]_i^a [B]_i^b} = \frac{[C]_{eq}^c [D]_{eq}^d}{[A]_{eq}^a [B]_{eq}^b}$$

Substituting 0 for ΔG and K_{eq} for $[C]_i^c [D]_i^d / [A]_i^a [B]_i^b$ in Equation 1–1, we obtain the relationship

$$\Delta G^\circ = -RT \ln K_{eq}$$

from which we see that ΔG° is simply a second way (besides K_{eq}) of expressing the driving force on a reaction. Because K_{eq} is experimentally measurable, we have a way of determining ΔG°, the thermodynamic constant characteristic of each reaction.

The units of ΔG° and ΔG are joules per mole (or calories per mole). When $K_{eq} \gg 1$, ΔG° is large and negative; when $K_{eq} \ll 1$, ΔG° is large and positive. From a table of experimentally determined values of either K_{eq} or ΔG°, we can see at a glance which reactions tend to go to completion and which do not.

One caution about the interpretation of ΔG°: *thermodynamic* constants such as this show where the final equilibrium for a reaction lies but tell us nothing about how fast that equilibrium will be achieved. The rates of reactions are governed by the parameters of *kinetics*, a topic we consider in detail in Chapter 6.

In biological organisms, just as in the mechanical example in Figure 1–27a, an exergonic reaction can be coupled to an endergonic reaction to drive otherwise unfavorable reactions. Figure 1–27b (a type of graph called a reaction coordinate diagram) illustrates this principle for the conversion of glucose to glucose 6-phosphate, the first step in the pathway for oxidation of glucose. The simplest way to produce glucose 6-phosphate would be:

Reaction 1:   Glucose + P_i ⟶ glucose 6-phosphate

(endergonic; ΔG_1 is positive)

(Here, P_i is an abbreviation for inorganic phosphate, HPO_4^{2-}. Don't be concerned about the structure of these compounds now; we describe them in detail later in the book.) This reaction does not occur spontaneously; ΔG_1 is positive. A second, very exergonic reaction can occur in all cells:

Reaction 2:   ATP ⟶ ADP + P_i

(exergonic; ΔG_2 is negative)

These two chemical reactions share a common intermediate, P_i, which is consumed in reaction 1 and produced in reaction 2. The two reactions can therefore be coupled in the form of a third reaction, which we can write as the sum of reactions 1 and 2, with the common intermediate, P_i, omitted from both sides of the equation:

Reaction 3:   Glucose + ATP ⟶

glucose 6-phosphate + ADP

Because more energy is released in reaction 2 than is consumed in reaction 1, the free-energy change for reaction 3, ΔG_3, is negative, and the synthesis of glucose 6-phosphate can therefore occur by reaction 3.

WORKED EXAMPLE 1–3   **Standard Free-Energy Changes Are Additive**

Given that the standard free-energy change for the reaction glucose + P_i ⟶ glucose 6-phosphate is 13.8 kJ/mol, and the standard free-energy change for the reaction ATP ⟶ ADP + P_i is −30.5 kJ/mol, what is the free-energy change for the reaction glucose + ATP ⟶ glucose 6-phosphate + ADP?

Solution: We can write the equation for this reaction as the sum of two other reactions:

(1) Glucose + P_i ⟶ glucose 6-phosphate $\Delta G_1^\circ = 13.8$ kJ/mol

(2) ATP ⟶ ADP + P_i $\Delta G_2^\circ = -30.5$ kJ/mol

Sum: Glucose + ATP ⟶ glucose 6-phosphate + ADP

$\Delta G_{Sum}^\circ = -16.7$ kJ/mol

The standard free-energy change for two reactions that sum to a third is simply the sum of the two individual reactions. A negative value for ΔG° (−16.7 kJ/mol) indicates that the reaction will tend to occur spontaneously.

The coupling of exergonic and endergonic reactions through a shared intermediate is central to the energy exchanges in living systems. As we shall see, reactions that break down ATP (such as reaction 2 in Fig. 1–27b) release energy that drives many endergonic processes in cells. ATP breakdown in cells is exergonic because *all living cells maintain a concentration of ATP far above its equilibrium concentration*. It is this disequilibrium that allows ATP to serve as the major carrier of chemical energy in all cells. As we shall see in more detail in Chapter 13, it is not the mere breakdown of ATP that provides energy to drive endergonic reactions; rather, it is the *transfer of a phosphoryl group* from ATP to another small molecule (glucose in the case above) that conserves some of the chemical potential originally in ATP.

FIGURE 1–28 Energy changes during a chemical reaction. An activation barrier, representing the transition state (see Chapter 6), must be overcome in the conversion of reactants (A) into products (B), even though the products are more stable than the reactants, as indicated by a large, negative free-energy change (ΔG). The energy required to overcome the activation barrier is the activation energy ($\Delta G^{\ddagger}$). Enzymes catalyze reactions by lowering the activation barrier. They bind the transition-state intermediates tightly, and the binding energy of this interaction effectively reduces the activation energy from $\Delta G^{\ddagger}_{uncat}$ (blue curve) to $\Delta G^{\ddagger}_{cat}$ (red curve). (Note that activation energy is *not* related to free-energy change, ΔG.)

WORKED EXAMPLE 1–4   Energetic Cost of ATP Synthesis

If the equilibrium constant, K_{eq}, for the reaction

$$\text{ATP} \longrightarrow \text{ADP} + \text{P}_i$$

is 2.22×10^5 M, calculate the standard free-energy change, $\Delta G°$, for the *synthesis* of ATP from ADP and P_i at 25 °C.

Solution: First calculate $\Delta G°$ for the reaction above.

$$\begin{aligned}
\Delta G° &= -RT \ln K_{eq} \\
&= -(8.315 \text{ J/mol} \cdot \text{K})(298 \text{ K})(\ln 2.22 \times 10^5) \\
&= -30.5 \text{ kJ/mol}
\end{aligned}$$

This is the standard free-energy change for the *breakdown* of ATP to ADP and P_i. The standard free-energy change for the *reverse* of a reaction has the same absolute value, but the opposite sign. The standard free-energy change for the reverse of the above reaction is therefore 30.5 kJ/mol. So, to synthesize 1 mol of ATP under standard conditions (25 °C, 1 M concentrations of ATP, ADP, P_i), at least 30.5 kJ of energy must be supplied. The actual free-energy change in cells—approximately 50 kJ/mol—is greater than this because the concentrations of ATP, ADP, and P_i in cells are not the standard 1 M (see Worked Example 13–2, p. 519).

WORKED EXAMPLE 1–5   Standard Free-Energy Change for Synthesis of Glucose 6-Phosphate

What is the standard free-energy change, $\Delta G°$, under physiological conditions (*E. coli* grows in the human gut, at 37 °C) for the following reaction?

$$\text{Glucose} + \text{ATP} \longrightarrow \text{glucose 6-phosphate} + \text{ADP}$$

Solution: We have the relationship $\Delta G° = -RT \ln K_{eq}$ and the value of K_{eq} for this reaction, 7.8×10^2. Substituting the values of R, T, and K_{eq} into this equation gives:

$$\Delta G° = -(8.315 \text{ J/mol} \cdot \text{K})(310 \text{ K})(\ln 7.8 \times 10^2) = -17 \text{ kJ/mol}$$

Notice that this value is slightly different from that in Worked Example 1–3. In that calculation we assumed a temperature of 25 °C (298 K), whereas in this calculation we used the physiological temperature of 37 °C (310 K).

Enzymes Promote Sequences of Chemical Reactions

All biological macromolecules are much less thermodynamically stable than their monomeric subunits, yet they are *kinetically stable*: their *uncatalyzed* breakdown occurs so slowly (over years rather than seconds) that, on a time scale that matters for the organism, these molecules are stable. Virtually every chemical reaction in a cell occurs at a significant rate only because of the presence of **enzymes**—biocatalysts that, like all other catalysts, greatly enhance the rate of specific chemical reactions without being consumed in the process.

The path from reactant(s) to product(s) almost invariably involves an energy barrier, called the activation barrier **(Fig. 1–28)**, that must be surmounted for any reaction to proceed. The breaking of existing bonds and formation of new ones generally requires, first, a distortion of the existing bonds to create a **transition state** of higher free energy than either reactant or product. The highest point in the reaction coordinate diagram represents the transition state, and the difference in energy between the reactant in its ground state and in its transition state is the **activation energy**, $\Delta G^{\ddagger}$. An enzyme catalyzes a reaction by providing a more comfortable fit for the transition state: a surface that complements the transition state in stereochemistry, polarity, and charge. The binding of enzyme to the transition state is exergonic, and the energy released by this binding reduces the activation energy for the reaction and greatly increases the reaction rate.

A further contribution to catalysis occurs when two or more reactants bind to the enzyme's surface close to each other and with stereospecific orientations that favor the reaction. This increases by orders of magnitude the probability of productive collisions between reactants. As a result of these factors and several others, discussed in Chapter 6, enzyme-catalyzed reactions commonly proceed at rates greater than 10^{12} times faster than the uncatalyzed reactions. (That is a *million million* times faster!)

Cellular catalysts are, with some notable exceptions, proteins. (Some RNA molecules have enzymatic activity, as discussed in Chapters 26 and 27.) Again with a few exceptions, each enzyme catalyzes a specific reaction, and each reaction in a cell is catalyzed by a different enzyme. Thousands of different enzymes are therefore required by each cell. The multiplicity of enzymes, their specificity (the ability to discriminate between reactants), and their susceptibility to regulation give cells the capacity to lower activation barriers selectively. This selectivity is crucial for the effective regulation of cellular processes. By allowing specific reactions to proceed at significant rates at particular times, enzymes determine how matter and energy are channeled into cellular activities.

The thousands of enzyme-catalyzed chemical reactions in cells are functionally organized into many sequences of consecutive reactions, called **pathways**, in which the product of one reaction becomes the reactant in the next. Some pathways degrade organic nutrients into simple end products in order to extract chemical energy and convert it into a form useful to the cell; together these degradative, free-energy-yielding reactions are designated **catabolism**. The energy released by catabolic reactions drives the synthesis of ATP. As a result, the cellular concentration of ATP is far above its equilibrium concentration, so that ΔG for ATP breakdown is large and negative. Similarly, catabolism results in the production of the reduced electron carriers NADH and NADPH, both of which can donate electrons in processes that generate ATP or drive reductive steps in biosynthetic pathways.

Other pathways start with small precursor molecules and convert them to progressively larger and more complex molecules, including proteins and nucleic acids. Such synthetic pathways, which invariably require the input of energy, are collectively designated **anabolism**. The overall network of enzyme-catalyzed pathways, both catabolic and anabolic, constitutes cellular **metabolism**. ATP (and the energetically equivalent nucleoside triphosphates cytidine triphosphate (CTP), uridine triphosphate (UTP), and guanosine triphosphate (GTP)) is the connecting link between the catabolic and anabolic components of this network (shown schematically in **Fig. 1–29**). The pathways of enzyme-catalyzed reactions that act on the main constituents of cells—proteins, fats, sugars, and nucleic acids—are virtually identical in all living organisms.

Metabolism Is Regulated to Achieve Balance and Economy

Not only do living cells simultaneously synthesize thousands of different kinds of carbohydrate, fat, protein, and nucleic acid molecules and their simpler subunits, but they do so in the precise proportions required by the cell under any given circumstance. For example, during rapid cell growth the precursors of proteins and nucleic acids must be made in large quantities, whereas in nongrowing cells the requirement for these precur-

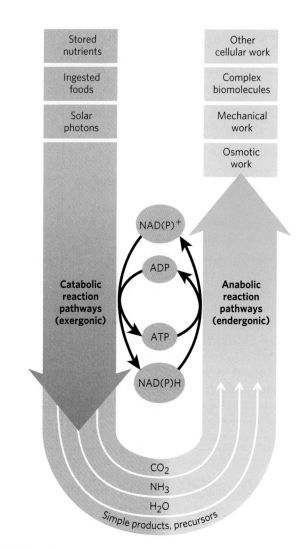

FIGURE 1–29 The central roles of ATP and NAD(P)H in metabolism. ATP is the shared chemical intermediate linking energy-releasing and energy-consuming cellular processes. Its role in the cell is analogous to that of money in an economy: it is "earned/produced" in exergonic reactions and "spent/consumed" in endergonic ones. NAD(P)H (nicotinamide adenine dinucleotide (phosphate)) is an electron-carrying cofactor that collects electrons from oxidative reactions and then donates them in a wide variety of reduction reactions in biosynthesis. Present in relatively low concentrations, these cofactors essential to anabolic reactions must be constantly regenerated by catabolic reactions.

sors is much reduced. Key enzymes in each metabolic pathway are regulated so that each type of precursor molecule is produced in a quantity appropriate to the current requirements of the cell.

Consider the pathway in *E. coli* that leads to the synthesis of the amino acid isoleucine, a constituent of proteins. The pathway has five steps catalyzed by five different enzymes (A through F represent the intermediates in the pathway):

If a cell begins to produce more isoleucine than it needs for protein synthesis, the unused isoleucine accumulates and the increased concentration inhibits the catalytic activity of the first enzyme in the pathway, immediately slowing the production of isoleucine. Such **feedback inhibition** keeps the production and utilization of each metabolic intermediate in balance. (Throughout the book, we will use ⊗ to indicate inhibition of an enzymatic reaction.)

Although the concept of discrete pathways is an important tool for organizing our understanding of metabolism, it is an oversimplification. There are thousands of metabolic intermediates in a cell, many of which are part of more than one pathway. Metabolism would be better represented as a web of interconnected and interdependent pathways. A change in the concentration of any one metabolite would start a ripple effect, influencing the flow of materials through other pathways. The task of understanding these complex interactions among intermediates and pathways in quantitative terms is daunting, but the new emphasis on **systems biology**, discussed in Chapter 15, has begun to offer important insights into the overall regulation of metabolism.

Cells also regulate the synthesis of their own catalysts, the enzymes, in response to increased or diminished need for a metabolic product; this is the substance of Chapter 28. The expression of genes (the translation from information in DNA to active protein in the cell) and synthesis of enzymes are other layers of metabolic control in the cell. All layers must be taken into account when describing the overall control of cellular metabolism.

SUMMARY 1.3 Physical Foundations

▶ Living cells are open systems, exchanging matter and energy with their surroundings, extracting and channeling energy to maintain themselves in a dynamic steady state distant from equilibrium. Energy is obtained from sunlight or chemical fuels by converting the energy from electron flow into the chemical bonds of ATP.

▶ The tendency for a chemical reaction to proceed toward equilibrium can be expressed as the free-energy change, ΔG, which has two components: enthalpy change, ΔH, and entropy change, ΔS. These variables are related by the equation $\Delta G = \Delta H - T\Delta S$.

▶ When ΔG of a reaction is negative, the reaction is exergonic and tends to go toward completion; when ΔG is positive, the reaction is endergonic and tends to go in the reverse direction. When two reactions can be summed to yield a third reaction, the ΔG for this overall reaction is the sum of the ΔGs of the two separate reactions.

▶ The reactions converting ATP to P_i and ADP or to AMP and PP_i are highly exergonic (large negative ΔG). Many endergonic cellular reactions are driven by coupling them, through a common intermediate, to these highly exergonic reactions.

▶ The standard free-energy change for a reaction, $\Delta G°$, is a physical constant that is related to the equilibrium constant by the equation $\Delta G° = -RT \ln K_{eq}$.

▶ Most cellular reactions proceed at useful rates only because enzymes are present to catalyze them. Enzymes act in part by stabilizing the transition state, reducing the activation energy, $\Delta G^{\ddagger}$, and increasing the reaction rate by many orders of magnitude. The catalytic activity of enzymes in cells is regulated.

▶ Metabolism is the sum of many interconnected reaction sequences that interconvert cellular metabolites. Each sequence is regulated to provide what the cell needs at a given time and to expend energy only when necessary.

1.4 Genetic Foundations

Perhaps the most remarkable property of living cells and organisms is their ability to reproduce themselves for countless generations with nearly perfect fidelity. This continuity of inherited traits implies constancy, over millions of years, in the structure of the molecules that contain the genetic information. Very few historical records of civilization, even those etched in copper or carved in stone **(Fig. 1–30)**, have survived for a thousand years. But there is good evidence that the genetic instructions in living organisms have remained nearly unchanged over very much longer periods; many bacteria have nearly the same size, shape, and internal structure and contain the same kinds of precursor molecules and enzymes as bacteria that lived nearly four billion years ago. This continuity of structure and composition is the result of continuity in the structure of the genetic material.

Among the seminal discoveries in biology in the twentieth century were the chemical nature and the three-dimensional structure of the genetic material, **deoxyribonucleic acid, DNA**. The sequence of the monomeric subunits, the nucleotides (strictly, deoxyribonucleotides, as discussed below), in this linear polymer encodes the instructions for forming all other cellular components and provides a template for the production of identical DNA molecules to be distributed to progeny when a cell divides. The perpetuation of a biological species requires that its genetic information be maintained in a stable form, expressed accurately in the form of gene products, and reproduced with a minimum of errors. Effective storage, expression, and reproduction of the genetic message define individual species, distinguish them from one another, and assure their continuity over successive generations.

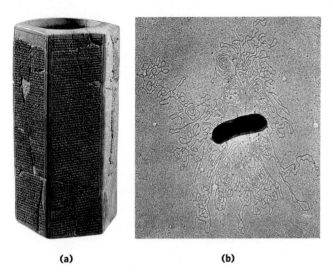

(a)   **(b)**

FIGURE 1–30 Two ancient scripts. (a) The Prism of Sennacherib, inscribed in about 700 BCE, describes in characters of the Assyrian language some historical events during the reign of King Sennacherib. The Prism contains about 20,000 characters, weighs about 50 kg, and has survived almost intact for about 2,700 years. **(b)** The single DNA molecule of the bacterium *E. coli*, leaking out of a disrupted cell, is hundreds of times longer than the cell itself and contains all the encoded information necessary to specify the cell's structure and functions. The bacterial DNA contains about 4.6 million characters (nucleotides), weighs less than 10^{-10} g, and has undergone only relatively minor changes during the past several million years. (The yellow spots and dark specks in this colorized electron micrograph are artifacts of the preparation.)

Genetic Continuity Is Vested in Single DNA Molecules

DNA is a long, thin, organic polymer, the rare molecule that is constructed on the atomic scale in one dimension (width) and the human scale in another (length: a molecule of DNA can be many centimeters long). A human sperm or egg, carrying the accumulated hereditary information of billions of years of evolution, transmits this inheritance in the form of DNA molecules, in which the linear sequence of covalently linked nucleotide subunits encodes the genetic message.

Usually when we describe the properties of a chemical species, we describe the average behavior of a very large number of identical molecules. While it is difficult to predict the behavior of any single molecule in a collection of, say, a picomole (about 6×10^{11} molecules) of a compound, the *average* behavior of the molecules is predictable because so many molecules enter into the average. Cellular DNA is a remarkable exception. The DNA that is the entire genetic material of *E. coli* is a *single molecule* containing 4.64 million nucleotide pairs. That single molecule must be replicated perfectly in every detail if an *E. coli* cell is to give rise to identical progeny by cell division; there is no room for averaging in this process! The same is true for all cells. A human sperm brings to the egg that it fertilizes just one molecule of DNA in each of its 23 different chromosomes, to combine with just one DNA molecule in each corresponding chromosome in the egg.

The result of this union is very highly predictable: an embryo with all of its ~25,000 genes, constructed of 3 billion nucleotide pairs, intact. An amazing chemical feat.

WORKED EXAMPLE 1–6   **Fidelity of DNA Replication**

Calculate the number of times the DNA of a modern *E. coli* cell has been copied accurately since its earliest bacterial precursor cell arose about 3.5 billion years ago. Assume for simplicity that over this time period *E. coli* has undergone, on average, one cell division every 12 hours (this is an overestimate for modern bacteria, but probably an underestimate for ancient bacteria).

Solution:

$$(1 \text{ generation}/12 \text{ hr})(24 \text{ hr/d})(365 \text{ d/yr})(3.5 \times 10^9 \text{ yr})$$
$$= 2.6 \times 10^{12} \text{ generations.}$$

A single page of this book contains about 5,000 characters, so the entire book contains about 5 million characters. The chromosome of *E. coli* also contains about 5 million characters (nucleotide pairs). If you made a handwritten copy of this book and then passed on the copy to a classmate to copy by hand, and this copy were then copied by a third classmate, and so on, how closely would each successive copy of the book resemble the original? Now, imagine the textbook that would result from hand-copying this one a few trillion times!

The Structure of DNA Allows for Its Replication and Repair with Near-Perfect Fidelity

The capacity of living cells to preserve their genetic material and to duplicate it for the next generation results from the structural complementarity between the two strands of the DNA molecule **(Fig. 1–31)**. The basic unit of DNA is a linear polymer of four different monomeric subunits, **deoxyribonucleotides**, arranged in a precise linear sequence. It is this linear sequence that encodes the genetic information. Two of these polymeric strands are twisted about each other to form the DNA double helix, in which each deoxyribonucleotide in one strand pairs specifically with a complementary deoxyribonucleotide in the opposite strand. Before a cell divides, the two DNA strands separate and each serves as a template for the synthesis of a new, complementary strand, generating two identical double-helical molecules, one for each daughter cell. If either strand is damaged at any time, continuity of information is assured by the information present in the other strand, which can act as a template for repair of the damage.

The Linear Sequence in DNA Encodes Proteins with Three-Dimensional Structures

The information in DNA is encoded in its linear (one-dimensional) sequence of deoxyribonucleotide subunits,

FIGURE 1–32 DNA to RNA to protein to enzyme (hexokinase). The linear sequence of deoxyribonucleotides in the DNA (the gene) that encodes the protein hexokinase is first transcribed into a ribonucleic acid (RNA) molecule with the complementary ribonucleotide sequence. The RNA sequence (messenger RNA) is then translated into the linear protein chain of hexokinase, which folds into its native three-dimensional shape, most likely aided by molecular chaperones. Once in its native form, hexokinase acquires its catalytic activity: it can catalyze the phosphorylation of glucose, using ATP as the phosphoryl group donor.

FIGURE 1–31 Complementarity between the two strands of DNA. DNA is a linear polymer of covalently joined deoxyribonucleotides of four types: deoxyadenylate (A), deoxyguanylate (G), deoxycytidylate (C), and deoxythymidylate (T). Each nucleotide, with its unique three-dimensional structure, can associate very specifically but noncovalently with one other nucleotide in the complementary chain: A always associates with T, and G with C. Thus, in the double-stranded DNA molecule, the entire sequence of nucleotides in one strand is *complementary* to the sequence in the other. The two strands, held together by hydrogen bonds (represented here by vertical light blue lines) between each pair of complementary nucleotides, twist about each other to form the DNA double helix. In DNA replication, the two strands (blue) separate and two new strands (pink) are synthesized, each with a sequence complementary to one of the original strands. The result is two double-helical molecules, each identical to the original DNA.

but the expression of this information results in a three-dimensional cell. This change from one to three dimensions occurs in two phases. A linear sequence of deoxyribonucleotides in DNA codes (through an intermediary, RNA) for the production of a protein with a corresponding linear sequence of amino acids **(Fig. 1–32)**. The protein folds into a particular three-dimensional shape,

determined by its amino acid sequence and stabilized primarily by noncovalent interactions. Although the final shape of the folded protein is dictated by its amino acid sequence, the folding is aided by "molecular chaperones" (see Fig. 4–30). The precise three-dimensional structure, or **native conformation**, of the protein is crucial to its function.

Once in its native conformation, a protein may associate noncovalently with other macromolecules (other proteins, nucleic acids, or lipids) to form supramolecular complexes such as chromosomes, ribosomes, and membranes. The individual molecules of these complexes have specific, high-affinity binding sites for each other, and within the cell they spontaneously self-assemble into functional complexes.

Although the amino acid sequences of proteins carry all necessary information for achieving the proteins' native conformation, accurate folding and self-assembly also require the right cellular environment—pH, ionic strength, metal ion concentrations, and so forth. Thus DNA sequence alone is not enough to form and maintain a fully functioning cell.

SUMMARY 1.4 Genetic Foundations

▶ Genetic information is encoded in the linear sequence of four types of deoxyribonucleotides in DNA.

▶ The double-helical DNA molecule contains an internal template for its own replication and repair.

▶ DNA molecules are extraordinarily large, with molecular weights in the millions or billions.

▶ Despite the enormous size of DNA, the sequence of nucleotides in it is very precise, and the maintenance of this precise sequence over very long times is the basis for genetic continuity in organisms.

▶ The linear sequence of amino acids in a protein, which is encoded in the DNA of the gene for that protein, produces a protein's unique three-dimensional structure—a process also dependent on environmental conditions.

▶ Individual macromolecules with specific affinity for other macromolecules self-assemble into supramolecular complexes.

1.5 Evolutionary Foundations

Nothing in biology makes sense except in the light of evolution.

—*Theodosius Dobzhansky,* The American Biology Teacher,
March 1973

Great progress in biochemistry and molecular biology in recent decades has amply confirmed the validity of Dobzhansky's striking generalization. The remarkable similarity of metabolic pathways and gene sequences across the three domains of life argues strongly that all modern organisms are derived from a common evolutionary progenitor by a series of small changes (mutations), each of which conferred a selective advantage to some organism in some ecological niche.

Changes in the Hereditary Instructions Allow Evolution

Despite the near-perfect fidelity of genetic replication, infrequent unrepaired mistakes in the DNA replication process lead to changes in the nucleotide sequence of DNA, producing a genetic **mutation** and changing the instructions for a cellular component. Incorrectly repaired damage to one of the DNA strands has the same effect. Mutations in the DNA handed down to offspring—that is, mutations carried in the reproductive cells—may be harmful or even lethal to the new organism or cell; they may, for example, cause the synthesis of a defective enzyme that is not able to catalyze an essential metabolic reaction. Occasionally, however, a mutation *better* equips an organism or cell to survive in its environment **(Fig. 1–33)**. The mutant enzyme might have acquired a slightly different specificity, for example, so that it is now able to use some compound that the cell was previously

FIGURE 1–33 **Gene duplication and mutation: one path to generate new enzymatic activities.** In this example, the single hexokinase gene in a hypothetical organism might occasionally, by accident, be copied twice during DNA replication, such that the organism has two full copies of the gene, one of which is superfluous. Over many generations, as the DNA with two hexokinase genes is repeatedly duplicated, rare mistakes occur, leading to changes in the nucleotide sequence of the superfluous gene and thus of the protein that it encodes. In a few very rare cases, the altered protein produced from this mutant gene can bind a new substrate—galactose in our hypothetical case. The cell containing the mutant gene has acquired a new capability (metabolism of galactose), which may allow it to survive in an ecological niche that provides galactose but not glucose. If no gene duplication precedes mutation, the original function of the gene product is lost.

unable to metabolize. If a population of cells were to find itself in an environment where that compound was the only or the most abundant available source of fuel, the mutant cell would have a selective advantage over the other, unmutated (**wild-type**) cells in the population. The mutant cell and its progeny would survive and prosper in the new environment, whereas wild-type cells would starve and be eliminated. This is what Darwin meant by natural selection—what is sometimes summarized as "survival of the fittest."

Occasionally, a second copy of a whole gene is introduced into the chromosome as a result of defective replication of the chromosome. The second copy is superfluous, and mutations in this gene will not be deleterious; it becomes a means by which the cell may evolve, by producing a new gene with a new function while retaining the original gene and gene function. Seen in this light, the DNA molecules of modern organisms are historical documents, records of the long journey from the earliest cells to modern organisms. The historical accounts in DNA are not complete, however; in the course of evolution, many mutations must have been erased or written over. But DNA molecules are the best source of biological history that we have. The frequency of errors in DNA replication represents a balance between too many errors, which would yield nonviable daughter cells, and too few, which would prevent the genetic variation that allows survival of mutant cells in new ecological niches.

Several billion years of natural selection have refined cellular systems to take maximum advantage of the chemical and physical properties of available raw materials. Chance genetic mutations occurring in individuals in a population, combined with natural selection, have resulted in the evolution of the enormous variety of species we see today, each adapted to its particular ecological niche.

Biomolecules First Arose by Chemical Evolution

In our account thus far we have passed over the first chapter of the story of evolution: the appearance of the first living cell. Apart from their occurrence in living organisms, organic compounds, including the basic biomolecules such as amino acids and carbohydrates, are found in only trace amounts in the Earth's crust, the sea, and the atmosphere. How did the first living organisms acquire their characteristic organic building blocks? According to one hypothesis, these compounds were created by the effects of powerful environmental forces—ultraviolet irradiation, lightning, or volcanic eruptions—on the gases in the prebiotic Earth's atmosphere, and on inorganic solutes in superheated thermal vents deep in the ocean.

This hypothesis was tested in a classic experiment on the abiotic (nonbiological) origin of organic biomolecules carried out in 1953 by Stanley Miller in the laboratory of Harold Urey. Miller subjected gaseous mixtures such as those presumed to exist on the prebiotic Earth,

including NH_3, CH_4, H_2O, and H_2, to electrical sparks produced across a pair of electrodes (to simulate lightning) for periods of a week or more, then analyzed the contents of the closed reaction vessel **(Fig. 1–34)**. The gas phase of the resulting mixture contained CO and CO_2 as well as the starting materials. The water phase contained a variety of organic compounds, including some amino acids, hydroxy acids, aldehydes, and hydrogen cyanide (HCN).

(a)

(b)

FIGURE 1–34 Abiotic production of biomolecules. (a) Spark-discharge apparatus of the type used by Miller and Urey in experiments demonstrating abiotic formation of organic compounds under primitive atmospheric conditions. After subjection of the gaseous contents of the system to electrical sparks, products were collected by condensation. Biomolecules such as amino acids were among the products. **(b)** Stanley L. Miller (1930–2007) using his spark-discharge apparatus.

This experiment established the possibility of abiotic production of biomolecules in relatively short times under relatively mild conditions. When Miller's carefully stored samples were rediscovered in 2010 and examined with much more sensitive and discriminating techniques (high-performance liquid chromatography and mass spectrometry), his original observations were confirmed and greatly broadened. Previously unpublished experiments by Miller that included H_2S in the gas mixture (mimicking the "smoking" volcanic plumes at the sea bottom; **Fig. 1–35**) showed the formation of 23 amino acids and 7 organosulfur compounds, as well as a large number of other simple compounds that might have served as building blocks in prebiotic evolution.

More-refined laboratory experiments have provided good evidence that many of the chemical components of living cells, including polypeptides and RNA-like molecules, can form under these conditions. Polymers of RNA can act as catalysts in biologically significant reactions (see Chapters 26 and 27), and RNA probably played a crucial role in prebiotic evolution, both as catalyst and as information repository.

RNA or Related Precursors May Have Been the First Genes and Catalysts

In modern organisms, nucleic acids encode the genetic information that specifies the structure of enzymes, and enzymes catalyze the replication and repair of nucleic acids. The mutual dependence of these two classes of biomolecules brings up the perplexing question: which came first, DNA or protein?

The answer may be that they appeared about the same time, and RNA preceded them both. The discovery that RNA molecules can act as catalysts in their own formation suggests that RNA or a similar molecule may have been the first gene *and* the first catalyst. According to this scenario **(Fig. 1–36)**, one of the earliest stages of biological evolution was the chance formation of an RNA molecule that could catalyze the formation of other RNA molecules of the same sequence—a self-replicating, self-perpetuating RNA. The concentration of a self-replicating RNA molecule would increase exponentially, as one molecule formed several, several formed many, and so on. The fidelity of self-replication was presumably less than perfect, so the process would generate variants of the RNA, some of which might be even better able to self-replicate. In the competition for nucleotides, the most efficient of the self-replicating sequences would win, and less efficient replicators would fade from the population.

FIGURE 1–35 Black smokers. Hydrothermal vents in the sea floor emit superheated water rich in dissolved minerals. Black "smoke" is formed when the vented solution meets cold sea water and dissolved sulfides precipitate. Diverse life forms, including a variety of archaea and some remarkably complex multicellular animals, are found in the immediate vicinity of such vents, which may have been the sites of early biogenesis.

Prebiotic formation of simple compounds, including nucleotides, from components of Earth's primitive atmosphere or gases in undersea volcanic vents

↓

Production of short RNA molecules with random sequences

↓

Selective replication of self-duplicating catalytic RNA segments

↓

Synthesis of specific peptides, catalyzed by RNA

↓

Increasing role of peptides in RNA replication; coevolution of RNA and protein

↓

Primitive translation system develops, with RNA genome and RNA-protein catalysts

↓

Genomic RNA begins to be copied into DNA

↓

DNA genome, translated on RNA-protein complex (ribosome) with RNA and protein catalysts

FIGURE 1–36 A possible "RNA world" scenario.

The division of function between DNA (genetic information storage) and protein (catalysis) was, according to the "RNA world" hypothesis, a later development. New variants of self-replicating RNA molecules developed, with the additional ability to catalyze the condensation of amino acids into peptides. Occasionally, the peptide(s) thus formed would reinforce the self-replicating ability of the RNA, and the pair—RNA molecule and helping peptide—could undergo further modifications in sequence, generating increasingly efficient self-replicating systems. The remarkable discovery that in the protein-synthesizing machinery of modern cells (ribosomes), RNA molecules, not proteins, catalyze the formation of peptide bonds is consistent with the RNA world hypothesis.

Some time after the evolution of this primitive protein-synthesizing system, there was a further development: DNA molecules with sequences complementary to the self-replicating RNA molecules took over the function of conserving the "genetic" information, and RNA molecules evolved to play roles in protein synthesis. (We explain in Chapter 8 why DNA is a more stable molecule than RNA and thus a better repository of inheritable information.) Proteins proved to be versatile catalysts and, over time, took over most of that function. Lipidlike compounds in the primordial mixture formed relatively impermeable layers around self-replicating collections of molecules. The concentration of proteins and nucleic acids within these lipid enclosures favored the molecular interactions required in self-replication.

The RNA world scenario is intellectually satisfying, but it leaves unanswered a vexing question: where did the nucleotides needed to make the initial RNA molecules come from? An alternative to this RNA world scenario supposes that simple metabolic pathways evolved first, perhaps at the hot vents in the ocean floor. A set of linked chemical reactions there might have produced precursors, including nucleotides, before the advent of lipid membranes or RNA. Without more experimental evidence, neither of these hypotheses can be disproved.

Biological Evolution Began More Than Three and a Half Billion Years Ago

Earth was formed about 4.6 billion years ago, and the first evidence of life dates to more than 3.5 billion years ago. In 1996, scientists working in Greenland found chemical evidence of life ("fossil molecules") from as far back as 3.85 billion years ago, forms of carbon embedded in rock that seem to have a distinctly biological origin. Somewhere on Earth during its first billion years the first simple organism arose, capable of replicating its own structure from a template (RNA?) that was the first genetic material. Because the terrestrial atmosphere at the dawn of life was nearly devoid of oxygen, and because there were few microorganisms to scavenge organic compounds formed by natural processes, these compounds were relatively stable. Given this stability and eons of time, the improbable became inevitable: lipid vesicles containing organic compounds and self-replicating RNA gave rise to the first cells (protocells), and those protocells with the greatest capacity for self-replication became more numerous. The process of biological evolution had begun.

The First Cell Probably Used Inorganic Fuels

The earliest cells arose in a reducing atmosphere (there was no oxygen) and probably obtained energy from inorganic fuels, such as ferrous sulfide and ferrous carbonate, both abundant on the early Earth. For example, the reaction

$$FeS + H_2S \longrightarrow FeS_2 + H_2$$

yields enough energy to drive the synthesis of ATP or similar compounds. The organic compounds these early cells required may have arisen by the nonbiological actions of lightning or of heat from volcanoes or thermal vents in the sea on components of the early atmosphere: CO, CO_2, N_2, NH_3, CH_4, and such. An alternative source of organic compounds has been proposed: extraterrestrial space. In 2006, the Stardust space mission brought back tiny particles of dust from the tail of a comet; the dust contained a variety of organic compounds, including the simple amino acid glycine.

Early unicellular organisms gradually acquired the ability to derive energy from compounds in their environment and to use that energy to synthesize more of their own precursor molecules, thereby becoming less dependent on outside sources. A very significant evolutionary event was the development of pigments capable of capturing the energy of light from the sun, which could be used to reduce, or "fix," CO_2 to form more complex, organic compounds. The original electron donor for these **photosynthetic** processes was probably H_2S, yielding elemental sulfur or sulfate (SO_4^{2-}) as the byproduct; later cells developed the enzymatic capacity to use H_2O as the electron donor in photosynthetic reactions, producing O_2 as waste. Cyanobacteria are the modern descendants of these early photosynthetic oxygen-producers.

Because the atmosphere of Earth in the earliest stages of biological evolution was nearly devoid of oxygen, the earliest cells were anaerobic. Under these conditions, chemotrophs could oxidize organic compounds to CO_2 by passing electrons not to O_2 but to acceptors such as SO_4^{2-}, in this case yielding H_2S as the product. With the rise of O_2-producing photosynthetic bacteria, the atmosphere became progressively richer in oxygen—a powerful oxidant and deadly poison to anaerobes. Responding to the evolutionary pressure of what Lynn Margulis and Dorion Sagan called the "oxygen holocaust," some lineages of microorganisms gave rise to aerobes that obtained energy by passing electrons from fuel molecules to oxygen.

Because the transfer of electrons from organic molecules to O_2 releases a great deal of energy, aerobic organisms had an energetic advantage over their anaerobic counterparts when both competed in an environment containing oxygen. This advantage translated into the predominance of aerobic organisms in O_2-rich environments.

Modern bacteria and archaea inhabit almost every ecological niche in the biosphere, and there are organisms capable of using virtually every type of organic compound as a source of carbon and energy. Photosynthetic microbes in both fresh and marine waters trap solar energy and use it to generate carbohydrates and all other cell constituents, which are in turn used as food by other forms of life. The process of evolution continues—and, in rapidly reproducing bacterial cells, on a time scale that allows us to witness it in the laboratory. One interesting line of research into evolutionary mechanisms aims at producing a "synthetic" cell in the laboratory (one in which the experimenter has provided every component from known, purified components). The first step in this direction involves determining the minimum number of genes necessary for life by examining the genomes of the simplest bacteria. The smallest known genome for a free-living bacterium is that of *Mycobacterium genitalium*, which comprises 580,000 base pairs encoding 483 genes. In 2010, scientists at the Craig Venter Institute succeeded in synthesizing the full chromosome of the mycobacterium in vitro, then incorporating that synthetic chromosome into a living bacterial cell of another species, which thereby acquired the properties of *Mycobacterium genitalium*. This technology opens the way to producing a synthetic cell, with the bare minimum of genes essential to life. With such a cell, one could hope to study in the laboratory the evolutionary processes by which protocells gradually diversified and became more complex.

Eukaryotic Cells Evolved from Simpler Precursors in Several Stages

Starting about 1.5 billion years ago, the fossil record begins to show evidence of larger and more complex organisms, probably the earliest eukaryotic cells **(Fig. 1–37)**. Details of the evolutionary path from non-nucleated to nucleated cells cannot be deduced from the fossil record alone, but morphological and biochemical comparisons of modern organisms have suggested a sequence of events consistent with the fossil evidence.

Three major changes must have occurred. First, as cells acquired more DNA, the mechanisms required to fold it compactly into discrete complexes with specific proteins and to divide it equally between daughter cells at cell division became more elaborate. Specialized proteins were required to stabilize folded DNA and to pull the resulting DNA-protein complexes (*chromosomes*) apart during cell division. Second, as cells became

FIGURE 1–37 **Landmarks in the evolution of life on Earth.**

Lynn Margulis, 1938–2011

larger, a system of intracellular membranes developed, including a double membrane surrounding the DNA. This membrane segregated the nuclear process of RNA synthesis on a DNA template from the cytoplasmic process of protein synthesis on ribosomes. Finally, according to a now widely accepted hypothesis advanced (initially, to much resistance) by Lynn Margulis, early eukaryotic cells, which were incapable of photosynthesis or aerobic metabolism, enveloped aerobic bacteria or photosynthetic bacteria to form **endosymbiotic** associations that eventually became permanent **(Fig. 1–38)**. Some aerobic bacteria evolved into the mitochondria of modern eukaryotes, and some photosynthetic cyanobacteria became the plastids, such as the chloroplasts of green algae, the likely ancestors of modern plant cells.

Anaerobic metabolism is inefficient because fuel is not completely oxidized.

Bacterium is engulfed by ancestral eukaryote, and multiplies within it.

Symbiotic system can now carry out aerobic catabolism. Some bacterial genes move to the nucleus, and the bacterial endosymbionts become **mitochondria**.

Nonphotosynthetic eukaryote

Nucleus

Ancestral anaerobic eukaryote

Bacterial genome

Aerobic bacterium

Aerobic metabolism is efficient because fuel is oxidized to CO_2.

Aerobic eukaryote

Mitochondrion

Photosynthetic eukaryote

Chloroplast

Cyanobacterial genome

Photosynthetic cyanobacterium

Light energy is used to synthesize biomolecules from CO_2.

Engulfed cyanobacterium becomes an endosymbiont and multiplies; new cell can make ATP using energy from sunlight.

In time, some cyanobacterial genes move to the nucleus, and endosymbionts become **chloroplasts**.

FIGURE 1–38 Evolution of eukaryotes through endosymbiosis. The earliest eukaryote, an anaerobe, acquired endosymbiotic purple bacteria, which carried with them their capacity for aerobic catabolism and became, over time, mitochondria. When photosynthetic cyanobacteria subsequently became endosymbionts of some aerobic eukaryotes, these cells became the photosynthetic precursors of modern green algae and plants.

At some later stage of evolution, unicellular organisms found it advantageous to cluster together, thereby acquiring greater motility, efficiency, or reproductive success than their free-living single-celled competitors. Further evolution of such clustered organisms led to permanent associations among individual cells and eventually to specialization within the colony—to cellular differentiation.

The advantages of cellular specialization led to the evolution of increasingly complex and highly differentiated organisms, in which some cells carried out the sensory functions, others the digestive, photosynthetic, or reproductive functions, and so forth. Many modern multicellular organisms contain hundreds of different cell types, each specialized for a function that supports the entire organism. Fundamental mechanisms that evolved early have been further refined and embellished through evolution. The same basic structures and mechanisms that underlie the beating motion of cilia in *Paramecium* and of flagella in *Chlamydomonas* are employed by the highly differentiated vertebrate sperm cell, for example.

Molecular Anatomy Reveals Evolutionary Relationships

Biochemists now have an enormously rich, ever increasing treasury of information on the molecular anatomy of cells that they can use to analyze evolutionary relationships and refine evolutionary theory. The sequence of the **genome**, the complete genetic endowment of an organism, has been determined for hundreds of bacteria and more than 40 archaea and for growing numbers of eukaryotic microorganisms, including *Saccharomyces cerevisiae* and *Plasmodium* species; plants, including *Arabidopsis thaliana* and rice; and multicellular animals, including *Caenorhabditis elegans* (a roundworm), *Drosophila melanogaster* (the fruit fly), mouse, rat, dog, chimpanzee, and *Homo sapiens* (Table 1–2). With such sequences in hand, detailed and quantitative comparisons among species can provide deep insight into the evolutionary process. Thus far, the molecular phylogeny derived from gene sequences is consistent with, but in many cases more precise than, the classical phylogeny based on macroscopic structures. Although organisms have continuously diverged at the level of gross anatomy, at the molecular level the basic unity of life is readily apparent; molecular structures and mechanisms are remarkably similar from the simplest to the most complex organisms. These similarities are most easily seen at the level of sequences, either the DNA sequences that encode proteins or the protein sequences themselves.

When two genes share readily detectable sequence similarities (nucleotide sequence in DNA or amino acid

TABLE 1–2   A Few of the Many Organisms Whose Genomes Have Been Completely Sequenced

Organism	Genome size (nucleotide pairs)	Biological interest
Mycoplasma genitalium	5.8×10^5	Smallest true organism
Helicobacter pylori	1.6×10^6	Causes gastric ulcers
Methanocaldococcus jannaschii	1.7×10^6	Archaeon; grows at 85 °C
Haemophilus influenzae	1.9×10^6	Causes bacterial influenza
Synechocystis sp.	3.9×10^6	Cyanobacterium
Bacillus subtilis	4.2×10^6	Common soil bacterium
Escherichia coli	4.6×10^6	Some strains are human pathogens
Saccharomyces cerevisiae	1.2×10^7	Unicellular eukaryote
Caenorhabditis elegans	1.0×10^8	Multicellular roundworm
Arabidopsis thaliana	1.2×10^8	Model plant
Drosophila melanogaster	1.8×10^8	Laboratory fly ("fruit fly")
Mus musculus	2.7×10^9	Laboratory mouse
Homo sapiens	3.0×10^9	Human

Source: www.ncbi.nlm.nih.gov/genome.

sequence in the proteins they encode), their sequences are said to be homologous and the proteins they encode are **homologs**. If two homologous genes occur in the *same* species, they are said to be paralogous and their protein products are **paralogs**. Paralogous genes are presumed to have been derived by gene duplication followed by gradual changes in the sequences of both copies. Typically, paralogous proteins are similar not only in sequence but also in three-dimensional structure, although they commonly have acquired different functions during their evolution.

Two homologous genes (or proteins) found in *different* species are said to be orthologous, and their protein products are **orthologs**. Orthologs are commonly found to have the same function in both organisms, and when a newly sequenced gene in one species is found to be strongly orthologous with a gene in another, this gene is presumed to encode a protein with the same function in both species. By this means, the function of gene products (proteins or RNA molecules) can be deduced from the genomic sequence, without any biochemical characterization of the molecules themselves. An **annotated genome** includes, in addition to the DNA sequence itself, a description of the likely function of each gene product, deduced from comparisons with other genomic sequences and established protein functions. Sometimes, by identifying the pathways (sets of enzymes) encoded in a genome, we can deduce from the genomic sequence alone the organism's metabolic capabilities.

The sequence differences between homologous genes may be taken as a rough measure of the degree to which the two species have diverged during evolution—

of how long ago their common evolutionary precursor gave rise to two lines with different evolutionary fates. The larger the number of sequence differences, the earlier the divergence in evolutionary history. One can construct a phylogeny (family tree) in which the evolutionary distance between any two species is represented by their proximity on the tree (Fig. 1–4 is an example).

In the course of evolution, new structures, processes, or regulatory mechanisms are acquired, reflections of the changing genomes of the evolving organisms. The genome of a simple eukaryote such as yeast should have genes related to formation of the nuclear membrane, genes not present in bacteria or archaea. The genome of an insect should contain genes that encode proteins involved in specifying insects' characteristic segmented body plan, genes not present in yeast. The genomes of all vertebrate animals should share genes that specify the development of a spinal column, and those of mammals should have unique genes necessary for the development of the placenta, a characteristic of mammals—and so on. Comparisons of the whole genomes of species in each phylum are leading to the identification of genes critical to fundamental evolutionary changes in body plan and development.

Functional Genomics Shows the Allocations of Genes to Specific Cellular Processes

When the sequence of a genome is fully determined and each gene is assigned a function, molecular geneticists can group genes according to the processes (DNA synthesis, protein synthesis, generation of ATP, and so forth) in which they function and thus find what fraction of the genome is allocated to each of a cell's activities. The largest category of genes in *E. coli*, *A. thaliana*, and *H. sapiens* consists of genes of (as yet) unknown function, which make up more than 40% of the genes in each species. The transporters that move ions and small molecules across plasma membranes take up a significant proportion of the genes in all three species, more in the bacterium and plant than in the mammal (10% of the ~4,400 genes of *E. coli*, ~8% of the ~32,000 genes of *A. thaliana*, and ~4% of the ~25,000 genes of *H. sapiens*). Genes that encode the proteins and RNA required for protein synthesis make up 3% to 4% of the *E. coli* genome, but in the more complex cells of *A. thaliana*, more genes are needed for targeting proteins to their final location in the cell than are needed to synthesize those proteins (about 6% and 2% of the genome, respectively). In general, the more complex the organism, the greater the proportion of its genome that encodes genes involved in the *regulation* of cellular processes and the smaller the proportion dedicated to the basic processes themselves, such as ATP generation and protein synthesis.

Genomic Comparisons Have Increasing Importance in Human Biology and Medicine

The genomes of chimpanzees and humans are 99.9% identical, yet the differences between the two species are vast. The relatively few differences in genetic endowment must explain the possession of language by humans, the extraordinary athleticism of chimpanzees, and myriad other differences. Genomic comparison is allowing researchers to identify candidate genes linked to divergences in the developmental programs of humans and the other primates and to the emergence of complex functions such as language. The picture will become clearer only as more primate genomes become available for comparison with the human genome.

Similarly, the differences in genetic endowment among humans are vanishingly small compared with the differences between humans and chimpanzees, yet these differences account for the variety among us— including differences in health and in susceptibility to chronic diseases. We have much to learn about the variability in sequence among humans, and the availability of genomic information will almost certainly transform medical diagnosis and treatment. We may expect that for some genetic diseases, palliatives will be replaced by cures; and that for disease susceptibilities associated with particular genetic markers, forewarning and perhaps increased preventive measures will prevail. Today's "medical history" may be replaced by a "medical forecast." ■

SUMMARY 1.5 Evolutionary Foundations

▶ Occasional inheritable mutations yield organisms that are better suited for survival and reproduction in an ecological niche, and their progeny come to dominate the population in that niche. This process of mutation and selection is the basis for the Darwinian evolution that led from the first cell to all modern organisms. The large number of genes shared by all living organisms explains their fundamental similarities.

▶ Life originated about 3.5 billion years ago, most likely with the formation of a membrane-enclosed compartment containing a self-replicating RNA molecule. The components for the first cell may have been produced near thermal vents at the bottom of the sea or by the action of lightning and high temperature on simple atmospheric molecules such as CO_2 and NH_3.

▶ The catalytic and genetic roles played by the early RNA genome were, over time, taken over by proteins and DNA, respectively.

▶ Eukaryotic cells acquired the capacity for photosynthesis and oxidative phosphorylation from endosymbiotic bacteria. In multicellular organisms, differentiated cell types specialize in one or more of the functions essential to the organism's survival.

▶ Knowledge of the complete genomic nucleotide sequences of organisms from different branches of the phylogenetic tree provides insights into evolution and offers great opportunities in human medicine.

Key Terms

All terms are defined in the glossary.

metabolite 3	endergonic reaction 23
nucleus 3	exergonic reaction 23
genome 3	equilibrium 25
eukaryotes 3	standard free-energy
bacteria 4	change, $\Delta G°$ 26
archaea 4	activation energy,
cytoskeleton 8	$\Delta G^{\ddagger}$ 27
stereoisomers 16	catabolism 28
configuration 16	anabolism 28
chiral center 17	metabolism 28
conformation 19	systems biology 29
entropy, S 23	mutation 32
enthalpy, H 23	
free-energy change,	
ΔG 23	

Further Reading

General

Fruton, J.S. (1999) *Proteins, Enzymes, Genes: The Interplay of Chemistry and Biochemistry*, Yale University Press, New Haven.

A distinguished historian of biochemistry traces the development of this science and discusses its impact on medicine, pharmacy, and agriculture.

Harold, F.M. (2001) *The Way of the Cell: Molecules, Organisms, and the Order of Life*, Oxford University Press, Oxford.

Judson, H.F. (1996) *The Eighth Day of Creation: The Makers of the Revolution in Biology*, expanded edn, Cold Spring Harbor Laboratory Press, Cold Spring Harbor, NY.

A highly readable and authoritative account of the rise of biochemistry and molecular biology in the twentieth century.

Kornberg, A. (1987) The two cultures: chemistry and biology. *Biochemistry* **26**, 6888–6891.

The importance of applying chemical tools to biological problems, described by an eminent practitioner.

Monod, J. (1971) *Chance and Necessity*, Alfred A. Knopf, Inc., New York. [Paperback edition, Vintage Books, 1972.] Originally published (1970) as *Le hasard et la nécessité*, Editions du Seuil, Paris.

An exploration of the philosophical implications of biological knowledge.

Morowitz, H.J. (2002) *The Emergence of Everything (How the World Became Complex)*, Oxford University Press, Oxford.

Short, beautifully written discussion of the emergence of complex organisms from simple beginnings.

Pace, N.R. (2001) The universal nature of biochemistry. *Proc. Natl. Acad. Sci. USA* **98**, 805–808.

A short discussion of the minimal definition of life, on Earth and elsewhere.

Cellular Foundations

Hardin, J., Bertoni, G.P., & Kleinsmith, L.J. (2011) *Becker's World of the Cell*, 8th edn, The Benjamin/Cummings Publishing Company, Redwood City, CA.

An excellent introductory textbook of cell biology.

Hosking, C.R. & Schwartz, J.L. (2009) The future's bright: imaging cell biology in the 21st century. *Trends Cell Biol.* **19**, 553–554.

Short editorial introduction to this whole issue of *Trends in Cell Biology*, about new methods for imaging cells.

Lodish, H., Berk, A., Kaiser, C.A., Krieger, M., Scott, M.R., Bretscher, A., Ploegh, H., & Amon, A. (2012) *Molecular Cell Biology*, 7th edn, W. H. Freeman and Company, New York.

A superb text, useful for this and later chapters.

Sadava, D., Hillis, D.M., Heller, H.C., & Berenbaum, M. (2010) *Life: The Science of Biology*, 9th edn, W. H. Freeman and Company, New York.

Wilson, C., Venditti, R., Rega, L.R., Colanzi, A., D'Angelo, G., & DeMatteis, M.A. (2011) The Golgi apparatus: an organelle with multiple complex functions. *Biochem. J.* **433**, 1–9.

Excellent intermediate-level review of the roles of the Golgi complex.

Chemical Foundations

Barta, N.S. & Stille, J.R. (1994) Grasping the concepts of stereochemistry. *J. Chem. Educ.* **71**, 20–23.

A clear description of the RS system for naming stereoisomers, with practical suggestions for determining and remembering the "handedness" of isomers.

Thall, E. (1996) When drug molecules look in the mirror. *J. Chem. Educ.* **73**, 481–484.

Biological difference in (R) and (S) isomers of drugs.

Vollhardt, K.P.C. & Shore, N.E. (2011) *Organic Chemistry: Structure and Function*, 6th edn, W. H. Freeman and Company, New York.

Up-to-date discussions of stereochemistry, functional groups, reactivity, and the chemistry of the principal classes of biomolecules.

Physical Foundations

Atkins, P.W. & de Paula, J. (2012) *Physical Chemistry for the Life Sciences*, 2nd edn, W. H. Freeman and Company, New York.

Blum, H.F. (1968) *Time's Arrow and Evolution*, 3rd edn, Princeton University Press, Princeton.

An excellent discussion of the way the second law of thermodynamics has influenced biological evolution.

Genetic Foundations

Griffiths, A.J.F., Wessler, S.R., Lewinton, R.C. & Carroll, S. (2008) *An Introduction to Genetic Analysis*, 9th edn, W. H. Freeman and Company, New York.

Jacob, F. (1973) *The Logic of Life: A History of Heredity*, Pantheon Books, Inc., New York. Originally published (1970) as *La logique du vivant: une histoire de l'hérédité*, Editions Gallimard, Paris.

A fascinating historical and philosophical account of the route to our present molecular understanding of life.

Pierce, B. (2012) *Genetics: A Conceptual Approach*, 4th edn, W. H. Freeman and Company, New York.

Evolutionary Foundations

Adams, J.C. (2009) Molecular and cellular evolution: a celebration of the 200th anniversary of the birth of Charles Darwin. *Int. J. Biochem. Cell Biol.* **41**, 249.

An editorial introduction to 14 essays on various aspects of molecular and cellular evolution that make up this issue of the journal.

Brow, J.R. & Doolittle, W.F. (1997) Archaea and the prokaryote-to-eukaryote transition. *Microbiol. Mol. Biol. Rev.* **61**, 456–502.

A very thorough discussion of the arguments for placing the Archaea on the phylogenetic branch that led to multicellular organisms.

Budin, I. & Szostak, J.W. (2010) Expanding roles for diverse physical phenomena during the origin of life. *Annu. Rev. Biophys.* **39**, 45–63.

A thoughtful discussion of the physical factors that may have influenced the course of prebiotic evolution.

Carroll, S.B. (2006) *The Making of the Fittest: DNA and the Ultimate Forensic Record of Evolution*, W.W. Norton & Company, Inc., New York.

Cavicchioli, R. (2011) Archaea—timeline of the third domain. *Nat. Rev. Microbiol.* **9**, 51–61.

Intermediate-level description of the discovery and investigation of the Archaea.

de Duve, C. (1995) The beginnings of life on earth. *Am. Sci.* **83**, 428–437.

One scenario for the succession of chemical steps that led to the first living organism.

de Duve, C. (1996) The birth of complex cells. *Sci. Am.* **274** (April), 50–57.

Evolution of Catalytic Function. (1987) *Cold Spring Harb. Symp. Quant. Biol.* **52**.

A collection of almost 100 articles on all aspects of prebiotic and early biological evolution; excellent source on molecular evolution.

Gesteland, R.F., Atkins, J.F., & Cech, T.R. (eds). (2006) *The RNA World*, Cold Spring Harbor Laboratory Press, Cold Spring Harbor, NY.

A collection of stimulating reviews on a wide range of topics related to the RNA world scenario.

Knoll, A.H. (2003) *Life on a Young Planet: The First Three Billion Years of Evolution on Earth*, Princeton University Press, Princeton.

Discussion of the early stages of evolution, with emphasis on the geochemistry of the environment in which it occurred.

Lazcano, A. & Miller, S.L. (1996) The origin and early evolution of life: prebiotic chemistry, the pre-RNA world, and time. *Cell* **85**, 793–798.

Brief review of developments in studies of the origin of life: primitive atmospheres, submarine vents, autotrophic versus heterotrophic origin, the RNA and pre-RNA worlds, and the time required for life to arise.

Martin, W., Baross, J., Kelley, D., & Russell, M.J. (2008) Hydrothermal vents and the origin of life. *Nat. Rev. Microbiol.* **6**, 805–814.

Miller, S.L. (1987) Which organic compounds could have occurred on the prebiotic earth? *Cold Spring Harb. Symp. Quant. Biol.* **52**, 17–27.

Summary of laboratory experiments on chemical evolution, by the person who did the original Miller-Urey experiment.

Noireaux, V., Maeda, Y.T., & Libchaber, A. (2011) Development of an artificial cell, from self-organization to computation and self-reproduction. *Proc. Natl. Acad. Sci. USA* **108**, 3473–3480.

Interesting discussion of the steps that must be taken to produce a living cell from scratch.

Parker, E.T., Cleaves, H.J., Dworkin, J.P., Glavin, D.P., Callahan, M., Aubrey, A., Lazcano, A., & Bada, J.L. (2011) Primordial synthesis of amines and amino acids in a 1958 Miller H_2S-rich spark discharge experiment. *Proc. Natl. Acad. Sci. USA* **108**, 5526–5531.

A modern reexamination of samples generated in the 1958 experiments of Miller, using H_2S as one of the atmospheric gases.

Stiller, J.W. (2007) Plastid endosymbiosis, genome evolution and the origin of green plants. *Trends Plant Sci.* **12**, 391–396.

Woese, C.R. (2002) On the evolution of cells. *Proc. Natl. Acad. Sci. USA* **99**, 8742–8747.

Short, clear review.

Woese, C.R. (2004) A new biology for a new century. *Microbiol. Mol. Biol. Rev.* **68**, 173–186.

Development of current thinking about cellular evolution by one of the seminal thinkers in the field.

Woese, C.R., Kandler, O., & Wheelis, M.L. (1990) Towards a natural system of organisms: proposal for the domains Archaea, Bacteria, and Eucarya. *Proc. Natl. Acad. Sci. USA* **87**, 4576–4579.

The arguments for dividing all living organisms into three domains.

Problems

Some problems related to the contents of the chapter follow. (In solving end-of-chapter problems, you may wish to refer to the tables on the inside of the back cover.) Each problem has a title for easy reference and discussion. For all numerical problems, keep in mind that answers should be expressed with the correct number of significant figures. Brief solutions are provided in Appendix B; expanded solutions are published in the *Absolute Ultimate Study Guide to Accompany Principles of Biochemistry*.

1. **The Size of Cells and Their Components**

(a) If you were to magnify a cell 10,000 fold (typical of the magnification achieved using an electron microscope), how big would it appear? Assume you are viewing a "typical" eukaryotic cell with a cellular diameter of 50 μm.

(b) If this cell were a muscle cell (myocyte), how many molecules of actin could it hold? Assume the cell is spherical and no other cellular components are present; actin molecules are spherical, with a diameter of 3.6 nm. (The volume of a sphere is $4/3\ \pi r^3$.)

(c) If this were a liver cell (hepatocyte) of the same dimensions, how many mitochondria could it hold? Assume the cell is spherical; no other cellular components are present; and the mitochondria are spherical, with a diameter of 1.5 μm.

(d) Glucose is the major energy-yielding nutrient for most cells. Assuming a cellular concentration of 1 mM (that is, 1 millimole/L), calculate how many molecules of glucose would be present in our hypothetical (and spherical) eukaryotic cell. (Avogadro's number, the number of molecules in 1 mol of a nonionized substance, is 6.02×10^{23}.)

(e) Hexokinase is an important enzyme in the metabolism of glucose. If the concentration of hexokinase in our eukaryotic cell is 20 μM, how many glucose molecules are present per hexokinase molecule?

2. **Components of *E. coli*** *E. coli* cells are rod-shaped, about 2 μm long and 0.8 μm in diameter. The volume of a cylinder is $\pi r^2 h$, where h is the height of the cylinder.

(a) If the average density of *E. coli* (mostly water) is 1.1×10^3 g/L, what is the mass of a single cell?

(b) *E. coli* has a protective cell envelope 10 nm thick. What percentage of the total volume of the bacterium does the cell envelope occupy?

(c) *E. coli* is capable of growing and multiplying rapidly because it contains some 15,000 spherical ribosomes (diameter 18 nm), which carry out protein synthesis. What percentage of the cell volume do the ribosomes occupy?

3. **Genetic Information in *E. coli* DNA** The genetic information contained in DNA consists of a linear sequence of coding units, known as codons. Each codon is a specific sequence of three deoxyribonucleotides (three deoxyribonucleotide pairs in double-stranded DNA), and each codon codes for a single amino acid unit in a protein. The molecular weight of an *E. coli* DNA molecule is about 3.1×10^9 g/mol. The average molecular weight of a nucleotide pair is 660 g/mol, and each nucleotide pair contributes 0.34 nm to the length of DNA.

(a) Calculate the length of an *E. coli* DNA molecule. Compare the length of the DNA molecule with the cell dimensions (see Problem 2). How does the DNA molecule fit into the cell?

(b) Assume that the average protein in *E. coli* consists of a chain of 400 amino acids. What is the maximum number of proteins that can be coded by an *E. coli* DNA molecule?

4. **The High Rate of Bacterial Metabolism** Bacterial cells have a much higher rate of metabolism than animal cells. Under ideal conditions some bacteria double in size and divide every 20 min, whereas most animal cells under rapid growth conditions require 24 hours. The high rate of bacterial metabolism requires a high ratio of surface area to cell volume.

(a) Why does surface-to-volume ratio affect the maximum rate of metabolism?

(b) Calculate the surface-to-volume ratio for the spherical bacterium *Neisseria gonorrhoeae* (diameter 0.5 μm), responsible for the disease gonorrhea. Compare it with the surface-to-volume ratio for a globular amoeba, a large eukaryotic cell (diameter 150 μm). The surface area of a sphere is $4\pi r^2$.

5. Fast Axonal Transport Neurons have long thin processes called axons, structures specialized for conducting signals throughout the organism's nervous system. Some axonal processes can be as long as 2 m—for example, the axons that originate in your spinal cord and terminate in the muscles of your toes. Small membrane-enclosed vesicles carrying materials essential to axonal function move along microtubules of the cytoskeleton, from the cell body to the tips of the axons. If the average velocity of a vesicle is 1 μm/s, how long does it take a vesicle to move from a cell body in the spinal cord to the axonal tip in the toes?

6. Is Synthetic Vitamin C as Good as the Natural Vitamin? A claim put forth by some purveyors of health foods is that vitamins obtained from natural sources are more healthful than those obtained by chemical synthesis. For example, pure L-ascorbic acid (vitamin C) extracted from rose hips is better than pure L-ascorbic acid manufactured in a chemical plant. Are the vitamins from the two sources different? Can the body distinguish a vitamin's source?

7. Identification of Functional Groups Figures 1–16 and 1–17 show some common functional groups of biomolecules. Because the properties and biological activities of biomolecules are largely determined by their functional groups, it is important to be able to identify them. In each of the compounds below, circle and identify by name each functional group.

Ethanolamine
(a)

Glycerol
(b)

Phosphoenolpyruvate, an intermediate in glucose metabolism
(c)

Threonine, an amino acid
(d)

Pantothenate, a vitamin
(e)

D-Glucosamine
(f)

8. Drug Activity and Stereochemistry The quantitative differences in biological activity between the two enantiomers of a compound are sometimes quite large.

For example, the D isomer of the drug isoproterenol, used to treat mild asthma, is 50 to 80 times more effective as a bronchodilator than the L isomer. Identify the chiral center in isoproterenol. Why do the two enantiomers have such radically different bioactivity?

Isoproterenol

9. Separating Biomolecules In studying a particular biomolecule (a protein, nucleic acid, carbohydrate, or lipid) in the laboratory, the biochemist first needs to separate it from other biomolecules in the sample—that is, to *purify* it. Specific purification techniques are described later in the text. However, by looking at the monomeric subunits of a biomolecule, you should have some ideas about the characteristics of the molecule that would allow you to separate it from other molecules. For example, how would you separate (a) amino acids from fatty acids and (b) nucleotides from glucose?

10. Silicon-Based Life? Silicon is in the same group of the periodic table as carbon and, like carbon, can form up to four single bonds. Many science fiction stories have been based on the premise of silicon-based life. Is this realistic? What characteristics of silicon make it *less* well adapted than carbon as the central organizing element for life? To answer this question, consider what you have learned about carbon's bonding versatility, and refer to a beginning inorganic chemistry textbook for silicon's bonding properties.

11. Drug Action and Shape of Molecules Some years ago two drug companies marketed a drug under the trade names Dexedrine and Benzedrine. The structure of the drug is shown below.

The physical properties (C, H, and N analysis, melting point, solubility, etc.) of Dexedrine and Benzedrine were identical. The recommended oral dosage of Dexedrine (which is still available) was 5 mg/day, but the recommended dosage of Benzedrine (no longer available) was twice that. Apparently it required considerably more Benzedrine than Dexedrine to yield the same physiological response. Explain this apparent contradiction.

12. Components of Complex Biomolecules Figure 1–10 shows the major components of complex biomolecules. For each of the three important biomolecules below (shown in their ionized forms at physiological pH), identify the constituents.

(a) Guanosine triphosphate (GTP), an energy-rich nucleotide that serves as a precursor to RNA:

(b) Methionine enkephalin, the brain's own opiate:

(c) Phosphatidylcholine, a component of many membranes:

13. Determination of the Structure of a Biomolecule An unknown substance, X, was isolated from rabbit muscle. Its structure was determined from the following observations and experiments. Qualitative analysis showed that X was composed entirely of C, H, and O. A weighed sample of X was completely oxidized, and the H_2O and CO_2 produced were measured; this quantitative analysis revealed that X contained 40.00% C, 6.71% H, and 53.29% O by weight. The molecular mass of X, determined by mass spectrometry, was 90.00 u (atomic mass units; see Box 1–1). Infrared spectroscopy showed that X contained one double bond. X dissolved readily in water to give an acidic solution; the solution demonstrated optical activity when tested in a polarimeter.

(a) Determine the empirical and molecular formula of X.

(b) Draw the possible structures of X that fit the molecular formula and contain one double bond. Consider *only* linear or branched structures and disregard cyclic structures. Note that oxygen makes very poor bonds to itself.

(c) What is the structural significance of the observed optical activity? Which structures in (b) are consistent with the observation?

(d) What is the structural significance of the observation that a solution of X was acidic? Which structures in (b) are consistent with the observation?

(e) What is the structure of X? Is more than one structure consistent with all the data?

14. Naming Stereoisomers with One Chiral Carbon Using the RS System Propranolol is a chiral compound. (R)-Propranolol is used as a contraceptive; (S)-propranolol is used to treat hypertension. Identify the chiral carbon in the structure below. Is this the (R) or the (S) isomer? Draw the other isomer.

15. Naming Stereoisomers with Two Chiral Carbons Using the RS System The (R,R) isomer of methylphenidate (Ritalin) is used to treat attention deficit hyperactivity disorder (ADHD). The (S,S) isomer is an antidepressant. Identify the two chiral carbons in the structure below. Is this the (R,R) or the (S,S) isomer? Draw the other isomer.

Data Analysis Problem

16. Interaction of Sweet-Tasting Molecules with Taste Receptors Many compounds taste sweet to humans. Sweet taste results when a molecule binds to the sweet receptor, one type of taste receptor, on the surface of certain tongue cells. The stronger the binding, the lower the concentration required to saturate the receptor and the sweeter a given concentration of that substance tastes. The standard free-energy change, $\Delta G°$, of the binding reaction between a sweet molecule and a sweet receptor can be measured in kilojoules or kilocalories per mole.

Sweet taste can be quantified in units of "molar relative sweetness" (MRS), a measure that compares the sweetness of a substance to the sweetness of sucrose. For example, saccharin has an MRS of 161; this means that saccharin is 161 times sweeter than sucrose. In practical terms, this is measured by asking human subjects to compare the sweetness of solutions containing different concentrations of each compound. Sucrose and saccharin taste equally sweet when sucrose is at a concentration 161 times higher than that of saccharin.

(a) What is the relationship between MRS and the $\Delta G°$ of the binding reaction? Specifically, would a more negative $\Delta G°$ correspond to a higher or lower MRS? Explain your reasoning.

Shown on the next page are the structures of 10 compounds, all of which taste sweet to humans. The MRS and $\Delta G°$ for binding to the sweet receptor are given for each substance.

Deoxysucrose
MRS = 0.95
$\Delta G° = -6.67$ kcal/mol

Sucrose
MRS = 1
$\Delta G° = -6.71$ kcal/mol

D-Tryptophan
MRS = 21
$\Delta G° = -8.5$ kcal/mol

Saccharin
MRS = 161
$\Delta G° = -9.7$ kcal/mol

Aspartame
MRS = 172
$\Delta G° = -9.7$ kcal/mol

6-Chloro-D-tryptophan
MRS = 906
$\Delta G° = -10.7$ kcal/mol

Alitame
MRS = 1,937
$\Delta G° = -11.1$ kcal/mol

Neotame
MRS = 11,057
$\Delta G° = -12.1$ kcal/mol

Tetrabromosucrose
MRS = 13,012
$\Delta G° = -12.2$ kcal/mol

Sucronic acid
MRS = 200,000
$\Delta G° = -13.8$ kcal/mol

Morini, Bassoli, and Temussi (2005) used computer-based methods (often referred to as "in silico" methods) to model the binding of sweet molecules to the sweet receptor.

(b) Why is it useful to have a computer model to predict the sweetness of molecules, instead of a human- or animal-based taste assay?

In earlier work, Schallenberger and Acree (1967) had suggested that all sweet molecules include an "AH-B" structural group, in which "A and B are electronegative atoms separated by a distance of greater than 2.5 Å [0.25 nm] but less than 4 Å [0.4 nm]. H is a hydrogen atom attached to one of the electronegative atoms by a covalent bond."

(c) Given that the length of a "typical" single bond is about 0.15 nm, identify the AH-B group(s) in each of the molecules shown on the left.

(d) Based on your findings from (c), give two objections to the statement that "molecules containing an AH-B structure will taste sweet."

(e) For two of the molecules shown here, the AH-B model *can* be used to explain the difference in MRS and $\Delta G°$. Which two molecules are these, and how would you use them to support the AH-B model?

(f) Several of the molecules have closely related structures but very different MRS and $\Delta G°$ values. Give two such examples, and use these to argue that the AH-B model is unable to explain the observed differences in sweetness.

In their computer-modeling study, Morini and coauthors used the three-dimensional structure of the sweet receptor and a molecular dynamics modeling program called GRAMM to predict the $\Delta G°$ of binding of sweet molecules to the sweet receptor. First, they "trained" their model—that is, they refined the parameters so that the $\Delta G°$ values predicted by the model matched the known $\Delta G°$ values for one set of sweet molecules (the "training set"). They then "tested" the model by asking it to predict the $\Delta G°$ values for a new set of molecules (the "test set").

(g) Why did Morini and colleagues need to test their model against a different set of molecules from the set it was trained on?

(h) The researchers found that the predicted $\Delta G°$ values for the test set differed from the actual values by, on average, 1.3 kcal/mol. Using the values given with the molecular structures, estimate the resulting error in MRS values.

References

Morini, G., Bassoli, A., & Temussi, P.A. (2005) From small sweeteners to sweet proteins: anatomy of the binding sites of the human T1R2_T1R3 receptor. *J. Med. Chem.* **48**, 5520–5529.

Schallenberger, R.S. & Acree, T.E. (1967) Molecular theory of sweet taste. *Nature* **216**, 480–482.

PART I

STRUCTURE AND CATALYSIS

Biochemistry is nothing less than the chemistry of life, and, yes, life can be investigated, analyzed, and understood. To begin, every student of biochemistry needs both a language and some fundamentals; these are provided in Part I.

The chapters of Part I are devoted to the structure and function of the major classes of cellular constituents: water (Chapter 2), amino acids and proteins (Chapters 3 through 6), sugars and polysaccharides (Chapter 7), nucleotides and nucleic acids (Chapter 8), fatty acids and lipids (Chapter 10), and, finally, membranes and membrane signaling proteins (Chapters 11 and 12). We also discuss, in the context of structure and function, the technologies used to study each class of biomolecules. One whole chapter (Chapter 9) is devoted entirely to biotechnologies associated with cloning and genomics.

We begin, in Chapter 2, with water, because its properties affect the structure and function of all other cellular constituents. For each class of organic molecules, we first consider the covalent chemistry of the monomeric units (amino acids, monosaccharides, nucleotides, and fatty acids) and then describe the structure of the macromolecules and supramolecular complexes derived from

them. An overarching theme is that the polymeric macromolecules in living systems, though large, are highly ordered chemical entities, with specific sequences of monomeric subunits giving rise to discrete structures and functions. This fundamental theme can be broken down into three interrelated principles: (1) the unique structure of each macromolecule determines its function; (2) noncovalent interactions play a critical role in the structure and thus the function of macromolecules; and (3) the monomeric subunits in polymeric macromolecules occur in specific sequences, representing a form of information on which the ordered living state depends.

The relationship between structure and function is especially evident in proteins, which exhibit an extraordinary diversity of functions. One particular polymeric sequence of amino acids produces a strong, fibrous structure found in hair and wool; another produces a protein that transports oxygen in the blood; a third binds other proteins and catalyzes the cleavage of the bonds between their amino acids. Similarly, the special functions of polysaccharides, nucleic acids, and lipids can be understood as resulting directly from their chemical structure, with their characteristic monomeric subunits precisely linked to form functional polymers.

Sugars linked together become energy stores, structural fibers, and points of specific molecular recognition; nucleotides strung together in DNA or RNA provide the blueprint for an entire organism; and aggregated lipids form membranes. Chapter 12 unifies the discussion of biomolecule function, describing how specific signaling systems regulate the activities of biomolecules—within a cell, within an organ, and among organs—to keep an organism in homeostasis.

As we move from monomeric units to larger and larger polymers, the chemical focus shifts from covalent bonds to noncovalent interactions. Covalent bonds, at the monomeric and macromolecular level, place constraints on the shapes assumed by large biomolecules. It is the numerous noncovalent interactions, however, that dictate the stable, native conformations of large molecules while permitting the flexibility necessary for their biological function. As we shall see, noncovalent interactions are essential to the catalytic power of enzymes, the critical interaction of complementary base pairs in nucleic acids, and the arrangement and properties of lipids in membranes. The principle that sequences of monomeric subunits are rich in information emerges most fully in the discussion of nucleic acids (Chapter 8). However, proteins and some short polymers of sugars (oligosaccharides) are also information-rich molecules. The amino acid sequence is a form of information that directs the folding of the protein into its unique three-dimensional structure and ultimately determines the function of the protein. Some oligosaccharides also have unique sequences and three-dimensional structures that are recognized by other macromolecules.

Each class of molecules has a similar structural hierarchy: subunits of fixed structure are connected by bonds of limited flexibility to form macromolecules with three-dimensional structures determined by noncovalent interactions. These macromolecules then interact to form the supramolecular structures and organelles that allow a cell to carry out its many metabolic functions. Together, the molecules described in Part I are the stuff of life.

Water

Water is the most abundant substance in living systems, making up 70% or more of the weight of most organisms. The first living organisms on Earth doubtless arose in an aqueous environment, and the course of evolution has been shaped by the properties of the aqueous medium in which life began.

This chapter begins with descriptions of the physical and chemical properties of water, to which all aspects of cell structure and function are adapted. The attractive forces between water molecules and the slight tendency of water to ionize are of crucial importance to the structure and function of biomolecules. We review the topic of ionization in terms of equilibrium constants, pH, and titration curves, and consider how aqueous solutions of weak acids or bases and their salts act as buffers against pH changes in biological systems. The water molecule and its ionization products, H^+ and OH^-, profoundly influence the structure, self-assembly, and properties of all cellular components, including proteins, nucleic acids, and lipids. The noncovalent interactions responsible for the strength and specificity of "recognition" among biomolecules are decisively influenced by water's properties as a solvent, including its ability to form hydrogen bonds with itself and with solutes.

2.1 Weak Interactions in Aqueous Systems

Hydrogen bonds between water molecules provide the cohesive forces that make water a liquid at room temperature and a crystalline solid (ice) with a highly ordered arrangement of molecules at cold tempera-tures. Polar biomolecules dissolve readily in water because they can replace water-water interactions with more energetically favorable water-solute interactions. In contrast, nonpolar biomolecules are poorly soluble in water because they interfere with water-water interactions but are unable to form water-solute interactions. In aqueous solutions, nonpolar molecules tend to cluster together. Hydrogen bonds and ionic, hydrophobic (Greek, "water-fearing"), and van der Waals interactions are individually weak, but collectively they have a very significant influence on the three-dimensional structures of proteins, nucleic acids, polysaccharides, and membrane lipids.

Hydrogen Bonding Gives Water Its Unusual Properties

Water has a higher melting point, boiling point, and heat of vaporization than most other common solvents (Table 2–1). These unusual properties are a consequence of attractions between adjacent water molecules that give liquid water great internal cohesion. A look at the electron structure of the H_2O molecule reveals the cause of these intermolecular attractions.

Each hydrogen atom of a water molecule shares an electron pair with the central oxygen atom. The geometry of the molecule is dictated by the shapes of the outer electron orbitals of the oxygen atom, which are similar to the sp^3 bonding orbitals of carbon (see Fig. 1–15). These orbitals describe a rough tetrahedron, with a hydrogen atom at each of two corners and unshared electron pairs at the other two corners (**Fig. 2–1a**). The H—O—H bond angle is 104.5°, slightly less than the 109.5° of a perfect tetrahedron because of crowding by the nonbonding orbitals of the oxygen atom.

The oxygen nucleus attracts electrons more strongly than does the hydrogen nucleus (a proton); that is, oxygen is more electronegative. This means that the shared electrons are more often in the vicinity of the oxygen atom than of the hydrogen. The result of this unequal electron sharing is two electric dipoles in the water molecule, one along each of the H—O bonds;

TABLE 2–1   Melting Point, Boiling Point, and Heat of Vaporization of Some Common Solvents

	Melting point (°C)	Boiling point (°C)	Heat of vaporization (J/g)*
Water	0	100	2,260
Methanol (CH_3OH)	−98	65	1,100
Ethanol (CH_3CH_2OH)	−117	78	854
Propanol ($CH_3CH_2CH_2OH$)	−127	97	687
Butanol ($CH_3(CH_2)_2CH_2OH$)	−90	117	590
Acetone (CH_3COCH_3)	−95	56	523
Hexane ($CH_3(CH_2)_4CH_3$)	−98	69	423
Benzene (C_6H_6)	6	80	394
Butane ($CH_3(CH_2)_2CH_3$)	−135	−0.5	381
Chloroform ($CHCl_3$)	−63	61	247

*The heat energy required to convert 1.0 g of a liquid at its boiling point and at atmospheric pressure into its gaseous state at the same temperature. It is a direct measure of the energy required to overcome attractive forces between molecules in the liquid phase.

each hydrogen atom bears a partial positive charge (δ+), and the oxygen atom bears a partial negative charge equal in magnitude to the sum of the two partial positives (2δ−). As a result, there is an electrostatic attraction between the oxygen atom of one water molecule and the hydrogen of another (Fig. 2–1b), called a **hydrogen bond**. Throughout this book, we represent hydrogen bonds with three parallel blue lines, as in Figure 2–1b.

Hydrogen bonds are relatively weak. Those in liquid water have a **bond dissociation energy** (the energy required to break a bond) of about 23 kJ/mol, compared with 470 kJ/mol for the covalent O—H bond in water or 348 kJ/mol for a covalent C—C bond. The hydrogen

bond is about 10% covalent, due to overlaps in the bonding orbitals, and about 90% electrostatic. At room temperature, the thermal energy of an aqueous solution (the kinetic energy of motion of the individual atoms and molecules) is of the same order of magnitude as that required to break hydrogen bonds. When water is heated, the increase in temperature reflects the faster motion of individual water molecules. At any given time, most of the molecules in liquid water are hydrogen bonded, but the lifetime of each hydrogen bond is just 1 to 20 picoseconds (1 ps = 10^{-12} s); when one hydrogen bond breaks, another hydrogen bond forms, with the same partner or a new one, within 0.1 ps. The apt phrase "flickering clusters" has been applied to the short-lived groups of water molecules interlinked by hydrogen bonds in liquid water. The sum of all the hydrogen bonds between H_2O molecules confers great internal cohesion on liquid water. Extended networks of hydrogen-bonded water molecules also form bridges between solutes (proteins and nucleic acids, for example) that allow the larger molecules to interact with each other over distances of several nanometers without physically touching.

The nearly tetrahedral arrangement of the orbitals about the oxygen atom (Fig. 2–1a) allows each water molecule to form hydrogen bonds with as many as four neighboring water molecules. In liquid water at room temperature and atmospheric pressure, however, water molecules are disorganized and in continuous motion, so that each molecule forms hydrogen bonds with an average of only 3.4 other molecules. In ice, on the other hand, each water molecule is fixed in space and forms hydrogen bonds with a full complement of four other water molecules to yield a regular lattice structure **(Fig. 2–2)**. Hydrogen bonds account for the relatively high melting point of water, because much thermal energy is required to break a sufficient proportion of hydrogen bonds to destabilize the crystal lattice of ice

104.5°

Hydrogen bond
0.177 nm

Covalent bond
0.0965 nm

(a)   **(b)**

FIGURE 2–1 Structure of the water molecule. (a) The dipolar nature of the H_2O molecule is shown in a ball-and-stick model; the dashed lines represent the nonbonding orbitals. There is a nearly tetrahedral arrangement of the outer-shell electron pairs around the oxygen atom; the two hydrogen atoms have localized partial positive charges (δ+) and the oxygen atom has a partial negative charge (δ−). **(b)** Two H_2O molecules joined by a hydrogen bond (designated here, and throughout this book, by three blue lines) between the oxygen atom of the upper molecule and a hydrogen atom of the lower one. Hydrogen bonds are longer and weaker than covalent O—H bonds.

FIGURE 2-2 Hydrogen bonding in ice. In ice, each water molecule forms four hydrogen bonds, the maximum possible for a water molecule, creating a regular crystal lattice. By contrast, in liquid water at room temperature and atmospheric pressure, each water molecule hydrogen-bonds with an average of 3.4 other water molecules. This crystal lattice structure makes ice less dense than liquid water, and thus ice floats on liquid water.

(Table 2–1). When ice melts or water evaporates, heat is taken up by the system:

$$H_2O\,(\text{solid}) \longrightarrow H_2O\,(\text{liquid}) \qquad \Delta H = +5.9 \text{ kJ/mol}$$
$$H_2O\,(\text{liquid}) \longrightarrow H_2O\,(\text{gas}) \qquad \Delta H = +44.0 \text{ kJ/mol}$$

During melting or evaporation, the entropy of the aqueous system increases as the highly ordered arrays of water molecules in ice relax into the less orderly hydrogen-bonded arrays in liquid water or into the wholly disordered gaseous state. At room temperature, both the melting of ice and the evaporation of water occur spontaneously; the tendency of the water molecules to associate through hydrogen bonds is outweighed by the energetic push toward randomness. Recall that the free-energy change (ΔG) must have a negative value for a process to occur spontaneously: $\Delta G = \Delta H - T\Delta S$, where ΔG represents the driving force, ΔH the enthalpy change from making and breaking bonds, and ΔS the change in randomness. Because ΔH is positive for melting and evaporation, it is clearly the increase in entropy (ΔS) that makes ΔG negative and drives these changes.

Water Forms Hydrogen Bonds with Polar Solutes

Hydrogen bonds are not unique to water. They readily form between an electronegative atom (the hydrogen acceptor, usually oxygen or nitrogen) and a hydrogen atom covalently bonded to another electronegative atom (the hydrogen donor) in the same or another molecule (**Fig. 2–3**). Hydrogen atoms covalently bonded to car-

FIGURE 2-3 Common hydrogen bonds in biological systems. The hydrogen acceptor is usually oxygen or nitrogen; the hydrogen donor is another electronegative atom.

bon atoms do not participate in hydrogen bonding, because carbon is only slightly more electronegative than hydrogen and thus the C—H bond is only very weakly polar. The distinction explains why butanol ($CH_3(CH_2)_2CH_2OH$) has a relatively high boiling point of 117 °C, whereas butane ($CH_3(CH_2)_2CH_3$) has a boiling point of only −0.5 °C. Butanol has a polar hydroxyl group and thus can form intermolecular hydrogen bonds. Uncharged but polar biomolecules such as sugars dissolve readily in water because of the stabilizing effect of hydrogen bonds between the hydroxyl groups or carbonyl oxygen of the sugar and the polar water molecules. Alcohols, aldehydes, ketones, and compounds containing N—H bonds all form hydrogen bonds with water molecules (**Fig. 2–4**) and tend to be soluble in water.

FIGURE 2-4 Some biologically important hydrogen bonds.

FIGURE 2–5 Directionality of the hydrogen bond. The attraction between the partial electric charges (see Fig. 2-1) is greatest when the three atoms involved in the bond (in this case O, H, and O) lie in a straight line. When the hydrogen-bonded moieties are structurally constrained (when they are parts of a single protein molecule, for example), this ideal geometry may not be possible and the resulting hydrogen bond is weaker.

Hydrogen bonds are strongest when the bonded molecules are oriented to maximize electrostatic interaction, which occurs when the hydrogen atom and the two atoms that share it are in a straight line—that is, when the acceptor atom is in line with the covalent bond between the donor atom and H **(Fig. 2–5).** This arrangement puts the positive charge of the hydrogen ion directly between the two partial negative charges. Hydrogen bonds are thus highly directional and capable of holding two hydrogen-bonded molecules or groups in a specific geometric arrangement. As we shall see later, this property of hydrogen bonds confers very precise three-dimensional structures on protein and nucleic acid molecules, which have many intramolecular hydrogen bonds.

Water Interacts Electrostatically with Charged Solutes

Water is a polar solvent. It readily dissolves most biomolecules, which are generally charged or polar compounds (Table 2–2); compounds that dissolve easily in water are **hydrophilic** (Greek, "water-loving"). In contrast, nonpolar solvents such as chloroform and benzene are poor solvents for polar biomolecules but easily dissolve those that are **hydrophobic**—nonpolar molecules such as lipids and waxes.

Water dissolves salts such as NaCl by hydrating and stabilizing the Na^+ and Cl^- ions, weakening the electrostatic interactions between them and thus counteracting their tendency to associate in a crystalline lattice **(Fig. 2–6).** Water also readily dissolves charged biomolecules, including compounds with functional groups such as ionized carboxylic acids ($-COO^-$), protonated amines ($-NH_3^+$), and phosphate esters or anhydrides. Water replaces the solute-solute hydrogen bonds linking these biomolecules to each other with solute-water hydrogen bonds, thus screening the electrostatic interactions between solute molecules.

Water is effective in screening the electrostatic interactions between dissolved ions because it has a high dielectric constant, a physical property that reflects the number of dipoles in a solvent. The strength, or force (F), of ionic interactions in a solution depends on the magnitude of the charges (Q), the distance between the charged groups (r), and the dielectric constant (ε, which is dimensionless) of the solvent in which the interactions occur:

$$F = \frac{Q_1 Q_2}{\varepsilon r^2}$$

For water at 25 °C, ε is 78.5, and for the very nonpolar solvent benzene, ε is 4.6. Thus, ionic interactions between dissolved ions are much stronger in less polar environments. The dependence on r^2 is such that ionic attractions or repulsions operate only over short distances—in the range of 10 to 40 nm (depending on the electrolyte concentration) when the solvent is water.

TABLE 2–2 **Some Examples of Polar, Nonpolar, and Amphipathic Biomolecules (Shown as Ionic Forms at pH 7)**

Polar	**Nonpolar**
Glucose	Typical wax
Glycine $^+NH_3-CH_2-COO^-$	
Aspartate $^-OOC-CH_2-\overset{\overset{+NH_3}{\mid}}{C}H-COO^-$	**Amphipathic**
	Phenylalanine
Lactate $CH_3-\overset{\overset{}{\mid}}{C}H-COO^-$, OH	Phosphatidylcholine
Glycerol $HOCH_2-\overset{\overset{OH}{\mid}}{C}H-CH_2OH$	

Wax structures:
$CH_3(CH_2)_7-CH=CH-(CH_2)_6-CH_2-C\overset{\overset{O}{\parallel}}{} O$
$CH_3(CH_2)_7-CH=CH-(CH_2)_7-CH_2$

Phenylalanine: $\bigcirc-CH_2-\overset{\overset{+NH_3}{\mid}}{C}H-COO^-$

Phosphatidylcholine:
$CH_3(CH_2)_{15}CH_2-\overset{\overset{O}{\parallel}}{C}-O-CH_2$
$CH_3(CH_2)_{15}CH_2-\overset{\overset{}{\mid}}{C}-O-\overset{\overset{}{\mid}}{C}H$
$\overset{\overset{}{\parallel}}{O}$
$CH_2-O-\overset{\overset{O}{\mid}}{\underset{\underset{O^-}{\mid}}{P}}-O-CH_2-\overset{\overset{+N(CH_3)_3}{\mid}}{C}H_2$

☐ Polar groups   ☐ Nonpolar groups

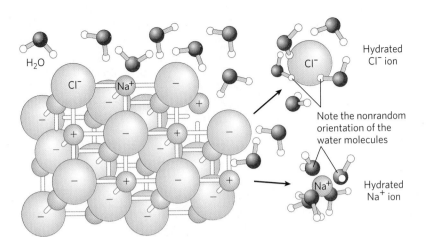

FIGURE 2–6 Water as solvent. Water dissolves many crystalline salts by hydrating their component ions. The NaCl crystal lattice is disrupted as water molecules cluster about the Cl^- and Na^+ ions. The ionic charges are partially neutralized, and the electrostatic attractions necessary for lattice formation are weakened.

Entropy Increases as Crystalline Substances Dissolve

As a salt such as NaCl dissolves, the Na^+ and Cl^- ions leaving the crystal lattice acquire far greater freedom of motion (Fig. 2–6). The resulting increase in entropy (randomness) of the system is largely responsible for the ease of dissolving salts such as NaCl in water. In thermodynamic terms, formation of the solution occurs with a favorable free-energy change: $\Delta G = \Delta H - T\Delta S$, where ΔH has a small positive value and $T\Delta S$ a large positive value; thus ΔG is negative.

Nonpolar Gases Are Poorly Soluble in Water

The molecules of the biologically important gases CO_2, O_2, and N_2 are nonpolar. In O_2 and N_2, electrons are shared equally by both atoms. In CO_2, each C=O bond is polar, but the two dipoles are oppositely directed and cancel each other (Table 2–3). The movement of molecules from the disordered gas phase into aqueous solution constrains their motion and the motion of water molecules and therefore represents a decrease in entropy. The nonpolar nature of these gases and the decrease in entropy when they enter solution combine to make

them very poorly soluble in water (Table 2–3). Some organisms have water-soluble "carrier proteins" (hemoglobin and myoglobin, for example) that facilitate the transport of O_2. Carbon dioxide forms carbonic acid (H_2CO_3) in aqueous solution and is transported as the HCO_3^- (bicarbonate) ion, either free—bicarbonate is very soluble in water (~100 g/L at 25 °C)—or bound to hemoglobin. Three other gases, NH_3, NO, and H_2S, also have biological roles in some organisms; these gases are polar, dissolve readily in water, and ionize in aqueous solution.

Nonpolar Compounds Force Energetically Unfavorable Changes in the Structure of Water

When water is mixed with benzene or hexane, two phases form; neither liquid is soluble in the other. Nonpolar compounds such as benzene and hexane are hydrophobic—they are unable to undergo energetically favorable interactions with water molecules, and they interfere with the hydrogen bonding among water molecules. All molecules or ions in aqueous solution interfere with the hydrogen bonding of some water

TABLE 2–3	Solubilities of Some Gases in Water		
Gas	Structure*	Polarity	Solubility in water (g/L)†
Nitrogen	N≡N	Nonpolar	0.018 (40 °C)
Oxygen	O=O	Nonpolar	0.035 (50 °C)
Carbon dioxide	O=C=O	Nonpolar	0.97 (45 °C)
Ammonia	H—N(H)—H	Polar	900 (10 °C)
Hydrogen sulfide	H—S—H	Polar	1,860 (40 °C)

*The arrows represent electric dipoles; there is a partial negative charge ($\delta-$) at the head of the arrow, a partial positive charge ($\delta+$; not shown here) at the tail.

†Note that polar molecules dissolve far better even at low temperatures than do nonpolar molecules at relatively high temperatures.

Hydrophilic "head group"

Hydrophobic alkyl group

"Flickering clusters" of H_2O molecules in bulk phase

Highly ordered H_2O molecules form "cages" around the hydrophobic alkyl chains

(a)

FIGURE 2–7 Amphipathic compounds in aqueous solution. (a) Long-chain fatty acids have very hydrophobic alkyl chains, each of which is surrounded by a layer of highly ordered water molecules. **(b)** By clustering together in micelles, the fatty acid molecules expose the smallest possible hydrophobic surface area to the water, and fewer water molecules are required in the shell of ordered water. The energy gained by freeing immobilized water molecules stabilizes the micelle.

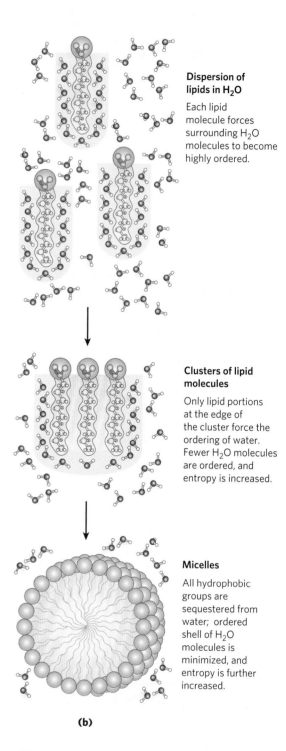

Dispersion of lipids in H_2O

Each lipid molecule forces surrounding H_2O molecules to become highly ordered.

Clusters of lipid molecules

Only lipid portions at the edge of the cluster force the ordering of water. Fewer H_2O molecules are ordered, and entropy is increased.

Micelles

All hydrophobic groups are sequestered from water; ordered shell of H_2O molecules is minimized, and entropy is further increased.

(b)

molecules in their immediate vicinity, but polar or charged solutes (such as NaCl) compensate for lost water-water hydrogen bonds by forming new solute-water interactions. The net change in enthalpy (ΔH) for dissolving these solutes is generally small. Hydrophobic solutes, however, offer no such compensation, and their addition to water may therefore result in a small gain of enthalpy; the breaking of hydrogen bonds between water molecules takes up energy from the system, requiring the input of energy from the surroundings. In addition to requiring this input of energy, dissolving hydrophobic compounds in water produces a measurable decrease in entropy. Water molecules in the immediate vicinity of a nonpolar solute are constrained in their possible orientations as they form a highly ordered cagelike shell around each solute molecule. These water molecules are not as highly oriented as those in **clathrates**, crystalline compounds of nonpolar solutes and water, but the effect is the same in both cases: the ordering of water molecules reduces entropy. The number of ordered water molecules, and therefore the magnitude of the entropy decrease, is proportional to the surface area of the hydrophobic solute enclosed within the cage of water molecules. The free-energy change for dissolving a nonpolar solute in water is thus unfavorable: $\Delta G = \Delta H - T\Delta S$, where ΔH has a positive value, ΔS has a negative value, and ΔG is positive.

Amphipathic compounds contain regions that are polar (or charged) and regions that are nonpolar (Table 2–2). When an amphipathic compound is mixed with water, the polar, hydrophilic region interacts favorably with the water and tends to dissolve, but the nonpolar, hydrophobic region tends to avoid contact with the water **(Fig. 2–7a)**. The nonpolar regions of the molecules cluster together to present the smallest hydrophobic area to the aqueous solvent, and the polar regions are arranged to maximize their interaction with the solvent (Fig. 2–7b). These stable structures of amphipathic compounds in water, called **micelles**, may contain hundreds or thousands of molecules. The forces that

Ordered water interacting with substrate and enzyme

Enzyme

Substrate

Disordered water displaced by enzyme-substrate interaction

Enzyme-substrate interaction stabilized by hydrogen-bonding, ionic, and hydrophobic interactions

FIGURE 2–8 Release of ordered water favors formation of an enzyme-substrate complex. While separate, both enzyme and substrate force neighboring water molecules into an ordered shell. Binding of substrate to enzyme releases some of the ordered water, and the resulting increase in entropy provides a thermodynamic push toward formation of the enzyme-substrate complex (see p. 198).

hold the nonpolar regions of the molecules together are called **hydrophobic interactions.** The strength of hydrophobic interactions is not due to any intrinsic attraction between nonpolar moieties. Rather, it results from the system's achieving the greatest thermodynamic stability by minimizing the number of ordered water molecules required to surround hydrophobic portions of the solute molecules.

Many biomolecules are amphipathic; proteins, pigments, certain vitamins, and the sterols and phospholipids of membranes all have both polar and nonpolar surface regions. Structures composed of these molecules are stabilized by hydrophobic interactions among the nonpolar regions. Hydrophobic interactions among lipids, and between lipids and proteins, are the most important determinants of structure in biological membranes. Hydrophobic interactions between nonpolar amino acids also stabilize the three-dimensional structures of proteins.

Hydrogen bonding between water and polar solutes also causes an ordering of water molecules, but the energetic effect is less significant than with nonpolar solutes. Disruption of ordered water molecules is part of the driving force for binding of a polar substrate (reactant) to the complementary polar surface of an enzyme: entropy increases as the enzyme displaces ordered water from the substrate, and as the substrate displaces ordered water from the enzyme surface **(Fig. 2–8)**.

van der Waals Interactions Are Weak Interatomic Attractions

When two uncharged atoms are brought very close together, their surrounding electron clouds influence each other. Random variations in the positions of the electrons around one nucleus may create a transient electric dipole, which induces a transient, opposite electric dipole in the nearby atom. The two dipoles weakly attract each other, bringing the two nuclei closer. These weak attractions are called **van der Waals interactions** (also known as London forces). As the two nuclei draw closer together, their electron clouds begin to repel each other. At the point where the net attraction is maximal, the nuclei are said to be in van der Waals contact. Each atom has a characteristic **van der Waals radius**, a measure of how close that atom will allow another to approach (Table 2–4). In the "space-filling" molecular models shown throughout this book, the atoms are depicted in sizes proportional to their van der Waals radii.

TABLE 2–4 van der Waals Radii and Covalent (Single-Bond) Radii of Some Elements

Element	van der Waals radius (nm)	Covalent radius for single bond (nm)
H	0.11	0.030
O	0.15	0.066
N	0.15	0.070
C	0.17	0.077
S	0.18	0.104
P	0.19	0.110
I	0.21	0.133

Sources: For van der Waals radii, Chauvin, R. (1992) Explicit periodic trend of van der Waals radii. *J. Phys. Chem.* 96, 9194–9197. For covalent radii, Pauling, L. (1960) *Nature of the Chemical Bond,* 3rd edn, Cornell University Press, Ithaca, NY.

Note: van der Waals radii describe the space-filling dimensions of atoms. When two atoms are joined covalently, the atomic radii at the point of bonding are less than the van der Waals radii, because the joined atoms are pulled together by the shared electron pair. The distance between nuclei in a van der Waals interaction or a covalent bond is about equal to the sum of the van der Waals or covalent radii, respectively, for the two atoms. Thus the length of a carbon–carbon single bond is about 0.077 nm + 0.077 nm = 0.154 nm.

Weak Interactions Are Crucial to Macromolecular Structure and Function

> I believe that as the methods of structural chemistry are further applied to physiological problems, it will be found that the significance of the hydrogen bond for physiology is greater than that of any other single structural feature.
>
> —*Linus Pauling,*
> The Nature of the Chemical Bond, *1939*

The noncovalent interactions we have described—hydrogen bonds and ionic, hydrophobic, and van der Waals interactions (Table 2–5)—are much weaker than covalent bonds. An input of about 350 kJ of energy is required to break a mole of (6×10^{23}) C—C single bonds, and about 410 kJ to break a mole of C—H bonds, but as little as 4 kJ is sufficient to disrupt a mole of typical van der Waals interactions. Hydrophobic interactions are also much weaker than covalent bonds, although they are substantially strengthened by a highly polar solvent (a concentrated salt solution, for example). Ionic interactions and hydrogen bonds are variable in strength, depending on the polarity of the solvent and the alignment of the hydrogen-bonded atoms, but they are always significantly weaker than covalent bonds. In aqueous solvent at 25 °C, the available thermal energy can be of the same order of magnitude as the strength of these weak interactions, and the interaction between solute and solvent (water) molecules is nearly as favorable as solute-solute interactions. Consequently, hydrogen bonds and ionic, hydrophobic, and van der Waals interactions are continually forming and breaking.

Although these four types of interactions are individually weak relative to covalent bonds, the cumulative effect of many such interactions can be very significant. For example, the noncovalent binding of an enzyme to its substrate may involve several hydrogen bonds and one or more ionic interactions, as well as hydrophobic and van der Waals interactions. The formation of each of these weak bonds contributes to a net decrease in the free energy of the system. We can calculate the stability of a noncovalent interaction, such as the hydrogen bonding of a small molecule to its macromolecular partner, from the binding energy, the reduction in the energy of the system when binding occurs. Stability, as measured by the equilibrium constant (see below) of the binding reaction, varies *exponentially* with binding energy. In order to dissociate two biomolecules (such as an enzyme and its bound substrate) that are associated noncovalently through multiple weak interactions, all these interactions must be disrupted at the same time. Because the interactions fluctuate randomly, such simultaneous disruptions are very unlikely. Therefore, 5 or 20 weak interactions bestow much greater molecular stability than would be expected intuitively from a simple summation of small binding energies.

Macromolecules such as proteins, DNA, and RNA contain so many sites of potential hydrogen bonding or ionic, van der Waals, or hydrophobic interactions that the cumulative effect of the many small binding forces can be enormous. For macromolecules, the most stable (that is, the native) structure is usually that in which weak interactions are maximized. The folding of a single polypeptide or polynucleotide chain into its three-dimensional shape is determined by this principle. The binding of an antigen to a specific antibody depends on the cumulative effects of many weak interactions. As noted earlier, the energy released when an enzyme binds noncovalently to its substrate is the main source of the enzyme's catalytic power. The binding of a hormone or a neurotransmitter to its cellular receptor protein is the result of multiple weak interactions. One consequence of the large size of enzymes and receptors (relative to their substrates or ligands) is that their extensive surfaces provide many opportunities for weak interactions. At the molecular level, the complementarity between interacting biomolecules reflects the complementarity and weak interactions between polar, charged, and hydrophobic groups on the surfaces of the molecules.

When the structure of a protein such as hemoglobin **(Fig. 2–9)** is determined by x-ray crystallography (see Box 4–5), water molecules are often found to be bound so tightly that they are part of the crystal structure; the same is true for water in crystals of RNA or DNA. These bound water molecules, which can also be detected in aqueous solutions by nuclear magnetic resonance, have distinctly different properties from

TABLE 2–5   Four Types of Noncovalent ("Weak") Interactions among Biomolecules in Aqueous Solvent

Hydrogen bonds Between neutral groups	
Between peptide bonds	
Ionic interactions Attraction	
Repulsion	$-^+NH_3 \longleftrightarrow H_3N^+ -$
Hydrophobic interactions	
van der Waals interactions	Any two atoms in close proximity

(a) **(b)**

FIGURE 2–9 Water binding in hemoglobin. (PDB ID 1A3N) The crystal structure of hemoglobin, shown **(a)** with bound water molecules (red spheres) and **(b)** without the water molecules. The water molecules are so firmly bound to the protein that they affect the x-ray diffraction pattern as though they were fixed parts of the crystal. The two α subunits of hemoglobin are shown in gray, the two β subunits in blue. Each subunit has a bound heme group (red stick structure), visible only in the β subunits in this view. The structure and function of hemoglobin are discussed in detail in Chapter 5.

those of the "bulk" water of the solvent. They are, for example, not osmotically active (see below). For many proteins, tightly bound water molecules are essential to their function. In a key reaction in photosynthesis, for example, protons flow across a biological membrane as light drives the flow of electrons through a series of electron-carrying proteins (see Fig. 19–62). One of these proteins, cytochrome *f*, has a chain of five bound water molecules **(Fig. 2–10)** that may provide a path for protons to move through the membrane by a process known as "proton hopping" (described below). Another such light-driven proton pump, bacteriorhodopsin, almost certainly uses a chain of precisely oriented bound water molecules in the transmembrane movement of protons (see Fig. 19–69b). Tightly bound water molecules can also form an essential part of the binding site of a protein for its ligand. In a bacterial arabinose-binding protein, for example, five water molecules form hydrogen bonds that provide critical cross-links between the sugar (arabinose) and the amino acid residues in the sugar-binding site **(Fig. 2–11)**.

Solutes Affect the Colligative Properties of Aqueous Solutions

Solutes of all kinds alter certain physical properties of the solvent, water: its vapor pressure, boiling point, melting point (freezing point), and osmotic pressure. These are called **colligative properties** (colligative meaning "tied together"), because the effect of solutes on all four properties has the same basis: the concentration of water is lower in solutions than in pure water. The effect of solute concentration on the colligative properties of water is independent of the chemical properties of the solute; it depends only on the *number* of solute particles (molecules or ions) in a given amount of water. For example, a compound such as NaCl, which dissociates in solution, has an effect on osmotic pressure

FIGURE 2–10 Water chain in cytochrome *f*. Water is bound in a proton channel of the membrane protein cytochrome *f*, which is part of the energy-trapping machinery of photosynthesis in chloroplasts (see Fig. 19–61). Five water molecules are hydrogen-bonded to each other and to functional groups of the protein: the peptide backbone atoms of valine, proline, arginine, and alanine residues, and the side chains of three asparagine and two glutamine residues. The protein has a bound heme (see Fig. 5–1), its iron ion facilitating electron flow during photosynthesis. Electron flow is coupled to the movement of protons across the membrane, which probably involves "proton hopping" (see Fig. 2–14) through this chain of bound water molecules.

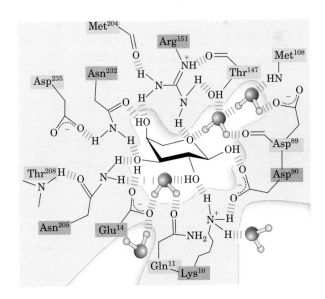

FIGURE 2–11 Hydrogen-bonded water as part of a protein's sugar-binding site. In the L-arabinose-binding protein of the bacterium *E. coli*, five water molecules are essential components of the hydrogen-bonded network of interactions between the sugar arabinose (center) and at least 13 amino acid residues in the sugar-binding site. Viewed in three dimensions, these interacting groups constitute two layers of binding moieties; amino acid residues in the first layer are screened in red, those in the second layer in green. Some of the hydrogen bonds are drawn longer than others for clarity; they are not actually longer than the others.

that is twice that of an equal number of moles of a non-dissociating solute such as glucose.

Water molecules tend to move from a region of higher water concentration to one of lower water concentration, in accordance with the tendency in nature for a system to become disordered. When two different aqueous solutions are separated by a semipermeable membrane (one that allows the passage of water but not solute molecules), water molecules diffusing from the region of higher water concentration to the region of lower water concentration produce osmotic pressure **(Fig. 2–12)**. Osmotic pressure, Π, measured as the force necessary to resist water movement (Fig. 2–12c), is approximated by the van't Hoff equation:

$$\Pi = icRT$$

in which R is the gas constant and T is the absolute temperature. The symbol i is the van't Hoff factor, which is a measure of the extent to which the solute dissociates into two or more ionic species. The term ic is the **osmolarity** of the solution, the product of the van't Hoff factor i and the solute's molar concentration c. In dilute NaCl solutions, the solute completely dissociates into Na^+ and Cl^-, doubling the number of solute particles, and thus $i = 2$. For all nonionizing solutes, $i = 1$. For solutions of several (n) solutes, Π is the sum of the contributions of each species:

$$\Pi = RT(i_1c_1 + i_2c_2 + \cdots + i_nc_n)$$

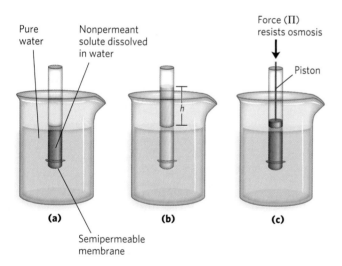

FIGURE 2–12 Osmosis and the measurement of osmotic pressure. **(a)** The initial state. The tube contains an aqueous solution, the beaker contains pure water, and the semipermeable membrane allows the passage of water but not solute. Water flows from the beaker into the tube to equalize its concentration across the membrane. **(b)** The final state. Water has moved into the solution of the nonpermeant compound, diluting it and raising the column of solution within the tube. At equilibrium, the force of gravity operating on the solution in the tube exactly balances the tendency of water to move into the tube, where its concentration is lower. **(c)** Osmotic pressure (Π) is measured as the force that must be applied to return the solution in the tube to the level of the water in the beaker. This force is proportional to the height, h, of the column in (b).

Osmosis, water movement across a semipermeable membrane driven by differences in osmotic pressure, is an important factor in the life of most cells. Plasma membranes are more permeable to water than to most other small molecules, ions, and macromolecules because protein channels (aquaporins; see Fig. 11–45) in the membrane selectively permit the passage of water. Solutions of osmolarity equal to that of a cell's cytosol are said to be **isotonic** relative to that cell. Surrounded by an isotonic solution, a cell neither gains nor loses water **(Fig. 2–13)**. In a **hypertonic** solution, one with higher osmolarity than that of the cytosol, the cell shrinks as water moves out. In a **hypotonic** solution, one with a lower osmolarity than the cytosol, the cell swells as water enters. In their natural environments, cells generally contain higher concentrations of biomolecules and ions than their surroundings, so osmotic pressure tends to drive water into cells. If not somehow counterbalanced, this inward movement of water would distend the plasma membrane and eventually cause bursting of the cell (osmotic lysis).

Several mechanisms have evolved to prevent this catastrophe. In bacteria and plants, the plasma membrane is surrounded by a nonexpandable cell wall of sufficient rigidity and strength to resist osmotic pressure

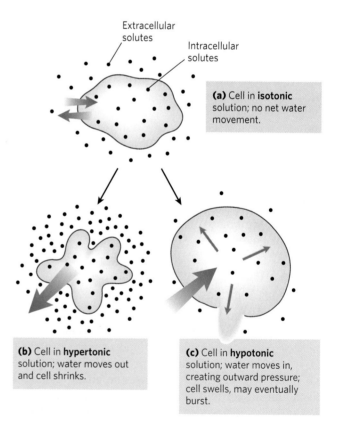

(a) Cell in **isotonic** solution; no net water movement.

(b) Cell in **hypertonic** solution; water moves out and cell shrinks.

(c) Cell in **hypotonic** solution; water moves in, creating outward pressure; cell swells, may eventually burst.

FIGURE 2–13 Effect of extracellular osmolarity on water movement across a plasma membrane. When a cell in osmotic balance with its surrounding medium—that is, a cell in **(a)** an isotonic medium—is transferred into **(b)** a hypertonic solution or **(c)** a hypotonic solution, water moves across the plasma membrane in the direction that tends to equalize osmolarity outside and inside the cell.

and prevent osmotic lysis. Certain freshwater protists that live in a highly hypotonic medium have an organelle (contractile vacuole) that pumps water out of the cell. In multicellular animals, blood plasma and interstitial fluid (the extracellular fluid of tissues) are maintained at an osmolarity close to that of the cytosol. The high concentration of albumin and other proteins in blood plasma contributes to its osmolarity. Cells also actively pump out Na^+ and other ions into the interstitial fluid to stay in osmotic balance with their surroundings.

Because the effect of solutes on osmolarity depends on the *number* of dissolved particles, not their *mass*, macromolecules (proteins, nucleic acids, polysaccharides) have far less effect on the osmolarity of a solution than would an equal mass of their monomeric components. For example, a *gram* of a polysaccharide composed of 1,000 glucose units has the same effect on osmolarity as a *milligram* of glucose. Storing fuel as polysaccharides (starch or glycogen) rather than as glucose or other simple sugars avoids an enormous increase in osmotic pressure in the storage cell.

Plants use osmotic pressure to achieve mechanical rigidity. The very high solute concentration in the plant cell vacuole draws water into the cell (Fig. 2–13), but the nonexpandable cell wall prevents swelling; instead, the pressure exerted against the cell wall (turgor pressure) increases, stiffening the cell, the tissue, and the plant body. When the lettuce in your salad wilts, it is because loss of water has reduced turgor pressure. Osmosis also has consequences for laboratory protocols. Mitochondria, chloroplasts, and lysosomes, for example, are enclosed by semipermeable membranes. In isolating these organelles from broken cells, biochemists must perform the fractionations in isotonic solutions (see Fig. 1–8) to prevent excessive entry of water into the organelles and the swelling and bursting that would follow. Buffers used in cellular fractionations commonly contain sufficient concentrations of sucrose or some other inert solute to protect the organelles from osmotic lysis.

WORKED EXAMPLE 2–1 Osmotic Strength of an Organelle I

Suppose the major solutes in intact lysosomes are KCl (~0.1 M) and NaCl (~0.03 M). When isolating lysosomes, what concentration of sucrose is required in the extracting solution at room temperature (25 °C) to prevent swelling and lysis?

Solution: We want to find a concentration of sucrose that gives an osmotic strength equal to that produced by the KCl and NaCl in the lysosomes. The equation for calculating osmotic strength (the van't Hoff equation) is

$$\Pi = RT(i_1c_1 + i_2c_2 + i_3c_3 + \cdots + i_nc_n)$$

where R is the gas constant 8.315 J/mol · K, T is the absolute temperature (Kelvin), c_1, c_2, and c_3 are the molar concentrations of each solute, and i_1, i_2, and i_3

are the numbers of particles each solute yields in solution ($i = 2$ for KCl and NaCl).

The osmotic strength of the lysosomal contents is

$$\begin{aligned}\Pi_{\text{lysosome}} &= RT(i_{\text{KCl}}c_{\text{KCl}} + i_{\text{NaCl}}c_{\text{NaCl}}) \\ &= RT[(2)(0.1 \text{ mol/L}) + (2)(0.03 \text{ mol/L})] \\ &= RT(0.26 \text{ mol/L})\end{aligned}$$

The osmotic strength of a sucrose solution is given by

$$\Pi_{\text{sucrose}} = RT(i_{\text{sucrose}}c_{\text{sucrose}})$$

In this case, $i_{\text{sucrose}} = 1$, because sucrose does not ionize. Thus,

$$\Pi_{\text{sucrose}} = RT(c_{\text{sucrose}})$$

The osmotic strength of the lysosomal contents equals that of the sucrose solution when

$$\begin{aligned}\Pi_{\text{sucrose}} &= \Pi_{\text{lysosome}} \\ RT(c_{\text{sucrose}}) &= RT(0.26 \text{ mol/L}) \\ c_{\text{sucrose}} &= 0.26 \text{ mol/L}\end{aligned}$$

So the required concentration of sucrose (FW 342) is (0.26 mol/L)(342 g/mol) = 88.92 g/L. Because the solute concentrations are only accurate to one significant figure, $c_{\text{sucrose}} = 0.09$ kg/L.

WORKED EXAMPLE 2–2 Osmotic Strength of an Organelle II

Suppose we decided to use a solution of a polysaccharide, say glycogen (p. 255), to balance the osmotic strength of the lysosomes (described in Worked Example 2–1). Assuming a linear polymer of 100 glucose units, calculate the amount of this polymer needed to achieve the same osmotic strength as the sucrose solution in Worked Example 2–1. The M_r of the glucose polymer is ~18,000, and, like sucrose, it does not ionize in solution.

Solution: As derived in Worked Example 2–1,

$$\Pi_{\text{sucrose}} = RT(0.26 \text{ mol/L})$$

Similarly,

$$\Pi_{\text{glycogen}} = RT(i_{\text{glycogen}}c_{\text{glycogen}}) = RT(c_{\text{glycogen}})$$

For a glycogen solution with the same osmotic strength as the sucrose solution,

$$\begin{aligned}\Pi_{\text{glycogen}} &= \Pi_{\text{sucrose}} \\ RT(c_{\text{glycogen}}) &= RT(0.26 \text{ mol/L}) \\ c_{\text{glycogen}} &= 0.26 \text{ mol/L} = (0.26 \text{ mol/L})(18,000 \text{ g/mol}) \\ &= 4.68 \text{ kg/L}\end{aligned}$$

Or, when significant figures are taken into account, $c_{\text{glycogen}} = 5$ kg/L, an absurdly high concentration.

As we'll see later (p. 256), cells of liver and muscle store carbohydrate not as low molecular weight sugars such as glucose or sucrose but as the high molecular weight polymer glycogen. This allows the cell to contain a large mass of glycogen with a minimal effect on the osmolarity of the cytosol.

SUMMARY 2.1 Weak Interactions in Aqueous Systems

▶ The very different electronegativities of H and O make water a highly polar molecule, capable of forming hydrogen bonds with itself and with solutes. Hydrogen bonds are fleeting, primarily electrostatic, and weaker than covalent bonds. Water is a good solvent for polar (hydrophilic) solutes, with which it forms hydrogen bonds, and for charged solutes, with which it interacts electrostatically.

▶ Nonpolar (hydrophobic) compounds dissolve poorly in water; they cannot hydrogen-bond with the solvent, and their presence forces an energetically unfavorable ordering of water molecules at their hydrophobic surfaces. To minimize the surface exposed to water, nonpolar compounds such as lipids form aggregates (micelles) in which the hydrophobic moieties are sequestered in the interior, associating through hydrophobic interactions, and only the more polar moieties interact with water.

▶ Weak, noncovalent interactions, in large numbers, decisively influence the folding of macromolecules such as proteins and nucleic acids. The most stable macromolecular conformations are those in which hydrogen bonding is maximized within the molecule and between the molecule and the solvent, and in which hydrophobic moieties cluster in the interior of the molecule away from the aqueous solvent.

▶ The physical properties of aqueous solutions are strongly influenced by the concentrations of solutes. When two aqueous compartments are separated by a semipermeable membrane (such as the plasma membrane separating a cell from its surroundings), water moves across that membrane to equalize the osmolarity in the two compartments. This tendency for water to move across a semipermeable membrane produces the osmotic pressure.

2.2 Ionization of Water, Weak Acids, and Weak Bases

Although many of the solvent properties of water can be explained in terms of the uncharged H_2O molecule, the small degree of ionization of water to hydrogen ions (H^+) and hydroxide ions (OH^-) must also be taken into account. Like all reversible reactions, the ionization of water can be described by an equilibrium constant. When weak acids are dissolved in water, they contribute H^+ by ionizing; weak bases consume H^+ by becoming protonated. These processes are also governed by equilibrium constants. The total hydrogen ion concentration from all sources is experimentally measurable and is expressed as the pH of the solution. To predict the state of ionization of solutes in water, we must take into account the relevant equilibrium constants for each ionization reaction. We therefore turn now to a brief

discussion of the ionization of water and of weak acids and bases dissolved in water.

Pure Water Is Slightly Ionized

Water molecules have a slight tendency to undergo reversible ionization to yield a hydrogen ion (a proton) and a hydroxide ion, giving the equilibrium

$$H_2O \rightleftharpoons H^+ + OH^- \qquad (2\text{--}1)$$

Although we commonly show the dissociation product of water as H^+, free protons do not exist in solution; hydrogen ions formed in water are immediately hydrated to form **hydronium ions** (H_3O^+). Hydrogen bonding between water molecules makes the hydration of dissociating protons virtually instantaneous:

$$H{-}O{\cdots}H{-}O \rightleftharpoons H{-}O^+{-}H + OH^-$$

The ionization of water can be measured by its electrical conductivity; pure water carries electrical current as H_3O^+ migrates toward the cathode and OH^- toward the anode. The movement of hydronium and hydroxide ions in the electric field is extremely fast compared with that of other ions such as Na^+, K^+, and Cl^-. This high ionic mobility results from the kind of "proton hopping" shown in **Figure 2–14**. No individual proton moves very

FIGURE 2–14 Proton hopping. Short "hops" of protons between a series of hydrogen-bonded water molecules result in an extremely rapid net movement of a proton over a long distance. As a hydronium ion (upper left) gives up a proton, a water molecule some distance away (bottom) acquires one, becoming a hydronium ion. Proton hopping is much faster than true diffusion and explains the remarkably high ionic mobility of H^+ ions compared with other monovalent cations such as Na^+ and K^+.

far through the bulk solution, but a series of proton hops between hydrogen-bonded water molecules causes the *net* movement of a proton over a long distance in a remarkably short time. (OH^- also moves rapidly by proton hopping, but in the opposite direction.) As a result of the high ionic mobility of H^+, acid-base reactions in aqueous solutions are exceptionally fast. As noted above, proton hopping very likely also plays a role in biological proton-transfer reactions (Fig. 2–10; see also Fig. 19–69b).

Because reversible ionization is crucial to the role of water in cellular function, we must have a means of expressing the extent of ionization of water in quantitative terms. A brief review of some properties of reversible chemical reactions shows how this can be done.

The position of equilibrium of any chemical reaction is given by its **equilibrium constant, K_{eq}** (sometimes expressed simply as K). For the generalized reaction

$$A + B \rightleftharpoons C + D \qquad (2\text{–}2)$$

the equilibrium constant K_{eq} can be defined in terms of the concentrations of reactants (A and B) and products (C and D) at equilibrium:

$$K_{eq} = \frac{[C]_{eq}[D]_{eq}}{[A]_{eq}[B]_{eq}}$$

Strictly speaking, the concentration terms should be the *activities*, or effective concentrations in nonideal solutions, of each species. Except in very accurate work, however, the equilibrium constant may be approximated by measuring the *concentrations* at equilibrium. For reasons beyond the scope of this discussion, equilibrium constants are dimensionless. Nonetheless, we have generally retained the concentration units (M) in the equilibrium expressions used in this book to remind you that molarity is the unit of concentration used in calculating K_{eq}.

The equilibrium constant is fixed and characteristic for any given chemical reaction at a specified temperature. It defines the composition of the final equilibrium mixture, regardless of the starting amounts of reactants and products. Conversely, we can calculate the equilibrium constant for a given reaction at a given temperature if the equilibrium concentrations of all its reactants and products are known. As we showed in Chapter 1 (p. 26), the standard free-energy change ($\Delta G°$) is directly related to $\ln K_{eq}$.

The Ionization of Water Is Expressed by an Equilibrium Constant

The degree of ionization of water at equilibrium (Eqn 2–1) is small; at 25 °C only about two of every 10^9 molecules in pure water are ionized at any instant. The equilibrium constant for the reversible ionization of water is

$$K_{eq} = \frac{[H^+][OH^-]}{[H_2O]} \qquad (2\text{–}3)$$

In pure water at 25 °C, the concentration of water is 55.5 M—grams of H_2O in 1 L divided by its gram molecular weight: $(1{,}000 \text{ g/L})/(18.015 \text{ g/mol})$—and is essentially constant in relation to the very low concentrations of H^+ and OH^-, namely 1×10^{-7} M. Accordingly, we can substitute 55.5 M in the equilibrium constant expression (Eqn 2–3) to yield

$$K_{eq} = \frac{[H^+][OH^-]}{[55.5 \text{ M}]}$$

On rearranging, this becomes

$$(55.5 \text{ M})(K_{eq}) = [H^+][OH^-] = K_w \qquad (2\text{–}4)$$

where K_w designates the product $(55.5 \text{ M})(K_{eq})$, the **ion product of water** at 25 °C.

The value for K_{eq}, determined by electrical-conductivity measurements of pure water, is 1.8×10^{-16} M at 25 °C. Substituting this value for K_{eq} in Equation 2–4 gives the value of the ion product of water:

$$K_w = [H^+][OH^-] = (55.5 \text{ M})(1.8 \times 10^{-16} \text{ M})$$
$$= 1.0 \times 10^{-14} \text{ M}^2$$

Thus the product $[H^+][OH^-]$ in aqueous solutions at 25 °C always equals 1×10^{-14} M^2. When there are exactly equal concentrations of H^+ and OH^-, as in pure water, the solution is said to be at **neutral pH**. At this pH, the concentration of H^+ and OH^- can be calculated from the ion product of water as follows:

$$K_w = [H^+][OH^-] = [H^+]^2 = [OH^-]^2$$

Solving for $[H^+]$ gives

$$[H^+] = \sqrt{K_w} = \sqrt{1 \times 10^{-14} \text{ M}^2}$$
$$[H^+] = [OH^-] = 10^{-7} \text{ M}$$

As the ion product of water is constant, whenever $[H^+]$ is greater than 1×10^{-7} M, $[OH^-]$ must be less than 1×10^{-7} M, and vice versa. When $[H^+]$ is very high, as in a solution of hydrochloric acid, $[OH^-]$ must be very low. From the ion product of water we can calculate $[H^+]$ if we know $[OH^-]$, and vice versa.

WORKED EXAMPLE 2–3 Calculation of $[H^+]$

What is the concentration of H^+ in a solution of 0.1 M NaOH?

Solution: We begin with the equation for the ion product of water:

$$K_w = [H^+][OH^-]$$

With $[OH^-] = 0.1$ M, solving for $[H^+]$ gives

$$[H^+] = \frac{K_w}{[OH^-]} = \frac{1 \times 10^{-14} \text{ M}^2}{0.1 \text{ M}} = \frac{10^{-14} \text{ M}^2}{10^{-1} \text{ M}}$$
$$= 10^{-13} \text{ M}$$

Calculation of [OH⁻]

What is the concentration of OH^- in a solution with an H^+ concentration of 1.3×10^{-4} M?

Solution: We begin with the equation for the ion product of water:

$$K_w = [H^+][OH^-]$$

With $[H^+] = 1.3 \times 10^{-4}$ M, solving for $[OH^-]$ gives

$$[OH^-] = \frac{K_w}{[H^+]} = \frac{1 \times 10^{-14} \text{ M}^2}{0.00013 \text{ M}} = \frac{10^{-14} \text{ M}^2}{1.3 \times 10^{-4} \text{ M}}$$
$$= 7.7 \times 10^{-11} \text{ M}$$

In all calculations be sure to round your answer to the correct number of significant figures, as here.

The pH Scale Designates the H⁺ and OH⁻ Concentrations

The ion product of water, K_w, is the basis for the **pH scale** (Table 2–6). It is a convenient means of designating the concentration of H^+ (and thus of OH^-) in any aqueous solution in the range between 1.0 M H^+ and 1.0 M OH^-. The term **pH** is defined by the expression

$$pH = \log \frac{1}{[H^+]} = -\log [H^+]$$

The symbol p denotes "negative logarithm of." For a precisely neutral solution at 25 °C, in which the concentration of hydrogen ions is 1.0×10^{-7} M, the pH can be calculated as follows:

$$pH = \log \frac{1}{1.0 \times 10^{-7}} = 7.0$$

Note that the concentration of H^+ must be expressed in molar (M) terms.

The value of 7 for the pH of a precisely neutral solution is not an arbitrarily chosen figure; it is derived from the absolute value of the ion product of water at 25 °C, which by convenient coincidence is a round number. Solutions having a pH greater than 7 are alkaline or basic; the concentration of OH^- is greater than that of H^+. Conversely, solutions having a pH less than 7 are acidic.

Keep in mind that the pH scale is logarithmic, not arithmetic. To say that two solutions differ in pH by 1 pH unit means that one solution has ten times the H^+ concentration of the other, but it does not tell us the absolute magnitude of the difference. **Figure 2–15** gives the pH values of some common aqueous fluids. A cola drink (pH 3.0) or red wine (pH 3.7) has an H^+ concentration approximately 10,000 times that of blood (pH 7.4).

The pH of an aqueous solution can be approximately measured with various indicator dyes, including litmus, phenolphthalein, and phenol red. These dyes undergo color changes as a proton dissociates from the dye

TABLE 2–6	The pH Scale		
[H⁺] (M)	**pH**	**[OH⁻] (M)**	**pOH***
10^0 (1)	0	10^{-14}	14
10^{-1}	1	10^{-13}	13
10^{-2}	2	10^{-12}	12
10^{-3}	3	10^{-11}	11
10^{-4}	4	10^{-10}	10
10^{-5}	5	10^{-9}	9
10^{-6}	6	10^{-8}	8
10^{-7}	7	10^{-7}	7
10^{-8}	8	10^{-6}	6
10^{-9}	9	10^{-5}	5
10^{-10}	10	10^{-4}	4
10^{-11}	11	10^{-3}	3
10^{-12}	12	10^{-2}	2
10^{-13}	13	10^{-1}	1
10^{-14}	14	10^0 (1)	0

*The expression pOH is sometimes used to describe the basicity, or OH^- concentration, of a solution; pOH is defined by the expression $pOH = -\log[OH^-]$, which is analogous to the expression for pH. Note that in all cases, $pH + pOH = 14$.

FIGURE 2–15   The pH of some aqueous fluids.

molecule. Accurate determinations of pH in the chemical or clinical laboratory are made with a glass electrode that is selectively sensitive to H^+ concentration but insensitive to Na^+, K^+, and other cations. In a pH meter, the signal from the glass electrode placed in a test solution is amplified and compared with the signal generated by a solution of accurately known pH.

Measurement of pH is one of the most important and frequently used procedures in biochemistry. The pH affects the structure and activity of biological macromolecules; for example, the catalytic activity of enzymes is strongly dependent on pH (see Fig. 2–22). Measurements of the pH of blood and urine are commonly used in medical diagnoses. The pH of the blood plasma of people with severe, uncontrolled diabetes, for example, is often below the normal value of 7.4; this condition is called **acidosis** (described in more detail below). In certain other diseases the pH of the blood is higher than normal, a condition known as **alkalosis**. Extreme acidosis or alkalosis can be life-threatening. ∎

Weak Acids and Bases Have Characteristic Acid Dissociation Constants

Hydrochloric, sulfuric, and nitric acids, commonly called strong acids, are completely ionized in dilute aqueous solutions; the strong bases NaOH and KOH are also completely ionized. Of more interest to biochemists is the behavior of weak acids and bases—those not completely ionized when dissolved in water. These are ubiquitous in biological systems and play important roles in metabolism and its regulation. The behavior of aqueous solutions of weak acids and bases is best understood if we first define some terms.

Acids may be defined as proton donors and bases as proton acceptors. When a proton donor such as acetic acid (CH_3COOH) loses a proton, it becomes the corresponding proton acceptor, in this case the acetate anion (CH_3COO^-). A proton donor and its corresponding proton acceptor make up a **conjugate acid-base pair** **(Fig. 2–16)**, related by the reversible reaction

$$CH_3COOH \rightleftharpoons CH_3COO^- + H^+$$

Each acid has a characteristic tendency to lose its proton in an aqueous solution. The stronger the acid, the greater its tendency to lose its proton. The tendency of any acid (HA) to lose a proton and form its conjugate base (A^-) is defined by the equilibrium constant (K_{eq}) for the reversible reaction

$$HA \rightleftharpoons H^+ + A^-$$

for which

$$K_{eq} = \frac{[H^+][A^-]}{[HA]} = K_a$$

FIGURE 2–16 Conjugate acid-base pairs consist of a proton donor and a proton acceptor. Some compounds, such as acetic acid and ammonium ion, are monoprotic; they can give up only one proton. Others are diprotic (carbonic acid and glycine) or triprotic (phosphoric acid). The dissociation reactions for each pair are shown where they occur along a pH gradient. The equilibrium or dissociation constant (K_a) and its negative logarithm, the pK_a, are shown for each reaction. *For an explanation of apparent discrepancies in pK_a values for carbonic acid (H_2CO_3), see p. 67.

Equilibrium constants for ionization reactions are usually called **ionization constants** or **acid dissociation constants**, often designated K_a. The dissociation constants of some acids are given in Figure 2–16. Stronger acids, such as phosphoric and carbonic acids, have larger ionization constants; weaker acids, such as monohydrogen phosphate (HPO_4^{2-}), have smaller ionization constants.

Also included in Figure 2–16 are values of pK_a, which is analogous to pH and is defined by the equation

$$pK_a = \log \frac{1}{K_a} = -\log K_a$$

The stronger the tendency to dissociate a proton, the stronger is the acid and the lower its pK_a. As we shall now see, the pK_a of any weak acid can be determined quite easily.

Titration Curves Reveal the pK_a of Weak Acids

Titration is used to determine the amount of an acid in a given solution. A measured volume of the acid is titrated with a solution of a strong base, usually sodium hydroxide (NaOH), of known concentration. The NaOH is added in small increments until the acid is consumed (neutralized), as determined with an indicator dye or a pH meter. The concentration of the acid in the original solution can be calculated from the volume and concentration of NaOH added. The amounts of acid and base in titrations are often expressed in terms of equivalents, where one equivalent is the amount of a substance that will react with, or supply, one mole of hydrogen ions in an acid-base reaction.

A plot of pH against the amount of NaOH added (a **titration curve**), reveals the pK_a of the weak acid. Consider the titration of a 0.1 M solution of acetic acid with 0.1 M NaOH at 25 °C (**Fig. 2–17**). Two reversible equilibria are involved in the process (here, for simplicity, acetic acid is denoted HAc):

$$H_2O \rightleftharpoons H^+ + OH^- \qquad (2\text{–}5)$$

$$HAc \rightleftharpoons H^+ + Ac^- \qquad (2\text{–}6)$$

The equilibria must simultaneously conform to their characteristic equilibrium constants, which are, respectively,

$$K_w = [H^+][OH^-] = 1 \times 10^{-14}\ \text{M}^2 \qquad (2\text{–}7)$$

$$K_a = \frac{[H^+][Ac^-]}{[HAc]} = 1.74 \times 10^{-5}\ \text{M} \qquad (2\text{–}8)$$

At the beginning of the titration, before any NaOH is added, the acetic acid is already slightly ionized, to an extent that can be calculated from its ionization constant (Eqn 2–8).

As NaOH is gradually introduced, the added OH^- combines with the free H^+ in the solution to form H_2O, to an extent that satisfies the equilibrium relationship in Equation 2–7. As free H^+ is removed, HAc dissociates

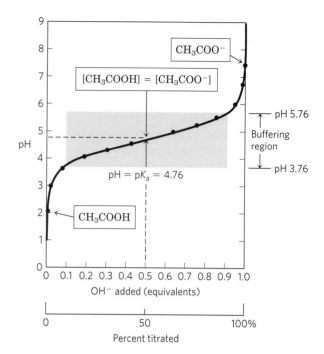

FIGURE 2–17 The titration curve of acetic acid. After addition of each increment of NaOH to the acetic acid solution, the pH of the mixture is measured. This value is plotted against the amount of NaOH added, expressed as a fraction of the total NaOH required to convert all the acetic acid (CH_3COOH) to its deprotonated form, acetate (CH_3COO^-). The points so obtained yield the titration curve. Shown in the boxes are the predominant ionic forms at the points designated. At the midpoint of the titration, the concentrations of the proton donor and proton acceptor are equal, and the pH is numerically equal to the pK_a. The shaded zone is the useful region of buffering power, generally between 10% and 90% titration of the weak acid.

further to satisfy its own equilibrium constant (Eqn 2–8). The net result as the titration proceeds is that more and more HAc ionizes, forming Ac^-, as the NaOH is added. At the midpoint of the titration, at which exactly 0.5 equivalent of NaOH has been added per equivalent of the acid, one-half of the original acetic acid has undergone dissociation, so that the concentration of the proton donor, [HAc], now equals that of the proton acceptor, [Ac^-]. At this midpoint a very important relationship holds: the pH of the equimolar solution of acetic acid and acetate is exactly equal to the pK_a of acetic acid ($pK_a = 4.76$; Figs 2–16, 2–17). The basis for this relationship, which holds for all weak acids, will soon become clear.

As the titration is continued by adding further increments of NaOH, the remaining nondissociated acetic acid is gradually converted into acetate. The end point of the titration occurs at about pH 7.0: all the acetic acid has lost its protons to OH^-, to form H_2O and acetate. Throughout the titration the two equilibria (Eqns 2–5, 2–6) coexist, each always conforming to its equilibrium constant.

Figure 2–18 compares the titration curves of three weak acids with very different ionization constants: acetic acid ($pK_a = 4.76$); dihydrogen phosphate,

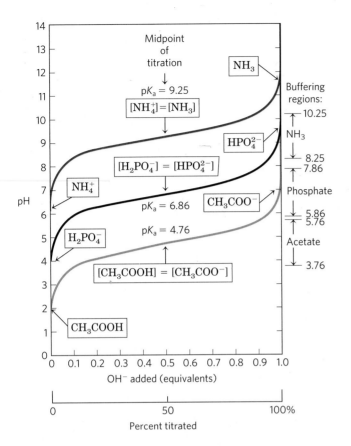

FIGURE 2–18 Comparison of the titration curves of three weak acids. Shown here are the titration curves for CH_3COOH, $H_2PO_4^-$, and NH_4^+. The predominant ionic forms at designated points in the titration are given in boxes. The regions of buffering capacity are indicated at the right. Conjugate acid-base pairs are effective buffers between approximately 10% and 90% neutralization of the proton-donor species.

$H_2PO_4^-$ ($pK_a = 6.86$); and ammonium ion, NH_4^+ ($pK_a = 9.25$). Although the titration curves of these acids have the same shape, they are displaced along the pH axis because the three acids have different strengths. Acetic acid, with the highest K_a (lowest pK_a) of the three, is the strongest of the three weak acids (loses its proton most readily); it is already half dissociated at pH 4.76. Dihydrogen phosphate loses a proton less readily, being half dissociated at pH 6.86. Ammonium ion is the weakest acid of the three and does not become half dissociated until pH 9.25.

The titration curve of a weak acid shows graphically that a weak acid and its anion—a conjugate acid-base pair—can act as a buffer, as we describe in the next section.

SUMMARY 2.2 Ionization of Water, Weak Acids, and Weak Bases

▶ Pure water ionizes slightly, forming equal numbers of hydrogen ions (hydronium ions, H_3O^+) and hydroxide ions. The extent of ionization is described by an equilibrium constant, $K_{eq} = \dfrac{[H^+][OH^-]}{[H_2O]}$,

from which the ion product of water, K_w, is derived. At 25 °C, $K_w = [H^+][OH^-] = (55.5 \text{ M})(K_{eq}) = 10^{-14} \text{ M}^2$.

▶ The pH of an aqueous solution reflects, on a logarithmic scale, the concentration of hydrogen ions:

$$pH = \log \frac{1}{[H^+]} = -\log [H^+].$$

▶ The greater the acidity of a solution, the lower its pH. Weak acids partially ionize to release a hydrogen ion, thus lowering the pH of the aqueous solution. Weak bases accept a hydrogen ion, increasing the pH. The extent of these processes is characteristic of each particular weak acid or base and is expressed as an acid dissociation constant:

$$K_{eq} = \frac{[H^+][A^-]}{[HA]} = K_a.$$

▶ The pK_a expresses, on a logarithmic scale, the relative strength of a weak acid or base:

$$pK_a = \log \frac{1}{K_a} = -\log K_a.$$

▶ The stronger the acid, the smaller its pK_a; the stronger the base, the larger its pK_a. The pK_a can be determined experimentally; it is the pH at the midpoint of the titration curve for the acid or base.

2.3 Buffering against pH Changes in Biological Systems

Almost every biological process is pH-dependent; a small change in pH produces a large change in the rate of the process. This is true not only for the many reactions in which the H^+ ion is a direct participant, but also for those reactions in which there is no apparent role for H^+ ions. The enzymes that catalyze cellular reactions, and many of the molecules on which they act, contain ionizable groups with characteristic pK_a values. The protonated amino and carboxyl groups of amino acids and the phosphate groups of nucleotides, for example, function as weak acids; their ionic state is determined by the pH of the surrounding medium. (When an ionizable group is sequestered in the middle of a protein, away from the aqueous solvent, its pK_a, or apparent pK_a, can be significantly different from its pK_a in water.) As we noted above, ionic interactions are among the forces that stabilize a protein molecule and allow an enzyme to recognize and bind its substrate.

Cells and organisms maintain a specific and constant cytosolic pH, usually near pH 7, keeping biomolecules in their optimal ionic state. In multicellular organisms, the pH of extracellular fluids is also tightly regulated. Constancy of pH is achieved primarily by biological buffers: mixtures of weak acids and their conjugate bases.

Buffers Are Mixtures of Weak Acids and Their Conjugate Bases

Buffers are aqueous systems that tend to resist changes in pH when small amounts of acid (H^+) or base (OH^-) are added. A buffer system consists of a weak acid (the proton donor) and its conjugate base (the proton acceptor). As an example, a mixture of equal concentrations of acetic acid and acetate ion, found at the midpoint of the titration curve in Figure 2–17, is a buffer system. Notice that the titration curve of acetic acid has a relatively flat zone extending about 1 pH unit on either side of its midpoint pH of 4.76. In this zone, a given amount of H^+ or OH^- added to the system has much less effect on pH than the same amount added outside the zone. This relatively flat zone is the **buffering region** of the acetic acid–acetate buffer pair. At the midpoint of the buffering region, where the concentration of the proton donor (acetic acid) exactly equals that of the proton acceptor (acetate), the buffering power of the system is maximal; that is, its pH changes least on addition of H^+ or OH^-. The pH at this point in the titration curve of acetic acid is equal to its pK_a. The pH of the acetate buffer system does change slightly when a small amount of H^+ or OH^- is added, but this change is very small compared with the pH change that would result if the same amount of H^+ or OH^- were added to pure water or to a solution of the salt of a strong acid and strong base, such as NaCl, which has no buffering power.

Buffering results from two reversible reaction equilibria occurring in a solution of nearly equal concentrations of a proton donor and its conjugate proton acceptor. **Figure 2–19** explains how a buffer system works. Whenever H^+ or OH^- is added to a buffer, the result is a small change in the ratio of the relative concentrations of the weak acid and its anion and thus a small change in pH. The decrease in concentration of one component of the system is balanced exactly by an increase in the other. The sum of the buffer components does not change, only their ratio.

Each conjugate acid-base pair has a characteristic pH zone in which it is an effective buffer (Fig. 2–18). The $H_2PO_4^-/HPO_4^{2-}$ pair has a pK_a of 6.86 and thus can serve as an effective buffer system between approximately pH 5.9 and pH 7.9; the NH_4^+/NH_3 pair, with a pK_a of 9.25, can act as a buffer between approximately pH 8.3 and pH 10.3.

The Henderson-Hasselbalch Equation Relates pH, pK_a, and Buffer Concentration

The titration curves of acetic acid, $H_2PO_4^-$, and NH_4^+ (Fig. 2–18) have nearly identical shapes, suggesting that these curves reflect a fundamental law or relationship. This is indeed the case. The shape of the titration curve of any weak acid is described by the Henderson-Hasselbalch equation, which is important for understanding buffer action and acid-base balance in the blood and tis-

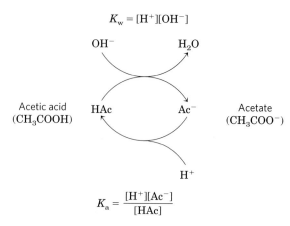

FIGURE 2–19 The acetic acid-acetate pair as a buffer system. The system is capable of absorbing either H^+ or OH^- through the reversibility of the dissociation of acetic acid. The proton donor, acetic acid (HAc), contains a reserve of bound H^+, which can be released to neutralize an addition of OH^- to the system, forming H_2O. This happens because the product $[H^+][OH^-]$ transiently exceeds K_w ($1 \times 10^{-14} M^2$). The equilibrium quickly adjusts to restore the product to $1 \times 10^{-14} M^2$ (at 25 °C), thus transiently reducing the concentration of H^+. But now the quotient $[H^+][Ac^-]/[HAc]$ is less than K_a, so HAc dissociates further to restore equilibrium. Similarly, the conjugate base, Ac^-, can react with H^+ ions added to the system; again, the two ionization reactions simultaneously come to equilibrium. Thus a conjugate acid-base pair, such as acetic acid and acetate ion, tends to resist a change in pH when small amounts of acid or base are added. Buffering action is simply the consequence of two reversible reactions taking place simultaneously and reaching their points of equilibrium as governed by their equilibrium constants, K_w and K_a.

sues of vertebrates. This equation is simply a useful way of restating the expression for the ionization constant of an acid. For the ionization of a weak acid HA, the Henderson-Hasselbalch equation can be derived as follows:

$$K_a = \frac{[H^+][A^-]}{[HA]}$$

First solve for $[H^+]$:

$$[H^+] = K_a \frac{[HA]}{[A^-]}$$

Then take the negative logarithm of both sides:

$$-\log[H^+] = -\log K_a - \log \frac{[HA]}{[A^-]}$$

Substitute pH for $-\log[H^+]$ and pK_a for $-\log K_a$:

$$pH = pK_a - \log \frac{[HA]}{[A^-]}$$

Now invert $-\log [HA]/[A^-]$, which involves changing its sign, to obtain the **Henderson-Hasselbalch equation**:

$$pH = pK_a + \log \frac{[A^-]}{[HA]} \qquad (2\text{–}9)$$

This equation fits the titration curve of all weak acids and enables us to deduce some important quantitative relationships. For example, it shows why the pK_a of a

FIGURE 2–20 Ionization of histidine. The amino acid histidine, a component of proteins, is a weak acid. The pK_a of the protonated nitrogen of the side chain is 6.0.

weak acid is equal to the pH of the solution at the midpoint of its titration. At that point, [HA] = [A$^-$], and

$$pH = pK_a + \log 1 = pK_a + 0 = pK_a$$

The Henderson-Hasselbalch equation also allows us to (1) calculate pK_a, given pH and the molar ratio of proton donor and acceptor; (2) calculate pH, given pK_a and the molar ratio of proton donor and acceptor; and (3) calculate the molar ratio of proton donor and acceptor, given pH and pK_a.

Weak Acids or Bases Buffer Cells and Tissues against pH Changes

The intracellular and extracellular fluids of multicellular organisms have a characteristic and nearly constant pH. The organism's first line of defense against changes in internal pH is provided by buffer systems. The cytoplasm of most cells contains high concentrations of proteins, and these proteins contain many amino acids with functional groups that are weak acids or weak bases. For example, the side chain of histidine **(Fig. 2–20)** has a pK_a of 6.0 and thus can exist in either the protonated or unprotonated form near neutral pH. Proteins containing histidine residues therefore buffer effectively near neutral pH.

Ionization of Histidine

Calculate the fraction of histidine that has its imidazole side chain protonated at pH 7.3. The pK_a values for histidine are pK_1 = 1.8, pK_2 (imidazole) = 6.0, and pK_3 = 9.2 (see Fig. 3–12b).

Solution: The three ionizable groups in histidine have sufficiently different pK_a values that the first acid (—COOH) is completely ionized before the second (protonated imidazole) begins to dissociate a proton, and the second ionizes completely before the third (—NH$_3^+$) begins to dissociate its proton. (With the Henderson-Hasselbalch equation, we can easily show that a weak acid goes from 1% ionized at 2 pH units below its pK_a to 99% ionized at 2 pH units above its pK_a; see also Fig. 3–12b.) At pH 7.3, the carboxyl group of histidine is entirely deprotonated (—COO$^-$) and the α-amino group is fully protonated (—NH$_3^+$). We can therefore assume that at pH 7.3, the only group that is partially

dissociated is the imidazole group, which can be protonated (we'll abbreviate as HisH$^+$) or not (His).

We use the Henderson-Hasselbalch equation:

$$pH = pK_a + \log \frac{[A^-]}{[HA]}$$

Substituting pK_2 = 6.0 and pH = 7.3:

$$7.3 = 6.0 + \log \frac{[His]}{[HisH^+]}$$
$$1.3 = \log \frac{[His]}{[HisH^+]}$$
$$\text{antilog } 1.3 = \frac{[His]}{[HisH^+]} = 2.0 \times 10^1$$

This gives us the *ratio* of [His] to [HisH$^+$] (20 to 1 in this case). We want to convert this ratio to the *fraction* of total histidine that is in the unprotonated form His at pH 7.3. That fraction is 20/21 (20 parts His per 1 part HisH$^+$, *in a total of 21 parts histidine* in either form), or about 95.2%; the remainder (100% minus 95.2%) is protonated—about 5%.

Nucleotides such as ATP, as well as many metabolites of low molecular weight, contain ionizable groups that can contribute buffering power to the cytoplasm. Some highly specialized organelles and extracellular compartments have high concentrations of compounds that contribute buffering capacity: organic acids buffer the vacuoles of plant cells; ammonia buffers urine.

Two especially important biological buffers are the phosphate and bicarbonate systems. The phosphate buffer system, which acts in the cytoplasm of all cells, consists of H$_2$PO$_4^-$ as proton donor and HPO$_4^{2-}$ as proton acceptor:

$$H_2PO_4^- \rightleftharpoons H^+ + HPO_4^{2-}$$

The phosphate buffer system is maximally effective at a pH close to its pK_a of 6.86 (Figs 2–16, 2–18) and thus tends to resist pH changes in the range between about 5.9 and 7.9. It is therefore an effective buffer in biological fluids; in mammals, for example, extracellular fluids and most cytoplasmic compartments have a pH in the range of 6.9 to 7.4.

Phosphate Buffers

(a) What is the pH of a mixture of 0.042 M NaH$_2$PO$_4$ and 0.058 M Na$_2$HPO$_4$?

Solution: We use the Henderson-Hasselbalch equation, which we'll express here as

$$pH = pK_a + \log \frac{[\text{conjugate base}]}{[\text{acid}]}$$

In this case, the acid (the species that gives up a proton) is H$_2$PO$_4^-$, and the conjugate base (the species that gains a proton) is HPO$_4^{2-}$. Substituting the given concentrations of acid and conjugate base and the pK_a (6.86),

$$pH = 6.86 + \log \frac{0.058}{0.042} = 6.86 + 0.14 = 7.0$$

We can roughly check this answer. When more conjugate base than acid is present, the acid is more than 50% titrated and thus the pH is above the pK_a (6.86), where the acid is exactly 50% titrated.

(b) If 1.0 mL of 10.0 M NaOH is added to a liter of the buffer prepared in (a), how much will the pH change?

Solution: A liter of the buffer contains 0.042 mol of NaH_2PO_4. Adding 1.0 mL of 10.0 M NaOH (0.010 mol) would titrate an equivalent amount (0.010 mol) of NaH_2PO_4 to Na_2HPO_4, resulting in 0.032 mol of NaH_2PO_4 and 0.068 mol of Na_2HPO_4. The new pH is

$$pH = pK_a + \log \frac{[HPO_4^{2-}]}{[H_2PO_4^-]}$$
$$= 6.86 + \log \frac{0.068}{0.032} = 6.86 + 0.33 = 7.2$$

(c) If 1.0 mL of 10.0 M NaOH is added to a liter of pure water at pH 7.0, what is the final pH? Compare this with the answer in (b).

Solution: The NaOH dissociates completely into Na^+ and OH^-, giving $[OH^-] = 0.010$ mol/L $= 1.0 \times 10^{-2}$ M. The pOH is the negative logarithm of $[OH^-]$, so pOH $= 2.0$. Given that in all solutions, pH + pOH = 14, the pH of the solution is 12.

So, an amount of NaOH that increases the pH of water from 7 to 12 increases the pH of a buffered solution, as in (b), from 7.0 to just 7.2. Such is the power of buffering!

Blood plasma is buffered in part by the bicarbonate system, consisting of carbonic acid (H_2CO_3) as proton donor and bicarbonate (HCO_3^-) as proton acceptor (K_1 is the first of several equilibrium constants in the bicarbonate buffering system):

$$H_2CO_3 \rightleftharpoons H^+ + HCO_3^-$$
$$K_1 = \frac{[H^+][HCO_3^-]}{[H_2CO_3]}$$

This buffer system is more complex than other conjugate acid-base pairs because one of its components, carbonic acid (H_2CO_3), is formed from dissolved (d) carbon dioxide and water, in a reversible reaction:

$$CO_2(d) + H_2O \rightleftharpoons H_2CO_3$$
$$K_2 = \frac{[H_2CO_3]}{[CO_2(d)][H_2O]}$$

Carbon dioxide is a gas under normal conditions, and CO_2 dissolved in an aqueous solution is in equilibrium with CO_2 in the gas (g) phase:

$$CO_2(g) \rightleftharpoons CO_2(d)$$
$$K_3 = \frac{[CO_2(d)]}{[CO_2(g)]}$$

The pH of a bicarbonate buffer system depends on the concentration of H_2CO_3 and HCO_3^-, the proton donor and acceptor components. The concentration of H_2CO_3 in turn depends on the concentration of dissolved CO_2, which in turn depends on the concentration of CO_2 in the gas phase, or the **partial pressure** of CO_2, denoted pCO_2. Thus the pH of a bicarbonate buffer exposed to a gas phase is ultimately determined by the concentration of HCO_3^- in the aqueous phase and by pCO_2 in the gas phase.

The bicarbonate buffer system is an effective physiological buffer near pH 7.4, because the H_2CO_3 of blood plasma is in equilibrium with a large reserve capacity of $CO_2(g)$ in the air space of the lungs. As noted above, this buffer system involves three reversible equilibria, in this case between gaseous CO_2 in the lungs and bicarbonate (HCO_3^-) in the blood plasma **(Fig. 2–21)**.

Blood can pick up H^+, such as from the lactic acid produced in muscle tissue during vigorous exercise. Alternatively, it can lose H^+, such as by protonation of the NH_3 produced during protein catabolism. When H^+ is added to blood as it passes through the tissues, reaction 1 in Figure 2–21 proceeds toward a new equilibrium, in which $[H_2CO_3]$ is increased. This in turn increases $[CO_2(d)]$ in the blood (reaction 2) and thus increases the partial pressure of $CO_2(g)$ in the air space of the lungs (reaction 3); the extra CO_2 is exhaled. Conversely, when H^+ is lost from the blood, the opposite events occur: more H_2CO_3 dissociates into H^+ and HCO_3^- and thus more $CO_2(g)$ from the lungs dissolves in blood plasma. The rate of respiration—that is, the rate of inhaling and exhaling—can quickly adjust these equilibria to keep the blood pH nearly constant. The rate of respiration is controlled by the brain stem, where detection of an increased blood pCO_2 or decreased blood pH triggers deeper and more frequent breathing.

At the pH of blood plasma (7.4) very little H_2CO_3 is present in comparison with HCO_3^-, and the addition of a small amount of base (NH_3 or OH^-) would titrate this H_2CO_3, exhausting the buffering capacity. The important

FIGURE 2–21 The bicarbonate buffer system. CO_2 in the air space of the lungs is in equilibrium with the bicarbonate buffer in the blood plasma passing through the lung capillaries. Because the concentration of dissolved CO_2 can be adjusted rapidly through changes in the rate of breathing, the bicarbonate buffer system of the blood is in near-equilibrium with a large potential reservoir of CO_2.

role for H_2CO_3 ($pK_a = 3.57$ at 37 °C) in buffering blood plasma (~pH 7.4) seems inconsistent with our earlier statement that a buffer is most effective in the range of 1 pH unit above and below its pK_a. The explanation for this paradox is the large reservoir of $CO_2(d)$ in blood. Its rapid equilibration with H_2CO_3 results in the formation of additional H_2CO_3:

$$CO_2(d) + H_2O \rightleftharpoons H_2CO_3$$

It is useful in clinical medicine to have a simple expression for blood pH in terms of dissolved CO_2, which is commonly monitored along with other blood gases. We can define a constant, K_h, which is the equilibrium constant for the hydration of CO_2 to form H_2CO_3:

$$K_h = \frac{[H_2CO_3]}{[CO_2(d)]}$$

Then, to take the $CO_2(d)$ reservoir into account, we can express $[H_2CO_3]$ as $K_h[CO_2(d)]$, and substitute this expression for $[H_2CO_3]$ in the equation for the acid dissociation of H_2CO_3:

$$K_a = \frac{[H^+][HCO_3^-]}{[H_2CO_3]} = \frac{[H^+][HCO_3^-]}{K_h[CO_2(d)]}$$

Now, the overall equilibrium for dissociation of H_2CO_3 can be expressed in these terms:

$$K_hK_a = K_{combined} = \frac{[H^+][HCO_3^-]}{[CO_2(d)]}$$

We can calculate the value of the new constant, $K_{combined}$, and the corresponding apparent pK, or $pK_{combined}$, from the experimentally determined values of K_h (3.0×10^{-3} M) and K_a (2.7×10^{-4} M) at 37 °C:

$$K_{combined} = (3.0 \times 10^{-3} \text{ M})(2.7 \times 10^{-4} \text{ M})$$
$$= 8.1 \times 10^{-7} \text{ M}^2$$
$$pK_{combined} = 6.1$$

In clinical medicine, it is common to refer to $CO_2(d)$ as the conjugate acid and to use the apparent, or combined, pK_a of 6.1 to simplify calculation of pH from $[CO_2(d)]$. In this convention,

$$pH = 6.1 + \log \frac{[HCO_3^-]}{(0.23 \times pCO_2)}$$

where pCO_2 is expressed in kilopascals (kPa; typically, pCO_2 is 4.6 to 6.7 kPa) and 0.23 is the corresponding solubility coefficient for CO_2 in water; thus the term $0.23 \times pCO_2 \approx 1.2$ kPa. Plasma $[HCO_3^-]$ is normally about 24 mM. ■

Untreated Diabetes Produces Life-Threatening Acidosis

Human blood plasma normally has a pH between 7.35 and 7.45, and many of the enzymes that function in the blood have evolved to have maximal activity in that pH range. Enzymes typically show maximal

catalytic activity at a characteristic pH, called the **pH optimum** (**Fig. 2–22**). On either side of this optimum pH, catalytic activity often declines sharply. Thus, a small change in pH can make a large difference in the rate of some crucial enzyme-catalyzed reactions. Biological control of the pH of cells and body fluids is therefore of central importance in all aspects of metabolism and cellular activities, and changes in blood pH have marked physiological consequences (described with gusto in Box 2–1!).

In individuals with untreated diabetes mellitus, the lack of insulin, or insensitivity to insulin (depending on the type of diabetes), disrupts the uptake of glucose from blood into the tissues and forces the tissues to use stored fatty acids as their primary fuel. For reasons we will describe in detail later (see Fig. 24–30), this dependence on fatty acids results in the accumulation of high concentrations of two carboxylic acids, β-hydroxybutyric acid and acetoacetic acid (blood plasma level of 90 mg/100 mL, compared with <3 mg/100 mL in control (healthy) individuals; urinary excretion of 5 g/24 hr, compared with <125 mg/24 hr in controls). Dissociation of these acids lowers the pH of blood plasma to less than 7.35, causing acidosis. Severe acidosis leads to headache, drowsiness, nausea, vomiting, and diarrhea, followed by stupor, coma, and convulsions, presumably because at the lower pH, some enzyme(s) do not function optimally. When a patient is found to have high blood glucose, low plasma pH, and high levels of β-hydroxybutyric acid and acetoacetic acid in blood and urine, diabetes mellitus is the likely diagnosis.

Other conditions can also produce acidosis. Fasting and starvation force the use of stored fatty acids as fuel,

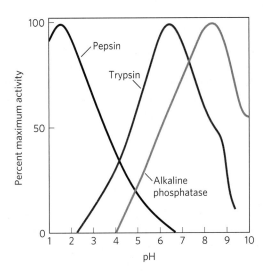

FIGURE 2–22 The pH optima of some enzymes. Pepsin is a digestive enzyme secreted into gastric juice, which has a pH of ~1.5, allowing pepsin to act optimally. Trypsin, a digestive enzyme that acts in the small intestine, has a pH optimum that matches the neutral pH in the lumen of the small intestine. Alkaline phosphatase of bone tissue is a hydrolytic enzyme thought to aid in bone mineralization.

BOX 2–1   ⚕ MEDICINE   On Being One's Own Rabbit (Don't Try This at Home!)

This is an account by J.B.S. Haldane of physiological experiments on controlling blood pH, from his book *Possible Worlds* (Harper and Brothers, 1928).

"I wanted to find out what happened to a man when one made him more acid or more alkaline . . . One might, of course, have tried experiments on a rabbit first, and some work had been done along these lines; but it is difficult to be sure how a rabbit feels at any time. Indeed, some rabbits make no serious attempt to cooperate with one.

". . . A human colleague and I therefore began experiments on one another . . . My colleague Dr. H.W. Davies and I made ourselves alkaline by over-breathing and by eating anything up to three ounces of bicarbonate of soda. We made ourselves acid by sitting in an airtight room with between six and seven per cent of carbon dioxide in the air. This makes one breathe as if one had just completed a boat-race, and also gives one a rather violent headache . . . Two hours was as long as any one wanted to stay in the carbon dioxide, even if the gas chamber at our disposal had not retained an ineradicable odour of 'yellow cross gas' from some wartime experiments, which made one weep gently every time one entered it. The most obvious thing to try was drinking hydrochloric acid. If one takes it strong it dissolves one's teeth and burns one's throat, whereas I wanted to let it diffuse gently all through my body. The strongest I ever cared to drink was about one part of the commercial strong acid in a hundred of water, but a pint of that was enough for me, as it irritated my throat and stomach, while my calculations showed that I needed a gallon and a half to get the effect I wanted . . . I argued that if one ate ammonium chloride, it would partly break up in the body, liberating hydrochloric acid. This proved to be correct . . . the liver turns ammonia into a harmless substance called urea before it reaches the heart and brain on absorption from the gut. The hydrochloric

acid is left behind and combines with sodium bicarbonate, which exists in all the tissues, producing sodium chloride and carbon dioxide. I have had this gas produced in me in this way at the rate of six quarts an hour (though not for an hour on end at that rate) . . .

"I was quite satisfied to have reproduced in myself the type of shortness of breath which occurs in the terminal stages of kidney disease and diabetes. This had long been known to be due to acid poisoning, but in each case the acid poisoning is complicated by other chemical abnormalities, and it had been rather uncertain which of the symptoms were due to the acid as such.

"The scene now shifts to Heidelberg, where Freudenberg and György were studying tetany in babies . . . it occurred to them that it would be well worth trying the effect of making the body unusually acid. For tetany had occasionally been observed in patients who had been treated for other complaints by very large doses of sodium bicarbonate, or had lost large amounts of hydrochloric acid by constant vomiting; and if alkalinity of the tissues will produce tetany, acidity may be expected to cure it. Unfortunately, one could hardly try to cure a dying baby by shutting it up in a room full of carbonic acid, and still less would one give it hydrochloric acid to drink; so nothing had come of their idea, and they were using lime salts, which are not very easily absorbed, and which upset the digestion, but certainly benefit many cases of tetany.

"However, the moment they read my paper on the effects of ammonium chloride, they began giving it to babies, and were delighted to find that the tetany cleared up in a few hours. Since then it has been used with effect both in England and America, both on children and adults. It does not remove the cause, but it brings the patient into a condition from which he has a very fair chance of recovering."

---

with the same consequences as for diabetes. Very heavy exertion, such as a sprint by runners or cyclists, leads to temporary accumulation of lactic acid in the blood. Kidney failure results in a diminished capacity to regulate bicarbonate levels. Lung diseases (such as emphysema, pneumonia, and asthma) reduce the capacity to dispose of the CO_2 produced by fuel oxidation in the tissues, with the resulting accumulation of H_2CO_3. Acidosis is treated by dealing with the underlying condition—administering insulin to people with diabetes, and steroids or antibiotics to people with lung disease. Severe acidosis can be reversed by administering bicarbonate solution intravenously. ∎

WORKED EXAMPLE 2–7   Treatment of Acidosis with Bicarbonate

Why does intravenous administration of a bicarbonate solution raise the plasma pH?

Solution: The ratio of $[HCO_3^-]$ to $[CO_2(d)]$ determines the pH of the bicarbonate buffer, according to the equation

$$pH = 6.1 + \log \frac{[HCO_3^-]}{(0.23 \times pCO_2)}$$

If $[HCO_3^-]$ is increased with no change in pCO_2, the pH will rise.

SUMMARY 2.3 Buffering against pH Changes in Biological Systems

▶ A mixture of a weak acid (or base) and its salt resists changes in pH caused by the addition of H^+ or OH^-. The mixture thus functions as a buffer.

▶ The pH of a solution of a weak acid (or base) and its salt is given by the Henderson-Hasselbalch equation: $pH = pK_a + \log \dfrac{[A^-]}{[HA]}$.

▶ In cells and tissues, phosphate and bicarbonate buffer systems maintain intracellular and extracellular fluids at their optimum (physiological) pH, which is usually close to pH 7. Enzymes generally work optimally at this pH.

▶ Medical conditions that lower the pH of blood, causing acidosis, or raise it, causing alkalosis, can be life threatening.

2.4 Water as a Reactant

Water is not just the solvent in which the chemical reactions of living cells occur; it is very often a direct participant in those reactions. The formation of ATP from ADP and inorganic phosphate is an example of a **condensation reaction** in which the elements of water are eliminated **(Fig. 2–23)**. The reverse of this reaction—cleavage accompanied by the addition of the elements of water—is a **hydrolysis reaction**. Hydrolysis reactions are also responsible for the enzymatic depolymerization of proteins, carbohydrates, and nucleic acids. Hydrolysis reactions, catalyzed by enzymes called **hydrolases**, are almost invariably exergonic; by producing two molecules from one, they lead to an increase in the randomness of the system. The formation of cellular polymers from their subunits by simple reversal of hydrolysis (that is, by condensation reactions) would be endergonic and therefore does not occur. As we shall see, cells circumvent this thermodynamic obstacle by coupling endergonic condensation reactions to exergonic processes, such as breakage of the anhydride bond in ATP.

FIGURE 2–23 Participation of water in biological reactions. ATP is a phosphoanhydride formed by a condensation reaction (loss of the elements of water) between ADP and phosphate. R represents adenosine monophosphate (AMP). This condensation reaction requires energy. The hydrolysis of (addition of the elements of water to) ATP to form ADP and phosphate releases an equivalent amount of energy. These condensation and hydrolysis reactions of ATP are just one example of the role of water as a reactant in biological processes.

You are (we hope!) consuming oxygen as you read. Water and carbon dioxide are the end products of the oxidation of fuels such as glucose. The overall reaction can be summarized as

$$C_6H_{12}O_6 + 6O_2 \longrightarrow 6CO_2 + 6H_2O$$
$$\text{Glucose}$$

The "metabolic water" formed by oxidation of foods and stored fats is actually enough to allow some animals in very dry habitats (gerbils, kangaroo rats, camels) to survive for extended periods without drinking water.

The CO_2 produced by glucose oxidation is converted in erythrocytes to the more soluble HCO_3^-, in a reaction catalyzed by the enzyme carbonic anhydrase:

$$CO_2 + H_2O \rightleftharpoons HCO_3^- + H^+$$

In this reaction, water not only is a substrate but also functions in proton transfer by forming a network of hydrogen-bonded water molecules through which proton hopping occurs (Fig. 2–14).

Green plants and algae use the energy of sunlight to split water in the process of photosynthesis:

$$2H_2O + 2A \xrightarrow{\text{light}} O_2 + 2AH_2$$

In this reaction, A is an electron-accepting species, which varies with the type of photosynthetic organism, and water serves as the electron donor in an oxidation-reduction sequence (see Fig. 19–59) that is fundamental to all life.

SUMMARY 2.4 Water as a Reactant

▶ Water is both the solvent in which metabolic reactions occur and a reactant in many biochemical processes, including hydrolysis, condensation, and oxidation-reduction reactions.

2.5 The Fitness of the Aqueous Environment for Living Organisms

Organisms have effectively adapted to their aqueous environment and have evolved means of exploiting the unusual properties of water. The high specific heat of water (the heat energy required to raise the temperature of 1 g of water by 1 °C) is useful to cells and organisms because it allows water to act as a "heat buffer," keeping the temperature of an organism relatively constant as the temperature of the surroundings fluctuates and as heat is generated as a byproduct of metabolism. Furthermore, some vertebrates exploit the high heat of vaporization of water (Table 2–1) by using (thus losing) excess body heat to evaporate sweat. The high degree of internal cohesion of liquid water, due to hydrogen bonding, is exploited by plants as a means of transporting dissolved nutrients from the roots to the leaves during the process of transpiration. Even the density of ice, lower than that of liquid water, has important biological consequences in the life cycles of aquatic organisms. Ponds freeze from the top down, and the layer of ice at the top insulates the water below from frigid air, preventing the pond (and the organisms in it) from freezing solid. Most fundamental to all living organisms is the fact that many physical and biological properties of cell macromolecules, particularly the proteins and nucleic acids, derive from their interactions with water molecules of the surrounding medium. The influence of water on the course of biological evolution has been profound and determinative. If life forms have evolved elsewhere in the universe, they are unlikely to resemble those of Earth unless liquid water is plentiful in their planet of origin.

Key Terms

Terms in bold are defined in the glossary.

hydrogen bond 48	**ion product of water**
bond energy 48	**(K_w)** 59
hydrophilic 50	**pH** 60
hydrophobic 50	**acidosis** 61
amphipathic 52	**alkalosis** 61
micelle 52	**conjugate acid-base**
hydrophobic	**pair** 61
interactions 53	**acid dissociation constant**
van der Waals	**(K_a)** 62
interactions 53	**pK_a** 62
osmolarity 56	**titration curve** 62
osmosis 56	**buffer** 64
isotonic 56	buffering region 64
hypertonic 56	**Henderson-Hasselbalch**
hypotonic 56	**equation** 64
equilibrium constant	**condensation** 69
(K_{eq}) 59	**hydrolysis** 69

Further Reading

General

Ball, P. (2001) *Life's Matrix: A Biography of Water*, University of California Press, Berkeley, CA.

A highly accessible and entertaining description of water, from the Big Bang to its many roles in the chemistry of life.

Denny, M.W. (1993) *Air and Water: The Biology and Physics of Life's Media*, Princeton University Press, Princeton, NJ.

A wonderful investigation of the biological relevance of the properties of water.

Eisenberg, D. & Kauzmann, W. (1969) *The Structure and Properties of Water*, Oxford University Press, New York.

An advanced, classic treatment of the physical chemistry of water and hydrophobic interactions.

Franks, F. & Mathias, S.F. (eds). (1982) *Biophysics of Water*, John Wiley & Sons, Inc., New York.

A large collection of papers on the structure of pure water and of the cytoplasm.

Gerstein, M. & Levitt, M. (1998) Simulating water and the molecules of life. *Sci. Am.* **279** (November), 100–105.

A well-illustrated description of the use of computer simulation to study the biologically important association of water with proteins and nucleic acids.

Kandori, H. (2000) Role of internal water molecules in bacteriorhodopsin. *Biochim. Biophys. Acta* 14__, 177–19_.

Intermediate-level review of the role of an internal chain of water molecules in proton movement through this protein.

Kornblatt, J. & Kornblatt, J. (1997) The role of water in recognition and catalysis by enzymes. *Biochemist* **19** (3), 14–17.

A short, useful summary of the ways in which bound water influences the structure and activity of proteins.

Lemieux, R.U. (1996) How water provides the impetus for molecular recognition in aqueous solution. *Acc. Chem. Res.* **29**, 373–380.

A study of the role of water in the binding of a sugar to its binding protein.

Luecke, H. (2000) Atomic resolution structures of bacteriorhodopsin photocycle intermediates: the role of discrete water molecules in the function of this light-driven ion pump. *Biochim. Biophys. Acta* **1460**, 133–156.

Advanced review of a proton pump that employs an internal chain of water molecules.

Nicolls, P. (2000) Introduction: the biology of the water molecule. *Cell. Mol. Life Sci.* **57**, 987–992.

A short review of the properties of water, introducing several excellent advanced reviews published in the same issue (see especially Pocker, 2000, and Rand et al., 2000, below).

Symons, M.C. (2000) Spectroscopy of aqueous solutions: protein and DNA interactions with water. *Cell. Mol. Life Sci.* **57**, 999–1007.

Wiggins, P.M. (1990) Role of water in some biological processes. *Microbiol. Rev.* **54**, 432–449.

A review of water in biology, including discussion of the physical structure of liquid water, its interaction with biomolecules, and the state of water in living cells.

Osmosis

Cayley, D.S., Guttman, H.J., & Record, M.T., Jr. (2000) Biophysical characterization of changes in amounts and activity of *Escherichia coli* cell and compartment water and turgor pressure in response to osmotic stress. *Biophys. J.* **78**, 1748–1764.

An advanced physical investigation of the cytoplasmic water fraction of the bacterium *Escherichia coli* grown in media of different osmolarities. (See also Record et al., 1998, below.)

Rand, R.P., Parsegian, V.A., & Rau, D.C. (2000) Intracellular osmotic action. *Cell. Mol. Life Sci.* **57**, 1018–1032.

Review of the roles of water in enzyme catalysis as revealed by studies in water-poor solutes.

Record, M.T., Jr., Courtenay, E.S., Cayley, D.S., & Guttman, H.J. (1998) Responses of *E. coli* to osmotic stress: large changes in amounts of cytoplasmic solutes and water. *Trends Biochem. Sci.* **23**, 143–148.

Intermediate-level review of the ways in which a bacterial cell counters changes in the osmolarity of its surroundings. (See also Cayley et al., 2000, above.)

Zonia, L. & Munnik, T. (2007) Life under pressure: hydrostatic pressure in cell growth and function. *Trends Plant Sci.* **12**, 90–97.

Weak Interactions in Aqueous Systems

Baldwin, R.L. (2007) Energetics of protein folding. *J. Mol. Biol.* **371**, 283–301.

Advanced discussion of the thermodynamic factors, including weak interactions, that determine the course of protein folding.

Ball, P. (2008) Water as an active constituent in cell biology. *Chem. Rev.* **108**, 74–108.

An advanced discussion of the role of water in biological structure and function.

Blokzijl, W. & Engberts, J.B.F.N. (1993) Hydrophobic effects. Opinions and facts. *Angew. Chem. Int. Ed. Engl.* **32**, 1545–1579.

Advanced, monumental, and critical review.

Chaplin, M. (2006) Do we underestimate the importance of water in cell biology? *Nat. Rev. Mol. Cell Biol.* **7**, 861–866.

Fersht, A.R. (198?) The hydrogen bond in molecular recognition. *Trends Biochem. Sci.* **12**, 301–304.

A clear, brief, quantitative discussion of the contribution of hydrogen bonding to molecular recognition and enzyme catalysis.

Jeffrey, G.A. (1997) *An Introduction to Hydrogen Bonding*, Oxford University Press, New York.

A detailed, advanced discussion of the structure and properties of hydrogen bonds, including those in water and biomolecules.

Kauzmann, W. (1959) Some factors in the interpretation of protein denaturation. *Adv. Protein Chem.* **14**, 1–63.

Remains the classic statement of the importance of hydrophobic interactions in the stability of proteins.

Ladbury, J. (1996) Just add water! The effect of water on the specificity of protein-ligand binding sites and its potential application to drug design. *Chem. Biol.* **3**, 973–980.

Levy, Y. & Onuchic, J.N. (2006) Water mediation in protein folding and molecular recognition. *Annu. Rev. Biophys. Biomol. Struct.* **35**, 389–415.

An advanced discussion of the role of water in protein structure.

Martin, T.W. & Derewenda, Z.S. (1999) The name is bond—H bond. *Nat. Struct. Biol.* **6**, 403–406.

Brief review of the evidence that hydrogen bonds have some covalent character.

Pace, C.N. (2009) Energetics of protein hydrogen bonds. *Nat. Struct. Mol. Biol.* **16**, 681–682.

Brief account of the historical contributions to understanding the strength of hydrogen bonds in proteins.

Pocker, Y. (2000) Water in enzyme reactions: biophysical aspects of hydration-dehydration processes. *Cell. Mol. Life Sci.* **57**, 1008–1017.

Review of the role of water in enzyme catalysis, with carbonic anhydrase as the featured example.

Schwabe, J.W.R. (1997) The role of water in protein-DNA interactions. *Curr. Opin. Struct. Biol.* **7**, 126–134.

An examination of the important role of water in both the specificity and the affinity of protein-DNA interactions.

Stillinger, F.H. (1980) Water revisited. *Science* **209**, 451–457.

A short review of the physical structure of water, including the importance of hydrogen bonding and the nature of hydrophobic interactions.

Tanford, C. (1978) The hydrophobic effect and the organization of living matter. *Science* **200**, 1012–1018.

A classic review of the chemical and energetic bases for hydrophobic interactions between biomolecules in aqueous solutions.

Weak Acids, Weak Bases, and Buffers: Problems for Practice

Segel, I.H. (1976) *Biochemical Calculations*, 2nd edn, John Wiley & Sons, Inc., New York.

Problems

1. Solubility of Ethanol in Water Explain why ethanol (CH_3CH_2OH) is more soluble in water than is ethane (CH_3CH_3).

2. Calculation of pH from Hydrogen Ion Concentration What is the pH of a solution that has an H^+ concentration of (a) 1.75×10^{-5} mol/L; (b) 6.50×10^{-10} mol/L; (c) 1.0×10^{-4} mol/L; (d) 1.50×10^{-5} mol/L?

3. Calculation of Hydrogen Ion Concentration from pH What is the H^+ concentration of a solution with pH of (a) 3.82; (b) 6.52; (c) 11.11?

 4. Acidity of Gastric HCl In a hospital laboratory, a 10.0 mL sample of gastric juice, obtained several hours after a meal, was titrated with 0.1 M NaOH to neutrality; 7.2 mL of NaOH was required. The patient's stomach contained no ingested food or drink; thus assume that no buffers were present. What was the pH of the gastric juice?

5. Calculation of the pH of a Strong Acid or Base (a) Write out the acid dissociation reaction for hydrochloric acid. (b) Calculate the pH of a solution of 5.0×10^{-4} M HCl. (c) Write out the acid dissociation reaction for sodium hydroxide. (d) Calculate the pH of a solution of 7.0×10^{-5} M NaOH.

6. Calculation of pH from Concentration of Strong Acid Calculate the pH of a solution prepared by diluting 3.0 mL of 2.5 M HCl to a final volume of 100 mL with H_2O.

7. Measurement of Acetylcholine Levels by pH Changes The concentration of acetylcholine (a neurotransmitter) in a sample can be determined from the pH changes that accompany its hydrolysis. When the sample is incubated with the enzyme acetylcholinesterase, acetylcholine is converted to choline and acetic acid, which dissociates to yield acetate and a hydrogen ion:

In a typical analysis, 15 mL of an aqueous solution containing an unknown amount of acetylcholine had a pH of 7.65. When incubated with acetylcholinesterase, the pH of the solution decreased to 6.87. Assuming there was no buffer in the assay mixture, determine the number of moles of acetylcholine in the 15 mL sample.

8. Physical Meaning of pK_a Which of the following aqueous solutions has the lowest pH: 0.1 M HCl; 0.1 M acetic acid ($pK_a = 4.86$); 0.1 M formic acid ($pK_a = 3.75$)?

9. Meanings of K_a and pK_a (a) Does a strong acid have a greater or lesser tendency to lose its proton than a weak acid? (b) Does the strong acid have a higher or lower K_a than the weak acid? (c) Does the strong acid have a higher or lower pK_a than the weak acid?

10. Simulated Vinegar One way to make vinegar (*not* the preferred way) is to prepare a solution of acetic acid, the sole acid component of vinegar, at the proper pH (see Fig. 2–15) and add appropriate flavoring agents. Acetic acid (M_r 60) is a liquid at 25 °C, with a density of 1.049 g/mL. Calculate the volume that must be added to distilled water to make 1 L of simulated vinegar (see Fig. 2–16).

11. Identifying the Conjugate Base Which is the conjugate base in each of the pairs below?

(a) RCOOH, RCOO⁻

(b) RNH₂, RNH₃⁺

(c) $H_2PO_4^-$, H_3PO_4

(d) H_2CO_3, HCO_3^-

12. Calculation of the pH of a Mixture of a Weak Acid and Its Conjugate Base Calculate the pH of a dilute solution that contains a molar ratio of potassium acetate to acetic acid (pK_a = 4.76) of (a) 2:1; (b) 1:3; (c) 5:1; (d) 1:1; (e) 1:10.

13. Effect of pH on Solubility The strongly polar, hydrogen-bonding properties of water make it an excellent solvent for ionic (charged) species. By contrast, nonionized, nonpolar organic molecules, such as benzene, are relatively insoluble in water. In principle, the aqueous solubility of any organic acid or base can be increased by converting the molecules to charged species. For example, the solubility of benzoic acid in water is low. The addition of sodium bicarbonate to a mixture of water and benzoic acid raises the pH and deprotonates the benzoic acid to form benzoate ion, which is quite soluble in water.

Benzoic acid
$pK_a \approx 5$

Benzoate ion

Are the following compounds more soluble in an aqueous solution of 0.1 M NaOH or 0.1 M HCl? (The dissociable protons are shown in red.)

Pyridine ion
$pK_a \approx 5$

(a)

β-Naphthol
$pK_a \approx 10$

(b)

N-Acetyltyrosine methyl ester
$pK_a \approx 10$

(c)

14. Treatment of Poison Ivy Rash The components of poison ivy and poison oak that produce the characteristic itchy rash are catechols substituted with long-chain alkyl groups.

$pK_a \approx 8$

If you were exposed to poison ivy, which of the treatments below would you apply to the affected area? Justify your choice.

(a) Wash the area with cold water.

(b) Wash the area with dilute vinegar or lemon juice.

(c) Wash the area with soap and water.

(d) Wash the area with soap, water, and baking soda (sodium bicarbonate).

15. pH and Drug Absorption Aspirin is a weak acid with a pK_a of 3.5 (the ionizable H is shown in red):

It is absorbed into the blood through the cells lining the stomach and the small intestine. Absorption requires passage through the plasma membrane, the rate of which is determined by the polarity of the molecule: charged and highly polar molecules pass slowly, whereas neutral hydrophobic ones pass rapidly. The pH of the stomach contents is about 1.5, and the pH of the contents of the small intestine is about 6. Is more aspirin absorbed into the bloodstream from the stomach or from the small intestine? Clearly justify your choice.

16. Calculation of pH from Molar Concentrations What is the pH of a solution containing 0.12 mol/L of NH_4Cl and 0.03 mol/L of NaOH (pK_a of NH_4^+/NH_3 is 9.25)?

17. Calculation of pH after Titration of Weak Acid A compound has a pK_a of 7.4. To 100 mL of a 1.0 M solution of this compound at pH 8.0 is added 30 mL of 1.0 M hydrochloric acid. What is the pH of the resulting solution?

18. Properties of a Buffer The amino acid glycine is often used as the main ingredient of a buffer in biochemical experiments. The amino group of glycine, which has a pK_a of 9.6, can exist either in the protonated form ($-NH_3^+$) or as the free base ($-NH_2$), because of the reversible equilibrium

$$R-NH_3^+ \rightleftharpoons R-NH_2 + H^+$$

(a) In what pH range can glycine be used as an effective buffer due to its amino group?

(b) In a 0.1 M solution of glycine at pH 9.0, what fraction of glycine has its amino group in the $-NH_3^+$ form?

(c) How much 5 M KOH must be added to 1.0 L of 0.1 M glycine at pH 9.0 to bring its pH to exactly 10.0?

(d) When 99% of the glycine is in its $-NH_3^+$ form, what is the numerical relation between the pH of the solution and the pK_a of the amino group?

19. Calculation of the pK_a of an Ionizable Group by Titration The pK_a values of a compound with two ionizable groups are $pK_1 = 4.10$ and pK_2 between 7 and 10. A biochemist has 10 mL of a 1.0 M solution of this compound at a pH of 8.00. She adds 10.0 mL of 1.00 M HCl, which changes the pH to 3.20. What is pK_2?

20. Calculation of the pH of a Solution of a Polyprotic Acid Histidine has ionizable groups with pK_a values of 1.8, 6.0, and 9.2, as shown below (His = imidazole group). A biochemist makes up 100 mL of a 0.100 M solution of histidine at a pH of 5.40. She then adds 40 mL of 0.10 M HCl. What is the pH of the resulting solution?

Ionizable group $-COOH \rightleftharpoons -COO^-$
$-HisH^+ \rightleftharpoons -His$
$-NH_3^+ \rightleftharpoons -NH_2$

21. Calculation of the Original pH from the Final pH after Titration A biochemist has 100 mL of a 0.10 M solution of a weak acid with a pK_a of 6.3. She adds 6.0 mL of 1.0 M HCl, which changes the pH to 5.7. What was the pH of the original solution?

22. Preparation of a Phosphate Buffer What molar ratio of HPO_4^{2-} to $H_2PO_4^-$ in solution would produce a pH of 7.0? Phosphoric acid (H_3PO_4), a triprotic acid, has three pK_a values: 2.14, 6.86, and 12.4. Hint: Only one of the pK_a values is relevant here.

23. Preparation of Standard Buffer for Calibration of a pH Meter The glass electrode used in commercial pH meters gives an electrical response proportional to the concentration of hydrogen ion. To convert these responses to a pH reading, the electrode must be calibrated against standard solutions of known H^+ concentration. Determine the weight in grams of sodium dihydrogen phosphate ($NaH_2PO_4 \cdot H_2O$; FW 138) and disodium hydrogen phosphate (Na_2HPO_4; FW 142) needed to prepare 1 L of a standard buffer at pH 7.00 with a total phosphate concentration of 0.100 M (see Fig. 2–16). See problem 22 for the pK_a values of phosphoric acid.

24. Calculation of Molar Ratios of Conjugate Base to Weak Acid from pH For a weak acid with a pK_a of 6.0, calculate the ratio of conjugate base to acid at a pH of 5.0.

25. Preparation of Buffer of Known pH and Strength Given 0.10 M solutions of acetic acid ($pK_a = 4.76$) and sodium acetate, describe how you would go about preparing 1.0 L of 0.10 M acetate buffer of pH 4.00.

26. Choice of Weak Acid for a Buffer Which of these compounds would be the best buffer at pH 5.0: formic acid ($pK_a = 3.8$), acetic acid ($pK_a = 4.76$), or ethylamine ($pK_a = 9.0$)? Briefly justify your answer.

27. Working with Buffers A buffer contains 0.010 mol of lactic acid ($pK_a = 3.86$) and 0.050 mol of sodium lactate per liter. (a) Calculate the pH of the buffer. (b) Calculate the change in pH when 5 mL of 0.5 M HCl is added to 1 L of the buffer. (c) What pH change would you expect if you added the same quantity of HCl to 1 L of pure water?

28. Use of Molar Concentrations to Calculate pH What is the pH of a solution that contains 0.20 M sodium acetate and 0.60 M acetic acid ($pK_a = 4.76$)?

29. Preparation of an Acetate Buffer Calculate the concentrations of acetic acid ($pK_a = 4.76$) and sodium acetate necessary to prepare a 0.2 M buffer solution at pH 5.0.

30. pH of Insect Defensive Secretion You have been observing an insect that defends itself from enemies by secreting a caustic liquid. Analysis of the liquid shows it to have a total concentration of formate plus formic acid ($K_a = 1.8 \times 10^{-4}$) of 1.45 M; the concentration of formate ion is 0.015 M. What is the pH of the secretion?

31. Calculation of pK_a An unknown compound, X, is thought to have a carboxyl group with a pK_a of 2.0 and another ionizable group with a pK_a between 5 and 8. When 75 mL of 0.1 M NaOH is added to 100 mL of a 0.1 M solution of X at pH 2.0, the pH increases to 6.72. Calculate the pK_a of the second ionizable group of X.

32. Ionic Forms of Alanine Alanine is a diprotic acid that can undergo two dissociation reactions (see Table 3–1 for pK_a values). (a) Given the structure of the partially protonated form (or zwitterion; see Fig. 3–9) below, draw the chemical structures of the other two forms of alanine that predominate in aqueous solution: the fully protonated form and the fully deprotonated form.

Alanine

Of the three possible forms of alanine, which would be present at the highest concentration in solutions of the following pH: (b) 1.0; (c) 6.2; (d) 8.02; (e) 11.9. Explain your answers in terms of pH relative to the two pK_a values.

33. Control of Blood pH by Respiratory Rate (a) The partial pressure of CO_2 in the lungs can be varied rapidly by the rate and depth of breathing. For example, a common remedy to alleviate hiccups is to increase the concentration of CO_2 in the lungs. This can be achieved by holding one's breath, by very slow and shallow breathing (hypoventilation), or by breathing in and out of a paper bag. Under such conditions, pCO_2 in the air space of the lungs rises above normal.

Qualitatively explain the effect of these procedures on the blood pH.

(b) A common practice of competitive short-distance runners is to breathe rapidly and deeply (hyperventilate) for about half a minute to remove CO_2 from their lungs just before the race begins. Blood pH may rise to 7.60. Explain why the blood pH increases.

(c) During a short-distance run, the muscles produce a large amount of lactic acid ($CH_3CH(OH)COOH$; $K_a = 1.38 \times 10^{-4}$ M) from their glucose stores. In view of this fact, why might hyperventilation before a dash be useful?

34. Calculation of Blood pH from CO_2 and Bicarbonate Levels Calculate the pH of a blood plasma sample with a total CO_2 concentration of 26.9 mM and bicarbonate concentration of 25.6 mM. Recall from page 67 that the relevant pK_a of carbonic acid is 6.1.

35. Effect of Holding One's Breath on Blood pH The pH of the extracellular fluid is buffered by the bicarbonate/carbonic acid system. Holding your breath can increase the concentration of $CO_2(g)$ in the blood. What effect might this have on the pH of the extracellular fluid? Explain by showing the relevant equilibrium equation(s) for this buffer system.

Data Analysis Problem

36. "Switchable" Surfactants Hydrophobic molecules do not dissolve well in water. Given that water is a very commonly used solvent, this makes certain processes very difficult: washing oily food residue off dishes, cleaning up spilled oil, keeping the oil and water phases of salad dressings well mixed, and carrying out chemical reactions that involve both hydrophobic and hydrophilic components.

Surfactants are a class of amphipathic compounds that includes soaps, detergents, and emulsifiers. With the use of surfactants, hydrophobic compounds can be suspended in aqueous solution by forming micelles (see Fig. 2–7). A micelle has a hydrophobic core consisting of the hydrophobic compound and the hydrophobic "tails" of the surfactant; the hydrophilic "heads" of the surfactant cover the surface of the micelle. A suspension of micelles is called an emulsion. The more hydrophilic the head group of the surfactant, the more powerful it is—that is, the greater its capacity to emulsify hydrophobic material.

When you use soap to remove grease from dirty dishes, the soap forms an emulsion with the grease that is easily removed by water through interaction with the hydrophilic head of the soap molecules. Likewise, a detergent can be used to emulsify spilled oil for removal by water. And emulsifiers in commercial salad dressings keep the oil suspended evenly throughout the water-based mixture.

There are some situations in which it would be very useful to have a "switchable" surfactant: a molecule that could be reversibly converted between a surfactant and a nonsurfactant.

(a) Imagine such a "switchable" surfactant existed. How would you use it to clean up and then recover the oil from an oil spill?

Liu et al. describe a prototypical switchable surfactant in their 2006 article "Switchable Surfactants." The switching is based on the following reaction:

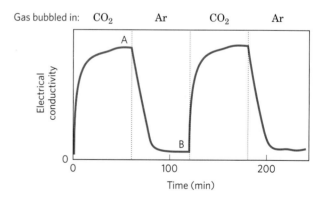

Amidine form   Amidinium form

(b) Given that the pK_a of a typical amidinium ion is 12.4, in which direction (left or right) would you expect the equilibrium of the above reaction to lie? (See Fig. 2–16 for relevant pK_a values.) Justify your answer. Hint: Remember the reaction $H_2O + CO_2 \rightleftharpoons H_2CO_3$.

Liu and colleagues produced a switchable surfactant for which R = $C_{16}H_{33}$. They do not name the molecule in their article; for brevity, we'll call it s-surf.

(c) The amidinium form of s-surf is a powerful surfactant; the amidine form is not. Explain this observation.

Liu and colleagues found that they could switch between the two forms of s-surf by changing the gas that they bubbled through a solution of the surfactant. They demonstrated this switch by measuring the electrical conductivity of the s-surf solution; aqueous solutions of ionic compounds have higher conductivity than solutions of nonionic compounds. They started with a solution of the amidine form of s-surf in water. Their results are shown below; dotted lines indicate the switch from one gas to another.

(d) In which form is the majority of s-surf at point A? At point B?

(e) Why does the electrical conductivity rise from time 0 to point A?

(f) Why does the electrical conductivity fall from point A to point B?

(g) Explain how you would use s-surf to clean up and recover the oil from an oil spill.

Reference

Liu, Y., Jessop, P.G., Cunningham, M., Eckert, C.A., & Liotta, C.L. (2006) Switchable surfactants. *Science* **313**, 958–960.

3

Amino Acids, Peptides, and Proteins

Proteins mediate virtually every process that takes place in a cell, exhibiting an almost endless diversity of functions. To explore the molecular mechanism of a biological process, a biochemist almost inevitably studies one or more proteins. Proteins are the most abundant biological macromolecules, occurring in all cells and all parts of cells. Proteins also occur in great variety; thousands of different kinds may be found in a single cell. As the arbiters of molecular function, proteins are the most important final products of the information pathways discussed in Part III of this book. Proteins are the molecular instruments through which genetic information is expressed.

Relatively simple monomeric subunits provide the key to the structure of the thousands of different proteins. The proteins of every organism, from the simplest of bacteria to human beings, are constructed from the same ubiquitous set of 20 amino acids. Because each of these amino acids has a side chain with distinctive chemical properties, this group of 20 precursor molecules may be regarded as the alphabet in which the language of protein structure is written.

To generate a particular protein, amino acids are covalently linked in a characteristic linear sequence. What is most remarkable is that cells can produce proteins with strikingly different properties and activities by joining the same 20 amino acids in many different combinations and sequences. From these building blocks different organisms can make such widely diverse products as enzymes, hormones, antibodies, transporters, muscle fibers, the lens protein of the eye, feathers, spider webs, rhinoceros horn, milk proteins, antibiotics, mushroom poisons, and myriad other substances having distinct biological activities **(Fig. 3–1)**. Among these protein products, the enzymes are the most varied and specialized. As the catalysts of virtually all cellular reactions, enzymes are one of the keys to understanding the chemistry of life and thus provide a focal point for any course in biochemistry.

Protein structure and function are the topics of this and the next three chapters. Here, we begin with a description of the fundamental chemical properties of amino acids, peptides, and proteins. We also consider how a biochemist works with proteins.

(a)

(b)

(c)

FIGURE 3–1 Some functions of proteins. (a) The light produced by fireflies is the result of a reaction involving the protein luciferin and ATP, catalyzed by the enzyme luciferase (see Box 13-1). **(b)** Erythrocytes contain large amounts of the oxygen-transporting protein hemoglobin. **(c)** The protein keratin, formed by all vertebrates, is the chief structural component of hair, scales, horn, wool, nails, and feathers. The black rhinoceros is nearing extinction in the wild because of the belief prevalent in some parts of the world that a powder derived from its horn has aphrodisiac properties. In reality, the chemical properties of powdered rhinoceros horn are no different from those of powdered bovine hooves or human fingernails.

3.1 Amino Acids

🖱 **Protein Architecture–Amino Acids** Proteins are polymers of amino acids, with each **amino acid residue** joined to its neighbor by a specific type of covalent bond. (The term "residue" reflects the loss of the elements of water when one amino acid is joined to another.) Proteins can be broken down (hydrolyzed) to their constituent amino acids by a variety of methods, and the earliest studies of proteins naturally focused on the free amino acids derived from them. Twenty different amino acids are commonly found in proteins. The first to be discovered was asparagine, in 1806. The last of the 20 to be found, threonine, was not identified until 1938. All the amino acids have trivial or common names, in some cases derived from the source from which they were first isolated. Asparagine was first found in asparagus, and glutamate in wheat gluten; tyrosine was first isolated from cheese (its name is derived from the Greek *tyros*, "cheese"); and glycine (Greek *glykos*, "sweet") was so named because of its sweet taste.

Amino Acids Share Common Structural Features

All 20 of the common amino acids are α-amino acids. They have a carboxyl group and an amino group bonded to the same carbon atom (the α carbon) **(Fig. 3–2)**. They differ from each other in their side chains, or **R groups**, which vary in structure, size, and electric charge, and which influence the solubility of the amino acids in water. In addition to these 20 amino acids there are many less common ones. Some are residues modified after a protein has been synthesized; others are amino acids present in living organisms but not as constituents of proteins. The common amino acids of proteins have been assigned three-letter abbreviations and one-letter symbols (Table 3–1), which are used as shorthand to indicate the composition and sequence of amino acids polymerized in proteins.

KEY CONVENTION: The three-letter code is transparent, the abbreviations generally consisting of the first three letters of the amino acid name. The one-letter code was devised by Margaret Oakley Dayhoff, considered by many to be the founder of the field of bioinformatics. The one-letter code reflects an attempt to reduce the size of the data files (in an era of punch-card computing) used to describe amino acid sequences. It was designed to be easily memorized, and understanding its origin can help students do just that. For six amino acids (CHIMSV), the first letter of the amino acid name is unique and thus is used as the symbol. For five others (AGLPT), the first letter is not

Margaret Oakley Dayhoff, 1925–1983

unique but is assigned to the amino acid that is most common in proteins (for example, leucine is more common than lysine). For another four, the letter used is phonetically suggestive (RFYW: aRginine, Fenyl-alanine, tYrosine, tWiptophan). The rest were harder to assign. Four (DNEQ) were assigned letters found within or suggested by their names (asparDic, asparagiNe, glutamEke, Q-tamine). That left lysine. Only a few letters were left in the alphabet, and K was chosen because it was the closest to L. ∎

For all the common amino acids except glycine, the α carbon is bonded to four different groups: a carboxyl group, an amino group, an R group, and a hydrogen atom (Fig. 3–2; in glycine, the R group is another hydrogen atom). The α-carbon atom is thus a **chiral center** (p. 17). Because of the tetrahedral arrangement of the bonding orbitals around the α-carbon atom, the four different groups can occupy two unique spatial arrangements, and thus amino acids have two possible stereoisomers. Since they are nonsuperposable mirror images of each other **(Fig. 3–3)**, the two forms

FIGURE 3–3 Stereoisomerism in α-amino acids. (a) The two stereoisomers of alanine, L- and D-alanine, are nonsuperposable mirror images of each other (enantiomers). **(b, c)** Two different conventions for showing the configurations in space of stereoisomers. In perspective formulas **(b)**, the solid wedge-shaped bonds project out of the plane of the paper, the dashed bonds behind it. In projection formulas **(c)**, the horizontal bonds are assumed to project out of the plane of the paper, the vertical bonds behind. However, projection formulas are often used casually and are not always intended to portray a specific stereochemical configuration.

FIGURE 3–2 General structure of an amino acid. This structure is common to all but one of the α-amino acids. (Proline, a cyclic amino acid, is the exception.) The R group, or side chain (purple), attached to the α carbon (gray) is different in each amino acid.

TABLE 3–1 Properties and Conventions Associated with the Common Amino Acids Found in Proteins

Amino acid	Abbreviation/ symbol	M_r*	pK_1 (—COOH)	pK_2 (—NH$_3^+$)	pK_R (R group)	pI	Hydropathy index[†]	Occurrence in proteins (%)[‡]
Nonpolar, aliphatic R groups								
Glycine	Gly G	75	2.34	9.60		5.97	−0.4	7.2
Alanine	Ala A	89	2.34	9.69		6.01	1.8	7.8
Proline	Pro P	115	1.99	10.96		6.48	−1.6	5.2
Valine	Val V	117	2.32	9.62		5.97	4.2	6.6
Leucine	Leu L	131	2.36	9.60		5.98	3.8	9.1
Isoleucine	Ile I	131	2.36	9.68		6.02	4.5	5.3
Methionine	Met M	149	2.28	9.21		5.74	1.9	2.3
Aromatic R groups								
Phenylalanine	Phe F	165	1.83	9.13		5.48	2.8	3.9
Tyrosine	Tyr Y	181	2.20	9.11	10.07	5.66	−1.3	3.2
Tryptophan	Trp W	204	2.38	9.39		5.89	−0.9	1.4
Polar, uncharged R groups								
Serine	Ser S	105	2.21	9.15		5.68	−0.8	6.8
Threonine	Thr T	119	2.11	9.62		5.87	−0.7	5.9
Cysteine[¶]	Cys C	121	1.96	10.28	8.18	5.07	2.5	1.9
Asparagine	Asn N	132	2.02	8.80		5.41	−3.5	4.3
Glutamine	Gln Q	146	2.17	9.13		5.65	−3.5	4.2
Positively charged R groups								
Lysine	Lys K	146	2.18	8.95	10.53	9.74	−3.9	5.9
Histidine	His H	155	1.82	9.17	6.00	7.59	−3.2	2.3
Arginine	Arg R	174	2.17	9.04	12.48	10.76	−4.5	5.1
Negatively charged R groups								
Aspartate	Asp D	133	1.88	9.60	3.65	2.77	−3.5	5.3
Glutamate	Glu E	147	2.19	9.67	4.25	3.22	−3.5	6.3

*M_r values reflect the structures as shown in Figure 3–5. The elements of water (M_r 18) are deleted when the amino acid is incorporated into a polypeptide.

[†]A scale combining hydrophobicity and hydrophilicity of R groups. The values reflect the free energy (ΔG) of transfer of the amino acid side chain from a hydrophobic solvent to water. This transfer is favorable ($\Delta G < 0$; negative value in the index) for charged or polar amino acid side chains, and unfavorable ($\Delta G > 0$; positive value in the index) for amino acids with nonpolar or more hydrophobic side chains. See Chapter 11. From Kyte, J. & Doolittle, R.F. (1982) A simple method for displaying the hydropathic character of a protein. *J. Mol. Biol.* **157**, 105–132.

[‡]Average occurrence in more than 1,150 proteins. From Doolittle, R.F. (1989) Redundancies in protein sequences. In *Prediction of Protein Structure and the Principles of Protein Conformation* (Fasman, G.D., ed.), pp. 599–623, Plenum Press, New York.

[¶]Cysteine is generally classified as polar despite having a positive hydropathy index. This reflects the ability of the sulfhydryl group to act as a weak acid and to form a weak hydrogen bond with oxygen or nitrogen.

represent a class of stereoisomers called **enantiomers** (see Fig. 1–20). All molecules with a chiral center are also **optically active**—that is, they rotate plane-polarized light (see Box 1–2).

KEY CONVENTION: Two conventions are used to identify the carbons in an amino acid—a practice that can be confusing. The additional carbons in an R group are commonly designated β, γ, δ, ε, and so forth, proceeding out from the α carbon. For most other organic molecules, carbon atoms are simply numbered from one end, giving highest priority (C-1) to the carbon with the substituent containing the atom of highest atomic number. Within this latter convention, the carboxyl carbon of an amino acid would be C-1 and the α carbon would be C-2.

$$^-OOC-\overset{1}{C}H-\overset{\overset{\beta}{3}}{C}H_2-\overset{\overset{\gamma}{4}}{C}H_2-\overset{\overset{\delta}{5}}{C}H_2-\overset{\overset{\epsilon}{6}}{C}H_2$$
$$\underset{^+NH_3}{|} \qquad\qquad\qquad \underset{^+NH_3}{|}$$

Lysine

In some cases, such as amino acids with heterocyclic R groups (such as histidine), the Greek lettering system is ambiguous and the numbering convention is therefore used. For branched amino acid side chains, equivalent carbons are given numbers after the Greek letters. Leucine thus has $\delta1$ and $\delta2$ carbons (see the structure in Fig. 3–5). ■

Special nomenclature has been developed to specify the **absolute configuration** of the four substituents of asymmetric carbon atoms. The absolute configurations of simple sugars and amino acids are specified by the **D, L system (Fig. 3–4)**, based on the absolute configuration of the three-carbon sugar glyceraldehyde, a convention proposed by Emil Fischer in 1891. (Fischer knew what groups surrounded the asymmetric carbon of glyceraldehyde but had to guess at their absolute configuration; he guessed right, as was later confirmed by x-ray diffraction analysis.) For all chiral compounds, stereoisomers having a configuration related to that of L-glyceraldehyde are designated L, and stereoisomers related to D-glyceraldehyde are designated D. The functional groups of L-alanine are matched with those of L-glyceraldehyde by aligning those that can be interconverted by simple, one-step chemical reactions. Thus the carboxyl group of L-alanine occupies the same position about the chiral carbon as does the aldehyde group of L-glyceraldehyde, because an aldehyde is readily converted to a carboxyl group via a one-step oxidation. Historically, the similar L and

D designations were used for levorotatory (rotating plane-polarized light to the left) and dextrorotatory (rotating light to the right). However, not all L-amino acids are levorotatory, and the convention shown in Figure 3–4 was needed to avoid potential ambiguities about absolute configuration. By Fischer's convention, L and D refer *only* to the absolute configuration of the four substituents around the chiral carbon, not to optical properties of the molecule.

Another system of specifying configuration around a chiral center is the **RS system**, which is used in the systematic nomenclature of organic chemistry and describes more precisely the configuration of molecules with more than one chiral center (p. 18).

The Amino Acid Residues in Proteins Are L Stereoisomers

Nearly all biological compounds with a chiral center occur naturally in only one stereoisomeric form, either D or L. The amino acid residues in protein molecules are exclusively L stereoisomers. D-Amino acid residues have been found in only a few, generally small peptides, including some peptides of bacterial cell walls and certain peptide antibiotics.

It is remarkable that virtually all amino acid residues in proteins are L stereoisomers. When chiral compounds are formed by ordinary chemical reactions, the result is a racemic mixture of D and L isomers, which are difficult for a chemist to distinguish and separate. But to a living system, D and L isomers are as different as the right hand and the left. The formation of stable, repeating substructures in proteins (Chapter 4) generally requires that their constituent amino acids be of one stereochemical series. Cells are able to specifically synthesize the L isomers of amino acids because the active sites of enzymes are asymmetric, causing the reactions they catalyze to be stereospecific.

Amino Acids Can Be Classified by R Group

Knowledge of the chemical properties of the common amino acids is central to an understanding of biochemistry. The topic can be simplified by grouping the amino acids into five main classes based on the properties of their R groups (Table 3–1), particularly their **polarity**, or tendency to interact with water at biological pH (near pH 7.0). The polarity of the R groups varies widely, from nonpolar and hydrophobic (water-insoluble) to highly polar and hydrophilic (water-soluble). A few amino acids are somewhat difficult to characterize or do not fit perfectly in any one group, particularly glycine, histidine, and cysteine. Their assignments to particular groupings are the results of considered judgments rather than absolutes.

The structures of the 20 common amino acids are shown in **Figure 3–5**, and some of their properties are listed in Table 3–1. Within each class there are gradations of polarity, size, and shape of the R groups.

Nonpolar, Aliphatic R Groups The R groups in this class of amino acids are nonpolar and hydrophobic. The side

$$^1CHO$$
$$HO-^2C-H$$
$$^3CH_2OH$$
L-Glyceraldehyde

$$CHO$$
$$H-C-OH$$
$$CH_2OH$$
D-Glyceraldehyde

$$COO^-$$
$$H_3\overset{+}{N}-C-H$$
$$CH_3$$
L-Alanine

$$COO^-$$
$$H-C-\overset{+}{N}H_3$$
$$CH_3$$
D-Alanine

FIGURE 3–4 Steric relationship of the stereoisomers of alanine to the absolute configuration of L- and D-glyceraldehyde. In these perspective formulas, the carbons are lined up vertically, with the chiral atom in the center. The carbons in these molecules are numbered beginning with the terminal aldehyde or carboxyl carbon (red), 1 to 3 from top to bottom as shown. When presented in this way, the R group of the amino acid (in this case the methyl group of alanine) is always below the α carbon. L-Amino acids are those with the α-amino group on the left, and D-amino acids have the α-amino group on the right.

FIGURE 3–5 The 20 common amino acids of proteins. The structural formulas show the state of ionization that would predominate at pH 7.0. The unshaded portions are those common to all the amino acids; the shaded portions are the R groups. Although the R group of histidine is shown uncharged, its pK_a (see Table 3–1) is such that a small but significant fraction of these groups are positively charged at pH 7.0. The protonated form of histidine is shown above the graph in Figure 3–12b.

chains of **alanine, valine, leucine,** and **isoleucine** tend to cluster together within proteins, stabilizing protein structure by means of hydrophobic interactions. **Glycine** has the simplest structure. Although it is most easily grouped with the nonpolar amino acids, its very small side chain makes no real contribution to hydrophobic interactions. **Methionine,** one of the two sulfur-containing amino acids, has a slightly nonpolar thioether group in its side chain. **Proline** has an aliphatic side chain with a distinctive cyclic structure. The secondary amino (imino) group of proline residues is held in a rigid conformation that reduces the structural flexibility of polypeptide regions containing proline.

Aromatic R Groups **Phenylalanine, tyrosine,** and **tryptophan,** with their aromatic side chains, are relatively nonpolar (hydrophobic). All can participate in hydrophobic interactions. The hydroxyl group of tyrosine can form hydrogen bonds, and it is an important functional group in some enzymes. Tyrosine and tryptophan are significantly more polar than phenylalanine, because of the tyrosine hydroxyl group and the nitrogen of the tryptophan indole ring.

Tryptophan and tyrosine, and to a much lesser extent phenylalanine, absorb ultraviolet light (**Fig. 3–6;** see also Box 3–1). This accounts for the characteristic strong absorbance of light by most proteins at a wavelength of

280 nm, a property exploited by researchers in the characterization of proteins.

Polar, Uncharged R Groups The R groups of these amino acids are more soluble in water, or more hydrophilic, than those of the nonpolar amino acids, because they contain functional groups that form hydrogen bonds

FIGURE 3–6 Absorption of ultraviolet light by aromatic amino acids. Comparison of the light absorption spectra of the aromatic amino acids tryptophan, tyrosine, and phenylalanine at pH 6.0. The amino acids are present in equimolar amounts (10^{-3} M) under identical conditions. The measured absorbance of tryptophan is more than four times that of tyrosine at a wavelength of 280 nm. Note that the maximum light absorption for both tryptophan and tyrosine occurs near 280 nm. Light absorption by phenylalanine generally contributes little to the spectroscopic properties of proteins.

BOX 3–1 | METHODS | Absorption of Light by Molecules: The Lambert–Beer Law

A wide range of biomolecules absorb light at characteristic wavelengths, just as tryptophan absorbs light at 280 nm (see Fig. 3–6). Measurement of light absorption by a spectrophotometer is used to detect and identify molecules and to measure their concentration in solution. The fraction of the incident light absorbed by a solution at a given wavelength is related to the thickness of the absorbing layer (path length) and the concentration of the absorbing species (Fig. 1). These two relationships are combined into the Lambert-Beer law,

$$\log \frac{I_0}{I} = \varepsilon c l$$

where I_0 is the intensity of the incident light, I is the intensity of the transmitted light, the ratio I/I_0 (the inverse of the ratio in the equation) is the transmittance, ε is the molar extinction coefficient (in units of liters per mole-centimeter), c is the concentration of the absorbing species (in moles per liter), and l is the

path length of the light-absorbing sample (in centimeters). The Lambert-Beer law assumes that the incident light is parallel and monochromatic (of a single wavelength) and that the solvent and solute molecules are randomly oriented. The expression log (I_0/I) is called the **absorbance**, designated A.

It is important to note that each successive millimeter of path length of absorbing solution in a 1.0 cm cell absorbs not a constant amount but a constant fraction of the light that is incident upon it. However, with an absorbing layer of fixed path length, *the absorbance, A, is directly proportional to the concentration of the absorbing solute.*

The molar extinction coefficient varies with the nature of the absorbing compound, the solvent, and the wavelength, and also with pH if the light-absorbing species is in equilibrium with an ionization state that has different absorbance properties.

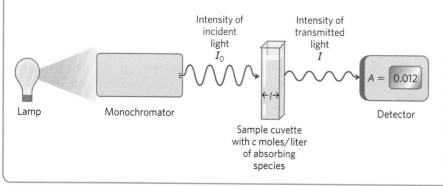

FIGURE 1 The principal components of a spectrophotometer. A light source emits light along a broad spectrum, then the monochromator selects and transmits light of a particular wavelength. The monochromatic light passes through the sample in a cuvette of path length l. The absorbance of the sample, log (I_0/I), is proportional to the concentration of the absorbing species. The transmitted light is measured by a detector.

FIGURE 3–7 Reversible formation of a disulfide bond by the oxidation of two molecules of cysteine. Disulfide bonds between Cys residues stabilize the structures of many proteins.

with water. This class of amino acids includes **serine**, **threonine**, **cysteine**, **asparagine**, and **glutamine**. The polarity of serine and threonine is contributed by their hydroxyl groups, and that of asparagine and glutamine by their amide groups. Cysteine is an outlier here because its polarity, contributed by its sulfhydryl group, is quite modest. Cysteine is a weak acid and can make weak hydrogen bonds with oxygen or nitrogen.

Asparagine and glutamine are the amides of two other amino acids also found in proteins—aspartate and glutamate, respectively—to which asparagine and glutamine are easily hydrolyzed by acid or base. Cysteine is readily oxidized to form a covalently linked dimeric amino acid called **cystine**, in which two cysteine molecules or residues are joined by a disulfide bond **(Fig. 3–7)**. The disulfide-linked residues are strongly hydrophobic (nonpolar). Disulfide bonds play a special role in the structures of many proteins by forming covalent links between parts of a polypeptide molecule or between two different polypeptide chains.

Positively Charged (Basic) R Groups The most hydrophilic R groups are those that are either positively or negatively charged. The amino acids in which the R groups have significant positive charge at pH 7.0 are **lysine**, which has a second primary amino group at the ε position on its aliphatic chain; **arginine**, which has a positively charged guanidinium group; and **histidine**, which has an aromatic imidazole group. As the only common amino acid having an ionizable side chain with pK_a near neutrality, histidine may be positively charged (protonated form) or uncharged at pH 7.0. His residues facilitate many enzyme-catalyzed reactions by serving as proton donors/acceptors.

Negatively Charged (Acidic) R Groups The two amino acids having R groups with a net negative charge at pH 7.0 are **aspartate** and **glutamate**, each of which has a second carboxyl group.

Uncommon Amino Acids Also Have Important Functions

In addition to the 20 common amino acids, proteins may contain residues created by modification of common residues already incorporated into a polypeptide **(Fig. 3–8a)**. Among these uncommon amino acids are **4-hydroxyproline**, a derivative of proline, and **5-hydroxylysine**, derived from lysine. The former is found in plant cell wall proteins, and both are found in collagen, a fibrous protein of connective tissues. **6-*N*-Methyllysine** is a constituent of myosin, a contractile protein of muscle. Another important uncommon amino acid is **γ-carboxyglutamate**, found in the blood-clotting protein prothrombin and in certain other proteins that bind Ca^{2+} as part of their biological function. More complex is **desmosine**, a derivative of four Lys residues, which is found in the fibrous protein elastin.

Selenocysteine is a special case. This rare amino acid residue is introduced during protein synthesis rather than created through a postsynthetic modification. It contains selenium rather than the sulfur of cysteine. Actually derived from serine, selenocysteine is a constituent of just a few known proteins.

Some amino acid residues in a protein may be modified transiently to alter the protein's function. The addition of phosphoryl, methyl, acetyl, adenylyl, ADP-ribosyl, or other groups to particular amino acid residues can increase or decrease a protein's activity (Fig. 3–8b). Phosphorylation is a particularly common regulatory modification. Covalent modification as a protein regulatory strategy is discussed in more detail in Chapter 6.

Some 300 additional amino acids have been found in cells. They have a variety of functions but are not all constituents of proteins. **Ornithine** and **citrulline** (Fig. 3–8c) deserve special note because they are key intermediates (metabolites) in the biosynthesis of arginine (Chapter 22) and in the urea cycle (Chapter 18).

Amino Acids Can Act as Acids and Bases

The amino and carboxyl groups of amino acids, along with the ionizable R groups of some amino acids, function as weak acids and bases. When an amino acid lacking an ionizable R group is dissolved in water at neutral pH, it exists in solution as the dipolar ion, or **zwitterion** (German for "hybrid ion"), which can act as either an acid or a base **(Fig. 3–9)**. Substances having this dual (acid-base) nature are **amphoteric** and are often called **ampholytes** (from "amphoteric electrolytes"). A simple monoamino monocarboxylic α-amino acid, such as alanine, is a diprotic acid when fully protonated; it has two groups, the —COOH group and the —NH_3^+ group, that can yield protons:

(a)

FIGURE 3–8 Uncommon amino acids. (a) Some uncommon amino acids found in proteins. All are derived from common amino acids. Extra functional groups added by modification reactions are shown in red. Desmosine is formed from four Lys residues (the carbon backbones are shaded in light red). Note the use of either numbers or Greek letters in the names of these structures to identify the altered carbon atoms. **(b)** Reversible amino acid modifications involved in regulation of protein activity. Phosphorylation is the most common type of regulatory modification. **(c)** Ornithine and citrulline, which are not found in proteins, are intermediates in the biosynthesis of arginine and in the urea cycle.

Phosphoserine

Phosphothreonine

Phosphotyrosine

σ-N-Methylarginine

6-N-Acetyllysine

Glutamate γ-methyl ester

(b) Adenylyltyrosine

(c)
Ornithine

Citrulline

Amino Acids Have Characteristic Titration Curves

Acid-base titration involves the gradual addition or removal of protons (Chapter 2). **Figure 3–10** shows the titration curve of the diprotic form of glycine. The two ionizable groups of glycine, the carboxyl group and the amino group, are titrated with a strong base such as NaOH. The plot has two distinct stages, corresponding to deprotonation of two different groups on glycine. Each of the two stages resembles in shape the titration

H
|
R—C—COOH
|
NH₂

Nonionic form

H
|
R—C—COO⁻
|
⁺NH₃

Zwitterionic form

$$H \atop R—C—COO⁻ \atop ⁺NH₃ \quad \rightleftharpoons \quad H \atop R—C—COO⁻ + H⁺ \atop NH₂$$

Zwitterion
as acid

$$H \atop R—C—COO⁻ + H⁺ \atop ⁺NH₃ \quad \rightleftharpoons \quad H \atop R—C—COOH \atop ⁺NH₃$$

Zwitterion
as base

FIGURE 3–9 Nonionic and zwitterionic forms of amino acids. The nonionic form does not occur in significant amounts in aqueous solutions. The zwitterion predominates at neutral pH. A zwitterion can act as either an acid (proton donor) or a base (proton acceptor).

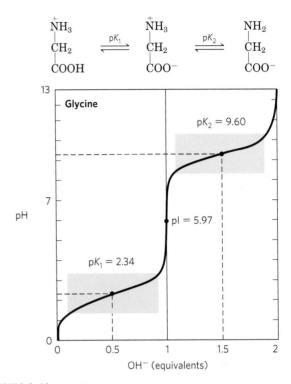

FIGURE 3–10 Titration of an amino acid. Shown here is the titration curve of 0.1 M glycine at 25°C. The ionic species predominating at key points in the titration are shown above the graph. The shaded boxes, centered at about pK_1 = 2.34 and pK_2 = 9.60, indicate the regions of greatest buffering power. Note that 1 equivalent of OH⁻ = 0.1 M NaOH added.

curve of a monoprotic acid, such as acetic acid (see Fig. 2–17), and can be analyzed in the same way. At very low pH, the predominant ionic species of glycine is the fully protonated form, $^+H_3N—CH_2—COOH$. In the first stage of the titration, the —COOH group of glycine loses its proton. At the midpoint of this stage, equimolar concentrations of the proton-donor ($^+H_3N—CH_2—COOH$) and proton-acceptor ($^+H_3N—CH_2—COO^-$) species are present. As in the titration of any weak acid, a point of inflection is reached at this midpoint where the pH is equal to the pK_a of the protonated group being titrated (see Fig. 2–18). For glycine, the pH at the midpoint is 2.34, thus its —COOH group has a pK_a (labeled pK_1 in Fig. 3–10) of 2.34. (Recall from Chapter 2 that pH and pK_a are simply convenient notations for proton concentration and the equilibrium constant for ionization, respectively. The pK_a is a measure of the tendency of a group to give up a proton, with that tendency decreasing tenfold as the pK_a increases by one unit.) As the titration of glycine proceeds, another important point is reached at pH 5.97. Here there is another point of inflection, at which removal of the first proton is essentially complete and removal of the second has just begun. At this pH glycine is present largely as the dipolar ion (zwitterion) $^+H_3N—CH_2—COO^-$. We shall return to the significance of this inflection point in the titration curve (labeled pI in Fig. 3–10) shortly.

The second stage of the titration corresponds to the removal of a proton from the —NH₃⁺ group of glycine. The pH at the midpoint of this stage is 9.60, equal to the pK_a (labeled pK_2 in Fig. 3–10) for the —NH₃⁺ group. The titration is essentially complete at a pH of about 12,

at which point the predominant form of glycine is $H_2N—CH_2—COO^-$.

From the titration curve of glycine we can derive several important pieces of information. First, it gives a quantitative measure of the pK_a of each of the two ionizing groups: 2.34 for the —COOH group and 9.60 for the —NH₃⁺ group. Note that the carboxyl group of glycine is over 100 times more acidic (more easily ionized) than the carboxyl group of acetic acid, which, as we saw in Chapter 2, has a pK_a of 4.76—about average for a carboxyl group attached to an otherwise unsubstituted aliphatic hydrocarbon. The perturbed pK_a of glycine is caused by repulsion between the departing proton and the nearby positively charged amino group on the α-carbon atom, as described in **Figure 3–11**. The opposite charges on the resulting zwitterion are stabilizing. Similarly, the pK_a of the amino group in glycine is perturbed downward relative to the average pK_a of an amino group. This effect is due partly to the electronegative oxygen atoms in the carboxyl groups, which tend to pull electrons toward them, increasing the tendency of the amino group to give up a proton. Hence, the α-amino group has a pK_a that is lower than that of an aliphatic amine such as methylamine (Fig. 3–11). In short, the pK_a of any functional group is greatly affected by its chemical environment, a phenomenon sometimes exploited in the active sites of enzymes to promote exquisitely adapted reaction mechanisms that depend

FIGURE 3–11 Effect of the chemical environment on pK_a. The pK_a values for the ionizable groups in glycine are lower than those for simple, methyl-substituted amino and carboxyl groups. These downward perturbations of pK$_a$ are due to intramolecular interactions. Similar effects can be caused by chemical groups that happen to be positioned nearby—for example, in the active site of an enzyme.

on the perturbed pK_a values of proton donor/acceptor groups of specific residues.

The second piece of information provided by the titration curve of glycine is that this amino acid has two regions of buffering power. One of these is the relatively flat portion of the curve, extending for approximately 1 pH unit on either side of the first pK_a of 2.34, indicating that glycine is a good buffer near this pH. The other buffering zone is centered around pH 9.60. (Note that glycine is not a good buffer at the pH of intracellular fluid or blood, about 7.4.) Within the buffering ranges of glycine, the Henderson-Hasselbalch equation (p. 64) can be used to calculate the proportions of proton-donor and proton-acceptor species of glycine required to make a buffer at a given pH.

Titration Curves Predict the Electric Charge of Amino Acids

Another important piece of information derived from the titration curve of an amino acid is the relationship between its net charge and the pH of the solution. At pH 5.97, the point of inflection between the two stages in its titration curve, glycine is present predominantly as its dipolar form, fully ionized but with no *net* electric charge (Fig. 3–10). The characteristic pH at which the *net* electric charge is zero is called the **isoelectric point** or **isoelectric pH**, designated **pI**. For glycine, which has no ionizable group in its side chain, the iso-

electric point is simply the arithmetic mean of the two pK_a values:

$$pI = \frac{1}{2}(pK_1 + pK_2) = \frac{1}{2}(2.34 + 9.60) = 5.97$$

As is evident in Figure 3–10, glycine has a net negative charge at any pH above its pI and will thus move toward the positive electrode (the anode) when placed in an electric field. At any pH below its pI, glycine has a net positive charge and will move toward the negative electrode (the cathode). The farther the pH of a glycine solution is from its isoelectric point, the greater the net electric charge of the population of glycine molecules. At pH 1.0, for example, glycine exists almost entirely as the form $^+H_3N—CH_2—COOH$ with a net positive charge of 1.0. At pH 2.34, where there is an equal mixture of $^+H_3N—CH_2—COOH$ and $^+H_3N—CH_2—COO^-$, the average or net positive charge is 0.5. The sign and the magnitude of the net charge of any amino acid at any pH can be predicted in the same way.

Amino Acids Differ in Their Acid-Base Properties

The shared properties of many amino acids permit some simplifying generalizations about their acid-base behaviors. First, all amino acids with a single α-amino group, a single α-carboxyl group, and an R group that does not ionize have titration curves resembling that of glycine (Fig. 3–10). These amino acids have very similar, although not identical, pK_a values: pK_a of the —COOH group in the range of 1.8 to 2.4, and pK_a of the —NH$_3^+$

FIGURE 3–12 Titration curves for **(a)** glutamate and **(b)** histidine. The pK_a of the R group is designated here as pK_R.

group in the range of 8.8 to 11.0 (Table 3–1). The differences in these pK_a values reflect the chemical environments imposed by their R groups. Second, amino acids with an ionizable R group have more complex titration curves, with *three* stages corresponding to the three possible ionization steps; thus they have three pK_a values. The additional stage for the titration of the ionizable R group merges to some extent with that for the titration of the α-carboxyl group, the titration of the α-amino group, or both. The titration curves for two amino acids of this type, glutamate and histidine, are shown in **Figure 3–12**. The isoelectric points reflect the nature of the ionizing R groups present. For example, glutamate has a pI of 3.22, considerably lower than that of glycine. This is due to the presence of two carboxyl groups, which, at the average of their pK_a values (3.22), contribute a net charge of −1 that balances the +1 contributed by the amino group. Similarly, the pI of histidine, with two groups that are positively charged when protonated, is 7.59 (the average of the pK_a values of the amino and imidazole groups), much higher than that of glycine.

Finally, as pointed out earlier, under the general condition of free and open exposure to the aqueous environment, only histidine has an R group (pK_a = 6.0) providing significant buffering power near the neutral pH usually found in the intracellular and extracellular fluids of most animals and bacteria (Table 3–1).

SUMMARY 3.1 Amino Acids

▶ The 20 amino acids commonly found as residues in proteins contain an α-carboxyl group, an α-amino group, and a distinctive R group substituted on the α-carbon atom. The α-carbon atom of all amino

acids except glycine is asymmetric, and thus amino acids can exist in at least two stereoisomeric forms. Only the L stereoisomers, with a configuration related to the absolute configuration of the reference molecule L-glyceraldehyde, are found in proteins.

▶ Other, less common amino acids also occur, either as constituents of proteins (through modification of common amino acid residues after protein synthesis) or as free metabolites.

▶ Amino acids can be classified into five types on the basis of the polarity and charge (at pH 7) of their R groups.

▶ Amino acids vary in their acid-base properties and have characteristic titration curves. Monoamino monocarboxylic amino acids (with nonionizable R groups) are diprotic acids ($^+H_3NCH(R)COOH$) at low pH and exist in several different ionic forms as the pH is increased. Amino acids with ionizable R groups have additional ionic species, depending on the pH of the medium and the pK_a of the R group.

3.2 Peptides and Proteins

We now turn to polymers of amino acids, the **peptides** and **proteins**. Biologically occurring polypeptides range in size from small to very large, consisting of two or three to thousands of linked amino acid residues. Our focus is on the fundamental chemical properties of these polymers.

Peptides Are Chains of Amino Acids

Two amino acid molecules can be covalently joined through a substituted amide linkage, termed a **peptide**

$$H_3\overset{+}{N}-\underset{\underset{\displaystyle O}{\|}}{\overset{\displaystyle R^1}{\underset{\displaystyle}{C}H-C}}-OH \;+\; H-\underset{\displaystyle H}{\overset{\displaystyle R^2}{N-CH}}-COO^-$$

$$H_2O \;\rightleftharpoons\; H_2O$$

$$H_3\overset{+}{N}-\overset{\displaystyle R^1}{CH}-\underset{\underset{\displaystyle O}{\|}}{C}-\overset{\displaystyle H}{N}-\overset{\displaystyle R^2}{CH}-COO^-$$

FIGURE 3–13 Formation of a peptide bond by condensation. The α-amino group of one amino acid (with R^2 group) acts as a nucleophile to displace the hydroxyl group of another amino acid (with R^1 group), forming a peptide bond (shaded in light red). Amino groups are good nucleophiles, but the hydroxyl group is a poor leaving group and is not readily displaced. At physiological pH, the reaction shown here does not occur to any appreciable extent.

bond, to yield a dipeptide. Such a linkage is formed by removal of the elements of water (dehydration) from the α-carboxyl group of one amino acid and the α-amino group of another **(Fig. 3–13)**. Peptide bond formation is an example of a condensation reaction, a common class of reactions in living cells. Under standard biochemical conditions, the equilibrium for the reaction shown in Figure 3–13 favors the amino acids over the dipeptide. To make the reaction thermodynamically more favorable, the carboxyl group must be chemically modified or activated so that the hydroxyl group can be more readily eliminated. A chemical approach to this problem is outlined later in this chapter. The biological approach to peptide bond formation is a major topic of Chapter 27.

Three amino acids can be joined by two peptide bonds to form a tripeptide; similarly, four amino acids can be linked to form a tetrapeptide, five to form a pentapeptide, and so forth. When a few amino acids are joined in this fashion, the structure is called an **oligopeptide.** When many amino acids are joined, the product is called a **polypeptide.** Proteins may have thousands of amino acid residues. Although the terms "protein" and "polypeptide" are sometimes used interchangeably, molecules referred to as polypeptides generally have molecular weights below 10,000, and those called proteins have higher molecular weights.

Figure 3–14 shows the structure of a pentapeptide. As already noted, an amino acid unit in a peptide is often called a residue (the part left over after losing the elements of water—a hydrogen atom from its amino group and the hydroxyl moiety from its carboxyl group). In a peptide, the amino acid residue at the end with a free α-amino group is the **amino-terminal** (or *N*-terminal) residue; the residue at the other end, which has a free carboxyl group, is the **carboxyl-terminal** (*C*-terminal) residue.

FIGURE 3–14 The pentapeptide serylglycyltyrosylalanylleucine, Ser-Gly-Tyr-Ala-Leu, or SGYAL. Peptides are named beginning with the amino-terminal residue, which by convention is placed at the left. The peptide bonds are shaded in light red; the R groups are in red.

KEY CONVENTION: When an amino acid sequence of a peptide, polypeptide, or protein is displayed, the amino-terminal end is placed on the left, the carboxyl-terminal end on the right. The sequence is read left to right, beginning with the amino-terminal end. ∎

Although hydrolysis of a peptide bond is an exergonic reaction, it occurs only slowly because it has a high activation energy (p. 27). As a result, the peptide bonds in proteins are quite stable, with an average half-life ($t_{1/2}$) of about 7 years under most intracellular conditions.

Peptides Can Be Distinguished by Their Ionization Behavior

Peptides contain only one free α-amino group and one free α-carboxyl group, at opposite ends of the chain **(Fig. 3–15)**. These groups ionize as they do in free amino acids, although the ionization constants are different because an oppositely charged group is no longer

$$
\begin{array}{ll}
 & \overset{+}{N}H_3 \\
\text{Ala} & CH-CH_3 \\
 & O{=}C \\
 & NH \\
\text{Glu} & CH-CH_2-CH_2-COO^- \\
 & O{=}C \\
 & NH \\
\text{Gly} & CH_2 \\
 & O{=}C \\
 & NH \\
\text{Lys} & CH-CH_2-CH_2-CH_2-CH_2-\overset{+}{N}H_3 \\
 & COO^-
\end{array}
$$

FIGURE 3–15 Alanylglutamylglycyllysine. This tetrapeptide has one free α-amino group, one free α-carboxyl group, and two ionizable R groups. The groups ionized at pH 7.0 are in red.

linked to the α carbon. The α-amino and α-carboxyl groups of all nonterminal amino acids are covalently joined in the peptide bonds, which do not ionize and thus do not contribute to the total acid-base behavior of peptides. However, the R groups of some amino acids can ionize (Table 3–1), and in a peptide these contribute to the overall acid-base properties of the molecule (Fig. 3–15). Thus the acid-base behavior of a peptide can be predicted from its free α-amino and α-carboxyl groups combined with the nature and number of its ionizable R groups.

Like free amino acids, peptides have characteristic titration curves and a characteristic isoelectric pH (pI) at which they do not move in an electric field. These properties are exploited in some of the techniques used to separate peptides and proteins, as we shall see later in the chapter. It should be emphasized that the pK_a value for an ionizable R group can change somewhat when an amino acid becomes a residue in a peptide. The loss of charge in the α-carboxyl and α-amino groups, the interactions with other peptide R groups, and other environmental factors can affect the pK_a. The pK_a values for R groups listed in Table 3–1 can be a useful guide to the pH range in which a given group will ionize, but they cannot be strictly applied to peptides.

Biologically Active Peptides and Polypeptides Occur in a Vast Range of Sizes and Compositions

No generalizations can be made about the molecular weights of biologically active peptides and proteins in relation to their functions. Naturally occurring peptides range in length from two to many thousands of amino acid residues. Even the smallest peptides can have biologically important effects. Consider the commercially synthesized dipeptide L-aspartyl-L-phenylalanine

methyl ester, the artificial sweetener better known as aspartame or NutraSweet.

L-Aspartyl-L-phenylalanine methyl ester
(aspartame)

Many small peptides exert their effects at very low concentrations. For example, a number of vertebrate hormones (Chapter 23) are small peptides. These include oxytocin (nine amino acid residues), which is secreted by the posterior pituitary gland and stimulates uterine contractions, and thyrotropin-releasing factor (three residues), which is formed in the hypothalamus and stimulates the release of another hormone, thyrotropin, from the anterior pituitary gland. Some extremely toxic mushroom poisons, such as amanitin, are also small peptides, as are many antibiotics.

How long are the polypeptide chains in proteins? As Table 3–2 shows, lengths vary considerably. Human cytochrome c has 104 amino acid residues linked in a single chain; bovine chymotrypsinogen has 245 residues. At the extreme is titin, a constituent of vertebrate muscle, which has nearly 27,000 amino acid residues and a molecular weight of about 3,000,000. The vast majority of naturally occurring proteins are much smaller than this, containing fewer than 2,000 amino acid residues.

Some proteins consist of a single polypeptide chain, but others, called **multisubunit** proteins, have two or more polypeptides associated noncovalently (Table 3–2). The individual polypeptide chains in a multisubunit

TABLE 3–2 Molecular Data on Some Proteins

	Molecular weight	Number of residues	Number of polypeptide chains
Cytochrome c (human)	12,400	104	1
Ribonuclease A (bovine pancreas)	13,700	124	1
Lysozyme (chicken egg white)	14,300	129	1
Myoglobin (equine heart)	16,700	153	1
Chymotrypsin (bovine pancreas)	25,200	241	3
Chymotrypsinogen (bovine)	25,700	245	1
Hemoglobin (human)	64,500	574	4
Serum albumin (human)	66,000	609	1
Hexokinase (yeast)	107,900	972	2
RNA polymerase (E. coli)	450,000	4,158	5
Apolipoprotein B (human)	513,000	4,536	1
Glutamine synthetase (E. coli)	619,000	5,628	12
Titin (human)	2,993,000	26,926	1

protein may be identical or different. If at least two are identical the protein is said to be **oligomeric**, and the identical units (consisting of one or more polypeptide chains) are referred to as **protomers**. Hemoglobin, for example, has four polypeptide subunits: two identical α chains and two identical β chains, all four held together by noncovalent interactions. Each α subunit is paired in an identical way with a β subunit within the structure of this multisubunit protein, so that hemoglobin can be considered either a tetramer of four polypeptide subunits or a dimer of $\alpha\beta$ protomers.

A few proteins contain two or more polypeptide chains linked covalently. For example, the two polypeptide chains of insulin are linked by disulfide bonds. In such cases, the individual polypeptides are not considered subunits but are commonly referred to simply as chains.

The amino acid composition of proteins is also highly variable. The 20 common amino acids almost never occur in equal amounts in a protein. Some amino acids may occur only once or not at all in a given type of protein; others may occur in large numbers. Table 3–3

shows the amino acid composition of bovine cytochrome c and chymotrypsinogen, the inactive precursor of the digestive enzyme chymotrypsin. These two proteins, with very different functions, also differ significantly in the relative numbers of each kind of amino acid residue.

We can calculate the approximate number of amino acid residues in a simple protein containing no other chemical constituents by dividing its molecular weight by 110. Although the average molecular weight of the 20 common amino acids is about 138, the smaller amino acids predominate in most proteins. If we take into account the proportions in which the various amino acids occur in an average protein (Table 3–1; the averages are determined by surveying the amino acid compositions of more than 1,000 different proteins), the average molecular weight of protein amino acids is nearer to 128. Because a molecule of water (M_r 18) is removed to create each peptide bond, the average molecular weight of an amino acid residue in a protein is about $128 - 18 = 110$.

TABLE 3–3 Amino Acid Composition of Two Proteins

Amino acid	Bovine cytochrome c		Bovine chymotrypsinogen	
	Number of residues per molecule	Percentage of total*	Number of residues per molecule	Percentage of total*
Ala	6	6	22	9
Arg	2	2	4	1.6
Asn	5	5	14	5.7
Asp	3	3	9	3.7
Cys	2	2	10	4
Gln	3	3	10	4
Glu	9	9	5	2
Gly	14	13	23	9.4
His	3	3	2	0.8
Ile	6	6	10	4
Leu	6	6	19	7.8
Lys	18	17	14	5.7
Met	2	2	2	0.8
Phe	4	4	6	2.4
Pro	4	4	9	3.7
Ser	1	1	28	11.4
Thr	8	8	23	9.4
Trp	1	1	8	3.3
Tyr	4	4	4	1.6
Val	3	3	23	9.4
Total	104	102	245	99.7

Note: In some common analyses, such as acid hydrolysis, Asp and Asn are not readily distinguished from each other and are together designated Asx (or B). Similarly, when Glu and Gln cannot be distinguished, they are together designated Glx (or Z). In addition, Trp is destroyed by acid hydrolysis. Additional procedures must be employed to obtain an accurate assessment of complete amino acid content.

*Percentages do not total to 100%, due to rounding.

TABLE 3–4 Conjugated Proteins

Class	Prosthetic group	Example
Lipoproteins	Lipids	β_1-Lipoprotein of blood
Glycoproteins	Carbohydrates	Immunoglobulin G
Phosphoproteins	Phosphate groups	Casein of milk
Hemoproteins	Heme (iron porphyrin)	Hemoglobin
Flavoproteins	Flavin nucleotides	Succinate dehydrogenase
Metalloproteins	Iron	Ferritin
	Zinc	Alcohol dehydrogenase
	Calcium	Calmodulin
	Molybdenum	Dinitrogenase
	Copper	Plastocyanin

Some Proteins Contain Chemical Groups Other Than Amino Acids

Many proteins, for example the enzymes ribonuclease A and chymotrypsin, contain only amino acid residues and no other chemical constituents; these are considered simple proteins. However, some proteins contain permanently associated chemical components in addition to amino acids; these are called **conjugated proteins**. The non–amino acid part of a conjugated protein is usually called its **prosthetic group**. Conjugated proteins are classified on the basis of the chemical nature of their prosthetic groups (Table 3–4); for example, **lipoproteins** contain lipids, **glycoproteins** contain sugar groups, and **metalloproteins** contain a specific metal. Some proteins contain more than one prosthetic group. Usually the prosthetic group plays an important role in the protein's biological function.

SUMMARY 3.2 Peptides and Proteins

▶   Amino acids can be joined covalently through peptide bonds to form peptides and proteins. Cells generally contain thousands of different proteins, each with a different biological activity.

▶   Proteins can be very long polypeptide chains of 100 to several thousand amino acid residues. However, some naturally occurring peptides have only a few amino acid residues. Some proteins are composed of several noncovalently associated polypeptide chains, called subunits.

▶   Simple proteins yield only amino acids on hydrolysis; conjugated proteins contain in addition some other component, such as a metal or organic prosthetic group.

3.3 Working with Proteins

Biochemists' understanding of protein structure and function has been derived from the study of many individual proteins. To study a protein in detail, the researcher must be able to separate it from other proteins in pure form and must have the techniques to determine its properties. The necessary methods come from protein chemistry, a discipline as old as biochemistry itself and one that retains a central position in biochemical research.

Proteins Can Be Separated and Purified

A pure preparation is essential before a protein's properties and activities can be determined. Given that cells contain thousands of different kinds of proteins, how can one protein be purified? Classical methods for separating proteins take advantage of properties that vary from one protein to the next, including size, charge, and binding properties. These have been supplemented in recent decades by methods involving DNA cloning and genome sequencing that can simplify the process of protein purification. The newer methods, presented in Chapter 9, often artificially modify the protein being purified, adding a few or many amino acid residues to one or both ends. Convenience thus comes at the price of potentially altering the activity of the purified protein. The purification of proteins in their native state (the form in which they function in the cell) usually relies on methods described here.

The source of a protein is generally tissue or microbial cells. The first step in any protein purification procedure is to break open these cells, releasing their proteins into a solution called a **crude extract**. If necessary, differential centrifugation can be used to prepare subcellular fractions or to isolate specific organelles (see Fig. 1–8).

Once the extract or organelle preparation is ready, various methods are available for purifying one or more of the proteins it contains. Commonly, the extract is subjected to treatments that separate the proteins into different **fractions** based on a property such as size or charge, a process referred to as **fractionation**. Early fractionation steps in a purification utilize differences in protein solubility, which is a complex function of pH, temperature, salt concentration, and other factors.

The solubility of proteins is lowered in the presence of some salts, an effect called "salting out." The addition of certain salts in the right amount can selectively precipitate some proteins, while others remain in solution. Ammonium sulfate ($(NH_4)_2SO_4$) is particularly effective and is often used to salt out proteins. The proteins thus precipitated are removed from those remaining in solution by low-speed centrifugation.

A solution containing the protein of interest usually must be further altered before subsequent purification steps are possible. For example, **dialysis** is a procedure that separates proteins from small solutes by taking advantage of the proteins' larger size. The partially purified extract is placed in a bag or tube made of a semipermeable membrane. When this is suspended in a much larger volume of buffered solution of appropriate ionic strength, the membrane allows the exchange of salt and buffer but not proteins. Thus dialysis retains large proteins within the membranous bag or tube while allowing the concentration of other solutes in the protein preparation to change until they come into equilibrium with the solution outside the membrane. Dialysis might be used, for example, to remove ammonium sulfate from the protein preparation.

The most powerful methods for fractionating proteins make use of **column chromatography**, which takes advantage of differences in protein charge, size, binding affinity, and other properties **(Fig. 3–16)**. A porous solid material with appropriate chemical properties (the stationary phase) is held in a column, and a buffered solution (the mobile phase) migrates through it. The protein, dissolved in the same buffered solution that was used to establish the mobile phase, is layered on the top of the column. The protein then percolates through the solid matrix as an ever-expanding band within the larger mobile phase. Individual proteins migrate faster or more slowly through the column depending on their properties.

Ion-exchange chromatography exploits differences in the sign and magnitude of the net electric charge of proteins at a given pH **(Fig. 3–17a)**. The column matrix is a synthetic polymer (resin) containing bound charged groups; those with bound anionic groups are called **cation exchangers**, and those with bound cationic groups are called **anion exchangers**. The affinity of each protein for the charged groups on the column is affected by the pH (which determines the ionization state of the molecule) and the concentration of competing free salt ions in the surrounding solution. Separation can be optimized by gradually changing the pH and/or salt concentration of the mobile phase so as to create a pH or salt gradient. In **cation-exchange chromatography**, the solid matrix has negatively charged groups. In the mobile phase, proteins with a net positive charge migrate through the matrix more slowly than those with a net negative charge, because the migration of the former is retarded more by interaction with the stationary phase.

In ion-exchange columns, the expansion of the protein band in the mobile phase (the protein solution) is

FIGURE 3–16 Column chromatography. The standard elements of a chromatographic column include a solid, porous material (matrix) supported inside a column, generally made of plastic or glass. A solution, the mobile phase, flows through the matrix, the stationary phase. The solution that passes out of the column at the bottom (the effluent) is constantly replaced by solution supplied from a reservoir at the top. The protein solution to be separated is layered on top of the column and allowed to percolate into the solid matrix. Additional solution is added on top. The protein solution forms a band within the mobile phase that is initially the depth of the protein solution applied to the column. As proteins migrate through the column (shown here at five different times), they are retarded to different degrees by their different interactions with the matrix material. The overall protein band thus widens as it moves through the column. Individual types of proteins (such as A, B, and C, shown in blue, red, and green) gradually separate from each other, forming bands within the broader protein band. Separation improves (i.e., resolution increases) as the length of the column increases. However, each individual protein band also broadens with time due to diffusional spreading, a process that decreases resolution. In this example, protein A is well separated from B and C, but diffusional spreading prevents complete separation of B and C under these conditions.

caused both by separation of proteins with different properties and by diffusional spreading. As the length of the column increases, the resolution of two types of protein with different net charges generally improves. However, the rate at which the protein solution can flow through

FIGURE 3–17 Three chromatographic methods used in protein purification. (a) Ion-exchange chromatography exploits differences in the sign and magnitude of the net electric charges of proteins at a given pH. **(b)** Size-exclusion chromatography, also called gel filtration, separates proteins according to size. **(c)** Affinity chromatography separates proteins by their binding specificities. Further details of these methods are given in the text.

the column usually decreases with column length. And as the length of time spent on the column increases, the resolution can decline as a result of diffusional spreading within each protein band. As the protein-containing solution exits a column, successive portions (fractions) of this effluent are collected in test tubes. Each fraction can be tested for the presence of the protein of interest as well as other properties, such as ionic strength or total protein concentration. All fractions positive for the protein of interest can be combined as the product of this chromatographic step of the protein purification.

WORKED EXAMPLE 3–1 Ion Exchange of Peptides

A biochemist wants to separate two peptides by ion-exchange chromatography. At the pH of the mobile phase to be used on the column, one peptide (A) has a net charge of −3, due to the presence of more Glu and Asp residues than Arg, Lys, and His residues. Peptide B has a net charge of +1. Which peptide would elute first from a cation-exchange resin? Which would elute first from an anion-exchange resin?

Solution: A cation-exchange resin has negative charges and binds positively charged molecules, retarding their progress through the column. Peptide B, with its net positive charge, will interact more strongly than peptide A with the cation-exchange resin, and thus peptide A will elute first. On the anion-exchange resin, peptide B will elute first. Peptide A, being negatively charged, will be retarded by its interaction with the positively charged resin.

Figure 3–17 shows two other variations of column chromatography in addition to ion exchange. **Size-exclusion chromatography**, also called gel filtration (Fig. 3–17b), separates proteins according to size. In this method, large proteins emerge from the column sooner than small ones—a somewhat counterintuitive result. The solid phase consists of cross-linked polymer beads with engineered pores or cavities of a particular size. Large proteins cannot enter the cavities and so take a shorter (and more rapid) path through the column, around the beads. Small proteins enter the cavities and are slowed by their more labyrinthine path through the column. Size-exclusion chromatography can also be used to approximate the size of a protein being purified, using methods similar to those described in Figure 3–19.

Affinity chromatography is based on binding affinity (Fig. 3–17c). The beads in the column have a covalently attached chemical group called a ligand—a group or molecule that binds to a macromolecule such as a protein. When a protein mixture is added to the column, any protein with affinity for this ligand binds to the beads, and its migration through the matrix is retarded. For example, if the biological function of a protein involves binding to ATP, then attaching a mol-ecule that resembles ATP to the beads in the column creates an affinity matrix that can help purify the protein. As the protein solution moves through the column, ATP-binding proteins (including the protein of interest) bind to the matrix. After proteins that do not bind are washed through the column, the bound protein is eluted by a solution containing either a high concentration of salt or free ligand—in this case, ATP or an analog of ATP. Salt weakens the binding of the protein to the immobilized ligand, interfering with ionic interactions. Free ligand competes with the ligand attached to the beads, releasing the protein from the matrix; the protein product that elutes from the column is often bound to the ligand used to elute it.

Chromatographic methods are typically enhanced by the use of **HPLC**, or **high-performance liquid chromatography**. HPLC makes use of high-pressure pumps that speed the movement of the protein molecules down the column, as well as higher-quality chromatographic materials that can withstand the crushing force of the pressurized flow. By reducing the transit time on the column, HPLC can limit diffusional spreading of protein bands and thus greatly improve resolution.

The approach to purification of a protein that has not previously been isolated is guided both by established precedents and by common sense. In most cases, several different methods must be used sequentially to purify a protein completely, each method separating proteins on the basis of different properties. For example, if one step separates ATP-binding proteins from those that do not bind ATP, then the next step must separate the various ATP-binding proteins on the basis of size or charge to isolate the particular protein that is wanted. The choice of methods is somewhat empirical, and many strategies may be tried before the most effective one is found. Trial and error can often be minimized by basing the new procedure on purification techniques developed for similar proteins. Published purification protocols are available for many thousands of proteins. Common sense dictates that inexpensive procedures such as salting out be used first, when the total volume and the number of contaminants are greatest. Chromatographic methods are often impractical at early stages, because the amount of chromatographic medium needed increases with sample size. As each purification step is completed, the sample size generally becomes smaller (Table 3–5), making it feasible to use more sophisticated (and expensive) chromatographic procedures at later stages.

Proteins Can Be Separated and Characterized by Electrophoresis

Another important technique for the separation of proteins is based on the migration of charged proteins in an electric field, a process called **electrophoresis**. These procedures are not generally used to purify proteins, because simpler alternatives are usually available and electrophoretic methods often adversely affect the

TABLE 3-5   A Purification Table for a Hypothetical Enzyme

Procedure or step	Fraction volume (mL)	Total protein (mg)	Activity (units)	Specific activity (units/mg)
1. Crude cellular extract	1,400	10,000	100,000	10
2. Precipitation with ammonium sulfate	280	3,000	96,000	32
3. Ion-exchange chromatography	90	400	80,000	200
4. Size-exclusion chromatography	80	100	60,000	600
5. Affinity chromatography	6	3	45,000	15,000

Note: All data represent the status of the sample *after* the designated procedure has been carried out. Activity and specific activity are defined on page 95.

structure and thus the function of proteins. However, as an analytical method, electrophoresis is extremely important. Its advantage is that proteins can be visualized as well as separated, permitting a researcher to estimate quickly the number of different proteins in a mixture or the degree of purity of a particular protein preparation. Also, electrophoresis can be used to determine crucial properties of a protein such as its isoelectric point and approximate molecular weight.

Electrophoresis of proteins is generally carried out in gels made up of the cross-linked polymer polyacrylamide (**Fig. 3–18**). The polyacrylamide gel acts as a molecular sieve, slowing the migration of proteins approximately in proportion to their charge-to-mass ratio. Migration may also be affected by protein shape. In electrophoresis, the force moving the macromolecule is the electrical potential, E. The electrophoretic mobility, μ, of a molecule is the ratio of its velocity, V,

(a)   **(b)**

FIGURE 3–18 Electrophoresis. (a) Different samples are loaded in wells or depressions at the top of the SDS–polyacrylamide gel. The proteins move into the gel when an electric field is applied. The gel minimizes convection currents caused by small temperature gradients, as well as protein movements other than those induced by the electric field. **(b)** Proteins can be visualized after electrophoresis by treating the gel with a stain such as Coomassie blue, which binds to the proteins but not to the gel itself. Each band on the gel represents a different protein (or protein subunit); smaller proteins move through the gel more rapidly than larger proteins and therefore are found nearer the bottom of the gel. This gel illustrates purification of the RecA protein of *Escherichia coli* (described in Chapter 25). The gene for the RecA protein was cloned (Chapter 9) so that its expression (synthesis of the protein) could be controlled. The first lane shows a set of standard proteins (of known M_r), serving as molecular weight markers. The next two lanes show proteins from *E. coli* cells before and after synthesis of RecA protein was induced. The fourth lane shows the proteins in a crude cellular extract. Subsequent lanes (left to right) show the proteins present after successive purification steps. The purified protein is a single polypeptide chain (M_r ~38,000), as seen in the rightmost lane.

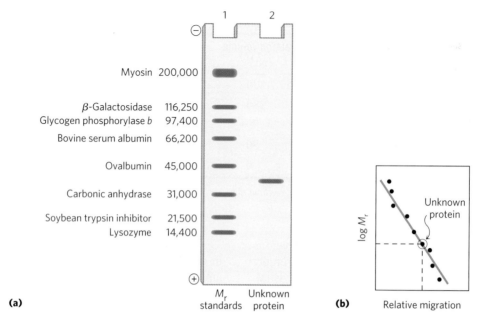

FIGURE 3–19 Estimating the molecular weight of a protein. The electrophoretic mobility of a protein on an SDS polyacrylamide gel is related to its molecular weight, M_r. **(a)** Standard proteins of known molecular weight are subjected to electrophoresis (lane 1). These marker proteins can be used to estimate the molecular weight of an unknown protein (lane 2). **(b)** A plot of log M_r of the marker proteins versus relative migration during electrophoresis is linear, which allows the molecular weight of the unknown protein to be read from the graph. (In similar fashion, a set of standard proteins with reproducible retention times on a size-exclusion column can be used to create a standard curve of retention time versus log M_r. The retention time of an unknown substance on the column can be compared with this standard curve to obtain an approximate M_r.)

to the electrical potential. Electrophoretic mobility is also equal to the net charge, Z, of the molecule divided by the frictional coefficient, f, which reflects in part a protein's shape. Thus:

$$\mu = \frac{V}{E} = \frac{Z}{f}$$

The migration of a protein in a gel during electrophoresis is therefore a function of its size and its shape.

An electrophoretic method commonly employed for estimation of purity and molecular weight makes use of the detergent **sodium dodecyl sulfate (SDS)** ("dodecyl" denoting a 12-carbon chain).

$$Na^+ \ {}^-O-\overset{\displaystyle O}{\underset{\displaystyle O}{\overset{\displaystyle \|}{\underset{\displaystyle \|}{S}}}}-O-(CH_2)_{11}CH_3$$

Sodium dodecyl sulfate
(SDS)

A protein will bind about 1.4 times its weight of SDS, nearly one molecule of SDS for each amino acid residue. The bound SDS contributes a large net negative charge, rendering the intrinsic charge of the protein insignificant and conferring on each protein a similar charge-to-mass ratio. In addition, SDS binding partially unfolds proteins, such that most SDS-bound proteins assume a similar rodlike shape. Electrophoresis in the presence of SDS therefore separates proteins almost exclusively on the basis of mass (molecular weight), with smaller polypeptides migrating more rapidly. After electrophoresis,

the proteins are visualized by adding a dye such as Coomassie blue, which binds to proteins but not to the gel itself (Fig. 3–18b). Thus, a researcher can monitor the progress of a protein purification procedure as the number of protein bands visible on the gel decreases after each new fractionation step. When compared with the positions to which proteins of known molecular weight migrate in the gel, the position of an unidentified protein can provide a good approximation of its molecular weight **(Fig. 3–19)**. If the protein has two or more different subunits, the subunits are generally separated by the SDS treatment, and a separate band appears for each. ◉ **SDS Gel Electrophoresis**

Isoelectric focusing is a procedure used to determine the isoelectric point (pI) of a protein **(Fig. 3–20)**. A pH gradient is established by allowing a mixture of low molecular weight organic acids and bases (ampholytes; p. 81) to distribute themselves in an electric field generated across the gel. When a protein mixture is applied, each protein migrates until it reaches the pH that matches its pI. Proteins with different isoelectric points are thus distributed differently throughout the gel.

Combining isoelectric focusing and SDS electrophoresis sequentially in a process called **two-dimensional electrophoresis** permits the resolution of complex mixtures of proteins **(Fig. 3–21)**. This is a more sensitive analytical method than either electrophoretic method alone. Two-dimensional electrophoresis separates proteins of identical molecular weight that differ

A protein sample may be applied to one end of a gel strip with an immobilized pH gradient. Or, a protein sample in a solution of ampholytes may be used to rehydrate a dehydrated gel strip.

An electric field is applied

⊖   ⊕

pH 9 ——— Decreasing pI ———→ pH 3

After staining, proteins are shown to be distributed along pH gradient according to their pI values.

FIGURE 3–20 Isoelectric focusing. This technique separates proteins according to their isoelectric points. A protein mixture is placed on a gel strip containing an immobilized pH gradient. With an applied electric field, proteins enter the gel and migrate until each reaches a pH equivalent to its pI. Remember that when pH = pI, the net charge of a protein is zero.

in pI, or proteins with similar pI values but different molecular weights.

Unseparated Proteins Can Be Quantified

To purify a protein, it is essential to have a way of detecting and quantifying that protein in the presence of many other proteins at each stage of the procedure. Often, purification must proceed in the absence of any information about the size and physical properties of the protein or about the fraction of the total protein mass it represents in the extract. For proteins that are enzymes, the amount in a given solution or tissue extract can be measured, or assayed, in terms of the catalytic effect the enzyme produces—that is, the *increase* in the rate at which its substrate is converted to reaction products when the enzyme is present. For this purpose the researcher must know (1) the overall equation of the reaction catalyzed, (2) an analytical procedure for determining the disappearance of the substrate or the appearance of a reaction product, (3) whether the enzyme requires cofactors such as metal ions or coenzymes, (4) the dependence of the enzyme activity on substrate concentration, (5) the optimum pH, and (6) a temperature zone in which the enzyme is stable and has high activity. Enzymes are usually assayed at their optimum pH and at some convenient temperature within the range 25 to 38 °C. Also, very high substrate concentrations are generally used so that the initial reaction rate, measured experimentally, is proportional to enzyme concentration (Chapter 6).

By international agreement, 1.0 unit of enzyme activity for most enzymes is defined as the amount of enzyme causing transformation of 1.0 μmol of substrate to product per minute at 25 °C under optimal conditions of measurement (for some enzymes, this definition is

Protein sample

Separate proteins in first dimension on gel strip with isoelectric focusing.

Gel strip   pH 9 ————————————→ pH 3

Separate proteins in second dimension on SDS–polyacrylamide gel.

(−)

(+)

Decreasing M_r

——— Decreasing pI ———→

FIGURE 3–21 Two-dimensional electrophoresis. Proteins are first separated by isoelectric focusing in a thin strip gel. The gel is then laid horizontally on a second, slab-shaped gel, and the proteins are separated by SDS polyacrylamide gel electrophoresis. Horizontal separation reflects differences in pI; vertical separation reflects differences in molecular weight. The original protein complement is thus spread in two dimensions. Thousands of cellular proteins can be resolved using this technique. Individual protein spots can be cut out of the gel and identified by mass spectrometry (see Figs 3–30 and 3–31).

inconvenient, and a unit may be defined differently). The term **activity** refers to the total units of enzyme in a solution. The **specific activity** is the number of enzyme units per milligram of total protein **(Fig. 3–22)**. The specific activity is a measure of enzyme purity: it increases during purification of an enzyme and becomes maximal and constant when the enzyme is pure (Table 3–5, p. 93).

FIGURE 3–22 Activity versus specific activity. The difference between these terms can be illustrated by considering two flasks containing marbles. The flasks contain the same number of red marbles, but different numbers of marbles of other colors. If the marbles represent proteins, both flasks contain the same *activity* of the protein represented by the red marbles. The second flask, however, has the higher *specific activity* because red marbles represent a higher fraction of the total.

After each purification step, the activity of the preparation (in units of enzyme activity) is assayed, the total amount of protein is determined independently, and the ratio of the two gives the specific activity. Activity and total protein generally decrease with each step. Activity decreases because there is always some loss due to inactivation or nonideal interactions with chromatographic materials or other molecules in the solution. Total protein decreases because the objective is to remove as much unwanted or nonspecific protein as possible. In a successful step, the loss of nonspecific protein is much greater than the loss of activity; therefore, specific activity increases even as total activity falls. The data are assembled in a purification table similar to Table 3–5. A protein is generally considered pure when further purification steps fail to increase specific activity and when only a single protein species can be detected (for example, by electrophoresis).

For proteins that are not enzymes, other quantification methods are required. Transport proteins can be assayed by their binding to the molecule they transport, and hormones and toxins by the biological effect they produce; for example, growth hormones will stimulate the growth of certain cultured cells. Some structural proteins represent such a large fraction of a tissue mass that they can be readily extracted and purified without a functional assay. The approaches are as varied as the proteins themselves.

SUMMARY 3.3 Working with Proteins

▶ Proteins are separated and purified on the basis of differences in their properties. Proteins can be selectively precipitated by changes in pH or temperature, and particularly by the addition of certain salts. A wide range of chromatographic procedures makes use of differences in size, binding affinities, charge, and other properties. These include ion-exchange, size-exclusion, affinity, and high-performance liquid chromatography.

▶ Electrophoresis separates proteins on the basis of mass or charge. SDS gel electrophoresis and isoelectric focusing can be used separately or in combination for higher resolution.

▶ All purification procedures require a method for quantifying or assaying the protein of interest in the presence of other proteins. Purification can be monitored by assaying specific activity.

3.4 The Structure of Proteins: Primary Structure

Purification of a protein is usually only a prelude to a detailed biochemical dissection of its structure and function. What is it that makes one protein an enzyme, another a hormone, another a structural protein, and still another an antibody? How do they differ chemically? The most obvious distinctions are structural, and to protein structure we now turn.

The structure of large molecules such as proteins can be described at several levels of complexity, arranged in a kind of conceptual hierarchy. Four levels of protein structure are commonly defined (**Fig. 3–23**). A description

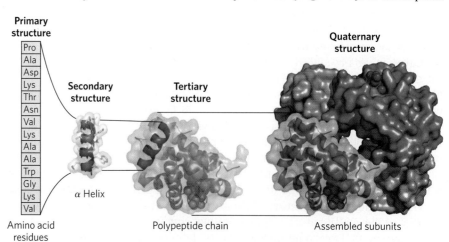

FIGURE 3–23 Levels of structure in proteins. The *primary structure* consists of a sequence of amino acids linked together by peptide bonds and includes any disulfide bonds. The resulting polypeptide can be arranged into units of *secondary structure*, such as an α helix. The helix is a part of the *tertiary structure* of the folded polypeptide, which is itself one of the subunits that make up the *quaternary structure* of the multisubunit protein, in this case hemoglobin.

of all covalent bonds (mainly peptide bonds and disulfide bonds) linking amino acid residues in a polypeptide chain is its **primary structure**. The most important element of primary structure is the *sequence* of amino acid residues. **Secondary structure** refers to particularly stable arrangements of amino acid residues giving rise to recurring structural patterns. **Tertiary structure** describes all aspects of the three-dimensional folding of a polypeptide. When a protein has two or more polypeptide subunits, their arrangement in space is referred to as **quaternary structure**. Our exploration of proteins will eventually include complex protein machines consisting of dozens to thousands of subunits. Primary structure is the focus of the remainder of this chapter; the higher levels of structure are discussed in Chapter 4.

Differences in primary structure can be especially informative. Each protein has a distinctive number and sequence of amino acid residues. As we shall see in Chapter 4, the primary structure of a protein determines how it folds up into its unique three-dimensional structure, and this in turn determines the function of the protein. We first consider empirical clues that amino acid sequence and protein function are closely linked, then describe how amino acid sequence is determined; finally, we outline the many uses to which this information can be put.

The Function of a Protein Depends on Its Amino Acid Sequence

The bacterium *Escherichia coli* produces more than 3,000 different proteins; a human has ~25,000 genes encoding a much larger number of proteins (through genetic processes discussed in Part III of this text). In both cases, each type of protein has a unique amino acid sequence that confers a particular three-dimensional structure. This structure in turn confers a unique function.

Some simple observations illustrate the importance of primary structure, or the amino acid sequence of a protein. First, as we have already noted, proteins with different functions always have different amino acid sequences. Second, thousands of human genetic diseases have been traced to the production of defective proteins. The defect can range from a single change in the amino acid sequence (as in sickle cell anemia, described in Chapter 5) to deletion of a larger portion of the polypeptide chain (as in most cases of Duchenne muscular dystrophy: a large deletion in the gene encoding the protein dystrophin leads to production of a shortened, inactive protein). Finally, on comparing functionally similar proteins from different species, we find that these proteins often have similar amino acid sequences. Thus, a close link between protein primary structure and function is evident.

Is the amino acid sequence absolutely fixed, or invariant, for a particular protein? No; some flexibility is possible. An estimated 20% to 30% of the proteins in humans are **polymorphic**, having amino acid sequence variants in the human population. Many of these variations in sequence have little or no effect on the function of the protein. Furthermore, proteins that carry out a broadly similar function in distantly related species can differ greatly in overall size and amino acid sequence.

Although the amino acid sequence in some regions of the primary structure might vary considerably without affecting biological function, most proteins contain crucial regions that are essential to their function and whose sequence is therefore conserved. The fraction of the overall sequence that is critical varies from protein to protein, complicating the task of relating sequence to three-dimensional structure, and structure to function. Before we can consider this problem further, however, we must examine how sequence information is obtained.

The Amino Acid Sequences of Millions of Proteins Have Been Determined

Two major discoveries in 1953 were of crucial importance in the history of biochemistry. In that year, James D. Watson and Francis Crick deduced the double-helical structure of DNA and proposed a structural basis for its precise replication (Chapter 8). Their proposal illuminated the molecular reality behind the idea of a gene. In the same year, Frederick Sanger worked out the sequence of amino acid residues in the polypeptide chains of the hormone insulin (**Fig. 3–24**), surprising many researchers who had long thought that determining the amino acid sequence of a polypeptide would be a hopelessly difficult task. It quickly became evident that the nucleotide sequence in DNA and the amino acid sequence in proteins were somehow related. Barely a decade after these discoveries, the genetic code relating the nucleotide sequence of DNA to the amino acid sequence of protein molecules was elucidated (Chapter 27). The amino acid sequences of proteins are now most often derived indirectly from the DNA sequences in genome databases. However, an array of techniques derived from traditional methods of polypeptide sequencing still command an important place in protein chemistry. Below, we summarize the traditional method and mention a few of the techniques derived from it.

FIGURE 3–24 Amino acid sequence of bovine insulin. The two polypeptide chains are joined by disulfide cross-linkages (yellow). The A chain of insulin is identical in human, pig, dog, rabbit, and sperm whale insulins. The B chains of the cow, pig, dog, goat, and horse are identical.

Protein Chemistry Is Enriched by Methods Derived from Classical Polypeptide Sequencing

The methods used in the 1950s by Fred Sanger to determine the sequence of the protein insulin are summarized, in their modern form, in **Figure 3–25**. Few proteins are sequenced in this way now, at least in their entirety. However, these traditional sequencing protocols have provided a rich array of tools for biochemists, and almost every step in Figure 3–25 makes use of methods that are widely used, sometimes in quite different contexts.

Frederick Sanger

In the traditional scheme for sequencing large proteins, the amino-terminal amino acid residue was first labeled and its identity determined. The amino-terminal α-amino group can be labeled with 1-fluoro-2,4-dinitrobenzene (FDNB), dansyl chloride, or dabsyl chloride (**Fig. 3–26**).

The chemical sequencing process itself is based on a two-step process developed by Pehr Edman (**Fig. 3–27**). The **Edman degradation** procedure labels and removes only the amino-terminal residue from a peptide, leaving all other peptide bonds intact. The peptide is reacted with phenylisothiocyanate under mildly alkaline conditions, which converts the amino-terminal amino acid to a phenylthiocarbamoyl (PTC) adduct. The peptide bond next to the PTC adduct is then cleaved in a step carried out in anhydrous trifluoroacetic acid, with removal of the amino-terminal amino acid as an anilinothiazolinone derivative. The derivatized amino acid is extracted with organic solvents, converted to the more stable phenylthiohydantoin derivative by treatment with aqueous acid, and then identified. The use of sequential reactions carried out under first basic and then acidic conditions provides a means of controlling the entire process. Each reaction with the

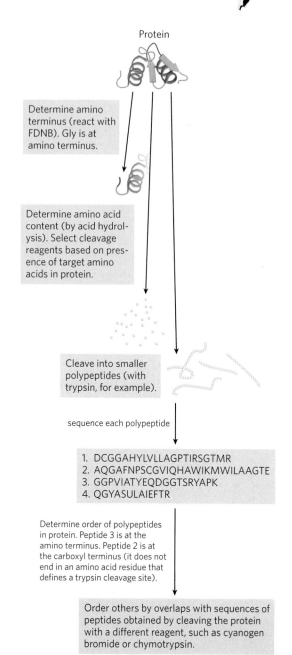

Protein

Determine amino terminus (react with FDNB). Gly is at amino terminus.

Determine amino acid content (by acid hydrolysis). Select cleavage reagents based on presence of target amino acids in protein.

Cleave into smaller polypeptides (with trypsin, for example).

sequence each polypeptide

1. DCGGAHYLVLLAGPTIRSGTMR
2. AQGAFNPSCGVIQHAWIKMWILAAGTE
3. GGPVIATYEQDGGTSRYAPK
4. QGYASULAIEFTR

Determine order of polypeptides in protein. Peptide 3 is at the amino terminus. Peptide 2 is at the carboxyl terminus (it does not end in an amino acid residue that defines a trypsin cleavage site).

Order others by overlaps with sequences of peptides obtained by cleaving the protein with a different reagent, such as cyanogen bromide or chymotrypsin.

FIGURE 3–25 Direct protein sequencing. The procedures shown here are those developed by Fred Sanger to sequence insulin, and they have been used subsequently for many additional proteins. FDNB is 1-fluoro-2,4-dinitrobenzene (see text and Fig. 3-26).

amino-terminal amino acid can go essentially to completion without affecting any of the other peptide bonds in the peptide. The process is repeated until, typically, as many as 40 sequential amino acid residues are identified. The reactions of the Edman degradation have been automated.

FDNB   Dansyl chloride   Dabsyl chloride

FIGURE 3–26 Reagents used to modify the α-amino group at the amino terminus.

FIGURE 3–27 The protein sequencing chemistry devised by Pehr Edman. The peptide bond nearest to the amino terminus of the protein or polypeptide is cleaved in two steps. The two steps are carried out under very different reaction conditions (basic conditions in step ❶, acidic in step ❷), allowing one step to proceed to completion before the second is initiated.

To determine the sequence of large proteins, early developers of sequencing protocols had to devise methods to eliminate disulfide bonds and to cleave proteins precisely into smaller polypeptides. Two approaches to irreversible breakage of disulfide bonds are outlined in **Figure 3–28.** Enzymes called **proteases** catalyze the hydrolytic cleavage of peptide bonds. Some proteases cleave only the peptide bond adjacent to particular amino acid residues (Table 3–6) and thus fragment a polypeptide chain in a predictable and reproducible way. A few chemical reagents also cleave the peptide bond adjacent to specific residues. Among proteases, the digestive enzyme trypsin catalyzes the hydrolysis of only those peptide bonds in

FIGURE 3–28 Breaking disulfide bonds in proteins. Two common methods are illustrated. Oxidation of a cystine residue with performic acid produces two cysteic acid residues. Reduction by dithiothreitol (or β-mercaptoethanol) to form Cys residues must be followed by further modification of the reactive —SH groups to prevent re-formation of the disulfide bond. Carboxymethylation by iodoacetate serves this purpose.

TABLE 3–6	The Specificity of Some Common Methods for Fragmenting Polypeptide Chains
Reagent (biological source)*	Cleavage points[†]
Trypsin (bovine pancreas)	Lys, Arg (C)
Submaxillary protease (mouse submaxillary gland)	Arg (C)
Chymotrypsin (bovine pancreas)	Phe, Trp, Tyr (C)
Staphylococcus aureus V8 protease (bacterium *S. aureus*)	Asp, Glu (C)
Asp-*N*-protease (bacterium *Pseudomonas fragi*)	Asp, Glu (N)
Pepsin (porcine stomach)	Leu, Phe, Trp, Tyr (N)
Endoproteinase Lys C (bacterium *Lysobacter enzymogenes*)	Lys (C)
Cyanogen bromide	Met (C)

*All reagents except cyanogen bromide are proteases. All are available from commercial sources.

[†]Residues furnishing the primary recognition point for the protease or reagent; peptide bond cleavage occurs on either the carbonyl (C) or the amino (N) side of the indicated amino acid residues.

which the carbonyl group is contributed by either a Lys or an Arg residue, regardless of the length or amino acid sequence of the chain. A polypeptide with three Lys and/or Arg residues will usually yield four smaller peptides on cleavage with trypsin. Moreover, all except one of these will have a carboxyl-terminal Lys or Arg. The choice of a reagent to cleave the protein into smaller peptides can be aided by first determining the amino acid content of the entire protein, employing acid to reduce the protein to its constituent amino acids. Trypsin would be used only on proteins that have an appropriate number of Lys or Arg residues.

In classical sequencing, a large protein would be cleaved into fragments twice, using a different protease or cleavage reagent each time so that the fragment endpoints would be different. Both sets of fragments would be purified and sequenced. The order in which the fragments appeared in the original protein could then be determined by examining the overlaps in sequence between the two sets of fragments.

Even if no longer used to sequence entire proteins, the traditional sequencing methods are still valuable in the lab. The sequencing of some amino acids from the amino terminus using the Edman chemistry is often sufficient to confirm the identity of a known protein that has just been purified, or to identify an unknown protein purified on the basis of an unusual activity. Techniques employed in individual steps of the traditional sequencing method are also useful for other purposes. For example, the methods used to break disulfide bonds can also be used to denature proteins when that is required. Furthermore, the effort to label the amino-terminal amino acid residue led eventually to the development of an array of reagents that could react with specific groups on a protein. The same reagents used to label the amino-terminal α-amino group can be used to label the primary-amines of Lys residues (Fig. 3–26). The sulfhydryl group on Cys residues can be modified with iodoacetamides, maleimides, benzyl halides, and bromomethyl ketones **(Fig. 3–29)**. Other amino acid residues can be modified

FIGURE 3–29 Reagents used to modify the sulfhydryl groups of Cys residues. (See also Fig. 3–28.)

by reagents linked to a dye or other molecule to aid in protein detection or functional studies.

Mass Spectrometry Offers an Alternative Method to Determine Amino Acid Sequences

Modern adaptations of **mass spectrometry** provide an important alternative to the sequencing methods described above. Mass spectrometry can provide a highly accurate measure of the molecular weight of a protein, but can also do much more. In particular, some variants of mass spectrometry can provide the sequences of multiple short polypeptide segments (20 to 30 amino acid residues) in a protein sample quite rapidly.

The mass spectrometer has long been an indispensable tool in chemistry. Molecules to be analyzed, referred to as **analytes**, are first ionized in a vacuum. When the newly charged molecules are introduced into an electric and/or magnetic field, their paths through the field are a function of their mass-to-charge ratio, m/z. This measured property of the ionized species can be used to deduce the mass (m) of the analyte with very high precision.

Although mass spectrometry has been in use for many years, it could not be applied to macromolecules such as proteins and nucleic acids. The m/z measurements are made on molecules in the gas phase, and the heating or other treatment needed to transfer a macromolecule to the gas phase usually caused its rapid decomposition. In 1988, two different techniques were developed to overcome this problem. In one, proteins are placed in a light-absorbing matrix. With a short pulse of laser light, the proteins are ionized and then desorbed from the matrix into the vacuum system. This process, known as **matrix-assisted laser desorption/ionization mass spectrometry**, or **MALDI MS**, has been successfully used to measure the mass of a wide range of macromolecules. In a second and equally successful method, macromolecules in solution are forced directly from the liquid to gas phase. A solution of analytes is passed through a charged needle that is kept at a high electrical potential, dispersing the solution into a fine mist of charged microdroplets. The solvent surrounding the macromolecules rapidly evaporates, leaving multiply charged macromolecular ions in the gas phase. This technique is called **electrospray ionization mass spectrometry**, or **ESI MS**. Protons added during passage through the needle give additional charge to the macromolecule. The m/z of the molecule can be analyzed in the vacuum chamber.

Mass spectrometry provides a wealth of information for proteomics research, enzymology, and protein chemistry in general. The techniques require only miniscule amounts of sample, so they can be readily applied to the small amounts of protein that can be extracted from a two-dimensional electrophoretic gel. The accurately measured molecular mass of a protein is critical to its identification. Once the mass of a protein is accurately known, mass spectrometry is a convenient and accurate method for detecting changes in mass due to the presence of bound cofactors, bound metal ions, covalent modifications, and so on.

The process for determining the molecular mass of a protein with ESI MS is illustrated in **Figure 3–30**. As it is injected into the gas phase, a protein acquires a variable number of protons, and thus positive charges, from the solvent. The variable addition of these charges creates a spectrum of species with different mass-to-charge ratios. Each successive peak corresponds to a species that differs from that of its neighboring peak by a charge difference of 1 and a mass difference of 1 (1 proton). The mass of the protein can be determined from any two neighboring peaks.

Mass spectrometry can also be used to sequence short stretches of polypeptide, an application that has emerged as an invaluable tool for quickly identifying unknown proteins. Sequence information is extracted using a technique called **tandem MS**, or **MS/MS**. A solution containing the protein under investigation is first treated with a protease or chemical reagent to hydrolyze it to a mixture of shorter peptides. The mixture is then injected into a device that is essentially two mass spectrometers in tandem (**Fig. 3–31a**, top). In the first, the peptide mixture is sorted so that only one of the several types of peptides produced by cleavage emerges at the other end. The sample of the selected peptide, each molecule of which has a charge somewhere along its length, then travels through a vacuum chamber between the two mass spectrometers. In this collision cell, the peptide is further fragmented by high-energy impact with a "collision gas" such as helium or argon that is bled into the vacuum chamber. Each individual peptide is broken in only one place, on average. Although the breaks are not hydrolytic, most occur at the peptide bonds.

The second mass spectrometer then measures the m/z ratios of all the charged fragments. This process generates one or more sets of peaks. A given set of peaks (Fig. 3–31b) consists of all the charged fragments that were generated by breaking the same type of bond (but at different points in the peptide). One set of peaks includes only the fragments in which the charge was retained on the amino-terminal side of the broken bonds; another includes only the fragments in which the charge was retained on the carboxyl-terminal side of the broken bonds. Each successive peak in a given set has one less amino acid than the peak before. The difference in mass from peak to peak identifies the amino

(a)

(b)

FIGURE 3–30 Electrospray ionization mass spectrometry of a protein. **(a)** A protein solution is dispersed into highly charged droplets by passage through a needle under the influence of a high-voltage electric field. The droplets evaporate, and the ions (with added protons in this case) enter the mass spectrometer for m/z measurement. The spectrum generated **(b)** is a family of peaks, with each successive peak (from right to left) corresponding to a charged species increased by 1 in both mass and charge. The inset shows a computer-generated transformation of this spectrum.

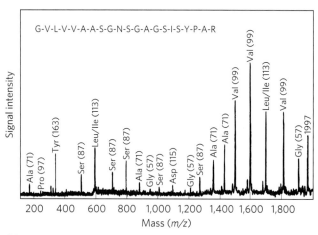

(a)

(b)

FIGURE 3–31 Obtaining protein sequence information with tandem MS. (a) After proteolytic hydrolysis, a protein solution is injected into a mass spectrometer (MS-1). The different peptides are sorted so that only one type is selected for further analysis. The selected peptide is further fragmented in a chamber between the two mass spectrometers, and *m/z* for each fragment is measured in the second mass spectrometer (MS-2). Many of the ions generated during this second fragmentation result from breakage of the peptide bond, as shown. These are called b-type or y-type ions, depending on whether the charge is retained on the amino- or carboxyl-terminal side, respectively. **(b)** A typical spectrum with peaks representing the peptide fragments generated from a sample of one small peptide (21 residues). The labeled peaks are y-type ions derived from amino acid residues. The number in parentheses over each peak is the molecular weight of the amino acid ion. The successive peaks differ by the mass of a particular amino acid in the original peptide. The deduced sequence is shown at the top.

acid that was lost in each case, thus revealing the sequence of the peptide. The only ambiguities involve leucine and isoleucine, which have the same mass. Although multiple sets of peaks are usually generated, the two most prominent sets generally consist of charged fragments derived from breakage of the peptide bonds. The amino acid sequence derived from one

set can be confirmed by the other, improving the confidence in the sequence information obtained.

The various methods for obtaining protein sequence information complement one another. The Edman degradation procedure is sometimes convenient to get sequence information uniquely from the amino terminus of a protein or peptide. However, it is relatively slow and requires a larger sample than does mass spectrometry. Mass spectrometry can be used for small amounts of sample and for mixed samples. It provides sequence information, but the fragmentation processes can leave unpredictable sequence gaps. Although most protein sequences are now extracted from genomic DNA sequences (Chapter 9) by employing our understanding of the genetic code (Chapter 27), direct protein sequencing is often necessary to identify unknown protein samples. Both protein sequencing methods permit the unambiguous identification of newly purified proteins. Mass spectrometry is the method of choice to identify proteins that are present in small amounts. For example, the technique is sensitive enough to analyze the few hundred nanograms of protein that might be extracted from a single protein band on a polyacrylamide gel. Direct sequencing by mass spectrometry also can reveal the addition of phosphoryl groups or other modifications (Chapter 6). Sequencing by either method can reveal changes in protein sequence that result from the editing of messenger RNA in eukaryotes (Chapter 26). Thus, these methods are all part of a robust toolbox used to investigate proteins and their functions.

Small Peptides and Proteins Can Be Chemically Synthesized

Many peptides are potentially useful as pharmacologic agents, and their production is of considerable commercial importance. There are three ways to obtain a peptide: (1) purification from tissue, a task often made difficult by the vanishingly low concentrations of some peptides; (2) genetic engineering (Chapter 9); or (3) direct chemical synthesis. Powerful techniques now make direct chemical synthesis an attractive option in many cases. In addition to commercial applications, the synthesis of specific peptide portions of larger proteins is an increasingly important tool for the study of protein structure and function.

The complexity of proteins makes the traditional synthetic approaches of organic chemistry impractical for peptides with more than four or five amino acid residues. One problem is the difficulty of purifying the product after each step.

The major breakthrough in this technology was provided by R. Bruce Merrifield in 1962. His innovation was to synthesize a peptide while keeping it attached at one end to a solid support. The support is an insoluble polymer (resin) contained within a column, similar to that used for chromatographic procedures. The peptide is built up on this support one amino acid at a time, through a standard set of reactions in a repeating cycle **(Fig. 3–32)**. At each successive step in the cycle, protective chemical groups block unwanted reactions.

The technology for chemical peptide synthesis is now automated. An important limitation of the process (a limitation shared by the Edman degradation sequencing process) is the efficiency of each chemical cycle, as can be seen by calculating the overall yields of peptides of various lengths when the yield for addition of each new amino acid is 96.0% versus 99.8% (Table 3–7).

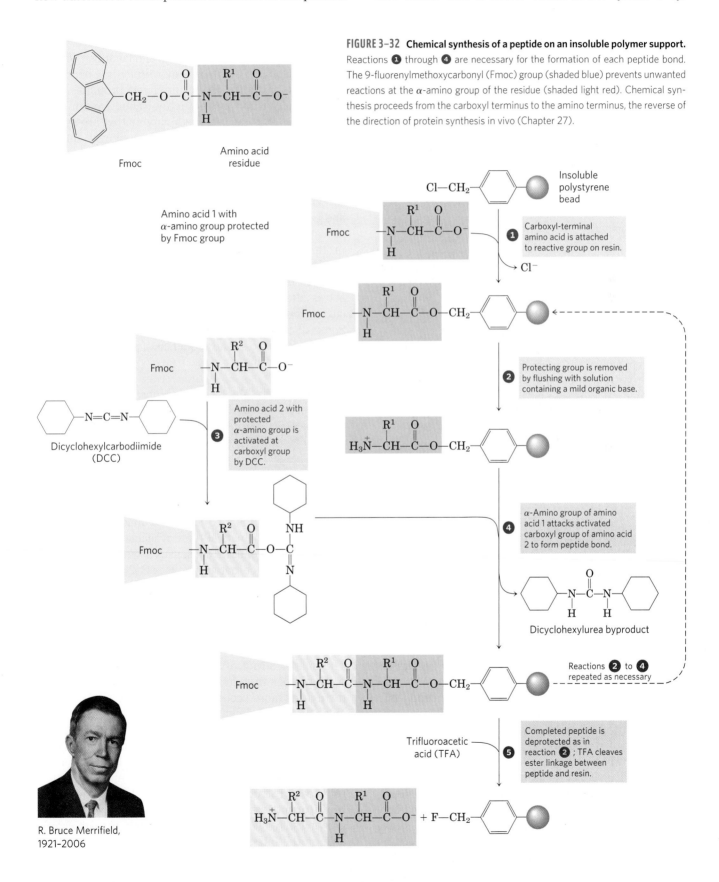

FIGURE 3–32 Chemical synthesis of a peptide on an insoluble polymer support. Reactions ❶ through ❹ are necessary for the formation of each peptide bond. The 9-fluorenylmethoxycarbonyl (Fmoc) group (shaded blue) prevents unwanted reactions at the α-amino group of the residue (shaded light red). Chemical synthesis proceeds from the carboxyl terminus to the amino terminus, the reverse of the direction of protein synthesis in vivo (Chapter 27).

R. Bruce Merrifield, 1921–2006

TABLE 3–7	Effect of Stepwise Yield on Overall Yield in Peptide Synthesis	
		Overall yield of final peptide (%) when the yield of each step is:	
Number of residues in the final polypeptide	96.0%	99.8%
11	66	98
21	44	96
31	29	94
51	13	90
100	1.8	82

Incomplete reaction at one stage can lead to formation of an impurity (in the form of a shorter peptide) in the next. The chemistry has been optimized to permit the synthesis of proteins of 100 amino acid residues in a few days in reasonable yield. A very similar approach is used to synthesize nucleic acids (see Fig. 8–35). It is worth noting that this technology, impressive as it is, still pales when compared with biological processes. The same 100-residue protein would be synthesized with exquisite fidelity in about 5 seconds in a bacterial cell.

A variety of new methods for the efficient ligation (joining together) of peptides has made possible the assembly of synthetic peptides into larger polypeptides and proteins. With these methods, novel forms of proteins can be created with precisely positioned chemical groups, including those that might not normally be found in a cellular protein. These novel forms provide new ways to test theories of enzyme catalysis, to create proteins with new chemical properties, and to design protein sequences that will fold into particular structures. This last application provides the ultimate test of our increasing ability to relate the primary structure of a peptide to the three-dimensional structure that it takes up in solution.

Amino Acid Sequences Provide Important Biochemical Information

Knowledge of the sequence of amino acids in a protein can offer insights into its three-dimensional structure and its function, cellular location, and evolution. Most of these insights are derived by searching for similarities between a protein of interest and previously studied proteins. Thousands of sequences are known and available in databases accessible through the Internet. A comparison of a newly obtained sequence with this large bank of stored sequences often reveals relationships both surprising and enlightening.

Exactly how the amino acid sequence determines three-dimensional structure is not understood in detail, nor can we always predict function from sequence. However, protein families that have some shared structural or functional features can be readily identified on the basis of amino acid sequence similarities. Individual proteins are assigned to families based on the degree of similarity in amino acid sequence. Members of a family are usually identical across 25% or more of their sequences, and proteins in these families generally share at least some structural and functional characteristics. Some families are defined, however, by identities involving only a few amino acid residues that are critical to a certain function. A number of similar substructures, or "domains" (to be defined more fully in Chapter 4), occur in many functionally unrelated proteins. These domains often fold into structural configurations that have an unusual degree of stability or that are specialized for a certain environment. Evolutionary relationships can also be inferred from the structural and functional similarities within protein families.

Certain amino acid sequences serve as signals that determine the cellular location, chemical modification, and half-life of a protein. Special signal sequences, usually at the amino terminus, are used to target certain proteins for export from the cell; other proteins are targeted for distribution to the nucleus, the cell surface, the cytosol, or other cellular locations. Other sequences act as attachment sites for prosthetic groups, such as sugar groups in glycoproteins and lipids in lipoproteins. Some of these signals are well characterized and are easily recognized in the sequence of a newly characterized protein (Chapter 27).

KEY CONVENTION: Much of the functional information encapsulated in protein sequences comes in the form of **consensus sequences**. This term is applied to such sequences in DNA, RNA, or protein. When a series of related nucleic acid or protein sequences are compared, a consensus sequence is the one that reflects the most common base or amino acid at each position. Parts of the sequence that have particularly good agreement often represent evolutionarily conserved functional domains. A range of mathematical tools available on the Internet can be used to generate consensus sequences or identify them in sequence databases. Box 3–2 illustrates common conventions for displaying consensus sequences. ■

Protein Sequences Can Elucidate the History of Life on Earth

The simple string of letters denoting the amino acid sequence of a protein holds a surprising wealth of information. As more protein sequences have become available, the development of more powerful methods for extracting information from them has become a major biochemical enterprise. Analysis of the information available in the many, ever-expanding biological databases, including gene and protein sequences and macromolecular structures, has given rise to the new field of **bioinformatics**. One outcome of this discipline is a growing suite of computer programs, many readily available on the Internet, that can be used by any scientist,

BOX 3–2 Consensus Sequences and Sequence Logos

Consensus sequences can be represented in several ways. To illustrate two types of conventions, we use two examples of consensus sequences, shown in Figure 1: (a) an ATP-binding structure called a P loop (see Box 12–2) and (b) a Ca^{2+}-binding structure called an EF hand (see Fig. 12–11). The rules described here are adapted from those used by the sequence comparison website PROSITE (expasy.org/prosite); they use the standard one-letter codes for the amino acids.

[AG]-x(4)-G-K-[ST].

(a)

D-{W}-[DNS]-{ILVFYW}-[DENSTG]-[DNQGHRK]-{GP}-
[LIVMC]-[DENQSTAGC]-x(2)-[DE]-[LIVMFYW].

(b)

FIGURE 1 **Representations of two consensus sequences. (a)** P loop, an ATP-binding structure; **(b)** EF hand, a Ca^{2+}-binding structure.

In one type of consensus sequence designation (shown at the top of (a) and (b)), each position is separated from its neighbor by a hyphen. A position where any amino acid is allowed is designated x. Ambiguities are indicated by listing the acceptable amino acids for a given position between square brackets. For example, in (a) [AG] means Ala or Gly. If all but a few amino acids are allowed at one position, the amino acids that are *not* allowed are listed between curly brackets. For example, in (b) {W} means any amino acid except Trp. Repetition of an element of the pattern is indicated by following that element with a number or range of numbers between parentheses. In (a), for example, x(4) means x-x-x-x; x(2,4) would mean x-x, or x-x-x, or x-x-x-x. When a pattern is restricted to either the amino or carboxyl terminus of a sequence, that pattern starts with < or ends with >, respectively (not so for either example here). A period ends the pattern. Applying these rules to the consensus sequence in (a), either A or G can be found at the first position. Any amino acid can occupy the next four positions, followed by an invariant G and an invariant K. The last position is either S or T.

Sequence logos provide a more informative and graphic representation of an amino acid (or nucleic acid) multiple sequence alignment. Each logo consists of a stack of symbols for each position in the sequence. The overall height of the stack (in bits) indicates the degree of sequence conservation at that position, while the height of each symbol in the stack indicates the relative frequency of that amino acid (or nucleotide). For amino acid sequences, the colors denote the characteristics of the amino acid: polar (G, S, T, Y, C, Q, N) green; basic (K, R, H) blue; acidic (D, E) red; and hydrophobic (A, V, L, I, P, W, F, M) black. The classification of amino acids in this scheme is somewhat different from that in Table 3–1 and Figure 3–5. The amino acids with aromatic side chains are subsumed into the nonpolar (F, W) and polar (Y) classifications. Glycine, always hard to group, is assigned to the polar group. Note that when multiple amino acids are acceptable at a particular position, they rarely occur with equal probability. One or a few usually predominate. The logo representation makes the predominance clear, and a conserved sequence in a protein is made obvious. However, the logo obscures some amino acid residues that may be allowed at a position, such as the Cys that occasionally occurs at position 8 of the EF hand in (b).

student, or knowledgeable layperson. Each protein's function relies on its three-dimensional structure, which in turn is determined largely by its primary structure. Thus, the biochemical information conveyed by a protein sequence is limited only by our own understanding of structural and functional principles. The constantly evolving tools of bioinformatics make it possible to identify functional segments in new proteins and help establish both their sequence and their structural relationships to proteins already in the databases. On a different level of inquiry, protein sequences are beginning to tell us how the proteins evolved and, ultimately, how life evolved on this planet.

The field of molecular evolution is often traced to Emile Zuckerkandl and Linus Pauling, whose work in the mid-1960s advanced the use of nucleotide and protein sequences to explore evolution. The premise is deceptively straightforward. If two organisms are closely related, the sequences of their genes and proteins should be similar. The sequences increasingly diverge as the evolutionary distance between two organisms increases. The promise of this approach began to be realized in the 1970s, when Carl Woese used ribosomal RNA sequences to define the Archaea as a group of living organisms distinct from the Bacteria and Eukarya (see Fig. 1–4). Protein sequences offer an opportunity to greatly refine the available information. With the advent of genome projects investigating organisms from bacteria to humans, the number of available sequences is growing at an enormous rate. This information can be used to trace biological history. The challenge is in learning to read the genetic hieroglyphics.

Evolution has not taken a simple linear path. Complexities abound in any attempt to mine the evolutionary information stored in protein sequences. For a given protein, the amino acid residues essential for the activity of the protein are conserved over evolutionary time. The residues that are less important to function may vary over time—that is, one amino acid may substitute for another—and these variable residues can provide the information to trace evolution. Amino acid substitutions are not always random, however. At some positions in the primary structure, the need to maintain protein function may mean that only particular amino acid substitutions can be tolerated. Some proteins have more variable amino acid residues than others. For these and other reasons, different proteins can evolve at different rates.

Another complicating factor in tracing evolutionary history is the rare transfer of a gene or group of genes from one organism to another, a process called **horizontal gene transfer**. The transferred genes may be quite similar to the genes they were derived from in the original organism, whereas most other genes in the same two organisms may be quite distantly related. An example of horizontal gene transfer is the recent rapid spread of antibiotic-resistance genes in bacterial populations. The proteins derived from these transferred genes would not be good candidates for the study of bacterial evolution, because they share only a very limited evolutionary history with their "host" organisms.

The study of molecular evolution generally focuses on families of closely related proteins. In most cases, the families chosen for analysis have essential functions in cellular metabolism that must have been present in the earliest viable cells, thus greatly reducing the chance that they were introduced relatively recently by horizontal gene transfer. For example, a protein called EF-1α (elongation factor 1α) is involved in the synthesis of proteins in all eukaryotes. A similar protein, EF-Tu, with the same function, is found in bacteria. Similarities in sequence and function indicate that EF-1α and EF-Tu are members of a family of proteins that share a common ancestor. The members of protein families are called **homologous proteins**, or **homologs**. The concept of a homolog can be further refined. If two proteins in a family (that is, two homologs) are present in the same species, they are referred to as **paralogs**. Homologs from different species are called **orthologs**. The process of tracing evolution involves

first identifying suitable families of homologous proteins and then using them to reconstruct evolutionary paths.

Homologs are identified through the use of increasingly powerful computer programs that can directly compare two or more chosen protein sequences, or can search vast databases to find the evolutionary relatives of one selected protein sequence. The electronic search process can be thought of as sliding one sequence past the other until a section with a good match is found. Within this sequence alignment, a positive score is assigned for each position where the amino acid residues in the two sequences are identical—the value of the score varying from one program to the next—to provide a measure of the quality of the alignment. The process has some complications. Sometimes the proteins being compared match well at, say, two sequence segments, and these segments are connected by less related sequences of different lengths. Thus the two matching segments cannot be aligned at the same time. To handle this, the computer program introduces "gaps" in one of the sequences to bring the matching segments into register **(Fig. 3–33)**. Of course, if a sufficient number of gaps are introduced, almost any two sequences could be brought into some sort of alignment. To avoid uninformative alignments, the programs include penalties for each gap introduced, thus lowering the overall alignment score. With electronic trial and error, the program selects the alignment with the optimal score that maximizes identical amino acid residues while minimizing the introduction of gaps.

Finding identical amino acids is often inadequate to identify related proteins or, more importantly, to determine how closely related the proteins are on an evolutionary time scale. A more useful analysis also considers the chemical properties of substituted amino acids. Many of the amino acid differences within a protein family may be conservative—that is, an amino acid residue is replaced by a residue having similar chemical properties. For example, a Glu residue may substitute in one family member for the Asp residue found in another; both amino acids are negatively charged. Such a conservative substitution should logically receive a higher score in a sequence alignment than does a nonconservative substitution, such as the replacement of the Asp residue with a hydrophobic Phe residue.

For most efforts to find homologies and explore evolutionary relationships, protein sequences (derived either directly from protein sequencing or from the

Escherichia coli	TGNRTIAVYDLGGGTFDISIIEIDEVDGEKTFEVLATNGDTHLGGEDFDSRLIHYL
Bacillus subtilis	DEDQTILLYDLGGGTFDVSILELGDG TFEVRSTAGDNRLGGDDFDQVIIDHL

Gap

FIGURE 3–33 Aligning protein sequences with the use of gaps. Shown here is the sequence alignment of a short section of the Hsp70 proteins (a widespread class of protein-folding chaperones) from two well-studied bacterial species, *E. coli* and *Bacillus subtilis*. Introduction of a gap in the *B. subtilis* sequence allows a better alignment of amino acid residues on either side of the gap. Identical amino acid residues are shaded.

Signature sequence

Archaea {	*Halobacterium halobium*	I G H V D H G K S T M V G R <u>L L Y E T G S V P E H V</u> I E Q H
	Sulfolobus solfataricus	I G H V D H G K S T L V G R <u>L L M D R G F I D E K T</u> V K E A
Eukaryotes {	*Saccharomyces cerevisiae*	I G H V D S G K S T T T G H <u>L I Y K C G G I D K R T</u> I E K F
	Homo sapiens	I G H V D S G K S T T T G H <u>L I Y K C G G I D K R T</u> I E K F
Gram-positive bacterium	*Bacillus subtilis*	I G H V D H G K S T M V G R I T T V
Gram-negative bacterium	*Escherichia coli*	I G H V D H G K T T L T A A I T T V

FIGURE 3–34 A signature sequence in the EF-1α/EF-Tu protein family. The signature sequence (boxed) is a 12-residue insertion near the amino terminus of the sequence. Residues that align in all species are shaded. Both archaea and eukaryotes have the signature, although the sequences of the insertions are quite distinct for the two groups. The variation in the signature sequence reflects the significant evolutionary divergence that has occurred at this site since it first appeared in a common ancestor of both groups.

sequencing of the DNA encoding the protein) are superior to nongenic nucleic acid sequences (those that do not encode a protein or functional RNA). For a nucleic acid, with its four different types of residues, random alignment of nonhomologous sequences will generally yield matches for at least 25% of the positions. Introduction of a few gaps can often increase the fraction of matched residues to 40% or more, and the probability of chance alignment of unrelated sequences becomes quite high. The 20 different amino acid residues in proteins greatly lower the probability of uninformative chance alignments of this type.

The programs used to generate a sequence alignment are complemented by methods that test the reliability of the alignments. A common computerized test is to shuffle the amino acid sequence of one of the proteins being compared to produce a random sequence, then to instruct the program to align the shuffled sequence with the other, unshuffled one. Scores are assigned to the new alignment, and the shuffling and alignment process is repeated many times. The original alignment, before shuffling, should have a score significantly higher than any of those within the distribution of scores generated by the random alignments; this increases the confidence that the sequence alignment has identified a pair of homologs. Note that the absence of a significant alignment score does not necessarily mean that no evolutionary relationship exists between two proteins. As we shall see in Chapter 4, three-dimensional structural similarities sometimes reveal evolutionary relationships where sequence homology has been wiped away by time.

To use a protein family to explore evolution, researchers identify family members with similar molecular functions in the widest possible range of organisms. Information from the family can then be used to trace the evolution of those organisms. By analyzing the sequence divergence in selected protein families, investigators can segregate organisms into classes based on their evolutionary relationships. This information must be reconciled with more classical examinations of the physiology and biochemistry of the organisms.

Certain segments of a protein sequence may be found in the organisms of one taxonomic group but not in other groups; these segments can be used as **signature sequences** for the group in which they are found.

An example of a signature sequence is an insertion of 12 amino acids near the amino terminus of the EF-1α/EF-Tu proteins in all archaea and eukaryotes but not in bacteria (**Fig. 3–34**). This particular signature is one of many biochemical clues that can help establish the evolutionary relatedness of eukaryotes and archaea. Signature sequences have been used to establish evolutionary relationships among groups of organisms at many different taxonomic levels.

By considering the entire sequence of a protein, researchers can now construct more elaborate evolutionary trees with many species in each taxonomic group. **Figure 3–35** presents one such tree for bacteria, based on sequence divergence in the protein GroEL (a protein present in all bacteria that assists in the proper folding of proteins). The tree can be refined by basing it on the sequences of multiple proteins and by supplementing the sequence information with data on the unique biochemical and physiological properties of each species. There are many methods for generating trees, each method with its own advantages and shortcomings, and many ways to represent the resulting evolutionary relationships. In Figure 3–35, the free end points of lines are called "external nodes"; each represents an extant species, and each is so labeled. The points where two lines come together, the "internal nodes," represent extinct ancestor species. In most representations (including Fig. 3–35), the lengths of the lines connecting the nodes are proportional to the number of amino acid substitutions separating one species from another. If we trace two extant species to a common internal node (representing the common ancestor of the two species), the length of the branch connecting each external node to the internal node represents the number of amino acid substitutions separating one extant species from this ancestor. The sum of the lengths of all the line segments that connect an extant species to another extant species through a common ancestor reflects the number of substitutions separating the two extant species. To determine how much time was needed for the various species to diverge, the tree must be calibrated by comparing it with information from the fossil record and other sources.

As more sequence information is made available in databases, we can generate evolutionary trees based on multiple proteins. And we can refine these trees as

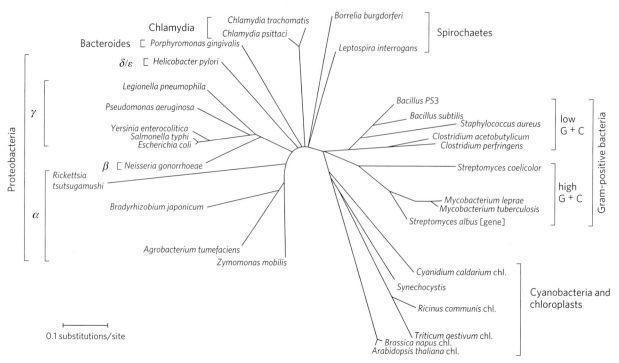

FIGURE 3–35 Evolutionary tree derived from amino acid sequence comparisons. A bacterial evolutionary tree, based on the sequence divergence observed in the GroEL family of proteins. Also included in this tree (lower right) are the chloroplasts (chl.) of some nonbacterial species.

additional genomic information emerges from increasingly sophisticated methods of analysis. All of this work moves us toward the goal of creating a detailed tree of life that describes the evolution and relationship of every organism on Earth. The story is a work in progress, of course **(Fig. 3–36)**. The questions being asked and answered are fundamental to how humans view themselves and the world around them. The field of molecular evolution promises to be among the most vibrant of the scientific frontiers in the twenty-first century.

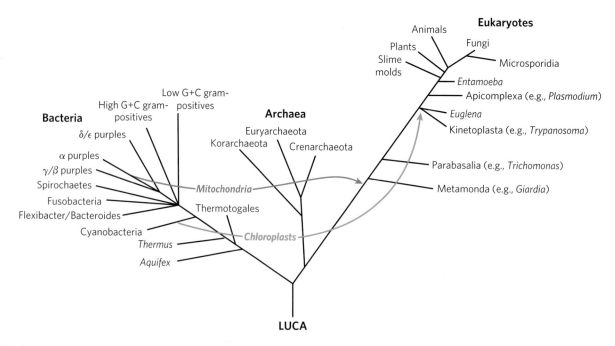

FIGURE 3–36 A consensus tree of life. The tree shown here is based on analyses of many different protein sequences and additional genomic features. The tree presents only a fraction of the available information, as well as only a fraction of the issues remaining to be resolved. Each extant group shown is a complex evolutionary story unto itself. LUCA is the last universal common ancestor from which all other life forms evolved. The blue and green arrows indicate the endosymbiotic assimilation of particular types of bacteria into eukaryotic cells to become mitochondria and chloroplasts, respectively (see Fig. 1–38).

SUMMARY 3.4 The Structure of Proteins: Primary Structure

▶ Differences in protein function result from differences in amino acid composition and sequence. Some variations in sequence may occur in a particular protein, with little or no effect on its function.

▶ Amino acid sequences are deduced by fragmenting polypeptides into smaller peptides with reagents known to cleave specific peptide bonds; determining the amino acid sequence of each fragment by the automated Edman degradation procedure; then ordering the peptide fragments by finding sequence overlaps between fragments generated by different reagents. A protein sequence can also be deduced from the nucleotide sequence of its corresponding gene in DNA, or by mass spectrometry.

▶ Short proteins and peptides (up to about 100 residues) can be chemically synthesized. The peptide is built up, one amino acid residue at a time, while tethered to a solid support.

▶ Protein sequences are a rich source of information about protein structure and function, as well as the evolution of life on Earth. Sophisticated methods are being developed to trace evolution by analyzing the resultant slow changes in amino acid sequences of homologous proteins.

Key Terms

Terms in bold are defined in the glossary.

Further Reading

Amino Acids

Dougherty, D.A. (2000) Unnatural amino acids as probes of protein structure and function. *Curr. Opin. Chem. Biol.* **4**, 645–652.

Kreil, G. (1997) D-Amino acids in animal peptides. *Annu. Rev. Biochem.* **66**, 337–345.
> Details the occurrence of these unusual stereoisomers of amino acids.

Meister, A. (1965) *Biochemistry of the Amino Acids*, 2nd edn, Vols 1 and 2, Academic Press, Inc., New York.
> Encyclopedic treatment of the properties, occurrence, and metabolism of amino acids.

Peptides and Proteins

Creighton, T.E. (1992) *Proteins: Structures and Molecular Properties*, 2nd edn, W. H. Freeman and Company, New York.
> Very useful general source.

Working with Proteins

Dunn, M.J. & Corbett, J.M. (1996) Two-dimensional polyacrylamide gel electrophoresis. *Methods Enzymol.* **271**, 177–203.
> A detailed description of the technology.

Kornberg, A. (1990) Why purify enzymes? *Methods Enzymol.* **182**, 1–5.
> The critical role of classical biochemical methods in a new age.

Scopes, R.K. (1994) *Protein Purification: Principles and Practice*, 3rd edn, Springer-Verlag, New York.
> A good source for more complete descriptions of the principles underlying chromatography and other methods.

Protein Primary Structure and Evolution

Andersson, L., Blomberg, L., Flegel, M., Lepsa, L., Nilsson, B., & Verlander, M. (2000) Large-scale synthesis of peptides. *Biopolymers* **55**, 227–250.
> A discussion of approaches to manufacturing peptides as pharmaceuticals.

Dell, A. & Morris, H.R. (2001) Glycoprotein structure determination by mass spectrometry. *Science* **291**, 2351–2356.
> Glycoproteins can be complex; mass spectrometry is a preferred method for sorting things out.

Delsuc, F., Brinkmann, H., & Philippe, H. (2005) Phylogenomics and the reconstruction of the tree of life. *Nat. Rev. Genet.* **6**, 361–375.

Gogarten, J.P. & Townsend, J.P. (2005) Horizontal gene transfer, genome innovation and evolution. *Nat. Rev. Microbiol.* **3**, 679–687.

Gygi, S.P. & Aebersold, R. (2000) Mass spectrometry and proteomics. *Curr. Opin. Chem. Biol.* **4**, 489–494.
> Uses of mass spectrometry to identify and study cellular proteins.

Koonin, E.V., Tatusov, R.L., & Galperin, M.Y. (1998) Beyond complete genomes: from sequence to structure and function. *Curr. Opin. Struct. Biol.* **8**, 355–363.
> A good discussion about the possible uses of the increasing amount of information on protein sequences.

Li, W.-H. & Graur, D. (2000) *Fundamentals of Molecular Evolution*, 2nd edn, Sinauer Associates, Inc., Sunderland, MA.
> A very readable text describing methods used to analyze protein and nucleic acid sequences. Chapter 5 provides one of the best

available descriptions of how evolutionary trees are constructed from sequence data.

Mayo, K.H. (2000) Recent advances in the design and construction of synthetic peptides: for the love of basics or just for the technology of it. *Trends Biotechnol.* **18**, 212–217.

Miranda, L.P. & Alewood, P.F. (2000) Challenges for protein chemical synthesis in the 21st century: bridging genomics and proteomics. *Biopolymers* **55**, 217–226.

This and the article by Mayo (above) describe how to make peptides and splice them together to address a wide range of problems in protein biochemistry.

Ramisetty, S.R. & Washburn, M.P. (2011) Unraveling the dynamics of protein interactions with quantitative mass spectrometry. *Crit. Rev. Biochem. Mol. Biol.* **46**, 216–228.

Rokas, A., Williams, B.L., King, N., & Carroll, S.B. (2003) Genome-scale approaches to resolving incongruence in molecular phylogenies. *Nature* **425**, 798–804.

How sequence comparisons of multiple proteins can yield accurate evolutionary information.

Sanger, F. (1988) Sequences, sequences, sequences. *Annu. Rev. Biochem.* **57**, 1–28.

A nice historical account of the development of sequencing methods.

Snel, B., Huynen, M.A., & Dutilh, B.E. (2005) Genome trees and the nature of genome evolution. *Annu. Rev. Microbiol.* **59**, 191–209.

Steen, H. & Mann, M. (2004) The ABC's (and XYZ's) of peptide sequencing. *Nat. Rev. Mol. Cell Biol.* **5**, 699–711.

Zuckerkandl, E. & Pauling, L. (1965) Molecules as documents of evolutionary history. *J. Theor. Biol.* **8**, 357–366.

Many consider this the founding paper in the field of molecular evolution.

Problems

1. Absolute Configuration of Citrulline The citrulline isolated from watermelons has the structure shown below. Is it a D- or L-amino acid? Explain.

$$\begin{array}{c} CH_2(CH_2)_2NH-C-NH_2 \\ | \quad\quad\quad\quad\quad || \\ H-C-\overset{+}{N}H_3 \quad\quad O \\ | \\ COO^- \end{array}$$

2. Relationship between the Titration Curve and the Acid-Base Properties of Glycine A 100 mL solution of 0.1 M glycine at pH 1.72 was titrated with 2 M NaOH solution. The pH was monitored and the results were plotted as shown in the graph. The key points in the titration are designated I to V. For each of the statements (a) to (o), *identify* the appropriate key point in the titration and *justify* your choice.

(a) Glycine is present predominantly as the species $^+H_3N-CH_2-COOH$.

(b) The *average* net charge of glycine is $+\frac{1}{2}$.

(c) Half of the amino groups are ionized.

(d) The pH is equal to the pK_a of the carboxyl group.

(e) The pH is equal to the pK_a of the protonated amino group.

(f) Glycine has its maximum buffering capacity.

(g) The *average* net charge of glycine is zero.

(h) The carboxyl group has been completely titrated (first equivalence point).

(i) Glycine is completely titrated (second equivalence point).

(j) The predominant species is $^+H_3N-CH_2-COO^-$.

(k) The *average* net charge of glycine is -1.

(l) Glycine is present predominantly as a 50:50 mixture of $^+H_3N-CH_2-COOH$ and $^+H_3N-CH_2-COO^-$.

(m) This is the isoelectric point.

(n) This is the end of the titration.

(o) These are the *worst* pH regions for buffering power.

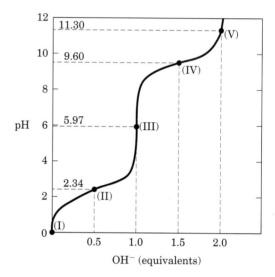

3. How Much Alanine Is Present as the Completely Uncharged Species? At a pH equal to the isoelectric point of alanine, the *net* charge on alanine is zero. Two structures can be drawn that have a net charge of zero, but the predominant form of alanine at its pI is zwitterionic.

$$\begin{array}{cc}
\overset{\displaystyle CH_3}{\underset{\displaystyle H}{H_3\overset{+}{N}-C-C}}\overset{\displaystyle O}{\underset{\displaystyle O^-}{\diagup\diagdown}} & \overset{\displaystyle CH_3}{\underset{\displaystyle H}{H_2N-C-C}}\overset{\displaystyle O}{\underset{\displaystyle OH}{\diagup\diagdown}} \\
\text{Zwitterionic} & \text{Uncharged}
\end{array}$$

(a) Why is alanine predominantly zwitterionic rather than completely uncharged at its pI?

(b) What fraction of alanine is in the completely uncharged form at its pI? Justify your assumptions.

4. Ionization State of Histidine Each ionizable group of an amino acid can exist in one of two states, charged or neutral. The electric charge on the functional group is determined by the relationship between its pK_a and the pH of the solution. This relationship is described by the Henderson-Hasselbalch equation.

(a) Histidine has three ionizable functional groups. Write the equilibrium equations for its three ionizations and assign the proper pK_a for each ionization. Draw the structure of histidine in each ionization state. What is the net charge on the histidine molecule in each ionization state?

(b) Draw the structures of the predominant ionization state of histidine at pH 1, 4, 8, and 12. Note that the ionization state can be approximated by treating each ionizable group independently.

(c) What is the net charge of histidine at pH 1, 4, 8, and 12? For each pH, will histidine migrate toward the anode (+) or cathode (–) when placed in an electric field?

5. Separation of Amino Acids by Ion-Exchange Chromatography Mixtures of amino acids can be analyzed by first separating the mixture into its components through ion-exchange chromatography. Amino acids placed on a cation-exchange resin (see Fig. 3–17a) containing sulfonate ($-SO_3^-$) groups flow down the column at different rates because of two factors that influence their movement: (1) ionic attraction between the sulfonate residues on the column and positively charged functional groups on the amino acids, and (2) hydrophobic interactions between amino acid side chains and the strongly hydrophobic backbone of the polystyrene resin. For each pair of amino acids listed, determine which will be eluted first from the cation-exchange column by a pH 7.0 buffer.

(a) Asp and Lys

(b) Arg and Met

(c) Glu and Val

(d) Gly and Leu

(e) Ser and Ala

6. Naming the Stereoisomers of Isoleucine The structure of the amino acid isoleucine is

$$
\begin{array}{c}
COO^- \\
| \\
H_3\overset{+}{N}-C-H \\
| \\
H-C-CH_3 \\
| \\
CH_2 \\
| \\
CH_3
\end{array}
$$

(a) How many chiral centers does it have?

(b) How many optical isomers?

(c) Draw perspective formulas for all the optical isomers of isoleucine.

7. Comparing the pK_a Values of Alanine and Polyalanine The titration curve of alanine shows the ionization of two functional groups with pK_a values of 2.34 and 9.69, corresponding to the ionization of the carboxyl and the protonated amino groups, respectively. The titration of di-, tri-, and larger oligopeptides of alanine also shows the ionization of only two functional groups, although the experimental pK_a values are different. The trend in pK_a values is summarized in the table.

Amino acid or peptide	pK_1	pK_2
Ala	2.34	9.69
Ala–Ala	3.12	8.30
Ala–Ala–Ala	3.39	8.03
Ala–(Ala)$_n$–Ala, $n \geq 4$	3.42	7.94

(a) Draw the structure of Ala–Ala–Ala. Identify the functional groups associated with pK_1 and pK_2.

(b) Why does the value of pK_1 *increase* with each additional Ala residue in the oligopeptide?

(c) Why does the value of pK_2 *decrease* with each additional Ala residue in the oligopeptide?

8. The Size of Proteins What is the approximate molecular weight of a protein with 682 amino acid residues in a single polypeptide chain?

9. The Number of Tryptophan Residues in Bovine Serum Albumin A quantitative amino acid analysis reveals that bovine serum albumin (BSA) contains 0.58% tryptophan (M_r 204) by weight.

(a) Calculate the *minimum* molecular weight of BSA (i.e., assume there is only one Trp residue per protein molecule).

(b) Size-exclusion chromatography of BSA gives a molecular weight estimate of 70,000. How many Trp residues are present in a molecule of serum albumin?

10. Subunit Composition of a Protein A protein has a molecular mass of 400 kDa when measured by size-exclusion chromatography. When subjected to gel electrophoresis in the presence of sodium dodecyl sulfate (SDS), the protein gives three bands with molecular masses of 180, 160, and 60 kDa. When electrophoresis is carried out in the presence of SDS and dithiothreitol, three bands are again formed, this time with molecular masses of 160, 90, and 60 kDa. Determine the subunit composition of the protein.

11. Net Electric Charge of Peptides A peptide has the sequence

Glu–His–Trp–Ser–Gly–Leu–Arg–Pro–Gly

(a) What is the net charge of the molecule at pH 3, 8, and 11? (Use pK_a values for side chains and terminal amino and carboxyl groups as given in Table 3–1.)

(b) Estimate the pI for this peptide.

12. Isoelectric Point of Pepsin Pepsin is the name given to a mix of several digestive enzymes secreted (as larger precursor proteins) by glands that line the stomach. These glands also secrete hydrochloric acid, which dissolves the particulate matter in food, allowing pepsin to enzymatically cleave individual protein molecules. The resulting mixture of food, HCl, and digestive enzymes is known as chyme and has a pH near 1.5. What pI would you predict for the pepsin proteins? What functional groups must be present to confer this pI on pepsin? Which amino acids in the proteins would contribute such groups?

13. Isoelectric Point of Histones Histones are proteins found in eukaryotic cell nuclei, tightly bound to DNA, which has many phosphate groups. The pI of histones is very high, about 10.8. What amino acid residues must be present in relatively large numbers in histones? In what way do these residues contribute to the strong binding of histones to DNA?

14. Solubility of Polypeptides One method for separating polypeptides makes use of their different solubilities. The solubility of large polypeptides in water depends on the relative polarity of their R groups, particularly on the number of ionized groups: the more ionized groups there are, the more soluble the polypeptide. Which of each pair of polypeptides that follow is more soluble at the indicated pH?

(a) (Gly)$_{20}$ or (Glu)$_{20}$ at pH 7.0

(b) (Lys–Ala)$_3$ or (Phe–Met)$_3$ at pH 7.0

(c) (Ala–Ser–Gly)$_5$ or (Asn–Ser–His)$_5$ at pH 6.0

(d) (Ala–Asp–Gly)$_5$ or (Asn–Ser–His)$_5$ at pH 3.0

15. Purification of an Enzyme A biochemist discovers and purifies a new enzyme, generating the purification table below.

Procedure	Total protein (mg)	Activity (units)
1. Crude extract	20,000	4,000,000
2. Precipitation (salt)	5,000	3,000,000
3. Precipitation (pH)	4,000	1,000,000
4. Ion-exchange chromatography	200	800,000
5. Affinity chromatography	50	750,000
6. Size-exclusion chromatography	45	675,000

(a) From the information given in the table, calculate the specific activity of the enzyme after each purification procedure.

(b) Which of the purification procedures used for this enzyme is most effective (i.e., gives the greatest relative increase in purity)?

(c) Which of the purification procedures is least effective?

(d) Is there any indication based on the results shown in the table that the enzyme after step 6 is now pure? What else could be done to estimate the purity of the enzyme preparation?

16. Dialysis A purified protein is in a Hepes (*N*-(2-hydroxyethyl)piperazine-*N'*-(2-ethanesulfonic acid)) buffer at pH 7 with 500 mM NaCl. A sample (1 mL) of the protein solution is placed in a tube made of dialysis membrane and dialyzed against 1 L of the same Hepes buffer with 0 mM NaCl. Small molecules and ions (such as Na^+, Cl^-, and Hepes) can diffuse across the dialysis membrane, but the protein cannot.

(a) Once the dialysis has come to equilibrium, what is the concentration of NaCl in the protein sample? Assume no volume changes occur in the sample during the dialysis.

(b) If the original 1 mL sample were dialyzed twice, successively, against 100 mL of the same Hepes buffer with 0 mM NaCl, what would be the final NaCl concentration in the sample?

17. Peptide Purification At pH 7.0, in what order would the following three peptides be eluted from a column filled with a cation-exchange polymer? Their amino acid compositions are:

Peptide A: Ala 10%, Glu 5%, Ser 5%, Leu 10%, Arg 10%, His 5%, Ile 10%, Phe 5%, Tyr 5%, Lys 10%, Gly 10%, Pro 5%, and Trp 10%.

Peptide B: Ala 5%, Val 5%, Gly 10%, Asp 5%, Leu 5%, Arg 5%, Ile 5%, Phe 5%, Tyr 5%, Lys 5%, Trp 5%, Ser 5%, Thr 5%, Glu 5%, Asn 5%, Pro 10%, Met 5%, and Cys 5%.

Peptide C: Ala 10%, Glu 10%, Gly 5%, Leu 5%, Asp 10%, Arg 5%, Met 5%, Cys 5%, Tyr 5%, Phe 5%, His 5%, Val 5%, Pro 5%, Thr 5%, Ser 5%, Asn 5%, and Gln 5%.

18. Sequence Determination of the Brain Peptide Leucine Enkephalin A group of peptides that influence nerve transmission in certain parts of the brain has been isolated from normal brain tissue. These peptides are known as opioids, because they bind to specific receptors that also bind opiate drugs, such as morphine and naloxone. Opioids thus mimic some of the properties of opiates. Some researchers consider these peptides to be the brain's own painkillers. Using the information below, determine the amino acid sequence of the opioid leucine enkephalin. Explain how your structure is consistent with each piece of information.

(a) Complete hydrolysis by 6 M HCl at 110 °C followed by amino acid analysis indicated the presence of Gly, Leu, Phe, and Tyr, in a 2:1:1:1 molar ratio.

(b) Treatment of the peptide with 1-fluoro-2,4-dinitrobenzene followed by complete hydrolysis and chromatography indicated the presence of the 2,4-dinitrophenyl derivative of tyrosine. No free tyrosine could be found.

(c) Complete digestion of the peptide with chymotrypsin followed by chromatography yielded free tyrosine and leucine, plus a tripeptide containing Phe and Gly in a 1:2 ratio.

19. Structure of a Peptide Antibiotic from *Bacillus brevis* Extracts from the bacterium *Bacillus brevis* contain a peptide with antibiotic properties. This peptide forms complexes with metal ions and seems to disrupt ion transport across the cell membranes of other bacterial species, killing them. The structure of the peptide has been determined from the following observations.

(a) Complete acid hydrolysis of the peptide followed by amino acid analysis yielded equimolar amounts of Leu, Orn, Phe, Pro, and Val. Orn is ornithine, an amino acid not present in proteins but present in some peptides. It has the structure

$$\overset{+}{H_3N}-CH_2-CH_2-CH_2-\underset{\overset{|}{\overset{+}{N}H_3}}{\overset{\overset{H}{|}}{C}}-COO^-$$

(b) The molecular weight of the peptide was estimated as about 1,200.

(c) The peptide failed to undergo hydrolysis when treated with the enzyme carboxypeptidase. This enzyme catalyzes the hydrolysis of the carboxyl-terminal residue of a polypeptide unless the residue is Pro or, for some reason, does not contain a free carboxyl group.

(d) Treatment of the intact peptide with 1-fluoro-2,4-dinitrobenzene, followed by complete hydrolysis and chromatography, yielded only free amino acids and the following derivative:

$$O_2N-\underset{NO_2}{\underset{|}{\bigcirc}}-NH-CH_2-CH_2-CH_2-\underset{\overset{+}{N}H_3}{\overset{\overset{H}{|}}{C}}-COO^-$$

(Hint: The 2,4-dinitrophenyl derivative involves the amino group of a side chain rather than the α-amino group.)

(e) Partial hydrolysis of the peptide followed by chromatographic separation and sequence analysis yielded the following di- and tripeptides (the amino-terminal amino acid is always at the left):

Leu–Phe   Phe–Pro   Orn–Leu   Val–Orn

Val–Orn–Leu   Phe–Pro–Val   Pro–Val–Orn

Given the above information, deduce the amino acid sequence of the peptide antibiotic. Show your reasoning. When you have arrived at a structure, demonstrate that it is consistent with *each* experimental observation.

20. Efficiency in Peptide Sequencing A peptide with the primary structure Lys–Arg–Pro–Leu–Ile–Asp–Gly–Ala is sequenced by the Edman procedure. If each Edman cycle is 96% efficient, what percentage of the amino acids liberated in the fourth cycle will be leucine? Do the calculation a second time, but assume a 99% efficiency for each cycle.

21. Sequence Comparisons Proteins called molecular chaperones (described in Chapter 4) assist in the process of protein folding. One class of chaperone found in organisms from bacteria to mammals is heat shock protein 90 (Hsp90). All Hsp90 chaperones contain a 10 amino acid "signature sequence," which allows for ready identification of these proteins in sequence databases. Two representations of this signature sequence are shown below.

Y-x-[NQHD]-[KHR]-[DE]-[IVA]-F-[LM]-R-[ED].

(a) In this sequence, which amino acid residues are invariant (conserved across all species)?

(b) At which position(s) are amino acids limited to those with positively charged side chains? For each position, which amino acid is more commonly found?

(c) At which positions are substitutions restricted to amino acids with negatively charged side chains? For each position, which amino acid predominates?

(d) There is one position that can be any amino acid, although one amino acid appears much more often than any other. What position is this, and which amino acid appears most often?

22. Chromatographic Methods Three polypeptides, the sequences of which are represented below using the one-letter code for their amino acids, are present in a mixture:

1. ATKNRASCLVPKHGALMFWRHKQLVSDPILQKR-QHILVCRNAAG
2. GPYFGDEPLDVHDEPEEG
3. PHLLSAWKGMEGVGKSQSFAALIVILA

Of the three, which one would migrate most slowly during chromatography through:

(a) an ion-exchange resin; beads coated with positively charged groups?

(b) an ion-exchange resin; beads coated with negatively charged groups?

(c) a size-exclusion (gel-filtration) column designed to separate small peptides such as these?

(d) Which peptide contains the ATP-binding motif shown in the following sequence logo?

Data Analysis Problem

23. Determining the Amino Acid Sequence of Insulin Figure 3–24 shows the amino acid sequence of bovine insulin. This structure was determined by Frederick Sanger and his coworkers. Most of this work is described in a series of articles published in the *Biochemical Journal* from 1945 to 1955.

When Sanger and colleagues began their work in 1945, it was known that insulin was a small protein consisting of two or four polypeptide chains linked by disulfide bonds. Sanger and his coworkers had developed a few simple methods for studying protein sequences.

Treatment with FDNB. FDNB (1-fluoro-2,4-dinitrobenzene) reacted with free amino (but not amido or guanidino) groups in proteins to produce dinitrophenyl (DNP) derivatives of amino acids:

$$R-NH_2 + F-\text{(FDNB)}-NO_2 \longrightarrow R-\underset{H}{N}-\text{(DNP-amine)}-NO_2 + HF$$

Amine   FDNB   DNP-amine

Acid Hydrolysis. Boiling a protein with 10% HCl for several hours hydrolyzed all of its peptide and amide bonds. Short treatments produced short polypeptides; the longer the treatment, the more complete the breakdown of the protein into its amino acids.

Oxidation of Cysteines. Treatment of a protein with performic acid cleaved all the disulfide bonds and converted all Cys residues to cysteic acid residues (see Fig. 3–28).

Paper Chromatography. This more primitive version of thin-layer chromatography (see Fig. 10–25) separated compounds based on their chemical properties, allowing identification of single amino acids and, in some cases, dipeptides. Thin-layer chromatography also separates larger peptides.

As reported in his first paper (1945), Sanger reacted insulin with FDNB and hydrolyzed the resulting protein. He found many free amino acids, but only three DNP–amino acids: α-DNP-glycine (DNP group attached to the α-amino group); α-DNP-phenylalanine; and ε-DNP-lysine (DNP attached to the ε-amino group). Sanger interpreted these results as showing that insulin had two protein chains: one with Gly at its amino terminus and one with Phe at its amino terminus. One of the two chains also contained a Lys residue, not at the amino terminus. He named the chain beginning with a Gly residue "A" and the chain beginning with Phe "B."

(a) Explain how Sanger's results support his conclusions.

(b) Are the results consistent with the known structure of bovine insulin (see Fig. 3–24)?

In a later paper (1949), Sanger described how he used these techniques to determine the first few amino acids (amino-terminal end) of each insulin chain. To analyze the B chain, for example, he carried out the following steps:

1. Oxidized insulin to separate the A and B chains.
2. Prepared a sample of pure B chain with paper chromatography.
3. Reacted the B chain with FDNB.
4. Gently acid-hydrolyzed the protein so that some small peptides would be produced.
5. Separated the DNP-peptides from the peptides that did not contain DNP groups.
6. Isolated four of the DNP-peptides, which were named B1 through B4.
7. Strongly hydrolyzed each DNP-peptide to give free amino acids.
8. Identified the amino acids in each peptide with paper chromatography.

The results were as follows:

B1: α-DNP-phenylalanine only

B2: α-DNP-phenylalanine; valine

B3: aspartic acid; α-DNP-phenylalanine; valine

B4: aspartic acid; glutamic acid; α-DNP-phenylalanine; valine

(c) Based on these data, what are the first four (amino-terminal) amino acids of the B chain? Explain your reasoning.

(d) Does this result match the known sequence of bovine insulin (see Fig. 3–24)? Explain any discrepancies.

Sanger and colleagues used these and related methods to determine the entire sequence of the A and B chains. Their sequence for the A chain was as follows (amino terminus on left):

$$
\overset{1}{\text{Gly}}-\text{Ile}-\text{Val}-\overset{5}{\text{Glx}}-\text{Glx}-\text{Cys}-\text{Cys}-\text{Ala}-\text{Ser}-\overset{10}{\text{Val}}-
$$
$$
\overset{15}{\text{Cys}}-\text{Ser}-\text{Leu}-\text{Tyr}-\text{Glx}-\text{Leu}-\overset{20}{\text{Glx}}-\text{Asx}-\text{Tyr}-\text{Cys}-\text{Asx}
$$

Because acid hydrolysis had converted all Asn to Asp and all Gln to Glu, these residues had to be designated Asx and Glx, respectively (exact identity in the peptide unknown). Sanger solved this problem by using protease enzymes that cleave peptide bonds, but not the amide bonds in Asn and Gln residues, to prepare short peptides. He then determined the number of amide groups present in each peptide by measuring the NH_4^+ released when the peptide was acid-hydrolyzed. Some of the results for the A chain are shown below. The peptides may not have been completely pure, so the numbers were approximate—but good enough for Sanger's purposes.

Peptide name	Peptide sequence	Number of amide groups in peptide
Ac1	Cys–Asx	0.7
Ap15	Tyr–Glx–Leu	0.98
Ap14	Tyr–Glx–Leu–Glx	1.06
Ap3	Asx–Tyr–Cys–Asx	2.10
Ap1	Glx–Asx–Tyr–Cys–Asx	1.94
Ap5pa1	Gly–Ile–Val–Glx	0.15
Ap5	Gly–Ile–Val–Glx–Glx–Cys–Cys– Ala–Ser–Val–Cys–Ser–Leu	1.16

(e) Based on these data, determine the amino acid sequence of the A chain. Explain how you reached your answer. Compare it with Figure 3–24.

References

Sanger, F. (1945) The free amino groups of insulin. *Biochem. J.* **39**, 507–515.

Sanger, F. (1949) The terminal peptides of insulin. *Biochem. J.* **45**, 563–574.

The Three-Dimensional Structure of Proteins

Proteins are big molecules. The covalent backbone of a typical protein contains hundreds of individual bonds. Because free rotation is possible around many of these bonds, the protein can in principle assume a virtually uncountable number of conformations. However, each protein has a specific chemical or structural function, suggesting that each has a unique three-dimensional structure **(Fig. 4–1)**. How stable is this structure, what factors guide its formation, and what holds it together? By the late 1920s, several proteins had been crystallized, including hemoglobin (M_r 64,500) and the enzyme urease (M_r 483,000). Given that, generally, the ordered array of molecules in a crystal can form only if the molecular units are identical, the finding that many proteins could be crystallized was evidence that even very large proteins are discrete chemical entities with unique structures. This conclusion revolutionized thinking about proteins and their functions, but the insight it provided was incomplete. Protein structure is always malleable in sometimes surprising ways. Changes in structure can be as important to a protein's function as the structure itself.

In this chapter, we examine the structure of proteins. We emphasize six themes. First, the three-dimensional structure or structures taken up by a protein are determined by its amino acid sequence. Second, the function of a typical protein depends on its structure. Third, most isolated proteins exist in one or a small number of stable structural forms. Fourth, the most important forces stabilizing the specific structures maintained by a given protein are noncovalent interactions. Fifth, amid the huge number of unique protein structures, we can recognize some common structural patterns that help to organize our understanding of protein architecture. Sixth, protein structures are not static. All proteins undergo changes in conformation ranging from subtle to quite dramatic. Parts of many proteins have no discernible structure. For some proteins, a lack of definable structure is critical to their function.

4.1 Overview of Protein Structure

The spatial arrangement of atoms in a protein or any part of a protein is called its **conformation**. The possible conformations of a protein or protein segment include any structural state it can achieve without breaking covalent bonds. A change in conformation could occur, for example, by rotation about single bonds. Of the many conformations that are theoretically possible in a protein containing hundreds of single bonds, one or (more commonly) a few generally predominate under

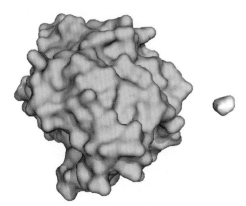

FIGURE 4–1 Structure of the enzyme chymotrypsin, a globular protein. A molecule of glycine (gray) is shown for size comparison. The known three-dimensional structures of proteins are archived in the Protein Data Bank, or PDB (see Box 4-4). The image shown here was made using data from the entry with PDB ID 6GCH.

biological conditions. The need for multiple stable conformations reflects the changes that must take place in most proteins as they bind to other molecules or catalyze reactions. The conformations existing under a given set of conditions are usually the ones that are thermodynamically the most stable—that is, having the lowest Gibbs free energy (G). Proteins in any of their functional, folded conformations are called **native** proteins.

For the vast majority of proteins, a particular structure or small set of structures is critical to function. However, in many cases, parts of proteins lack discernible structure. These protein segments are intrinsically disordered. In a few cases, entire proteins are intrinsically disordered, yet fully functional.

What principles determine the most stable conformations of a typical protein? An understanding of protein conformation can be built stepwise from the discussion of primary structure in Chapter 3 through a consideration of secondary, tertiary, and quaternary structures. To this traditional approach we must add the newer emphasis on common and classifiable folding patterns, variously called supersecondary structures, folds, or motifs, which provide an important organizational context to this complex endeavor. We begin by introducing some guiding principles.

A Protein's Conformation Is Stabilized Largely by Weak Interactions

In the context of protein structure, the term **stability** can be defined as the tendency to maintain a native conformation. Native proteins are only marginally stable; the ΔG separating the folded and unfolded states in typical proteins under physiological conditions is in the range of only 20 to 65 kJ/mol. A given polypeptide chain can theoretically assume countless conformations, and as a result the unfolded state of a protein is characterized by a high degree of conformational entropy. This entropy, and the hydrogen-bonding interactions of many groups in the polypeptide chain with the solvent (water), tend to maintain the unfolded state. The chemical interactions that counteract these effects and stabilize the native conformation include disulfide (covalent) bonds and the weak (noncovalent) interactions described in Chapter 2: hydrogen bonds and hydrophobic and ionic interactions.

Many proteins do not have disulfide bonds. The environment within most cells is highly reducing due to high concentrations of reductants such as glutathione, and most sulfhydryls will thus remain in the reduced state. Outside the cell, the environment is often more oxidizing, and disulfide formation is more likely to occur. In eukaryotes, disulfide bonds are found primarily in secreted, extracellular proteins (for example, the hormone insulin). Disulfide bonds are also uncommon in bacterial proteins. However, thermophilic bacteria, as well as the archaea, typically have many proteins with disulfide bonds, which stabilize proteins; this is presumably an adaptation to life at high temperatures.

For all proteins of all organisms, weak interactions are especially important in the folding of polypeptide chains into their secondary and tertiary structures. The association of multiple polypeptides to form quaternary structures also relies on these weak interactions.

About 200 to 460 kJ/mol are required to break a single covalent bond, whereas weak interactions can be disrupted by a mere 0.4 to 30 kJ/mol. Individual covalent bonds, such as disulfide bonds linking separate parts of a single polypeptide chain, are clearly much stronger than individual weak interactions. Yet, because they are so numerous, it is weak interactions that predominate as a stabilizing force in protein structure. In general, the protein conformation with the lowest free energy (that is, the most stable conformation) is the one with the maximum number of weak interactions.

The stability of a protein is not simply the sum of the free energies of formation of the many weak interactions within it. For every hydrogen bond formed in a protein during folding, a hydrogen bond (of similar strength) between the same group and water was broken. The net stability contributed by a given hydrogen bond, or the *difference* in free energies of the folded and unfolded states, may be close to zero. Ionic interactions may be either stabilizing or destabilizing. We must therefore look elsewhere to understand why a particular native conformation is favored.

On carefully examining the contribution of weak interactions to protein stability, we find that **hydrophobic interactions** generally predominate. Pure water contains a network of hydrogen-bonded H_2O molecules. No other molecule has the hydrogen-bonding potential of water, and the presence of other molecules in an aqueous solution disrupts the hydrogen bonding of water. When water surrounds a hydrophobic molecule, the optimal arrangement of hydrogen bonds results in a highly structured shell, or **solvation layer**, of water around the molecule (see Fig. 2–7). The increased order of the water molecules in the solvation layer correlates with an unfavorable decrease in the entropy of the water. However, when nonpolar groups cluster together, the extent of the solvation layer decreases, because each group no longer presents its entire surface to the solution. The result is a favorable increase in entropy. As described in Chapter 2, this increase in entropy is the major thermodynamic driving force for the association of hydrophobic groups in aqueous solution. Hydrophobic amino acid side chains therefore tend to cluster in a protein's interior, away from water (think of an oil droplet in water). The amino acid sequences of most proteins thus feature a significant content of hydrophobic amino acid side chains (especially Leu, Ile, Val, Phe, and Trp). These are positioned so that they are clustered when the protein is folded, forming a hydrophobic protein core.

Under physiological conditions, the formation of hydrogen bonds in a protein is driven largely by this same entropic effect. Polar groups can generally form hydrogen bonds with water and hence are soluble in water. However, the number of hydrogen bonds per unit mass is generally greater for pure water than for any other liquid or solution, and there are limits to the solubility of even the most polar molecules as their presence causes a net decrease in hydrogen bonding per unit mass. Therefore, a solvation layer also forms to some extent around polar molecules. Even though the energy of formation of an intramolecular hydrogen bond between two polar groups in a macromolecule is largely canceled by the elimination of such interactions between these polar groups and water, the release of structured water as intramolecular interactions form provides an entropic driving force for folding. Most of the net change in free energy as weak interactions form within a protein is therefore derived from the increased entropy in the surrounding aqueous solution resulting from the burial of hydrophobic surfaces. This more than counterbalances the large loss of conformational entropy as a polypeptide is constrained into its folded conformation.

Hydrophobic interactions are clearly important in stabilizing conformation; the interior of a structured protein is generally a densely packed core of hydrophobic amino acid side chains. It is also important that any polar or charged groups in the protein interior have suitable partners for hydrogen bonding or ionic interactions. One hydrogen bond seems to contribute little to the stability of a native structure, but the presence of hydrogen-bonding groups without partners in the hydrophobic core of a protein can be so *destabilizing* that conformations containing these groups are often thermodynamically untenable. The favorable free-energy change resulting from the combination of several such groups with partners in the surrounding solution can be greater than the free-energy difference between the folded and unfolded states. In addition, hydrogen bonds between groups in a protein form cooperatively (formation of one makes the next one more likely) in repeating secondary structures that optimize hydrogen bonding, as described below. In this way, hydrogen bonds often have an important role in guiding the protein-folding process.

The interaction of oppositely charged groups that form an ion pair, or salt bridge, can have either a stabilizing or destabilizing effect on protein structure. As in the case of hydrogen bonds, charged amino acid side chains interact with water and salts when the protein is unfolded, and the loss of those interactions must be considered when evaluating the effect of a salt bridge on the overall stability of a folded protein. However, the strength of a salt bridge increases as it moves to an environment of lower dielectric constant, ε (p. 50): from the polar aqueous solvent (ε near 80) to the nonpolar protein interior (ε near 4). Salt bridges, especially those that are partly or entirely buried, can thus provide significant stabilization to a protein structure. This trend explains the increased occurrence of buried salt bridges in the proteins of thermophilic organisms. Ionic interactions also limit structural flexibility and confer a uniqueness to a particular protein structure that nonspecific hydrophobic interactions cannot provide.

In the tightly packed atomic environment of a protein, one more type of weak interaction can have a significant effect—van der Waals interactions (p. 54). Van der Waals interactions are dipole-dipole interactions involving the permanent electric dipoles in groups such as carbonyls, transient dipoles derived from fluctuations of the electron cloud surrounding any atom, and dipoles induced by interaction of an atom with another that has a permanent or transient dipole. As atoms approach each other, these dipole-dipole interactions provide an attractive intermolecular force that operates only over a limited intermolecular distance (0.3 to 0.6 nm). Van der Waals interactions are weak and individually contribute little to overall protein stability. However, in a well-packed protein, or in an interaction between a protein and another protein or other molecule at a complementary surface, the number of such interactions can be substantial.

Most of the structural patterns outlined in this chapter reflect two simple rules: (1) hydrophobic residues are largely buried in the protein interior, away from water, and (2) the number of hydrogen bonds and ionic interactions within the protein is maximized, thus reducing the number of hydrogen-bonding and ionic groups that are not paired with a suitable partner. Proteins within membranes (which we examine in Chapter 11) and proteins that are intrinsically disordered or have intrinsically disordered segments follow different rules. This reflects their particular function or environment, but weak interactions are still critical structural elements. For example, soluble but intrinsically disordered protein segments are enriched in amino acid side chains that are charged (especially Arg, Lys, Glu) or small (Gly, Ala), providing little or no opportunity for the formation of a stable hydrophobic core.

The Peptide Bond Is Rigid and Planar

Protein Architecture–Primary Structure Covalent bonds, too, place important constraints on the conformation of a polypeptide. In the late 1930s, Linus Pauling and Robert Corey embarked on a series of studies that laid the foundation for our current understanding of protein structure. They began with a careful analysis of the peptide bond.

Linus Pauling, 1901–1994   Robert Corey, 1897–1971

The α carbons of adjacent amino acid residues are separated by three covalent bonds, arranged as C_α—C—N—C_α. X-ray diffraction studies of crystals of amino acids and of simple dipeptides and tripeptides showed that the peptide C—N bond is somewhat shorter than the C—N bond in a simple amine and that the atoms associated with the peptide bond are coplanar. This indicated a resonance or partial sharing of two pairs of electrons between the carbonyl oxygen and the amide nitrogen **(Fig. 4–2a)**. The oxygen has a partial negative charge and the hydrogen bonded to the nitrogen has a net partial positive charge, setting up a small electric dipole. The six atoms of the **peptide group** lie in a single plane, with the oxygen atom of the carbonyl group trans to the hydrogen atom of the amide nitrogen. From these findings Pauling and Corey concluded that the peptide C—N bonds, because of their partial double-bond character, cannot rotate freely. Rotation is permitted about the N—C_α and the C_α—C bonds. The backbone of a polypeptide chain can thus be pictured as a series of rigid planes, with consecutive planes sharing a common point of rotation at C_α (Fig. 4–2b). The rigid peptide bonds limit the range of conformations possible for a polypeptide chain.

Peptide conformation is defined by three dihedral angles (also known as torsion angles) called ϕ (phi), ψ (psi), and ω (omega), reflecting rotation about each of the three repeating bonds in the peptide backbone. A dihedral angle is the angle at the intersection of two planes. In the case of peptides, the planes are defined by bond vectors in the peptide backbone. Two successive bond vectors describe a plane. Three successive bond vectors describe two planes (the central bond vector is common to both; Fig. 4–2c), and the angle between these two planes is what we measure to describe protein conformation.

KEY CONVENTION: The important dihedral angles in a peptide are defined by the three bond vectors connecting four consecutive main-chain (peptide backbone) atoms (Fig. 4–2c): ϕ involves the C—N—C_α—C bonds (with the rotation occurring about the N—C_α bond), and ψ involves the N—C_α—C—N bonds. Both ϕ and ψ are defined as $\pm180°$ when the polypeptide is fully extended and all peptide groups are in the same plane (Fig. 4–2d). As one looks down the central bond vector in the direction of the vector arrow (as depicted in Fig. 4–2c for ψ), the dihedral angles increase as the distal

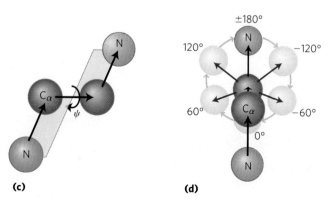

(a)

(b)

(c)

(d)

FIGURE 4–2 **The planar peptide group. (a)** Each peptide bond has some double-bond character due to resonance and cannot rotate. Although the N atom in a peptide bond is often represented with a partial positive charge, careful consideration of bond orbitals and quantum mechanics indicates that the N has a net charge that is neutral or slightly negative. **(b)** Three bonds separate sequential α carbons in a polypeptide chain. The N—C_α and C_α—C bonds can rotate, described by dihedral angles designated ϕ and ψ, respectively. The peptide C—N bond is not free to rotate. Other single bonds in the backbone may also be rotationally hindered, depending on the size and charge of the R groups. **(c)** The atoms and planes defining ψ. **(d)** By convention, ϕ and ψ are 180° (or −180°) when the first and fourth atoms are farthest apart and the peptide is fully extended. As the viewer looks out along the bond undergoing rotation (from either direction), the ϕ and ψ angles increase as the fourth atom rotates clockwise relative to the first. In a protein, some of the conformations shown here (e.g., 0°) are prohibited by steric overlap of atoms. In (b) through (d), the balls representing atoms are smaller than the van der Waals radii for this scale.

(fourth) atom is rotated clockwise (Fig. 4–2d). From the ±180° position, the dihedral angle increases from –180° to 0°, at which point the first and fourth atoms are eclipsed. The rotation can be continued from 0° to +180° (same position as –180°) to bring the structure back to the starting point. The third dihedral angle, ω, is not often considered. It involves the C_α—C—N—C_α bonds. The central bond in this case is the peptide bond, where rotation is constrained. The peptide bond is normally (99.6% of the time) in the trans configuration, constraining ω to a value of ±180°. For a rare cis peptide bond, ω = 0°. ■

In principle, ϕ and ψ can have any value between –180° and +180°, but many values are prohibited by steric interference between atoms in the polypeptide backbone and amino acid side chains. The conformation in which both ϕ and ψ are 0° (Fig. 4–2d) is prohibited for this reason; this conformation is merely a reference point for describing the dihedral angles. Allowed values for ϕ and ψ become evident when ψ is plotted versus ϕ in a **Ramachandran plot (Fig. 4–3)**, introduced by G. N. Ramachandran. We will see that Ramachandran

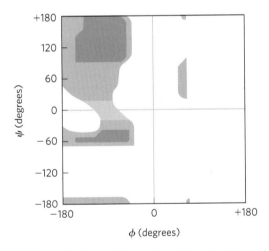

FIGURE 4–3 Ramachandran plot for L-Ala residues. Peptide conformations are defined by the values of ϕ and ψ. Conformations deemed possible are those that involve little or no steric interference, based on calculations using known van der Waals radii and dihedral angles. The areas shaded dark blue represent conformations that involve no steric overlap if the van der Waals radii of each atom are modeled as a hard sphere and thus are fully allowed; medium blue indicates conformations permitted if atoms are allowed to approach each other by an additional 0.1 nm, a slight clash; the lightest blue indicates conformations that are permissible if a very modest flexibility (a few degrees) is allowed in the ω dihedral angle that describes the peptide bond itself (generally constrained to 180°). The white regions are conformations that are not allowed. The asymmetry of the plot results from the L stereochemistry of the amino acid residues. The plots for other L residues with unbranched side chains are nearly identical. Allowed ranges for branched residues such as Val, Ile, and Thr are somewhat smaller than for Ala. The Gly residue, which is less sterically hindered, has a much broader range of allowed conformations. The range for Pro residues is greatly restricted because ϕ is limited by the cyclic side chain to the range of –35° to –85°.

plots are very useful tools that are often used to test the quality of three-dimensional protein structures that are deposited in international databases.

SUMMARY 4.1 Overview of Protein Structure

▶ A typical protein usually has one or more stable three-dimensional structures, or conformations, that reflect its function. Some proteins have segments that are intrinsically disordered.

▶ Protein structure is stabilized largely by multiple weak interactions. Hydrophobic interactions, derived from the increase in entropy of the surrounding water when nonpolar molecules or groups are clustered together, are the major contributors to stabilizing the globular form of most soluble proteins. Van der Waals interactions also contribute. Hydrogen bonds and ionic interactions are optimized in the thermodynamically most stable structures.

▶ Nonpeptide covalent bonds, particularly disulfide bonds, play a role in the stabilization of structure in some proteins.

▶ The nature of the covalent bonds in the polypeptide backbone places constraints on structure. The peptide bond has a partial double-bond character that keeps the entire six-atom peptide group in a rigid planar configuration. The N—C_α and C_α—C bonds can rotate to define the dihedral angles ϕ and ψ, respectively.

▶ The Ramachandran plot is a visual description of the combinations of ϕ and ψ dihedral angles that are permitted in a peptide backbone or that are not permitted due to steric constraints.

4.2 Protein Secondary Structure

The term **secondary structure** refers to any chosen segment of a polypeptide chain and describes the local spatial arrangement of its main-chain atoms, without regard to the positioning of its side chains or its relationship to other segments. A *regular* secondary structure occurs when each dihedral angle, ϕ and ψ, remains the same or nearly the same throughout the segment. There are a few types of secondary structure that are particularly stable and occur widely in proteins. The most prominent are the α helix and β conformations; another common type is the β turn. Where a regular pattern is not found, the secondary structure is sometimes referred to as undefined or as a random coil. This last designation, however, does not properly describe the structure of these segments. The path of most of the polypeptide backbone in a typical protein is not random; rather, it is unchanging and highly specific to the structure and function of that particular protein. Our discussion here focuses on the regular structures that are most common.

The α Helix Is a Common Protein Secondary Structure

🖱 **Protein Architecture–α Helix** Pauling and Corey were aware of the importance of hydrogen bonds in orienting polar chemical groups such as the C=O and N—H groups of the peptide bond. They also had the experimental results of William Astbury, who in the 1930s had conducted pioneering x-ray studies of proteins. Astbury demonstrated that the protein that makes up hair and porcupine quills (the fibrous protein α-keratin) has a regular structure that repeats every 5.15 to 5.2 Å. (The angstrom, Å, named after the physicist Anders J. Ångström, is equal to 0.1 nm. Although not an SI unit, it is used universally by structural biologists to describe atomic distances—it is approximately the length of a typical C—H bond.) With this information and their data on the peptide bond, and with the help of precisely constructed models, Pauling and Corey set out to determine the likely conformations of protein molecules.

The first breakthrough came in 1948. Pauling was a visiting lecturer at Oxford University, became ill, and retired to his apartment for several days of rest. Bored with the reading available, Pauling grabbed some paper and pencils to work out a plausible stable structure that could be taken up by a polypeptide chain. The model he developed, and later confirmed in work with Corey and

coworker Herman Branson, was the simplest arrangement the polypeptide chain can assume that maximizes the use of internal hydrogen bonding. It is a helical structure, and Pauling and Corey called it the **α helix (Fig. 4–4)**. In this structure, the polypeptide backbone is tightly wound around an imaginary axis drawn longitudinally through the middle of the helix, and the R groups of the amino acid residues protrude outward from the helical backbone. The repeating unit is a single turn of the helix, which extends about 5.4 Å along the long axis, slightly greater than the periodicity Astbury observed on x-ray analysis of hair keratin. The backbone atoms of the amino acid residues in the prototypical α helix have a characteristic set of dihedral angles that define the α-helix conformation (Table 4–1), and each helical turn includes 3.6 amino acid residues. The α-helical segments in proteins often deviate slightly from these dihedral angles, and even vary somewhat within a single contiguous segment to produce subtle bends or kinks in the helical axis. Pauling and Corey considered both right- and left-handed variants of the α helix. The subsequent elucidation of the three-dimensional structure of myoglobin and other proteins showed that the right-handed α helix is the common form (Box 4–1). Extended left-handed α helices are theoretically less stable and have not been observed in proteins. The α helix proved to be the predominant structure in

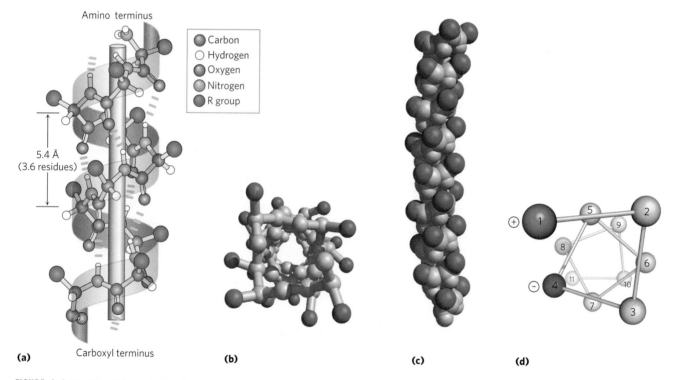

(a) Amino terminus / Carboxyl terminus
5.4 Å (3.6 residues)

● Carbon
○ Hydrogen
● Oxygen
● Nitrogen
● R group

(b)

(c)

(d)

FIGURE 4–4 Models of the α helix, showing different aspects of its structure. (a) Ball-and-stick model showing the intrachain hydrogen bonds. The repeat unit is a single turn of the helix, 3.6 residues. **(b)** The α helix viewed from one end, looking down the longitudinal axis (derived from PDB ID 4TNC). Note the positions of the R groups, represented by purple spheres. This ball-and-stick model, which emphasizes the helical arrangement, gives the false impression that the helix is hollow, because the balls do not represent the van der Waals radii of the individual atoms. **(c)** As this space-filling model shows, the atoms in the center of the α helix are in very close contact. **(d)** Helical wheel projection of an α helix. This representation can be colored to identify surfaces with particular properties. The yellow residues, for example, could be hydrophobic and conform to an interface between the helix shown here and another part of the same or another polypeptide. The red (negative) and blue (positive) residues illustrate the potential for interaction of oppositely charged side chains separated by two residues in the helix.

TABLE 4–1 Idealized ϕ and ψ Angles for Common Secondary Structures in Proteins

Structure	ϕ	ψ
α Helix	−57°	−47°
β Conformation		
Antiparallel	−139°	+135°
Parallel	−119°	+113°
Collagen triple helix	−51°	+153°
β Turn type I		
i + 1*	−60°	−30°
i + 2*	−90°	0°
β Turn type II		
i + 1	−60°	+120°
i + 2	+80°	0°

Note: In real proteins, the dihedral angles often vary somewhat from these idealized values.

*The i + 1 and i + 2 angles are those for the second and third amino acid residues in the β turn, respectively.

α-keratins. More generally, about one-fourth of all amino acid residues in proteins are found in α helices, the exact fraction varying greatly from one protein to another.

Why does the α helix form more readily than many other possible conformations? The answer lies in part in its optimal use of internal hydrogen bonds. The structure is stabilized by a hydrogen bond between the hydrogen atom attached to the electronegative nitrogen atom of a peptide linkage and the electronegative carbonyl oxygen atom of the fourth amino acid on the amino-terminal side of that peptide bond (Fig. 4–4a). Within the α helix, every peptide bond (except those close to each end of the helix) participates in such hydrogen bonding. Each successive turn of the α helix is held to adjacent turns by three to four hydrogen bonds, conferring significant stability on the overall structure. At the ends of an α-helical segment, there are always three or four amide carbonyl or amino groups that cannot participate in this helical pattern of hydrogen bonding. These may be exposed to the surrounding solvent, where they hydrogen-bond with water, or other parts of the protein may cap the helix to provide the needed hydrogen-bonding partners.

Further experiments have shown that an α helix can form in polypeptides consisting of either L- or D-amino acids. However, all residues must be of one stereoisomeric series; a D-amino acid will disrupt a regular structure consisting of L-amino acids, and vice versa. The most stable form of an α helix consisting of D-amino acids is left-handed.

WORKED EXAMPLE 4–1 Secondary Structure and Protein Dimensions

What is the length of a polypeptide with 80 amino acid residues in a single contiguous α helix?

Solution: An idealized α helix has 3.6 residues per turn and the rise along the helical axis is 5.4 Å. Thus, the rise along the axis for each amino acid residue is 1.5 Å. The length of the polypeptide is therefore 80 residues × 1.5 Å/residue = 120 Å.

Amino Acid Sequence Affects Stability of the α Helix

Not all polypeptides can form a stable α helix. Each amino acid residue in a polypeptide has an intrinsic propensity to form an α helix (Table 4–2), reflecting the properties of the R group and how they affect the capacity of the adjoining main-chain atoms to take up the characteristic ϕ and ψ angles. Alanine shows the greatest tendency to form α helices in most experimental model systems.

The position of an amino acid residue relative to its neighbors is also important. Interactions between amino

BOX 4–1 METHODS Knowing the Right Hand from the Left

There is a simple method for determining whether a helical structure is right-handed or left-handed. Make fists of your two hands with thumbs outstretched and pointing away from you. Looking at your right hand, think of a helix spiraling up your right thumb in the direction in which the other four fingers are curled as shown (clockwise). The resulting helix is right-handed. Your left hand will demonstrate a left-handed helix, which rotates in the counterclockwise direction as it spirals up your thumb.

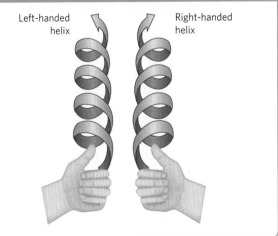

Left-handed helix   Right-handed helix

TABLE 4–2 Propensity of Amino Acid Residues to Take Up an α-Helical Conformation

Amino acid	$\Delta\Delta G°$ (kJ/mol)*	Amino acid	$\Delta\Delta G°$ (kJ/mol)*
Ala	0	Leu	0.79
Arg	0.3	Lys	0.63
Asn	3	Met	0.88
Asp	2.5	Phe	2.0
Cys	3	Pro	>4
Gln	1.3	Ser	2.2
Glu	1.4	Thr	2.4
Gly	4.6	Tyr	2.0
His	2.6	Trp	2.0
Ile	1.4	Val	2.1

Sources: Data (except proline) from Bryson, J.W., Betz, S.F., Lu, H.S., Suich, D.J., Zhou, H.X., O'Neil, K.T., & DeGrado, W.F. (1995) Protein design: a hierarchic approach. *Science* 270, 935. Proline data from Myers, J.K., Pace, C.N., & Scholtz, J.M. (1997) Helix propensities are identical in proteins and peptides. *Biochemistry* 36, 10,926.

*$\Delta\Delta G°$ is the difference in free-energy change, relative to that for alanine, required for the amino acid residue to take up the α-helical conformation. Larger numbers reflect greater difficulty taking up the α-helical structure. Data are a composite derived from multiple experiments and experimental systems.

acid side chains can stabilize or destabilize the α-helical structure. For example, if a polypeptide chain has a long block of Glu residues, this segment of the chain will not form an α helix at pH 7.0. The negatively charged carboxyl groups of adjacent Glu residues repel each other so strongly that they prevent formation of the α helix. For the same reason, if there are many adjacent Lys and/or Arg residues, with positively charged R groups at pH 7.0, they also repel each other and prevent formation of the α helix. The bulk and shape of Asn, Ser, Thr, and Cys residues can also destabilize an α helix if they are close together in the chain.

The twist of an α helix ensures that critical interactions occur between an amino acid side chain and the side chain three (and sometimes four) residues away on either side of it. This is clear when the α helix is depicted as a helical wheel (Fig. 4–4d). Positively charged amino acids are often found three residues away from negatively charged amino acids, permitting the formation of an ion pair. Two aromatic amino acid residues are often similarly spaced, resulting in a hydrophobic interaction.

A constraint on the formation of the α helix is the presence of Pro or Gly residues, which have the least proclivity to form α helices. In proline, the nitrogen atom is part of a rigid ring (see Fig. 4–8), and rotation about the N—C_α bond is not possible. Thus, a Pro residue introduces a destabilizing kink in an α helix. In addition, the nitrogen atom of a Pro residue in a peptide linkage has no substituent hydrogen to participate in hydrogen bonds with other residues. For these reasons,

proline is only rarely found in an α helix. Glycine occurs infrequently in α helices for a different reason: it has more conformational flexibility than the other amino acid residues. Polymers of glycine tend to take up coiled structures quite different from an α helix.

A final factor affecting the stability of an α helix is the identity of the amino acid residues near the ends of the α-helical segment of the polypeptide. A small electric dipole exists in each peptide bond (Fig. 4–2a). These dipoles are aligned through the hydrogen bonds of the helix, resulting in a net dipole along the helical axis that increases with helix length (**Fig. 4–5**). The partial positive and negative charges of the helix dipole reside on the peptide amino and carbonyl groups near the amino-terminal and carboxyl-terminal ends, respectively. For this reason, negatively charged amino acids are often found near the amino terminus of the helical segment, where they have a stabilizing interaction with the positive charge of the helix dipole; a positively charged amino acid at the amino-terminal end is destabilizing. The opposite is true at the carboxyl-terminal end of the helical segment.

In summary, five types of constraints affect the stability of an α helix: (1) the intrinsic propensity of an amino acid residue to form an α helix; (2) the interactions between R groups, particularly those spaced three (or four) residues apart; (3) the bulkiness of adjacent R groups; (4) the occurrence of Pro and Gly residues; and (5) interactions between amino acid residues at the ends of the helical segment and the electric dipole inherent to the α helix. The tendency of a given segment of a polypeptide chain to form an α helix therefore depends on the identity and sequence of amino acid residues within the segment.

Amino terminus

Carboxyl terminus

FIGURE 4–5 Helix dipole. The electric dipole of a peptide bond (see Fig. 4–2a) is transmitted along an α-helical segment through the intrachain hydrogen bonds, resulting in an overall helix dipole. In this illustration, the amino and carbonyl constituents of each peptide bond are indicated by + and − symbols, respectively. Non-hydrogen-bonded amino and carbonyl constituents of the peptide bonds near each end of the α-helical region are circled and shown in color.

The β Conformation Organizes Polypeptide Chains into Sheets

Protein Architecture–β Sheet In 1951, Pauling and Corey predicted a second type of repetitive structure, the **β conformation**. This is a more extended conformation of polypeptide chains, and its structure is again defined by backbone atoms arranged according to a characteristic set of dihedral angles (Table 4–1). In the β conformation, the backbone of the polypeptide chain is extended into a zigzag rather than helical structure **(Fig. 4–6)**. The arrangement of several segments side by side, all of which are in the β conformation, is called a **β sheet**. The zigzag structure of the individual polypeptide segments gives rise to a pleated appearance of the overall sheet. Hydrogen bonds form between adjacent segments of polypeptide chain within the sheet. The individual segments that form a β sheet are usually nearby on the polypeptide chain but can also be quite distant from each other in the linear sequence of the polypeptide;

(a) β strand
Side view

Side chains (above)

Side chains (below)

(b) Antiparallel β sheet
Top view

7 Å

(c) Parallel β sheet
Top view

6.5 Å

FIGURE 4–6 The β conformation of polypeptide chains. These **(a)** side and **(b, c)** top views reveal the R groups extending out from the β sheet and emphasize the pleated shape formed by the planes of the peptide bonds. (An alternative name for this structure is β-pleated sheet.) Hydrogen-bond cross-links between adjacent chains are also shown. The amino-terminal to carboxyl-terminal orientations of adjacent chains (arrows) can be the same or opposite, forming (b) an antiparallel β sheet or (c) a parallel β sheet.

they may even be in different polypeptide chains. The R groups of adjacent amino acids protrude from the zigzag structure in opposite directions, creating the alternating pattern seen in the side view in Figure 4–6.

The adjacent polypeptide chains in a β sheet can be either parallel or antiparallel (having the same or opposite amino-to-carboxyl orientations, respectively). The structures are somewhat similar, although the repeat period is shorter for the parallel conformation (6.5 Å, vs. 7 Å for antiparallel) and the hydrogen-bonding patterns are different. The interstrand hydrogen bonds are essentially in-line (see Fig. 2–5) in the antiparallel β sheet, whereas they are distorted or not in-line for the parallel variant. The idealized structures exhibit the bond angles given in Table 4–1; these values vary somewhat in real proteins, resulting in structural variation, as seen above for α helices.

β Turns Are Common in Proteins

Protein Architecture–β Turn In globular proteins, which have a compact folded structure, some amino acid residues are in turns or loops where the polypeptide chain reverses direction **(Fig. 4–7)**. These are the connecting elements that link successive runs of α helix or β conformation. Particularly common are **β turns** that connect the ends of two adjacent segments of an antiparallel β sheet. The structure is a 180° turn involving four amino acid residues, with the carbonyl oxygen of the first residue forming a hydrogen bond with the amino-group hydrogen of the fourth. The peptide groups of the central two residues do not participate in any inter-residue hydrogen bonding. Several types of β turns have been described, each defined by the ϕ and ψ angles of the bonds that link the four amino acid residues that make up the particular turn (Table 4–1). Gly and Pro residues often occur in β turns, the former because it is small and flexible, the latter because peptide bonds involving the imino nitrogen of proline readily assume the cis configuration **(Fig. 4–8)**, a form that is particularly amenable to a tight turn. The two types of β turns shown in Figure 4–7 are the most common. Beta turns are often found near the surface of a protein, where the peptide groups of the central two amino acid residues in the turn can hydrogen-bond with water. Considerably less common is the γ turn, a three-residue turn with a hydrogen bond between the first and third residues.

Common Secondary Structures Have Characteristic Dihedral Angles

The α helix and the β conformation are the major repetitive secondary structures in a wide variety of proteins, although other repetitive structures exist in some specialized proteins (an example is collagen; see Fig. 4–13). Every type of secondary structure can be completely described by the dihedral angles ϕ and ψ associated with each residue. As shown by a Ramachandran plot, the dihedral angles that define the α helix and β conformation fall within a relatively restricted range of

Type I β turn

Type II β turn

FIGURE 4–7 Structures of β turns. Type I and type II β turns are most common, distinguished by the φ and ψ angles taken up by the peptide backbone in the turn (see Table 4-1). Type I turns occur more than twice as frequently as type II. Type II β turns usually have Gly as the third residue. Note the hydrogen bond between the peptide groups of the first and fourth residues of the bends. (Individual amino acid residues are framed by large blue circles. Not all H atoms are shown in these depictions.)

trans   cis

Proline isomers

FIGURE 4–8 Trans and cis isomers of a peptide bond involving the imino nitrogen of proline. Of the peptide bonds between amino acid residues other than Pro, more than 99.95% are in the trans configuration. For peptide bonds involving the imino nitrogen of proline, however, about 6% are in the cis configuration; many of these occur at β turns.

sterically allowed structures **(Fig. 4–9a)**. Most values of φ and ψ taken from known protein structures fall into the expected regions, with high concentrations near the α helix and β conformation values as predicted (Fig. 4–9b). The only amino acid residue often found in a conformation outside these regions is glycine. Because its side chain is small, a Gly residue can take part in many conformations that are sterically forbidden for other amino acids.

Common Secondary Structures Can Be Assessed by Circular Dichroism

Any form of structural asymmetry in a molecule gives rise to differences in absorption of left-handed versus right-handed circularly polarized light. Measurement of this difference is called **circular dichroism (CD)**

(a)

(b)

FIGURE 4–9 Ramachandran plots showing a variety of structures.
(a) The values of φ and ψ for various allowed secondary structures are overlaid on the plot from Figure 4-3. Although left-handed α helices extending over several amino acid residues are theoretically possible, they have not been observed in proteins. **(b)** The values of φ and ψ for all the amino acid residues except Gly in the enzyme pyruvate kinase (isolated from rabbit) are overlaid on the plot of theoretically allowed conformations (Fig. 4-3). The small, flexible Gly residues were excluded because they frequently fall outside the expected (blue) ranges.

spectroscopy. An ordered structure, such as a folded protein, gives rise to an absorption spectrum that can have peaks or regions with both positive and negative values. For proteins, spectra are obtained in the far UV region (190 to 250 nm). The light-absorbing entity, or chromophore, in this region is the peptide bond; a signal is obtained when the peptide bond is in a folded environment. The difference in molar extinction coefficients (see Box 3–1) for left- and right-handed, circularly polarized light ($\Delta\varepsilon$) is plotted as a function of wavelength. The α helix and β conformations have characteristic CD spectra **(Fig. 4–10)**. Using CD spectra, biochemists can determine whether proteins are properly folded, estimate the fraction of the protein that is folded in either of the common secondary structures, and monitor transitions between the folded and unfolded states.

SUMMARY 4.2 Protein Secondary Structure

▶ Secondary structure is the local spatial arrangement of the main-chain atoms in a selected segment of a polypeptide chain.

▶ The most common regular secondary structures are the α helix, the β conformation, and β turns.

▶ The secondary structure of a polypeptide segment can be completely defined if the ϕ and ψ angles are known for all amino acid residues in that segment.

▶ Circular dichroism spectroscopy is a method for assessing common secondary structure and monitoring folding in proteins.

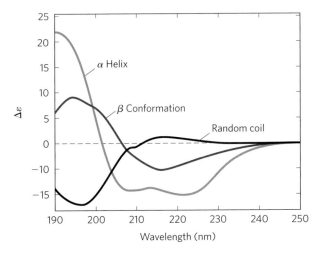

FIGURE 4–10 Circular dichroism spectroscopy. These spectra show polylysine entirely as α helix, as β conformation, or as a denatured, random coil. The y axis unit is a simplified version of the units most commonly used in CD experiments. Since the curves are different for α helix, β conformation, and random coil, the CD spectrum for a given protein can provide a rough estimate for the fraction of the protein made up of the two most common secondary structures. The CD spectrum of the native protein can serve as a benchmark for the folded state, useful for monitoring denaturation or conformational changes brought about by changes in solution conditions.

4.3 Protein Tertiary and Quaternary Structures

🖱 **Protein Architecture–Introduction to Tertiary Structure** The overall three-dimensional arrangement of all atoms in a protein is referred to as the protein's **tertiary structure**. Whereas the term "secondary structure" refers to the spatial arrangement of amino acid residues that are adjacent in a segment of a polypeptide, tertiary structure includes *longer-range* aspects of amino acid sequence. Amino acids that are far apart in the polypeptide sequence and are in different types of secondary structure may interact within the completely folded structure of a protein. The location of bends (including β turns) in the polypeptide chain and the direction and angle of these bends are determined by the number and location of specific bend-producing residues, such as Pro, Thr, Ser, and Gly. Interacting segments of polypeptide chains are held in their characteristic tertiary positions by several kinds of weak interactions (and sometimes by covalent bonds such as disulfide cross-links) between the segments.

Some proteins contain two or more separate polypeptide chains, or subunits, which may be identical or different. The arrangement of these protein subunits in three-dimensional complexes constitutes **quaternary structure**.

In considering these higher levels of structure, it is useful to designate two major groups into which many proteins can be classified: **fibrous proteins**, with polypeptide chains arranged in long strands or sheets, and **globular proteins**, with polypeptide chains folded into a spherical or globular shape. The two groups are structurally distinct. Fibrous proteins usually consist largely of a single type of secondary structure, and their tertiary structure is relatively simple. Globular proteins often contain several types of secondary structure. The two groups also differ functionally: the structures that provide support, shape, and external protection to vertebrates are made of fibrous proteins, whereas most enzymes and regulatory proteins are globular proteins.

Fibrous Proteins Are Adapted for a Structural Function

🖱 **Protein Architecture–Tertiary Structure of Fibrous Proteins** α-Keratin, collagen, and silk fibroin nicely illustrate the relationship between protein structure and biological function (Table 4–3). Fibrous proteins share properties that give strength and/or flexibility to the structures in which they occur. In each case, the fundamental structural unit is a simple repeating element of secondary structure. All fibrous proteins are insoluble in water, a property conferred by a high concentration of hydrophobic amino acid residues both in the interior of the protein and on its surface. These hydrophobic surfaces are largely buried as many similar polypeptide chains are packed together to form elaborate supramolecular complexes. The underlying structural simplicity of fibrous proteins makes them particularly useful for illustrating some of the fundamental principles of protein structure discussed above.

α-Keratin The α-keratins have evolved for strength. Found only in mammals, these proteins constitute almost the entire dry weight of hair, wool, nails, claws, quills, horns, hooves, and much of the outer layer of skin. The α-keratins are part of a broader family of proteins called intermediate filament (IF) proteins. Other IF proteins are found in the cytoskeletons of animal cells. All IF proteins have a structural function and share the structural features exemplified by the α-keratins.

The α-keratin helix is a right-handed α helix, the same helix found in many other proteins. Francis Crick and Linus Pauling in the early 1950s independently suggested that the α helices of keratin were arranged as a coiled coil. Two strands of α-keratin, oriented in parallel (with their amino termini at the same end), are wrapped about each other to form a supertwisted coiled coil. The supertwisting amplifies the strength of the overall structure, just as strands are twisted to make a strong rope **(Fig. 4–11)**. The twisting of the axis of an α helix to form a coiled coil explains the discrepancy between the 5.4 Å per turn predicted for an α helix by Pauling and Corey and the 5.15 to 5.2 Å repeating structure observed in the x-ray diffraction of hair (p. 120). The helical path of the supertwists is left-handed, opposite in sense to the α helix. The surfaces where the two α helices touch are made up of hydrophobic amino acid residues, their R groups meshed together in a regular interlocking pattern. This permits a close packing of the polypeptide chains within the left-handed supertwist. Not surprisingly, α-keratin is rich in the hydrophobic residues Ala, Val, Leu, Ile, Met, and Phe.

An individual polypeptide in the α-keratin coiled coil has a relatively simple tertiary structure, dominated by an α-helical secondary structure with its helical axis twisted in a left-handed superhelix. The intertwining of the two α-helical polypeptides is an example of quaternary structure. Coiled coils of this type are common structural elements in filamentous proteins and in the muscle protein myosin (see Fig. 5–27). The quaternary structure of α-keratin can be quite complex. Many coiled coils can be assembled into large supramolecular complexes, such as the arrangement of α-keratin to form the intermediate filament of hair (Fig. 4–11b).

The strength of fibrous proteins is enhanced by covalent cross-links between polypeptide chains in the multihelical "ropes" and between adjacent chains in a supramolecular assembly. In α-keratins, the cross-links stabilizing quaternary structure are disulfide bonds (Box 4–2). In the hardest and toughest α-keratins, such

(a)

(b) Cross section of a hair

FIGURE 4–11 Structure of hair. (a) Hair α-keratin is an elongated α helix with somewhat thicker elements near the amino and carboxyl termini. Pairs of these helices are interwound in a left-handed sense to form two-chain coiled coils. These then combine in higher-order structures called protofilaments and protofibrils. About four protofibrils—32 strands of α-keratin in all—combine to form an intermediate filament. The individual two-chain coiled coils in the various substructures also seem to be interwound, but the handedness of the interwinding and other structural details are unknown. **(b)** A hair is an array of many α-keratin filaments, made up of the substructures shown in (a).

TABLE 4–3	Secondary Structures and Properties of Some Fibrous Proteins	
Structure	**Characteristics**	**Examples of occurrence**
α Helix, cross-linked by disulfide bonds	Tough, insoluble protective structures of varying hardness and flexibility	α-Keratin of hair, feathers, nails
β Conformation	Soft, flexible filaments	Silk fibroin
Collagen triple helix	High tensile strength, without stretch	Collagen of tendons, bone matrix

BOX 4-2   Permanent Waving Is Biochemical Engineering

When hair is exposed to moist heat, it can be stretched. At the molecular level, the α helices in the α-keratin of hair are stretched out until they arrive at the fully extended β conformation. On cooling they spontaneously revert to the α-helical conformation. The characteristic "stretchability" of α-keratins, and their numerous disulfide cross-linkages, are the basis of permanent waving. The hair to be waved or curled is first bent around a form of appropriate shape. A solution of a reducing agent, usually a compound containing a thiol or sulfhydryl group (—SH), is then applied with heat. The reducing agent cleaves the cross-linkages by reducing each disulfide bond to form two Cys residues. The moist heat breaks hydrogen bonds and causes the α-helical structure of the polypeptide chains to uncoil. After a time the reducing solution is removed, and an oxidizing agent is added to establish *new* disulfide bonds between pairs of Cys residues of adjacent polypeptide chains, but not the same pairs as before the treatment. After the hair is washed and cooled, the polypeptide chains

revert to their α-helical conformation. The hair fibers now curl in the desired fashion because the new disulfide cross-linkages exert some torsion or twist on the bundles of α-helical coils in the hair fibers. The same process can be used to straighten hair that is naturally curly. A permanent wave (or hair straightening) is not truly permanent, because the hair grows; in the new hair replacing the old, the α-keratin has the natural pattern of disulfide bonds.

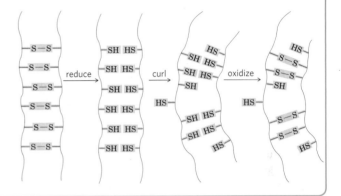

---

as those of rhinoceros horn, up to 18% of the residues are cysteines involved in disulfide bonds.

Collagen Like the α-keratins, collagen has evolved to provide strength. It is found in connective tissue such as tendons, cartilage, the organic matrix of bone, and the cornea of the eye. The collagen helix is a unique secondary structure, quite distinct from the α helix. It is left-handed and has three amino acid residues per turn (**Fig. 4–12** and Table 4–1). Collagen is also a coiled coil, but one with distinct tertiary and quaternary structures: three separate polypeptides, called α chains (not to be confused with α helices), are supertwisted about each other (Fig. 4–12c). The superhelical twisting is right-handed in collagen, opposite in sense to the left-handed helix of the α chains.

There are many types of vertebrate collagen. Typically they contain about 35% Gly, 11% Ala, and 21% Pro and 4-Hyp (4-hydroxyproline, an uncommon amino acid; see Fig. 3–8a). The food product gelatin is derived from collagen. It has little nutritional value as a protein, because collagen is extremely low in many amino acids that are essential in the human diet. The unusual amino acid content of collagen is related to structural constraints unique to the collagen helix. The amino acid sequence in collagen is generally a repeating tripeptide unit, Gly–X–Y, where X is often Pro, and Y is often 4-Hyp. Only Gly residues can be accommodated at the very tight junctions between the individual α chains (Fig. 4–12d). The Pro and 4-Hyp residues permit the sharp twisting of the collagen helix. The amino acid sequence and the supertwisted quaternary structure of

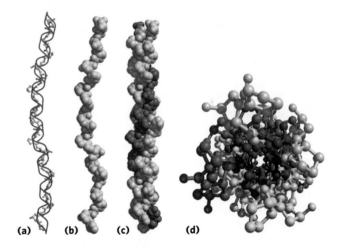

FIGURE 4–12   Structure of collagen. (Derived from PDB ID 1CGD) **(a)** The α chain of collagen has a repeating secondary structure unique to this protein. The repeating tripeptide sequence Gly–X–Pro or Gly–X–4-Hyp adopts a left-handed helical structure with three residues per turn. The repeating sequence used to generate this model is Gly–Pro–4-Hyp. **(b)** Space-filling model of the same α chain. **(c)** Three of these helices (shown here in gray, blue, and purple) wrap around one another with a right-handed twist. **(d)** The three-stranded collagen superhelix shown from one end, in a ball-and-stick representation. Gly residues are shown in red. Glycine, because of its small size, is required at the tight junction where the three chains are in contact. The balls in this illustration do not represent the van der Waals radii of the individual atoms. The center of the three-stranded superhelix is not hollow, as it appears here, but very tightly packed.

collagen allow a very close packing of its three polypeptides. 4-Hydroxyproline has a special role in the structure of collagen—and in human history (Box 4–3).

BOX 4-3 🜚 MEDICINE Why Sailors, Explorers, and College Students Should Eat Their Fresh Fruits and Vegetables

. . . from this misfortune, together with the unhealthiness of the country, where there never falls a drop of rain, we were stricken with the "camp-sickness," which was such that the flesh of our limbs all shrivelled up, and the skin of our legs became all blotched with black, mouldy patches, like an old jack-boot, and proud flesh came upon the gums of those of us who had the sickness, and none escaped from this sickness save through the jaws of death. The signal was this: when the nose began to bleed, then death was at hand . . .

—The Memoirs of the Lord of Joinville, *ca. 1300*

This excerpt describes the plight of Louis IX's army toward the end of the Seventh Crusade (1248–1254), when the scurvy-weakened Crusader army was destroyed by the Egyptians. What was the nature of the malady afflicting these thirteenth-century soldiers?

Scurvy is caused by lack of vitamin C, or ascorbic acid (ascorbate). Vitamin C is required for, among other things, the hydroxylation of proline and lysine in collagen; scurvy is a deficiency disease characterized by general degeneration of connective tissue. Manifestations of advanced scurvy include numerous small hemorrhages caused by fragile blood vessels, tooth loss, poor wound healing and the reopening of old wounds, bone pain and degeneration, and eventually heart failure. Milder cases of vitamin C deficiency are accompanied by fatigue, irritability, and an increased severity of respiratory tract infections. Most animals make large amounts of vitamin C, converting glucose to ascorbate in four enzymatic steps. But in the course of evolution, humans and some other animals—gorillas, guinea pigs, and fruit bats—have lost the last enzyme in this pathway and must obtain ascorbate in their diet. Vitamin C is available in a wide range of fruits and vegetables. Until 1800, however, it was often absent in the dried foods and other food supplies stored for winter or for extended travel.

Scurvy was recorded by the Egyptians in 1500 BCE, and it is described in the fifth century BCE writings of Hippocrates. Yet it did not come to wide public notice until the European voyages of discovery from 1500 to 1800. The first circumnavigation of the globe (1519–1522), led by Ferdinand Magellan, was accomplished only with the loss of more than 80% of his crew to scurvy. During Jacques Cartier's second voyage to explore the St. Lawrence River (1535–1536), his band was threatened with complete disaster until the native Americans taught the men to make a cedar tea that cured and prevented scurvy (it contained vitamin C). Winter outbreaks of scurvy in Europe were gradually eliminated in the nineteenth century as the cultivation of the potato, introduced from South America, became widespread.

In 1747, James Lind, a Scottish surgeon in the Royal Navy, carried out the first controlled clinical study in recorded history. During an extended voyage on the 50-gun warship HMS *Salisbury*, Lind selected 12 sailors suffering from scurvy and separated them into groups of two. All 12 received the same diet, except that each group was given a different remedy for scurvy from among those recommended at the time. The sailors given lemons and oranges recovered and returned to duty. The sailors given boiled apple juice improved slightly. The remainder continued to deteriorate. Lind's *Treatise on the Scurvy* was published in 1753, but inaction persisted in the Royal Navy for another 40 years. In 1795 the British admiralty finally mandated a ration of concentrated lime or lemon juice for all British sailors (hence the name "limeys"). Scurvy continued to be a problem in some other parts of the world until 1932, when Hungarian scientist

James Lind, 1716–1794; naval surgeon, 1739–1748

Albert Szent-Györgyi, and W. A. Waugh and C. G. King at the University of Pittsburgh, isolated and synthesized ascorbic acid.

L-Ascorbic acid (vitamin C) is a white, odorless, crystalline powder. It is freely soluble in water and relatively insoluble in organic solvents. In a dry state, away from light, it is stable for a considerable length of time. The appropriate daily intake of this vitamin is still in dispute. The recommended value in the United States is 90 mg (Australia and the United Kingdom recommend 60 mg; Russia recommends 125 mg). Along with citrus fruits and almost all other fresh fruits, good sources of vitamin C include peppers, tomatoes, potatoes, and broccoli. The vitamin C of fruits and vegetables is destroyed by overcooking or prolonged storage.

So why is ascorbate so necessary to good health? Of particular interest to us here is its role in the formation of collagen. As noted in the text, collagen is constructed of the repeating tripeptide unit Gly–X–Y, where X and Y are generally Pro or 4-Hyp—the proline derivative $(4R)$-L-hydroxyproline, which plays an essential role in the folding of collagen and in maintaining its structure. The proline ring is normally found as a mixture of two puckered conformations, called C_γ-endo and C_γ-exo (Fig. 1). The collagen helix structure requires the Pro residue in the Y positions to

FIGURE 1 The C_γ-endo conformation of proline and the C_γ-exo conformation of 4-hydroxyproline.

be in the C_γ-exo conformation, and it is this conformation that is enforced by the hydroxyl substitution at C-4 in 4-Hyp. The collagen structure also requires that the Pro residue in the X positions have the C_γ-endo conformation, and introduction of 4-Hyp here can destabilize the helix. In the absence of vitamin C, cells cannot hydroxylate the Pro at the Y positions. This leads to collagen instability and the connective tissue problems seen in scurvy.

The hydroxylation of specific Pro residues in procollagen, the precursor of collagen, requires the action of the enzyme prolyl 4-hydroxylase. This enzyme (M_r 240,000) is an $\alpha_2\beta_2$ tetramer in all vertebrates. The proline-hydroxylating activity is found in the α subunits. Each α subunit contains one atom of nonheme iron (Fe^{2+}), and the enzyme is one of a class of hydroxylases that require α-ketoglutarate in their reactions.

In the normal prolyl 4-hydroxylase reaction (Fig. 2a), one molecule of α-ketoglutarate and one of O_2 bind to the enzyme. The α-ketoglutarate is oxidatively decarboxylated to form CO_2 and succinate. The remaining oxygen atom is then used to hydroxylate an appropriate Pro residue in procollagen. No ascorbate is needed in this reaction. However, prolyl 4-hydroxylase also catalyzes an oxidative decarboxylation of α-ketoglutarate that is not coupled to proline hydroxylation (Fig. 2b). During this reaction the heme Fe^{2+} becomes oxidized, inactivating the enzyme and preventing the proline hydroxylation. The ascorbate consumed in the reaction is needed to restore enzyme activity—by reducing the heme iron.

Scurvy remains a problem today, not only in remote regions where nutritious food is scarce but, surprisingly, on U.S. college campuses. The only vegetables consumed by some students are those in tossed salads, and days go by without these young adults consuming fruit. A 1998 study of 230 students at Arizona State University revealed that 10% had serious vitamin C deficiencies, and 2 students had vitamin C levels so low that they probably had scurvy. Only half the students in the study consumed the recommended daily allowance of vitamin C.

Eat your fresh fruits and vegetables.

FIGURE 2 Reactions catalyzed by prolyl 4-hydroxylase. **(a)** The normal reaction, coupled to proline hydroxylation, which does not require ascorbate. The fate of the two oxygen atoms from O_2 is shown in red. **(b)** The uncoupled reaction, in which α-ketoglutarate is oxidatively decarboxylated without hydroxylation of proline. Ascorbate is consumed stoichiometrically in this process as it is converted to dehydroascorbate, preventing Fe^{2+} oxidation.

The tight wrapping of the α chains in the collagen triple helix provides tensile strength greater than that of a steel wire of equal cross section. Collagen fibrils **(Fig. 4–13)** are supramolecular assemblies consisting of triple-helical collagen molecules (sometimes referred to as tropocollagen molecules) associated in a variety of ways to provide different degrees of tensile strength. The α chains of collagen molecules and the collagen molecules of fibrils are cross-linked by unusual types of covalent bonds involving Lys, HyLys (5-hydroxylysine; see Fig. 3–8a), or His residues that are present at a few of the X and Y positions. These links create uncommon amino acid residues such as dehydrohydroxylysinonorleucine. The increasingly rigid and brittle character of aging connective tissue results from accumulated covalent cross-links in collagen fibrils.

H—N
CH—CH$_2$—CH$_2$—CH$_2$—CH=N—CH$_2$—CH—CH$_2$—CH$_2$—CH
O=C   OH   C=O
N—H

| Polypeptide chain | Lys residue minus ε-amino group (norleucine) | HyLys residue | Polypeptide chain |

Dehydrohydroxylysinonorleucine

A typical mammal has more than 30 structural variants of collagen, particular to certain tissues and each somewhat different in sequence and function. Some human genetic defects in collagen structure illustrate the close relationship between amino acid sequence and three-dimensional structure in this protein.

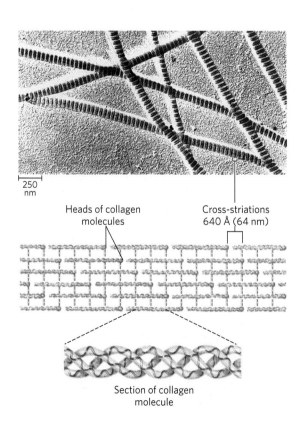

250 nm

Heads of collagen molecules

Cross-striations 640 Å (64 nm)

Section of collagen molecule

Osteogenesis imperfecta is characterized by abnormal bone formation in babies; at least eight variants of this condition, with different degrees of severity, occur in the human population. Ehlers-Danlos syndrome is characterized by loose joints, and at least six variants occur in humans. The composer Niccolò Paganini (1782–1840) was famed for his seemingly impossible dexterity in playing the violin. He suffered from a variant of Ehlers-Danlos syndrome that rendered him effectively double-jointed. In both disorders, some variants can be lethal, whereas others cause lifelong problems.

All of the variants of both conditions result from the substitution of an amino acid residue with a larger R group (such as Cys or Ser) for a single Gly residue in an α chain in one or another collagen protein (a different Gly residue in each disorder). These single-residue substitutions have a catastrophic effect on collagen function because they disrupt the Gly–X–Y repeat that gives collagen its unique helical structure. Given its role in the collagen triple helix (Fig. 4–12d), Gly cannot be replaced by another amino acid residue without substantial deleterious effects on collagen structure. ∎

Silk Fibroin Fibroin, the protein of silk, is produced by insects and spiders. Its polypeptide chains are predominantly in the β conformation. Fibroin is rich in Ala and Gly residues, permitting a close packing of β sheets and an interlocking arrangement of R groups **(Fig. 4–14)**. The overall structure is stabilized by extensive hydrogen bonding between all peptide linkages in the polypeptides of each β sheet and by the optimization of van der Waals interactions between sheets. Silk does not stretch, because the β conformation is already highly extended (Fig. 4–6). However, the structure is flexible, because the sheets are held together by numerous weak interactions rather than by covalent bonds such as the disulfide bonds in α-keratins.

Structural Diversity Reflects Functional Diversity in Globular Proteins

In a globular protein, different segments of the polypeptide chain (or multiple polypeptide chains) fold back on each other, generating a more compact shape than is seen in the fibrous proteins **(Fig. 4–15)**. The folding also provides the structural diversity necessary for proteins to carry out a wide array of biological functions.

FIGURE 4–13 Structure of collagen fibrils. Collagen (M_r 300,000) is a rod-shaped molecule, about 3,000 Å long and only 15 Å thick. Its three helically intertwined α chains may have different sequences; each chain has about 1,000 amino acid residues. Collagen fibrils are made up of collagen molecules aligned in a staggered fashion and cross-linked for strength. The specific alignment and degree of cross-linking vary with the tissue and produce characteristic cross-striations in an electron micrograph. In the example shown here, alignment of the head groups of every fourth molecule produces striations 640 Å (64 nm) apart.

Antiparallel β sheet

Ala side chains

Gly side chains

(a)

(b)

70 μm

FIGURE 4–14 Structure of silk. The fibers in silk cloth and in a spiderweb are made up primarily of the protein fibroin. **(a)** Fibroin consists of layers of antiparallel β sheets rich in Ala and Gly residues. The small side chains interdigitate and allow close packing of the sheets, as shown in the ball-and-stick view. The segments shown would be just a small part of the fibroin strand. **(b)** Strands of silk (blue) emerge from the spinnerets of a spider in this colorized scanning electron micrograph.

Globular proteins include enzymes, transport proteins, motor proteins, regulatory proteins, immunoglobulins, and proteins with many other functions.

Our discussion of globular proteins begins with the principles gleaned from the first protein structures to be elucidated. This is followed by a detailed description of protein substructure and comparative categorization. Such discussions are possible only because of the vast amount of information available over the Internet from

β Conformation
2,000 × 5 Å

α Helix
900 × 11 Å

Native globular form
100 × 60 Å

FIGURE 4–15 Globular protein structures are compact and varied. Human serum albumin (M_r 64,500) has 585 residues in a single chain. Given here are the approximate dimensions its single polypeptide chain would have if it occurred entirely in extended β conformation or as an α helix. Also shown is the size of the protein in its native globular form, as determined by x-ray crystallography; the polypeptide chain must be very compactly folded to fit into these dimensions.

publicly accessible databases, particularly the Protein Data Bank (Box 4–4).

Myoglobin Provided Early Clues about the Complexity of Globular Protein Structure

Protein Architecture–Tertiary Structure of Small Globular Proteins, II. Myoglobin The first breakthrough in understanding the three-dimensional structure of a globular protein came from x-ray diffraction studies of myoglobin carried out by John Kendrew and his colleagues in the 1950s. Myoglobin is a relatively small (M_r 16,700), oxygen-binding protein of muscle cells. It functions both to store oxygen and to facilitate oxygen diffusion in rapidly contracting muscle tissue. Myoglobin contains a single polypeptide chain of 153 amino acid residues of known sequence and a single iron protoporphyrin, or heme, group. The same heme group that is found in myoglobin is found in hemoglobin, the oxygen-binding protein of erythrocytes, and is responsible for the deep red-brown color of both myoglobin and hemoglobin. Myoglobin is particularly abundant in the muscles of diving mammals such as the whale, seal, and porpoise—so abundant that the muscles of these animals are brown. Storage and distribution of oxygen by muscle myoglobin permits diving mammals to remain submerged for long periods. The activities of

BOX 4–4   The Protein Data Bank

The number of known three-dimensional protein structures is now in the tens of thousands and more than doubles every couple of years. This wealth of information is revolutionizing our understanding of protein structure, the relation of structure to function, and the evolutionary paths by which proteins arrived at their present state, which can be seen in the family resemblances that come to light as protein databases are sifted and sorted. One of the most important resources available to biochemists is the **Protein Data Bank** (**PDB**; www.pdb.org).

The PDB is an archive of experimentally determined three-dimensional structures of biological macromolecules, containing virtually all of the macromolecular structures (proteins, RNAs, DNAs, etc.) elucidated to date. Each structure is assigned an identifying label (a four-character identifier called the PDB ID). Such labels are provided in the figure legends for every PDB-derived structure illustrated in this text so that students and instructors can explore the same structures on their own. The data files in the PDB describe the spatial coordinates of each atom whose position has been determined (many of the cataloged structures are not complete). Additional data files provide information on how the structure was determined and its accuracy. The atomic coordinates can be converted into an image of the macromolecule by using structure visualization software. Students are encouraged to access the PDB and explore structures, using visualization software linked to the database. Macromolecular structure files can also be downloaded and explored on the desktop, using free software such as Jmol.

---

myoglobin and other globin molecules are investigated in greater detail in Chapter 5.

Figure 4–16 shows several structural representations of myoglobin, illustrating how the polypeptide chain is folded in three dimensions—its tertiary structure. The red group surrounded by protein is heme. The backbone of the myoglobin molecule consists of eight relatively straight segments of α helix interrupted by bends, some of which are β turns. The longest α helix has 23 amino acid residues and the shortest only 7; all helices are right-handed. More than 70% of the residues in myoglobin are in these α-helical regions. X-ray analysis has revealed the precise position of each of the R groups, which fill up nearly all the space within the folded chain that is not occupied by backbone atoms.

Many important conclusions were drawn from the structure of myoglobin. The positioning of amino acid side chains reflects a structure that derives much of its stability from hydrophobic interactions. Most of the hydrophobic R groups are in the interior of the molecule, hidden from exposure to water. All but two of the polar R groups are located on the outer surface of the molecule, and all are hydrated. The myoglobin molecule is so compact that its interior has room for only four molecules of water. This dense hydrophobic core is typical of globular proteins. The fraction of space occupied by atoms in an organic liquid is 0.4 to 0.6. In a globular protein the fraction is about 0.75, comparable to that in a crystal (in a typical crystal the fraction is 0.70 to 0.78, near the theoretical maximum). In this packed environment, weak interactions strengthen and reinforce each other. For example, the nonpolar side chains in the core are so close together that short-range van der Waals interactions make a significant contribution to stabilizing hydrophobic interactions.

(a)   (b)   (c)   (d)

FIGURE 4–16 Tertiary structure of sperm whale myoglobin. (PDB ID 1MBO) Orientation of the protein is similar in (a) through (d); the heme group is shown in red. In addition to illustrating the myoglobin structure, this figure provides examples of several different ways to display protein structure. **(a)** The polypeptide backbone in a ribbon representation of a type introduced by Jane Richardson, which highlights regions of secondary structure. The α-helical regions are evident. **(b)** Surface contour image; this is useful for visualizing pockets in the protein where other molecules might bind. **(c)** Ribbon representation including side chains (yellow) for the hydrophobic residues Leu, Ile, Val, and Phe. **(d)** Space-filling model with all amino acid side chains. Each atom is represented by a sphere encompassing its van der Waals radius. The hydrophobic residues are again shown in yellow; most are buried in the interior of the protein and thus not visible.

Deduction of the structure of myoglobin confirmed some expectations and introduced some new elements of secondary structure. As predicted by Pauling and Corey, all the peptide bonds are in the planar trans configuration. The α helices in myoglobin provided the first direct experimental evidence for the existence of this type of secondary structure. Three of the four Pro residues are found at bends. The fourth Pro residue occurs within an α helix, where it creates a kink necessary for tight helix packing.

The flat heme group rests in a crevice, or pocket, in the myoglobin molecule. The iron atom in the center of the heme group has two bonding (coordination) positions perpendicular to the plane of the heme (**Fig. 4–17**). One of these is bound to the R group of the His residue at position 93; the other is the site at which an O_2 molecule binds. Within this pocket, the accessibility of the heme group to solvent is highly restricted. This is important for function, because free heme groups in an oxygenated solution are rapidly oxidized from the ferrous (Fe^{2+}) form, which is active in the reversible binding of O_2, to the ferric (Fe^{3+}) form, which does not bind O_2.

As many different myoglobin structures were resolved, investigators were able to observe the structural changes that accompany the binding of oxygen or other molecules and thus, for the first time, to understand the correlation between protein structure and function. Hundreds of proteins have now been subjected to similar analysis. Today, nuclear magnetic resonance (NMR) spectroscopy and other techniques supplement x-ray diffraction data, providing more information on a protein's structure (Box 4–5). In addition, the sequencing of the genomic DNA of many organisms (Chapter 9) has identified thousands of genes that encode proteins of known sequence but, as yet, unknown function; this work continues apace.

Globular Proteins Have a Variety of Tertiary Structures

From what we now know about the tertiary structures of hundreds of globular proteins, it is clear that myoglobin illustrates just one of many ways in which a polypeptide chain can fold. Table 4–4 shows the proportions of α helix and β conformations (expressed as percentage of residues in each type) in several small, single-chain, globular proteins. Each of these proteins has a distinct structure, adapted for its particular biological function, but together they share several important properties with myoglobin. Each is folded compactly, and in each case the hydrophobic amino acid side chains are oriented toward the interior (away from water) and the hydrophilic side chains are on the surface. The structures are also stabilized by a multitude of hydrogen bonds and some ionic interactions.

For the beginning student, the very complex tertiary structures of globular proteins—some much larger than myoglobin—are best approached by focusing on common structural patterns, recurring in different and often unrelated proteins. The three-dimensional structure of a typical globular protein can be considered an assemblage of polypeptide segments in the α-helical and β conformations, linked by connecting segments. The structure can then be defined by how these segments stack on one another and how the segments that connect them are arranged.

To understand a complete three-dimensional structure, we need to analyze its folding patterns. We begin by defining two important terms that describe protein structural patterns or elements in a polypeptide chain and then turn to the folding rules.

FIGURE 4–17 The heme group. This group is present in myoglobin, hemoglobin, cytochromes, and many other proteins (the heme proteins). **(a)** Heme consists of a complex organic ring structure, protoporphyrin, which binds an iron atom in its ferrous (Fe^{2+}) state. The iron atom has six coordination bonds, four in the plane of, and bonded to, the flat porphyrin molecule and two perpendicular to it. **(b)** In myoglobin and hemoglobin, one of the perpendicular coordination bonds is bound to a nitrogen atom of a His residue. The other is "open" and serves as the binding site for an O_2 molecule.

TABLE 4–4 Approximate Proportion of α Helix and β Conformation in Some Single-Chain Proteins

Protein (total residues)	Residues (%)*	
	α Helix	β Conformation
Chymotrypsin (247)	14	45
Ribonuclease (124)	26	35
Carboxypeptidase (307)	38	17
Cytochrome c (104)	39	0
Lysozyme (129)	40	12
Myoglobin (153)	78	0

Source: Data from Cantor, C.R. & Schimmel, P.R. (1980) *Biophysical Chemistry*, Part I: *The Conformation of Biological Macromolecules*, p. 100, W. H. Freeman and Company, New York.

*Portions of the polypeptide chains not accounted for by α helix or β conformation consist of bends and irregularly coiled or extended stretches. Segments of α helix and β conformation sometimes deviate slightly from their normal dimensions and geometry.

BOX 4–5 | METHODS | Methods for Determining the Three-Dimensional Structure of a Protein

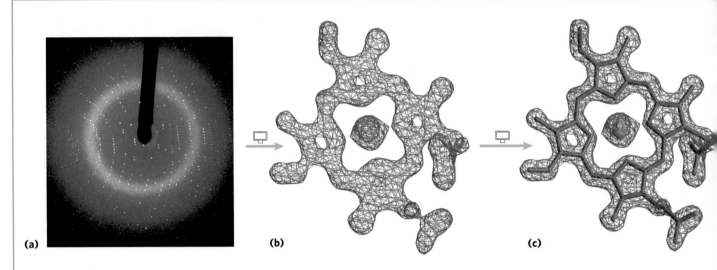

(a)   (b)   (c)

X-Ray Diffraction

The spacing of atoms in a crystal lattice can be determined by measuring the locations and intensities of spots produced on photographic film by a beam of x rays of given wavelength, after the beam has been diffracted by the electrons of the atoms. For example, x-ray analysis of sodium chloride crystals shows that Na^+ and Cl^- ions are arranged in a simple cubic lattice. The spacing of the different kinds of atoms in complex organic molecules, even very large ones such as proteins, can also be analyzed by x-ray diffraction methods. However, the technique for analyzing crystals of complex molecules is far more laborious than for simple salt crystals. When the repeating pattern of the crystal is a molecule as large as, say, a protein, the numerous atoms in the molecule yield thousands of diffraction spots that must be analyzed by computer.

Consider how images are generated in a light microscope. Light from a point source is focused on an object. The object scatters the light waves, and these scattered waves are recombined by a series of lenses to generate an enlarged image of the object. The smallest object whose structure can be determined by such a system—that is, the resolving power of the microscope—is determined by the wavelength of the light, in this case visible light, with wavelengths in the range of 400 to 700 nm. Objects smaller than half the wavelength of the incident light cannot be resolved. To resolve objects as small as proteins we must use x rays, with wavelengths in the range of 0.7 to 1.5 Å (0.07 to 0.15 nm). However, there are no lenses that can recombine x rays to form an image; instead, the pattern of diffracted x rays is collected directly and an image is reconstructed by mathematical techniques.

The amount of information obtained from x-ray crystallography depends on the degree of structural order in the sample. Some important structural parameters were obtained from early studies of the diffraction patterns of the fibrous proteins arranged in regular arrays in hair and wool. However, the orderly bundles formed by fibrous proteins are not crystals—the molecules are aligned side by side, but not all are oriented in the same direction. More detailed three-dimensional structural information about proteins requires a highly ordered protein crystal. The structures of many proteins are not yet known, simply because they have proved difficult to crystallize. Practitioners have compared making protein crystals to holding together a stack of bowling balls with cellophane tape.

Operationally, there are several steps in x-ray structural analysis (Fig. 1). A crystal is placed in an x-ray beam between the x-ray source and a detector, and a regular array of spots called reflections is generated. The spots are created by the diffracted x-ray beam, and each atom in a molecule makes a contribution to each spot. An electron-density map of the protein is reconstructed from the overall diffraction pattern of spots by a mathematical technique called a Fourier transform. In effect, the computer acts as a "computational lens." A model for the structure is then built that is consistent with the electron-density map.

John Kendrew found that the x-ray diffraction pattern of crystalline myoglobin (isolated from muscles of the sperm whale) is very complex, with nearly 25,000 reflections. Computer analysis of these reflections took place in stages. The resolution improved at each stage, until in 1959 the positions of virtually all the non-hydrogen atoms in the protein had been determined. The amino acid sequence of the protein, obtained by chemical analysis, was consistent with the molecular model. The structures of thousands of proteins, many of them much more complex than

(d)

FIGURE 1 Steps in determining the structure of sperm whale myoglobin by x-ray crystallography. **(a)** X-ray diffraction patterns are generated from a crystal of the protein. **(b)** Data extracted from the diffraction patterns are used to calculate a three-dimensional electron-density map. The electron density of only part of the structure, the heme, is shown here. **(c)** Regions of greatest electron density reveal the location of atomic nuclei, and this information is used to piece together the final structure. Here, the heme structure is modeled into its electron-density map. **(d)** The completed structure of sperm whale myoglobin, including the heme (PDB ID 2MBW).

myoglobin, have since been determined to a similar level of resolution.

The physical environment in a crystal, of course, is not identical to that in solution or in a living cell. A crystal imposes a space and time average on the structure deduced from its analysis, and x-ray diffraction studies provide little information about molecular motion within the protein. The conformation of proteins in a crystal could in principle also be affected by nonphysiological factors such as incidental protein-protein contacts within the crystal. However, when structures derived from the analysis of crystals are compared with structural information obtained by other means (such as NMR, as described below), the crystal-derived structure almost always represents a functional conformation of the protein. X-ray crystallography can be applied successfully to proteins too large to be structurally analyzed by NMR.

Nuclear Magnetic Resonance

An advantage of nuclear magnetic resonance (NMR) studies is that they are carried out on macromolecules in solution, whereas x-ray crystallography is limited to molecules that can be crystallized. NMR can also illuminate the dynamic side of protein structure, including conformational changes, protein folding, and interactions with other molecules.

NMR is a manifestation of nuclear spin angular momentum, a quantum mechanical property of atomic nuclei. Only certain atoms, including ^1H, ^{13}C, ^{15}N, ^{19}F, and ^{31}P, have the kind of nuclear spin that gives rise to an NMR signal. Nuclear spin generates a magnetic dipole. When a strong, static magnetic field is applied to a solution containing a single type of macromolecule, the magnetic dipoles are aligned in the field in one of two orientations, parallel (low energy) or antiparallel

(high energy). A short ($\sim 10\ \mu s$) pulse of electromagnetic energy of suitable frequency (the resonant frequency, which is in the radio frequency range) is applied at right angles to the nuclei aligned in the magnetic field. Some energy is absorbed as nuclei switch to the high-energy state, and the absorption spectrum that results contains information about the identity of the nuclei and their immediate chemical environment. The data from many such experiments on a sample are averaged, increasing the signal-to-noise ratio, and an NMR spectrum such as that in Figure 2 is generated.

^1H is particularly important in NMR experiments because of its high sensitivity and natural abundance. For macromolecules, ^1H NMR spectra can become quite complicated. Even a small protein has hundreds of ^1H atoms, typically resulting in a one-dimensional NMR spectrum too complex for analysis. Structural analysis of proteins became possible with the advent of two-dimensional NMR techniques (Fig. 3). These methods allow measurement of distance-dependent coupling of nuclear spins in nearby atoms through space (the nuclear Overhauser effect (NOE), in a method dubbed NOESY) or the coupling of nuclear spins in atoms connected by covalent bonds (total correlation spectroscopy, or TOCSY).

(Continued on next page)

FIGURE 2 One-dimensional NMR spectrum of a globin from a marine blood worm. This protein and sperm whale myoglobin are very close structural analogs, belonging to the same protein structural family and sharing an oxygen-transport function.

BOX 4–5 | **METHODS** | Methods for Determining the Three-Dimensional Structure of a Protein (*Continued*)

Translating a two-dimensional NMR spectrum into a complete three-dimensional structure can be a laborious process. The NOE signals provide some information about the distances between individual atoms, but for these distance constraints to be useful, the atoms giving rise to each signal must be identified. Complementary TOCSY experiments can help identify which NOE signals reflect atoms that are linked by covalent bonds. Certain patterns of NOE signals have been associated with secondary structures such as α helices. Genetic engineering (Chapter 9) can be used to prepare proteins that contain the rare isotopes ^{13}C or ^{15}N. The new NMR signals produced by these atoms, and the coupling with ^{1}H signals resulting from these substitutions, help in the assignment of individual ^{1}H NOE signals. The process is also aided by a knowledge of the amino acid sequence of the polypeptide.

To generate a three-dimensional structure, researchers feed the distance constraints into a computer along with known geometric constraints such as chirality, van der Waals radii, and bond lengths and angles. The computer generates a family of closely related structures that represent the range of conformations consistent with the NOE distance constraints (Fig. 3c). The uncertainty in structures generated by NMR is in part a reflection of the molecular vibrations (known as breathing) within a protein structure in solution, discussed in more detail in Chapter 5. Normal experimental uncertainty can also play a role.

Protein structures determined by both x-ray crystallography and NMR generally agree well. In some cases, the precise locations of particular amino acid side chains on the protein exterior are different, often because of effects related to the packing of adjacent protein molecules in a crystal. The two techniques together are at the heart of the rapid increase in the availability of structural information about the macromolecules of living cells.

FIGURE 3 Use of two-dimensional NMR to generate a three-dimensional structure of a globin, the same protein used to generate the data in Figure 2. The diagonal in a two-dimensional NMR spectrum is equivalent to a one-dimensional spectrum. The off-diagonal peaks are NOE signals generated by close-range interactions of ^{1}H atoms that may generate signals quite distant in the one-dimensional spectrum. Two such interactions are identified in **(a)**, and their identities are shown with blue lines in **(b)** (PDB ID 1VRF). Three lines are drawn for interaction 2 between a methyl group in the protein and a hydrogen on the heme. The methyl group rotates rapidly such that each of its three hydrogens contributes equally to the interaction and the NMR signal. Such information is used to determine the complete three-dimensional structure (PDB ID 1VRE), as in **(c)**. The multiple lines shown for the protein backbone in (c) represent the family of structures consistent with the distance constraints in the NMR data. The structural similarity with myoglobin (Fig. 1) is evident. The proteins are oriented in the same way in both figures.

(a)

(b)

(c)

The first term is **motif**, also called a **fold** or (more rarely) **supersecondary structure**. *A motif or fold is a recognizable folding pattern involving two or more elements of secondary structure and the connection(s) between them.* A motif can be very simple, such as two elements of secondary structure folded against each other, and represent only a small part of a protein. An example is a **β-α-β loop** (**Fig. 4–18a**). A motif can also be a very elaborate structure involving scores of protein segments folded together, such as the β barrel (Fig. 4–18b). In some cases, a single large motif may comprise the entire protein. The terms "motif" and "fold" are often used interchangeably, although "fold" is applied more commonly to somewhat more complex folding patterns. The terms encompass any advantageous folding pattern and are useful for describing such patterns. The segment defined as a motif or fold may or may not be independently stable. We have already encountered a well-studied motif, the coiled coil of α-keratin, which is also found in some other proteins. The distinctive arrangement of eight α helices in myoglobin is replicated in all globins and is called the globin fold. Note that a motif is not a hierarchical structural element falling between secondary and tertiary structure. It is simply a folding pattern. The synonymous term "supersecondary structure" is thus somewhat misleading because it suggests hierarchy.

The second term for describing structural patterns is **domain**. A domain, as defined by Jane Richardson in 1981, is a part of a polypeptide chain that is independently stable or could undergo movements as a single entity with respect to the entire protein. Polypeptides with more than a few hundred amino acid residues often fold into two or more domains, sometimes with different functions. In many cases, a domain from a large protein will retain its native three-dimensional structure even when separated (for example, by proteolytic cleavage) from the remainder of the polypeptide chain. In a protein with multiple domains, each domain may appear as

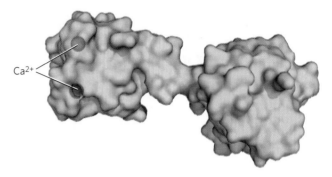

Ca²⁺

FIGURE 4–19 Structural domains in the polypeptide troponin C. (PDB ID 4TNC) This calcium-binding protein, associated with muscle, has two separate calcium-binding domains, shown here in brown and blue.

a distinct globular lobe (**Fig. 4–19**); more commonly, extensive contacts between domains make individual domains hard to discern. Different domains often have distinct functions, such as the binding of small molecules or interaction with other proteins. Small proteins usually have only one domain (the domain *is* the protein).

Folding of polypeptides is subject to an array of physical and chemical constraints, and several rules have emerged from studies of common protein folding patterns.

1. Hydrophobic interactions make a large contribution to the stability of protein structures. Burial of hydrophobic amino acid R groups so as to exclude water requires at least two layers of secondary structure. Simple motifs, such as the β-α-β loop (Fig. 4–18a), create two such layers.

2. Where they occur together in a protein, α helices and β sheets generally are found in different structural layers. This is because the backbone of a polypeptide segment in the β conformation (Fig. 4–6) cannot readily hydrogen-bond to an α helix that is adjacent to it.

3. Segments adjacent to each other in the amino acid sequence are usually stacked adjacent to each other in the folded structure. Distant segments of a polypeptide may come together in the tertiary structure, but this is not the norm.

4. The β conformation is most stable when the individual segments are twisted slightly in a right-handed sense. This influences both the arrangement of β sheets derived from the twisted segments and the path of the polypeptide connections between them. Two parallel β strands, for example, must be connected by a crossover strand (**Fig. 4–20a**). In principle, this crossover could have a right- or left-handed conformation, but in proteins it is almost always right-handed. Right-handed connections tend to be shorter than left-handed connections and tend to bend through smaller angles, making them easier to

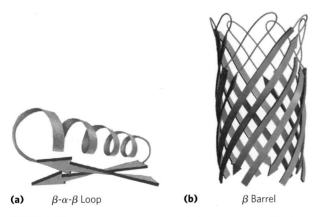

(a)   β-α-β Loop   **(b)**   β Barrel

FIGURE 4–18 Motifs. (a) A simple motif, the β-α-β loop. **(b)** A more elaborate motif, the β barrel. This β barrel is a single domain of α-hemolysin (a toxin that kills a cell by creating a hole in its membrane) from the bacterium *Staphylococcus aureus* (derived from PDB ID 7AHL).

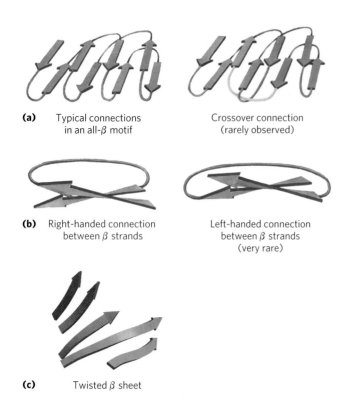

(a) Typical connections in an all-β motif   Crossover connection (rarely observed)

(b) Right-handed connection between β strands   Left-handed connection between β strands (very rare)

(c) Twisted β sheet

FIGURE 4–20 Stable folding patterns in proteins. (a) Connections between β strands in layered β sheets. The strands here are viewed from one end, with no twisting. Thick lines represent connections at the ends nearest the viewer; thin lines are connections at the far ends of the β strands. The connections at a given end (e.g., near the viewer) rarely cross one another. An example of such a rare crossover is illustrated by the yellow strand in the structure on the right. **(b)** Because of the right-handed twist in β strands, connections between strands are generally right-handed. Left-handed connections must traverse sharper angles and are harder to form. **(c)** This twisted β sheet is from a domain of photolyase (a protein that repairs certain types of DNA damage) from *E. coli* (derived from PDB ID 1DNP). Connecting loops have been removed so as to focus on the folding of the β sheet.

form. The twisting of β sheets also leads to a characteristic twisting of the structure formed by many such segments together, as seen in the β barrel (Fig. 4–18b) and twisted β sheet (Fig. 4–20c), which form the core of many larger structures.

Following these rules, complex motifs can be built up from simple ones. For example, a series of β-α-β loops arranged so that the β strands form a barrel creates a particularly stable and common motif, the ***α/β* barrel** (**Fig. 4–21**). In this structure, each parallel β segment is attached to its neighbor by an α-helical segment. All connections are right-handed. The α/β barrel is found in many enzymes, often with a binding site (for a cofactor or substrate) in the form of a pocket near one end of the barrel. Note that domains with similar folding patterns are said to have the same motif even though their constituent α helices and β sheets may differ in length.

β-α-β Loop   *α/β* Barrel

FIGURE 4–21 Constructing large motifs from smaller ones. The *α/β* barrel is a commonly occurring motif constructed from repetitions of the *β-α-β* loop motif. This *α/β* barrel is a domain of pyruvate kinase (a glycolytic enzyme) from rabbit (derived from PDB ID 1PKN).

Protein Motifs Are the Basis for Protein Structural Classification

Protein Architecture–Tertiary Structure of Large Globular Proteins, IV. Structural Classification of Proteins As we have seen, understanding the complexities of tertiary structure is made easier by considering substructures. Taking this idea further, researchers have organized the complete contents of protein databases according to hierarchical levels of structure. All of these databases rely on data and information deposited in the Protein Data Bank. The Structural Classification of Proteins (SCOP) database is a good example of this important trend in biochemistry. At the highest level of classification, the SCOP database (http://scop.mrc-lmb.cam.ac.uk/scop) borrows a scheme already in common use, with four classes of protein structure: **all α**, **all β**, **α/β** (with α and β segments interspersed or alternating), and **α + β** (with α and β regions somewhat segregated). Each class includes tens to hundreds of different folding arrangements (motifs), built up from increasingly identifiable substructures. Some of the substructure arrangements are very common; others have been found in just one protein. **Figure 4–22** shows a variety of motifs arrayed among the four classes of protein structure; this is just a minute sample of the hundreds of known motifs. The number of folding patterns is not infinite, however. As the rate at which new protein structures are elucidated has increased, the fraction of those structures containing a new motif has steadily declined. Fewer than 1,000 different folds or motifs may exist in all. Figure 4–22 also shows how proteins can be organized based on the presence of the various motifs. The top two levels of organization, **class** and **fold**, are purely structural. Below the fold level (see color key in Fig. 4–22), categorization is based on evolutionary relationships.

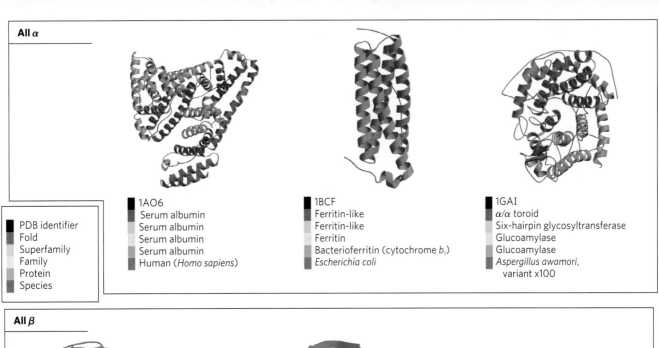

All α

1AO6
Serum albumin
Serum albumin
Serum albumin
Serum albumin
Human (*Homo sapiens*)

1BCF
Ferritin-like
Ferritin-like
Ferritin
Bacterioferritin (cytochrome b_1)
Escherichia coli

1GAI
α/α toroid
Six-hairpin glycosyltransferase
Glucoamylase
Glucoamylase
Aspergillus awamori,
  variant x100

PDB identifier
Fold
Superfamily
Family
Protein
Species

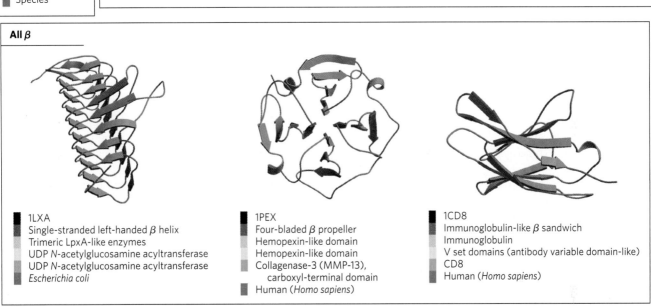

All β

1LXA
Single-stranded left-handed β helix
Trimeric LpxA-like enzymes
UDP *N*-acetylglucosamine acyltransferase
UDP *N*-acetylglucosamine acyltransferase
Escherichia coli

1PEX
Four-bladed β propeller
Hemopexin-like domain
Hemopexin-like domain
Collagenase-3 (MMP-13),
  carboxyl-terminal domain
Human (*Homo sapiens*)

1CD8
Immunoglobulin-like β sandwich
Immunoglobulin
V set domains (antibody variable domain-like)
CD8
Human (*Homo sapiens*)

α/β

1DEH
NAD(P)-binding Rossmann-fold domains
NAD(P)-binding Rossmann-fold domains
Alcohol/glucose dehydrogenases,
  carboxyl-terminal domain
Alcohol dehydrogenase
Human (*Homo sapiens*)

1DUB
ClpP/crotonase
ClpP/crotonase
Crotonase-like
Enoyl-CoA hydratase (crotonase)
Rat (*Rattus norvegicus*)

1PFK
Phosphofructokinase
Phosphofructokinase
Phosphofructokinase
ATP-dependent
  phosphofructokinase
Escherichia coli

FIGURE 4–22 Organization of proteins based on motifs. Shown here are a few of the hundreds of known stable motifs divided into four classes: all α, all β, α/β, and α + β. Structural classification data from the SCOP (Structural Classification of Proteins) database (http://scop.mrc-lmb.cam.ac.uk/scop) are also provided (see the color key). The PDB identifier (listed first for each structure) is the unique accession code given to each structure archived in the Protein Data Bank (www.pdb.org). The α/β, barrel (see Fig. 4–21) is another particularly common α/β motif. **(Continued on next page)**

$\alpha + \beta$

PDB identifier	■ 2PIL	■ 1SYN	■ 1EMA
Fold	Pilin	Thymidylate synthase/dCMP hydroxymethylase	GFP-like
Superfamily	Pilin	Thymidylate synthase/dCMP hydroxymethylase	GFP-like
Family	Pilin	Thymidylate synthase/dCMP hydroxymethylase	Fluorescent proteins
Protein	Pilin	Thymidylate synthase	Green fluorescent protein, GFP
Species	*Neisseria gonorrhoeae*	*Escherichia coli*	Jellyfish (*Aequorea victoria*)

FIGURE 4–22 *(Continued)*

Many examples of recurring domain or motif structures are available, and these reveal that protein tertiary structure is more reliably conserved than amino acid sequence. The comparison of protein structures can thus provide much information about evolution. Proteins with significant similarity in primary structure and/or with similar tertiary structure and function are said to be in the same **protein family**. A strong evolutionary relationship is usually evident within a protein family. For example, the globin family has many different proteins with both structural and sequence similarity to myoglobin (as seen in the proteins used as examples in Box 4–5 and in Chapter 5). Two or more families with little similarity in amino acid sequence sometimes make use of the same major structural motif and have functional similarities; these families are grouped as **superfamilies**. An evolutionary relationship among families in a superfamily is considered probable, even though time and functional distinctions—that is, different adaptive pressures—may have erased many of the telltale sequence relationships. A protein family may be widespread in all three domains of cellular life, the Bacteria, Archaea, and Eukarya, suggesting an ancient origin. Many proteins involved in intermediary metabolism and the metabolism of nucleic acids and proteins fall into this category. Other families may be present in only a small group of organisms, indicating that the structure arose more recently. Tracing the natural history of structural motifs, using structural classifications in databases such as SCOP, provides a powerful complement to sequence analyses in tracing evolutionary relationships. The SCOP database is curated manually, with the objective of placing proteins in the correct evolutionary framework based on conserved structural features.

Structural motifs become especially important in defining protein families and superfamilies. Improved classification and comparison systems for proteins lead inevitably to the elucidation of new functional relationships. Given the central role of proteins in living systems, these structural comparisons can help illuminate every aspect of biochemistry, from the evolution of individual proteins to the evolutionary history of complete metabolic pathways.

Protein Quaternary Structures Range from Simple Dimers to Large Complexes

Protein Architecture–Quaternary Structure Many proteins have multiple polypeptide subunits (from two to hundreds). The association of polypeptide chains can serve a variety of functions. Many multisubunit proteins have regulatory roles; the binding of small molecules may affect the interaction between subunits, causing large changes in the protein's activity in response to small changes in the concentration of substrate or regulatory molecules (Chapter 6). In other cases, separate subunits take on separate but related functions, such as catalysis and regulation. Some associations, such as the fibrous proteins considered earlier in this chapter and the coat proteins of viruses, serve primarily structural roles. Some very large protein assemblies are the site of complex, multistep reactions. For example, each ribosome, the site of protein synthesis, incorporates dozens of protein subunits along with a number of RNA molecules.

A multisubunit protein is also referred to as a **multimer**. A multimer with just a few subunits is often called an **oligomer**. If a multimer has nonidentical subunits, the overall structure of the protein can be asymmetric and quite complicated. However, most multimers have identical subunits or repeating groups of nonidentical subunits, usually in symmetric arrangements. As noted in Chapter 3, the repeating structural unit in such a multimeric protein, whether a single subunit or a group of subunits, is called a

protomer. Greek letters are sometimes used to distinguish the individual subunits that make up a protomer.

The first oligomeric protein to have its three-dimensional structure determined was hemoglobin (M_r 64,500), which contains four polypeptide chains and four heme prosthetic groups, in which the iron atoms are in the ferrous (Fe^{2+}) state (Fig. 4–17). The protein portion, the globin, consists of two α chains (141 residues each) and two β chains (146 residues each). Note that in this case, α and β do not refer to secondary structures. In a practice that can be confusing to the beginning student, the Greek letters α and β (and γ and δ, and others) are often used to distinguish two different kinds of subunits within a multisubunit protein, regardless of what kinds of secondary structure may predominate in the subunits. Because hemoglobin is four times as large as myoglobin, much more time and effort were required to solve its three-dimensional structure by x-ray analysis, finally achieved by Max Perutz, John Kendrew, and their colleagues in 1959. The subunits of hemoglobin are arranged in symmetric pairs **(Fig. 4–23)**, each pair having one α and one β subunit. Hemoglobin can therefore be described either as a tetramer or as a dimer of $\alpha\beta$ protomers. The role these distinct subunits play in hemoglobin function is discussed extensively in Chapter 5.

Max Perutz, 1914–2002 (left), and John Kendrew, 1917–1997

Some Proteins or Protein Segments Are Intrinsically Disordered

In spite of decades of progress in the understanding of protein structure, many proteins cannot be crystallized, making it difficult to determine their three-dimensional structure by methods now considered classical (see Box 4–5). Even where crystallization succeeds, parts of the protein are often sufficiently disordered within the crystal that the determined structure does not include those parts. Sometimes, this is due to subtle features of the structure that render crystallization difficult. However, the reason can be more straightforward: some proteins or protein segments lack an ordered structure in solution.

The concept that some proteins function in the absence of a definable structure is a product of the reassessment of data involving many different proteins. As many as a third of all human proteins may be unstructured or have significant unstructured segments. All organisms have some proteins that fall into this category. **Intrinsically disordered proteins** have properties that are distinct from classical structured proteins. They lack a hydrophobic core, and instead are characterized by high densities of charged amino acid residues such as Lys, Arg, and Glu. Pro residues are also prominent, as they tend to disrupt ordered structures.

Structural disorder and high charge density can facilitate the function of some proteins as spacers, insulators, or linkers in larger structures. Other disordered proteins are scavengers, binding up ions and small molecules in solution and serving as reservoirs or garbage dumps. However, many intrinsically disordered proteins are at the heart of important protein interaction networks. The lack of an ordered structure can facilitate a kind of functional promiscuity, allowing one protein to interact with multiple partners. Some intrinsically disordered proteins act to inhibit the action of other proteins by an unusual mechanism: wrapping around their protein targets. One disordered protein may have several or even dozens of protein partners. The structural disorder allows the inhibitor protein to wrap around the multiple

(a)   **(b)**

FIGURE 4–23 Quaternary structure of deoxyhemoglobin. (PDB ID 2HHB) X-ray diffraction analysis of deoxyhemoglobin (hemoglobin without oxygen molecules bound to the heme groups) shows how the four polypeptide subunits are packed together. **(a)** A ribbon representation reveals the secondary structural elements of the structure and the positioning of all the heme cofactors. **(b)** A surface contour model shows the pockets in which the heme cofactors are bound and helps to visualize subunit packing. The α subunits are shown in shades of gray; the β subunits in shades of blue. Note that the heme groups (red) are relatively far apart.

targets in different ways. The intrinsically disordered protein p27 plays a key role in controlling mammalian cell division. This protein lacks definable structure in solution. It wraps around and thus inhibits the action of several enzymes called protein kinases (see Chapter 6) that facilitate cell division. The flexible structure of p27 allows it to accommodate itself to its different target proteins. Human tumor cells, which are simply cells that have lost the capacity to control cell division normally, generally have reduced levels of p27; the lower the levels of p27, the poorer the prognosis for the cancer patient. Similarly, intrinsically disordered proteins are often present as hubs or scaffolds at the center of protein networks that constitute signaling pathways. These proteins, or parts of them, may interact with many different binding partners. They often take on a structure when they interact with other proteins, but the structure they assume may vary with different binding partners. The mammalian protein p53 is also critical in the control of cell division. It features both structured and unstructured segments, and the different segments interact with dozens of other proteins. An unstructured region of p53 at the carboxyl terminus interacts with at least four different binding partners and assumes a different structure in each of the complexes **(Fig. 4–24)**.

SUMMARY 4.3 Protein Tertiary and Quaternary Structures

▶ Tertiary structure is the complete three-dimensional structure of a polypeptide chain. Many proteins fall into one of two general classes of proteins based on tertiary structure: fibrous and globular.

▶ Fibrous proteins, which serve mainly structural roles, have simple repeating elements of secondary structure.

▶ Globular proteins have more complicated tertiary structures, often containing several types of secondary structure in the same polypeptide chain. The first globular protein structure to be determined, by x-ray diffraction methods, was that of myoglobin.

▶ The complex structures of globular proteins can be analyzed by examining folding patterns called motifs (also called folds or supersecondary structures). The thousands of known protein structures are generally assembled from a repertoire of only a few hundred motifs. Domains are regions of a polypeptide chain that can fold stably and independently.

FIGURE 4–24 Binding of the intrinsically disordered carboxyl terminus of p53 protein to its binding partners. (a) The p53 protein is made up of several different segments (PDB ID 1XQH). Only the central domain is well ordered. **(b)** The linear sequence of the p53 protein is depicted as a colored bar. The overlaid graph presents a plot of the PONDR (Predictor of Natural Disordered Regions) score versus the protein sequence. PONDR is one of the best available algorithms for predicting the likelihood that a given amino acid residue is in a region of intrinsic disorder, based on the surrounding amino acid sequence and amino acid composition. A score of 1.0 indicates a probability of 100% that a protein will be disordered. In the actual protein structure, the tan central domain is ordered. The amino-terminal (blue) and carboxyl-terminal (red) regions are disordered. The very end of the carboxyl-terminal region has multiple binding partners and folds when it binds to each of them; however, the three-dimensional structure that is assumed when binding occurs is different for each of the interactions shown, and thus the color of this carboxyl-terminal segment (11 to 20 residues) is shown in a different color in each complex (cyclin A, PDB ID 1H26; sirtuin, PDB ID 1MA3; CBP bromo domain, PDB ID 1JSP; s100B($\beta\beta$), PDB ID 1DT7).

▶ Quaternary structure results from interactions between the subunits of multisubunit (multimeric) proteins or large protein assemblies. Some multimeric proteins have a repeated unit consisting of a single subunit or a group of subunits, each unit called a protomer.

▶ Some proteins or protein segments are intrinsically disordered, lacking definable structure. These proteins have distinctive amino acid compositions that allow a more flexible structure. Some of these disordered proteins function as structural components or scavengers; others can interact with many different protein partners, serving as versatile inhibitors or as central components of protein interaction networks.

4.4 Protein Denaturation and Folding

Proteins lead a surprisingly precarious existence. As we have seen, a native protein conformation is only marginally stable. In addition, most proteins must maintain conformational flexibility to function. The continual maintenance of the active set of cellular proteins required under a given set of conditions is called **proteostasis**. Cellular proteostasis requires the coordinated function of pathways for protein synthesis and folding, the refolding of proteins that are partially unfolded, and the sequestration and degradation of proteins that have been irreversibly unfolded. In all cells, these networks involve hundreds of enzymes and specialized proteins.

As seen in **Figure 4–25**, the life of a protein encompasses much more than its synthesis and later degradation. The marginal stability of most proteins can produce a tenuous balance between folded and unfolded states. As proteins are synthesized on ribosomes (Chapter 27), they must fold into their native conformations. Sometimes this occurs spontaneously, but more often it occurs with the assistance of specialized enzymes and complexes called chaperones. Many of these same folding helpers function to refold proteins that become transiently unfolded. Proteins that are not properly folded often have exposed hydrophobic surfaces that render them "sticky," leading to the formation of inactive aggregates. These aggregates may lack their normal function but are not inert; their accumulation in cells lies at the heart of diseases ranging from diabetes to Parkinson and Alzheimer diseases. Not surprisingly, all cells have elaborate pathways for recycling and/or degrading proteins that are irreversibly misfolded.

The transitions between the folded and unfolded states, and the network of pathways that control these transitions, now become our focus.

Loss of Protein Structure Results in Loss of Function

Protein structures have evolved to function in particular cellular environments. Conditions different from those in the cell can result in protein structural changes, large

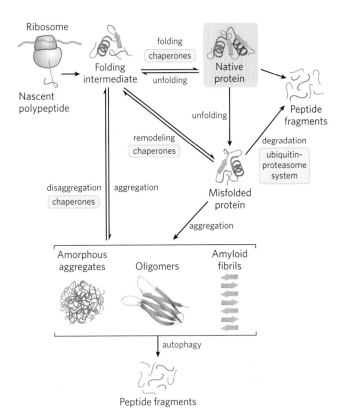

FIGURE 4–25 Pathways that contribute to proteostasis. Three kinds of processes contribute to proteostasis, in some cases with multiple contributing pathways. First, proteins are synthesized on a ribosome. Second, multiple pathways contribute to protein folding, many of which involve the activity of complexes called chaperones. Chaperones (including chaperonins) also contribute to the refolding of proteins that are partially and transiently unfolded. Finally, proteins that are irreversibly unfolded are subject to sequestration and degradation by several additional pathways. Partially unfolded proteins and protein-folding intermediates that escape the quality-control activities of the chaperones and degradative pathways may aggregate, forming both disordered aggregates and ordered amyloid-like aggregates that contribute to disease and aging processes.

and small. A loss of three-dimensional structure sufficient to cause loss of function is called **denaturation**. The denatured state does not necessarily equate with complete unfolding of the protein and randomization of conformation. Under most conditions, denatured proteins exist in a set of partially folded states.

Most proteins can be denatured by heat, which has complex effects on many weak interactions in a protein (primarily on the hydrogen bonds). If the temperature is increased slowly, a protein's conformation generally remains intact until an abrupt loss of structure (and function) occurs over a narrow temperature range **(Fig. 4–26)**. The abruptness of the change suggests that unfolding is a cooperative process: loss of structure in one part of the protein destabilizes other parts. The effects of heat on proteins are not readily predictable. The very heat-stable proteins of thermophilic bacteria and archaea have evolved to function at the temperature of hot springs (~100 °C). Yet the structures of these proteins often differ only slightly from those of

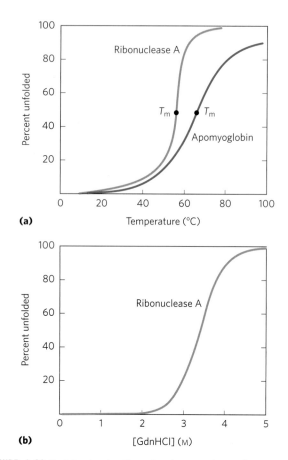

(a)

(b)

FIGURE 4–26 Protein denaturation. Results are shown for proteins denatured by two different environmental changes. In each case, the transition from the folded to the unfolded state is abrupt, suggesting cooperativity in the unfolding process. **(a)** Thermal denaturation of horse apomyoglobin (myoglobin without the heme prosthetic group) and ribonuclease A (with its disulfide bonds intact; see Fig. 4–27). The midpoint of the temperature range over which denaturation occurs is called the melting temperature, or T_m. Denaturation of apomyoglobin was monitored by circular dichroism (see Fig. 4–10), which measures the amount of helical structure in the protein. Denaturation of ribonuclease A was tracked by monitoring changes in the intrinsic fluorescence of the protein, which is affected by changes in the environment of Trp residues. **(b)** Denaturation of disulfide-intact ribonuclease A by guanidine hydrochloride (GdnHCl), monitored by circular dichroism.

homologous proteins derived from bacteria such as *Escherichia coli*. How these small differences promote structural stability at high temperatures is imperfectly understood.

Proteins can also be denatured by extremes of pH, by certain miscible organic solvents such as alcohol or acetone, by certain solutes such as urea and guanidine hydrochloride, or by detergents. Each of these denaturing agents represents a relatively mild treatment in the sense that no covalent bonds in the polypeptide chain are broken. Organic solvents, urea, and detergents act primarily by disrupting the hydrophobic interactions that make up the stable core of globular proteins; urea also disrupts hydrogen bonds; extremes of pH alter the net charge on the protein, causing electrostatic repulsion and the disruption of some hydrogen bonding. The

denatured structures resulting from these various treatments are not necessarily the same.

Denaturation often leads to protein precipitation, a consequence of protein aggregate formation as exposed hydrophobic surfaces associate. The aggregates are often highly disordered. The protein precipitate that is seen after boiling an egg white is one example. More-ordered aggregates are also observed in some proteins, as we shall see.

Amino Acid Sequence Determines Tertiary Structure

The tertiary structure of a globular protein is determined by its amino acid sequence. The most important proof of this came from experiments showing that denaturation of some proteins is reversible. Certain globular proteins denatured by heat, extremes of pH, or denaturing reagents will regain their native structure and their biological activity if returned to conditions in which the native conformation is stable. This process is called **renaturation**.

A classic example is the denaturation and renaturation of ribonuclease A, demonstrated by Christian Anfinsen in the 1950s. Purified ribonuclease A denatures completely in a concentrated urea solution in the presence of a reducing agent. The reducing agent cleaves the four disulfide bonds to yield eight Cys residues, and the urea disrupts the stabilizing hydrophobic interactions, thus freeing the entire polypeptide from its folded conformation. Denaturation of ribonuclease is accompanied by a complete loss of catalytic activity. When the urea and the reducing agent are removed, the randomly coiled, denatured ribonuclease spontaneously refolds into its correct tertiary structure, with full restoration of its catalytic activity (**Fig. 4–27**). The refolding of ribonuclease is so accurate that the four intrachain disulfide bonds are re-formed in the same positions in the renatured molecule as in the native ribonuclease. Later, similar results were obtained using chemically synthesized, catalytically active ribonuclease A. This eliminated the possibility that some minor contaminant in Anfinsen's purified ribonuclease preparation might have contributed to the renaturation of the enzyme, thus dispelling any remaining doubt that this enzyme folds spontaneously.

The Anfinsen experiment provided the first evidence that the amino acid sequence of a polypeptide chain contains all the information required to fold the chain into its native, three-dimensional structure. Subsequent work has shown that only a minority of proteins, many of them small and inherently stable, will fold spontaneously into their native form. Even though all proteins have the potential to fold into their native structure, many require some assistance.

Polypeptides Fold Rapidly by a Stepwise Process

In living cells, proteins are assembled from amino acids at a very high rate. For example, *E. coli* cells can make a complete, biologically active protein molecule containing 100 amino acid residues in about 5 seconds at 37 °C.

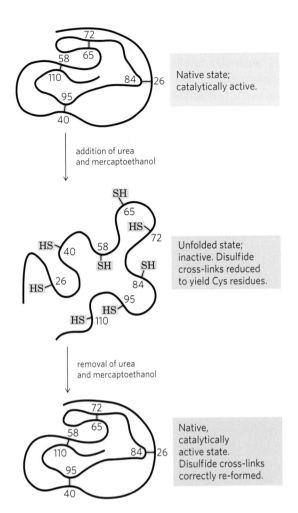

FIGURE 4–27 Renaturation of unfolded, denatured ribonuclease.
Urea denatures the ribonuclease, and mercaptoethanol (HOCH₂CH₂SH) reduces and thus cleaves the disulfide bonds to yield eight Cys residues. Renaturation involves reestablishing the correct disulfide cross-links.

However, the synthesis of peptide bonds on the ribosome is not enough; the protein must fold.

How does the polypeptide chain arrive at its native conformation? Let's assume conservatively that each of the amino acid residues could take up 10 different conformations on average, giving 10^{100} different conformations for the polypeptide. Let's also assume that the protein folds spontaneously by a random process in which it tries out all possible conformations around every single bond in its backbone until it finds its native, biologically active form. If each conformation were sampled in the shortest possible time ($\sim 10^{-13}$ second, or the time required for a single molecular vibration), it would take about 10^{77} years to sample all possible conformations. Clearly, protein folding is not a completely random, trial-and-error process. There must be shortcuts. This problem was first pointed out by Cyrus Levinthal in 1968 and is sometimes called Levinthal's paradox.

The folding pathway of a large polypeptide chain is unquestionably complicated. However, rapid progress has been made in this field, sufficient to produce robust algorithms that can often predict the structure of smaller

proteins on the basis of their amino acid sequences. The major folding pathways are hierarchical. Local secondary structures form first. Certain amino acid sequences fold readily into α helices or β sheets, guided by constraints such as those reviewed in our discussion of secondary structure. Ionic interactions, involving charged groups that are often near one another in the linear sequence of the polypeptide chain, can play an important role in guiding these early folding steps. Assembly of local structures is followed by longer-range interactions between, say, two elements of secondary structure that come together to form stable folded structures. Hydrophobic interactions play a significant role throughout the process, as the aggregation of nonpolar amino acid side chains provides an entropic stabilization to intermediates and, eventually, to the final folded structure. The process continues until complete domains form and the entire polypeptide is folded **(Fig. 4–28)**. Notably, proteins dominated by close-range interactions (between pairs of residues generally located near each other in the polypeptide sequence) tend to fold faster than proteins with more complex folding patterns and many long-range interactions between different segments. As larger proteins with multiple domains are synthesized, domains near the amino terminus (which are synthesized

Amino acid sequence of a 56-residue peptide
MTYKLILNGKTLKGETTTEAVDAATAEKVFKQYANDNGVDGEWTYDDATKTFTVTE

FIGURE 4–28 A protein-folding pathway as defined for a small protein. A hierarchical pathway is shown, based on computer modeling. Small regions of secondary structure are assembled first and then gradually incorporated into larger structures. The program used for this model has been highly successful in predicting the three-dimensional structure of small proteins from their amino acid sequence. The numbers indicate the amino acid residues in this 56 residue peptide that have acquired their final structure in each of the steps shown.

first) may fold before the entire polypeptide has been assembled.

Thermodynamically, the folding process can be viewed as a kind of free-energy funnel (**Fig. 4–29**). The unfolded states are characterized by a high degree of conformational entropy and relatively high free energy. As folding proceeds, the narrowing of the funnel reflects the decrease in the conformational space that must be searched as the protein approaches its native state. Small depressions along the sides of the free-energy funnel represent semistable intermediates that can briefly slow the folding process. At the bottom of the funnel, an ensemble of folding intermediates has been reduced to a single native conformation (or one of a small set of native conformations). The funnels can have a variety of shapes depending on the complexity of the folding pathway, the existence of semistable intermediates, and the potential for particular intermediates to assemble into aggregates of misfolded proteins (Fig. 4–29).

Thermodynamic stability is not evenly distributed over the structure of a protein—the molecule has regions of relatively high stability and others of low or negligible stability. For example, a protein may have two stable domains joined by a segment that is entirely disordered. Regions of low stability may allow a protein to alter its conformation between two or more states. As we shall see in the next two chapters, variations in the stability of regions within a protein are often essential to protein function. Intrinsically disordered proteins or protein segments do not fold at all.

As our understanding of protein folding and protein structure improves, increasingly sophisticated computer programs for predicting the structure of proteins from their amino acid sequence are being developed. Prediction of protein structure is a specialty field of bioinformatics, and progress in this area is monitored with a biennial test called the CASP (Critical Assessment of Structural Prediction) competition. Entrants from around the world vie to predict the structure of an assigned protein (whose structure has been determined but not yet published). The most successful teams are invited to present their results at a CASP conference. The success of these efforts is improving rapidly.

Some Proteins Undergo Assisted Folding

Not all proteins fold spontaneously as they are synthesized in the cell. Folding for many proteins requires **chaperones**, proteins that interact with partially folded or improperly folded polypeptides, facilitating correct folding pathways or providing microenvironments in which folding can occur. Several types of molecular chaperones are found in organisms ranging from bacteria to humans. Two major families of chaperones, both well studied, are the **Hsp70** family and the **chaperonins**.

The Hsp70 family of proteins generally have a molecular weight near 70,000 and are more abundant in

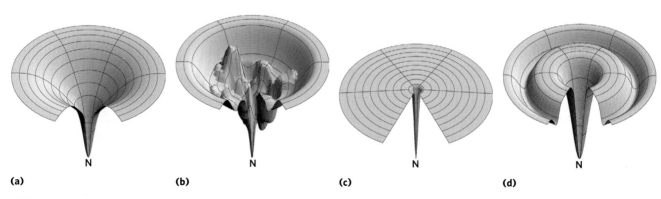

(a)   **(b)**   **(c)**   **(d)**

FIGURE 4–29 The thermodynamics of protein folding depicted as free-energy funnels. As proteins fold, the conformational space that can be explored by the structure is constrained. This is modeled as a three-dimensional thermodynamic funnel, with ΔG represented as depth and with the native structure (N) at the bottom (lowest free-energy point) of the funnel. The funnel for a given protein can have a variety of shapes, depending on the number and types of folding intermediates in the folding pathways. Any folding intermediate with significant stability and a finite lifetime would be represented as a local free-energy minimum—a depression on the surface of the funnel. **(a)** A simple but relatively wide and smooth funnel represents a protein that has multiple folding pathways (that is, the order in which different parts of the protein fold would be somewhat random), but it assumes its three-dimensional structure with no folding intermediates that have significant stability. **(b)** This funnel represents a more typical protein that has multiple possible folding intermediates with significant stability on the multiple pathways leading to the native structure. **(c)** A protein with one stable native structure, essentially no other folded intermediates with significant stability, and only one or a very few productive folding pathways is shown as a funnel with one narrow depression leading to the native form. **(d)** A protein with folding intermediates of substantial stability on virtually every pathway leading to the native state (that is, a protein in which a particular motif or domain always folds quickly, but other parts of the protein fold more slowly and in a random order) is depicted by a funnel with a major depression surrounding the depression leading to the native form.

cells stressed by elevated temperatures (hence, *heat shock* proteins of M_r 70,000, or Hsp70). Hsp70 proteins bind to regions of unfolded polypeptides that are rich in hydrophobic residues. These chaperones thus "protect" both proteins subject to denaturation by heat and new peptide molecules being synthesized (and not yet folded). Hsp70 proteins also block the folding of certain proteins that must remain unfolded until they have been translocated across a membrane (as described in Chapter 27). Some chaperones also facilitate the quaternary assembly of oligomeric proteins. The Hsp70 proteins bind to and release polypeptides in a cycle that uses energy from ATP hydrolysis and involves several other proteins (including a class called Hsp40). **Figure 4–30** illustrates chaperone-assisted folding as elucidated for the eukaryotic Hsp70 and Hsp40

chaperones. The binding of an unfolded polypeptide by an Hsp70 chaperone may break up a protein aggregate or prevent the formation of a new one. When the bound polypeptide is released, it has a chance to resume folding to its native structure. If folding does not occur rapidly enough, the polypeptide may be bound again and the process repeated. Alternatively, the Hsp70-bound polypeptide may be delivered to a chaperonin.

Chaperonins are elaborate protein complexes required for the folding of some cellular proteins that do not fold spontaneously. In *E. coli*, an estimated 10% to 15% of cellular proteins require the resident chaperonin system, called GroEL/GroES, for folding under normal conditions (up to 30% require this assistance when the cells are heat stressed). The analogous chaperonin system in eukaryotes is called Hsp60. The chaperonins first became known when they were found to be necessary for the growth of certain bacterial viruses (hence the designation "Gro"). This family of proteins is structured as a series of multisubunit rings, forming two chambers oriented back to back. An unfolded protein is first bound to an exposed hydrophobic surface near the apical end of one GroEL chamber. The protein is then trapped within the chamber when it is capped transiently by the GroES "lid" **(Fig. 4–31)**. GroEL undergoes substantial conformational changes, coupled to slow ATP hydrolysis, which also regulates the binding and release of GroES. Inside the chamber, a protein has about 10 seconds to fold—the time required for the bound ATP to hydrolyze. Constraining a protein within the chamber prevents inappropriate protein aggregation and also restricts the conformational space that a polypeptide chain can explore as it folds. The protein is released when the GroES cap dissociates but can rebound rapidly for another round if folding has not been completed. The two chambers in a GroEL complex alternate in binding and releasing unfolded polypeptide substrates. In eukaryotes, the Hsp60 system utilizes a similar process to fold proteins. However, in place of the GroES lid, protrusions from the apical domains of the subunits flex and close over the chamber. The ATP hydrolytic cycle is also slower in the Hsp60 complexes, giving the proteins constrained inside more time to fold.

Finally, the folding pathways of some proteins require two enzymes that catalyze isomerization reactions. **Protein disulfide isomerase (PDI)** is a widely distributed enzyme that catalyzes the interchange, or shuffling, of disulfide bonds until the bonds of the native conformation are formed. Among its functions, PDI catalyzes the elimination of folding intermediates with inappropriate disulfide cross-links. **Peptide prolyl cis-trans isomerase (PPI)** catalyzes the interconversion of the cis and trans isomers of Pro residue peptide bonds (Fig. 4–8), which can be a slow step in the folding of proteins that contain some Pro peptide bonds in the cis conformation.

Aggregates

Protein

Folding intermediate

Native protein

ATP

Hsp40

P_i

ATP ADP

NEF

Open Hsp70-ATP
Low affinity

Closed Hsp70-ADP
High affinity

FIGURE 4–30 Chaperones in protein folding. The pathway by which Hsp70-class chaperones bind and release polypeptides is illustrated for the eukaryotic chaperones Hsp70 and Hsp40. The chaperones do not actively promote the folding of the substrate protein, but instead prevent aggregation of unfolded peptides. The unfolded or partly folded proteins bind first to the open, ATP-bound form of Hsp70 (PDB ID 2QXL). Hsp40 then interacts with this complex and triggers ATP hydrolysis that produces the closed form of the complex (derived from PDB IDs 2KHO and 1DKZ), where the domains colored orange and yellow come together like the two parts of a jaw, trapping parts of the unfolded protein inside. Dissociation of ADP and recycling of the Hsp70 requires interaction with another protein, nucleotide-exchange factor (NEF). For a population of polypeptide molecules, some fraction of the molecules released after the transient binding of partially folded proteins by Hsp70 will take up the native conformation. The remainder are rebound by Hsp70 or diverted to the chaperonin system (Hsp60; see Fig. 4–31). In bacteria, the Hsp70 and Hsp40 chaperones are called DnaK and DnaJ, respectively. DnaK and DnaJ were first identified as proteins required for in vitro replication of certain viral DNA molecules (hence the "Dna" designation).

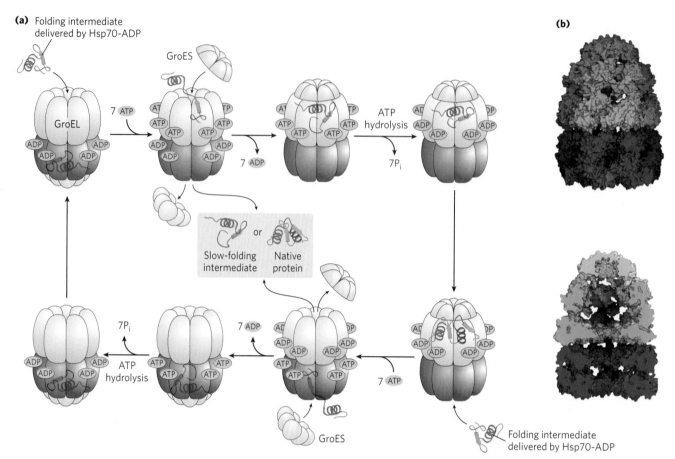

FIGURE 4–31 Chaperonins in protein folding. (a) A proposed pathway for the action of the *E. coli* chaperonins GroEL (a member of the Hsp60 protein family) and GroES. Each GroEL complex consists of two large chambers formed by two heptameric rings (each subunit M_r 57,000). GroES, also a heptamer (subunit M_r 10,000), blocks one of the GroEL chambers after an unfolded protein is bound inside. The chamber with the unfolded protein is referred to as cis; the opposite one is trans. Folding occurs within the cis chamber, during the time it takes to hydrolyze the 7 ATP bound to the subunits in the heptameric ring. The GroES and the ADP molecules then dissociate, and the protein is released. The two chambers of the GroEL/Hsp60 systems alternate in the binding and facilitated folding of client proteins. **(b)** Surface and cutaway images of the GroEL/GroES complex (PDB ID 1AON). The cutaway (below) illustrates the large interior space within which other proteins are bound.

Defects in Protein Folding Provide the Molecular Basis for a Wide Range of Human Genetic Disorders

Despite the many processes that assist in protein folding, misfolding does occur. In fact, protein misfolding is a substantial problem in all cells, and a quarter or more of all polypeptides synthesized may be destroyed because they do not fold correctly. In some cases, the misfolding causes or contributes to the development of serious disease.

Many conditions, including type 2 diabetes, Alzheimer disease, Huntington disease, and Parkinson disease, are associated with a misfolding mechanism: a soluble protein that is normally secreted from the cell is secreted in a misfolded state and converted into an insoluble extracellular **amyloid** fiber. The diseases are collectively referred to as **amyloidoses**. The fibers are highly ordered and unbranched, with a diameter of 7 to 10 nm and a high degree of β-sheet structure. The β segments are oriented perpendicular to the axis of the fiber. In some amyloid fibers the overall structure features two layers of β sheet, such as that shown for amyloid-β peptide in **Figure 4–32**.

Many proteins can take on the amyloid fibril structure as an alternative to their normal folded conformations, and most of these proteins have a concentration of aromatic amino acid residues in a core region of β sheet or α helix. The proteins are secreted in an incompletely folded conformation. The core (or some part of it) folds into a β sheet before the rest of the protein folds correctly, and the β sheets from two or more incompletely folded protein molecules associate to begin forming an amyloid fibril. The fibril grows in the extracellular space. Other parts of the protein then fold differently, remaining on the outside of the β-sheet core in the growing fibril. The effect of aromatic residues in stabilizing the structure is shown in Figure 4–32c. Because most of the protein molecules fold normally, the onset of symptoms in the amyloidoses is often very slow. If a person inherits a mutation such as substitution with an aromatic residue at a position that favors formation of amyloid fibrils, disease symptoms may begin at an earlier age.

Native   Misfolded or partially unfolded   Denatured

Self-association

Amyloid fibril core structure

(a)   Further assembly of protofilaments

Phe

(b)   Amyloid-β peptide   **(c)**   Amyloid fibrils

FIGURE 4–32 Formation of disease-causing amyloid fibrils. (a) Protein molecules whose normal structure includes regions of β sheet undergo partial folding. In a small number of the molecules, before folding is complete, the β-sheet regions of one polypeptide associate with the same region in another polypeptide, forming the nucleus of an amyloid. Additional protein molecules slowly associate with the amyloid and extend it to form a fibril. **(b)** The amyloid-β peptide begins as two α-helical segments of a larger protein. Proteolytic cleavage of this larger protein leaves the relatively unstable amyloid-β peptide, which loses its α-helical structure. It can then assemble slowly into amyloid fibrils **(c)**, which contribute to the characteristic plaques on the exterior of nervous tissue in people with Alzheimer disease. The aromatic side chains shown here play a significant role in stabilizing the amyloid structure. Amyloid is rich in β sheet, with the β strands arranged perpendicular to the axis of the amyloid fibril. Amyloid-β peptide takes the form of two layers of extended parallel β sheet. Some amyloid-forming peptides may fold to form left-handed β-helices (see Fig. 4-22).

In eukaryotes, proteins destined for secretion undergo their initial folding in the endoplasmic reticulum (ER; see pathway in Chapter 27). When stress conditions arise, or when protein synthesis threatens to overwhelm the protein-folding capacity of the ER, unfolded proteins can accumulate. These conditions

trigger the unfolded protein response (UPR). A set of transcriptional regulators that constitute the UPR bring the various systems into alignment by increasing the concentration of chaperones in the ER or decreasing the rate of overall protein synthesis, or both. Amyloid aggregates that form before the UPR can come into play may be removed. Some are degraded by **autophagy**. In this process, they are first encapsulated in a membrane, then the contents of the resulting vesicle are degraded after the vesicle docks with a cytosolic lysosome. Alternatively, misfolded proteins can be degraded by a system of proteases called the ubiquitin-proteasome system (described in Chapter 27). Defects in any of these systems decrease the capacity to deal with misfolded proteins and increase the propensity for development of amyloid-related diseases.

Some amyloidoses are systemic, involving many tissues. Primary systemic amyloidosis is caused by deposition of fibrils consisting of misfolded immunoglobulin light chains (see Chapter 5), or fragments of light chains derived from proteolytic degradation. The mean age of onset is about 65 years. Patients have symptoms including fatigue, hoarseness, swelling, and weight loss, and many die within the first year after diagnosis. The kidneys or heart are often most affected. Some amyloidoses are associated with other types of disease. People with certain chronic infectious or inflammatory diseases such as rheumatoid arthritis, tuberculosis, cystic fibrosis, and some cancers can experience a sharp increase in secretion of an amyloid-prone polypeptide called serum amyloid A (SAA) protein. This protein, or fragments of it, deposits in the connective tissue of the spleen, kidney, and liver, and around the heart. People with this condition, known as secondary systemic amyloidosis, have a wide range of symptoms, depending on the organs initially affected. The disease is generally fatal within a few years. More than 80 amyloidoses are associated with mutations in transthyretin (a protein that binds to and transports thyroid hormones, distributing them throughout the body and brain). A variety of mutations in this protein lead to amyloid deposition concentrated around different tissues, thus producing different symptoms. Amyloidoses are also associated with inherited mutations in the proteins lysozyme, fibrinogen A α chain, and apolipoproteins A-I and A-II; all of these proteins are described in later chapters.

Some amyloid diseases are associated with particular organs. The amyloid-prone protein is generally secreted only by the affected tissue, and its locally high concentration leads to amyloid deposition around that tissue (although some of the protein may be distributed systemically). One common site of amyloid deposition is near the pancreatic islet β cells, responsible for insulin secretion and regulation of glucose metabolism (see Fig. 23–26). Secretion by β cells of a small (37 amino acid) peptide called islet amyloid polypeptide (IAPP), or amylin, can lead to amyloid deposits around the islets, gradually destroying the cells. A healthy human adult has 1 to 1.5 million

BOX 4–6 🝊 MEDICINE Death by Misfolding: The Prion Diseases

A misfolded brain protein seems to be the causative agent of several rare degenerative brain diseases in mammals. Perhaps the best known of these is bovine spongiform encephalopathy (BSE; also known as mad cow disease). Related diseases include kuru and Creutzfeldt-Jakob disease in humans, scrapie in sheep, and chronic wasting disease in deer and elk. These diseases are also referred to as spongiform encephalopathies, because the diseased brain frequently becomes riddled with holes (Fig. 1). Progressive deterioration of the brain leads to a spectrum of neurological symptoms, including weight loss, erratic behavior, problems with posture, balance, and coordination, and loss of cognitive function. The diseases are fatal.

In the 1960s, investigators found that preparations of the disease-causing agents seemed to lack nucleic acids. At this time, Tikvah Alper suggested that the agent was a protein. Initially, the idea seemed heretical. All disease-causing agents known up to that time—viruses, bacteria, fungi, and so on—contained nucleic acids, and their virulence was related to genetic reproduction and propagation. However, four decades of investigations, pursued most notably by Stanley Prusiner, have provided evidence that spongiform encephalopathies are different.

The infectious agent has been traced to a single protein (M_r 28,000), which Prusiner dubbed **prion** protein (PrP). The name was derived from *proteinaceous infectious*, but Prusiner thought that "prion"

FIGURE 1 Stained section of cerebral cortex from autopsy of a patient with Creutzfeldt-Jakob disease shows spongiform (vacuolar) degeneration, the most characteristic neurohistological feature. The yellowish vacuoles are intracellular and occur mostly in pre- and postsynaptic processes of neurons. The vacuoles in this section vary in diameter from 20 to 100 μm.

sounded better than "proin." Prion protein is a normal constituent of brain tissue in all mammals. Its role is not known in detail, but it may have a molecular signaling function. Strains of mice lacking the gene for PrP (and thus the protein itself) suffer no obvious ill effects. Illness occurs only when the normal cellular PrP, or PrP^C, occurs in an altered conformation called PrP^{Sc} (Sc denotes scrapie). The structure

pancreatic β cells. With progressive loss of these cells, glucose homeostasis is affected and eventually, when 50% or more of the cells are lost, the condition matures into type 2 (non–insulin-dependent) diabetes mellitus.

The amyloid deposition diseases that trigger neurodegeneration, particularly in older adults, are a special class of localized amyloidoses. Alzheimer disease is associated with extracellular amyloid deposition by neurons, involving the amyloid-β peptide (Fig. 4–32b), derived from a larger transmembrane protein (amyloid-β precursor protein) found in most human tissues. When it is part of the larger protein, the peptide is composed of two α-helical segments spanning the membrane. When the external and internal domains are cleaved off by dedicated proteases, the relatively unstable amyloid-β peptide leaves the membrane and loses its α-helical structure. It can then take the form of two layers of extended parallel β sheet, which can slowly assemble into amyloid fibrils (Fig. 4–32c). Deposits of these amyloid fibers seem to be the primary cause of Alzheimer disease, but a second type of amyloidlike

aggregation, involving a protein called tau, also occurs intracellularly (in neurons) in people with Alzheimer disease. Inherited mutations in the tau protein do not result in Alzheimer disease, but they cause a frontotemporal dementia and parkinsonism (a condition with symptoms resembling Parkinson disease) that can be equally devastating.

Several other neurodegenerative conditions involve intracellular aggregation of misfolded proteins. In Parkinson disease, the misfolded form of the protein α-synuclein aggregates into spherical filamentous masses called Lewy bodies. Huntington disease involves the protein huntingtin, which has a long polyglutamine repeat. In some individuals, the polyglutamine repeat is longer than normal and a more subtle type of intracellular aggregation occurs. Notably, when the mutant human proteins involved in Parkinson and Huntington diseases are expressed in *Drosophila melanogaster*, the flies display neurodegeneration expressed as eye deterioration, tremors, and early death. All of these symptoms are highly suppressed if expression of the Hsp70 chaperone is also increased.

of PrP^C features two α helices. The structure of PrP^{Sc} is very different, with much of the structure converted to amyloid-like β sheets (Fig. 2). The interaction of PrP^{Sc} with PrP^C converts the latter to PrP^{Sc}, initiating a domino effect in which more and more of the brain protein converts to the disease-causing form. The mechanism by which the presence of PrP^{Sc} leads to spongiform encephalopathy is not understood.

In inherited forms of prion diseases, a mutation in the gene encoding PrP produces a change in one amino acid residue that is believed to make the conversion of PrP^C to PrP^{Sc} more likely. A complete understanding of prion diseases awaits new information on how prion protein affects brain function. Structural information about PrP is beginning to provide insights into the molecular process that allows the prion proteins to interact so as to alter their conformation (Fig. 2).

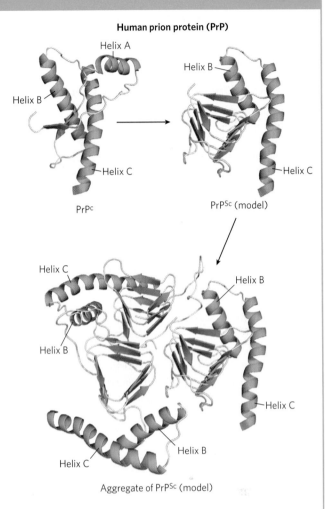

Human prion protein (PrP)

FIGURE 2 Structure of the globular domain of human PrP (PDB ID 1QLX) and models of the misfolded, disease-causing conformation PrP^{Sc}, and an aggregate of PrP^{Sc}. The α helices are labeled to help illustrate the conformation change. Helix A is incorporated into the β-sheet structure of the misfolded conformation.

Protein misfolding need not lead to amyloid formation to cause serious disease. For example, cystic fibrosis is caused by defects in a membrane-bound protein called *cystic fibrosis transmembrane conductance regulator* (CFTR), which acts as a channel for chloride ions. The most common cystic fibrosis–causing mutation is the deletion of a Phe residue at position 508 in CFTR, which causes improper protein folding. Most of this protein is then degraded and its normal function is lost (see Box 11–2). Many of the disease-related mutations in collagen (p. 130) also cause defective folding. A particularly remarkable type of protein misfolding is seen in the prion diseases (Box 4–6). ■

SUMMARY 4.4 Protein Denaturation and Folding

▶ The maintenance of the steady-state collection of active cellular proteins required under a particular set of conditions—called proteostasis—involves an elaborate set of pathways and processes that fold, refold, and degrade polypeptide chains.

▶ The three-dimensional structure and the function of most proteins can be destroyed by denaturation, demonstrating a relationship between structure and function. Some denatured proteins can renature spontaneously to form biologically active protein, showing that tertiary structure is determined by amino acid sequence.

▶ Protein folding in cells is generally hierarchical. Initially, regions of secondary structure may form, followed by folding into motifs and domains. Large ensembles of folding intermediates are rapidly brought to a single native conformation.

▶ For many proteins, folding is facilitated by Hsp70 chaperones and by chaperonins. Disulfide bond formation and the cis-trans isomerization of Pro peptide bonds are catalyzed by specific enzymes.

▶ Protein misfolding is the molecular basis of a wide range of human diseases, including the amyloidoses.

Key Terms

Terms in bold are defined in the glossary.

conformation 115	**motif** 137
native conformation 116	**fold** 137
hydrophobic	**domain** 137
interactions 116	protein family 140
solvation layer 116	multimer 140
peptide group 118	**oligomer** 140
Ramachandran plot 119	**protomer** 141
secondary structure 119	**intrinsically disordered**
α helix 120	**proteins** 141
β **conformation** 123	**proteostasis** 143
β sheet 123	**denaturation** 143
β turn 123	**renaturation** 144
circular dichroism (CD)	**chaperone** 146
spectroscopy 124	Hsp70 146
tertiary structure 125	**chaperonin** 146
quaternary structure 125	protein disulfide isomerase
fibrous proteins 125	(PDI) 147
globular proteins 125	peptide prolyl cis-trans
α-keratin 126	isomerase (PPI) 147
collagen 127	amyloid 148
silk fibroin 130	**amyloidoses** 148
Protein Data Bank	**autophagy** 149
(PDB) 132	prion 150

Further Reading

General

Anfinsen, C.B. (1973) Principles that govern the folding of protein chains. *Science* **181**, 223–230.

The author reviews his classic work on ribonuclease.

Creighton, T.E. (1993) *Proteins: Structures and Molecular Properties,* 2nd edn, W. H. Freeman and Company, New York.

A comprehensive and authoritative source.

Kendrew, J.C. (1961) The three-dimensional structure of a protein molecule. *Sci. Am.* **205** (December), 96–111.

Describes how the structure of myoglobin was determined and what was learned from it.

Richardson, J.S. (1981) The anatomy and taxonomy of protein structure. *Adv. Protein Chem.* **34**, 167–339.

An outstanding summary of protein structural patterns and principles; the author originated the very useful "ribbon" representations of protein structure.

Secondary, Tertiary, and Quaternary Structures

Beeby, M., O'Connor, B.D., Ryttersgaard, C., Boutz, D.R., Perry, L.J., & Yeates, T.O. (2005) The genomics of disulfide bonding and protein stabilization in thermophiles. *PLoS Biol.* **3**, e309.

Brown, J.H. (2006) Breaking symmetry in protein dimers: designs and function. *Protein Sci.* **15**, 1–13.

Dunker, A.K. & Kriwacki, R.W. (2011) The orderly chaos of proteins. *Sci. Am.* **304** (April), 68–73.

A nice summary of the work on proteins that lack intrinsic structure.

Herráez, A. (2006) Biomolecules in the computer. *Biochem. Mol. Biol. Educ.* **34**, 255–261.

McPherson, A. (1989) Macromolecular crystals. *Sci. Am.* **260** (March), 62–69.

A description of how macromolecules such as proteins are crystallized.

Milner-White, E.J. (1997) The partial charge of the nitrogen atom in peptide bonds. *Protein Sci.* **6**, 2477–2482.

Ponting, C.P. & Russell, R.R. (2002) The natural history of protein domains. *Annu. Rev. Biophys. Biomol. Struct.* **31**, 45–71.

An explanation of how structural databases can be used to explore evolution.

Protein Denaturation and Folding

Chiti, F. & Dobson, C.M. (2006) Protein misfolding, functional amyloid, and human disease. *Annu. Rev. Biochem.* **75**, 333–366.

Dill, K.A., Ozkan, S.B., Shell, M.S., & Weikl, T.R. (2008) The protein folding problem. *Annu. Rev. Biophys.* **37**, 289–316.

Gazit, E. (2005) Mechanisms of amyloid fibril self-assembly and inhibition. *FEBS J.* **272**, 5971–5978.

Hartl, F.U., Bracher, A., & Hayer-Hartl, M. (2011) Molecular chaperones in protein folding and proteostasis. *Nature* **475**, 324–332.

Hoppener, J.W.M. & Lips, C.J.M. (2006) Role of islet amyloid in type 2 diabetes mellitus. *Int. J. Biochem. Cell Biol.* **38**, 726–736.

Kapinga, H.H. & Craig, E.A. (2010) The HSP70 chaperone machinery: J proteins as drivers of functional specificity. *Nat. Rev. Mol. Cell. Biol.* **11**, 579–592.

Norrby, E. (2011) Prions and protein-folding diseases. *J. Intern. Med.* **270**, 1–14.

Prusiner, S.B. (1995) The prion diseases. *Sci. Am.* **272** (January), 48–57.

A good summary of the evidence leading to the prion hypothesis.

Selkoe, D.J. (2003) Folding proteins in fatal ways. *Nature* **426**, 900–904.

A good summary of amyloidoses.

Tang, Y., Chang, H., Roeben, A., Wischnewski, D., Wischnewski, N., Kerner, M., Hartl, F., & Hayer-Hartl, M. (2006) Structural features of the GroEL-GroES nanocage required for rapid folding of encapsulated protein. *Cell* **125**, 903–914.

Tyedmers, J., Mogk, A., & Bukau, B. (2010) Cellular strategies for controlling protein aggregation. *Nat. Rev. Mol. Cell Biol.* **11**, 777–788.

Problems

1. Properties of the Peptide Bond In x-ray studies of crystalline peptides, Linus Pauling and Robert Corey found that the C—N bond in the peptide link is intermediate in length (1.32 Å) between a typical C—N single bond (1.49 Å) and a C=N double bond (1.27 Å). They also found that the peptide bond is planar (all four atoms attached to the C—N group are located in the same plane) and that the two *α*-carbon atoms attached to the C—N are always trans to each other (on opposite sides of the peptide bond).

(a) What does the length of the C—N bond in the peptide linkage indicate about its strength and its bond order (i.e., whether it is single, double, or triple)?

(b) What do the observations of Pauling and Corey tell us about the ease of rotation about the C—N peptide bond?

2.   Structural and Functional Relationships in Fibrous Proteins William Astbury discovered that the x-ray diffraction pattern of wool shows a repeating structural unit spaced about 5.2 Å along the length of the wool fiber. When he steamed and stretched the wool, the x-ray pattern showed a new repeating structural unit at a spacing of 7.0 Å. Steaming and stretching the wool and then letting it shrink gave an x-ray pattern consistent with the original spacing of about 5.2 Å. Although these observations provided important clues to the molecular structure of wool, Astbury was unable to interpret them at the time.

(a) Given our current understanding of the structure of wool, interpret Astbury's observations.

(b) When wool sweaters or socks are washed in hot water or heated in a dryer, they shrink. Silk, on the other hand, does not shrink under the same conditions. Explain.

3.   Rate of Synthesis of Hair α-Keratin Hair grows at a rate of 15 to 20 cm/yr. All this growth is concentrated at the base of the hair fiber, where α-keratin filaments are synthesized inside living epidermal cells and assembled into ropelike structures (see Fig. 4–11). The fundamental structural element of α-keratin is the α helix, which has 3.6 amino acid residues per turn and a rise of 5.4 Å per turn (see Fig. 4–4a). Assuming that the biosynthesis of α-helical keratin chains is the rate-limiting factor in the growth of hair, calculate the rate at which peptide bonds of α-keratin chains must be synthesized (peptide bonds per second) to account for the observed yearly growth of hair.

4.   Effect of pH on the Conformation of α-Helical Secondary Structures The unfolding of the α helix of a polypeptide to a randomly coiled conformation is accompanied by a large decrease in a property called specific rotation, a measure of a solution's capacity to rotate circularly polarized light. Polyglutamate, a polypeptide made up of only L-Glu residues, has the α-helical conformation at pH 3. When the pH is raised to 7, there is a large decrease in the specific rotation of the solution. Similarly, polylysine (L-Lys residues) is an α helix at pH 10, but when the pH is lowered to 7 the specific rotation also decreases, as shown by the following graph.

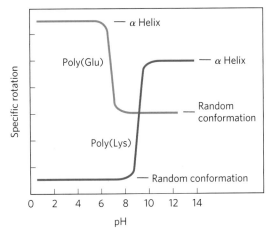

What is the explanation for the effect of the pH changes on the conformations of poly(Glu) and poly(Lys)? Why does the transition occur over such a narrow range of pH?

5.   Disulfide Bonds Determine the Properties of Many Proteins Some natural proteins are rich in disulfide bonds, and their mechanical properties (tensile strength, viscosity, hardness, etc.) are correlated with the degree of disulfide bonding.

(a) Glutenin, a wheat protein rich in disulfide bonds, is responsible for the cohesive and elastic character of dough made from wheat flour. Similarly, the hard, tough nature of tortoise shell is due to the extensive disulfide bonding in its α-keratin. What is the molecular basis for the correlation between disulfide-bond content and mechanical properties of the protein?

(b) Most globular proteins are denatured and lose their activity when briefly heated to 65 °C. However, globular proteins that contain multiple disulfide bonds often must be heated longer at higher temperatures to denature them. One such protein is bovine pancreatic trypsin inhibitor (BPTI), which has 58 amino acid residues in a single chain and contains three disulfide bonds. On cooling a solution of denatured BPTI, the activity of the protein is restored. What is the molecular basis for this property?

6.   Dihedral Angles A series of torsion angles, ϕ and ψ, that might be taken up by the peptide backbone is shown below. Which of these closely correspond to ϕ and ψ for an idealized collagen triple helix? Refer to Figure 4–9 as a guide.

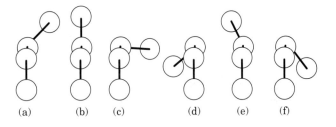

7.   Amino Acid Sequence and Protein Structure Our growing understanding of how proteins fold allows researchers to make predictions about protein structure based on primary amino acid sequence data. Consider the following amino acid sequence.

1	2	3	4	5	6	7	8	9	10
Ile–	Ala –	His –	Thr –	Tyr –	Gly –	Pro –	Phe –	Glu –	Ala –

11	12	13	14	15	16	17	18	19	20
Ala –	Met –	Cys –	Lys –	Trp –	Glu –	Ala –	Gln –	Pro –	Asp –

21	22	23	24	25	26	27	28
Gly –	Met –	Glu –	Cys –	Ala –	Phe –	His –	Arg

(a) Where might bends or β turns occur?

(b) Where might intrachain disulfide cross-linkages be formed?

(c) Assuming that this sequence is part of a larger globular protein, indicate the probable location (the external surface or interior of the protein) of the following amino acid residues: Asp, Ile, Thr, Ala, Gln, Lys. Explain your reasoning. (Hint: See the hydropathy index in Table 3–1.)

8. Bacteriorhodopsin in Purple Membrane Proteins Under the proper environmental conditions, the salt-loving archaeon *Halobacterium halobium* synthesizes a membrane protein (M_r 26,000) known as bacteriorhodopsin, which is purple because it contains retinal (see Fig. 10–21). Molecules of this protein aggregate into "purple patches" in the cell membrane. Bacteriorhodopsin acts as a light-activated proton pump that provides energy for cell functions. X-ray analysis of this protein reveals that it consists of seven parallel α-helical segments, each of which traverses the bacterial cell membrane (thickness 45 Å). Calculate the minimum number of amino acid residues necessary for one segment of α helix to traverse the membrane completely. Estimate the fraction of the bacteriorhodopsin protein that is involved in membrane-spanning helices. (Use an average amino acid residue weight of 110.)

9. Protein Structure Terminology Is myoglobin a motif, a domain, or a complete three-dimensional structure?

10. Interpreting Ramachandran Plots Examine the two proteins labeled (a) and (b) below. Which of the two Ramachandran plots, labeled (c) and (d), is more likely to be derived from which protein? Why?

11. Pathogenic Action of Bacteria That Cause Gas Gangrene The highly pathogenic anaerobic bacterium *Clostrid-*

ium perfringens is responsible for gas gangrene, a condition in which animal tissue structure is destroyed. This bacterium secretes an enzyme that efficiently catalyzes the hydrolysis of the peptide bond indicated in red:

$$-X—Gly—Pro—Y— \xrightarrow{H_2O}$$
$$-X—COO^- + H_3\overset{+}{N}—Gly—Pro—Y—$$

where X and Y are any of the 20 common amino acids. How does the secretion of this enzyme contribute to the invasiveness of this bacterium in human tissues? Why does this enzyme not affect the bacterium itself?

12. Number of Polypeptide Chains in a Multisubunit Protein A sample (660 mg) of an oligomeric protein of M_r 132,000 was treated with an excess of 1-fluoro-2,4-dinitrobenzene (Sanger's reagent) under slightly alkaline conditions until the chemical reaction was complete. The peptide bonds of the protein were then completely hydrolyzed by heating it with concentrated HCl. The hydrolysate was found to contain 5.5 mg of the following compound:

(a)

(b)

(c)

(d)

2,4-Dinitrophenyl derivatives of the α-amino groups of other amino acids could not be found.

(a) Explain how this information can be used to determine the number of polypeptide chains in an oligomeric protein.

(b) Calculate the number of polypeptide chains in this protein.

(c) What other protein analysis technique could you employ to determine whether the polypeptide chains in this protein are similar or different?

13. Predicting Secondary Structure Which of the following peptides is more likely to take up an α-helical structure, and why?

(a) LKAENDEAARAMSEA

(b) CRAGGFPWDQPGTSN

14. Amyloid Fibers in Disease Several small aromatic molecules, such as phenol red (used as a nontoxic drug model), have been shown to inhibit the formation of amyloid in laboratory model systems. A goal of the research on these small aromatic compounds is to find a drug that would efficiently inhibit the formation of amyloid in the brain in people with incipient Alzheimer disease.

(a) Suggest why molecules with aromatic substituents would disrupt the formation of amyloid.

(b) Some researchers have suggested that a drug used to treat Alzheimer disease may also be effective in treating type 2 (non–insulin-dependent) diabetes mellitus. Why might a single drug be effective in treating these two different conditions?

Using the Web

15. Protein Modeling on the Internet A group of patients with Crohn disease (an inflammatory bowel disease) underwent biopsies of their intestinal mucosa in an attempt to identify the causative agent. Researchers identified a protein that was present at higher levels in patients with Crohn disease than in patients with an unrelated inflammatory bowel disease or in unaffected controls. The protein was isolated, and the following *partial* amino acid sequence was obtained (reads left to right):

EAELCPDRCI	HSFQNLGIQC	VKKRDLEQAI
SQRIQTNNNP	FQVPIEEQRG	DYDLNAVRLC
FQVTVRDPSG	RPLRLPPVLP	HPIFDNRAPN
TAELKICRVN	RNSGSCLGGD	EIFLLCDKVQ
KEDIEVYFTG	PGWEARGSFS	QADVHRQVAI
VFRTPPYADP	SLQAPVRVSM	QLRRPSDREL
SEPMEFQYLP	DTDDRHRIEE	KRKRTYETFK
SIMKKSPFSG	PTDPRPPPRR	IAVPSRSSAS
VPKPAPQPYP		

(a) You can identify this protein using a protein database on the Internet. Some good places to start include Protein Information Resource (PIR; http://pir.georgetown.edu), Structural Classification of Proteins (SCOP; http://scop.mrc-lmb.cam.ac.uk/scop), and Prosite (http://prosite.expasy.org).

At your selected database site, follow links to the sequence comparison engine. Enter about 30 residues from the protein sequence in the appropriate search field and submit it for analysis. What does this analysis tell you about the identity of the protein?

(b) Try using different portions of the amino acid sequence. Do you always get the same result?

(c) A variety of websites provide information about the three-dimensional structure of proteins. Find information about the protein's secondary, tertiary, and quaternary structure using database sites such as the Protein Data Bank (PDB; www.pdb.org) or SCOP.

(d) In the course of your Web searches, what did you learn about the cellular function of the protein?

Data Analysis Problem

16. Mirror-Image Proteins As noted in Chapter 3, "The amino acid residues in protein molecules are exclusively L stereoisomers." It is not clear whether this selectivity is necessary for proper protein function or is an accident of evolution. To explore this question, Milton and colleagues (1992) published a study of an enzyme made entirely of D stereoisomers. The enzyme they chose was HIV protease, a proteolytic enzyme made by HIV that converts inactive viral preproteins to their active forms.

Previously, Wlodawer and coworkers (1989) had reported the complete chemical synthesis of HIV protease from L-amino acids (the L-enzyme), using the process shown in Figure 3–32. Normal HIV protease contains two Cys residues at positions 67 and 95. Because chemical synthesis of proteins containing Cys is technically difficult, Wlodawer and colleagues substituted the synthetic amino acid L-α-amino-n-butyric acid (Aba) for the two Cys residues in the protein. In the authors' words, this was done to "reduce synthetic difficulties associated with Cys deprotection and ease product handling."

(a) The structure of Aba is shown below. Why was this a suitable substitution for a Cys residue? Under what circumstances would it not be suitable?

$$\overset{\displaystyle -O \quad\quad O}{\underset{\displaystyle \quad}{\overset{\displaystyle \diagdown \;\; \diagup}{C}}}$$

$$H-\underset{\displaystyle \underset{+}{N H_3}}{\overset{\displaystyle |}{\underset{\displaystyle |}{C}}}-CH_2-CH_3$$

L-α-Amino-n-butyric acid

Wlodawer and coworkers denatured the newly synthesized protein by dissolving it in 6 M guanidine HCl and then allowed it to fold slowly by dialyzing away the guanidine against a neutral buffer (10% glycerol, 25 mM $NaPO_4$, pH 7).

(b) There are many reasons to predict that a protein synthesized, denatured, and folded in this manner would not be active. Give three such reasons.

(c) Interestingly, the resulting L-protease was active. What does this finding tell you about the role of disulfide bonds in the native HIV protease molecule?

In their new study, Milton and coworkers synthesized HIV protease from D-amino acids, using the same protocol as the earlier study (Wlodawer et al.). Formally, there are three possibilities for the folding of the D-protease: it would give (1) the same shape as the L-protease, (2) the mirror image of the L-protease, or (3) something else, possibly inactive.

(d) For each possibility, decide whether or not it is a likely outcome and defend your position.

In fact, the D-protease was active: it cleaved a particular synthetic substrate and was inhibited by specific inhibitors. To examine the structure of the D- and L-enzymes, Milton and coworkers tested both forms for activity with D and L forms of a chiral peptide substrate and for inhibition by D and L forms of a chiral peptide-analog inhibitor. Both forms were also tested for inhibition by the achiral inhibitor Evans blue. The findings are given in the table.

HIV Protease	Substrate hydrolysis		Inhibition		
			Peptide inhibitor		Evans blue
	D-substrate	L-substrate	D-inhibitor	L-inhibitor	(achiral)
L-protease	−	+	−	+	+
D-protease	+	−	+	−	+

(e) Which of the three models proposed above is supported by these data? Explain your reasoning.

(f) Why does Evans blue inhibit both forms of the protease?

(g) Would you expect chymotrypsin to digest the D-protease? Explain your reasoning.

(h) Would you expect total synthesis from D-amino acids followed by renaturation to yield active enzyme for any enzyme? Explain your reasoning.

References

Milton, R.C., Milton, S.C., & Kent, S.B. (1992) Total chemical synthesis of a D-enzyme: the enantiomers of HIV-1 protease show demonstration of reciprocal chiral substrate specificity. *Science* **256**, 1445–1448.

Wlodawer, A., Miller, M., Jaskólski, M., Sathyanarayana, B.K., Baldwin, E., Weber, I.T., Selk, L.M., Clawson, L., Schneider, J., & Kent, S.B. (1989) Conserved folding in retroviral proteases: crystal structure of a synthetic HIV-1 protease. *Science* **245**, 616–621.

Protein Function

Knowing the three-dimensional structure of a protein is an important part of understanding how the protein functions. However, the structure shown in two dimensions on a page is deceptively static. Proteins are dynamic molecules whose functions almost invariably depend on interactions with other molecules, and these interactions are affected in physiologically important ways by sometimes subtle, sometimes striking changes in protein conformation. In this chapter, we explore how proteins interact with other molecules and how their interactions are related to dynamic protein structure. The importance of molecular interactions to a protein's function can hardly be overemphasized. In Chapter 4, we saw that the function of fibrous proteins as structural elements of cells and tissues depends on stable, long-term quaternary interactions between identical polypeptide chains. As we shall see in this chapter, the functions of many other proteins involve interactions with a variety of different molecules. Most of these interactions are fleeting, though they may be the basis of complex physiological processes such as oxygen transport, immune function, and muscle contraction—the topics we examine in this chapter. The proteins that carry out these processes illustrate the following key principles of protein function, some of which will be familiar from the previous chapter:

The functions of many proteins involve the reversible binding of other molecules. A molecule bound reversibly by a protein is called a **ligand**. A ligand may be any kind of molecule, including another protein. The transient nature of protein-ligand interactions is critical to life, allowing an organism to respond rapidly and reversibly to changing environmental and metabolic circumstances.

A ligand binds at a site on the protein called the **binding site**, which is complementary to the ligand in size, shape, charge, and hydrophobic or hydrophilic character. Furthermore, the interaction is specific: the protein can discriminate among the thousands of different molecules in its environment and selectively bind only one or a few. A given protein may have separate binding sites for several different ligands. These specific molecular interactions are crucial in maintaining the high degree of order in a living system. (This discussion excludes the binding of water, which may interact weakly and nonspecifically with many parts of a protein. In Chapter 6, we consider water as a specific ligand for many enzymes.)

Proteins are flexible. Changes in conformation may be subtle, reflecting molecular vibrations and small movements of amino acid residues throughout the protein. A protein flexing in this way is sometimes said to "breathe." Changes in conformation may also be quite dramatic, with major segments of the protein structure moving as much as several nanometers. Specific conformational changes are frequently essential to a protein's function.

The binding of a protein and ligand is often coupled to a conformational change in the protein that makes the binding site more complementary to the ligand, permitting tighter binding. The structural adaptation that occurs between protein and ligand is called **induced fit**.

In a multisubunit protein, a conformational change in one subunit often affects the conformation of other subunits.

Interactions between ligands and proteins may be regulated, usually through specific interactions with one or more additional ligands. These other ligands may cause conformational changes in the protein that affect the binding of the first ligand.

Enzymes represent a special case of protein function. Enzymes bind and chemically transform other molecules—they catalyze reactions. The molecules acted upon by enzymes are called reaction **substrates** rather than ligands, and the ligand-binding site is called the **catalytic site** or **active site**. In this chapter we emphasize the noncatalytic functions of proteins. In Chapter 6 we consider catalysis by enzymes, a central topic in biochemistry. You will see that the themes of this chapter—binding, specificity, and conformational change—are continued in the next chapter, with the added element of proteins participating in chemical transformations.

5.1 Reversible Binding of a Protein to a Ligand: Oxygen-Binding Proteins

Myoglobin and hemoglobin may be the most-studied and best-understood proteins. They were the first proteins for which three-dimensional structures were determined, and these two molecules illustrate almost every aspect of that most central of biochemical processes: the reversible binding of a ligand to a protein. This classic model of protein function tells us a great deal about how proteins work. 🖱 **Oxygen-Binding Proteins—Myoglobin: Oxygen Storage**

Oxygen Can Bind to a Heme Prosthetic Group

Oxygen is poorly soluble in aqueous solutions (see Table 2–3) and cannot be carried to tissues in sufficient quantity if it is simply dissolved in blood serum. Diffusion of oxygen through tissues is also ineffective over distances greater than a few millimeters. The evolution of larger, multicellular animals depended on the evolution of proteins that could transport and store oxygen. However, none of the amino acid side chains in proteins are suited for the reversible binding of oxygen molecules. This role is filled by certain transition metals, among them iron and copper, that have a strong tendency to bind oxygen. Multicellular organisms exploit the properties of metals, most commonly iron, for oxygen transport. However, free iron promotes the formation of highly reactive oxygen species such as hydroxyl radicals that can damage DNA and other macromolecules. Iron used in cells is therefore bound in forms that sequester it and/or make it less reactive. In multicellular organisms—especially those in which iron, in its oxygen-carrying capacity, must be transported over large distances—iron is often incorporated into a protein-bound prosthetic group called **heme** (or haem). (Recall from Chapter 3 that a prosthetic group is a compound permanently associated with a protein that contributes to the protein's function.)

Heme consists of a complex organic ring structure, **protoporphyrin**, to which is bound a single iron atom in its ferrous (Fe^{2+}) state **(Fig. 5–1)**. The iron atom has six coordination bonds, four to nitrogen atoms that are part of the flat **porphyrin ring** system and two perpendicular to the porphyrin. The coordinated nitrogen atoms (which have an electron-donating character) help prevent conversion of the heme iron to the ferric (Fe^{3+}) state. Iron in the Fe^{2+} state binds oxygen reversibly; in the Fe^{3+} state it does not bind oxygen. Heme is found in many oxygen-transporting proteins, as well as in some proteins, such as the cytochromes, that participate in oxidation-reduction (electron-transfer) reactions (Chapter 19).

Free heme molecules (heme not bound to protein) leave Fe^{2+} with two "open" coordination bonds. Simultaneous reaction of one O_2 molecule with two free heme molecules (or two free Fe^{2+}) can result in irreversible conversion of Fe^{2+} to Fe^{3+}. In heme-containing proteins, this reaction is prevented by sequestering each

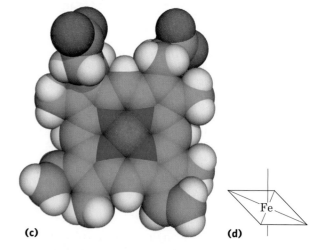

FIGURE 5–1 Heme. The heme group is present in myoglobin, hemoglobin, and many other proteins, designated heme proteins. Heme consists of a complex organic ring structure, protoporphyrin IX, with a bound iron atom in its ferrous (Fe^{2+}) state. **(a)** Porphyrins, of which protoporphyrin IX is only one example, consist of four pyrrole rings linked by methene bridges, with substitutions at one or more of the positions denoted X. **(b, c)** Two representations of heme (derived from PDB ID 1CCR). The iron atom of heme has six coordination bonds: four in the plane of, and bonded to, the flat porphyrin ring system, and **(d)** two perpendicular to it.

Edge view

HN—CH
N—Fe—O$_2$
C
CH$_2$
CH

Histidine residue   Plane of porphyrin ring system

FIGURE 5–2 The heme group viewed from the side. This view shows the two coordination bonds to Fe^{2+} that are perpendicular to the porphyrin ring system. One is occupied by a His residue, sometimes called the proximal His; the other is the binding site for oxygen. The remaining four coordination bonds are in the plane of, and bonded to, the flat porphyrin ring system.

heme deep within the protein structure. Thus, access to the two open coordination bonds is restricted. One of these two coordination bonds is occupied by a side-chain nitrogen of a His residue. The other is the binding site for molecular oxygen (O$_2$) **(Fig. 5–2)**. When oxygen binds, the electronic properties of heme iron change; this accounts for the change in color from the dark purple of oxygen-depleted venous blood to the bright red of oxygen-rich arterial blood. Some small molecules, such as carbon monoxide (CO) and nitric oxide (NO), coordinate to heme iron with greater affinity than does O$_2$. When a molecule of CO is bound to heme, O$_2$ is excluded, which is why CO is highly toxic to aerobic organisms (a topic explored later, in Box 5–1). By surrounding and sequestering heme, oxygen-binding proteins regulate the access of CO and other small molecules to the heme iron.

Globins Are a Family of Oxygen-Binding Proteins

The **globins** are a widespread family of proteins, all having similar primary and tertiary structures. Globins are commonly found in eukaryotes of all classes and even in some bacteria. Most function in oxygen transport or storage, although some play a role in the sensing of oxygen, nitric oxide, or carbon monoxide. The simple nematode worm *Caenorhabditis elegans* has genes encoding 33 different globins. In humans and other mammals, there are at least four kinds of globins. The monomeric myoglobin facilitates oxygen diffusion in muscle tissue. Myoglobin is particularly abundant in the muscles of diving marine mammals such as seals and whales, where it also has an oxygen-storage function for prolonged excursions undersea. The tetrameric hemoglobin is responsible for oxygen transport in the blood stream. The monomeric neuroglobin is expressed largely in neurons and helps to protect the brain from hypoxia (low oxygen) or ischemia (restricted blood supply). Cytoglobin, another monomeric globin, is found at high levels in a range of tissues, but its function is unknown.

Myoglobin Has a Single Binding Site for Oxygen

Myoglobin (M_r 16,700; abbreviated Mb) is a single polypeptide of 153 amino acid residues with one molecule of heme. As is typical for a globin polypeptide, myoglobin is made up of eight α-helical segments connected by bends **(Fig. 5–3)**. About 78% of the amino acid residues in the protein are found in these α helices.

Any detailed discussion of protein function inevitably involves protein structure. In the case of myoglobin, we first introduce some structural conventions peculiar to globins. As seen in Figure 5–3, the helical segments are named A through H. An individual amino acid residue is designated either by its position in the amino acid sequence or by its location in the sequence of a particular α-helical segment. For example, the His residue coordinated to the heme in myoglobin, His^{93} (the 93rd residue from the amino-terminal end of the myoglobin polypeptide sequence), is also called His F8 (the 8th residue in α helix F). The bends in the structure are designated AB, CD, EF, FG, and so forth, reflecting the α-helical segments they connect.

Protein-Ligand Interactions Can Be Described Quantitatively

The function of myoglobin depends on the protein's ability not only to bind oxygen but also to release it when and where it is needed. Function in biochemistry

FIGURE 5–3 Myoglobin. (PDB ID 1MBO) The eight α-helical segments (shown here as cylinders) are labeled A through H. Nonhelical residues in the bends that connect them are labeled AB, CD, EF, and so forth, indicating the segments they interconnect. A few bends, including BC and DE, are abrupt and do not contain any residues; these are not normally labeled. (The short segment visible between D and E is an artifact of the computer representation.) The heme is bound in a pocket made up largely of the E and F helices, although amino acid residues from other segments of the protein also participate.

often revolves around a reversible protein-ligand interaction of this type. A quantitative description of this interaction is therefore a central part of many biochemical investigations.

In general, the reversible binding of a protein (P) to a ligand (L) can be described by a simple **equilibrium expression**:

$$P + L \rightleftharpoons PL \qquad (5\text{--}1)$$

The reaction is characterized by an equilibrium constant, K_a, such that

$$K_a = \frac{[PL]}{[P][L]} = \frac{k_a}{k_d} \qquad (5\text{--}2)$$

where k_a and k_d are rate constants (more on these below). The term $\boldsymbol{K_a}$ is an **association constant** (not to be confused with the K_a that denotes an acid dissociation constant; p. 62) that describes the equilibrium between the complex and the unbound components of the complex. The association constant provides a measure of the affinity of the ligand L for the protein. K_a has units of M^{-1}; a higher value of K_a corresponds to a higher affinity of the ligand for the protein.

The equilibrium term K_a is also equivalent to the ratio of the rates of the forward (association) and reverse (dissociation) reactions that form the PL complex. The association rate is described by a rate constant k_a, and dissociation by the rate constant k_d. As discussed further in the next chapter, rate constants are proportionality constants, describing the fraction of a pool of reactant that reacts in a given amount of time. When the reaction involves one molecule, such as the dissociation reaction $PL \longrightarrow P + L$, the reaction is *first order* and the rate constant (k_d) has units of reciprocal time (s^{-1}). When the reaction involves two molecules, such as the association reaction $P + L \longrightarrow PL$, it is called *second order*, and its rate constant (k_a) has units of $\text{M}^{-1}\,\text{s}^{-1}$.

KEY CONVENTION: Equilibrium constants are denoted with a capital K and rate constants with a lower case k. ∎

A rearrangement of the first part of Equation 5–2 shows that the ratio of bound to free protein is directly proportional to the concentration of free ligand:

$$K_a[L] = \frac{[PL]}{[P]} \qquad (5\text{--}3)$$

When the concentration of the ligand is much greater than the concentration of ligand-binding sites, the binding of the ligand by the protein does not appreciably change the concentration of free (unbound) ligand—that is, [L] remains constant. This condition is broadly applicable to most ligands that bind to proteins in cells and simplifies our description of the binding equilibrium.

We can now consider the binding equilibrium from the standpoint of the fraction, θ (theta), of ligand-binding sites on the protein that are occupied by ligand:

$$\theta = \frac{\text{binding sites occupied}}{\text{total binding sites}} = \frac{[PL]}{[PL] + [P]} \qquad (5\text{--}4)$$

Substituting $K_a[L][P]$ for [PL] (see Eqn 5–3) and rearranging terms gives

$$\theta = \frac{K_a[L][P]}{K_a[L][P] + [P]} = \frac{K_a[L]}{K_a[L] + 1} = \frac{[L]}{[L] + \dfrac{1}{K_a}} \qquad (5\text{--}5)$$

The value of K_a can be determined from a plot of θ versus the concentration of free ligand, [L] (**Fig. 5–4a**). Any equation of the form $x = y/(y + z)$ describes a hyperbola, and θ is thus found to be a hyperbolic function of [L]. The fraction of ligand-binding sites occupied approaches saturation asymptotically as [L] increases. The [L] at which half of the available ligand-binding sites are occupied (that is, $\theta = 0.5$) corresponds to $1/K_a$.

It is more common (and intuitively simpler), however, to consider the **dissociation constant**, $\boldsymbol{K_d}$, which is the reciprocal of K_a ($K_d = 1/K_a$) and is given in units of molar concentration (M). K_d is the equilibrium constant

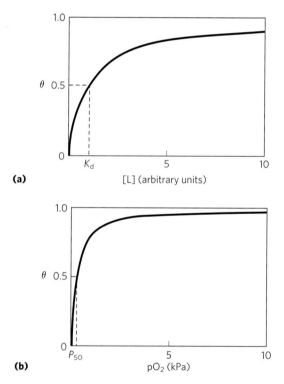

(a)

(b)

FIGURE 5–4 Graphical representations of ligand binding. The fraction of ligand-binding sites occupied, θ, is plotted against the concentration of free ligand. Both curves are rectangular hyperbolas. **(a)** A hypothetical binding curve for a ligand L. The [L] at which half of the available ligand-binding sites are occupied is equivalent to $1/K_a$, or K_d. The curve has a horizontal asymptote at $\theta = 1$ and a vertical asymptote (not shown) at $[L] = -1/K_a$. **(b)** A curve describing the binding of oxygen to myoglobin. The partial pressure of O_2 in the air above the solution is expressed in kilopascals (kPa). Oxygen binds tightly to myoglobin, with a P_{50} of only 0.26 kPa.

TABLE 5–1 Some Protein Dissociation Constants

Protein	Ligand	K_d (M)*
Avidin (egg white)	Biotin	1×10^{-15}
Insulin receptor (human)	Insulin	1×10^{-10}
Anti-HIV immunoglobulin (human)[†]	gp41 (HIV-1 surface protein)	4×10^{-10}
Nickel-binding protein (E. coli)	Ni^{2+}	1×10^{-7}
Calmodulin (rat)[‡]	Ca^{2+}	3×10^{-6}
		2×10^{-5}

Color bars indicate the range of dissociation constants typical of various classes of interactions in biological systems. A few interactions, such as that between the protein avidin and the enzyme cofactor biotin, fall outside the normal ranges. The avidin-biotin interaction is so tight it may be considered irreversible. Sequence-specific protein-DNA interactions reflect proteins that bind to a particular sequence of nucleotides in DNA, as opposed to general binding to any DNA site.

*A reported dissociation constant is valid only for the particular solution conditions under which it was measured. K_d values for a protein–ligand interaction can be altered, sometimes by several orders of magnitude, by changes in the solution's salt concentration, pH, or other variables.

[†]This immunoglobulin was isolated as part of an effort to develop a vaccine against HIV. Immunoglobulins (described later in the chapter) are highly variable, and the K_d reported here should not be considered characteristic of all immunoglobulins.

[‡]Calmodulin has four binding sites for calcium. The values shown reflect the highest- and lowest-affinity binding sites observed in one set of measurements.

for the release of ligand. The relevant expressions change to

$$K_d = \frac{[P][L]}{[PL]} = \frac{k_d}{k_a} \qquad (5–6)$$

$$[PL] = \frac{[P][L]}{K_d} \qquad (5–7)$$

$$\theta = \frac{[L]}{[L] + K_d} \qquad (5–8)$$

When [L] equals K_d, half of the ligand-binding sites are occupied. As [L] falls below K_d, progressively less of the protein has ligand bound to it. In order for 90% of the available ligand-binding sites to be occupied, [L] must be nine times greater than K_d.

In practice, K_d is used much more often than K_a to express the affinity of a protein for a ligand. Note that a lower value of K_d corresponds to a higher affinity of ligand for the protein. The mathematics can be reduced to simple statements: K_d is equivalent to the molar concentration of ligand at which half of the available ligand-binding sites are occupied. At this point, the protein is said to have reached half-saturation with respect to ligand binding. The more tightly a protein binds a ligand, the lower the concentration of ligand required for half

the binding sites to be occupied, and thus the lower the value of K_d. Some representative dissociation constants are given in Table 5–1; the scale shows typical ranges for dissociation constants found in biological systems.

WORKED EXAMPLE 5–1 Receptor-Ligand Dissociation Constants

Two proteins, X and Y, bind to the same ligand, A, with the binding curves shown below.

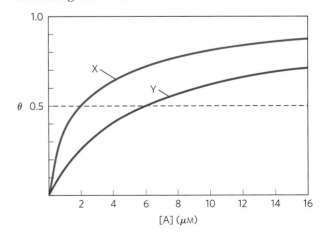

What is the dissociation constant, K_d, for each protein? Which protein (X or Y) has a greater affinity for ligand A?

Solution: We can determine the dissociation constants by inspecting the graph. Since θ represents the fraction of binding sites occupied by ligand, the concentration of ligand at which half the binding sites are occupied—that is, the point where the binding curve crosses the line where $\theta = 0.5$—is the dissociation constant. For X, $K_d = 2\ \mu M$; for Y, $K_d = 6\ \mu M$. Because X is half-saturated at a lower [A], it has a higher affinity for the ligand.

The binding of oxygen to myoglobin follows the patterns discussed above. However, because oxygen is a gas, we must make some minor adjustments to the equations so that laboratory experiments can be carried out more conveniently. We first substitute the concentration of dissolved oxygen for [L] in Equation 5–8 to give

$$\theta = \frac{[O_2]}{[O_2] + K_d} \qquad (5\text{–}9)$$

As for any ligand, K_d equals the $[O_2]$ at which half of the available ligand-binding sites are occupied, or $[O_2]_{0.5}$. Equation 5–9 thus becomes

$$\theta = \frac{[O_2]}{[O_2] + [O_2]_{0.5}} \qquad (5\text{–}10)$$

In experiments using oxygen as a ligand, it is the partial pressure of oxygen (pO_2) in the gas phase above the solution that is varied, because this is easier to measure than the concentration of oxygen dissolved in the solution. The concentration of a volatile substance in solution is always proportional to the local partial pressure of the gas. So, if we define the partial pressure of oxygen at $[O_2]_{0.5}$ as P_{50}, substitution in Equation 5–10 gives

$$\theta = \frac{pO_2}{pO_2 + P_{50}} \qquad (5\text{–}11)$$

A binding curve for myoglobin that relates θ to pO_2 is shown in Figure 5–4b.

Protein Structure Affects How Ligands Bind

The binding of a ligand to a protein is rarely as simple as the above equations would suggest. The interaction is greatly affected by protein structure and is often accompanied by conformational changes. For example, the specificity with which heme binds its various ligands is altered when the heme is a component of myoglobin. Carbon monoxide binds to free heme molecules more than 20,000 times better than does O_2 (that is, the K_d or P_{50} for CO binding to free heme is more than 20,000 times lower than that for O_2), but it binds only about 200 times better than O_2 when the heme is bound in myoglobin. The difference may be partly explained by steric hindrance. When O_2 binds to free heme, the axis of the oxygen molecule is positioned at an angle to the Fe—O bond **(Fig. 5–5a)**. In contrast, when CO binds to free heme, the Fe, C, and O atoms lie in a straight line

FIGURE 5–5 Steric effects caused by ligand binding to the heme of myoglobin. (a) Oxygen binds to heme with the O_2 axis at an angle, a binding conformation readily accommodated by myoglobin. **(b)** Carbon monoxide binds to free heme with the CO axis perpendicular to the plane of the porphyrin ring. When binding to the heme in myoglobin, CO is forced to adopt a slight angle because the perpendicular arrangement is sterically blocked by His E7, the distal His. This effect weakens the binding of CO to myoglobin. **(c)** Another view of the heme of myoglobin (derived from PDB ID 1MBO), showing the arrangement of key amino acid residues around the heme. The bound O_2 is hydrogen-bonded to the distal His, His E7 (His[64]), further facilitating the binding of O_2.

(Fig. 5–5b). In both cases, the binding reflects the geometry of hybrid orbitals in each ligand. In myoglobin, His[64] (His E7), on the O_2-binding side of the heme, is too far away to coordinate with the heme iron, but it does interact with a ligand bound to heme. This residue, called the *distal His* (as distinct from the *proximal His*, His F8), forms a hydrogen bond with O_2 (Fig. 5–5c) but may help preclude the linear binding of CO, providing one explanation for the selectively diminished binding of CO to heme in myoglobin (and hemoglobin). A reduction in CO binding is physiologically important, because CO is a low-level byproduct of cellular metabolism. Other factors, not yet well-defined, also may modulate the interaction of heme with CO in these proteins.

The binding of O_2 to the heme in myoglobin also depends on molecular motions, or "breathing," in the protein structure. The heme molecule is deeply buried in the folded polypeptide, with no direct path for oxygen to move from the surrounding solution to the ligand-binding site. If the protein were rigid, O_2 could not enter or leave the heme pocket at a measurable rate. However, rapid molecular flexing of the amino acid side chains produces transient cavities in the protein structure, and O_2 makes its way in and out by moving through these cavities. Computer simulations of rapid structural fluctuations in myoglobin suggest that there are many such pathways. One major route is provided by rotation of the side chain of the distal His (His^{64}), which occurs on a nanosecond (10^{-9} s) time scale. Even subtle conformational changes can be critical for protein activity.

In neuroglobin, cytoglobin, and some globins found in plants and invertebrates, the distal His (His E7) is directly coordinated with the heme iron. In these globins, the oxygen or other ligand must displace the distal His in the process of binding.

Hemoglobin Transports Oxygen in Blood

🔊 **Oxygen-Binding Proteins—Hemoglobin: Oxygen Transport** Nearly all the oxygen carried by whole blood in animals is bound and transported by hemoglobin in erythrocytes (red blood cells). Normal human erythrocytes are small (6 to 9 μm in diameter), biconcave disks. They are formed from precursor stem cells called **hemocytoblasts**. In the maturation process, the stem cell produces daughter cells that form large amounts of hemoglobin and then lose their intracellular organelles—nucleus, mitochondria, and endoplasmic reticulum. Erythrocytes are thus incomplete, vestigial cells, unable to reproduce and, in humans, destined to survive for only about 120 days. Their main function is to carry hemoglobin, which is dissolved in the cytosol at a very high concentration (~34% by weight).

In arterial blood passing from the lungs through the heart to the peripheral tissues, hemoglobin is about 96% saturated with oxygen. In the venous blood returning to the heart, hemoglobin is only about 64% saturated. Thus, each 100 mL of blood passing through a tissue releases about one-third of the oxygen it carries, or 6.5 mL of O_2 gas at atmospheric pressure and body temperature.

Myoglobin, with its hyperbolic binding curve for oxygen (Fig. 5–4b), is relatively insensitive to small changes in the concentration of dissolved oxygen and so functions well as an oxygen-storage protein. Hemoglobin, with its multiple subunits and O_2-binding sites, is better suited to oxygen transport. As we shall see, interactions between the subunits of a multimeric protein can permit a highly sensitive response to small changes in ligand concentration. Interactions among the subunits in hemoglobin cause conformational changes that alter the affinity of the protein for oxygen. The modulation of oxygen binding allows the O_2-transport protein to respond to changes in oxygen demand by tissues.

Hemoglobin Subunits Are Structurally Similar to Myoglobin

Hemoglobin (M_r 64,500; abbreviated Hb) is roughly spherical, with a diameter of nearly 5.5 nm. It is a tetrameric protein containing four heme prosthetic groups,

FIGURE 5–6 Comparison of the structures of myoglobin (PDB ID 1MBO) and the β subunit of hemoglobin (derived from PDB ID 1HGA).

one associated with each polypeptide chain. Adult hemoglobin contains two types of globin, two α chains (141 residues each) and two β chains (146 residues each). Although fewer than half of the amino acid residues are identical in the polypeptide sequences of the α and β subunits, the three-dimensional structures of the two types of subunits are very similar. Furthermore, their structures are very similar to that of myoglobin **(Fig. 5–6)**, even though the amino acid sequences of the three polypeptides are identical at only 27 positions **(Fig. 5–7)**. All three polypeptides are members of the globin family of proteins. The helix-naming convention described for myoglobin is also applied to the hemoglobin polypeptides, except that the α subunit lacks the short D helix. The heme-binding pocket is made up largely of the E and F helices in each of the subunits.

The quaternary structure of hemoglobin features strong interactions between unlike subunits. The $\alpha_1\beta_1$ interface (and its $\alpha_2\beta_2$ counterpart) involves more than 30 residues, and its interaction is sufficiently strong that although mild treatment of hemoglobin with urea tends to disassemble the tetramer into $\alpha\beta$ dimers, these dimers remain intact. The $\alpha_1\beta_2$ (and $\alpha_2\beta_1$) interface involves 19 residues **(Fig. 5–8)**. Hydrophobic interactions predominate at all the interfaces, but there are also many hydrogen bonds and a few ion pairs (or salt bridges), whose importance is discussed below.

Hemoglobin Undergoes a Structural Change on Binding Oxygen

X-ray analysis has revealed two major conformations of hemoglobin: the **R state** and the **T state**. Although oxygen binds to hemoglobin in either state, it has a significantly higher affinity for hemoglobin in the R state. Oxygen binding stabilizes the R state. When oxygen is absent experimentally, the T state is more stable and is thus the predominant conformation of **deoxyhemoglobin**. T and R originally denoted "tense" and "relaxed," respectively, because the T state is stabilized by a greater number of ion pairs, many of which lie at the $\alpha_1\beta_2$ (and $\alpha_2\beta_1$) interface **(Fig. 5–9)**. The binding of O_2

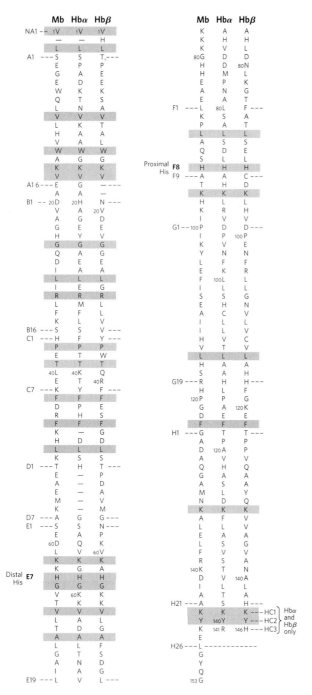

FIGURE 5–7 The amino acid sequences of whale myoglobin and the α and β chains of human hemoglobin. Dashed lines mark helix boundaries. To align the sequences optimally, short gaps must be introduced into both Hb sequences where a few amino acids are present in the other, compared sequences. With the exception of the missing D helix in Hbα, this alignment permits the use of the helix lettering convention that emphasizes the common positioning of amino acid residues that are identical in all three structures (shaded). Residues shaded in light red are conserved in all known globins. Note that the common helix-letter-and-number designation for amino acids does not necessarily correspond to a common position in the linear sequence of amino acids in the polypeptides. For example, the distal His residue is His E7 in all three structures, but corresponds to His[64], His[58], and His[63] in the linear sequences of Mb, Hbα, and Hbβ, respectively. Nonhelical residues at the amino and carboxyl termini, beyond the first (A) and last (H) α-helical segments, are labeled NA and HC, respectively.

FIGURE 5–8 Dominant interactions between hemoglobin subunits. (PDB ID 1HGA) In this representation, α subunits are light and β subunits are dark. The strongest subunit interactions (highlighted) occur between unlike subunits. When oxygen binds, the $\alpha_1\beta_1$ contact changes little, but there is a large change at the $\alpha_1\beta_2$ contact, with several ion pairs broken.

FIGURE 5–9 Some ion pairs that stabilize the T state of deoxyhemoglobin. (a) Close-up view of a portion of a deoxyhemoglobin molecule in the T state (PDB ID 1HGA). Interactions between the ion pairs His HC3 and Asp FG1 of the β subunit (blue) and between Lys C5 of the α subunit (gray) and His HC3 (its α-carboxyl group) of the β subunit are shown with dashed lines. (Recall that HC3 is the carboxyl-terminal residue of the β subunit.) **(b)** Interactions between these ion pairs, and between others not shown in (a), are schematized in this representation of the extended polypeptide chains of hemoglobin.

T state   **R state**

FIGURE 5–10 The T → R transition. (PDB ID 1HGA and 1BBB) In these depictions of deoxyhemoglobin, as in Figure 5-9, the β subunits are blue and the α subunits are gray. Positively charged side chains and chain termini involved in ion pairs are shown in blue, their negatively charged partners in red. The Lys C5 of each α subunit and Asp FG1 of each β subunit are visible but not labeled (compare Fig. 5-9a). Note that the molecule is oriented slightly differently than in Figure 5-9. The transition from the T state to the R state shifts the subunit pairs substantially, affecting certain ion pairs. Most noticeably, the His HC3 residues at the carboxyl termini of the β subunits, which are involved in ion pairs in the T state, rotate in the R state toward the center of the molecule, where they are no longer in ion pairs. Another dramatic result of the T → R transition is a narrowing of the pocket between the β subunits.

to a hemoglobin subunit in the T state triggers a change in conformation to the R state. When the entire protein undergoes this transition, the structures of the individual subunits change little, but the $\alpha\beta$ subunit pairs slide past each other and rotate, narrowing the pocket between the β subunits **(Fig. 5–10).** In this process, some of the ion pairs that stabilize the T state are broken and some new ones are formed.

Max Perutz proposed that the T → R transition is triggered by changes in the positions of key amino acid side chains surrounding the heme. In the T state, the porphyrin is slightly puckered, causing the heme iron to protrude somewhat on the proximal His (His F8) side. The binding of O_2 causes the heme to assume a more planar conformation, shifting the position of the proximal His and the attached F helix **(Fig. 5–11).** These changes lead to adjustments in the ion pairs at the $\alpha_1\beta_2$ interface.

Hemoglobin Binds Oxygen Cooperatively

Hemoglobin must bind oxygen efficiently in the lungs, where the pO_2 is about 13.3 kPa, and release oxygen in the tissues, where the pO_2 is about 4 kPa. Myoglobin, or any protein that binds oxygen with a hyperbolic binding curve, would be ill-suited to this function, for the reason illustrated in **Figure 5–12.** A protein that bound O_2 with high affinity would bind it efficiently in the lungs but would not release much of it in the tissues. If the protein bound oxygen with a sufficiently low affinity to release it in the tissues, it would not pick up much oxygen in the lungs.

Hemoglobin solves the problem by undergoing a transition from a low-affinity state (the T state) to a high-affinity state (the R state) as more O_2 molecules are bound. As a result, hemoglobin has a hybrid S-shaped, or sigmoid, binding curve for oxygen (Fig. 5–12). A single-subunit protein with a single ligand-binding site cannot produce a sigmoid binding curve—even if binding elicits a conformational change—because each molecule of ligand binds independently and cannot affect ligand binding to another molecule. In contrast, O_2 binding to individual subunits of hemoglobin can alter the affinity for O_2 in adjacent subunits. The first molecule of O_2 that interacts with deoxyhemoglobin binds weakly, because it binds to a subunit in the T state. Its binding, however,

T state   **R state**

FIGURE 5–11 Changes in conformation near heme on O_2 binding to deoxyhemoglobin. (Derived from PDB ID 1HGA and 1BBB) The shift in the position of helix F when heme binds O_2 is thought to be one of the adjustments that triggers the T → R transition.

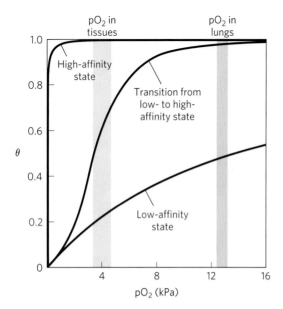

FIGURE 5–12 A sigmoid (cooperative) binding curve. A sigmoid binding curve can be viewed as a hybrid curve reflecting a transition from a low-affinity to a high-affinity state. Because of its cooperative binding, as manifested by a sigmoid binding curve, hemoglobin is more sensitive to the small differences in O_2 concentration between the tissues and the lungs, allowing it to bind oxygen in the lungs (where pO_2 is high) and release it in the tissues (where pO_2 is low).

leads to conformational changes that are communicated to adjacent subunits, making it easier for additional molecules of O_2 to bind. In effect, the T → R transition occurs more readily in the second subunit once O_2 is bound to the first subunit. The last (fourth) O_2 molecule binds to a heme in a subunit that is already in the R state, and hence it binds with much higher affinity than the first molecule.

An **allosteric protein** is one in which the binding of a ligand to one site affects the binding properties of another site on the same protein. The term "allosteric" derives from the Greek *allos*, "other," and *stereos*, "solid" or "shape." Allosteric proteins are those having "other shapes," or conformations, induced by the binding of ligands referred to as modulators. The conformational changes induced by the modulator(s) interconvert more-active and less-active forms of the protein. The modulators for allosteric proteins may be either inhibitors or activators. When the normal ligand and modulator are identical, the interaction is termed **homotropic**. When the modulator is a molecule other than the normal ligand, the interaction is **heterotropic**. Some proteins have two or more modulators and therefore can have both homotropic and heterotropic interactions.

Cooperative binding of a ligand to a multimeric protein, such as we observe with the binding of O_2 to hemoglobin, is a form of allosteric binding. The binding of one ligand affects the affinities of any remaining unfilled binding sites, and O_2 can be considered as both a ligand and an activating homotropic modulator. There is only one binding site for O_2 on each subunit, so the allosteric

effects giving rise to cooperativity are mediated by conformational changes transmitted from one subunit to another by subunit-subunit interactions. A sigmoid binding curve is diagnostic of cooperative binding. It permits a much more sensitive response to ligand concentration and is important to the function of many multisubunit proteins. The principle of allostery extends readily to regulatory enzymes, as we shall see in Chapter 6.

Cooperative conformational changes depend on variations in the structural stability of different parts of a protein, as described in Chapter 4. The binding sites of an allosteric protein typically consist of stable segments in proximity to relatively unstable segments, with the latter capable of frequent changes in conformation or intrinsic disorder (**Fig. 5–13**). When a ligand binds, the moving parts of the protein's binding site may be stabilized in a particular conformation, affecting the conformation of adjacent polypeptide

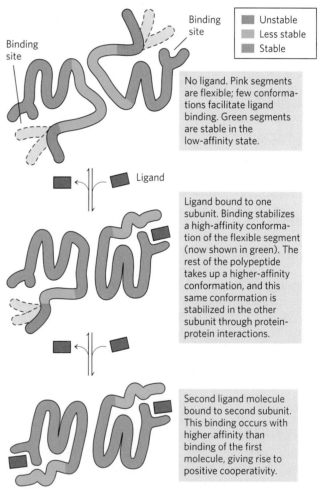

FIGURE 5–13 Structural changes in a multisubunit protein undergoing cooperative binding to ligand. Structural stability is not uniform throughout a protein molecule. Shown here is a hypothetical dimeric protein, with regions of high (blue), medium (green), and low (red) stability. The ligand-binding sites are composed of both high- and low-stability segments, so affinity for ligand is relatively low. The conformational changes that occur as ligand binds convert the protein from a low- to a high-affinity state, a form of induced fit.

subunits. If the entire binding site were highly stable, then few structural changes could occur in this site or be propagated to other parts of the protein when a ligand binds.

As is the case with myoglobin, ligands other than oxygen can bind to hemoglobin. An important example is carbon monoxide, which binds to hemoglobin about 250 times better than does oxygen. Human exposure to CO can have tragic consequences (Box 5–1).

Cooperative Ligand Binding Can Be Described Quantitatively

Cooperative binding of oxygen by hemoglobin was first analyzed by Archibald Hill in 1910. From this work came a general approach to the study of cooperative ligand binding to multisubunit proteins.

For a protein with n binding sites, the equilibrium of Equation 5–1 becomes

$$\text{P} + n\text{L} \rightleftharpoons \text{PL}_n \qquad (5\text{–}12)$$

and the expression for the association constant becomes

$$K_a = \frac{[\text{PL}_n]}{[\text{P}][\text{L}]^n} \qquad (5\text{–}13)$$

The expression for θ (see Eqn 5–8) is

$$\theta = \frac{[\text{L}]^n}{[\text{L}]^n + K_d} \qquad (5\text{–}14)$$

Rearranging, then taking the log of both sides, yields

$$\frac{\theta}{1 - \theta} = \frac{[\text{L}]^n}{K_d} \qquad (5\text{–}15)$$

$$\log\left(\frac{\theta}{1 - \theta}\right) = n \log [\text{L}] - \log K_d \qquad (5\text{–}16)$$

where $K_d = [L]_{0.5}^n$.

Equation 5–16 is the **Hill equation**, and a plot of log $[\theta/(1 - \theta)]$ versus log [L] is called a **Hill plot**. Based on the equation, the Hill plot should have a slope of n. However, the experimentally determined slope actually reflects not the number of binding sites but the degree of interaction between them. The slope of a Hill plot is therefore denoted by n_H, the **Hill coefficient**, which is a measure of the degree of cooperativity. If n_H equals 1, ligand binding is not cooperative, a situation that can arise even in a multisubunit protein if the subunits do not communicate. An n_H of greater than 1 indicates positive cooperativity in ligand binding. This is the situation observed in hemoglobin, in which the binding of one molecule of ligand facilitates the binding of others. The theoretical upper limit for n_H is reached when $n_H = n$. In this case the

binding would be completely cooperative: all binding sites on the protein would bind ligand simultaneously, and no protein molecules partially saturated with ligand would be present under any conditions. This limit is never reached in practice, and the measured value of n_H is always less than the actual number of ligand-binding sites in the protein.

An n_H of less than 1 indicates negative cooperativity, in which the binding of one molecule of ligand *impedes* the binding of others. Well-documented cases of negative cooperativity are rare.

To adapt the Hill equation to the binding of oxygen to hemoglobin we must again substitute pO_2 for [L] and P_{50}^n for K_d:

$$\log\left(\frac{\theta}{1 - \theta}\right) = n \log pO_2 - n \log P_{50} \qquad (5\text{–}17)$$

Hill plots for myoglobin and hemoglobin are given in **Figure 5–14**.

Two Models Suggest Mechanisms for Cooperative Binding

Biochemists now know a great deal about the T and R states of hemoglobin, but much remains to be learned about how the T → R transition occurs. Two models for the cooperative binding of ligands to proteins with multiple binding sites have greatly influenced thinking about this problem.

The first model was proposed by Jacques Monod, Jeffries Wyman, and Jean-Pierre Changeux in 1965, and is called the **MWC model** or the **concerted**

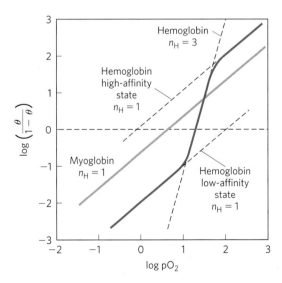

FIGURE 5–14 Hill plots for oxygen binding to myoglobin and hemoglobin. When $n_H = 1$, there is no evident cooperativity. The maximum degree of cooperativity observed for hemoglobin corresponds approximately to $n_H = 3$. Note that while this indicates a high level of cooperativity, n_H is less than n, the number of O_2-binding sites in hemoglobin. This is normal for a protein that exhibits allosteric binding behavior.

BOX 5-1 ⚕ MEDICINE Carbon Monoxide: A Stealthy Killer

Lake Powell, Arizona, August 2000. A family was vacationing in a rented houseboat. They turned on the electrical generator to power an air conditioner and a television. About 15 minutes later, two brothers, aged 8 and 11, jumped off the swim deck at the stern. Situated immediately below the deck was the exhaust port for the generator. Within two minutes, both boys were overcome by the carbon monoxide in the exhaust, which had become concentrated in the space under the deck. Both drowned. These deaths, along with a series of deaths in the 1990s that were linked to houseboats of similar design, eventually led to the recall and redesign of the generator exhaust assembly.

Carbon monoxide (CO), a colorless, odorless gas, is responsible for more than half of yearly deaths due to poisoning worldwide. CO has an approximately 250-fold greater affinity for hemoglobin than does oxygen. Consequently, relatively low levels of CO can have substantial and tragic effects. When CO combines with hemoglobin, the complex is referred to as carboxyhemoglobin, or COHb.

Some CO is produced by natural processes, but locally high levels generally result only from human activities. Engine and furnace exhausts are important sources, as CO is a byproduct of the incomplete combustion of fossil fuels. In the United States alone, nearly 4,000 people succumb to CO poisoning each year, both accidentally and intentionally. Many of the accidental deaths involve undetected CO buildup in enclosed spaces, such as when a household furnace malfunctions or leaks, venting CO into a home. However, CO poisoning can also occur in open spaces, as unsuspecting people at work or play inhale the exhaust from generators, outboard motors, tractor engines, recreational vehicles, or lawn mowers.

Carbon monoxide levels in the atmosphere are rarely dangerous, ranging from less than 0.05 part per million (ppm) in remote and uninhabited areas to 3 to 4 ppm in some cities of the northern hemisphere. In the United States, the government-mandated (Occupational Safety and Health Administration, OSHA) limit for CO at worksites is 50 ppm for people working an eight-hour shift. The tight binding of CO to hemoglobin

means that COHb can accumulate over time as people are exposed to a constant low-level source of CO.

In an average, healthy individual, 1% or less of the total hemoglobin is complexed as COHb. Since CO is a product of tobacco smoke, many smokers have COHb levels in the range of 3% to 8% of total hemoglobin, and the levels can rise to 15% for chain-smokers. COHb levels equilibrate at 50% in people who breathe air containing 570 ppm of CO for several hours. Reliable methods have been developed that relate CO content in the atmosphere to COHb levels in the blood (Fig. 1). In tests of houseboats with a generator exhaust like the one responsible for the Lake Powell deaths, CO levels reached 6,000 to 30,000 ppm under the swim deck, and atmospheric O_2 levels under the deck declined from 21% to 12%. Even above the swim deck, CO levels of up to 7,200 ppm were detected, high enough to cause death within a few minutes.

How is a human affected by COHb? At levels of less than 10% of total hemoglobin, symptoms are rarely observed. At 15%, the individual experiences mild headaches. At 20% to 30%, the headache is severe and is generally accompanied by nausea, dizziness, confusion,

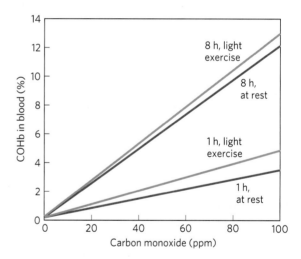

FIGURE 1 Relationship between levels of COHb in blood and concentration of CO in the surrounding air. Four different conditions of exposure are shown, comparing the effects of short versus extended exposure, and exposure at rest versus exposure during light exercise.

---

model (Fig. 5–15a). The concerted model assumes that the subunits of a cooperatively binding protein are functionally identical, that each subunit can exist in (at least) two conformations, and that all subunits undergo the transition from one conformation to the other simultaneously. In this model, no protein has individual subunits in different conformations. The two conformations are in equilibrium. The ligand can bind to either

conformation, but binds each with different affinity. Successive binding of ligand molecules to the low-affinity conformation (which is more stable in the absence of ligand) makes a transition to the high-affinity conformation more likely.

In the second model, the **sequential model** (Fig. 5–15b), proposed in 1966 by Daniel Koshland and colleagues, ligand binding can induce a change of disorientation, and some visual disturbances; these symptoms are generally reversed if the individual is treated with oxygen. At COHb levels of 30% to 50%, the neurological symptoms become more severe, and at levels near 50%, the individual loses consciousness and can sink into coma. Respiratory failure may follow. With prolonged exposure, some damage becomes permanent. Death normally occurs when COHb levels rise above 60%. Autopsy on the boys who died at Lake Powell revealed COHb levels of 59% and 52%.

Binding of CO to hemoglobin is affected by many factors, including exercise (Fig. 1) and changes in air pressure related to altitude. Because of their higher base levels of COHb, smokers exposed to a source of CO often develop symptoms faster than nonsmokers. Individuals with heart, lung, or blood diseases that reduce the availability of oxygen to tissues may also experience symptoms at lower levels of CO exposure. Fetuses are at particular risk for CO poisoning, because fetal hemoglobin has a somewhat higher affinity for CO than adult hemoglobin. Cases of CO exposure have been recorded in which the fetus died but the mother recovered.

It may seem surprising that the loss of half of one's hemoglobin to COHb can prove fatal—we know that people with any of several anemic conditions manage to function reasonably well with half the usual complement of active hemoglobin. However, the binding of CO to hemoglobin does more than remove protein from the pool available to bind oxygen. It also affects the affinity of the remaining hemoglobin subunits for oxygen. As CO binds to one or two subunits of a hemoglobin tetramer, the affinity for O_2 is increased substantially in the remaining subunits (Fig. 2). Thus, a hemoglobin tetramer with two bound CO molecules can efficiently bind O_2 in the lungs—but it releases very little of it in the tissues. Oxygen deprivation in the tissues rapidly becomes severe. To add to the problem, the effects of CO are not limited to interference with hemoglobin function. CO binds to other heme proteins and a variety of metalloproteins. The effects of these interactions are not yet well understood, but they may be responsible for some of the longer-term effects of acute but nonfatal CO poisoning.

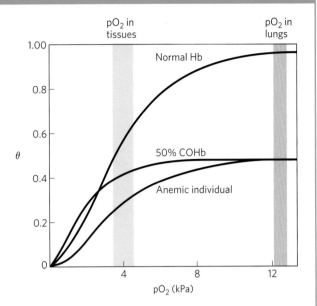

FIGURE 2 Several oxygen-binding curves: for normal hemoglobin, hemoglobin from an anemic individual with only 50% of her hemoglobin functional, and hemoglobin from an individual with 50% of his hemoglobin subunits complexed with CO. The pO_2 in human lungs and tissues is indicated.

When CO poisoning is suspected, rapid evacuation of the person away from the CO source is essential, but this does not always result in rapid recovery. When an individual is moved from the CO-polluted site to a normal, outdoor atmosphere, O_2 begins to replace the CO in hemoglobin—but the COHb levels drop only slowly. The half-time is 2 to 6.5 hours, depending on individual and environmental factors. If 100% oxygen is administered with a mask, the rate of exchange can be increased about fourfold; the half-time for O_2-CO exchange can be reduced to tens of minutes if 100% oxygen at a pressure of 3 atm (303 kPa) is supplied. Thus, rapid treatment by a properly equipped medical team is critical.

Carbon monoxide detectors in all homes are highly recommended. This is a simple and inexpensive measure to avoid possible tragedy. After completing the research for this box, we immediately purchased several new CO detectors for our homes.

conformation in an individual subunit. A conformational change in one subunit makes a similar change in an adjacent subunit, as well as the binding of a second ligand molecule, more likely. There are more potential intermediate states in this model than in the concerted model. The two models are not mutually exclusive; the concerted model may be viewed as the "all-or-none" limiting case of the sequential model. In

Chapter 6 we use these models to investigate allosteric enzymes.

Hemoglobin Also Transports H⁺ and CO₂

In addition to carrying nearly all the oxygen required by cells from the lungs to the tissues, hemoglobin carries two end products of cellular respiration—H^+ and CO_2—from the tissues to the lungs and the kidneys, where

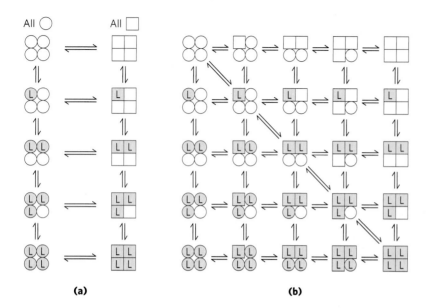

FIGURE 5–15 Two general models for the interconversion of inactive and active forms of a protein during cooperative ligand binding. Although the models may be applied to any protein—including any enzyme (Chapter 6)—that exhibits cooperative binding, we show here four subunits because the model was originally proposed for hemoglobin. **(a)** In the concerted, or all-or-none, model (MWC model), all subunits are postulated to be in the same conformation, either all ○ (low affinity or inactive) or all □ (high affinity or active). Depending on the equilibrium, K_{eq}, between ○ and □ forms, the binding of one or more ligand molecules (L) will pull the equilibrium toward the □ form. Subunits with bound L are shaded. **(b)** In the sequential model, each individual subunit can be in either the ○ or □ form. A very large number of conformations is thus possible.

they are excreted. The CO_2, produced by oxidation of organic fuels in mitochondria, is hydrated to form bicarbonate:

$$CO_2 + H_2O \rightleftharpoons H^+ + HCO_3^-$$

This reaction is catalyzed by **carbonic anhydrase**, an enzyme particularly abundant in erythrocytes. Carbon dioxide is not very soluble in aqueous solution, and bubbles of CO_2 would form in the tissues and blood if it were not converted to bicarbonate. As you can see from the reaction catalyzed by carbonic anhydrase, the hydration of CO_2 results in an increase in the H^+ concentration (a decrease in pH) in the tissues. The binding of oxygen by hemoglobin is profoundly influenced by pH and CO_2 concentration, so the interconversion of CO_2 and bicarbonate is of great importance to the regulation of oxygen binding and release in the blood.

Hemoglobin transports about 40% of the total H^+ and 15% to 20% of the CO_2 formed in the tissues to the lungs and kidneys. (The remainder of the H^+ is absorbed by the plasma's bicarbonate buffer; the remainder of the CO_2 is transported as dissolved HCO_3^- and CO_2.) The binding of H^+ and CO_2 is inversely related to the binding of oxygen. At the relatively low pH and high CO_2 concentration of peripheral tissues, the affinity of hemoglobin for oxygen decreases as H^+ and CO_2 are bound, and O_2 is released to the tissues. Conversely, in the capillaries of the lung, as CO_2 is excreted and the blood pH consequently rises, the affinity of hemoglobin for oxygen increases and the protein binds more O_2 for transport to the peripheral tissues. This effect of pH and CO_2 concentration on the binding and release of oxygen by hemoglobin is called the **Bohr effect**, after Christian Bohr, the Danish physiologist (and father of physicist Niels Bohr) who discovered it in 1904.

The binding equilibrium for hemoglobin and one molecule of oxygen can be designated by the reaction

$$Hb + O_2 \rightleftharpoons HbO_2$$

but this is not a complete statement. To account for the effect of H^+ concentration on this binding equilibrium, we rewrite the reaction as

$$HHb^+ + O_2 \rightleftharpoons HbO_2 + H^+$$

where HHb^+ denotes a protonated form of hemoglobin. This equation tells us that the O_2-saturation curve of hemoglobin is influenced by the H^+ concentration **(Fig. 5–16)**. Both O_2 and H^+ are bound by hemoglobin, but with inverse affinity. When the oxygen concentration is high, as in the lungs, hemoglobin binds O_2 and releases protons. When the oxygen concentration is

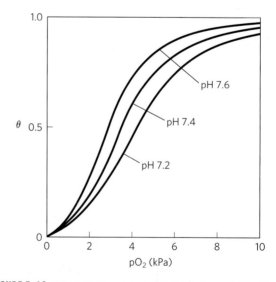

FIGURE 5–16 Effect of pH on oxygen binding to hemoglobin. The pH of blood is 7.6 in the lungs and 7.2 in the tissues. Experimental measurements on hemoglobin binding are often performed at pH 7.4.

low, as in the peripheral tissues, H^+ is bound and O_2 is released.

Oxygen and H^+ are not bound at the same sites in hemoglobin. Oxygen binds to the iron atoms of the hemes, whereas H^+ binds to any of several amino acid residues in the protein. A major contribution to the Bohr effect is made by His^{146} (His HC3) of the β subunits. When protonated, this residue forms one of the ion pairs—to Asp^{94} (Asp FG1)—that helps stabilize deoxyhemoglobin in the T state (Fig. 5–9). The ion pair stabilizes the protonated form of His HC3, giving this residue an abnormally high pK_a in the T state. The pK_a falls to its normal value of 6.0 in the R state because the ion pair cannot form, and this residue is largely unprotonated in oxyhemoglobin at pH 7.6, the blood pH in the lungs. As the concentration of H^+ rises, protonation of His HC3 promotes release of oxygen by favoring a transition to the T state. Protonation of the amino-terminal residues of the α subunits, certain other His residues, and perhaps other groups has a similar effect.

Thus we see that the four polypeptide chains of hemoglobin communicate with each other not only about O_2 binding to their heme groups but also about H^+ binding to specific amino acid residues. And there is still more to the story. Hemoglobin also binds CO_2, again in a manner inversely related to the binding of oxygen. Carbon dioxide binds as a carbamate group to the α-amino group at the amino-terminal end of each globin chain, forming carbaminohemoglobin:

Amino-terminal residue → Carbamino-terminal residue

This reaction produces H^+, contributing to the Bohr effect. The bound carbamates also form additional salt bridges (not shown in Fig. 5–9) that help to stabilize the T state and promote the release of oxygen.

When the concentration of carbon dioxide is high, as in peripheral tissues, some CO_2 binds to hemoglobin and the affinity for O_2 decreases, causing its release. Conversely, when hemoglobin reaches the lungs, the high oxygen concentration promotes binding of O_2 and release of CO_2. It is the capacity to communicate ligand-binding information from one polypeptide subunit to the others that makes the hemoglobin molecule so beautifully adapted to integrating the transport of O_2, CO_2, and H^+ by erythrocytes.

Oxygen Binding to Hemoglobin Is Regulated by 2,3-Bisphosphoglycerate

The interaction of **2,3-bisphosphoglycerate (BPG)** with hemoglobin molecules further refines the function of hemoglobin, and provides an example of heterotropic allosteric modulation.

2,3-Bisphosphoglycerate

BPG is present in relatively high concentrations in erythrocytes. When hemoglobin is isolated, it contains substantial amounts of bound BPG, which can be difficult to remove completely. In fact, the O_2-binding curves for hemoglobin that we have examined to this point were obtained in the presence of bound BPG. 2,3-Bisphosphoglycerate is known to greatly reduce the affinity of hemoglobin for oxygen—there is an inverse relationship between the binding of O_2 and the binding of BPG. We can therefore describe another binding process for hemoglobin:

$$HbBPG + O_2 \rightleftharpoons HbO_2 + BPG$$

BPG binds at a site distant from the oxygen-binding site and regulates the O_2-binding affinity of hemoglobin in relation to the pO_2 in the lungs. BPG is important in the physiological adaptation to the lower pO_2 at high altitudes. For a healthy human at sea level, the binding of O_2 to hemoglobin is regulated such that the amount of O_2 delivered to the tissues is nearly 40% of the maximum that could be carried by the blood **(Fig. 5–17)**. Imagine that such a person is suddenly transported from sea level to an altitude of 4,500 meters, where the pO_2 is considerably lower. The delivery of O_2 to the tissues is now reduced. However, after just a few hours at the higher altitude, the BPG concentration in the blood has begun to rise, leading to a decrease in the affinity of hemoglobin for oxygen. This adjustment in the BPG level has only a small effect on the binding of O_2 in the lungs but a considerable effect on the release of O_2 in the tissues. As a result, the delivery of oxygen to the tissues is restored to nearly 40% of the O_2 that can be transported by the blood. The situation is reversed when the person returns to sea level. The BPG concentration in erythrocytes also increases in people suffering from **hypoxia**, lowered oxygenation of peripheral tissues due to inadequate functioning of the lungs or circulatory system.

The site of BPG binding to hemoglobin is the cavity between the β subunits in the T state **(Fig. 5–18)**. This cavity is lined with positively charged amino acid residues that interact with the negatively charged groups of BPG. Unlike O_2, only one molecule of BPG is bound to each hemoglobin tetramer. BPG lowers hemoglobin's affinity for oxygen by stabilizing the T state. The transition to the R state narrows the binding pocket for BPG, precluding BPG binding. In the absence

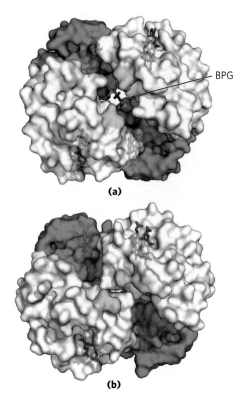

FIGURE 5–17 **Effect of BPG on oxygen binding to hemoglobin.** The BPG concentration in normal human blood is about 5 mM at sea level and about 8 mM at high altitudes. Note that hemoglobin binds to oxygen quite tightly when BPG is entirely absent, and the binding curve seems to be hyperbolic. In reality, the measured Hill coefficient for O_2-binding cooperativity decreases only slightly (from 3 to about 2.5) when BPG is removed from hemoglobin, but the rising part of the sigmoid curve is confined to a very small region close to the origin. At sea level, hemoglobin is nearly saturated with O_2 in the lungs, but just over 60% saturated in the tissues, so the amount of O_2 released in the tissues is about 38% of the maximum that can be carried in the blood. At high altitudes, O_2 delivery declines by about one-fourth, to 30% of maximum. An increase in BPG concentration, however, decreases the affinity of hemoglobin for O_2, so approximately 37% of what can be carried is again delivered to the tissues.

FIGURE 5–18 **Binding of BPG to deoxyhemoglobin. (a)** BPG binding stabilizes the T state of deoxyhemoglobin (PDB ID 1B86). The negative charges of BPG interact with several positively charged groups (shown in blue in this surface contour image) that surround the pocket between the β subunits on the surface of deoxyhemoglobin in the T state. **(b)** The binding pocket for BPG disappears on oxygenation, following transition to the R state (PDB ID 1BBB). (Compare with Fig. 5–10.)

of BPG, hemoglobin is converted to the R state more easily.

Regulation of oxygen binding to hemoglobin by BPG has an important role in fetal development. Because a fetus must extract oxygen from its mother's blood, fetal hemoglobin must have greater affinity than the maternal hemoglobin for O_2. The fetus synthesizes γ subunits rather than β subunits, forming $\alpha_2\gamma_2$ hemoglobin. This tetramer has a much lower affinity for BPG than normal adult hemoglobin, and a correspondingly higher affinity for O_2. 🖱 **Oxygen-Binding Proteins–Hemoglobin Is Susceptible to Allosteric Regulation**

Sickle-Cell Anemia Is a Molecular Disease of Hemoglobin

 The hereditary human disease sickle-cell anemia demonstrates strikingly the importance of amino acid sequence in determining the secondary, tertiary, and quaternary structures of globular proteins, and thus their biological functions. Almost 500 genetic variants of hemoglobin are known to occur in the human population; all but a few are quite rare. Most variations consist of differences in a single amino acid residue. The effects on hemoglobin structure and function are often minor but can sometimes be extraordinary. Each hemoglobin variation is the product of an altered gene. The variant genes are called alleles. Because humans generally have two copies of each gene, an individual may have two copies of one allele (thus being homozygous for that gene) or one copy of each of two different alleles (thus heterozygous).

Sickle-cell anemia occurs in individuals who inherit the allele for sickle-cell hemoglobin from both parents. The erythrocytes of these individuals are fewer and also abnormal. In addition to an unusually large number of immature cells, the blood contains many long, thin, sickle-shaped erythrocytes **(Fig. 5–19)**. When hemoglobin from sickle cells (called hemoglobin S) is deoxygenated, it becomes insoluble and forms polymers that

(a)   2 μm   **(b)**

FIGURE 5–19 A comparison of **(a)** uniform, cup-shaped, normal erythrocytes with **(b)** the variably shaped erythrocytes seen in sickle-cell anemia, which range from normal to spiny or sickle-shaped.

aggregate into tubular fibers **(Fig. 5–20)**. Normal hemoglobin (hemoglobin A) remains soluble on deoxygenation. The insoluble fibers of deoxygenated hemoglobin S cause the deformed, sickle shape of the erythrocytes, and the proportion of sickled cells increases greatly as blood is deoxygenated.

The altered properties of hemoglobin S result from a single amino acid substitution, a Val instead of a Glu residue at position 6 in the two β chains. The R group of valine has no electric charge, whereas glutamate has a negative charge at pH 7.4. Hemoglobin S therefore has two fewer negative charges than hemoglobin A (one fewer on each β chain). Replacement of the Glu residue by Val creates a "sticky" hydrophobic contact point at position 6 of the β chain, which is on the outer surface of the molecule. These sticky spots cause deoxyhemoglobin S molecules to associate abnormally with each other, forming the long, fibrous aggregates characteristic of this disorder. 🖱 Oxygen-Binding Proteins—Defects in Hb Lead to Serious Genetic Disease

Sickle-cell anemia, as we have noted, occurs in individuals homozygous for the sickle-cell allele of the gene encoding the β subunit of hemoglobin. Individuals who receive the sickle-cell allele from only one parent and are thus heterozygous experience a milder condition called sickle-cell trait; only about 1% of their erythrocytes become sickled on deoxygenation. These individuals may live completely normal lives if they avoid vigorous exercise and other stresses on the circulatory system.

Sickle-cell anemia is life-threatening and painful. People with this disease suffer repeated crises brought on by physical exertion. They become weak, dizzy, and short of breath, and they also experience heart murmurs and an increased pulse rate. The hemoglobin content of their blood is only about half the normal value of 15 to 16 g/100 mL, because sickled cells are very fragile and

rupture easily; this results in anemia ("lack of blood"). An even more serious consequence is that capillaries become blocked by the long, abnormally shaped cells, causing severe pain and interfering with normal organ function—a major factor in the early death of many people with the disease.

Without medical treatment, people with sickle-cell anemia usually die in childhood. Curiously, the frequency of the sickle-cell allele in populations is unusually high in

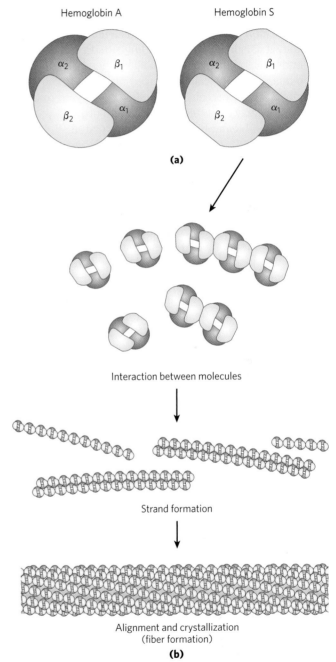

FIGURE 5–20 Normal and sickle-cell hemoglobin. (a) Subtle differences between the conformations of hemoglobin A and hemoglobin S result from a single amino acid change in the β chains. **(b)** As a result of this change, deoxyhemoglobin S has a hydrophobic patch on its surface, which causes the molecules to aggregate into strands that align into insoluble fibers.

certain parts of Africa. Investigation into this matter led to the finding that in heterozygous individuals, the allele confers a small but significant resistance to lethal forms of malaria. Natural selection has resulted in an allele population that balances the deleterious effects of the homozygous condition against the resistance to malaria afforded by the heterozygous condition. ■

SUMMARY 5.1 Reversible Binding of a Protein to a Ligand: Oxygen-Binding Proteins

▶   Protein function often entails interactions with other molecules. A protein binds a molecule, known as a ligand, at its binding site. Proteins may undergo conformational changes when a ligand binds, a process called induced fit. In a multisubunit protein, the binding of a ligand to one subunit may affect ligand binding to other subunits. Ligand binding can be regulated.

▶   Myoglobin contains a heme prosthetic group, which binds oxygen. Heme consists of a single atom of Fe^{2+} coordinated within a porphyrin. Oxygen binds to myoglobin reversibly; this simple reversible binding can be described by an association constant K_a or a dissociation constant K_d. For a monomeric protein such as myoglobin, the fraction of binding sites occupied by a ligand is a hyperbolic function of ligand concentration.

▶   Normal adult hemoglobin has four heme-containing subunits, two α and two β, similar in structure to each other and to myoglobin. Hemoglobin exists in two interchangeable structural states, T and R. The T state is most stable when oxygen is not bound. Oxygen binding promotes transition to the R state.

▶   Oxygen binding to hemoglobin is both allosteric and cooperative. As O_2 binds to one binding site, the hemoglobin undergoes conformational changes that affect the other binding sites—an example of allosteric behavior. Conformational changes between the T and R states, mediated by subunit-subunit interactions, result in cooperative binding; this is described by a sigmoid binding curve and can be analyzed by a Hill plot.

▶   Two major models have been proposed to explain the cooperative binding of ligands to multisubunit proteins: the concerted model and the sequential model.

▶   Hemoglobin also binds H^+ and CO_2, resulting in the formation of ion pairs that stabilize the T state and lessen the protein's affinity for O_2 (the Bohr effect). Oxygen binding to hemoglobin is also modulated by 2,3-bisphosphoglycerate, which binds to and stabilizes the T state.

▶   Sickle-cell anemia is a genetic disease caused by a single amino acid substitution (Glu^6 to Val^6) in each β chain of hemoglobin. The change produces a hydrophobic patch on the surface of the hemoglobin that causes the molecules to aggregate into bundles of fibers. This homozygous condition results in serious medical complications.

5.2 Complementary Interactions between Proteins and Ligands: The Immune System and Immunoglobulins

We have seen how the conformations of oxygen-binding proteins affect and are affected by the binding of small ligands (O_2 or CO) to the heme group. However, most protein-ligand interactions do not involve a prosthetic group. Instead, the binding site for a ligand is more often like the hemoglobin binding site for BPG—a cleft in the protein lined with amino acid residues, arranged to make the binding interaction highly specific. Effective discrimination between ligands is the norm at binding sites, even when the ligands have only minor structural differences.

All vertebrates have an immune system capable of distinguishing molecular "self" from "nonself" and then destroying what is identified as nonself. In this way, the immune system eliminates viruses, bacteria, and other pathogens and molecules that may pose a threat to the organism. On a physiological level, the immune response is an intricate and coordinated set of interactions among many classes of proteins, molecules, and cell types. At the level of individual proteins, the immune response demonstrates how an acutely sensitive and specific biochemical system is built upon the reversible binding of ligands to proteins.

The Immune Response Features a Specialized Array of Cells and Proteins

Immunity is brought about by a variety of **leukocytes** (white blood cells), including **macrophages** and **lymphocytes**, all of which develop from undifferentiated stem cells in the bone marrow. Leukocytes can leave the bloodstream and patrol the tissues, each cell producing one or more proteins capable of recognizing and binding to molecules that might signal an infection.

The immune response consists of two complementary systems, the humoral and cellular immune systems. The **humoral immune system** (Latin *humor*, "fluid") is directed at bacterial infections and extracellular viruses (those found in the body fluids), but can also respond to individual foreign proteins. The **cellular immune system** destroys host cells infected by viruses and also destroys some parasites and foreign tissues.

At the heart of the humoral immune response are soluble proteins called **antibodies** or **immunoglobulins**, often abbreviated **Ig**. Immunoglobulins bind bacteria, viruses, or large molecules identified as foreign and target them for destruction. Making up 20% of blood protein, the immunoglobulins are produced by **B lymphocytes**, or **B cells**, so named because they complete their development in the *bone* marrow.

The agents at the heart of the cellular immune response are a class of **T lymphocytes**, or **T cells** (so called because the latter stages of their development occur in the *thymus*), known as **cytotoxic T cells** (**T_C cells**, also called killer T cells). Recognition of infected cells or parasites involves proteins called **T-cell receptors** on the surface of T_C cells. Receptors are proteins, usually found on the outer surface of cells and extending through the plasma membrane; they recognize and bind extracellular ligands, triggering changes inside the cell.

In addition to cytotoxic T cells, there are **helper T cells (T_H cells)**, whose function it is to produce soluble signaling proteins called cytokines, which include the interleukins. T_H cells interact with macrophages. The T_H cells participate only indirectly in the destruction of infected cells and pathogens, stimulating the selective proliferation of those T_C and B cells that can bind to a particular antigen. This process, called **clonal selection**, increases the number of immune system cells that can respond to a particular pathogen. The importance of T_H cells is dramatically illustrated by the epidemic produced by HIV (human immunodeficiency virus), the virus that causes AIDS (acquired immune deficiency syndrome). The primary targets of HIV infection are T_H cells. Elimination of these cells progressively incapacitates the entire immune system. Table 5–2 summarizes the functions of some leukocytes of the immune system.

Each recognition protein of the immune system, either a T-cell receptor or an antibody produced by a B cell, specifically binds some particular chemical structure, distinguishing it from virtually all others. Humans are capable of producing more than 10^8 different antibodies with distinct binding specificities. Given this extraordinary diversity, any chemical structure on the surface of a virus or invading cell will most likely be recognized and bound by one or more antibodies. Antibody diversity is derived from random reassembly of a set of immunoglobulin gene segments through genetic recombination mechanisms that are discussed in Chapter 25 (see Fig. 25–42).

A specialized lexicon is used to describe the unique interactions between antibodies or T-cell receptors and the molecules they bind. Any molecule or pathogen capable of eliciting an immune response is called an **antigen**. An antigen may be a virus, a bacterial cell wall, or an individual protein or other macromolecule. A complex antigen may be bound by several different antibodies. An individual antibody or T-cell receptor binds only a particular molecular structure within the antigen, called its **antigenic determinant** or **epitope**.

It would be unproductive for the immune system to respond to small molecules that are common intermediates and products of cellular metabolism. Molecules of M_r <5,000 are generally not antigenic. However, when small molecules are covalently attached to large proteins in the laboratory, they can be used to elicit an immune response. These small molecules are called **haptens**. The antibodies produced in response to protein-linked haptens will then bind to the same small molecules in their free form. Such antibodies are sometimes used in the development of analytical tests described later in this chapter or as a specific ligand in affinity chromatography (see Fig. 3–18c). We now turn to a more detailed description of antibodies and their binding properties.

Antibodies Have Two Identical Antigen-Binding Sites

Immunoglobulin G (IgG) is the major class of antibody molecule and one of the most abundant proteins in the blood serum. IgG has four polypeptide chains: two large ones, called heavy chains, and two light chains, linked by noncovalent and disulfide bonds into a complex of M_r 150,000. The heavy chains of an IgG molecule interact at one end, then branch to interact separately with the light chains, forming a Y-shaped molecule (**Fig. 5–21**). At the "hinges" separating the base of an IgG molecule from its branches, the immunoglobulin can be cleaved with proteases. Cleavage with the protease papain liberates the basal fragment, called **Fc** because it usually *c*rystallizes readily, and the two branches, called **Fab**, the *a*ntigen-*b*inding fragments. Each branch has a single antigen-binding site.

The fundamental structure of immunoglobulins was first established by Gerald Edelman and Rodney Porter. Each chain is made up of identifiable domains; some are constant in sequence and structure from one IgG to the next, others are variable. The constant domains have a characteristic structure known as the **immunoglobulin fold**, a well-conserved structural motif in the all-β class of proteins (Chapter 4).

TABLE 5–2	Some Types of Leukocytes Associated with the Immune System

Cell type	Function
Macrophages	Ingest large particles and cells by phagocytosis
B lymphocytes (B cells)	Produce and secrete antibodies
T lymphocytes (T cells)	
Cytotoxic (killer) T cells (T_C)	Interact with infected host cells through receptors on T-cell surface
Helper T cells (T_H)	Interact with macrophages and secrete cytokines (interleukins) that stimulate T_C, T_H, and B cells to proliferate.

FIGURE 5-21 **Immunoglobulin G. (a)** Pairs of heavy and light chains combine to form a Y-shaped molecule. Two antigen-binding sites are formed by the combination of variable domains from one light (V_L) and one heavy (V_H) chain. Cleavage with papain separates the Fab and Fc portions of the protein in the hinge region. The Fc portion of the molecule also contains bound carbohydrate (shown in (b)). **(b)** A ribbon model of the first complete IgG molecule to be crystallized and structurally analyzed (PDB ID 1IGT). Although the molecule has two identical heavy chains (two shades of blue) and two identical light chains (two shades of red), it crystallized in the asymmetric conformation shown here. Conformational flexibility may be important to the function of immunoglobulins.

There are three of these constant domains in each heavy chain and one in each light chain. The heavy and light chains also have one variable domain each, in which most of the variability in amino acid sequence is found. The variable domains associate to create the antigen-binding site (Fig. 5–21), allowing formation of an antigen-antibody complex **(Fig. 5–22)**.

In many vertebrates, IgG is but one of five classes of immunoglobulins. Each class has a characteristic type of heavy chain, denoted α, δ, ε, γ, and μ for IgA, IgD, IgE, IgG, and IgM, respectively. Two types of light chain, κ and λ, occur in all classes of immunoglobulins. The overall structures of **IgD** and **IgE** are similar to that of IgG. **IgM** occurs either in a monomeric, membrane-bound form or in a secreted form that is a cross-linked pentamer of this basic structure **(Fig. 5–23)**. **IgA**, found principally in secretions such as saliva, tears, and milk, can be a monomer, dimer, or trimer. IgM is the first antibody to

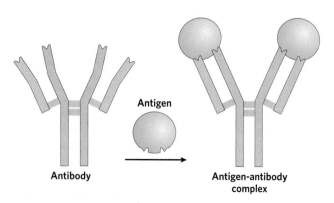

FIGURE 5-22 **Binding of IgG to an antigen.** To generate an optimal fit for the antigen, the binding sites of IgG often undergo slight conformational changes. Such induced fit is common to many protein-ligand interactions.

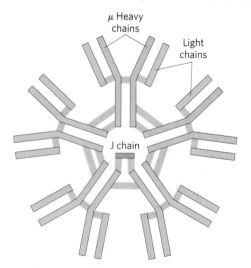

FIGURE 5-23 **IgM pentamer of immunoglobulin units.** The pentamer is cross-linked with disulfide bonds (yellow). The J chain is a polypeptide of M_r 20,000 found in both IgA and IgM.

be made by B lymphocytes and the major antibody in the early stages of a primary immune response. Some B cells soon begin to produce IgD (with the same antigen-binding site as the IgM produced by the same cell), but the particular function of IgD is less clear.

The IgG described above is the major antibody in secondary immune responses, which are initiated by a class of B cells called memory B cells. As part of the organism's ongoing immunity to antigens already encountered and dealt with, IgG is the most abundant immunoglobulin in the blood. When IgG binds to an invading bacterium or virus, it activates certain leukocytes such as macrophages to engulf and destroy the invader, and also activates some other parts of the immune response. Receptors on the macrophage surface recognize and bind the Fc region of IgG. When these Fc receptors bind an antibody-pathogen complex, the macrophage engulfs the complex by phagocytosis (**Fig. 5–24**).

IgE plays an important role in the allergic response, interacting with basophils (phagocytic leukocytes) in the blood and with histamine-secreting cells called mast cells, which are widely distributed in tissues. This immunoglobulin binds, through its Fc region, to special Fc receptors on the basophils or mast cells. In this form, IgE serves as a receptor for antigen. If antigen is bound, the cells are induced to secrete histamine and other biologically active amines that cause dilation and increased permeability of blood vessels. These effects on the blood vessels are thought to facilitate the movement of immune system cells and proteins to sites of inflammation. They also produce the symptoms normally associated with allergies. Pollen or other allergens are recognized as foreign, triggering an immune response normally reserved for pathogens. ∎

Antibodies Bind Tightly and Specifically to Antigen

The binding specificity of an antibody is determined by the amino acid residues in the variable domains of its

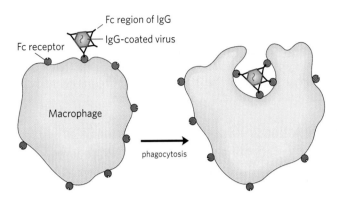

FIGURE 5–24 Phagocytosis of an antibody-bound virus by a macrophage. The Fc regions of antibodies bound to the virus now bind to Fc receptors on the surface of a macrophage, triggering the macrophage to engulf and destroy the virus.

heavy and light chains. Many residues in these domains are variable, but not equally so. Some, particularly those lining the antigen-binding site, are hypervariable—especially likely to differ. Specificity is conferred by chemical complementarity between the antigen and its specific binding site, in terms of shape and the location of charged, nonpolar, and hydrogen-bonding groups. For example, a binding site with a negatively charged group may bind an antigen with a positive charge in the complementary position. In many instances, complementarity is achieved interactively as the structures of antigen and binding site influence each other as they come closer together. Conformational changes in the antibody and/or the antigen then allow the complementary groups to interact fully. This is an example of induced fit. The complex of a peptide derived from HIV (a model antigen) and an Fab molecule, shown in **Figure 5–25**, illustrates some of these properties. The changes in structure observed on antigen binding are particularly striking in this example.

A typical antibody-antigen interaction is quite strong, characterized by K_d values as low as 10^{-10} M

(a) Conformation with no antigen bound

(b) Antigen bound (but not shown)

(c) Antigen bound (shown)

FIGURE 5–25 Induced fit in the binding of an antigen to IgG. The molecule here, shown in surface contour, is the Fab fragment of an IgG. The antigen is a small peptide derived from HIV. Two residues in the heavy chain (blue) and one in the light chain (red) are colored to provide visual points of reference. **(a)** View of the Fab fragment in the absence of antigen, looking down on the antigen-binding site (PDB ID 1GGC). **(b)** The same view, but with the Fab fragment in the "bound" conformation (PDB ID 1GGI); the antigen is omitted to provide an unobstructed view of the altered binding site. Note how the binding cavity has enlarged and several groups have shifted position. **(c)** The same view as (b), but with the antigen in the binding site, pictured as a red stick structure.

(recall that a lower K_d corresponds to a stronger binding interaction; see Table 5–1). The K_d reflects the energy derived from the various ionic, hydrogen-bonding, hydrophobic, and van der Waals interactions that stabilize the binding. The binding energy required to produce a K_d of 10^{-10} M is about 65 kJ/mol.

The Antibody-Antigen Interaction Is the Basis for a Variety of Important Analytical Procedures

The extraordinary binding affinity and specificity of antibodies make them valuable analytical reagents. Two types of antibody preparations are in use: polyclonal and monoclonal. **Polyclonal antibodies** are those produced by many different B lymphocytes responding to one antigen, such as a protein injected into an animal. Cells in the population of B lymphocytes produce antibodies that bind specific, different epitopes within the antigen. Thus, polyclonal preparations contain a mixture of antibodies that recognize different parts of the protein. **Monoclonal antibodies**, in contrast, are synthesized by a population of identical B cells (a **clone**) grown in cell culture. These antibodies are homogeneous, all recognizing the same epitope. The techniques for producing monoclonal antibodies were developed by Georges Köhler and Cesar Milstein.

The specificity of antibodies has practical uses. A selected antibody can be covalently attached to a resin

Georges Köhler, 1946–1995   Cesar Milstein, 1927–2002

and used in a chromatography column of the type shown in Figure 3–17c . When a mixture of proteins is added to the column, the antibody specifically binds its target protein and retains it on the column while other proteins are washed through. The target protein can then be eluted from the resin by a salt solution or some other agent. This is a powerful protein analytical tool.

In another versatile analytical technique, an antibody is attached to a radioactive label or some other reagent that makes it easy to detect. When the antibody binds the target protein, the label reveals the presence of the protein in a solution or its location in a gel or even a living cell. Several variations of this procedure are illustrated in **Figure 5–26**.

FIGURE 5–26 Antibody techniques. The specific reaction of an antibody with its antigen is the basis of several techniques that identify and quantify a specific protein in a complex sample. **(a)** A schematic representation of the general method. **(b)** An ELISA to test for the presence of herpes simplex virus (HSV) antibodies in blood samples. Wells were coated with an HSV antigen, to which antibodies against HSV will bind. The second antibody is anti-human IgG linked to horseradish peroxidase. Following completion of the steps shown in (a), blood samples with greater amounts of HSV antibody turn brighter yellow. **(c)** An immunoblot. Lanes 1 to 3 are from an SDS gel; samples from successive stages in the purification of a protein kinase were separated and stained with Coomassie blue. Lanes 4 to 6 show the same samples, but these were electrophoretically transferred to a nitrocellulose membrane after separation on an SDS gel. The membrane was then "probed" with antibody against the protein kinase. The numbers between the SDS gel and the immunoblot indicate M_r in thousands.

① Coat surface with sample (antigens).

② Block unoccupied sites with nonspecific protein.

③ Incubate with primary antibody against specific antigen.

④ Incubate with secondary antibody–enzyme complex that binds primary antibody.

⑤ Add substrate.

⑥ Formation of colored product indicates presence of specific antigen.

(a)

ELISA
(b)

SDS gel   Immunoblot
(c)

An **ELISA** (enzyme-linked immunosorbent assay) can be used to rapidly screen for and quantify an antigen in a sample (Fig. 5–26b). Proteins in the sample are adsorbed to an inert surface, usually a 96-well polystyrene plate. The surface is washed with a solution of an inexpensive nonspecific protein (often casein from nonfat dry milk powder) to block proteins introduced in subsequent steps from adsorbing to unoccupied sites. The surface is then treated with a solution containing the primary antibody—an antibody against the protein of interest. Unbound antibody is washed away, and the surface is treated with a solution containing a secondary antibody—antibody against the primary antibody—linked to an enzyme that catalyzes a reaction that forms a colored product. After unbound secondary antibody is washed away, the substrate of the antibody-linked enzyme is added. Product formation (monitored as color intensity) is proportional to the concentration of the protein of interest in the sample.

In an **immunoblot assay**, also called a **Western blot** (Fig. 5–26c), proteins that have been separated by gel electrophoresis are transferred electrophoretically to a nitrocellulose membrane. The membrane is blocked (as described above for ELISA), then treated successively with primary antibody, secondary antibody linked to enzyme, and substrate. A colored precipitate forms only along the band containing the protein of interest. Immunoblotting allows the detection of a minor component in a sample and provides an approximation of its molecular weight. 🖱 **Immunoblotting**

We will encounter other aspects of antibodies in later chapters. They are extremely important in medicine and can tell us much about the structure of proteins and the action of genes.

SUMMARY 5.2 Complementary Interactions between Proteins and Ligands: The Immune System and Immunoglobulins

▶ The immune response is mediated by interactions among an array of specialized leukocytes and their associated proteins. T lymphocytes produce T-cell receptors. B lymphocytes produce immunoglobulins. In a process called clonal selection, helper T cells induce the proliferation of B cells and cytotoxic T cells that produce immunoglobulins or of T-cell receptors that bind to a specific antigen.

▶ Humans have five classes of immunoglobulins, each with different biological functions. The most abundant class is IgG, a Y-shaped protein with two heavy and two light chains. The domains near the upper ends of the Y are hypervariable within the broad population of IgGs and form two antigen-binding sites.

▶ A given immunoglobulin generally binds to only a part, called the epitope, of a large antigen. Binding often involves a conformational change in the IgG, an induced fit to the antigen.

▶ The exquisite binding specificity of immunoglobins is exploited in analytical techniques such as the ELISA and the immunoblot.

5.3 Protein Interactions Modulated by Chemical Energy: Actin, Myosin, and Molecular Motors

Organisms move. Cells move. Organelles and macromolecules within cells move. Most of these movements arise from the activity of a fascinating class of protein-based molecular motors. Fueled by chemical energy, usually derived from ATP, large aggregates of motor proteins undergo cyclic conformational changes that accumulate into a unified, directional force—the tiny force that pulls apart chromosomes in a dividing cell, and the immense force that levers a pouncing, quarter-ton jungle cat into the air.

The interactions among motor proteins, as you might predict, feature complementary arrangements of ionic, hydrogen-bonding, hydrophobic, and van der Waals interactions at protein binding sites. In motor proteins, however, these interactions achieve exceptionally high levels of spatial and temporal organization.

Motor proteins underlie the contraction of muscles, the migration of organelles along microtubules, the rotation of bacterial flagella, and the movement of some proteins along DNA. Proteins called kinesins and dyneins move along microtubules in cells, pulling along organelles or reorganizing chromosomes during cell division. An interaction of dynein with microtubules brings about the motion of eukaryotic flagella and cilia. Flagellar motion in bacteria involves a complex rotational motor at the base of the flagellum (see Fig. 19–41). Helicases, polymerases, and other proteins move along DNA as they carry out their functions in DNA metabolism (Chapter 25). Here, we focus on the well-studied example of the contractile proteins of vertebrate skeletal muscle as a paradigm for how proteins translate chemical energy into motion.

The Major Proteins of Muscle Are Myosin and Actin

The contractile force of muscle is generated by the interaction of two proteins, myosin and actin. These proteins are arranged in filaments that undergo transient interactions and slide past each other to bring about contraction. Together, actin and myosin make up more than 80% of the protein mass of muscle.

Myosin (M_r 520,000) has six subunits: two heavy chains (each of M_r 220,000) and four light chains (each of M_r 20,000). The heavy chains account for much of the overall structure. At their carboxyl termini, they are arranged as extended α helices, wrapped around each other in a fibrous, left-handed coiled coil similar to that of α-keratin **(Fig. 5–27a)**. At its amino terminus, each heavy chain has a large globular domain containing a site where ATP is hydrolyzed. The light chains are associated with the globular domains. When myosin is treated briefly with the pro-tease trypsin, much of the fibrous tail is cleaved off, dividing the protein into components called light and heavy meromyosin (Fig. 5–27b). The globular domain—called myosin subfragment 1, or S1, or simply the myosin head group—is liberated from heavy meromyosin by cleavage with papain. The S1 fragment is the motor domain that makes muscle contraction possible. S1 fragments can be crystallized, and their overall structure as determined by Ivan Rayment and Hazel Holden is shown in Figure 5–27c.

In muscle cells, molecules of myosin aggregate to form structures called **thick filaments (Fig. 5–28a)**.

FIGURE 5–27 Myosin. (a) Myosin has two heavy chains (in two shades of pink), the carboxyl termini forming an extended coiled coil (tail) and the amino termini having globular domains (heads). Two light chains (blue) are associated with each myosin head. **(b)** Cleavage with trypsin and papain separates the myosin heads (S1 fragments) from the tails. **(c)** Ribbon representation of the myosin S1 fragment (from coordinates supplied by Ivan Rayment). The heavy chain is in gray, the two light chains in two shades of blue.

FIGURE 5–28 The major components of muscle. (a) Myosin aggregates to form a bipolar structure called a thick filament. **(b)** F-actin is a fila-mentous assemblage of G-actin monomers that polymerize two by two, giving the appearance of two filaments spiraling about one another in a right-handed fashion. **(c)** Space-filling model of an actin filament (shades of red) with one myosin head (gray and two shades of blue) bound to an actin monomer within the filament (from coordinates sup-plied by Ivan Rayment).

These rodlike structures are the core of the contractile unit. Within a thick filament, several hundred myosin molecules are arranged with their fibrous "tails" associated to form a long bipolar structure. The globular domains project from either end of this structure, in regular stacked arrays.

The second major muscle protein, **actin**, is abundant in almost all eukaryotic cells. In muscle, molecules of monomeric actin, called G-actin (*g*lobular actin; M_r 42,000), associate to form a long polymer called F-actin (*f*ilamentous actin). The **thin filament** consists of F-actin (Fig. 5–28b), along with the proteins troponin and tropomyosin (discussed below). The filamentous parts of thin filaments assemble as successive monomeric actin molecules add to one end. On addition, each monomer binds ATP, then hydrolyzes it to ADP, so every actin molecule in the filament is complexed to ADP. This ATP hydrolysis by actin functions only in the assembly of the filaments; it does not contribute directly to the energy expended in muscle contraction. Each actin monomer in the thin filament can bind tightly and specifically to one myosin head group (Fig. 5–28c).

Additional Proteins Organize the Thin and Thick Filaments into Ordered Structures

Skeletal muscle consists of parallel bundles of **muscle fibers**, each fiber a single, very large, multinucleated cell, 20 to 100 μm in diameter, formed from many cells fused together; a single fiber often spans the length of the muscle. Each fiber contains about 1,000 **myofibrils**, 2 μm in diameter, each consisting of a vast number of regularly arrayed thick and thin filaments complexed to other proteins (**Fig. 5–29**). A system of flat membranous vesicles called the **sarcoplasmic reticulum** surrounds each myofibril. Examined under the electron microscope, muscle fibers reveal alternating regions of high and low electron density, called the **A bands** and **I bands** (Fig. 5–29b, c). The A and I bands arise from the arrangement of thick and thin filaments, which are aligned and partially overlapping. The I band is the region of the bundle that in cross section would contain only thin filaments. The darker A band stretches the length of the thick filament and includes the region where parallel thick and thin filaments overlap. Bisecting the I band is a thin structure called the **Z disk**, perpendicular to the thin filaments and serving as an anchor to which the thin filaments are attached. The A band, too, is bisected by a thin line, the **M line** or M disk, a region of high electron density in the middle of the thick filaments. The entire contractile unit, consisting of bundles of thick filaments interleaved at either end with bundles of thin filaments, is called the **sarcomere**. The arrangement of interleaved bundles allows the thick and thin filaments to slide past each other (by a mechanism discussed below), causing a progressive shortening of each sarcomere (**Fig. 5–30**).

The thin actin filaments are attached at one end to the Z disk in a regular pattern. The assembly includes the minor muscle proteins α-**actinin**, **desmin**, and

(a)

Muscle

Bundle of muscle fibers

Myofibrils

Nuclei

Capillaries

Muscle fiber

Sarcoplasmic reticulum

Sarcomere

Myofibril

I band   A band

Z disk   M line

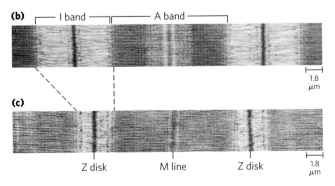

(b) — I band — — A band —

1.8 μm

(c)

Z disk   M line   Z disk

1.8 μm

FIGURE 5–29 Skeletal muscle. (a) Muscle fibers consist of single, elongated, multinucleated cells that arise from the fusion of many precursor cells. The fibers are made up of many myofibrils (only six are shown here for simplicity) surrounded by the membranous sarcoplasmic reticulum. The organization of thick and thin filaments in a myofibril gives it a striated appearance. When muscle contracts, the I bands narrow and the Z disks come closer together, as seen in electron micrographs of **(b)** relaxed and **(c)** contracted muscle.

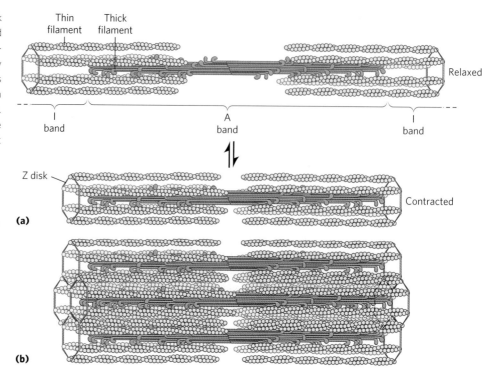

FIGURE 5–30 Muscle contraction. Thick filaments are bipolar structures created by the association of many myosin molecules. **(a)** Muscle contraction occurs by the sliding of the thick and thin filaments past each other so that the Z disks in neighboring I bands draw closer together. **(b)** The thick and thin filaments are interleaved such that each thick filament is surrounded by six thin filaments.

vimentin. Thin filaments also contain a large protein called **nebulin** (~7,000 amino acid residues), thought to be structured as an α helix that is long enough to span the length of the filament. The M line similarly organizes the thick filaments. It contains the proteins **paramyosin**, **C-protein**, and **M-protein**. Another class of proteins called **titins**, the largest single polypeptide chains discovered thus far (the titin of human cardiac muscle has 26,926 amino acid residues), link the thick filaments to the Z disk, providing additional organization to the overall structure. Among their structural functions, the proteins nebulin and titin are believed to act as "molecular rulers," regulating the length of the thin and thick filaments, respectively. Titin extends from the Z disk to the M line, regulating the length of the sarcomere itself and preventing overextension of the muscle. The characteristic sarcomere length varies from one muscle tissue to the next in a vertebrate, largely due to the different titin variants in the tissues.

Myosin Thick Filaments Slide along Actin Thin Filaments

The interaction between actin and myosin, like that between all proteins and ligands, involves weak bonds. When ATP is not bound to myosin, a face on the myosin head group binds tightly to actin (**Fig. 5–31**). When ATP binds to myosin and is hydrolyzed to ADP and phosphate, a coordinated and cyclic series of conformational changes occurs in which myosin releases the F-actin subunit and binds another subunit farther along the thin filament.

The cycle has four major steps (Fig. 5–31). In step **1**, ATP binds to myosin and a cleft in the myosin molecule opens, disrupting the actin-myosin interaction so that the bound actin is released. ATP is then hydrolyzed in step **2**, causing a conformational change in the protein to a "high-energy" state that moves the myosin head and changes its orientation in relation to the actin thin filament. Myosin then binds weakly to an F-actin subunit closer to the Z disk than the one just released. As the phosphate product of ATP hydrolysis is released from myosin in step **3**, another conformational change occurs in which the myosin cleft closes, strengthening the myosin-actin binding. This is followed quickly by step **4**, a "power stroke" during which the conformation of the myosin head returns to the original resting state, its orientation relative to the bound actin changing so as to pull the tail of the myosin toward the Z disk. ADP is then released to complete the cycle. Each cycle generates about 3 to 4 pN (piconewtons) of force and moves the thick filament 5 to 10 nm relative to the thin filament.

Because there are many myosin heads in a thick filament, at any given moment some (probably 1% to 3%) are bound to thin filaments. This prevents thick filaments from slipping backward when an individual myosin head releases the actin subunit to which it was bound. The thick filament thus actively slides forward past the adjacent thin filaments. This process, coordinated among the many sarcomeres in a muscle fiber, brings about muscle contraction.

The interaction between actin and myosin must be regulated so that contraction occurs only in response to appropriate signals from the nervous system. The

FIGURE 5–31 Molecular mechanism of muscle contraction. Conformational changes in the myosin head that are coupled to stages in the ATP hydrolytic cycle cause myosin to successively dissociate from one actin subunit, then associate with another farther along the actin filament. In this way the myosin heads slide along the thin filaments, drawing the thick filament array into the thin filament array (see Fig. 5-30).

(1) ATP binds to myosin head, causing dissociation from actin.

(2) As tightly bound ATP is hydrolyzed, a conformational change occurs. ADP and P_i remain associated with the myosin head.

(3) Myosin head attaches to actin filament, causing release of P_i.

(4) P_i release triggers a "power stroke," a conformational change in the myosin head that moves actin and myosin filaments relative to one another. ADP is released in the process.

FIGURE 5–32 Regulation of muscle contraction by tropomyosin and troponin. Tropomyosin and troponin are bound to F-actin in the thin filaments. In the relaxed muscle, these two proteins are arranged around the actin filaments so as to block the binding sites for myosin. Tropomyosin is a two-stranded coiled coil of α helices, the same structural motif as in α-keratin (see Fig. 4-11). It forms head-to-tail polymers twisting around the two actin chains. Troponin is attached to the actin-tropomyosin complex at regular intervals of 38.5 nm. Troponin consists of three different subunits: I, C, and T. Troponin I prevents binding of the myosin head to actin; troponin C has a binding site for Ca^{2+}; and troponin T links the entire troponin complex to tropomyosin. When the muscle receives a neural signal to initiate contraction, Ca^{2+} is released from the sarcoplasmic reticulum (see Fig. 5-29a) and binds to troponin C. This causes a conformational change in troponin C, which alters the positions of troponin I and tropomyosin so as to relieve the inhibition by troponin I and allow muscle contraction.

regulation is mediated by a complex of two proteins, **tropomyosin** and **troponin** (**Fig. 5–32**). Tropomyosin binds to the thin filament, blocking the attachment sites for the myosin head groups. Troponin is a Ca^{2+}-binding protein. A nerve impulse causes release of Ca^{2+} from the sarcoplasmic reticulum. The released Ca^{2+} binds to troponin (another protein-ligand interaction) and causes a conformational change in the tropomyosin-troponin complexes, exposing the myosin-binding sites on the thin filaments. Contraction follows.

Working skeletal muscle requires two types of molecular functions that are common in proteins—binding and catalysis. The actin-myosin interaction, a protein-ligand interaction like that of immunoglobulins with antigens, is reversible and leaves the participants unchanged. When ATP binds myosin, however, it is hydrolyzed to ADP and P_i. Myosin is not only an actin-binding protein, it is also an ATPase—an enzyme. The function of enzymes in catalyzing chemical transformations is the topic of the next chapter.

SUMMARY 5.3 Protein Interactions Modulated by Chemical Energy: Actin, Myosin, and Molecular Motors

▶ Protein-ligand interactions achieve a special degree of spatial and temporal organization in motor proteins. Muscle contraction results from choreographed interactions between myosin and actin, coupled to the hydrolysis of ATP by myosin.

▶ Myosin consists of two heavy and four light chains, forming a fibrous coiled coil (tail) domain and a globular (head) domain. Myosin molecules are organized into thick filaments, which slide past thin filaments composed largely of actin. ATP hydrolysis in myosin is coupled to a series of conformational changes in the myosin head, leading to dissociation of myosin from one F-actin subunit and its eventual reassociation with another, farther along the thin filament. The myosin thus slides along the actin filaments.

▶ Muscle contraction is stimulated by the release of Ca^{2+} from the sarcoplasmic reticulum. The Ca^{2+} binds to the protein troponin, leading to a conformational change in a troponin-tropomyosin complex that triggers the cycle of actin-myosin interactions.

Key Terms

Terms in bold are defined in the glossary.

ligand 157	**T lymphocyte** or
binding site 157	**T cell** 175
induced fit 157	**antigen** 175
heme 158	**epitope** 175
porphyrin 158	**hapten** 175
globins 159	immunoglobulin fold 175
equilibrium expression 160	**polyclonal**
association constant, K_a 160	**antibodies** 178
dissociation constant,	**monoclonal**
K_d 160	**antibodies** 178
allosteric protein 166	ELISA 179
Hill equation 167	**immunoblotting** 179
Bohr effect 170	**Western blotting** 179
lymphocytes 174	**myosin** 179
antibody 174	**actin** 181
immunoglobulin 174	**sarcomere** 181
B lymphocyte or	
B cell 175	

Further Reading

Oxygen-Binding Proteins

Changeux, J.P. & Edelstein, S.J. (2005) Allosteric mechanisms of signal transduction *Science* **308**, 1424–1428.

Koder, R.L., Anderson, J.L.R., Solomon, L.A., Reddy, K.S., Moser, C.C., & Dutton, P.L. (2009) Design and engineering of an O_2 transport protein. *Nature* **458**, 305–309.

Koshland, D.E., Jr., Nemethy, G., & Filmer, D. (1966) Comparison of experimental binding data and theoretical models in proteins containing subunits. *Biochemistry* **5**, 365–385.

The paper that introduced the sequential model.

Laberge, M. & Yonetani, T. (2007) Common dynamics of globin family proteins. *IUBMB Life* **59**, 528–534.

Monod, J., Wyman, J., & Changeux, J.-P. (1965) On the nature of allosteric transitions: a plausible model. *J. Mol. Biol.* **12**, 88–118.

The concerted model was first proposed in this landmark paper.

Perutz, M.F., Wilkinson, A.J., Paoli, M., & Dodson, G.G. (1998) The stereochemical mechanism of the cooperative effects in hemoglobin revisited. *Annu. Rev. Biophys. Biomol. Struct.* **27**, 1–34.

Immune System Proteins

Cooper, M.D. & Alder, M.N. (2006) The evolution of adaptive immune systems. *Cell* **124**, 815–822.

An interesting essay tracing the origins of our immune system.

Flajnik, M.F. & Kasahara, M. (2010) Origin and evolution of the adaptive immune system: genetic events and selective pressures. *Nat. Rev. Genet.* **11**, 47–59.

Kindt, T.J., Osborne, B.A., & Goldsby, R.A. (2007) *Kuby Immunology*, 6th edn, W. H. Freeman and Company, New York.

Ploegh, H.L. (1998) Viral strategies of immune evasion. *Science* **280**, 248–253.

Yewdell, J.W. & Haeryfar, S.M.M. (2005) Understanding presentation of viral antigens to CD8(+) T cells in vivo: the key to rational vaccine design. *Annu. Rev. Immunol.* **23**, 651–682.

Molecular Motors

Geeves, M.A. & Holmes, K.C. (1999) Structural mechanism of muscle contraction. *Annu. Rev. Biochem.* **68**, 687–728.

Geigel, C. & Schmidt, C.F. (2011) Moving into the cell: single-molecule studies of molecular motors in complex environments. *Nat. Rev. Mol. Cell Biol.* **12**, 163–176.

Huxley, H.E. (1998) Getting to grips with contraction: the interplay of structure and biochemistry. *Trends Biochem. Sci.* **23**, 84–87.

An interesting historical perspective on deciphering the mechanism of muscle contraction.

Molloy, J.E. & Veigel, C. (2003) Myosin motors walk the walk. *Science* **300**, 2045–2046.

Rayment, I. (1996) The structural basis of the myosin ATPase activity. *J. Biol. Chem.* **271**, 15,850–15,853.

Examines the muscle contraction mechanism from a structural perspective.

Rayment, I. & Holden, H.M. (1994) The three-dimensional structure of a molecular motor. *Trends Biochem. Sci.* **19**, 129–134.

Vale, R.D. (2003) The molecular motor toolbox for intracellular transport. *Cell* **112**, 467–480.

Problems

1. Relationship between Affinity and Dissociation Constant Protein A has a binding site for ligand X with a K_d of 10^{-6} M. Protein B has a binding site for ligand X with a K_d of 10^{-9} M. Which protein has a higher affinity for ligand X? Explain your reasoning. Convert the K_d to K_a for both proteins.

2. Negative Cooperativity Which of the following situations would produce a Hill plot with $n_H < 1.0$? Explain your reasoning in each case.

(a) The protein has multiple subunits, each with a single ligand-binding site. Binding of ligand to one site decreases the binding affinity of other sites for the ligand.

(b) The protein is a single polypeptide with two ligand-binding sites, each having a different affinity for the ligand.

(c) The protein is a single polypeptide with a single ligand-binding site. As purified, the protein preparation is heterogeneous, containing some protein molecules that are partially denatured and thus have a lower binding affinity for the ligand.

3. Hemoglobin's Affinity for Oxygen What is the effect of the following changes on the O_2 affinity of hemoglobin? (a) A drop in the pH of blood plasma from 7.4 to 7.2. (b) A decrease in the partial pressure of CO_2 in the lungs from 6 kPa (holding one's breath) to 2 kPa (normal). (c) An increase in the BPG level from 5 mM (normal altitudes) to 8 mM (high altitudes). (d) An increase in CO from 1.0 part per million (ppm) in a normal indoor atmosphere to 30 ppm in a home that has a malfunctioning or leaking furnace.

4. Reversible Ligand Binding I The protein calcineurin binds to the protein calmodulin with an association rate of $8.9 \times 10^3 \, M^{-1}s^{-1}$ and an overall dissociation constant, K_d, of 10 nM. Calculate the dissociation rate, k_d, including appropriate units.

5. Reversible Ligand Binding II Three membrane receptor proteins bind tightly to a hormone. Based on the data in the table below, (a) what is the K_d for hormone binding by protein 2? (Include appropriate units.) (b) Which of these proteins binds *most* tightly to this hormone?

Hormone concentration (nM)	θ		
	Protein 1	Protein 2	Protein 3
0.2	0.048	0.29	0.17
0.5	0.11	0.5	0.33
1	0.2	0.67	0.5
4	0.5	0.89	0.8
10	0.71	0.95	0.91
20	0.83	0.97	0.95
50	0.93	0.99	0.98

6. Cooperativity in Hemoglobin Under appropriate conditions, hemoglobin dissociates into its four subunits. The isolated α subunit binds oxygen, but the O_2-saturation curve is hyperbolic rather than sigmoid. In addition, the binding of oxygen to the isolated α subunit is not affected by the presence of H^+, CO_2, or BPG. What do these observations indicate about the source of the cooperativity in hemoglobin?

7. Comparison of Fetal and Maternal Hemoglobins Studies of oxygen transport in pregnant mammals show that the O_2-saturation curves of fetal and maternal blood are markedly different when measured under the same conditions. Fetal erythrocytes contain a structural variant of hemoglobin, HbF, consisting of two α and two γ subunits ($\alpha_2\gamma_2$), whereas maternal erythrocytes contain HbA ($\alpha_2\beta_2$).

(a) Which hemoglobin has a higher affinity for oxygen under physiological conditions, HbA or HbF? Explain.

(b) What is the physiological significance of the different O_2 affinities?

(c) When all the BPG is carefully removed from samples of HbA and HbF, the measured O_2-saturation curves (and consequently the O_2 affinities) are displaced to the left. However, HbA now has a greater affinity for oxygen than does HbF. When BPG is reintroduced, the O_2-saturation curves return to normal, as shown in the graph. What is the effect of BPG on the O_2 affinity of hemoglobin? How can the above information be used to explain the different O_2 affinities of fetal and maternal hemoglobin?

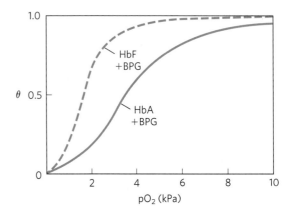

8. Hemoglobin Variants There are almost 500 naturally occurring variants of hemoglobin. Most are the result of a single amino acid substitution in a globin polypeptide chain. Some variants produce clinical illness, though not all variants have deleterious effects. A brief sample follows.

HbS (sickle-cell Hb): substitutes a Val for a Glu on the surface

Hb Cowtown: eliminates an ion pair involved in T-state stabilization

Hb Memphis: substitutes one uncharged polar residue for another of similar size on the surface

Hb Bibba: substitutes a Pro for a Leu involved in an α helix

Hb Milwaukee: substitutes a Glu for a Val

Hb Providence: substitutes an Asn for a Lys that normally projects into the central cavity of the tetramer

Hb Philly: substitutes a Phe for a Tyr, disrupting hydrogen bonding at the $\alpha_1\beta_1$ interface

Explain your choices for each of the following:

(a) The Hb variant *least* likely to cause pathological symptoms.

(b) The variant(s) most likely to show pI values different from that of HbA on an isoelectric focusing gel.

(c) The variant(s) most likely to show a decrease in BPG binding and an increase in the overall affinity of the hemoglobin for oxygen.

9. Oxygen Binding and Hemoglobin Structure A team of biochemists uses genetic engineering to modify the interface region between hemoglobin subunits. The resulting hemoglobin variants exist in solution primarily as $\alpha\beta$ dimers (few, if any, $\alpha_2\beta_2$ tetramers form). Are these variants likely to bind oxygen more weakly or more tightly? Explain your answer.

10. Reversible (but Tight) Binding to an Antibody An antibody binds to an antigen with a K_d of 5×10^{-8} M. At what concentration of antigen will θ be (a) 0.2, (b) 0.5, (c) 0.6, (d) 0.8?

11. Using Antibodies to Probe Structure-Function Relationships in Proteins A monoclonal antibody binds to G-actin but not to F-actin. What does this tell you about the epitope recognized by the antibody?

12. The Immune System and Vaccines A host organism needs time, often days, to mount an immune response against a new antigen, but memory cells permit a rapid response to pathogens previously encountered. A vaccine to protect against a particular viral infection often consists of weakened or killed virus or isolated proteins from a viral protein coat. When injected into a human patient, the vaccine generally does not cause an infection and illness, but it effectively "teaches" the immune system what the viral particles look like, stimulating the production of memory cells. On subsequent infection, these cells can bind to the virus and trigger a rapid immune response. Some pathogens, including HIV, have developed mechanisms to evade the immune system, making it difficult or impossible to develop effective vaccines against them. What strategy could a pathogen use to evade the immune system? Assume that a host's antibodies and/or T-cell receptors are available to bind to any structure that might appear on the surface of a pathogen and that, once bound, the pathogen is destroyed.

13. How We Become a "Stiff" When a vertebrate dies, its muscles stiffen as they are deprived of ATP, a state called rigor mortis. Explain the molecular basis of the rigor state.

14. Sarcomeres from Another Point of View The symmetry of thick and thin filaments in a sarcomere is such that six thin filaments ordinarily surround each thick filament in a hexagonal array. Draw a cross section (transverse cut) of a myofibril at the following points: (a) at the M line; (b) through the I band; (c) through the dense region of the A band; (d) through the less dense region of the A band, adjacent to the M line (see Fig. 5–29b, c).

Using the Web

15. Lysozyme and Antibodies To fully appreciate how proteins function in a cell, it is helpful to have a three-dimensional view of how proteins interact with other cellular components. Fortunately, this is possible using Web-based protein databases and three-dimensional molecular viewing utilities such as Jmol, a free and user-friendly molecular viewer that is compatible with most browsers and operating systems.

In this exercise you will examine the interactions between the enzyme lysozyme (Chapter 4) and the Fab portion of the anti-lysozyme antibody. Use the PDB identifier 1FDL to explore the structure of the IgG1 Fab fragment–lysozyme complex (antibody-antigen complex). To answer the following questions, use the information on the Structure Summary page at the Protein Data Bank (www.rcsb.org), and view the structure using Jmol or a similar viewer.

(a) Which chains in the three-dimensional model correspond to the antibody fragment and which correspond to the antigen, lysozyme?

(b) What type of secondary structure predominates in this Fab fragment?

(c) How many amino acid residues are in the heavy and light chains of the Fab fragment? In lysozyme? Estimate the percentage of the lysozyme that interacts with the antigen-binding site of the antibody fragment.

(d) Identify the specific amino acid residues in lysozyme and in the variable regions of the Fab heavy and light chains that are situated at the antigen-antibody interface. Are the residues contiguous in the primary sequence of the polypeptide chains?

16. Exploring Reversible Interactions of Proteins and Ligands with Living Graphs 🔘 Use the living graphs for Equations 5–8, 5–11, 5–14, and 5–16 to work through the following exercises.

(a) Reversible binding of a ligand to a simple protein, without cooperativity. For Equation 5–8, set up a plot of θ versus [L] (vertical and horizontal axes, respectively). Examine the plots generated when K_d is set at 5, 10, 20, and 100 μM. Higher affinity of the protein for the ligand means more binding at lower ligand concentrations. Suppose that four different proteins exhibit these four different K_d values for ligand L. Which protein would have the highest affinity for L?

Examine the plot generated when $K_d = 10$ μM. How much does θ increase when [L] increases from 0.2 to 0.4 μM? How much does θ increase when [L] increases from 40 to 80 μM?

You can do the same exercise for Equation 5–11. Convert [L] to pO_2 and K_d to P_{50}. Examine the curves generated when P_{50} is set at 0.5, 1, 2, and 10 kPa. For the curve generated when $P_{50} = 1$ kPa, how much does θ change when the pO_2 increases from 0.02 to 0.04 kPa? From 4 to 8 kPa?

(b) Cooperative binding of a ligand to a multisubunit protein. Using Equation 5–14, generate a binding curve for a protein and ligand with $K_d = 10$ μM and $n = 3$. Note the altered definition of K_d in Equation 5–16. On the same plot, add a curve for a protein with $K_d = 20$ μM and $n = 3$. Now see how both curves change when you change to $n = 4$. Generate Hill plots (Eqn 5–16) for each of these cases. For $K_d = 10$ μM and $n = 3$, what is θ when [L] = 20 μM?

(c) Explore these equations further by varying all the parameters used above.

Data Analysis Problem

17. Protein Function During the 1980s, the structures of actin and myosin were known only at the resolution shown in Figure 5–28a, b. Although researchers knew that the S1 portion of myosin binds to actin and hydrolyzes ATP, there was a substantial debate about where in the myosin molecule the contractile force was generated. At the time, two competing models were proposed for the mechanism of force generation in myosin.

In the "hinge" model, S1 bound to actin, but the pulling force was generated by contraction of the "hinge region" in the myosin tail. The hinge region is in the heavy meromyosin portion of the myosin molecule, near where trypsin cleaves off light meromyosin (see Fig. 5–27b). This is roughly the point labeled "Two supercoiled α helices" in Figure 5–27a. In the "S1" model, the pulling force was generated in the S1 "head" itself and the tail was just for structural support.

Many experiments had been performed but provided no conclusive evidence. In 1987, James Spudich and his colleagues at Stanford University published a study that, although not conclusive, went a long way toward resolving this controversy.

Recombinant DNA techniques were not sufficiently developed to address this issue in vivo, so Spudich and colleagues used an interesting in vitro motility assay. The alga *Nitella* has extremely long cells, often several centimeters in length and about 1 mm in diameter. These cells have actin fibers that run along their long axes, and the cells can be cut open along their length to expose the actin fibers. Spudich and his group had observed that plastic beads coated with myosin would "walk" along these fibers in the presence of ATP, just as myosin would do in contracting muscle.

For these experiments, they used a more well-defined method for attaching the myosin to the beads. The "beads" were clumps of killed bacterial (*Staphylococcus aureus*) cells. These cells have a protein on their surface that binds to the Fc region of antibody molecules (Fig. 5–21a). The antibodies, in turn, bind to several (unknown) places along the tail of the myosin molecule. When bead-antibody-myosin complexes were prepared with intact myosin molecules, they would move along *Nitella* actin fibers in the presence of ATP.

(a) Sketch a diagram showing what a bead-antibody-myosin complex might look like at the molecular level.

(b) Why was ATP required for the beads to move along the actin fibers?

(c) Spudich and coworkers used antibodies that bound to the myosin tail. Why would this experiment have failed if they had used an antibody that bound to the part of S1 that normally binds to actin? Why would this experiment have failed if they had used an antibody that bound to actin?

To help focus in on the part of myosin responsible for force production, Spudich and his colleagues used trypsin to produce two partial myosin molecules (see Fig. 5–27b): (1) heavy meromyosin (HMM), made by briefly digesting myosin with trypsin; HMM consists of S1 and the part of the tail that includes the hinge; and (2) short heavy meromyosin (SHMM), made from a more extensive digestion of HMM with trypsin; SHMM consists of S1 and a shorter part of the tail that does not include the hinge. Brief digestion of myosin with trypsin produces HMM and light meromyosin, by cleavage of a single specific peptide bond in the myosin molecule.

(d) Why might trypsin attack this peptide bond first rather than other peptide bonds in myosin?

Spudich and colleagues prepared bead-antibody-myosin complexes with varying amounts of myosin, HMM, and SHMM and measured their speeds along *Nitella* actin fibers in the presence of ATP. The graph below sketches their results.

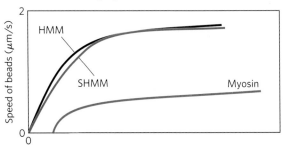

(e) Which model ("S1" or "hinge") is consistent with these results? Explain your reasoning.

(f) Provide a plausible explanation for why the speed of the beads increased with increasing myosin density.

(g) Provide a plausible explanation for why the speed of the beads reached a plateau at high myosin density.

The more extensive trypsin digestion required to produce SHMM had a side effect: another specific cleavage of the myosin polypeptide backbone in addition to the cleavage in the tail. This second cleavage was in the S1 head.

(h) Based on this information, why is it surprising that SHMM was still capable of moving beads along actin fibers?

(i) As it turns out, the tertiary structure of the S1 head remains intact in SHMM. Provide a plausible explanation of how the protein remains intact and functional even though the polypeptide backbone has been cleaved and is no longer continuous.

Reference

Hynes, T.R., Block, S.M., White, B.T., & Spudich, J.A. (1987) Movement of myosin fragments in vitro: domains involved in force production. *Cell* **48**, 953–963.

6

Enzymes

There are two fundamental conditions for life. First, the organism must be able to self-replicate (a topic considered in Part III); second, it must be able to catalyze chemical reactions efficiently and selectively. The central importance of catalysis may seem surprising, but it is easy to demonstrate. As described in Chapter 1, living systems make use of energy from the environment. Many of us, for example, consume substantial amounts of sucrose—common table sugar—as a kind of fuel, usually in the form of sweetened foods and drinks. The conversion of sucrose to CO_2 and H_2O in the presence of oxygen is a highly exergonic process, releasing free energy that we can use to think, move, taste, and see. However, a bag of sugar can remain on the shelf for years without any obvious conversion to CO_2 and H_2O. Although this chemical process is thermodynamically favorable, it is very slow. Yet when sucrose is consumed by a human (or almost any other organism), it releases its chemical energy in seconds. The difference is catalysis. Without catalysis, chemical reactions such as sucrose oxidation could not occur on a useful time scale, and thus could not sustain life.

In this chapter, then, we turn our attention to the reaction catalysts of biological systems: enzymes, the most remarkable and highly specialized proteins. Enzymes have extraordinary catalytic power, often far greater than that of synthetic or inorganic catalysts. They have a high degree of specificity for their substrates, they accelerate chemical reactions tremendously, and they function in aqueous solutions under very mild conditions of temperature and pH. Few nonbiological catalysts have all these properties.

Enzymes are central to every biochemical process. Acting in organized sequences, they catalyze the hundreds of stepwise reactions that degrade nutrient molecules, conserve and transform chemical energy, and make biological macromolecules from simple precursors.

The study of enzymes has immense practical importance. In some diseases, especially inheritable genetic disorders, there may be a deficiency or even a total absence of one or more enzymes. Other disease conditions may be caused by excessive activity of an enzyme. Measurements of the activities of enzymes in blood plasma, erythrocytes, or tissue samples are important in diagnosing certain illnesses. Many drugs act through interactions with enzymes. Enzymes are also important practical tools in chemical engineering, food technology, and agriculture.

We begin with descriptions of the properties of enzymes and the principles underlying their catalytic power, then introduce enzyme kinetics, a discipline that provides much of the framework for any discussion of enzymes. Specific examples of enzyme mechanisms are then provided, illustrating principles introduced earlier in the chapter. We end with a discussion of how enzyme activity is regulated.

6.1 An Introduction to Enzymes

Much of the history of biochemistry is the history of enzyme research. Biological catalysis was first recognized and described in the late 1700s, in studies on the digestion of meat by secretions of the stomach. Research continued in the 1800s with examinations of the conversion of starch to sugar by saliva and various plant extracts. In the 1850s, Louis Pasteur concluded that fermentation of sugar into alcohol by yeast is catalyzed by "ferments." He postulated that these ferments were inseparable from the structure of living yeast cells; this view, called vitalism, prevailed for decades. Then in 1897 Eduard Buchner discovered that yeast extracts could ferment sugar to alcohol, proving that fermentation was promoted by molecules that continued to function when removed from cells. Buchner's experiment at once marked the

Eduard Buchner, 1860–1917   James Sumner, 1887–1955   J. B. S. Haldane, 1892–1964

end of vitalistic notions and the dawn of the science of biochemistry. Frederick W. Kühne later gave the name **enzymes** (from the Greek *enzymos*, "leavened") to the molecules detected by Buchner.

The isolation and crystallization of urease by James Sumner in 1926 was a breakthrough in early enzyme studies. Sumner found that urease crystals consisted entirely of protein, and he postulated that all enzymes are proteins. In the absence of other examples, this idea remained controversial for some time. Only in the 1930s was Sumner's conclusion widely accepted, after John Northrop and Moses Kunitz crystallized pepsin, trypsin, and other digestive enzymes and found them also to be proteins. During this period, J. B. S. Haldane wrote a treatise titled *Enzymes*. Although the molecular nature of enzymes was not yet fully appreciated, Haldane made the remarkable suggestion that weak bonding interactions between an enzyme and its substrate might be used to catalyze a reaction. This insight lies at the heart of our current understanding of enzymatic catalysis.

Since the latter part of the twentieth century, thousands of enzymes have been purified, their structures elucidated, and their mechanisms explained.

Most Enzymes Are Proteins

With the exception of a small group of catalytic RNA molecules (Chapter 26), all enzymes are proteins. Their catalytic activity depends on the integrity of their native protein conformation. If an enzyme is denatured or dissociated into its subunits, catalytic activity is usually lost. If an enzyme is broken down into its component amino acids, its catalytic activity is always destroyed. Thus the primary, secondary, tertiary, and quaternary structures of protein enzymes are essential to their catalytic activity.

Enzymes, like other proteins, have molecular weights ranging from about 12,000 to more than 1 million. Some enzymes require no chemical groups for activity other than their amino acid residues. Others require an additional chemical component called a **cofactor**—either one or more inorganic ions, such as Fe^{2+}, Mg^{2+}, Mn^{2+}, or Zn^{2+} (Table 6–1), or a complex organic or metalloorganic molecule called a **coenzyme**. Coenzymes act as transient carriers of specific functional groups (Table 6–2). Most are derived from vitamins, organic nutrients required in small amounts in the diet. We consider coenzymes in more detail as we encounter them in the metabolic pathways discussed in Part II. Some enzymes require *both* a coenzyme and one or more metal ions for activity. A coenzyme or metal ion that is very tightly or even covalently bound to the enzyme protein is called a **prosthetic group**. A complete, catalytically active enzyme together with its bound coenzyme and/or metal ions is called a **holoenzyme**. The protein part of such an enzyme is called the **apoenzyme** or **apoprotein**. Finally, some enzyme proteins are modified covalently by phosphorylation, glycosylation, and other processes. Many of these alterations are involved in the regulation of enzyme activity.

Enzymes Are Classified by the Reactions They Catalyze

Many enzymes have been named by adding the suffix "-ase" to the name of their substrate or to a word or phrase describing their activity. Thus urease catalyzes hydrolysis of urea, and DNA polymerase catalyzes the polymerization of nucleotides to form DNA. Other enzymes were named by their discoverers for a broad function, before the specific reaction catalyzed was known. For example, an enzyme known to act in the digestion of foods was named pepsin, from the Greek *pepsis*, "digestion," and lysozyme was named for its ability to lyse (break down) bacterial cell walls. Still others were named for their source: trypsin, named in part

TABLE 6–1 Some Inorganic Ions That Serve as Cofactors for Enzymes

Ions	Enzymes
Cu^{2+}	Cytochrome oxidase
Fe^{2+} or Fe^{3+}	Cytochrome oxidase, catalase, peroxidase
K^+	Pyruvate kinase
Mg^{2+}	Hexokinase, glucose 6-phosphatase, pyruvate kinase
Mn^{2+}	Arginase, ribonucleotide reductase
Mo	Dinitrogenase
Ni^{2+}	Urease
Zn^{2+}	Carbonic anhydrase, alcohol dehydrogenase, carboxypeptidases A and B

TABLE 6-2 Some Coenzymes That Serve as Transient Carriers of Specific Atoms or Functional Groups

Coenzyme	Examples of chemical groups transferred	Dietary precursor in mammals
Biocytin	CO_2	Biotin
Coenzyme A	Acyl groups	Pantothenic acid and other compounds
5′-Deoxyadenosylcobalamin (coenzyme B_{12})	H atoms and alkyl groups	Vitamin B_{12}
Flavin adenine dinucleotide	Electrons	Riboflavin (vitamin B_2)
Lipoate	Electrons and acyl groups	Not required in diet
Nicotinamide adenine dinucleotide	Hydride ion ($:H^-$)	Nicotinic acid (niacin)
Pyridoxal phosphate	Amino groups	Pyridoxine (vitamin B_6)
Tetrahydrofolate	One-carbon groups	Folate
Thiamine pyrophosphate	Aldehydes	Thiamine (vitamin B_1)

Note: The structures and modes of action of these coenzymes are described in Part II.

from the Greek *tryein*, "to wear down," was obtained by rubbing pancreatic tissue with glycerin. Sometimes the same enzyme has two or more names, or two different enzymes have the same name. Because of such ambiguities, and the ever-increasing number of newly discovered enzymes, biochemists, by international agreement, have adopted a system for naming and classifying enzymes. This system divides enzymes into six classes, each with subclasses, based on the type of reaction catalyzed (Table 6–3). Each enzyme is assigned a four-part classification number and a systematic name, which identifies the reaction it catalyzes. As an example, the formal systematic name of the enzyme catalyzing the reaction

$$\text{ATP} + \text{D-glucose} \longrightarrow \text{ADP} + \text{D-glucose 6-phosphate}$$

is ATP:glucose phosphotransferase, which indicates that it catalyzes the transfer of a phosphoryl group from ATP to glucose. Its Enzyme Commission number (E.C. number) is 2.7.1.1. The first number (2) denotes the class name (transferase); the second number (7), the subclass (phosphotransferase); the third number (1), a phosphotransferase with a hydroxyl group as acceptor;

and the fourth number (1), D-glucose as the phosphoryl group acceptor. For many enzymes, a trivial name is more frequently used—in this case, hexokinase. A complete list and description of the thousands of known enzymes is maintained by the Nomenclature Committee of the International Union of Biochemistry and Molecular Biology (www.chem.qmul.ac.uk/iubmb/enzyme). This chapter is devoted primarily to principles and properties common to all enzymes.

SUMMARY 6.1 An Introduction to Enzymes

▶ Life depends on powerful and specific catalysts: enzymes. Almost every biochemical reaction is catalyzed by an enzyme.

▶ With the exception of a few catalytic RNAs, all known enzymes are proteins. Many require nonprotein coenzymes or cofactors for their catalytic function.

▶ Enzymes are classified according to the type of reaction they catalyze. All enzymes have formal E.C. numbers and names, and most have trivial names.

TABLE 6-3 International Classification of Enzymes

Class no.	Class name	Type of reaction catalyzed
1	Oxidoreductases	Transfer of electrons (hydride ions or H atoms)
2	Transferases	Group transfer reactions
3	Hydrolases	Hydrolysis reactions (transfer of functional groups to water)
4	Lyases	Cleavage of C—C, C—O, C—N, or other bonds by elimination, leaving double bonds or rings, or addition of groups to double bonds
5	Isomerases	Transfer of groups within molecules to yield isomeric forms
6	Ligases	Formation of C—C, C—S, C—O, and C—N bonds by condensation reactions coupled to cleavage of ATP or similar cofactor

6.2 How Enzymes Work

The enzymatic catalysis of reactions is essential to living systems. Under biologically relevant conditions, uncatalyzed reactions tend to be slow—most biological molecules are quite stable in the neutral-pH, mild-temperature, aqueous environment inside cells. Furthermore, many common chemical processes are unfavorable or unlikely in the cellular environment, such as the transient formation of unstable charged intermediates or the collision of two or more molecules in the precise orientation required for reaction. Reactions required to digest food, send nerve signals, or contract a muscle simply do not occur at a useful rate without catalysis.

An enzyme circumvents these problems by providing a specific environment within which a given reaction can occur more rapidly. The distinguishing feature of an enzyme-catalyzed reaction is that it takes place within the confines of a pocket on the enzyme called the **active site** (**Fig. 6–1**). The molecule that is bound in the active site and acted upon by the enzyme is called the **substrate**. The surface of the active site is lined with amino acid residues with substituent groups that bind the substrate and catalyze its chemical transformation. Often, the active site encloses a substrate, sequestering it completely from solution. The enzyme-substrate complex, whose existence was first proposed by Charles-Adolphe Wurtz in 1880, is central to the action of enzymes. It is also the starting point for mathematical treatments that define the kinetic behavior of enzyme-catalyzed reactions and for theoretical descriptions of enzyme mechanisms.

Enzymes Affect Reaction Rates, Not Equilibria

A simple enzymatic reaction might be written

$$E + S \rightleftharpoons ES \rightleftharpoons EP \rightleftharpoons E + P \quad (6\text{–}1)$$

FIGURE 6–1 Binding of a substrate to an enzyme at the active site. The enzyme chymotrypsin with bound substrate (PDB ID 7GCH). Some key active-site amino acid residues appear as a red splotch on the enzyme surface.

where E, S, and P represent the enzyme, substrate, and product; ES and EP are transient complexes of the enzyme with the substrate and with the product.

To understand catalysis, we must first appreciate the important distinction between reaction equilibria and reaction rates. The function of a catalyst is to increase the *rate* of a reaction. Catalysts do not affect reaction *equilibria*. (Recall that a reaction is at equilibrium when there is no net change in the concentrations of reactants or products.) Any reaction, such as $S \rightleftharpoons P$, can be described by a reaction coordinate diagram (**Fig. 6–2**), a picture of the energy changes during the reaction. As discussed in Chapter 1, energy in biological systems is described in terms of free energy, G. In the coordinate diagram, the free energy of the system is plotted against the progress of the reaction (the reaction coordinate). The starting point for either the forward or the reverse reaction is called the **ground state**, the contribution to the free energy of the system by an average molecule (S or P) under a given set of conditions.

KEY CONVENTION: To describe the free-energy changes for reactions, chemists define a standard set of conditions (temperature 298 K; partial pressure of each gas 1 atm, or 101.3 kPa; concentration of each solute 1 M) and express the free-energy change for a reacting system under these conditions as $\Delta G°$, the **standard free-energy change**. Because biochemical systems commonly involve H^+ concentrations far below 1 M, biochemists define a **biochemical standard free-energy change**, $\Delta G'°$, the standard free-energy change *at pH 7.0*; we employ this definition throughout the book. A more complete definition of $\Delta G'°$ is given in Chapter 13. ■

The equilibrium between S and P reflects the difference in the free energies of their ground states. In the example shown in Figure 6–2, the free energy of the ground state of P is lower than that of S, so $\Delta G'°$ for

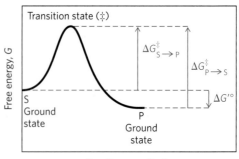

FIGURE 6–2 Reaction coordinate diagram. The free energy of the system is plotted against the progress of the reaction $S \rightarrow P$. A diagram of this kind is a description of the energy changes during the reaction, and the horizontal axis (reaction coordinate) reflects the progressive chemical changes (e.g., bond breakage or formation) as S is converted to P. The activation energies, $\Delta G^‡$, for the $S \rightarrow P$ and $P \rightarrow S$ reactions are indicated. $\Delta G'°$ is the overall standard free-energy change in the direction $S \rightarrow P$.

the reaction is negative (the reaction is exergonic) and at equilibrium there is more P than S (the equilibrium favors P). The position and direction of equilibrium are *not* affected by any catalyst.

A favorable equilibrium does not mean that the S → P conversion will occur at a detectable rate. The *rate* of a reaction is dependent on an entirely different parameter. There is an energy barrier between S and P: the energy required for alignment of reacting groups, formation of transient unstable charges, bond rearrangements, and other transformations required for the reaction to proceed in either direction. This is illustrated by the energy "hill" in Figures 6–2 and 6–3. To undergo reaction, the molecules must overcome this barrier and therefore must be raised to a higher energy level. At the top of the energy hill is a point at which decay to the S or P state is equally probable (it is downhill either way). This is called the **transition state**. The transition state is not a chemical species with any significant stability and should not be confused with a reaction intermediate (such as ES or EP). It is simply a fleeting molecular moment in which events such as bond breakage, bond formation, and charge development have proceeded to the precise point at which decay to either substrate or product is equally likely. The difference between the energy levels of the ground state and the transition state is the **activation energy, $\Delta G^{\ddagger}$**. The rate of a reaction reflects this activation energy: a higher activation energy corresponds to a slower reaction. Reaction rates can be increased by raising the temperature and/or pressure, thereby increasing the number of molecules with sufficient energy to overcome the energy barrier. Alternatively, the activation energy can be lowered by adding a catalyst **(Fig. 6–3)**. *Catalysts enhance reaction rates by lowering activation energies.*

Enzymes are no exception to the rule that catalysts do not affect reaction equilibria. The bidirectional arrows in Equation 6–1 make this point: any enzyme that catalyzes the reaction S → P also catalyzes the reaction P → S. The role of enzymes is to *accelerate* the interconversion of S and P. The enzyme is not used up in the process, and the equilibrium point is unaffected. However, the reaction reaches equilibrium much faster when the appropriate enzyme is present, because the rate of the reaction is increased.

This general principle is illustrated in the conversion of sucrose and oxygen to carbon dioxide and water:

$$C_{12}H_{22}O_{11} + 12O_2 \rightleftharpoons 12CO_2 + 11H_2O$$

This conversion, which takes place through a series of separate reactions, has a very large and negative $\Delta G'^{\circ}$, and at equilibrium the amount of sucrose present is negligible. Yet sucrose is a stable compound, because the activation energy barrier that must be overcome before sucrose reacts with oxygen is quite high. Sucrose can be stored in a container with oxygen almost indefinitely without reacting. In cells, however, sucrose is readily broken down to CO_2 and H_2O in a series of reactions catalyzed by enzymes. These enzymes not only accelerate the reactions, they organize and control them so that much of the energy released is recovered in other chemical forms and made available to the cell for other tasks. The reaction pathway by which sucrose (and other sugars) is broken down is the primary energy-yielding pathway for cells, and the enzymes of this pathway allow the reaction sequence to proceed on a biologically useful time scale.

Any reaction may have several steps, involving the formation and decay of transient chemical species called **reaction intermediates**.* A reaction intermediate is any species on the reaction pathway that has a finite chemical lifetime (longer than a molecular vibration, ~10^{-13} second). When the S $\rightleftharpoons$ P reaction is catalyzed by an enzyme, the ES and EP complexes can be considered intermediates, even though S and P are stable chemical species (Eqn 6–1); the ES and EP complexes occupy valleys in the reaction coordinate diagram (Fig. 6–3). Additional, less stable chemical intermediates often exist in the course of an enzyme-catalyzed reaction. The interconversion of two sequential reaction intermediates thus constitutes a reaction step. When several steps occur in a reaction, the overall rate is determined by the step (or steps) with the highest activation energy; this is called the **rate-limiting step**. In a simple case, the rate-limiting step is the highest-energy point in the diagram for interconversion of S and P. In practice, the rate-limiting step can vary with reaction conditions, and for many enzymes several steps may have similar activation energies, which means they are all partially rate-limiting.

FIGURE 6–3 Reaction coordinate diagram comparing enzyme-catalyzed and uncatalyzed reactions. In the reaction S → P, the ES and EP intermediates occupy minima in the energy progress curve of the enzyme-catalyzed reaction. The terms $\Delta G^{\ddagger}_{uncat}$ and $\Delta G^{\ddagger}_{cat}$ correspond to the activation energy for the uncatalyzed reaction and the overall activation energy for the catalyzed reaction, respectively. The activation energy is lower when the enzyme catalyzes the reaction.

*In this chapter, *step* and *intermediate* refer to chemical species in the reaction pathway of a single enzyme-catalyzed reaction. In the context of metabolic pathways involving many enzymes (discussed in Part II), these terms are used somewhat differently. An entire enzymatic reaction is often referred to as a "step" in a pathway, and the product of one enzymatic reaction (which is the substrate for the next enzyme in the pathway) is referred to as a pathway "intermediate."

Activation energies are energy barriers to chemical reactions. These barriers are crucial to life itself. The rate at which a molecule undergoes a particular reaction decreases as the activation barrier for that reaction increases. Without such energy barriers, complex macromolecules would revert spontaneously to much simpler molecular forms, and the complex and highly ordered structures and metabolic processes of cells could not exist. Over the course of evolution, enzymes have developed to lower activation energies *selectively* for reactions that are needed for cell survival.

Reaction Rates and Equilibria Have Precise Thermodynamic Definitions

Reaction *equilibria* are inextricably linked to the standard free-energy change for the reaction, $\Delta G'^{\circ}$, and reaction *rates* are linked to the activation energy, $\Delta G^{\ddagger}$. A basic introduction to these thermodynamic relationships is the next step in understanding how enzymes work.

An equilibrium such as $S \rightleftharpoons P$ is described by an **equilibrium constant**, K_{eq}, or simply K (p. 25). Under the standard conditions used to compare biochemical processes, an equilibrium constant is denoted K'_{eq} (or K'):

$$K'_{eq} = \frac{[P]}{[S]} \qquad (6\text{–}2)$$

From thermodynamics, the relationship between K'_{eq} and $\Delta G'^{\circ}$ can be described by the expression

$$\Delta G'^{\circ} = -RT \ln K'_{eq} \qquad (6\text{–}3)$$

where R is the gas constant, 8.315 J/mol $\cdot$ K, and T is the absolute temperature, 298 K (25 °C). Equation 6–3 is developed and discussed in more detail in Chapter 13. The important point here is that the equilibrium constant is directly related to the overall standard free-energy change for the reaction (Table 6–4). A large negative value for $\Delta G'^{\circ}$ reflects a favorable reaction equilibrium

(one in which there is much more product than substrate at equilibrium)—but as already noted, this does not mean the reaction will proceed at a rapid rate.

The rate of any reaction is determined by the concentration of the reactant (or reactants) and by a **rate constant**, usually denoted by k. For the unimolecular reaction $S \rightarrow P$, the rate (or velocity) of the reaction, V—representing the amount of S that reacts per unit time—is expressed by a **rate equation**:

$$V = k[S] \qquad (6\text{–}4)$$

In this reaction, the rate depends only on the concentration of S. This is called a first-order reaction. The factor k is a proportionality constant that reflects the probability of reaction under a given set of conditions (pH, temperature, and so forth). Here, k is a first-order rate constant and has units of reciprocal time, such as s^{-1}. If a first-order reaction has a rate constant k of 0.03 s^{-1}, this may be interpreted (qualitatively) to mean that 3% of the available S will be converted to P in 1 s. A reaction with a rate constant of 2,000 s^{-1} will be over in a small fraction of a second. If a reaction rate depends on the concentration of two different compounds, or if the reaction is between two molecules of the same compound, the reaction is second order and k is a second-order rate constant, with units of $M^{-1}s^{-1}$. The rate equation then becomes

$$V = k[S_1][S_2] \qquad (6\text{–}5)$$

From transition-state theory we can derive an expression that relates the magnitude of a rate constant to the activation energy:

$$k = \frac{\mathbf{k}T}{h} e^{-\Delta G^{\ddagger}/RT} \qquad (6\text{–}6)$$

where $\mathbf{k}$ is the Boltzmann constant and h is Planck's constant. The important point here is that the relationship between the rate constant k and the activation energy $\Delta G^{\ddagger}$ is inverse and exponential. In simplified terms, this is the basis for the statement that a lower activation energy means a faster reaction rate.

Now we turn from *what* enzymes do to *how* they do it.

A Few Principles Explain the Catalytic Power and Specificity of Enzymes

Enzymes are extraordinary catalysts. The rate enhancements they bring about are in the range of 5 to 17 orders of magnitude (Table 6–5). Enzymes are also very specific, readily discriminating between substrates with quite similar structures. How can these enormous and highly selective rate enhancements be explained? What is the source of the energy for the dramatic lowering of the activation energies for specific reactions?

The answer to these questions has two distinct but interwoven parts. The first lies in the rearrangement of covalent bonds during an enzyme-catalyzed reaction. Chemical reactions of many types take place between

TABLE 6–4	Relationship between K'_{eq} and $\Delta G'^{\circ}$
K'_{eq}	$\Delta G'^{\circ}$ (kJ/mol)
10^{-6}	34.2
10^{-5}	28.5
10^{-4}	22.8
10^{-3}	17.1
10^{-2}	11.4
10^{-1}	5.7
1	0.0
10^{1}	−5.7
10^{2}	−11.4
10^{3}	−17.1

Note: The relationship is calculated from $\Delta G'^{\circ} = -RT \ln K'_{eq}$ (Eqn 6–3).

TABLE 6–5 Some Rate Enhancements Produced by Enzymes

Cyclophilin	10^5
Carbonic anhydrase	10^7
Triose phosphate isomerase	10^9
Carboxypeptidase A	10^{11}
Phosphoglucomutase	10^{12}
Succinyl-CoA transferase	10^{13}
Urease	10^{14}
Orotidine monophosphate decarboxylase	10^{17}

substrates and enzymes' functional groups (specific amino acid side chains, metal ions, and coenzymes). Catalytic functional groups on an enzyme may form a transient covalent bond with a substrate and activate it for reaction, or a group may be transiently transferred from the substrate to the enzyme. In many cases, these reactions occur only in the enzyme active site. Covalent interactions between enzymes and substrates lower the activation energy (and thereby accelerate the reaction) by providing an alternative, lower-energy reaction path. The specific types of rearrangements that occur are described in Section 6.4.

The second part of the explanation lies in the *noncovalent* interactions between enzyme and substrate. Recall from Chapter 4 that weak, noncovalent interactions help stabilize protein structure and protein-protein interactions. These same interactions are critical to the formation of complexes between proteins and small molecules, including enzyme substrates. Much of the energy required to lower activation energies is derived from weak, noncovalent interactions between substrate and enzyme. What really sets enzymes apart from most other catalysts is the formation of a specific ES complex. The interaction between substrate and enzyme in this complex is mediated by the same forces that stabilize protein structure, including hydrogen bonds and hydrophobic and ionic interactions (Chapter 4). Formation of each weak interaction in the ES complex is accompanied by release of a small amount of free energy that stabilizes the interaction. The energy derived from enzyme-substrate interaction is called **binding energy**, ΔG_B. Its significance extends beyond a simple stabilization of the enzyme-substrate interaction. *Binding energy is a major source of free energy used by enzymes to lower the activation energies of reactions.*

Two fundamental and interrelated principles provide a general explanation for how enzymes use noncovalent binding energy:

1. Much of the catalytic power of enzymes is ultimately derived from the free energy released in forming many weak bonds and interactions between an enzyme and its substrate. This binding energy contributes to specificity as well as to catalysis.

2. Weak interactions are optimized in the reaction transition state; enzyme active sites are complementary not to the substrates per se but to the transition states through which substrates pass as they are converted to products during an enzymatic reaction.

These themes are critical to an understanding of enzymes, and they now become our primary focus.

Weak Interactions between Enzyme and Substrate Are Optimized in the Transition State

How does an enzyme use binding energy to lower the activation energy for a reaction? Formation of the ES complex is not the explanation in itself, although some of the earliest considerations of enzyme mechanisms began with this idea. Studies on enzyme specificity carried out by Emil Fischer led him to propose, in 1894, that enzymes were structurally complementary to their substrates, so that they fit together like a lock and key **(Fig. 6–4)**. This elegant idea, that a specific (exclusive) interaction between two biological molecules is mediated

FIGURE 6–4 Complementary shapes of a substrate and its binding site on an enzyme. The enzyme dihydrofolate reductase with its substrate NADP⁺, unbound and bound; another bound substrate, tetrahydrofolate, is also visible (PDB ID 1RA2). In this model, the NADP⁺ binds to a pocket that is complementary to it in shape and ionic properties, an illustration of Emil Fischer's "lock and key" hypothesis of enzyme action. In reality, the complementarity between protein and ligand (in this case, substrate) is rarely perfect, as we saw in Chapter 5.

by molecular surfaces with complementary shapes, has greatly influenced the development of biochemistry, and such interactions lie at the heart of many biochemical processes. However, the "lock and key" hypothesis can be misleading when applied to enzymatic catalysis. An enzyme completely complementary to its substrate would be a very poor enzyme, as we can demonstrate.

Consider an imaginary reaction, the breaking of a magnetized metal stick. The uncatalyzed reaction is shown in **Figure 6–5a**. Let's examine two imaginary enzymes—two "stickases"—that could catalyze this reaction, both of which employ magnetic forces as a paradigm for the binding energy used by real enzymes. We first design an enzyme perfectly complementary to

the substrate (Fig. 6–5b). The active site of this stickase is a pocket lined with magnets. To react (break), the stick must reach the transition state of the reaction, but the stick fits so tightly in the active site that it cannot bend, because bending would eliminate some of the magnetic interactions between stick and enzyme. Such an enzyme *impedes* the reaction, stabilizing the substrate instead. In a reaction coordinate diagram (Fig. 6–5b), this kind of ES complex would correspond to an energy trough from which the substrate would have difficulty escaping. Such an enzyme would be useless.

The modern notion of enzymatic catalysis, first proposed by Michael Polanyi (1921) and Haldane (1930), was elaborated by Linus Pauling in 1946 and by

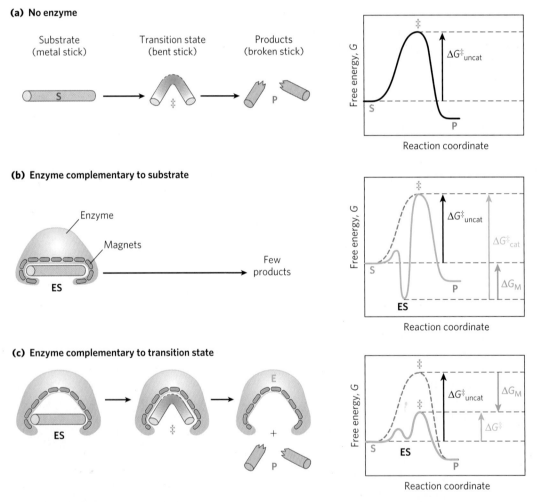

FIGURE 6–5 An imaginary enzyme (stickase) designed to catalyze breakage of a metal stick. (a) Before the stick is broken, it must first be bent (the transition state). In both stickase examples, magnetic interactions take the place of weak bonding interactions between enzyme and substrate. **(b)** A stickase with a magnet-lined pocket complementary in structure to the stick (the substrate) stabilizes the substrate. Bending is impeded by the magnetic attraction between stick and stickase. **(c)** An enzyme with a pocket complementary to the reaction transition state helps to destabilize the stick, contributing to catalysis of the reaction. The binding energy of the magnetic interactions compensates for the

increase in free energy required to bend the stick. Reaction coordinate diagrams (right) show the energy consequences of complementarity to substrate versus complementarity to transition state (EP complexes are omitted). ΔG_M, the difference between the transition-state energies of the uncatalyzed and catalyzed reactions, is contributed by the magnetic interactions between the stick and stickase. When the enzyme is complementary to the substrate (b), the ES complex is more stable and has less free energy in the ground state than substrate alone. The result is an *increase* in the activation energy.

William P. Jencks in the 1970s: in order to catalyze reactions, an enzyme must be complementary to the *reaction transition state*. This means that optimal interactions between substrate and enzyme occur only in the transition state. Figure 6–5c demonstrates how such an enzyme can work. The metal stick binds to the stickase, but only a subset of the possible magnetic interactions are used in forming the ES complex. The bound substrate must still undergo the increase in free energy needed to reach the transition state. Now, however, the increase in free energy required to draw the stick into a bent and partially broken conformation is offset, or "paid for," by the magnetic interactions (binding energy) that form between the enzyme and substrate in the transition state. Many of these interactions involve parts of the stick that are distant from the point of breakage; thus interactions between the stickase and nonreacting parts of the stick provide some of the energy needed to catalyze stick breakage. This "energy payment" translates into a lower net activation energy and a faster reaction rate.

Real enzymes work on an analogous principle. Some weak interactions are formed in the ES complex, but the full complement of such interactions between substrate and enzyme is formed only when the substrate reaches the transition state. The free energy (binding energy) released by the formation of these interactions partially offsets the energy required to reach the top of the energy hill. The summation of the unfavorable (positive) activation energy $\Delta G^{\ddagger}$ and the favorable (negative) binding energy ΔG_B results in a lower *net activation energy* **(Fig. 6–6)**. Even on the enzyme, the transition state is not a stable species but a brief point in time that the substrate spends atop an energy hill. The enzyme-catalyzed reaction is much faster than the uncatalyzed process, however, because the hill is much smaller. The important principle is that *weak binding interactions between the enzyme and the substrate provide a substantial driving force for enzymatic catalysis*. The groups on the substrate that are involved in these weak interactions can be at some distance from the bonds that are broken or changed. The weak interactions formed only in the transition state are those that make the primary contribution to catalysis.

The requirement for multiple weak interactions to drive catalysis is one reason why enzymes (and some coenzymes) are so large. An enzyme must provide functional groups for ionic, hydrogen-bond, and other interactions, and also must precisely position these groups so that binding energy is optimized in the transition state. Adequate binding is accomplished most readily by positioning a substrate in a cavity (the active site) where it is effectively removed from water. The size of proteins reflects the need for superstructure to keep interacting groups properly positioned and to keep the cavity from collapsing.

Binding Energy Contributes to Reaction Specificity and Catalysis

Can we demonstrate quantitatively that binding energy accounts for the huge rate accelerations brought about by enzymes? Yes. As a point of reference, Equation 6–6 allows us to calculate that $\Delta G^{\ddagger}$ must be lowered by about 5.7 kJ/mol to accelerate a first-order reaction by a factor of 10, under conditions commonly found in cells. The energy available from formation of a single weak interaction is generally estimated to be 4 to 30 kJ/mol. The overall energy available from a number of such interactions is therefore sufficient to lower activation energies by the 60 to 100 kJ/mol required to explain the large rate enhancements observed for many enzymes.

The same binding energy that provides energy for catalysis also gives an enzyme its **specificity**, the ability to discriminate between a substrate and a competing molecule. Conceptually, specificity is easy to distinguish from catalysis, but this distinction is much more difficult to make experimentally, because catalysis and specificity arise from the same phenomenon. If an enzyme active site has functional groups arranged optimally to form a variety of weak interactions with a particular substrate in the transition state, the enzyme will not be able to interact to the same degree with any other molecule. For example, if the substrate has a hydroxyl group that forms a hydrogen bond with a specific Glu residue on the enzyme, any molecule lacking a hydroxyl group at that particular position will be a poorer substrate for the enzyme. In addition, any molecule with an extra functional group for which the enzyme has no pocket or binding site is likely to be excluded from the enzyme. In general, *specificity* is derived from the formation of many weak interactions between the enzyme and its specific substrate molecule.

The importance of binding energy to catalysis can be readily demonstrated. For example, the glycolytic enzyme triose phosphate isomerase catalyzes the interconversion

FIGURE 6–6 Role of binding energy in catalysis. To lower the activation energy for a reaction, the system must acquire an amount of energy equivalent to the amount by which $\Delta G^{\ddagger}$ is lowered. Much of this energy comes from binding energy (ΔG_B) contributed by formation of weak noncovalent interactions between substrate and enzyme in the transition state. The role of ΔG_B is analogous to that of ΔG_M in Figure 6-5.

of glyceraldehyde 3-phosphate and dihydroxyacetone phosphate:

$$HC=O \quad \underset{\text{triose}}{\overset{}{\rightleftharpoons}} \quad H_2C-OH$$

Glyceraldehyde 3-phosphate

triose phosphate isomerase

Dihydroxyacetone phosphate

This reaction rearranges the carbonyl and hydroxyl groups on carbons 1 and 2. However, more than 80% of the enzymatic rate acceleration has been traced to enzyme-substrate interactions involving the phosphate group on carbon 3 of the substrate. This was determined by comparing the enzyme-catalyzed reactions with glyceraldehyde 3-phosphate and with glyceraldehyde (no phosphate group at position 3) as substrate.

The general principles outlined above can be illustrated by a variety of recognized catalytic mechanisms. These mechanisms are not mutually exclusive, and a given enzyme might incorporate several types in its overall mechanism of action.

Consider what needs to occur for a reaction to take place. Prominent physical and thermodynamic factors contributing to $\Delta G^{\ddagger}$, the barrier to reaction, might include: (1) the entropy (freedom of motion) of molecules in solution, which reduces the possibility that they will react together, (2) the solvation shell of hydrogen-bonded water that surrounds and helps to stabilize most biomolecules in aqueous solution, (3) the distortion of substrates that must occur in many reactions, and (4) the need for proper alignment of catalytic functional groups on the enzyme. Binding energy can be used to overcome all these barriers.

First, a large restriction in the relative motions of two substrates that are to react, or **entropy reduction**, is one obvious benefit of binding them to an enzyme. Binding energy holds the substrates in the proper orientation to react—a substantial contribution to catalysis, because productive collisions between molecules in solution can be exceedingly rare. Substrates can be precisely aligned on the enzyme, with many weak interactions between each substrate and strategically located groups on the enzyme clamping the substrate molecules into the proper positions. Studies have shown that constraining the motion of two reactants can produce rate enhancements of many orders of magnitude **(Fig. 6–7)**.

Second, formation of weak bonds between substrate and enzyme results in **desolvation** of the substrate. Enzyme-substrate interactions replace most or all of the hydrogen bonds between the substrate and water that would otherwise impede reaction. Third, binding energy involving weak interactions formed only in the reaction transition state helps to compensate thermodynamically for any distortion, primarily electron redistribution, that the substrate must undergo to react.

FIGURE 6–7 Rate enhancement by entropy reduction. Shown here are reactions of an ester with a carboxylate group to form an anhydride. The R group is the same in each case. **(a)** For this bimolecular reaction, the rate constant k is second-order, with units of $M^{-1}s^{-1}$. **(b)** When the two reacting groups are in a single molecule, and thus have less freedom of motion, the reaction is much faster. For this unimolecular reaction, k has units of s^{-1}. Dividing the rate constant for (b) by the rate constant for (a) gives a rate enhancement of about 10^5 M. (The enhancement has units of molarity because we are comparing a unimolecular and a bimolecular reaction.) Put another way, if the reactant in (b) were present at a concentration of 1 M, the reacting groups would *behave* as though they were present at a concentration of 10^5 M. Note that the reactant in (b) has freedom of rotation about three bonds (shown with curved arrows), but this still represents a substantial reduction of entropy over (a). If the bonds that rotate in (b) are constrained as in **(c)**, the entropy is reduced further and the reaction exhibits a rate enhancement of 10^8 M relative to (a).

Finally, the enzyme itself usually undergoes a change in conformation when the substrate binds, induced by multiple weak interactions with the substrate. This is referred to as **induced fit**, a mechanism postulated by Daniel Koshland in 1958. The motions can affect a small part of the enzyme near the active site or can involve changes in the positioning of entire domains. Typically, a network of coupled motions occurs throughout the enzyme that ultimately brings about the required changes in the active site. Induced fit serves to bring specific functional groups on the enzyme into the proper position to catalyze the reaction. The conformational change also permits formation of additional weak bonding interactions in the transition state. In either case, the new enzyme conformation has enhanced catalytic properties. As we have seen, induced fit is a common feature of the reversible binding of ligands to proteins (Chapter 5). Induced fit is also important in the interaction of almost every enzyme with its substrate.

Specific Catalytic Groups Contribute to Catalysis

In most enzymes, the binding energy used to form the ES complex is just one of several contributors to the overall catalytic mechanism. Once a substrate is bound to an enzyme, properly positioned catalytic functional groups aid in the cleavage and formation of bonds by a variety of mechanisms, including general acid-base catalysis, covalent catalysis, and metal ion catalysis. These are distinct from mechanisms based on binding energy, because they generally involve transient *covalent* interaction with a substrate or group transfer to or from a substrate.

General Acid-Base Catalysis A proton transfer is the single most common reaction in biochemistry. One or, often, many proton transfers occur in the course of most reactions that take place in cells. Many biochemical reactions involve the formation of unstable charged intermediates that tend to break down rapidly to their constituent reactant species, thus impeding the reaction **(Fig. 6–8)**. Charged intermediates can often be stabilized by the transfer of protons to or from the substrate or intermediate to form a species that breaks down more readily to products. The effects of catalysis by acids and bases are often studied using nonenzymatic model reactions, in which the proton transfers can involve either the constituents of water alone or other weak proton donors or acceptors. Catalysis of the type that uses only the H^+ (H_3O^+) or OH^- ions present in water is referred to as **specific acid-base catalysis**. If protons are transferred between the intermediate and water faster than the intermediate breaks down to reactants, the intermediate is effectively stabilized every time it forms. No additional catalysis mediated by other proton acceptors or donors will occur. In many cases, however, water is not enough. The term **general acid-base catalysis** refers to proton transfers mediated by weak acids and bases other than water. For nonenzymatic reactions in aqueous solutions, added weak acids and bases provide an observed rate acceleration only when the unstable reaction intermediate breaks down to reactants faster than protons can be transferred to or from water alone. Many weak organic acids can supplement water as proton donors in this situation, or weak organic bases can serve as proton acceptors.

In the active site of an enzyme, where water may not be available as a proton donor or acceptor, general acid-base catalysis becomes crucial. A number of amino acid side chains can and do take on the role of proton donors and acceptors **(Fig. 6–9)**. These groups can be precisely positioned in an enzyme active site to allow proton transfers, providing rate enhancements of the order of 10^2 to 10^5. This type of catalysis occurs on the vast majority of enzymes.

FIGURE 6–8 How a catalyst circumvents unfavorable charge development during cleavage of an amide. The hydrolysis of an amide bond, shown here, is the same reaction as that catalyzed by chymotrypsin and other proteases. Charge development is unfavorable and can be circumvented by donation of a proton by H_3O^+ (specific acid catalysis) or HA (general acid catalysis), where HA represents any acid. Similarly, charge can be neutralized by proton abstraction by OH^- (specific base catalysis) or B: (general base catalysis), where B: represents any base.

FIGURE 6–9 Amino acids in general acid-base catalysis. Many organic reactions that are used to model biochemical processes are promoted by proton donors (general acids) or proton acceptors (general bases). The active sites of some enzymes contain amino acid functional groups, such as those shown here, that can participate in the catalytic process as proton donors or proton acceptors.

Covalent Catalysis In covalent catalysis, a transient covalent bond is formed between the enzyme and the substrate. Consider the hydrolysis of a bond between groups A and B:

$$A{-}B \xrightarrow{\;H_2O\;} A + B$$

In the presence of a covalent catalyst (an enzyme with a nucleophilic group X:) the reaction becomes

$$A{-}B + X: \longrightarrow A{-}X + B \xrightarrow{\;H_2O\;} A + X: + B$$

This alters the pathway of the reaction, and it results in catalysis *only* when the new pathway has a lower activation energy than the uncatalyzed pathway. Both of the new steps must be faster than the uncatalyzed reaction. A number of amino acid side chains, including all those in Figure 6–9, and the functional groups of some enzyme cofactors can serve as nucleophiles in the formation of covalent bonds with substrates. These covalent complexes always undergo further reaction to regenerate the free enzyme. The covalent bond formed between the enzyme and the substrate can activate a substrate for further reaction in a manner that is usually specific to the particular group or coenzyme.

Metal Ion Catalysis Metals, whether tightly bound to the enzyme or taken up from solution along with the substrate, can participate in catalysis in several ways. Ionic interactions between an enzyme-bound metal and a substrate can help orient the substrate for reaction or stabilize charged reaction transition states. This use of weak bonding interactions between metal and substrate is similar to some of the uses of enzyme-substrate binding energy described earlier. Metals can also mediate oxidation-reduction reactions by reversible changes in the metal ion's oxidation state. Nearly a third of all known enzymes require one or more metal ions for catalytic activity.

Most enzymes combine several catalytic strategies to bring about a rate enhancement. A good example is the use of covalent catalysis, general acid-base catalysis, and transition-state stabilization in the reaction catalyzed by chymotrypsin, detailed in Section 6.4.

SUMMARY 6.2 How Enzymes Work

▶ Enzymes are highly effective catalysts, commonly enhancing reaction rates by a factor of 10^5 to 10^{17}.

▶ Enzyme-catalyzed reactions are characterized by the formation of a complex between substrate and enzyme (an ES complex). Substrate binding occurs in a pocket on the enzyme called the active site.

▶ The function of enzymes and other catalysts is to lower the activation energy, $\Delta G^{\ddagger}$, for a reaction and thereby enhance the reaction rate. The equilibrium of a reaction is unaffected by the enzyme.

▶ A significant part of the energy used for enzymatic rate enhancements is derived from weak interactions (hydrogen bonds and hydrophobic and ionic interactions) between substrate and enzyme. The enzyme active site is structured so that some of these weak interactions occur preferentially in the reaction transition state, thus stabilizing the transition state. The need for multiple interactions is one reason for the large size of enzymes. The binding energy, ΔG_B, is used to offset the energy required for activation, $\Delta G^{\ddagger}$, in several ways. It can be used, for example, to lower substrate entropy, for substrate desolvation, or to cause a conformational change in the enzyme (induced fit). Binding energy also accounts for the exquisite specificity of enzymes for their substrates.

▶ Additional catalytic mechanisms employed by enzymes include general acid-base catalysis, covalent catalysis, and metal ion catalysis. Catalysis often involves transient covalent interactions between the substrate and the enzyme, or group transfers to and from the enzyme, so as to provide a new, lower-energy reaction path. In all cases, the enzyme reverts to the unbound state once the reaction is complete.

6.3 Enzyme Kinetics as an Approach to Understanding Mechanism

Biochemists commonly use several approaches to study the mechanism of action of purified enzymes. The three-dimensional structure of the protein provides important information, which is enhanced by classical protein chemistry and modern methods of site-directed mutagenesis (changing the amino acid sequence of a protein by genetic engineering; see Fig. 9–10). These technologies permit enzymologists to examine the role of individual amino acids in enzyme structure and action. However, the oldest approach to understanding enzyme mechanisms, and the one that remains most important, is to determine the *rate* of a reaction and how it changes in response to changes in experimental parameters, a discipline known as **enzyme kinetics**. We provide here a basic introduction to the kinetics of enzyme-catalyzed reactions. More advanced treatments are available in the sources cited at the end of the chapter.

Substrate Concentration Affects the Rate of Enzyme-Catalyzed Reactions

A key factor affecting the rate of a reaction catalyzed by an enzyme is the concentration of substrate, [S]. However, studying the effects of substrate concentration is complicated by the fact that [S] changes during the course of an in vitro reaction as substrate is converted to product. One simplifying approach in kinetics experiments is to measure the **initial rate** (or **initial**

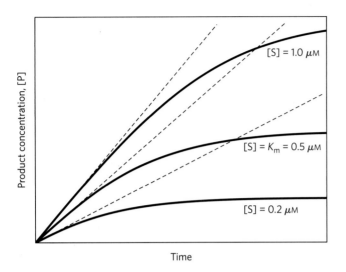

FIGURE 6–10 Initial velocities of enzyme-catalyzed reactions. A theoretical enzyme catalyzes the reaction S ⇌ P, and is present at a concentration sufficient to catalyze the reaction at a maximum velocity, V_{max}, of 1 μM/min. The Michaelis constant, K_m (explained in the text), is 0.5 μM. Progress curves are shown for substrate concentrations below, at, and above the K_m. The rate of an enzyme-catalyzed reaction declines as substrate is converted to product. A tangent to each curve taken at time = 0 defines the initial velocity, V_0, of each reaction.

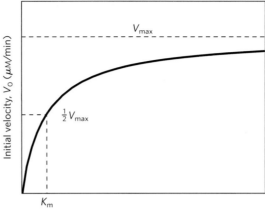

FIGURE 6–11 Effect of substrate concentration on the initial velocity of an enzyme-catalyzed reaction. The maximum velocity, V_{max}, is extrapolated from the plot, because V_0 approaches but never quite reaches V_{max}. The substrate concentration at which V_0 is half maximal is K_m, the Michaelis constant. The concentration of enzyme in an experiment such as this is generally so low that [S] >> [E] even when [S] is described as low or relatively low. The units shown are typical for enzyme-catalyzed reactions and are given only to help illustrate the meaning of V_0 and [S]. (Note that the curve describes *part* of a rectangular hyperbola, with one asymptote at V_{max}. If the curve were continued below [S] = 0, it would approach a vertical asymptote at [S] = $-K_m$.)

velocity), designated V_0 **(Fig. 6–10)**. In a typical reaction, the enzyme may be present in nanomolar quantities, whereas [S] may be five or six orders of magnitude higher. If only the beginning of the reaction is monitored (often the first 60 seconds or less), changes in [S] can be limited to a few percent, and [S] can be regarded as constant. V_0 can then be explored as a function of [S], which is adjusted by the investigator. The effect on V_0 of varying [S] when the enzyme concentration is held constant is shown in **Figure 6–11**. At relatively low concentrations of substrate, V_0 increases almost linearly with an increase in [S]. At higher substrate concentrations, V_0 increases by smaller and smaller amounts in response to increases in [S]. Finally, a point is reached beyond which increases in V_0 are vanishingly small as [S] increases. This plateau-like V_0 region is close to the **maximum velocity**, V_{max}.

The ES complex is the key to understanding this kinetic behavior, just as it was a starting point for our discussion of catalysis. The kinetic pattern in Figure 6–11 led Victor Henri, following the lead of Wurtz, to propose in 1903 that the combination of an enzyme with its substrate molecule to form an ES complex is a necessary step in enzymatic catalysis. This idea was expanded into a general theory of enzyme action, particularly by Leonor Michaelis and Maud Menten in 1913. They postulated that the enzyme first combines reversibly with its substrate to form an enzyme-substrate complex in a relatively fast reversible step:

$$E + S \underset{k_{-1}}{\overset{k_1}{\rightleftharpoons}} ES \qquad (6–7)$$

The ES complex then breaks down in a slower second step to yield the free enzyme and the reaction product P:

$$ES \underset{k_{-2}}{\overset{k_2}{\rightleftharpoons}} E + P \qquad (6–8)$$

Because the slower second reaction (Eqn 6–8) must limit the rate of the overall reaction, the overall rate must be proportional to the concentration of the species that reacts in the second step, that is, ES.

At any given instant in an enzyme-catalyzed reaction, the enzyme exists in two forms, the free or uncombined form E and the combined form ES. At low [S], most of the enzyme is in the uncombined form E. Here, the rate is proportional to [S] because the equilibrium of Equation 6–7 is pushed toward formation of more ES as [S] increases. The maximum initial rate of the catalyzed reaction (V_{max}) is observed when virtually all the enzyme is present as the ES complex and [E] is vanishingly

Leonor Michaelis, 1875–1949   Maud Menten, 1879–1960

small. Under these conditions, the enzyme is "saturated" with its substrate, so that further increases in [S] have no effect on rate. This condition exists when [S] is sufficiently high that essentially all the free enzyme has been converted to the ES form. After the ES complex breaks down to yield the product P, the enzyme is free to catalyze reaction of another molecule of substrate (and will do so rapidly under saturating conditions). The saturation effect is a distinguishing characteristic of enzymatic catalysts and is responsible for the plateau observed in Figure 6–11. The pattern seen in Figure 6–11 is sometimes referred to as saturation kinetics.

When the enzyme is first mixed with a large excess of substrate, there is an initial period, the **pre–steady state**, during which the concentration of ES builds up. This period is usually too short to be easily observed, lasting just microseconds, and is not evident in Figure 6–10. The reaction quickly achieves a **steady state** in which [ES] (and the concentrations of any other intermediates) remains approximately constant over time. The concept of a steady state was introduced by G. E. Briggs and Haldane in 1925. The measured V_0 generally reflects the steady state, even though V_0 is limited to the early part of the reaction, and analysis of these initial rates is referred to as **steady-state kinetics**.

The Relationship between Substrate Concentration and Reaction Rate Can Be Expressed Quantitatively

The curve expressing the relationship between [S] and V_0 (Fig. 6–11) has the same general shape for most enzymes (it approaches a rectangular hyperbola), which can be expressed algebraically by the Michaelis-Menten equation. Michaelis and Menten derived this equation starting from their basic hypothesis that the rate-limiting step in enzymatic reactions is the breakdown of the ES complex to product and free enzyme. The equation is

$$V_0 = \frac{V_{max}[S]}{K_m + [S]} \qquad \text{🖱} \quad (6\text{–}9)$$

All these terms—[S], V_0, V_{max}, and a constant called the Michaelis constant, K_m—are readily measured experimentally.

Here we develop the basic logic and the algebraic steps in a modern derivation of the Michaelis-Menten equation, which includes the steady-state assumption introduced by Briggs and Haldane. The derivation starts with the two basic steps of the formation and breakdown of ES (Eqns 6–7 and 6–8). Early in the reaction, the concentration of the product, [P], is negligible, and we make the simplifying assumption that the reverse reaction, $P \rightarrow S$ (described by k_{-2}), can be ignored. This assumption is not critical but it simplifies our task. The overall reaction then reduces to

$$E + S \underset{k_{-1}}{\overset{k_1}{\rightleftharpoons}} ES \overset{k_2}{\longrightarrow} E + P \qquad (6\text{–}10)$$

V_0 is determined by the breakdown of ES to form product, which is determined by [ES]:

$$V_0 = k_2[ES] \qquad (6\text{–}11)$$

Because [ES] in Equation 6–11 is not easily measured experimentally, we must begin by finding an alternative expression for this term. First, we introduce the term $[E_t]$, representing the total enzyme concentration (the sum of free and substrate-bound enzyme). Free or unbound enzyme [E] can then be represented by $[E_t] - [ES]$. Also, because [S] is ordinarily far greater than $[E_t]$, the amount of substrate bound by the enzyme at any given time is negligible compared with the total [S]. With these conditions in mind, the following steps lead us to an expression for V_0 in terms of easily measurable parameters.

Step 1 The rates of formation and breakdown of ES are determined by the steps governed by the rate constants k_1 (formation) and $k_{-1} + k_2$ (breakdown to reactants and products, respectively), according to the expressions

$$\text{Rate of ES formation} = k_1([E_t] - [ES])[S] \quad (6\text{–}12)$$

$$\text{Rate of ES breakdown} = k_{-1}[ES] + k_2[ES] \quad (6\text{–}13)$$

Step 2 We now make an important assumption: that the initial rate of reaction reflects a steady state in which [ES] is constant—that is, the rate of formation of ES is equal to the rate of its breakdown. This is called the **steady-state assumption**. The expressions in Equations 6–12 and 6–13 can be equated for the steady state, giving

$$k_1([E_t] - [ES])[S] = k_{-1}[ES] + k_2[ES] \quad (6\text{–}14)$$

Step 3 In a series of algebraic steps, we now solve Equation 6–14 for [ES]. First, the left side is multiplied out and the right side simplified to give

$$k_1[E_t][S] - k_1[ES][S] = (k_{-1} + k_2)[ES] \quad (6\text{–}15)$$

Adding the term $k_1[ES][S]$ to both sides of the equation and simplifying gives

$$k_1[E_t][S] = (k_1[S] + k_{-1} + k_2)[ES] \quad (6\text{–}16)$$

We then solve this equation for [ES]:

$$[ES] = \frac{k_1[E_t][S]}{k_1[S] + k_{-1} + k_2} \qquad (6\text{–}17)$$

This can now be simplified further, combining the rate constants into one expression:

$$[ES] = \frac{[E_t][S]}{[S] + (k_{-1} + k_2)/k_1} \qquad (6\text{–}18)$$

The term $(k_{-1} + k_2)/k_1$ is defined as the **Michaelis constant**, K_m. Substituting this into Equation 6–18 simplifies the expression to

$$[ES] = \frac{[E_t][S]}{K_m + [S]} \qquad (6\text{–}19)$$

Step 4 We can now express V_0 in terms of [ES]. Substituting the right side of Equation 6–19 for [ES] in Equation 6–11 gives

$$V_0 = \frac{k_2[E_t][S]}{K_m + [S]} \qquad (6-20)$$

This equation can be further simplified. Because the maximum velocity occurs when the enzyme is saturated (that is, with $[ES] = [E_t]$), V_{max} can be defined as $k_2[E_t]$. Substituting this in Equation 6–20 gives Equation 6–9:

$$V_0 = \frac{V_{max}[S]}{K_m + [S]}$$

This is the **Michaelis-Menten equation**, the **rate equation** for a one-substrate enzyme-catalyzed reaction. It is a statement of the quantitative relationship between the initial velocity V_0, the maximum velocity V_{max}, and the initial substrate concentration [S], all related through the Michaelis constant K_m. Note that K_m has units of molar concentration. Does the equation fit experimental observations? Yes; we can confirm this by considering the limiting situations where [S] is very high or very low, as shown in **Figure 6–12**.

An important numerical relationship emerges from the Michaelis-Menten equation in the special case when V_0 is exactly one-half V_{max} (Fig. 6–12). Then

$$\frac{V_{max}}{2} = \frac{V_{max}[S]}{K_m + [S]} \qquad (6-21)$$

On dividing by V_{max}, we obtain

$$\frac{1}{2} = \frac{[S]}{K_m + [S]} \qquad (6-22)$$

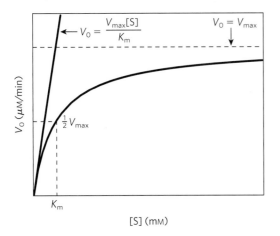

FIGURE 6–12 **Dependence of initial velocity on substrate concentration.** This graph shows the kinetic parameters that define the limits of the curve at high and low [S]. At low [S], $K_m \gg [S]$ and the [S] term in the denominator of the Michaelis-Menten equation (Eqn 6–9) becomes insignificant. The equation simplifies to $V_0 = V_{max}[S]/K_m$ and V_0 exhibits a linear dependence on [S], as observed here. At high [S], where $[S] \gg K_m$, the K_m term in the denominator of the Michaelis-Menten equation becomes insignificant and the equation simplifies to $V_0 = V_{max}$; this is consistent with the plateau observed at high [S]. The Michaelis-Menten equation is therefore consistent with the observed dependence of V_0 on [S], and the shape of the curve is defined by the terms V_{max}/K_m at low [S] and V_{max} at high [S].

Solving for K_m, we get $K_m + [S] = 2[S]$, or

$$K_m = [S], \quad \text{when} \quad V_0 = \tfrac{1}{2}V_{max} \qquad (6-23)$$

This is a very useful, practical definition of K_m: K_m is equivalent to the substrate concentration at which V_0 is one-half V_{max}.

The Michaelis-Menten equation (Eqn 6–9) can be algebraically transformed into versions that are useful in the practical determination of K_m and V_{max} (Box 6–1) and, as we describe later, in the analysis of inhibitor action (see Box 6–2 on p. 209).

Kinetic Parameters Are Used to Compare Enzyme Activities

It is important to distinguish between the Michaelis-Menten equation and the specific kinetic mechanism on which it was originally based. The equation describes the kinetic behavior of a great many enzymes, and all enzymes that exhibit a hyperbolic dependence of V_0 on [S] are said to follow **Michaelis-Menten kinetics**. The practical rule that $K_m = [S]$ when $V_0 = \tfrac{1}{2}V_{max}$ (Eqn 6–23) holds for all enzymes that follow Michaelis-Menten kinetics. (The most important exceptions to Michaelis-Menten kinetics are the regulatory enzymes, discussed in Section 6.5.) However, the Michaelis-Menten equation does not depend on the relatively simple two-step reaction mechanism proposed by Michaelis and Menten (Eqn 6–10). Many enzymes that follow Michaelis-Menten kinetics have quite different reaction mechanisms, and enzymes that catalyze reactions with six or eight identifiable steps often exhibit the same steady-state kinetic behavior. Even though Equation 6–23 holds true for many enzymes, both the magnitude and the real meaning of V_{max} and K_m can differ from one enzyme to the next. This is an important limitation of the steady-state approach to enzyme kinetics. The parameters V_{max} and K_m can be obtained experimentally for any given enzyme, but by themselves they provide little information about the number, rates, or chemical nature of discrete steps in the reaction. Steady-state kinetics nevertheless is the standard language by which biochemists compare and characterize the catalytic efficiencies of enzymes.

Interpreting V_{max} and K_m Figure 6–12 shows a simple graphical method for obtaining an approximate value for K_m. A more convenient procedure, using a **double-reciprocal plot**, is presented in Box 6–1. The K_m can vary greatly from enzyme to enzyme, and even for different substrates of the same enzyme (Table 6–6). The term is sometimes used (often inappropriately) as an indicator of the affinity of an enzyme for its substrate. The actual meaning of K_m depends on specific aspects of the reaction mechanism such as the number and relative rates of the individual steps. For reactions with two steps,

BOX 6–1 Transformations of the Michaelis-Menten Equation: The Double-Reciprocal Plot

The Michaelis-Menten equation

$$V_0 = \frac{V_{max}[S]}{K_m + [S]}$$

can be algebraically transformed into equations that are more useful in plotting experimental data. One common transformation is derived simply by taking the reciprocal of both sides of the Michaelis-Menten equation:

$$\frac{1}{V_0} = \frac{K_m + [S]}{V_{max}[S]}$$

Separating the components of the numerator on the right side of the equation gives

$$\frac{1}{V_0} = \frac{K_m}{V_{max}[S]} + \frac{[S]}{V_{max}[S]}$$

which simplifies to

$$\frac{1}{V_0} = \frac{K_m}{V_{max}[S]} + \frac{1}{V_{max}}$$

This form of the Michaelis-Menten equation is called the **Lineweaver-Burk equation**. For enzymes obeying the Michaelis-Menten relationship, a plot of $1/V_0$ versus $1/[S]$ (the "double reciprocal" of the V_0 versus [S] plot we have been using to this point) yields a straight line (Fig. 1). This line has a slope of K_m/V_{max}, an intercept of $1/V_{max}$ on the $1/V_0$ axis, and an intercept of $-1/K_m$ on the $1/[S]$ axis. The double-reciprocal presentation, also called a Lineweaver-Burk plot, has the great advantage of allowing a more accurate determination of V_{max}, which can only be *approximated* from a simple plot of V_0 versus [S] (see Fig. 6–12).

Other transformations of the Michaelis-Menten equation have been derived, each with some particular advantage in analyzing enzyme kinetic data. (See Problem 14 at the end of this chapter.)

The double-reciprocal plot of enzyme reaction rates is very useful in distinguishing between certain types of enzymatic reaction mechanisms (see Fig. 6–14) and in analyzing enzyme inhibition (see Box 6–2).

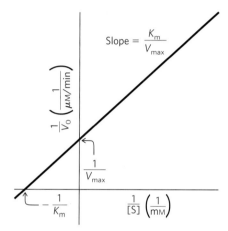

FIGURE 1 A double-reciprocal or Lineweaver-Burk plot.

$$K_m = \frac{k_2 + k_{-1}}{k_1} \quad (6\text{--}24)$$

When k_2 is rate-limiting, $k_2 \ll k_{-1}$ and K_m reduces to k_{-1}/k_1, which is defined as the **dissociation constant**, K_d, of the ES complex. Where these conditions hold, K_m does represent a measure of the affinity of the enzyme for its substrate in the ES complex. However, this scenario does not apply for most enzymes. Sometimes $k_2 \gg k_{-1}$, and then $K_m = k_2/k_1$. In other cases, k_2 and k_{-1} are comparable and K_m remains a more complex function of all three rate constants (Eqn 6–24). The Michaelis-Menten equation and the characteristic saturation behavior of the enzyme still apply, but K_m cannot be considered a simple measure of substrate affinity. Even more common are cases in which the reaction goes through several steps after formation of ES; K_m can then become a very complex function of many rate constants.

The quantity V_{max} also varies greatly from one enzyme to the next. If an enzyme reacts by the two-step Michaelis-Menten mechanism, $V_{max} = k_2[E_t]$, where k_2 is rate limiting. However, the number of reaction steps and the identity of the rate-limiting step(s) can vary from enzyme to enzyme. For example, consider the quite common situation where product release, EP → E + P, is rate-limiting. Early in the reaction (when [P] is low), the overall reaction can be described by the scheme

$$E + S \underset{k_{-1}}{\overset{k_1}{\rightleftharpoons}} ES \underset{k_{-2}}{\overset{k_2}{\rightleftharpoons}} EP \overset{k_3}{\rightleftharpoons} E + P \quad (6\text{--}25)$$

In this case, most of the enzyme is in the EP form at saturation, and $V_{max} = k_3[E_t]$. It is useful to define a more general rate constant,

TABLE 6–6 K_m for Some Enzymes and Substrates

Enzyme	Substrate	K_m (mM)
Hexokinase (brain)	ATP	0.4
	D-Glucose	0.05
	D-Fructose	1.5
Carbonic anhydrase	HCO_3^-	26
Chymotrypsin	Glycyltyrosinylglycine	108
	N-Benzoyltyrosinamide	2.5
β-Galactosidase	D-Lactose	4.0
Threonine dehydratase	L-Threonine	5.0

k_{cat}, to describe the limiting rate of any enzyme-catalyzed reaction at saturation. If the reaction has several steps and one is clearly rate limiting, k_{cat} is equivalent to the rate constant for that limiting step. For the simple reaction of Equation 6–10, $k_{cat} = k_2$. For the reaction of Equation 6–25, when product release is clearly rate-limiting, $k_{cat} = k_3$. When several steps are partially rate-limiting, k_{cat} can become a complex function of several of the rate constants that define each individual reaction step. In the Michaelis-Menten equation, $k_{cat} = V_{max}/[E_t]$, and Equation 6–9 becomes

$$V_0 = \frac{k_{cat}[E_t][S]}{K_m + [S]} \qquad (6\text{–}26)$$

The constant k_{cat} is a first-order rate constant and hence has units of reciprocal time. It is also called the **turnover number**. It is equivalent to the number of substrate molecules converted to product in a given unit of time on a single enzyme molecule when the enzyme is saturated with substrate. The turnover numbers of several enzymes are given in Table 6–7.

Comparing Catalytic Mechanisms and Efficiencies The kinetic parameters k_{cat} and K_m are useful for the study and comparison of different enzymes, whether their reaction mechanisms are simple or complex. Each enzyme has values of k_{cat} and K_m that reflect the cellular environment, the concentration of substrate normally encountered in vivo by the enzyme, and the chemistry of the reaction being catalyzed.

The parameters k_{cat} and K_m also allow us to evaluate the kinetic efficiency of enzymes, but either parameter alone is insufficient for this task. Two enzymes catalyzing different reactions may have the same k_{cat} (turnover number), yet the rates of the uncatalyzed reactions may be different and thus the rate enhancements brought about by the enzymes may differ greatly. Experimentally,

TABLE 6–7 Turnover Number, k_{cat}, of Some Enzymes

Enzyme	Substrate	k_{cat} (s^{-1})
Catalase	H_2O_2	40,000,000
Carbonic anhydrase	HCO_3^-	400,000
Acetylcholinesterase	Acetylcholine	14,000
β-Lactamase	Benzylpenicillin	2,000
Fumarase	Fumarate	800
RecA protein (an ATPase)	ATP	0.5

the K_m for an enzyme tends to be similar to the cellular concentration of its substrate. An enzyme that acts on a substrate present at a very low concentration in the cell usually has a lower K_m than an enzyme that acts on a substrate that is more abundant.

The best way to compare the catalytic efficiencies of different enzymes or the turnover of different substrates by the same enzyme is to compare the ratio k_{cat}/K_m for the two reactions. This parameter, sometimes called the **specificity constant**, is the rate constant for the conversion of E + S to E + P. When $[S] << K_m$, Equation 6–26 reduces to the form

$$V_0 = \frac{k_{cat}}{K_m}[E_t][S] \qquad (6\text{–}27)$$

V_0 in this case depends on the concentration of two reactants, $[E_t]$ and $[S]$; therefore this is a second-order rate equation and the constant k_{cat}/K_m is a second-order rate constant with units of $M^{-1}s^{-1}$. There is an upper limit to k_{cat}/K_m, imposed by the rate at which E and S can diffuse together in an aqueous solution. This diffusion-controlled limit is 10^8 to 10^9 $M^{-1}s^{-1}$, and many enzymes have a k_{cat}/K_m near this range (Table 6–8). Such enzymes are said to have achieved catalytic perfection. Note that different values of k_{cat} and K_m can produce the maximum ratio.

TABLE 6–8 Enzymes for Which k_{cat}/K_m Is Close to the Diffusion-Controlled Limit (10^8 to 10^9 $M^{-1}s^{-1}$)

Enzyme	Substrate	K_{cat} (s^{-1})	K_m (M)	K_{cat}/K_m $(M^{-1}s^{-1})$
Acetylcholinesterase	Acetylcholine	1.4×10^4	9×10^{-5}	1.6×10^8
Carbonic anhydrase	CO_2	1×10^6	1.2×10^{-2}	8.3×10^7
	HCO_3^-	4×10^5	2.6×10^{-2}	1.5×10^7
Catalase	H_2O_2	4×10^7	1.1×10^0	4×10^7
Crotonase	Crotonyl-CoA	5.7×10^3	2×10^{-5}	2.8×10^8
Fumarase	Fumarate	8×10^2	5×10^{-6}	1.6×10^8
	Malate	9×10^2	2.5×10^{-5}	3.6×10^7
β-Lactamase	Benzylpenicillin	2.0×10^3	2×10^{-5}	1×10^8

Source: Fersht, A. (1999) *Structure and Mechanism in Protein Science*, p. 166, W. H. Freeman and Company, New York.

WORKED EXAMPLE 6–1   Determination of K_m

An enzyme is discovered that catalyzes the chemical reaction

$$SAD \rightleftharpoons HAPPY$$

A team of motivated researchers sets out to study the enzyme, which they call happyase. They find that the k_{cat} for happyase is 600 s^{-1}. They carry out several experiments.

When $[E_t] = 20$ nM and $[SAD] = 40$ μM, the reaction velocity, V_0, is 9.6 μM s^{-1}. Calculate K_m for the substrate SAD.

Solution: We know k_{cat}, $[E_t]$, $[S]$, and V_0. We want to solve for K_m. Equation 6–26, in which we substitute $k_{cat}[E_t]$ for V_{max} in the Michaelis-Menten equation, is most useful here. Substituting our known values in Equation 6–26 allows us to solve for K_m.

$$V_0 = \frac{k_{cat}[E_t][S]}{K_m + [S]}$$

$$9.6 \ \mu\text{M s}^{-1} = \frac{(600 \text{ s}^{-1})(0.020 \ \mu\text{M})(40 \ \mu\text{M})}{K_m + 40 \ \mu\text{M}}$$

$$9.6 \ \mu\text{M s}^{-1} = \frac{480 \ \mu\text{M}^2 \text{ s}^{-1}}{K_m + 40 \ \mu\text{M}}$$

$$9.6 \ \mu\text{M s}^{-1}(K_m + 40 \ \mu\text{M}) = 480 \ \mu\text{M}^2 \text{ s}^{-1}$$

$$K_m + 40 \ \mu\text{M} = \frac{480 \ \mu\text{M}^2 \text{ s}^{-1}}{9.6 \ \mu\text{M s}^{-1}}$$

$$K_m + 40 \ \mu\text{M} = 50 \ \mu\text{M}$$

$$K_m = 50 \ \mu\text{M} - 40 \ \mu\text{M}$$

$$K_m = 10 \ \mu\text{M}$$

Once you have worked with this equation, you will recognize shortcuts to solve problems like this. For example, one can calculate V_{max} knowing that $k_{cat}[E_t] = V_{max}$ (600 s^{-1} × 0.020 μM = 12 μM s^{-1} in this case). A simple rearrangement of Equation 6–26 by dividing both sides by V_{max} gives

$$\frac{V_0}{V_{max}} = \frac{[S]}{K_m + [S]}$$

Thus, the ratio $V_0/V_{max} = 9.6$ μM s^{-1}/12 μM s^{-1} = $[S]/(K_m + [S])$. This simplifies the process of solving for K_m, giving 0.25[S] or 10 μM.

WORKED EXAMPLE 6–2   Determination of [S]

In a separate happyase experiment using $[E_t] = 10$ nM, the reaction velocity, V_0, is measured as 3 μM s^{-1}. What is the [S] used in this experiment?

Solution: Using the same logic as in Worked Example 6–1, we see that the V_{max} for this enzyme concentration is 6 μM s^{-1}. Note that the V_0 is exactly half of the V_{max}.

Recall that K_m is by definition equal to the [S] where $V_0 = \frac{1}{2}V_{max}$. Thus, the [S] in this problem must be the same as the K_m, or 10 μM. If V_0 were anything other than $\frac{1}{2}V_{max}$, it would be simplest to use the expression $V_0/V_{max} = [S]/(K_m + [S])$ to solve for [S].

Many Enzymes Catalyze Reactions with Two or More Substrates

We have seen how [S] affects the rate of a simple enzymatic reaction (S → P) with only one substrate molecule. In most enzymatic reactions, however, two (and sometimes more) different substrate molecules bind to the enzyme and participate in the reaction. For example, in the reaction catalyzed by hexokinase, ATP and glucose are the substrate molecules, and ADP and glucose 6-phosphate are the products:

$$\text{ATP} + \text{glucose} \longrightarrow \text{ADP} + \text{glucose 6-phosphate}$$

The rates of such bisubstrate reactions can also be analyzed by the Michaelis-Menten approach. Hexokinase has a characteristic K_m for each of its substrates (Table 6–6).

Enzymatic reactions with two substrates usually involve transfer of an atom or a functional group from one substrate to the other. These reactions proceed by one of several different pathways. In some cases, both substrates are bound to the enzyme concurrently at some point in the course of the reaction, forming a noncovalent ternary complex (**Fig. 6–13a**); the substrates bind in a random sequence or in a specific order. In

(a) Enzyme reaction involving a ternary complex

Random order

E → ES$_1$, ES$_2$ → ES$_1$S$_2$ → E + P$_1$ + P$_2$

Ordered

E + S$_1$ ⇌ ES$_1$ $\xrightarrow{S_2}$ ES$_1$S$_2$ → E + P$_1$ + P$_2$

(b) Enzyme reaction in which no ternary complex is formed

E + S$_1$ ⇌ ES$_1$ ⇌ E′P$_1$ $\xrightarrow{P_1}$ E′ $\xrightarrow{S_2}$ E′S$_2$ → E + P$_2$

FIGURE 6–13 Common mechanisms for enzyme-catalyzed bisubstrate reactions. (a) The enzyme and both substrates come together to form a ternary complex. In ordered binding, substrate 1 must bind before substrate 2 can bind productively. In random binding, the substrates can bind in either order. **(b)** An enzyme-substrate complex forms, a product leaves the complex, the altered enzyme forms a second complex with another substrate molecule, and the second product leaves, regenerating the enzyme. Substrate 1 may transfer a functional group to the enzyme (to form the covalently modified E′), which is subsequently transferred to substrate 2. This is called a Ping-Pong or double-displacement mechanism.

other cases, the first substrate is converted to product and dissociates before the second substrate binds, so no ternary complex is formed. An example of this is the Ping-Pong, or double-displacement, mechanism (Fig. 6–13b). Steady-state kinetics can often help distinguish among these possibilities **(Fig. 6–14)**.

Pre–Steady State Kinetics Can Provide Evidence for Specific Reaction Steps

We have introduced kinetics as the primary method for studying the steps in an enzymatic reaction, and we have also outlined the limitations of the most common kinetic parameters in providing such information. The two most important experimental parameters obtained from steady-state kinetics are k_{cat} and k_{cat}/K_m. Variation in k_{cat} and k_{cat}/K_m with changes in pH or temperature can provide additional information about steps in a reaction pathway. In the case of bisubstrate reactions, steady-state kinetics can help determine whether a ternary complex is formed during the reaction (Fig. 6–14). A more complete picture generally requires more sophisticated kinetic methods that go beyond the scope of an introductory text. Here, we briefly introduce one of the most important kinetic approaches for studying reaction mechanisms, pre–steady state kinetics.

A complete description of an enzyme-catalyzed reaction requires direct measurement of the rates of individual reaction steps—for example, the association of enzyme and substrate to form the ES complex. It is during the pre–steady state (p. 202) that the rates of many reaction steps can be measured independently and events during reaction of a single substrate molecule can be observed. Because the pre–steady state phase is generally very short, the experiments often require specialized techniques for very rapid mixing and sampling. One objective is to gain a complete and quantitative picture of the energy changes during the reaction. As we have already noted, reaction rates and equilibria are related to the free-energy changes during a reaction. Another objective is to measure the rate of individual reaction steps. In a number of cases, investigators have been able to record the rates of every individual step in a multistep enzymatic reaction. Some examples of the application of pre–steady state kinetics are included in the descriptions of specific enzymes in Section 6.4.

Enzymes Are Subject to Reversible or Irreversible Inhibition

Enzyme inhibitors are molecules that interfere with catalysis, slowing or halting enzymatic reactions. Enzymes catalyze virtually all cellular processes, so it should not be surprising that enzyme inhibitors are among the most important pharmaceutical agents known. For example, aspirin (acetylsalicylate) inhibits the enzyme that catalyzes the first step in the synthesis of prostaglandins, compounds involved in many processes, including some that produce pain. The study of enzyme inhibitors also has provided valuable information about enzyme mechanisms and has helped define some metabolic pathways. There are two broad classes of enzyme inhibitors: reversible and irreversible.

Reversible Inhibition One common type of **reversible inhibition** is called competitive **(Fig. 6–15a)**. A **competitive inhibitor** competes with the substrate for the active site of an enzyme. While the inhibitor (I) occupies the active site, it prevents binding of the substrate to the enzyme. Many competitive inhibitors are structurally similar to the substrate and combine with the enzyme to form an EI complex, but without leading to catalysis. Even fleeting combinations of this type will reduce the efficiency of the enzyme. By taking into account the molecular geometry of inhibitors, we can reach conclusions about which parts of the normal substrate bind to the enzyme. Competitive inhibition can be analyzed quantitatively by steady-state kinetics. In the presence of a competitive inhibitor, the Michaelis-Menten equation (Eqn 6–9) becomes

$$V_0 = \frac{V_{max}[S]}{\alpha K_m + [S]} \qquad (6\text{–}28)$$

FIGURE 6–14 Steady-state kinetic analysis of bisubstrate reactions. In these double-reciprocal plots (see Box 6-1), the concentration of substrate 1 is varied while the concentration of substrate 2 is held constant. This is repeated for several values of [S₂], generating several separate lines. **(a)** Intersecting lines indicate that a ternary complex is formed in the reaction; **(b)** parallel lines indicate a Ping-Pong (double-displacement) pathway.

(a) Competitive inhibition

$$E + S \rightleftharpoons ES \longrightarrow E + P$$
$$+$$
$$I$$
$$\Big\Vert K_I'$$
$$EI$$

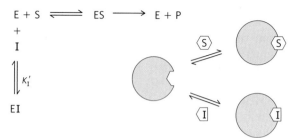

(b) Uncompetitive inhibition

$$E + S \rightleftharpoons ES \longrightarrow E + P$$
$$+$$
$$I$$
$$\Big\Vert K_I'$$
$$ESI$$

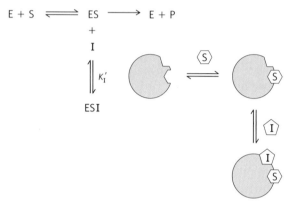

(c) Mixed inhibition

$$E + S \rightleftharpoons ES \longrightarrow E + P$$
$$+ \qquad +$$
$$I \qquad I$$
$$\Big\Vert K_I \qquad \Big\Vert K_I'$$
$$EI + S \rightleftharpoons ESI$$

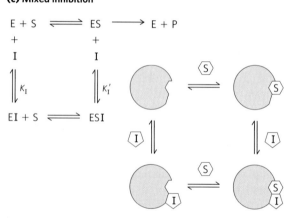

FIGURE 6–15 Three types of reversible inhibition. (a) Competitive inhibitors bind to the enzyme's active site; K_I is the equilibrium constant for inhibitor binding to E. **(b)** Uncompetitive inhibitors bind at a separate site, but bind only to the ES complex; K_I' is the equilibrium constant for inhibitor binding to ES. **(c)** Mixed inhibitors bind at a separate site, but may bind to either E or ES.

where

$$\alpha = 1 + \frac{[I]}{K_I} \quad \text{and} \quad K_I = \frac{[E][I]}{[EI]}$$

Equation 6–28 describes the important features of competitive inhibition. The experimentally determined variable αK_m, the K_m observed in the presence of the inhibitor, is often called the "apparent" K_m.

Because the inhibitor binds reversibly to the enzyme, the competition can be biased to favor the substrate simply by adding more substrate. When [S] far exceeds [I], the probability that an inhibitor molecule will bind to the enzyme is minimized and the reaction exhibits a normal

V_{max}. However, the [S] at which $V_0 = \frac{1}{2}V_{max}$, the apparent K_m, increases in the presence of inhibitor by the factor α. This effect on apparent K_m, combined with the absence of an effect on V_{max}, is diagnostic of competitive inhibition and is readily revealed in a double-reciprocal plot (Box 6–2). The equilibrium constant for inhibitor binding, K_I, can be obtained from the same plot.

A medical therapy based on competition at the active site is used to treat patients who have ingested methanol, a solvent found in gas-line antifreeze. The liver enzyme alcohol dehydrogenase converts methanol to formaldehyde, which is damaging to many tissues. Blindness is a common result of methanol ingestion, because the eyes are particularly sensitive to formaldehyde. Ethanol competes effectively with methanol as an alternative substrate for alcohol dehydrogenase. The effect of ethanol is much like that of a competitive inhibitor, with the distinction that ethanol is also a substrate for alcohol dehydrogenase and its concentration will decrease over time as the enzyme converts it to acetaldehyde. The therapy for methanol poisoning is slow intravenous infusion of ethanol, at a rate that maintains a controlled concentration in the bloodstream for several hours. This slows the formation of formaldehyde, lessening the danger while the kidneys filter out the methanol to be excreted harmlessly in the urine. ■

Two other types of reversible inhibition, uncompetitive and mixed, can be defined in terms of one-substrate enzymes, but are in practice observed only with enzymes having two or more substrates. An **uncompetitive inhibitor** (Fig. 6–15b) binds at a site distinct from the substrate active site and, unlike a competitive inhibitor, binds only to the ES complex. In the presence of an uncompetitive inhibitor, the Michaelis-Menten equation is altered to

$$V_0 = \frac{V_{max}[S]}{K_m + \alpha'[S]} \qquad (6\text{–}29)$$

where

$$\alpha' = 1 + \frac{[I]}{K_I'} \quad \text{and} \quad K_I' = \frac{[ES][I]}{[ESI]}$$

As described by Equation 6–29, at high concentrations of substrate, V_0 approaches V_{max}/α'. Thus, an uncompetitive inhibitor lowers the measured V_{max}. Apparent K_m also decreases, because the [S] required to reach one-half V_{max} decreases by the factor α'.

A **mixed inhibitor** (Fig. 6–15c) also binds at a site distinct from the substrate active site, but it binds to either E or ES. The rate equation describing mixed inhibition is

$$V_0 = \frac{V_{max}[S]}{\alpha K_m + \alpha'[S]} \qquad (6\text{–}30)$$

where α and α' are defined as above. A mixed inhibitor usually affects both K_m and V_{max}. The special case of $\alpha = \alpha'$, rarely encountered in experiments, classically has been defined as **noncompetitive inhibition**.

BOX 6–2   Kinetic Tests for Determining Inhibition Mechanisms

The double-reciprocal plot (see Box 6–1) offers an easy way of determining whether an enzyme inhibitor is competitive, uncompetitive, or mixed. Two sets of rate experiments are carried out, with the enzyme concentration held constant in each set. In the first set, [S] is also held constant, permitting measurement of the effect of increasing inhibitor concentration [I] on the initial rate V_0 (not shown). In the second set, [I] is held constant but [S] is varied. The results are plotted as $1/V_0$ versus $1/[S]$.

Figure 1 shows a set of double-reciprocal plots, one obtained in the absence of inhibitor and two at different concentrations of a competitive inhibitor. Increasing [I] results in a family of lines with a common intercept on the $1/V_0$ axis but with different slopes. Because the intercept on the $1/V_0$ axis equals

$1/V_{max}$, we know that V_{max} is unchanged by the presence of a competitive inhibitor. That is, regardless of the concentration of a competitive inhibitor, a sufficiently high substrate concentration will always displace the inhibitor from the enzyme's active site. Above the graph is the rearrangement of Equation 6–28 on which the plot is based. The value of α can be calculated from the change in slope at any given [I]. Knowing [I] and α, we can calculate K_I from the expression

$$\alpha = 1 + \frac{[I]}{K_I}$$

For uncompetitive and mixed inhibition, similar plots of rate data give the families of lines shown in Figures 2 and 3. Changes in axis intercepts signal changes in V_{max} and K_m.

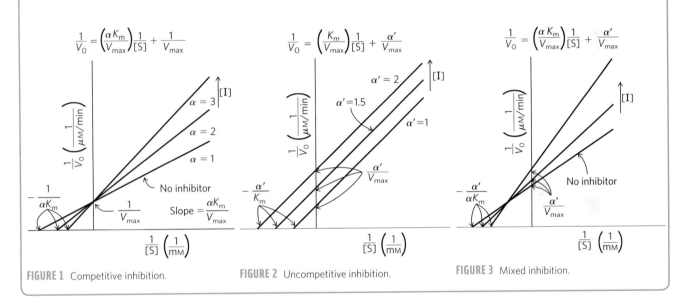

FIGURE 1 Competitive inhibition.

FIGURE 2 Uncompetitive inhibition.

FIGURE 3 Mixed inhibition.

Examine Equation 6–30 to see why a noncompetitive inhibitor would affect the V_{max} but not the K_m.

Equation 6–30 is a general expression for the effects of reversible inhibitors, simplifying to the expressions for competitive and uncompetitive inhibition when $\alpha' = 1.0$ or $\alpha = 1.0$, respectively. From this expression we can summarize the effects of inhibitors on individual kinetic parameters. For all reversible inhibitors, the apparent $V_{max} = V_{max}/\alpha'$, because the right side of Equation 6–30 always simplifies to V_{max}/α' at sufficiently high substrate concentrations. For competitive inhibitors, $\alpha' = 1.0$ and can thus be ignored. Taking this expression for apparent V_{max}, we can also derive a general expression for apparent K_m to show how this parameter changes in the presence of reversible inhibitors. Apparent K_m, as always, equals the [S] at which V_0 is one-half apparent V_{max} or, more generally,

when $V_0 = V_{max}/2\alpha'$. This condition is met when $[S] = \alpha K_m/\alpha'$. Thus, apparent $K_m = \alpha K_m/\alpha'$. This expression is simpler when either α or α' is 1.0 (for uncompetitive or competitive inhibitors), as summarized in Table 6–9.

In practice, uncompetitive and mixed inhibition are observed only for enzymes with two or more substrates—

TABLE 6–9 Effects of Reversible Inhibitors on Apparent V_{max} and Apparent K_m

Inhibitor type	Apparent V_{max}	Apparent K_m
None	V_{max}	K_m
Competitive	V_{max}	αK_m
Uncompetitive	V_{max}/α'	K_m/α'
Mixed	V_{max}/α'	$\alpha K_m/\alpha'$

say, S_1 and S_2—and are very important in the experimental analysis of such enzymes. If an inhibitor binds to the site normally occupied by S_1, it may act as a competitive inhibitor in experiments in which $[S_1]$ is varied. If an inhibitor binds to the site normally occupied by S_2, it may act as a mixed or uncompetitive inhibitor of S_1. The actual inhibition patterns observed depend on whether the S_1- and S_2-binding events are ordered or random, and thus the order in which substrates bind and products leave the active site can be determined. Use of one of the reaction products as an inhibitor is often particularly informative. If only one of two reaction products is present, no reverse reaction can take place. However, a product generally binds to some part of the active site, thus serving as an inhibitor. Enzymologists can use elaborate kinetic studies involving different combinations and amounts of products and inhibitors to develop a detailed picture of the mechanism of a bisubstrate reaction.

WORKED EXAMPLE 6–3   Effect of Inhibitor on K_m

The researchers working on happyase (see Worked Examples 6–1 and 6–2) discover that the compound STRESS is a potent competitive inhibitor of happyase. Addition of 1 nM STRESS increases the measured K_m for SAD by a factor of 2. What are the values for α and α' under these conditions?

Solution: Recall that the apparent K_m, the K_m measured in the presence of a competitive inhibitor, is defined as αK_m. Because K_m for SAD increases by a factor of 2 in the presence of 1 nM STRESS, the value of α must be 2. The value of α' for a competitive inhibitor is 1, by definition.

Irreversible Inhibition The **irreversible inhibitors** bind covalently with or destroy a functional group on an enzyme that is essential for the enzyme's activity, or form a particularly stable noncovalent association. Formation of a covalent link between an irreversible inhibitor and an enzyme is one approach. Irreversible inhibitors are another useful tool for studying reaction mechanisms. Amino acids with key catalytic functions in the active site can sometimes be identified by determining which residue is covalently linked to an inhibitor after the enzyme is inactivated. An example is shown in **Figure 6–16**.

A special class of irreversible inhibitors is the **suicide inactivators**. These compounds are relatively unreactive until they bind to the active site of a specific enzyme. A suicide inactivator undergoes the first few chemical steps of the normal enzymatic reaction, but instead of being transformed into the normal product, the inactivator is converted to a very reactive compound that combines irreversibly with the enzyme. These compounds are also called **mechanism-based inactivators**,

FIGURE 6–16 Irreversible inhibition. Reaction of chymotrypsin with diisopropylfluorophosphate (DIFP), which modifies Ser[195], irreversibly inhibits the enzyme. This has led to the conclusion that Ser[195] is the key active-site Ser residue in chymotrypsin.

because they hijack the normal enzyme reaction mechanism to inactivate the enzyme. Suicide inactivators play a significant role in *rational drug design*, a modern approach to obtaining new pharmaceutical agents in which chemists synthesize novel substrates based on knowledge of substrates and reaction mechanisms. A well-designed suicide inactivator is specific for a single enzyme and is unreactive until it is within that enzyme's active site, so drugs based on this approach can offer the important advantage of few side effects (Box 6–3). Some additional examples of irreversible inhibitors of medical importance are described in Section 6.4.

An irreversible inhibitor need not bind covalently to the enzyme. Noncovalent binding is enough, if that binding is so tight that the inhibitor dissociates only rarely. How does one develop a tight-binding inhibitor? Recall that enzymes evolve to bind most tightly to the transition states of the reactions that they catalyze. In principle, if one can design a molecule that looks like that reaction transition state, it should bind tightly to the enzyme. Even though transition states cannot be observed directly, chemists can often predict the approximate structure of a transition state based on accumulated knowledge about reaction mechanisms. The transition state is by definition transient and so unstable that direct measurement of the binding interaction between this species and the enzyme is impossible. In some cases, however, stable molecules can be designed that resemble transition states. These are called **transition-state analogs**. They bind to an enzyme more tightly than does the substrate in the ES complex, because they fit into the active site better (that is, form a greater number of weak interactions) than the substrate itself. The idea of transition-state analogs was suggested by Pauling in the 1940s, and it has been explored using a number of enzymes. For example, transition-state analogs designed to inhibit the glycolytic enzyme aldolase bind to that enzyme more than four orders of magnitude more

BOX 6–3 ⚕ MEDICINE Curing African Sleeping Sickness with a Biochemical Trojan Horse

African sleeping sickness, or African trypanosomiasis, is caused by protists (single-celled eukaryotes) called trypanosomes (Fig. 1). This disease (and related trypanosome-caused diseases) is medically and economically significant in many developing nations. Until the late twentieth century, the disease was virtually incurable. Vaccines are ineffective because the parasite has a novel mechanism to evade the host immune system.

The cell coat of trypanosomes is covered with a single protein, which is the antigen to which the immune system responds. Every so often, however, by a process of genetic recombination (see Table 28–1), a few cells in the population of infecting trypanosomes switch to a new protein coat, not recognized by the immune system. This process of "changing coats" can occur hundreds of times. The result is a chronic cyclic infection: the human host develops a fever, which subsides as the immune system beats back the first infection; trypanosomes with changed coats then become the seed for a second infection, and the fever recurs. This cycle can repeat for weeks, and the weakened person eventually dies.

Some modern approaches to treating African sleeping sickness have been based on an understanding of enzymology and metabolism. In at least one such approach, this involves pharmaceutical agents designed as mechanism-based enzyme inactivators (suicide inactivators). A vulnerable point in trypanosome metabolism is the pathway of polyamine biosynthesis. The polyamines spermine and spermidine, involved in DNA packaging, are required in large amounts in rapidly dividing cells. The first step in their synthesis is catalyzed by ornithine decarboxylase, an enzyme that requires a coenzyme called pyridoxal

FIGURE 1 *Trypanosoma brucei rhodesiense*, one of several trypanosomes known to cause African sleeping sickness.

phosphate for its function. Pyridoxal phosphate (PLP), derived from vitamin B_6, makes a covalent bond to the amino acid substrates of the reactions it is involved in and acts as an electron sink to facilitate a variety of reactions (see Fig. 22–32). In mammalian cells, ornithine decarboxylase undergoes rapid turnover—that is, a constant round of enzyme degradation and synthesis. In some trypanosomes, however, the enzyme (for reasons not well understood) is stable, not readily replaced by newly synthesized enzyme. An inhibitor of ornithine decarboxylase that binds permanently to the enzyme would thus have little effect on human cells, which could rapidly replace inactivated enzyme, but would adversely affect the parasite.

The first few steps of the normal reaction catalyzed by ornithine decarboxylase are shown in Figure 2. Once CO_2 is released, the electron movement is reversed and putrescine is produced (see Fig. 22–32). Based on this mechanism, several suicide inactivators have been designed, one of which is difluoromethylornithine (DFMO). DFMO is relatively inert in solution. When it

(continued on next page)

FIGURE 2 Mechanism of ornithine decarboxylase reaction.

BOX 6–3 🩺 MEDICINE Curing African Sleeping Sickness with a Biochemical Trojan Horse (Continued)

binds to ornithine decarboxylase, however, the enzyme is quickly inactivated (Fig. 3). The inhibitor acts by providing an alternative electron sink in the form of two strategically placed fluorine atoms, which are excellent leaving groups. Instead of electrons moving into the ring structure of PLP, the reaction results in displacement of a fluorine atom. The —S of a Cys residue at the enzyme's active site then forms a covalent complex with the highly reactive PLP-inhibitor adduct in an essentially irreversible reaction. In this way, the inhibitor makes use of the enzyme's own reaction mechanisms to kill it.

DFMO has proved highly effective against African sleeping sickness in clinical trials and is now used to treat African sleeping sickness caused by *Trypanosoma brucei gambiense*. Approaches such as this show great promise for treating a wide range of diseases. The design of drugs based on enzyme mechanism and structure can complement the more traditional trial-and-error methods of developing pharmaceuticals.

FIGURE 3 Inhibition of ornithine decarboxylase by DFMO.

tightly than substrates **(Fig. 6–17)**. These experiments have the limitation that a transition-state analog cannot perfectly mimic a transition state. Some analogs, however, bind to a target enzyme 10^2 to 10^8 times more tightly than does the normal substrate, providing good evidence that enzyme active sites are indeed complementary to transition states. The concept of transition-state analogs is important to the design of new pharmaceutical agents. As we shall see, the powerful anti-HIV drugs called protease inhibitors were designed in part as tight-binding transition-state analogs.

Enzyme Activity Depends on pH

Enzymes have an optimum pH (or pH range) at which their activity is maximal **(Fig. 6–18)**; at higher or lower pH, activity decreases. This is not surprising. Amino acid side chains in the active site may act as weak acids and bases with critical functions that depend on their maintaining a certain state of ionization, and elsewhere in the protein ionized side chains may play an essential role in the interactions that maintain protein structure. Removing a proton from a His residue, for example, might eliminate an ionic interaction essential for stabilizing the active conformation of the enzyme. A less common cause of pH sensitivity is titration of a group on the substrate.

The pH range over which an enzyme undergoes changes in activity can provide a clue to the type of amino acid residue involved (see Table 3–1). A change in activity near pH 7.0, for example, often reflects titration of a His residue. The effects of pH must be interpreted with some caution, however. In the closely packed environment of a protein, the pK_a of amino acid side chains can be significantly altered. For example, a nearby positive charge can lower the pK_a of a Lys residue, and a nearby negative charge can increase it. Such effects

Fructose 1,6-bisphosphate

Glyceraldehyde 3-phosphate

Proposed transition state

Transition-state analog, phosphoglycolohydroxamate

Dihydroxyacetone phosphate

FIGURE 6–17 A transition-state analog. In glycolysis, a class II aldolase (found in bacteria and fungi) catalyzes the cleavage of fructose 1,6-bisphosphate to form glyceraldehyde 3-phosphate and dihydroxyacetone phosphate (see Fig. 14-6 for an example of a class I aldolase, found in animals and higher plants). The reaction proceeds via a reverse aldol condensation–like mechanism. The compound phosphoglycolohydroxamate resembles the proposed enediolate transition state, and binds to the enzyme nearly 10,000 times better than does the dihydroxyacetone phosphate product.

FIGURE 6–18 The pH-activity profiles of two enzymes. These curves are constructed from measurements of initial velocities when the reaction is carried out in buffers of different pH. Because pH is a logarithmic scale reflecting 10-fold changes in $[H^+]$, the changes in V_0 are also plotted on a logarithmic scale. The pH optimum for the activity of an enzyme is generally close to the pH of the environment in which the enzyme is normally found. Pepsin, a peptidase found in the stomach, has a pH optimum of about 1.6. The pH of gastric juice is between 1 and 2. Glucose 6-phosphatase of hepatocytes (liver cells), with a pH optimum of about 7.8, is responsible for releasing glucose into the blood. The normal pH of the cytosol of hepatocytes is about 7.2.

sometimes result in a pK_a that is shifted by several pH units from its value in the free amino acid. In the enzyme acetoacetate decarboxylase, for example, one Lys residue has a pK_a of 6.6 (compared with 10.5 in free lysine) due to electrostatic effects of nearby positive charges.

SUMMARY 6.3 Enzyme Kinetics as an Approach to Understanding Mechanism

▶ Most enzymes have certain kinetic properties in common. When substrate is added to an enzyme, the reaction rapidly achieves a steady state in which the rate at which the ES complex forms balances the rate at which it breaks down. As [S] increases, the steady-state activity of a fixed concentration of enzyme increases in a hyperbolic fashion to approach a characteristic maximum rate, V_{max}, at which essentially all the enzyme has formed a complex with substrate.

▶ The substrate concentration that results in a reaction rate equal to one-half V_{max} is the Michaelis

constant K_m, which is characteristic for each enzyme acting on a given substrate. The Michaelis-Menten equation

$$V_0 = \frac{V_{max}[S]}{K_m + [S]}$$

relates initial velocity to [S] and V_{max} through the constant K_m. Michaelis-Menten kinetics is also called steady-state kinetics.

▶ K_m and V_{max} have different meanings for different enzymes. The limiting rate of an enzyme-catalyzed reaction at saturation is described by the constant k_{cat}, the turnover number. The ratio k_{cat}/K_m provides a good measure of catalytic efficiency. The Michaelis-Menten equation is also applicable to bisubstrate reactions, which occur by ternary complex or Ping-Pong (double-displacement) pathways.

▶ Reversible inhibition of an enzyme may be competitive, uncompetitive, or mixed. Competitive inhibitors compete with substrate by binding reversibly to the active site, but they are not transformed by the enzyme. Uncompetitive inhibitors bind only to the ES complex, at a site distinct from the active site. Mixed inhibitors bind to either E or ES, again at a site distinct from the active site. In irreversible inhibition an inhibitor binds permanently to an active site by forming a covalent bond or a very stable noncovalent interaction.

▶ Every enzyme has an optimum pH (or pH range) at which it has maximal activity.

6.4 Examples of Enzymatic Reactions

Thus far we have focused on the general principles of catalysis and on introducing some of the kinetic parameters used to describe enzyme action. We now turn to several examples of specific enzyme reaction mechanisms.

An understanding of the complete mechanism of action of a purified enzyme requires identification of all substrates, cofactors, products, and regulators. Moreover, it requires a knowledge of (1) the temporal sequence in which enzyme-bound reaction intermediates form, (2) the structure of each intermediate and each transition state, (3) the rates of interconversion between intermediates, (4) the structural relationship of the enzyme to each intermediate, and (5) the energy contributed by all reacting and interacting groups to intermediate complexes and transition states. As yet, there is probably no enzyme for which we have an understanding that meets all these requirements.

We present here the mechanisms for four enzymes: chymotrypsin, hexokinase, enolase, and lysozyme. These examples are not intended to cover all possible classes of enzyme chemistry. They are chosen in part because they are among the best-understood enzymes, and in part because they clearly illustrate some general principles outlined in this chapter. The discussion concentrates on selected principles, along with some key experiments that have helped to bring these principles into focus. We use the chymotrypsin example to review some of the conventions used to depict enzyme mechanisms. Much mechanistic detail and experimental evidence is necessarily omitted; no one book could completely document the rich experimental history of these enzymes. In addition, the special contribution of coenzymes to the catalytic activity of many enzymes is considered only briefly. The function of coenzymes is chemically varied, and we describe each coenzyme in detail as it is encountered in Part II.

The Chymotrypsin Mechanism Involves Acylation and Deacylation of a Ser Residue

Bovine pancreatic chymotrypsin (M_r 25,191) is a protease, an enzyme that catalyzes the hydrolytic cleavage of peptide bonds. This protease is specific for peptide bonds adjacent to aromatic amino acid residues (Trp, Phe, Tyr). The three-dimensional structure of chymotrypsin is shown in **Figure 6–19**, with functional groups in the active site emphasized. The reaction catalyzed by this enzyme illustrates the principle of transition-state stabilization and also provides a classic example of general acid-base catalysis and covalent catalysis.

Chymotrypsin enhances the rate of peptide bond hydrolysis by a factor of at least 10^9. It does not catalyze a direct attack of water on the peptide bond; instead, a transient covalent acyl-enzyme intermediate is formed. The reaction thus has two distinct phases. In the acylation phase, the peptide bond is cleaved and an ester linkage is

FIGURE 6–19 Structure of chymotrypsin. (PDB ID 7GCH) **(a)** A representation of primary structure, showing disulfide bonds and the amino acid residues crucial to catalysis. The protein consists of three polypeptide chains linked by disulfide bonds. (The numbering of residues in chymotrypsin, with "missing" residues 14, 15, 147, and 148, is explained in Fig. 6–38.) The active-site amino acid residues are grouped together in the three-dimensional structure. **(b)** A depiction of the enzyme emphasizing its surface. The hydrophobic pocket in which the aromatic amino acid side chain of the substrate is bound is shown in yellow. Key active-site residues, including Ser[195], His[57], and Asp[102], are red. The roles of these residues in catalysis are illustrated in Figure 6-22. **(c)** The polypeptide backbone as a ribbon structure. Disulfide bonds are yellow; the three chains are colored as in part (a). **(d)** A close-up of the active site with a substrate (white and yellow) bound. The hydroxyl of Ser[195] attacks the carbonyl group of the substrate (the oxygens are red); the developing negative charge on the oxygen is stabilized by the oxyanion hole (amide nitrogens from Ser[195] and Gly[193], in blue), as explained in Figure 6-22. The aromatic amino acid side chain of the substrate (yellow) sits in the hydrophobic pocket. The amide nitrogen of the peptide bond to be cleaved (protruding toward the viewer and projecting the path of the rest of the substrate polypeptide chain) is shown in white.

formed between the peptide carbonyl carbon and the enzyme. In the deacylation phase, the ester linkage is hydrolyzed and the nonacylated enzyme regenerated.

The first evidence for a covalent acyl-enzyme intermediate came from a classic application of pre–steady state kinetics. In addition to its action on polypeptides, chymotrypsin also catalyzes the hydrolysis of small esters and amides. These reactions are much slower than hydrolysis of peptides because less binding energy is available with smaller substrates, and they are therefore easier to study. Investigations by B. S. Hartley and B. A. Kilby in 1954 found that chymotrypsin hydrolysis of the ester p-nitrophenylacetate, as measured by release of p-nitrophenol, proceeds with a rapid burst before leveling off to a slower rate **(Fig. 6–20)**. By extrapolating back to zero time, they concluded that the burst phase corresponded to just under one molecule of p-nitrophenol released for every enzyme molecule present. Hartley and Kilby suggested that this reflected a rapid acylation of all the enzyme molecules (with release of p-nitrophenol), with the rate for subsequent turnover of the enzyme limited by a slow deacylation step. Similar results have since been obtained with many other enzymes. The observation of a burst phase provides yet another example of the use of kinetics to break down a reaction into its constituent steps.

Additional features of the chymotrypsin mechanism have been discovered by analyzing the dependence of the reaction on pH. The rate of chymotrypsin-catalyzed cleavage generally exhibits a bell-shaped pH-rate profile **(Fig. 6–21)**. The rates plotted in Figure 6–21a are obtained at low (subsaturating) substrate concentrations and therefore represent k_{cat}/K_m (see Eqn 6–27, p. 205). A more complete analysis of the rates at different substrate concentrations at each pH allows researchers to determine the individual contributions of the k_{cat} and K_m terms. After obtaining the maximum rates at each pH, one can plot the k_{cat} alone versus pH (Fig. 6–21b); after obtaining the K_m at each pH, researchers can then plot $1/K_m$ (Fig. 6–21c). Kinetic and structural analyses have revealed that the change in k_{cat} reflects the ionization state of His[57]. The decline in k_{cat} at low pH results from protonation of His[57] (so that it cannot extract a proton from Ser[195] in step ❷ of the reaction; see **Fig. 6–22**). This rate reduction illustrates the

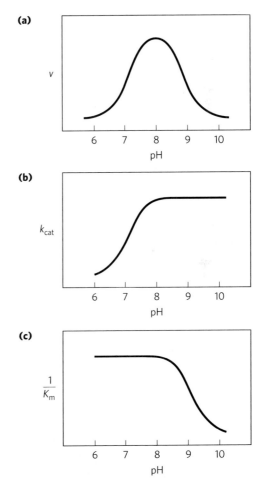

FIGURE 6–21 The pH dependence of chymotrypsin-catalyzed reactions. **(a)** The rates of chymotrypsin-mediated cleavage produce a bell-shaped pH-rate profile with an optimum at pH 8.0. The rate (v) being plotted is that at low substrate concentrations and thus reflects the term k_{cat}/K_m. The plot can be broken down to its components by using kinetic methods to determine the terms k_{cat} and K_m separately at each pH. When this is done **(b, c)**, it becomes clear that the transition just above pH 7 is due to changes in k_{cat}, whereas the transition above pH 8.5 is due to changes in $1/K_m$. Kinetic and structural studies have shown that the transitions illustrated in (b) and (c) reflect the ionization states of the His[57] side chain (when substrate is not bound) and the α-amino group of Ile[16] (at the amino terminus of the B chain), respectively. For optimal activity, His[57] must be unprotonated and Ile[16] must be protonated.

FIGURE 6–20 Pre-steady state kinetic evidence for an acyl-enzyme intermediate. The hydrolysis of p-nitrophenylacetate by chymotrypsin is measured by release of p-nitrophenol (a colored product). Initially, the reaction releases a rapid burst of p-nitrophenol nearly stoichiometric with the amount of enzyme present. This reflects the fast acylation phase of the reaction. The subsequent rate is slower, because enzyme turnover is limited by the rate of the slower deacylation phase.

How to Read Reaction Mechanisms—A Refresher

Chemical reaction mechanisms, which trace the formation and breakage of covalent bonds, are communicated with dots and curved arrows, a convention known informally as "electron pushing." A covalent bond consists of a shared pair of electrons. Nonbonded electrons important to the reaction mechanism are designated by dots ($^-\ddot{O}H$). Curved arrows ($\frown$) represent the movement of electron pairs. For movement of a single electron (as in a free radical reaction), a single-headed (fishhook-type) arrow is used ($\frown$). Most reaction steps involve an unshared electron pair (as in the chymotrypsin mechanism).

Some atoms are more electronegative than others; that is, they more strongly attract electrons. The relative electronegativities of atoms encountered in this text are F > O > N > C≈S > P≈H. For example, the two electron pairs making up a C=O (carbonyl) bond are not shared equally; the carbon is relatively electron-deficient as the oxygen draws away the electrons. Many reactions involve an electron-rich atom (a nucleophile) reacting with an electron-deficient atom (an electrophile). Some common nucleophiles and electrophiles in biochemistry are shown at right.

In general, a reaction mechanism is initiated at an unshared electron pair of a nucleophile. In mechanism diagrams, the base of the electron-pushing arrow originates near the electron-pair dots, and the head of the arrow points directly at the electrophilic center being attacked. Where the unshared electron pair confers a formal negative charge on the nucleophile, the negative charge symbol itself can represent the unshared electron pair and serves as the base of the arrow. In the chymotrypsin mechanism, the nucleophilic electron pair in the ES complex between steps ❶ and ❷ is provided by the oxygen of the Ser^{195} hydroxyl group. This electron pair (2 of the 8 valence electrons of the hydroxyl oxygen) provides the base of the curved arrow. The electrophilic center under attack is the carbonyl carbon of the peptide bond to be cleaved. The C, O, and N atoms have a maximum of 8 valence electrons, and H has a maximum of 2. These atoms are occasionally found in unstable states with less than their maximum allotment of electrons, but C, O, and N cannot have more than 8. Thus, when the electron pair from chymotrypsin's Ser^{195} attacks the substrate's carbonyl carbon, an electron pair is displaced from the carbon valence shell (you cannot have 5 bonds to carbon!). These electrons move toward the more electronegative carbonyl oxygen. The oxygen has 8 valence electrons both before and after this chemical process, but the number shared with the carbon is reduced from 4 to 2, and the carbonyl oxygen acquires a negative charge. In the next step, the electron pair conferring the negative charge on the oxygen moves back to re-form a bond with carbon and reestablish the carbonyl linkage. Again, an electron pair must be displaced from the carbon, and this time it is the electron pair shared with the amino group of the peptide linkage. This breaks the peptide bond. The remaining steps follow a similar pattern.

Chymotrypsin (free enzyme)

When substrate binds, the side chain of the residue adjacent to the peptide bond to be cleaved nestles in a hydrophobic pocket on the enzyme, positioning the peptide bond for attack.

❶

Substrate (a polypeptide)

Product 2

❼

Dissociation of the second product from the active site regenerates free enzyme.

Enzyme-product 2 complex

Nucleophiles	Electrophiles
$-O^-$ Negatively charged oxygen (as in an unprotonated hydroxyl group or an ionized carboxylic acid)	:R Carbon atom of a carbonyl group (the more electronegative oxygen of the carbonyl group pulls electrons away from the carbon)
$-S^-$ Negatively charged sulfhydryl	:R Protonated imine group (activated for nucleophilic attack at the carbon by protonation of the imine)
Carbanion $-C^-$	
$-N-$ Uncharged amine group	
Imidazole	Phosphorus of a phosphate group
$H-O^-$ Hydroxide ion	:R Proton H^+

Enzyme-substrate complex

Interaction of Ser[195] and His[57] generates a strongly nucleophilic alkoxide ion on Ser[195]; the ion attacks the peptide carbonyl group, forming a tetrahedral acyl-enzyme. This is accompanied by formation of a short-lived negative charge on the carbonyl oxygen of the substrate, which is stabilized by hydrogen bonding in the oxyanion hole.

Short-lived intermediate* (acylation)

Instability of the negative charge on the substrate carbonyl oxygen leads to collapse of the tetrahedral intermediate; re-formation of a double bond with carbon displaces the bond between carbon and the amino group of the peptide linkage, breaking the peptide bond. The amino leaving group is protonated by His[57], facilitating its displacement.

❷

Product 1

❸

MECHANISM FIGURE 6–22 Hydrolytic cleavage of a peptide bond by chymotrypsin. The reaction has two phases. In the acylation phase (steps ❶ to ❹), formation of a covalent acyl-enzyme intermediate is coupled to cleavage of the peptide bond. In the deacylation phase (steps ❺ to ❼), deacylation regenerates the free enzyme; this is essentially the reverse of the acylation phase, with water mirroring, in reverse, the role of the amine component of the substrate.

🖱 **Chymotrypsin Mechanism**

*The tetrahedral intermediate in the chymotrypsin reaction pathway, and the second tetrahedral intermediate that forms later, are sometimes referred to as transition states, which can lead to confusion. An *intermediate* is any chemical species with a finite lifetime, "finite" being defined as longer than the time required for a molecular vibration ($\sim10^{-13}$ second). A *transition state* is simply the maximum-energy species formed on the reaction coordinate and does not have a finite lifetime. The tetrahedral intermediates formed in the chymotrypsin reaction closely resemble, both energetically and structurally, the transition states leading to their formation and breakdown. However, the intermediate represents a committed stage of completed bond formation, whereas the transition state is part of the process of reaction. In the case of chymotrypsin, given the close relationship between the intermediate and the actual transition state the distinction between them is routinely glossed over. Furthermore, the interaction of the negatively charged oxygen with the amide nitrogens in the oxyanion hole, often referred to as transition-state stabilization, also serves to stabilize the intermediate in this case. Not all intermediates are so short-lived that they resemble transition states. The chymotrypsin acyl-enzyme intermediate is much more stable and more readily detected and studied, and it is never confused with a transition state.

Acyl-enzyme intermediate

❹

An incoming water molecule is deprotonated by general base catalysis, generating a strongly nucleophilic hydroxide ion. Attack of hydroxide on the ester linkage of the acyl-enzyme generates a second tetrahedral intermediate, with oxygen in the oxyanion hole again taking on a negative charge.

Acyl-enzyme intermediate

❺

Short-lived intermediate* (deacylation)

❻

Collapse of the tetrahedral intermediate forms the second product, a carboxylate anion, and displaces Ser[195].

importance of general acid and general base catalysis in the mechanism for chymotrypsin. The changes in the $1/K_m$ term reflect the ionization of the α-amino group of Ile[16] (at the amino-terminal end of one of chymotrypsin's three polypeptide chains). This group forms a salt bridge to Asp[194], stabilizing the active conformation of the enzyme. When this group loses its proton at high pH, the salt bridge is eliminated and a conformational change closes the hydrophobic pocket where the aromatic amino acid side chain of the substrate inserts (Fig. 6–19). Substrates can no longer bind properly, which is measured kinetically as an increase in K_m.

As can be seen in Figure 6-22, the nucleophile in the acylation phase is the oxygen of Ser[195]. (Proteases with a Ser residue that plays this role in reaction mechanisms are called **serine proteases**.) The pK_a of a Ser hydroxyl group is generally too high for the unprotonated form to be present in significant concentrations at physiological pH. However, in chymotrypsin, Ser[195] is linked to His[57] and Asp[102] in a hydrogen-bonding network referred to as the **catalytic triad**. When a peptide substrate binds to chymotrypsin, a subtle change in conformation compresses the hydrogen bond between His[57] and Asp[102], resulting in a stronger interaction, called a low-barrier hydrogen bond. This enhanced interaction increases the pK_a of His[57] from ~7 (for free histidine) to >12, allowing the His residue to act as an enhanced general base that can remove the proton from the Ser[195] hydroxyl group. Deprotonation prevents development of a very unstable positive charge on the Ser[195] hydroxyl and makes the Ser side chain a stronger nucleophile. At later reaction stages, His[57] also acts as a proton donor, protonating the amino group in the displaced portion of the substrate (the leaving group).

As the Ser[195] oxygen attacks the carbonyl group of the substrate, a very short-lived tetrahedral intermediate* is formed in which the carbonyl oxygen acquires a negative charge (Fig. 6–22, step ❷). This charge, forming within a pocket on the enzyme called the oxyanion hole, is stabilized by hydrogen bonds contributed by the amide groups of two peptide bonds in the chymotrypsin backbone. One of these hydrogen bonds (contributed by Gly[193]) is present only in this intermediate and in the transition states for its formation and breakdown; it reduces the energy required to reach these states. This is an example of the use of binding energy in catalysis.

An Understanding of Protease Mechanisms Leads to New Treatments for HIV Infections

New pharmaceutical agents are almost always designed to inhibit an enzyme. The very successful therapies developed to treat HIV infections provide a case in point. The human immunodeficiency virus (HIV) is the causative agent of acquired immune deficiency syndrome (AIDS). In 2005, an estimated 37 to 45 million people worldwide were living with HIV infections, with 3.9 to 6.6 million new infections that year

and more than 2.4 million fatalities. AIDS first surfaced as a world epidemic in the 1980s; HIV was discovered soon after and identified as a **retrovirus**. Retroviruses possess an RNA genome and an enzyme, reverse transcriptase, capable of using RNA to direct the synthesis of a complementary DNA. Efforts to understand HIV and develop therapies for HIV infection benefited from decades of basic research, both on enzyme mechanisms and on the properties of other retroviruses. A retrovirus such as HIV has a relatively simple life cycle (see Fig. 26–32). Its RNA genome is converted to duplex DNA in several steps catalyzed by a reverse transcriptase (described in Chapter 26). The duplex DNA is then inserted into a chromosome in the nucleus of the host cell by the enzyme integrase (described in Chapter 25). The integrated copy of the viral genome can remain dormant indefinitely. Alternatively, it can be transcribed back into RNA, which can then be translated into proteins to construct new virus particles. Most of the viral genes are translated into large polyproteins, which are cut by the HIV protease into the individual proteins needed to make the virus (see Fig. 26–33). There are only three key enzymes in this cycle—the reverse transcriptase, the integrase, and the protease. These enzymes thus represent the most promising drug targets.

There are four major subclasses of proteases. Serine proteases, such as chymotrypsin and trypsin, and cysteine proteases (in which Cys serves a catalytic role similar to that of Ser in the active site) feature covalent enzyme-substrate complexes; aspartyl proteases and metalloproteases do not. The HIV protease is an aspartyl protease. Two active-site Asp residues facilitate a direct attack of water on the peptide bond to be cleaved **(Fig. 6–23)**. The initial product of the attack of water on the carbonyl group of the peptide bond is an unstable tetrahedral intermediate, much as we have seen for the chymotrypsin reaction. This intermediate is close in structure and energy to the reaction transition state. The drugs that have been developed as HIV protease inhibitors form noncovalent complexes with the enzyme, but they bind to it so tightly that they can be considered irreversible inhibitors. The tight binding is derived in part from their design as transition-state analogs. The success of these drugs makes a point worth emphasizing. The catalytic principles we have studied in this chapter are not simply abstruse ideas to be memorized—their application saves lives.

The HIV protease cleaves peptide bonds between Phe and Pro residues most efficiently. The active site thus has a pocket to bind aromatic groups next to the bond to be cleaved. The structures of several HIV protease inhibitors are shown in **Figure 6–24**. Although the structures appear varied, they all share a core structure—a main chain with a hydroxyl group positioned next to a branch containing a benzyl group. This arrangement targets the benzyl group to an aromatic (hydrophobic) binding pocket. The adjacent hydroxyl group mimics the negatively charged oxygen in the

FIGURE 6–23 Mechanism of action of HIV protease. Two active-site Asp residues (from different subunits) act as general acid-base cata- lysts, facilitating the attack of water on the peptide bond. The unstable tetrahedral intermediate in the reaction pathway is shaded light red.

Indinavir

Nelfinavir

Lopinavir

Saquinavir

FIGURE 6–24 HIV protease inhibitors. The hydroxyl group (red) acts as a transition-state analog, mimicking the oxygen of the tetrahedral inter- mediate. The adjacent benzyl group (blue) helps to properly position the drug in the active site.

tetrahedral intermediate in the normal reaction, provid- ing a transition-state analog. The remainder of each inhibitor structure was designed to fit into and bind to various crevices along the surface of the enzyme, enhancing overall binding. Availability of these effective drugs has vastly increased the lifespan and quality of life of millions of people with HIV and AIDS. ■

Hexokinase Undergoes Induced Fit on Substrate Binding

Yeast hexokinase (M_r 107,862) is a bisubstrate enzyme that catalyzes the reversible reaction

β-D-Glucose   Glucose 6-phosphate

ATP and ADP always bind to enzymes as a complex with the metal ion Mg^{2+}.

The hydroxyl at C-6 of glucose (to which the γ-phosphoryl of ATP is transferred in the hexokinase reaction) is similar in chemical reactivity to water, and water freely enters the enzyme active site. Yet hexoki- nase favors the reaction with glucose by a factor of 10^6. The enzyme can discriminate between glucose and water because of a conformational change in the enzyme when the correct substrates binds **(Fig. 6–25).** Hexokinase thus provides a good example of induced fit. When glucose is not present, the enzyme is in an inactive conformation, with the active-site amino acid side chains out of position for reaction. When glucose (but not water) and Mg · ATP bind, the binding energy derived from this interaction induces a conformational change in hexokinase to the catalytically active form.

This model has been reinforced by kinetic studies. The five-carbon sugar xylose, stereochemically similar to glucose but one carbon shorter, binds to hexokinase

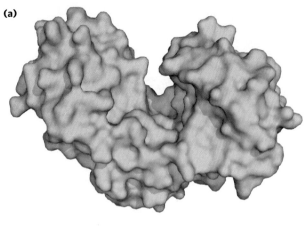

(a)

(b)

ADP   Glucose

FIGURE 6–25 Induced fit in hexokinase. (a) Hexokinase has a U-shaped structure (PDB ID 2YHX). **(b)** The ends pinch toward each other in a conformational change induced by binding of D-glucose (derived from PDB ID 1HKG and PDB ID 1GLK).

but in a position where it cannot be phosphorylated. Nevertheless, addition of xylose to the reaction mixture increases the rate of ATP hydrolysis. Evidently, the binding of xylose is sufficient to induce a change in hexokinase to its active conformation, and the enzyme is thereby "tricked" into phosphorylating water. The hexokinase reaction also illustrates that enzyme specificity is not always a simple matter of binding one compound but not another. In the case of hexokinase, specificity is observed not in the formation of the ES complex but in the relative rates of subsequent catalytic steps. Water is not excluded from the active site, but reaction rates increase greatly in the presence of the functional phosphoryl group acceptor (glucose).

Xylose   Glucose

Induced fit is only one aspect of the catalytic mechanism of hexokinase—like chymotrypsin, hexokinase uses several catalytic strategies. For example, the active-site amino acid residues (those brought into position by the conformational change that follows substrate binding) participate in general acid-base catalysis and transition-state stabilization.

The Enolase Reaction Mechanism Requires Metal Ions

Another glycolytic enzyme, enolase, catalyzes the reversible dehydration of 2-phosphoglycerate to phosphoenolpyruvate:

2-Phosphoglycerate → (enolase) → Phosphoenolpyruvate + H_2O

The reaction provides an example of the use of an enzymatic cofactor, in this case a metal ion (an example of coenzyme function is provided in Box 6–3). Yeast enolase (M_r 93,316) is a dimer with 436 amino acid residues per subunit. The enolase reaction illustrates one type of metal ion catalysis and provides an additional example of general acid-base catalysis and transition-state stabilization. The reaction occurs in two steps (**Fig. 6–26a**). First, Lys[345] acts as a general base catalyst, abstracting a proton from C-2 of 2-phosphoglycerate; then Glu[211] acts as a general acid catalyst, donating a proton to the —OH leaving group. The proton at C-2 of 2-phosphoglycerate is not acidic and thus is quite resistant to its removal by Lys[345]. However, the electronegative oxygen atoms of the adjacent carboxyl group pull electrons away from C-2, making the attached protons somewhat more labile. In the active site, the carboxyl group of 2-phosphoglycerate undergoes strong ionic interactions with two bound Mg^{2+} ions (Fig. 6–26b), strongly enhancing the electron withdrawal by the carboxyl. Together, these effects render the C-2 protons sufficiently acidic (lowering the pK_a) so that one can be abstracted to initiate the reaction. As the unstable enolate intermediate is formed, the metal ions further act to shield the two negative charges (on the carboxyl oxygen atoms) that transiently exist in close proximity to each other. Hydrogen bonding to other active-site amino acid residues also contributes to the overall mechanism. The various interactions effectively stabilize both the enolate intermediate and the transition state preceding its formation.

Lysozyme Uses Two Successive Nucleophilic Displacement Reactions

Lysozyme is a natural antibacterial agent found in tears and egg whites. The hen egg white lysozyme (M_r 14,296) is a monomer with 129 amino acid residues.

(a)

2-Phosphoglycerate bound to enzyme

Lys³⁴⁵ abstracts a proton by general base catalysis. Two Mg²⁺ ions stabilize the resulting enolate intermediate.

Enolate intermediate

Glu²¹¹ facilitates elimination of the –OH group by general acid catalysis.

Phosphoenolpyruvate

(b)

MECHANISM FIGURE 6–26 Two-step reaction catalyzed by enolase. (a) The mechanism by which enolase converts 2-phosphoglycerate (2-PGA) to phosphoenolpyruvate. The carboxyl group of 2-PGA is coordinated by two magnesium ions at the active site. **(b)** The substrate, 2-PGA, in relation to the Mg^{2+}, Lys^{345}, and Glu^{211} in the enolase active site (gray outline). Nitrogen is shown in blue, phosphorus in orange; hydrogen atoms are not shown (PDB ID 1ONE).

This was the first enzyme to have its three-dimensional structure determined, by David Phillips and colleagues in 1965. The structure revealed four stabilizing disulfide bonds and a cleft containing the active site **(Fig. 6–27a)**. More than five decades of investigations have provided a detailed picture of the structure and activity of the enzyme, and an interesting story of how biochemical science progresses.

The substrate of lysozyme is peptidoglycan, a carbohydrate found in many bacterial cell walls (see Fig. 20–30). Lysozyme cleaves the $(\beta 1 \rightarrow 4)$ glycosidic C—O bond (p. 252) between the two types of sugar residues in the molecule, N-acetylmuramic acid (Mur2Ac) and N-acetylglucosamine (GlcNAc) **(Fig. 6–27b)**, often referred to as NAM and NAG, respectively, in the research literature on enzymology. Six residues of the alternating Mur2Ac and GlcNAc in peptidoglycan bind in the active site, in binding sites labeled A through F. Model building has shown that the lactyl side chain of Mur2Ac cannot be accommodated in sites C and E, restricting Mur2Ac binding to sites B, D, and F. Only one of the bound glycosidic bonds is cleaved, that between a Mur2Ac residue in site D and a GlcNAc residue in site E. The key catalytic amino acid residues in the active site are Glu^{35} and Asp^{52} **(Fig. 6–28a)**. The reaction is a nucleophilic substitution, with —OH from water replacing the GlcNAc at C-1 of Mur2Ac.

With the active-site residues identified and a detailed structure of the enzyme available, the path to understanding the reaction mechanism seemed open in the 1960s. However, definitive evidence for a particular mechanism eluded investigators for nearly four decades.

There are two chemically reasonable mechanisms that could generate the observed product of lysozyme-mediated cleavage of the glycosidic bond. Phillips and colleagues proposed a dissociative (S_N1-type) mechanism (Fig. 6–28a, left), in which the GlcNAc initially dissociates in step ① to leave behind a glycosyl cation (a carbocation) intermediate. In this mechanism, the departing GlcNAc is protonated by general acid catalysis by Glu^{35}, located in a hydrophobic pocket that gives its carboxyl group an unusually high pK_a. The carbocation is stabilized by resonance involving the adjacent ring oxygen, as well as by electrostatic interaction with the negative charge on the nearby Asp^{52}. In step ② water attacks at C-1 of Mur2Ac to yield the product. The alternative mechanism (Fig. 6–28a, right) involves two consecutive direct displacement (S_N2-type) steps. In step ❶, Asp^{52} attacks C-1 of Mur2Ac to displace the GlcNAc. As in the first mechanism, Glu^{35} acts as a general acid to protonate the departing GlcNAc. In step ❷, water attacks at C-1 of Mur2Ac to displace the Asp^{52} and generate product.

The Phillips mechanism (S_N1), was widely accepted for more than three decades. However, some controversy persisted and tests continued. The scientific method sometimes advances an issue slowly, and a truly insightful experiment can be difficult to design. Some early arguments against the Phillips mechanism were suggestive but not completely persuasive. For example, the half-life of the proposed glycosyl cation was estimated to be 10^{-12} second, just longer than a molecular vibration and not long enough for the needed diffusion of other molecules. More important, lysozyme is a member

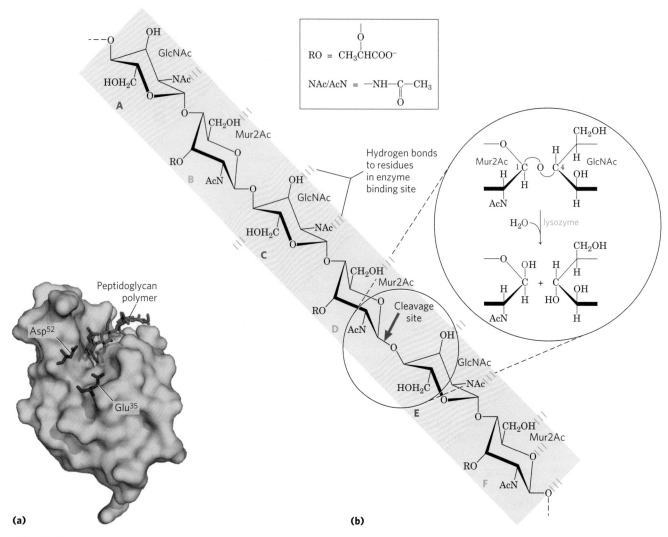

FIGURE 6–27 Hen egg white lysozyme and the reaction it catalyzes.
(a) Surface representation of the enzyme, with the active-site residues Glu[35] and Asp[52] shown as black stick structures and bound substrate shown as a red stick structure (PDB ID 1LZE). **(b)** Reaction catalyzed by hen egg white lysozyme. A segment of a peptidoglycan polymer is shown, with the lysozyme binding sites A through F shaded. The glyco- sidic C—O bond between sugar residues bound to sites D and E is cleaved, as indicated by the red arrow. The hydrolytic reaction is shown in the inset, with the fate of the oxygen in the H_2O traced in red. Mur2Ac is N-acetylmuramic acid; GlcNAc, N-acetylglucosamine. RO— represents a lactyl (lactic acid) group; —NAc and AcN—, an N-acetyl group (see key).

of a family of enzymes called "retaining glycosidases," all of which catalyze reactions in which the product has the same anomeric configuration as the substrate (anomeric configurations of carbohydrates are examined in Chapter 7), and all of which are known to have reactive covalent intermediates like that envisioned in the alternative (S_N2) pathway. Hence, the Phillips mechanism ran counter to experimental findings for closely related enzymes.

A compelling experiment tipped the scales decidedly in favor of the S_N2 pathway, as reported by Stephen Withers and colleagues in 2001. Making use of a mutant enzyme (with residue 35 changed from Glu to Gln) and artificial substrates, which combined to slow the rate of key steps in the reaction, these workers were able to stabilize the elusive covalent intermediate. This in turn allowed them to observe the intermediate directly, using both mass spectrometry and x-ray crystallography (Fig. 6–28b).

Is the lysozyme mechanism now proven? No. A key feature of the scientific method, as Albert Einstein once summarized it, is "No amount of experimentation can ever prove me right; a single experiment can prove me wrong." In the case of the lysozyme mechanism, one might argue (and some have) that the artificial substrates, with fluorine substitutions at C-1 and C-2, as were used to stabilize the covalent intermediate, might have altered the reaction pathway. The highly electronegative fluorine could destabilize an already electron-deficient oxocarbenium ion in the glycosyl cation intermediate that might occur in an S_N1 pathway. However, the S_N2 pathway is now the mechanism most in concert with available data.

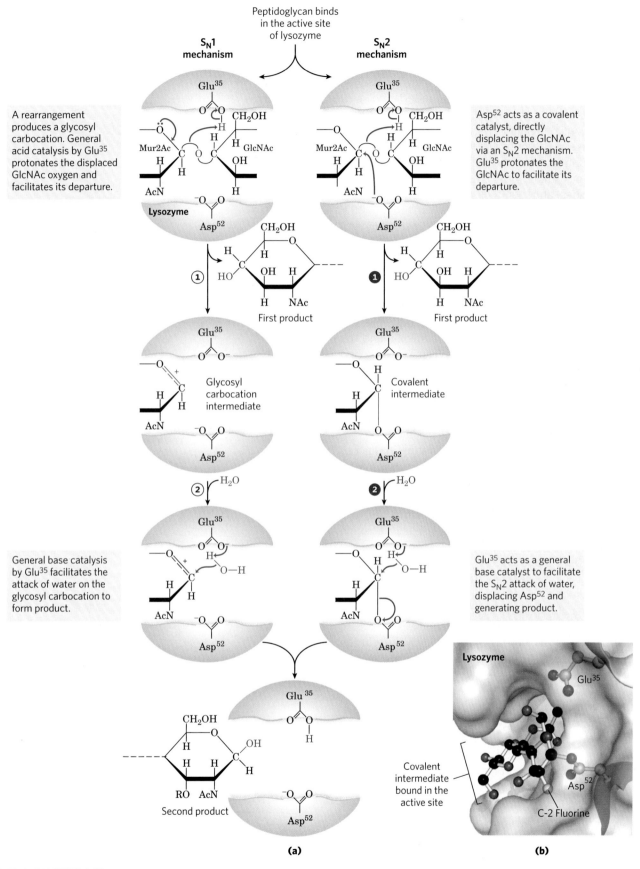

MECHANISM FIGURE 6–28 Lysozyme reaction. In this reaction (described in the text), the water introduced into the product at C-1 of Mur2Ac is in the same configuration as the original glycosidic bond. The reaction is thus a molecular substitution with retention of configuration. (a) Two proposed pathways potentially explain the overall reaction and its properties. The S_N1 pathway (left) is the original Phillips mechanism. The S_N2 pathway (right) is the mechanism most consistent with current data. (b) A surface rendering of the lysozyme active site, with the covalent enzyme-substrate intermediate shown as a ball-and-stick structure. Side chains of active-site residues are shown as ball-and-stick structures protruding from ribbons (PDB ID 1H6M).

An Understanding of Enzyme Mechanism Produces Useful Antibiotics

Penicillin was discovered in 1928 by Alexander Fleming, but it took another 15 years before this relatively unstable compound was understood well enough to use it as a pharmaceutical agent to treat bacterial infections. Penicillin interferes with the synthesis of peptidoglycan (described in Chapter 20, Fig. 20–31), the major component of the rigid cell wall that protects bacteria from osmotic lysis. Peptidoglycan consists of polysaccharides and peptides cross-linked in several steps that include a transpeptidase reaction (**Fig. 6–29**). It is this reaction that is inhibited by penicillin and related compounds (**Fig. 6–30a**), all of which mimic one conformation of the D-Ala–D-Ala segment of the peptidoglycan precursor. The peptide bond in the precursor is replaced by a highly reactive β-lactam ring. When penicillin binds to the transpeptidase, an active-site Ser attacks the carbonyl of the β-lactam ring and generates a covalent adduct between penicillin and the enzyme. However, the leaving group remains attached because it is linked by the remnant of the β-lactam ring (Fig. 6–30b). The covalent complex irreversibly inactivates the enzyme. This, in turn, blocks synthesis of the bacterial cell wall, and most bacteria die as the fragile inner membrane bursts under osmotic pressure.

Human use of penicillin and its derivatives has led to the evolution of strains of pathogenic bacteria that express **β-lactamases** (**Fig. 6–31a**), enzymes that cleave β-lactam antibiotics, rendering them inactive. The bacteria thereby become resistant to the antibiotics. The genes for these enzymes have spread rapidly through bacterial populations under the selective pressure imposed by the use (and often overuse) of β-lactam antibiotics. Human medicine responded with the development of compounds such as clavulanic acid, a suicide inactivator, which irreversibly inactivates the β-lactamases (Fig. 6–31b). Clavulanic acid mimics the structure of a β-lactam antibiotic, and forms a covalent adduct with a Ser in the β-lactamase active site. This leads to a rearrangement that creates a much more reactive derivative, which is subsequently attacked by another nucleophile in the active site to irreversibly acylate the enzyme and inactivate it. Amoxicillin and clavulanic acid are combined in a widely used pharmaceutical formulation with the trade name Augmentin. The cycle of chemical warfare between humans and bacteria continues unabated. Strains of disease-causing bacteria that are resistant to both amoxicillin and clavulanic acid (reflecting mutations in β-lactamase that render it unreactive to clavulanic acid) have been discovered. The development of new antibiotics promises to be a growth industry for the foreseeable future. ∎

SUMMARY 6.4 Examples of Enzymatic Reactions

▶ Chymotrypsin is a serine protease with a well-understood mechanism, featuring general acid-base catalysis, covalent catalysis, and transition-state stabilization.

FIGURE 6–29 The transpeptidase reaction. This reaction, which links two peptidoglycan precursors into a larger polymer, is facilitated by an active-site Ser and a covalent catalytic mechanism similar to that of chymotrypsin. Note that peptidoglycan is one of the few places in nature where D-amino acid residues are found. The active-site Ser attacks the carbonyl of the peptide bond between the two D-Ala residues, creating a covalent ester linkage between the substrate and the enzyme, with release of the terminal D-Ala residue. An amino group from the second peptidoglycan precursor then attacks the ester linkage, displacing the enzyme and cross-linking the two precursors.

FIGURE 6–30 Transpeptidase inhibition by β-lactam antibiotics.
(a) β-Lactam antibiotics feature a five-membered thiazolidine ring fused to a four-membered β-lactam ring. The latter ring is strained and includes an amide moiety that plays a critical role in the inactivation of peptidoglycan synthesis. The R group differs in different penicillins. Penicillin G was the first to be isolated and remains one of the most effective, but it is degraded by stomach acid and must be administered by injection. Penicillin V is nearly as effective and is acid stable, so it can be administered orally. Amoxicillin has a broad range of effectiveness, is readily administered orally, and is thus the most widely prescribed β-lactam antibiotic. **(b)** Attack on the amide moiety of the β-lactam ring by a transpeptidase active-site Ser results in a covalent acyl-enzyme product. This is hydrolyzed so slowly that adduct formation is practically irreversible, and the transpeptidase is inactivated.

FIGURE 6–31 β-Lactamases and β-lactamase inhibition. (a) β-Lactamases promote cleavage of the β-lactam ring in β-lactam antibiotics, inactivating them. **(b)** Clavulanic acid is a suicide inhibitor, making use of the normal chemical mechanism of β-lactamases to create a reactive species at the active site. This reactive species is attacked by a nucleophilic group (Nu:) in the active site to irreversibly acylate the enzyme.

▶ Hexokinase provides an excellent example of induced fit as a means of using substrate binding energy.

▶ The enolase reaction proceeds via metal ion catalysis.

▶ Lysozyme makes use of covalent catalysis and general acid catalysis as it promotes two successive nucleophilic displacement reactions.

▶ Understanding enzyme mechanism allows the development of drugs to inhibit enzyme action.

6.5 Regulatory Enzymes

In cellular metabolism, groups of enzymes work together in sequential pathways to carry out a given metabolic process, such as the multireaction breakdown of glucose to lactate or the multireaction synthesis of an amino acid from simpler precursors. In such enzyme systems, the reaction product of one enzyme becomes the substrate of the next.

Most of the enzymes in each metabolic pathway follow the kinetic patterns we have already described. Each pathway, however, includes one or more enzymes that have a greater effect on the rate of the overall sequence. These **regulatory enzymes** exhibit increased or decreased catalytic activity in response to certain signals. Adjustments in the rate of reactions catalyzed by regulatory enzymes, and therefore in the rate of entire metabolic sequences, allow the cell to meet changing needs for energy and for biomolecules required in growth and repair.

The activities of regulatory enzymes are modulated in a variety of ways. **Allosteric enzymes** function through reversible, noncovalent binding of regulatory compounds called **allosteric modulators** or **allosteric effectors**, which are generally small metabolites or cofactors. Other enzymes are regulated by reversible **covalent modification**. Both classes of regulatory enzymes tend to be multisubunit proteins, and in some cases the regulatory site(s) and the active site are on separate subunits. Metabolic systems have at least two other mechanisms of enzyme regulation. Some enzymes are stimulated or inhibited when they are bound by separate **regulatory proteins**. Others are activated when peptide segments are removed by **proteolytic cleavage**; unlike effector-mediated regulation, regulation by proteolytic cleavage is irreversible. Important examples of both mechanisms are found in physiological processes such as digestion, blood clotting, hormone action, and vision.

Cell growth and survival depend on efficient use of resources, and this efficiency is made possible by regulatory enzymes. No single rule governs the occurrence of different types of regulation in different systems. To a degree, allosteric (noncovalent) regulation may permit fine-tuning of metabolic pathways that are required continuously but at different levels of activity as cellular conditions change. Regulation by covalent modification may be all or none—usually the case with proteolytic cleavage—or it may allow for subtle changes in activity. Several types of regulation may occur in a single regulatory enzyme. The remainder of this chapter is devoted to a discussion of these methods of enzyme regulation.

Allosteric Enzymes Undergo Conformational Changes in Response to Modulator Binding

As we saw in Chapter 5, allosteric proteins are those having "other shapes" or conformations induced by the binding of modulators. The same concept applies to certain regulatory enzymes, as conformational changes induced by one or more modulators interconvert more-active and less-active forms of the enzyme. The modulators for allosteric enzymes may be inhibitory or stimulatory. Often the modulator is the substrate itself; regulation in which substrate and modulator are identical is called homotropic. The effect is similar to that of O_2 binding to hemoglobin (Chapter 5): binding of the ligand—or substrate, in the case of enzymes—causes conformational changes that affect the subsequent activity of other sites on the protein. In most cases, the conformational change converts a relatively inactive conformation (often referred to as a T state) to a more active conformation (an R state). When the modulator is a molecule other than the substrate, the enzyme is said to be heterotropic. Note that allosteric modulators should not be confused with uncompetitive and mixed inhibitors. Although the latter bind at a second site on the enzyme, they do not necessarily mediate conformational changes between active and inactive forms, and the kinetic effects are distinct.

The properties of allosteric enzymes are significantly different from those of simple nonregulatory enzymes. Some of the differences are structural. In addition to active sites, allosteric enzymes generally have one or more regulatory, or allosteric, sites for binding the modulator (**Fig. 6–32**). Just as an enzyme's active site is specific for its substrate, each regulatory site is specific for its modulator. Enzymes with several modulators generally have different specific binding sites for each. In homotropic enzymes, the active site and regulatory site are the same.

Allosteric enzymes are typically larger and more complex than nonallosteric enzymes, with two or more subunits. A classic example is aspartate transcarbamoylase (often abbreviated ATCase), which catalyzes an early step in the biosynthesis of pyrimidine nucleotides, the reaction of carbamoyl phosphate and aspartate to form carbamoyl aspartate.

Carbamoyl phosphate   Aspartate   N-Carbamoylaspartate

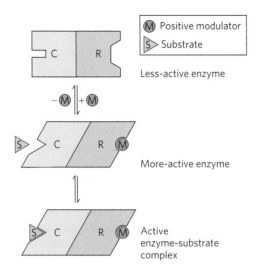

Less-active enzyme

More-active enzyme

Active enzyme-substrate complex

FIGURE 6–32 Subunit interactions in an allosteric enzyme, and interactions with inhibitors and activators. In many allosteric enzymes the substrate-binding site and the modulator-binding site(s) are on different subunits, the catalytic (C) and regulatory (R) subunits, respectively. Binding of the positive (stimulatory) modulator (M) to its specific site on the regulatory subunit is communicated to the catalytic subunit through a conformational change. This change renders the catalytic subunit active and capable of binding the substrate (S) with higher affinity. On dissociation of the modulator from the regulatory subunit, the enzyme reverts to its inactive or less active form.

ATCase has 12 polypeptide chains organized into 6 catalytic (organized as 2 trimeric complexes) and 6 regulatory (organized as 3 dimeric complexes) subunits. **Figure 6–33** shows the quaternary structure of this enzyme, deduced from x-ray analysis. The enzyme exhibits allosteric behavior as detailed below, as the catalytic subunits function cooperatively. The regulatory subunits have binding sites for ATP and CTP, which function as positive and negative regulators, respectively. CTP is one of the end products of the pathway, and negative regulation by CTP serves to limit ATCase action under conditions when CTP is abundant. On the other hand, high concentrations of ATP indicate that cellular metabolism is robust, the cell is growing, and additional pyrimidine nucleotides may be needed to support RNA transcription and DNA replication.

The Kinetic Properties of Allosteric Enzymes Diverge from Michaelis-Menten Behavior

Allosteric enzymes show relationships between V_0 and [S] that differ from Michaelis-Menten kinetics. They do exhibit saturation with the substrate when [S] is sufficiently high, but for allosteric enzymes, plots of V_0 versus [S] **(Fig. 6–34)** usually produce a sigmoid saturation curve, rather than the hyperbolic curve typical of nonregulatory enzymes. On the sigmoid saturation curve we can find a value of [S] at which V_0 is half-maximal, but we cannot refer to it with the designation K_m, because the enzyme does not follow the hyperbolic Michaelis-Menten relationship. Instead, the symbol $[S]_{0.5}$ or $K_{0.5}$ is often used to represent the substrate concentration giving half-maximal velocity of the reaction catalyzed by an allosteric enzyme (Fig. 6–34).

Sigmoid kinetic behavior generally reflects cooperative interactions between multiple protein subunits. In other words, changes in the structure of one subunit are translated into structural changes in adjacent subunits, an effect mediated by noncovalent interactions at the

(a) Inactive T state

(b) Active R state

FIGURE 6–33 The regulatory enzyme aspartate transcarbamoylase. **(a)** The inactive T state (PDB ID 1RAB) and **(b)** active R state (PDB ID 1F1B) of the enzyme are shown. This allosteric regulatory enzyme has two stacked catalytic clusters, each with three catalytic polypeptide chains (in shades of blue and purple), and three regulatory clusters, each with two regulatory polypeptide chains (in beige and yellow). The regulatory clusters form the points of a triangle (not evident in this side view) surrounding the catalytic subunits. Binding sites for allosteric modulators (including CTP) are on the regulatory subunits. Modulator binding produces large changes in enzyme conformation and activity. The role of this enzyme in nucleotide synthesis, and details of its regulation, are discussed in Chapter 22.

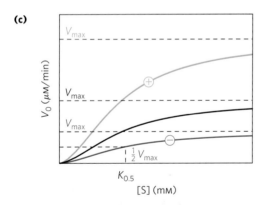

FIGURE 6–34 Substrate-activity curves for representative allosteric enzymes. Three examples of complex responses of allosteric enzymes to their modulators. **(a)** The sigmoid curve of a homotropic enzyme, in which the substrate also serves as a positive (stimulatory) modulator, or activator. Note the resemblance to the oxygen-saturation curve of hemoglobin (see Fig. 5-12). The sigmoidal curve is a hybrid curve in which the enzyme is present primarily in the relatively inactive T state at low substrate concentration, and primarily in the more active R state at high substrate concentration. The curves for the pure T and R states are plotted separately in color. ATCase exhibits a kinetic pattern similar to this. **(b)** The effects of several different concentrations of a positive modulator (+) or a negative modulator (−) on an allosteric enzyme in which $K_{0.5}$ is altered without a change in V_{max}. The central curve shows the substrate-activity relationship without a modulator. For ATCase, CTP is a negative modulator and ATP is a positive modulator. **(c)** A less common type of modulation, in which V_{max} is altered and $K_{0.5}$ is nearly constant.

interface between subunits. The principles are particularly well illustrated by a nonenzyme: O_2 binding to hemoglobin. Sigmoid kinetic behavior is explained by the concerted and sequential models for subunit interactions (see Fig. 5–15).

ATCase effectively illustrates both homotropic and heterotropic allosteric kinetic behavior. The binding of the substrates, aspartate and carbamoyl phosphate, to the enzyme gradually bring about a transition from the relatively inactive T state to the more active R state. This accounts for the sigmoid rather than hyperbolic change in V_0 with increasing [S]. One characteristic of sigmoid kinetics is that small changes in the concentration of a modulator can be associated with large changes in activity. As exemplified in Figure 6–34a, a relatively small increase in [S] in the steep part of the curve causes a comparatively large increase in V_0.

The heterotropic allosteric regulation of ATCase is brought about by its interactions with ATP and CTP. For heterotropic allosteric enzymes, an activator may cause the curve to become more nearly hyperbolic, with a decrease in $K_{0.5}$ but no change in V_{max}, resulting in an increased reaction velocity at a fixed substrate concentration. For ATCase, the interaction with ATP brings this about, and the enzyme exhibits a V_0 versus [S] curve that is characteristic of the active R state at sufficiently high ATP concentrations (V_0 is higher for any value of [S]; Fig. 6–34b). A negative modulator (an inhibitor) may produce a *more* sigmoid substrate-saturation curve, with an increase in $K_{0.5}$, as illustrated by the effects of CTP on ATCase kinetics (see curves for negative modulater, Fig. 6–34b). Other heterotropic allosteric enzymes respond to an activator by an increase in V_{max} with little change in $K_{0.5}$ (Fig. 6–34c). Heterotropic allosteric enzymes therefore show different kinds of responses in their substrate-activity curves because some have inhibitory modulators, some have activating modulators, and some (like ATCase) have both.

Some Enzymes Are Regulated by Reversible Covalent Modification

In another important class of regulatory enzymes, activity is modulated by covalent modification of one or more of the amino acid residues in the enzyme molecule. Over 500 different types of covalent modification have been found in proteins. Common modifying groups include phosphoryl, acetyl, adenylyl, uridylyl, methyl, amide, carboxyl, myristoyl, palmitoyl, prenyl, hydroxyl, sulfate, and adenosine diphosphate ribosyl groups **(Fig. 6–35)**. There are even entire proteins that are used as specialized modifying groups, including ubiquitin and sumo. These varied groups are generally linked to and removed from a regulated enzyme by separate enzymes. When an amino acid residue in an enzyme is modified, a novel amino acid with altered properties has effectively been introduced into the enzyme. Introduction of a charge can alter the local properties of the enzyme and induce

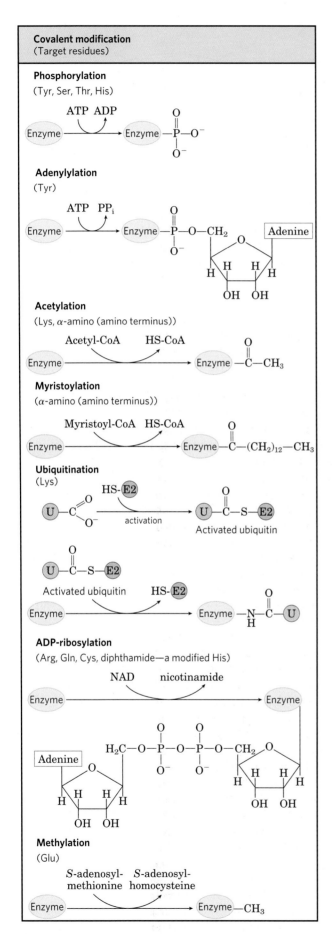

FIGURE 6–35 Some enzyme modification reactions.

a change in conformation. Introduction of a hydrophobic group can trigger association with a membrane. The changes are often substantial and can be critical to the function of the altered enzyme.

The variety of enzyme modifications is too great to cover in detail, but some examples can be offered. An example of an enzyme regulated by methylation is the methyl-accepting chemotaxis protein of bacteria. This protein is part of a system that permits a bacterium to swim toward an attractant (such as a sugar) in solution and away from repellent chemicals. The methylating agent is *S*-adenosylmethionine (adoMet) (see Fig. 18–18). Acetylation is another common modification, with approximately 80% of the soluble proteins in eukaryotes, including many enzymes, acetylated at their amino terminus. Ubiquitin is added to proteins as a tag that predestines them for proteolytic degradation (see Fig. 27–47). Ubiquitination can also have a regulatory function. Sumo is found attached to many eukaryotic nuclear proteins with roles in the regulation of transcription, chromatin structure, and DNA repair.

ADP-ribosylation is an especially interesting reaction, observed in a number of proteins; the ADP-ribose is derived from nicotinamide adenine dinucleotide (NAD) (see Fig. 8–38). This type of modification occurs for the bacterial enzyme dinitrogenase reductase, resulting in regulation of the important process of biological nitrogen fixation. Diphtheria toxin and cholera toxin are enzymes that catalyze the ADP-ribosylation (and inactivation) of key cellular enzymes or proteins.

Phosphorylation is the most important type of regulatory modification. It is estimated that one-third of all proteins in a eukaryotic cell are phosphorylated, and one or (often) many phosphorylation events are part of virtually every regulatory process. Some proteins have only one phosphorylated residue, others have several, and a few have dozens of sites for phosphorylation. This mode of covalent modification is central to a large number of regulatory pathways, and we therefore discuss it in some detail here, and again in Chapter 12.

We will encounter all of these types of modification again in later chapters.

Phosphoryl Groups Affect the Structure and Catalytic Activity of Enzymes

The attachment of phosphoryl groups to specific amino acid residues of a protein is catalyzed by **protein kinases**. In the reactions, the γ-phosphoryl group derived from a nucleoside triphosphate (usually ATP) is transferred to a particular Ser, Thr, or Tyr residue (occasionally His as well) on the target protein. This introduces a bulky, charged group into a region of the target protein that was only moderately polar. The oxygen atoms of a phosphoryl group can hydrogen-bond with one or several groups in a protein, commonly the amide groups of the peptide backbone at the start of an α helix or the charged guanidinium group of an Arg residue. The two negative charges on a phosphorylated side chain can also repel neighboring negatively charged (Asp or Glu) residues. When the modified side chain is located in a region of an enzyme critical to its three-dimensional structure, phosphorylation can have dramatic effects on enzyme conformation and thus on substrate binding and catalysis. Removal of phosphoryl groups from these same target proteins is catalyzed by **protein phosphatases**.

An important example of enzyme regulation by phosphorylation is seen in glycogen phosphorylase (M_r 94,500) of muscle and liver (Chapter 15), which catalyzes the reaction

$$(\text{Glucose})_n + P_i \longrightarrow (\text{glucose})_{n-1} + \text{glucose 1-phosphate}$$

$$\text{Glycogen} \qquad\qquad \text{Shortened} \\ \text{glycogen} \\ \text{chain}$$

The glucose 1-phosphate so formed can be used for ATP synthesis in muscle or converted to free glucose in the liver. Note that glycogen phosphorylase, though it adds a phosphate to a substrate, is not itself a kinase, because it does not utilize ATP or any other nucleotide triphosphate as a phosphoryl donor in its catalyzed reaction. It is, however, the substrate for a protein kinase that phosphorylates it. In the discussion below, the phosphoryl groups we are concerned with are those involved in regulation of the enzyme, as distinguished from its catalytic function.

Glycogen phosphorylase occurs in two forms: the more active phosphorylase *a* and the less active phosphorylase *b* **(Fig. 6–36)**. Phosphorylase *a* has two subunits, each with a specific Ser residue that is phosphorylated at its hydroxyl group. These serine phosphate residues are required for maximal activity of the enzyme. The phosphoryl groups can be hydrolytically removed by a separate enzyme called phosphorylase phosphatase:

$$\text{Phosphorylase } a + 2H_2O \longrightarrow \text{Phosphorylase } b + 2P_i$$

$$\text{(more active)} \qquad\qquad\qquad \text{(less active)}$$

In this reaction, phosphorylase *a* is converted to phosphorylase *b* by the cleavage of two serine phosphate covalent bonds, one on each subunit of glycogen phosphorylase.

Phosphorylase *b* can in turn be reactivated—covalently transformed back into active phosphorylase *a*—by another enzyme, phosphorylase kinase, which catalyzes the transfer of phosphoryl groups from ATP to the hydroxyl groups of the two specific Ser residues in phosphorylase *b*:

$$2ATP + \text{phosphorylase } b \longrightarrow 2ADP + \text{phosphorylase } a$$

$$\text{(less active)} \qquad\qquad\qquad \text{(more active)}$$

The breakdown of glycogen in skeletal muscles and the liver is regulated by variations in the ratio of the two forms of glycogen phosphorylase. The *a* and *b* forms differ in their secondary, tertiary, and quaternary structures; the active site undergoes changes in structure

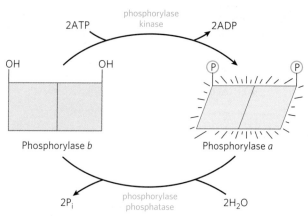

FIGURE 6–36 Regulation of muscle glycogen phosphorylase activity by phosphorylation. In the more active form of the enzyme, phosphorylase *a*, specific Ser residues, one on each subunit, are phosphorylated. Phosphorylase *a* is converted to the less active phosphorylase *b* by enzymatic loss of these phosphoryl groups, promoted by phosphoprotein phosphatase 1 (PP1). Phosphorylase *b* can be reconverted (reactivated) to phosphorylase *a* by the action of phosphorylase kinase.

and, consequently, changes in catalytic activity as the two forms are interconverted.

The regulation of glycogen phosphorylase by phosphorylation illustrates the effects on both structure and catalytic activity of adding a phosphoryl group. In the unphosphorylated state, each subunit of this enzyme is folded so as to bring the 20 residues at its amino terminus, including a number of basic residues, into a region containing several acidic amino acids; this produces an electrostatic interaction that stabilizes the conformation. Phosphorylation of Ser[14] interferes with this interaction, forcing the amino-terminal domain out of the acidic environment and into a conformation that allows interaction between the $\circledP$-Ser and several Arg side chains. In this conformation, the enzyme is much more active.

Phosphorylation of an enzyme can affect catalysis in another way: by altering substrate-binding affinity. For example, when isocitrate dehydrogenase (an enzyme of the citric acid cycle; Chapter 16) is phosphorylated, electrostatic repulsion by the phosphoryl group inhibits the binding of citrate (a tricarboxylic acid) at the active site.

Multiple Phosphorylations Allow Exquisite Regulatory Control

The Ser, Thr, or Tyr residues that are typically phosphorylated in regulated proteins occur within common structural motifs, called consensus sequences, that are recognized by specific protein kinases (Table 6–10). Some kinases are basophilic, preferentially phosphorylating a residue having basic neighbors; others have different substrate preferences, such as for a residue near a Pro residue. Amino acid sequence is not the only important factor in determining whether a given residue will be phosphorylated, however. Protein folding brings together

TABLE 6–10 Consensus Sequences for Protein Kinases

Protein kinase	Consensus sequence and phosphorylated residue
Protein kinase A	-x-R-[RK]-x-[ST]-B-
Protein kinase G	-x-R-[RK]-x-[ST]-x-
Protein kinase C	-[RK](2)-x-[ST]-B-[RK](2)-
Protein kinase B	-x-R-x-[ST]-x-K-
Ca^{2+}/calmodulin kinase I	-B-x-R-x(2)-[ST]-x(3)-B-
Ca^{2+}/calmodulin kinase II	-B-x-[RK]-x(2)-[ST]-x(2)-
Myosin light chain kinase (smooth muscle)	-K(2)-R-x(2)-S-x-B(2)-
Phosphorylase *b* kinase	*-K-R-K-Q-I-S-V-R-*
Extracellular signal–regulated kinase (ERK)	-P-x-[ST]-P(2)-
Cyclin-dependent protein kinase (cdc2)	-x-[ST]-P-x-[KR]-
Casein kinase I	-[SpTp]-x(2)-[ST]-B*
Casein kinase II	-x-[ST]-x(2)-[ED]-x-
β-Adrenergic receptor kinase	-[DE](n)-[ST]-x(3)
Rhodopsin kinase	-x(2)-[ST]-E(n)-
Insulin receptor kinase	-x-E(3)-Y-M(4)-*K(2)-S-R-G-D-Y-M-T-M-Q-I-* *G-K(3)-L-P-A-T-G-D-Y-M-N-M-S-P-V-G-D-*
Epidermal growth factor (EGF) receptor kinase	*-E(4)-Y-F-E-L-V-*

Sources: Pinna, L.A. & Ruzzene, M.H. (1996) How do protein kinases recognize their substrates? *Biochim. Biophys. Acta* 1314, 191–225; Kemp, B.E. & Pearson, R.B. (1990) Protein kinase recognition sequence motifs. *Trends Biochem. Sci.* 15, 342–346; Kennelly, P.J. & Krebs, E.G. (1991) Consensus sequences as substrate specificity determinants for protein kinases and protein phosphatases. *J. Biol. Chem.* 266, 15,555–15,558.

Note: Shown here are deduced consensus sequences (in roman type) and actual sequences from known substrates (italic). The Ser (S), Thr (T), or Tyr (Y) residue that undergoes phosphorylation is in red; all amino acid residues are shown as their one-letter abbreviations (see Table 3–1). x represents any amino acid; B, any hydrophobic amino acid. Sp, Tp, and Yp are Ser, Thr, and Tyr residues that must already be phosphorylated for the kinase to recognize the site.

*The best target site has two amino acid residues separating the phosphorylated and target Ser/Thr residues; target sites with one or three intervening residues function at a reduced level.

residues that are distant in the primary sequence; the resulting three-dimensional structure can determine whether a protein kinase has access to a given residue and can recognize it as a substrate. Another factor influencing the substrate specificity of certain protein kinases is the proximity of other phosphorylated residues.

Regulation by phosphorylation is often complicated. Some proteins have consensus sequences recognized by several different protein kinases, each of which can phosphorylate the protein and alter its enzymatic activity. In some cases, phosphorylation is hierarchical: a certain residue can be phosphorylated only if a neighboring residue has already been phosphorylated. For example, glycogen synthase, the enzyme that catalyzes the condensation of glucose monomers to form glycogen (Chapter 15), is inactivated by phosphorylation of specific Ser residues and is also modulated by at least four other protein kinases that phosphorylate four other sites in the enzyme (Fig. 6–37). The enzyme is not a substrate for glycogen synthase kinase 3, for example, until one site has been phosphorylated by casein kinase II. Some phosphorylations inhibit glycogen synthase more than others, and some combinations of phosphorylations are cumulative. These multiple regulatory phosphorylations provide the potential for extremely subtle modulation of enzyme activity.

To serve as an effective regulatory mechanism, phosphorylation must be reversible. In general, phosphoryl groups are added and removed by different enzymes, and the processes can therefore be separately regulated. Cells contain a family of phosphoprotein phosphatases that hydrolyze specific Ⓟ–Ser, Ⓟ–Thr, and Ⓟ–Tyr esters, releasing P_i. The phosphoprotein phosphatases we know of thus far act only on a subset of phosphoproteins, but they show less substrate specificity than protein kinases.

Some Enzymes and Other Proteins Are Regulated by Proteolytic Cleavage of an Enzyme Precursor

For some enzymes, an inactive precursor called a **zymogen** is cleaved to form the active enzyme. Many proteolytic enzymes (proteases) of the stomach and pancreas are regulated in this way. Chymotrypsin and trypsin are initially synthesized as chymotrypsinogen and trypsinogen (Fig. 6–38). Specific cleavage causes conformational changes that expose the enzyme active site.

Kinase	Phosphorylation sites	Degree of synthase inactivation
Protein kinase A	1A, 1B, 2, 4	+
Protein kinase G	1A, 1B, 2	+
Protein kinase C	1A	+
Ca²⁺/calmodulin kinase	1B, 2	+
Phosphorylase *b* kinase	2	+
Casein kinase I	At least nine	+ + + +
Casein kinase II	5	0
Glycogen synthase kinase 3	3A, 3B, 3C	+ + +
Glycogen synthase kinase 4	2	+

FIGURE 6–37 Multiple regulatory phosphorylations. The enzyme glycogen synthase has at least nine separate sites in five designated regions susceptible to phosphorylation by one of the cellular protein kinases. Thus, regulation of this enzyme is a matter not of binary (on/off) switching but of finely tuned modulation of activity over a wide range in response to a variety of signals.

Because this type of activation is irreversible, other mechanisms are needed to inactivate these enzymes. Proteases are inactivated by inhibitor proteins that bind very tightly to the enzyme active site. For example, pancreatic trypsin inhibitor (M_r 6,000) binds to and inhibits trypsin. α_1-Antiproteinase (M_r 53,000) primarily inhibits neutrophil elastase (neutrophils are a type of leukocyte, or white blood cell; elastase is a protease acting on elastin, a component of some connective tissues). An insufficiency of α_1-antiproteinase, which can be caused by exposure to cigarette smoke, has been associated with lung damage, including emphysema.

Proteases are not the only proteins activated by proteolysis. In other cases, however, the precursors are called not zymogens but, more generally, **proproteins** or **proenzymes**, as appropriate. For example, the connective tissue protein collagen is initially synthesized as the soluble precursor procollagen.

A Cascade of Proteolytically Activated Zymogens Leads to Blood Coagulation

A blood clot is an aggregate of cell fragments called platelets, cross-linked and stabilized by proteinaceous fibers consisting mainly of fibrin (**Fig. 6–39a**). Fibrin is derived from a soluble zymogen called fibrinogen. After albumins and globulins, fibrinogen is generally the third most abundant type of protein in blood plasma. The formation of a blood clot provides a well-studied example of a **regulatory cascade**, a mechanism that allows a very sensitive response to—and amplification of—a molecular signal. The pathways also bring together several other types of regulation.

In a regulatory cascade, a signal leads to the activation of protein X. Protein X catalyzes the activation of protein Y. Protein Y catalyzes the activation of protein Z, and so on. Since proteins X, Y, and Z are catalysts and activate multiple copies of the next protein in the chain, the signal is amplified in each step. In some cases, the

FIGURE 6–38 Activation of zymogens by proteolytic cleavage. Shown here is the formation of chymotrypsin and trypsin from their zymogens, chymotrypsinogen and trypsinogen. The bars represent the amino acid sequences of the polypeptide chains, with numbers indicating the positions of the residues (the amino-terminal residue is number 1). Residues at the termini of the polypeptide fragments generated by cleavage are indicated below the bars. Note that in the final active forms, some numbered residues are missing. Recall that the three polypeptide chains (A, B, and C) of chymotrypsin are linked by disulfide bonds (see Fig. 6-19).

Some Regulatory Enzymes Use Several Regulatory Mechanisms

Glycogen phosphorylase catalyzes the first reaction in a pathway that feeds stored glucose into energy-yielding carbohydrate metabolism (Chapters 14 and 15). This is an important metabolic pathway, and its regulation is correspondingly complex. Although the primary regulation of glycogen phosphorylase is through covalent modification, as outlined in Figure 6–36, glycogen phosphorylase is also modulated allosterically by AMP, which is an activator of phosphorylase b, and by glucose 6-phosphate and ATP, both inhibitors. In addition, the enzymes that add and remove the phosphoryl groups are themselves regulated by—and so the entire system is sensitive to—the levels of hormones that regulate blood sugar (**Fig. 6–42**; see also Chapters 15 and 23).

Other complex regulatory enzymes are found at key metabolic crossroads. Bacterial glutamine synthetase, which catalyzes a reaction that introduces reduced nitrogen into cellular metabolism (Chapter 22), is among the most complex regulatory enzymes known. It is regulated allosterically (with at least eight different modulators); by reversible covalent modification; and by the association of other regulatory proteins, a mechanism examined in detail when we consider the regulation of specific metabolic pathways.

What is the advantage of such complexity in the regulation of enzymatic activity? We began this chapter by stressing the central importance of catalysis to the very existence of life. The *control* of catalysis is also critical to life. If all possible reactions in a cell were catalyzed simultaneously, macromolecules and metabolites would quickly be broken down to much simpler chemical forms. Instead, cells catalyze only the reactions they need at a given moment. When chemical resources are plentiful, cells synthesize and store glucose and other metabolites. When chemical resources are scarce, cells use these stores to fuel cellular metabolism. Chemical energy is used economically, parceled out to various metabolic pathways as cellular needs dictate. The availability of powerful catalysts, each specific for a given reaction, makes the regulation of these reactions possible. This in turn gives rise to the complex, highly regulated symphony we call life.

SUMMARY 6.5 Regulatory Enzymes

▶ The activities of metabolic pathways in cells are regulated by control of the activities of certain enzymes.

▶ The activity of an allosteric enzyme is adjusted by reversible binding of a specific modulator to a regulatory site. A modulator may be the substrate itself or some other metabolite, and the effect of the modulator may be inhibitory or stimulatory. The kinetic behavior of allosteric enzymes reflects cooperative interactions among enzyme subunits.

▶ Other regulatory enzymes are modulated by covalent modification of a specific functional group necessary for activity. The phosphorylation of specific amino acid residues is a particularly common way to regulate enzyme activity.

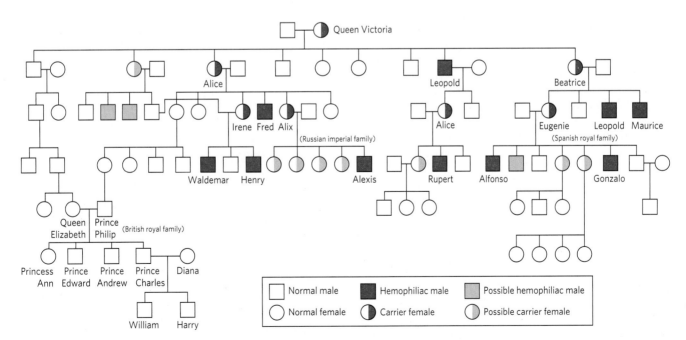

FIGURE 6–41 The royal families of Europe and inheritance of hemophilia A. Males are indicated by squares and females by circles. Males who suffered from hemophilia are represented by red squares, and presumed female carriers by half-red circles.

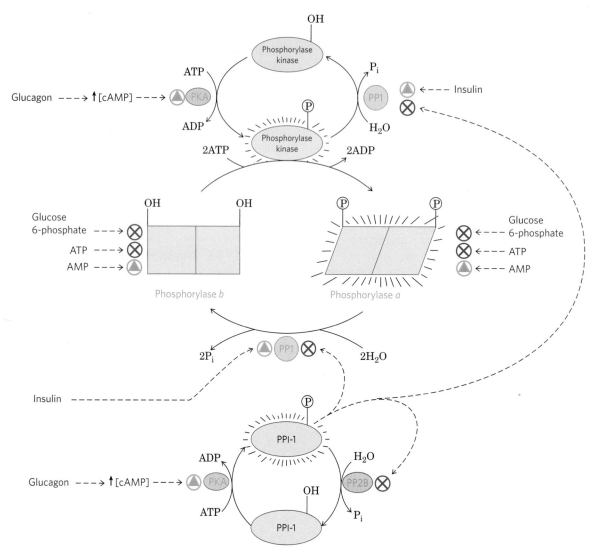

FIGURE 6–42 Regulation of muscle glycogen phosphorylase activity by phosphorylation. The activity of glycogen phosphorylase in muscle is subjected to a multilevel system of regulation involving much more than the covalent modification (phosphorylation) shown in Figure 6–36. Allosteric regulation, and a regulatory cascade sensitive to hormonal status that acts on the enzymes involved in phosphorylation and dephosphorylation, also play important roles. The activity of both forms of the enzyme is allosterically regulated by an activator (AMP) and by inhibitors (glucose 6-phosphate and ATP) that bind to separate sites on the enzyme. The activities of phosphorylase kinase and phosphorylase phosphatase 1 (PP1) are also regulated by covalent modification, via a short pathway that responds to the hormones glucagon and epinephrine.

One path leads to the phosphorylation of phosphorylase kinase and phosphoprotein phosphatase inhibitor 1 (PPI-1). The phosphorylated phosphorylase kinase is activated and in turn phosphorylates and activates glycogen phosphorylase. At the same time, the phosphorylated PPI-1 interacts with and inhibits PP1. PPI-1 also keeps itself active (phosphorylated) by inhibiting phosphoprotein phosphatase 2B (PP2B), the enzyme that dephosphorylates (inactivates) it. In this way, the equilibrium between the a and b forms of glycogen phosphorylase is shifted decisively toward the more active glycogen phosphorylase a. Note that the two forms of phosphorylase kinase are both activated to a degree by Ca^{2+} ion (not shown). This pathway is discussed in more detail in Chapters 14, 15, and 23.

▶ Many proteolytic enzymes are synthesized as inactive precursors called zymogens, which are activated by cleavage of small peptide fragments.

▶ Blood clotting is mediated by two interlinked regulatory cascades of proteolytically activated zymogens.

▶ Enzymes at important metabolic intersections may be regulated by complex combinations of effectors, allowing coordination of the activities of interconnected pathways.

Key Terms

Terms in bold are defined in the glossary.

enzyme 190	**active site** 192
cofactor 190	**substrate** 192
coenzyme 190	**ground state** 192
prosthetic group 190	**transition state** 193
holoenzyme 190	**activation energy**
apoenzyme 190	**($\Delta G^{\ddagger}$)** 193
apoprotein 190	**reaction intermediate** 193

Further Reading

General

Evolution of Catalytic Function. (1987) *Cold Spring Harb. Symp. Quant. Biol.* **52**.
   A collection of excellent papers on fundamentals; continues to be very useful.

Fersht, A. (1999) *Structure and Mechanism in Protein Science: A Guide to Enzyme Catalysis and Protein Folding*, W. H. Freeman and Company, New York.
   A clearly written, concise introduction. More advanced.

Frey, P.A. & Hegeman, A.D. (2006) *Enzymatic Reaction Mechanisms*, Oxford University Press, New York.
   An authoritative and up-to-date resource on the reactions that occur in living systems.

Jencks, W.P. (1987) *Catalysis in Chemistry and Enzymology*, Dover Publications, Inc., New York.
   An outstanding book on the subject. More advanced.

Kornberg, A. (1989) *For the Love of Enzymes: The Odyssey of a Biochemist*, Harvard University Press, Cambridge, MA.

Principles of Catalysis

Amyes, T.L., O'Donoghue, A.C., & Richard, J.P. (2001) Contribution of phosphate intrinsic binding energy to the enzymatic rate acceleration for triosephosphate isomerase. *J. Am. Chem. Soc.* **123**, 11,325–11,326.

Gutteridge, A. & Thornton, J.M. (2005) Understanding nature's catalytic toolkit. *Trends Biochem. Sci.* **30**, 622–629.
   A nice discussion of where binding energy comes from and how it is used.

Hammes-Schiffer, S. & Benkovic, S.J. (2006) Relating protein motion to catalysis. *Annu. Rev. Biochem.* **75**, 519–541.
   A good description of the importance of protein motions in catalysis.

Hansen, D.E. & Raines, R.T. (1990) Binding energy and enzymatic catalysis. *J. Chem. Educ.* **67**, 483–489.
   A good place for the beginning student to acquire a better understanding of principles.

Harris, T.K. & Turner, G.J. (2002) Structural basis of perturbed p*K*~a~ values of catalytic groups in enzyme active sites. *IUBMB Life* **53**, 85–98.

Kraut, D.A., Carroll, K.S., & Herschlag, D. (2003) Challenges in enzyme mechanism and energetics. *Annu. Rev. Biochem.* **72**, 517–571.
   A good summary of the principles of enzymatic catalysis as currently understood, and of what we still do not understand.

Schramm, V.L. (2011) Enzymatic transition states, transition-state analogs, dynamics, thermodynamics, and lifetimes. *Annu. Rev. Biochem.* **80**, 703–732.

Williams, D.H. (2010) Enzyme catalysis from improved packing in their transition-state structures. *Curr. Opin. Chem. Biol.* **14**, 666–670.

Wolfenden, R. (2011) Benchmark reaction rates, the stability of biological molecules in water, and the evolution of catalytic power in enzymes. *Annu. Rev. Biochem.* **80**, 645–667.

Kinetics

Cleland, W.W. (2002) Enzyme kinetics: steady state. In *Encyclopedia of Life Sciences*, John Wiley & Sons, Inc./Wiley Interscience, www.els.net.
   A clear and concise presentation of the basics.

Raines, R.T. & Hansen, D.E. (1988) An intuitive approach to steady-state kinetics. *J. Chem. Educ.* **65**, 757–759.

Enzyme Examples

Babbitt, P.C. & Gerlt, J.A. (1997) Understanding enzyme superfamilies: chemistry as the fundamental determinant in the evolution of new catalytic activities. *J. Biol. Chem.* **27**, 30,591–30,594.
   An interesting description of the evolution of enzymes with different catalytic specificities, and the use of a limited repertoire of protein structural motifs.

Babbitt, P.C., Hasson, M.S., Wedekind, J.E., Palmer, D.R.J., Barrett, W.C., Reed, G.H., Rayment, I., Ringe, D., Kenyon, G.L., & Gerlt, J.A. (1996) The enolase superfamily: a general strategy for enzyme-catalyzed abstraction of the α-protons of carboxylic acids. *Biochemistry* **35**, 16,489–16,501.

Kirby, A.J. (2001) The lysozyme mechanism sorted—after 50 years. *Nat. Struct. Biol.* **8**, 737–739.
   A nice discussion of the catalytic power of enzymes and the principles underlying it.

Regulatory Enzymes

Changeux, J.-P. (2012) Allostery and the Monod-Wyman-Changeux model after 50 years. *Annu. Rev. Biophys.* **41**, 83–113.

Ehrmann, M. & Clausen, T. (2004) Proteolysis as a regulatory mechanism. *Annu. Rev. Genet.* **38**, 709–724.

Hunter, T. & Plowman, G.D. (1997) The protein kinases of budding yeast: six score and more. *Trends Biochem. Sci.* **22**, 18–22.

Details of the variety of these important enzymes in a model eukaryote.

Johnson, L.N. & Barford, D. (1993) The effects of phosphorylation on the structure and function of proteins. *Annu. Rev. Biophys. Biomol. Struct.* **22**, 199–232.

Problems

1. Keeping the Sweet Taste of Corn The sweet taste of freshly picked corn (maize) is due to the high level of sugar in the kernels. Store-bought corn (several days after picking) is not as sweet, because about 50% of the free sugar is converted to starch within one day of picking. To preserve the sweetness of fresh corn, the husked ears can be immersed in boiling water for a few minutes ("blanched"), then cooled in cold water. Corn processed in this way and stored in a freezer maintains its sweetness. What is the biochemical basis for this procedure?

2. Intracellular Concentration of Enzymes To approximate the actual concentration of enzymes in a bacterial cell, assume that the cell contains equal concentrations of 1,000 different enzymes in solution in the cytosol and that each protein has a molecular weight of 100,000. Assume also that the bacterial cell is a cylinder (diameter 1.0 μm, height 2.0 μm), that the cytosol (specific gravity 1.20) is 20% soluble protein by weight, and that the soluble protein consists entirely of enzymes. Calculate the *average* molar concentration of each enzyme in this hypothetical cell.

3. Rate Enhancement by Urease The enzyme urease enhances the rate of urea hydrolysis at pH 8.0 and 20°C by a factor of 10^{14}. If a given quantity of urease can completely hydrolyze a given quantity of urea in 5.0 min at 20°C and pH 8.0, how long would it take for this amount of urea to be hydrolyzed under the same conditions in the absence of urease? Assume that both reactions take place in sterile systems so that bacteria cannot attack the urea.

4. Protection of an Enzyme against Denaturation by Heat When enzyme solutions are heated, there is a progressive loss of catalytic activity over time due to denaturation of the enzyme. A solution of the enzyme hexokinase incubated at 45°C lost 50% of its activity in 12 min, but when incubated at 45°C in the presence of a very large concentration of one of its substrates, it lost only 3% of its activity in 12 min. Suggest why thermal denaturation of hexokinase was retarded in the presence of one of its substrates.

5. Requirements of Active Sites in Enzymes Carboxypeptidase, which sequentially removes carboxyl-terminal amino acid residues from its peptide substrates, is a single polypeptide of 307 amino acids. The two essential catalytic groups in the active site are furnished by Arg^{145} and Glu^{270}.

(a) If the carboxypeptidase chain were a perfect α helix, how far apart (in Å) would Arg^{145} and Glu^{270} be? (Hint: See Fig. 4–4a.)

(b) Explain how the two amino acid residues can catalyze a reaction occurring in the space of a few angstroms.

6. Quantitative Assay for Lactate Dehydrogenase The muscle enzyme lactate dehydrogenase catalyzes the reaction

$$CH_3-\overset{O}{\overset{\|}{C}}-COO^- + NADH + H^+ \longrightarrow CH_3-\underset{\underset{H}{|}}{\overset{OH}{\overset{|}{C}}}-COO^- + NAD^+$$

Pyruvate   Lactate

NADH and NAD^+ are the reduced and oxidized forms, respectively, of the coenzyme NAD. Solutions of NADH, but *not* NAD^+, absorb light at 340 nm. This property is used to determine the concentration of NADH in solution by measuring spectrophotometrically the amount of light absorbed at 340 nm by the solution. Explain how these properties of NADH can be used to design a quantitative assay for lactate dehydrogenase.

7. Effect of Enzymes on Reactions Which of the listed effects would be brought about by any enzyme catalyzing the following simple reaction?

$$S \underset{k_2}{\overset{k_1}{\rightleftharpoons}} P \quad \text{where} \quad K'_{eq} = \frac{[P]}{[S]}$$

(a) Decreased K'_{eq}; (b) increased k_1; (c) increased K'_{eq}; (d) increased $\Delta G^{\ddagger}$; (e) decreased $\Delta G^{\ddagger}$; (f) more negative $\Delta G'^{\circ}$; (g) increased k_2.

8. Relation between Reaction Velocity and Substrate Concentration: Michaelis-Menten Equation

(a) At what substrate concentration would an enzyme with a k_{cat} of 30.0 s^{-1} and a K_m of 0.0050 M operate at one-quarter of its maximum rate?

(b) Determine the fraction of V_{max} that would be obtained at the following substrate concentrations [S]: $\frac{1}{2}K_m$, $2K_m$, and $10K_m$.

(c) An enzyme that catalyzes the reaction $X \rightleftharpoons Y$ is isolated from two bacterial species. The enzymes have the same V_{max}, but different K_m values for the substrate X. Enzyme A has a K_m of 2.0 μM, while enzyme B has a K_m of 0.5 μM. The plot below shows the kinetics of reactions carried out with the same concentration of each enzyme and with [X] = 1 μM. Which curve corresponds to which enzyme?

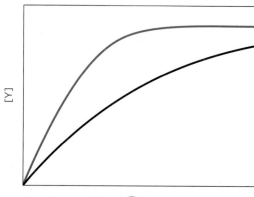

9. Applying the Michaelis-Menten Equation I A research group discovers a new version of happyase, which they call happyase*, that catalyzes the chemical reaction

$$\text{HAPPY} \rightleftharpoons \text{SAD}$$

The researchers begin to characterize the enzyme.

(a) In the first experiment, with $[E_t]$ at 4 nM, they find that the V_{max} is 1.6 $\mu M \ s^{-1}$. Based on this experiment, what is the k_{cat} for happyase*? (Include appropriate units.)

(b) In another experiment, with $[E_t]$ at 1 nM and [HAPPY] at 30 μM, the researchers find that $V_0 = 300$ nM s^{-1}. What is the measured K_m of happyase* for its substrate HAPPY? (Include appropriate units.)

(c) Further research shows that the purified happyase* used in the first two experiments was actually contaminated with a reversible inhibitor called ANGER. When ANGER is carefully removed from the happyase* preparation and the two experiments repeated, the measured V_{max} in (a) is increased to 4.8 $\mu M \ s^{-1}$, and the measured K_m in (b) is now 15 μM. For the inhibitor ANGER, calculate the values of α and α'.

(d) Based on the information given above, what type of inhibitor is ANGER?

10. Applying the Michaelis-Menten Equation II An enzyme is found that catalyzes the reaction

$$A \rightleftharpoons B$$

Researchers find that the K_m for the substrate A is 4 μM, and the k_{cat} is 20 min^{-1}.

(a) In an experiment, [A] = 6 mM, and $V_0 = 480$ nM min^{-1}. What was the $[E_t]$ used in the experiment?

(b) In another experiment, $[E_t] = 0.5 \ \mu M$, and the measured $V_0 = 5 \ \mu M \ min^{-1}$. What was the [A] used in the experiment?

(c) The compound Z is found to be a very strong competitive inhibitor of the enzyme, with an α of 10. In an experiment with the same $[E_t]$ as in (a), but a different [A], an amount of Z is added that reduces V_0 to 240 nM min^{-1}. What is the [A] in this experiment?

(d) Based on the kinetic parameters given above, has this enzyme evolved to achieve catalytic perfection? Explain your answer briefly, using the kinetic parameter(s) that define catalytic perfection.

11. Estimation of V_{max} and K_m by Inspection Although graphical methods are available for accurate determination of the V_{max} and K_m of an enzyme-catalyzed reaction (see Box 6–1), sometimes these quantities can be quickly estimated by inspecting values of V_0 at increasing [S]. Estimate the V_{max} and K_m of the enzyme-catalyzed reaction for which the following data were obtained.

[S] (M)	V_0 (μM/min)
2.5×10^{-6}	28
4.0×10^{-6}	40
1×10^{-5}	70
2×10^{-5}	95
4×10^{-5}	112
1×10^{-4}	128
2×10^{-3}	139
1×10^{-2}	140

12. Properties of an Enzyme of Prostaglandin Synthesis Prostaglandins are a class of eicosanoids, fatty acid derivatives with a variety of extremely potent actions on vertebrate tissues. They are responsible for producing fever and inflammation and its associated pain. Prostaglandins are derived from the 20-carbon fatty acid arachidonic acid in a reaction catalyzed by the enzyme prostaglandin endoperoxide synthase. This enzyme, a cyclooxygenase, uses oxygen to convert arachidonic acid to PGG_2, the immediate precursor of many different prostaglandins (prostaglandin synthesis is described in Chapter 21).

(a) The kinetic data given below are for the reaction catalyzed by prostaglandin endoperoxide synthase. Focusing here on the first two columns, determine the V_{max} and K_m of the enzyme.

[Arachidonic acid] (mM)	Rate of formation of PGG_2 (mM/min)	Rate of formation of PGG_2 with 10 mg/mL ibuprofen (mM/min)
0.5	23.5	16.67
1.0	32.2	25.25
1.5	36.9	30.49
2.5	41.8	37.04
3.5	44.0	38.91

(b) Ibuprofen is an inhibitor of prostaglandin endoperoxide synthase. By inhibiting the synthesis of prostaglandins, ibuprofen reduces inflammation and pain. Using the data in the first and third columns of the table, determine the type of inhibition that ibuprofen exerts on prostaglandin endoperoxide synthase.

13. Graphical Analysis of V_{max} and K_m The following experimental data were collected during a study of the catalytic activity of an intestinal peptidase with the substrate glycylglycine:

$$\text{Glycylglycine} + H_2O \longrightarrow 2 \text{ glycine}$$

[S] (mM)	Product formed (μmol/min)
1.5	0.21
2.0	0.24
3.0	0.28
4.0	0.33
8.0	0.40
16.0	0.45

Use graphical analysis (see Box 6–1 and its associated Living Graph) to determine the K_m and V_{max} for this enzyme preparation and substrate.

14. The Eadie-Hofstee Equation There are several ways to transform the Michaelis-Menten equation so as to plot data and derive kinetic parameters, each with different advantages depending on the data set being analyzed. One transformation of the Michaelis-Menten equation is the Lineweaver-Burk, or double-reciprocal, equation. Multiplying both sides of the Lineweaver-Burk equation by V_{max} and rearranging gives the Eadie-Hofstee equation:

$$V_0 = (-K_m)\frac{V_0}{[S]} + V_{max}$$

A plot of V_0 versus $V_0/[S]$ for an enzyme-catalyzed reaction is shown below. The blue curve was obtained in the absence of inhibitor. Which of the other curves (A, B, or C) shows the enzyme activity when a competitive inhibitor is added to the reaction mixture? Hint: See Equation 6–30.

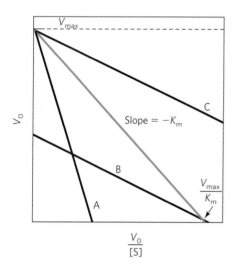

15. The Turnover Number of Carbonic Anhydrase Carbonic anhydrase of erythrocytes (M_r 30,000) has one of the highest turnover numbers known. It catalyzes the reversible hydration of CO_2:

$$H_2O + CO_2 \rightleftharpoons H_2CO_3$$

This is an important process in the transport of CO_2 from the tissues to the lungs. If 10.0 μg of pure carbonic anhydrase catalyzes the hydration of 0.30 g of CO_2 in 1 min at 37 °C at V_{max}, what is the turnover number (k_{cat}) of carbonic anhydrase (in units of min^{-1})?

16. Deriving a Rate Equation for Competitive Inhibition The rate equation for an enzyme subject to competitive inhibition is

$$V_0 = \frac{V_{max}[S]}{\alpha K_m + [S]}$$

Beginning with a new definition of total enzyme as

$$[E_t] = [E] + [ES] + [EI]$$

and the definitions of α and K_I provided in the text, derive the rate equation above. Use the derivation of the Michaelis-Menten equation as a guide.

17. Irreversible Inhibition of an Enzyme Many enzymes are inhibited irreversibly by heavy metal ions such as Hg^{2+}, Cu^{2+}, or Ag^+, which can react with essential sulfhydryl groups to form mercaptides:

$$Enz–SH + Ag^+ \longrightarrow Enz–S–Ag + H^+$$

The affinity of Ag^+ for sulfhydryl groups is so great that Ag^+ can be used to titrate —SH groups quantitatively. To 10.0 mL of a solution containing 1.0 mg/mL of a pure enzyme, an investigator added just enough $AgNO_3$ to completely inactivate the enzyme. A total of 0.342 μmol of $AgNO_3$ was required. Calculate the minimum molecular weight of the enzyme. Why does the value obtained in this way give only the *minimum* molecular weight?

18. Clinical Application of Differential Enzyme Inhibition Human blood serum contains a class of enzymes known as acid phosphatases, which hydrolyze biological phosphate esters under slightly acidic conditions (pH 5.0):

$$R—O—\overset{\displaystyle O^-}{\underset{\displaystyle O}{\overset{|}{\underset{||}{P}}}}—O^- + H_2O \longrightarrow R—OH + HO—\overset{\displaystyle O^-}{\underset{\displaystyle O}{\overset{|}{\underset{||}{P}}}}—O^-$$

Acid phosphatases are produced by erythrocytes, the liver, kidney, spleen, and prostate gland. The enzyme of the prostate gland is clinically important, because its increased activity in the blood can be an indication of prostate cancer. The phosphatase from the prostate gland is strongly inhibited by tartrate ion, but acid phosphatases from other tissues are not. How can this information be used to develop a specific procedure for measuring the activity of the acid phosphatase of the prostate gland in human blood serum?

19. Inhibition of Carbonic Anhydrase by Acetazolamide Carbonic anhydrase is strongly inhibited by the drug acetazolamide, which is used as a diuretic (i.e., to increase the production of urine) and to lower excessively high pressure in the eye (due to accumulation of intraocular fluid) in glaucoma. Carbonic anhydrase plays an important role in these and other secretory processes, because it participates in regulating the pH and bicarbonate content of several body fluids. The experimental curve of initial reaction velocity (as percentage of V_{max}) versus [S] for the carbonic anhydrase reaction is illustrated below (upper curve). When the experiment is repeated in the presence of acetazolamide, the lower curve is obtained. From an inspection of the curves and your knowledge of the kinetic properties of competitive and mixed enzyme inhibitors, determine the nature of the inhibition by acetazolamide. Explain your reasoning.

20. The Effects of Reversible Inhibitors Derive the expression for the effect of a reversible inhibitor on observed K_m (apparent $K_m = \alpha K_m/\alpha'$). Start with Equation 6–30 and the statement that apparent K_m is equivalent to the [S] at which $V_0 = V_{max}/2\alpha'$.

21. pH Optimum of Lysozyme The active site of lysozyme contains two amino acid residues essential for catalysis: Glu[35] and Asp[52]. The pK_a values of the carboxyl side chains of these residues are 5.9 and 4.5, respectively. What is the ionization state (protonated or deprotonated) of each residue at pH 5.2, the pH optimum of lysozyme? How can the ionization states of these residues explain the pH-activity profile of lysozyme shown below?

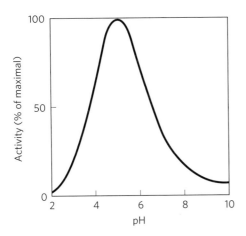

22. Working with Kinetics Go to the Living Graphs for Chapter 6.

(a) Using the Living Graph for Equation 6–9, create a V versus [S] plot. Use $V_{max} = 100\ \mu M\ s^{-1}$, and $K_m = 10\ \mu M$. How much does V_0 increase when [S] is doubled, from 0.2 to 0.4 μM? What is V_0 when [S] = 10 μM? How much does the V_0 increase when [S] increases from 100 to 200 μM? Observe how the graph changes when the values for V_{max} or K_m are halved or doubled.

(b) Using the Living Graph for Equation 6–30 and the kinetic parameters in (a), create a plot in which both α and α' are 1.0. Now observe how the plot changes when $\alpha = 2.0$, when $\alpha' = 3.0$, and when $\alpha = 2.0$ and $\alpha' = 3.0$.

(c) Using the Living Graphs for Equation 6–30 and the Lineweaver-Burk equation in Box 6–1, create Lineweaver-Burk (double-reciprocal) plots for all the cases in (a) and (b). When $\alpha = 2.0$, does the x intercept move to the right or to the left? If $\alpha = 2.0$ and $\alpha' = 3.0$, does the x intercept move to the right or to the left?

Data Analysis Problem

23. Exploring and Engineering Lactate Dehydrogenase Examining the structure of an enzyme results in hypotheses about the relationship between different amino acids in the protein's structure and the protein's function. One way to test these hypotheses is to use recombinant DNA technology to generate mutant versions of the enzyme and then examine the structure and function of these altered forms. The technology used to do this is described in Chapter 9.

One example of this kind of analysis is the work of A. R. Clarke and colleagues on the enzyme lactate dehydrogenase, published in 1989. Lactate dehydrogenase (LDH) catalyzes the reduction of pyruvate with NADH to form lactate (see Section 14.3). A schematic of the enzyme's active site is shown below; the pyruvate is in the center:

The reaction mechanism is similar to that of many NADH reductions (see Fig. 13–24); it is approximately the reverse of steps ❷ and ❸ of Figure 14–8. The transition state involves a strongly polarized carbonyl group of the pyruvate molecule, as shown below:

(a) A mutant form of LDH in which Arg[109] is replaced with Gln shows only 5% of the pyruvate binding and 0.07% of the activity of wild-type enzyme. Provide a plausible explanation for the effects of this mutation.

(b) A mutant form of LDH in which Arg[171] is replaced with Lys shows only 0.05% of the wild-type level of substrate binding. Why is this dramatic effect surprising?

(c) In the crystal structure of LDH, the guanidinium group of Arg[171] and the carboxyl group of pyruvate are aligned as shown above in a co-planar "forked" configuration. Based on this, explain the dramatic effect of substituting Arg[171] with Lys.

(d) A mutant form of LDH in which Ile[250] is replaced with Gln shows reduced binding of NADH. Provide a plausible explanation for this result.

Clarke and colleagues also set out to engineer a mutant version of LDH that would bind and reduce oxaloacetate rather than pyruvate. They made a single substitution, replacing Gln[102] with Arg; the resulting enzyme would reduce oxaloacetate to malate and would no longer reduce pyruvate to lactate. They had therefore converted LDH to malate dehydrogenase.

(e) Sketch the active site of this mutant LDH with oxaloacetate bound.

(f) Why does this mutant enzyme now use oxaloacetate as a substrate instead of pyruvate?

(g) The authors were surprised that substituting a larger amino acid in the active site allowed a larger substrate to bind. Explain this result.

References

Clarke, A.R., Atkinson, T., & Holbrook, J.J. (1989) From analysis to synthesis: new ligand binding sites on the lactate dehydrogenase framework, Part I. *Trends Biochem. Sci.* **14**, 101–105.

Clarke, A.R., Atkinson, T., & Holbrook, J.J. (1989) From analysis to synthesis: new ligand binding sites on the lactate dehydrogenase framework, Part II. *Trends Biochem. Sci.* **14**, 145–148.

Carbohydrates and Glycobiology

Carbohydrates are the most abundant biomolecules on Earth. Each year, photosynthesis converts more than 100 billion metric tons of CO_2 and H_2O into cellulose and other plant products. Certain carbohydrates (sugar and starch) are a dietary staple in most parts of the world, and the oxidation of carbohydrates is the central energy-yielding pathway in most nonphotosynthetic cells. Carbohydrate polymers (also called glycans) serve as structural and protective elements in the cell walls of bacteria and plants and in the connective tissues of animals. Other carbohydrate polymers lubricate skeletal joints and participate in recognition and adhesion between cells. Complex carbohydrate polymers covalently attached to proteins or lipids act as signals that determine the intracellular destination or metabolic fate of these hybrid molecules, called **glycoconjugates**. This chapter introduces the major classes of carbohydrates and glycoconjugates and provides a few examples of their many structural and functional roles.

Carbohydrates are polyhydroxy aldehydes or ketones, or substances that yield such compounds on hydrolysis. Many, but not all, carbohydrates have the empirical formula $(CH_2O)_n$; some also contain nitrogen, phosphorus, or sulfur. There are three major size classes of carbohydrates: monosaccharides, oligosaccharides, and polysaccharides (the word "saccharide" is derived from the Greek *sakcharon*, meaning "sugar"). **Monosaccharides**, or simple sugars, consist of a single polyhydroxy aldehyde or ketone unit. The most abundant monosaccharide in nature is the six-carbon sugar D-glucose, sometimes referred to as dextrose. Monosaccharides of four or more carbons tend to have cyclic structures.

Oligosaccharides consist of short chains of monosaccharide units, or residues, joined by characteristic linkages called glycosidic bonds. The most abundant are the **disaccharides**, with two monosaccharide units. Typical is sucrose (cane sugar), which consists of the six-carbon sugars D-glucose and D-fructose. All common monosaccharides and disaccharides have names ending with the suffix "-ose." In cells, most oligosaccharides consisting of three or more units do not occur as free entities but are joined to nonsugar molecules (lipids or proteins) in glycoconjugates.

The **polysaccharides** are sugar polymers containing more than 20 or so monosaccharide units; some have hundreds or thousands of units. Some polysaccharides, such as cellulose, are linear chains; others, such as glycogen, are branched. Both glycogen and cellulose consist of recurring units of D-glucose, but they differ in the type of glycosidic linkage and consequently have strikingly different properties and biological roles.

7.1 Monosaccharides and Disaccharides

The simplest of the carbohydrates, the monosaccharides, are either aldehydes or ketones with two or more hydroxyl groups; the six-carbon monosaccharides glucose and fructose have five hydroxyl groups. Many of the carbon atoms to which hydroxyl groups are attached are chiral centers, which give rise to the many sugar stereoisomers found in nature. Stereoisomerism in sugars is biologically significant because the enzymes that act on sugars are strictly stereospecific, typically preferring one stereoisomer to another by three or more orders of magnitude, as reflected in K_m values or binding constants. It is as difficult to fit the wrong sugar stereoisomer into an enzyme's binding site as it is to put your left glove on your right hand.

We begin by describing the families of monosaccharides with backbones of three to seven carbons—their structure and stereoisomeric forms, and the means of representing their three-dimensional structures on paper. We then discuss several chemical reactions of the carbonyl groups of monosaccharides. One such reaction, the addition of a hydroxyl group from within the same molecule, generates cyclic forms having four or more backbone carbons (the forms that predominate in aqueous solution). This ring closure creates a new chiral center, adding further stereochemical complexity to this class of compounds. The nomenclature for unambiguously specifying the configuration about each carbon atom in a cyclic form and the means of representing these structures on paper are therefore described in some detail; this information will be useful as we discuss the metabolism of monosaccharides in Part II. We also introduce here some important monosaccharide derivatives encountered in later chapters.

The Two Families of Monosaccharides Are Aldoses and Ketoses

Monosaccharides are colorless, crystalline solids that are freely soluble in water but insoluble in nonpolar solvents. Most have a sweet taste (see Box 7–2, p. 254). The backbones of common monosaccharides are unbranched carbon chains in which all the carbon atoms are linked by single bonds. In this open-chain form, one of the carbon atoms is double-bonded to an oxygen atom to form a carbonyl group; each of the other carbon atoms has a hydroxyl group. If the carbonyl group is at an end of the carbon chain (that is, in an aldehyde group) the monosaccharide is an **aldose**; if the carbonyl group is at any other position (in a ketone group) the monosaccharide is a **ketose**. The simplest monosaccharides are the two three-carbon trioses: glyceraldehyde, an aldotriose, and dihydroxyacetone, a ketotriose **(Fig. 7–1a)**.

Monosaccharides with four, five, six, and seven carbon atoms in their backbones are called, respectively, tetroses, pentoses, hexoses, and heptoses. There are aldoses and ketoses of each of these chain lengths: aldotetroses and ketotetroses, aldopentoses and ketopentoses, and so on. The hexoses, which include the aldohexose D-glucose and the ketohexose D-fructose (Fig. 7–1b), are the most common monosaccharides in nature—the products of photosynthesis, and key intermediates in the central energy-yielding reaction sequence in most organisms. The aldopentoses D-ribose and 2-deoxy-D-ribose (Fig. 7–1c) are components of nucleotides and nucleic acids (Chapter 8).

Monosaccharides Have Asymmetric Centers

All the monosaccharides except dihydroxyacetone contain one or more asymmetric (chiral) carbon atoms and thus occur in optically active isomeric forms (pp. 17–18). The simplest aldose, glyceraldehyde, contains one chiral center (the middle carbon atom) and therefore has two different optical isomers, or **enantiomers** (Fig. 7–2).

KEY CONVENTION: One of the two enantiomers of glyceraldehyde is, by convention, designated the D isomer, the other the L isomer. As for other biomolecules with chiral centers, the absolute configurations of sugars are known from x-ray crystallography. To represent three-dimensional sugar structures on paper, we often use **Fischer projection formulas** (Fig. 7–2). In Fischer projection formulas, horizontal bonds project out of the plane of the paper, toward the reader; vertical bonds project behind the plane of the paper, away from the reader. ■

In general, a molecule with n chiral centers can have 2^n stereoisomers. Glyceraldehyde has $2^1 = 2$; the aldohexoses, with four chiral centers, have $2^4 = 16$. The stereoisomers of monosaccharides of each carbon-chain

D-Glyceraldehyde, an aldotriose

Dihydroxyacetone, a ketotriose

(a)

D-Glucose, an aldohexose

(b)

D-Fructose, a ketohexose

D-Ribose, an aldopentose

2-Deoxy-D-ribose, an aldopentose

(c)

FIGURE 7–1 Representative monosaccharides. (a) Two trioses, an aldose and a ketose. The carbonyl group in each is shaded. **(b)** Two common hexoses. **(c)** The pentose components of nucleic acids. D-Ribose is a component of ribonucleic acid (RNA), and 2-deoxy-D-ribose is a component of deoxyribonucleic acid (DNA).

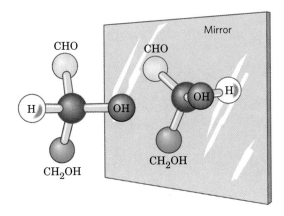

Ball-and-stick models

CHO
H—C—OH
CH₂OH
ᴅ-Glyceraldehyde

CHO
HO—C—H
CH₂OH
ʟ-Glyceraldehyde

Fischer projection formulas

CHO
H—C—OH
CH₂OH
ᴅ-Glyceraldehyde

CHO
HO—C—H
CH₂OH
ʟ-Glyceraldehyde

Perspective formulas

FIGURE 7-2 Three ways to represent the two enantiomers of glyceraldehyde. The enantiomers are mirror images of each other. Ball-and-stick models show the actual configuration of molecules. Recall (see Fig. 1-18) that in perspective formulas, the wide end of a solid wedge projects out of the plane of the paper, toward the reader; a dashed wedge extends behind.

length can be divided into two groups that differ in the configuration about the chiral center *most distant* from the carbonyl carbon. Those in which the configuration at this reference carbon is the same as that of ᴅ-glyceraldehyde are designated ᴅ isomers, and those with the same configuration as ʟ-glyceraldehyde are ʟ isomers. In other words, when the hydroxyl group on the reference carbon is on the right (*dextro*) in a projection formula that has the carbonyl carbon at the top, the sugar is the ᴅ isomer; when on the left (*levo*), it is the ʟ isomer. Of the 16 possible aldohexoses, eight are ᴅ forms and eight are ʟ. Most of the hexoses of living organisms are ᴅ isomers. Why ᴅ isomers? An interesting and unanswered question. Recall that all of the amino acids found in protein are exclusively one of two possible stereoisomers, ʟ. The basis for this initial preference for one isomer during evolution is also unknown; however, once one isomer had been selected, it was likely that evolving enzymes would retain their preference for that stereoisomer (p. 78).

Figure 7–3 shows the structures of the ᴅ stereoisomers of all the aldoses and ketoses having three to six carbon atoms. The carbons of a sugar are numbered beginning at the end of the chain nearest the carbonyl group. Each of the eight ᴅ-aldohexoses, which differ in the stereochemistry at C-2, C-3, or C-4, has its own name: ᴅ-glucose, ᴅ-galactose, ᴅ-mannose, and so forth (Fig. 7–3a). The four- and five-carbon ketoses are designated by inserting "ul" into the name of a corresponding aldose; for example, ᴅ-ribulose is the keto-pentose corresponding to the aldopentose ᴅ-ribose. (We will see the importance of ribulose when we discuss the fixation of atmospheric CO_2 by green plants, in Chapter 20.) The ketohexoses are named otherwise: for example, fructose (from the Latin *fructus*, "fruit"; fruits are one source of this sugar) and sorbose (from *Sorbus*, the genus of mountain ash, which has berries rich in the related sugar alcohol sorbitol). Two sugars that differ only in the configuration around one carbon atom are called **epimers**; ᴅ-glucose and ᴅ-mannose, which differ only in the stereochemistry at C-2, are epimers, as are ᴅ-glucose and ᴅ-galactose (which differ at C-4) **(Fig. 7–4)**.

Some sugars occur naturally in their ʟ form; examples are ʟ-arabinose and the ʟ isomers of some sugar derivatives that are common components of glycoconjugates (Section 7.3).

H   O
\\ //
C
H—C—OH
HO—C—H
HO—C—H
CH₂OH
ʟ-Arabinose

The Common Monosaccharides Have Cyclic Structures

For simplicity, we have thus far represented the structures of aldoses and ketoses as straight-chain molecules (Figs 7–3, 7–4). In fact, in aqueous solution, aldotetroses and all monosaccharides with five or more carbon atoms in the backbone occur predominantly as cyclic (ring) structures in which the carbonyl group has formed a covalent bond with the oxygen of a hydroxyl group along the chain. The formation of these ring structures is the result of a general reaction between alcohols and aldehydes or ketones to form derivatives called **hemiacetals** or **hemiketals**. Two molecules of an alcohol can add to a carbonyl carbon; the product of the first addition is a hemiacetal (for addition to an aldose) or a hemiketal (for addition to a ketose). If the —OH and carbonyl groups are from the same molecule,

(a) D-Aldoses

Three carbons

H—C=O
|
H—C—OH
|
CH₂OH

D-Glyceraldehyde

Four carbons

H—C=O
|
H—C—OH
|
H—C—OH
|
CH₂OH

D-Erythrose

H—C=O
|
HO—C—H
|
H—C—OH
|
CH₂OH

D-Threose

Five carbons

H—C=O
|
H—C—OH
|
H—C—OH
|
H—C—OH
|
CH₂OH

D-Ribose

H—C=O
|
HO—C—H
|
H—C—OH
|
H—C—OH
|
CH₂OH

D-Arabinose

H—C=O
|
H—C—OH
|
HO—C—H
|
H—C—OH
|
CH₂OH

D-Xylose

H—C=O
|
HO—C—H
|
HO—C—H
|
H—C—OH
|
CH₂OH

D-Lyxose

Six carbons

H—C=O
|
H—C—OH
|
H—C—OH
|
H—C—OH
|
H—C—OH
|
CH₂OH

D-Allose

H—C=O
|
HO—C—H
|
H—C—OH
|
H—C—OH
|
H—C—OH
|
CH₂OH

D-Altrose

H—C=O
|
H—C—OH
|
HO—C—H
|
H—C—OH
|
H—C—OH
|
CH₂OH

D-Glucose

H—C=O
|
HO—C—H
|
HO—C—H
|
H—C—OH
|
H—C—OH
|
CH₂OH

D-Mannose

H—C=O
|
H—C—OH
|
H—C—OH
|
HO—C—H
|
H—C—OH
|
CH₂OH

D-Gulose

H—C=O
|
HO—C—H
|
H—C—OH
|
HO—C—H
|
H—C—OH
|
CH₂OH

D-Idose

H—C=O
|
H—C—OH
|
HO—C—H
|
HO—C—H
|
H—C—OH
|
CH₂OH

D-Galactose

H—C=O
|
HO—C—H
|
HO—C—H
|
HO—C—H
|
H—C—OH
|
CH₂OH

D-Talose

FIGURE 7–3 Aldoses and ketoses. The series of **(a)** D-aldoses and **(b)** D-ketoses having from three to six carbon atoms, shown as projection formulas. The carbon atoms in red are chiral centers. In all these D isomers, the chiral carbon *most distant from the carbonyl carbon* has the same configuration as the chiral carbon in D-glyceraldehyde. The sugars named in boxes are the most common in nature; you will encounter these again in this and later chapters.

(b) D-Ketoses

Three carbons

CH₂OH
|
C=O
|
CH₂OH

Dihydroxyacetone

Four carbons

CH₂OH
|
C=O
|
H—C—OH
|
CH₂OH

D-Erythrulose

Five carbons

CH₂OH
|
C=O
|
H—C—OH
|
H—C—OH
|
CH₂OH

D-Ribulose

CH₂OH
|
C=O
|
HO—C—H
|
H—C—OH
|
CH₂OH

D-Xylulose

Six carbons

CH₂OH
|
C=O
|
H—C—OH
|
H—C—OH
|
H—C—OH
|
CH₂OH

D-Psicose

CH₂OH
|
C=O
|
HO—C—H
|
H—C—OH
|
H—C—OH
|
CH₂OH

D-Fructose

CH₂OH
|
C=O
|
H—C—OH
|
HO—C—H
|
H—C—OH
|
CH₂OH

D-Sorbose

CH₂OH
|
C=O
|
HO—C—H
|
HO—C—H
|
H—C—OH
|
CH₂OH

D-Tagatose

a five- or six-membered ring results. Addition of the second molecule of alcohol produces the full acetal or ketal (**Fig. 7–5**), and the bond formed is a glycosidic linkage. When the two molecules that react are both monosaccharides, the acetal or ketal formed is a disaccharide.

The reaction with the first molecule of alcohol creates an additional chiral center (the carbonyl carbon). Because the alcohol can add in either of two ways, attacking either the "front" or the "back" of the carbonyl carbon, the reaction can produce either of two stereoisomeric configurations, denoted α and β. For example, D-glucose exists in solution as an intramolecular hemiacetal in which the free hydroxyl group at C-5 has reacted with the aldehydic C-1, rendering the latter carbon asymmetric and producing two possible stereoisomers, designated α and β (**Fig. 7–6**). Isomeric forms of monosaccharides that differ only in their configuration about the hemiacetal or hemiketal carbon atom are called **anomers**, and the carbonyl carbon atom is called the **anomeric carbon**.

Six-membered ring compounds are called **pyranoses** because they resemble the six-membered ring compound

FIGURE 7-4 **Epimers.** D-Glucose and two of its epimers are shown as projection formulas. Each epimer differs from D-glucose in the configuration at one chiral center (shaded light red or blue).

FIGURE 7-5 **Formation of hemiacetals and hemiketals.** An aldehyde or ketone can react with an alcohol in a 1:1 ratio to yield a hemiacetal or hemiketal, respectively, creating a new chiral center at the carbonyl carbon. Substitution of a second alcohol molecule produces an acetal or ketal. When the second alcohol is part of another sugar molecule, the bond produced is a glycosidic bond (p. 252).

FIGURE 7-6 **Formation of the two cyclic forms of D-glucose.** Reaction between the aldehyde group at C-1 and the hydroxyl group at C-5 forms a hemiacetal linkage, producing either of two stereoisomers, the α and β anomers, which differ only in the stereochemistry around the hemiacetal carbon. This reaction is reversible. The interconversion of α and β anomers is called mutarotation.

pyran (**Fig. 7-7**). The systematic names for the two ring forms of D-glucose are therefore α-D-glucopyranose and β-D-glucopyranose. Ketohexoses (such as fructose) also occur as cyclic compounds with α and β anomeric forms. In these compounds the hydroxyl group at C-5 (or C-6) reacts with the keto group at C-2, forming a **furanose** (or pyranose) ring containing a hemiketal linkage (Fig. 7-5). D-Fructose readily forms the furanose ring (Fig. 7-7); the more common anomer of this sugar in combined forms or in derivatives is β-D-fructofuranose.

Cyclic sugar structures are more accurately represented in **Haworth perspective formulas** than in the Fischer projections commonly used for linear sugar structures. In Haworth projections the six-membered ring is tilted to make its plane almost perpendicular to that of the paper, with the bonds closest to the reader drawn thicker than those farther away, as in Figure 7-7.

KEY CONVENTION: To convert the Fischer projection formula of any linear D-hexose to a Haworth perspective formula showing the molecule's cyclic structure, draw the six-membered ring (five carbons and one oxygen, at the upper right), number the carbons in a clockwise direction beginning with the anomeric carbon, then

place the hydroxyl groups. If a hydroxyl group is to the right in the Fischer projection, it is placed pointing down (i.e., below the plane of the ring) in the Haworth perspective; if it is to the left in the Fischer projection, it is placed pointing up (i.e., above the plane) in the Haworth perspective. The terminal —CH$_2$OH group projects upward for the D-enantiomer, downward for the L-enantiomer. The hydroxyl on the anomeric carbon can point up or down. When the anomeric hydroxyl of a D-hexose is on the same side of the ring as C-6, the structure is by definition β; when it is on the opposite side from C-6, the structure is α. ∎

FIGURE 7-7 Pyranoses and furanoses. The pyranose forms of D-glucose and the furanose forms of D-fructose are shown here as Haworth perspective formulas. The edges of the ring nearest the reader are represented by bold lines. Hydroxyl groups below the plane of the ring in these Haworth perspectives would appear at the right side of a Fischer projection (compare with Fig. 7-6). Pyran and furan are shown for comparison.

WORKED EXAMPLE 7-1 Conversion of Fischer Projection to Haworth Perspective Formulas

Draw the Haworth perspective formulas for D-mannose and D-galactose.

Solution: Pyranoses are six-membered rings, so start with six-membered Haworth structures with the oxygen atom at the top right. Number the carbon atoms clockwise, starting with the aldose carbon. For mannose, place the hydroxyls on C-2, C-3, and C-4 above, above, and below the ring, respectively (because in the Fischer projection they are on the left, left, and right sides of the mannose structure). For D-galactose, the hydroxyls are oriented below, above, and above for C-2, C-3, and C-4, respectively. The hydroxyl at C-1 can be either up or down; there are two possible configurations, α and β, at this carbon.

WORKED EXAMPLE 7-2 Drawing Haworth Perspective Formulas of Sugar Isomers

Draw the Haworth perspective formulas for α-D-mannose and β-L-galactose.

Solution: The Haworth perspective formula of D-mannose from Worked Example 7-1 can have the hydroxyl group at C-1 pointing either up or down. According to the Key Convention, for the α form, the C-1 hydroxyl is pointing down when C-6 is up, as it is in D-mannose.

For β-L-galactose, use the Fischer representation of D-galactose (see Worked Example 7-1) to draw the correct Fischer representation of L-galactose, which is its mirror image: the hydroxyls at C-2, C-3, C-4, and C-5 are on the left, right, right, and left sides, respectively. Now draw the Haworth perspective, a six-membered ring in which the —OH groups on C-2, C-3, and C-4 are oriented up, down, and down, respectively, because in the Fischer representation they are on the left, right, and right sides. Because it is the β form, the —OH on the anomeric carbon points down (same side as C-5).

The α and β anomers of D-glucose interconvert in aqueous solution by a process called **mutarotation**, in which one ring form (say, the α anomer) opens briefly into the linear form, then closes again to produce the β anomer (Fig. 7-6). Thus, a solution of β-D-glucose and a solution of α-D-glucose eventually form identical equilibrium mixtures having identical optical properties. This mixture consists of about one-third α-D-glucose, two-thirds β-D-glucose, and very small amounts of the linear and five-membered ring (glucofuranose) forms.

Haworth perspective formulas like those in Figure 7-7 are commonly used to show the stereochemistry of ring forms of monosaccharides. However, the six-membered pyranose ring is not planar, as Haworth perspectives suggest, but tends to assume either of two "chair" conformations **(Fig. 7-8)**. Recall from Chapter 1 (pp. 18–19) that two *conformations* of a molecule are interconvertible without the breakage of covalent bonds, whereas two *configurations* can be interconverted only by breaking a covalent bond. To interconvert α and β configurations, the bond involving the ring oxygen atom would have to be broken, but interconversion of the two chair forms (which are *conformers*) does not require bond breakage and does not change configurations at any of the ring carbons. The specific three-dimensional structures of the monosaccharide units are important in determining the biological properties and functions of some polysaccharides, as we shall see.

Axis

Two possible chair forms of β-D-glucopyranose
(a)

Axis

α-D-Glucopyranose
(b)

FIGURE 7–8 Conformational formulas of pyranoses. (a) Two chair forms of the pyranose ring of β-D-glucopyranose. Two *conformers* such as these are not readily interconvertible; an input of about 46 kJ of energy per mole of sugar is required to force the interconversion of chair forms. Another conformation, the "boat" (not shown), is seen only in derivatives with very bulky substituents. **(b)** The preferred chair conformation of α-D-glucopyranose.

Organisms Contain a Variety of Hexose Derivatives

In addition to simple hexoses such as glucose, galactose, and mannose, there are a number of sugar derivatives in which a hydroxyl group in the parent compound is replaced with another substituent, or a carbon atom is oxidized to a carboxyl group **(Fig. 7–9)**. In glucosamine, galactosamine, and mannosamine, the hydroxyl at C-2 of the parent compound is replaced with an amino group. The amino group is commonly condensed with acetic

FIGURE 7–9 Some hexose derivatives important in biology. In amino sugars, an —NH_2 group replaces one of the —OH groups in the parent hexose. Substitution of —H for —OH produces a deoxy sugar; note that the deoxy sugars shown here occur in nature as the L isomers. The acidic sugars contain a carboxylate group, which confers a negative charge at neutral pH. D-Glucono-δ-lactone results from formation of an ester linkage between the C-1 carboxylate group and the C-5 (also known as the δ carbon) hydroxyl group of D-gluconate.

acid, as in *N*-acetylglucosamine. This glucosamine derivative is part of many structural polymers, including those of the bacterial cell wall. The substitution of a hydrogen for the hydroxyl group at C-6 of L-galactose or L-mannose produces L-fucose or L-rhamnose, respectively. L-Fucose is found in the complex oligosaccharide components of glycoproteins and glycolipids; L-rhamnose is found in plant polysaccharides.

Oxidation of the carbonyl (aldehyde) carbon of glucose to the carboxyl level produces gluconic acid, used in medicine as an innocuous counterion with which to administer positively charged drugs (such as quinine) or ions (such as Ca^{2+}). Other aldoses yield other **aldonic acids**. Oxidation of the carbon at the other end of the carbon chain—C-6 of glucose, galactose, or mannose—forms the

corresponding **uronic acid**: glucuronic, galacturonic, or mannuronic acid. Both aldonic and uronic acids form stable intramolecular esters called lactones (Fig. 7–9, lower left). The sialic acids are a family of sugars with the same nine-carbon backbone. One of them, *N*-acetylneuraminic acid (often referred to simply as "sialic acid"), is a derivative of *N*-acetylmannosamine that occurs in many glycoproteins and glycolipids on animal cell surfaces, providing sites of recognition by other cells or extracellular carbohydrate-binding proteins. The carboxylic acid groups of the acidic sugar derivatives are ionized at pH 7, and the compounds are therefore correctly named as the carboxylates—glucuronate, galacturonate, and so forth.

In the synthesis and metabolism of carbohydrates, the intermediates are very often not the sugars

---

BOX 7–1 🩺 MEDICINE   Blood Glucose Measurements in the Diagnosis and Treatment of Diabetes

Glucose is the principal fuel for the brain. When the amount of glucose reaching the brain is too low, the consequences can be dire: lethargy, coma, permanent brain damage, and death (see Fig. 23–24). Animals have evolved complex hormonal mechanisms to ensure that the concentration of glucose in the blood remains high enough (about 5 mM) to satisfy the brain's needs, but not too high, because elevated blood glucose can also have serious physiological consequences.

Individuals with insulin-dependent diabetes mellitus do not produce sufficient insulin, the hormone that normally serves to reduce blood glucose concentration, and if the diabetes is untreated their blood glucose levels may rise to severalfold higher than normal. These high glucose levels are believed to be at least one cause of the serious long-term consequences of untreated diabetes—kidney failure, cardiovascular disease, blindness, and impaired wound healing—so one goal of therapy is to provide just enough insulin (by injection) to keep blood glucose levels near normal. To maintain the correct balance of exercise, diet, and insulin for the individual, blood glucose concentration needs to be measured several times a day, and the amount of insulin injected adjusted appropriately.

The concentrations of glucose in blood and urine can be determined by a simple assay for reducing sugar, such as Fehling's reaction, which for many years was used as a diagnostic test for diabetes. Modern measurements require just a drop of blood, added to a test strip containing the enzyme glucose oxidase, which catalyzes the following reaction:

D-Glucose + O_2 $\xrightarrow{\text{glucose oxidase}}$

D-Glucono-δ-lactone + H_2O_2

A second enzyme, a peroxidase, catalyzes the reaction of the H_2O_2 with colorless compound to create a colored product, which is quantified with a simple photometer that reads out the blood glucose concentration.

Because blood glucose levels change with the timing of meals and exercise, single-time measurements do not reflect the *average* blood glucose over hours and days, so dangerous increases may go undetected. The average glucose concentration can be assessed by looking at its effect on hemoglobin, the oxygen-carrying protein in erythrocytes (p. 163). Transporters in the erythrocyte membrane equilibrate intracellular and plasma glucose concentrations, so hemoglobin is constantly exposed to glucose at whatever concentration is present in the blood. A nonenzymatic reaction occurs between glucose and primary amino groups in hemoglobin (either the amino-terminal Val or the ε-amino groups of Lys residues) (Fig. 1). The rate of this process is proportional to the concentration of glucose, so the reaction can be used as the basis for estimating the average blood glucose level over weeks. The amount of glycated hemoglobin (GHB) present at any time reflects the average blood glucose concentration over the circulating "lifetime" of the erythrocyte (about 120 days), although the concentration in the last two weeks is the most important in setting the level of GHB.

The extent of **hemoglobin glycation** (so named to distinguish it from glycosylation, the *enzymatic* transfer of glucose to a protein) is measured clinically by extracting hemoglobin from a small sample of blood and separating GHB from unmodified hemoglobin electrophoretically, taking advantage of the charge difference resulting from modification of the amino group(s). Normal GHB values are about 5% of total hemoglobin (corresponding to blood glucose of themselves but their phosphorylated derivatives. Condensation of phosphoric acid with one of the hydroxyl groups of a sugar forms a phosphate ester, as in glucose 6-phosphate (Fig. 7–9), the first metabolite in the pathway by which most organisms oxidize glucose for energy. Sugar phosphates are relatively stable at neutral pH and bear a negative charge. One effect of sugar phosphorylation within cells is to trap the sugar inside the cell; most cells do not have plasma membrane transporters for phosphorylated sugars. Phosphorylation also activates sugars for subsequent chemical transformation. Several important phosphorylated derivatives of sugars are components of nucleotides (discussed in the next chapter).

Monosaccharides Are Reducing Agents

Monosaccharides can be oxidized by relatively mild oxidizing agents such as cupric (Cu^{2+}) ion. The carbonyl carbon is oxidized to a carboxyl group. Glucose and other sugars capable of reducing cupric ion are called **reducing sugars**. Cupric ion oxidizes glucose and certain other sugars to a complex mixture of carboxylic acids. This is the basis of Fehling's reaction, a semi-quantitative test for the presence of reducing sugar that for many years was used to detect and measure elevated glucose levels in people with diabetes mellitus. Today, more sensitive methods that involve an immobilized enzyme on a test strip are used; they require only a single drop of blood (Box 7–1). ■

120 mg/100 mL). In people with untreated diabetes, however, this value may be as high as 13%, indicating an average blood glucose level of about 300 mg/100 mL—dangerously high. One criterion for success in an individual program of insulin therapy (the timing, frequency, and amount of insulin injected) is maintaining GHB values at about 7%.

In the hemoglobin glycation reaction, the first step (formation of a Schiff base) is followed by a series of rearrangements, oxidations, and dehydrations of the carbohydrate moiety to produce a heterogeneous mixture of AGEs, advanced glycation end products. These products can leave the erythrocyte and form covalent cross-links between proteins, interfering with normal protein function (Fig. 1). The accumulation of relatively high concentrations of AGEs in people with diabetes may, by cross-linking critical proteins, cause the damage to the kidneys, retinas, and cardiovascular system that characterizes the disease. This pathogenic process is a potential target for drug action.

FIGURE 1 The nonenzymatic reaction of glucose with a primary amino group in hemoglobin begins with ❶ formation of a Schiff base, which ❷ undergoes a rearrangement to generate a stable product; ❸ this ketoamine can further cyclize to yield GHB. ❹ Subsequent reactions generate advanced glycation end products (AGEs), such as ε-N-carboxymethyllysine and methylglyoxal, compounds that ❺ can damage other proteins by cross-linking them, causing pathological changes.

Disaccharides Contain a Glycosidic Bond

Disaccharides (such as maltose, lactose, and sucrose) consist of two monosaccharides joined covalently by an **O-glycosidic bond**, which is formed when a hydroxyl group of one sugar molecule, typically cyclic, reacts with the anomeric carbon of the other **(Fig. 7–10)**. This reaction represents the formation of an acetal from a hemiacetal (such as glucopyranose) and an alcohol (a hydroxyl group of the second sugar molecule) (Fig. 7–5), and the resulting compound is called a glycoside. Glycosidic bonds are readily hydrolyzed by acid but resist cleavage by base. Thus disaccharides can be hydrolyzed to yield their free monosaccharide components by boiling with dilute acid. **N-glycosyl bonds** join the anomeric carbon of a sugar to a nitrogen atom in glycoproteins (see Fig. 7–30) and nucleotides (see Fig. 8–1).

The oxidation of a sugar by cupric ion (the reaction that defines a reducing sugar) occurs only with the linear form, which exists in equilibrium with the cyclic form(s). When the anomeric carbon is involved in a glycosidic bond (that is, when the compound is a full acetal or ketal; see Fig. 7–5), the easy interconversion of linear and cyclic forms shown in Figure 7–6 is prevented. Because the carbonyl carbon can be oxidized only when the sugar is in its linear form, formation of a glycosidic bond renders a sugar nonreducing. In describing disaccharides or polysaccharides, the end of a chain with a free anomeric carbon (one not involved in a glycosidic bond) is commonly called the **reducing end**.

The disaccharide maltose (Fig. 7–10) contains two D-glucose residues joined by a glycosidic linkage between C-1 (the anomeric carbon) of one glucose residue and C-4 of the other. Because the disaccharide retains a free anomeric carbon (C-1 of the glucose residue on the right in Fig. 7–10), maltose is a reducing sugar. The configuration of the anomeric carbon atom in the glycosidic linkage is α. The glucose residue with the free anomeric carbon is capable of existing in α- and β-pyranose forms.

KEY CONVENTION: To name reducing disaccharides such as maltose unambiguously, and especially to name more complex oligosaccharides, several rules are followed. By convention, the name describes the compound written with its nonreducing end to the left, and we can "build up" the name in the following order. (1) Give the configuration (α or β) at the anomeric carbon joining the first monosaccharide unit (on the left) to the second. (2) Name the nonreducing residue; to distinguish five- and six-membered ring structures, insert "furano" or "pyrano" into the name. (3) Indicate in parentheses the two carbon atoms joined by the glycosidic bond, with an arrow connecting the two numbers; for example, (1→4) shows that C-1 of the first-named sugar residue is joined to C-4 of the second. (4) Name the second residue. If there is a third residue, describe the second glycosidic bond by the same conventions. (To shorten the description of complex polysaccharides, three-letter abbreviations or colored symbols for the monosaccharides are often used, as given in Table 7–1.) Following this convention for naming oligosaccharides, maltose is α-D-glucopyranosyl-(1→4)-D-glucopyranose. Because

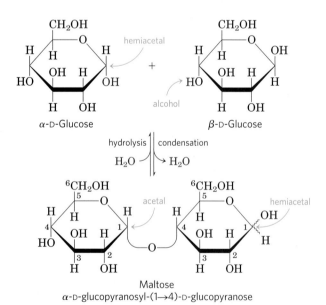

FIGURE 7–10 Formation of maltose. A disaccharide is formed from two monosaccharides (here, two molecules of D-glucose) when an —OH (alcohol) of one monosaccharide molecule (right) condenses with the intramolecular hemiacetal of the other (left), with elimination of H_2O and formation of a glycosidic bond. The reversal of this reaction is hydrolysis—attack by H_2O on the glycosidic bond. The maltose molecule shown here retains a reducing hemiacetal at the C-1 not involved in the glycosidic bond. Because mutarotation interconverts the α and β forms of the hemiacetal, the bonds at this position are sometimes depicted with wavy lines, as shown here, to indicate that the structure may be either α or β.

TABLE 7–1 Symbols and Abbreviations for Common Monosaccharides and Some of Their Derivatives

Abequose	Abe	Glucuronic acid	◆ GlcA
Arabinose	Ara	Galactosamine	□ GalN
Fructose	Fru	Glucosamine	◣ GlcN
Fucose	▲ Fuc	N-Acetylgalactosamine	▢ GalNAc
Galactose	○ Gal	N-Acetylglucosamine	■ GlcNAc
Glucose	● Glc	Iduronic acid	◇ IdoA
Mannose	● Man	Muramic acid	Mur
Rhamnose	Rha	N-Acetylmuramic acid	Mur2Ac
Ribose	Rib	N-Acetylneuraminic	
Xylose	★ Xyl	acid (a sialic acid)	◆ Neu5Ac

Note: In a commonly used convention, hexoses are represented as circles, N-acetylhexosamines as squares, and hexosamines as squares divided diagonally. All sugars with the "gluco" configuration are blue, those with the "galacto" configuration are yellow, and "manno" sugars are green. Other substituents can be added as needed: sulfate (S), phosphate (P), O-acetyl (OAc), or O-methyl (OMe).

most sugars encountered in this book are the D enantiomers and the pyranose form of hexoses predominates, we generally use a shortened version of the formal name of such compounds, giving the configuration of the anomeric carbon and naming the carbons joined by the glycosidic bond. In this abbreviated nomenclature, maltose is Glc($\alpha1\rightarrow4$)Glc. ■

The disaccharide lactose **(Fig. 7–11)**, which yields D-galactose and D-glucose on hydrolysis, occurs naturally in milk. The anomeric carbon of the glucose residue is available for oxidation, and thus lactose is a reducing disaccharide. Its abbreviated name is Gal($\beta1\rightarrow4$)Glc. Sucrose (table sugar) is a disaccharide of glucose and fructose. It is formed by plants but not by animals. In contrast to maltose and lactose, sucrose contains no free anomeric carbon atom; the anomeric carbons of both monosaccharide units are involved in the glycosidic bond (Fig. 7–11). Sucrose is therefore a nonreducing sugar, and its stability toward oxidation makes it a suitable molecule for the storage and transport of energy in plants. In the abbreviated nomenclature, a double-headed arrow connects the symbols

specifying the anomeric carbons and their configurations. For example, the abbreviated name of sucrose is either Glc($\alpha1\leftrightarrow2\beta$)Fru or Fru($\beta2\leftrightarrow1\alpha$)Glc. Sucrose is a major intermediate product of photosynthesis; in many plants it is the principal form in which sugar is transported from the leaves to other parts of the plant body. Trehalose, Glc($\alpha1\leftrightarrow1\alpha$)Glc (Fig. 7–11)—a disaccharide of D-glucose that, like sucrose, is a nonreducing sugar—is a major constituent of the circulating fluid (hemolymph) of insects, serving as an energy-storage compound. Lactose gives milk its sweetness, and sucrose, of course, is table sugar. Trehalose is also used commercially as a sweetener. Box 7–2 explains how humans detect sweetness, and how artificial sweeteners such as aspartame act.

SUMMARY 7.1 Monosaccharides and Disaccharides

▶ Sugars (also called saccharides) are compounds containing an aldehyde or ketone group and two or more hydroxyl groups.

▶ Monosaccharides generally contain several chiral carbons and therefore exist in a variety of stereochemical forms, which may be represented on paper as Fischer projections. Epimers are sugars that differ in configuration at only one carbon atom.

▶ Monosaccharides commonly form internal hemiacetals or hemiketals, in which the aldehyde or ketone group joins with a hydroxyl group of the same molecule, creating a cyclic structure; this can be represented as a Haworth perspective formula. The carbon atom originally found in the aldehyde or ketone group (the anomeric carbon) can assume either of two configurations, α and β, which are interconvertible by mutarotation. In the linear form of the monosaccharide, which is in equilibrium with the cyclic forms, the anomeric carbon is easily oxidized, making the compound a reducing sugar.

▶ A hydroxyl group of one monosaccharide can add to the anomeric carbon of a second monosaccharide to form an acetal called a glycoside. In this disaccharide, the glycosidic bond protects the anomeric carbon from oxidation, making it a nonreducing sugar.

▶ Oligosaccharides are short polymers of several monosaccharides joined by glycosidic bonds. At one end of the chain, the reducing end, is a monosaccharide unit with its anomeric carbon not involved in a glycosidic bond.

▶ The common nomenclature for di- or oligosaccharides specifies the order of monosaccharide units, the configuration at each anomeric carbon, and the carbon atoms involved in the glycosidic linkage(s).

FIGURE 7–11 Two common disaccharides. Like maltose in Figure 7–10, these are shown as Haworth perspectives. The common name, full systematic name, and abbreviation are given for each disaccharide. Formal nomenclature for sucrose names glucose as the parent glycoside, although it is typically depicted as shown, with glucose on the left. The two abbreviated symbols shown for sucrose are equivalent ($\equiv$).

BOX 7–2   Sugar Is Sweet, and So Are . . . a Few Other Things

Sweetness is one of the five basic flavors that humans can taste (Fig. 1); the others are sour, bitter, salty, and umami. Sweet taste is detected by protein receptors in the plasma membranes of gustatory cells in the taste buds on the surface of the tongue. In humans, two closely related genes (*T1R2* and *T1R3*) encode sweetness receptors (Fig. 2). When a molecule with a compatible structure binds these receptors on a gustatory cell's extracellular domain, it triggers a series of events in the cell (including activation of a GTP-binding protein; see Fig. 12–42) that lead to an electrical signal being sent to the brain that is interpreted there as "sweet." During evolution, there has probably been selection for the ability to taste compounds found in foods containing important nutrients, such as the carbohy-

drates that are major fuels for most organisms. Most simple sugars, including sucrose, glucose, and fructose,

FIGURE 2 The receptor for sweet-tasting substances, with its regions of interaction (short arrows) with various sweet-tasting compounds indicated. Each receptor has an extracellular domain, a cysteine-rich domain (CRD), and a membrane domain with seven transmembrane helices, a common feature of signaling receptors. Artificial sweeteners bind to only one of the two receptor subunits; natural sugars bind to both. See Chapter 1, Problem 14, for the structures of many of these artificial sweeteners.

FIGURE 1 A strong stimulus for the sweetness receptors.

7.2 Polysaccharides

Most carbohydrates found in nature occur as polysaccharides, polymers of medium to high molecular weight (M_r >20,000). Polysaccharides, also called **glycans**, differ from each other in the identity of their recurring monosaccharide units, in the length of their chains, in the types of bonds linking the units, and in the degree of branching. **Homopolysaccharides** contain only a single monomeric species; **heteropolysaccharides** contain two or more different kinds **(Fig. 7–12)**. Some homopolysaccharides serve as storage forms of monosaccharides that are used as fuels; starch and glycogen are homopolysaccharides of this type. Other homopolysaccharides (cellulose and chitin, for example) serve as structural elements in plant cell walls and animal exoskeletons. Heteropolysaccharides provide extracellular support for organisms of all kingdoms. For example, the rigid layer of the bacterial cell envelope (the peptidoglycan) is composed in part of a heteropolysaccharide built from two alternating monosaccharide units (see Fig. 20–30). In animal tissues, the extracellular space is

FIGURE 7–12 Homo- and heteropolysaccharides. Polysaccharides may be composed of one, two, or several different monosaccharides, in straight or branched chains of varying length.

taste sweet, but there are other classes of compounds that also bind the sweet receptors: the amino acids glycine, alanine, and serine are mildly sweet and harmless; nitrobenzene and ethylene glycol have a strong sweet taste, but are toxic. (See Box 18–2 for a remarkable medical mystery involving ethylene glycol poisoning.) Several natural products are extraordinarily sweet: stevioside, a sugar derivative isolated from the leaves of the stevia plant (*Stevia rebaudiana* Berton), is several hundred times sweeter than an equivalent amount of sucrose (table sugar), and the small (54 amino acids) protein brazzein, isolated from berries of the Oubli vine (*Pentadiplandra brazzeana* Baillon) in Gabon and Cameroon, is 17,000 times as sweet as sucrose on a molar basis. Presumably the sweet taste of the berries encourages their consumption by animals, which then disperse the seeds geographically so new plants may be established.

There is great interest in the development of artificial sweeteners as weight-reduction aids—compounds that give foods a sweet taste without adding the calories found in sugars. The artificial sweetener aspartame demonstrates the importance of stereochemistry in biology (Fig. 3). According to one simple model of sweetness receptor binding, binding involves three sites on the receptor: AH^+, B^-, and X. Site AH^+ contains some group (an alcohol or amine) that can form a hydrogen bond with a partial negative charge, such as a carbonyl oxygen, on the sweetener molecule; the carboxylic acid of aspartame contains such an oxy-

gen. Site B^- contains a group with a partially negative oxygen available to hydrogen-bond with some partially positive atom on the sweetener molecule, such as the amine group of aspartame. Site X is oriented perpendicular to the other two groups and is capable of interacting with a hydrophobic patch on the sweetener molecule, such as the benzene ring of aspartame.

When the steric match is correct, as on the left in Figure 3, the sweet receptor is stimulated and the signal "sweet" is conducted to the brain. When the match is not correct, as on the right in Figure 3, the sweet receptor is not stimulated; in fact, in this case, another receptor (for bitterness) is stimulated by the "wrong" stereoisomer of aspartame. Stereoisomerism really matters!

FIGURE 3 Stereochemical basis for the taste of two isomers of aspartame.

occupied by several types of heteropolysaccharides, which form a matrix that holds individual cells together and provides protection, shape, and support to cells, tissues, and organs.

Unlike proteins, polysaccharides generally do not have defining molecular weights. This difference is a consequence of the mechanisms of assembly of the two types of polymer. As we shall see in Chapter 27, proteins are synthesized on a template (messenger RNA) of defined sequence and length, by enzymes that follow the template exactly. For polysaccharide synthesis there is no template; rather, the program for polysaccharide synthesis is intrinsic to the enzymes that catalyze the polymerization of the monomeric units, and there is no specific stopping point in the synthetic process; the products thus vary in length.

Some Homopolysaccharides Are Stored Forms of Fuel

The most important storage polysaccharides are starch in plant cells and glycogen in animal cells. Both polysaccharides occur intracellularly as large clusters or granules.

Starch and glycogen molecules are heavily hydrated, because they have many exposed hydroxyl groups available to hydrogen-bond with water. Most plant cells have the ability to form starch (see Fig. 20–2), and starch storage is especially abundant in tubers (underground stems), such as potatoes, and in seeds.

Starch contains two types of glucose polymer, amylose and amylopectin **(Fig. 7–13)**. Amylose consists of long, unbranched chains of D-glucose residues connected by $(\alpha1{\rightarrow}4)$ linkages (as in maltose). Such chains vary in molecular weight from a few thousand to more than a million. Amylopectin also has a high molecular weight (up to 200 million) but unlike amylose is highly branched. The glycosidic linkages joining successive glucose residues in amylopectin chains are $(\alpha1{\rightarrow}4)$; the branch points (occurring every 24 to 30 residues) are $(\alpha1{\rightarrow}6)$ linkages.

Glycogen is the main storage polysaccharide of animal cells. Like amylopectin, glycogen is a polymer of $(\alpha1{\rightarrow}4)$-linked subunits of glucose, with $(\alpha1{\rightarrow}6)$-linked branches, but glycogen is more extensively branched (on average, every 8 to 12 residues) and more compact than

(a) Amylose

(b)

(c)

FIGURE 7–13 **Glycogen and starch.** **(a)** A short segment of amylose, a linear polymer of D-glucose residues in $(\alpha 1 \rightarrow 4)$ linkage. A single chain can contain several thousand glucose residues. Amylopectin has stretches of similarly linked residues between branch points. Glycogen has the same basic structure, but has more branching than amylopectin. **(b)** An $(\alpha 1 \rightarrow 6)$ branch point of glycogen or amylopectin. **(c)** A cluster of amylose and amylopectin like that believed to occur in starch granules. Strands of amylopectin (black) form double-helical structures with each other or with amylose strands (blue). Amylopectin has frequent $(\alpha 1 \rightarrow 6)$ branch points (red). Glucose residues at the nonreducing ends of the outer branches are removed enzymatically during the mobilization of starch for energy production. Glycogen has a similar structure but is more highly branched and more compact.

starch. Glycogen is especially abundant in the liver, where it may constitute as much as 7% of the wet weight; it is also present in skeletal muscle. In hepatocytes glycogen is found in large granules, which are themselves clusters of smaller granules composed of single, highly branched glycogen molecules with an average molecular weight of several million. Such glycogen granules also contain, in tightly bound form, the enzymes responsible for the synthesis and degradation of glycogen.

Because each branch in glycogen ends with a nonreducing sugar unit, a glycogen molecule with n branches has $n + 1$ nonreducing ends, but only one reducing end. When glycogen is used as an energy source, glucose units are removed one at a time from the nonreducing ends. Degradative enzymes that act only at nonreducing ends can work simultaneously on the many branches, speeding the conversion of the polymer to monosaccharides.

Why not store glucose in its monomeric form? It has been calculated that hepatocytes store glycogen equivalent to a glucose concentration of 0.4 M. The actual concentration of glycogen, which is insoluble and contributes little to the osmolarity of the cytosol, is about 0.01 μM. If the cytosol contained 0.4 M glucose, the osmolarity would be threateningly elevated, leading to osmotic entry of water that might rupture the cell (see Fig. 2–13). Furthermore, with an intracellular glucose

concentration of 0.4 M and an external concentration of about 5 mM (the concentration in the blood of a mammal), the free-energy change for glucose uptake into cells against this very high concentration gradient would be prohibitively large.

Dextrans are bacterial and yeast polysaccharides made up of $(\alpha 1 \rightarrow 6)$-linked poly-D-glucose; all have $(\alpha 1 \rightarrow 3)$ branches, and some also have $(\alpha 1 \rightarrow 2)$ or $(\alpha 1 \rightarrow 4)$ branches. Dental plaque, formed by bacteria growing on the surface of teeth, is rich in dextrans, which are adhesive and allow the bacteria to stick to teeth and to each other. Dextrans also provide a source of glucose for bacterial metabolism. Synthetic dextrans are used in several commercial products (for example, Sephadex) that serve in the fractionation of proteins by size-exclusion chromatography (see Fig. 3–17b). The dextrans in these products are chemically cross-linked to form insoluble materials of various sizes.

Some Homopolysaccharides Serve Structural Roles

Cellulose, a fibrous, tough, water-insoluble substance, is found in the cell walls of plants, particularly in stalks, stems, trunks, and all the woody portions of the plant body. Cellulose constitutes much of the mass of wood, and cotton is almost pure cellulose. Like amylose, the

(β1→4)-linked D-glucose units

FIGURE 7–14 Cellulose. Two units of a cellulose chain; the D-glucose residues are in (β1→4) linkage. The rigid chair structures can rotate relative to one another.

cellulose molecule is a linear, unbranched homopolysaccharide, consisting of 10,000 to 15,000 D-glucose units. But there is a very important difference: in cellulose the glucose residues have the β configuration (**Fig. 7–14**), whereas in amylose the glucose is in the α configuration. The glucose residues in cellulose are linked by (β1→4) glycosidic bonds, in contrast to the (α1→4) bonds of amylose. This difference causes individual molecules of cellulose and amylose to fold differently in space, giving them very different macroscopic structures and physical properties (see below). The tough, fibrous nature of cellulose makes it useful in such commercial products as cardboard and insulation material, and it is a major constituent of cotton and linen fabrics. Cellulose is also the starting material for the commercial production of cellophane and rayon.

Glycogen and starch ingested in the diet are hydrolyzed by α-amylases and glycosidases, enzymes in saliva and the intestine that break (α1→4) glycosidic bonds between glucose units. Most vertebrate animals cannot use cellulose as a fuel source, because they lack an enzyme to hydrolyze the (β1→4) linkages. Termites readily digest cellulose (and therefore wood), but only because their intestinal tract harbors a symbiotic microorganism, *Trichonympha*, that secretes cellulase, which hydrolyzes the (β1→4) linkages (**Fig. 7–15**). Molecular

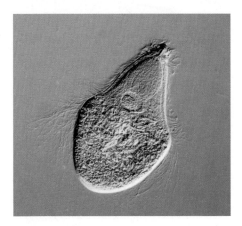

FIGURE 7–15 Cellulose breakdown by *Trichonympha*, a protist in the gut of a wood-eating termite. *Trichonympha* produces the enzyme cellulase, which breaks the (β1→4) glycosidic bonds in cellulose, making wood a source of metabolizable sugar (glucose) for the protist and the termite. A number of invertebrates can digest cellulose, but only a few vertebrates (the ruminants, such as cattle, sheep, and goats); the ruminants are able to use cellulose as food because the first of their four stomach compartments (rumen) teems with bacteria and protists that secrete cellulase.

genetic studies have revealed that genes encoding cellulose-degrading enzymes are present in the genomes of a wide range of invertebrate animals, including arthropods and nematodes. There is one important exception to the absence of cellulase in vertebrates: ruminant animals such as cattle, sheep, and goats harbor symbiotic microorganisms in the rumen (the first of their four stomach compartments) that can hydrolyze cellulose, allowing the animal to degrade dietary cellulose from soft grasses, but not from woody plants. Fermentation in the rumen yields acetate, propionate, and β-hydroxybutyrate, which the animal uses to synthesize the sugars in milk (p. 560).

Biomass (such as switch grass) that is rich in cellulose can be used as starting material for the fermentation of carbohydrates to ethanol, to be used as a gasoline additive. The annual production of biomass on Earth (accomplished primarily by photosynthetic organisms) is the energetic equivalent of nearly a trillion barrels of crude oil, when converted to ethanol by fermentation. Because of their potential use in biomass conversion to bioenergy, cellulose-degrading enzymes such as cellulase are under vigorous investigation. Supramolecular complexes called cellulosomes, found on the outside surface of the bacterium *Clostridium cellulolyticum*, include the catalytic subunit of cellulase, along with proteins that hold one or more cellulase molecules to the bacterial surface, and a subunit that binds cellulose and positions it in the catalytic site.

A major fraction of photosynthetic biomass is the woody portion of plants and trees, which consists of cellulose plus several other polymers derived from carbohydrates that are not easily digestible, either chemically or biologically. Lignins, for example, make up some 30% of the mass of wood. Synthesized from precursors that include phenylalanine and glucose, lignins are complex polymers with covalent cross-links to cellulose that complicate the digestion of cellulose by cellulase. If woody plants are to be used in the production of ethanol from biomass, better means of digesting wood components will need to be found.

Chitin is a linear homopolysaccharide composed of *N*-acetylglucosamine residues in (β1→4) linkage (**Fig. 7–16**). The only chemical difference from cellulose is the replacement of the hydroxyl group at C-2 with an acetylated amino group. Chitin forms extended fibers similar to those of cellulose, and like cellulose cannot be digested by vertebrates. Chitin is the principal component of the hard exoskeletons of nearly a million species of arthropods—insects, lobsters, and crabs, for example—and is probably the second most abundant polysaccharide, next to cellulose, in nature; an estimated 1 billion tons of chitin are produced each year in the biosphere.

Steric Factors and Hydrogen Bonding Influence Homopolysaccharide Folding

The folding of polysaccharides in three dimensions follows the same principles as those governing polypeptide

(a)

(b)

FIGURE 7-16 Chitin. (a) A short segment of chitin, a homopolymer of N-acetyl-D-glucosamine units in (β1→4) linkage. **(b)** A spotted June beetle (*Pelidnota punctata*), showing its surface armor (exoskeleton) of chitin.

structure: subunits with a more-or-less rigid structure dictated by covalent bonds form three-dimensional macromolecular structures that are stabilized by weak interactions within or between molecules, such as hydrogen bonds and hydrophobic and van der Waals interactions, and, for polymers with charged subunits, electrostatic interactions. Because polysaccharides have so many hydroxyl groups, hydrogen bonding has an especially important influence on their structure. Glycogen, starch, and cellulose are composed of pyranoside subunits (having six-membered rings), as are the oligosaccharides of glycoproteins and glycolipids, to be discussed later. Such molecules can be represented as a series of rigid pyranose rings connected by an oxygen atom bridging two carbon atoms (the glycosidic bond). There is, in principle, free rotation about both C—O bonds linking the residues (Fig. 7–14), but as in polypeptides (see Figs 4–2, 4–9), rotation about each bond is limited by steric hindrance by substituents. The three-dimensional structures of these molecules can be described in terms of the dihedral angles, ϕ and ψ, about the glycosidic bond **(Fig. 7–17)**, analogous to angles ϕ and ψ made by the peptide bond (see Fig. 4–2).

The bulkiness of the pyranose ring and its substituents, and electronic effects at the anomeric carbon, place constraints on the angles ϕ and ψ; thus certain conformations are much more stable than others, as can be shown on a map of energy as a function of ϕ and ψ **(Fig. 7–18)**.

The most stable three-dimensional structure for the (α1→4)-linked chains of starch and glycogen is a tightly coiled helix **(Fig. 7–19)**, stabilized by interchain hydrogen bonds. In amylose (with no branches) this structure is regular enough to allow crystallization and

thus determination of the structure by x-ray diffraction. The average plane of each residue along the amylose chain forms a 60° angle with the average plane of the

FIGURE 7-17 Conformation at the glycosidic bonds of cellulose, amylose, and dextran. The polymers are depicted as rigid pyranose rings joined by glycosidic bonds, with free rotation about these bonds. Note that in dextran there is also free rotation about the bond between C-5 and C-6 (torsion angle ω (omega)).

$\phi,\psi = 30°, -40°$

(a)

(b) $\phi,\psi = -170°, -170°$

FIGURE 7–18 **A map of favored conformations for oligosaccharides and polysaccharides.** The torsion angles ψ and ϕ (see Fig. 7–17), which define the spatial relationship between adjacent rings, can in principle have any value from 0° to 360°. In fact, some of the torsion angles would give conformations that are sterically hindered, whereas others give conformations that maximize hydrogen bonding. **(a)** When the relative energy (Σ) is plotted for each value of ϕ and ψ, with isoenergy ("same energy") contours drawn at intervals of 1 kcal/mol above the minimum energy state, the result is a map of preferred conformations. This is analogous to the Ramachandran plot for peptides (see Figs 4-3, 4-9). **(b)** Two energetic extremes for the disaccharide Gal($\beta1 \rightarrow 3$)Gal; these values fall on the energy diagram (a) as shown by the red and blue dots. The red dot indicates the least favored conformation; the blue dot, the most favored conformation. The known conformations of the three polysaccharides shown in Figure 7-17 have been determined by x-ray crystallography, and all fall within the lowest-energy regions of the map.

preceding residue, so the helical structure has six residues per turn. For amylose, the core of the helix is of precisely the right dimensions to accommodate iodine as complex ions (I_3^- and I_5^-), giving an intensely blue complex. This interaction is a common qualitative test for amylose.

For cellulose, the most stable conformation is that in which each chair is turned 180° relative to its neighbors, yielding a straight, extended chain. All —OH groups are available for hydrogen bonding with neighboring chains. With several chains lying side by side, a stabilizing network of interchain and intrachain hydrogen bonds produces straight, stable supramolecular fibers of great tensile strength **(Fig. 7–20)**. This property of cellulose has made it a useful substance to civilizations for millennia. Many manufactured products, including papyrus, paper, cardboard, rayon, insulating tiles, and a variety of other useful materials, are derived from cellulose. The water content of these materials is low because extensive interchain hydrogen bonding between cellulose molecules satisfies their capacity for hydrogen-bond formation.

Bacterial and Algal Cell Walls Contain Structural Heteropolysaccharides

The rigid component of bacterial cell walls (peptidoglycan) is a heteropolymer of alternating ($\beta1 \rightarrow 4$)-linked *N*-acetylglucosamine and *N*-acetylmuramic acid residues (see Fig. 20–30). The linear polymers lie side by side in the cell wall, cross-linked by short peptides, the exact structure of which depends on the bacterial species. The peptide cross-links weld the polysaccharide chains into a strong sheath (peptidoglycan) that envelops the entire cell and prevents cellular swelling and lysis due to the osmotic entry of water. The enzyme

($\alpha1 \rightarrow 4$)-linked
D-glucose units

(a)

(b)

FIGURE 7–19 **Helical structure of starch (amylose).** **(a)** In the most stable conformation, with adjacent rigid chairs, the polysaccharide chain is curved, rather than linear as in cellulose (see Fig. 7-14). **(b)** A model of a segment of amylose; for clarity, the hydroxyl groups have been omitted from all but one of the glucose residues. Compare the two residues shaded in pink with the chemical structures in (a). The conformation of ($\alpha1 \rightarrow 4$) linkages in amylose, amylopectin, and glycogen causes these polymers to assume tightly coiled helical structures. These compact structures produce the dense granules of stored starch or glycogen seen in many cells (see Fig. 20-2).

FIGURE 7–20 Cellulose chains. Scale drawing of segments of two parallel cellulose chains, showing the conformation of the D-glucose residues and the hydrogen-bond cross-links. In the hexose unit at the lower left, all hydrogen atoms are shown; in the other three hexose units, the hydrogens attached to carbon have been omitted for clarity, as they do not participate in hydrogen bonding.

lysozyme kills bacteria by hydrolyzing the ($\beta1\rightarrow4$) glycosidic bond between *N*-acetylglucosamine and *N*-acetylmuramic acid (see Fig. 6–27); the enzyme is found in human tears, where it is presumably a defense against bacterial infections of the eye, and is also produced by certain bacterial viruses to ensure their release from the host bacterial cell, an essential step of the viral infection cycle. Penicillin and related antibiotics kill bacteria by preventing synthesis of the cross-links, leaving the cell wall too weak to resist osmotic lysis (p. 224).

Certain marine red algae, including some of the seaweeds, have cell walls that contain **agar**, a mixture of sulfated heteropolysaccharides made up of D-galactose and an L-galactose derivative ether-linked between C-3 and C-6. Agar is a complex mixture of polysaccharides, all with the same backbone structure but substituted to varying degrees with sulfate and pyruvate. **Agarose** (M_r ~150,000) is the agar component with the fewest charged groups (sulfates, pyruvates) **(Fig. 7–21)**. The remarkable gel-forming property of agarose makes it useful in the biochemistry laboratory. When a suspension of agarose in water is heated and cooled, the agarose forms a double helix: two molecules in parallel orientation twist together with a helix repeat of three residues; water molecules are trapped in the central cavity. These structures in turn associate with each other to form a gel—a three-dimensional matrix that traps large amounts of water. Agarose gels are used as inert supports for the electrophoretic separation of nucleic acids, an essential part of the DNA-sequencing process (p. 302). Agar is also used to form a surface for the growth of bacterial colonies. Another commercial use of agar is for the capsules in which some vitamins and drugs are packaged; the dried agar material dissolves readily in the stomach and is metabolically inert.

Agarose
3)D-Gal($\beta1\rightarrow4$)3,6-anhydro-L-Gal2S ($\alpha1$
repeating units

FIGURE 7–21 Agarose. The repeating unit consists of D-galactose ($\beta1\rightarrow4$)-linked to 3,6-anhydro-L-galactose (in which an ether bridge connects C-3 and C-6). These units are joined by ($\alpha1\rightarrow3$) glycosidic links to form a polymer 600 to 700 residues long. A small fraction of the 3,6-anhydrogalactose residues have a sulfate ester at C-2 (as shown here). The open parentheses in the systematic name indicate that the repeating unit extends from both ends.

Glycosaminoglycans Are Heteropolysaccharides of the Extracellular Matrix

The extracellular space in the tissues of multicellular animals is filled with a gel-like material, the **extracellular matrix (ECM)**, also called ground substance, which holds the cells together and provides a porous pathway for the diffusion of nutrients and oxygen to individual cells. The ECM that surrounds fibroblasts and other connective tissue cells is composed of an interlocking meshwork of heteropolysaccharides and fibrous proteins such as fibrillar collagens, elastins, and fibronectins. Basement membrane is a specialized ECM that underlies epithelial cells; it consists of specialized collagens, laminins, and heteropolysaccharides. These heteropolysaccharides, the **glycosaminoglycans**, are a family of linear polymers composed of repeating disaccharide units **(Fig. 7–22)**. They are unique to animals and bacteria and are not found in plants. One of the two monosaccharides is always either *N*-acetylglucosamine or *N*-acetylgalactosamine; the other is in most cases a uronic acid, usually D-glucuronic or L-iduronic acid. Some glycosaminoglycans contain esterified sulfate groups. The combination of sulfate groups and the carboxylate groups of the uronic acid residues gives glycosaminoglycans a very high density of negative charge. To minimize the repulsive forces among neighboring charged groups, these molecules assume an extended conformation in solution, forming a rodlike helix in which the negatively charged carboxylate groups occur on alternate sides of the helix (as shown for heparin in Fig. 7–22). The extended rod form also provides maximum separation between the negatively charged sulfate groups. The specific patterns of sulfated and nonsulfated sugar residues in glycosaminoglycans provide for specific recognition by a variety of protein ligands that bind electrostatically to these molecules. The sulfated glycosaminoglycans are attached to extracellular proteins to form proteoglycans (Section 7.3).

The glycosaminoglycan **hyaluronan** (hyaluronic acid) contains alternating residues of D-glucuronic acid and *N*-acetylglucosamine (Fig. 7–22). With up to 50,000 repeats of the basic disaccharide unit, hyaluronan has a

Glycosaminoglycan	Repeating disaccharide
Number of disaccharides per chain	

Hyaluronan
~50,000

GlcA   GlcNAc

Chondroitin 4-sulfate
20–60

GlcA   GalNAc4S

Keratan sulfate
~25

Gal   GlcNAc6S

Heparin
15–90

IdoA2S   GlcNS3S6S

Heparin segment

FIGURE 7–22 Repeating units of some common glycosaminoglycans of extracellular matrix. The molecules are copolymers of alternating uronic acid and amino sugar residues (keratan sulfate is the exception), with sulfate esters in any of several positions, except in hyaluronan. The ionized carboxylate and sulfate groups (red in the perspective formulas) give these polymers their characteristic high negative charge. Therapeutic heparin contains primarily iduronic acid (IdoA) and a smaller proportion of glucuronic acid (GlcA, not shown), and is generally highly sulfated and heterogeneous in length. The space-filling model shows a heparin segment as its solution structure, as determined by NMR spectroscopy (PDB ID 1HPN). The carbons in the iduronic acid sulfate are colored blue; those in glucosamine sulfate are green. Oxygen and sulfur atoms are shown in their standard colors of red and yellow, respectively. The hydrogen atoms are not shown (for clarity). Heparan sulfate (not shown) is similar to heparin but has a higher proportion of GlcA and fewer sulfate groups, arranged in a less regular pattern.

some pathogenic bacteria, can hydrolyze the glycosidic linkages of hyaluronan, rendering tissues more susceptible to bacterial invasion. In many animal species, a similar enzyme in sperm hydrolyzes an outer glycosaminoglycan coat around the ovum, allowing sperm penetration.

Other glycosaminoglycans differ from hyaluronan in three respects: they are generally much shorter polymers, they are covalently linked to specific proteins (proteoglycans), and one or both monomeric units differ from those of hyaluronan. **Chondroitin sulfate** (Greek *chondros*, "cartilage") contributes to the tensile strength of cartilage, tendons, ligaments, and the walls of the aorta. Dermatan sulfate (Greek *derma*, "skin") contributes to the pliability of skin and is also present in blood vessels and heart valves. In this polymer, many of the glucuronate residues present in chondroitin sulfate are replaced by their 5-epimer, L-iduronate (IdoA).

α-L-Iduronate
(IdoA)

β-D-Glucuronate
(GlcA)

Keratan sulfates (Greek *keras*, "horn") have no uronic acid and their sulfate content is variable. They are present in cornea, cartilage, bone, and a variety of horny structures formed of dead cells: horn, hair, hoofs, nails, and claws. **Heparan sulfate** (Greek *hēpar*, "liver"; it was originally isolated from dog liver) is produced by all animal cells and contains variable arrangements of sulfated and nonsulfated sugars. The sulfated segments of the chain allow it to interact with a large number of proteins, including growth factors and ECM components, as well as various enzymes and factors present in plasma. Heparin is a fractionated form of heparan sulfate derived mostly from mast cells (a type of leukocyte). Heparin is a therapeutic agent used to inhibit coagulation through its capacity to bind the protease inhibitor antithrombin.

molecular weight of several million; it forms clear, highly viscous solutions that serve as lubricants in the synovial fluid of joints and give the vitreous humor of the vertebrate eye its jellylike consistency (the Greek *hyalos* means "glass"; hyaluronan can have a glassy or translucent appearance). Hyaluronan is also a component of the extracellular matrix of cartilage and tendons, to which it contributes tensile strength and elasticity as a result of its strong noncovalent interactions with other components of the matrix. Hyaluronidase, an enzyme secreted by

Glycosaminoglycan
(heparin)

FIGURE 7–23 Interaction between a glycosaminoglycan and its binding protein. Fibroblast growth factor 1 (FGF1), its cell surface receptor (FGFR), and a short segment of a glycosaminoglycan (heparin) were co-crystallized to yield the structure shown here (PDB ID 1EOO). The proteins are represented as surface contour images, with color to represent surface electrostatic potential: red, predominantly negative charge; blue, predominantly positive charge. Heparin is shown in a ball-and-stick representation, with the negative charges ($-SO_3^-$ and $-COO^-$) attracted to the positive (blue) surface of the FGF1 protein. Heparin was used in this experiment, but the glycosaminoglycan that binds FGF1 in vivo is heparan sulfate on the cell surface.

Heparin binding causes antithrombin to bind to and inhibit thrombin, a protease essential to blood clotting. The interaction is strongly electrostatic; heparin has the highest negative charge density of any known biological macromolecule **(Fig. 7–23)**. Purified heparin is routinely added to blood samples obtained for clinical analysis, and to blood donated for transfusion, to prevent clotting.

Table 7–2 summarizes the composition, properties, roles, and occurrence of the polysaccharides described in Section 7.2.

TABLE 7–2   Structures and Roles of Some Polysaccharides

Polymer	Type*	Repeating unit†	Size (number of monosaccharide units)	Roles/significance
Starch				Energy storage: in plants
Amylose	Homo-	$(\alpha1{\rightarrow}4)$Glc, linear	50–5,000	
Amylopectin	Homo-	$(\alpha1{\rightarrow}4)$Glc, with $(\alpha1{\rightarrow}6)$Glc branches every 24–30 residues	Up to 10^6	
Glycogen	Homo-	$(\alpha1{\rightarrow}4)$Glc, with $(\alpha1{\rightarrow}6)$Glc branches every 8–12 residues	Up to 50,000	Energy storage: in bacteria and animal cells
Cellulose	Homo-	$(\beta1{\rightarrow}4)$Glc	Up to 15,000	Structural: in plants, gives rigidity and strength to cell walls
Chitin	Homo-	$(\beta1{\rightarrow}4)$GlcNAc	Very large	Structural: in insects, spiders, crustaceans, gives rigidity and strength to exoskeletons
Dextran	Homo-	$(\alpha1{\rightarrow}6)$Glc, with $(\alpha1{\rightarrow}3)$ branches	Wide range	Structural: in bacteria, extracellular adhesive
Peptidoglycan	Hetero-; peptides attached	4)Mur2Ac$(\beta1{\rightarrow}4)$ GlcNAc$(\beta1$	Very large	Structural: in bacteria, gives rigidity and strength to cell envelope
Agarose	Hetero-	3)D-Gal$(\beta1{\rightarrow}4)$3,6-anhydro-L-Gal$(\alpha1$	1,000	Structural: in algae, cell wall material
Hyaluronan (a glycosaminoglycan)	Hetero-; acidic	4)GlcA$(\beta1{\rightarrow}3)$ GlcNAc$(\beta1$	Up to 100,000	Structural: in vertebrates, extracellular matrix of skin and connective tissue; viscosity and lubrication in joints

*Each polymer is classified as a homopolysaccharide (homo-) or heteropolysaccharide (hetero-).

†The abbreviated names for the peptidoglycan, agarose, and hyaluronan repeating units indicate that the polymer contains repeats of this disaccharide unit. For example, in peptidoglycan, the GlcNAc of one disaccharide unit is $(\beta1{\rightarrow}4)$-linked to the first residue of the next disaccharide unit.

SUMMARY 7.2 Polysaccharides

▸ Polysaccharides (glycans) serve as stored fuel and as structural components of cell walls and extracellular matrix.

▸ The homopolysaccharides starch and glycogen are stored fuels in plant, animal, and bacterial cells. They consist of D-glucose with ($\alpha1\rightarrow4$) linkages, and both contain some branches.

▸ The homopolysaccharides cellulose, chitin, and dextran serve structural roles. Cellulose, composed of ($\beta1\rightarrow4$)-linked D-glucose residues, lends strength and rigidity to plant cell walls. Chitin, a polymer of ($\beta1\rightarrow4$)-linked N-acetylglucosamine, strengthens the exoskeletons of arthropods. Dextran forms an adhesive coat around certain bacteria.

▸ Homopolysaccharides fold in three dimensions. The chair form of the pyranose ring is essentially rigid, so the conformation of the polymers is determined by rotation about the bonds from the rings to the oxygen atom in the glycosidic linkage. Starch and glycogen form helical structures with intrachain hydrogen bonding; cellulose and chitin form long, straight strands that interact with neighboring strands.

▸ Bacterial and algal cell walls are strengthened by heteropolysaccharides—peptidoglycan in bacteria, agar in red algae. The repeating disaccharide in peptidoglycan is GlcNAc($\beta1\rightarrow4$)Mur2Ac; in agar, it is D-Gal($\beta1\rightarrow4$)3,6-anhydro-L-Gal.

▸ Glycosaminoglycans are extracellular heteropolysaccharides in which one of the two monosaccharide units is a uronic acid (keratin sulfate is an exception) and the other an N-acetylated amino sugar. Sulfate esters on some of the hydroxyl groups and on the amino group of some glucosamine residues in heparin and in heparan sulfate give these polymers a high density of negative charge, forcing them to assume extended conformations. These polymers (hyaluronan, chondroitin sulfate, dermatan sulfate, and keratan sulfate) provide viscosity, adhesiveness, and tensile strength to the extracellular matrix.

7.3 Glycoconjugates: Proteoglycans, Glycoproteins, and Glycosphingolipids

In addition to their important roles as stored fuels (starch, glycogen, dextran) and as structural materials (cellulose, chitin, peptidoglycans), polysaccharides and oligosaccharides are information carriers. Some provide communication between cells and their extracellular surroundings; others label proteins for transport to and localization in specific organelles, or for destruction when the protein is malformed or superfluous; and others serve as recognition sites for extracellular signal molecules (growth factors, for example) or extracellular parasites (bacteria or viruses). On almost every eukaryotic cell, specific oligosaccharide chains attached to components of the plasma membrane form a carbohydrate layer (the glycocalyx), several nanometers thick, that serves as an information-rich surface that the cell shows to its surroundings. These oligosaccharides are central players in cell-cell recognition and adhesion, cell migration during development, blood clotting, the immune response, wound healing, and other cellular processes. In most of these cases, the informational carbohydrate is covalently joined to a protein or a lipid to form a **glycoconjugate**, which is the biologically active molecule **(Fig. 7–24)**.

Proteoglycans are macromolecules of the cell surface or extracellular matrix in which one or more sulfated glycosaminoglycan chains are joined covalently to a membrane protein or a secreted protein. The glycosaminoglycan chain can bind to extracellular proteins through electrostatic interactions between the protein and the negatively charged sugar moieties on the proteoglycan. Proteoglycans are major components of all extracellular matrices.

Glycoproteins have one or several oligosaccharides of varying complexity joined covalently to a protein. They are usually found on the outer face of the plasma membrane (as part of the glycocalyx), in the extracellular matrix, and in the blood. Inside cells they are found in specific organelles such as Golgi complexes,

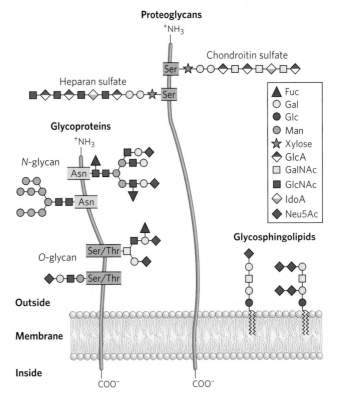

FIGURE 7–24 Glycoconjugates. The structures of some typical proteoglycans, glycoproteins, and glycosphingolipids described in the text.

secretory granules, and lysosomes. The oligosaccharide portions of glycoproteins are very heterogeneous and, like glycosaminoglycans, they are rich in information, forming highly specific sites for recognition and high-affinity binding by carbohydrate-binding proteins called lectins. Some cytosolic and nuclear proteins can be glycosylated as well.

Glycosphingolipids are plasma membrane components in which the hydrophilic head groups are oligosaccharides. As in glycoproteins, the oligosaccharides act as specific sites for recognition by lectins. The brain and neurons are rich in glycosphingolipids, which help in nerve conduction and myelin formation. Glycosphingolipids also play a role in signal transduction in cells. Sphingolipids are considered in more detail in Chapters 10 and 11.

Proteoglycans Are Glycosaminoglycan-Containing Macromolecules of the Cell Surface and Extracellular Matrix

Mammalian cells can produce 40 types of proteoglycans. These molecules act as tissue organizers, and they influence various cellular activities, such as growth factor activation and adhesion. The basic proteoglycan unit consists of a "core protein" with covalently attached glycosaminoglycan(s). The point of attachment is a Ser residue, to which the glycosaminoglycan is joined through a tetrasaccharide bridge (**Fig. 7–25**). The Ser residue is generally in the sequence –Ser–Gly–X–Gly– (where X is any amino acid residue), although not every protein with this sequence has an attached glycosaminoglycan.

Many proteoglycans are secreted into the extracellular matrix, but some are integral membrane proteins (see Fig. 11–7). For example, the sheetlike extracellular matrix (basal lamina) that separates organized groups of cells from other groups contains a family of core proteins (M_r 20,000 to 40,000), each with several covalently attached heparan sulfate chains. There are two major families of membrane heparan sulfate proteoglycans. **Syndecans** have a single transmembrane domain and an extracellular domain bearing three to five chains of heparan sulfate and in some cases chondroitin sulfate (**Fig. 7–26a**). **Glypicans** are attached to the membrane by a lipid anchor, a derivative of the membrane lipid phosphatidylinositol (see Fig. 11–15). Both syndecans and glypicans can be shed into the extracellular space. A protease in the ECM that cuts close to the membrane surface releases syndecan ectodomains (those domains outside the plasma membrane), and a phospholipase that breaks the connection

FIGURE 7–26 Two families of membrane proteoglycans. (a) Schematic diagrams of a syndecan and a glypican in the plasma membrane. Syndecans are held in the membrane by hydrophobic interactions between a sequence of nonpolar amino acid residues and plasma membrane lipids; they can be released by a single proteolytic cut near the membrane surface. In a typical syndecan, the extracellular amino-terminal domain is covalently attached (by tetrasaccharide linkers such as those in Fig. 7-25) to three heparan sulfate chains and two chondroitin sulfate chains. Glypicans are held in the membrane by a covalently attached membrane lipid (GPI anchor; see Fig. 11-15), but are shed if the bond between the lipid portion of the GPI anchor (phosphatidylinositol) and the oligosaccharide linked to the protein is cleaved by a phospholipase. All glypicans have 14 conserved Cys residues, which form disulfide bonds to stabilize the protein moiety, and either two or three glycosaminoglycan chains attached near the carboxyl terminus, close to the membrane surface. **(b)** Along a heparan sulfate chain, regions rich in sulfated sugars, the NS domains (green), alternate with regions with chiefly unmodified residues of GlcNAc and GlcA, the NA domains (gray). One of the NS domains is shown in more detail, revealing a high density of modified residues: GlcNS (N-sulfoglucosamine), with a sulfate ester at C-6; and both GlcA and IdoA, with a sulfate ester at C-2. The exact pattern of sulfation in the NS domain differs among proteoglycans.

FIGURE 7–25 Proteoglycan structure, showing the tetrasaccharide bridge. A typical tetrasaccharide linker (blue) connects a glycosaminoglycan—in this case chondroitin 4-sulfate (orange)—to a Ser residue in the core protein. The xylose residue at the reducing end of the linker is joined by its anomeric carbon to the hydroxyl of the Ser residue.

to the membrane lipid releases glypicans. These mechanisms provide a way for a cell to change its surface features quickly. Shedding is highly regulated and is activated in proliferating cells, such as cancer cells. Proteoglycan shedding is involved in cell-cell recognition and adhesion, and in the proliferation and differentiation of cells. Numerous chondroitin sulfate and dermatan sulfate proteoglycans also exist, some as membrane-bound entities, others as secreted products in the ECM.

The glycosaminoglycan chains can bind to a variety of extracellular ligands and thereby modulate the ligands' interaction with specific receptors of the cell surface. Detailed studies of heparan sulfate demonstrate a domain structure that is not random; some domains (typically 3 to 8 disaccharide units long) differ from neighboring domains in sequence and in ability to bind to specific proteins. Highly sulfated domains (called NS domains) alternate with domains having unmodified GlcNAc and GlcA residues (*N*-acetylated, or NA, domains) (Fig. 7–26b). The exact pattern of sulfation in the NS domain depends on the particular proteoglycan; given the number of possible modifications of the GlcNAc–IdoA (iduronic acid) dimer, at least 32 different disaccharide units are possible. Furthermore, the same core protein can display different heparan sulfate structures when synthesized in different cell types.

Heparan sulfate molecules with precisely organized NS domains bind specifically to extracellular proteins and signaling molecules to alter their activities. The change in activity may result from a conformational change in the protein that is induced by the binding **(Fig. 7–27a)**, or it may be due to the ability of adjacent domains of heparan sulfate to bind to two different proteins, bringing them into close proximity and enhancing protein-protein interactions (Fig. 7–27b). A third general mechanism of action is the binding of extracellular signal molecules (growth factors, for example) to heparan sulfate, which increases their local concentrations and enhances their interaction with growth factor receptors in the cell surface; in this case, the heparan sulfate acts as a coreceptor (Fig. 7–27c). For example, fibroblast growth factor (FGF), an extracellular protein signal that stimulates cell division, first binds to heparan sulfate moieties of syndecan molecules in the target cell's plasma membrane. Syndecan presents FGF to the FGF plasma membrane receptor, and only then can FGF interact productively with its receptor to trigger cell division. Finally, in another type of mechanism, the NS domains interact—electrostatically and otherwise—with a variety of soluble molecules outside the cell, maintaining high local concentrations at the cell surface (Fig. 7–27d).

The importance of correctly synthesizing sulfated domains in heparan sulfate is demonstrated in mutant

(a) Conformational activation

A conformational change induced in the protein antithrombin (AT) on binding a specific pentasaccharide NS domain allows its interaction with blood clotting factor Xa, preventing clotting.

(b) Enhanced protein-protein interaction

Binding of AT and thrombin to two adjacent NS domains brings the two proteins into close proximity, favoring their interaction, which inhibits blood clotting.

(c) Coreceptor for extracellular ligands

NS domains interact with both the fibroblast growth factor (FGF) and its receptor, bringing the oligomeric complex together and increasing the effectiveness of a low concentration of FGF.

(d) Cell surface localization/concentration

The high density of negative charges in heparan sulfate attracts positively charged lipoprotein lipase molecules and holds them by electrostatic and sequence-specific interactions with NS domains.

FIGURE 7–27 Four types of protein interactions with NS domains of heparan sulfate.

("knockout") mice lacking the enzyme that sulfates the C-2 hydroxyl of iduronate (IdoA). These animals are born without kidneys and with very severe developmental abnormalities of the skeleton and eyes. Other studies demonstrate that membrane proteoglycans are important in lipoprotein clearance in the liver. There is growing evidence that the path taken by developing axons in the nervous system, and thus the wiring circuitry, is influenced by proteoglycans containing heparan sulfates and chondroitin sulfate, which provide directional cues for axon outgrowth.

Some proteoglycans can form **proteoglycan aggregates**, enormous supramolecular assemblies of many core proteins all bound to a single molecule of hyaluronan. Aggrecan core protein (M_r ~250,000) has multiple chains of chondroitin sulfate and keratan sulfate, joined to Ser residues in the core protein through trisaccharide linkers, to give an aggrecan monomer of M_r ~2×10^6. When a hundred or more of these "decorated" core proteins bind a single, extended molecule of hyaluronate **(Fig. 7–28)**, the resulting proteoglycan aggregate (M_r >2×10^8) and its associated water of hydration occupy a volume about equal to that of a bacterial cell! Aggrecan interacts strongly with collagen in the extracellular matrix of cartilage, contributing to the

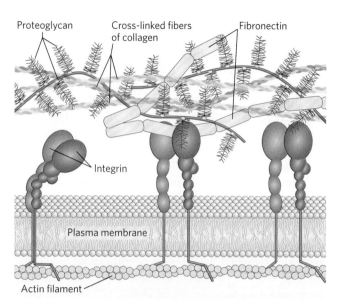

FIGURE 7–29 Interactions between cells and the extracellular matrix. The association between cells and the proteoglycan of the extracellular matrix is mediated by a membrane protein (integrin) and by an extracellular protein (fibronectin in this example) with binding sites for both integrin and the proteoglycan. Note the close association of collagen fibers with the fibronectin and proteoglycan.

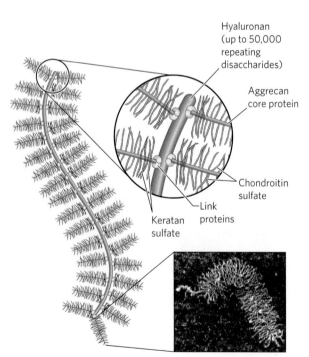

FIGURE 7–28 Proteoglycan aggregate of the extracellular matrix. Schematic drawing of a proteoglycan with many aggrecan molecules. One very long molecule of hyaluronan is associated noncovalently with about 100 molecules of the core protein aggrecan. Each aggrecan molecule contains many covalently bound chondroitin sulfate and keratan sulfate chains. Link proteins at the junction between each core protein and the hyaluronan backbone mediate the core protein–hyaluronan interaction. The micrograph shows a single molecule of aggrecan, viewed with the atomic force microscope (see Box 19–2).

development, tensile strength, and resilience of this connective tissue.

Interwoven with these enormous extracellular proteoglycans are fibrous matrix proteins such as collagen, elastin, and fibronectin, forming a cross-linked meshwork that gives the whole extracellular matrix strength and resilience. Some of these proteins are multiadhesive, a single protein having binding sites for several different matrix molecules. Fibronectin, for example, has separate domains that bind fibrin, heparan sulfate, collagen, and a family of plasma membrane proteins called integrins that mediate signaling between the cell interior and the extracellular matrix (see Fig. 12–29). The overall picture of cell-matrix interactions that emerges **(Fig. 7–29)** shows an array of interactions between cellular and extracellular molecules. These interactions serve not merely to anchor cells to the extracellular matrix but also to provide paths that direct the migration of cells in developing tissue and to convey information in both directions across the plasma membrane.

Glycoproteins Have Covalently Attached Oligosaccharides

Glycoproteins are carbohydrate-protein conjugates in which the glycans are smaller, branched, and more structurally diverse than the huge glycosaminoglycans of proteoglycans. The carbohydrate is attached at its anomeric carbon through a glycosidic link to the —OH of a Ser or Thr residue (O-linked), or through an *N*-glycosyl link to the amide nitrogen of an Asn residue (*N*-linked) **(Fig. 7–30)**. Some glycoproteins have a single oligosaccharide chain, but many have more than one; the carbohydrate may constitute from 1% to 70% or more of the glycoprotein by mass. About half of all proteins of mammals are glycosylated, and about 1% of all mammalian genes encode enzymes involved in the synthesis and attachment of these oligosaccharide chains. *N*-linked oligosaccharides are generally found in the consensus sequence N-{P}-[ST]; not all potential sites are used. (See Box 3–2 for the conventions on representing consensus sequences.) There appears to be no specific consensus sequence for *O*-linked oligosaccharides, although regions bearing *O*-linked chains tend to be rich in Gly, Val, and Pro residues.

One class of glycoproteins found in the cytoplasm and the nucleus is unique in that the glycosylated positions in the protein carry only single residues of *N*-acetylglucosamine, in *O*-glycosidic linkage to the hydroxyl group of Ser side chains. This modification is reversible and often occurs on the same Ser residues that are phosphorylated at some stage in the protein's activity. The two modifications are mutually exclusive, and this type of glycosylation is important in the regulation of protein activity. We discuss protein phosphorylation at length in Chapter 12.

As we shall see in Chapter 11, the external surface of the plasma membrane has many membrane glycoproteins with arrays of covalently attached oligosaccharides of varying complexity. **Mucins** are secreted or membrane glycoproteins that can contain large numbers of *O*-linked oligosaccharide chains. Mucins are present in most secretions; they are what gives mucus its characteristic slipperiness.

Glycomics is the systematic characterization of all of the carbohydrate components of a given cell or tissue, including those attached to proteins and to lipids. For glycoproteins, this also means determining which proteins are glycosylated and where in the amino acid sequence each oligosaccharide is attached. This is a challenging undertaking, but worthwhile because of the potential insights it offers into normal patterns of glycosylation and the ways in which they are altered during development or in genetic diseases or cancer. Current methods of characterizing the whole carbohydrate complement of cells depend heavily on sophisticated application of mass spectrometry (see Fig. 7–39).

The structures of a large number of *O*- and *N*-linked oligosaccharides from a variety of glycoproteins are known; Figures 7–24 and 7–30 show a few typical examples. We consider the mechanisms by which specific proteins acquire specific oligosaccharide moieties in Chapter 27.

Many of the proteins secreted by eukaryotic cells are glycoproteins, including most of the proteins of blood. For example, immunoglobulins (antibodies) and certain hormones, such as follicle-stimulating hormone, luteinizing hormone, and thyroid-stimulating hormone, are glycoproteins. Many milk proteins, including the major whey protein *α*-lactalbumin, and some of the proteins secreted by the pancreas (such as ribonuclease) are glycosylated, as are most of the proteins contained in lysosomes.

The biological advantages of adding oligosaccharides to proteins are slowly being uncovered. The very hydrophilic clusters of carbohydrate alter the polarity and solubility of the proteins with which they are conjugated. Oligosaccharide chains that are attached to newly synthesized proteins in the endoplasmic reticulum (ER) and elaborated in the Golgi complex serve as destination labels (see Fig. 27–39) and also act in protein quality control, targeting misfolded proteins for degradation (see Fig. 27–40). When numerous negatively charged oligosaccharide chains are clustered in a single region of a protein, the charge repulsion among them favors the formation of an extended,

FIGURE 7–30 Oligosaccharide linkages in glycoproteins. (a) *O*-linked oligosaccharides have a glycosidic bond to the hydroxyl group of Ser or Thr residues (light red), illustrated here with GalNAc as the sugar at the reducing end of the oligosaccharide. One simple chain and one complex chain are shown. **(b)** *N*-linked oligosaccharides have an *N*-glycosyl bond to the amide nitrogen of an Asn residue (green), illustrated here with GlcNAc as the terminal sugar. Three common types of oligosaccharide chains that are *N*-linked in glycoproteins are shown. A complete description of oligosaccharide structure requires specification of the position and stereochemistry (*α* or *β*) of each glycosidic linkage.

rodlike structure in that region. The bulkiness and negative charge of oligosaccharide chains also protect some proteins from attack by proteolytic enzymes. Beyond these global physical effects on protein structure, there are also more specific biological effects of oligosaccharide chains in glycoproteins (Section 7.4). The importance of normal protein glycosylation is clear from the finding of at least 18 different genetic disorders of glycosylation in humans, all causing severely defective physical or mental development; some of these disorders are fatal.

Glycolipids and Lipopolysaccharides Are Membrane Components

Glycoproteins are not the only cellular components that bear complex oligosaccharide chains; some lipids, too, have covalently bound oligosaccharides. **Gangliosides** are membrane lipids of eukaryotic cells in which the polar head group, the part of the lipid that forms the outer surface of the membrane, is a complex oligosaccharide containing a sialic acid (Fig. 7–9) and other monosaccharide residues. Some of the oligosaccharide moieties of gangliosides, such as those that determine human blood groups (see Fig. 10–15), are identical with those found in certain glycoproteins, which therefore also contribute to blood group type. Like the oligosaccharide moieties of glycoproteins, those of membrane lipids are generally, perhaps always, found on the outer face of the plasma membrane.

Lipopolysaccharides are the dominant surface feature of the outer membrane of gramnegative bacteria such as *Escherichia coli* and *Salmonella typhimurium*. These molecules are prime targets of the antibodies produced by the vertebrate immune system in response to bacterial infection and are therefore important determinants of the serotype of bacterial strains (serotypes are strains that are distinguished on the basis of antigenic properties). The lipopolysaccharides of *S. typhimurium* contain six fatty acids bound to two glucosamine residues, one of which is the point of attachment for a complex oligosaccharide (**Fig. 7–31**). *E. coli* has similar but unique lipopolysaccharides. The lipid A portion of the lipopolysaccharides of some bacteria is called endotoxin; its toxicity to humans and other animals is responsible for the dangerously lowered blood pressure that occurs in toxic shock syndrome resulting from gram-negative bacterial infections. ■

SUMMARY 7.3 Glycoconjugates: Proteoglycans, Glycoproteins, and Glycosphingolipids

▶ Proteoglycans are glycoconjugates in which one or more large glycans, called sulfated

FIGURE 7–31 Bacterial lipopolysaccharides. Schematic diagram of the lipopolysaccharide of the outer membrane of *Salmonella typhimurium*. Kdo is 3-deoxy-D-*manno*-octulosonic acid (previously called ketodeoxyoctonic acid); Hep is L-*glycero*-D-*manno*-heptose; AbeOAc is abequose (a 3,6-dideoxyhexose) acetylated on one of its hydroxyls. There are six fatty acid residues in the lipid A portion of the molecule. Different bacterial species have subtly different lipopolysaccharide structures, but they have in common a lipid region (lipid A), a core oligosaccharide also known as endotoxin, and an "O-specific" chain, which is the principal determinant of the serotype (immunological reactivity) of the bacterium. The outer membranes of the gram-negative bacteria *S. typhimurium* and *E. coli* contain so many lipopolysaccharide molecules that the cell surface is virtually covered with O-specific chains.

glycosaminoglycans (heparan sulfate, chondroitin sulfate, dermatan sulfate, or keratan sulfate) are covalently attached to a core protein. Bound to the outside of the plasma membrane by a transmembrane peptide or a covalently attached lipid, proteoglycans provide points of adhesion, recognition, and information transfer between cells, or between the cell and the extracellular matrix.

▶ Glycoproteins contain oligosaccharides covalently linked to Asp or Ser/Thr residues. The glycans are typically branched and smaller than glycosaminoglycans. Many cell surface or extracellular proteins are glycoproteins, as are most secreted proteins. The covalently attached oligosaccharides influence the folding and stability of the proteins, provide critical information about the targeting of newly synthesized proteins, and allow for specific recognition by other proteins.

▶ Glycomics is the determination of the full complement of sugar-containing molecules in a cell or tissue, and the determination of the function of each such molecule.

▶ Glycolipids and glycosphingolipids in plants and animals and lipopolysaccharides in bacteria are components of the cell envelope, with covalently attached oligosaccharide chains exposed on the cell's outer surface.

7.4 Carbohydrates as Informational Molecules: The Sugar Code

Glycobiology, the study of the structure and function of glycoconjugates, is one of the most active and exciting areas of biochemistry and cell biology. It is becoming increasingly clear that cells use specific oligosaccharides to encode important information about intracellular targeting of proteins, cell-cell interactions, cell differentiation and tissue development, and extracellular signals. Our discussion uses just a few examples to illustrate the diversity of structure and the range of biological activity of the glycoconjugates. In Chapter 20 we discuss the biosynthesis of polysaccharides, including peptidoglycan; and in Chapter 27, the assembly of oligosaccharide chains on glycoproteins.

Improved methods for the analysis of oligosaccharide and polysaccharide structure have revealed remarkable complexity and diversity in the oligosaccharides of glycoproteins and glycolipids. Consider the oligosaccharide chains in Figure 7–30, typical of those found in many glycoproteins. The most complex of those shown contains 14 monosaccharide residues of four different kinds, variously linked as (1→2), (1→3), (1→4), (1→6), (2→3), and (2→6), some with the α and some with the β configuration. Branched structures, not found in nucleic acids or proteins, are common in oligosaccharides. With the reasonable assumption that 20 different monosaccharide subunits are available for construction of oligosaccharides, we can calculate that many billions of different hexameric oligosaccharides are possible; this compares with 6.4×10^7 (20^6) different hexapeptides possible with the 20 common amino acids, and 4,096 (4^6) different hexanucleotides with the four nucleotide subunits. If we also allow for variations in oligosaccharides resulting from sulfation of one or more residues, the number of possible oligosaccharides increases by two orders of magnitude. In reality, only a subset of possible combinations is found, given the restrictions imposed by the biosynthetic enzymes and the availability of precursors. Nevertheless, the enormously rich structural information in glycans does not merely rival but far surpasses that of nucleic acids in the density of information contained in a molecule of modest size. Each of the oligosaccharides represented in Figures 7–24 and 7–30 presents a unique, three-dimensional face—a word in the sugar code—readable by the proteins that interact with it.

Lectins Are Proteins That Read the Sugar Code and Mediate Many Biological Processes

Lectins, found in all organisms, are proteins that bind carbohydrates with high specificity and with moderate to high affinity. Lectins serve in a wide variety of cell-cell recognition, signaling, and adhesion processes and in intracellular targeting of newly synthesized proteins. Plant lectins, abundant in seeds, probably serve as deterrents to insects and other predators. In the laboratory, purified plant lectins are useful reagents for detecting and separating glycans and glycoproteins with different oligosaccharide moieties. Here we discuss just a few examples of the roles of lectins in animal cells.

Some peptide hormones that circulate in the blood have oligosaccharide moieties that strongly influence their circulatory half-life. Luteinizing hormone and thyrotropin (polypeptide hormones produced in the pituitary) have *N*-linked oligosaccharides that end with the disaccharide GalNAc4S(β1→4)GlcNAc, which is recognized by a lectin (receptor) of hepatocytes. (GalNAc4S is *N*-acetylgalactosamine sulfated on the —OH group at C-4.) Receptor-hormone interaction mediates the uptake and destruction of luteinizing hormone and thyrotropin, reducing their concentration in the blood. Thus the blood levels of these hormones undergo a periodic rise (due to pulsatile secretion by the pituitary) and fall (due to continual destruction by hepatocytes).

The residues of Neu5Ac (a sialic acid) situated at the ends of the oligosaccharide chains of many plasma glycoproteins (Fig. 7–24) protect those proteins from uptake and degradation in the liver. For example, ceruloplasmin, a copper-containing serum glycoprotein, has several oligosaccharide chains ending in Neu5Ac. The mechanism that removes sialic acid residues from serum glycoproteins is unclear. It may be due to the activity of the enzyme neuraminidase (also called sialidase) produced by invading organisms or to a steady, slow release by extracellular enzymes. The plasma membrane of hepatocytes has lectin molecules (asialoglycoprotein receptors; "asialo-" indicating "without sialic acid") that specifically bind oligosaccharide chains with galactose residues no longer "protected" by a terminal Neu5Ac residue. Receptor-ceruloplasmin interaction triggers endocytosis and destruction of the ceruloplasmin.

OH
HOH₂C | H
C
H—C H COO⁻
HO H O OH
HN H H
H₃C—C H OH H
O

N-Acetylneuraminic acid (Neu5Ac)
(a sialic acid)

A similar mechanism is apparently responsible for removing "old" erythrocytes from the mammalian bloodstream. Newly synthesized erythrocytes have several membrane glycoproteins with oligosaccharide chains that end in Neu5Ac. When the sialic acid residues are removed by withdrawing a sample of blood from experimental animals, treating it with neuraminidase in vitro, and reintroducing it into the circulation, the treated erythrocytes disappear from the bloodstream within a few hours; erythrocytes with intact oligosaccharides (withdrawn and reintroduced without neuraminidase treatment) continue to circulate for days.

Cell surface lectins are important in the development of some human diseases—both human lectins and the lectins of infectious agents. **Selectins** are a family of plasma membrane lectins that mediate cell-cell recognition and adhesion in a wide range of cellular processes. One such process is the movement of immune cells (leukocytes) through the capillary wall, from blood to tissues, at sites of infection or inflammation **(Fig. 7–32)**. At an infection site, P-selectin on the surface of capillary endothelial cells interacts with a specific oligosaccharide of the surface glycoproteins of circulating leukocytes. This interaction slows the leukocytes as they roll along the endothelial lining of the capillaries. A second interaction, between integrin molecules (p. 470) in the leukocyte plasma membrane and an adhesion protein on the endothelial cell surface, now stops the leukocyte and allows it to move through the capillary wall into the infected tissues to initiate the immune attack. Two other selectins participate in this "lymphocyte homing": E-selectin on the endothelial cell and L-selectin on the leukocyte bind their cognate oligosaccharides on the leukocyte and endothelial cell, respectively.

Human selectins mediate the inflammatory responses in rheumatoid arthritis, asthma, psoriasis, multiple sclerosis, and the rejection of transplanted organs, and thus there is great interest in developing drugs that inhibit selectin-mediated cell adhesion. Many carcinomas express an antigen normally present only in fetal cells (sialyl Lewis x, or sialyl Le^x) that, when shed into the circulation, facilitates tumor cell survival and metastasis. Carbohydrate derivatives that mimic the sialyl Le^x portion of sialoglycoproteins or that alter the biosynthesis of the oligosaccharide might prove effective as selectin-specific drugs for treating chronic inflammation or metastatic disease.

Several animal viruses, including the influenza virus, attach to their host cells through interactions with oligosaccharides displayed on the host cell surface. The

FIGURE 7–32 Role of lectin-ligand interactions in leukocyte movement to the site of an infection or injury. A leukocyte circulating through a capillary is slowed by transient interactions between P-selectin molecules in the plasma membrane of the capillary endothelial cells and glycoprotein ligands for P-selectin on the leukocyte surface. As it interacts with successive P-selectin molecules, the leukocyte rolls along the capillary surface. Near a site of inflammation, stronger interactions between integrin in the leukocyte surface and its ligand in the capillary surface lead to tight adhesion. The leukocyte stops rolling and, under the influence of signals sent out from the site of inflammation, begins extravasation—escape through the capillary wall—as it moves toward the site of inflammation.

N-Acetylneuraminic acid   Oseltamivir   Zanamivir

(a)

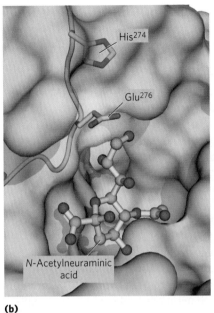

His²⁷⁴

Glu²⁷⁶

N-Acetylneuraminic acid

(b)

His²⁷⁴

Glu²⁷⁶

Oseltamivir

(c)

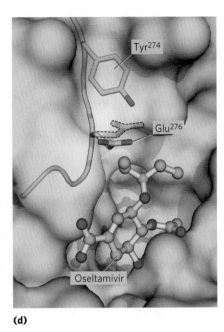

Tyr²⁷⁴

Glu²⁷⁶

Oseltamivir

(d)

FIGURE 7–33 Binding site on influenza neuraminidase for *N*-acetyl-neuraminic acid and an antiviral drug, oseltamivir. (a) The normal binding ligand for this enzyme is a sialic acid, *N*-acetylneuraminic acid. The drugs oseltamivir and zanamivir occupy the same site on the enzyme, competitively inhibiting it and blocking viral release from the host cell. **(b)** The normal interaction with *N*-acetylneuraminic acid in the binding site (PDB ID 2BAT). **(c)** Oseltamivir can fit into this site by pushing a nearby Glu residue out of the way (PDB ID 2HU4). **(d)** A mutation in the influenza virus's gene for neuraminidase replaces a His near this Glu residue with the larger side chain of a Tyr (PDB ID 3CLO). Now, oseltamivir is not as effective at pushing the Glu out of its way, and binds much less well to the binding site, making the mutant virus effectively resistant to oseltamivir.

lectin of the influenza virus, known as the HA (hemagglutinin) protein, is essential for viral entry and infection. After the virus has entered a host cell and has been replicated, the newly synthesized viral particles bud out of the cell, wrapped in a portion of its plasma membrane. A viral sialidase (neuraminidase) trims the terminal sialic acid residue from the host cell's oligosaccharides, releasing the viral particles from their interaction with the cell and preventing their aggregation with one another. Another round of infection can now begin. The antiviral drugs oseltamivir (Tamiflu) and zanamivir (Relenza) are used clinically in the treatment of influenza. These drugs are sugar analogs; they inhibit the viral sialidase by competing with the host cell's oligosaccharides for binding **(Fig. 7–33)**. This prevents the release of viruses from the infected cell and also causes viral particles to aggregate, both of which block another cycle of infection.

Some microbial pathogens have lectins that mediate bacterial adhesion to host cells or the entry of toxin into cells. For example, *Helicobacter pylori* has a surface lectin that adheres to oligosaccharides on the surface of epithelial cells that line the inner surface of the stomach **(Fig. 7–34)**. Among the binding sites

FIGURE 7–34 An ulcer in the making. *Helicobacter pylori* cells adhering to the gastric surface. This bacterium causes ulcers through interactions between a bacterial surface lectin and the Le^b oligosaccharide (a blood group antigen) of the epithelial cells lining the inside surface of the stomach.

recognized by the *H. pylori* lectin is the oligosaccharide Lewis b (Le[b]), which is present in the glycoproteins and glycolipids that define the type O blood group determinant (see Fig. 10–15). This observation helps to explain the severalfold greater incidence of gastric ulcers in people of blood type O than in those of type A or B; *H. pylori* attacks their epithelial cells more effectively. Chemically synthesized analogs of the Le[b] oligosaccharide may prove useful in treating this type of ulcer. Administered orally, they could prevent bacterial adhesion (and thus infection) by competing with the gastric glycoproteins for binding to the bacterial lectin.

Some of the most devastating of the human parasitic diseases, widespread in much of the developing world, are caused by eukaryotic microorganisms that display unusual surface oligosaccharides, which in some cases are known to be protective for the parasites. These organisms include the trypanosomes, responsible for African sleeping sickness and Chagas disease (see Box 6–3); *Plasmodium falciparum*, the malaria parasite; and *Entamoeba histolytica*, the causative agent of amoebic dysentery. The prospect of finding drugs that interfere with the synthesis of these unusual oligosaccharide chains, and therefore with the replication of the parasites, has inspired much recent work on the biosynthetic pathways of these oligosaccharides. ∎

Lectins also act intracellularly, in sorting proteins for transportation to specific cellular compartments (see Chapter 27). For example, an oligosaccharide containing mannose 6-phosphate, recognized by a lectin, marks newly synthesized proteins in the Golgi complex for transfer to the lysosome (see Fig. 27–39).

Lectin-Carbohydrate Interactions Are Highly Specific and Often Multivalent

The high density of information in the structure of oligosaccharides provides a sugar code with an essentially unlimited number of unique "words" small enough to be read by a single protein. In their carbohydrate-binding sites, lectins have a subtle molecular complementarity that allows interaction only with their correct carbohydrate cognates. The result is an extraordinarily high specificity in these interactions. The affinity between an oligosaccharide and an individual carbohydrate-binding domain (CBD) of a lectin is sometimes modest (micromolar to millimolar K_d values), but the effective affinity is in many cases greatly increased by lectin multivalency, in which a single lectin molecule has multiple CBDs. In a cluster of oligosaccharides—as is commonly found on a membrane surface, for example—each oligosaccharide can engage one of the lectin's CBDs, strengthening the interaction. When cells express multiple lectin receptors, the avidity of the interaction can be very high, enabling highly cooperative events such as cell attachment and rolling (Fig. 7–32).

X-ray crystallographic studies of the structure of the mannose 6-phosphate receptor/lectin reveal details of its interaction with mannose 6-phosphate that explain the specificity of the binding and the role for a divalent cation in the lectin-sugar interaction (**Fig. 7–35a**). His[105] is hydrogen-bonded to one of the oxygen atoms of

(a)   **(b)**

FIGURE 7–35 Details of a lectin-carbohydrate interaction. Structure of the bovine mannose 6-phosphate receptor complexed with mannose 6-phosphate (PDB ID 1M6P). The protein is represented as a surface contour image, showing the surface as predominantly negatively charged (red) or positively charged (blue). Mannose 6-phosphate is shown as a stick structure; a manganese ion is shown as a violet sphere. **(b)** An enlarged view of the binding site. Mannose 6-phosphate is hydrogen-bonded to Arg[111] and coordinated with the manganese ion (shown smaller than its van der Waals radius for clarity). Each hydroxyl group of mannose is hydrogen-bonded to the protein. The His[105] hydrogen-bonded to a phosphate oxygen of mannose 6-phosphate may be the residue that, when protonated at low pH, causes the receptor to release mannose 6-phosphate into the lysosome.

the phosphate (Fig. 7–35b). When the protein tagged with mannose 6-phosphate reaches the lysosome (which has a lower internal pH than the Golgi complex), the receptor loses its affinity for mannose 6-phosphate. Protonation of His[105] may be responsible for this change in binding.

In addition to such very specific interactions, there are more general interactions that contribute to the binding of many carbohydrates to their lectins. For example, many sugars have a more polar and a less polar side **(Fig. 7–36)**; the more polar side hydrogen-bonds with the lectin, while the less polar undergoes hydrophobic interactions with nonpolar amino acid residues. The sum of all these interactions produces high-affinity binding and high specificity of lectins for their carbohydrates. This represents a kind of information transfer that is clearly central in many processes within and between cells. **Figure 7–37** summarizes some of the biological interactions mediated by the sugar code.

SUMMARY 7.4 Carbohydrates as Informational Molecules: The Sugar Code

▶ Monosaccharides can be assembled into an almost limitless variety of oligosaccharides, which differ in the stereochemistry and position of glycosidic bonds, the type and orientation of substituent groups, and the number and type of branches. Glycans are far more information-dense than nucleic acids or proteins.

FIGURE 7–36 Hydrophobic interactions of sugar residues. Sugar units such as galactose have a more polar side (the top of the chair as shown here, with the ring oxygen and several hydroxyls) that is available to hydrogen-bond with the lectin, and a less polar side that can have hydrophobic interactions with nonpolar side chains in the protein, such as the indole ring of Trp residues.

▶ Lectins, proteins with highly specific carbohydrate-binding domains, are commonly found on the outer surface of cells, where they initiate interaction with other cells. In vertebrates, oligosaccharide tags "read" by lectins govern the rate of degradation of certain peptide hormones, circulating proteins, and blood cells.

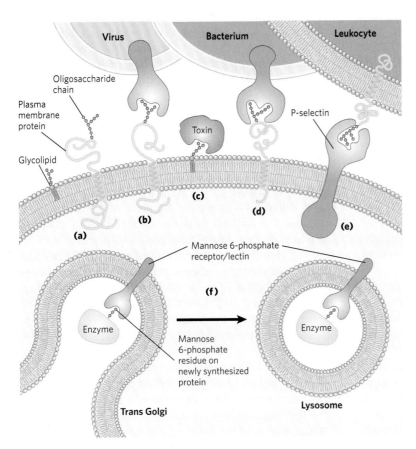

FIGURE 7–37 Role of oligosaccharides in recognition events at the cell surface and in the endomembrane system. (a) Oligosaccharides with unique structures (represented as strings of hexagons) are components of a variety of glycoproteins or glycolipids on the outer surface of plasma membranes. Their oligosaccharide moieties are bound by extracellular lectins with high specificity and affinity. **(b)** Viruses that infect animal cells, such as the influenza virus, bind to cell surface glycoproteins as the first step in infection. **(c)** Bacterial toxins, such as the cholera and pertussis toxins, bind to a surface glycolipid before entering a cell. **(d)** Some bacteria, such as *H. pylori*, adhere to and then colonize or infect animal cells. **(e)** Selectins (lectins) in the plasma membrane of certain cells mediate cell-cell interactions, such as those of leukocytes with the endothelial cells of the capillary wall at an infection site. **(f)** The mannose 6-phosphate receptor/lectin of the trans Golgi complex binds to the oligosaccharide of lysosomal enzymes, targeting them for transfer into the lysosome.

▶ Bacterial and viral pathogens and some eukaryotic parasites adhere to their animal cell targets by the binding of lectins in the pathogens to oligosaccharides on the target cell surface.

▶ X-ray crystallography of lectin-sugar complexes shows the detailed complementarity between the two molecules, which accounts for the strength and specificity of lectin interactions with carbohydrates.

7.5 Working with Carbohydrates

The growing appreciation of the importance of oligosaccharide structure in biological signaling and recognition has been the driving force behind the development of methods for analyzing the structure and stereochemistry of complex oligosaccharides. Oligosaccharide analysis is complicated by the fact that, unlike nucleic acids and proteins, oligosaccharides can be branched and are joined by a variety of linkages. The high charge density of many oligosaccharides and polysaccharides, and the relative lability of the sulfate esters in glycosaminoglycans, present further difficulties.

For simple, linear polymers such as amylose, the positions of the glycosidic bonds are determined by the classical method of exhaustive methylation: treating the intact polysaccharide with methyl iodide in a strongly basic medium to convert all free hydroxyls to acid-stable methyl ethers, then hydrolyzing the methylated polysaccharide in acid. The only free hydroxyls present in the monosaccharide derivatives so produced are those that were involved in glycosidic bonds. To determine the sequence of monosaccharide residues, including any branches that are present, exoglycosidases of known specificity are used to remove residues one at a time from the nonreducing end(s). The known specificity of these exoglycosidases often allows deduction of the position and stereochemistry of the linkages.

For analysis of the oligosaccharide moieties of glycoproteins and glycolipids, the oligosaccharides are released by purified enzymes—glycosidases that specifically cleave O- or N-linked oligosaccharides or lipases that remove lipid head groups. Alternatively, O-linked glycans can be released from glycoproteins by treatment with hydrazine.

The resulting mixtures of carbohydrates are resolved into their individual components by a variety of methods (Fig. 7–38), including the same techniques used in protein and amino acid separation: fractional precipitation by solvents, and ion-exchange and size-exclusion chromatography (see Fig. 3–17). Highly purified lectins, attached covalently to an insoluble support, are commonly used in affinity chromatography of carbohydrates (see Fig. 3–17c).

Hydrolysis of oligosaccharides and polysaccharides in strong acid yields a mixture of monosaccharides, which may be identified and quantified by chromatographic techniques to yield the overall composition of the polymer.

Oligosaccharide analysis relies increasingly on mass spectrometry and high-resolution NMR spectroscopy. Matrix-assisted laser desorption/ionization mass spectrometry (MALDI MS) and tandem mass spectrometry (MS/MS), both described in Chapter 3, are readily applicable to polar compounds such as oligosaccharides. MALDI MS is a very sensitive method for determining the mass of a molecular ion (in this case, the entire oligosaccharide chain; Fig. 7–39). MS/MS reveals the mass of the molecular ion and many of its fragments, which are usually the result of breakage of the glycosidic bonds. NMR analysis alone (see Box 4–5), especially for oligosaccharides of moderate size, can yield much information about sequence, linkage position, and anomeric carbon configuration. For example, the structure of the heparin segment shown as a space-filling model in Figure 7–22 was obtained entirely by NMR spectroscopy. Automated procedures and commercial instruments are used for the routine determination of oligosaccharide structure, but the sequencing of branched oligosaccharides joined by more than one type of bond remains a far more formidable task than determining the linear sequences of proteins and nucleic acids.

Another important tool in working with carbohydrates is chemical synthesis, which has proved to be a powerful approach to understanding the biological functions of glycosaminoglycans and oligosaccharides. The chemistry involved in such syntheses is difficult, but carbohydrate chemists can now synthesize short segments of almost any glycosaminoglycan, with correct stereochemistry, chain length, and sulfation pattern, and oligosaccharides significantly more complex than those shown in Figure 7–30. Solid-phase oligosaccharide synthesis is based on the same principles (and has the same advantages) as peptide synthesis (see Fig. 3–32), but requires a set of tools unique to carbohydrate chemistry: blocking groups and activating groups that allow the synthesis of glycosidic linkages with the correct hydroxyl group. Synthetic approaches of this type currently represent an area of great interest, because it is difficult to purify defined oligosaccharides in adequate quantities from natural sources.

To identify proteins with specific affinity for particular oligosaccharides, **oligosaccharide microarrays** are used. The principle is the same as for DNA microarrays (Figs 9–22 and 9–23), but the technical problems are more challenging. Pure oligosaccharides are attached to a glass slide in microdroplets, and the slide is exposed to a potential lectin (glycan-binding protein) that has been tagged with a fluorescent molecule (Fig. 7–40). After all the unadsorbed protein is washed away, observation of the microarrays with a fluorescence microscope identifies the oligosaccharides that the lectin recognizes, and quantification of the fluorescence gives a rough measure of lectin-oligosaccharide affinity.

FIGURE 7–38 Methods of carbohydrate analysis. A carbohydrate purified in the first stage of the analysis often requires all four analytical routes for its complete characterization.

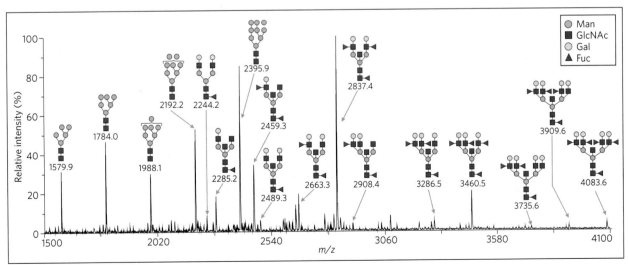

FIGURE 7–39 Separation and quantification of the oligosaccharides in a group of glycoproteins. In this experiment, the mixture of proteins extracted from kidney tissue was treated to release oligosaccharides from glycoproteins, and the oligosaccharides were analyzed by matrix-assisted laser desorption/ionization mass spectrometry (MALDI MS). Each distinct oligosaccharide produces a peak at its molecular mass, and the area under the curve reflects the quantity of that oligosaccharide. The most prominent oligosaccharide here (mass 2837.4 u) is composed of 13 sugar residues; other oligosaccharides, containing as few as 7 and as many as 19 residues, were also resolved by this method.

Oligosaccharide microarray

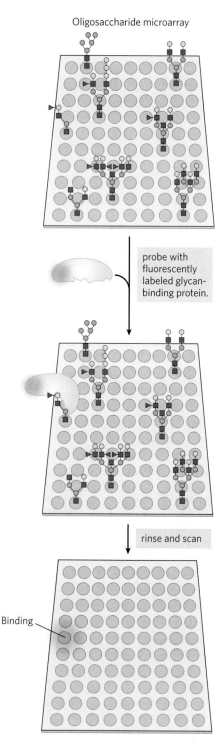

probe with fluorescently labeled glycan-binding protein.

rinse and scan

Binding

FIGURE 7–40 Oligosaccharide microarrays to determine the specificity and affinity of carbohydrate binding by lectins. Solutions of pure samples of oligosaccharides, synthesized or isolated from nature, are placed in microscopic droplets on a glass slide and attached to the glass through an inert spacer. Each spot represents a different oligosaccharide. The protein sample to be tested for its affinity for oligosaccharides is first conjugated with a fluorescent marker, then the sample is poured over the slide and allowed to equilibrate, and any nonadsorbed protein is washed away. Observation of the microarray with a fluorescence microscope shows which spots have adsorbed protein (they glow green), and assessment of the fluorescence intensity gives a rough measure of protein-oligosaccharide binding affinity.

SUMMARY 7.5 Working with Carbohydrates

▶ Establishing the complete structure of oligosaccharides and polysaccharides requires determination of linear sequence, branching positions, the configuration of each monosaccharide unit, and the positions of the glycosidic linkages—a more complex problem than protein and nucleic acid analysis.

▶ The structures of oligosaccharides and polysaccharides are usually determined by a combination of methods: specific enzymatic hydrolysis to determine stereochemistry at the glycosidic bond and to produce smaller fragments for further analysis; methylation to locate glycosidic bonds; and stepwise degradation to determine sequence and configuration of anomeric carbons.

▶ Mass spectrometry and high-resolution NMR spectroscopy, applicable to small samples of carbohydrate, yield essential information about sequence, configuration at anomeric and other carbons, and positions of glycosidic bonds.

▶ Solid-phase synthetic methods yield defined oligosaccharides that are of great value in exploring lectin-oligosaccharide interactions and may prove clinically useful.

▶ Microarrays of pure oligosaccharides are useful in determining the specificity and affinity of lectin binding to specific oligosaccharides.

Key Terms

Terms in bold are defined in the glossary.

glycoconjugate 243
carbohydrate 243
monosaccharide 243
oligosaccharide 243
disaccharide 243
polysaccharide 243
aldose 244
ketose 244
Fischer projection formulas 244
epimers 245
hemiacetal 245
hemiketal 245
anomers 246
anomeric carbon 246
pyranose 246
furanose 247
Haworth perspective formulas 247
mutarotation 248
reducing sugar 250
hemoglobin glycation 251

O-**glycosidic bonds** 252
reducing end 252
glycan 254
starch 255
glycogen 255
cellulose 256
extracellular matrix (ECM) 260
glycosaminoglycan 260
hyaluronan 260
chondroitin sulfate 261
heparan sulfate 261
proteoglycan 263
glycoprotein 263
glycosphingolipid 264
syndecan 264
glypican 264
glycomics 267
lectin 269
selectins 270
oligosaccharide microarrays 274

Further Reading

General

Assadi-Porter, F.M., Maillet, E.L., Radek, J.T., Quijaada, J., Markley, J.L., & Max, M. (2010) Key amino acid residues involved in multi-point binding interactions between brazzein, a sweet protein, and the T1R2–T1R3 human sweet receptor. *J. Mol. Biol.* **398**, 584–599.

Hayes, J.E. (2007) Transdisciplinary perspectives on sweetness. *Chemosens. Percept.* **1**, 48–57.

Description of the theory of sweetness discussed in Box 7–2.

Varki, A., Cummings, R.D., Esko, J.D., Freeze, H.H., Stanley, P., Bertozzi, C.R., Hart, G.W., & Etzler, M.E. (eds). (2009) *Essentials of Glycobiology*, 2nd edn, Cold Spring Harbor Laboratory Press, Cold Spring Harbor, NY.

Structure, biosynthesis, metabolism, and function of glycosaminoglycans, proteoglycans, glycoproteins, and glycolipids, all presented at an intermediate level and very well illustrated. The book is available free online (www.ncbi.nlm.nih.gov/books/NBK1908).

Watanabe, H. & Tokuda, G. (2010) Cellulolytic systems in insects. *Annu. Rev. Entomol.* **55**, 609–632.

Glycosaminoglycans and Proteoglycans

Bishop, J.R., Schuksz, M., & Esko, J.D. (2007) Heparan sulfate proteoglycans fine-tune mammalian physiology. *Nature* **446**, 1030–1037.

Couchman, J.R. (2010) Transmembrane signaling proteoglycans. *Annu. Rev. Cell Dev. Biol.* **26**, 89–114.

Advanced review of the role of proteoglycans in signal transduction in vertebrates.

Fears, C.Y. & Woods, A. (2006) The role of syndecans in disease and wound healing. *Matrix Biol.* **25**, 443–456.

Intermediate-level review.

Frantz, C., Stewart, K.M., & Weaver, V.M. (2010) The extracellular matrix at a glance. *J. Cell Sci.* **123**, 4195–4200.

Poster-style summary of roles of glycans and glycoconjugates in the extracellular matrix.

Kirkpatrick, C.A. & Selleck, S.B. (2007) Heparan sulfate proteoglycans at a glance. *J. Cell Sci.* **120**, 1829–1832.

Poster-style summary of much useful information about proteoglycans.

Manon-Jensen, T., Itoh, Y., & Couchman, J.R. (2010) Proteoglycans in health and disease: the multiple roles of syndecan shedding. *FEBS J.* **277**, 3876–3889.

Roseman, S. (2001) Reflections on glycobiology. *J. Biol. Chem.* **276**, 41,527–41,542.

A masterful review of the history of carbohydrate and glycosaminoglycan studies, by one of the major contributors to this field.

Glycoproteins

Boraston, A. & Mulloy, B. (2010) Structural glycobiology: biosynthesis, recognition events, and new methods. *Curr. Opin. Struct. Biol.* **20**, 533–535.

Editorial introduction to a series of excellent reviews on these subjects published in this issue.

Luac, G. & Zoldos, V. (2010) Protein glycosylation—an evolutionary crossroad between genes and environment. *Mol. Biol. Syst.* **6**, 2372–2379.

A detailed discussion of the factors that determine whether and where a protein will be glycosylated.

Molinaro, M. (2007) *N*-glycan structure dictates extension of protein folding or onset of disposal. *Nat. Chem. Biol.* **3**, 313–320.

Intermediate-level review of the importance of protein glycosylation in the Golgi complex.

Sharon, N. & Gallagher, J. (2009) *Curr. Opin. Struct. Biol.* **19**, 495–497.

Editorial introduction to a series of good reviews on glycoproteins and glycolipids in this issue of the journal.

Weerapana, E. & Imperiali, B. (2006) Asparagine-linked protein glycosylation: from eukaryotic to prokaryotic systems. *Glycobiology* **16**, 91R–101R.

Intermediate-level review of the biosynthetic process of protein glycosylation.

Glycobiology and the Sugar Code

Boraston, A.B., Bolam, D.N., Gilbert, H.J., & Davies, G.J. (2004) Carbohydrate-binding modules: fine-tuning polysaccharide recognition. *Biochem. J.* **382**, 769–781.

Excellent review of the structural basis for the specificity of sugar-binding proteins.

Gabius, H.-J., Andre, S., Jimenez-Barbero, J., Romero, A., & Solis, D. (2011) From lectin structure to functional glycomics: principles of the sugar code. *Trends Biochem. Sci.* **36**, 298–313.

Intermediate-level review of structural basis for lectin-sugar recognition.

Ghosh, P., Dahms, N.M., & Kornfeld, S. (2003) Mannose 6-phosphate receptors: new twists in the tale. *Nat. Rev. Mol. Cell Biol.* **4**, 202–212.

Hebert, D.N., Garman, S.C., & Molinari, M. (2005) The glycan code of the endoplasmic reticulum: asparagine-linked carbohydrates as protein maturation and quality control tags. *Trends Cell Biol.* **15**, 364–370.

Intermediate-level review.

Helenius, A. & Aebi, M. (2004) Roles of *N*-linked glycans in the endoplasmic reticulum. *Annu. Rev. Biochem.* **73**, 1019–1049.

Lütteke, T., Bohne-Lang, A., Loss, A., Goetz, T., Frank, M., & von der Lieth, C.-W. (2006) Glycosciences.de: an internet portal to support glycomics and glycobiology research. *Glycobiology* **16**, 71R–81R.

McEver, R.P. & Zhu, C. (2010) Rolling cell adhesion. *Annu. Rev. Cell Dev. Biol.* **26**, 363–396.

Taylor, M.E. & Drickamer, K. (2006) *Introduction to Glycobiology*, 2nd edn, Oxford University Press, Oxford.

Working with Carbohydrates

Fukuda, M. (ed.). (2006) *Functional Glycomics*, Methods in Enzymology, Vol. 417, Academic Press, Inc., New York.

Fukuda, M. (ed.). (2006) *Glycobiology*, Methods in Enzymology, Vol. 415, Academic Press, Inc., New York.

Fukuda, M. (ed.). (2006) *Glycomics*, Methods in Enzymology, Vol. 416, Academic Press, Inc., New York.

Jay, A. (1996) The methylation reaction in carbohydrate analysis. *J. Carbohydr. Chem.* **15**, 897–923.

Paulson, J.C., Blixt, O., & Collins, B.E. (2006) Sweet spots in functional glycomics. *Nat. Chem. Biol.* **2**, 238–248.

Intermediate-level review of newly developed tools in glycobiology.

Zaia, J. (2008) Mass spectrometry and the emerging field of glycomics. *Chem. Biol.* **15**, 881–892.

Excellent introduction to the use of mass spectrometry in studies of glycan structure and function.

Problems

1. Sugar Alcohols In the monosaccharide derivatives known as sugar alcohols, the carbonyl oxygen is reduced to a hydroxyl group. For example, D-glyceraldehyde can be reduced to glycerol. However, this sugar alcohol is no longer designated D or L. Why?

2. Recognizing Epimers Using Figure 7–3, identify the epimers of (a) D-allose, (b) D-gulose, and (c) D-ribose at C-2, C-3, and C-4.

3. Melting Points of Monosaccharide Osazone Derivatives Many carbohydrates react with phenylhydrazine ($C_6H_5NHNH_2$) to form bright yellow crystalline derivatives known as osazones:

Glucose   Osazone derivative
                 of glucose

The melting temperatures of these derivatives are easily determined and are characteristic for each osazone. This information was used to help identify monosaccharides before the development of HPLC or gas-liquid chromatography. Listed below are the melting points (MPs) of some aldose-osazone derivatives.

Monosaccharide	MP of anhydrous monosaccharide (°C)	MP of osazone derivative (°C)
Glucose	146	205
Mannose	132	205
Galactose	165–168	201
Talose	128–130	201

As the table shows, certain pairs of derivatives have the same melting points, although the underivatized monosaccharides do not. Why do glucose and mannose, and similarly galactose and talose, form osazone derivatives with the same melting points?

4. Configuration and Conformation Which bond(s) in α-D-glucose must be broken to change its configuration to β-D-glucose? Which bond(s) to convert D-glucose to D-mannose? Which bond(s) to convert one "chair" form of D-glucose to the other?

5. Deoxysugars Is D-2-deoxygalactose the same chemical as D-2-deoxyglucose? Explain.

6. Sugar Structures Describe the common structural features and the differences for each pair: (a) cellulose and glycogen; (b) D-glucose and D-fructose; (c) maltose and sucrose.

7. Reducing Sugars Draw the structural formula for α-D-glucosyl-(1→6)-D-mannosamine and circle the part of this structure that makes the compound a reducing sugar.

8. Hemiacetal and Glycosidic Linkages Explain the difference between a hemiacetal and a glycoside.

9. A Taste of Honey The fructose in honey is mainly in the β-D-pyranose form. This is one of the sweetest carbohydrates known, about twice as sweet as glucose; the β-D-furanose form of fructose is much less sweet. The sweetness of honey gradually decreases at a high temperature. Also, high-fructose corn syrup (a commercial product in which much of the glucose in corn syrup is converted to fructose) is used for sweetening *cold* but not *hot* drinks. What chemical property of fructose could account for both these observations?

10. Glucose Oxidase in Determination of Blood Glucose The enzyme glucose oxidase isolated from the mold *Penicillium notatum* catalyzes the oxidation of β-D-glucose to D-glucono-δ-lactone. This enzyme is highly specific for the β anomer of glucose and does not affect the α anomer. In spite of this specificity, the reaction catalyzed by glucose oxidase is commonly used in a clinical assay for total blood glucose—that is, for solutions consisting of a mixture of β- and α-D-glucose. What are the circumstances required to make this possible? Aside from allowing the detection of smaller quantities of glucose, what advantage does glucose oxidase offer over Fehling's reagent for measuring blood glucose?

11. Invertase "Inverts" Sucrose The hydrolysis of sucrose (specific rotation +66.5°) yields an equimolar mixture of D-glucose (specific rotation +52.5°) and D-fructose (specific rotation −92°). (See Problem 4 for details of specific rotation.)

(a) Suggest a convenient way to determine the rate of hydrolysis of sucrose by an enzyme preparation extracted from the lining of the small intestine.

(b) Explain why, in the food industry, an equimolar mixture of D-glucose and D-fructose formed by hydrolysis of sucrose is called invert sugar.

(c) The enzyme invertase (now commonly called sucrase) is allowed to act on a 10% (0.1 g/mL) solution of sucrose until hydrolysis is complete. What will be the observed optical rotation of the solution in a 10 cm cell? (Ignore a possible small contribution from the enzyme.)

12. Manufacture of Liquid-Filled Chocolates The manufacture of chocolates containing a liquid center is an interesting application of enzyme engineering. The flavored liquid center consists largely of an aqueous solution of sugars rich in fructose to provide sweetness. The technical dilemma is the following: the chocolate coating must be prepared by pouring hot melted chocolate over a solid (or almost solid) core, yet the final product must have a liquid, fructose-rich center. Suggest a way to solve this problem. (Hint: Sucrose is much less soluble than a mixture of glucose and fructose.)

13. Anomers of Sucrose? Lactose exists in two anomeric forms, but no anomeric forms of sucrose have been reported. Why?

14. Gentiobiose Gentiobiose (D-Glc(β1→6)D-Glc) is a disaccharide found in some plant glycosides. Draw the structure of gentiobiose based on its abbreviated name. Is it a reducing sugar? Does it undergo mutarotation?

15. Identifying Reducing Sugars Is *N*-acetyl-β-D-glucosamine (Fig. 7–9) a reducing sugar? What about D-gluconate? Is the disaccharide GlcN(α1↔1α)Glc a reducing sugar?

16. Cellulose Digestion Cellulose could provide a widely available and cheap form of glucose, but humans cannot digest it. Why not? If you were offered a procedure that allowed you to acquire this ability, would you accept? Why or why not?

17. Physical Properties of Cellulose and Glycogen The almost pure cellulose obtained from the seed threads of *Gossypium* (cotton) is tough, fibrous, and completely insoluble in water. In contrast, glycogen obtained from muscle or liver disperses readily in hot water to make a turbid solution. Despite their markedly different physical properties, both substances are (1→4)-linked D-glucose polymers of comparable molecular weight. What structural features of these two polysaccharides underlie their different physical properties? Explain the biological advantages of their respective properties.

18. Dimensions of a Polysaccharide Compare the dimensions of a molecule of cellulose and a molecule of amylose, each of M_r 200,000.

19. Growth Rate of Bamboo The stems of bamboo, a tropical grass, can grow at the phenomenal rate of 0.3 m/day under optimal conditions. Given that the stems are composed almost entirely of cellulose fibers oriented in the direction of growth, calculate the number of sugar residues per second that must be added enzymatically to growing cellulose chains to account for the growth rate. Each D-glucose unit contributes ~0.5 nm to the length of a cellulose molecule.

20. Glycogen as Energy Storage: How Long Can a Game Bird Fly? Since ancient times it has been observed that certain game birds, such as grouse, quail, and pheasants, are easily fatigued. The Greek historian Xenophon wrote, "The bustards . . . can be caught if one is quick in starting them up, for they will fly only a short distance, like partridges, and soon tire; and their flesh is delicious." The flight muscles of game birds rely almost entirely on the use of glucose 1-phosphate for energy, in the form of ATP (Chapter 14). The glucose 1-phosphate is formed by the breakdown of stored muscle glycogen, catalyzed by the enzyme glycogen phosphorylase. The rate of ATP production is limited by the rate at which glycogen can be broken down. During a "panic flight," the game bird's rate of glycogen breakdown is quite high, approximately 120 μmol/min of glucose 1-phosphate produced per gram of fresh tissue. Given that the flight muscles usually contain about 0.35% glycogen by weight, calculate how long a game bird can fly. (Assume the average molecular weight of a glucose residue in glycogen is 162 g/mol.)

21. Relative Stability of Two Conformers Explain why the two structures shown in Figure 7–18b are so different in energy (stability). Hint: See Figure 1–22.

22. Volume of Chondroitin Sulfate in Solution One critical function of chondroitin sulfate is to act as a lubricant in skeletal joints by creating a gel-like medium that is resilient to friction and shock. This function seems to be related to a distinctive property of chondroitin sulfate: the volume occupied by the molecule is much greater in solution than in the dehydrated solid. Why is the volume so much larger in solution?

23. Heparin Interactions Heparin, a highly negatively charged glycosaminoglycan, is used clinically as an anticoagulant. It acts by binding several plasma proteins, including antithrombin III, an inhibitor of blood clotting. The 1:1 binding of heparin to antithrombin III seems to cause a conformational change in the protein that greatly increases its ability to inhibit clotting. What amino acid residues of antithrombin III are likely to interact with heparin?

24. Permutations of a Trisaccharide Think about how one might estimate the number of possible trisaccharides composed of *N*-acetylglucosamine 4-sulfate (GlcNAc4S) and glucuronic acid (GlcA), and draw 10 of them.

25. Effect of Sialic Acid on SDS Polyacrylamide Gel Electrophoresis Suppose you have four forms of a protein, all with identical amino acid sequence but containing zero, one, two, or three oligosaccharide chains, each ending in a single sialic acid residue. Draw the gel pattern you would expect when a mixture of these four glycoproteins is subjected to SDS polyacrylamide gel electrophoresis (see Fig. 3–18) and stained for protein. Identify any bands in your drawing.

26. Information Content of Oligosaccharides The carbohydrate portion of some glycoproteins may serve as a cellular recognition site. To perform this function, the oligosaccharide moiety must have the potential to exist in a large variety of forms. Which can produce a greater variety of structures: oligopeptides composed of five different amino acid residues, or oligosaccharides composed of five different monosaccharide residues? Explain.

27. Determination of the Extent of Branching in Amylopectin The amount of branching (number of (α1→6) glycosidic bonds) in amylopectin can be determined by the following procedure. A sample of amylopectin is exhaustively methylated—treated with a methylating agent (methyl iodide) that replaces the hydrogen of every sugar hydroxyl with a methyl group, converting —OH to —OCH$_3$. All the glycosidic bonds in the treated sample are then hydrolyzed in aqueous acid, and the amount of 2,3-di-*O*-methylglucose so formed is determined.

2,3-Di-*O*-methylglucose

(a) Explain the basis of this procedure for determining the number of (α1→6) branch points in amylopectin. What happens to the unbranched glucose residues in amylopectin during the methylation and hydrolysis procedure?

(b) A 258 mg sample of amylopectin treated as described above yielded 12.4 mg of 2,3-di-*O*-methylglucose. Determine what percentage of the glucose residues in the amylopectin contained an $(\alpha 1 \rightarrow 6)$ branch. (Assume that the average molecular weight of a glucose residue in amylopectin is 162 g/mol.)

28. Structural Analysis of a Polysaccharide A polysaccharide of unknown structure was isolated, subjected to exhaustive methylation, and hydrolyzed. Analysis of the products revealed three methylated sugars: 2,3,4-tri-O-methyl-D-glucose, 2,4-di-O-methyl-D-glucose, and 2,3,4,6-tetra-O-methyl-D-glucose, in the ratio 20:1:1. What is the structure of the polysaccharide?

Data Analysis Problem

29. Determining the Structure of ABO Blood Group Antigens The human ABO blood group system was first discovered in 1901, and in 1924 this trait was shown to be inherited at a single gene locus with three alleles. In 1960, W. T. J. Morgan published a paper summarizing what was known at that time about the structure of the ABO antigen molecules. When the paper was published, the complete structures of the A, B, and O antigens were not yet known; this paper is an example of what scientific knowledge looks like "in the making."

In any attempt to determine the structure of an unknown biological compound, researchers must deal with two fundamental problems: (1) If you don't know what *it* is, how do you know if *it* is pure? (2) If you don't know what *it* is, how do you know that your extraction and purification conditions have not changed *its* structure? Morgan addressed problem 1 through several methods. One method is described in his paper (p. 312) as observing "constant analytical values after fractional solubility tests." In this case, "analytical values" are measurements of chemical composition, melting point, and so forth.

(a) Based on your understanding of chemical techniques, what could Morgan mean by "fractional solubility tests"?

(b) Why would the analytical values obtained from fractional solubility tests of a *pure* substance be constant, and those of an *impure* substance not be constant?

Morgan addressed problem 2 by using an assay to measure the immunological activity of the substance present in different samples.

(c) Why was it important for Morgan's studies, and especially for addressing problem 2, that this activity assay be quantitative (measuring a level of activity) rather than simply qualitative (measuring only the presence or absence of a substance)?

The structure of the blood group antigens is shown in Figure 10–15. In his paper (p. 314), Morgan listed several properties of the three antigens, A, B, and O, that were known at that time:

1. Type B antigen has a higher content of galactose than A or O.
2. Type A antigen contains more total amino sugars than B or O.
3. The glucosamine/galactosamine ratio for the A antigen is roughly 1.2; for B, it is roughly 2.5.

(d) Which of these findings is (are) consistent with the known structures of the blood group antigens?

(e) How do you explain the discrepancies between Morgan's data and the known structures?

In later work, Morgan and his colleagues used a clever technique to obtain structural information about the blood group antigens. Enzymes had been found that would specifically degrade the antigens. However, these were available only as crude enzyme preparations, perhaps containing more than one enzyme of unknown specificity. Degradation of the blood type antigens by these crude enzymes could be inhibited by the addition of particular sugar molecules to the reaction. Only sugars found in the blood type antigens would cause this inhibition. One enzyme preparation, isolated from the protozoan *Trichomonas foetus*, would degrade all three antigens and was inhibited by the addition of particular sugars. The results of these studies are summarized in the table below, showing the percentage of substrate remaining unchanged when the *T. foetus* enzyme acted on the blood group antigens in the presence of sugars.

	Unchanged substrate (%)		
Sugar added	A antigen	B antigen	O antigen
Control—no sugar	3	1	1
L-Fucose	3	1	100
D-Fucose	3	1	1
L-Galactose	3	1	3
D-Galactose	6	100	1
N-Acetylglucosamine	3	1	1
N-Acetylgalactosamine	100	6	1

For the O antigen, a comparison of the control and L-fucose results shows that L-fucose inhibits the degradation of the antigen. This is an example of product inhibition, in which an excess of reaction product shifts the equilibrium of the reaction, preventing further breakdown of substrate.

(f) Although the O antigen contains galactose, N-acetylglucosamine, and N-acetylgalactosamine, none of these sugars inhibited the degradation of this antigen. Based on these data, is the enzyme preparation from *T. foetus* an endo- or exoglycosidase? (Endoglycosidases cut bonds between interior residues; exoglycosidases remove one residue at a time from the end of a polymer.) Explain your reasoning.

(g) Fucose is also present in the A and B antigens. Based on the structure of these antigens, why does fucose fail to prevent their degradation by the *T. foetus* enzyme? What structure would be produced?

(h) Which of the results in (f) and (g) are consistent with the structures shown in Figure 10–15? Explain your reasoning.

Reference

Morgan, W.T.J. (1960) The Croonian Lecture: a contribution to human biochemical genetics; the chemical basis of blood-group specificity. *Proc. R. Soc. Lond. B Biol. Sci.* **151**, 308–347.

Nucleotides and Nucleic Acids

Nucleotides have a variety of roles in cellular metabolism. They are the energy currency in metabolic transactions, the essential chemical links in the response of cells to hormones and other extracellular stimuli, and the structural components of an array of enzyme cofactors and metabolic intermediates. And, last but certainly not least, they are the constituents of nucleic acids: deoxyribonucleic acid (DNA) and ribonucleic acid (RNA), the molecular repositories of genetic information. The structure of every protein, and ultimately of every biomolecule and cellular component, is a product of information programmed into the nucleotide sequence of a cell's nucleic acids. The ability to store and transmit genetic information from one generation to the next is a fundamental condition for life.

This chapter provides an overview of the chemical nature of the nucleotides and nucleic acids found in most cells; a more detailed examination of the function of nucleic acids is the focus of Part III of this text.

8.1 Some Basics

Nucleotides, Building Blocks of Nucleic Acids The amino acid sequence of every protein in a cell, and the nucleotide sequence of every RNA, is specified by a nucleotide sequence in the cell's DNA. A segment of a DNA molecule that contains the information required for the synthesis of a functional biological product, whether protein or RNA, is referred to as a **gene**. A cell typically has many thousands of genes, and DNA molecules, not surprisingly, tend to be very large. The storage and transmission of biological information are the only known functions of DNA.

RNAs have a broader range of functions, and several classes are found in cells. **Ribosomal RNAs**

(**rRNAs**) are components of ribosomes, the complexes that carry out the synthesis of proteins. **Messenger RNAs (mRNAs)** are intermediaries, carrying genetic information from one or a few genes to a ribosome, where the corresponding proteins can be synthesized. **Transfer RNAs (tRNAs)** are adapter molecules that faithfully translate the information in mRNA into a specific sequence of amino acids. In addition to these major classes there is a wide variety of RNAs with special functions, described in depth in Part III.

Nucleotides and Nucleic Acids Have Characteristic Bases and Pentoses

Nucleotides have three characteristic components: (1) a nitrogenous (nitrogen-containing) base, (2) a pentose, and (3) one or more phosphates (**Fig. 8–1**). The molecule without a phosphate group is called a **nucleoside**. The nitrogenous bases are derivatives of two parent

(a)

(b) Pyrimidine Purine

FIGURE 8–1 Structure of nucleotides. (a) General structure showing the numbering convention for the pentose ring. This is a ribonucleotide. In deoxyribonucleotides the —OH group on the 2′ carbon (in red) is replaced with H. **(b)** The parent compounds of the pyrimidine and purine bases of nucleotides and nucleic acids, showing the numbering conventions.

compounds, **pyrimidine** and **purine**. The bases and pentoses of the common nucleotides are heterocyclic compounds.

KEY CONVENTION: The carbon and nitrogen atoms in the parent structures are conventionally numbered to facilitate the naming and identification of the many derivative compounds. The convention for the pentose ring follows rules outlined in Chapter 7, but in the pentoses of nucleotides and nucleosides the carbon numbers are given a prime (′) designation to distinguish them from the numbered atoms of the nitrogenous bases. ∎

The base of a nucleotide is joined covalently (at N-1 of pyrimidines and N-9 of purines) in an *N*-β-glycosyl bond to the 1′ carbon of the pentose, and the phosphate is esterified to the 5′ carbon. The *N*-β-glycosyl bond is formed by removal of the elements of water (a hydroxyl group from the pentose and hydrogen from the base), as in *O*-glycosidic bond formation (see Fig. 7–30).

Both DNA and RNA contain two major purine bases, **adenine** (A) and **guanine** (G), and two major pyrimidines. In both DNA and RNA one of the pyrimidines is **cytosine** (C), but the second common pyrimidine is not the same in both: it is **thymine** (T) in DNA and **uracil** (U) in RNA. Only occasionally does thymine occur in RNA or uracil in DNA. The structures of the five major bases are shown in **Figure 8–2**, and the nomenclature of their corresponding nucleotides and nucleosides is summarized in Table 8–1.

Nucleic acids have two kinds of pentoses. The recurring deoxyribonucleotide units of DNA contain 2′-deoxy-D-ribose, and the ribonucleotide units of RNA contain D-ribose. In nucleotides, both types of pentoses are in their β-furanose (closed five-membered ring)

FIGURE 8–2 Major purine and pyrimidine bases of nucleic acids. Some of the common names of these bases reflect the circumstances of their discovery. Guanine, for example, was first isolated from guano (bird manure), and thymine was first isolated from thymus tissue.

form. As **Figure 8–3** shows, the pentose ring is not planar but occurs in one of a variety of conformations generally described as "puckered."

KEY CONVENTION: Although DNA and RNA seem to have two distinctions—different pentoses and the presence of uracil in RNA and thymine in DNA—it is the pentoses that define the identity of a nucleic acid. If the nucleic acid contains 2′-deoxy-D-ribose, it is DNA by definition even though it may contain uracil. Similarly, if the nucleic acid contains D-ribose, it is RNA regardless of its base composition. ∎

TABLE 8–1	Nucleotide and Nucleic Acid Nomenclature		
Base	Nucleoside	Nucleotide	Nucleic acid
Purines			
Adenine	Adenosine	Adenylate	RNA
	Deoxyadenosine	Deoxyadenylate	DNA
Guanine	Guanosine	Guanylate	RNA
	Deoxyguanosine	Deoxyguanylate	DNA
Pyrimidines			
Cytosine	Cytidine	Cytidylate	RNA
	Deoxycytidine	Deoxycytidylate	DNA
Thymine	Thymidine or deoxythymidine	Thymidylate or deoxythymidylate	DNA
Uracil	Uridine	Uridylate	RNA

Note: "Nucleoside" and "nucleotide" are generic terms that include both ribo- and deoxyribo- forms. Also, ribonucleosides and ribonucleotides are here designated simply as nucleosides and nucleotides (e.g., riboadenosine as adenosine), and deoxyribonucleosides and deoxyribonucleotides as deoxynucleosides and deoxynucleotides (e.g., deoxyriboadenosine as deoxyadenosine). Both forms of naming are acceptable, but the shortened names are more commonly used. Thymine is an exception; "ribothymidine" is used to describe its unusual occurrence in RNA.

FIGURE 8–3 Conformations of ribose. (a) In solution, the straight-chain (aldehyde) and ring (β-furanose) forms of free ribose are in equilibrium. RNA contains only the ring form, β-D-ribofuranose. Deoxyribose undergoes a similar interconversion in solution, but in DNA exists solely as β-2'-deoxy-D-ribofuranose. **(b)** Ribofuranose rings in nucleotides can exist in four different puckered conformations. In all cases, four of the five atoms are nearly in a single plane. The fifth atom (C-2' or C-3') is on either the same (endo) or the opposite (exo) side of the plane relative to the C-5' atom.

Figure 8–4 gives the structures and names of the four major **deoxyribonucleotides** (deoxyribonucleoside 5'-monophosphates), the structural units of DNAs, and the four major **ribonucleotides** (ribonucleoside 5'-monophosphates), the structural units of RNAs.

Nucleotide:	Deoxyadenylate (deoxyadenosine 5'-monophosphate)	Deoxyguanylate (deoxyguanosine 5'-monophosphate)	Deoxythymidylate (deoxythymidine 5'-monophosphate)	Deoxycytidylate (deoxycytidine 5'-monophosphate)
Symbols:	A, dA, dAMP	G, dG, dGMP	T, dT, dTMP	C, dC, dCMP
Nucleoside:	Deoxyadenosine	Deoxyguanosine	Deoxythymidine	Deoxycytidine

(a) Deoxyribonucleotides

Nucleotide:	Adenylate (adenosine 5'-monophosphate)	Guanylate (guanosine 5'-monophosphate)	Uridylate (uridine 5'-monophosphate)	Cytidylate (cytidine 5'-monophosphate)
Symbols:	A, AMP	G, GMP	U, UMP	C, CMP
Nucleoside:	Adenosine	Guanosine	Uridine	Cytidine

(b) Ribonucleotides

FIGURE 8–4 Deoxyribonucleotides and ribonucleotides of nucleic acids. All nucleotides are shown in their free form at pH 7.0. The nucleotide units of DNA **(a)** are usually symbolized as A, G, T, and C, sometimes as dA, dG, dT, and dC; those of RNA **(b)** as A, G, U, and C. In their free form the deoxyribonucleotides are commonly abbreviated dAMP, dGMP, dTMP, and dCMP; the ribonucleotides, AMP, GMP, UMP, and CMP. For each nucleotide in the figure, the more common name is followed by the complete name in parentheses. All abbreviations assume that the phosphate group is at the 5' position. The nucleoside portion of each molecule is shaded in light red. In this and the following illustrations, the ring carbons are not shown.

FIGURE 8–5 Some minor purine and pyrimidine bases, shown as the nucleosides. (a) Minor bases of DNA. 5-Methylcytidine occurs in the DNA of animals and higher plants, N^6-methyladenosine in bacterial DNA, and 5-hydroxymethylcytidine in the DNA of animals and of bacteria infected with certain bacteriophages. **(b)** Some minor bases of tRNAs. Inosine contains the base hypoxanthine. Note that pseudouridine, like uridine, contains uracil; they are distinct in the point of attachment to the ribose—in uridine, uracil is attached through N-1, the usual attachment point for pyrimidines; in pseudouridine, through C-5.

Although nucleotides bearing the major purines and pyrimidines are most common, both DNA and RNA also contain some minor bases **(Fig. 8–5)**. In DNA the most common of these are methylated forms of the major bases; in some viral DNAs, certain bases may be hydroxymethylated or glucosylated. Altered or unusual bases in DNA molecules often have roles in regulating or protecting the genetic information. Minor bases of many types are also found in RNAs, especially in tRNAs (see Fig. 8–25 and Fig. 26–22).

KEY CONVENTION: The nomenclature for the minor bases can be confusing. Like the major bases, many have common names—hypoxanthine, for example, shown as its nucleoside inosine in Figure 8–5. When an atom in the purine or pyrimidine ring is substituted, the usual

FIGURE 8–6 Some adenosine monophosphates. Adenosine 2′-monophosphate, 3′-monophosphate, and 2′,3′-cyclic monophosphate are formed by enzymatic and alkaline hydrolysis of RNA.

convention (used here) is simply to indicate the ring position of the substituent by its number—for example, 5-methylcytosine, 7-methylguanine, and 5-hydroxymethylcytosine (shown as the nucleosides in Fig. 8–5). The element to which the substituent is attached (N, C, O) is not identified. The convention changes when the substituted atom is exocyclic (not within the ring structure), in which case the type of atom is identified and the ring position to which it is attached is denoted with a superscript. The amino nitrogen attached to C-6 of adenine is N^6; similarly, the carbonyl oxygen and amino nitrogen at C-6 and C-2 of guanine are O^6 and N^2, respectively. Examples of this nomenclature are N^6-methyladenosine and N^2-methylguanosine (Fig. 8–5). ■

Cells also contain nucleotides with phosphate groups in positions other than on the 5′ carbon **(Fig. 8–6)**. **Ribonucleoside 2′,3′-cyclic monophosphates** are isolatable intermediates, and **ribonucleoside 3′-monophosphates** are end products of the hydrolysis of RNA by certain ribonucleases. Other variations are adenosine 3′,5′-cyclic monophosphate (cAMP) and guanosine 3′,5′-cyclic monophosphate (cGMP), considered at the end of this chapter.

Phosphodiester Bonds Link Successive Nucleotides in Nucleic Acids

The successive nucleotides of both DNA and RNA are covalently linked through phosphate-group "bridges," in which the 5′-phosphate group of one nucleotide unit is joined to the 3′-hydroxyl group of the next nucleotide, creating a **phosphodiester linkage (Fig. 8–7)**. Thus the covalent backbones of nucleic acids consist of alternating phosphate and pentose residues, and the nitrogenous bases may be regarded as side groups joined to the backbone at regular intervals. The backbones of both DNA and RNA are hydrophilic. The hydroxyl groups of the sugar residues form hydrogen bonds with water. The phosphate groups, with a pK_a near 0, are completely ionized and negatively charged at pH 7, and the negative charges are generally neutralized by ionic interactions with positive charges on proteins, metal ions, and polyamines.

KEY CONVENTION: All the phosphodiester linkages in DNA and RNA have the same orientation along the chain (Fig. 8–7), giving each linear nucleic acid strand a specific polarity and distinct 5′ and 3′ ends. By definition, the **5′ end** lacks a nucleotide at the 5′ position and the **3′ end** lacks a nucleotide at the 3′ position. Other groups (most often one or more phosphates) may be present on one or both ends. The 5′ to 3′ orientation of a strand of nucleic acid refers to the *ends* of the strand, not the orientation of the individual phosphodiester bonds linking its constituent nucleotides. ■

The covalent backbone of DNA and RNA is subject to slow, nonenzymatic hydrolysis of the phosphodiester bonds. In the test tube, RNA is hydrolyzed rapidly under alkaline conditions, but DNA is not; the 2′-hydroxyl groups in RNA (absent in DNA) are directly involved in the process. Cyclic 2′,3′-monophosphate nucleotides are the first products of the action of alkali on RNA and are rapidly hydrolyzed further to yield a mixture of 2′- and 3′-nucleoside monophosphates **(Fig. 8–8)**.

The nucleotide sequences of nucleic acids can be represented schematically, as illustrated below by a segment of DNA with five nucleotide units. The phosphate

FIGURE 8–7 Phosphodiester linkages in the covalent backbone of DNA and RNA. The phosphodiester bonds (one of which is shaded in the DNA) link successive nucleotide units. The backbone of alternating pentose and phosphate groups in both types of nucleic acid is highly polar. The 5′ and 3′ ends of the macromolecule may be free or may have an attached phosphoryl group.

FIGURE 8–8 Hydrolysis of RNA under alkaline conditions. The 2′ hydroxyl acts as a nucleophile in an intramolecular displacement. The 2′,3′-cyclic monophosphate derivative is further hydrolyzed to a mixture of 2′- and 3′-monophosphates. DNA, which lacks 2′ hydroxyls, is stable under similar conditions.

groups are symbolized by Ⓟ, and each deoxyribose is symbolized by a vertical line, from C-1′ at the top to C-5′ at the bottom (but keep in mind that the sugar is always in its closed-ring β-furanose form in nucleic acids). The connecting lines between nucleotides (which pass through Ⓟ) are drawn diagonally from the middle (C-3′) of the deoxyribose of one nucleotide to the bottom (C-5′) of the next.

Some simpler representations of this pentadeoxyribonucleotide are pA-C-G-T-A$_{OH}$, pApCpGpTpA, and pACGTA.

KEY CONVENTION: The sequence of a single strand of nucleic acid is always written with the 5′ end at the left and the 3′ end at the right—that is, in the 5′→3′ direction. ∎

A short nucleic acid is referred to as an **oligonucleotide**. The definition of "short" is somewhat arbitrary, but polymers containing 50 or fewer nucleotides are generally called oligonucleotides. A longer nucleic acid is called a **polynucleotide**.

The Properties of Nucleotide Bases Affect the Three-Dimensional Structure of Nucleic Acids

Free pyrimidines and purines are weakly basic compounds and thus are called bases. The purines and pyrimidines common in DNA and RNA are aromatic molecules (Fig. 8–2), a property with important consequences for the structure, electron distribution, and light absorption of nucleic acids. Electron delocalization among atoms in the ring gives most of the bonds partial double-bond character. One result is that pyrimidines are planar molecules and purines are very nearly planar, with a slight pucker. Free pyrimidine and purine bases

FIGURE 8–9 Tautomeric forms of uracil. The lactam form predominates at pH 7.0; the other forms become more prominent as pH decreases. The other free pyrimidines and the free purines also have tautomeric forms, but they are more rarely encountered.

may exist in two or more tautomeric forms depending on the pH. Uracil, for example, occurs in lactam, lactim, and double lactim forms **(Fig. 8–9)**. The structures shown in Figure 8–2 are the tautomers that predominate at pH 7.0. All nucleotide bases absorb UV light, and nucleic acids are characterized by a strong absorption at wavelengths near 260 nm **(Fig. 8–10)**.

The purine and pyrimidine bases are hydrophobic and relatively insoluble in water at the near-neutral pH of the cell. At acidic or alkaline pH the bases become charged and their solubility in water increases. Hydrophobic stacking interactions in which two or more bases are positioned with the planes of their rings parallel (like a stack of coins) are one of two important modes of interaction between bases in nucleic acids. The stacking also involves a combination of van der Waals and dipole-dipole interactions between the bases. Base stacking helps to minimize contact of the bases with water, and base-stacking interactions are very important in stabilizing the three-dimensional structure of nucleic acids, as described later.

The functional groups of pyrimidines and purines are ring nitrogens, carbonyl groups, and exocyclic amino groups. Hydrogen bonds involving the amino and carbonyl groups are the most important mode of interaction between two (and occasionally three or four) complementary strands of nucleic acid. The most common

FIGURE 8–10 Absorption spectra of the common nucleotides. The spectra are shown as the variation in molar extinction coefficient with wavelength. The molar extinction coefficients at 260 nm and pH 7.0 (ε_{260}) are listed in the table. The spectra of corresponding ribonucleotides and deoxyribonucleotides, as well as the nucleosides, are essentially identical. For mixtures of nucleotides, a wavelength of 260 nm (dashed vertical line) is used for absorption measurements.

Molar extinction coefficient at 260 nm, ε_{260} ($M^{-1}cm^{-1}$)	
AMP	15,400
GMP	11,700
UMP	9,900
dTMP	9,200
CMP	7,500

FIGURE 8–11 **Hydrogen-bonding patterns in the base pairs defined by Watson and Crick.** Here as elsewhere, hydrogen bonds are represented by three blue lines.

hydrogen-bonding patterns are those defined by James D. Watson and Francis Crick in 1953, in which A bonds specifically to T (or U) and G bonds to C **(Fig. 8–11)**. These two types of **base pairs** predominate in double-stranded DNA and RNA, and the tautomers shown in Figure 8–2 are responsible for these patterns. It is this specific pairing of bases that permits the duplication of genetic information, as we shall discuss later in this chapter.

James D. Watson

Francis Crick, 1916–2004

SUMMARY 8.1 Some Basics

▶ A nucleotide consists of a nitrogenous base (purine or pyrimidine), a pentose sugar, and one or more phosphate groups. Nucleic acids are polymers of nucleotides, joined together by phosphodiester linkages between the 5′-hydroxyl group of one pentose and the 3′-hydroxyl group of the next.

▶ There are two types of nucleic acid: RNA and DNA. The nucleotides in RNA contain ribose, and the common pyrimidine bases are uracil and cytosine. In DNA, the nucleotides contain 2′-deoxyribose, and the common pyrimidine bases are thymine and cytosine. The primary purines are adenine and guanine in both RNA and DNA.

8.2 Nucleic Acid Structure

The discovery of the structure of DNA by Watson and Crick in 1953 gave rise to entirely new disciplines and influenced the course of many established ones. In this section we focus on DNA structure, some of the events that led to its discovery, and more recent refinements in our understanding of DNA. RNA structure is also introduced.

As in the case of protein structure (Chapter 4), it is sometimes useful to describe nucleic acid structure in terms of hierarchical levels of complexity (primary, secondary, tertiary). The primary structure of a nucleic acid is its covalent structure and nucleotide sequence. Any regular, stable structure taken up by some or all of the nucleotides in a nucleic acid can be referred to as secondary structure. All structures considered in the remainder of this chapter fall under the heading of secondary structure. The complex folding of large chromosomes within eukaryotic chromatin and bacterial nucleoids, or the elaborate folding of large tRNA or rRNA molecules, is generally considered tertiary structure. DNA tertiary structure is discussed in Chapter 24, and RNA tertiary structure is considered in Chapter 26.

DNA Is a Double Helix That Stores Genetic Information

DNA was first isolated and characterized by Friedrich Miescher in 1868. He called the phosphorus-containing substance "nuclein." Not until the 1940s, with the work of Oswald T. Avery, Colin MacLeod, and Maclyn McCarty, was there any compelling evidence that DNA was the genetic material. Avery and his colleagues found that DNA extracted from a virulent (disease-causing) strain of the bacterium *Streptococcus pneumoniae* and injected into a nonvirulent strain of the same bacterium transformed the nonvirulent strain into a virulent strain. They concluded that the DNA from the virulent strain carried the genetic information for virulence. Then in 1952, experiments by Alfred D. Hershey and Martha Chase, in which they studied the infection of bacterial cells by a virus (bacteriophage) with radioactively labeled DNA or protein, removed any remaining doubt that DNA, not protein, carried the genetic information.

Another important clue to the structure of DNA came from the work of Erwin Chargaff and his colleagues in the late 1940s. They found that the four nucleotide bases of DNA occur in different ratios in the DNAs of different organisms and that the amounts of certain bases are closely related. These data, collected from DNAs of a great many different species, led Chargaff to the following conclusions:

1. The base composition of DNA generally varies from one species to another.

2. DNA specimens isolated from different tissues of the same species have the same base composition.

3. The base composition of DNA in a given species does not change with an organism's age, nutritional state, or changing environment.

4. In all cellular DNAs, regardless of the species, the number of adenosine residues is equal to the number of thymidine residues (that is, A = T), and the number of guanosine residues is equal to the number of cytidine residues (G = C). From these relationships it follows that the sum of the purine residues equals the sum of the pyrimidine residues; that is, A + G = T + C.

These quantitative relationships, sometimes called "Chargaff's rules," were confirmed by many subsequent researchers. They were a key to establishing the three-dimensional structure of DNA and yielded clues to how genetic information is encoded in DNA and passed from one generation to the next.

To shed more light on the structure of DNA, Rosalind Franklin and Maurice Wilkins used the powerful method

FIGURE 8–12 X-ray diffraction pattern of DNA fibers. The spots forming a cross in the center denote a helical structure. The heavy bands at the left and right arise from the recurring bases.

of x-ray diffraction (see Box 4–5) to analyze DNA fibers. They showed in the early 1950s that DNA produces a characteristic x-ray diffraction pattern **(Fig. 8–12)**. From this pattern it was deduced that DNA molecules are helical with two periodicities along their long axis, a primary one of 3.4 Å and a secondary one of 34 Å. The problem then was to formulate a three-dimensional model of the DNA molecule that could account not only for the x-ray diffraction data but also for the specific A = T and G = C base equivalences discovered by Chargaff and for the other chemical properties of DNA.

Rosalind Franklin, 1920–1958

Maurice Wilkins, 1916–2004

James Watson and Francis Crick relied on this accumulated information about DNA to set about deducing its structure. In 1953 they postulated a three-dimensional model of DNA structure that accounted for all the available data. It consists of two helical DNA chains wound around the same axis to form a right-handed double helix (see Box 4–1 for an explanation of the right- or left-handed sense of a helical structure). The hydrophilic backbones of alternating deoxyribose and phosphate groups are on the outside of the double helix, facing the surrounding water. The furanose ring of each deoxyribose is in the C-2' endo conformation. The purine and pyrimidine bases of both strands are stacked inside the double helix, with their hydrophobic and nearly planar

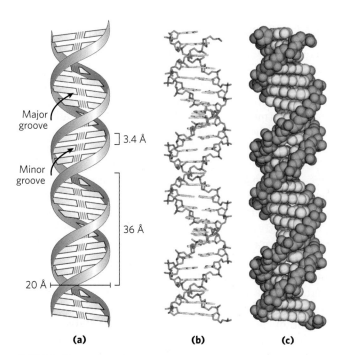

FIGURE 8–13 Watson-Crick model for the structure of DNA. The original model proposed by Watson and Crick had 10 base pairs, or 34 Å (3.4 nm), per turn of the helix; subsequent measurements revealed 10.5 base pairs, or 36 Å (3.6 nm), per turn. **(a)** Schematic representation, showing dimensions of the helix. **(b)** Stick representation showing the backbone and stacking of the bases. **(c)** Space-filling model.

ring structures very close together and perpendicular to the long axis. The offset pairing of the two strands creates a **major groove** and **minor groove** on the surface of the duplex **(Fig. 8–13)**. Each nucleotide base of one strand is paired in the same plane with a base of the other strand. Watson and Crick found that the hydrogen-bonded base pairs illustrated in Figure 8–11, G with C and A with T, are those that fit best within the structure, providing a rationale for Chargaff's rule that in any DNA, G = C and A = T. It is important to note that three hydrogen bonds can form between G and C, symbolized G≡C, but only two can form between A and T, symbolized A=T. This is one reason for the finding that separation of paired DNA strands is more difficult the higher the ratio of G≡C to A=T base pairs. Other pairings of bases tend (to varying degrees) to destabilize the double-helical structure.

When Watson and Crick constructed their model, they had to decide at the outset whether the strands of DNA should be **parallel** or **antiparallel**—whether their 3′,5′-phosphodiester bonds should run in the same or opposite directions. An antiparallel orientation produced the most convincing model, and later work with DNA polymerases (Chapter 25) provided experimental evidence that the strands are indeed antiparallel, a finding ultimately confirmed by x-ray analysis.

To account for the periodicities observed in the x-ray diffraction patterns of DNA fibers, Watson and Crick manipulated molecular models to arrive at a structure in which the vertically stacked bases inside the double helix would be 3.4 Å apart; the secondary repeat

distance of about 34 Å was accounted for by the presence of 10 base pairs in each complete turn of the double helix. In aqueous solution the structure differs slightly from that in fibers, having 10.5 base pairs per helical turn (Fig. 8–13).

As **Figure 8–14** shows, the two antiparallel polynucleotide chains of double-helical DNA are not identical in either base sequence or composition. Instead they are **complementary** to each other. Wherever adenine occurs in one chain, thymine is found in the other; similarly, wherever guanine occurs in one chain, cytosine is found in the other.

The DNA double helix, or duplex, is held together by two forces, as described earlier: hydrogen bonding between complementary base pairs (Fig. 8–11) and base-stacking interactions. The complementarity between the DNA strands is attributable to the hydrogen bonding between base pairs. The base-stacking interactions, which are largely nonspecific with respect to the identity of the stacked bases, make the major contribution to the stability of the double helix.

The important features of the double-helical model of DNA structure are supported by much chemical and biological evidence. Moreover, the model immediately suggested a mechanism for the transmission of genetic information. The essential feature of the model is the complementarity of the two DNA strands. As Watson and Crick were able to see, well before confirmatory data became available, this structure could logically be replicated by (1) separating the two strands and (2) synthesizing a complementary strand for each. Because nucleotides in each new strand are joined in a sequence specified by the base-pairing rules stated above, each preexisting

FIGURE 8–14 Complementarity of strands in the DNA double helix. The complementary antiparallel strands of DNA follow the pairing rules proposed by Watson and Crick. The base-paired antiparallel strands differ in base composition: the left strand has the composition $A_3T_2G_1C_3$; the right, $A_2T_3G_3C_1$. They also differ in sequence when each chain is read in the 5′→3′ direction. Note the base equivalences: A = T and G = C in the duplex.

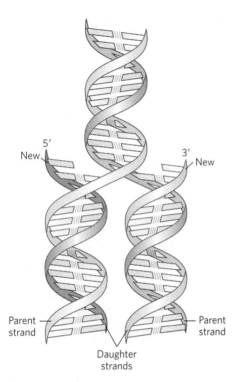

FIGURE 8–15 Replication of DNA as suggested by Watson and Crick. The preexisting or "parent" strands become separated, and each is the template for biosynthesis of a complementary "daughter" strand (in pink).

strand functions as a template to guide the synthesis of one complementary strand **(Fig. 8–15)**. These expectations were experimentally confirmed, inaugurating a revolution in our understanding of biological inheritance.

WORKED EXAMPLE 8–1 Base Pairing in DNA

In samples of DNA isolated from two unidentified species of bacteria, X and Y, adenine makes up 32% and 17%, respectively, of the total bases. What relative proportions of adenine, guanine, thymine, and cytosine would you expect to find in the two DNA samples? What assumptions have you made? One of these species was isolated from a hot spring (64 °C). Which species is most likely the thermophilic bacterium, and why?

Solution: For any double-helical DNA, A = T and G = C. The DNA from species X has 32% A and therefore must contain 32% T. This accounts for 64% of the bases and leaves 36% as G≡C pairs: 18% G and 18% C. The sample from species Y, with 17% A, must contain 17% T, accounting for 34% of the base pairs. The remaining 66% of the bases are thus equally distributed as 33% G and 33% C. This calculation is based on the assumption that both DNA molecules are double-stranded.

The higher the G + C content of a DNA molecule, the higher the melting temperature. Species Y, having the DNA with the higher G + C content (66%), most likely is the thermophilic bacterium; its DNA has a higher melting temperature and thus is more stable at the temperature of the hot spring.

DNA Can Occur in Different Three-Dimensional Forms

DNA is a remarkably flexible molecule. Considerable rotation is possible around several types of bonds in the sugar–phosphate (phosphodeoxyribose) backbone, and thermal fluctuation can produce bending, stretching, and unpairing (melting) of the strands. Many significant deviations from the Watson-Crick DNA structure are found in cellular DNA, some or all of which may be important in DNA metabolism. These structural variations generally do not affect the key properties of DNA defined by Watson and Crick: strand complementarity, antiparallel strands, and the requirement for A=T and G≡C base pairs.

Structural variation in DNA reflects three things: the different possible conformations of the deoxyribose, rotation about the contiguous bonds that make up the phosphodeoxyribose backbone **(Fig. 8–16a)**, and free rotation about the C-1′–N-glycosyl bond (Fig. 8–16b). Because of steric constraints, purines in purine nucleotides are restricted to two stable conformations with respect to deoxyribose, called syn and anti (Fig. 8–16b). Pyrimidines are generally restricted to the anti conformation because of steric interference between the sugar and the carbonyl oxygen at C-2 of the pyrimidine.

FIGURE 8–16 Structural variation in DNA. (a) The conformation of a nucleotide in DNA is affected by rotation about seven different bonds. Six of the bonds rotate freely. The limited rotation about bond 4 gives rise to ring pucker. This conformation is endo or exo, depending on whether the atom is displaced to the same side of the plane as C-5′ or to the opposite side (see Fig. 8–3b). **(b)** For purine bases in nucleotides, only two conformations with respect to the attached ribose units are sterically permitted, anti or syn. Pyrimidines occur in the anti conformation.

The Watson-Crick structure is also referred to as **B-form DNA**, or B-DNA. The B form is the most stable structure for a random-sequence DNA molecule under physiological conditions and is therefore the standard point of reference in any study of the properties of DNA. Two structural variants that have been well characterized in crystal structures are the **A** and **Z forms**. These three DNA conformations are shown in **Figure 8–17**, with a summary of their properties. The A form is favored in many solutions that are relatively devoid of water. The DNA is still arranged in a right-handed double helix, but the helix is wider and the number of base pairs per helical turn is 11, rather than 10.5 as in B-DNA. The plane of the base pairs in A-DNA is tilted about 20° relative to B-DNA base pairs, thus the base pairs in A-DNA are not perfectly perpendicular to the helix axis. These structural changes deepen the major groove while making the minor groove shallower. The reagents used to promote crystallization of DNA tend to dehydrate it, and thus most short DNA molecules tend to crystallize in the A form.

Z-form DNA is a more radical departure from the B structure; the most obvious distinction is the left-handed helical rotation. There are 12 base pairs per helical turn, and the structure appears more slender and elongated. The DNA backbone takes on a zigzag appearance. Certain nucleotide sequences fold into left-handed Z helices much more readily than others. Prominent examples are sequences in which pyrimidines alternate with purines, especially alternating C and G or 5-methyl-C and G residues. To form the left-handed helix in Z-DNA, the purine residues flip to the syn conformation, alternating with pyrimidines in the anti conformation. The major groove is barely apparent in Z-DNA, and the minor groove is narrow and deep.

Whether A-DNA occurs in cells is uncertain, but there is evidence for some short stretches (tracts) of Z-DNA in both bacteria and eukaryotes. These Z-DNA tracts may play a role (as yet undefined) in regulating the expression of some genes or in genetic recombination.

Certain DNA Sequences Adopt Unusual Structures

Other sequence-dependent structural variations found in larger chromosomes may affect the function and metabolism of the DNA segments in their immediate vicinity. For example, bends occur in the DNA helix wherever four or more adenosine residues appear sequentially in one strand. Six adenosines in a row produce a bend of about 18°. The bending observed with this and other sequences may be important in the binding of some proteins to DNA.

A rather common type of DNA sequence is a **palindrome**. A palindrome is a word, phrase, or sentence that is spelled identically read either forward or backward; two examples are ROTATOR and NURSES RUN. The term is applied to regions of DNA with **inverted repeats** of base sequence having twofold symmetry

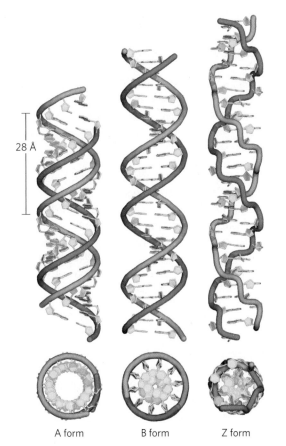

A form   B form   Z form

FIGURE 8–17 Comparison of A, B, and Z forms of DNA. Each structure shown here has 36 base pairs. The riboses and bases are shown in yellow. The phosphodiester backbone is represented as a blue rope. Blue is the color used to represent DNA strands in later chapters. The table summarizes some properties of the three forms of DNA.

	A form	B form	Z form
Helical sense	Right handed	Right handed	Left handed
Diameter	~26 Å	~20 Å	~18 Å
Base pairs per helical turn	11	10.5	12
Helix rise per base pair	2.6 Å	3.4 Å	3.7 Å
Base tilt normal to the helix axis	20°	6°	7°
Sugar pucker conformation	C-3′ endo	C-2′ endo	C-2′ endo for pyrimidines; C-3′ endo for purines
Glycosyl bond conformation	Anti	Anti	Anti for pyrimidines; syn for purines

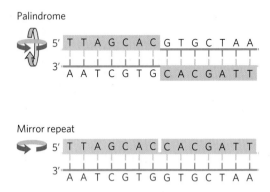

FIGURE 8–18 Palindromes and mirror repeats. Palindromes are sequences of double-stranded nucleic acids with twofold symmetry. In order to superimpose one repeat (shaded sequence) on the other, it must be rotated 180° about the horizontal axis then 180° about the vertical axis, as shown by the colored arrows. A mirror repeat, on the other hand, has a symmetric sequence within each strand. Superimposing one repeat on the other requires only a single 180° rotation about the vertical axis.

over two strands of DNA **(Fig. 8–18)**. Such sequences are self-complementary within each strand and therefore have the potential to form **hairpin** or **cruciform** (cross-shaped) structures **(Fig. 8–19)**. When the inverted repeat occurs within each individual strand of the DNA, the sequence is called a **mirror repeat**. Mirror repeats do not have complementary sequences within the same strand and cannot form hairpin or cruciform structures. Sequences of these types are found in virtually every large DNA molecule and can encompass a few base pairs or thousands. The extent to which palindromes occur as cruciforms in cells is not known, although some cruciform structures have been demonstrated in vivo in *Escherichia coli*. Self-complementary sequences cause isolated single strands of DNA (or RNA) in solution to fold into complex structures containing multiple hairpins.

Several unusual DNA structures involve three or even four DNA strands. Nucleotides participating in a Watson-Crick base pair (Fig. 8–11) can form additional hydrogen bonds, particularly with functional groups arrayed in the major groove. For example, a cytidine residue (if protonated) can pair with the guanosine residue of a G≡C nucleotide pair **(Fig. 8–20)**; a thymidine can pair with the adenosine of an A=T pair. The N-7, O^6, and N^6 of purines, the atoms that participate in the hydrogen bonding of triplex DNA, are often referred to as **Hoogsteen positions**, and the non-Watson-Crick pairing is called **Hoogsteen pairing**, after Karst Hoogsteen, who in 1963 first recognized the potential for these unusual pairings. Hoogsteen pairing allows the formation of **triplex DNAs**. The triplexes shown in Figure 8–20 (a, b) are most stable at low pH because the C≡G·C$^+$ triplet requires a protonated cytosine. In the triplex, the pK_a of this cytosine is >7.5, altered from its normal value of 4.2. The triplexes also form most readily within long sequences containing only pyrimidines or only purines in a given strand. Some triplex DNAs contain two pyrimidine

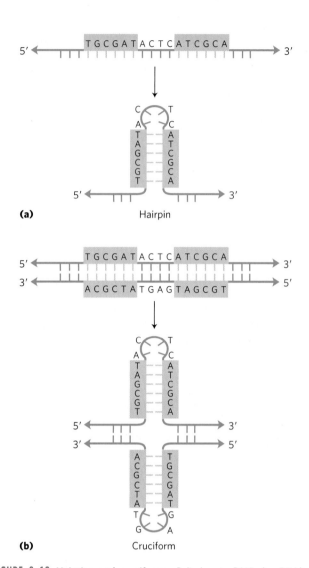

(a) Hairpin

(b) Cruciform

FIGURE 8–19 Hairpins and cruciforms. Palindromic DNA (or RNA) sequences can form alternative structures with intrastrand base pairing. **(a)** When only a single DNA (or RNA) strand is involved, the structure is called a hairpin. **(b)** When both strands of a duplex DNA are involved, it is called a cruciform. Blue shading highlights asymmetric sequences that can pair with the complementary sequence either in the same strand or in the complementary strand.

strands and one purine strand; others contain two purine strands and one pyrimidine strand.

Four DNA strands can also pair to form a tetraplex (quadruplex), but this occurs readily only for DNA sequences with a very high proportion of guanosine residues (Fig. 8–20c, d). The guanosine tetraplex, or **G tetraplex**, is quite stable over a wide range of conditions. The orientation of strands in the tetraplex can vary as shown in Figure 8–20e.

In the DNA of living cells, sites recognized by many sequence-specific DNA-binding proteins (Chapter 28) are arranged as palindromes, and polypyrimidine or polypurine sequences that can form triple helices are found within regions involved in the regulation of expression of some eukaryotic genes. In principle, synthetic DNA strands designed to pair with these sequences to form triplex DNA could disrupt gene

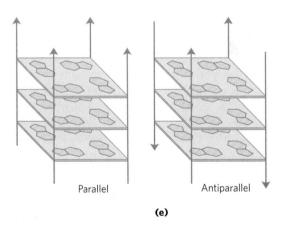

$T=A \cdot T$

$C \equiv G \cdot C^+$

(a)

(b)

Guanosine tetraplex

(c)

(d)

FIGURE 8-20 DNA structures containing three or four DNA strands.
(a) Base-pairing patterns in one well-characterized form of triplex DNA. The Hoogsteen pair in each case is shown in red. **(b)** Triple-helical DNA containing two pyrimidine strands (red and white; sequence TTCCT) and one purine strand (blue; sequence AAGGAA) (derived from PDB ID 1BCE). The blue and white strands are antiparallel and paired by normal Watson-Crick base-pairing patterns. The third (all-pyrimidine) strand (red) is parallel to the purine strand and paired through non-Watson-Crick hydrogen bonds. The triplex is viewed from the side, with six triplets shown. **(c)** Base-pairing pattern in the guanosine tetraplex structure. **(d)** Four successive tetraplets from a G tetraplex structure (PDB ID 244D). **(e)** Possible variants in the orientation of strands in a G tetraplex.

Parallel   Antiparallel

(e)

expression. This approach to controlling cellular metabolism is of commercial interest for its potential application in medicine and agriculture.

Messenger RNAs Code for Polypeptide Chains

We now turn our attention to the expression of the genetic information that DNA contains. RNA, the second major form of nucleic acid in cells, has many functions. In gene expression, RNA acts as an intermediary by using the information encoded in DNA to specify the amino acid sequence of a functional protein.

Given that the DNA of eukaryotes is largely confined to the nucleus whereas protein synthesis occurs on ribosomes in the cytoplasm, some molecule other than DNA must carry the genetic message from the nucleus to the cytoplasm. As early as the 1950s, RNA was considered the logical candidate: RNA is found in both the nucleus and the cytoplasm, and an increase in protein synthesis is accompanied by an increase in the amount of cytoplasmic RNA and an increase in its rate of turnover. These and other observations led several researchers to suggest that RNA carries genetic information from DNA to the protein biosynthetic machinery of the ribosome. In 1961 François Jacob and Jacques Monod presented a unified (and essentially correct) picture of many aspects of this process. They proposed

5′ ▬▬▬▬▬▬▬▬ 3′
Gene
(a) Monocistronic

5′ ▬▬▬▬▬▬▬▬▬▬▬▬▬ 3′
Gene 1   Gene 2   Gene 3
(b) Polycistronic

FIGURE 8–21 Bacterial mRNA. Schematic diagrams show **(a)** monocistronic and **(b)** polycistronic mRNAs of bacteria. Red segments represent RNA coding for a gene product; gray segments represent noncoding RNA. In the polycistronic transcript, noncoding RNA separates the three genes.

the name "messenger RNA" (mRNA) for that portion of the total cellular RNA carrying the genetic information from DNA to the ribosomes, where the messengers provide the templates that specify amino acid sequences in polypeptide chains. Although mRNAs from different genes can vary greatly in length, the mRNAs from a particular gene generally have a defined size. The process of forming mRNA on a DNA template is known as **transcription**.

In bacteria and archaea, a single mRNA molecule may code for one or several polypeptide chains. If it carries the code for only one polypeptide, the mRNA is **monocistronic**; if it codes for two or more different polypeptides, the mRNA is **polycistronic**. In eukaryotes, most mRNAs are monocistronic. (For the purposes of this discussion, "cistron" refers to a gene. The term itself has historical roots in the science of genetics, and its formal genetic definition is beyond the scope of this text.) The minimum length of an mRNA is set by the length of the polypeptide chain for which it codes. For example, a polypeptide chain of 100 amino acid residues requires an RNA coding sequence of at least 300 nucleotides, because each amino acid is coded by a nucleotide triplet (this and other details of protein synthesis are discussed in Chapter 27). However, mRNAs transcribed from DNA are always somewhat longer than the length needed simply to code for a polypeptide sequence (or sequences). The additional, noncoding RNA includes sequences that regulate protein synthesis. **Figure 8–21** summarizes the general structure of bacterial mRNAs.

Many RNAs Have More Complex Three-Dimensional Structures

Messenger RNA is only one of several classes of cellular RNA. Transfer RNAs are adapter molecules in protein synthesis; covalently linked to an amino acid at one end, they pair with the mRNA in such a way that amino acids are joined to a growing polypeptide in the correct sequence. Ribosomal RNAs are components of ribosomes. There is also a wide variety of special-function RNAs, including some (called ribozymes) that have enzymatic activity. All the RNAs are considered in detail

in Chapter 26. The diverse and often complex functions of these RNAs reflect a diversity of structure much richer than that observed in DNA molecules.

The product of transcription of DNA is always single-stranded RNA. The single strand tends to assume a right-handed helical conformation dominated by base-stacking interactions **(Fig. 8–22)**, which are stronger between two purines than between a purine and pyrimidine or between two pyrimidines. The purine-purine interaction is so strong that a pyrimidine separating two purines is often displaced from the stacking pattern so that the purines can interact. Any self-complementary sequences in the molecule produce more complex structures. RNA can base-pair with complementary regions of either RNA or DNA. Base pairing matches the pattern for DNA: G pairs with C and A pairs with U (or with the occasional T residue in some RNAs). One difference is that base pairing between G and U residues—unusual in DNA—is allowed in RNA (see Fig. 8–24) when complementary sequences in two single strands of RNA pair with each other. The paired strands in RNA or RNA-DNA duplexes are antiparallel, as in DNA.

When two strands of RNA with perfectly complementary sequences are paired, the predominant double-stranded structure is an A-form right-handed double helix. However, strands of RNA that are perfectly paired over long regions of sequence are uncommon. The three-dimensional structures of many RNAs, like those of proteins, are complex and unique. Weak

FIGURE 8–22 Typical right-handed stacking pattern of single-stranded RNA. The bases are shown in yellow, the phosphorus atoms in orange, and the riboses and phosphate oxygens in green. Green is used to represent RNA strands in succeeding chapters, just as blue is used for DNA.

interactions, especially base-stacking interactions, help stabilize RNA structures, just as they do in DNA. Z-form helices have been made in the laboratory (under very high-salt or high-temperature conditions). The B form of RNA has not been observed. Breaks in the regular A-form helix caused by mismatched or unmatched bases in one or both strands are common and result in bulges or internal loops (**Fig. 8–23**). Hairpin loops form between nearby self-complementary (palindromic) sequences. The potential for base-paired helical segments in many RNAs is extensive (**Fig. 8–24**), and the resulting hairpins are the most common type of secondary structure in RNA. Specific short base sequences (such as UUCG) are often found at the ends of RNA hairpins and are known to form particularly tight and stable loops. Such sequences may act as starting points for the folding of an RNA molecule into its precise three-dimensional structure. Other contributions are made by hydrogen bonds that are not part of standard Watson-Crick base pairs. For example, the 2'-hydroxyl group of ribose can hydrogen-bond with other groups. Some of these properties are evident in the structure of the phenylalanine transfer RNA of yeast—the tRNA responsible for inserting Phe residues into polypeptides—and in two RNA enzymes, or ribozymes, whose functions, like those of

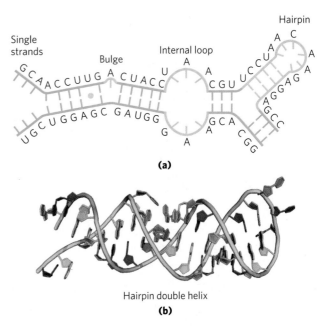

FIGURE 8–23 Secondary structure of RNAs. (a) Bulge, internal loop, and hairpin loop. **(b)** The paired regions generally have an A-form right-handed helix, as shown for a hairpin (derived from PDB ID 1GID).

FIGURE 8–24 Base-paired helical structures in an RNA. Shown here is the possible secondary structure of the M1 RNA component of the enzyme RNase P of *E. coli*, with many hairpins. RNase P, which also contains a protein component (not shown), functions in the processing of transfer RNAs (see Fig. 26-26). The two brackets indicate additional complementary sequences that may be paired in the three dimensional structure. The blue dots indicate non-Watson-Crick G≡U base pairs (boxed inset). Note that G≡U base pairs are allowed only when presynthesized strands of RNA fold up or anneal with each other. There are no RNA polymerases (the enzymes that synthesize RNAs on a DNA template) that insert a U opposite a template G, or vice versa, during RNA synthesis.

FIGURE 8–25 Three-dimensional structure in RNA. (a) Three-dimensional structure of phenylalanine tRNA of yeast (PDB ID 1TRA). Some unusual base-pairing patterns found in this tRNA are shown. Note also the involvement of the oxygen of a ribose phosphodiester bond in one hydrogen-bonding arrangement, and a ribose 2'-hydroxyl group in another (both in red). **(b)** A hammerhead ribozyme (so named because the secondary structure at the active site looks like the head of a hammer), derived from certain plant viruses (derived from PDB ID 1MME).

Ribozymes, or RNA enzymes, catalyze a variety of reactions, primarily in RNA metabolism and protein synthesis. The complex three-dimensional structures of these RNAs reflect the complexity inherent in catalysis, as described for protein enzymes in Chapter 6. **(c)** A segment of mRNA known as an intron, from the ciliated protozoan *Tetrahymena thermophila* (derived from PDB ID 1GRZ). This intron (a ribozyme) catalyzes its own excision from between exons in an mRNA strand (discussed in Chapter 26).

protein enzymes, depend on their three-dimensional structures **(Fig. 8–25)**.

The analysis of RNA structure and the relationship between its structure and its function is an emerging field of inquiry that has many of the same complexities as the analysis of protein structure. The importance of understanding RNA structure grows as we become increasingly aware of the large number of functional roles for RNA molecules.

SUMMARY 8.2 Nucleic Acid Structure

▶ Many lines of evidence show that DNA bears genetic information. Some of the earliest evidence came from the Avery-MacLeod-McCarty experiment, which showed that DNA isolated from one bacterial strain can enter and transform the cells of another strain, endowing it with some of the inheritable characteristics of the donor. The

Hershey-Chase experiment showed that the DNA of a bacterial virus, but not its protein coat, carries the genetic message for replication of the virus in a host cell.

▶ Putting together the available data, Watson and Crick postulated that native DNA consists of two antiparallel chains in a right-handed double-helical arrangement. Complementary base pairs, A=T and G≡C, are formed by hydrogen bonding within the helix. The base pairs are stacked perpendicular to the long axis of the double helix, 3.4 Å apart, with 10.5 base pairs per turn.

▶ DNA can exist in several structural forms. Two variations of the Watson-Crick form, or B-DNA, are A- and Z-DNA. Some sequence-dependent structural variations cause bends in the DNA molecule. DNA strands with appropriate sequences can form hairpin or cruciform structures or triplex or tetraplex DNA.

▶ Messenger RNA transfers genetic information from DNA to ribosomes for protein synthesis. Transfer RNA and ribosomal RNA are also involved in protein synthesis. RNA can be structurally complex; single RNA strands can fold into hairpins, double-stranded regions, or complex loops.

8.3 Nucleic Acid Chemistry

The role of DNA as a repository of genetic information depends in part on its inherent stability. The chemical transformations that do occur are generally very slow in the absence of an enzyme catalyst. The long-term storage of information without alteration is so important to a cell, however, that even very slow reactions that alter DNA structure can be physiologically significant. Processes such as carcinogenesis and aging may be intimately linked to slowly accumulating, irreversible alterations of DNA. Other, nondestructive alterations also occur and are essential to function, such as the strand separation that must precede DNA replication or transcription. In addition to providing insights into physiological processes, our understanding of nucleic acid chemistry has given us a powerful array of technologies that have applications in molecular biology, medicine, and forensic science. We now examine the chemical properties of DNA and some of these technologies.

Double-Helical DNA and RNA Can Be Denatured

Solutions of carefully isolated, native DNA are highly viscous at pH 7.0 and room temperature (25 °C). When such a solution is subjected to extremes of pH or to temperatures above 80 °C, its viscosity decreases sharply, indicating that the DNA has undergone a physical change. Just as heat and extremes of pH denature globular proteins, they also cause denaturation, or melting, of double-helical DNA. Disruption of the hydrogen

bonds between paired bases and of base stacking causes unwinding of the double helix to form two single strands, completely separate from each other along the entire length or part of the length (partial denaturation) of the molecule. No covalent bonds in the DNA are broken **(Fig. 8–26)**.

Renaturation of a DNA molecule is a rapid one-step process, as long as a double-helical segment of a dozen or more residues still unites the two strands. When the temperature or pH is returned to the range in which most organisms live, the unwound segments of the two strands spontaneously rewind, or **anneal**, to yield the intact duplex (Fig. 8–26). However, if the two strands are completely separated, renaturation occurs in two steps. In the first, relatively slow step, the two strands "find" each other by random collisions and form a short segment of complementary double helix. The second step is much faster: the remaining unpaired bases successively come into register as base pairs, and the two strands "zipper" themselves together to form the double helix.

The close interaction between stacked bases in a nucleic acid has the effect of decreasing its absorption of UV light relative to that of a solution with the same concentration of free nucleotides, and the absorption is decreased further when two complementary nucleic acid strands are paired. This is called the hypochromic

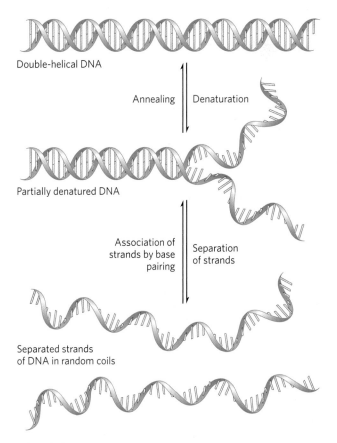

Double-helical DNA

Annealing ⇅ Denaturation

Partially denatured DNA

Association of strands by base pairing ⇅ Separation of strands

Separated strands of DNA in random coils

FIGURE 8–26 **Reversible denaturation and annealing (renaturation) of DNA.**

effect. Denaturation of a double-stranded nucleic acid produces the opposite result: an increase in absorption called the hyperchromic effect. The transition from double-stranded DNA to the single-stranded, denatured form can thus be detected by monitoring UV absorption at 260 nm.

Viral or bacterial DNA molecules in solution denature when they are heated slowly **(Fig. 8–27)**. Each species of DNA has a characteristic denaturation temperature, or melting point (t_m; formally, the temperature at which half the DNA is present as separated single strands): the higher its content of G≡C base pairs, the higher the melting point of the DNA. This is because G≡C base pairs, with three hydrogen bonds, require more heat energy to dissociate than A=T base pairs. Thus the melting point of a DNA molecule, determined under fixed conditions of pH and ionic strength, can yield an estimate of its base composition. If denaturation conditions are carefully controlled, regions that are rich in A=T base pairs will specifically denature while most of the DNA remains double-stranded. Such

FIGURE 8–27 Heat denaturation of DNA. (a) The denaturation, or melting, curves of two DNA specimens. The temperature at the midpoint of the transition (t_m) is the melting point; it depends on pH and ionic strength and on the size and base composition of the DNA. **(b)** Relationship between t_m and the G+C content of a DNA.

FIGURE 8–28 Partially denatured DNA. This DNA was partially denatured, then fixed to prevent renaturation during sample preparation. The shadowing method used to visualize the DNA in this electron micrograph increases its diameter approximately fivefold and obliterates most details of the helix. However, length measurements can be obtained, and single-stranded regions are readily distinguishable from double-stranded regions. The arrows point to some single-stranded bubbles where denaturation has occurred. The regions that denature are highly reproducible and are rich in A=T base pairs.

denatured regions (called bubbles) can be visualized with electron microscopy **(Fig. 8–28)**. Note that in the strand separation of DNA that occurs in vivo during processes such as DNA replication and transcription, the sites where these processes are initiated are often rich in A=T base pairs, as we shall see.

Duplexes of two RNA strands or one RNA strand and one DNA strand (RNA-DNA hybrids) can also be denatured. Notably, RNA duplexes are more stable to heat denaturation than DNA duplexes. At neutral pH, denaturation of a double-helical RNA often requires temperatures 20 °C or more higher than those required for denaturation of a DNA molecule with a comparable sequence, assuming the strands in each molecule are perfectly complementary. The stability of an RNA-DNA hybrid is generally intermediate between that of RNA and DNA duplexes. The physical basis for these differences in thermal stability is not known.

Nucleic Acids from Different Species Can Form Hybrids

The ability of two complementary DNA strands to pair with one another can be used to detect similar DNA sequences in two different species or within the genome of a single species. If duplex DNAs isolated from human cells and from mouse cells are completely denatured by heating, then mixed and kept at about 25 °C below their t_m for many hours, much of the DNA will anneal. The rate of DNA annealing is affected by temperature, the length and concentration of the DNA fragments being annealed, the concentration of salts in the reaction mixture, and properties of the sequence itself (e.g., complexity and G≡C content). Temperature is especially important. If the temperature is too low, short sequences with coincidental similarity from distant, heterologous parts of the DNA molecules will anneal unproductively and interfere with the more general alignment of complementary DNA strands. Temperatures that are too high will favor denaturation. Most of the reannealing occurs between complementary mouse DNA strands to form mouse duplex DNA; similarly, most human DNA strands anneal with complementary human DNA strands. However, some strands of the mouse DNA will associate with human DNA strands to yield **hybrid duplexes**, in which segments of a mouse DNA strand form base-paired regions with segments of a human DNA strand **(Fig. 8–29)**. This reflects a common evolutionary heritage; different organisms generally have many proteins and RNAs with similar functions and, often, similar structures. In many cases, the DNAs encoding these proteins and RNAs have similar sequences. The closer the evolutionary relationship between two species, the more extensively their DNAs will hybridize. For example, human DNA hybridizes much more extensively with mouse DNA than with DNA from yeast.

The hybridization of DNA strands from different sources forms the basis for a powerful set of techniques essential to the practice of modern molecular genetics. A specific DNA sequence or gene can be detected in the presence of many other sequences if one already has an appropriate complementary DNA strand (usually labeled in some way) to hybridize with it (Chapter 9). The complementary DNA can be from a different species or from the same species, or it can be synthesized chemically in the laboratory using techniques described later in this chapter. Hybridization techniques can be varied to detect a specific RNA rather than DNA. The isolation and identification of specific genes and RNAs rely on these hybridization techniques. Applications of this technology make possible the identification of an individual on the basis of a single hair left at the scene of a crime or the prediction of the onset of a disease decades before symptoms appear (see Box 9–1).

Nucleotides and Nucleic Acids Undergo Nonenzymatic Transformations

Purines and pyrimidines, along with the nucleotides of which they are a part, undergo spontaneous alterations in their covalent structure. The rate of these reactions is generally *very slow*, but they are physiologically significant because of the cell's very low tolerance for alterations in its genetic information. Alterations in DNA structure that produce permanent changes in the genetic information encoded therein are called **mutations**, and much evidence suggests an intimate link between the accumulation of mutations in an individual organism and the process of aging and carcinogenesis.

Several nucleotide bases undergo spontaneous loss of their exocyclic amino groups (deamination) **(Fig. 8–30a)**. For example, under typical cellular conditions, deamination of cytosine (in DNA) to uracil occurs in about one of every 10^7 cytidine residues in 24 hours. This corresponds to about 100 spontaneous events per day, on average, in a mammalian cell. Deamination of adenine and guanine occurs at about 1/100th this rate.

The slow cytosine deamination reaction seems innocuous enough, but is almost certainly the reason why DNA contains thymine rather than uracil. The product of cytosine deamination (uracil) is readily recognized as foreign in DNA and is removed by a repair system (Chapter 25). If DNA normally contained uracil, recognition of uracils resulting from cytosine deamination would be more difficult, and unrepaired uracils would lead to permanent sequence changes as they were paired with adenines during replication. Cytosine deamination would gradually lead to a decrease in G≡C base pairs and an increase in A=U base pairs in the DNA of all cells. Over the millennia, cytosine deamination could eliminate G≡C base pairs and the genetic code that depends on them. Establishing thymine as one of the four bases in DNA may well

FIGURE 8–29 DNA hybridization. Two DNA samples to be compared are completely denatured by heating. When the two solutions are mixed and slowly cooled, DNA strands of each sample associate with their normal complementary partner and anneal to form duplexes. If the two DNAs have significant sequence similarity, they also tend to form partial duplexes or hybrids with each other: the greater the sequence similarity between the two DNAs, the greater the number of hybrids formed. Hybrid formation can be measured in several ways. One of the DNAs is usually labeled with a radioactive isotope to simplify their detection and measurement.

(a) Deamination

(b) Depuration

FIGURE 8–30 Some well-characterized nonenzymatic reactions of nucleotides. **(a)** Deamination reactions. Only the base is shown. **(b)** Depurination, in which a purine is lost by hydrolysis of the N-β-glycosyl bond. Loss of pyrimidines via a similar reaction occurs, but much more slowly. The resulting lesion, in which the deoxyribose is present but the base is not, is called an abasic site or an AP site (apurinic site or, rarely, apyrimidinic site). The deoxyribose remaining after depurination is readily converted from the β-furanose to the aldehyde form (see Fig. 8–3), further destabilizing the DNA at this position. More nonenzymatic reactions are illustrated in Figures 8–31 and 8–32.

have been one of the crucial turning points in evolution, making the long-term storage of genetic information possible.

Another important reaction in deoxyribonucleotides is the hydrolysis of the N-β-glycosyl bond between the base and the pentose, to create a DNA lesion called an AP (apurinic, apyrimidinic) site or abasic site (Fig. 8–30b). This occurs at a higher rate for purines than for pyrimidines. As many as one in 10^5 purines (10,000 per mammalian cell) are lost from DNA every 24 hours under typical cellular conditions. Depurination of ribonucleotides and RNA is much slower and generally is not considered physiologically significant. In the test tube, loss of purines can be accelerated by dilute acid. Incubation of DNA at pH 3 causes selective removal of the purine bases, resulting in a derivative called apurinic acid.

Other reactions are promoted by radiation. UV light induces the condensation of two ethylene groups to form a cyclobutane ring. In the cell, the same reaction between adjacent pyrimidine bases in nucleic acids forms cyclobutane pyrimidine dimers. This happens most frequently between adjacent thymidine residues

on the same DNA strand **(Fig. 8–31)**. A second type of pyrimidine dimer, called a 6-4 photoproduct, is also formed during UV irradiation. Ionizing radiation (x rays and gamma rays) can cause ring opening and fragmentation of bases as well as breaks in the covalent backbone of nucleic acids.

Virtually all forms of life are exposed to energy-rich radiation capable of causing chemical changes in DNA. Near-UV radiation (with wavelengths of 200 to 400 nm), which makes up a significant portion of the solar spectrum, is known to cause pyrimidine dimer formation and other chemical changes in the DNA of bacteria and of human skin cells. We are subject to a constant field of ionizing radiation in the form of cosmic rays, which can penetrate deep into the earth, as well as radiation emitted from radioactive elements, such as radium, plutonium, uranium, radon, ^{14}C, and ^{3}H. X rays used in medical and dental examinations and in radiation therapy of cancer and other diseases are another form of ionizing radiation. It is estimated that UV and ionizing radiations are responsible for about 10% of all DNA damage caused by environmental agents.

FIGURE 8–31 Formation of pyrimidine dimers induced by UV light.
(a) One type of reaction (on the left) results in the formation of a cyclo-butyl ring involving C-5 and C-6 of adjacent pyrimidine residues. An alternative reaction (on the right) results in a 6-4 photoproduct, with a linkage between C-6 of one pyrimidine and C-4 of its neighbor.
(b) Formation of a cyclobutane pyrimidine dimer introduces a bend or kink into the DNA (PDB ID 1TTD).

DNA also may be damaged by reactive chemicals introduced into the environment as products of industrial activity. Such products may not be injurious per se but may be metabolized by cells into forms that are. There are two prominent classes of such agents **(Fig. 8–32)**: (1) deaminating agents, particularly nitrous acid (HNO_2) or compounds that can be metabolized to nitrous acid or nitrites, and (2) alkylating agents.

Nitrous acid, formed from organic precursors such as nitrosamines and from nitrite and nitrate salts, is a potent accelerator of the deamination of bases. Bisulfite has similar effects. Both agents are used as preservatives in processed foods to prevent the growth of toxic bacteria. They do not seem to increase cancer risks significantly when used in this way, perhaps because they are used in small amounts and make only a minor contribution to the overall levels of DNA damage. (The potential health risk from food spoilage if these preservatives were not used is much greater.)

FIGURE 8–32 Chemical agents that cause DNA damage. (a) Precursors of nitrous acid, which promotes deamination reactions. **(b)** Alkylating agents. Only *S*-adenosylmethionine acts enzymatically.

(a) Nitrous acid precursors

(b) Alkylating agents

Alkylating agents can alter certain bases of DNA. For example, the highly reactive chemical dimethylsulfate (Fig. 8–32b) can methylate a guanine to yield O^6-methylguanine, which cannot base-pair with cytosine.

Guanine tautomers

O^6-Methylguanine

Many similar reactions are brought about by alkylating agents normally present in cells, such as *S*-adenosyl methionine.

The most important source of mutagenic alterations in DNA is oxidative damage. Excited-oxygen species such as hydrogen peroxide, hydroxyl radicals, and superoxide radicals arise during irradiation or as a byproduct of aerobic metabolism. Of these species, the hydroxyl radicals are responsible for most oxidative DNA damage. Cells have an elaborate defense system to destroy reactive oxygen species, including enzymes such as catalase and superoxide dismutase that convert reactive oxygen species to harmless products. A fraction of these oxidants inevitably escape cellular defenses, however, and damage to DNA occurs through any of a large, complex group of reactions ranging from oxidation of deoxyribose and base moieties to strand breaks. Accurate estimates for the extent of this damage are not yet available, but every day the DNA of each human cell is subjected to thousands of damaging oxidative reactions.

This is merely a sampling of the best-understood reactions that damage DNA. Many carcinogenic compounds in food, water, or air exert their cancer-causing effects by modifying bases in DNA. Nevertheless, the integrity of DNA as a polymer is better maintained than that of either RNA or protein, because DNA is the only macromolecule that has the benefit of extensive biochemical repair systems. These repair processes (described in Chapter 25) greatly lessen the impact of damage to DNA. ■

Some Bases of DNA Are Methylated

Certain nucleotide bases in DNA molecules are enzymatically methylated. Adenine and cytosine are methylated more often than guanine and thymine. Methylation is generally confined to certain sequences or regions of a DNA molecule. In some cases the function of methylation is well understood; in others the function remains unclear. All known DNA methylases use *S*-adenosylmethionine as a methyl group donor (Fig. 8–32b). *E. coli*

has two prominent methylation systems. One serves as part of a defense mechanism that helps the cell to distinguish its DNA from foreign DNA by marking its own DNA with methyl groups and destroying (foreign) DNA without the methyl groups (this is known as a restriction-modification system; see p. 314). The other system methylates adenosine residues within the sequence (5′)GATC(3′) to N^6-methyladenosine (Fig. 8–5a). This is mediated by the Dam (*DNA adenine methylation*) methylase, a component of a system that repairs mismatched base pairs formed occasionally during DNA replication (see Fig. 25–21).

In eukaryotic cells, about 5% of cytidine residues in DNA are methylated to 5-methylcytidine (Fig. 8–5a). Methylation is most common at CpG sequences, producing methyl-CpG symmetrically on both strands of the DNA. The extent of methylation of CpG sequences varies by molecular region in large eukaryotic DNA molecules.

The Sequences of Long DNA Strands Can Be Determined

In its capacity as a repository of information, a DNA molecule's most important property is its nucleotide sequence. Until the late 1970s, determining the sequence of a nucleic acid containing even five or ten nucleotides was very laborious. The development of two new techniques in 1977, one by Alan Maxam and Walter Gilbert and the other by Frederick Sanger, made possible the sequencing of larger DNA molecules with an ease unimagined just a few years before. The techniques depend on an improved understanding of nucleotide chemistry and DNA metabolism, and on electrophoretic methods for separating DNA strands differing in size by only one nucleotide. Electrophoresis of DNA is similar to that of proteins (see Fig. 3–18). Polyacrylamide is often used as the gel matrix in work with short DNA molecules (up to a few hundred nucleotides); agarose is generally used for longer pieces of DNA.

In both Sanger and Maxam-Gilbert sequencing, the general principle is to reduce the DNA to four sets of labeled fragments. The reaction producing each set is base-specific, so the lengths of the fragments correspond to positions in the DNA sequence where a certain base occurs. For example, for an oligonucleotide with the sequence pAATCGACT, labeled at the 5′ end (the left end), a reaction that breaks the DNA after each C residue will generate two labeled fragments: a four-nucleotide and a seven-nucleotide fragment; a reaction that breaks the DNA after each G will produce only one labeled, five-nucleotide fragment. Because the fragments are radioactively labeled at their 5′ ends, only the fragment to the 5′ side of the break is visualized. The fragment sizes correspond to the relative positions of C and G residues in the sequence. When the sets of fragments corresponding to each of the four bases are electrophoretically separated side by side, they produce a ladder of bands from which the sequence can be read directly **(Fig. 8–33)**. We illustrate only the Sanger

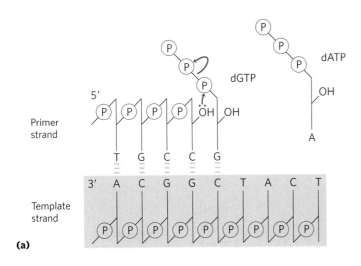

(a)

(b) ddNTP analog

FIGURE 8–33 DNA sequencing by the Sanger method. This method makes use of the mechanism of DNA synthesis by DNA polymerases (Chapter 25). **(a)** DNA polymerases require both a primer (a short oligonucleotide strand), to which nucleotides are added, and a template strand to guide selection of each new nucleotide. In cells, the 3′-hydroxyl group of the primer reacts with an incoming deoxynucleoside triphosphate (dNTP) to form a new phosphodiester bond. **(b)** The Sanger sequencing procedure uses dideoxynucleoside triphosphate (ddNTP) analogs to interrupt DNA synthesis. (The Sanger method is also known as the dideoxy method.) When a ddNTP is inserted in place of a dNTP, strand elongation is halted after the analog is added, because it lacks the 3′-hydroxyl group needed for the next step. **(c)** The DNA to be sequenced is used as the template strand, and a short primer, radioactively or fluorescently labeled, is annealed to it. By addition of small amounts of a single ddNTP, for example ddCTP, to an otherwise normal reaction system, the synthesized strands will be prematurely terminated at some locations where dC normally occurs. Given the excess of dCTP over ddCTP, the chance that the analog will be incorporated whenever a dC is to be added is small. However, ddCTP is present in sufficient amounts to ensure that each new strand has a high probability of acquiring at least one ddC at some point during synthesis. The result is a solution containing a mixture of labeled fragments, each ending with a C residue. Each C residue in the sequence generates a set of fragments of a particular length, such that the different-sized fragments, separated by electrophoresis, reveal the location of C residues. This procedure is repeated separately for each of the four ddNTPs, and the sequence can be read directly from an autoradiogram of the gel. Because shorter DNA fragments migrate faster, the fragments near the bottom of the gel represent the nucleotide positions closest to the primer (the 5′ end), and the sequence is read (in the 5′→3′ direction) from bottom to top. Note that the sequence obtained is that of the strand *complementary* to the strand being analyzed.

(c)

method, because it has proved to be technically easier and is in more widespread use. It requires the enzymatic synthesis of a DNA strand complementary to the strand under analysis, using a radioactively labeled "primer" and dideoxynucleotides.

Since these first practical DNA-sequencing methods appeared, methodology has improved rapidly. Much of the advance has been fueled by the Human Genome Project, described in Chapter 9. A variation of Sanger's sequencing method, in which the dideoxynucleotides used for each reaction are labeled with a differently colored fluorescent tag **(Fig. 8–34)**, was used in early efforts to automate large DNA-sequencing efforts. With this technology, researchers can sequence DNA molecules containing thousands of nucleotides in a few hours. This approach was heavily used in the initial efforts to sequence entire organism genomes, and is still used for routine sequencing of genes or DNA segments. However, modern genomic sequencing now makes use of vastly more efficient methods, sometimes referred to as next-generation or **next-gen sequencing**. These are described in Chapter 9. ▣ **Dideoxy Sequencing of DNA**

The Chemical Synthesis of DNA Has Been Automated

An important practical advance in nucleic acid chemistry was the rapid and accurate synthesis of short oligonucleotides of known sequence. The methods were pioneered by H. Gobind Khorana and his colleagues in the 1970s. Refinements by Robert Letsinger and Marvin Caruthers led to the chemistry now in widest use, called the phosphoramidite method **(Fig. 8–35)**. The synthesis is carried out with the growing strand attached to a solid support, using principles similar to those used by Merrifield for peptide synthesis (see Fig. 3–32), and is readily automated. The efficiency of each addition step is very high, allowing the routine synthesis of polymers containing 70 or 80 nucleotides and, in some laboratories, much longer strands. The availability of relatively inexpensive DNA polymers with predesigned sequences is having a powerful impact on all areas of biochemistry (Chapter 9).

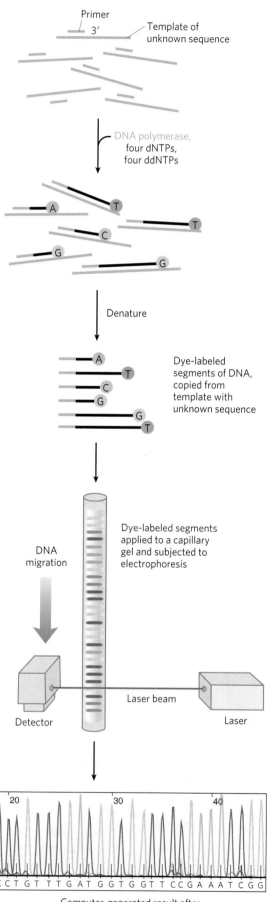

FIGURE 8–34 Strategy for automating DNA-sequencing reactions. Each dideoxynucleotide used in the Sanger method can be linked to a fluorescent molecule that gives all the fragments terminating in that nucleotide a particular color. All four labeled ddNTPs are added to a single tube. The resulting colored DNA fragments are then separated by size in a single electrophoretic gel contained in a capillary tube (a refinement of gel electrophoresis that allows for faster separations). All fragments of a given length migrate through the capillary gel in a single peak, and the color associated with each peak is detected using a laser beam. The DNA sequence is read by determining the sequence of colors in the peaks as they pass the detector. This information is fed directly to a computer, which determines the sequence.

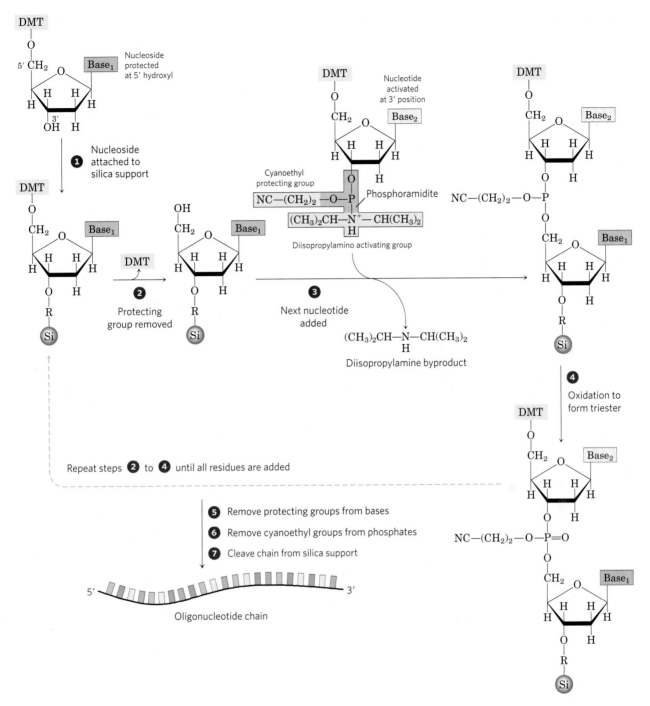

FIGURE 8–35 Chemical synthesis of DNA by the phosphoramidite method. Automated DNA synthesis is conceptually similar to the synthesis of polypeptides on a solid support. The oligonucleotide is built up on the solid support (silica), one nucleotide at a time, in a repeated series of chemical reactions with suitably protected nucleotide precursors. ❶ The first nucleoside (which will be the 3′ end) is attached to the silica support at the 3′ hydroxyl (through a linking group, R) and is protected at the 5′ hydroxyl with an acid-labile dimethoxytrityl group (DMT). The reactive groups on all bases are also chemically protected. ❷ The protecting DMT group is removed by washing the column with acid (the DMT group is colored, so this reaction can be followed spectrophotometrically). ❸ The next nucleotide has a reactive phosphoramidite at its 3′ position: a trivalent phosphite (as opposed to the more oxidized pentavalent phosphate normally present in nucleic acids) with one linked oxygen replaced by an amino group or substituted amine. In the common variant shown, one of the phosphoramidite oxygens is bonded to the deoxyribose, the other is protected by a cyanoethyl group, and the third position is occupied by a readily displaced diisopropylamino group. Reaction with the immobilized nucleotide forms a 5′,3′ linkage, and the diisopropylamino group is eliminated. In step ❹, the phosphite linkage is oxidized with iodine to produce a phosphotriester linkage. Reactions ❷ through ❹ are repeated until all nucleotides are added. At each step, excess nucleotide is removed before addition of the next nucleotide. In steps ❺ and ❻ the remaining protecting groups on the bases and the phosphates are removed, and in ❼ the oligonucleotide is separated from the solid support and purified. The chemical synthesis of RNA is somewhat more complicated because of the need to protect the 2′ hydroxyl of ribose without adversely affecting the reactivity of the 3′ hydroxyl.

SUMMARY 8.3 Nucleic Acid Chemistry

▶ Native DNA undergoes reversible unwinding and separation of strands (melting) on heating or at extremes of pH. DNAs rich in G≡C pairs have higher melting points than DNAs rich in A=T pairs.

▶ Denatured single-stranded DNAs from two species can form a hybrid duplex, the degree of hybridization depending on the extent of sequence similarity. Hybridization is the basis for important techniques used to study and isolate specific genes and RNAs.

▶ DNA is a relatively stable polymer. Spontaneous reactions such as deamination of certain bases, hydrolysis of base-sugar N-glycosyl bonds, radiation-induced formation of pyrimidine dimers, and oxidative damage occur at very low rates, yet are important because of a cell's very low tolerance for changes in genetic material.

▶ DNA sequences can be determined with a range of modern methods.

▶ Oligonucleotides of known sequence can be synthesized rapidly and accurately.

8.4 Other Functions of Nucleotides

In addition to their roles as the subunits of nucleic acids, nucleotides have a variety of other functions in every cell: as energy carriers, components of enzyme cofactors, and chemical messengers.

Nucleotides Carry Chemical Energy in Cells

The phosphate group covalently linked at the 5′ hydroxyl of a ribonucleotide may have one or two additional phosphates attached. The resulting molecules are referred to as nucleoside mono-, di-, and triphosphates (Fig. 8–36). Starting from the ribose, the three phosphates are generally labeled α, β, and γ. Hydrolysis of nucleoside triphosphates provides the chemical energy to drive many cellular reactions. Adenosine 5′-triphosphate, ATP, is by far the most widely used for this purpose, but UTP, GTP, and CTP are also used in some reactions. Nucleoside triphosphates also serve as the activated precursors of DNA and RNA synthesis, as described in Chapters 25 and 26.

The energy released by hydrolysis of ATP and the other nucleoside triphosphates is accounted for by the structure of the triphosphate group. The bond between the ribose and the α phosphate is an ester linkage. The α,β and β,γ linkages are phosphoanhydrides (Fig. 8–37). Hydrolysis of the ester linkage yields about 14 kJ/mol under standard conditions, whereas hydrolysis of each anhydride bond yields about 30 kJ/mol. ATP hydrolysis often plays an important thermodynamic role in biosynthesis. When coupled to a reaction with a positive free-energy change, ATP hydrolysis shifts the equilibrium of the overall process to favor product formation (recall the relationship between equilibrium constant and free-energy change described by Eqn 6–3 on p. 194).

Adenine Nucleotides Are Components of Many Enzyme Cofactors

A variety of enzyme cofactors serving a wide range of chemical functions include adenosine as part of their structure (Fig. 8–38). They are unrelated structurally except for the presence of adenosine. In none of these cofactors does the adenosine portion participate directly in the primary function, but removal of adenosine generally results in a drastic reduction of cofactor activities. For example, removal of the adenine nucleotide (3′-phosphoadenosine diphosphate) from acetoacetyl-CoA, the coenzyme A derivative of acetoacetate, reduces its reactivity as a substrate for β-ketoacyl-CoA transferase (an enzyme of lipid metabolism) by a factor of 10^6. Although this requirement for adenosine has not been investigated in detail, it must involve the binding energy between

	Abbreviations of ribonucleoside 5′-phosphates				Abbreviations of deoxyribonucleoside 5′-phosphates		
Base	Mono-	Di-	Tri-	Base	Mono-	Di-	Tri-
Adenine	AMP	ADP	ATP	Adenine	dAMP	dADP	dATP
Guanine	GMP	GDP	GTP	Guanine	dGMP	dGDP	dGTP
Cytosine	CMP	CDP	CTP	Cytosine	dCMP	dCDP	dCTP
Uracil	UMP	UDP	UTP	Thymine	dTMP	dTDP	dTTP

FIGURE 8–36 Nucleoside phosphates. General structure of the nucleoside 5′-mono-, di-, and triphosphates (NMPs, NDPs, and NTPs) and their standard abbreviations. In the deoxyribonucleoside phosphates (dNMPs, dNDPs, and dNTPs), the pentose is 2′-deoxy-D-ribose.

FIGURE 8–37 The phosphate ester and phosphoanhydride bonds of ATP. Hydrolysis of an anhydride bond yields more energy than hydrolysis of the ester. A carboxylic acid anhydride and carboxylic acid ester are shown for comparison.

enzyme and substrate (or cofactor) that is used both in catalysis and in stabilizing the initial enzyme-substrate complex (Chapter 6). In the case of β-ketoacyl-CoA transferase, the nucleotide moiety of coenzyme A seems to be a binding "handle" that helps to pull the substrate (acetoacetyl-CoA) into the active site. Similar roles may be found for the nucleoside portion of other nucleotide cofactors.

Why is adenosine, rather than some other large molecule, used in these structures? The answer here may involve a form of evolutionary economy. Adenosine is certainly not unique in the amount of potential binding energy it can contribute. The importance of adenosine probably lies not so much in some special chemical characteristic as in the evolutionary advantage of using one compound for multiple roles. Once ATP became the universal source of chemical energy, systems developed

FIGURE 8–38 Some coenzymes containing adenosine. The adenosine portion is shaded in light red. Coenzyme A (CoA) functions in acyl group transfer reactions; the acyl group (such as the acetyl or acetoacetyl group) is attached to the CoA through a thioester linkage to the β-mercaptoethylamine moiety. NAD^+ functions in hydride transfers, and FAD, the active form of vitamin B_2 (riboflavin), in electron transfers. Another coenzyme incorporating adenosine is 5'-deoxyadenosylcobalamin, the active form of vitamin B_{12} (see Box 17–2), which participates in intramolecular group transfers between adjacent carbons.

to synthesize ATP in greater abundance than the other nucleotides; because it is abundant, it becomes the logical choice for incorporation into a wide variety of structures. The economy extends to protein structure. A single protein domain that binds adenosine can be used in different enzymes. Such a domain, called a **nucleotide-binding fold**, is found in many enzymes that bind ATP and nucleotide cofactors.

Some Nucleotides Are Regulatory Molecules

Cells respond to their environment by taking cues from hormones or other external chemical signals. The interaction of these extracellular chemical signals ("first messengers") with receptors on the cell surface often leads to the production of **second messengers** inside the cell, which in turn leads to adaptive changes in the cell interior (Chapter 12). Often, the second messenger is a nucleotide **(Fig. 8–39)**. One of the most common is **adenosine 3′,5′-cyclic monophosphate (cyclic AMP, or cAMP)**, formed from ATP in a reaction catalyzed by adenylyl cyclase, an enzyme associated with the inner face of the plasma membrane. Cyclic AMP serves regulatory functions in virtually every cell outside the plant kingdom. Guanosine 3′,5′-cyclic monophosphate (cGMP) occurs in many cells and also has regulatory functions.

Another regulatory nucleotide, ppGpp (Fig. 8–39), is produced in bacteria in response to a slowdown in protein synthesis during amino acid starvation. This nucleotide inhibits the synthesis of the rRNA and tRNA molecules (see Fig. 28–22) needed for protein synthesis, preventing the unnecessary production of nucleic acids.

SUMMARY 8.4 Other Functions of Nucleotides

▶ ATP is the central carrier of chemical energy in cells. The presence of an adenosine moiety in a variety of enzyme cofactors may be related to binding-energy requirements.

▶ Cyclic AMP, formed from ATP in a reaction catalyzed by adenylyl cyclase, is a common second messenger produced in response to hormones and other chemical signals.

Adenosine 3′,5′-cyclic monophosphate
(cyclic AMP; cAMP)

Guanosine 3′,5′-cyclic monophosphate
(cyclic GMP; cGMP)

Guanosine 5′-diphosphate,3′-diphosphate
(guanosine tetraphosphate)
(ppGpp)

FIGURE 8–39 **Three regulatory nucleotides.**

Key Terms

Terms in bold are defined in the glossary.

gene 281	major groove 289
ribosomal RNA	minor groove 289
(rRNA) 281	B-form DNA 291
messenger RNA	A-form DNA 291
(mRNA) 281	Z-form DNA 291
transfer RNA	**palindrome** 291
(tRNA) 281	**hairpin** 292
nucleotide 281	**cruciform** 292
nucleoside 281	triplex DNA 292
pyrimidine 282	G tetraplex 292
purine 282	**transcription** 294
deoxyribonucleotides 283	**monocistronic**
ribonucleotide 283	**mRNA** 294
phosphodiester	**polycistronic mRNA** 294
linkage 285	**mutation** 299
5′ end 285	**second messenger** 308
3′ end 285	**adenosine 3′,5′-cyclic**
oligonucleotide 286	**monophosphate (cyclic**
polynucleotide 286	**AMP, cAMP)** 308
base pair 287	

Further Reading

General

Cox, M.M., Doudna, J.A., & O'Donnell, M. (2012) *Molecular Biology: Principles and Practice*, W. H. Freeman and Company, New York.

The best place to start to learn more about nucleic acid structure and function.

Friedberg, E.C., Walker, G.C., Siede, W., Wood, R.D., Schultz, R.A., & Ellenberger, T. (2006) *DNA Repair and Mutagenesis*, 2nd edn, ASM Press, Washington, DC.

A good source for more information on the chemistry of nucleotides and nucleic acids.

Historical

Judson, H.F. (1996) *The Eighth Day of Creation: Makers of the Revolution in Biology*, expanded edn, Cold Spring Harbor Laboratory Press, Cold Spring Harbor, NY.

Olby, R.C. (1994) *The Path to the Double Helix: The Discovery of DNA*, Dover Publications, Inc., New York.

Sayre, A. (1978) *Rosalind Franklin and DNA*, W. W. Norton & Co., Inc., New York.

Watson, J.D. (1968) *The Double Helix: A Personal Account of the Discovery of the Structure of DNA*, Atheneum, New York. [Paperback edition, Touchstone Books, 2001.]

Nucleic Acid Structure

Frank-Kamenetskii, M.D. & Mirkin, S.M. (1995) Triplex DNA structures. *Annu. Rev. Biochem.* **64**, 65–95.

Holbrook, S.R. (2008) Structural principles from large RNAs. *Annu. Rev. Biophys.* **37**, 445–464.

Keniry, M.A. (2000) Quadruplex structures in nucleic acids. *Biopolymers* **56**, 123–146.

Good summary of the structural properties of quadruplexes.

Nucleic Acid Chemistry

Bonetta, L. (2006) Genome sequencing in the fast lane. *Nat. Methods* **3**, 141–147.

This paper introduces a newer generation of sequencing methods that are described in Chapter 9.

Collins, A.R. (1999) Oxidative DNA damage, antioxidants, and cancer. *Bioessays* **21**, 238–246.

Cooke, M.S., Evans, M.D., Dizdaroglu, M., & Lunt J. (2003) Oxidative DNA damage: mechanisms, mutation, and disease. *FASEB J.* **17**, 1195–1214.

Imlay, J.A. (2008) Cellular defenses against superoxide and hydrogen peroxide. *Annu. Rev. Biochem.* **77**, 755–776.

Marnett, L.J. & Plastaras, J.P. (2001) Endogenous DNA damage and mutation. *Trends Genet.* **17**, 214–221.

ATP as Energy Carrier

Jencks, W.P. (1987) Economics of enzyme catalysis. *Cold Spring Harb. Symp. Quant. Biol.* **52**, 65–73.

A relatively short article, full of insights.

Problems

1. Nucleotide Structure Which positions in the purine ring of a purine nucleotide in DNA have the potential to form hydrogen bonds but are not involved in Watson-Crick base pairing?

2. Base Sequence of Complementary DNA Strands One strand of a double-helical DNA has the sequence (5′)GCGCAATATTTCTCAAAATATTGCGC(3′). Write the base sequence of the complementary strand. What special type of sequence is contained in this DNA segment? Does the double-stranded DNA have the potential to form any alternative structures?

3. DNA of the Human Body Calculate the weight in grams of a double-helical DNA molecule stretching from the Earth to the moon (~320,000 km). The DNA double helix weighs about 1×10^{-18} g per 1,000 nucleotide pairs; each base pair extends 3.4 Å. For an interesting comparison, your body contains about 0.5 g of DNA!

4. DNA Bending Assume that a poly(A) tract five base pairs long produces a 20° bend in a DNA strand. Calculate the total (net) bend produced in a DNA if the center base pairs (the third of five) of two successive $(dA)_5$ tracts are located (a) 10 base pairs apart; (b) 15 base pairs apart. Assume 10 base pairs per turn in the DNA double helix.

5. Distinction between DNA Structure and RNA Structure Hairpins may form at palindromic sequences in single strands of either RNA or DNA. How is the helical structure of a long and fully base-paired (except at the end) hairpin in RNA different from that of a similar hairpin in DNA?

6. Nucleotide Chemistry The cells of many eukaryotic organisms have highly specialized systems that specifically repair G–T mismatches in DNA. The mismatch is repaired to form a G≡C (not A=T) base pair. This G–T mismatch repair mechanism occurs in addition to a more general system that repairs virtually all mismatches. Suggest why cells might require a specialized system to repair G–T mismatches.

7. Denaturation of Nucleic Acids A duplex DNA oligonucleotide in which one of the strands has the sequence TAATACGACTCACTATAGGG has a melting temperature (t_m) of 59 °C. If an RNA duplex oligonucleotide of identical sequence (substituting U for T) is constructed, will its melting temperature be higher or lower?

8. Spontaneous DNA Damage Hydrolysis of the *N*-glycosyl bond between deoxyribose and a purine in DNA creates an AP site. An AP site generates a thermodynamic destabilization greater than that created by any DNA mismatched base pair. This effect is not completely understood. Examine the structure of an AP site (see Fig. 8–30b) and describe some chemical consequences of base loss.

9. Prediction of Nucleic Acid Structure from Its Sequence A part of a sequenced chromosome has the sequence (on one strand) ATTGCATCCGCGCGTGCGCGCGCG-ATCCCGTTACTTTCCG. Which part of this sequence is most likely to take up the Z conformation?

10. Nucleic Acid Structure Explain why the absorption of UV light by double-stranded DNA increases (the hyperchromic effect) when the DNA is denatured.

11. Determination of Protein Concentration in a Solution Containing Proteins and Nucleic Acids The concentration of protein or nucleic acid in a solution containing both can be estimated by using their different light absorption properties: proteins absorb most strongly at 280 nm and nucleic acids at 260 nm. Estimates of their respective concentrations in a mixture can be made by measuring the absorbance (A) of the solution at 280 and 260 nm and using the table on the next page, which gives $R_{280/260}$, the ratio of absorbances at 280 and 260 nm; the percentage of total mass that is nucleic acid; and a factor, F, that corrects the A_{280} reading and gives a more accurate protein estimate. The protein concentration (in mg/mL) = $F \times A_{280}$ (assuming the cuvette is 1 cm wide). Calculate the protein concentration in a solution of $A_{280} = 0.69$ and $A_{260} = 0.94$.

$R_{280/260}$	Proportion of nucleic acid (%)	F
1.75	0.00	1.116
1.63	0.25	1.081
1.52	0.50	1.054
1.40	0.75	1.023
1.36	1.00	0.994
1.30	1.25	0.970
1.25	1.50	0.944
1.16	2.00	0.899
1.09	2.50	0.852
1.03	3.00	0.814
0.979	3.50	0.776
0.939	4.00	0.743
0.874	5.00	0.682
0.846	5.50	0.656
0.822	6.00	0.632
0.804	6.50	0.607
0.784	7.00	0.585
0.767	7.50	0.565
0.753	8.00	0.545
0.730	9.00	0.508
0.705	10.00	0.478
0.671	12.00	0.422
0.644	14.00	0.377
0.615	17.00	0.322
0.595	20.00	0.278

12. Solubility of the Components of DNA Draw the following structures and rate their relative solubilities in water (most soluble to least soluble): deoxyribose, guanine, phosphate. How are these solubilities consistent with the three-dimensional structure of double-stranded DNA?

13. Sanger Sequencing Logic In the Sanger (dideoxy) method for DNA sequencing, a small amount of a dideoxynucleotide triphosphate—say, ddCTP—is added to the sequencing reaction along with a larger amount of the corresponding dCTP. What result would be observed if the dCTP were omitted?

14. DNA Sequencing The following DNA fragment was sequenced by the Sanger method. The red asterisk indicates a fluorescent label.

*5′ ━━━━━ 3′-OH
3′ ━━━━━ ATTACGCAAGGACATTAGAC---5′

A sample of the DNA was reacted with DNA polymerase and each of the nucleotide mixtures (in an appropriate buffer) listed below. Dideoxynucleotides (ddNTPs) were added in relatively small amounts.

1. dATP, dTTP, dCTP, dGTP, ddTTP
2. dATP, dTTP, dCTP, dGTP, ddGTP
3. dATP, dCTP, dGTP, ddTTP
4. dATP, dTTP, dCTP, dGTP

The resulting DNA was separated by electrophoresis on an agarose gel, and the fluorescent bands on the gel were located. The band pattern resulting from nucleotide mixture 1 is shown below. Assuming that all mixtures were run on the same gel, what did the remaining lanes of the gel look like?

15. Snake Venom Phosphodiesterase An exonuclease is an enzyme that sequentially cleaves nucleotides from the end of a polynucleotide strand. Snake venom phosphodiesterase, which hydrolyzes nucleotides from the 3′ end of any oligonucleotide with a free 3′-hydroxyl group, cleaves between the 3′ hydroxyl of the ribose or deoxyribose and the phosphoryl group of the next nucleotide. It acts on single-stranded DNA or RNA and has no base specificity. This enzyme was used in sequence determination experiments before the development of modern nucleic acid sequencing techniques. What are the products of partial digestion by snake venom phosphodiesterase of an oligonucleotide with the following sequence?

(5′)GCGCCAUUGC(3′)—OH

16. Preserving DNA in Bacterial Endospores Bacterial endospores form when the environment is no longer conducive to active cell metabolism. The soil bacterium *Bacillus subtilis*, for example, begins the process of sporulation when one or more nutrients are depleted. The end product is a small, metabolically dormant structure that can survive almost indefinitely with no detectable metabolism. Spores have mechanisms to prevent accumulation of potentially lethal mutations in their DNA over periods of dormancy that can exceed 1,000 years. *B. subtilis* spores are much more resistant than are the organism's growing cells to heat, UV radiation, and oxidizing agents, all of which promote mutations.

(a) One factor that prevents potential DNA damage in spores is their greatly decreased water content. How would this affect some types of mutations?

(b) Endospores have a category of proteins called small acid-soluble proteins (SASPs) that bind to their DNA, preventing formation of cyclobutane-type dimers. What causes cyclobutane dimers, and why do bacterial endospores need mechanisms to prevent their formation?

17. Oligonucleotide Synthesis In the scheme of Figure 8–35, each new base to be added to the growing oligonucleotide is modified so that its 3′ hydroxyl is activated and the 5′ hydroxyl has a dimethoxytrityl (DMT) group attached. What is the function of the DMT group on the incoming base?

Using the Web

18. The Structure of DNA Elucidation of the three-dimensional structure of DNA helped researchers understand how this molecule conveys information that can be faithfully replicated from one generation to the next. To see the secondary structure of double-stranded DNA, go to the Protein Data Bank website (www.pdb.org). Use the PDB identifiers listed below to retrieve the structure summaries for the two forms of DNA. Open the structures using Jmol, and use the controls in the Jmol menu (accessed with a control-click or by clicking on the Jmol logo in the lower right corner of the image screen) to complete the following exercises. Refer to the Jmol help links as needed.

(a) Obtain the file for 141D, a highly conserved, repeated DNA sequence from the end of the HIV-1 (the virus that causes AIDS) genome. Display the molecule as a ball-and-stick structure and color by element. Identify the sugar–phosphate backbone for each strand of the DNA duplex. Locate and identify individual bases. Identify the 5′ end of each strand. Locate the major and minor grooves. Is this a right- or left-handed helix?

(b) Obtain the file for 145D, a DNA with the Z conformation. Display the molecule as a ball-and-stick structure and color by element. Identify the sugar–phosphate backbone for each strand of the DNA duplex. Is this a right- or left-handed helix?

(c) To fully appreciate the secondary structure of DNA, view the molecules in stereo. On the control menu, Select > All, then Style > Stereographic > Cross-eyed viewing or Wall-eyed viewing. (If you have stereographic glasses available, select the appropriate option.) You will see two images of the DNA molecule. Sit with your nose approximately 10 inches from the monitor and focus on the tip of your nose (cross-eyed) or the opposite edges of the screen (wall-eyed). In the background you should see three images of the DNA helix. Shift your focus to the middle image, which should appear three-dimensional. (Note that only one of the two authors can make this work.)

Data Analysis Problem

19. Chargaff's Studies of DNA Structure The chapter section "DNA Is a Double Helix That Stores Genetic Information" includes a summary of the main findings of Erwin Chargaff and his coworkers, listed as four conclusions ("Chargaff's rules"; p. 288). In this problem, you will examine the data Chargaff collected in support of these conclusions.

In one paper, Chargaff (1950) described his analytical methods and some early results. Briefly, he treated DNA samples with acid to remove the bases, separated the bases by paper chromatography, and measured the amount of each base with UV spectroscopy. His results are shown in the three tables below. The *molar ratio* is the ratio of the number of moles of each base in the sample to the number of moles of phosphate in the sample—this gives the fraction of the total number of bases represented by each particular base. The *recovery* is the sum of all four bases (the sum of the molar ratios); full recovery of all bases in the DNA would give a recovery of 1.0.

Molar ratios in ox DNA

Base	Thymus Prep. 1	Prep. 2	Prep. 3	Spleen Prep. 1	Prep. 2	Liver Prep. 1
Adenine	0.26	0.28	0.30	0.25	0.26	0.26
Guanine	0.21	0.24	0.22	0.20	0.21	0.20
Cytosine	0.16	0.18	0.17	0.15	0.17	
Thymine	0.25	0.24	0.25	0.24	0.24	
Recovery	*0.88*	*0.94*	*0.94*	*0.84*	*0.88*	

Molar ratios in human DNA

Base	Sperm Prep. 1	Prep. 2	Thymus Prep. 1	Liver Normal	Carcinoma
Adenine	0.29	0.27	0.28	0.27	0.27
Guanine	0.18	0.17	0.19	0.19	0.18
Cytosine	0.18	0.18	0.16		0.15
Thymine	0.31	0.30	0.28		0.27
Recovery	*0.96*	*0.92*	*0.91*		*0.87*

Molar ratios in DNA of microorganisms

Base	Yeast Prep. 1	Prep. 2	Avian tubercle bacilli Prep. 1
Adenine	0.24	0.30	0.12
Guanine	0.14	0.18	0.28
Cytosine	0.13	0.15	0.26
Thymine	0.25	0.29	0.11
Recovery	*0.76*	*0.92*	*0.77*

(a) Based on these data, Chargaff concluded that "no differences in composition have so far been found in DNA from different tissues of the same species." This corresponds to conclusion 2 in this chapter. However, a skeptic looking at the data above might say, "They certainly look different to me!" If you were Chargaff, how would you use the data to convince the skeptic to change her mind?

(b) The base composition of DNA from normal and cancerous liver cells (hepatocarcinoma) was not distinguishably different. Would you expect Chargaff's technique to be capable of detecting a difference between the DNA of normal and cancerous cells? Explain your reasoning.

As you might expect, Chargaff's data were not completely convincing. He went on to improve his techniques, as described in his 1951 paper, in which he reported molar ratios of bases in DNA from a variety of organisms:

Source	A:G	T:C	A:T	G:C	Purine:pyrimidine
Ox	1.29	1.43	1.04	1.00	1.1
Human	1.56	1.75	1.00	1.00	1.0
Hen	1.45	1.29	1.06	0.91	0.99
Salmon	1.43	1.43	1.02	1.02	1.02
Wheat	1.22	1.18	1.00	0.97	0.99
Yeast	1.67	1.92	1.03	1.20	1.0
Haemophilus influenzae type c	1.74	1.54	1.07	0.91	1.0
E. coli K-12	1.05	0.95	1.09	0.99	1.0
Avian tubercle bacillus	0.4	0.4	1.09	1.08	1.1
Serratia marcescens	0.7	0.7	0.95	0.86	0.9
Bacillus schatz	0.7	0.6	1.12	0.89	1.0

(c) According to Chargaff, as stated in conclusion 1 in this chapter, "The base composition of DNA generally varies from one species to another." Provide an argument, based on the data presented so far, that supports this conclusion.

(d) According to conclusion 4, "In all cellular DNAs, regardless of the species . . . A + G = T + C." Provide an argument, based on the data presented so far, that supports this conclusion.

Part of Chargaff's intent was to disprove the "tetranucleotide hypothesis"; this was the idea that DNA was a monotonous tetranucleotide polymer $(AGCT)_n$ and therefore not capable of containing sequence information. Although the data presented above show that DNA cannot be simply a tetranucleotide—if so, all samples would have molar ratios of 0.25 for each base—it was still possible that the DNA from different organisms was a slightly more complex, but still monotonous, repeating sequence.

To address this issue, Chargaff took DNA from wheat germ and treated it with the enzyme deoxyribonuclease for different time intervals. At each time interval, some of the DNA was converted to small fragments; the remaining, larger fragments he called the "core." In the table below, the "19% core" corresponds to the larger fragments left behind when 81% of the DNA was degraded; the "8% core" corresponds to the larger fragments left after 92% degradation.

Base	Intact DNA	19% Core	8% Core
Adenine	0.27	0.33	0.35
Guanine	0.22	0.20	0.20
Cytosine	0.22	0.16	0.14
Thymine	0.27	0.26	0.23
Recovery	*0.98*	*0.95*	*0.92*

(e) How would you use these data to argue that wheat germ DNA is not a monotonous repeating sequence?

References

Chargaff, E. (1950) Chemical specificity of nucleic acids and mechanism of their enzymic degradation. *Experientia* **6**, 201–209.

Chargaff, E. (1951) Structure and function of nucleic acids as cell constituents. *Fed. Proc.* **10**, 654–659.

9

DNA-Based Information Technologies

The complexity of the molecules and systems revealed in this book can sometimes conceal a biochemical reality—what we have learned is just a beginning. Novel proteins and lipids and carbohydrates and nucleic acids are discovered every day, and we often have no clue as to their functions. How many have yet to be encountered, and what might they do? Even well-characterized biomolecules continue to challenge researchers with countless unresolved mechanistic and functional questions. The thousands of completed genome sequences in hand have provided one look at the immensity of the task ahead. Simply put, we do not know the function of most of the DNA—often including half or more of the genes—in a typical genome. Those same genomic sequences also provide an unprecedented opportunity. There is no greater source of information about a cell or organism than that buried in its own DNA. The task of extracting that information has given rise to sophisticated DNA-based technologies that are now at the heart of almost every biochemical story. The technologies we now turn to touch every topic we explore in subsequent chapters.

As objects of study, DNA molecules present a special problem—their size. Chromosomes are far and away the largest biomolecules in any cell. How does a researcher find the information he or she seeks when it is just a small part of a chromosome that can include millions or even billions of contiguous base pairs? Solutions to these problems began to emerge in the 1970s.

Decades of advances by thousands of scientists working in genetics, biochemistry, cell biology, and physical chemistry came together in the laboratories of Paul Berg, Herbert Boyer, and Stanley Cohen to yield techniques for locating, isolating, preparing, and studying small segments of DNA derived from much larger chromosomes. Techniques for DNA cloning paved the way to the modern field of **genomics** and more broadly to many of the technologies that contribute to **systems biology**, the study of biochemistry on the scale of whole cells and organisms.

Every student and instructor, when considering the topics we present in this chapter, encounters a conflict. First, the methods we describe were made possible by advances in our understanding of DNA and RNA metabolism. Hence, one must understand some fundamental concepts of DNA replication, RNA transcription, protein synthesis, and gene regulation to appreciate how these methods work. At the same time, however, modern biochemistry relies on these same methods to such an extent that a current treatment of any aspect of the discipline becomes very difficult without a proper introduction to them. By presenting these methods early in the book, we acknowledge that they are inextricably interwoven with both the advances that gave rise to them and the newer discoveries they

Paul Berg

Herbert Boyer

Stanley N. Cohen

now make possible. The background we necessarily provide makes the discussion here not just an introduction to technology, but also a preview of many of the fundamentals of DNA and RNA biochemistry encountered in later chapters.

We begin by outlining the principles of DNA cloning, then illustrate the range of applications and the potential of many newer technologies that support and accelerate the advance of biochemistry.

9.1 Studying Genes and Their Products

A researcher has isolated a new enzyme that she knows is the key to a human disease. She hopes to isolate large amounts of the protein to crystallize it for structural analysis and to study it. She wants to alter amino acid residues at its active site to understand the reaction it catalyzes. She plans an elaborate research program to elucidate how this enzyme interacts with, and is regulated by, other proteins in the cell. All of this, and much more, becomes possible if she can obtain the gene encoding her enzyme. Unfortunately, that gene is just a few thousand base pairs within a human chromosome with a size measured in hundreds of millions of base pairs. How does she isolate the small segment that she needs and then study it? The answer lies in DNA cloning and methods developed to manipulate cloned genes.

Genes Can Be Isolated by DNA Cloning

A *clone* is an identical copy. This term originally applied to cells of a single type, isolated and allowed to reproduce to create a population of identical cells. When applied to DNA, a clone represents many identical copies of a particular gene segment. In brief, our researcher must cut the gene out of the larger chromosome, attach it to a much smaller piece of carrier DNA, and allow microorganisms to make many copies of it for her. This is the process of **DNA cloning**. The result is selective amplification of a particular gene or DNA segment, facilitating its isolation and study. Classically, the cloning of DNA from any organism entails five general procedures:

1. *Cutting target DNA at precise locations.* Sequence-specific endonucleases (restriction endonucleases) provide the necessary molecular scissors.

2. *Selecting a small carrier molecule of DNA capable of self-replication.* These DNAs are called **cloning vectors** (a vector is a delivery agent). They are typically plasmids or viral DNAs.

3. *Joining two DNA fragments covalently.* The enzyme DNA ligase links the cloning vector and the DNA to be cloned. Composite DNA molecules comprising covalently linked segments from two or more sources are called **recombinant DNAs**.

4. *Moving recombinant DNA from the test tube to a host cell* that will provide the enzymatic machinery for DNA replication.

5. *Selecting or identifying host cells that contain recombinant DNA.*

The methods used to accomplish these and related tasks are collectively referred to as **recombinant DNA technology** or, more informally, **genetic engineering**.

Much of our initial discussion will focus on DNA cloning in the bacterium *Escherichia coli*, the first organism used for recombinant DNA work and still the most common host cell. *E. coli* has many advantages: its DNA metabolism (like many other of its biochemical processes) is well understood; many naturally occurring cloning vectors associated with *E. coli*, such as plasmids and bacteriophages (bacterial viruses; also called phages), are well characterized; and techniques are available for moving DNA expeditiously from one bacterial cell to another. The principles discussed here are broadly applicable to DNA cloning in other organisms, a topic discussed more fully later in the section.

Restriction Endonucleases and DNA Ligases Yield Recombinant DNA

Particularly important to recombinant DNA technology is a set of enzymes (Table 9–1) made available through decades of research on nucleic acid metabolism. Two classes of enzymes lie at the heart of the classic approach to generating and propagating a recombinant DNA molecule (**Fig. 9–1**). First, **restriction endonucleases** (also called restriction enzymes) recognize and cleave DNA at specific sequences (recognition sequences or restriction sites) to generate a set of smaller fragments. Second, the DNA fragment to be cloned is joined to a suitable cloning vector by using **DNA ligases** to link the DNA molecules together. The recombinant vector is then introduced into a host cell, which amplifies the fragment in the course of many generations of cell division.

Restriction endonucleases are found in a wide range of bacterial species. Werner Arber discovered in the early 1960s that their biological function is to recognize and cleave foreign DNA (the DNA of an infecting virus, for example); such DNA is said to be *restricted*. In the host cell's DNA, the sequence that would be recognized by its own restriction endonuclease is protected from digestion by methylation of the DNA, catalyzed by a specific DNA methylase. The restriction endonuclease and the corresponding methylase are sometimes referred to as a **restriction-modification system**.

There are three types of restriction endonucleases, designated I, II, and III. Types I and III are generally large, multisubunit complexes containing both the

TABLE 9–1	Some Enzymes Used in Recombinant DNA Technology
Enzyme(s)	Function
Type II restriction endonucleases	Cleave DNAs at specific base sequences
DNA ligase	Joins two DNA molecules or fragments
DNA polymerase I (*E. coli*)	Fills gaps in duplexes by stepwise addition of nucleotides to 3′ ends
Reverse transcriptase	Makes a DNA copy of an RNA molecule
Polynucleotide kinase	Adds a phosphate to the 5′-OH end of a polynucleotide to label it or permit ligation
Terminal transferase	Adds homopolymer tails to the 3′-OH ends of a linear duplex
Exonuclease III	Removes nucleotide residues from the 3′ ends of a DNA strand
Bacteriophage λ exonuclease	Removes nucleotides from the 5′ ends of a duplex to expose single-stranded 3′ ends
Alkaline phosphatase	Removes terminal phosphates from either the 5′ or 3′ end (or both)

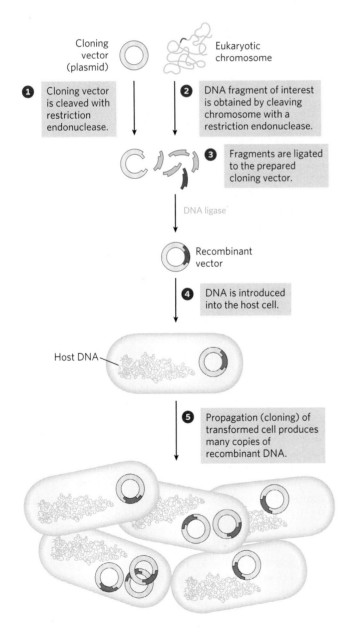

endonuclease and methylase activities. Type I restriction endonucleases cleave DNA at random sites that can be more than 1,000 base pairs (bp) from the recognition sequence. Type III restriction endonucleases cleave the DNA about 25 bp from the recognition sequence. Both types move along the DNA in a reaction that requires the energy of ATP. **Type II restriction endonucleases**, first isolated by Hamilton Smith in 1970, are simpler, require no ATP, and catalyze the hydrolytic cleavage of particular phosphodiester bonds in the DNA within the recognition sequence itself. The extraordinary utility of this group of restriction endonucleases was demonstrated by Daniel Nathans, who first used them to develop novel methods for mapping and analyzing genes and genomes.

Thousands of type II restriction endonucleases have been discovered in different bacterial species, and more than 100 different DNA sequences are recognized by one or more of these enzymes. The recognition sequences are usually 4 to 6 bp long and palindromic (see Fig. 8–18). Table 9–2 lists sequences recognized by a few type II restriction endonucleases.

Some restriction endonucleases make staggered cuts on the two DNA strands, leaving two to four nucleotides of one strand unpaired at each resulting end.

① Cloning vector is cleaved with restriction endonuclease.

② DNA fragment of interest is obtained by cleaving chromosome with a restriction endonuclease.

③ Fragments are ligated to the prepared cloning vector.

DNA ligase

Recombinant vector

④ DNA is introduced into the host cell.

Host DNA

⑤ Propagation (cloning) of transformed cell produces many copies of recombinant DNA.

Cloning vector (plasmid)

Eukaryotic chromosome

FIGURE 9–1 Schematic illustration of DNA cloning. A cloning vector and eukaryotic chromosomes are separately cleaved with the same restriction endonuclease. (A single chromosome is shown here for simplicity.) The fragments to be cloned are then ligated to the cloning vector. The resulting recombinant DNA (only one recombinant vector is shown here) is introduced into a host cell, where it can be propagated (cloned). Note that this drawing is not to scale: the size of the *E. coli* chromosome relative to that of a typical cloning vector (such as a plasmid) is much greater than depicted here.

TABLE 9–2	Recognition Sequences for Some Type II Restriction Endonucleases

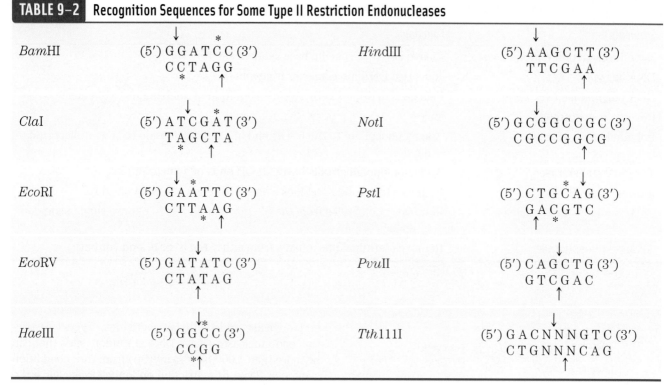

Note: Arrows indicate the phosphodiester bonds cleaved by each restriction endonuclease. Asterisks indicate bases that are methylated by the corresponding methylase (where known). N denotes any base. Note that the name of each enzyme consists of a three-letter abbreviation (in italics) of the bacterial species from which it is derived, sometimes followed by a strain designation and Roman numerals to distinguish different restriction endonucleases isolated from the same bacterial species. Thus *Bam*HI is the first (I) restriction endonuclease characterized from *Bacillus amyloliquefaciens*, strain H.

These unpaired strands are referred to as **sticky ends (Fig. 9–2a)** because they can base-pair with each other or with complementary sticky ends of other DNA fragments. Other restriction endonucleases cleave both strands of DNA at the opposing phosphodiester bonds, leaving no unpaired bases on the ends, often called **blunt ends** (Fig. 9–2b).

The average size of the DNA fragments produced by cleaving genomic DNA with a restriction endonuclease depends on the frequency with which a particular restriction site occurs in the DNA molecule; this in turn depends largely on the size of the recognition sequence.

In a DNA molecule with a random sequence in which all four nucleotides were equally abundant, a 6 bp sequence recognized by a restriction endonuclease such as *Bam*HI would occur on average once every 4^6 (4,096) bp. Enzymes that recognize a 4 bp sequence would produce smaller DNA fragments from a random-sequence DNA molecule; a recognition sequence of this size would be expected to occur about once every 4^4 (256) bp. In natural DNA molecules, particular recognition sequences tend to occur less frequently than this because nucleotide sequences in DNA are not random and the four nucleotides are not equally abundant. In laboratory experiments, the average size of the fragments produced by restriction endonuclease cleavage of a large DNA can be increased by simply terminating the reaction before completion; the result is called a partial digest. Average fragment size can also be increased by using a special class of endonucleases called homing endonucleases (see Fig. 26–37). These recognize and cleave much longer DNA sequences (14 to 20 bp).

Once a DNA molecule has been cleaved into fragments, a particular fragment of known size can be partially purified by agarose or acrylamide gel electrophoresis (p. 302) or by HPLC (p. 92). For a typical mammalian genome, however, cleavage by a restriction endonuclease usually yields too many different DNA fragments to permit convenient isolation of a particular fragment. A common intermediate step in the cloning of a specific gene or DNA segment is the construction of a DNA library (described in Section 9.2).

After the target DNA fragment is isolated, DNA ligase can be used to join it to a similarly digested cloning vector—that is, a vector digested by the *same* restriction endonuclease; a fragment generated by *Eco*RI, for example, generally will not link to a fragment generated by *Bam*HI. As described in more detail in Chapter 25 (see Fig. 25–16), DNA ligase catalyzes the formation of new phosphodiester bonds in a reaction that uses ATP or a similar cofactor. The base pairing of complementary sticky ends greatly facilitates the ligation reaction (Fig. 9–2a). Blunt ends can also be ligated, albeit less efficiently. Researchers can create new DNA sequences by inserting synthetic DNA fragments (called **linkers**) between the ends that are being ligated.

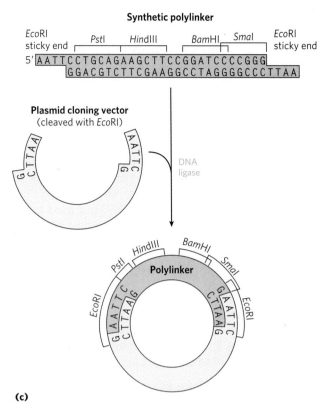

FIGURE 9–2 Cleavage of DNA molecules by restriction endonucleases. Restriction endonucleases recognize and cleave only specific sequences, leaving either **(a)** sticky ends (with protruding single strands) or **(b)** blunt ends. Fragments can be ligated to other DNAs, such as the cleaved cloning vector (a plasmid) shown here. This reaction is facilitated by the annealing of complementary sticky ends. Ligation is less efficient for DNA fragments with blunt ends than for those with complementary sticky ends, and DNA fragments with different (noncomplementary) sticky ends generally are not ligated. **(c)** A synthetic DNA fragment with recognition sequences for several restriction endonucleases can be inserted into a plasmid that has been cleaved by a restriction endonuclease. The insert is called a linker; an insert with multiple restriction sites is called a polylinker. 🖱 Restriction Endonucleases

Inserted DNA fragments with multiple recognition sequences for restriction endonucleases (often useful later as points for inserting additional DNA by cleavage and ligation) are called **polylinkers** (Fig. 9–2c).

The effectiveness of sticky ends in selectively joining two DNA fragments was apparent in the earliest recombinant DNA experiments. Before restriction endonucleases were widely available, some workers found they could generate sticky ends by the combined action of the bacteriophage λ exonuclease and terminal transferase (Table 9–1). The fragments to be joined were given complementary homopolymeric tails. Peter Lobban and Dale Kaiser used this method in 1971 in the first experiments to join naturally occurring DNA fragments. Similar methods were used soon after in the laboratory of Paul Berg to join DNA segments from simian virus 40 (SV40) to DNA derived from bacteriophage λ, thereby creating the first recombinant DNA molecule with DNA segments from different species.

Cloning Vectors Allow Amplification of Inserted DNA Segments

The principles that govern the delivery of recombinant DNA in clonable form to a host cell, and its subsequent amplification in the host, are well illustrated by considering three popular cloning vectors—plasmids and bacterial artificial chromosomes, used in experiments with *E. coli*, and a vector used to clone large DNA segments in yeast.

Plasmids A plasmid is a circular DNA molecule that replicates separately from the host chromosome. The wide variety of naturally occurring bacterial plasmids range in size from 5,000 to 400,000 bp. Many of the plasmids found in bacterial populations are little more than molecular parasites, similar to viruses but with a more limited capacity to transfer from one cell to another. To survive in the host cell, plasmids incorporate several specialized sequences that enable them to make use of the cell's resources for their own replication and gene expression.

Naturally occurring plasmids usually have a symbiotic role in the cell. They may provide genes that confer resistance to antibiotics or that perform new functions for the cell. For example, the Ti plasmid of *Agrobacterium tumefaciens* allows the host bacterium to colonize the cells of plants and make use of the plant's resources. The same properties that enable plasmids to grow and survive in a bacterial or eukaryotic host are useful to molecular biologists who want to engineer a vector for cloning a specific DNA segment. The classic *E. coli* plasmid pBR322, constructed in 1977, is a good example of a plasmid with features useful in almost all cloning vectors (**Fig. 9–3**):

1. The plasmid pBR322 has an **origin of replication**, or **ori**, a sequence where replication is initiated by cellular enzymes (see Chapter 25). This sequence is required to propagate the plasmid. An associated regulatory system is present that limits replication to maintain pBR322 at a level of 10 to 20 copies per cell.

2. The plasmid contains genes that confer resistance to the antibiotics tetracycline (Tet^R) and ampicillin (Amp^R), allowing the selection of cells that contain the intact plasmid or a recombinant version of the plasmid (discussed below).

3. Several unique recognition sequences in pBR322 are targets for restriction endonucleases (*Pst*I, *Eco*RI, *Bam*HI, *Sal*I, and *Pvu*II), providing sites where the plasmid can be cut to insert foreign DNA.

FIGURE 9–3 The constructed *E. coli* plasmid pBR322. Note the location of some important restriction sites—for *Pst*I, *Eco*RI, *Bam*HI, *Sal*I, and *Pvu*II; ampicillin- and tetracycline-resistance genes; and the replication origin (ori). Constructed in 1977, this was one of the early plasmids designed expressly for cloning in *E. coli*.

4. The small size of the plasmid (4,361 bp) facilitates its entry into cells and the biochemical manipulation of the DNA. This small size was generated simply by trimming away many DNA segments from a larger, parent plasmid—sequences that the molecular biologist does not need.

The replication origins inserted in common plasmid vectors were originally derived from naturally occurring plasmids. As in pBR322, each of these origins is regulated to maintain a particular plasmid copy number. Depending on the origin used, the plasmid copy number can vary from one to hundreds or thousands per cell, providing many options for investigators. Two different plasmids cannot function in the same cell if they use the same origin of replication, because the regulation of one will interfere with the replication of the other. Such plasmids are said to be incompatible. When a researcher wants to introduce two or more different plasmids into a bacterial cell, each plasmid must have a different replication origin.

In the laboratory, small plasmids can be introduced into bacterial cells by a process called **transformation**. The cells (often *E. coli*, but other bacterial species are also used) and plasmid DNA are incubated together at 0°C in a calcium chloride solution, then subjected to heat shock by rapidly shifting the temperature to between 37°C and 43°C. For reasons not well understood, some of the cells treated in this way take up the plasmid DNA. Some species of bacteria, such as *Acinetobacter baylyi*, are naturally competent for DNA uptake and do not require the calcium chloride–heat shock treatment. In an alternative method, cells incubated with the plasmid DNA are subjected to a high-voltage pulse. This approach, called **electroporation**, transiently renders the bacterial membrane permeable to large molecules.

Regardless of the approach, relatively few cells take up the plasmid DNA, so a method is needed to identify those that do. The usual strategy is to utilize one of two types of genes in the plasmid, referred to as selectable and screenable markers. **Selectable markers** either permit the growth of a cell (positive selection) or kill the cell (negative selection) under a defined set of conditions. The plasmid pBR322 provides examples of both positive and negative selection (**Fig. 9–4**). A **screenable marker** is a gene encoding a protein that causes the cell to produce a colored or fluorescent molecule. Cells are not harmed when the gene is present, and the cells that carry the plasmid are easily identified by the colored or fluorescent colonies they produce.

Transformation of typical bacterial cells with purified DNA (never a very efficient process) becomes less successful as plasmid size increases, and it is difficult to clone DNA segments longer than about 15,000 bp when plasmids are used as the vector.

To illustrate the use of a plasmid as a cloning vector, consider a typical bacterial gene, that encoding a

pBR322 plasmids

amp^R

tet^R

1 pBR322 is cleaved at the *amp*^R element by *Pst*I.

*Pst*I restriction endonuclease

Foreign DNA

2 DNA fragments to be cloned are ligated to cleaved pBR322. Where ligation is successful, the *amp*^R element is disrupted. The *tet*^R element remains intact.

DNA ligase

3 *E. coli* cells are transformed, then grown on agar plates containing tetracycline to select for those that have taken up plasmid.

transformation of *E. coli* cells

Incorporated plasmid

selection of transformed cells

All colonies have plasmids

Agar containing tetracycline

4 Individual colonies are transferred to matching positions on additional plates. One plate contains tetracycline, the other tetracycline and ampicillin.

colonies transferred for testing

Colonies with recombinant plasmids

Agar containing tetracycline (control)

Agar containing ampicillin + tetracycline

5 Cells that grow on tetracycline but not on tetracycline + ampicillin contain recombinant plasmids with disrupted ampicillin resistance, hence the foreign DNA. Cells with pBR322 without foreign DNA retain ampicillin resistance and grow on both plates.

FIGURE 9–4 Use of pBR322 to clone foreign DNA in *E. coli* and identify cells containing it. Plasmid Cloning

recombinase called the RecA protein (see Chapter 25). In most bacteria, the gene encoding RecA is one of thousands of other genes on a chromosome millions of base pairs long. The *recA* gene is just over 1,000 bp long. A plasmid would be a good choice for cloning a gene of this size. As described later, the cloned gene can be altered in a variety of ways, and the gene variants can be expressed at high levels to enable purification of the encoded protein.

Bacterial Artificial Chromosomes Large-genome sequencing projects often require the cloning of much longer DNA segments than can typically be incorporated into standard plasmid cloning vectors such as pBR322. To meet this need, plasmid vectors have been developed with special features that allow the cloning of very long segments (typically 100,000 to 300,000 bp) of DNA. Once such large segments of cloned DNA have been added, these vectors are large enough to be thought of as chromosomes, and are known as **bacterial artificial chromosomes**, or **BACs (Fig. 9–5)**.

A BAC vector (without any cloned DNA inserted) is a relatively simple plasmid, generally not much larger than other plasmid vectors. To accommodate very long segments of cloned DNA, BAC vectors have stable origins of replication that maintain the plasmid at one or two copies per cell. The low copy number is useful in cloning large segments of DNA, because it limits the opportunities for unwanted recombination reactions that can unpredictably alter large cloned DNAs over time. BACs also include *par* genes, which encode proteins that direct the reliable distribution of the recombinant chromosomes to daughter cells at cell division, thereby increasing the likelihood of each daughter cell carrying one copy, even when few copies are present. The BAC vector includes both selectable and screenable markers. The BAC vector shown in Figure 9–5 contains a gene that confers resistance to the antibiotic chloramphenicol (Cm^R). Positive selection for vector-containing cells occurs on agar plates containing this antibiotic. A *lacZ* gene, required for production of the enzyme β-galactosidase, is a screenable marker that can reveal which cells contain plasmids—now chromosomes—that incorporate the cloned DNA segments. The β-galactosidase catalyzes the conversion of the colorless molecule 5-bromo-4-chloro-3-indolyl-β-D-galactopyranoside (X-gal) to a blue product. If the gene is intact and expressed, the colony containing it will be blue. If gene expression is disrupted by the introduction of a cloned DNA segment, the colony will be white.

Yeast Artificial Chromosomes As with *E. coli*, yeast genetics is a well-developed discipline. The genome of *Saccharomyces cerevisiae* contains only 14×10^6 bp (less than four times the size of the *E. coli* chromosome), and its entire sequence is known. Yeast is also very easy to maintain and grow on a large scale in the laboratory. Plasmid vectors have been constructed for yeast,

Colonies with recombinant BACs are white.

FIGURE 9–5 Bacterial artificial chromosomes (BACs) as cloning vectors. The vector is a relatively simple plasmid, with a replication origin (ori) that directs replication. The *par* genes, derived from a type of plasmid called an F plasmid, assist in the even distribution of plasmids to daughter cells at cell division. This increases the likelihood of each daughter cell carrying one copy of the plasmid, even when few copies are present. The low number of copies is useful in cloning large segments of DNA because it limits the opportunities for unwanted recombination reactions that can unpredictably alter large cloned DNAs over time. The BAC includes selectable markers. A *lacZ* gene (required for the production of the enzyme β-galactosidase) is situated in the cloning region such that it is inactivated by cloned DNA inserts. Introduction of recombinant BACs into cells by electroporation is promoted by the use of cells with an altered (more porous) cell wall. Recombinant DNAs are screened for resistance to the antibiotic chloramphenicol (Cm^R). Plates also contain X-gal, a substrate for β-galactosidase that yields a blue product. Colonies with active β-galactosidase and hence no DNA insert in the BAC vector turn blue; colonies without β-galactosidase activity—and thus with the desired DNA inserts—are white.

employing the same principles that govern the use of *E. coli* vectors. Convenient methods for moving DNA into and out of yeast cells permit the study of many aspects of eukaryotic cell biochemistry. Some recombinant plasmids incorporate multiple replication origins and other elements that allow them to be used in more than one species (e.g., in yeast and in *E. coli*). Such plasmids that can be propagated in cells of two or more species are called **shuttle vectors**.

Research on large genomes and the associated need for high-capacity cloning vectors led to the development of **yeast artificial chromosomes**, or **YACs (Fig. 9–6)**. YAC vectors contain all the elements needed to maintain a eukaryotic chromosome in the yeast nucleus: a yeast origin of replication, two selectable markers, and specialized sequences (derived from the

telomeres and centromere) needed for stability and proper segregation of the chromosomes at cell division (see Chapter 24). In preparation for its use in cloning, the vector is propagated as a circular bacterial plasmid and then isolated and purified. Cleavage with a restriction endonuclease (*Bam*HI in Fig. 9–6) removes a length of DNA between two telomere sequences (TEL), leaving the telomeres at the ends of the linearized DNA. Cleavage at another internal site (by *Eco*RI in Fig. 9–6) divides the vector into two DNA segments, referred to as vector arms, each with a different selectable marker.

The genomic DNA to be cloned is prepared by partial digestion with restriction endonucleases to obtain a suitable fragment size. Genomic fragments are then separated by **pulsed field gel electrophoresis**, a variation of gel electrophoresis (see Fig. 3–18) that segregates very large DNA segments. DNA fragments of appropriate size (up to about 2×10^6 bp) are mixed with the prepared vector arms and ligated. The ligation mixture is then used to transform yeast cells (pretreated to partially degrade their cell walls) with these very large DNA molecules—which now have the structure and size to be considered yeast chromosomes. Culture on a medium that requires the presence of both selectable marker genes ensures the growth of only those yeast cells that contain an artificial chromosome with a large insert sandwiched between the two vector arms (Fig. 9–6). The stability of YAC clones increases with the length of the cloned DNA segment (up to a point). Those with inserts of more than 150,000 bp are nearly as stable as normal cellular chromosomes, whereas those with inserts less than 100,000 bp long are gradually lost during mitosis (so, generally, there are no yeast cell clones carrying only the two vector ends ligated together or vectors with only short inserts). YACs that lack a telomere at either end are rapidly degraded.

As with BACs, YAC vectors can be used to clone very long segments of DNA. In addition, the DNA cloned

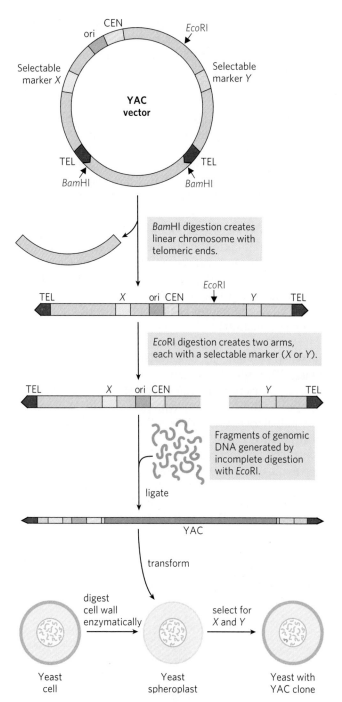

FIGURE 9–6 Construction of a yeast artificial chromosome (YAC). A YAC vector includes an origin of replication (ori), a centromere (CEN), two telomeres (TEL), and selectable markers (X and Y). Digestion with BamHI and EcoRI generates two separate DNA arms, each with a telomeric end and one selectable marker. A large segment of DNA (e.g., up to 2×10^6 bp from the human genome) is ligated to the two arms to create a yeast artificial chromosome. The YAC transforms yeast cells (prepared by removal of the cell wall to form spheroplasts), and the cells are selected for X and Y; the surviving cells propagate the DNA insert.

*Labels within figure:*
CEN
ori
EcoRI
Selectable marker X
Selectable marker Y
YAC vector
TEL
BamHI
TEL
BamHI

BamHI digestion creates linear chromosome with telomeric ends.

TEL   X   ori CEN   EcoRI   Y   TEL

EcoRI digestion creates two arms, each with a selectable marker (X or Y).

TEL   X   ori CEN   Y   TEL

Fragments of genomic DNA generated by incomplete digestion with EcoRI.

ligate

YAC

transform

digest cell wall enzymatically

select for X and Y

Yeast cell   Yeast spheroplast   Yeast with YAC clone

in a YAC can be altered to study the function of specialized sequences in chromosome metabolism, mechanisms of gene regulation and expression, and many other problems in eukaryotic molecular biology.

Cloned Genes Can Be Expressed to Amplify Protein Production

Frequently, the product of a cloned gene, rather than the gene itself, is of primary interest—particularly when the protein has commercial, therapeutic, or research value. Biochemists use purified proteins for many purposes, including to elucidate protein function, study reaction mechanisms, generate antibodies to the proteins, reconstitute complex cellular activities in the test tube with purified components, and examine protein binding partners. With an increased understanding of the fundamentals of DNA, RNA, and protein metabolism and their regulation in a host organism such as *E. coli* or yeast, investigators can manipulate cells to express cloned genes in order to study their protein products. The general goal is to alter the sequences around a cloned gene to trick the host organism into producing the protein product of the gene, often at very high levels. This overexpression of a protein can make its subsequent purification much easier.

We'll use the expression of a eukaryotic protein in a bacterium as an example. Eukaryotic genes have surrounding sequences needed for their transcription and regulation in the cells they are derived from, but these sequences do not function in bacteria. Thus, eukaryotic genes lack the DNA sequence elements required for their controlled expression in bacterial cells—promoters (sequences that instruct RNA polymerase where to bind to initiate mRNA synthesis), ribosome-binding sites (sequences that allow translation of the mRNA to protein), and additional regulatory sequences. Therefore, appropriate bacterial regulatory sequences for transcription and translation must be inserted in the vector DNA at the correct positions relative to the eukaryotic gene. In some cases, cloned genes are so efficiently expressed that their protein product represents 10% or more of the cellular protein. At these concentrations, some foreign proteins can kill the host cell (usually *E. coli*), so expression of the cloned gene must be limited to the few hours before the planned harvesting of the cells.

Cloning vectors with the transcription and translation signals needed for the regulated expression of a cloned gene are called **expression vectors**. The rate of expression of the cloned gene is controlled by replacing the gene's normal promoter and regulatory sequences with more efficient and convenient versions supplied by the vector. Generally, a well-characterized promoter and its regulatory elements are positioned near several unique restriction sites for cloning, so that genes inserted at the restriction sites will be expressed from the regulated promoter elements **(Fig. 9–7)**. Some of these vectors incorporate other features, such as a bacterial ribosome-binding site to enhance translation of the mRNA derived from the gene (Chapter 27) or a transcription termination sequence (Chapter 26).

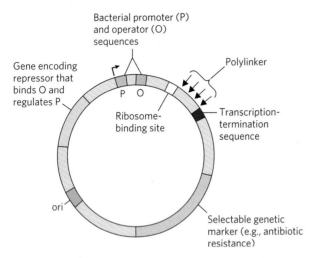

FIGURE 9-7 DNA sequences in a typical *E. coli* expression vector. The gene to be expressed is inserted into one of the restriction sites in the polylinker, near the promoter (P), with the end of the gene encoding the amino terminus of the protein positioned closest to the promoter. The promoter allows efficient transcription of the inserted gene, and the transcription-termination sequence sometimes improves the amount and stability of the mRNA produced. The operator (O) permits regulation by a repressor that binds to it. The ribosome-binding site provides sequence signals for the efficient translation of the mRNA derived from the gene. The selectable marker allows the selection of cells containing the recombinant DNA.

Many Different Systems Are Used to Express Recombinant Proteins

Every living organism has the capacity to express genes in its genomic DNA; thus, in principle, any organism can serve as a host to express proteins from a different (heterologous) species. Almost every sort of organism has, indeed, been used for this purpose, and each host type has a particular set of advantages and disadvantages.

Bacteria Bacteria, especially *E. coli*, remain the most common hosts for protein expression. The regulatory sequences that govern gene expression in *E. coli* and many other bacteria are well understood and can be harnessed to express cloned proteins at high levels. Bacteria are easy to store and grow in the laboratory, on inexpensive growth media. Efficient methods also exist to get DNA into bacteria and extract DNA from them. Bacteria can be grown in huge amounts in commercial fermenters, providing a rich source of the cloned protein. Problems do exist, however. When expressed in bacteria, some heterologous proteins do not fold correctly, and many do not undergo the covalent modifications or proteolytic cleavage that may be necessary for their activity. Certain features of a gene sequence also can make a particular gene difficult to express in bacteria. For example, intrinsically disordered regions are more common in eukaryotic proteins. Expressed in bacteria, many eukaryotic proteins aggregate into

insoluble cellular precipitates called inclusion bodies. For these and many other reasons, some eukaryotic proteins are inactive when purified from bacteria or cannot be expressed at all. To help address some of these problems, new bacterial host strains are regularly being developed that include enhancements such as the engineered presence of eukaryotic protein chaperones or enzymes that modify eukaryotic proteins.

There are many specialized systems for expressing proteins in bacteria. The promoter and regulatory sequences associated with the lactose operon (see Chapter 28) are often fused to the gene of interest to direct transcription. The cloned gene will be transcribed when lactose is added to the growth medium. However, regulation in the lactose system is "leaky"; it is not turned off completely when lactose is absent—a potential problem if the product of the cloned gene is toxic to the host cells. Transcription from the Lac promoter is also not efficient enough for some applications.

An alternative system uses a promoter and RNA polymerase found in a bacterial virus called bacteriophage T7. If the cloned gene is fused to a T7 promoter, it is not transcribed by the *E. coli* RNA polymerase, but instead by the T7 RNA polymerase. The gene encoding this polymerase is separately cloned into the same cell in a construct that affords tight regulation (allowing controlled production of the T7 RNA polymerase). The polymerase is also very efficient and directs high levels of expression of most genes fused to the T7 promoter. This system has been used to express the RecA protein in bacterial cells **(Fig. 9-8)**.

FIGURE 9-8 Regulated expression of RecA protein in a bacterial cell. The gene encoding the RecA protein, fused to a bacteriophage T7 promoter, is cloned into an expression vector. Under normal growth conditions (uninduced), no RecA protein appears. When the T7 RNA polymerase is induced in the cell, the *recA* gene is expressed, and large amounts of RecA protein are produced. The positions of standard molecular weight markers run on the same gel are indicated.

Yeast The yeast *Saccharomyces cerevisiae* is probably the best understood eukaryotic organism and one of the easiest to grow and manipulate in the laboratory. Like bacteria, this yeast can be grown on inexpensive media. Yeast have tough cell walls that are difficult to breach in order to introduce DNA vectors, so bacteria are more convenient for doing much of the genetic engineering and vector maintenance. This is why the yeast vector was first propagated in bacteria. Several excellent shuttle vectors exist for this purpose.

The principles underlying the expression of a protein in yeast are the same as those in bacteria. Cloned genes must be linked to promoters that can direct high-level expression in yeast. For example, the yeast *GAL1* and *GAL10* genes are under cellular regulation such that they are expressed when yeast cells are grown in media with galactose but shut down when the cells are grown in glucose. Thus, if a heterologous gene is expressed using the same regulatory sequences, the expression of that gene can be controlled simply by choosing an appropriate medium for cell growth.

Some of the same problems that accompany protein expression in bacteria also occur with yeast. Heterologous proteins may not fold properly, yeast may lack the enzymes needed to modify the proteins to their active forms, or the expression of proteins may be made difficult by certain features of the gene sequence. However, because *S. cerevisiae* is a eukaryote, the expression of eukaryotic genes (especially yeast genes) is sometimes more efficient in this host than in bacteria. Folding and modification of the products may also be more accurate than for proteins expressed in bacteria.

Insects and Insect Viruses Baculoviruses are insect viruses with double-stranded DNA genomes. When they infect their insect larval hosts, they act as parasites, killing the larvae and turning them into factories for virus production. Late in the infection process, the viruses produce large amounts of two proteins (p10 and polyhedrin), neither of which is needed for virus production in cultured insect cells. The genes for both of these proteins can be replaced with the gene of a heterologous protein. When the resulting recombinant virus is used to infect insect cells or larvae, the heterologous protein is often produced at very high levels—up to 25% of the total protein present at the end of the infection cycle.

Autographa californica multicapsid nucleopolyhedrovirus (AcMNPV) is the **baculovirus** most often used for protein expression. It has a large genome (134,000 bp), too large for direct cloning. Virus purification is also cumbersome. These problems have been solved by the creation of **bacmids**, large circular DNAs that include the entire baculovirus genome along with sequences that allow replication of the bacmid in *E. coli*

(Fig. 9–9). The gene of interest is cloned into a smaller plasmid and combined with the larger plasmid by site-specific recombination in vivo (see Fig. 25–37). The recombinant bacmid is then isolated and transfected into insect cells (the term **transfection** is used when the DNA used for transformation includes viral sequences and leads to viral replication), followed by recovery of the protein once the infection cycle is finished. A wide range of bacmid systems is available commercially. Baculovirus systems are not successful with all proteins. However, with these systems, insect cells sometimes successfully replicate the protein-modification patterns of higher eukaryotes and produce active, correctly modified eukaryotic proteins.

Mammalian Cells in Culture The most convenient way to introduce cloned genes into a mammalian cell is with viruses. This method takes advantage of the natural capacity of a virus to insert its DNA or RNA into a cell, and sometimes into the cellular chromosome. A variety of engineered mammalian viruses are available as vectors, including human adenoviruses and retroviruses. The gene of interest is cloned so that its expression is controlled by a virus promoter. The virus uses its natural infection mechanisms to introduce the recombinant genome into cells, where the cloned protein is expressed. These systems have the advantage that proteins can be expressed either transiently (if the viral DNA is maintained separately from the host cell genome and eventually degraded) or permanently (if the viral DNA is integrated into the host cell genome). With the correct choice of host cell, the proper post-translational modification of the protein to its active form can be ensured. However, the growth of mammalian cells in tissue culture is very expensive, and this technology is generally used to test the function of a protein in vivo rather than to produce a protein in large amounts.

Alteration of Cloned Genes Produces Altered Proteins

Cloning techniques can be used not only to overproduce proteins but to produce protein products altered, subtly or dramatically, from their native forms. Specific amino acids may be replaced individually by **site-directed mutagenesis**. This technique has greatly enhanced research on proteins by allowing investigators to make specific changes in the primary structure and examine the effects of these changes on the protein's folding, three-dimensional structure, and activity. This powerful approach to studying protein structure and function changes the amino acid sequence by altering the DNA sequence of the cloned gene. If appropriate restriction sites flank the sequence to be altered, researchers can simply remove a DNA segment and replace it with a

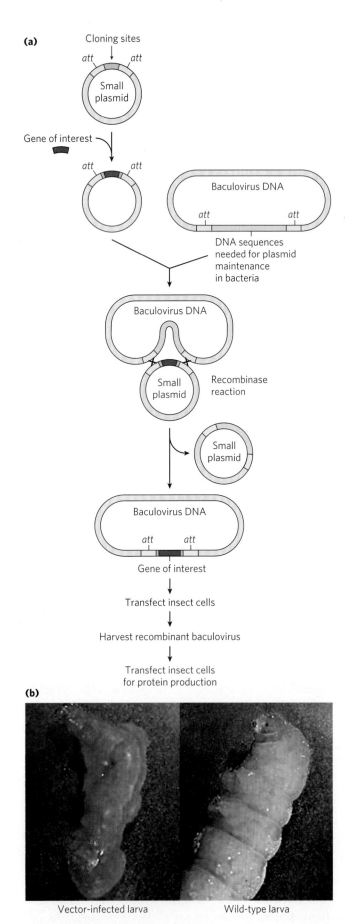

(a)

Cloning sites

att ↓ att

Small plasmid

Gene of interest

att att

Baculovirus DNA

att att

DNA sequences needed for plasmid maintenance in bacteria

Baculovirus DNA

Small plasmid

Recombinase reaction

Small plasmid

Baculovirus DNA

att att

Gene of interest

Transfect insect cells

Harvest recombinant baculovirus

Transfect insect cells for protein production

(b)

Vector-infected larva   Wild-type larva

FIGURE 9–9 Cloning with baculoviruses. (a) Shown here is the construction of a typical vector used for protein expression in baculoviruses. The gene of interest is cloned into a small plasmid (left) between two sites (*att*) recognized by a site-specific recombinase, then introduced into the baculovirus vector by site-specific recombination (see Fig. 25-37). This generates a circular DNA product that is used to infect the cells of an insect larva. The gene of interest is expressed during the infection cycle, downstream of a promoter that normally expresses a baculovirus coat protein at very high levels. **(b)** The photographs show (left) an insect larva infected with a recombinant baculovirus vector expressing a protein that produces a red color and (right) an uninfected larva.

synthetic one, identical to the original except for the desired change **(Fig. 9–10a)**.

When suitably located restriction sites are not present, **oligonucleotide-directed mutagenesis** can create a specific DNA sequence change (Fig. 9–10b). Two short, complementary synthetic DNA strands, each with the desired base change, are annealed to opposite strands of the cloned gene within a suitable circular DNA vector. The mismatch of a single base pair in 30 to 40 bp does not prevent annealing. The two annealed oligonucleotides serve to prime DNA synthesis in both directions around the plasmid vector, creating two complementary strands that contain the mutation. After several cycles of selective amplification, using a technique called polymerase chain reaction (PCR; described on p. 327), the mutation-containing DNA predominates in the population and can be used to transform bacteria. Most of the transformed bacteria will have plasmids carrying the mutation. If necessary, the nonmutant template plasmid DNA can be selectively eliminated by cleavage with the restriction enzyme *Dpn*I. The template plasmid, usually isolated from wild-type *E. coli*, has a methylated A residue in every copy of the four-nucleotide palindrome GATC (called a dam site; see Fig. 25–21). The new DNA containing the mutation does not have methylated A residues because the replication is done in vitro. *Dpn*I selectively cleaves DNA at the sequence GATC only if the A residue in one or both strands is methylated—that is, it breaks down only the template.

For an example, we go back to the bacterial *recA* gene. The product of this gene, the RecA protein, has several activities (see Section 25.3). It binds to and forms a filamentous structure on DNA, aligns two DNAs of similar sequence, and hydrolyzes ATP. A particular amino acid residue in RecA (a 352 residue polypeptide), the Lys residue at position 72, is involved in ATP hydrolysis. By changing Lys[72] to an Arg, a variant of RecA protein is created that will bind, but not hydrolyze, ATP (Fig. 9–10c). The engineering and purification of this variant RecA protein has facilitated research into the roles of ATP hydrolysis in the functioning of this protein.

Changes can be introduced into a gene that involve far more than one base pair. Large parts of a gene can be deleted by cutting out a segment with restriction

(a) Site-directed mutagenesis

Recombinant vector

Cleave with restriction endonuclease.

Mutation

Insert synthetic DNA segment containing mutation.

(b) Oligonucleotide-directed mutagenesis

Gene
Target site for mutation
Plasmid

Primers

Denature and anneal oligonucleotide primers with mutation.

Use DNA polymerase to extend and incorporate the mutagenic primers.

Digest nonmutated parental DNA template with methylation-specific nuclease and anneal newly synthesized strands.

Mutated plasmid — with nicked strands

Transform dsDNA into cells. Cell repairs nicks in mutated plasmid.

(c)

Wild-type recA

T C T T C C G G T A A A A C C A C G C

Mutant recA K72R

T C T T C C G G T C G C A C C A C G C

FIGURE 9–10 Two approaches to site-directed mutagenesis. (a) A synthetic DNA segment replaces a fragment removed by a restriction endonuclease. **(b)** A pair of synthetic and complementary oligonucleotides with a specific sequence change at one position is hybridized to a circular plasmid with a cloned copy of the gene to be altered. The mutated oligonucleotides act as primers for the synthesis of full-length duplex DNA copies of the plasmid that contain the specified sequence change. These plasmid copies are then used to transform cells. **(c)** Results from an automated sequencer (see Fig. 8–34), showing sequences from the wild-type recA gene (top) and an altered recA gene (bottom) with the triplet (codon) at position 72 changed from AAA to CGC, specifying an Arg (R) instead of a Lys (K) residue.

endonucleases and ligating the remaining portions to form a smaller gene. For example, if a protein has two domains, the gene segment encoding one of the domains can be removed to produce a protein that now has only one of the original two. Parts of two different genes can be ligated to create new combinations; the product of such a fused gene is called a **fusion protein**. Researchers have ingenious methods to bring about virtually any genetic alteration in vitro. After reintroducing the altered DNA into the cell, they can investigate the consequences of the alteration.

Terminal Tags Provide Handles for Affinity Purification

Affinity chromatography is one of the most efficient methods for purifying proteins (see Fig. 3–17c). Unfortunately, many proteins do not bind a ligand that can be conveniently immobilized on a column matrix. However, the gene for almost any protein can be altered to express a fusion protein that can be purified by affinity chromatography. The gene encoding the target protein is fused to a gene encoding a peptide or protein that binds a simple, stable ligand with high affinity and specificity. The peptide or protein used for this purpose is referred to as a **tag**. Tag sequences can be added to genes such that the resulting proteins have tags at their amino or carboxyl terminus. Table 9–3 lists some of the peptides or proteins commonly used as tags.

TABLE 9–3   Commonly Used Protein Tags

Tag protein/ peptide	Molecular mass (kDa)	Immobilized ligand
Protein A	59	Fc portion of IgG
$(His)_6$	0.8	Ni^{2+}
Glutathione-S-transferase (GST)	26	Glutathione
Maltose-binding protein	41	Maltose
β-Galactosidase	116	p-Aminophenyl-β-D-thiogalactoside (TPEG)
Chitin-binding domain	5.7	Chitin

The general procedure can be illustrated by focusing on a system that uses the glutathione-*S*-transferase (GST) tag **(Fig. 9–11)**. GST is a small enzyme (M_r 26,000) that binds tightly and specifically to glutathione. When the GST gene sequence is fused to a target gene, the fusion protein acquires the capacity to bind glutathione. The fusion protein is expressed in a host organism such as a bacterium, and a crude extract is prepared. A column is filled with a porous matrix consisting of the ligand (glutathione) immobilized on microscopic beads of a stable polymer such as cross-linked agarose. As the crude extract percolates through this matrix, the fusion protein becomes immobilized by binding the glutathione. The other proteins in the extract are washed through the column and discarded. The interaction between GST and glutathione is tight but noncovalent, allowing the fusion protein to be gently eluted from the column with a solution containing either a higher concentration of salts or free glutathione to compete with the immobilized ligand for GST binding. The fusion protein is often obtained with good yield and high purity. In some commercially available systems, the tag can be entirely or largely removed from the purified fusion protein by a protease that cleaves a sequence near the junction between the target protein and its tag.

(a)

γ-Glu–Cys–Gly Glutathione (GSH)

Glutathione-*S*-transferase (GST)

(b)

Gene for target protein

Gene for GST

Transcription

Gene for fusion protein

Express fusion protein in cell.

Prepare cell extract containing fusion protein as part of the cell protein mixture.

Pump

Pump

Protein mixture is added to column.

Solution of free glutathione is added to column.

GST tag

Glutathione anchored to medium binds GST tag.

1 2 3 4 5 6 7 8 9

1 2 3 4 5 6 7 8 9

Other proteins flow through column.

Fusion protein is eluted by glutathione solution.

FIGURE 9–11 Use of tagged proteins in protein purification. (a) Glutathione-*S*-transferase (GST) is a small enzyme that binds glutathione (a glutamate residue to which a Cys–Gly dipeptide is attached at the carboxyl carbon of the Glu side chain, hence the abbreviation GSH). **(b)** The GST tag is fused to the carboxyl terminus of the protein by genetic engineering. The tagged protein is expressed in the cell and is present in the crude extract when the cells are lysed. The extract is subjected to affinity chromatography (see Fig. 3-17c) through a matrix with immobilized glutathione. The GST-tagged protein binds to the glutathione, retarding its migration through the column, while the other proteins are washed through rapidly. The tagged protein is subsequently eluted with a solution containing elevated salt concentration or free glutathione.

A shorter tag with widespread application consists of a simple sequence of six or more His residues. These histidine tags, or His tags, bind tightly and specifically to nickel ions. A chromatography matrix with immobilized Ni^{2+} can be used to quickly separate a His-tagged protein from other proteins in an extract. Some of the larger tags, such as maltose-binding protein, provide added stability and solubility, allowing the purification of cloned proteins that are otherwise inactive due to improper folding or insolubility.

Affinity chromatography using terminal tags is powerful and convenient. The tags have been successfully used in thousands of published studies; in many cases, the protein would be impossible to purify and study without the tag. However, even very small tags can affect the properties of the proteins they are attached to, thereby influencing the study results. For example, the tag may adversely affect protein folding. Even if the tag is removed by a protease, one or a few extra amino acid residues can remain behind on the target protein, which may or may not affect the protein's activity. The types of experiments to be carried out, and the results obtained from them, should always be evaluated with the aid of well-designed controls to assess any effect of a tag on protein function.

Gene Sequences Can Be Amplified with the Polymerase Chain Reaction

Genome projects worldwide generate international databases containing the complete genome sequences of hundreds of organisms and provide unprecedented access to gene sequence information. This, in turn, simplifies the process of cloning individual genes for more detailed analysis. If the sequence of at least the end portions of a DNA segment of interest is known, the number of copies of that DNA segment can be hugely amplified with the **polymerase chain reaction (PCR)**, a process conceived by Kary Mullis in 1983. The amplified DNA can then be cloned by the methods described earlier or can be used in a variety of analytical procedures.

The PCR procedure has an elegant simplicity. It relies on DNA polymerases, which synthesize DNA strands from deoxyribonucleotides, using a DNA template (Chapter 25). All DNA polymerases synthesize DNA in the $5'{\rightarrow}3'$ direction (see Fig. 8–33a). Further, DNA polymerases do not synthesize DNA de novo, but instead must add nucleotides to preexisting strands, referred to as primers. Two synthetic oligonucleotides are prepared, complementary to sequences on opposite strands of the target DNA at positions defining the ends of the segment to be amplified. The oligonucleotides serve as replication primers that can be extended by a DNA polymerase. The 3' ends of the hybridized primers are oriented toward each other and positioned to prime DNA synthesis across the DNA segment **(Fig. 9–12a)**. Basic PCR requires four components: a DNA sample containing the segment to be amplified, the pair of synthetic oligonucleotide primers, deoxynucleoside triphosphates (dNTPs), and DNA polymerase. The reaction mixture is heated briefly to denature the DNA, separating the two strands. The mixture is cooled so that the primers can anneal to the DNA. The high concentration of primers increases the likelihood that they will anneal to each strand of the denatured DNA before the two DNA strands (present at a much lower concentration) can reanneal to each other. The primed segment is then replicated selectively by the DNA polymerase, using the pool of dNTPs. The cycle of heating, cooling, and replication is repeated 25 to 30 times over a few hours in an automated process, amplifying the DNA segment between the primers until it can be readily analyzed or cloned. Each cycle increases the amount of the DNA segment by a factor of 2, so the concentration of this DNA grows exponentially. After 20 cycles, the DNA segment has been amplified more than a millionfold (2^{20}); after 30 cycles, more than a billionfold. All other DNA in the sample remains unamplified. PCR uses a heat-stable DNA polymerase that remains active after every heating step and does not have to be replenished.

By careful design of the primers used for PCR, the amplified segment can be altered by the inclusion, at each end, of additional DNA not present in the chromosome that is being targeted. For example, restriction endonuclease cleavage sites can be included to facilitate the subsequent cloning of the amplified DNA (Fig. 9–12b).

This technology is highly sensitive: PCR can detect and amplify as little as one DNA molecule in almost any type of sample. The double-helical structure of DNA makes it a highly stable molecule, but DNA does degrade slowly over time (see Chapter 8). However, PCR has allowed the successful cloning of rare, undegraded DNA segments from samples more than 40,000 years old. Investigators have used the technique to clone DNA fragments from the mummified remains of humans and extinct animals, such as the woolly mammoth, creating the new fields of molecular archaeology and molecular paleontology. DNA from burial sites has been amplified by PCR and used to trace ancient human migrations. Epidemiologists can use PCR-enhanced DNA samples from human remains to trace the evolution of human pathogenic viruses. Thus, in addition to its usefulness for cloning DNA, PCR is a potent tool in forensic medicine (Box 9–1). It can be used to detect viral infections before they cause symptoms. For prospective parents from families with inherited genetic conditions of concern, PCR is an important tool for genetic counseling.

Given the extreme sensitivity of PCR methods, contamination of samples is a serious issue. In many applications, including forensic and ancient DNA tests, control samples ensure that the amplified DNA is not derived from the researcher or from contaminating bacteria.

Many specialized adaptations of PCR have increased the utility of the method. For example, sequences in RNA can be amplified if the first PCR cycle uses reverse transcriptase, an enzyme that works like DNA polymerase (see Fig. 8–33a), but uses RNA as a template

(a)

After 20 cycles, the target sequence has been amplified about 10^6-fold.

FIGURE 9–12 Amplification of a DNA segment by the polymerase chain reaction (PCR). (a) The PCR procedure has three steps. DNA strands are ❶ separated by heating, then ❷ annealed to an excess of short synthetic DNA primers (orange) that flank the region to be amplified (dark blue); ❸ new DNA is synthesized by polymerization catalyzed by DNA polymerase. The three steps are repeated for 25 or 30 cycles. The thermostable *Taq* DNA polymerase (from *Thermus aquaticus*, a bacterial species that grows in hot springs) is not denatured by the heating steps. **(b)** DNA amplified by PCR can be cloned. The primers can include noncomplementary ends that have a site for cleavage by a restriction endonuclease. Although these parts of the primers do not anneal to the target DNA, the PCR process incorporates them into the DNA that is amplified. Cleavage of the amplified fragments at these sites creates sticky ends, used in ligation of the amplified DNA to a cloning vector. 🖱 **Polymerase Chain Reaction**

(b)

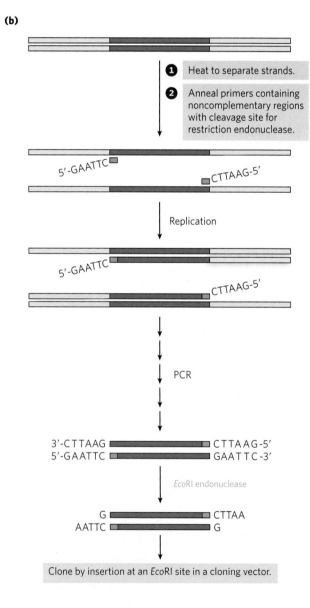

Clone by insertion at an *Eco*RI site in a cloning vector.

BOX 9–1 | METHODS | A Powerful Tool in Forensic Medicine

One of the most accurate methods for placing an individual at the scene of a crime is a fingerprint. The advent of recombinant DNA technology has made available a much more powerful tool: **DNA genotyping** (also called DNA fingerprinting or DNA profiling). The method is based on sequence polymorphisms—slight sequence differences among individuals: 1 in every 1,000 bp, on average. The use of these differences for forensic identification was first described by English geneticist Alec Jeffreys in 1985. Each difference from the prototype human genome sequence (the first one obtained) occurs in some fraction of the human population; every person has some differences from this prototype.

Modern forensic work focuses on differences in the lengths of **short tandem repeat (STR)** sequences. An STR is a short DNA sequence, repeated many times in tandem at a specific location in a chromosome;

usually, the repeated sequence is 4 bp long. The STR loci most often used in DNA genotyping are short—4 to 50 repeats long (16 to 200 bp for tetranucleotide repeats)—and have multiple length variants in the human population. More than 20,000 tetranucleotide STR loci have been characterized and more than a million STRs of all types may be present in the human genome, accounting for about 3% of all human DNA.

The length of a particular STR in a given individual can be determined with the aid of the polymerase chain reaction (see Fig. 9–12). The use of PCR also makes the procedure sensitive enough to be applied to the very small samples of DNA often collected at crime scenes. The DNA sequences flanking STRs are unique to each STR locus and are identical (except for extremely rare mutations) in all humans. PCR primers are targeted to this flanking DNA and are designed to amplify the DNA across the STR (Fig. 1a). The length

(continued on next page)

FIGURE 1 **(a)** STR loci can be analyzed by PCR. Suitable PCR primers (with an attached dye to aid in subsequent detection) are targeted to sequences on either side of the STR, and the region between them is amplified. Individuals usually have two different alleles at a particular locus (one inherited from each parent). If the STR sequences have different lengths on the two chromosomes of an individual, two PCR products of different lengths will result. **(b)** The PCR products from amplification of 16 STR loci run on a single capillary acrylamide gel are shown. Particular loci are targeted with primers labeled with only one of three different fluorescent dyes (six loci with a green dye, five with a blue dye, and five with a yellow dye, plotted here with black ink for visibility). Determination of which locus corresponds to which signal depends on the color of the fluorescent dye attached to the primers used in the process and to the size range in which the signal appears (the size range can be controlled by which sequences—closer to or more distant from the STR—are targeted by the designed PCR primers).

BOX 9–1 **METHODS** A Powerful Tool in Forensic Medicine (*Continued*)

of the PCR product then reflects the length of the STR in that sample. Because each human inherits one chromosome of each chromosome pair from each parent, the STR lengths on the two chromosomes are often different, generating two different STR lengths from one individual. The PCR products are subjected to electrophoresis on a very thin polyacrylamide gel in a capillary tube. The resulting bands are converted into a set of peaks that accurately reveal the size of each PCR fragment and so the length of the STR in the corresponding allele. Analysis of multiple STR loci can yield a profile that is unique to an individual (Fig. 1b). This is typically done with a commercially available kit that includes PCR primers unique to each locus, linked to colored dyes to help distinguish the different PCR products. PCR amplification enables investigators to obtain STR genotypes from less than 1 ng of partially degraded DNA, an amount that can be obtained from a single hair follicle, a drop of blood, a small semen sample, or samples that might be months or even many years old. When good STR genotypes are obtained, the chance of misidentification is less than 1 in 10^{18} (quintillion).

The successful forensic use of STR analysis required standardization, first attempted in the United Kingdom in 1995. The U.S. standard, called the Combined DNA Index System (CODIS), established in 1998, is based on 13 well-studied STR loci, which must be present in any DNA-typing experiment carried out in the United States (Table 1). The amelogenin gene is also used as a marker in the analyses. Present on the human sex chromosomes, this gene has a slightly different length on the X and Y chromosomes. PCR amplification across this gene thus generates different-size products that can reveal the sex of the DNA donor. By the beginning of 2010, the CODIS database contained more than 7 million STR genotypes and had assisted more than 100,000 forensic investigations.

DNA genotyping has been used to both convict and acquit suspects and to establish paternity with an extraordinary degree of certainty. The impact of these procedures on court cases will continue to grow as standards are refined and as international STR genotyping databases grow. Even very old mysteries can be solved. In 1996, DNA fingerprinting helped confirm the identification of the bones of the last Russian czar and his family, who were assassinated in 1918.

TABLE 1 Properties of the Loci Used for the CODIS Database

Locus	Chromosome	Repeat motif	Repeat length (range)*	Number of alleles seen[†]
CSF1PO	5	TAGA	5–16	20
FGA	4	CTTT	12.2–51.2	80
TH01	11	TCAT	3–14	20
TPOX	2	GAAT	4–16	15
VWA	12	[TCTG][TCTA]	10–25	28
D3S1358	3	[TCTG][TCTA]	8–21	24
D5S818	5	AGAT	7–18	15
D7S820	7	GATA	5–16	30
D8S1179	8	[TCTA][TCTG]	7–20	17
D13S317	13	TATC	5–16	17
D16S539	16	GATA	5–16	19
D18S51	18	AGAA	7–39.2	51
D21S11	21	[TCTA][TCTG]	12–41.2	82
Amelogenin[‡]	X, Y	Not applicable		

Source: Adapted from Butler, J.M. (2005) *Forensic DNA Typing*, 2nd edn, Academic Press, San Diego, p. 96.

*Repeat lengths observed in the human population. Partial or imperfect repeats can be included in some alleles.

[†]Number of different alleles observed as of 2005 in the human population. Careful analysis of a locus in many individuals is a prerequisite to its use in forensic DNA typing.

[‡]Amelogenin is a gene, of slightly different size on the X and Y chromosomes, that is used to establish gender.

(Fig. 9–12). After the DNA strand is made from the RNA template, the remaining cycles can be carried out with DNA polymerases, using standard PCR protocols. This **reverse transcriptase PCR (RT-PCR)** can be used, for example, to detect sequences derived from living cells (which are transcribing their DNA into RNA) as opposed to dead tissues.

PCR protocols can also be made quantitative for estimating the relative copy numbers of particular sequences in a sample. The approach is called **quantitative PCR**, or **qPCR**. If a DNA sequence is present in higher than usual amounts in a sample—for example, if certain genes are amplified in tumor cells—qPCR can reveal the increased representation of that sequence. In brief, the PCR is carried out in the presence of a probe that emits a fluorescent signal when the PCR product is present (**Fig. 9–13**). If the sequence of interest is present at higher levels than other sequences in the sample, the PCR signal will reach a predetermined threshold faster. Reverse transcriptase PCR and real-time PCR can be combined to determine the relative concentrations of a particular RNA molecule in a cell under different environmental conditions.

Finally, PCR has facilitated the development of powerful new DNA sequencing procedures, as we shall see.

SUMMARY 9.1 Studying Genes and Their Products

▸ DNA cloning and genetic engineering involve the cleavage of DNA and assembly of DNA segments in new combinations—recombinant DNA.

▸ Cloning entails cutting DNA into fragments with enzymes; selecting and possibly modifying a fragment of interest; inserting the DNA fragment into a suitable cloning vector; transferring the vector with the DNA insert into a host cell for replication; and identifying and selecting cells that contain the DNA fragment.

▸ Key enzymes in gene cloning include restriction endonucleases (especially the type II enzymes) and DNA ligase.

▸ Cloning vectors include plasmids, and, for the longest DNA inserts, bacterial artificial chromosomes (BACs) and yeast artificial chromosomes (YACs).

▸ Genetic engineering techniques manipulate cells to express and/or alter cloned genes.

▸ Proteins or peptides can be attached to a protein of interest by altering its cloned gene, creating a fusion protein. The additional peptide segments can be used to detect the protein or to purify it using convenient affinity chromatography methods.

▸ The polymerase chain reaction (PCR) permits the amplification of chosen segments of DNA or RNA for detailed study or cloning.

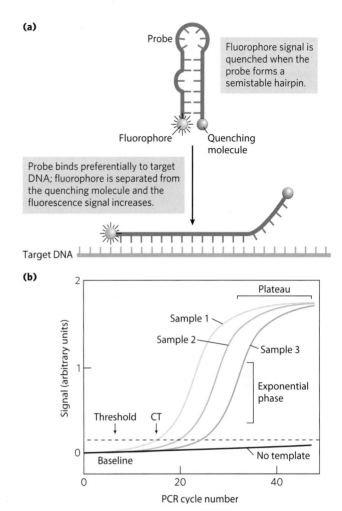

FIGURE 9–13 Quantitative PCR. PCR can be used quantitatively, by carefully monitoring the progress of a PCR amplification and determining when a DNA segment has been amplified to a specific threshold level. **(a)** The amount of PCR product present is determined by measuring the level of a fluorescent probe attached to a reporter oligonucleotide complementary to the DNA segment that is being amplified. Probe fluorescence is initially not detectable due to a fluorescence quencher attached to the same oligonucleotide. When the reporter oligonucleotide pairs with its complement in a copy of the amplified DNA segment, the fluorophore is separated from the quenching molecule and fluorescence results. **(b)** As the PCR reaction proceeds, the amount of the targeted DNA segment increases exponentially, and the fluorescent signal also increases exponentially as the oligonucleotide probes anneal to the amplified segments. After many PCR cycles, the signal reaches a plateau as one or more reaction components become exhausted. When a segment is present in greater amounts in one sample than another, its amplification reaches a defined threshold level earlier. The "No template" line follows the slow increase in background signal observed in a control that does not include added sample DNA. CT is the cycle number at which the threshold is first surpassed.

9.2 Using DNA-Based Methods to Understand Protein Function

Protein function can be described on three levels. **Phenotypic function** describes the effects of a protein on the entire organism. For example, the loss of the protein may lead to slower growth of the organism, an altered development pattern, or even death. **Cellular function** is a description of the network of interactions a protein engages in at the cellular level. Identifying interactions with other proteins in the cell can help define the kinds of metabolic processes in which the protein participates. Finally, **molecular function** refers to the precise biochemical activity of a protein, including details such as the reactions an enzyme catalyzes or the ligands a receptor binds. The challenge of understanding the functions of the thousands of uncharacterized or poorly characterized proteins found in a typical cell has given rise to a wide variety of techniques. DNA-based methods make a critical contribution to this effort and can provide information on all three levels.

DNA Libraries Are Specialized Catalogs of Genetic Information

A **DNA library** is a collection of DNA clones, gathered for purposes of genome sequencing, gene discovery, or determination of gene/protein function. The library can take a variety of forms, depending on the source of the DNA and the ultimate purpose of the library.

The largest is a **genomic library**, produced when the complete genome of an organism is cleaved into thousands of fragments. All the fragments are cloned by insertion of each fragment into a cloning vector. This creates a complex mixture of recombinant vectors, each one with a different cloned fragment. The first step is *partial* digestion of the DNA by restriction endonucleases, such that any given sequence will appear in fragments of a limited range of sizes—a range compatible with the cloning vector to ensure that virtually all sequences are represented among the clones in the library. Fragments that are too large or too small for cloning are removed by centrifugation or electrophoresis. The cloning vector, such as a BAC or YAC, is cleaved with the same restriction endonuclease used to digest the DNA and ligated to the genomic DNA fragments. The ligated DNA mixture is then used to transform bacteria or yeast cells to produce a library of cells, each harboring a different recombinant DNA molecule. Ideally, all the DNA in the genome under study is represented in the library. Each transformed bacterium or yeast cell grows into a colony, or clone, of identical cells, each cell bearing the same recombinant plasmid, one of many represented in the overall library.

Efforts to define gene or protein function often make use of more specialized libraries. An example is a library that includes only those sequences of DNA that are *expressed*—that is, transcribed into RNA—in a given organism, or even just in certain cells or tissues. Such a library lacks the noncoding DNA that makes up a large portion of many eukaryotic genomes. The researcher first extracts mRNA from an organism, or from specific cells of an organism, and then prepares the **complementary DNAs (cDNAs)**. This multistep

reaction, shown in **Figure 9–14**, relies on reverse transcriptase, which synthesizes DNA from a template RNA. The resulting double-stranded DNA fragments are inserted into a suitable vector and cloned, creating a population of clones called a **cDNA library**. The appearance of a gene for a particular protein in such a library implies that it is expressed in the cells and under the conditions used to generate the library.

FIGURE 9–14 Building a cDNA library from mRNA. A cell's total mRNA content includes transcripts from thousands of genes, and the cDNAs generated from this mRNA are correspondingly heterogeneous. Reverse transcriptase can synthesize DNA on an RNA or a DNA template (see Fig. 26–32). To prime the synthesis of a second DNA strand, oligonucleotides of known sequence are ligated to the 3′ end of the first strand, and the double-stranded cDNA so produced is cloned into a plasmid.

mRNA
5′ [AAAAAAAA]

mRNA template is annealed to synthetic oligonucleotide (oligo-dT) primer.

5′ [AAAAAAAA]
3′ [TTTTTTT]

Reverse transcriptase and dNTPs yield a complementary DNA strand.

mRNA-DNA hybrid
5′ [AAAAAAAA]
3′ [TTTTTTT]

mRNA is degraded with alkali.

3′ [TTTTTTT]

To prime synthesis of a second strand, an oligonucleotide of known sequence is often ligated to the 3′ end of the cDNA.

5′ [ ]
3′ [TTTTTTT]

DNA polymerase I and dNTPs extend the primer to yield double-stranded DNA.

Duplex DNA
5′ [AAAAAAAA]
3′ [TTTTTTT]

Sequence or Structural Relationships Provide Information on Protein Function

One important reason to sequence many genomes is to provide a database that can be used to assign gene functions by genome comparisons, an enterprise referred to as **comparative genomics**. Sometimes a newly discovered gene is related by sequence homologies to a gene previously studied in another or the same species, and its function can be entirely or partly defined by that relationship. Genes that occur in different species but have a clear sequence and functional relationship to each other are called **orthologs**. Genes similarly related to each other within a single species are called **paralogs**. If the function of a gene has been characterized for one species, this information can be used to at least tentatively assign gene function to the ortholog found in the second species. The correlation is easiest to make when comparing genomes from relatively closely related species, such as mouse and human, although many clearly orthologous genes have been identified in species as distant as bacteria and humans. Sometimes even the order of genes on a chromosome is conserved over large segments of the genomes of closely related species **(Fig. 9–15)**. Conserved gene order, called **synteny**, provides additional evidence for an orthologous relationship between genes at identical locations within the related segments.

Alternatively, certain sequences associated with particular structural motifs (Chapter 4) may be identified within a protein. The presence of a structural motif may help to define molecular function by suggesting that a protein, say, catalyzes ATP hydrolysis, binds to DNA, or forms a complex with zinc ions. These relationships are determined with the aid of sophisticated computer programs, limited only by the current information on gene and protein structure and by our capacity to associate sequences with particular structural motifs.

FIGURE 9–15 Synteny in the human and mouse genomes. Large segments of the two genomes have closely related genes aligned in the same order on the chromosomes. In these short segments of human chromosome 9 and mouse chromosome 2, the genes exhibit a very high degree of homology, as well as the same gene order. The different lettering schemes for the gene names simply reflect the different naming conventions in the two species.

Fusion Proteins and Immunofluorescence Can Localize Proteins in Cells

Often, an important clue to a gene product's function comes from determining its location within the cell. For example, a protein found exclusively in the nucleus could be involved in processes that are unique to that organelle, such as transcription, replication, or chromatin condensation. Researchers often engineer fusion proteins for the purpose of locating a protein in the cell or organism. Some of the most useful fusions involve the attachment of marker proteins that allow the investigator to determine the location by direct visualization or by immunofluorescence.

A particularly useful marker is the **green fluorescent protein (GFP)**. A target gene (coding the protein of interest) fused to the GFP gene generates a fusion protein that is highly fluorescent—it literally lights up when exposed to blue light—and can be visualized directly in a living cell. GFP is a protein derived from the jellyfish *Aequorea victoria* (see Box 12–3, Fig. 1). It has a β-barrel structure, and the fluorophore (the fluorescent component of the protein) is in the center of the barrel **(Fig. 9–16a)**. The fluorophore is derived from a rearrangement and oxidation of several amino acid residues. Because this reaction is autocatalytic and requires no other proteins or cofactors other than molecular oxygen (see Box 12–3, Fig. 3), GFP is readily cloned in an active form in almost any cell. Just a few molecules of this protein can be observed microscopically, allowing the study of its location and movements in a cell. Careful protein engineering, coupled with the isolation of related fluorescent proteins from other marine coelenterates, has made a wide range of these proteins available, in an array of colors (Fig. 9–16b) and other characteristics (brightness, stability). With this technology, for example, the protein GLR1 (a glutamate receptor of nervous tissue) has been visualized as a GLR1-GFP fusion protein in the nematode *Caenorhabditis elegans* (Fig. 9–16c).

In many cases, visualization of a GFP fusion protein in a live cell is not possible or practical or desirable. The GFP fusion protein may be inactive or may not be expressed at sufficient levels to allow visualization. In this case, **immunofluorescence** is an alternative approach for visualizing the endogenous (unaltered) protein. This approach requires the fixation (and thus death) of the cell. The protein of interest is sometimes expressed as a fusion protein with an **epitope tag**, a short protein sequence that is bound tightly by a well-characterized, commercially available antibody. Fluorescent molecules (fluorochromes) are attached to this antibody. More commonly, the target protein is unaltered. This protein is bound by an antibody that is specific for it. Next, a second antibody is added that binds specifically to the first one, and it is the second antibody that has the attached fluorochrome(s) **(Fig. 9–17a)**. A variation of this indirect visualization approach is to attach biotin molecules to the first antibody, then add

FIGURE 9–16 Green fluorescent protein (GFP). (a) The GFP protein (PDB ID 1GFL), derived from the jellyfish *Aequorea victoria*, has a β-barrel structure; the fluorophore (shown as a space-filling model) is in the center of the barrel. **(b)** Variants of GFP are now available in almost any color of the visible spectrum. **(c)** A GLR1-GFP fusion protein fluoresces bright green in *Caenorhabditis elegans*, a nematode worm (left). GLR1 is a glutamate receptor of nervous tissue. (Autofluorescing fat droplets are false colored in magenta.) The membranes of *E. coli* cells (right) are stained with a red fluorescent dye. The cells are expressing a protein that binds to a resident plasmid, fused to GFP. The green spots indicate the locations of plasmids.

streptavidin (a bacterial protein closely related to avidin, a protein that binds to biotin; see Table 5–1) complexed with fluorochromes. The interaction between biotin and streptavidin is one of the strongest and most specific known, and the potential to add multiple fluorochromes to each target protein gives this method great sensitivity.

Highly specialized cDNA libraries (Fig. 9–14) can be made by cloning cDNAs or cDNA fragments into a vector that fuses each cDNA sequence with the sequence for a marker called a reporter gene. The fused gene is often called a reporter construct. For example, all the genes in the library may be fused to the GFP gene **(Fig. 9–18)**. Each cell in the library expresses one of these fused genes. The cellular location of the product of any gene represented in the library will be revealed as foci of light in cells that express the appropriate fused gene at sufficient levels—assuming that the gene retains its normal function and location.

Protein-Protein Interactions Can Help Elucidate Protein Function

Another key to defining the function of a particular protein is to determine what it binds to. In the case of

(a)

Secondary antibody
Primary antibody
Fluorescent dye
Protein of interest

(b)

| DNA polymerase ε | PCNA | BrdU | Merge of DNA polymerase ε and PCNA |

FIGURE 9–17 Indirect immunofluorescence. (a) The protein of interest is bound to a primary antibody, and a secondary antibody is added; this second antibody, with one or more attached fluorescent groups, binds to the first. Multiple secondary antibodies can bind the primary antibody, amplifying the signal. If the protein of interest is in the interior of the cell, the cell is fixed and permeabilized, and the two antibodies are added in succession. **(b)** The end result is an image in which bright spots indicate the location of the protein or proteins of interest in the cell. The images show a nucleus from a human fibroblast, successively stained with antibodies and fluorescent labels for DNA polymerase ε, for PCNA, an important polymerase accessory protein, and for bromo-deoxyuridine (BrdU), a nucleotide analog. The BrdU, added as a brief pulse, identifies regions undergoing active DNA replication. The patterns of staining show that DNA polymerase ε and PCNA co-localize to regions of active DNA synthesis. One such region is visible in the white box.

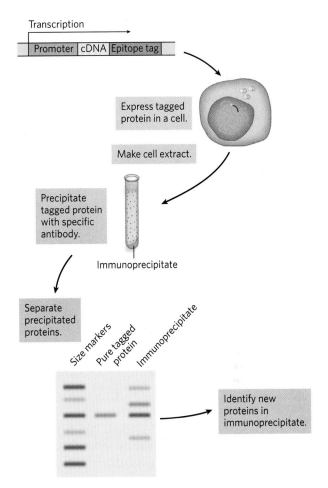

FIGURE 9–18 Specialized DNA libraries. Cloning of a cDNA next to the GFP gene creates a reporter construct. Transcription proceeds through the gene of interest (the inserted cDNA) and the reporter gene (here, GFP), and the mRNA transcript is expressed as a fusion protein. The GFP part of the protein is visible with the fluorescence microscope. Although only one example is shown, thousands of genes can be fused to GFP in similar constructs and stored in libraries in which each cell or organism in the library expresses a different protein fused to GFP. If the fusion protein is properly expressed, its location in the cell or organism can be assessed. The photograph shows a nematode worm containing a GFP fusion protein expressed only in the four "touch" neurons that run the length of its body.

protein-protein interactions, the association of a protein of unknown function with one whose function is known can provide useful and compelling "guilt by association." The techniques used in this effort are quite varied.

Purification of Protein Complexes With the construction of cDNA libraries in which each gene is fused to an epitope tag, investigators can precipitate the protein product of a gene by complexing it with the antibody that binds the epitope, a process called **immunoprecipitation (Fig. 9–19).** If the tagged protein is expressed in cells, other proteins that bind to it precipitate with it. Identifying the associated proteins reveals some of the intracellular protein-protein interactions of the tagged protein. There are many variations of this process. For example, a crude extract of cells that express a tagged protein is added to a column containing immobilized antibody (see Fig. 3–17c for a description of affinity chromatography). The tagged protein binds to the antibody, and proteins that interact with the tagged protein are sometimes also retained on the column. The connection between the protein and the tag is cleaved with a specific protease, and the protein complexes are eluted from the column and analyzed. Researchers can use these methods to define complex networks of interactions within a cell. In principle, the chromatographic approach to analyzing protein-protein interactions can be used with any type of protein tag (His tag, GST, etc.) that can be immobilized on a suitable chromatographic medium.

The selectivity of this approach has been enhanced with **tandem affinity purification (TAP) tags**. Two consecutive tags are fused to a target protein, and the fusion protein is expressed in a cell **(Fig. 9–20).** The first tag is protein A, a protein found at the surface of the bacterium *Staphylococcus aureus* that binds tightly to mammalian immunoglobulin G (IgG). The second tag is often a calmodulin-binding peptide. A crude extract containing the TAP-tagged fusion protein is passed through a column matrix with attached IgG antibodies that bind to protein A. Most of the unbound cellular proteins are

FIGURE 9–19 The use of epitope tags to study protein-protein interactions. The gene of interest is cloned next to a gene for an epitope tag, and the resulting fusion protein is precipitated by antibodies to the epitope. Any other proteins that interact with the tagged protein also precipitate, thereby helping to elucidate protein-protein interactions.

washed through the column, but proteins that normally interact with the target protein in the cell are retained. The first tag is then cleaved from the fusion protein with a highly specific protease, TEV protease, and the shortened fusion target protein and any proteins associated noncovalently with the target protein are eluted from the column. The eluent is then passed through a second column containing a matrix with attached calmodulin that binds the second tag. Loosely bound proteins are again washed from the column. After the second tag is cleaved, the target protein is eluted from the column with its associated proteins. The two consecutive purification steps eliminate any weakly bound contaminants. False positives are minimized, and protein interactions that persist through both steps are likely to be functionally significant.

Yeast Two-Hybrid Analysis A sophisticated genetic approach to defining protein-protein interactions is based on the properties of the Gal4 protein (Gal4p; see Fig. 28–30), which activates the transcription of *GAL* genes in yeast (genes encoding the enzymes of galactose metabolism). Gal4p has two domains: one that binds

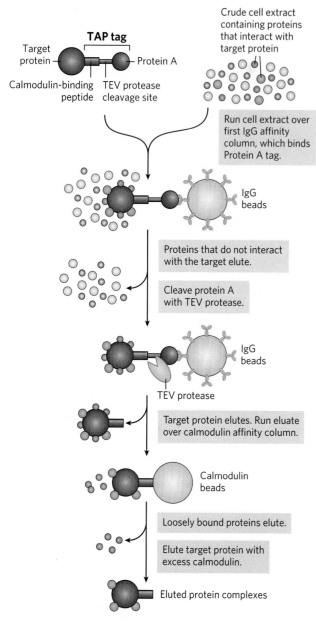

FIGURE 9–20 Tandem affinity purification (TAP) tags. A TAP-tagged protein and associated proteins are isolated by two consecutive affinity purifications, as described in the text.

a specific DNA sequence and another that activates RNA polymerase to synthesize mRNA from an adjacent gene. The two domains of Gal4p are stable when separated, but activation of RNA polymerase requires interaction with the activation domain, which in turn requires positioning by the DNA-binding domain. Hence, the domains must be brought together to function correctly.

In **yeast two-hybrid analysis**, the protein-coding regions of the genes to be analyzed are fused to the yeast gene for either the DNA-binding domain or the activation domain of Gal4p, and the resulting genes express a

series of fusion proteins (**Fig. 9–21**). If a protein fused to the DNA-binding domain interacts with a protein fused to the activation domain, transcription is activated. The reporter gene transcribed by this activation is generally one that yields a protein required for growth or an enzyme that catalyzes a reaction with a colored product. Thus, when grown on the proper medium, cells that contain a pair of interacting proteins are easily distinguished from those that do not. A library can be set up

FIGURE 9–21 Yeast two-hybrid analysis. (a) The goal is to bring together the DNA-binding domain and the activation domain of the yeast Gal4 protein (Gal4p) through the interaction of two proteins, X and Y, to which one or other of the domains is fused. This interaction is accompanied by the expression of a reporter gene. **(b)** The two gene fusions are created in separate yeast strains, which are then mated. The mated mixture is plated on a medium on which the yeast cannot survive unless the reporter gene is expressed. Thus, all surviving colonies have interacting fusion proteins. Sequencing of the fusion proteins in the survivors reveals which proteins are interacting.

with a particular yeast strain in which each cell in the library has a gene fused to the Gal4p DNA-binding domain gene, and many such genes are represented in the library. In a second yeast strain, a gene of interest is fused to the gene for the Gal4p activation domain. The yeast strains are mated, and individual diploid cells are grown into colonies. The only cells that grow on the selective medium, or that produce the appropriate color, are those in which the gene of interest is binding to a partner, allowing transcription of the reporter gene. This allows for large-scale screening for cellular proteins that interact with the target protein. The interacting protein that is fused to the Gal4p DNA-binding domain present in a particular selected colony can be quickly identified by DNA sequencing of the fusion protein's gene. Some false positive results occur, due to the formation of multiprotein complexes.

These techniques for determining cellular localization and molecular interactions provide important clues to protein function. However, they do not replace classical biochemistry. They simply give researchers an expedited entrée into important new biological problems. When paired with the simultaneously evolving tools of biochem-istry and molecular biology, the techniques described here are speeding the discovery not only of new proteins, but of new biological processes and mechanisms.

DNA Microarrays Reveal RNA Expression Patterns and Other Information

Major refinements of the technology underlying DNA libraries, PCR, and hybridization have come together in the development of **DNA microarrays**, which allow the rapid and simultaneous screening of many thousands of genes. DNA segments from genes of known sequence, a few dozen to hundreds of base pairs long, are amplified by PCR. Robotic devices then accurately deposit nano-liter quantities of the DNA solutions on a solid surface of just a few square centimeters, in a predesigned array, with each of the thousands of spots containing sequences derived from a particular gene. An alternative and increasingly common strategy is to synthesize DNA directly on the solid surface, using photolithography **(Fig. 9–22)**. The resulting array, often called a chip, may include sequences derived from every gene of a bacterial or yeast genome, or selected families of genes

FIGURE 9–22 Photolithography to create a DNA microarray. ❶ A computer is programmed with the desired oligonucleotide sequences. **❷** The reactive groups, attached to a solid surface, are initially rendered inactive by photoactive blocking groups, which can be removed by a flash of light. An opaque screen blocks the light from some areas of the surface, preventing their activation. Other areas or "spots" are exposed. **❸** A solution containing one activated nucleotide (e.g., A*) is washed over the spots. The 5' hydroxyl of the nucleotide is blocked to prevent unwanted reactions, and the nucleotide links to the surface groups at the appropriate spots through its 3' hydroxyl. The surface is washed successively with solutions containing each remaining activated nucleotide (G*, C*, T*). The 5'-blocking groups on each nucleotide limit the reactions to addition of one nucleotide at a time, and these groups can also be removed by light. Once each spot has one nucleotide, a second nucleotide can be added to extend the nascent oligonucleotide at each spot, using screens and light to ensure that the correct nucleotides are added at each spot in the correct sequence. This continues until the required sequences are built up on each of the thousands of spots in a DNA microarray.

from a larger genome. Once constructed, the microarray can be probed with mRNAs or cDNAs from a particular cell type or cell culture to identify the genes being expressed in those cells.

A microarray can provide a snapshot of all the genes in an organism, informing the researcher about the genes that are expressed at a given stage in the organism's development or under a particular set of environmental conditions. For example, the total complement of mRNA can be isolated from cells at two different stages of development and converted to cDNA with reverse transcriptase. Fluorescently labeled deoxyribonucleotides can be used to make one cDNA sample fluoresce red, the other green (**Fig. 9–23**). The cDNA

from the two samples is mixed and used to probe the microarray. Each cDNA anneals to only one spot on the microarray, corresponding to the gene encoding the mRNA that gave rise to that cDNA. Spots that fluoresce green represent genes that produce mRNAs at higher levels at one developmental stage; those that fluoresce red represent genes expressed at higher levels at another stage. If a gene produces mRNAs that are equally abundant at both stages of development, the corresponding spot fluoresces yellow. By using a mixture of two samples to measure relative rather than absolute sequence abundance, the method corrects for variations in the amount of DNA originally deposited in each spot on the grid, as well as other possible inconsistencies among spots in the microarray. The spots that fluoresce provide a snapshot of all the genes being expressed in the cells at the moment they were harvested—gene expression examined on a genome-wide scale. For a gene of unknown function, the time and circumstances of its expression can provide important clues about its role in the cell. These technologies also reveal expression of many types of specialized RNAs, such as microRNAs (miRNAs; Chapter 26) and small interfering RNAs (siRNAs; Chapter 28).

An example of this technique is illustrated in **Figure 9–24**, which shows the dramatic results that microarray experiments can produce. Segments from each of the roughly 6,500 genes in the completely sequenced yeast genome were separately amplified by PCR, and each segment was deposited in a defined pattern to create the microarray. In a sense, this array provides a snapshot of the functioning of the entire yeast genome under one set of conditions.

SUMMARY 9.2 Using DNA-Based Methods to Understand Protein Function

▶ Proteins can be studied at the level of phenotypic, cellular, or molecular function.

▶ DNA libraries can be a prelude to many types of investigations that yield information about protein function.

▶ By fusing a gene of interest with genes that encode green fluorescent protein or epitope tags, researchers can visualize the cellular location of the gene product, either directly or by immunofluorescence.

▶ The interactions of a protein with other proteins or RNA can be investigated with epitope tags and immunoprecipitation or affinity chromatography.

FIGURE 9–23 **A DNA microarray experiment.** A microarray can be prepared from any known DNA sequence, from any source. Once the DNA is attached to a solid support, the microarray can be probed with other fluorescently labeled nucleic acids. Here, mRNA samples are collected from cells of a frog at two different stages of development.

FIGURE 9–24 Enlarged image of a DNA microarray. Each glowing spot contains DNA from one of the roughly 6,500 genes of the yeast (*S. cerevisiae*) genome, with every gene represented in the array. The microarray has been probed with fluorescently labeled nucleic acid derived from mRNAs obtained when the cells were growing in culture (green) and 5 hours after they began to form spores (red). The green spots represent genes expressed at higher levels during growth; the red spots, genes expressed at higher levels during sporulation. The yellow spots represent genes that do not change their level of expression during sporulation. This image is enlarged; the microarray actually measures only 1.8 × 1.8 cm.

Yeast two-hybrid analysis probes molecular interactions in vivo.

▶ Microarrays can reveal expression patterns of genes that change in response to cellular stimuli, developmental stage, or conditions.

9.3 Genomics and the Human Story

DNA sequencing methods described in Chapter 8 led to the first complete genomic sequences of bacterial species in the 1990s. Two complete human genome sequences appeared in 2001. One of these was generated by a publicly funded effort led first by James Watson and later by

Francis S. Collins

J. Craig Venter

Francis Collins. A parallel, private effort was led by Craig Venter. These accomplishments reflected more than a decade of intense effort coordinated in dozens of laboratories around the world, but they were just a beginning. DNA sequence databases now contain the complete genomic sequences of thousands of organisms of all types. We have only begun to effectively mine this vast informational resource.

Genomic Sequencing Is Aided by New Generations of DNA-Sequencing Methods

DNA-sequencing technologies continue to evolve. A complete human genome can now be sequenced in a day or two, a bacterial genome in a few hours. The day when a personal genomic sequence might be a routine part of one's medical record is within reach (Box 9–2). These advances have been made possible by approaches sometimes referred to as next-generation, or "next-gen," sequencing. The sequencing strategy is sometimes similar to and sometimes quite different from that used in the Sanger method described in Chapter 8. Innovations have allowed a miniaturization of the procedure, a massive increase in scale, and a corresponding decrease in cost.

A genomic sequence is generated in several steps. First, genomic sequences are broken at random locations by shearing to generate fragments that are a few hundred base pairs long. Synthetic oligonucleotides are ligated to the ends of all of the fragments, providing a known point of reference on every DNA molecule. The individual fragments are then immobilized on a solid surface, and each is amplified using PCR (Fig. 9–12). The solid surface is part of a channel that allows liquid solutions to flow over the samples. The result is a solid surface just a few centimeters across, with millions of DNA clusters attached, each cluster containing multiple copies of a single DNA sequence derived from one or another random genomic DNA fragment. The efficiency comes from sequencing all of these millions of clusters at the same time, with the data from each cluster captured and stored in a computer.

Two widely utilized next-generation sequencers use different strategies to accomplish the sequencing reactions. One of these, 454 sequencing (the numbers refer to a code used in the developmental phase of the technology and have no scientific meaning), uses a strategy called pyrosequencing in which the addition of nucleotides is detected with flashes of light **(Fig. 9–25)**. The four deoxynucleoside triphosphates (unaltered) are pulsed onto the reacting surface one at a time, in a repeating sequence. The nucleotide solution is retained on the surface just long enough for DNA polymerase to add that nucleotide to any cluster where it is complementary to the next template base in the sequence. Excess nucleotide is destroyed quickly with the enzyme apyrase before the next nucleotide pulse. When a specific nucleotide is successfully added to the strands of a cluster, pyrophosphate is released as a byproduct.

BOX 9–2 🜲 MEDICINE Personalized Genomic Medicine

When twins Noah and Alexis Beery were born in California, they exhibited symptoms that elicited a diagnosis of cerebral palsy. Treatments seemed to have no effect. Not satisfied with the diagnosis or the treatment, the twins' parents, Joe and Retta, took the twins, then age 5, to see a specialist in Michigan, who diagnosed them with a rare genetic condition called DOPA-responsive dystonia. A treatment regimen was devised that successfully suppressed the symptoms and allowed the twins to assume normal lives. However, at age 12, Alexis developed a severe cough and breathing difficulties that again seemed to threaten the child's survival. In one episode, paramedics had to revive her twice. The symptoms did not seem to be related to the dystonia. Might Noah be next? Frustrated and deeply worried, the twins' parents sought a complete genome sequence of both Noah and Alexis. This seemingly unusual step was a natural one for the Beery family. Joe was the chief information officer at Life Technologies, developers of sequencing technologies in use by many large DNA-sequencing centers. The cases of Noah and Alexis were taken up by Matthew Bainbridge and his team at the Baylor College of Medicine Human Genome Sequencing Center in Houston, Texas. The results proved decisive. The twins had mutations in their genomes that produced not only a deficiency in DOPA but also a potential deficiency in production of the hormone serotonin. A small adjustment in Alexis's therapy brought her life-threatening symptoms to an end, and the same therapy was given to her brother. Both siblings now lead normal lives.

The first draft human genome sequence was completed in 2001, after 12 years, at a cost of $3 billion. That cost has plummeted (Fig. 1), and newly completed human genomes are commonplace. The goal of a $1,000 human genome is on the horizon and promises to make this technology widely available. Since most genomic changes that affect human health are thought to be in protein-coding genes (an assumption that may be challenged in years to come), a cheaper alternative is simply to sequence the 1% of the genome that represents the coding regions (exons) of genes, or the **exome**.

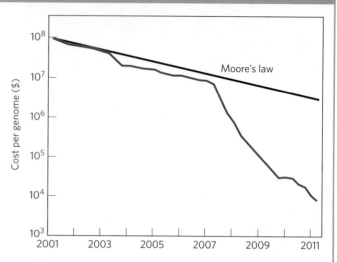

FIGURE 1 Since January 2008, the cost of human genome sequencing has been declining faster than the projected decline in the cost of processing data on computers (Moore's law).

That first human genome sequence came from a haploid genome, derived from a DNA amalgam from several different humans. Completion of a high-quality sequence in 2004 established this as the reference human genome. Subsequent completed human genome sequences, many from individual diploid genomes, have demonstrated how much individual genetic variation exists. Relative to the reference sequence, a typical human has about 3.5 million SNPs (about 60% of these are heterozygous, present on only one of two chromosomes) and another few hundred thousand differences in the form of small insertions and deletions and changes in repeat copy numbers. Only a small portion (5,000 to 10,000) of the SNPs affect the amino acid sequences of proteins encoded by genes.

This complexity ensures that, at least in the short term, successful diagnosis of a condition by whole genome sequencing will be the exception rather than the rule. However, few disciplines are advancing as rapidly as human genomics. The number of success stories is increasing rapidly as the technology becomes more widely available and the capacity of genomic analysis to recognize causative genetic changes improves.

Another enzyme in the solution bathing the surface is sulfurylase, which converts the pyrophosphate to ATP.

The appearance of ATP ultimately provides the signal that a nucleotide has been added to the DNA. Also present in the medium is an enzyme called luciferase and a substrate molecule, luciferin (luciferase is the enzyme that generates the flash of light produced by fireflies; see Box 13–1). When ATP is generated, luciferase catalyzes a reaction with luciferin that results in a tiny flash of light.

When many tiny flashes occur in a cluster, the emitted light can be recorded in a captured image. For example, when dCTP is added to the solution, flashes will occur only at clusters where G is present in the template and C is the next nucleotide to be added to the growing DNA chain. If there is a string of 2, 3, or 4 G residues in the template, a similar number of C residues will be added to the growing strand in one cycle. This is recorded as a "flash" amplitude at that cluster that is 2, 3, or 4 times

(a)

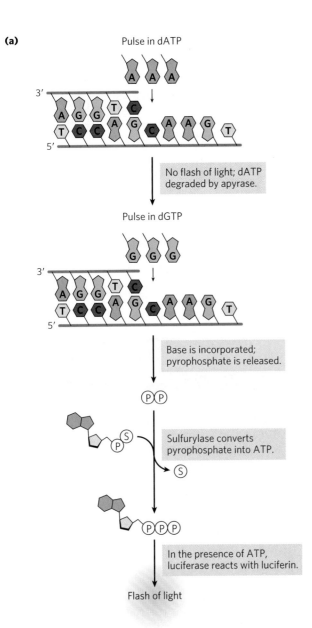

Pulse in dATP

3′

5′

No flash of light; dATP degraded by apyrase.

Pulse in dGTP

3′

5′

Base is incorporated; pyrophosphate is released.

Sulfurylase converts pyrophosphate into ATP.

In the presence of ATP, luciferase reacts with luciferin.

Flash of light

Nucleotide sequence

A GG T C G TT C A G

Relative light intensity

A G T C A G T C A G

Nucleotide added

greater than when only one C residue is added. Similarly, when dGTP is added, flashes occur at a different set of clusters, marking those as clusters where G is the next nucleotide added to the sequence. The length of DNA that can be reliably sequenced in a single cluster by this method—often referred to as the read length or "read"—is typically 400 to 500 nucleotides, as of early 2012, and is constantly improving.

The alternative method is the Illumina sequencer, which uses a technique known as reversible terminator sequencing **(Fig. 9–26)**. A special sequencing primer is added that is complementary to the oligonucleotides of known sequence that were ligated to the ends of the DNA fragments in each cluster (as described above). In addition, fluorescently labeled terminator nucleotides and DNA polymerase are added. The polymerase adds the appropriate nucleotide to the strands in each cluster, each type of nucleotide (A, T, G, or C) carrying a different fluorescent label. These terminator nucleotides have blocking groups attached to the 3′ ends that permit only one nucleotide addition to each strand. Next, lasers excite all the fluorescent labels, and an image of the entire surface reveals the color (and thus the identity of the base) added to each cluster. The fluorescent label and the blocking groups are then chemically or photolytically removed, in preparation for adding a new nucleotide to each cluster. The sequencing proceeds stepwise. Read lengths are shorter for this method, typically 100 to 200 nucleotides per cluster.

These technologies are modern manifestations of an approach to genomic sequencing that is sometimes called shotgun sequencing. Many copies of the genomic DNA are sheared to generate each set of fragments. Thus, a particular short segment of the genome may be present in dozens or even hundreds of different

FIGURE 9–25 Next-generation pyrosequencing. (a) Pyrosequencing uses the enzymes sulfurylase (see Fig. 22–15) and luciferase (see Box 13–1) to detect nucleotide addition with flashes of light. The plot shows the light intensities observed during successive sequencing cycles for a DNA segment immobilized at a particular spot of a picotiter plate and (at top) the DNA nucleotide sequence derived from them. **(b)** An image of a very small part of one cycle of a 454 sequencing run. Each individual segment of DNA to be sequenced is attached to a tiny DNA capture bead, then amplified on the bead by PCR. Each bead is immersed in an emulsion and placed in a tiny (~29 μm) well on a picotiter plate. The reaction of luciferin and ATP with luciferase produces light flashes when a nucleotide is added to a particular DNA cluster in a particular well. Circles represent the same cluster over multiple cycles. In this case, reading the top (or bottom) circle from left to right across each row gives the sequence for that cluster.

(b)

dNTP added

DNA sequence at circled clusters

A G T C A G T C A G

AGGTCGTTCAG:

_ _ _CAGTTCAG:

Duplicate dNTP incorporation results in 2X brighter flash

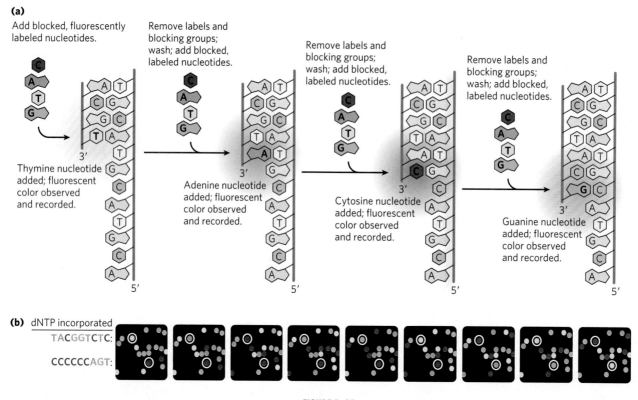

(a)

Add blocked, fluorescently labeled nucleotides.

Thymine nucleotide added; fluorescent color observed and recorded.

Remove labels and blocking groups; wash; add blocked, labeled nucleotides.

Adenine nucleotide added; fluorescent color observed and recorded.

Remove labels and blocking groups; wash; add blocked, labeled nucleotides.

Cytosine nucleotide added; fluorescent color observed and recorded.

Remove labels and blocking groups; wash; add blocked, labeled nucleotides.

Guanine nucleotide added; fluorescent color observed and recorded.

(b) dNTP incorporated

TACGGTCTC:

CCCCCCAGT:

(c)

FIGURE 9–26 Next-generation reversible terminator sequencing. (a) The reversible terminator method of sequencing uses fluorescent tags to identify nucleotides. Blocking groups on each fluorescently labeled nucleotide prevent multiple nucleotides from being added per cycle. **(b)** Six successive cycles from one very small part of an Illumina sequencing run. Each colored spot represents the location of an immobilized DNA oligonucleotide affixed to the surface of the flow cell. The circled clusters represent the same spot on the surface over successive cycles and give the sequences indicated. Data are automatically recorded and analyzed digitally. **(c)** Typical flow cell used for a next-generation sequencer. Millions of DNA fragments can be sequenced simultaneously in each of the eight channels.

sequenced clusters. However, there is no landmark on an individual fragment to tell where in the genome it came from. Translating the sequences of these millions of fragments into a genomic sequence requires the computerized alignment of overlapping fragment sequences **(Fig. 9–27)**. The overlaps allow the computer to trace the sequence through a chromosome, from one fragment to another. This allows the assembly of long contiguous sequences called **contigs**. In a successful genomic sequencing exercise, many contigs can extend over millions of base pairs. Special strategies are needed to fill in the inevitable gaps and to deal with repetitive sequences.

The Human Genome Contains Genes and Many Other Types of Sequences

New and more-efficient DNA sequencing methods have produced an explosion of new genomic sequences. These are stored and made available in a range of publicly accessible databases. A good entrée to these resources can be found at the National Center for Biotechnology Information (NCBI; www.ncbi.nlm.nih.gov). The analysis of that growing wealth of genomic information is ongoing. What does our own genome, and its comparison with those of other organisms, tell us?

In some ways, we are not as complicated as we once imagined. Decades-old estimates that humans had about 100,000 genes within the approximately 3.2×10^9 bp of the human genome have been supplanted by the discovery that we have only about 25,000 protein-coding genes—less than twice the number in a fruit fly (13,600), not many more than in a nematode worm (20,000), and fewer than in a rice plant (38,000).

In other ways, however, we are more complex than we previously realized. The study of eukaryotic chromosome structure and genome sequences has revealed that many, if not most, eukaryotic genes contain one or more intervening segments of DNA that do not code for the amino acid sequence of the polypeptide product. These nontranslated inserts interrupt the otherwise colinear relationship between the gene's nucleotide

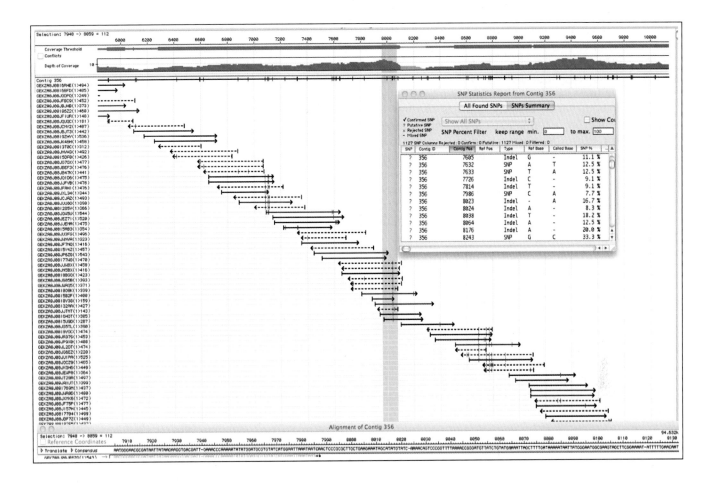

FIGURE 9–27 Sequence assembly. In a genomic sequence, each base pair of the genome is usually represented in several, often dozens, of the sequenced fragments, referred to as reads. Shown is a small part of the sequence of a new variant species of *E. coli*, with the reads generated by a 454 sequencer. The numbers at the top represent genomic base-pair positions, relative to an arbitrarily defined "0." The sequences all come from a particular long contig designated 356. The reads themselves are represented by horizontal arrows, with computer-assigned identifiers listed for each one at the left. DNA strand segments are sequenced at random, with sequences obtained from one strand (5′→3′, left to right) represented by solid arrows and sequences obtained from the other strand (5′→3′, right to left) represented by dashed lines. The latter sequences are automatically reported as

their complement when they are merged with the overall dataset. The coverage threshold at the top is a measure of sequence quality. The wider green bar indicates sequences that have been obtained enough times to generate high confidence in the results. The depth of the coverage line indicates how many times a given base pair appears in a sequenced read. The vertical blue bar denotes a part of the sequence that is highlighted in the sequence line at the bottom of the figure. The SNP statistics report (inset) is a listing of positions where single nucleotide polymorphisms appear to be present in some of the reads. These putative SNPs are often checked by additional sequencing. They are indicated within the reads by thin, blue vertical slash marks within the horizontal lines for each read.

sequence and the amino acid sequence of the encoded polypeptide. Such nontranslated DNA segments are called **intervening sequences**, or **introns**, and the coding segments are called **exons (Fig. 9–28).** Few bacterial genes contain introns. Introns are removed from a primary RNA transcript and the exons are spliced together to generate a transcript that can be translated contiguously into a protein product (see Chapter 26). An exon often (but not always) encodes a single domain of a larger, multidomain protein. Alternative modes of gene expression and RNA splicing permit the production of various combinations of exons, leading to the production of more than one protein from a single gene. Humans share many protein domain types with plants, worms, and flies, but we use these domains

FIGURE 9–28 Introns and exons. This gene transcript contains five exons and four introns, along with 5′ and 3′ untranslated regions (5′UTR and 3′UTR). Splicing removes the introns to create an mRNA product for translation into protein.

in more complex arrangements and thereby generate more complex proteins.

In mammals and some other eukaryotes, the typical gene has a much higher proportion of intron DNA than exon DNA; in most cases, the function of introns is not clear. Only about 1.5% of human DNA is "coding" or exon DNA, carrying information for protein products **(Fig. 9–29a)**. However, when the much larger introns are included in the count, as much as 30% of the human genome consists of genes. Several efforts are underway to categorize the protein-coding genes by function (Fig. 9–29b).

The relative paucity of genes in the human genome leaves a lot of DNA unaccounted for. Much of the nongene DNA is in the form of repeated sequences of several kinds. Perhaps most surprising, about half the human genome is made up of moderately repeated sequences that are derived from transposable elements—segments of DNA, ranging from a few hundred to several thousand base pairs long, that can move from one location to another in the genome. Originally discovered in corn by Barbara McClintock, transposable elements, or **transposons**, are a kind of molecular parasite. They efficiently, and largely passively, make

their home in the genomes of essentially every organism. Many transposons contain genes encoding the proteins that catalyze the transposition process itself, as described in more detail in Chapters 25 and 26. There are multiple classes of transposons in the human genome. Some are active, moving at a low frequency, but most are inactive, evolutionary relics altered by mutations.

Once the protein-coding genes (including exons and introns) and transposons are accounted for, perhaps 25% of the total DNA remains. The largest portion of this consists of unique sequences found between protein-coding genes. As described in Chapter 26, virtually all of these DNA segments are transcribed into RNA in at least some human cells. New classes of functional RNAs—encoded by genes whose existence was previously unsuspected—are being discovered at a rapid pace. Many genes encoding functional RNA are difficult to identify by automated methods, particularly when the RNA products have not been characterized. However, the RNA-coding genes are clearly a prominent feature of these otherwise uncharted genomic regions.

Another 3% or so of the human genome consists of highly repetitive sequences referred to as **simple-sequence repeats (SSRs)**. Generally less than 10 bp long, an SSR is sometimes repeated millions of times per cell and has identifiable functional importance in human cellular metabolism. The most prominent examples of SSR DNA occur in centromeres and telomeres (see Chapter 24). However, long repeats of simple sequences also occur throughout the genome.

What does all this information tell us about the similarities and differences among individual humans? Within the human population there are millions of single-base variations, called **single nucleotide polymorphisms**, or **SNPs** (pronounced "snips"). Each human differs from the next by, on average, 1 in every 1,000 bp. Many of these variations are in the form of SNPs, but a wide range of larger deletions, insertions, and small rearrangements occur in the human population as well. From these often subtle genetic differences comes the human variety we are all aware of—differences in hair color, stature, eyesight, allergies, foot size, and (to some unknown degree) behavior.

The process of genetic recombination and chromosomal segregation during meiosis tends to mix and match these small genetic variations so that different combinations of genes are inherited (see Chapter 25). When two such genetic variations are on different chromosomes, the particular variants that might be inherited by a given individual are a result of random chance. If the genetic variants lie on the same chromosome, the chance that they will be inherited together is an inverse function of the distance between them. Groups of SNPs and other genetic differences that are close together on the same chromosome are rarely affected by recombination and are usually inherited together; these groupings are known as **haplotypes**. Haplotypes provide

(a) Human genome: DNA sequence types

Other transposons (11.2%)
SINEs (13.1%)
LINEs (20.4%)
Simple-sequence repeats (3%)
Segment duplications (5%)
Long repetitive sequences (centromeres, telomeres) (8%)
Miscellaneous unique sequences (11.6%)
Introns (25.9%)
Protein-coding genes (1.5%)

(b) Human genome: Protein-coding genes

Extracellular structures (3%)
Cytoskeleton (3%)
Immune proteins (1.3%)
Enzymes (16.2%)
Unknown (37.4%)
Membrane proteins (2.4%)
Miscellaneous (4.5%)
Nucleic acid binding (9.8%)
Receptors (5.2%)
Transport (3.4%)
Regulation (4.1%)
Signaling (2.7%)
Transcription factor (7.0%)

FIGURE 9–29 A snapshot of the human genome. (a) This pie chart shows the proportions of various types of sequences in our genome. Transposons, including SINEs and LINEs, are described in Chapters 25 and 26. **(b)** The approximately 25,000 protein-coding genes in the human genome can be classified by the type of protein encoded.

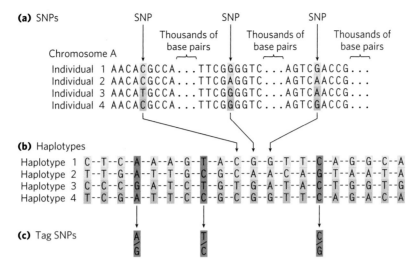

FIGURE 9-30 Haplotype identification. (a) Single nucleotide polymorphisms (SNPs), or positions in the human genome where the sequence varies from one individual to another, are identified in genomic samples, and **(b)** groups of SNPs that are relatively close to each other on a chromosome (within a few tens of thousands of base pairs) are compiled into a haplotype. The SNPs will vary in the overall human population, such as in the four fictitious individuals shown in panels (a) and (b). However, the SNPs chosen to define a haplotype are often the same in most individuals of a particular population. **(c)** Haplotype-defining SNPs (tag SNPs) can be used to simplify the process of identifying an individual's haplotype (by sequencing 3 defining SNPs instead of all 20). If the tag SNPs shown in light red are sequenced, the nucleotides A, T, and G at the three successive tag SNP positions might be characteristic of a population native to one location in northern Europe, whereas the nucleotides G, T, and C at these same three genomic positions might be found in a population in Asia. Multiple haplotypes of this kind are used to trace prehistoric human migrations (see Fig. 9-35).

convenient markers for certain human populations and individuals within populations.

Defining a haplotype requires several steps. First, positions that contain SNPs in the human population are identified in genomic DNA samples from multiple individuals **(Fig. 9-30a)**. Each SNP may be separated from the next by many thousands of base pairs. Second, SNPs that are relatively close to each other on a chromosome and thus usually inherited together are compiled into haplotypes (Fig. 9-30b). Each haplotype consists of the particular bases found at the various SNP positions in the defined haplotype. Finally, tag SNPs—a subset of the SNPs that define the entire haplotype—are chosen to uniquely identify each haplotype (Fig. 9-30c). By sequencing just these tag positions in genomic samples from human populations, researchers can quickly identify which haplotypes are present in each individual. Especially stable haplotypes exist in the mitochondrial genome (which, being inherited maternally, never undergoes meiotic recombination) and on the male Y chromosome (only 3% of which is homologous to the X chromosome and thus subject to recombination).

Genome Sequencing Informs Us about Our Humanity

A primary purpose of most genome sequencing projects is to identify conserved genetic elements of functional significance, such as conserved exon sequences, regulatory regions, and other genomic features such as

centromeres and telomeres. One major purpose of sequencing the human genome is in identifying the differences between our genome and those of other organisms. Although the human genome is very closely related to other mammalian genomes over large segments of every chromosome, differences of a few percent in billions of base pairs add up to millions of genetic distinctions. Searching among these differences using comparative genomics techniques can reveal the molecular basis of human genetic diseases and can help identify genes, gene alterations, and other genomic features that are uniquely human and thus likely to contribute to definably human characteristics such as our large brain, language skills, tool-making ability, or bipedalism.

The genome sequences of our closest biological relatives, the chimpanzees and bonobos, offer some important clues and can illustrate the comparative process. Humans and chimpanzees shared a common ancestor about 7 million years ago. Genomic differences between the two species fall into two types: base-pair changes (SNPs) and larger genomic rearrangements of many types. SNPs in the protein-coding regions often result in amino acid changes that can be used to construct a phylogenetic tree **(Fig. 9-31a)**. Segments of chromosomes may become inverted over evolutionary time. The processes leading to such inversions are complicated and rare, but the human lineage has long segments of DNA that are inverted (relative to other primates) due to this process on chromosomes 1, 12, 15,

FIGURE 9–31 Genomic alterations in the human lineage. (a) This evolutionary tree for primates is derived from sequences for the progesterone receptor, which helps regulate many events in reproduction. The gene encoding this protein has undergone more evolutionary alterations than most. Amino acid changes associated uniquely with humans, chimpanzees, and bonobos are listed by residue number beside each branch. **(b)** The genes on chimpanzee chromosomes 2p and 2q are homologous to those on human chromosome 2, implying that two chromosomes fused into one at some point in the line leading to humans.

16, and 18. Chromosome fusions can also occur. In the human lineage, two chromosomes found in other primate lineages have been fused to form the single human chromosome 2 (Fig. 9–31b). The human lineage thus has 23 chromosome pairs rather than the 24 pairs typical of other primates. Once this fusion appeared in the line leading to humans, it would have represented a major barrier to interbreeding with other primates that lacked it.

If we ignore transposons and large chromosomal rearrangements, the published human and chimpanzee genomes differ by only 1.23% at the level of base pairs (compared with the 0.1% variance from one human to another). If we further ignore variations at positions where there is a known polymorphism in either the human or the chimpanzee population (these are unlikely to reflect a species-defining evolutionary change), the differences amount to about 1.06%, or about 1 in 100 bp. This small percentage translates into more than 30 million base-pair changes, some of which affect protein function and gene regulation. The genome rearrangements that help distinguish chimpanzees and humans include 5 million short insertions or deletions involving a few base pairs each, as well as a substantial number of larger insertions, deletions, inversions, or duplications that can involve many thousands of base pairs. When transposon insertions—a major source of genomic variance—are included, the differences between the human and

chimpanzee genomes increase by about 90 million bp, representing another 3% of these genomes. In effect, each species has segments of DNA, constituting 40 to 45 million bp, that are entirely unique to that particular genome, with larger chromosomal insertions, duplications, and other rearrangements affecting more base pairs than do single-nucleotide changes. Thus, the total genomic difference between chimpanzees and humans amounts to about 4% of their genomes.

Sorting out which genomic distinctions are relevant to features that are uniquely human is a daunting task. If the two species share a common ancestor, then, assuming a similar rate of evolution in both lines, half the changes represent chimpanzee lineage changes and half represent human lineage changes. When you see a difference, how do you tell which variant was the one present in the common ancestor? One way is to compare both genome sequences with those of more distantly related organisms referred to as **outgroups**. Consider a locus, X, where there is a difference between the human and chimpanzee genomes **(Fig. 9–32)**. The lineage of the orangutan, an outgroup, diverged from that of chimps and humans prior to the chimpanzee-human common ancestor. If the sequence at locus X is identical in orangutans and chimpanzees, this sequence was probably present in the chimpanzee-human ancestor, and the sequence seen in humans is specific to the human lineage. Sequences that are identical in humans and orangutans can be eliminated as candidates for

FIGURE 9–32 Determination of sequence alterations unique to one ancestral line. (a) Sequences from the same hypothetical gene in humans and chimpanzees are compared. The sequence of this gene in their last common ancestor is unknown. **(b)** The orangutan genome is used as an outgroup. The sequence of the orangutan gene is identical to the chimpanzee gene. This means that the mutation causing the difference between humans and chimpanzees almost certainly occurred in the line leading to modern humans, and the common ancestor of humans and chimpanzees (and orangutans) had the sequence now found in chimpanzees.

human-specific genomic features. The importance of comparisons with closely related outgroups has given rise to new efforts to sequence the genomes of orangutans, macaques, and many other primate species.

The search for the genetic underpinnings of special human characteristics, such as our enhanced brain function, can benefit from two complementary approaches. The first searches for genomic regions showing extreme changes from other primates. These could include multiple gene duplications or the addition of large genomic segments not present in other primates. The second approach looks at genes known to be involved in relevant human conditions. For brain function, for example, one would examine genes that, when mutated, contribute to cognitive or other mental disorders.

Notably, analyses of the human lineage have not detected an enrichment of genetic changes in protein-coding genes involved in brain development or size. In primates, most genes that function uniquely in the brain are even more highly conserved than genes functioning in other tissues. However, some differences in gene expression are observed. When changes in genomic regions related to gene regulation are analyzed, genes involved in neural development and nutrition are disproportionately affected. A variety of RNA-coding genes, some with expression concentrated in the brain, also show evidence of accelerated evolution **(Fig. 9–33)**. The many new classes of RNA that are being discovered (see Chapter 26) are likely to radically change our perspective on how evolution alters the workings of living systems. It is increasingly evident that which genes are expressed may not be as important as when, where, and how much they are expressed.

Genome Comparisons Help Locate Genes Involved in Disease

The Human Genome Project has fulfilled its potential for accelerating the discovery of genes underlying genetic diseases: over 1,600 human genetic diseases have now been mapped to particular genes. With use of a method called **linkage analysis**, the gene involved in a disease condition is mapped relative to well-characterized genetic polymorphisms that occur throughout the human genome. The search often begins with one or more large families that include several individuals affected by a particular disease over several generations. The most common approach is basically an exercise in phylogenetics (the study of evolutionary relatedness between groups of organisms) and is deeply rooted in concepts derived from evolutionary biology. We can illustrate by describing the search for one gene involved in Alzheimer disease. About 10% of all cases of this condition in the United States result from an inherited predisposition. Several different genes have been discovered that, when mutated, can lead to early onset of Alzheimer disease. One such gene (*PS1*) encodes the protein presenilin-1, and its discovery made heavy use of linkage analysis.

Just as haplotypes rely on SNPs that lie close together on a chromosome, linkage analysis involves a search for SNPs that lie close to a gene of interest. In studies of this type, researchers focus on families affected by the disease, and look for families in which DNA samples can be obtained from individuals of multiple generations. DNA samples are collected from both affected and unaffected family members. Researchers first localize the region associated with the disease to a specific chromosome, using sets (called panels) of genomic locations where common SNPs or other mapped genomic alterations occur in a significant proportion of the human population. Thus, many but not all humans will differ in the genomic sequence at these locations. Using a panel that includes several well-characterized SNP loci mapped to each chromosome, investigators compare the genotypes of individuals with and without the disease, focusing especially on close family members. For Alzheimer disease, two of the many family pedigrees used to search for this gene in the early 1990s are shown in **Figure 9–34**. Using such pedigrees, researchers look for the particular SNP variants that are inherited in the same or nearly the same pattern as the disease-causing gene. The responsible gene can gradually be localized to a single chromosome, if the inheritance of particular SNP variants on that chromosome is mirrored closely by the inheritance of the disease condition. More-detailed localization of a disease-causing gene on a chromosome relies on statistical methods to correlate the inheritance of additional, more closely spaced polymorphisms with the occurrence of the disease, focusing on a denser panel of polymorphisms known to occur on the chromosome of interest. The more closely a marker is located to a disease gene, the more likely it is

FIGURE 9–33 Accelerated evolution in some human genes. The HAR1F locus specifies a noncoding RNA that is highly conserved in vertebrates. In humans, the HAR1F gene exhibits an unusual number of mutations (shaded light red), providing evidence of accelerated evolution.

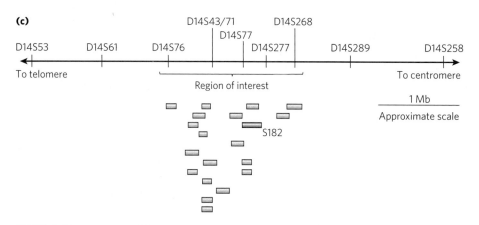

(c)

D14S53 — D14S61 — D14S76 — D14S43/71 / D14S77 — D14S268 / D14S277 — D14S289 — D14S258

To telomere

Region of interest

To centromere

1 Mb
Approximate scale

S182

FIGURE 9–34 Linkage analysis in the discovery of disease genes.
(a) These pedigrees for two families affected by early-onset Alzheimer disease are based on the data available at the time of the study. Red symbols represent affected individuals; slashes indicate deaths either before or soon after the study. The number above each symbol is the person's age, either at the time of the study or at time of death (indicated with a D). To protect family privacy, gender is not indicated. **(b)** Chromosome 14, with bands created by certain dyes. Chromosome marker positions are shown at the right, with the genetic distance between them presented in a genetic distance measurement called centimorgans, reflecting the frequency of recombination between them. *TCRD* (T-cell receptor delta) and *PI* (AACT (α1-antichymotrypsin)) are genes with alterations in the human population that were used as markers, along with SNPs, in chromosome mapping. **(c)** By comparisons of DNA from affected and unaffected family members, a region of interest that contains 19 expressed genes was eventually defined near marker D14S43. The gene labeled *S182* (red) encodes presenilin-1.

to be inherited along with that gene. This process can point to a region of the chromosome that contains the gene. In this example of Alzheimer disease, linkage analysis indicated that the disease-causing gene was somewhere near a SNP locus called D14S43 (Fig. 9–34c).

The final steps of the search for the disease gene again use the human genome databases. The local region containing the gene is examined and the genes within it are identified. DNA from many individuals, some who have the disease and some who do not, is sequenced over this region. This process gradually leads, with an increasing number of individuals analyzed, to the identification of gene variants consistently present in individuals with the disease state and not in unaffected individuals. The search can be aided by an understanding of the function of the genes in the target region, because particular metabolic pathways may be more likely than others to produce the disease state. In 1995, the chromosome 14 gene associated with Alzheimer disease was identified as gene *S182*. The product of this gene was given the name presenilin-1, and the gene itself was subsequently renamed *PS1*.

More complex are cases where a disease condition is caused by the presence of mutations in two different genes (neither of which, alone, causes the disease), or where a particular condition is enhanced by an otherwise innocuous mutation in another gene. Identifying the genes and mutations responsible for such digenic diseases is exceedingly difficult, and these diseases are sometimes possible to document only within small, isolated, and highly inbred populations.

Genome databases open alternative paths to the identification of disease genes, especially when biochemical information about the disease is known. In the case of Alzheimer disease, an accumulation of the amyloid β-protein in limbic and association cortices of the brain is at least partly responsible for the symptoms. Defects in presenilin-1 (and in a related protein, presenilin-2, encoded by a gene on chromosome 1) lead to elevated cortical levels of amyloid β-protein. Focused databases catalog such functional information on the protein products of genes and on protein-interaction networks, SNP locations, and other data, providing a streamlined path to the identification of candidate genes for a particular disease. A researcher with some knowledge about the kinds of enzymes or other proteins likely to contribute to a disease can use these databases to generate a list of genes known to encode proteins with relevant functions, additional uncharacterized genes with orthologous or paralogous relationships to the genes in this list, a list of proteins known to interact with the target proteins or orthologs in other organisms, and a map of gene positions. With data from selected family pedigrees, a short list of potentially relevant genes can often be determined rapidly.

These approaches are not limited to human diseases. The same methods can be used to identify the genes involved in diseases—or genes that produce desirable characteristics—in other animals and in plants. ■

Genome Sequences Inform Us about Our Past and Provide Opportunities for the Future

About 70,000 years ago, a small group of humans in Africa looked out across the Red Sea to Asia. Perhaps encouraged by some innovation in small boat construction, or driven by conflict or famine, or simply curious, they crossed the water barrier. That initial colonization, involving maybe 1,000 individuals, began a journey that did not stop until humans reached Tierra del Fuego (at the southern tip of South America) many thousands of years later. In the process, an established population from a previous hominid expansion into Eurasia, including *Homo neanderthalensis*, was displaced. The Neanderthals disappeared, just as *Homo erectus* and other hominid lines disappeared before them.

The story of how modern humans first appeared in Africa a few hundred thousand years ago, and their migrations as they eventually radiated out of Africa, is written in our DNA. Through the use of genomic sequences from multiple species, both primate and hominid evolution have come into sharper focus. Using haplotypes present in extant human populations, we can trace the migrations of our intrepid ancestors across the planet (**Fig. 9–35**).

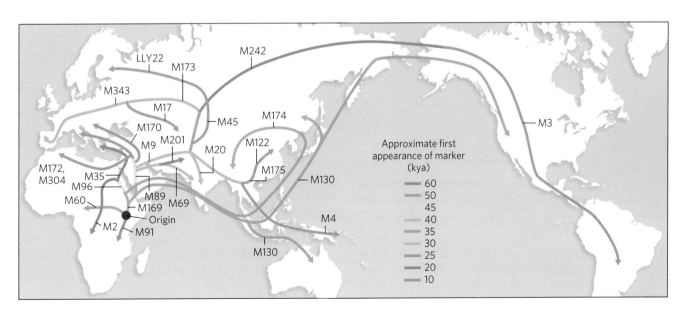

FIGURE 9–35 The paths of human migrations. When a small part of a human population migrates away from a larger group, it takes only part of the overall genetic diversity with it. Thus, some haplotypes are present in the migrating group but many are not. At the same time, mutations can create novel haplotypes over time. This map was generated from an analysis of genetic markers (defined haplotypes with M or LLY numbers) on the Y chromosome. The genetic samples were taken from indigenous populations long established at geographic points along the routes shown. Haplotypes that appear suddenly along a migration path, reflecting new changes (mutations) in particular SNP genomic locations in certain isolated populations, are called "founder events." These enable researchers to trace migrations from that point, as other populations that possess the new haplotype were probably descended from the founder population. The abbreviation *kya* means "thousand years ago."

BOX 9–3   Getting to Know the Neanderthals

Modern humans and Neanderthals coexisted in Europe and Asia as recently as 30,000 years ago. The human and Neanderthal ancestral populations diverged about 370,000 years ago, before the appearance of anatomically modern humans. Neanderthals used tools, lived in small groups, and buried their dead. Of the known hominid relatives of modern humans, Neanderthals are the closest. For hundreds of millennia, they inhabited large parts of Europe and western Asia (Fig. 1). If the chimpanzee genome can tell us something about what it is to be human, perhaps the Neanderthal genome can tell us more. Buried in the bones and remains taken from burial sites are fragments of Neanderthal genomic DNA. Technologies developed for use in forensic science (see Box 9–1) and ancient DNA studies have been combined to initiate a Neanderthal genome project.

This endeavor is unlike the genome projects aimed at extant species. The Neanderthal DNA is present in small amounts, and it is contaminated with DNA from other animals and bacteria. How does one get at it, and how can one be certain that the sequences really came from Neanderthals? The answers have been revealed by innovative applications of biotechnology. In essence, the small quantities of DNA fragments found in a Neanderthal bone or other remains are cloned into a library, and the cloned DNA segments are sequenced at random, contaminants and all. The sequencing results are compared with the existing human genome and chimpanzee genome databases. Segments derived from Neanderthal DNA are readily distinguished from segments derived from bacteria or insects by computerized analysis, because they have sequences closely related to human and chimpanzee DNA. Once a collection of Neanderthal DNA segments is sequenced, they can be used as probes to identify sequence fragments in ancient samples that overlap with these known fragments.

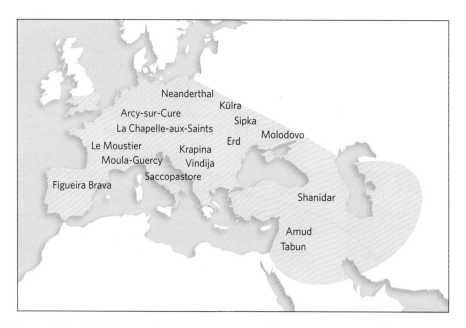

FIGURE 1 Neanderthals occupied much of Europe and western Asia until about 30,000 years ago. Major Neanderthal archaeological sites are shown here. (Note that the group was named for the site at Neanderthal in Germany.)

The Neanderthals were not simply displaced. Some mingling occurred. Using sensitive PCR-based methods, we now have a nearly complete sequence of the Neanderthal genome (Box 9–3). We know that about 4% of the human genomes of non-Africans are derived from Neanderthals. Some human populations also acquired genomic DNA from another recently discovered group, the Denisovans. Neanderthal DNA gave humans a more complex immune system, making us more resistant to infection but also a little more susceptible to autoimmune diseases. The story of our past is gradually taking shape as more human genomes, of those alive today and those who lived in past millennia, are being assembled.

The medical promise of personal genomic sequences grows as more genes underlying inherited diseases are defined. Knowledge of genomic sequences also provides the prospect of altering them. It is now commonplace to engineer the DNA sequences of organisms from bacteria and yeast to plants and mammals for research and commercial purposes. Efforts to cure inherited human diseases by human gene therapy have not yet lived up

The potential problem of contamination with the closely related modern human DNA can be controlled for by examining mitochondrial DNA. Human populations have readily identifiable haplotypes (distinctive sets of genomic differences; see Fig. 9–30) in their mitochondrial DNA, and analysis of Neanderthal samples has shown that Neanderthals' mitochondrial DNA has its own distinct haplotypes. The presence in the Neanderthal samples of some base-pair differences that are found in the chimpanzee database but not in the human database is more evidence that non-human hominid sequences are being found.

Completion of this challenging endeavor is on the horizon. The draft sequence of the Neanderthal genome unveiled in early 2009 covered more than 60% of the genomic sequences. A finished sequence will require just a little more time. The data provide evidence that modern humans and the Neanderthals who were the source of this DNA shared a common ancestor about 700,000 years ago (Fig. 2). Analysis of mitochondrial DNA suggests that the two groups continued on the same track, with some gene flow between them, for about 300,000 more years. The lines split with the appearance of anatomically modern humans, although evidence now exists for some intermingling of the lines somewhat later.

Expanded libraries of Neanderthal DNA from different sets of remains should eventually allow an analysis of Neanderthal genetic diversity, and perhaps Neanderthal migrations, providing a fascinating look at our hominid past.

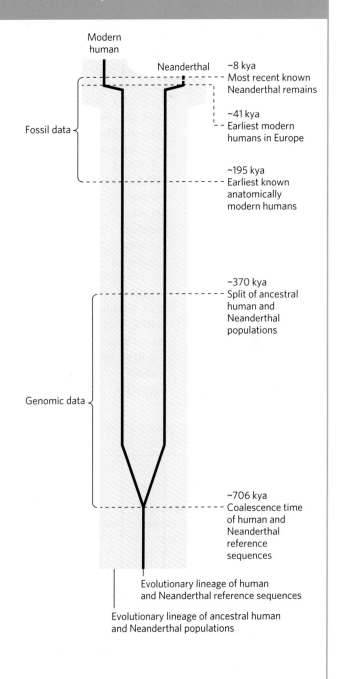

FIGURE 2 This timeline shows the divergence of human and Neanderthal genome sequences (black lines) and of ancestral human and Neanderthal populations (yellow screen). Genomic data provide evidence for some intermingling of the populations up to about 45,000 years ago. Key events in human evolution are noted.

to their potential, but technologies for gene delivery are constantly being improved. Few scientific disciplines will affect the future of our species more than modern genomics.

SUMMARY 9.3 Genomics and the Human Story

▶ Next-generation sequencing methods have vastly reduced the time required to generate complete genomic sequences.

▶ About 30% of the DNA in the human genome is in the exons and introns of genes encoding proteins.

Nearly half of the DNA is derived from parasitic transposons. Much of the rest encodes RNAs of many types. Simple-sequence repeats make up the centromere and telomeres.

▶ The gene alterations that define humanity can be discerned in part through comparative genomics using other primates.

▶ Comparative genomics is also used to locate the gene alterations that define inherited diseases and can be used to study the evolution and migration of our human ancestors over millennia.

Key Terms

Terms in bold are defined in the glossary.

genomics 313
systems biology 313
cloning 314
vector 314
recombinant DNA 314
genetic engineering 314
restriction endonucleases 314
DNA ligases 314
plasmid 317
bacterial artificial chromosome (BAC) 319
yeast artificial chromosome (YAC) 320
expression vector 321
baculovirus 323
bacmid 323
site-directed mutagenesis 323
fusion protein 325
tag 325
polymerase chain reaction (PCR) 327

short tandem repeat (STR) 329
quantitative PCR (qPCR) 331
DNA library 332
genomic library 332
complementary DNA (cDNA) 332
cDNA library 332
comparative genomics 333
orthologs 333
paralogs 333
synteny 333
epitope tag 333
yeast two-hybrid analysis 336
DNA microarray 337
contig 342
single nucleotide polymorphism (SNP) 344
haplotype 344

Further Reading

General

Jackson, D.A., Symons, R.H., & Berg, P. (1972) Biochemical method for inserting new genetic information into DNA of simian virus 40: circular SV40 DNA molecules containing lambda phage genes and the galactose operon of *Escherichia coli. Proc. Natl. Acad. Sci. USA* **69**, 2904–2909.

The first recombinant DNA experiment linking DNA from two species.

Lobban, P.E. & Kaiser, A.D. (1973) Enzymatic end-to-end joining of DNA molecules. *J. Mol. Biol.* **78**, 453–471.

Report of the first recombinant DNA experiment.

Studying Genes and Their Products

Arnheim, N. & Erlich, H. (1992) Polymerase chain reaction strategy. *Annu. Rev. Biochem.* **61**, 131–156.

Foster, E.A., Jobling, M.A., Taylor, P.G., Donnelly, P., de Knijff, P., Mieremet, R., Zerjal, T., & Tyler-Smith, C. (1999) The Thomas Jefferson paternity case. *Nature* **397**, 32.

Last article of a series in an interesting case study of the uses of biotechnology to address historical questions.

Giepmans, B.N.G., Adams, S.R., Ellisman, M.H., & Tsien, R.Y. (2006) The fluorescent toolbox for assessing protein location and function. *Science* **312**, 217–224.

Kayser, M. & de Knijff, P. (2011) Improving human forensics through advances in genetics, genomics, and molecular biology. *Nat. Rev. Genet.* **12**, 179–192.

Ståhl, P.L. & Lundeber, J. (2012) Toward the single-hour high-quality genome. *Annu. Rev. Biochem.* **81**, 359–378.

Zhao, J. & Grant, S.F.A. (2011) Advances in whole genome sequencing technology. *Curr. Pharm. Biotechnol.* **12**, 293–305.

Using DNA-Based Methods to Understand Protein Function

Budowle, B., Johnson, M.D., Fraser, C.M., Leighton, T.J., Murch, R.S., & Chakraborty, R. (2005) Genetic analysis and attribution of microbial forensics evidence. *Crit. Rev. Microbiol.* **31**, 233–254.

How biotechnology is used to fight bioterrorism.

Koonin, E.V. (2005) Orthologs, paralogs, and evolutionary genomics. *Annu. Rev. Genet.* **39**, 309–338.

Good description of some comparative genomics basics.

Stoughton, R.B. (2005) Applications of DNA microarrays in biology. *Annu. Rev. Biochem.* **74**, 53–82.

Terpe, K. (2006) Overview of bacterial expression systems for heterologous protein production: from molecular and biochemical fundamentals to commercial systems. *App. Microbiol. Biotechnol.* **72**, 211–222.

Yooseph, S., Sutton, G., Rusch, D.B., Halpern, A.L., Williamson, S.J., Remington, K., Eisen, J.A., Heidelberg, K.B., Manning, G., Li, W., et al. (2007) The *Sorcerer II* global ocean sampling expedition: expanding the universe of protein families. *PLoS Biol.* **5**, e16.

One of a series of articles arising from an ambitious effort to sample microbial biodiversity throughout the world's oceans.

Genomics and the Human Story

Alkan, C., Sajjadian, S., & Eichler, E.E. (2011) Limitations of next-generation genome sequence assembly. *Nat. Methods.* **8**, 61–65.

Callaway, E. (2011) Ancient DNA reveals secrets of human history. *Nature* **476**, 136–137.

Carr, P.A. & Church, G.M. (2009) Genome engineering. *Nat. Biotechnol.* **27**, 1151–1162.

Carroll, S.B. (2003) Genetics and the making of *Homo sapiens. Nature* **422**, 849–857.

Gonzaga-Jauregui, C., Lupski, J.R., & Gibbs, R.A. (2012) Human genome sequencing in health and disease. *Annu. Rev. Med.* **63**, 35–61.

Green, E.D. & Guyer, M.S. (2011) Charting a course for genomic medicine from base pairs to bedside. *Nature* **470**, 204–213.

Kay, M.A. (2011) State-of-the-art gene-based therapies: the road ahead. *Nat. Rev. Genet.* **12**, 316–328.

Lander, E.S. (2011) Initial impact of the sequencing of the human genome. *Nature* **470**, 187–197.

Metzker, M.L. (2010) Sequencing technologies: the next generation. *Nat. Rev. Genet.* **11**, 31–46.

Perkel, J.M. (2011) Synthetic genomics: building a better bacterium. *Science* **331**, 1628–1630.

Reich, D., Green, R.E., Kircher, M., Krause, J., Patterson, N., Durand, E.Y., Viola, B., Briggs, A.W., Stenzel, U., Johnson, P.L., et al. (2010) Genetic history of an archaic hominin group from Denisova cave in Siberia. *Nature* **468**, 1053–1060.

Neanderthals were not the only other hominid group that humans encountered in their trek across Asia.

Stoneking, M. & Krause, J. (2011) Learning about human population history from ancient and modern genomes. *Nat. Rev. Genet.* **12**, 603–614.

Problems

1. Engineering Cloned DNA When joining two or more DNA fragments, a researcher can adjust the sequence at the junction in a variety of subtle ways, as seen in the following exercises.

(a) Draw the structure of each end of a linear DNA fragment produced by an *Eco*RI restriction digest (include those sequences remaining from the *Eco*RI recognition sequence).

(b) Draw the structure resulting from the reaction of this end sequence with DNA polymerase I and the four deoxynucleoside triphosphates (see Fig. 8–33).

(c) Draw the sequence produced at the junction that arises if two ends with the structure derived in (b) are ligated (see Fig. 25–16).

(d) Draw the structure produced if the structure derived in (a) is treated with a nuclease that degrades only single-stranded DNA.

(e) Draw the sequence of the junction produced if an end with structure (b) is ligated to an end with structure (d).

(f) Draw the structure of the end of a linear DNA fragment that was produced by a *Pvu*II restriction digest (include those sequences remaining from the *Pvu*II recognition sequence).

(g) Draw the sequence of the junction produced if an end with structure (b) is ligated to an end with structure (f).

(h) Suppose you can synthesize a short duplex DNA fragment with any sequence you desire. With this synthetic fragment and the procedures described in (a) through (g), design a protocol that would remove an *Eco*RI restriction site from a DNA molecule and incorporate a new *Bam*HI restriction site at approximately the same location. (See Fig. 9–2.)

(i) Design four different short synthetic double-stranded DNA fragments that would permit ligation of structure (a) with a DNA fragment produced by a *Pst*I restriction digest. In one of these fragments, design the sequence so that the final junction contains the recognition sequences for both *Eco*RI and *Pst*I. In the second and third fragments, design the sequence so that the junction contains only the *Eco*RI and only the *Pst*I recognition sequence, respectively. Design the sequence of the fourth fragment so that neither the *Eco*RI nor the *Pst*I sequence appears in the junction.

2. Selecting for Recombinant Plasmids When cloning a foreign DNA fragment into a plasmid, it is often useful to insert the fragment at a site that interrupts a selectable marker (such as the tetracycline-resistance gene of pBR322). The loss of function of the interrupted gene can be used to identify clones containing recombinant plasmids with foreign DNA. With a bacteriophage λ vector it is not necessary to do this, yet one can easily distinguish vectors that incorporate large foreign DNA fragments from those that do not. How are these recombinant vectors identified?

3. DNA Cloning The plasmid cloning vector pBR322 (see Fig. 9–3) is cleaved with the restriction endonuclease *Pst*I. An isolated DNA fragment from a eukaryotic genome (also produced by *Pst*I cleavage) is added to the prepared vector and ligated. The mixture of ligated DNAs is then used to transform bacteria, and plasmid-containing bacteria are selected by growth in the presence of tetracycline.

(a) In addition to the desired recombinant plasmid, what other types of plasmids might be found among the transformed bacteria that are tetracycline-resistant? How can the types be distinguished?

(b) The cloned DNA fragment is 1,000 bp long and has an *Eco*RI site 250 bp from one end. Three different recombinant plasmids are cleaved with *Eco*RI and analyzed by gel electrophoresis, giving the patterns shown below. What does each pattern say about the cloned DNA? Note that in pBR322, the *Pst*I and *Eco*RI restriction sites are about 750 bp apart. The entire plasmid with no cloned insert is 4,361 bp. Size markers in lane 4 have the number of nucleotides noted.

4. Restriction Enzymes The partial sequence of one strand of a double-stranded DNA molecule is

5′ – – – GACGAAGTGCTGCAGAAAGTCCGCGTTATAGGCAT GAATTCCTGAGG – – – 3′

The cleavage sites for the restriction enzymes *Eco*RI and *Pst*I are shown below.

*Eco*RI

(5′) G A A T T C (3′)
    C T T A A G

*Pst*I

(5′) C T G C A G (3′)
    G A C G T C

Write the sequence of *both strands* of the DNA fragment created when this DNA is cleaved with both *Eco*RI and *Pst*I. The top strand of your duplex DNA fragment should be derived from the strand sequence given above.

5. Designing a Diagnostic Test for a Genetic Disease Huntington disease (HD) is an inherited neurodegenerative disorder, characterized by the gradual, irreversible impairment of psychological, motor, and cognitive functions. Symptoms typically appear in middle age, but onset can occur at almost any age. The course of the disease can last 15 to 20 years. The molecular basis of the disease is becoming better understood. The genetic mutation underlying HD has been traced to a gene encoding a protein (M_r 350,000) of unknown function. In individuals who will not develop HD, a region of the gene that encodes the amino terminus of the protein has a sequence of CAG codons (for glutamine) that is repeated 6 to 39 times in succession. In individuals with adult-onset HD, this codon is typically repeated 40 to 55 times. In individuals with childhood-onset HD, this codon is repeated more than 70 times.

The length of this simple trinucleotide repeat indicates whether an individual will develop HD, and at approximately what age the first symptoms will occur.

A small portion of the amino-terminal coding sequence of the 3,143-codon HD gene is given below. The nucleotide sequence of the DNA is shown in black, the amino acid sequence corresponding to the gene is shown in blue, and the CAG repeat is shaded. Using Figure 27–7 to translate the genetic code, outline a PCR-based test for HD that could be carried out using a blood sample. Assume the PCR primer must be 25 nucleotides long. By convention, unless otherwise specified a DNA sequence encoding a protein is displayed with the coding strand (the sequence identical to the mRNA transcribed from the gene) on top such that it is read 5′ to 3′, left to right.

```
307  ATGGCGACCCTGGAAAAGCTGATGAAGGCCTTCGAGTCCCTCAAGTCCTTC
1    M   A   T   L   E   K   L   M   K   A   F   E   S   L   K   S   F

358  CAGCAGTTCCAGCAGCAGCAGCAGCAGCAGCAGCAGCAGCAGCAGCAGCAG
18   Q   Q   F   Q   Q   Q   Q   Q   Q   Q   Q   Q   Q   Q   Q   Q

409  CAGCAGCAGCAGCAGCAGCAGCAACAGCCGCCACCGCCGCCGCCGCCGCCG
35   Q   Q   Q   Q   Q   Q   Q   Q   Q   P   P   P   P   P   P   P

460  CCGCCTCCTCAGCTTCCTCAGCCGCCGCCG
52   P   P   P   Q   L   P   Q   P   P   P
```

Source: The Huntington's Disease Collaborative Research Group. (1993) A novel gene containing a trinucleotide repeat that is expanded and unstable on Huntington's disease chromosomes. *Cell* 72, 971–983.

6. Using PCR to Detect Circular DNA Molecules In a species of ciliated protist, a segment of genomic DNA is sometimes deleted. The deletion is a genetically programmed reaction associated with cellular mating. A researcher proposes that the DNA is deleted in a type of recombination called site-specific recombination, with the DNA on either end of the segment joined together and the deleted DNA ending up as a circular DNA reaction product.

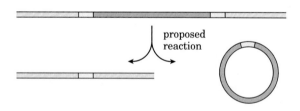

Suggest how the researcher might use the polymerase chain reaction (PCR) to detect the presence of the circular form of the deleted DNA in an extract of the protist.

7. Glowing Plants When grown in ordinary garden soil and watered normally, a plant engineered to express green fluorescent protein (see Fig. 9–16) will glow in the dark, whereas a plant engineered to express firefly luciferase will not. Explain these observations.

8. Designing PCR Primers One strand of a chromosomal DNA sequence is shown at the top of the next column. An investigator wants to amplify and isolate a DNA fragment defined by the segment shown in red, using the polymerase chain reaction. Design two PCR primers, each 20 nucleotides long, that can be used to amplify this DNA segment.

```
5′ – – – AATGCCGTCAGCCGATCTGCCTCGAGTCAATCGA
TGCTGGTAACTTGGGGTATAAAGCTTACCCATGGTATCG
TAGTTAGATTGATTGTTAGGTTCTTAGGTTTAGGTTTC
TGGTATTGGTTTAGGGTCTTTGATGCTATTAATTGTTTGG
TTTTGATTTGGTCTTTATATGGTTTATGTTTTAAGCCGGGT
TTTGTCTGGGATGGTTCGTCTGATGTGCGCGTAGCGTGCG
GCG – – – 3′
```

9. Mapping a Chromosome Segment A group of overlapping clones, designated A through F, is isolated from one region of a chromosome. Each of the clones is separately cleaved by a restriction enzyme and the pieces resolved by agarose gel electrophoresis, with the results shown below. There are nine different restriction fragments in this chromosomal region, with a subset appearing in each clone. Using this information, deduce the order of the restriction fragments in the chromosome.

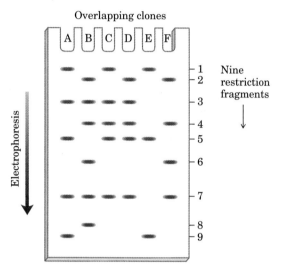

10. Immunofluorescence In the more common protocol for immunofluorescence detection of cellular proteins, an investigator uses two antibodies. The first binds specifically to the protein of interest. The second is labeled with fluorochromes for easy visualization, and it binds to the first antibody. In principle, one could simply label the first antibody and skip one step. Why use two successive antibodies?

11. Yeast Two-Hybrid Analysis You are a researcher who has just discovered a new protein in a fungus. Design a yeast two-hybrid experiment to identify the other proteins in the fungal cell with which your protein interacts and explain how this could help you determine the function of your protein.

12. Use of Photolithography to Make a DNA Microarray Figure 9–22 shows the first steps in the process of making a DNA microarray, or DNA chip, using photolithography. Describe the remaining steps needed to obtain the desired sequences (a different four-nucleotide sequence on each of the four spots) shown in the first panel of the figure. After each step, give the resulting nucleotide sequence attached at each spot.

13. Genomic Sequencing In large-genome sequencing projects, the initial data usually reveal gaps where no sequence information has been obtained. To close the gaps, DNA primers complementary to the 5′-ending strand (i.e., identical to the sequence of the 3′-ending strand) at the end of each contig are especially useful. Explain how these primers might be used.

14. Use of Outgroups in Comparative Genomics A hypothetical protein is found in orangutans, chimpanzees, and humans that has the following sequences (red indicates the amino acid residue differences):

Human: ATSAAG**Y**DEWEGGKV**L**IHL – – **KLQ**NRGALL ELDIGAV

Orangutan: ATSAAG**W**DEWEGGKV**L**IHL**DG**KLQNRGALL ELDIGAV

Chimpanzee: ATSAAG**W**DEWEGGK**I**LIHL**DG**KLQNRGALL ELDIGAV

(Dashes indicate a deletion—the residues are missing in that sequence.)

What is the most likely sequence of the protein present in the last common ancestor of chimpanzees and humans?

15. Finding Disease Genes You are a gene hunter, trying to find the genetic basis for a rare inherited disease. Examination of six pedigrees of families affected by the disease provides inconsistent results. For two of the families, the disease is co-inherited with markers on chromosome 7. For the other four families, the disease is co-inherited with markers on chromosome 12. Explain how this might occur.

Data Analysis Problem

16. *Hinc*II: The First Restriction Endonuclease Discovery of the first restriction endonuclease to be of practical use was reported in two papers published in 1970. In the first paper, Smith and Wilcox described the isolation of an enzyme that cleaved double-stranded DNA. They initially demonstrated the enzyme's nuclease activity by measuring the decrease in viscosity of DNA samples treated with the enzyme.

(a) Why does treatment with a nuclease decrease the viscosity of a solution of DNA?

The authors determined whether the enzyme was an endo- or an exonuclease by treating ^{32}P-labeled DNA with the enzyme, then adding trichloroacetic acid (TCA). Under the conditions used in their experiment, single nucleotides would be TCA-soluble and oligonucleotides would precipitate.

(b) No TCA-soluble ^{32}P-labeled material formed on treatment of ^{32}P-labeled DNA with the nuclease. Based on this finding, is the enzyme an endo- or exonuclease? Explain your reasoning.

When a polynucleotide is cleaved, the phosphate usually is not removed but remains attached to the 5′ or 3′ end of the resulting DNA fragment. Smith and Wilcox determined the location of the phosphate on the fragment formed by the nuclease in the following steps:

1. Treat unlabeled DNA with the nuclease.
2. Treat a sample (A) of the product with γ-^{32}P-labeled ATP and polynucleotide kinase (which can attach the γ-phos-

phate of ATP to a 5′ OH but not to a 5′ phosphate or to a 3′ OH or 3′ phosphate). Measure the amount of ^{32}P incorporated into the DNA.
3. Treat another sample (B) of the product of step 1 with alkaline phosphatase (which removes phosphate groups from free 5′ and 3′ ends), followed by polynucleotide kinase and γ-^{32}P-labeled ATP. Measure the amount of ^{32}P incorporated into the DNA.

(c) Smith and Wilcox found that sample A had 136 counts/ min of ^{32}P; sample B had 3,740 counts/min. Did the nuclease cleavage leave the phosphate on the 5′ or the 3′ end of the DNA fragments? Explain your reasoning.

(d) Treatment of bacteriophage T7 DNA with the nuclease gave approximately 40 specific fragments of various lengths. How is this result consistent with the enzyme's recognizing a specific sequence in the DNA as opposed to making random double-strand breaks?

At this point, there were two possibilities for the site-specific cleavage: the cleavage occurred either (1) at the site of recognition or (2) near the site of recognition but not within the sequence recognized. To address this issue, Kelly and Smith determined the sequence of the 5′ ends of the DNA fragments generated by the nuclease, in the following steps:

1. Treat phage T7 DNA with the enzyme.
2. Treat the resulting fragments with alkaline phosphatase to remove the 5′ phosphates.
3. Treat the dephosphorylated fragments with polynucleotide kinase and γ-^{32}P-labeled ATP to label the 5′ ends.
4. Treat the labeled molecules with DNases to break them into a mixture of mono-, di-, and trinucleotides.
5. Determine the sequence of the labeled mono-, di-, and trinucleotides by comparing them with oligonucleotides of known sequence on thin-layer chromatography.

The labeled products were identified as follows: mononucleotides: A and G; dinucleotides: (5′)ApA(3′) and (5′)GpA(3′); trinucleotides: (5′)ApApC(3′) and (5′)GpApC(3′).

(e) Which model of cleavage is consistent with these results? Explain your reasoning.

Kelly and Smith went on to determine the sequence of the 3′ ends of the fragments. They found a mixture of (5′)TpC(3′) and (5′)TpT(3′). They did not determine the sequence of any trinucleotides at the 3′ end.

(f) Based on these data, what is the recognition sequence for the nuclease and where in the sequence is the DNA backbone cleaved? Use Table 9–2 as a model for your answer.

References

Kelly, T.J. & Smith, H.O. (1970) A restriction enzyme from *Haemophilus influenzae*: II. Base sequence of the recognition site. *J. Mol. Biol.* **51**, 393–409.

Smith, H.O. & Wilcox, K.W. (1970) A restriction enzyme from *Haemophilus influenzae*: I. Purification and general properties. *J. Mol. Biol.* **51**, 379–391.

Lipids

Biological lipids are a chemically diverse group of compounds, the common and defining feature of which is their insolubility in water. The biological functions of the lipids are as diverse as their chemistry. Fats and oils are the principal stored forms of energy in many organisms. Phospholipids and sterols are major structural elements of biological membranes. Other lipids, although present in relatively small quantities, play crucial roles as enzyme cofactors, electron carriers, light-absorbing pigments, hydrophobic anchors for proteins, "chaperones" to help membrane proteins fold, emulsifying agents in the digestive tract, hormones, and intracellular messengers. This chapter introduces representative lipids of each type, organized according to their functional roles, with emphasis on their chemical structure and physical properties. Although we follow a functional organization for our discussion, the literally thousands of different lipids can also be organized into eight general categories of chemical structure (see Table 10–3). We discuss the energy-yielding oxidation of lipids in Chapter 17 and their synthesis in Chapter 21.

10.1 Storage Lipids

The fats and oils used almost universally as stored forms of energy in living organisms are derivatives of **fatty acids**. The fatty acids are hydrocarbon derivatives, at about the same low oxidation state (that is, as highly reduced) as the hydrocarbons in fossil fuels. The cellular oxidation of fatty acids (to CO_2 and H_2O), like the controlled, rapid burning of fossil fuels in internal combustion engines, is highly exergonic.

We introduce here the structures and nomenclature of the fatty acids most commonly found in living organisms. Two types of fatty acid–containing compounds, triacylglycerols and waxes, are described to illustrate the diversity of structure and physical properties in this family of compounds.

Fatty Acids Are Hydrocarbon Derivatives

Fatty acids are carboxylic acids with hydrocarbon chains ranging from 4 to 36 carbons long (C_4 to C_{36}). In some fatty acids, this chain is unbranched and fully saturated (contains no double bonds); in others the chain contains one or more double bonds (Table 10–1). A few contain three-carbon rings, hydroxyl groups, or methyl-group branches.

KEY CONVENTION: A simplified nomenclature for unbranched fatty acids specifies the chain length and number of double bonds, separated by a colon **(Fig. 10–1a)**; for example, the 16-carbon saturated palmitic acid is abbreviated 16:0, and the 18-carbon oleic acid, with one double bond, is 18:1. The positions of any double bonds are

(a) 18:1(Δ^9) *cis*-9-Octadecenoic acid

(b) 20:5($\Delta^{5,8,11,14,17}$) Eicosapentaenoic acid (EPA), an omega-3 fatty acid

FIGURE 10–1 Two conventions for naming fatty acids. (a) Standard nomenclature assigns the number 1 to the carboxyl carbon (C-1), and α to the carbon next to it. Each line segment of the zigzag represents a single bond between adjacent carbons. The position of any double bond(s) is indicated by Δ followed by a superscript number indicating the lower-numbered carbon in the double bond. **(b)** For polyunsaturated fatty acids (PUFAs), an alternative convention numbers the carbons in the opposite direction, assigning the number 1 to the methyl carbon at the other end of the chain; this carbon is also designated ω (omega; the last letter in the Greek alphabet). The positions of the double bonds are indicated relative to the ω carbon.

TABLE 10-1 Some Naturally Occurring Fatty Acids: Structure, Properties, and Nomenclature

Carbon skeleton	Structure*	Systematic name[†]	Common name (derivation)	Melting point (°C)	Solubility at 30 °C (mg/g solvent) Water	Benzene
12:0	$CH_3(CH_2)_{10}COOH$	*n*-Dodecanoic acid	Lauric acid (Latin *laurus*, "laurel plant")	44.2	0.063	2,600
14:0	$CH_3(CH_2)_{12}COOH$	*n*-Tetradecanoic acid	Myristic acid (Latin *Myristica*, nutmeg genus)	53.9	0.024	874
16:0	$CH_3(CH_2)_{14}COOH$	*n*-Hexadecanoic acid	Palmitic acid (Latin *palma*, "palm tree")	63.1	0.0083	348
18:0	$CH_3(CH_2)_{16}COOH$	*n*-Octadecanoic acid	Stearic acid (Greek *stear*, "hard fat")	69.6	0.0034	124
20:0	$CH_3(CH_2)_{18}COOH$	*n*-Eicosanoic acid	Arachidic acid (Latin *Arachis*, legume genus)	76.5		
24:0	$CH_3(CH_2)_{22}COOH$	*n*-Tetracosanoic acid	Lignoceric acid (Latin *lignum*, "wood" + *cera*, "wax")	86.0		
16:1(Δ^9)	$CH_3(CH_2)_5CH{=}CH(CH_2)_7COOH$	*cis*-9-Hexadecenoic acid	Palmitoleic acid	1 to –0.5		
18:1(Δ^9)	$CH_3(CH_2)_7CH{=}CH(CH_2)_7COOH$	*cis*-9-Octadecenoic acid	Oleic acid (Latin *oleum*, "oil")	13.4		
18:2($\Delta^{9,12}$)	$CH_3(CH_2)_4CH{=}CHCH_2CH{=}CH(CH_2)_7COOH$	*cis*-,*cis*-9,12-Octadecadienoic acid	Linoleic acid (Greek *linon*, "flax")	1–5		
18:3($\Delta^{9,12,15}$)	$CH_3CH_2CH{=}CHCH_2CH{=}CHCH_2CH{=}CH(CH_2)_7COOH$	*cis*-,*cis*-,*cis*-9,12,15-Octadecatrienoic acid	α-Linolenic acid	–11		
20:4($\Delta^{5,8,11,14}$)	$CH_3(CH_2)_4CH{=}CHCH_2CH{=}CHCH_2CH{=}CHCH_2CH{=}CH(CH_2)_3COOH$	*cis*-,*cis*-,*cis*-,*cis*-5,8,11,14-Icosatetraenoic acid	Arachidonic acid	–49.5		

*All acids are shown in their nonionized form. At pH 7, all free fatty acids have an ionized carboxylate. Note that numbering of carbon atoms begins at the carboxyl carbon.

[†]The prefix *n*- indicates the "normal" unbranched structure. For instance, "dodecanoic" simply indicates 12 carbon atoms, which could be arranged in a variety of branched forms; "*n*-dodecanoic" specifies the linear, unbranched form. For unsaturated fatty acids, the configuration of each double bond is indicated; in biological fatty acids the configuration is almost always cis.

specified relative to the carboxyl carbon, numbered 1, by superscript numbers following Δ (delta); a 20-carbon fatty acid with one double bond between C-9 and C-10 (C-1 being the carboxyl carbon) and another between C-12 and C-13 is designated 20:2($\Delta^{9,12}$). ■

The most commonly occurring fatty acids have even numbers of carbon atoms in an unbranched chain of 12 to 24 carbons (Table 10–1). As we shall see in Chapter 21, the even number of carbons results from the mode of synthesis of these compounds, which involves successive condensations of two-carbon (acetate) units.

There is also a common pattern in the location of double bonds; in most monounsaturated fatty acids the double bond is between C-9 and C-10 (Δ^9), and the other double bonds of polyunsaturated fatty acids are generally Δ^{12} and Δ^{15}. (Arachidonic acid is an exception to this generalization.) The double bonds of polyunsaturated fatty acids are almost never conjugated (alternating single and double bonds, as in

—CH=CH—CH=CH—), but are separated by a methylene group: —CH=CH—CH$_2$—CH=CH— (Fig. 10–1b). In nearly all naturally occurring unsaturated fatty acids, the double bonds are in the cis configuration. Trans fatty acids are produced by fermentation in the rumen of dairy animals and are obtained from dairy products and meat.

KEY CONVENTION: The family of **polyunsaturated fatty acids (PUFAs)** with a double bond between the third and fourth carbon from the methyl end of the chain are of special importance in human nutrition. Because the physiological role of PUFAs is related more to the position of the first double bond near the *methyl* end of the chain than to the carboxyl end, an alternative nomenclature is sometimes used for these fatty acids. The carbon of the methyl group—that is, the carbon most distant from the carboxyl group—is called the ω (omega) carbon and is given the number 1 (Fig. 10–1b). In this convention, PUFAs with a double bond between C-3 and C-4 are called **omega-3 (ω-3) fatty acids**, and those with a double bond between C-6 and C-7 are **omega-6 (ω-6) fatty acids**. ∎

Humans require but do not have the enzymatic capacity to synthesize the omega-3 PUFA α-linolenic acid (ALA; 18:3($\Delta^{9,12,15}$), in the standard convention), and must therefore obtain it in the diet. From ALA, humans can synthesize two other omega-3 PUFAs important in cellular function: eicosapentaenoic acid (EPA; 20:5($\Delta^{5,8,11,14,17}$), shown in Fig. 10–1b) and docosahexaenoic acid (DHA; 22:6($\Delta^{4,7,10,13,16,19}$)). An imbalance of omega-6 and omega-3 PUFAs in the diet is associated with an increased risk of cardiovascular disease. The optimal dietary ratio of omega-6 to omega-3 PUFAs is between 1:1 and 4:1, but the ratio in the diets of most North Americans is closer to 10:1 to 30:1. The "Mediterranean diet," which has been associated with lowered cardiovascular risk, is richer in omega-3 PUFAs, obtained in leafy vegetables (salads) and fish oils. The latter oils are especially rich in EPA and DHA, and fish oil supplements are often prescribed for individuals with a history of cardiovascular disease. ∎

The physical properties of the fatty acids, and of compounds that contain them, are largely determined by the length and degree of unsaturation of the hydrocarbon chain. The nonpolar hydrocarbon chain accounts for the poor solubility of fatty acids in water. Lauric acid (12:0, M_r 200), for example, has a solubility in water of 0.063 mg/g—much less than that of glucose (M_r 180), which is 1,100 mg/g. The longer the fatty acyl chain and the fewer the double bonds, the lower is the solubility in water. The carboxylic acid group is polar (and ionized at neutral pH) and accounts for the slight solubility of short-chain fatty acids in water.

Melting points are also strongly influenced by the length and degree of unsaturation of the hydrocarbon chain. At room temperature (25 °C), the saturated fatty acids from 12:0 to 24:0 have a waxy consistency, whereas

unsaturated fatty acids of these lengths are oily liquids. This difference in melting points is due to different degrees of packing of the fatty acid molecules **(Fig. 10–2)**. In the fully saturated compounds, free rotation around each carbon–carbon bond gives the hydrocarbon chain great flexibility; the most stable conformation is the fully extended form, in which the steric hindrance of neighboring atoms is minimized. These molecules can pack together tightly in nearly crystalline arrays, with atoms all along their lengths in van der Waals contact with the atoms of neighboring molecules. In unsaturated fatty acids, a cis double bond forces a kink in the hydrocarbon chain. Fatty acids with one or several such kinks cannot pack together as tightly as fully saturated fatty acids, and their interactions with each other are therefore weaker. Because less thermal energy is needed to disorder these poorly ordered arrays of unsaturated fatty acids, they have markedly lower melting points than saturated fatty acids of the same chain length (Table 10–1).

In vertebrates, free fatty acids (unesterified fatty acids, with a free carboxylate group) circulate in the blood bound noncovalently to a protein carrier, serum albumin. However, fatty acids are present in blood plasma mostly as carboxylic acid derivatives such as

FIGURE 10–2 The packing of fatty acids into stable aggregates. The extent of packing depends on the degree of saturation. **(a)** Two representations of the fully saturated acid stearic acid, 18:0 (stearate at pH 7), in its usual extended conformation. **(b)** The cis double bond (red) in oleic acid, 18:1(Δ^9) (oleate), restricts rotation and introduces a rigid bend in the hydrocarbon tail. All other bonds in the chain are free to rotate. **(c)** Fully saturated fatty acids in the extended form pack into nearly crystalline arrays, stabilized by many hydrophobic interactions. **(d)** The presence of one or more fatty acids with cis double bonds (red) interferes with this tight packing and results in less stable aggregates.

esters or amides. Lacking the charged carboxylate group, these fatty acid derivatives are generally even less soluble in water than are the free fatty acids.

Triacylglycerols Are Fatty Acid Esters of Glycerol

The simplest lipids constructed from fatty acids are the **triacylglycerols**, also referred to as triglycerides, fats, or neutral fats. Triacylglycerols are composed of three fatty acids each in ester linkage with a single glycerol **(Fig. 10–3)**. Those containing the same kind of fatty acid in all three positions are called simple triacylglycerols and are named after the fatty acid they contain. Simple triacylglycerols of 16:0, 18:0, and 18:1, for example, are tripalmitin, tristearin, and triolein, respectively. Most naturally occurring triacylglycerols are mixed; they contain two or three different fatty acids. To name these compounds unambiguously, the name and position of each fatty acid must be specified.

Because the polar hydroxyls of glycerol and the polar carboxylates of the fatty acids are bound in ester linkages, triacylglycerols are nonpolar, hydrophobic molecules, essentially insoluble in water. Lipids have lower specific gravities than water, which explains why mixtures of oil and water (oil-and-vinegar salad dressing, for example) have two phases: oil, with the lower specific gravity, floats on the aqueous phase.

Triacylglycerols Provide Stored Energy and Insulation

In most eukaryotic cells, triacylglycerols form a separate phase of microscopic, oily droplets in the aqueous cytosol, serving as depots of metabolic fuel. In vertebrates, specialized cells called adipocytes, or fat cells, store large amounts of triacylglycerols as fat droplets that nearly fill the cell **(Fig. 10–4a)**. Triacylglycerols are also stored as oils in the seeds of many types of plants, providing energy and biosynthetic precursors during seed germination (Fig. 10–4b). Adipocytes and germinating seeds contain **lipases**, enzymes that catalyze the hydrolysis of stored triacylglycerols, releasing fatty acids for export to sites where they are required as fuel.

There are two significant advantages to using triacylglycerols as stored fuels, rather than polysaccharides such as glycogen and starch. First, the carbon atoms of fatty acids are more reduced than those of

1-Stearoyl, 2-linoleoyl, 3-palmitoyl glycerol,
a mixed triacylglycerol

FIGURE 10–3 Glycerol and a triacylglycerol. The mixed triacylglycerol shown here has three different fatty acids attached to the glycerol backbone. When glycerol has different fatty acids at C-1 and C-3, C-2 is a chiral center (p. 17).

FIGURE 10–4 Fat stores in cells. (a) Cross section of human white adipose tissue. Each cell contains a fat droplet (white) so large that it squeezes the nucleus (stained red) against the plasma membrane. **(b)** Cross section of a cotyledon cell from a seed of the plant *Arabidopsis*. The large dark structures are protein bodies, which are surrounded by stored oils in the light-colored oil bodies.

sugars, and oxidation of triacylglycerols yields more than twice as much energy, gram for gram, as the oxidation of carbohydrates. Second, because triacylglycerols are hydrophobic and therefore unhydrated, the organism that carries fat as fuel does not have to carry the extra weight of water of hydration that is associated with stored polysaccharides (2 g per gram of polysaccharide). Humans have fat tissue (composed primarily of adipocytes) under the skin, in the abdominal cavity, and in the mammary glands. Moderately obese people with 15 to 20 kg of triacylglycerols deposited in their adipocytes could meet their energy needs for months by drawing on their fat stores. In contrast, the human body can store less than a day's energy supply in the form of glycogen. Carbohydrates such as glucose do offer certain advantages as quick sources of metabolic energy, one of which is their ready solubility in water.

In some animals, triacylglycerols stored under the skin serve not only as energy stores but as insulation against low temperatures. Seals, walruses, penguins, and other warm-blooded polar animals are amply padded with triacylglycerols. In hibernating animals (bears, for example), the huge fat reserves accumulated before hibernation serve the dual purposes of insulation and energy storage (see Box 17–1).

Partial Hydrogenation of Cooking Oils Produces Trans Fatty Acids

Most natural fats, such as those in vegetable oils, dairy products, and animal fat, are complex mixtures of simple and mixed triacylglycerols. These contain a variety of fatty acids differing in chain length and degree of saturation **(Fig. 10–5)**. Vegetable oils such as corn (maize) and olive oil are composed largely of triacylglycerols with unsaturated fatty acids and thus are liquids at room temperature. Triacylglycerols containing only saturated fatty acids, such as tristearin, the major component of beef fat, are white, greasy solids at room temperature.

When lipid-rich foods are exposed too long to the oxygen in air, they may spoil and become rancid. The unpleasant taste and smell associated with rancidity result from the oxidative cleavage of double bonds in unsaturated fatty acids, which produces aldehydes and carboxylic acids of shorter chain length and therefore higher volatility; these compounds pass readily through the air to your nose. To improve the shelf life of vegetable oils used in cooking, and to increase their stability at the high temperatures used in deep-frying, commercial vegetable oils are prepared by partial hydrogenation. This process converts many of the cis double bonds in the fatty acids to single bonds and increases the melting temperature of the oils so that they are more nearly solid at room temperature (margarine is produced from vegetable oil in this way). Partial hydrogenation has another, undesirable, effect: some cis double bonds are converted to trans double bonds. There is now strong evidence that dietary intake of trans fatty acids (often referred to simply as "trans fats") leads to a higher incidence of cardiovascular disease, and that avoiding these fats in the diet substantially reduces the risk of coronary heart disease. Dietary trans fatty acids raise the level of triacylglycerols and of LDL ("bad") cholesterol in the blood, and lower the level of HDL ("good") cholesterol, and these changes alone are enough to increase the risk of coronary heart disease. But trans fatty acids may have further adverse effects. They seem, for example, to increase the body's inflammatory response, which is another risk factor for heart disease. (See Chapter 21 for a description of LDL and HDL—low-density and high-density lipoprotein—cholesterol and their health effects.)

Many fast foods are deep-fried in partially hydrogenated vegetable oils and therefore contain high levels of trans fatty acids (Table 10–2). In view of the detrimental effects of these fats, some countries (Denmark, for example) and some cities (New York City and Philadelphia) severely restrict the use of partially hydrogenated oils in restaurants. French fries prepared in a chain fast-food restaurant in Denmark now contain almost no detectable trans fatty acids, whereas the same product prepared in the United States contains 5 to 10 g of trans fatty acids per serving (Table 10–2). The deleterious effects of trans fats occur at intakes of 2 to 7 g/day (20 to 60 kcal in a daily caloric intake of 2,000 kcal; note that a nutritional Calorie is the equivalent of the kilocalorie used by chemists and biochemists, so a 2,000 Calorie diet is the equivalent of a 2,000 kcal diet). A single serving of french fries in a U.S. restaurant may contain this amount of trans fatty acid! Many other prepared foods, baked goods, and snacks on the shelves of supermarkets have comparably high levels of trans fats. ■

Natural fats at 25 °C

FIGURE 10–5 Fatty acid composition of three food fats. Olive oil, butter, and beef fat consist of mixtures of triacylglycerols, differing in their fatty acid composition. The melting points of these fats—and hence their physical state at room temperature (25 °C)—are a direct function of their fatty acid composition. Olive oil has a high proportion of long-chain (C_{16} and C_{18}) unsaturated fatty acids, which accounts for its liquid state at 25 °C. The higher proportion of long-chain (C_{16} and C_{18}) saturated fatty acids in butter increases its melting point, so butter is a soft solid at room temperature. Beef fat, with an even higher proportion of long-chain saturated fatty acids, is a hard solid.

TABLE 10–2 Trans Fatty Acids in Some Typical Fast Foods and Snacks

	Trans fatty acid content	
	In a typical serving (g)	As % of total fatty acids
French fries	4.7–6.1	28–36
Breaded fish burger	5.6	28
Breaded chicken nuggets	5.0	25
Pizza	1.1	9
Corn tortilla chips	1.6	22
Doughnut	2.7	25
Muffin	0.7	14
Chocolate bar	0.2	2

Source: Adapted from Table 1 in Mozaffarian, D., Katan, M.B., Ascherio, P.H., Stampfer, M.J., & Willet, W.C. (2006). Trans fatty acids and cardiovascular disease. *N. Engl. J. Med.* **354,** 1604–1605.

Note: All data for foods prepared with partially hydrogenated vegetable oil in the United States in 2002.

Waxes Serve as Energy Stores and Water Repellents

Biological waxes are esters of long-chain (C_{14} to C_{36}) saturated and unsaturated fatty acids with long-chain (C_{16} to C_{30}) alcohols **(Fig. 10–6)**. Their melting points (60 to 100 °C) are generally higher than those of triacylglycerols. In plankton, the free-floating microorganisms at the bottom of the food chain for marine animals, waxes are the chief storage form of metabolic fuel.

Waxes also serve a diversity of other functions related to their water-repellent properties and their firm consistency. Certain skin glands of vertebrates secrete waxes to protect hair and skin and keep it pliable, lubricated, and waterproof. Birds, particularly waterfowl, secrete waxes from their preen glands to keep their feathers water-repellent. The shiny leaves of holly, rhododendrons, poison ivy, and many tropical plants are coated with a thick layer of waxes, which prevents excessive evaporation of water and protects against parasites.

Biological waxes find a variety of applications in the pharmaceutical, cosmetic, and other industries. Lanolin (from lamb's wool), beeswax (Fig. 10–6), carnauba wax (from a Brazilian palm tree), and wax extracted from spermaceti oil (from whales) are widely used in the manufacture of lotions, ointments, and polishes.

SUMMARY 10.1 Storage Lipids

- Lipids are water-insoluble cellular components, of diverse structure, that can be extracted from tissues by nonpolar solvents.

- Almost all fatty acids, the hydrocarbon components of many lipids, have an even number of carbon

Palmitic acid 1-Triacontanol

(a)

(b)

FIGURE 10–6 Biological wax. (a) Triacontanoylpalmitate, the major component of beeswax, is an ester of palmitic acid with the alcohol triacontanol. **(b)** A honeycomb, constructed of beeswax, is firm at 25 °C and completely impervious to water. The term "wax" originates in the Old English *weax*, meaning "the material of the honeycomb."

atoms (usually 12 to 24); they are either saturated or unsaturated, with double bonds almost always in the cis configuration.

- Triacylglycerols contain three fatty acid molecules esterified to the three hydroxyl groups of glycerol. Simple triacylglycerols contain only one type of fatty acid; mixed triacylglycerols, two or three types. Triacylglycerols are primarily storage fats; they are present in many foods.

- Partial hydrogenation of vegetable oils in the food industry converts some cis double bonds to the trans configuration. Trans fatty acids in the diet are an important risk factor for coronary heart disease.

10.2 Structural Lipids in Membranes

The central architectural feature of biological membranes is a double layer of lipids, which acts as a barrier to the passage of polar molecules and ions. Membrane lipids are amphipathic: one end of the molecule is hydrophobic, the other hydrophilic. Their hydrophobic interactions with each other and their hydrophilic interactions with water direct their packing into sheets called membrane bilayers. In this section we describe five general types of membrane lipids: glycerophospholipids, in which the hydrophobic regions are composed

FIGURE 10–7 Some common types of storage and membrane lipids. All the lipid types shown here have either glycerol or sphingosine as the backbone (light red screen), to which are attached one or more long-chain alkyl groups (yellow) and a polar head group (blue). In triacylglycerols, glycerophospholipids, galactolipids, and sulfolipids, the alkyl groups are fatty acids in ester linkage. Sphingolipids contain a single fatty acid, in amide linkage to the sphingosine backbone. The membrane lipids of archaea are variable; that shown here has two very long, branched alkyl chains, each end in ether linkage with a glycerol moiety. In phospholipids the polar head group is joined through a phosphodiester, whereas glycolipids have a direct glycosidic linkage between the head-group sugar and the backbone glycerol.

of two fatty acids joined to glycerol; galactolipids and sulfolipids, which also contain two fatty acids esterified to glycerol, but lack the characteristic phosphate of phospholipids; archaeal tetraether lipids, in which two very long alkyl chains are ether-linked to glycerol at both ends; sphingolipids, in which a single fatty acid is joined to a fatty amine, sphingosine; and sterols, compounds characterized by a rigid system of four fused hydrocarbon rings.

The hydrophilic moieties in these amphipathic compounds may be as simple as a single —OH group at one end of the sterol ring system, or they may be much more complex. In glycerophospholipids and some sphingolipids, a polar head group is joined to the hydrophobic moiety by a phosphodiester linkage; these are the **phospholipids**. Other sphingolipids lack phosphate but have a simple sugar or complex oligosaccharide at their polar ends; these are the **glycolipids (Fig. 10–7)**. Within these groups of membrane lipids, enormous diversity results from various combinations of fatty acid "tails" and polar "heads." The arrangement of these lipids in membranes, and their structural and functional roles therein, are considered in the next chapter.

Glycerophospholipids Are Derivatives of Phosphatidic Acid

Glycerophospholipids, also called phosphoglycerides, are membrane lipids in which two fatty acids are attached in ester linkage to the first and second carbons of glycerol, and a highly polar or charged group is attached through a phosphodiester linkage to the third carbon. Glycerol is prochiral; it has no asymmetric carbons, but attachment of phosphate at one end converts it into a chiral compound, which can be correctly named either L-glycerol 3-phosphate, D-glycerol 1-phosphate, or *sn*-glycerol 3-phosphate **(Fig. 10–8)**. Glycerophospholipids are named as derivatives of the parent compound, phospha-

tidic acid **(Fig. 10–9)**, according to the polar alcohol in the head group. Phosphatidylcholine and phosphatidylethanolamine have choline and ethanolamine as their polar head groups, for example. In all these compounds, the head group is joined to glycerol through a phosphodiester bond, in which the phosphate group bears a negative charge at neutral pH. The polar alcohol may be negatively charged (as in phosphatidylinositol 4,5-bisphosphate), neutral (phosphatidylserine), or positively charged (phosphatidylcholine, phosphatidylethanolamine). As we shall see in Chapter 11, these charges contribute greatly to the surface properties of membranes.

The fatty acids in glycerophospholipids can be any of a wide variety, so a given phospholipid (phosphatidylcholine,

$$^1CH_2OH$$
$$HO - {}^2C - H$$
$$^3CH_2 - O - \overset{O}{\underset{O^-}{\overset{\|}{P}}} - O^-$$

L-Glycerol 3-phosphate (*sn*-glycerol 3-phosphate)

FIGURE 10–8 L-Glycerol 3-phosphate, the backbone of phospholipids. Glycerol itself is not chiral, as it has a plane of symmetry through C-2. However, glycerol is prochiral—it can be converted to a chiral compound by adding a substituent such as phosphate to either of the —CH₂OH groups. One unambiguous nomenclature for glycerol phosphate is the D, L system (described on p. 78), in which the isomers are named according to their stereochemical relationships to glyceraldehyde isomers. By this system, the stereoisomer of glycerol phosphate found in most lipids is correctly named either L-glycerol 3-phosphate or D-glycerol 1-phosphate. Another way to specify stereoisomers is the *sn* (stereospecific numbering) system, in which C-1 is, by definition, the group of the prochiral compound that occupies the pro-S position. The common form of glycerol phosphate in phospholipids is, by this system, *sn*-glycerol 3-phosphate (in which C-2 has the R configuration). In archaea, the glycerol in lipids has the other configuration; it is D-glycerol 3-phosphate.

Name of glycerophospholipid	Name of X—O	Formula of X	Net charge (at pH 7)
Phosphatidic acid	—	—H	−2
Phosphatidylethanolamine	Ethanolamine	$\overset{+}{N}H_3$	0
Phosphatidylcholine	Choline	$\overset{+}{N}$	0
Phosphatidylserine	Serine		−1
Phosphatidylglycerol	Glycerol		−1
Phosphatidylinositol 4,5-bisphosphate	myo-Inositol 4,5-bisphosphate		−4*
Cardiolipin	Phosphatidyl-glycerol		−2

FIGURE 10–9 Glycerophospholipids. The common glycerophospholipids are diacylglycerols linked to head-group alcohols through a phosphodiester bond. Phosphatidic acid, a phosphomonoester, is the parent compound. Each derivative is named for the head-group alcohol (X), with the prefix "phosphatidyl-." In cardiolipin, two phosphatidic acids share a single glycerol (R^1 and R^2 are fatty acyl groups). *Note that the phosphate esters in phosphatidylinositol 4,5-bisphosphate each have a charge of about −1.5; one of their —OH groups is only partially ionized at pH 7.

for example) may consist of several molecular species, each with its unique complement of fatty acids. The distribution of molecular species is specific for different organisms, different tissues of the same organism, and different glycerophospholipids in the same cell or tissue. In general, glycerophospholipids contain a C_{16} or C_{18} saturated fatty acid at C-1 and a C_{18} or C_{20} unsaturated fatty acid at C-2. With few exceptions, the biological significance of the variation in fatty acids and head groups is not yet understood.

Some Glycerophospholipids Have Ether-Linked Fatty Acids

Some animal tissues and some unicellular organisms are rich in **ether lipids**, in which one of the two acyl chains is attached to glycerol in ether, rather than ester, linkage. The ether-linked chain may be saturated, as in the alkyl ether lipids, or may contain a double bond between C-1 and C-2, as in **plasmalogens (Fig. 10–10)**. Vertebrate heart tissue is uniquely enriched in ether lipids; about

FIGURE 10-10 Ether lipids. Plasmalogens have an ether-linked alkenyl chain where most glycerophospholipids have an ester-linked fatty acid (compare Fig. 10-9). Platelet-activating factor has a long ether-linked alkyl chain at C-1 of glycerol, but C-2 is ester-linked to acetic acid, which makes the compound much more water-soluble than most glycerophospholipids and plasmalogens. The head-group alcohol is ethanolamine in plasmalogens and choline in platelet-activating factor.

half of the heart phospholipids are plasmalogens. The membranes of halophilic bacteria, ciliated protists, and certain invertebrates also contain high proportions of ether lipids. The functional significance of ether lipids in these membranes is unknown; perhaps their resistance to the phospholipases that cleave ester-linked fatty acids from membrane lipids is important in some roles.

At least one ether lipid, **platelet-activating factor**, is a potent molecular signal. It is released from leukocytes called basophils and stimulates platelet aggregation and the release of serotonin (a vasoconstrictor) from platelets. It also exerts a variety of effects on liver, smooth muscle, heart, uterine, and lung tissues and plays an important role in inflammation and the allergic response. ∎

Chloroplasts Contain Galactolipids and Sulfolipids

The second group of membrane lipids are those that predominate in plant cells: the **galactolipids**, in which one or two galactose residues are connected by a glycosidic linkage to C-3 of a 1,2-diacylglycerol **(Fig. 10-11**; see also Fig. 10-7). Galactolipids are localized in the thylakoid membranes (internal membranes) of chloroplasts; they make up 70% to 80% of the total membrane lipids of a

vascular plant, and are therefore probably the most abundant membrane lipids in the biosphere. Phosphate is often the limiting plant nutrient in soil, and perhaps the evolutionary pressure to conserve phosphate for more critical roles favored plants that made phosphate-free lipids. Plant membranes also contain sulfolipids, in which a sulfonated glucose residue is joined to a diacylglycerol in glycosidic linkage. The sulfonate group bears a negative charge like that of the phosphate group in phospholipids.

Archaea Contain Unique Membrane Lipids

Some archaea that live in ecological niches with extreme conditions—high temperatures (boiling water), low pH, high ionic strength, for example—have membrane lipids containing long-chain (32 carbons) branched hydrocarbons linked at each end to glycerol **(Fig. 10-12)**. These linkages are through ether bonds, which are much more stable to hydrolysis at low pH and high temperature than are the ester bonds found in the lipids of bacteria and eukaryotes. In their fully extended form, these archaeal lipids are twice the length of phospholipids and sphingolipids, and can span the full width of the plasma membrane. At each end of the extended molecule is a polar head consisting of glycerol linked to either phosphate or

Monogalactosyldiacylglycerol (MGDG)

Digalactosyldiacylglycerol (DGDG)

FIGURE 10-11 Two galactolipids of chloroplast thylakoid membranes. In monogalactosyldiacylglycerols (MGDGs) and digalactosyldiacylglycerols (DGDGs) the acyl groups are both polyunsaturated and the head groups are uncharged.

Glycerol phosphate

Diphytanyl groups

Glycerol

αGlc(β1→2)Gal-1

FIGURE 10-12 An unusual membrane lipid found only in some archaea. In this diphytanyl tetraether lipid, the diphytanyl moieties (yellow) are long hydrocarbons composed of eight five-carbon isoprene groups condensed end-to-end (on the condensation of isoprene units, see Fig. 21–36; also, compare the diphytanyl groups with the 20-carbon phytol side chain of chlorophylls in Fig. 19–49a). In this extended form, the diphytanyl groups are about twice the length of a 16-carbon fatty acid typically found in the membrane lipids of bacteria and eukaryotes. The glycerol moieties in the archaeal lipids are in the R configuration, in contrast to those of bacteria and eukaryotes, which have the S configuration. Archaeal lipids differ in the substituents on the glycerols. In the molecule shown here, one glycerol is linked to the disaccharide α-glucopyranosyl-(1→2)-β-galactofuranose; the other glycerol is linked to a glycerol phosphate head group.

sugar residues. The general name for these compounds, glycerol dialkyl glycerol tetraethers (GDGTs), reflects their unique structure. The glycerol moiety of the archaeal lipids is not the same stereoisomer as that in the lipids of bacteria and eukaryotes; the central carbon is in the R configuration in archaea, in the S configuration in bacteria and eukaryotes (Fig. 10–8).

Sphingolipids Are Derivatives of Sphingosine

Sphingolipids, the fourth large class of membrane lipids, also have a polar head group and two nonpolar tails, but unlike glycerophospholipids and galactolipids they contain no glycerol. Sphingolipids are composed of one molecule of the long-chain amino alcohol sphingosine (also called 4-sphingenine) or one of its derivatives, one molecule of a long-chain fatty acid, and a polar head group that is joined by a glycosidic linkage in some cases and a phosphodiester in others (Fig. 10–13).

Carbons C-1, C-2, and C-3 of the sphingosine molecule are structurally analogous to the three carbons of glycerol in glycerophospholipids. When a fatty acid is attached in amide linkage to the —NH₂ on C-2, the resulting compound is a **ceramide**, which is structurally similar to a diacylglycerol. Ceramide is the structural parent of all sphingolipids.

There are three subclasses of sphingolipids, all derivatives of ceramide but differing in their head groups: sphingomyelins, neutral (uncharged) glycolipids, and gangliosides. **Sphingomyelins** contain phosphocholine or phosphoethanolamine as their polar head group and are therefore classified along with glycerophospholipids as phospholipids (Fig. 10–7). Indeed, sphingomyelins resemble phosphatidylcholines in their general properties and three-dimensional structure, and in having no net charge on their head groups (Fig. 10–14). Sphingomyelins are present in the plasma membranes of animal cells and are especially prominent in myelin, a membranous sheath that surrounds and insulates the axons of some neurons—thus the name "sphingomyelins."

Glycosphingolipids, which occur largely in the outer face of plasma membranes, have head groups with one or more sugars connected directly to the —OH at C-1 of the ceramide moiety; they do not contain phosphate. **Cerebrosides** have a single sugar linked to ceramide; those with galactose are characteristically found in the plasma membranes of cells in neural tissue, and those with glucose in the plasma membranes of cells in nonneural tissues. **Globosides** are glycosphingolipids with two or more sugars, usually D-glucose, D-galactose, or N-acetyl-D-galactosamine. Cerebrosides and globosides are sometimes called **neutral glycolipids**, as they have no charge at pH 7.

Gangliosides, the most complex sphingolipids, have oligosaccharides as their polar head groups and one or more residues of N-acetylneuraminic acid (Neu5Ac), a sialic acid (often simply called "sialic acid"), at the termini. Sialic acid gives gangliosides the negative charge at pH 7 that distinguishes them from globosides. Gangliosides with one sialic acid residue are in the GM (M for mono-) series, those with two are in the GD (D for di-) series, and so on (GT, three sialic acid residues; GQ, four).

N-Acetylneuraminic acid (a sialic acid)
(Neu5Ac)

Johann Thudichum,
1829-1901

FIGURE 10-13 Sphingolipids. The first three carbons at the polar end of sphingosine are analogous to the three carbons of glycerol in glycerophospholipids. The amino group at C-2 bears a fatty acid in amide linkage. The fatty acid is usually saturated or monounsaturated, with 16, 18, 22, or 24 carbon atoms. Ceramide is the parent compound for this group. Other sphingolipids differ in the polar head group (X) attached at C-1. Gangliosides have very complex oligosaccharide head groups. Standard symbols for sugars are used in this figure, as shown in Table 7-1.

Sphingolipids at Cell Surfaces Are Sites of Biological Recognition

When sphingolipids were discovered more than a century ago by the physician-chemist Johann Thudichum, their biological role seemed as enigmatic as the Sphinx, for which he therefore named them. In humans, at least 60 different sphingolipids have been identified in cellular membranes. Many of these are especially prominent in the plasma membranes of neurons, and some are clearly recognition sites on the cell surface, but a specific function for only a few sphingolipids has been

FIGURE 10-14 The molecular structures of two types of membrane lipid classes are similar. Phosphatidylcholine (a glycerophospholipid) and sphingomyelin (a sphingolipid) have similar dimensions and physical properties, but presumably play different roles in membranes.

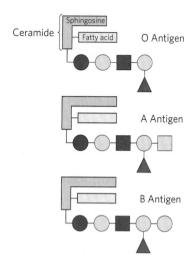

discovered thus far. The carbohydrate moieties of certain sphingolipids define the human blood groups and therefore determine the type of blood that individuals can safely receive in blood transfusions **(Fig. 10–15)**.

Gangliosides are concentrated in the outer surface of cells, where they present points of recognition for extracellular molecules or surfaces of neighboring cells. The kinds and amounts of gangliosides in the plasma membrane change dramatically during embryonic development. Tumor formation induces the synthesis of a new complement of gangliosides, and very low concentrations of a specific ganglioside have been found to induce differentiation of cultured neuronal tumor cells. Investigation of the biological roles of diverse gangliosides remains fertile ground for future research.

Phospholipids and Sphingolipids Are Degraded in Lysosomes

Most cells continually degrade and replace their membrane lipids. For each hydrolyzable bond in a glycerophospholipid, there is a specific hydrolytic enzyme in the lysosome **(Fig. 10–16)**. Phospholipases of the A type remove one of the two fatty acids, producing a lysophospholipid. (These esterases do not attack the ether link of plasmalogens.) Lysophospholipases remove the remaining fatty acid.

Gangliosides are degraded by a set of lysosomal enzymes that catalyze the stepwise removal of sugar units, finally yielding a ceramide. A genetic defect in any of these hydrolytic enzymes leads to the accumulation of gangliosides in the cell, with severe medical consequences (Box 10–1).

Sterols Have Four Fused Carbon Rings

Sterols are structural lipids present in the membranes of most eukaryotic cells. The characteristic structure of

FIGURE 10–16 **The specificities of phospholipases.** Phospholipases A_1 and A_2 hydrolyze the ester bonds of intact glycerophospholipids at C-1 and C-2 of glycerol, respectively. When one of the fatty acids has been removed by a type A phospholipase, the second fatty acid is removed by a lysophospholipase (not shown). Phospholipases C and D each split one of the phosphodiester bonds in the head group. Some phospholipases act on only one type of glycerophospholipid, such as phosphatidylinositol 4,5-bisphosphate (shown here) or phosphatidylcholine; others are less specific.

this fifth group of membrane lipids is the steroid nucleus, consisting of four fused rings, three with six carbons and one with five **(Fig. 10–17)**. The steroid nucleus is almost planar and is relatively rigid; the fused rings do not allow rotation about C—C bonds. **Cholesterol**, the major sterol in animal tissues, is amphipathic, with a polar head group (the hydroxyl group at C-3) and a nonpolar hydrocarbon body (the steroid nucleus and the hydrocarbon side chain at C-17), about as long as a 16-carbon fatty acid in its extended form. Similar sterols are found in other eukaryotes: stigmasterol in plants and ergosterol in fungi, for example. Bacteria cannot synthesize sterols; a few bacterial species, however, can

FIGURE 10–17 **Cholesterol.** In this chemical structure of cholesterol, the rings are labeled A through D to simplify reference to derivatives of the steroid nucleus; the carbon atoms are numbered in blue. The C-3 hydroxyl group (shaded blue) is the polar head group. For storage and transport of the sterol, this hydroxyl group condenses with a fatty acid to form a sterol ester.

BOX 10–1 MEDICINE   Abnormal Accumulations of Membrane Lipids: Some Inherited Human Diseases

The polar lipids of membranes undergo constant metabolic turnover, the rate of their synthesis normally counterbalanced by the rate of breakdown. The breakdown of lipids is promoted by hydrolytic enzymes in lysosomes, each enzyme capable of hydrolyzing a specific bond. When sphingolipid degradation is impaired by a defect in one of these enzymes (Fig. 1), partial breakdown products accumulate in the tissues, causing serious disease.

For example, Niemann-Pick disease is caused by a rare genetic defect in the enzyme sphingomyelinase, which cleaves phosphocholine from sphingomyelin. Sphingomyelin accumulates in the brain, spleen, and liver. The disease becomes evident in infants and causes mental retardation and early death. More common is Tay-Sachs disease, in which ganglioside GM2 accumulates in the brain and spleen (Fig. 2) owing to lack of the enzyme hexosaminidase A. The symptoms of Tay-Sachs disease are progressive developmental retardation, paralysis, blindness, and death by the age of 3 or 4 years.

Genetic counseling can predict and avert many inheritable diseases. Tests on prospective parents can detect abnormal enzymes, then DNA testing can determine the exact nature of the defect and the risk it poses for offspring. Once a pregnancy occurs, fetal cells obtained by sampling a part of the placenta (chorionic villus sampling) or the fluid surrounding the fetus (amniocentesis) can be tested in the same way.

FIGURE 1 Pathways for the breakdown of GM1, globoside, and sphingomyelin to ceramide. A defect in the enzyme hydrolyzing a particular step is indicated by ⊗; the disease that results from accumulation of the partial breakdown product is noted.

1 μm

FIGURE 2 Electron micrograph of a portion of a brain cell from an infant with Tay-Sachs disease, obtained post mortem, showing abnormal ganglioside deposits in the lysosomes.

incorporate exogenous sterols into their membranes. The sterols of all eukaryotes are synthesized from simple five-carbon isoprene subunits, as are the fat-soluble vitamins, quinones, and dolichols described in Section 10.3.

In addition to their roles as membrane constituents, the sterols serve as precursors for a variety of products with specific biological activities. Steroid hormones, for example, are potent biological signals that regulate gene expression. **Bile acids** are polar derivatives of cholesterol that act as detergents in the intestine, emulsifying dietary fats to make them more readily accessible to digestive lipases.

Taurocholic acid
(a bile acid)

We return to cholesterol and other sterols in later chapters, to consider the structural role of cholesterol in biological membranes (Chapter 11), signaling by steroid hormones (Chapter 12), and the remarkable biosynthetic pathway to cholesterol and transport of cholesterol by lipoprotein carriers (Chapter 21).

SUMMARY 10.2 Structural Lipids in Membranes

▶ The polar lipids, with polar heads and nonpolar tails, are major components of membranes. The most abundant are the glycerophospholipids, which contain fatty acids esterified to two of the hydroxyl groups of glycerol, and a second alcohol, the head group, esterified to the third hydroxyl of glycerol via a phosphodiester bond. Other polar lipids are the sterols.

▶ Glycerophospholipids differ in the structure of their head group; common glycerophospholipids are phosphatidylethanolamine and phosphatidylcholine. The polar heads of the glycerophospholipids are charged at pH near 7.

▶ Chloroplast membranes are rich in galactolipids, composed of a diacylglycerol with one or two linked galactose residues, and sulfolipids, diacylglycerols with a linked sulfonated sugar residue and thus a negatively charged head group.

▶ Some archaea have unique membrane lipids, with long-chain alkyl groups ether-linked to glycerol at both ends and with sugar residues and/or phosphate joined to the glycerol to provide a polar or charged head group. These lipids are stable under the harsh conditions in which these archaea live.

▶ The sphingolipids contain sphingosine, a long-chain aliphatic amino alcohol, but no glycerol. Sphingomyelin has, in addition to phosphoric acid and choline, two long hydrocarbon chains, one contributed by a fatty acid and the other by sphingosine. Three other classes of sphingolipids are cerebrosides, globosides, and gangliosides, which contain sugar components.

▶ Sterols have four fused rings and a hydroxyl group. Cholesterol, the major sterol in animals, is both a structural component of membranes and precursor to a wide variety of steroids.

10.3 Lipids as Signals, Cofactors, and Pigments

The two functional classes of lipids considered thus far (storage lipids and structural lipids) are major cellular components; membrane lipids make up 5% to 10% of the dry mass of most cells, and storage lipids more than 80% of the mass of an adipocyte. With some important exceptions, these lipids play a *passive* role in the cell; lipid fuels are stored until oxidized by enzymes, and membrane lipids form impermeable barriers around cells and cellular compartments. Another group of lipids, present in much smaller amounts, have *active* roles in the metabolic traffic as metabolites and messengers. Some serve as potent signals—as hormones, carried in the blood from one tissue to another, or as intracellular messengers generated in response to an extracellular signal (hormone or growth factor). Others function as enzyme cofactors in electron-transfer reactions in chloroplasts and mitochondria, or in the transfer of sugar moieties in a variety of glycosylation reactions. A third group consists of lipids with a system of conjugated double bonds: pigment molecules that absorb visible light. Some of these act as light-capturing pigments in vision and photosynthesis; others produce natural colorations, such as the orange of pumpkins and carrots and the yellow of canary feathers. Finally, a very large group of volatile lipids produced in plants serve as signals that pass through the air, allowing plants to communicate with each other, and to invite animal friends and deter foes. We describe in this section a few representatives of these biologically active lipids. In later chapters, their synthesis and biological roles are considered in more detail.

Phosphatidylinositols and Sphingosine Derivatives Act as Intracellular Signals

Phosphatidylinositol and its phosphorylated derivatives act at several levels to regulate cell structure and metabolism. Phosphatidylinositol 4,5-bisphosphate (Fig. 10–16) in the cytoplasmic (inner) face of plasma membranes serves as a reservoir of messenger molecules that are released inside the cell in response to extracellular signals interacting with specific surface receptors. Extracellular signals such as the hormone vasopressin activate a specific phospholipase C in the membrane, which hydrolyzes phosphatidylinositol 4,5-bisphosphate to release two products that act as intracellular messengers: inositol 1,4,5-trisphosphate (IP_3), which is water-soluble, and diacylglycerol, which remains associated with the plasma membrane. IP_3 triggers release of Ca^{2+} from the endoplasmic reticulum, and the combination of diacylglycerol and elevated cytosolic Ca^{2+} activates the enzyme protein kinase C. By phosphorylating specific proteins, this enzyme brings about the cell's response to the extracellular signal. This signaling mechanism is described more fully in Chapter 12 (see Fig. 12–10).

Inositol phospholipids also serve as points of nucleation for supramolecular complexes involved in signaling or in exocytosis. Certain signaling proteins bind specifically to phosphatidylinositol 3,4,5-trisphosphate in the plasma membrane, initiating the formation of multienzyme complexes at the membrane's cytosolic surface. Formation of phosphatidylinositol 3,4,5-trisphosphate in response to extracellular signals therefore brings the proteins together in signaling complexes at the surface of the plasma membrane (see Fig. 12–16).

Membrane sphingolipids also can serve as sources of intracellular messengers. Both ceramide and sphingomyelin (Fig. 10–13) are potent regulators of protein kinases, and ceramide or its derivatives are involved in the regulation of cell division, differentiation, migration, and programmed cell death (also called apoptosis; see Chapter 12).

Eicosanoids Carry Messages to Nearby Cells

Eicosanoids are paracrine hormones, substances that act only on cells near the point of hormone synthesis instead of being transported in the blood to act on cells in other tissues or organs. These fatty acid derivatives have a variety of dramatic effects on vertebrate tissues. They are involved in reproductive function; in the inflammation, fever, and pain associated with injury or disease; in the formation of blood clots and the regulation of blood pressure; in gastric acid secretion; and in various other processes important in human health or disease.

All eicosanoids are derived from arachidonic acid $(20:4(\Delta^{5,8,11,14}))$ **(Fig. 10–18)**, the 20-carbon polyunsaturated fatty acid from which they take their general name (Greek *eikosi*, "twenty"). There are three classes of eicosanoids: prostaglandins, thromboxanes, and leukotrienes.

Prostaglandins (PG) contain a five-carbon ring originating from the chain of arachidonic acid. Their name derives from the prostate gland, the tissue from which they were first isolated by Bengt Samuelsson and Sune Bergström. Two groups of prostaglandins were originally defined: PGE (*ether*-soluble) and PGF (*fosfat* (Swedish for phosphate) buffer–soluble). Each group contains numerous subtypes, named PGE_1, PGE_2, PGF_1, and so forth. Prostaglandins have an array of functions. Some stimulate contraction of the smooth muscle of the uterus during menstruation and labor. Others affect blood flow to specific organs, the wake-sleep cycle, and the responsiveness of certain tissues to hormones such as epinephrine and glucagon. Prostaglandins in a third group elevate body temperature (producing fever) and cause inflammation and pain.

The **thromboxanes** have a six-membered ring containing an ether. They are produced by platelets (also called thrombocytes) and act in the formation of blood clots and the reduction of blood flow to the site of a clot. As shown by John Vane, the nonsteroidal antiinflammatory drugs (NSAIDs)—aspirin, ibuprofen, and

FIGURE 10–18 Arachidonic acid and some eicosanoid derivatives. Arachidonic acid (arachidonate at pH 7) is the precursor of eicosanoids, including the prostaglandins, thromboxanes, and leukotrienes. In prostaglandin E_1, C-8 and C-12 of arachidonate are joined to form the characteristic five-membered ring. In thromboxane A_2, the C-8 and C-12 are joined and an oxygen atom is added to form the six-membered ring. Leukotriene A_4 has a series of three conjugated double bonds. Nonsteroidal antiinflammatory drugs (NSAIDs) such as aspirin and ibuprofen block the formation of prostaglandins and thromboxanes from arachidonate by inhibiting the enzyme cyclooxygenase (prostaglandin H_2 synthase).

John Vane, Sune Bergström, and Bengt Samuelsson

meclofenamate, for example—inhibit the enzyme prostaglandin H_2 synthase (also called cyclooxygenase, or COX), which catalyzes an early step in the pathway from arachidonate to prostaglandins and thromboxanes (Fig. 10–18; see also Fig. 21–15).

Leukotrienes, first found in leukocytes, contain three conjugated double bonds. They are powerful biological signals. For example, leukotriene D_4, derived from leukotriene A_4, induces contraction of the smooth muscle lining the airways to the lung. Overproduction of leukotrienes causes asthmatic attacks, and leukotriene synthesis is one target of antiasthmatic drugs such as prednisone. The strong contraction of the smooth muscle of the lungs that occurs during anaphylactic shock is part of the potentially fatal allergic reaction in individuals hypersensitive to bee stings, penicillin, or other agents. ■

Steroid Hormones Carry Messages between Tissues

Steroids are oxidized derivatives of sterols; they have the sterol nucleus but lack the alkyl chain attached to ring D of cholesterol, and they are more polar than cholesterol. Steroid hormones move through the bloodstream (on protein carriers) from their site of production to target tissues, where they enter cells, bind to highly specific receptor proteins in the nucleus, and trigger changes in gene expression and thus metabolism. Because hormones have very high affinity for their recep-

tors, very low concentrations of hormones (nanomolar or less) are sufficient to produce responses in target tissues. The major groups of steroid hormones are the male and female sex hormones and the hormones produced by the adrenal cortex, cortisol and aldosterone **(Fig. 10–19)**. Prednisone and prednisolone are steroid drugs with potent antiinflammatory activities, mediated in part by the inhibition of arachidonate release by phospholipase A_2 and consequent inhibition of the synthesis of leukotrienes, prostaglandins, and thromboxanes. These drugs have a variety of medical applications, including the treatment of asthma and rheumatoid arthritis. ■

Vascular plants contain the steroidlike brassinolide (Fig. 10–19), a potent growth regulator that increases the rate of stem elongation and affects the orientation of cellulose microfibrils in the cell wall during growth.

Vascular Plants Produce Thousands of Volatile Signals

Plants produce literally thousands of different lipophilic compounds, volatile substances that are used to attract pollinators, to repel herbivores, to attract organisms that defend the plant against herbivores, and to communicate with other plants. Jasmonate, for example (see Fig. 12–33), derived from the fatty acid $18:3(\Delta^{9,12,15})$ in membrane lipids, triggers the plant's defenses in response to insect-inflicted damage. The methyl ester of jasmonate gives the characteristic fragrance of jasmine oil, which is

Testosterone

Cortisol

Prednisone

β-Estradiol

Aldosterone

Prednisolone

Brassinolide
(a brassinosteroid)

FIGURE 10–19 Steroids derived from cholesterol. Testosterone, the male sex hormone, is produced in the testes. Estradiol, one of the female sex hormones, is produced in the ovaries and placenta. Cortisol and aldosterone are hormones synthesized in the cortex of the adrenal gland; they regulate glucose metabolism and salt excretion, respectively. Prednisone and prednisolone are synthetic steroids used as antiinflammatory agents. Brassinolide is a growth regulator found in vascular plants.

widely used in the perfume industry. Many of the plant volatiles are derived from fatty acids, or from compounds made by the condensation of five-carbon isoprene units; these include geraniol (the characteristic scent of geraniums), β-pinene (pine trees), limonene (limes), menthol, and carvone (see Fig. 1–24a), to name but a few.

$$CH_2=\overset{\overset{\displaystyle CH_3}{|}}{C}-CH=CH_2$$

Isoprene

Vitamins A and D Are Hormone Precursors

During the first third of the twentieth century, a major focus of research in physiological chemistry was the identification of **vitamins**, compounds that are essential to the health of humans and other vertebrates but cannot be synthesized by these animals and must therefore be obtained in the diet. Early nutritional studies identified two general classes of such compounds: those soluble in nonpolar organic solvents (fat-soluble vitamins) and those that could be extracted from foods with aqueous solvents (water-soluble vitamins). Eventually the fat-soluble group was resolved into the four vitamin groups A, D, E, and K, all of which are isoprenoid compounds synthesized by the condensation of multiple isoprene units. Two of these (D and A) serve as hormone precursors.

Vitamin D_3, also called **cholecalciferol**, is normally formed in the skin from 7-dehydrocholesterol in a photochemical reaction driven by the UV component of sunlight **(Fig. 10–20a)**. Vitamin D_3 is not itself biologically active, but it is converted by enzymes in the liver and kidney to $1\alpha,25$-dihydroxyvitamin D_3 (calcitriol), a hormone that regulates calcium uptake in the intestine and calcium levels in kidney and bone. Deficiency of vitamin D leads to defective bone formation and the disease rickets, for which administration of vitamin D produces a dramatic cure (Fig. 10–20b). Vitamin D_2 (ergocalciferol) is a commercial product formed by UV irradiation of the ergosterol of yeast. Vitamin D_2 is structurally similar to D_3, with slight modification to the side chain attached to the sterol D ring. Both have the same biological effects, and D_2 is commonly added to milk and butter as a dietary supplement. Like steroid hormones, the product of vitamin D metabolism, $1\alpha,25$-dihydroxyvitamin D_3, regulates gene expression by interacting with specific nuclear receptor proteins (pp. 1182–1183).

Vitamin A (retinol), in its various forms, functions as a hormone and as the visual pigment of the vertebrate eye **(Fig. 10–21)**. Acting through receptor proteins in the cell nucleus, the vitamin A derivative retinoic acid regulates gene expression in the development of epithelial tissue, including skin. Retinoic acid is the active ingredient in the drug tretinoin (Retin-A), used in the treatment of severe acne and wrinkled skin. Retinal, another vitamin A derivative, is the pigment that initiates the response of rod and cone cells of the retina to light, producing a

(a)

7-Dehydrocholesterol → UV light 2 steps (in skin) → Cholecalciferol (vitamin D_3) → 1 step in the liver 1 step in the kidney → $1\alpha,25$-Dihydroxyvitamin D_3 (calcitriol)

FIGURE 10–20 Vitamin D_3 production and metabolism. (a) Cholecalciferol (vitamin D_3) is produced in the skin by UV irradiation of 7-dehydrocholesterol, which breaks the bond shaded light red. In the liver, a hydroxyl group is added at C-25; in the kidney, a second hydroxylation at C-1 produces the active hormone, $1\alpha,25$-dihydroxyvitamin D_3. This hormone regulates the metabolism of Ca^{2+} in kidney, intestine, and bone. **(b)** Dietary vitamin D prevents rickets, a disease once common in cold climates where heavy clothing blocks the UV component of sunlight necessary for the production of vitamin D_3 in skin. In this detail from a large mural by John Steuart Curry, *The Social Benefits of Biochemical Research* (1943), the people and animals on the left show the effects of poor nutrition, including the bowed legs of a boy with classical rickets. On the right are the people and animals made healthier with the "social benefits of research," including the use of vitamin D to prevent and treat rickets. This mural is in the Department of Biochemistry at the University of Wisconsin–Madison.

(b)

FIGURE 10–21 Vitamin A_1 and its precursor and derivatives. (a) β-Carotene is the precursor of vitamin A_1. Isoprene structural units are set off by dashed red lines (see p. 373). Cleavage of β-carotene yields two molecules of vitamin A_1 (retinol) **(b)**. Oxidation at C-15 converts retinol to the aldehyde, retinal **(c)**, and further oxidation produces retinoic acid **(d)**, a hormone that regulates gene expression. Retinal combines with the protein opsin to form rhodopsin (not shown), a visual pigment wide-spread in nature. In the dark, retinal of rhodopsin is in the 11-cis form (c). When a rhodopsin molecule is excited by visible light, the 11-cis-retinal undergoes a series of photochemical reactions that convert it to all-trans-retinal **(e)**, forcing a change in the shape of the entire rhodopsin molecule. This transformation in the rod cell of the vertebrate retina sends an electrical signal to the brain that is the basis of visual transduction, a topic we address in more detail in Chapter 12.

neuronal signal to the brain. This role of retinal is described in detail in Chapter 12.

Vitamin A was first isolated from fish liver oils; liver, eggs, whole milk, and butter are also good dietary sources. In vertebrates, β-carotene, the pigment that gives carrots, sweet potatoes, and other yellow vegetables their characteristic color, can be enzymatically converted to vitamin A. Deficiency of vitamin A leads to a variety of symptoms in humans, including dryness of the skin, eyes, and mucous membranes; retarded development and growth; and night blindness, an early symptom commonly used in diagnosing vitamin A deficiency. ∎

Vitamins E and K and the Lipid Quinones Are Oxidation-Reduction Cofactors

Vitamin E is the collective name for a group of closely related lipids called **tocopherols**, all of which contain a substituted aromatic ring and a long isoprenoid side chain **(Fig. 10–22a)**. Because they are hydrophobic, tocopherols associate with cell membranes, lipid deposits, and lipoproteins in the blood. Tocopherols are biological antioxidants. The aromatic ring reacts with and destroys the most reactive forms of oxygen radicals and other free radicals, protecting unsaturated fatty acids from oxidation and preventing oxidative damage to membrane lipids, which can cause cell fragility. Tocopherols are found in eggs and vegetable oils and are especially abundant in wheat germ. Laboratory animals fed diets depleted of vitamin E develop scaly skin, muscular weakness and wasting, and sterility. Vitamin E deficiency in humans is very rare; the principal symptom is fragile erythrocytes.

The aromatic ring of **vitamin K** (Fig. 10–22b) undergoes a cycle of oxidation and reduction during the formation of active prothrombin, a blood plasma protein essential in blood clotting. Prothrombin is a proteolytic enzyme that splits peptide bonds in the blood protein fibrinogen to convert it to fibrin, the insoluble fibrous protein that holds blood clots together (see Fig. 6–39). Henrik Dam and Edward A. Doisy independently discovered that vitamin K deficiency slows blood clotting, which can be fatal. Vitamin K deficiency is very uncommon in humans, aside from a small percentage of infants who suffer from hemorrhagic disease of the newborn, a potentially fatal disorder. In the United States, newborns are routinely given a 1 mg injection of vitamin K. Vitamin K_1 (phylloquinone) is found in green plant leaves; a related form, vitamin K_2 (menaquinone), is formed by bacteria living in the vertebrate intestine.

(a)
Vitamin E: an antioxidant

(b)
Vitamin K₁: a blood-clotting
cofactor (phylloquinone)

(c)
Warfarin: a blood
anticoagulant

(d)
Ubiquinone: a mitochondrial
electron carrier (coenzyme Q)
(n = 4 to 8)

(e)
Plastoquinone: a chloroplast
electron carrier (n = 4 to 8)

(f)
Dolichol: a sugar carrier
(n = 9 to 22)

FIGURE 10–22 Some other biologically active isoprenoid compounds or derivatives. Units derived from isoprene are set off by dashed red lines. In most mammalian tissues, ubiquinone (also called coenzyme Q) has 10 isoprene units. Dolichols of animals have 17 to 21 isoprene units (85 to 105 carbon atoms), bacterial dolichols have 11, and those of plants and fungi have 14 to 24.

Ubiquinone (also called coenzyme Q) and plastoquinone (Fig. 10–22d, e) are isoprenoids that function as lipophilic electron carriers in the oxidation-reduction reactions that drive ATP synthesis in mitochondria and chloroplasts, respectively. Both ubiquinone and plastoquinone can accept either one or two electrons and either one or two protons (see Fig. 19–3).

Dolichols Activate Sugar Precursors for Biosynthesis

During assembly of the complex carbohydrates of bacterial cell walls, and during the addition of polysaccharide units to certain proteins (glycoproteins) and lipids (glycolipids) in eukaryotes, the sugar units to be added are chemically activated by attachment to isoprenoid alcohols called **dolichols** (Fig. 10–22f). These compounds have strong hydrophobic interactions with membrane lipids, anchoring the attached sugars to the membrane, where they participate in sugar-transfer reactions.

Henrik Dam,
1895–1976

Edward A. Doisy,
1893–1986

Warfarin (Fig. 10–22c) is a synthetic compound that inhibits the formation of active prothrombin. It is particularly poisonous to rats, causing death by internal bleeding. Ironically, this potent rodenticide is also an invaluable anticoagulant drug for treating humans at risk for excessive blood clotting, such as surgical patients and those with coronary thrombosis. ■

Canthaxanthin
(bright red)

Zeaxanthin
(bright yellow)

FIGURE 10–23 Lipids as pigments in plants and bird feathers. Compounds with long conjugated systems absorb light in the visible region of the spectrum. Subtle differences in the chemistry of these compounds produce pigments of strikingly different colors. Birds acquire the pigments that color their feathers red or yellow by eating plant materials that contain carotenoid pigments, such as canthaxanthin and zeaxanthin. The differences in pigmentation between male and female birds are the result of differences in intestinal uptake and processing of carotenoids.

Many Natural Pigments Are Lipidic Conjugated Dienes

Conjugated dienes have carbon chains with alternating single and double bonds. Because this structural arrangement allows the delocalization of electrons, the compounds can be excited by low-energy electromagnetic radiation (visible light), giving them colors visible to humans and other animals. Carotene (Fig. 10–21) is yellow-orange; similar compounds give bird feathers their striking reds, oranges, and yellows **(Fig. 10–23)**. Like sterols, steroids, dolichols, vitamins A, E, D, and K, ubiquinone, and plastoquinone, these pigments are synthesized from five-carbon isoprene derivatives; the biosynthetic pathway is described in detail in Chapter 21.

Polyketides Are Natural Products with Potent Biological Activities

Polyketides are a diverse group of lipids with biosynthetic pathways (Claisen condensations) similar to those for fatty acids. They are **secondary metabolites**, compounds that are not central to an organism's metabolism but that serve some subsidiary function that gives their producers an advantage in some ecological niche. Many polyketides find use in medicine as antibiotics (erythromycin), antifungals (amphotericin B), or inhibitors of cholesterol synthesis (lovastatin) **(Fig. 10–24)**.

Erythromycin (antibiotic)

Amphotericin B (antifungal)

Lovastatin (statin)

 FIGURE 10–24 Three polyketide natural products used in human medicine.

SUMMARY 10.3 Lipids as Signals, Cofactors, and Pigments

▶ Some types of lipids, although present in relatively small quantities, play critical roles as cofactors or signals.

▶ Phosphatidylinositol bisphosphate is hydrolyzed to yield two intracellular messengers, diacylglycerol and inositol 1,4,5-trisphosphate. Phosphatidylinositol 3,4,5-trisphosphate is a nucleation point for supramolecular protein complexes involved in biological signaling.

▶ Prostaglandins, thromboxanes, and leukotrienes (the eicosanoids), derived from arachidonate, are extremely potent hormones.

▶ Steroid hormones, such as the sex hormones, are derived from sterols. They serve as powerful biological signals, altering gene expression in target cells.

▶ Vitamins D, A, E, and K are fat-soluble compounds made up of isoprene units. All play essential roles in the metabolism or physiology of animals. Vitamin D is precursor to a hormone that regulates calcium metabolism. Vitamin A furnishes the visual pigment of the vertebrate eye and is a regulator of gene expression during epithelial cell growth. Vitamin E functions in the protection of membrane lipids from oxidative damage, and vitamin K is essential in the blood-clotting process.

▶ Ubiquinones and plastoquinones, also isoprenoid derivatives, are electron carriers in mitochondria and chloroplasts, respectively.

▶ Dolichols activate and anchor sugars to cellular membranes; the sugar groups are then used in the synthesis of complex carbohydrates, glycolipids, and glycoproteins.

▶ Lipidic conjugated dienes serve as pigments in flowers and fruits and give bird feathers their striking colors.

▶ Polyketides are natural products widely used in medicine.

10.4 Working with Lipids

Because lipids are insoluble in water, their extraction and subsequent fractionation require the use of organic solvents and some techniques not commonly used in the purification of water-soluble molecules such as proteins and carbohydrates. In general, complex mixtures of lipids are separated by differences in polarity or solubility in nonpolar solvents. Lipids that contain ester- or amide-linked fatty acids can be hydrolyzed by treatment with acid or alkali or with specific hydrolytic enzymes (phospholipases, glycosidases) to yield their components for analysis. Some methods commonly used in lipid analysis are shown in **Figure 10–25** and discussed below.

Lipid Extraction Requires Organic Solvents

Neutral lipids (triacylglycerols, waxes, pigments, and so forth) are readily extracted from tissues with ethyl ether, chloroform, or benzene, solvents that do not permit lipid clustering driven by hydrophobic interactions. Membrane lipids are more effectively extracted

FIGURE 10-25 Common procedures in the extraction, separation, and identification of cellular lipids. (a) Tissue is homogenized in a chloroform/methanol/water mixture, which on addition of water and removal of unextractable sediment by centrifugation yields two phases. **(b)** Major classes of extracted lipids in the chloroform phase may first be separated by thin-layer chromatography (TLC), in which lipids are carried up a silica gel–coated plate by a rising solvent front, less-polar lipids traveling farther than more-polar or charged lipids, or by adsorption chromatography on a column of silica gel, through which solvents of increasing polarity are passed. For example, column chromatography with appropriate solvents can be used to separate closely related lipid species such as phosphatidylserine, phosphatidylglycerol, and phosphatidylinositol. Once separated, each lipid's complement of fatty acids can be determined by mass spectrometry. **(c)** Alternatively, in the "shotgun" approach, an unfractionated extract of lipids can be directly subjected to high-resolution mass spectrometry of different types and under different conditions to determine the total composition of all the lipids: the lipidome.

by more polar organic solvents, such as ethanol or methanol, which reduce the hydrophobic interactions among lipid molecules while also weakening the hydrogen bonds and electrostatic interactions that bind membrane lipids to membrane proteins. A commonly used extractant is a mixture of chloroform, methanol, and water, initially in volume proportions (1:2:0.8) that are miscible, producing a single phase. After tissue is homogenized in this solvent to extract all lipids, more water is added to the resulting extract and the mixture separates into two phases, methanol/water (top phase) and chloroform (bottom phase). The lipids remain in the chloroform layer, and the more polar molecules such as proteins and sugars partition into the methanol/water layer (Fig. 10–25a).

Adsorption Chromatography Separates Lipids of Different Polarity

Complex mixtures of tissue lipids can be fractionated by chromatographic procedures based on the different polarities of each class of lipid (Fig. 10–25b). In adsorption chromatography, an insoluble, polar material such as silica gel (a form of silicic acid, $Si(OH)_4$) is packed into a glass column, and the lipid mixture (in chloroform solution) is applied to the top of the column. (In high-performance liquid chromatography, the column is of smaller diameter and solvents are forced through the column under high pressure.) The polar lipids bind tightly to the polar silicic acid, but the neutral lipids pass directly through the column and emerge in the first chloroform wash. The polar lipids are then eluted, in order of increasing polarity, by washing the column with solvents of progressively higher polarity. Uncharged but polar lipids (cerebrosides, for example) are eluted with acetone, and very polar or charged lipids (such as glycerophospholipids) are eluted with methanol.

Thin-layer chromatography on silicic acid employs the same principle (Fig. 10–25b). A thin layer of silica gel is spread onto a glass plate, to which it adheres. A small sample of lipids dissolved in chloroform is applied near one edge of the plate, which is dipped in a shallow container of an organic solvent or solvent mixture; the entire setup is enclosed in a chamber saturated with the solvent vapor. As the solvent rises on the plate by capillary action, it carries lipids with it. The less polar lipids move farthest, as they have less tendency to bind to the silicic acid. The separated lipids can be detected by spraying the plate with a dye (rhodamine) that fluoresces when associated with lipids, or by exposing the plate to iodine fumes. Iodine reacts reversibly with the double bonds in fatty acids, such that lipids containing unsaturated fatty acids develop a yellow or brown color. Several other spray reagents are also useful in detecting specific lipids. For subsequent analysis, regions containing separated lipids can be scraped from the plate and the lipids recovered by extraction with an organic solvent.

Gas-Liquid Chromatography Resolves Mixtures of Volatile Lipid Derivatives

Gas-liquid chromatography separates volatile components of a mixture according to their relative tendencies to dissolve in the inert material packed in the chromatography column or to volatilize and move through the column, carried by a current of an inert gas such as helium. Some lipids are naturally volatile, but most must first be derivatized to increase their volatility (that is, lower their boiling point). For an analysis of the fatty acids in a sample of phospholipids, the lipids are first transesterified: heated in a methanol/HCl or methanol/NaOH mixture to convert fatty acids esterified to glycerol into their methyl esters. These fatty acyl methyl esters are then loaded onto the gas-liquid chromatography column, and the column is heated to volatilize the compounds. Those fatty acyl esters most soluble in the column material partition into (dissolve in) that material; the less soluble lipids are carried by the stream of inert gas and emerge first from the column. The order of elution depends on the nature of the solid adsorbant in the column and on the boiling point of the components of the lipid mixture. Using these techniques, mixtures of fatty acids of various chain lengths and various degrees of unsaturation can be completely resolved.

Specific Hydrolysis Aids in Determination of Lipid Structure

Certain classes of lipids are susceptible to degradation under specific conditions. For example, all ester-linked fatty acids in triacylglycerols, phospholipids, and sterol esters are released by mild acid or alkaline treatment, and somewhat harsher hydrolysis conditions release amide-bound fatty acids from sphingolipids. Enzymes that specifically hydrolyze certain lipids are also useful in the determination of lipid structure. Phospholipases A, C, and D (Fig. 10–16) each split particular bonds in phospholipids and yield products with characteristic solubilities and chromatographic behaviors. Phospholipase C, for example, releases a water-soluble phosphoryl alcohol (such as phosphocholine from phosphatidylcholine) and a chloroform-soluble diacylglycerol, each of which can be characterized separately to determine the structure of the intact phospholipid. The combination of specific hydrolysis with characterization of the products by thin-layer, gas-liquid, or high-performance liquid chromatography often allows determination of a lipid structure.

Mass Spectrometry Reveals Complete Lipid Structure

To establish unambiguously the length of a hydrocarbon chain or the position of double bonds, mass spectrometric analysis of lipids or their volatile derivatives is invaluable. The chemical properties of similar lipids (for example, two fatty acids of similar length unsaturated at different positions, or two isoprenoids with different numbers of isoprene units) are very much alike, and their order of

FIGURE 10–26 Determination of fatty acid structure by mass spectrometry. The fatty acid is first converted to a derivative that minimizes migration of the double bonds when the molecule is fragmented by electron bombardment. The derivative shown here is a picolinyl ester of linoleic acid—18:2($\Delta^{9,12}$) (M_r, 371)—in which the alcohol is picolinol (red). When bombarded with a stream of electrons, this molecule is volatilized and converted to a parent ion (M$^+$; M_r, 371), in which the N atom bears the positive charge, and a series of smaller fragments produced by breakage of C—C bonds in the fatty acid. The mass spectrometer separates these charged fragments according to their mass/charge ratio (m/z). (To review the principles of mass spectrometry, see pp. 100–102.)

The prominent ions at m/z = 92, 108, 151, and 164 contain the pyridine ring of the picolinol and various fragments of the carboxyl group, showing that the compound is indeed a picolinyl ester. The molecular ion, M$^+$ (m/z = 371), confirms the presence of a C_{18} fatty acid with two double bonds. The uniform series of ions 14 atomic mass units (u) apart represents loss of each successive methyl and methylene group from the methyl end of the acyl chain (beginning at C-18; the right end of the molecule as shown here), until the ion at m/z = 300 is reached. This is followed by a gap of 26 u for the carbons of the terminal double bond, at m/z = 274; a further gap of 14 u for the C-11 methylene group, at m/z = 260; and so forth. By this means the entire structure is determined, although these data alone do not reveal the configuration (cis or trans) of the double bonds.

elution from the various chromatographic procedures often does not distinguish between them. When the eluate from a chromatography column is sampled by mass spectrometry, however, the components of a lipid mixture can be simultaneously separated and identified by their unique pattern of fragmentation **(Fig. 10–26)**. With the increased resolution of mass spectrometry, it is possible to identify individual lipids in very complex mixtures without first fractionating the lipids in a crude extract. This "shotgun" method (Fig. 10–25c) avoids losses during preliminary separation of lipid subclasses, and it is faster.

Lipidomics Seeks to Catalog All Lipids and Their Functions

In exploring the biological role of lipids in cells and tissues, it is important to know which lipids are present and in what proportions, and to know how this lipid composition changes with embryonic development, disease, or drug treatment. As lipid biochemists have become aware of the thousands of different naturally occurring lipids, they have proposed a new nomenclature system, with the aim of making it easier to compile and search databases of lipid composition. The system places each lipid in one of eight chemical groups (Table 10–3) designated by two letters. Within these groups, finer distinctions are indicated by numbered classes and subclasses. For

example, all glycerophosphocholines are GP01; the subgroup of glycerophosphocholines with two fatty acids in ester linkage is designated GP0101; with one fatty acid ether-linked at position 1 and one in ester linkage at position 2, this becomes GP0102. Specific fatty acids are designated by numbers that give every lipid its own unique identifier, so that each individual lipid, including lipid types not yet discovered, can be unambiguously described in terms of a 12-character identifier. One factor used in this classification is the nature of the biosynthetic precursor. For example, prenol lipids (dolichols and vitamins E and K, for example) are formed from isoprenyl precursors. Polyketides include some natural products, many toxic, with biosynthetic pathways related to those for fatty acids. The eight chemical categories in Table 10–3 do not coincide perfectly with the divisions according to biological function that we have used in this chapter. For example, the structural lipids of membranes include both glycerophospholipids and sphingolipids, separate categories in Table 10–3. Each method of categorization has its advantages.

The application of mass spectrometric techniques with high throughput and high resolution can provide quantitative catalogs of all the lipids present in a specific cell type under particular conditions—the **lipidome**—and of the ways in which the lipidome changes with differentiation, disease such as cancer, or drug treatment.

TABLE 10-3	Eight Major Categories of Biological Lipids	
Category	Category code	Examples
Fatty acids	FA	Oleate, stearoyl-CoA, palmitoylcarnitine
Glycerolipids	GL	Di- and triacylglycerols
Glycerophospholipids	GP	Phosphatidylcholine, phosphatidylserine, phosphatidylethanolamine
Sphingolipids	SP	Sphingomyelin, ganglioside GM2
Sterol lipids	ST	Cholesterol, progesterone, bile acids
Prenol lipids	PR	Farnesol, geraniol, retinol, ubiquinone
Saccharolipids	SL	Lipopolysaccharide
Polyketides	PK	Tetracycline, erythromycin, aflatoxin B_1

An animal cell contains more than a thousand different lipid species, each presumably having a specific function. These functions are known for a growing number of lipids, but the still largely unexplored lipidome offers a rich source of new problems for the next generation of biochemists and cell biologists to solve.

SUMMARY 10.4 Working with Lipids

▶ In the determination of lipid composition, the lipids are first extracted from tissues with organic solvents and separated by thin-layer, gas-liquid, or high-performance liquid chromatography.

▶ Phospholipases specific for one of the bonds in a phospholipid can be used to generate simpler compounds for subsequent analysis.

▶ Individual lipids are identified by their chromatographic behavior, their susceptibility to hydrolysis by specific enzymes, or mass spectrometry.

▶ High-resolution mass spectrometry allows the analysis of crude mixtures of lipids without prefractionation—the "shotgun" approach.

▶ Lipidomics combines powerful analytical techniques to determine the full complement of lipids in a cell or tissue (the lipidome) and to assemble annotated databases that allow comparisons between lipids of different cell types and under different conditions.

Key Terms

Terms in bold are defined in the glossary.

fatty acid 357
polyunsaturated fatty acid (PUFA) 359
triacylglycerol 360
lipases 360
phospholipid 363
glycolipid 363
glycerophospholipid 363
ether lipid 364
plasmalogen 364
galactolipid 365
sphingolipid 366
ceramide 366
sphingomyelin 366
glycosphingolipid 366
cerebroside 366
globoside 366
ganglioside 366

sterol 368
cholesterol 368
prostaglandin 371
thromboxane 371
leukotriene 372
vitamin 373
vitamin D_3 373
cholecalciferol 373

vitamin A (retinol) 373
vitamin E 374
tocopherol 374
vitamin K 374
dolichol 375
polyketide 376
lipidome 379

Further Reading

General

Fahy, E., Subramaniam, S., Brown, H.A., Glass, C.K., Merrill, A.H., Jr., Murphy, R.C., Raetz, C.R.H., Russell, D.W., Seyama, Y., Shaw, W., et al. (2005) A comprehensive classification system for lipids. *J. Lipid Res.* **46**, 839–862.

A new system of nomenclature for biological lipids, separating them into eight major categories. The definitive reference on lipid classification.

Gurr, M.I., Harwood, J.L., & Frayn, K.N. (2002) *Lipid Biochemistry: An Introduction*, 5th edn, Blackwell Science Ltd., Oxford.

A good general resource on lipid structure and metabolism, at the intermediate level.

Lipid Maps. *Nature* Lipidomics Gateway, www.lipidmaps.org.

Tutorials and lectures on lipid structure and function, lipidomics methods, and databases on lipid structures and properties.

Vance, J.E. & Vance, D.E. (eds). (2008) *Biochemistry of Lipids, Lipoproteins, and Membranes*, 5th edn, Elsevier Science Publishing Co., Inc., New York.

An excellent collection of reviews on various aspects of lipid structure, biosynthesis, and function.

Lipids as Nutrients

Angerer, P. & von Schacky, C. (2000) Omega-3 polyunsaturated fatty acids and the cardiovascular system. *Curr. Opin. Lipidol.* **11**, 57–63.

Covington, M.B. (2004) Omega-3 fatty acids. *Am. Fam. Physician* **70**, 133–140.

Succinct statement of the findings that omega-3 fatty acids reduce the risk of cardiovascular disease.

de Logeril, M., Salen, P., Martin, J.L., Monjaud, I., Delaye, J., & Mamelle, N. (1999) Mediterranean diet, traditional risk factors, and the rate of cardiovascular complications after myocardial infarction: final report of the Lyon Diet Heart Study. *Circulation* **99**, 779–785.

Lavie, C.J., Milani, R.V., Mehra, M.R., & Ventura, H.O. (2009) Omega-3 polyunsaturated fatty acids and cardiovascular diseases. *J. Am. Coll. Cardiol.* **54**, 585–594.

Mello, M.M. (2009) New York City's war on fat. *N. Engl. J. Med.* **360**, 2015–2020.

Legal and ethical aspects of the ban on trans fatty acids in restaurants.

Mozaffarian, D., Katan, M.B., Ascherio, P.H., Stampfer, M.J., & Willet, W.C. (2006) Trans fatty acids and cardiovascular disease. *N. Engl. J. Med.* **354**, 1601–1613.

A summary of the evidence that dietary trans fatty acids predispose to coronary heart disease.

Structural Lipids in Membranes

Bogdanov, M. & Dowhan, W. (1999) Lipid-assisted protein folding. *J. Biol. Chem.* **274**, 36,827–36,830.

A minireview of the role of membrane lipids in the folding of membrane proteins.

Dowhan, W. (1997) Molecular basis for membrane phospholipid diversity: why are there so many lipids? *Annu. Rev. Biochem.* **66**, 199–232.

Jacquemet, A., Barbeau, J., Lemiegre, L., & Benvegnu, T. (2009) Archaeal tetraether bipolar lipids: structures, functions and applications. *Biochimie* **91**, 711–717.

Valle, D., Beaudet, A.L., Vogelstein, B., Kinzler, K.W., Antonarakis, S.E., & Ballabio. A. (eds). (2006) *Scriver's Online Metabolic & Molecular Bases of Inherited Disease*, www.ommbid.com.

This classic medical encyclopedia, last published in 2001 as a four-volume set, is now maintained and kept updated online. It contains definitive descriptions of the clinical, biochemical, and genetic aspects of hundreds of human metabolic diseases—an authoritative source and fascinating reading. Part 16: Lysosomal Disorders includes 24 articles on various disorders of lipid metabolism.

Lipids as Signals, Cofactors, and Pigments

Bell, R.M., Exton, J.H., & Prescott, S.M. (eds). (1996) *Lipid Second Messengers*, Handbook of Lipid Research, Vol. 8, Plenum Press, New York.

Berkner, K.L. & Runge, K.W. (2004) The physiology of vitamin K nutriture and vitamin K-dependent protein function in atherosclerosis. *J. Thromb. Haemost.* **2**, 2118–2132.

Binkley, N.C. & Suttie, J.W. (1995) Vitamin K nutrition and osteoporosis. *J. Nutr.* **125**, 1812–1821.

Brigelius-Flohé, R. & Traber, M.G. (1999) Vitamin E: function and metabolism. *FASEB J.* **13**, 1145–1155.

Chojnacki, T. & Dallner, G. (1988) The biological role of dolichol. *Biochem. J.* **251**, 1–9.

Clouse, S.D. (2002) Brassinosteroid signal transduction: clarifying the pathway from ligand perception to gene expression. *Mol. Cell* **10**, 973–982.

DeLuca, H.F. (2008) Evolution of our understanding of vitamin D. *Nutr. Rev.* **66**, S73–S87.

Dicke, M., van Loon, J.J.A., & Soler, R. (2009) Chemical complexity of volatiles from plants induced by multiple attack. *Nat. Chem. Biol.* **5**, 317–324.

Holick, M.F. (2007) Vitamin D deficiency. *N. Engl. J. Med.* **357**, 266–281.

James, D.J., Khodthong, C., Kowalchyk, J.A., & Martin, T.F. (2010) Phosphatidylinositol 4,5-bisphosphate regulation of SNARE function in membrane fusion mediated by CAPS. *Adv. Enzyme Regul.* **50**, 62–70.

Jones M.B., Rosenberg, J.N., Betenbaugh, M.J., & Krag, S.S. (2009) Structure and synthesis of polyisoprenoids used in *N*-glycosylation across the three domains of life. *Biochim. Biophys. Acta* **1790**, 485–494.

Intermediate-level review of role of dolichols in glycosylation.

Lee, D. (2007) *Nature's Palette: The Science of Plant Color*, University of Chicago Press, Chicago, IL.

Fascinating intermediate-level book about lipids as biological pigments.

Rosen, H., Gonzalez-Cabrera, P.J., Sanna, M.G., & Brown, S. (2009) Sphingosine 1-phosphate receptor signaling. *Annu. Rev. Biochem.* **78**, 743–768.

Suttie, J.W. (1993) Synthesis of vitamin K-dependent proteins. *FASEB J.* **7**, 445–452.

Describes the biochemical basis for the requirement of vitamin K in blood clotting and the importance of carboxylation in the synthesis of the blood-clotting protein thrombin.

Weber, H. (2002) Fatty acid-derived signals in plants. *Trends Plant Sci.* **7**, 217–224.

Wymann, M.P. & Schneiter, R. (2008) Lipid signaling in disease. *Nat. Rev. Mol. Cell Biol.* **9**, 162–176.

Working with Lipids

Christie, W.W. (1998) Gas chromatography–mass spectrometry methods for structural analysis of fatty acids. *Lipids* **33**, 343–353.

A detailed description of the methods used to obtain data such as those presented in Figure 10–26.

Christie, W.W. (2003) *Lipid Analysis*, 3rd edn, The Oily Press, Bridgwater, England.

Dennis, E.A., Deems, R.A., Harkewicz, R., Quehenberger, O., Brown, H.A., Milne, S.B., Myers, D.S., Glass, C.K., Hardiman, G., Reichart, D., et al. (2010) A mouse macrophage lipidome. *J. Biol. Chem.* **285**, 39,976–39,985.

Research paper describing variations in the total macrophage lipidome.

Harkewicz, R. & Dennis, E.A. (2011) Applications of mass spectrometry to lipids and membranes. *Annu. Rev. Biochem.* **80**, 301–325.

Advanced discussion of mass spectrometry and lipidomics.

Murphy, R.C. & Gasskell, S.J. (2011) New applications of mass spectrometry in lipid analysis. *J. Biol. Chem.* **286**, 25,427–25,433.

Watson, A.D. (2006) Lipidomics: a global approach to lipid analysis in biological systems. *J. Lipid Res.* **47**, 2101–2111.

A short, intermediate-level review of the classes of lipids, the methods for extracting and separating them, and mass spectrometric means for identifying and quantifying all lipids in a given cell, tissue, or organelle.

Problems

1. Operational Definition of Lipids How is the definition of "lipid" different from the types of definitions used for other biomolecules that we have considered, such as amino acids, nucleic acids, and proteins?

2. Melting Points of Lipids The melting points of a series of 18-carbon fatty acids are: stearic acid, 69.6 °C; oleic acid, 13.4 °C; linoleic acid, –5 °C ; and linolenic acid, –11 °C.

(a) What structural aspect of these 18-carbon fatty acids can be correlated with the melting point?

(b) Draw all the possible triacylglycerols that can be constructed from glycerol, palmitic acid, and oleic acid. Rank them in order of increasing melting point.

(c) Branched-chain fatty acids are found in some bacterial membrane lipids. Would their presence increase or decrease the fluidity of the membranes (that is, give them a lower or higher melting point)? Why?

3. Preparation of Béarnaise Sauce During the preparation of béarnaise sauce, egg yolks are incorporated into melted butter to stabilize the sauce and avoid separation. The stabilizing agent in the egg yolks is lecithin (phosphatidylcholine). Suggest why this works.

4. Isoprene Units in Isoprenoids Geraniol, farnesol, and squalene are called isoprenoids, because they are synthesized from five-carbon isoprene units. In each compound, circle the five-carbon units representing isoprene units (see Fig. 10–22).

Geraniol

Farnesol

Squalene

5. Naming Lipid Stereoisomers The two compounds below are stereoisomers of carvone with quite different properties; the one on the left smells like spearmint, and that on the right, like caraway. Name the compounds using the RS system.

Spearmint   Caraway

6. RS Designations for Alanine and Lactate Draw (using wedge-bond notation) and label the (R) and (S) isomers of 2-aminopropanoic acid (alanine) and 2-hydroxypropanoic acid (lactic acid).

2-Aminopropanoic acid   2-Hydroxypropanoic acid
        (alanine)   (lactic acid)

7. Hydrophobic and Hydrophilic Components of Membrane Lipids A common structural feature of membrane lipids is their amphipathic nature. For example, in phosphatidylcholine, the two fatty acid chains are hydrophobic and the phosphocholine head group is hydrophilic. For each of the following membrane lipids, name the components that serve as the hydrophobic and hydrophilic units: (a) phosphatidylethanolamine; (b) sphingomyelin; (c) galactosylcerebroside; (d) ganglioside; (e) cholesterol.

8. Structure of Omega-6 Fatty Acid Draw the structure of the omega-6 fatty acid 16:1.

9. Catalytic Hydrogenation of Vegetable Oils Catalytic hydrogenation, used in the food industry, converts double bonds in the fatty acids of the oil triacylglycerols to $—CH_2—CH_2—$. How does this affect the physical properties of the oils?

10. Alkali Lability of Triacylglycerols A common procedure for cleaning the grease trap in a sink is to add a product that contains sodium hydroxide. Explain why this works.

11. Deducing Lipid Structure from Composition Compositional analysis of a certain lipid shows that it has exactly one mole of fatty acid per mole of inorganic phosphate. Could this be a glycerophospholipid? A ganglioside? A sphingomyelin?

12. Deducing Lipid Structure from Molar Ratio of Components Complete hydrolysis of a glycerophospholipid yields glycerol, two fatty acids ($16:1(\Delta^9)$ and $16:0$), phosphoric acid, and serine in the molar ratio 1:1:1:1:1. Name this lipid and draw its structure.

13. Impermeability of Waxes What property of the waxy cuticles that cover plant leaves makes the cuticles impermeable to water?

14. The Action of Phospholipases The venom of the Eastern diamondback rattler and the Indian cobra contains phospholipase A_2, which catalyzes the hydrolysis of fatty acids at the C-2 position of glycerophospholipids. The phospholipid breakdown product of this reaction is lysolecithin (lecithin is phosphatidylcholine). At high concentrations, this and other lysophospholipids act as detergents, dissolving the membranes of erythrocytes and lysing the cells. Extensive hemolysis may be life-threatening.

(a) All detergents are amphipathic. What are the hydrophilic and hydrophobic portions of lysolecithin?

(b) The pain and inflammation caused by a snake bite can be treated with certain steroids. What is the basis of this treatment?

(c) Though the high levels of phospholipase A_2 in venom can be deadly, this enzyme is necessary for a variety of normal metabolic processes. What are these processes?

15. Lipids in Blood Group Determination We note in Figure 10–15 that the structure of glycosphingolipids determines the blood groups A, B, and O in humans. It is also true that glycoproteins determine blood groups. How can both statements be true?

16. Intracellular Messengers from Phosphatidylinositols When the hormone vasopressin stimulates cleavage of phosphatidylinositol 4,5-bisphosphate by hormone-sensitive phospholipase C, two products are formed. What are they? Compare their properties and their solubilities in water, and predict whether either would diffuse readily through the cytosol.

17. Storage of Fat-Soluble Vitamins In contrast to water-soluble vitamins, which must be part of our daily diet, fat-soluble vitamins can be stored in the body in amounts sufficient for many months. Suggest an explanation for this difference.

18. Hydrolysis of Lipids Name the products of mild hydrolysis with dilute NaOH of (a) 1-stearoyl-2,3-dipalmitoylglycerol; (b) 1-palmitoyl-2-oleoylphosphatidylcholine.

19. Effect of Polarity on Solubility Rank the following in order of increasing solubility in water: a triacylglycerol, a diacylglycerol, and a monoacylglycerol, all containing only palmitic acid.

20. Chromatographic Separation of Lipids A mixture of lipids is applied to a silica gel column, and the column is then washed with increasingly polar solvents. The mixture consists of phosphatidylserine, phosphatidylethanolamine, phosphatidylcholine, cholesteryl palmitate (a sterol ester), sphingomyelin, palmitate, n-tetradecanol, triacylglycerol, and cholesterol. In what order will the lipids elute from the column? Explain your reasoning.

21. Identification of Unknown Lipids Johann Thudichum, who practiced medicine in London about 100 years ago, also dabbled in lipid chemistry in his spare time. He isolated a variety of lipids from neural tissue, and characterized and named many of them. His carefully sealed and labeled vials of isolated lipids were rediscovered many years later.

(a) How would you confirm, using techniques not available to Thudichum, that the vials labeled "sphingomyelin" and "cerebroside" actually contain these compounds?

(b) How would you distinguish sphingomyelin from phosphatidylcholine by chemical, physical, or enzymatic tests?

22. Ninhydrin to Detect Lipids on TLC Plates Ninhydrin reacts specifically with primary amines to form a purplish-blue product. A thin-layer chromatogram of rat liver phospholipids is sprayed with ninhydrin, and the color is allowed to develop. Which phospholipids can be detected in this way?

Data Analysis Problem

23. Determining the Structure of the Abnormal Lipid in Tay-Sachs Disease Box 10–1, Figure 1, shows the pathway of breakdown of gangliosides in healthy (normal) individuals and individuals with certain genetic diseases. Some of the data on which the figure is based were presented in a paper by Lars Svennerholm (1962). Note that the sugar Neu5Ac, N-acetylneuraminic acid, represented in the Box 10–1 figure as ◆, is a sialic acid.

Svennerholm reported that "about 90% of the monosialiogangliosides isolated from normal human brain" consisted of a compound with ceramide, hexose, N-acetylgalactosamine, and N-acetylneuraminic acid in the molar ratio 1:3:1:1.

(a) Which of the gangliosides (GM1 through GM3 and globoside) in Box 10–1, Figure 1, fits this description? Explain your reasoning.

(b) Svennerholm reported that 90% of the gangliosides from a patient with Tay-Sachs had a molar ratio (of the same four components given above) of 1:2:1:1. Is this consistent with the Box 10–1 figure? Explain your reasoning.

To determine the structure in more detail, Svennerholm treated the gangliosides with neuraminidase to remove the N-acetylneuraminic acid. This resulted in an asialoganglioside that was much easier to analyze. He hydrolyzed it with acid, collected the ceramide-containing products, and determined the molar ratio of the sugars in each product. He did this for both the normal and the Tay-Sachs gangliosides. His results are shown below.

Ganglioside	Ceramide	Glucose	Galactose	Galactosamine
Normal				
Fragment 1	1	1	0	0
Fragment 2	1	1	1	0
Fragment 3	1	1	1	1
Fragment 4	1	1	2	1
Tay-Sachs				
Fragment 1	1	1	0	0
Fragment 2	1	1	1	0
Fragment 3	1	1	1	1

(c) Based on these data, what can you conclude about the structure of the normal ganglioside? Is this consistent with the structure in Box 10–1? Explain your reasoning.

(d) What can you conclude about the structure of the Tay-Sachs ganglioside? Is this consistent with the structure in Box 10–1? Explain your reasoning.

Svennerholm also reported the work of other researchers who "permethylated" the normal asialoganglioside. Permethylation is the same as exhaustive methylation: a methyl group is added to every free hydroxyl group on a sugar. They found the following permethylated sugars: 2,3,6-trimethylglycopyranose; 2,3,4,6-tetramethylgalactopyranose; 2,4,6-trimethylgalactopyranose; and 4,6-dimethyl-2-deoxy-2-aminogalactopyranose.

(e) To which sugar of GM1 does each of the permethylated sugars correspond? Explain your reasoning.

(f) Based on all the data presented so far, what pieces of information about normal ganglioside structure are missing?

Reference

Svennerholm, L. (1962) The chemical structure of normal human brain and Tay-Sachs gangliosides. *Biochem. Biophys. Res. Comm.* **9,** 436–441.

Biological Membranes and Transport

The first cell probably came into being when a membrane formed, enclosing a small volume of aqueous solution and separating it from the rest of the universe. Membranes define the external boundaries of cells and control the molecular traffic across that boundary **(Fig. 11–1)**; in eukaryotic cells, they divide the internal space into discrete compartments to segregate processes and components. They organize complex reaction sequences and are central to both biological energy conservation and cell-to-cell communication. The biological activities of membranes flow from their remarkable physical properties. Membranes are flexible, self-sealing, and selectively permeable to polar solutes. Their flexibility permits the shape changes that accompany cell growth and movement (such as amoeboid movement). With their ability to break and reseal, two membranes can fuse, as in exocytosis, or a single membrane-enclosed compartment can undergo fission to yield two sealed compartments, as in endocytosis or cell division, without creating gross leaks through cellular surfaces. Because membranes are selectively permeable, they retain certain compounds and ions within cells and within specific cellular compartments while excluding others.

Membranes are not merely passive barriers. They include an array of proteins specialized for promoting or catalyzing various cellular processes. At the cell surface, transporters move specific organic solutes and inorganic ions across the membrane; receptors sense extracellular signals and trigger molecular changes in the cell; adhesion molecules hold neighboring cells together. Within the cell, membranes organize cellular processes such as the synthesis of lipids and certain proteins, and the energy transductions in mitochondria and chloroplasts. Because membranes consist of just two layers of molecules, they are very thin—essentially two-dimensional. Intermolecular collisions are far more probable in this two-dimensional space than in three-dimensional space, so the efficiency of enzyme-catalyzed processes organized within membranes is vastly increased.

In this chapter we first describe the composition of cellular membranes and their chemical architecture—the molecular structures that underlie their biological functions. Next, we consider the remarkable dynamic features of membranes, in which lipids and proteins move relative to each other. Cell adhesion, endocytosis, and the membrane fusion accompanying neurotransmitter secretion illustrate the dynamic roles of membrane proteins. We then turn to the protein-mediated passage of solutes across membranes via transporters and ion channels. In later chapters we discuss the roles of membranes in signal transduction (Chapters 12 and 23), energy transduction (Chapter 19), lipid synthesis (Chapter 21), and protein synthesis (Chapter 27).

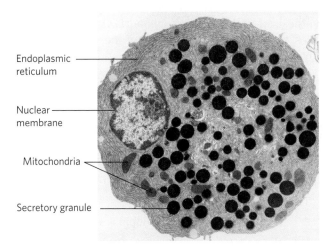

Endoplasmic
reticulum

Nuclear
membrane

Mitochondria

Secretory granule

FIGURE 11–1 Biological membranes. This electron micrograph of a thin-sectioned exocrine pancreas cell shows several compartments made of or bounded by membranes: the endoplasmic reticulum, mitochondria, secretory granules, and the nuclear membrane.

11.1 The Composition and Architecture of Membranes

One approach to understanding membrane function is to study membrane composition—to determine, for example, which components are common to all membranes and which are unique to membranes with specific functions. So before describing membrane structure and function, we consider the molecular components of membranes: proteins and polar lipids, which account for almost all the mass of biological membranes, and carbohydrates, present as part of glycoproteins and glycolipids.

Each Type of Membrane Has Characteristic Lipids and Proteins

The relative proportions of protein and lipid vary with the type of membrane (Table 11–1), reflecting the diversity of biological roles. For example, certain neurons have a myelin sheath—an extended plasma membrane that wraps around the cell many times and acts as a passive electrical insulator. The myelin sheath consists primarily of lipids, whereas the plasma membranes of bacteria and the membranes of mitochondria and chloroplasts, the sites of many enzyme-catalyzed processes, contain more protein than lipid (in mass per total mass).

For studies of membrane composition, the first task is to isolate a selected membrane. When eukaryotic cells are subjected to mechanical shear, their plasma membranes are torn and fragmented, releasing cytoplasmic components and membrane-bounded organelles such as mitochondria, chloroplasts, lysosomes, and nuclei. Plasma membrane fragments and intact organelles can be isolated by techniques described in Chapter 1 (see Fig. 1–8) and in Worked Example 2–1 (p. 57).

Cells clearly have mechanisms to control the kinds and amounts of membrane lipid they synthesize and to target specific lipids to particular organelles. Each kingdom, each species, each tissue or cell type, and the organelles of each cell type have a characteristic set of membrane lipids. Plasma membranes, for example, are enriched in cholesterol and contain no detectable car-

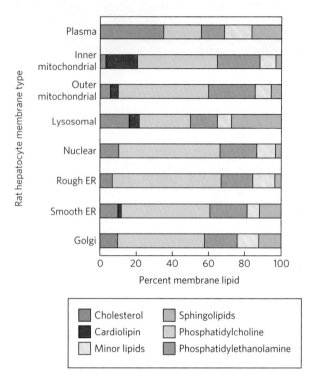

FIGURE 11–2 Lipid composition of the plasma membrane and organelle membranes of a rat hepatocyte. The functional specialization of each membrane type is reflected in its unique lipid composition. Cholesterol is prominent in plasma membranes but barely detectable in mitochondrial membranes. Cardiolipin is a major component of the inner mitochondrial membrane but not of the plasma membrane. Phosphatidylserine, phosphatidylinositol, and phosphatidylglycerol are relatively minor components of most membranes but serve critical functions; phosphatidylinositol and its derivatives, for example, are important in signal transductions triggered by hormones. Sphingolipids, phosphatidylcholine, and phosphatidylethanolamine are present in most membranes but in varying proportions. Glycolipids, which are major components of the chloroplast membranes of plants, are virtually absent from animal cells.

diolipin **(Fig. 11–2)**; mitochondrial membranes are very low in cholesterol and sphingolipids, but they contain phosphatidylglycerol and cardiolipin, which are synthesized within the mitochondria. In all but a few cases, the functional significance of these combinations is not yet known.

TABLE 11–1	Major Components of Plasma Membranes in Various Organisms				
	Components (% by weight)				
	Protein	Phospholipid	Sterol	Sterol type	Other lipids
Human myelin sheath	30	30	19	Cholesterol	Galactolipids, plasmalogens
Mouse liver	45	27	25	Cholesterol	—
Maize leaf	47	26	7	Sitosterol	Galactolipids
Yeast	52	7	4	Ergosterol	Triacylglycerols, steryl esters
Paramecium (ciliated protist)	56	40	4	Stigmasterol	—
E. coli	75	25	0	—	—

Note: Values do not add up to 100% in every case because there are components other than protein, phospholipids, and sterol; plants, for example, have high levels of glycolipids.

The protein composition of membranes from different sources varies even more widely than their lipid composition, reflecting functional specialization. In addition, some membrane proteins are covalently linked to oligosaccharides. For example, in glycophorin, a glycoprotein of the erythrocyte plasma membrane, 60% of the mass consists of complex oligosaccharides covalently attached to specific amino acid residues. Ser, Thr, and Asn residues are the most common points of attachment (see Fig. 7–30). The sugar moieties of surface glycoproteins influence the folding of the proteins as well as their stability and intracellular destination, and they play a significant role in the specific binding of ligands to glycoprotein surface receptors (see Fig. 7–37).

Some membrane proteins are covalently attached to one or more lipids, which serve as hydrophobic anchors that hold the proteins to the membrane, as we shall see.

All Biological Membranes Share Some Fundamental Properties

Membranes are impermeable to most polar or charged solutes, but permeable to nonpolar compounds. They are 5 to 8 nm (50 to 80 Å) thick when proteins protruding on both sides are included and appear trilaminar when viewed in cross section with the electron microscope. The combined evidence from electron microscopy and studies of chemical composition, as well as physical studies of permeability and the motion of individual protein and lipid molecules within membranes, led to the development of the **fluid mosaic model** for the structure of biological membranes **(Fig. 11–3)**. Phospholipids form a bilayer in which the nonpolar regions of the lipid molecules in each layer face the core

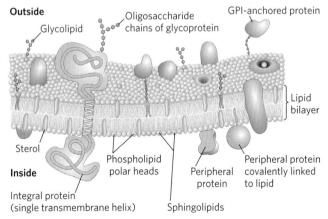

FIGURE 11–3 Fluid mosaic model for plasma membrane structure. The fatty acyl chains in the interior of the membrane form a fluid, hydrophobic region. Integral proteins float in this sea of lipid, held by hydrophobic interactions with their nonpolar amino acid side chains. Both proteins and lipids are free to move laterally in the plane of the bilayer, but movement of either from one leaflet of the bilayer to the other is restricted. The carbohydrate moieties attached to some proteins and lipids of the plasma membrane are exposed on the extracellular surface.

of the bilayer and their polar head groups face outward, interacting with the aqueous phase on either side. Proteins are embedded in this bilayer sheet, held by hydrophobic interactions between the membrane lipids and hydrophobic domains in the proteins. Some proteins protrude from only one side of the membrane; others have domains exposed on both sides. The orientation of proteins in the bilayer is asymmetric, giving the membrane "sidedness": the protein domains exposed on one side of the bilayer are different from those exposed on the other side, reflecting functional asymmetry. The individual lipid and protein units in a membrane form a fluid mosaic with a pattern that, unlike a mosaic of ceramic tile and mortar, is free to change constantly. The membrane mosaic is fluid because most of the interactions among its components are noncovalent, leaving individual lipid and protein molecules free to move laterally in the plane of the membrane.

We now look at some of these features of the fluid mosaic model in more detail and consider the experimental evidence that supports the basic model but has necessitated its refinement in several ways.

A Lipid Bilayer Is the Basic Structural Element of Membranes

Glycerophospholipids, sphingolipids, and sterols are virtually insoluble in water. When mixed with water, they spontaneously form microscopic lipid aggregates, clustering together, with their hydrophobic moieties in contact with each other and their hydrophilic groups interacting with the surrounding water. This clustering reduces the amount of hydrophobic surface exposed to water and thus minimizes the number of molecules in the shell of ordered water at the lipid-water interface (see Fig. 2–7), resulting in an increase in entropy. Hydrophobic interactions among lipid molecules provide the thermodynamic driving force for the formation and maintenance of these clusters.

Depending on the precise conditions and the nature of the lipids, three types of lipid aggregate can form when amphipathic lipids are mixed with water **(Fig. 11–4)**. **Micelles** are spherical structures that contain anywhere from a few dozen to a few thousand amphipathic molecules. These molecules are arranged with their hydrophobic regions aggregated in the interior, where water is excluded, and their hydrophilic head groups at the surface, in contact with water. Micelle formation is favored when the cross-sectional area of the head group is greater than that of the acyl side chain(s), as in free fatty acids, lysophospholipids (phospholipids lacking one fatty acid), and detergents such as sodium dodecyl sulfate (SDS; p. 94).

A second type of lipid aggregate in water is the **bilayer**, in which two lipid monolayers (leaflets) form a two-dimensional sheet. Bilayer formation is favored if the cross-sectional areas of the head group and acyl side chain(s) are similar, as in glycerophospholipids and

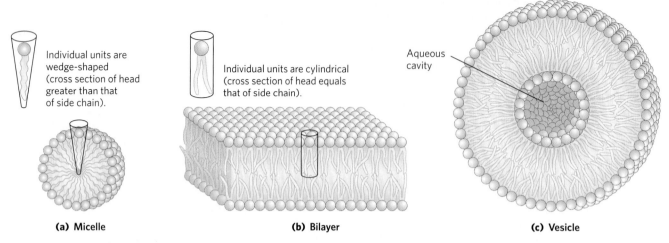

(a) Micelle **(b) Bilayer** **(c) Vesicle**

FIGURE 11–4 Amphipathic lipid aggregates that form in water. (a) In micelles, the hydrophobic chains of the fatty acids are sequestered at the core of the sphere. There is virtually no water in the hydrophobic interior. **(b)** In an open bilayer, all acyl side chains except those at the edges of the sheet are protected from interaction with water. **(c)** When a two-dimensional bilayer folds on itself, it forms a closed bilayer, a three-dimensional hollow vesicle (liposome) enclosing an aqueous cavity.

sphingolipids. The hydrophobic portions in each monolayer, excluded from water, interact with each other. The hydrophilic head groups interact with water at each surface of the bilayer. Because the hydrophobic regions at its edges (Fig. 11–4b) are in contact with water, the bilayer sheet is relatively unstable and spontaneously folds back on itself to form a hollow sphere, a **vesicle** (Fig. 11–4c). The continuous surface of vesicles eliminates exposed hydrophobic regions, allowing bilayers to achieve maximal stability in their aqueous environment. Vesicle formation also creates a separate aqueous compartment. It is likely that the precursors to the first living cells resembled lipid vesicles, their aqueous contents segregated from their surroundings by a hydrophobic shell.

The lipid bilayer is 3 nm (30 Å) thick. The hydrocarbon core, made up of the —CH_2— and —CH_3 of the fatty acyl groups, is about as nonpolar as decane, and vesicles formed in the laboratory from pure lipids (liposomes) are essentially impermeable to polar solutes, as is the lipid bilayer of biological membranes (although biological membranes, as we shall see, are permeable to solutes for which they have specific transporters).

Plasma membrane lipids are asymmetrically distributed between the two monolayers of the bilayer, although the asymmetry, unlike that of membrane proteins, is not absolute. In the plasma membrane of the erythrocyte, for example, choline-containing lipids (phosphatidylcholine and sphingomyelin) are typically found in the outer (extracellular, or exoplasmic) leaflet **(Fig. 11–5)**, whereas phosphatidylserine, phosphatidylethanolamine, and the phosphatidylinositols are much more common in the inner (cytoplasmic) leaflet. The flow of membrane components from the endoplasmic reticulum through the Golgi apparatus and to the plasma membrane via transport vesicles is accompanied by changes in lipid composition and disposition across

the bilayer **(Fig. 11–6)**. Phosphatidylcholine is the principal phospholipid in the lumenal monolayer of the Golgi membrane, but in transport vesicles phosphatidylcholine has been largely replaced by sphingolipids and cholesterol, which, on fusion of transport vesicles with the plasma membrane, make up the majority of the lipids in the outer monolayer of the plasma membrane.

FIGURE 11–5 Asymmetric distribution of phospholipids between the inner and outer monolayers of the erythrocyte plasma membrane. The distribution of a specific phospholipid is determined by treating the intact cell with phospholipase C, which cannot reach lipids in the inner monolayer (leaflet) but removes the head groups of lipids in the outer monolayer. The proportion of each head group released provides an estimate of the fraction of each lipid in the outer monolayer.

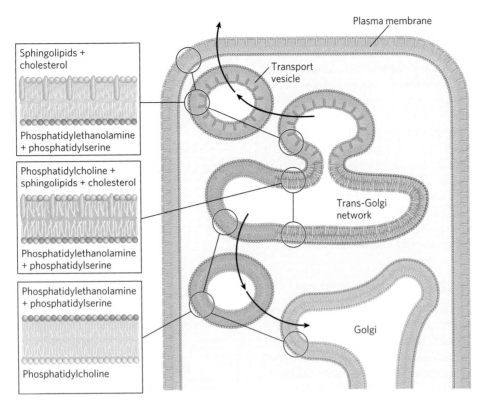

FIGURE 11–6 The distribution of lipids in the membranes of a typical cell. Each membrane has its own characteristic composition, and the two monolayers of a given membrane may differ in composition as well.

Changes in the distribution of lipids between plasma membrane leaflets have biological consequences. For example, only when the phosphatidylserine in the plasma membrane moves into the outer leaflet is a platelet able to play its role in formation of a blood clot. For many other cell types, phosphatidylserine exposure on the outer surface marks a cell for destruction by programmed cell death. The transbilayer movement of phospholipid molecules is catalyzed and regulated by specific proteins (see Fig. 11–17).

Three Types of Membrane Proteins Differ in Their Association with the Membrane

Integral membrane proteins are very firmly associated with the lipid bilayer and are removable only by agents that interfere with hydrophobic interactions, such as detergents, organic solvents, or denaturants **(Fig. 11–7)**. **Peripheral membrane proteins** associate with the membrane through electrostatic interactions and hydrogen bonding with the hydrophilic domains of integral proteins and with the polar head

FIGURE 11–7 Peripheral, integral, and amphitropic proteins. Membrane proteins can be operationally distinguished by the conditions required to release them from the membrane. Most peripheral proteins are released by changes in pH or ionic strength, removal of Ca^{2+} by a chelating agent, or addition of urea or carbonate. Integral proteins are extractable with detergents, which disrupt the hydrophobic interactions with the lipid bilayer and form micelle-like clusters around individual protein molecules. Integral proteins covalently attached to a membrane lipid, such as a glycosyl phosphatidylinositol (GPI; see Fig. 11–15), can be released by treatment with phospholipase C. Amphitropic proteins are sometimes associated with membranes and sometimes not, depending on some type of regulatory process such as reversible palmitoylation.

groups of membrane lipids. They can be released by relatively mild treatments that interfere with electrostatic interactions or break hydrogen bonds; a commonly used agent is carbonate at high pH. **Amphitropic proteins** are found both in the cytosol and in association with membranes. Their affinity for membranes results in some cases from the protein's noncovalent interaction with a membrane protein or lipid, and in other cases from the presence of one or more lipids covalently attached to the amphitropic protein (see Fig. 11–15). Generally, the reversible association of amphitropic proteins with the membrane is regulated; for example, phosphorylation or ligand binding can force a conformational change in the protein, exposing a membrane-binding site that was previously inaccessible.

Many Membrane Proteins Span the Lipid Bilayer

Membrane protein topology (the localization of protein domains relative to the lipid bilayer) can be determined with reagents that react with protein side chains but cannot cross membranes—polar chemical reagents that react with primary amines of Lys residues, for example, or enzymes such as trypsin that cleave proteins but cannot cross the membrane. The human erythrocyte is convenient for such studies because it has no membrane-bounded organelles; the plasma membrane is the only membrane present. If a membrane protein in an intact erythrocyte reacts with a membrane-impermeant reagent, that protein must have at least one domain exposed on the outer (extracellular) face of the membrane. Trypsin cleaves extracellular domains but does not affect domains buried within the bilayer or exposed on the inner surface only, unless the plasma membrane is broken to make these domains accessible to the enzyme.

Experiments with such topology-specific reagents show that the erythrocyte glycoprotein **glycophorin** spans the plasma membrane. Its amino-terminal domain (bearing the carbohydrate chains) is on the outer surface and is cleaved by trypsin. The carboxyl terminus protrudes on the inside of the cell, where it cannot react with impermeant reagents. Both the amino-terminal and carboxyl-terminal domains contain many polar or charged amino acid residues and are therefore hydrophilic. However, a segment in the center of the protein (residues 75 to 93) contains mainly hydrophobic amino acid residues, suggesting that glycophorin has a transmembrane segment arranged as shown in **Figure 11–8**.

These noncrystallographic experiments also revealed that the orientation of glycophorin in the membrane is asymmetric: its amino-terminal segment is always on the outside. Similar studies of other membrane proteins show that each has a specific orientation in the bilayer, giving the membrane a distinct sidedness. For glycophorin, and for all other glycoproteins of the plasma membrane, the glycosylated domains

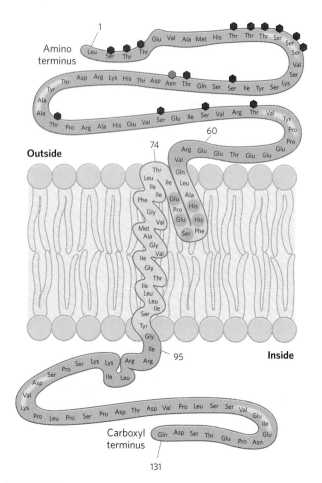

FIGURE 11–8 Transbilayer disposition of glycophorin in an erythrocyte. One hydrophilic domain, containing all the sugar residues, is on the outer surface, and another hydrophilic domain protrudes from the inner face of the membrane. Each red hexagon represents a tetrasaccharide (containing two Neu5Ac (sialic acid), Gal, and GalNAc) O-linked to a Ser or Thr residue; the blue hexagon represents an oligosaccharide N-linked to an Asn residue. The relative size of the oligosaccharide units is larger than shown here. A segment of 19 hydrophobic residues (residues 75 to 93) forms an α helix that traverses the membrane bilayer (see Fig. 11-12a). The segment from residues 64 to 74 has some hydrophobic residues and probably penetrates the outer face of the lipid bilayer, as shown.

are invariably found on the extracellular face of the bilayer. As we shall see, the asymmetric arrangement of membrane proteins results in functional asymmetry. All the molecules of a given ion pump, for example, have the same orientation in the membrane and pump ions in the same direction.

Integral Proteins Are Held in the Membrane by Hydrophobic Interactions with Lipids

The firm attachment of integral proteins to membranes is the result of hydrophobic interactions between membrane lipids and hydrophobic domains of the protein. Some proteins have a single hydrophobic sequence in the middle (as in glycophorin) or at the amino or carboxyl terminus. Others have multiple hydrophobic sequences,

Inside

Outside

Type VI   Type IV   Type V   Type III   Type II   Type I

H_3N^+ COO^-

^-OOC

NH_3^+

FIGURE 11–9 Integral membrane proteins. For known proteins of the plasma membrane, the spatial relationships of protein domains to the lipid bilayer fall into six categories. Types I and II have a single transmembrane helix; the amino-terminal domain is outside the cell in type I proteins and inside in type II. Type III proteins have multiple transmembrane helices in a single polypeptide. In type IV proteins, transmembrane domains of several different polypeptides assemble to form a channel through the membrane. Type V proteins are held to the bilayer primarily by covalently linked lipids (see Fig. 11–15), and type VI proteins have both transmembrane helices and lipid anchors.

In this figure, and in figures throughout the book, we represent transmembrane protein segments in their most likely conformations: as α helices of six to seven turns. Sometimes these helices are shown simply as cylinders. As relatively few membrane protein structures have been deduced by x-ray crystallography, our representation of the extramembrane domains is arbitrary and not necessarily to scale.

each of which, when in the α-helical conformation, is long enough to span the lipid bilayer **(Fig. 11–9)**.

One of the best-studied membrane-spanning proteins, bacteriorhodopsin, has seven very hydrophobic internal sequences and crosses the lipid bilayer seven times. Bacteriorhodopsin is a light-driven proton pump densely packed in regular arrays in the purple membrane of the bacterium *Halobacterium salinarum*. X-ray crystallography reveals a structure with seven α-helical segments, each traversing the lipid bilayer, connected by nonhelical loops at the inner and outer face of the membrane **(Fig. 11–10)**. In the amino acid sequence of bacteriorhodopsin, seven segments of about 20 hydrophobic residues can be identified, each forming an α helix that spans the bilayer. The seven helices are clustered together and oriented not quite perpendicular to the bilayer plane, a pattern that (as we shall see in Chapter 12) is a common motif in membrane proteins involved in signal reception. Hydrophobic interactions between the nonpolar amino acids and the fatty acyl groups of the membrane lipids firmly anchor the protein in the membrane.

Crystallized membrane proteins solved (i.e., their molecular structure deduced) by crystallography often include molecules of phospholipids, which are presumed to be positioned in the crystals as they are in the native membranes. Many of these phospholipid molecules lie on the protein surface, their head groups interacting with polar amino acid residues at the inner

Outside

Amino terminus

Inside

Carboxyl terminus

FIGURE 11–10 Bacteriorhodopsin, a membrane-spanning protein. (PDB ID 2AT9) The single polypeptide chain folds into seven hydrophobic α helices, each of which traverses the lipid bilayer roughly perpendicular to the plane of the membrane. The seven transmembrane helices are clustered, and the space around and between them is filled with the acyl chains of membrane lipids. The light-absorbing pigment retinal (see Fig. 10–21) is buried deep in the membrane in contact with several of the helical segments (not shown). The helices are colored to correspond with the hydropathy plot in Figure 11–12b.

and outer membrane–water interfaces and their side chains associated with nonpolar residues. These **annular lipids** form a bilayer shell (annulus) around the protein, oriented roughly as expected for phospholipids in a bilayer **(Fig. 11–11)**. Other phospholipids are found at the interfaces between monomers of multisubunit membrane proteins, where they form a "grease seal." Yet others are embedded deep within a membrane protein, often with their head groups well below the plane of the bilayer. For example, succinate dehydrogenase (Complex II, found in mitochondria; see Fig. 19–10) has several deeply embedded phospholipid molecules.

The Topology of an Integral Membrane Protein Can Sometimes Be Predicted from Its Sequence

Determination of the three-dimensional structure of a membrane protein—that is, its topology—is generally much more difficult than determining its amino acid sequence, either directly or by gene sequencing. The amino acid sequences are known for thousands of membrane proteins, but relatively few three-dimensional structures have been established by crystallography or NMR spectroscopy. The presence of unbroken sequences of more than 20 hydrophobic residues in a membrane protein is commonly taken as evidence that these sequences traverse the lipid bilayer, acting as hydrophobic anchors or forming transmembrane channels. Virtually all integral proteins have at least one such sequence. Application of this logic to entire genomic sequences leads to the conclusion that in many species, 20% to 30% of all proteins are integral membrane proteins.

(a) Sheep aquaporin

Phospholipids

(b) V$_o$ from V-type ATPase

FIGURE 11–11 Lipid annuli associated with two integral membrane proteins. (a) The crystal structure of sheep aquaporin (PDB ID 2B6O), a transmembrane water channel, includes a shell of phospholipids positioned with their head groups (blue) at the expected positions on the inner and outer membrane surfaces and their hydrophobic acyl chains (gold) intimately associated with the surface of the protein exposed to the bilayer. The lipid forms a "grease seal" around the protein, which is depicted as a dark blue surface representation. **(b)** The crystal structure of the V$_o$ integral protein complex of the V-type Na$^+$ ATPase from *Enterococcus hirae* (PDB ID 2BL2) has 10 identical subunits, each with four transmembrane helices, surrounding a central cavity filled with phosphatidylglycerol (PG). Here five of the subunits have been cut away to reveal the PG molecules associated with each subunit around the interior of this structure.

What can we predict about the secondary structure of the membrane-spanning portions of integral proteins? An α-helical sequence of 20 to 25 residues is just long enough to span the thickness (30 Å) of the lipid bilayer (recall that the length of an α helix is 1.5 Å (0.15 nm) per amino acid residue). A polypeptide chain surrounded by lipids, having no water molecules with which to hydrogen-bond, will tend to form α helices or β sheets, in which intrachain hydrogen bonding is maximized. If the side chains of all amino acids in a helix are nonpolar, hydrophobic interactions with the surrounding lipids further stabilize the helix.

Several simple methods of analyzing amino acid sequences yield reasonably accurate predictions of secondary structure for transmembrane proteins. The relative polarity of each amino acid has been determined experimentally by measuring the free-energy change accompanying the movement of that amino acid side chain from a hydrophobic solvent into water. This free energy of transfer, which can be expressed as a **hydropathy index** (see Table 3–1), ranges from very exergonic for charged or polar residues to very endergonic for amino acids with aromatic or aliphatic hydrocarbon side chains. The overall hydropathy index (hydrophobicity) of a sequence of amino acids is estimated by summing the free energies of transfer for the residues in the sequence. To scan a polypeptide sequence for potential membrane-spanning segments, an investigator calculates the hydropathy index for successive segments (called windows) of a given size, from 7 to 20 residues. For a window of seven residues, for example, the average indices for residues 1 to 7, 2 to 8, 3 to 9, and so on are plotted as in **Figure 11–12** (plotted for the

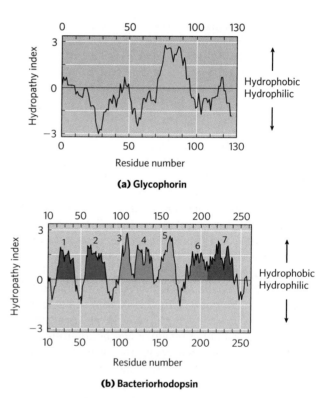

(a) Glycophorin

(b) Bacteriorhodopsin

FIGURE 11–12 Hydropathy plots. Average hydropathy index (see Table 3-1) is plotted against residue number for two integral membrane proteins. The hydropathy index for each amino acid residue in a sequence of defined length, or "window," is used to calculate the average hydropathy for that window. The horizontal axis shows the residue number in the middle of the window. **(a)** Glycophorin from human erythrocytes has a single hydrophobic sequence between residues 75 and 93 (yellow); compare this with Figure 11-8. **(b)** Bacteriorhodopsin, known from independent physical studies to have seven transmembrane helices (see Fig. 11-10), has seven hydrophobic regions. Note, however, that the hydropathy plot is ambiguous in the region of segments 6 and 7. X-ray crystallography has confirmed that this region has two transmembrane segments.

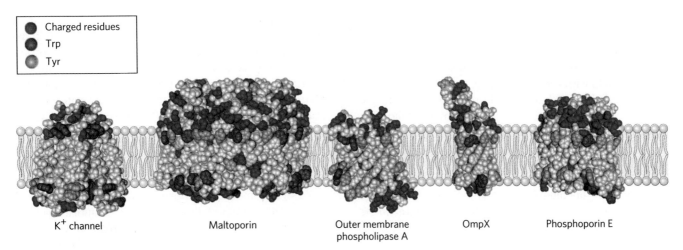

Charged residues
Trp
Tyr

K⁺ channel   Maltoporin   Outer membrane phospholipase A   OmpX   Phosphoporin E

FIGURE 11–13 Tyr and Trp residues of membrane proteins clustering at the water-lipid interface. The detailed structures of these five integral membrane proteins are known from crystallographic studies. The K⁺ channel (PDB ID 1BL8) is from the bacterium *Streptomyces lividans* (see Fig. 11–47); maltoporin (PDB ID 1AF6), outer membrane phospholipase A (OmpLA, PDB ID 1QD5), OmpX (PDB ID 1QJ9), and phosphoporin E (PDB ID 1PHO) are proteins of the outer membrane of *E. coli*. Residues of Tyr and Trp are found predominantly where the nonpolar region of acyl chains meets the polar head group region. Charged residues (Lys, Arg, Glu, Asp) are found almost exclusively in the aqueous phases.

middle residue in each window—residue 4 for residues 1 to 7, for example). A region with more than 20 residues of high hydropathy index is presumed to be a transmembrane segment. When the sequences of membrane proteins of known three-dimensional structure are scanned in this way, we find a reasonably good correspondence between predicted and known membrane-spanning segments. Hydropathy analysis predicts a single hydrophobic helix for glycophorin (Fig. 11–12a) and seven transmembrane segments for bacteriorhodopsin (Fig. 11–12b)—in agreement with experimental studies.

On the basis of their amino acid sequences and hydropathy plots, many of the transport proteins described in this chapter are believed to have multiple membrane-spanning helical regions—that is, they are type III or type IV integral proteins (Fig. 11–9). When predictions are consistent with chemical studies of protein localization (such as those described above for glycophorin and bacteriorhodopsin), the assumption that hydrophobic regions correspond to membrane-spanning domains is much better justified.

A further remarkable feature of many transmembrane proteins of known structure is the presence of Tyr and Trp residues at the interface between lipid and water (**Fig. 11–13**). The side chains of these residues apparently serve as membrane interface anchors, able to interact simultaneously with the central lipid phase and the aqueous phases on either side of the membrane. Another generalization about amino acid location relative to the bilayer is described by the **positive-inside rule**: the positively charged Lys, His, and Arg residues of membrane proteins occur more commonly on the cytoplasmic face of membranes.

Not all integral membrane proteins are composed of transmembrane α helices. Another structural motif common in bacterial membrane proteins is the β **barrel**

(see Fig. 4–18b), in which 20 or more transmembrane segments form β sheets that line a cylinder (**Fig. 11–14**). The same factors that favor α-helix formation in the hydrophobic interior of a lipid bilayer also stabilize β barrels: when no water molecules are available to hydrogen-bond with the carbonyl oxygen and nitrogen of the peptide bond, maximal intrachain hydrogen bonding gives the most stable conformation. Planar β sheets do not maximize these interactions and are generally not found in the membrane interior; β barrels allow all possible hydrogen bonds and are apparently common among membrane proteins. **Porins**, proteins that allow certain polar solutes to cross the outer membrane of gram-negative bacteria such as *E. coli*, have many-stranded β barrels lining the polar transmembrane passage. The outer membranes of mitochondria and chloroplasts also contain a variety of β barrels.

A polypeptide is more extended in the β conformation than in an α helix; just seven to nine residues of β conformation are needed to span a membrane.

FepA   OmpLA   Maltoporin

FIGURE 11–14 Membrane proteins with β-barrel structure. Three proteins of the *E. coli* outer membrane are shown, viewed in the plane of the membrane. FepA (PDB ID 1FEP), involved in iron uptake, has 22 membrane-spanning β strands. OmpLA (derived from PDB ID 1QD5), a phospholipase, is a 12-stranded β barrel that exists as a dimer in the membrane. Maltoporin (derived from PDB ID 1MAL), a maltose transporter, is a trimer; each monomer consists of 16 β strands.

Recall that in the β conformation, alternating side chains project above and below the sheet (see Fig. 4–6). In β strands of membrane proteins, every second residue in the membrane-spanning segment is hydrophobic and interacts with the lipid bilayer; aromatic side chains are commonly found at the lipid-protein interface. The other residues may or may not be hydrophilic. The hydropathy plot is not useful in predicting transmembrane segments for proteins with β barrel motifs, but as the database of known β-barrel motifs increases, sequence-based predictions of transmembrane β conformations have become feasible. For example, sequence analysis has correctly predicted that some outer membrane proteins of gram-negative bacteria (Fig. 11–14) contain β barrels.

Covalently Attached Lipids Anchor Some Membrane Proteins

Some membrane proteins contain one or more covalently linked lipids, which may be of several types: long-chain fatty acids, isoprenoids, sterols, or glycosylated derivatives of *phosphatidylinositol* (GPIs; **Fig. 11–15**). The attached lipid provides a hydrophobic anchor that inserts into the lipid bilayer and holds the protein at the membrane surface. The strength of the hydrophobic interaction between a bilayer and a single hydro-

carbon chain linked to a protein is barely enough to anchor the protein securely, but many proteins have more than one attached lipid moiety. Other interactions, such as ionic attractions between positively charged Lys residues in the protein and negatively charged lipid head groups, probably contribute to the stability of the attachment. The association of these lipid-linked proteins with the membrane is certainly weaker than that for integral membrane proteins and is, at least in the case of cysteine palmitoylation, reversible.

Beyond merely anchoring a protein to the membrane, the attached lipid may have a more specific role. In the plasma membrane, proteins with GPI anchors are exclusively on the outer face and are clustered in certain regions, as discussed later in the chapter (p. 399), whereas other types of lipid-linked proteins (with farnesyl or geranylgeranyl groups attached; Fig. 11–15) are exclusively on the inner face. In polarized epithelial cells (such as intestinal epithelial cells; see Fig. 11–43), in which apical and basal surfaces have different roles, GPI-anchored proteins are directed specifically to the apical surface. Attachment of a specific lipid to a newly synthesized membrane protein therefore has a targeting function, directing the protein to its correct membrane location.

FIGURE 11–15 Lipid-linked membrane proteins. Covalently attached lipids anchor membrane proteins to the lipid bilayer. A palmitoyl group is shown attached by thioester linkage to a Cys residue; an *N*-myristoyl group is generally attached to an amino-terminal Gly, typically of a protein that also has a hydrophobic transmembrane segment; the farnesyl and geranylgeranyl groups attached to carboxyl-terminal Cys residues are isoprenoids of 15 and 20 carbons, respectively. The carboxyl-terminal Cys residue is invariably methylated. Glycosyl phosphatidylinositol (GPI) anchors are derivatives of phosphatidylinositol in which the inositol bears a short oligosaccharide covalently joined to the carboxyl-terminal residue of a protein through phosphoethanolamine. GPI-anchored proteins are always on the extracellular face of the plasma membrane. Farnesylated and palmitoylated membrane proteins are found on the inner surface of the plasma membrane, and myristoylated proteins have domains both inside and outside the plasma membrane.

SUMMARY 11.1 The Composition and Architecture of Membranes

▶ Biological membranes define cellular boundaries, divide cells into discrete compartments, organize complex reaction sequences, and act in signal reception and energy transformations.

▶ Membranes are composed of lipids and proteins in varying combinations particular to each species, cell type, and organelle. The lipid bilayer is the basic structural unit.

▶ Peripheral membrane proteins are loosely associated with the membrane through electrostatic interactions and hydrogen bonds or by covalently attached lipid anchors. Integral proteins associate firmly with membranes by hydrophobic interactions between the lipid bilayer and their nonpolar amino acid side chains, which are oriented toward the outside of the protein molecule. Amphitropic proteins associate reversibly with membranes.

▶ Many membrane proteins span the lipid bilayer several times, with hydrophobic sequences of about 20 amino acid residues forming transmembrane α helices. Multistranded β barrels are also common in integral proteins in bacterial membranes. Tyr and Trp residues of transmembrane proteins are commonly found at the lipid-water interface.

▶ The lipids and proteins of membranes are inserted into the bilayer with specific sidedness; thus membranes are structurally and functionally asymmetric. Plasma membrane glycoproteins are always oriented with the oligosaccharide-bearing domain on the extracellular surface.

11.2 Membrane Dynamics

One remarkable feature of all biological membranes is their flexibility—their ability to change shape without losing their integrity and becoming leaky. The basis for this property is the noncovalent interactions among lipids in the bilayer and the mobility allowed to individual lipids because they are not covalently anchored to one another. We turn now to the dynamics of membranes: the motions that occur and the transient structures allowed by these motions.

Acyl Groups in the Bilayer Interior Are Ordered to Varying Degrees

Although the lipid bilayer structure is stable, its individual phospholipid molecules have much freedom of motion **(Fig. 11–16)**, depending on the temperature and the lipid composition. Below normal physiological temperatures, the lipids in a bilayer form a semisolid **liquid-ordered (L_o) state**, in which all types of motion of individual lipid molecules are strongly constrained;

(a) Liquid-ordered state L_o

Heat produces thermal motion of side chains ($L_o \rightarrow L_d$ transition).

(b) Liquid-disordered state L_d

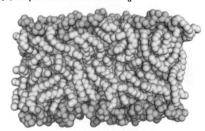

FIGURE 11–16 Two extreme states of bilayer lipids. (a) In the liquid-ordered (L_o) state, polar head groups are uniformly arrayed at the surface, and the acyl chains are nearly motionless and packed with regular geometry. **(b)** In the liquid-disordered (L_d) state, or fluid state, acyl chains undergo much thermal motion and have no regular organization. The state of membrane lipids in biological membranes is maintained somewhere between these extremes.

the bilayer is paracrystalline (Fig. 11–16a). Above physiological temperatures, individual hydrocarbon chains of fatty acids are in constant motion produced by rotation about the carbon–carbon bonds of the long acyl side chains and by lateral diffusion of individual lipid molecules in the plane of the bilayer. This is the **liquid-disordered (L_d) state** (Fig. 11–16b). In the transition from the L_o state to the L_d state, the general shape and dimensions of the bilayer are maintained; what changes is the degree of motion (lateral and rotational) allowed to individual lipid molecules.

At temperatures in the physiological range for a mammal (about 20 to 40 °C), long-chain saturated fatty acids (such as 16:0 and 18:0) tend to pack into an L_o gel phase, but the kinks in unsaturated fatty acids (see Fig. 10–2) interfere with packing, favoring the L_d state. Shorter-chain fatty acyl groups have the same effect. The sterol content of a membrane (which varies greatly with organism and organelle; Table 11–1) is another important determinant of lipid state. Sterols (such as cholesterol) have paradoxical effects on bilayer fluidity: they interact with phospholipids containing unsaturated fatty acyl chains, compacting them and constraining their motion in bilayers. Sterol association with sphingolipids and phospholipids with long, saturated fatty acyl chains tends, rather, to fluidize the bilayer, which, without

TABLE 11–2	Fatty Acid Composition of *E. coli* Cells Cultured at Different Temperatures			
	Percentage of total fatty acids*			
	10 °C	**20 °C**	**30 °C**	**40 °C**
Myristic acid (14:0)	4	4	4	8
Palmitic acid (16:0)	18	25	29	48
Palmitoleic acid (16:1)	26	24	23	9
Oleic acid (18:1)	38	34	30	12
Hydroxymyristic acid	13	10	10	8
Ratio of unsaturated to saturated[†]	2.9	2.0	1.6	0.38

Source: Data from Marr, A.G. & Ingraham, J.L. (1962) Effect of temperature on the composition of fatty acids in *Escherichia coli*. *J. Bacteriol.* 84, 1260.

*The exact fatty acid composition depends not only on growth temperature but on growth stage and growth medium composition.

[†]Ratios calculated as the total percentage of 16:1 plus 18:1 divided by the total percentage of 14:0 plus 16:0. Hydroxymyristic acid was omitted from this calculation.

cholesterol, would adopt the L_o state. In biological membranes composed of a variety of phospholipids and sphingolipids, cholesterol tends to associate with sphingolipids and to form regions in the L_o state surrounded by cholesterol-poor regions in the L_d state (see the discussion of membrane rafts below).

Cells regulate their lipid composition to achieve a constant membrane fluidity under various growth conditions. For example, bacteria synthesize more unsaturated fatty acids and fewer saturated ones when cultured at low temperatures than when cultured at higher temperatures (Table 11–2). As a result of this adjustment in lipid composition, membranes of bacteria cultured at high or low temperatures have about the same degree of fluidity. This is presumably essential for the function of many proteins—enzymes, transporters, and receptors—that act within the lipid bilayer.

Transbilayer Movement of Lipids Requires Catalysis

At physiological temperatures, transbilayer—or "flip-flop"—diffusion of a lipid molecule from one leaflet of the bilayer to the other **(Fig. 11–17a)** occurs very slowly if at all in most membranes, although lateral diffusion *in the plane* of the bilayer is very rapid (Fig. 11–17b). Transbilayer movement requires that a polar or charged head group leave its aqueous environment and move into the hydrophobic interior of the bilayer, a process with a large, positive free-energy change. There are, however, situations in which such movement is essential. For example, in the ER, membrane glycerophospholipids are synthesized on the cytosolic surface, whereas sphingolipids are synthesized or modified on the lumenal surface. To get from their site of synthesis to their eventual point of deposition, these lipids must undergo flip-flop diffusion.

The asymmetric disposition of lipid types in the bilayer predicts the existence of flippases, floppases, and scramblases (Fig. 11–17c), which facilitate the transbilayer movement of lipids, providing a path that is energetically more favorable and much faster than the uncatalyzed movement. The combination of asymmetric

biosynthesis of membrane lipids, very slow uncatalyzed flip-flop diffusion, and the presence of selective, energy-dependent lipid translocators could account for the transbilayer asymmetry in lipid composition shown in Figure 11–5. Besides contributing to this asymmetry of composition, the energy-dependent transport of lipids to one bilayer leaflet may, by creating a larger surface on one side of the bilayer, be important in generating the membrane curvature essential in the budding of vesicles.

(a) Uncatalyzed transbilayer ("flip-flop") diffusion

very slow
($t_{1/2}$ in days)

(b) Uncatalyzed lateral diffusion

very fast
(1 µm/s)

(c) Catalyzed transbilayer translocations

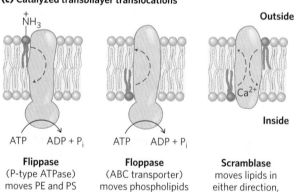

| **Flippase**
(P-type ATPase)
moves PE and PS
from outer to
cytosolic leaflet | **Floppase**
(ABC transporter)
moves phospholipids
from cytosolic to
outer leaflet | **Scramblase**
moves lipids in
either direction,
toward equilibrium |

FIGURE 11–17 Motion of single phospholipids in a bilayer. (a) Uncatalyzed movement from one leaflet to the other is very slow, but **(b)** lateral diffusion within the leaflet is very rapid, requiring no catalysis. **(c)** Three types of phospholipid translocaters in the plasma membrane. PE is phosphatidylethanolamine; PS is phosphatidylserine.

Flippases catalyze translocation of the *amino*-phospholipids phosphatidylethanolamine and phosphatidylserine from the extracellular to the cytosolic leaflet of the plasma membrane, contributing to the asymmetric distribution of phospholipids: phosphatidylethanolamine and phosphatidylserine primarily in the cytosolic leaflet, and the sphingolipids and phosphatidylcholine in the outer leaflet. Keeping phosphatidylserine out of the extracellular leaflet is important: its exposure on the outer surface triggers apoptosis (programmed cell death; see Chapter 12) and engulfment by macrophages that carry phosphatidylserine receptors. Flippases also act in the ER, where they move newly synthesized phospholipids from their site of synthesis in the cytosolic leaflet to the lumenal leaflet. Flippases consume about one ATP per molecule of phospholipid translocated, and they are structurally and functionally related to the P-type ATPases (active transporters) described on page 410.

Two other types of lipid-translocating activities are known but less well characterized. **Floppases** move plasma membrane phospholipids from the cytosolic to the extracellular leaflet and like flippases are ATP-dependent. Floppases are members of the ABC transporter family described on page 413, all of which actively transport hydrophobic substrates outward across the plasma membrane. **Scramblases** are proteins that move any membrane phospholipid across the bilayer down its concentration gradient (from the leaflet where it has a higher concentration to the leaflet where it has a lower concentration); their activity is not dependent on ATP. Scramblase activity leads to controlled randomization of the head-group composition on the two faces of the bilayer. The activity rises sharply with an increase in cytosolic Ca^{2+} concentration, which may result from cell activation, cell injury, or apoptosis; as noted above, exposure of phosphatidylserine on the outer surface marks a cell for apoptosis and engulfment by macrophages. Finally, a group of proteins that act primarily to move phosphatidylinositol lipids across lipid bilayers, the phosphatidylinositol transfer proteins, are believed to have important roles in lipid signaling and membrane trafficking.

Lipids and Proteins Diffuse Laterally in the Bilayer

Individual lipid molecules can move laterally in the plane of the membrane by changing places with neighboring lipid molecules; that is, they undergo Brownian movement within the bilayer (Fig. 11–17b), which can be quite rapid. A molecule in the outer leaflet of the erythrocyte plasma membrane, for example, can diffuse laterally so fast that it circumnavigates the erythrocyte in seconds. This rapid lateral diffusion in the plane of the bilayer tends to randomize the positions of individual molecules in a few seconds.

Lateral diffusion can be shown experimentally by attaching fluorescent probes to the head groups of lipids and using fluorescence microscopy to follow the probes over time **(Fig. 11–18)**. In one technique, a small region (5 μm^2) of a cell surface with fluorescence-tagged lipids

FIGURE 11–18 Measurement of lateral diffusion rates of lipids by fluorescence recovery after photobleaching (FRAP). Lipids in the outer leaflet of the plasma membrane are labeled by reaction with a membrane-impermeant fluorescent probe (red) so that the surface is uniformly labeled when viewed with a fluorescence microscope. A small area is bleached by irradiation with an intense laser beam and becomes nonfluorescent. With the passage of time, labeled lipid molecules diffuse into the bleached region, and it again becomes fluorescent. Researchers can track the time course of fluorescence return and determine a diffusion coefficient for the labeled lipid. The diffusion rates are typically high; a lipid moving at this speed could circumnavigate an *E. coli* cell in one second. (The FRAP method can also be used to measure lateral diffusion of membrane proteins.)

Cell

Fluorescent probe on lipids

React cell with fluorescent probe to label lipids

View surface with fluorescence microscope

Intense laser beam bleaches small area

With time, unbleached phospholipids diffuse into bleached area

Measure rate of fluorescence return

is bleached by intense laser radiation so that the irradiated patch no longer fluoresces when viewed with less-intense (nonbleaching) light in the fluorescence microscope. However, within milliseconds, the region recovers its fluorescence as unbleached lipid molecules diffuse into the bleached patch and bleached lipid molecules diffuse away from it. The rate of *fluorescence recovery after photobleaching*, or **FRAP**, is a measure of the rate of lateral diffusion of the lipids. Using the FRAP technique, researchers have shown that some membrane lipids diffuse laterally at rates of up to 1 μm/s.

Another technique, single particle tracking, allows one to follow the movement of a *single* lipid molecule in the plasma membrane on a much shorter time scale. Results from these studies confirm rapid lateral diffusion within small, discrete regions of the cell surface and show that movement from one such region to a nearby region ("hop diffusion") is inhibited; membrane lipids behave as though corralled by fences that they can occasionally cross by hop diffusion **(Fig. 11–19)**.

Many membrane proteins move as if afloat in a sea of lipids. Like membrane lipids, these proteins are free to diffuse laterally in the plane of the bilayer and are in constant motion, as shown by the FRAP technique with fluorescence-tagged surface proteins. Some membrane proteins associate to form large aggregates ("patches") on the surface of a cell or organelle in which individual protein molecules do not move relative to one another; for example, acetylcholine receptors form dense, near-crystalline patches on neuronal plasma membranes at synapses. Other membrane proteins are anchored to

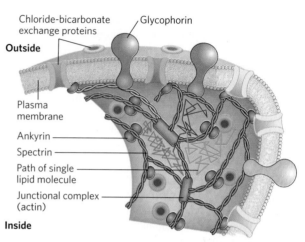

FIGURE 11–20 Restricted motion of the erythrocyte chloride-bicarbonate exchanger and glycophorin. The proteins span the membrane and are tethered to spectrin, a cytoskeletal protein, by another protein, ankyrin, limiting their lateral mobility. Ankyrin is anchored in the membrane by a covalently bound palmitoyl side chain (see Fig. 11–15). Spectrin, a long, filamentous protein, is cross-linked at junctional complexes containing actin. A network of cross-linked spectrin molecules attached to the cytoplasmic face of the plasma membrane stabilizes the membrane, making it resistant to deformation. This network of anchored membrane proteins may form the "corral" suggested by the experiment shown in Figure 11–19; the lipid tracks shown here are confined to different regions defined by the tethered membrane proteins. Occasionally a lipid molecule (green track) jumps from one corral to another (blue track), then another (red track).

internal structures that prevent their free diffusion. In the erythrocyte membrane, both glycophorin and the chloride-bicarbonate exchanger (p. 407) are tethered to spectrin, a filamentous cytoskeletal protein **(Fig. 11–20)**. One possible explanation for the pattern of lateral diffusion of lipid molecules shown in Figure 11–19 is that membrane proteins immobilized by their association with spectrin form the "fences" that define the regions of relatively unrestricted lipid motion.

Sphingolipids and Cholesterol Cluster Together in Membrane Rafts

We have seen that diffusion of membrane lipids from one bilayer leaflet to the other is very slow unless catalyzed and that the different lipid species of the plasma membrane are asymmetrically distributed in the two leaflets of the bilayer (Fig. 11–5). Even within a single leaflet, the lipid distribution is not uniform. Glycosphingolipids (cerebrosides and gangliosides), which typically contain long-chain saturated fatty acids, form transient clusters in the outer leaflet that largely exclude glycerophospholipids, which typically contain one unsaturated fatty acyl group and a shorter saturated acyl group. The long, saturated acyl groups of sphingolipids can form more compact, more stable associations with the long ring system of cholesterol than can the shorter, often unsaturated, chains of phospholipids. The cholesterol-sphingolipid **microdomains** in the outer monolayer of the plasma

Finish   Start

0.1 μm

FIGURE 11–19 Hop diffusion of individual lipid molecules. The motion of a single fluorescently labeled lipid molecule in a cell surface is recorded on video by fluorescence microscopy, with a time resolution of 25 μs (equivalent to 40,000 frames/s). The track shown here represents a molecule followed for 56 ms (2,250 frames); the trace begins in the purple area and continues through blue, green, and orange. The pattern of movement indicates rapid diffusion within a confined region (about 250 nm in diameter, shown by a single color), with occasional hops into an adjoining region. This finding suggests that the lipids are corralled by molecular fences that they occasionally jump.

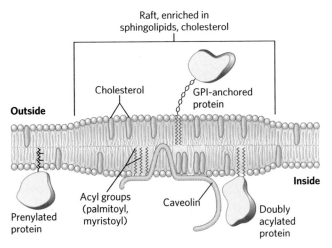

FIGURE 11–21 Membrane microdomains (rafts). Stable associations of sphingolipids and cholesterol in the outer leaflet produce a microdomain, slightly thicker than other membrane regions, that is enriched with specific types of membrane proteins. GPI-anchored proteins are prominent in the outer leaflet of these rafts, and proteins with one or several covalently attached long-chain acyl groups are common in the inner leaflet. Inwardly curved rafts called caveolae are especially enriched in the protein caveolin (see Fig. 11–22). Proteins with attached prenyl groups (such as Ras; see Box 12-2) tend to be excluded from rafts.

membrane are slightly thicker and more ordered (less fluid) than neighboring microdomains rich in phospholipids and are more difficult to dissolve with nonionic detergents; they behave like liquid-ordered sphingolipid **rafts** adrift on an ocean of liquid-disordered phospholipids **(Fig. 11–21)**.

These lipid rafts are remarkably enriched in two classes of integral membrane proteins: those anchored to the membrane by two covalently attached long-chain saturated fatty acids attached through Cys residues (two palmitoyl groups or a palmitoyl and a myristoyl group) and **GPI-anchored proteins** (Fig. 11–15). Presumably these lipid anchors, like the long, saturated acyl chains of sphingolipids, form more stable associations with the cholesterol and long acyl groups in rafts than with the surrounding phospholipids. (It is notable that other lipid-linked proteins, those with covalently attached isoprenyl groups such as farnesyl, are *not* preferentially associated with the outer leaflet of sphingolipid/cholesterol rafts (Fig. 11–21).) The "raft" and "sea" domains of the plasma membrane are not rigidly separated; membrane proteins can move into and out of lipid rafts on a time scale of seconds. But in the shorter time scale (microseconds) more relevant to many membrane-mediated biochemical processes, many of these proteins reside primarily in a raft.

We can estimate the fraction of the cell surface occupied by rafts from the fraction of the plasma membrane that resists detergent solubilization, which can be as high as 50% in some cases: the rafts cover half of the ocean. Indirect measurements in cultured fibroblasts suggest a diameter of roughly 50 nm for an individual raft, which corresponds to a patch containing a few

thousand sphingolipids and perhaps 10 to 50 membrane proteins. Because most cells express more than 50 different kinds of plasma membrane proteins, it is likely that a single raft contains only a subset of membrane proteins and that this segregation of membrane proteins is functionally significant. For a process that involves interaction of two membrane proteins, their presence in a single raft would hugely increase the likelihood of their collision. Certain membrane receptors and signaling proteins, for example, seem to be segregated together in membrane rafts. Experiments show that signaling through these proteins can be disrupted by manipulations that deplete the plasma membrane of cholesterol and destroy lipid rafts.

Caveolin is an integral membrane protein with two globular domains connected by a hairpin-shaped hydrophobic domain, which binds the protein to the cytoplasmic leaflet of the plasma membrane. Three palmitoyl groups attached to the carboxyl-terminal globular domain further anchor it to the membrane. Caveolin (actually, a family of related caveolins) forms dimers and associates with cholesterol-rich regions in the membrane, and the presence of caveolin dimers forces the associated lipid bilayer to curve inward, forming **caveolae** ("little caves") in the surface of the cell **(Fig. 11–22)**. Caveolae are unusual rafts: they involve *both* leaflets of the bilayer—the cytoplasmic leaflet, from which the caveolin globular domains project, and the extracellular leaflet, a typical sphingolipid/cholesterol raft with associated GPI-anchored proteins. Caveolae are implicated in a variety of cellular functions, including membrane trafficking within cells and the transduction of external signals into cellular responses. The receptors for insulin and other growth factors, as well as certain GTP-binding proteins and protein kinases associated with transmembrane signaling, seem to be localized in rafts and perhaps in caveolae. We discuss some possible roles of rafts in signaling in Chapter 12.

Membrane Curvature and Fusion Are Central to Many Biological Processes

Caveolin is not unique in its ability to induce curvature in membranes. Changes of curvature are central to one of the most remarkable features of biological membranes: their ability to undergo fusion with other membranes without losing their continuity. Although membranes are stable, they are by no means static. Within the eukaryotic endomembrane system (which includes the nuclear membrane, endoplasmic reticulum, Golgi complex, and various small vesicles), the membranous compartments constantly reorganize. Vesicles bud from the ER to carry newly synthesized lipids and proteins to other organelles and to the plasma membrane. Exocytosis, endocytosis, cell division, fusion of egg and sperm cells, and entry of a membrane-enveloped virus into its host cell all involve membrane reorganization in which the fundamental operation is fusion of two membrane

(a)

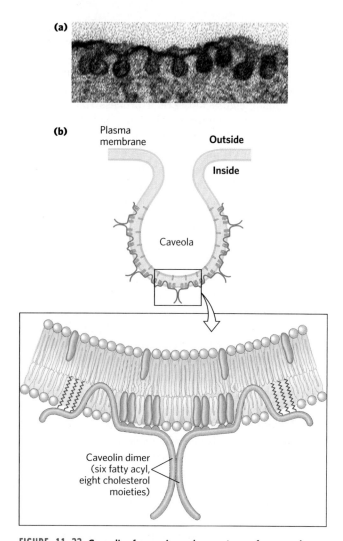

(b)

FIGURE 11-22 **Caveolin forces inward curvature of a membrane.** Caveolae are small invaginations in the plasma membrane, as seen in **(a)** an electron micrograph of an adipocyte that is surface-labeled with an electron-dense marker. **(b)** Cartoon showing the location and role of caveolin in causing inward membrane curvature. Each caveolin monomer has a central hydrophobic domain and three long-chain acyl groups (red), which hold the molecule to the inside of the plasma membrane. When several caveolin dimers are concentrated in a small region (a raft), they force a curvature in the lipid bilayer, forming a caveola. Cholesterol molecules in the bilayer are shown in orange.

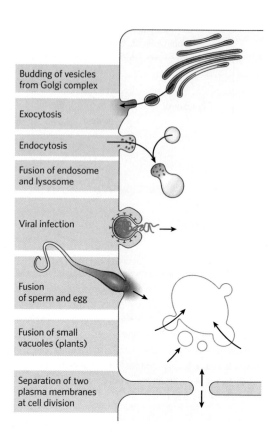

FIGURE 11-23 **Membrane fusion.** The fusion of two membranes is central to a variety of cellular processes involving organelles and the plasma membrane.

segments without loss of continuity **(Fig. 11-23).** Most of these processes begin with a local increase in membrane curvature. A protein that is intrinsically curved may force curvature in a bilayer by binding to it **(Fig. 11-24)**; the binding energy provides the driving force for the increase in bilayer curvature. Alternatively, multiple subunits of a scaffold protein may assemble into curved supramolecular complexes and stabilize curves that spontaneously form in the bilayer. For example, a superfamily of proteins containing **BAR domains** (named for the first three members of the family to be identified: *B*IN1, *a*mphiphysin, and *R*VS167) can assemble into a crescent-shaped scaffold that binds to the membrane surface, forcing or favoring membrane curvature.

BAR domains consist of coiled coils that form long, thin, curved dimers with a positively charged concave surface that tends to form ionic interactions with the negatively charged head groups of membrane phospholipids (Fig. 11-24). Some of these BAR proteins also have a helical region that inserts into one leaflet of the bilayer, expanding its area relative to the other leaflet and thereby forcing curvature.

Specific fusion of two membranes requires that (1) they recognize each other; (2) their surfaces become closely apposed, which requires the removal of water molecules normally associated with the polar head groups of lipids; (3) their bilayer structures become locally disrupted, resulting in fusion of the outer leaflet of each membrane (hemifusion); and (4) their bilayers fuse to form a single continuous bilayer. The fusion occurring in receptor-mediated endocytosis, or regulated secretion, also requires that (5) the process is triggered at the appropriate time or in response to a specific signal. Integral proteins called **fusion proteins** mediate these events, bringing about specific recognition and a transient local distortion of the bilayer structure that favors membrane fusion. (Note that these fusion proteins are unrelated to the products encoded by two fused genes, also called fusion proteins, discussed in Chapter 9.)

A well-studied example of membrane fusion is that occurring at synapses, when intracellular vesicles loaded

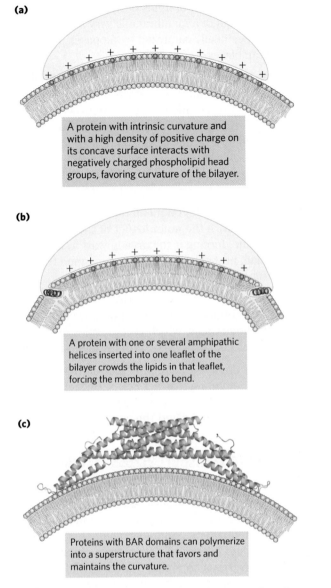

(a)

A protein with intrinsic curvature and with a high density of positive charge on its concave surface interacts with negatively charged phospholipid head groups, favoring curvature of the bilayer.

(b)

A protein with one or several amphipathic helices inserted into one leaflet of the bilayer crowds the lipids in that leaflet, forcing the membrane to bend.

(c)

Proteins with BAR domains can polymerize into a superstructure that favors and maintains the curvature.

FIGURE 11–24 Three models for protein-induced curvature of membranes.

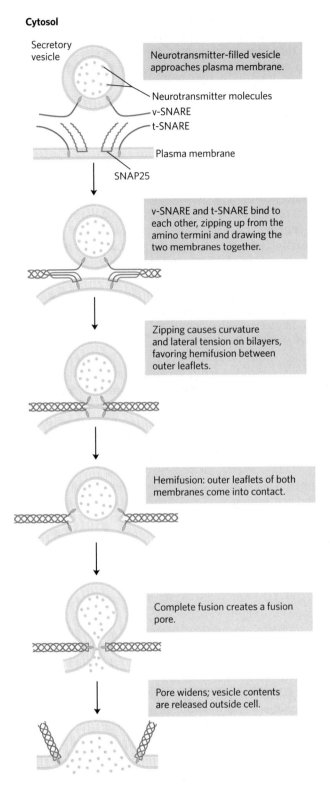

Cytosol

Secretory vesicle

Neurotransmitter-filled vesicle approaches plasma membrane.

Neurotransmitter molecules
v-SNARE
t-SNARE
Plasma membrane
SNAP25

v-SNARE and t-SNARE bind to each other, zipping up from the amino termini and drawing the two membranes together.

Zipping causes curvature and lateral tension on bilayers, favoring hemifusion between outer leaflets.

Hemifusion: outer leaflets of both membranes come into contact.

Complete fusion creates a fusion pore.

Pore widens; vesicle contents are released outside cell.

FIGURE 11–25 Membrane fusion during neurotransmitter release at a synapse. The secretory vesicle membrane contains the v-SNARE synaptobrevin (red). The target (plasma) membrane contains the t-SNAREs syntaxin (blue) and SNAP25 (violet). When a local increase in [Ca^{2+}] signals release of neurotransmitter, the v-SNARE, SNAP25, and t-SNARE interact, forming a coiled bundle of four α helices, pulling the two membranes together and disrupting the bilayer locally. This leads first to hemifusion, joining the outer leaflets of the two membranes, then to complete membrane fusion and neurotransmitter release. NSF (*N*-ethylmaleimide-sensitive *f*usion factor) acts in disassembly of the SNARE complex when fusion is complete.

with neurotransmitter fuse with the plasma membrane. This process involves a family of proteins called SNARES **(Fig. 11–25)**. SNAREs in the cytoplasmic face of the intracellular vesicle are called **v-SNAREs**; those in the target membrane with which the vesicle fuses (the plasma membrane during exocytosis) are **t-SNAREs**. Two other proteins, SNAP25 and NSF, are also involved. During fusion, a v-SNARE and t-SNARE bind to each other and undergo a structural change that produces a bundle of long, thin rods made up of helices from both SNARES and two helices from SNAP25 (Fig. 11–25). The two SNAREs initially interact at their ends, then zip up into the bundle of helices. This structural change pulls the two membranes into contact and initiates the fusion of their lipid bilayers.

The complex of SNAREs and SNAP25 is the target of the powerful *Clostridium botulinum* toxin, a protease that cleaves specific bonds in these proteins, preventing neurotransmission and thereby causing the death of the organism. Because of its very high specificity for these proteins, purified botulinum toxin has served as a powerful tool for dissecting the mechanism of neurotransmitter release in vivo and in vitro.

Integral Proteins of the Plasma Membrane Are Involved in Surface Adhesion, Signaling, and Other Cellular Processes

Several families of integral proteins in the plasma membrane provide specific points of attachment between cells or between a cell and extracellular matrix proteins. **Integrins** are surface adhesion proteins that mediate a cell's interaction with the extracellular matrix and with other cells, including some pathogens. Integrins also carry signals in both directions across the plasma membrane, integrating information about the extracellular and intracellular environments. All integrins are heterodimeric proteins composed of two unlike subunits, α and β, each anchored to the plasma membrane by a single transmembrane helix. The large extracellular domains of the α and β subunits combine to form a specific binding site for extracellular proteins such as collagen and fibronectin, which contain a common determinant of integrin binding, the sequence Arg–Gly–Asp (RGD). We discuss the signaling functions of integrins in more detail in Chapter 12 (p. 470).

Other plasma membrane proteins involved in surface adhesion are the **cadherins**, which undergo homophilic ("with same kind") interactions with identical cadherins in an adjacent cell. **Selectins** have extracellular domains that, in the presence of Ca^{2+}, bind specific polysaccharides on the surface of an adjacent cell. Selectins are present primarily in the various types of blood cells and in the endothelial cells that line blood vessels (see Fig. 7–32). They are an essential part of the blood-clotting process.

Integral membrane proteins play roles in many other cellular processes. They serve as transporters and ion channels (discussed in Section 11.3) and as receptors for hormones, neurotransmitters, and growth factors (Chapter 12). They are central to oxidative phosphorylation and photophosphorylation (Chapter 19) and to cell-cell and cell-antigen recognition in the immune system (Chapter 5). Integral proteins are also important players in the membrane fusion that accompanies exocytosis, endocytosis, and the entry of many types of viruses into host cells.

SUMMARY 11.2 Membrane Dynamics

▶ Lipids in a biological membrane can exist in liquid-ordered or liquid-disordered states; in the latter state, thermal motion of acyl chains makes the interior of the bilayer fluid. Fluidity is affected by temperature, fatty acid composition, and sterol content.

▶ Flip-flop diffusion of lipids between the inner and outer leaflets of a membrane is very slow except when specifically catalyzed by flippases, floppases, or scramblases.

▶ Lipids and proteins can diffuse laterally within the plane of the membrane, but this mobility is limited by interactions of membrane proteins with internal cytoskeletal structures and interactions of lipids with lipid rafts. One class of lipid rafts consists of sphingolipids and cholesterol with a subset of membrane proteins that are GPI-linked or attached to several long-chain fatty acyl moieties.

▶ Caveolin is an integral membrane protein that associates with the inner leaflet of the plasma membrane, forcing it to curve inward to form caveolae, probably involved in membrane transport and signaling.

▶ Specific proteins containing BAR domains cause local membrane curvature and mediate the fusion of two membranes, which accompanies processes such as endocytosis, exocytosis, and viral invasion.

▶ Integrins are transmembrane proteins of the plasma membrane that act both to attach cells to each other and to carry messages between the extracellular matrix and the cytoplasm.

11.3 Solute Transport across Membranes

Every living cell must acquire from its surroundings the raw materials for biosynthesis and for energy production, and must release the byproducts of metabolism to its environment. A few nonpolar compounds can dissolve in the lipid bilayer and cross the membrane unassisted, but for transmembrane movement of any polar compound or ion, a membrane protein is essential. In some cases a membrane protein simply facilitates the diffusion of a solute down its concentration gradient, but transport can also occur against a gradient of concentration, electric charge, or both, in which case the process requires energy **(Fig. 11–26)**. The energy may come directly from ATP hydrolysis or may be supplied in the form of one solute moving down its electrochemical gradient, which provides sufficient energy to drive another solute up its gradient. Ions may also move across membranes via ion channels formed by proteins, or they may be carried across by ionophores, small molecules that mask the charge of ions and allow them to diffuse through the lipid bilayer. With very few exceptions, the traffic of small molecules across the plasma membrane is mediated by proteins such as transmembrane channels, carriers, or pumps. Within the eukaryotic cell, different compartments have different concentrations of ions and of metabolic intermediates and products, and these, too, must move across intracellular membranes in tightly regulated, protein-mediated processes.

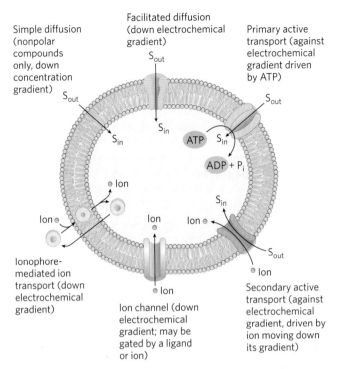

Simple diffusion (nonpolar compounds only, down concentration gradient)

Facilitated diffusion (down electrochemical gradient)

Primary active transport (against electrochemical gradient driven by ATP)

S_{out} S_{out} S_{in} S_{in} S_{in} ATP S_{in} ADP + P_i S_{in} S_{out}

Ion

Ion

Ion Ion Ion Ion Ion

Ionophore-mediated ion transport (down electrochemical gradient)

Ion channel (down electrochemical gradient; may be gated by a ligand or ion)

Secondary active transport (against electrochemical gradient, driven by ion moving down its gradient)

FIGURE 11–26 Summary of transporter types. Some types (ionophores, ion channels, and passive transporters) simply speed transmembrane movement of solutes down their electrochemical gradients, whereas others (active transporters) can pump solutes against a gradient, using ATP or a gradient of a second solute to provide the energy.

Passive Transport Is Facilitated by Membrane Proteins

When two aqueous compartments containing unequal concentrations of a soluble compound or ion are separated by a permeable divider (membrane), the solute moves by **simple diffusion** from the region of higher concentration, through the membrane, to the region of lower concentration, until the two compartments have equal solute concentrations **(Fig. 11–27a)**. When ions of opposite charge are separated by a permeable membrane, there is a transmembrane electrical gradient, a **membrane potential, V_m** (expressed in millivolts). This membrane potential produces a force opposing ion movements that increase V_m and driving ion movements that reduce V_m (Fig. 11–27b). Thus, the direction in which a charged solute tends to move spontaneously across a membrane depends on both the chemical gradient (the difference in solute concentration) and the electrical gradient (V_m) across the membrane. Together these two factors are referred to as the **electrochemical gradient** or **electrochemical potential**. This behavior of solutes is in accord with the second law of thermodynamics: molecules tend to spontaneously assume the distribution of greatest randomness and lowest energy.

To pass through a lipid bilayer, a polar or charged solute must first give up its interactions with the water molecules in its hydration shell, then diffuse about 3 nm (30 Å) through a substance (lipid) in which it is poorly soluble **(Fig. 11–28)**. The energy used to strip away the hydration shell and to move the polar compound from water into lipid, then through the lipid bilayer, is regained as the compound leaves the membrane on the other side and is rehydrated. However, the intermediate stage of transmembrane passage is a high-energy state comparable to the transition state in an enzyme-catalyzed chemical reaction. In both cases, an activation barrier must be overcome to reach the intermediate stage (Fig. 11–28; compare with Fig. 6–3). The energy of activation ($\Delta G^{\ddagger}$) for translocation of a polar solute across the bilayer is so large that pure lipid bilayers are virtually impermeable to polar and charged species over periods relevant to cell growth and division.

Membrane proteins lower the activation energy for transport of polar compounds and ions by providing an alternative path across the membrane for specific solutes. Proteins that bring about this **facilitated diffusion**, or

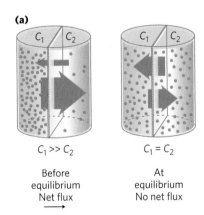

(a)

C_1 C_2 C_1 C_2

$C_1 \gg C_2$ $C_1 = C_2$

Before equilibrium Net flux

At equilibrium No net flux

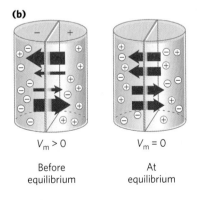

(b)

$V_m > 0$ $V_m = 0$

Before equilibrium

At equilibrium

FIGURE 11–27 Movement of solutes across a permeable membrane. **(a)** Net movement of an electrically neutral solute is toward the side of lower solute concentration until equilibrium is achieved. The solute concentrations on the left and right sides of the membrane are designated C_1 and C_2. The rate of transmembrane solute movement (indicated by the arrows) is proportional to the concentration ratio. **(b)** Net movement of an electrically charged solute is dictated by a combination of the electrical potential (V_m) and the ratio of chemical concentrations (C_2/C_1) across the membrane; net ion movement continues until this electrochemical potential reaches zero.

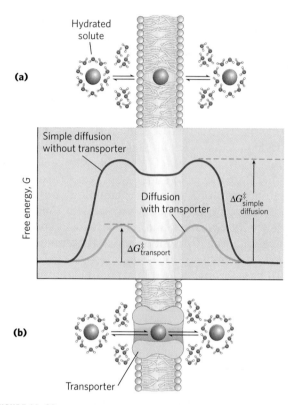

(a)

Hydrated solute

(b)

Transporter

FIGURE 11–28 Energy changes accompanying passage of a hydrophilic solute through the lipid bilayer of a biological membrane. (a) In simple diffusion, removal of the hydration shell is highly endergonic, and the energy of activation ($\Delta G^{\ddagger}$) for diffusion through the bilayer is very high. **(b)** A transporter protein reduces the $\Delta G^{\ddagger}$ for transmembrane diffusion of the solute. It does this by forming noncovalent interactions with the dehydrated solute to replace the hydrogen bonding with water and by providing a hydrophilic transmembrane pathway.

passive transport, are not enzymes in the usual sense; their "substrates" are moved from one compartment to another but are not chemically altered. Membrane proteins that speed the movement of a solute across a membrane by facilitating diffusion are called **transporters** or **permeases**.

Like enzymes, transporters bind their substrates with stereochemical specificity through multiple weak, noncovalent interactions. The negative free-energy change associated with these weak interactions, $\Delta G_{binding}$, counterbalances the positive free-energy change that accompanies loss of the water of hydration from the substrate, $\Delta G_{dehydration}$, thereby lowering $\Delta G^{\ddagger}$ for transmembrane passage (Fig. 11–28). Transporters span the lipid bilayer several times, forming a transmembrane pathway lined with hydrophilic amino acid side chains. The pathway provides an alternative route for a specific substrate to move across the lipid bilayer without its having to dissolve in the bilayer, further lowering $\Delta G^{\ddagger}$ for transmembrane diffusion. The result is an increase of several to many orders of magnitude in the rate of transmembrane passage of the substrate.

Transporters and Ion Channels Are Fundamentally Different

We know from genomic studies that transporters constitute a significant fraction of all proteins encoded in the genomes of both simple and complex organisms. There are probably a thousand or more different genes in the human genome encoding proteins that allow molecules and ions to cross membranes. These proteins fall within two very broad categories: transporters and channels (**Fig. 11–29**). **Transporters** for molecules and ions bind their substrates with high specificity, catalyze transport at rates well below the limits of free diffusion, and are saturable in the same sense as are enzymes: there is some substrate concentration above which further increases will not produce a greater rate of transport. **Channels** generally allow transmembrane movement of ions at rates that are orders of magnitude greater than those typical of transporters, approaching the limit of unhindered diffusion (tens of millions of ions per second per channel). Channels typically show some specificity for an ion, but are not saturable with the ion substrate, in contrast to the saturation kinetics seen with transporters. The direction of ion movement through an ion channel is dictated by the ion's charge and the electrochemical

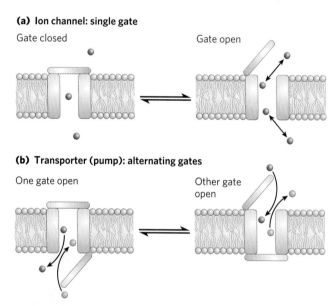

(a) Ion channel: single gate

Gate closed   Gate open

(b) Transporter (pump): alternating gates

One gate open   Other gate open

FIGURE 11–29 Differences between channels and transporters. (a) In an ion channel, a transmembrane pore is either open or closed, depending on the position of the single gate. When it is open, ions move through at a rate limited only by the maximum rate of diffusion. **(b)** Transporters (pumps) have two gates, and they are never both open. Movement of a substrate (an ion or a small molecule) through the membrane is therefore limited by the time needed for one gate to open and close (on one side of the membrane) and for the second gate to open. Rates of movement through ion channels can be orders of magnitude greater than rates through pumps, but channels simply allow the ion to flow down the electrochemical gradient, whereas pumps can move a substrate against a concentration gradient.

gradient across the membrane. Within each of these categories are families of various types, defined not only by their primary sequences but by their secondary structures. Among the transporters, some simply facilitate diffusion down a concentration gradient; they are the **passive transporters**. **Active transporters** can drive substrates across the membrane against a concentration gradient, some using energy provided directly by a chemical reaction (primary active transporters) and some coupling uphill transport of one substrate with downhill transport of another (secondary active transporters). We now consider some well-studied representatives of the main transporter and channel families. You will encounter some of these in Chapter 12 when we discuss transmembrane signaling and again in later chapters in the context of the metabolic pathways in which they participate.

The Glucose Transporter of Erythrocytes Mediates Passive Transport

Energy-yielding metabolism in erythrocytes depends on a constant supply of glucose from the blood plasma, where the glucose concentration is maintained at about 5 mM. Glucose enters the erythrocyte by facilitated diffusion via a specific glucose transporter, at a rate about 50,000 times greater than uncatalyzed transmembrane diffusion. The glucose transporter of erythrocytes (called GLUT1 to distinguish it from related glucose transporters in other tissues) is a type III integral protein

(M_r ~45,000) with 12 hydrophobic segments, each of which is believed to form a membrane-spanning helix. The detailed structure of GLUT1 is not yet known, but one plausible model suggests that the side-by-side assembly of several helices produces a transmembrane channel lined with hydrophilic residues that can hydrogen-bond with glucose as it moves through the aqueous pore **(Fig. 11–30)**.

The process of glucose transport can be described by its analogy with an enzymatic reaction in which the "substrate" is glucose outside the cell (S_{out}), the "product" is glucose inside (S_{in}), and the "enzyme" is the transporter, T. When the initial rate of glucose uptake is measured as a function of external glucose concentration **(Fig. 11–31)**, the resulting plot is hyperbolic: at high external glucose concentrations the rate of uptake approaches V_{max}. Formally, such a transport process can be described by the equations

$$S_{out} + T_1 \underset{k_{-1}}{\overset{k_1}{\rightleftharpoons}} S_{out} \bullet T_1$$

$$k_{-4} {\Big\Vert} k_4 \qquad\qquad k_{-2} {\Big\Vert} k_2$$

$$S_{in} + T_2 \underset{k_{-3}}{\overset{k_3}{\rightleftharpoons}} S_{in} \bullet T_2$$

in which k_1, k_{-1}, and so forth are the forward and reverse rate constants for each step; T_1 is the transporter conformation in which the glucose-binding site faces out, and T_2 is the conformation in which it faces in.

(a)

(b)

−Ser −Leu −Val −Thr −Asn−Phe −Ile −

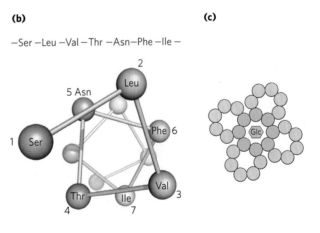

(c)

FIGURE 11–30 Membrane topology of the glucose transporter GLUT1. **(a)** Transmembrane helices are represented here as oblique (angled) rows of three or four amino acid residues, each row depicting one turn of the α helix. Nine of the 12 helices contain three or more polar or charged residues (blue or red), often separated by several hydrophobic residues (yellow). This representation of topology is not intended to depict three-dimensional structure. **(b)** A helical wheel diagram shows the distribution of polar and nonpolar residues on the surface of a helical segment. The helix is diagrammed as though observed along its axis from the amino terminus. Adjacent residues in the linear sequence are connected, and each residue is placed around the wheel in the position it occupies in the helix; recall that 3.6 residues are required to make one complete turn of the α helix. In this example, the polar residues (blue) are on one side of the helix and the hydrophobic residues (yellow) on the other. This is, by definition, an amphipathic helix. **(c)** Side-by-side association of four amphipathic helices, each with its polar face oriented toward the central cavity, can produce a transmembrane channel lined with polar (and charged) residues. This channel provides many opportunities for hydrogen bonding with glucose as it moves through.

(a)

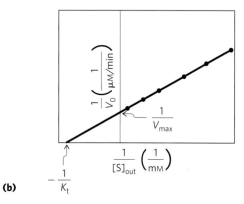

(b)

FIGURE 11–31 Kinetics of glucose transport into erythrocytes. (a) The initial rate of glucose entry into an erythrocyte, V_0, depends on the initial concentration of glucose on the outside, $[S]_{out}$. **(b)** Double-reciprocal plot of the data in (a). The kinetics of facilitated diffusion is analogous to the kinetics of an enzyme-catalyzed reaction. Compare these plots with Figure 6-11 and with Figure 1 in Box 6-1. Note that K_t is analogous to K_m, the Michaelis constant.

The steps are summarized in **Figure 11–32**. Given that every step in this sequence is reversible, the transporter is, in principle, equally able to move glucose into or out of the cell. However, with GLUT1, glucose

FIGURE 11–32 Model of glucose transport into erythrocytes by GLUT1. The transporter exists in two conformations: T_1, with the glucose-binding site exposed on the outer surface of the plasma membrane, and T_2, with the binding site exposed on the inner surface. Glucose transport occurs in four steps. ❶ Glucose in blood plasma binds to a stereospecific site on T_1; this lowers the activation energy for ❷ a conformational change from $glucose_{out} \cdot T_1$ to $glucose_{in} \cdot T_2$, effecting the transmembrane passage of the glucose. ❸ Glucose is released from T_2 into the cytoplasm, and ❹ the transporter returns to the T_1 conformation, ready to transport another glucose molecule.

always moves down its concentration gradient, which normally means *into* the cell. Glucose that enters a cell is generally metabolized immediately, and the intracellular glucose concentration is thereby kept low relative to its concentration in the blood.

The rate equations for glucose transport can be derived exactly as for enzyme-catalyzed reactions (Chapter 6), yielding an expression analogous to the Michaelis-Menten equation:

$$V_0 = \frac{V_{max}[S]_{out}}{K_t + [S]_{out}} \qquad (11\text{--}1)$$

in which V_0 is the initial velocity of accumulation of glucose inside the cell when its concentration in the surrounding medium is $[S]_{out}$ and $\boldsymbol{K_t}$ $(\boldsymbol{K_{transport}})$ is a constant analogous to the Michaelis constant, a combination of rate constants that is characteristic of each transport system. This equation describes the *initial* velocity, the rate observed when $[S]_{in} = 0$. As is the case for enzyme-catalyzed reactions, the slope-intercept form of the equation describes a linear plot of $1/V_0$ against $1/[S]_{out}$, from which we can obtain values of K_t and V_{max} (Fig. 11–31b). When $[S]_{out} = K_t$, the rate of uptake is $\frac{1}{2}V_{max}$; the transport process is half-saturated. The concentration of glucose in blood is 4.5 to 5 mM, close to the K_t, which ensures that GLUT1 is nearly saturated with substrate and operates near V_{max}.

Because no chemical bonds are made or broken in the conversion of S_{out} to S_{in}, neither "substrate" nor "product" is intrinsically more stable, and the process of entry is therefore fully reversible. As $[S]_{in}$ approaches $[S]_{out}$, the rates of entry and exit become equal. Such a system is therefore incapable of accumulating glucose within a cell at concentrations above that in the surrounding medium; it simply equilibrates glucose on the two sides of the membrane much faster than would occur in the absence of a specific transporter. GLUT1 is specific for D-glucose, with a measured K_t of about 6 mM. For the close analogs D-mannose and D-galactose, which differ only in the position of one hydroxyl group, the values of K_t are 20 and 30 mM, respectively, and for L-glucose, K_t exceeds 3,000 mM. Thus, GLUT1 shows the three hallmarks of passive transport: high rates of diffusion down a concentration gradient, saturability, and specificity.

Twelve passive glucose transporters are encoded in the human genome, each with its unique kinetic properties, patterns of tissue distribution, and function (Table 11–3). In the liver, GLUT2 transports glucose out of hepatocytes when liver glycogen is broken down to replenish blood glucose. GLUT2 has a large K_t (17 mM or greater) and can therefore respond to increased levels of intracellular glucose (produced by glycogen breakdown) by increasing outward transport. Skeletal and heart muscle and adipose tissue have yet another glucose transporter, GLUT4

TABLE 11-3	Glucose Transporters in Humans		
Transporter	Tissue(s) where expressed	K_t (mM)*	Role[†]
GLUT1	Ubiquitous	3	Basal glucose uptake
GLUT2	Liver, pancreatic islets, intestine	17	In liver and kidney, removal of excess glucose from blood; in pancreas, regulation of insulin release
GLUT3	Brain (neuronal), testis (sperm)	1.4	Basal glucose uptake
GLUT4	Muscle, fat, heart	5	Activity increased by insulin
GLUT5	Intestine (primarily), testis, kidney	6[‡]	Primarily fructose transport
GLUT6	Spleen, leukocytes, brain	>5	Possibly no transporter function
GLUT7	Small intestine, colon	0.3	—
GLUT8	Testis	~2	—
GLUT9	Liver, kidney	0.6	—
GLUT10	Heart, lung, brain, liver, muscle, pancreas, kidney	0.3[§]	—
GLUT11	Heart, skeletal muscle, kidney	0.16	—
GLUT12	Skeletal muscle, heart, prostate, small intestine	—	—

*K_t for glucose, except as noted, from Augustin, R. (2010) The protein family of glucose transport facilitators: it's not only about glucose after all. *IUBMB Life* 62, 315–333.

[†] Dash indicates role uncertain.

[‡] K_m for fructose.

[§] K_m for 2-deoxyglucose.

(K_t = 5 mM), which is distinguished by its response to insulin: its activity increases when insulin signals a high blood glucose concentration, thus increasing the rate of glucose uptake into muscle and adipose tissue (Box 11–1 describes the effect of insulin on this transporter).

The Chloride-Bicarbonate Exchanger Catalyzes Electroneutral Cotransport of Anions across the Plasma Membrane

The erythrocyte contains another facilitated diffusion system, an anion exchanger that is essential in CO_2 transport to the lungs from tissues such as skeletal muscle and liver. Waste CO_2 released from respiring tissues into the blood plasma enters the erythrocyte, where it is converted to bicarbonate (HCO_3^-) by the enzyme carbonic anhydrase. (Recall that HCO_3^- is the primary buffer of blood pH; see Fig. 2–21.) The HCO_3^- reenters the blood plasma for transport to the lungs (**Fig. 11–33**). Because HCO_3^- is much more soluble in blood plasma than is CO_2, this roundabout route increases the capacity of the blood to carry carbon dioxide from the tissues to the lungs. In the lungs, HCO_3^- reenters the erythrocyte and is converted to CO_2, which is eventually released into the lung space and exhaled. To be effective, this shuttle requires very rapid movement of HCO_3^- across the erythrocyte membrane.

The **chloride-bicarbonate exchanger**, also called the **anion exchange (AE) protein**, increases the rate of HCO_3^- transport across the erythrocyte membrane more than a millionfold. Like the glucose transporter, it is an integral protein that probably

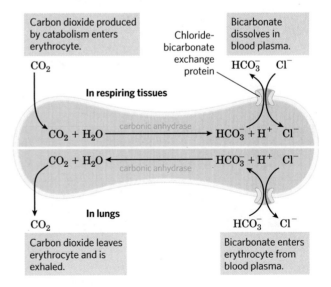

FIGURE 11-33 Chloride-bicarbonate exchanger of the erythrocyte membrane. This cotransport system allows the entry and exit of HCO_3^- without changing the membrane potential. Its role is to increase the CO_2-carrying capacity of the blood. The top half of the figure illustrates the events that take place in respiring tissues; the bottom half, the events in the lungs.

BOX 11–1 🜍 MEDICINE Defective Glucose and Water Transport in Two Forms of Diabetes

When ingestion of a carbohydrate-rich meal causes blood glucose to exceed the usual concentration between meals (about 5 mM), excess glucose is taken up by the myocytes of cardiac and skeletal muscle (which store it as glycogen) and by adipocytes (which convert it to triacylglycerols). Glucose uptake into myocytes and adipocytes is mediated by the glucose transporter GLUT4. Between meals, some GLUT4 is present in the plasma membrane, but most is sequestered in the membranes of small intracellular vesicles (Fig. 1). Insulin released from the pancreas in response to high blood glucose triggers the movement of these intracellular vesicles to the plasma membrane, with which they fuse, bringing GLUT4 molecules to the plasma membrane (see Fig. 12–16). With more GLUT4 molecules in action, the rate of glucose uptake increases 15-fold or more. When blood glucose levels return to normal, insulin release slows and most GLUT4 molecules are removed from the plasma membrane and stored in vesicles.

In type 1 (insulin-dependent) diabetes mellitus, the inability to release insulin (and thus to mobilize glucose transporters) results in low rates of glucose uptake into muscle and adipose tissue. One consequence is a prolonged period of high blood glucose after a carbohydrate-rich meal. This condition is the basis for the glucose tolerance test used to diagnose diabetes (Chapter 23).

The water permeability of epithelial cells lining the renal collecting duct in the kidney is due to the presence of an aquaporin (AQP2) in their apical plasma membranes (facing the lumen of the duct). Vasopressin (antidiuretic hormone, ADH) regulates the retention of water by mobilizing AQP2 molecules stored in vesicle membranes within the epithelial cells, much as insulin mobilizes GLUT4 in muscle and adipose tissue. When the vesicles fuse with the epithelial cell plasma membrane, water permeability greatly increases and more water is reabsorbed from the collecting duct and returned to the blood. When the vasopressin level drops, AQP2 is resequestered within vesicles, reducing water retention. In the relatively rare human disease diabetes insipidus, a genetic defect in AQP2 leads to impaired water reabsorption by the kidney. The result is excretion of copious volumes of very dilute urine. If the individual drinks enough water to replace that lost in the urine, there are no serious medical consequences, but insufficient water intake leads to dehydration and imbalances in blood electrolytes, which can lead to fatigue, headache, muscle pain, or even death.

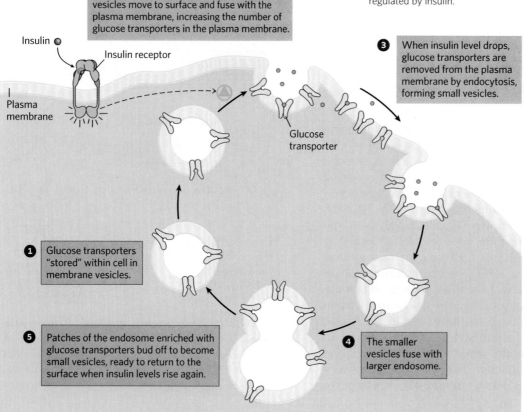

FIGURE 1 Transport of glucose into a myocyte by GLUT4 is regulated by insulin.

2 When insulin interacts with its receptor, vesicles move to surface and fuse with the plasma membrane, increasing the number of glucose transporters in the plasma membrane.

3 When insulin level drops, glucose transporters are removed from the plasma membrane by endocytosis, forming small vesicles.

Insulin

Insulin receptor

Plasma membrane

Glucose transporter

1 Glucose transporters "stored" within cell in membrane vesicles.

5 Patches of the endosome enriched with glucose transporters bud off to become small vesicles, ready to return to the surface when insulin levels rise again.

4 The smaller vesicles fuse with larger endosome.

FIGURE 11–34 Three general classes of transport systems. Transporters differ in the number of solutes (substrates) transported and the direction in which each solute moves. Examples of all three types of transporter are discussed in the text. Note that this classification tells us nothing about whether these are energy-requiring (active transport) or energy-independent (passive transport) processes.

FIGURE 11–35 Two types of active transport. (a) In primary active transport, the energy released by ATP hydrolysis drives solute (S_1) movement against an electrochemical gradient. **(b)** In secondary active transport, a gradient of ion X (S_1) (often Na^+) has been established by primary active transport. Movement of X (S_1) down its electrochemical gradient now provides the energy to drive cotransport of a second solute (S_2) against its electrochemical gradient.

spans the membrane at least 12 times. This protein mediates the simultaneous movement of two anions: for each HCO_3^- ion that moves in one direction, one Cl^- ion moves in the opposite direction, with no net transfer of charge; the exchange is **electroneutral**. The coupling of Cl^- and HCO_3^- movements is obligatory; in the absence of chloride, bicarbonate transport stops. In this respect, the anion exchanger is typical of those systems, called **cotransport systems**, that simultaneously carry two solutes across a membrane **(Fig. 11–34)**. When, as in this case, the two substrates move in opposite directions, the process is **antiport**. In **symport**, two substrates are moved simultaneously in the same direction. Transporters that carry only one substrate, such as the erythrocyte glucose transporter, are known as **uniport** systems.

The human genome has genes for three closely related chloride-bicarbonate exchangers, all with the same predicted transmembrane topology. Erythrocytes contain the AE1 transporter, AE2 is prominent in the liver, and AE3 is present in plasma membranes of the brain, heart, and retina. Similar anion exchangers are also found in plants and microorganisms.

Active Transport Results in Solute Movement against a Concentration or Electrochemical Gradient

In passive transport, the transported species always moves down its electrochemical gradient and is not accumulated above the equilibrium concentration. **Active transport**, by contrast, results in the accumulation of a solute above the equilibrium point. Active transport is thermodynamically unfavorable (endergonic) and takes place only when coupled (directly or indirectly) to an exergonic process such as the absorp-

tion of sunlight, an oxidation reaction, the breakdown of ATP, or the concomitant flow of some other chemical species down its electrochemical gradient. In **primary active transport**, solute accumulation is coupled directly to an exergonic chemical reaction, such as conversion of ATP to ADP + P_i **(Fig. 11–35)**. **Secondary active transport** occurs when endergonic (uphill) transport of one solute is coupled to the exergonic (downhill) flow of a different solute that was originally pumped uphill by primary active transport.

The amount of energy needed for the transport of a solute against a gradient can be calculated from the initial concentration gradient. The general equation for the free-energy change in the chemical process that converts S to P is

$$\Delta G = \Delta G'^\circ + RT \ln ([P]/[S]) \qquad (11\text{–}2)$$

where $\Delta G'^\circ$ is the standard free-energy change, R is the gas constant, $8.315\ J/mol \cdot K$, and T is the absolute temperature. When the "reaction" is simply transport of a solute from a region where its concentration is C_1 to a region where its concentration is C_2, no bonds are made or broken and $\Delta G'^\circ$ is zero. The free-energy change for transport, ΔG_t, is then

$$\Delta G_t = RT \ln (C_2/C_1) \qquad (11\text{–}3)$$

If there is a 10-fold difference in concentration between two compartments, the cost of moving 1 mol of an uncharged solute at 25 °C uphill across a membrane separating the compartments is

$$\Delta G_t = (8.315\ J/mol \cdot K)(298\ K) \ln (10/1) = 5{,}700\ J/mol$$
$$= 5.7\ kJ/mol$$

Equation 11–3 holds for all uncharged solutes.

Energy Cost of Pumping an Uncharged Solute

Calculate the energy cost (free-energy change) of pumping an uncharged solute against a 1.0×10^4–fold concentration gradient at 25 °C.

Solution: Begin with Equation 11–3. Substitute 1.0×10^4 for (C_2/C_1), 8.315 J/mol·K for R, and 298 K for T:

$$\Delta G_t = RT \ln(C_2/C_1)$$
$$= (8.315 \text{ J/mol·K})(298 \text{ K}) \ln(1.0 \times 10^4)$$
$$= 23 \text{ kJ/mol}$$

When the solute is an *ion*, its movement without an accompanying counterion results in the endergonic separation of positive and negative charges, producing an electrical potential; such a transport process is said to be **electrogenic**. The energetic cost of moving an ion depends on the electrochemical potential (Fig 11–27), the sum of the chemical and electrical gradients:

$$\Delta G_t = RT \ln(C_2/C_1) + Z\mathcal{F}\Delta\psi \quad (11\text{–}4)$$

where Z is the charge on the ion, $\mathcal{F}$ is the Faraday constant (96,480 J/V·mol), and $\Delta\psi$ is the transmembrane electrical potential (in volts). Eukaryotic cells typically have plasma membrane potentials of about 0.05 V (with the inside negative relative to the outside), so the second term of Equation 11–4 can make a significant contribution to the total free-energy change for transporting an ion. Most cells maintain more than a 10-fold difference in ion concentrations across their plasma or intracellular membranes, and for many cells and tissues active transport is therefore a major energy-consuming process.

Energy Cost of Pumping a Charged Solute

Calculate the energy cost (free-energy change) of pumping Ca^{2+} from the cytosol, where its concentration is about 1.0×10^{-7} M, to the extracellular fluid, where its concentration is about 1.0 mM. Assume a temperature of 37 °C (body temperature in a mammal) and a standard transmembrane potential of 50 mV (inside negative) for the plasma membrane.

Solution: In this calculation, both the concentration gradient and the electrical potential must be taken into account. In Equation 11–4, substitute 8.315 J/mol · K for R, 310 K for T, 1.0×10^{-3} for C_2, 1.0×10^{-7} for C_1, 96,500 J/V · mol for $\mathcal{F}$, +2 (the charge on a Ca^{2+} ion) for Z, and 0.050 V for $\Delta\psi$. Note that the transmembrane potential is 50 mV (inside negative), so the change in potential when an ion moves from inside to outside is 50 mV.

$$\Delta G_t = RT \ln(C_2/C_1) + Z\mathcal{F}\Delta\psi$$
$$= (8.315 \text{ J/mol·K})(310 \text{ K}) \ln \frac{1.0 \times 10^{-3}}{1.0 \times 10^{-7}} +$$
$$2(96{,}500 \text{ J/V·mol})(0.050 \text{ V})$$
$$= 33 \text{ kJ/mol}$$

The mechanism of active transport is of fundamental importance in biology. As we shall see in Chapter 19, ATP is formed in mitochondria and chloroplasts by a mechanism that is essentially ATP-driven ion transport operating in reverse. The energy made available by the spontaneous flow of protons across a membrane is calculable from Equation 11–4; remember that ΔG for flow *down* an electrochemical gradient has a negative value and ΔG for transport of ions *against* an electrochemical gradient has a positive value.

P-Type ATPases Undergo Phosphorylation during Their Catalytic Cycles

The family of active transporters called **P-type ATPases** are cation transporters that are reversibly phosphorylated by ATP (thus the name P-type) as part of the transport cycle. Phosphorylation forces a conformational change that is central to movement of the cation across the membrane. The human genome encodes at least 70 P-type ATPases that share similarities in amino acid sequence and topology, especially near the Asp residue that undergoes phosphorylation. All are integral proteins with 8 or 10 predicted membrane-spanning regions in a single polypeptide (type III in Fig. 11–9), and all are sensitive to inhibition by the phosphate analog **vanadate**.

Phosphate   Vanadate

The P-type ATPases are widespread in eukaryotes and bacteria. The Na^+K^+ ATPase of animal cells (an antiporter for Na^+ and K^+ ions) and the plasma membrane H^+ ATPase of plants and fungi set the transmembrane electrochemical potential in cells by establishing ion gradients across the plasma membrane. These gradients provide the driving force for secondary active transport and are also the basis for electrical signaling in neurons. In animal tissues, the **sarcoplasmic/endoplasmic reticulum Ca^{2+} ATPase (SERCA) pump** and the plasma membrane Ca^{2+} ATPase pump are uniporters for Ca^{2+} ions, which together maintain the cytosolic level of Ca^{2+} below 1 μM. Parietal cells in the lining of the mammalian stomach have a P-type ATPase that pumps H^+ and K^+ across the plasma membrane, thereby acidifying the stomach contents. Lipid flippases, as we noted earlier, are structurally and functionally related to P-type transporters. Bacteria and eukaryotes use P-type ATPases to pump out toxic heavy metal ions such as Cd^{2+} and Cu^{2+}.

The P-type pumps have similar structures **(Fig. 11–36)** and similar mechanisms. The mechanism postulated for P-type ATPases takes into account the large conformational changes and the phosphorylation-dephosphorylation of the critical Asp residue in the P domain that is known to occur during a catalytic cycle. For the SERCA pump **(Fig. 11–37)**, each catalytic cycle moves two Ca^{2+} ions across the membrane and converts an ATP to ADP and P_i. ATP has two roles in this mechanism, one catalytic and one modulatory. The

Jens Skou

role of ATP binding and phosphoryl transfer to the enzyme is to bring about the interconversion of two conformations (E1 and E2) of the transporter. In the E1 conformation, the two Ca^{2+}-binding sites are exposed on the cytosolic side of the ER or sarcoplasmic reticulum and bind Ca^{2+} with high affinity. ATP binding and Asp phosphorylation drive a conformational change from E1 to E2 in which the Ca^{2+}-binding sites are now exposed on the lumenal side of the membrane and their affinity for Ca^{2+} is greatly reduced, causing Ca^{2+} release into the lumen. By this mechanism, the energy released by hydrolysis of ATP during one phosphorylation-dephosphorylation cycle drives Ca^{2+} across the membrane against a large electrochemical gradient.

A variation on this basic mechanism is seen in the **Na^+K^+ ATPase** of the plasma membrane, discovered by Jens Skou in 1957. This cotransporter couples phosphorylation-dephosphorylation of the critical Asp residue to the simultaneous movement of both Na^+ and K^+ against their electrochemical gradients. The Na^+K^+ ATPase is responsible for maintaining low Na^+ and high K^+ concentrations in the cell relative to the extracellular fluid **(Fig. 11–38)**. For each molecule of ATP converted to ADP and P_i, the transporter moves two K^+ ions inward

(a)

(b) Ca^{2+} ATPase (SERCA pump)   **(c)** $Na^+ K^+$ ATPase   Plasma membrane H^+ ATPase   Gastric $H^+ K^+$ ATPase

FIGURE 11–36 The general structure of the P-type ATPases. (a) P-type ATPases have three cytoplasmic domains (A, N, and P) and two transmembrane domains (T and S) consisting of multiple helices. The N (nucleotide) domain binds ATP and Mg^{2+}, and it has protein kinase activity that phosphorylates the specific Asp residue found in the P (phosphorylated) domain of all P-type ATPases. The A (actuator) domain has protein phosphatase activity and removes the phosphoryl group from the Asp residue with each catalytic cycle of the pump. A transport domain with six transmembrane helices (T) includes the ion-transporting structure, and four more transmembrane helices make up the support (S) domain, which provides physical support to the transport domain and may have other specialized function in

certain P-type ATPases. The binding sites for the ions to be transported are near the middle of the membrane, 40 to 50 Å from the phosphorylated Asp residue—thus Asp phosphorylation-dephosphorylation does not *directly* affect ion binding. The A domain communicates movements of the N and P domains to the ion-binding sites. **(b)** A ribbon representation of the Ca^{2+} ATPase (SERCA pump) (PDB 1T5S). ATP binds to the N domain and the Ca^{2+} ions to be transported bind to the T domain. **(c)** Other P-type ATPases have domain structures, and presumably mechanisms, like the SERCA pump: Na^+K^+ ATPase (PDB ID 3KDP), the plasma membrane H^+ ATPase (PDB ID 3B8C), and the gastric H^+K^+ ATPase (derived from PDB ID 3IXZ).

FIGURE 11–37 Postulated mechanism of the SERCA pump. The transport cycle begins with the protein in the E1 conformation, with the Ca^{2+}-binding sites facing the cytosol. Two Ca^{2+} ions bind, then ATP binds to the transporter and phosphorylates Asp^{351}, forming E1-P. Phosphorylation favors the second conformation, E2-P, in which the Ca^{2+}-binding sites, now with a reduced affinity for Ca^{2+}, are accessible on the other side of the membrane (the lumen or extracellular space), and the released Ca^{2+} diffuses away. Finally, E2-P is dephosphorylated, returning the protein to the E1 conformation for another round of transport.

and three Na^+ ions outward across the plasma membrane. Cotransport is therefore electrogenic—it creates a net separation of charge across the membrane; in animals, this produces the membrane potential of −50 to −70 mV (inside negative relative to outside) that is characteristic of most cells and is essential to the conduction of action potentials in neurons. The central role of the Na^+K^+ ATPase is reflected in the energy invested in this single reaction: about 25% of the total energy consumption of a human at rest!

V-Type and F-Type ATPases Are ATP-Driven Proton Pumps

V-type ATPases, a class of proton-transporting ATPases, are responsible for acidifying intracellular compartments in many organisms (thus V for $v$acuolar). Proton pumps of this type maintain the vacuoles of fungi and higher plants at a pH between 3 and 6, well below that of the surrounding cytosol (pH 7.5). V-type ATPases are also responsible for the acidification of lysosomes, endosomes, the Golgi complex, and secretory vesicles in animal cells. All V-type ATPases have a similar complex structure, with an integral (transmembrane) domain (V_o) that serves as a proton channel and a peripheral domain (V_1) that contains the ATP-binding site and the ATPase activity **(Fig 11–39a)**. The structure is similar to that of the well-characterized F-type ATPases.

F-type ATPase active transporters catalyze the uphill transmembrane passage of protons driven by ATP hydrolysis. The "F-type" designation derives from the identification of these ATPases as energy-coupling *f*actors. The F_o integral membrane protein complex (Fig. 11–39b; subscript o denotes its inhibition by the drug *oli*gomycin) provides a transmembrane pathway for protons, and the peripheral protein F_1 (subscript 1 indicating this was the first of several factors isolated from mitochondria) uses the energy of ATP to drive protons uphill (into a region of higher H^+ concentration). The

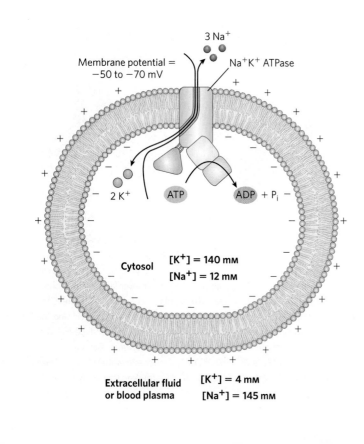

FIGURE 11–38 Role of the Na^+K^+ ATPase in animal cells. This active transport system is primarily responsible for setting and maintaining the intracellular concentrations of Na^+ and K^+ in animal cells and for generating the membrane potential. It does this by moving three Na^+ out of the cell for every two K^+ it moves in. The electrical potential across the plasma membrane is central to electrical signaling in neurons, and the gradient of Na^+ is used to drive the uphill cotransport of solutes in many cell types.

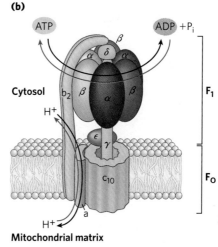

FIGURE 11–39 Two proton pumps with similar structures. (a) The V_oV_1 H^+ ATPase uses ATP to pump protons into vacuoles and lysosomes, creating their low internal pH. It has an integral (membrane-embedded) domain, V_o (orange), that includes multiple identical c subunits, and a peripheral domain that projects into the cytosol and contains the ATP hydrolysis sites, located on three identical B subunits (purple). **(b)** The F_oF_1 ATPase/ATP synthase of mitochondria has an integral domain, F_o (orange), with multiple copies of the c subunit, and a peripheral domain, F_1, consisting of three α subunits, three β subunits, and a central shaft joined to the integral domain. F_o, and presumably V_o, provides a transmembrane channel through which protons are pumped as ATP is hydrolyzed on the

β subunits of F_1 (B subunits of V_1). The remarkable mechanism by which ATP hydrolysis is coupled to proton movement is described in detail in Chapter 19. It involves rotation of F_o in the plane of the membrane. The structures of the V_oV_1 ATPase and its analogs A_oA_1 ATPase (of archaea) and CF_oCF_1 ATPase (of chloroplasts) are essentially similar to that of F_oF_1, and the mechanisms are also conserved. An ATP-driven proton transporter also can catalyze ATP synthesis (red arrows) as protons flow *down* their electrochemical gradient. This is the central reaction in the processes of oxidative phosphorylation and photophosphorylation, both described in detail in Chapter 19.

F_oF_1 organization of proton-pumping transporters must have developed very early in evolution. Bacteria such as *E. coli* use an F_oF_1 ATPase complex in their plasma membrane to pump protons outward, and archaea have a closely homologous proton pump, the A_oA_1 ATPase.

Like all enzymes, F-type ATPases catalyze their reactions in both directions. Therefore, a sufficiently large proton gradient can supply the energy to drive the reverse reaction, ATP synthesis (Fig. 11–39b). When functioning in this direction, the F-type ATPases are more appropriately named **ATP synthases**. ATP synthases are central to ATP production in mitochondria during oxidative phosphorylation and in chloroplasts during photophosphorylation, as well as in bacteria and archaea. The proton gradient needed to drive ATP synthesis is produced by other types of proton pumps powered by substrate oxidation or sunlight. We provide a detailed description of these processes in Chapter 19 (p. 750).

ABC Transporters Use ATP to Drive the Active Transport of a Wide Variety of Substrates

ABC transporters (Fig. 11–40) constitute a large family of ATP-dependent transporters that pump amino acids, peptides, proteins, metal ions, various lipids, bile salts, and many hydrophobic compounds, including drugs, out of cells against a concentration gradient. One ABC transporter in humans, the **multidrug transporter** (**MDR1**; also called **P glycoprotein**), is

responsible for the striking resistance of certain tumors to some generally effective antitumor drugs. MDR1 has broad substrate specificity for hydrophobic compounds, including, for example, the chemotherapeutic drugs adriamycin, doxorubicin, and vinblastine. By pumping these drugs out of the cell, the transporter prevents their accumulation within a tumor and thus blocks their therapeutic effects. MDR1 (Fig. 11–40a) is an integral membrane protein (M_r 170,000) with two homologous halves, each with six transmembrane helices and a cytoplasmic ATP-binding domain ("cassette"), which give the family its name: *A*TP-*b*inding *c*assette transporters. Overexpression of MDR1 is associated with treatment failure in cancers of the liver, kidney, and colon. A related ABC transporter, ABCC1, is overexpressed in drug-resistant prostate, lung, and breast cancer cells. Highly selective inhibitors of multidrug transporters, which would be expected to enhance the effectiveness of antitumor drugs that are otherwise pumped out of tumor cells, are the objects of current drug discovery and design. Why have multidrug transporters been conserved in evolution? MDR1 in the placental membrane and in the blood-brain barrier keep out toxic compounds that would damage the fetus or the brain.

ABC transporters are also present in simpler animals and in plants and microorganisms. Yeast has 31 genes that encode ABC transporters, *Drosophila* has 56, and *E. coli* has 80, representing 2% of its entire genome. ABC transporters, used by *E. coli* and other bacteria to import

(a)

Transmembrane domains (TMDs)

Extracellular space

Cytoplasm

Nucleotide-binding domains (NBDs)

(b)

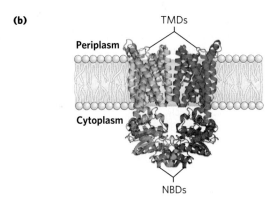

TMDs

Periplasm

Cytoplasm

NBDs

(c)

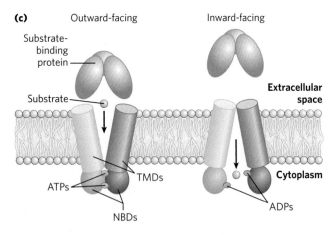

Outward-facing   Inward-facing

Substrate-binding protein

Substrate

Extracellular space

ATPs

TMDs

Cytoplasm

NBDs

ADPs

FIGURE 11–40 Two ABC transporters. (a) The multidrug transporter of animal cells (MDR1, also called P glycoprotein; PDB ID 3G60), responsible for pumping a variety of antitumor drugs out of human cells, has two homologous halves (blue and light blue), each with six transmembrane helices in two transmembrane domains (TMDs; blue), and a cytoplasmic nucleotide-binding domain (NBD; red). **(b)** The vitamin B_{12} importer BtuCD (PDB ID 1L7V) of *E. coli* is a homodimer with 10 transmembrane helices (blue, light blue) in each monomer and two NBDs (red) that extend into the cytoplasm. **(c)** Mechanisms proposed for the *E. coli* vitamin B_{12} ABC transporter coupling of ATP hydrolysis to transport. Substrate is brought to the transporter on the periplasmic side by a substrate-specific binding protein. With ATP bound to the NBD sites, the transporter is open to the outside (periplasm), but on substrate binding and ATP hydrolysis to ADP, a conformational change exposes the substrate to the inside surface, and it diffuses away from the transporter and into the cytosol.

essentials such as vitamin B_{12} (Fig. 11–40b), are the presumed evolutionary precursors of the MDRs of animal cells. The presence of ABC transporters that confer antibiotic resistance in pathogenic microbes (*Pseudomonas aeruginosa, Staphylococcus aureus, Candida albicans, Neisseria gonorrhoeae,* and *Plasmodium falciparum*) is a serious public health concern and makes these transporters attractive targets for drug design. ∎

All ABC transporters have two nucleotide-binding domains (NBDs) and two transmembrane domains containing multiple transmembrane helices. In some cases, all these domains are in a single long polypeptide; other ABC transporters have two subunits, each contributing an NBD and a domain with six transmembrane helices. Many ABC transporters are in the plasma membrane, but some types are also found in the endoplasmic reticulum and in the membranes of mitochondria and lysosomes. The CFTR protein (see Box 11–2) is an interesting case of an ion channel (for Cl⁻), operated by ATP hydrolysis, that is apparently derived from an ABC transporter in which evolution has eliminated the pumping function but left a functional channel.

The NBDs of all ABC proteins are similar in sequence and presumably in three-dimensional structure; they are the conserved molecular motor that can be coupled to a wide variety of pumps and channels. When coupled with a pump, the ATP-driven motor moves solutes against a concentration gradient. The stoichiometry of ABC pumps is approximately one ATP hydrolyzed per molecule of substrate transported, but neither the mechanism of coupling nor the site of substrate binding is fully understood (Fig. 11–40c).

Some ABC transporters have very high specificity for a single substrate; others are more promiscuous. The human genome contains at least 48 genes that encode ABC transporters, many of which are involved in maintaining the lipid bilayer and in transporting sterols, sterol derivatives, and fatty acids throughout the body. The flippases that move membrane lipids from one leaflet of the bilayer to the other are ABC transporters, and the cellular machinery for exporting excess cholesterol includes an ABC transporter. Mutations in the genes that encode some of these proteins contribute to several genetic diseases, including cystic fibrosis (Box 11–2), Tangier disease (p. 874), retinal degeneration, anemia, and liver failure. ∎

Ion Gradients Provide the Energy for Secondary Active Transport

The ion gradients formed by primary transport of Na^+ or H^+ can in turn provide the driving force for cotransport of other solutes. Many cell types contain transport systems that couple the spontaneous, downhill flow of these ions to the simultaneous uphill pumping of another ion, sugar, or amino acid (Table 11–4).

BOX 11-2 ⚕ **MEDICINE** A Defective Ion Channel in Cystic Fibrosis

Cystic fibrosis (CF) is a serious and relatively common hereditary human disease. About 5% of white Americans are carriers, having one defective and one normal copy of the gene. Only individuals with two defective copies show the severe symptoms of the disease: obstruction of the gastrointestinal and respiratory tracts, commonly leading to bacterial infection of the airways and death due to respiratory insufficiency before the age of 30. In CF, the thin layer of mucus that normally coats the internal surfaces of the lungs is abnormally thick, obstructing air flow and providing a haven for pathogenic bacteria, particularly *Staphylococcus aureus* and *Pseudomonas aeruginosa*.

The defective gene in CF patients was discovered in 1989. It encodes a membrane protein called *cystic fibrosis transmembrane conductance regulator*, or CFTR. This protein has two segments, each containing six transmembrane helices, two nucleotide-binding domains (NBDs), and a regulatory region (Fig. 1). CFTR is therefore very similar to other ABC transporter proteins except that it functions as an *ion channel* (for Cl⁻ ion), not as a pump. The channel conducts Cl⁻ across the plasma membrane when both NBDs have bound ATP, and it closes when the ATP

FIGURE 2 Mucus lining the surface of the lungs traps bacteria. In healthy lungs (shown here), these bacteria are killed and swept away by the action of cilia. In CF, this mechanism is impaired, resulting in recurring infections and progressive damage to the lungs.

on one of the NBDs is broken down to ADP and P_i. The Cl⁻ channel is further regulated by phosphorylation of several Ser residues in the regulatory domain, catalyzed by cAMP-dependent protein kinase (Chapter 12). When the regulatory domain is not phosphorylated, the Cl⁻ channel is closed. The mutation responsible for CF in 70% of cases results in deletion of a Phe residue at position 508. The mutant protein folds incorrectly, which interferes with its insertion in the plasma membrane, resulting in reduced Cl⁻ and H_2O movement across the plasma membranes of epithelial cells that line the airways (Fig. 2), the digestive tract, and exocrine glands (pancreas, sweat glands, bile ducts, and vas deferens). Liquid secretion is essential to keep the mucus on the surface of alveoli of the lung at just the right viscosity to trap and clear microorganisms that are inhaled.

Diminished export of Cl⁻ is accompanied by diminished export of water from cells, causing the mucus on the surfaces of the cells to become dehydrated, thick, and excessively sticky. In normal circumstances, cilia on the epithelial cells that line the inner surface of the lungs constantly sweep away bacteria that settle in this mucus, but the thick mucus in individuals with CF hinders this process. Frequent infections by bacteria such as *S. aureus* and *P. aeruginosa* result, causing progressive damage to the lungs and reduced respiratory efficiency. Respiratory failure is commonly the cause of death in people with CF.

FIGURE 1 Three states of the cystic fibrosis transmembrane conductance regulator, CFTR. The protein has two segments, each with six transmembrane helices, and three functionally significant domains extend from the cytoplasmic surface: NBD_1 and NBD_2 (green) are nucleotide-binding domains that bind ATP, and a regulatory R domain (blue) is the site of phosphorylation by cAMP-dependent protein kinase. When this R domain is phosphorylated but no ATP is bound to the NBDs (left), the channel is closed. The binding of ATP opens the channel (middle) until the bound ATP is hydrolyzed. When the regulatory domain is unphosphorylated (right), it binds the NBD domains and prevents ATP binding and channel opening. The most commonly occurring mutation leading to CF is the deletion of Phe^{508} in the NBD_1 domain (left). CFTR is a typical ABC transporter in all but two respects: most ABC transporters lack the regulatory domain, and CFTR acts as an ion channel (for Cl⁻), not as a typical transporter.

Channel closed
R domain phosphorylated
No ATP bound to NBDs

Channel open
R domain phosphorylated
ATP bound to NBDs

Channel closed
R domain unphosphorylated
No ATP bound to NBDs

TABLE 11–4 Cotransport Systems Driven by Gradients of Na⁺ or H⁺

Organism/ tissue/cell type	Transported solute (moving against its gradient)	Cotransported solute (moving down its gradient)	Type of transport
E. coli	Lactose	H⁺	Symport
	Proline	H⁺	Symport
	Dicarboxylic acids	H⁺	Symport
Intestine, kidney (vertebrates)	Glucose	Na⁺	Symport
	Amino acids	Na⁺	Symport
Vertebrate cells (many types)	Ca²⁺	Na⁺	Antiport
Higher plants	K⁺	H⁺	Antiport
Fungi (*Neurospora*)	K⁺	H⁺	Antiport

The **lactose transporter** (**lactose permease**, or **galactoside permease**) of *E. coli* is the well-studied prototype for proton-driven cotransporters. This protein consists of a single polypeptide chain (417 residues) that functions as a monomer to transport one proton and one lactose molecule into the cell, with the net accumulation of lactose (**Fig. 11–41**). *E. coli* normally produces a gradient of protons and charge across its plasma membrane by oxidizing fuels and using the energy of oxidation to pump protons outward. (This mechanism is discussed in detail in Chapter 19.) The lipid bilayer is impermeable to protons, but the lactose transporter provides a route for proton reentry, and lactose is simultaneously carried into the cell by

symport. The endergonic accumulation of lactose is thereby coupled to the exergonic flow of protons into the cell, with a negative overall free-energy change.

The lactose transporter is one member of the **major facilitator superfamily (MFS)** of transporters, which comprises 28 families. Almost all proteins in this superfamily have 12 transmembrane domains (the few exceptions have 14). The proteins share relatively little sequence homology, but the similarity of their secondary structures and topology suggests a common tertiary structure. The crystallographic solution of the *E. coli* lactose transporter provides a glimpse of this general structure (**Fig. 11–42a**). The protein has 12 transmembrane helices, and connecting loops protrude into the cytoplasm or the periplasmic space (between the plasma membrane and outer membrane or cell wall). The six amino-terminal and six carboxyl-terminal helices form very similar domains to produce a structure with a rough twofold symmetry. In the crystallized form of the protein, a large aqueous cavity is exposed on the cytoplasmic side of the membrane. The substrate-binding site is in this cavity, more or less in the middle of the membrane. The side of the transporter facing outward (the periplasmic face) is closed tightly, with no channel big enough for lactose to enter. The proposed mechanism for transmembrane passage of the substrate (Fig. 11–42b) involves a rocking motion between the two domains, driven by substrate binding and proton movement, alternately exposing the substrate-binding domain to the cytoplasm and to the periplasm. This "rocking banana" model is similar to that shown in Figure 11–32 for GLUT1.

How is proton movement into the cell coupled with lactose uptake? Extensive genetic studies of the lactose transporter have established that of the 417 residues in the protein, only 6 are absolutely essential for cotransport of H⁺ and lactose—some for lactose binding, others for

(a)

(b)

FIGURE 11–41 Lactose uptake in *E. coli*. (a) The primary transport of H⁺ out of the cell, driven by the oxidation of a variety of fuels, establishes both a proton gradient and an electrical potential (inside negative) across the membrane. Secondary active transport of lactose into the cell involves symport of H⁺ and lactose by the lactose transporter. The uptake of lactose against its concentration gradient is entirely dependent on this inflow of protons driven by the electrochemical gradient. **(b)** When the energy-yielding oxidation reactions of metabolism are blocked by cyanide (CN⁻), the lactose transporter allows equilibration of lactose across the membrane via passive transport. Mutations that affect Glu³²⁵ or Arg³⁰² have the same effect as cyanide. The dashed line represents the concentration of lactose in the surrounding medium.

Cytoplasm

Periplasmic space

(a)   **(b)**

FIGURE 11-42 The lactose transporter (lactose permease) of *E. coli*.
(a) Ribbon representation viewed parallel to the plane of the membrane shows the 12 transmembrane helices arranged in two nearly symmetric domains, shown in different shades of purple. In the form of the protein for which the crystal structure was determined, the substrate sugar (red) is bound near the middle of the membrane where the sugar is exposed to the cytoplasm (derived from PDB ID 1PV7). **(b)** The postulated second conformation of the transporter (PDB ID 2CFQ), related to the first by a large, reversible conformational change in which the substrate-binding site is exposed first to the periplasm, where lactose is picked up, then to the cytoplasm, where the lactose is released. The interconversion of the two forms is driven by changes in the pairing of charged (protonatable) side chains such as those of Glu[325] and Arg[302] (green), which is affected by the transmembrane proton gradient.

proton transport. Mutation in either of two residues (Glu[325] and Arg[302]; Fig. 11–42) results in a protein still able to catalyze facilitated diffusion of lactose but incapable of coupling H^+ flow to uphill lactose transport. A similar effect is seen in wild-type (unmutated) cells when their ability to generate a proton gradient is blocked with CN^-: the transporter carries out facilitated diffusion normally, but it cannot pump lactose against a concentration gradient (Fig. 11–41b). The balance between the two conformations of the lactose transporter is affected by changes in charge pairing between the side chains of Glu[325] and Arg[302].

In intestinal epithelial cells, glucose and certain amino acids are accumulated by symport with Na^+, down the Na^+ gradient established by the Na^+K^+ ATPase of the plasma membrane **(Fig. 11–43)**. The apical surface of the intestinal epithelial cell is covered with microvilli, long, thin projections of the plasma membrane that greatly increase the surface area exposed to the intestinal contents. **Na^+-glucose symporters** in the apical plasma membrane take up glucose from the intestine in a process driven by the downhill flow of Na^+:

$$2Na^+_{out} + glucose_{out} \longrightarrow 2Na^+_{in} + glucose_{in}$$

The energy required for this process comes from two sources: the greater concentration of Na^+ outside than inside (the chemical potential) and the membrane (electrical) potential, which is inside negative and therefore draws Na^+ inward.

WORKED EXAMPLE 11-3 Energetics of Pumping by Symport

Calculate the maximum $\dfrac{[glucose]_{in}}{[glucose]_{out}}$ ratio that can be achieved by the plasma membrane Na^+-glucose symporter of an epithelial cell when $[Na^+]_{in}$ is 12 mM, $[Na^+]_{out}$ is 145 mM, the membrane potential is −50 mV (inside negative), and the temperature is 37 °C.

Solution: Using Equation 11–4 (p. 412), we can calculate the energy inherent in an electrochemical Na^+ gradient—that is, the cost of moving one Na^+ ion up this gradient:

$$\Delta G_t = RT \ln \frac{[Na^+]_{out}}{[Na^+]_{in}} + Z\mathcal{F}\Delta\psi$$

We then substitute standard values for R, T, and $\mathcal{F}$; the given values for $[Na^+]$ (expressed as molar concentrations); +1 for Z (because Na^+ has a positive charge); and 0.050 V for $\Delta\psi$. Note that the membrane potential

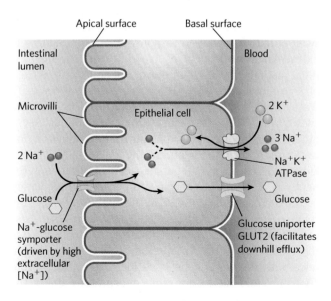

Apical surface   **Basal surface**

Intestinal lumen

Blood

Microvilli

Epithelial cell

2 K$^+$

3 Na$^+$

2 Na$^+$

Na$^+$K$^+$
ATPase

Glucose

Glucose

Na$^+$-glucose
symporter
(driven by high
extracellular
[Na$^+$])

Glucose uniporter
GLUT2 (facilitates
downhill efflux)

FIGURE 11-43 Glucose transport in intestinal epithelial cells. Glucose is cotransported with Na^+ across the apical plasma membrane into the epithelial cell. It moves through the cell to the basal surface, where it passes into the blood via GLUT2, a passive glucose uniporter. The Na^+K^+ ATPase continues to pump Na^+ outward to maintain the Na^+ gradient that drives glucose uptake.

is -50 mV (inside negative), so the change in potential when an ion moves from inside to outside is 50 mV.

$$\Delta G_t = (8.315 \text{ J/mol·K})(310 \text{ K}) \ln \frac{1.45 \times 10^{-1}}{1.2 \times 10^{-2}} +$$
$$1(96{,}500 \text{ J/V·mol})(0.050 \text{ V})$$
$$= 11.2 \text{ kJ/mol}$$

When Na^+ reenters the cell, it releases the electrochemical potential created by pumping it out; ΔG for reentry is -11.2 kJ/mol of Na^+. This is the potential energy per mole of Na^+ that is available to pump glucose. Given that two Na^+ ions pass down their electrochemical gradient and into the cell for each glucose carried in by symport, the energy available to pump 1 mole of glucose is 2×11.2 kJ/mol $= 22.4$ kJ/mol. We can now calculate the maximum concentration ratio of glucose that can be achieved by this pump (from Equation 11–3, p. 411):

$$\Delta G_t = RT \ln \frac{[\text{glucose}]_{in}}{[\text{glucose}]_{out}}$$

Rearranging, then substituting the values of ΔG_t, R, and T, gives

$$\ln \frac{[\text{glucose}]_{in}}{[\text{glucose}]_{out}} = \frac{\Delta G_t}{RT} = \frac{22.4 \text{ kJ/mol}}{(8.315 \text{ J/mol·K})(310 \text{ K})} = 8.69$$

$$\frac{[\text{glucose}]_{in}}{[\text{glucose}]_{out}} = e^{8.69}$$

$$= 5.94 \times 10^3$$

Thus the cotransporter can pump glucose inward until its concentration inside the epithelial cell is about 6,000 times that outside (in the intestine). (This is the maximum theoretical ratio, assuming a perfectly efficient coupling of Na^+ reentry and glucose uptake.)

As glucose is pumped from the intestine into the epithelial cell at the apical surface, it is simultaneously moved from the cell into the blood by passive transport through a glucose transporter (GLUT2) in the basal surface (Fig. 11–43). The crucial role of Na^+ in symport and antiport systems such as this requires the continued outward pumping of Na^+ to maintain the transmembrane Na^+ gradient.

Because of the essential role of ion gradients in active transport and energy conservation, compounds that collapse ion gradients across cellular membranes are effective poisons, and those that are specific for infectious microorganisms can serve as antibiotics. One such substance is valinomycin, a small cyclic peptide that neutralizes the K^+ charge by surrounding it with six carbonyl oxygens **(Fig. 11–44)**. The hydrophobic peptide then acts as a shuttle, carrying K^+ across membranes down its concentration gradient and deflating that gradient. Compounds that shuttle ions across membranes in this way are called **ionophores** ("ion bearers"). Both valinomycin and monensin (a Na^+-carrying ionophore)

FIGURE 11–44 Valinomycin, a peptide ionophore that binds K⁺. In this image, the surface contours are shown as a yellow envelope, through which a stick structure of the peptide and a K⁺ ion (green) are visible. The oxygen atoms (red) that bind K⁺ are part of a central hydrophilic cavity. Hydrophobic amino acid side chains (yellow) coat the outside of the molecule. Because the exterior of the K⁺-valinomycin complex is hydrophobic, the complex readily diffuses through membranes, carrying K⁺ down its concentration gradient. The resulting dissipation of the transmembrane ion gradient kills microbial cells, making valinomycin a potent antibiotic.

are antibiotics; they kill microbial cells by disrupting secondary transport processes and energy-conserving reactions. Monensin is widely used as an antifungal and antiparasitic agent. ■

Aquaporins Form Hydrophilic Transmembrane Channels for the Passage of Water

Peter Agre

A family of integral membrane proteins discovered by Peter Agre, the **aquaporins (AQPs)**, provide channels for rapid movement of water molecules across all plasma membranes. Aquaporins are found in all organisms, and multiple aquaporin genes are generally present, encoding similar but not identical proteins. Eleven aquaporins are known in mammals, each with a specific localization and role (Table 11–5). Erythrocytes, which swell or shrink rapidly in response to abrupt changes in extracellular osmolarity as blood travels through the renal medulla, have a high density of aquaporin in their plasma membrane (2×10^5 copies of AQP1 per cell). Water secretion by the exocrine glands that produce sweat, saliva, and tears occurs through aquaporins. Seven different aquaporins play roles in urine production and water retention in the nephron (the functional unit of the kidney). Each renal AQP has a specific localization in the nephron, and each has specific properties and regulatory features. For example, AQP2 in the epithelial cells of the renal collecting duct is regulated by vasopressin (also called antidiuretic hormone): more water is reabsorbed in the kidney when the vasopressin level is high. Mutant mice with no AQP2 gene have increased urine output (polyuria) and decreased urine-concentrating ability, the

TABLE 11–5 Permeability Characteristics and Predominant Distribution of Known Mammalian Aquaporins

Aquaporin	Permeant (permeability)	Tissue distribution	Subcellular distribution*
AQP0	Water (low)	Lens	Plasma membrane
AQP1	Water (high)	Erythrocyte, kidney, lung, vascular endothelium, brain, eye	Plasma membrane
AQP2	Water (high)	Kidney, vas deferens	Apical plasma membrane, intracellular vesicles
AQP3	Water (high), glycerol (high), urea (moderate)	Kidney, skin, lung, eye, colon	Basolateral plasma membrane
AQP4	Water (high)	Brain, muscle, kidney, lung, stomach, small intestine	Basolateral plasma membrane
AQP5	Water (high)	Salivary gland, lacrimal gland, sweat gland, lung, cornea	Apical plasma membrane
AQP6	Water (low), anions ($NO_3^- > Cl^-$)	Kidney	Intracellular vesicles
AQP7	Water (high), glycerol (high), urea (high)	Adipose tissue, kidney, testis	Plasma membrane
AQP8[†]	Water (high)	Testis, kidney, liver, pancreas, small intestine, colon	Plasma membrane, intracellular vesicles
AQP9	Water (low), glycerol (high), urea (high), arsenite	Liver, leukocyte, brain, testis	Plasma membrane
AQP10	Water (low), glycerol (high), urea (high)	Small intestine	Intracellular vesicles

Source: Data from King, L.S., Kozono, D., & Agre, P. (2004) From structure to disease: the evolving tale of aquaporin biology. *Nat. Rev. Mol. Cell Biol.* 5, 688.

*Aquaporins that are present primarily in the apical or in the basolateral membrane are noted as localized in one of these membranes; those present in both membranes are described as localized in the plasma membrane.

[†]AQP8 might also be permeated by urea.

result of decreased water permeability of the proximal tubule. In humans, genetically defective AQPs are known to be responsible for a variety of diseases, including a relatively rare form of diabetes that is accompanied by polyuria (Box 11–1).

Water molecules flow through an AQP1 channel at a rate of about 10^9 s^{-1}. For comparison, the highest known turnover number for an enzyme is that for catalase, $4 \times 10^7 \text{ s}^{-1}$, and many enzymes have turnover numbers between 1 s^{-1} and 10^4 s^{-1} (see Table 6–7). The low activation energy for passage of water through aquaporin channels ($\Delta G^{\ddagger} < 15 \text{ kJ/mol}$) suggests that water moves through the channels in a continuous stream, in the direction dictated by the osmotic gradient. (For a discussion of osmosis, see p. 56.) Aquaporins do not allow passage of protons (hydronium ions, H_3O^+), which would collapse membrane electrochemical gradients. What is the basis for this extraordinary selectivity?

We find an answer in the structure of AQP1, as determined by x-ray crystallography. AQP1 **(Fig. 11–45a)** consists of four identical monomers (each M_r 28,000), each of which forms a transmembrane pore with a diameter sufficient to allow passage of water molecules in single file. Each monomer has six transmembrane helical segments and two shorter helices, both of which contain the sequence Asn–Pro–Ala (NPA). The six transmembrane helices form the pore through the monomer, and the two short loops containing the NPA sequences extend toward the middle of the bilayer from opposite sides. Their NPA regions overlap in the middle of the membrane to form part of the specificity filter—the structure that allows only water to pass (Fig. 11–45b).

The water channel narrows to a diameter of 2.8 Å near the center of the membrane, severely restricting the size of molecules that can travel through. The positive charge of a highly conserved Arg residue at this bottleneck discourages the passage of cations such as H_3O^+. The residues that line the channel of each AQP1 monomer are generally nonpolar, but carbonyl oxygens in the peptide backbone, projecting into the narrow part of the channel at intervals, can hydrogen-bond with individual water molecules as they pass through; the two Asn residues (Asn^{76} and Asn^{192}) in the NPA loops also form hydrogen bonds with the water. The structure of the channel does not permit formation of a chain of water molecules close enough to allow proton hopping (see Fig. 2–14), which would effectively move protons across the membrane. Critical Arg and His residues and

Outside

Inside

(a)

(b)

FIGURE 11–45 Aquaporin. The protein is a tetramer of identical subunits, each with a transmembrane pore. **(a)** A monomer of spinach aquaporin SoPIP2;1 (derived from PDB ID 2B5F), viewed in the plane of the membrane. The helices form a central pore, and two short helical segments (green) contain the Asn–Pro–Ala (NPA) sequences, found in all aquaporins, that form part of the water channel. **(b)** This cartoon of bovine aquaporin 1 (derived from PDB ID 1J4N) shows that the pore (brown; filled with water molecules shown in red and white) narrows at His^{180} to a diameter of 2.8 Å (about the size of a water molecule), limiting passage of molecules larger than H_2O. The positive charge of Arg^{195} repels cations, including H_3O^+, preventing their passage through the pore. The two short helices shown in green are oriented with their positively charged dipoles pointed at the pore in such a way as to force a water molecule to reorient as it passes through; this breaks up hydrogen-bonded chains of water molecules, preventing proton passage by "proton hopping."

electric dipoles formed by the short helices of the NPA loops provide positive charges in positions that repel any protons that might leak through the pore and prevent hydrogen bonding between adjacent water molecules.

An aquaporin isolated from spinach is known to be "gated"—open when two critical Ser residues near the intracellular end of the channel are phosphorylated, and closed when they are dephosphorylated. Both the open and closed structures have been determined by crystallography. Phosphorylation favors a conformation that presses two nearby Leu residues and a His residue into the channel, blocking the movement of water past that point and effectively closing the channel. Other aquaporins are regulated in other ways, allowing rapid changes in membrane permeability to water.

Although generally highly specific for water, some AQPs also allow glycerol or urea to pass at high rates (Table 11–5); these AQPs are believed to be important in the metabolism of glycerol. AQP7, for example, found in the plasma membranes of adipocytes (fat cells), transports glycerol efficiently. Mice with defective AQP7 develop obesity and non–insulin-dependent diabetes, presumably as a result of their inability to move glycerol into or out of adipocytes as triacylglycerols are converted to free fatty acids and glycerol, and as glycerol is acylated to triacylglycerol.

Ion-Selective Channels Allow Rapid Movement of Ions across Membranes

Ion-selective channels—first recognized in neurons and now known to be present in the plasma membranes of all cells, as well as in the intracellular membranes of eukaryotes—provide another mechanism for moving inorganic ions across membranes. Ion channels, together with ion pumps such as the Na^+K^+ ATPase, determine a plasma membrane's permeability to specific ions and regulate the cytosolic concentration of ions and the membrane potential. In neurons, very rapid changes in the activity of ion channels cause the changes in membrane potential (action potentials) that carry signals from one end of a neuron to the other. In myocytes, rapid opening of Ca^{2+} channels in the sarcoplasmic reticulum releases the Ca^{2+} that triggers muscle contraction. We discuss the signaling functions of ion channels in Chapter 12. Here we describe the structural basis for ion-channel function, using as examples a voltage-gated K^+ channel, the neuronal Na^+ channel, and the acetylcholine receptor ion channel.

Ion channels are distinct from ion transporters in at least three ways. First, the rate of flux through channels can be several orders of magnitude greater than the turnover number for a transporter—10^7 to 10^8 ions/s for an ion channel, approaching the theoretical maximum for unrestricted diffusion. By contrast, the turnover rate of the Na^+K^+ ATPase is about $100\ s^{-1}$. Second, ion channels are not saturable: rates do not approach a maximum at high substrate concentration. Third, they are gated in response to some cellular event. In **ligand-gated channels** (which are generally oligomeric), binding of an extracellular or intracellular small molecule forces an allosteric transition in the protein, which opens or closes the channel. In **voltage-gated ion channels**, a change in transmembrane electrical potential (V_m) causes a charged protein domain to move relative to the membrane, opening or closing the channel. Both types of gating can be very fast. A channel typically opens in a fraction of a millisecond and may remain open for only milliseconds, making these molecular devices effective for very fast signal transmission in the nervous system.

Ion-Channel Function Is Measured Electrically

Because a single ion channel typically remains open for only a few milliseconds, monitoring this process is beyond the limit of most biochemical measurements.

Ion fluxes must therefore be measured electrically, either as changes in V_m (in the millivolt range) or as electric current I (in the microampere or picoampere range), using microelectrodes and appropriate amplifiers. In **patch-clamping**, a technique developed by Erwin Neher and Bert Sakmann in 1976, very small currents are measured through a tiny region of the membrane surface containing only one or a few ion-channel molecules **(Fig. 11–46)**. The researcher can measure the size and duration of the current that flows during one opening of an ion channel and can determine how often a channel opens and how that frequency is affected by membrane potential, regulatory ligands, toxins, and other agents. Patch-clamp studies have revealed that as many as 10^4 ions can move through a single ion channel in 1 ms. Such an ion flux represents a huge amplification of the initial signal; for example, only two acetylcholine molecules are needed to open an acetylcholine receptor channel (as described below).

Erwin Neher

Bert Sakmann

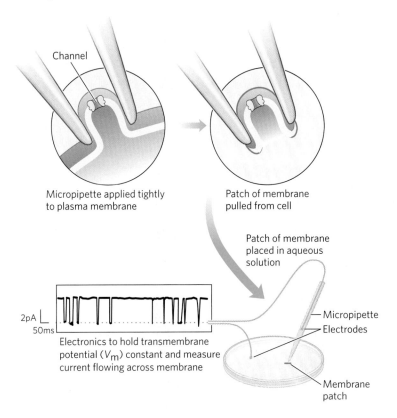

2pA
50ms

Micropipette applied tightly to plasma membrane

Patch of membrane pulled from cell

Patch of membrane placed in aqueous solution

Electronics to hold transmembrane potential (V_m) constant and measure current flowing across membrane

Micropipette
Electrodes

Membrane patch

FIGURE 11–46 Electrical measurements of ion-channel function. The "activity" of an ion channel is estimated by measuring the flow of ions through it, using the patch-clamp technique. A finely drawn-out pipette (micropipette) is pressed against the cell surface, and negative pressure in the pipette forms a pressure seal between pipette and membrane. As the pipette is pulled away from the cell, it pulls off a tiny patch of membrane (which may contain one or a few ion channels). After placing the pipette and attached patch in an aqueous solution, the researcher can measure channel activity as the electric current that flows between the contents of the pipette and the aqueous solution. In practice, a circuit is set up that "clamps" the transmembrane potential at a given value and measures the current that must flow to maintain this voltage. With highly sensitive current detectors, researchers can measure the current flowing through a single ion channel, typically a few picoamperes. The trace shows the current through a single acetylcholine receptor channel as a function of time (in milliseconds), revealing how fast the channel opens and closes, how frequently it opens, and how long it stays open. Downward deflection represents channel opening. Clamping the V_m at different values permits determination of the effect of membrane potential on these parameters of channel function.

The Structure of a K⁺ Channel Reveals the Basis for Its Specificity

Roderick MacKinnon

The structure of a potassium channel from the bacterium *Streptomyces lividans*, determined crystallographically by Roderick MacKinnon in 1998, provides important insight into the way ion channels work. This bacterial ion channel is related in sequence to all other known K⁺ channels and serves as the prototype for such channels, including the voltage-gated K⁺ channel of neurons. Among the members of this protein family, the similarities in sequence are greatest in the "pore region," which contains the ion selectivity filter that allows K⁺ (radius 1.33 Å) to pass 10^4 times more readily than Na⁺ (radius 0.95 Å)—at a rate (about 10^8 ions/s) approaching the theoretical limit for unrestricted diffusion.

The K⁺ channel consists of four identical subunits that span the membrane and form a cone within a cone surrounding the ion channel, with the wide end of the double cone facing the extracellular space **(Fig. 11–47a)**. Each subunit has two transmembrane α helices as well as a third, shorter helix that contributes to the pore region. The outer cone is formed by one of the transmembrane helices of each subunit. The inner cone, formed by the other four transmembrane helices, surrounds the ion channel and cradles the ion selectivity filter. Viewed perpendicular to the plane of the membrane, the central channel is seen to be just wide enough to accommodate an unhydrated metal ion such as potassium (Fig. 11–47b).

Both the ion specificity and the high flux through the channel are understandable from what we know of the channel's structure (Fig. 11–47c). At the inner and outer plasma membrane surfaces, the entryways to the channel have several negatively charged amino acid

residues, which presumably increase the local concentration of cations such as K⁺ and Na⁺. The ion path through the membrane begins (on the inner surface) as a wide, water-filled channel in which the ion can retain its hydration sphere. Further stabilization is provided by the short helices in the pore region of each subunit, with the partial negative charges of their electric dipoles pointed at K⁺ in the channel. About two-thirds of the way through the membrane, this channel narrows in the region of the selectivity filter, forcing the ion to give up its hydrating water molecules. Carbonyl oxygen atoms in the backbone of the selectivity filter replace the water molecules in the hydration sphere, forming a series of perfect coordination shells through which the K⁺ moves. This favorable interaction with the filter is not possible for Na⁺, which is too small to make contact with all the potential oxygen ligands. The preferential stabilization of K⁺ is the basis for the ion selectivity of the filter, and mutations that change residues in this part of the protein eliminate the channel's ion selectivity. The K⁺-binding sites of the filter are flexible enough to collapse to fit any

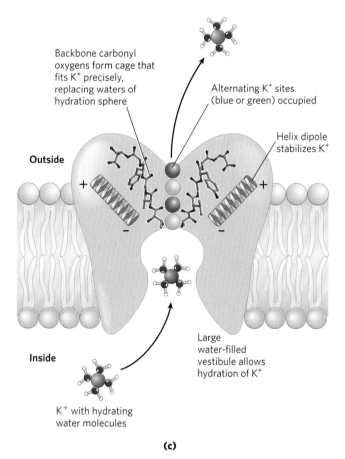

Backbone carbonyl oxygens form cage that fits K⁺ precisely, replacing waters of hydration sphere

Alternating K⁺ sites (blue or green) occupied

Helix dipole stabilizes K⁺

Outside

Inside

Large water-filled vestibule allows hydration of K⁺

K⁺ with hydrating water molecules

Outside

Inside

(a)   (b)   (c)

FIGURE 11–47 The K⁺ channel of *Streptomyces lividans*. (PDB ID 1BL8) **(a)** Viewed in the plane of the membrane, the channel consists of eight transmembrane helices (two from each of four identical subunits), forming a cone with its wide end toward the extracellular space. The inner helices of the cone (lighter colored) line the transmembrane channel, and the outer helices interact with the lipid bilayer. Short segments of each subunit converge in the open end of the cone to make a selectivity filter. **(b)** This view, perpendicular to the plane of the membrane, shows the four subunits

arranged around a central channel just wide enough for a single K⁺ ion to pass. **(c)** Diagram of a K⁺ channel in cross section, showing the structural features critical to function. Carbonyl oxygens (red) of the peptide backbone in the selectivity filter protrude into the channel, interacting with and stabilizing a K⁺ ion passing through. These ligands are perfectly positioned to interact with each of four K⁺ ions but not with the smaller Na⁺ ions. This preferential interaction with K⁺ is the basis for the ion selectivity.

Na$^+$ that enters the channel, and this conformational change closes the channel.

There are four potential K$^+$-binding sites along the selectivity filter, each composed of an oxygen "cage" that provides ligands for the K$^+$ ions (Fig. 11–47c). In the crystal structure, two K$^+$ ions are visible within the selectivity filter, about 7.5 Å apart, and two water molecules occupy the unfilled positions. K$^+$ ions pass through the filter in single file; their mutual electrostatic repulsion most likely just balances the interaction of each ion with the selectivity filter and keeps them moving. Movement of the two K$^+$ ions is concerted: first they occupy positions 1 and 3, then they hop to positions 2 and 4. The energetic difference between these two configurations (1, 3 and 2, 4) is very small;

energetically, the selectivity pore is not a series of hills and valleys but a flat surface, which is ideal for rapid ion movement through the channel. The structure of the channel seems to have been optimized during evolution to give maximal flow rates and high specificity.

Voltage-gated K$^+$ channels are more complex structures than that illustrated in Figure 11–47, but they are variations on the same theme. For example, the mammalian voltage-gated K$^+$ channels in the *Shaker* family have an ion channel like that of the bacterial channel shown in Figure 11–47, but with additional protein domains that sense the membrane potential, move in response to a change in potential, and in moving trigger the opening or closing of the K$^+$ channel **(Fig. 11–48)**. The critical transmembrane helix in the voltage-sensing domain of

FIGURE 11–48 Structural basis for voltage gating in the K$^+$ channel. (PDB ID 2A79) This crystal structure of the Kv1.2-β2 subunit complex from rat brain shows the basic K$^+$ channel (corresponding to that shown in Fig. 11–47) with the extra machinery necessary to make the channel sensitive to gating by membrane potential: four transmembrane helical extensions of each subunit and four β subunits. The entire complex, viewed **(a)** in the plane of the membrane and **(b)** perpendicular to the plane (as viewed from outside the membrane), is represented as in Figure 11–47, with each subunit in a different color; each of the four β subunits is the same color as the subunit with which it associates. In (b), each transmembrane helix of one subunit (red) is numbered, S1 to S6. S5 and S6 from each of four subunits form the channel itself and are comparable to the two transmembrane helices of each subunit in Figure 11–47. S1 to S4 are four transmembrane helices. The

S4 helix contains the highly conserved Arg residues and is believed to be the chief moving part of the voltage-sensing mechanism. **(c)** A schematic diagram of the voltage-gated channel, showing the basic pore structure (center) and the extra structures that make the channel voltage-sensitive; S4, the Arg-containing helix, is orange. For clarity, the β subunits are not shown in this view. In the resting membrane, the transmembrane electrical potential (inside negative) exerts a pull on positively charged Arg side chains in S4, toward the cytosolic side. When the membrane is depolarized, the pull is lessened, and with complete reversal of the membrane potential, S4 is drawn toward the extracellular side. **(d)** This movement of S4 is physically coupled to opening and closing of the K$^+$ channel, which is shown here in its open and closed conformations. Although K$^+$ is present in the closed channel, the pore closes on the bottom, near the cytosol, preventing K$^+$ passage.

Shaker K$^+$ channels contains four Arg residues; the positive charges on these residues cause the helix to move relative to the membrane in response to changes in the transmembrane electric field (the membrane potential).

Cells also have channels that specifically conduct Na$^+$ or Ca^{2+} and exclude K$^+$. In each case, the ability to discriminate among cations requires both a cavity in the binding site of just the right size (neither too large nor too small) to accommodate the ion and the precise positioning within the cavity of carbonyl oxygens that can replace the ion's hydration shell. This fit can be achieved with molecules smaller than proteins; for example, valinomycin (Fig. 11–44) can provide the precise fit that gives high specificity for the binding of one ion rather than another. Chemists have designed small molecules with very high specificity for binding of Li$^+$ (radius 0.60 Å), Na$^+$ (radius 0.95 Å), K$^+$ (radius 1.33 Å), or Rb$^+$ (radius 1.48 Å). The biological versions, however—the channel proteins—not only *bind* specifically but *conduct* ions across membranes in a *gated* fashion.

Gated Ion Channels Are Central in Neuronal Function

Virtually all rapid signaling between neurons and their target tissues (such as muscle) is mediated by the rapid opening and closing of ion channels in plasma membranes. For example, Na$^+$ channels in neuronal plasma membranes sense the transmembrane electrical gradient and respond to changes by opening or closing. These voltage-gated ion channels are typically very selective for Na$^+$ over other monovalent or divalent cations (by factors of 100 or more) and have very high flux rates ($>10^7$ ions/s). Closed in the resting state, Na$^+$ channels are opened—activated—by a reduction in the membrane potential; they then undergo very rapid inactivation. Within milliseconds of opening, a channel closes and remains inactive for many milliseconds. Activation followed by inactivation of Na$^+$ channels is the basis for signaling by neurons (see Fig. 12–26).

Another very well-studied ion channel is the **nicotinic acetylcholine receptor**, which functions in the passage of an electric signal from a motor neuron to a muscle fiber at the neuromuscular junction (signaling the muscle to contract). Acetylcholine released by the motor neuron diffuses a few micrometers to the plasma membrane of a myocyte, where it binds to an acetylcholine receptor. This forces a conformational change in the receptor, causing its ion channel to open. The resulting inward movement of positively charged ions into the myocyte depolarizes its plasma membrane and triggers contraction. The acetylcholine receptor allows Na$^+$, Ca^{2+}, and K$^+$ to pass through its channel with equal ease, but other cations and all anions are unable to pass. Movement of Na$^+$ through an acetylcholine receptor ion channel is unsaturable (its rate is linear with respect to extracellular [Na$^+$]) and very fast—about 2×10^7 ions/s under physiological conditions.

Acetylcholine

The acetylcholine receptor channel is typical of many other ion channels that produce or respond to electric signals: it has a "gate" that opens in response to stimulation by a signal molecule (in this case acetylcholine) and an intrinsic timing mechanism that closes the gate after a split second. Thus the acetylcholine signal is transient—an essential feature of all electric signal conduction.

Based on similarities between the amino acid sequences of other ligand-gated ion channels and the acetylcholine receptor, neuronal receptor channels that respond to the extracellular signals γ-aminobutyric acid (GABA), glycine, and serotonin are grouped in the acetylcholine receptor superfamily and probably share three-dimensional structure and gating mechanisms. The GABA$_A$ and glycine receptors are anion channels specific for Cl$^-$ or HCO$_3^-$, whereas the serotonin receptor, like the acetylcholine receptor, is cation-specific.

Another class of ligand-gated ion channels respond to *intracellular* ligands: 3′,5′-cyclic guanosine mononucleotide (cGMP) in the vertebrate eye, cGMP and cAMP in olfactory neurons, and ATP and inositol 1,4,5-trisphosphate (IP$_3$) in many cell types. These channels are composed of multiple subunits, each with six transmembrane helical domains. We discuss the signaling functions of these ion channels in Chapter 12.

Table 11–6 shows some transporters discussed in other chapters in the context of the pathways in which they act.

Defective Ion Channels Can Have Severe Physiological Consequences

The importance of ion channels to physiological processes is clear from the effects of mutations in specific ion-channel proteins (Table 11–7, Box 11–2). Genetic defects in the voltage-gated Na$^+$ channel of the myocyte plasma membrane result in diseases in which muscles are periodically either paralyzed (as in hyperkalemic periodic paralysis) or stiff (as in paramyotonia congenita). Cystic fibrosis is the result of a mutation that changes one amino acid in the protein CFTR, a Cl$^-$ ion channel; the defective process here is not neurotransmission but secretion by various exocrine gland cells with activities tied to Cl$^-$ ion fluxes.

Many naturally occurring toxins act on ion channels, and the potency of these toxins further illustrates the importance of normal ion-channel function. Tetrodotoxin (produced by the puffer fish, *Sphaeroides rubripes*) and saxitoxin (produced by the marine

TABLE 11–6 Transport Systems Described Elsewhere in This Text

Transport system and location	Figure	Role
Adenine nucleotide antiporter of mitochondrial inner membrane	19–30	Imports substrate ADP for oxidative phosphorylation and exports product ATP
Acetylcholine receptor/channel	12–28	Signals muscle contraction
Acyl-carnitine/carnitine transporter of mitochondrial inner membrane	17–6	Imports fatty acids into matrix for β oxidation
P_i-H^+ symporter of mitochondrial inner membrane	19–30	Supplies P_i for oxidative phosphorylation
Malate–α-ketoglutarate transporter of mitochondrial inner membrane	19–31	Shuttles reducing equivalents (as malate) from matrix to cytosol
Glutamate-aspartate transporter of mitochondrial inner membrane	19–31	Completes shuttling begun by malate–α-ketoglutarate shuttle
Citrate transporter of mitochondrial inner membrane	21–10	Provides cytosolic citrate as source of acetyl-CoA for lipid synthesis
Pyruvate transporter of mitochondrial inner membrane	21–10	Is part of mechanism for shuttling citrate from matrix to cytosol
Fatty acid transporter of myocyte plasma membrane	17–3	Imports fatty acids for fuel
Complex I, III, and IV proton transporters of mitochondrial inner membrane	19–16	Act as energy-conserving mechanism in oxidative phosphorylation, converting electron flow into proton gradient
Thermogenin (uncoupling protein 1), a proton pore of mitochondrial inner membrane	19–36, 23–34	Allows dissipation of proton gradient in mitochondria as means of thermogenesis and/or disposal of excess fuel
Cytochrome bf complex, a proton transporter of chloroplast thylakoid	19–61	Acts as proton pump, driven by electron flow through the Z scheme; source of proton gradient for photosynthetic ATP synthesis
Bacteriorhodopsin, a light-driven proton pump	19–69	Is light-driven source of proton gradient for ATP synthesis in halophilic bacterium
F_oF_1 ATPase/ATP synthase of mitochondrial inner membrane, chloroplast thylakoid, and bacterial plasma membrane	19–25, 19–62a, 19–66	Interconverts energy of proton gradient and ATP during oxidative phosphorylation and photophosphorylation
P_i–triose phosphate antiporter of chloroplast inner membrane	20–15, 20–16	Exports photosynthetic product from stroma; imports P_i for ATP synthesis
Bacterial protein transporter	27–44	Exports secreted proteins through plasma membrane
Protein translocase of ER	27–38	Transports into ER proteins destined for plasma membrane, secretion, or organelles
Nuclear pore protein translocase	27–42	Shuttles proteins between nucleus and cytoplasm
LDL receptor in animal cell plasma membrane	21–41	Imports, by receptor-mediated endocytosis, lipid-carrying particles
Glucose transporter of animal cell plasma membrane; regulated by insulin	12–16	Increases capacity of muscle and adipose tissue to take up excess glucose from blood
IP_3-gated Ca^{2+} channel of ER	12–10	Allows signaling via changes in cytosolic $[Ca^{2+}]$
cGMP-gated Ca^{2+} channel of retinal rod and cone cells	12–37	Allows signaling via rhodopsin linked to cAMP-dependent phosphodiesterase in vertebrate eye
Voltage-gated Na^+ channel of neuron	12–26	Creates action potentials in neuronal signal transmission

TABLE 11–7   Some Diseases Resulting from Ion Channel Defects

Ion channel	Affected gene	Disease
Na$^+$ (voltage-gated, skeletal muscle)	SCN4A	Hyperkalemic periodic paralysis (or paramyotonia congenita)
Na$^+$ (voltage-gated, neuronal)	SCN1A	Generalized epilepsy with febrile seizures
Na$^+$ (voltage-gated, cardiac muscle)	SCN5A	Long QT syndrome 3
Ca^{2+} (neuronal)	CACNA1A	Familial hemiplegic migraine
Ca^{2+} (voltage-gated, retina)	CACNA1F	Congenital stationary night blindness
Ca^{2+} (polycystin-1)	PKD1	Polycystic kidney disease
K$^+$ (neuronal)	KCNQ4	Dominant deafness
K$^+$ (voltage-gated, neuronal)	KCNQ2	Benign familial neonatal convulsions
Nonspecific cation (cGMP-gated, retinal)	CNCG1	Retinitis pigmentosa
Acetylcholine receptor (skeletal muscle)	CHRNA1	Congenital myasthenic syndrome
Cl$^-$	CFTR	Cystic fibrosis

dinoflagellate *Gonyaulax*, which causes "red tides") act by binding to the voltage-gated Na$^+$ channels of neurons and preventing normal action potentials. Puffer fish is an ingredient of the Japanese delicacy fugu, which may be prepared only by chefs specially trained to separate succulent morsel from deadly poison. Eating shellfish that have fed on *Gonyaulax* can also be fatal; shellfish are not sensitive to saxitoxin, but they concentrate it in their muscles, which become highly poisonous to organisms higher up the food chain. The venom of the black mamba snake contains dendrotoxin, which interferes with voltage-gated K$^+$ channels. Tubocurarine, the active component of curare (used as an arrow poison in the Amazon region), and two other toxins from snake venoms, cobrotoxin and bungarotoxin, block the acetylcholine receptor or prevent the opening of its ion channel. By blocking signals from nerves to muscles, all these toxins cause paralysis and possibly death. On the positive side, the extremely high affinity of bungarotoxin for the acetylcholine receptor ($K_d = 10^{-15}$ M) has proved useful experimentally: the radiolabeled toxin was used to quantify the receptor during its purification. ∎

SUMMARY 11.3 Solute Transport across Membranes

- Movement of polar compounds and ions across biological membranes requires transporter proteins. Some transporters simply facilitate passive diffusion across the membrane from the side with higher concentration to the side with lower. Others transport solutes against an electrochemical gradient; this requires a source of metabolic energy.

- Carriers, like enzymes, show saturation and stereospecificity for their substrates. Transport via these systems may be passive or active. Primary active transport is driven by ATP or electron-transfer reactions; secondary active transport is driven by coupled flow of two solutes, one of which (often H$^+$ or Na$^+$) flows down its electrochemical gradient as the other is pulled up its gradient.

- The GLUT transporters, such as GLUT1 of erythrocytes, carry glucose into cells by facilitated diffusion. These transporters are uniporters, carrying only one substrate. Symporters permit simultaneous passage of two substances in the same direction; examples are the lactose transporter of *E. coli*, driven by the energy of a proton gradient (lactose-H$^+$ symport), and the glucose transporter of intestinal epithelial cells, driven by a Na$^+$ gradient (glucose-Na$^+$ symport). Antiporters mediate simultaneous passage of two substances in opposite directions; examples are the chloride-bicarbonate exchanger of erythrocytes and the ubiquitous Na$^+$K$^+$ ATPase.

- In animal cells, Na$^+$K$^+$ ATPase maintains the differences in cytosolic and extracellular concentrations of Na$^+$ and K$^+$, and the resulting Na$^+$ gradient is used as the energy source for a variety of secondary active transport processes.

- The Na$^+$K$^+$ ATPase of the plasma membrane and the Ca^{2+} transporters of the sarcoplasmic and endoplasmic reticulum (the SERCA pumps) are examples of P-type ATPases; they undergo reversible phosphorylation during their catalytic cycle. F-type ATPase proton pumps (ATP synthases) are central to energy-conserving mechanisms in mitochondria and chloroplasts. V-type ATPases produce gradients of protons across some intracellular membranes, including plant vacuolar membranes.

- ABC transporters carry a variety of substrates (including many drugs) out of cells, using ATP as energy source.

▶ Ionophores are lipid-soluble molecules that bind specific ions and carry them passively across membranes, dissipating the energy of electrochemical ion gradients.

▶ Water moves across membranes through aquaporins. Some aquaporins are regulated; some also transport glycerol or urea.

▶ Ion channels provide hydrophilic pores through which select ions can diffuse, moving down their electrical or chemical concentration gradients; they characteristically are unsaturable, have very high flux rates, and are highly specific for one ion. Most are voltage- or ligand-gated. The neuronal Na^+ channel is voltage-gated, and the acetylcholine receptor ion channel is gated by acetylcholine, which triggers conformational changes that open and close the transmembrane path.

Key Terms

Terms in bold are defined in the glossary.

fluid mosaic model 387
micelle 387
bilayer 387
vesicle 388
integral proteins 389
peripheral proteins 389
amphitropic proteins 390
annular lipid 391
hydropathy index 392
positive-inside rule 393
β barrel 393
porin 393
liquid-disordered
   state (l_d) 395
liquid-ordered
   state (l_o) 395
flippases 397
floppases 397
scramblases 397
FRAP 398
microdomains 398
rafts 399
GPI-anchored
   protein 399
caveolin 399
caveolae 399
BAR domain 400
fusion protein 400
v-SNAREs 401
t-SNAREs 401
selectins 402
simple diffusion 403
membrane potential
   (V_m) 403
electrochemical
   gradient 403

electrochemical
   potential 403
facilitated diffusion 403
passive transport 404
transporters 404
permeases 404
channels 404
K_t ($K_{transport}$) 406
electroneutral 409
cotransport 409
antiport 409
symport 409
uniport 409
active transport 409
electrogenic 410
P-type ATPases 410
SERCA pump 410
Na^+K^+ ATPase 411
V-type ATPases 412
F-type ATPases 412
ATP synthase 413
ABC transporters 413
multidrug
   transporters 413
lactose transporter 416
major facilitator superfamily
   (MFS) 416
Na^+-glucose symporters 417
ionophore 418
aquaporins (AQPs) 418
ion channel 420
ligand-gated channel 421
voltage-gated channel 421
patch-clamping 421
nicotinic acetylcholine
   receptor 424

Further Reading

Composition and Architecture of Membranes

Dowhan, W. (1997) Molecular basis for membrane phospholipids diversity: why are there so many lipids? *Annu. Rev. Biochem.* **66**, 199–232.

Ediden, M. (2002) Lipids on the frontier: a century of cell-membrane bilayers. *Nat. Rev. Mol. Cell Biol.* **4**, 414–418.
   Short review of how the notion of a lipid bilayer membrane was developed and confirmed.

Leventis, P.A. & Grinstein, S. (2010) The distribution and function of phosphatidylserine in cellular membranes. *Annu. Rev. Biophys.* **39**, 407–427.

Maxfield, R.R. & van Meer, G. (2010) Cholesterol, the central lipid of mammalian cells *Curr. Opin. Cell Biol.* **22**, 422–429.

Von Heijne, G. (2006) Membrane protein topology. *Nat. Rev. Mol. Cell Biol.* **7**, 909–918.

White, S.H., Ladokhin, A.S., Jayasinghe, S., & Hristova, K. (2001) How membranes shape protein structure. *J. Biol. Chem.* **276**, 32,395–32,398.
   Brief, intermediate-level review of the forces that shape transmembrane helices.

Wimley, W.C. (2003) The versatile β barrel membrane protein. *Curr. Opin. Struct. Biol.* **13**, 1–8.
   Intermediate-level review.

Zeth, K. & Thein, M. (2010) Porins in prokaryotes and eukaryotes: common themes and variations. *Biochem. J.* **431**, 13–22.
   Intermediate-level review of the β-barrel porins.

Membrane Dynamics

Daleke, D.L. (2007) Phospholipid flippases. *J. Biol. Chem.* **282**, 821–825.
   Intermediate-level review.

Deveaux, P.F., Lopez-Montero, I., & Bryde, S. (2006) Proteins involved in lipid translocation in eukaryotic cells. *Chem. Phys. Lipids* **141**, 119–132.

Didier, M., Lenne, P.-F., Rigneault, H., & He, H.-T. (2006) Dynamics in the plasma membrane: how to combine fluidity and order. *EMBO J.* **25**, 3446–3457.
   Intermediate-level review of studies of membrane dynamics, with fluorescent and other probes.

Frost, A., Unger, V.M., & De Camilli, P. (2009) The BAR domain superfamily: membrane-molding macromolecules. *Cell* **137**, 191–196.

Frye, L.D. & Ediden, M. (1970) The rapid intermixing of cell-surface antigens after formation of mouse-human heterokaryons. *J. Cell Sci.* **7**, 319–335.
   The classic demonstration of membrane protein mobility.

Graham, T.R. (2004) Flippases and vesicle-mediated protein transport. *Trends Cell Biol.* **14**, 670–677.
   Intermediate-level review of flippase function.

Graham, T.R. & Kozlov, M.M. (2010) Interplay of proteins and lipids in generating membrane curvature. *Curr. Opin. Cell Biol.* **22**, 430–436.

Hannich, J.T., Umebayashi, K., & Riezman, H. (2011) Distribution and functions of sterols and sphingolipids. In *The Biology of Lipids: Trafficking, Regulation, and Function* (Simons, K., ed.), Cold Spring Harbor Laboratory Press, Cold Spring Harbor, NY; also in *CSH Perspect. Cell Biol.* doi:10.1101/cshperspect.a004762.

Jahn, R. & Scheller, R.H. (2006) SNAREs—engines for membrane fusion. *Nat. Rev. Cell Mol. Biol.* **7**, 631–643.
   Excellent intermediate-level review of the role of SNAREs in membrane fusion and the fusion mechanism itself.

Janmey, P.A. & Kunnunen, P.K.J. (2006) Biophysical properties of lipids and dynamic membranes. *Trends Cell Biol.* **16**, 538–546.

Leventis, P.A. & Grinstein, S. (2010) The distribution and function of phosphatidylserine in cellular membranes. *Annu. Rev. Biophys.* **39**, 407–427.
  Advanced review that includes discussion of flippases.

Lingwood, D. & Simons, K. (2010) Lipid rafts as a membrane-organizing principle. *Science* **327**, 46–50.

MacCallum, J.L. & Tieleman, D.P. (2011) Hydrophobicity scales: a thermodynamic looking glass into lipid-protein interactions. *Trends Biochem. Sci.* **36**, 653–662.
  Intermediate-level review of several methods for determining an amino acid's hydrophobicity.

Marguet, D., Lenne, P.-F., Rigneault, H., & He, H.-T. (2006) Dynamics in the plasma membrane: how to combine fluidity and order. *EMBO J.* **25**, 3446–3457.
  Intermediate-level review of the methods and results of studies on molecular motions in the membrane.

Martens, S. & McMahon, H.T. (2008) Mechanisms of membrane fusion: disparate players and common principles. *Nat. Rev. Mol. Cell Biol.* **9**, 543–566.

Niessen, C.M., Leckband, D., & Yap, A.S. (2011) Tissue organization by cadherin adhesion molecules: dynamic molecular and cellular mechanisms of morphogenetic regulation. *Physiol. Rev.* **91**, 691–731.

Palmgren, M.G. & Nissen, P. (2011) P-type ATPases. *Annu. Rev. Biophys.* **40**, 243–266.
  Advanced review of transporters that include P4-type flippases.

Palsdottir, H. & Hunte, C. (2004) Lipids in membrane protein structures. *Biochim. Biophys. Acta* **1666**, 2–18.

Parton, R.G. & Simons, K. (2007) The multiple faces of caveolae. *Nat. Rev. Mol. Cell Biol.* **8**, 185–194.

Phillips, R., Ursell, T., Wiggins, P., & Sens, P. (2009) Emerging roles for lipids in shaping membrane-protein function. *Nature* **459**, 379–385.
  Intermediate-level review.

Qualman, B., Koch, D., & Kessels, M.M. (2011) Let's go bananas: revisiting the endocytic BAR code. *EMBO J.* **30**, 3501–3515.
  Intermediate-level review of the action of BAR domains in causing membrane curvature.

Quinn, P.J. & Wolf, C. (2009) The liquid ordered phase in membranes. *Biochim. Biophys. Acta* **1788**, 33–46.
  Advanced review of the state of membrane lipids. This is one of 26 excellent reviews that appear in this issue of the journal on all aspects of the physical state of lipids in biological membranes.

Sanyal, S. & Menon, A.K. (2009) Flipping lipids: why an' what's the reason for? *ACS Chem. Biol.* **4**, 895–909.
  Intermediate-level review of flippases in membrane biogenesis.

Sezgin, E. & Schwille, P. (2011) Fluorescence techniques to study lipid dynamics. In *The Biology of Lipids: Trafficking, Regulation, and Function* (Simons, K., ed.), Cold Spring Harbor Laboratory Press, Cold Spring Harbor, NY; also in *CSH Perspect. Cell Biol.*, doi:10.1101/cshperspect.a009803.

Simons, K. & Sampaio, J.L. (2011) Membrane organization and lipid rafts. In *The Biology of Lipids: Trafficking, Regulation, and Function* (Simons, K., ed.), Cold Spring Harbor Laboratory Press, Cold Spring Harbor, NY; also in *CSH Perspect. Cell Biol.*, doi:10.1101/cshperspect.a004697.

Tanaka, K., Fujimura-Kamada, K., & Yamamoto, T. (2011) Functions of phospholipid flippases. *J. Biochem.* **149**, 131–143.
  Intermediate-level review of flippase structure and function.

van der Velden, L.M., van de Graaf, S.F.J., & Klomp, L.W.J. (2010) Biochemical and cellular functions of P$_4$ ATPases. *Biochem. J.* **431**, 1–11.

Intermediate-level review of the P$_4$ ATPase flippases.

van Deurs, B., Roepstorff, K., Hommelgaard, A.M., & Sandpit, K. (2003) Caveolae: anchored, multifunctional platforms in the lipid ocean. *Trends Cell Biol.* **13**, 92–100.

Van Meer, G. (2011) Dynamic transbilayer lipid asymmetry. In *The Biology of Lipids: Trafficking, Regulation, and Function* (Simons, K., ed.), Cold Spring Harbor Laboratory Press, Cold Spring Harbor, NY; also in *CSH Perspect. Cell Biol.*, doi:10.1101/cshperspect.a004671.

Wickner, W. & Shekmana, R. (2010) Membrane fusion *Nat. Struct. Mol. Biol.* **15**, 658–664.
  Brief and accessible review of the molecules involved in membrane fusion.

Zhang, Y.-M. & Rock, C.O. (2008) Membrane lipid homeostasis in bacteria. *Nat. Rev. Microbiol.* **6**, 222–233.

Zimmerberg, J. & Kozlov, M.M. (2006) How proteins produce cellular membrane curvature. *Nat. Rev. Mol. Cell Biol.* **7**, 9–19.

Transporters

Augustin, R. (2010) The protein family of glucose transport facilitators: it's not only about glucose after all. *IUBMB Life* **62**, 315–333.
  Advanced review of the GLUT proteins' structure and function.

Brini, M. & Carafoli, E. (2009) Calcium pumps in health and disease *Physiol. Rev.* **89**, 1341–1378.

Bublitz, M., Poulson, H., Preben Morth, J., & Nissen, P. (2010) In and out of the cation pumps: P-type ATPase structure revisited. *Curr. Opin. Struct. Biol.* **20**, 431–439.

Bublitz, M., Preben Morth, J., & Nissen, P. (2011) P-type ATPases at a glance. *J. Cell Sci.* **124**, 2515–2519.

Fujiyoshi, Y., Mitsuoka, K., de Groot, B.L., Philippsen, A., Grubmüller, H., Agre, P., & Engel, A. (2002) Structure and function of water channels. *Curr. Opin. Struct. Biol.* **12**, 509–515.

Guan, L. & Kaback, H.R. (2006) Lessons from lactose permease. *Annu. Rev. Biophys. Biomol. Struct.* **35**, 67–91.

Hoffman, N.J. & Elmendorf, J.S. (2011) Signaling, cytoskeletal and membrane mechanisms regulating GLUT4 exocytosis. *Trends Endocrinol. Metab.* **22**, 110–116.

Jones, P.M., O'Mara, M.L., & George, A.M. (2001) ABC transporters: a riddle wrapped in a mystery inside an enigma. *Trends Biochem. Sci.* **34**, 520–531.
  Intermediate-level review of ATP transporter structure and function.

Kjellbom, P., Larsson, C., Johansson, I., Karlsson, M., & Johanson, U. (1999) Aquaporins and water homeostasis in plants. *Trends Plant Sci.* **4**, 308–314.
  Intermediate-level review.

Krishnamurthy, H., Piscitelli, C.L., & Gouaus, E. (2009) Unlocking the molecular secrets of sodium-coupled transporters. *Nature* **459**, 347–355.
  Intermediate-level review of the structure and function of the Na$^+$ cotransporters.

Preben Morth, J., Pedersen, B.P., Buch-Pedersen, M.J., Andersen, J.P., Vilsen, B., Palmgren, M.G., & Nissen, P. (2011) A structural overview of the plasma membrane Na$^+$,K$^+$-ATPase and H$^+$-ATPase ion pumps. *Nat. Rev. Mol. Cell Biol.* **12**, 60–70.

Rees, D.C., Johnson, E., & Lewinson, O. (2009) ABC transporters: the power to change. *Nat. Rev. Mol. Cell Biol.* **10**, 218–227.

Sui, H., Han, B.-G., Lee, J.K., Walian, P., & Jap, B.K. (2001) Structural basis of water-specific transport through the AQP1 water channel. *Nature* **414**, 872–878.
  High-resolution solution of the aquaporin structure by x-ray crystallography.

Thorens, B. & Mueckler, M. (2010) Glucose transporters in the 21st century. *Am. J. Physiol. Endocrinol. Metab.* **298**, E141–E145.
  Intermediate-level review of glucose facilitators.

Toyoshima, C. & Mizutani, T. (2004) Crystal structure of the calcium pump with a bound ATP analogue. *Nature* **430**, 529–535.

Toyoshima, C., Nomura, H., & Tsuda, T. (2004) Lumenal gating mechanism revealed in calcium pump crystal structures with phosphate analogs. *Nature* **432**, 361–368.

The supplementary materials available with the online version of this article include an excellent movie of the putative gating mechanism.

Watson, R.T. & Pessin, J.E. (2006) Bridging the GAP between insulin signaling and GLUT4 translocation. *Trends Biochem. Sci.* **31**, 215–222.

Intermediate-level review of the regulation of glucose transport through GLUT4.

Wright, E.M., Loo, D.D.F., & Hirayama, B.A. (2011) Biology of human sodium glucose transporters. *Physiol. Rev.* **91**, 733–794.

Advanced review of the Na⁺-glucose cotransporter's biochemistry and physiology.

Zeth, K. & Thein, M. (2010) Porins in prokaryotes and eukaryotes: common themes and variations. *Biochem. J.* **431**, 13–22.

Ion Channels

Ashcroft, F.M. (2006). From molecule to malady. *Nature* **440**, 440–447.

A short review of the many known cases in which genetic defects in ion channels lead to disease in humans.

Doyle, D.A., Cabral, K.M., Pfuetzner, R.A., Kuo, A., Gulbis, J.M., Cohen, S.L., Chait, B.T., & MacKinnon, R. (1998) The structure of the potassium channel: molecular basis of K⁺ conduction and selectivity. *Science* **280**, 69–77.

The first crystal structure of an ion channel is described.

Gadsby, D.C., Vergani, P., & Csanady, L. (2006) The ABC protein turned chloride channel whose failure causes cystic fibrosis. *Nature* **440**, 477–483.

This is one of seven excellent reviews of ion channels published together in this issue of *Nature*.

Gouaux, E. & MacKinnon, R. (2005) Principles of selective ion transport in channels and pumps. *Science* **310**, 1461–1465.

Short review of the architectural features of channels and pumps that give each protein its ion specificity.

Guggino, W.B. & Stanton, B.A. (2006) New insights into cystic fibrosis: molecular switches that regulate CFTR. *Nat. Rev. Mol. Cell Biol.* **7**, 426–436.

Hille, B. (2001) *Ion Channels of Excitable Membranes*, 3rd edn, Sinauer Associates, Sunderland, MA.

Intermediate-level text emphasizing the function of ion channels.

Jiang, Y., Lee, A., Chen, J., Ruta, V., Cadene, M., Chait, B.T., & MacKinnon, R. (2003) X-ray structure of a voltage-dependent K⁺ channel. *Nature* **423**, 33–41.

King, L.S., Kozono, D., & Agre, P. (2004) From structure to disease: the evolving tale of aquaporin biology. *Nat. Rev. Mol. Cell Biol.* **5**, 687–698.

Intermediate-level review of the localization of aquaporins in mammalian tissues and the effects of aquaporin defects on physiology.

Lee, A.G. & East, J.M. (2001) What the structure of a calcium pump tells us about its mechanism. *Biochem J.* **356**, 665–683.

Long, S.B., Campbell, E.B., & MacKinnon, R. (2005) Crystal structure of a mammalian voltage-dependent *Shaker* family K⁺ channel. *Science* **309**, 897–902.

Neher, E. & Sakmann, B. (1992) The patch clamp technique. *Sci. Am.* (March) **266**, 44–51.

Clear description of the electrophysiological methods used to measure the activity of single ion channels, by the Nobel Prize–winning developers of this technique.

Tombola, F., Pathak, M.M., & Isacoff, E.Y. (2006) How does voltage open an ion channel? *Annu. Rev. Cell Dev. Biol.* **22**, 23–52.

Advanced review of the mechanisms of voltage gating of ion channels.

Verkman, A.S. (2011) Aquaporins at a glance. *J. Cell Sci.* **124**, 2107–2112.

Intermediate-level, poster-style review of aquaporin structure and function.

Problems

1. Determining the Cross-Sectional Area of a Lipid Molecule When phospholipids are layered gently onto the surface of water, they orient at the air-water interface with their head groups in the water and their hydrophobic tails in the air. An experimental apparatus **(a)** has been devised that reduces the surface area available to a layer of lipids. By measuring the force necessary to push the lipids together, it is possible to determine when the molecules are packed tightly in a continuous monolayer; as that area is approached, the force needed to further reduce the surface area increases sharply **(b)**. How would you use this apparatus to determine the average area occupied by a single lipid molecule in the monolayer?

(a) Force applied here to compress monolayer

(b)

2. Evidence for a Lipid Bilayer In 1925, E. Gorter and F. Grendel used an apparatus like that described in Problem 1 to determine the surface area of a lipid monolayer formed by lipids extracted from erythrocytes of several animal species. They used a microscope to measure the dimensions of individual cells, from which they calculated the average surface area of one erythrocyte. They obtained the data shown in the table. Were these investigators justified in concluding that "chromocytes [erythrocytes] are covered by a layer of fatty substances that is two molecules thick" (i.e., a lipid bilayer)?

Animal	Volume of packed cells (mL)	Number of cells (per mm³)	Total surface area of lipid monolayer from cells (m²)	Total surface area of one cell (μm²)
Dog	40	8,000,000	62	98
Sheep	10	9,900,000	6.0	29.8
Human	1	4,740,000	0.92	99.4

Source: Data from Gorter, E. & Grendel, F. (1925) On bimolecular layers of lipoids on the chromocytes of the blood. *J. Exp. Med.* 41, 439–443.

3. Number of Detergent Molecules per Micelle When a small amount of the detergent sodium dodecyl sulfate (SDS; $Na^+CH_3(CH_2)_{11}OSO_3^-$) is dissolved in water, the detergent ions enter the solution as monomeric species. As more detergent is added, a concentration is reached (the critical micelle concentration) at which the monomers associate to form micelles. The critical micelle concentration of SDS is 8.2 mM. The micelles have an average particle weight (the sum of the molecular weights of the constituent monomers) of 18,000. Calculate the number of detergent molecules in the average micelle.

4. Properties of Lipids and Lipid Bilayers Lipid bilayers formed between two aqueous phases have this important property: they form two-dimensional sheets, the edges of which close on each other and undergo self-sealing to form vesicles (liposomes).

(a) What properties of lipids are responsible for this property of bilayers? Explain.

(b) What are the consequences of this property for the structure of biological membranes?

5. Length of a Fatty Acid Molecule The carbon–carbon bond distance for single-bonded carbons such as those in a saturated fatty acyl chain is about 1.5 Å. Estimate the length of a single molecule of palmitate in its fully extended form. If two molecules of palmitate were placed end to end, how would their total length compare with the thickness of the lipid bilayer in a biological membrane?

6. Temperature Dependence of Lateral Diffusion The experiment described in Figure 11–18 was performed at 37 °C. If the experiment were carried out at 10 °C, what effect would you expect on the rate of diffusion? Why?

7. Synthesis of Gastric Juice: Energetics Gastric juice (pH 1.5) is produced by pumping HCl from blood plasma (pH 7.4) into the stomach. Calculate the amount of free energy required to concentrate the H^+ in 1 L of gastric juice at 37 °C. Under cellular conditions, how many moles of ATP must be hydrolyzed to provide this amount of free energy? The free-energy change for ATP hydrolysis under cellular conditions is about −58 kJ/mol (as explained in Chapter 13). Ignore the effects of the transmembrane electrical potential.

8. Energetics of the Na^+K^+ ATPase For a typical vertebrate cell with a membrane potential of −0.070 V (inside negative), what is the free-energy change for transporting 1 mol of Na^+ from the cell into the blood at 37 °C? Assume the concentration of Na^+ inside the cell is 12 mM and that in blood plasma is 145 mM.

9. Action of Ouabain on Kidney Tissue Ouabain specifically inhibits the Na^+K^+ ATPase activity of animal tissues but is not known to inhibit any other enzyme. When ouabain is added to thin slices of living kidney tissue, it inhibits oxygen consumption by 66%. Why? What does this observation tell us about the use of respiratory energy by kidney tissue?

10. Energetics of Symport Suppose you determined experimentally that a cellular transport system for glucose, driven by symport of Na^+, could accumulate glucose to concentrations 25 times greater than in the external medium, while the external $[Na^+]$ was only 10 times greater than the intracellular $[Na^+]$. Would this violate the laws of thermodynamics? If not, how could you explain this observation?

11. Location of a Membrane Protein The following observations are made on an unknown membrane protein, X. It can be extracted from disrupted erythrocyte membranes into a concentrated salt solution, and it can be cleaved into fragments by proteolytic enzymes. Treatment of erythrocytes with proteolytic enzymes followed by disruption and extraction of membrane components yields intact X. However, treatment of erythrocyte "ghosts" (which consist of just plasma membranes, produced by disrupting the cells and washing out the hemoglobin) with proteolytic enzymes followed by disruption and extraction yields extensively fragmented X. What do these observations indicate about the location of X in the plasma membrane? Do the properties of X resemble those of an integral or peripheral membrane protein?

12. Membrane Self-Sealing Cellular membranes are self-sealing—if they are punctured or disrupted mechanically, they quickly and automatically reseal. What properties of membranes are responsible for this important feature?

13. Lipid Melting Temperatures Membrane lipids in tissue samples obtained from different parts of a reindeer's leg have different fatty acid compositions. Membrane lipids from tissue near the hooves contain a larger proportion of unsaturated fatty acids than those from tissue in the upper leg. What is the significance of this observation?

14. Flip-Flop Diffusion The inner leaflet (monolayer) of the human erythrocyte membrane consists predominantly of phosphatidylethanolamine and phosphatidylserine. The outer leaflet consists predominantly of phosphatidylcholine and sphingomyelin. Although the phospholipid components of the membrane can diffuse in the fluid bilayer, this sidedness is preserved at all times. How?

15. Membrane Permeability At pH 7, tryptophan crosses a lipid bilayer at about one-thousandth the rate of indole, a closely related compound:

Suggest an explanation for this observation.

16. Water Flow through an Aquaporin A human erythrocyte has about 2×10^5 AQP1 monomers. If water molecules flow through the plasma membrane at a rate of 5×10^8 per AQP1 tetramer per second and the volume of an erythrocyte is 5×10^{-11} mL, how rapidly could an erythrocyte halve its volume as it encountered the high osmolarity (1 M) in the interstitial fluid of the renal medulla? Assume that the erythrocyte consists entirely of water.

17. Labeling the Lactose Transporter A bacterial lactose transporter, which is highly specific for lactose, contains a Cys residue that is essential to its transport activity. Covalent reaction of *N*-ethylmaleimide (NEM) with this Cys residue irreversibly inactivates the transporter. A high concentration of lactose in the medium prevents inactivation by NEM, presumably by sterically protecting the Cys residue, which is in or near the lactose-binding site. You know nothing else about the transporter protein. Suggest an experiment that might allow you to determine the M_r of the Cys-containing transporter polypeptide.

18. Predicting Membrane Protein Topology from Sequence You have cloned the gene for a human erythrocyte protein, which you suspect is a membrane protein. From the nucleotide sequence of the gene, you know the amino acid sequence. From this sequence alone, how would you evaluate the possibility that the protein is an integral protein? Suppose the protein proves to be an integral protein, either type I or type II. Suggest biochemical or chemical experiments that might allow you to determine which type it is.

19. Intestinal Uptake of Leucine You are studying the uptake of L-leucine by epithelial cells of the mouse intestine. Measurements of the rate of uptake of L-leucine and several of its analogs, with and without Na^+ in the assay buffer, yield the results given in the table. What can you conclude about the properties and mechanism of the leucine transporter? Would you expect L-leucine uptake to be inhibited by ouabain?

Substrate	Uptake in presence of Na^+		Uptake in absence of Na^+	
	V_{max}	K_t (mM)	V_{max}	K_t (mM)
L-Leucine	420	0.24	23	0.2
D-Leucine	310	4.7	5	4.7
L-Valine	225	0.31	19	0.31

20. Effect of an Ionophore on Active Transport Consider the leucine transporter described in Problem 19. Would V_{max} and/or K_t change if you added a Na^+ ionophore to the assay solution containing Na^+? Explain.

21. Surface Density of a Membrane Protein *E. coli* can be induced to make about 10,000 copies of the lactose transporter (M_r 31,000) per cell. Assume that *E. coli* is a cylinder 1 μm in diameter and 2 μm long. What fraction of the plasma membrane surface is occupied by the lactose transporter molecules? Explain how you arrived at this conclusion.

22. Use of the Helical Wheel Diagram A helical wheel is a two-dimensional representation of a helix, a view along its central axis (see Fig. 11–30b; see also Fig. 4–4d). Use the helical wheel diagram shown here to determine the distribution of amino acid residues in a helical segment with the sequence –Val–Asp–Arg–Val–Phe–Ser–Asn–Val–Cys–Thr–His–Leu–Lys–Thr–Leu–Gln–Asp–Lys–

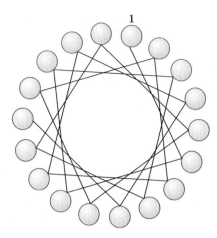

What can you say about the surface properties of this helix? How would you expect the helix to be oriented in the tertiary structure of an integral membrane protein?

23. Molecular Species in the *E. coli* Membrane The plasma membrane of *E. coli* is about 75% protein and 25% phospholipid by weight. How many molecules of membrane lipid are present for each molecule of membrane protein? Assume an average protein M_r of 50,000 and an average phospholipid M_r of 750. What more would you need to know to estimate the fraction of the membrane surface that is covered by lipids?

Using the Web

24. Membrane Protein Topology The receptor for the hormone epinephrine in animal cells is an integral membrane protein (M_r 64,000) that is believed to have seven membrane-spanning regions.

(a) Show that a protein of this size is capable of spanning the membrane seven times.

(b) Given the amino acid sequence of this protein, how would you predict which regions of the protein form the membrane-spanning helices?

(c) Go to the Protein Data Bank (www.pdb.org). Use the PDB identifier 1DEP to retrieve the data page for a portion of the β-adrenergic receptor (one type of epinephrine receptor) isolated from a turkey. Using Jmol to explore the structure, predict whether this portion of the receptor is located within the membrane or at the membrane surface. Explain.

(d) Retrieve the data for a portion of another receptor, the acetylcholine receptor of neurons and myocytes, using PDB identifier 1A11. As in (c), predict where this portion of the receptor is located and explain your answer.

If you have not used the PDB, see Box 4–4 (p. 132) for more information.

Data Analysis Problem

25. The Fluid Mosaic Model of Biological Membrane Structure Figure 11–3 shows the currently accepted fluid mosaic model of biological membrane structure. This model was presented in detail in a review article by S. J. Singer in 1971. In the article, Singer presented the three models of membrane structure that had been proposed by that time:

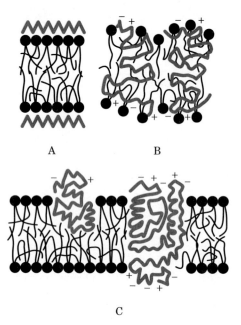

A   B

C

A. The Davson-Danielli-Robertson Model. This was the most widely accepted model in 1971, when Singer's review was published. In this model, the phospholipids are arranged as a bilayer. Proteins are found on both surfaces of the bilayer, attached to it by ionic interactions between the charged head groups of the phospholipids and charged groups in the proteins. Crucially, there is no protein in the interior of the bilayer.

B. The Benson Lipoprotein Subunit Model. Here the proteins are globular and the membrane is a protein-lipid mixture. The hydrophobic tails of the lipids are embedded in the hydrophobic parts of the proteins. The lipid head groups are exposed to the solvent. There is no lipid bilayer.

C. The Lipid-Globular Protein Mosaic Model. This is the model shown in Figure 11–3. The lipids form a bilayer and proteins are embedded in it, some extending through the bilayer and others not. Proteins are anchored in the bilayer by hydrophobic interactions between the hydrophobic tails of the lipids and hydrophobic portions of the protein.

For the data given below, consider how each piece of information aligns with each of the three models of membrane structure. Which model(s) are supported, which are not supported, and what reservations do you have about the data or their interpretation? Explain your reasoning.

(a) When cells were fixed, stained with osmium tetroxide, and examined in the electron microscope, the membranes showed a "railroad track" appearance, with two dark-staining lines separated by a light space.

(b) The thickness of membranes in cells fixed and stained in the same way was found to be 5 to 9 nm. The thickness of a "naked" phospholipid bilayer, without proteins, was 4 to 4.5 nm. The thickness of a single monolayer of proteins was about 1 nm.

(c) Singer wrote in his article: "The average amino acid composition of membrane proteins is not distinguishable from that of soluble proteins. In particular, a substantial fraction of the residues is hydrophobic" (p. 165).

(d) As described in Problems 1 and 2 of this chapter, researchers had extracted membranes from cells, extracted the lipids, and compared the area of the lipid monolayer with the area of the original cell membrane. The interpretation of the results was complicated by the issue illustrated in the graph of Problem 1: the area of the monolayer depended on how hard it was pushed. With very light pressures, the ratio of monolayer area to cell membrane area was about 2.0. At higher pressures—thought to be more like those found in cells—the ratio was substantially lower.

(e) Circular dichroism spectroscopy uses changes in polarization of UV light to make inferences about protein secondary structure (see Fig. 4–10). On average, this technique showed that membrane proteins have a large amount of α helix and little or no β sheet. This finding was consistent with most membrane proteins having a globular structure.

(f) Phospholipase C is an enzyme that removes the polar head group (including the phosphate) from phospholipids. In several studies, treatment of intact membranes with phospholipase C removed about 70% of the head groups without disrupting the "railroad track" structure of the membrane.

(g) Singer described in his article a study in which "a glycoprotein of molecular weight about 31,000 in human red blood cell membranes is cleaved by tryptic treatment of the membranes into soluble glycopeptides of about 10,000 molecular weight, while the remaining portions are quite hydrophobic" (p. 199). Trypsin treatment did not cause gross changes in the membranes, which remained intact.

Singer's review also included many more studies in this area. In the end, though, the data available in 1971 did not conclusively prove Model C was correct. As more data have accumulated, this model of membrane structure has been accepted by the scientific community.

Reference

Singer, S.J. (1971) The molecular organization of biological membranes. In *Structure and Function of Biological Membranes* (Rothfield, L.I., ed.), pp. 145–222, Academic Press, Inc., New York.

Biosignaling

The ability of cells to receive and act on signals from beyond the plasma membrane is fundamental to life. Bacterial cells receive constant input from membrane proteins that act as information receptors, sampling the surrounding medium for pH, osmotic strength, the availability of food, oxygen, and light, and the presence of noxious chemicals, predators, or competitors for food. These signals elicit appropriate responses, such as motion toward food or away from toxic substances or the formation of dormant spores in a nutrient-depleted medium. In multicellular organisms, cells with different functions exchange a wide variety of signals. Plant cells respond to growth hormones and to variations in sunlight. Animal cells exchange information about the concentrations of ions and glucose in extracellular fluids, the interdependent metabolic activities taking place in different tissues, and, in an embryo, the correct placement of cells during development. In all these cases, the signal represents *information* that is detected by specific receptors and converted to a cellular response, which always involves a *chemical* process. This conversion of information into a chemical change, **signal transduction**, is a universal property of living cells.

12.1 General Features of Signal Transduction

Signal transductions are remarkably specific and exquisitely sensitive. **Specificity** is achieved by precise molecular complementarity between the signal and receptor molecules (**Fig. 12–1a**), mediated by the same kinds of weak (noncovalent) forces that mediate enzyme-substrate and antigen-antibody interactions. Multicellular organisms have an additional level of specificity, because the receptors for a given signal, or the intracellular targets of a given signal pathway, are present only in certain cell types. Thyrotropin-releasing hormone, for example, triggers responses in the cells of the anterior pituitary but not in hepatocytes, which lack receptors for this hormone. Epinephrine alters glycogen metabolism in hepatocytes but not in adipocytes; in this case, both cell types have receptors for the hormone, but whereas hepatocytes contain glycogen and the glycogen-metabolizing enzyme that is stimulated by epinephrine, adipocytes contain neither. Adipocytes respond to epinephrine by releasing fatty acids from triacylglycerols and exporting them to other tissues.

Three factors account for the extraordinary sensitivity of signal transduction: the high affinity of receptors for signal molecules, cooperativity (often but not always) in the ligand-receptor interaction, and amplification of the signal by enzyme cascades. The **affinity** between signal (ligand) and receptor can be expressed as the dissociation constant K_d, commonly 10^{-10} M or less—meaning that the receptor detects picomolar concentrations of a signal molecule. Receptor-ligand interactions are quantified by Scatchard analysis, which

(a) Specificity
Signal molecule fits
binding site on its
complementary receptor;
other signals do not fit.

(b) Amplification
When enzymes activate
enzymes, the number of
affected molecules
increases geometrically
in an enzyme cascade.

(c) Modularity
Proteins with multivalent
affinities form diverse
signaling complexes from
interchangeable parts.
Phosphorylation provides
reversible points of
interaction.

(d) Desensitization/Adaptation
Receptor activation triggers
a feedback circuit that shuts
off the receptor or removes
it from the cell surface.

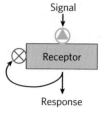

(e) Integration
When two signals have
opposite effects on a
metabolic characteristic
such as the concentration
of a second messenger X,
or the membrane potential
V_m, the regulatory outcome
results from the integrated
input from both receptors.

FIGURE 12–1 Five features of signal-transducing systems.

yields a quantitative measure of affinity (K_d) and the number of ligand-binding sites in a receptor sample (Box 12–1).

Cooperativity in receptor-ligand interactions results in large changes in receptor activation with small changes in ligand concentration (recall the effect of cooperativity on oxygen binding to hemoglobin; see Fig. 5–12). **Amplification** results when an enzyme associated with a signal receptor is activated and, in turn, catalyzes the activation of many molecules of a second enzyme, each of which activates many molecules of a third enzyme, and so on, in a so-called **enzyme cascade** (Fig. 12–1b). Such cascades can produce amplifications of several orders of magnitude within milliseconds. The response to a signal must also be terminated such that the downstream effects are in proportion to the strength of the original stimulus.

Modularity of interacting signaling proteins allows a cell to mix and match a set of signaling molecules to create complexes with different functions or cellular locations. Many signaling proteins have multiple domains that recognize specific features in other proteins, or in the cytoskeleton or plasma membrane, and the resulting multivalency of individual modules allows their assembly into a wide variety of multienzyme complexes. One common theme in such interactions is the binding of one modular signaling protein to phosphorylated residues in another protein; the resulting interaction can be regulated by phosphorylation or dephosphorylation of the protein partner (Fig. 12–1c). Nonenzymatic **scaffold proteins** with affinity for several enzymes that interact in cascades bring those proteins together, ensuring their interaction at specific cellular locations and at specific times.

The sensitivity of receptor systems is subject to modification. When a signal is present continuously, **desensitization** of the receptor system results (Fig. 12–1d); when the stimulus falls below a certain threshold, the system again becomes sensitive. Think of what happens to your visual transduction system when you walk from bright sunlight into a darkened room or from darkness into the light.

A final noteworthy feature of signal-transducing systems is **integration** (Fig. 12–1e), the ability of the system to receive multiple signals and produce a unified response appropriate to the needs of the cell or organism. Different signaling pathways converse with each other at several levels, generating complex cross talk that maintains homeostasis in the cell and the organism.

BOX 12–1 | METHODS | Scatchard Analysis Quantifies the Receptor-Ligand Interaction

The cellular actions of a hormone begin when the hormone (ligand, L) binds specifically and tightly to its protein receptor (R) on or in the target cell. Binding is mediated by noncovalent interactions (hydrogen-bonding, hydrophobic, and electrostatic) between the complementary surfaces of ligand and receptor. Receptor-ligand interaction brings about a conformational change that alters the biological activity of the receptor, which may be an enzyme, an enzyme regulator, an ion channel, or a regulator of gene expression.

Receptor-ligand binding is described by the equation

$$\underset{\text{Receptor}}{R} + \underset{\text{Ligand}}{L} \rightleftharpoons \underset{\text{Receptor-ligand complex}}{RL}$$

This binding, like that of an enzyme to its substrate, depends on the concentrations of the interacting components and can be described by an equilibrium constant:

$$\underset{\text{Receptor}}{R} + \underset{\text{Ligand}}{L} \underset{k_{-1}}{\overset{k_{+1}}{\rightleftharpoons}} \underset{\text{Receptor-ligand complex}}{RL}$$

$$K_a = \frac{[RL]}{[R][L]} = \frac{k_{+1}}{k_{-1}} = 1/K_d$$

where K_a is the association constant and K_d is the dissociation constant.

Like enzyme-substrate binding, receptor-ligand binding is saturable. As more ligand is added to a fixed amount of receptor, an increasing fraction of receptor molecules is occupied by ligand (Fig. 1a). A rough measure of receptor-ligand affinity is given by the concentration of ligand needed to give half-saturation of the receptor. Using **Scatchard analysis** of receptor-ligand binding, we can estimate both the dissociation constant K_d and the number of receptor-binding sites in a given preparation. When binding has reached equilibrium, the total number of possible binding sites, B_{max}, equals the number of unoccupied sites, represented by [R], plus the number of occupied or ligand-bound sites, [RL]; that is, $B_{max} = [R] + [RL]$. The number of unbound sites can be expressed in terms of total sites minus occupied sites: $[R] = B_{max} - [RL]$. The equilibrium expression can now be written

$$K_a = \frac{[RL]}{[L](B_{max} - [RL])}$$

Rearranging to obtain the ratio of receptor-bound ligand to free (unbound) ligand, we get

$$\frac{[\text{Bound}]}{[\text{Free}]} = \frac{[RL]}{[L]} = K_a(B_{max} - [RL])$$
$$= \frac{1}{K_d}(B_{max} - [RL])$$

From this slope-intercept form of the equation, we can see that a plot of [bound ligand]/[free ligand] versus

(a) Total hormone added, [L] + [RL]

(b) Bound hormone, [RL]

FIGURE 1 Scatchard analysis of a receptor-ligand interaction. A radio-labeled ligand (L)—a hormone, for example—is added at several concentrations to a fixed amount of receptor (R), and the fraction of the hormone bound to receptor is determined by separating the receptor-hormone complex (RL) from free hormone.

(a) A plot of [RL] versus [L] + [RL] (total hormone added) is hyperbolic, rising toward a maximum for [RL] as the receptor sites become saturated. To control for nonsaturable, nonspecific binding sites (eicosanoid hormones bind nonspecifically to the lipid bilayer, for example), a separate series of binding experiments is also necessary. A large excess of unlabeled hormone is added along with the dilute solution of labeled hormone. The unlabeled molecules compete with the labeled molecules for specific binding to the saturable site on the receptor, but not for the nonspecific binding. The true value for specific binding is obtained by subtracting nonspecific binding from total binding.

(b) A linear plot of [RL]/[L] versus [RL] gives K_d and B_{max} for the receptor-hormone complex. Compare these plots with those of V_0 versus [S] and $1/V_0$ versus $1/[S]$ for an enzyme-substrate complex (see Fig. 6–12, Box 6-1).

[bound ligand] should give a straight line with a slope of $-K_a$ $(-1/K_d)$ and an intercept on the abscissa of B_{max}, the total number of binding sites (Fig. 1b). Hormone-ligand interactions typically have K_d values of 10^{-9} to 10^{-11} M, corresponding to very tight binding.

Scatchard analysis is reliable for the simplest cases, but as with Lineweaver-Burk plots for enzymes, when the receptor is an allosteric protein, the plots deviate from linearity.

TABLE 12–1   Some Signals to Which Cells Respond

Antigens	Light
Cell surface glycoproteins/ oligosaccharides	Mechanical touch
	Microbial, insect pathogens
Developmental signals	Neurotransmitters
Extracellular matrix components	Nutrients
Growth factors	Odorants
Hormones	Pheromones
Hypoxia	Tastants

One of the revelations of research on signaling is the remarkable degree to which signaling mechanisms have been conserved during evolution. Although the number of different biological signals (Table 12–1) is probably in the thousands, and the kinds of response elicited by these signals are comparably numerous, the machinery for transducing all of these signals is built from about 10 basic types of protein components.

In this chapter we examine some examples of the major classes of signaling mechanisms, looking at how they are integrated in specific biological functions such as the transmission of nerve signals; responses to hormones and growth factors; the senses of sight, smell, and taste; and control of the cell cycle. Often, the end result of a signaling pathway is the phosphorylation of a few specific target-cell proteins, which changes their activities and thus the activities of the cell. Throughout our discussion we emphasize the conservation of fundamental mechanisms for the transduction of biological signals and the adaptation of these basic mechanisms to a wide range of signaling pathways.

We consider the molecular details of several representative signal-transduction systems, classified according to the type of receptor. The trigger for each system is different, but the general features of signal transduction are common to all: a signal interacts with a receptor; the activated receptor interacts with cellular machinery, producing a second signal or a change in the activity of a cellular protein; the metabolic activity of the target cell undergoes a change; and finally, the transduction event ends. To illustrate these general features of signaling systems, we will look at examples of six basic receptor types **(Fig. 12–2)**.

1. *G protein–coupled receptors* that *indirectly* activate (through GTP-binding proteins, or G proteins) enzymes that generate intracellular second messengers. This type of receptor is illustrated by the β-adrenergic receptor system that detects epinephrine (adrenaline) (Section 12.2).

2. *Receptor tyrosine kinases*, plasma membrane receptors that are also enzymes. When one of these receptors is activated by its extracellular

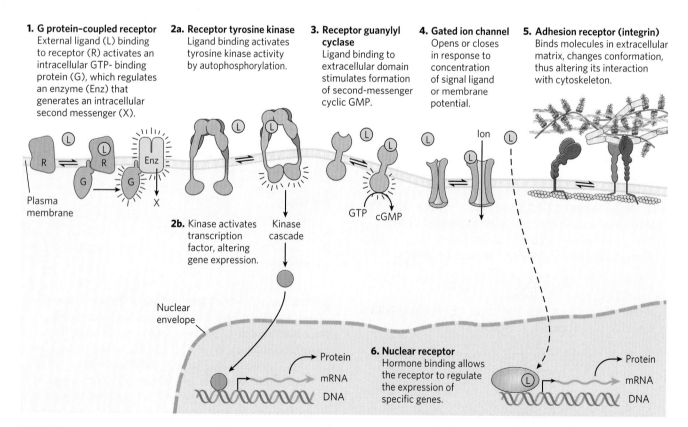

1. G protein–coupled receptor
External ligand (L) binding to receptor (R) activates an intracellular GTP- binding protein (G), which regulates an enzyme (Enz) that generates an intracellular second messenger (X).

2a. Receptor tyrosine kinase
Ligand binding activates tyrosine kinase activity by autophosphorylation.

2b. Kinase activates transcription factor, altering gene expression.

Kinase cascade

3. Receptor guanylyl cyclase
Ligand binding to extracellular domain stimulates formation of second-messenger cyclic GMP.

4. Gated ion channel
Opens or closes in response to concentration of signal ligand or membrane potential.

5. Adhesion receptor (integrin)
Binds molecules in extracellular matrix, changes conformation, thus altering its interaction with cytoskeleton.

6. Nuclear receptor
Hormone binding allows the receptor to regulate the expression of specific genes.

FIGURE 12–2 Six general types of signal transducers.

ligand, it catalyzes the phosphorylation of several cytosolic or plasma membrane proteins. The insulin receptor is one example (Section 12.3); the receptor for epidermal growth factor (EGFR) is another.

3. *Receptor guanylyl cyclases*, which are also plasma membrane receptors with an enzymatic cytoplasmic domain. The intracellular second messenger for these receptors, cyclic guanosine monophosphate (cGMP), activates a cytosolic protein kinase that phosphorylates cellular proteins and thereby changes their activities (Section 12.4).

4. *Gated ion channels* of the plasma membrane that open and close (hence the term "gated") in response to the binding of chemical ligands or changes in transmembrane potential. These are the simplest signal transducers. The acetylcholine receptor ion channel is an example of this mechanism (Section 12.6).

5. *Adhesion receptors* that interact with macromolecular components of the extracellular matrix (such as collagen) and convey instructions to the cytoskeletal system about cell migration or adherence to the matrix. Integrins illustrate this general type of transduction mechanism (Section 12.7).

6. *Nuclear receptors* that bind specific ligands (such as the hormone estrogen) and alter the rate at which specific genes are transcribed and translated into cellular proteins. Because steroid hormones function through mechanisms intimately related to the regulation of gene expression, we consider them here only briefly (Section 12.8) and defer a detailed discussion of their action until Chapter 28.

As we begin this discussion of biological signaling, a word about the nomenclature of signaling proteins is in order. These proteins are typically discovered in one context and named accordingly, then prove to be involved in a broader range of biological functions for which the original name is not helpful. For example, the retinoblastoma protein, pRb, was initially identified as the site of a mutation that contributes to cancer of the retina (retinoblastoma), but it is now known to function in many pathways essential to cell division in all cells, not just those of the retina. Some genes and proteins are given noncommittal names: the tumor suppressor protein p53, for example, is a *protein* of *53* kDa, but its name gives no clue to its great importance in the regulation of cell division and the development of cancer. In this chapter we generally define these protein names as we encounter them, introducing the names commonly used by researchers in the field. Don't be discouraged if you can't get them all straight the first time you encounter them!

SUMMARY 12.1 General Features of Signal Transduction

▶ All cells have specific and highly sensitive signal-transducing mechanisms, which have been conserved during evolution.

▶ A wide variety of stimuli act through specific protein receptors in the plasma membrane.

▶ The receptors bind the signal molecule and initiate a process that amplifies the signal, integrates it with input from other receptors, and transmits the information throughout the cell. If the signal persists, receptor desensitization reduces or ends the response.

▶ Multicellular organisms have six general types of signaling mechanisms: plasma membrane proteins that act through G proteins, receptor tyrosine kinases, receptor guanylyl cyclases that act through a protein kinase, gated ion channels, adhesion receptors that carry information between the extracellular matrix and the cytoskeleton, and nuclear receptors that bind steroids and alter gene expression.

12.2 G Protein–Coupled Receptors and Second Messengers

As their name implies, **G protein–coupled receptors (GPCRs)** are receptors that are closely associated with a member of the **guanosine nucleotide–binding protein (G protein)** family. Three essential components define signal transduction through GPCRs: a plasma membrane receptor with seven transmembrane helical segments, a G protein that cycles between active (GTP-bound) and inactive (GDP-bound) forms, and an effector enzyme (or ion channel) in the plasma membrane that is regulated by the activated G protein. The G protein, stimulated by the activated receptor, exchanges bound GDP for GTP, then dissociates from the occupied receptor and binds to the nearby effector enzyme, altering its activity. The activated enzyme then generates a **second messenger** that affects downstream targets. The human genome encodes about 350 GPCRs for detecting hormones, growth factors, and other endogenous ligands, and perhaps 500 that serve as olfactory (smell) and gustatory (taste) receptors.

GPCRs have been implicated in many common human diseases, including allergies, depression, blindness, diabetes, and various cardiovascular defects with serious health consequences. Close to half of *all* drugs on the market target one GPCR or another. For example, the β-adrenergic receptor, which mediates the effects of epinephrine, is the target of the "beta blockers," prescribed for such diverse conditions as hypertension, cardiac arrhythmia, glaucoma, anxiety, and migraine headache. At least 150 of the GPCRs found in the human genome are still "orphan receptors": their natural ligands are not yet identified, and so we know nothing about their biology. The β-adrenergic receptor, with well-understood biology and pharmacology, is the prototype for all GPCRs, and our discussion of signal-transducing systems begins there. ■

The β-Adrenergic Receptor System Acts through the Second Messenger cAMP

Epinephrine sounds the alarm when some threat requires the organism to mobilize its energy-generating machinery; it signals the need to fight or flee. Epinephrine action begins when the hormone binds to a protein receptor in the plasma membrane of an epinephrine-sensitive cell. **Adrenergic receptors** ("adrenergic" reflects the alternative name for epinephrine, adrenaline) are of four general types, α_1, α_2, β_1, and β_2, defined by differences in their affinities and responses to a group of agonists and antagonists. **Agonists** are structural analogs that bind to a receptor and mimic the effects of its natural ligand; **antagonists** are analogs that bind the receptor without triggering the normal effect and thereby block the effects of agonists, including the biological ligand. In some cases, the affinity of the synthetic agonist or antagonist for the receptor is greater than that of the natural agonist **(Fig. 12–3)**. The four types of adrenergic receptors are found in different target tissues and mediate different responses to epinephrine. Here we focus on the **β-adrenergic receptors** of muscle, liver, and adipose tissue. These receptors mediate changes in fuel metabolism, as described in Chapter 23, including

the increased breakdown of glycogen and fat. Adrenergic receptors of the β_1 and β_2 subtypes act through the same mechanism, so in our discussion, "β-adrenergic" applies to both types.

Like all GPCRs, the β-adrenergic receptor is an integral protein with seven hydrophobic, helical regions of 20 to 28 amino acid residues that span the plasma membrane seven times, thus the alternative name for GPCRs: **heptahelical receptors**. The binding of epinephrine to a site on the receptor deep within the plasma membrane **(Fig. 12–4a**, step **❶**) promotes a conformational change in the receptor's intracellular domain that affects its interaction with an associated G protein, promoting the dissociation of GDP and the binding of GTP (step **❷**). For all GPCRs, the G protein is heterotrimeric, composed of three different subunits: α, β, and γ. Such G proteins are therefore known as **trimeric G proteins**. In this case, it is the α subunit that binds GDP or GTP and transmits the signal from the activated receptor to the effector protein. Because this G protein activates its effector, it is referred to as a **stimulatory G protein**, or **G_s**. Like other G proteins (Box 12–2), G_s functions as a biological "switch": when the nucleotide-binding site of G_s (on the α subunit) is occupied by GTP, G_s is turned on and can activate its effector protein (adenylyl cyclase in the present case); with GDP bound to the site, G_s is switched off. In the active form, the β and γ subunits of G_s dissociate from the α subunit as a $\beta\gamma$ dimer, and $G_{s\alpha}$, with its bound GTP, moves in the plane of the membrane from the receptor to a nearby molecule of adenylyl cyclase (step **❸**). $G_{s\alpha}$ is held to the membrane by a covalently attached palmitoyl group (see Fig. 11–15).

Adenylyl cyclase is an integral protein of the plasma membrane, with its active site on the cytoplasmic face. The association of active $G_{s\alpha}$ with adenylyl cyclase stimulates the cyclase to catalyze cAMP synthesis from ATP (Fig. 12–4a, step **❹**, and Fig. 12–4b), raising the cytosolic [cAMP]. The interaction between $G_{s\alpha}$ and adenylyl cyclase is possible only when $G_{s\alpha}$ is bound to GTP.

The stimulation by $G_{s\alpha}$ is self-limiting; $G_{s\alpha}$ *has intrinsic GTPase activity that inactivates* $G_{s\alpha}$ by converting its bound GTP to GDP **(Fig. 12–5)**. The now inactive $G_{s\alpha}$ dissociates from adenylyl cyclase, rendering the cyclase inactive. $G_{s\alpha}$ reassociates with the $\beta\gamma$ dimer ($G_{s\beta\gamma}$), and inactive G_s is again available to interact with a hormone-bound receptor.

The role of $G_{s\alpha}$ in serving as a biological "switch" protein is not unique. A variety of G proteins act as binary switches in signaling systems with GPCRs and in many processes that involve membrane fusion or fission (Box 12–2). ⬚ **Trimeric G Proteins: Molecular On/Off Switches**

Epinephrine exerts its downstream effects through the increase in [cAMP] that results from the activation of adenylyl cyclase. Cyclic AMP, in turn, allosterically activates **cAMP-dependent protein kinase**, also called **protein kinase A** or **PKA** (Fig. 12–4a, step **❺**),

FIGURE 12–3 Epinephrine and its synthetic analogs. Epinephrine, also called adrenaline, is released from the adrenal gland and regulates energy-yielding metabolism in muscle, liver, and adipose tissue. It also serves as a neurotransmitter in adrenergic neurons. Its affinity for its receptor is expressed as a dissociation constant for the receptor-ligand complex. Isoproterenol and propranolol are synthetic analogs, one an agonist with an affinity for the receptor that is higher than that of epinephrine, and the other an antagonist with extremely high affinity.

(a)

1 Epinephrine binds to its specific receptor.

β-adrenergic receptor

N

Outside

Adenylyl cyclase

C

Inside

$G_{s\beta}$

GDP

GTP

GTP

$G_{s\alpha}$

$G_{s\alpha}$

ATP

$G_{s\gamma}$

GDP

2 Hormone-receptor complex causes the GDP bound to $G_{s\alpha}$ to be replaced by GTP, activating $G_{s\alpha}$.

3 Activated $G_{s\alpha}$ separates from $G_{s\beta\gamma}$, moves to adenylyl cyclase, and activates it. Many $G_{s\alpha}$ subunits may be activated by one occupied receptor.

4 Adenylyl cyclase catalyzes the formation of cAMP.

cAMP

cyclic nucleotide phosphodiesterase

5 cAMP activates PKA.

6 Phosphorylation of cellular proteins by PKA causes the cellular response to epinephrine.

5'-AMP **7** cAMP is degraded, reversing the activation of PKA.

NH_2

PP_i

adenylyl cyclase

H_2O

cyclic nucleotide phosphodiesterase

NH_2

NH_2

ATP

Adenosine 3',5'-cyclic monophosphate (cAMP)

Adenosine 5'-monophosphate (AMP)

(b)

FIGURE 12–4 Transduction of the epinephrine signal: the β-adrenergic pathway. (a) The mechanism that couples binding of epinephrine to its receptor with activation of adenylyl cyclase; the seven steps are discussed further in the text. The same adenylyl cyclase molecule in the plasma membrane may be regulated by a stimulatory G protein (G_s), as shown, or an inhibitory G protein (G_i, not shown). G_s and G_i are under the influence of different hormones. Hormones that induce GTP binding to G_i cause *inhibition* of adenylyl cyclase, resulting in lower cellular [cAMP]. **(b)** The combined action of the enzymes that catalyze steps **4** and **7**, synthesis and hydrolysis of cAMP by adenylyl cyclase and cAMP phosphodiesterase, respectively.

which catalyzes the phosphorylation of specific Ser or Thr residues of targeted proteins, including glycogen phosphorylase *b* kinase. This enzyme is active when phosphorylated and can begin the process of mobilizing glycogen stores in muscle and liver in anticipation of the need for energy, as signaled by epinephrine.

The inactive form of PKA contains two identical catalytic subunits (C) and two identical regulatory subunits (R) **(Fig. 12–6a)**. The tetrameric R_2C_2 complex is catalytically inactive, because an autoinhibitory domain of each R subunit occupies the substrate-binding cleft of each C subunit. When cAMP binds to the R subunits,

① G_s with GDP bound is turned off; it cannot activate adenylyl cyclase.

② Contact of G_s with hormone-receptor complex causes displacement of bound GDP by GTP.

③ G_s with GTP bound dissociates into α and $\beta\gamma$ subunits. $G_{s\alpha}$-GTP is turned on; it can activate adenylyl cyclase.

④ GTP bound to $G_{s\alpha}$ is hydrolyzed by the protein's intrinsic GTPase; $G_{s\alpha}$ thereby turns itself off. The inactive α subunit reassociates with the $\beta\gamma$ subunit.

FIGURE 12–5 The GTPase switch. G proteins cycle between GDP-bound (off) and GTP-bound (on). The protein's intrinsic GTPase activity, in many cases stimulated by RGS proteins (*regulators of G-protein signaling*; see Box 12-2), determines how quickly bound GTP is hydrolyzed to GDP and thus how long the G protein remains active.

they undergo a conformational change that moves the autoinhibitory domain of R out of the catalytic domain of C, and the R₂C₂ complex dissociates to yield two free, catalytically active C subunits. This same basic mechanism—displacement of an autoinhibitory domain—mediates the allosteric activation of many types of protein kinases by their second messengers (as in Figs 12–14 and 12–22, for example). The structure of the substrate-binding cleft in PKA is the prototype for all known protein kinases (Fig. 12–6b); certain residues in this cleft region have identical counterparts in all of the more than 1,000 known protein kinases. The ATP-binding site of each catalytic subunit positions ATP perfectly for the transfer of its terminal (γ) phosphoryl group to the —OH in the side chain of a Ser or Thr residue.

FIGURE 12–6 Activation of cAMP-dependent protein kinase (PKA).
(a) When [cAMP] is low, the two identical regulatory subunits (R; red) associate with the two identical catalytic subunits (C). In this R₂C₂ complex, the inhibitor sequences of the R subunits lie in the substrate-binding cleft of the C subunits and prevent binding of protein substrates; the complex is therefore catalytically inactive. The amino-terminal sequences of the R subunits interact to form an R₂ dimer, the site of binding to an A kinase anchoring protein (AKAP), described later in the text. When [cAMP] rises in response to a hormonal signal, each R subunit binds two cAMP molecules and undergoes a dramatic reorganization that

pulls its inhibitory sequence away from the C subunit, opening up the substrate-binding cleft and releasing each C subunit in its catalytically active form. **(b)** A crystal structure showing part of the R₂C₂ complex (PDB ID 1U7E)—one C subunit and part of one R subunit. The amino-terminal dimerization region of the R subunit is omitted for simplicity. The small lobe of C contains the ATP-binding site, and the large lobe surrounds and defines the cleft where the protein substrate binds and undergoes phosphorylation at a Ser or Thr residue, with the phosphoryl group transferred from ATP. In this inactive form, the inhibitor sequence of R blocks the substrate-binding cleft of C, inactivating it.

BOX 12–2 ⚕ MEDICINE G Proteins: Binary Switches in Health and Disease

Alfred G. Gilman and Martin Rodbell (Fig. 1) discovered the critical roles of guanosine nucleotide–binding proteins (G proteins) in a wide variety of cellular processes, including sensory perception, signaling for cell division, growth and differentiation, intracellular movements of proteins and membrane vesicles, and protein synthesis. The human genome encodes nearly 200 of these proteins, which differ in size and subunit structure, intracellular location, and function. But all G proteins share a common feature: they can become activated and then, after a brief period, can inactivate themselves, thereby serving as molecular binary switches with built-in timers. This superfamily of proteins includes the trimeric G proteins involved in adrenergic signaling (G_s and G_i) and vision (transducin); small G proteins such as that involved in insulin signaling (Ras) and others that function in vesicle trafficking (ARF and Rab), transport into and out of the nucleus (Ran; see Fig. 27–42), and timing of the cell cycle (Rho); and several proteins involved in protein synthesis (initiation factor IF2 and elongation factors EF-Tu and EF-G; see Chapter 26). Many G proteins have covalently bound lipids, which give them an affinity for membranes and dictate their locations in the cell.

All G proteins have the same core structure and use the same mechanism for switching between an inactive conformation, favored when GDP is bound, and an active conformation, favored when GTP is bound. We can use the Ras protein (~20 kDa), a minimal signaling unit, as a prototype for all members of this superfamily (Fig. 2).

In the GTP-bound conformation, the G protein exposes previously buried regions (called **switch I** and **switch II**) that interact with proteins downstream in the signaling pathway, until the G protein inactivates itself by hydrolyzing its bound GTP to GDP. The critical determinant of G-protein conformation is the γ phosphate of GTP, which interacts with a

FIGURE 2 The Ras protein, the prototype for all G proteins (PDB ID 5P21). Mg^{2+}-GTP is held by critical residues in the phosphate-binding P loop (blue), and by Thr^{35} in the switch I (red) and Gly^{60} in the switch II (green) regions. Ala^{146} gives specificity for GTP over ATP. In this structure, the nonhydrolyzable GTP analog Gpp(NH)p is in the GTP-binding site.

region called the **P loop** (*p*hosphate-binding; Fig. 3). In Ras, the γ phosphate of GTP binds to a Lys residue in the P loop and to two critical residues, Thr^{35} in switch I and Gly^{60} in switch II, that hydrogen-bond with the oxygens of the γ phosphate of GTP. These

(Continued on next page)

FIGURE 1 Alfred G. Gilman (left) and Martin Rodbell (1925–1998). Their Nobel lectures on the discovery and exploration of G proteins are available at www.nobelprize.org.

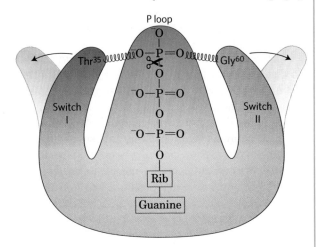

FIGURE 3 When bound GTP is hydrolyzed by the GTPase activities of Ras and its GAP, loss of hydrogen bonds to Thr^{35} and Gly^{60} allows the switch I and switch II regions to relax into a conformation in which they are no longer available to interact with downstream targets such as Raf.

BOX 12–2 MEDICINE G Proteins: Binary Switches in Health and Disease (Continued)

hydrogen bonds act like a pair of springs holding the protein in its active conformation. When GTP is cleaved to GDP and P_i is released, these hydrogen bonds are lost; the protein relaxes into its inactive conformation, burying the sites that interact with other partners in its active state. Ala[146] hydrogen-bonds to the guanine oxygen, allowing GTP, but not ATP, to bind.

The intrinsic GTPase activity of G proteins is increased up to 10^5-fold by **GTPase activator proteins (GAPs)**, also called, in the case of heterotrimeric G proteins, **regulators of G protein signaling (RGSs**; Fig. 4). GAPs (and RGSs) thus determine how long the switch remains "on." They contribute a critical Arg residue that reaches into the G-protein GTPase active site and assists in catalysis. The intrinsically slow process of replacing bound GDP with GTP, switching the protein on, is catalyzed by **guanosine nucleotide–exchange factors (GEFs)** associated with the G protein (Fig. 4).

Because G proteins play crucial roles in so many signaling processes, it is not surprising that defects in G proteins lead to a variety of diseases. In about 25% of all human cancers (and in a much higher proportion of certain types of cancer), there is a mutation in a Ras protein—typically in one of the critical residues around the GTP-binding site or in the P loop—that virtually eliminates its GTPase activity. Once activated by GTP binding, this Ras protein remains constitutively active, promoting cell division in cells that should not divide. The tumor suppressor gene *NF1* encodes a GAP that enhances the GTPase activity of normal Ras. Mutations in *NF1* that result in a nonfunctioning GAP leave Ras with only its intrinsic GTPase activity, which is very weak (has a very low turnover number); once activated by GTP binding, Ras stays active for an extended period, continuing to send the signal: divide.

Defective heterotrimeric G proteins can also lead to disease. Mutations in the gene that encodes the α subunit of G_s (which mediates changes in [cAMP] in response to hormonal stimuli) may result in a G_α that is permanently active or permanently inactive. "Activating" mutations generally occur in residues crucial to GTPase activity; they lead to a continuously elevated [cAMP], with significant downstream consequences, including undesirable cell proliferation. For example,

FIGURE 4 Factors that regulate the activity of G proteins (green). Inactive G proteins (both small G proteins such as Ras and heterotrimeric G proteins such as G_s) interact with upstream GTP-GDP exchange factors, GEFs (red; often these are activated (*) receptors such as rhodopsin, β-adrenergic receptors, and Sos), and are activated by GTP binding. In this form, the G proteins activate downstream effector enzymes (blue; enzymes such as cGMP phosphodiesterase, adenylyl cyclase, and Raf). GTPase activator proteins (GAPs, in the case of small G proteins) and regulators of G protein signaling (RGSs) (yellow), by modulating the GTPase activity of G proteins, determine how long the G protein will remain active.

such mutations are found in about 40% of pituitary tumors (adenomas). Individuals with "inactivating" mutations in G_α are unresponsive to hormones (such as thyroid hormone) that act through cAMP. Mutation in the gene for the transducin α subunit (T_α), which is involved in visual signaling, leads to a type of night blindness, apparently due to defective interaction between the activated T_α subunit and the phosphodiesterase of the rod outer segment (see Fig. 12–39). A sequence variation in the gene encoding the β subunit of a heterotrimeric G protein is commonly found in individuals with hypertension (high blood pressure), and this variant gene is suspected of involvement in obesity and atherosclerosis.

The pathogenic bacteria that cause cholera and pertussis (whooping cough) produce toxins that target G proteins, interfering with normal signaling in host cells. **Cholera toxin**, secreted by *Vibrio cholerae* in the intestine of an infected person, is a heterodimeric protein. Subunit B recognizes and binds to specific

---

As indicated in Figure 12–4a (step ❻), PKA regulates several enzymes downstream in the signaling pathway (Table 12–2). Although these downstream targets have diverse functions, they share a region of sequence similarity around the Ser or Thr residue that

undergoes phosphorylation, a sequence that marks them for regulation by PKA. The substrate-binding cleft of PKA recognizes these sequences and phosphorylates their Thr or Ser residue. Comparison of the sequences of various protein substrates for PKA has

Normal G_s: GTPase activity terminates the signal from receptor to adenylyl cyclase.

cholera toxin

ADP-ribosylated G_s: GTPase activity is inactivated; G_s constantly activates adenylyl cyclase.

NAD^+

ADP-ribose

FIGURE 5 The bacterial toxins that cause cholera and whooping cough (pertussis) are enzymes that catalyze transfer of the ADP-ribose moiety of NAD^+ to an Arg residue of G_s (in the case of cholera toxin, as shown here) or a Cys residue of G_i (pertussis toxin). The G proteins thus modi-fied fail to respond to normal hormonal stimuli. The pathology of both diseases results from defective regulation of adenylyl cyclase and overproduction of cAMP.

gangliosides on the surface of intestinal epithelial cells and provides a route for subunit A to enter these cells. After entry, subunit A is broken into two pieces: the A1 fragment and the A2 fragment. A1 then associates with the ADP-ribosylation factor ARF6, a small G protein in host cells, through residues in its switch I and switch II regions—which are accessible only when ARF6 is in its active (GTP-bound) form. This association with ARF6 activates A1, which catalyzes the transfer of ADP-ribose from NAD^+ to the critical Arg residue in the P loop of the α subunit of G_s (Fig. 5). ADP-ribosylation blocks the GTPase activity of G_s and thereby renders G_s perma-nently active. This results in continuous activation of the adenylyl cyclase of intestinal epithelial cells, chroni-cally high [cAMP], and chronically active PKA. PKA phosphorylates the CFTR Cl^- channel (see Box 11–2) and a Na^+-H^+ exchanger in the intestinal epithelial cells. The resultant efflux of NaCl triggers massive water loss through the intestine as cells respond to the ensuing osmotic imbalance. Severe dehydration and electrolyte loss are the major pathologies in cholera. These can be fatal in the absence of prompt rehydra-tion therapy.

The **pertussis toxin**, produced by *Bordetella pertussis*, catalyzes ADP-ribosylation of the α subunit of G_i, in this case preventing GDP-GTP exchange and blocking inhibition of adenylyl cyclase by G_i. The bac-terium infects the respiratory tract, where it destroys the ciliated epithelial cells that normally sweep away mucus. Without this ciliary action, vigorous coughing is needed to clear the tract; this is the gasping cough that gives the disease its name (and spreads the bac-terium to others). How the defect in G-protein signal-ing kills ciliated epithelial cells is not yet clear.

Given the large number of G protein–coupled receptors in the human genome, it seems likely that future studies will reveal many more examples of how defective G-protein signaling affects human health.

yielded the **consensus sequence**—the neighboring residues needed to mark a Ser or Thr residue for phosphorylation (see Table 12–2).

As in many signaling pathways, signal transduction by adenylyl cyclase entails several steps that *amplify* the original hormone signal (**Fig. 12–7**). First, the binding of one hormone molecule to one receptor mol-ecule catalytically activates many G_s molecules that associate with the activated receptor, one after the other. Next, by activating one molecule of adenylyl

TABLE 12–2 Some Enzymes and Other Proteins Regulated by cAMP-Dependent Phosphorylation (by PKA)

Enzyme/protein	Sequence phosphorylated*	Pathway/process regulated
Glycogen synthase	RASCTSSS	Glycogen synthesis

Phosphorylase b kinase
  α subunit
  β subunit
	VEFRRLSI ⎫
	RTKRSGSV ⎭
		Glycogen breakdown

Pyruvate kinase (rat liver)	GVLRRASVAZL	Glycolysis
Pyruvate dehydrogenase complex (type L)	GYLRRASV	Pyruvate to acetyl-CoA
Hormone-sensitive lipase	PMRRSV	Triacylglycerol mobilization and fatty acid oxidation
Phosphofructokinase-2/fructose 2,6-bisphosphatase	LQRRRGSSIPQ	Glycolysis/gluconeogenesis
Tyrosine hydroxylase	FIGRRQSL	Synthesis of L-dopa, dopamine, norepinephrine, and epinephrine
Histone H1	AKRKASGPPVS	DNA condensation
Histone H2B	KKAKASRKESYSVYVYK	DNA condensation
Cardiac phospholamban (cardiac pump regulator)	AIRRAST	Intracellular $[Ca^{2+}]$
Protein phosphatase-1 inhibitor-1	IRRRPTP	Protein dephosphorylation
PKA consensus sequence†	xR[RK]x[ST]B	Many

*The phosphorylated S or T residue is shown in red. All residues are given as their one-letter abbreviations (see Table 3–1).

†x is any amino acid; B is any hydrophobic amino acid. See Box 3–2 for conventions used in displaying consensus sequences.

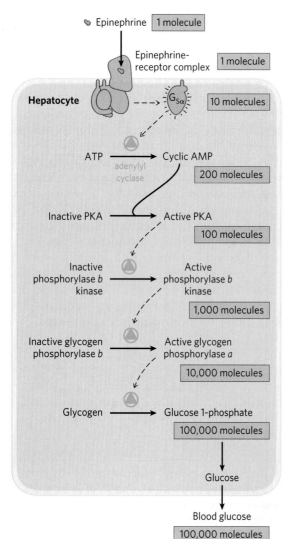

cyclase, each active $G_{s\alpha}$ molecule stimulates the catalytic synthesis of *many* molecules of cAMP. The second messenger cAMP now activates PKA, each molecule of which catalyzes the phosphorylation of *many* molecules of the target protein—phosphorylase b kinase in Figure 12–7. This kinase activates glycogen phosphorylase b, which leads to the rapid mobilization of glucose from glycogen. The net effect of the cascade is amplification of the hormonal signal by several orders of magnitude, which accounts for the very low concentration of epinephrine (or any other hormone) required for hormone activity.

Several Mechanisms Cause Termination of the β-Adrenergic Response

To be useful, a signal-transducing system has to *turn off* after the hormonal or other stimulus has ended, and mechanisms for shutting off the signal are intrinsic to

FIGURE 12–7 Epinephrine cascade. Epinephrine triggers a series of reactions in hepatocytes in which catalysts activate catalysts, resulting in great amplification of the original hormone signal. The numbers of molecules shown are simply to illustrate amplification and are almost certainly gross underestimates. Binding of one molecule of epinephrine to one β-adrenergic receptor on the cell surface activates a number (possibly hundreds) of G proteins, one after another, each of which goes on to activate a molecule of the enzyme adenylyl cyclase. Adenylyl cyclase acts catalytically, producing many molecules of cAMP for each activated adenylyl cyclase. (Because two molecules of cAMP are required to activate one PKA catalytic subunit, this step does not amplify the signal.)

all signaling systems. Most systems also adapt to the continued presence of the signal by becoming less sensitive to it, in the process of desensitization. The β-adrenergic system illustrates both. When the concentration of epinephrine in the blood drops below the K_d for its receptor, the hormone dissociates from the receptor and the latter reassumes the inactive conformation, in which it can no longer activate G_s.

A second means of ending the response to β-adrenergic stimulation is the hydrolysis of GTP bound to the G_α subunit, catalyzed by the intrinsic GTPase activity of the G protein. Conversion of bound GTP to GDP favors the return of G_α to the conformation in which it binds the $G_{\beta\gamma}$ subunits—the conformation in which the G protein is unable to interact with or stimulate adenylyl cyclase. This ends the production of cAMP. The rate of inactivation of G_s depends on the GTPase activity, which for G_α alone is very feeble. However, GTPase activator proteins (GAPs) strongly stimulate this GTPase activity, causing more rapid inactivation of the G protein (see Box 12–2). GAPs can themselves be regulated by other factors, providing a fine-tuning of the response to β-adrenergic stimulation. A third mechanism for terminating the response is to remove the second messenger: hydrolysis of cAMP to 5′-AMP (not active as a second messenger) by **cyclic nucleotide phosphodiesterase** (Fig. 12–4a, step ❼; 12–4b).

Finally, at the end of the signaling pathway, the metabolic effects that result from enzyme phosphorylation are reversed by the action of phosphoprotein phosphatases, which hydrolyze phosphorylated Ser, Thr, or Tyr residues, releasing inorganic phosphate (P_i). About 150 genes in the human genome encode phosphoprotein phosphatases, fewer than the number encoding protein kinases (~500). Some of these phosphatases are known to be regulated; others may act constitutively. When [cAMP] drops and PKA returns to its inactive form (step ❼ in Fig. 12–4a), the balance between phosphorylation and dephosphorylation is tipped toward dephosphorylation by these phosphatases.

The β-Adrenergic Receptor Is Desensitized by Phosphorylation and by Association with Arrestin

The mechanisms for signal termination described above take effect when the stimulus ends. A different mechanism, desensitization, damps the response *even while the signal persists*. Desensitization of the β-adrenergic receptor is mediated by a protein kinase that phosphorylates the receptor on the intracellular domain that normally interacts with G_s **(Fig. 12–8)**. When the receptor is occupied by epinephrine, **β-adrenergic receptor kinase**, or **βARK** (also commonly called **GRK2**; see below), phosphorylates several Ser residues

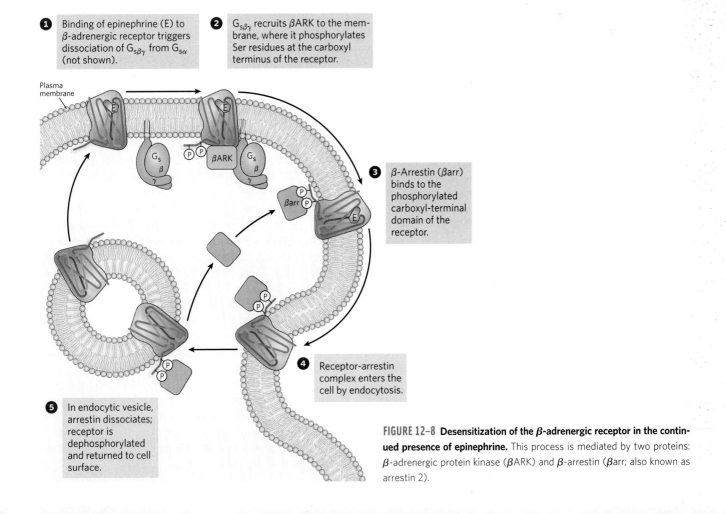

❶ Binding of epinephrine (E) to β-adrenergic receptor triggers dissociation of $G_{s\beta\gamma}$ from $G_{s\alpha}$ (not shown).

❷ $G_{s\beta\gamma}$ recruits βARK to the membrane, where it phosphorylates Ser residues at the carboxyl terminus of the receptor.

❸ β-Arrestin (βarr) binds to the phosphorylated carboxyl-terminal domain of the receptor.

❹ Receptor-arrestin complex enters the cell by endocytosis.

❺ In endocytic vesicle, arrestin dissociates; receptor is dephosphorylated and returned to cell surface.

Plasma membrane

FIGURE 12–8 Desensitization of the β-adrenergic receptor in the continued presence of epinephrine. This process is mediated by two proteins: β-adrenergic protein kinase (βARK) and β-arrestin (βarr; also known as arrestin 2).

near the carboxyl terminus of the receptor, which is on the cytoplasmic side of the plasma membrane. Usually located in the cytosol, βARK is drawn to the plasma membrane by its association with the $G_{s\beta\gamma}$ subunits and is thus positioned to phosphorylate the receptor. Receptor phosphorylation creates a binding site for the protein **β-arrestin**, or **βarr** (also called arrestin 2), and binding of β-arrestin effectively prevents further interaction between the receptor and the G protein. The binding of β-arrestin also facilitates receptor sequestration, the removal of receptor molecules from the plasma membrane by endocytosis into small intracellular vesicles. The arrestin-receptor complex recruits two proteins involved in vesicle formation (see Fig. 27–45), the AP-2 complex and clathrin, which initiate membrane invagination, leading to the formation of endosomes containing the adrenergic receptor. In this state the receptors are inaccessible to epinephrine and therefore inactive. Receptors in the endocytic vesicles are eventually dephosphorylated and returned to the plasma membrane, completing the circuit and resensitizing the system to epinephrine. β-Adrenergic receptor kinase is a member of a family of **G protein–coupled receptor kinases (GRKs)**, all of which phosphorylate GPCRs on their carboxyl-terminal cytoplasmic domains and play roles similar to that of βARK in desensitization and resensitization of their receptors. At least five different GRKs and four different arrestins are encoded in the human genome; each GRK is capable of desensitizing a particular subset of GPCRs, and each arrestin can interact with many different types of phosphorylated receptors.

Cyclic AMP Acts as a Second Messenger for Many Regulatory Molecules

Epinephrine is just one of many hormones, growth factors, and other regulatory molecules that act by changing the intracellular [cAMP] and thus the activity of PKA (Table 12–3). For example, glucagon binds to its receptors in the plasma membrane of adipocytes, activating (via a G_s protein) adenylyl cyclase. PKA, stimulated by the resulting rise in [cAMP], phosphorylates and activates two proteins critical to the mobilization of the fatty acids of stored fats (see Fig. 17–3). Similarly, the peptide hormone ACTH (adrenocorticotropic hormone, also called corticotropin), produced by the anterior pituitary, binds to specific receptors in the adrenal cortex, activating adenylyl cyclase and raising the intracellular [cAMP]. PKA then phosphorylates and activates several of the enzymes required for the synthesis of cortisol and other steroid hormones. In many cell types, the catalytic subunit of PKA can also move into the nucleus, where it phosphorylates the **cAMP response element binding protein (CREB)**, which alters the expression of specific genes regulated by cAMP.

Some hormones act by *inhibiting* adenylyl cyclase, thus *lowering* [cAMP] and *suppressing* protein phos-

TABLE 12–3	Some Signals That Use cAMP as Second Messenger

Corticotropin (ACTH)

Corticotropin-releasing hormone (CRH)

Dopamine [D_1, D_2]

Epinephrine (β-adrenergic)

Follicle-stimulating hormone (FSH)

Glucagon

Histamine [H_2]

Luteinizing hormone (LH)

Melanocyte-stimulating hormone (MSH)

Odorants (many)

Parathyroid hormone

Prostaglandins E_1, E_2 (PGE_1, PGE_2)

Serotonin [5-HT-1a, 5-HT-2]

Somatostatin

Tastants (sweet, bitter)

Thyroid-stimulating hormone (TSH)

Note: Receptor subtypes in square brackets. Subtypes may have different transduction mechanisms. For example, serotonin is detected in some tissues by receptor subtypes 5-HT-1a and 5-HT-1b, which act through adenylyl cyclase and cAMP, and in other tissues by receptor subtype 5-HT-1c, acting through the phospholipase C–IP$_3$ mechanism (see Table 12–4).

phorylation. For example, the binding of somatostatin to its receptor leads to activation of an **inhibitory G protein**, or G_i, structurally homologous to G_s, that inhibits adenylyl cyclase and lowers [cAMP]. Somatostatin therefore counterbalances the effects of glucagon. In adipose tissue, prostaglandin E_1 (PGE_1; see Fig. 10–18) inhibits adenylyl cyclase, thus lowering [cAMP] and slowing the mobilization of lipid reserves triggered by epinephrine and glucagon. In certain other tissues PGE_1 stimulates cAMP synthesis: its receptors are coupled to adenylyl cyclase through a stimulatory G protein, G_s. In tissues with α_2-adrenergic receptors, epinephrine lowers [cAMP]; in this case, the receptors are coupled to adenylyl cyclase through an inhibitory G protein, G_i. In short, an extracellular signal such as epinephrine or PGE_1 can have quite different effects on different tissues or cell types, depending on three factors: the type of receptor in the tissue, the type of G protein (G_s or G_i) with which the receptor is coupled, and the set of PKA target enzymes in the cells. By summing the influences that tend to increase and decrease [cAMP], a cell achieves the integration of signals that we noted as a general feature of signal-transducing mechanisms (Fig. 12–1e).

Another factor that explains how so many types of signals can be mediated by a single second messenger (cAMP) is the confinement of the signaling process to a specific region of the cell by **adaptor proteins**—noncatalytic proteins that hold together other protein molecules that function in concert (further described below). **AKAPs (A kinase anchoring proteins)** are multivalent adaptor proteins; one part binds to the R subunits of PKA (see Fig. 12–6a) and another to a specific structure in the cell, confining the PKA to the vicinity of that structure. For example, specific AKAPs bind PKA to microtubules, actin filaments, ion channels, mitochondria, or the nucleus. Different types of cells have different complements of AKAPs, so cAMP might stimulate phosphorylation of mitochondrial proteins in one cell and phosphorylation of actin filaments in another. In some cases, an AKAP connects PKA with the enzyme that triggers PKA activation (adenylyl cyclase) or terminates PKA action (cAMP phosphodiesterase or phosphoprotein phosphatase) **(Fig. 12–9)**. The very close proximity of these activating and inactivating enzymes presumably achieves a highly localized, and very brief, response.

As is now clear, to fully understand cellular signaling, researchers need tools precise enough to detect and study the spatiotemporal aspects of signaling processes at the subcellular level and in real time. In studies of the intracellular localization of biochemical changes, biochemistry meets cell biology, and techniques that cross this boundary have become essential in understanding signaling pathways. Fluorescent probes have found wide application in signaling studies. Labeling of functional proteins with a fluorescent tag such as the green fluorescent protein (GFP) reveals their subcellular localizations (see Fig. 9–16c). Changes in the state of association of two proteins (such as the R and C subunits of PKA) can be seen by measuring the nonradiative transfer of energy between fluorescent probes attached to each protein, a technique called fluorescence resonance energy transfer (FRET; Box 12–3).

Diacylglycerol, Inositol Trisphosphate, and Ca²⁺ Have Related Roles as Second Messengers

A second broad class of GPCRs are coupled through a G protein to a plasma membrane **phospholipase C (PLC)** that is specific for the membrane phospholipids phosphatidylinositol 4,5-bisphosphate, or PIP_2 (see Fig. 10–16). When one of the hormones that acts by this mechanism (Table 12–4) binds its specific receptor in the plasma membrane **(Fig. 12–10**, step **❶)**, the receptor-hormone complex catalyzes GTP-GDP exchange on an associated

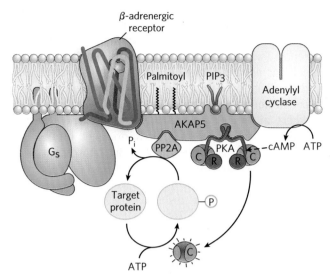

FIGURE 12–9 Nucleation of supramolecular complexes by A kinase anchoring proteins (AKAPs). AKAP5 is one of a family of proteins that act as multivalent scaffolds, holding PKA catalytic subunits—through the AKAP's interaction with the PKA regulatory subunits—in proximity to a particular region or structure in the cell. AKAP5 is targeted to rafts in the cytoplasmic surface of the plasma membrane by two covalently attached palmitoyl groups and a site that binds phosphatidylinositol 3,4,5-trisphosphate (PIP_3) in the membrane. AKAP5 also has binding sites for the β-adrenergic receptor, adenylyl cyclase, PKA, and a phosphoprotein phosphatase (PP2A), bringing them all together in the plane of the membrane. When epinephrine binds to the β-adrenergic receptor, $G_{s\alpha}$ triggers adenylyl cyclase produces cAMP, which reaches the nearby PKA quickly and with very little dilution. PKA phosphorylates its target protein, altering its activity, until the phosphoprotein phosphatase removes the phosphoryl group and returns the target protein to its pre-stimulus state. The AKAPs in this and other cases bring about a high local concentration of enzymes and second messengers, so that the signaling circuit remains highly localized, and the duration of the signal is limited.

G protein, G_q (step **❷**), activating it in much the same way that the β-adrenergic receptor activates G_s (Fig. 12–4). The activated G_q activates the PIP_2-specific PLC (Fig. 12–10, step **❸**), which catalyzes (step **❹**) the production of two potent second messengers, **diacylglycerol** and **inositol 1,4,5-trisphosphate**, or **IP₃** (not to be confused with PIP_3, p. 456).

TABLE 12–4	Some Signals That Act through Phospholipase C, IP₃, and Ca²⁺		
Acetylcholine [muscarinic M₁]	Gastrin-releasing peptide		Platelet-derived growth factor (PDGF)
α_1-Adrenergic agonists	Glutamate		Serotonin [5-HT-1c]
Angiogenin	Gonadotropin-releasing hormone (GRH)		Thyrotropin-releasing hormone (TRH)
Angiotensin II	Histamine [H₁]		Vasopressin
ATP [P₂ₓ, P₂y]	Light (*Drosophila*)		
Auxin	Oxytocin		

Note: Receptor subtypes are in square brackets; see footnote to Table 12–3.

BOX 12–3 METHODS FRET: Biochemistry Visualized in a Living Cell

Fluorescent probes are commonly used to detect rapid biochemical changes in single living cells. They can be designed to give an essentially instantaneous report (within nanoseconds) on the changes in intracellular concentration of a second messenger or in the activity of a protein kinase. Furthermore, fluorescence microscopy has sufficient resolution to reveal where in the cell such changes are occurring. In one widely used procedure, the fluorescent probes are derived from a naturally occurring fluorescent protein, the **green fluorescent protein (GFP)** of the jellyfish *Aequorea victoria* (Fig. 1).

When excited by absorption of a photon of light, GFP emits a photon (that is, it fluoresces) in the green region of the spectrum. The light-absorbing/emitting center of GFP (its chromophore) comprises an oxidized form of the tripeptide –Ser^{65}–Tyr^{66}–Gly^{67}– (Fig. 2). Oxidation of the tripeptide is catalyzed by the GFP protein itself (Fig. 3), so it is possible to clone the protein into virtually any cell, where it can serve as a fluorescent marker for any protein to which it is fused (see Fig. 9–18). Variants of GFP, with different fluorescence spectra, are produced by genetic engineering. For example, in the yellow fluorescent protein (YFP), Ala^{206} in GFP is replaced by a Lys residue, changing the wavelength of light absorption and fluorescence. Other variants of GFP fluoresce blue (BFP) or cyan (CFP) light, and a related protein (mRFP1) fluoresces red light (Fig. 4). GFP and its variants are compact structures that retain their ability to fold into their native β-barrel conformation even when fused with another protein. These fluorescent hybrid proteins act as spectroscopic rulers for measuring distances between interacting proteins within

FIGURE 2 Green fluorescent protein (GFP), with the fluorescent chromophore shown in ball-and-stick form (derived from PDB ID 1GFL).

Chromophore
(Ser^{65}–Tyr^{66}–Gly^{67})

a cell and, indirectly, to measure local concentrations of compounds that change the distance between two proteins.

An excited fluorescent molecule such as GFP or YFP can dispose of the energy from the absorbed photon in either of two ways: (1) by fluorescence, emitting a photon of slightly longer wavelength (lower energy) than the exciting light, or (2) by nonradiative **fluorescence resonance energy transfer (FRET)**, in which the energy of the excited molecule (the donor) passes directly to a nearby molecule (the acceptor) *without emission of a photon*, exciting the acceptor (Fig. 5). The acceptor can now decay to its ground state by fluorescence; the emitted photon has a longer wavelength (lower energy) than both the original exciting light and the fluorescence emission of the donor. This second mode of decay (FRET) is possible only when donor and acceptor are close to each other (within 1 to 50 Å); the efficiency of FRET is inversely proportional to the *sixth power* of the distance between donor and acceptor. Thus very small changes in the distance between donor and acceptor register as very large changes in FRET, measured as the fluorescence of the acceptor molecule when the donor is excited. With sufficiently sensitive light detectors, this fluorescence signal can be located to specific regions of a single, living cell.

FRET has been used to measure [cAMP] in living cells. The gene for GFP is fused with that for the regulatory subunit (R) of cAMP-dependent protein kinase (PKA), and the gene for BFP is fused with that for the catalytic subunit (C) (Fig. 6). When these two hybrid

FIGURE 1 *Aequorea victoria*, a jellyfish abundant in Puget Sound, Washington State.

FIGURE 3 The chromophore in GFP is derived from a series of three amino acids: -Ser^{65}-Tyr^{66}-Gly^{67}-. Maturation of the chromophore involves an internal rearrangement, coupled to an oxidation reaction that takes place in multiple steps. An abbreviated mechanism is shown here.

Mature chromophore

FIGURE 4 Emission spectra of GFP variants.

proteins are expressed in a cell, BFP (donor; excitation at 380 nm, emission at 460 nm) and GFP (acceptor; excitation at 475 nm, emission at 545 nm) in the inactive PKA (R_2C_2 tetramer) are close enough to undergo FRET. Wherever in the cell [cAMP] increases, the R_2C_2 complex dissociates into R_2 and 2C and the FRET signal is lost, because donor and acceptor are now too far apart for efficient FRET. Viewed in the fluorescence microscope, the region of higher [cAMP] has a minimal GFP signal and higher BFP signal. Measuring the ratio of emission at 460 nm and 545 nm gives a sensitive measure of the change in [cAMP]. By determining this ratio for all regions of the cell, the investigator can generate a false color image of the cell in which the ratio, or relative [cAMP], is represented by the intensity of the color. Images recorded at timed intervals reveal changes in [cAMP] over time.

A variation of this technology has been used to measure the activity of PKA in a living cell (Fig. 7). Researchers create a phosphorylation target for PKA by producing a hybrid protein containing four elements: YFP (acceptor); a short peptide with a Ser residue surrounded by the consensus sequence for PKA; a Ⓟ–Ser-binding domain (called 14-3-3); and

FIGURE 5 When the donor protein (CFP) is excited with monochromatic light of wavelength 433 nm, it emits fluorescent light at 476 nm (left). When the (red) protein fused with CFP interacts with the (purple) protein fused with YFP, that interaction brings CFP and YFP close enough to allow fluorescence resonance energy transfer (FRET) between them. Now, when CFP absorbs light of 433 nm, instead of fluorescing at 476 nm, it transfers energy directly to YFP, which then fluoresces at its characteristic emission wavelength, 527 nm. The ratio of light emission at 527 and 476 nm is therefore a measure of the interaction of the red and purple proteins.

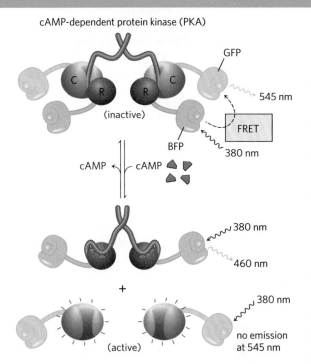

FIGURE 6 Measuring [cAMP] with FRET. Gene fusion creates hybrid proteins that exhibit FRET when the PKA regulatory (R) and catalytic (C) subunits are associated (low [cAMP]). When [cAMP] rises, the subunits dissociate and FRET ceases. The ratio of emission at 460 nm (dissociated) and 545 nm (complexed) thus offers a sensitive measure of [cAMP].

CFP (donor). When the Ser residue is not phosphorylated, 14-3-3 has no affinity for the Ser residue and the hybrid protein exists in an extended form, with the donor and acceptor too far apart to generate a FRET signal. Wherever PKA is active in the cell, it phosphorylates the Ser residue of the hybrid protein, and 14-3-3 binds to the Ⓟ–Ser. In doing so, it draws YFP and CFP together and a FRET signal is detected with the fluorescence microscope, revealing the presence of active PKA.

FIGURE 7 Measuring the activity of PKA with FRET. An engineered protein links YFP and CFP via a peptide that contains a Ser residue surrounded by the consensus sequence for phosphorylation by PKA, and the 14-3-3 Ⓟ-Ser-binding domain. Active PKA phosphorylates the Ser residue, which docks with the 14-3-3 binding domain, bringing the fluorescence proteins close enough to allow FRET to occur, revealing the presence of active PKA.

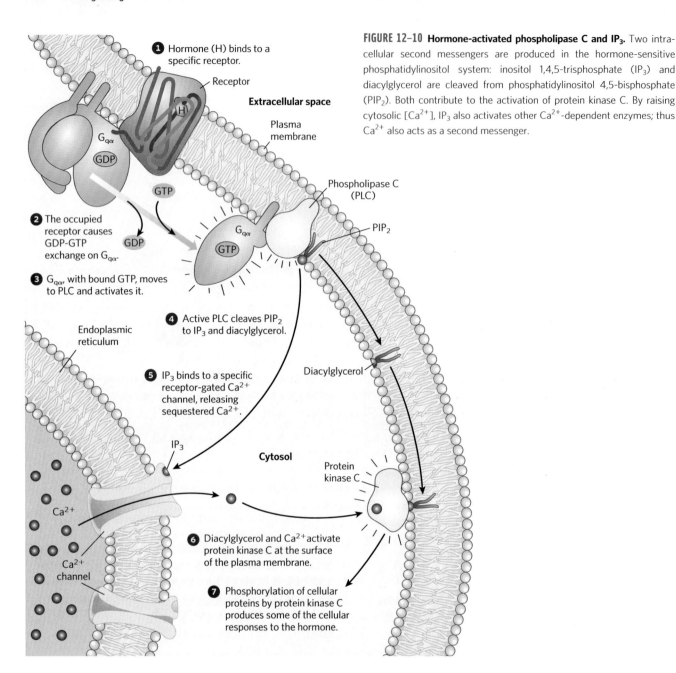

① Hormone (H) binds to a specific receptor.

Receptor

Extracellular space

Plasma membrane

② The occupied receptor causes GDP-GTP exchange on $G_{q\alpha}$.

③ $G_{q\alpha}$, with bound GTP, moves to PLC and activates it.

Phospholipase C (PLC)

PIP_2

Endoplasmic reticulum

④ Active PLC cleaves PIP_2 to IP_3 and diacylglycerol.

Diacylglycerol

⑤ IP_3 binds to a specific receptor-gated Ca^{2+} channel, releasing sequestered Ca^{2+}.

IP_3

Cytosol

Protein kinase C

Ca^{2+}

Ca^{2+} channel

⑥ Diacylglycerol and Ca^{2+} activate protein kinase C at the surface of the plasma membrane.

⑦ Phosphorylation of cellular proteins by protein kinase C produces some of the cellular responses to the hormone.

FIGURE 12–10 Hormone-activated phospholipase C and IP₃. Two intracellular second messengers are produced in the hormone-sensitive phosphatidylinositol system: inositol 1,4,5-trisphosphate (IP_3) and diacylglycerol are cleaved from phosphatidylinositol 4,5-bisphosphate (PIP_2). Both contribute to the activation of protein kinase C. By raising cytosolic $[Ca^{2+}]$, IP_3 also activates other Ca^{2+}-dependent enzymes; thus Ca^{2+} also acts as a second messenger.

Inositol 1,4,5-trisphosphate (IP_3)

Inositol trisphosphate, a water-soluble compound, diffuses from the plasma membrane to the endoplasmic reticulum (ER), where it binds to specific IP_3-gated Ca^{2+} channels, causing them to open. The action of the SERCA pump (see p. 410) ensures that $[Ca^{2+}]$ in the ER is orders of magnitude higher than that in the cytosol, so when these gated Ca^{2+} channels open, Ca^{2+} rushes into the cytosol (Fig. 12–10, step ⑤), and the cytosolic $[Ca^{2+}]$ rises sharply to about 10^{-6} M. One effect of elevated $[Ca^{2+}]$ is the activation of **protein kinase C (PKC)**. Diacylglycerol cooperates with Ca^{2+} in activating PKC, thus also acting as a second messenger (step ⑥). Activation involves the movement of a PKC domain (the pseudosubstrate domain) away from its location in the substrate-binding region of the enzyme, allowing the enzyme to bind and phosphorylate proteins that contain a PKC consensus sequence—Ser or Thr residues embedded in an amino acid sequence recognized by PKC (step ⑦). There are several isozymes of PKC, each with a characteristic tissue distribution, target protein specificity, and role. Their targets include cytoskeletal proteins, enzymes, and nuclear proteins that regulate gene expression. Taken together, this family of enzymes has a wide range of cellular actions, affecting neuronal and immune function and the regulation of cell division, for example.

Calcium Is a Second Messenger That May Be Localized in Space and Time

There are many variations on this basic scheme for Ca^{2+} signaling. In many cell types that respond to extracellular signals, Ca^{2+} serves as a second messenger that triggers intracellular responses, such as exocytosis in neurons and endocrine cells, contraction in muscle, and cytoskeletal rearrangements during amoeboid movement. In unstimulated cells, cytosolic $[Ca^{2+}]$ is kept very low ($<10^{-7}$ M) by the action of Ca^{2+} pumps in the ER, mitochondria, and plasma membrane (as further discussed below). Hormonal, neural, or other stimuli cause either an influx of Ca^{2+} into the cell through specific Ca^{2+} channels in the plasma membrane or the release of sequestered Ca^{2+} from the ER or mitochondria, in either case raising the cytosolic $[Ca^{2+}]$ and triggering a cellular response.

Changes in intracellular $[Ca^{2+}]$ are detected by Ca^{2+}-binding proteins that regulate a variety of Ca^{2+}-dependent enzymes. **Calmodulin** (**CaM**; M_r 17,000) is an acidic protein with four high-affinity Ca^{2+}-binding sites. When intracellular $[Ca^{2+}]$ rises to about 10^{-6} M (1 μM), the binding of Ca^{2+} to calmodulin drives a conformational change in the protein **(Fig. 12–11a)**. Calmodulin associates with a variety of proteins and, in its Ca^{2+}-bound state, modulates their activities (Fig. 12–11b). It is a member of a family of Ca^{2+}-binding proteins that also includes troponin (see Fig. 5–32), which triggers skeletal muscle contraction in response to increased $[Ca^{2+}]$. This family shares a characteristic Ca^{2+}-binding structure, the EF hand (Fig. 12–11c).

Calmodulin is an integral subunit of the **Ca^{2+}/ calmodulin-dependent protein kinases** (**CaM kinases**, types I through IV). When intracellular $[Ca^{2+}]$ increases in response to a stimulus, calmodulin binds Ca^{2+}, undergoes a change in conformation, and activates the CaM kinase. The kinase then phosphorylates target enzymes, regulating their activities. Calmodulin is also a regulatory subunit of phosphorylase b kinase of muscle, which is activated by Ca^{2+}. Thus Ca^{2+} triggers ATP-requiring muscle contractions while also activating glycogen breakdown, providing fuel for ATP synthesis. Many other enzymes are also known to be modulated by Ca^{2+} through calmodulin (Table 12–5). The activity of the second messenger Ca^{2+}, like that of cAMP, can be spatially restricted; after its release triggers a local response, Ca^{2+} is generally removed before it can diffuse to distant parts of the cell.

Very commonly, Ca^{2+} level does not simply rise and then decrease, but rather oscillates with a period of a few seconds **(Fig. 12–12)**—even when the extracellular concentration of the triggering hormone remains constant. The mechanism underlying $[Ca^{2+}]$ oscillations presumably entails feedback regulation by Ca^{2+} on

(a)

(b)

(c)

FIGURE 12–11 Calmodulin. This is the protein mediator of many Ca^{2+}-stimulated enzymatic reactions. Calmodulin has four high-affinity Ca^{2+}-binding sites ($K_d \approx 0.1$ to 1 μM). **(a)** A ribbon model of the crystal structure of calmodulin (PDB ID 1CLL). The four Ca^{2+}-binding sites are occupied by Ca^{2+} (purple). The amino-terminal domain is on the left; the carboxyl-terminal domain on the right. **(b)** Calmodulin associated with a helical domain (red) of one of the many enzymes it regulates, calmodulin-dependent protein kinase II (PDB ID 1CDL). Notice that the long central α helix of calmodulin visible in (a) has bent back on itself in binding to the helical substrate domain. The central helix of calmodulin is clearly more flexible in solution than in the crystal. **(c)** Each of the four Ca^{2+}-binding sites occurs in a helix-loop-helix motif called the EF hand, also found in many other Ca^{2+}-binding proteins.

TABLE 12–5	Some Proteins Regulated by Ca^{2+} and Calmodulin

Adenylyl cyclase (brain)

Ca^{2+}/calmodulin-dependent protein kinases (CaM kinases I to IV)

Ca^{2+}-dependent Na^+ channel (*Paramecium*)

Ca^{2+}-release channel of sarcoplasmic reticulum

Calcineurin (phosphoprotein phosphatase 2B)

cAMP phosphodiesterase

cAMP-gated olfactory channel

cGMP-gated Na^+, Ca^{2+} channels (rod and cone cells)

Glutamate decarboxylase

Myosin light-chain kinases

NAD^+ kinase

Nitric oxide synthase

Phosphatidylinositol 3-kinase

Plasma membrane Ca^{2+} ATPase (Ca^{2+} pump)

RNA helicase (p68)

(a)

0   0.5   1.0

$[Ca^{2+}]$ (μM)

(b)

FIGURE 12–12 Triggering of oscillations in intracellular $[Ca^{2+}]$ by extracellular signals. (a) A dye (fura) that undergoes fluorescence changes when it binds Ca^{2+} is allowed to diffuse into cells, and its instantaneous light output is measured by fluorescence microscopy. Fluorescence intensity is represented by color; the color scale relates intensity of color to $[Ca^{2+}]$, allowing determination of the absolute $[Ca^{2+}]$. In this case, thymocytes (cells of the thymus) have been stimulated with extracellular ATP, which raises their internal $[Ca^{2+}]$. The cells are heterogeneous in their responses; some have high intracellular $[Ca^{2+}]$ (red), others much lower (blue). **(b)** When such a probe is used in a single hepatocyte, the agonist norepinephrine (added at the arrow) causes oscillations of $[Ca^{2+}]$ from 200 to 500 nM. Similar oscillations are induced in other cell types by other extracellular signals.

some part of the Ca^{2+}-release process. Whatever the mechanism, the effect is that one kind of signal (hormone concentration, for example) is converted into another (frequency and amplitude of intracellular $[Ca^{2+}]$ "spikes"). The Ca^{2+} signal diminishes as Ca^{2+} diffuses away from the initial source (the Ca^{2+} channel), is sequestered in the ER, or is pumped out of the cell.

There is significant cross talk between the Ca^{2+} and cAMP signaling systems. In some tissues, both the enzyme that produces cAMP (adenylyl cyclase) and the enzyme that degrades cAMP (phosphodiesterase) are stimulated by Ca^{2+}. Temporal and spatial changes in $[Ca^{2+}]$ can therefore produce transient, localized changes in [cAMP]. We have noted already that PKA, the enzyme that responds to cAMP, is often part of a highly localized supramolecular

complex assembled on scaffold proteins such as AKAPs. This subcellular localization of target enzymes, combined with temporal and spatial gradients in $[Ca^{2+}]$ and [cAMP], allows a cell to respond to one or several signals with subtly nuanced metabolic changes, localized in space and time.

GPCRs Mediate the Actions of a Wide Variety of Signals

The human genome encodes about 1,000 G protein–coupled receptors, recognizable by their seven transmembrane helical segments and certain highly conserved residues. Each is expressed selectively, in certain cell types or under certain conditions. Together, they allow cells and tissues to respond to a wide array of different stimuli, including various low molecular weight amines, peptides, proteins, eicosanoids and other lipids, as well as light and compounds detected by olfaction and gustation. The determination of several GCPR structures by crystallography **(Fig. 12–13)**, including the β-adrenergic receptor with its G protein, and the histamine receptor, has stimulated great interest in both the transduction mechanism(s) and the possibilities of altering receptor activity with drugs. These two receptors are the targets of a variety of widely used beta-blocker and antihistamine medications, respectively. The similarities in GPCR structures go beyond the common seven–transmembrane helix pattern; the structures of five different GPCRs are almost superimposable (Fig. 12–13c). Clearly, something about this three-dimensional structure makes it effective as a transducer of many disparate signals.

SUMMARY 12.2 G Protein–Coupled Receptors and Second Messengers

▶ G protein–coupled receptors (GPCRs) share a common structural arrangement of seven transmembrane helices and act through heterotrimeric G proteins. On ligand binding, GPCRs catalyze the exchange of GTP for GDP on the G protein, causing dissociation of the G_α subunit; G_α then stimulates or inhibits the activity of an effector enzyme, changing the level of its second-messenger product.

▶ The β-adrenergic receptor activates a stimulatory G protein, G_s, thereby activating adenylyl cyclase and raising the concentration of the second messenger cAMP. Cyclic AMP stimulates cAMP-dependent protein kinase to phosphorylate key target enzymes, changing their activities.

▶ Enzyme cascades, in which a single molecule of hormone activates a catalyst to activate another catalyst, and so on, result in the large signal amplification that is characteristic of hormone receptor systems.

(a) β_2-adrenergic receptor | Ligand-binding site | **(b)** μ-opioid receptor | Morphine analog in ligand-binding site | **(c)** Histamine H1 receptor | Doxepin in ligand-binding site | **(d)**

Outside

Inside

$G_{s\alpha}$ $G_{s\beta}$ $G_{s\gamma}$

FIGURE 12–13 The β-adrenergic receptor and several other GPCRs. (a) The β_2-adrenergic receptor and its associated trimeric G protein, G_s (PDB ID 3SN6). The bound agonist (epinephrine) is shown in yellow, and the subunits of G_s are three shades of green. **(b)** The μ opioid receptor (PDB ID 4DKL), the target of morphine and codeine, with a morphine analog in the ligand-binding site. **(c)** The histamine H1 receptor with the bound drug doxepin (PDB ID 3RZE). **(d)** Five GPCR structures superimposed to show the remarkable conservation of structure. Shown are the human A2A adenosine receptor (orange; PDB ID 3EML); turkey β_1-adrenergic receptor (blue; PDB ID 2VT4), human β_2-adrenergic receptor (green; PDB ID 2RH1), rhodopsin from squid (yellow; PDB ID 2Z73); and bovine rhodopsin, (red; PDB ID 1U19).

- ▶ Cyclic AMP concentration is eventually reduced by cAMP phosphodiesterase, and G_s turns itself off by hydrolysis of its bound GTP to GDP, acting as a self-limiting binary switch.

- ▶ When the epinephrine signal persists, β-adrenergic receptor–specific protein kinase and β-arrestin temporarily desensitize the receptor and cause it to move into intracellular vesicles.

- ▶ Some receptors stimulate adenylyl cyclase through G_s; others inhibit it through G_i. Thus cellular [cAMP] reflects the integrated input of two (or more) signals.

- ▶ Noncatalytic adaptor proteins such as AKAPs hold together proteins involved in a signaling process, increasing the efficiency of their interactions and in some cases confining the process to a specific subcellular location.

- ▶ Some GPCRs act via a plasma membrane phospholipase C that cleaves PIP_2 to diacylglycerol and IP_3. By opening Ca^{2+} channels in the endoplasmic reticulum, IP_3 raises cytosolic $[Ca^{2+}]$. Diacylglycerol and Ca^{2+} act together to activate protein kinase C, which phosphorylates and changes the activity of specific cellular proteins. Cellular $[Ca^{2+}]$ also regulates (often through calmodulin) many other enzymes and proteins involved in secretion, cytoskeletal rearrangements, or contraction.

12.3 Receptor Tyrosine Kinases

The **receptor tyrosine kinases (RTKs)**, a large family of plasma membrane receptors with intrinsic protein kinase activity, transduce extracellular signals by a mechanism fundamentally different from that of GPCRs. RTKs have a ligand-binding domain on the extracellular face of the plasma membrane and an enzyme active site

on the cytoplasmic face, connected by a single transmembrane segment. The cytoplasmic domain is a protein kinase that phosphorylates Tyr residues in specific target proteins—a Tyr kinase. The receptors for insulin and epidermal growth factor are prototypes for this group.

Stimulation of the Insulin Receptor Initiates a Cascade of Protein Phosphorylation Reactions

Insulin regulates both metabolic enzymes and gene expression. Insulin does not enter cells, but initiates a signal that travels a branched pathway from the plasma membrane receptor to insulin-sensitive enzymes in the cytosol and to the nucleus, where it stimulates the transcription of specific genes. The active insulin receptor protein (INSR) consists of two identical α subunits protruding from the outer face of the plasma membrane and two transmembrane β subunits with their carboxyl termini protruding into the cytosol—a dimer of $\alpha\beta$ monomers **(Fig. 12–14)**. The α subunits contain the insulin-binding domain, and the intracellular domains of the β subunits contain the protein kinase activity that transfers a phosphoryl group from ATP to the hydroxyl group of Tyr residues in specific target proteins. Signaling through INSR begins when the binding of one insulin molecule between the two subunits of the dimer activates the Tyr kinase activity, and each β subunit phosphorylates three critical Tyr residues near the carboxyl terminus of the other β subunit. This **autophosphorylation** opens up the active site so that the enzyme can phosphorylate Tyr residues of other target proteins.

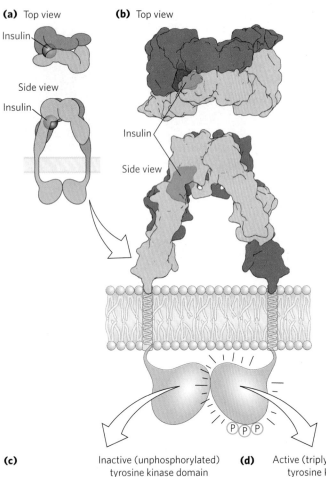

(a) Top view

Insulin

Side view

Insulin

(b) Top view

Insulin

Side view

FIGURE 12–14 Activation of the insulin-receptor tyrosine kinase by auto-phosphorylation. (a) The insulin-binding region of the insulin receptor lies outside the cell and comprises **(b)** two α subunits and the extracellular portions of two β subunits, intertwined to form the insulin-binding site (shown as a surface contour model of the crystal structure, derived from PDB ID 2DTG). (The structure of the transmembrane domain has not been solved by crystallography.) The binding of insulin (red; PDB ID 2CEU) is communicated through the single transmembrane helix of each β subunit to the paired Tyr kinase domains inside the cell, activating them to phosphorylate each other on three Tyr residues. **(c)** (PDB ID 1IRK) In the inactive form of the Tyr kinase domain (before Tyr phosphorylations), the activation loop (teal) sits in the active site, and none of the critical Tyr residues (white and red ball-and-stick structures) are phosphorylated. This conformation is stabilized by hydrogen bonding between Tyr[1162] and Asp[1132]. **(d)** (PDB ID 1IR3) Activation of the Tyr kinase allows each β subunit to phosphorylate three Tyr residues (Tyr[1158], Tyr[1162], Tyr[1163]) on the other β subunit. (Phosphoryl groups are depicted in red and orange.) The introduction of three highly charged ℗-Tyr residues forces a 30 Å change in the position of the activation loop, away from the substrate-binding site, which becomes available to bind and phosphorylate a target protein.

(c) Inactive (unphosphorylated) tyrosine kinase domain

Tyr[1158]
Tyr[1162]
Tyr[1163]
Asp[1132]

Activation loop blocks substrate-binding site

(d) Active (triply phosphorylated) tyrosine kinase domain

Tyr[1163]
Tyr[1158]
Asp[1132]
Target protein
Tyr[1162]

Triply phosphorylated activation loop moves dramatically, making room for the target protein in the substrate-binding site.

The mechanism of activation of the INSR protein kinase is similar to that described for PKA and PKC: a region of the cytoplasmic domain (an autoinhibitory sequence) that usually occludes the active site moves out of the active site after being phosphorylated, opening up the site for the binding of target proteins (Fig. 12–14).

When INSR is autophosphorylated (**Fig. 12–15,** step ❶), one of its targets is insulin receptor substrate-1 (IRS-1; step ❷). Once phosphorylated on several of its Tyr residues, IRS-1 becomes the point of nucleation for a complex of proteins (step ❸) that carry the message from the insulin receptor to end targets in the cytosol and nucleus, through a long series of intermediate proteins. First, a ℗-Tyr residue of IRS-1 binds to the **SH2 domain** of the protein Grb2. (SH2 is an abbreviation of *Src homology 2,* so named because the sequence of an SH2 domain is similar to that of a domain in Src (pronounced *sark*), another protein Tyr kinase.) Several signaling proteins contain SH2 domains, all of which bind ℗-Tyr residues in a protein partner. Grb2 is an adaptor protein, with no intrinsic enzymatic activity. Its function is to bring together two proteins (in this case, IRS-1 and the protein Sos) that must interact to enable signal transduction. In addition to its SH2 (℗-Tyr-binding) domain, Grb2 also contains a second protein-binding domain, SH3, that binds to a proline-rich region of Sos, recruiting Sos to the growing receptor complex. When bound to Grb2, Sos acts as a guanosine nucleotide–exchange factor

① Insulin receptor binds insulin and undergoes autophosphorylation on its carboxyl-terminal Tyr residues.

② Insulin receptor phosphorylates IRS-1 on its Tyr residues.

③ SH2 domain of Grb2 binds to Ⓟ-Tyr of IRS-1. Sos binds to Grb2, then to Ras, causing GDP release and GTP binding to Ras.

④ Activated Ras binds and activates Raf-1.

⑤ Raf-1 phosphorylates MEK on two Ser residues, activating it. MEK phosphorylates ERK on a Thr and a Tyr residue, activating it.

⑥ ERK moves into the nucleus and phosphorylates nuclear transcription factors such as Elk1, activating them.

⑦ Phosphorylated Elk1 joins SRF to stimulate the transcription and translation of a set of genes needed for cell division.

FIGURE 12–15 Regulation of gene expression by insulin through a MAP kinase cascade. The insulin receptor (INSR) consists of two α subunits on the outer face of the plasma membrane and two β subunits that traverse the membrane and protrude from the cytosolic face. Binding of insulin to the α subunits triggers a conformational change that allows the autophosphorylation of Tyr residues in the carboxyl-terminal domain of the β subunits. Autophosphorylation further activates the Tyr kinase domain, which then catalyzes phosphorylation of other target proteins. The signaling pathway by which insulin regulates the expression of specific genes consists of a cascade of protein kinases, each of which activates the next. INSR is a Tyr-specific kinase; the other kinases (all shown in blue) phosphorylate Ser or Thr residues. MEK is a dual-specificity kinase, which phosphorylates both a Thr and a Tyr residue in ERK (extracellular regulated kinase); MEK is mitogen-activated, ERK-activating kinase; SRF is serum response factor.

(GEF), catalyzing the replacement of bound GDP with GTP on Ras, a G protein.

Ras is the prototype of a family of **small G proteins** that mediate a wide variety of signal transductions (see Box 12–2). Like the trimeric G protein that functions with the β-adrenergic system (Fig. 12–5), Ras can exist in either the GTP-bound (active) or GDP-bound (inactive) conformation, but Ras (~20 kDa) acts as a monomer. When GTP binds, Ras can activate a protein kinase, Raf-1 (Fig. 12–15, step ④), the first of three protein kinases—Raf-1, MEK, and ERK—that form a cascade in which each kinase activates the next by phosphorylation (step ⑤). The protein kinases MEK and ERK are activated by phosphorylation of both a Thr and a Tyr residue. When activated, ERK mediates some of the biological effects of insulin by entering the nucleus and phosphorylating transcription factors, such as Elk1 (step ⑥), that modulate the transcription of about 100 insulin-regulated genes (step ⑦), some of which encode proteins essential for cell division. Thus, insulin acts as a growth factor.

The proteins Raf-1, MEK, and ERK are members of three larger families, for which several nomenclatures are employed. ERK is in the **MAPK** family (*m*itogen-*a*ctivated *p*rotein *k*inases; mitogens are extracellular signals that induce mitosis and cell division). Soon after discovery of the first MAPK enzyme, that enzyme was found to be activated by another protein kinase, which was named MAP kinase kinase (MEK belongs to this family), and when a third kinase that activated MAP kinase kinase was discovered, it was given the slightly ludicrous family name MAP kinase kinase kinase (Raf-1 is in this family). Somewhat less cumbersome are the abbreviations for these three families: MAPK, MAPKK, and MAPKKK. Kinases in the MAPK and MAPKKK families are specific for Ser or Thr residues, and MAPKKs (here, MEK) phosphorylate both a Ser and a Tyr residue in their substrate, a MAPK (here, ERK).

Biochemists now recognize this insulin pathway as but one instance of a more general scheme in which hormone signals, via pathways similar to that shown in Figure 12–15, result in phosphorylation of target enzymes by protein kinases. The target of phosphorylation is often another protein kinase, which then phosphorylates a third protein kinase, and so on. The result is a cascade of reactions that amplifies the initial signal by many orders of magnitude (see Fig. 12–1b). **MAPK cascades** (Fig. 12–15) mediate signaling initiated by a variety of growth factors, such as platelet-derived growth factor (PDGF) and epidermal growth factor (EGF). Another general scheme exemplified by the insulin receptor pathway is the use of nonenzymatic adaptor proteins to bring together the components of a branched signaling pathway, to which we now turn.

The Membrane Phospholipid PIP₃ Functions at a Branch in Insulin Signaling

The signaling pathway from insulin branches at IRS-1 (Fig. 12–15, step ❷). Grb2 is not the only protein that associates with phosphorylated IRS-1. The enzyme phosphoinositide 3-kinase (PI3K) binds IRS-1 through PI3K's SH2 domain **(Fig. 12–16)**. Thus activated, PI3K converts the membrane lipid phosphatidylinositol 4,5-bisphosphate (PIP₂) to phosphatidylinositol 3,4,5-trisphosphate (PIP₃). The multiply charged head group of PIP₃, protruding on the cytoplasmic side of the plasma membrane, is the starting point for a second signaling branch involving another cascade of protein kinases. When bound to PIP₃, protein kinase B (PKB; also called Akt) is phosphorylated and activated by yet another protein kinase, PDK1. The activated PKB then phosphorylates Ser or Thr residues in its target proteins, one of which is glycogen synthase kinase 3 (GSK3). In its active, nonphosphorylated form, GSK3 phosphorylates glycogen synthase, inactivating it and thereby contributing to the slowing of glycogen synthesis. (This mechanism is only part of the explanation for the effects of insulin on glycogen metabolism.) When phosphorylated by PKB, GSK3 is inactivated. By thus preventing inactivation of glycogen synthase in liver and muscle, the cascade of protein phosphorylations initiated by insulin stimulates glycogen synthesis (Fig. 12–16). In a third signaling branch in muscle and fat tissue, PKB triggers the clathrin-aided movement of glucose transporters (GLUT4) from internal vesicles to the plasma membrane, stimulating glucose uptake from the blood (Fig. 12–16, step ❺; see also Box 11–1).

As in all signaling pathways, there is a mechanism for terminating the activity of the PI3K–PKB pathway. A PIP₃-specific phosphatase (PTEN in humans) removes the phosphoryl group at the 3 position of PIP₃ to produce PIP₂, which no longer serves as a binding site for PKB, and the signaling chain is broken. In various types of cancer, it is often found that the PTEN gene has undergone mutation, resulting in a defective regulatory circuit and abnormally high levels of PIP₃ and of PKB activity. The result seems to be a continuing signal for cell division and thus tumor growth. ■

The insulin receptor is the prototype for several receptor enzymes with a similar structure and RTK activity **(Fig. 12–17)**. The receptors for EGF and PDGF, for example, have structural and sequence similarities to INSR, and both have a protein Tyr kinase activity that phosphorylates IRS-1. Many of these receptors dimerize after binding ligand; INSR is the exception, as it is already an $(\alpha\beta)_2$ dimer before insulin binds.

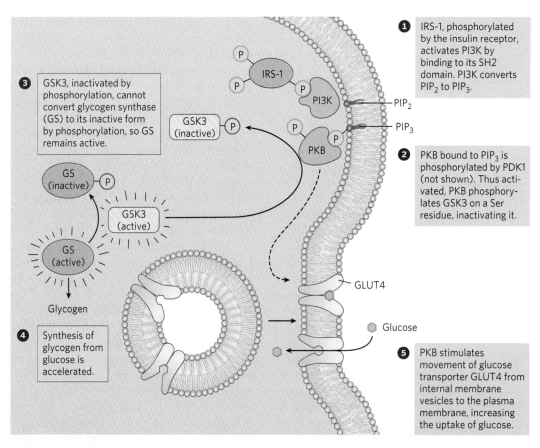

FIGURE 12–16 Insulin action on glycogen synthesis and GLUT4 movement to the plasma membrane. The activation of PI3 kinase (PI3K) by phosphorylated IRS-1 signals (through protein kinase B, PKB) movement of the glucose transporter GLUT4 to the plasma membrane, and the activation of glycogen synthase.

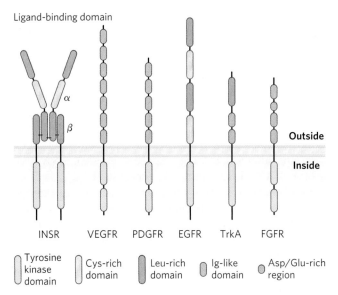

Ligand-binding domain

INSR   VEGFR   PDGFR   EGFR   TrkA   FGFR

Outside

Inside

- Tyrosine kinase domain
- Cys-rich domain
- Leu-rich domain
- Ig-like domain
- Asp/Glu-rich region

FIGURE 12–17 Receptor tyrosine kinases. Growth factor receptors that signal through Tyr kinase activity include those for insulin (INSR), vascular epidermal growth factor (VEGFR), platelet-derived growth factor (PDGFR), epidermal growth factor (EGFR), high-affinity nerve growth factor (TrkA), and fibroblast growth factor (FGFR). All these receptors have a Tyr kinase domain on the cytoplasmic side of the plasma membrane (blue). The extracellular domain is unique to each type of receptor, reflecting the different growth-factor specificities. These extracellular domains are typically combinations of structural motifs such as cysteine- or leucine-rich segments and segments containing one of several motifs common to immunoglobulins (Ig-like domains; see Fig. 4-22). Many other receptors of this type are encoded in the human genome, each with a different extracellular domain and ligand specificity.

(The protomer of the insulin receptor is one $\alpha\beta$ unit.) The binding of adaptor proteins such as Grb2 to ℗–Tyr residues is a common mechanism for promoting protein-protein interactions initiated by RTKs, a subject to which we return in Section 12.5.

In addition to the many receptors that act as protein Tyr kinases (the RTKs), several receptorlike plasma membrane proteins have protein Tyr phosphatase activity. Based on the structures of these proteins, we can surmise that their ligands are components of the extracellular matrix or are surface molecules on other cells. Although their signaling roles are not yet as well understood as those of the RTKs, they clearly have the potential to reverse the actions of signals that stimulate RTKs.

What spurred the evolution of such complicated regulatory machinery? This system allows one activated receptor to activate several IRS-1 molecules, amplifying the insulin signal, and it provides for the integration of signals from different receptors such as EGFR and PDGFR, each of which can phosphorylate IRS-1. Furthermore, because IRS-1 can activate any of several proteins that contain SH2 domains, a single receptor acting through IRS-1 can trigger two or more signaling pathways; insulin affects gene expression through the Grb2-Sos-Ras-MAPK pathway and affects glycogen metabolism and glucose transport through the PI3K–PKB pathway. Finally, there are several closely

related IRS proteins (IRS2, IRS3), each with its own characteristic tissue distribution and function, further enriching the signaling possibilities in pathways initiated by RTKs.

The JAK-STAT Signaling System Also Involves Tyrosine Kinase Activity

A variation on the basic theme of receptor Tyr kinases is receptors that have no intrinsic protein kinase activity but, when occupied by their ligand, bind a *cytosolic* Tyr kinase. One example is the system that regulates the formation of erythrocytes in mammals. The developmental signal, or **cytokine**, for this system is erythropoietin (EPO), a 165 amino acid protein produced in the kidneys. When EPO binds to its plasma membrane receptor **(Fig. 12–18)**, the receptor dimerizes, and the dimer can bind and activate the soluble protein kinase JAK (*Ja*nus *k*inase). The activated JAK phosphorylates

FIGURE 12–18 The JAK-STAT transduction mechanism for the erythropoietin receptor. Binding of erythropoietin (EPO) causes dimerization of the EPO receptor, which allows JAK, a soluble Tyr kinase, to bind to the internal domain of the receptor and phosphorylate it on several Tyr residues. **(a)** In one signaling pathway, the SH2 domain of the STAT protein STAT5 binds to ℗-Tyr residues on the receptor, bringing it into proximity with JAK. Following phosphorylation of STAT5 by JAK, two STAT5 molecules dimerize, each binding the other's ℗-Tyr residue, thus exposing a nuclear localization sequence (NLS) that targets the dimer for transport into the nucleus. In the nucleus, STAT5 turns on the expression of EPO-controlled genes. **(b)** In a second signaling pathway, following EPO binding and autophosphorylation of JAK, the adaptor protein SHC binds the ℗-Tyr of the receptor, then Grb2 binds SHC and triggers the MAPK cascade, as in the insulin system (see Fig. 12-15).

several Tyr residues in the cytoplasmic domain of the EPO receptor. A family of transcription factors, collectively called STATs (signal *t*ransducers and *a*ctivators of *t*ranscription), are also targets of JAK. An SH2 domain in STAT5 binds ⓟ–Tyr residues in the EPO receptor, positioning the STAT for phosphorylation by JAK in response to EPO. The phosphorylated STAT5 forms dimers, exposing a signal that causes it to be transported into the nucleus. There, STAT5 induces the expression (transcription) of specific genes essential for erythrocyte maturation. This JAK-STAT system also operates in other signaling pathways, including that for the hormone leptin, described in detail in Chapter 23 (see Fig. 23–36). Activated JAK can also trigger, through Grb2, the MAPK cascade (Fig. 12–18b), which leads to altered expression of specific genes.

Src is another soluble protein Tyr kinase that associates with certain receptors when they bind their ligands. Src was the first protein found to have the characteristic ⓟ–Tyr-binding domain that was subsequently named the Src homology (SH2) domain.

Cross Talk among Signaling Systems Is Common and Complex

Although, for simplicity, we have treated individual signaling pathways as separate sequences of events leading to separate metabolic consequences, there is in fact extensive cross talk among signaling systems. The regulatory circuitry that governs metabolism is richly interwoven and multilayered. We have discussed the signaling pathways for insulin and epinephrine separately, but they do not operate independently. Insulin opposes the metabolic effects of epinephrine in most tissues, and activation of the insulin signaling pathway directly attenuates signaling through the β-adrenergic signaling system. For example, the INSR kinase directly phosphorylates two Tyr residues in the cytoplasmic tail

of a β_2-adrenergic receptor, and PKB, activated by insulin **(Fig. 12–19)**, phosphorylates two Ser residues in the same region. Phosphorylation of these four residues triggers clathrin-aided internalization of the β_2-adrenergic receptor, taking it out of service and lowering the cell's sensitivity to epinephrine. A second type of cross talk between these receptors occurs when ⓟ–Tyr residues on the β_2-adrenergic receptor, phosphorylated by INSR, serve as nucleation points for SH2 domain–containing proteins such as Grb2 (Fig. 12–19, left side). Activation of the MAPK ERK by insulin (see Fig. 12–15) is 5- to 10-fold greater in the presence of the β_2-adrenergic receptor, presumably because of this cross talk. Signaling systems that use cAMP and Ca^{2+} also show extensive interaction; each second messenger affects the generation and concentration of the other. One of the major challenges of systems biology is to sort out the effects of such interactions on the overall metabolic patterns in each tissue—a daunting task.

SUMMARY 12.3 Receptor Tyrosine Kinases

▶ The insulin receptor, INSR, is the prototype of receptor enzymes with Tyr kinase activity. When insulin binds, each $\alpha\beta$ unit of INSR phosphorylates the β subunit of its partner, activating the receptor's Tyr kinase activity. The kinase catalyzes the phosphorylation of Tyr residues on other proteins, such as IRS-1.

▶ Phosphotyrosine residues in IRS-1 serve as binding sites for proteins with SH2 domains. Some of these proteins, such as Grb2, have two or more protein-binding domains and can serve as adaptors that bring two proteins into proximity.

▶ Sos bound to Grb2 catalyzes GDP-GTP exchange on Ras (a small G protein), which in turn activates a MAPK cascade that ends with the phosphorylation of target proteins in the cytosol

FIGURE 12–19 Cross talk between the insulin receptor and the β_2-adrenergic receptor (or other GPCR). When INSR is activated by insulin binding, its Tyr kinase directly phosphorylates the β_2-adrenergic receptor (right side) on two Tyr residues (Tyr[350] and Tyr[364]) near its carboxyl terminus, and indirectly (through activation of protein kinase B (PKB); see Fig. 12–16) causes phosphorylation of two Ser residues in the same region. The effect of these phosphorylations is internalization of the adrenergic receptor, reducing the response to the adrenergic stimulus. Alternatively (left side), INSR-catalyzed phosphorylation of a GPCR (an adrenergic or other receptor) on a carboxyl-terminal Tyr creates the point of nucleation for activating the MAPK cascade (see Fig. 12–15), with Grb2 serving as the adaptor protein. In this case, INSR has used the GPCR to enhance its own signaling.

and nucleus. The result is specific metabolic changes and altered gene expression.

▶ The enzyme PI3K, activated by interaction with IRS-1, converts the membrane lipid PIP$_2$ to PIP$_3$, which becomes the point of nucleation for proteins in a second and third branch of insulin signaling.

▶ In the JAK-STAT signaling system, a soluble protein Tyr kinase (JAK) is activated by association with a receptor, and then phosphorylates the transcription factor STAT, which enters the nucleus and alters the expression of a set of genes.

▶ There are extensive interconnections among signaling pathways, allowing integration and fine-tuning of multiple hormonal effects.

12.4 Receptor Guanylyl Cyclases, cGMP, and Protein Kinase G

Guanylyl cyclases (**Fig. 12–20**) are receptor enzymes that, when activated, convert GTP to the second messenger **guanosine 3′,5′-cyclic monophosphate (cyclic GMP, cGMP)**:

GTP

↘ PP$_i$

Guanosine 3′,5′-cyclic monophosphate (cGMP)

Many of the actions of cGMP in animals are mediated by **cGMP-dependent protein kinase**, also called **protein kinase G (PKG)**. On activation by cGMP, PKG phosphorylates Ser and Thr residues in target proteins.

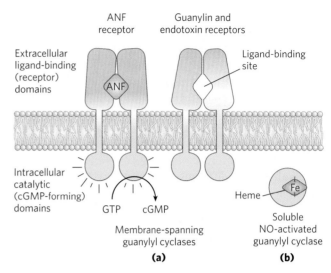

FIGURE 12–20 Two types of guanylyl cyclase that participate in signal transduction. (a) One type is a homodimer with a single membrane-spanning segment in each monomer, connecting the extracellular ligand-binding domain and the intracellular guanylyl cyclase domain. Receptors of this type are used to detect two extracellular ligands: atrial natriuretic factor (ANF; receptors in cells of the renal collecting ducts and vascular smooth muscle) and guanylin (peptide hormone produced in the intestine, with receptors in intestinal epithelial cells). The guanylin receptor is also the target of a bacterial endotoxin that triggers severe diarrhea. **(b)** The other type is a soluble heme-containing enzyme that is activated by intracellular nitric oxide (NO); this form is present in many tissues, including smooth muscle of the heart and blood vessels.

The catalytic and regulatory domains of this enzyme are in a single polypeptide ($M_r \sim 80,000$). Part of the regulatory domain fits snugly in the substrate-binding cleft. Binding of cGMP forces this pseudosubstrate out of the binding site, opening the site to target proteins containing the PKG consensus sequence.

Cyclic GMP carries different messages in different tissues. In the kidney and intestine it triggers changes in ion transport and water retention; in cardiac muscle (a type of smooth muscle) it signals relaxation; in the brain it may be involved both in development and in adult brain function. Guanylyl cyclase in the kidney is activated by the peptide hormone **atrial natriuretic factor (ANF)**, which is released by cells in the cardiac atrium when the heart is stretched by increased blood volume. Carried in the blood to the kidney, ANF activates guanylyl cyclase in cells of the collecting ducts (Fig. 12–20a). The resulting rise in [cGMP] triggers increased renal excretion of Na$^+$ and consequently of water, driven by the change in osmotic pressure. Water loss reduces the blood volume, countering the stimulus that initially led to ANF secretion. Vascular smooth muscle also has an ANF receptor–guanylyl cyclase; on binding to this receptor, ANF causes relaxation (vasodilation) of the blood vessels, which increases blood flow while decreasing blood pressure.

A similar receptor guanylyl cyclase in the plasma membrane of epithelial cells lining the intestine is activated by the peptide **guanylin** (Fig. 12–20a), which regulates Cl^- secretion in the intestine. This receptor is also the target of a heat-stable peptide endotoxin produced by *Escherichia coli* and other gram-negative bacteria. The elevation in [cGMP] caused by the endotoxin increases Cl^- secretion and consequently decreases reabsorption of water by the intestinal epithelium, producing diarrhea.

A distinctly different type of guanylyl cyclase is a cytosolic protein with a tightly associated heme group (Fig. 12–20b), an enzyme activated by nitric oxide (NO). Nitric oxide is produced from arginine by Ca^{2+}-dependent **NO synthase**, present in many mammalian tissues, and diffuses from its cell of origin into nearby cells.

$$\begin{array}{c}
NH_2 \\
| \\
C{=}\overset{+}{N}H_2 \\
| \\
NH \\
| \\
(CH_2)_3 \\
| \\
CH{-}COO^- \\
| \\
{}^+NH_3 \\
\text{Arginine}
\end{array}
\quad
\xrightarrow[\text{NO synthase}]{}
\quad
\begin{array}{c}
NH_2 \\
| \\
C{=}O \\
| \\
NH \\
| \\
(CH_2)_3 \\
| \\
CH{-}COO^- \\
| \\
{}^+NH_3 \\
\text{Citrulline}
\end{array}
\quad + \; NO$$

(with O_2, NADPH → NADP$^+$, Ca^{2+} above the arrow)

NO is sufficiently nonpolar to cross plasma membranes without a carrier. In the target cell, it binds to the heme group of guanylyl cyclase and activates cGMP production. In the heart, cGMP-dependent protein kinase reduces the forcefulness of contractions by stimulating the ion pump(s) that remove Ca^{2+} from the cytosol.

NO-induced relaxation of cardiac muscle is the same response brought about by nitroglycerin and other nitrovasodilators taken to relieve **angina pectoris**, the pain caused by contraction of a heart deprived of O_2 because of blocked coronary arteries. Nitric oxide is unstable and its action is brief; within seconds of its formation, it undergoes oxidation to nitrite or nitrate. Nitrovasodilators produce long-lasting relaxation of cardiac muscle because they break down over several hours, yielding a steady stream of NO. The value of nitroglycerin as a treatment for angina was discovered serendipitously in factories producing nitroglycerin as an explosive in the 1860s. Workers with angina reported that their condition was much improved during the workweek but worsened on weekends. The physicians treating these workers heard this story so often that they made the connection, and a drug was born.

The effects of increased cGMP synthesis diminish after the stimulus ceases, because a specific phosphodiesterase (cGMP PDE) converts cGMP to the inactive 5′-GMP. Humans have several isoforms of cGMP PDE, with different tissue distributions. The isoform in the blood vessels of the penis is inhibited by the drug sildenafil (Viagra), which therefore causes [cGMP] to remain elevated once raised by an appropriate stimulus,

accounting for the usefulness of this drug in the treatment of erectile dysfunction.

Sildenafil (Viagra) ∎

Cyclic GMP has another mode of action in the vertebrate eye: it causes ion-specific channels to open in the retinal rod and cone cells. We return to this role of cGMP in the discussion of vision in Section 12.10.

SUMMARY 12.4 Receptor Guanylyl Cyclases, cGMP, and Protein Kinase G

▶ Several signals, including atrial natriuretic factor and guanylin, act through receptor enzymes with guanylyl cyclase activity. The cGMP so produced is a second messenger that activates cGMP-dependent protein kinase (PKG). This enzyme alters metabolism by phosphorylating specific enzyme targets.

▶ Nitric oxide is a short-lived messenger that stimulates a soluble guanylyl cyclase, raising [cGMP] and stimulating PKG.

12.5 Multivalent Adaptor Proteins and Membrane Rafts

Two generalizations have emerged from studies of signaling systems such as those we have discussed so far: (1) protein kinases that phosphorylate Tyr, Ser, and Thr residues are central to signaling, *directly* affecting the activities of a large number of protein substrates by phosphorylation, and (2) protein-protein interactions brought about by the reversible phosphorylation of Tyr, Ser, and Thr residues in signaling proteins create *docking sites* for other proteins that bring about *indirect* effects on proteins downstream in the signaling pathway. In fact, many signaling proteins are *multivalent*—they can interact with several different proteins simultaneously to form multiprotein signaling complexes. In this section we present a few examples to illustrate the general principles of phosphorylation-dependent protein interactions in signaling pathways.

Protein Modules Bind Phosphorylated Tyr, Ser, or Thr Residues in Partner Proteins

The protein Grb2 in the insulin signaling pathway (Figs 12–15 and 12–19) binds through its SH2 domain to other proteins that have exposed ℗–Tyr residues. The human genome encodes at least 87 SH2-containing proteins, many already known to participate in signaling. The Ⓟ–Tyr residue is bound in a deep pocket in an SH2 domain, with each of its phosphate oxygens participating in hydrogen bonding or electrostatic interactions; the positive charges on two Arg residues figure prominently in the binding. Subtle differences in the structure of SH2 domains account for the specificities of the interactions of SH2-containing proteins with various Ⓟ–Tyr-containing proteins. The SH2 domain typically interacts with a Ⓟ–Tyr (which is assigned the index position 0) and the next three residues toward the carboxyl terminus (designated +1, +2, +3). Some SH2 domains (Src, Fyn, Hck, Nck) favor negatively charged residues in the +1 and +2 positions; others (PLCγ1, SHP2) have a long hydrophobic groove that binds to aliphatic residues in positions +1 to +5. These differences define subclasses of SH2 domains that have different partner specificities.

Phosphotyrosine-binding domains (**PTB domains**; **Fig. 12–21**) are another binding partner for Ⓟ–Tyr proteins, but their critical sequences and three-dimensional structure distinguish them from SH2 domains. The human genome encodes 24 proteins that contain PTB domains, including IRS-1, which we have already encountered in its role as an adaptor protein in insulin-signal transduction (Fig. 12–15). The Ⓟ–Tyr binding sites for SH2 and PTB domains on partner proteins are created by Tyr kinases and eliminated by protein tyrosine phosphatases (PTPases).

Other signaling protein kinases, including PKA, PKC, PKG, and members of the MAPK cascade, phosphorylate Ser or Thr residues in their target proteins, which in some cases acquire the ability to interact with partner proteins through the phosphorylated residue, triggering a downstream process. An alphabet soup of domains that bind Ⓟ–Ser or Ⓟ–Thr residues has been identified, and more are sure to be found. Each domain favors a certain sequence around the phosphorylated residue, so the domains represent families of highly specific recognition sites, able to bind to a specific subset of phosphorylated proteins.

In some cases, the region on a protein that binds Ⓟ–Tyr of a substrate protein is masked by its interaction with a Ⓟ–Tyr in the same protein. For example, the soluble protein Tyr kinase Src, when phosphorylated on a critical Tyr residue, is rendered inactive; an SH2 domain needed to bind to the substrate protein instead binds to the internal Ⓟ–Tyr. When this Ⓟ–Tyr residue is hydrolyzed by a phosphoprotein phosphatase, the Tyr kinase activity of Src is activated (**Fig. 12–22a**). Similarly, glycogen synthase kinase 3 (GSK3) is inactive

FIGURE 12–22 Mechanism of autoinhibition of Src and GSK3. (a) In the active form of the Tyr kinase Src, an SH2 domain binds a Ⓟ-Tyr in the protein substrate, and an SH3 domain binds a proline-rich region of the substrate, lining up the active site of the kinase with several target Tyr residues in the substrate (top). When Src is phosphorylated on a specific Tyr residue (bottom), the SH2 domain binds the internal Ⓟ-Tyr instead of the Ⓟ-Tyr of the substrate, and the SH3 domain binds an internal proline-rich region, preventing productive enzyme-substrate binding; the enzyme is thus autoinhibited. **(b)** In the active form of glycogen synthase kinase 3 (GSK3), an internal Ⓟ-Ser-binding domain is available to bind Ⓟ-Ser in its substrate (glycogen synthase), and thus to position the kinase to phosphorylate neighboring Ser residues (top). Phosphorylation of an internal Ser residue allows this internal kinase segment to occupy the Ⓟ-Ser-binding site, blocking substrate binding (bottom).

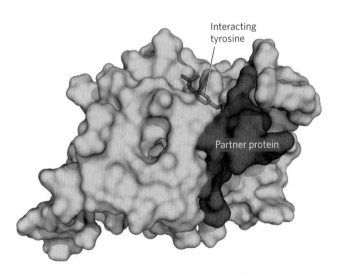

FIGURE 12–21 Interaction of a PTB domain with a Ⓟ-Tyr residue in a partner protein. (PDB ID 1SHC) The PTB domain is represented as a blue surface contour. The Ⓟ-Tyr residue of the partner protein (red) projects into a binding pocket in the PTB domain and is held firmly by multiple noncovalent interactions.

when phosphorylated on a Ser residue in its autoinhibitory domain (Fig. 12–22b). Dephosphorylation of that domain frees the enzyme to bind (and then phosphorylate) its target proteins.

In addition to the three commonly phosphorylated residues in proteins, there is a fourth structure that nucleates the formation of supramolecular complexes of signaling proteins: the phosphorylated head group of the membrane phosphatidylinositols. Many signaling proteins contain domains such as SH3 and PH (plextrin homology domain) that bind tightly to PIP_3 protruding from the inner leaflet of the plasma membrane. Wherever the enzyme PI3K creates this head group (as it does in response to the insulin signal), proteins that bind it will congregate at the membrane surface.

Most of the proteins involved in signaling at the plasma membrane have one or more protein- or phospholipid-binding domains; many have three or more, and thus are multivalent in their interactions with other signaling proteins. **Figure 12–23** shows just a few of the multivalent proteins known to participate in

signaling. Many of the complexes include components with membrane-binding domains. Given the location of so many signaling processes at the inner surface of the plasma membrane, the molecules that must collide to produce the signaling response are effectively confined to two-dimensional space—the membrane surface; collisions here are far more likely than in the three-dimensional space of the cytosol.

In summary, a remarkable picture of signaling pathways has emerged from studies of many signaling proteins and their multiple binding domains. An initial signal results in phosphorylation of the receptor or a target protein, triggering the assembly of large multiprotein complexes, held together on scaffolds with multivalent binding capacities. Some of these complexes contain several protein kinases that activate each other in turn, producing a cascade of phosphorylation and a great amplification of the initial signal. The interactions between cascade kinases are not left to the vagaries of random collisions in three-dimensional space. In the MAPK cascade, for example, a scaffold protein, KSR,

FIGURE 12–23 Some binding modules of signaling proteins. Each protein is represented by a line (with the amino terminus to the left); symbols indicate the location of conserved binding domains (with specificities as listed in the key; abbreviations are explained in the text); green boxes indicate catalytic activities. The name of each protein is given at its carboxyl-terminal end. These signaling proteins interact with phosphorylated proteins or phospholipids in many permutations and combinations to form integrated signaling complexes.

binds all three kinases (MAPK, MAPKK, and MAPKKK), assuring their proximity and correct orientation and even conferring allosteric properties on the interactions among the kinases, which makes their serial phosphorylation sensitive to very small stimuli **(Fig. 12–24)**.

Phosphotyrosine phosphatases remove the phosphate from Ⓟ–Tyr residues, reversing the effect of phosphorylation. Some of these are receptorlike membrane proteins, presumably controlled by extracellular factors not yet identified; other PTPases are soluble and contain SH2 domains. In addition, animal cells have protein Ⓟ–Ser and Ⓟ–Thr phosphatases, which reverse the effects of Ser- and Thr-specific protein kinases. We can see, then, that signaling occurs in *protein circuits*, which are effectively hardwired from signal receptor to response effector and can be switched off instantly by the hydrolysis of a single upstream phosphate ester bond.

The multivalency of signaling proteins allows for the assembly of many different combinations of signaling modules, each combination suited to particular signals, cell types, and metabolic circumstances, yielding diverse signaling circuits of extraordinary complexity.

Membrane Rafts and Caveolae May Segregate Signaling Proteins

Membrane rafts (Chapter 11) are regions of the membrane bilayer enriched in sphingolipids, sterols, and certain proteins, including many attached to the bilayer by GPI anchors. The β-adrenergic receptor is segregated in rafts that contain G proteins, adenylyl cyclase, PKA, and a specific protein phosphatase, PP2, which together provide a highly integrated signaling unit. By segregating in a small region of the plasma membrane all of the elements required for responding to and ending the signal, the cell is able to produce a highly localized and brief "puff" of second messenger.

Some RTKs (EGFR and PDGFR) seem to be localized in rafts, and this sequestration is very probably functionally significant. When cholesterol is removed from rafts by treatment of the membrane with cyclodextrin (which binds and removes cholesterol), the rafts are disrupted and the RTK signaling pathways become defective.

If an RTK in a raft is phosphorylated, and the only locally available PTPase that reverses this phosphorylation is in another raft, then dephosphorylation of the RTK is slowed or prevented. Interactions between adaptor proteins might be strong enough to recruit into a raft a signaling protein not usually located there, or might even be strong enough to pull receptors out of a raft. For example, the EGFR in isolated fibroblasts is usually concentrated in specialized rafts called caveolae (see Fig. 11–22), but treatment with EGF causes the receptor to leave the raft. This migration depends on the receptor's protein kinase activity; mutant receptors lacking this activity remain in the raft during treatment with EGF. Caveolin, an integral membrane protein localized in caveolae, is phosphorylated on Tyr residues in response to insulin, and the now-activated EGFR may be able to draw its binding partners into the raft. Spatial segregation of signaling proteins in rafts adds yet another dimension to the already complex processes initiated by extracellular signals.

FIGURE 12–24 A scaffold protein from yeast that organizes and regulates a protein kinase cascade. (a) The scaffold protein KSR has binding sites for all three of the kinases in the Raf/MEK/Erk cascade. By binding all three in appropriate orientations, the scaffold makes interactions among the proteins rapid and efficient. When Erk has been activated (left), it phosphorylates the binding site for Raf (right), forcing a conformational change that displaces Raf and thereby prevents the phosphorylation of MEK. The result of this feedback regulation is that MEK phosphorylation is temporary. **(b)** In yeast cells with mutant KSR lacking the phosphorylation sites (red curve), no feedback occurs, producing a different time course of signaling.

SUMMARY 12.5 Multivalent Adaptor Proteins and Membrane Rafts

▶ Many signaling proteins have domains that bind phosphorylated Tyr, Ser, or Thr residues in other proteins; the binding specificity for each domain is determined by sequences that adjoin the phosphorylated residue in the substrate.

▶ SH2 and PTB domains bind to proteins containing Ⓟ–Tyr residues; other domains bind Ⓟ–Ser and Ⓟ–Thr residues in various contexts.

▶ SH3 and PH domains bind the membrane phospholipid PIP$_3$.

▶ Many signaling proteins are multivalent, with several different binding modules. By combining the substrate specificities of various protein kinases with the specificities of domains that bind phosphorylated Ser, Thr, or Tyr residues, and with phosphatases that can rapidly inactivate a signaling pathway, cells create a large number of multiprotein signaling complexes.

▶ Membrane rafts and caveolae sequester groups of signaling proteins in small regions of the plasma membrane, enhancing their interactions and making signaling more efficient.

12.6 Gated Ion Channels

Ion Channels Underlie Electrical Signaling in Excitable Cells

Certain cells in multicellular organisms are "excitable": they can detect an external signal, convert it into an electrical signal (specifically, a change in membrane potential), and pass it on. Excitable cells play central roles in nerve conduction, muscle contraction, hormone secretion, sensory processes, and learning and memory. The excitability of sensory cells, neurons, and myocytes depends on ion channels, signal transducers that provide a regulated path for the movement of inorganic ions such as Na$^+$, K$^+$, Ca^{2+}, and Cl$^-$ across the plasma membrane in response to various stimuli. Recall from Chapter 11 that these ion channels are "gated": they may be open or closed, depending on whether the associated receptor has been activated by the binding of its specific ligand (a neurotransmitter, for example) or by a change in the transmembrane electrical potential, V_m. The Na$^+$K$^+$ ATPase is electrogenic; it creates a charge imbalance across the plasma membrane by carrying 3 Na$^+$ out of the cell for every 2 K$^+$ carried in (**Fig. 12–25a**), making the inside negative relative to the outside. The membrane is said to be polarized.

KEY CONVENTION: V_m is negative when the inside of the cell is negative relative to the outside. For a typical animal cell, $V_m = -50$ to -70 mV. ∎

Because ion channels generally allow passage of either anions or cations but not both, ion flux through a channel causes a redistribution of charge on the two sides of the membrane, changing V_m. Influx of a positively charged ion such as Na$^+$, or efflux of a negatively

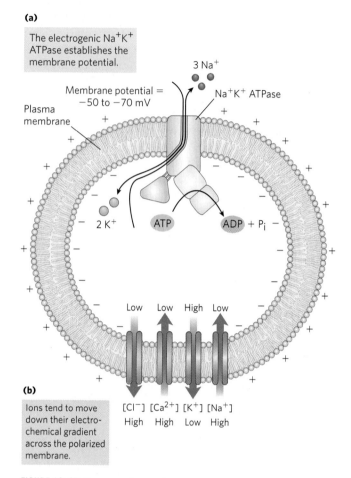

(a)

The electrogenic Na$^+$K$^+$ ATPase establishes the membrane potential.

Membrane potential = -50 to -70 mV

3 Na$^+$

Na$^+$K$^+$ ATPase

Plasma membrane

2 K$^+$ ATP   ADP + P$_i$

Low   Low   High   Low

(b)

Ions tend to move down their electro-chemical gradient across the polarized membrane.

[Cl$^-$]	[Ca^{2+}]	[K$^+$]	[Na$^+$]
High	High	Low	High

FIGURE 12–25 Transmembrane electrical potential. (a) The electrogenic Na$^+$K$^+$ ATPase produces a transmembrane electrical potential of about -60 mV (inside negative). **(b)** Blue arrows show the direction in which ions tend to move spontaneously across the plasma membrane in an animal cell, driven by the combination of chemical and electrical gradients. The chemical gradient drives Na$^+$ and Ca^{2+} inward (producing depolarization) and K$^+$ outward (producing hyperpolarization). The electrical gradient drives Cl$^-$ outward, against its concentration gradient (producing depolarization).

charged ion such as Cl$^-$, *depolarizes* the membrane and brings V_m closer to zero. Conversely, efflux of K$^+$ *hyperpolarizes* the membrane and V_m becomes more negative. These ion fluxes through channels are passive, in contrast to active transport by the Na$^+$K$^+$ ATPase.

The direction of spontaneous ion flow across a polarized membrane is dictated by the electrochemical potential of that ion across the membrane, which has two components: the difference in concentration (C) of the ion on the two sides of the membrane, and the difference in electrical potential, typically expressed in millivolts. The force (ΔG) that causes a cation (say, Na$^+$) to pass spontaneously inward through an ion channel is a function of the ratio of its concentrations on the two sides of the membrane (C_{in}/C_{out}) and of the difference in electrical potential (V_m or $\Delta\psi$):

$$\Delta G = RT \ln (C_{in}/C_{out}) + Z\mathcal{F}V_m \qquad (12\text{–}1)$$

where R is the gas constant, T the absolute temperature, Z the charge on the ion, and $\mathcal{J}$ the Faraday constant. (Note that the sign of the charge on the ion determines the sign of the second term in Eqn 12–1.) In a typical neuron or myocyte, the concentrations of Na^+, K^+, Ca^{2+}, and Cl^- in the cytosol are very different from those in the extracellular fluid (Table 12–6). Given these concentration differences, the resting V_m of about -60 mV, and the relationship shown in Equation 12–1, the opening of a Na^+ or Ca^{2+} channel will result in a spontaneous inward flow of Na^+ or Ca^{2+} (and depolarization), whereas opening of a K^+ channel will result in a spontaneous outward flux of K^+ (and hyperpolarization) (Fig. 12–25b). In this case, K^+ moves outward, against the electrical gradient, because the large concentration difference inside and outside the cell produces a more powerful, outward chemical force on the ion.

A given ionic species continues to flow through a channel only as long as the combination of concentration gradient and electrical potential provides a driving force. For example, as Na^+ flows down its concentration gradient, it depolarizes the membrane. When the membrane potential reaches $+70$ mV, the effect of this membrane potential (resistance to further entry of Na^+) exactly equals the effect of the $[Na^+]$ gradient (promotion of Na^+ flow inward). At this equilibrium potential (E), the driving force (ΔG) tending to move a Na^+ ion is zero. The equilibrium potential is different for each ionic species, because the concentration gradients differ.

The number of ions that must flow to produce a physiologically significant change in the membrane potential is negligible relative to the concentrations of Na^+, K^+, and Cl^- in cells and extracellular fluid, so the ion fluxes that occur during signaling in excitable cells have essentially no effect on the concentrations of these ions. With Ca^{2+}, the situation is different; because the intracellular $[Ca^{2+}]$ is generally very low ($\sim 10^{-7}$ M), inward flow of Ca^{2+} can significantly alter the cytosolic $[Ca^{2+}]$.

The membrane potential of a cell at a given time is the result of the types and numbers of ion channels open at that instant. In most cells at rest, more K^+ channels than Na^+, Cl^-, or Ca^{2+} channels are open and thus the resting potential is closer to the E for K^+ (-98 mV) than that for any other ion. When channels for Na^+, Ca^{2+}, or Cl^- open, the membrane potential moves toward the E for that ion. The precisely timed opening and closing of ion channels and the resulting transient changes in membrane potential underlie the electrical signaling by which the nervous system stimulates the skeletal muscles to contract, the heart to beat, or secretory cells to release their contents. Moreover, many hormones exert their effects by altering the membrane potential of their target cells. These mechanisms are not limited to animals; ion channels play important roles in the responses of bacteria, protists, and plants to environmental signals.

To illustrate the action of ion channels in cell-to-cell signaling, we describe the mechanisms by which a neuron passes a signal along its length and across a synapse to the next neuron (or to a myocyte) in a cellular circuit, using acetylcholine as the neurotransmitter.

Voltage-Gated Ion Channels Produce Neuronal Action Potentials

Signaling in the nervous system is accomplished by networks of neurons, specialized cells that carry an electrical impulse (action potential) from one end of the cell (the cell body) through an elongated cytoplasmic extension (the axon). The electrical signal triggers release of neurotransmitter molecules at the synapse, carrying the signal to the next cell in the circuit. Three types of **voltage-gated ion channels** are essential to this signaling mechanism. Along the entire length of the axon are **voltage-gated Na^+ channels (Fig. 12–26)**, which are closed when the membrane is at rest ($V_m = -60$ mV) but open briefly when the membrane is depolarized locally in response to acetylcholine (or some other neurotransmitter). Also distributed along the axon are **voltage-gated K^+ channels**, which open, a split second later, in response to the depolarization when nearby Na^+ channels open. The depolarizing flow of Na^+ into the axon (influx) is thus rapidly countered by a repolarizing flow of K^+ out (efflux). At the distal end of the axon are **voltage-gated Ca^{2+} channels**, which open when the wave of depolarization (step ❶) and repolarization (step ❷) caused by the activity of Na^+ and K^+ channels arrives, triggering release of the neurotransmitter acetylcholine—which carries the signal to another neuron (fire an action potential!) or to a muscle fiber (contract!).

TABLE 12–6	Ion Concentrations in Cells and Extracellular Fluids (mM)							
	K^+		Na^+		Ca^{2+}		Cl^-	
Cell type	In	Out	In	Out	In	Out	In	Out
Squid axon	400	20	50	440	≤ 0.4	10	40–150	560
Frog muscle	124	2.3	10.4	109	< 0.1	2.1	1.5	78

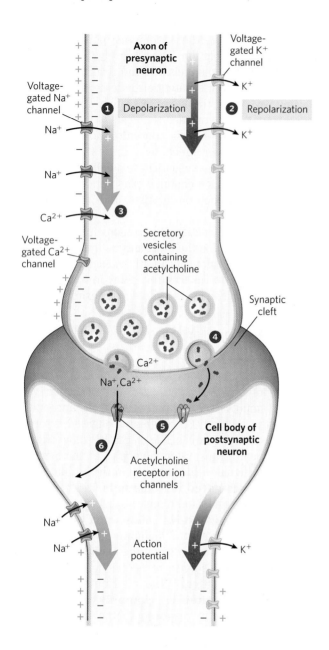

Initially, the plasma membrane of the presynaptic neuron is polarized (inside negative) through the action of the electrogenic Na^+K^+ ATPase, which pumps out 3 Na^+ for every 2 K^+ pumped in (see Fig. 12-25). ❶ A stimulus to this neuron (not shown) causes an action potential to move along the axon (blue arrow), away from the cell body. The opening of a voltage-gated Na^+ channel allows Na^+ entry, and the resulting local depolarization causes the adjacent Na^+ channel to open, and so on. The directionality of movement of the action potential is ensured by the brief refractory period that follows the opening of each voltage-gated Na^+ channel. ❷ A split second after the action potential passes a point in the axon, voltage-operated K^+ channels open, allowing K^+ exit that brings about repolarization of the membrane (red arrow), to make it ready for the next action potential. (For clarity, Na^+ channels and K^+ channels are drawn on opposite sides of the axon; both types of channels are uniformly distributed in the axonal membrane.) ❸ When the wave of depolarization reaches the axon tip, voltage-gated Ca^{2+} channels open, allowing Ca^{2+} entry. ❹ The resulting increase in internal $[Ca^{2+}]$ triggers exocytic release of the neurotransmitter acetylcholine into the synaptic cleft. ❺ Acetylcholine binds to a receptor on the postsynaptic neuron (or myocyte), causing its ligand-gated ion channel to open. ❻ Extracellular Na^+ and Ca^{2+} enter through this channel, depolarizing the postsynaptic cell. The electrical signal has thus passed to the cell body of the postsynaptic neuron (or myocyte) and will move along its axon to a third neuron (or a myocyte) by this same sequence of events.

unidirectional wave of depolarization—the action potential—sweeps from the nerve cell body toward the end of the axon.

When the wave of depolarization reaches the voltage-gated Ca^{2+} channels, they open (step ❸), and Ca^{2+} enters from the extracellular space. The rise in cytoplasmic $[Ca^{2+}]$ then triggers release of acetylcholine by exocytosis into the synaptic cleft (step ❹). Acetylcholine diffuses to the postsynaptic cell (another neuron or a myocyte), where it binds to acetylcholine receptors and triggers depolarization. Thus the message is passed to the next cell in the circuit. We see, then, that gated ion channels convey signals in either of two ways: by changing the cytoplasmic concentration of an ion (such as Ca^{2+}), which then serves as an intracellular second messenger, or by changing V_m and affecting other membrane proteins that are sensitive to V_m. The passage of an electrical signal through one neuron and on to the next illustrates both types of mechanism.

We discussed the structure and mechanism of voltage-gated K^+ channels in some detail in Section 11.3 (see Figs 11–47 and 11–48). Here we take a closer look at Na^+ channels. The essential component of a Na^+ channel is a single, large polypeptide (1,840 amino acid residues) organized into four domains clustered around a central channel **(Fig. 12–27a, b)**, providing a path for Na^+ through the membrane. The path is made Na^+-specific by a "pore region" composed of the segments between transmembrane helices 5 and 6 of each

The voltage-gated Na^+ channels are very selective for Na^+ over other cations (by a factor of 100 or more) and have a very high flux rate ($>10^7$ ions/s). After being opened—activated—by a reduction in transmembrane electrical potential, a Na^+ channel undergoes very rapid inactivation—within milliseconds, the channel closes and remains inactive for many milliseconds. As voltage-gated K^+ channels open in response to the depolarization induced by the opening of Na^+ channels (step ❶ in Fig. 12–26), the resulting efflux of K^+ repolarizes the membrane locally (it reestablishes the inside-negative membrane potential; step ❷). A brief pulse of depolarization thus traverses the axon as local depolarization triggers the brief opening of neighboring Na^+ channels, then K^+ channels. The short refractory period that follows the opening of each Na^+ channel, during which it cannot open again, ensures that a

FIGURE 12–27 Voltage-gated Na⁺ channels of neurons. Sodium channels of different tissues and organisms have a variety of subunits, but only the principal subunit (α) is essential. **(a)** The α subunit is a large protein with four homologous domains (I to IV, shown spread out here to illustrate the parts), each containing six transmembrane helices (1 to 6). Helix 4 in each domain (blue) is the voltage sensor; helix 6 (orange) is thought to be the activation gate. The segments between helices 5 and 6, the pore region (red), form the selectivity filter, and the segment connecting domains III and IV (green) is the inactivation gate. **(b)** (PDB ID 3RW0) Structure of a voltage-gated Na⁺ channel (from the bacterium *Arcobacter butzleri*, but probably similar to channels of vertebrate neurons), with a funnel-shaped opening on the extracellular side, an ion selectivity filter, a central aqueous cavity, and the activation domain on the cytoplasmic side. **(c)** A schematic view of the Na⁺ channel. The four domains are wrapped about a central transmembrane channel lined with polar amino acid residues. The four pore regions (red) come together near the extracellular surface to form the selectivity filter, which is conserved in all Na⁺ channels. The filter gives the channel its ability to discriminate between Na⁺ and other ions of similar size. The inactivation gate (green) closes (dotted lines) soon after the activation gate opens. **(d)** The voltage-sensing mechanism involves movement of helix 4 (blue) perpendicular to the plane of the membrane in response to a change in transmembrane potential. As shown at the top, the strong positive charge on helix 4 allows it to be pulled inward in response to the inside-negative membrane potential (V_m). Depolarization lessens this pull, and helix 4 relaxes by moving outward (bottom). This movement is communicated to the activation gate (orange), inducing conformational changes that open the channel.

domain, which fold into the channel. Helix 4 of each domain has a high density of positively charged Arg residues; this segment is believed to move within the membrane in response to changes in the transmembrane voltage, from the resting potential of about -60 mV to about $+30$ mV. The movement of helix 4 triggers opening of the channel, and this is the basis for the voltage gating (Fig. 12–27c).

Inactivation of the channel (during the refractory period) is thought to occur by a ball-and-chain mechanism. A protein domain on the cytoplasmic surface of the Na⁺ channel, the inactivation gate (the ball), is tethered to the channel by a short segment of the polypeptide (the chain; Fig. 12–27b). This domain is free to move about when the channel is closed, but when it opens, a site on the inner face of the channel becomes available for the tethered ball to bind, blocking the channel. The length of the tether seems to determine how long an ion channel stays open: the longer the tether, the longer the open period. Other gated ion channels may be inactivated by a similar mechanism.

The Acetylcholine Receptor Is a Ligand-Gated Ion Channel

The **nicotinic acetylcholine receptor** mediates the passage of the signal from an electrically excited neuron at some types of synapses and at neuromuscular junctions (between motor neuron and muscle fiber), triggering muscle contraction. (Nicotinic acetylcholine receptors were originally distinguished from muscarinic acetylcholine receptors by the sensitivity of the former to nicotine, the latter to the mushroom alkaloid muscarine. They are structurally and functionally different.) Acetylcholine released by the presynaptic neuron or motor neuron diffuses a few micrometers to the plasma membrane of the postsynaptic neuron or myocyte, where it binds to the acetylcholine receptor. This forces a conformational change in the receptor, causing its ion channel to open. The resulting inward movement of cations depolarizes the plasma membrane. In a muscle fiber, this triggers contraction. The acetylcholine receptor allows ready passage of Na^+, Ca^{2+}, and K^+ ions, but other cations and all anions are unable to pass. Movement of Na^+ through an acetylcholine receptor ion channel is unsaturable (its rate is linear with respect to extracellular $[Na^+]$) and very fast—about 2×10^7 ions/s under physiological conditions.

H_3C—C (=O) —O—CH_2—CH_2—$\overset{+}{N}$(—CH_3)(—CH_3)—CH_3

Acetylcholine

Like other gated ion channels, the acetylcholine receptor opens in response to stimulation by its signal molecule and has an intrinsic timing mechanism that closes the gate milliseconds later. Thus the acetylcholine signal is transient—as we have seen, an essential feature of electrical signal conduction. We understand the structural changes underlying gating in the acetylcholine receptor, but not the exact mechanism of "desensitization," in which the gate remains closed even in the continued presence of acetylcholine.

The nicotinic acetylcholine receptor has five subunits ($\alpha_2\beta\gamma\delta$), each having four transmembrane helical segments (M1 to M4) **(Fig. 12–28)**. The five subunits, which are related in sequence and tertiary structure, surround a central pore, which is lined with their M2 helices. The pore is about 20 Å wide in the parts of the channel that protrude on the cytoplasmic and extracellular surfaces, but narrows as it passes through the lipid bilayer. Near the center of the bilayer is a ring of bulky hydrophobic side chains of Leu residues in the M2 helices, positioned so close together that they prevent ions from passing through the channel (Fig. 12–28d). Binding of acetylcholine to sites on each α subunit forces all M2 helices to rotate slightly, moving the bulky Leu residues aside and replacing them with smaller, polar residues. The widening of the pore allows passage of ions (Na^+ and Ca^{2+}).

Neurons Have Receptor Channels That Respond to Different Neurotransmitters

Animal cells, especially those of the nervous system, contain a variety of ion channels gated by ligands, voltage, or both. Receptors that are themselves ion channels are classified as **ionotropic** to distinguish them from receptors that generate a second messenger (metabotropic receptors). We have so far focused on acetylcholine as neurotransmitter, but there are many others. 5-Hydroxytryptamine (serotonin), glutamate, and glycine all can act through receptor channels that are structurally related to the acetylcholine receptor. Serotonin and glutamate trigger the opening of cation (K^+, Na^+, Ca^{2+}) channels, whereas glycine opens Cl^--specific channels. Cation and anion channels are distinguished by subtle differences in the amino acid residues that line the hydrophilic channel. Cation channels have negatively charged Glu and Asp side chains at crucial positions. When a few of these acidic residues are experimentally replaced with basic residues, the cation channel is converted to an anion channel.

Depending on which ion passes through a channel, binding of the ligand (neurotransmitter) for that channel results in either depolarization or hyperpolarization of the target cell. A single neuron normally receives input from many other neurons, each releasing its own characteristic neurotransmitter with its characteristic depolarizing or hyperpolarizing effect. The target cell's V_m therefore reflects the *integrated* input (Fig. 12–1e) from multiple neurons. The cell responds with an action potential only if the integrated input adds up to a net depolarization of sufficient size.

The receptor channels for acetylcholine, glycine, glutamate, and γ-aminobutyric acid (GABA) are gated by *extracellular* ligands. *Intracellular* second messengers—such as cAMP, cGMP, IP_3, Ca^{2+}, and ATP—regulate ion channels of another class, which, as we shall see in Section 12.10, participate in the sensory transductions of vision, olfaction, and gustation.

Toxins Target Ion Channels

Many of the most potent toxins found in nature act on ion channels. For example, dendrotoxin (from the black mamba snake) blocks the action of voltage-gated K^+ channels, tetrodotoxin (produced by puffer fish) acts on voltage-gated Na^+ channels, and cobrotoxin disables acetylcholine receptor ion channels. Why, in the course of evolution, have ion channels become the preferred target of toxins, rather than some critical metabolic target such as an enzyme essential in energy metabolism?

Ion channels are extraordinary amplifiers; opening of a single channel can allow the flow of 10 million ions per second. Consequently, relatively few molecules of

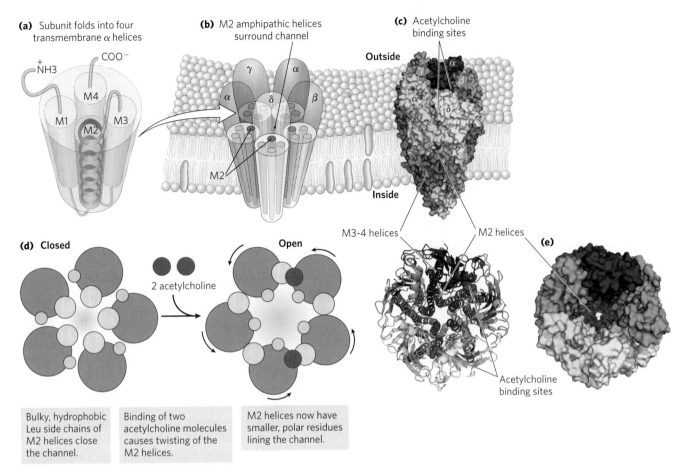

(a) Subunit folds into four transmembrane α helices

(b) M2 amphipathic helices surround channel

(c) Acetylcholine binding sites

Outside

Inside

M3-4 helices   M2 helices   **(e)**

Acetylcholine binding sites

(d) Closed   Open

2 acetylcholine

| Bulky, hydrophobic Leu side chains of M2 helices close the channel. | Binding of two acetylcholine molecules causes twisting of the M2 helices. | M2 helices now have smaller, polar residues lining the channel. |

FIGURE 12–28 The acetylcholine receptor ion channel. (a) Each of the five homologous subunits ($\alpha_2\beta\gamma\delta$) has four transmembrane helices, M1 to M4. The M2 helices are amphipathic; the others have mainly hydrophobic residues. **(b)** The five subunits are arranged around a central transmembrane channel, which is lined with the polar sides of the M2 helices. At the top and bottom of the channel are rings of negatively charged amino acid residues. **(c)** A molecular model of the acetylcholine receptor, based on x-ray structure determination of a related protein (the acetylcholine-binding protein from a mollusk; PDB ID 1UV6). **(d)** This cartoon view of a cross section through the center of the M2 helices shows five Leu side chains (yellow), one from each M2 helix, protruding into the channel and constricting it to a diameter too small to allow passage of Ca^{2+}, Na^+, or K^+. When both acetylcholine receptor sites (one on each α subunit) are occupied, a conformational change occurs. As the M2 helices twist slightly, the five Leu residues rotate away from the channel and are replaced by smaller, polar residues (blue). This gating mechanism opens the channel, allowing the passage of Ca^{2+}, Na^+, or K^+. **(e)** Molecular model of the acetylcholine receptor viewed perpendicular to the membrane, showing the small central pore that allows ion passage.

an ion channel protein are needed per neuron for signaling functions. This means that a relatively small number of toxin molecules with high affinity for ion channels, acting from outside the cell, can have a very pronounced effect on neurosignaling throughout the body. A comparable effect by way of a metabolic enzyme, typically present in cells at much higher concentrations than ion channels, would require far more copies of the toxin molecule.

SUMMARY 12.6 Gated Ion Channels

▶ Ion channels gated by membrane potential or ligands are central to signaling in neurons and other cells.

▶ The voltage-gated Na^+ and K^+ channels of neuronal membranes carry the action potential along the axon as a wave of depolarization (Na^+ influx) followed by repolarization (K^+ efflux).

▶ The gating mechanism for voltage-sensitive channels involves the movement, perpendicular to the plane of the membrane, of a transmembrane peptide with a high charge density, due to the presence of Arg or other charged residues.

▶ Arrival of an action potential at the distal end of a presynaptic neuron triggers neurotransmitter release. The neurotransmitter (acetylcholine, for example) diffuses to the postsynaptic neuron (or the myocyte, at a neuromuscular junction), binds to specific receptors in the plasma membrane, and triggers a change in V_m.

▶ The acetylcholine receptor of neurons and myocytes is a ligand-gated ion channel; acetylcholine binding triggers a conformational change that opens the channel to Na^+ and Ca^{2+} ions.

▶ Neurotoxins produced by many organisms attack neuronal ion channels, and are therefore fast-acting and deadly.

12.7 Integrins: Bidirectional Cell Adhesion Receptors

Integrins are proteins of the plasma membrane that mediate the adhesion of cells to each other and to the extracellular matrix, and carry signals in both directions across the membrane **(Fig. 12–29)**. The mammalian genome encodes 18 different α subunits and 8 different β subunits, which are found in a range of combinations with various ligand-binding specificities in various tissues. Each of the 24 different integrins found thus far seems to have a unique function. Because they can inform cells about the extracellular neighborhood, integrins play crucial roles in processes that require selective cell-cell interactions, such as embryonic development, blood clotting, immune cell function, normal differentiation, and tumor growth and metastasis.

The extracellular ligands that interact with integrins include collagen, fibrinogen, fibronectin, and many other proteins that have the sequence recognized by integrins: −Arg–Gly–Asp− (RGD, using single-letter amino acid abbreviations). The short, cytoplasmic extensions of the α and β subunits interact with cytoskeletal proteins just beneath the plasma membrane—talin, α-actinin, vinculin, paxillin, and others—modulating the assembly of actin-based cytoskeletal structures. The dual association of integrins with the extracellular matrix and the cytoskeleton allows the cell to integrate information about its extracellular and intracellular environments and to coordinate cytoskeletal positioning with extracellular adhesion sites. In this capacity, integrins govern the shape, motility, polarity, and differentiation of many cell types. In "outside-in" signaling, the extracellular domains of an integrin undergo dramatic, global conformational changes when ligand binds at a site many angstroms from the transmembrane helices. These changes somehow alter the dispositions of the cytoplasmic tails of the α and β subunits, changing their interactions with intracellular proteins and thereby conducting the signal inward.

The conformation and adhesiveness of integrin extracellular domains are also dramatically altered by "inside-out" signaling initiated by signals from *inside* the cell. In one conformation, the extracellular domains have no affinity for the proteins of the extracellular

FIGURE 12–29 Two-way signaling by integrins. All integrins have one α and one β subunit, each with a short cytoplasmic extension, a single transmembrane helix, and a large extracellular domain with the ligand-binding site. The β subunit (purple) is rich in Cys residues and has extensive intrachain disulfide bonding. The α subunit (pink) in many integrins has several binding sites for divalent cations such as Ca^{2+}, which are intrinsic to the ligand-binding activity. In its inactive state, integrin's extracellular domain is folded upon itself (center). Contact with an extracellular ligand (collagen or heparan sulfate, for example) straightens the extracel-

lular domain and moves the cytosolic tails of the α and β subunits apart (left), altering their interactions with intracellular proteins such as talin, which in turn connect the integrin to actin filaments in the cytoskeleton. In inside-out signaling, contact of the cytosolic domain with talin produces a dramatic unbending of the extracellular domain (right) and an increase in its affinity for extracellular binding partners, allowing interactions with extracellular proteins or proteoglycans and changing the cell's adhesion to the extracellular matrix. Protein ligands in the extracellular matrix have the RGD sequence recognized by integrins.

matrix, but signals from the cell can favor another conformation in which integrins adhere tightly to extracellular proteins (Fig. 12–29).

Regulation of adhesiveness is central to leukocyte homing to the site of an infection (see Fig. 7–32), interactions between immune cells, and phagocytosis by macrophages. During an immune response, for example, leukocyte integrins are activated (exposing their extracellular ligand-binding sites) from inside the cell via a signaling pathway *triggered* by cytokines (extracellular developmental signals). Thus activated, the integrins can mediate the attachment of leukocytes to other immune cells or can target cells for phagocytosis. Mutation in an integrin gene encoding the β subunit known as CD18 is the cause of leukocyte adhesion deficiency, a rare human genetic disease in which leukocytes fail to pass out of blood vessels to reach sites of infection. Infants with a severe defect in CD18 commonly die of infections before the age of two.

An integrin specific to platelets ($\alpha_{\text{IIb}}\beta_3$) is involved in both normal and pathological blood clotting. Local damage to blood vessels at a site of injury exposes high-affinity binding sites (RGD sequences in thrombin and collagen, for example) for the integrins of platelets, which attach themselves to the lesion, to other platelets, and to the clotting protein fibrinogen, leading to clot formation that prevents further bleeding. Mutations in the α or β subunit of platelet integrin $\alpha_{\text{IIb}}\beta_3$ lead to a bleeding disorder known as Glanzmann thrombasthenia, in which individuals bleed excessively after a relatively minor injury. Overly effective blood coagulation is also undesirable. Dysregulation of platelet adhesion can lead to pathological blood clot formation, resulting in blockage of the arteries that supply blood to the heart and brain and increasing the risk of heart attack and stroke. Drugs such as tirofiban and eptifibatide that block the external ligand-binding sites of platelet integrin reduce clot formation and are useful in treating and preventing heart attacks and strokes.

When tumors metastasize, tumor cells lose their adhesion to the originating tissue and invade new locations. Both the changes in tumor cell adhesion and the development of new blood vessels (angiogenesis) to support the tumor at a new location are modulated by specific integrins. These proteins are therefore potential targets for drugs that suppress the migration and relocation of tumor cells. ∎

SUMMARY 12.7 Integrins: Bidirectional Cell Adhesion Receptors

▶ Integrins are a family of dimeric ($\alpha\beta$) plasma membrane receptors that interact with extracellular macromolecules and the cytoskeleton, carrying signals in and out of the cell.

▶ The active and inactive forms of an integrin differ in the conformation of their extracellular domains. Intracellular events and signals can interconvert the active and inactive forms.

▶ Integrins mediate various aspects of the immune response, blood clotting, and angiogenesis, and they play a role in tumor metastasis.

12.8 Regulation of Transcription by Nuclear Hormone Receptors

The steroid, retinoic acid (retinoid), and thyroid hormones form a large group of hormones (receptor ligands) that exert at least part of their effects by a mechanism fundamentally different from that of other hormones: they act in the nucleus to alter gene expression. We discuss their mode of action in detail in Chapter 28, along with other mechanisms for regulating gene expression. Here we give a brief overview.

Steroid hormones (estrogen, progesterone, and cortisol, for example), too hydrophobic to dissolve readily in the blood, are transported on specific carrier proteins from their point of release to their target tissues. In target cells, these hormones pass through the plasma membrane by simple diffusion and bind to specific receptor proteins in the nucleus (**Fig. 12–30**). Steroid hormone receptors with no bound ligand (aporeceptors) often act to suppress the transcription of target genes. Hormone binding triggers changes in the conformation of a receptor protein so that it becomes capable of interacting with specific regulatory sequences in DNA called **hormone response elements (HREs)**, thus altering gene expression (see Fig. 28–33). The bound receptor-hormone complex enhances the expression of specific genes adjacent to HREs, with the help of several other proteins essential for transcription. Hours or days are required for these regulators to have their full effect—the time required for the changes in RNA synthesis and subsequent protein synthesis to become evident in altered metabolism.

The specificity of the steroid-receptor interaction is exploited in the use of the drug **tamoxifen** to treat breast cancer. In some types of breast cancer, division of the cancerous cells depends on the continued presence of estrogen. Tamoxifen is an estrogen antagonist; it competes with estrogen for binding to the estrogen receptor, but the tamoxifen-receptor complex has little or no effect on gene expression. Consequently, tamoxifen administered after surgery or during chemotherapy for hormone-dependent breast cancer slows or stops the growth of remaining cancerous cells. Another steroid analog, the drug **mifepristone (RU486)**, binds to the progesterone receptor and blocks hormone actions essential to implantation of the fertilized ovum in the uterus, and thus functions as a contraceptive.

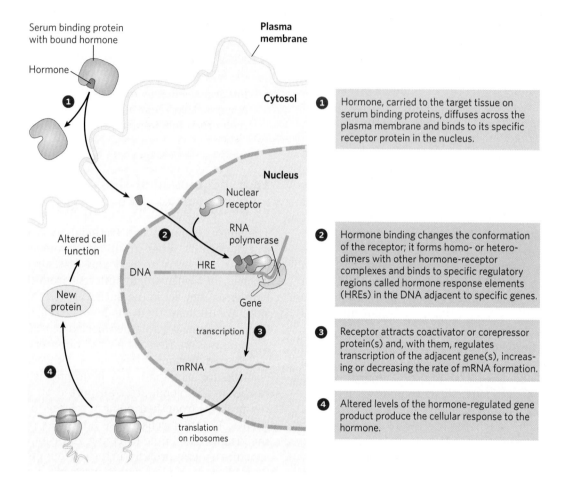

FIGURE 12–30 General mechanism by which steroid and thyroid hormones, retinoids, and vitamin D regulate gene expression. The details of transcription and protein synthesis are discussed in Chapters 26 and 27. Some steroids also act through plasma membrane receptors by a completely different mechanism.

Tamoxifen

Mifepristone
(RU486)

Certain effects of steroids seem to occur too fast to be the result of altered protein synthesis via the classic

mechanism of steroid hormone action through nuclear receptors. For example, the estrogen-mediated dilation of blood vessels is known to be independent of gene transcription or protein synthesis, as is the steroid-induced decrease in cellular [cAMP]. Another transduction mechanism involving plasma membrane receptors is believed to be responsible for some of these effects. ∎

SUMMARY 12.8 Regulation of Transcription by Nuclear Hormone Receptors

▶ Steroid hormones enter cells and bind to specific receptor proteins.

▶ The hormone-receptor complex binds specific regions of DNA, the hormone response elements, and interacts with other proteins to regulate the expression of nearby genes.

▶ Certain effects of steroid hormones may occur through a different, faster signaling pathway.

12.9 Signaling in Microorganisms and Plants

Much of what we have said here about signaling relates to mammalian tissues or cultured cells from such tissues. Bacteria, archaea, eukaryotic microorganisms, and vascular plants must also respond to a variety of external signals—O_2, nutrients, light, noxious chemicals, and so on. We turn here to a brief consideration of the kinds of signaling machinery used by microorganisms and plants.

Bacterial Signaling Entails Phosphorylation in a Two-Component System

In pioneering studies of chemotaxis in bacteria, Julius Adler showed that *Escherichia coli* responds to nutri-

Julius Adler

ents in its environment, including sugars and amino acids, by swimming toward them, propelled by one or a few flagella. A family of membrane proteins have binding domains on the outside of the plasma membrane to which specific **attractants** (sugars or amino acids) bind **(Fig. 12–31)**. Ligand binding causes an intrinsic kinase activity of the receptor to phosphorylate a His residue in its cytosolic domain. This first component of the **two-component system**, the **receptor histidine kinase**, then catalyzes transfer of the phosphoryl group from the His residue to an Asp residue on a second, soluble protein, the **response regulator**. This phosphoprotein moves to the base of the flagellum, carrying the signal from the membrane receptor. The flagellum is driven by a rotary motor that can propel the cell through its medium or cause it to stall, depending on the direction of motor rotation. The change in attractant concentration over time, signaled through the receptor, allows the cell to determine whether it is moving toward or away from the source of the attractant. If its motion is toward the attractant, the response regulator signals the cell to continue in a straight line (a run); if away from it, the cell tumbles momentarily, acquiring a new direction. Repetition of this behavior results in a random path, biased toward movement in the direction of increasing attractant concentration.

E. coli detects not only sugars and amino acids but also O_2, extremes of temperature, and other environmental factors, using this basic two-component system. Two-component systems have been detected in many other bacteria, both gram-positive and gram-negative, and in archaea, as well as in protists and fungi. Clearly, this signaling mechanism developed early in the course of cellular evolution and has been conserved.

Various signaling systems used by animal cells also have analogs in bacteria. As the full genomic sequences of

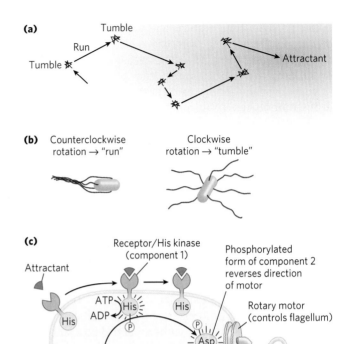

FIGURE 12–31 The two-component signaling mechanism in bacterial chemotaxis. (a) When placed near a source of an attractant solute, *E. coli* performs a random walk, biased toward the attractant. **(b)** Flagella have intrinsic helical structure, and when all flagella rotate counterclockwise, the flagellar helices twist together and move in concert to propel the cell forward in a "run." When the flagella rotate clockwise, the flagellar bundles fly apart, and the cell tumbles briefly until counterclockwise rotation resumes and the cell begins to swim forward again in a new, random direction. When moving toward the attractant, the cell has fewer tumbles, and therefore longer runs; when moving away, the frequent tumbles eventually result in movement toward the attractant. **(c)** Flagellar rotation is controlled by a two-component system consisting of a receptor–histidine kinase and an effector protein. When an attractant ligand binds to the receptor domain of the membrane-bound receptor, a protein kinase in the cytosolic domain (component 1) is activated and autophosphorylates a His residue. This phosphoryl group is then transferred to an Asp residue on component 2. After phosphorylation, component 2 moves to the base of the flagellum, where it causes counterclockwise rotation of the flagella, producing a run.

more, and more diverse, bacteria become known, researchers have discovered genes that encode proteins similar to protein Ser or Thr kinases, Ras-like proteins regulated by GTP binding, and proteins with SH3 domains. Receptor Tyr kinases have not been detected in bacteria, but Ⓟ–Tyr residues do occur in some bacteria.

Signaling Systems of Plants Have Some of the Same Components Used by Microbes and Mammals

Like animals, vascular plants must have a means of communication between tissues to coordinate and direct growth and development; to adapt to conditions of O_2, nutrients, light, temperature, and water availability; and to warn of the presence of noxious chemicals and damaging pathogens **(Fig. 12–32)**. At least a billion years of evolution have passed since the plant and animal branches of the eukaryotes diverged, which is reflected in the differences in signaling mechanisms: some plant mechanisms are conserved—that is, are similar to those in animals (protein kinases, adaptor proteins, cyclic nucleotides, electrogenic ion pumps, and gated ion channels); some are similar to bacterial two-component systems; and some are unique to plants (light-sensing mechanisms that reflect seasonal changes in the angle, and hence color, of sunlight, for example) (Table 12–7). The genome of the plant *Arabidopsis thaliana* encodes about 1,000 protein Ser/Thr kinases, including about 60 MAPKs and nearly 400 membrane-associated receptor kinases that phosphorylate Ser or Thr residues; a variety of protein phosphatases; adaptor proteins that form scaffolds on which proteins assemble in signaling complexes; enzymes for the synthesis and degradation of cyclic nucleotides; and 100 or more ion channels, including about 20 gated by cyclic nucleotides. Inositol phospholipids are present, as are kinases that interconvert them by phosphorylation of inositol head groups. Even given the fact that *Arabidopsis* has multiple copies of many genes, the presence of this many genes certainly reflects a wide array of signaling potential.

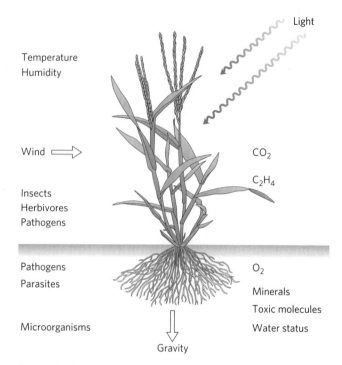

FIGURE 12–32 Some stimuli that produce responses in plants.

TABLE 12–7	Signaling Components Present in Mammals, Plants, or Bacteria		
Signaling component	Mammals	Plants	Bacteria
Ion channels	+	+	+
Electrogenic ion pumps	+	+	+
Two-component His kinases	+	+	+
Adenylyl cyclase	+	+	+
Guanylyl cyclase	+	+	?
Receptor protein kinases (Ser/Thr)	+	+	?
Ca^{2+} as second messenger	+	+	?
Ca^{2+} channels	+	+	?
Calmodulin, CaM-binding protein	+	+	−
MAPK cascade	+	+	−
Cyclic nucleotide–gated channels	+	+	−
IP_3-gated Ca^{2+} channels	+	+	−
Phosphatidylinositol kinases	+	+	−
GPCRs	+	+/−	+
Trimeric G proteins	+	+/−	−
PI-specific phospholipase C	+	?	−
Tyrosine kinase receptors	+	?	−
SH2 domains	+	?	?
Nuclear steroid receptors	+	−	−
Protein kinase A	+	−	−
Protein kinase G	+	−	−

However, some types of signaling proteins common in animal tissues are not present in plants, or are represented by only a few genes. Cyclic nucleotide–dependent protein kinases (PKA and PKG) seem to be absent, for example. Heterotrimeric G proteins and protein Tyr kinase genes are much less prominent in the plant genome, and genes for GPCRs, the largest family of proteins in the human genome (~1,000 genes), are very sparsely represented in the plant genome. DNA-binding nuclear steroid receptors are certainly not prominent, and may be absent from plants. Although plants lack the most widely conserved light-sensing mechanism present in animals (rhodopsin, with retinal as pigment), they have a rich collection of other light-detecting mechanisms not found in animal tissues—phytochromes and cryptochromes, for example (Chapter 19).

The kinds of compounds that elicit signals in plants are similar to certain signaling molecules in animals (Fig. 12–33). Instead of prostaglandins, plants have jasmonate; instead of steroid hormones, brassinosteroids. About 100 different small peptides serve as plant signals, and both plants and animals use compounds derived from aromatic amino acids as signals.

Plants Detect Ethylene through a Two-Component System and a MAPK Cascade

The gaseous plant hormone ethylene ($CH_2{=}CH_2$), which stimulates the ripening of fruits (among other func-

tions), acts through receptors that are related in primary sequence to the receptor His kinases of the bacterial two-component systems and probably evolved from them. In *Arabidopsis*, the two-component signaling system is contained within a single integral membrane protein of the endoplasmic reticulum (*not* the plasma membrane). Ethylene diffuses into the cell through the plasma membrane and into the ER. The first downstream component affected by ethylene signaling is a protein Ser/Thr kinase (CTR1; **Fig. 12–34**) with sequence homology to Raf, the protein kinase that begins the MAPK cascade in the mammalian response to insulin (see Fig. 12–15). In plants, in the absence of ethylene, the CTR1 kinase is active and *inhibits* the MAPK cascade, preventing transcription of ethylene-responsive genes. Exposure to ethylene *inactivates* the CTR1 kinase, thereby activating the MAPK cascade that leads to activation of the transcription factor EIN3. Active EIN3 stimulates the synthesis of a second transcription factor (ERF1), which in turn activates transcription of

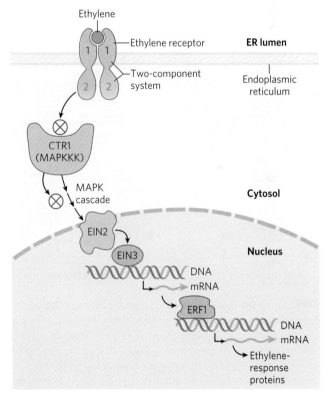

FIGURE 12–34 Transduction mechanism for detection of ethylene by plants. The ethylene receptor (pink) in the endoplasmic reticulum is a two-component system contained in a single protein, with a receptor domain (component 1) and a response regulator domain (component 2). The receptor controls (in ways we do not yet understand) the activity of CTR1, a protein kinase similar to MAPKKKs and therefore presumed to be part of a MAPK cascade. CTR1 is a negative regulator of the ethylene response; when CTR1 is *inactive*, the ethylene signal is transmitted through the gene product EIN2 (thought to be a nuclear envelope protein), which causes increased synthesis of ERF1, a transcription factor. ERF1 stimulates expression of proteins specific to the ethylene response.

Plants

Jasmonate

Animals

Prostaglandin E₁

Indole-3-acetate
(an auxin)

Serotonin
(5-hydroxytryptamine)

Brassinolide
(a brassinosteroid)

Estradiol

FIGURE 12–33 Structural similarities between plant and animal signals.

ethylene-responsive genes; the gene products affect processes ranging from seedling development to fruit ripening. Although apparently derived from the bacterial two-component signaling system, the ethylene system in *Arabidopsis* is different in that the His kinase activity that defines component 1 in bacteria is not essential to signal transduction in *Arabidopsis*.

Receptorlike Protein Kinases Transduce Signals from Peptides

One common motif in plant signaling involves **receptorlike kinases (RLKs)**, which have a single helical segment in the plasma membrane that connects a receptor domain on the outside with a protein Ser/Thr kinase on the cytoplasmic side. This type of receptor participates in the defense mechanism triggered by infection with a bacterial pathogen **(Fig. 12–35a)**. The signal to turn on the genes needed for defense against infection is a peptide

(flg22) released by breakdown of flagellin, the major protein of the bacterial flagellum. Binding of flg22 to the FLS2 receptor of *Arabidopsis* induces receptor dimerization and autophosphorylation on Ser and Thr residues, and the downstream effect is activation of a MAPK cascade like that described above for insulin action. The final kinase in this cascade activates a specific transcription factor, triggering synthesis of the proteins that defend against the bacterial infection. The steps between receptor phosphorylation and the MAPK cascade are not yet known. A phosphoprotein phosphatase (KAPP) associates with the active receptor protein and inactivates it by dephosphorylation to end the response.

The MAPK cascade in the plant's defense against bacterial pathogens is remarkably similar to the innate immune response in mammals (Fig. 12–35b) that is triggered by bacterial lipopolysaccharide and mediated by the Toll-like receptors (TLRs, a name derived from a *Drosophila* mutant originally called Toll (German for "mad"); TLRs were subsequently found in many other organisms and were shown to function in embryonic development). Other membrane receptors use similar mechanisms to activate a MAPK cascade, ultimately activating transcription factors and turning on the genes essential to the defense response.

Most of the several hundred RLKs in plants are presumed to act in similar ways: ligand binding induces dimerization and autophosphorylation, and the activated receptor kinase triggers downstream responses by phosphorylating key proteins at Ser or Thr residues.

FIGURE 12–35 Similarities between the signaling pathways that trigger immune responses in plants and animals. (a) In *Arabidopsis thaliana*, the peptide flg22, derived from the flagella of a bacterial pathogen, binds to its receptor (FLS) in the plasma membrane, causing the receptor to form dimers and triggering autophosphorylation of the cytosolic protein kinase domain on a Ser or Thr residue (*not* a Tyr). Thus activated, the protein kinase phosphorylates downstream proteins (not shown). The activated receptor also activates (by means unknown) a MAPK cascade, which leads to phosphorylation of a nuclear protein that normally inhibits the transcription factors WRKY22 and 29; this phosphorylation triggers proteolytic degradation of the inhibitor and frees the transcription factors to stimulate gene expression related to the immune response. **(b)** In mammals, a toxic bacterial lipopolysaccharide (LPS; see Fig. 7–31) is detected by plasma membrane receptors, which then associate with and activate a soluble protein kinase (IRAK). The major flagellar protein of pathogenic bacteria acts through a similar receptor, also activating IRAK. The activated IRAK initiates two distinct MAPK cascades that end in the nucleus, causing the synthesis of proteins needed in the immune response. Jun, Fos, and NFκB are transcription factors.

SUMMARY 12.9 Signaling in Microorganisms and Plants

▶ Bacteria and eukaryotic microorganisms have a variety of sensory systems that allow them to sample and respond to their environment. In the two-component system, a receptor His kinase senses the signal and autophosphorylates a His residue, then phosphorylates an Asp residue of the response regulator.

▶ Plants respond to many environmental stimuli and employ hormones and growth factors to coordinate the development and metabolic activities of their tissues. Plant genomes encode hundreds of signaling proteins, including some very similar to those of mammals.

▶ Two-component signaling mechanisms common in bacteria are found in modified forms in plants, used in the detection of chemical signals and light.

▶ Plant receptorlike kinases (RLKs) participate in detecting a wide variety of stimuli, including brassinosteroids, peptides that originate from pathogens, and developmental signals. RLKs autophosphorylate Ser/Thr residues, then activate downstream proteins, which in some cases are MAPK cascades. The end result is increased transcription of specific genes.

12.10 Sensory Transduction in Vision, Olfaction, and Gustation

The detection of light, odors, and tastes (vision, olfaction, and gustation, respectively) in animals is accomplished by specialized sensory neurons that use signal-transduction mechanisms fundamentally similar to those that detect hormones, neurotransmitters, and growth factors. An initial sensory signal is amplified greatly by mechanisms that include gated ion channels and intracellular second messengers; the system adapts to continued stimulation by changing its sensitivity to the stimulus (desensitization); and sensory input from several receptors is integrated before the final signal goes to the brain.

The Visual System Uses Classic GPCR Mechanisms

In the vertebrate eye, light entering through the pupil is focused on a highly organized collection of light-sensitive neurons (Fig. 12–36). The light-sensing neurons are of two types: rods (about 10^9 per retina), which sense low levels of light but cannot discriminate colors, and cones (about 3×10^6 per retina), which are less sensitive to light but can discriminate colors. Both cell types are long, narrow, specialized sensory neurons with two distinct cellular compartments: the outer segment contains dozens of membranous disks loaded with receptor proteins and their photosensitive chromophore **retinal**; the inner segment contains the nucleus and many mitochondria, which produce the ATP essential to phototransduction.

Like other neurons, rods and cones have a transmembrane electrical potential (V_m), produced by the electrogenic pumping of the Na^+K^+ ATPase in the plasma membrane of the inner segment (**Fig. 12–37**). Also contributing to the membrane potential is an ion channel in the outer segment that permits passage of either Na^+ or Ca^{2+} and is gated (opened) by cGMP. In the dark, rod cells contain enough cGMP to keep this channel open. The membrane potential is therefore determined by the difference between the amount of Na^+ and K^+ pumped by the inner segment (which polarizes the membrane) and the influx of Na^+ through the ion channels of the outer segment (which tends to depolarize the membrane).

The essence of signaling in the rod or cone cell is a light-induced decrease in [cGMP], which causes the cGMP-gated ion channel to close. The plasma membrane then becomes hyperpolarized by the Na^+K^+ ATPase. Rod and cone cells synapse with interconnecting neurons (Fig. 12–36) that carry information about the electrical activity to ganglion neurons near the inner surface of the retina. The ganglion neurons integrate the output from many rod or cone cells and send the resulting signal through the optic nerve to the visual cortex of the brain.

Visual transduction begins when light falls on rhodopsin, many thousands of molecules of which are present in each disk of the outer segments of rod and cone cells. **Rhodopsin** (M_r 40,000) is an integral protein with seven membrane-spanning α helices (**Fig. 12–38**), the characteristic GPCR architecture. The light-absorbing pigment (chromophore) 11-*cis*-retinal is covalently attached to **opsin**, the protein component of rhodopsin, through a Schiff base to a Lys residue. The retinal molecule lies near the middle of the bilayer (Fig. 12–38), oriented with its long axis approximately in the plane of the membrane. When a photon is absorbed by the retinal component of rhodopsin, the energy causes a photochemical change; 11-*cis*-retinal is converted to all-*trans*-retinal (see Figs 1–19b and 10–21). This change in the structure of the chromophore forces conformational changes in the rhodopsin molecule—the first stage in visual transduction.

Retinal is derived from vitamin A_1 (retinol), which is produced from β-carotene (see Fig. 10–21). Dietary deficiency of vitamin A leads to night blindness (the inability to adapt to low light levels), which is relatively common in some developing countries. Vitamin A supplements or vegetables rich in carotene (such as carrots) supply the vitamin and reverse the night blindness. ■

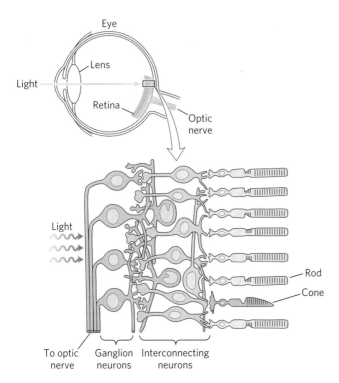

FIGURE 12–36 Light reception in the vertebrate eye. The lens focuses light on the retina, which is composed of layers of neurons. The primary photosensory neurons are rod cells (yellow), which are responsible for high-resolution and night vision, and cone cells of three subtypes (pink), which initiate color vision. The rods and cones form synapses with several ranks of interconnecting neurons that convey and integrate the electrical signals. The signals eventually pass from ganglion neurons through the optic nerve to the brain. Note that light must pass through the layers of ganglion neurons and interconnecting neurons before reaching the rod and cone cells.

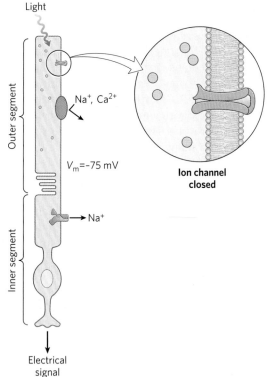

Light

Outer segment

Inner segment

cGMP

Ca^{2+}

Na^+

$V_m = -45$ mV

Na^+

Na^+K^+
ATPase

**Ion channel
open**

Na^+, Ca^{2+}

$V_m = -75$ mV

**Ion channel
closed**

Na^+

Electrical
signal

FIGURE 12–37 Light-induced hyperpolarization of rod cells. The rod cell consists of an outer segment, filled with stacks of membranous disks (not shown) containing the photoreceptor rhodopsin, and an inner segment that contains the nucleus and other organelles (not shown). The inner segment forms a synapse with interconnecting neurons (Fig. 12–36). Cones have a similar structure. ATP in the inner segment powers the Na^+K^+ ATPase, which creates a transmembrane electrical potential by pumping 3 Na^+ out for every 2 K^+ pumped in. The membrane potential is reduced by the inflow of Na^+ and Ca^{2+} through cGMP-gated cation channels in the outer-segment plasma membrane. When rhodopsin absorbs light, it triggers degradation of cGMP (green dots) in the outer segment, causing closure of the ion channel. Without cation influx through this channel, the cell becomes hyperpolarized. This electrical signal is passed to the brain through the ranks of neurons shown in Figure 12–36.

FIGURE 12–38 Complex of rhodopsin with the G protein transducin. (PDB ID 1BAC) Rhodopsin (red) has seven transmembrane helices embedded in the disk membranes of rod outer segments and is oriented with its carboxyl terminus on the cytosolic side and its amino terminus inside the disk. The chromophore 11-*cis*-retinal (yellow space-filling structure), attached through a Schiff base linkage to Lys^{256} of the seventh helix, lies near the center of the bilayer. (This location is similar to that of the epinephrine-binding site in the β-adrenergic receptor.) Several Ser and Thr residues near the carboxyl terminus are substrates for phosphorylations that are part of the desensitization mechanism for rhodopsin. Cytosolic loops that interact with the G protein transducin are shown in orange; their exact positions are not yet known. The three subunits of transducin (green) are shown in their likely arrangement. Rhodopsin is palmitoylated at its carboxyl terminus, and both the α and γ subunits of transducin have attached lipids (yellow) that assist in anchoring them to the membrane.

Excited Rhodopsin Acts through the G Protein Transducin to Reduce the cGMP Concentration

In its excited conformation, rhodopsin interacts with a second protein, **transducin**, which hovers nearby on the cytoplasmic face of the disk membrane (Fig. 12–38). Transducin (T) belongs to the same family of heterotrimeric GTP-binding proteins as G_s and G_i. Although specialized for visual transduction, transducin shares many functional features with G_s and G_i. It can bind either GDP or GTP. In the dark, GDP is bound, all three subunits of the protein (T_α, T_β, and T_γ) remain together, and no signal is sent. When rhodopsin is excited by light, it interacts with transducin, catalyzing the replacement of bound GDP by GTP from the cytosol (**Fig. 12–39**, steps ❶ and ❷). Transducin then dissociates into T_α and $T_{\beta\gamma}$, and the T_α-GTP carries the signal from the excited receptor to the next element in the transduction pathway, a cGMP phosphodiesterase; this enzyme converts cGMP to 5′-GMP (steps ❸ and ❹). Note that this is not the same cyclic nucleotide phosphodiesterase that hydrolyzes cAMP to terminate the β-adrenergic response. One isoform of the cGMP-specific PDE is unique to the visual cells of the retina.

The PDE of the retina is a peripheral protein with its active site on the cytoplasmic side of the disk membrane. In the dark, a tightly bound inhibitory subunit very effectively suppresses the PDE activity. When T_α-GTP encounters the PDE, the inhibitory subunit leaves the enzyme and instead binds T_α, and the enzyme's

activity immediately increases by several orders of magnitude. Each molecule of the active PDE degrades many molecules of cGMP to the biologically inactive 5′-GMP, lowering [cGMP] in the outer segment within a fraction of a second. At the new, lower [cGMP], the cGMP-gated ion channels close, blocking reentry of Na^+ and Ca^{2+} into the outer segment and hyperpolarizing the membrane of the rod or cone cell (step ❺). Through this process, the initial stimulus—a photon—changes the V_m of the cell. The brighter the illumination of the rod cell, the greater the hyperpolarization. This hyperpolarization is perceived by the integrating neurons of the retina, which pass the integrated signal on to the ganglion cells, which send axons via the optic nerve to the brain.

Several steps in the visual-transduction process result in a huge amplification of the signal. Each excited rhodopsin molecule activates at least 500 molecules of transducin, each of which can activate a molecule of

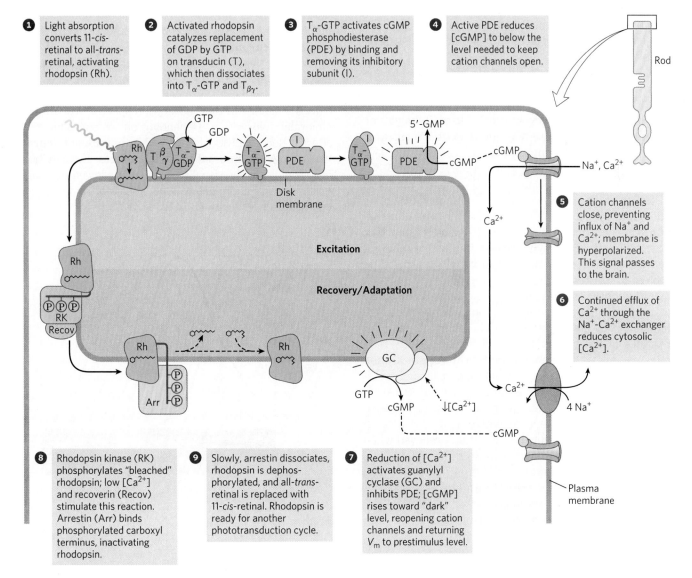

❶ Light absorption converts 11-*cis*-retinal to all-*trans*-retinal, activating rhodopsin (Rh).

❷ Activated rhodopsin catalyzes replacement of GDP by GTP on transducin (T), which then dissociates into T_α-GTP and $T_{\beta\gamma}$.

❸ T_α-GTP activates cGMP phosphodiesterase (PDE) by binding and removing its inhibitory subunit (I).

❹ Active PDE reduces [cGMP] to below the level needed to keep cation channels open.

❺ Cation channels close, preventing influx of Na^+ and Ca^{2+}; membrane is hyperpolarized. This signal passes to the brain.

❻ Continued efflux of Ca^{2+} through the Na^+-Ca^{2+} exchanger reduces cytosolic [Ca^{2+}].

❼ Reduction of [Ca^{2+}] activates guanylyl cyclase (GC) and inhibits PDE; [cGMP] rises toward "dark" level, reopening cation channels and returning V_m to prestimulus level.

❽ Rhodopsin kinase (RK) phosphorylates "bleached" rhodopsin; low [Ca^{2+}] and recoverin (Recov) stimulate this reaction. Arrestin (Arr) binds phosphorylated carboxyl terminus, inactivating rhodopsin.

❾ Slowly, arrestin dissociates, rhodopsin is dephosphorylated, and all-*trans*-retinal is replaced with 11-*cis*-retinal. Rhodopsin is ready for another phototransduction cycle.

FIGURE 12–39 Molecular consequences of photon absorption by rhodopsin in the rod outer segment. The top half of the figure (steps ❶ to ❺) describes excitation; the bottom shows post-illumination steps: recovery (steps ❻ and ❼) and adaptation (steps ❽ and ❾).

the PDE. This phosphodiesterase has a remarkably high turnover number, each activated molecule hydrolyzing 4,200 molecules of cGMP per second. The binding of cGMP to cGMP-gated ion channels is cooperative, and a relatively small change in [cGMP] therefore registers as a large change in ion conductance. The result of these amplifications is exquisite sensitivity to light. Absorption of a single photon closes 1,000 or more ion channels and changes the cell's membrane potential by about 1 mV.

The Visual Signal Is Quickly Terminated

As your eyes move across this line, the retinal images of the first words disappear rapidly—before you see the next series of words. In that short interval, a great deal of biochemistry has taken place. Very shortly after illumination of the rod or cone cells stops, the photosensory system shuts off. The α subunit of transducin (with bound GTP) has intrinsic GTPase activity. Within milliseconds after the decrease in light intensity, GTP is hydrolyzed and T_α reassociates with $T_{\beta\gamma}$. The inhibitory subunit of the PDE, which had been bound to T_α-GTP, is released and reassociates with the enzyme, strongly inhibiting its activity.

To return [cGMP] to its "dark" level, the enzyme guanylyl cyclase converts GTP to cGMP (step ❼ in Fig. 12–39) in a reaction that is inhibited by high [Ca^{2+}] (>100 nM). Calcium levels drop during illumination, because the steady-state [Ca^{2+}] in the outer segment is the result of outward pumping of Ca^{2+} through the Na^+-Ca^{2+} exchanger of the plasma membrane (see Fig. 12–37) and influx of Ca^{2+} through open cGMP-gated channels. In the dark, this produces a [Ca^{2+}] of about 500 nM—enough to inhibit cGMP synthesis. After brief illumination, Ca^{2+} entry slows and [Ca^{2+}] declines (step ❻). The inhibition of guanylyl cyclase by Ca^{2+} is relieved, and the cyclase converts GTP to cGMP to return the system to its prestimulus state (step ❼).

Rhodopsin itself also undergoes changes in response to prolonged illumination. The conformational change induced by light absorption exposes several Thr and Ser residues in the carboxyl-terminal domain. These residues are quickly phosphorylated by **rhodopsin kinase** (step ❽ in Fig. 12–39), which is functionally and structurally homologous to the β-adrenergic kinase (βARK) that desensitizes the β-adrenergic receptor (Fig. 12–8). The Ca^{2+}-binding protein **recoverin** inhibits rhodopsin kinase at high [Ca^{2+}], but the inhibition is relieved when [Ca^{2+}] drops after illumination, as described above. The phosphorylated carboxyl-terminal domain of rhodopsin is bound by the protein **arrestin 1**, preventing further interaction between activated rhodopsin and transducin. Arrestin 1 is a close homolog of arrestin 2 (βarr; Fig. 12–8). On a relatively long time scale (seconds to minutes), the all-*trans*-retinal of an excited rhodopsin molecule is removed and replaced by 11-*cis*-retinal, to produce rhodopsin that is ready for another round of excitation (step ❾ in Fig. 12–39).

Cone Cells Specialize in Color Vision

Color vision involves a path of sensory transduction in cone cells essentially identical to that described above, but triggered by slightly different light receptors. Three types of cone cells are specialized to detect light from different regions of the spectrum, using three related photoreceptor proteins (opsins). Each cone cell expresses only one kind of opsin, but each type is closely related to rhodopsin in size, amino acid sequence, and presumably three-dimensional structure. The differences among the opsins, however, are great enough to place the chromophore, 11-*cis*-retinal, in three slightly different environments, with the result that the three photoreceptors have different absorption spectra **(Fig. 12–40)**. We discriminate colors and hues by integrating the output from the three types of cone cells, each containing one of the three photoreceptors.

Color blindness, such as the inability to distinguish red from green, is a fairly common, genetically inherited trait in humans. The various types of color blindness result from different opsin mutations. One form is due to loss of the red photoreceptor; affected individuals are **red⁻ dichromats** (they see only two primary colors). Others lack the green pigment and are **green⁻ dichromats**. In some cases, the red and green photoreceptors are present but have a changed amino acid sequence that causes a change in their absorption spectra, resulting in abnormal color vision. Depending on which pigment is altered, such individuals are **red-anomalous trichromats** or **green-anomalous trichromats**. Examination of the genes for the visual receptors has allowed the diagnosis of color blindness in a famous "patient" more than a century after his death (Box 12–4). ■

FIGURE 12–40 Absorption spectra of purified rhodopsin and the red, green, and blue receptors of cone cells. The receptor spectra, obtained from individual cone cells isolated from cadavers, peak at about 420, 530, and 560 nm, and the maximum absorption for rhodopsin is at about 500 nm. For reference, the visible spectrum for humans is about 380 to 750 nm.

BOX 12–4 ⚕ MEDICINE Color Blindness: John Dalton's Experiment from the Grave

The chemist John Dalton (of atomic theory fame) was color-blind. He thought it probable that the vitreous humor of his eyes (the fluid that fills the eyeball behind the lens) was tinted blue, unlike the colorless fluid of normal eyes. He proposed that after his death, his eyes should be dissected and the color of the vitreous humor determined. His wish was honored. The day after Dalton's death in July 1844, Joseph Ransome dissected his eyes and found the vitreous humor to be perfectly colorless. Ransome, like many scientists, was reluctant to throw samples away. He placed Dalton's eyes in a jar of preservative, where they stayed for a century and a half (Fig. 1).

Then, in the mid-1990s, molecular biologists in England took small samples of Dalton's retinas and extracted DNA. Using the known gene sequences for the opsins of the red and green light receptors, they amplified the relevant sequences (using techniques described in Chapter 9) and determined that Dalton had the opsin gene for the red photopigment but lacked the opsin gene for the green photopigment. Dalton was a green-dichromat. So, 150 years after his death, the experiment Dalton started—by hypothesizing about the cause of his color blindness—was finally finished.

FIGURE 1 Dalton's eyes.

Vertebrate Olfaction and Gustation Use Mechanisms Similar to the Visual System

The sensory cells that detect odors and tastes have much in common with the rod and cone cells. Olfactory neurons have long thin cilia extending from one end of the cell into a mucous layer that overlays the cell. These cilia present a large surface area for interaction with olfactory signals. The receptors for olfactory stimuli are ciliary membrane proteins with the familiar GPCR structure of seven transmembrane α helices. The olfactory signal can be any one of the many volatile compounds for which there are specific receptor proteins. Our ability to discriminate odors stems from hundreds of different olfactory receptors in the tongue and nasal passages and from the brain's ability to integrate input from different types of olfactory receptors to recognize a "hybrid" pattern, extending our range of discrimination far beyond the number of receptors.

The olfactory stimulus arrives at the sensory cells by diffusion through the air. In the mucous layer covering the olfactory neurons, the odorant molecule binds directly to an olfactory receptor or to a specific binding protein that carries the odorant to a receptor **(Fig. 12–41)**. Interaction between odorant and receptor triggers a change in receptor conformation that results in the replacement of bound GDP by GTP on a G protein, G_{olf}, analogous to transducin and to G_s of the β-adrenergic system. The activated G_{olf} then activates adenylyl cyclase of the ciliary membrane, which synthesizes cAMP from ATP, raising the local [cAMP]. The cAMP-gated Na^+ and Ca^{2+} channels of the ciliary membrane open, and the influx of Na^+ and Ca^{2+} produces a small depolarization called the **receptor potential**. If a sufficient number of odorant molecules encounter receptors, the receptor potential is strong enough to cause the neuron to fire an action potential. This is relayed to the brain in several stages and registers as a specific smell. All these events occur within 100 to 200 ms.

When the olfactory stimulus is no longer present, the transducing machinery shuts itself off in several ways. A cAMP phosphodiesterase returns [cAMP] to the prestimulus level. G_{olf} hydrolyzes its bound GTP to GDP, thereby inactivating itself. Phosphorylation of the receptor by a specific kinase prevents its interaction with G_{olf}, by a mechanism analogous to that used to desensitize the β-adrenergic receptor and rhodopsin. And lastly, some odorants are enzymatically destroyed by oxidases.

The sense of taste in vertebrates reflects the activity of gustatory neurons clustered in taste buds on the surface of the tongue. In these sensory neurons, GPCRs are coupled to the heterotrimeric G protein **gustducin** (very similar to the transducin of rod and cone cells). Sweet-tasting molecules are those that bind receptors in "sweet" taste buds. When the molecule (tastant) binds, gustducin is activated by replacement of bound GDP with GTP and then stimulates cAMP production by adenylyl cyclase. The resulting elevation of [cAMP] activates PKA, which phosphorylates K^+ channels in the plasma membrane, causing them to close. Reduced efflux of K^+ depolarizes the cell **(Fig. 12–42)** sending an electrical signal to the brain. Other taste buds specialize in detecting bitter, sour, salty, or umami (savory) tastants, using various combinations of second messengers and ion channels in the transduction mechanisms.

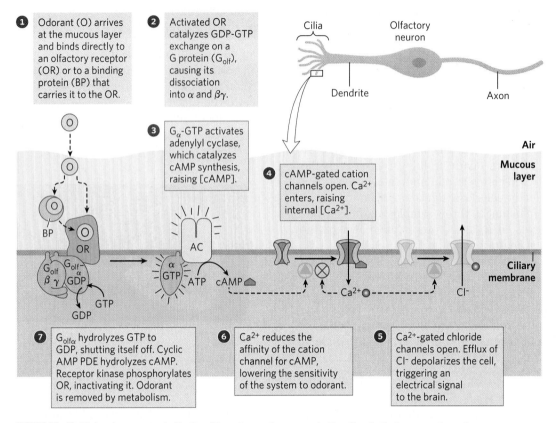

FIGURE 12–41 Molecular events of olfaction. These interactions occur in the cilia of olfactory receptor cells.

GPCRs of the Sensory Systems Share Several Features with GPCRs of Hormone Signaling Systems

We have now looked at several types of signaling systems (hormone signaling, vision, olfaction, and gustation) in which membrane receptors are coupled to second messenger–generating enzymes through G proteins. As we have intimated, signaling mechanisms must have arisen early in evolution; genomic studies have revealed hundreds of genes encoding GPCRs in vertebrates, arthropods (*Drosophila* and mosquito), and the roundworm *Caenorhabditis elegans*. Even the common baker's yeast *Saccharomyces* uses GPCRs and G proteins to detect the opposite mating type. Overall patterns have been conserved, and the introduction of variety has given modern organisms the ability to

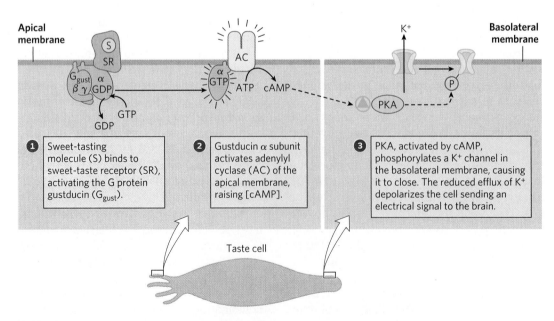

FIGURE 12–42 Transduction mechanism for sweet tastants.

respond to a wide range of stimuli (Table 12–8). Of the approximately 29,000 genes in the human genome, as many as 1,000 encode GPCRs, including hundreds for olfactory stimuli and many "orphan receptors" for which the natural ligand is not yet known.

All well-studied signal-transducing systems that act through heterotrimeric G proteins share some common features, which reflect their evolutionary relatedness **(Fig. 12–43)**. The receptors have seven transmembrane segments, a domain (generally the loop between transmembrane helices 6 and 7) that interacts with a G protein, and a carboxyl-terminal cytoplasmic domain that undergoes reversible phosphorylation on several Ser or Thr residues. The ligand-binding site (or, in the case of light reception, the light receptor) is buried deep in the membrane and includes residues from several of the transmembrane segments. Ligand binding (or light) induces a conformational change in the receptor, exposing a domain that can interact with a G protein. Heterotrimeric G proteins activate or inhibit effector enzymes (adenylyl cyclase, PDE, or PLC), which change the concentration of a second messenger (cAMP, cGMP, IP_3, or Ca^{2+}). In the hormone-detecting systems, the final output is an activated protein kinase that regulates some cellular process by phosphorylating a protein critical to that process. In sensory neurons, the output is a change in membrane potential and a consequent electrical signal that passes to another neuron in the pathway connecting the sensory cell to the brain.

All these systems self-inactivate. Bound GTP is converted to GDP by the intrinsic GTPase activity of G proteins, often augmented by GTPase-activating proteins (GAPs) or RGS proteins (regulators of G-protein

TABLE 12–8	Some Signals That Act through GPCRs
Amines	Somatostatin
Acetylcholine (muscarinic)	Tachykinin
Dopamine	Thyrotropin-releasing hormone
Epinephrine	Urotensin II
Histamine	**Protein hormones**
Serotonin	Follicle-stimulating hormone
Peptides	
Angiotensin	Gonadotropin
Bombesin	Lutropin-choriogonadotropic hormone
Bradykinin	
Chemokine	
Colecystokinin (CCK)	Thyrotropin
	Prostanoids
Endothelin	Prostacyclin
Gonadotropin-releasing hormone	Prostaglandin
	Thromboxane
Interleukin-8	**Others**
Melanocortin	Cannabinoids
Neuropeptide Y	Lysosphingolipids
Neurotensin	Melatonin
Opioid	Olfactory stimuli
Orexin	Rhodopsin

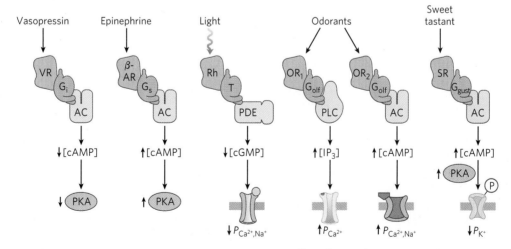

FIGURE 12–43 Common features of signaling systems that detect hormones, light, smells, and tastes. GPCRs provide signal specificity, and their interaction with G proteins provides signal amplification. Heterotrimeric G proteins activate effector enzymes: adenylyl cyclase (AC), phospholipase C (PLC), and phosphodiesterases (PDEs) that degrade cAMP or cGMP. Changes in concentration of the second messengers (cAMP, cGMP, IP_3) result in alterations of enzymatic activities by phosphorylation or alterations in the permeability (*P*) of surface membranes to Ca^{2+}, Na^+, and K^+. The resulting depolarization or hyperpolarization of the sensory cell (the signal) passes through relay neurons to sensory centers in the brain. In the best-studied cases, desensitization includes phosphorylation of the receptor and binding of a protein (arrestin) that interrupts receptor–G protein interactions. VR is the vasopressin receptor; β-AR is the β-adrenergic receptor. Other receptor and G-protein abbreviations are as used in earlier illustrations.

signaling; see Fig. 12–5 and Box 12–2, Fig. 4). In some cases, the effector enzymes that are the targets of modulation by G proteins also serve as GAPs. The desensitization mechanism involving phosphorylation of the carboxyl-terminal region followed by arrestin binding is widespread, and may be universal.

SUMMARY 12.10 Sensory Transduction in Vision, Olfaction, and Gustation

▶ Vision, olfaction, and gustation in vertebrates employ GPCRs, which act through heterotrimeric G proteins to change the V_m of a sensory neuron.

▶ In rod and cone cells of the retina, light activates rhodopsin, which activates the G protein transducin. The freed α subunit of transducin activates a cGMP phosphodiesterase, which lowers [cGMP] and thus closes cGMP-dependent ion channels in the outer segment of the neuron. The resulting hyperpolarization of the rod or cone cell carries the signal to the next neuron in the pathway, and eventually to the brain.

▶ In olfactory neurons, olfactory stimuli, acting through GPCRs and G proteins, trigger either an increase in [cAMP] (by activating adenylyl cyclase) or an increase in [Ca²⁺] (by activating PLC). These second messengers affect ion channels and thus the V_m.

▶ Gustatory neurons have GPCRs that respond to tastants by altering levels of cAMP, which changes V_m by gating ion channels.

▶ There is a high degree of conservation of signaling proteins and transduction mechanisms across signaling systems and across species.

12.11 Regulation of the Cell Cycle by Protein Kinases

One of the most dramatic manifestations of signaling pathways is the regulation of the eukaryotic cell cycle. During embryonic growth and later development, cell division occurs in virtually every tissue. In the adult organism most tissues become quiescent. A cell's "decision" to divide or not is of crucial importance to the organism. When the regulatory mechanisms that limit cell division are defective and cells undergo unregulated division, the result is catastrophic—cancer. Proper cell division requires a precisely ordered sequence of biochemical events that assures every daughter cell a full complement of the molecules required for life. Investigations into the control of cell division in diverse eukaryotic cells have revealed universal regulatory mechanisms. Signaling mechanisms much like those discussed above are central in determining whether and when a cell undergoes cell division, and they also ensure orderly passage through the stages of the cell cycle.

The Cell Cycle Has Four Stages

Cell division accompanying mitosis in eukaryotes occurs in four well-defined stages **(Fig. 12–44)**. In the S (synthesis) phase, the DNA is replicated to produce copies for both daughter cells. In the G2 phase (G indicates the gap between divisions), new proteins are synthesized and the cell approximately doubles in size. In the M phase (mitosis), the maternal nuclear envelope breaks down, paired chromosomes are pulled to opposite poles of the cell, each set of daughter chromosomes is surrounded by a newly formed nuclear envelope, and cytokinesis pinches the cell in half, producing two daughter cells (see Fig. 24–24). In embryonic or rapidly proliferating tissue, each daughter cell divides again, but only after a waiting period (G1). In cultured animal cells the entire process takes about 24 hours.

After passing through mitosis and into G1, a cell either continues through another division or ceases to divide, entering a quiescent phase (G0) that may last hours, days, or the lifetime of the cell. When a cell in G0 begins to divide again, it reenters the division cycle through the G1 phase. Differentiated cells such as hepatocytes or adipocytes have acquired their specialized function and form; they remain in the G0 phase. Stem cells retain their potential to divide and to differentiate into any of a number of cell types.

Levels of Cyclin-Dependent Protein Kinases Oscillate

The timing of the cell cycle is controlled by a family of protein kinases with activities that change in response

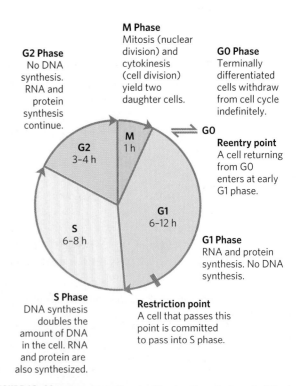

FIGURE 12–44 Eukaryotic cell cycle. The durations (in hours) of the four stages vary, but those shown are typical.

to cellular signals. By phosphorylating specific proteins at precisely timed intervals, these protein kinases orchestrate the metabolic activities of the cell to produce orderly cell division. The kinases are heterodimers with a regulatory subunit, **cyclin**, and a catalytic subunit, **cyclin-dependent protein kinase (CDK)**. In the absence of cyclin, the catalytic subunit is virtually inactive. When cyclin binds, the catalytic site opens up, a residue essential to catalysis becomes accessible **(Fig. 12–45)**, and the protein kinase activity of the catalytic subunit increases 10,000-fold. Animal cells have at least 10 different cyclins (designated A, B, and so forth) and at least 8 CDKs (CDK1 through CDK8), which act in various combinations at specific points in the cell cycle. Plants also use a family of CDKs to regulate their cell division in root and shoot meristems, the principal tissues in which division occurs.

In a population of animal cells undergoing synchronous division, some CDK activities show striking oscillations **(Fig. 12–46)**. These oscillations are the result of four mechanisms for regulating CDK activity: phosphorylation or dephosphorylation of the CDK, controlled degradation of the cyclin subunit, periodic synthesis of CDKs and cyclins, and the action of specific CDK-inhibiting proteins. The precisely timed activation and inactivation of a series of CDKs produce signals serving as a master clock that orchestrates the events in normal cell division and ensures that one stage is completed before the next begins.

Regulation of CDKs by Phosphorylation The activity of a CDK is strikingly affected by phosphorylation and dephosphorylation of two critical residues in the protein **(Fig. 12–47a)**. Phosphorylation of Tyr^{15} near the amino terminus by another protein kinase renders CDK2 inactive; the ⓟ–Tyr residue is in the ATP-binding site of the kinase, and the negatively charged phosphate group blocks the entry of ATP. A specific phosphatase (a PTPase) dephosphorylates this ⓟ–Tyr residue, permitting the binding of ATP. Phosphorylation of Thr^{160} in the "T loop" of CDK, catalyzed by yet another protein kinase, forces the T loop out of the substrate-binding cleft, permitting the binding of a specific downstream target protein and its phosphorylation by CDK (Fig. 12–45c).

(a)

(b)

(c)

FIGURE 12–45 Activation of cyclin-dependent protein kinases (CDKs) by cyclin and phosphorylation. CDKs, a family of related enzymes, are active only when associated with cyclins, another protein family. The crystal structure of CDK2 with and without cyclin reveals the basis for this activation. **(a)** Without cyclin (PDB ID 1HCK), CDK2 folds so that one segment, the T loop, obstructs the binding site for protein substrates and thus inhibits protein kinase activity. The binding site for ATP is also near the T loop. **(b)** When cyclin binds (PDB ID 1FIN), it forces conformational changes that move the T loop away from the active site and reorient an amino-terminal helix, bringing a residue critical to catalysis (Glu^{51}) into the active site. **(c)** Phosphorylation of a Thr residue in the T loop produces a negatively charged residue that is stabilized by interaction with three Arg residues, holding CDK in its active conformation (PDB ID 1JST).

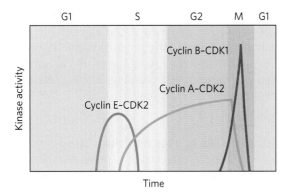

FIGURE 12–46 Variations in the activities of specific CDKs during the cell cycle in animals. Cyclin E–CDK2 activity peaks near the G1 phase–S phase boundary, when the active enzyme triggers synthesis of enzymes required for DNA synthesis (see Fig. 12-49). Cyclin A–CDK2 activity rises during the S and G2 phases, then drops sharply in the M phase, as cyclin B–CDK1 peaks.

The presence of single-strand breaks in DNA leads to arrest of the cell cycle in G2 by regulating a particular CDK. A specific protein kinase (called Rad3 in yeast), which is activated by single-strand breaks, triggers a cascade leading to the inactivation of the PTPase that dephosphorylates Tyr^{15} of CDK. The CDK remains inactive and the cell is arrested in G2, unable to divide until the DNA is repaired and the effects of the cascade are reversed.

Controlled Degradation of Cyclin Highly specific and precisely timed proteolytic breakdown of mitotic cyclins regulates CDK activity throughout the cell cycle. Progress through mitosis requires first the activation then the destruction of cyclins A and B, which activate the catalytic subunit of the M-phase CDK. These cyclins contain near their amino terminus the sequence –Arg–Thr–Ala–Leu–Gly–Asp–Ile–Gly–Asn–, the "destruction box," which targets them for degradation.

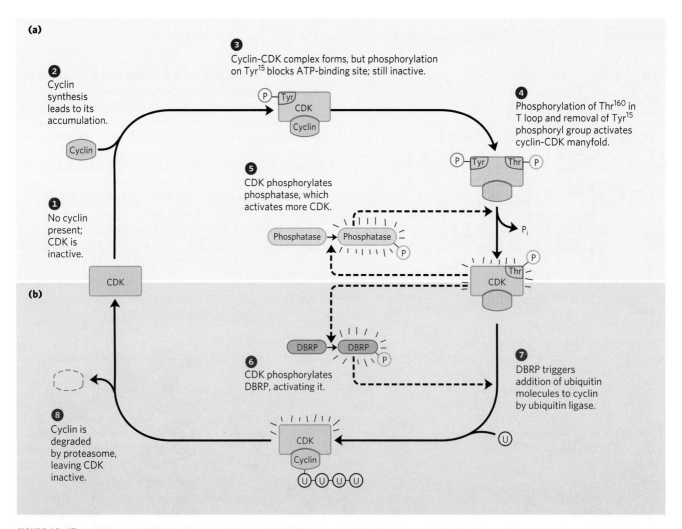

FIGURE 12–47 Regulation of CDK by phosphorylation and proteolysis.
(a) The cyclin-dependent protein kinase activated at the time of mitosis (the M-phase CDK) has a "T loop" that can fold into the substrate-binding site. When Thr^{160} in the T loop is phosphorylated, the loop moves out of the substrate-binding site, activating the CDK manyfold (step ❹).

(b) The active cyclin-CDK complex triggers its own inactivation by phosphorylation of DBRP (destruction box recognizing protein; step ❻). DBRP and ubiquitin ligase then attach several molecules of ubiquitin (U) to cyclin (step ❼), targeting it for destruction by proteasomes, proteolytic enzyme complexes (step ❽).

(This usage of "box" derives from the common practice, in diagramming the sequence of a nucleic acid or protein, of enclosing within a box a short sequence of nucleotide or amino acid residues with some specific function. It does not imply any three-dimensional structure.) The protein DBRP (*destruction box recognizing protein*) recognizes this sequence and initiates the process of cyclin degradation by bringing together the cyclin and another protein, **ubiquitin**. Cyclin and activated ubiquitin are covalently joined by the enzyme ubiquitin ligase (Fig. 12–47b). Several more ubiquitin molecules are then appended, providing the signal for a proteolytic enzyme complex, or **proteasome**, to degrade cyclin.

What controls the timing of cyclin breakdown? A feedback loop occurs in the overall process shown in Figure 12–47. Increased CDK activity (step ❹) leads eventually to cyclin proteolysis (step ❽). Newly synthesized cyclin associates with and activates CDK, which phosphorylates and activates DBRP. Active DBRP then causes proteolysis of cyclin. The lowered cyclin level causes a decline in CDK activity, and the activity of DBRP also drops through slow, constant dephosphorylation and inactivation by a DBRP phosphatase. The cyclin level is ultimately restored by synthesis of new cyclin molecules.

The role of ubiquitin and proteasomes is not limited to the regulation of cyclin; as we shall see in Chapter 27, both also take part in the turnover of cellular proteins, a process fundamental to cellular housekeeping.

Regulated Synthesis of CDKs and Cyclins The third mechanism for changing CDK activity is regulation of the rate of synthesis of cyclin or CDK or both. For example, cyclin D, cyclin E, CDK2, and CDK4 are synthesized only when a specific transcription factor, E2F, is present in the nucleus to activate transcription of their genes. Synthesis of E2F is in turn regulated by extracellular signals such as **growth factors** and cytokines (developmental signals that induce cell division), compounds found to be essential for the division of mammalian cells in culture. They induce the synthesis of specific nuclear transcription factors essential to the production of the enzymes of DNA synthesis. Growth factors trigger phosphorylation of the nuclear proteins Jun and Fos, transcription factors that promote the synthesis of a variety of gene products, including cyclins, CDKs, and E2F. In turn, E2F controls production of several enzymes essential for the synthesis of deoxynucleotides and DNA, enabling cells to enter the S phase **(Fig. 12–48)**.

Inhibition of CDKs Finally, specific protein inhibitors bind to and inactivate specific CDKs. One such protein is p21, which we discuss below.

These four control mechanisms modulate the activity of specific CDKs that, in turn, control whether a cell will divide, differentiate, become permanently quiescent, or begin a new cycle of division after a period of quiescence. The details of cell cycle regulation, such as the number

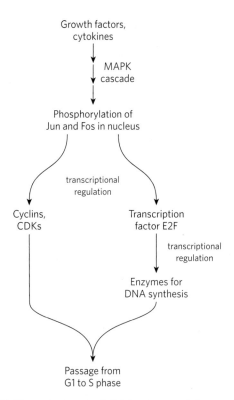

FIGURE 12–48 Regulation of cell division by growth factors. The path from growth factors to cell division leads through the enzyme cascade that activates MAPK; phosphorylation of the nuclear transcription factors Jun and Fos; and the activity of the transcription factor E2F, which promotes synthesis of several enzymes essential for DNA synthesis.

of different cyclins and kinases and the combinations in which they act, differ from species to species, but the basic mechanism has been conserved in the evolution of all eukaryotic cells.

CDKs Regulate Cell Division by Phosphorylating Critical Proteins

We have examined how cells maintain close control of CDK activity, but how does the activity of CDK control the cell cycle? The list of target proteins that CDKs are known to act upon continues to grow, and much remains to be learned. But we can see a general pattern behind CDK regulation by inspecting the effect of CDKs on the structures of lamin and myosin and on the activity of retinoblastoma protein.

The structure of the nuclear envelope is maintained in part by highly organized meshworks of intermediate filaments composed of the protein lamin. Breakdown of the nuclear envelope before segregation of the sister chromatids in mitosis is partly due to the phosphorylation of lamin by a CDK, which causes lamin filaments to depolymerize.

A second kinase target is the ATP-driven contractile machinery (actin and myosin) that pinches a dividing cell into two equal parts during cytokinesis. After the division, CDK phosphorylates a small regulatory subunit of myosin, causing dissociation of myosin from actin filaments and inactivating the contractile machinery. Subsequent dephosphorylation allows reassembly of the contractile apparatus for the next round of cytokinesis.

A third and very important CDK substrate is the **retinoblastoma protein, pRb**; when DNA damage is detected, this protein participates in a mechanism that arrests cell division in G1 **(Fig. 12–49)**. Named for the retinal tumor cell line in which it was discovered, pRb functions in most, perhaps all, cell types to regulate cell division in response to a variety of stimuli. Unphosphorylated pRb binds the transcription factor E2F;

FIGURE 12–49 Regulation of passage from G1 to S by phosphorylation of pRb. Transcription factor E2F promotes transcription of genes for certain enzymes essential to DNA synthesis. The retinoblastoma protein, pRb, can bind E2F (lower left), inactivating it and preventing transcription of these genes. Phosphorylation of pRb by CDK2 prevents it from binding and inactivating E2F, and the genes are transcribed, allowing cell division. Damage to the cell's DNA (upper left) triggers a series of events that inactivate CDK2, blocking cell division. When the protein MRN detects damage to the DNA, it activates two protein kinases, ATM and ATR, and they phosphorylate and activate the transcription factor p53. Active p53 promotes the synthesis of another protein, p21, an inhibitor of CDK2. Inhibition of CDK2 stops the phosphorylation of pRb, which therefore continues to bind and inhibit E2F. With E2F inactivated, genes essential to cell division are not transcribed and cell division is blocked. When DNA has been repaired, this inhibition is released, and the cell divides.

while bound to pRb, E2F cannot promote transcription of a group of genes necessary for DNA synthesis (the genes for DNA polymerase α, ribonucleotide reductase, and other proteins; see Chapter 25). In this state, the cell cycle cannot proceed from the G1 to the S phase, the step that commits a cell to mitosis and cell division. The pRb-E2F blocking mechanism is relieved when pRb is phosphorylated by cyclin E–CDK2, which occurs in response to a signal for cell division to proceed.

When the protein kinases ATM and ATR detect damage to DNA (signaled by the presence of the protein MRN at a double-strand break site), they phosphorylate p53, activating it to serve as a transcription factor that stimulates the synthesis of the protein p21 (Fig. 12–49). This protein inhibits the protein kinase activity of cyclin E–CDK2. In the presence of p21, pRb remains unphosphorylated and bound to E2F, blocking the activity of this transcription factor, and the cell cycle is arrested in G1. This gives the cell time to repair its DNA before entering the S phase, thereby avoiding the potentially disastrous transfer of a defective genome to one or both daughter cells. When the damage is too severe to allow effective repair, this same machinery triggers a process (apoptosis, described below) that leads to the death of the cell, preventing the possible development of a cancer.

SUMMARY 12.11 Regulation of the Cell Cycle by Protein Kinases

▶ Progression through the cell cycle is regulated by the cyclin-dependent protein kinases (CDKs), which act at specific points in the cycle, phosphorylating key proteins and modulating their activities. The catalytic subunit of CDKs is inactive unless associated with the regulatory cyclin subunit.

▶ The activity of a cyclin-CDK complex changes during the cell cycle through differential synthesis of CDKs, specific degradation of cyclin, phosphorylation and dephosphorylation of critical residues in CDKs, and binding of inhibitory proteins to specific cyclin-CDKs.

▶ Among the targets phosphorylated by cyclin-CDKs are proteins of the nuclear envelope and proteins required for cytokinesis and DNA repair.

12.12 Oncogenes, Tumor Suppressor Genes, and Programmed Cell Death

Tumors and cancer are the result of uncontrolled cell division. Normally, cell division is regulated by a family of extracellular growth factors, proteins that cause resting cells to divide and, in some cases, differentiate. The result is a precise balance between the formation of new cells (such as skin cells that die and are replaced every few weeks, or white blood cells that are replaced every few days) and cell destruction. When this balance is disturbed by defects in regulatory proteins, the result is sometimes the formation of a clone of cells that divide repeatedly and without regulation (a tumor) until their presence interferes with the function of normal tissues—cancer. The direct cause is almost always a genetic defect in one or more of the proteins that regulate cell division. In some cases, a defective gene is inherited from one parent; in other cases, the mutation occurs when a toxic compound from the environment (a mutagen or carcinogen) or high-energy radiation interacts with the DNA of a single cell to damage it and introduce a mutation. In most cases there is both an inherited and an environmental contribution, and in most cases, more than one mutation is required to cause completely unregulated division and full-blown cancer.

Oncogenes Are Mutant Forms of the Genes for Proteins That Regulate the Cell Cycle

Oncogenes were originally discovered in tumor-causing viruses, then later found to be derived from genes in the animal host cells, **proto-oncogenes**, which encode growth-regulating proteins. During a viral infection, the host DNA sequence of a proto-oncogene is sometimes copied into the viral genome, where it proliferates with the virus. In subsequent viral infection cycles, the proto-oncogenes can become defective by truncation or mutation. Viruses, unlike animal cells, do not have effective mechanisms for correcting mistakes during DNA replication, so they accumulate mutations rapidly. When a virus carrying an oncogene infects a new host cell, the viral DNA (and oncogene) can be incorporated into the host cell's DNA, where it can now interfere with the regulation of cell division in the host cell. In an alternative, nonviral mechanism, a single cell in a tissue exposed to carcinogens may suffer DNA damage that renders one of its regulatory proteins defective, with the same effect as the oncogenic mechanism: failed regulation of cell division.

The mutations that produce oncogenes are genetically dominant; if either of a pair of chromosomes contains a defective gene, that gene product sends the signal "divide" and a tumor may result. The oncogenic defect can be in any of the proteins involved in communicating the "divide" signal. Oncogenes discovered thus far include those that encode secreted proteins, growth factors, transmembrane proteins (receptors), cytoplasmic proteins (G proteins and protein kinases), and the nuclear transcription factors that control the expression of genes essential for cell division (Jun, Fos).

Some oncogenes encode surface receptors with defective or missing signal-binding sites, such that their intrinsic Tyr kinase activity is unregulated. For example, the oncoprotein ErbB is essentially identical to the normal receptor for epidermal growth factor, except that ErbB lacks the amino-terminal domain that normally binds EGF **(Fig. 12–50)** and as a result sends the "divide" signal whether EGF is present or not. Mutations

FIGURE 12–50 Oncogene-encoded defective EGF receptor. The product of the *erbB* oncogene (the ErbB protein) is a truncated version of the normal receptor for epidermal growth factor (EGF). Its intracellular domain has the structure normally induced by EGF binding, but the protein lacks the extracellular binding site for EGF. Unregulated by EGF, ErbB continuously signals cell division.

in *erbB2*, the gene for a receptor Tyr kinase related to ErbB, are commonly associated with cancers of the glandular epithelium in breast, stomach, and ovary. (For an explanation of the use of abbreviations in naming genes and their products, see Chapter 25.)

The prominent role played by protein kinases in signaling processes related to normal and abnormal cell division has made them a prime target in the development of drugs for the treatment of cancer (Box 12–5). Mutant forms of the G protein Ras are common in tumor cells. The *ras* oncogene encodes a protein with normal GTP binding but no GTPase activity. The mutant Ras protein is therefore always in its activated (GTP-bound) form, regardless of the signals arriving through normal receptors. The result can be unregulated growth. Mutations in *ras* are associated with 30% to 50% of lung and colon carcinomas and more than 90% of pancreatic carcinomas. ■

Defects in Certain Genes Remove Normal Restraints on Cell Division

Tumor suppressor genes encode proteins that normally restrain cell division. Mutation in one or more of these genes can lead to tumor formation. Unregulated growth due to defective tumor suppressor genes, unlike that due to oncogenes, is genetically recessive; tumors form only if *both* chromosomes of a pair contain a defective gene. This is because the function of these genes is to prevent cell division, and if either copy of the gene for such a protein is normal, the normal inhibition of division will take place. In a person who inherits one correct copy and one defective copy, every cell begins with one defective copy of the gene. If any one of those 10^{12} somatic cells undergoes mutation in the one good copy, a tumor may grow from that doubly

BOX 12–5 ⚕ MEDICINE   Development of Protein Kinase Inhibitors for Cancer Treatment

When a single cell divides without any regulatory limitation, it eventually gives rise to a clone of cells so large that it interferes with normal physiological functions (Fig. 1). This is cancer, a leading cause of death in the developed world, and increasingly so in the developing world. In all types of cancer, the normal regulation of cell division has become dysfunctional due to defects in one or more genes. For example, genes encoding proteins that normally send intermittent signals for cell division become oncogenes, producing constitutively active signaling proteins, or genes encoding proteins that normally restrain cell division (tumor suppressor genes) mutate to produce proteins that lack this braking function. In many tumors, both kinds of mutation have occurred.

Many oncogenes and tumor suppressor genes encode protein kinases or proteins that act in pathways upstream from protein kinases. It is therefore reasonable to hope that specific inhibitors of protein kinases could prove valuable in the treatment of cancer. For example, a mutant form of the EGF receptor is a constantly active receptor Tyr kinase (RTK), signaling cell division whether EGF is present or not (see Fig. 12–50). In about 30% of all women with invasive breast cancer, a mutation in the receptor gene HER2/neu yields an RTK with activity increased up to 100-fold. Another RTK, **vascular endothelial growth factor receptor (VEGFR)**, must be activated for the formation of new blood vessels (angiogenesis) to provide a solid tumor with its own blood supply, and inhibition of VEGFR might starve a tumor of essential nutrients. Nonreceptor Tyr kinases can also mutate, resulting in constant signaling and unregulated cell division. For example, the oncogene *Abl* (from the *Ab*elson *l*eukemia virus) is associated with

acute myeloid leukemia, a relatively rare blood disease (~5,000 cases a year in the United States). Another group of oncogenes encode unregulated cyclin-dependent protein kinases. In each of these cases, specific protein kinase inhibitors might be valuable chemotherapeutic agents in the treatment of disease. Not surprisingly, huge efforts are under way to develop such inhibitors. How should one approach this challenge?

Protein kinases of all types show striking conservation of structure at the active site. All share with the prototypical PKA structure the features shown in Figure 2: two lobes that enclose the active site, with a P loop that helps to align and bind the phosphoryl groups of ATP, an activation loop that moves to open the active site to the protein substrate, and a C helix that changes position as the enzyme is activated, bringing the residues in the substrate-binding cleft into their binding positions.

The simplest protein kinase inhibitors are ATP analogs that occupy the ATP-binding site but cannot serve as phosphoryl group donors. Many such compounds are known, but their clinical usefulness is limited by their lack of selectivity—they inhibit virtually all protein kinases and would produce unacceptable side effects. More selectivity

FIGURE 1 Unregulated division of a single cell in the colon led to a primary cancer that metastasized to the liver. Secondary cancers are seen as white patches in this liver obtained at autopsy.

FIGURE 2 Conserved features of the active site of protein kinases (PDB ID 1S9I). The amino-terminal and carboxyl-terminal lobes surround the active site of the enzyme, near the catalytic loop and the site where ATP binds. The activation loop of this and many other kinases undergoes phosphorylation, then moves away from the active site to expose the substrate-binding cleft, which in this image is occupied by a specific inhibitor of this enzyme, PD318088. The P loop is essential in the binding of ATP, and the C helix must also be correctly aligned for ATP binding and kinase activity.

is seen with compounds that fill part of the ATP-binding site but also interact outside this site, with parts of the protein unique to the target protein kinase. A third possible strategy is based on the fact that although the active conformations of all protein kinases are similar, their inactive conformations are not. Drugs that target the inactive conformation of a specific protein kinase and prevent its conversion to the active form may have a higher specificity of action. A fourth approach employs the great specificity of antibodies. For example, monoclonal antibodies (p. 178) that bind the extracellular portions of specific RTKs could eliminate the receptors' kinase activity by preventing dimerization or by causing their removal from the cell surface. In some cases, an antibody selectively binding to the surface of cancer cells could cause the immune system to attack those cells.

The search for drugs active against specific protein kinases has yielded encouraging results. For example, imatinib mesylate (Gleevec; Fig. 3a), one of the small-molecule inhibitors, has proved nearly 100% effective in bringing about remission in patients with early-stage chronic myeloid leukemia. Erlotinib (Tarceva; Fig. 3b), which targets EGFR, is effective against advanced non-small-cell lung cancer (NSCLC). Because many cell-division signaling systems involve more than one protein kinase, inhibitors that act on several protein kinases may be useful in the treatment of cancer. Sunitinib (Sutent) and sorafenib (Nexavar) target several protein kinases, including VEGFR and PDGFR. These two drugs are in clinical use for patients with gastrointestinal stromal tumors and advanced renal cell carcinoma, respectively. Trastuzumab (Herceptin), cetuximab (Erbitux), and bevacizumab (Avastin) are monoclonal antibodies that target HER2/neu, EGFR, and VEGFR, respectively; all three drugs are in clinical use for certain types of cancer. Detailed knowledge of the structure around the ATP-binding site makes it possible to design drugs that inhibit a *specific* protein kinase by (1) blocking the critical ATP-binding site, while (2) interacting with residues around that site that are *unique* to that particular protein kinase.

At least a hundred more compounds are in preclinical trials. Among the drugs being evaluated are some obtained from natural sources and some produced by synthetic chemistry. Indirubin is a component of a Chinese herbal preparation traditionally used to treat certain leukemias; it inhibits CDK2 and CDK5. Roscovatine (Fig. 3d), a substituted adenine, has a benzyl ring that makes it highly specific as an inhibitor of CDK2. With several hundred potential anti-

(a) Imatinib (Gleevec) bound to Abl

(b) Erlotinib (Tarceva) bound to EGF-R

Imatinib (Gleevec)

Erlotinib (Tarceva)

(c) ATP bound to CDK2

(d) Roscovatine bound to CDK2

Roscovatine

FIGURE 3 Some protein kinase inhibitors now in clinical trials or clinical use, showing their binding to the target protein. **(a)** Imatinib binds to the Abl oncogene kinase active site (PDB ID 1IEP); it occupies both the ATP-binding site and a region adjacent to that site. **(b)** Erlotinib binds to the active site of EGFR (PDB ID 1M17). **(c), (d)** Roscovatine is an inhibitor of the cyclin-dependent kinase CDK2; shown here are normal Mg-ATP binding (c) at the active site (PDB ID 1S9I) and roscovatine binding (d), which prevents the binding of ATP (PDB ID 2A4L).

cancer drugs heading toward clinical testing, it is realistic to hope that some will prove more effective or more target-specific than those now in use.

mutant cell. Mutations in both copies of the genes for pRb, p53, or p21 yield cells in which the normal restraint on cell division is lost and a tumor forms.

Retinoblastoma occurs in children and causes blindness if not surgically treated. The cells of a retinoblastoma have two defective versions of the *Rb* gene (two defective alleles). Very young children who develop retinoblastoma commonly have multiple tumors in both eyes. These children have inherited one defective copy of the *Rb* gene, which is present in every cell; each tumor is derived from a single retinal cell that has undergone a mutation in its one good copy of the *Rb* gene. (A fetus with two mutant alleles in every cell is nonviable.) People with retinoblastoma who survive childhood also have a high incidence of cancers of the lung, prostate, and breast later in life.

A far less likely event is that a person born with two good copies of the *Rb* gene will have independent mutations in both copies in the *same* cell. Some individuals do develop retinoblastomas later in childhood, usually with only one tumor in one eye. These individuals were presumably born with two good copies (alleles) of *Rb* in every cell, but both *Rb* alleles in a single retinal cell have undergone mutation, leading to a tumor. After about age three, retinal cells stop dividing, and retinoblastomas at later ages are quite rare.

Stability genes (also called caretaker genes) encode proteins that function in the repair of major genetic defects that result from aberrant DNA replication, ionizing radiation, or environmental carcinogens. Mutations in these genes lead to a high frequency of unrepaired damage (mutations) in other genes, including proto-oncogenes and tumor suppressor genes, and thus to cancer. Among the stability genes are *ATM* (see Fig. 12–49); the *XP* gene family, in which mutations lead to xeroderma pigmentosum; and the *BRCA1* genes associated with some types of breast cancer (see Box 25–1). Mutations in the gene for p53 also cause tumors; in more than 90% of human cutaneous squamous cell carcinomas (skin cancers) and in about 50% of all other human cancers, *p53* is defective. Those very rare individuals who *inherit* one defective copy of *p53* commonly have the Li-Fraumeni cancer syndrome, with multiple cancers (of the breast, brain, bone, blood, lung, and skin) occurring at high frequency and at an early age. The explanation for multiple tumors in this case is the same as that for *Rb* mutations: an individual born with one defective copy of *p53* in every somatic cell is likely to suffer a second *p53* mutation in more than one cell during his or her lifetime.

In summary, then, three classes of defects can contribute to the development of cancer: oncogenes, in which the defect is the equivalent of a car's accelerator pedal being stuck down, with the engine racing; mutated tumor suppressor genes, in which the defect leads to the equivalent of brake failure; and mutated stability genes, with the defect leading to unrepaired damage to the cell's replication machinery, the equivalent of an unskilled car mechanic.

Mutations in oncogenes and tumor suppressor genes do not have an all-or-none effect. In some cancers, perhaps in all, the progression from a normal cell to a malignant tumor requires an accumulation of mutations (sometimes over several decades), none of which, alone, is responsible for the end effect. For example, the development of colorectal cancer has several recognizable stages, each associated with a mutation **(Fig. 12–51)**. If an epithelial cell in the colon undergoes mutation of both copies of the tumor suppressor gene *APC* (adenomatous polyposis coli), it begins to divide faster than normal and produces a clone of itself, a benign polyp (early adenoma). For reasons not yet known, the *APC* mutation results in chromosomal instability, and whole regions of a chromosome are lost or rearranged during cell division. This instability can lead to another mutation, commonly in *ras*, that converts the clone into an intermediate adenoma. A third mutation (often in the tumor suppressor gene *DCC*) leads to a late adenoma. Only when both copies of *p53* become defective does this cell mass become a carcinoma—a malignant, life-threatening tumor. The full sequence therefore requires at least seven genetic "hits": two on each of three tumor suppressor genes (*APC*, *DCC*, and *p53*) and one on the proto-oncogene *ras*. There are probably several other routes to colorectal cancer as well, but the principle that full malignancy results only from multiple mutations is likely to hold true for all of them. When a polyp is detected in the early adenoma stage and the cells containing the first mutations are removed surgically, late adenomas and carcinomas will not develop; hence the importance of early detection. Cells and organisms, too, have their early detection systems. For example, the ATM and ATR proteins described in Section 12.11 can detect DNA damage too extensive to be repaired effectively. They then trigger, through a pathway that includes p53, the process of apoptosis, in which a cell that has become dangerous to the organism kills itself. ∎

Apoptosis Is Programmed Cell Suicide

Many cells can precisely control the time of their own death by the process of **programmed cell death**, or **apoptosis** (app′-a-toe′-sis; from the Greek for "dropping off," as in leaves dropping in the fall). One trigger for apoptosis is irreparable damage to DNA. Programmed cell death also occurs during the development of an embryo, when some cells must die to give a tissue or organ its final shape. Carving fingers from stubby limb buds requires the precisely timed death of cells between developing finger bones. During development of the nematode *C. elegans* from a fertilized egg, exactly 131 cells (of a total of 1,090 somatic cells in the embryo) must undergo programmed death in order to construct the adult body.

Apoptosis also has roles in processes other than development. If a developing antibody-producing cell

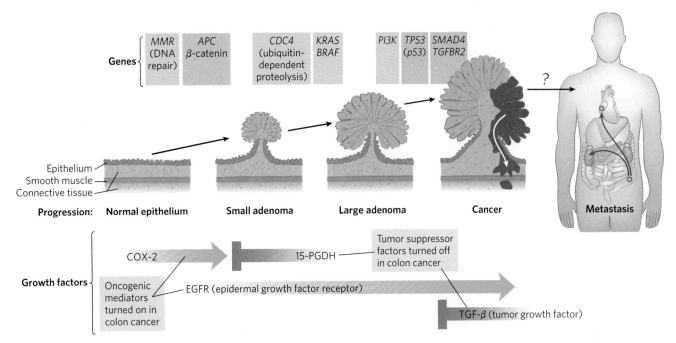

FIGURE 12–51 Multistep transition from normal epithelial cell to colorectal cancer. Serial mutations in oncogenes (green) or tumor suppressor genes (red) lead to progressively less control of cell division, until finally an active tumor forms, which can sometimes metastasize (spread from the initial site to other regions of the body). Mutation of the *MMR* gene leads to defective DNA repair and consequently to a higher rate of mutation. Mutations in both copies of the tumor suppressor gene *APC* lead to benign clusters of epithelial cells that multiply too rapidly (early adenoma). The *CDC4* oncogene results in defective ubiquitination, which is essential to the regulation of cyclin-dependent kinases (see Fig. 12–47). The oncogenes *KRAS* and *BRAF* encode ras and raf proteins (see Fig. 12–15), and this further disruption of signaling leads to the formation of a large adenoma, which may be detected by colonoscopy as a benign polyp. Oncogenic mutations in the *PI3K* gene that encodes the enzyme phosphoinositide-3 kinase, or in *PTEN*, which regulates the synthesis of this enzyme, lead to a further strengthening of the signal: divide now. When a cell in one of the polyps undergoes further mutations, in the tumor suppressor genes *DCC* and *p53* (see Fig. 12–49) for example, increasingly aggressive tumors form. Finally, mutations in other tumor suppressor genes such as *SMAD4* lead to a malignant tumor and sometimes to a metastatic tumor that can spread to other tissues. A second type of mutation that can add to the deleterious effects is one that affects the production or action of growth factors or their receptors (bottom). Mutations in EGFR (epidermal growth factor receptor) or TGF-β (transforming growth factor-β) favor uncontrolled growth, as do mutations in the enzymes that produce certain prostaglandins (COX-2; cyclooxygenase; see pp. 845–846) or 15-PGDH (15-hydroxyprostaglandin dehydrogenase). Most malignant tumors of other tissues probably result from a series of mutations such as this, although not necessarily these particular genes, or in this order.

generates antibodies against a protein or glycoprotein normally present in the body, that cell undergoes programmed death in the thymus gland—an essential mechanism for eliminating anti-self antibodies (the cause of many autoimmune diseases). The monthly sloughing of cells of the uterine wall (menstruation) is another case of apoptosis mediating normal cell death. The dropping of leaves in the fall is the result of apoptosis in specific cells of the stem. Sometimes cell suicide is not programmed but occurs in response to biological circumstances that threaten the rest of the organism. For example, a virus-infected cell that dies before completion of the infection cycle prevents spread of the virus to nearby cells. Severe stresses such as heat, hyperosmolarity, UV light, and gamma irradiation also trigger cell suicide; presumably the organism is better off with any aberrant, potentially mutated cells dead.

The regulatory mechanisms that trigger apoptosis involve some of the same proteins that regulate the cell cycle. The signal for suicide often comes from outside, through a surface receptor. Tumor necrosis factor (TNF), produced by cells of the immune system, interacts with cells through specific TNF receptors. These receptors have TNF-binding sites on the outer face of the plasma membrane and a "death domain" (~80 amino acid residues) that carries the self-destruct signal through the membrane to cytosolic proteins such as TRADD (*TNF receptor–associated death domain*) **(Fig. 12–52)**. Another receptor, Fas, has a similar death domain that allows it to interact with the cytosolic protein FADD (*Fas-associated death domain*), which activates the cytosolic protease caspase 8. This enzyme belongs to a family of proteases that participate in apoptosis; all are synthesized as inactive proenzymes, all have a critical Cys residue at the active site, and all hydrolyze their target proteins on the carboxyl-terminal side of specific Asp residues (hence the name caspase, from *Cys* and *Asp*).

When caspase 8, an "initiator" caspase, is activated by an apoptotic signal carried through FADD, it further self-activates by cleaving its own proenzyme form. Mitochondria are one target of active caspase 8. The protease

FIGURE 12–52 Initial events of apoptosis. Apoptosis-triggering signals from outside the cell (Fas and TNFα) bind to their specific receptors in the plasma membrane (FasR, TNFαR). The occupied receptors interact with the cytosolic proteins FADD and TRADD through an 80 amino acid sequence called the "death domain" on both the receptors and these cytosolic targets, which become activated. Activation of FADD and TRADD initiates a proteolytic cascade that leads to apoptosis. FADD and TRADD activate caspase-8, which acts to release cytochrome c from mitochondria, which, in concert with protein Apaf-1, activates caspase-9, triggering apoptosis.

causes the release of certain proteins contained between the inner and outer mitochondrial membranes: cytochrome *c* (Chapter 19) and several "effector" caspases. Cytochrome *c* binds to the proenzyme form of the effector enzyme caspase 9 and stimulates its proteolytic activation. The activated caspase 9 in turn catalyzes wholesale destruction of cellular proteins—a major cause of apoptotic cell death. One specific target of caspase action is a caspase-activated deoxyribonuclease.

In apoptosis, the monomeric products of protein and DNA degradation (amino acids and nucleotides) are released in a controlled process that allows them to be taken up and reused by neighboring cells. Apoptosis thus allows the organism to eliminate a cell that is unneeded or potentially dangerous without wasting its components.

SUMMARY 12.12 Oncogenes, Tumor Suppressor Genes, and Programmed Cell Death

▶ Oncogenes encode defective signaling proteins. By continually giving the signal for cell division, they lead to tumor formation. Oncogenes are genetically dominant and may encode defective growth factors, receptors, G proteins, protein kinases, or nuclear regulators of transcription.

▶ Tumor suppressor genes encode regulatory proteins that normally inhibit cell division; mutations in these genes are genetically recessive but can lead to tumor formation.

▶ Cancer is generally the result of an accumulation of mutations in oncogenes and tumor suppressor genes.

▶ When stability genes, which encode proteins necessary for the repair of genetic damage, are mutated, other mutations go unrepaired, including mutations in proto-oncogenes and tumor suppressor genes that can lead to cancer.

▶ Apoptosis is programmed and controlled cell death that functions during normal development and adulthood to get rid of unnecessary, damaged, or infected cells. Apoptosis can be triggered by extracellular signals such as TNF, acting through plasma membrane receptors.

Key Terms

Terms in bold are defined in the glossary.

signal transduction   433
specificity   433
cooperativity   434
amplification   434
enzyme cascade   434
modularity   434
scaffold proteins   434
desensitization   434
integration   434
Scatchard analysis   435
G protein–coupled
   receptors (GPCRs)   437
guanosine nucleotide–
   binding proteins   437
G proteins   437
second messenger   437
agonist   438
antagonist   438
β-adrenergic
   receptors   438
heptahelical
   receptors   438
stimulatory G protein
   (G_s)   438
adenylyl cyclase   438
cAMP-dependent protein
   kinase (protein kinase A;
   PKA)   438
P loop   441
GTPase activator protein
   (GAP)   442
regulator of G protein
   signaling (RGS)   442
guanosine nucleotide–
   exchange factor
   (GEF)   442
consensus sequence   443
β-arrestin (βarr; arrestin 2)
   446
G protein–coupled receptor
   kinases (GRKs)   446
cAMP response element
   binding protein
   (CREB)   446
inhibitory G protein
   (G_i)   446
adaptor proteins   446
AKAPs (A kinase anchoring
   proteins)   447
phospholipase C (PLC)   447
inositol 1,4,5-trisphosphate
   (IP_3)   447
green fluorescent protein
   (GFP)   448

fluorescence resonance
   energy transfer
   (FRET)   448
protein kinase C (PKC)   450
calmodulin (CaM)   451
Ca^{2+}/calmodulin-dependent
   protein kinases (CaM
   kinases)   451
receptor Tyr kinase
   (RTK)   453
autophosphorylation
   453
SH2 domain   454
small G proteins   455
MAPKs   455
cytokine   457
guanosine 3′,5′-cyclic
   monophosphate (cyclic
   GMP; cGMP)   459
cGMP-dependent protein
   kinase (protein kinase G;
   PKG)   459
atrial natriuretic factor
   (ANF)   459
NO synthase   460
angina pectoris   460
PTB domains   461
voltage-gated ion
   channels   465
nicotinic acetylcholine
   receptor   467
ionotropic   468
integrin   470
hormone response
   element (HRE)   471
two-component signaling
   systems   473
receptor histidine
   kinase   473
response regulator   473
receptorlike kinase
   (RLK)   476
retinal   477
rhodopsin   477
opsin   477
transducin   478
rhodopsin kinase   480
receptor potential   481
gustducin   481
cyclin   485
cyclin-dependent protein
   kinase (CDK)   485
ubiquitin   487
proteasome   487

growth factors   487
retinoblastoma protein
   (pRb)   488
oncogene   489
proto-oncogene   489

tumor suppressor
   gene   489
programmed cell death   492
apoptosis   492

Further Reading

General

Cohen, P. (2000) The regulation of protein function by multisite phosphorylation—a 25 year update. *Trends Biochem. Sci.* **25**, 596–601.
   Historical account of protein phosphorylation.

Giepmans, B.N.G., Adams, S.R., Ellisman, M.H., & Tsien, R.Y. (2006) The fluorescent toolbox for assessing protein location and function. *Science* **312**, 217–224.
   A short, intermediate-level review of FRET.

Gomperts, B., Kramer, I.M., & Tatham, P.E.R. (2009) *Signal Transduction*, 2nd edn, Academic Press, New York.
   A beautifully illustrated and clear description of all of the signaling systems discussed in this chapter.

Marks, F., Klingmüller, U., & Müller-Decker, K. (2009) *Cellular Signal Processing: An Introduction to the Molecular Mechanisms of Signal Transduction*, Garland Science, New York.

Pawson, T. & Scott, J.D. (2005) Protein phosphorylation in signaling—50 years and counting. *Trends Biochem. Sci.* **30**, 286–290.

G Protein–Coupled Receptors (GPCRs)

Aktories, K. (2011) Bacterial protein toxins that modify host regulatory GTPases. *Nat. Rev. Microbiol.* **9**, 487–498.

Awais, M. & Ozawa, T. (2011) Illuminating intracellular signaling and molecules for single cell analyisis. *Mol. Biosyst.* **7**, 1376–1387.
   Intermediate-level review of methods for using fluorescent probes in cell biology and biochemistry.

Beene, D.L. & Scott, J.D. (2007) A-kinase anchoring proteins take shape. *Curr. Opin. Cell Biol.* **19**, 192–198.

Birnbaumer, L. (2007) The discovery of signal transduction by G proteins: a personal account and an overview of the initial findings and contributions that led to our present understanding. *Biochim. Biophys. Acta Biomembr.* **1768**, 756–771.

Escribá, P.V. (2007) G protein–coupled receptors, signaling mechanisms and pathophysiological relevance. *Biochim. Biophys. Acta Biomembr.* **1768**, 747.
   The editorial introduction to a series of 20 papers on GPCRs.

Francis, S.H., Blount, M.A., & Corbin, J.D. (2011) Mammalian cyclic nucleotide phosphodiesterases: molecular mechanisms and physiological function. *Physiol. Rev.* **91**, 651–690.

Kobilka, B.K. (2011) Structural insights into adrenergic receptor function and pharmacology. *Trends Pharmacol. Sci.* **32**, 213–218.
   Intermediate-level review.

Kremers, G.-J., Gilbert, S.G., Cranfill, P.J., Davidson, M.W., & Piston, D.W. (2011) Fluorescent proteins at a glance. *J. Cell Sci.* **124**, 157–160.
   Brief review of methodology for using fluorescently labeled proteins.

Lefkowitz, R.J. (2007) Introduction to special section on β-arrestins. *Annu. Rev. Physiol.* **69**.
   This introduces five excellent advanced reviews on the roles of arrestin.

Malumbres, M. (2011) Physiological relevance of cell cycle kinases. *Physiol. Rev.* **91**, 973–1007.

Advanced and extensive review of the roles of protein kinases in regulating the cell cycle and their roles in cancer.

Pearce, L.R., Komander, D., & Alessi, D.R. (2010) The nuts and bolts of AGC protein kinases. *Nat. Rev. Mol. Cell Biol.* **11**, 9–22.

Advanced review of the PKA, PKG, PKC family of kinases.

Pinna, L.A. & Ruzzene, M. (1996) How do protein kinases recognize their substrates? *Biochim. Biophys. Acta* **1314**, 191–225.

Advanced review of the factors, including consensus sequences, that give protein kinases their specificity.

Rasmussen, S.G.F., DeVree, B.T., Zou, Y., Kruse, A.C., Chung, K.Y., Kobilka, T.S., Thian, F.S., Chae, P.S., Pardon, E., Calinski, D., et al. (2011) Crystal structure of the β_2 adrenergic receptor-Gs protein complex. *Nature* **477**, 549–555.

The definitive determination of the structure and interaction of receptor and G protein.

Receptor Enzymes

Garbers, D.L., Chrisman, T.D., Wiegn, P., Katafuchi, T., Albanesi, J.P., Bielinski, V., Barylko, B., Redfield, M.M., & Burnett, J.C., Jr. (2006) Membrane guanylyl cyclase receptors: an update. *Trends Endocrinol. Metab.* **17**, 251–258.

Karnoub, A.E. & Weinberg, R.A. (2008) Ras oncogenes: split personalities. *Nat. Rev. Mol. Cell Biol.* **9**, 517–531.

Lemmon, M.A. & Schlessinger, J. (2010) Cell signaling by receptor tyrosine kinases. *Cell* **141**, 1117–1134.

Misono, K.S., Philo, J.S., Arakawa, T., Ogata, C.M., Qiu, Y., Ogawa, H., & Young, H.S. (2011) Structure, signaling mechanism and regulation of the natriuretic peptide receptor guanylate cyclase. *FEBS J.* **278**, 1818–1829.

Adaptor Proteins and Membrane Rafts

Dehmelt, L. & Bastiaens, P.I.H. (2010) Spatial organization of intracellular communication: insights from imaging. *Nat. Rev. Mol. Cell Biol.* **11**, 440–452.

Good, M.C., Zalatan, J.G., & Lim, W.A. (2011) Scaffold proteins: hubs for controlling the flow of cellular information. *Science* **332**, 680–686.

Intermediate-level review of scaffold structure and function.

Schwarz-Romond, T. & Gorski, S.A. (2010) Focus on the spatial organization of signaling. *EMBO J.* **29**, 2675–2676.

Editorial introduction to a collection of eight excellent reviews of spatial aspects of signaling.

Smith, F.D. & Scott, J.D. (2006) Anchored cAMP signaling: onward and upward. A short history of compartmentalized cAMP signal transduction. *Eur. J. Cell Biol.* **85**, 582–592.

Short, intermediate-level review introducing an entire journal issue on the subject of AKAPs and cAMP signaling.

Receptor Ion Channels

See also Chapter 11, Further Reading, Ion Channels.

Ashcroft, F.M. (2006) From molecule to malady. *Nature* **440**, 440–447.

Short, intermediate-level review of human diseases associated with defects in ion channels.

Changeux, J.-P. (2010) Allosteric receptors: from electric organ to cognition. *Annu. Rev. Pharmacol. Toxicol.* **50**, 1–38.

Grigoryan, G., Moore, D.T., & DeGrado, W.F. (2011) Transmembrane communication: general principles and lessons from the structure and function of the M2 proton channel, K^+ channels, and integrin receptors. *Annu. Rev. Biochem.* **80**, 211–237.

Advanced, rewarding review.

Tombola, F., Pathak, M.M., & Isacoff, E.Y. (2006) How does voltage open an ion channel? *Annu. Rev. Cell Dev. Biol.* **22**, 23–52.

Calcium Ions in Signaling

Berridge, M.J. (2009) Inositol trisphosphate and calcium signalling mechanisms. *Biochim. Biophys. Acta Mol. Cell Res.* **1793**, 933–940.

Bunney, T.D. & Katan, M. (2011) PLC regulation: emerging pictures for molecular mechanisms. *Trends Biochem. Sci.* **36**, 88–96.

Intermediate review of phospholipase C in Ca^{2+}/IP_3 signaling.

Chazin, W.J. (2011) Relating form and function of EF-hand calcium binding proteins. *Acc. Chem. Res.* **44**, 171–179.

Chin, D. & Means, A.R. (2000) Calmodulin: a prototypical calcium receptor. *Trends Cell Biol.* **10**, 322–328.

Haiech, J., Heizmann, C., & Krebs, J. (eds). (2009) 10th European Symposium on Calcium. *Biochim. Biophys. Acta Mol. Cell Res.* **1793**, 931–1114.

All the papers in this volume deal with calcium signaling.

Parekh, A.B. (2011) Decoding cytosolic Ca^{2+} oscillations. *Trends Biochem. Sci.* **36**, 78–87.

Integrins

Harburger, D.S. & Calderwood, D.A. (2009) Integrin signaling at a glance. *J. Cell Sci.* **122**, 159–163.

Poster format review of integrin signaling.

Valdembri, D., Sandri, C., Santambrogio, M., & Serini, G. (2011) Regulation of integrins by conformation and traffic: it takes two to tango. *Mol. Biosyst.* **7**, 2539–2546.

Steroid Hormone Receptors and Action

Biggins, J.B. & Koh, J.T. (2007) Chemical biology of steroid and nuclear hormone receptors. *Curr. Opin. Chem. Biol.* **11**, 99–110.

Huang, P., Chandra, V., & Rastinejad, F. (2010) Structural overview of the nuclear receptor superfamily: insights into physiology and therapeutics *Annu. Rev. Physiol.* **72**, 247–272.

Signaling in Plants and Bacteria

Chen, Y.F., Etheridge, N., & Schaller, G.E. (2005) Ethylene signal transduction. *Ann. Botany* **95**, 901–915.

Intermediate-level review.

Clouse, S.D. (2011) Brassinosteroid signal transduction: from receptor kinase activation to transcriptional networks regulating plant development. *Plant Cell* **23**, 1219–1230.

Cutler, S.R., Rodriguez, P.L., Finkelstein, R.R., & Abrams, S. (2010) Abscisic acid: emergence of a core signaling network. *Annu. Rev. Plant Biol.* **61**, 651–679.

Dodd, A.N., Kudla, J., & Sanders, D. (2010) The language of calcium signaling. *Annu. Rev. Plant Biol.* **61**, 593–620.

Ferreira, F.J. & Kieber, J.J. (2005) Cytokinin signaling. *Curr. Opin. Plant Biol.* **8**, 518–525.

Paciorek, T. & Friml, J. (2006) Auxin signaling. *J. Cell Sci.* **119**, 1199–1202.

Pauwels, L. & Goossens, A. (2011) The JAZ proteins: a crucial interface in the jasmonate signaling cascade. *Plant Cell* **23**, 3089–3100.

Perry, J., Koteva, K., & Wright, G. (2011) Receptor domains of two-component signal transduction systems. *Mol. Biosyst.* **7**, 1388–1398.

Rodriguez, M.C.S., Petersen, M., & Mundy, J. (2010) Mitogen-activated protein kinase signaling in plants. *Annu. Rev. Plant Biol.* **61**, 621–649.

Stepanova, A.N. & Alonso, J.M. (2009) Ethylene signaling and response: where different regulatory modules meet. *Curr. Opin. Plant Biol.* **12**, 548–555.

Wang, Z. (2010) From receptors to responses (in plants). *Curr. Opin. Plant Biol.* **13**, 485–488.

Editorial introduction to a series of papers on plant signaling, all published in this issue.

Vision, Olfaction, and Gustation

Kaupp, U.B. (2010) Olfactory signalling in vertebrates and insects: differences and commonalities. *Nat. Rev. Neurosci.* **11**, 188–200.

Smith, S.O. (2010) Structure and activation of the visual pigment rhodopsin. *Annu. Rev. Biophys.* **39**, 309–328.

Advanced review.

Yarmolinsky, D.A., Zuker, C.S., & Ryba, N.J.P. (2009) Common sense about taste: from mammals to insects. *Cell* **139**, 234–244.

Cell Cycle and Cancer

Bublil, E.M. & Yarden, Y. (2007) The EGF receptor family: spearheading a merger of signaling and therapeutics. *Curr. Opin. Cell Biol.* **19**, 124–134.

Clarke, P.R. & Allan, L.A. (2009) Cell-cycle control in the face of damage—a matter of life or death. *Trends Cell Biol.* **19**, 89–98.

Intermediate review of roles of ATP, p53, and other regulatory proteins in the cell cycle.

Dorsam, R.T. & Gutkind, J.S. (2007) G-protein-coupled receptors and cancer. *Nat. Rev. Cancer* **7**, 79–94.

Levine, A.J. & Oren, M. (2009) The first 30 years of p53: growing ever more complex. *Nat. Rev. Cancer* **9**, 749–758.

Historical perspective on the studies of p53 and cancer.

Ma, H.T. & Poon, R.Y.C. (2011) How protein kinases co-ordinate mitosis in animal cells. *Biochem. J.* **435**, 17–31.

Malumbres, M. (2011) Physiological relevance of cell cycle kinases *Physiol. Rev.* **91**, 973–1007.

Advanced review of roles of CDKs and other protein kinases in controlling the cell cycle.

Markowitz, S.D. & Bertagnolli, M.M. (2009) Molecular basis of colorectal cancer. *N. Engl. J. Med.* **361**, 2449–2460.

Novak, B., Kapuy, O., Domingo-Sananes, M.R., & Tyson, J.J. (2010) Regulated protein kinases and phosphatases in cell cycle decisions. *Curr. Opin. Cell Biol.* **22**, 801–808.

Pylayeva-Gupta, Y., Grabocka, E., & Bar-Sagi, D. (2011) RAS oncogenes: weaving a tumorigenic web. *Nat. Rev. Cancer* **11**, 761–774.

Apoptosis

Wylie, A.H. (2010) "Where, O death, is thy sting?" A brief review of apoptosis biology. *Mol. Neurobiol.* **42**, 4–9.

Problems

1. Hormone Experiments in Cell-Free Systems In the 1950s, Earl W. Sutherland, Jr., and his colleagues carried out pioneering experiments to elucidate the mechanism of action of epinephrine and glucagon. Given what you have learned in this chapter about hormone action, interpret each of the experiments described below. Identify substance X and indicate the significance of the results.

(a) Addition of epinephrine to a homogenate of normal liver resulted in an increase in the activity of glycogen phosphorylase. However, when the homogenate was first centrifuged at a high speed and epinephrine or glucagon was added to the clear supernatant fraction that contains phosphorylase, no increase in the phosphorylase activity occurred.

(b) When the particulate fraction from the centrifugation in (a) was treated with epinephrine, substance X was produced. The substance was isolated and purified. Unlike epinephrine, substance X activated glycogen phosphorylase when added to the clear supernatant fraction of the centrifuged homogenate.

(c) Substance X was heat stable; that is, heat treatment did not affect its capacity to activate phosphorylase. (Hint: Would this be the case if substance X were a protein?) Substance X was nearly identical to a compound obtained when pure ATP was treated with barium hydroxide. (Fig. 8–6 will be helpful.)

2. Effect of Dibutyryl cAMP versus cAMP on Intact Cells The physiological effects of epinephrine should in principle be mimicked by addition of cAMP to the target cells. In practice, addition of cAMP to intact target cells elicits only a minimal physiological response. Why? When the structurally related derivative dibutyryl cAMP (shown below) is added to intact cells, the expected physiological response is readily apparent. Explain the basis for the difference in cellular response to these two substances. Dibutyryl cAMP is widely used in studies of cAMP function.

Dibutyryl cAMP
($N^6,O^{2'}$-Dibutyryl adenosine 3',5'-cyclic monophosphate)

3. Effect of Cholera Toxin on Adenylyl Cyclase The gram-negative bacterium *Vibrio cholerae* produces a protein, cholera toxin (M_r 90,000), that is responsible for the characteristic symptoms of cholera: extensive loss of body water and Na^+ through continuous, debilitating diarrhea. If body fluids and Na^+ are not replaced, severe dehydration results; untreated, the disease is often fatal. When the cholera toxin gains access to the human intestinal tract, it binds tightly to specific sites in the plasma membrane of the epithelial cells lining the small intestine, causing adenylyl cyclase to undergo prolonged activation (hours or days).

(a) What is the effect of cholera toxin on [cAMP] in the intestinal cells?

(b) Based on the information above, suggest how cAMP normally functions in intestinal epithelial cells.

(c) Suggest a possible treatment for cholera.

4. Mutations in PKA Explain how mutations in the R or C subunit of cAMP-dependent protein kinase (PKA) might lead to (a) a constantly active PKA or (b) a constantly inactive PKA.

5. Therapeutic Effects of Albuterol The respiratory symptoms of asthma result from constriction of the bronchi and bronchioles of the lungs, caused by contraction of the smooth muscle of their walls. This constriction can be reversed by raising [cAMP] in the smooth muscle. Explain the therapeutic effects of albuterol, a β-adrenergic agonist taken (by inhalation) for asthma. Would you expect this drug to have any side effects? How might one design a better drug that does not have these effects?

6. Termination of Hormonal Signals Signals carried by hormones must eventually be terminated. Describe several different mechanisms for signal termination.

7. Using FRET to Explore Protein-Protein Interactions in Vivo Figure 12–8 shows the interaction between β-arrestin and the β-adrenergic receptor. How would you use FRET (see Box 12–3) to demonstrate this interaction in living cells? Which proteins would you fuse? Which wavelengths would you use to illuminate the cells, and which would you monitor? What would you expect to observe if the interaction occurred? If it did not occur? How might you explain the failure of this approach to demonstrate this interaction?

8. EGTA Injection EGTA (ethylene glycol-bis(β-amino-ethyl ether)-N,N,N',N'-tetraacetic acid) is a chelating agent with high affinity and specificity for Ca^{2+}. By microinjecting a cell with an appropriate Ca^{2+}-EGTA solution, an experimenter can prevent cytosolic $[Ca^{2+}]$ from rising above 10^{-7} M. How would EGTA microinjection affect a cell's response to vasopressin (see Table 12–4)? To glucagon?

9. Amplification of Hormonal Signals Describe all the sources of amplification in the insulin receptor system.

10. Mutations in *ras* How would a mutation in *ras* that leads to formation of a Ras protein with no GTPase activity affect a cell's response to insulin?

11. Differences among G Proteins Compare the G proteins G_s, which acts in transducing the signal from β-adrenergic receptors, and Ras. What properties do they share? How do they differ? What is the functional difference between G_s and G_i?

12. Mechanisms for Regulating Protein Kinases Identify eight general types of protein kinases found in eukaryotic cells, and explain what factor is *directly* responsible for activating each type.

13. Nonhydrolyzable GTP Analogs Many enzymes can hydrolyze GTP between the β and γ phosphates. The GTP analog β,γ-imidoguanosine 5'-triphosphate (Gpp(NH)p), shown below, cannot be hydrolyzed between the β and γ phosphates.

Gpp(NH)p
(β,γ-imidoguanosine 5'-triphosphate)

Predict the effect of microinjection of Gpp(NH)p into a myocyte on the cell's response to β-adrenergic stimulation.

14. Use of Toxin Binding to Purify a Channel Protein α-Bungarotoxin is a powerful neurotoxin found in the venom of a poisonous snake (*Bungarus multicinctus*). It binds with high specificity to the nicotinic acetylcholine receptor (AChR) protein and prevents the ion channel from opening. This interaction was used to purify AChR from the electric organ of torpedo fish.

(a) Outline a strategy for using α-bungarotoxin covalently bound to chromatography beads to purify the AChR protein. (Hint: See Fig. 3–17c.)

(b) Outline a strategy for the use of $[^{125}I]\alpha$-bungarotoxin to purify the AChR protein.

15. Resting Membrane Potential A variety of unusual invertebrates, including giant clams, mussels, and polychaete worms, live on the fringes of deep-sea hydrothermal vents, where the temperature is 60 °C.

(a) The adductor muscle of a giant clam has a resting membrane potential of −95 mV. Given the intracellular and extracellular ionic compositions shown below, would you have predicted this membrane potential? Why or why not?

Ion	Concentration (mM)	
	Intracellular	**Extracellular**
Na^+	50	440
K^+	400	20
Cl^-	21	560
Ca^{2+}	0.4	10

(b) Assume that the adductor muscle membrane is permeable to only one of the ions listed above. Which ion could determine the V_m?

16. Membrane Potentials in Frog Eggs Fertilization of a frog oocyte by a sperm cell triggers ionic changes similar to those observed in neurons (during movement of the action potential) and initiates the events that result in cell division and development of the embryo. Oocytes can be stimulated to divide without fertilization, by suspending them in 80 mM KCl (normal pond water contains 9 mM KCl).

(a) Calculate how much the change in extracellular [KCl] changes the resting membrane potential of the oocyte. (Hint: Assume the oocyte contains 120 mM K^+ and is permeable *only* to K^+.) Assume a temperature of 20 °C.

(b) When the experiment is repeated in Ca^{2+}-free water, elevated [KCl] has no effect. What does this suggest about the mechanism of the KCl effect?

17. Excitation Triggered by Hyperpolarization In most neurons, membrane *depolarization* leads to the opening of voltage-dependent ion channels, generation of an action potential, and ultimately an influx of Ca^{2+}, which causes release of neurotransmitter at the axon terminus. Devise a cellular strategy by which *hyperpolarization* in rod cells could produce excitation of the visual pathway and passage of visual signals to the brain. (Hint: The neuronal signaling pathway in higher organisms consists of a *series* of neurons that relay information to the brain (see Fig. 12–36). The signal released by one neuron can be either excitatory or inhibitory to the following, postsynaptic neuron.)

18. Genetic "Channelopathies" There are many genetic diseases that result from defects in ion channels. For each of the following, explain how the molecular defect might lead to the symptoms described.

(a) A loss-of-function mutation in the gene encoding the α subunit of the cGMP-gated cation channel of retinal cone cells leads to a complete inability to distinguish colors.

(b) Loss-of-function alleles of the gene encoding the α subunit of the ATP-gated K^+ channel shown in Figure 23–28 lead to a condition known as congenital hyperinsulinism—persistently high levels of insulin in the blood.

(c) Mutations affecting the β subunit of the ATP-gated K^+ channel that prevent ATP binding lead to neonatal diabetes—persistently low levels of insulin in the blood in newborn babies.

19. Visual Desensitization Oguchi disease is an inherited form of night blindness. Affected individuals are slow to recover vision after a flash of bright light against a dark background, such as the headlights of a car on the freeway. Suggest what the molecular defect(s) might be in Oguchi disease. Explain in molecular terms how this defect would account for night blindness.

20. Effect of a Permeant cGMP Analog on Rod Cells An analog of cGMP, 8-Br-cGMP, will permeate cellular membranes, is only slowly degraded by a rod cell's PDE activity, and is as effective as cGMP in opening the gated channel in the cell's outer segment. If you suspended rod cells in a buffer containing a relatively high [8-Br-cGMP], then illuminated the cells while measuring their membrane potential, what would you observe?

21. Hot and Cool Taste Sensations The sensations of heat and cold are transduced by a group of temperature-gated cation channels. For example, TRPV1, TRPV3, and TRPM8 are usually closed, but open under the following conditions: TRPV1 at $\geq 43\,°C$; TRPV3 at $\geq 33\,°C$; and TRPM8 at $<25\,°C$. These channels are expressed in sensory neurons known to be responsible for temperature sensation.

(a) Propose a reasonable model to explain how exposing a sensory neuron containing TRPV1 to high temperature leads to a sensation of heat.

(b) Capsaicin, one of the active ingredients in "hot" peppers, is an agonist of TRPV1. Capsaicin shows 50% activation of the TRPV1 response at a concentration (i.e., it has an EC_{50}) of 32 nM. Explain why even a very few drops of hot pepper sauce can taste very "hot" without actually burning you.

(c) Menthol, one of the active ingredients in mint, is an agonist of TRPM8 ($EC_{50} = 30\ \mu M$) and TRPV3 ($EC_{50} = 20$ mM). What sensation would you expect from contact with low levels of menthol? With high levels?

22. Oncogenes, Tumor-Suppressor Genes, and Tumors For each of the following situations, provide a plausible explanation for how it could lead to unrestricted cell division.

(a) Colon cancer cells often contain mutations in the gene encoding the prostaglandin E_2 receptor. PGE_2 is a growth factor required for the division of cells in the gastrointestinal tract.

(b) Kaposi sarcoma, a common tumor in people with untreated AIDS, is caused by a virus carrying a gene for a protein similar to the chemokine receptors CXCR1 and CXCR2. Chemokines are cell-specific growth factors.

(c) Adenovirus, a tumor virus, carries a gene for the protein E1A, which binds to the retinoblastoma protein, pRb. (Hint: See Fig. 12–49.)

(d) An important feature of many oncogenes and tumor suppressor genes is their cell-type specificity. For example, mutations in the PGE_2 receptor are not typically found in lung tumors. Explain this observation. (Note that PGE_2 acts through a GPCR in the plasma membrane.)

23. Mutations in Tumor-Suppressor Genes and Oncogenes Explain why mutations in tumor-suppressor genes are recessive (both copies of the gene must be defective for the regulation of cell division to be defective) whereas mutations in oncogenes are dominant.

24. Retinoblastoma in Children Explain why some children with retinoblastoma develop multiple tumors of the retina in both eyes, whereas others have a single tumor in only one eye.

25. Specificity of a Signal for a Single Cell Type Discuss the validity of the following proposition. A signaling molecule (hormone, growth factor, or neurotransmitter) elicits identical responses in different types of target cells if they contain identical receptors.

Data Analysis Problem

26. Exploring Taste Sensation in Mice Figure 12–42 shows the signal-transduction pathway for sweet taste in mammals. Pleasing tastes are an evolutionary adaptation to encourage animals to consume nutritious foods. Zhao and coauthors (2003) examined the two major pleasurable taste sensations: sweet and umami. Umami is a "distinct savory taste" triggered by amino acids, especially aspartate and glutamate, and probably encourages animals to consume protein-rich foods. Monosodium glutamate (MSG) is a flavor enhancer that exploits this sensitivity.

At the time the article was published, specific taste receptor proteins (labeled SR in Fig. 12–42) for sweet and umami had been tentatively characterized. Three such proteins were known—T1R1, T1R2, and T1R3—which function as heterodimeric receptor complexes: T1R1-T1R3 was tentatively identified as the umami receptor, and T1R2-T1R3 as the sweet receptor. It was not clear how taste sensation was encoded and sent to the brain, and two possible models had been suggested. In the cell-based model, individual taste-sensing cells express only one kind of receptor; that is, there are "sweet cells," "bitter cells," "umami cells," and so on, and each type of cell sends its information to the brain via a different nerve. The brain "knows" which taste is detected by the identity of the nerve fiber that transmits the message. In the receptor-based model, individual taste-sensing cells have several kinds of receptors and send different messages along the same nerve fiber to the brain, the message depending on which receptor is activated. Also unclear at the time was whether there was any interaction between the different taste sensations, or whether parts of one taste-sensing system were required for other taste sensations.

(a) Previous work had shown that different taste receptor proteins are expressed in nonoverlapping sets of taste receptor cells. Which model does this support? Explain your reasoning.

Zhao and colleagues constructed a set of "knockout mice"—mice homozygous for loss-of-function alleles for one of the three receptor proteins, T1R1, T1R2, or T1R3—and double-knockout mice with nonfunctioning T1R2 and T1R3. The researchers measured the taste perception of these mice by measuring their "lick rate" of solutions containing different taste molecules. Mice will lick the spout of a feeding bottle with a pleasant-tasting solution more often than one with an unpleasant-tasting solution. The researchers measured relative lick rates: how often the mice licked a sample solution compared with water. A relative lick rate of 1 indicated no preference; <1, an aversion; and >1, a preference.

(b) All four types of knockout strains had the same responses to salt and bitter tastes as did wild-type mice. Which of the above issues did this experiment address? What do you conclude from these results?

The researchers then studied umami taste reception by measuring the relative lick rates of the different mouse strains with different quantities of MSG in the feeding solution. Note that the solutions also contained inosine monophosphate (IMP), a strong potentiator of umami taste reception (and a common ingredient in ramen soups, along with MSG), and ameloride, which suppresses the pleasant salty taste imparted by the sodium of MSG. The results are shown in the graph.

MSG + IMP + ameloride (mM)

(c) Are these data consistent with the umami taste receptor consisting of a heterodimer of T1R1 and T1R3? Why or why not?

(d) Which model(s) of taste encoding does this result support? Explain your reasoning.

Zhao and coworkers then performed a series of similar experiments using sucrose as a sweet taste. These results are shown below.

Sucrose (mM)

(e) Are these data consistent with the sweet taste receptor consisting of a heterodimer of T1R2 and T1R3? Why or why not?

(f) There were some unexpected responses at very high sucrose concentrations. How do these complicate the idea of a heterodimeric system as presented above?

In addition to sugars, humans also taste other compounds (e.g., the peptides monellin and aspartame) as sweet; mice do not taste these as sweet. Zhao and coworkers inserted into T1R2-knockout mice a copy of the human T1R2 gene under the control of the mouse T1R2 promoter. These modified mice now tasted monellin and saccharin as sweet. The researchers then went further, adding to T1R1-knockout mice the RASSL protein—a G protein–linked receptor for the synthetic opiate spiradoline; the RASSL gene was under the control of a promoter that could be induced by feeding the mice tetracycline. These mice did not prefer spiradoline in the absence of tetracycline; in the presence of tetracycline, they showed a strong preference for nanomolar concentrations of spiradoline.

(g) How do these results strengthen Zhao and coauthors' conclusions about the mechanism of taste sensation?

Reference

Zhao, G.Q., Zhang, Y., Hoon, M.A., Chandrashekar, J., Erlenbach, I., Ryba, N.J.P., & Zuker, C. (2003) The receptors for mammalian sweet and umami taste. *Cell* **115,** 255–266.

PART II

BIOENERGETICS AND METABOLISM

Metabolism is a highly coordinated cellular activity in which many multienzyme systems (metabolic pathways) cooperate to (1) obtain chemical energy by capturing solar energy or degrading energy-rich nutrients from the environment; (2) convert nutrient molecules into the cell's own characteristic molecules, including precursors of macromolecules; (3) polymerize monomeric precursors into macromolecules: proteins, nucleic acids, and polysaccharides; and (4) synthesize and degrade biomolecules required for specialized cellular functions, such as membrane lipids, intracellular messengers, and pigments.

Although metabolism embraces hundreds of different enzyme-catalyzed reactions, our major concern in Part II is the central metabolic pathways, which are few in number and remarkably similar in all forms of life. Living organisms can be divided into two large groups according to the chemical form in which they obtain carbon from the environment. **Autotrophs** (such as photosynthetic bacteria, green algae, and vascular plants) can use carbon dioxide from the atmosphere as their sole source of carbon, from which they construct all their carbon-containing biomolecules (see Fig. 1–5). Some autotrophic organisms, such as cyanobacteria, can also use atmospheric nitrogen to generate all their nitrogenous components. **Heterotrophs** cannot use atmospheric carbon dioxide and must obtain carbon from their environment in the form of relatively complex organic molecules such as glucose. Multicellular animals and most microorganisms are heterotrophic. Autotrophic cells and organisms are relatively self-sufficient, whereas heterotrophic cells and organisms, with their requirements for carbon in more complex forms, must subsist on the products of other organisms.

Many autotrophic organisms are photosynthetic and obtain their energy from sunlight, whereas heterotrophic organisms obtain their energy from the degradation of organic nutrients produced by autotrophs. In our biosphere, autotrophs and heterotrophs live together in a vast, interdependent cycle in which autotrophic organisms use atmospheric carbon dioxide to build their organic biomolecules, some of them generating

501

oxygen from water in the process. Heterotrophs in turn use the organic products of autotrophs as nutrients and return carbon dioxide to the atmosphere. Some of the oxidation reactions that produce carbon dioxide also consume oxygen, converting it to water. Thus carbon, oxygen, and water are constantly cycled between the heterotrophic and autotrophic worlds, with solar energy as the driving force for this global process **(Fig. 1)**.

All living organisms also require a source of nitrogen, which is necessary for the synthesis of amino acids, nucleotides, and other compounds. Bacteria and plants can generally use either ammonia or nitrate as their sole source of nitrogen, but vertebrates must obtain nitrogen in the form of amino acids or other organic compounds. Only a few organisms—the cyanobacteria and many species of soil bacteria that live symbiotically on the roots of some plants—are capable of converting ("fixing") atmospheric nitrogen (N_2) into ammonia. Other bacteria (the nitrifying bacteria) oxidize ammonia to nitrites and nitrates; yet others convert nitrate to N_2. The anammox bacteria convert ammonia and nitrite to N_2. Thus, in addition to the global carbon and oxygen cycles, a nitrogen cycle operates in the biosphere, turning over huge amounts of nitrogen **(Fig. 2)**. The cycling of carbon, oxygen, and nitrogen, which ultimately involves all species, depends on a proper balance between the activities of the producers (autotrophs) and consumers (heterotrophs) in our biosphere.

These cycles of matter are driven by an enormous flow of energy into and through the biosphere, beginning with the capture of solar energy by photosynthetic organisms and use of this energy to generate energy-rich carbohydrates and other organic nutrients; these nutrients are then used as energy sources by

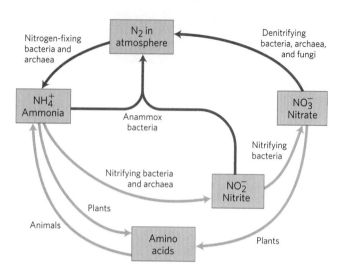

FIGURE 2 Cycling of nitrogen in the biosphere. Gaseous nitrogen (N_2) makes up 80% of the earth's atmosphere.

heterotrophic organisms. In metabolic processes, and in all energy transformations, there is a loss of useful energy (free energy) and an inevitable increase in the amount of unusable energy (heat and entropy). In contrast to the cycling of matter, therefore, energy flows one way through the biosphere; organisms cannot regenerate useful energy from energy dissipated as heat and entropy. Carbon, oxygen, and nitrogen recycle continuously, but energy is constantly transformed into unusable forms such as heat.

Metabolism, the sum of all the chemical transformations taking place in a cell or organism, occurs through a series of enzyme-catalyzed reactions that constitute **metabolic pathways**. Each of the consecutive steps in a metabolic pathway brings about a specific, small chemical change, usually the removal, transfer, or addition of a particular atom or functional group. The precursor is converted into a product through a series of metabolic intermediates called **metabolites**. The term **intermediary metabolism** is often applied to the combined activities of all the metabolic pathways that interconvert precursors, metabolites, and products of low molecular weight (generally, $M_r < 1,000$).

Catabolism is the degradative phase of metabolism in which organic nutrient molecules (carbohydrates, fats, and proteins) are converted into smaller, simpler end products (such as lactic acid, CO_2, and NH_3). Catabolic pathways release energy, some of which is conserved in the formation of ATP and reduced electron carriers (NADH, NADPH, and $FADH_2$); the rest is lost as heat. In **anabolism**, also called biosynthesis, small, simple precursors are built up into larger and more complex molecules, including lipids, polysaccharides, proteins,

FIGURE 1 Cycling of carbon dioxide and oxygen between the autotrophic (photosynthetic) and heterotrophic domains in the biosphere. The flow of mass through this cycle is enormous; about 4×10^{11} metric tons of carbon are turned over in the biosphere annually.

and nucleic acids. Anabolic reactions require an input of energy, generally in the form of the phosphoryl group transfer potential of ATP and the reducing power of NADH, NADPH, and $FADH_2$ **(Fig. 3)**.

Some metabolic pathways are linear, and some are branched, yielding multiple useful end products from a single precursor or converting several starting materials into a single product. In general, catabolic pathways are *convergent* and anabolic pathways *divergent* **(Fig. 4)**. Some pathways are cyclic: one starting component of the pathway is regenerated in a series of reactions that converts another starting component into a product. We shall see examples of each type of pathway in the following chapters.

Most cells have the enzymes to carry out both the degradation and the synthesis of the important categories of biomolecules—fatty acids, for example. The simultaneous synthesis and degradation of fatty acids would be wasteful, however, and this is prevented by reciprocally regulating the anabolic and catabolic

reaction sequences: when one sequence is active, the other is suppressed. Such regulation could not occur if anabolic and catabolic pathways were catalyzed by exactly the same set of enzymes, operating in one direction for anabolism, the opposite direction for catabolism: inhibition of an enzyme involved in catabolism would also inhibit the reaction sequence in the anabolic direction. Catabolic and anabolic pathways that connect the same two end points (glucose $\rightarrow \rightarrow$ pyruvate, and pyruvate $\rightarrow \rightarrow$ glucose, for example) may employ many of the same enzymes, but invariably at least one of the steps is catalyzed by different enzymes in the catabolic and anabolic directions, and these enzymes are the sites of separate regulation. Moreover, for both anabolic and catabolic pathways to be essentially irreversible, the reactions unique to each direction must include at least one that is thermodynamically very favorable—in other words, a reaction for which the reverse reaction is very unfavorable. As a further contribution to the separate regulation of catabolic and anabolic reaction sequences, paired catabolic and anabolic pathways commonly take place in different cellular compartments: for example, fatty acid catabolism in mitochondria, fatty acid synthesis in the cytosol. The concentrations of intermediates, enzymes, and regulators can be maintained at different levels in these different compartments. Because metabolic pathways are subject to kinetic control by substrate concentration, separate pools of anabolic and catabolic intermediates also contribute to the control of metabolic rates. Devices that separate anabolic and catabolic processes will be of particular interest in our discussions of metabolism.

Metabolic pathways are regulated at several levels, from within the cell and from outside. The most immediate regulation is by the availability of substrate; when the intracellular concentration of an enzyme's substrate is near or below K_m (as is commonly the case), the rate of the reaction depends strongly on substrate concentration (see Fig. 6–11). A second type of rapid control from within is allosteric regulation (p. 226) by a metabolic intermediate or coenzyme—an amino acid or ATP, for example—that signals the cell's internal metabolic state. When the cell contains an amount of, say, aspartate sufficient for its immediate needs or when the cellular level of ATP indicates that further fuel consumption is unnecessary at the moment, these signals allosterically inhibit the activity of one or more enzymes in the relevant pathway. In multicellular organisms, the metabolic activities of different tissues are regulated and integrated by growth factors and hormones that act from outside

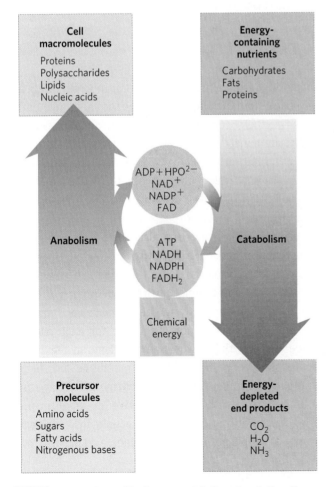

FIGURE 3 Energy relationships between catabolic and anabolic pathways. Catabolic pathways deliver chemical energy in the form of ATP, NADH, NADPH, and $FADH_2$. These energy carriers are used in anabolic pathways to convert small precursor molecules into cellular macromolecules.

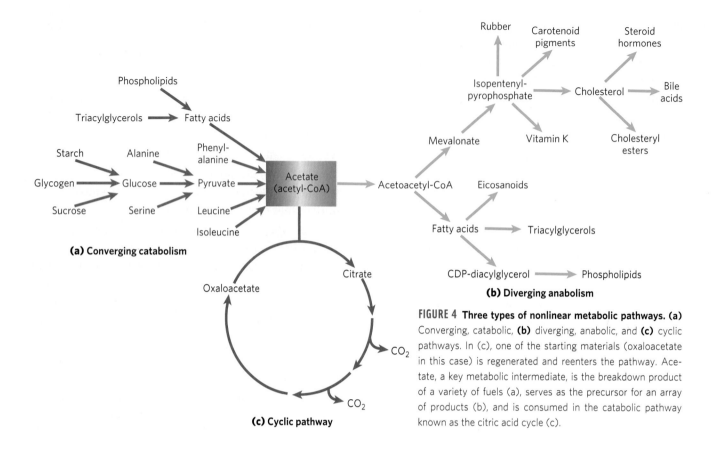

FIGURE 4 Three types of nonlinear metabolic pathways. (a) Converging, catabolic, **(b)** diverging, anabolic, and **(c)** cyclic pathways. In (c), one of the starting materials (oxaloacetate in this case) is regenerated and reenters the pathway. Acetate, a key metabolic intermediate, is the breakdown product of a variety of fuels (a), serves as the precursor for an array of products (b), and is consumed in the catabolic pathway known as the citric acid cycle (c).

the cell. In some cases, this regulation occurs virtually instantaneously (sometimes in less than a millisecond) through changes in the levels of intracellular messengers that modify the activity of existing enzyme molecules by allosteric mechanisms or by covalent modification such as phosphorylation. In other cases, the extracellular signal changes the cellular concentration of an enzyme by altering the rate of its synthesis or degradation, so the effect is seen only after minutes or hours.

We begin Part II with a discussion of the basic energetic principles that govern all metabolism (Chapter 13). We then consider the major catabolic pathways by which cells obtain energy from the oxidation of various fuels (Chapters 14 through 19). Chapter 19 is the pivotal point of our discussion of metabolism; it concerns chemiosmotic energy coupling, a universal mechanism in which a transmembrane electrochemical potential, produced either by substrate oxidation or by light absorption, drives the synthesis of ATP.

Chapters 20 through 22 describe the major anabolic pathways by which cells use the energy in ATP to produce carbohydrates, lipids, amino acids, and nucleotides from simpler precursors. In Chapter 23 we step back from our detailed look at the metabolic pathways—as they occur in all organisms, from *Escherichia coli* to humans—and consider how they are regulated and integrated in mammals by hormonal mechanisms.

As we undertake our study of intermediary metabolism, a final word. Keep in mind that the myriad reactions described in these pages take place in, and play crucial roles in, living organisms. As you encounter each reaction and each pathway, ask, What does this chemical transformation do for the organism? How does this pathway interconnect with the other pathways operating simultaneously in the same cell to produce the energy and products required for cell maintenance and growth? How do the multilayered regulatory mechanisms cooperate to balance metabolic and energy inputs and outputs, achieving the dynamic steady state of life? Studied with this perspective, metabolism provides fascinating and revealing insights into life, with countless applications in medicine, agriculture, and biotechnology.

Bioenergetics and Biochemical Reaction Types

iving cells and organisms must perform work to stay alive, to grow, and to reproduce. The ability to harness energy and to channel it into biological work is a fundamental property of all living organisms; it must have been acquired very early in cellular evolution. Modern organisms carry out a remarkable variety of energy transductions, conversions of one form of energy to another. They use the chemical energy in fuels to bring about the synthesis of complex, highly ordered macromolecules from simple precursors. They also convert the chemical energy of fuels into concentration gradients and electrical gradients, into motion and heat, and, in a few organisms such as fireflies and deep-sea fish, into light. Photosynthetic organisms transduce light energy into all these other forms of energy.

The chemical mechanisms that underlie biological energy transductions have fascinated and challenged biologists for centuries. The French chemist Antoine Lavoisier recognized that animals somehow transform chemical fuels (foods) into heat and that this process of respiration is essential to life. He observed that

. . . in general, respiration is nothing but a slow combustion of carbon and hydrogen, which is entirely similar to that which occurs in a lighted lamp or

Antoine Lavoisier, 1743-1794

candle, and that, from this point of view, animals that respire are true combustible bodies that burn and consume themselves. . . . One may say that this analogy between combustion and respiration has not escaped the notice of the poets, or rather the philosophers of antiquity, and which they had expounded and interpreted. This fire stolen from heaven, this torch of Prometheus, does not only represent an ingenious and poetic idea, it is a faithful picture of the operations of nature, at least for animals that breathe; one may therefore say, with the ancients, that the torch of life lights itself at the moment the infant breathes for the first time, and it does not extinguish itself except at death.*

In the twentieth century, we began to understand much of the chemistry underlying that "torch of life." Biological energy transductions obey the same chemical and physical laws that govern all other natural processes. It is therefore essential for a student of biochemistry to understand these laws and how they apply to the flow of energy in the biosphere.

In this chapter we first review the laws of thermodynamics and the quantitative relationships among free energy, enthalpy, and entropy. We then review the common types of biochemical reactions that occur in living cells, reactions that harness, store, transfer, and release the energy taken up by organisms from their surroundings. Our focus then shifts to reactions that have special roles in biological energy exchanges, particularly those involving ATP. We finish by considering the importance of oxidation-reduction reactions in living cells, the energetics of biological electron transfers, and the electron carriers commonly employed as cofactors in these processes.

*From a memoir by Armand Seguin and Antoine Lavoisier, dated 1789, quoted in Lavoisier, A. (1862) *Oeuvres de Lavoisier*, Imprimerie Impériale, Paris.

13.1 Bioenergetics and Thermodynamics

Bioenergetics is the quantitative study of **energy transductions**—changes of one form of energy into another—that occur in living cells, and of the nature and function of the chemical processes underlying these transductions. Although many of the principles of thermodynamics have been introduced in earlier chapters and may be familiar to you, a review of the quantitative aspects of these principles is useful here.

Biological Energy Transformations Obey the Laws of Thermodynamics

Many quantitative observations made by physicists and chemists on the interconversion of different forms of energy led, in the nineteenth century, to the formulation of two fundamental laws of thermodynamics. The first law is the principle of the conservation of energy: *for any physical or chemical change, the total amount of energy in the universe remains constant; energy may change form or it may be transported from one region to another, but it cannot be created or destroyed.* The second law of thermodynamics, which can be stated in several forms, says that the universe always tends toward increasing disorder: *in all natural processes, the entropy of the universe increases.*

"Now, in the *second* law of thermodynamics . . ."

Living organisms consist of collections of molecules much more highly organized than the surrounding materials from which they are constructed, and organisms maintain and produce order, seemingly immune to the second law of thermodynamics. But living organisms do not violate the second law; they operate strictly within it. To discuss the application of the second law to biological systems, we must first define those systems and their surroundings.

The reacting system is the collection of matter that is undergoing a particular chemical or physical process; it may be an organism, a cell, or two reacting compounds. The reacting system and its surroundings together constitute the universe. In the laboratory, some chemical or physical processes can be carried out in isolated or closed systems, in which no material or energy is exchanged with the surroundings. Living cells and organisms, however, are open systems, exchanging both material and energy with their surroundings; living systems are never at equilibrium with their surroundings, and the constant transactions between system and surroundings explain how organisms can create order within themselves while operating within the second law of thermodynamics.

In Chapter 1 (p. 23) we defined three thermodynamic quantities that describe the energy changes occurring in a chemical reaction:

Gibbs free energy, **G**, expresses the amount of an energy capable of doing work during a reaction at constant temperature and pressure. When a reaction proceeds with the release of free energy (that is, when the system changes so as to possess less free energy), the free-energy change, ΔG, has a negative value and the reaction is said to be exergonic. In endergonic reactions, the system gains free energy and ΔG is positive.

Enthalpy, **H**, is the heat content of the reacting system. It reflects the number and kinds of chemical bonds in the reactants and products. When a chemical reaction releases heat, it is said to be exothermic; the heat content of the products is less than that of the reactants and ΔH has, by convention, a negative value. Reacting systems that take up heat from their surroundings are endothermic and have positive values of ΔH.

Entropy, **S**, is a quantitative expression for the randomness or disorder in a system (see Box 1–3). When the products of a reaction are less complex and more disordered than the reactants, the reaction is said to proceed with a gain in entropy.

The units of ΔG and ΔH are joules/mole or calories/mole (recall that 1 cal = 4.184 J); units of entropy are joules/mole·Kelvin (J/mol·K) (Table 13–1).

Under the conditions existing in biological systems (including constant temperature and pressure), changes in free energy, enthalpy, and entropy are related to each other quantitatively by the equation

$$\Delta G = \Delta H - T\,\Delta S \tag{13–1}$$

in which ΔG is the change in Gibbs free energy of the reacting system, ΔH is the change in enthalpy of the system, T is the absolute temperature, and ΔS is the change in entropy of the system. By convention, ΔS has a positive sign when entropy increases and ΔH, as noted above, has a negative sign when heat is released by the system to its surroundings. Either of these conditions,

TABLE 13–1	Some Physical Constants and Units Used in Thermodynamics

Boltzmann constant, $\mathbf{k} = 1.381 \times 10^{-23}$ J/K

Avogadro's number, $N = 6.022 \times 10^{23}$ mol^{-1}

Faraday constant, $\mathcal{F} = 96{,}480$ J/V·mol

Gas constant, $R = 8.315$ J/mol·K

($= 1.987$ cal/mol·K)

Units of ΔG and ΔH are J/mol (or cal/mol)

Units of ΔS are J/mol·K (or cal/mol·K)

1 cal $= 4.184$ J

Units of absolute temperature, T, are Kelvin, K

25°C = 298 K

At 25°C, $RT = 2.478$ kJ/mol

($= 0.592$ kcal/mol)

which are typical of energetically favorable processes, tend to make ΔG negative. In fact, ΔG of a spontaneously reacting system is always negative.

The second law of thermodynamics states that the entropy *of the universe* increases during all chemical and physical processes, but it does not require that the entropy increase take place *in the reacting system* itself. The order produced within cells as they grow and divide is more than compensated for by the disorder they create in their surroundings in the course of growth and division (see Box 1–3, case 2). In short, living organisms preserve their internal order by taking from the surroundings free energy in the form of nutrients or sunlight, and returning to their surroundings an equal amount of energy as heat and entropy.

Cells Require Sources of Free Energy

Cells are isothermal systems—they function at essentially constant temperature (and also function at constant pressure). Heat flow is not a source of energy for cells, because heat can do work only as it passes to a zone or object at a lower temperature. The energy that cells can and must use is free energy, described by the Gibbs free-energy function G, which allows prediction of the direction of chemical reactions, their exact equilibrium position, and the amount of work they can (in theory) perform at constant temperature and pressure. Heterotrophic cells acquire free energy from nutrient molecules, and photosynthetic cells acquire it from absorbed solar radiation. Both kinds of cells transform this free energy into ATP and other energy-rich compounds capable of providing energy for biological work at constant temperature.

Standard Free-Energy Change Is Directly Related to the Equilibrium Constant

The composition of a reacting system (a mixture of chemical reactants and products) tends to continue changing until equilibrium is reached. At the equilibrium concentration of reactants and products, the rates of the forward and reverse reactions are exactly equal and no further net change occurs in the system. The concentrations of reactants and products *at equilibrium* define the equilibrium constant, K_{eq} (p. 25). In the general reaction aA + bB $\rightleftharpoons$ cC + dD, where a, b, c, and d are the number of molecules of A, B, C, and D participating, the equilibrium constant is given by

$$K_{eq} = \frac{[\text{C}]^{\text{c}}[\text{D}]^{\text{d}}}{[\text{A}]^{\text{a}}[\text{B}]^{\text{b}}} \qquad (13-2)$$

where [A], [B], [C], and [D] are the molar concentrations of the reaction components at the point of equilibrium.

When a reacting system is not at equilibrium, the tendency to move toward equilibrium represents a driving force, the magnitude of which can be expressed as the free-energy change for the reaction, ΔG. Under standard conditions (298 K = 25°C), when reactants and products are initially present at 1 M concentrations or, for gases, at partial pressures of 101.3 kilopascals (kPa), or 1 atm, the force driving the system toward equilibrium is defined as the standard free-energy change, $\Delta G°$. By this definition, the standard state for reactions that involve hydrogen ions is [H$^+$] = 1 M, or pH 0. Most biochemical reactions, however, occur in well-buffered aqueous solutions near pH 7; both the pH and the concentration of water (55.5 M) are essentially constant.

KEY CONVENTION: For convenience of calculations, biochemists define a standard state different from that used in chemistry and physics: in the biochemical standard state, [H$^+$] is 10^{-7} M (pH 7) and [H$_2$O] is 55.5 M. For reactions that involve Mg^{2+} (which include most of those with ATP as a reactant), [Mg^{2+}] in solution is commonly taken to be constant at 1 mM. ■

Physical constants based on this biochemical standard state are called **standard transformed constants** and are written with a prime (such as $\Delta G'°$ and K'_{eq}) to distinguish them from the untransformed constants used by chemists and physicists. (Note that most other textbooks use the symbol $\Delta G°'$ rather than $\Delta G'°$. Our use of $\Delta G'°$, recommended by an international committee of chemists and biochemists, is intended to emphasize that the transformed free energy, $\Delta G'$, is the criterion for equilibrium.) For simplicity, we will hereafter refer to these transformed constants as **standard free-energy changes**.

KEY CONVENTION: In another simplifying convention used by biochemists, when H$_2$O, H$^+$, and/or Mg^{2+} are reactants or products, their concentrations are not included in equations such as Equation 13–2 but are instead incorporated into the constants K'_{eq} and $\Delta G'°$. ■

Just as K'_{eq} is a physical constant characteristic for each reaction, so too is $\Delta G'°$ a constant. As we noted in

Chapter 6, there is a simple relationship between K'_{eq} and $\Delta G'^{\circ}$:

$$\Delta G'^{\circ} = -RT \ln K'_{eq} \qquad (13\text{–}3)$$

The standard free-energy change of a chemical reaction is simply an alternative mathematical way of expressing its equilibrium constant. Table 13–2 shows the relationship between $\Delta G'^{\circ}$ and K'_{eq}. If the equilibrium constant for a given chemical reaction is 1.0, the standard free-energy change of that reaction is 0.0 (the natural logarithm of 1.0 is zero). If K'_{eq} of a reaction is greater than 1.0, its $\Delta G'^{\circ}$ is negative. If K'_{eq} is less than 1.0, $\Delta G'^{\circ}$ is positive. Because the relationship between $\Delta G'^{\circ}$ and K'_{eq} is exponential, relatively small changes in $\Delta G'^{\circ}$ correspond to large changes in K'_{eq}.

It may be helpful to think of the standard free-energy change in another way. $\Delta G'^{\circ}$ is the difference between the free-energy content of the products and the free-energy content of the reactants, under standard conditions. When $\Delta G'^{\circ}$ is negative, the products contain less free energy than the reactants and the reaction will proceed spontaneously under standard conditions; all chemical reactions tend to go in the direction that results in a decrease in the free energy of the system. A positive value of $\Delta G'^{\circ}$ means that the products of the reaction contain more free energy than the reactants, and this reaction will tend to go in the reverse direction if we start with 1.0 M concentrations of all components (standard conditions). Table 13–3 summarizes these points.

TABLE 13–2 Relationship between Equilibrium Constants and Standard Free-Energy Changes of Chemical Reactions

K'_{eq}	$\Delta G'^{\circ}$ (kJ/mol)	$\Delta G'^{\circ}$ (kcal/mol)*
10^3	−17.1	−4.1
10^2	−11.4	−2.7
10^1	−5.7	−1.4
1	0.0	0.0
10^{-1}	5.7	1.4
10^{-2}	11.4	2.7
10^{-3}	17.1	4.1
10^{-4}	22.8	5.5
10^{-5}	28.5	6.8
10^{-6}	34.2	8.2

*Although joules and kilojoules are the standard units of energy and are used throughout this text, biochemists and nutritionists sometimes express $\Delta G'^{\circ}$ values in kilocalories per mole. We have therefore included values in both kilojoules and kilocalories in this table and in Tables 13–4 and 13–6. To convert kilojoules to kilocalories, divide the number of kilojoules by 4.184.

TABLE 13–3 Relationships among K'_{eq}, $\Delta G'^{\circ}$, and the Direction of Chemical Reactions

When K'_{eq} is ...	$\Delta G'^{\circ}$ is ...	Starting with all components at 1 M, the reaction ...
>1.0	negative	proceeds forward
1.0	zero	is at equilibrium
<1.0	positive	proceeds in reverse

WORKED EXAMPLE 13–1 Calculation of $\Delta G'^{\circ}$

Calculate the standard free-energy change of the reaction catalyzed by the enzyme phosphoglucomutase

$$\text{Glucose 1-phosphate} \rightleftharpoons \text{glucose 6-phosphate}$$

given that, starting with 20 mM glucose 1-phosphate and no glucose 6-phosphate, the final equilibrium mixture at 25°C and pH 7.0 contains 1.0 mM glucose 1-phosphate and 19 mM glucose 6-phosphate. Does the reaction in the direction of glucose 6-phosphate formation proceed with a loss or a gain of free energy?

Solution: First we calculate the equilibrium constant:

$$K'_{eq} = \frac{[\text{glucose 6-phosphate}]}{[\text{glucose 1-phosphate}]} = \frac{19 \text{ mM}}{1.0 \text{ mM}} = 19$$

We can now calculate the standard free-energy change:

$$\begin{aligned} \Delta G'^{\circ} &= -RT \ln K'_{eq} \\ &= -(8.315 \text{ J/mol} \cdot \text{K})(298 \text{ K})(\ln 19) \\ &= -7.3 \text{ kJ/mol} \end{aligned}$$

Because the standard free-energy change is negative, the conversion of glucose 1-phosphate to glucose 6-phosphate proceeds with a loss (release) of free energy. (For the reverse reaction, $\Delta G'^{\circ}$ has the same magnitude but the *opposite* sign.)

Table 13–4 gives the standard free-energy changes for some representative chemical reactions. Note that hydrolysis of simple esters, amides, peptides, and glycosides, as well as rearrangements and eliminations, proceed with relatively small standard free-energy changes, whereas hydrolysis of acid anhydrides is accompanied by relatively large decreases in standard free energy. The complete oxidation of organic compounds such as glucose or palmitate to CO_2 and H_2O, which in cells requires many steps, results in very large decreases in standard free energy. However, standard free-energy changes such as those in Table 13–4 indicate how much free energy is available from a reaction under *standard conditions.* To describe the energy released under the conditions existing in cells, an expression for the *actual* free-energy change is essential.

TABLE 13-4 Standard Free-Energy Changes of Some Chemical Reactions

Reaction type	$\Delta G'^{\circ}$ (kJ/mol)	$\Delta G'^{\circ}$ (kcal/mol)
Hydrolysis reactions		
Acid anhydrides		
Acetic anhydride + $H_2O \longrightarrow$ 2 acetate	−91.1	−21.8
ATP + $H_2O \longrightarrow$ ADP + P_i	−30.5	−7.3
ATP + $H_2O \longrightarrow$ AMP + PP_i	−45.6	−10.9
PP_i + $H_2O \longrightarrow 2P_i$	−19.2	−4.6
UDP-glucose + $H_2O \longrightarrow$ UMP + glucose 1-phosphate	−43.0	−10.3
Esters		
Ethyl acetate + $H_2O \longrightarrow$ ethanol + acetate	−19.6	−4.7
Glucose 6-phosphate + $H_2O \longrightarrow$ glucose + P_i	−13.8	−3.3
Amides and peptides		
Glutamine + $H_2O \longrightarrow$ glutamate + NH_4^+	−14.2	−3.4
Glycylglycine + $H_2O \longrightarrow$ 2 glycine	−9.2	−2.2
Glycosides		
Maltose + $H_2O \longrightarrow$ 2 glucose	−15.5	−3.7
Lactose + $H_2O \longrightarrow$ glucose + galactose	−15.9	−3.8
Rearrangements		
Glucose 1-phosphate $\longrightarrow$ glucose 6-phosphate	−7.3	−1.7
Fructose 6-phosphate $\longrightarrow$ glucose 6-phosphate	−1.7	−0.4
Elimination of water		
Malate $\longrightarrow$ fumarate + H_2O	3.1	0.8
Oxidations with molecular oxygen		
Glucose + $6O_2 \longrightarrow 6CO_2$ + $6H_2O$	−2,840	−686
Palmitate + $23O_2 \longrightarrow 16CO_2$ + $16H_2O$	−9,770	−2,338

Actual Free-Energy Changes Depend on Reactant and Product Concentrations

We must be careful to distinguish between two different quantities: the actual free-energy change, ΔG, and the standard free-energy change, $\Delta G'^{\circ}$. Each chemical reaction has a characteristic standard free-energy change, which may be positive, negative, or zero, depending on the equilibrium constant of the reaction. The standard free-energy change tells us in which direction and how far a given reaction must go to reach equilibrium *when the initial concentration of each component is 1.0 M, the pH is 7.0, the temperature is 25 °C, and the pressure is 101.3 kPa (1 atm)*. Thus $\Delta G'^{\circ}$ is a constant: it has a characteristic, unchanging value for a given reaction. But the *actual* free-energy change, ΔG, is a function of reactant and product concentrations and of the temperature prevailing during the reaction, none of which

will necessarily match the standard conditions as defined above. Moreover, the ΔG of any reaction proceeding spontaneously toward its equilibrium is always negative, becomes less negative as the reaction proceeds, and is zero at the point of equilibrium, indicating that no more work can be done by the reaction.

ΔG and $\Delta G'^{\circ}$ for any reaction aA + bB $\rightleftharpoons$ cC + dD are related by the equation

$$\Delta G = \Delta G'^{\circ} + RT \ln \frac{[C]^c[D]^d}{[A]^a[B]^b} \qquad (13\text{-}4)$$

in which the terms in red are those *actually prevailing* in the system under observation. The concentration terms in this equation express the effects commonly called mass action, and the term $[C]^c[D]^d/[A]^a[B]^b$ is called the **mass-action ratio**, Q. Thus Equation 13–4 can be expressed as $\Delta G = \Delta G'^{\circ} + RT \ln Q$. As an example, let us suppose that the reaction A + B $\rightleftharpoons$ C + D is taking place under the standard conditions of temperature (25 °C) and pressure (101.3 kPa) but that the concentrations of A, B, C, and D are *not* equal and none of the components is present at the standard concentration of 1.0 M. To determine the actual free-energy change, ΔG, under these nonstandard conditions of concentration as the reaction proceeds from left to right, we simply enter the *actual* concentrations of A, B, C, and D in Equation 13–4; the values of R, T, and $\Delta G'^\circ$ are the standard values. ΔG is negative and approaches zero as the reaction proceeds, because the actual concentrations of A and B decrease and the concentrations of C and D increase. Notice that when a reaction is at equilibrium—when there is no force driving the reaction in either direction and ΔG is zero—Equation 13–4 reduces to

$$0 = \Delta G = \Delta G'^\circ + RT \ln \frac{[C]_{eq}[D]_{eq}}{[A]_{eq}[B]_{eq}}$$

or

$$\Delta G'^\circ = -RT \ln K'_{eq}$$

which is the equation relating the standard free-energy change and equilibrium constant (Eqn 13–3).

The criterion for spontaneity of a reaction is the value of ΔG, not $\Delta G'^\circ$. A reaction with a positive $\Delta G'^\circ$ can go in the forward direction *if ΔG is negative*. This is possible if the term $RT \ln$ ([products]/[reactants]) in Equation 13–4 is negative and has a larger absolute value than $\Delta G'^\circ$. For example, the immediate removal of the products of a reaction can keep the ratio [products]/[reactants] well below 1, such that the term $RT \ln$ ([products]/[reactants]) has a large, negative value. $\Delta G'^\circ$ and ΔG are expressions of the *maximum* amount of free energy that a given reaction can *theoretically* deliver—an amount of energy that could be realized only if a perfectly efficient device were available to trap or harness it. Given that no such device is possible (some energy is always lost to entropy during any process), the amount of work done by the reaction at constant temperature and pressure is always less than the theoretical amount.

Another important point is that some thermodynamically favorable reactions (that is, reactions for which $\Delta G'^\circ$ is large and negative) do not occur at measurable rates. For example, combustion of firewood to CO_2 and H_2O is very favorable thermodynamically, but firewood remains stable for years because the activation energy (see Figs 6–2 and 6–3) for the combustion reaction is higher than the energy available at room temperature. If the necessary activation energy is provided (with a lighted match, for example), combustion will begin, converting the wood to the more stable products CO_2 and H_2O and releasing energy as heat and light. The heat released by this exothermic reaction provides the activation energy for combustion of neighboring regions of the firewood; the process is self-perpetuating.

In living cells, reactions that would be extremely slow *if uncatalyzed* are caused to proceed not by supplying additional heat but by lowering the activation energy through use of an enzyme. An enzyme provides an alternative reaction pathway with a lower activation energy than the uncatalyzed reaction, so that at room temperature a large fraction of the substrate molecules have enough thermal energy to overcome the activation barrier, and the reaction rate increases dramatically. *The free-energy change for a reaction is independent of the pathway by which the reaction occurs*; it depends only on the nature and concentration of the initial reactants and the final products. *Enzymes cannot, therefore, change equilibrium constants*; but they can and do increase the *rate* at which a reaction proceeds in the direction dictated by thermodynamics (see Section 6.2).

Standard Free-Energy Changes Are Additive

In the case of two sequential chemical reactions, A $\rightleftharpoons$ B and B $\rightleftharpoons$ C, each reaction has its own equilibrium constant and each has its characteristic standard free-energy change, $\Delta G_1'^\circ$ and $\Delta G_2'^\circ$. As the two reactions are sequential, B cancels out to give the overall reaction A $\rightleftharpoons$ C, which has its own equilibrium constant and thus its own standard free-energy change, $\Delta G_{total}'^\circ$. *The $\Delta G'^\circ$ values of sequential chemical reactions are additive.* For the overall reaction A $\rightleftharpoons$ C, $\Delta G_{total}'^\circ$ is the sum of the individual standard free-energy changes, $\Delta G_1'^\circ$ and $\Delta G_2'^\circ$, of the two reactions: $\Delta G_{total}'^\circ = \Delta G_1'^\circ + \Delta G_2'^\circ$.

(1)	A $\longrightarrow$ B	$\Delta G_1'^\circ$
(2)	B $\longrightarrow$ C	$\Delta G_2'^\circ$
Sum:	A $\longrightarrow$ C	$\Delta G_1'^\circ + \Delta G_2'^\circ$

This principle of bioenergetics explains how a thermodynamically unfavorable (endergonic) reaction can be driven in the forward direction by coupling it to a highly exergonic reaction through a common intermediate. For example, the synthesis of glucose 6-phosphate is the first step in the utilization of glucose by many organisms:

Glucose + P_i $\longrightarrow$ glucose 6-phosphate + H_2O
$$\Delta G'^\circ = 13.8 \text{ kJ/mol}$$

The positive value of $\Delta G'^\circ$ predicts that under standard conditions the reaction will tend not to proceed spontaneously in the direction written. Another cellular reaction, the hydrolysis of ATP to ADP and P_i, is very exergonic:

ATP + H_2O $\longrightarrow$ ADP + P_i $\Delta G'^\circ = -30.5 \text{ kJ/mol}$

These two reactions share the common intermediates P_i and H_2O and may be expressed as sequential reactions:

(1)	Glucose + P_i $\longrightarrow$ glucose 6-phosphate + H_2O
(2)	ATP + H_2O $\longrightarrow$ ADP + P_i
Sum:	ATP + glucose $\longrightarrow$ ADP + glucose 6-phosphate

The overall standard free-energy change is obtained by adding the $\Delta G'^{\circ}$ values for individual reactions:

$$\Delta G'^{\circ} = 13.8 \text{ kJ/mol} + (-30.5 \text{ kJ/mol}) = -16.7 \text{ kJ/mol}$$

The overall reaction is exergonic. In this case, energy stored in ATP is used to drive the synthesis of glucose 6-phosphate, even though its formation from glucose and inorganic phosphate (P_i) is endergonic. The *pathway* of glucose 6-phosphate formation from glucose by phosphoryl transfer from ATP is different from reactions (1) and (2) above, but the net result is the same as the sum of the two reactions. In thermodynamic calculations, all that matters is the state of the system at the beginning of the process and its state at the end; the route between the initial and final states is immaterial.

We have said that $\Delta G'^{\circ}$ is a way of expressing the equilibrium constant for a reaction. For reaction (1) above,

$$K'_{eq_1} = \frac{[\text{glucose 6-phosphate}]}{[\text{glucose}][P_i]} = 3.9 \times 10^{-3} \text{ M}^{-1}$$

Notice that H_2O is not included in this expression, as its concentration (55.5 M) is assumed to remain unchanged by the reaction. The equilibrium constant for the hydrolysis of ATP is

$$K'_{eq_2} = \frac{[\text{ADP}][P_i]}{[\text{ATP}]} = 2.0 \times 10^5 \text{ M}$$

The equilibrium constant for the two coupled reactions is

$$\begin{aligned} K'_{eq_3} &= \frac{[\text{glucose 6-phosphate}][\text{ADP}][P_i]}{[\text{glucose}][P_i][\text{ATP}]} \\ &= (K'_{eq_1})(K'_{eq_2}) = (3.9 \times 10^{-3} \text{ M}^{-1})(2.0 \times 10^5 \text{ M}) \\ &= 7.8 \times 10^2 \end{aligned}$$

This calculation illustrates an important point about equilibrium constants: although the $\Delta G'^{\circ}$ values for two reactions that sum to a third, overall reaction are *additive*, the K'_{eq} for the overall reaction is the *product* of the individual K'_{eq} values for the two reactions. Equilibrium constants are *multiplicative*. By coupling ATP hydrolysis to glucose 6-phosphate synthesis, the K'_{eq} for formation of glucose 6-phosphate from glucose has been raised by a factor of about 2×10^5.

This common-intermediate strategy is employed by all living cells in the synthesis of metabolic intermediates and cellular components. Obviously, the strategy works only if compounds such as ATP are continuously available. In the following chapters we consider several of the most important cellular pathways for producing ATP.

SUMMARY 13.1 Bioenergetics and Thermodynamics

▶ Living cells constantly perform work. They require energy for maintaining their highly organized structures, synthesizing cellular components, generating electric currents, and many other processes.

▶ Bioenergetics is the quantitative study of energy relationships and energy conversions in biological systems. Biological energy transformations obey the laws of thermodynamics.

▶ All chemical reactions are influenced by two forces: the tendency to achieve the most stable bonding state (for which enthalpy, H, is a useful expression) and the tendency to achieve the highest degree of randomness, expressed as entropy, S. The net driving force in a reaction is ΔG, the free-energy change, which represents the net effect of these two factors: $\Delta G = \Delta H - T\Delta S$.

▶ The standard transformed free-energy change, $\Delta G'^{\circ}$, is a physical constant that is characteristic for a given reaction and can be calculated from the equilibrium constant for the reaction: $\Delta G'^{\circ} = -RT \ln K'_{eq}$.

▶ The actual free-energy change, ΔG, is a variable that depends on $\Delta G'^{\circ}$ and on the concentrations of reactants and products: $\Delta G = \Delta G'^{\circ} + RT \ln ([\text{products}]/[\text{reactants}])$.

▶ When ΔG is large and negative, the reaction tends to go in the forward direction; when ΔG is large and positive, the reaction tends to go in the reverse direction; and when $\Delta G = 0$, the system is at equilibrium.

▶ The free-energy change for a reaction is independent of the pathway by which the reaction occurs. Free-energy changes are additive; the net chemical reaction that results from successive reactions sharing a common intermediate has an overall free-energy change that is the sum of the ΔG values for the individual reactions.

13.2 Chemical Logic and Common Biochemical Reactions

The biological energy transductions we are concerned with in this book are chemical reactions. Cellular chemistry does not encompass every kind of reaction learned in a typical organic chemistry course. Which reactions take place in biological systems and which do not is determined by (1) their relevance to that particular metabolic system and (2) their rates. Both considerations play major roles in shaping the metabolic pathways we consider throughout the rest of the book. A relevant reaction is one that makes use of an available substrate and converts it to a useful product. However, even a potentially relevant reaction may not occur. Some chemical transformations are too slow (have activation energies that are too high) to contribute to living systems even with the aid of powerful enzyme catalysts. The reactions that do occur in cells represent a toolbox that evolution has used to construct metabolic pathways that circumvent the "impossible" reactions. Learning to recognize the plausible reactions can be a great aid in developing a command of biochemistry.

Even so, the number of metabolic transformations taking place in a typical cell can seem overwhelming. Most cells have the capacity to carry out thousands of specific, enzyme-catalyzed reactions: for example, transformation of a simple nutrient such as glucose into amino acids, nucleotides, or lipids; extraction of energy from fuels by oxidation; and polymerization of monomeric subunits into macromolecules.

To study these reactions, some organization is essential. There are patterns within the chemistry of life; you do not need to learn every individual reaction to comprehend the molecular logic of biochemistry. Most of the reactions in living cells fall into one of five general categories: (1) reactions that make or break carbon–carbon bonds; (2) internal rearrangements, isomerizations, and eliminations; (3) free-radical reactions; (4) group transfers; and (5) oxidation-reductions. We discuss each of these in more detail below and refer to some examples of each type in later chapters. Note that the five reaction types are not mutually exclusive; for example, an isomerization reaction may involve a free-radical intermediate.

Before proceeding, however, we should review two basic chemical principles. First, a covalent bond consists of a shared pair of electrons, and the bond can be broken in two general ways **(Fig. 13–1)**. In **homolytic cleavage**, each atom leaves the bond as a **radical**, carrying one unpaired electron. In **heterolytic cleavage**,

FIGURE 13–1 Two mechanisms for cleavage of a C—C or C—H bond. In a homolytic cleavage, each atom keeps one of the bonding electrons, resulting in the formation of carbon radicals (carbons having unpaired electrons) or uncharged hydrogen atoms. In a heterolytic cleavage, one of the atoms retains both bonding electrons. This can result in the formation of carbanions, carbocations, protons, or hydride ions.

FIGURE 13–2 Common nucleophiles and electrophiles in biochemical reactions. Chemical reaction mechanisms, which trace the formation and breakage of covalent bonds, are communicated with dots and curved arrows, a convention known informally as "electron pushing." A covalent bond consists of a shared pair of electrons. Nonbonded electrons important to the reaction mechanism are designated by dots (:). Curved arrows (⌢) represent the movement of electron pairs. For movement of a single electron (as in a free radical reaction), a single-headed (fishhook-type) arrow is used (⌢). Most reaction steps involve an unshared electron pair.

which is more common, one atom retains both bonding electrons. The species most often generated when C—C and C—H bonds are cleaved are illustrated in Figure 13–1. Carbanions, carbocations, and hydride ions are highly unstable; this instability shapes the chemistry of these ions, as we shall see.

The second basic principle is that many biochemical reactions involve interactions between **nucleophiles** (functional groups rich in and capable of donating electrons) and **electrophiles** (electron-deficient functional groups that seek electrons). Nucleophiles combine with and give up electrons to electrophiles. Common biological nucleophiles and electrophiles are shown in **Figure 13–2**. Note that a carbon atom can act as either a nucleophile or an electrophile, depending on which bonds and functional groups surround it.

Reactions That Make or Break Carbon–Carbon Bonds Heterolytic cleavage of a C—C bond yields a **carbanion** and a **carbocation** (Fig. 13–1). Conversely, the formation of a C—C bond involves the combination of a nucleophilic carbanion and an electrophilic carbocation. Carbanions and carbocations are generally so unstable that their formation as reaction intermediates can be energetically inaccessible even with enzyme catalysts. For the purpose of cellular biochemistry they are impossible reactions—unless chemical assistance is provided in the form of functional groups containing electronegative atoms (O and N) that can alter the electronic structure of adjacent carbon atoms so as to stabilize and facilitate the formation of carbanion and carbocation intermediates.

Carbonyl groups are particularly important in the chemical transformations of metabolic pathways. The carbon of a carbonyl group has a partial positive charge due to the electron-withdrawing property of the carbonyl oxygen, and thus is an electrophilic carbon (**Fig. 13–3a**). A carbonyl group can thus facilitate the formation of a carbanion on an adjoining carbon by delocalizing the carbanion's negative charge (Fig. 13–3b). An imine group (see Fig. 1–16) can serve a similar function (Fig. 13–3c). The capacity of carbonyl and imine groups to delocalize electrons can be further enhanced by a general acid catalyst or by a metal ion such as Mg^{2+} (Fig. 13–3d).

The importance of a carbonyl group is evident in three major classes of reactions in which C—C bonds are formed or broken (**Fig. 13–4**): aldol condensations, Claisen ester condensations, and decarboxylations. In each type of reaction, a carbanion intermediate is stabilized by a carbonyl group, and in many cases another carbonyl provides the electrophile with which the nucleophilic carbanion reacts.

An **aldol condensation** is a common route to the formation of a C—C bond; the aldolase reaction, which converts a six-carbon compound to two three-carbon compounds in glycolysis, is an aldol condensation in reverse (see Fig. 14–6). In a **Claisen condensation**, the carbanion is stabilized by the carbonyl of an adjacent thioester; an example is the synthesis of citrate in the

Aldol condensation

Claisen ester condensation

Decarboxylation of a β-keto acid

FIGURE 13–4 Some common reactions that form and break C—C bonds in biological systems. For both the aldol condensation and the Claisen condensation, a carbanion serves as nucleophile and the carbon of a carbonyl group serves as electrophile. The carbanion is stabilized in each case by another carbonyl at the adjoining carbon. In the decarboxylation reaction, a carbanion is formed on the carbon shaded blue as the CO_2 leaves. The reaction would not occur at an appreciable rate without the stabilizing effect of the carbonyl adjacent to the carbanion carbon. Wherever a carbanion is shown, a stabilizing resonance with the adjacent carbonyl, as shown in Figure 13-3b, is assumed. An imine (Fig. 13-3c) or other electron-withdrawing group (including certain enzymatic cofactors such as pyridoxal) can replace the carbonyl group in the stabilization of carbanions.

citric acid cycle (see Fig. 16–9). Decarboxylation also commonly involves the formation of a carbanion stabilized by a carbonyl group; the acetoacetate decarboxylase reaction that occurs in the formation of ketone bodies during fatty acid catabolism provides an example (see Fig. 17–19). Entire metabolic pathways are organized around the introduction of a carbonyl group in a particular location so that a nearby carbon–carbon bond can be formed or cleaved. In some reactions, an imine or a specialized cofactor such as pyridoxal phosphate plays the electron-withdrawing role of the carbonyl group.

The carbocation intermediate occurring in some reactions that form or cleave C—C bonds is generated by the elimination of a very good leaving group, such as pyrophosphate (see Group Transfer Reactions below). An example is the prenyltransferase reaction (**Fig. 13–5**), an early step in the pathway of cholesterol biosynthesis.

Internal Rearrangements, Isomerizations, and Eliminations

Another common type of cellular reaction is an intramolecular rearrangement in which redistribution of electrons results in alterations of many different types without a change in the overall oxidation state of the molecule. For example, different groups in a molecule may undergo oxidation-reduction, with no net change in oxidation state of the molecule; groups at a double bond may undergo a cis-trans rearrangement; or the positions of double bonds may be transposed. An example of an isomerization entailing oxidation-reduction is the formation of

FIGURE 13–3 Chemical properties of carbonyl groups. (a) The carbon atom of a carbonyl group is an electrophile by virtue of the electron-withdrawing capacity of the electronegative oxygen atom, which results in a structure in which the carbon has a partial positive charge. **(b)** Within a molecule, delocalization of electrons into a carbonyl group stabilizes a carbanion on an adjacent carbon, facilitating its formation. **(c)** Imines function much like carbonyl groups in facilitating electron withdrawal. **(d)** Carbonyl groups do not always function alone; their capacity as electron sinks often is augmented by interaction with either a metal ion (Me^{2+}, such as Mg^{2+}) or a general acid (HA).

Dimethylallyl pyrophosphate

Isopentenyl → PP$_i$
pyrophosphate

Isopentenyl pyrophosphate   **Dimethylallylic carbocation**

→ H$^+$

Geranyl pyrophosphate

FIGURE 13–5 Carbocations in carbon-carbon bond formation. In one of the early steps in cholesterol biosynthesis, the enzyme prenyltransferase catalyzes condensation of isopentenyl pyrophosphate and dimethylallyl pyrophosphate to form geranyl pyrophosphate (see Fig. 21-36). The reaction is initiated by elimination of pyrophosphate from the dimethylallyl pyrophosphate to generate a carbocation, stabilized by resonance with the adjacent C=C bond.

fructose 6-phosphate from glucose 6-phosphate in glycolysis **(Fig. 13–6)**; this reaction is discussed in detail in Chapter 14): C-1 is reduced (aldehyde to alcohol) and C-2 is oxidized (alcohol to ketone). Figure 13–6b shows the details of the electron movements in this type of isomerization. A cis-trans rearrangement is illustrated by the prolyl cis-trans isomerase reaction in the folding of certain proteins (see Fig. 4–8). A simple transposition of a C=C bond occurs during metabolism of oleic acid, a common fatty acid (see Fig. 17–10). Some spectacular examples of double-bond repositioning occur in the biosynthesis of cholesterol (see Fig. 21–33).

An example of an elimination reaction that does not affect overall oxidation state is the loss of water from an alcohol, resulting in the introduction of a C=C bond:

Similar reactions can result from eliminations in amines.

Free-Radical Reactions Once thought to be rare, the homolytic cleavage of covalent bonds to generate free radicals has now been found in a wide range of biochemical processes. These include: isomerizations that make use of adenosylcobalamin (vitamin B$_{12}$) or S-adenosylmethionine, which are initiated with a 5'-deoxyadenosyl radical (see the methylmalonyl-CoA mutase reaction in Box 17–2); certain radical-initiated decarboxylation reactions **(Fig. 13–7)**; some reductase reactions, such as that catalyzed by ribonucleotide reductase (see Fig. 22–41);

(a)

Glucose 6-phosphate   phosphohexose isomerase   **Fructose 6-phosphate**

(b)

① B$_1$ abstracts a proton.

② This allows the formation of a C=C double bond.

③ Electrons from carbonyl form an O—H bond with the hydrogen ion donated by B$_2$.

⑤ An electron pair is displaced from the C=C bond to form a C—H bond with the proton donated by B$_1$.

④ B$_2$ abstracts a proton, allowing the formation of a C=O bond.

Enediol intermediate

FIGURE 13–6 Isomerization and elimination reactions. (a) The conversion of glucose 6-phosphate to fructose 6-phosphate, a reaction of sugar metabolism catalyzed by phosphohexose isomerase. **(b)** This reaction proceeds through an enediol intermediate. Light red screens

follow the path of oxidation from left to right. B$_1$ and B$_2$ are ionizable groups on the enzyme; they are capable of donating and accepting protons (acting as general acids or general bases) as the reaction proceeds.

Coproporphyrinogen III   Coproporphyrinogenyl   Protoporphyrinogen IX
                                 III radical

FIGURE 13–7 A free radical-initiated decarboxylation reaction. The biosynthesis of heme (see Fig. 22-26) in *Escherichia coli* includes a decarboxylation step in which propionyl side chains on the coproporphyrinogen III intermediate are converted to the vinyl side chains of protoporphyrinogen IX. When the bacteria are grown anaerobically the enzyme oxygen-independent coproporphyrinogen III oxidase, also called HemN protein, promotes decarboxylation via the free-radical mechanism shown here. The acceptor of the released electron is not known. For simplicity, only the relevant portions of the large coproporphyrinogen III and protoporphyrinogen molecules are shown; the entire structures are given in Figure 22-26. When *E. coli* are grown in the presence of oxygen, this reaction is an oxidative decarboxylation and is catalyzed by a different enzyme.

and some rearrangement reactions, such as that catalyzed by DNA photolyase (see Fig. 25–26).

Group Transfer Reactions The transfer of acyl, glycosyl, and phosphoryl groups from one nucleophile to another is common in living cells. Acyl group transfer generally involves the addition of a nucleophile to the carbonyl carbon of an acyl group to form a tetrahedral intermediate:

Tetrahedral
intermediate

The chymotrypsin reaction is one example of acyl group transfer (see Fig. 6–22). Glycosyl group transfers involve nucleophilic substitution at C-1 of a sugar ring, which is the central atom of an acetal. In principle, the substitution could proceed by an S_N1 or S_N2 pathway, as described in Figure 6–28 for the enzyme lysozyme.

Phosphoryl group transfers play a special role in metabolic pathways, and these transfer reactions are discussed in detail in Section 13.3. A general theme in metabolism is the attachment of a good leaving group to a metabolic intermediate to "activate" the intermediate for subsequent reaction. Among the better leaving groups in nucleophilic substitution reactions are inorganic orthophosphate (the ionized form of H_3PO_4 at neutral pH, a mixture of $H_2PO_4^-$ and HPO_4^{2-}, commonly abbreviated P_i) and inorganic pyrophosphate ($P_2O_7^{4-}$, abbreviated PP_i); esters and anhydrides of phosphoric acid are effectively activated for reaction. Nucleophilic substitution is made more favorable by the attachment of a phosphoryl group to an otherwise poor leaving group such as —OH. Nucleophilic substitutions in which the phosphoryl group ($-PO_3^{2-}$) serves as a leaving group occur in hundreds of metabolic reactions.

Phosphorus can form five covalent bonds. The conventional representation of P_i **(Fig. 13–8a)**, with three P—O bonds and one P=O bond, is a convenient

(a)

(b)

(c)

ATP

ADP   Glucose 6-phosphate,
             a phosphate ester

(d)

$Z = R—OH$
$W = ADP$

FIGURE 13–8 Phosphoryl group transfers: some of the participants.
(a) In one (inadequate) representation of P_i, three oxygens are single-bonded to phosphorus, and the fourth is double-bonded, allowing the four different resonance structures shown here. **(b)** The resonance structures of P_i can be represented more accurately by showing all four phosphorus–oxygen bonds with some double-bond character; the hybrid orbitals so represented are arranged in a tetrahedron with P at its center. **(c)** When a nucleophile Z (in this case, the —OH on C-6 of glucose) attacks ATP, it displaces ADP (W). In this S_N2 reaction, a pentacovalent intermediate **(d)** forms transiently.

but inaccurate picture. In P_i, four equivalent phosphorus–oxygen bonds share some double-bond character, and the anion has a tetrahedral structure (Fig. 13–8b). Because oxygen is more electronegative than phosphorus, the sharing of electrons is unequal: the central phosphorus bears a partial positive charge and can therefore act as an electrophile. In a great many metabolic reactions, a phosphoryl group ($-PO_3^{2-}$) is transferred from ATP to an alcohol, forming a phosphate ester (Fig. 13–8c), or to a carboxylic acid, forming a mixed anhydride. When a nucleophile attacks the electrophilic phosphorus atom in ATP, a relatively stable pentacovalent structure forms as a reaction intermediate (Fig. 13–8d). With departure of the leaving group (ADP), the transfer of a phosphoryl group is complete. The large family of enzymes that catalyze phosphoryl group transfers with ATP as donor are called **kinases** (Greek *kinein*, "to move"). Hexokinase, for example, "moves" a phosphoryl group from ATP to glucose.

Phosphoryl groups are not the only groups that activate molecules for reaction. Thioalcohols (thiols), in which the oxygen atom of an alcohol is replaced with a sulfur atom, are also good leaving groups. Thiols activate carboxylic acids by forming thioesters (thiol esters). In later chapters we discuss several reactions, including those catalyzed by the fatty acyl synthases in lipid synthesis (see Fig. 21–2), in which nucleophilic substitution at the carbonyl carbon of a thioester results in transfer of the acyl group to another moiety.

Oxidation-Reduction Reactions Carbon atoms can exist in five oxidation states, depending on the elements with which they share electrons **(Fig. 13–9)**, and transitions between these states are of crucial importance in metabolism (oxidation-reduction reactions are the topic of Section 13.4). In many biological oxidations, a

FIGURE 13–10 **An oxidation-reduction reaction.** Shown here is the oxidation of lactate to pyruvate. In this dehydrogenation, two electrons and two hydrogen ions (the equivalent of two hydrogen atoms) are removed from C-2 of lactate, an alcohol, to form pyruvate, a ketone. In cells the reaction is catalyzed by lactate dehydrogenase and the electrons are transferred to the cofactor nicotinamide adenine dinucleotide (NAD). This reaction is fully reversible; pyruvate can be reduced by electrons transferred from the cofactor.

compound loses two electrons and two hydrogen ions (that is, two hydrogen atoms); these reactions are commonly called dehydrogenations and the enzymes that catalyze them are called dehydrogenases **(Fig. 13–10)**. In some, but not all, biological oxidations, a carbon atom becomes covalently bonded to an oxygen atom. The enzymes that catalyze these oxidations are generally called oxidases or, if the oxygen atom is derived directly from molecular oxygen (O_2) oxygenases.

Every oxidation must be accompanied by a reduction, in which an electron acceptor acquires the electrons removed by oxidation. Oxidation reactions generally release energy (think of camp fires: the compounds in wood are oxidized by oxygen molecules in the air). Most living cells obtain the energy needed for cellular work by oxidizing metabolic fuels such as carbohydrates or fat (photosynthetic organisms can also trap and use the energy of sunlight). The catabolic (energy-yielding) pathways described in Chapters 14 through 19 are oxidative reaction sequences that result in the transfer of electrons from fuel molecules, through a series of electron carriers, to oxygen. The high affinity of O_2 for electrons makes the overall electron-transfer process highly exergonic, providing the energy that drives ATP synthesis—the central goal of catabolism.

Many of the reactions within these five classes are facilitated by cofactors, in the form of coenzymes and metals (vitamin B_{12}, *S*-adenosylmethionine, folate, nicotinamide, and iron are some examples). Cofactors bind to enzymes—in some cases reversibly, in other cases almost irreversibly—and confer on them the capacity to promote a particular kind of chemistry (p. 190). Most cofactors participate in a narrow range of closely related reactions. In the following chapters, we will introduce and discuss each important cofactor at the point where we first encounter it. The cofactors provide another way to organize the study of biochemical processes, since the reactions facilitated by a given cofactor generally are mechanistically related.

FIGURE 13–9 **The oxidation levels of carbon in biomolecules.** Each compound is formed by oxidation of the red carbon in the compound shown immediately above. Carbon dioxide is the most highly oxidized form of carbon found in living systems.

Biochemical and Chemical Equations Are Not Identical

Biochemists write metabolic equations in a simplified way, and this is particularly evident for reactions involving ATP. Phosphorylated compounds can exist in several ionization states and, as we have noted, the different species can bind Mg^{2+}. For example, at pH 7 and 2 mM Mg^{2+}, ATP exists in the forms ATP^{4-}, $HATP^{3-}$, H_2ATP^{2-}, $MgHATP^-$, and Mg_2ATP. In thinking about the biological role of ATP, however, we are not always interested in all this detail, and so we consider ATP as an entity made up of a sum of species, and we write its hydrolysis as the biochemical equation

$$ATP + H_2O \longrightarrow ADP + P_i$$

where ATP, ADP, and P_i are sums of species. The corresponding standard transformed equilibrium constant, $K'_{eq} = [ADP][P_i]/[ATP]$, depends on the pH and the concentration of free Mg^{2+}. Note that H^+ and Mg^{2+} do not appear in the biochemical equation because they are held constant. Thus a biochemical equation does not necessarily balance H, Mg, or charge, although it does balance all other elements involved in the reaction (C, N, O, and P in the equation above).

We can write a chemical equation that *does* balance for all elements and for charge. For example, when ATP is hydrolyzed at a pH above 8.5 in the absence of Mg^{2+}, the chemical reaction is represented by

$$ATP^{4-} + H_2O \longrightarrow ADP^{3-} + HPO_4^{2-} + H^+$$

The corresponding equilibrium constant, $K'_{eq} = [ADP^{3-}][HPO_4^{2-}][H^+]/[ATP^{4-}]$, depends only on temperature, pressure, and ionic strength.

Both ways of writing a metabolic reaction have value in biochemistry. Chemical equations are needed when we want to account for all atoms and charges in a reaction, as when we are considering the mechanism of a chemical reaction. Biochemical equations are used to determine in which direction a reaction will proceed spontaneously, given a specified pH and $[Mg^{2+}]$, or to calculate the equilibrium constant of such a reaction.

Throughout this book we use biochemical equations, unless the focus is on chemical mechanism, and we use values of $\Delta G'^{\circ}$ and K'_{eq} as determined at pH 7 and 1 mM Mg^{2+}.

SUMMARY 13.2 Chemical Logic and Common Biochemical Reactions

▶ Living systems make use of a large number of chemical reactions that can be classified into five general types.

▶ Carbonyl groups play a special role in reactions that form or cleave C—C bonds. Carbanion intermediates are common and are stabilized by adjacent carbonyl groups or, less often, by imines or certain cofactors.

▶ A redistribution of electrons can produce internal rearrangements, isomerizations, and eliminations. Such reactions include intramolecular oxidation-reduction, change in cis-trans arrangement at a double bond, and transposition of double bonds.

▶ Homolytic cleavage of covalent bonds to generate free radicals occurs in some pathways, such as in certain isomerization, decarboxylation, reductase, and rearrangement reactions.

▶ Phosphoryl transfer reactions are an especially important type of group transfer in cells, required for the activation of molecules for reactions that would otherwise be highly unfavorable.

▶ Oxidation-reduction reactions involve the loss or gain of electrons: one reactant gains electrons and is reduced, while the other loses electrons and is oxidized. Oxidation reactions generally release energy and are important in catabolism.

13.3 Phosphoryl Group Transfers and ATP

Having developed some fundamental principles of energy changes in chemical systems and reviewed the common classes of reactions, we can now examine the energy cycle in cells and the special role of ATP as the energy currency that links catabolism and anabolism (see Fig. 1–29). Heterotrophic cells obtain free energy in a chemical form by the catabolism of nutrient molecules, and they use that energy to make ATP from ADP and P_i. ATP then donates some of its chemical energy to endergonic processes such as the synthesis of metabolic intermediates and macromolecules from smaller precursors, the transport of substances across membranes against concentration gradients, and mechanical motion. This donation of energy from ATP generally involves the covalent participation of ATP in the reaction that is to be driven, with the eventual result that ATP is converted to ADP and P_i or, in some reactions, to AMP and 2 P_i. We discuss here the chemical basis for the large free-energy changes that accompany hydrolysis of ATP and other high-energy phosphate compounds, and we show that most cases of energy donation by ATP involve group transfer, not simple hydrolysis of ATP. To illustrate the range of energy transductions in which ATP provides the energy, we consider the synthesis of information-rich macromolecules, the transport of solutes across membranes, and motion produced by muscle contraction.

FIGURE 13–11 Chemical basis for the large free-energy change associated with ATP hydrolysis. ❶ The charge separation that results from hydrolysis relieves electrostatic repulsion among the four negative charges on ATP. ❷ The product inorganic phosphate (P_i) is stabilized by formation of a resonance hybrid, in which each of the four phosphorus-oxygen bonds has the same degree of double-bond character and the hydrogen ion is not permanently associated with any one of the oxygens. (Some degree of resonance stabilization also occurs in phosphates involved in ester or anhydride linkages, but fewer resonance forms are possible than for P_i.) A third factor (not shown) that favors ATP hydrolysis is the greater degree of solvation (hydration) of the products P_i and ADP relative to ATP, which further stabilizes the products relative to the reactants.

The Free-Energy Change for ATP Hydrolysis Is Large and Negative

Figure 13–11 summarizes the chemical basis for the relatively large, negative, standard free energy of hydrolysis of ATP. The hydrolytic cleavage of the terminal phosphoric acid anhydride (phosphoanhydride) bond in ATP separates one of the three negatively charged phosphates and thus relieves some of the electrostatic repulsion in ATP; the P_i released is stabilized by the formation of several resonance forms not possible in ATP.

The free-energy change for ATP hydrolysis is -30.5 kJ/mol under standard conditions, but the *actual* free energy of hydrolysis (ΔG) of ATP in living cells is very different: the cellular concentrations of ATP, ADP, and P_i are not identical and are much lower than the 1.0 M of standard conditions (Table 13–5). Furthermore, Mg^{2+} in the cytosol binds to ATP and ADP **(Fig. 13–12)**, and for most enzymatic reactions that involve ATP as phosphoryl group donor, the true substrate is $MgATP^{2-}$. The relevant $\Delta G'^{\circ}$ is therefore that for $MgATP^{2-}$ hydrolysis. We can calculate ΔG for ATP hydrolysis using data such as those in Table 13–5. The actual free energy of hydrolysis of ATP under intracellular conditions is often called its **phosphorylation potential**, ΔG_p.

TABLE 13–5 Adenine Nucleotide, Inorganic Phosphate, and Phosphocreatine Concentrations in Some Cells

	Concentration (mM)*				
	ATP	ADP[†]	AMP	P_i	PCr
Rat hepatocyte	3.38	1.32	0.29	4.8	0
Rat myocyte	8.05	0.93	0.04	8.05	28
Rat neuron	2.59	0.73	0.06	2.72	4.7
Human erythrocyte	2.25	0.25	0.02	1.65	0
E. coli cell	7.90	1.04	0.82	7.9	0

*For erythrocytes the concentrations are those of the cytosol (human erythrocytes lack a nucleus and mitochondria). In the other types of cells the data are for the entire cell contents, although the cytosol and the mitochondria have very different concentrations of ADP. PCr is phosphocreatine, discussed on p. 526.

[†]This value reflects total concentration; the true value for free ADP may be much lower (p. 519).

FIGURE 13–12 Mg^{2+} and ATP. Formation of Mg^{2+} complexes partially shields the negative charges and influences the conformation of the phosphate groups in nucleotides such as ATP and ADP.

WORKED EXAMPLE 13–2   **Calculation of ΔG_p**

Calculate the actual free energy of hydrolysis of ATP, ΔG_p, in human erythrocytes. The standard free energy of hydrolysis of ATP is -30.5 kJ/mol, and the concentrations of ATP, ADP, and P_i in erythrocytes are as shown in Table 13–5. Assume that the pH is 7.0 and the temperature is 37 °C (body temperature). What does this reveal about the amount of energy required to *synthesize* ATP under the same cellular conditions?

Solution: The concentrations of ATP, ADP, and P_i in human erythrocytes are 2.25, 0.25, and 1.65 mM, respectively. The actual free energy of hydrolysis of ATP under these conditions is given by the relationship (see Eqn 13–4)

$$\Delta G_p = \Delta G'^\circ + RT \ln \frac{[ADP][P_i]}{[ATP]}$$

Substituting the appropriate values we get

$$\Delta G_p = -30.5 \text{ kJ/mol} + \left[(8.315 \text{ J/mol} \cdot \text{K})(310 \text{ K}) \ln \frac{(0.25 \times 10^{-3})(1.65 \times 10^{-3})}{(2.25 \times 10^{-3})} \right]$$

$$= -30.5 \text{ kJ/mol} + (2.58 \text{ kJ/mol}) \ln 1.8 \times 10^{-4}$$

$$= -30.5 \text{ kJ/mol} + (2.58 \text{ kJ/mol})(-8.6)$$

$$= -30.5 \text{ kJ/mol} - 22 \text{ kJ/mol}$$

$$= -52 \text{ kJ/mol}$$

(Note that the final answer has been rounded to the correct number of significant figures (52.5 rounded to 52), following rules for rounding a number that ends in a 5 to the nearest even number.) Thus ΔG_p, the actual free-energy change for ATP hydrolysis in the intact erythrocyte (-52 kJ/mol), is much larger than the standard free-energy change (-30.5 kJ/mol). By the same token, the free energy required to *synthesize* ATP from ADP and P_i under the conditions prevailing in the erythrocyte would be 52 kJ/mol.

Because the concentrations of ATP, ADP, and P_i differ from one cell type to another, ΔG_p for ATP likewise differs among cells. Moreover, in any given cell, ΔG_p can vary from time to time, depending on the metabolic conditions and how they influence the concentrations of ATP, ADP, P_i, and H^+ (pH). We can calculate the actual free-energy change for any given metabolic reaction as it occurs in a cell, providing we know the concentrations of all the reactants and products and other factors (such as pH, temperature, and $[Mg^{2+}]$) that may affect the actual free-energy change.

To further complicate the issue, the *total* concentrations of ATP, ADP, P_i, and H^+ in a cell may be substantially higher than the *free* concentrations, which are the thermodynamically relevant values. The difference is due to tight binding of ATP, ADP, and P_i to cellular proteins. For example, the free [ADP] in resting muscle has been variously estimated at between 1 and 37 μM. Using the value 25 μM in Worked Example 13–2, we would get a ΔG_p of -64 kJ/mol. Calculation of the exact value of ΔG_p, however, is perhaps less instructive than the generalization we can make about actual free-energy changes: in vivo, the energy released by ATP hydrolysis is greater than the standard free-energy change, $\Delta G'^\circ$.

In the following discussions we use the $\Delta G'^\circ$ value for ATP hydrolysis because this allows comparison, on the same basis, with the energetics of other cellular reactions. Always keep in mind, however, that in living cells ΔG is the relevant quantity—for ATP hydrolysis and all other reactions—and may be quite different from $\Delta G'^\circ$.

Here we must make an important point about cellular ATP levels. We have shown (and will discuss further) how the chemical properties of ATP make it a suitable form of energy currency in cells. But it is not merely the molecule's intrinsic chemical properties that give it this ability to drive metabolic reactions and other energy-requiring processes. Even more important is that, in the course of evolution, there has been a very strong selective pressure for regulatory mechanisms that *hold cellular ATP concentrations far above the equilibrium concentrations* for the hydrolysis reaction. When the ATP level drops, not only does the *amount* of fuel decrease, but the fuel itself *loses its potency*: ΔG for its hydrolysis (that is, its phosphorylation potential, ΔG_p) is diminished. As our discussions of the metabolic pathways that produce and consume ATP will show, living cells have developed elaborate mechanisms—often at what might seem to us the expense of efficiency and common sense—to maintain high concentrations of ATP.

FIGURE 13-13 Hydrolysis of phosphoenolpyruvate (PEP). Catalyzed by pyruvate kinase, this reaction is followed by spontaneous tautomerization of the product, pyruvate. Tautomerization is not possible in PEP, and thus the products of hydrolysis are stabilized relative to the reactants. Resonance stabilization of P_i also occurs, as shown in Figure 13-11.

PEP

Pyruvate
(enol form)

Pyruvate
(keto form)

$$PEP^{3-} + H_2O \longrightarrow pyruvate^- + HPO_4^{2-}$$
$$\Delta G'^\circ = -61.9 \text{ kJ/mol}$$

Other Phosphorylated Compounds and Thioesters Also Have Large Free Energies of Hydrolysis

Phosphoenolpyruvate (PEP; **Fig. 13-13**) contains a phosphate ester bond that undergoes hydrolysis to yield the enol form of pyruvate, and this direct product can tautomerize to the more stable keto form. Because the reactant (PEP) has only one form (enol) and the product (pyruvate) has two possible forms, the product is stabilized relative to the reactant. This is the greatest contributing factor to the high standard free energy of hydrolysis of phosphoenolpyruvate: $\Delta G'^\circ = -61.9$ kJ/mol.

Another three-carbon compound, 1,3-bisphosphoglycerate **(Fig. 13-14)**, contains an anhydride bond between the C-1 carboxyl group and a number that ends in a phosphoric acid. Hydrolysis of this acyl phosphate is accompanied by a large, negative, standard free-energy change ($\Delta G'^\circ = -49.3$ kJ/mol), which

can, again, be explained in terms of the structure of reactant and products. When H_2O is added across the anhydride bond of 1,3-bisphosphoglycerate, one of the direct products, 3-phosphoglyceric acid, can lose a proton to give the carboxylate ion, 3-phosphoglycerate, which has two equally probable resonance forms (Fig. 13-14). Removal of the direct product (3-phosphoglyceric acid) and formation of the resonance-stabilized ion favor the forward reaction.

In phosphocreatine **(Fig. 13-15)**, the P—N bond can be hydrolyzed to generate free creatine and P_i. The release of P_i and the resonance stabilization of creatine favor the forward reaction. The standard free-energy change of phosphocreatine hydrolysis is again large, -43.0 kJ/mol.

In all these phosphate-releasing reactions, the several resonance forms available to P_i (Fig. 13-11) stabilize this product relative to the reactant, contributing to an already negative free-energy change. Table 13-6 lists

FIGURE 13-14 Hydrolysis of 1,3-bisphosphoglycerate. The direct product of hydrolysis is 3-phosphoglyceric acid, with an undissociated carboxylic acid. Its dissociation allows resonance structures that stabilize the product relative to the reactants. Resonance stabilization of P_i further contributes to the negative free-energy change.

1,3-Bisphosphoglycerate

3-Phosphoglyceric acid

3-Phosphoglycerate

$$\text{1,3-Bisphosphoglycerate}^{4-} + H_2O \longrightarrow \text{3-phosphoglycerate}^{3-} + HPO_4^{2-} + H^+$$
$$\Delta G'^\circ = -49.3 \text{ kJ/mol}$$

FIGURE 13-15 Hydrolysis of phosphocreatine. Breakage of the P—N bond in phosphocreatine produces creatine, which is stabilized by formation of a resonance hybrid. The other product, P_i, is also resonance stabilized.

Phosphocreatine

Creatine

$$\text{Phosphocreatine}^{2-} + H_2O \longrightarrow creatine + HPO_4^{2-}$$
$$\Delta G'^\circ = -43.0 \text{ kJ/mol}$$

TABLE 13-6 Standard Free Energies of Hydrolysis of Some Phosphorylated Compounds and Acetyl-CoA (a Thioester)

	$\Delta G'^{\circ}$	
	(kJ/mol)	(kcal/mol)
Phosphoenolpyruvate	−61.9	−14.8
1,3-Bisphosphoglycerate ($\rightarrow$ 3-phosphoglycerate + P_i)	−49.3	−11.8
Phosphocreatine	−43.0	−10.3
ADP ($\rightarrow$ AMP + P_i)	−32.8	−7.8
ATP ($\rightarrow$ ADP + P_i)	−30.5	−7.3
ATP ($\rightarrow$ AMP + PP_i)	−45.6	−10.9
AMP ($\rightarrow$ adenosine + P_i)	−14.2	−3.4
PP_i ($\rightarrow$ 2P_i)	−19.2	−4.0
Glucose 3-phosphate	−20.9	−5.0
Fructose 6-phosphate	−15.9	−3.8
Glucose 6-phosphate	−13.8	−3.3
Glycerol 3-phosphate	−9.2	−2.2
Acetyl-CoA	−31.4	−7.5

Source: Data mostly from Jencks, W.P. (1976) in *Handbook of Biochemistry and Molecular Biology*, 3rd edn (Fasman, G.D., ed.), *Physical and Chemical Data*, Vol. 1, pp. 296–304, CRC Press, Boca Raton, FL. The value for the free energy of hydrolysis of PP_i is from Frey, P.A. & Arabshahi, A. (1995) Standard free-energy change for the hydrolysis of the α-β-phosphoanhydride bridge in ATP. *Biochemistry* 34, 11,307-11,310.

the standard free energies of hydrolysis for some biologically important phosphorylated compounds.

Thioesters, in which a sulfur atom replaces the usual oxygen in the ester bond, also have large, negative, standard free energies of hydrolysis. Acetyl-coenzyme A, or acetyl-CoA **(Fig. 13–16)**, is one of many thioesters important in metabolism. The acyl group in these compounds is activated for transacylation, condensation, or oxidation-reduction reactions. Thioesters undergo much less resonance stabilization than do oxygen esters; consequently, the difference in free energy between the

$$\text{Acetyl-CoA} + H_2O \longrightarrow \text{acetate}^- + \text{CoA} + H^+$$
$$\Delta G'^{\circ} = -31.4 \text{ kJ/mol}$$

FIGURE 13–16 Hydrolysis of acetyl-coenzyme A. Acetyl-CoA is a thioester with a large, negative, standard free energy of hydrolysis. Thioesters contain a sulfur atom in the position occupied by an oxygen atom in oxygen esters. The complete structure of coenzyme A (CoA, or CoASH) is shown in Figure 8-38.

reactant and its hydrolysis products, which *are* resonance-stabilized, is greater for thioesters than for comparable oxygen esters **(Fig. 13–17)**. In both cases, hydrolysis of the ester generates a carboxylic acid, which can ionize and assume several resonance forms. Together, these factors result in the large, negative $\Delta G'^{\circ}$ (−31.4 kJ/mol) for acetyl-CoA hydrolysis.

To summarize, for hydrolysis reactions with large, negative, standard free-energy changes, the products are more stable than the reactants for one or more of the following reasons: (1) the bond strain in reactants due to electrostatic repulsion is relieved by charge separation, as for ATP; (2) the products are stabilized by ionization, as for ATP, acyl phosphates, and thioesters; (3) the

FIGURE 13–17 Free energy of hydrolysis for thioesters and oxygen esters. The *products* of both types of hydrolysis reaction have about the same free-energy content (G), but the thioester has a higher free-energy content than the oxygen ester. Orbital overlap between the O and C atoms allows resonance stabilization in oxygen esters; orbital overlap between S and C atoms is poorer and provides little resonance stabilization.

products are stabilized by isomerization (tautomerization), as for PEP; and/or (4) the products are stabilized by resonance, as for creatine released from phosphocreatine, carboxylate ion released from acyl phosphates and thioesters, and phosphate (P_i) released from anhydride or ester linkages.

ATP Provides Energy by Group Transfers, Not by Simple Hydrolysis

Throughout this book you will encounter reactions or processes for which ATP supplies energy, and the contribution of ATP to these reactions is commonly indicated as in **Figure 13–18a**, with a single arrow showing the conversion of ATP to ADP and P_i (or, in some cases, of ATP to AMP and pyrophosphate, PP_i). When written this way, these reactions of ATP seem to be simple hydrolysis reactions in which water displaces P_i (or PP_i), and one is tempted to say that an ATP-dependent reaction is "driven by the hydrolysis of ATP." This is *not* the case. ATP hydrolysis per se usually accomplishes nothing but the liberation of heat, which cannot drive a chemical process in an isothermal system. A single reaction arrow such as that in Figure 13–18a almost invariably represents a two-step process (Fig. 13–18b) in which part of the ATP molecule, a phosphoryl or pyrophosphoryl group or the adenylate moiety (AMP), is first transferred to a substrate molecule or to an amino

acid residue in an enzyme, becoming covalently attached to the substrate or the enzyme and raising its free-energy content. Then, in a second step, the phosphate-containing moiety transferred in the first step is displaced, generating P_i, PP_i, or AMP. Thus ATP participates *covalently* in the enzyme-catalyzed reaction to which it contributes free energy.

Some processes *do* involve direct hydrolysis of ATP (or GTP), however. For example, noncovalent binding of ATP (or GTP), followed by its hydrolysis to ADP (or GDP) and P_i, can provide the energy to cycle some proteins between two conformations, producing mechanical motion. This occurs in muscle contraction (see Fig. 5–31), and in the movement of enzymes along DNA (see Fig. 25–31) or of ribosomes along messenger RNA (see Fig. 27–31). The energy-dependent reactions catalyzed by helicases, RecA protein, and some topoisomerases (Chapter 25) also involve direct hydrolysis of phosphoanhydride bonds. The AAA+ ATPases involved in DNA replication and other processes described in Chapter 25 use ATP hydrolysis to cycle associated proteins between active and inactive forms. GTP-binding proteins that act in signaling pathways directly hydrolyze GTP to drive conformational changes that terminate signals triggered by hormones or by other extracellular factors (Chapter 12).

The phosphate compounds found in living organisms can be divided somewhat arbitrarily into two groups, based on their standard free energies of hydrolysis **(Fig. 13–19)**. "High-energy" compounds have a $\Delta G'^\circ$ of hydrolysis more negative than -25 kJ/mol; "low-energy" compounds have a less negative $\Delta G'^\circ$. Based on this criterion, ATP, with a $\Delta G'^\circ$ of hydrolysis of -30.5 kJ/mol (-7.3 kcal/mol), is a high-energy compound; glucose 6-phosphate, with a $\Delta G'^\circ$ of hydrolysis of -13.8 kJ/mol (-3.3 kcal/mol), is a low-energy compound.

The term "high-energy phosphate bond," long used by biochemists to describe the P—O bond broken in hydrolysis reactions, is incorrect and misleading as it wrongly suggests that the bond itself contains the energy. In fact, the breaking of all chemical bonds requires an *input* of energy. The free energy released by hydrolysis of phosphate compounds does not come from the specific bond that is broken; it results from the products of the reaction having a lower free-energy content than the reactants. For simplicity, we will sometimes use the term "high-energy phosphate compound" when referring to ATP or other phosphate compounds with a large, negative, standard free energy of hydrolysis.

As is evident from the additivity of free-energy changes of sequential reactions (see Section 13.1), any phosphorylated compound can be synthesized by coupling the synthesis to the breakdown of another phosphorylated compound with a more negative free energy of hydrolysis. For example, because cleavage of P_i from phosphoenolpyruvate releases more energy than is needed to drive the condensation of P_i with ADP, the

FIGURE 13–18 ATP hydrolysis in two steps. (a) The contribution of ATP to a reaction is often shown as a single step, but is almost always a two-step process. **(b)** Shown here is the reaction catalyzed by ATP-dependent glutamine synthetase. ❶ A phosphoryl group is transferred from ATP to glutamate, then ❷ the phosphoryl group is displaced by NH_3 and released as P_i.

FIGURE 13–19 Ranking of biological phosphate compounds by standard free energies of hydrolysis. This shows the flow of phosphoryl groups, represented by Ⓟ, from high-energy phosphoryl group donors via ATP to acceptor molecules (such as glucose and glycerol) to form their low-energy phosphate derivatives. (The location of each compound's donor phosphoryl group along the scale approximately indicates the $\Delta G'^{\circ}$ of hydrolysis.) This flow of phosphoryl groups, catalyzed by kinases, proceeds with an overall loss of free energy under intracellular conditions. Hydrolysis of low-energy phosphate compounds releases P_i, which has an even lower phosphoryl group transfer potential (as defined in the text).

direct donation of a phosphoryl group from PEP to ADP is thermodynamically feasible:

$$\Delta G'^{\circ} \text{ (kJ/mol)}$$

(1)	$PEP + H_2O \longrightarrow$ pyruvate $+ P_i$	−61.9
(2)	$ADP + P_i \longrightarrow ATP + H_2O$	+30.5
Sum:	$PEP + ADP \longrightarrow$ pyruvate $+ ATP$	−31.4

Notice that while the overall reaction is represented as the algebraic sum of the first two reactions, the overall reaction is actually a third, distinct reaction that does not involve P_i; PEP donates a *phosphoryl* group *directly* to ADP. We can describe phosphorylated compounds as having a high or low phosphoryl group transfer potential, on the basis of their standard free energies of hydrolysis (as listed in Table 13–6). The phosphoryl group transfer potential of PEP is very high, that of ATP is high, and that of glucose 6-phosphate is low (Fig. 13–19).

Much of catabolism is directed toward the synthesis of high-energy phosphate compounds, but their formation is not an end in itself; they are the means of activating a very wide variety of compounds for further chemical transformation. The transfer of a phosphoryl group to a compound effectively puts free energy into that compound, so that it has more free energy to give up during subsequent metabolic transformations. We described above how the synthesis of glucose 6-phosphate is accomplished by phosphoryl group transfer from ATP. In the next chapter we see how this phosphorylation of glucose activates, or "primes," the glucose for catabolic reactions that occur in nearly every living cell. Because of its intermediate position on the scale of group transfer potential, ATP can carry energy from

high-energy phosphate compounds produced by catabolism to compounds such as glucose, converting them into more reactive species. ATP thus serves as the universal energy currency in all living cells.

One more chemical feature of ATP is crucial to its role in metabolism: although in aqueous solution ATP is thermodynamically unstable and is therefore a good phosphoryl group donor, it is *kinetically* stable. Because of the huge activation energies (200 to 400 kJ/mol) required for uncatalyzed cleavage of its phosphoanhydride bonds, ATP does not spontaneously donate phosphoryl groups to water or to the hundreds of other potential acceptors in the cell. Only when specific enzymes are present to lower the energy of activation does phosphoryl group transfer from ATP proceed. The cell is therefore able to regulate the disposition of the energy carried by ATP by regulating the various enzymes that act on it.

ATP Donates Phosphoryl, Pyrophosphoryl, and Adenylyl Groups

The reactions of ATP are generally S_N2 nucleophilic displacements (see Section 13.2) in which the nucleophile may be, for example, the oxygen of an alcohol or carboxylate, or a nitrogen of creatine or of the side chain of arginine or histidine. Each of the three phosphates of ATP is susceptible to nucleophilic attack **(Fig. 13–20)**, and each position of attack yields a different type of product.

Nucleophilic attack by an alcohol on the γ phosphate (Fig. 13–20a) displaces ADP and produces a new phosphate ester. Studies with ^{18}O-labeled reactants have shown that the bridge oxygen in the new compound is

FIGURE 13–20 Nucleophilic displacement reactions of ATP. Any of the three P atoms (α, β, or γ) may serve as the electrophilic target for nucleophilic attack—in this case, by the labeled nucleophile R—^{18}O:. The nucleophile may be an alcohol (ROH), a carboxyl group (RCOO$^-$), or a phosphoanhydride (a nucleoside mono- or diphosphate, for example). **(a)** When the oxygen of the nucleophile attacks the γ position, the bridge oxygen of the product is labeled, indicating that the group transferred from ATP is a phosphoryl ($-$PO$_3^{2-}$), not a phosphate ($-$OPO$_3^{2-}$). **(b)** Attack on the β position displaces AMP and leads to the transfer of a pyrophosphoryl (not pyrophosphate) group to the nucleophile. **(c)** Attack on the α position displaces PP$_i$ and transfers the adenylyl group to the nucleophile.

Three positions on ATP for attack by the nucleophile R$^{18}\ddot{O}$

(a) Phosphoryl transfer — ADP

(b) Pyrophosphoryl transfer — AMP

(c) Adenylyl transfer — PP$_i$

derived from the alcohol, not from ATP; the group transferred from ATP is therefore a phosphoryl ($-$PO$_3^{2-}$), not a phosphate ($-$OPO$_3^{2-}$). Phosphoryl group transfer from ATP to glutamate (Fig. 13–18) or to glucose (p. 219) involves attack at the γ position of the ATP molecule.

Attack at the β phosphate of ATP displaces AMP and transfers a pyrophosphoryl (not pyrophosphate) group to the attacking nucleophile (Fig. 13–20b). For example, the formation of 5-phosphoribosyl-1-pyrophosphate (p. 892), a key intermediate in nucleotide synthesis, results from attack of an —OH of the ribose on the β phosphate.

Nucleophilic attack at the α position of ATP displaces PP$_i$ and transfers adenylate (5'-AMP) as an adenylyl group (Fig. 13–20c); the reaction is an **adenylylation** (a-den'-i-li-la'-shun, one of the most ungainly words in the biochemical language). Notice that hydrolysis of the α–β phosphoanhydride bond releases considerably more energy (~46 kJ/mol) than hydrolysis of the β–γ bond (~31 kJ/mol) (Table 13–6). Furthermore, the PP$_i$ formed as a byproduct of the adenylylation is hydrolyzed to two P$_i$ by the ubiquitous enzyme **inorganic pyrophosphatase**, releasing 19 kJ/mol and thereby providing a further energy "push" for the adenylylation reaction. In effect, both phosphoanhydride bonds of ATP are split in the overall reaction. Adenylylation reactions are therefore thermodynamically very favorable. When the energy of ATP is used to drive a particularly unfavorable metabolic reaction, adenylylation is often the mechanism of energy coupling. Fatty acid activation is a good example of this energy-coupling strategy.

The first step in the activation of a fatty acid—either for energy-yielding oxidation or for use in the synthesis of more complex lipids—is the formation of its thiol ester (see Fig. 17–5). The direct condensation of a fatty acid with coenzyme A is endergonic, but the formation of fatty acyl–CoA is made exergonic by stepwise removal of *two* phosphoryl groups from ATP. First, adenylate (AMP) is transferred from ATP to the carboxyl group of the fatty acid, forming a mixed anhydride

(fatty acyl adenylate) and liberating PP$_i$. The thiol group of coenzyme A then displaces the adenylyl group and forms a thioester with the fatty acid. The sum of these two reactions is energetically equivalent to the exergonic hydrolysis of ATP to AMP and PP$_i$ ($\Delta G'^\circ = -45.6$ kJ/mol) and the endergonic formation of fatty acyl–CoA ($\Delta G'^\circ = 31.4$ kJ/mol). The formation of fatty acyl–CoA is made energetically favorable by hydrolysis of the PP$_i$ by inorganic pyrophosphatase. Thus, in the activation of a fatty acid, both phosphoanhydride bonds of ATP are broken. The resulting $\Delta G'^\circ$ is the sum of the $\Delta G'^\circ$ values for the breakage of these bonds, or -45.6 kJ/mol $+ (-19.2)$ kJ/mol:

$$\text{ATP} + 2\text{H}_2\text{O} \longrightarrow \text{AMP} + 2\text{P}_i \qquad \Delta G'^\circ = -64.8 \text{ kJ/mol}$$

The activation of amino acids before their polymerization into proteins (see Fig. 27–19) is accomplished by an analogous set of reactions in which a transfer RNA molecule takes the place of coenzyme A. An interesting use of the cleavage of ATP to AMP and PP$_i$ occurs in the firefly, which uses ATP as an energy source to produce light flashes (Box 13–1).

Assembly of Informational Macromolecules Requires Energy

When simple precursors are assembled into high molecular weight polymers with defined sequences (DNA, RNA, proteins), as described in detail in Part III, energy is required both for the condensation of monomeric units and for the creation of *ordered* sequences. The precursors for DNA and RNA synthesis are nucleoside triphosphates, and polymerization is accompanied by cleavage of the phosphoanhydride linkage between the α and β phosphates, with the release of PP$_i$ (Fig. 13–20). The moieties transferred to the growing polymer in these reactions are adenylate (AMP), guanylate (GMP), cytidylate (CMP), or uridylate (UMP) for RNA synthesis, and their deoxy analogs (with TMP in place

BOX 13–1 Firefly Flashes: Glowing Reports of ATP

Bioluminescence requires considerable amounts of energy. In the firefly, ATP is used in a set of reactions that converts chemical energy into light energy. In the 1950s, from many thousands of fireflies collected by children in and around Baltimore, William McElroy and his colleagues at the Johns Hopkins University isolated the principal biochemical components: luciferin, a complex carboxylic acid, and luciferase, an enzyme. The generation of a light flash requires activation of luciferin by an enzymatic reaction involving pyrophosphate cleavage of ATP to form luciferyl adenylate (Fig. 1). In the presence of molecular oxygen and luciferase, the luciferin undergoes a multistep oxidative decarboxylation to oxyluciferin. This process is accompanied by emission of light. The color of the light flash differs with the firefly species and seems to be determined by differences in the structure of the luciferase. Luciferin is regenerated from oxyluciferin in a subsequent series of reactions.

In the laboratory, pure firefly luciferin and luciferase are used to measure minute quantities of ATP by the intensity of the light flash produced. As little as a few picomoles (10^{-12} mol) of ATP can be measured in this way. Next-gen pyrosequencing of DNA relies on flashes of light from the luciferin-luciferase reaction to detect the presence of ATP after addition of nucleotides to a growing strand of DNA (see Fig. 9–25).

The firefly, a beetle of the *Lampyridae* family.

FIGURE 1 Important components in the firefly bioluminescence cycle.

of UMP) for DNA synthesis. As noted above, the activation of amino acids for protein synthesis involves the donation of adenylyl groups from ATP, and we shall see in Chapter 27 that several steps of protein synthesis on the ribosome are also accompanied by GTP hydrolysis. In all these cases, the exergonic breakdown of a nucleoside triphosphate is coupled to the endergonic process of synthesizing a polymer of a specific sequence.

ATP Energizes Active Transport and Muscle Contraction

ATP can supply the energy for transporting an ion or a molecule across a membrane into another aqueous compartment where its concentration is higher (see Fig. 11–38). Transport processes are major consumers of energy; in human kidney and brain, for example, as much as two-thirds of the energy consumed at rest is used to pump Na^+ and K^+ across plasma membranes via the Na^+K^+ ATPase. The transport of Na^+ and K^+ is driven by cyclic phosphorylation and dephosphorylation of the transporter protein, with ATP as the phosphoryl group donor. Na^+-dependent phosphorylation of the Na^+K^+ ATPase forces a change in the protein's conformation, and K^+-dependent dephosphorylation favors return to the original conformation. Each cycle in the transport process results in the conversion of ATP to ADP and P_i, and it is the free-energy change of ATP hydrolysis that drives the cyclic changes in protein conformation that result in the electrogenic pumping of Na^+ and K^+. Note that in this case ATP interacts covalently by phosphoryl group transfer to the enzyme, not the substrate.

In the contractile system of skeletal muscle cells, myosin and actin are specialized to transduce the chemical energy of ATP into motion (see Fig. 5–31). ATP binds tightly but noncovalently to one conformation of myosin, holding the protein in that conformation. When myosin catalyzes the hydrolysis of its bound ATP, the ADP and P_i dissociate from the protein, allowing it to relax into a second conformation until another molecule of ATP binds. The binding and subsequent hydrolysis of ATP (by myosin ATPase) provide the energy that forces cyclic changes in the conformation of the myosin head. The change in conformation of many individual myosin molecules results in the sliding of myosin fibrils along actin filaments (see Fig. 5–30), which translates into macroscopic contraction of the muscle fiber. As we noted earlier, this production of mechanical motion at the expense of ATP is one of the few cases in which ATP hydrolysis per se, rather than group transfer from ATP, is the source of the chemical energy in a coupled process.

Transphosphorylations between Nucleotides Occur in All Cell Types

Although we have focused on ATP as the cell's energy currency and donor of phosphoryl groups, all other nucleoside triphosphates (GTP, UTP, and CTP) and all deoxynucleoside triphosphates (dATP, dGTP, dTTP, and dCTP) are energetically equivalent to ATP. The standard free-energy changes associated with hydrolysis of their phosphoanhydride linkages are very nearly identical with those shown in Table 13–6 for ATP. In preparation for their various biological roles, these other nucleotides are generated and maintained as the nucleoside triphosphate (NTP) forms by phosphoryl group transfer to the corresponding nucleoside diphosphates (NDPs) and monophosphates (NMPs).

ATP is the primary high-energy phosphate compound produced by catabolism, in the processes of glycolysis, oxidative phosphorylation, and, in photosynthetic cells, photophosphorylation. Several enzymes then carry phosphoryl groups from ATP to the other nucleotides. **Nucleoside diphosphate kinase**, found in all cells, catalyzes the reaction

$$ATP + NDP \text{ (or dNDP)} \xrightleftharpoons{Mg^{2+}} ADP + NTP \text{ (or dNTP)}$$
$$\Delta G'^{\circ} \approx 0$$

Although this reaction is fully reversible, the relatively high [ATP]/[ADP] ratio in cells normally drives the reac-

tion to the right, with the net formation of NTPs and dNTPs. The enzyme actually catalyzes a two-step phosphoryl group transfer, which is a classic case of a double-displacement (Ping-Pong) mechanism (**Fig. 13–21**; see also Fig. 6–13b). First, phosphoryl group transfer from ATP to an active-site His residue produces a phosphoenzyme intermediate; then the phosphoryl group is transferred from the (P)–His residue to an NDP acceptor. Because the enzyme is nonspecific for the base in the NDP and works equally well on dNDPs and NDPs, it can synthesize all NTPs and dNTPs, given the corresponding NDPs and a supply of ATP.

Phosphoryl group transfers from ATP result in an accumulation of ADP; for example, when muscle is contracting vigorously, ADP accumulates and interferes with ATP-dependent contraction. During periods of intense demand for ATP, the cell lowers the ADP concentration, and at the same time replenishes ATP, by the action of **adenylate kinase**:

$$2ADP \xrightleftharpoons{Mg^{2+}} ATP + AMP \qquad \Delta G'^{\circ} \approx 0$$

This reaction is fully reversible, so after the intense demand for ATP ends, the enzyme can recycle AMP by converting it to ADP, which can then be phosphorylated to ATP in mitochondria. A similar enzyme, guanylate kinase, converts GMP to GDP at the expense of ATP. By pathways such as these, energy conserved in the catabolic production of ATP is used to supply the cell with all required NTPs and dNTPs.

Phosphocreatine (PCr; Fig. 13–15), also called creatine phosphate, serves as a ready source of phosphoryl groups for the quick synthesis of ATP from ADP. The PCr concentration in skeletal muscle is approximately 30 mM, nearly 10 times the concentration of ATP, and in other tissues such as smooth muscle, brain, and kidney [PCr] is 5 to 10 mM. The enzyme **creatine kinase** catalyzes the reversible reaction

$$ADP + PCr \xrightleftharpoons{Mg^{2+}} ATP + Cr \qquad \Delta G'^{\circ} = -12.5 \text{ kJ/mol}$$

When a sudden demand for energy depletes ATP, the PCr reservoir is used to replenish ATP at a rate considerably faster than ATP can be synthesized by catabolic pathways. When the demand for energy slackens, ATP produced by catabolism is used to replenish the PCr

FIGURE 13–21 Ping-Pong mechanism of nucleoside diphosphate kinase. The enzyme binds its first substrate (ATP in our example), and a phosphoryl group is transferred to the side chain of a His residue. ADP departs, and another nucleoside (or deoxynucleoside) diphosphate replaces it, and this is converted to the corresponding triphosphate by transfer of the phosphoryl group from the phosphohistidine residue.

reservoir by reversal of the creatine kinase reaction (see Box 23–2). Organisms in the lower phyla employ other PCr-like molecules (collectively called **phosphagens**) as phosphoryl reservoirs.

Inorganic Polyphosphate Is a Potential Phosphoryl Group Donor

Inorganic polyphosphate, polyP (or $(polyP)_n$, where n is the number of orthophosphate residues), is a linear polymer composed of many tens or hundreds of P_i residues linked through phosphoanhydride bonds. This polymer, present in all organisms, may accumulate to high levels in some cells. In yeast, for example, the amount of polyP that accumulates in the vacuoles would represent, if distributed uniformly throughout the cell, a concentration of 200 mM! (Compare this with the concentrations of other phosphoryl group donors listed in Table 13–5.)

Inorganic polyphosphate (polyP)

One potential role for polyP is to serve as a phosphagen, a reservoir of phosphoryl groups that can be used to generate ATP, as creatine phosphate is used in muscle. PolyP has about the same phosphoryl group transfer potential as PP_i. The shortest polyphosphate, PP_i ($n = 2$), can serve as the energy source for active transport of H^+ across the vacuolar membrane in plant cells. For at least one form of the enzyme phosphofructokinase in plants, PP_i is the phosphoryl group donor, a role played by ATP in animals and microbes (p. 550). The finding of high concentrations of polyP in volcanic condensates and steam vents suggests that it could have served as an energy source in prebiotic and early cellular evolution.

In bacteria, the enzyme **polyphosphate kinase-1** (PPK-1) catalyzes the reversible reaction

$$\text{ATP} + (polyP)_n \underset{}{\overset{Mg^{2+}}{\rightleftharpoons}} \text{ADP} + (polyP)_{n+1}$$
$$\Delta G'^{\circ} = -20 \text{ kJ/mol}$$

by a mechanism involving an enzyme-bound Ⓟ–His intermediate (recall the mechanism of nucleoside diphosphate kinase, described in Fig. 13–21). A second enzyme, **polyphosphate kinase-2** (PPK-2), catalyzes the reversible synthesis of GTP (or ATP) from polyphosphate and GDP (or ADP):

$$\text{GDP} + (polyP)_{n+1} \underset{}{\overset{Mn^{2+}}{\rightleftharpoons}} \text{GTP} + (polyP)_n$$

PPK-2 is believed to act primarily in the direction of GTP and ATP synthesis, and PPK-1 in the direction of polyphosphate synthesis. PPK-1 and PPK-2 are present

in a wide variety of bacteria, including many pathogenic species.

In bacteria, elevated levels of polyP have been shown to promote expression of genes involved in adaptation of the organism to conditions of starvation or other threats to survival. In *Escherichia coli*, for example, polyP accumulates when cells are starved for amino acids or P_i, and this accumulation confers a survival advantage. Deletion of the genes for polyphosphate kinases diminishes the ability of certain pathogenic bacteria to invade animal tissues. The enzymes may therefore prove to be suitable targets in the development of new antimicrobial drugs.

No yeast gene encodes a PPK-like protein, but four genes—unrelated to bacterial PPK genes—are necessary for the synthesis of polyphosphate. The mechanism for polyphosphate synthesis in eukaryotes seems to be quite different from that in bacteria.

SUMMARY 13.3 Phosphoryl Group Transfers and ATP

▶ ATP is the chemical link between catabolism and anabolism. It is the energy currency of the living cell. The exergonic conversion of ATP to ADP and P_i, or to AMP and PP_i, is coupled to many endergonic reactions and processes.

▶ Direct hydrolysis of ATP is the source of energy in some processes driven by conformational changes, but in general it is not ATP hydrolysis but the transfer of a phosphoryl, pyrophosphoryl, or adenylyl group from ATP to a substrate or enzyme that couples the energy of ATP breakdown to endergonic transformations of substrates.

▶ Through these group transfer reactions, ATP provides the energy for anabolic reactions, including the synthesis of informational macromolecules, and for the transport of molecules and ions across membranes against concentration gradients and electrical potential gradients.

▶ To maintain its high group transfer potential, ATP concentration must be held far above the equilibrium concentration by energy-yielding reactions of catabolism.

▶ Cells contain other metabolites with large, negative, free energies of hydrolysis, including phosphoenolpyruvate, 1,3-bisphosphoglycerate, and phosphocreatine. These high-energy compounds, like ATP, have a high phosphoryl group transfer potential. Thioesters also have high free energies of hydrolysis.

▶ Inorganic polyphosphate, present in all cells, may serve as a reservoir of phosphoryl groups with high group transfer potential.

13.4 Biological Oxidation-Reduction Reactions

The transfer of phosphoryl groups is a central feature of metabolism. Equally important is another kind of transfer, electron transfer in oxidation-reduction reactions. These reactions involve the loss of electrons by one chemical species, which is thereby oxidized, and the gain of electrons by another, which is reduced. The flow of electrons in oxidation-reduction reactions is responsible, directly or indirectly, for all work done by living organisms. In nonphotosynthetic organisms, the sources of electrons are reduced compounds (foods); in photosynthetic organisms, the initial electron donor is a chemical species excited by the absorption of light. The path of electron flow in metabolism is complex. Electrons move from various metabolic intermediates to specialized electron carriers in enzyme-catalyzed reactions. The carriers in turn donate electrons to acceptors with higher electron affinities, with the release of energy. Cells contain a variety of molecular energy transducers, which convert the energy of electron flow into useful work.

We begin by discussing how work can be accomplished by an electromotive force (emf), then consider the theoretical and experimental basis for measuring energy changes in oxidation reactions in terms of emf and the relationship between this force, expressed in volts, and the free-energy change, expressed in joules. We conclude by describing the structures and oxidation-reduction chemistry of the most common of the specialized electron carriers, which you will encounter repeatedly in later chapters.

The Flow of Electrons Can Do Biological Work

Every time we use a motor, an electric light or heater, or a spark to ignite gasoline in a car engine, we use the flow of electrons to accomplish work. In the circuit that powers a motor, the source of electrons can be a battery containing two chemical species that differ in affinity for electrons. Electrical wires provide a pathway for electron flow from the chemical species at one pole of the battery, through the motor, to the chemical species at the other pole of the battery. Because the two chemical species differ in their affinity for electrons, electrons flow spontaneously through the circuit, driven by a force proportional to the difference in electron affinity, the **electromotive force**, **emf**. The emf (typically a few volts) can accomplish work if an appropriate energy transducer—in this case a motor—is placed in the circuit. The motor can be coupled to a variety of mechanical devices to do useful work.

Living cells have an analogous biological "circuit," with a relatively reduced compound such as glucose as the source of electrons. As glucose is enzymatically oxidized, the released electrons flow spontaneously through a series of electron-carrier intermediates to another chemical species, such as O_2. This electron flow is exergonic, because O_2 has a higher affinity for electrons than do the electron-carrier intermediates. The resulting emf

provides energy to a variety of molecular energy transducers (enzymes and other proteins) that do biological work. In the mitochondrion, for example, membrane-bound enzymes couple electron flow to the production of a transmembrane pH difference and a transmembrane electrical potential, accomplishing osmotic and electrical work. The proton gradient thus formed has potential energy, sometimes called the proton-motive force by analogy with electromotive force. Another enzyme, ATP synthase in the inner mitochondrial membrane, uses the proton-motive force to do chemical work: synthesis of ATP from ADP and P_i as protons flow spontaneously across the membrane. Similarly, membrane-localized enzymes in *E. coli* convert emf to proton-motive force, which is then used to power flagellar motion. The principles of electrochemistry that govern energy changes in the macroscopic circuit with a motor and battery apply with equal validity to the molecular processes accompanying electron flow in living cells.

Oxidation-Reductions Can Be Described as Half-Reactions

Although oxidation and reduction must occur together, it is convenient when describing electron transfers to consider the two halves of an oxidation-reduction reaction separately. For example, the oxidation of ferrous ion by cupric ion,

$$Fe^{2+} + Cu^{2+} \rightleftharpoons Fe^{3+} + Cu^+$$

can be described in terms of two half-reactions:

$$(1) \qquad Fe^{2+} \rightleftharpoons Fe^{3+} + e^-$$
$$(2) \qquad Cu^{2+} + e^- \rightleftharpoons Cu^+$$

The electron-donating molecule in an oxidation-reduction reaction is called the reducing agent or reductant; the electron-accepting molecule is the oxidizing agent or oxidant. A given agent, such as an iron cation existing in the ferrous (Fe^{2+}) or ferric (Fe^{3+}) state, functions as a conjugate reductant-oxidant pair (redox pair), just as an acid and corresponding base function as a conjugate acid-base pair. Recall from Chapter 2 that in acid-base reactions we can write a general equation: proton donor $\rightleftharpoons H^+$ + proton acceptor. In redox reactions we can write a similar general equation: electron donor (reductant) $\rightleftharpoons e^-$ + electron acceptor (oxidant). In the reversible half-reaction (1) above, Fe^{2+} is the electron donor and Fe^{3+} is the electron acceptor; together, Fe^{2+} and Fe^{3+} constitute a **conjugate redox pair**.

The electron transfers in the oxidation-reduction reactions of organic compounds are not fundamentally different from those of inorganic species. Consider the oxidation of a reducing sugar (an aldehyde or ketone) by cupric ion:

$$R\overset{O}{\underset{H}{-C}} + 4OH^- + 2Cu^{2+} \rightleftharpoons R\overset{O}{\underset{OH}{-C}} + Cu_2O + 2H_2O$$

This overall reaction can be expressed as two half-reactions:

$$(1) \quad R-C\!\!\!\overset{\displaystyle O}{\underset{\displaystyle H}{\big\langle}} + 2OH^- \rightleftharpoons R-C\!\!\!\overset{\displaystyle O}{\underset{\displaystyle OH}{\big\langle}} + 2e^- + H_2O$$

$$(2) \quad 2Cu^{2+} + 2e^- + 2OH^- \rightleftharpoons Cu_2O + H_2O$$

Because two electrons are removed from the aldehyde carbon, the second half-reaction (the one-electron reduction of cupric to cuprous ion) must be doubled to balance the overall equation.

Biological Oxidations Often Involve Dehydrogenation

The carbon in living cells exists in a range of oxidation states **(Fig. 13–22)**. When a carbon atom shares an electron pair with another atom (typically H, C, S, N, or O), the sharing is unequal in favor of the more electronegative atom. The order of increasing electronegativity is H < C < S < N < O. In oversimplified but useful terms, the more electronegative atom "owns" the bonding electrons it shares with another atom. For example, in methane (CH_4), carbon is more electronegative than the four hydrogens bonded to it, and the C atom therefore "owns" all eight bonding electrons (Fig. 13–22). In ethane, the electrons in the C—C bond are shared equally, so each C atom "owns" only seven of its eight bonding electrons. In ethanol, C-1 is less electronegative than the oxygen to which it is bonded, and the O atom therefore "owns" both electrons of the C—O bond, leaving C-1 with only five bonding electrons. With each formal loss of "owned" electrons, the carbon atom has undergone oxidation—even when no oxygen is involved, as in the conversion of an alkane ($-CH_2-CH_2-$) to an alkene ($-CH=CH-$). In this case, oxidation (loss of electrons) is coincident with the loss of hydrogen. In biological systems, as we noted earlier in the chapter, oxidation is often synonymous with **dehydrogenation** and many enzymes that catalyze oxidation reactions are **dehydrogenases**. Notice that the more reduced compounds in Figure 13–22 (top) are richer in hydrogen than in oxygen, whereas the more oxidized compounds (bottom) have more oxygen and less hydrogen.

Not all biological oxidation-reduction reactions involve carbon. For example, in the conversion of molecular nitrogen to ammonia, $6H^+ + 6e^- + N_2 \rightarrow 2NH_3$, the nitrogen atoms are reduced.

Electrons are transferred from one molecule (electron donor) to another (electron acceptor) in one of four ways:

1. Directly as *electrons*. For example, the Fe^{2+}/Fe^{3+} redox pair can transfer an electron to the Cu^+/Cu^{2+} redox pair:

$$Fe^{2+} + Cu^{2+} \rightleftharpoons Fe^{3+} + Cu^+$$

FIGURE 13–22 Different levels of oxidation of carbon compounds in the biosphere. To approximate the level of oxidation of these compounds, focus on the red carbon atom and its bonding electrons. When this carbon is bonded to the less electronegative H atom, both bonding electrons (red) are assigned to the carbon. When carbon is bonded to another carbon, bonding electrons are shared equally, so one of the two electrons is assigned to the red carbon. When the red carbon is bonded to the more electronegative O atom, the bonding electrons are assigned to the oxygen. The number to the right of each compound is the number of electrons "owned" by the red carbon, a rough expression of the degree of oxidation of that compound. As the red carbon undergoes oxidation (loses electrons), the number gets smaller.

2. As *hydrogen atoms*. Recall that a hydrogen atom consists of a proton (H^+) and a single electron (e^-). In this case we can write the general equation

$$AH_2 \rightleftharpoons A + 2e^- + 2H^+$$

where AH_2 is the hydrogen/electron donor. (Do not mistake the above reaction for an acid dissociation, which involves a proton and no electron.) AH_2 and A together constitute a conjugate redox pair (A/AH_2), which can reduce another compound B (or redox pair, B/BH_2) by transfer of hydrogen atoms:

$$AH_2 + B \rightleftharpoons A + BH_2$$

3. As a *hydride ion* (:H^-), which has two electrons. This occurs in the case of NAD-linked dehydrogenases, described below.

4. Through direct *combination with oxygen*. In this case, oxygen combines with an organic reductant and is covalently incorporated in the product, as in the oxidation of a hydrocarbon to an alcohol:

$$R{-}CH_3 + \tfrac{1}{2}O_2 \longrightarrow R{-}CH_2{-}OH$$

The hydrocarbon is the electron donor and the oxygen atom is the electron acceptor.

All four types of electron transfer occur in cells. The neutral term **reducing equivalent** is commonly used to designate a single electron equivalent participating in an oxidation-reduction reaction, no matter whether this equivalent is an electron per se or part of a hydrogen atom or a hydride ion, or whether the electron transfer takes place in a reaction with oxygen to yield an oxygenated product. Because biological fuel molecules are usually enzymatically dehydrogenated to lose *two* reducing equivalents at a time, and because each oxygen atom can accept two reducing equivalents, biochemists by convention regard the unit of biological oxidations as two reducing equivalents passing from substrate to oxygen.

Reduction Potentials Measure Affinity for Electrons

When two conjugate redox pairs are together in solution, electron transfer from the electron donor of one pair to the electron acceptor of the other may proceed spontaneously. The tendency for such a reaction depends on the relative affinity of the electron acceptor of each redox pair for electrons. The **standard reduction potential, $E°$**, a measure (in volts) of this affinity, can be determined in an experiment such as that described in **Figure 13–23**. Electrochemists have chosen as a standard of reference the half-reaction

$$H^+ + e^- \longrightarrow \tfrac{1}{2}H_2$$

The electrode at which this half-reaction occurs (called a half-cell) is arbitrarily assigned an $E°$ of 0.00 V.

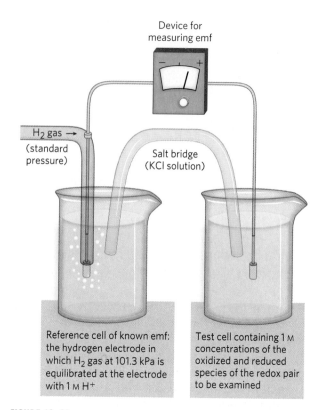

Device for measuring emf

H_2 gas → (standard pressure)

Salt bridge (KCl solution)

Reference cell of known emf: the hydrogen electrode in which H_2 gas at 101.3 kPa is equilibrated at the electrode with 1 M H^+

Test cell containing 1 M concentrations of the oxidized and reduced species of the redox pair to be examined

FIGURE 13–23 Measurement of the standard reduction potential ($E'°$) of a redox pair. Electrons flow from the test electrode to the reference electrode, or vice versa. The ultimate reference half-cell is the hydrogen electrode, as shown here, at pH 0. The electromotive force (emf) of this electrode is designated 0.00 V. At pH 7 in the test cell (and 25 °C), $E'°$ for the hydrogen electrode is −0.414 V. The direction of electron flow depends on the relative electron "pressure" or potential of the two cells. A salt bridge containing a saturated KCl solution provides a path for counter-ion movement between the test cell and the reference cell. From the observed emf and the known emf of the reference cell, the experimenter can find the emf of the test cell containing the redox pair. The cell that gains electrons has, by convention, the more positive reduction potential.

When this hydrogen electrode is connected through an external circuit to another half-cell in which an oxidized species and its corresponding reduced species are present at standard concentrations (25 °C, each solute at 1 M, each gas at 101.3 kPa), electrons tend to flow through the external circuit from the half-cell of lower $E°$ to the half-cell of higher $E°$. By convention, a half-cell that takes electrons from the standard hydrogen cell is assigned a positive value of $E°$, and one that donates electrons to the hydrogen cell, a negative value. When any two half-cells are connected, that with the larger (more positive) $E°$ will get reduced; it has the greater reduction potential.

The reduction potential of a half-cell depends not only on the chemical species present but also on their activities, approximated by their concentrations. About a century ago, Walther Nernst derived an equation that relates standard reduction potential ($E°$) to the actual reduction potential (E) at any concentration of oxidized and reduced species in a living cell:

$$E = E° + \frac{RT}{n\mathcal{F}} \ln \frac{[\text{electron acceptor}]}{[\text{electron donor}]} \quad (13\text{–}5)$$

where R and T have their usual meanings, n is the number of electrons transferred per molecule, and $\mathcal{F}$ is the Faraday constant (Table 13–1). At 298 K (25 °C), this expression reduces to

$$E = E° + \frac{0.026\text{V}}{n} \ln \frac{[\text{electron acceptor}]}{[\text{electron donor}]} \quad (13\text{–}6)$$

KEY CONVENTION: Many half-reactions of interest to biochemists involve protons. As in the definition of $\Delta G'°$, biochemists define the standard state for oxidation-reduction reactions as pH 7 and express a standard transformed reduction potential, $E'°$, the standard reduction potential at pH 7 and 25 °C. By convention, $\Delta E'°$ for any redox reaction is given as $E'°$ of the electron acceptor minus $E'°$ of the electron donor. ∎

The standard reduction potentials given in Table 13–7 and used throughout this book are values for $E'°$ and are therefore valid only for systems at neutral pH. Each value represents the potential difference when the conjugate redox pair, at 1 M concentrations, 25 °C, and pH 7, is connected with the standard (pH 0) hydrogen electrode. Notice in Table 13–7 that when the conjugate pair $2H^+/H_2$ at pH 7 is connected with the standard hydrogen electrode (pH 0), electrons tend to flow from the pH 7 cell to the standard (pH 0) cell; the measured $E'°$ for the $2H^+/H_2$ pair is -0.414 V.

Standard Reduction Potentials Can Be Used to Calculate Free-Energy Change

Why are reduction potentials so useful to the biochemist? When E values have been determined for any two half-cells, relative to the standard hydrogen electrode, we also know their reduction potentials relative to each other. We can then predict the direction in which electrons will tend to flow when the two half-cells are connected through an external circuit or when components of both half-cells are present in the same solution. Electrons tend to flow to the half-cell with the more positive E, and the strength of that tendency is proportional to ΔE, the difference in reduction potential. The energy made available by this spontaneous electron flow (the free-energy change, ΔG, for the oxidation-reduction reaction) is proportional to ΔE:

$$\Delta G = -n\mathcal{F}\Delta E \quad \text{or} \quad \Delta G'° = -n\mathcal{F}\Delta E'° \quad (13\text{–}7)$$

where n is the number of electrons transferred in the reaction. With this equation we can calculate the actual free-energy change for any oxidation-reduction

TABLE 13–7	Standard Reduction Potentials of Some Biologically Important Half-Reactions

Half-reaction	$E'°$ (V)
$\frac{1}{2}O_2 + 2H^+ + 2e^- \longrightarrow H_2O$	0.816
$Fe^{3+} + e^- \longrightarrow Fe^{2+}$	0.771
$NO_3^- + 2H^+ + 2e^- \longrightarrow NO_2^- + H_2O$	0.421
Cytochrome f (Fe^{3+}) $+ e^- \longrightarrow$ cytochrome f (Fe^{2+})	0.365
$Fe(CN)_6^{3-}$ (ferricyanide) $+ e^- \longrightarrow Fe(CN)_6^{4-}$	0.36
Cytochrome a_3 (Fe^{3+}) $+ e^- \longrightarrow$ cytochrome a_3 (Fe^{2+})	0.35
$O_2 + 2H^+ + 2e^- \longrightarrow H_2O_2$	0.295
Cytochrome a (Fe^{3+}) $+ e^- \longrightarrow$ cytochrome a (Fe^{2+})	0.29
Cytochrome c (Fe^{3+}) $+ e^- \longrightarrow$ cytochrome c (Fe^{2+})	0.254
Cytochrome c_1 (Fe^{3+}) $+ e^- \longrightarrow$ cytochrome c_1 (Fe^{2+})	0.22
Cytochrome b (Fe^{3+}) $+ e^- \longrightarrow$ cytochrome b (Fe^{2+})	0.077
Ubiquinone $+ 2H^+ + 2e^- \longrightarrow$ ubiquinol $+ H_2$	0.045
Fumarate^{2-} $+ 2H^+ + 2e^- \longrightarrow$ succinate^{2-}	0.031
$2H^+ + 2e^- \longrightarrow H_2$ (at standard conditions, pH 0)	0.000
Crotonyl-CoA $+ 2H^+ + 2e^- \longrightarrow$ butyryl-CoA	-0.015
Oxaloacetate^{2-} $+ 2H^+ + 2e^- \longrightarrow$ malate^{2-}	-0.166
Pyruvate$^-$ $+ 2H^+ + 2e^- \longrightarrow$ lactate$^-$	-0.185
Acetaldehyde $+ 2H^+ + 2e^- \longrightarrow$ ethanol	-0.197
FAD $+ 2H^+ + 2e^- \longrightarrow$ FADH$_2$	-0.219*
Glutathione $+ 2H^+ + 2e^- \longrightarrow$ 2 reduced glutathione	-0.23
S $+ 2H^+ + 2e^- \longrightarrow H_2S$	-0.243
Lipoic acid $+ 2H^+ + 2e^- \longrightarrow$ dihydrolipoic acid	-0.29
NAD$^+$ $+ H^+ + 2e^- \longrightarrow$ NADH	-0.320
NADP$^+$ $+ H^+ + 2e^- \longrightarrow$ NADPH	-0.324
Acetoacetate $+ 2H^+ + 2e^- \longrightarrow$ β-hydroxybutyrate	-0.346
α-Ketoglutarate $+ CO_2 + 2H^+ + 2e^- \longrightarrow$ isocitrate	-0.38
$2H^+ + 2e^- \longrightarrow H_2$ (at pH 7)	-0.414
Ferredoxin (Fe^{3+}) $+ e^- \longrightarrow$ ferredoxin (Fe^{2+})	-0.432

Source: Data mostly from Loach, R.A. (1976) in *Handbook of Biochemistry and Molecular Biology*, 3rd edn (Fasman, G.D., ed.), *Physical and Chemical Data*, Vol. 1, pp. 122–130, CRC Press, Boca Raton, FL.

* This is the value for free FAD; FAD bound to a specific flavoprotein (e.g., succinate dehydrogenase) has a different $E'°$ that depends on its protein environment.

reaction from the values of $E'°$ in a table of reduction potentials (Table 13–7) and the concentrations of reacting species.

WORKED EXAMPLE 13–3 Calculation of $\Delta G'^\circ$ and ΔG of a Redox Reaction

Calculate the standard free-energy change, $\Delta G'^\circ$, for the reaction in which acetaldehyde is reduced by the biological electron carrier NADH:

$$\text{Acetaldehyde} + \text{NADH} + \text{H}^+ \longrightarrow \text{ethanol} + \text{NAD}^+$$

Then calculate the *actual* free-energy change, ΔG, when [acetaldehyde] and [NADH] are 1.00 M, and [ethanol] and [NAD$^+$] are 0.100 M. The relevant half-reactions and their E'° values are:

(1) Acetaldehyde $+ 2\text{H}^+ + 2e^- \longrightarrow$ ethanol
$$E'^\circ = -0.197 \text{ V}$$

(2) $\text{NAD}^+ + 2\text{H}^+ + 2e^- \longrightarrow \text{NADH} + \text{H}^+$
$$E'^\circ = -0.320 \text{ V}$$

Remember that, by convention, $\Delta E'^\circ$ is E'° of the electron acceptor minus E'° of the electron donor.

Solution: Because acetaldehyde is accepting electrons ($n = 2$) from NADH, $\Delta E'^\circ = -0.197 \text{ V} - (-0.320 \text{ V}) = 0.123$ V. Therefore,

$$\Delta G'^\circ = -n\mathcal{F}\,\Delta E'^\circ = -2(96.5 \text{ kJ/V} \cdot \text{mol})(0.123 \text{ V})$$
$$= -23.7 \text{ kJ/mol}$$

This is the free-energy change for the oxidation-reduction reaction at 25 °C and pH 7, when acetaldehyde, ethanol, NAD$^+$, and NADH are all present at 1.00 M concentrations.

To calculate ΔG when [acetaldehyde] and [NADH] are 1.00 M, and [ethanol] and [NAD$^+$] are 0.100 M, we can use Equation 13–4 and the standard free-energy change we calculated above:

$$\Delta G = \Delta G'^\circ + RT \ln \frac{[\text{ethanol}][\text{NAD}^+]}{[\text{acetaldehyde}][\text{NADH}]}$$
$$= -23.7 \text{ kJ/mol} +$$
$$(8.315 \text{ J/mol} \cdot \text{K})(298 \text{ K}) \ln \frac{(0.100 \text{ M})(0.100 \text{ M})}{(1.00 \text{ M})(1.00 \text{ M})}$$
$$= -23.7 \text{ kJ/mol} + (2.48 \text{ J/mol}) \ln 0.01$$
$$= -35.1 \text{ kJ/mol}$$

This is the actual free-energy change at the specified concentrations of the redox pairs.

Cellular Oxidation of Glucose to Carbon Dioxide Requires Specialized Electron Carriers

The principles of oxidation-reduction energetics described above apply to the many metabolic reactions that involve electron transfers. For example, in many organisms, the oxidation of glucose supplies energy for the production of ATP. The complete oxidation of glucose:

$$\text{C}_6\text{H}_{12}\text{O}_6 + 6\text{O}_2 \longrightarrow 6\text{CO}_2 + 6\text{H}_2\text{O}$$

has a $\Delta G'^\circ$ of $-2,840$ kJ/mol. This is a much larger release of free energy than is required for ATP synthesis in cells (50 to 60 kJ/mol; see Worked Example 13–2).

Cells convert glucose to CO_2 not in a single, high-energy-releasing reaction but rather in a series of controlled reactions, some of which are oxidations. The free energy released in these oxidation steps is of the same order of magnitude as that required for ATP synthesis from ADP, with some energy to spare. Electrons removed in these oxidation steps are transferred to coenzymes specialized for carrying electrons, such as NAD$^+$ and FAD (described below).

A Few Types of Coenzymes and Proteins Serve as Universal Electron Carriers

The multitude of enzymes that catalyze cellular oxidations channel electrons from their hundreds of different substrates into just a few types of universal electron carriers. The reduction of these carriers in catabolic processes results in the conservation of free energy released by substrate oxidation. NAD, NADP, FMN, and FAD are water-soluble coenzymes that undergo reversible oxidation and reduction in many of the electron-transfer reactions of metabolism. The nucleotides NAD and NADP move readily from one enzyme to another; the flavin nucleotides FMN and FAD are usually very tightly bound to the enzymes, called flavoproteins, for which they serve as prosthetic groups. Lipid-soluble quinones such as ubiquinone and plastoquinone act as electron carriers and proton donors in the nonaqueous environment of membranes. Iron-sulfur proteins and cytochromes, which have tightly bound prosthetic groups that undergo reversible oxidation and reduction, also serve as electron carriers in many oxidation-reduction reactions. Some of these proteins are water-soluble, but others are peripheral or integral membrane proteins (see Fig. 11–7).

We conclude this chapter by describing some chemical features of nucleotide coenzymes and some of the enzymes (dehydrogenases and flavoproteins) that use them. The oxidation-reduction chemistry of quinones, iron-sulfur proteins, and cytochromes is discussed in Chapter 19.

NADH and NADPH Act with Dehydrogenases as Soluble Electron Carriers

Nicotinamide adenine dinucleotide (NAD; NAD$^+$ in its oxidized form) and its close analog nicotinamide adenine dinucleotide phosphate (NADP; NADP$^+$ when oxidized) are composed of two nucleotides joined through their phosphate groups by a phosphoanhydride bond **(Fig. 13–24a)**. Because the nicotinamide ring resembles pyridine, these compounds are sometimes called **pyridine nucleotides**. The vitamin niacin is the source of the nicotinamide moiety in nicotinamide nucleotides.

Both coenzymes undergo reversible reduction of the nicotinamide ring (Fig. 13–24). As a substrate molecule undergoes oxidation (dehydrogenation), giving up two hydrogen atoms, the oxidized form of the nucleotide

FIGURE 13-24 NAD and NADP. (a) Nicotinamide adenine dinucleotide, NAD⁺, and its phosphorylated analog NADP⁺ undergo reduction to NADH and NADPH, accepting a hydride ion (two electrons and one proton) from an oxidizable substrate. The hydride ion is added to either the front (the A side) or the back (the B side) of the planar nicotinamide ring (see Table 13-8). **(b)** The UV absorption spectra of NAD⁺ and NADH. Reduction of the nicotinamide ring produces a new, broad absorption band with a maximum at 340 nm. The production of NADH during an enzyme-catalyzed reaction can be conveniently followed by observing the appearance of the absorbance at 340 nm (molar extinction coefficient $\varepsilon_{340} = 6{,}200\ M^{-1}\ cm^{-1}$).

(NAD⁺ or NADP⁺) accepts a hydride ion (:H⁻, the equivalent of a proton and two electrons) and is reduced (to NADH or NADPH). The second proton removed from the substrate is released to the aqueous solvent. The half-reactions for these nucleotide cofactors are

$$NAD^+ + 2e^- + 2H^+ \longrightarrow NADH + H^+$$
$$NADP^+ + 2e^- + 2H^+ \longrightarrow NADPH + H^+$$

Reduction of NAD⁺ or NADP⁺ converts the benzenoid ring of the nicotinamide moiety (with a fixed positive charge on the ring nitrogen) to the quinonoid form (with no charge on the nitrogen). The reduced nucleotides absorb light at 340 nm; the oxidized forms do not (Fig. 13–24b); this difference in absorption is used by biochemists to assay reactions involving these coenzymes. Note that the plus sign in the abbreviations NAD⁺ and NADP⁺ does *not* indicate the net charge on these molecules (in fact, both are negatively charged); rather, it indicates that the nicotinamide ring is in its oxidized form, with a positive charge on the nitrogen atom. In the abbreviations NADH and NADPH, the "H" denotes the added hydride ion. To refer to these nucleotides without specifying their oxidation state, we use NAD and NADP.

The total concentration of NAD⁺ + NADH in most tissues is about 10^{-5} M; that of NADP⁺ + NADPH is about 10^{-6} M. In many cells and tissues, the ratio of NAD⁺ (oxidized) to NADH (reduced) is high, favoring hydride transfer from a substrate *to* NAD⁺ to form NADH. By contrast, NADPH is generally present at a higher concentration than NADP⁺, favoring hydride transfer *from* NADPH to a substrate. This reflects the specialized metabolic roles of the two coenzymes: NAD⁺ generally functions in oxidations—usually as part of a catabolic reaction; NADPH is the usual coenzyme in reductions—nearly always as part of an anabolic reaction. A few enzymes can use either coenzyme, but most show a strong preference for one over the other. Also, the processes in which these two cofactors function are segregated in eukaryotic cells: for example, oxidations of fuels such as pyruvate, fatty acids, and α-keto acids derived from amino acids occur in the mitochondrial matrix, whereas reductive biosynthetic processes such as fatty acid synthesis take place in the cytosol. This functional and spatial specialization allows a cell to maintain two distinct pools of electron carriers, with two distinct functions.

More than 200 enzymes are known to catalyze reactions in which NAD⁺ (or NADP⁺) accepts a hydride ion from a reduced substrate, or NADPH (or NADH) donates a hydride ion to an oxidized substrate. The general reactions are

$$AH_2 + NAD^+ \longrightarrow A + NADH + H^+$$
$$A + NADPH + H^+ \longrightarrow AH_2 + NADP^+$$

where AH_2 is the reduced substrate and A the oxidized substrate. The general name for an enzyme of this type is **oxidoreductase**; they are also commonly called dehydrogenases. For example, alcohol dehydrogenase catalyzes the first step in the catabolism of ethanol, in which ethanol is oxidized to acetaldehyde:

$$CH_3CH_2OH + NAD^+ \longrightarrow CH_3CHO + NADH + H^+$$

Ethanol   Acetaldehyde

Notice that one of the carbon atoms in ethanol has lost a hydrogen; the compound has been oxidized from an alcohol to an aldehyde (refer again to Fig. 13–22 for the oxidation states of carbon).

When NAD^+ or $NADP^+$ is reduced, the hydride ion could in principle be transferred to either side of the nicotinamide ring: the front (A side) or the back (B side), as represented in Figure 13–24a. Studies with isotopically labeled substrates have shown that a given enzyme catalyzes either an A-type or a B-type transfer, but not both. For example, yeast alcohol dehydrogenase and lactate dehydrogenase of vertebrate heart transfer a hydride ion to (or remove a hydride ion from) the A side of the nicotinamide ring; they are classed as type A dehydrogenases to distinguish them from another group of enzymes that transfer a hydride ion to (or remove a hydride ion from) the B side of the nicotinamide ring (Table 13–8). The specificity for one side or another can be very striking; lactate dehydrogenase, for example, prefers the A side over the B side by a factor of 5×10^7! The basis for this preference lies in the exact positioning of the enzyme groups involved in hydrogen bonding with the —$CONH_2$ group of the nicotinamide.

Most dehydrogenases that use NAD or NADP bind the cofactor in a conserved protein domain called the Rossmann fold (named for Michael Rossmann, who deduced the structure of lactate dehydrogenase and first described this structural motif). The Rossmann fold typically consists of a six-stranded parallel β sheet and four associated α helices (**Fig. 13–25**).

(a)

(b)

NAD

Rossmann fold 1   Rossmann fold 2

FIGURE 13–25 The Rossmann fold. This structural motif is found in the NAD-binding site of many dehydrogenases. **(a)** It consists of a pair of structurally similar motifs (only one of which is shown here), each having three parallel β sheets and two α helices (β-α-β-α-β). **(b)** The nucleotide-binding domain of the enzyme lactate dehydrogenase (derived from PDB ID 3LDH) with NAD (ball-and-stick structure) bound in an extended conformation through hydrogen bonds and salt bridges to the paired β-α-β-α-β motifs of the Rossmann fold (shades of red and blue).

The association between a dehydrogenase and NAD or NADP is relatively loose; the coenzyme readily diffuses from one enzyme to another, acting as a water-soluble

TABLE 13–8   Stereospecificity of Dehydrogenases That Employ NAD^+ or $NADP^+$ as Coenzymes

Enzyme	Coenzyme	Stereochemical specificity for nicotinamide ring (A or B)	Text page
Isocitrate dehydrogenase	NAD^+	A	643
α-Ketoglutarate dehydrogenase	NAD^+	B	644
Glucose 6-phosphate dehydrogenase	$NADP^+$	B	577
Malate dehydrogenase	NAD^+	A	647
Glutamate dehydrogenase	NAD^+ or $NADP^+$	B	702
Glyceraldehyde 3-phosphate dehydrogenase	NAD^+	B	553
Lactate dehydrogenase	NAD^+	A	563
Alcohol dehydrogenase	NAD^+	A	565

carrier of electrons from one metabolite to another. For example, in the production of alcohol during fermentation of glucose by yeast cells, a hydride ion is removed from glyceraldehyde 3-phosphate by one enzyme (glyceraldehyde 3-phosphate dehydrogenase, a type B enzyme) and transferred to NAD$^+$. The NADH produced then leaves the enzyme surface and diffuses to another enzyme (alcohol dehydrogenase, a type A enzyme), which transfers a hydride ion to acetaldehyde, producing ethanol:

(1) Glyceraldehyde 3-phosphate + NAD$^+$ $\longrightarrow$
$\qquad\qquad\qquad$ 3-phosphoglycerate + NADH + H$^+$

(2) Acetaldehyde + NADH + H$^+$ $\longrightarrow$ ethanol + NAD$^+$

Sum: Glyceraldehyde 3-phosphate + acetaldehyde $\longrightarrow$
$\qquad\qquad\qquad$ 3-phosphoglycerate + ethanol

Notice that in the overall reaction there is no net production or consumption of NAD$^+$ or NADH; the coenzymes function catalytically and are recycled repeatedly without a net change in the concentration of NAD$^+$ + NADH.

Dietary Deficiency of Niacin, the Vitamin Form of NAD and NADP, Causes Pellagra

As we noted in Chapter 6, and will discuss further in the chapters to follow, most coenzymes are derived from the substances we call vitamins. The pyridine-like rings of NAD and NADP are derived from the vitamin **niacin** (nicotinic acid; **Fig. 13–26**), which is synthesized from tryptophan. Humans generally cannot synthesize sufficient quantities of niacin, and this is especially so for individuals with diets low in tryptophan (maize, for example, has a low tryptophan content). Niacin deficiency, which affects all the NAD(P)-dependent dehydrogenases, causes the serious human disease pellagra (Italian for "rough skin") and a related disease in dogs, blacktongue. These diseases are characterized by the "three Ds": dermatitis, diarrhea, and dementia, followed in many cases by death. A century ago, pellagra was a common human disease; in the southern United States, where maize was a dietary staple, about 100,000 people were afflicted and about 10,000 died as a result of this disease between 1912 and 1916. In 1920 Joseph Goldberger showed pellagra to be caused by a dietary insufficiency, and in 1937 Frank Strong, D. Wayne Woolley, and Conrad Elvehjem identified niacin as the curative agent for blacktongue. Supplementation of the human diet with this inexpensive compound has eradicated pellagra in the populations of the developed world, with one significant exception: people with alcoholism, or who drink excessive amounts of alcohol. In these individuals, intestinal absorption of niacin is much reduced, and caloric needs are often met with distilled spirits that are virtually devoid of vitamins, including niacin. In some parts of the world, including the Deccan Plateau in India, pellagra still occurs in the general population, especially among people living in poverty. ■

Frank Strong, 1908-1993

D. Wayne Woolley, 1914-1966

Conrad Elvehjem, 1901-1962

FIGURE 13–26 Niacin (nicotinic acid) and its derivative nicotinamide. The biosynthetic precursor of these compounds is tryptophan. In the laboratory, nicotinic acid was first produced by oxidation of the natural product nicotine—thus the name. Both nicotinic acid and nicotinamide cure pellagra, but nicotine (from cigarettes or elsewhere) has no curative activity.

Flavin Nucleotides Are Tightly Bound in Flavoproteins

Flavoproteins (Table 13–9) are enzymes that catalyze oxidation-reduction reactions using either flavin mononucleotide (FMN) or flavin adenine dinucleotide (FAD) as coenzyme (**Fig. 13–27**). These coenzymes, the **flavin nucleotides**, are derived from the vitamin riboflavin. The fused ring structure of flavin nucleotides (the isoalloxazine ring) undergoes reversible reduction, accepting either one or two electrons in the form of one or two hydrogen atoms (each atom an electron plus a proton) from a reduced substrate. The fully reduced forms are abbreviated FADH$_2$ and FMNH$_2$. When a fully oxidized flavin nucleotide accepts only one electron (one hydrogen atom), the semiquinone form of the isoalloxazine ring is produced, abbreviated FADH$^\bullet$ and FMNH$^\bullet$. Because flavin nucleotides have a slightly different chemical specialty from that of the nicotinamide

TABLE 13–9 Some Enzymes (Flavoproteins) That Employ Flavin Nucleotide Coenzymes

Enzyme	Flavin nucleotide	Text page(s)
Acyl-CoA dehydrogenase	FAD	673
Dihydrolipoyl dehydrogenase	FAD	637
Succinate dehydrogenase	FAD	646
Glycerol 3-phosphate dehydrogenase	FAD	759
Thioredoxin reductase	FAD	917
NADH dehydrogenase (Complex I)	FMN	738–739
Glycolate oxidase	FMN	813

coenzymes—the ability to participate in either one- or two-electron transfers—flavoproteins are involved in a greater diversity of reactions than the NAD(P)-linked dehydrogenases.

Like the nicotinamide coenzymes (Fig. 13–24), the flavin nucleotides undergo a shift in a major absorption band on reduction (again, useful to biochemists who want to monitor reactions involving these coenzymes). Flavoproteins that are fully reduced (two electrons accepted) generally have an absorption maximum near 360 nm. When partially reduced (one electron), they acquire another absorption maximum at about 450 nm; when fully oxidized, the flavin has maxima at 370 and 440 nm.

The flavin nucleotide in most flavoproteins is bound rather tightly to the protein, and in some enzymes, such as succinate dehydrogenase, it is bound covalently. Such tightly bound coenzymes are properly called prosthetic groups. They do not transfer electrons by diffusing from one enzyme to another; rather, they provide a means by which the flavoprotein can temporarily hold electrons while it catalyzes electron transfer from a reduced substrate to an electron acceptor. One important feature of the flavoproteins is the variability in the standard reduction potential (E'°) of the bound flavin nucleotide. Tight association between the enzyme and prosthetic group confers on the flavin ring a reduction potential typical of that particular flavoprotein, sometimes quite different from the reduction potential of the free flavin nucleotide. FAD bound to succinate dehydrogenase, for example, has an E'° close to 0.0 V, compared with -0.219 V for free FAD; E'° for other flavoproteins ranges from -0.40 V to $+0.06$ V. Flavoproteins are often very complex; some have, in addition to a flavin nucleotide, tightly bound inorganic ions (iron or molybdenum, for example) capable of participating in electron transfers.

Certain flavoproteins act in a quite different role, as light receptors. **Cryptochromes** are a family of flavoproteins, widely distributed in the eukaryotic phyla, that mediate the effects of blue light on plant development and the effects of light on mammalian circadian rhythms (oscillations in physiology and biochemistry, with a 24-hour period). The cryptochromes are homologs of another family of flavoproteins, the photolyases. Found in both bacteria and eukaryotes, **photolyases** use the energy of absorbed light to repair chemical defects in DNA.

FIGURE 13–27 Oxidized and reduced FAD and FMN. FMN consists of the structure above the dashed line on the FAD (oxidized form). The flavin nucleotides accept two hydrogen atoms (two electrons and two protons), both of which appear in the flavin ring system. When FAD or FMN accepts only one hydrogen atom, the semiquinone, a stable free radical, forms.

We examine the function of flavoproteins as electron carriers in Chapter 19, when we consider their roles in oxidative phosphorylation (in mitochondria) and photophosphorylation (in chloroplasts), and we describe the photolyase reactions in Chapter 25.

SUMMARY 13.4 Biological Oxidation-Reduction Reactions

▶ In many organisms, a central energy-conserving process is the stepwise oxidation of glucose to CO_2, in which some of the energy of oxidation is conserved in ATP as electrons are passed to O_2.

▶ Biological oxidation-reduction reactions can be described in terms of two half-reactions, each with a characteristic standard reduction potential, E'°.

▶ When two electrochemical half-cells, each containing the components of a half-reaction, are connected, electrons tend to flow to the half-cell with the higher reduction potential. The strength of this tendency is proportional to the difference between the two reduction potentials (ΔE) and is a function of the concentrations of oxidized and reduced species.

▶ The standard free-energy change for an oxidation-reduction reaction is directly proportional to the difference in standard reduction potentials of the two half-cells: $\Delta G'^{\circ} = -n\,\mathcal{F}\,\Delta E'^{\circ}$.

▶ Many biological oxidation reactions are dehydrogenations in which one or two hydrogen atoms ($H^+ + e^-$) are transferred from a substrate to a hydrogen acceptor. Oxidation-reduction reactions in living cells involve specialized electron carriers.

▶ NAD and NADP are the freely diffusible coenzymes of many dehydrogenases. Both NAD^+ and $NADP^+$ accept two electrons and one proton.

▶ FAD and FMN, the flavin nucleotides, serve as tightly bound prosthetic groups of flavoproteins. They can accept either one or two electrons and one or two protons. Flavoproteins also serve as light receptors in cryptochromes and photolyases.

Key Terms

Terms in bold are defined in the glossary.

Further Reading

Bioenergetics and Thermodynamics

Atkins, P.W. (1984) *The Second Law*, Scientific American Books, Inc., New York.

A well-illustrated and elementary discussion of the second law and its implications.

Atkinson, D.E. (1977) *Cellular Energy Metabolism and Its Regulation*, Academic Press, Inc., New York.

A classic treatment of the roles of ATP, ADP, and AMP in controlling the rate of catabolism.

Bergethon, P.R. (1998) *The Physical Basis of Biochemistry*, Springer Verlag, New York.

Chapters 11 through 13 of this book, and the books by Tinoco et al. and van Holde et al. (below), are excellent general references for physical biochemistry, with good discussions of the applications of thermodynamics to biochemistry.

Edsall, J.T. & Gutfreund, H. (1983) *Biothermodynamics: The Study of Biochemical Processes at Equilibrium*, John Wiley & Sons, Inc., New York.

Hammes, G. (2000) *Thermodynamics and Kinetics for the Biological Sciences*, John Wiley & Sons, Inc., New York.

Clearly written, well illustrated, with excellent examples and problems.

Harold, F.M. (1986) *The Vital Force: A Study of Bioenergetics*, W.H. Freeman and Company, New York.

A beautifully clear discussion of thermodynamics in biological processes.

Harris, D.A. (1995) *Bioenergetics at a Glance*, Blackwell Science, Oxford.

A short, clearly written account of cellular energetics, including introductory chapters on thermodynamics.

Haynie, D.T. (2001) *Biological Thermodynamics*, Cambridge University Press, Cambridge.

Highly accessible discussions of thermodynamics and kinetics in biological systems.

Loewenstein, W.R. (1999) *The Touchstone of Life: Molecular Information, Cell Communication, and the Foundations of Life*, Oxford University Press, New York.

Beautifully written discussion of the relationship between entropy and information.

Nicholls, D.G. & Ferguson, S.J. (2002) *Bioenergetics 3*, Academic Press, Inc., New York.

Clear, well-illustrated, intermediate-level discussion of the theory of bioenergetics and the mechanisms of energy transductions.

Tinoco, I., Jr., Sauer, K., Wang, J.C., & Puglisi, J.D. (2002) *Physical Chemistry: Principles and Applications in Biological Sciences*, 4th edn, Prentice-Hall, Inc., Upper Saddle River, NJ.

Chapters 2 through 5 cover thermodynamics.

van Holde, K.E., Johnson, C., & Ho, P.S. (2006) *Principles of Physical Biochemistry*, 2nd edn, Prentice-Hall, Inc., Upper Saddle River, NJ.

Chapters 2 and 3 are especially relevant.

Chemical Logic and Common Biochemical Reactions

Frey, P.A. (2001) Radical mechanisms of enzymatic catalysis. *Annu. Rev. Biochem.* **70**, 121–148.

A very useful survey of reactions that proceed by free-radical mechanisms.

Frey, P.A. & Hegeman, A.D. (2006) *Enzymatic Reaction Mechanisms*, Oxford University Press, New York.

An authoritative and up-to-date resource on the reactions that occur in living systems.

Gutteridge, A. & Thornton, J.M. (2005) Understanding nature's catalytic toolkit. *Trends Biochem. Sci.* **11**, 622–629.

Kraut, D.A., Carroll, K.S., & Herschlag, D. (2003) Challenges in enzyme mechanism and energetics. *Annu. Rev. Biochem.* **72**, 517–571.

A good summary of the principles of enzyme catalysis as currently understood, and what we still do not understand.

Phosphoryl Group Transfers and ATP

Alberty, R.A. (1994) Biochemical thermodynamics. *Biochim. Biophys. Acta* **1207**, 1–11.

Explains the distinction between biochemical and chemical equations, and the calculation and meaning of transformed thermodynamic properties for ATP and other phosphorylated compounds.

Bridger, W.A. & Henderson, J.F. (1983) *Cell ATP,* John Wiley & Sons, Inc., New York.

The chemistry of ATP, its place in metabolic regulation, and its catabolic and anabolic roles.

Brown, M.R.W. & Kornberg, A. (2004) Inorganic polyphosphate in the origin and survival of species. *Proc. Natl. Acad. Sci. USA* **101**, 16,085–16,087.

Fraley, C.D., Rashid, M.H., Lee, S.S.K., Gottschalk, R., Harrison, J., Wood, P.J., Brown, M.R.W, & Kornberg, A. (2007) A polyphosphate kinase 1 (ppk1) mutant of *Pseudomonas aeruginosa* exhibits multiple ultrastructural and functional defects. *Proc. Natl. Acad. Sci. USA* **104**, 3526–3531.

Frey, P.A. & Arabshahi, A. (1995) Standard free-energy change for the hydrolysis of the α–β-phosphoanhydride bridge in ATP. *Biochemistry* **34**, 11,307–11,310.

Hanson, R.W. (1989) The role of ATP in metabolism. *Biochem. Educ.* **17**, 86–92.

Excellent summary of the chemistry and biology of ATP.

Kalckar, H.M. (1991) Fifty years of biological research: from oxidative phosphorylation to energy requiring transport regulation. *Annu. Rev. Biochem.* **60**, 1–37.

Intermediate-level discussion of the history of ATP studies, in which the author was a major player.

Kornberg, A. (1999) Inorganic polyphosphate: a molecule of many functions. *Annu. Rev. Biochem.* **68**, 89–125.

Lipmann, F. (1941) Metabolic generation and utilization of phosphate bond energy. *Adv. Enzymol.* **11**, 99–162.

The classic description of the role of high-energy phosphate compounds in biology.

Pullman, B. & Pullman, A. (1960) Electronic structure of energy-rich phosphates. *Radiat. Res.,* Suppl. 2, 160–181.

An advanced discussion of the chemistry of ATP and other "energy-rich" compounds.

Rees, D.C. & Howard, J.B. (1999) Structural bioenergetics and energy transduction mechanisms. *J. Mol. Biol.* **293**, 343–350.

Discussion of the structural basis for the efficient coupling of two energetic processes by way of changes in conformational states.

Veech, R.L., Lawson, J.W.R., Cornell, N.W., & Krebs, H.A. (1979) Cytosolic phosphorylation potential. *J. Biol. Chem.* **254**, 6538–6547.

Experimental determination of ATP, ADP, and P_i concentrations in brain, muscle, and liver, and a discussion of the difficulties in determining the real free-energy change for ATP synthesis in cells.

Westheimer, F.H. (1987) Why nature chose phosphates. *Science* **235**, 1173–1178.

A chemist's description of the unique suitability of phosphate esters and anhydrides for metabolic transformations.

Biological Oxidation-Reduction Reactions

Cashmore, A.R., Jarillo, J.A., Wu, Y.J., & Liu, D. (1999) Cryptochromes: blue light receptors for plants and animals. *Science* **284**, 760–765.

Dolphin, D., Avramović, O., & Poulson, R. (eds). (1987) *Pyridine Nucleotide Coenzymes: Chemical, Biochemical, and Medical Aspects*, John Wiley & Sons, Inc., New York.

An excellent two-volume collection of authoritative reviews. Among the most useful are the chapters by Kaplan, Westheimer, Veech, and Ohno and Ushio.

Fraaije, M.W. & Mattevi, A. (2000) Flavoenzymes: diverse catalysts with recurrent features. *Trends Biochem. Sci.* **25**, 126–132.

Hosler, J.P., Ferguson-Miller, S., & Mills, D.S. (2006) Energy transduction: proton transfer through the respiratory complexes. *Annu. Rev. Biochem.* **75**, 165–187.

Massey, V. (1994) Activation of molecular oxygen by flavins and flavoproteins. *J. Biol. Chem.* **269**, 22,459–22,462.

A short review of the chemistry of flavin-oxygen interactions in flavoproteins.

Rees, D.C. (2002) Great metalloclusters in enzymology. *Annu. Rev. Biochem.* **71**, 221–246.

Advanced review of the types of metal ion clusters found in enzymes and their modes of action.

Roehm, K.-H. (2001) Electron carriers: proteins and cofactors in oxidative phosphorylation. In *Encyclopedia of Life Sciences*, John Wiley & Sons, Inc./Wiley InterScience, www.els.net.

Good overview of the different classes of electron carriers that participate in respiration.

Williams, R.E. & Bruce, N.C. (2002) New uses for an old enzyme—the old yellow enzyme family of flavoenzymes. *Microbiology* **148**, 1607–1614.

Problems

1. Entropy Changes during Egg Development Consider a system consisting of an egg in an incubator. The white and yolk of the egg contain proteins, carbohydrates, and lipids. If fertilized, the egg is transformed from a single cell to a complex organism. Discuss this irreversible process in terms of the entropy changes in the system, surroundings, and universe. Be sure that you first clearly define the system and surroundings.

2. Calculation of $\Delta G'^\circ$ from an Equilibrium Constant Calculate the standard free-energy change for each of the following metabolically important enzyme-catalyzed reactions, using the equilibrium constants given for the reactions at 25 °C and pH 7.0.

(a) Glutamate + oxaloacetate $\xrightleftharpoons{\text{aspartate aminotransferase}}$

aspartate + α-ketoglutarate $\quad K'_{eq} = 6.8$

(b) Dihydroxyacetone phosphate $\xrightleftharpoons{\text{triose phosphate isomerase}}$

glyceraldehyde 3-phosphate $\quad K'_{eq} = 0.0475$

(c) Fructose 6-phosphate + ATP $\xrightleftharpoons{\text{phosphofructokinase}}$

fructose 1,6-bisphosphate + ADP $\quad K'_{eq} = 254$

3. Calculation of the Equilibrium Constant from $\Delta G'^\circ$ Calculate the equilibrium constant K'_{eq} for each of the following reactions at pH 7.0 and 25 °C, using the $\Delta G'^\circ$ values in Table 13–4.

(a) Glucose 6-phosphate + H_2O $\xrightleftharpoons{\text{glucose 6-phosphatase}}$

glucose + P_i

(b) Lactose + H_2O $\xrightleftharpoons{\text{β-galactosidase}}$ glucose + galactose

(c) Malate $\xrightleftharpoons{\text{fumarase}}$ fumarate + H_2O

4. Experimental Determination of K'_{eq} and $\Delta G'^\circ$ If a 0.1 M solution of glucose 1-phosphate at 25 °C is incubated with a catalytic amount of phosphoglucomutase, the glucose 1-phosphate is transformed to glucose 6-phosphate. At equilibrium, the concentrations of the reaction components are

Glucose 1-phosphate $\rightleftharpoons$ glucose 6-phosphate
4.5×10^{-3} M $\qquad\quad$ 9.6×10^{-2} M

Calculate K'_{eq} and $\Delta G'^\circ$ for this reaction.

5. Experimental Determination of $\Delta G'^\circ$ for ATP Hydrolysis A direct measurement of the standard free-energy change associated with the hydrolysis of ATP is technically demanding because the minute amount of ATP remaining at equilibrium is difficult to measure accurately. The value of $\Delta G'^\circ$ can be calculated indirectly, however, from the equilibrium constants of two other enzymatic reactions having less favorable equilibrium constants:

Glucose 6-phosphate + H_2O $\longrightarrow$ glucose + P_i $\quad K'_{eq} = 270$

ATP + glucose $\longrightarrow$ ADP + glucose 6-phosphate $\quad K'_{eq} = 890$

Using this information for equilibrium constants determined at 25 °C, calculate the standard free energy of hydrolysis of ATP.

6. Difference between $\Delta G'^\circ$ and ΔG Consider the following interconversion, which occurs in glycolysis (Chapter 14):

Fructose 6-phosphate $\rightleftharpoons$ glucose 6-phosphate
$K'_{eq} = 1.97$

(a) What is $\Delta G'^\circ$ for the reaction (K'_{eq} measured at 25 °C)?

(b) If the concentration of fructose 6-phosphate is adjusted to 1.5 M and that of glucose 6-phosphate is adjusted to 0.50 M, what is ΔG?

(c) Why are $\Delta G'^\circ$ and ΔG different?

7. Free Energy of Hydrolysis of CTP Compare the structure of the nucleoside triphosphate CTP with the structure of ATP.

Cytidine triphosphate (CTP)

Adenosine triphosphate (ATP)

Now predict the K'_{eq} and $\Delta G'^\circ$ for the following reaction:

ATP + CDP $\longrightarrow$ ADP + CTP

8. Dependence of ΔG on pH The free energy released by the hydrolysis of ATP under standard conditions is -30.5 kJ/mol. If ATP is hydrolyzed under standard conditions except at pH 5.0, is more or less free energy released? Explain. Use the Living Graph to explore this relationship. 🖱

9. The $\Delta G'^\circ$ for Coupled Reactions Glucose 1-phosphate is converted into fructose 6-phosphate in two successive reactions:

Glucose 1-phosphate $\longrightarrow$ glucose 6-phosphate

Glucose 6-phosphate $\longrightarrow$ fructose 6-phosphate

Using the $\Delta G'^\circ$ values in Table 13–4, calculate the equilibrium constant, K'_{eq}, for the sum of the two reactions:

Glucose 1-phosphate $\longrightarrow$ fructose 6-phosphate

10. Effect of [ATP]/[ADP] Ratio on Free Energy of Hydrolysis of ATP Using Equation 13–4, plot ΔG against ln Q (mass-action ratio) at 25 °C for the concentrations of ATP, ADP, and P_i in the table below. $\Delta G'^\circ$ for the reaction is -30.5 kJ/mol. Use the resulting plot to explain why metabolism is regulated to keep the ratio [ATP]/[ADP] high.

	Concentration (mM)				
ATP	5	3	1	0.2	5
ADP	0.2	2.2	4.2	5.0	25
P_i	10	12.1	14.1	14.9	10

11. Strategy for Overcoming an Unfavorable Reaction: ATP-Dependent Chemical Coupling The phosphorylation of glucose to glucose 6-phosphate is the initial step in the catabolism of glucose. The direct phosphorylation of glucose by P_i is described by the equation

$$\text{Glucose} + P_i \longrightarrow \text{glucose 6-phosphate} + H_2O$$

$$\Delta G'^{\circ} = 13.8 \text{ kJ/mol}$$

(a) Calculate the equilibrium constant for the above reaction at 37 °C. In the rat hepatocyte the physiological concentrations of glucose and P_i are maintained at approximately 4.8 mM. What is the equilibrium concentration of glucose 6-phosphate obtained by the direct phosphorylation of glucose by P_i? Does this reaction represent a reasonable metabolic step for the catabolism of glucose? Explain.

(b) In principle, at least, one way to increase the concentration of glucose 6-phosphate is to drive the equilibrium reaction to the right by increasing the intracellular concentrations of glucose and P_i. Assuming a fixed concentration of P_i at 4.8 mM, how high would the intracellular concentration of glucose have to be to give an equilibrium concentration of glucose 6-phosphate of 250 μM (the normal physiological concentration)? Would this route be physiologically reasonable, given that the maximum solubility of glucose is less than 1 M?

(c) The phosphorylation of glucose in the cell is coupled to the hydrolysis of ATP; that is, part of the free energy of ATP hydrolysis is used to phosphorylate glucose:

(1) $\text{Glucose} + P_i \longrightarrow \text{glucose 6-phosphate} + H_2O$

$$\Delta G'^{\circ} = 13.8 \text{ kJ/mol}$$

(2) $\text{ATP} + H_2O \longrightarrow \text{ADP} + P_i$ $\Delta G'^{\circ} = -30.5 \text{ kJ/mol}$

Sum: $\text{Glucose} + \text{ATP} \longrightarrow \text{glucose 6-phosphate} + \text{ADP}$

Calculate K'_{eq} at 37 °C for the overall reaction. For the ATP-dependent phosphorylation of glucose, what concentration of glucose is needed to achieve a 250 μM intracellular concentration of glucose 6-phosphate when the concentrations of ATP and ADP are 3.38 mM and 1.32 mM, respectively? Does this coupling process provide a feasible route, at least in principle, for the phosphorylation of glucose in the cell? Explain.

(d) Although coupling ATP hydrolysis to glucose phosphorylation makes thermodynamic sense, we have not yet specified how this coupling is to take place. Given that coupling requires a common intermediate, one conceivable route is to use ATP hydrolysis to raise the intracellular concentration of P_i and thus drive the unfavorable phosphorylation of glucose by P_i. Is this a reasonable route? (Think about the solubility products of metabolic intermediates.)

(e) The ATP-coupled phosphorylation of glucose is catalyzed in hepatocytes by the enzyme glucokinase. This enzyme binds ATP and glucose to form a glucose-ATP-enzyme complex, and the phosphoryl group is transferred directly from ATP to glucose. Explain the advantages of this route.

12. Calculations of $\Delta G'^{\circ}$ for ATP-Coupled Reactions
From data in Table 13–6 calculate the $\Delta G'^{\circ}$ value for the following reactions

(a) Phosphocreatine + ADP $\longrightarrow$ creatine + ATP

(b) ATP + fructose $\longrightarrow$ ADP + fructose 6-phosphate

13. Coupling ATP Cleavage to an Unfavorable Reaction
To explore the consequences of coupling ATP hydrolysis

under physiological conditions to a thermodynamically unfavorable biochemical reaction, consider the hypothetical transformation X → Y, for which $\Delta G'^{\circ} = 20.0$ kJ/mol.

(a) What is the ratio [Y]/[X] at equilibrium?

(b) Suppose X and Y participate in a sequence of reactions during which ATP is hydrolyzed to ADP and P_i. The overall reaction is

$$\text{X} + \text{ATP} + H_2O \longrightarrow \text{Y} + \text{ADP} + P_i$$

Calculate [Y]/[X] for this reaction at equilibrium. Assume that the temperature is 25 °C and the equilibrium concentrations of ATP, ADP, and P_i are 1 M.

(c) We know that [ATP], [ADP], and [P_i] are *not* 1 M under physiological conditions. Calculate [Y]/[X] for the ATP-coupled reaction when the values of [ATP], [ADP], and [P_i] are those found in rat myocytes (Table 13–5).

14. Calculations of ΔG at Physiological Concentrations
Calculate the actual, physiological ΔG for the reaction

$$\text{Phosphocreatine} + \text{ADP} \longrightarrow \text{creatine} + \text{ATP}$$

at 37 °C, as it occurs in the cytosol of neurons, with phosphocreatine at 4.7 mM, creatine at 1.0 mM, ADP at 0.73 mM, and ATP at 2.6 mM.

15. Free Energy Required for ATP Synthesis under Physiological Conditions In the cytosol of rat hepatocytes, the temperature is 37 °C and the mass-action ratio, Q, is

$$\frac{[\text{ATP}]}{[\text{ADP}][P_i]} = 5.33 \times 10^2 \text{ M}^{-1}$$

Calculate the free energy required to synthesize ATP in a rat hepatocyte.

16. Chemical Logic In the glycolytic pathway, a six-carbon sugar (fructose 1,6-bisphosphate) is cleaved to form two three-carbon sugars, which undergo further metabolism (see Fig. 14–6). In this pathway, an isomerization of glucose 6-phosphate to fructose 6-phosphate (shown below) occurs two steps before the cleavage reaction (the intervening step is phosphorylation of fructose 6-phosphate to fructose 1,6-bisphosphate (p. 549)).

Glucose 6-phosphate   Fructose 6-phosphate

What does the isomerization step accomplish from a chemical perspective? (Hint: Consider what might happen if the C—C bond cleavage were to proceed without the preceding isomerization.)

17. Enzymatic Reaction Mechanisms I Lactate dehydrogenase is one of the many enzymes that require NADH as coenzyme. It catalyzes the conversion of pyruvate to lactate:

Draw the mechanism of this reaction (show electron-pushing arrows). (Hint: This is a common reaction throughout metabolism; the mechanism is similar to that catalyzed by other dehydrogenases that use NADH, such as alcohol dehydrogenase.)

18. Enzymatic Reaction Mechanisms II Biochemical reactions often look more complex than they really are. In the pentose phosphate pathway (Chapter 14), sedoheptulose 7-phosphate and glyceraldehyde 3-phosphate react to form erythrose 4-phosphate and fructose 6-phosphate in a reaction catalyzed by transaldolase.

Draw a mechanism for this reaction (show electron-pushing arrows). (Hint: Take another look at aldol condensations, then consider the name of this enzyme.)

19. Daily ATP Utilization by Human Adults

(a) A total of 30.5 kJ/mol of free energy is needed to synthesize ATP from ADP and P_i when the reactants and products are at 1 M concentrations and the temperature is 25 °C (standard state). Because the actual physiological concentrations of ATP, ADP, and P_i are not 1 M, and the temperature is 37 °C, the free energy required to synthesize ATP under physiological conditions is different from $\Delta G'^\circ$. Calculate the free energy required to synthesize ATP in the human hepatocyte when the

physiological concentrations of ATP, ADP, and P_i are 3.5, 1.50, and 5.0 mM, respectively.

(b) A 68 kg (150 lb) adult requires a caloric intake of 2,000 kcal (8,360 kJ) of food per day (24 hours). The food is metabolized and the free energy is used to synthesize ATP, which then provides energy for the body's daily chemical and mechanical work. Assuming that the efficiency of converting food energy into ATP is 50%, calculate the weight of ATP used by a human adult in 24 hours. What percentage of the body weight does this represent?

(c) Although adults synthesize large amounts of ATP daily, their body weight, structure, and composition do not change significantly during this period. Explain this apparent contradiction.

20. Rates of Turnover of γ and β Phosphates of ATP If a small amount of ATP labeled with radioactive phosphorus in the terminal position, $[\gamma\text{-}^{32}P]ATP$, is added to a yeast extract, about half of the ^{32}P activity is found in P_i within a few minutes, but the concentration of ATP remains unchanged. Explain. If the same experiment is carried out using ATP labeled with ^{32}P in the central position, $[\beta\text{-}^{32}P]ATP$, the ^{32}P does not appear in P_i within such a short time. Why?

21. Cleavage of ATP to AMP and PP_i during Metabolism Synthesis of the activated form of acetate (acetyl-CoA) is carried out in an ATP-dependent process:

$$\text{Acetate} + \text{CoA} + \text{ATP} \longrightarrow \text{acetyl-CoA} + \text{AMP} + PP_i$$

(a) The $\Delta G'^\circ$ for hydrolysis of acetyl-CoA to acetate and CoA is −32.2 kJ/mol and that for hydrolysis of ATP to AMP and PP_i is −30.5 kJ/mol. Calculate $\Delta G'^\circ$ for the ATP-dependent synthesis of acetyl-CoA.

(b) Almost all cells contain the enzyme inorganic pyrophosphatase, which catalyzes the hydrolysis of PP_i to P_i. What effect does the presence of this enzyme have on the synthesis of acetyl-CoA? Explain.

22. Energy for H$^+$ Pumping The parietal cells of the stomach lining contain membrane "pumps" that transport hydrogen ions from the cytosol (pH 7.0) into the stomach, contributing to the acidity of gastric juice (pH 1.0). Calculate the free energy required to transport 1 mol of hydrogen ions through these pumps. (Hint: See Chapter 11.) Assume a temperature of 37 °C.

23. Standard Reduction Potentials The standard reduction potential, E'°, of any redox pair is defined for the half-cell reaction:

$$\text{Oxidizing agent} + n \text{ electrons} \longrightarrow \text{reducing agent}$$

The E'° values for the NAD$^+$/NADH and pyruvate/lactate conjugate redox pairs are −0.32 V and −0.19 V, respectively.

(a) Which redox pair has the greater tendency to lose electrons? Explain.

(b) Which pair is the stronger oxidizing agent? Explain.

(c) Beginning with 1 M concentrations of each reactant and product at pH 7 and 25 °C, in which direction will the following reaction proceed?

$$\text{Pyruvate} + \text{NADH} + \text{H}^+ \rightleftharpoons \text{lactate} + \text{NAD}^+$$

(d) What is the standard free-energy change ($\Delta G'^\circ$) for the conversion of pyruvate to lactate?

(e) What is the equilibrium constant (K'_{eq}) for this reaction?

24. Energy Span of the Respiratory Chain Electron transfer in the mitochondrial respiratory chain may be represented by the net reaction equation

$$NADH + H^+ + \tfrac{1}{2}O_2 \rightleftharpoons H_2O + NAD^+$$

(a) Calculate $\Delta E'^\circ$ for the net reaction of mitochondrial electron transfer. Use E'° values from Table 13–7.

(b) Calculate $\Delta G'^\circ$ for this reaction.

(c) How many ATP molecules can *theoretically* be generated by this reaction if the free energy of ATP synthesis under cellular conditions is 52 kJ/mol?

25. Dependence of Electromotive Force on Concentrations Calculate the electromotive force (in volts) registered by an electrode immersed in a solution containing the following mixtures of NAD^+ and NADH at pH 7.0 and 25 °C, with reference to a half-cell of E'° 0.00 V.

(a) 1.0 mM NAD^+ and 10 mM NADH

(b) 1.0 mM NAD^+ and 1.0 mM NADH

(c) 10 mM NAD^+ and 1.0 mM NADH

26. Electron Affinity of Compounds List the following in order of increasing tendency to accept electrons: (a) α-ketoglutarate + CO_2 (yielding isocitrate); (b) oxaloacetate; (c) O_2; (d) $NADP^+$.

27. Direction of Oxidation-Reduction Reactions Which of the following reactions would you expect to proceed in the direction shown, under standard conditions, in the presence of the appropriate enzymes?

(a) Malate + NAD^+ $\longrightarrow$ oxaloacetate + NADH + H^+

(b) Acetoacetate + NADH + H^+ $\longrightarrow$
β-hydroxybutyrate + NAD^+

(c) Pyruvate + NADH + H^+ $\longrightarrow$ lactate + NAD^+

(d) Pyruvate + β-hydroxybutyrate $\longrightarrow$
lactate + acetoacetate

(e) Malate + pyruvate $\longrightarrow$ oxaloacetate + lactate

(f) Acetaldehyde + succinate $\longrightarrow$ ethanol + fumarate

Data Analysis Problem

28. Thermodynamics Can Be Tricky Thermodynamics is a challenging area of study and one with many opportunities for confusion. An interesting example is found in an article by Robinson, Hampson, Munro, and Vaney, published in *Science* in 1993. Robinson and colleagues studied the movement of small molecules between neighboring cells of the nervous system through cell-to-cell channels (gap junctions). They found that the dyes Lucifer yellow (a small, negatively charged molecule) and biocytin (a small zwitterionic molecule) moved in

only one direction between two particular types of glia (non-neuronal cells of the nervous system). Dye injected into astrocytes would rapidly pass into adjacent astrocytes, oligodendrocytes, or Müller cells, but dye injected into oligodendrocytes or Müller cells passed slowly if at all into astrocytes. All of these cell types are connected by gap junctions.

Although it was not a central point of their article, the authors presented a molecular model for how this unidirectional transport might occur, as shown in their Figure 3:

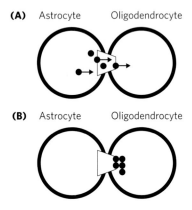

The figure legend reads: "Model of the unidirectional diffusion of dye between coupled oligodendrocytes and astrocytes, based on differences in connection pore diameter. Like a fish in a fish trap, dye molecules (black circles) can pass from an astrocyte to an oligodendrocyte (A) but not back in the other direction (B)."

Although this article clearly passed review at a well-respected journal, several letters to the editor (1994) followed, showing that Robinson and coauthors' model violated the second law of thermodynamics.

(a) Explain how the model violates the second law. Hint: Consider what would happen to the entropy of the system if one started with equal concentrations of dye in the astrocyte and oligodendrocyte connected by the "fish trap" type of gap junctions.

(b) Explain why this model cannot work for small molecules, although it may allow one to catch fish.

(c) Explain why a fish trap *does* work for fish.

(d) Provide two plausible mechanisms for the unidirectional transport of dye molecules between the cells that do not violate the second law of thermodynamics.

References

Letters to the editor. (1994) *Science* **265**, 1017–1019.

Robinson, S.R., Hampson, E.C.G.M., Munro, M.N., & Vaney, D.I. (1993) Unidirectional coupling of gap junctions between neuroglia. *Science* **262**, 1072–1074.

Glycolysis, Gluconeogenesis, and the Pentose Phosphate Pathway

Glucose occupies a central position in the metabolism of plants, animals, and many microorganisms. It is relatively rich in potential energy, and thus a good fuel; the complete oxidation of glucose to carbon dioxide and water proceeds with a standard free-energy change of −2,840 kJ/mol. By storing glucose as a high molecular weight polymer such as starch or glycogen, a cell can stockpile large quantities of hexose units while maintaining a relatively low cytosolic osmolarity. When energy demands increase, glucose can be released from these intracellular storage polymers and used to produce ATP either aerobically or anaerobically.

Glucose is not only an excellent fuel, it is also a remarkably versatile precursor, capable of supplying a huge array of metabolic intermediates for biosynthetic reactions. A bacterium such as *Escherichia coli* can obtain from glucose the carbon skeletons for every amino acid, nucleotide, coenzyme, fatty acid, or other metabolic intermediate it needs for growth. A comprehensive study of the metabolic fates of glucose would encompass hundreds or thousands of transformations. In animals and vascular plants, glucose has four major fates: it may be used in the synthesis of complex polysaccharides destined for the extracellular space; stored in cells (as a polysaccharide or as sucrose); oxidized to a three-carbon compound (pyruvate) via glycolysis to provide ATP and metabolic intermediates; or oxidized via the pentose phosphate (phosphogluconate) pathway to yield ribose 5-phosphate for

nucleic acid synthesis and NADPH for reductive biosynthetic processes **(Fig. 14–1)**.

Organisms that do not have access to glucose from other sources must make it. Photosynthetic organisms make glucose by first reducing atmospheric CO_2 to trioses, then converting the trioses to glucose. Nonphotosynthetic cells make glucose from simpler three- and four-carbon precursors by the process of gluconeogenesis, effectively reversing glycolysis in a pathway that uses many of the glycolytic enzymes.

In this chapter we describe the individual reactions of glycolysis, gluconeogenesis, and the pentose phosphate pathway and the functional significance of each pathway. We also describe the various metabolic fates of the pyruvate produced by glycolysis. They include the fermentations that are used by many organisms in anaerobic niches to produce ATP and that are exploited industrially as sources of ethanol, lactic acid, and other commercially useful products. And we look at the pathways that feed various sugars from mono-, di-, and polysaccharides into the glycolytic pathway. The discussion of glucose metabolism continues in Chapter 15, where we

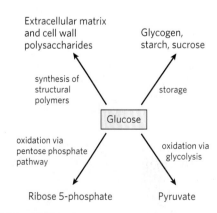

FIGURE 14–1 Major pathways of glucose utilization. Although not the only possible fates for glucose, these four pathways are the most significant in terms of the amount of glucose that flows through them in most cells.

use the processes of carbohydrate synthesis and degradation to illustrate the many mechanisms by which organisms regulate metabolic pathways. The biosynthetic pathways from glucose to extracellular matrix and cell wall polysaccharides and storage polysaccharides are discussed in Chapter 20.

14.1 Glycolysis

In **glycolysis** (from the Greek *glykys*, "sweet" or "sugar," and *lysis*, "splitting"), a molecule of glucose is degraded in a series of enzyme-catalyzed reactions to yield two molecules of the three-carbon compound pyruvate. During the sequential reactions of glycolysis, some of the free energy released from glucose is conserved in the form of ATP and NADH. Glycolysis was the first metabolic pathway to be elucidated and is probably the best understood. From Eduard Buchner's discovery in 1897 of fermentation in broken extracts of yeast cells until the elucidation of the whole pathway in yeast (by Otto Warburg and Hans von Euler-Chelpin) and in muscle (by Gustav Embden and Otto Meyerhof) in the 1930s, the reactions of glycolysis in extracts of yeast and muscle were a major focus of biochemical research. The philosophical shift that accompanied these discoveries was announced by Jacques Loeb in 1906:

> Through the discovery of Buchner, Biology was relieved of another fragment of mysticism. The splitting up of sugar into CO₂ and alcohol is no more the effect of a "vital principle" than the splitting up of cane sugar by invertase. The history of this problem is instructive, as it warns us against considering problems as beyond our reach because they have not yet found their solution.

The development of methods of enzyme purification, the discovery and recognition of the importance of coenzymes such as NAD, and the discovery of the pivotal metabolic role of ATP and other phosphorylated compounds all came out of studies of glycolysis. The glycolytic enzymes of many species have long since been purified and thoroughly studied.

Glycolysis is an almost universal central pathway of glucose catabolism, the pathway with the largest flux of carbon in most cells. The glycolytic breakdown of glucose is the sole source of metabolic energy in some mammalian tissues and cell types (erythrocytes, renal medulla, brain, and sperm, for example). Some plant tissues that are modified to store starch (such as potato tubers) and some aquatic plants (watercress, for example) derive most of their energy from glycolysis; many anaerobic microorganisms are entirely dependent on glycolysis.

Fermentation is a general term for the *anaerobic* degradation of glucose or other organic nutrients to obtain energy, conserved as ATP. Because living organisms first arose in an atmosphere without oxygen, anaerobic breakdown of glucose is probably the most ancient biological mechanism for obtaining energy from organic fuel molecules. And as genome sequencing of a wide variety of organisms has revealed, some archaea and some parasitic microorganisms lack one or more of the enzymes of glycolysis but retain the core of the pathway; they presumably carry out variant forms of glycolysis. In the course of evolution, the chemistry of this reaction sequence has been completely conserved; the glycolytic enzymes of vertebrates are closely similar, in amino acid sequence and three-dimensional structure, to their homologs in yeast and spinach. Glycolysis differs among species only in the details of its regulation and in the subsequent metabolic fate of the pyruvate formed. The thermodynamic principles and the types of regulatory mechanisms that govern glycolysis are common to all pathways of cell metabolism. The glycolytic pathway, of central importance in itself, can also serve as a model for many aspects of the pathways discussed throughout this book.

Before examining each step of the pathway in some detail, we take a look at glycolysis as a whole.

An Overview: Glycolysis Has Two Phases

The breakdown of the six-carbon glucose into two molecules of the three-carbon pyruvate occurs in 10 steps, the first 5 of which constitute the *preparatory phase* **(Fig. 14–2a)**. In these reactions, glucose is first phosphorylated at the hydroxyl group on C-6 (step ❶). The D-glucose 6-phosphate thus formed is converted to D-fructose 6-phosphate (step ❷), which is again phosphorylated, this time at C-1, to yield D-fructose 1,6-bisphosphate (step ❸). For both phosphorylations, ATP is the phosphoryl group donor. As all sugar derivatives in glycolysis are the D isomers, we will usually omit the D designation except when emphasizing stereochemistry.

Fructose 1,6-bisphosphate is split to yield two three-carbon molecules, dihydroxyacetone phosphate and glyceraldehyde 3-phosphate (step ❹); this is the "lysis" step that gives the pathway its name. The dihydroxyacetone phosphate is isomerized to a second molecule of glyceraldehyde 3-phosphate (step ❺),

Hans Von Euler-Chelpin, 1873-1964

Gustav Embden, 1874-1933

Otto Meyerhof, 1884-1951

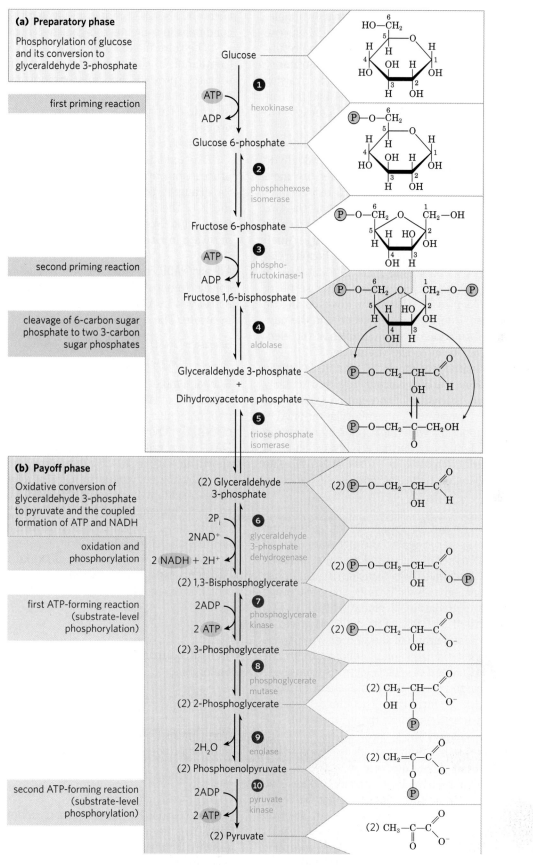

FIGURE 14–2 The two phases of glycolysis. For each molecule of glucose that passes through the preparatory phase **(a)**, two molecules of glyceraldehyde 3-phosphate are formed; both pass through the payoff phase **(b)**. Pyruvate is the end product of the second phase of glycolysis. For each glucose molecule, two ATP are consumed in the preparatory phase and four ATP are produced in the payoff phase, giving a net yield of two ATP per molecule of glucose converted to pyruvate. The numbered reaction steps correspond to the numbered headings in the text discussion. Keep in mind that each phosphoryl group, represented here as P, has two negative charges $(-PO_3^{2-})$.

ending the first phase of glycolysis. Note that two molecules of ATP are invested before the cleavage of glucose into two three-carbon pieces; there will be a good return on this investment. To summarize: in the preparatory phase of glycolysis the energy of ATP is invested, raising the free-energy content of the intermediates, and the carbon chains of all the metabolized hexoses are converted to a common product, glyceraldehyde 3-phosphate.

The energy gain comes in the *payoff phase* of glycolysis (Fig. 14–2b). Each molecule of glyceraldehyde 3-phosphate is oxidized and phosphorylated by inorganic phosphate (*not* by ATP) to form 1,3-bisphosphoglycerate (step ❻). Energy is then released as the two molecules of 1,3-bisphosphoglycerate are converted to two molecules of pyruvate (steps ❼ through ❿). Much of this energy is conserved by the coupled phosphorylation of four molecules of ADP to ATP. The net yield is two molecules of ATP per molecule of glucose used, because two molecules of ATP were invested in the preparatory phase. Energy is also conserved in the payoff phase in the formation of two molecules of the electron carrier NADH per molecule of glucose.

In the sequential reactions of glycolysis, three types of chemical transformations are particularly noteworthy: (1) degradation of the carbon skeleton of glucose to yield pyruvate; (2) phosphorylation of ADP to ATP by compounds with high phosphoryl group transfer potential, formed during glycolysis; and (3) transfer of a hydride ion to NAD^+, forming NADH. The overall chemical logic of the pathway is described in **Figure 14–3**.

Fates of Pyruvate

With the exception of some interesting variations in the bacterial realm, the pyruvate formed by glycolysis is further metabolized via one of three catabolic routes. In aerobic organisms or tissues, under aerobic conditions, glycolysis is only the first stage in the complete degradation of glucose **(Fig. 14–4)**. Pyruvate is oxidized, with loss of its carboxyl group as CO_2, to yield the acetyl group of acetyl-coenzyme A; the acetyl group is then oxidized completely to CO_2 by the citric acid cycle (Chapter 16). The electrons from these oxidations are passed to O_2 through a chain of carriers in mitochondria, to form H_2O. The energy from the electron-transfer reactions drives the synthesis of ATP in mitochondria (Chapter 19).

The second route for pyruvate is its reduction to lactate via **lactic acid fermentation**. When vigorously contracting skeletal muscle must function under low-oxygen conditions (**hypoxia**), NADH cannot be reoxidized to NAD^+, but NAD^+ is required as an electron acceptor for the further oxidation of pyruvate. Under these conditions pyruvate is reduced to lactate, accepting electrons from NADH and thereby regenerating the NAD^+ necessary for glycolysis to continue. Certain tissues and cell types (retina and erythrocytes, for example) convert glucose to lactate even under aerobic conditions, and

lactate is also the product of glycolysis under anaerobic conditions in some microorganisms (Fig. 14–4).

The third major route of pyruvate catabolism leads to ethanol. In some plant tissues and in certain invertebrates, protists, and microorganisms such as brewer's or baker's yeast, pyruvate is converted under hypoxic or anaerobic conditions to ethanol and CO_2, a process called **ethanol (alcohol) fermentation** (Fig. 14–4).

The oxidation of pyruvate is an important catabolic process, but pyruvate has anabolic fates as well. It can, for example, provide the carbon skeleton for the synthesis of the amino acid alanine or for the synthesis of fatty acids. We return to these anabolic reactions of pyruvate in later chapters.

ATP and NADH Formation Coupled to Glycolysis

During glycolysis some of the energy of the glucose molecule is conserved in ATP, while much remains in the product, pyruvate. The overall equation for glycolysis is

$$\text{Glucose} + 2NAD^+ + 2ADP + 2P_i \longrightarrow$$
$$2 \text{ pyruvate} + 2NADH + 2H^+ + 2ATP + 2H_2O \quad (14\text{–}1)$$

For each molecule of glucose degraded to pyruvate, two molecules of ATP are generated from ADP and P_i, and two molecules of NADH are produced by the reduction of NAD^+. The hydrogen acceptor in this reaction is NAD^+ (see Fig. 13–24), bound to a Rossmann fold as shown in Figure 13–25. The reduction of NAD^+ proceeds by the enzymatic transfer of a hydride ion $(:H^-)$ from the aldehyde group of glyceraldehyde 3-phosphate to the nicotinamide ring of NAD^+, yielding the reduced coenzyme NADH. The other hydrogen atom of the substrate molecule is released to the solution as H^+.

We can now resolve the equation of glycolysis into two processes—the conversion of glucose to pyruvate, which is exergonic:

$$\text{Glucose} + 2NAD^+ \longrightarrow 2 \text{ pyruvate} + 2NADH + 2H^+ \quad (14\text{–}2)$$
$$\Delta G_1'^\circ = -146 \text{ kJ/mol}$$

and the formation of ATP from ADP and P_i, which is endergonic:

$$2ADP + 2P_i \longrightarrow 2ATP + 2H_2O \quad (14\text{–}3)$$
$$\Delta G_2'^\circ = 2(30.5 \text{ kJ/mol}) = 61.0 \text{ kJ/mol}$$

The sum of Equations 14–2 and 14–3 gives the overall standard free-energy change of glycolysis, $\Delta G_s'^\circ$:

$$\Delta G_s'^\circ = \Delta G_1'^\circ + \Delta G_2'^\circ = -146 \text{ kJ/mol} + 61.0 \text{ kJ/mol}$$
$$= -85 \text{ kJ/mol}$$

Under standard conditions, and under the (nonstandard) conditions that prevail in a cell, glycolysis is an essentially irreversible process, driven to completion by a large net decrease in free energy.

Energy Remaining in Pyruvate

Glycolysis releases only a small fraction of the total available energy of the glucose

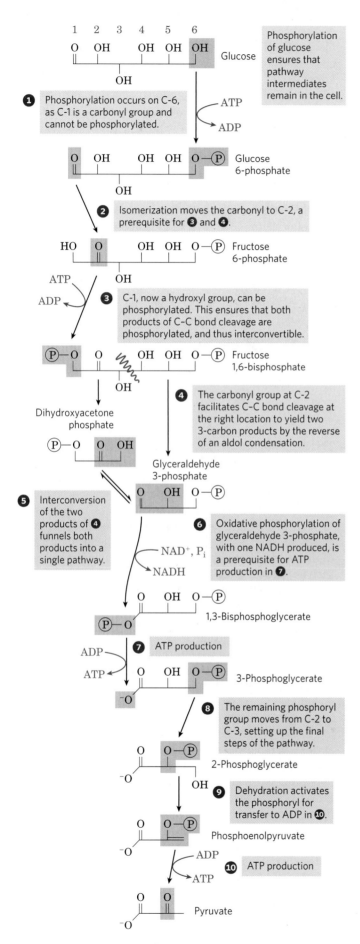

1 Phosphorylation occurs on C-6, as C-1 is a carbonyl group and cannot be phosphorylated.

Phosphorylation of glucose ensures that pathway intermediates remain in the cell.

2 Isomerization moves the carbonyl to C-2, a prerequisite for **3** and **4**.

3 C-1, now a hydroxyl group, can be phosphorylated. This ensures that both products of C–C bond cleavage are phosphorylated, and thus interconvertible.

4 The carbonyl group at C-2 facilitates C–C bond cleavage at the right location to yield two 3-carbon products by the reverse of an aldol condensation.

5 Interconversion of the two products of **4** funnels both products into a single pathway.

6 Oxidative phosphorylation of glyceraldehyde 3-phosphate, with one NADH produced, is a prerequisite for ATP production in **7**.

7 ATP production

8 The remaining phosphoryl group moves from C-2 to C-3, setting up the final steps of the pathway.

9 Dehydration activates the phosphoryl for transfer to ADP in **10**.

10 ATP production

molecule; the two molecules of pyruvate formed by glycolysis still contain most of the chemical potential energy of glucose, energy that can be extracted by oxidative reactions in the citric acid cycle (Chapter 16) and oxidative phosphorylation (Chapter 19).

Importance of Phosphorylated Intermediates Each of the nine glycolytic intermediates between glucose and pyruvate is phosphorylated (Fig. 14–2). The phosphoryl groups seem to have three functions.

1. Because the plasma membrane generally lacks transporters for phosphorylated sugars, the phosphorylated glycolytic intermediates cannot leave the cell. After the initial phosphorylation, no further energy is necessary to retain phosphorylated intermediates in the cell, despite the large difference in their intracellular and extracellular concentrations.

2. Phosphoryl groups are essential components in the enzymatic conservation of metabolic energy. Energy released in the breakage of phosphoanhydride bonds (such as those in ATP) is partially conserved in the formation of phosphate esters such as glucose 6-phosphate. High-energy phosphate compounds formed in glycolysis (1,3-bisphosphoglycerate and phosphoenolpyruvate) donate phosphoryl groups to ADP to form ATP.

3. Binding energy resulting from the binding of phosphate groups to the active sites of enzymes lowers the activation energy and increases the specificity of the enzymatic reactions (Chapter 6). The phosphate groups of ADP, ATP, and the glycolytic intermediates form complexes with Mg^{2+}, and the substrate binding sites of many glycolytic enzymes are specific for these Mg^{2+} complexes. Most glycolytic enzymes require Mg^{2+} for activity.

FIGURE 14–3 The chemical logic of the glycolytic pathway. In this simplified version of the pathway, each molecule is shown in a linear form, with carbon and hydrogen atoms not depicted, in order to highlight chemical transformations. Remember that glucose and fructose are present mostly in their cyclized forms in solution, although they are transiently present in linear form at the active sites of some of the enzymes in this pathway.

The preparatory phase, steps **1** to **5**, converts the six-carbon glucose into two three-carbon units, each of them phosphorylated. Oxidation of the three-carbon units is initiated in the payoff phase. To produce pyruvate, the chemical steps must occur in the order shown.

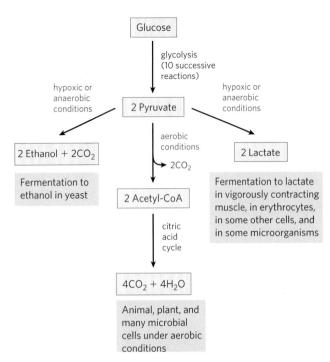

FIGURE 14–4 Three possible catabolic fates of the pyruvate formed in glycolysis. Pyruvate also serves as a precursor in many anabolic reactions, not shown here.

The Preparatory Phase of Glycolysis Requires ATP

In the preparatory phase of glycolysis, two molecules of ATP are invested and the hexose chain is cleaved into two triose phosphates. The realization that *phosphorylated* hexoses were intermediates in glycolysis came slowly and serendipitously. In 1906, Arthur Harden and William Young tested their hypothesis that inhibitors of proteolytic enzymes would stabilize the glucose-fermenting enzymes in yeast extract. They added blood serum (known to contain inhibitors of proteolytic enzymes) to yeast extracts and observed the predicted stimulation of glucose metabolism. However, in a control experiment intended to show that boiling the serum destroyed the stimulatory activity, they discovered that boiled serum was just as effective at stimulating glycolysis! Careful examination and testing of the contents of the boiled serum revealed that inorganic phosphate was

Arthur Harden,
1865–1940

William Young,
1878–1942

responsible for the stimulation. Harden and Young soon discovered that glucose added to their yeast extract was converted to a hexose bisphosphate (the "Harden-Young ester," eventually identified as fructose 1,6-bisphosphate). This was the beginning of a long series of investigations on the role of organic esters and anhydrides of phosphate in biochemistry, which has led to our current understanding of the central role of phosphoryl group transfer in biology.

❶ Phosphorylation of Glucose In the first step of glycolysis, glucose is activated for subsequent reactions by its phosphorylation at C-6 to yield **glucose 6-phosphate**, with ATP as the phosphoryl donor:

$$\Delta G'^\circ = -16.7 \text{ kJ/mol}$$

This reaction, which is irreversible under intracellular conditions, is catalyzed by **hexokinase**. Recall that kinases are enzymes that catalyze the transfer of the terminal phosphoryl group from ATP to an acceptor nucleophile (see Fig. 13–20). Kinases are a subclass of transferases (see Table 6–3). The acceptor in the case of hexokinase is a hexose, normally D-glucose, although hexokinase also catalyzes the phosphorylation of other common hexoses, such as D-fructose and D-mannose, in some tissues.

Hexokinase, like many other kinases, requires Mg^{2+} for its activity, because the true substrate of the enzyme is not ATP^{4-} but the $MgATP^{2-}$ complex (see Fig. 13–12). Mg^{2+} shields the negative charges of the phosphoryl groups in ATP, making the terminal phosphorus atom an easier target for nucleophilic attack by an —OH of glucose. Hexokinase undergoes a profound change in shape, an induced fit, when it binds glucose; two domains of the protein move about 8 Å closer to each other when ATP binds (see Fig. 6–25). This movement brings bound ATP closer to a molecule of glucose also bound to the enzyme and blocks the access of water (from the solvent), which might otherwise enter the active site and attack (hydrolyze) the phosphoanhydride bonds of ATP. Like the other nine enzymes of glycolysis, hexokinase is a soluble, cytosolic protein.

Hexokinase is present in nearly all organisms. The human genome encodes four different hexokinases (I to IV), all of which catalyze the same reaction. Two or more enzymes that catalyze the same reaction but are encoded by different genes are called **isozymes** (see Box 15–2). One of the isozymes present in hepatocytes, hexokinase IV (also called glucokinase), differs from

MECHANISM FIGURE 14–5 The phosphohexose isomerase reaction. The ring opening and closing reactions (steps ❶ and ❹) are catalyzed by an active-site His residue, by mechanisms omitted here for simplicity. The proton (light red) initially at C-2 is made more easily abstractable by electron withdrawal by the adjacent carbonyl and nearby hydroxyl groups. After its transfer from C-2 to the active-site Glu residue (a weak acid), the proton is freely exchanged with the surrounding solution; that is, the proton abstracted from C-2 in step ❷ is not necessarily the same one that is added to C-1 in step ❸.

🖱 Phosphohexose Isomerase Mechanism

other forms of hexokinase in kinetic and regulatory properties, with important physiological consequences that are described in Section 15.3.

❷ **Conversion of Glucose 6-Phosphate to Fructose 6-Phosphate** The enzyme **phosphohexose isomerase (phosphoglucose isomerase)** catalyzes the reversible isomerization of glucose 6-phosphate, an aldose, to **fructose 6-phosphate**, a ketose:

$$\Delta G'^{\circ} = 1.7 \text{ kJ/mol}$$

The mechanism for this reaction involves an enediol intermediate **(Fig. 14–5)**. The reaction proceeds readily in either direction, as might be expected from the relatively small change in standard free energy.

❸ **Phosphorylation of Fructose 6-Phosphate to Fructose 1,6-Bisphosphate** In the second of the two priming reactions of glycolysis, **phosphofructokinase-1 (PFK-1)** catalyzes the transfer of a phosphoryl group from ATP to fructose 6-phosphate to yield **fructose 1,6-bisphosphate**:

$$\Delta G'^{\circ} = -14.2 \text{ kJ/mol}$$

KEY CONVENTION: Compounds that contain two phosphate or phosphoryl groups attached at different positions in the molecule are named *bisphosphates* (or *bisphospho* compounds); for example, fructose 1,6-bisphosphate and 1,3-bisphosphoglycerate. Compounds with two phosphates linked together as a pyrophosphoryl group are named *diphosphates*; for example, adenosine diphosphate (ADP). Similar rules apply for the naming of *trisphosphates* (such as inositol 1,4,5-trisphosphate; see p. 450) and *triphosphates* (such as adenosine triphosphate, ATP). ■

The enzyme that forms fructose 1,6-bisphosphate is called PFK-1 to distinguish it from a second enzyme (PFK-2) that catalyzes the formation of fructose 2,6-bisphosphate from fructose 6-phosphate in a separate pathway (the roles of PFK-2 and fructose 2,6-bisphosphate are discussed in Chapter 15). The PFK-1 reaction is essentially irreversible under cellular conditions, and it is the first "committed" step in the glycolytic pathway; glucose 6-phosphate and fructose 6-phosphate have other possible fates, but fructose 1,6-bisphosphate is targeted for glycolysis.

Some bacteria and protists and perhaps all plants have a phosphofructokinase that uses pyrophosphate (PP_i), not ATP, as the phosphoryl group donor in the synthesis of fructose 1,6-bisphosphate:

$$\text{Fructose 6-phosphate} + \xrightarrow{\text{Mg}^{2+}}$$
$$\text{fructose 1,6-bisphosphate} + P_i$$
$$\Delta G'^{\circ} = -2.9 \text{ kJ/mol}$$

Phosphofructokinase-1 is subject to complex allosteric regulation; its activity is increased whenever the cell's ATP supply is depleted or when the ATP breakdown products, ADP and AMP (particularly the latter), accumulate. The enzyme is inhibited whenever the cell has ample ATP and is well supplied by other fuels such as fatty acids. In some organisms, fructose 2,6-bisphosphate (not to be confused with the PFK-1 reaction product, fructose 1,6-bisphosphate) is a potent allosteric activator of PFK-1. Ribulose 5-phosphate, an intermediate in the pentose phosphate pathway discussed later in this chapter, also activates phosphofructokinase indirectly. The multiple layers of regulation of this step in glycolysis are discussed in greater detail in Chapter 15.

❹ Cleavage of Fructose 1,6-Bisphosphate The enzyme **fructose 1,6-bisphosphate aldolase**, often called simply **aldolase**, catalyzes a reversible aldol condensation (see Fig. 13–4). Fructose 1,6-bisphosphate is cleaved to yield two different triose phosphates, **glyceraldehyde 3-phosphate**, an aldose, and **dihydroxyacetone phosphate**, a ketose:

$$\Delta G'^{\circ} = 23.8 \text{ kJ/mol}$$

There are two classes of aldolases. Class I aldolases, found in animals and plants, use the mechanism shown in **Figure 14–6**. Class II enzymes, in fungi and bacteria, do not form the Schiff base intermediate. Instead, a zinc ion at the active site is coordinated with the carbonyl oxygen at C-2; the Zn^{2+} polarizes the carbonyl group and stabilizes the enolate intermediate created in the C—C bond cleavage step (see Fig. 6–17).

Although the aldolase reaction has a strongly positive standard free-energy change in the direction of fructose 1,6-bisphosphate cleavage, at the lower concentrations of reactants present in cells the actual free-energy change is small and the aldolase reaction is readily reversible. We shall see later that aldolase acts in the reverse direction during the process of gluconeogenesis (see Fig. 14–17).

❺ Interconversion of the Triose Phosphates Only one of the two triose phosphates formed by aldolase, glyceraldehyde 3-phosphate, can be directly degraded in the subsequent steps of glycolysis. The other product, dihydroxyacetone phosphate, is rapidly and reversibly converted to glyceraldehyde 3-phosphate by the fifth enzyme of the glycolytic sequence, **triose phosphate isomerase**:

The reaction mechanism is similar to the reaction promoted by phosphohexose isomerase in step ❷ of glycolysis (Fig. 14–5). After the triose phosphate isomerase reaction, the carbon atoms derived from C-1, C-2, and C-3 of the starting glucose are chemically indistinguishable from C-6, C-5, and C-4, respectively (**Fig. 14–7**); the two "halves" of glucose have both yielded glyceraldehyde 3-phosphate.

This reaction completes the preparatory phase of glycolysis. The hexose molecule has been phosphorylated at C-1 and C-6 and then cleaved to form two molecules of glyceraldehyde 3-phosphate.

The Payoff Phase of Glycolysis Yields ATP and NADH

The payoff phase of glycolysis (Fig. 14–2b) includes the energy-conserving phosphorylation steps in which some of the chemical energy of the glucose molecule is conserved in the form of ATP and NADH. Remember that one molecule of glucose yields two molecules of glyceraldehyde 3-phosphate, and both halves of the

MECHANISM FIGURE 14–6 The class I aldolase reaction. The reaction shown here is the reverse of an aldol condensation. Note that cleavage between C-3 and C-4 depends on the presence of the carbonyl group at C-2, which is converted to an imine on the enzyme. A and B represent amino acid residues that serve as general acid (A) or base (B).

glucose molecule follow the same pathway in the second phase of glycolysis. The conversion of two molecules of glyceraldehyde 3-phosphate to two molecules of pyruvate is accompanied by the formation of four molecules of ATP from ADP. However, the net yield of ATP per molecule of glucose degraded is only two, because two ATP were invested in the preparatory phase of glycolysis to phosphorylate the two ends of the hexose molecule.

❻ Oxidation of Glyceraldehyde 3-Phosphate to 1,3-Bisphosphoglycerate The first step in the payoff phase is the oxidation of glyceraldehyde 3-phosphate to **1,3-bisphosphoglycerate**, catalyzed by **glyceraldehyde 3-phosphate dehydrogenase**:

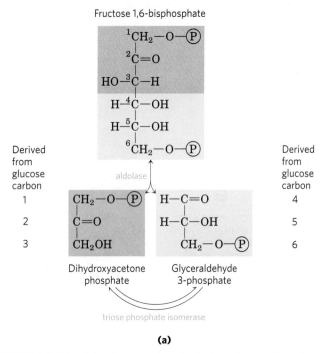

(a)

(b)

FIGURE 14–7 Fate of the glucose carbons in the formation of glyceraldehyde 3-phosphate. (a) The origin of the carbons in the two three-carbon products of the aldolase and triose phosphate isomerase reactions. The end product of the two reactions is glyceraldehyde 3-phosphate (two molecules). **(b)** Each carbon of glyceraldehyde 3-phosphate is derived from either of two specific carbons of glucose. Note that the numbering of the carbon atoms of glyceraldehyde 3-phosphate differs from that of the glucose from which it is derived. In glyceraldehyde 3-phosphate, the most complex functional group (the carbonyl) is specified as C-1. This numbering change is important for interpreting experiments with glucose in which a single carbon is labeled with a radioisotope. (See Problems 6 and 9 at the end of this chapter.)

This is the first of the two energy-conserving reactions of glycolysis that eventually lead to the formation of ATP. The aldehyde group of glyceraldehyde 3-phosphate is oxidized, not to a free carboxyl group but to a carboxylic acid anhydride with phosphoric acid. This type of anhydride, called an **acyl phosphate**, has a very high standard free energy of hydrolysis ($\Delta G'^{\circ} = -49.3$ kJ/mol; see Fig. 13–14, Table 13–6). Much of the free energy of oxidation of the aldehyde group of glyceraldehyde 3-phosphate is conserved by formation of the acyl phosphate group at C-1 of 1,3-bisphosphoglycerate.

Glyceraldehyde 3-phosphate is covalently bound to the dehydrogenase during the reaction **(Fig. 14–8)**. The aldehyde group of glyceraldehyde 3-phosphate reacts with the —SH group of an essential Cys residue in the active site, in a reaction analogous to the formation of a hemiacetal (see Fig. 7–5), in this case producing a *thio*hemiacetal. Reaction of the essential Cys residue with a heavy metal such as Hg^{2+} irreversibly inhibits the enzyme.

The amount of NAD^+ in a cell ($\leq 10^{-5}$ M) is far smaller than the amount of glucose metabolized in a few minutes. Glycolysis would soon come to a halt if the NADH formed in this step of glycolysis were not continuously reoxidized and recycled. We return to a discussion of this recycling of NAD^+ later in the chapter.

➐ Phosphoryl Transfer from 1,3-Bisphosphoglycerate to ADP The enzyme **phosphoglycerate kinase** transfers the high-energy phosphoryl group from the carboxyl group of 1,3-bisphosphoglycerate to ADP, forming ATP and **3-phosphoglycerate**:

1,3-Bisphosphoglycerate   ADP

Mg^{2+} phosphoglycerate kinase

3-Phosphoglycerate   ATP

$\Delta G'^{\circ} = -18.5$ kJ/mol

Notice that phosphoglycerate kinase is named for the reverse reaction, in which it transfers a phosphoryl

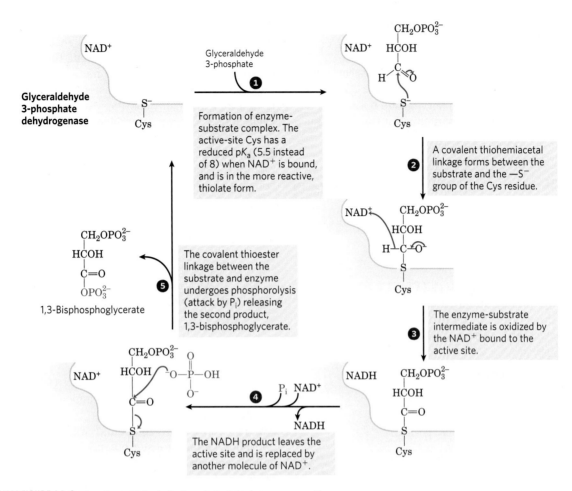

MECHANISM FIGURE 14–8 The glyceraldehyde 3-phosphate dehydrogenase reaction.

group from ATP to 3-phosphoglycerate. Like all enzymes, it catalyzes the reaction in both directions. This enzyme acts in the direction suggested by its name during gluconeogenesis (see Fig. 14–17) and during photosynthetic CO_2 assimilation (see Fig. 20–4). In glycolysis, the reaction it catalyzes proceeds as shown above, in the direction of ATP synthesis.

Steps ❻ and ❼ of glycolysis together constitute an energy-coupling process in which 1,3-bisphosphoglycerate is the common intermediate; it is formed in the first reaction (which would be endergonic in isolation), and its acyl phosphate group is transferred to ADP in the second reaction (which is strongly exergonic). The sum of these two reactions is

Glyceraldehyde 3-phosphate + ADP + P_i + NAD^+ $\rightleftharpoons$
$$3\text{-phosphoglycerate} + ATP + NADH + H^+$$
$$\Delta G'^\circ = -12.2 \text{ kJ/mol}$$

Thus the overall reaction is exergonic.

Recall from Chapter 13 that the actual free-energy change, ΔG, is determined by the standard free-energy change, $\Delta G'^\circ$, and the mass-action ratio, Q, which is the ratio [products]/[reactants] (see Eqn 13–4). For step ❻

$$\Delta G = \Delta G'^\circ + RT \ln Q$$

$$= \Delta G'^\circ + RT \ln \frac{[\text{1,3-bisphosphoglycerate}][\text{NADH}]}{[\text{glyceraldehyde 3-phosphate}][P_i][\text{NAD}^+]}$$

Notice that $[H^+]$ is not included in Q. In biochemical calculations, $[H^+]$ is assumed to be a constant (10^{-7} M), and this constant is included in the definition of $\Delta G'^\circ$ (p. 507).

When the mass-action ratio is less than 1.0, its natural logarithm has a negative sign. In the cytosol, where these reactions are taking place, the ratio [NADH]/[NAD$^+$] is a small fraction, contributing to a low Q. Step ❼, by consuming the product of step ❻ (1,3-bisphosphoglycerate), keeps [1,3-bisphosphoglycerate] relatively low in the steady state and thereby keeps Q for the overall energy-coupling process small. When Q is small, the contribution of $\ln Q$ can make ΔG strongly negative. This is simply another way of showing how the two reactions, steps ❻ and ❼, are coupled through a common intermediate.

The outcome of these coupled reactions, both reversible under cellular conditions, is that the energy released on oxidation of an aldehyde to a carboxylate group is conserved by the coupled formation of ATP from ADP and P_i. The formation of ATP by phosphoryl group transfer from a substrate such as 1,3-bisphosphoglycerate is referred to as a **substrate-level phosphorylation**, to distinguish this mechanism from **respiration-linked phosphorylation**. Substrate-level phosphorylations involve soluble enzymes and chemical intermediates (1,3-bisphosphoglycerate in this case). Respiration-linked phosphorylations, on the other hand, involve membrane-bound enzymes and transmembrane gradients of protons (Chapter 19).

⑧ Conversion of 3-Phosphoglycerate to 2-Phosphoglycerate The enzyme **phosphoglycerate mutase** catalyzes a reversible shift of the phosphoryl group between C-2 and C-3 of glycerate; Mg^{2+} is essential for this reaction:

$$\Delta G'^{\circ} = 4.4 \text{ kJ/mol}$$

The reaction occurs in two steps **(Fig. 14–9)**. A phosphoryl group initially attached to a His residue of the mutase is transferred to the hydroxyl group at C-2 of 3-phosphoglycerate, forming 2,3-bisphosphoglycerate

(2,3-BPG). The phosphoryl group at C-3 of 2,3-BPG is then transferred to the same His residue, producing 2-phosphoglycerate and regenerating the phosphorylated enzyme. Phosphoglycerate mutase is initially phosphorylated by phosphoryl transfer from 2,3-BPG, which is required in small quantities to initiate the catalytic cycle and is continuously regenerated by that cycle.

⑨ Dehydration of 2-Phosphoglycerate to Phosphoenolpyruvate In the second glycolytic reaction that generates a compound with high phosphoryl group transfer potential (the first was step ⑥), **enolase** promotes reversible removal of a molecule of water from 2-phosphoglycerate to yield **phosphoenolpyruvate (PEP)**:

$$\Delta G'^{\circ} = 7.5 \text{ kJ/mol}$$

The mechanism of the enolase reaction involves an enolic intermediate stabilized by Mg^{2+} (see Fig. 6–26). The reaction converts a compound with a relatively low phosphoryl group transfer potential ($\Delta G'^{\circ}$ for hydrolysis of 2-phosphoglycerate is -17.6 kJ/mol) to one with high phosphoryl group transfer potential ($\Delta G'^{\circ}$ for PEP hydrolysis is -61.9 kJ/mol) (see Fig. 13–13, Table 13–6).

⑩ Transfer of the Phosphoryl Group from Phosphoenolpyruvate to ADP The last step in glycolysis is the transfer of the phosphoryl group from phosphoenolpyruvate to ADP, catalyzed by **pyruvate kinase**, which requires K^{+} and either Mg^{2+} or Mn^{2+}:

$$\Delta G'^{\circ} = -31.4 \text{ kJ/mol}$$

Phosphoglycerate mutase

① Phosphoryl transfer occurs between an active-site His and C-2 (OH) of the substrate. A second active-site His acts as general base catalyst.

② Phosphoryl transfer from C-3 of the substrate to the first active-site His. The second active-site His acts as general acid catalyst.

MECHANISM FIGURE 14–9 The phosphoglycerate mutase reaction.

In this substrate-level phosphorylation, the product **pyruvate** first appears in its enol form, then tautomerizes rapidly and nonenzymatically to its keto form, which predominates at pH 7:

Pyruvate
(enol form)

Pyruvate
(keto form)

The overall reaction has a large, negative standard free-energy change, due in large part to the spontaneous conversion of the enol form of pyruvate to the keto form (see Fig. 13–13). About half of the energy released by PEP hydrolysis ($\Delta G'^{\circ} = -61.9$ kJ/mol) is conserved in the formation of the phosphoanhydride bond of ATP ($\Delta G'^{\circ} = -30.5$ kJ/mol), and the rest (-31.4 kJ/mol) constitutes a large driving force pushing the reaction toward ATP synthesis. We discuss the regulation of pyruvate kinase in Chapter 15.

The Overall Balance Sheet Shows a Net Gain of ATP

We can now construct a balance sheet for glycolysis to account for (1) the fate of the carbon skeleton of glucose, (2) the input of P_i and ADP and output of ATP, and (3) the pathway of electrons in the oxidation-reduction reactions. The left-hand side of the following equation shows all the inputs of ATP, NAD^+, ADP, and P_i (consult Fig. 14–2), and the right-hand side shows all the outputs (keep in mind that each molecule of glucose yields two molecules of pyruvate):

Glucose + 2ATP + $2NAD^+$ + 4ADP + $2P_i \longrightarrow$
    2 pyruvate + 2ADP + 2NADH + $2H^+$ + 4ATP + $2H_2O$

Canceling out common terms on both sides of the equation gives the overall equation for glycolysis under aerobic conditions:

Glucose + $2NAD^+$ + 2ADP + $2P_i \longrightarrow$
    2 pyruvate + 2NADH + $2H^+$ + 2ATP + $2H_2O$

The two molecules of NADH formed by glycolysis in the cytosol are, under aerobic conditions, reoxidized to NAD^+ by transfer of their electrons to the electron-transfer chain, which in eukaryotic cells is located in the mitochondria. The electron-transfer chain passes these electrons to their ultimate destination, O_2:

2NADH + $2H^+$ + $O_2 \longrightarrow 2NAD^+$ + $2H_2O$

Electron transfer from NADH to O_2 in mitochondria provides the energy for synthesis of ATP by respiration-linked phosphorylation (Chapter 19).

In the overall glycolytic process, one molecule of glucose is converted to two molecules of pyruvate (the pathway of carbon). Two molecules of ADP and two of P_i are converted to two molecules of ATP (the pathway of phosphoryl groups). Four electrons, as two hydride ions, are transferred from two molecules of glyceraldehyde 3-phosphate to two of NAD^+ (the pathway of electrons).

Glycolysis Is under Tight Regulation

During his studies on the fermentation of glucose by yeast, Louis Pasteur discovered that both the rate and the total amount of glucose consumption were many times greater under anaerobic than aerobic conditions. Later studies of muscle showed the same large difference in the rates of anaerobic and aerobic glycolysis. The biochemical basis of this "Pasteur effect" is now clear. The ATP yield from glycolysis under anaerobic conditions (2 ATP per molecule of glucose) is much smaller than that from the complete oxidation of glucose to CO_2 under aerobic conditions (30 or 32 ATP per glucose; see Table 19–5). About 15 times as much glucose must therefore be consumed anaerobically as aerobically to yield the same amount of ATP.

The flux of glucose through the glycolytic pathway is regulated to maintain nearly constant ATP levels (as well as adequate supplies of glycolytic intermediates that serve biosynthetic roles). The required adjustment in the rate of glycolysis is achieved by a complex interplay among ATP consumption, NADH regeneration, and allosteric regulation of several glycolytic enzymes—including hexokinase, PFK-1, and pyruvate kinase—and by second-to-second fluctuations in the concentration of key metabolites that reflect the cellular balance between ATP production and consumption. On a slightly longer time scale, glycolysis is regulated by the hormones glucagon, epinephrine, and insulin, and by changes in the expression of the genes for several glycolytic enzymes. An especially interesting case of abnormal regulation of glycolysis is seen in cancer. The German biochemist Otto Warburg first observed in 1928 that tumors of nearly all types carry out glycolysis at a much higher rate than normal tissue, *even when oxygen is available*. This "Warburg effect" is the basis for several methods of detecting and treating cancer (Box 14–1).

Warburg is generally considered the preeminent biochemist of the first half of the twentieth century. He

Otto Warburg, 1883–1970

made seminal contributions to many other areas of biochemistry, including respiration, photosynthesis, and the enzymology of intermediary metabolism. Beginning in 1930, Warburg and his associates purified and crystallized seven of the enzymes of glycolysis. They developed an experimental tool that revolutionized biochemical studies of oxidative metabolism: the Warburg manometer, which directly measured the oxygen consumption of tissues by monitoring changes in gas volume, and thus allowed quantitative measurement of any enzyme with oxidase activity.

BOX 14–1 MEDICINE High Rate of Glycolysis in Tumors Suggests Targets for Chemotherapy and Facilitates Diagnosis

In many types of tumors found in humans and other animals, glucose uptake and glycolysis proceed about 10 times faster than in normal, noncancerous tissues. Most tumor cells grow under hypoxic conditions (i.e., with limited oxygen supply) because, at least initially, they lack the capillary network to supply sufficient oxygen. Cancer cells located more than 100 to 200 μm from the nearest capillaries must depend on glycolysis alone (without further oxidation of pyruvate) for much of their ATP production. The energy yield (2 ATP per glucose) is far lower than can be obtained by the complete oxidation of pyruvate to CO_2 in mitochondria (about 30 ATP per glucose; Chapter 19). So, to make the same amount of ATP, tumor cells must take up much more glucose than do normal cells, converting it to pyruvate and then to lactate as they recycle NADH. It is likely that two early steps in the transformation of a normal cell into a tumor cell are (1) the change to dependence on glycolysis for ATP production, and (2) the development of tolerance to a low pH in the extracellular fluid (caused by release of the end product of glycolysis, lactic acid). In general, the more aggressive the tumor, the greater is its rate of glycolysis.

This increase in glycolysis is achieved at least in part by increased synthesis of the glycolytic enzymes and of the plasma membrane transporters GLUT1 and GLUT3 (see Table 11–3) that carry glucose into cells. (Recall that GLUT1 and GLUT3 are not dependent on insulin.) The **hypoxia-inducible transcription factor (HIF-1)** is a protein that acts at the level of mRNA synthesis to stimulate the production of at least eight glycolytic enzymes and the glucose transporters when oxygen supply is limited (Fig. 1). With the resulting high rate of glycolysis, the tumor cell can survive anaerobic conditions until the supply of blood vessels has caught up with tumor growth. Another protein induced by HIF-1 is the peptide hormone VEGF (vascular endothelial growth factor), which stimulates the outgrowth of blood vessels (angiogenesis) toward the tumor.

There is also evidence that the tumor suppressor protein p53, which is mutated in most types of cancer (see Section 12.12), controls the synthesis and assembly of mitochondrial proteins essential to the passage of electrons to O_2. Cells with mutant p53 are defective in mitochondrial electron transport and are forced to rely more heavily on glycolysis for ATP production (Fig. 1).

This heavier reliance of tumors than of normal tissue on glycolysis suggests a possibility for anticancer therapy: inhibitors of glycolysis might target and kill tumors by depleting their supply of ATP. Three inhibitors of hexokinase have shown promise as chemotherapeutic agents: 2-deoxyglucose, lonidamine, and 3-bromopyruvate. By preventing the formation

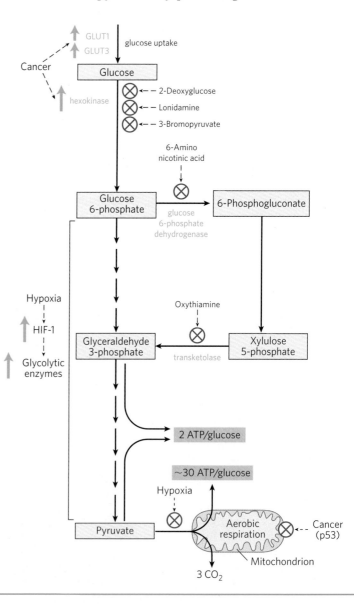

FIGURE 1 The anaerobic metabolism of glucose in tumor cells yields far less ATP (2 per glucose) than the complete oxidation to CO_2 that takes place in healthy cells under aerobic conditions (~30 ATP per glucose), so a tumor cell must consume much more glucose to produce the same amount of ATP. Glucose transporters and most of the glycolytic enzymes are overproduced in tumors. Compounds that inhibit hexokinase, glucose 6-phosphate dehydrogenase, or transketolase block ATP production by glycolysis, thus depriving the cancer cell of energy and killing it.

[^{18}F]2-Fluoro-2-deoxyglucose
(FdG)

$\xrightarrow[\text{hexokinase}]{\text{ATP} \quad \text{ADP}}$

[^{18}F]6-Phospho-2-fluoro-2-deoxyglucose
(6-Phospho-FdG)

FIGURE 2 Phosphorylation of ^{18}F-labeled 2-fluoro-2-deoxyglucose by hexokinase traps the FdG in cells (as 6-phospho-FdG), where its presence can be detected by positron emission from ^{18}F.

of glucose 6-phosphate, these compounds not only deprive tumor cells of glycolytically produced ATP but also prevent the formation of pentose phosphates via the pentose phosphate pathway, which also begins with glucose 6-phosphate. Without pentose phosphates, a cell cannot synthesize the nucleotides essential to DNA and RNA synthesis and thus cannot grow or divide. Another anticancer drug already approved for clinical use is imatinib (Gleevec), described in Box 12–5. It inhibits a specific tyrosine kinase, preventing the increased synthesis of hexokinase normally triggered by that kinase. The thiamine analog oxythiamine, which blocks the action of a transketolase-like enzyme that converts xylulose 5-phosphate to glyceraldehyde 3-phosphate (Fig. 1), is in preclinical trials as an antitumor drug.

The high glycolytic rate in tumor cells also has diagnostic usefulness. The relative rates at which tissues take up glucose can be used in some cases to pinpoint the location of tumors. In positron emission tomography (PET), individuals are injected with a harmless, isotopically labeled glucose analog that is taken up but not metabolized by tissues. The labeled compound is 2-fluoro-2-deoxyglucose (FdG), in which the hydroxyl group at the C-2 of glucose is replaced with ^{18}F (Fig. 2). This compound is taken up via GLUT transporters and is a good substrate for hexokinase, but it cannot be converted to the enediol intermediate in the phosphohexose isomerase reaction (see Fig. 14–5) and therefore accumulates as 6-phospho-FdG. The extent of its accumulation depends on its rate of uptake and phosphorylation, which as noted above is typically 10 or more times higher in tumors than in normal tissue. Decay of ^{18}F yields positrons (two per ^{18}F atom) that can be detected by a series of sensitive detectors positioned around the body, which allows accurate localization of accumulated 6-phospho-FdG (Fig. 3).

FIGURE 3 Detection of cancerous tissue by positron emission tomography (PET). The adult male patient had undergone surgical removal of a primary skin cancer (malignant melanoma). The image on the left, obtained by whole-body computed tomography (CT scan), shows the location of the soft tissues and bones. The central panel is a PET scan after the patient had ingested ^{18}F-labeled 2-fluoro-2-deoxyglucose (FdG). Dark spots indicate regions of high glucose utilization. As expected, the brain and bladder are heavily labeled—the brain because it uses most of the glucose consumed in the body, and the bladder because the ^{18}F-labeled 6-phospho-FdG is excreted in the urine. When the intensity of the label in the PET scan is translated into false color (the intensity increases from green to yellow to red) and the image is superimposed on the CT scan, the fused image (right) reveals cancer in the bones of the upper spine, in the liver, and in some regions of muscle, all the result of cancer spreading from the primary malignant melanoma.

Trained in carbohydrate chemistry in the laboratory of the great Emil Fischer (who won the Nobel Prize in Chemistry in 1902), Warburg himself won the Nobel Prize in Physiology or Medicine in 1931. Several of Warburg's students and colleagues also were awarded Nobel Prizes: Otto Meyerhof in 1922, Hans Krebs and Fritz Lipmann in 1953, and Hugo Theorell in 1955. Meyerhof's laboratory provided training for Lipmann, and for several other Nobel Prize winners: Severo Ochoa (1959), Andre Lwoff (1965), and George Wald (1967).

Glucose Uptake Is Deficient in Type 1 Diabetes Mellitus

The metabolism of glucose in mammals is limited by the rate of glucose uptake into cells and its phosphorylation by hexokinase. Glucose uptake from the blood is mediated by the GLUT family of glucose transporters (see Table 11–3). The transporters of hepatocytes (GLUT1, GLUT2) and of brain neurons (GLUT3) are always present in plasma membranes. In contrast, the main glucose transporter in the cells of skeletal muscle, cardiac muscle, and adipose tissue (GLUT4) is sequestered in small intracellular vesicles and moves into the plasma membrane only in response to an insulin signal **(Fig. 14–10)**. We discussed this insulin signaling mechanism in Chapter 12 (see Fig. 12–16). Thus in skeletal muscle, heart, and adipose tissue, glucose uptake and metabolism depend on the normal release of insulin by pancreatic β cells in response to elevated blood glucose (see Fig. 23–26).

Individuals with type 1 diabetes mellitus (also called insulin-dependent diabetes) have too few β cells and cannot release sufficient insulin to trigger glucose uptake by the cells of skeletal muscle, heart, or adipose tissue. Thus, after a meal containing carbohydrates, glucose accumulates to abnormally high levels in the blood, a condition known as hyperglycemia. Unable to take up glucose, muscle and fat tissue use the fatty acids of stored triacylglycerols as their principal fuel. In the liver, acetyl-CoA derived from this fatty acid breakdown is converted to "ketone bodies"—acetoacetate and β-hydroxybutyrate—which are exported and carried to other tissues to be used as fuel (Chapter 17). These compounds are especially critical to the brain, which uses ketone bodies as alternative fuel when glucose is unavailable. (Fatty acids cannot pass through the blood-brain barrier and thus are not a fuel for brain neurons.)

In untreated type 1 diabetes, overproduction of acetoacetate and β-hydroxybutyrate leads to their accumulation in the blood, and the consequent lowering of blood pH produces **ketoacidosis**, a life-threatening condition. Insulin injection reverses this sequence of events: GLUT4 moves into the plasma membranes of hepatocytes and adipocytes, glucose is taken up into the cells and phosphorylated, and the blood glucose level falls, greatly reducing the production of ketone bodies.

Diabetes mellitus has profound effects on the metabolism of both carbohydrates and fats. We return to this topic in Chapter 23, after considering lipid metabolism (Chapters 17 and 21). ∎

SUMMARY 14.1 Glycolysis

▶ Glycolysis is a near-universal pathway by which a glucose molecule is oxidized to two molecules of pyruvate, with energy conserved as ATP and NADH.

▶ All 10 glycolytic enzymes are in the cytosol, and all 10 intermediates are phosphorylated compounds of three or six carbons.

▶ In the preparatory phase of glycolysis, ATP is invested to convert glucose to fructose 1,6-bisphosphate. The bond between C-3 and C-4 is then broken to yield two molecules of triose phosphate.

▶ In the payoff phase, each of the two molecules of glyceraldehyde 3-phosphate derived from glucose undergoes oxidation at C-1; the energy of this oxidation reaction is conserved in the form of one NADH and two ATP per triose phosphate oxidized. The net equation for the overall process is

$$\text{Glucose} + 2\text{NAD}^+ + 2\text{ADP} + 2\text{P}_i \longrightarrow$$
$$2 \text{ pyruvate} + 2\text{NADH} + 2\text{H}^+ + 2\text{ATP} + 2\text{H}_2\text{O}$$

▶ Glycolysis is tightly regulated in coordination with other energy-yielding pathways to ensure a steady supply of ATP.

▶ In type 1 diabetes, defective uptake of glucose by muscle and adipose tissue has profound effects on the metabolism of carbohydrates and fats.

14.2 Feeder Pathways for Glycolysis

Many carbohydrates besides glucose meet their catabolic fate in glycolysis, after being transformed into one of the glycolytic intermediates. The most significant are the storage polysaccharides glycogen and starch, either within cells (endogenous) or obtained in the diet; the disaccharides maltose, lactose, trehalose, and sucrose; and the monosaccharides fructose, mannose, and galactose **(Fig. 14–11)**.

Dietary Polysaccharides and Disaccharides Undergo Hydrolysis to Monosaccharides

For most humans, starch is the major source of carbohydrates in the diet (Fig. 14–11). Digestion begins in the mouth, where salivary α-**amylase** hydrolyzes the internal ($\alpha1\rightarrow4$) glycosidic linkages of starch, producing short polysaccharide fragments or oligosaccharides. (Note that in this *hydrolysis* reaction, water, not P_i, is the attacking species.) In the stomach, salivary α-amylase is inactivated

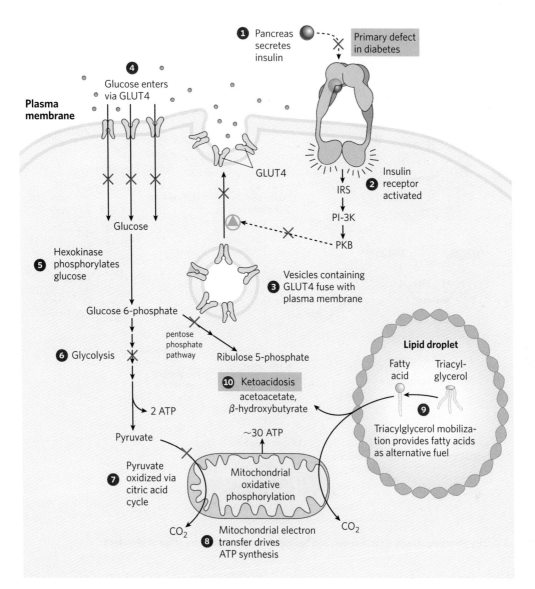

FIGURE 14–10 Effect of type 1 diabetes on carbohydrate and fat metabolism in an adipocyte. Normally, insulin triggers the insertion of GLUT4 transporters into the plasma membrane by the fusion of GLUT4-containing vesicles with the membrane, allowing glucose uptake from the blood. When blood levels of insulin drop, GLUT4 is resequestered in vesicles by endocytosis. In type 1 (insulin-dependent) diabetes mellitus, the insertion of GLUT4 into membranes, as well as other processes normally stimulated by insulin, are inhibited as indicated by **X**. The lack of insulin prevents glucose uptake via GLUT4; as a consequence, cells are deprived of glucose, and blood glucose is elevated.

Lacking glucose for energy supply, adipocytes break down triacylglycerols stored in fat droplets and supply the resulting fatty acids to other tissues for mitochondrial ATP production. Two byproducts of fatty acid oxidation in the liver (acetoacetate and β-hydroxybutyrate, see p. 686) accumulate and are released into the blood, providing fuel for the brain but also decreasing blood pH, causing ketoacidosis. The same sequence of events takes place in muscle, except that myocytes do not store triacylglycerols and instead take up fatty acids that are released into the blood by adipocytes.

by the low pH, but a second form of α-amylase, secreted by the pancreas into the small intestine, continues the breakdown process. Pancreatic α-amylase yields mainly maltose and maltotriose (the di- and trisaccharides of glucose) and oligosaccharides called limit dextrins, fragments of amylopectin containing ($\alpha1\rightarrow6$) branch points. Maltose and dextrins are degraded to glucose by enzymes of the intestinal brush border (the fingerlike microvilli of intestinal epithelial cells, which greatly increase the area of the intestinal surface). Dietary glycogen has essentially the

same structure as starch, and its digestion proceeds by the same pathway.

As we noted in Chapter 7, most animals cannot digest cellulose for lack of the enzyme cellulase, which attacks the ($\beta1\rightarrow4$) glycosidic bonds of cellulose. In ruminant animals, the extended stomach includes a chamber in which symbiotic microorganisms that produce cellulase break down cellulose into glucose molecules. These microorganisms use the resulting glucose in an anaerobic fermentation that produces large quantities

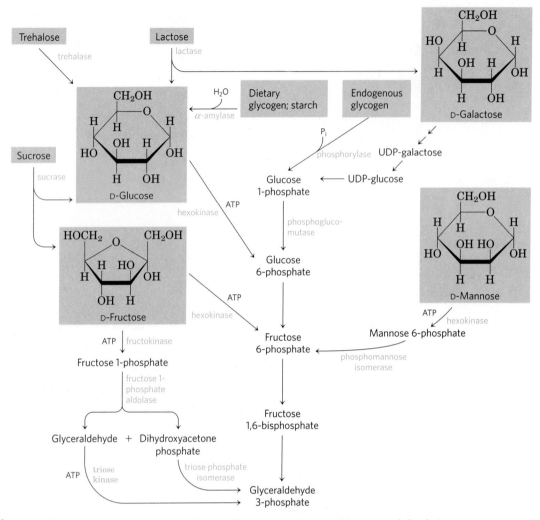

FIGURE 14–11 Entry of dietary glycogen, starch, disaccharides, and hexoses into the preparatory stage of glycolysis.

of propionate. This propionate serves as the starting material for gluconeogenesis, which produces much of the lactose in milk.

Endogenous Glycogen and Starch Are Degraded by Phosphorolysis

Glycogen stored in animal tissues (primarily liver and skeletal muscle), in microorganisms, or in plant tissues can be mobilized for use within the same cell by a *phosphorolytic* reaction catalyzed by **glycogen phosphorylase** (**starch phosphorylase** in plants) **(Fig. 14–12)**. These enzymes catalyze an attack by P_i on the ($\alpha1{\rightarrow}4$) glycosidic linkage that joins the last two glucose residues at a nonreducing end, generating glucose 1-phosphate and a polymer one glucose unit shorter. *Phosphorolysis* preserves some of the energy of the glycosidic bond in the phosphate ester glucose 1-phosphate. Glycogen phosphorylase (or starch phosphorylase) acts repetitively until it approaches an ($\alpha1{\rightarrow}6$) branch point (see Fig. 7–13), where its action stops. A **debranching**

enzyme removes the branches. The mechanisms and control of glycogen degradation are described in greater detail in Chapter 15.

Glucose 1-phosphate produced by glycogen phosphorylase is converted to glucose 6-phosphate by **phosphoglucomutase**, which catalyzes the reversible reaction

$$\text{Glucose 1-phosphate} \rightleftharpoons \text{glucose 6-phosphate}$$

Phosphoglucomutase employs essentially the same mechanism as phosphoglycerate mutase (Fig. 14–9): both entail a bisphosphate intermediate, and the enzyme is transiently phosphorylated in each catalytic cycle. The general name **mutase** is given to enzymes that catalyze the transfer of a functional group from one position to another in the same molecule. Mutases are a subclass of **isomerases**, enzymes that interconvert stereoisomers or structural or positional isomers (see Table 6–3). The glucose 6-phosphate formed in the phosphoglucomutase reaction can enter glycolysis or another pathway such as the pentose phosphate pathway, described in Section 14.5.

Nonreducing end

Glycogen (starch)
n glucose units

glycogen (starch)
phosphorylase

Glucose
1-phosphate

Glycogen (starch)
(n−1) glucose units

FIGURE 14–12 Breakdown of intracellular glycogen by glycogen phosphorylase. The enzyme catalyzes attack by inorganic phosphate (pink) on the terminal glucosyl residue (blue) at the nonreducing end of a glycogen molecule, releasing glucose 1-phosphate and generating a glycogen molecule shortened by one glucose residue. The reaction is a *phosphorolysis* (not hydrolysis).

WORKED EXAMPLE 14–1   Energy Savings for Glycogen Breakdown by Phosphorolysis

---

Calculate the energy savings (in ATP molecules per glucose monomer) achieved by breaking down glycogen by *phosphorolysis* rather than *hydrolysis* to begin the process of glycolysis.

Solution: Phosphorolysis produces a phosphorylated glucose (glucose 1-phosphate), which is then converted to glucose 6-phosphate—without expenditure of the cellular energy (1 ATP) needed for formation of glucose 6-phosphate from free glucose. Thus only 1 ATP is consumed per glucose monomer in the preparatory phase, compared with 2 ATP when glycolysis starts with free glucose. The cell therefore gains 3 ATP per glucose monomer (4 ATP produced in the payoff phase minus 1 ATP used in the preparatory phase), rather than 2—a saving of 1 ATP per glucose monomer.

Breakdown of dietary polysaccharides such as glycogen and starch in the gastrointestinal tract by phosphorolysis rather than hydrolysis would produce no energy gain: sugar phosphates are not transported into the cells that line the intestine, but must first be dephosphorylated to the free sugar.

Disaccharides must be hydrolyzed to monosaccharides before entering cells. Intestinal disaccharides and dextrins are hydrolyzed by enzymes attached to the outer surface of the intestinal epithelial cells:

$$\text{Dextrin} + n\text{H}_2\text{O} \xrightarrow{\text{dextrinase}} n \text{ D-glucose}$$

$$\text{Maltose} + \text{H}_2\text{O} \xrightarrow{\text{maltase}} 2 \text{ D-glucose}$$

$$\text{Lactose} + \text{H}_2\text{O} \xrightarrow{\text{lactase}} \text{D-galactose} + \text{D-glucose}$$

$$\text{Sucrose} + \text{H}_2\text{O} \xrightarrow{\text{sucrase}} \text{D-fructose} + \text{D-glucose}$$

$$\text{Trehalose} + \text{H}_2\text{O} \xrightarrow{\text{trehalase}} 2 \text{ D-glucose}$$

The monosaccharides so formed are actively transported into the epithelial cells (see Fig. 11–43), then passed into the blood to be carried to various tissues, where they are phosphorylated and funneled into the glycolytic sequence.

Lactose intolerance, common among adults of most human populations except those originating in Northern Europe and some parts of Africa, is due to the disappearance after childhood of most or all of the lactase activity of the intestinal epithelial cells. Without intestinal lactase, lactose cannot be completely digested and absorbed in the small intestine, and it passes into the large intestine, where bacteria convert it to toxic products that cause abdominal cramps and diarrhea. The problem is further complicated because undigested lactose and its metabolites increase the osmolarity of the intestinal contents, favoring retention of water in the intestine. In most parts of the world where lactose intolerance is prevalent, milk is not used as a food by adults, although milk products predigested with lactase are commercially available in some countries. In certain human disorders, several or all of the intestinal disaccharidases are missing. In these cases, the digestive disturbances triggered by dietary disaccharides can sometimes be minimized by a controlled diet. ■

Other Monosaccharides Enter the Glycolytic Pathway at Several Points

In most organisms, hexoses other than glucose can undergo glycolysis after conversion to a phosphorylated derivative. D-Fructose, present in free form in many fruits and formed by hydrolysis of sucrose in the small intestine of vertebrates, is phosphorylated by hexokinase:

$$\text{Fructose} + \text{ATP} \xrightarrow{\text{Mg}^{2+}} \text{fructose 6-phosphate} + \text{ADP}$$

This is a major pathway of fructose entry into glycolysis in the muscles and kidney. In the liver, fructose enters by a different pathway. The liver enzyme **fructokinase** catalyzes the phosphorylation of fructose at C-1 rather than C-6:

$$\text{Fructose} + \text{ATP} \xrightarrow{\text{Mg}^{2+}} \text{fructose 1-phosphate} + \text{ADP}$$

The fructose 1-phosphate is then cleaved to glyceraldehyde and dihydroxyacetone phosphate by **fructose 1-phosphate aldolase**:

$$
\begin{array}{l}
^{1}CH_2OPO_3^{2-} \\
^{2}C{=}O \\
^{3}HOCH \\
^{4}HCOH \\
^{5}HCOH \\
^{6}CH_2OH
\end{array}
\quad \underset{\text{aldolase}}{\overset{\text{fructose 1-phosphate}}{\rightleftharpoons}} \quad
$$

Fructose 1-phosphate

$$
\begin{array}{l}
CH_2OPO_3^{2-} \\
C{=}O \\
CH_2OH
\end{array}
$$
Dihydroxyacetone phosphate

+

$$
\begin{array}{l}
H \\
C{=}O \\
HCOH \\
CH_2OH
\end{array}
$$
Glyceraldehyde

Dihydroxyacetone phosphate is converted to glyceraldehyde 3-phosphate by the glycolytic enzyme triose phosphate isomerase. Glyceraldehyde is phosphorylated by ATP and **triose kinase** to glyceraldehyde 3-phosphate:

$$\text{Glyceradehyde} + \text{ATP} \xrightarrow{\text{Mg}^{2+}} \text{glyceraldehyde 3-phosphate} + \text{ADP}$$

Thus both products of fructose 1-phosphate hydrolysis enter the glycolytic pathway as glyceraldehyde 3-phosphate.

D-Galactose, a product of the hydrolysis of lactose (milk sugar), passes in the blood from the intestine to the liver, where it is first phosphorylated at C-1, at the expense of ATP, by the enzyme **galactokinase**:

$$\text{Galactose} + \text{ATP} \xrightarrow{\text{Mg}^{2+}} \text{galactose 1-phosphate} + \text{ADP}$$

The galactose 1-phosphate is then converted to its epimer at C-4, glucose 1-phosphate, by a set of reactions in which **uridine diphosphate** (UDP) functions as a coenzyme-like carrier of hexose groups **(Fig. 14–13)**. The epimerization involves first the oxidation of the C-4 —OH group to a ketone, then reduction of the ketone to an —OH, with inversion of the configuration at C-4. NAD is the cofactor for both the oxidation and the reduction.

A defect in any of the three enzymes in this pathway causes **galactosemia** in humans. In galactokinase-deficiency galactosemia, high galactose concentrations are found in blood and urine. Affected individuals develop cataracts in infancy, caused by deposition of the galactose metabolite galactitol in the lens.

$$
\begin{array}{l}
CH_2OH \\
H{-}C{-}OH \\
HO{-}C{-}H \\
HO{-}C{-}H \\
H{-}C{-}OH \\
CH_2OH
\end{array}
$$
D-Galactitol

The other symptoms in this disorder are relatively mild, and strict limitation of galactose in the diet greatly diminishes their severity.

Transferase-deficiency galactosemia is more serious; it is characterized by poor growth in childhood, speech abnormality, mental deficiency, and liver damage

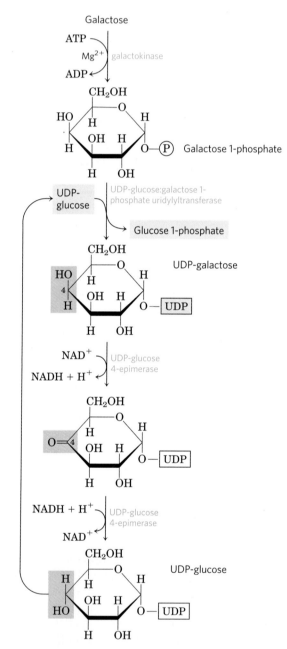

FIGURE 14–13 Conversion of galactose to glucose 1-phosphate. The conversion proceeds through a sugar-nucleotide derivative, UDP-galactose, which is formed when galactose 1-phosphate displaces glucose 1-phosphate from UDP-glucose. UDP-galactose is then converted by UDP-glucose 4-epimerase to UDP-glucose, in a reaction that involves oxidation of C-4 (light red) by NAD⁺, then reduction of C-4 by NADH; the result is inversion of the configuration at C-4. The UDP-glucose is recycled through another round of the same reaction. The net effect of this cycle is the conversion of galactose 1-phosphate to glucose 1-phosphate; there is no net production or consumption of UDP-galactose or UDP-glucose.

that may be fatal, even when galactose is withheld from the diet. Epimerase-deficiency galactosemia leads to similar symptoms, but is less severe when dietary galactose is carefully controlled. ■

D-Mannose, released in the digestion of various polysaccharides and glycoproteins of foods, can be phosphorylated at C-6 by hexokinase:

$$\text{Mannose} + \text{ATP} \xrightarrow{\text{Mg}^{2+}} \text{mannose 6-phosphate} + \text{ADP}$$

Mannose 6-phosphate is isomerized by **phosphomannose isomerase** to yield fructose 6-phosphate, an intermediate of glycolysis.

SUMMARY 14.2 Feeder Pathways for Glycolysis

▶ Endogenous glycogen and starch, storage forms of glucose, enter glycolysis in a two-step process. Phosphorolytic cleavage of a glucose residue from an end of the polymer, forming glucose 1-phosphate, is catalyzed by glycogen phosphorylase or starch phosphorylase. Phosphoglucomutase then converts the glucose 1-phosphate to glucose 6-phosphate, which can enter glycolysis.

▶ Ingested polysaccharides and disaccharides are converted to monosaccharides by intestinal hydrolytic enzymes, and the monosaccharides then enter intestinal cells and are transported to the liver or other tissues.

▶ A variety of D-hexoses, including fructose, galactose, and mannose, can be funneled into glycolysis. Each is phosphorylated and converted to glucose 6-phosphate, fructose 6-phosphate, or fructose 1-phosphate.

▶ Conversion of galactose 1-phosphate to glucose 1-phosphate involves two nucleotide derivatives: UDP-galactose and UDP-glucose. Genetic defects in any of the three enzymes that catalyze conversion of galactose to glucose 1-phosphate result in galactosemias of varying severity.

14.3 Fates of Pyruvate under Anaerobic Conditions: Fermentation

Under aerobic conditions, the pyruvate formed in the final step of glycolysis is oxidized to acetate (acetyl-CoA), which enters the citric acid cycle and is oxidized to CO_2 and H_2O. The NADH formed by dehydrogenation of glyceraldehyde 3-phosphate is ultimately reoxidized to NAD^+ by passage of its electrons to O_2 in mitochondrial respiration. Under hypoxic (low-oxygen) conditions, however—as in very active skeletal muscle, in submerged plant tissues, in solid tumors, or in lactic acid bacteria—NADH generated by glycolysis cannot be reoxidized by O_2. Failure to regenerate NAD^+ would leave the cell with no electron acceptor for the oxidation

of glyceraldehyde 3-phosphate, and the energy-yielding reactions of glycolysis would stop. NAD^+ must therefore be regenerated in some other way.

The earliest cells lived in an atmosphere almost devoid of oxygen and had to develop strategies for deriving energy from fuel molecules under anaerobic conditions. Most modern organisms have retained the ability to continually regenerate NAD^+ during anaerobic glycolysis by transferring electrons from NADH to form a reduced end product such as lactate or ethanol.

Pyruvate Is the Terminal Electron Acceptor in Lactic Acid Fermentation

When animal tissues cannot be supplied with sufficient oxygen to support aerobic oxidation of the pyruvate and NADH produced in glycolysis, NAD^+ is regenerated from NADH by the reduction of pyruvate to **lactate**. As mentioned earlier, some tissues and cell types (such as erythrocytes, which have no mitochondria and thus cannot oxidize pyruvate to CO_2) produce lactate from glucose even under aerobic conditions. The reduction of pyruvate in this pathway is catalyzed by **lactate dehydrogenase**, which forms the L isomer of lactate at pH 7:

$$\Delta G'^{\circ} = -25.1 \text{ kJ/mol}$$

The overall equilibrium of the reaction strongly favors lactate formation, as shown by the large negative standard free-energy change.

In glycolysis, dehydrogenation of the two molecules of glyceraldehyde 3-phosphate derived from each molecule of glucose converts two molecules of NAD^+ to two of NADH. Because the reduction of two molecules of pyruvate to two of lactate regenerates two molecules of NAD^+, there is no net change in NAD^+ or NADH:

The lactate formed by active skeletal muscles (or by erythrocytes) can be recycled; it is carried in the blood to the liver, where it is converted to glucose during the recovery from strenuous muscular activity. When lactate is produced in large quantities during vigorous muscle contraction (during a sprint, for example), the acidification that results from ionization of lactic acid in muscle and blood limits the period of vigorous activity. The best-conditioned athletes can sprint at top speed for no more than a minute (Box 14–2).

Although conversion of glucose to lactate includes two oxidation-reduction steps, there is no net change in the oxidation state of carbon; in glucose ($C_6H_{12}O_6$) and lactic acid ($C_3H_6O_3$), the H:C ratio is the same.

BOX 14–2 Athletes, Alligators, and Coelacanths: Glycolysis at Limiting Concentrations of Oxygen

Most vertebrates are essentially aerobic organisms; they convert glucose to pyruvate by glycolysis, then use molecular oxygen to oxidize the pyruvate completely to CO_2 and H_2O. Anaerobic catabolism of glucose to lactate occurs during short bursts of extreme muscular activity, for example in a 100 m sprint, during which oxygen cannot be carried to the muscles fast enough to oxidize pyruvate. Instead, the muscles use their stored glucose (glycogen) as fuel to generate ATP by fermentation, with lactate as the end product. In a sprint, lactate in the blood builds up to high concentrations. It is slowly converted back to glucose by gluconeogenesis in the liver in the subsequent rest or recovery period, during which oxygen is consumed at a gradually diminishing rate until the breathing rate returns to normal. The excess oxygen consumed in the recovery period represents a repayment of the oxygen debt. This is the amount of oxygen required to supply ATP for gluconeogenesis during recovery respiration, in order to regenerate the glycogen "borrowed" from liver and muscle to carry out intense muscular activity in the sprint. The cycle of reactions that includes glucose conversion to lactate in muscle and lactate conversion to glucose in liver is called the Cori cycle, for Carl and Gerty Cori, whose studies in the 1930s and 1940s clarified the pathway and its role (see Box 15–4).

The circulatory systems of most small vertebrates can carry oxygen to their muscles fast enough to avoid having to use muscle glycogen anaerobically. For example, migrating birds often fly great distances at high speeds without rest and without incurring an oxygen debt. Many running animals of moderate size also maintain an essentially aerobic metabolism in their skeletal muscle. However, the circulatory systems of larger animals, including humans, cannot completely sustain aerobic metabolism in skeletal muscles over long periods of intense muscular activity. These animals generally are slow-moving under normal circumstances and engage in intense muscular activity only in the gravest emergencies, because such bursts of activity require long recovery periods to repay the oxygen debt.

Alligators and crocodiles, for example, are normally sluggish animals. Yet when provoked they are capable of lightning-fast charges and dangerous lashings of their powerful tails. Such intense bursts of activity are short and must be followed by long periods of recovery. The fast emergency movements require lactic acid fermentation to generate ATP in skeletal muscles. The stores of muscle glycogen are rapidly expended in intense muscular activity, and lactate reaches very high concentrations in myocytes and extracellular fluid. Whereas a trained athlete can recover from a 100 m sprint in 30 min or less, an alligator may require many hours of rest and extra oxygen consumption to clear the excess lactate from its blood and regenerate muscle glycogen after a burst of activity.

Other large animals, such as the elephant and rhinoceros, have similar metabolic characteristics, as do diving mammals such as whales and seals. Dinosaurs and other huge, now-extinct animals probably had to depend on lactic acid fermentation to supply energy for muscular activity, followed by very long recovery periods during which they were vulnerable to attack by smaller predators better able to use oxygen and thus better adapted to continuous, sustained muscular activity.

Deep-sea explorations have revealed many species of marine life at great ocean depths, where the oxygen concentration is near zero. For example, the primitive coelacanth, a large fish recovered from depths of 4,000 m or more off the coast of South Africa, has an essentially anaerobic metabolism in virtually all its tissues. It converts carbohydrates to lactate and other products, most of which must be excreted. Some marine vertebrates ferment glucose to ethanol and CO_2 in order to generate ATP.

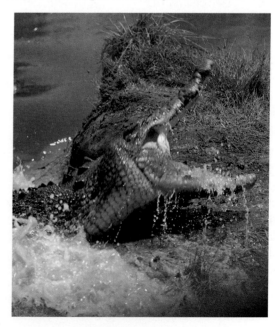

Nevertheless, some of the energy of the glucose molecule has been extracted by its conversion to lactate—enough to give a net yield of two molecules of ATP for every glucose molecule consumed. **Fermentation** is the general term for such processes, which extract energy (as ATP) but do not consume oxygen or change the concentrations of NAD^+ or NADH. Fermentations are carried out by a wide range of organisms, many of which occupy anaerobic niches, and they yield a variety of end products, some of which find commercial uses.

Ethanol Is the Reduced Product in Ethanol Fermentation

Yeast and other microorganisms ferment glucose to ethanol and CO_2, rather than to lactate. Glucose is converted to pyruvate by glycolysis, and the pyruvate is converted to ethanol and CO_2 in a two-step process:

In the first step, pyruvate is decarboxylated in an irreversible reaction catalyzed by **pyruvate decarboxylase**. This reaction is a simple decarboxylation and does not involve the net oxidation of pyruvate. Pyruvate decarboxylase requires Mg^{2+} and has a tightly bound coenzyme, thiamine pyrophosphate, which is discussed below. In the second step, acetaldehyde is reduced to ethanol through the action of **alcohol dehydrogenase**, with the reducing power furnished by NADH derived from the dehydrogenation of glyceraldehyde 3-phosphate. This reaction is a well-studied case of hydride transfer from NADH **(Fig. 14–14)**. Ethanol and CO_2 are thus the end products of ethanol fermentation, and the overall equation is

$$\text{Glucose} + 2ADP + 2P_i \longrightarrow 2 \text{ ethanol} + 2CO_2 + 2ATP + 2H_2O$$

As in lactic acid fermentation, there is no net change in the ratio of hydrogen to carbon atoms when glucose (H:C ratio = 12/6 = 2) is fermented to two ethanol and two CO_2 (combined H:C ratio = 12/6 = 2). In all fermentations, the H:C ratio of the reactants and products remains the same.

Pyruvate decarboxylase is present in brewer's and baker's yeast (*Saccharomyces cerevisiae*) and in all other organisms that ferment glucose to ethanol, including some plants. The CO_2 produced by pyruvate decarboxylation in brewer's yeast is responsible for the characteristic carbonation of champagne. The ancient art of brewing beer involves several enzymatic processes in addition to the reactions of ethanol fermentation (Box 14–3). In baking, CO_2 released by pyruvate decarboxylase when yeast is mixed with a fermentable sugar causes dough to rise. The enzyme is absent in

MECHANISM FIGURE 14–14 The alcohol dehydrogenase reaction.
🖱 Alcohol Dehydrogenase Mechanism

vertebrate tissues and in other organisms that carry out lactic acid fermentation.

Alcohol dehydrogenase is present in many organisms that metabolize ethanol, including humans. In the liver it catalyzes the oxidation of ethanol, either ingested or produced by intestinal microorganisms, with the concomitant reduction of NAD^+ to NADH. In this case, the reaction proceeds in the direction opposite to that involved in the production of ethanol by fermentation.

Thiamine Pyrophosphate Carries "Active Acetaldehyde" Groups

💊 The pyruvate decarboxylase reaction provides our first encounter with **thiamine pyrophosphate (TPP)** (Fig. 14–15), a coenzyme derived from vitamin B_1. Lack of vitamin B_1 in the human diet leads to the condition known as beriberi, characterized by an accumulation of body fluids (swelling), pain, paralysis, and ultimately death. ∎

Thiamine pyrophosphate plays an important role in the cleavage of bonds adjacent to a carbonyl group, such as the decarboxylation of α-keto acids, and in chemical rearrangements in which an activated acetaldehyde group is transferred from one carbon atom to another (Table 14–1). The functional part of TPP, the thiazolium ring, has a relatively acidic proton at C-2. Loss of this proton produces a carbanion that is the active species in TPP-dependent reactions (Fig. 14–15). The carbanion readily adds to carbonyl groups, and the thiazolium ring is thereby positioned to act as an "electron sink" that greatly facilitates reactions such as the decarboxylation catalyzed by pyruvate decarboxylase.

BOX 14–3 Ethanol Fermentations: Brewing Beer and Producing Biofuels

Beer brewing was a science learned early in human history, and later refined for larger-scale production. Brewers prepare beer by ethanol fermentation of the carbohydrates in cereal grains (seeds) such as barley, carried out by yeast glycolytic enzymes. The carbohydrates, largely polysaccharides, must first be degraded to disaccharides and monosaccharides. In a process called malting, the barley seeds are allowed to germinate until they form the hydrolytic enzymes required to break down their polysaccharides, at which point germination is stopped by controlled heating. The product is malt, which contains enzymes that catalyze the hydrolysis of the β linkages of cellulose and other cell wall polysaccharides of the barley husks, and enzymes such as α-amylase and maltase.

The brewer next prepares the wort, the nutrient medium required for fermentation by yeast cells. The malt is mixed with water and then mashed or crushed. This allows the enzymes formed in the malting process to act on the cereal polysaccharides to form maltose, glucose, and other simple sugars, which are soluble in the aqueous medium. The remaining cell matter is then separated, and the liquid wort is boiled with hops to give flavor. The wort is cooled and then aerated.

Now the yeast cells are added. In the aerobic wort the yeast grows and reproduces very rapidly, using energy obtained from available sugars. No ethanol forms during this stage, because the yeast, amply supplied with oxygen, oxidizes the pyruvate formed by glycolysis to CO_2 and H_2O via the citric acid cycle. When all the dissolved oxygen in the vat of wort has been consumed, the yeast cells switch to anaerobic metabolism, and from this point they ferment the sugars into ethanol and CO_2. The fermentation process is controlled in part by the concentration of the ethanol formed, by the pH, and by the amount of remaining sugar. After fermentation has been stopped, the cells are removed and the "raw" beer is ready for final processing.

In the final steps of brewing, the amount of foam (or head) on the beer, which results from dissolved proteins, is adjusted. Normally this is controlled by proteolytic enzymes that arise in the malting process. If these enzymes act on the proteins too long, the beer will have very little head and will be flat; if they do not act long enough, the beer will not be clear when it is cold. Sometimes proteolytic enzymes from other sources are added to control the head.

Much of the technology developed for large-scale production of alcoholic beverages is now finding application to a wholly different problem: the production of ethanol as a renewable fuel. With the continuing depletion of the known stores of fossil fuels and the rising cost of fuel for internal combustion engines, there is increased interest in the use of ethanol as a fuel substitute or extender. The principal advantage of ethanol as a fuel is that it can be produced from relatively *inexpensive* and *renewable* resources rich in sucrose, starch, or cellulose—starch from corn or wheat, sucrose from beets or cane, and cellulose from straw, forest industry waste, or municipal solid waste. Typically, the raw material (feedstock) is first converted chemically to monosaccharides, then fed to a hardy strain of yeast in an industrial-scale fermenter (Fig. 1). The fermentation can yield not only ethanol for fuel but also side products such as proteins that can be used as animal feed.

FIGURE 1 Industrial-scale fermentations to produce biofuel and other products are typically carried out in tanks that hold thousands of liters of medium.

Fermentations Are Used to Produce Some Common Foods and Industrial Chemicals

Our progenitors learned millennia ago to use fermentation in the production and preservation of foods. Certain microorganisms present in raw food products ferment the carbohydrates and yield metabolic products that give the foods their characteristic forms, textures, and tastes. Yogurt, already known in biblical times, is produced when the bacterium *Lactobacillus bulgaricus* ferments the carbohydrate in milk, producing lactic acid; the resulting drop in pH causes the milk proteins to precipitate, producing the thick texture and sour taste of unsweetened yogurt. Another bacterium,

(a)

Thiamine pyrophosphate (TPP)

(b)

Hydroxyethyl thiamine pyrophosphate

(c)

MECHANISM FIGURE 14–15 Thiamine pyrophosphate (TPP) and its role in pyruvate decarboxylation. (a) TPP is the coenzyme form of vitamin B_1 (thiamine). The reactive carbon atom in the thiazolium ring of TPP is shown in red. In the reaction catalyzed by pyruvate decarboxylase, two of the three carbons of pyruvate are carried transiently on TPP in the form of a hydroxyethyl, or "active acetaldehyde," group **(b)**, which is subsequently released as acetaldehyde. **(c)** The thiazolium ring of TPP stabilizes carbanion intermediates by providing an electrophilic (electron-deficient) structure into which the carbanion electrons can be delocalized by resonance. Structures with this property, often called "electron sinks," play a role in many biochemical reactions—here, facilitating carbon–carbon bond cleavage. 🖱 Thiamine Pyrophosphate Mechanism

Propionibacterium freudenreichii, ferments milk to produce propionic acid and CO_2; the propionic acid precipitates milk proteins, and bubbles of CO_2 cause the holes characteristic of Swiss cheese. Many other food products are the result of fermentations: pickles, sauerkraut, sausage, soy sauce, and a variety of national favorites, such as kimchi (Korea), tempoyak (Indonesia), kefir (Russia), dahi (India), and pozol (Mexico). The drop in pH associated with fermentation also helps to preserve foods, because most of the microorganisms that cause food spoilage cannot grow at low pH. In agriculture, plant byproducts such as corn stalks are preserved for use as animal feed by packing them into a large container (a silo) with limited access to air; microbial fermentation produces acids that lower the pH. The silage that results from this fermentation process can be kept as animal feed for long periods without spoilage.

In 1910 Chaim Weizmann (later to become the first president of Israel) discovered that the bacterium *Clostridium acetobutyricum* ferments starch to butanol and acetone. This discovery opened the field of industrial fermentations, in which some readily available material rich in carbohydrate (corn starch or molasses, for example) is supplied to a pure culture of a specific microorganism, which ferments it into a product of greater commercial value. The ethanol used to make "gasohol" is produced by microbial fermentation, as are formic, acetic, propionic, butyric, and succinic acids,

TABLE 14–1   Some TPP-Dependent Reactions

Enzyme	Pathway(s)	Bond cleaved	Bond formed
Pyruvate decarboxylase	Ethanol fermentation	$R^1-\overset{O}{\overset{\|}{C}}-C\overset{O}{\underset{O^-}{\diagdown}}$	$R^1-\overset{O}{C}\diagdown H$
Pyruvate dehydrogenase α-Ketoglutarate dehydrogenase	Synthesis of acetyl-CoA Citric acid cycle	$R^2-\overset{O}{\overset{\|}{C}}-C\overset{O}{\underset{O^-}{\diagdown}}$	$R^2-\overset{O}{C}\diagdown S\text{-CoA}$
Transketolase	Carbon-assimilation reactions Pentose phosphate pathway	$R^3-\overset{O}{\overset{\|}{C}}-\overset{OH}{\underset{H}{\overset{\|}{C}}}-R^4$	$R^3-\overset{O}{\overset{\|}{C}}-\overset{OH}{\underset{H}{\overset{\|}{C}}}-R^5$

and glycerol, methanol, isopropanol, butanol, and butanediol. These fermentations are generally carried out in huge closed vats in which temperature and access to air are controlled to favor the multiplication of the desired microorganism and to exclude contaminating organisms. The beauty of industrial fermentations is that complicated, multistep chemical transformations are carried out in high yields and with few side products by chemical factories that reproduce themselves—microbial cells. For some industrial fermentations, technology has been developed to immobilize the cells in an inert support, to pass the starting material continuously through the bed of immobilized cells, and to collect the desired product in the effluent—an engineer's dream!

SUMMARY 14.3   Fates of Pyruvate under Anaerobic Conditions: Fermentation

▶   The NADH formed in glycolysis must be recycled to regenerate NAD^+, which is required as an electron acceptor in the first step of the payoff phase. Under aerobic conditions, electrons pass from NADH to O_2 in mitochondrial respiration.

▶   Under anaerobic or hypoxic conditions, many organisms regenerate NAD^+ by transferring electrons from NADH to pyruvate, forming lactate. Other organisms, such as yeast, regenerate NAD^+ by reducing pyruvate to ethanol and CO_2. In these anaerobic processes (fermentations), there is no *net* oxidation or reduction of the carbons of glucose.

▶   A variety of microorganisms can ferment sugar in fresh foods, resulting in changes in pH, taste, and texture, and preserving food from spoilage. Fermentations are used in industry to produce a wide variety of commercially valuable organic compounds from inexpensive starting materials.

14.4 Gluconeogenesis

The central role of glucose in metabolism arose early in evolution, and this sugar remains the nearly universal fuel and building block in modern organisms, from microbes to humans. In mammals, some tissues depend almost completely on glucose for their metabolic energy. For the human brain and nervous system, as well as the erythrocytes, testes, renal medulla, and embryonic tissues, glucose from the blood is the sole or major fuel source. The brain alone requires about 120 g of glucose each day—more than half of all the glucose stored as glycogen in muscle and liver. However, the supply of glucose from these stores is not always sufficient; between meals and during longer fasts, or after vigorous exercise, glycogen is depleted. For these times, organisms need a method for synthesizing glucose from noncarbohydrate precursors. This is accomplished by a pathway called **gluconeogenesis** ("new formation of sugar"), which converts pyruvate and related three- and four-carbon compounds to glucose.

Gluconeogenesis occurs in all animals, plants, fungi, and microorganisms. The reactions are essentially the same in all tissues and all species. The important precursors of glucose in animals are three-carbon compounds such as lactate, pyruvate, and glycerol, as well as certain amino acids **(Fig. 14–16)**. In mammals, gluconeogenesis takes place mainly in the liver, and to a lesser extent in renal cortex and in the epithelial cells that line the inside of the small intestine. The glucose produced passes into the blood to supply other tissues. After vigorous exercise, lactate produced by anaerobic glycolysis in skeletal muscle returns to the liver and is converted to glucose, which moves back to muscle and is converted to glycogen—a circuit called the Cori cycle (Box 14–2; see also Fig. 23–19). In plant seedlings, stored fats and proteins are converted, via paths that include gluconeogenesis, to the disaccharide sucrose for transport throughout the developing

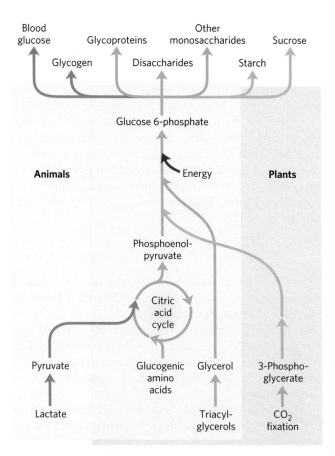

FIGURE 14–16 Carbohydrate synthesis from simple precursors. The pathway from phosphoenolpyruvate to glucose 6-phosphate is common to the biosynthetic conversion of many different precursors of carbohydrates in animals and plants. The path from pyruvate to phosphoenolpyruvate leads through oxaloacetate, an intermediate of the citric acid cycle, which we discuss in Chapter 16. Any compound that can be converted to either pyruvate or oxaloacetate can therefore serve as starting material for gluconeogenesis. This includes alanine and aspartate, which are convertible to pyruvate and oxaloacetate, respectively, and other amino acids that can also yield three- or four-carbon fragments, the so-called glucogenic amino acids (see Table 14-4; see also Fig. 18-15). Plants and photosynthetic bacteria are uniquely able to convert CO_2 to carbohydrates, using the Calvin cycle (see Section 20.1).

plant. Glucose and its derivatives are precursors for the synthesis of plant cell walls, nucleotides and coenzymes, and a variety of other essential metabolites. In many microorganisms, gluconeogenesis starts from simple organic compounds of two or three carbons, such as acetate, lactate, and propionate, in their growth medium.

Although the reactions of gluconeogenesis are the same in all organisms, the metabolic context and the regulation of the pathway differ from one species to another and from tissue to tissue. In this section we focus on gluconeogenesis as it occurs in the mammalian liver. In Chapter 20 we show how photosynthetic organisms use this pathway to convert the primary products of photosynthesis into glucose, to be stored as sucrose or starch.

Gluconeogenesis and glycolysis are not identical pathways running in opposite directions, although they do share several steps **(Fig. 14–17)**; 7 of the 10 enzymatic reactions of gluconeogenesis are the reverse of glycolytic reactions. However, three reactions of glycolysis are essentially irreversible in vivo and cannot

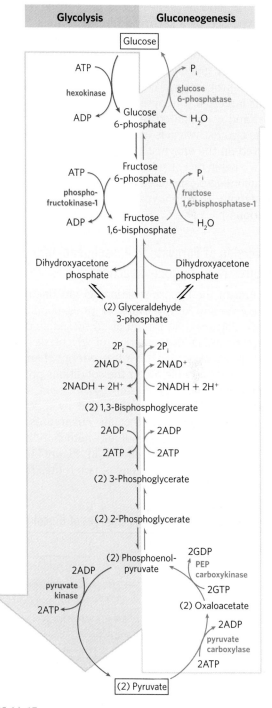

FIGURE 14–17 Opposing pathways of glycolysis and gluconeogenesis in rat liver. The reactions of glycolysis are on the left side, in red; the opposing pathway of gluconeogenesis is on the right, in blue. The major sites of regulation of gluconeogenesis shown here are discussed later in this chapter, and in detail in Chapter 15. Figure 14–20 illustrates an alternative route for oxaloacetate produced in mitochondria.

be used in gluconeogenesis: the conversion of glucose to glucose 6-phosphate by hexokinase, the phosphorylation of fructose 6-phosphate to fructose 1,6-bisphosphate by phosphofructokinase-1, and the conversion of phosphoenolpyruvate to pyruvate by pyruvate kinase (Fig. 14–17). In cells, these three reactions are characterized by a large negative free-energy change, whereas other glycolytic reactions have a ΔG near 0 (Table 14–2). In gluconeogenesis, the three irreversible steps are bypassed by a separate set of enzymes, catalyzing reactions that are sufficiently exergonic to be effectively irreversible in the direction of glucose synthesis. Thus, both glycolysis and gluconeogenesis are irreversible processes in cells. In animals, both pathways occur largely in the cytosol, necessitating their reciprocal and coordinated regulation. Separate regulation of the two pathways is brought about through controls exerted on the enzymatic steps unique to each.

We begin by considering the three bypass reactions of gluconeogenesis. (Keep in mind that "bypass" refers throughout to the bypass of irreversible glycolytic reactions.)

Conversion of Pyruvate to Phosphoenolpyruvate Requires Two Exergonic Reactions

The first of the bypass reactions in gluconeogenesis is the conversion of pyruvate to phosphoenolpyruvate (PEP). This reaction cannot occur by simple reversal of the pyruvate kinase reaction of glycolysis (p. 554), which has a large, negative free-energy change and is therefore irreversible under the conditions prevailing in intact cells (Table 14–2, step **10**). Instead, the phosphorylation of pyruvate is achieved by a roundabout sequence of reactions that in eukaryotes requires enzymes in both the cytosol and mitochondria. As we shall see, the pathway shown in Figure 14–17 and described in detail here is one of two routes from pyruvate to PEP; it is the predominant path when pyruvate or alanine is the glucogenic precursor. A second pathway, described later, predominates when lactate is the glucogenic precursor.

Pyruvate is first transported from the cytosol into mitochondria or is generated from alanine within mitochondria by transamination, in which the α-amino group is transferred from alanine (leaving pyruvate) to an α-keto carboxylic acid (transamination reactions are discussed in detail in Chapter 18). Then **pyruvate carboxylase**, a mitochondrial enzyme that requires the coenzyme **biotin**, converts the pyruvate to oxaloacetate (**Fig. 14–18**):

$$\text{Pyruvate} + \text{HCO}_3^- + \text{ATP} \longrightarrow$$
$$\text{oxaloacetate} + \text{ADP} + \text{P}_i \quad (14\text{–}4)$$

The carboxylation reaction involves biotin as a carrier of activated bicarbonate, as shown in **Figure 14–19**; the reaction mechanism is shown in Figure 16–17. (Note that HCO_3^- is formed by ionization of carbonic acid formed from $\text{CO}_2 + \text{H}_2\text{O}$.) HCO_3^- is phosphorylated by ATP to form a mixed anhydride (a carboxyphosphate); then biotin displaces the phosphate in the formation of carboxybiotin.

Pyruvate carboxylase is the first regulatory enzyme in the gluconeogenic pathway, requiring acetyl-CoA as a positive effector. (Acetyl-CoA is produced by fatty acid oxidation (Chapter 17), and its accumulation signals the availability of fatty acids as fuel.) As we shall see in Chapter 16 (see Fig. 16–16), the pyruvate carboxylase reaction can replenish intermediates in another central metabolic pathway, the citric acid cycle.

Because the mitochondrial membrane has no transporter for oxaloacetate, before export to the cytosol the oxaloacetate formed from pyruvate must be reduced to malate by mitochondrial **malate dehydrogenase**, at the expense of NADH:

TABLE 14–2 Free-Energy Changes of Glycolytic Reactions in Erythrocytes		
Glycolytic reaction step	$\Delta G'^{\circ}$ (kJ/mol)	ΔG (kJ/mol)
1 Glucose + ATP $\longrightarrow$ glucose 6-phosphate + ADP	−16.7	−33.4
2 Glucose 6-phosphate $\rightleftharpoons$ fructose 6-phosphate	1.7	0 to 25
3 Fructose 6-phosphate + ATP $\longrightarrow$ fructose 1,6-bisphosphate + ADP	−14.2	−22.2
4 Fructose 1,6-bisphosphate $\rightleftharpoons$ dihydroxyacetone phosphate + glyceraldehyde 3-phosphate	23.8	−6 to 0
5 Dihydroxyacetone phosphate $\rightleftharpoons$ glyceraldehyde 3-phosphate	7.5	0 to 4
6 Glyceraldehyde 3-phosphate + P$_i$ + NAD$^+$ $\rightleftharpoons$ 1,3-bisphosphoglycerate + NADH + H$^+$	6.3	−2 to 2
7 1,3-Bisphosphoglycerate + ADP $\rightleftharpoons$ 3-phosphoglycerate + ATP	−18.8	0 to 2
8 3-Phosphoglycerate $\rightleftharpoons$ 2-phosphoglycerate	4.4	0 to 0.8
9 2-Phosphoglycerate $\rightleftharpoons$ phosphoenolpyruvate + H$_2$O	7.5	0 to 3.3
10 Phosphoenolpyruvate + ADP $\longrightarrow$ pyruvate + ATP	−31.4	−16.7

Note: $\Delta G'^{\circ}$ is the standard free-energy change, as defined in Chapter 13 (pp. 507–508). ΔG is the free-energy change calculated from the actual concentrations of glycolytic intermediates present under physiological conditions in erythrocytes, at pH 7. The glycolytic reactions bypassed in gluconeogenesis are shown in red. Biochemical equations are not necessarily balanced for H or charge (p. 517).

(a)

(b)

FIGURE 14–18 **Synthesis of phosphoenolpyruvate from pyruvate. (a)** In mitochondria, pyruvate is converted to oxaloacetate in a biotin-requiring reaction catalyzed by pyruvate carboxylase. **(b)** In the cytosol, oxaloacetate is converted to phosphoenolpyruvate by PEP carboxykinase. The CO_2 incorporated in the pyruvate carboxylase reaction is lost here as CO_2. The decarboxylation leads to a rearrangement of electrons that facilitates attack of the carbonyl oxygen of the pyruvate moiety on the γ phosphate of GTP.

$$\text{Oxaloacetate} + \text{NADH} + \text{H}^+ \rightleftharpoons \text{L-malate} + \text{NAD}^+ \quad (14\text{–}5)$$

The standard free-energy change for this reaction is quite high, but under physiological conditions (including a very low concentration of oxaloacetate) $\Delta G \approx 0$ and the reaction is readily reversible. Mitochondrial malate dehydrogenase functions in both gluconeogenesis and the citric acid cycle, but the overall flow of metabolites in the two processes is in opposite directions.

Malate leaves the mitochondrion through a specific transporter in the inner mitochondrial membrane (see Fig. 19–31), and in the cytosol it is reoxidized to oxaloacetate, with the production of cytosolic NADH:

$$\text{Malate} + \text{NAD}^+ \longrightarrow \text{oxaloacetate} + \text{NADH} + \text{H}^+ \quad (14\text{–}6)$$

The oxaloacetate is then converted to PEP by **phosphoenolpyruvate carboxykinase** (Fig. 14–18). This Mg^{2+}-dependent reaction requires GTP as the phosphoryl group donor:

$$\text{Oxaloacetate} + \text{GTP} \rightleftharpoons \text{PEP} + CO_2 + \text{GDP} \quad (14\text{–}7)$$

The reaction is reversible under intracellular conditions; the formation of one high-energy phosphate compound (PEP) is balanced by the hydrolysis of another (GTP).

Pyruvate carboxylase

FIGURE 14–19 **Role of biotin in the pyruvate carboxylase reaction.** The cofactor biotin is covalently attached to the enzyme through an amide linkage to the ε-amino group of a Lys residue, forming a biotinyl-enzyme. The reaction occurs in two phases, which occur at two different sites in the enzyme. At catalytic site 1, bicarbonate ion is converted to CO_2 at the expense of ATP. Then CO_2 reacts with biotin, forming carboxybiotinyl-enzyme. The long arm composed of biotin and the Lys side chain to which it is attached then carry the CO_2 of carboxybiotinyl-enzyme to catalytic site 2 on the enzyme surface, where CO_2 is released and reacts with the pyruvate, forming oxaloacetate and regenerating the biotinyl-enzyme. The general role of flexible arms in carrying reaction intermediates between enzyme active sites is described in Figure 16–18 and the mechanistic details of the pyruvate carboxylase reaction are shown in Figure 16–17. Similar mechanisms occur in other biotin-dependent carboxylation reactions, such as those catalyzed by propionyl-CoA carboxylase (see Fig. 17–12) and acetyl-CoA carboxylase (see Fig. 21–1).

The overall equation for this set of bypass reactions, the sum of Equations 14–4 through 14–7, is

$$\text{Pyruvate} + \text{ATP} + \text{GTP} + \text{HCO}_3^- \longrightarrow$$
$$\text{PEP} + \text{ADP} + \text{GDP} + \text{P}_i + \text{CO}_2 \quad (14\text{–}8)$$
$$\Delta G'^\circ = 0.9 \text{ kJ/mol}$$

Two high-energy phosphate equivalents (one from ATP and one from GTP), each yielding about 50 kJ/mol under cellular conditions, must be expended to phosphorylate one molecule of pyruvate to PEP. In contrast, when PEP is converted to pyruvate during glycolysis, only one ATP is generated from ADP. Although the standard free-energy change ($\Delta G'^\circ$) of the two-step path from pyruvate to PEP is 0.9 kJ/mol, the actual free-energy change (ΔG), calculated from measured cellular concentrations of intermediates, is very strongly negative (–25 kJ/mol); this results from the ready consumption of PEP in other reactions such that its concentration remains relatively low. The reaction is thus effectively irreversible in the cell.

Note that the CO_2 added to pyruvate in the pyruvate carboxylase step is the same molecule that is lost in the PEP carboxykinase reaction (Fig. 14–18b). This carboxylation-decarboxylation sequence represents a way of "activating" pyruvate, in that the decarboxylation of oxaloacetate facilitates PEP formation. In Chapter 21 we shall see how a similar carboxylation-decarboxylation sequence is used to activate acetyl-CoA for fatty acid biosynthesis (see Fig. 21–1).

There is a logic to the route of these reactions through the mitochondrion. The [NADH]/[NAD⁺] ratio in the cytosol is 8×10^{-4}, about 10^5 times lower than in mitochondria. Because cytosolic NADH is consumed in gluconeogenesis (in the conversion of 1,3-bisphosphoglycerate to glyceraldehyde 3-phosphate; Fig. 14–17), glucose biosynthesis cannot proceed unless NADH is available. The transport of malate from the mitochondrion to the cytosol and its reconversion there to oxaloacetate effectively moves reducing equivalents to the cytosol, where they are scarce. This path from pyruvate to PEP therefore provides an important balance between NADH produced and consumed in the cytosol during gluconeogenesis.

A second pyruvate → PEP bypass predominates when lactate is the glucogenic precursor (**Fig. 14–20**). This pathway makes use of lactate produced by glycolysis in erythrocytes or anaerobic muscle, for example, and it is particularly important in large vertebrates after vigorous exercise (Box 14–2). The conversion of lactate to pyruvate in the cytosol of hepatocytes yields NADH, and the export of reducing equivalents (as malate) from mitochondria is therefore unnecessary. After the pyruvate produced by the lactate dehydrogenase reaction is transported into the mitochondrion, it is converted to oxaloacetate by pyruvate carboxylase, as described above. This oxaloacetate, however, is converted directly to PEP by a mitochondrial isozyme of PEP carboxykinase, and the PEP is transported out of the mitochon-

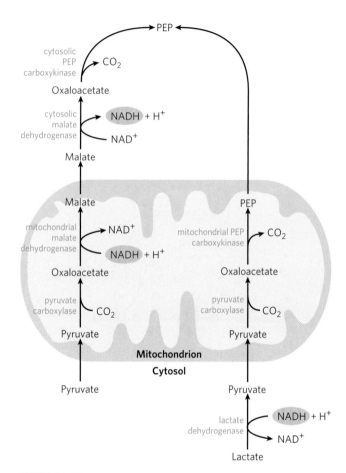

FIGURE 14–20 Alternative paths from pyruvate to phosphoenolpyruvate. The relative importance of the two pathways depends on the availability of lactate or pyruvate and the cytosolic requirements for NADH for gluconeogenesis. The path on the right predominates when lactate is the precursor, because cytosolic NADH is generated in the lactate dehydrogenase reaction and does not have to be shuttled out of the mitochondrion (see text). The requirements of ATP for pyruvate carboxylase and GTP for PEP carboxykinase (see Fig. 14–17) are omitted for simplicity.

drion to continue on the gluconeogenic path. The mitochondrial and cytosolic isozymes of PEP carboxykinase are encoded by separate genes in the nuclear chromosomes, providing another example of two distinct enzymes catalyzing the same reaction but having different cellular locations or metabolic roles (recall the isozymes of hexokinase).

Conversion of Fructose 1,6-Bisphosphate to Fructose 6-Phosphate Is the Second Bypass

The second glycolytic reaction that cannot participate in gluconeogenesis is the phosphorylation of fructose 6-phosphate by PFK-1 (Table 14–2, step ❸). Because this reaction is highly exergonic and therefore irreversible in intact cells, the generation of fructose 6-phosphate from fructose 1,6-bisphosphate (Fig. 14–17) is catalyzed by a different enzyme, Mg²⁺-dependent **fructose 1,6-bisphosphatase (FBPase-1)**, which promotes the essentially irreversible *hydrolysis* of the C-1 phosphate (*not* phosphoryl group transfer to ADP):

Fructose 1,6-bisphosphate + H_2O $\longrightarrow$
$$\text{fructose 6-phosphate} + P_i$$
$$\Delta G'^\circ = -16.3 \text{ kJ/mol}$$

FBPase-1 is so named to distinguish it from another, similar enzyme (FBPase-2) with a regulatory role, which we discuss in Chapter 15.

Conversion of Glucose 6-Phosphate to Glucose Is the Third Bypass

The third bypass is the final reaction of gluconeogenesis, the dephosphorylation of glucose 6-phosphate to yield glucose (Fig. 14–17). Reversal of the hexokinase reaction (p. 548) would require phosphoryl group transfer from glucose 6-phosphate to ADP, forming ATP, an energetically unfavorable reaction (Table 14–2, step ❶). The reaction catalyzed by **glucose 6-phosphatase** does not require synthesis of ATP; it is a simple hydrolysis of a phosphate ester:

Glucose 6-phosphate + H_2O $\longrightarrow$ glucose + P_i
$$\Delta G'^\circ = -13.8 \text{ kJ/mol}$$

This Mg^{2+}-activated enzyme is found on the lumenal side of the endoplasmic reticulum of hepatocytes, renal cells, and epithelial cells of the small intestine (see Fig. 15–30), but not in other tissues, which are therefore unable to supply glucose to the blood. If other tissues had glucose 6-phosphatase, this enzyme's activity would hydrolyze the glucose 6-phosphate needed within those tissues for glycolysis. Glucose produced by gluconeogenesis in the liver or kidney or ingested in the diet is delivered to these other tissues, including brain and muscle, through the bloodstream.

Gluconeogenesis Is Energetically Expensive, but Essential

The sum of the biosynthetic reactions leading from pyruvate to free blood glucose (Table 14–3) is

2 Pyruvate + 4ATP + 2GTP + 2NADH + $2H^+$ + $4H_2O$ $\longrightarrow$
$$\text{glucose} + 4ADP + 2GDP + 6P_i + 2NAD^+ \quad (14\text{-}9)$$

For each molecule of glucose formed from pyruvate, six high-energy phosphate groups are required, four from ATP and two from GTP. In addition, two molecules of NADH are required for the reduction of two molecules of 1,3-bisphosphoglycerate. Clearly, Equation 14–9 is not simply the reverse of the equation for conversion of glucose to pyruvate by glycolysis, which would require only two molecules of ATP:

Glucose + 2ADP + $2P_i$ + NAD^+ $\longrightarrow$
$$2 \text{ pyruvate} + 2ATP + 2NADH + 2H^+ + 2H_2O$$

The synthesis of glucose from pyruvate is a relatively expensive process. Much of this high energy cost is necessary to ensure the irreversibility of gluconeogenesis. Under intracellular conditions, the overall free-energy change of glycolysis is at least −63 kJ/mol. Under the same conditions the overall ΔG of gluconeogenesis is −16 kJ/mol. Thus both glycolysis and gluco-neogenesis are essentially irreversible processes in cells. A second advantage to investing energy to convert pyruvate to glucose is that if pyruvate were instead excreted, its considerable potential for ATP production by complete, aerobic oxidation would be lost (more than 10 ATP are produced per pyruvate, as we shall see in Chapter 16).

TABLE 14–3   Sequential Reactions in Gluconeogenesis Starting from Pyruvate

Pyruvate + HCO_3^- + ATP $\longrightarrow$ oxaloacetate + ADP + P_i	×2
Oxaloacetate + GTP $\rightleftharpoons$ phosphoenolpyruvate + CO_2 + GDP	×2
Phosphoenolpyruvate + H_2O $\rightleftharpoons$ 2-phosphoglycerate	×2
2-Phosphoglycerate $\rightleftharpoons$ 3-phosphoglycerate	×2
3-Phosphoglycerate + ATP $\rightleftharpoons$ 1,3-bisphosphoglycerate + ADP	×2
1,3-Bisphosphoglycerate + NADH + H^+ $\rightleftharpoons$ glyceraldehyde 3-phosphate + NAD^+ + P_i	×2
Glyceraldehyde 3-phosphate $\rightleftharpoons$ dihydroxyacetone phosphate	
Glyceraldehyde 3-phosphate + dihydroxyacetone phosphate $\rightleftharpoons$ fructose 1,6-bisphosphate	
Fructose 1,6-bisphosphate $\longrightarrow$ fructose 6-phosphate + P_i	
Fructose 6-phosphate $\rightleftharpoons$ glucose 6-phosphate	
Glucose 6-phosphate + H_2O $\longrightarrow$ glucose + P_i	
Sum: 2 Pyruvate + 4ATP + 2GTP + 2NADH + $2H^+$ + $4H_2O$ $\longrightarrow$ glucose + 4ADP + 2GDP + 6P_i + 2NAD^+	

Note: The bypass reactions are in red; all other reactions are reversible steps of glycolysis. The figures at the right indicate that the reaction is to be counted twice, because two three-carbon precursors are required to make a molecule of glucose. The reactions required to replace the cytosolic NADH consumed in the glyceraldehyde 3-phosphate dehydrogenase reaction (the conversion of lactate to pyruvate in the cytosol or the transport of reducing equivalents from mitochondria to the cytosol in the form of malate) are not considered in this summary. Biochemical equations are not necessarily balanced for H and charge (p. 517).

Citric Acid Cycle Intermediates and Some Amino Acids Are Glucogenic

The biosynthetic pathway to glucose described above allows the net synthesis of glucose not only from pyruvate but also from the four-, five-, and six-carbon intermediates of the citric acid cycle (Chapter 16). Citrate, isocitrate, α-ketoglutarate, succinyl-CoA, succinate, fumarate, and malate—all are citric acid cycle intermediates that can undergo oxidation to oxaloacetate (see Fig. 16–7). Some or all of the carbon atoms of most amino acids derived from proteins are ultimately catabolized to pyruvate or to intermediates of the citric acid cycle. Such amino acids can therefore undergo net conversion to glucose and are said to be **glucogenic** (Table 14–4). Alanine and glutamine, the principal molecules that transport amino groups from extrahepatic tissues to the liver (see Fig. 18–9), are particularly important glucogenic amino acids in mammals. After removal of their amino groups in liver mitochondria, the carbon skeletons remaining (pyruvate and α-ketoglutarate, respectively) are readily funneled into gluconeogenesis.

Mammals Cannot Convert Fatty Acids to Glucose

No net conversion of fatty acids to glucose occurs in mammals. As we shall see in Chapter 17, the catabolism of most fatty acids yields only acetyl-CoA. Mammals cannot use acetyl-CoA as a precursor of glucose, because the pyruvate dehydrogenase reaction is irreversible and cells have no other pathway to convert acetyl-CoA to pyruvate. Plants, yeast, and many bacteria do have a pathway (the glyoxylate cycle; see Fig. 16–22) for converting acetyl-CoA to oxaloacetate, so these organisms can use fatty acids as the starting material for gluconeogenesis. This is important during the germination of seedlings, for example; before leaves develop and photosynthesis can provide energy and carbohydrates, the seedling relies on stored seed oils for energy production and cell wall biosynthesis.

Although mammals cannot convert fatty acids to carbohydrate, they can use the small amount of glycerol produced from the breakdown of fats (triacyl*glycerols*) for gluconeogenesis. Phosphorylation of glycerol by glycerol kinase, followed by oxidation of the central carbon, yields dihydroxyacetone phosphate, an intermediate in gluconeogenesis in liver.

As we shall see in Chapter 21, glycerol phosphate is an essential intermediate in triacylglycerol synthesis in adipocytes, but these cells lack glycerol kinase and so cannot simply phosphorylate glycerol. Instead, adipocytes carry out a truncated version of gluconeogenesis, known as **glyceroneogenesis**: the conversion of pyruvate to dihydroxyacetone phosphate via the early reactions of gluconeogenesis, followed by reduction of the dihydroxyacetone phosphate to glycerol phosphate (see Fig. 21–21).

Glycolysis and Gluconeogenesis Are Reciprocally Regulated

If glycolysis (the conversion of glucose to pyruvate) and gluconeogenesis (the conversion of pyruvate to glucose) were allowed to proceed simultaneously at high rates, the result would be the consumption of ATP and the production of heat. For example, PFK-1 and FBPase-1 catalyze opposing reactions:

$$\text{ATP} + \text{fructose 6-phosphate} \xrightarrow{\text{PFK-1}}$$
$$\text{ADP} + \text{fructose 1,6-bisphosphate}$$

$$\text{Fructose 1,6-bisphosphate} + \text{H}_2\text{O} \xrightarrow{\text{FBPase-1}}$$
$$\text{fructose 6-phosphate} + \text{P}_i$$

The sum of these two reactions is

$$\text{ATP} + \text{H}_2\text{O} \longrightarrow \text{ADP} + \text{P}_i + \text{heat}$$

These two enzymatic reactions, and several others in the two pathways, are regulated allosterically and by covalent modification (phosphorylation). In Chapter 15 we take up the mechanisms of this regulation in detail. For now, suffice it to say that the pathways are regulated so that when the flux of glucose through glycolysis goes up, the flux of pyruvate toward glucose goes down, and vice versa.

TABLE 14–4	Glucogenic Amino Acids, Grouped by Site of Entry

Pyruvate	**Succinyl-CoA**
Alanine	Isoleucine*
Cysteine	Methionine
Glycine	Threonine
Serine	Valine
Threonine	**Fumarate**
Tryptophan*	Phenylalanine*
α-Ketoglutarate	Tyrosine*
Arginine	**Oxaloacetate**
Glutamate	Asparagine
Glutamine	Aspartate
Histidine	
Proline	

Note: All these amino acids are precursors of blood glucose or liver glycogen, because they can be converted to pyruvate or citric acid cycle intermediates. Of the 20 common amino acids, only leucine and lysine are unable to furnish carbon for net glucose synthesis.

*These amino acids are also ketogenic (see Fig. 18–15).

SUMMARY 14.4 Gluconeogenesis

▶ Gluconeogenesis is a ubiquitous multistep process in which glucose is produced from lactate, pyruvate, or oxaloacetate, or any compound (including citric acid cycle intermediates) that can be converted to one of these intermediates. Seven of the steps in gluconeogenesis are catalyzed by the same enzymes used in glycolysis; these are the reversible reactions.

▶ Three irreversible steps in glycolysis are bypassed by reactions catalyzed by gluconeogenic enzymes: (1) conversion of pyruvate to PEP via oxaloacetate, catalyzed by pyruvate carboxylase and PEP carboxykinase; (2) dephosphorylation of fructose 1,6-bisphosphate by FBPase-1; and (3) dephosphorylation of glucose 6-phosphate by glucose 6-phosphatase.

▶ Formation of one molecule of glucose from pyruvate requires 4 ATP, 2 GTP, and 2 NADH; it is expensive.

▶ In mammals, gluconeogenesis in the liver, kidney, and small intestine provides glucose for use by the brain, muscles, and erythrocytes.

▶ Pyruvate carboxylase is stimulated by acetyl-CoA, increasing the rate of gluconeogenesis when the cell has adequate supplies of other substrates (fatty acids) for energy production.

▶ Animals cannot convert acetyl-CoA derived from fatty acids into glucose; plants and microorganisms can.

▶ Glycolysis and gluconeogenesis are reciprocally regulated to prevent wasteful operation of both pathways at the same time.

FIGURE 14–21 General scheme of the pentose phosphate pathway. NADPH formed in the oxidative phase is used to reduce glutathione, GSSG (see Box 14–4) and to support reductive biosynthesis. The other product of the oxidative phase is ribose 5-phosphate, which serves as a precursor for nucleotides, coenzymes, and nucleic acids. In cells that are not using ribose 5-phosphate for biosynthesis, the nonoxidative phase recycles six molecules of the pentose into five molecules of the hexose glucose 6-phosphate, allowing continued production of NADPH and converting glucose 6-phosphate (in six cycles) to CO_2.

14.5 Pentose Phosphate Pathway of Glucose Oxidation

In most animal tissues, the major catabolic fate of glucose 6-phosphate is glycolytic breakdown to pyruvate, much of which is then oxidized via the citric acid cycle, ultimately leading to the formation of ATP. Glucose 6-phosphate does have other catabolic fates, however, which lead to specialized products needed by the cell. Of particular importance in some tissues is the oxidation of glucose 6-phosphate to pentose phosphates by the **pentose phosphate pathway** (also called the **phosphogluconate pathway** or the **hexose monophosphate pathway**; **Fig. 14–21**). In this oxidative pathway, $NADP^+$ is the electron acceptor, yielding NADPH. Rapidly dividing cells, such as those of bone marrow, skin, and intestinal mucosa, and those of tumors, use the pentose ribose 5-phosphate to make RNA, DNA, and such coenzymes as ATP, NADH, $FADH_2$, and coenzyme A.

In other tissues, the essential product of the pentose phosphate pathway is not the pentoses but the electron donor NADPH, needed for reductive biosynthesis or to counter the damaging effects of oxygen radicals. Tissues that carry out extensive fatty acid syn-

thesis (liver, adipose, lactating mammary gland) or very active synthesis of cholesterol and steroid hormones (liver, adrenal glands, gonads) require the NADPH provided by this pathway. Erythrocytes and the cells of the lens and cornea are directly exposed to oxygen and thus to the damaging free radicals generated by oxygen. By maintaining a reducing atmosphere (a high ratio of NADPH to $NADP^+$ and a high ratio of reduced to oxidized glutathione), such cells can prevent or undo oxidative damage to proteins, lipids, and other sensitive molecules. In erythrocytes, the NADPH produced by the pentose phosphate pathway is so important in preventing oxidative damage that a genetic defect in glucose 6-phosphate dehydrogenase, the first enzyme of the pathway, can have serious medical consequences (Box 14–4). ■

The Oxidative Phase Produces Pentose Phosphates and NADPH

The first reaction of the pentose phosphate pathway **(Fig. 14–22)** is the oxidation of glucose 6-phosphate by **glucose 6-phosphate dehydrogenase (G6PD)** to form 6-phosphoglucono-δ-lactone, an intramolecular ester. $NADP^+$ is the electron acceptor, and the

BOX 14–4 ⚕ MEDICINE   Why Pythagoras Wouldn't Eat Falafel: Glucose 6-Phosphate Dehydrogenase Deficiency

Fava beans, an ingredient of falafel, have been an important food source in the Mediterranean and Middle East since antiquity. The Greek philosopher and mathematician Pythagoras prohibited his followers from dining on fava beans, perhaps because they make many people sick with a condition called favism, which can be fatal. In favism, erythrocytes begin to lyse 24 to 48 hours after ingestion of the beans, releasing free hemoglobin into the blood. Jaundice and sometimes kidney failure can result. Similar symptoms can occur with ingestion of the antimalarial drug primaquine or of sulfa antibiotics, or following exposure to certain herbicides. These symptoms have a genetic basis: glucose 6-phosphate dehydrogenase (G6PD) deficiency, which affects about 400 million people worldwide. Most G6PD-deficient individuals are asymptomatic; only the combination of G6PD deficiency and certain environmental factors produces the clinical manifestations.

Glucose 6-phosphate dehydrogenase catalyzes the first step in the pentose phosphate pathway (see Fig. 14–22), which produces NADPH. This reductant, essential in many biosynthetic pathways, also protects cells from oxidative damage by hydrogen peroxide (H_2O_2) and superoxide free radicals, highly reactive oxidants generated as metabolic byproducts and through the actions of drugs such as primaquine and natural products such as divicine—the toxic ingredient of fava beans. During normal detoxification, H_2O_2 is converted to H_2O by reduced glutathione and glutathione peroxidase, and the oxidized glutathione is converted back to the reduced form by glutathione reductase and NADPH (Fig. 1). H_2O_2 is also broken down to H_2O and O_2 by catalase, which also requires NADPH. In G6PD-deficient individuals, the NADPH production is diminished and detoxification of H_2O_2 is inhibited. Cellular damage results: lipid peroxidation leading to breakdown of erythrocyte membranes and oxidation of proteins and DNA.

The geographic distribution of G6PD deficiency is instructive. Frequencies as high as 25% occur in tropical Africa, parts of the Middle East, and Southeast Asia, areas where malaria is most prevalent. In addition to such epidemiological observations, in vitro studies show that growth of one malaria parasite, *Plasmodium falciparum*, is inhibited in G6PD-deficient eryth-

rocytes. The parasite is very sensitive to oxidative damage and is killed by a level of oxidative stress that is tolerable to a G6PD-deficient human host. Because the advantage of resistance to malaria balances the disadvantage of lowered resistance to oxidative damage, natural selection sustains the G6PD-deficient genotype in human populations where malaria is prevalent. Only under overwhelming oxidative stress, caused by drugs, herbicides, or divicine, does G6PD deficiency cause serious medical problems.

An antimalarial drug such as primaquine is believed to act by causing oxidative stress to the parasite. It is ironic that antimalarial drugs can cause human illness through the same biochemical mechanism that provides resistance to malaria. Divicine also acts as an antimalarial drug, and ingestion of fava beans may protect against malaria. By refusing to eat falafel, many Pythagoreans with normal G6PD activity may have unwittingly increased their risk of malaria!

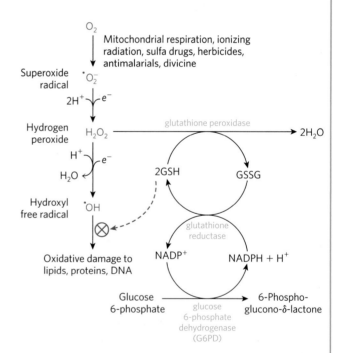

FIGURE 1 Role of NADPH and glutathione in protecting cells against highly reactive oxygen derivatives. Reduced glutathione (GSH) protects the cell by destroying hydrogen peroxide and hydroxyl free radicals. Regeneration of GSH from its oxidized form (GSSG) requires the NADPH produced in the glucose 6-phosphate dehydrogenase reaction.

overall equilibrium lies far in the direction of NADPH formation. The lactone is hydrolyzed to the free acid 6-phosphogluconate by a specific **lactonase**, then 6-phosphogluconate undergoes oxidation and decarboxylation by **6-phosphogluconate dehydrogenase**

to form the ketopentose ribulose 5-phosphate; the reaction generates a second molecule of NADPH. (This ribulose 5-phosphate is important in the regulation of glycolysis and gluconeogenesis, as we shall see in Chapter 15.) **Phosphopentose isomerase** converts

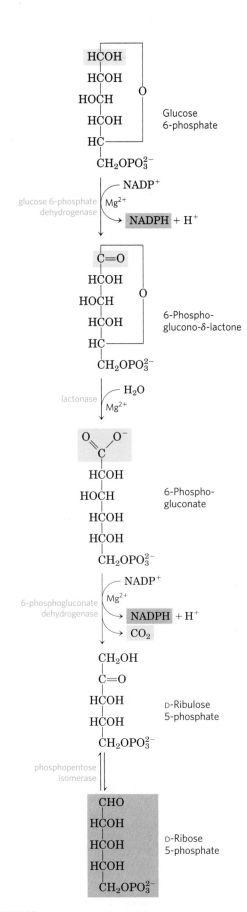

FIGURE 14-22 Oxidative reactions of the pentose phosphate pathway.
The end products are ribose 5-phosphate, CO_2, and NADPH.

ribulose 5-phosphate to its aldose isomer, ribose 5-phosphate. In some tissues, the pentose phosphate pathway ends at this point, and its overall equation is

$$\text{Glucose 6-phosphate} + 2NADP^+ + H_2O \longrightarrow$$
$$\text{ribose 5-phosphate} + CO_2 + 2NADPH + 2H^+$$

The net result is the production of NADPH, a reductant for biosynthetic reactions, and ribose 5-phosphate, a precursor for nucleotide synthesis.

The Nonoxidative Phase Recycles Pentose Phosphates to Glucose 6-Phosphate

In tissues that require primarily NADPH, the pentose phosphates produced in the oxidative phase of the pathway are recycled into glucose 6-phosphate. In this nonoxidative phase, ribulose 5-phosphate is first epimerized to xylulose 5-phosphate:

$$
\begin{array}{ccc}
CH_2OH & & CH_2OH \\
| & & | \\
C{=}O & & C{=}O \\
| & & | \\
H{-}C{-}OH & \rightleftharpoons & HO{-}C{-}H \\
| & & | \\
H{-}C{-}OH & & H{-}C{-}OH \\
| & & | \\
CH_2OPO_3^{2-} & & CH_2OPO_3^{2-} \\
\text{Ribulose} & & \text{Xylulose 5-phosphate} \\
\text{5-phosphate} & &
\end{array}
$$

ribulose 5-phosphate epimerase

Then, in a series of rearrangements of the carbon skeletons **(Fig. 14–23)**, six five-carbon sugar phosphates are converted to five six-carbon sugar phosphates, completing the cycle and allowing continued oxidation of glucose 6-phosphate with production of NADPH. Continued recycling leads ultimately to the conversion of glucose 6-phosphate to six CO_2. Two enzymes unique to the pentose phosphate pathway act in these interconversions of sugars: transketolase and transaldolase. **Transketolase** catalyzes the transfer of a two-carbon fragment from a ketose donor to an aldose acceptor **(Fig. 14–24a)**. In its first appearance in the pentose phosphate pathway, transketolase transfers C-1 and C-2 of xylulose 5-phosphate to ribose 5-phosphate, forming the seven-carbon product sedoheptulose 7-phosphate (Fig. 14–24b). The remaining three-carbon fragment from xylulose is glyceraldehyde 3-phosphate.

Next, **transaldolase** catalyzes a reaction similar to the aldolase reaction of glycolysis: a three-carbon fragment is removed from sedoheptulose 7-phosphate and condensed with glyceraldehyde 3-phosphate, forming fructose 6-phosphate and the tetrose erythrose 4-phosphate **(Fig. 14–25)**. Now transketolase acts again, forming fructose 6-phosphate and glyceraldehyde 3-phosphate from erythrose 4-phosphate and xylulose 5-phosphate **(Fig. 14–26)**. Two molecules of glyceraldehyde 3-phosphate formed by two iterations of these

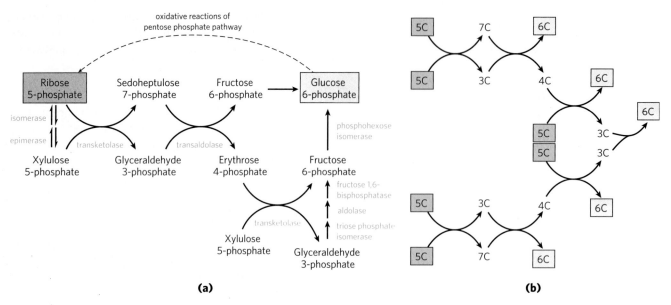

(a)   **(b)**

FIGURE 14–23 Nonoxidative reactions of the pentose phosphate pathway. (a) These reactions convert pentose phosphates to hexose phosphates, allowing the oxidative reactions (see Fig. 14-22) to continue. Transketolase and transaldolase are specific to this pathway; the other enzymes also serve in the glycolytic or gluconeogenic pathways. **(b)** A schematic diagram showing the pathway from six pentoses (5C) to five hexoses (6C). Note that this involves two sets of the interconversions shown in **(a)**. Every reaction shown here is reversible; unidirectional arrows are used only to make clear the direction of the reactions during continuous oxidation of glucose 6-phosphate. In the light-independent reactions of photosynthesis, the direction of these reactions is reversed (see Fig. 20-10).

(a)

(b)

FIGURE 14–24 The first reaction catalyzed by transketolase. (a) The general reaction catalyzed by transketolase is the transfer of a two-carbon group, carried temporarily on enzyme-bound TPP, from a ketose donor to an aldose acceptor. **(b)** Conversion of two pentose phosphates to a triose phosphate and a seven-carbon sugar phosphate, sedoheptulose 7-phosphate.

FIGURE 14–25 The reaction catalyzed by transaldolase.

Sedoheptulose 7-phosphate + Glyceraldehyde 3-phosphate ⇌ (transaldolase) Erythrose 4-phosphate + Fructose 6-phosphate

FIGURE 14–26 The second reaction catalyzed by transketolase.

Xylulose 5-phosphate + Erythrose 4-phosphate ⇌ (TPP, transketolase) Glyceraldehyde 3-phosphate + Fructose 6-phosphate

reactions can be converted to a molecule of fructose 1,6-bisphosphate as in gluconeogenesis (Fig. 14–17), and finally FBPase-1 and phosphohexose isomerase convert fructose 1,6-bisphosphate to glucose 6-phosphate. Overall, six pentose phosphates have been converted to five hexose phosphates (Fig. 14–23b)—the cycle is now complete!

Transketolase requires the cofactor thiamine pyrophosphate (TPP), which stabilizes a two-carbon carbanion in this reaction **(Fig. 14–27a)**, just as it does in the pyruvate decarboxylase reaction (Fig. 14–15). Transaldolase uses a Lys side chain to form a Schiff base with the carbonyl group of its substrate, a ketose, thereby stabilizing a carbanion (Fig. 14–27b) that is central to the reaction mechanism.

The process described in Figure 14–22 is known as the **oxidative pentose phosphate pathway**. The first and third steps are oxidations with large, negative standard free-energy changes and are essentially irreversible in the cell. The reactions of the nonoxidative part of the pentose phosphate pathway (Fig. 14–23) are readily reversible and thus also provide a means of converting hexose phosphates to pentose phosphates. As we shall see in Chapter 20, a process that converts hexose phosphates to pentose phosphates is crucial to the photosynthetic assimilation of CO_2 by plants. That pathway, the **reductive pentose phosphate pathway**, is essentially the reversal of the reactions shown in Figure 14–23 and employs many of the same enzymes.

All the enzymes in the pentose phosphate pathway are located in the cytosol, like those of glycolysis and most of those of gluconeogenesis. In fact, these three pathways are connected through several shared intermediates and enzymes. The glyceraldehyde 3-phosphate formed by the action of transketolase is readily

(a) Transketolase

TPP ⇌ (resonance stabilization)

(b) Transaldolase

Lys—N Protonated Schiff base ⇌ (resonance stabilization) Lys—N

FIGURE 14–27 Carbanion intermediates stabilized by covalent interactions with transketolase and transaldolase. (a) The ring of TPP stabilizes the carbanion in the dihydroxyethyl group carried by transketolase; see Fig. 14–15 for the chemistry of TPP action. **(b)** In the transaldolase reaction, the protonated Schiff base formed between the ε-amino group of a Lys side chain and the substrate stabilizes the C-3 carbanion formed after aldol cleavage.

converted to dihydroxyacetone phosphate by the glycolytic enzyme triose phosphate isomerase, and these two trioses can be joined by the aldolase as in gluconeogenesis, forming fructose 1,6-bisphosphate. Alternatively, the triose phosphates can be oxidized to pyruvate by the glycolytic reactions. The fate of the trioses is determined by the cell's relative needs for pentose phosphates, NADPH, and ATP.

Wernicke-Korsakoff Syndrome Is Exacerbated by a Defect in Transketolase

Wernicke-Korsakoff syndrome is a disorder caused by a severe deficiency of thiamine, a component of TPP. The syndrome is more common among people with alcoholism than in the general population, because chronic, heavy alcohol consumption interferes with the intestinal absorption of thiamine. The syndrome can be exacerbated by a mutation in the gene for transketolase that results in an enzyme with a lowered affinity for TPP—an affinity one-tenth that of the normal enzyme. This defect makes individuals much more sensitive to a thiamine deficiency: even a moderate thiamine deficiency (tolerable in individuals with an unmutated transketolase) can drop the level of TPP below that needed to saturate the enzyme. The result is a slowing down of the whole pentose phosphate pathway. In people with Wernicke-Korsakoff syndrome this results in a worsening of symptoms, which can include severe memory loss, mental confusion, and partial paralysis. ■

Glucose 6-Phosphate Is Partitioned between Glycolysis and the Pentose Phosphate Pathway

Whether glucose 6-phosphate enters glycolysis or the pentose phosphate pathway depends on the current needs of the cell and on the concentration of $NADP^+$ in the cytosol. Without this electron acceptor, the first reaction of the pentose phosphate pathway (catalyzed by G6PD) cannot proceed. When a cell is rapidly converting NADPH to $NADP^+$ in biosynthetic reductions, the level of $NADP^+$ rises, allosterically stimulating G6PD and thereby increasing the flux of glucose 6-phosphate through the pentose phosphate pathway **(Fig. 14–28)**. When the demand for NADPH slows, the level of $NADP^+$ drops, the pentose phosphate pathway slows, and glucose 6-phosphate is instead used to fuel glycolysis.

SUMMARY 14.5 Pentose Phosphate Pathway of Glucose Oxidation

▶ The *oxidative* pentose phosphate pathway (phosphogluconate pathway, or hexose monophosphate pathway) brings about oxidation and decarboxylation at C-1 of glucose 6-phosphate, reducing $NADP^+$ to NADPH and producing pentose phosphates.

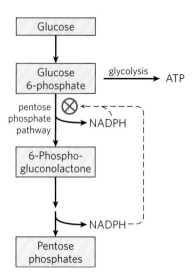

FIGURE 14–28 Role of NADPH in regulating the partitioning of glucose 6-phosphate between glycolysis and the pentose phosphate pathway. When NADPH is forming faster than it is being used for biosynthesis and glutathione reduction (see Fig. 14-21), [NADPH] rises and inhibits the first enzyme in the pentose phosphate pathway. As a result, more glucose 6-phosphate is available for glycolysis.

▶ NADPH provides reducing power for biosynthetic reactions, and ribose 5-phosphate is a precursor for nucleotide and nucleic acid synthesis. Rapidly growing tissues and tissues carrying out active biosynthesis of fatty acids, cholesterol, or steroid hormones send more glucose 6-phosphate through the pentose phosphate pathway than do tissues with less demand for pentose phosphates and reducing power.

▶ The first phase of the pentose phosphate pathway consists of two oxidations that convert glucose 6-phosphate to ribulose 5-phosphate and reduce $NADP^+$ to NADPH. The second phase comprises nonoxidative steps that convert pentose phosphates to glucose 6-phosphate, which begins the cycle again.

▶ In the second phase, transketolase (with TPP as cofactor) and transaldolase catalyze the interconversion of three-, four-, five-, six-, and seven-carbon sugars, with the reversible conversion of six pentose phosphates to five hexose phosphates. In the carbon-assimilating reactions of photosynthesis, the same enzymes catalyze the reverse process, the *reductive* pentose phosphate pathway: conversion of five hexose phosphates to six pentose phosphates.

▶ A genetic defect in transketolase that lowers its affinity for TPP exacerbates the Wernicke-Korsakoff syndrome.

▶ Entry of glucose 6-phosphate either into glycolysis or into the pentose phosphate pathway is largely determined by the relative concentrations of $NADP^+$ and NADPH.

Key Terms

Terms in bold are defined in the glossary.

glycolysis 544
fermentation 544
lactic acid
    fermentation 546
hypoxia 546
ethanol (alcohol)
    fermentation 546
isozymes 549
acyl phosphate 552
substrate-level
    phosphorylation 553
respiration-linked
    phosphorylation 553
phosphoenolpyruvate
    (PEP) 554
mutases 560

isomerases 560
lactose intolerance 561
galactosemia 562
thiamine pyrophosphate
    (TPP) 565
gluconeogenesis 568
biotin 570
pentose phosphate
    pathway 575
phosphogluconate
    pathway 575
hexose monophosphate
    pathway 575

Further Reading

General

Fruton, J.S. (1999) *Proteins, Genes, and Enzymes: The Interplay of Chemistry and Biology*, Yale University Press, New Haven.
    This text includes a detailed historical account of research on glycolysis.

Glycolysis

Boiteux, A. & Hess, B. (1981) Design of glycolysis. *Philos. Trans. R. Soc. Lond. Ser. B Biol. Sci.* **293**, 5–22.
    Intermediate-level review of the pathway and the classic view of its control.

Dandekar, T., Schuster, S., Snel, B., Huynen, M., & Bork, P. (1999) Pathway alignment: application to the comparative analysis of glycolytic enzymes. *Biochem. J.* **343**, 115–124.
    Intermediate-level review of the bioinformatic view of the evolution of glycolysis.

Dang, C.V. & Semenza, G.L. (1999) Oncogenic alterations of metabolism. *Trends Biochem. Sci.* **24**, 68–72.
    Brief review of the molecular basis for increased glycolysis in tumors.

Erlandsen, H., Abola, E.E., & Stevens, R.C. (2000) Combining structural genomics and enzymology: completing the picture in metabolic pathways and enzyme active sites. *Curr. Opin. Struct. Biol.* **10**, 719–730.
    Intermediate-level review of the structures of the glycolytic enzymes.

Gatenby, R.A. & Gillies, R.J. (2004) Why do cancers have high aerobic glycolysis? *Nat. Rev. Cancer* **4**, 891–899.

Hardie, D.G. (2000) Metabolic control: a new solution to an old problem. *Curr. Biol.* **10**, R757–R759.

Harris, A.L. (2002) Hypoxia—a key regulatory factor in tumour growth. *Nat. Rev. Cancer* **2**, 38–47.

Heinrich, R., Melendez-Hevia, E., Montero, F., Nuno, J.C., Stephani, A., & Waddell, T.D. (1999) The structural design of glycolysis: an evolutionary approach. *Biochem. Soc. Trans.* **27**, 294–298.

Herling, A., König, M., Bulik, S., & Holzhütter, H.G. (2011) Enzymatic features of the glucose metabolism in tumor cells. *FEBS J.* **278**, 2436–2459.

Keith, B. & Simon, M.C. (2007) Hypoxia-inducible factors, stem cells, and cancer. *Cell* **129**, 465–472.
    Intermediate-level review.

Knowles, J. & Albery, W.J. (1977) Perfection in enzyme catalysis: the energetics of triose phosphate isomerase. *Acc. Chem. Res.* **10**, 105–111.

Kresge, N., Simoni, R.D., & Hill, R.L. (2005) Otto Fritz Meyerhof and the elucidation of the glycolytic pathway. *J. Biol. Chem.* **280**, e3.
    Brief review of classic papers, which are also available online.

Kritikou, E. (2006) p53 turns on the energy switch. *Nat. Rev. Mol. Cell Biol.* **7**, 552–553.

Pelicano, H., Martin, D.S., Zu, R-H., & Huang, P. (2006) Glycolysis inhibition for anticancer treatment. *Oncogene* **25**, 4633–4646.
    Intermediate-level review.

Phillips, D., Blake, C.C.F., & Watson, H.C. (eds). (1981) The Enzymes of Glycolysis: Structure, Activity and Evolution. *Philos. Trans. R. Soc. Lond. Ser. B Biol. Sci.* **293**, 1–214.
    A collection of excellent reviews on the enzymes of glycolysis, written at a level challenging but comprehensible to a beginning student of biochemistry.

Plaxton, W.C. (1996) The organization and regulation of plant glycolysis. *Annu. Rev. Plant Physiol. Plant Mol. Biol.* **47**, 185–214.
    Very helpful review of the subcellular localization of glycolytic enzymes and the regulation of glycolysis in plants.

Rose, I. (1981) Chemistry of proton abstraction by glycolytic enzymes (aldolase, isomerases, and pyruvate kinase). *Philos. Trans. R. Soc. Lond. Ser. B Biol. Sci.* **293**, 131–144.
    Intermediate-level review of the mechanisms of these enzymes.

Shirmer, T. & Evans, P.R. (1990) Structural basis for the allosteric behavior of phosphofructokinase. *Nature* **343**, 140–145.

Smith, T.A. (2000) Mammalian hexokinases and their abnormal expression in cancer. *Br. J. Biomed. Sci.* **57**, 170–178.
    A review of the four hexokinase isozymes of mammals: their properties and tissue distributions and their expression during the development of tumors.

Feeder Pathways for Glycolysis

Elsas, L.J. & Lai, K. (1998) The molecular biology of galactosemia. *Genet. Med.* **1**, 40–48.

Novelli, G. & Reichardt, J.K. (2000) Molecular basis of disorders of human galactose metabolism: past, present, and future. *Mol. Genet. Metab.* **71**, 62–65.

Petry, K.G. & Reichardt, J.K. (1998) The fundamental importance of human galactose metabolism: lessons from genetics and biochemistry. *Trends Genet.* **14**, 98–102.

Van Beers, E.H., Buller, H.A., Grand, R.J., Einerhand, A.W.C., & Dekker, J. (1995) Intestinal brush border glycohydrolases: structure, function, and development. *Crit. Rev. Biochem. Mol. Biol.* **30**, 197–262.

Fermentations

Demain, A.L., Davies, J.E., Atlas, R.M., Cohen, G., Hershberger, C.L., Hu, W.-S., Sherman, D.H., Willson, R.C., & Wu, J.H.D. (eds). (1999) *Manual of Industrial Microbiology and Biotechnology*, American Society for Microbiology, Washington, DC.
    Classic introduction to all aspects of industrial fermentations.

Liese, A., Seelbach, K., & Wandrey, C. (eds). (2006) *Industrial Biotransformations*, John Wiley & Sons, New York.
    The use of microorganisms in industry for the synthesis of valuable products from inexpensive starting materials.

Sticklen, M.B. (2008) Plant genetic engineering for biofuel production: towards affordable cellulosic ethanol. *Nat. Rev. Genet.* **9**, 433–443.

Gluconeogenesis

Aschenback, J.R., Kristensen, N.B., Donkin, S.S., Hammon, H.M., & Penner, G.B. (2010) Gluconeogenesis in dairy cows: the secret of making sweet milk from sour dough. *IUBMB Life* **62**, 869–877.

Gerich, J.E., Meyer, C., Woerle, H.J., & Stumvoll, M. (2001) Renal gluconeogenesis: its importance in human glucose homeostasis. *Diabetes Care* **24**, 382–391.

Intermediate-level review of the contribution of kidney tissue to gluconeogenesis.

Gleeson, T. (1996) Post-exercise lactate metabolism: a comparative review of sites, pathways, and regulation. *Annu. Rev. Physiol.* **58**, 565–581.

Hers, H.G. & Hue, L. (1983) Gluconeogenesis and related aspects of glycolysis. *Annu. Rev. Biochem.* **52**, 617–653.

Matte, A., Tari, L.W., Goldie, H., & Delbaere, L.T.J. (1997) Structure and mechanism of phosphoenolpyruvate carboxykinase. *J. Biol. Chem.* **272**, 8105–8108.

Oxidative Pentose Phosphate Pathway

Chayen, J., Howat, D.W., & Bitensky, L. (1986) Cellular biochemistry of glucose 6-phosphate and 6-phosphogluconate dehydrogenase activities. *Cell Biochem. Funct.* **4**, 249–253.

Horecker, B.L. (1976) Unraveling the pentose phosphate pathway. In *Reflections on Biochemistry* (Kornberg, A., Cornudella, L., Horecker, B.L., & Oro, J., eds), pp. 65–72, Pergamon Press, Inc., Oxford.

Kletzien, R.F., Harris, P.K., & Foellmi, L.A. (1994) Glucose 6-phosphate dehydrogenase: a "housekeeping" enzyme subject to tissue-specific regulation by hormones, nutrients, and oxidant stress. *FASEB J.* **8**, 174–181.

An intermediate-level review.

Kresge, N., Simoni, R.D., & Hill, R.L. (2005) Bernard L. Horecker's contributions to elucidating the pentose phosphate pathway. *J. Biol. Chem.* **280**, e26.

Brief review of classic papers, which are also available online.

Martini, G. & Ursini, M.V. (1996) A new lease on life for an old enzyme. *BioEssays* **18**, 631–637.

An intermediate-level review of glucose 6-phosphate dehydrogenase, the effects of mutations in this enzyme in humans, and the effects of knock-out mutations in mice.

Notaro, R., Afolayan, A., & Luzzatto, L. (2000) Human mutations in glucose 6-phosphate dehydrogenase reflect evolutionary history. *FASEB J.* **14**, 485–494.

Perl, A., Hanczko, R., Relarico, T., Oaks, Z., & Landas, S. (2011) Oxidative stress, inflammation and carcinogenesis are controlled through the pentose phosphate pathway by transaldolase. *Trends Mol. Med.* **17**, 395–403.

Saggerson, D. (2009) Getting to grips with the pentose phosphate pathway in 1953. *Biochem J.* (doi:10.1042/BJ20081961).

Vulliamy, T., Mehta, A., Luzzatto, L. (2006) Glucose 6-Phosphate Dehydrogenas Deficiency. In *Scriver's Online Metabolic and Molecular Bases of Inherited Disease* (Valle, D., Beaudet, A.L., Vogelstein, B., Kinzler, K.W., Antonarakis, S.E., Ballabio, A., eds) (http://dx.doi.org/10.1036/ommbid.212).

This classic medical encyclopedia, last published in 2001 as a four-volume set, is now maintained online (www.ommbid.com). It contains definitive descriptions of the clinical, biochemical, and genetic aspects of hundreds of human metabolic diseases—an authoritative source and fascinating reading.

Wood, T. (1985) *The Pentose Phosphate Pathway*, Academic Press, Inc., Orlando, FL.

Wood, T. (1986) Physiological functions of the pentose phosphate pathway. *Cell Biochem. Funct.* **4**, 241–247.

Problems

1. Equation for the Preparatory Phase of Glycolysis
Write balanced biochemical equations for all the reactions in the catabolism of glucose to two molecules of glyceraldehyde 3-phosphate (the preparatory phase of glycolysis), including the standard free-energy change for each reaction. Then write the overall or net equation for the preparatory phase of glycolysis, with the net standard free-energy change.

2. The Payoff Phase of Glycolysis in Skeletal Muscle
In working skeletal muscle under anaerobic conditions, glyceraldehyde 3-phosphate is converted to pyruvate (the payoff phase of glycolysis), and the pyruvate is reduced to lactate. Write balanced biochemical equations for all the reactions in this process, with the standard free-energy change for each reaction. Then write the overall or net equation for the payoff phase of glycolysis (with lactate as the end product), including the net standard free-energy change.

3. GLUT Transporters Compare the localization of GLUT4 with that of GLUT2 and GLUT3, and explain why these localizations are important in the response of muscle, adipose tissue, brain, and liver to insulin.

4. Ethanol Production in Yeast When grown anaerobically on glucose, yeast (*S. cerevisiae*) converts pyruvate to acetaldehyde, then reduces acetaldehyde to ethanol using electrons from NADH. Write the equation for the second reaction, and calculate its equilibrium constant at 25 °C, given the standard reduction potentials in Table 13–7.

5. Energetics of the Aldolase Reaction Aldolase catalyzes the glycolytic reaction

Fructose 1,6-bisphosphate $\longrightarrow$
glyceraldehyde 3-phosphate + dihydroxyacetone phosphate

The standard free-energy change for this reaction in the direction written is +23.8 kJ/mol. The concentrations of the three intermediates in the hepatocyte of a mammal are: fructose 1,6-bisphosphate, 1.4×10^{-5} M; glyceraldehyde 3-phosphate, 3×10^{-6} M; and dihydroxyacetone phosphate, 1.6×10^{-5} M. At body temperature (37 °C), what is the actual free-energy change for the reaction?

6. Pathway of Atoms in Fermentation A "pulse-chase" experiment using ^{14}C-labeled carbon sources is carried out on a yeast extract maintained under strictly anaerobic conditions to produce ethanol. The experiment consists of incubating a small amount of ^{14}C-labeled substrate (the pulse) with the yeast extract just long enough for each intermediate in the fermentation pathway to become labeled. The label is then "chased" through the pathway by the addition of excess unlabeled glucose. The chase effectively prevents any further entry of labeled glucose into the pathway.

(a) If [1-^{14}C]glucose (glucose labeled at C-1 with ^{14}C) is used as a substrate, what is the location of ^{14}C in the product ethanol? Explain.

(b) Where would ^{14}C have to be located in the starting glucose to ensure that all the ^{14}C activity is liberated as ^{14}CO$_2$ during fermentation to ethanol? Explain.

7. Heat from Fermentations Large-scale industrial fermenters generally require constant, vigorous cooling. Why?

8. Fermentation to Produce Soy Sauce Soy sauce is prepared by fermenting a salted mixture of soybeans and wheat with several microorganisms, including yeast, over a period of 8 to 12 months. The resulting sauce (after solids are removed) is rich in lactate and ethanol. How are these two compounds produced? To prevent the soy sauce from having a strong vinegary taste (vinegar is dilute acetic acid), oxygen must be kept out of the fermentation tank. Why?

9. Equivalence of Triose Phosphates ^{14}C-Labeled glyceraldehyde 3-phosphate was added to a yeast extract. After a short time, fructose 1,6-bisphosphate labeled with ^{14}C at C-3 and C-4 was isolated. What was the location of the ^{14}C label in the starting glyceraldehyde 3-phosphate? Where did the second ^{14}C label in fructose 1,6-bisphosphate come from? Explain.

10. Glycolysis Shortcut Suppose you discovered a mutant yeast whose glycolytic pathway was shorter because of the presence of a new enzyme catalyzing the reaction

$$\text{Glyceraldehyde 3-phosphate} + \text{H}_2\text{O} \xrightarrow[\text{NAD}^+ \quad \text{NADH} + \text{H}^+]{} \text{3-phosphoglycerate}$$

Would shortening the glycolytic pathway in this way benefit the cell? Explain.

11. Role of Lactate Dehydrogenase During strenuous activity, the demand for ATP in muscle tissue is vastly increased. In rabbit leg muscle or turkey flight muscle, the ATP is produced almost exclusively by lactic acid fermentation. ATP is formed in the payoff phase of glycolysis by two reactions, promoted by phosphoglycerate kinase and pyruvate kinase. Suppose skeletal muscle were devoid of lactate dehydrogenase. Could it carry out strenuous physical activity; that is, could it generate ATP at a high rate by glycolysis? Explain.

12. Efficiency of ATP Production in Muscle The transformation of glucose to lactate in myocytes releases only about 7% of the free energy released when glucose is completely oxidized to CO$_2$ and H$_2$O. Does this mean that anaerobic glycolysis in muscle is a wasteful use of glucose? Explain.

13. Free-Energy Change for Triose Phosphate Oxidation The oxidation of glyceraldehyde 3-phosphate to 1,3-bisphos-

phoglycerate, catalyzed by glyceraldehyde 3-phosphate dehydrogenase, proceeds with an unfavorable equilibrium constant ($K'_{eq} = 0.08$; $\Delta G'^\circ = 6.3$ kJ/mol), yet the flow through this point in the glycolytic pathway proceeds smoothly. How does the cell overcome the unfavorable equilibrium?

14. Arsenate Poisoning Arsenate is structurally and chemically similar to inorganic phosphate (P$_i$), and many enzymes that require phosphate will also use arsenate. Organic compounds of arsenate are less stable than analogous phosphate compounds, however. For example, acyl *arsenates* decompose rapidly by hydrolysis:

$$\text{R}-\overset{\overset{\displaystyle O}{\|}}{\text{C}}-\text{O}-\overset{\overset{\displaystyle O}{\|}}{\underset{\underset{\displaystyle O^-}{|}}{\text{As}}}-\text{O}^- + \text{H}_2\text{O} \longrightarrow$$

$$\text{R}-\overset{\overset{\displaystyle O}{\|}}{\text{C}}-\text{O}^- + \text{HO}-\overset{\overset{\displaystyle O}{\|}}{\underset{\underset{\displaystyle O^-}{|}}{\text{As}}}-\text{O}^- + \text{H}^+$$

On the other hand, acyl *phosphates*, such as 1,3-bisphosphoglycerate, are more stable and undergo further enzyme-catalyzed transformation in cells.

(a) Predict the effect on the net reaction catalyzed by glyceraldehyde 3-phosphate dehydrogenase if phosphate were replaced by arsenate.

(b) What would be the consequence to an organism if arsenate were substituted for phosphate? Arsenate is very toxic to most organisms. Explain why.

15. Requirement for Phosphate in Ethanol Fermentation In 1906 Harden and Young, in a series of classic studies on the fermentation of glucose to ethanol and CO$_2$ by extracts of brewer's yeast, made the following observations. (1) Inorganic phosphate was essential to fermentation; when the supply of phosphate was exhausted, fermentation ceased before all the glucose was used. (2) During fermentation under these conditions, ethanol, CO$_2$, and a hexose bisphosphate accumulated. (3) When arsenate was substituted for phosphate, no hexose bisphosphate accumulated, but the fermentation proceeded until all the glucose was converted to ethanol and CO$_2$.

(a) Why did fermentation cease when the supply of phosphate was exhausted?

(b) Why did ethanol and CO$_2$ accumulate? Was the conversion of pyruvate to ethanol and CO$_2$ essential? Why? Identify the hexose bisphosphate that accumulated. Why did it accumulate?

(c) Why did the substitution of arsenate for phosphate prevent the accumulation of the hexose bisphosphate yet allow fermentation to ethanol and CO$_2$ to go to completion? (See Problem 14.)

16. Role of the Vitamin Niacin Adults engaged in strenuous physical activity require an intake of about 160 g of carbohydrate daily but only about 20 mg of niacin for optimal nutrition. Given the role of niacin in glycolysis, how do you explain the observation?

17. Synthesis of Glycerol Phosphate The glycerol 3-phosphate required for the synthesis of glycerophospholipids can be synthesized from a glycolytic intermediate. Propose a reaction sequence for this conversion.

18. Severity of Clinical Symptoms Due to Enzyme Deficiency The clinical symptoms of two forms of galactosemia—deficiency of galactokinase or of UDP-glucose:galactose 1-phosphate uridylyltransferase—show radically different severity. Although both types produce gastric discomfort after milk ingestion, deficiency of the transferase also leads to liver, kidney, spleen, and brain dysfunction and eventual death. What products accumulate in the blood and tissues with each type of enzyme deficiency? Estimate the relative toxicities of these products from the above information.

19. Muscle Wasting in Starvation One consequence of starvation is a reduction in muscle mass. What happens to the muscle proteins?

20. Pathway of Atoms in Gluconeogenesis A liver extract capable of carrying out all the normal metabolic reactions of the liver is briefly incubated in separate experiments with the following ^{14}C-labeled precursors.

(a) [^{14}C]Bicarbonate, $HO-^{14}C\overset{O^-}{\underset{O}{\diagdown}}$

(b) [1-^{14}C]Pyruvate, $CH_3-\overset{O}{\underset{\|}{C}}-^{14}COO^-$

Trace the pathway of each precursor through gluconeogenesis. Indicate the location of ^{14}C in all intermediates and in the product, glucose.

21. Energy Cost of a Cycle of Glycolysis and Gluconeogenesis What is the cost (in ATP equivalents) of transforming glucose to pyruvate via glycolysis and back again to glucose via gluconeogenesis?

22. Relationship between Gluconeogenesis and Glycolysis Why is it important that gluconeogenesis is not the exact reversal of glycolysis?

23. Energetics of the Pyruvate Kinase Reaction Explain in bioenergetic terms how the conversion of pyruvate to phosphoenolpyruvate in gluconeogenesis overcomes the large, negative, standard free-energy change of the pyruvate kinase reaction in glycolysis.

24. Glucogenic Substrates A common procedure for determining the effectiveness of compounds as precursors of glucose in mammals is to starve the animal until the liver glycogen stores are depleted and then administer the compound in question. A substrate that leads to a *net* increase in liver glycogen is termed glucogenic, because it must first be converted to glucose 6-phosphate. Show by means of known enzymatic reactions which of the following substances are glucogenic.

(a) Succinate, $^-OOC-CH_2-CH_2-COO^-$

(b) Glycerol, $\underset{\underset{H}{|}}{CH_2}\overset{OH}{\underset{\ }{-}}\overset{OHOH}{\underset{\ }{C}}-CH_2$

(c) Acetyl-CoA, $CH_3-\overset{O}{\underset{\|}{C}}-S\text{-CoA}$

(d) Pyruvate, $CH_3-\overset{O}{\underset{\|}{C}}-COO^-$

(e) Butyrate, $CH_3-CH_2-CH_2-COO^-$

25. Ethanol Affects Blood Glucose Levels The consumption of alcohol (ethanol), especially after periods of strenuous activity or after not eating for several hours, results in a deficiency of glucose in the blood, a condition known as hypoglycemia. The first step in the metabolism of ethanol by the liver is oxidation to acetaldehyde, catalyzed by liver alcohol dehydrogenase:

$$CH_3CH_2OH + NAD^+ \longrightarrow CH_3CHO + NADH + H^+$$

Explain how this reaction inhibits the transformation of lactate to pyruvate. Why does this lead to hypoglycemia?

26. Blood Lactate Levels during Vigorous Exercise The concentrations of lactate in blood plasma before, during, and after a 400 m sprint are shown in the graph.

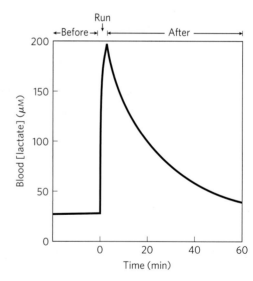

(a) What causes the rapid rise in lactate concentration?

(b) What causes the decline in lactate concentration after completion of the sprint? Why does the decline occur more slowly than the increase?

(c) Why is the concentration of lactate not zero during the resting state?

27. Relationship between Fructose 1,6-Bisphosphatase and Blood Lactate Levels A congenital defect in the liver enzyme fructose 1,6-bisphosphatase results in abnormally high levels of lactate in the blood plasma. Explain.

28. Effect of Phloridzin on Carbohydrate Metabolism Phloridzin, a toxic glycoside from the bark of the pear tree, blocks the normal reabsorption of glucose from the kidney tubule, thus causing blood glucose to be almost completely excreted in the urine. In an experiment, rats fed phloridzin and sodium succinate excreted about 0.5 mol of glucose (made by gluconeogenesis) for every 1 mol of sodium succinate ingested. How is the succinate transformed to glucose? Explain the stoichiometry.

Phloridzin

29. Excess O_2 Uptake during Gluconeogenesis Lactate absorbed by the liver is converted to glucose, with the input of 6 mol of ATP for every mole of glucose produced. The extent of this process in a rat liver preparation can be monitored by administering [^{14}C]lactate and measuring the amount of [^{14}C]glucose produced. Because the stoichiometry of O_2 consumption and ATP production is known (about 5 ATP per O_2), we can predict the extra O_2 consumption above the normal rate when a given amount of lactate is administered. However, when the extra O_2 used in the synthesis of glucose from lactate is actually measured, it is always higher than predicted by known stoichiometric relationships. Suggest a possible explanation for this observation.

30. Role of the Pentose Phosphate Pathway If the oxidation of glucose 6-phosphate via the pentose phosphate pathway were being used primarily to generate NADPH for biosynthesis, the other product, ribose 5-phosphate, would accumulate. What problems might this cause?

Data Analysis Problem

31. Engineering a Fermentation System Fermentation of plant matter to produce ethanol for fuel is one potential method for reducing the use of fossil fuels and thus the CO_2 emissions that lead to global warming. Many microorganisms can break down cellulose then ferment the glucose to ethanol. However, many potential cellulose sources, including agricultural residues and switchgrass, also contain substantial amounts of arabinose, which is not as easily fermented.

D-Arabinose

Escherichia coli is capable of fermenting arabinose to ethanol, but it is not naturally tolerant of high ethanol levels,

thus limiting its utility for commercial ethanol production. Another bacterium, *Zymomonas mobilis,* is naturally tolerant of high levels of ethanol but cannot ferment arabinose. Deanda, Zhang, Eddy, and Picataggio (1996) described their efforts to combine the most useful features of these two organisms by introducing the *E. coli* genes for the arabinose-metabolizing enzymes into *Z. mobilis.*

(a) Why is this a simpler strategy than the reverse: engineering *E. coli* to be more ethanol-tolerant?

Deanda and colleagues inserted five *E. coli* genes into the *Z. mobilis* genome: *araA,* coding for L-arabinose isomerase, which interconverts L-arabinose and L-ribulose; *araB,* L-ribulokinase, which uses ATP to phosphorylate L-ribulose at C-5; *araD,* L-ribulose 5-phosphate epimerase, which interconverts L-ribulose 5-phosphate and L-xylulose 5-phosphate; *talB,* transaldolase; and *tktA,* transketolase.

(b) For each of the three *ara* enzymes, briefly describe the chemical transformation it catalyzes and, where possible, name an enzyme discussed in this chapter that carries out an analogous reaction.

The five *E. coli* genes inserted in *Z. mobilis* allowed the entry of arabinose into the nonoxidative phase of the pentose phosphate pathway (Fig. 14–23), where it was converted to glucose 6-phosphate and fermented to ethanol.

(c) The three *ara* enzymes eventually converted arabinose into which sugar?

(d) The product from part (c) feeds into the pathway shown in Figure 14–23. Combining the five *E. coli* enzymes listed above with the enzymes of this pathway, describe the overall pathway for the fermentation of six molecules of arabinose to ethanol.

(e) What is the stoichiometry of the fermentation of six molecules of arabinose to ethanol and CO_2? How many ATP molecules would you expect this reaction to generate?

(f) *Zymomonas mobilis* uses a slightly different pathway for ethanol fermentation from the one described in this chapter. As a result, the expected ATP yield is only 1 ATP per molecule of arabinose. Although this is less beneficial for the bacterium, it is better for ethanol production. Why?

Another sugar commonly found in plant matter is xylose.

D-Xylose

(g) What additional enzymes would you need to introduce into the modified *Z. mobilis* strain described above to enable it to use xylose as well as arabinose to produce ethanol? You don't need to name the enzymes (they may not even exist in the real world!); just give the reactions they would need to catalyze.

Reference

Deanda, K., Zhang, M., Eddy, C., & Picataggio, S. (1996) Development of an arabinose-fermenting *Zymomonas mobilis* strain by metabolic pathway engineering. *Appl. Environ. Microbiol.* **62,** 4465–4470.

Principles of Metabolic Regulation

Metabolic regulation, a central theme in biochemistry, is one of the most remarkable features of living organisms. Of the thousands of enzyme-catalyzed reactions that can take place in a cell, there is probably not one that escapes some form of regulation. This need to regulate every aspect of cellular metabolism becomes clear as one examines the complexity of metabolic reaction sequences. Although it is convenient for the student of biochemistry to divide metabolic processes into "pathways" that play discrete roles in the cell's economy, no such separation exists in the living cell. Rather, every pathway we discuss in this book is inextricably intertwined with all the other cellular pathways in a multidimensional network of reactions **(Fig. 15–1)**. For example, in Chapter 14 we discussed four possible fates for **glucose 6-phosphate** in a hepatocyte: breakdown by glycolysis for the production of ATP, breakdown in the pentose phosphate pathway for the production of NADPH and pentose phosphates, use in the synthesis of complex polysaccharides of the extracellular matrix, or hydrolysis to glucose and phosphate to replenish blood glucose. In fact, glucose 6-phosphate has other possible fates in hepatocytes, too; it may, for example, be used to synthesize other sugars, such as glucosamine, galactose, galactosamine, fucose, and neuraminic acid, for use in protein glycosylation, or it may be partially degraded to provide acetyl-CoA for fatty acid and sterol synthesis. And the bacterium *Escherichia coli* can use glucose to produce the carbon skeleton of *every one* of its several thousand types of molecules. When any cell uses glucose 6-phosphate for one purpose, that "decision" affects all the other pathways for which glucose 6-phosphate is a precursor or intermediate: any change in the allocation of glucose 6-phosphate to one pathway affects, directly or indirectly, the flow of metabolites through all the others.

Such changes in allocation are common in the life of a cell. Louis Pasteur was the first to describe the more than 10-fold increase in glucose consumption by a yeast culture when it was shifted from aerobic to anaerobic conditions. This "Pasteur effect" occurs without a significant change in the concentrations of ATP or most of the hundreds of metabolic intermediates and products derived from glucose. A similar effect occurs in the cells of skeletal muscle when a sprinter leaves the starting blocks. The ability of a cell to carry out all these interlocking metabolic processes simultaneously—obtaining every product in the amount needed and at the right time, in the face of major perturbations from outside, and without generating leftovers—is an *astounding* accomplishment.

In this chapter we use the metabolism of glucose to illustrate some general principles of metabolic regulation. First we look at the general roles of regulation in achieving metabolic homeostasis and introduce metabolic control analysis, a system for analyzing complex metabolic interactions quantitatively. We then describe the specific regulatory properties of the individual enzymes of glucose metabolism; for glycolysis and gluconeogenesis, we described the catalytic activities of the enzymes in Chapter 14. Here we also discuss both the catalytic and regulatory properties of the enzymes of glycogen synthesis and breakdown, one of the best-studied cases of metabolic regulation. Note that in selecting carbohydrate metabolism to illustrate the principles of metabolic regulation, we have artificially separated the metabolism of fats and carbohydrates. In fact, these two activities are very tightly integrated, as we shall see in Chapter 23.

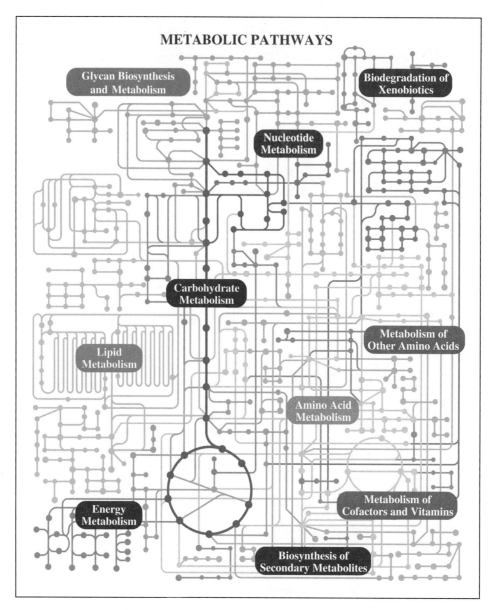

FIGURE 15–1 Metabolism as a three-dimensional meshwork. A typical eukaryotic cell has the capacity to make about 30,000 different proteins, which catalyze thousands of different reactions involving many hundreds of metabolites, most shared by more than one "pathway." In this much-simplified overview of metabolic pathways, each dot represents an intermediate compound and each connecting line represents an enzymatic reaction. For a more realistic and far more complex diagram of metabolism, see the online KEGG PATHWAY database (www.genome.ad.jp/kegg/pathway/map/map01100.html); in this interactive map, each dot can be clicked to obtain extensive data about the compound and the enzymes for which it is a substrate. The cover of this book shows the interlocking reactions that occur in the mitochondrion.

15.1 Regulation of Metabolic Pathways

The pathways of glucose metabolism provide, in the catabolic direction, the energy essential to oppose the forces of entropy and, in the anabolic direction, biosynthetic precursors and a storage form of metabolic energy. These reactions are so important to survival that very complex regulatory mechanisms have evolved to ensure that metabolites move through each pathway in the correct direction and at the correct rate to match exactly the cell's or the organism's changing circumstances. By a variety of mechanisms operating on different time scales, adjustments are made in the rate of metabolite flow through an entire pathway when external circumstances change.

Circumstances do change, sometimes dramatically. For example, the demand for ATP in insect flight muscle increases 100-fold in a few seconds when the insect takes flight. In humans, the availability of oxygen may decrease due to hypoxia (diminished delivery of oxygen to tissues) or ischemia (diminished flow of blood to tissues). The relative proportions of carbohydrate, fat, and protein in the diet vary from meal to meal, and the supply of fuels obtained in the diet is intermittent, requiring metabolic adjustments between meals and during periods of starvation. Wound healing requires huge amounts of energy and biosynthetic precursors.

Cells and Organisms Maintain a Dynamic Steady State

Fuels such as glucose enter a cell, and waste products such as CO_2 leave, but the mass and the gross composition of a typical cell, organ, or adult animal do not change appreciably over time; cells and organisms exist in a dynamic steady state. For each metabolic reaction in a pathway, the substrate is provided by the preceding reaction at the same rate at which it is converted to product. Thus, although the rate (v) of metabolite flow, or **flux**, through this step of the pathway may be high and variable, the concentration of substrate, S, remains constant. So, for the two-step reaction

$$A \xrightarrow{v_1} S \xrightarrow{v_2} P$$

when $v_1 = v_2$, [S] is constant. For example, changes in v_1 for the entry of glucose from various sources into the blood are balanced by changes in v_2 for the uptake of glucose from the blood into various tissues, so the concentration of glucose in the blood ([S]) is held nearly constant at 5 mM. This is **homeostasis** at the molecular level. The failure of homeostatic mechanisms is often at the root of human disease. In diabetes mellitus, for example, the regulation of blood glucose concentration is defective as a result of the lack of or insensitivity to insulin, with profound medical consequences.

When the external perturbation is not merely transient, or when one kind of cell develops into another, the adjustments in cell composition and metabolism can be more dramatic and may require significant and lasting changes in the allocation of energy and synthetic precursors to bring about a new dynamic steady state. Consider, for example, the differentiation of stem cells in the bone marrow into erythrocytes. The precursor cell contains a nucleus, mitochondria, and little or no hemoglobin, whereas the fully differentiated erythrocyte contains prodigious amounts of hemoglobin but has neither nucleus nor mitochondria; the cell's composition has permanently changed in response to external developmental signals, with accompanying changes in metabolism. This **cellular differentiation** requires precise regulation of the levels of cellular proteins.

In the course of evolution, organisms have acquired a remarkable collection of regulatory mechanisms for maintaining homeostasis at the molecular, cellular, and organismal levels, as reflected in the proportion of genes that encode regulatory machinery. In humans, about 4,000 genes (~12% of all genes) encode regulatory proteins, including a variety of receptors, regulators of gene expression, and more than 500 different protein kinases! In many cases, the regulatory mechanisms overlap: one enzyme is subject to regulation by several different mechanisms.

Both the Amount and the Catalytic Activity of an Enzyme Can Be Regulated

The flux through an enzyme-catalyzed reaction can be modulated by changes in the *number* of enzyme molecules or by changes in the *catalytic activity* of each enzyme molecule already present. Such changes occur on time scales from milliseconds to many hours, in response to signals from within or outside the cell. Very rapid allosteric changes in enzyme activity are generally triggered locally, by changes in the local concentration of a small molecule—a substrate of the pathway in which that reaction is a step (say, glucose for glycolysis), a product of the pathway (ATP from glycolysis), or a key metabolite or cofactor (such as NADH) that indicates the cell's metabolic state. Second messengers (such as cyclic AMP and Ca^{2+}) generated intracellularly in response to extracellular signals (hormones, cytokines, and so forth) also mediate allosteric regulation, on a slightly slower time scale set by the rate of the signal-transduction mechanism (see Chapter 12).

Extracellular signals (**Fig. 15–2, ❶**) may be hormonal (insulin or epinephrine, for example) or neuronal (acetylcholine), or may be growth factors or cytokines. The number of molecules of a given enzyme in a cell is a function of the relative rates of synthesis and degradation of that enzyme. The rate of synthesis can be adjusted by the activation (in response to some outside signal) of a transcription factor (Fig. 15–2, ❷; described in more detail in Chapter 28). **Transcription factors** are nuclear proteins that, when activated, bind specific DNA regions (**response elements**) near a gene's promoter (its transcriptional starting point) and activate or repress the transcription of that gene, leading to increased or decreased synthesis of the encoded protein. Activation of a transcription factor is sometimes the result of its binding of a specific ligand and sometimes the result of its phosphorylation or dephosphorylation. Each gene is controlled by one or more response elements that are recognized by specific transcription factors. Genes that have several response elements are therefore controlled by several different transcription factors responding to several different signals. Groups of genes encoding proteins that act together, such as the enzymes of glycolysis or gluconeogenesis, often share common response element sequences, so that a single signal, acting through a particular transcription factor, turns all of these genes on and off together. The regulation of carbohydrate metabolism by specific transcription factors is described in Section 15.3.

The stability of messenger RNAs—their resistance to degradation by cellular ribonucleases (Fig. 15–2, ❸)— varies, and the amount of a given mRNA in the cell is a function of its rates of synthesis and degradation (Chapter 26). The rate at which an mRNA is translated into a protein by ribosomes (Fig. 15–2, ❹) is also regulated, and depends on several factors described in detail in Chapter 27. Note that an n-fold increase in an mRNA

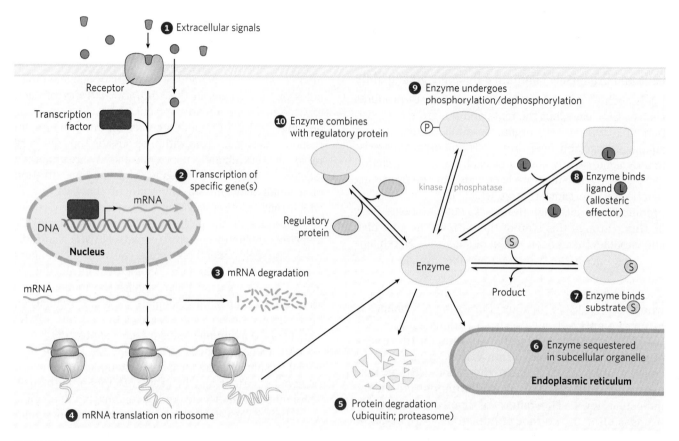

FIGURE 15–2 Factors affecting the activity of enzymes. The total activity of an enzyme can be changed by altering the *number* of its molecules in the cell, or its *effective* activity in a subcellular compartment (❶ through ❻), or by modulating the *activity* of existing molecules (❼ through ❿), as detailed in the text. An enzyme may be influenced by a combination of such factors.

does not always mean an n-fold increase in its protein product.

Once synthesized, protein molecules have a finite lifetime, which may range from minutes to many days (Table 15–1). The rate of protein degradation (Fig. 15–2, ❺) differs from one enzyme to another and depends on the conditions in the cell. Some proteins are tagged by the covalent attachment of ubiquitin for degradation in proteasomes, as discussed in Chapter 27 (see, for example, the case of cyclin, in Fig. 12–47). Rapid **turnover** (synthesis followed by degradation) is energetically expensive, but proteins with a short half-life can reach new steady state levels much faster than those with a long half-life, and the benefit of this quick responsiveness must balance or outweigh the cost to the cell.

TABLE 15–1	Average Half-Life of Proteins in Mammalian Tissues

Tissue	Average half-life (days)
Liver	0.9
Kidney	1.7
Heart	4.1
Brain	4.6
Muscle	10.7

Yet another way to alter the *effective* activity of an enzyme is to sequester the enzyme and its substrate in different compartments (Fig. 15–2, ❻). In muscle, for example, hexokinase cannot act on glucose until the sugar enters the myocyte from the blood, and the rate at which it enters depends on the activity of glucose transporters (see Table 11–3) in the plasma membrane. Within cells, membrane-bounded compartments segregate certain enzymes and enzyme systems, and the transport of substrate across these intracellular membranes may be the limiting factor in enzyme action.

By these several mechanisms for regulating enzyme level, cells can dramatically change their complement of enzymes in response to changes in metabolic circumstances. In vertebrates, liver is the most adaptable tissue; a change from a high-carbohydrate to a high-lipid diet, for example, affects the transcription of hundreds of genes and thus the levels of hundreds of proteins. These global changes in gene expression can be quantified by the use of DNA microarrays (see Fig. 9–23) that display the entire complement of mRNAs present in a given cell type or organ (the **transcriptome**) or by two-dimensional gel electrophoresis (see Fig. 3–21) that displays the protein complement of a cell type or organ (its **proteome**). Both techniques offer great insights into metabolic regulation. The effect of changes in the proteome is often a change in the total ensemble of low molecular weight metabolites, the **metabolome** (**Fig. 15–3**). The metabolome of *E. coli* growing on glucose is dominated by a few classes of metabolites: glutamate (49%); nucleotides (mainly ribonucleoside triphosphates) (15%); intermediates of glycolysis, the citric acid cycle, and the pentose phosphate pathway (central pathways of carbon metabolism) (15%); and redox cofactors and glutathiones (9%).

Once the regulatory mechanisms that involve protein synthesis and degradation have produced a certain number of molecules of each enzyme in a cell, the activity of those enzymes can be further regulated in several other ways: by the concentration of substrate, the presence of allosteric effectors, covalent modifications, or binding of regulatory proteins—all of which can change the activity of an individual enzyme molecule (Fig. 15–2, ❼ to ❿).

All enzymes are sensitive to the concentration of their substrate(s) (Fig. 15–2, ❼). Recall that in the simplest case (an enzyme that follows Michaelis-Menten kinetics), the initial rate of the reaction is half-maximal when the substrate is present at a concentration equal to K_m (that is, when the enzyme is half-saturated with substrate). Activity drops off at lower [S], and when $[S] \ll K_m$, the reaction rate is linearly dependent on [S].

The relationship between [S] and K_m is important because intracellular concentrations of substrate are often in the same range as, or lower than, K_m. The activity of hexokinase, for example, changes with [glucose], and intracellular [glucose] varies with the concentration of glucose in the blood. As we will see, the different forms (isozymes) of hexokinase have different K_m values and are therefore affected differently by changes in

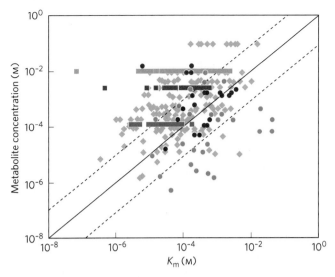

■ NAD+   ■ NADPH   ● Degradation reactions
■ ATP   ◆ Other   ● Central pathways of carbon metabolism

FIGURE 15–4 Comparison of K_m and substrate concentration for some metabolic enzymes. Measured metabolite concentrations for *E. coli* growing on glucose are plotted against the known K_m for enzymes that consume that metabolite. The solid line is the line of unity (where metabolite concentration = K_m), and the dashed lines each denote a tenfold deviation from the line of unity.

intracellular [glucose], in ways that make sense physiologically. For a number of phosphoryl transfers from ATP, and for redox reactions using NADPH or NAD$^+$, the metabolite concentration is well above the K_m (**Fig. 15–4**); these cofactors are not likely to be the limiting factors in such reactions.

WORKED EXAMPLE 15–1   Activity of a Glucose Transporter

If K_t (the equivalent of K_m) for the glucose transporter in liver (GLUT2) is 40 mM, calculate the effect on the rate of glucose flux into a hepatocyte of increasing the blood glucose concentration from 3 mM to 10 mM.

Solution: We use Equation 11–1 (p. 406) to find the initial velocity (flux) of glucose uptake.

$$V_0 = \frac{V_{max}[S]_{out}}{K_t + [S]_{out}}$$

At 3 mM glucose

$$V_0 = V_{max}\,(3\ \text{mM})/(40\ \text{mM} + 3\ \text{mM})$$
$$= V_{max}\,(3\ \text{mM}/43\ \text{mM}) = 0.07\ V_{max}$$

At 10 mM glucose

$$V_0 = V_{max}\,(10\ \text{mM})/(40\ \text{mM} + 10\ \text{mM})$$
$$= V_{max}\,(10\ \text{mM}/50\ \text{mM}) = 0.20\ V_{max}$$

So a rise in blood glucose from 3 mM to 10 mM increases the rate of glucose influx into a hepatocyte by a factor of $0.20/0.07 \approx 3$.

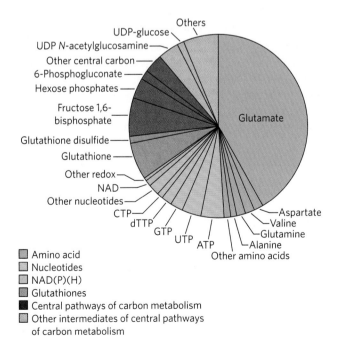

□ Amino acid
□ Nucleotides
□ NAD(P)(H)
▨ Glutathiones
■ Central pathways of carbon metabolism
□ Other intermediates of central pathways of carbon metabolism

FIGURE 15–3 The metabolome of *E. coli* growing on glucose. Summary of the amounts of 103 metabolites measured by a combination of liquid chromatography and tandem mass spectrometry (LC-MS/MS).

Enzyme activity can be either increased or decreased by an allosteric effector (Fig. 15–2, ❽; see Fig. 6–34). Allosteric effectors typically convert hyperbolic kinetics to sigmoid kinetics, or vice versa (see Fig. 15–16b, for example). In the steepest part of the sigmoid curve, a small change in the concentration of substrate, or of allosteric effector, can have a large impact on reaction rate. Recall from Chapter 5 (p. 167) that the cooperativity of an allosteric enzyme can be expressed as a Hill coefficient, with higher coefficients meaning greater cooperativity. For an allosteric enzyme with a Hill coefficient of 4, activity increases from 10% V_{max} to 90% V_{max} with only a 3-fold increase in [S], compared with the 81-fold rise in [S] needed by an enzyme with no cooperative effects (Hill coefficient of 1; Table 15–2).

Covalent modifications of enzymes or other proteins (Fig. 15–2, ❾) occur within seconds or minutes of a regulatory signal, typically an extracellular signal. By far the most common modifications are phosphorylation and dephosphorylation **(Fig. 15–5)**; up to half the proteins in a eukaryotic cell are phosphorylated under some circumstances. Phosphorylation by a specific protein kinase may alter the electrostatic features of an enzyme's active site, cause movement of an inhibitory region of the enzyme out of the active site, alter the enzyme's interaction with other proteins, or force conformational changes that translate into changes in V_{max} or K_m. For covalent modification to be useful in regulation, the cell must be able to restore the altered enzyme to its original activity state. A family of phosphoprotein phosphatases, at least some of which are themselves under regulation, catalyzes the dephosphorylation of proteins.

Finally, many enzymes are regulated by association with and dissociation from another, regulatory protein (Fig. 15–2, ❿). For example, the cyclic AMP–dependent protein kinase (PKA; see Fig. 12–6) is inactive until cAMP binding separates catalytic from regulatory (inhibitory) subunits of the enzyme.

These several mechanisms for altering the flux through a step in a metabolic pathway are not mutually exclusive. It is very common for a single enzyme to be regulated at the level of transcription and by both allosteric

FIGURE 15–5 Protein phosphorylation and dephosphorylation. Protein kinases transfer a phosphoryl group from ATP to a Ser, Thr, or Tyr residue in an enzyme or other protein substrate. Protein phosphatases remove the phosphoryl group as P_i.

and covalent mechanisms. The combination provides fast, smooth, effective regulation in response to a very wide array of perturbations and signals.

In the discussions that follow, it is useful to think of changes in enzymatic activity as serving two distinct though complementary roles. We use the term **metabolic regulation** to refer to processes that serve to maintain homeostasis at the molecular level—to hold some cellular parameter (concentration of a metabolite, for example) at a steady level over time, even as the flow of metabolites through the pathway changes. The term **metabolic control** refers to a process that leads to a change in the output of a metabolic pathway over time, in response to some outside signal or change in circumstances. The distinction, although useful, is not always easy to make.

Reactions Far from Equilibrium in Cells Are Common Points of Regulation

For some steps in a metabolic pathway the reaction is close to equilibrium, with the cell in its dynamic steady state **(Fig. 15–6)**. The net flow of metabolites through these steps is the small difference between the rates of the forward and reverse reactions, rates that are very similar when a reaction is near equilibrium. Small changes in

TABLE 15–2	Relationship between Hill Coefficient and the Effect of Substrate Concentration on Reaction Rate for Allosteric Enzymes
Hill coefficient (n_H)	**Required change in [S] to increase V_0 from 10% to 90% V_{max}**
0.5	×6,600
1.0	×81
2.0	×9
3.0	×4.3
4.0	×3

$$\longrightarrow A \underset{V = 0.01}{\overset{V = 10.01}{\rightleftharpoons}} B \underset{V = 190}{\overset{V = 200}{\rightleftharpoons}} C \underset{V = 490}{\overset{V = 500}{\rightleftharpoons}} D$$

net rate:　　10　　　　10　　　　10

FIGURE 15–6 Near-equilibrium and nonequilibrium steps in a metabolic pathway. Steps ❷ and ❸ of this pathway are near equilibrium in the cell; for each step, the rate (V) of the forward reaction is only slightly greater than the reverse rate, so the net forward rate (10) is relatively low and the free-energy change, ΔG, is close to zero. An increase in [C] or [D] can reverse the direction of these steps. Step ❶ is maintained in the cell far from equilibrium; its forward rate greatly exceeds its reverse rate. The net rate of step ❶ (10) is much larger than the reverse rate (0.01) and is identical to the net rates of steps ❷ and ❸ when the pathway is operating in the steady state. Step ❶ has a large, negative ΔG.

| **TABLE 15-3** | Equilibrium Constants, Mass-Action Coefficients, and Free-Energy Changes for Enzymes of Carbohydrate Metabolism | | | | | |

| Enzyme | K'_{eq} | Mass-action ratio, Q | | Reaction near equilibrium in vivo?* | $\Delta G'^{\circ}$ (kJ/mol) | ΔG (kJ/mol) in heart |
		Liver	Heart			
Hexokinase	1×10^3	2×10^{-2}	8×10^{-2}	No	-17	-27
PFK-1	1.0×10^3	9×10^{-2}	3×10^{-2}	No	-14	-23
Aldolase	1.0×10^{-4}	1.2×10^{-6}	9×10^{-6}	Yes	$+24$	-6.0
Triose phosphate isomerase	4×10^{-2}	—†	2.4×10^{-1}	Yes	$+7.5$	$+3.8$
Glyceraldehyde 3-phosphate dehydrogenase + phosphoglycerate kinase	2×10^3	6×10^2	9.0	Yes	-13	$+3.5$
Phosphoglycerate mutase	1×10^{-1}	1×10^{-1}	1.2×10^{-1}	Yes	$+4.4$	$+0.6$
Enolase	3	2.9	1.4	Yes	-3.2	-0.5
Pyruvate kinase	2×10^4	7×10^{-1}	40	No	-31	-17
Phosphoglucose isomerase	4×10^{-1}	3.1×10^{-1}	2.4×10^{-1}	Yes	$+2.2$	-1.4
Pyruvate carboxylase + PEP carboxykinase	7	1×10^{-3}	—†	No	-5.0	-23
Glucose 6-phosphatase	8.5×10^2	1.2×10^2	—†	Yes	-17	-5.0

Source: K'_{eq} and Q from Newsholme, E.A. & Start, C. (1973) *Regulation in Metabolism*, Wiley Press, New York, pp. 97, 263. ΔG and $\Delta G'^{\circ}$ were calculated from these data.

*For simplicity, any reaction for which the absolute value of the calculated ΔG is less than 6 is considered near equilibrium.

†Data not available.

substrate or product concentration can produce large changes in the net rate, and can even change the direction of the net flow. We can identify these near-equilibrium reactions in a cell by comparing the **mass-action ratio**, Q, with the equilibrium constant for the reaction, K'_{eq}. Recall that for the reaction A + B → C + D, $Q = [C][D]/[A][B]$. When Q and K'_{eq} are within 1 to 2 orders of magnitude of each other, the reaction is near equilibrium. This is the case for 6 of the 10 steps in the glycolytic pathway (Table 15–3).

Other reactions are far from equilibrium in the cell. For example, K'_{eq} for the phosphofructokinase-1 (PFK-1) reaction is about 1,000, but Q ([fructose 1,6-bisphosphate][ADP]/[fructose 6-phosphate][ATP]) in a hepatocyte in the steady state is about 0.1 (Table 15–3). It is *because* the reaction is so far from equilibrium that the process is exergonic under cellular conditions and tends to go in the forward direction. The reaction is held far from equilibrium because, under prevailing cellular conditions of substrate, product, and effector concentrations, the rate of conversion of fructose 6-phosphate to fructose 1,6-bisphosphate is limited by the activity of PFK-1, which is itself limited by the number of PFK-1 molecules present and by the actions of allosteric effectors. Thus the net forward rate of the enzyme-catalyzed reaction is equal to the net flow of glycolytic intermediates through other steps in the pathway, and the reverse flow through PFK-1 remains near zero.

The cell *cannot* allow reactions with large equilibrium constants to reach equilibrium. If [fructose 6-phos-

phate], [ATP], and [ADP] in the cell were held at typical levels (low millimolar concentrations) and the PFK-1 reaction were allowed to reach equilibrium by an increase in [fructose 1,6-bisphosphate], the concentration of fructose 1,6-bisphosphate would rise into the molar range, wreaking osmotic havoc on the cell. Consider another case: if the reaction ATP → ADP + P$_i$ were allowed to approach equilibrium in the cell, the actual free-energy change ($\Delta G'$) for that reaction (ΔG_p; see Worked Example 13–2, p. 519) would approach zero, and ATP would lose the high phosphoryl group transfer potential that makes it valuable to the cell. It is therefore essential that enzymes catalyzing ATP breakdown and other highly exergonic reactions in a cell be sensitive to regulation, so that when metabolic changes are forced by external circumstances, the flow through these enzymes will be adjusted to ensure that [ATP] remains far above its equilibrium level. When such metabolic changes occur, the activities of enzymes in all interconnected pathways adjust to keep these critical steps away from equilibrium. Thus, not surprisingly, many enzymes (such as PFK-1) that catalyze highly exergonic reactions are subject to a variety of subtle regulatory mechanisms. The multiplicity of these adjustments is so great that we cannot predict by examining the properties of any one enzyme in a pathway whether that enzyme has a strong influence on net flow through the entire pathway. This complex problem can be approached by metabolic control analysis, as described in Section 15.2.

FIGURE 15–7 Effect of ATP concentration on the initial reaction velocity of a typical ATP-dependent enzyme. These experimental data yield a K_m for ATP of 5 mM. The concentration of ATP in animal tissues is ~5 mM.

Adenine Nucleotides Play Special Roles in Metabolic Regulation

After the protection of its DNA from damage, perhaps nothing is more important to a cell than maintaining a constant supply and concentration of ATP. Many ATP-using enzymes have K_m values between 0.1 and 1 mM, and the ATP concentration in a typical cell is about 5 to 10 mM (Fig. 15–4). If [ATP] were to drop significantly, these enzymes would be less than fully saturated by their substrate (ATP), and the rates of hundreds of reactions that involve ATP would decrease **(Fig. 15–7)**; the cell would probably not survive this *kinetic* effect on so many reactions.

There is also an important *thermodynamic* effect of lowered [ATP]. Because ATP is converted to ADP or AMP when "spent" to accomplish cellular work, the [ATP]/[ADP] ratio profoundly affects all reactions that employ these cofactors. (The same is true for other important cofactors, such as NADH/NAD$^+$ and NADPH/NADP$^+$.) For example, consider the reaction catalyzed by hexokinase:

$$\text{ATP} + \text{glucose} \longrightarrow \text{ADP} + \text{glucose 6-phosphate}$$

$$K'_{eq} = \frac{[\text{ADP}]_{eq}[\text{glucose 6-phosphate}]_{eq}}{[\text{ATP}]_{eq}[\text{glucose}]_{eq}} = 2 \times 10^3$$

Note that this expression holds true *only* when reactants and products are at their *equilibrium* concentra-

tions, where $\Delta G' = 0$. At any other set of concentrations, $\Delta G'$ is not zero. Recall (from Chapter 13) that the ratio of products to substrates (the mass action ratio, Q) determines the magnitude and sign of $\Delta G'$ and therefore the driving force, $\Delta G'$, of the reaction:

$$\Delta G' = \Delta G'^{\circ} + RT \ln \frac{[\text{ADP}][\text{glucose 6-phosphate}]}{[\text{ATP}][\text{glucose}]}$$

Because an alteration of this driving force profoundly influences every reaction that involves ATP, organisms have evolved under strong pressure to develop regulatory mechanisms responsive to the [ATP]/[ADP] ratio.

AMP concentration is an even more sensitive indicator of a cell's energetic state than is [ATP]. Normally cells have a far higher concentration of ATP (5 to 10 mM) than of AMP (<0.1 mM). When some process (say, muscle contraction) consumes ATP, AMP is produced in two steps. First, hydrolysis of ATP produces ADP, then the reaction catalyzed by **adenylate kinase** produces AMP:

$$2\text{ADP} \longrightarrow \text{AMP} + \text{ATP}$$

If ATP is consumed such that its concentration drops 10%, the *relative* increase in [AMP] is much greater than that of [ADP] (Table 15–4). It is not surprising, therefore, that many regulatory processes are keyed to changes in [AMP]. Probably the most important mediator of regulation by AMP is **AMP-activated protein kinase (AMPK)**, which responds to an increase in [AMP] by phosphorylating key proteins and thus regulating their activities. The rise in [AMP] may be caused by a reduced nutrient supply or by increased exercise. The action of AMPK (not to be confused with the *cyclic* AMP–dependent protein kinase; see Section 15.5) increases glucose transport and activates glycolysis and fatty acid oxidation, while suppressing energy-requiring processes such as the synthesis of fatty acids, cholesterol, and protein **(Fig. 15–8)**. We discuss AMPK further, and the detailed mechanisms by which it effects these changes, in Chapter 23.

In addition to ATP, hundreds of metabolic intermediates also must be present at appropriate concentrations in the cell. To take just one example: the glycolytic intermediates dihydroxyacetone phosphate and 3-phosphoglycerate are precursors of triacylglycerols and serine, respectively. When these products are needed, the rate of glycolysis must be adjusted to provide them without reducing the glycolytic production of ATP. The same

TABLE 15–4	Relative Changes in [ATP] and [AMP] When ATP Is Consumed		
Adenine nucleotide	Concentration before ATP depletion (mM)	Concentration after ATP depletion (mM)	Relative change
ATP	5.0	4.5	10%
ADP	1.0	1.0	0
AMP	0.1	0.6	600%

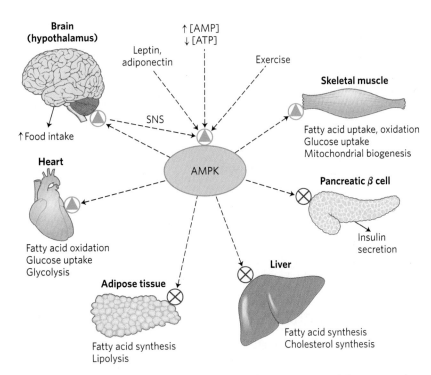

FIGURE 15–8 Role of AMP-activated protein kinase (AMPK) in carbohydrate and fat metabolism. AMPK is activated by elevated [AMP] or decreased [ATP], by exercise, by the sympathetic nervous system (SNS), or by peptide hormones produced in adipose tissue (leptin and adiponectin, described in more detail in Chapter 23). When activated, AMPK phosphorylates target proteins and shifts metabolism in a variety of tissues away from energy-consuming processes such as the synthesis of glycogen, fatty acids, and cholesterol; shifts metabolism in extrahepatic tissues to the use of fatty acids as a fuel; and triggers gluconeogenesis in the liver to provide glucose for the brain. In the hypothalamus, AMPK stimulates feeding behavior to provide more dietary fuel.

is true for maintaining the levels of other important cofactors, such as NADH and NADPH: changes in their mass action ratios (that is, in the ratio of reduced to oxidized cofactor) have global effects on metabolism.

Of course, priorities at the *organismal* level have also driven the evolution of regulatory mechanisms. In mammals, the brain has virtually no stored source of energy, depending instead on a constant supply of glucose from the blood. If blood glucose drops from its normal concentration of 4 to 5 mM to half that level, mental confusion results, and a fivefold reduction in blood glucose can lead to coma and death. To buffer against changes in blood glucose concentration, release of the hormones insulin and glucagon, elicited by high or low blood glucose, respectively, triggers metabolic changes that tend to return the blood glucose concentration to normal.

Other selective pressures must also have operated throughout evolution, selecting for regulatory mechanisms that accomplish the following:

1. Maximize the efficiency of fuel utilization by preventing the simultaneous operation of pathways in opposite directions (such as glycolysis and gluconeogenesis).

2. Partition metabolites appropriately between alternative pathways (such as glycolysis and the pentose phosphate pathway).

3. Draw on the fuel best suited for the immediate needs of the organism (glucose, fatty acids, glycogen, or amino acids).

4. Slow down biosynthetic pathways when their products accumulate.

The remaining chapters of this book present many examples of each kind of regulatory mechanism.

SUMMARY 15.1 Regulation of Metabolic Pathways

▶ In a metabolically active cell in a steady state, intermediates are formed and consumed at equal rates. When a transient perturbation alters the rate of formation or consumption of a metabolite, compensating changes in enzyme activities return the system to the steady state.

▶ Cells regulate their metabolism by a variety of mechanisms over a time scale ranging from less than a millisecond to days, either by changing the activity of existing enzyme molecules or by changing the number of molecules of a specific enzyme.

▶ Various signals activate or inactivate transcription factors, which act in the nucleus to regulate gene expression. Changes in the transcriptome lead to changes in the proteome, and ultimately in the metabolome of a cell or tissue.

▶ In multistep processes such as glycolysis, certain reactions are essentially at equilibrium in the steady state; the rates of these reactions rise and fall with substrate concentration. Other reactions are far from equilibrium; these steps are typically the points of regulation of the overall pathway.

▶ Regulatory mechanisms maintain nearly constant levels of key metabolites such as ATP and NADH in cells and glucose in the blood, while matching the use or production of glucose to the organism's changing needs.

▶ The levels of ATP and AMP are a sensitive reflection of a cell's energy status, and when the [ATP]/[AMP] ratio decreases, the AMP-activated protein kinase (AMPK) triggers a variety of cellular responses to raise [ATP] and lower [AMP].

15.2 Analysis of Metabolic Control

Eduard Buchner, 1860–1917

Detailed studies of metabolic regulation were not feasible until the basic chemical steps in a pathway had been clarified and the responsible enzymes characterized. Beginning with Eduard Buchner's discovery (c. 1900) that an extract of broken yeast cells could convert glucose to ethanol and CO_2, a major thrust of biochemical research was to deduce the steps by which this transformation occurred and to purify and characterize the enzymes that catalyzed each step. By the middle of the twentieth century, all 10 enzymes of the glycolytic pathway had been purified and characterized. In the next 50 years much was learned about the regulation of these enzymes by intracellular and extracellular signals, through the kinds of allosteric and covalent mechanisms described in this chapter. The conventional wisdom was that in a linear pathway such as glycolysis, catalysis by one enzyme must be the slowest and must therefore determine the rate of metabolite flow, or flux, through the whole pathway. For glycolysis, PFK-1 was considered the rate-limiting enzyme, because it was known to be closely regulated by fructose 2,6-bisphosphate and other allosteric effectors.

With the advent of genetic engineering technology, it became possible to test this "single rate-determining step" hypothesis by increasing the concentration of the enzyme that catalyzes the "rate-limiting step" in a pathway and determining whether flux through the pathway increases proportionally. Most often it does not; the simple solution (a single rate-determining step) is wrong. It has now become clear that in most pathways the control of flux is distributed among several enzymes, and the extent to which each contributes to the control

varies with metabolic circumstances—the supply of the starting material (say, glucose), the supply of oxygen, the need for other products derived from intermediates of the pathway (say, glucose 6-phosphate for the pentose phosphate pathway in cells synthesizing large amounts of nucleotides), the effects of metabolites with regulatory roles, and the hormonal status of the organism (such as the levels of insulin and glucagon), among other factors.

Why are we interested in what limits the flux through a pathway? To understand the action of hormones or drugs, or the pathology that results from a failure of metabolic regulation, we must know where control is exercised. If researchers wish to develop a drug that stimulates or inhibits a pathway, the logical target is the enzyme that has the greatest impact on the flux through that pathway. And the bioengineering of a microorganism to overproduce a product of commercial value (p. 321) requires a knowledge of what limits the flux of metabolites toward that product.

The Contribution of Each Enzyme to Flux through a Pathway Is Experimentally Measurable

There are several ways to determine experimentally how a change in the activity of one enzyme in a pathway affects metabolite flux through that pathway. Consider the experimental results shown in **Figure 15–9**. When a sample of rat liver was homogenized to release all soluble enzymes, the extract carried out the glycolytic conversion of glucose to fructose 1,6-bisphosphate at a measurable rate. (This experiment, for simplicity, focused on just the first part of the glycolytic pathway.) When increasing amounts of purified hexokinase IV

FIGURE 15–9 Dependence of glycolytic flux in a rat liver homogenate on added enzymes. Purified enzymes in the amounts shown on the x axis were added to an extract of liver carrying out glycolysis in vitro. The flux through the pathway is shown on the y axis.

(glucokinase) were added to the extract, the rate of glycolysis progressively increased. The addition of purified PFK-1 to the extract also increased the rate of glycolysis, but not as dramatically as did hexokinase. Addition of purified phosphohexose isomerase was without effect. These results suggest that hexokinase and PFK-1 both contribute to setting the flux through the pathway (hexokinase more than PFK-1), and that phosphohexose isomerase does not.

Similar experiments can be done on intact cells or organisms, using specific inhibitors or activators to change the activity of one enzyme while observing the effect on flux through the pathway. The amount of an enzyme can also be altered genetically; bioengineering can produce a cell that makes extra copies of the enzyme under investigation or has a version of the enzyme that is less active than the normal enzyme. Increasing the concentration of an enzyme genetically sometimes has significant effects on flux; sometimes it has no effect.

Three critical parameters, which together describe the responsiveness of a pathway to changes in metabolic circumstances, lie at the center of **metabolic control analysis**. We turn now to a qualitative description of these parameters and their meaning in the context of a living cell. Box 15–1 provides a more rigorous quantitative discussion.

The Flux Control Coefficient Quantifies the Effect of a Change in Enzyme Activity on Metabolite Flux through a Pathway

Quantitative data on metabolic flux, obtained as described in Figure 15–9, can be used to calculate a **flux control coefficient**, C, for each enzyme in a pathway. This coefficient expresses the relative contribution of each enzyme to setting the rate at which metabolites flow through the pathway—that is, the **flux, J**. C can have any value from 0.0 (for an enzyme with no impact on the flux) to 1.0 (for an enzyme that wholly determines the flux). An enzyme can also have a *negative* flux control coefficient. In a branched pathway, an enzyme in one branch, by drawing intermediates away from the other branch, can have a negative impact on the flux through that other branch **(Fig. 15–10)**. C is not a constant, and it is not intrinsic to a single enzyme; it is a function of the whole system of enzymes, and its value depends on the concentrations of substrates and effectors.

When real data from the experiment on glycolysis in a rat liver extract (Fig. 15–9) were subjected to this kind of analysis, investigators found flux control coefficients (for enzymes at the concentrations found in the extract) of 0.79 for hexokinase, 0.21 for PFK-1, and 0.0 for phosphohexose isomerase. It is not just fortuitous that these values add up to 1.0; we can show that for any complete pathway, the sum of the flux control coefficients must equal unity.

$$C_4 = -0.2$$

$$A \xrightleftharpoons[C_1 = 0.3]{} B \xrightleftharpoons[C_2 = 0.0]{} C \xrightleftharpoons[C_3 = 0.9]{} D$$

FIGURE 15–10 Flux control coefficient, C, in a branched metabolic pathway. In this simple pathway, the intermediate B has two alternative fates. To the extent that reaction B → E draws B away from the pathway A → D, it controls that pathway, which will result in a *negative* flux control coefficient for the enzyme that catalyzes step B → E. Note that the sum of all four coefficients equals 1.0, as it must for any defined system of enzymes.

The Elasticity Coefficient Is Related to an Enzyme's Responsiveness to Changes in Metabolite or Regulator Concentrations

A second parameter, the **elasticity coefficient, ε,** expresses quantitatively the responsiveness of a single enzyme to changes in the concentration of a metabolite or regulator; it is a function of the enzyme's intrinsic kinetic properties. For example, an enzyme with typical Michaelis-Menten kinetics shows a hyperbolic response to increasing substrate concentration **(Fig. 15–11)**. At low concentrations of substrate (say, $0.1\ K_m$), each increment in substrate concentration results in a comparable increase in enzymatic activity, yielding an ε near 1.0. At relatively high substrate concentrations (say, $10\ K_m$), increasing the substrate concentration has little effect on the reaction rate, because the enzyme is already saturated with substrate. The elasticity in this case approaches zero. For allosteric enzymes that show positive cooperativity, ε may exceed 1.0, but it cannot exceed the Hill coefficient, which is typically between 1.0 and 4.0.

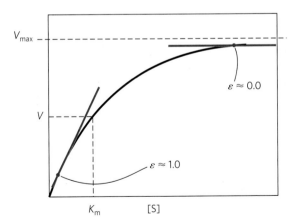

FIGURE 15–11 Elasticity coefficient, ε, of an enzyme with typical Michaelis-Menten kinetics. At substrate concentrations far below the K_m, each increase in [S] produces a correspondingly large increase in the reaction velocity, V. For this region of the curve, the enzyme has an ε of about 1.0. At [S] $\gg K_m$, increasing [S] has little effect on V; ε here is close to 0.0.

BOX 15–1 | METHODS | Metabolic Control Analysis: Quantitative Aspects

The factors that influence the flow of intermediates (flux) through a pathway may be determined quantitatively by experiment and expressed in terms useful for predicting the change in flux when some factor involved in the pathway changes. Consider the simple reaction sequence in Figure 1, in which a substrate X (say, glucose) is converted in several steps to a product Z (perhaps pyruvate, formed glycolytically). An enzyme late in the pathway is a dehydrogenase (ydh) that acts on substrate Y. Because the action of a dehydrogenase is easily measured (see Fig. 13–24), we can use the flux (J) through this step (J_{ydh}) to measure the flux through the whole path. We manipulate experimentally the level of an early enzyme in the pathway (xase, which acts on the substrate X) and measure the flux through the path (J_{ydh}) for several levels of the enzyme xase.

$$X \xrightarrow[J_{xase}]{xase} S_1 \xrightarrow{multistep} Y \xrightarrow[J_{ydh}]{ydh} S_6 \xrightarrow{multistep} Z$$

FIGURE 1 Flux through a hypothetical multienzyme pathway.

The relationship between the flux through the pathway from X to Z in the intact cell and the concentration of each enzyme in the path should be hyperbolic, with virtually no flux at infinitely low enzyme activity and near-maximum flux at very high enzyme activity. In a plot of J_{ydh} against the concentration of xase, E_{xase}, the change in flux with a small change in enzyme level is $\partial J_{ydh}/\partial E_{xase}$, which is simply the slope of the tangent to the curve at any concentration of enzyme, E_{xase}, and which tends toward zero at saturating E_{xase}. At low E_{xase}, the slope is steep; the flux increases with each incremental increase in enzyme activity. At very high E_{xase}, the slope is much smaller; the system is less responsive to added xase because it is already present in excess over the other enzymes in the pathway.

To show quantitatively the dependence of flux through the pathway, ∂J_{ydh}, on ∂E_{xase}, we could use the ratio $\partial J_{ydh}/\partial E_{xase}$. However, its usefulness is limited because its value depends on the units used to express flux and enzyme activity. By expressing the *fractional* changes in flux and enzyme activity, $\partial J_{ydh}/J_{ydh}$ and $\partial E_{xase}/E_{xase}$, we obtain a unitless expression for the **flux control coefficient**, C, in this case $C_{xase}^{J_{ydh}}$:

$$C_{xase}^{J_{ydh}} \approx \frac{\partial J_{ydh}}{J_{ydh}} \bigg/ \frac{\partial E_{xase}}{E_{xase}} \qquad (1)$$

This can be rearranged to

$$C_{xase}^{J_{ydh}} \approx \frac{\partial J_{ydh}}{\partial E_{xase}} \cdot \frac{E_{xase}}{J_{ydh}}$$

which is mathematically identical to

$$C_{xase}^{J_{ydh}} \approx \frac{\partial \ln J_{ydh}}{\partial \ln E_{xase}}$$

This equation suggests a simple graphical means for determining the flux control coefficient: $C_{xase}^{J_{ydh}}$ is the slope of the tangent to the plot of $\ln J_{ydh}$ versus $\ln E_{xase}$, which can be obtained by replotting the experimental data in Figure 2a to obtain Figure 2b. Notice that $C_{xase}^{J_{ydh}}$ is not a constant; it depends on the starting E_{xase} from which the change in enzyme level takes place. For the cases shown in Figure 2, $C_{xase}^{J_{ydh}}$ is about 1.0 at the lowest E_{xase}, but only about 0.2 at high E_{xase}. A value near 1.0 for $C_{xase}^{J_{ydh}}$ means that the enzyme's concentration wholly determines the flux through the pathway; a value near 0.0 means that the enzyme's concentration does not limit the flux through the path. Unless the flux control coefficient is greater than about 0.5, changes in the activity of the enzyme will not have a strong effect on the flux.

The **elasticity**, ε, of an enzyme is a measure of how that enzyme's catalytic activity changes when the concentration of a metabolite—substrate, product, or effector—changes. It is obtained from an experimental plot of the rate of the reaction catalyzed by the enzyme versus the concentration of the metabolite, at metabolite concentrations that prevail in the cell. By arguments analogous to those used to

The Response Coefficient Expresses the Effect of an Outside Controller on Flux through a Pathway

We can also derive a quantitative expression for the relative impact of an outside factor (such as a hormone or growth factor), which is neither a metabolite nor an enzyme in the pathway, on the flux through the pathway. The experiment would measure the flux through the pathway (glycolysis, in this case) at various levels of the parameter P (the insulin concentration, for example) to obtain the **response coefficient**, R, which expresses the change in pathway flux when P ([insulin]) changes.

The three coefficients C, ε, and R are related in a simple way: the responsiveness (R) of a pathway to an outside factor that affects a certain enzyme is a function of (1) how sensitive the pathway is to changes in the activity of that enzyme (the flux control coefficient, C) and (2) how sensitive that specific enzyme is to changes in the outside controlling factor (the elasticity, ε):

$$R = C \cdot \varepsilon$$

Each enzyme in the pathway can be examined in this way, and the effects of any of several outside factors on flux through the pathway can be separately determined.

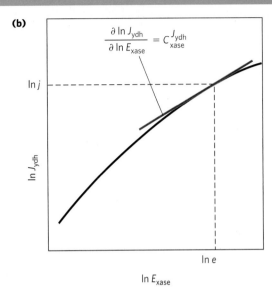

FIGURE 2 The flux control coefficient. (a) Typical variation of the pathway flux, J_{ydh}, measured at the step catalyzed by the enzyme ydh, as a function of the amount of the enzyme xase, E_{xase}, which catalyzes an earlier step in the pathway. The flux control coefficient at (e,j) is the product of the slope of the tangent to the curve, $\partial J_{ydh}/\partial E_{xase}$, and the ratio (scaling factor) e/j. (b) On a double-logarithmic plot of the same curve, the flux control coefficient is the slope of the tangent to the curve.

derive C, we can show ε to be the slope of the tangent to a plot of ln V versus ln [substrate, or product, or effector]:

$$\varepsilon_S^{xase} = \frac{\partial V_{xase}}{\partial S} \cdot \frac{S}{V_{xase}}$$

$$= \frac{\partial \ln |V_{xase}|}{\partial \ln S}$$

For an enzyme with typical Michaelis-Menten kinetics, the value of ε ranges from about 1 at substrate concentrations far below K_m to near 0 as V_{max} is approached. Allosteric enzymes can have elasticities greater than 1.0, but not larger than their Hill coefficient (p. 167).

Finally, the effect of controllers outside the pathway itself (that is, not metabolites) can be measured and expressed as the **response coefficient**, R. The change in flux through the pathway is measured for changes in the concentration of the controlling param-

eter P, and R is defined in a form analogous to that of Equation 1, yielding the expression

$$R_P^{J_{ydh}} = \frac{\partial J_{ydh}}{\partial P} \cdot \frac{P}{J_{ydh}}$$

Using the same logic and graphical methods as described above for determining C, we can obtain R as the slope of the tangent to the plot of ln J versus ln P.

The three coefficients we have described are related in this simple way:

$$R_P^{J_{ydh}} = C_{xase}^{J_{ydh}} \cdot \varepsilon_P^{xase}$$

Thus the responsiveness of each enzyme in a pathway to a change in an outside controlling factor is a simple function of two things: the control coefficient, a variable that expresses the extent to which that enzyme influences the flux under a given set of conditions, and the elasticity, an intrinsic property of the enzyme that reflects its sensitivity to substrate and effector concentrations.

Thus, in principle, we can predict how the flux of substrate through a series of enzymatic steps will change when there is a change in one or more controlling factors external to the pathway. Box 15–1 shows how these qualitative concepts are treated quantitatively.

Metabolic Control Analysis Has Been Applied to Carbohydrate Metabolism, with Surprising Results

Metabolic control analysis provides a framework within which we can think quantitatively about regulation, interpret the significance of the regulatory properties of

each enzyme in a pathway, identify the steps that most affect the flux through the pathway, and distinguish between *regulatory* mechanisms that act to maintain metabolite concentrations and *control* mechanisms that actually alter the flux through the pathway. Analysis of the glycolytic pathway in yeast, for example, has revealed an unexpectedly low flux control coefficient for PFK-1, which, as we have noted, has been viewed as the main point of flux control—the "rate-determining step"—in glycolysis. Experimentally raising the level of PFK-1 fivefold led to a change in flux through glycolysis of less than 10%, suggesting that the real role of PFK-1 regulation is not to control flux through glycolysis but to mediate metabolite homeostasis—to prevent large changes in metabolite concentrations when the flux through glycolysis increases in response to elevated blood glucose or insulin. Recall that the study of glycolysis in a liver extract (Fig. 15–9) also yielded a flux control coefficient that contradicted the conventional wisdom; it showed that hexokinase, not PFK-1, is most influential in setting the flux through glycolysis. We must note here that a liver extract is far from equivalent to a hepatocyte; the ideal way to study flux control is by manipulating one enzyme at a time in the living cell. This is already feasible in many cases.

Investigators have used nuclear magnetic resonance (NMR) as a noninvasive means to determine the concentrations of glycogen and metabolites in the five-step pathway from glucose in the blood to glycogen in myocytes **(Fig. 15–12)** in rat and human muscle. They found that the flux control coefficient for glycogen synthase was smaller than that for either the glucose transporter (GLUT4) or hexokinase. (We discuss glycogen synthase and other enzymes of glycogen metabolism in Sections 15.4 and 15.5.) This finding contradicts the conventional wisdom that glycogen synthase is the locus of flux control and suggests that the importance of the phosphorylation/dephosphorylation of glycogen synthase is related instead to the maintenance of metabolite homeostasis—that is, *regulation*, not *control*. Two metabolites in this pathway, glucose and glucose 6-phosphate, are key intermediates in other pathways, including glycolysis, the pentose phosphate pathway, and the synthesis of glucosamine. Metabolic

control analysis suggests that when the blood glucose level rises, insulin acts in muscle to (1) increase glucose transport into cells by conveying GLUT4 to the plasma membrane, (2) induce the synthesis of hexokinase, and (3) activate glycogen synthase by covalent alteration (see Fig. 15–41). The first two effects of insulin increase glucose flux through the pathway (control), and the third serves to adapt the activity of glycogen synthase so that metabolite levels (glucose 6-phosphate, for example) will not change dramatically with the increased flux (regulation).

Metabolic Control Analysis Suggests a General Method for Increasing Flux through a Pathway

How could an investigator engineer a cell to increase the flux through one pathway without altering the concentrations of other metabolites or the fluxes through other pathways? More than three decades ago Henrik Kacser predicted, on the basis of metabolic control analysis, that this could be accomplished by increasing the concentrations of every enzyme in a pathway. The prediction has been confirmed in several experimental tests, and it also fits with the way cells normally control fluxes through a pathway. For example, rats fed a high-protein diet dispose of excess amino groups by converting them to urea in the urea cycle (Chapter 18). After such a dietary shift, the urea output increases fourfold, and the amount of all eight enzymes in the urea cycle increases two- to threefold. Similarly, when increased fatty acid oxidation is triggered by activation of peroxisome proliferator-activated receptor γ (PPARγ, a ligand-activated transcription factor; see Fig. 21–22), synthesis of the *whole set* of fatty acid oxidative enzymes is increased. With the growing use of DNA microarrays to study the expression of whole sets of genes in response to various perturbations, we should soon learn whether this is the general mechanism by which cells make long-term adjustments in flux through specific pathways.

SUMMARY 15.2 Analysis of Metabolic Control

▸ Metabolic control analysis shows that control of the rate of metabolite flux through a pathway is distributed among several of the enzymes in that path.

▸ The flux control coefficient, C, is an experimentally determined measure of the effect of an enzyme's concentration on flux through a multienzyme pathway. It is characteristic of the whole system, not intrinsic to the enzyme.

▸ The elasticity coefficient, ε, of an enzyme is an experimentally determined measure of its responsiveness to changes in the concentration of a metabolite or regulator molecule.

▸ The response coefficient, R, is a measure of the experimentally determined change in flux through

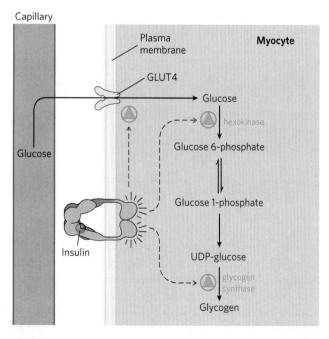

FIGURE 15–12 Control of glycogen synthesis from blood glucose in muscle. Insulin affects three of the five steps in this pathway, but it is the effects on transport and hexokinase activity, not the change in glycogen synthase activity, that increase the flux toward glycogen.

a pathway in response to a regulatory hormone or second messenger. It is a function of C and ε: $R = C \cdot \varepsilon$.

▶ Some regulated enzymes control the flux through a pathway, while others rebalance the level of metabolites in response to the change in flux. The first activity is *control*; the second, rebalancing activity is *regulation.*

▶ Metabolic control analysis predicts, and experiments have confirmed, that flux toward a specific product is most effectively increased by raising the concentration of all enzymes in the pathway.

15.3 Coordinated Regulation of Glycolysis and Gluconeogenesis

In mammals, **gluconeogenesis** occurs primarily in the liver, where its role is to provide glucose for export to other tissues when glycogen stores are exhausted and when no dietary glucose is available. As we discussed in Chapter 14, gluconeogenesis employs several of the enzymes that act in glycolysis, but it is not simply the reversal of glycolysis. Seven of the glycolytic reactions are freely reversible, and the enzymes that catalyze these reactions also function in gluconeogenesis (**Fig. 15–13**). Three reactions of glycolysis are so exergonic as to be essentially irreversible: those catalyzed by hexokinase, PFK-1, and pyruvate kinase. All three reactions have a large, negative $\Delta G'$ (Table 15–3 shows the values in heart muscle). Gluconeogenesis uses detours around each of these irreversible steps; for example, the conversion of fructose 1,6-bisphosphate to fructose 6-phosphate is catalyzed by fructose 1,6-bisphosphatase (FBPase-1). Each of these bypass reactions also has a large, negative $\Delta G'$.

At each of the three points where glycolytic reactions are bypassed by alternative, gluconeogenic reactions, simultaneous operation of both pathways would consume ATP without accomplishing any chemical or biological work. For example, PFK-1 and FBPase-1 catalyze opposing reactions:

$$\text{ATP} + \text{fructose 6-phosphate} \xrightarrow{\text{PFK-1}}$$
$$\text{ADP} + \text{fructose 1,6-bisphosphate}$$

$$\text{Fructose 1,6-bisphosphate} + \text{H}_2\text{O} \xrightarrow{\text{FBPase-1}}$$
$$\text{fructose 6-phosphate} + \text{P}_i$$

The sum of these two reactions is:

$$\text{ATP} + \text{H}_2\text{O} \longrightarrow \text{ADP} + \text{P}_i + \text{heat}$$

that is, hydrolysis of ATP without any useful metabolic work being done. Clearly, if these two reactions were allowed to proceed simultaneously at a high rate in the same cell, a large amount of chemical energy would be dissipated as heat. This uneconomical process has been called a **futile cycle**. However, as we shall see later,

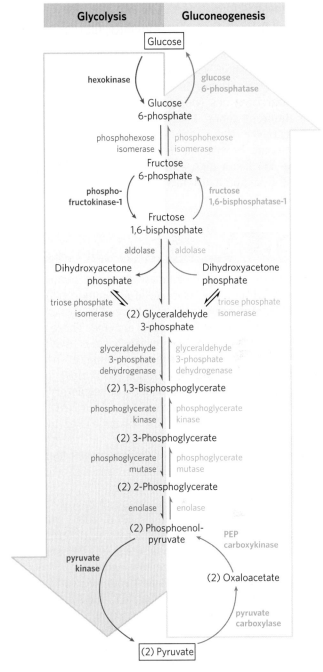

FIGURE 15–13 Glycolysis and gluconeogenesis. Opposing pathways of glycolysis (pink) and gluconeogenesis (blue) in rat liver. Three steps are catalyzed by different enzymes in gluconeogenesis (the "bypass reactions") and glycolysis; seven steps are catalyzed by the same enzymes in the two pathways. Cofactors have been omitted for simplicity.

such cycles may provide advantages for controlling pathways, and the term **substrate cycle** is a better description. Similar substrate cycles also occur with the other two sets of bypass reactions of gluconeogenesis (Fig. 15–13).

We look now in some detail at the mechanisms that regulate glycolysis and gluconeogenesis at the three points where these pathways diverge.

Hexokinase Isozymes of Muscle and Liver Are Affected Differently by Their Product, Glucose 6-Phosphate

Hexokinase, which catalyzes the entry of glucose into the glycolytic pathway, is a regulatory enzyme. Humans have four isozymes (designated I to IV), encoded by four different genes. **Isozymes** are different proteins that catalyze the same reaction (Box 15–2). The predominant hexokinase isozyme of myocytes (**hexokinase II**) has a high affinity for glucose—it is half-saturated at about 0.1 mM. Because glucose entering myocytes from the blood (where the glucose concentration is 4 to 5 mM) produces an intracellular glucose concentration high enough to saturate hexokinase II, the enzyme normally acts at or near its maximal rate. Muscle **hexokinase I** and hexokinase II are allosterically inhibited by their product, glucose 6-phosphate, so whenever the cellular concentration of glucose 6-phosphate rises above its normal level, these isozymes are temporarily and reversibly inhibited, bringing the rate of glucose 6-phosphate formation into balance with the rate of its utilization and reestablishing the steady state.

BOX 15–2   Isozymes: Different Proteins That Catalyze the Same Reaction

The four forms of hexokinase found in mammalian tissues are but one example of a common biological situation: the same reaction catalyzed by two or more different molecular forms of an enzyme. These multiple forms, called isozymes or isoenzymes, may occur in the same species, in the same tissue, even in the same cell. The different forms (isoforms) of the enzyme generally differ in kinetic or regulatory properties, in the cofactor they use (NADH or NADPH for dehydrogenase isozymes, for example), or in their subcellular distribution (soluble or membrane-bound). Isozymes may have similar, but not identical, amino acid sequences, and in many cases they clearly share a common evolutionary origin.

One of the first enzymes found to have isozymes was lactate dehydrogenase (LDH; p. 563), which in vertebrate tissues exists as at least five different isozymes separable by electrophoresis. All LDH isozymes contain four polypeptide chains (each of M_r 33,500), each type containing a different ratio of two kinds of polypeptides. The M (for muscle) chain and the H (for heart) chain are encoded by two different genes.

In skeletal muscle the predominant isozyme contains four M chains, and in heart the predominant isozyme contains four H chains. Other tissues have some combination of the five possible types of LDH isozymes:

Type	Composition	Location
LDH$_1$	HHHH	Heart and erythrocyte
LDH$_2$	HHHM	Heart and erythrocyte
LDH$_3$	HHMM	Brain and kidney
LDH$_4$	HMMM	Skeletal muscle and liver
LDH$_5$	MMMM	Skeletal muscle and liver

The differences in the isozyme content of tissues can be used to assess the timing and extent of heart damage due to myocardial infarction (heart attack). Damage to heart tissue results in the release of heart LDH into the blood. Shortly after a heart attack, the blood level of total LDH increases, and there is more LDH$_2$ than LDH$_1$. After 12 hours the amounts of LDH$_1$ and LDH$_2$ are very similar, and after 24 hours there is more LDH$_1$ than LDH$_2$. This switch in the [LDH$_1$]/[LDH$_2$] ratio, combined with increased concentrations in the blood of another heart enzyme, creatine kinase, is very strong evidence of a recent myocardial infarction. ■

The different LDH isozymes have significantly different values of V_{max} and K_m, particularly for pyruvate. The properties of LDH$_4$ favor rapid reduction of very low concentrations of pyruvate to lactate in skeletal muscle, whereas those of isozyme LDH$_1$ favor rapid oxidation of lactate to pyruvate in the heart.

In general, the distribution of different isozymes of a given enzyme reflects at least four factors:

1. *Different metabolic patterns in different organs.* For glycogen phosphorylase, the isozymes in skeletal muscle and liver have different regulatory properties, reflecting the different roles of glycogen breakdown in these two tissues.

2. *Different locations and metabolic roles for isozymes in the same cell.* The isocitrate dehydrogenase isozymes of the cytosol and the mitochondrion are an example (Chapter 16).

3. *Different stages of development in embryonic or fetal tissues and in adult tissues.* For example, the fetal liver has a characteristic isozyme distribution of LDH, which changes as the organ develops into its adult form. Some enzymes of glucose catabolism in malignant (cancer) cells occur as their fetal, not adult, isozymes.

4. *Different responses of isozymes to allosteric modulators.* This difference is useful in fine-tuning metabolic rates. Hexokinase IV (glucokinase) of liver and the hexokinase isozymes of other tissues differ in their sensitivity to inhibition by glucose 6-phosphate.

The different hexokinase isozymes of liver and muscle reflect the different roles of these organs in carbohydrate metabolism: muscle consumes glucose, using it for energy production, whereas liver maintains blood glucose homeostasis by consuming or producing glucose, depending on the prevailing blood glucose concentration. The predominant hexokinase isozyme of liver is **hexokinase IV** (glucokinase), which differs in three important respects from hexokinases I–III of muscle. First, the glucose concentration at which hexokinase IV is half-saturated (about 10 mM) is higher than the usual concentration of glucose in the blood. Because an efficient glucose transporter in hepatocytes (**GLUT2**) rapidly equilibrates the glucose concentrations in cytosol and blood (see Fig. 11–31 for the kinetics of the same transporter, GLUT1, in erythrocytes), the high K_m of hexokinase IV allows its direct regulation by the level of blood glucose (**Fig. 15–14**). When blood glucose is high, as it is after a meal rich in carbohydrates, excess glucose is transported into hepatocytes, where hexokinase IV converts it to glucose 6-phosphate. Because hexokinase IV is not saturated at 10 mM glucose, its activity continues to increase as the glucose concentration rises to 10 mM or more. Under conditions of low blood glucose, the glucose concentration in a hepatocyte is low relative to the K_m of hexokinase IV, and the glucose generated by gluconeogenesis leaves the cell before being trapped by phosphorylation.

Second, hexokinase IV is not inhibited by glucose 6-phosphate, and it can therefore continue to operate when the accumulation of glucose 6-phosphate completely inhibits hexokinases I–III. Finally, hexokinase IV is subject to inhibition by the reversible binding of a regulatory protein specific to liver (**Fig. 15–15**). The binding is much tighter in the presence of the allosteric effector fructose 6-phosphate. Glucose competes with fructose 6-phosphate for binding and causes dissociation of the regulatory protein from the hexokinase, relieving the inhibition. Immediately after a carbohydrate-rich meal, when blood glucose is high, glucose enters the hepatocyte via GLUT2 and activates hexokinase IV by this mechanism. During a fast, when blood glucose drops

FIGURE 15–14 Comparison of the kinetic properties of hexokinase IV (glucokinase) and hexokinase I. Note the sigmoidicity for hexokinase IV and the much lower K_m for hexokinase I. When blood glucose rises above 5 mM, hexokinase IV activity increases, but hexokinase I is already operating near V_{max} and cannot respond to an increase in glucose concentration. Hexokinases I, II, and III have similar kinetic properties.

below 5 mM, fructose 6-phosphate triggers the inhibition of hexokinase IV by the regulatory protein, so the liver does not compete with other organs for the scarce glucose. The mechanism of inhibition by the regulatory protein is interesting: the protein anchors hexokinase IV inside the nucleus, where it is segregated from the other enzymes of glycolysis in the cytosol (Fig. 15–15). When the glucose concentration in the cytosol rises, it equilibrates with glucose in the nucleus by transport through the nuclear pores. Glucose causes dissociation of the regulatory protein, and hexokinase IV enters the cytosol and begins to phosphorylate glucose.

Hexokinase IV (Glucokinase) and Glucose 6-Phosphatase Are Transcriptionally Regulated

Hexokinase IV is also regulated at the level of protein synthesis. Circumstances that call for greater energy production (low [ATP], high [AMP], vigorous muscle contraction) or for greater glucose consumption (high blood glucose, for example) cause increased transcription of the hexokinase IV gene. Glucose 6-phosphatase, the

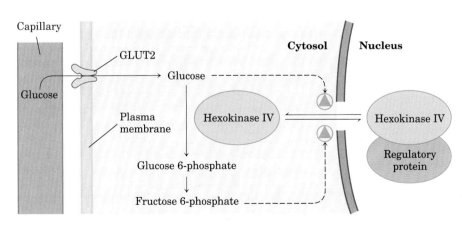

FIGURE 15–15 Regulation of hexokinase IV (glucokinase) by sequestration in the nucleus. The protein inhibitor of hexokinase IV is a nuclear binding protein that draws hexokinase IV into the nucleus when the fructose 6-phosphate concentration in liver is high and releases it to the cytosol when the glucose concentration is high.

gluconeogenic enzyme that bypasses the hexokinase step of glycolysis, is transcriptionally regulated by factors that call for increased production of glucose (low blood glucose, glucagon signaling). The transcriptional regulation of these two enzymes (along with other enzymes of glycolysis and gluconeogenesis) is described below.

Phosphofructokinase-1 and Fructose 1,6-Bisphosphatase Are Reciprocally Regulated

As we have noted, glucose 6-phosphate can flow either into glycolysis or through any of several other pathways, including glycogen synthesis and the pentose phosphate pathway. The metabolically irreversible reaction catalyzed by PFK-1 is the step that commits glucose to glycolysis. In addition to its substrate-binding sites, this complex enzyme has several regulatory sites at which allosteric activators or inhibitors bind.

ATP is not only a substrate for PFK-1 but also an end product of the glycolytic pathway. When high cellular [ATP] signals that ATP is being produced faster than it is being consumed, ATP inhibits PFK-1 by binding to an allosteric site and lowering the affinity of the enzyme for its substrate fructose 6-phosphate **(Fig. 15–16)**. ADP and AMP, which increase in concentration as consumption of ATP outpaces production, act allosterically to relieve this inhibition by ATP. These effects combine to produce higher enzyme activity when ADP or AMP accumulates and lower activity when ATP accumulates.

Citrate (the ionized form of citric acid), a key intermediate in the aerobic oxidation of pyruvate, fatty acids, and amino acids, is also an allosteric regulator of PFK-1; high citrate concentration increases the inhibitory effect of ATP, further reducing the flow of glucose through glycolysis. In this case, as in several others encountered later, citrate serves as an intracellular signal that the cell is meeting its current needs for energy-yielding metabolism by the oxidation of fats and proteins.

The corresponding step in gluconeogenesis is the conversion of fructose 1,6-bisphosphate to fructose 6-phosphate **(Fig. 15–17)**. The enzyme that catalyzes this reaction, FBPase-1, is strongly inhibited (allosterically) by AMP; when the cell's supply of ATP is low (corresponding to high [AMP]), the ATP-requiring synthesis of glucose slows.

Thus these opposing steps in the glycolytic and gluconeogenic pathways—those catalyzed by PFK-1 and FBPase-1—are regulated in a coordinated and reciprocal manner. In general, when sufficient concentrations of acetyl-CoA or citrate (the product of acetyl-CoA condensation with oxaloacetate) are present, or when a high proportion of the cell's adenylate is in the form of ATP, gluconeogenesis is favored. When the level of AMP increases, it promotes glycolysis by stimulating PFK-1 (and, as we shall see in Section 15.5, promotes glycogen degradation by activating glycogen phosphorylase).

(a)

(b)

(c)

FIGURE 15–16 Phosphofructokinase-1 (PFK-1) and its regulation.
(a) Surface contour image of *E. coli* PFK-1, showing portions of its four identical subunits (PDB ID 1PFK). Each subunit has its own catalytic site, where the products ADP and fructose 1,6-bisphosphate (red and yellow stick structures, respectively) are almost in contact, and its own binding sites for the allosteric regulator ATP, buried in the protein in the positions indicated. **(b)** Allosteric regulation of muscle PFK-1 by ATP, shown by a substrate-activity curve. At low [ATP], the $K_{0.5}$ for fructose 6-phosphate is relatively low, enabling the enzyme to function at a high rate at relatively low [fructose 6-phosphate]. (Recall from Chapter 6 that $K_{0.5}$ is the K_m term for regulatory enzymes.) When [ATP] is high, $K_{0.5}$ for fructose 6-phosphate is greatly increased, as indicated by the sigmoid relationship between substrate concentration and enzyme activity. **(c)** Summary of the regulators affecting PFK-1 activity.

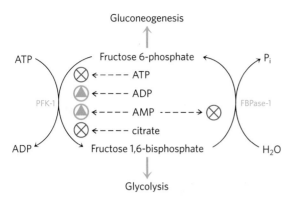

FIGURE 15–17 Regulation of fructose 1,6-bisphosphatase (FBPase-1) and phosphofructokinase-1 (PFK-1). The important role of fructose 2,6-bisphosphate in the regulation of this substrate cycle is detailed in subsequent figures.

Fructose 2,6-Bisphosphate Is a Potent Allosteric Regulator of PFK-1 and FBPase-1

The special role of the liver in maintaining a constant blood glucose level requires additional regulatory mechanisms to coordinate glucose production and consumption. When the blood glucose level decreases, the hormone **glucagon** signals the liver to produce and release more glucose and to stop consuming it for its own

needs. One source of glucose is glycogen stored in the liver; another source is gluconeogenesis, using pyruvate, lactate, glycerol, or certain amino acids as starting material. When blood glucose is high, insulin signals the liver to use glucose as a fuel and as a precursor for the synthesis and storage of glycogen and triacylglycerol.

The rapid hormonal regulation of glycolysis and gluconeogenesis is mediated by **fructose 2,6-bisphosphate**, an allosteric effector for the enzymes PFK-1 and FBPase-1:

Fructose 2,6-bisphosphate

When fructose 2,6-bisphosphate binds to its allosteric site on PFK-1, it increases the enzyme's affinity for its substrate fructose 6-phosphate and reduces its affinity for the allosteric inhibitors ATP and citrate **(Fig. 15–18)**. At the physiological concentrations of its substrates, ATP and fructose 6-phosphate, and of its other positive and

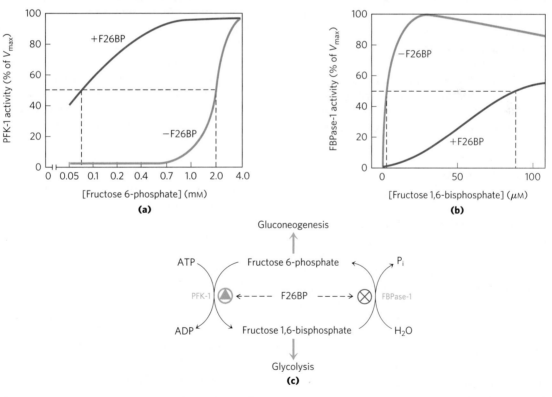

(a)

(b)

(c)

FIGURE 15–18 Role of fructose 2,6-bisphosphate in regulation of glycolysis and gluconeogenesis. Fructose 2,6-bisphosphate (F26BP) has opposite effects on the enzymatic activities of phosphofructokinase-1 (PFK-1, a glycolytic enzyme) and fructose 1,6-bisphosphatase (FBPase-1, a gluconeogenic enzyme). **(a)** PFK-1 activity in the absence of F26BP (blue curve) is half-maximal when the concentration of fructose 6-phosphate is 2 mM (that is, $K_{0.5}$ = 2 mM). When 0.13 μM F26BP is present (red curve), the $K_{0.5}$ for fructose 6-phosphate is only 0.08 mM. Thus

F26BP activates PFK-1 by increasing its apparent affinity for fructose 6-phosphate (see Fig. 15-16b). **(b)** FBPase-1 activity is inhibited by as little as 1 μM F26BP and is strongly inhibited by 25 μM. In the absence of this inhibitor (blue curve) the $K_{0.5}$ for fructose 1,6-bisphosphate is 5 μM, but in the presence of 25 μM F26BP (red curve) the $K_{0.5}$ is >70 μM. Fructose 2,6-bisphosphate also makes FBPase-1 more sensitive to inhibition by another allosteric regulator, AMP. **(c)** Summary of regulation by F26BP.

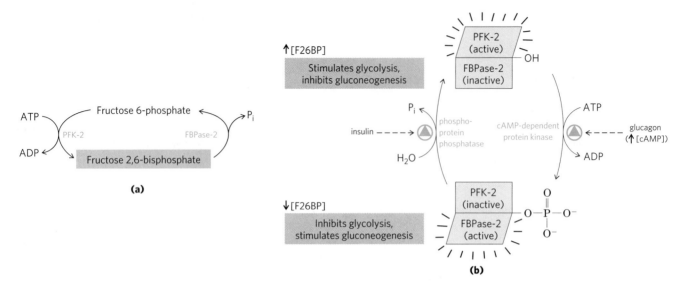

FIGURE 15–19 Regulation of fructose 2,6-bisphosphate level. (a) The cellular concentration of the regulator fructose 2,6-bisphosphate (F26BP) is determined by the rates of its synthesis by phosphofructoki- nase-2 (PFK-2) and its breakdown by fructose 2,6-bisphosphatase (FBPase-2). **(b)** Both enzyme activities are part of the same polypeptide chain, and they are reciprocally regulated by insulin and glucagon.

negative effectors (ATP, AMP, citrate), PFK-1 is virtually inactive in the absence of fructose 2,6-bisphosphate. Fructose 2,6-bisphosphate has the opposite effect on FBPase-1: it reduces its affinity for its substrate (Fig. 15–18b), thereby slowing gluconeogenesis.

The cellular concentration of the allosteric regulator fructose 2,6-bisphosphate is set by the relative rates of its formation and breakdown **(Fig. 15–19a)**. It is formed by phosphorylation of fructose 6-phosphate, catalyzed by **phosphofructokinase-2 (PFK-2)**, and is broken down by **fructose 2,6-bisphosphatase (FBPase-2)**. (Note that these enzymes are distinct from PFK-1 and FBPase-1, which catalyze the formation and breakdown, respectively, of fructose 1,6-bisphosphate.) PFK-2 and FBPase-2 are two separate enzymatic activities of a single, bifunctional protein. The balance of these two activities in the liver, which determines the cellular level of fructose 2,6-bisphosphate, is regulated by glucagon and insulin (Fig. 15–19b).

As we saw in Chapter 12 (p. 446), glucagon stimulates the adenylyl cyclase of liver to synthesize 3′,5′-cyclic AMP (cAMP) from ATP. Cyclic AMP then activates cAMP-dependent protein kinase, which transfers a phosphoryl group from ATP to the bifunctional protein PFK-2/FBPase-2. Phosphorylation of this protein enhances its FBPase-2 activity and inhibits its PFK-2 activity. Glucagon thereby lowers the cellular level of fructose 2,6-bisphosphate, inhibiting glycolysis and stimulating gluconeogenesis. The resulting production of more glucose enables the liver to replenish blood glucose in response to glucagon. Insulin has the opposite effect, stimulating the activity of a phosphoprotein phosphatase that catalyzes removal of the phosphoryl group from the bifunctional protein PFK-2/FBPase-2, activating its PFK-2 activity, increasing the level of fructose 2,6-bisphosphate, stimulating glycolysis, and inhibiting gluconeogenesis.

Xylulose 5-Phosphate Is a Key Regulator of Carbohydrate and Fat Metabolism

Another regulatory mechanism also acts by controlling the level of fructose 2,6-bisphosphate. In the mammalian liver, xylulose 5-phosphate (p. 577), a product of the pentose phosphate pathway (hexose monophosphate pathway), mediates the increase in glycolysis that follows ingestion of a high-carbohydrate meal. The xylulose 5-phosphate concentration rises as glucose entering the liver is converted to glucose 6-phosphate and enters both the glycolytic and pentose phosphate pathways. Xylulose 5-phosphate activates phosphoprotein phosphatase 2A (PP2A; **Fig. 15–20**), which dephosphorylates the bifunctional PFK-2/FBPase-2 enzyme (Fig. 15–19). Dephosphorylation activates PFK-2 and inhibits FBPase-2, and the resulting rise in fructose 2,6-bisphosphate concentration stimulates glycolysis and inhibits gluconeogenesis. The increased glycolysis boosts the production of acetyl-CoA, while the increased flow of hexose through the pentose phosphate pathway generates NADPH. Acetyl-CoA and NADPH are the starting materials for fatty acid synthesis, which has long been known to increase dramatically in response to intake of a high-carbohydrate meal. Xylulose 5-phosphate also increases the synthesis of *all* the enzymes required for fatty acid synthesis, meeting the prediction from metabolic control analysis. We return to this effect in our discussion of the integration of carbohydrate and lipid metabolism in Chapter 23.

The Glycolytic Enzyme Pyruvate Kinase Is Allosterically Inhibited by ATP

At least three isozymes of pyruvate kinase are found in vertebrates, differing in their tissue distribution and their response to modulators. High concentrations of

(a)

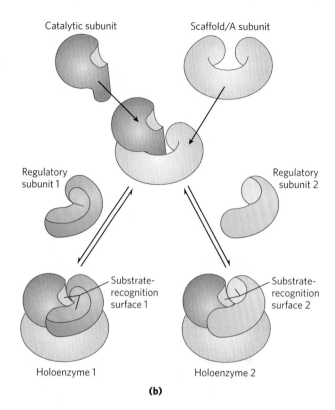

(b)

FIGURE 15–20 Structure and action of phosphoprotein phosphatase 2A (PP2A). (a) The catalytic subunit has two Mn^{2+} ions in its active site, positioned close to the substrate-recognition surface formed by the interface between the catalytic subunit and the regulatory subunit (PDB ID 2NPP). Microcystin-LR, shown here in red, is a specific inhibitor of PP2A. The catalytic and regulatory subunits rest in a scaffold (the A subunit) that positions them relative to each other and shapes the substrate-recognition site. **(b)** PP2A recognizes several target proteins, its specificity provided by the regulatory subunit. Each of several regulatory subunits fits the scaffold containing the catalytic subunit, and each regulatory subunit creates its unique substrate-binding site.

ATP, acetyl-CoA, and long-chain fatty acids (signs of abundant energy supply) allosterically inhibit all isozymes of pyruvate kinase **(Fig. 15–21)**. The liver isozyme (L form), but not the muscle isozyme (M form), is subject to further regulation by phosphorylation. When low blood glucose causes glucagon release, cAMP-dependent protein kinase phosphorylates the L isozyme of pyruvate kinase, inactivating it. This slows the use of

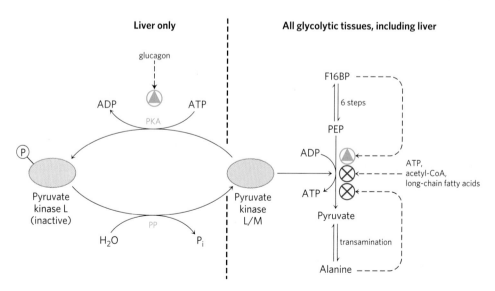

FIGURE 15–21 Regulation of pyruvate kinase. The enzyme is allosterically inhibited by ATP, acetyl-CoA, and long-chain fatty acids (all signs of an abundant energy supply), and the accumulation of fructose 1,6-bisphosphate triggers its activation. Accumulation of alanine, which can be synthesized from pyruvate in one step, allosterically inhibits pyruvate kinase, slowing the production of pyruvate by glycolysis. The liver isozyme (L form) is also regulated hormonally. Glucagon activates

cAMP-dependent protein kinase (PKA; see Fig. 15–37), which phosphorylates the pyruvate kinase L isozyme, inactivating it. When the glucagon level drops, a protein phosphatase (PP) dephosphorylates pyruvate kinase, activating it. This mechanism prevents the liver from consuming glucose by glycolysis when blood glucose is low; instead, the liver exports glucose. The muscle isozyme (M form) is not affected by this phosphorylation mechanism.

glucose as a fuel in liver, sparing it for export to the brain and other organs. In muscle, the effect of increased [cAMP] is quite different. In response to epinephrine, cAMP activates glycogen breakdown and glycolysis and provides the fuel needed for the fight-or-flight response.

The Gluconeogenic Conversion of Pyruvate to Phosphoenolpyruvate Is under Multiple Types of Regulation

In the pathway leading from pyruvate to glucose, the first control point determines the fate of pyruvate in the mitochondrion: its conversion either to acetyl-CoA (by the pyruvate dehydrogenase complex) to fuel the citric acid cycle (Chapter 16) or to oxaloacetate (by pyruvate carboxylase) to start the process of gluconeogenesis **(Fig. 15–22)**. When fatty acids are readily available as fuels, their breakdown in liver mitochondria yields acetyl-CoA, a signal that further oxidation of glucose for fuel is not necessary. Acetyl-CoA is a positive allosteric modulator of pyruvate carboxylase and a negative modulator of pyruvate dehydrogenase, through stimulation of a protein kinase that inactivates the dehydrogenase. When the cell's energy needs are being met, oxidative phosphorylation slows, [NADH] rises relative to [NAD$^+$] and inhibits the citric acid cycle, and acetyl-CoA

accumulates. The increased concentration of acetyl-CoA inhibits the pyruvate dehydrogenase complex, slowing the formation of acetyl-CoA from pyruvate, and stimulates gluconeogenesis by activating pyruvate carboxylase, allowing conversion of excess pyruvate to oxaloacetate (and, eventually, glucose).

Oxaloacetate formed in this way is converted to phosphoenolpyruvate (PEP) in the reaction catalyzed by PEP carboxykinase (Fig. 15–13). In mammals, the regulation of this key enzyme occurs primarily at the level of its synthesis and breakdown, in response to dietary and hormonal signals. Fasting or high glucagon levels act through cAMP to increase the rate of transcription and to stabilize the mRNA. Insulin, or high blood glucose, has the opposite effects. We discuss this transcriptional regulation in more detail below. Generally triggered by a signal from outside the cell (diet, hormones), these changes take place on a time scale of minutes to hours.

Transcriptional Regulation of Glycolysis and Gluconeogenesis Changes the Number of Enzyme Molecules

Most of the regulatory actions discussed thus far are mediated by fast, quickly reversible mechanisms: allosteric effects, covalent alteration (phosphorylation) of the enzyme, or binding of a regulatory protein. Another set of regulatory processes involves changes in the number of molecules of an enzyme in the cell, through changes in the balance of enzyme synthesis and breakdown, and our discussion now turns to regulation of transcription through signal-activated transcription factors.

In Chapter 12 we encountered nuclear receptors and transcription factors in the context of insulin signaling. Insulin acts through its receptor in the plasma membrane to turn on at least two distinct signaling pathways, each involving activation of a protein kinase. The MAP kinase ERK, for example, phosphorylates the transcription factors SRF and Elk1 (see Fig. 12–15), which then stimulate the synthesis of enzymes needed for cell growth and division. Protein kinase B (PKB; also called Akt) phosphorylates another set of transcription factors (PDX1, for example), and these stimulate the synthesis of enzymes that metabolize carbohydrates and the fats formed and stored following excess carbohydrate intake in the diet. In pancreatic β cells, PDX1 also stimulates the synthesis of insulin itself.

More than 150 genes are transcriptionally regulated by insulin; humans have at least seven general types of insulin response elements, each recognized by a subset of transcription factors activated by insulin under various conditions. Insulin stimulates the transcription of the genes that encode hexokinases II and IV, PFK-1, pyruvate kinase, and PFK-2/FBPase-2 (all involved in glycolysis and its regulation); several enzymes of fatty acid synthesis; and glucose 6-phosphate dehydrogenase and 6-phosphogluconate dehydrogenase, enzymes of the pentose phosphate pathway that generate the NADPH

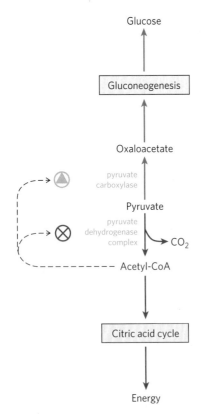

FIGURE 15–22 Two alternative fates for pyruvate. Pyruvate can be converted to glucose and glycogen via gluconeogenesis or oxidized to acetyl-CoA for energy production. The first enzyme in each path is regulated allosterically; acetyl-CoA, produced either by fatty acid oxidation or by the pyruvate dehydrogenase complex, stimulates pyruvate carboxylase and inhibits pyruvate dehydrogenase.

TABLE 15–5 Some of the Genes Regulated by Insulin

Change in gene expression	Pathway
Increased expression	
Hexokinase II	Glycolysis
Hexokinase IV	Glycolysis
Phosphofructokinase-1 (PFK-1)	Glycolysis
Pyruvate kinase	Glycolysis
PFK-2/FBPase-2	Regulation of glycolysis/gluconeogenesis
Glucose 6-phosphate dehydrogenase	Pentose phosphate pathway (NADPH)
6-Phosphogluconate dehydrogenase	Pentose phosphate pathway (NADPH)
Pyruvate dehydrogenase	Fatty acid synthesis
Acetyl-CoA carboxylase	Fatty acid synthesis
Malic enzyme	Fatty acid synthesis (NADPH)
ATP-citrate lyase	Fatty acid synthesis (provides acetyl-CoA)
Fatty acid synthase complex	Fatty acid synthesis
Stearoyl-CoA dehydrogenase	Fatty acid desaturation
Acyl-CoA–glycerol transferases	Triacylglycerol synthesis
Decreased expression	
PEP carboxykinase	Gluconeogenesis
Glucose 6-phosphatase (catalytic subunit)	Glucose release to blood

required for fatty acid synthesis. Insulin also slows the expression of the genes for two enzymes of gluconeogenesis: PEP carboxykinase and glucose 6-phosphatase (Table 15–5).

One transcription factor important to carbohydrate metabolism is **ChREBP (carbohydrate response element binding protein; Fig. 15–23)**, which is expressed primarily in liver, adipose tissue, and kidney. It serves to coordinate the synthesis of enzymes needed for carbohydrate and fat synthesis. ChREBP in its inactive state is phosphorylated, and is located in the cytosol. When the phosphoprotein phosphatase PP2A (Fig. 15–20) removes a phosphoryl group from ChREBP, the transcription factor can enter the nucleus. Here, nuclear PP2A removes another phosphoryl group, and ChREBP now joins with a partner protein, Mlx, and turns on the synthesis of several enzymes: pyruvate kinase, fatty acid synthase, and acetyl-CoA carboxylase, the first enzyme in the path to fatty acid synthesis.

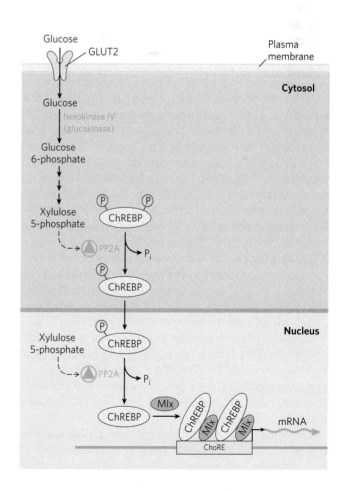

FIGURE 15–23 Mechanism of gene regulation by the transcription factor ChREBP. When ChREBP in the cytosol of a hepatocyte is phosphorylated on a Ser and a Thr residue, it cannot enter the nucleus. Dephosphorylation of (P)-Ser by protein phosphatase PP2A allows ChREBP to enter the nucleus, where a second dephosphorylation, of (P)-Thr, activates ChREBP so that it can associate with its partner protein, Mlx. ChREBP-Mlx now binds to the carbohydrate response element (ChoRE) in the promoter and stimulates transcription. PP2A is allosterically activated by xylulose 5-phosphate, an intermediate in the pentose phosphate pathway.

Controlling the activity of PP2A—and thus, ultimately, the synthesis of this group of metabolic enzymes—is xylulose 5-phosphate, an intermediate of the pentose phosphate pathway (Fig. 14–23). When blood glucose concentration is high, glucose enters the liver and is phosphorylated by hexokinase IV. The glucose 6-phosphate thus formed can enter either the glycolytic pathway or the pentose phosphate pathway. If the latter, two initial oxidations produce xylulose 5-phosphate, which serves as a signal that the glucose-utilizing pathways are well-supplied with substrate. It accomplishes this by allosterically activating PP2A, which then dephosphorylates ChREBP, allowing the transcription factor to turn on the expression of genes for enzymes of glycolysis and fat synthesis (Fig. 15–23). Glycolysis yields pyruvate, and conversion of pyruvate to acetyl-CoA provides the starting material for fatty acid synthesis: acetyl-CoA carboxylase converts acetyl-CoA to malonyl-CoA, the first committed intermediate in the path to fatty acids. The fatty acid synthase complex produces fatty acids for export to adipose tissue and storage as triacylglycerols (Chapter 21). In this way, excess dietary carbohydrate is stored as fat.

Another transcription factor in the liver, **SREBP-1c**, a member of the family of **sterol regulatory element binding proteins** (see Fig. 21–44), turns on the synthesis of pyruvate kinase, hexokinase IV, lipoprotein lipase, acetyl-CoA carboxylase, and the fatty acid synthase complex that will convert acetyl-CoA (produced from pyruvate) into fatty acids for storage in adipocytes. The synthesis of SREBP-1c is stimulated by insulin and depressed by glucagon. SREBP-1c also suppresses the expression of several gluconeogenic enzymes: glucose 6-phosphatase, PEP carboxykinase, and FBPase-1.

The transcription factor **CREB** (**cyclic AMP response element binding protein**) turns on the synthesis of glucose 6-phosphatase and PEP carboxykinase in response to the increase in [cAMP] triggered by glucagon. In contrast, insulin-stimulated *inactivation* of other transcription factors turns off several gluconeogenic enzymes in the liver: PEP carboxykinase, fructose 1,6-bisphosphatase, the glucose 6-phosphate transporter of the endoplasmic reticulum, and glucose 6-phosphatase. For example, **FOXO1 (forkhead box other)** stimulates the synthesis of gluconeogenic enzymes and suppresses the synthesis of the enzymes of glycolysis, the pentose phosphate pathway, and triacylglycerol synthesis **(Fig. 15–24)**. In its unphosphorylated form, FOXO1 acts as a nuclear transcription factor. In response to insulin, FOXO1 leaves the nucleus and in the cytosol is phosphorylated by PKB, then tagged with ubiquitin and degraded by the proteasome. Glucagon prevents this phosphorylation by PKB, and FOXO1 remains active in the nucleus.

Complicated though the processes outlined above may seem, regulation of the genes encoding enzymes of carbohydrate and fat metabolism is proving far more

FIGURE 15–24 Mechanism of gene regulation by the transcription factor FOXO1. Insulin activates the signaling cascade shown in Figure 12–16, leading to activation of protein kinase B (PKB). FOXO1 in the cytosol is phosphorylated by PKB, and the phosphorylated transcription factor is tagged by the attachment of ubiquitin for degradation by proteasomes. FOXO1 that remains unphosphorylated or is dephosphorylated can enter the nucleus, bind to a response element, and trigger transcription of the associated genes. Insulin therefore has the effect of turning *off* the expression of these genes, which include PEP carboxykinase and glucose 6-phosphatase.

complex and more subtle than we have shown here. Multiple transcription factors can act on the same gene promoter; multiple protein kinases and phosphatases can activate or inactivate these transcription factors; and a variety of protein accessory factors modulate the action of the transcription factors. This complexity is apparent, for example, in the gene encoding PEP carboxykinase, a very well-studied case of transcriptional control. Its promoter region **(Fig. 15–25)** has 15 or more response elements that are recognized by at least a dozen known transcription factors, with more likely to be discovered. The transcription factors act in combination on this promoter region, and on hundreds of other gene promoters, to fine-tune the levels of hundreds of metabolic enzymes, coordinating their activity in the metabolism of carbohydrates and fats. The critical importance of transcription factors in metabolic regulation is made clear by observing the effects of mutations in their genes. For example, at least five different types of maturity-onset diabetes of the young (MODY) are associated with mutations in specific transcription factors (Box 15–3).

Transcription factors	
FOXO1	forkhead box other 1
PPARγ2	peroxisome proliferator-activated receptor γ2
HNF-3β	hepatic nuclear factor-3β
SREBP-1	sterol regulatory element binding protein-1
HNF-4α	hepatic nuclear factor-4α
COUP-TF	chicken ovalbumin upstream promoter-transcription factor
RAR	retinoic acid receptor
GR	glucocorticoid receptor
T₃R	thyroid hormone receptor
C/EBP	CAAT/enhance binding protein
HNF-1	hepatic nuclear factor-1
NF1	nuclear factor 1
ATF3	activating transcription factor 3
CREB	cAMP response element binding protein
NFκB	nuclear factor κB
TBP	TATA-box binding protein
TFIIH	transcription factor IIH

Response elements and regulatory binding sites in promoter	
dAF2	distal accessory factor 2
dAF1	distal accessory factor 1
SRE	sterol regulatory element
AF1	accessory factor 1
AF2	accessory factor 2
GRE	glucocorticoid regulatory element
TRE	thyroid hormone regulatory element
CRE	cAMP regulatory element

FIGURE 15–25 The PEP carboxykinase promoter region, showing the complexity of regulatory input to this gene. This diagram shows the transcription factors (smaller icons, bound to the DNA) known to regulate the transcription of the PEP carboxykinase gene. The extent to which this gene is expressed depends on the combined input affecting all of these factors, which can reflect the availability of nutrients, blood glucose level, and other circumstances that affect the cell's need for this enzyme at any particular time. P1, P2, P3I, P3II, and P4 are protein-binding sites identified by DNase I footprinting (see Box 26–1). The TATA box is the assembly point for the RNA polymerase II (Pol II) transcription complex.

BOX 15-3 MEDICINE Genetic Mutations That Lead to Rare Forms of Diabetes

The term "diabetes" describes a variety of medical conditions that have in common an excessive production of urine. In Box 11–1 we described diabetes insipidus, in which defective water reabsorption in the kidney results from a mutation in the gene for aquaporin. "Diabetes mellitus" refers specifically to disease in which the ability to metabolize glucose is defective, due either to the failure of the pancreas to produce insulin or to tissue resistance to the actions of insulin.

There are two common types of diabetes mellitus. Type 1, also called insulin-dependent diabetes mellitus (IDDM), is caused by autoimmune attack on the insulin-producing β cells of the pancreas. Individuals with IDDM must take insulin by injection or inhalation to compensate for their missing β cells. IDDM develops in childhood or in the teen years; an older name for the disease is juvenile diabetes. Type 2, also called non–insulin-dependent diabetes mellitus (NIDDM), typically develops in adults over 40 years old. It is far more common than IDDM, and its occurrence in the population is strongly correlated with obesity. The current epidemic of obesity in the more developed countries brings with it the promise of an epidemic of NIDDM, providing a strong incentive to understand the relationship between obesity and the onset of NIDDM at the genetic and biochemical levels. After completing our look at the metabolism of fats and proteins in later chapters, we will return (in Chapter 23) to the discussion of diabetes, which has a broad effect on metabolism: of carbohydrates, fats, and proteins.

Here we consider another type of diabetes in which carbohydrate and fat metabolism is deranged: mature onset diabetes of the young (MODY), in which genetic mutation affects a transcription factor important in carrying the insulin signal into the nucleus, or affects an enzyme that responds to insulin. In MODY2, a mutation in the hexokinase IV (glucokinase) gene affects the liver and pancreas, tissues in which this is the main isoform of hexokinase. The glucokinase of pancreatic β cells functions as a glucose sensor. Normally, when blood glucose

(Continued on the next page)

rises, so does the glucose level in β cells, and because glucokinase has a relatively high K_m for glucose, its activity increases with rising blood glucose levels. Metabolism of the glucose 6-phosphate formed in this reaction raises the ATP level in β cells, and this triggers insulin release by the mechanism shown in Fig. 23–27. In healthy individuals, blood glucose concentrations of ~5 mM trigger this insulin release. But individuals with inactivating mutations in both copies of the glucokinase gene have very high thresholds for insulin release, and consequently, from birth, they have severe hyperglycemia—permanent neonatal diabetes. In individuals with one mutated and one normal copy of the glucokinase gene, the glucose threshold for insulin release rises to about 7 mM. As a result these individuals have blood glucose levels only slightly above normal: they generally have only mild hyperglycemia and no symptoms. This condition

(MODY2) is generally discovered by accident during routine blood glucose analysis.

There are at least five other types of MODY, each the result of an inactivating mutation in one or another of the transcription factors essential to the normal development and function of pancreatic β cells. Individuals with these mutations have varying degrees of reduced insulin production and the associated defects in blood glucose homeostasis. In MODY1 and MODY3, the defects are severe enough to produce the long-term complications associated with IDDM and NIDDM—cardiovascular problems, kidney failure, and blindness. MODY4, 5, and 6 are less severe forms of the disease. Altogether, MODY disorders represent a small percentage of NIDDM cases. Also very rare are individuals with mutations in the insulin gene itself; they have defects in insulin signaling of varying severity.

SUMMARY 15.3 Coordinated Regulation of Glycolysis and Gluconeogenesis

▶ Gluconeogenesis and glycolysis share seven enzymes, catalyzing the freely reversible reactions of the pathways. For the other three steps, the forward and reverse reactions are catalyzed by different enzymes, and these are the points of regulation of the two pathways.

▶ Hexokinase IV (glucokinase) has kinetic properties related to its special role in the liver: releasing glucose to the blood when blood glucose is low, and taking up and metabolizing glucose when blood glucose is high.

▶ PFK-1 is allosterically inhibited by ATP and citrate. In most mammalian tissues, including liver, fructose 2,6-bisphosphate is an allosteric activator of this enzyme.

▶ Pyruvate kinase is allosterically inhibited by ATP, and the liver isozyme also is inhibited by cAMP-dependent phosphorylation.

▶ Gluconeogenesis is regulated at the level of pyruvate carboxylase (which is activated by acetyl-CoA) and FBPase-1 (which is inhibited by fructose 2,6-bisphosphate and AMP).

▶ To limit substrate cycling between glycolysis and gluconeogenesis, the two pathways are under reciprocal allosteric control, mainly achieved by the opposing effects of fructose 2,6-bisphosphate on PFK-1 and FBPase-1.

▶ Glucagon or epinephrine decreases [fructose 2,6-bisphosphate], by raising [cAMP] and bringing

about phosphorylation of the bifunctional enzyme PFK-2/FBPase-2. Insulin increases [fructose 2,6-bisphosphate] by activating a phosphoprotein phosphatase that dephosphorylates and thus activates PFK-2.

▶ Xylulose 5-phosphate, an intermediate of the pentose phosphate pathway, activates phosphoprotein phosphatase PP2A, which dephosphorylates several target proteins, including PFK-2/FBPase-2, tilting the balance toward glucose uptake, glycogen synthesis, and lipid synthesis in the liver.

▶ Transcription factors including ChREBP, CREB, SREBP, and FOXO1 act in the nucleus to regulate the expression of specific genes coding for enzymes of the glycolytic and gluconeogenic pathways. Insulin and glucagon act antagonistically in activating these transcription factors, thus turning on and off large numbers of genes.

15.4 The Metabolism of Glycogen in Animals

Our discussion of metabolic regulation, using carbohydrate metabolism as the primary example, now turns to the synthesis and breakdown of glycogen. In this section we focus on the metabolic pathways; in Section 15.5 we turn to the regulatory mechanisms.

In organisms from bacteria to plants to vertebrates, excess glucose is converted to polymeric forms for storage—glycogen in vertebrates and many microorganisms, starch in plants. In vertebrates, glycogen is found primarily in the liver and skeletal muscle; it may repre-

FIGURE 15-26 Glycogen granules in a hepatocyte. Glycogen, a storage form of carbohydrate, appears as electron-dense particles, often in aggregates or rosettes. In hepatocytes glycogen is closely associated with tubules of the smooth endoplasmic reticulum. Many mitochondria are also evident in this micrograph.

sent up to 10% of the weight of liver and 1% to 2% of the weight of muscle. If this much glucose were dissolved in the cytosol of a hepatocyte, its concentration would be about 0.4 M, enough to dominate the osmotic properties of the cell. When stored as a large polymer (glycogen), however, the same mass of glucose has a concentration of only 0.01 μM. Glycogen is stored in large cytosolic granules. The elementary particle of glycogen, the β-particle, is about 21 nm in diameter and consists of up to 55,000 glucose residues with about 2,000 nonreducing ends. Twenty to 40 of these particles cluster together to form α-rosettes, easily seen with the microscope in tissue samples from well-fed animals **(Fig. 15–26)** but essentially absent after a 24-hour fast.

The glycogen in muscle is there to provide a quick source of energy for either aerobic or anaerobic metabolism. Muscle glycogen can be exhausted in less than an

hour during vigorous activity. Liver glycogen serves as a reservoir of glucose for other tissues when dietary glucose is not available (between meals or during a fast); this is especially important for the neurons of the brain, which cannot use fatty acids as fuel. Liver glycogen can be depleted in 12 to 24 hours. In humans, the total amount of energy stored as glycogen is far less than the amount stored as fat (triacylglycerol) (see Table 23–5), but fats cannot be converted to glucose in mammals and cannot be catabolized anaerobically.

Glycogen granules are complex aggregates of glycogen and the enzymes that synthesize it and degrade it, as well as the machinery for regulating these enzymes. The general mechanisms for storing and mobilizing glycogen are the same in muscle and liver, but the enzymes differ in subtle yet important ways that reflect the different roles of glycogen in the two tissues. Glycogen is also obtained in the diet and broken down in the gut, and this involves a separate set of hydrolytic enzymes that convert glycogen to free glucose. (Dietary starch is hydrolyzed in a similar way.) We begin our discussion with the breakdown of glycogen to glucose 1-phosphate (**glycogenolysis**), then turn to synthesis of glycogen (**glycogenesis**).

Glycogen Breakdown Is Catalyzed by Glycogen Phosphorylase

In skeletal muscle and liver, the glucose units of the outer branches of glycogen enter the glycolytic pathway through the action of three enzymes: glycogen phosphorylase, glycogen debranching enzyme, and phosphoglucomutase. Glycogen phosphorylase catalyzes the reaction in which an ($\alpha 1\rightarrow 4$) glycosidic linkage between two glucose residues at a nonreducing end of glycogen undergoes attack by inorganic phosphate (P_i), removing the terminal glucose residue as **α-D-glucose 1-phosphate (Fig. 15–27)**. This *phosphorolysis* reaction is

FIGURE 15-27 Removal of a glucose residue from the nonreducing end of a glycogen chain by glycogen phosphorylase. This process is repetitive; the enzyme removes successive glucose residues until it reaches the fourth glucose unit from a branch point (see Fig. 15–28).

Nonreducing end

Glycogen chain (glucose)$_n$

P_i glycogen phosphorylase

Glucose 1-phosphate

Nonreducing end

Glycogen shortened by one residue (glucose)$_{n-1}$

different from the *hydrolysis* of glycosidic bonds by amylase during intestinal degradation of dietary glycogen and starch. In phosphorolysis, some of the energy of the glycosidic bond is preserved in the formation of the phosphate ester, glucose 1-phosphate (see Section 14.2).

Pyridoxal phosphate is an essential cofactor in the glycogen phosphorylase reaction; its phosphate group acts as a general acid catalyst, promoting attack by P_i on the glycosidic bond. (This is an unusual role for pyridoxal phosphate; its more typical role is as a cofactor in amino acid metabolism; see Fig. 18–6.)

Glycogen phosphorylase acts repetitively on the nonreducing ends of glycogen branches until it reaches a point four glucose residues away from an ($\alpha1{\rightarrow}6$) branch point (see Fig. 7–13), where its action stops. Further degradation by glycogen phosphorylase can occur only after the **debranching enzyme**, formally known as **oligo ($\alpha1{\rightarrow}6$) to ($\alpha1{\rightarrow}4$) glucantransferase**, catalyzes two successive reactions that transfer branches **(Fig. 15–28)**. Once these branches are transferred and the glucosyl residue at C-6 is hydrolyzed, glycogen phosphorylase activity can continue.

Glucose 1-Phosphate Can Enter Glycolysis or, in Liver, Replenish Blood Glucose

Glucose 1-phosphate, the end product of the glycogen phosphorylase reaction, is converted to glucose 6-phosphate by **phosphoglucomutase**, which catalyzes the reversible reaction

$$\text{Glucose 1-phosphate} \rightleftharpoons \text{glucose 6-phosphate}$$

Initially phosphorylated at a Ser residue, the enzyme donates a phosphoryl group to C-6 of the substrate, then accepts a phosphoryl group from C-1 **(Fig. 15–29)**.

The glucose 6-phosphate formed from glycogen in skeletal muscle can enter glycolysis and serve as an energy source to support muscle contraction. In liver, glycogen breakdown serves a different purpose: to release glucose into the blood when the blood glucose

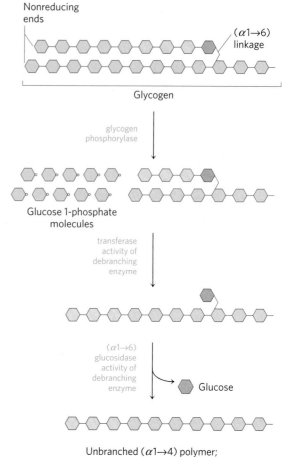

Glucose 1-phosphate molecules

FIGURE 15–28 Glycogen breakdown near an ($\alpha1{\rightarrow}6$) branch point. Following sequential removal of terminal glucose residues by glycogen phosphorylase (see Fig. 15–27), glucose residues near a branch are removed in a two-step process that requires a bifunctional debranching enzyme. First, the transferase activity of the enzyme shifts a block of three glucose residues from the branch to a nearby nonreducing end, to which they are reattached in ($\alpha1{\rightarrow}4$) linkage. The single glucose residue remaining at the branch point, in ($\alpha1{\rightarrow}6$) linkage, is then released as free glucose by the debranching enzyme's ($\alpha1{\rightarrow}6$) glucosidase activity. The glucose residues are shown in shorthand form, which omits the —H, —OH, and —CH₂OH groups from the pyranose rings.

Phosphoglucomutase

FIGURE 15–29 Reaction catalyzed by phosphoglucomutase. The reaction begins with the enzyme phosphorylated on a Ser residue. In step **1**, the enzyme donates its phosphoryl group (blue) to glucose 1-phosphate, producing glucose 1,6-bisphosphate. In step **2**, the phosphoryl group at C-1 of glucose 1,6-bisphosphate (red) is transferred back to the enzyme, reforming the phosphoenzyme and producing glucose 6-phosphate.

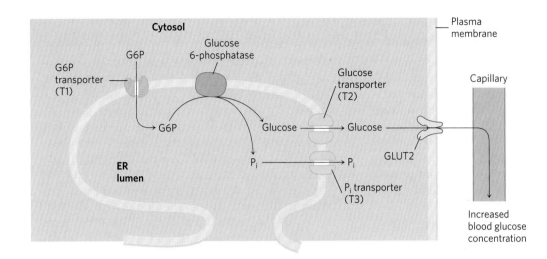

FIGURE 15–30 Hydrolysis of glucose 6-phosphate by glucose 6-phosphatase of the ER. The catalytic site of glucose 6-phosphatase faces the lumen of the ER. A glucose 6-phosphate (G6P) transporter (T1) carries the substrate from the cytosol to the lumen, and the products glucose and P$_i$ pass to the cytosol on specific transporters (T2 and T3). Glucose leaves the cell via the GLUT2 transporter in the plasma membrane.

level drops, as it does between meals. This requires the enzyme glucose 6-phosphatase, present in liver and kidney but not in other tissues. The enzyme is an integral membrane protein of the endoplasmic reticulum, predicted to contain nine transmembrane helices, with its active site on the lumenal side of the ER. Glucose 6-phosphate formed in the cytosol is transported into the ER lumen by a specific transporter (T1) **(Fig. 15–30)** and hydrolyzed at the lumenal surface by the glucose 6-phosphatase. The resulting P$_i$ and glucose are thought to be carried back into the cytosol by two different transporters (T2 and T3), and the glucose leaves the hepatocyte via the plasma membrane transporter, GLUT2. Notice that by having the active site of glucose 6-phosphatase inside the ER lumen, the cell separates this reaction from the process of glycolysis, which takes place in the cytosol and would be aborted by the action of glucose 6-phosphatase. Genetic defects in either glucose 6-phosphatase or T1 lead to serious derangement of glycogen metabolism, resulting in type Ia glycogen storage disease (Box 15–4).

Because muscle and adipose tissue lack glucose 6-phosphatase, they cannot convert the glucose 6-phosphate formed by glycogen breakdown to glucose, and these tissues therefore do not contribute glucose to the blood.

The Sugar Nucleotide UDP-Glucose Donates Glucose for Glycogen Synthesis

Luis Leloir, 1906–1987

Many of the reactions in which hexoses are transformed or polymerized involve **sugar nucleotides**, compounds in which the anomeric carbon of a sugar is activated by attachment to a nucleotide through a phosphate ester linkage. Sugar nucleotides are the substrates for polymerization of monosaccharides into disaccharides, glycogen, starch, cellulose, and more complex extracellular polysaccharides. They are also key intermediates in the production of the aminohexoses and deoxyhexoses found in some of these polysaccharides, and in the synthesis of vitamin C (l-ascorbic acid). The role of sugar nucleotides in the biosynthesis of glycogen and many other carbohydrate derivatives was discovered in 1953 by the Argentine biochemist Luis Leloir.

D-Glucosyl group

Uridine

UDP-glucose
(a sugar nucleotide)

The suitability of sugar nucleotides for biosynthetic reactions stems from several properties:

1. Their formation is metabolically irreversible, contributing to the irreversibility of the synthetic pathways in which they are intermediates. The condensation of a nucleoside triphosphate with a hexose 1-phosphate to form a sugar nucleotide has a small positive free-energy change, but the reaction releases PP$_i$, which is rapidly hydrolyzed by inorganic pyrophosphatase **(Fig. 15–31)**, in a reaction that is strongly exergonic ($\Delta G'^{\circ} = -19.2$ kJ/mol). This keeps the cellular concentration of PP$_i$ low, ensuring that the actual free-energy change in the cell is favorable. In effect, rapid removal of the product, driven by the large, negative free-energy change of PP$_i$

BOX 15–4   Carl and Gerty Cori: Pioneers in Glycogen Metabolism and Disease

Much of what is written in present-day biochemistry textbooks about the metabolism of glycogen was discovered between about 1925 and 1950 by the remarkable husband and wife team of Carl F. Cori and Gerty T. Cori. Both trained in medicine in Europe at the end of World War I (she completed premedical studies and medical school in one year!). They left Europe together in 1922 to establish research laboratories in the United States, first for nine years in Buffalo, New York, at what is now the Roswell Park Memorial Institute, then from 1931 until the end of their lives at Washington University in St. Louis.

In their early physiological studies of the origin and fate of glycogen in animal muscle, the Coris demonstrated the conversion of glycogen to lactate in tissues, movement of lactate in the blood to the liver, and, in the liver, reconversion of lactate to glycogen—a pathway that came to be known as the Cori cycle (see Fig. 23–19). Pursuing these observations at the biochemical level, they showed that glycogen was mobilized in a phosphorolysis reaction catalyzed by the enzyme they discovered, glycogen phosphorylase. They identified the product of this reaction (the "Cori ester") as glucose 1-phosphate and showed that it could be reincorporated into glycogen in the reverse reaction. Although this did not prove to be the reaction by which glycogen is synthesized in cells, it was the first in vitro demonstration of the synthesis of a macromolecule from simple monomeric subunits, and it inspired others to search for polymerizing enzymes. Arthur Kornberg, discoverer of the first DNA polymerase, said of his experience in the Coris' lab, "Glycogen phosphorylase, not base pairing, was what led me to DNA polymerase."

Gerty Cori became interested in human genetic diseases in which too much glycogen is stored in the liver. She was able to identify the biochemical defect in several of these diseases and to show that the diseases could be diagnosed by assays of the enzymes of glycogen metabolism in small samples of tissue obtained by biopsy. Table 1 summarizes what we now know about 13 genetic diseases of this sort. ∎

The Coris in Gerty Cori's laboratory, around 1947.

Carl and Gerty Cori shared the Nobel Prize in Physiology or Medicine in 1947 with Bernardo Houssay of Argentina, who was cited for his studies of hormonal regulation of carbohydrate metabolism. The Cori laboratories in St. Louis became an international center of biochemical research in the 1940s and 1950s, and at least six scientists who trained with the Coris became Nobel laureates: Arthur Kornberg (for DNA synthesis, 1959), Severo Ochoa (for RNA synthesis, 1959), Luis Leloir (for the role of sugar nucleotides in polysaccharide synthesis, 1970), Earl Sutherland (for the discovery of cAMP in the regulation of carbohydrate metabolism, 1971), Christian de Duve (for subcellular fractionation, 1974), and Edwin Krebs (for the discovery of phosphorylase kinase, 1991).

hydrolysis, pulls the synthetic reaction forward, a common strategy in biological polymerization reactions.

2. Although the chemical transformations of sugar nucleotides do not involve the atoms of the nucleotide itself, the nucleotide moiety has many groups that can undergo noncovalent interactions with enzymes; the additional free energy of binding can contribute significantly to catalytic activity (Chapter 6; see also pp. 306–307).

3. Like phosphate, the nucleotidyl group (UMP or AMP, for example) is an excellent leaving group, facilitating nucleophilic attack by activating the sugar carbon to which it is attached.

4. By "tagging" some hexoses with nucleotidyl groups, cells can set them aside in a pool for one purpose (glycogen synthesis, for example), separate from hexose phosphates destined for another purpose (such as glycolysis).

TABLE 1 ⚕ Glycogen Storage Diseases of Humans

Type (name)	Enzyme affected	Primary organ affected	Symptoms
Type 0	Glycogen synthase	Liver	Low blood glucose, high ketone bodies, early death
Type Ia (von Gierke)	Glucose 6-phosphatase	Liver	Enlarged liver, kidney failure
Type Ib	Microsomal glucose 6-phosphate translocase	Liver	As in type Ia; also high susceptibility to bacterial infections
Type Ic	Microsomal P_i transporter	Liver	As in type Ia
Type II (Pompe)	Lysosomal glucosidase	Skeletal and cardiac muscle	Infantile form: death by age 2; juvenile form: muscle defects (myopathy); adult form: as in muscular dystrophy
Type IIIa (Cori or Forbes)	Debranching enzyme	Liver, skeletal and cardiac muscle	Enlarged liver in infants; myopathy
Type IIIb	Liver debranching enzyme (muscle enzyme normal)	Liver	Enlarged liver in infants
Type IV (Andersen)	Branching enzyme	Liver, skeletal muscle	Enlarged liver and spleen, myoglobin in urine
Type V (McArdle)	Muscle phosphorylase	Skeletal muscle	Exercise-induced cramps and pain; myoglobin in urine
Type VI (Hers)	Liver phosphorylase	Liver	Enlarged liver
Type VII (Tarui)	Muscle PFK-1	Muscle, erythrocytes	As in type V; also hemolytic anemia
Type VIb, VIII, or IX	Phosphorylase kinase	Liver, leukocytes, muscle	Enlarged liver
Type XI (Fanconi-Bickel)	Glucose transporter (GLUT2)	Liver	Failure to thrive, enlarged liver, rickets, kidney dysfunction

Glycogen synthesis takes place in virtually all animal tissues but is especially prominent in the liver and skeletal muscles. The starting point for synthesis of glycogen is glucose 6-phosphate. As we have seen, this can be derived from free glucose in a reaction catalyzed by the isozymes hexokinase I and hexokinase II in muscle and hexokinase IV (glucokinase) in liver:

$$\text{D-Glucose} + \text{ATP} \longrightarrow \text{D-glucose 6-phosphate} + \text{ADP}$$

However, some ingested glucose takes a more roundabout path to glycogen. It is first taken up by erythrocytes and converted to lactate glycolytically; the lactate is then taken up by the liver and converted to glucose 6-phosphate by gluconeogenesis.

To initiate glycogen synthesis, the glucose 6-phosphate is converted to glucose 1-phosphate in the phosphoglucomutase reaction:

$$\text{Glucose 6-phosphate} \rightleftharpoons \text{glucose 1-phosphate}$$

The product of this reaction is converted to UDP-glucose by the action of **UDP-glucose pyrophosphorylase**, in a key step of glycogen biosynthesis:

$$\text{Glucose 1-phosphate} + \text{UTP} \longrightarrow \text{UDP-glucose} + \text{PP}_i$$

FIGURE 15–31 Formation of a sugar nucleotide. A condensation reaction occurs between a nucleoside triphosphate (NTP) and a sugar phosphate. The negatively charged oxygen on the sugar phosphate serves as a nucleophile, attacking the α phosphate of the nucleoside triphosphate and displacing pyrophosphate. The reaction is pulled in the forward direction by the hydrolysis of PP_i by inorganic pyrophosphatase.

Net reaction: Sugar phosphate + NTP $\longrightarrow$ NDP-sugar + $2P_i$

Notice that this enzyme is named for the reverse reaction; in the cell, the reaction proceeds in the direction of UDP-glucose formation, because pyrophosphate is rapidly hydrolyzed by inorganic pyrophosphatase (Fig. 15–31).

UDP-glucose is the immediate donor of glucose residues in the reaction catalyzed by **glycogen synthase**, which promotes the transfer of the glucose resi-

due from UDP-glucose to a nonreducing end of a branched glycogen molecule **(Fig. 15–32)**. The overall equilibrium of the path from glucose 6-phosphate to glycogen lengthened by one glucose unit greatly favors synthesis of glycogen.

Glycogen synthase cannot make the ($\alpha1{\to}6$) bonds found at the branch points of glycogen; these are

FIGURE 15–32 Glycogen synthesis. A glycogen chain is elongated by glycogen synthase. The enzyme transfers the glucose residue of UDP-glucose to the nonreducing end of a glycogen branch (see Fig. 7–13) to make a new ($\alpha1{\to}4$) linkage.

FIGURE 15–33 Branch synthesis in glycogen. The glycogen-branching enzyme (also called amylo (1→4) to (1→6) transglycosylase, or glycosyl-(4→6) transferase) forms a new branch point during glycogen synthesis.

formed by the glycogen-branching enzyme, also called **amylo (1→4) to (1→6) transglycosylase**, or glycosyl-(4→6) transferase. The glycogen-branching enzyme catalyzes transfer of a terminal fragment of 6 or 7 glucose residues from the nonreducing end of a glycogen branch having at least 11 residues to the C-6 hydroxyl group of a glucose residue at a more interior position of the same or another glycogen chain, thus creating a new branch **(Fig. 15–33)**. Further glucose residues may be added to the new branch by glycogen synthase. The biological effect of branching is to make the glycogen molecule more soluble and to increase the number of nonreducing ends. This increases the number of sites accessible to glycogen phosphorylase and glycogen synthase, both of which act only at nonreducing ends.

Glycogenin Primes the Initial Sugar Residues in Glycogen

Glycogen synthase cannot initiate a new glycogen chain de novo. It requires a primer, usually a preformed (α1→4) polyglucose chain or branch having at least eight glucose residues. So, how is a *new* glycogen molecule initiated? The intriguing protein **glycogenin**

(Fig. 15–34) is both the primer on which new chains are assembled and the enzyme that catalyzes their assembly. The first step in the synthesis of a new glycogen molecule is the transfer of a glucose residue from UDP-glucose to the hydroxyl group of Tyr[194] of glycogenin, catalyzed by the protein's intrinsic glucosyltransferase activity **(Fig. 15–35)**. The nascent chain is extended by the sequential addition of seven more glucose residues, each derived from UDP-glucose; the reactions are catalyzed by the chain-extending activity of glycogenin. At this point, glycogen synthase takes over, further extending the glycogen chain. Glycogenin remains buried within the β-particle, covalently attached to the single reducing end of the glycogen molecule (Fig. 15–35b). Medical consequences of a mutation in the gene for glycogenin that knocks out that protein's polymerizing activity include muscle weakness and fatigue, depleted glycogen in the liver, and an irregular heartbeat (cardiac arrhythmia).

SUMMARY 15.4 The Metabolism of Glycogen in Animals

▶ Glycogen is stored in muscle and liver as large particles. Contained within the particles are the

FIGURE 15–34 Glycogenin structure. (PDB 1D 1LL2) Muscle glycogenin (M_r 37,000) forms dimers in solution. Humans have a second isoform in liver, glycogenin-2. The substrate, UDP-glucose, is bound to a Rossmann fold near the amino terminus and is some distance from the Tyr[194] residues—15 Å from the Tyr in the same monomer, 12 Å from the Tyr in the dimeric partner. Each UDP-glucose is bound through its phosphates to a Mn^{2+} ion, which is essential to catalysis. Mn^{2+} is believed to function as an electron-pair acceptor (Lewis acid) to stabilize the leaving group, UDP. The glycosidic bond in the product has the same configuration about the C-1 of glucose as the substrate UDP-glucose, suggesting that the transfer of glucose from UDP to Tyr[194] occurs in two steps. The first step is probably a nucleophilic attack by Asp[162], forming a temporary intermediate with inverted configuration. A second nucleophilic attack by Tyr[194] then restores the starting configuration.

(a)

(b)

FIGURE 15–35 Glycogenin and the structure of the glycogen particle. (a) Glycogenin catalyzes two distinct reactions. Initial attack by the hydroxyl group of Tyr^{194} on C-1 of the glucosyl moiety of UDP-glucose results in a glucosylated Tyr residue. The C-1 of another UDP-glucose molecule is now attacked by the C-4 hydroxyl group of the terminal glucose, and this sequence repeats to form a nascent glycogen molecule of eight glucose residues attached by (α1→4) glycosidic linkages. **(b)** Structure of the glycogen particle. Starting at a central glycogenin molecule, glycogen chains (12 to 14 residues) extend in tiers. Inner chains have two (α1→6) branches each. Chains in the outer tier are unbranched. There are 12 tiers in a mature glycogen particle (only 5 are shown here), consisting of about 55,000 glucose residues in a molecule of about 21 nm diameter and $M_r \sim 1 \times 10^7$.

enzymes that metabolize glycogen, as well as regulatory enzymes.

▶ Glycogen phosphorylase catalyzes phosphorolytic cleavage at the nonreducing ends of glycogen chains, producing glucose 1-phosphate. The debranching enzyme transfers branches onto main chains and releases the residue at the (α1→6) branch as free glucose.

▶ Phosphoglucomutase interconverts glucose 1-phosphate and glucose 6-phosphate. Glucose 6-phosphate can enter glycolysis or, in liver, can be converted to free glucose by glucose 6-phosphatase in the endoplasmic reticulum, then released to replenish blood glucose.

▶ The sugar nucleotide UDP-glucose donates glucose residues to the nonreducing end of glycogen in the reaction catalyzed by glycogen synthase. A separate branching enzyme produces the (α1→6) linkages at branch points.

▶ New glycogen particles begin with the autocatalytic formation of a glycosidic bond between the glucose of UDP-glucose and a Tyr residue in the protein glycogenin, followed by addition of several glucose residues to form a primer that can be acted on by glycogen synthase.

15.5 Coordinated Regulation of Glycogen Synthesis and Breakdown

As we have seen, the mobilization of stored glycogen is brought about by glycogen phosphorylase, which degrades glycogen to glucose 1-phosphate (Fig. 15–27). Glycogen phosphorylase provides an especially instructive case of enzyme regulation. It was one of the first known examples of an allosterically regulated enzyme and the first enzyme shown to be controlled by reversible phosphorylation. It was also one of the first allosteric enzymes for which the detailed three-dimensional structures of the active and inactive forms were revealed by x-ray crystallographic studies. Glycogen phosphorylase is also another illustration of how isozymes play their tissue-specific roles.

Glycogen Phosphorylase Is Regulated Allosterically and Hormonally

Earl W. Sutherland, Jr., 1915-1974

In the late 1930s, Carl and Gerty Cori (Box 15–4) discovered that the glycogen phosphorylase of skeletal muscle exists in two interconvertible forms: **glycogen phosphorylase *a***, which is catalytically active, and **glycogen phosphorylase *b***, which is less active **(Fig. 15–36)**. Subsequent studies by Earl Sutherland showed that phosphorylase *b* predominates in resting muscle, but during vigorous muscular activity epinephrine triggers phosphorylation of a specific Ser residue in phosphorylase *b*, converting it to its more active form, phosphorylase *a*. (Note that glycogen phosphorylase is often

referred to simply as phosphorylase—so honored because it was the first phosphorylase to be discovered; the shortened name has persisted in common usage and in the literature.)

The enzyme (phosphorylase *b* kinase) responsible for activating phosphorylase by transferring a phosphoryl group to its Ser residue is itself activated by epinephrine or glucagon through a series of steps shown in **Figure 15–37**. Sutherland discovered the second messenger cAMP, which increases in concentration in response to stimulation by epinephrine (in muscle) or glucagon (in liver). Elevated [cAMP] initiates an **enzyme cascade**, in which a catalyst activates a catalyst, which activates a catalyst (see Section 12.1). Such cascades allow for large amplification of the initial signal (see pink boxes in Fig. 15–37). The rise in [cAMP] activates cAMP-dependent protein kinase, also called protein kinase A (PKA). PKA then phosphorylates and activates **phosphorylase *b* kinase**, which catalyzes the phosphorylation of Ser residues in each of the two identical subunits of glycogen phosphorylase, activating it and thus stimulating glycogen breakdown. In muscle, this provides fuel for glycolysis to sustain muscle contraction for the fight-or-flight response signaled by epinephrine. In liver, glycogen breakdown counters the low blood glucose signaled by glucagon, releasing glucose. These different roles are reflected in subtle differences in the regulatory mechanisms in muscle and liver. The glycogen phosphorylases of liver and muscle are isozymes, encoded by different genes and differing in their regulatory properties.

In muscle, superimposed on the regulation of phosphorylase by covalent modification are two allosteric control mechanisms (Fig. 15–37). Ca^{2+}, the signal for muscle contraction, binds to and activates phosphorylase *b* kinase, promoting conversion of phosphorylase *b* to the active *a* form. Ca^{2+} binds to phosphorylase *b* kinase through its δ subunit, which is calmodulin (see Fig. 12–11). AMP, which accumulates in vigorously contracting muscle as a result of ATP breakdown, binds to and activates phosphorylase, speeding the release of glucose 1-phosphate from glycogen. When ATP levels are adequate, ATP blocks the allosteric site to which AMP binds, inactivating phosphorylase.

When the muscle returns to rest, a second enzyme, **phosphorylase *a* phosphatase**, also called **phosphoprotein phosphatase 1 (PP1)**, removes the phosphoryl groups from phosphorylase *a*, converting it to the less active form, phosphorylase *b*.

Like the enzyme of muscle, the glycogen phosphorylase of liver is regulated hormonally (by phosphorylation/dephosphorylation) and allosterically. The dephosphorylated form is essentially inactive. When the blood glucose level is too low, glucagon (acting through the cascade mechanism shown in Fig. 15–37) activates phosphorylase *b* kinase, which in turn converts phosphorylase *b* to its active *a* form, initiating the release of

FIGURE 15–36 Regulation of muscle glycogen phosphorylase by covalent modification. In the more active form of the enzyme, phosphorylase *a*, Ser[14] residues, one on each subunit, are phosphorylated. Phosphorylase *a* is converted to the less active form, phosphorylase *b*, by enzymatic loss of these phosphoryl groups, catalyzed by phosphorylase *a* phosphatase (also known as phosphoprotein phosphatase 1, PP1). Phosphorylase *b* can be reconverted (reactivated) to phosphorylase *a* by the action of phosphorylase *b* kinase. (See also Fig. 6–42 on glycogen phosphorylase regulation.)

FIGURE 15–37 Cascade mechanism of epinephrine and glucagon action. By binding to specific surface receptors, either epinephrine acting on a myocyte (left) or glucagon acting on a hepatocyte (right) activates a GTP-binding protein, $G_{s\alpha}$ (see Fig. 12–4). Active $G_{s\alpha}$ triggers a rise in [cAMP], activating PKA. This sets off a cascade of phosphorylations; PKA activates phosphorylase *b* kinase, which then activates glycogen phosphorylase. Such cascades effect a large amplification of the initial signal; the figures in pink boxes are probably low estimates of the actual increase in number of molecules at each stage of the cascade. The resulting breakdown of glycogen provides glucose, which in the myocyte can supply ATP (via glycolysis) for muscle contraction and in the hepatocyte is released into the blood to counter the low blood glucose.

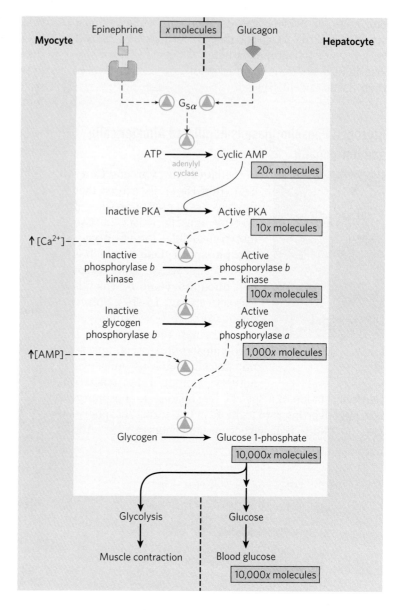

glucose into the blood. When blood glucose levels return to normal, glucose enters hepatocytes and binds to an inhibitory allosteric site on phosphorylase *a*. This binding also produces a conformational change that exposes the phosphorylated Ser residues to PP1, which catalyzes their dephosphorylation and inactivates the phosphorylase **(Fig. 15–38)**. The allosteric site for glucose allows liver glycogen phosphorylase to act as its own glucose sensor and to respond appropriately to changes in blood glucose.

FIGURE 15–38 Glycogen phosphorylase of liver as a glucose sensor. Glucose binding to an allosteric site of the phosphorylase *a* isozyme of liver induces a conformational change that exposes its phosphorylated Ser residues to the action of phosphorylase *a* phosphatase (PP1). This phosphatase converts phosphorylase *a* to phosphorylase *b*, sharply reducing the activity of phosphorylase and slowing glycogen breakdown in response to high blood glucose. Insulin also acts indirectly to stimulate PP1 and slow glycogen breakdown.

FIGURE 15–39 Effects of GSK3 on glycogen synthase activity. Glycogen synthase *a*, the active form, has three Ser residues near its carboxyl terminus, which are phosphorylated by glycogen synthase kinase 3 (GSK3). This converts glycogen synthase to the inactive (*b*) form. GSK3 action requires prior phosphorylation (priming) by casein kinase (CKII). Insulin triggers activation of glycogen synthase *b* by blocking the activity of GSK3 (see the pathway for this action in Fig. 12–16) and activating a phosphoprotein phosphatase (PP1 in muscle, another phosphatase in liver). In muscle, epinephrine activates PKA, which phosphorylates the glycogen-targeting protein GM (see Fig. 15–42) on a site that causes dissociation of PP1 from glycogen. Glucose 6-phosphate favors dephosphorylation of glycogen synthase by binding to it and promoting a conformation that is a good substrate for PP1. Glucose also promotes dephosphorylation; the binding of glucose to glycogen phosphorylase *a* forces a conformational change that favors dephosphorylation to glycogen phosphorylase *b*, thus allowing the action of PP1 (see Fig. 15–41).

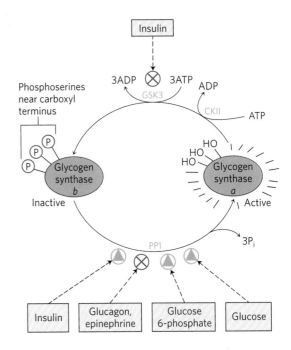

Glycogen Synthase Is Also Regulated by Phosphorylation and Dephosphorylation

Like glycogen phosphorylase, glycogen synthase can exist in phosphorylated and dephosphorylated forms **(Fig. 15–39)**. Its active form, **glycogen synthase *a***, is unphosphorylated. Phosphorylation of the hydroxyl side chains of several Ser residues of both subunits converts glycogen synthase *a* to **glycogen synthase *b***, which is inactive unless its allosteric activator, glucose 6-phosphate, is present. Glycogen synthase is remarkable for its ability to be phosphorylated on various residues by at least 11 different protein kinases. The most important regulatory kinase is **glycogen synthase kinase 3 (GSK3)**, which adds phosphoryl groups to three Ser residues near the carboxyl termi-

nus of glycogen synthase, strongly inactivating it. The action of GSK3 is hierarchical; it cannot phosphorylate glycogen synthase until another protein kinase, **casein kinase II (CKII)**, has first phosphorylated the glycogen synthase on a nearby residue, an event called **priming (Fig. 15–40a)**.

In liver, conversion of glycogen synthase *b* to the active form is promoted by PP1, which is bound to the glycogen particle. PP1 removes the phosphoryl groups from the three Ser residues phosphorylated by GSK3. Glucose 6-phosphate binds to an allosteric site on

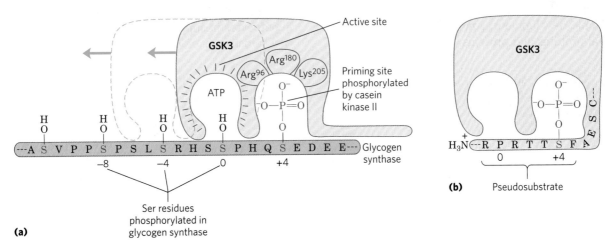

FIGURE 15–40 Priming of GSK3 phosphorylation of glycogen synthase. **(a)** Glycogen synthase kinase 3 first associates with its substrate (glycogen synthase) by interaction between three positively charged residues (Arg^{96}, Arg^{180}, Lys^{205}) and a phosphoserine residue at position +4 in the substrate. (For orientation, the Ser or Thr residue to be phosphorylated in the substrate is assigned the index 0. Residues on the amino-terminal side of this residue are numbered −1, −2, and so forth; residues on the carboxyl-terminal side are numbered +1, +2, and so forth.) This association aligns the active site of the enzyme with a Ser residue at position 0, which it phosphorylates. This creates a new priming site,

and the enzyme moves down the protein to phosphorylate the Ser residue at position −4, and then the Ser at −8. **(b)** GSK3 has a Ser residue near its amino terminus that can be phosphorylated by PKA or PKB (see Fig. 15–41). This produces a "pseudosubstrate" region in GSK3 that folds into the priming site and makes the active site inaccessible to another protein substrate, inhibiting GSK3 until the priming phosphoryl group of its pseudosubstrate region is removed by PP1. Other proteins that are substrates for GSK3 also have a priming site at position +4, which must be phosphorylated by another protein kinase before GSK3 can act on them. (See also Figs 6–37 and 12–22b on glycogen synthase regulation.)

glycogen synthase *b*, making the enzyme a better substrate for dephosphorylation by PP1 and causing its activation. By analogy with glycogen phosphorylase, which acts as a glucose sensor, glycogen synthase can be regarded as a glucose 6-phosphate sensor. In muscle, a different phosphatase may have the role played by PP1 in liver, activating glycogen synthase by dephosphorylating it.

Glycogen Synthase Kinase 3 Mediates Some of the Actions of Insulin

As we saw in Chapter 12, one way in which insulin triggers intracellular changes is by activating a protein kinase (PKB) that in turn phosphorylates and inactivates GSK3 (**Fig. 15–41**; see also Fig. 12–16). Phosphorylation of a Ser residue near the amino terminus of GSK3 converts that region of the protein to a pseudosubstrate, which folds into the site at which the priming phosphorylated Ser residue normally binds (Fig. 15–40b). This prevents GSK3 from binding the priming site of a real substrate, thereby inactivating the enzyme and tipping the balance in favor of dephosphorylation of glycogen synthase by PP1. Glycogen phosphorylase can also affect the phosphorylation of glycogen synthase: active glycogen phosphorylase directly inhibits PP1, preventing it from activating glycogen synthase (Fig. 15–39).

Although first discovered in its role in glycogen metabolism (hence the name glycogen synthase kinase), GSK3 clearly has a much broader role than the regulation of glycogen synthase. It mediates signaling by insulin and other growth factors and nutrients, and it acts in the specification of cell fates during embryonic development. Among its targets are cytoskeletal proteins and proteins essential for mRNA and protein synthesis. These targets, like glycogen synthase, must first undergo a priming phosphorylation by another protein kinase before they can be phosphorylated by GSK3.

Phosphoprotein Phosphatase 1 Is Central to Glycogen Metabolism

A single enzyme, PP1, can remove phosphoryl groups from all three of the enzymes phosphorylated in response to glucagon (liver) and epinephrine (liver and muscle): phosphorylase kinase, glycogen phosphorylase, and glycogen synthase. Insulin stimulates glycogen synthesis by activating PP1 and by inactivating GSK3.

Phosphoprotein phosphatase 1 does not exist free in the cytosol, but is tightly bound to its target proteins by one of a family of **glycogen-targeting proteins** that bind glycogen and each of the three enzymes, glycogen phosphorylase, phosphorylase kinase, and glycogen synthase (**Fig. 15–42**). PP1 is itself subject to covalent and allosteric regulation: it is inactivated when phosphorylated by PKA and is allosterically activated by glucose 6-phosphate.

Allosteric and Hormonal Signals Coordinate Carbohydrate Metabolism Globally

Having looked at the mechanisms that regulate individual enzymes, we can now consider the overall shifts in carbohydrate metabolism that occur in the well-fed state, during fasting, and in the fight-or-flight response—signaled by insulin, glucagon, and epinephrine, respectively. We need to contrast two cases in which regulation serves different ends: (1) the role of hepatocytes in supplying glucose to the blood, and (2) the selfish use of carbohydrate fuels by nonhepatic

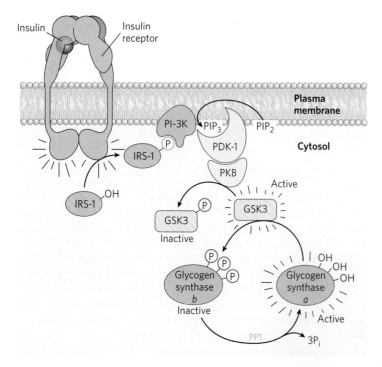

FIGURE 15–41 The path from insulin to GSK3 and glycogen synthase. Insulin binding to its receptor activates a tyrosine protein kinase in the receptor, which phosphorylates insulin receptor substrate-1 (IRS-1). The phosphotyrosine in this protein is then bound by phosphatidylinositol 3-kinase (PI-3K), which converts phosphatidylinositol 4,5-bisphosphate (PIP$_2$) in the membrane to phosphatidylinositol 3,4,5-trisphosphate (PIP$_3$). A protein kinase (PDK-1) that is activated when bound to PIP$_3$ activates a second protein kinase (PKB), which phosphorylates glycogen synthase kinase 3 (GSK3) in its pseudosubstrate region, inactivating it by the mechanism shown in Figure 15–40b. The inactivation of GSK3 allows phosphoprotein phosphatase 1 (PP1) to dephosphorylate and thus activate glycogen synthase. In this way, insulin stimulates glycogen synthesis. (See Fig. 12–16 for more details on insulin action.)

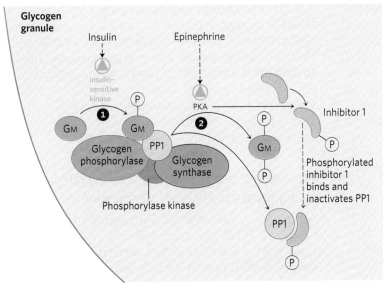

FIGURE 15–42 Glycogen-targeting protein Gм. The glycogen-targeting protein Gм is one of a family of proteins that bind other proteins (including PP1) to glycogen particles. Gм can be phosphorylated at two different sites in response to insulin or epinephrine. ❶ Insulin-stimulated phosphorylation of Gм site 1 activates PP1, which dephosphorylates phosphorylase kinase, glycogen phosphorylase, and glycogen synthase. ❷ Epinephrine-stimulated phosphorylation of Gм site 2 causes dissociation of PP1 from the glycogen particle, preventing its access to glycogen phosphorylase and glycogen synthase. PKA also phosphorylates a protein (inhibitor 1) that, when phosphorylated, inhibits PP1. By these means, insulin inhibits glycogen breakdown and stimulates glycogen synthesis, and epinephrine (or glucagon in the liver) has the opposite effects.

tissues, typified by skeletal muscle (myocytes), to support their own activities.

After ingestion of a carbohydrate-rich meal, the elevation of blood glucose triggers insulin release (**Fig. 15–43**, top). In a hepatocyte, insulin has two immediate effects: it inactivates GSK3, acting through the cascade shown in Figure 15–41, and activates a protein phosphatase, perhaps PP1. These two actions fully activate glycogen synthase. PP1 also inactivates glycogen phosphorylase a and phosphorylase kinase by dephosphorylating both, effectively stopping glycogen breakdown. Glucose enters the hepatocyte through the high-capacity transporter GLUT2, always present in the plasma membrane, and the elevated intracellular glucose leads to dissociation of hexokinase IV (glucokinase) from its nuclear regulatory protein (Fig. 15–15). Hexokinase IV enters the cytosol and phosphorylates glucose, stimulating glycolysis and supplying the precursor for glycogen synthesis. Under these conditions, hepatocytes use the excess glucose in the blood to synthesize glycogen, up to the limit of about 10% of the total weight of the liver.

Between meals, or during an extended fast, the drop in blood glucose triggers the release of glucagon, which, acting through the cascade shown in Figure 15–37, activates PKA. PKA mediates all the effects of glucagon (Fig. 15–43, bottom). It phosphorylates phosphorylase kinase, activating it and leading to the activation of glycogen phosphorylase. It phosphorylates glycogen synthase, inactivating it and blocking glycogen synthesis. It phosphorylates PFK-2/FBPase-2, leading to a

FIGURE 15–43 Regulation of carbohydrate metabolism in the liver. Arrows indicate causal relationships between the changes they connect. For example, an arrow from ↓A to ↑B means that a decrease in A causes an increase in B. Red arrows connect events that result from high blood glucose; blue arrows connect events that result from low blood glucose.

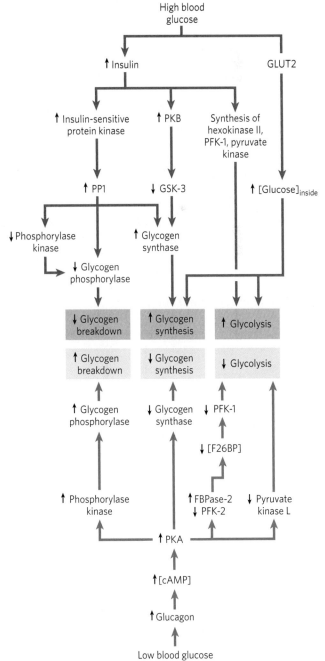

drop in the concentration of the regulator fructose 2,6-bisphosphate, which has the effect of inactivating the glycolytic enzyme PFK-1 and activating the gluconeogenic enzyme FBPase-1. And it phosphorylates and inactivates the glycolytic enzyme pyruvate kinase. Under these conditions, the liver produces glucose 6-phosphate by glycogen breakdown and by gluconeogenesis, and it stops using glucose to fuel glycolysis or make glycogen, maximizing the amount of glucose it can release to the blood. This release of glucose is possible only in liver and kidney, because other tissues lack glucose 6-phosphatase (Fig. 15–30).

The physiology of skeletal muscle differs from that of liver in three ways important to our discussion of metabolic regulation (**Fig. 15–44**): (1) muscle uses its stored glycogen only for its own needs; (2) as it goes from rest to vigorous contraction, muscle undergoes very large changes in its demand for ATP, which is supported by glycolysis; (3) muscle lacks the enzymatic machinery for gluconeogenesis. The regulation of carbohydrate metabolism in muscle reflects these differences from liver. First, myocytes lack receptors for glucagon. Second, the muscle isozyme of pyruvate kinase is not phosphorylated by PKA, so glycolysis is not turned off when [cAMP] is high. In fact, cAMP *increases* the rate of glycolysis in muscle, probably by activating glycogen phosphorylase. When epinephrine is released into the blood in a fight-or-flight situation, PKA is activated by the rise in [cAMP] and phosphorylates and activates glycogen phosphorylase kinase. The resulting phosphorylation and activation of glycogen phosphorylase results in faster glycogen breakdown. Epinephrine is not released under low-stress conditions, but with each neuronal stimulation of muscle contraction, cytosolic [Ca^{2+}] rises briefly and activates phosphorylase kinase through its calmodulin subunit.

Elevated insulin triggers increased glycogen synthesis in myocytes by activating PP1 and inactivating GSK3. Unlike hepatocytes, myocytes have a reserve of GLUT4 sequestered in intracellular vesicles. Insulin triggers their movement to the plasma membrane (see Fig. 12–16), where they allow increased glucose uptake. In response to insulin, therefore, myocytes help to lower blood glucose by increasing their rates of glucose uptake, glycogen synthesis, and glycolysis.

Carbohydrate and Lipid Metabolism Are Integrated by Hormonal and Allosteric Mechanisms

As complex as the regulation of carbohydrate metabolism is, it is far from the whole story of fuel metabolism. The metabolism of fats and fatty acids is very closely tied to that of carbohydrates. Hormonal signals such as insulin and changes in diet or exercise are equally important in regulating fat metabolism and integrating it with that of carbohydrates. We return to this overall metabolic integration in mammals in Chapter 23, after first considering the metabolic pathways for fats and amino acids (Chapters 17 and 18). The message we wish to convey here is that metabolic pathways are overlaid with complex regulatory controls that are exquisitely sensitive to changes in metabolic circumstances. These mechanisms act to adjust the flow of metabolites through various metabolic pathways, as needed by the cell and organism, and to do so without causing major changes in the concentrations of intermediates shared with other pathways.

SUMMARY 15.5 Coordinated Regulation of Glycogen Synthesis and Breakdown

▶ Glycogen phosphorylase is activated in response to glucagon or epinephrine, which raise [cAMP] and activate PKA. PKA phosphorylates and activates phosphorylase kinase, which converts glycogen phosphorylase *b* to its active *a* form. Phosphoprotein phosphatase 1 (PP1) reverses the phosphorylation of glycogen phosphorylase *a*, inactivating it. Glucose binds to the liver isozyme of glycogen phosphorylase *a*, favoring its dephosphorylation and inactivation.

▶ Glycogen synthase *a* is inactivated by phosphorylation catalyzed by GSK3. Insulin blocks GSK3. PP1, which is activated by insulin, reverses the inhibition by dephosphorylating glycogen synthase *b*.

▶ Insulin increases glucose uptake into myocytes and adipocytes by triggering movement of the glucose transporter GLUT4 to the plasma membrane.

▶ Insulin stimulates the synthesis of hexokinases II and IV, PFK-1, pyruvate kinase, and several

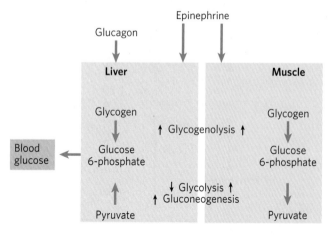

FIGURE 15–44 Difference in the regulation of carbohydrate metabolism in liver and muscle. In liver, either glucagon (indicating low blood glucose) or epinephrine (signaling the need to fight or flee) has the effect of maximizing the output of glucose into the bloodstream. In muscle, epinephrine increases glycogen breakdown and glycolysis, which together provide fuel to produce the ATP needed for muscle contraction.

enzymes involved in lipid synthesis. Insulin stimulates glycogen synthesis in muscle and liver.

▶ In liver, glucagon stimulates glycogen breakdown and gluconeogenesis while blocking glycolysis, thereby sparing glucose for export to the brain and other tissues.

▶ In muscle, epinephrine stimulates glycogen breakdown and glycolysis, providing ATP to support contraction.

Key Terms

Terms in bold are defined in the glossary.

glucose 6-phosphate   587
flux, *J*   589
homeostasis   589
cellular
   differentiation   589
transcription factor   589
response element   589
turnover   590
transcriptome   590
proteome   590
metabolome   591
metabolic regulation   592
metabolic control   592
mass-action ratio, Q   593
adenylate kinase   594
AMP-activated protein
   kinase (AMPK)   594
flux control
   coefficient, *C*   597, 598
elasticity coefficient, *ε*
   597, 598
response coefficient, *R*   598
gluconeogenesis   601
futile cycle   601
substrate cycle   601
hexokinase II   602
hexokinase I   602
hexokinase IV   603
GLUT2   603
glucagon   605
fructose
   2,6-bisphosphate   605
phosphofructokinase-2
   (PFK-2)   606
fructose 2,6-bisphosphatase
   (FBPase-2)   606
carbohydrate response
   element binding protein
   (ChREBP)   609

sterol regulatory element
   binding protein
   (SREBP)   610
cyclic AMP response element
   binding protein
   (CREB)   610
forkhead box other
   (FOXO1)   610
glycogenolysis   613
glycogenesis   613
glucose 1-phosphate   613
debranching enzyme   614
oligo (α1→6) to (α1→4)
   glucantransferase   614
phosphoglucomutase   614
sugar nucleotides   615
UDP-glucose
   pyrophosphorylase   617
amylo (1→4) to (1→6)
   transglycosylase   619
glycogenin   619
glycogen
   phosphorylase *a*   621
glycogen
   phosphorylase *b*   621
enzyme cascade   621
phosphorylase *b*
   kinase   621
phosphoprotein
   phosphatase 1 (PP1)   621
glycogen synthase *a*   623
glycogen synthase *b*   623
glycogen synthase
   kinase 3 (GSK3)   623
casein kinase II
   (CKII)   623
priming   623
glycogen-targeting
   proteins   624

Further Reading

Regulation of Metabolic Pathways

Desvergne, B., Michalik, L., & Wahli, W. (2006) Transcriptional regulation of metabolism. *Physiol. Rev.* **86**, 465–514.
   Advanced and comprehensive review.

Gibson, D. & Harris, R.A. (2001) *Metabolic Regulation in Mammals*, Taylor & Francis, New York.
   Excellent, readable account of metabolic regulation.

Li, X. & Snyder, M. (2011) Metabolites as global regulators: a new view of protein regulation. *Bioessays* **33**, 485–489.

Naïmi, M., Arous, C., & Obberghen, E. (2010) Energetic cell sensors: a key to metabolic homeostasis. *Trends Endocrinol. Med.* **21**, 75–82.

Storey, K.B. (ed.). (2004) *Functional Metabolism: Regulation and Adaptation*, Wiley-Liss, Inc., Hoboken, NJ.
   Excellent discussion of the principles of metabolic regulation, signal transduction, transcriptional control, and energy metabolism in health and disease.

Analysis of Metabolic Control

Castrillo, J.I. & Oliver, S.G. (2007) Metabolic control in the eukaryotic cell, a systems biology perspective. *Meth. Microbiol.* **36**, 527–549. doi:10.1016/S0580-9517(06)36021-7

Fell, D.A. (1992) Metabolic control analysis: a survey of its theoretical and experimental development. *Biochem. J.* **286**, 313–330.
   Clear statement of the principles of metabolic control analysis.

Fell, D.A. (1997) *Understanding the Control of Metabolism*, Portland Press, Ltd., London.
   An excellent, clear exposition of metabolic regulation, from the point of view of metabolic control analysis. If you read only one treatment on metabolic control analysis, this should be it.

Heinrich, R. & Rapoport, T.A. (1974) A linear steady-state treatment of enzymatic chains: general properties, control and effector strength. *Eur. J. Biochem.* **42**, 89–95.
   Early statement of principles of metabolic control analysis. See also the paper by Kacser & Burns, listed below.

Jeffrey, F.M.H., Rajagopal, A., Maloy, C.R., & Sherry, A.D. (1991) ^{13}C-NMR: a simple yet comprehensive method for analysis of intermediary metabolism. *Trends Biochem. Sci.* **16**, 5–10.
   Brief, intermediate-level review.

Kacser, H. & Burns, J.A. (1973) The control of flux. *Symp. Soc. Exp. Biol.* **32**, 65–104.
   A classic paper in the field. See also the paper by Heinrich & Rapoport, listed above.

Kacser, H., Burns, J.A., & Fell, D.A. (1995) The control of flux: 21 years on. *Biochem. Soc. Trans.* **23**, 341–366.

Saavedra, E., Rodriguez-Enriquesz, S., Quezada, H., Jasso-Chavez, R., & Moreno-Sanchez, R. (2011) Rational design of strategies based on metabolic control analysis. In *Comprehensive Biotechnology*, 2nd edn, Vol. 1 (Moo-Young, M., Butler, M., Webb. C., Moreira, A., Grodzinski, B., Cui, Z.E., and Agathos, S., eds), pp. 511–524. Pergamon Press, Elmsford, NY.

Schilling, C.H., Schuster, S., Palsson, B.O., & Heinrich, R. (1999) Metabolic pathway analysis: basic concepts and scientific applications in the post-genomic era. *Biotechnol. Prog.* **15**, 296–303.
   Short, advanced discussion of theoretical treatments that attempt to find ways of manipulating metabolism to optimize the formation of metabolic products.

Schuster, S., Fell, D.A., & Dandekar, T. (2000) A general definition of metabolic pathways useful for systematic organization and analysis of complex metabolic networks. *Nat. Biotechnol.* **18**, 326–332.
   An interesting and provocative analysis of the interplay between the pentose phosphate pathway and glycolysis, from a theoretical standpoint.

Sohn, S.B., Kim, T.Y., Kim, J.M., Park, J.M., & Lee, S.Y. (2011) Metabolic control. In *Comprehensive Biotechnology*, 2nd edn, Vol. 2 (Moo-Young, M., Butler, M., Webb, C., Moreira, A., Grodzinski, B., Cui, Z.F., and Agathos, S., eds), pp. 853–861. Pergamon Press, Elmsford, NY.

Varma, A. & Palsson, B.O. (1994) Metabolic flux balancing: basic concepts, scientific and practical use. *Bio/Technology* **12**, 994–98.

Westerhoff, H.V., Hofmeyr, J.-H.S., & Kholodenko, B.N. (1994) Getting to the inside of cells using metabolic control analysis. *Biophys. Chem.* **50**, 273–283.

Coordinated Regulation of Glycolysis and Gluconeogenesis

Armoni, M., Harel, C., & Karnieli, E. (2007) Transcriptional regulation of the GLUT4 gene: from PPAR-gamma and FOXO1 to FFA and inflammation. *Trends Endocrinol. Metab.* **18**, 100–107.

Barthel, A., Schmoll, D., & Unterman, T.G. (2005) FoxO proteins in insulin action and metabolism. *Trends Endocrinol. Metab.* **16**, 183–189.
   Intermediate-level review of the transcription factor's effects on carbohydrate metabolism.

Brady, M.J., Pessin, J.E., & Saltiel, A.R. (1999) Spatial compartmentalization in the regulation of glucose metabolism by insulin. *Trends Endocrinol. Metab.* **10**, 408–413.
   Intermediate-level review.

Carling, D. (2004) The AMP-activated protein kinase cascade—a unifying system for energy control. *Trends Biochem. Sci.* **29**, 18–24.
   Intermediate-level review of AMPK and its role in energy metabolism.

de la Iglesia, N., Mukhtar, M., Seoane, J., Guinovart, J.J., & Agius, L. (2000) The role of the regulatory protein of glucokinase in the glucose sensory mechanism of the hepatocyte. *J. Biol. Chem.* **275**, 10,597–10,603.
   Report of the experimental determination of the flux control coefficients for glucokinase and the glucokinase regulatory protein in hepatocytes.

Dean, L. & McEntyre, J. (2004) *The Genetic Landscape of Diabetes*, National Center for Biotechnology Information, www.ncbi.nlm.nih.gov/books/bv.fcgi?rid=diabetes.
   An excellent, highly readable, downloadable book (free). It includes an introduction to diabetes, a history of studies of diabetes, and chapters on the genetic factors in IDDM, NIDDM, and MODY.

Desverne, B., Michalik, L., & Wahli, W. (2006) Transcriptional regulation of metabolism. *Physiol. Rev.* **86**, 465–514.
   Extensive, advanced review of transcription factors, including those that regulate carbohydrate and fat metabolism.

Grüning, N.-M., Lehrach, H., & Ralser, M. (2010) Regulatory crosstalk of the metabolic network. *Trends Biochem. Sci.* **35**, 220–227.

Hardie, D.G. (2007) AMP-activated protein kinase as a drug target. *Annu. Rev. Pharmacol. Toxicol.* **47**, 185–210.
   Advanced review, with emphasis on the possible role of this enzyme in type II diabetes.

Herman, M.A. & Kahn, B.B. (2006) Glucose transport and sensing in the maintenance of glucose homeostasis and metabolic harmony. *J. Clin. Invest.* **116**, 1767–1775.
   Beautifully illustrated, intermediate-level review.

Hers, H.G. & Van Schaftingen, E. (1982) Fructose 2,6-bisphosphate 2 years after its discovery. *Biochem. J.* **206**, 1–12.
   Classic description of this regulatory molecule and its role in regulating carbohydrate metabolism.

Hue, L. & Rider, M.H. (1987) Role of fructose 2,6-bisphosphate in the control of glycolysis in mammalian tissues. *Biochem. J.* **245**, 313–324.

Jorgensen, S.B., Richter, E.A., & Wojtaszewski, J.F.P. (2006) Role of AMPK in skeletal muscle metabolic regulation and adaptation in relation to exercise. *J. Physiol.* **574**, 17–31.

Kahn, B.B., Alquier, T., Carling, D., & Hardie, D.G. (2005) AMP-activated protein kinase: ancient energy gauge provides clues to modern understanding of metabolism. *Cell Metab.* **1**, 15–25.
   Well-illustrated, intermediate-level review.

Long, Y.C. & Zierath, J.R. (2006) AMP-activated protein kinase signaling in metabolic regulation. *J. Clin. Invest.* **116**, 1776–1783.
   Advanced, short review of AMPK role in metabolism, including data on knockout mice.

Nordlie, R.C., Foster, J.D., & Lange, A.J. (1999) Regulation of glucose production by the liver. *Annu. Rev. Nutr.* **19**, 379–406.
   Advanced review.

Okar, D.A., Manzano, A., Navarro-Sabate, A., Riera, L., Bartrons, R., & Lange, A.J. (2001) PFK-2/FBPase-2: maker and breaker of the essential biofactor fructose-2,6-bisphosphate. *Trends Biochem. Sci.* **26**, 30–35.

Pilkis, S.J. & Granner, D.K. (1992) Molecular physiology of the regulation of hepatic gluconeogenesis and glycolysis. *Annu. Rev. Physiol.* **54**, 885–909.

Postic, C., Dentin, R., Denechaud, P.-D., & Girard, J. (2007) ChREBP, a transcriptional regulator of glucose and lipid metabolism. *Annu. Rev. Nutr.* **27**, 179–192.
   Advanced review of the role of transcription factor ChREBP in carbohydrate metabolism.

Rider, M.H. & Bartrons, R. (2010) Fructose 2,6-bisphosphate: the last milestone of the 20th century in metabolic control? *Biochem. J.* doi:10.1042/BJ20091921.

Schirmer, T. & Evans, P.R. (1990) Structural basis of the allosteric behavior of phosphofructokinase. *Nature* **343**, 140–145.

Sola-Penna, M., Da Silva, D., Coelho, W.S., Marinho-Carvalho, M.M., & Zancan, P. (2010) Regulation of mammalian muscle type 6-phosphofructo-1-kinase and its implication for the control of metabolism. *IUBMB Life* **62**, 791–796.

Towle, H.C. (2005) Glucose as a regulator of eukaryotic gene transcription. *Trends Endocrinol. Metab.* **16**, 489–494.
   Intermediate-level review.

Towler, M.C. & Hardie, D.G. (2007) AMP-activated protein kinase in metabolic control and insulin signaling. *Circ. Res.* **100**, 328–341.

van Shaftingen, E. & Gerin, I. (2002) The glucose-6-phosphatase system. *Biochem. J.* **362**, 513–532.

Veech, R.L. (2003) A humble hexose monophosphate pathway metabolite regulates short- and long-term control of lipogenesis. *Proc. Natl. Acad. Sci. USA* **100**, 5578–5580.
   Short review of the work from K. Uyeda's laboratory on the role of xylulose 5-phosphate in carbohydrate and fat metabolism; Uyeda's papers are cited in this review.

Yamada, K. & Noguchi, T. (1999) Nutrient and hormonal regulation of pyruvate kinase gene expression. *Biochem. J.* **337**, 1–11.

The Metabolism of Glycogen in Animals

Moslemi, A.-R., Lindberg, C., Nilsson, J., Tajsharghi, H., Andersson, B., & Oldfors, A. (2010) Glycogenin-1 deficiency and inactivated priming of glycogen synthesis. *N. Engl. J. Med.* **362**, 1203–1210.

Whelan, W.J. (1976) On the origin of primer for glycogen synthesis. *Trends Biochem. Sci.* **1**, 13–15.
   Intermediate review of the discovery, properties, and role of glycogenin.

Coordinated Regulation of Glycogen Synthesis and Breakdown

Aiston, S., Hampson, L., Gomez-Foix, A.M., Guinovart, J.J., & Agius, L. (2001) Hepatic glycogen synthesis is highly sensitive to phosphorylase activity: evidence from metabolic control analysis. *J. Biol. Chem.* **276**, 23,858–23,866.

Jope, R.S. & Johnson, G.V.W. (2004) The glamour and gloom of glycogen synthase kinase-3. *Trends Biochem. Sci.* **29**, 95–102.
Intermediate-level, well-illustrated review.

Problems

1. Measurement of Intracellular Metabolite Concentrations Measuring the concentrations of metabolic intermediates in a living cell presents great experimental difficulties—usually a cell must be destroyed before metabolite concentrations can be measured. Yet enzymes catalyze metabolic interconversions very rapidly, so a common problem associated with these types of measurements is that the findings reflect not the physiological concentrations of metabolites but the equilibrium concentrations. A reliable experimental technique requires all enzyme-catalyzed reactions to be instantaneously stopped in the intact tissue so that the metabolic intermediates do not undergo change. This objective is accomplished by rapidly compressing the tissue between large aluminum plates cooled with liquid nitrogen (−190 °C), a process called **freeze-clamping**. After freezing, which stops enzyme action instantly, the tissue is powdered and the enzymes are inactivated by precipitation with perchloric acid. The precipitate is removed by centrifugation, and the clear supernatant extract is analyzed for metabolites. To calculate intracellular concentrations, the intracellular volume is determined from the total water content of the tissue and a measurement of the extracellular volume.

The intracellular concentrations of the substrates and products of the phosphofructokinase-1 reaction in isolated rat heart tissue are given in the table below.

Metabolite	Concentration (μM)*
Fructose 6-phosphate	87.0
Fructose 1,6-bisphosphate	22.0
ATP	11,400
ADP	1,320

Source: From Williamson, J.R. (1965) Glycolytic control mechanisms I: inhibition of glycolysis by acetate and pyruvate in the isolated, perfused rat heart. *J. Biol. Chem.* **240**, 2308–2321.

*Calculated as μmol/mL of intracellular water.

(a) Calculate Q, [fructose 1,6-bisphosphate][ADP]/[fructose 6-phosphate][ATP], for the PFK-1 reaction under physiological conditions.

(b) Given a $\Delta G'°$ for the PFK-1 reaction of −14.2 kJ/mol, calculate the equilibrium constant for this reaction.

(c) Compare the values of Q and K'_{eq}. Is the physiological reaction near or far from equilibrium? Explain. What does this experiment suggest about the role of PFK-1 as a regulatory enzyme?

2. Are All Metabolic Reactions at Equilibrium?

(a) Phosphoenolpyruvate (PEP) is one of the two phosphoryl group donors in the synthesis of ATP during glycolysis. In human erythrocytes, the steady-state concentration of ATP is 2.24 mM, that of ADP is 0.25 mM, and that of pyruvate is 0.051 mM. Calculate the concentration of PEP at 25 °C, assuming that the pyruvate kinase reaction (see Fig. 13–13) is at equilibrium in the cell.

(b) The physiological concentration of PEP in human erythrocytes is 0.023 mM. Compare this with the value obtained in (a). Explain the significance of this difference.

3. Effect of O_2 Supply on Glycolytic Rates The regulated steps of glycolysis in intact cells can be identified by studying the catabolism of glucose in whole tissues or organs. For example, the glucose consumption by heart muscle can be measured by artificially circulating blood through an isolated intact heart and measuring the concentration of glucose before and after the blood passes through the heart. If the circulating blood is deoxygenated, heart muscle consumes glucose at a steady rate. When oxygen is added to the blood, the rate of glucose consumption drops dramatically, then is maintained at the new, lower rate. Explain.

4. Regulation of PFK-1 The effect of ATP on the allosteric enzyme PFK-1 is shown below. For a given concentration of fructose 6-phosphate, the PFK-1 activity increases with increasing concentrations of ATP, but a point is reached beyond which increasing the concentration of ATP inhibits the enzyme.

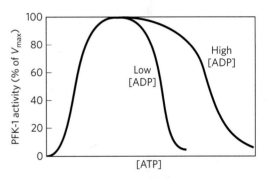

(a) Explain how ATP can be both a substrate and an inhibitor of PFK-1. How is the enzyme regulated by ATP?

(b) In what ways is glycolysis regulated by ATP levels?

(c) The inhibition of PFK-1 by ATP is diminished when the ADP concentration is high, as shown in the illustration. How can this observation be explained?

5. Cellular Glucose Concentration The concentration of glucose in human blood plasma is maintained at about 5 mM. The concentration of free glucose inside a myocyte is much lower. Why is the concentration so low in the cell? What happens to glucose after entry into the cell? Glucose is administered intravenously as a food source in certain clinical situations. Given that the transformation of glucose to glucose 6-phosphate consumes ATP, why not administer intravenous glucose 6-phosphate instead?

6. Enzyme Activity and Physiological Function The V_{max} of the glycogen phosphorylase from skeletal muscle is much greater than the V_{max} of the same enzyme from liver tissue.

(a) What is the physiological function of glycogen phosphorylase in skeletal muscle? In liver tissue?

(b) Why does the V_{max} of the muscle enzyme need to be greater than that of the liver enzyme?

7. Glycogen Phosphorylase Equilibrium Glycogen phosphorylase catalyzes the removal of glucose from glycogen. The $\Delta G'^{\circ}$ for this reaction is 3.1 kJ/mol.

(a) Calculate the ratio of $[P_i]$ to [glucose 1-phosphate] when the reaction is at equilibrium. (Hint: The removal of glucose units from glycogen does not change the glycogen concentration.)

(b) The measured ratio $[P_i]$/[glucose 1-phosphate] in myocytes under physiological conditions is more than 100:1. What does this indicate about the direction of metabolite flow through the glycogen phosphorylase reaction in muscle?

(c) Why are the equilibrium and physiological ratios different? What is the possible significance of this difference?

8. Regulation of Glycogen Phosphorylase In muscle tissue, the rate of conversion of glycogen to glucose 6-phosphate is determined by the ratio of phosphorylase a (active) to phosphorylase b (less active). Determine what happens to the rate of glycogen breakdown if a muscle preparation containing glycogen phosphorylase is treated with (a) phosphorylase kinase and ATP; (b) PP1; (c) epinephrine.

9. Glycogen Breakdown in Rabbit Muscle The intracellular use of glucose and glycogen is tightly regulated at four points. To compare the regulation of glycolysis when oxygen is plentiful and when it is depleted, consider the utilization of glucose and glycogen by rabbit leg muscle in two physiological settings: a resting rabbit, with low ATP demands, and a rabbit that sights its mortal enemy, the coyote, and dashes into its burrow. For each setting, determine the relative levels (high, intermediate, or low) of AMP, ATP, citrate, and acetyl-CoA and describe how these levels affect the flow of metabolites through glycolysis by regulating specific enzymes. In periods of stress, rabbit leg muscle produces much of its ATP by anaerobic glycolysis (lactate fermentation) and very little by oxidation of acetyl-CoA derived from fat breakdown.

10. Glycogen Breakdown in Migrating Birds Unlike the rabbit with its short dash, migratory birds require energy for extended periods of time. For example, ducks generally fly several thousand miles during their annual migration. The flight muscles of migratory birds have a high oxidative capacity and obtain the necessary ATP through the oxidation of acetyl-CoA (obtained from fats) via the citric acid cycle. Compare the regulation of muscle glycolysis during short-term intense activity, as in the fleeing rabbit, and during extended activity, as in the migrating duck. Why must the regulation in these two settings be different?

11. Enzyme Defects in Carbohydrate Metabolism Summaries of four clinical case studies follow. For each case determine which enzyme is defective and designate the appropriate treatment, from the lists provided at the end of the problem. Justify your choices. Answer the questions contained in each case study. (You may need to refer to information in Chapter 14.)

Case A The patient develops vomiting and diarrhea shortly after milk ingestion. A lactose tolerance test is administered. (The patient ingests a standard amount of lactose, and the glucose and galactose concentrations of blood plasma are measured at intervals. In individuals with normal carbohydrate metabolism, the levels increase to a maximum in about 1 hour, then decline.) The patient's blood glucose and galactose concentrations do not increase during the test. Why do blood glucose and galactose increase and then decrease during the test in healthy individuals? Why do they fail to rise in the patient?

Case B The patient develops vomiting and diarrhea after ingestion of milk. His blood is found to have a low concentration of glucose but a much higher than normal concentration of reducing sugars. The urine tests positive for galactose. Why is the concentration of reducing sugar in the blood high? Why does galactose appear in the urine?

Case C The patient complains of painful muscle cramps when performing strenuous physical exercise but has no other symptoms. A muscle biopsy indicates a muscle glycogen concentration much higher than normal. Why does glycogen accumulate?

Case D The patient is lethargic, her liver is enlarged, and a biopsy of the liver shows large amounts of excess glycogen. She also has a lower than normal blood glucose level. What is the reason for the low blood glucose in this patient?

Defective Enzyme
(a) Muscle PFK-1
(b) Phosphomannose isomerase
(c) Galactose 1-phosphate uridylyltransferase
(d) Liver glycogen phosphorylase
(e) Triose kinase
(f) Lactase in intestinal mucosa
(g) Maltase in intestinal mucosa
(h) Muscle debranching enzyme

Treatment
1. Jogging 5 km each day
2. Fat-free diet
3. Low-lactose diet
4. Avoiding strenuous exercise
5. Large doses of niacin (the precursor of NAD^+)
6. Frequent feedings (smaller portions) of a normal diet

12. Effects of Insufficient Insulin in a Person with Diabetes A man with insulin-dependent diabetes is brought to the emergency room in a near-comatose state. While vacationing in an isolated place, he lost his insulin medication and has not taken any insulin for two days.

(a) For each tissue listed below, is each pathway faster, slower, or unchanged in this patient, compared with the normal level when he is getting appropriate amounts of insulin?

(b) For each pathway, describe at least one control mechanism responsible for the change you predict.

Tissue and Pathways
1. Adipose: fatty acid synthesis
2. Muscle: glycolysis; fatty acid synthesis; glycogen synthesis
3. Liver: glycolysis; gluconeogenesis; glycogen synthesis; fatty acid synthesis; pentose phosphate pathway

13. Blood Metabolites in Insulin Insufficiency For the patient described in Problem 12, predict the levels of the following metabolites in his blood *before* treatment in the emergency room, relative to levels maintained during adequate insulin treatment: (a) glucose; (b) ketone bodies; (c) free fatty acids.

14. Metabolic Effects of Mutant Enzymes Predict and explain the effect on glycogen metabolism of each of the following defects caused by mutation: (a) loss of the cAMP-binding site on the regulatory subunit of protein kinase A (PKA); (b) loss of the protein phosphatase inhibitor (inhibitor 1 in Fig. 15–42); (c) overexpression of phosphorylase *b* kinase in liver; (d) defective glucagon receptors in liver.

15. Hormonal Control of Metabolic Fuel Between your evening meal and breakfast, your blood glucose drops and your liver becomes a net producer rather than consumer of glucose. Describe the hormonal basis for this switch, and explain how the hormonal change triggers glucose production by the liver.

16. Altered Metabolism in Genetically Manipulated Mice Researchers can manipulate the genes of a mouse so that a single gene in a single tissue either produces an inactive protein (a "knockout" mouse) or produces a protein that is always (constitutively) active. What effects on metabolism would you predict for mice with the following genetic changes: (a) knockout of glycogen debranching enzyme in the liver; (b) knockout of hexokinase IV in liver; (c) knockout of FBPase-2 in liver; (d) constitutively active FBPase-2 in liver; (e) constitutively active AMPK in muscle; (f) constitutively active ChREBP in liver?

Data Analysis Problem

17. Optimal Glycogen Structure Muscle cells need rapid access to large amounts of glucose during heavy exercise. This glucose is stored in liver and skeletal muscle in polymeric form as particles of glycogen. The typical glycogen particle contains about 55,000 glucose residues (see Fig. 15–35b). Meléndez-Hevia, Waddell, and Shelton (1993) explored some theoretical aspects of the structure of glycogen, as described in this problem.

(a) The cellular concentration of glycogen in liver is about 0.01 μM. What cellular concentration of free glucose would be required to store an equivalent amount of glucose? Why would this concentration of free glucose present a problem for the cell?

Glucose is released from glycogen by glycogen phosphorylase, an enzyme that can remove glucose molecules, one at a time, from one end of a glycogen chain. Glycogen chains are branched (see Figs 15–28 and 15–35b), and the degree of branching—the number of branches per chain—has a powerful influence on the rate at which glycogen phosphorylase can release glucose.

(b) Why would a degree of branching that was too low (i.e., below an optimum level) reduce the rate of glucose release? (Hint: Consider the extreme case of no branches in a chain of 55,000 glucose residues.)

(c) Why would a degree of branching that was too high also reduce the rate of glucose release? (Hint: Think of the physical constraints.)

Meléndez-Hevia and colleagues did a series of calculations and found that two branches per chain (see Fig. 15–35b) was optimal for the constraints described above. This is what is found in glycogen stored in muscle and liver.

To determine the optimum number of glucose residues per chain, Meléndez-Hevia and coauthors considered two key parameters that define the structure of a glycogen particle: t = the number of tiers of glucose chains in a particle (the molecule in Fig. 15–35b has five tiers); g_c = the number of glucose residues in each chain. They set out to find the values of t and g_c that would maximize three quantities: (1) the amount of glucose stored in the particle (G_T) per unit volume; (2) the number of unbranched glucose chains (C_A) per unit volume (i.e., number of chains in the outermost tier, readily accessible to glycogen phosphorylase); and (3) the amount of glucose available to phosphorylase in these unbranched chains (G_{PT}).

(d) Show that $C_A = 2^{t-1}$. This is the number of chains available to glycogen phosphorylase before the action of the debranching enzyme.

(e) Show that C_T, the total number of chains in the particle, is given by $C_T = 2^t - 1$. Thus $G_T = g_c(C_T) = g_c(2^t - 1)$, the total number of glucose residues in the particle.

(f) Glycogen phosphorylase cannot remove glucose from glycogen chains that are shorter than five glucose residues. Show that $G_{PT} = (g_c - 4)(2^{t-1})$. This is the amount of glucose readily available to glycogen phosphorylase.

(g) Based on the size of a glucose residue and the location of branches, the thickness of one tier of glycogen is $0.12\, g_c$ nm + 0.35 nm. Show that the volume of a particle, V_s, is given by the equation $V_s = \frac{4}{3}\pi t^3(0.12 g_c + 0.35)^3$ nm³.

Meléndez-Hevia and coauthors then determined the optimum values of t and g_c—those that gave the maximum value of a quality function, f, that maximizes G_T, C_A, and G_{PT}, while minimizing V_s: $f = \dfrac{G_T C_A G_{PT}}{V_S}$. They found that the optimum value of g_c is independent of t.

(h) Choose a value of t between 5 and 15 and find the optimum value of g_c. How does this compare with the g_c found in liver glycogen (see Fig. 15–35b)? (Hint: You may find it useful to use a spreadsheet program.)

Reference

Meléndez-Hevia, E., Waddell, T.G., & Shelton, E.D. (1993) Optimization of molecular design in the evolution of metabolism: the glycogen molecule. *Biochem. J.* **295**, 477–483.

The Citric Acid Cycle

As we saw in Chapter 14, some cells obtain energy (ATP) by fermentation, breaking down glucose in the absence of oxygen. For most eukaryotic cells and many bacteria, which live under aerobic conditions and oxidize their organic fuels to carbon dioxide and water, glycolysis is but the first stage in the complete oxidation of glucose. Rather than being reduced to lactate, ethanol, or some other fermentation product, the pyruvate produced by glycolysis is further oxidized to H_2O and CO_2. This aerobic phase of catabolism is called **respiration**. In the broader physiological or macroscopic sense, respiration refers to a multicellular organism's uptake of O_2 and release of CO_2. Biochemists and cell biologists, however, use the term in a narrower sense to refer to the molecular processes by which *cells* consume O_2 and produce CO_2—processes more precisely termed **cellular respiration**.

Cellular respiration occurs in three major stages **(Fig. 16–1)**. In the first, organic fuel molecules—glucose, fatty acids, and some amino acids—are oxidized to yield two-carbon fragments in the form of the acetyl group of acetyl-coenzyme A (acetyl-CoA). In the second stage, the acetyl groups are fed into the citric acid cycle, which enzymatically oxidizes them to CO_2; the energy released is conserved in the reduced electron carriers NADH and $FADH_2$. In the third stage of respiration, these reduced coenzymes are themselves oxidized, giving up protons (H^+) and electrons. The electrons are transferred to O_2—the final electron acceptor—via a chain of electron-carrying molecules known as the respiratory chain. In the course of electron transfer, the large amount of energy released is conserved in the form of ATP, by a process called oxidative phosphorylation (Chapter 19). Respiration is more complex than

glycolysis and is believed to have evolved much later, after the appearance of cyanobacteria. The metabolic activities of cyanobacteria account for the rise of oxygen levels in the earth's atmosphere, a dramatic turning point in evolutionary history.

We consider first the conversion of pyruvate to acetyl groups, then the entry of those groups into the **citric acid cycle**, also called the **tricarboxylic acid (TCA) cycle** or the **Krebs cycle** (after its discoverer, Hans Krebs). We next examine the cycle reactions and the enzymes that catalyze them. Because intermediates of the citric acid cycle are also siphoned off as biosynthetic precursors, we go on to consider some ways in which these intermediates are replenished. The citric acid cycle is a hub in metabolism, with degradative pathways leading in and anabolic pathways leading out, and it is closely regulated in coordination with other pathways. The chapter ends with a description of the glyoxylate pathway, a metabolic sequence in some organisms that employs several of the same enzymes and reactions used in the citric acid cycle, bringing about the net synthesis of glucose from stored triacylglycerols.

Hans Krebs, 1900–1981

16.1 Production of Acetyl-CoA (Activated Acetate)

In aerobic organisms, glucose and other sugars, fatty acids, and most amino acids are ultimately oxidized to CO_2 and H_2O via the citric acid cycle and the respiratory chain. Before entering the citric acid cycle, the carbon skeletons of sugars and fatty acids are degraded to the acetyl group of acetyl-CoA, the form in which the cycle accepts most of its fuel input. Many amino acid carbons also enter the cycle this way, although several amino acids are degraded to other cycle intermediates. Here

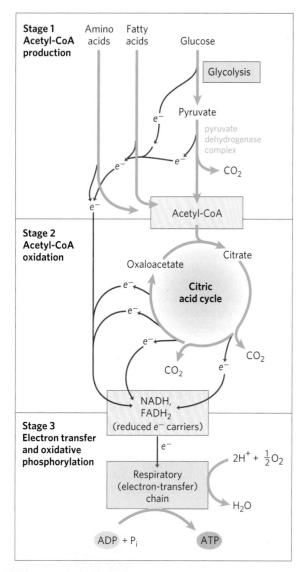

FIGURE 16–1 Catabolism of proteins, fats, and carbohydrates in the three stages of cellular respiration. Stage 1: oxidation of fatty acids, glucose, and some amino acids yields acetyl-CoA. Stage 2: oxidation of acetyl groups in the citric acid cycle includes four steps in which electrons are abstracted. Stage 3: electrons carried by NADH and $FADH_2$ are funneled into a chain of mitochondrial (or, in bacteria, plasma membrane–bound) electron carriers—the respiratory chain—ultimately reducing O_2 to H_2O. This electron flow drives the production of ATP.

we focus on how pyruvate, derived from glucose and other sugars by glycolysis, is oxidized to acetyl-CoA and CO_2 by the **pyruvate dehydrogenase (PDH) complex**, a cluster of enzymes—multiple copies of each of three enzymes—located in the mitochondria of eukaryotic cells and in the cytosol of bacteria.

A careful examination of this enzyme complex is rewarding in several respects. The PDH complex is a classic, much-studied example of a multienzyme complex in which a series of chemical intermediates remain bound to the enzyme molecules as a substrate is transformed into the final product. Five cofactors, four derived from vitamins, participate in the reaction mechanism. The regulation of this enzyme complex also illus-

trates how a combination of covalent modification and allosteric mechanism results in precisely regulated flux through a metabolic step. Finally, the PDH complex is the prototype for two other important enzyme complexes: α-ketoglutarate dehydrogenase, of the citric acid cycle, and the branched-chain α-keto acid dehydrogenase, of the oxidative pathways of several amino acids (see Fig. 18–28). The remarkable similarity in the protein structure, cofactor requirements, and reaction mechanisms of these three complexes doubtless reflects a common evolutionary origin.

Pyruvate Is Oxidized to Acetyl-CoA and CO_2

The overall reaction catalyzed by the pyruvate dehydrogenase complex is an **oxidative decarboxylation**, an irreversible oxidation process in which the carboxyl group is removed from pyruvate as a molecule of CO_2 and the two remaining carbons become the acetyl group of acetyl-CoA **(Fig. 16–2)**. The NADH formed in this reaction gives up a hydride ion ($:H^-$) to the respiratory chain (Fig. 16–1), which carries the two electrons to oxygen or, in anaerobic microorganisms, to an alternative electron acceptor such as nitrate or sulfate. The transfer of electrons from NADH to oxygen ultimately generates 2.5 molecules of ATP per pair of electrons. The irreversibility of the PDH complex reaction has been demonstrated by isotopic labeling experiments: the complex cannot reattach radioactively labeled CO_2 to acetyl-CoA to yield carboxyl-labeled pyruvate.

The Pyruvate Dehydrogenase Complex Requires Five Coenzymes

The combined dehydrogenation and decarboxylation of pyruvate to the acetyl group of acetyl-CoA (Fig. 16–2) requires the sequential action of three different enzymes and five different coenzymes or prosthetic groups—thiamine pyrophosphate (TPP), flavin adenine dinucleotide (FAD), coenzyme A (CoA, sometimes denoted CoA-SH, to emphasize the role of the —SH group), nicotinamide adenine dinucleotide (NAD), and lipoate. Four different vitamins required in human nutrition are vital components of this system: thiamine (in TPP), riboflavin (in FAD), niacin (in NAD), and pantothenate (in CoA). We have

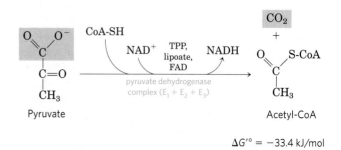

FIGURE 16–2 Overall reaction catalyzed by the pyruvate dehydrogenase complex. The five coenzymes participating in this reaction, and the three enzymes that make up the enzyme complex, are discussed in the text.

FIGURE 16–3 Coenzyme A (CoA). A hydroxyl group of pantothenic acid is joined to a modified ADP moiety by a phosphate ester bond, and its carboxyl group is attached to β-mercaptoethylamine in amide linkage. The hydroxyl group at the 3' position of the ADP moiety has a phosphoryl group not present in free ADP. The —SH group of the mercaptoethylamine moiety forms a thioester with acetate in acetyl-coenzyme A (acetyl-CoA) (lower left).

already described the roles of FAD and NAD as electron carriers (Chapter 13), and we have encountered TPP as the coenzyme of pyruvate decarboxylase (see Fig. 14–15).

Coenzyme A **(Fig. 16–3)** has a reactive thiol (—SH) group that is critical to the role of CoA as an acyl carrier in a number of metabolic reactions. Acyl groups are covalently linked to the thiol group, forming **thioesters**. Because of their relatively high standard free energies of hydrolysis (see Figs 13–16, 13–17), thioesters have a high acyl group transfer potential and can donate their acyl groups to a variety of acceptor molecules. The acyl group attached to coenzyme A may thus be thought of as "activated" for group transfer.

The fifth cofactor of the PDH complex, **lipoate (Fig. 16–4)**, has two thiol groups that can undergo reversible oxidation to a disulfide bond (—S—S—), similar to that between two Cys residues in a protein. Because of its capacity to undergo oxidation-reduction reactions, lipoate can serve both as an electron (hydrogen) carrier and as an acyl carrier, as we shall see.

The Pyruvate Dehydrogenase Complex Consists of Three Distinct Enzymes

The PDH complex contains three enzymes—**pyruvate dehydrogenase** (E_1), **dihydrolipoyl transacetylase** (E_2), and **dihydrolipoyl dehydrogenase** (E_3)—each present in multiple copies. The number of copies of each enzyme and therefore the size of the complex varies among species. The PDH complex isolated from mammals is about 50 nm in diameter—more than five times the size of an entire ribosome and big enough to be visualized with the electron microscope **(Fig. 16–5a)**. In the bovine enzyme, 60 identical copies of E_2 form a pentagonal dodecahedron (the core) with a diameter of about 25 nm (Fig. 16–5b). (The core of the

Escherichia coli enzyme contains 24 copies of E_2.) E_2 is the point of connection for the prosthetic group lipoate, attached through an amide bond to the ε-amino group of a Lys residue (Fig. 16–4). E_2 has three functionally distinct domains (Fig. 16–5c): the amino-terminal *lipoyl domain*, containing the lipoyl-Lys residue(s); the central E_1- and E_3-*binding domain*; and the inner-core *acyltransferase domain*, which contains the acyltransferase active site. The yeast PDH complex has a single lipoyl domain with a lipoate attached, but the

FIGURE 16–4 Lipoic acid (lipoate) in amide linkage with a Lys residue. The lipoyllysyl moiety is the prosthetic group of dihydrolipoyl transacetylase (E_2 of the PDH complex). The lipoyl group occurs in oxidized (disulfide) and reduced (dithiol) forms and acts as a carrier of both hydrogen and an acetyl (or other acyl) group.

(a)

50 nm

(b)

10 nm

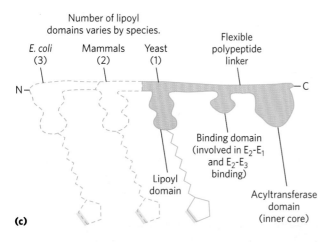

(c)

FIGURE 16–5 **The pyruvate dehydrogenase complex. (a)** Cryoelectron micrograph of PDH complexes isolated from bovine kidney. In cryoelectron microscopy, biological samples are viewed at extremely low temperatures; this avoids potential artifacts introduced by the usual process of dehydrating, fixing, and staining. **(b)** Three-dimensional image of PDH complex, showing the subunit structure: E_1, pyruvate dehydrogenase; E_2, dihydrolipoyl transacetylase; and E_3, dihydrolipoyl dehydrogenase. This image is reconstructed by analysis of a large number of images such as those in (a), combined with crystallographic studies of individual subunits. The core (green) consists of 60 molecules of E_2, arranged in 20 trimers to form a pentagonal dodecahedron. The lipoyl domain of E_2 (blue) reaches outward to touch the active sites of E_1 molecules (yellow) arranged on the E_2 core. Several E_3 subunits (red) are also bound to the core, where the swinging arm on E_2 can reach their active sites. An asterisk marks the site where a lipoyl group is attached to the lipoyl domain of E_2. To make the structure clearer, about half of the complex has been cut away from the front. This model was prepared by Z. H. Zhou and colleagues (2001); in another model, proposed by J. L. S. Milne and colleagues (2002), the E_3 subunits are located more toward the periphery (see Further Reading). **(c)** E_2 consists of three types of domains linked by short polypeptide linkers: a catalytic acyltransferase domain; a binding domain, involved in the binding of E_2 to E_1 and E_3; and one or more (depending on the species) lipoyl domains.

mammalian complex has two, and *E. coli* has three (Fig. 16–5c). The domains of E_2 are separated by linkers, sequences of 20 to 30 amino acid residues, rich in Ala and Pro and interspersed with charged residues; these linkers tend to assume their extended forms, holding the three domains apart.

The active site of E_1 has bound TPP, and that of E_3 has bound FAD. Also part of the complex are two regulatory proteins, a protein kinase and a phosphoprotein phosphatase, discussed below. This basic E_1–E_2–E_3 structure has been conserved during evolution and used in a number of similar metabolic reactions, including the oxidation of α-ketoglutarate in the citric acid cycle (described below) and the oxidation of α-keto acids derived from the breakdown of the branched-chain amino acids valine, isoleucine, and leucine (see Fig. 18–28). Within a given species, E_3 of PDH is identical to E_3 of the other two enzyme complexes. The attachment of lipoate to the end of a Lys side chain in E_2 produces a long, flexible arm that can move from the active site of E_1 to the active sites of E_2 and E_3, a distance of perhaps 5 nm or more.

In Substrate Channeling, Intermediates Never Leave the Enzyme Surface

Figure 16–6 shows schematically how the pyruvate dehydrogenase complex carries out the five consecutive reactions in the decarboxylation and dehydrogenation of pyruvate. Step **❶** is essentially identical to the reaction catalyzed by pyruvate decarboxylase (see Fig. 14–15c); C-1 of pyruvate is released as CO_2, and C-2, which in pyruvate has the oxidation state of an aldehyde, is attached to TPP as a hydroxyethyl group. This first step is the slowest and therefore limits the rate of

FIGURE 16–6 Oxidative decarboxylation of pyruvate to acetyl-CoA by the PDH complex. The fate of pyruvate is traced in red. In step ❶ pyruvate reacts with the bound thiamine pyrophosphate (TPP) of pyruvate dehydrogenase (E₁), undergoing decarboxylation to the hydroxyethyl derivative (see Fig. 14–15). Pyruvate dehydrogenase also carries out step ❷, the transfer of two electrons and the acetyl group from TPP to the oxidized form of the lipoyllysyl group of the core enzyme, dihydrolipoyl transacetylase (E₂), to form the acetyl thioester of the reduced lipoyl group. Step ❸ is a transesterification in which the —SH group of CoA replaces the —SH group of E₂ to yield acetyl-CoA and the fully reduced (dithiol) form of the lipoyl group. In step ❹ dihydrolipoyl dehydrogenase (E₃) promotes transfer of two hydrogen atoms from the reduced lipoyl groups of E₂ to the FAD prosthetic group of E₃, restoring the oxidized form of the lipoyllysyl group of E₂. In step ❺ the reduced FADH₂ of E₃ transfers a hydride ion to NAD⁺, forming NADH. The enzyme complex is now ready for another catalytic cycle. (Subunit colors correspond to those in Fig. 16–5b.)

the overall reaction. It is also the point at which the PDH complex exercises its substrate specificity. In step ❷ the hydroxyethyl group is oxidized to the level of a carboxylic acid (acetate). The two electrons removed in this reaction reduce the —S—S— of a lipoyl group on E₂ to two thiol (—SH) groups. The acetyl moiety produced in this oxidation-reduction reaction is first esterified to one of the lipoyl —SH groups, then transesterified to CoA to form acetyl-CoA (step ❸). Thus the energy of oxidation drives the formation of a high-energy thioester of acetate. The remaining reactions catalyzed by the PDH complex (by E₃, in steps ❹ and ❺) are electron transfers necessary to regenerate the oxidized (disulfide) form of the lipoyl group of E₂ to prepare the enzyme complex for another round of oxidation. The electrons removed from the hydroxyethyl group derived from pyruvate pass through FAD to NAD⁺.

Central to the mechanism of the PDH complex are the swinging lipoyllysyl arms of E₂, which accept from E₁ the two electrons and the acetyl group derived from pyruvate, passing them to E₃. All these enzymes and coenzymes are clustered, allowing the intermediates to react quickly without diffusing away from the surface of the enzyme complex. The five-reaction sequence shown in Figure 16–6 is thus an example of **substrate channeling.** The intermediates of the multistep sequence never leave the complex, and the local concentration of the substrate of E₂ is kept very high. Channeling also prevents theft of the activated acetyl group by other enzymes that use this group as substrate. As we shall see, a similar tethering mechanism for the channeling of substrate between active sites is used in some other enzymes, with lipoate, biotin, or a CoA-like moiety serving as cofactors.

As one might predict, mutations in the genes for the subunits of the PDH complex, or a dietary thiamine deficiency, can have severe consequences. Thiamine-deficient animals are unable to oxidize pyruvate normally. This is of particular importance to the brain, which usually obtains all its energy from the aerobic oxidation of glucose in a pathway that necessarily includes the oxidation of pyruvate. Beriberi, a disease that results from thiamine deficiency, is characterized by loss of neural function. This disease occurs primarily in populations that rely on a diet consisting mainly of white (polished) rice, which lacks the hulls in which most of the thiamine of rice is found. People who habitually consume large amounts of alcohol can also develop thiamine deficiency, because much of their dietary intake consists of the vitamin-free "empty calories" of distilled spirits. An elevated level of pyruvate in the blood is often an indicator of defects in pyruvate oxidation due to one of these causes. ∎

SUMMARY 16.1 Production of Acetyl-CoA (Activated Acetate)

▶ Pyruvate, the product of glycolysis, is converted to acetyl-CoA, the starting material for the citric acid cycle, by the pyruvate dehydrogenase complex.

▶ The PDH complex is composed of multiple copies of three enzymes: pyruvate dehydrogenase, E₁ (with its bound cofactor TPP); dihydrolipoyltransacetylase, E₂ (with its covalently bound lipoyl group); and dihydrolipoyl dehydrogenase, E_3 (with its cofactors FAD and NAD).

▶ E_1 catalyzes first the decarboxylation of pyruvate, producing hydroxyethyl-TPP, and then the oxidation of the hydroxyethyl group to an acetyl group. The electrons from this oxidation reduce the disulfide of lipoate bound to E_2, and the acetyl group is transferred into thioester linkage with one —SH group of reduced lipoate.

▶ E_2 catalyzes the transfer of the acetyl group to coenzyme A, forming acetyl-CoA.

▶ E_3 catalyzes the regeneration of the disulfide (oxidized) form of lipoate; electrons pass first to FAD, then to NAD^+.

▶ The long lipoyllysyl arm swings from the active site of E_1 to E_2 to E_3, tethering the intermediates to the enzyme complex to allow substrate channeling.

▶ The organization of the PDH complex is very similar to that of the enzyme complexes that catalyze the oxidation of α-ketoglutarate and the branched-chain α-keto acids.

16.2 Reactions of the Citric Acid Cycle

We are now ready to trace the process by which acetyl-CoA undergoes oxidation. This chemical transformation is carried out by the citric acid cycle, the first *cyclic* pathway we have encountered **(Fig. 16–7)**. To begin a turn of the cycle, acetyl-CoA donates its acetyl group to the four-carbon compound oxaloacetate to form the six-carbon citrate. Citrate is then transformed into isocitrate, also a six-carbon molecule, which is dehydrogenated with loss of CO_2 to yield the five-carbon compound α-ketoglutarate (also called oxoglutarate). α-Ketoglutarate undergoes loss of a second molecule of CO_2 and ultimately yields the four-carbon compound succinate. Succinate is then enzymatically converted in three steps into the four-carbon oxaloacetate—which is then ready to react with another molecule of acetyl-CoA. In each turn of the cycle, one acetyl group (two carbons) enters as acetyl-CoA and two molecules of CO_2 leave; one molecule of oxaloacetate is used to form citrate and one molecule of oxaloacetate is regenerated. No net removal of oxaloacetate occurs; one molecule of oxaloacetate can theoretically bring about oxidation of an infinite number of acetyl groups, and, in fact, oxaloacetate is present in cells in very low concentrations. Four of the eight steps in this process are oxidations, in which the energy of oxidation is very efficiently conserved in the form of the reduced coenzymes NADH and $FADH_2$.

As noted earlier, although the citric acid cycle is central to energy-yielding metabolism its role is not limited to energy conservation. Four- and five-carbon intermediates of the cycle serve as precursors for a wide variety of products. To replace intermediates removed for this purpose, cells employ anaplerotic (replenishing) reactions, which are described below.

Eugene Kennedy and Albert Lehninger showed in 1948 that, in eukaryotes, the entire set of reactions of the citric acid cycle takes place in mitochondria. Isolated mitochondria were found to contain not only all the enzymes and coenzymes required for the citric acid cycle, but also all the enzymes and proteins necessary for the last stage of respiration—electron transfer and ATP synthesis by oxidative phosphorylation. As we shall see in later chapters, mitochondria also contain the enzymes for the oxidation of fatty acids and some amino acids to acetyl-CoA, and the oxidative degradation of other amino acids to α-ketoglutarate, succinyl-CoA, or oxaloacetate. Thus, in nonphotosynthetic eukaryotes, the mitochondrion is the site of most energy-yielding oxidative reactions and of the coupled synthesis of ATP. In photosynthetic eukaryotes, mitochondria are the major site of ATP production in the dark, but in daylight chloroplasts produce most of the organism's ATP. In most bacteria, the enzymes of the citric acid cycle are in the cytosol, and the plasma membrane plays a role analogous to that of the inner mitochondrial membrane in ATP synthesis (Chapter 19).

The Sequence of Reactions in the Citric Acid Cycle Makes Chemical Sense

Acetyl-CoA produced in the breakdown of carbohydrates, fats, and proteins must be completely oxidized to CO_2 if the maximum potential energy is to be extracted from these fuels. However, the direct oxidation of acetate (or acetyl-CoA) to CO_2 is not biochemically feasible. Decarboxylation of this two-carbon acid would yield CO_2 and methane (CH_4). Methane is chemically rather stable, and except for certain methanotrophic bacteria that grow in methane-rich niches, organisms do not have the cofactors and enzymes needed to oxidize methane. Methylene groups (—CH_2—), however, are readily metabolized by enzyme systems present in most organisms. In typical oxidation sequences, two adjacent methylene groups (—CH_2—CH_2—) are involved, at least one of which is adjacent to a carbonyl group. As we noted in Chapter 13 (p. 513), carbonyl groups are particularly important in the chemical transformations of metabolic pathways. The carbon of the carbonyl group has a partial positive charge due to the electron-withdrawing property of the carbonyl oxygen and is therefore an electrophilic center. A carbonyl group can facilitate the formation of a carbanion on an adjoining carbon by delocalizing the carbanion's negative charge. We see in the citric acid cycle an example of the oxidation of a methylene group as succinate is oxidized (steps ❻ to ❽ in Fig. 16–7), forming a carbonyl (in oxaloacetate) that is more chemically reactive than either a methylene group or methane.

In short, if acetyl-CoA is to be oxidized efficiently, the methyl group of the acetyl-CoA must be attached to

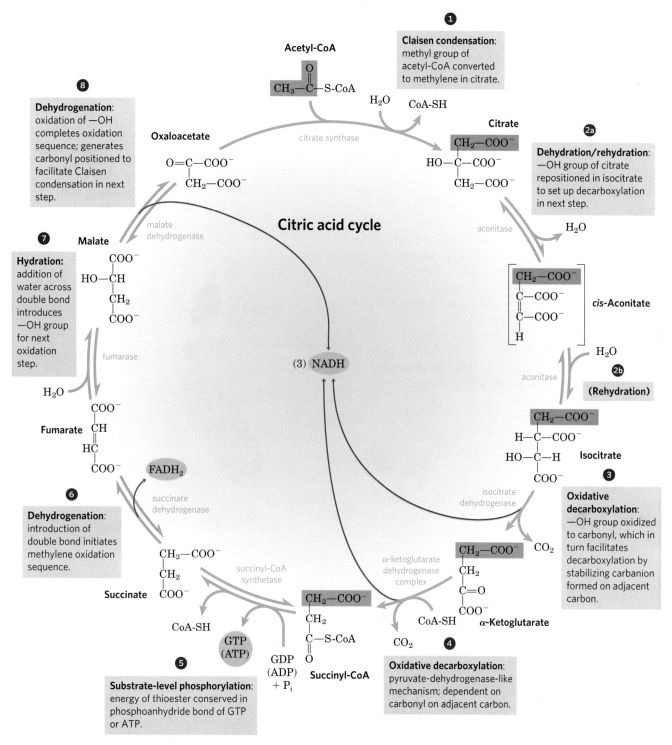

FIGURE 16–7 Reactions of the citric acid cycle. The carbon atoms shaded in pink are those derived from the acetate of acetyl-CoA in the first turn of the cycle; these are *not* the carbons released as CO_2 in the first turn. Note that in succinate and fumarate, the two-carbon group derived from acetate can no longer be specifically denoted; because succinate and fumarate are symmetric molecules, C-1 and C-2 are indistinguishable from C-4 and C-3. The num-ber beside each reaction step corresponds to a numbered heading on pages 640–647). The red arrows show where energy is conserved by electron transfer to FAD or NAD^+, forming $FADH_2$ or $NADH + H^+$. Steps **1**, **3**, and **4** are essentially irreversible in the cell; all other steps are reversible. The nucleoside triphosphate product of step **5** may be either ATP or GTP, depending on which succinyl-CoA synthetase isozyme is the catalyst.

something. The first step of the citric acid cycle neatly solves the problem of the unreactive methyl group by means of the condensation of acetyl-CoA with oxaloacetate. The carbonyl of oxaloacetate acts as an electrophilic center, which is attacked by the methyl carbon of

acetyl-CoA in a Claisen condensation (p. 513) to form citrate (step **1** in Fig. 16–7). The methyl group of acetate has been converted into a methylene in citric acid. This tricarboxylic acid then readily undergoes a series of oxidations that eliminate two carbons as CO_2. Note that all steps featuring the breakage or formation of carbon–carbon bonds (steps ❶, ❸, and ❹) rely on properly positioned carbonyl groups. As in all metabolic pathways, there is a chemical logic to the sequence of steps in the citric acid cycle: each step either involves an energy-conserving oxidation or is a necessary prelude to the oxidation, placing functional groups in position to facilitate oxidation or oxidative decarboxylation. As you learn the steps of the cycle, keep in mind the chemical rationale for each; it will make the process easier to understand and remember.

The Citric Acid Cycle Has Eight Steps

In examining the eight successive reaction steps of the citric acid cycle, we place special emphasis on the chemical transformations taking place as citrate formed from acetyl-CoA and oxaloacetate is oxidized to yield CO_2 and the energy of this oxidation is conserved in the reduced coenzymes NADH and $FADH_2$.

❶ **Formation of Citrate** The first reaction of the cycle is the condensation of acetyl-CoA with **oxaloacetate** to form **citrate**, catalyzed by **citrate synthase**:

$$\Delta G'^\circ = -32.2 \text{ kJ/mol}$$

In this reaction the methyl carbon of the acetyl group is joined to the carbonyl group (C-2) of oxaloacetate. Citroyl-CoA is a transient intermediate formed on the active site of the enzyme (see Fig. 16–9). It rapidly undergoes hydrolysis to free CoA and citrate, which are released from the active site. The hydrolysis of this high-energy thioester intermediate makes the forward reaction highly exergonic. The large, negative standard free-energy change of the citrate synthase reaction is essential to the operation of the cycle because, as noted earlier, the concentration of oxaloacetate is normally very low. The CoA liberated in this reaction is recycled to participate in the oxidative decarboxylation of another molecule of pyruvate by the PDH complex.

Citrate synthase from mitochondria has been crystallized and visualized by x-ray diffraction in the presence and absence of its substrates and inhibitors **(Fig. 16–8)**. Each subunit of the homodimeric enzyme is a single polypeptide with two domains, one large and rigid, the other smaller and more flexible, with the active site between

FIGURE 16–8 Structure of citrate synthase. The flexible domain of each subunit undergoes a large conformational change on binding oxaloacetate, creating a binding site for acetyl-CoA. **(a)** Open form of the enzyme alone (PDB ID 5CSC); **(b)** closed form with bound oxaloacetate and a stable analog of acetyl-CoA (carboxymethyl-CoA) (derived from PDB ID 5CTS). In these representations one subunit is colored tan and one green.

them. Oxaloacetate, the first substrate to bind to the enzyme, induces a large conformational change in the flexible domain, creating a binding site for the second substrate, acetyl-CoA. When citroyl-CoA has formed in the enzyme active site, another conformational change brings about thioester hydrolysis, releasing CoA-SH. This induced fit of the enzyme first to its substrate and then to its reaction intermediate decreases the likelihood of premature and unproductive cleavage of the thioester bond of acetyl-CoA. Kinetic studies of the enzyme are consistent with this ordered bisubstrate mechanism (see Fig. 6–13). The reaction catalyzed by citrate synthase is essentially a Claisen condensation (p. 513), involving a thioester (acetyl-CoA) and a ketone (oxaloacetate) **(Fig. 16–9)**.

2 Formation of Isocitrate via *cis*-Aconitate The enzyme **aconitase** (more formally, **aconitate hydratase**) catalyzes the reversible transformation of citrate to **isocitrate**, through the intermediary formation of the tricarboxylic acid ***cis*-aconitate**, which normally does not dissociate from the active site. Aconitase can promote the reversible addition of H_2O to the double bond of enzyme-bound *cis*-aconitate in two different ways, one leading to citrate and the other to isocitrate:

$$
\begin{array}{c}
CH_2-COO^- \\
HO-C-COO^- \\
H-C-COO^- \\
H
\end{array}
\underset{\text{aconitase}}{\overset{H_2O}{\rightleftharpoons}}
\left[
\begin{array}{c}
CH_2-COO^- \\
C-COO^- \\
C-COO^- \\
H
\end{array}
\right]
\underset{\text{aconitase}}{\overset{H_2O}{\rightleftharpoons}}
$$

Citrate   *cis*-Aconitate

$$
\begin{array}{c}
CH_2-COO^- \\
H-C-COO^- \\
HO-C-H \\
COO^-
\end{array}
$$

Isocitrate

$\Delta G'^{\circ} = 13.3 \text{ kJ/mol}$

Although the equilibrium mixture at pH 7.4 and 25 °C contains less than 10% isocitrate, in the cell the reaction is pulled to the right because isocitrate is rapidly consumed in the next step of the cycle, lowering its steady-state concentration. Aconitase contains an **iron-sulfur center (Fig. 16–10)**, which acts both in the binding of the substrate at the active site and in the catalytic addition or removal of H_2O. In iron-depleted cells, aconitase loses its iron-sulfur center and acquires a new role in the regulation of iron homeostasis. Aconitase is one of many enzymes known to "moonlight" in a second role (Box 16–1).

3 Oxidation of Isocitrate to α-Ketoglutarate and CO_2 In the next step, **isocitrate dehydrogenase** catalyzes oxidative decarboxylation of isocitrate to form **α-keto-glutarate (Fig. 16–11)**. Mn^{2+} in the active site interacts with the carbonyl group of the intermediate

The thioester linkage in acetyl-CoA activates the methyl hydrogens. Asp375 abstracts a proton from the methyl group, forming an enolate intermediate. The intermediate is stabilized by hydrogen bonding to and/or protonation by His274 (full protonation is shown).

The enol(ate) rearranges to attack the carbonyl carbon of oxaloacetate, with His274 positioned to abstract the proton it had previously donated. His320 acts as a general acid. The resulting condensation generates citroyl-CoA.

The thioester is subsequently hydrolyzed, regenerating CoA-SH and producing citrate.

MECHANISM FIGURE 16–9 Citrate synthase. In the mammalian citrate synthase reaction, oxaloacetate binds first, in a strictly ordered reaction sequence. This binding triggers a conformation change that opens up the binding site for acetyl-CoA. Oxaloacetate is specifically oriented in the active site of citrate synthase by interaction of its two carboxylates with two positively charged Arg residues (not shown here). 🖱 **Citrate Synthase Mechanism**

BOX 16–1 Moonlighting Enzymes: Proteins with More Than One Job

The "one gene–one enzyme" dictum, put forward by George Beadle and Edward Tatum in 1940 (see Chapter 24), went unchallenged for much of the twentieth century, as did the associated assumption that each protein had only one role. But in recent years, many striking exceptions to this simple formula have been discovered—cases in which a single protein encoded by a single gene clearly is **"moonlighting,"** doing more than one job in the cell. Aconitase is one such protein: it acts both as an enzyme and as a regulator of protein synthesis.

Eukaryotic cells have two isozymes of aconitase. The mitochondrial isozyme converts citrate to isocitrate in the citric acid cycle. The cytosolic isozyme has two distinct functions. It catalyzes the conversion of citrate to isocitrate, providing the substrate for a cytosolic isocitrate dehydrogenase that generates NADPH as reducing power for fatty acid synthesis and other anabolic processes in the cytosol. It also has a role in cellular iron homeostasis.

All cells must obtain iron for the activity of the many proteins that require it as a cofactor. In humans, severe iron deficiency results in anemia, an insufficient supply of erythrocytes and a reduced oxygen-carrying capacity that can be life-threatening. Too much iron is also harmful: it accumulates in and damages the liver in hemochromatosis and other diseases. Iron obtained in the diet is carried in the blood by the protein **transferrin** and enters cells via endocytosis mediated by the **transferrin receptor**. Once inside cells, iron is used in the synthesis of hemes, cytochromes, Fe-S proteins, and other Fe-dependent proteins, and excess iron is stored bound to the protein **ferritin**. The levels of transferrin, transferrin receptor, and ferritin are therefore crucial to cellular iron homeostasis. The synthesis of these three proteins is regulated in response to iron availability—and aconitase, in its moonlighting job, plays a key regulatory role.

Aconitase has an essential Fe-S cluster at its active site (see Fig. 16–10). When a cell is depleted of iron, this Fe-S cluster is disassembled and the enzyme loses its aconitase activity. But the apoenzyme (apoaconitase, lacking its Fe-S cluster) so formed has now acquired its second activity—the ability to bind to specific sequences in the mRNAs for the transferrin receptor and ferritin, thus regulating protein synthesis at the translational level. Two **iron regulatory proteins**, **IRP1** and **IRP2**, were independently discovered as regulators of iron metabolism. As it turned out, IRP1 is identical to cytosolic apoaconitase, and IRP2 is very closely related to IRP1 in structure and function, but unlike IRP1 it cannot be converted to enzymatically active aconitase. Both IRP1 and IRP2 bind to regions in the mRNAs encoding ferritin and the transferrin receptor, with effects on iron mobilization and iron uptake. These mRNA sequences are part of hairpin structures (p. 292) called **iron response elements (IREs),**

IRP bound to iron response element (IRE)?		Yes	No
Ferritin mRNA translation	Repressed	Activated	
Ferritin synthesis	Decreased	Increased	
TfR mRNA stability	Increased	Decreased	
TfR synthesis	Increased	Decreased	

FIGURE 1 Effect of IRP1 and IRP2 on the mRNAs for ferritin and the transferrin receptor.

FIGURE 16–10 Iron-sulfur center in aconitase. The iron-sulfur center is in red, the citrate molecule in blue. Three Cys residues of the enzyme bind three iron atoms; the fourth iron is bound to one of the carboxyl groups of citrate and also interacts noncovalently with a hydroxyl group of citrate (dashed bond). A basic residue (:B) in the enzyme helps to position the citrate in the active site. The iron-sulfur center acts in both substrate binding and catalysis. The general properties of iron-sulfur proteins are discussed in Chapter 19 (see Fig. 19–5).

located at the 5′ and 3′ ends of the mRNAs (Fig. 1). When bound to the 5′-untranslated IRE sequence in the ferritin mRNA, IRPs block ferritin synthesis; when bound to the 3′-untranslated IRE sequences in the transferrin receptor mRNA, they stabilize the mRNA, preventing its degradation and thus allowing the synthesis of more copies of the receptor protein per mRNA molecule. So, in iron-deficient cells, iron uptake becomes more efficient and iron storage (bound to ferritin) is reduced. When cellular iron concentrations return to normal levels, IRP1 is converted to aconitase, and IRP2 undergoes proteolytic degradation, ending the low-iron response.

The enzymatically active aconitase and the moonlighting, regulatory apoaconitase have different structures. As the active aconitase, the protein has two lobes that close around the Fe-S cluster; as IRP1, the two lobes open, exposing the mRNA-binding site (Fig. 2).

Aconitase is just one of a growing list of enzymes known (or believed) to moonlight in a second role. Many of the glycolytic enzymes are included in this group. Pyruvate kinase acts in the nucleus to regulate the transcription of genes that respond to thyroid hormone. Glyceraldehyde 3-phosphate dehydrogenase moonlights both as uracil DNA glycosylase, effecting the repair of damaged DNA, and as a regulator of histone H2B transcription. The crystallins in the lens of the vertebrate eye are several moonlighting glycolytic enzymes, including phosphoglycerate kinase, triose phosphate isomerase, and lactate dehydrogenase.

Until recently, the discovery that a protein has more than one function was largely a matter of serendipity: two groups of investigators studying two unrelated questions discovered that "their" proteins had similar properties, compared them carefully, and found them to be identical. With the growth of annotated protein and DNA sequence databases, researchers can now deliberately look for moonlighting proteins by searching the databases for any other protein with the same sequence as the one under study, but with a different function. This also means that in the databases, a protein annotated as having a given function doesn't necessarily have *only* that function. Protein moonlighting may also explain some puzzling findings: experiments in which a protein with a known function is made inactive by a mutation, and the resulting mutant organisms show a phenotype with no obvious relation to that function.

FIGURE 2 Two forms of cytosolic aconitase/IRP1 with two distinct functions. **(a)** In aconitase, the two major lobes are closed and the Fe-S cluster is buried; the protein has been made transparent here to show the Fe-S cluster (PDB ID 2B3Y). **(b)** In IRP1, the lobes open up, exposing a binding site for the mRNA hairpin of the substrate (PDB ID 2IPY).

MECHANISM FIGURE 16–11 Isocitrate dehydrogenase. In this reaction, the substrate, isocitrate, loses one carbon by oxidative decarboxylation.

See Figure 14-14 for more information on hydride transfer reactions involving NAD^+ and $NADP^+$.

① Isocitrate is oxidized by hydride transfer to NAD^+ or $NADP^+$ (depending on the isocitrate dehydrogenase isozyme).

② Decarboxylation is facilitated by electron withdrawal by the adjacent carbonyl and coordinated Mn^{2+}.

③ Rearrangement of the enol intermediate generates α-ketoglutarate.

FIGURE 16–12 A conserved mechanism for oxidative decarboxylation. The pathways shown employ the same five cofactors (thiamine pyrophosphate, coenzyme A, lipoate, FAD, and NAD^+), closely similar multienzyme complexes, and the same enzymatic mechanism to carry out oxidative decarboxylations of pyruvate (by the pyruvate dehydrogenase complex), α-ketoglutarate (in the citric acid cycle), and the carbon skeletons of the three branched-chain amino acids, isoleucine (shown here), leucine, and valine. A fourth reaction, catalyzed by glycine decarboxylase, involves a very similar mechanism (see Fig. 20–22).

oxalosuccinate, which is formed transiently but does not leave the binding site until decarboxylation converts it to α-ketoglutarate. Mn^{2+} also stabilizes the enol formed transiently by decarboxylation.

There are two different forms of isocitrate dehydrogenase in all cells, one requiring NAD^+ as electron acceptor and the other requiring $NADP^+$. The overall reactions are otherwise identical. In eukaryotic cells, the NAD-dependent enzyme occurs in the mitochondrial matrix and serves in the citric acid cycle. The main function of the NADP-dependent enzyme, found in both the mitochondrial matrix and the cytosol, may be the generation of NADPH, which is essential for reductive anabolic reactions.

④ Oxidation of α-Ketoglutarate to Succinyl-CoA and CO_2 The next step is another oxidative decarboxylation, in which α-ketoglutarate is converted to **succinyl-CoA** and CO_2 by the action of the **α-ketoglutarate dehydrogenase complex**; NAD^+ serves as electron acceptor and CoA as the carrier of the succinyl group. The energy of oxidation of α-ketoglutarate is conserved in the formation of the thioester bond of succinyl-CoA:

This reaction is virtually identical to the pyruvate dehydrogenase reaction discussed above and to the reaction sequence responsible for the breakdown of branched-chain amino acids **(Fig. 16–12)**. The α-ketoglutarate dehydrogenase complex closely resembles the PDH complex in both structure and function. It includes three enzymes, homologous to E_1, E_2, and E_3 of the PDH complex, as well as enzyme-bound TPP, bound lipoate, FAD, NAD, and coenzyme A. Both complexes are certainly derived from a common evolutionary ancestor. Although the E_1 components of the two complexes are structurally similar, their amino acid sequences differ and, of course, they have different binding specificities: E_1 of the PDH complex binds pyruvate, and E_1 of the α-ketoglutarate dehydrogenase complex binds α-ketoglutarate. The E_2 components of the two complexes are also very similar, both having covalently bound lipoyl moieties. The subunits of E_3 are identical in the two enzyme complexes. The complex that degrades branched-chain α-keto acids (see Fig. 18–28) catalyzes the same reaction sequence using the same five cofactors. This is a clear case of **divergent evolution**, in which the genes for an enzyme with one substrate specificity give rise, during evolution, to closely related enzymes with different substrate specificities but the same enzymatic mechanism.

⑤ Conversion of Succinyl-CoA to Succinate Succinyl-CoA, like acetyl-CoA, has a thioester bond with a strongly negative standard free energy of hydrolysis ($\Delta G'^\circ \approx -36$ kJ/mol). In the next step of the citric acid cycle, energy released in the breakage of this bond is used to drive the synthesis of a phosphoanhydride bond in

$\Delta G'^\circ = -33.5$ kJ/mol

either GTP or ATP, with a net $\Delta G'^\circ$ of only -2.9 kJ/mol. **Succinate** is formed in the process:

$$\text{Succinyl-CoA} \xrightarrow[\text{synthetase}]{\text{succinyl-CoA}} \text{Succinate}$$

$$\Delta G'^\circ = -2.9 \text{ kJ/mol}$$

The enzyme that catalyzes this reversible reaction is called **succinyl-CoA synthetase** or **succinic thiokinase**; both names indicate the participation of a nucleoside triphosphate in the reaction (Box 16–2).

This energy-conserving reaction involves an intermediate step in which the enzyme molecule itself becomes phosphorylated at a His residue in the active site **(Fig. 16–13a)**. This phosphoryl group, which has a high group transfer potential, is transferred to ADP (or GDP) to form ATP (or GTP). Animal cells have two isozymes of succinyl-CoA synthetase, one specific for ADP and the other for GDP. The enzyme has two subunits, α (M_r 32,000), which has the Ⓟ–His residue (His²⁴⁶) and the binding site for CoA, and β (M_r 42,000), which confers specificity for either ADP or GDP. The active site is at the interface between subunits. The crystal structure of succinyl-CoA synthetase reveals two "power helices" (one from each subunit), oriented so that their electric dipoles situate partial positive charges close to the negatively charged Ⓟ–His (Fig. 16–13b), stabilizing the phosphoenzyme intermediate. (Recall the similar role of helix dipoles in stabilizing K⁺ ions in the K⁺ channel; see Fig. 11–47.)

The formation of ATP (or GTP) at the expense of the energy released by the oxidative decarboxylation of α-ketoglutarate is a substrate-level phosphorylation, like the synthesis of ATP in the glycolytic reactions catalyzed by glyceraldehyde 3-phosphate dehydrogenase and pyruvate kinase (see Fig. 14–2). The GTP formed by succinyl-CoA synthetase can donate its terminal phosphoryl group to ADP to form ATP, in a reversible reaction catalyzed by **nucleoside diphosphate kinase** (p. 526):

$$\text{GTP} + \text{ADP} \rightleftharpoons \text{GDP} + \text{ATP} \qquad \Delta G'^\circ = 0 \text{ kJ/mol}$$

Thus the net result of the activity of either isozyme of succinyl-CoA synthetase is the conservation of energy as ATP. There is no change in free energy for the nucleoside diphosphate kinase reaction; ATP and GTP are energetically equivalent.

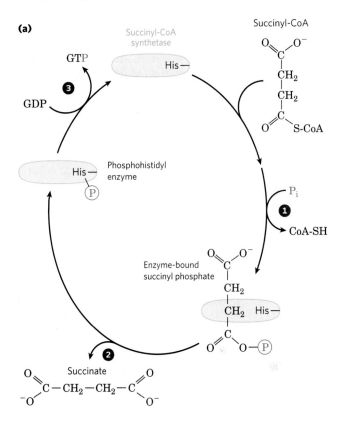

FIGURE 16–13 The succinyl-CoA synthetase reaction. (a) In step ❶ a phosphoryl group replaces the CoA of succinyl-CoA bound to the enzyme, forming a high-energy acyl phosphate. In step ❷ the succinyl phosphate donates its phosphoryl group to a His residue of the enzyme, forming a high-energy phosphohistidyl enzyme. In step ❸ the phosphoryl group is transferred from the His residue to the terminal phosphate of GDP (or ADP), forming GTP (or ATP). **(b)** Active site of succinyl-CoA synthetase of *E. coli* (derived from PDB ID 1SCU). The active site includes part of both the α (blue) and the β (brown) subunits. The power helices (blue, brown) place the partial positive charges of the helix dipole near the phosphate group of Ⓟ–His²⁴⁶ in the α chain, stabilizing the phosphohistidyl enzyme. The bacterial and mammalian enzymes have similar amino acid sequences and three-dimensional structures.

BOX 16–2 Synthases and Synthetases; Ligases and Lyases; Kinases, Phosphatases, and Phosphorylases: Yes, the Names Are Confusing!

Citrate synthase is one of many enzymes that catalyze condensation reactions, yielding a product more chemically complex than its precursors. **Synthases** catalyze condensation reactions in which no nucleoside triphosphate (ATP, GTP, and so forth) is required as an energy source. **Synthetases** catalyze condensations that *do* use ATP or another nucleoside triphosphate as a source of energy for the synthetic reaction. Succinyl-CoA synthetase is such an enzyme. **Ligases** (from the Latin *ligare*, "to tie together") are enzymes that catalyze condensation reactions in which two atoms are joined, using ATP or another energy source. (Thus synthetases are ligases.) DNA ligase, for example, closes breaks in DNA molecules, using energy supplied by either ATP or NAD^+; it is widely used in joining DNA pieces for genetic engineering. Ligases are not to be confused with **lyases**, enzymes that catalyze cleavages (or, in the reverse direction, additions) in which electronic rearrangements occur. The PDH complex, which oxidatively cleaves CO_2 from pyruvate, is a member of the large class of lyases.

The name **kinase** is applied to enzymes that transfer a phosphoryl group from a nucleoside triphosphate such as ATP to an acceptor molecule—a sugar (as in hexokinase and glucokinase), a protein (as in glycogen phosphorylase kinase), another nucleotide (as in nucleoside diphosphate kinase), or a metabolic intermediate such as oxaloacetate (as in PEP carboxykinase). The reaction catalyzed by a kinase is a *phosphorylation*. On the other hand, *phosphorolysis* is a displacement reaction in which phosphate is the attacking species and becomes covalently attached at the point of bond breakage. Such reactions are catalyzed by **phosphorylases**. Glycogen phosphorylase, for example, catalyzes the phosphorolysis of glycogen, producing glucose 1-phosphate. *Dephosphorylation*, the removal of a phosphoryl group from a phosphate

ester, is catalyzed by **phosphatases**, with water as the attacking species. Fructose bisphosphatase-1 converts fructose 1,6-bisphosphate to fructose 6-phosphate in gluconeogenesis, and phosphorylase *a* phosphatase removes phosphoryl groups from phosphoserine in phosphorylated glycogen phosphorylase. Whew!

Unfortunately, these descriptions of enzyme types overlap, and many enzymes are commonly called by two or more names. Succinyl-CoA synthetase, for example, is also called succinate thiokinase; the enzyme is both a synthetase in the citric acid cycle and a kinase when acting in the direction of succinyl-CoA synthesis. This raises another source of confusion in the naming of enzymes. An enzyme may have been discovered by the use of an assay in which, say, A is converted to B. The enzyme is then named for that reaction. Later work may show, however, that in the cell, the enzyme functions primarily in converting B to A. Commonly, the first name continues to be used, although the metabolic role of the enzyme would be better described by naming it for the reverse reaction. The glycolytic enzyme pyruvate kinase illustrates this situation (p. 554). To a beginner in biochemistry, this duplication in nomenclature can be bewildering. International committees have made heroic efforts to systematize the nomenclature of enzymes (see Table 6–3 for a brief summary of the system), but some systematic names have proved too long and cumbersome and are not frequently used in biochemical conversation.

We have tried throughout this book to use the enzyme name most commonly used by working biochemists and to point out cases in which an enzyme has more than one widely used name. For current information on enzyme nomenclature, refer to the recommendations of the Nomenclature Committee of the International Union of Biochemistry and Molecular Biology (www.chem.qmw.ac.uk/iubmb/nomenclature/).

⑥ Oxidation of Succinate to Fumarate The succinate formed from succinyl-CoA is oxidized to **fumarate** by the flavoprotein **succinate dehydrogenase**:

$$\Delta G'^{\circ} = 0 \text{ kJ/mol}$$

In eukaryotes, succinate dehydrogenase is tightly bound to the mitochondrial inner membrane; in bacteria, to the plasma membrane. The enzyme contains three different iron-sulfur clusters and one molecule of covalently bound FAD (see Fig. 19–10). Electrons pass from succinate through the FAD and iron-sulfur centers before entering the chain of electron carriers in the mitochondrial inner membrane (the plasma membrane in bacteria). Electron flow from succinate through these carriers to the final electron acceptor, O_2, is coupled to the synthesis of about 1.5 ATP molecules per pair of electrons (respiration-linked phosphorylation). Malonate, an analog of succinate not normally present in cells, is a strong competitive inhibitor of succinate dehydrogenase, and its addition to mitochondria blocks the activity of the citric acid cycle.

Malonate   Succinate

7 Hydration of Fumarate to Malate The reversible hydration of fumarate to L-**malate** is catalyzed by **fumarase** (formally, **fumarate hydratase**). The transition state in this reaction is a carbanion:

Fumarate

Carbanion transition state

L-Malate

$\Delta G'^{\circ} = -3.8$ kJ/mol

This enzyme is highly stereospecific; it catalyzes hydration of the trans double bond of fumarate but not the cis double bond of maleate (the cis isomer of fumarate). In the reverse direction (from L-malate to fumarate), fumarase is equally stereospecific: D-malate is not a substrate.

Fumarate   Maleate

L-Malate   D-Malate

8 Oxidation of Malate to Oxaloacetate In the last reaction of the citric acid cycle, NAD-linked L-**malate dehydrogenase** catalyzes the oxidation of L-malate to oxaloacetate:

L-Malate   Oxaloacetate

$\Delta G'^{\circ} = 29.7$ kJ/mol

The equilibrium of this reaction lies far to the left under standard thermodynamic conditions, but in intact cells oxaloacetate is continually removed by the highly exergonic citrate synthase reaction (step **2** of Fig. 16–7). This keeps the concentration of oxaloacetate in the cell extremely low ($<10^{-6}$ M), pulling the malate dehydrogenase reaction toward the formation of oxaloacetate.

Although the individual reactions of the citric acid cycle were initially worked out in vitro, using minced muscle tissue, the pathway and its regulation have also been studied extensively in vivo. By using radioactively labeled precursors such as [14C]pyruvate and [14C]acetate, researchers have traced the fate of individual carbon atoms through the citric acid cycle. Some of the earliest experiments with isotopes produced an unexpected result, however, which aroused considerable controversy about the pathway and mechanism of the citric acid cycle. In fact, these experiments at first seemed to show that citrate was not the first tricarboxylic acid to be formed. Box 16–3 gives some details of this episode in the history of citric acid cycle research. Metabolic flux through the cycle can now be monitored in living tissue by using 13C-labeled precursors and whole-tissue NMR spectroscopy. Because the NMR signal is unique to the compound containing the 13C, biochemists can trace the movement of precursor carbons into each cycle intermediate and into compounds derived from the intermediates. This technique has great promise for studies of regulation of the citric acid cycle and its interconnections with other metabolic pathways such as glycolysis.

The Energy of Oxidations in the Cycle Is Efficiently Conserved

We have now covered one complete turn of the citric acid cycle (**Fig. 16–14**). A two-carbon acetyl group entered the cycle by combining with oxaloacetate. Two carbon atoms emerged from the cycle as CO_2 from the oxidation of isocitrate and α-ketoglutarate. The energy released by these oxidations was conserved in the reduction of three NAD+ and one FAD and the production of one ATP or GTP. At the end of the cycle a molecule of oxaloacetate was regenerated. Note that the two carbon atoms appearing as CO_2 are not the same two carbons that entered in the form of the acetyl group; additional turns around the cycle are required to release these carbons as CO_2 (Fig. 16–7).

Although the citric acid cycle directly generates only one ATP per turn (in the conversion of succinyl-CoA

BOX 16–3 Citrate: A Symmetric Molecule That Reacts Asymmetrically

When compounds enriched in the heavy-carbon isotope ^{13}C and the radioactive carbon isotopes ^{11}C and ^{14}C became available about 60 years ago, they were soon put to use in tracing the pathway of carbon atoms through the citric acid cycle. One such experiment initiated the controversy over the role of citrate. Acetate labeled in the carboxyl group (designated [1-^{14}C] acetate) was incubated aerobically with an animal tissue preparation. Acetate is enzymatically converted to acetyl-CoA in animal tissues, and the pathway of the labeled carboxyl carbon of the acetyl group in the cycle reactions could thus be traced. α-Ketoglutarate was isolated from the tissue after incubation, then degraded by known chemical reactions to establish the position(s) of the isotopic carbon.

Condensation of unlabeled oxaloacetate with carboxyl-labeled acetate would be expected to produce citrate labeled in one of the two primary carboxyl groups. Citrate is a symmetric molecule, its two terminal carboxyl groups being chemically indistinguishable. Therefore, half the labeled citrate molecules were expected to yield α-ketoglutarate labeled in the α-carboxyl group and the other half to yield α-ketoglutarate labeled in the γ-carboxyl group; that is, the α-ketoglutarate isolated was expected to be a mixture of the two types of labeled molecules (Fig. 1, pathways ❶ and ❷). Contrary to this expectation, the labeled α-ketoglutarate isolated from the tissue suspension contained ^{14}C only in the γ-carboxyl group (Fig. 1, pathway ❶). The investigators concluded that citrate (or any other symmetric molecule) could not be an intermediate in the pathway from acetate to α-ketoglutarate. Rather, an asymmetric tricarboxylic acid, presumably *cis*-aconitate or isocitrate, must be the first product formed from condensation of acetate and oxaloacetate.

In 1948, however, Alexander Ogston pointed out that although citrate has no chiral center (see Fig. 1–20), it has the *potential* to react asymmetrically if an enzyme with which it interacts has an active site that is asymmetric. He suggested that the active site of aconitase may have three points to which the citrate must be bound and that the citrate must undergo a specific three-point attachment to these binding points. As seen in Figure 2, the binding of citrate to three such points could happen in only one way, and this would account for the formation of only one type of labeled α-ketoglutarate. Organic molecules such as citrate that have no chiral center but are potentially capable of reacting asymmetrically with an asymmetric active site are now called **prochiral molecules**.

FIGURE 1 Incorporation of the isotopic carbon (^{14}C) of the labeled acetyl group into α-ketoglutarate by the citric acid cycle. The carbon atoms of the entering acetyl group are shown in red.

FIGURE 2 The prochiral nature of citrate. **(a)** Structure of citrate; **(b)** schematic representation of citrate: X = —OH; Y = COO⁻; Z = —CH₂COO⁻. **(c)** Correct complementary fit of citrate to the binding site of aconitase. There is only one way in which the three specified groups of citrate can fit on the three points of the binding site. Thus only one of the two —CH₂COO⁻ groups is bound by aconitase.

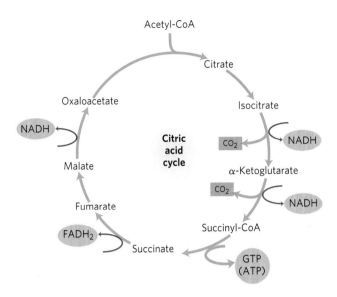

Acetyl-CoA

Citrate

Oxaloacetate

Isocitrate

NADH

Citric acid cycle

CO_2 NADH

Malate

α-Ketoglutarate

CO_2

Fumarate

NADH

$FADH_2$

Succinyl-CoA

Succinate

GTP (ATP)

FIGURE 16–14 Products of one turn of the citric acid cycle. At each turn of the cycle, three NADH, one $FADH_2$, one GTP (or ATP), and two CO_2 are released in oxidative decarboxylation reactions. Here and in several following figures, all cycle reactions are shown as proceeding in one direction only, but keep in mind that most of the reactions are reversible (see Fig. 16–7).

to succinate), the four oxidation steps in the cycle provide a large flow of electrons into the respiratory chain via NADH and $FADH_2$ and thus lead to formation of a large number of ATP molecules during oxidative phosphorylation.

We saw in Chapter 14 that the energy yield from the production of two molecules of pyruvate from one molecule of glucose in glycolysis is 2 ATP and 2 NADH. In oxidative phosphorylation (Chapter 19), passage of two electrons from NADH to O_2 drives the formation of about 2.5 ATP, and passage of two electrons from $FADH_2$ to O_2 yields about 1.5 ATP. This stoichiometry allows us to calculate the overall yield of ATP from the complete oxidation of glucose. When both pyruvate molecules are oxidized to 6 CO_2 via the pyruvate dehydrogenase complex and the citric acid cycle, and the electrons are transferred to O_2 via oxidative phosphorylation, as many as 32 ATP are obtained per glucose (Table 16–1). In round numbers, this represents the conservation of 32×30.5 kJ/mol = 976 kJ/mol or 34% of the theoretical maximum of about 2,840 kJ/mol available from the complete oxidation of glucose. These calculations employ the standard free-energy changes; when corrected for the actual free energy required to form ATP within cells (see Worked Example 13–2, p. 519), the calculated efficiency of the process is closer to 65%.

Why Is the Oxidation of Acetate So Complicated?

The eight-step cyclic process for oxidation of simple two-carbon acetyl groups to CO_2 may seem unnecessarily cumbersome and not in keeping with the biological principle of maximum economy. The role of the citric acid cycle is not confined to the oxidation of acetate, however. This pathway is the hub of intermediary metabolism. Four- and five-carbon end products of many catabolic processes feed into the cycle to serve as fuels. Oxaloacetate and α-ketoglutarate, for example, are produced from aspartate and glutamate, respectively, when proteins are degraded. Under some metabolic circumstances, intermediates are drawn out of the

TABLE 16–1	Stoichiometry of Coenzyme Reduction and ATP Formation in the Aerobic Oxidation of Glucose via Glycolysis, the Pyruvate Dehydrogenase Complex Reaction, the Citric Acid Cycle, and Oxidative Phosphorylation		
Reaction	**Number of ATP or reduced coenzyme directly formed**	**Number of ATP ultimately formed***
Glucose ⟶ glucose 6-phosphate	−1 ATP	−1
Fructose 6-phosphate ⟶ fructose 1,6-bisphosphate	−1 ATP	−1
2 Glyceraldehyde 3-phosphate ⟶ 2 1,3-bisphosphoglycerate	2 NADH	3 or 5[†]
2 1,3-Bisphosphoglycerate ⟶ 2 3-phosphoglycerate	2 ATP	2
2 Phosphoenolpyruvate ⟶ 2 pyruvate	2 ATP	2
2 Pyruvate ⟶ 2 acetyl-CoA	2 NADH	5
2 Isocitrate ⟶ 2 α-ketoglutarate	2 NADH	5
2 α-Ketoglutarate ⟶ 2 succinyl-CoA	2 NADH	5
2 Succinyl-CoA ⟶ 2 succinate	2 ATP (or 2 GTP)	2
2 Succinate ⟶ 2 fumarate	2 $FADH_2$	3
2 Malate ⟶ 2 oxaloacetate	2 NADH	5
Total		30–32

*This is calculated as 2.5 ATP per NADH and 1.5 ATP per $FADH_2$. A negative value indicates consumption.

[†]This number is either 3 or 5, depending on the mechanism used to shuttle NADH equivalents from the cytosol to the mitochondrial matrix; see Figures 19–30 and 19–31.

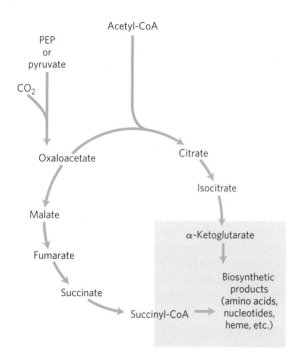

FIGURE 16–15 Biosynthetic precursors produced by an incomplete citric acid cycle in anaerobic bacteria. These anaerobes lack α-ketoglutarate dehydrogenase and therefore cannot carry out the complete citric acid cycle. α-Ketoglutarate and succinyl-CoA serve as precursors in a variety of biosynthetic pathways. (See Fig. 16–14 for the "normal" direction of these reactions in the citric acid cycle.)

cycle to be used as precursors in a variety of biosynthetic pathways.

The citric acid cycle, like all other metabolic pathways, is the product of evolution, and much of this evolution occurred before the advent of aerobic organisms. It does not necessarily represent the *shortest* pathway from acetate to CO_2, but it is the pathway that has, over time, conferred the greatest selective advantage. Early anaerobes most probably used some of the reactions of the citric acid cycle in linear biosynthetic processes. In fact, some modern anaerobic microorganisms use an incomplete citric acid cycle as a source of, not energy, but biosynthetic precursors **(Fig. 16–15)**. These organisms use the first three reactions of the cycle to make α-ketoglutarate but, lacking α-ketoglutarate dehydrogenase, they cannot carry out the complete set of citric acid cycle reactions. They do have the four enzymes that catalyze the reversible conversion of oxaloacetate to succinyl-CoA and can produce malate, fumarate, succinate, and succinyl-CoA from oxaloacetate in a reversal of the "normal" (oxidative) direction of flow through the cycle. This pathway is a fermentation, with the NADH produced by isocitrate oxidation recycled to NAD^+ by reduction of oxaloacetate to succinate.

With the evolution of cyanobacteria that produced O_2 from water, the earth's atmosphere became aerobic and organisms were under selective pressure to develop aerobic metabolism, which, as we have seen, is much more efficient than anaerobic fermentation.

Citric Acid Cycle Components Are Important Biosynthetic Intermediates

In aerobic organisms, the citric acid cycle is an **amphibolic pathway**, one that serves in both catabolic and anabolic processes. Besides its role in the oxidative catabolism of carbohydrates, fatty acids, and amino acids, the cycle provides precursors for many biosynthetic pathways **(Fig. 16–16)**, through reactions that served the same purpose in anaerobic ancestors. α-Ketoglutarate and oxaloacetate can, for example, serve as precursors of the amino acids aspartate and glutamate by simple transamination (Chapter 22). Through aspartate and glutamate, the carbons of oxaloacetate and α-ketoglutarate are then used to build other amino acids, as well as purine and pyrimidine nucleotides. Oxaloacetate is converted to glucose in gluconeogenesis (see Fig. 15–13). Succinyl-CoA is a central intermediate in the synthesis of the porphyrin ring of heme groups, which serve as oxygen carriers (in hemoglobin and myoglobin) and electron carriers (in cytochromes) (see Fig. 22–25). And the citrate produced in some organisms is used commercially for a variety of purposes.

Anaplerotic Reactions Replenish Citric Acid Cycle Intermediates

As intermediates of the citric acid cycle are removed to serve as biosynthetic precursors, they are replenished by **anaplerotic reactions** (Fig. 16–16; Table 16–2). Under normal circumstances, the reactions by which cycle intermediates are siphoned off into other pathways and those by which they are replenished are in dynamic balance, so that the concentrations of the citric acid cycle intermediates remain almost constant.

Table 16–2 shows the most common anaplerotic reactions, all of which, in various tissues and organisms, convert either pyruvate or phosphoenolpyruvate to oxaloacetate or malate. The most important anaplerotic reaction in mammalian liver and kidney is the reversible carboxylation of pyruvate by CO_2 to form oxaloacetate, catalyzed by **pyruvate carboxylase.** When the citric acid cycle is deficient in oxaloacetate or any other intermediates, pyruvate is carboxylated to produce more oxaloacetate. The enzymatic addition of a carboxyl group to pyruvate requires energy, which is supplied by ATP—the free energy required to attach a carboxyl group to pyruvate is about equal to the free energy available from ATP.

Pyruvate carboxylase is a regulatory enzyme and is virtually inactive in the absence of acetyl-CoA, its positive allosteric modulator. Whenever acetyl-CoA, the fuel for the citric acid cycle, is present in excess, it stimulates the pyruvate carboxylase reaction to produce more oxaloacetate, enabling the cycle to use more acetyl-CoA in the citrate synthase reaction.

The other anaplerotic reactions shown in Table 16–2 are also regulated to keep the level of intermediates

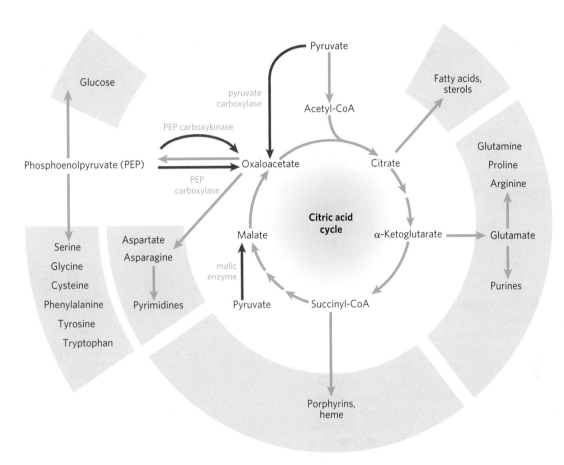

FIGURE 16–16 Role of the citric acid cycle in anabolism. Intermediates of the citric acid cycle are drawn off as precursors in many biosynthetic pathways. Shown in red are four anaplerotic reactions that replenish depleted cycle intermediates (see Table 16–2).

high enough to support the activity of the citric acid cycle. Phosphoenolpyruvate (PEP) carboxylase, for example, is activated by the glycolytic intermediate fructose 1,6-bisphosphate, which accumulates when the citric acid cycle operates too slowly to process the pyruvate generated by glycolysis.

Biotin in Pyruvate Carboxylase Carries CO_2 Groups

The pyruvate carboxylase reaction requires the vitamin **biotin (Fig. 16–17),** which is the prosthetic group of the enzyme. Biotin plays a key role in many carboxyl-

ation reactions. It is a specialized carrier of one-carbon groups in their most oxidized form: CO_2. (The transfer of one-carbon groups in more reduced forms is mediated by other cofactors, notably tetrahydrofolate and S-adenosylmethionine, as described in Chapter 18.) Carboxyl groups are activated in a reaction that consumes ATP and joins CO_2 to enzyme-bound biotin. This "activated" CO_2 is then passed to an acceptor (pyruvate in this case) in a carboxylation reaction.

Pyruvate carboxylase has four identical subunits, each containing a molecule of biotin covalently attached through an amide linkage to the ε-amino

TABLE 16–2	**Anaplerotic Reactions**
Reaction	**Tissue(s)/organism(s)**
Pyruvate + HCO_3^- + ATP $\xrightleftharpoons[]{\text{pyruvate carboxylase}}$ oxaloacetate + ADP + P_i	Liver, kidney
Phosphoenolpyruvate + CO_2 + GDP $\xrightleftharpoons[]{\text{PEP carboxykinase}}$ oxaloacetate + GTP	Heart, skeletal muscle
Phosphoenolpyruvate + HCO_3^- $\xrightleftharpoons[]{\text{PEP carboxylase}}$ oxaloacetate + P_i	Higher plants, yeast, bacteria
Pyruvate + HCO_3^- + NAD(P)H $\xrightleftharpoons[]{\text{malic enzyme}}$ malate + NAD(P)$^+$	Widely distributed in eukaryotes and bacteria

MECHANISM FIGURE 16–17 The role of biotin in the reaction catalyzed by pyruvate carboxylase. Biotin is attached to the enzyme through an amide bond with the ε-amino group of a Lys residue, forming biotinyl-enzyme. Biotin-mediated carboxylation reactions occur in two phases, generally catalyzed in separate active sites on the enzyme, as exemplified by the pyruvate carboxylase reaction. In the first phase (steps ❶ to ❸), bicarbonate is converted to the more activated CO_2, and then used to carboxylate biotin. The biotin acts as a carrier to transport the CO_2 from one active site to another on an adjacent monomer of the tetrameric enzyme (step ❹). In the second phase (steps ❺ to ❼), catalyzed in this second active site, the CO_2 reacts with pyruvate to form oxaloacetate.

group of a specific Lys residue in the enzyme active site. Carboxylation of pyruvate proceeds in two steps (Fig. 16–17): first, a carboxyl group derived from HCO_3^- is attached to biotin, then the carboxyl group is transferred to pyruvate to form oxaloacetate. These two steps occur at separate active sites; the long flex-ible arm of biotin transfers activated carboxyl groups from the first active site (on one monomer of the tetramer) to the second (on the adjacent monomer), functioning much like the long lipoyllysyl arm of E_2 in the PDH complex (Fig. 16–6) and the long arm of the CoA-like moiety in the acyl carrier protein involved in

FIGURE 16–18 **Biological tethers.** The cofactors lipoate, biotin, and the combination of β-mercaptoethylamine and pantothenate form long, flexible arms (blue) on the enzymes to which they are covalently bound, acting as tethers that move intermediates from one active site to the next. The group shaded light red is in each case the point of attachment of the activated intermediate to the tether.

fatty acid synthesis (see Fig. 21–5); these are compared in **Figure 16–18**. Lipoate, biotin, and pantothenate all enter cells on the same transporter; all become covalently attached to proteins by similar reactions; and all provide a flexible tether that allows bound reaction intermediates to move from one active site to another in an enzyme complex, without dissociating from it—all, that is, participate in substrate channeling.

Biotin is a vitamin required in the human diet; it is abundant in many foods and is synthesized by intestinal bacteria. Biotin deficiency is rare, but can sometimes be caused by a diet rich in raw eggs. Egg whites contain a large amount of the protein **avidin** (M_r 70,000), which binds very tightly to biotin and prevents its absorption in the intestine. The avidin of egg whites may be a defense mechanism for the potential chick embryo, inhibiting the growth of bacteria. When eggs are cooked, avidin is denatured (and thereby inactivated) along with all other egg white proteins. Purified avidin is a useful reagent in biochemistry and cell biology. A protein that contains covalently bound biotin (derived experimentally or produced in vivo) can be recovered by affinity chromatography (see Fig. 3–17c) based on biotin's strong affinity for avidin. The protein is then eluted from the column with an excess of free biotin. The very high affinity of biotin for avidin is also used in the laboratory in the form of a molecular glue that can hold two structures together (see Fig. 19–27).

SUMMARY 16.2 Reactions of the Citric Acid Cycle

▶ The citric acid cycle (Krebs cycle, TCA cycle) is a nearly universal central catabolic pathway in which compounds derived from the breakdown of carbohydrates, fats, and proteins are oxidized to CO_2, with most of the energy of oxidation temporarily held in the electron carriers $FADH_2$ and NADH. During aerobic metabolism, these electrons are transferred to O_2 and the energy of electron flow is trapped as ATP.

▶ Acetyl-CoA enters the citric acid cycle (in the mitochondria of eukaryotes, the cytosol of bacteria) as citrate synthase catalyzes its condensation with oxaloacetate to form citrate.

▶ In seven sequential reactions, including two decarboxylations, the citric acid cycle converts citrate to oxaloacetate and releases two CO_2. The pathway is cyclic in that the intermediates of the cycle are not used up; for each oxaloacetate consumed in the path, one is produced.

▶ For each acetyl-CoA oxidized by the citric acid cycle, the energy gain consists of three molecules of NADH, one $FADH_2$, and one nucleoside triphosphate (either ATP or GTP).

▶ Besides acetyl-CoA, any compound that gives rise to a four- or five-carbon intermediate of the citric acid cycle—for example, the breakdown products of many amino acids—can be oxidized by the cycle.

▶ The citric acid cycle is amphibolic, serving in both catabolism and anabolism; cycle intermediates can be drawn off and used as the starting material for a variety of biosynthetic products.

▶ When intermediates are shunted from the citric acid cycle to other pathways, they are replenished by several anaplerotic reactions, which produce four-carbon intermediates by carboxylation of three-carbon compounds; these reactions are catalyzed by pyruvate carboxylase, PEP carboxykinase, PEP carboxylase, and malic enzyme. Enzymes that catalyze carboxylations commonly employ biotin to activate CO_2 and to carry it to acceptors such as pyruvate or phosphoenolpyruvate.

16.3 Regulation of the Citric Acid Cycle

As we have seen in Chapter 15, the regulation of key enzymes in metabolic pathways, by allosteric effectors and by covalent modification, ensures the production of intermediates at the rates required to keep the cell in a stable steady state while avoiding wasteful overproduction. The flow of carbon atoms from pyruvate into and through the citric acid cycle is under tight regulation at two levels: the conversion of pyruvate to acetyl-CoA, the starting material for the cycle (the pyruvate dehydrogenase complex reaction), and the entry of acetyl-CoA into the cycle (the citrate synthase reaction). Acetyl-CoA is also produced by pathways other than the PDH complex reaction—most cells produce acetyl-CoA from the oxidation of fatty acids and certain amino acids—and the availability of intermediates from these other pathways is important in the regulation of pyruvate oxidation and of the citric acid cycle. The cycle is also regulated at the isocitrate dehydrogenase and α-ketoglutarate dehydrogenase reactions.

Production of Acetyl-CoA by the Pyruvate Dehydrogenase Complex Is Regulated by Allosteric and Covalent Mechanisms

The PDH complex of mammals is strongly inhibited by ATP and by acetyl-CoA and NADH, the products of the reaction catalyzed by the complex **(Fig. 16–19)**. The allosteric inhibition of pyruvate oxidation is greatly enhanced when long-chain fatty acids are available. AMP, CoA, and NAD^+, all of which accumulate when too little acetate flows into the citric acid cycle, allosterically activate the PDH complex. Thus, this enzyme

activity is turned off when ample fuel is available in the form of fatty acids and acetyl-CoA and when the cell's [ATP]/[ADP] and [NADH]/[NAD$^+$] ratios are high, and it is turned on again when energy demands are high and the cell requires greater flux of acetyl-CoA into the citric acid cycle.

In mammals, these allosteric regulatory mechanisms are complemented by a second level of regulation: covalent protein modification. The PDH complex is inhibited by reversible phosphorylation of a specific Ser residue on one of the two subunits of E_1. As noted earlier, in addition to the enzymes E_1, E_2, and E_3, the mammalian PDH complex contains two regulatory proteins whose sole purpose is to regulate the activity of the complex. Pyruvate dehydrogenase kinase phosphorylates and thereby inactivates E_1, and a specific phosphoprotein phosphatase removes the phosphoryl group by hydrolysis and thereby activates E_1. The kinase is allosterically activated by ATP: when [ATP] is high (reflecting a sufficient supply of energy), the PDH complex is inactivated by phosphorylation of E_1. When [ATP] declines, kinase activity decreases and phosphatase action removes the phosphoryl groups from E_1, activating the complex.

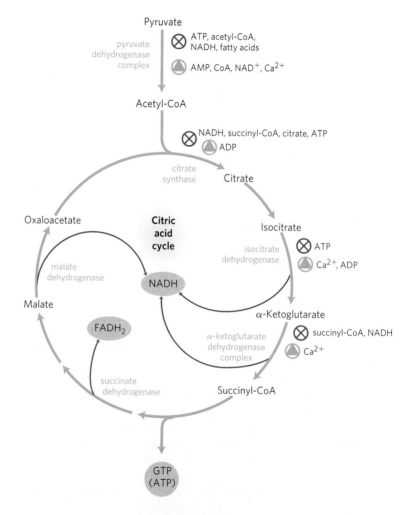

FIGURE 16–19 Regulation of metabolite flow from the PDH complex through the citric acid cycle in mammals. The PDH complex is allosterically inhibited when [ATP]/[ADP], [NADH]/[NAD$^+$], and [acetyl-CoA]/[CoA] ratios are high, indicating an energy-sufficient metabolic state. When these ratios decrease, allosteric activation of pyruvate oxidation results. The rate of flow through the citric acid cycle can be limited by the availability of the citrate synthase substrates, oxaloacetate and acetyl-CoA, or of NAD^+, which is depleted by its conversion to NADH, slowing the three NAD-dependent oxidation steps. Feedback inhibition by succinyl-CoA, citrate, and ATP also slows the cycle by inhibiting early steps. In muscle tissue, Ca^{2+} signals contraction and, as shown here, stimulates energy-yielding metabolism to replace the ATP consumed by contraction.

The PDH complex of plants, located in the mitochondrial matrix and in plastids, is inhibited by its products, NADH and acetyl-CoA. The plant mitochondrial enzyme is also regulated by reversible phosphorylation; pyruvate inhibits the kinase, thus activating the PDH complex, and NH_4^+ stimulates the kinase, causing inactivation of the complex. The PDH complex of *E. coli* is under allosteric regulation similar to that of the mammalian enzyme, but it does not seem to be regulated by phosphorylation.

The Citric Acid Cycle Is Regulated at Its Three Exergonic Steps

The flow of metabolites through the citric acid cycle is under stringent regulation. Three factors govern the rate of flux through the cycle: substrate availability, inhibition by accumulating products, and allosteric feedback inhibition of the enzymes that catalyze early steps in the cycle.

Each of the three strongly exergonic steps in the cycle—those catalyzed by citrate synthase, isocitrate dehydrogenase, and α-ketoglutarate dehydrogenase (Fig. 16–19)—can become the rate-limiting step under some circumstances. The availability of the substrates for citrate synthase (acetyl-CoA and oxaloacetate) varies with the metabolic state of the cell and sometimes limits the rate of citrate formation. NADH, a product of isocitrate and α-ketoglutarate oxidation, accumulates under some conditions, and at high [NADH]/[NAD$^+$] both dehydrogenase reactions are severely inhibited by mass action. Similarly, in the cell, the malate dehydrogenase reaction is essentially at equilibrium (that is, it is substrate-limited), and when [NADH]/[NAD$^+$] is high the concentration of oxaloacetate is low, slowing the first step in the cycle. Product accumulation inhibits all three limiting steps of the cycle: succinyl-CoA inhibits α-ketoglutarate dehydrogenase (and also citrate synthase); citrate blocks citrate synthase; and the end product, ATP, inhibits both citrate synthase and isocitrate dehydrogenase. The inhibition of citrate synthase by ATP is relieved by ADP, an allosteric activator of this enzyme. In vertebrate muscle, Ca^{2+}, the signal for contraction and for a concomitant increase in demand for ATP, activates both isocitrate dehydrogenase and α-ketoglutarate dehydrogenase, as well as the PDH complex. In short, the concentrations of substrates and intermediates in the citric acid cycle set the flux through this pathway at a rate that provides optimal concentrations of ATP and NADH.

Under normal conditions, the rates of glycolysis and of the citric acid cycle are integrated so that only as much glucose is metabolized to pyruvate as is needed to supply the citric acid cycle with its fuel, the acetyl groups of acetyl-CoA. Pyruvate, lactate, and acetyl-CoA are normally maintained at steady-state concentrations. The rate of glycolysis is matched to the rate of the citric acid cycle not only through its inhibition by high levels of ATP and NADH, which are common to both the glycolytic and respiratory stages of glucose oxidation, but also by the concentration of citrate. Citrate, the product of the first step of the citric acid cycle, is an important allosteric inhibitor of phosphofructokinase-1 in the glycolytic pathway (see Fig. 15–16).

Substrate Channeling through Multienzyme Complexes May Occur in the Citric Acid Cycle

Although the enzymes of the citric acid cycle are usually described as soluble components of the mitochondrial matrix (except for succinate dehydrogenase, which is membrane-bound), growing evidence suggests that within the mitochondrion these enzymes exist as multienzyme complexes. The classic approach of enzymology—purification of individual proteins from extracts of broken cells—was applied with great success to the citric acid cycle enzymes. However, the first casualty of cell breakage is higher-level organization within the cell—the noncovalent, reversible interaction of one protein with another, or of an enzyme with some structural component such as a membrane, microtubule, or microfilament. When cells are broken open, their contents, including enzymes, are diluted 100- or 1,000-fold **(Fig. 16–20)**.

In the cytosol, high concentrations of enzymes 1, 2, and 3 favor their association.

In extract of broken cells, dilution by buffer reduces the concentrations of enzymes 1, 2, and 3, favoring their dissociation.

FIGURE 16–20 Dilution of a solution containing a noncovalently bound protein complex—such as one consisting of three enzymes (illustrated here in red, blue, and green)—favors dissociation of the complex into its constituents.

Several types of evidence suggest that, in cells, multienzyme complexes ensure efficient passage of the product of one enzyme reaction to the next enzyme in the pathway. Such complexes are called **metabolons**. Certain enzymes of the citric acid cycle have been isolated together as supramolecular complexes, or have been found associated with the inner mitochondrial membrane, or have been shown to diffuse in the mitochondrial matrix more slowly than expected for the individual protein in solution. There is strong evidence for substrate channeling through multienzyme complexes in other metabolic pathways, and many enzymes thought of as "soluble" probably function in the cell as highly organized complexes that channel intermediates. We will encounter other examples of channeling when we discuss the biosynthesis of amino acids and nucleotides in Chapter 22.

Some Mutations in Enzymes of the Citric Acid Cycle Lead to Cancer

When the mechanisms for regulating a pathway such as the citric acid cycle are overwhelmed by a major metabolic perturbation, the result can be serious disease. Mutations in citric acid cycle enzymes are very rare in humans and other mammals, but those that do occur are devastating. Genetic defects in the fumarase gene lead to tumors of smooth muscle (leiomas) and kidney; mutations in succinate dehydrogenase lead to tumors of the adrenal gland (pheochromocytomas). In cultured cells with these mutations, fumarate (in the case of fumarase mutations) and, to a lesser extent, succinate (in the case of succinate dehydrogenase mutations) accumulate, and this accumulation induces the hypoxia-inducible transcription factor HIF-1α (see Box 14–1). The mechanism of tumor formation may be the production of a pseudohypoxic state. In cells with these mutations, there is an up-regulation of genes normally regulated by HIF-1α. These effects of mutations in the fumarase and succinate dehydrogenase genes define them as tumor suppressor genes (p. 489).

Another remarkable connection between citric acid cycle intermediates and cancer is the finding that in many glial cell tumors (gliomas), the NADPH-dependent isocitrate dehydrogenase has an unusual genetic defect. The mutant enzyme loses its normal activity (converting isocitrate to α-ketoglutarate) but *gains* a new activity: it converts α-ketoglutarate to 2-hydroxyglutarate **(Fig. 16–21)**, which accumulates in the tumor cells. α-Ketoglutarate and Fe^{3+} are essential cofactors for a family of histone demethylases that alter gene expression by removing methyl groups from Arg and Lys residues in the histones that organize nuclear DNA. By competing with α-ketoglutarate for binding to the histone demethylases, 2-hydroxyglutarate inhibits their activity. The inhibition of the histone demethylases in turn interferes with normal gene regulation, leading to unrestricted glial cell growth. ■

FIGURE 16-21 A mutant isocitrate dehydrogenase acquires a new activity. Wild-type isocitrate dehydrogenase catalyzes the conversion of isocitrate to α-ketoglutarate, but mutations that alter the binding site for isocitrate cause loss of the normal enzymatic activity and gain of a new activity: conversion of α-ketoglutarate to 2-hydroxyglutarate. Accumulation of this product inhibits histone demethylase, altering gene regulation and leading to glial cell tumors in the brain.

SUMMARY 16.3 Regulation of the Citric Acid Cycle

▶ The overall rate of the citric acid cycle is controlled by the rate of conversion of pyruvate to acetyl-CoA and by the flux through citrate synthase, isocitrate dehydrogenase, and α-ketoglutarate dehydrogenase. These fluxes are largely determined by the concentrations of substrates and products: the end products ATP and NADH are inhibitory, and the substrates NAD^+ and ADP are stimulatory.

▶ The production of acetyl-CoA for the citric acid cycle by the PDH complex is inhibited allosterically by metabolites that signal a sufficiency of metabolic energy (ATP, acetyl-CoA, NADH, and fatty acids) and stimulated by metabolites that indicate a reduced energy supply (AMP, NAD^+, CoA).

▶ Complexes of consecutive enzymes in a pathway allow substrate channeling between them.

16.4 The Glyoxylate Cycle

Vertebrates cannot convert fatty acids, or the acetate derived from them, to carbohydrates. Conversion of phosphoenolpyruvate to pyruvate (p. 554) and of pyruvate to acetyl-CoA (Fig. 16–2) are so exergonic as to be essentially irreversible. If a cell cannot convert acetate into phosphoenolpyruvate, acetate cannot serve as the starting material for the gluconeogenic pathway, which leads from phosphoenolpyruvate to glucose (see Fig. 15–13). Without this capacity, then, a cell or organism is unable to convert fuels or metabolites that are degraded to acetate (fatty acids and certain amino acids) into carbohydrates.

As noted in the discussion of anaplerotic reactions (Table 16–2), phosphoenolpyruvate can be synthesized from oxaloacetate in the reversible reaction catalyzed by PEP carboxykinase:

$$\text{Oxaloacetate} + \text{GTP} \rightleftharpoons$$
$$\text{phosphoenolpyruvate} + CO_2 + \text{GDP}$$

Because the carbon atoms of acetate molecules that enter the citric acid cycle appear eight steps later in oxaloacetate, it might seem that this pathway could generate oxaloacetate from acetate and thus generate phosphoenolpyruvate for gluconeogenesis. However, as an examination of the stoichiometry of the citric acid cycle shows, there is no *net* conversion of acetate to oxaloacetate; in vertebrates, for every two carbons that enter the cycle as acetyl-CoA, two leave as CO_2. In many organisms other than vertebrates, the glyoxylate cycle serves as a mechanism for converting acetate to carbohydrate.

The Glyoxylate Cycle Produces Four-Carbon Compounds from Acetate

In plants, certain invertebrates, and some microorganisms (including *E. coli* and yeast) acetate can serve both as an energy-rich fuel and as a source of phosphoenolpyruvate for carbohydrate synthesis. In these organisms, enzymes of the **glyoxylate cycle** catalyze the net conversion of acetate to succinate or other four-carbon intermediates of the citric acid cycle:

$$2\,\text{Acetyl-CoA} + NAD^+ + 2H_2O \longrightarrow$$
$$\text{succinate} + 2\text{CoA} + \text{NADH} + H^+$$

In the glyoxylate cycle, acetyl-CoA condenses with oxaloacetate to form citrate, and citrate is converted to isocitrate, exactly as in the citric acid cycle. The next step, however, is not the breakdown of isocitrate by isocitrate dehydrogenase but the cleavage of isocitrate by **isocitrate lyase**, forming succinate and **glyoxylate**. The glyoxylate then condenses with a second molecule of acetyl-CoA to yield malate, in a reaction catalyzed by **malate synthase**. The malate is subsequently oxidized to oxaloacetate, which can condense with another molecule of acetyl-CoA to start another turn of the cycle **(Fig. 16–22)**. Each turn of the glyoxylate cycle consumes two molecules of acetyl-CoA and produces one molecule of succinate, which is then available for biosynthetic purposes. The succinate may be converted through fumarate and malate into oxaloacetate, which can then be converted to phosphoenolpyruvate by PEP carboxykinase, and

FIGURE 16–22 Glyoxylate cycle. The citrate synthase, aconitase, and malate dehydrogenase of the glyoxylate cycle are isozymes of the citric acid cycle enzymes; isocitrate lyase and malate synthase are unique to the glyoxylate cycle. Notice that two acetyl groups (light red) enter the cycle and four carbons leave as succinate (blue). The glyoxylate cycle was elucidated by Hans Kornberg and Neil Madsen in the laboratory of Hans Krebs.

thus to glucose by gluconeogenesis. Vertebrates do not have the enzymes specific to the glyoxylate cycle (isocitrate lyase and malate synthase) and therefore cannot bring about the net synthesis of glucose from lipids.

In plants, the enzymes of the glyoxylate cycle are sequestered in membrane-bounded organelles called glyoxysomes, which are specialized peroxisomes **(Fig. 16–23)**. Those enzymes common to the citric acid and glyoxylate cycles have two isozymes, one specific to mitochondria, the other to glyoxysomes. Glyoxysomes are not present in all plant tissues at all times. They develop in lipid-rich seeds during germination, before the developing plant acquires the ability to make glucose by photosynthesis. In addition to glyoxylate cycle enzymes, glyoxysomes contain all the enzymes needed for the degradation of the fatty acids stored in seed oils (see Fig. 17–14). Acetyl-CoA formed from lipid breakdown is converted to succinate via the glyoxylate cycle, and the succinate is exported to mitochondria, where citric acid cycle enzymes transform it to malate. A cytosolic isozyme

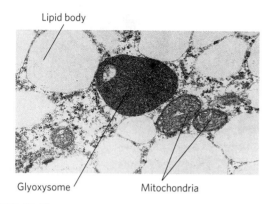

FIGURE 16–23 Electron micrograph of a germinating cucumber seed, showing a glyoxysome, mitochondria, and surrounding lipid bodies.

of malate dehydrogenase oxidizes malate to oxaloacetate, a precursor for gluconeogenesis. Germinating seeds can therefore convert the carbon of stored lipids into glucose.

The Citric Acid and Glyoxylate Cycles Are Coordinately Regulated

In germinating seeds, the enzymatic transformations of dicarboxylic and tricarboxylic acids occur in three intracellular compartments: mitochondria, glyoxysomes, and the cytosol. There is a continuous interchange of metabolites among these compartments (**Fig. 16–24**).

The carbon skeleton of oxaloacetate from the citric acid cycle (in the mitochondrion) is carried to the glyoxysome in the form of aspartate. Aspartate is converted to oxaloacetate, which condenses with acetyl-CoA derived from fatty acid breakdown. The citrate thus formed is converted to isocitrate by aconitase, then split into glyoxylate and succinate by isocitrate lyase. The succinate returns to the mitochondrion, where it reenters the citric acid cycle and is transformed into malate, which enters the cytosol and is oxidized (by cytosolic malate dehydrogenase) to oxaloacetate. Oxaloacetate is converted via gluconeogenesis into hexoses and sucrose, which can be transported to the growing roots and shoot. Four distinct pathways participate in these conversions: fatty acid breakdown to acetyl-CoA (in glyoxysomes), the glyoxylate cycle (in glyoxysomes), the citric acid cycle (in mitochondria), and gluconeogenesis (in the cytosol).

The sharing of common intermediates requires that these pathways be coordinately regulated. Isocitrate is a crucial intermediate, at the branch point between the glyoxylate and citric acid cycles (**Fig. 16–25**). Isocitrate dehydrogenase is regulated by covalent modification: a specific protein kinase phosphorylates and thereby inactivates the dehydrogenase. This inactivation shunts isocitrate to the glyoxylate cycle, where it begins the synthetic route toward glucose. A phosphoprotein phosphatase removes the phosphoryl group from isocitrate dehydrogenase, reactivating the enzyme and sending more isocitrate through the energy-yielding

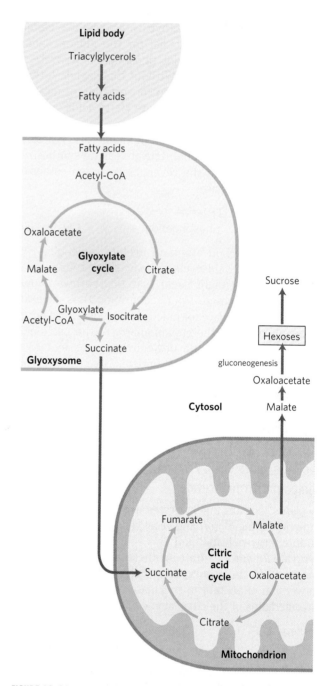

FIGURE 16–24 Relationship between the glyoxylate and citric acid cycles. The reactions of the glyoxylate cycle (in glyoxysomes) proceed simultaneously with, and mesh with, those of the citric acid cycle (in mitochondria), as intermediates pass between these compartments. The conversion of succinate to oxaloacetate is catalyzed by citric acid cycle enzymes. The oxidation of fatty acids to acetyl-CoA is described in Chapter 17; the synthesis of hexoses from oxaloacetate is described in Chapter 20.

citric acid cycle. The regulatory protein kinase and phosphoprotein phosphatase are separate enzymatic activities of a single polypeptide.

Some bacteria, including *E. coli*, have the full complement of enzymes for the glyoxylate and citric acid cycles in the cytosol and can therefore grow on acetate as their sole source of carbon and energy. The phosphoprotein phosphatase that activates isocitrate dehydrogenase

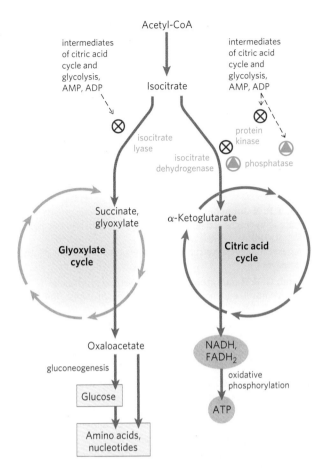

FIGURE 16–25 Coordinated regulation of glyoxylate and citric acid cycles. Regulation of isocitrate dehydrogenase activity determines the partitioning of isocitrate between the glyoxylate and citric acid cycles. When the enzyme is inactivated by phosphorylation (by a specific protein kinase), isocitrate is directed into biosynthetic reactions via the glyoxylate cycle. When the enzyme is activated by dephosphorylation (by a specific phosphatase), isocitrate enters the citric acid cycle and ATP is produced.

is stimulated by intermediates of the citric acid cycle and glycolysis and by indicators of reduced cellular energy supply (Fig. 16–25). The same metabolites *inhibit* the protein kinase activity of the bifunctional polypeptide. Thus, the accumulation of intermediates of the central energy-yielding pathways—indicating energy depletion—results in the activation of isocitrate dehydrogenase. When the concentration of these regulators falls, signaling a sufficient flux through the energy-yielding citric acid cycle, isocitrate dehydrogenase is inactivated by the protein kinase.

The same intermediates of glycolysis and the citric acid cycle that activate isocitrate dehydrogenase are allosteric inhibitors of isocitrate lyase. When energy-yielding metabolism is sufficiently fast to keep the concentrations of glycolytic and citric acid cycle intermediates low, isocitrate dehydrogenase is inactivated, the inhibition of isocitrate lyase is relieved, and isocitrate flows into the glyoxylate pathway, to be used in the biosynthesis of carbohydrates, amino acids, and other cellular components.

SUMMARY 16.4 The Glyoxylate Cycle

▶ The glyoxylate cycle is active in the germinating seeds of some plants and in certain microorganisms that can live on acetate as the sole carbon source. In plants, the pathway takes place in glyoxysomes in seedlings. It involves several citric acid cycle enzymes and two additional enzymes: isocitrate lyase and malate synthase.

▶ In the glyoxylate cycle, the bypassing of the two decarboxylation steps of the citric acid cycle makes possible the *net* formation of succinate, oxaloacetate, and other cycle intermediates from acetyl-CoA. Oxaloacetate thus formed can be used to synthesize glucose via gluconeogenesis.

▶ Vertebrates lack the glyoxylate cycle and cannot synthesize glucose from acetate or the fatty acids that give rise to acetyl-CoA.

▶ The partitioning of isocitrate between the citric acid cycle and the glyoxylate cycle is controlled at the level of isocitrate dehydrogenase, which is regulated by reversible phosphorylation.

Key Terms

Terms in bold are defined in the glossary.

respiration 633
cellular respiration 633
citric acid cycle 633
tricarboxylic acid (TCA) cycle 633
Krebs cycle 633
pyruvate dehydrogenase (PDH) complex 634
oxidative decarboxylation 634
thioester 635
lipoate 635
substrate channeling 637
iron-sulfur center 641
moonlighting enzymes 642
α-ketoglutarate dehydrogenase complex 644

nucleoside diphosphate kinase 645
synthases 646
synthetases 646
ligases 646
lyases 646
kinases 646
phosphorylases 646
phosphatases 646
prochiral molecule 648
amphibolic pathway 650
anaplerotic reaction 650
biotin 651
avidin 653
metabolon 656
glyoxylate cycle 657

Further Reading

General

Holmes, F.L. (1990, 1993) *Hans Krebs*, Vol 1: *Formation of a Scientific Life, 1900–1933*; Vol. 2: *Architect of Intermediary Metabolism, 1933–1937*, Oxford University Press, Oxford.

A scientific and personal biography of Krebs by an eminent historian of science, with a thorough description of the work that revealed the urea and citric acid cycles.

Kay, J. & Weitzman, P.D.J. (eds). (1987) *Krebs' Citric Acid Cycle: Half a Century and Still Turning*, Biochemical Society Symposium **54**, The Biochemical Society, London.

A multiauthor book on the citric acid cycle, including molecular genetics, regulatory mechanisms, variations on the cycle in microorganisms from unusual ecological niches, and evolution of the pathway. Especially relevant are the chapters by H. Gest (Evolutionary Roots of the Citric Acid Cycle in Prokaryotes), W. H. Holms (Control of Flux through the Citric Acid Cycle and the Glyoxylate Bypass in *Escherichia coli*), and R. N. Perham et al. (α-Keto Acid Dehydrogenase Complexes).

Pyruvate Dehydrogenase Complex

Harris, R.A., Bowker-Kinley, M.M., Huang, B., & Wu, P. (2002) Regulation of the activity of the pyruvate dehydrogenase complex. *Adv. Enzyme Regul.* **42**, 249–259.

Milne, J.L.S., Shi, D., Rosenthal, P.B., Sunshine, J.S., Domingo, G.J., Wu, X., Brooks, B.R., Perham, R.N., Henderson, R., & Subramaniam, S. (2002) Molecular architecture and mechanism of an icosahedral pyruvate dehydrogenase complex: a multifunctional catalytic machine. *EMBO J.* **21**, 5587–5598.

Beautiful illustration of the power of image reconstruction methodology with cryoelectron microscopy, here used to develop a plausible model for the structure of the PDH complex. Compare this model with that in the paper by Zhou et al. (below).

Perham, R.N. (2000) Swinging arms and swinging domains in multifunctional enzymes: catalytic machines for multistep reactions. *Annu. Rev. Biochem.* **69**, 961–1004.

Review of the roles of swinging arms containing lipoate, biotin, and pantothenate in substrate channeling through multienzyme complexes.

Zhou, Z.H., McCarthy, D.B., O'Conner, C.M., Reed, L.J., & Stoops, J.K. (2001) The remarkable structural and functional organization of the eukaryotic pyruvate dehydrogenase complexes. *Proc. Natl. Acad. Sci. USA* **98**, 14,802–14,807.

Another striking paper in which image reconstruction with cryoelectron microscopy yields a model of the PDH complex. Compare this model with that in the paper by Milne et al. (above).

Citric Acid Cycle Enzymes

de la Fuente, J.M., Ramírez-Rodríguez, V., Cabrera-Ponce, J.L., & Herrera-Estrella, L. (1997) Aluminum tolerance in transgenic plants by alteration of citrate synthesis. *Science* **276**, 1566–1568.

Fraser, M.D., James, M.N., Bridger, W.A., & Wolodko, W.T. (1999) A detailed structural description of *Escherichia coli* succinyl-CoA synthetase. *J. Mol. Biol.* **285**, 1633–1653. (See also the erratum in *J. Mol. Biol.* **288**, 501 (1998).)

Goward, C.R. & Nicholls, D.J. (1994) Malate dehydrogenase: a model for structure, evolution, and catalysis. *Protein Sci.* **3**, 1883–1888.

A good, short review.

Hagerhall, C. (1997) Succinate:quinone oxidoreductases: variations on a conserved theme. *Biochim. Biophys. Acta* **1320**, 107–141.

A review of the structure and function of succinate dehydrogenases.

Hanson, R.W. (2009) Thematic minireview series: a perspective on the biology of phosphoenolpyruvate carboxykinase 55 years after its discovery. *J. Biol. Chem.* **284**, 27,021–27,023.

The editorial introduction to a series of minireviews in this journal issue on PEP carboxykinase.

Jitrapakdee, S., St. Maurice, M., Rayment, I., Cleland, W.W., Wallace, J.C., & Attwood, P.V. (2008) Structure, mechanism and regulation of pyruvate carboxylase. *Biochem. J.* **413**, 369–387.

Ma, J.F., Ryan, P.R., & Delhaize, E. (2001) Aluminium tolerance in plants and the complexing role of organic acids. *Trends Plant Sci.* **6**, 273–278.

Matte, A., Tari, L.W., Goldie, H., & Delbaere, L.T.J. (1997) Structure and mechanism of phosphoenolpyruvate carboxykinase. *J. Biol. Chem.* **272**, 8105–8108.

Ovadi, J. & Srere, P. (2000) Macromolecular compartmentation and channeling. *Int. Rev. Cytol.* **192**, 255–280.

Advanced review of the evidence for channeling and metabolons.

Prensner, J.R. & Chinnaiyan, A.M. (2011) Metabolism unhinged: IDH mutations in cancer. *Nat. Med.* **17**, 291–293.

Brief review of the mutations of isocitrate dehydrogenase associated with cancer.

Remington, S.J. (1992) Structure and mechanism of citrate synthase. *Curr. Top. Cell. Regul.* **33**, 209–229.

A thorough review of this enzyme.

Singer, T.P. & Johnson, M.K. (1985) The prosthetic groups of succinate dehydrogenase: 30 years from discovery to identification. *FEBS Lett.* **190**, 189–198.

A description of the structure and role of the iron-sulfur centers in this enzyme.

Weigand, G. & Remington, S.J. (1986) Citrate synthase: structure, control, and mechanism. *Annu. Rev. Biophys. Biophys. Chem.* **15**, 97–117.

Wolodko, W.T., Fraser, M.E., James, M.N.G., & Bridger, W.A. (1994) The crystal structure of succinyl-CoA synthetase from *Escherichia coli* at 2.5-Å resolution. *J. Biol. Chem.* **269**, 10,883–10,890.

Yang, J., Kalhan, S.C., & Hanson, R.W. (2009) What is the metabolic role of phosphenolpyruvate carboxykinase? *J. Biol. Chem.* **284**, 27,025–27,029.

A minireview of the several roles played by this enzyme.

Moonlighting Enzymes

Eisenstein, R.S. (2000) Iron regulatory proteins and the molecular control of mammalian iron metabolism. *Annu. Rev. Nutr.* **20**, 627–662.

Flores, C.-L. & Gancedo, C. (2011) Unraveling moonlighting functions with yeasts. *IUBMB Life* **63**, 457–462.

Jeffery, C.J. (1999) Moonlighting proteins. *Trends Biochem. Sci.* **24**, 8–11.

Kim, J.-W. & Dang, C.V. (2006) Multifaceted roles of glycolytic enzymes. *Trends Biochem. Sci.* **30**, 142–150.

Intermediate-level review of moonlighting enzymes.

Rouault, T.A. (2006) The role of iron regulatory proteins in mammalian iron homeostasis and disease. *Nat. Chem. Biol.* **2**, 406–414.

An advanced review.

Regulation of the Citric Acid Cycle

Briere, J.-J., Favier, J., Gimenez-Roqueplo, A.-P., & Rustin, P. (2006) Tricarboxylic acid cycle dysfunction as a cause of human diseases and tumor formation. *Am. J. Physiol. Cell Physiol.* **291**, 1114–1120.

Intermediate-level review of clinical effects of mutations in succinate dehydrogenase, fumarase, and α-ketoglutarate dehydrogenase.

Hansford, R.G. (1980) Control of mitochondrial substrate oxidation. *Curr. Top. Bioenerget.* **10**, 217–278.

A detailed review of the regulation of the citric acid cycle.

Kaplan, N.O. (1985) The role of pyridine nucleotides in regulating cellular metabolism. *Curr. Top. Cell. Regul.* **26**, 371–381.

An excellent general discussion of the importance of the [NADH]/[NAD$^+$] ratio in cellular regulation.

King, A., Selak, M.A., & Gottlieb, E. (2006) Succinate dehydrogenase and fumarate hydratase: linking mitochondrial dysfunction and cancer. *Oncogene* **25**, 4675–4682.

Reed, L.J., Damuni, Z., & Merryfield, M.L. (1985) Regulation of mammalian pyruvate and branched-chain α-keto acid dehydrogenase complexes by phosphorylation-dephosphorylation. *Curr. Top. Cell. Regul.* **27**, 41–49.

Glyoxylate Cycle

Eastmond, P.J. & Graham, I.A. (2001) Re-examining the role of the glyoxylate cycle in oilseeds. *Trends Plant Sci.* **6**, 72–77.

Intermediate-level review of studies of the glyoxylate cycle in *Arabidopsis*.

Holms, W.H. (1986) The central metabolic pathways of *Escherichia coli*: relationship between flux and control at a branch point, efficiency of conversion to biomass, and excretion of acetate. *Curr. Top. Cell. Regul.* **28**, 69–106.

Problems

1. Balance Sheet for the Citric Acid Cycle The citric acid cycle has eight enzymes: citrate synthase, aconitase, isocitrate dehydrogenase, α-ketoglutarate dehydrogenase, succinyl-CoA synthetase, succinate dehydrogenase, fumarase, and malate dehydrogenase.

(a) Write a balanced equation for the reaction catalyzed by each enzyme.

(b) Name the cofactor(s) required by each enzyme reaction.

(c) For each enzyme determine which of the following describes the type of reaction(s) catalyzed: condensation (carbon–carbon bond formation); dehydration (loss of water); hydration (addition of water); decarboxylation (loss of CO_2); oxidation-reduction; substrate-level phosphorylation; isomerization.

(d) Write a balanced net equation for the catabolism of acetyl-CoA to CO_2.

2. Net Equation for Glycolysis and the Citric Acid Cycle Write the net biochemical equation for the metabolism of a molecule of glucose by glycolysis and the citric acid cycle, including all cofactors.

3. Recognizing Oxidation and Reduction Reactions One biochemical strategy of many living organisms is the step-wise oxidation of organic compounds to CO_2 and H_2O and the conservation of a major part of the energy thus produced in the form of ATP. It is important to be able to recognize oxidation-reduction processes in metabolism. Reduction of an organic molecule results from the hydrogenation of a double bond (Eqn 1, below) or of a single bond with accompanying cleavage (Eqn 2). Conversely, oxidation results from dehydrogenation. In biochemical redox reactions, the coenzymes NAD and FAD dehydrogenate/hydrogenate organic molecules in the presence of the proper enzymes.

(1)

(2)

For each of the metabolic transformations in (a) through (h), determine whether oxidation or reduction has occurred. Balance each transformation by inserting H—H and, where necessary, H_2O.

4. Relationship between Energy Release and the Oxidation State of Carbon A eukaryotic cell can use glucose ($C_6H_{12}O_6$) and hexanoic acid ($C_6H_{14}O_2$) as fuels for cellular respiration. On the basis of their structural formulas, which substance releases more energy per gram on complete combustion to CO_2 and H_2O?

5. Nicotinamide Coenzymes as Reversible Redox Carriers The nicotinamide coenzymes (see Fig. 13–24) can undergo reversible oxidation-reduction reactions with specific substrates in the presence of the appropriate dehydrogenase. In these reactions, NADH + H^+ serves as the hydrogen source, as described in Problem 3. Whenever the coenzyme is oxidized, a substrate must be simultaneously reduced:

$$\text{Substrate} + \text{NADH} + \text{H}^+ \rightleftharpoons \text{product} + \text{NAD}^+$$
$$\text{Oxidized} \qquad \text{Reduced} \qquad\qquad \text{Reduced} \qquad \text{Oxidized}$$

For each of the reactions in (a) through (f), determine whether the substrate has been oxidized or reduced or is unchanged in oxidation state (see Problem 3). If a redox change has occurred, balance the reaction with the necessary amount of NAD^+, NADH, H^+, and H_2O. The objective is to recognize when a redox coenzyme is necessary in a metabolic reaction.

(a) $CH_3CH_2OH \longrightarrow CH_3-C\overset{O}{\underset{H}{<}}$

     Ethanol   Acetaldehyde

(b) $\underset{\underset{H}{|}}{CH_2}\overset{\overset{OPO_3^{2-}}{|}}{-}\underset{}{C}\overset{\overset{OH}{|}}{-}C\overset{O}{\underset{OPO_3^{2-}}{<}} \longrightarrow \underset{\underset{H}{|}}{CH_2}\overset{\overset{OPO_3^{2-}}{|}}{-}\underset{}{C}\overset{\overset{OH}{|}}{-}C\overset{O}{\underset{H}{<}} + HPO_4^{2-}$

     1,3-Bisphosphoglycerate   Glyceraldehyde 3-phosphate

(c) $CH_3-\overset{\overset{O}{\|}}{C}-C\overset{O^-}{\underset{O}{<}} \longrightarrow CH_3-C\overset{O}{\underset{H}{<}} + CO_2$

     Pyruvate   Acetaldehyde

(d) $CH_3-\overset{\overset{O}{\|}}{C}-C\overset{O^-}{\underset{O}{<}} \longrightarrow CH_3-C\overset{O}{\underset{O^-}{<}} + CO_2$

     Pyruvate   Acetate

(e) $^-OOC-CH_2-\overset{\overset{O}{\|}}{C}-COO^- \longrightarrow {}^-OOC-CH_2-\underset{\underset{H}{|}}{\overset{\overset{OH}{|}}{C}}-COO^-$

     Oxaloacetate   Malate

(f) $CH_3-\overset{\overset{O}{\|}}{C}-CH_2-C\overset{O}{\underset{O^-}{<}} + H^+ \longrightarrow CH_3-\overset{\overset{O}{\|}}{C}-CH_3 + CO_2$

     Acetoacetate   Acetone

6. Pyruvate Dehydrogenase Cofactors and Mechanism Describe the role of each cofactor involved in the reaction catalyzed by the pyruvate dehydrogenase complex.

7. Thiamine Deficiency Individuals with a thiamine-deficient diet have relatively high levels of pyruvate in their blood. Explain this in biochemical terms.

8. Isocitrate Dehydrogenase Reaction What type of chemical reaction is involved in the conversion of isocitrate to α-ketoglutarate? Name and describe the role of any cofactors. What other reaction(s) of the citric acid cycle are of this same type?

9. Stimulation of Oxygen Consumption by Oxaloacetate and Malate In the early 1930s, Albert Szent-Györgyi reported the interesting observation that the addition of small amounts of oxaloacetate or malate to suspensions of minced pigeon breast muscle stimulated the oxygen consumption of the preparation. Surprisingly, the amount of oxygen consumed was about seven times more than the amount necessary for complete oxidation (to CO_2 and H_2O) of the added oxaloacetate or malate. Why did the addition of oxaloacetate or malate stimulate oxygen consumption? Why was the amount of oxygen consumed so much greater than the amount necessary to completely oxidize the added oxaloacetate or malate?

10. Formation of Oxaloacetate in a Mitochondrion In the last reaction of the citric acid cycle, malate is dehydrogenated to regenerate the oxaloacetate necessary for the entry of acetyl-CoA into the cycle:

$$\text{L-Malate} + \text{NAD}^+ \longrightarrow \text{oxaloacetate} + \text{NADH} + \text{H}^+$$
$$\Delta G'^\circ = 30.0 \text{ kJ/mol}$$

(a) Calculate the equilibrium constant for this reaction at 25 °C.

(b) Because $\Delta G'^\circ$ assumes a standard pH of 7, the equilibrium constant calculated in (a) corresponds to

$$K'_{eq} = \frac{[\text{oxaloacetate}][\text{NADH}]}{[\text{L-malate}][\text{NAD}^+]}$$

The measured concentration of L-malate in rat liver mitochondria is about 0.20 mM when [NAD$^+$]/[NADH] is 10. Calculate the concentration of oxaloacetate at pH 7 in these mitochondria.

(c) To appreciate the magnitude of the mitochondrial oxaloacetate concentration, calculate the number of oxaloacetate molecules in a single rat liver mitochondrion. Assume the mitochondrion is a sphere of diameter 2.0 μm.

11. Cofactors for the Citric Acid Cycle Suppose you have prepared a mitochondrial extract that contains all of the soluble enzymes of the matrix but has lost (by dialysis) all the low molecular weight cofactors. What must you add to the extract so that the preparation will oxidize acetyl-CoA to CO_2?

12. Riboflavin Deficiency How would a riboflavin deficiency affect the functioning of the citric acid cycle? Explain your answer.

13. Oxaloacetate Pool What factors might decrease the pool of oxaloacetate available for the activity of the citric acid cycle? How can the pool of oxaloacetate be replenished?

14. Energy Yield from the Citric Acid Cycle The reaction catalyzed by succinyl-CoA synthetase produces the high-energy compound GTP. How is the free energy contained in GTP incorporated into the cellular ATP pool?

15. Respiration Studies in Isolated Mitochondria Cellular respiration can be studied in isolated mitochondria by measuring oxygen consumption under different conditions. If 0.01 M sodium malonate is added to actively respiring mitochondria that are using pyruvate as fuel source, respiration soon stops and a metabolic intermediate accumulates.

(a) What is the structure of this intermediate?

(b) Explain why it accumulates.

(c) Explain why oxygen consumption stops.

(d) Aside from removal of the malonate, how can this inhibition of respiration be overcome? Explain.

16. Labeling Studies in Isolated Mitochondria The metabolic pathways of organic compounds have often been delineated by using a radioactively labeled substrate and following the fate of the label.

(a) How can you determine whether glucose added to a suspension of isolated mitochondria is metabolized to CO_2 and H_2O?

(b) Suppose you add a brief pulse of $[3-^{14}C]$pyruvate (labeled in the methyl position) to the mitochondria. After one turn of the citric acid cycle, what is the location of the ^{14}C in the oxaloacetate? Explain by tracing the ^{14}C label through the pathway. How many turns of the cycle are required to release all the $[3-^{14}C]$pyruvate as CO_2?

17. Pathway of CO_2 in Gluconeogenesis In the first bypass step of gluconeogenesis, the conversion of pyruvate to phosphoenolpyruvate (PEP), pyruvate is carboxylated by pyruvate carboxylase to oxaloacetate, which is subsequently decarboxylated to PEP by PEP carboxykinase (Chapter 14). Because the addition of CO_2 is directly followed by the loss of CO_2, you might expect that in tracer experiments, the ^{14}C of $^{14}CO_2$ would not be incorporated into PEP, glucose, or any intermediates in gluconeogenesis. However, investigators find that when a rat liver preparation synthesizes glucose in the presence of $^{14}CO_2$, ^{14}C slowly appears in PEP and eventually at C-3 and C-4 of glucose. How does the ^{14}C label get into the PEP and glucose? (Hint: During gluconeogenesis in the presence of $^{14}CO_2$, several of the four-carbon citric acid cycle intermediates also become labeled.)

18. $[1-^{14}C]$Glucose Catabolism An actively respiring bacterial culture is briefly incubated with $[1-^{14}C]$glucose, and the glycolytic and citric acid cycle intermediates are isolated. Where is the ^{14}C in each of the intermediates listed below? Consider only the initial incorporation of ^{14}C, in the first pass of labeled glucose through the pathways.

(a) Fructose 1,6-bisphosphate

(b) Glyceraldehyde 3-phosphate

(c) Phosphoenolpyruvate

(d) Acetyl-CoA

(e) Citrate

(f) α-Ketoglutarate

(g) Oxaloacetate

 19. Role of the Vitamin Thiamine People with beriberi, a disease caused by thiamine deficiency, have elevated levels of blood pyruvate and α-ketoglutarate, especially after consuming a meal rich in glucose. How are these effects related to a deficiency of thiamine?

20. Synthesis of Oxaloacetate by the Citric Acid Cycle Oxaloacetate is formed in the last step of the citric acid cycle by the NAD^+-dependent oxidation of L-malate. Can a net synthesis of oxaloacetate from acetyl-CoA occur using only the enzymes and cofactors of the citric acid cycle, without depleting the intermediates of the cycle? Explain. How is oxaloacetate that is lost from the cycle (to biosynthetic reactions) replenished?

21. Oxaloacetate Depletion Mammalian liver can carry out gluconeogenesis using oxaloacetate as the starting material (Chapter 14). Would the operation of the citric acid cycle be affected by extensive use of oxaloacetate for gluconeogenesis? Explain your answer.

22. Mode of Action of the Rodenticide Fluoroacetate Fluoroacetate, prepared commercially for rodent control, is also produced by a South African plant. After entering a cell, fluoroacetate is converted to fluoroacetyl-CoA in a reaction catalyzed by the enzyme acetate thiokinase:

$$F-CH_2COO^- + CoA\text{-}SH + ATP \longrightarrow$$
$$F-CH_2\underset{\underset{O}{\|}}{C}-S\text{-}CoA + AMP + PP_i$$

The toxic effect of fluoroacetate was studied in an experiment using intact isolated rat heart. After the heart was perfused with 0.22 mM fluoroacetate, the measured rate of glucose uptake and glycolysis decreased, and glucose 6-phosphate and fructose 6-phosphate accumulated. Examination of the citric acid cycle intermediates revealed that their concentrations were below normal, except for citrate, with a concentration 10 times higher than normal.

(a) Where did the block in the citric acid cycle occur? What caused citrate to accumulate and the other cycle intermediates to be depleted?

(b) Fluoroacetyl-CoA is enzymatically transformed in the citric acid cycle. What is the structure of the end product of fluoroacetate metabolism? Why does it block the citric acid cycle? How might the inhibition be overcome?

(c) In the heart perfusion experiments, why did glucose uptake and glycolysis decrease? Why did hexose monophosphates accumulate?

(d) Why is fluoroacetate poisoning fatal?

23. Synthesis of L-Malate in Wine Making The tartness of some wines is due to high concentrations of L-malate. Write a sequence of reactions showing how yeast cells synthesize L-malate from glucose under anaerobic conditions in the presence of dissolved CO_2 (HCO_3^-). Note that the overall reaction for this fermentation cannot involve the consumption of nicotinamide coenzymes or citric acid cycle intermediates.

24. Net Synthesis of α-Ketoglutarate α-Ketoglutarate plays a central role in the biosynthesis of several amino acids. Write a sequence of enzymatic reactions that could result in the net synthesis of α-ketoglutarate from pyruvate. Your proposed sequence must not involve the net consumption of other citric acid cycle intermediates. Write an equation for the overall reaction and identify the source of each reactant.

25. Amphibolic Pathways Explain, giving examples, what is meant by the statement that the citric acid cycle is amphibolic.

26. Regulation of the Pyruvate Dehydrogenase Complex In animal tissues, the rate of conversion of pyruvate to acetyl-CoA is regulated by the ratio of active, phosphorylated to inactive, unphosphorylated PDH complex. Determine what happens to the rate of this reaction when a preparation of rabbit muscle mitochondria containing the PDH complex is treated with (a) pyruvate dehydrogenase kinase, ATP, and NADH; (b) pyruvate dehydrogenase phosphatase and Ca^{2+}; (c) malonate.

27. Commercial Synthesis of Citric Acid Citric acid is used as a flavoring agent in soft drinks, fruit juices, and many other foods. Worldwide, the market for citric acid is valued at hundreds of millions of dollars per year. Commercial production uses the mold *Aspergillus niger*, which metabolizes sucrose under carefully controlled conditions.

(a) The yield of citric acid is strongly dependent on the concentration of $FeCl_3$ in the culture medium, as indicated in the graph. Why does the yield decrease when the concentration of Fe^{3+} is above or below the optimal value of 0.5 mg/L?

(b) Write the sequence of reactions by which *A. niger* synthesizes citric acid from sucrose. Write an equation for the overall reaction.

(c) Does the commercial process require the culture medium to be aerated—that is, is this a fermentation or an aerobic process? Explain.

28. Regulation of Citrate Synthase In the presence of saturating amounts of oxaloacetate, the activity of citrate synthase from pig heart tissue shows a sigmoid dependence on the concentration of acetyl-CoA, as shown in the graph below. When succinyl-CoA is added, the curve shifts to the right and the sigmoid dependence is more pronounced.

On the basis of these observations, suggest how succinyl-CoA regulates the activity of citrate synthase. (Hint: See Fig. 6–34.) Why is succinyl-CoA an appropriate signal for regulation of the citric acid cycle? How does the regulation of citrate synthase control the rate of cellular respiration in pig heart tissue?

29. Regulation of Pyruvate Carboxylase The carboxylation of pyruvate by pyruvate carboxylase occurs at a very low rate unless acetyl-CoA, a positive allosteric modulator, is present. If you have just eaten a meal rich in fatty acids (triacylglycerols) but low in carbohydrates (glucose), how does this regulatory property shut down the oxidation of glucose to CO_2 and H_2O but increase the oxidation of acetyl-CoA derived from fatty acids?

30. Relationship between Respiration and the Citric Acid Cycle Although oxygen does not participate directly in the citric acid cycle, the cycle operates only when O_2 is present. Why?

31. Effect of [NADH]/[NAD⁺] on the Citric Acid Cycle How would you expect the operation of the citric acid cycle to respond to a rapid increase in the $[NADH]/[NAD^+]$ ratio in the mitochondrial matrix? Why?

32. Thermodynamics of Citrate Synthase Reaction in Cells Citrate is formed by the condensation of acetyl-CoA with oxaloacetate, catalyzed by citrate synthase:

$$\text{Oxaloacetate} + \text{acetyl-CoA} + H_2O \rightleftharpoons \text{citrate} + \text{CoA} + H^+$$

In rat heart mitochondria at pH 7.0 and 25 °C, the concentrations of reactants and products are: oxaloacetate, 1 μM; acetyl-CoA, 1 μM; citrate, 220 μM; and CoA, 65 μM. The standard free-energy change for the citrate synthase reaction is −32.2 kJ/mol. What is the direction of metabolite flow through the citrate synthase reaction in rat heart cells? Explain.

33. Reactions of the Pyruvate Dehydrogenase Complex Two of the steps in the oxidative decarboxylation of pyruvate (steps ❹ and ❺ in Fig. 16–6) do not involve any of the three carbons of pyruvate yet are essential to the operation of the PDH complex. Explain.

34. Citric Acid Cycle Mutants There are many cases of human disease in which one or another enzyme activity is lacking due to genetic mutation. However, cases in which individuals lack one of the enzymes of the citric acid cycle are extremely rare. Why?

35. Partitioning between the Citric Acid and Glyoxylate Cycles In an organism (such as *E. coli*) that has both the citric acid cycle and the glyoxylate cycle, what determines which of these pathways isocitrate will enter?

Data Analysis Problem

36. How the Citric Acid Cycle Was Determined The detailed biochemistry of the citric acid cycle was determined by several researchers over a period of decades. In a 1937 article, Krebs and Johnson summarized their work and the work of others in the first published description of this pathway.

The methods used by these researchers were very different from those of modern biochemistry. Radioactive tracers were not commonly available until the 1940s, so Krebs and other researchers had to use nontracer techniques to work out the pathway. Using freshly prepared samples of pigeon breast muscle, they determined oxygen consumption by suspending minced muscle in buffer in a sealed flask and measuring the volume (in μL) of oxygen consumed under different conditions. They measured levels of substrates (intermediates) by treating samples with acid to remove contaminating proteins, then assaying the quantities of various small organic molecules. The two key observations that led Krebs and colleagues to propose a citric acid *cycle* as opposed to a *linear pathway* (like that of glycolysis) were made in the following experiments.

Experiment I. They incubated 460 mg of minced muscle in 3 mL of buffer at 40 °C for 150 minutes. Addition of *citrate* increased O_2 consumption by 893 μL compared with samples without added citrate. They calculated, based on the O_2 consumed during respiration of other carbon-containing compounds, that the expected O_2 consumption for complete respiration of this quantity of citrate was only 302 μL.

Experiment II. They measured O_2 consumption by 460 mg of minced muscle in 3 mL of buffer when incubated with *citrate* and/or with *1-phosphoglycerol* (glycerol 1-phosphate; this was known to be readily oxidized by cellular respiration) at 40 °C for 140 minutes. The results are shown in the table.

Sample	Substrate(s) added	μL O_2 absorbed
1	No extra	342
2	0.3 mL 0.2 M 1-phosphoglycerol	757
3	0.15 mL 0.02 M citrate	431
4	0.3 mL 0.2 M 1-phosphoglycerol and 0.15 mL 0.02 M citrate	1,385

(a) Why is O_2 consumption a good measure of cellular respiration?

(b) Why does sample 1 (unsupplemented muscle tissue) consume some oxygen?

(c) Based on the results for samples 2 and 3, can you conclude that 1-phosphoglycerol and citrate serve as substrates for cellular respiration in this system? Explain your reasoning.

(d) Krebs and colleagues used the results from these experiments to argue that citrate was "catalytic"—that it helped the muscle tissue samples metabolize 1-phosphoglycerol more completely. How would you use their data to make this argument?

(e) Krebs and colleagues further argued that citrate was not simply consumed by these reactions, but had to be *regenerated*. Therefore, the reactions had to be a *cycle* rather than a linear pathway. How would you make this argument?

Other researchers had found that *arsenate* (AsO_4^{3-}) inhibits α-ketoglutarate dehydrogenase and that *malonate* inhibits succinate dehydrogenase.

(f) Krebs and coworkers found that muscle tissue samples treated with arsenate and citrate would consume citrate only in the presence of oxygen; under these conditions, oxygen was consumed. Based on the pathway in Figure 16–7, what was the citrate converted to in this experiment, and why did the samples consume oxygen?

In their article, Krebs and Johnson further reported the following. (1) In the presence of arsenate, 5.48 mmol of citrate was converted to 5.07 mmol of α-ketoglutarate. (2) In the presence of malonate, citrate was quantitatively converted to large amounts of succinate and small amounts of α-ketoglutarate. (3) Addition of oxaloacetate in the absence of oxygen led to production of a large amount of citrate; the amount was increased if glucose was also added.

Other workers had found the following pathway in similar muscle tissue preparations:

$$\text{Succinate} \longrightarrow \text{fumarate} \longrightarrow \text{malate} \longrightarrow$$
$$\text{oxaloacetate} \longrightarrow \text{pyruvate}$$

(g) Based only on the data presented in this problem, what is the order of the intermediates in the citric acid cycle? How does this compare with Figure 16–7? Explain your reasoning.

(h) Why was it important to show the *quantitative* conversion of citrate to α-ketoglutarate?

The Krebs and Johnson article also contains other data that filled in most of the missing components of the cycle. The only component left unresolved was the molecule that reacted with oxaloacetate to form citrate.

Reference

Krebs, H.A. & Johnson, W.A. (1937) The role of citric acid in intermediate metabolism in animal tissues. *Enzymologia* **4**, 148–156. [Reprinted (1980) in *FEBS Lett.* **117** (Suppl.), K2–K10.]

Fatty Acid Catabolism

The oxidation of long-chain fatty acids to acetyl-CoA is a central energy-yielding pathway in many organisms and tissues. In mammalian heart and liver, for example, it provides as much as 80% of the energetic needs under all physiological circumstances. The electrons removed from fatty acids during oxidation pass through the respiratory chain, driving ATP synthesis; the acetyl-CoA produced from the fatty acids may be completely oxidized to CO_2 in the citric acid cycle, resulting in further energy conservation. In some species and in some tissues, the acetyl-CoA has alternative fates. In liver, acetyl-CoA may be converted to ketone bodies—water-soluble fuels exported to the brain and other tissues when glucose is not available. In higher plants, acetyl-CoA serves primarily as a biosynthetic precursor, only secondarily as fuel. Although the biological role of fatty acid oxidation differs from organism to organism, the mechanism is essentially the same. The repetitive four-step process, called **β oxidation**, by which fatty acids are converted into acetyl-CoA is the main topic of this chapter.

In Chapter 10 we described the properties of triacylglycerols (also called triglycerides or neutral fats) that make them especially suitable as storage fuels. The long alkyl chains of their constituent fatty acids are essentially hydrocarbons, highly reduced structures with an energy of complete oxidation ($\sim$38 kJ/g) more than twice that for the same weight of carbohydrate or protein. This advantage is compounded by the extreme insolubility of lipids in water; cellular triacylglycerols aggregate in lipid droplets, which do not raise the osmolarity of the cytosol, and they are unsolvated. (In storage polysaccharides, by contrast, water of solvation can account for two-thirds of the overall weight of the stored molecules.) And because of their relative chemical inertness, triacylglycerols can be stored in large quantity in cells without the risk of undesired chemical reactions with other cellular constituents.

The properties that make triacylglycerols good storage compounds, however, present problems in their role as fuels. Because they are insoluble in water, ingested triacylglycerols must be emulsified before they can be digested by water-soluble enzymes in the intestine, and triacylglycerols absorbed in the intestine or mobilized from storage tissues must be carried in the blood bound to proteins that counteract their insolubility. To overcome the relative stability of the C—C bonds in a fatty acid, the carboxyl group at C-1 is activated by attachment to coenzyme A, which allows stepwise oxidation of the fatty acyl group at the C-3, or β, position—hence the name β oxidation.

We begin this chapter with a brief discussion of the sources of fatty acids and the routes by which they travel to the site of their oxidation, with special emphasis on the process in vertebrates. We then describe the chemical steps of fatty acid oxidation in mitochondria. The complete oxidation of fatty acids to CO_2 and H_2O takes place in three stages: the oxidation of long-chain fatty acids to two-carbon fragments, in the form of acetyl-CoA (β oxidation); the oxidation of acetyl-CoA to CO_2 in the citric acid cycle (Chapter 16); and the transfer of electrons from reduced electron carriers to the mitochondrial respiratory chain (Chapter 19). In this chapter we focus on the first of these stages. We begin our discussion of β oxidation with the simple case in which a fully saturated fatty acid with an even number of carbon atoms is degraded to acetyl-CoA. We then look briefly at the extra transformations necessary for the degradation of unsaturated fatty acids and fatty acids with an odd number of carbons. Finally, we discuss variations on the β-oxidation theme in specialized organelles—peroxisomes and glyoxysomes—and two less common pathways of fatty acid catabolism, ω and α oxidation. The chapter concludes with a description of an alternative fate for the acetyl-CoA formed by β oxidation in vertebrates: the production of ketone bodies in the liver.

17.1 Digestion, Mobilization, and Transport of Fats

Cells can obtain fatty acid fuels from three sources: fats consumed in the diet, fats stored in cells as lipid droplets, and fats synthesized in one organ for export to another. Some species use all three sources under various circumstances, others use one or two. Vertebrates, for example, obtain fats in the diet, mobilize fats stored in specialized tissue (adipose tissue, consisting of cells called adipocytes), and, in the liver, convert excess dietary carbohydrates to fats for export to other tissues. On average, 40% or more of the daily energy requirement of humans in highly industrialized countries is supplied by dietary triacylglycerols (although most nutritional guidelines recommend no more than 30% of daily caloric intake from fats). Triacylglycerols provide more than half the energy requirements of some organs, particularly the liver, heart, and resting skeletal muscle. Stored triacylglycerols are virtually the sole source of energy in hibernating animals and migrating birds. Protists obtain fats by consuming organisms lower in the food chain, and some also store fats as cytosolic lipid droplets. Vascular plants mobilize fats stored in seeds during germination, but do not otherwise depend on fats for energy.

Dietary Fats Are Absorbed in the Small Intestine

In vertebrates, before ingested triacylglycerols can be absorbed through the intestinal wall they must be converted from insoluble macroscopic fat particles to finely dispersed microscopic micelles. This solubilization is carried out by bile salts, such as taurocholic acid (p. 370), which are synthesized from cholesterol in the liver, stored in the gallbladder, and released into the small intestine after ingestion of a fatty meal. Bile salts are amphipathic compounds that act as biological detergents, converting dietary fats into mixed micelles of bile salts and triacylglycerols (**Fig. 17–1**, step ❶). Micelle formation enormously increases the fraction of lipid molecules accessible to the action of water-soluble lipases in the intestine, and lipase action converts triacylglycerols to monoacylglycerols (monoglycerides) and diacylglycerols (diglycerides), free fatty acids, and glycerol (step ❷). These products of lipase action diffuse into the epithelial cells lining the intestinal surface (the intestinal mucosa) (step ❸), where they are reconverted to triacylglycerols and packaged

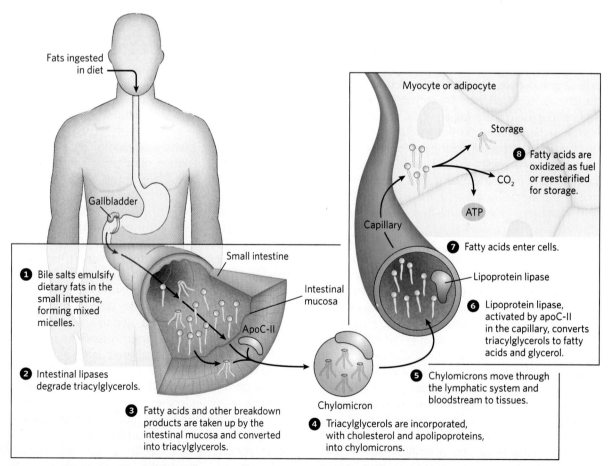

FIGURE 17–1 Processing of dietary lipids in vertebrates. Digestion and absorption of dietary lipids occur in the small intestine, and the fatty acids released from triacylglycerols are packaged and delivered to muscle and adipose tissues. The eight steps are discussed in the text.

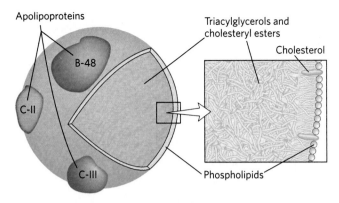

Apolipoproteins

Triacylglycerols and
cholesteryl esters

Cholesterol

B-48

C-II

C-III

Phospholipids

FIGURE 17–2 Molecular structure of a chylomicron. The surface is a layer of phospholipids, with head groups facing the aqueous phase. Triacylglycerols sequestered in the interior (yellow) make up more than 80% of the mass. Several apolipoproteins that protrude from the surface (B-48, C-III, C-II) act as signals in the uptake and metabolism of chylomicron contents. The diameter of chylomicrons ranges from about 100 to 500 nm.

with dietary cholesterol and specific proteins into lipoprotein aggregates called **chylomicrons** (**Fig. 17–2**; see also Fig. 17–1, step **❹**).

Apolipoproteins are lipid-binding proteins in the blood, responsible for the transport of triacylglycerols, phospholipids, cholesterol, and cholesteryl esters between organs. Apolipoproteins ("apo" means "detached" or "separate," designating the protein in its lipid-free form) combine with lipids to form several classes of **lipoprotein** particles, spherical aggregates with hydrophobic lipids at the core and hydrophilic protein side chains and lipid head groups at the surface. Various combinations of lipid and protein produce particles of different densities, ranging from chylomicrons and very-low-density lipoproteins (VLDL) to very-high-density lipoproteins (VHDL), which can be separated by ultracentrifugation. The structures of these lipoprotein particles and their roles in lipid transport are detailed in Chapter 21.

The protein moieties of lipoproteins are recognized by receptors on cell surfaces. In lipid uptake from the intestine, chylomicrons, which contain apolipoprotein C-II (apoC-II), move from the intestinal mucosa into the lymphatic system, and then enter the blood, which carries them to muscle and adipose tissue (Fig. 17–1, step **❺**). In the capillaries of these tissues, the extracellular enzyme **lipoprotein lipase**, activated by apoC-II, hydrolyzes triacylglycerols to fatty acids and glycerol (step **❻**), which are taken up by cells in the target tissues (step **❼**). In muscle, the fatty acids are oxidized for energy; in adipose tissue, they are reesterified for storage as triacylglycerols (step **❽**).

The remnants of chylomicrons, depleted of most of their triacylglycerols but still containing cholesterol and apolipoproteins, travel in the blood to the liver,

where they are taken up by endocytosis, mediated by receptors for their apolipoproteins. Triacylglycerols that enter the liver by this route may be oxidized to provide energy or to provide precursors for the synthesis of ketone bodies, as described in Section 17.3. When the diet contains more fatty acids than are needed immediately for fuel or as precursors, the liver converts them to triacylglycerols, which are packaged with specific apolipoproteins into VLDLs. The VLDLs are transported in the blood to adipose tissues, where the triacylglycerols are removed and stored in lipid droplets within adipocytes.

Hormones Trigger Mobilization of Stored Triacylglycerols

Neutral lipids are stored in adipocytes (and in steroid-synthesizing cells of the adrenal cortex, ovary, and testis) in the form of lipid droplets, with a core of sterol esters and triacylglycerols surrounded by a monolayer of phospholipids. The surface of these droplets is coated with **perilipins**, a family of proteins that restrict access to lipid droplets, preventing untimely lipid mobilization. When hormones signal the need for metabolic energy, triacylglycerols stored in adipose tissue are mobilized (brought out of storage) and transported to tissues (skeletal muscle, heart, and renal cortex) in which fatty acids can be oxidized for energy production. The hormones epinephrine and glucagon, secreted in response to low blood glucose levels or impending activity, stimulate the enzyme adenylyl cyclase in the adipocyte plasma membrane (**Fig. 17–3**), which produces the intracellular second messenger cyclic AMP (cAMP; see Fig. 12–4). Cyclic AMP–dependent protein kinase (PKA) triggers changes that open the lipid droplet up to the action of three lipases, which act on tri-, di-, and monoacylglycerols, releasing fatty acids and glycerol.

The fatty acids thus released (**free fatty acids, FFA**) pass from the adipocyte into the blood, where they bind to the blood protein **serum albumin** (Fig. 17–3). This protein (M_r 66,000), which makes up about half of the total serum protein, noncovalently binds as many as 10 fatty acids per protein monomer. Bound to this soluble protein, the otherwise insoluble fatty acids are carried to tissues such as skeletal muscle, heart, and renal cortex. In these target tissues, fatty acids dissociate from albumin and are moved by plasma membrane transporters into cells to serve as fuel. The glycerol liberated by lipase action is phosphorylated and oxidized to dihydroxyacetone phosphate, which can enter the glycolytic or gluconeogenic pathways. Alternatively, glycerol phosphate can be used in triacylglycerol or phospholipid synthesis.

About 95% of the biologically available energy of triacylglycerols resides in their three long-chain fatty acids; only 5% is contributed by the glycerol moiety. The glycerol released by lipase action is phosphorylated

FIGURE 17–3 Mobilization of triacylglycerols stored in adipose tissue.
When low levels of glucose in the blood trigger the release of glucagon,
❶ the hormone binds its receptor in the adipocyte membrane and thus
❷ stimulates adenylyl cyclase, via a G protein, to produce cAMP. This activates PKA, which phosphorylates ❸ the hormone-sensitive lipase (HSL)
and ❹ perilipin molecules on the surface of the lipid droplet. Phosphorylation of perilipin causes ❺ dissociation of the protein CGI from perilipin. CGI
then associates with the enzyme adipose triacylglycerol lipase (ATGL), activating it. Active ATGL ❻ converts triacylglycerols to diacylglycerols. The

phosphorylated perilipin associates with phosphorylated HSL, allowing it
access to the surface of the lipid droplet, where ❼ it converts diacylglycerols
to monoacylglycerols. A third lipase, monoacylglycerol lipase (MGL) ❽,
hydrolyzes monoacylglycerols. ❾ Fatty acids leave the adipocyte, bind
serum albumin in the blood, and are carried in the blood; they are released
from the albumin and ❿ enter a myocyte via a specific fatty acid transporter.
⓫ In the myocyte, fatty acids are oxidized to CO_2, and the energy of
oxidation is conserved in ATP, which fuels muscle contraction and other
energy-requiring metabolism in the myocyte.

by **glycerol kinase (Fig. 17–4)**, and the resulting
glycerol 3-phosphate is oxidized to dihydroxyacetone
phosphate. The glycolytic enzyme triose phosphate
isomerase converts this compound to glyceraldehyde
3-phosphate, which is oxidized via glycolysis.

Fatty Acids Are Activated and Transported into Mitochondria

The enzymes of fatty acid oxidation in animal cells are
located in the mitochondrial matrix, as demonstrated in
1948 by Eugene P. Kennedy and Albert Lehninger. The
fatty acids with chain lengths of 12 or fewer carbons
enter mitochondria without the help of membrane
transporters. Those with 14 or more carbons, which
constitute the majority of the FFA obtained in the diet
or released from adipose tissue, cannot pass directly
through the mitochondrial membranes—they must first
undergo the three enzymatic reactions of the **carnitine
shuttle**. The first reaction is catalyzed by a family of
isozymes (different isozymes specific for fatty acids
having short, intermediate, or long carbon chains) pres-

ent in the outer mitochondrial membrane, the **acyl-
CoA synthetases**, which promote the general reaction

$$\text{Fatty acid} + \text{CoA} + \text{ATP} \rightleftharpoons \text{fatty acyl–CoA} + \text{AMP} + \text{PP}_i$$

Thus, acyl–CoA synthetases catalyze the formation of a
thioester linkage between the fatty acid carboxyl group
and the thiol group of coenzyme A to yield a **fatty acyl–
CoA**, coupled to the cleavage of ATP to AMP and PP_i.
(Recall the description of this reaction in Chapter 13, to
illustrate how the free energy released by cleavage of
phosphoanhydride bonds in ATP could be coupled to
the formation of a high-energy compound; p. 524.) The
reaction occurs in two steps and involves a fatty acyl–
adenylate intermediate **(Fig. 17–5)**.

Fatty acyl–CoAs, like acetyl-CoA, are high-energy
compounds; their hydrolysis to FFA and CoA has a
large, negative standard free-energy change ($\Delta G'^{\circ} =$
-31 kJ/mol). The formation of a fatty acyl–CoA is made
more favorable by the hydrolysis of *two* high-energy
bonds in ATP; the pyrophosphate formed in the activation reaction is immediately hydrolyzed by inorganic
pyrophosphatase (left side of Fig. 17–5), which pulls

FIGURE 17-4 **Entry of glycerol into the glycolytic pathway.**

ATP, or they can be used in the cytosol to synthesize membrane lipids. Fatty acids destined for mitochondrial oxidation are transiently attached to the hydroxyl group of **carnitine** to form fatty acyl–carnitine—the second reaction of the shuttle.

Carnitine

This transesterification is catalyzed by **carnitine acyltransferase I**, in the outer membrane. Either the acyl-CoA passes through the outer membrane and is converted to the carnitine ester in the intermembrane space **(Fig. 17–6)**, or the carnitine ester is formed on the cytosolic face of the outer membrane, then moved across the outer membrane to the intermembrane space—the current evidence does not reveal which. In either case, passage into the intermembrane space (the space between the outer and inner membranes) occurs through large pores (formed by the protein porin) in the outer membrane. The fatty acyl–carnitine ester then enters the matrix by facilitated diffusion through the **acyl-carnitine/carnitine transporter** of the inner mitochondrial membrane (Fig. 17–6).

In the third and final step of the carnitine shuttle, the fatty acyl group is enzymatically transferred from carnitine to intramitochondrial coenzyme A by **carnitine acyltransferase II**. This isozyme, located on the inner face of the inner mitochondrial membrane, regenerates fatty acyl–CoA and releases it, along with free carnitine, into the matrix (Fig. 17–6). Carnitine reenters the intermembrane space via the acylcarnitine/carnitine transporter.

the preceding activation reaction in the direction of fatty acyl–CoA formation. The overall reaction is

Fatty acid + CoA + ATP $\longrightarrow$

$$\text{fatty acyl–CoA} + \text{AMP} + 2\text{P}_i \quad (17\text{–}1)$$

$$\Delta G'^{\circ} = -34 \text{ kJ/mol}$$

Fatty acyl–CoA esters formed at the cytosolic side of the outer mitochondrial membrane can be transported into the mitochondrion and oxidized to produce

MECHANISM FIGURE 17-5 **Conversion of a fatty acid to a fatty acyl-CoA.** The conversion is catalyzed by fatty acyl-CoA synthetase and inorganic pyrophosphatase. Fatty acid activation by formation of the fatty acyl-CoA derivative occurs in two steps. The overall reaction is highly exergonic. 🔵 **Fatty Acyl-CoA Synthetase Mechanism**

❶ The carboxylate ion is adenylylated by ATP, to form a fatty acyl-adenylate and PP$_i$. The PP$_i$ is immediately hydrolyzed to two molecules of P$_i$.

❷ The thiol group of coenzyme A attacks the acyl-adenylate (a mixed anhydride), displacing AMP and forming the thioester fatty acyl-CoA.

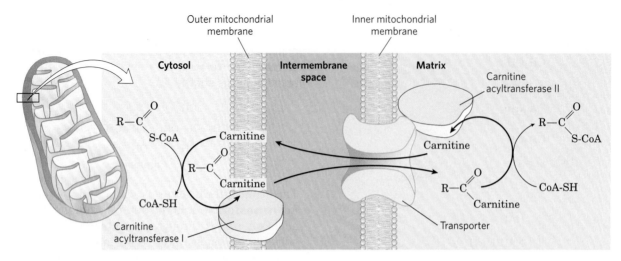

FIGURE 17–6 Fatty acid entry into mitochondria via the acyl-carnitine/carnitine transporter. After fatty acyl-carnitine is formed at the outer membrane or in the intermembrane space, it moves into the matrix by facilitated diffusion through the transporter in the inner membrane. In the matrix, the acyl group is transferred to mitochondrial coenzyme A, freeing carnitine to return to the intermembrane space through the same transporter. Acyltransferase I is inhibited by malonyl-CoA, the first intermediate in fatty acid synthesis (see Fig. 21-2). This inhibition prevents the simultaneous synthesis and degradation of fatty acids.

This three-step process for transferring fatty acids into the mitochondrion—esterification to CoA, transesterification to carnitine followed by transport, and transesterification back to CoA—links two separate pools of coenzyme A and of fatty acyl–CoA, one in the cytosol, the other in mitochondria. These pools have different functions. Coenzyme A in the mitochondrial matrix is largely used in oxidative degradation of pyruvate, fatty acids, and some amino acids, whereas cytosolic coenzyme A is used in the biosynthesis of fatty acids (see Fig. 21–10). Fatty acyl–CoA in the cytosolic pool can be used for membrane lipid synthesis or can be moved into the mitochondrial matrix for oxidation and ATP production. Conversion to the carnitine ester commits the fatty acyl moiety to the oxidative fate.

The carnitine-mediated entry process is the rate-limiting step for oxidation of fatty acids in mitochondria and, as discussed later, is a regulation point. Once inside the mitochondrion, the fatty acyl–CoA is acted upon by a set of enzymes in the matrix.

SUMMARY 17.1 Digestion, Mobilization, and Transport of Fats

▶ The fatty acids of triacylglycerols furnish a large fraction of the oxidative energy in animals. Dietary triacylglycerols are emulsified in the small intestine by bile salts, hydrolyzed by intestinal lipases, absorbed by intestinal epithelial cells, reconverted into triacylglycerols, then formed into chylomicrons by combination with specific apolipoproteins.

▶ Chylomicrons deliver triacylglycerols to tissues, where lipoprotein lipase releases free fatty acids for entry into cells. Triacylglycerols stored in adipose tissue are mobilized by a hormone-sensitive triacylglycerol lipase. The released fatty acids bind to serum albumin and are carried in the blood to the heart, skeletal muscle, and other tissues that use fatty acids for fuel.

▶ Once inside cells, fatty acids are activated at the outer mitochondrial membrane by conversion to fatty acyl–CoA thioesters. Fatty acyl–CoA that is to be oxidized enters mitochondria in three steps, via the carnitine shuttle.

17.2 Oxidation of Fatty Acids

As noted earlier, mitochondrial oxidation of fatty acids takes place in three stages (**Fig. 17–7**). In the first stage—β oxidation—fatty acids undergo oxidative removal of successive two-carbon units in the form of acetyl-CoA, starting from the carboxyl end of the fatty acyl chain. For example, the 16-carbon palmitic acid (palmitate at pH 7) undergoes seven passes through the oxidative sequence, in each pass losing two carbons as acetyl-CoA. At the end of seven cycles the last two carbons of palmitate (originally C-15 and C-16) remain as acetyl-CoA. The overall result is the conversion of the 16-carbon chain of palmitate to eight two-carbon acetyl groups of acetyl-CoA molecules. Formation of each acetyl-CoA requires removal of four hydrogen atoms (two pairs of electrons and four H^+) from the fatty acyl moiety by dehydrogenases.

In the second stage of fatty acid oxidation, the acetyl groups of acetyl-CoA are oxidized to CO_2 in the citric acid cycle, which also takes place in the mitochondrial matrix. Acetyl-CoA derived from fatty acids thus enters a final common pathway of oxidation with the acetyl-CoA derived from glucose via glycolysis and pyruvate oxidation (see Fig. 16–1). The first two stages of fatty acid oxidation produce the reduced electron carriers NADH and $FADH_2$, which in the third stage donate electrons to the mitochondrial respiratory chain, through which the

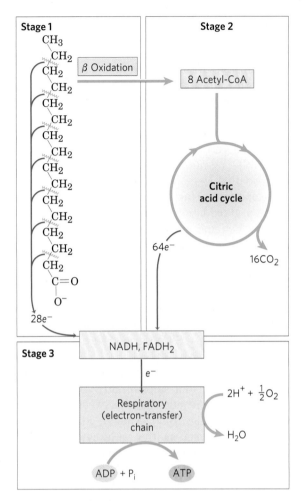

FIGURE 17-7 Stages of fatty acid oxidation. Stage 1: A long-chain fatty acid is oxidized to yield acetyl residues in the form of acetyl-CoA. This process is called β oxidation. Stage 2: The acetyl groups are oxidized to CO_2 via the citric acid cycle. Stage 3: Electrons derived from the oxidations of stages 1 and 2 pass to O_2 via the mitochondrial respiratory chain, providing the energy for ATP synthesis by oxidative phosphorylation.

electrons pass to oxygen with the concomitant phosphorylation of ADP to ATP (Fig. 17–7). The energy released by fatty acid oxidation is thus conserved as ATP.

We now take a closer look at the first stage of fatty acid oxidation, beginning with the simple case of a saturated fatty acyl chain with an even number of carbons, then turning to the slightly more complicated cases of unsaturated and odd-number chains. We also consider the regulation of fatty acid oxidation, the β-oxidative processes as they occur in organelles other than mitochondria, and, finally, two less-general modes of fatty acid catabolism, α oxidation and ω oxidation.

The β Oxidation of Saturated Fatty Acids Has Four Basic Steps

Four enzyme-catalyzed reactions make up the first stage of fatty acid oxidation **(Fig. 17–8a)**. First, dehydrogenation of fatty acyl–CoA produces a double bond between the α and β carbon atoms (C-2 and C-3), yielding a ***trans-Δ^2-enoyl-CoA*** (the symbol Δ^2 desig-

nates the position of the double bond; you may want to review fatty acid nomenclature, p. 357.) Note that the new double bond has the trans configuration, whereas the double bonds in naturally occurring unsaturated fatty acids are normally in the cis configuration. We consider the significance of this difference later.

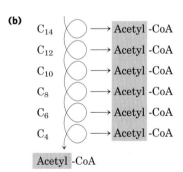

FIGURE 17-8 The β-oxidation pathway. (a) In each pass through this four-step sequence, one acetyl residue (shaded in pink) is removed in the form of acetyl-CoA from the carboxyl end of the fatty acyl chain—in this example palmitate (C_{16}), which enters as palmitoyl-CoA. **(b)** Six more passes through the pathway yield seven more molecules of acetyl-CoA, the seventh arising from the last two carbon atoms of the 16-carbon chain. Eight molecules of acetyl-CoA are formed in all.

This first step is catalyzed by three isozymes of **acyl-CoA dehydrogenase**, each specific for a range of fatty-acyl chain lengths: very-long-chain acyl-CoA dehydrogenase (VLCAD), acting on fatty acids of 12 to 18 carbons; medium-chain (MCAD), acting on fatty acids of 4 to 14 carbons; and short-chain (SCAD), acting on fatty acids of 4 to 8 carbons. All three isozymes are flavoproteins with FAD (see Fig. 13–27) as a prosthetic group. The electrons removed from the fatty acyl–CoA are transferred to FAD, and the reduced form of the dehydrogenase immediately donates its electrons to an electron carrier of the mitochondrial respiratory chain, the **electron-transferring flavoprotein (ETF)** (see Fig. 19–8). The oxidation catalyzed by an acyl-CoA dehydrogenase is analogous to succinate dehydrogenation in the citric acid cycle (p. 646); in both reactions the enzyme is bound to the inner membrane, a double bond is introduced into a carboxylic acid between the α and β carbons, FAD is the electron acceptor, and electrons from the reaction ultimately enter the respiratory chain and pass to O_2, with the concomitant synthesis of about 1.5 ATP molecules per electron pair.

In the second step of the β-oxidation cycle (Fig. 17–8a), water is added to the double bond of the *trans*-Δ^2-enoyl-CoA to form the L stereoisomer of **β-hydroxyacyl-CoA (3-hydroxyacyl-CoA)**. This reaction, catalyzed by **enoyl-CoA hydratase**, is formally analogous to the fumarase reaction in the citric acid cycle, in which H_2O adds across an α–β double bond (p. 647).

In the third step, L-β-hydroxyacyl-CoA is dehydrogenated to form **β-ketoacyl-CoA**, by the action of **β-hydroxyacyl-CoA dehydrogenase**; NAD^+ is the electron acceptor. This enzyme is absolutely specific for the L stereoisomer of hydroxyacyl-CoA. The NADH formed in the reaction donates its electrons to **NADH dehydrogenase**, an electron carrier of the respiratory chain, and ATP is formed from ADP as the electrons pass to O_2. The reaction catalyzed by β-hydroxyacyl-CoA dehydrogenase is closely analogous to the malate dehydrogenase reaction of the citric acid cycle (p. 647).

The fourth and last step of the β-oxidation cycle is catalyzed by **acyl-CoA acetyltransferase**, more commonly called **thiolase**, which promotes reaction of β-ketoacyl-CoA with a molecule of free coenzyme A to split off the carboxyl-terminal two-carbon fragment of the original fatty acid as acetyl-CoA. The other product is the coenzyme A thioester of the fatty acid, now shortened by two carbon atoms (Fig. 17–8a). This reaction is called thiolysis, by analogy with the process of hydrolysis, because the β-ketoacyl-CoA is cleaved by reaction with the thiol group of coenzyme A. The thiolase reaction is a reverse Claisen condensation (see Fig. 13–4).

The last three steps of this four-step sequence are catalyzed by either of two sets of enzymes, with the enzymes employed depending on the length of the fatty acyl chain. For fatty acyl chains of 12 or more carbons, the reactions are catalyzed by a multienzyme complex associated with the inner mitochondrial membrane, the **trifunctional protein (TFP)**. TFP is a heterooctamer of $\alpha_4\beta_4$ subunits. Each α subunit contains two activities,

the enoyl-CoA hydratase and the β-hydroxyacyl-CoA dehydrogenase; the β subunits contain the thiolase activity. This tight association of three enzymes may allow efficient substrate channeling from one active site to the next, without diffusion of the intermediates away from the enzyme surface. When TFP has shortened the fatty acyl chain to 12 or fewer carbons, further oxidations are catalyzed by a set of four soluble enzymes in the matrix.

As noted earlier, the single bond between methylene (—CH_2—) groups in fatty acids is relatively stable. The β-oxidation sequence is an elegant mechanism for destabilizing and breaking these bonds. The first three reactions of β oxidation create a much less stable C—C bond, in which the α carbon (C-2) is bonded to *two* carbonyl carbons (the β-ketoacyl-CoA intermediate). The ketone function on the β carbon (C-3) makes it a good target for nucleophilic attack by the —SH of coenzyme A, catalyzed by thiolase. The acidity of the α hydrogen and the resonance stabilization of the carbanion generated by the departure of this hydrogen make the terminal —CH_2—CO—S-CoA a good leaving group, facilitating breakage of the α–β bond.

We have already seen a reaction sequence nearly identical with these four steps of fatty acid oxidation, in the citric acid cycle reaction steps between succinate and oxaloacetate (see Fig. 16–7). A nearly identical reaction sequence occurs again in the pathways by which the branched-chain amino acids (isoleucine, leucine, and valine) are oxidized as fuels (see Fig. 18–28). **Figure 17–9** shows the common features of these three sequences, almost certainly an example of the conservation of a mechanism by gene duplication and evolution of a new specificity in the enzyme products of the duplicated genes.

The Four β-Oxidation Steps Are Repeated to Yield Acetyl-CoA and ATP

In one pass through the β-oxidation sequence, one molecule of acetyl-CoA, two pairs of electrons, and four protons (H^+) are removed from the long-chain fatty acyl–CoA, shortening it by two carbon atoms. The equation for one pass, beginning with the coenzyme A ester of our example, palmitate, is

$$\text{Palmitoyl-CoA} + \text{CoA} + \text{FAD} + NAD^+ + H_2O \longrightarrow$$
$$\text{myristoyl-CoA} + \text{acetyl-CoA} + FADH_2 + \text{NADH} + H^+$$

$$(17\text{–}2)$$

Following removal of one acetyl-CoA unit from palmitoyl-CoA, the coenzyme A thioester of the shortened fatty acid (now the 14-carbon myristate) remains. The myristoyl-CoA can now go through another set of four β-oxidation reactions, exactly analogous to the first, to yield a second molecule of acetyl-CoA and lauroyl-CoA, the coenzyme A thioester of the 12-carbon laurate. Altogether, seven passes through the β-oxidation sequence are required to oxidize one molecule of palmitoyl-CoA to eight molecules of acetyl-CoA (Fig. 17–8b). The overall equation is

$$\text{Palmitoyl-CoA} + 7\text{CoA} + 7\text{FAD} + 7NAD^+ + 7H_2O \longrightarrow$$
$$8 \text{ acetyl-CoA} + 7FADH_2 + 7\text{NADH} + 7H^+ \quad (17\text{–}3)$$

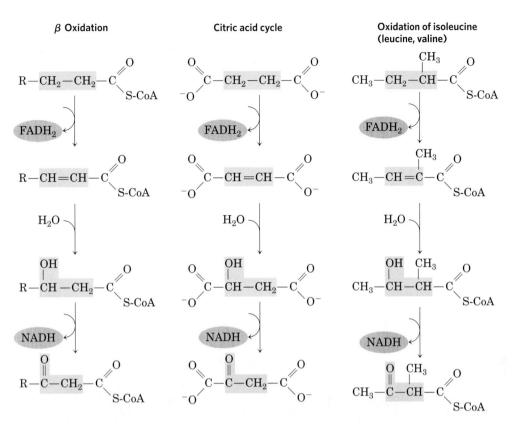

FIGURE 17–9 A conserved reaction sequence to introduce a carbonyl function on the carbon β to a carboxyl group. The β-oxidation pathway for fatty acyl-CoAs, the pathway from succinate to oxaloacetate in the citric acid cycle, and the pathway by which the deaminated carbon skeletons from isoleucine, leucine, and valine are oxidized as fuels, use the same reaction sequence.

Each molecule of $FADH_2$ formed during oxidation of the fatty acid donates a pair of electrons to ETF of the respiratory chain, and about 1.5 molecules of ATP are generated during the ensuing transfer of each electron pair to O_2. Similarly, each molecule of NADH formed delivers a pair of electrons to the mitochondrial NADH dehydrogenase, and the subsequent transfer of each pair of electrons to O_2 results in formation of about 2.5 molecules of ATP. Thus four molecules of ATP are formed for each two-carbon unit removed in one pass through the sequence. Note that water is also produced in this process. Transfer of electrons from NADH or $FADH_2$ to O_2 yields one H_2O per electron pair. Reduction of O_2 by NADH also consumes one H^+ per NADH molecule: $NADH + H^+ + \frac{1}{2}O_2 \longrightarrow NAD^+ + H_2O$. In hibernating animals, fatty acid oxidation provides metabolic energy, heat, and water—all essential for survival of an animal that neither eats nor drinks for long periods (Box 17–1). Camels obtain water to supplement the meager supply available in their natural environment by oxidation of fats stored in their hump.

The overall equation for the oxidation of palmitoyl-CoA to eight molecules of acetyl-CoA, including the electron transfers and oxidative phosphorylations, is

Palmitoyl-CoA + 7CoA + 7O$_2$ + 28P$_i$ + 28ADP $\longrightarrow$
$$8 \text{ acetyl-CoA} + 28\text{ATP} + 7\text{H}_2\text{O} \qquad (17\text{–}4)$$

Acetyl-CoA Can Be Further Oxidized in the Citric Acid Cycle

The acetyl-CoA produced from the oxidation of fatty acids can be oxidized to CO_2 and H_2O by the citric acid cycle. The following equation represents the balance sheet for the second stage in the oxidation of palmitoyl-CoA, together with the coupled phosphorylations of the third stage:

8 Acetyl-CoA + 16O$_2$ + 80P$_i$ + 80ADP $\longrightarrow$
$$8\text{CoA} + 80\text{ATP} + 16\text{CO}_2 + 16\text{H}_2\text{O} \qquad (17\text{–}5)$$

Combining Equations 17–4 and 17–5, we obtain the overall equation for the complete oxidation of palmitoyl-CoA to carbon dioxide and water:

Palmitoyl-CoA + 23O$_2$ + 108P$_i$ + 108ADP $\longrightarrow$
$$\text{CoA} + 108\text{ATP} + 16\text{CO}_2 + 23\text{H}_2\text{O} \qquad (17\text{–}6)$$

Table 17–1 summarizes the yields of NADH, $FADH_2$, and ATP in the successive steps of palmitoyl-CoA oxidation. Note that because the activation of palmitate to palmitoyl-CoA breaks both phosphoanhydride bonds in ATP (Fig. 17–5), the energetic cost of activating a fatty acid is equivalent to two ATP, and the net gain per molecule of palmitate is 106 ATP. The standard free-energy change for the oxidation of palmitate to CO_2 and H_2O is about 9,800 kJ/mol. Under standard conditions, the energy recovered as the phosphate bond energy of ATP is 106 × 30.5 kJ/mol = 3,230 kJ/mol, about 33% of the

BOX 17–1   Fat Bears Carry Out β Oxidation in Their Sleep

Many animals depend on fat stores for energy during hibernation, during migratory periods, and in other situations involving radical metabolic adjustments. One of the most pronounced adjustments of fat metabolism occurs in hibernating grizzly bears. These animals remain in a continuous state of dormancy for periods as long as seven months. Unlike most hibernating species, the bear maintains a body temperature of between 32 and 35 °C, close to the normal (nonhibernating) level. Although expending about 25,000 kJ/day (6,000 kcal/day), the bear does not eat, drink, urinate, or defecate for months at a time.

Experimental studies have shown that hibernating grizzly bears use body fat as their sole fuel. Fat oxidation yields sufficient energy for maintenance of body temperature, active synthesis of amino acids and proteins, and other energy-requiring activities, such as membrane transport. Fat oxidation also releases large amounts of water, as described in the text, which replenishes water lost in breathing. The glycerol released by degradation of triacylglycerols is converted into blood glucose by gluconeogenesis. Urea formed during breakdown of amino acids is reabsorbed in the kidneys and recycled, the amino groups reused to make new amino acids for maintaining body proteins.

Bears store an enormous amount of body fat in preparation for their long sleep. An adult grizzly consumes about 38,000 kJ/day during the late spring and summer, but as winter approaches it feeds 20 hours a day, consuming up to 84,000 kJ daily. This change in feeding is a response to a seasonal change in hormone secretion. Large amounts of triacylglycerols are formed from the huge intake of carbohydrates during the fattening-up period. Other hibernating species, including the tiny dormouse, also accumulate large amounts of body fat.

A grizzly bear prepares its hibernation nest near the McNeil River in Canada.

TABLE 17–1   Yield of ATP during Oxidation of One Molecule of Palmitoyl-CoA to CO_2 and H_2O

Enzyme catalyzing the oxidation step	Number of NADH or $FADH_2$ formed	Number of ATP ultimately formed*
Acyl-CoA dehydrogenase	7 $FADH_2$	10.5
β-Hydroxyacyl-CoA dehydrogenase	7 NADH	17.5
Isocitrate dehydrogenase	8 NADH	20
α-Ketoglutarate dehydrogenase	8 NADH	20
Succinyl-CoA synthetase		8†
Succinate dehydrogenase	8 $FADH_2$	12
Malate dehydrogenase	8 NADH	20
Total		108

*These calculations assume that mitochondrial oxidative phosphorylation produces 1.5 ATP per $FADH_2$ oxidized and 2.5 ATP per NADH oxidized.

†GTP produced directly in this step yields ATP in the reaction catalyzed by nucleoside diphosphate kinase (p. 526).

theoretical maximum. However, when the free-energy changes are calculated from actual concentrations of reactants and products under intracellular conditions (see Worked Example 13–2, p. 519), the free-energy recovery is more than 60%; the energy conservation is remarkably efficient.

Oxidation of Unsaturated Fatty Acids Requires Two Additional Reactions

The fatty acid oxidation sequence just described is typical when the incoming fatty acid is saturated (that is, has only single bonds in its carbon chain). However, most of the fatty acids in the triacylglycerols and phospholipids of animals and plants are unsaturated, having one or more double bonds. These bonds are in the cis configuration and cannot be acted upon by enoyl-CoA hydratase, the enzyme catalyzing the addition of H_2O to the trans double bond of the Δ^2-enoyl-CoA generated during β oxidation. Two auxiliary enzymes are needed for β oxidation of the common unsaturated fatty acids: an isomerase and a reductase. We illustrate these auxiliary reactions with two examples.

Oleate is an abundant 18-carbon monounsaturated fatty acid with a cis double bond between C-9 and C-10 (denoted Δ^9). In the first step of oxidation, oleate is converted to oleoyl-CoA and, like the saturated fatty acids, enters the mitochondrial matrix via the carnitine shuttle (Fig. 17–6). Oleoyl-CoA then undergoes three passes through the fatty acid oxidation cycle to yield three molecules of acetyl-CoA and the coenzyme A ester of a Δ^3, 12-carbon unsaturated fatty acid, cis-Δ^3-dodecenoyl-CoA **(Fig. 17–10)**. This product cannot serve as a substrate for enoyl-CoA hydratase, which acts only on trans double bonds. The auxiliary enzyme Δ^3,Δ^2-**enoyl-CoA isomerase** isomerizes the cis-Δ^3-enoyl-CoA to the $trans$-Δ^2-enoyl-CoA, which is converted by enoyl-CoA hydratase into the corresponding L-β-hydroxyacyl-CoA ($trans$-Δ^2-dodecenoyl-CoA). This intermediate is now acted upon by the remaining enzymes of β oxidation to yield acetyl-CoA and the coenzyme A ester of a 10-carbon saturated fatty acid, decanoyl-CoA. The latter undergoes four more passes through the β-oxidation pathway to yield five more molecules of acetyl-CoA. Altogether, nine acetyl-CoAs are produced from one molecule of the 18-carbon oleate.

The other auxiliary enzyme (a reductase) is required for oxidation of polyunsaturated fatty acids—for example, the 18-carbon linoleate, which has a cis-Δ^9,cis-Δ^{12} configuration **(Fig. 17–11)**. Linoleoyl-CoA undergoes three passes through the β-oxidation sequence to yield three molecules of acetyl-CoA and the coenzyme A ester of a 12-carbon unsaturated fatty acid with a cis-Δ^3,cis-Δ^6 configuration. This intermediate cannot be used by the enzymes of the β-oxidation pathway; its double bonds are in the wrong position

FIGURE 17–10 Oxidation of a monounsaturated fatty acid. Oleic acid, as oleoyl-CoA (Δ^9), is the example used here. Oxidation requires an additional enzyme, enoyl-CoA isomerase, to reposition the double bond, converting the cis isomer to a trans isomer, a normal intermediate in β oxidation.

and have the wrong configuration (cis, not trans). However, the combined action of enoyl-CoA isomerase and **2,4-dienoyl-CoA reductase**, as shown in Figure 17–11, allows reentry of this intermediate into the β-oxidation pathway and its degradation to six acetyl-CoAs. The overall result is conversion of linoleate to nine molecules of acetyl-CoA.

Complete Oxidation of Odd-Number Fatty Acids Requires Three Extra Reactions

Although most naturally occurring lipids contain fatty acids with an even number of carbon atoms, fatty acids with an odd number of carbons are common in the lipids of many plants and some marine organisms. Cattle and other ruminant animals form large amounts of the three-carbon **propionate** (CH_3—CH_2—COO^-) during fermentation of carbohydrates in the rumen. The propionate is absorbed into the blood and oxidized by the liver and other tissues. And small quantities of propionate are added as a mold inhibitor to some breads and cereals, thus entering the human diet.

Long-chain odd-number fatty acids are oxidized in the same pathway as the even-number acids, beginning at the carboxyl end of the chain. However, the substrate for the last pass through the β-oxidation sequence is a fatty acyl–CoA with a five-carbon fatty acid. When this

FIGURE 17–11 Oxidation of a polyunsaturated fatty acid. The example here is linoleic acid, as linoleoyl-CoA ($\Delta^{9,12}$). Oxidation requires a second auxiliary enzyme in addition to enoyl-CoA isomerase: NADPH-dependent 2,4-dienoyl-CoA reductase. The combined action of these two enzymes converts a *trans*-Δ^2,*cis*-Δ^4-dienoyl-CoA intermediate to the *trans*-Δ^2-enoyl-CoA substrate necessary for β oxidation.

FIGURE 17–12 Oxidation of propionyl-CoA produced by β oxidation of odd-number fatty acids. The sequence involves the carboxylation of propionyl-CoA to D-methylmalonyl-CoA and conversion of the latter to succinyl-CoA. This conversion requires epimerization of D- to L-methylmalonyl-CoA, followed by a remarkable reaction in which substituents on adjacent carbon atoms exchange positions (see Box 17–2).

is oxidized and cleaved, the products are acetyl-CoA and **propionyl-CoA**. The acetyl-CoA can be oxidized in the citric acid cycle, of course, but propionyl-CoA enters a different pathway having three enzymes.

Propionyl-CoA is first carboxylated to form the D stereoisomer of **methylmalonyl-CoA (Fig. 17–12)** by **propionyl-CoA carboxylase**, which contains the cofactor biotin. In this enzymatic reaction, as in the pyruvate carboxylase reaction (see Fig. 16–17), CO_2 (or its hydrated ion, HCO_3^-) is activated by attachment to biotin before its transfer to the substrate, in this case the propionate moiety. Formation of the carboxybiotin intermediate requires energy, which is provided by ATP. The D-methylmalonyl-CoA thus formed is enzymatically

epimerized to its L stereoisomer by **methylmalonyl-CoA epimerase** (Fig. 17–12). The L-methylmalonyl-CoA then undergoes an intramolecular rearrangement to form succinyl-CoA, which can enter the citric acid cycle. This rearrangement is catalyzed by **methylmalonyl-CoA mutase**, which requires as its coenzyme **5′-deoxyadenosylcobalamin**, or **coenzyme B$_{12}$**, which is derived from vitamin B$_{12}$ (cobalamin). Box 17–2 describes the role of coenzyme B$_{12}$ in this remarkable exchange reaction.

Fatty Acid Oxidation Is Tightly Regulated

Oxidation of fatty acids consumes a precious fuel, and it is regulated so as to occur only when the need for energy requires it. In the liver, fatty acyl–CoA formed in the cytosol has two major pathways open to it: (1) β oxidation by enzymes in mitochondria or (2) conversion into triacylglycerols and phospholipids by enzymes in the cytosol. The pathway taken depends on the rate of transfer of long-chain fatty acyl–CoA into mitochondria. The three-step process (carnitine shuttle) by which fatty acyl groups are carried from cytosolic fatty acyl–CoA into the mitochondrial matrix (Fig. 17–6) is rate-limiting for fatty acid oxidation and is an important point of regulation. Once fatty acyl groups have entered the mitochondrion, they are committed to oxidation to acetyl-CoA.

Malonyl-CoA, the first intermediate in the cytosolic biosynthesis of long-chain fatty acids from acetyl-CoA (see Fig. 21–2), increases in concentration whenever the animal is well supplied with carbohydrate; excess glucose that cannot be oxidized or stored as glycogen is converted in the cytosol into fatty acids for storage as triacylglycerol. The inhibition of carnitine acyltransferase I by malonyl-CoA **(Fig. 17–13)** ensures that the oxidation of fatty acids is inhibited whenever the liver is amply supplied with glucose as fuel and is actively making triacylglycerols from excess glucose.

$$\overset{O}{\overset{\|}{^-OOC{-}CH_2{-}C{-}S\text{-}CoA}}$$
$$\text{Malonyl-CoA}$$

Two of the enzymes of β oxidation are also regulated by metabolites that signal energy sufficiency. When the [NADH]/[NAD$^+$] ratio is high, β-hydroxyacyl-

CoA dehydrogenase is inhibited; in addition, high concentrations of acetyl-CoA inhibit thiolase.

Recall from Chapter 15 that during periods of vigorous muscle contraction or during fasting, the fall in [ATP] and the rise in [AMP] activate AMPK, the AMP-activated protein kinase. AMPK phosphorylates several target enzymes, including acetyl-CoA carboxylase, which catalyzes malonyl-CoA synthesis. This phosphorylation and thus inhibition of acetyl-CoA carboxylase lowers the concentration of malonyl-CoA, relieving the inhibition of fatty acyl–carnitine transport into mitochondria (Fig. 17–13) and allowing β oxidation to replenish the supply of ATP.

Transcription Factors Turn on the Synthesis of Proteins for Lipid Catabolism

In addition to the various short-term regulatory mechanisms that modulate the activity of existing enzymes, transcriptional regulation can change the number of molecules of the enzymes of fatty acid oxidation on a longer time scale, minutes to hours. The **PPAR** family of nuclear receptors are transcription factors that affect many metabolic processes in response to a variety of fatty acid–like ligands. (They were originally recognized as *p*eroxisome *p*roliferator-*a*ctivated *r*eceptors, then were found to function more broadly.) PPARα acts in muscle, adipose tissue, and liver to turn on a set of genes essential for fatty acid oxidation, including the

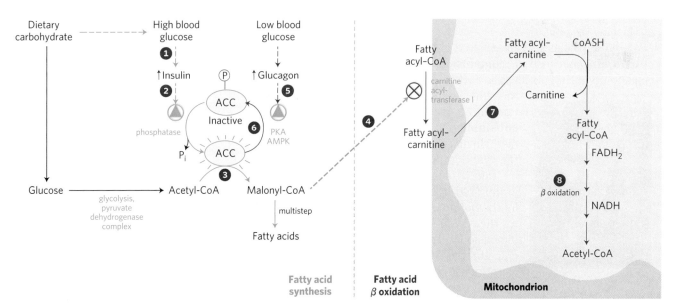

FIGURE 17–13 Coordinated regulation of fatty acid synthesis and breakdown. When the diet provides a ready source of carbohydrate as fuel, β oxidation of fatty acids is unnecessary and is therefore downregulated. Two enzymes are key to the coordination of fatty acid metabolism: acetyl-CoA carboxylase (ACC), the first enzyme in the synthesis of fatty acids (see Fig. 21-1), and carnitine acyltransferase I, which limits the transport of fatty acids into the mitochondrial matrix for β oxidation (see Fig. 17-6). Ingestion of a high-carbohydrate meal raises the blood glucose level and thus ❶ triggers the release of insulin. ❷ Insulin-dependent protein phosphatase dephosphorylates ACC, activating it.

❸ ACC catalyzes the formation of malonyl-CoA (the first intermediate of fatty acid synthesis), and ❹ malonyl-CoA inhibits carnitine acyltransferase I, thereby preventing fatty acid entry into the mitochondrial matrix. When blood glucose levels drop between meals, ❺ glucagon release activates cAMP-dependent protein kinase (PKA), which ❻ phosphorylates and inactivates ACC. The concentration of malonyl-CoA falls, the inhibition of fatty acid entry into mitochondria is relieved, and ❼ fatty acids enter the mitochondrial matrix and ❽ become the major fuel. Because glucagon also triggers the mobilization of fatty acids in adipose tissue, a supply of fatty acids begins arriving in the blood.

In the methylmalonyl-CoA mutase reaction (see Fig. 17–12), the group —CO—S-CoA at C-2 of the original propionate exchanges position with a hydrogen atom at C-3 of the original propionate (Fig. 1a). Coenzyme B$_{12}$ is the cofactor for this reaction, as it is for almost all enzymes that catalyze reactions of this general type (Fig. 1b). These coenzyme B$_{12}$–dependent processes are among the very few enzymatic reactions in biology in which there is an exchange of an alkyl or substituted alkyl group (X) with a hydrogen atom on an adjacent carbon, *with no mixing of the transferred hydrogen atom with the hydrogen of the solvent,* H$_2$O. How can the hydrogen atom move between two carbons without mixing with the enormous excess of hydrogen atoms in the solvent?

Coenzyme B$_{12}$ is the cofactor form of vitamin B$_{12}$, which is unique among all the vitamins in that it contains not only a complex organic molecule but an essential trace element, cobalt. The complex **corrin ring system** of vitamin B$_{12}$ (colored blue in Fig. 2), to which cobalt (as Co^{3+}) is coordinated, is chemically related to the porphyrin ring system of heme and heme proteins (see Fig. 5–1). A fifth coordination position of cobalt is filled by dimethylbenzimidazole ribonucleotide (shaded yellow), bound covalently by its 3′-phosphate group to a side chain of the corrin ring, through aminoisopropanol. The formation of this complex cofactor occurs in one of only two known reactions in which triphosphate is cleaved from ATP (Fig. 3); the other reaction is the formation of *S*-adenosylmethionine from ATP and methionine (see Fig. 18–18).

Vitamin B$_{12}$ as usually isolated is called **cyanocobalamin**, because it contains a cyano group (picked up during purification) attached to cobalt in the sixth coordination position. In **5′-deoxyadenosylcobalamin**, the cofactor for methylmalonyl-CoA mutase, the cyano group is replaced by the **5′-deoxyadenosyl** group (red in Fig. 2), covalently bound through C-5′ to the cobalt. The three-dimensional structure of the cofactor was determined by Dorothy Crowfoot Hodgkin in 1956, using x-ray crystallography.

Dorothy Crowfoot Hodgkin, 1910–1994

The key to understanding how coenzyme B$_{12}$ catalyzes hydrogen exchange lies in the properties of the covalent bond between cobalt and C-5′ of the deoxyadenosyl group (Fig. 2). This is a relatively weak bond; its bond dissociation energy is about 110 kJ/mol, compared with 348 kJ/mol for a typical C—C bond or 414 kJ/mol for a C—H bond. Merely illuminating the

(a)

L-Methylmalonyl-CoA ⇌ (coenzyme B$_{12}$ / methylmalonyl-CoA mutase) Succinyl-CoA

(b)

⇌ coenzyme B$_{12}$

FIGURE 1

compound with visible light is enough to break this Co—C bond. (This extreme photolability probably accounts for the absence of vitamin B$_{12}$ in plants.) Dissociation produces a 5′-deoxyadenosyl radical and the Co^{2+} form of the vitamin. The chemical function of 5′-deoxyadenosylcobalamin is to generate free radicals in this way, thus initiating a series of transformations such as that illustrated in Figure 4—a postulated mechanism for the reaction catalyzed by methylmalonyl-CoA

FIGURE 2

mutase and several other coenzyme B$_{12}$–dependent transformations. In this postulated mechanism, the migrating hydrogen atom never exists as a free species and is thus never free to exchange with the hydrogen of surrounding water molecules.

Vitamin B$_{12}$ deficiency results in serious disease. This vitamin is not made by plants or animals and can be synthesized only by a few species of microorganisms. It is required by healthy people in only minute amounts, about 3 μg/day. The serious disease **pernicious anemia** results from failure to absorb vitamin B$_{12}$ efficiently from the intestine, where it is synthesized by intestinal bacteria or obtained from digestion of meat. Individuals with this disease do not produce sufficient amounts of **intrinsic factor**, a glycoprotein essential to vitamin B$_{12}$ absorption. The pathology in pernicious anemia includes reduced production of erythrocytes, reduced levels of hemoglobin, and severe, progressive impairment of the central nervous system. Administration of large doses of vitamin B$_{12}$ alleviates these symptoms in at least some cases. ■

FIGURE 3

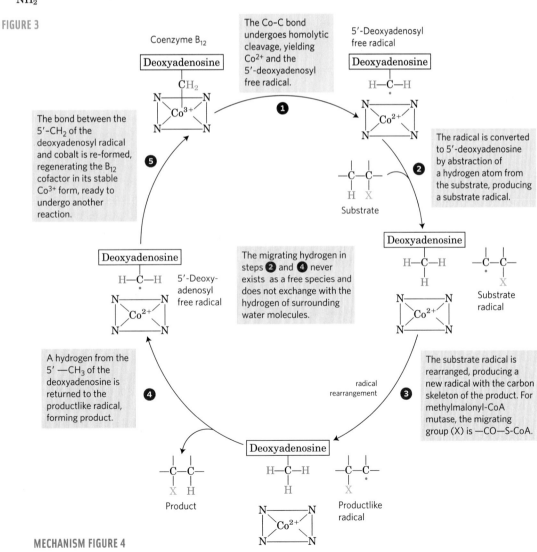

The Co-C bond undergoes homolytic cleavage, yielding Co^{2+} and the 5'-deoxyadenosyl free radical.

The radical is converted to 5'-deoxyadenosine by abstraction of a hydrogen atom from the substrate, producing a substrate radical.

The substrate radical is rearranged, producing a new radical with the carbon skeleton of the product. For methylmalonyl-CoA mutase, the migrating group (X) is —CO—S-CoA.

A hydrogen from the 5' —CH$_3$ of the deoxyadenosine is returned to the productlike radical, forming product.

The bond between the 5'-CH$_2$ of the deoxyadenosyl radical and cobalt is re-formed, regenerating the B$_{12}$ cofactor in its stable Co^{3+} form, ready to undergo another reaction.

The migrating hydrogen in steps ❷ and ❹ never exists as a free species and does not exchange with the hydrogen of surrounding water molecules.

Coenzyme B$_{12}$ · Deoxyadenosine · 5'-Deoxyadenosyl free radical · Substrate · Substrate radical · Product · Productlike radical · radical rearrangement · 5'-Deoxyadenosyl free radical

MECHANISM FIGURE 4

fatty acid transporter, carnitine acyltransferases I and II, fatty acyl–CoA dehydrogenases for short, medium, long, and very long acyl chains, and related enzymes. This response is triggered when a cell or organism has an increased demand for energy from fat catabolism, such as during a fast between meals or under conditions of longer-term starvation. Glucagon, released in response to low blood glucose, can act through cAMP and the transcription factor CREB to turn on certain genes for lipid catabolism.

Another situation that is accompanied by major changes in the expression of the enzymes of fatty acid oxidation is the transition from fetal to neonatal metabolism in the heart. In the fetus the principal fuels are glucose and lactate, but in the neonatal heart, fatty acids are the main fuel. At the time of this transition, PPARα is activated and in turn activates the genes essential for fatty acid metabolism. As we will see in Chapter 23, two other transcription factors in the PPAR family also play crucial roles in setting the enzyme complements—and therefore the metabolic activities—of specific tissues at particular times (see Fig. 23–42).

The major site of fatty acid oxidation, at rest and during exercise, is skeletal muscle. Endurance training increases PPARα expression in muscle, leading to increased levels of fatty acid–oxidizing enzymes and increased oxidative capacity of the muscle.

Genetic Defects in Fatty Acyl–CoA Dehydrogenases Cause Serious Disease

Stored triacylglycerols are typically the chief source of energy for muscle contraction, and an inability to oxidize fatty acids from triacylglycerols has serious consequences for health. The most common genetic defect in fatty acid catabolism in U.S. and northern European populations is due to a mutation in the gene encoding the **medium-chain acyl-CoA dehydrogenase (MCAD)**. Among northern Europeans, the frequency of carriers (individuals with this recessive mutation on one of the two homologous chromosomes) is about 1 in 40, and about 1 individual in 10,000 has the disease—that is, has two copies of the mutant MCAD allele and is unable to oxidize fatty acids of 6 to 12 carbons. The disease is characterized by recurring episodes of a syndrome that includes fat accumulation in the liver, high blood levels of octanoic acid (8:0), low blood glucose (hypoglycemia), sleepiness, vomiting, and coma. The pattern of organic acids in the urine helps in the diagnosis of this disease: the urine commonly contains high levels of 6-carbon to 10-carbon dicarboxylic acids (produced by ω oxidation) and low levels of urinary ketone bodies (we discuss ω oxidation below and ketone bodies in Section 17.3). Although individuals may have no symptoms between episodes, the episodes are very serious; mortality from this disease is 25% to 60% in early childhood. If the genetic defect is detected shortly after birth, the infant can be started on a low-fat, high-carbohydrate diet. With early detection and careful management of the diet—including avoiding long intervals between meals, to prevent the body from turning to its fat reserves for energy—the prognosis for these individuals is good.

More than 20 other human genetic defects in fatty acid transport or oxidation have been documented, most much less common than the defect in MCAD. One of the most severe disorders results from loss of the long-chain β-hydroxyacyl-CoA dehydrogenase activity of the trifunctional protein, TFP. Other disorders include defects in the α or β subunits that affect all three activities of TFP and cause serious heart disease and abnormal skeletal muscle. ∎

Peroxisomes Also Carry Out β Oxidation

The mitochondrial matrix is the major site of fatty acid oxidation in animal cells, but in certain cells other compartments also contain enzymes capable of oxidizing fatty acids to acetyl-CoA, by a pathway similar but not identical to that in mitochondria. In plant cells, the major site of β oxidation is not mitochondria but peroxisomes.

In **peroxisomes**, membrane-enclosed organelles of animal and plant cells, the intermediates for β oxidation of fatty acids are coenzyme A derivatives, and the process consists of four steps, as in mitochondrial β oxidation **(Fig. 17–14)**: (1) dehydrogenation, (2) addition of water to the resulting double bond, (3) oxidation of the β-hydroxyacyl-CoA to a ketone, and (4) thiolytic cleavage by coenzyme A. (The identical reactions also occur in glyoxysomes, as discussed below.)

One difference between the peroxisomal and mitochondrial pathways is in the chemistry of the first step. In peroxisomes, the flavoprotein acyl-CoA oxidase that introduces the double bond passes electrons directly to O_2, producing H_2O_2 (Fig. 17–14). (Thus the name peroxisomes.) This strong and potentially damaging oxidant is immediately cleaved to H_2O and O_2 by **catalase**. Recall that in mitochondria, the electrons removed in the first oxidation step pass through the respiratory chain to O_2 to produce H_2O, and this process is accompanied by ATP synthesis. In peroxisomes, the energy released in the first oxidative step of fatty acid breakdown is not conserved as ATP, but is dissipated as heat.

A second important difference between mitochondrial and peroxisomal β oxidation in mammals is in the specificity for fatty acyl–CoAs; the peroxisomal system is much more active on very-long-chain fatty acids such as hexacosanoic acid (26:0) and on branched-chain fatty acids such as phytanic acid and pristanic acid (see Fig. 17–18). These less-common fatty acids are obtained in the diet from dairy products, the fat of ruminant animals, meat, and fish. Their catabolism in the peroxisome involves several auxiliary enzymes unique to this organelle. The inability to oxidize these compounds is responsible for several serious human diseases. Individuals with **Zellweger syndrome** are unable to make peroxisomes and therefore lack all the

FIGURE 17–14 Comparison of β oxidation in mitochondria and in peroxisomes and glyoxysomes. The peroxisomal/glyoxysomal system differs from the mitochondrial system in three respects: (1) the peroxisomal system prefers very-long-chain fatty acids; (2) in the first oxidative step electrons pass directly to O_2, generating H_2O_2, and (3) the NADH formed in the second oxidative step cannot be reoxidized in the peroxisome or glyoxysome, so reducing equivalents are exported to the cytosol, eventually entering mitochondria. The acetyl-CoA produced by peroxisomes and glyoxysomes is also exported; the acetate from glyoxysomes (organelles found only in germinating seeds) serves as a biosynthetic precursor (see Fig. 17-15). Acetyl-CoA produced in mitochondria is further oxidized in the citric acid cycle.

metabolism unique to that organelle. In **X-linked adrenoleukodystrophy (XALD)**, peroxisomes fail to oxidize very-long-chain fatty acids, apparently for lack of a functional transporter for these fatty acids in the peroxisomal membrane. Both defects lead to accumulation in the blood of very-long-chain fatty acids, especially 26:0. XALD affects young boys before the age of 10 years, causing loss of vision, behavioral disturbances, and death within a few years. ∎

In mammals, high concentrations of fats in the diet result in increased synthesis of the enzymes of peroxisomal β oxidation in the liver. Liver peroxisomes do not contain the enzymes of the citric acid cycle and cannot catalyze the oxidation of acetyl-CoA to CO_2. Instead, long-chain or branched fatty acids are catabolized to shorter-chain products, such as hexanoyl-CoA, which are exported to mitochondria and completely oxidized.

Plant Peroxisomes and Glyoxysomes Use Acetyl-CoA from β Oxidation as a Biosynthetic Precursor

In plants, fatty acid oxidation does not occur primarily in mitochondria but in the peroxisomes of leaf tissue and in the glyoxysomes of germinating seeds. Plant peroxisomes and glyoxysomes are similar in structure and function; glyoxysomes, which occur only in germinating seeds, may be considered specialized peroxisomes. The biological role of β oxidation in these organelles is to use stored lipids primarily to provide biosynthetic precursors, not energy.

During seed germination, stored triacylglycerols are converted into glucose, sucrose, and a wide variety of essential metabolites **(Fig. 17–15)**. Fatty acids released from the triacylglycerols are first activated to their coenzyme A derivatives and oxidized in glyoxysomes by the same four-step process that takes place in peroxisomes (Fig. 17–14). The acetyl-CoA produced is converted via the glyoxylate cycle to four-carbon precursors for gluconeogenesis (see Fig. 16–24). Glyoxysomes, like peroxisomes, contain high concentrations of catalase, which converts the H_2O_2 produced by β oxidation to H_2O and O_2.

The β-Oxidation Enzymes of Different Organelles Have Diverged during Evolution

Although the β-oxidation reactions in mitochondria are essentially the same as those in peroxisomes and glyoxysomes, the enzymes (isozymes) differ significantly

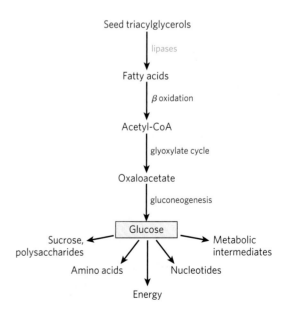

FIGURE 17–15 Triacylglycerols as glucose source in seeds. β Oxidation is one stage in a pathway that converts stored triacylglycerols to glucose in germinating seeds. For more detail, see Figure 16-24.

between the two types of organelles. The differences apparently reflect an evolutionary divergence that occurred very early, with the separation of gram-positive and gram-negative bacteria (see Fig. 1–6).

In mitochondria, the four β-oxidation enzymes that act on short-chain fatty acyl–CoAs are separate, soluble proteins (as noted earlier), similar in structure to the analogous enzymes of gram-positive bacteria **(Fig. 17–16a)**. The gram-negative bacteria have four activities in three soluble subunits (Fig. 17–16b), and the eukaryotic enzyme system that acts on long-chain fatty acids—the trifunctional protein, TFP—has three enzyme activities in two subunits that are membrane-associated (Fig. 17–16c). The β-oxidation enzymes of plant peroxisomes and glyoxysomes, however, form a complex of proteins, one of which contains four enzymatic activities in a single polypeptide chain (Fig. 17–16d). The first enzyme, acyl-CoA oxidase, is a single polypeptide chain; the **multifunctional protein (MFP)** contains the second and third enzyme activities (enoyl-CoA hydratase and hydroxyacyl-CoA dehydrogenase) as well as two auxiliary activities needed for the oxidation of unsaturated fatty acids (D-3-hydroxyacyl-CoA epimerase and Δ^3,Δ^2-enoyl-CoA isomerase); the fourth enzyme, thiolase, is a separate, soluble polypeptide.

It is interesting that the enzymes that catalyze essentially the reversal of β oxidation in the synthesis of fatty acids are also organized differently in bacteria and eukaryotes; in bacteria, the seven enzymes needed for fatty acid synthesis are separate polypeptides, but in mammals, all seven activities are part of a single, huge polypeptide chain. One advantage to the cell in having several enzymes of the same pathway encoded in a single polypeptide chain is that this solves the problem of regulating the synthesis of enzymes that must interact functionally; regulation of the expression of *one* gene ensures production of the same number of active sites for all enzymes in the path. When each enzyme activity is on a separate polypeptide, some mechanism is required to coordinate the synthesis of all the gene products. The *disadvantage* of having several activities on the same polypeptide is that the longer the polypeptide chain, the greater is the probability of a mistake in its synthesis: a single incorrect amino acid in the chain may make all the enzyme activities in that chain useless. Comparison of the gene structures for these proteins in many species may shed light on the reasons for the selection of one or the other strategy in evolution.

The ω Oxidation of Fatty Acids Occurs in the Endoplasmic Reticulum

Although mitochondrial β oxidation, in which enzymes act at the carboxyl end of a fatty acid, is by far the most important catabolic fate for fatty acids in animal cells, there is another pathway in some species, including vertebrates, that involves oxidation of the ω (omega) carbon—the carbon most distant from the carboxyl group. The enzymes unique to **ω oxidation** are located (in vertebrates) in the endoplasmic reticulum of liver and

(a) Gram-positive bacteria and mitochondrial short-chain-specific system

(b) Gram-negative bacteria

(c) Mitochondrial very-long-chain-specific system

(d) Peroxisomal and glyoxysomal system of plants

FIGURE 17–16 The enzymes of β oxidation. Shown here are the different subunit structures of the enzymes of β oxidation in gram-positive and gram-negative bacteria, mitochondria, and plant peroxisomes and glyoxysomes. Enz$_1$ is acyl-CoA dehydrogenase; Enz$_2$, enoyl-CoA hydratase; Enz$_3$, L-β-hydroxyacyl-CoA dehydrogenase; Enz$_4$, thiolase; Enz$_5$, D-3-hydroxyacyl-CoA epimerase, and Enz$_6$, Δ^3,Δ^2-enoyl-CoA isomerase. **(a)** The four enzymes of β oxidation in gram-positive bacteria are separate, soluble entities, as are those of the short-chain-specific system of mitochondria. **(b)** In gram-negative bacteria, the four enzyme activities reside in three polypeptides; Enz$_2$ and Enz$_3$ are parts of a single polypeptide chain. **(c)** The very-long-chain-specific system of mitochondria is also composed of three polypeptides, one of which includes the activities of Enz$_2$ and Enz$_3$; in this case, the system is bound to the inner mitochondrial membrane. **(d)** In the peroxisomal and glyoxysomal β-oxidation systems of plants, Enz$_1$ and Enz$_4$ are separate polypeptides, but Enz$_2$ and Enz$_3$, as well as two auxiliary enzymes (Enz$_5$ and Enz$_6$), are part of a single polypeptide chain: the multifunctional protein, MFP.

FIGURE 17–17 The ω oxidation of fatty acids in the endoplasmic reticulum. This alternative to β oxidation begins with oxidation of the carbon most distant from the β carbon—the ω (omega) carbon. The substrate is usually a medium-chain fatty acid; shown here is lauric acid (laurate). This pathway is generally not the major route for oxidative catabolism of fatty acids.

kidney, and the preferred substrates are fatty acids of 10 or 12 carbon atoms. In mammals ω oxidation is normally a minor pathway for fatty acid degradation, but when β oxidation is defective (because of mutation or a carnitine deficiency, for example) it becomes more important.

The first step introduces a hydroxyl group onto the ω carbon **(Fig. 17–17).** The oxygen for this group comes from molecular oxygen (O_2) in a complex reaction that involves cytochrome P450 and the electron donor NADPH. Reactions of this type are catalyzed by **mixed-function oxidases**, described in Box 21–1. Two more enzymes now act on the ω carbon: **alcohol dehydrogenase** oxidizes the hydroxyl group to an aldehyde, and **aldehyde dehydrogenase** oxidizes the aldehyde group to a carboxylic acid, producing a fatty acid with a carboxyl group at each end. At this point, either end can be attached to coenzyme A, and the molecule can enter the mitochondrion and undergo β oxidation by the normal route. In each pass through the β-oxidation pathway, the "double-ended" fatty acid yields dicarboxylic acids

such as succinic acid, which can enter the citric acid cycle, and adipic acid (Fig. 17–17).

Phytanic Acid Undergoes α Oxidation in Peroxisomes

The presence of a methyl group on the β carbon of a fatty acid makes β oxidation impossible, and these branched fatty acids are catabolized in peroxisomes of animal cells by **α oxidation**. In the oxidation of phytanic acid, for example **(Fig. 17–18),** phytanoyl-CoA is hydroxylated on its α carbon, in a reaction that

FIGURE 17–18 The α oxidation of a branched-chain fatty acid (phytanic acid) in peroxisomes. Phytanic acid has a methyl-substituted β carbon and therefore cannot undergo β oxidation. The combined action of the enzymes shown here removes the carboxyl carbon of phytanic acid to produce pristanic acid, in which the β carbon is unsubstituted, allowing β oxidation. Notice that β oxidation of pristanic acid releases propionyl-CoA, not acetyl-CoA. This is further catabolized as in Figure 17-12. (The details of the reaction that produces pristanal remain controversial.)

involves molecular oxygen; decarboxylated to form an aldehyde one carbon shorter; and then oxidized to the corresponding carboxylic acid, which now has no substituent on the β carbon and can be oxidized further by β oxidation. **Refsum disease**, resulting from a genetic defect in phytanoyl-CoA hydroxylase, leads to very high blood levels of phytanic acid and severe neurological problems, including blindness and deafness. ■

SUMMARY 17.2 Oxidation of Fatty Acids

▶ In the first stage of β oxidation, four reactions remove each acetyl-CoA unit from the carboxyl end of a saturated fatty acyl–CoA: (1) dehydrogenation of the α and β carbons (C-2 and C-3) by FAD-linked acyl-CoA dehydrogenases, (2) hydration of the resulting trans-Δ^2 double bond by enoyl-CoA hydratase, (3) dehydrogenation of the resulting L-β-hydroxyacyl-CoA by NAD-linked β-hydroxyacyl-CoA dehydrogenase, and (4) CoA-requiring cleavage of the resulting β-ketoacyl-CoA by thiolase, to form acetyl-CoA and a fatty acyl–CoA shortened by two carbons. The shortened fatty acyl–CoA then reenters the sequence.

▶ In the second stage of fatty acid oxidation, the acetyl-CoA is oxidized to CO_2 in the citric acid cycle. A large fraction of the theoretical yield of free energy from fatty acid oxidation is recovered as ATP by oxidative phosphorylation, the final stage of the oxidative pathway.

▶ Malonyl-CoA, an early intermediate of fatty acid synthesis, inhibits carnitine acyltransferase I, preventing fatty acid entry into mitochondria. This blocks fatty acid breakdown while synthesis is occurring.

▶ Genetic defects in the medium-chain acyl-CoA dehydrogenase result in serious human disease, as do mutations in other components of the β-oxidation system.

▶ Oxidation of unsaturated fatty acids requires two additional enzymes: enoyl-CoA isomerase and 2,4-dienoyl-CoA reductase. Odd-number fatty acids are oxidized by the β-oxidation pathway to yield acetyl-CoA and a molecule of propionyl-CoA. This is carboxylated to methylmalonyl-CoA, which is isomerized to succinyl-CoA in a reaction catalyzed by methylmalonyl-CoA mutase, an enzyme requiring coenzyme B_{12}.

▶ Peroxisomes of plants and animals, and glyoxysomes of plants, carry out β oxidation in four steps similar to those of the mitochondrial pathway in animals. The first oxidation step, however, transfers electrons directly to O_2, generating H_2O_2. Peroxisomes of animal tissues specialize in the oxidation of very-long-chain fatty acids and branched fatty acids. In glyoxysomes, in germinating seeds, β oxidation is one step in the conversion of stored lipids into a variety of intermediates and products.

▶ The reactions of ω oxidation, occurring in the endoplasmic reticulum, produce dicarboxylic fatty acyl intermediates, which can undergo β oxidation at either end to yield short dicarboxylic acids such as succinate.

▶ The reactions of α oxidation degrade branched fatty acids such as phytanic acid.

17.3 Ketone Bodies

In humans and most other mammals, acetyl-CoA formed in the liver during oxidation of fatty acids can either enter the citric acid cycle (stage 2 of Fig. 17–7) or undergo conversion to the "ketone bodies," **acetone**, **acetoacetate**, and D-β-**hydroxybutyrate**, for export to other tissues. (The term "bodies" is a historical artifact; the term is occasionally applied to insoluble particles, but these compounds are soluble in blood and urine.)

$$CH_3-\overset{\displaystyle O}{\underset{\displaystyle \|}{C}}-CH_3$$

Acetone

$$CH_3-\overset{\displaystyle O}{\underset{\displaystyle \|}{C}}-CH_2-\overset{\displaystyle O}{C}\diagdown_{O^-}$$

Acetoacetate

$$CH_3-\overset{\displaystyle OH}{\underset{\displaystyle |}{\underset{\displaystyle H}{C}}}-CH_2-\overset{\displaystyle O}{C}\diagdown_{O^-}$$

D-β-Hydroxybutyrate

Acetone, produced in smaller quantities than the other ketone bodies, is exhaled. Acetoacetate and D-β-hydroxybutyrate are transported by the blood to tissues other than the liver (extrahepatic tissues), where they are converted to acetyl-CoA and oxidized in the citric acid cycle, providing much of the energy required by tissues such as skeletal and heart muscle and the renal cortex. The brain, which preferentially uses glucose as fuel, can adapt to the use of acetoacetate or D-β-hydroxybutyrate under starvation conditions, when glucose is unavailable. The production and export of ketone bodies from the liver to extrahepatic tissues allows continued oxidation of fatty acids in the liver when acetyl-CoA is not being oxidized in the citric acid cycle.

Ketone Bodies, Formed in the Liver, Are Exported to Other Organs as Fuel

The first step in the formation of acetoacetate, occurring in the liver (**Fig. 17–19**), is the enzymatic condensation of two molecules of acetyl-CoA, catalyzed by thiolase;

FIGURE 17–19 **Formation of ketone bodies from acetyl-CoA.** Healthy, well-nourished individuals produce ketone bodies at a relatively low rate. When acetyl-CoA accumulates (as in starvation or untreated diabetes, for example), thiolase catalyzes the condensation of two acetyl-CoA molecules to acetoacetyl-CoA, the parent compound of the three ketone bodies. The reactions of ketone body formation occur in the matrix of liver mitochondria. The six-carbon compound β-hydroxy-β-methylglutaryl-CoA (HMG-CoA) is also an intermediate of sterol biosynthesis, but the enzyme that forms HMG-CoA in that pathway is cytosolic. HMG-CoA lyase is present only in the mitochondrial matrix.

this is simply the reversal of the last step of β oxidation. The acetoacetyl-CoA then condenses with acetyl-CoA to form **β-hydroxy-β-methylglutaryl-CoA (HMG-CoA)**, which is cleaved to free acetoacetate and acetyl-CoA. The acetoacetate is reversibly reduced by D-β-hydroxybutyrate dehydrogenase, a mitochondrial enzyme, to D-β-hydroxybutyrate. This enzyme is specific for the D stereoisomer; it does not act on L-β-hydroxyacyl-CoAs and is not to be confused with L-β-hydroxyacyl-CoA dehydrogenase of the β-oxidation pathway.

In healthy people, acetone is formed in very small amounts from acetoacetate, which is easily decarboxylated, either spontaneously or by the action of **acetoacetate decarboxylase** (Fig. 17–19). Because individuals with untreated diabetes produce large quantities of acetoacetate, their blood contains significant amounts of acetone, which is toxic. Acetone is volatile and imparts a characteristic odor to the breath, which is sometimes useful in diagnosing diabetes. ■

In extrahepatic tissues, D-β-hydroxybutyrate is oxidized to acetoacetate by D-β-hydroxybutyrate dehydrogenase **(Fig. 17–20)**. The acetoacetate is activated to its coenzyme A ester by transfer of CoA from succinyl-CoA, an intermediate of the citric acid cycle (see Fig. 16–7), in a reaction catalyzed by **β-ketoacyl-CoA transferase**, also called thiophorase. The acetoacetyl-CoA is then cleaved by thiolase to yield two acetyl-CoAs, which enter the citric acid cycle. Thus the ketone bodies are used as fuels in all tissues except liver, which lacks thiophorase. The liver is therefore a producer of ketone bodies for the other tissues, but not a consumer.

The production and export of ketone bodies by the liver allows continued oxidation of fatty acids with only minimal oxidation of acetyl-CoA. When intermediates of the citric acid cycle are being siphoned off for glucose synthesis by gluconeogenesis, for example, oxidation of

FIGURE 17–20 **D-β-Hydroxybutyrate as a fuel.** D-β-Hydroxybutyrate, synthesized in the liver, passes into the blood and thus to other tissues, where it is converted in three steps to acetyl-CoA. It is first oxidized to acetoacetate, which is activated with coenzyme A donated from succinyl-CoA, then split by thiolase. The acetyl-CoA thus formed is used for energy production.

cycle intermediates slows—and so does acetyl-CoA oxidation. Moreover, the liver contains only a limited amount of coenzyme A, and when most of it is tied up in acetyl-CoA, β oxidation slows for want of the free coenzyme. The production and export of ketone bodies frees coenzyme A, allowing continued fatty acid oxidation.

Ketone Bodies Are Overproduced in Diabetes and during Starvation

Starvation and untreated diabetes mellitus lead to overproduction of ketone bodies, with several associated medical problems. During starvation, gluconeogenesis depletes citric acid cycle intermediates, diverting acetyl-CoA to ketone body production (Fig. 17–21). In untreated diabetes, when the insulin level is insufficient, extrahepatic tissues cannot take up glucose efficiently from the blood, either for fuel or for conversion to fat. Under these conditions, levels of malonyl-CoA (the starting material for fatty acid synthesis) fall, inhibition of carnitine acyltransferase I is relieved, and fatty acids enter mitochondria to be degraded to acetyl-CoA—which cannot pass through the citric acid cycle because cycle intermediates have been drawn off for use as substrates in gluconeogenesis. The resulting accumulation of acetyl-CoA accelerates the formation of ketone bodies beyond the capacity of extrahepatic tissues to oxidize them. The increased blood levels of acetoacetate and D-β-hydroxybutyrate lower the blood

pH, causing the condition known as **acidosis**. Extreme acidosis can lead to coma and in some cases death. Ketone bodies in the blood and urine of individuals with untreated diabetes can reach extraordinary levels—a blood concentration of 90 mg/100 mL (compared with a normal level of <3 mg/100 mL) and urinary excretion of 5,000 mg/24 hr (compared with a normal rate of ≤125 mg/24 hr). This condition is called **ketosis**.

Individuals on very low-calorie diets, using the fats stored in adipose tissue as their major energy source, also have increased levels of ketone bodies in their blood and urine. These levels must be monitored to avoid the dangers of acidosis and ketosis (ketoacidosis). ∎

SUMMARY 17.3 Ketone Bodies

▶ The ketone bodies—acetone, acetoacetate, and D-β-hydroxybutyrate—are formed in the liver. The latter two compounds serve as fuel molecules in extrahepatic tissues, through oxidation to acetyl-CoA and entry into the citric acid cycle.

▶ Overproduction of ketone bodies in uncontrolled diabetes or severely reduced calorie intake can lead to acidosis or ketosis.

Key Terms

Terms in bold are defined in the glossary.

β oxidation 667
chylomicron 669
apolipoprotein 669
lipoprotein 669
perilipin 669
free fatty acids 669
serum albumin 669
carnitine shuttle 670
carnitine
  acyltransferase I 671
acyl-carnitine/carnitine
  transporter 671
carnitine
  acyltransferase II 671
trifunctional protein
  (TFP) 674
methylmalonyl-CoA
  mutase 678

coenzyme B₁₂ 678
malonyl-CoA 679
PPAR (peroxisome
  proliferator-activated
  receptor) 679
pernicious anemia 681
intrinsic factor 681
medium-chain acyl-CoA
  dehydrogenase
  (MCAD) 682
multifunctional
  protein (MFP) 684
ω oxidation 684
mixed-function
  oxidases 685
α oxidation 685
acidosis 688
ketosis 688

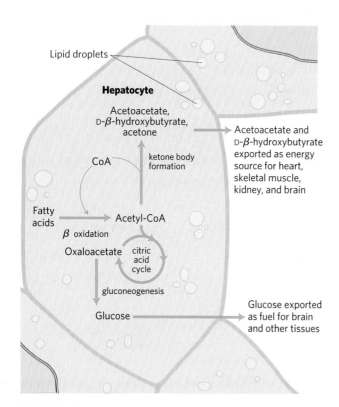

FIGURE 17–21 **Ketone body formation and export from the liver.** Conditions that promote gluconeogenesis (untreated diabetes, severely reduced food intake) slow the citric acid cycle (by drawing off oxaloacetate) and enhance the conversion of acetyl-CoA to acetoacetate. The released coenzyme A allows continued β oxidation of fatty acids.

Further Reading

General

Boyer, P.D. (1983) *The Enzymes*, 3rd edn, Vol. 16: *Lipid Enzymology*, Academic Press, Inc., San Diego, CA.

Ferry, G. (1998) *Dorothy Hodgkin: A Life*, Cold Spring Harbor Laboratory Press, Cold Spring Harbor, NY.
  Fascinating biography of an amazing woman.

Gurr, M.I., Harwood, J.L., & Frayn, K.N. (2002) *Lipid Biochemistry: An Introduction*, 5th edn, Blackwell Science, Oxford, UK.

Plutzky, J. (2009) The mighty mighty fatty acid. *Nat. Med.* **15**, 618–619.

Scheffler, I.E. (1999) *Mitochondria*, Wiley-Liss, New York.

   An excellent book on mitochondrial structure and function.

Wood, P.A. (2006) *How Fat Works*, Harvard University Press, Cambridge, MA.

   Very readable, intermediate-level account of the contributions of genetics and mouse models to the understanding of lipid metabolism and obesity.

Digestion, Mobilization, and Transport of Fats

Farese, R.V., Jr., & Walther, T.C. (2009) Lipid droplets finally get a little r-e-s-p-e-c-t. *Cell* **139**, 855–860.

Glatz, J.F.C., Luiken, J.J.F.P., & Bonen, A. (2010) Membrane fatty acid transporters as regulators of lipid metabolism: implications for metabolic disease. *Physiol. Rev.* **90**, 367–417.

Greenberg, A.S. & Coleman, R.A. (2011) Expanding roles for lipid droplets. *Trends Endocrinol. Metab.* **22**, 195–196.

   Editorial introduction to an issue of this journal devoted to lipid droplets.

Langin, D., Holm, C., & Lafontan, M. (1996) Adipocyte hormone-sensitive lipase: a major regulator of lipid metabolism. *Proc. Nutr. Soc.* **55**, 93–109.

Ramsay, T.G. (1996) Fat cells. *Endocrinol. Metab. Clin. N. Am.* **25**, 847–870.

   A review of all aspects of fat storage and mobilization in adipocytes.

Reue, K. (2011) A thematic review series: lipid droplet storage and metabolism: from yeast to man. *J. Lipid Res.* **52**, 1865–1868.

   Editorial introduction to a series of articles on lipid droplets published together in this issue.

Shaw, C.S., Clark, J., & Wagenmakers, A.J.M. (2010) The effect of exercise and nutrition on intramuscular fat metabolism and insulin sensitivity. *Annu. Rev. Nutr.* **30**, 13–34.

Steinberg, G.R. (2009) Role of the AMP-activated protein kinase in regulating fatty acid metabolism during exercise. *Appl. Physiol. Nutr. Metab.* **34**, 315–322.

Storch, J. & Rhumsey, A.D. (2010) Tissue-specific functions in the fatty acid-binding protein family (minireview) *J. Biol. Chem.* **285**, 32,679–32,683.

Wang, C.S., Hartsuck, J., & McConathy, W.J. (1992) Structure and functional properties of lipoprotein lipase. *Biochim. Biophys. Acta* **1123**, 1–17.

   Advanced-level discussion of the enzyme that releases fatty acids from lipoproteins in the capillaries of muscle and adipose tissue.

Watt, M.J. & Steinberg, G.R. (2008) Regulation and function of triacylglycerol lipases in cellular metabolism. *Biochem. J.* **414**, 313–325.

Zechner, R., Kienesberger, P.C., Haemmerle, G., Zimmermann R., & Lass, A. (2009) Adipose triglyceride lipase and the lipolytic catabolism of cellular fat stores. *J. Lipid Res.* **50**, 3–21.

Mitochondrial β Oxidation

Bannerjee, R. (1997) The yin-yang of cobalamin biochemistry. *Chem. Biol.* **4**, 175–186.

   A review of the biochemistry of coenzyme B_{12} reactions, including the methylmalonyl-CoA mutase reaction.

Carey, H.V., Andrews, M.T., & Martin, S.L. (2003) Mammalian hibernation: cellular and molecular responses to depressed metabolism and low temperature. *Physiol. Rev.* **83**, 1153–1181.

Desverne, B., Michalik, L., & Wahij, W. (2006) Transcriptional regulation of metabolism. *Physiol. Rev.* **86**, 465–514.

   An extensive review of the regulation of metabolism, including fat metabolism, by transcription factors.

Eaton, S., Bartlett, K., & Pourfarzam, M. (1996) Mammalian mitochondrial β-oxidation. *Biochem. J.* **320**, 345–357.

   A review of the enzymology of β oxidation, inherited defects in this pathway, and regulation of the process in mitochondria.

Eaton, S., Bursby, T., Middleton, B., Pourfarzam, M., Mills, K., Johnson, A.W., & Bartlett, K. (2000) The mitochondrial trifunctional protein: centre of a β-oxidation metabolon? *Biochem. Soc. Trans.* **28**, 177–182.

   Short, intermediate-level review.

Evans, R.M., Barish, G.D., & Wang, Y.-X. (2004) PPARs and the complex journey to obesity. *Nat. Med.* **10**, 1–7.

   A very readable, intermediate-level account of the discovery of the PPARs and their functions.

Harwood, J.L. (1988) Fatty acid metabolism. *Annu. Rev. Plant Physiol. Plant Mol. Biol.* **39**, 101–138.

Jeukendrup, A.E., Saris, W.H., & Wagenmakers, A.J. (1998) Fat metabolism during exercise: a review. Part III: effects of nutritional interventions. *Int. J. Sports Med.* **19**, 371–379.

   This paper is one of a series that reviews the factors that influence fat mobilization and utilization during exercise.

Kampf, J.P. & Kleinfeld, A.M. (2007) Is membrane transport of FFA mediated by lipid, protein, or both? *Physiology* **22**, 7–14.

Kerner, J. & Hoppel, C. (1998) Genetic disorders of carnitine metabolism and their nutritional management. *Annu. Rev. Nutr.* **18**, 179–206.

Kerner, J. & Hoppel, C. (2000) Fatty acid import into mitochondria. *Biochim. Biophys. Acta* **1486**, 1–17.

Kunau, W.H., Dommes, V., & Schulz, H. (1995) β-Oxidation of fatty acids in mitochondria, peroxisomes, and bacteria: a century of continued progress. *Prog. Lipid Res.* **34**, 267–342.

   A good historical account and a useful comparison of β oxidation in different systems.

Mandard, S., Muller, M., & Kersten, S. (2004) Peroxisome proliferator-activated receptor alpha target genes. *Cell. Mol. Life Sci.* **61**, 393–416.

   A review of the genes controlled by PPARα.

Rinaldo, P., Matern, D., & Bennett, M.J. (2002) Fatty acid oxidation disorders. *Annu. Rev. Physiol.* **64**, 477–502.

   Advanced review of metabolic defects in fat oxidation, including MCAD mutations.

Rufer, A.C., Thoma, R., Benz, J., Stihle, M., Gsell, B., De Roo, E., Banner, D.W., Mueller, F., Chomienne, O., & Hennig, M. (2006) The crystal structure of carnitine palmitoyltransferase 2 and implications for diabetes treatment. *Structure* **14**, 713–723.

Sherratt, H.S. (1994) Introduction: the regulation of fatty acid oxidation in cells. *Biochem. Soc. Trans.* **22**, 421–422.

   Introduction to reviews in this journal issue of various aspects of fatty acid oxidation and its regulation.

Thorpe, C. & Kim, J.J. (1995) Structure and mechanism of action of the acyl-CoA dehydrogenases. *FASEB J.* **9**, 718–725.

   Short, clear description of the three-dimensional structure and catalytic mechanism of these enzymes.

Peroxisomal β Oxidation

Graham, I.A. & Eastmond, P.J. (2002) Pathways of straight and branched chain fatty acid catabolism in higher plants. *Prog. Lipid Res.* **41**, 156–181.

Wanders, R.J.A., van Grunsven, E.G., & Jansen, G.A. (2000) Lipid metabolism in peroxisomes: enzymology, functions and dysfunctions of the fatty acid α- and β-oxidation systems in humans. *Biochem. Soc. Trans.* **28**, 141–148.

Ketone Bodies

Foster, D.W. & McGarry, J.D. (1983) The metabolic derangements and treatment of diabetic ketoacidosis. *N. Engl. J. Med.* **309**, 159–169.

McGarry, J.D. & Foster, D.W. (1980) Regulation of hepatic fatty acid oxidation and ketone body production. *Annu. Rev. Biochem.* **49**, 395–420.

Robinson, A.M. & Williamson, D.H. (1980) Physiological roles of ketone bodies as substrates and signals in mammalian tissues. *Physiol. Rev.* **60,** 143–187.

Problems

1. Energy in Triacylglycerols On a per-carbon basis, where does the largest amount of biologically available energy in triacylglycerols reside: in the fatty acid portions or the glycerol portion? Indicate how knowledge of the chemical structure of triacylglycerols provides the answer.

2. Fuel Reserves in Adipose Tissue Triacylglycerols, with their hydrocarbon-like fatty acids, have the highest energy content of the major nutrients.

(a) If 15% of the body mass of a 70.0 kg adult consists of triacylglycerols, what is the total available fuel reserve, in both kilojoules and kilocalories, in the form of triacylglycerols? Recall that 1.00 kcal = 4.18 kJ.

(b) If the basal energy requirement is approximately 8,400 kJ/day (2,000 kcal/day), how long could this person survive if the oxidation of fatty acids stored as triacylglycerols were the only source of energy?

(c) What would be the weight loss in pounds per day under such starvation conditions (1 lb = 0.454 kg)?

3. Common Reaction Steps in the Fatty Acid Oxidation Cycle and Citric Acid Cycle Cells often use the same enzyme reaction pattern for analogous metabolic conversions. For example, the steps in the oxidation of pyruvate to acetyl-CoA and of α-ketoglutarate to succinyl-CoA, although catalyzed by different enzymes, are very similar. The first stage of fatty acid oxidation follows a reaction sequence closely resembling a sequence in the citric acid cycle. Use equations to show the analogous reaction sequences in the two pathways.

4. β Oxidation: How Many Cycles? How many cycles of β oxidation are required for the complete oxidation of activated oleic acid, 18:1(Δ^9)?

5. Chemistry of the Acyl-CoA Synthetase Reaction Fatty acids are converted to their coenzyme A esters in a reversible reaction catalyzed by acyl-CoA synthetase:

$$R{-}COO^- + ATP + CoA \rightleftharpoons$$

$$\underset{\substack{\parallel \\ O}}{R{-}C}{-}CoA + AMP + PP_i$$

(a) The enzyme-bound intermediate in this reaction has been identified as the mixed anhydride of the fatty acid and adenosine monophosphate (AMP), acyl-AMP:

Write two equations corresponding to the two steps of the reaction catalyzed by acyl-CoA synthetase.

(b) The acyl-CoA synthetase reaction is readily reversible, with an equilibrium constant near 1. How can this reaction be made to favor formation of fatty acyl–CoA?

6. Intermediates in Oleic Acid Oxidation What is the structure of the partially oxidized fatty acyl group that is formed when oleic acid, 18:1(Δ^9), has undergone three cycles of β oxidation? What are the next two steps in the continued oxidation of this intermediate?

7. β Oxidation of an Odd-Chain Fatty Acid What are the direct products of β oxidation of a fully saturated, straight-chain fatty acid of 11 carbons?

8. Oxidation of Tritiated Palmitate Palmitate uniformly labeled with tritium (^{3}H) to a specific activity of 2.48×10^8 counts per minute (cpm) per micromole of palmitate is added to a mitochondrial preparation that oxidizes it to acetyl-CoA. The acetyl-CoA is isolated and hydrolyzed to acetate. The specific activity of the isolated acetate is 1.00×10^7 cpm/μmol. Is this result consistent with the β-oxidation pathway? Explain. What is the final fate of the removed tritium?

9. Compartmentation in β Oxidation Free palmitate is activated to its coenzyme A derivative (palmitoyl-CoA) in the cytosol before it can be oxidized in the mitochondrion. If palmitate and [^{14}C]coenzyme A are added to a liver homogenate, palmitoyl-CoA isolated from the cytosolic fraction is radioactive, but that isolated from the mitochondrial fraction is not. Explain.

10. Comparative Biochemistry: Energy-Generating Pathways in Birds One indication of the relative importance of various ATP-producing pathways is the V_{max} of certain enzymes of these pathways. The values of V_{max} of several enzymes from the pectoral muscles (chest muscles used for flying) of pigeon and pheasant are listed below.

Enzyme	V_{max} (μmol substrate/min/g tissue)	
	Pigeon	**Pheasant**
Hexokinase	3.0	2.3
Glycogen phosphorylase	18.0	120.0
Phosphofructokinase-1	24.0	143.0
Citrate synthase	100.0	15.0
Triacylglycerol lipase	0.07	0.01

(a) Discuss the relative importance of glycogen metabolism and fat metabolism in generating ATP in the pectoral muscles of these birds.

(b) Compare oxygen consumption in the two birds.

(c) Judging from the data in the table, which bird is the long-distance flyer? Justify your answer.

(d) Why were these particular enzymes selected for comparison? Would the activities of triose phosphate isomerase and malate dehydrogenase be equally good bases for comparison? Explain.

11. Mutant Carnitine Acyltransferase What changes in metabolic pattern would result from a mutation in the muscle carnitine acyltransferase I in which the mutant protein has lost its affinity for malonyl-CoA but not its catalytic activity?

12. Effect of Carnitine Deficiency An individual developed a condition characterized by progressive muscular weakness and aching muscle cramps. The symptoms were aggravated by fasting, exercise, and a high-fat diet. The homogenate of a skeletal muscle specimen from the patient oxidized added oleate more slowly than did control homogenates, consisting of muscle specimens from healthy individuals. When carnitine was added to the patient's muscle homogenate, the rate of oleate oxidation equaled that in the control homogenates. The patient was diagnosed as having a carnitine deficiency.

(a) Why did added carnitine increase the rate of oleate oxidation in the patient's muscle homogenate?

(b) Why were the patient's symptoms aggravated by fasting, exercise, and a high-fat diet?

(c) Suggest two possible reasons for the deficiency of muscle carnitine in this individual.

13. Fatty Acids as a Source of Water Contrary to legend, camels do not store water in their humps, which actually consist of large fat deposits. How can these fat deposits serve as a source of water? Calculate the amount of water (in liters) that a camel can produce from 1.0 kg of fat. Assume for simplicity that the fat consists entirely of tripalmitoylglycerol.

14. Petroleum as a Microbial Food Source Some microorganisms of the genera *Nocardia* and *Pseudomonas* can grow in an environment where hydrocarbons are the only food source. These bacteria oxidize straight-chain aliphatic hydrocarbons, such as octane, to their corresponding carboxylic acids:

$$CH_3(CH_2)_6CH_3 + NAD^+ + O_2 \rightleftharpoons$$
$$CH_3(CH_2)_6COOH + NADH + H^+$$

How could these bacteria be used to clean up oil spills? What would be some of the limiting factors in the efficiency of this process?

15. Metabolism of a Straight-Chain Phenylated Fatty Acid A crystalline metabolite was isolated from the urine of a rabbit that had been fed a straight-chain fatty acid containing a terminal phenyl group:

$$\langle\rangle-CH_2-(CH_2)_n-COO^-$$

A 302 mg sample of the metabolite in aqueous solution was completely neutralized by 22.2 mL of 0.100 M NaOH.

(a) What is the probable molecular weight and structure of the metabolite?

(b) Did the straight-chain fatty acid contain an even or an odd number of methylene ($-CH_2-$) groups (i.e., is n even or odd)? Explain.

16. Fatty Acid Oxidation in Uncontrolled Diabetes When the acetyl-CoA produced during β oxidation in the liver exceeds the capacity of the citric acid cycle, the excess acetyl-CoA forms ketone bodies—acetone, acetoacetate, and D-β-hydroxybutyrate. This occurs in severe,

uncontrolled diabetes: because the tissues cannot use glucose, they oxidize large amounts of fatty acids instead. Although acetyl-CoA is not toxic, the mitochondrion must divert the acetyl-CoA to ketone bodies. What problem would arise if acetyl-CoA were not converted to ketone bodies? How does the diversion to ketone bodies solve the problem?

17. Consequences of a High-Fat Diet with No Carbohydrates Suppose you had to subsist on a diet of whale blubber and seal blubber, with little or no carbohydrate.

(a) What would be the effect of carbohydrate deprivation on the utilization of fats for energy?

(b) If your diet were totally devoid of carbohydrate, would it be better to consume odd- or even-numbered fatty acids? Explain.

18. Even- and Odd-Chain Fatty Acids in the Diet In a laboratory experiment, two groups of rats are fed two different fatty acids as their sole source of carbon for a month. The first group gets heptanoic acid (7:0), and the second gets octanoic acid (8:0). After the experiment, a striking difference is seen between the two groups. Those in the first group are healthy and have gained weight, whereas those in the second group are weak and have lost weight as a result of losing muscle mass. What is the biochemical basis for this difference?

19. Metabolic Consequences of Ingesting ω-Fluorooleate The shrub *Dichapetalum toxicarium*, native to Sierra Leone, produces ω-fluorooleate, which is highly toxic to warm-blooded animals.

$$F-CH_2-(CH_2)_7-\overset{\displaystyle\overset{H}{|}}{C}=\overset{\displaystyle\overset{H}{|}}{C}-(CH_2)_7-COO^-$$
$$\omega\text{-Fluorooleate}$$

This substance has been used as an arrow poison, and powdered fruit from the plant is sometimes used as a rat poison (hence the plant's common name, ratsbane). Why is this substance so toxic? (Hint: Review Chapter 16, Problem 22.)

20. Mutant Acetyl-CoA Carboxylase What would be the consequences for fat metabolism of a mutation in acetyl-CoA carboxylase that replaced the Ser residue normally phosphorylated by AMPK with an Ala residue? What might happen if the same Ser were replaced by Asp? (Hint: See Fig. 17–13.)

21. Effect of PDE Inhibitor on Adipocytes How would an adipocyte's response to epinephrine be affected by the addition of an inhibitor of cAMP phosphodiesterase (PDE)? (Hint: See Fig. 12–4.)

22. Role of FAD as Electron Acceptor Acyl-CoA dehydrogenase uses enzyme-bound FAD as a prosthetic group to dehydrogenate the α and β carbons of fatty acyl–CoA. What is the advantage of using FAD as an electron acceptor rather than NAD$^+$? Explain in terms of the standard reduction potentials for the Enz-FAD/FADH$_2$ ($E'^\circ = -0.219$ V) and NAD$^+$/NADH ($E'^\circ = -0.320$ V) half-reactions.

23. β Oxidation of Arachidic Acid How many turns of the fatty acid oxidation cycle are required for complete oxidation of arachidic acid (see Table 10–1) to acetyl-CoA?

24. Fate of Labeled Propionate If [3-^{14}C]propionate (^{14}C in the methyl group) is added to a liver homogenate, ^{14}C-labeled oxaloacetate is rapidly produced. Draw a flow chart for the pathway by which propionate is transformed to oxaloacetate, and indicate the location of the ^{14}C in oxaloacetate.

25. Phytanic Acid Metabolism When phytanic acid uniformly labeled with ^{14}C is fed to a mouse, radioactivity can be detected in malate, a citric acid cycle intermediate, within minutes. Draw a metabolic pathway that could account for this. Which of the carbon atoms in malate would contain ^{14}C label?

26. Sources of H$_2$O Produced in β Oxidation The complete oxidation of palmitoyl-CoA to carbon dioxide and water is represented by the overall equation

$$\text{Palmitoyl-CoA} + 23O_2 + 108P_i + 108ADP \longrightarrow$$
$$CoA + 16CO_2 + 108ATP + 23H_2O$$

Water is also produced in the reaction

$$ADP + P_i \longrightarrow ATP + H_2O$$

but is not included as a product in the overall equation. Why?

27. Biological Importance of Cobalt In cattle, deer, sheep, and other ruminant animals, large amounts of propionate are produced in the rumen through the bacterial fermentation of ingested plant matter. Propionate is the principal source of glucose for these animals, via the route propionate $\longrightarrow$ oxaloacetate $\longrightarrow$ glucose. In some areas of the world, notably Australia, ruminant animals sometimes show symptoms of anemia with concomitant loss of appetite and retarded growth, resulting from an inability to transform propionate to oxaloacetate. This condition is due to a cobalt deficiency caused by very low cobalt levels in the soil and thus in plant matter. Explain.

28. Fat Loss during Hibernation Bears expend about 25×10^6 J/day during periods of hibernation, which may last as long as seven months. The energy required to sustain life is obtained from fatty acid oxidation. How much weight loss (in kilograms) has occurred after seven months? How might ketosis be minimized during hibernation? (Assume the oxidation of fat yields 38 kJ/g.)

Data Analysis Problem

29. β Oxidation of Trans Fats Unsaturated fats with trans double bonds are commonly referred to as "trans fats." There has been much discussion about the effects of dietary trans fats on health. In their investigations of the effects of trans fatty acid metabolism on health, Yu and colleagues (2004) showed that a model trans fatty acid was processed differently from its cis isomer. They used three related 18-carbon fatty acids to explore the difference in β oxidation between cis and trans isomers of the same-size fatty acid.

The researchers incubated the coenzyme A derivative of each acid with rat liver mitochondria for 5 minutes, then separated the remaining CoA derivatives in each mixture by HPLC (high-performance liquid chromatography). The results are shown below, with separate panels for the three experiments.

In the figure, IS indicates an internal standard (pentadecanoyl-CoA) added to the mixture, after the reaction, as a molecular marker. The researchers abbreviated the CoA derivatives as follows: stearoyl-CoA, C_{18}-CoA; *cis*-Δ^5-tetradecenoyl-CoA, cΔ^5C$_{14}$-CoA; oleoyl-CoA, cΔ^9C$_{18}$-CoA; *trans*-Δ^5-tetradecenoyl-CoA, tΔ^5C$_{14}$-CoA; and elaidoyl-CoA, tΔ^9C$_{18}$-CoA.

(a) Why did Yu and colleagues need to use CoA derivatives rather than the free fatty acids in these experiments?

(b) Why were no lower molecular weight CoA derivatives found in the reaction with stearoyl-CoA?

(c) How many rounds of β oxidation would be required to convert the oleoyl-CoA and the elaidoyl-CoA to cis-Δ^5-tetradecenoyl-CoA and $trans$-Δ^5-tetradecenoyl-CoA, respectively?

There are two forms of the enzyme acyl-CoA dehydrogenase (see Fig. 17–8a): long-chain acyl-CoA dehydrogenase (LCAD) and very-long-chain acyl-CoA dehydrogenase (VLCAD). Yu and coworkers measured the kinetic parameters of both enzymes. They used the CoA derivatives of three fatty acids: tetradecanoyl-CoA (C_{14}-CoA), cis-Δ^5-tetradecenoyl-CoA ($c\Delta^5C_{14}$-CoA), and $trans$-Δ^5-tetradecenoyl-CoA ($t\Delta^5C_{14}$-CoA). The results are shown below. (See Chapter 6 for definitions of the kinetic parameters.)

	LCAD			VLCAD		
	C_{14}-CoA	$c\Delta^5C_{14}$-CoA	$t\Delta^5C_{14}$-CoA	C_{14}-CoA	$c\Delta^5C_{14}$-CoA	$t\Delta^5C_{14}$-CoA
V_{max}	3.3	3.0	2.9	1.4	0.32	0.88
K_m	0.41	0.40	1.6	0.57	0.44	0.97
k_{cat}	9.9	8.9	8.5	2.0	0.42	1.12
k_{cat}/K_m	24	22	5	4	1	1

(d) For LCAD, the K_m differs dramatically for the cis and trans substrates. Provide a plausible explanation for this observation in terms of the structures of the substrate molecules. (Hint: You may want to refer to Fig. 10–2.)

(e) The kinetic parameters of the two enzymes are relevant to the differential processing of these fatty acids *only* if the LCAD or VLCAD reaction (or both) is the rate-limiting step in the pathway. What evidence is there to support this assumption?

(f) How do these different kinetic parameters explain the different levels of the CoA derivatives found after incubation of rat liver mitochondria with stearoyl-CoA, oleoyl-CoA, and elaidoyl-CoA (shown in the three-panel figure)?

Yu and coworkers measured the substrate specificity of rat liver mitochondrial thioesterase, which hydrolyzes acyl-CoA to CoA and free fatty acid (see Chapter 21). This enzyme was approximately twice as active with C_{14}-CoA thioesters as with C_{18}-CoA thioesters.

(g) Other research has suggested that free fatty acids can pass through membranes. In their experiments, Yu and colleagues found $trans$-Δ^5-tetradecenoic acid outside mitochondria (i.e., in the medium) that had been incubated with elaidoyl-CoA. Describe the pathway that led to this extramitochondrial $trans$-Δ^5-tetradecenoic acid. Be sure to indicate where in the cell the various transformations take place, as well as the enzymes that catalyze the transformations.

(h) It is often said in the popular press that "trans fats are not broken down by your cells and instead accumulate in your body." In what sense is this statement correct and in what sense is it an oversimplification?

Reference

Yu, W., Liang, X., Ensenauer, R., Vockley, J., Sweetman, L., & Schultz, H. (2004) Leaky β-oxidation of a *trans*-fatty acid. *J. Biol. Chem.* **279**, 52,160–52,167.

Amino Acid Oxidation and the Production of Urea

W e now turn our attention to the amino acids, the final class of biomolecules that, through their oxidative degradation, make a significant contribution to the generation of metabolic energy. The fraction of metabolic energy obtained from amino acids, whether they are derived from dietary protein or from tissue protein, varies greatly with the type of organism and with metabolic conditions. Carnivores can obtain (immediately following a meal) up to 90% of their energy requirements from amino acid oxidation, whereas herbivores may fill only a small fraction of their energy needs by this route. Most microorganisms can scavenge amino acids from their environment and use them as fuel when required by metabolic conditions. Plants, however, rarely if ever oxidize amino acids to provide energy; the carbohydrate produced from CO_2 and H_2O in photosynthesis is generally their sole energy source. Amino acid concentrations in plant tissues are carefully regulated to just meet the requirements for biosynthesis of proteins, nucleic acids, and other molecules needed to support growth. Amino acid catabolism does occur in plants, but its purpose is to produce metabolites for other biosynthetic pathways.

In animals, amino acids undergo oxidative degradation in three different metabolic circumstances:

1. During the normal synthesis and degradation of cellular proteins (protein turnover; Chapter 27), some amino acids that are released from protein breakdown and are not needed for new protein synthesis undergo oxidative degradation.

2. When a diet is rich in protein and the ingested amino acids exceed the body's needs for protein synthesis, the surplus is catabolized; amino acids cannot be stored.

3. During starvation or in uncontrolled diabetes mellitus, when carbohydrates are either unavailable or not properly utilized, cellular proteins are used as fuel.

Under all these metabolic conditions, amino acids lose their amino groups to form α-keto acids, the "carbon skeletons" of amino acids. The α-keto acids undergo oxidation to CO_2 and H_2O or, often more importantly, provide three- and four-carbon units that can be converted by gluconeogenesis into glucose, the fuel for brain, skeletal muscle, and other tissues.

The pathways of amino acid catabolism are quite similar in most organisms. The focus of this chapter is on the pathways in vertebrates, because these have received the most research attention. As in carbohydrate and fatty acid catabolism, the processes of amino acid degradation converge on the central catabolic pathways, with the carbon skeletons of most amino acids finding their way to the citric acid cycle. In some cases the reaction pathways of amino acid breakdown closely parallel steps in the catabolism of fatty acids (see Fig. 17–9).

One important feature distinguishes amino acid degradation from other catabolic processes described to this point: every amino acid contains an amino group, and the pathways for amino acid degradation therefore include a key step in which the α-amino group is separated from the carbon skeleton and shunted into the pathways of amino group metabolism (**Fig. 18–1**). We deal first with amino group metabolism and nitrogen excretion, then with the fate of the carbon skeletons derived from the amino acids; along the way we see how the pathways are interconnected.

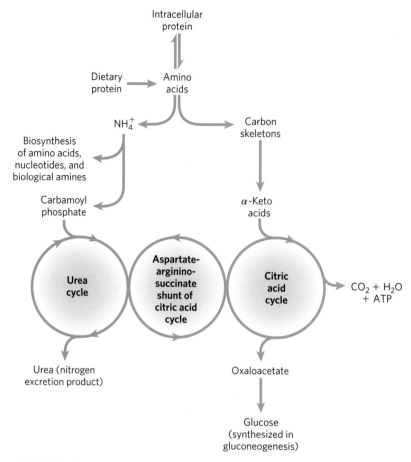

FIGURE 18–1 Overview of amino acid catabolism in mammals. The amino groups and the carbon skeleton take separate but interconnected pathways.

18.1 Metabolic Fates of Amino Groups

Nitrogen, N_2, is abundant in the atmosphere but is too inert for use in most biochemical processes. Because only a few microorganisms can convert N_2 to biologically useful forms such as NH_3 (Chapter 22), amino groups are carefully husbanded in biological systems.

Figure 18–2a provides an overview of the catabolic pathways of ammonia and amino groups in vertebrates. Amino acids derived from dietary protein are the source of most amino groups. Most amino acids are metabolized in the liver. Some of the ammonia generated in this process is recycled and used in a variety of biosynthetic pathways; the excess is either excreted directly or converted to urea or uric acid for excretion, depending on the organism (Fig. 18–2b). Excess ammonia generated in other (extrahepatic) tissues travels to the liver (in the form of amino groups, as described below) for conversion to the excretory form.

Four amino acids play central roles in nitrogen metabolism: glutamate, glutamine, alanine, and aspartate. The special place of these four amino acids in nitrogen metabolism is not an evolutionary accident. These particular amino acids are the ones most easily converted into citric acid cycle intermediates: glutamate and glutamine to α-ketoglutarate, alanine to pyruvate, and aspartate to oxaloacetate. Glutamate and glutamine are especially important, acting as a kind of general collection point for amino groups. In the cytosol of liver cells (hepatocytes), amino groups from most amino acids are transferred to α-ketoglutarate to form glutamate, which enters mitochondria and gives up its amino group to form NH_4^+. Excess ammonia generated in most other tissues is converted to the amide nitrogen of glutamine, which passes to the liver, then into liver mitochondria. Glutamine or glutamate or both are present in higher concentrations than other amino acids in most tissues.

In skeletal muscle, excess amino groups are generally transferred to pyruvate to form alanine, another important molecule in the transport of amino groups to the liver. We will see in Section 18.2 that aspartate comes into play in the metabolic processes that occur once the amino groups are delivered to the liver.

We begin with a discussion of the breakdown of dietary proteins, then give a general description of the metabolic fates of amino groups.

FIGURE 18–2 Amino group catabolism. (a) Overview of catabolism of amino groups (shaded) in vertebrate liver. **(b)** Excretory forms of nitrogen. Excess NH_4^+ is excreted as ammonia (microbes, bony fishes), urea (most terrestrial vertebrates), or uric acid (birds and terrestrial reptiles).

Notice that the carbon atoms of urea and uric acid are highly oxidized; the organism discards carbon only after extracting most of its available energy of oxidation.

Dietary Protein Is Enzymatically Degraded to Amino Acids

In humans, the degradation of ingested proteins to their constituent amino acids occurs in the gastrointestinal tract. Entry of dietary protein into the stomach stimulates the gastric mucosa to secrete the hormone **gastrin**, which in turn stimulates the secretion of hydrochloric acid by the parietal cells and pepsinogen by the chief cells of the gastric glands **(Fig. 18–3a)**. The acidic gastric juice (pH 1.0 to 2.5) is both an antiseptic, killing most bacteria and other foreign cells, and a denaturing agent, unfolding globular proteins and rendering their internal peptide bonds more accessible to enzymatic hydrolysis. **Pepsinogen** (M_r 40,554), an inactive precursor, or zymogen (p. 231), is converted to active pepsin (M_r 34,614) by an autocatalytic cleavage (a cleavage mediated by the pepsinogen itself) that occurs only at low pH. In the stomach, pepsin hydrolyzes ingested proteins at peptide bonds on the amino-terminal side of the aromatic amino acid residues Phe, Trp, and Tyr (see Table 3–6), cleaving long polypeptide chains into a mixture of smaller peptides.

As the acidic stomach contents pass into the small intestine, the low pH triggers secretion of the hormone **secretin** into the blood. Secretin stimulates the pancreas

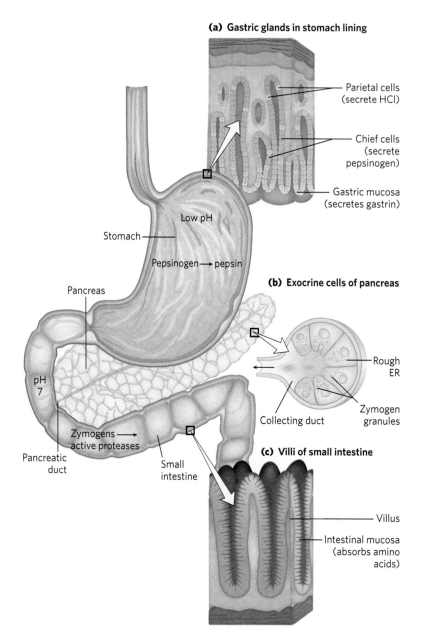

(a) Gastric glands in stomach lining

Parietal cells (secrete HCl)

Chief cells (secrete pepsinogen)

Gastric mucosa (secretes gastrin)

Low pH

Stomach

Pepsinogen → pepsin

Pancreas

pH 7

Zymogens → active proteases

Pancreatic duct

Small intestine

(b) Exocrine cells of pancreas

Rough ER

Collecting duct

Zymogen granules

(c) Villi of small intestine

Villus

Intestinal mucosa (absorbs amino acids)

FIGURE 18–3 Part of the human digestive (gastrointestinal) tract. (a) The parietal cells and chief cells of the gastric glands secrete their products in response to the hormone gastrin. Pepsin begins the process of protein degradation in the stomach. **(b)** The cytoplasm of exocrine cells is completely filled with rough endoplasmic reticulum, the site of synthesis of the zymogens of many digestive enzymes. The zymogens are concentrated in membrane-enclosed transport particles called zymogen granules. When an exocrine cell is stimulated, its plasma membrane fuses with the zymogen granule membrane and zymogens are released into the lumen of the collecting duct by exocytosis. The collecting ducts ultimately lead to the pancreatic duct and thence to the small intestine. **(c)** Amino acids are absorbed through the epithelial cell layer (intestinal mucosa) of the villi and enter the capillaries. Recall that the products of lipid hydrolysis in the small intestine enter the lymphatic system after their absorption by the intestinal mucosa (see Fig. 17–1).

to secrete bicarbonate into the small intestine to neutralize the gastric HCl, abruptly increasing the pH to about 7. (All pancreatic secretions pass into the small intestine through the pancreatic duct.) The digestion of proteins now continues in the small intestine. Arrival of amino acids in the upper part of the intestine (duodenum) causes release into the blood of the hormone **cholecystokinin**, which stimulates secretion of several pancreatic enzymes with activity optima at pH 7 to 8. **Trypsinogen**, **chymotrypsinogen**, and **procarboxypeptidases A** and **B**—the zymogens of **trypsin**, **chymotrypsin**, and **carboxypeptidases A** and **B**—are synthesized and secreted by the exocrine cells of the pancreas (Fig. 18–3b). Trypsinogen is converted to its active form, trypsin, by **enteropeptidase**, a proteolytic enzyme secreted by intestinal cells. Free trypsin then catalyzes the conversion of additional trypsinogen to trypsin (see Fig. 6–38). Trypsin also activates chymotrypsinogen, the procarboxypeptidases, and proelastase.

Why this elaborate mechanism for getting active digestive enzymes into the gastrointestinal tract? Synthesis of the enzymes as inactive precursors protects the exocrine cells from destructive proteolytic attack. The pancreas further protects itself against self-digestion by making a specific inhibitor, a protein called **pancreatic trypsin inhibitor** (p. 232), that effectively prevents premature production of active proteolytic enzymes within the pancreatic cells.

Trypsin and chymotrypsin further hydrolyze the peptides that were produced by pepsin in the stomach. This stage of protein digestion is accomplished very efficiently, because pepsin, trypsin, and chymotrypsin have different amino acid specificities (see Table 3–6). Degradation of the short peptides in the small intestine is then completed by other intestinal peptidases. These include carboxypeptidases A and B (both of which are zinc-containing enzymes), which remove successive carboxyl-terminal residues from peptides, and an **aminopeptidase** that hydrolyzes successive amino-terminal residues from short peptides. The resulting mixture of free amino acids is transported into the epithelial cells lining the small intestine (Fig. 18–3c), through which the amino acids enter the blood capillaries in the villi and travel to the liver. In humans, most globular proteins from animal sources are almost completely hydrolyzed to amino acids in the gastrointestinal tract, but some fibrous proteins, such as keratin, are only partly digested. In addition, the protein content of some plant foods is protected against breakdown by indigestible cellulose husks.

Acute pancreatitis is a disease caused by obstruction of the normal pathway by which pancreatic secretions enter the intestine. The zymogens of the proteolytic enzymes are converted to their catalytically active forms prematurely, *inside* the pancreatic cells, and attack the pancreatic tissue itself. This causes excruciating pain and damage to the organ that can prove fatal. ∎

Pyridoxal Phosphate Participates in the Transfer of α-Amino Groups to α-Ketoglutarate

The first step in the catabolism of most L-amino acids, once they have reached the liver, is removal of the α-amino groups, promoted by enzymes called **aminotransferases** or **transaminases**. In these **transamination** reactions, the α-amino group is transferred to the α-carbon atom of α-ketoglutarate, leaving behind the corresponding α-keto acid analog of the amino acid (**Fig. 18–4**). There is no net deamination (loss of amino groups) in these reactions, because the α-ketoglutarate becomes aminated as the α-amino acid is deaminated. The effect of transamination reactions is to collect the

FIGURE 18–4 Enzyme-catalyzed transaminations. In many aminotransferase reactions, α-ketoglutarate is the amino group acceptor. All aminotransferases have pyridoxal phosphate (PLP) as cofactor. Although the reaction is shown here in the direction of transfer of the amino group to α-ketoglutarate, it is readily reversible.

amino groups from many different amino acids in the form of L-glutamate. The glutamate then functions as an amino group donor for biosynthetic pathways or for excretion pathways that lead to the elimination of nitrogenous waste products.

Cells contain different types of aminotransferases. Many are specific for α-ketoglutarate as the amino group acceptor but differ in their specificity for the L-amino acid. The enzymes are named for the amino group donor (alanine aminotransferase, aspartate aminotransferase, for example). The reactions catalyzed by aminotransferases are freely reversible, having an equilibrium constant of about 1.0 ($\Delta G' \approx 0$ kJ/mol).

All aminotransferases have the same prosthetic group and the same reaction mechanism. The prosthetic group is **pyridoxal phosphate (PLP)**, the coenzyme form of pyridoxine, or vitamin B_6. We encountered pyridoxal phosphate in Chapter 15, as a coenzyme in the glycogen phosphorylase reaction, but its role in that reaction is not representative of its usual coenzyme function. Its primary role in cells is in the metabolism of molecules with amino groups.

Pyridoxal phosphate functions as an intermediate carrier of amino groups at the active site of aminotransferases. It undergoes reversible transformations between its aldehyde form, pyridoxal phosphate, which can accept an amino group, and its aminated form, pyridoxamine phosphate, which can donate its amino group to an α-keto acid (**Fig. 18–5a**). Pyridoxal phosphate is generally covalently bound to the enzyme's active site through an aldimine (Schiff base) linkage to the ε-amino group of a Lys residue (Fig. 18–5b, d).

FIGURE 18-5 Pyridoxal phosphate, the prosthetic group of aminotransferases. **(a)** Pyridoxal phosphate (PLP) and its aminated form, pyridoxamine phosphate, are the tightly bound coenzymes of aminotransferases. The functional groups are shaded. **(b)** Pyridoxal phosphate is bound to the enzyme through noncovalent interactions and a Schiff-base (aldimine) linkage to a Lys residue at the active site. The steps in the formation of a Schiff base from a primary amine and a carbonyl group are detailed in Figure 14-6. **(c)** PLP (red) bound to one of the two active sites of the dimeric enzyme aspartate aminotransferase, a typical aminotransferase; **(d)** close-up view of the active site, with PLP (red, with yellow phosphorus) in aldimine linkage with the side chain of Lys[258] (purple); **(e)** another close-up view of the active site, with PLP linked to the substrate analog 2-methylaspartate (green) via a Schiff base (PDB ID 1AJS).

Pyridoxal phosphate participates in a variety of reactions at the α, β, and γ carbons (C-2 to C-4) of amino acids. Reactions at the α carbon **(Fig. 18–6)** include racemizations (interconverting L- and D-amino acids) and decarboxylations, as well as transaminations. Pyridoxal phosphate plays the same chemical role in each of these reactions. A bond to the α carbon of the substrate is broken, removing either a proton or a carboxyl group. The electron pair left behind on the α carbon would form a highly unstable carbanion, but pyridoxal phosphate provides resonance stabilization of this intermediate (Fig. 18–6 inset). The highly conjugated structure of PLP (an electron sink) permits delocalization of the negative charge.

Aminotransferases (Fig. 18–5) are classic examples of enzymes catalyzing bimolecular Ping-Pong reactions (see Fig. 6–13b), in which the first substrate reacts and the product must leave the active site before the second substrate can bind. Thus the incoming amino acid binds to the active site, donates its amino group to pyridoxal phosphate, and departs in the form of an α-keto acid. The incoming α-keto acid then binds, accepts the amino group from pyridoxamine phosphate, and departs in the form of an amino acid.

Glutamate Releases Its Amino Group As Ammonia in the Liver

As we have seen, the amino groups from many of the α-amino acids are collected in the liver in the form of the amino group of L-glutamate molecules. These amino groups must next be removed from glutamate to prepare them for excretion. In hepatocytes, glutamate is transported from the cytosol into mitochondria, where it undergoes **oxidative deamination** catalyzed by **L-glutamate dehydrogenase** (M_r 330,000). In

MECHANISM FIGURE 18–6 Some amino acid transformations at the α carbon that are facilitated by pyridoxal phosphate. Pyridoxal phosphate is generally bonded to the enzyme through a Schiff base, also called an internal aldimine. This activated form of PLP readily undergoes transimination to form a new Schiff base (external aldimine) with the α-amino group of the substrate amino acid (see Fig. 18–5b, d). Three alternative fates for the external aldimine are shown: Ⓐ transamination, Ⓑ racemization, and Ⓒ decarboxylation. The PLP–amino acid Schiff base is in conjugation with the pyridine ring, an electron sink that permits delocalization of an electron pair to avoid formation of an unstable carbanion on the α carbon (inset). A quinonoid intermediate is involved in all three types of reactions. The transamination route Ⓐ is especially important in the pathways described in this chapter. The pathway highlighted in yellow (shown left to right) represents only part of the overall reaction catalyzed by aminotransferases. To complete the process, a second α-keto acid replaces the one that is released, and this is converted to an amino acid in a reversal of the reaction steps (right to left). Pyridoxal phosphate is also involved in certain reactions at the β and γ carbons of some amino acids (not shown). 🖱 **Pyridoxal Phosphate Reaction Mechanisms**

mammals, this enzyme is present in the mitochondrial matrix. It is the only enzyme that can use either NAD^+ or $NADP^+$ as the acceptor of reducing equivalents **(Fig. 18–7)**.

The combined action of an aminotransferase and glutamate dehydrogenase is referred to as **transdeamination**. A few amino acids bypass the transdeamination pathway and undergo direct oxidative deamination. The fate of the NH_4^+ produced by any of these deamination processes is discussed in detail in Section 18.2. The α-ketoglutarate formed from glutamate deamination can be used in the citric acid cycle and for glucose synthesis.

Glutamate dehydrogenase operates at an important intersection of carbon and nitrogen metabolism. An allosteric enzyme with six identical subunits, its activity

FIGURE 18-7 Reaction catalyzed by glutamate dehydrogenase. The glutamate dehydrogenase of mammalian liver has the unusual capacity to use either NAD⁺ or NADP⁺ as cofactor. The glutamate dehydrogenases of plants and microorganisms are generally specific for one or the other. The mammalian enzyme is allosterically regulated by GTP and ADP.

FIGURE 18-8 Ammonia transport in the form of glutamine. Excess ammonia in tissues is added to glutamate to form glutamine, a process catalyzed by glutamine synthetase. After transport in the bloodstream, the glutamine enters the liver and NH_4^+ is liberated in mitochondria by the enzyme glutaminase.

is influenced by a complicated array of allosteric modulators. The best-studied of these are the positive modulator ADP and the negative modulator GTP. The metabolic rationale for this regulatory pattern has not been elucidated in detail. Mutations that alter the allosteric binding site for GTP or otherwise cause permanent activation of glutamate dehydrogenase lead to a human genetic disorder called hyperinsulinism-hyperammonemia syndrome, characterized by elevated levels of ammonia in the bloodstream and hypoglycemia.

Glutamine Transports Ammonia in the Bloodstream

Ammonia is quite toxic to animal tissues (we examine some possible reasons for this toxicity later), and the levels present in blood are regulated. In many tissues, including the brain, some processes such as nucleotide degradation generate free ammonia. In most animals much of the free ammonia is converted to a nontoxic compound before export from the extrahepatic tissues into the blood and transport to the liver or kidneys. For this transport function, glutamate, critical to *intracellular* amino group metabolism, is supplanted by L-glutamine. The free ammonia produced in tissues is combined with glutamate to yield glutamine by the action of **glutamine synthetase**. This reaction requires ATP and occurs in two steps **(Fig. 18-8)**. First, glutamate and ATP react to form ADP and a γ-glutamyl phosphate

intermediate, which then reacts with ammonia to produce glutamine and inorganic phosphate. Glutamine is a nontoxic transport form of ammonia; it is normally present in blood in much higher concentrations than other amino acids. Glutamine also serves as a source of amino groups in a variety of biosynthetic reactions. Glutamine synthetase is found in all organisms, always playing a central metabolic role. In microorganisms, the enzyme serves as an essential portal for the entry of fixed nitrogen into biological systems. (The roles of glutamine and glutamine synthetase in metabolism are further discussed in Chapter 22.)

In most terrestrial animals, glutamine in excess of that required for biosynthesis is transported in the blood to the intestine, liver, and kidneys for processing. In these tissues, the amide nitrogen is released as ammonium ion in the mitochondria, where the enzyme

glutaminase converts glutamine to glutamate and NH_4^+ (Fig. 18–8). The NH_4^+ from intestine and kidney is transported in the blood to the liver. In the liver, the ammonia from all sources is disposed of by urea synthesis. Some of the glutamate produced in the glutaminase reaction may be further processed in the liver by glutamate dehydrogenase, releasing more ammonia and producing carbon skeletons for metabolic fuel. However, most glutamate enters the transamination reactions required for amino acid biosynthesis and other processes (Chapter 22).

In metabolic acidosis (p. 688) there is an increase in glutamine processing by the kidneys. Not all the excess NH_4^+ thus produced is released into the bloodstream or converted to urea; some is excreted directly into the urine. In the kidney, the NH_4^+ forms salts with metabolic acids, facilitating their removal in the urine. Bicarbonate produced by the decarboxylation of α-ketoglutarate in the citric acid cycle can also serve as a buffer in blood plasma. Taken together, these effects of glutamine metabolism in the kidney tend to counteract acidosis. ∎

Alanine Transports Ammonia from Skeletal Muscles to the Liver

Alanine also plays a special role in transporting amino groups to the liver in a nontoxic form, via a pathway called the **glucose-alanine cycle (Fig. 18–9).** In muscle and certain other tissues that degrade amino acids for fuel, amino groups are collected in the form of glutamate by transamination (Fig. 18–2a). Glutamate can be converted to glutamine for transport to the liver, as described above, or it can transfer its α-amino group to pyruvate, a readily available product of muscle glycolysis, by the action of **alanine aminotransferase** (Fig. 18–9). The alanine so formed passes into the blood and travels to the liver. In the cytosol of hepatocytes, alanine aminotransferase transfers the amino group from alanine to α-ketoglutarate, forming pyruvate and glutamate. Glutamate can then enter mitochondria, where the glutamate dehydrogenase reaction releases NH_4^+ (Fig. 18–7), or can undergo transamination with oxaloacetate to form aspartate, another nitrogen donor in urea synthesis, as we shall see.

The use of alanine to transport ammonia from skeletal muscles to the liver is another example of the intrinsic economy of living organisms. Vigorously contracting skeletal muscles operate anaerobically, producing pyruvate and lactate from glycolysis as well as ammonia from protein breakdown. These products must find their way to the liver, where pyruvate and lactate are incorporated into glucose, which is returned to the muscles, and ammonia is converted to urea for excretion. The glucose-alanine cycle, in concert with the Cori cycle (see Box 14–2 and Fig. 23–19), accomplishes this transaction. The energetic burden of gluconeogenesis is thus imposed

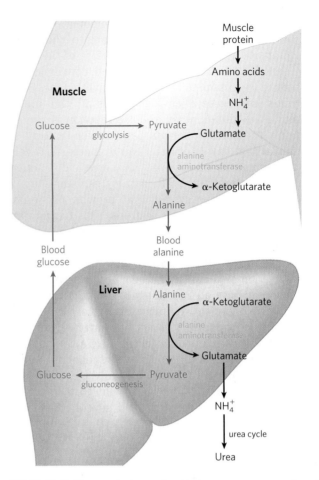

FIGURE 18–9 Glucose-alanine cycle. Alanine serves as a carrier of ammonia and of the carbon skeleton of pyruvate from skeletal muscle to liver. The ammonia is excreted and the pyruvate is used to produce glucose, which is returned to the muscle.

on the liver rather than the muscle, and all available ATP in muscle is devoted to muscle contraction.

Ammonia Is Toxic to Animals

The catabolic production of ammonia poses a serious biochemical problem, because ammonia is very toxic. The molecular basis for this toxicity is not entirely understood. The terminal stages of ammonia intoxication in humans are characterized by onset of a comatose state accompanied by cerebral edema (an increase in the brain's water content) and increased cranial pressure, so research and speculation on ammonia toxicity have focused on this tissue. Speculation centers on a potential depletion of ATP in brain cells.

Ammonia readily crosses the blood-brain barrier, so any condition that raises the level of ammonia in the bloodstream will expose the brain to high concentrations too. The developing brain is more susceptible to the deleterious effects of ammonium ion than the adult brain. The damage from ammonium toxicity includes loss of neurons, altered synapse formation, and a general defect in cellular energy metabolism.

Ridding the cytosol of excess ammonia requires reductive amination of α-ketoglutarate to glutamate by glutamate dehydrogenase (the reverse of the reaction described earlier; Fig. 18–7) and conversion of glutamate to glutamine by glutamine synthetase. Both enzymes are present at high levels in the brain, although the glutamine synthetase reaction is almost certainly the more important pathway for removal of ammonia. High levels of NH_4^+ lead to increased levels of glutamine, which acts as an osmotically active solute (osmolyte) in brain astrocytes, star-shaped cells of the nervous system that provide nutrients, support, and insulation for neurons. This triggers an uptake of water into the astrocytes to maintain osmotic balance, leading to swelling of the cells and the brain, which in turn can lead to coma.

Depletion of glutamate in the glutamine synthetase reaction may have additional effects on the brain. Glutamate and its derivative γ-aminobutyrate (GABA; see Fig. 22–31) are important neurotransmitters; the sensitivity of the brain to ammonia may reflect a depletion of neurotransmitters as well as changes in cellular osmotic balance. ■

As we close this discussion of amino group metabolism, note that we have described several processes that deposit excess ammonia in the mitochondria of hepatocytes (Fig. 18–2). We now look at the fate of that ammonia.

SUMMARY 18.1 Metabolic Fates of Amino Groups

▸ Humans derive a small fraction of their oxidative energy from the catabolism of amino acids. Amino acids are derived from the normal breakdown (recycling) of cellular proteins, degradation of ingested proteins, and breakdown of body proteins in lieu of other fuel sources during starvation or in uncontrolled diabetes mellitus.

▸ Proteases degrade ingested proteins in the stomach and small intestine. Most proteases are initially synthesized as inactive zymogens.

▸ An early step in the catabolism of amino acids is the separation of the amino group from the carbon skeleton. In most cases, the amino group is transferred to α-ketoglutarate to form glutamate. This transamination reaction requires the coenzyme pyridoxal phosphate.

▸ Glutamate is transported to liver mitochondria, where glutamate dehydrogenase liberates the amino group as ammonium ion (NH_4^+). Ammonia formed in other tissues is transported to the liver as the amide nitrogen of glutamine or, in transport from skeletal muscle, as the amino group of alanine.

▸ The pyruvate produced by deamination of alanine in the liver is converted to glucose, which is transported back to muscle as part of the glucose-alanine cycle.

18.2 Nitrogen Excretion and the Urea Cycle

If not reused for the synthesis of new amino acids or other nitrogenous products, amino groups are channeled into a single excretory end product **(Fig. 18–10)**. Most aquatic species, such as the bony fishes, are **ammonotelic**, excreting amino nitrogen as ammonia. The toxic ammonia is simply diluted in the surrounding water. Terrestrial animals require pathways for nitrogen excretion that minimize toxicity and water loss. Most terrestrial animals are **ureotelic**, excreting amino nitrogen in the form of urea; birds and reptiles are **uricotelic**, excreting amino nitrogen as uric acid. (The pathway of uric acid synthesis is described in Fig. 22–48.) Plants recycle virtually all amino groups—they excrete nitrogen only under very unusual circumstances.

In ureotelic organisms, the ammonia deposited in the mitochondria of hepatocytes is converted to urea in the **urea cycle**. This pathway was discovered in 1932 by Hans Krebs (who later also discovered the citric acid cycle) and a medical student associate, Kurt Henseleit. Urea production occurs almost exclusively in the liver and is the fate of most of the ammonia channeled there. The urea passes into the bloodstream and thus to the kidneys and is excreted into the urine. The production of urea now becomes the focus of our discussion.

Urea Is Produced from Ammonia in Five Enzymatic Steps

The urea cycle begins inside liver mitochondria, but three of the subsequent steps take place in the cytosol; the cycle thus spans two cellular compartments (Fig. 18–10). The first amino group to enter the urea cycle is derived from ammonia in the mitochondrial matrix—most of this NH_4^+ arises by the pathways described in the previous section. The liver also receives some ammonia via the portal vein from the intestine, from the bacterial oxidation of amino acids. Whatever its source, the NH_4^+ generated in liver mitochondria is immediately used, together with CO_2 (as HCO_3^-) produced by mitochondrial respiration, to form carbamoyl phosphate in

FIGURE 18–10 Urea cycle and reactions that feed amino groups into the cycle. The enzymes catalyzing these reactions (named in the text) are distributed between the mitochondrial matrix and the cytosol. One amino group enters the urea cycle as carbamoyl phosphate, formed in the matrix; the other enters as aspartate, formed in the matrix by transamination of oxaloacetate and glutamate, catalyzed by aspartate aminotransferase. The urea cycle consists of four steps. ❶ Formation of citrulline from ornithine and carbamoyl phosphate (entry of the first amino group); the citrulline passes into the cytosol. ❷ Formation of argininosuccinate through a citrullyl-AMP intermediate (entry of the second amino group). ❸ Formation of arginine from argininosuccinate; this reaction releases fumarate, which enters the citric acid cycle. ❹ Formation of urea; this reaction also regenerates ornithine. The pathways by which NH_4^+ arrives in the mitochondrial matrix of hepatocytes were discussed in Section 18.1.

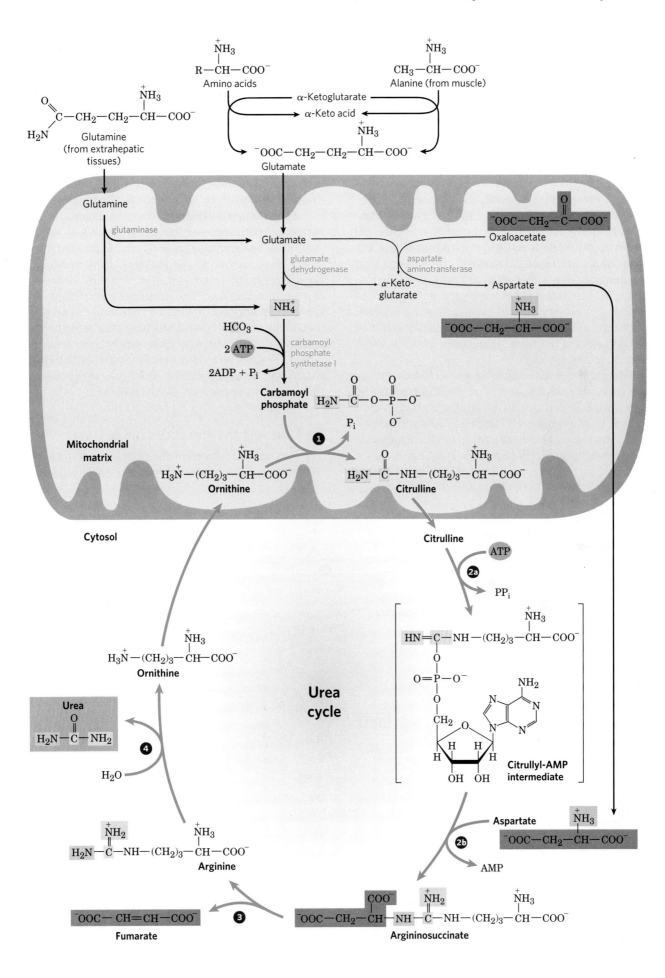

the matrix (**Fig. 18–11a**; see also Fig. 18–10). This ATP-dependent reaction is catalyzed by **carbamoyl phosphate synthetase I**, a regulatory enzyme (see below). The mitochondrial form of the enzyme is distinct from the cytosolic (II) form, which has a separate function in pyrimidine biosynthesis (Chapter 22).

The carbamoyl phosphate, which functions as an activated carbamoyl group donor, now enters the urea cycle. The cycle has only four enzymatic steps. First, carbamoyl phosphate donates its carbamoyl group to ornithine to form citrulline, with the release of P_i (Fig. 18–10, step **1**). The reaction is catalyzed by **ornithine transcarbamoylase**. Ornithine is not one of the 20 common amino acids found in proteins, but it is a key intermediate in nitrogen metabolism. It is synthesized from glutamate in a five-step pathway described in Chapter 22. Ornithine plays a role resembling that of oxaloacetate in the citric acid cycle, accepting material at each turn of the urea cycle. The citrulline produced in the first step of the urea cycle passes from the mitochondrion to the cytosol.

The next two steps bring in the second amino group. The source is aspartate generated in mitochondria by transamination and transported into the cytosol. A condensation reaction between the amino group of aspartate and the ureido (carbonyl) group of citrulline forms argininosuccinate (step **2** in Fig. 18–10). This cytosolic reaction, catalyzed by **argininosuccinate synthetase**,

requires ATP and proceeds through a citrullyl-AMP intermediate (Fig. 18–11b). The argininosuccinate is then cleaved by **argininosuccinase** (step **3** in Fig. 18–10) to form free arginine and fumarate, the latter being converted to malate before entering mitochondria to join the pool of citric acid cycle intermediates. This is the only reversible step in the urea cycle. In the last reaction of the urea cycle (step **4**), the cytosolic enzyme **arginase** cleaves arginine to yield **urea** and ornithine. Ornithine is transported into the mitochondrion to initiate another round of the urea cycle.

As we noted in Chapter 16, the enzymes of many metabolic pathways are clustered (p. 636), with the product of one enzyme reaction being channeled directly to the next enzyme in the pathway. In the urea cycle, the mitochondrial and cytosolic enzymes seem to be clustered in this way. The citrulline transported out of the mitochondrion is not diluted into the general pool of metabolites in the cytosol but is passed directly to the active site of argininosuccinate synthetase. This channeling between enzymes continues for argininosuccinate, arginine, and ornithine. Only urea is released into the general cytosolic pool of metabolites.

The Citric Acid and Urea Cycles Can Be Linked

The fumarate produced in the argininosuccinase reaction is also an intermediate of the citric acid cycle.

(a)

(b)

MECHANISM FIGURE 18–11 Nitrogen-acquiring reactions in the synthesis of urea. The urea nitrogens are acquired in two reactions, each requiring ATP. **(a)** In the reaction catalyzed by carbamoyl phosphate synthetase I, the first nitrogen enters from ammonia. The terminal phosphate groups of two molecules of ATP are used to form one molecule of carbamoyl phosphate. In other words, this reaction has two activation steps (**1** and **3**). 🔵 Carbamoyl Phosphate Synthetase I Mechanism **(b)** In the reaction catalyzed by argininosuccinate synthetase, the second nitrogen enters from aspartate. Activation of the ureido oxygen of citrulline in step **1** sets up the addition of aspartate in step **2**. 🔵 Argininosuccinate Synthetase Mechanism

Thus, the cycles are, in principle, interconnected—in a process dubbed the "Krebs bicycle" **(Fig. 18–12)**. However, each cycle can operate independently, and communication between them depends on the transport of key intermediates between the mitochondrion and cytosol. Major transporters in the inner mitochondrial membrane include the malate–α-ketoglutarate transporter, the glutamate-aspartate transporter, and the glutamate-OH$^-$ transporter. Together, these transporters facilitate the movement of malate and glutamate into the mitochondrial matrix and the movement of aspartate and α-ketoglutarate out to the cytosol.

Several enzymes of the citric acid cycle, including fumarase (fumarate hydratase) and malate dehydrogenase (p. 647), are also present as isozymes in the cytosol. There is no transporter to directly move the fumarate generated in cytosolic arginine synthesis back into the mitochondrial matrix. However, fumarate can be converted to malate in the cytosol. Fumarate and malate can be further metabolized in the cytosol or malate can be transported into mitochondria for use in the citric acid cycle. Aspartate formed in mitochondria by transamination between oxaloacetate and glutamate can be transported to the cytosol, where it serves as nitrogen donor in the urea cycle reaction catalyzed by argininosuccinate synthetase. These reactions, making up the **aspartate-argininosuccinate shunt**, provide metabolic links between the separate pathways by which the amino groups and carbon skeletons of amino acids are processed.

The use of aspartate as a nitrogen donor in the urea cycle may appear to be a relatively complicated way to introduce the second amino group into urea. However, we shall see in Chapter 22 that this pathway for nitrogen incorporation is one of the two common ways to introduce amino groups in the biosynthesis of molecules that contain them. In the urea cycle, additional pathway interconnections can help explain why aspartate is used as a nitrogen donor. The urea and citric acid cycles are closely tied to an additional process that brings NADH in the form of reducing equivalents into the mitochondrion. As detailed in the next chapter, the NADH produced by glycolysis, fatty acid oxidation, and other processes cannot be transported across

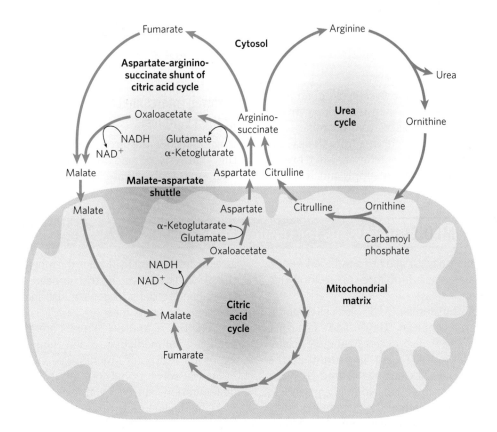

FIGURE 18–12 Links between the urea cycle and citric acid cycle. The interconnected cycles have been called the "Krebs bicycle." The pathways linking the citric acid and urea cycles are known as the aspartate-argininosuccinate shunt; these effectively link the fates of the amino groups and the carbon skeletons of amino acids. The interconnections are quite elaborate. For example, some citric acid cycle enzymes, such as fumarase and malate dehydrogenase, have both cytosolic and mitochondrial isozymes. Fumarate produced in the cytosol—whether by the urea cycle, purine biosynthesis, or other processes—can be converted to cytosolic malate, which is used in the cytosol, or transported into mitochondria to enter the citric acid cycle. These processes are further intertwined with the malate-aspartate shuttle, a set of reactions that brings reducing equivalents into the mitochondrion (see also Fig. 19–31). These different cycles and processes rely on a limited number of transporters in the inner mitochondrial membrane.

the mitochondrial inner membrane. Reducing equivalents are instead brought into the mitochondrion by converting aspartate to oxaloacetate in the cytosol, reducing the oxaloacetate to malate with NADH, and transporting the malate into the mitochondrial matrix via the malate–α-ketoglutarate transporter. Once inside the mitochondrion, the malate can be reconverted to oxaloacetate while generating NADH. The oxaloacetate is converted to aspartate in the matrix and transported out of the mitochondrion by the aspartate-glutamate transporter. This malate-aspartate shuttle completes yet another cycle that functions to keep the mitochondrion supplied with NADH (see Fig. 19–31).

These processes require that a balance be maintained in the cytosol between the concentrations of glutamate and aspartate. The enzyme that transfers amino groups between these key amino acids is aspartate aminotransferase, AAT (also called glutamate-oxaloacetate transaminase, GOT). This enzyme is among the most active enzymes in hepatocytes and other tissues. When tissue damage occurs, this easily assayed enzyme and others leak into the blood. Thus, measuring blood levels of liver enzymes is important in diagnosing a variety of medical conditions (Box 18–1).

The Activity of the Urea Cycle Is Regulated at Two Levels

The flux of nitrogen through the urea cycle in an individual animal varies with diet. When the dietary intake is primarily protein, the carbon skeletons of amino acids are used for fuel, producing much urea from the excess amino groups. During prolonged starvation, when breakdown of muscle protein begins to supply much of the organism's metabolic energy, urea production also increases substantially.

These changes in demand for urea cycle activity are met over the long term by regulation of the rates of synthesis of the four urea cycle enzymes and carbamoyl phosphate synthetase I in the liver. All five enzymes are synthesized at higher rates in starving animals and in animals on very-high-protein diets than in well-fed animals eating primarily carbohydrates and fats. Animals on protein-free diets produce lower levels of urea cycle enzymes.

On a shorter time scale, allosteric regulation of at least one key enzyme adjusts the flux through the urea cycle. The first enzyme in the pathway, carbamoyl phosphate synthetase I, is allosterically activated by **N-acetylglutamate**, which is synthesized from acetyl-CoA and glutamate by **N-acetylglutamate synthase (Fig. 18–13)**. In plants and microorganisms this enzyme catalyzes the first step in the de novo synthesis of arginine from glutamate (see Fig. 22–12), but in mammals N-acetylglutamate synthase activity in the liver has a purely regulatory function (mammals lack the other enzymes needed to convert glutamate to arginine). The steady-state levels of N-acetylglutamate are determined by the concentrations of glutamate and acetyl-CoA (the substrates for N-acetylglutamate synthase) and arginine (an activator of N-acetylglutamate synthase, and thus an activator of the urea cycle).

Pathway Interconnections Reduce the Energetic Cost of Urea Synthesis

If we consider the urea cycle in isolation, we see that the synthesis of one molecule of urea requires four high-energy phosphate groups (Fig. 18–10). Two ATP

BOX 18–1   ☤ MEDICINE   Assays for Tissue Damage

Analyses of certain enzyme activities in blood serum give valuable diagnostic information for several disease conditions.

Alanine aminotransferase (ALT; also called glutamate-pyruvate transaminase, GPT) and aspartate aminotransferase (AST; also called glutamate-oxaloacetate transaminase, GOT) are important in the diagnosis of heart and liver damage caused by heart attack, drug toxicity, or infection. After a heart attack, a variety of enzymes, including these aminotransferases, leak from injured heart cells into the bloodstream. Measurements of the blood serum concentrations of the two aminotransferases by the SGPT and SGOT tests (S for serum)—and of another enzyme, **creatine kinase**, by the SCK test—can provide information about the severity of the damage. Creatine kinase is the first heart enzyme to appear in the blood after a heart attack; it also disappears quickly from the blood. GOT is the next to appear, and GPT follows later. Lactate dehydrogenase also leaks from injured or anaerobic heart muscle.

The SGOT and SGPT tests are also important in occupational medicine, to determine whether people exposed to carbon tetrachloride, chloroform, or other industrial solvents have suffered liver damage. Liver degeneration caused by these solvents is accompanied by leakage of various enzymes from injured hepatocytes into the blood. Aminotransferases are most useful in the monitoring of people exposed to these chemicals, because these enzyme activities are high in liver and thus are likely to be among the proteins leaked from damaged hepatocytes; also, they can be detected in the bloodstream in very small amounts.

FIGURE 18–13 Synthesis of N-acetylglutamate and its activation of carbamoyl phosphate synthetase I.

molecules are required to make carbamoyl phosphate, and one ATP to make argininosuccinate—the latter ATP undergoing a pyrophosphate cleavage to AMP and PP$_i$, which is hydrolyzed to two P$_i$. The overall equation of the urea cycle is

$$2NH_4^+ + HCO_3^- + 3ATP^{4-} + H_2O \longrightarrow$$
$$urea + 2ADP^{3-} + 4P_i^{2-} + AMP^{2-} + 2H^+$$

However, this apparent cost is ameliorated by the pathway interconnections detailed above. The fumarate generated by the urea cycle is converted to malate, and the malate is transported into the mitochondrion (Fig. 18–12). Inside the mitochondrial matrix, NADH is generated in the malate dehydrogenase reaction. Each NADH molecule can generate up to 2.5 ATP during mitochondrial respiration (Chapter 19), greatly reducing the overall energetic cost of urea synthesis.

Genetic Defects in the Urea Cycle Can Be Life-Threatening

People with genetic defects in any enzyme involved in urea formation cannot tolerate protein-rich diets. Amino acids ingested in excess of the minimum daily requirements for protein synthesis are

deaminated in the liver, producing free ammonia that cannot be converted to urea and exported into the bloodstream, and, as we have seen, ammonia is highly toxic. The absence of a urea cycle enzyme can result in hyperammonemia or in the buildup of one or more urea cycle intermediates, depending on the enzyme that is missing. Given that most urea cycle steps are irreversible, the absent enzyme activity can often be identified by determining which cycle intermediate is present in especially elevated concentration in the blood and/or urine. Although the breakdown of amino acids can have serious health consequences in individuals with urea cycle deficiencies, a protein-free diet is not a treatment option. Humans are incapable of synthesizing half of the 20 common amino acids, and these **essential amino acids** (Table 18–1) must be provided in the diet.

A variety of treatments are available for individuals with urea cycle defects. Careful administration of the aromatic acids benzoate or phenylbutyrate in the diet can help lower the level of ammonia in the blood. Benzoate is converted to benzoyl-CoA, which combines with glycine to form hippurate (**Fig. 18–14**, left). The glycine used up in this reaction must be regenerated, and ammonia is thus taken up in the glycine synthase reaction. Phenylbutyrate is converted to phenylacetate by β oxidation. The phenylacetate is then converted to phenylacetyl-CoA, which combines with glutamine to form phenylacetylglutamine (Fig. 18–14, right). The resulting removal of glutamine triggers its further synthesis by glutamine synthetase (see Eqn 22–1) in a reaction that takes up ammonia. Both hippurate and phenylacetylglutamine are nontoxic compounds that are excreted in the urine. The pathways shown in Figure 18–14 make only minor contributions to normal metabolism, but they become prominent when aromatic acids are ingested.

TABLE 18–1   Nonessential and Essential Amino Acids for Humans and the Albino Rat

Nonessential	Conditionally essential*	Essential
Alanine	Arginine	Histidine
Asparagine	Cysteine	Isoleucine
Aspartate	Glutamine	Leucine
Glutamate	Glycine	Lysine
Serine	Proline	Methionine
	Tyrosine	Phenylalanine
		Threonine
		Tryptophan
		Valine

*Required to some degree in young, growing animals and/or sometimes during illness.

CH₂—CH₂—CH₂—COO⁻

Phenylbutyrate

CoA-SH
β oxidation
Acetyl-CoA

CH₂—COO⁻

COO⁻

+ CoA-SH

Benzoate

CH₂—COO⁻

+ CoA-SH

Phenylacetate

ATP

AMP + PP$_i$

ATP

AMP + PP$_i$

O S-CoA
‖
C

Benzoyl-CoA

O
‖
CH₂—C—S-CoA

Phenylacetyl-CoA

H₃N⁺—CH₂—COO⁻
Glycine

CoA-SH

COO⁻
H₃N⁺—CH
CH₂
CH₂
C=O
NH₂
Glutamine

CoA-SH

O NH—CH₂—COO⁻
‖
C

Hippurate
(benzoylglycine)

O COO⁻
‖
CH₂—C—NH—CH
CH₂
CH₂
C=O
NH₂

Phenylacetylglutamine

FIGURE 18–14 Treatment for deficiencies in urea cycle enzymes. The aromatic acids benzoate and phenylbutyrate, administered in the diet, are metabolized and combine with glycine and glutamine, respectively. The products are excreted in the urine. Subsequent synthesis of glycine and glutamine to replenish the pool of these intermediates removes ammonia from the bloodstream.

Other therapies are more specific to a particular enzyme deficiency. Deficiency of N-acetylglutamate synthase results in the absence of the normal activator of carbamoyl phosphate synthetase I (Fig. 18–13). This condition can be treated by administering carbamoyl glutamate, an analog of N-acetylglutamate that is effective in activating carbamoyl phosphate synthetase I.

O COO⁻
‖ |
H₂N—C—NH—C—H
CH₂
CH₂
COO⁻

Carbamoyl glutamate

Supplementing the diet with arginine is useful in treating deficiencies of ornithine transcarbamoylase, argininosuccinate synthetase, and argininosuccinase. Many of these treatments must be accompanied by strict dietary control and supplements of essential amino acids. In the rare cases of arginase deficiency, arginine, the substrate of the defective enzyme, must be excluded from the diet. ■

SUMMARY 18.2 Nitrogen Excretion and the Urea Cycle

▶ Ammonia is highly toxic to animal tissues. In the urea cycle, ornithine combines with ammonia, in the form of carbamoyl phosphate, to form citrulline. A second amino group is transferred to citrulline from aspartate to form arginine—the immediate precursor of urea. Arginase catalyzes hydrolysis of arginine to urea and ornithine; thus ornithine is regenerated in each turn of the cycle.

▶ The urea cycle results in a net conversion of oxaloacetate to fumarate, both of which are intermediates in the citric acid cycle. The two cycles are thus interconnected.

▶ The activity of the urea cycle is regulated at the level of enzyme synthesis and by allosteric regulation of the enzyme that catalyzes the formation of carbamoyl phosphate.

18.3 Pathways of Amino Acid Degradation

The pathways of amino acid catabolism, taken together, normally account for only 10% to 15% of the human body's energy production; these pathways are not nearly as active as glycolysis and fatty acid oxidation. Flux through these catabolic routes also varies greatly, depending on the balance between requirements for biosynthetic processes and the availability of a particular amino acid. The 20 catabolic pathways converge to form only six major products, all of which enter the citric acid cycle (**Fig. 18–15**). From here the carbon skeletons are diverted to gluconeogenesis or ketogenesis or are completely oxidized to CO_2 and H_2O.

All or part of the carbon skeletons of seven amino acids are ultimately broken down to acetyl-CoA. Five amino acids are converted to α-ketoglutarate, four to succinyl-CoA, two to fumarate, and two to oxaloacetate. Parts or all of six amino acids are converted to

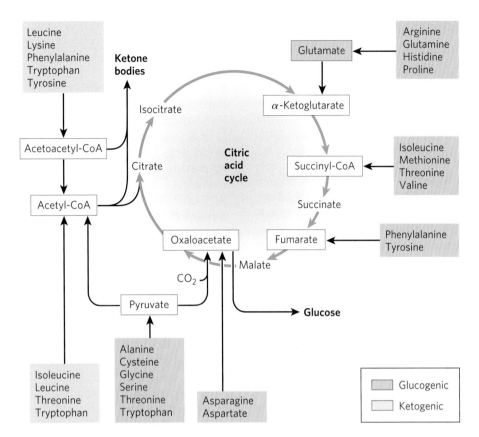

FIGURE 18–15 Summary of amino acid catabolism. Amino acids are grouped according to their major degradative end product. Some amino acids are listed more than once because different parts of their carbon skeletons are degraded to different end products. The figure shows the most important catabolic pathways in vertebrates, but there are minor variations among vertebrate species. Threonine, for instance, is degraded via at least two different pathways (see Figs 18–19, 18–27), and the importance of a given pathway can vary with the organism and its metabolic conditions. The glucogenic and ketogenic amino acids are also delineated in the figure, by color shading. Notice that five of the amino acids are both glucogenic and ketogenic. The amino acids degraded to pyruvate are also potentially ketogenic. Only two amino acids, leucine and lysine, are exclusively ketogenic.

pyruvate, which can be converted to either acetyl-CoA or oxaloacetate. We later summarize the individual pathways for the 20 amino acids in flow diagrams, each leading to a specific point of entry into the citric acid cycle. In these diagrams the carbon atoms that enter the citric acid cycle are shown in color. Note that some amino acids appear more than once, reflecting different fates for different parts of their carbon skeletons. Rather than examining every step of every pathway in amino acid catabolism, we single out for special discussion some enzymatic reactions that are particularly noteworthy for their mechanisms or their medical significance.

Some Amino Acids Are Converted to Glucose, Others to Ketone Bodies

The seven amino acids that are degraded entirely or in part to acetoacetyl-CoA and/or acetyl-CoA— phenylalanine, tyrosine, isoleucine, leucine, tryptophan, threonine, and lysine—can yield ketone bodies in the liver, where acetoacetyl-CoA is converted to acetoacetate and then to acetone and β-hydroxybutyrate (see Fig. 17–19). These are the **ketogenic** amino acids (Fig. 18–15). Their ability to form ketone bodies is particularly evident in uncontrolled diabetes mellitus, in which the liver produces large amounts of ketone bodies from both fatty acids and the ketogenic amino acids.

The amino acids that are degraded to pyruvate, α-ketoglutarate, succinyl-CoA, fumarate, and/or oxaloacetate can be converted to glucose and glycogen by pathways described in Chapters 14 and 15. They are the **glucogenic** amino acids. The division between ketogenic and glucogenic amino acids is not sharp; five amino acids—tryptophan, phenylalanine, tyrosine, threonine, and isoleucine—are both ketogenic and glucogenic. Catabolism of amino acids is particularly critical to the survival of animals with high-protein diets or during starvation. Leucine is an exclusively ketogenic amino acid that is very common in proteins. Its degradation makes a substantial contribution to ketosis under starvation conditions.

Several Enzyme Cofactors Play Important Roles in Amino Acid Catabolism

A variety of interesting chemical rearrangements occur in the catabolic pathways of amino acids. It is useful to begin our study of these pathways by noting the classes of reactions that recur and introducing their enzyme cofactors. We have already considered one important class: transamination reactions requiring pyridoxal phosphate. Another common type of reaction in amino acid catabolism is one-carbon transfers, which usually involve one of three cofactors: biotin, tetrahydrofolate, or S-adenosylmethionine (Fig. 18–16). These cofactors transfer one-carbon groups in different oxidation states: biotin transfers carbon in its most oxidized state, CO_2 (see Fig. 14–19); tetrahydrofolate transfers one-carbon groups in intermediate oxidation states and sometimes as methyl groups; and S-adenosylmethionine transfers methyl groups, the most reduced state of carbon. The latter two cofactors are especially important in amino acid and nucleotide metabolism.

Tetrahydrofolate (**H_4 folate**), synthesized in bacteria, consists of substituted pterin (6-methylpterin), p-aminobenzoate, and glutamate moieties (Fig. 18–16).

Pterin

The oxidized form, folate, is a vitamin for mammals; it is converted in two steps to tetrahydrofolate by the enzyme dihydrofolate reductase. The one-carbon group undergoing transfer, in any of three oxidation states, is bonded to N-5 or N-10 or both. The most reduced form of the cofactor carries a methyl group, a more oxidized form carries a methylene group, and the most oxidized forms carry a methenyl, formyl, or formimino group

(Fig. 18–17). Most forms of tetrahydrofolate are interconvertible and serve as donors of one-carbon units in a variety of metabolic reactions. The primary source of one-carbon units for tetrahydrofolate is the carbon removed in the conversion of serine to glycine, producing N^5,N^{10}-methylenetetrahydrofolate.

Although tetrahydrofolate can carry a methyl group at N-5, the transfer potential of this methyl group is insufficient for most biosynthetic reactions. **S-Adenosylmethionine (adoMet)** is the preferred cofactor for biological methyl group transfers. It is synthesized from ATP and methionine by the action of **methionine adenosyl transferase** (**Fig. 18–18**, step ❶). This reaction is unusual in that the nucleophilic sulfur atom of methionine attacks the 5′ carbon of the ribose moiety of ATP rather than one of the phosphorus atoms. Triphosphate is released and is cleaved to P_i and PP_i on the enzyme, and the PP_i is cleaved by inorganic pyrophosphatase; thus three bonds, including two bonds of high-energy phosphate groups, are broken in this reaction. The only other known reaction in which triphosphate is displaced from ATP occurs in the synthesis of coenzyme B_{12} (see Box 17–2, Fig. 3).

S-Adenosylmethionine is a potent alkylating agent by virtue of its destabilizing sulfonium ion. The methyl group is subject to attack by nucleophiles and is about 1,000 times more reactive than the methyl group of N^5-methyltetrahydrofolate.

Transfer of the methyl group from S-adenosylmethionine to an acceptor yields **S-adenosylhomocysteine** (Fig. 18–18, step ❷), which is subsequently broken down to homocysteine and adenosine (step ❸). Methionine is regenerated by transfer of a methyl group to homocysteine in a reaction catalyzed by methionine synthase (step ❹), and methionine is reconverted to S-adenosylmethionine to complete an activated-methyl cycle.

One form of methionine synthase common in bacteria uses N^5-methyltetrahydrofolate as a methyl donor.

FIGURE 18–16 Some enzyme cofactors important in one-carbon transfer reactions. The nitrogen atoms to which one-carbon groups are attached in tetrahydrofolate are shown in blue.

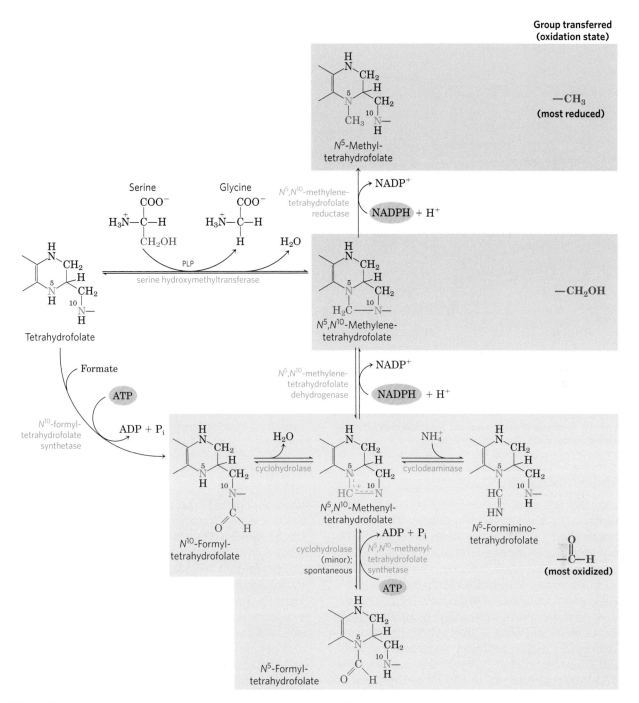

FIGURE 18–17 Conversions of one-carbon units on tetrahydrofolate. The different molecular species are grouped according to oxidation state, with the most reduced at the top and most oxidized at the bottom. All species within a single shaded box are at the same oxidation state. The conversion of N^5,N^{10}-methylenetetrahydrofolate to N^5-methyltetrahydrofolate is effectively irreversible. The enzymatic transfer of formyl groups, as in purine synthesis (see Fig. 22–35) and in the formation of formylmethionine in bacteria (Chapter 27), generally uses N^{10}-formyltetrahydrofolate rather than N^5-formyltetrahydrofolate. The latter species is significantly more stable and therefore a weaker donor of formyl groups. N^5-Formyltetrahydrofolate is a minor byproduct of the cyclohydrolase reaction, and can also form spontaneously. Conversion of N^5-formyltetrahydrofolate to N^5,N^{10}-methenyltetrahydrofolate requires ATP, because of an otherwise unfavorable equilibrium. Note that N^5-formiminotetrahydrofolate is derived from histidine in a pathway shown in Figure 18-26.

Another form of the enzyme present in some bacteria and mammals uses N^5-methyltetrahydrofolate, but the methyl group is first transferred to cobalamin, derived from coenzyme B_{12}, to form methylcobalamin as the methyl donor in methionine formation. This reaction and the rearrangement of L-methylmalonyl-CoA to succinyl-CoA (see Box 17–2, Fig. 1a) are the only known coenzyme B_{12}–dependent reactions in mammals. The vitamins B_{12} and folate are closely linked in these metabolic pathways. The B_{12} deficiency disease **pernicious anemia** is rare, seen only in individuals who have a defect in the intestinal absorption

FIGURE 18–18 Synthesis of methionine and S-adenosylmethionine in an activated-methyl cycle. The steps are described in the text. In the methionine synthase reaction (step ❹), the methyl group is transferred to cobalamin to form methylcobalamin, which in turn is the methyl donor in the formation of methionine. S-Adenosylmethionine, which has a positively charged sulfur (and is thus a sulfonium ion), is a powerful methylating agent in several biosynthetic reactions. The methyl group acceptor (step ❷) is designated R.

pathways for this vitamin (see Box 17–2) or in strict vegetarians (B_{12} is not present in plants). The disease progresses slowly, because only small amounts of vitamin B_{12} are required and normal stores of B_{12} in the liver can last three to five years. Symptoms include not only anemia but a variety of neurological disorders.

The anemia can be traced to the methionine synthase reaction. As noted above, the methyl group of methylcobalamin is derived from N^5-methyltetrahydrofolate, and this is the only reaction in mammals that uses N^5-methyltetrahydrofolate. The reaction converting the N^5,N^{10}-methylene form to the N^5-methyl form of tetrahydrofolate is irreversible (Fig. 18–17). Thus, if coenzyme B_{12} is not available for the synthesis of methylcobalamin, metabolic folates become trapped in the N^5-methyl form. The anemia associated with vitamin B_{12} deficiency is called **megaloblastic anemia**. It manifests as a decline in the production of mature erythrocytes (red blood cells) and the appearance in the bone marrow of immature precursor cells, or **megaloblasts**. Erythrocytes are gradually replaced in the blood by smaller numbers of abnormally large erythrocytes called **macrocytes**. The defect in erythrocyte development is a direct consequence of the depletion of the N^5,N^{10}-methylenetetrahydrofolate, which is required for synthesis of the thymidine nucleotides needed for DNA synthesis (see Chapter 22). Folate deficiency, in which all forms of tetrahydrofolate are depleted, also produces anemia, for much the same reasons. The anemia symptoms of B_{12} deficiency can be alleviated by administering either vitamin B_{12} or folate.

However, it is dangerous to treat pernicious anemia by folate supplementation alone, because the neurological symptoms of B_{12} deficiency will progress. These symptoms do not arise from the defect in the methionine synthase reaction. Instead, the impaired methylmalonyl-CoA mutase (see Box 17–2 and Fig. 17–12) causes accumulation of unusual, odd-number fatty acids in neuronal membranes. The anemia associated with folate deficiency is thus often treated by administering both folate and vitamin B_{12}, at least until the metabolic source of the anemia is unambiguously defined. Early diagnosis of B_{12} deficiency is important because some of its associated neurological conditions may be irreversible.

Folate deficiency also reduces the availability of the N^5-methyltetrahydrofolate required for methionine synthase function. This leads to a rise in homocysteine levels in blood, a condition linked to heart disease, hypertension, and stroke. High levels of homocysteine may be responsible for 10% of all cases of heart disease. The condition is treated with folate supplements. ∎

Tetrahydrobiopterin, another cofactor of amino acid catabolism, is similar to the pterin moiety of tetrahydrofolate, but it is not involved in one-carbon transfers; instead it participates in oxidation reactions. We consider its mode of action when we discuss phenylalanine degradation (see Fig. 18–24).

Six Amino Acids Are Degraded to Pyruvate

The carbon skeletons of six amino acids are converted in whole or in part to pyruvate. The pyruvate can then be converted to acetyl-CoA and eventually oxidized via the citric acid cycle, or to oxaloacetate and shunted into gluconeogenesis. The six amino acids are alanine, tryptophan, cysteine, serine, glycine, and threonine (**Fig. 18–19**). **Alanine** yields pyruvate directly on transamination with α-ketoglutarate, and the side chain of **tryptophan** is cleaved to yield alanine and thus pyruvate. **Cysteine** is converted to pyruvate in two steps; one removes the sulfur atom, the other is a transamination. **Serine** is converted to pyruvate by serine dehydratase. Both the β-hydroxyl and the α-amino groups of serine are removed in this single pyridoxal phosphate–dependent reaction (**Fig. 18–20a**).

   Glycine is degraded via three pathways, only one of which leads to pyruvate. Glycine is converted to serine by

enzymatic addition of a hydroxymethyl group (Figs 18–19, 18–20b). This reaction, catalyzed by **serine hydroxymethyltransferase**, requires the coenzymes tetrahydrofolate and pyridoxal phosphate. The serine is converted to pyruvate as described above. In the second pathway, which predominates in animals, glycine undergoes oxidative cleavage to CO_2, NH_4^+, and a methylene group (—CH_2—) (Figs 18–19, 18–20c). This readily reversible reaction, catalyzed by **glycine cleavage enzyme** (also called glycine synthase), also requires tetrahydrofolate, which accepts the methylene group. In this oxidative cleavage pathway, the two carbon atoms of glycine do not enter the citric acid cycle. One carbon is lost as CO_2 and the other becomes the methylene group of N^5,N^{10}-methylenetetrahydrofolate (Fig. 18–17), a one-carbon group donor in certain biosynthetic pathways.

   This second pathway for glycine degradation seems to be critical in mammals. Humans with serious defects in glycine cleavage enzyme activity suffer from a condition known as nonketotic hyperglycinemia. The condition is characterized by elevated serum levels of glycine, leading to severe mental deficiencies and death in

FIGURE 18–19 Catabolic pathways for alanine, glycine, serine, cysteine, tryptophan, and threonine. The fate of the indole group of tryptophan is shown in Figure 18–21. Details of most of the reactions involving serine and glycine are shown in Figure 18–20. The pathway for threonine degradation shown here accounts for only about a third of threonine catabolism (for the alternative pathway, see Fig. 18–27). Several pathways for cysteine degradation lead to pyruvate. The sulfur of cysteine has several alternative fates, one of which is shown in Figure 22–17. Carbon atoms here and in subsequent figures are color-coded as necessary to trace their fates.

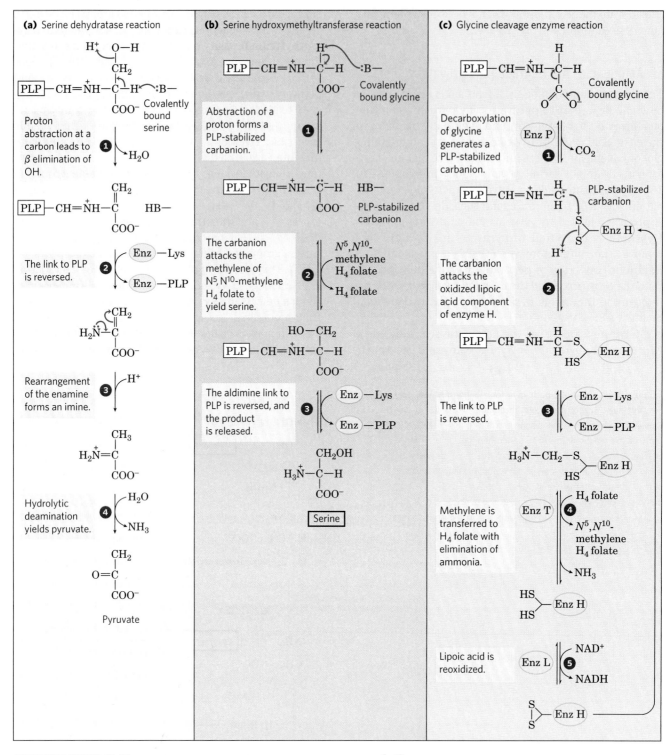

MECHANISM FIGURE 18–20 Interplay of the pyridoxal phosphate and tetrahydrofolate cofactors in serine and glycine metabolism. The first step in each of these reactions (not shown) involves the formation of a covalent imine linkage between enzyme-bound PLP and the substrate amino acid—serine in (a), glycine in (b) and (c). **(a)** A PLP-catalyzed elimination of water in the serine dehydratase reaction (step **1**) begins the pathway to pyruvate. **(b)** In the serine hydroxymethyltransferase reaction, a PLP-stabilized carbanion (product of step **1**) is a key intermediate in the reversible transfer of the methylene group (as —CH$_2$—OH)

from N^5,N^{10}-methylenetetrahydrofolate to form serine. **(c)** The glycine cleavage enzyme is a multienzyme complex, with components P, H, T, and L. The overall reaction, which is reversible, converts glycine to CO_2 and NH_4^+, with the second glycine carbon taken up by tetrahydrofolate to form N^5,N^{10}-methylenetetrahydrofolate. Pyridoxal phosphate activates the α carbon of amino acids at critical stages in all these reactions, and tetrahydrofolate carries one-carbon units in two of them (see Figs 18–6, 18–17).

very early childhood. At high levels, glycine is an inhibitory neurotransmitter, perhaps explaining the neurological effects of the disease. Many genetic defects of amino acid metabolism have been identified in humans (Table 18–2). We shall encounter several more in this chapter. ■

In the third and final pathway of glycine degradation, the achiral glycine molecule is a substrate for the enzyme D-amino acid oxidase. The glycine is converted to glyoxylate, an alternative substrate for hepatic lactate dehydrogenase (p. 563). Glyoxylate is oxidized in an NAD^+-dependent reaction to oxalate:

The primary function of D-amino acid oxidase, present at high levels in the kidney, is thought to be the detoxification of ingested D-amino acids derived from bacterial cell walls and from grilled foodstuffs (high heat causes some spontaneous racemization of the L-amino acids in proteins). Oxalate, whether obtained in foods or produced enzymatically in the kidneys, has medical significance. Crystals of calcium oxalate account for up to 75% of all kidney stones. ■

There are two significant pathways for **threonine** degradation. One pathway leads to pyruvate via glycine (Fig. 18–19). The conversion to glycine occurs in two steps, with threonine first converted to 2-amino-3-ketobutyrate by the action of threonine dehydrogenase. This is a relatively minor pathway in humans, accounting for 10% to 30% of threonine catabolism, but is more important in some other mammals. The major pathway in humans leads to succinyl-CoA and is described later.

In the laboratory, serine hydroxymethyltransferase will catalyze the conversion of threonine to glycine and acetaldehyde in one step, but this is not a significant pathway for threonine degradation in mammals.

Seven Amino Acids Are Degraded to Acetyl-CoA

Portions of the carbon skeletons of seven amino acids— **tryptophan, lysine, phenylalanine, tyrosine, leucine, isoleucine,** and **threonine**—yield acetyl-CoA and/or

---

TABLE 18–2   Some Human Genetic Disorders Affecting Amino Acid Catabolism

Medical condition	Approximate incidence (per 100,000 births)	Defective process	Defective enzyme	Symptoms and effects
Albinism	<3	Melanin synthesis from tyrosine	Tyrosine 3-monooxygenase (tyrosinase)	Lack of pigmentation; white hair, pink skin
Alkaptonuria	<0.4	Tyrosine degradation	Homogentisate 1,2-dioxygenase	Dark pigment in urine; late-developing arthritis
Argininemia	<0.5	Urea synthesis	Arginase	Mental retardation
Argininosuccinic acidemia	<1.5	Urea synthesis	Argininosuccinase	Vomiting; convulsions
Carbamoyl phosphate synthetase I deficiency	<0.5	Urea synthesis	Carbamoyl phosphate synthetase I	Lethargy; convulsions; early death
Homocystinuria	<0.5	Methionine degradation	Cystathionine β-synthase	Faulty bone development; mental retardation
Maple syrup urine disease (branched-chain ketoaciduria)	<0.4	Isoleucine, leucine, and valine degradation	Branched-chain α-keto acid dehydrogenase complex	Vomiting; convulsions; mental retardation; early death
Methylmalonic acidemia	<0.5	Conversion of propionyl-CoA to succinyl-CoA	Methylmalonyl-CoA mutase	Vomiting; convulsions; mental retardation; early death
Phenylketonuria	<8	Conversion of phenylalanine to tyrosine	Phenylalanine hydroxylase	Neonatal vomiting; mental retardation

acetoacetyl-CoA, the latter being converted to acetyl-CoA **(Fig. 18–21)**. Some of the final steps in the degradative pathways for leucine, lysine, and tryptophan resemble steps in the oxidation of fatty acids (see Fig. 17–9). Threonine (not shown in Fig. 18–21) yields some acetyl-CoA via the minor pathway illustrated in Figure 18–19.

The degradative pathways of two of these seven amino acids deserve special mention. Tryptophan breakdown is the most complex of all the pathways of amino acid catabolism in animal tissues; portions of tryptophan (four of its carbons) yield acetyl-CoA via acetoacetyl-CoA. Some of the intermediates in tryptophan catabolism are precursors for the synthesis of other biomolecules **(Fig. 18–22)**, including nicotinate, a precursor of NAD and NADP in animals; serotonin, a

neurotransmitter in vertebrates; and indoleacetate, a growth factor in plants. Some of these biosynthetic pathways are described in more detail in Chapter 22 (see Figs 22–30, 22–31).

The breakdown of phenylalanine is noteworthy because genetic defects in the enzymes of this pathway lead to several inheritable human diseases **(Fig. 18–23)**, as discussed below. Phenylalanine and its oxidation product tyrosine (both with nine carbons) are degraded into two fragments, both of which can enter the citric acid cycle: four of the nine carbon atoms yield free acetoacetate, which is converted to acetoacetyl-CoA and thus acetyl-CoA, and a second four-carbon fragment is recovered as fumarate. Eight of the nine carbons of these two amino acids thus enter the citric acid cycle; the

FIGURE 18–21 Catabolic pathways for tryptophan, lysine, phenylalanine, tyrosine, leucine, and isoleucine. These amino acids donate some of their carbons (red) to acetyl-CoA. Tryptophan, phenylalanine, tyrosine, and isoleucine also contribute carbons (blue) to pyruvate or citric acid cycle intermediates. The phenylalanine pathway is described in more detail in Figure 18-23. The fate of nitrogen atoms is not traced in this scheme; in most cases they are transferred to α-ketoglutarate to form glutamate.

FIGURE 18-22 Tryptophan as precursor. The aromatic rings of tryptophan give rise to nicotinate (niacin), indoleacetate, and serotonin. Colored atoms trace the source of the ring atoms in nicotinate.

remaining carbon is lost as CO_2. Phenylalanine, after its hydroxylation to tyrosine, is also the precursor of dopamine, a neurotransmitter, and of norepinephrine and epinephrine, hormones secreted by the adrenal medulla (see Fig. 22–31). Melanin, the black pigment of skin and hair, is also derived from tyrosine.

Phenylalanine Catabolism Is Genetically Defective in Some People

Given that many amino acids are either neurotransmitters or precursors or antagonists of neurotransmitters, it is not surprising that genetic defects of amino acid metabolism can cause defective neural development and intellectual deficits. In most such diseases specific intermediates accumulate. For example, a genetic defect in **phenylalanine hydroxylase**, the first enzyme in the catabolic pathway for phenylalanine (Fig. 18–23), is responsible for the disease **phenylketonuria (PKU)**, the most common cause of elevated levels of phenylalanine in the blood (hyperphenylalaninemia).

Phenylalanine hydroxylase (also called phenylalanine-4-monooxygenase) is one of a general class of enzymes

FIGURE 18-23 Catabolic pathways for phenylalanine and tyrosine. In humans these amino acids are normally converted to acetoacetyl-CoA and fumarate. Genetic defects in many of these enzymes cause inheritable human diseases (shaded yellow).

FIGURE 18–24 Role of tetrahydrobiopterin in the phenylalanine hydroxylase reaction. The H atom shaded pink is transferred directly from C-4 to C-3 in the reaction. This feature, discovered at the National Institutes of Health, is called the NIH shift.

called **mixed-function oxidases** (see Box 21–1), all of which catalyze simultaneous hydroxylation of a substrate by an oxygen atom of O_2 and reduction of the other oxygen atom to H_2O. Phenylalanine hydroxylase requires the cofactor tetrahydrobiopterin, which carries electrons from NADPH to O_2 and becomes oxidized to dihydrobiopterin in the process **(Fig. 18–24).** It is subsequently reduced by the enzyme **dihydrobiopterin reductase** in a reaction that requires NADPH.

In individuals with PKU, a secondary, normally little-used pathway of phenylalanine metabolism comes into play. In this pathway phenylalanine undergoes transamination with pyruvate to yield **phenylpyruvate (Fig. 18–25).** Phenylalanine and phenylpyruvate accumulate in the blood and tissues and are excreted in the urine—hence the name "phenylketonuria." Much of the phenylpyruvate, rather than being excreted as such, is either decarboxylated to phenylacetate or reduced to phenyllactate. Phenylacetate imparts a characteristic odor to the urine, which nurses have traditionally used to detect PKU in infants. The accumulation of phenylalanine or its metabolites in early life impairs normal development of the brain, causing severe intellectual deficits. This may be caused by excess phenylalanine competing with other amino acids for transport across the blood-brain barrier, resulting in a deficit of required metabolites.

Phenylketonuria was among the first inheritable metabolic defects discovered in humans. When this condition is recognized early in infancy, mental retardation can be prevented by rigid dietary control. The diet must supply only enough phenylalanine and tyrosine to meet the needs for protein synthesis. Consumption of protein-rich foods must be curtailed. Natural proteins, such as casein of milk, must first be hydrolyzed and much of the phenylalanine removed to provide an appropriate diet, at least through childhood. Because the artificial sweetener aspartame is a dipeptide of aspartate and the methyl ester of phenylalanine (see Fig. 1–24b), foods sweetened with aspartame bear warnings addressed to individuals on phenylalanine-controlled diets.

Phenylketonuria can also be caused by a defect in the enzyme that catalyzes the regeneration of tetrahydrobiopterin (Fig. 18–24). The treatment in this case is more complex than restricting the intake of phenylalanine and tyrosine. Tetrahydrobiopterin is also required for the formation of L-3,4-dihydroxyphenylalanine (L-dopa) and 5-hydroxytryptophan—precursors of the neurotransmitters norepinephrine and serotonin, respectively—and in phenylketonuria of this type, these precursors must be supplied in the diet. Supplementing the diet with tetrahydrobiopterin itself is ineffective because it is unstable and does not cross the blood-brain barrier.

FIGURE 18–25 Alternative pathways for catabolism of phenylalanine in phenylketonuria. In PKU, phenylpyruvate accumulates in the tissues, blood, and urine. The urine may also contain phenylacetate and phenyllactate.

Screening newborns for genetic diseases can be highly cost-effective, especially in the case of PKU. The tests (no longer relying on urine odor) are relatively inexpensive, and the detection and early treatment of PKU in infants (eight to ten cases per 100,000 newborns) saves millions of dollars in later health care costs each year. More importantly, the emotional trauma avoided by early detection with these simple tests is inestimable.

Another inheritable disease of phenylalanine catabolism is **alkaptonuria**, in which the defective enzyme is **homogentisate dioxygenase** (Fig. 18–23). Less serious than PKU, this condition produces few ill effects, although large amounts of homogentisate are excreted and its oxidation turns the urine black. Individuals with alkaptonuria are also prone to develop a form of arthritis. Alkaptonuria is of considerable historical interest. Archibald Garrod discovered in the early 1900s that this condition is inherited, and he traced the cause to the absence of a single enzyme. Garrod was the first to make a connection between an inheritable trait and an enzyme—a great advance on the path that ultimately led to our current understanding of genes and the information pathways described in Part III. ■

Five Amino Acids Are Converted to α-Ketoglutarate

The carbon skeletons of five amino acids (proline, glutamate, glutamine, arginine, and histidine) enter the citric acid cycle as α-ketoglutarate **(Fig. 18–26)**. **Proline**, **glutamate**, and **glutamine** have five-carbon skeletons. The cyclic structure of proline is opened by oxidation of the carbon most distant from the carboxyl group to create a Schiff base, then hydrolysis of the Schiff base to a linear semialdehyde, glutamate γ-semialdehyde. This intermediate is further oxidized at the same carbon to produce glutamate. The action of glutaminase, or any of several enzyme reactions in which glutamine donates its amide nitrogen to an acceptor, converts glutamine to glutamate. Transamination or deamination of glutamate produces α-ketoglutarate.

Arginine and **histidine** contain five adjacent carbons and a sixth carbon attached through a nitrogen atom. The catabolic conversion of these amino acids to glutamate is therefore slightly more complex than the path from proline or glutamine (Fig. 18–26). Arginine is

FIGURE 18–26 Catabolic pathways for arginine, histidine, glutamate, glutamine, and proline. These amino acids are converted to α-ketoglutarate. The numbered steps in the histidine pathway are catalyzed by ❶ histidine ammonia lyase, ❷ urocanate hydratase, ❸ imidazolonepropionase, and ❹ glutamate formimino transferase.

converted to the five-carbon skeleton of ornithine in the urea cycle (Fig. 18–10), and the ornithine is transaminated to glutamate γ-semialdehyde. Conversion of histidine to the five-carbon glutamate occurs in a multistep pathway; the extra carbon is removed in a step that uses tetrahydrofolate as cofactor.

Four Amino Acids Are Converted to Succinyl-CoA

The carbon skeletons of methionine, isoleucine, threonine, and valine are degraded by pathways that yield succinyl-CoA **(Fig. 18–27)**, an intermediate of the citric acid cycle. **Methionine** donates its methyl group to one of several possible acceptors through S-adenosylmethionine, and three of its four remaining carbon atoms are converted to the propionate of propionyl-CoA, a precursor of succinyl-CoA. **Isoleucine** undergoes transamination, followed by oxidative decarboxylation of the resulting α-keto acid. The remaining five-carbon skeleton is further oxidized to acetyl-CoA and propionyl-CoA. **Valine**

undergoes transamination and decarboxylation, then a series of oxidation reactions that convert the remaining four carbons to propionyl-CoA. Some parts of the valine and isoleucine degradative pathways closely parallel steps in fatty acid degradation (see Fig. 17–9). In human tissues, **threonine** is also converted in two steps to propionyl-CoA. This is the primary pathway for threonine degradation in humans (see Fig. 18–19 for the alternative pathway). The mechanism of the first step is analogous to that catalyzed by serine dehydratase, and the serine and threonine dehydratases may actually be the same enzyme.

The propionyl-CoA derived from these three amino acids is converted to succinyl-CoA by a pathway described in Chapter 17: carboxylation to methylmalonyl-CoA, epimerization of the methylmalonyl-CoA, and conversion to succinyl-CoA by the coenzyme B_{12}–dependent methylmalonyl-CoA mutase (see Fig. 17–12). In the rare genetic disease known as methylmalonic acidemia, methylmalonyl-CoA mutase is lacking—with serious metabolic consequences (Table 18–2; Box 18–2).

FIGURE 18–27 Catabolic pathways for methionine, isoleucine, threonine, and valine. These amino acids are converted to succinyl-CoA; isoleucine also contributes two of its carbon atoms to acetyl-CoA (see Fig. 18–21). The pathway of threonine degradation shown here occurs in humans; a pathway found in other organisms is shown in Figure 18–19. The route from methionine to homocysteine is described in more detail in Figure 18–18; the conversion of homocysteine to α-ketobutyrate in Figure 22–16; and the conversion of propionyl-CoA to succinyl-CoA in Figure 17–12.

Branched-Chain Amino Acids Are Not Degraded in the Liver

Although much of the catabolism of amino acids takes place in the liver, the three amino acids with branched side chains (leucine, isoleucine, and valine) are oxidized as fuels primarily in muscle, adipose, kidney, and brain tissue. These extrahepatic tissues contain an aminotransferase, absent in liver, that acts on all three branched-chain amino acids to produce the corresponding α-keto acids **(Fig. 18–28)**. The **branched-chain α-keto acid dehydrogenase complex** then catalyzes oxidative decarboxylation of all three α-keto acids, in each case releasing the carboxyl group as CO_2 and producing the acyl-CoA derivative. This reaction is formally analogous to two other oxidative decarboxylations encountered in Chapter 16: oxidation of pyruvate to acetyl-CoA by the pyruvate dehydrogenase complex (see Fig. 16–6) and oxidation of α-ketoglutarate to succinyl-CoA by the α-ketoglutarate dehydrogenase complex (p. 644). In fact, all three enzyme complexes are similar in structure and share essentially the same reaction mechanism. Five cofactors (thiamine pyrophosphate, FAD, NAD, lipoate, and coenzyme A) participate, and the three proteins in each complex catalyze homologous reactions. This is clearly a case in which enzymatic machinery that evolved to catalyze one reaction was "borrowed" by gene duplication and further evolved to catalyze similar reactions in other pathways.

Experiments with rats have shown that the branched-chain α-keto acid dehydrogenase complex is regulated by covalent modification in response to the content of branched-chain amino acids in the diet. With little or no excess dietary intake of branched-chain amino acids, the enzyme complex is phosphorylated and thereby inactivated by a protein kinase. Addition of excess branched-chain amino acids to the diet results in dephosphorylation and consequent activation of the enzyme. Recall that the pyruvate dehydrogenase complex is subject to similar regulation by phosphorylation and dephosphorylation (p. 654).

There is a relatively rare genetic disease in which the three branched-chain α-keto acids (as well as their precursor amino acids, especially leucine) accumulate in the blood and "spill over" into the urine. This condition, called **maple syrup urine disease** because of the characteristic odor imparted to the urine by the α-keto acids, results from a defective branched-chain α-keto acid dehydrogenase complex. Untreated, the disease results in abnormal development of the brain, mental retardation, and death in early infancy. Treatment entails rigid control of the diet, limiting the intake of valine, isoleucine, and leucine to the minimum required to permit normal growth. ■

FIGURE 18–28 Catabolic pathways for the three branched-chain amino acids: valine, isoleucine, and leucine. All three pathways occur in extrahepatic tissues and share the first two enzymes, as shown here. The branched-chain α-keto acid dehydrogenase complex is analogous to the pyruvate and α-ketoglutarate dehydrogenase complexes and requires the same five cofactors (some not shown here). This enzyme is defective in people with maple syrup urine disease.

BOX 18–2 🜋 MEDICINE   Scientific Sleuths Solve a Murder Mystery

Truth can sometimes be stranger than fiction—or at least as strange as a made-for-TV movie. Take, for example, the case of Patricia Stallings. Convicted of the murder of her infant son, she was sentenced to life in prison—but was later found innocent, thanks to the medical sleuthing of three persistent researchers.

The story began in the summer of 1989 when Stallings brought her three-month-old son, Ryan, to the emergency room of Cardinal Glennon Children's Hospital in St. Louis. The child had labored breathing, uncontrollable vomiting, and gastric distress. According to the attending physician, a toxicologist, the child's symptoms indicated that he had been poisoned with ethylene glycol, an ingredient of antifreeze, a conclusion apparently confirmed by analysis at a commercial lab.

After he recovered, the child was placed in a foster home, and Stallings and her husband, David, were allowed to see him in supervised visits. But when the infant became ill, and subsequently died, after a visit in which Stallings had been briefly left alone with him, she was charged with first-degree murder and held without bail. At the time, the evidence seemed compelling, as both the commercial lab and the hospital lab found large amounts of ethylene glycol in the boy's blood and traces of it in a bottle of milk Stallings had fed her son during the visit.

But without knowing it, Stallings had performed a brilliant experiment. While in custody, she learned she was pregnant; she subsequently gave birth to another son, David Stallings Jr., in February 1990. He was placed immediately in a foster home, but within

two weeks he started having symptoms similar to Ryan's. David was eventually diagnosed with a rare metabolic disorder called methylmalonic acidemia (MMA). A recessive genetic disorder of amino acid metabolism, MMA affects about 1 in 48,000 newborns and presents symptoms almost identical with those caused by ethylene glycol poisoning.

Stallings couldn't possibly have poisoned her second son, but the Missouri state prosecutor's office was not impressed by the new developments and pressed forward with her trial anyway. The court wouldn't allow the MMA diagnosis of the second child to be introduced as evidence, and in January 1991 Patricia Stallings was convicted of assault with a deadly weapon and sentenced to life in prison.

Fortunately for Stallings, however, William Sly, chairman of the Department of Biochemistry and Molecular Biology at St. Louis University, and James Shoemaker, head of a metabolic screening lab at the university, got interested in her case when they heard about it from a television broadcast. Shoemaker performed his own analysis of Ryan's blood and didn't detect ethylene glycol. He and Sly then contacted Piero Rinaldo, a metabolic disease expert at Yale University School of Medicine whose lab is equipped to diagnose MMA from blood samples.

When Rinaldo analyzed Ryan's blood serum, he found high concentrations of methylmalonic acid, a breakdown product of the branched-chain amino acids isoleucine and valine, which accumulates in MMA patients because the enzyme that should convert it to the next product in the metabolic pathway is defective (Fig. 1).

Asparagine and Aspartate Are Degraded to Oxaloacetate

The carbon skeletons of **asparagine** and **aspartate** ultimately enter the citric acid cycle as malate in mammals or oxaloacetate in bacteria. The enzyme **asparaginase** catalyzes the hydrolysis of asparagine to aspartate, which undergoes transamination with α-ketoglutarate to yield glutamate and oxaloacetate (**Fig. 18–29**). The oxaloacetate is converted to malate in the cytosol and then transported into the mitochondrial matrix through the malate–α-ketoglutarate transporter. In bacteria, the oxaloacetate produced in the transamination reaction can be used directly in the citric acid cycle.

We have now seen how the 20 common amino acids, after losing their nitrogen atoms, are degraded by dehydrogenation, decarboxylation, and other reactions to yield portions of their carbon backbones in the form of six central metabolites that can enter the citric acid cycle. Those portions degraded to acetyl-CoA are completely oxidized to carbon dioxide and water, with generation of ATP by oxidative phosphorylation.

FIGURE 18–29 Catabolic pathway for asparagine and aspartate. Both amino acids are converted to oxaloacetate.

Methylmalonic acid accumulates
in tissues, blood, urine

FIGURE 1 Children with a mutation (red X) that inactivates the enzyme methylmalony-CoA mutase cannot degrade isoleucine, methionine, threonine, and valine normally. Instead, a potentially fatal accumulation of methylmalonic acid occurs, with symptoms similar to those of ethylene glycol poisoning.

And particularly telling, he says, the child's blood and urine contained massive amounts of ketones, another metabolic consequence of the disease. Like Shoemaker, he did not find any ethylene glycol in a sample of the baby's bodily fluids. The bottle couldn't be tested, since it had mysteriously disappeared. Rinaldo's analyses convinced him that Ryan had died from MMA, but how to account for the results from two labs, indicating that the boy had ethylene glycol in his blood? Could they both be wrong?

When Rinaldo obtained the lab reports, what he saw was, he says, "scary." One lab said that Ryan Stallings' blood contained ethylene glycol, even though the blood sample analysis did not match the lab's own profile for a known sample containing ethylene glycol. "This was not just a matter of questionable interpretation. The quality of their analysis was unacceptable," Rinaldo says. And the second laboratory? According to Rinaldo, that lab detected an abnormal component in Ryan's blood and just "assumed it was ethylene glycol." Samples from the bottle had produced nothing unusual, says Rinaldo, yet the lab claimed evidence of ethylene glycol in that, too.

Rinaldo presented his findings to the case's prosecutor, George McElroy, who called a press conference the very next day. "I no longer believe the laboratory data," he told reporters. Having concluded that Ryan Stallings had died of MMA after all, McElroy dismissed all charges against Patricia Stallings on September 20, 1991.

By Michelle Hoffman (1991). *Science* **253**, 931. Copyright 1991 by the American Association for the Advancement of Science.

As was the case for carbohydrates and lipids, the degradation of amino acids results ultimately in the generation of reducing equivalents (NADH and $FADH_2$) through the action of the citric acid cycle. Our survey of catabolic processes concludes in the next chapter with a discussion of respiration, in which these reducing equivalents fuel the ultimate oxidative and energy-generating process in aerobic organisms.

SUMMARY 18.3 Pathways of Amino Acid Degradation

▶ After the removal of amino groups, the carbon skeletons of amino acids undergo oxidation to compounds that can enter the citric acid cycle for oxidation to CO_2 and H_2O. The reactions of these pathways require several cofactors, including tetrahydrofolate and S-adenosylmethionine in one-carbon transfer reactions and tetrahydrobiopterin in the oxidation of phenylalanine by phenylalanine hydroxylase.

▶ Depending on their degradative end product, some amino acids can be converted to ketone bodies, some to glucose, and some to both. Thus amino acid degradation is integrated into intermediary metabolism and can be critical to survival under conditions in which amino acids are a significant source of metabolic energy.

▶ The carbon skeletons of amino acids enter the citric acid cycle through five intermediates: acetyl-CoA, α-ketoglutarate, succinyl-CoA, fumarate, and oxaloacetate. Some are also degraded to pyruvate, which can be converted to either acetyl-CoA or oxaloacetate.

▶ The amino acids producing pyruvate are alanine, cysteine, glycine, serine, threonine, and tryptophan. Leucine, lysine, phenylalanine, and tryptophan yield acetyl-CoA via acetoacetyl-CoA. Isoleucine, leucine, threonine, and tryptophan also form acetyl-CoA directly.

▶ Arginine, glutamate, glutamine, histidine, and proline produce α-ketoglutarate; isoleucine, methionine, threonine, and valine produce succinyl-CoA; four carbon atoms of phenylalanine and tyrosine give rise to fumarate; and asparagine and aspartate produce oxaloacetate.

▶ The branched-chain amino acids (isoleucine, leucine, and valine), unlike the other amino acids, are degraded only in extrahepatic tissues.

▶ Several serious human diseases can be traced to genetic defects in the enzymes of amino acid catabolism.

Key Terms

Terms in bold are defined in the glossary.

aminotransferases 699
transaminases 699
transamination 699
pyridoxal phosphate (PLP) 699
oxidative deamination 700
L-glutamate dehydrogenase 700
glutamine synthetase 702
glutaminase 703
glucose-alanine cycle 703
ammonotelic 704
ureotelic 704
uricotelic 704
urea cycle 704
urea 704

creatine kinase 708
essential amino acids 709
ketogenic 711
glucogenic 711
tetrahydrofolate 712
S-adenosylmethionine **(adoMet)** 712
tetrahydrobiopterin 714
phenylketonuria (PKU) 719
mixed-function oxidases 720
alkaptonuria 721
maple syrup urine disease 723

Further Reading

General

Amon, J., Titgemeyer, F., & Burkovski, A. (2010) Common patterns—unique features: nitrogen metabolism and regulation in Gram-positive bacteria. *FEMS Microbiol. Rev.* **34**, 588–605.

Arias, I.M., Alter, H.J., Boyer, J.L., Cohen, D.E., Fausto, N., Shafritz, D.A., & Wolkof, A.W. (2009) *The Liver: Biology and Pathobiology*, 5th edn, John Wiley & Sons, Hoboken, New Jersey.

Brosnan, J.T. (2001) Amino acids, then and now—a reflection on Sir Hans Krebs' contribution to nitrogen metabolism. *IUBMB Life* **52**, 265–270.

An interesting tour through the life of this important biochemist.

Frey, P.A. & Hegeman, A.D. (2006) *Enzymatic Reaction Mechanisms*, Oxford University Press, New York.

A good source for in-depth discussion of the classes of enzymatic reaction mechanisms described in the chapter.

Amino Acid Metabolism

Christen, P. & Metzler, D.E. (1985) *Transaminases*, Wiley-Interscience, Inc., New York.

Curthoys, N.P. & Watford, M. (1995) Regulation of glutaminase activity and glutamine metabolism. *Annu. Rev. Nutr.* **15**, 133–159.

Eliot, A.C. & Kirsch, J.F. (2004) Pyridoxal phosphate enzymes: mechanistic, structural and evolutionary considerations. *Annu. Rev. Biochem.* **73**, 383–415.

Fitzpatrick, P.F. (1999) Tetrahydropterin-dependent amino acid hydroxylases. *Annu. Rev. Biochem.* **68**, 355–382.

Kalhan, S.C. & Bier, D.M. (2008) Protein and amino acid metabolism in the human newborn. *Annu. Rev. Nutr.* **28**, 389–410.

Pencharz, P.B. & Ball, R.O. (2003) Different approaches to define individual amino acid requirements. *Annu. Rev. Nutr.* **23**, 101–116.

Determination of which amino acids are essential in the human diet is not a trivial problem, as this review relates.

The Urea Cycle

Braissant, O. (2010) Current concepts in the pathogenesis of urea cycle diosorders. *Mol. Genet. Metab.* **100**, S3–S12.

Brusilow, S.W. & Horwich, A.L. (2006) Urea cycle enzymes. In *Scriver's Online Metabolic and Molecular Bases of Inherited Disease* (Valle, D., Beaudet, A.L., Vogelstein, B., Kinzler, K.W., Antonarakis, S.E., Ballabio A., eds), http:// dx.doi.org/10.10 36/ommbid, 108.

An authoritative source on this pathway.

Enns, G.M., Berry, S.A., Berry, G.T., Rhead, W.J., Brusilow, S.W., & Hamosh, A. (2007) Survival after treatment with phenylacetate and benzoate for urea-cycle disorders. *N. Engl. J. Med.* **356**, 2282–2292.

Kresge, N., Simoni R.D., & Hill, R.L. (2005) Four decades of research on the biosynthesis of urea: the work of Sarah Ratner. *J. Biol. Chem.* **280**, e34.

Morris, S.M. (2002) Regulation of enzymes of the urea cycle and arginine metabolism. *Annu. Rev. Nutr.* **22**, 87–105.

This review details what is known about some levels of regulation not covered in the chapter, such as hormonal and nutritional regulation.

Simoni, R.D., Hill, R.L., Vaughan, M., & Tabor, H. (2004) Transaminases: the work of Philip P. Cohen. *J. Biol. Chem.* **279**, e1.

Disorders of Amino Acid Degradation

Ledley, F.D., Levy, H.L., & Woo, S.L.C. (1986) Molecular analysis of the inheritance of phenylketonuria and mild hyperphenylalaninemia in families with both disorders. *N. Engl. J. Med.* **314**, 1276–1280.

Scriver, C.R., Kaufman, S., & Woo, S.L.C. (1988) Mendelian hyperphenylalaninemia. *Annu. Rev. Genet.* **22**, 301–321.

Valle, D., Beaudet, A.L., Vogelstein, B., Kinzler, K.W., Antonarakis, S.E., & Ballabio, A. (eds) (2006, updated 2011) *Scriver's Online Metabolic and Molecular Bases of Inherited Disease*, Part 8: Amino Acids, www.ommbid.com.

Werner, E.R., Blau, N., & Thöny, B. (2011) Tetrahydrobiopterin: biochemistry and pathophysiology. *Biochem. J.* **438**, 397–414.

Problems

1. Products of Amino Acid Transamination Name and draw the structure of the α-keto acid resulting when each of the following amino acids undergoes transamination with α-ketoglutarate: (a) aspartate, (b) glutamate, (c) alanine, (d) phenylalanine.

2. Measurement of Alanine Aminotransferase Activity The activity (reaction rate) of alanine aminotransferase is usually measured by including an excess of pure lactate dehydrogenase and NADH in the reaction system. The rate of alanine disappearance is equal to the rate of NADH disappearance measured spectrophotometrically. Explain how this assay works.

3. Alanine and Glutamine in the Blood Normal human blood plasma contains all the amino acids required for the synthesis of body proteins, but not in equal concentrations.

Alanine and glutamine are present in much higher concentrations than any other amino acids. Suggest why.

4. Distribution of Amino Nitrogen If your diet is rich in alanine but deficient in aspartate, will you show signs of aspartate deficiency? Explain.

5. Lactate versus Alanine as Metabolic Fuel: The Cost of Nitrogen Removal The three carbons in lactate and alanine have identical oxidation states, and animals can use either carbon source as a metabolic fuel. Compare the net ATP yield (moles of ATP per mole of substrate) for the complete oxidation (to CO_2 and H_2O) of lactate versus alanine when the cost of nitrogen excretion as urea is included.

$$
\begin{array}{cc}
COO^- & COO^- \\
| & | \\
HO-C-H & H_3\overset{+}{N}-C-H \\
| & | \\
H-C-H & H-C-H \\
| & | \\
H & H \\
\text{Lactate} & \text{Alanine}
\end{array}
$$

6. Ammonia Toxicity Resulting from an Arginine-Deficient Diet In a study conducted some years ago, cats were fasted overnight then given a single meal complete in all amino acids except arginine. Within 2 hours, blood ammonia levels increased from a normal level of 18 μg/L to 140 μg/L, and the cats showed the clinical symptoms of ammonia toxicity. A control group fed a complete amino acid diet or an amino acid diet in which arginine was replaced by ornithine showed no unusual clinical symptoms.

(a) What was the role of fasting in the experiment?

(b) What caused the ammonia levels to rise in the experimental group? Why did the absence of arginine lead to ammonia toxicity? Is arginine an essential amino acid in cats? Why or why not?

(c) Why can ornithine be substituted for arginine?

7. Oxidation of Glutamate Write a series of balanced equations, and an overall equation for the net reaction, describing the oxidation of 2 mol of glutamate to 2 mol of α-ketoglutarate and 1 mol of urea.

8. Transamination and the Urea Cycle Aspartate aminotransferase has the highest activity of all the mammalian liver aminotransferases. Why?

9. The Case against the Liquid Protein Diet A weight-reducing diet heavily promoted some years ago required the daily intake of "liquid protein" (soup of hydrolyzed gelatin), water, and an assortment of vitamins. All other food and drink were to be avoided. People on this diet typically lost 10 to 14 lb in the first week.

(a) Opponents argued that the weight loss was almost entirely due to water loss and would be regained very soon after a normal diet was resumed. What is the biochemical basis for this argument?

(b) A few people on this diet died. What are some of the dangers inherent in the diet, and how can they lead to death?

10. Ketogenic Amino Acids Which amino acids are exclusively ketogenic?

11. A Genetic Defect in Amino Acid Metabolism: A Case History A two-year-old child was taken to the hospital. His mother said that he vomited frequently, especially after feedings. The child's weight and physical development were below normal. His hair, although dark, contained patches of white. A urine sample treated with ferric chloride ($FeCl_3$) gave a green color characteristic of the presence of phenylpyruvate. Quantitative analysis of urine samples gave the results shown in the table.

| | Concentration (mM) | |
Substance	Patient's urine	Normal urine
Phenylalanine	7.0	0.01
Phenylpyruvate	4.8	0
Phenyllactate	10.3	0

(a) Suggest which enzyme might be deficient in this child. Propose a treatment.

(b) Why does phenylalanine appear in the urine in large amounts?

(c) What is the source of phenylpyruvate and phenyllactate? Why does this pathway (normally not functional) come into play when the concentration of phenylalanine rises?

(d) Why does the boy's hair contain patches of white?

12. Role of Cobalamin in Amino Acid Catabolism Pernicious anemia is caused by impaired absorption of vitamin B_{12}. What is the effect of this impairment on the catabolism of amino acids? Are all amino acids equally affected? (Hint: See Box 17–2.)

13. Vegetarian Diets Vegetarian diets can provide high levels of antioxidants and a lipid profile that can help prevent coronary disease. However, there can be some associated problems. Blood samples were taken from a large group of volunteer subjects who were vegans (strict vegetarians: no animal products), lactovegetarians (vegetarians who eat dairy products), or omnivores (individuals with a normal, varied diet including meat). In each case, the volunteers had followed the diet for several years. The blood levels of both homocysteine and methylmalonate were elevated in the vegan group, somewhat lower in the lactovegetarian group, and much lower in the omnivore group. Explain.

14. Pernicious Anemia Vitamin B_{12} deficiency can arise from a few rare genetic diseases that lead to low B_{12} levels despite a normal diet that includes B_{12}-rich meat and dairy sources. These conditions cannot be treated with dietary B_{12} supplements. Explain.

15. Pyridoxal Phosphate Reaction Mechanisms Threonine can be broken down by the enzyme threonine dehydratase, which catalyzes the conversion of threonine to α-ketobutyrate and ammonia. The enzyme uses PLP as a cofactor. Suggest a mechanism for this reaction, based on the mechanisms in Figure 18–6. Note that this reaction includes an elimination at the β carbon of threonine.

$$CH_3-\overset{OH}{\underset{}{CH}}-\overset{\overset{+}{NH_3}}{\underset{}{CH}}-COO^- \xrightarrow[\substack{\text{threonine} \\ \text{dehydratase}}]{\text{PLP}}$$

Threonine

$$CH_3-CH_2-\overset{O}{\overset{\|}{C}}-COO^- + NH_3 + H_2O$$

α-Ketobutyrate

16. Pathway of Carbon and Nitrogen in Glutamate Metabolism When [2-^{14}C,^{15}N] glutamate undergoes oxidative degradation in the liver of a rat, in which atoms of the following metabolites will each isotope be found: (a) urea, (b) succinate, (c) arginine, (d) citrulline, (e) ornithine, (f) aspartate?

$$\begin{array}{c} H \quad COO^- \\ H-\overset{15}{\underset{}{N}}\overset{+}{}-\overset{14}{\underset{}{C}}-H \\ H \quad CH_2 \\ CH_2 \\ COO^- \end{array}$$

Labeled glutamate

17. Chemical Strategy of Isoleucine Catabolism Isoleucine is degraded in six steps to propionyl-CoA and acetyl-CoA.

$$\begin{array}{c} O \quad O^- \\ \overset{\|}{C} \\ H_3\overset{+}{N}-C-H \\ H-C-CH_3 \\ CH_2 \\ CH_3 \end{array} \xrightarrow{\text{6 steps}} \begin{array}{c} O \quad S\text{-CoA} \\ \overset{\|}{C} \\ CH_2 \\ CH_3 \quad \text{Propionyl-CoA} \\ + \\ O \quad S\text{-CoA} \\ \overset{\|}{C} \\ CH_3 \quad \text{Acetyl-CoA} \end{array}$$

Isoleucine

(a) The chemical process of isoleucine degradation includes strategies analogous to those used in the citric acid cycle and the β oxidation of fatty acids. The intermediates of isoleucine degradation (I to V) shown below are not in the proper order. Use your knowledge and understanding of the citric acid cycle and β-oxidation pathway to arrange the intermediates in the proper metabolic sequence for isoleucine degradation.

I
$$\begin{array}{c} O \quad S\text{-CoA} \\ \overset{\|}{C} \\ C-CH_3 \\ H-C \\ CH_3 \end{array}$$

II
$$\begin{array}{c} O \quad O^- \\ \overset{\|}{C} \\ C=O \\ H-C-CH_3 \\ CH_2 \\ CH_3 \end{array}$$

III
$$\begin{array}{c} O \quad S\text{-CoA} \\ \overset{\|}{C} \\ H-C-CH_3 \\ C=O \\ CH_3 \end{array}$$

IV
$$\begin{array}{c} O \quad S\text{-CoA} \\ \overset{\|}{C} \\ H-C-CH_3 \\ CH_2 \\ CH_3 \end{array}$$

V
$$\begin{array}{c} O \quad S\text{-CoA} \\ \overset{\|}{C}\;H \\ H\;\;C \\ HO-C \quad CH_3 \\ CH_3 \end{array}$$

(b) For each step you propose, describe the chemical process, provide an analogous example from the citric acid cycle or β-oxidation pathway (where possible), and indicate any necessary cofactors.

18. Role of Pyridoxal Phosphate in Glycine Metabolism The enzyme serine hydroxymethyltransferase requires pyridoxal phosphate as cofactor. Propose a mechanism for the reaction catalyzed by this enzyme, in the direction of serine degradation (glycine production). (Hint: See Figs 18–19 and 18–20b.)

19. Parallel Pathways for Amino Acid and Fatty Acid Degradation The carbon skeleton of leucine is degraded by a series of reactions closely analogous to those of the citric acid cycle and β oxidation. For each reaction, (a) through (f), shown below, indicate its type, provide an analogous example from the citric acid cycle or β-oxidation pathway (where possible), and note any necessary cofactors.

$$\begin{array}{c} H \quad\quad \overset{+}{NH_3} \\ CH_3-\overset{}{\underset{}{C}}-CH_2-\overset{}{\underset{}{C}}-COO^- \quad \text{Leucine} \\ CH_3 \quad\quad H \end{array}$$

(a) ↓

$$\begin{array}{c} H \quad\quad O \\ CH_3-\overset{}{\underset{}{C}}-CH_2-\overset{\|}{C}-COO^- \quad \text{α-Ketoisocaproate} \\ CH_3 \end{array}$$

(b) ⟨ CoA-SH → CO$_2$

$$\begin{array}{c} H \quad\quad O \\ CH_3-\overset{}{\underset{}{C}}-CH_2-\overset{\|}{C}-S\text{-CoA} \quad \text{Isovaleryl-CoA} \\ CH_3 \end{array}$$

(c) ↓

$$\begin{array}{c} O \\ CH_3-C=C-\overset{\|}{C}-S\text{-CoA} \quad \text{β-Methylcrotonyl-CoA} \\ H_3C \quad H \end{array}$$

(d) ⟨ HCO$_3^-$

$$\begin{array}{c} O \\ {}^-OOC-CH_2-C=C-\overset{\|}{C}-S\text{-CoA} \quad \text{β-Methylglutaconyl-CoA} \\ H_3C \quad H \end{array}$$

(e) ⟨ H$_2$O

$$\begin{array}{c} OH \quad\quad O \\ {}^-OOC-CH_2-\overset{}{\underset{}{C}}-CH_2-\overset{\|}{C}-S\text{-CoA} \quad \text{β-Hydroxy-β-methylglutaryl-CoA} \\ CH_3 \end{array}$$

(f) ↓

$$\begin{array}{cc} O & O \\ {}^-OOC-CH_2-\overset{\|}{C}-CH_3 + CH_3-\overset{\|}{C}-S\text{-CoA} \end{array}$$

Acetoacetate Acetyl-CoA

Data Analysis Problem

20. Maple Syrup Urine Disease Figure 18–28 shows the pathway for the degradation of branched-chain amino acids and the site of the biochemical defect that causes maple syrup urine disease. The initial findings that eventually led to the discovery of the defect in this disease were presented in three papers published in the late 1950s and early 1960s. This problem traces the history of the findings from initial clinical observations to proposal of a biochemical mechanism.

Menkes, Hurst, and Craig (1954) presented the cases of four siblings, all of whom died following a similar course of symptoms. In all four cases, the mother's pregnancy and the birth had been normal. The first 3 to 5 days of each child's life were also normal. But soon thereafter each child began having convulsions, and the children died between the ages of 11 days and 3 months. Autopsy showed considerable swelling of the brain in all cases. The children's urine had a strong, unusual "maple syrup" odor, starting from about the third day of life.

Menkes (1959) reported data collected from six more children. All showed symptoms similar to those described above, and died within 15 days to 20 months of birth. In one case, Menkes was able to obtain urine samples during the last months of the infant's life. When he treated the urine with 2,4-dinitrophenylhydrazone, which forms colored precipitates with keto compounds, he found three α-keto acids in unusually large amounts:

α-Ketoisocaproate α-Ketoisovalerate α-Keto-β-methyl-n-valerate

(a) These α-keto acids are produced by the deamination of amino acids. For each of the α-keto acids above, draw and name the amino acid from which it was derived.

Dancis, Levitz, and Westall (1960) collected further data that led them to propose the biochemical defect shown in Figure 18–28. In one case, they examined a patient whose urine first showed the maple syrup odor when he was 4 months old. At the age of 10 months (March 1956), the child was admitted to the hospital because he had a fever, and he showed grossly retarded motor development. At the age of 20 months (January 1957), he was readmitted and was found to have the degenerative neurological symptoms seen in previous cases of

	Urine (mg/24 h)			Plasma (mg/mL)	
	Normal	**Patient**		**Normal**	**Patient**
Amino acid(s)		**Mar. 1956**	**Jan. 1957**		**Jan. 1957**
Alanine	5–15	0.2	0.4	3.0–4.8	0.6
Asparagine and glutamine	5–15	0.4	0	3.0–5.0	2.0
Aspartic acid	1–2	0.2	1.5	0.1–0.2	0.04
Arginine	1.5–3	0.3	0.7	0.8–1.4	0.8
Cystine	2–4	0.5	0.3	1.0–1.5	0
Glutamic acid	1.5–3	0.7	1.6	1.0–1.5	0.9
Glycine	20–40	4.6	20.7	1.0–2.0	1.5
Histidine	8–15	0.3	4.7	1.0–1.7	0.7
Isoleucine	2–5	2.0	13.5	0.8–1.5	2.2
Leucine	3–8	2.7	39.4	1.7–2.4	14.5
Lysine	2–12	1.6	4.3	1.5–2.7	1.1
Methionine	2–5	1.4	1.4	0.3–0.6	2.7
Ornithine	1–2	0	1.3	0.6–0.8	0.5
Phenylalanine	2–4	0.4	2.6	1.0–1.7	0.8
Proline	2–4	0.5	0.3	1.5–3.0	0.9
Serine	5–15	1.2	0	1.3–2.2	0.9
Taurine	1–10	0.2	18.7	0.9–1.8	0.4
Threonine	5–10	0.6	0	1.2–1.6	0.3
Tryptophan	3–8	0.9	2.3	Not measured	0
Tyrosine	4–8	0.3	3.7	1.5–2.3	0.7
Valine	2–4	1.6	15.4	2.0–3.0	13.1

maple syrup urine disease; he died soon after. Results of his blood and urine analyses are shown in the table on page 729, along with normal values for each component.

(b) The table includes taurine, an amino acid not normally found in proteins. Taurine is often produced as a byproduct of cell damage. Its structure is:

$$H_3\overset{+}{N}-CH_2-CH_2-\underset{\underset{O}{\|}}{\overset{\overset{O}{\|}}{S}}-O^-$$

Based on its structure and the information in this chapter, what is the most likely amino acid precursor of taurine? Explain your reasoning.

(c) Compared with the normal values given in the table, which amino acids showed significantly elevated levels in the patient's blood in January 1957? Which ones in the patient's urine?

Based on their results and their knowledge of the pathway shown in Figure 18–28, Dancis and coauthors concluded:

"although it appears most likely to the authors that the primary block is in the metabolic degradative pathway of the branched-chain amino acids, this cannot be considered established beyond question."

(d) How do the data presented here support this conclusion?

(e) Which data presented here do *not* fit this model of maple syrup urine disease? How do you explain these seemingly contradictory data?

(f) What data would you need to collect to be more secure in your conclusion?

References

Dancis, J., Levitz, M., & Westall, R. (1960) Maple syrup urine disease: branched-chain ketoaciduria. *Pediatrics* **25**, 72–79.

Menkes, J.H. (1959) Maple syrup disease: isolation and identification of organic acids in the urine. *Pediatrics* **23**, 348–353.

Menkes, J.H., Hurst, P.L., & Craig J.M. (1954) A new syndrome: progressive familial infantile cerebral dysfunction associated with an unusual urinary substance. *Pediatrics* **14**, 462–466.

Oxidative Phosphorylation and Photophosphorylation

Oxidative phosphorylation is the culmination of energy-yielding metabolism in aerobic organisms. All oxidative steps in the degradation of carbohydrates, fats, and amino acids converge at this final stage of cellular respiration, in which the energy of oxidation drives the synthesis of ATP. Photophosphorylation is the means by which photosynthetic organisms capture the energy of sunlight—the ultimate source of energy in the biosphere—and harness it to make ATP. Together, oxidative phosphorylation and photophosphorylation account for most of the ATP synthesized by most organisms most of the time. In eukaryotes, oxidative phosphorylation occurs in mitochondria, photophosphorylation in chloroplasts. The pathways to ATP synthesis in mitochondria and chloroplasts have challenged and fascinated biochemists

for more than half a century, and the fascination has grown with our deepening appreciation of these fundamental mechanisms in living organisms, their conservation in evolution, and their structural bases.

Our current understanding of ATP synthesis in mitochondria and chloroplasts is based on the hypothesis, introduced by Peter Mitchell in 1961, that transmembrane differences in proton concentration are the reservoir for the energy extracted from biological oxidation reactions. This **chemiosmotic theory** has been accepted as one of the great unifying principles of twentieth-century biology. It provides insight into the processes of oxidative phosphorylation and photophosphorylation, and into such apparently disparate energy transductions as active transport across membranes and the motion of bacterial flagella.

Oxidative phosphorylation and photophosphorylation are mechanistically similar in three respects (**Fig. 19–1**). (1) Both processes involve the flow of electrons through a chain of membrane-bound carriers. (2) The free energy made available by this "downhill" (exergonic) electron flow is coupled to the "uphill" transport of protons across a proton-impermeable membrane, conserving the free energy of fuel oxidation as a transmembrane electrochemical potential (p. 403). (3) The transmembrane flow of protons back down their concentration gradient through specific protein channels provides the free energy for synthesis of ATP, catalyzed by a membrane protein complex (ATP synthase) that couples proton flow to phosphorylation of ADP.

The chapter begins with mitochondrial oxidative phosphorylation. We first describe the components of the electron-transfer chain, their organization into large functional complexes in the inner mitochondrial membrane, the path of electron flow through them, and the proton movements that accompany this flow. We then consider the remarkable enzyme complex that, by "rota-

(a) Mitochondrion

❶ Reduced substrate (fuel) donates e^-.

❷ Electron carriers pump H^+ out as electrons flow to O_2.

❸ Energy of e^- flow stored as electrochemical potential.

❹ ATP synthase uses electrochemical potential to synthesize ATP.

(b) Chloroplast

❶ Light converts H_2O to a good e^- donor.

❷ Electron carriers pump H^+ in as electrons flow to $NADP^+$.

❸ Energy of e^- flow stored as electrochemical potential.

❹ ATP synthase uses electrochemical potential to synthesize ATP.

FIGURE 19–1 The chemiosmotic mechanism for ATP synthesis. (a) In mitochondria, electrons move through a chain of membrane-bound carriers (the respiratory chain) spontaneously, driven by the high reduction potential of oxygen and the relatively low reduction potentials of the various reduced substrates (fuels) that undergo oxidation in the mitochondrion. **(b)** In chloroplasts, the movement of electrons through a chain of membrane-bound carriers is driven by the energy of photons absorbed by the green pigment chlorophyll. In both organelles, electron flow creates an electrochemical potential by the transmembrane movement of protons and positive charge. In both cases this electrochemical potential drives ATP synthesis by a membrane-bound enzyme, ATP synthase, that is fundamentally similar in both mitochondria and chloroplasts, and in bacteria and archaea as well.

tional catalysis," captures the energy of proton flow in ATP, and the regulatory mechanisms that coordinate oxidative phosphorylation with the many catabolic pathways by which fuels are oxidized.

The metabolic role of mitochondria is so critical to cellular and organismal function that defects in mitochondrial function have very serious medical consequences. Mitochondria are central to neuronal and muscular function, and to the regulation of whole-body energy metabolism and body weight. Human neurodegenerative diseases, as well as cancer, diabetes, and obesity, are recognized as possible results of compromised mitochondrial function, and one theory of aging is based on gradual loss of mitochondrial integrity. ATP production is not the only important mitochondrial function; this organelle also acts in thermogenesis, steroid synthesis, and apoptosis (programmed cell death). The discovery of these diverse and important roles of mitochondria has stimulated much current research on the biochemistry of this organelle.

After discussing these various mitochondrial functions, we turn to photophosphorylation, looking first at the absorption of light by photosynthetic pigments, then at the light-driven flow of electrons from H_2O to $NADP^+$ and the molecular basis for coupling electron and proton flow. We also consider the similarities of structure and mechanism between the ATP synthases of chloroplasts and mitochondria, and the evolutionary basis for this conservation of mechanism. The remarkable ability of

chloroplasts to make ATP by oxidizing a compound of unlimited availability (water), while producing a compound essential to most animal life (oxygen), poses a set of challenges equally fascinating to biologist, biochemist, and chemist. Determination of the structures of supramolecular complexes that carry out these processes in chloroplasts has provided invaluable physical and chemical clues to understanding the process of photosynthesis.

OXIDATIVE PHOSPHORYLATION

19.1 Electron-Transfer Reactions in Mitochondria

The discovery in 1948 by Eugene Kennedy and Albert Lehninger that mitochondria are the site of oxidative phosphorylation in eukaryotes marked the beginning of the modern phase of studies in biological energy transductions. Mitochondria, like gramnegative bacteria, have two membranes **(Fig. 19–2a)**. The outer mitochondrial membrane is readily permeable to

Albert L. Lehninger, 1917–1986

(a)

ATP synthase (F_oF_1)

Cristae

Outer membrane
Freely permeable to small molecules and ions

Inner membrane
Impermeable to most small molecules and ions, including H^+
Contains:
- Respiratory electron carriers (Complexes I–IV)
- ADP-ATP translocase
- ATP synthase (F_oF_1)
- Other membrane transporters

Matrix
Contains:
- Pyruvate dehydrogenase complex
- Citric acid cycle enzymes
- Fatty acid β-oxidation enzymes
- Amino acid oxidation enzymes
- DNA, ribosomes
- Many other enzymes
- ATP, ADP, P_i, Mg^{2+}, Ca^{2+}, K^+
- Many soluble metabolic intermediates

Ribosomes

Porin channels

(b)

(c)

FIGURE 19–2 Biochemical anatomy of a mitochondrion. (a) The outer membrane has pores that make it permeable to small molecules and ions, but not to proteins. The convolutions (cristae) of the inner membrane provide a very large surface area. The inner membrane of a single liver mitochondrion may have more than 10,000 sets of electron-transfer systems (respiratory chains) and ATP synthase molecules, distributed over the membrane surface. **(b)** The mitochondria of heart muscle, which have more profuse cristae and thus a much larger area of inner membrane, contain more than three times as many sets of electron-transfer systems as **(c)** liver mitochondria. Muscle and liver mitochondria are about the size of a bacterium—1 to 2 μm long. The mitochondria of invertebrates, plants, and microbial eukaryotes are similar to those shown here, but with much variation in size, shape, and degree of convolution of the inner membrane.

small molecules (M_r <5,000) and ions, which move freely through transmembrane channels formed by a family of integral membrane proteins called porins. The inner membrane is impermeable to most small molecules and ions, including protons (H^+); the only species that cross this membrane do so through specific transporters. The inner membrane bears the components of the respiratory chain and the ATP synthase.

The mitochondrial matrix, enclosed by the inner membrane, contains the pyruvate dehydrogenase complex and the enzymes of the citric acid cycle, the fatty acid β-oxidation pathway, and the pathways of amino acid oxidation—all the pathways of fuel oxidation except glycolysis, which takes place in the cytosol. The selectively permeable inner membrane segregates the intermediates and enzymes of cytosolic metabolic path-

ways from those of metabolic processes occurring in the matrix. However, specific transporters carry pyruvate, fatty acids, and amino acids or their α-keto derivatives into the matrix for access to the machinery of the citric acid cycle. ADP and P_i are specifically transported into the matrix as newly synthesized ATP is transported out. The best current inventory of proteins in mammalian mitochondria lists about 1,100, at least 300 of which have unknown functions.

The bean-shaped representation of a mitochondrion in Figure 19–2 is an oversimplification, derived in part from early studies in which thin sections of cells were observed in the electron microscope. Three-dimensional images obtained either by reconstruction from serial sections or by confocal microscopy reveal greater variation in mitochondrial size and shape. In living cells stained with mitochondrion-specific fluorescent dyes, large numbers of variously shaped mitochondria are seen, clustered about the nucleus (see Fig. 19–42).

Tissues with a high demand for aerobic metabolism (brain, skeletal and heart muscle, and eye, for example) contain many hundreds or thousands of mitochondria per cell, and in general, mitochondria of cells with high metabolic activity have more, and more densely packed, cristae (Fig. 19–1b). In tissues with less-active metabolism (skin, for example) there are fewer mitochondria, each with fewer cristae. During cell growth and division, mitochondria divide by fission (like bacteria), and under some circumstances individual mitochondria fuse to form larger, more-extended structures.

Electrons Are Funneled to Universal Electron Acceptors

Oxidative phosphorylation begins with the entry of electrons into the chain of electron carriers called the **respiratory chain**. Most of these electrons arise from the action of dehydrogenases that collect electrons from catabolic pathways and funnel them into universal electron acceptors—nicotinamide nucleotides (NAD^+ or $NADP^+$) or flavin nucleotides (FMN or FAD) (see Figs 13–24, 13–27).

Nicotinamide nucleotide–linked dehydrogenases catalyze reversible reactions of the following general types:

$$\text{Reduced substrate} + NAD^+ \rightleftharpoons$$
$$\text{oxidized substrate} + NADH + H^+$$

$$\text{Reduced substrate} + NADP^+ \rightleftharpoons$$
$$\text{oxidized substrate} + NADPH + H^+$$

Most dehydrogenases that act in catabolism are specific for NAD^+ as electron acceptor (Table 19–1). Some are in the cytosol, others are in mitochondria, and still others have mitochondrial and cytosolic isozymes.

NAD-linked dehydrogenases remove two hydrogen atoms from their substrates. One of these is transferred as a hydride ion ($:H^-$) to NAD^+, the other is released as H^+ in the medium (see Fig. 13–24). NADH and NADPH are water-soluble electron carriers that associate *reversibly* with dehydrogenases. NADH carries electrons from catabolic reactions to their point of entry into the respiratory chain, the NADH dehydrogenase complex described below. NADPH generally supplies electrons to anabolic reactions. Cells maintain separate pools of NADPH and NADH, with different redox potentials. This is accomplished by holding the ratio of [reduced form]/[oxidized form] relatively high for NADPH and relatively low for NADH. Neither NADH nor NADPH can cross the inner mitochondrial membrane, but the electrons they carry can be shuttled across indirectly, as we shall see.

Flavoproteins contain a very tightly, sometimes covalently, bound flavin nucleotide, either FMN or FAD (see Fig. 13–27). The oxidized flavin nucleotide can accept either one electron (yielding the semiquinone form) or two (yielding $FADH_2$ or $FMNH_2$). Electron transfer occurs because the flavoprotein has a higher reduction potential than the compound oxidized. Recall that reduction potential is a quantitative measure of the relative tendency of a given chemical species to accept electrons in an oxidation-reduction reaction (p. 530). The standard reduction potential of a flavin nucleotide, unlike that of NAD or NADP, depends on the protein with which it is associated. Local interactions

TABLE 19–1 Some Important Reactions Catalyzed by NAD(P)H-Linked Dehydrogenases	
Reaction*	Location†
NAD-linked	
α-Ketoglutarate + CoA + NAD^+ $\rightleftharpoons$ succinyl-CoA + CO_2 + NADH + H^+	M
L-Malate + NAD^+ $\rightleftharpoons$ oxaloacetate + NADH + H^+	M and C
Pyruvate + CoA + NAD^+ $\rightleftharpoons$ acetyl-CoA + CO_2 + NADH + H^+	M
Glyceraldehyde 3-phosphate + P_i + NAD^+ $\rightleftharpoons$ 1,3-bisphosphoglycerate + NADH + H^+	C
Lactate + NAD^+ $\rightleftharpoons$ pyruvate + NADH + H^+	C
β-Hydroxyacyl-CoA + NAD^+ $\rightleftharpoons$ β-ketoacyl-CoA + NADH + H^+	M
NADP-linked	
Glucose 6-phosphate + $NADP^+$ $\rightleftharpoons$ 6-phosphogluconate + NADPH + H^+	C
L-Malate + $NADP^+$ $\rightleftharpoons$ pyruvate + CO_2 + NADPH + H^+	C
NAD- or NADP-linked	
L-Glutamate + H_2O + $NAD(P)^+$ $\rightleftharpoons$ α-ketoglutarate + NH_4^+ + NAD(P)H	M
Isocitrate + $NAD(P)^+$ $\rightleftharpoons$ α-ketoglutarate + CO_2 + NAD(P)H + H^+	M and C

*These reactions and their enzymes are discussed in Chapters 14 through 18.
†M designates mitochondria; C, cytosol.

with functional groups in the protein distort the electron orbitals in the flavin ring, changing the relative stabilities of oxidized and reduced forms. The relevant standard reduction potential is therefore that of the particular flavoprotein, not that of isolated FAD or FMN. The flavin nucleotide should be considered part of the flavoprotein's active site rather than a reactant or product in the electron-transfer reaction. Because flavoproteins can participate in either one- or two-electron transfers, they can serve as intermediates between reactions in which two electrons are donated (as in dehydrogenations) and those in which only one electron is accepted (as in the reduction of a quinone to a hydroquinone, described below).

Electrons Pass through a Series of Membrane-Bound Carriers

The mitochondrial respiratory chain consists of a series of sequentially acting electron carriers, most of which are integral proteins with prosthetic groups capable of accepting and donating either one or two electrons. Three types of electron transfers occur in oxidative phosphorylation: (1) direct transfer of electrons, as in the reduction of Fe^{3+} to Fe^{2+}, (2) transfer as a hydrogen atom ($H^+ + e^-$), and (3) transfer as a hydride ion ($:H^-$), which bears two electrons. The term **reducing equivalent** is used to designate a single electron equivalent transferred in an oxidation-reduction reaction.

In addition to NAD and flavoproteins, three other types of electron-carrying molecules function in the respiratory chain: a hydrophobic quinone (ubiquinone) and two different types of iron-containing proteins (cytochromes and iron-sulfur proteins). **Ubiquinone** (also called **coenzyme Q**, or simply **Q**) is a lipid-soluble benzoquinone with a long isoprenoid side chain **(Fig. 19–3)**. The closely related compounds plastoquinone (of plant chloroplasts) and menaquinone (of bacteria) play roles analogous to that of ubiquinone, carrying electrons in membrane-associated electron-transfer chains. Ubiquinone can accept one electron to become the semiquinone radical (·QH) or two electrons to form ubiquinol (QH_2) (Fig. 19–3) and, like flavoprotein carriers, it can act at the junction between a two-electron donor and a one-electron acceptor. Because ubiquinone is both small and hydrophobic, it is freely diffusible within the lipid bilayer of the inner mitochondrial membrane and can shuttle reducing equivalents between other, less mobile electron carriers in the membrane. And because it carries both electrons and protons, it plays a central role in coupling electron flow to proton movement.

The **cytochromes** are proteins with characteristic strong absorption of visible light, due to their iron-containing heme prosthetic groups **(Fig. 19–4a)**. Mitochondria contain three classes of cytochromes, designated a, b, and c, which are distinguished by differences in their light-absorption spectra. Each type of cytochrome in its reduced (Fe^{2+}) state has three absorption bands in the visible range (Fig. 19–4b). The longest-wavelength band

FIGURE 19–3 Ubiquinone (Q, or coenzyme Q). Complete reduction of ubiquinone requires two electrons and two protons, and occurs in two steps through the semiquinone radical intermediate.

is near 600 nm in type a cytochromes, near 560 nm in type b, and near 550 nm in type c. To distinguish among closely related cytochromes of one type, the exact absorption maximum is sometimes used in the names, as in cytochrome b_{562}.

The heme cofactors of a and b cytochromes are tightly, but not covalently, bound to their associated proteins; the hemes of c-type cytochromes are covalently attached through Cys residues (Fig. 19–4). As with the flavoproteins, the standard reduction potential of the heme iron atom of a cytochrome depends on its interaction with protein side chains and is therefore different for each cytochrome. The cytochromes of type a and b and some of type c are integral proteins of the inner mitochondrial membrane. One striking exception is the cytochrome c of mitochondria, a soluble protein that associates through electrostatic interactions with the outer surface of the inner membrane.

In **iron-sulfur proteins**, the iron is present not in heme but in association with inorganic sulfur atoms or with the sulfur atoms of Cys residues in the protein, or both. These iron-sulfur (Fe-S) centers range from simple structures with a single Fe atom coordinated to four Cys —SH groups to more complex Fe-S centers with two or four Fe atoms **(Fig. 19–5)**. **Rieske iron-sulfur proteins** (named after their discoverer, John S. Rieske) are a variation on this theme, in which one Fe atom is coordinated to two His residues rather than two Cys residues. All iron-sulfur proteins participate in one-electron transfers in which one iron atom of the iron-sulfur cluster is oxidized or reduced. At least eight Fe-S proteins function in mitochondrial electron transfer. The reduction potential

FIGURE 19–4 Prosthetic groups of cytochromes. (a) Each group consists of four five-membered, nitrogen-containing rings in a cyclic structure called a porphyrin. The four nitrogen atoms are coordinated with a central Fe ion, either Fe^{2+} or Fe^{3+}. Iron protoporphyrin IX is found in b-type cytochromes and in hemoglobin and myoglobin (see Fig. 4–17). Heme c is covalently bound to the protein of cytochrome c through thioether bonds to two Cys residues. Heme a, found in a-type cytochromes, has a long isoprenoid tail attached to one of the five-membered rings. The conjugated double-bond system (shaded light red) of the porphyrin ring has delocalized π electrons that are relatively easily excited by photons with the wavelengths of visible light, which accounts for the strong absorption by hemes (and related compounds) in the visible region of the spectrum. **(b)** Absorption spectra of cytochrome c (cyt c) in its oxidized (blue) and reduced (red) forms. The characteristic α, β, and γ bands of the reduced form are labeled.

of Fe-S proteins varies from -0.65 V to $+0.45$ V, depending on the microenvironment of the iron within the protein.

In the overall reaction catalyzed by the mitochondrial respiratory chain, electrons move from NADH, succinate, or some other primary electron donor through flavoproteins, ubiquinone, iron-sulfur proteins, and cytochromes, and finally to O_2. A look at the methods used to determine the sequence in which the carriers act is instructive, as the same general approaches have been used to study other electron-transfer chains, such as those of chloroplasts.

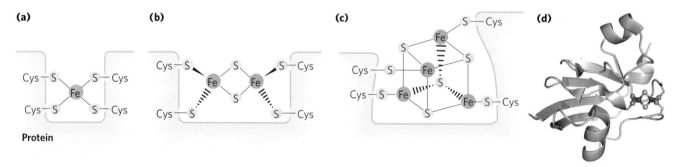

FIGURE 19–5 Iron-sulfur centers. The Fe-S centers of iron-sulfur proteins may be as simple as **(a)**, with a single Fe ion surrounded by the S atoms of four Cys residues. Other centers include both inorganic and Cys S atoms, as in **(b)** 2Fe-2S or **(c)** 4Fe-4S centers. **(d)** The ferredoxin of the cyanobacterium *Anabaena* 7120 has one 2Fe-2S center (PDB ID 1FRD); Fe is red, inorganic S is yellow, and the S of Cys is orange. (Note that in these designations only the inorganic S atoms are counted. For example, in the 2Fe-2S center (b), each Fe ion is actually surrounded by four S atoms.) The exact standard reduction potential of the iron in these centers depends on the type of center and its interaction with the associated protein.

TABLE 19–2 Standard Reduction Potentials of Respiratory Chain and Related Electron Carriers

Redox reaction (half-reaction)	E'° (V)
$2H^+ + 2e^- \longrightarrow H_2$	-0.414
$NAD^+ + H^+ + 2e^- \longrightarrow NADH$	-0.320
$NADP^+ + H^+ + 2e^- \longrightarrow NADPH$	-0.324
NADH dehydrogenase (FMN) $+ 2H^+ + 2e^- \longrightarrow$ NADH dehydrogenase (FMNH$_2$)	-0.30
Ubiquinone $+ 2H^+ + 2e^- \longrightarrow$ ubiquinol	0.045
Cytochrome b (Fe^{3+}) $+ e^- \longrightarrow$ cytochrome b (Fe^{2+})	0.077
Cytochrome c_1 (Fe^{3+}) $+ e^- \longrightarrow$ cytochrome c_1 (Fe^{2+})	0.22
Cytochrome c (Fe^{3+}) $+ e^- \longrightarrow$ cytochrome c (Fe^{2+})	0.254
Cytochrome a (Fe^{3+}) $+ e^- \longrightarrow$ cytochrome a (Fe^{2+})	0.29
Cytochrome a_3 (Fe^{3+}) $+ e^- \longrightarrow$ cytochrome a_3 (Fe^{2+})	0.35
$\frac{1}{2}O_2 + 2H^+ + 2e^- \longrightarrow H_2O$	0.8166

First, the standard reduction potentials of the individual electron carriers have been determined experimentally (Table 19–2). We would expect the carriers to function in order of increasing reduction potential, because electrons tend to flow spontaneously from carriers of lower E'° to carriers of higher E'°. The order of carriers deduced by this method is NADH $\rightarrow$ Q $\rightarrow$ cytochrome b $\rightarrow$ cytochrome c_1 $\rightarrow$ cytochrome c $\rightarrow$ cytochrome a $\rightarrow$ cytochrome a_3 $\rightarrow$ O$_2$. Note, however, that the order of standard reduction potentials is not necessarily the same as the order of *actual* reduction potentials under cellular conditions, which depend on the concentration of reduced and oxidized forms (see Eqn 13–5, p. 531). A second method for determining the sequence of electron carriers involves reducing the entire chain of carriers experimentally by providing an electron source but no electron acceptor (no O$_2$). When O$_2$ is suddenly introduced into the system, the rate at which each electron carrier becomes oxidized (measured spectroscopically) reveals the order in which the carriers function. The carrier nearest O$_2$ (at the end of the chain) gives up its

electrons first, the second carrier from the end is oxidized next, and so on. Such experiments have confirmed the sequence deduced from standard reduction potentials.

In a final confirmation, agents that inhibit the flow of electrons through the chain have been used in combination with measurements of the degree of oxidation of each carrier. In the presence of O$_2$ and an electron donor, carriers that function before the inhibited step become fully reduced, and those that function after this step are completely oxidized (**Fig. 19–6**). By using several inhibitors that block different steps in the chain, investigators have determined the entire sequence; it is the same as deduced in the first two approaches.

Electron Carriers Function in Multienzyme Complexes

The electron carriers of the respiratory chain are organized into membrane-embedded supramolecular complexes that can be physically separated. Gentle treatment of the inner mitochondrial membrane with detergents allows the resolution of four unique electron-carrier complexes, each capable of catalyzing

FIGURE 19–6 **Method for determining the sequence of electron carriers.** This method measures the effects of inhibitors of electron transfer on the oxidation state of each carrier. In the presence of an electron donor and O$_2$, each inhibitor causes a characteristic pattern of oxidized/reduced carriers: those before the block become reduced (blue), and those after the block become oxidized (light red).

TABLE 19-3 The Protein Components of the Mitochondrial Electron-Transfer Chain

Enzyme complex/protein	Mass (kDa)	Number of subunits*	Prosthetic group(s)
I NADH dehydrogenase	850	43 (14)	FMN, Fe-S
II Succinate dehydrogenase	140	4	FAD, Fe-S
III Ubiquinone:cytochrome c oxidoreductase	250	11	Hemes, Fe-S
Cytochrome $c^{\dagger}$	13	1	Heme
IV Cytochrome oxidase	160	13 (3–4)	Hemes; Cu_A, Cu_B

*Number of subunits in the bacterial equivalents in parentheses.

†Cytochrome c is not part of an enzyme complex; it moves between Complexes III and IV as a freely soluble protein.

electron transfer through a portion of the chain (Table 19–3; **Fig. 19–7**). Complexes I and II catalyze electron transfer to ubiquinone from two different electron donors: NADH (Complex I) and succinate (Complex II). Complex III carries electrons from reduced ubiquinone to cytochrome c, and Complex IV completes the sequence by transferring electrons from cytochrome c to O_2.

We now look in more detail at the structure and function of each complex of the mitochondrial respiratory chain.

Complex I: NADH to Ubiquinone **Figure 19–8** illustrates the relationship between Complexes I and II, and the enzymes of fatty acid β-oxidation, and ubiquinone. **Complex I**, also called **NADH:ubiquinone oxidoreductase** or **NADH dehydrogenase**, is a large enzyme composed of 42 different polypeptide chains, including an FMN-containing flavoprotein and at least six iron-sulfur centers. Complex I is L-shaped, with one arm of the L in the membrane and the other extending into the matrix. As shown in **Figure 19–9**, Complex I catalyzes two simultaneous and obligately coupled processes: (1) the exergonic transfer to ubiquinone of a hydride ion from NADH and a proton from the matrix, expressed by

$$NADH + H^+ + Q \longrightarrow NAD^+ + QH_2 \quad (19\text{–}1)$$

and (2) the endergonic transfer of four protons from the matrix to the intermembrane space. Complex I is therefore a proton pump driven by the energy of electron transfer, and the reaction it catalyzes is **vectorial**: it moves protons in a specific direction from one location (the matrix, which becomes negatively charged with the departure of protons) to another (the intermembrane space, which becomes positively charged). To emphasize the vectorial nature of the process, the overall reaction is often written with subscripts that indicate the location of the protons: P for the positive side of the inner membrane (the intermembrane space), N for the negative side (the matrix):

$$NADH + 5H_N^+ + Q \longrightarrow NAD^+ + QH_2 + 4H_P^+ \quad (19\text{–}2)$$

Amytal (a barbiturate drug), rotenone (a plant product commonly used as an insecticide), and piericidin A (an antibiotic) inhibit electron flow from the Fe-S centers of Complex I to ubiquinone (Table 19–4) and therefore block the overall process of oxidative phosphorylation.

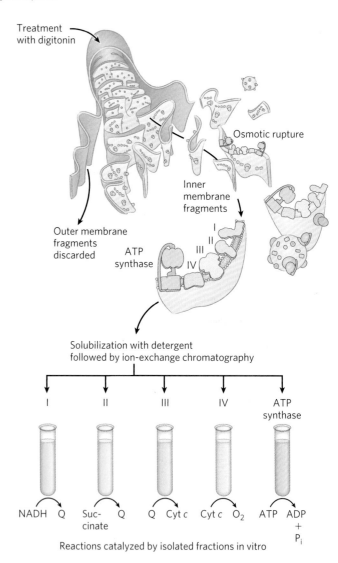

FIGURE 19-7 Separation of functional complexes of the respiratory chain. The outer mitochondrial membrane is first removed by treatment with the detergent digitonin. Fragments of inner membrane are then obtained by osmotic rupture of the mitochondria, and the fragments are gently dissolved in a second detergent. The resulting mixture of inner membrane proteins is resolved by ion-exchange chromatography into different complexes (I through IV) of the respiratory chain, each with its unique protein composition (see Table 19–3), and the enzyme ATP synthase (sometimes called Complex V). The isolated Complexes I through IV catalyze transfers between donors (NADH and succinate), intermediate carriers (Q and cytochrome c), and O_2, as shown. In vitro, isolated ATP synthase has only ATP-hydrolyzing (ATPase), not ATP-synthesizing, activity.

FIGURE 19–8 Path of electrons from NADH, succinate, fatty acyl-CoA, and glycerol 3-phosphate to ubiquinone. Ubiquinone (Q) is the point of entry for electrons derived from reactions in the cytosol, from fatty acid oxidation, and from succinate oxidation (in the citric acid cycle). Electrons from NADH pass through a flavoprotein with the cofactor FMN to a series of Fe-S centers (in Complex I) and then to Q. Electrons from succinate pass through a flavoprotein with the cofactor FAD and several Fe-S centers (in Complex II) on the way to Q. Glycerol 3-phosphate donates electrons to a flavoprotein (glycerol 3-phosphate dehydrogenase) on the outer face of the inner mitochondrial membrane, from which they pass to Q. Acyl-CoA dehydrogenase (the first enzyme of β oxidation) transfers electrons to electron-transferring flavoprotein (ETF), from which they pass to Q via ETF:ubiquinone oxidoreductase.

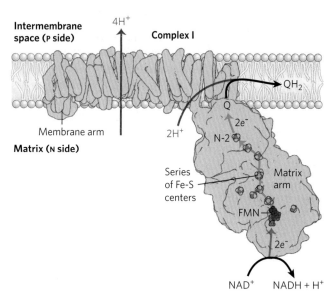

FIGURE 19–9 NADH:ubiquinone oxidoreductase (Complex I). (PDB ID 3M9S) Complex I (the crystal structure from the bacterium *Thermus thermophilus* is shown) catalyzes the transfer of a hydride ion from NADH to FMN, from which two electrons pass through a series of Fe-S centers to the Fe-S center N-2 in the matrix arm of the complex. Electron transfer from N-2 to ubiquinone on the membrane arm forms QH_2, which diffuses into the lipid bilayer. This electron transfer also drives the expulsion from the matrix of four protons per pair of electrons. The detailed mechanism that couples electron and proton transfer in Complex I is not yet known, but probably involves a Q cycle similar to that in Complex III in which QH_2 participates twice per electron pair (see Fig. 19–12). Proton flux produces an electrochemical potential across the inner mitochondrial membrane (N side negative, P side positive).

TABLE 19–4 Agents That Interfere with Oxidative Phosphorylation or Photophosphorylation

Type of interference	Compound*	Target/mode of action
Inhibition of electron transfer	Cyanide Carbon monoxide	Inhibit cytochrome oxidase
	Antimycin A	Blocks electron transfer from cytochrome b to cytochrome c_1
	Myxothiazol Rotenone Amytal Piericidin A	Prevent electron transfer from Fe-S center to ubiquinone
	DCMU	Competes with Q_B for binding site in PSII
Inhibition of ATP synthase	Aurovertin	Inhibits F_1
	Oligomycin Venturicidin	Inhibits F_o and CF_o
	DCCD	Blocks proton flow through F_o and CF_o
Uncoupling of phosphorylation from electron transfer	FCCP DNP	Hydrophobic proton carriers
	Valinomycin	K^+ ionophore
	Thermogenin	In brown adipose tissue, forms proton-conducting pores in inner mitochondrial membrane
Inhibition of ATP-ADP exchange	Atractyloside	Inhibits adenine nucleotide translocase

*DCMU is 3-(3,4-dichlorophenyl)-1,1-dimethylurea; DCCD, dicyclohexylcarbodiimide; FCCP, cyanide-*p*-trifluoromethoxyphenylhydrazone; DNP, 2,4-dinitrophenol.

Complex II: Succinate to Ubiquinone We encountered **Complex II** in Chapter 16 as **succinate dehydrogenase**, the only membrane-bound enzyme in the citric acid cycle (p. 646). Although smaller and simpler than Complex I, it contains five prosthetic groups of two types and four different protein subunits **(Fig. 19–10)**. Subunits C and D are integral membrane proteins, each with three transmembrane helices. They contain a heme group, heme b, and a binding site for ubiquinone, the final electron acceptor in the reaction catalyzed by Complex II. Subunits A and B extend into the matrix; they contain three 2Fe-2S centers, bound FAD, and a binding site for the substrate, succinate. The path of electron transfer from the succinate-binding site to FAD, then through the Fe-S centers to the Q-binding site, is more than 40 Å long, but none of the individual electron-transfer distances exceeds about 11 Å—a reasonable distance for rapid electron transfer (Fig. 19–10).

The heme b of Complex II is apparently not in the direct path of electron transfer; it may serve instead to reduce the frequency with which electrons "leak" out of the system, moving from succinate to molecular oxygen to produce the **reactive oxygen species (ROS)** hydrogen peroxide (H_2O_2) and the **superoxide radical** ($\cdot O_2^-$), as described below. Humans with point mutations in Complex II subunits near heme b or the quinone-binding site suffer from hereditary paraganglioma. This inherited condition is characterized by benign tumors of the head and neck, commonly in the carotid body, an organ that senses O_2 levels in the blood. These mutations result in greater production of ROS and perhaps greater tissue damage during succinate oxidation. ■

Other substrates for mitochondrial dehydrogenases pass electrons into the respiratory chain at the level of ubiquinone, but not through Complex II. The first step in the β oxidation of fatty acyl–CoA, catalyzed by the flavoprotein **acyl-CoA dehydrogenase** (see Fig. 17–8), involves transfer of electrons from the substrate to the FAD of the dehydrogenase, then to electron-transferring flavoprotein (ETF), which in turn passes its electrons to **ETF:ubiquinone oxidoreductase** (Fig. 19–8). This enzyme transfers electrons into the respiratory chain by reducing ubiquinone. Glycerol 3-phosphate, formed either from glycerol released by triacylglycerol breakdown or by the reduction of dihydroxyacetone phosphate from glycolysis, is oxidized by **glycerol 3-phosphate dehydrogenase** (see Fig. 17–4). This enzyme is a flavoprotein located on the outer face of the inner mitochondrial membrane, and like succinate dehydrogenase and acyl-CoA dehydrogenase, it channels electrons into the respiratory chain by reducing ubiquinone (Fig. 19–8). The important role of glycerol 3-phosphate dehydrogenase in shuttling reducing equivalents from cytosolic NADH into the mitochondrial matrix is described in Section 19.2 (see Fig. 19–32). The effect of each of these electron-transferring enzymes is to contribute to the pool of reduced ubiquinone. QH_2 from all these reactions is reoxidized by Complex III.

Intermembrane space (P side)

Phosphatidylethanolamine

QH₂

Ubiquinone

Heme b

Fe-S centers

Matrix (N side)

FAD

e^-

Succinate binding site

FIGURE 19–10 Structure of Complex II (succinate dehydrogenase). (PDB ID 1ZOY) This complex (shown here is the porcine heart enzyme) has two transmembrane subunits, C and D; the cytoplasmic extensions contain subunits A and B. Just behind the FAD in subunit A is the binding site for succinate. Subunit B has three Fe-S centers, ubiquinone is bound to subunit B, and heme b is sandwiched between subunits C and D. Two phosphatidylethanolamine molecules are so tightly bound to subunit D that they show up in the crystal structure. Electrons move (blue arrows) from succinate to FAD, then through the three Fe-S centers to ubiquinone. The heme b is not on the main path of electron transfer but protects against the formation of reactive oxygen species (ROS) by electrons that go astray.

Complex III: Ubiquinone to Cytochrome c The next respiratory complex, **Complex III**, also called **cytochrome bc_1 complex** or **ubiquinone:cytochrome c oxidoreductase**, couples the transfer of electrons from ubiquinol (QH_2) to cytochrome c with the vectorial transport of protons from the matrix to the intermembrane space. The determinations of the complete structure of this huge complex **(Fig. 19–11)** and of Complex IV (below) by x-ray crystallography, achieved between 1995 and 1998, were landmarks in the study of mitochondrial electron transfer, providing the structural framework to integrate the many biochemical observations on the functions of the respiratory complexes.

The functional unit of Complex III is a dimer, with the two monomeric units of cytochrome b surrounding a "cavern" in the middle of the membrane, in which ubiquinone is free to move from the matrix side of the membrane (site Q_N on one monomer) to the intermembrane space (site Q_P of the other monomer) as it shuttles electrons and protons across the inner mitochondrial membrane (Fig. 19–11b).

Based on the structure of Complex III and detailed biochemical studies of the redox reactions, a reasonable

FIGURE 19–11 Cytochrome bc_1 complex (Complex III). (PDB ID 1BGY) The complex is a dimer of identical monomers, each with 11 different subunits. The functional core of each monomer is three subunits: cytochrome b (green) with its two hemes (b_H and b_L), the Rieske iron-sulfur protein (purple) with its 2Fe-2S centers, and cytochrome c_1 (blue) with its heme. This cartoon view of the complex shows how cytochrome c_1 and the Rieske iron-sulfur protein project from the P surface and can interact with cytochrome c (not part of the functional complex) in the intermembrane space. The complex has two distinct binding sites for ubiquinone, Q_N and Q_P, which correspond to the sites of inhibition by two drugs that block oxidative phosphorylation. Antimycin A, which blocks electron flow from heme b_H to Q, binds at Q_N, close to heme b_H on the N (matrix) side of the membrane. Myxothiazol, which prevents electron flow from QH_2 to the Rieske iron-sulfur protein, binds at Q_P, near the 2Fe-2S center and heme b_L on the P side. The dimeric structure is essential to the function of Complex III. The interface between monomers forms two caverns, each containing a Q_P site from one monomer and a Q_N site from the other. The ubiquinone intermediates move within these sheltered caverns.

Complex III crystallizes in two distinct conformations (not shown). In one, the Rieske Fe-S center is close to its electron acceptor, the heme of cytochrome c_1, but relatively distant from cytochrome b and the QH_2-binding site at which the Rieske Fe-S center receives electrons. In the other, the Fe-S center has moved away from cytochrome c_1 and toward cytochrome b. The Rieske protein is thought to oscillate between these two conformations as it is first reduced, then oxidized.

model, the **Q cycle**, has been proposed for the passage of electrons and protons through the complex. The net equation for the redox reactions of the Q cycle **(Fig. 19–12)** is

$$QH_2 + 2 \text{ cyt } c_1 \text{ (oxidized)} + 2H_N^+ \longrightarrow$$
$$Q + 2 \text{ cyt } c_1 \text{ (reduced)} + 4H_P^+ \quad (19\text{–}3)$$

The Q cycle accommodates the switch between the two-electron carrier ubiquinol (the reduced form of ubiquinone) and the one-electron carriers—hemes b_L and b_H of cytochrome b and cytochromes c_1 and c—and results in

the uptake of two protons on the N side, and the release of four protons on the P side per pair of electrons passing through Complex III to cytochrome c. Two of the protons released on the P side are electrogenic; the other two are electroneutral, balanced by the two charges (electrons) passed to cytochrome c on the P side. Although the path of electrons through this segment of the respiratory chain is complicated, the net effect of the transfer is simple: QH_2 is

FIGURE 19–12 The Q cycle, shown in two stages. The path of electrons through Complex III is shown by blue arrows. The movement of various forms of ubiquinone is shown with black arrows. In the first stage (left), Q on the N side is reduced to the semiquinone radical, which moves back into position to accept another electron. In the second stage (right), the semiquinone radical is converted to QH_2. Meanwhile, on the P side of the membrane, two molecules of QH_2 are oxidized to Q, releasing two protons per Q molecule (four protons in all) into the intermembrane space. Each QH_2 donates one electron (via the Rieske Fe-S center) to cytochrome c_1, and one electron (via cytochrome b) to a molecule of Q near the N side, reducing it in two steps to QH_2. This reduction also consumes two protons per Q, which are taken up from the matrix (N side). Reduced cyt c_1 passes electrons one at a time to cyt c, which dissociates and carries electrons to Complex IV.

$$QH_2 + Q + \text{cyt } c_1 \text{ (oxidized)} \longrightarrow$$
$$Q + {}^\bullet Q^- + 2H_P^+ + \text{cyt } c_1 \text{ (reduced)}$$

$$QH_2 + {}^\bullet Q^- + 2H_N^+ + \text{cyt } c_1 \text{ (oxidized)} \longrightarrow$$
$$Q + 2H_P^+ + QH_2 + \text{cyt } c_1 \text{ (reduced)}$$

Net equation: $QH_2 + 2 \text{ cyt } c_1 \text{ (oxidized)} + 2H_N^+ \longrightarrow Q + 2 \text{ cyt } c_1 \text{ (reduced)} + 4H_P^+$

oxidized to Q, two molecules of cytochrome c are reduced, and protons are moved from the P side to the N side.

Cytochrome c is a soluble protein of the intermembrane space. After its single heme accepts an electron from Complex III, cytochrome c moves to Complex IV to donate the electron to a binuclear copper center.

Complex IV: Cytochrome c to O_2 In the final step of the respiratory chain, **Complex IV**, also called **cytochrome oxidase**, carries electrons from cytochrome c to molecular oxygen, reducing it to H_2O. Complex IV is a large enzyme (13 subunits; M_r 204,000) of the inner mitochondrial membrane. Bacteria contain a form that is much simpler, with only three or four subunits, but still capable of catalyzing both electron transfer and proton pumping. Comparison of the mitochondrial and bacterial complexes suggests that three subunits are critical to the function **(Fig. 19–13)**.

Mitochondrial subunit II contains two Cu ions complexed with the —SH groups of two Cys residues in a binuclear center (Cu_A; Fig. 19–13b) that resembles the 2Fe-2S centers of iron-sulfur proteins. Subunit I contains two heme groups, designated a and a_3, and another copper ion (Cu_B). Heme a_3 and Cu_B form a second binuclear center that accepts electrons from heme a and transfers them to O_2 bound to heme a_3.

Electron transfer through Complex IV is from cytochrome c to the Cu_A center, to heme a, to the heme a_3–Cu_B center, and finally to O_2 **(Fig. 19–14)**. For every four electrons passing through this complex, the enzyme consumes four "substrate" H^+ from the matrix (N side) in converting O_2 to $2H_2O$. It also uses the energy of this redox reaction to pump one proton outward into the intermembrane space (P side) for each electron that passes through, adding to the electrochemical potential

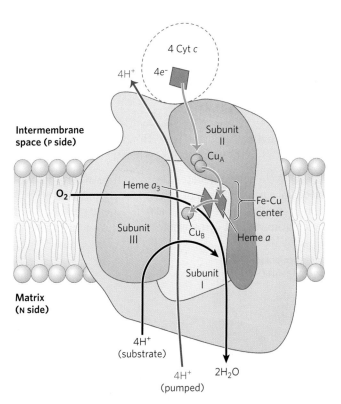

FIGURE 19–13 Structure of cytochrome oxidase (Complex IV). This complex from bovine mitochondria has 13 subunits, but only four core proteins are shown here (PDB ID 1OCC). **(a)** Complex IV, with four subunits in each of two identical units of a dimer. Subunit I (yellow) has two heme groups, a and a_3, near a single copper ion, Cu_B (not visible here). Heme a_3 and Cu_B form a binuclear Fe-Cu center. Subunit II (purple) contains two Cu ions complexed with the —SH groups of two Cys residues in a binuclear center, Cu_A, that resembles the 2Fe-2S centers of iron-sulfur proteins. This binuclear center and the cytochrome c-binding site are located in a domain of subunit II that protrudes from the P side of the inner membrane (into the intermembrane space). Subunit III (blue) is essential for rapid proton movement through subunit II. The role of subunit IV (green) is not yet known. **(b)** The binuclear center of Cu_A. The Cu ions (green spheres) share electrons equally. When the center is reduced, the ions have the formal charges $Cu^{1+}Cu^{1+}$; when oxidized, $Cu^{1.5+}Cu^{1.5+}$. Six amino acid residues are ligands around the Cu ions: two His, two Cys, Glu, and Met.

FIGURE 19–14 Path of electrons through Complex IV. The three proteins critical to electron flow are subunits I, II, and III. The larger green structure includes the other 10 proteins in the complex. Electron transfer through Complex IV begins with cytochrome c (top). Two molecules of reduced cytochrome c each donate an electron to the binuclear center Cu_A. From here electrons pass through heme a to the Fe-Cu center (heme a_3 and Cu_B). Oxygen now binds to heme a_3 and is reduced to its peroxy derivative (O_2^{2-}; not shown here) by two electrons from the Fe-Cu center. Delivery of two more electrons from cytochrome c (top, making four electrons in all) converts the O_2^{2-} to two molecules of water, with consumption of four "substrate" protons from the matrix. At the same time, four protons are pumped from the matrix by an as yet unknown mechanism.

produced by redox-driven proton transport through Complexes I and III. The overall reaction catalyzed by Complex IV is

$$4 \text{ cyt } c \text{ (reduced)} + 8H_N^+ + O_2 \longrightarrow$$
$$4 \text{ cyt } c \text{ (oxidized)} + 4H_P^+ + 2H_2O \quad (19\text{--}4)$$

This four-electron reduction of O_2 involves redox centers that carry only one electron at a time, and it must occur without the release of incompletely reduced intermediates such as hydrogen peroxide or hydroxyl free radicals—very reactive species that would damage cellular components. The intermediates remain tightly bound to the complex until completely converted to water.

Mitochondrial Complexes May Associate in Respirasomes

There is growing experimental evidence that in the intact mitochondrion, the respiratory complexes tightly associate with each other in the inner membrane to form **respirasomes**, functional combinations of two or more different electron-transfer complexes. For example, when Complex III is gently extracted from mitochondrial membranes, it is found to be associated with Complex I and remains associated during gentle electrophoresis. Supercomplexes of Complex III and IV can also be isolated, and when viewed with the electron microscope are of the right size and shape to accommodate the crystal structures of both complexes **(Fig. 19–15)**. Proteins of a family synthesized under the control of the hypoxia-inducible factor HIF are found associated with respirasomes and may be essential for their assembly or stability. HIF mediates changes in the composition of Complex IV under hypoxic conditions (see Fig. 19–34). The kinetics of electron flow through the series of respiratory complexes would be very different in the two extreme cases of tight versus no association: (1) if complexes were tightly associated, electron transfers would essentially occur through a solid state, and (2) if the complexes functioned separately, electrons would be carried between them by ubiquinone and cytochrome c. The kinetic evidence supports electron transfer through a solid state, and thus the respirasome model.

Cardiolipin, the lipid that is especially abundant in the inner mitochondrial membrane (see Figs 10–9 and 11–2), may be critical to the integrity of respirasomes; its removal with detergents, or its absence in certain yeast mutants, results in defective mitochondrial electron transfer and a loss of affinity between the respiratory complexes.

The Energy of Electron Transfer Is Efficiently Conserved in a Proton Gradient

The transfer of two electrons from NADH through the respiratory chain to molecular oxygen can be written as

$$NADH + H^+ + \tfrac{1}{2}O_2 \longrightarrow NAD^+ + H_2O \quad (19\text{--}5)$$

(a)

(b)

FIGURE 19–15 A putative respirasome composed of Complexes III and IV. (a) Purified supercomplexes containing Complexes III and IV, from yeast, visualized by electron microscopy after staining with uranyl acetate. The electron densities of hundreds of images were averaged to yield this composite view. **(b)** The x-ray–derived structures of one molecule of Complex III (red; from yeast) and two of Complex IV (green; from bovine heart) could be fitted to the electron-density map to suggest one possible mode of interaction of these complexes in a respirasome. This view is in the plane of the bilayer (yellow).

This net reaction is highly exergonic. For the redox pair $NAD^+/NADH$, E'° is -0.320 V, and for the pair O_2/H_2O, E'° is 0.816 V. The $\Delta E'^\circ$ for this reaction is therefore 1.14 V, and the standard free-energy change (see Eqn 13–7, p. 531) is

$$\Delta G'^\circ = -n\mathscr{F}\Delta E'^\circ \quad (19\text{--}6)$$
$$= -2(96.5 \text{ kJ/V} \cdot \text{mol})(1.14 \text{ V})$$
$$= -220 \text{ kJ/mol (of NADH)}$$

This *standard* free-energy change is based on the assumption of equal concentrations (1 M) of NADH and NAD^+. In actively respiring mitochondria, the actions of many dehydrogenases keep the actual $[NADH]/[NAD^+]$ ratio well above unity, and the real free-energy change for the reaction shown in Equation 19–5 is therefore substantially greater (more negative) than -220 kJ/mol. A similar calculation for the oxidation of succinate shows that electron transfer from succinate (E'° for fumarate/succinate $= 0.031$ V) to O_2 has a smaller, but still negative, standard free-energy change of about -150 kJ/mol.

Much of this energy is used to pump protons out of the matrix. For each pair of electrons transferred to O_2, four protons are pumped out by Complex I, four by

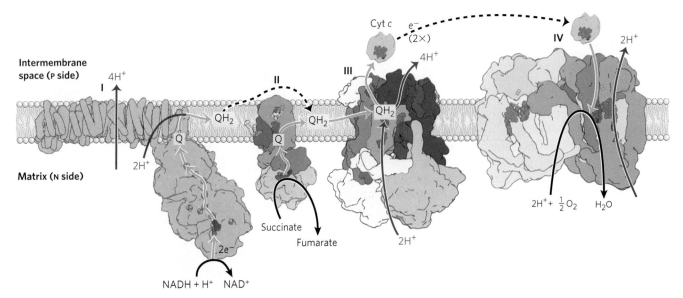

FIGURE 19–16 Summary of the flow of electrons and protons through the four complexes of the respiratory chain. Electrons reach Q through Complexes I and II. The reduced Q (QH_2) serves as a mobile carrier of electrons and protons. It passes electrons to Complex III, which passes them to another mobile connecting link, cytochrome c. Complex IV then transfers electrons from reduced cytochrome c to O_2. Electron flow through Complexes I, III, and IV is accompanied by proton flow from the matrix to the intermembrane space. Recall that electrons from β oxidation of fatty acids can also enter the respiratory chain through Q (see Fig. 19–8). The structures shown here are from several sources: Complex I, *Thermus thermophilus* (PDB ID 3M9S); Complex II, porcine heart (PDB ID 1ZOY); Complex III, bovine heart (PDB ID 1BGY); cytochrome c, equine heart (PDB ID 1HRC); Complex IV, bovine heart (PDB ID 1OCC).

Complex III, and two by Complex IV **(Fig. 19–16)**. The *vectorial* equation for the process is therefore

$$NADH + 11H_N^+ + \tfrac{1}{2}O_2 \longrightarrow$$
$$NAD^+ + 10H_P^+ + H_2O \qquad (19\text{–}7)$$

The electrochemical energy inherent in this difference in proton concentration and separation of charge represents a temporary conservation of much of the energy of electron transfer. The energy stored in such a gradient, termed the **proton-motive force**, has two components: (1) the *chemical potential energy* due to the difference in concentration of a chemical species (H^+) in the two regions separated by the membrane, and (2) the *electrical potential energy* that results from the separation of charge when a proton moves across the membrane without a counterion **(Fig. 19–17)**.

As we showed in Chapter 11, the free-energy change for the creation of an electrochemical gradient by an ion pump is

$$\Delta G = RT \ln (C_2/C_1) + Z \mathcal{F} \Delta\psi \qquad (19\text{–}8)$$

where C_2 and C_1 are the concentrations of an ion in two regions, and $C_2 > C_1$; Z is the absolute value of its electrical charge (1 for a proton); and $\Delta\psi$ is the transmembrane difference in electrical potential, measured in volts.

For protons at 25 °C,

$$\ln (C_2/C_1) = 2.3(\log [H^+]_P - \log [H^+]_N)$$
$$= 2.3(pH_N - pH_P) = 2.3 \, \Delta pH$$

and Equation 19–8 reduces to

$$\Delta G = 2.3RT \, \Delta pH + \mathcal{F}\Delta\psi \qquad (19\text{–}9)$$
$$= (5.70 \text{ kJ/mol})\Delta pH + (96.5 \text{ kJ/V·mol})\Delta\psi$$

In actively respiring mitochondria, the measured $\Delta\psi$ is 0.15 to 0.20 V and the pH of the matrix is about 0.75 unit more alkaline than that of the intermembrane space.

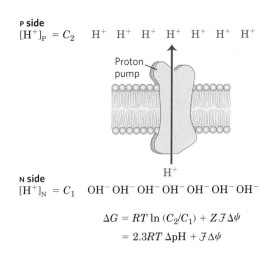

$$\Delta G = RT \ln (C_2/C_1) + Z\mathcal{F}\Delta\psi$$
$$= 2.3RT \, \Delta pH + \mathcal{F}\Delta\psi$$

FIGURE 19–17 Proton-motive force. The inner mitochondrial membrane separates two compartments of different [H^+], resulting in differences in chemical concentration (ΔpH) and charge distribution ($\Delta\psi$) across the membrane. The net effect is the proton-motive force (ΔG), which can be calculated as shown here. This is explained more fully in the text.

WORKED EXAMPLE 19–1 **Energetics of Electron Transfer**

Calculate the amount of energy conserved in the proton gradient across the inner mitochondrial membrane per pair of electrons transferred through the respiratory chain from NADH to oxygen. Assume $\Delta\psi$ is 0.15 V and the pH difference is 0.75 unit.

Solution: Equation 19–9 gives the free-energy change when *one* mole of protons moves across the inner membrane. Substituting the measured values for ΔpH (0.75 unit) and $\Delta\psi$ (0.15 V) in this equation gives $\Delta G = 19$ kJ/mol (of protons). Because the transfer of two electrons from NADH to O_2 is accompanied by the outward pumping of 10 protons (Eqn 19–7), roughly 200 kJ (of the 220 kJ released by oxidation of 1 mol of NADH) is conserved in the proton gradient.

When protons flow spontaneously *down* their electrochemical gradient, energy is made available to do work. In mitochondria, chloroplasts, and aerobic bacteria, the electrochemical energy in the proton gradient drives the synthesis of ATP from ADP and P_i. We return to the energetics and stoichiometry of ATP synthesis driven by the electrochemical potential of the proton gradient in Section 19.2.

Reactive Oxygen Species Are Generated during Oxidative Phosphorylation

Several steps in the path of oxygen reduction in mitochondria have the potential to produce highly reactive free radicals that can damage cells. The passage of electrons from QH_2 to Complex III and the passage of electrons from Complexes I and II to QH_2 involve the radical $^{\bullet}Q^-$ as an intermediate. The $^{\bullet}Q^-$ can, with a low probability, pass an electron to O_2 in the reaction

$$O_2 + e^- \longrightarrow {}^{\bullet}O_2^-$$

The superoxide free radical thus generated is highly reactive; its formation also leads to production of the even more reactive hydroxyl free radical, $^{\bullet}$OH **(Fig. 19–18)**.

Reactive oxygen species can wreak havoc, reacting with and damaging enzymes, membrane lipids, and nucleic acids. In actively respiring mitochondria, 0.1% to as much as 4% of the O_2 used in respiration forms $^{\bullet}O_2^-$—more than enough to have lethal effects unless the free radical is quickly disposed of. Factors that slow the flow of electrons through the respiratory chain increase the formation of superoxide, perhaps by prolonging the lifetime of $^{\bullet}O_2^-$ generated in the Q cycle. The formation of ROS is favored when two conditions are met: (1) mitochondria are not making ATP (for lack of ADP or O_2) and therefore have a large proton-motive force and a high ratio of QH_2/Q, and (2) there is a high $NADH/NAD^+$ ratio in the matrix. In these situations, the mitochondrion is under oxidative stress—more elec-

FIGURE 19–18 ROS formation in mitochondria and mitochondrial defenses. When the rate of electron entry into the respiratory chain and the rate of electron transfer through the chain are mismatched, superoxide radical ($^{\bullet}O_2^-$) production increases at Complexes I and III as the partially reduced ubiquinone radical ($^{\bullet}Q^-$) donates an electron to O_2. Superoxide acts on aconitase, a 4Fe-4S protein, to release Fe^{2+}. In the presence of Fe^{2+}, the Fenton reaction leads to formation of the highly reactive hydroxyl free radical ($^{\bullet}$OH). The reactions shown in blue defend the cell against the damaging effects of superoxide. Reduced glutathione (GSH; see Fig. 22–29) donates electrons for the reduction of H_2O_2 and of the oxidized Cys residues (—S—S—) of enzymes and other proteins, and GSH is regenerated from the oxidized form (GSSG) by reduction with NADPH.

trons are available to enter the respiratory chain than can be immediately passed through to oxygen. When the supply of electron donors (NADH) is matched with that of electron acceptors, there is less oxidative stress, and ROS production is much reduced. Although overproduction of ROS is clearly detrimental, *low* levels of ROS may be used by the cell as a signal reflecting the insufficient supply of oxygen (hypoxia), triggering metabolic adjustments (see Fig. 19–34).

To prevent oxidative damage by $^{\bullet}O_2^-$, cells have several forms of the enzyme **superoxide dismutase**, which catalyzes the reaction

$$2\,{}^{\bullet}O_2^- + 2H^+ \longrightarrow H_2O_2 + O_2$$

The hydrogen peroxide (H_2O_2) thus generated is rendered harmless by the action of **glutathione peroxidase** (Fig. 19–18). Glutathione reductase recycles the oxidized glutathione to its reduced form, using electrons from the NADPH generated by nicotinamide nucleotide transhydrogenase (in the mitochondrion) or by the pentose phosphate pathway (in the cytosol; see Fig. 14–21). Reduced glutathione also serves to keep protein sulfhydryl groups in their

BOX 19–1 Hot, Stinking Plants and Alternative Respiratory Pathways

Many flowering plants attract insect pollinators by releasing odorant molecules that mimic an insect's natural food sources or potential egg-laying sites. Plants pollinated by flies or beetles that normally feed on or lay their eggs in dung or carrion sometimes use foul-smelling compounds to attract these insects.

One family of stinking plants is the Araceae, which includes philodendrons, arum lilies, and skunk cabbages. These plants have tiny flowers densely packed on an erect structure, the spadix, surrounded by a modified leaf, the spathe. The spadix releases odors of rotting flesh or dung. Before pollination the spadix also heats up, in some species to as much as 20 to 40 °C above the ambient temperature. Heat production (thermogenesis) helps evaporate odorant molecules for better dispersal, and because rotting flesh and dung are usually warm from the hyperactive metabolism of scavenging microbes, the heat itself might also attract insects. In the case of the eastern skunk cabbage (Fig. 1), which flowers in late winter or early spring when snow still covers the ground, thermogenesis allows the spadix to grow up through the snow.

How does a skunk cabbage heat its spadix? The mitochondria of plants, fungi, and unicellular eukary-

FIGURE 1 Eastern skunk cabbage.

otes have electron-transfer systems that are essentially the same as those in animals, but they also have an alternative respiratory pathway. A cyanide-resistant QH_2 oxidase transfers electrons from the ubiquinone pool directly to oxygen, bypassing the two proton-translocating steps of Complexes III and IV (Fig. 2). Energy that might have been conserved as ATP is instead released as heat. Plant mitochondria also have an alternative NADH dehydrogenase, insensitive to the Complex I inhibitor rotenone (see

---

reduced state, preventing some of the deleterious effects of oxidative stress (Fig. 19–18). Nicotinamide nucleotide transhydrogenase is critical in this process: it produces the NADPH essential for glutathione reductase activity.

Plant Mitochondria Have Alternative Mechanisms for Oxidizing NADH

Plant mitochondria supply the cell with ATP during periods of low illumination or darkness by mechanisms entirely analogous to those used by nonphotosynthetic organisms. In the light, the principal source of mitochondrial NADH is a reaction in which glycine, produced by a process known as photorespiration, is converted to serine (see Fig. 20–21):

$$2 \text{ Glycine} + NAD^+ \longrightarrow$$
$$\text{serine} + CO_2 + NH_3 + NADH + H^+$$

For reasons discussed in Chapter 20, plants must carry out this reaction even when they do not need NADH for ATP production. To regenerate NAD^+ from unneeded NADH, mitochondria of plants (and of some fungi and protists) transfer electrons from NADH directly to ubiquinone and from ubiquinone directly to O_2, bypassing Complexes III and IV and their proton pumps. In this process the energy in NADH is dissi-

pated as heat, which can sometimes be of value to the plant (Box 19–1). Unlike cytochrome oxidase (Complex IV), the alternative QH_2 oxidase is not inhibited by cyanide. Cyanide-resistant NADH oxidation is therefore the hallmark of this unique plant electron-transfer pathway.

SUMMARY 19.1 Electron-Transfer Reactions in Mitochondria

▶ Chemiosmotic theory provides the intellectual framework for understanding many biological energy transductions, including oxidative phosphorylation and photophosphorylation. The mechanism of energy coupling is similar in both cases: the energy of electron flow is conserved by the concomitant pumping of protons across the membrane, producing an electrochemical gradient, the proton-motive force.

▶ In mitochondria, hydride ions removed from substrates (such as α-ketoglutarate and malate) by NAD-linked dehydrogenases donate electrons to the respiratory (electron-transfer) chain, which transfers the electrons to molecular O_2, reducing it to H_2O.

▶ Reducing equivalents from NADH are passed through a series of Fe-S centers to ubiquinone,

Table 19–4), that transfers electrons from NADH in the matrix directly to ubiquinone, bypassing Complex I and its associated proton pumping. And plant mitochondria have yet another NADH dehydrogenase, on the external face of the inner membrane, that transfers electrons from NADPH or NADH in the intermembrane space to ubiquinone, again bypassing Complex I. Thus when electrons enter the alternative respiratory pathway through the rotenone-insensitive NADH dehydrogenase, the external NADH dehydrogenase, or succinate dehydrogenase (Complex II), and pass to O_2 via the cyanide-resistant alternative oxidase, energy is not conserved as ATP but is released as heat. A skunk cabbage can use the heat to melt snow, produce a foul stench, or attract beetles or flies.

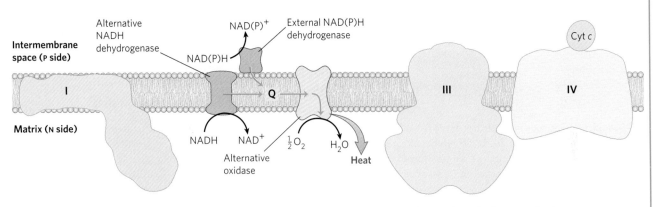

FIGURE 2 Electron carriers of the inner membrane of plant mitochondria. Electrons can flow through Complexes I, III, and IV, as in animal mitochondria, or through plant-specific alternative carriers by the paths shown with blue arrows.

which transfers the electrons to cytochrome b, the first carrier in Complex III. In this complex, electrons take two separate paths through two b-type cytochromes and cytochrome c_1 to an Fe-S center. The Fe-S center passes electrons, one at a time, through cytochrome c and into Complex IV, cytochrome oxidase. This copper-containing enzyme, which also contains cytochromes a and a_3, accumulates electrons, then passes them to O_2, reducing it to H_2O.

▶ Some electrons enter this chain of carriers through alternative paths. Succinate is oxidized by succinate dehydrogenase (Complex II), which contains a flavoprotein that passes electrons through several Fe-S centers to ubiquinone. Electrons derived from the oxidation of fatty acids pass to ubiquinone via the electron-transferring flavoprotein.

▶ Potentially harmful reactive oxygen species produced in mitochondria are inactivated by a set of protective enzymes, including superoxide dismutase and glutathione peroxidase.

▶ Plants, fungi, and unicellular eukaryotes have, in addition to the typical cyanide-sensitive path for electron transfer, an alternative, cyanide-resistant NADH oxidation pathway.

19.2 ATP Synthesis

How is a concentration gradient of protons transformed into ATP? We have seen that electron transfer releases, and the proton-motive force conserves, more than enough free energy (about 200 kJ) per "mole" of electron pairs to drive the formation of a mole of ATP, which requires about 50 kJ (p. 519). Mitochondrial oxidative phosphorylation therefore poses no thermodynamic problem. But what is the chemical mechanism that couples proton flux with phosphorylation?

The **chemiosmotic model**, proposed by Peter Mitchell, is the paradigm for this mechanism. According to the model **(Fig. 19–19)**, the electrochemical energy inherent in the difference in proton concentration and the separation of charge across the inner mitochondrial membrane—the proton-motive force—drives the synthesis of ATP as protons flow passively back into the matrix through a proton pore in **ATP synthase**. To emphasize this crucial role of the proton-motive force, the equation for ATP synthesis is sometimes written

$$\text{ADP} + \text{P}_i + n\text{H}_\text{P}^+ \longrightarrow \text{ATP} + \text{H}_2\text{O} + n\text{H}_\text{N}^+ \quad (19\text{-}10)$$

Mitchell used "chemiosmotic" to describe enzymatic reactions that involve, simultaneously, a chemical reaction

FIGURE 19–19 Chemiosmotic model. In this simple representation of the chemiosmotic theory applied to mitochondria, electrons from NADH and other oxidizable substrates pass through a chain of carriers arranged asymmetrically in the inner membrane. Electron flow is accompanied by proton transfer across the membrane, producing both a chemical gradient (ΔpH) and an electrical gradient ($\Delta\psi$) (combined, the proton-motive force). The inner mitochondrial membrane is impermeable to protons; protons can reenter the matrix only through proton-specific channels (F_o). The proton-motive force that drives protons back into the matrix provides the energy for ATP synthesis, catalyzed by the F_1 complex associated with F_o.

Peter Mitchell, 1920–1992

and a transport process, and the overall process is sometimes referred to as "chemiosmotic coupling." Coupling refers to the *obligate* connection between mitochondrial ATP synthesis and electron flow through the respiratory chain; neither of the two processes can proceed without the other. The operational definition of coupling is shown in **Figure 19–20**. When isolated mitochondria are suspended in a buffer containing ADP, P_i, and an oxidizable substrate such as succinate, three easily measured processes occur: (1) the substrate is oxidized (succinate yields fumarate), (2) O_2 is consumed, and (3) ATP is synthesized. Oxygen consumption and ATP synthesis depend on the presence of an oxidizable substrate (succinate in this case) as well as ADP and P_i.

Because the energy of substrate oxidation drives ATP synthesis in mitochondria, we would expect inhibitors of the passage of electrons to O_2 (such as cyanide, carbon monoxide, and antimycin A) to block ATP synthesis (Fig. 19–20a). More surprising is the finding that the converse is also true: inhibition of ATP synthesis blocks electron transfer in intact mitochondria. This obligatory coupling can be demonstrated in isolated mitochondria by providing O_2 and oxidizable substrates, but not ADP (Fig. 19–20b). Under these condi-

Henry Lardy, 1917–2010

tions, no ATP synthesis can occur and electron transfer to O_2 does not proceed. Henry Lardy, who pioneered the use of antibiotics to explore mitochondrial function, demonstrated coupling of oxidation and phosphorylation by using oligomycin and venturicidin, toxic antibiotics that bind to the ATP synthase in mitochondria. These compounds are potent inhibitors of both ATP synthesis *and* the transfer of electrons through the chain of carriers to O_2 (Fig. 19–20b). Because oligomycin is known to interact not directly with the electron carriers but with ATP synthase, it follows that electron transfer and ATP synthesis are obligately coupled: neither reaction occurs without the other.

Chemiosmotic theory readily explains the dependence of electron transfer on ATP synthesis in mitochondria. When the flow of protons into the matrix through the proton channel of ATP synthase is blocked (with oligomycin, for example), no path exists for the return of protons to the matrix, and the continued extrusion of protons driven by the activity of the respiratory chain generates a large proton gradient. The proton-motive force builds up until the cost (free energy) of pumping protons out of the matrix against this gradient equals or exceeds the energy released by the transfer of electrons from

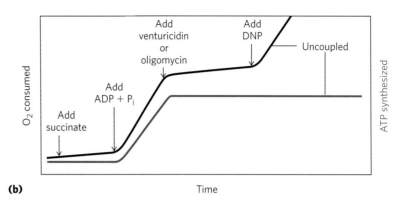

FIGURE 19–20 Coupling of electron transfer and ATP synthesis in mitochondria. In experiments to demonstrate coupling, mitochondria are suspended in a buffered medium and an O_2 electrode monitors O_2 consumption. At intervals, samples are removed and assayed for the presence of ATP. **(a)** Addition of ADP and P_i alone results in little or no increase in either respiration (O_2 consumption; black) or ATP synthesis (red). When succinate is added, respiration begins immediately and ATP is synthe-

sized. Addition of cyanide (CN^-), which blocks electron transfer between cytochrome oxidase (Complex IV) and O_2, inhibits both respiration and ATP synthesis. **(b)** Mitochondria provided with succinate respire and synthesize ATP only when ADP and P_i are added. Subsequent addition of venturicidin or oligomycin, inhibitors of ATP synthase, blocks both ATP synthesis and respiration. Dinitrophenol (DNP) is an uncoupler, allowing respiration to continue without ATP synthesis.

NADH to O_2. At this point electron flow must stop; the free energy for the overall process of electron flow coupled to proton pumping becomes zero, and the system is at equilibrium.

Certain conditions and reagents, however, can uncouple oxidation from phosphorylation. When intact mitochondria are disrupted by treatment with detergent or by physical shear, the resulting membrane fragments can still catalyze electron transfer from succinate or NADH to O_2, but no ATP synthesis is coupled to this respiration. Certain chemical compounds cause uncoupling without disrupting mitochondrial structure. Chemical uncouplers include 2,4-dinitrophenol (DNP) and carbonylcyanide-p-trifluoromethoxyphenylhydrazone (FCCP) (Table 19–4; **Fig. 19–21**), weak acids with hydrophobic properties that permit them to diffuse readily across mitochondrial membranes. After entering the matrix in the protonated form, they can release a proton, thus dissipating the proton gradient. Resonance stabilization delocalizes the charge on the anionic forms, making them sufficiently permeant to diffuse back across the membrane, where they can pick up a proton and repeat the process. Ionophores such as valinomycin (see Fig. 11–44) allow inorganic ions to pass easily through membranes. Ionophores uncouple electron transfer from oxidative phosphorylation by dissipating the electrical contribution to the electrochemical gradient across the mitochondrial membrane.

A prediction of the chemiosmotic theory is that, because the role of electron transfer in mitochondrial ATP synthesis is simply to pump protons to create the electrochemical potential of the proton-motive force, an artificially created proton gradient should be able to replace electron transfer in driving ATP synthesis. This has been experimentally confirmed

FIGURE 19–21 Two chemical uncouplers of oxidative phosphorylation. Both DNP and FCCP have a dissociable proton and are very hydrophobic. They carry protons across the inner mitochondrial membrane, dissipating the proton gradient. Both also uncouple photophosphorylation (see Fig. 19-65).

(**Fig. 19–22**). Mitochondria manipulated so as to impose a difference of proton concentration and a separation of charge across the inner membrane synthesize ATP *in the absence of an oxidizable substrate*; the proton-motive force alone suffices to drive ATP synthesis.

(a)

(b)

FIGURE 19–22 Evidence for the role of a proton gradient in ATP synthesis. An artificially imposed electrochemical gradient can drive ATP synthesis in the absence of an oxidizable substrate as electron donor. In this two-step experiment, **(a)** isolated mitochondria are first incubated in a pH 9 buffer containing 0.1 M KCl. Slow leakage of buffer and KCl into the mitochondria eventually brings the matrix into equilibrium with the surrounding medium. No oxidizable substrates are present. **(b)** Mitochondria are now separated from the pH 9 buffer and resuspended in pH 7 buffer containing valinomycin but no KCl. The change in buffer creates a difference of two pH units across the inner mitochondrial membrane. The outward flow of K^+, carried (by valinomycin) down its concentration gradient without a counter-ion, creates a charge imbalance across the membrane (matrix negative). The sum of the chemical potential provided by the pH difference and the electrical potential provided by the separation of charges is a proton-motive force large enough to support ATP synthesis in the absence of an oxidizable substrate.

ATP Synthase Has Two Functional Domains, F_o and F_1

Mitochondrial ATP synthase is an F-type ATPase (see Fig. 11–39) similar in structure and mechanism to the ATP synthases of chloroplasts and bacteria. This large enzyme complex of the inner mitochondrial membrane catalyzes the formation of ATP from ADP and P_i, driven by the flow of protons from the P to the N side of the membrane (Eqn 19–10). ATP synthase, also called Complex V, has two distinct components: F_1, a peripheral membrane protein, and F_o (o denoting oligomycin-

Efraim Racker, 1913–1991

sensitive), which is integral to the membrane. F_1, the first factor recognized as essential for oxidative phosphorylation, was identified and purified by Efraim Racker and his colleagues in the early 1960s.

In the laboratory, small membrane vesicles formed from inner mitochondrial membranes carry out ATP synthesis coupled to electron transfer. When F_1 is gently extracted, the "stripped" vesicles still contain intact respiratory chains and the F_o portion of ATP synthase. The vesicles can catalyze electron transfer from NADH to O_2 but cannot produce a proton gradient: F_o has a proton pore through which protons leak as fast as they are pumped by electron transfer, and without a proton gradient the F_1-depleted vesicles cannot make ATP. Isolated F_1 catalyzes ATP hydrolysis (the reversal of synthesis) and was therefore originally called **F_1 ATPase**. When purified F_1 is added back to the depleted vesicles, it reassociates with F_o, plugging its proton pore and restoring the membrane's capacity to couple electron transfer and ATP synthesis.

ATP Is Stabilized Relative to ADP on the Surface of F_1

Isotope exchange experiments with purified F_1 reveal a remarkable fact about the enzyme's catalytic mechanism: on the enzyme surface, the reaction ADP + P_i $\rightleftharpoons$ ATP + H_2O is readily reversible—the free-energy change for ATP synthesis is close to zero! When ATP is hydrolyzed by F_1 in the presence of ^{18}O-labeled water, the P_i released contains an ^{18}O atom. Careful measurement of the ^{18}O content of P_i formed in vitro by F_1-catalyzed hydrolysis of ATP reveals that the P_i has not one, but three or four ^{18}O atoms **(Fig. 19–23)**. This indicates that the terminal pyrophosphate bond in ATP is cleaved and re-formed repeatedly before P_i leaves the enzyme surface. With P_i free to tumble in its binding site, each hydrolysis inserts ^{18}O randomly at one of the four positions in the molecule. This exchange reaction occurs in unenergized F_oF_1 complexes (with no proton gradient) and with isolated F_1—the exchange does not require the input of energy.

Kinetic studies of the initial rates of ATP synthesis and hydrolysis confirm the conclusion that $\Delta G'^{\circ}$ for ATP synthesis on the enzyme is near zero. From the measured rates of hydrolysis ($k_1 = 10\ \text{s}^{-1}$) and synthesis ($k_{-1} = 24\ \text{s}^{-1}$), the calculated equilibrium constant for the reaction

$$\text{Enz-ATP} \rightleftharpoons \text{Enz-(ADP + }P_i)$$

is

$$K'_{eq} = \frac{k_{-1}}{k_1} = \frac{24\ \text{s}^{-1}}{10\ \text{s}^{-1}} = 2.4$$

From this K'_{eq}, the calculated apparent $\Delta G'^{\circ}$ is close to zero. This is much different from the K'_{eq} of about 10^5 ($\Delta G'^{\circ} = -30.5$ kJ/mol) for the hydrolysis of ATP free in solution (not on the enzyme surface).

What accounts for the huge difference? ATP synthase stabilizes ATP relative to ADP + P_i by binding

(a)

FIGURE 19–23 Catalytic mechanism of F_1. (a) ^{18}O-exchange experiment. F_1 solubilized from mitochondrial membranes is incubated with ATP in the presence of ^{18}O-labeled water. At intervals, a sample of the solution is withdrawn and analyzed for the incorporation of ^{18}O into the P_i produced from ATP hydrolysis. In minutes, the P_i contains three or four ^{18}O atoms, indicating that both ATP hydrolysis and ATP synthesis have occurred several times during the incubation. **(b)** The likely transition state complex for ATP hydrolysis and synthesis by ATP synthase (derived from PDB ID 1BMF). The α subunit is shown in gray, β in purple. The positively charged residues β-Arg^{182} and α-Arg^{376} coordinate two oxygens of the pentavalent phosphate intermediate; β-Lys^{155} interacts with a third oxygen, and the Mg^{2+} ion further stabilizes the intermediate. The blue sphere represents the leaving group (H_2O). These interactions result in the ready equilibration of ATP and ADP + P_i in the active site.

(b)

ATP more tightly, releasing enough energy to counterbalance the cost of making ATP. Careful measurements of the binding constants show that F_oF_1 binds ATP with very high affinity ($K_d \leq 10^{-12}$ M) and ADP with much lower affinity ($K_d \approx 10^{-5}$ M). The difference in K_d corresponds to a difference of about 40 kJ/mol in binding energy, and this binding energy drives the equilibrium toward formation of the product ATP.

The Proton Gradient Drives the Release of ATP from the Enzyme Surface

Although ATP synthase equilibrates ATP with ADP + P_i, in the absence of a proton gradient the newly synthesized ATP does not leave the surface of the enzyme. It is the proton gradient that causes the enzyme to release the ATP formed on its surface. The reaction coordinate diagram of the process **(Fig. 19–24)** illustrates the difference between the mechanism of ATP synthase and that of many other enzymes that catalyze endergonic reactions.

For the continued synthesis of ATP, the enzyme must cycle between a form that binds ATP very tightly and a form that releases ATP. Chemical and

FIGURE 19–24 Reaction coordinate diagrams for ATP synthase and for a more typical enzyme. In a typical enzyme-catalyzed reaction (left), reaching the transition state ($\ddagger$) between substrate and product is the major energy barrier to overcome. In the reaction catalyzed by ATP synthase (right), release of ATP from the enzyme, not formation of ATP, is the major energy barrier. The free-energy change for the formation of ATP from ADP and P_i in aqueous solution is large and positive, but on the enzyme surface, the very tight binding of ATP provides sufficient binding energy to bring the free energy of the enzyme-bound ATP close to that of ADP + P_i, so the reaction is readily reversible. The equilibrium constant is near 1. The free energy required for the release of ATP is provided by the proton-motive force.

crystallographic studies of the ATP synthase have revealed the structural basis for this alternation in function.

Each β Subunit of ATP Synthase Can Assume Three Different Conformations

Mitochondrial F_1 has nine subunits of five different types, with the composition $\alpha_3\beta_3\gamma\delta\varepsilon$. Each of the three β subunits has one catalytic site for ATP synthesis. The crystallographic determination of the F_1 structure by John E. Walker and colleagues revealed structural details very helpful in explaining the catalytic mechanism of the enzyme. The knoblike portion of F_1 is a flattened sphere, 8 nm high and 10 nm across, consisting of alternating α and β subunits arranged like the sections of an orange **(Fig. 19–25a, b, c)**. The polypeptides that make up the stalk in the F_1 crystal structure are asymmetrically arranged, with one domain of the single γ subunit making up a central shaft that passes through F_1, and another domain of γ associated primarily with one of the three β subunits, designated β-empty (Fig. 19–25b). Although the amino acid sequences of the three β subunits are identical, *their conformations differ*, in part because of the association of the γ subunit with just one of the three. The structures of the δ and ε subunits are not revealed in these crystallographic studies.

John E. Walker

The conformational differences among β subunits extend to differences in their ATP/ADP-binding sites. When researchers crystallized the protein in the presence of ADP and App(NH)p, a close structural analog of ATP that cannot be hydrolyzed by the ATPase activity of F_1, the binding site of one of the three β subunits was filled with App(NH)p, the second was filled with ADP, and the third was empty. The corresponding β subunit conformations are designated β-ATP, β-ADP, and β-empty (Fig. 19–25b). This difference in nucleotide binding among the three subunits is critical to the mechanism of the complex.

The F_o complex, with its proton pore, is composed of three subunits, a, b, and c, in the proportion ab_2c_{n}, where n ranges from 8 to 15 in various organisms; in yeast, it is 10. Subunit c is a small (M_r 8,000), very hydrophobic polypeptide, consisting almost entirely of two transmembrane helices, with a small loop extending from the matrix side of the membrane. The crystal structure of the yeast F_oF_1, solved in 1999, shows 10 c subunits, each with two transmembrane helices roughly perpendicular to the plane of the membrane and arranged in two concentric circles. The inner circle is made up of the amino-terminal helices of each c subunit; the outer circle, about 55 Å in diameter, is made up of the carboxyl-terminal helices. *The c subunits in this c ring rotate together as a unit around an axis perpendicular to the membrane.* The ε and γ subunits of F_1 form a leg-and-foot that projects from the bottom (membrane) side of F_1 and stands firmly on the ring of c subunits. The a subunit consists of several hydrophobic helices that span the membrane in close association with one of the c subunits in the c ring. The schematic drawing in Figure 19–25a combines the structural information from studies of bovine F_1, yeast F_oF_1, and the c ring of *Ilyobacter tartaricus*.

Rotational Catalysis Is Key to the Binding-Change Mechanism for ATP Synthesis

On the basis of detailed kinetic and binding studies of the reactions catalyzed by F_oF_1, Paul Boyer proposed a **rotational catalysis** mechanism in which the three active sites of F_1 take turns catalyzing ATP synthesis **(Fig. 19–26)**. A given β subunit starts in the β-ADP conformation, which binds ADP and P_i from the surrounding medium. The subunit now changes conformation, assuming the β-ATP form that tightly binds and stabilizes ATP, bringing about the ready equilibration of ADP + P_i with ATP on the enzyme surface. Finally, the subunit changes to the β-empty conformation, which has very low affinity for ATP, and the newly synthesized ATP leaves the enzyme surface. Another round of catalysis begins when this subunit again assumes the β-ADP form and binds ADP and P_i.

The conformational changes central to this mechanism are driven by the passage of protons through the F_o portion of ATP synthase. The streaming of protons through the F_o pore causes the cylinder of c subunits and the attached γ subunit to rotate about the long axis of γ, which is perpendicular to the plane of the membrane. The γ subunit passes through the center of the $\alpha_3\beta_3$ spheroid, which is held stationary relative to the

Paul Boyer

Nonhydrolyzable
β-γ bond

App(NH)p (β,γ-imidoadenosine 5'-triphosphate)

FIGURE 19-25 Mitochondrial ATP synthase complex. (a) A cartoon view of the F_oF_1 complex. **(b)** (PDB ID 1BMF and PDB ID 1JNV) F_1 viewed from above (that is, from the N side of the membrane), showing the three β (shades of purple) and three α (shades of gray) subunits and the central shaft (γ subunit, green). Each β subunit, near its interface with the neighboring α subunit, has a nucleotide-binding site critical to the catalytic activity. The single γ subunit associates primarily with one of the three $\alpha\beta$ pairs, forcing each of the three β subunits into slightly different conformations, with different nucleotide-binding sites. In the crystalline enzyme, one subunit (β-ADP) has ADP (yellow) in its binding site, the next (β-ATP) has ATP (red), and the third (β-empty) has no bound nucleotide.

(c) The entire enzyme viewed from the side (in the plane of the membrane). The F_1 portion (PDB IDs 1BMF, 1JNV, and 2A7U) has three α and three β subunits arranged like the segments of an orange around a central shaft, the γ subunit (green). (Two α subunits and one β subunit have been omitted to reveal the γ subunit and the binding sites for ATP and ADP on the β subunits.) The δ subunit confers oligomycin sensitivity on the ATP synthase, and the ε subunit may serve to inhibit its ATPase activity under some circumstances. The F_o subunit consists of one a subunit and two b subunits (PDB ID 2CLY and PDB ID 1B9U), which anchor the F_oF_1 complex in the membrane and act as a stator, holding the α and β subunits in place. F_o also includes the c ring, made up of a number (8 to 15, depending on the species) of identical c subunits, small, hydrophobic proteins. The c ring and the a subunit interact to provide a transmembrane path for protons. Each c subunit has an Asp residue near the middle of the membrane, which can bind or give up a proton. In this structure (PDB ID 1YCE), we have shown the homologous c_{11} ring of the Na^+-ATPase of *Ilyobacter tartaricus*, for which the structure is well established. The Na^+-binding sites in it, which correspond to the proton-binding sites of the F_oF_1 complex, are shown with their bound Na^+ ions (red spheres). **(d)** A view of F_o perpendicular to the membrane. Each of the c subunits in F_o has a critical Asp residue near the middle of the membrane, which undergoes protonation/deprotonation during the catalytic cycle of the ATP synthase. As in (c), red spheres represent the Na^+- or proton-binding sites in Asp residues.

(a)

(b) Top view of F_1

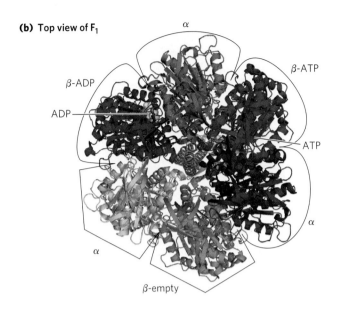

(c) Side view of F_oF_1

(d) Bottom view of F_o

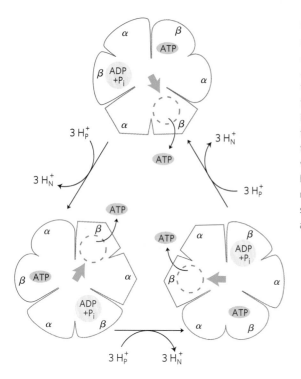

FIGURE 19–26 Binding-change model for ATP synthase. The F_1 complex has three nonequivalent adenine nucleotide-binding sites, one for each pair of α and β subunits. At any given moment, one of these sites is in the β-ATP conformation (which binds ATP tightly), a second is in the β-ADP (loose-binding) conformation, and a third is in the β-empty (very-loose-binding) conformation. The proton-motive force causes rotation of the central shaft—the γ subunit, shown as a green arrowhead—which comes into contact with each $\alpha\beta$ subunit pair in succession. This produces a cooperative conformational change in which the β-ATP site is converted to the β-empty conformation, and ATP dissociates; the β-ADP site is converted to the β-ATP conformation, which promotes condensation of bound ADP + P_i to form ATP; and the β-empty site becomes a β-ADP site, which loosely binds ADP + P_i entering from the solvent. This model, based on experimental findings, requires that at least two of the three catalytic sites alternate in activity; ATP cannot be released from one site unless and until ADP and P_i are bound at the other.

membrane surface by the b_2 and δ subunits (Fig. 19–25a). With each rotation of 120°, γ comes into contact with a different β subunit, and the contact forces that β subunit into the β-empty conformation.

The three β subunits interact in such a way that when one assumes the β-empty conformation, its neighbor to one side *must* assume the β-ADP form, and the other neighbor the β-ATP form. Thus one complete rotation of the γ subunit causes each β subunit to cycle through all three of its possible conformations, and for each rotation, three ATP are synthesized and released from the enzyme surface.

One strong prediction of this **binding-change model** is that the γ subunit should rotate in one direction when F_oF_1 is synthesizing ATP and in the opposite direction when the enzyme is hydrolyzing ATP. This prediction of rotation with ATP hydrolysis was confirmed in elegant experiments in the laboratories of Masasuke Yoshida and Kazuhiko Kinosita, Jr. The rotation of γ in a single F_1 molecule was observed microscopically by attaching a long, thin, fluorescent actin polymer to γ and watching it move relative to $\alpha_3\beta_3$ immobilized on a microscope slide as ATP was hydrolyzed. (The expected reversal of the rotation when ATP is being synthesized could not be tested in this experiment; there is no proton gradient to drive ATP synthesis.) When the entire F_oF_1 complex (not just F_1) was used in a similar experiment, the entire ring of c subunits rotated with γ **(Fig. 19–27)**. The "shaft" rotated in the predicted direction through 360°. The rotation was not smooth, but occurred in three discrete steps of 120°. As calculated from the known rate of ATP hydrolysis by one F_1 molecule and from the frictional drag on the long

FIGURE 19–27 Experimental demonstration of rotation of F_o and γ. F_1 genetically engineered to contain a run of His residues adheres tightly to a microscope slide coated with a Ni complex; biotin is covalently attached to a c subunit of F_o. The protein avidin, which binds biotin very tightly, is covalently attached to long filaments of actin labeled with a fluorescent probe. Biotin-avidin binding now attaches the actin filaments to the c subunit. When ATP is provided as substrate for the ATPase activity of F_1, the labeled filament is seen to rotate continuously in one direction, proving that the F_o cylinder of c subunits rotates. In another experiment, a fluorescent actin filament was attached directly to the γ subunit. The series of fluorescence micrographs (read left to right) shows the position of the actin filament at intervals of 133 ms. Note that as the filament rotates, it makes a discrete jump about every eleventh frame. Presumably the cylinder and shaft move as one unit.

actin polymer, the efficiency of this mechanism in converting chemical energy into motion is close to 100%. It is, in Boyer's words, "a splendid molecular machine!"

How Does Proton Flow through the F_o Complex Produce Rotary Motion?

One feasible model to explain how proton flow and rotary motion are coupled in the F_o complex is shown in **Figure 19–28**. The individual subunits in F_o are

4 Displaced H⁺ exits on N side H⁺

5 c_{10} ring rotates, Arg²¹⁰ returns to P-side half-channel

c_{10}

⁺H   H⁺

Arg²¹⁰

a

3 Arg²¹⁰ rotates, displacing H⁺ from Asp

6 Process repeats

Half-channel (N side)

c_{10}

⁺H   H⁺

Arg²¹⁰

H⁺

a

Half-channel (P side)

2 Proton displaces Arg²¹⁰ to adjacent c subunit

H⁺ **1** Proton enters half-channel on P side

FIGURE 19–28 A model for proton-driven rotation of the c ring. The a subunit of the F_o complex of the ATP synthase (see Fig. 19-25a) has two hydrophilic half-channels for protons, one leading from the P side to the middle of the membrane, and another leading from the middle of the membrane to the N side (matrix). The individual c subunits in F_o (10 in the yeast enzyme) are arranged in a circle about a central core. Each c subunit has a critical Asp residue about midway across the membrane, which can donate or accept a proton (red H⁺). The a subunit has a positively charged Arg side chain that forms an electrostatic interaction with the negatively charged carboxylate of Asp on the adjoining c subunit. This c subunit is initially positioned so that a proton that enters the half-channel on the P side (where the proton concentration is relatively high) encounters and protonates the Asp residue, weakening its interaction with Arg. The Arg side chain rotates toward the protonated Asp residue in the next c subunit, and displaces its carboxyl proton, as a new electrostatic Arg-Asp interaction forms. The displaced proton moves through the second half-channel in subunit a and is released on the N side. The c subunit with its unprotonated Asp residue moves so that its Arg-Asp pair faces the half-channel on the P side, and a second cycle begins: proton entry, protonation of Asp, Arg movement, and proton exit. Rotation of the ring occurs by thermal (Brownian) motion and is effectively ratcheted; the orientation of the proton gradient dictates the direction of proton flow and makes the rotation of the c ring essentially unidirectional.

arranged in a circle about a central core that is probably filled with membrane lipids. In each c subunit, there is a critical Asp (or Glu) residue located at about the middle of the membrane. Protons cross the membrane through a path made up of both a and c subunits. The a subunit has a proton half-channel leading from the cytosol (P side) to the middle of the membrane, where it ends near the Asp residue of the adjoining c subunit. A proton diffuses from the cytosolic side (where the proton concentration is relatively high) down this half-channel and binds to the Asp residue, displacing a positively charged Arg residue that had been associated with the Asp. This Arg residue swings aside, forming an interaction with the Asp on the adjacent c subunit in the ring and displacing the proton from that Asp; this adjacent proton exits through the second half-channel to the N side, where the proton concentration is relatively low, completing the movement of one proton equivalent from outside to inside the matrix. Now another proton has entered the half-channel on the cytosolic side, moved to the Asp of the next c subunit, and protonated it, again displacing Arg, which in turn displaces a proton from the next c subunit, and so forth. The rotary movement of the c ring is the result of thermal (Brownian) motion, made unidirectional by the large difference in proton concentration across the membrane. The number of protons that must be transferred to produce one complete rotation of the c ring is equal to the number of c subunits in the ring. Studies of the c ring with atomic force microscopy (Box 19–2) or x-ray diffraction have shown that the number of c subunits is different in different organisms **(Fig. 19–29)**. In animal mitochondria this number is 8, in yeast mitochondria and in *E. coli* it is 10, and the number of c subunits can range as high as 15 in the cyanobacterium *Spirulina platensis*. The rate of rotation in intact mitochondria has been estimated at about 6,000 rpm—100 rotations per second!

Chemiosmotic Coupling Allows Nonintegral Stoichiometries of O_2 Consumption and ATP Synthesis

Before the general acceptance of the chemiosmotic model for oxidative phosphorylation, the assumption was that the overall reaction equation would take the following form:

$$x\text{ADP} + x\text{P}_i + \tfrac{1}{2}\text{O}_2 + \text{H}^+ + \text{NADH} \longrightarrow$$
$$x\text{ATP} + \text{H}_2\text{O} + \text{NAD}^+ \qquad (19\text{–}11)$$

with the value of x—sometimes called the **P/O ratio** or the **P/2e^- ratio**—always an integer. When intact mitochondria are suspended in solution with an oxidizable substrate such as succinate or NADH and are provided with O_2, ATP synthesis is readily measurable, as is the decrease in O_2. In principle, these two measurements should yield the number of ATP synthesized per $\tfrac{1}{2}O_2$ consumed, the P/O ratio. Measurement of P/O, however,

(a) Top view   **Side view**

(b)

(c)   **(d)**

5 nm

FIGURE 19–29 Different organisms have different numbers of c subunits in the c ring of the F_o complex. The structure of the c ring from several species has been determined by x-ray crystallography. Each helix in the inner ring is half of a hairpin-shaped c subunit; the outer ring of helices is the other half of the hairpin structure. Views of the c ring perpendicular to the membrane show the number of c subunits for **(a)** bovine mitochondria (8) and **(b)** yeast mitochondria (10). Atomic force microscopy has been used to visualize the c rings of **(c)** the thermophilic bacterium *Bacillus* species TA2.A1 (13) and **(d)** spinach (14). According to the model in Fig. 19–28, different numbers of c subunits in the c ring should result in different ratios of ATP formed per pair of electrons passing through the respiratory chain (different P/O ratios).

is complicated by the fact that intact mitochondria *consume* ATP in many unrelated reactions taking place in the matrix, and they consume O_2 for purposes other than oxidative phosphorylation. The contribution of these interfering reactions must be carefully measured, and the calculation of P/O must be corrected for their contributions. Most experiments have yielded P/O (ATP to $\frac{1}{2}O_2$) ratios of between 2 and 3 when NADH was the electron donor, and between 1 and 2 when succinate was the donor. Given the assumption that P/O should have an integral value, most experimenters agreed that the P/O ratios must be 3 for NADH and 2 for succinate, and for years those values appeared in research papers and textbooks.

With introduction of the chemiosmotic paradigm for coupling ATP synthesis to electron transfer, there was no

---

BOX 19–2 | METHODS | Atomic Force Microscopy to Visualize Membrane Proteins

In atomic force microscopy (AFM), the sharp tip of a microscopic probe attached to a flexible cantilever is drawn across an uneven surface such as a membrane (Fig. 1). Electrostatic and van der Waals interactions between the tip and the sample produce a force that moves the probe up and down (in the z dimension) as it encounters hills and valleys in the sample. A laser beam reflected from the cantilever detects motions of as little as 1 Å. In one type of atomic force microscope, the force on the probe is held constant (relative to a standard force, on the order of piconewtons) by a feedback circuit that causes the platform holding the sample to rise or fall to keep the force constant. A series of scans in the x and y dimensions (the plane of the membrane) yields a three-dimensional contour map of the surface with resolution near the atomic scale—0.1 nm in the vertical dimension, 0.5 to 1.0 nm in the lateral dimensions.

In favorable cases, AFM can be used to study single membrane protein molecules, such as the c subunits of the F_o complex (Fig. 19–29c, d).

FIGURE 1 The principle of atomic force microscopy.

theoretical requirement for P/O to be integral. The relevant questions about stoichiometry became, How many protons are pumped outward by electron transfer from one NADH to O_2, and how many protons must flow inward through the F_oF_1 complex to drive the synthesis of one ATP? The measurement of proton fluxes is technically complicated; the investigator must take into account the buffering capacity of mitochondria, nonproductive leakage of protons across the inner membrane, and use of the proton gradient for functions other than ATP synthesis, such as driving the transport of substrates across the inner mitochondrial membrane (described below). The consensus experimental values for number of protons pumped out per pair of electrons are 10 for NADH and 6 for succinate (which sends electrons into the respiratory chain at the level of ubiquinone). The most widely accepted experimental value for number of protons required to drive the synthesis of an ATP molecule is 4, of which 1 is used in transporting P_i, ATP, and ADP across the mitochondrial membrane (see below). If 10 protons are pumped out per NADH and 4 must flow in to produce 1 ATP, the proton-based P/O ratio is 2.5 for NADH as the electron donor and 1.5 (6/4) for succinate.

The Proton-Motive Force Energizes Active Transport

Although the primary role of the proton gradient in mitochondria is to furnish energy for the synthesis of ATP, the proton-motive force also drives several transport processes essential to oxidative phosphorylation. The inner mitochondrial membrane is generally impermeable to charged species, but two specific systems transport ADP and P_i into the matrix and ATP out to the cytosol **(Fig. 19–30)**.

The **adenine nucleotide translocase**, integral to the inner membrane, binds ADP^{3-} in the intermembrane space and transports it into the matrix in exchange for an ATP^{4-} molecule simultaneously transported outward (see Fig. 13–11 for the ionic forms of ATP and ADP). Because this antiporter moves four negative charges out for every three moved in, its activity is favored by the transmembrane electrochemical gradient, which gives the matrix a net negative charge; the proton-motive force drives ATP-ADP exchange. Adenine nucleotide translocase is specifically inhibited by atractyloside, a toxic glycoside formed by a species of thistle. If the transport of ADP into and ATP out of mitochondria is inhibited, cytosolic ATP cannot be regenerated from ADP, explaining the toxicity of atractyloside.

A second membrane transport system essential to oxidative phosphorylation is the **phosphate translocase**, which promotes symport of one $H_2PO_4^-$ and one H^+ into the matrix. This transport process, too, is favored by the transmembrane proton gradient (Fig. 19–30). Notice that the process requires movement of one proton from the P to the N side of the inner membrane, consuming some of the energy of electron transfer. A complex of the

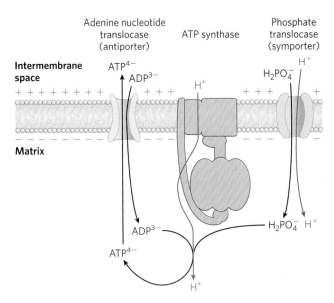

FIGURE 19–30 Adenine nucleotide and phosphate translocases. Transport systems of the inner mitochondrial membrane carry ADP and P_i into the matrix and newly synthesized ATP into the cytosol. The adenine nucleotide translocase is an antiporter; the same protein moves ADP into the matrix and ATP out. The effect of replacing ATP^{4-} with ADP^{3-} in the matrix is the net efflux of one negative charge, which is favored by the charge difference across the inner membrane (outside positive). At pH 7, P_i is present as both HPO_4^{2-} and $H_2PO_4^-$; the phosphate translocase is specific for $H_2PO_4^-$. There is no net flow of charge during symport of $H_2PO_4^-$ and H^+, but the relatively low proton concentration in the matrix favors the inward movement of H^+. Thus the proton-motive force is responsible both for providing the energy for ATP synthesis and for transporting substrates (ADP and P_i) into and product (ATP) out of the mitochondrial matrix. All three of these transport systems can be isolated as a single membrane-bound complex (ATP synthasome).

ATP synthase and both translocases, the **ATP synthasome**, can be isolated from mitochondria by gentle dissection with detergents, suggesting that the functions of these three proteins are very tightly integrated.

WORKED EXAMPLE 19–2 **Stoichiometry of ATP Production: Effect of c Ring Size**

(a) If *bovine* mitochondria have 8 c subunits per c ring, what is the predicted ratio of ATP formed per NADH oxidized? (b) What is the predicted value for *yeast* mitochondria, with 10 c subunits? (c) What are the comparable values for electrons entering the respiratory chain from $FADH_2$?

Solution: (a) The question asks us to determine how many ATP are produced per NADH. This is another way of asking us to calculate the P/O ratio, or x in Equation 19–11. If the c ring has 8 c subunits, then one full rotation will transfer 8 protons to the matrix and produce 3 ATP molecules. But this synthesis also requires the transport of 3 P_i into the matrix, at a cost of 1 proton each, adding 3 more protons to the total number required. This brings the total cost to (11 protons)/

(3 ATP) = 3.7 protons/ATP. The consensus value for the number of protons pumped out per pair of electrons transferred from NADH is 10 (see Fig. 19–19). So, oxidizing 1 NADH produces (10 protons)/(3.7 protons/ATP) = 2.7 ATP.

(b) If the c ring has 10 c subunits, then one full rotation will transfer 10 protons to the matrix and produce 3 ATP molecules. Adding in the 3 protons to transport the 3 P_i into the matrix brings the total cost to (13 protons)/(3 ATP) = 4.3 protons/ATP. Oxidizing 1 NADH produces (10 protons)/(4.3 protons/ATP) = 2.3 ATP.

(c) When electrons enter the respiratory chain from $FADH_2$ (at ubiquinone), only 6 protons are available to drive ATP synthesis. This changes the calculation for bovine mitochondria to (6 protons)/(3.7 protons/ATP) = 1.6 ATP per pair of electrons from $FADH_2$. For yeast mitochondria, the calculation is (6 protons)/(4.3 protons/ATP) = 1.4 ATP per pair of electrons from $FADH_2$.

These calculated values of x or the P/O ratio define a range that includes the experimental values of 2.5

ATP/NADH and 1.5 $ATP/FADH_2$, and we therefore use these values throughout this book.

Shuttle Systems Indirectly Convey Cytosolic NADH into Mitochondria for Oxidation

The NADH dehydrogenase of the inner mitochondrial membrane of animal cells can accept electrons only from NADH in the matrix. Given that the inner membrane is not permeable to NADH, how can the NADH generated by glycolysis in the cytosol be reoxidized to NAD^+ by O_2 via the respiratory chain? Special shuttle systems carry reducing equivalents from cytosolic NADH into mitochondria by an indirect route. The most active NADH shuttle, which functions in liver, kidney, and heart mitochondria, is the **malate-aspartate shuttle (Fig. 19–31)**. The reducing equivalents of cytosolic NADH are first transferred to cytosolic oxaloacetate to yield malate, catalyzed by cytosolic malate dehydrogenase. The malate thus formed passes through

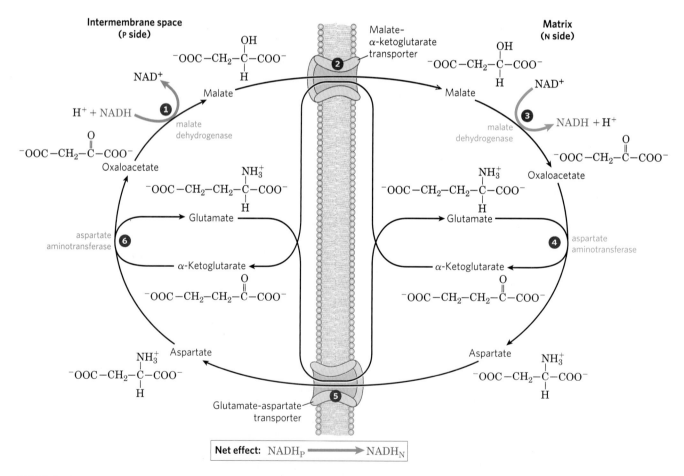

FIGURE 19–31 Malate-aspartate shuttle. This shuttle for transporting reducing equivalents from cytosolic NADH into the mitochondrial matrix is used in liver, kidney, and heart. ❶ NADH in the cytosol enters the intermembrane space through openings in the outer membrane (porins), then passes two reducing equivalents to oxaloacetate, producing malate. ❷ Malate crosses the inner membrane via the malate-α-ketoglutarate transporter. ❸ In the matrix, malate passes two reducing equivalents to NAD^+, and the resulting NADH is oxidized by the respiratory chain; the oxaloacetate formed from malate cannot pass directly into the cytosol. ❹ Oxaloacetate is first transaminated to aspartate, and ❺ aspartate can leave via the glutamate-aspartate transporter. ❻ Oxaloacetate is regenerated in the cytosol, completing the cycle.

the inner membrane via the malate–α-ketoglutarate transporter. Within the matrix the reducing equivalents are passed to NAD^+ by the action of matrix malate dehydrogenase, forming NADH; this NADH can pass electrons directly to the respiratory chain. About 2.5 molecules of ATP are generated as this pair of electrons passes to O_2. Cytosolic oxaloacetate must be regenerated by transamination reactions and the activity of membrane transporters to start another cycle of the shuttle.

Skeletal muscle and brain use a different NADH shuttle, the **glycerol 3-phosphate shuttle (Fig. 19–32)**. It differs from the malate-aspartate shuttle in that it delivers the reducing equivalents from NADH to ubiquinone and thus into Complex III, not Complex I (Fig. 19–8), providing only enough energy to synthesize 1.5 ATP molecules per pair of electrons.

The mitochondria of plants have an *externally* oriented NADH dehydrogenase that can transfer electrons directly from cytosolic NADH into the respiratory chain at the level of ubiquinone. Because this pathway bypasses the NADH dehydrogenase of Complex I and the associated proton movement, the yield of ATP from cytosolic NADH is less than that from NADH generated in the matrix (Box 19–1).

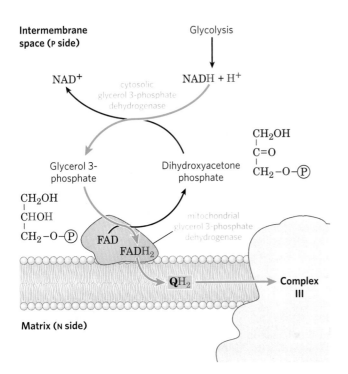

FIGURE 19–32 Glycerol 3-phosphate shuttle. This alternative means of moving reducing equivalents from the cytosol to the mitochondrial matrix operates in skeletal muscle and the brain. In the cytosol, dihydroxyacetone phosphate accepts two reducing equivalents from NADH in a reaction catalyzed by cytosolic glycerol 3-phosphate dehydrogenase. An isozyme of glycerol 3-phosphate dehydrogenase bound to the outer face of the inner membrane then transfers two reducing equivalents from glycerol 3-phosphate in the intermembrane space to ubiquinone. Note that this shuttle does not involve membrane transport systems.

SUMMARY 19.2 ATP Synthesis

▶ The flow of electrons through Complexes I, III, and IV results in pumping of protons across the inner mitochondrial membrane, making the matrix alkaline relative to the intermembrane space. This proton gradient provides the energy (in the form of the proton-motive force) for ATP synthesis from ADP and P_i by ATP synthase (F_oF_1 complex) in the inner membrane.

▶ ATP synthase carries out "rotational catalysis," in which the flow of protons through F_o causes each of three nucleotide-binding sites in F_1 to cycle from (ADP + P_i)–bound to ATP-bound to empty conformations.

▶ ATP formation on the enzyme requires little energy; the role of the proton-motive force is to push ATP from its binding site on the synthase.

▶ The ratio of ATP synthesized per $\frac{1}{2}O_2$ reduced to H_2O (the P/O ratio) is about 2.5 when electrons enter the respiratory chain at Complex I, and 1.5 when electrons enter at ubiquinone. This ratio may vary somewhat in different organisms based on the number of c subunits in the F_o complex.

▶ Energy conserved in a proton gradient can drive solute transport uphill across a membrane.

▶ The inner mitochondrial membrane is impermeable to NADH and NAD^+, but NADH equivalents are moved from the cytosol to the matrix by either of two shuttles. NADH equivalents moved in by the malate-aspartate shuttle enter the respiratory chain at Complex I and yield a P/O ratio of 2.5; those moved in by the glycerol 3-phosphate shuttle enter at ubiquinone and give a P/O ratio of 1.5.

19.3 Regulation of Oxidative Phosphorylation

Oxidative phosphorylation produces most of the ATP made in aerobic cells. Complete oxidation of a molecule of glucose to CO_2 yields 30 or 32 ATP (Table 19–5). By comparison, glycolysis under anaerobic conditions (lactate fermentation) yields only 2 ATP per glucose. Clearly, the evolution of oxidative phosphorylation provided a tremendous increase in the energy efficiency of catabolism. Complete oxidation to CO_2 of the coenzyme A derivative of palmitate (16:0), which also occurs in the mitochondrial matrix, yields 108 ATP per palmitoyl-CoA (see Table 17–1). A similar calculation can be made for the ATP yield from oxidation of each of the amino acids (Chapter 18). Aerobic oxidative pathways that result in electron transfer to O_2 accompanied by oxidative phosphorylation therefore account for the vast majority of the ATP produced in catabolism, so the regulation of ATP production by oxidative phosphorylation to match the cell's fluctuating needs for ATP is absolutely essential.

TABLE 19–5 ATP Yield from Complete Oxidation of Glucose

Process	Direct product	Final ATP
Glycolysis	2 NADH (cytosolic)	3 or 5*
	2 ATP	2
Pyruvate oxidation (two per glucose)	2 NADH (mitochondrial matrix)	5
Acetyl-CoA oxidation in citric acid cycle (two per glucose)	6 NADH (mitochondrial matrix)	15
	2 FADH$_2$	3
	2 ATP or 2 GTP	2
Total yield per glucose		30 or 32

*The number depends on which shuttle system transfers reducing equivalents into the mitochondrion.

Oxidative Phosphorylation Is Regulated by Cellular Energy Needs

The rate of respiration (O$_2$ consumption) in mitochondria is tightly regulated; it is generally limited by the availability of ADP as a substrate for phosphorylation. Dependence of the rate of O$_2$ consumption on the availability of the P$_i$ acceptor, ADP (Fig. 19–20b), the **acceptor control** of respiration, can be remarkable. In some animal tissues, the **acceptor control ratio**, the ratio of the maximal rate of ADP-induced O$_2$ consumption to the basal rate in the absence of ADP, is at least 10.

The intracellular concentration of ADP is one measure of the energy status of cells. Another, related measure is the **mass-action ratio** of the ATP-ADP system, [ATP]/([ADP][P$_i$]). Usually this ratio is very high, so the ATP-ADP system is almost fully phosphorylated. When the rate of some energy-requiring process (protein synthesis, for example) increases, the rate of breakdown of ATP to ADP and P$_i$ increases, lowering the mass-action ratio. With more ADP available for oxidative phosphorylation, the rate of respiration increases, causing regeneration of ATP. This continues until the mass-action ratio returns to its normal high level, at which point respiration slows again. The rate of oxidation of cellular fuels is regulated with such sensitivity and precision that the [ATP]/([ADP][P$_i$]) ratio fluctuates only slightly in most tissues, even during extreme variations in energy demand. In short, ATP is formed only as fast as it is used in energy-requiring cellular activities.

An Inhibitory Protein Prevents ATP Hydrolysis during Hypoxia

We have already encountered ATP synthase as an ATP-driven proton pump (see Fig. 11–39), catalyzing the reverse of ATP synthesis. When a cell is hypoxic (deprived of oxygen), as in a heart attack or stroke, electron transfer to oxygen slows, and so does the pumping of protons. The proton-motive force soon collapses. Under these conditions, the ATP synthase could operate in reverse, hydrolyzing ATP to pump protons outward and causing a disastrous drop in ATP levels. This is prevented by a small (84 amino acids) protein

inhibitor, IF$_1$, which simultaneously binds to two ATP synthase molecules, inhibiting their ATPase activity **(Fig. 19–33)**. IF$_1$ is inhibitory only in its dimeric form, which is favored at pH lower than 6.5. In a cell starved for oxygen, the main source of ATP becomes glycolysis, and the pyruvic or lactic acid thus formed lowers the pH in the cytosol and the mitochondrial matrix. This favors IF$_1$ dimerization, leading to inhibition of the ATPase activity of ATP synthase and thereby preventing wasteful hydrolysis of ATP. When aerobic metabolism resumes, production of pyruvic acid slows, the pH of the cytosol rises, the IF$_1$ dimer is destabilized, and the inhibition of ATP synthase is lifted. IF$_1$ is one of a growing number of proteins known to be intrinsically disordered (p. 141); it acquires a favored conformation only upon interaction with the ATP synthase.

Hypoxia Leads to ROS Production and Several Adaptive Responses

In hypoxic cells there is an imbalance between the input of electrons from fuel oxidation in the mitochondrial matrix and transfer of electrons to molecular oxygen,

FIGURE 19–33 Structure of bovine F$_1$-ATPase in a complex with its regulatory protein IF$_1$. (Derived from PDB ID 1OHH) Two F$_1$ molecules are viewed here as in Figure 19–25c. The inhibitor IF$_1$ (red) binds to the $\alpha\beta$ interface of the subunits in the diphosphate (ADP) conformation (α-ADP and β-ADP), freezing the two F$_1$ complexes and thereby blocking ATP hydrolysis (and synthesis). (Parts of IF$_1$ that failed to resolve in the crystals of F$_1$ are shown in light red as they occur in crystals of isolated IF$_1$.) This complex is stable only at the low cytosolic pH characteristic of cells that are producing ATP by glycolysis; when aerobic metabolism resumes, the cytosolic pH rises, the inhibitor is destabilized, and ATP synthase becomes active.

FIGURE 19–34 Hypoxia-inducible factor (HIF-1) regulates gene expression to reduce ROS formation. Under conditions of low oxygen (hypoxia), HIF-1 is synthesized in greater amounts and acts as a transcription factor, increasing the synthesis of glucose transporter, glycolytic enzymes, pyruvate dehydrogenase kinase (PDH kinase), lactate dehydrogenase, a protease that degrades the cytochrome oxidase subunit COX4-1, and cytochrome oxidase subunit COX4-2. These changes counter the formation of ROS by decreasing the supply of NADH and FADH$_2$ and making cytochrome oxidase of Complex IV more effective.

leading to increased formation of reactive oxygen species. In addition to the glutathione peroxidase system (Fig. 19–18), cells have two other lines of defense against ROS **(Fig. 19–34)**. One is regulation of pyruvate dehydrogenase (PDH), the enzyme that delivers acetyl-CoA to the citric acid cycle (Chapter 16). Under hypoxic conditions, PDH kinase phosphorylates mitochondrial PDH, inactivating it and slowing the delivery of FADH$_2$ and NADH from the citric acid cycle to the respiratory chain. A second means of preventing ROS formation is the replacement of one subunit of Complex IV, known as COX4-1, with another subunit, COX4-2, that is better suited to hypoxic conditions. With COX4-1, the catalytic properties of Complex IV are optimal for respiration at normal oxygen concentrations; with COX4-2, Complex IV is optimized for operation under hypoxic conditions.

The changes in PDH activity and the COX4-2 content of Complex IV are both mediated by HIF-1, the hypoxia-inducible factor. HIF-1 (another intrinsically disordered protein) accumulates in hypoxic cells and, acting as a transcription factor, triggers increased synthesis of PDH kinase, COX4-2, and a protease that

degrades COX4-1. Recall that HIF-1 also mediates the changes in glucose transport and glycolytic enzymes that produce the Pasteur effect (see Box 14–1).

When these mechanisms for dealing with ROS are insufficient, due to genetic mutation affecting one of the protective proteins or under conditions of very high rates of ROS production, mitochondrial function is compromised. Mitochondrial damage is thought to be involved in aging, heart failure, certain rare cases of diabetes (described below), and several maternally inherited genetic diseases that affect the nervous system. ■

ATP-Producing Pathways Are Coordinately Regulated

The major catabolic pathways have interlocking and concerted regulatory mechanisms that allow them to function together in an economical and self-regulating manner to produce ATP and biosynthetic precursors. The relative concentrations of ATP and ADP control not only the rates of electron transfer and oxidative phosphorylation but also the rates of the citric acid cycle, pyruvate oxidation, and glycolysis **(Fig. 19–35)**. Whenever ATP consumption increases, the rate of electron

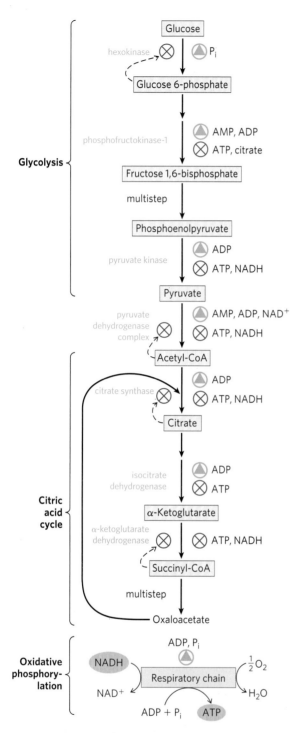

FIGURE 19–35 Regulation of the ATP-producing pathways. This diagram shows the interlocking regulation of glycolysis, pyruvate oxidation, the citric acid cycle, and oxidative phosphorylation by the relative concentrations of ATP, ADP, and AMP, and by NADH. High [ATP] (or low [ADP] and [AMP]) produces low rates of glycolysis, pyruvate oxidation, acetate oxidation via the citric acid cycle, and oxidative phosphorylation. All four pathways are accelerated when the use of ATP and the formation of ADP, AMP, and P$_i$ increase. The interlocking of glycolysis and the citric acid cycle by citrate, which inhibits glycolysis, supplements the action of the adenine nucleotide system. In addition, increased levels of NADH and acetyl-CoA also inhibit the oxidation of pyruvate to acetyl-CoA, and a high [NADH]/[NAD$^+$] ratio inhibits the dehydrogenase reactions of the citric acid cycle (see Fig. 16–19).

transfer and oxidative phosphorylation increases. Simultaneously, the rate of pyruvate oxidation via the citric acid cycle increases, increasing the flow of electrons into the respiratory chain. These events can in turn evoke an increase in the rate of glycolysis, increasing the rate of pyruvate formation. When conversion of ADP to ATP lowers the ADP concentration, acceptor control slows electron transfer and thus oxidative phosphorylation. Glycolysis and the citric acid cycle are also slowed, because ATP is an allosteric inhibitor of the glycolytic enzyme phosphofructokinase-1 (see Fig. 15–16) and of pyruvate dehydrogenase (see Fig. 16–19).

Phosphofructokinase-1 is also inhibited by citrate, the first intermediate of the citric acid cycle. When the cycle is "idling," citrate accumulates within mitochondria, then is transported into the cytosol. When the concentrations of both ATP and citrate rise, they produce a concerted allosteric inhibition of phosphofructokinase-1 that is greater than the sum of their individual effects, slowing glycolysis.

SUMMARY 19.3 Regulation of Oxidative Phosphorylation

▶ Oxidative phosphorylation is regulated by cellular energy demands. The intracellular [ADP] and the mass-action ratio [ATP]/([ADP][P$_i$]) are measures of a cell's energy status.

▶ In hypoxic (oxygen-deprived) cells, a protein inhibitor blocks ATP hydrolysis by the reverse activity of ATP synthase, preventing a drastic drop in [ATP].

▶ The adaptive responses to hypoxia, mediated by HIF-1, slow electron transfer into the respiratory chain and modify Complex IV to act more efficiently under low-oxygen conditions.

▶ ATP and ADP concentrations set the rate of electron transfer through the respiratory chain via a series of interlocking controls on respiration, glycolysis, and the citric acid cycle.

19.4 Mitochondria in Thermogenesis, Steroid Synthesis, and Apoptosis

Although ATP production is a central role for the mitochondrion, this organelle has other functions that, in specific tissues or under specific circumstances, are also crucial. In adipose tissue, mitochondria generate heat to protect vital organs from low ambient temperature; in the adrenal glands and the gonads, mitochondria are the sites of steroid hormone synthesis; and in most or all tissues they are key participants in apoptosis (programmed cell death).

Uncoupled Mitochondria in Brown Adipose Tissue Produce Heat

We noted above that respiration slows when the cell is adequately supplied with ATP. There is a remarkable and instructive exception to this general rule. Most newborn mammals, including humans, have a type of adipose tissue called **brown adipose tissue** (**BAT**; p. 944) in which fuel oxidation serves, not to produce ATP, but to generate heat to keep the newborn warm. This specialized adipose tissue is brown because of the presence of large numbers of mitochondria and thus high concentrations of cytochromes, with heme groups that are strong absorbers of visible light.

The mitochondria of brown adipocytes are much like those of other mammalian cells, except in having a unique protein in their inner membrane. **Thermogenin**, also called **uncoupling protein 1** (the product of the *UCP1* gene), provides a path for protons to return to the matrix without passing through the F_oF_1 complex (**Fig. 19–36**). As a result of this short-circuiting of protons, the energy of oxidation is not conserved by ATP formation but is dissipated as heat, which contributes to maintaining the body temperature (see Fig. 23–16). Hibernating animals also depend on the activity of uncoupled BAT mitochondria to generate heat during their long dormancy (see Box 17–1). We will return to the role of thermogenin when we discuss the regulation of body mass in Chapter 23 (pp. 961–962).

Mitochondrial P-450 Oxygenases Catalyze Steroid Hydroxylations

Mitochondria are the site of biosynthetic reactions that produce steroid hormones, including the sex hormones, glucocorticoids, mineralocorticoids, and vitamin D hormone. These compounds are synthesized from cholesterol or a related sterol in a series of hydroxylations catalyzed by enzymes of the **cytochrome P-450** family, all of which have a critical heme group (its absorption at 450 nm gives this family its name). In the hydroxylation reactions, one atom of

FIGURE 19–37 Mitochondria of adrenal gland, specialized for steroid synthesis. As seen in this electron micrograph of a thin section of adrenal gland, the mitochondria are profuse and have extensive cristae, providing a large surface for the P-450 enzymes of the inner membrane.

molecular oxygen is incorporated into the substrate and the second is reduced to H_2O:

$$R\text{—}H + O_2 + NADPH + H^+ \longrightarrow R\text{—}OH + H_2O + NADP^+$$

There are dozens of P-450 enzymes, all situated in the inner mitochondrial membrane with their catalytic site exposed to the matrix. Steroidogenic cells are packed with mitochondria specialized for steroid synthesis; the mitochondria are generally larger than those in other tissues and have more extensive and highly convoluted inner membranes (**Fig. 19–37**).

The path of electron flow in the mitochondrial P-450 system is complex, involving a flavoprotein and an iron-sulfur protein that carry electrons from NADPH to the P-450 heme (**Fig. 19–38**). All P-450 enzymes have a heme that interacts with O_2 and a substrate-binding site that confers specificity.

Another large family of P-450 enzymes is found in the endoplasmic reticulum of hepatocytes. These enzymes catalyze reactions similar to the mitochondrial P-450 reactions, but their substrates include a wide variety of hydrophobic compounds, many of which are **xenobiotics**—compounds not found in nature but synthesized industrially. The P-450 enzymes of the ER have very broad and overlapping substrate specificities. Hydroxylation of the hydrophobic compounds makes them more water soluble, and they can then be cleared by the kidneys and excreted in urine. Among the substrates for these P-450 oxygenases are many commonly used prescription drugs. Metabolism by P-450 enzymes limits the drugs' lifetime in the bloodstream and their therapeutic effects. Humans differ in their genetic complement of P-450 enzymes in the ER, and in the extent to which certain P-450 enzymes have been induced, such as by a history of ethanol ingestion. In principle, therefore, an individual's genetics and personal history could figure into determinations of therapeutic

Intermembrane space (P side)

Uncoupling protein UCP1 (thermogenin)

H^+

Cyt c

F_o

I   II   III   IV

F_1

Heat

H^+

Matrix (N side)

FIGURE 19–36 Heat generation by uncoupled mitochondria. The uncoupling protein (thermogenin) in the mitochondria of brown adipose tissue, by providing an alternative route for protons to reenter the mitochondrial matrix, causes the energy conserved by proton pumping to be dissipated as heat.

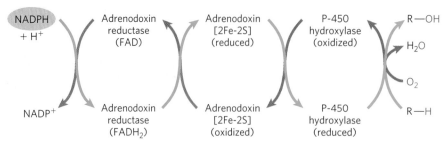

FIGURE 19–38 Path of electron flow in mitochondrial cytochrome P-450 reactions in adrenal gland. Two electrons are transferred from NADPH to the FAD-containing flavoprotein adrenodoxin reductase, which passes the electrons, one at a time, to adrenodoxin, a small, soluble 2Fe-2S protein. Adrenodoxin passes single electrons to the cytochrome P-450 hydroxylase, which interacts directly with O_2 and the substrate (R—H) to form the products, H_2O and R—OH.

drug dose; in practice, this precise tailoring of dosage is not yet economically feasible, but it may become so. ■

Mitochondria Are Central to the Initiation of Apoptosis

Apoptosis, also called **programmed cell death,** is a process in which individual cells die for the good of the organism (for example, in the course of normal embryonic development), and the organism conserves the cells' molecular components (amino acids, nucleotides, and so forth). Apoptosis may be triggered by an external signal, acting at a plasma membrane receptor, or by internal events such as DNA damage, viral infection, oxidative stress from the accumulation of ROS, or another stress such as a heat shock.

Mitochondria play a critical role in triggering apoptosis. When a stressor gives the signal for cell death, one early consequence is an increase in the permeability of the outer mitochondrial membrane, allowing cytochrome c to escape from the intermembrane space into the cytosol **(Fig. 19–39).** The increased permeability is due to the opening of the **permeability transition pore complex (PTPC),** a multisubunit protein in the outer membrane; its opening and closing are affected by several proteins that stimulate or suppress apoptosis. When released into the cytosol, cytochrome c interacts with monomers of the protein **Apaf-1 (apoptosis protease activating factor-1),** causing the formation of an **apoptosome** composed of seven Apaf-1 and seven cytochrome c molecules. The apoptosome provides the platform on which the protease procaspase-9 is activated to caspase-9, a member of a family of highly specific proteases (the **caspases**) involved in apoptosis. They share a critical Cys residue at their active site, and all cleave proteins only on the carboxyl-terminal

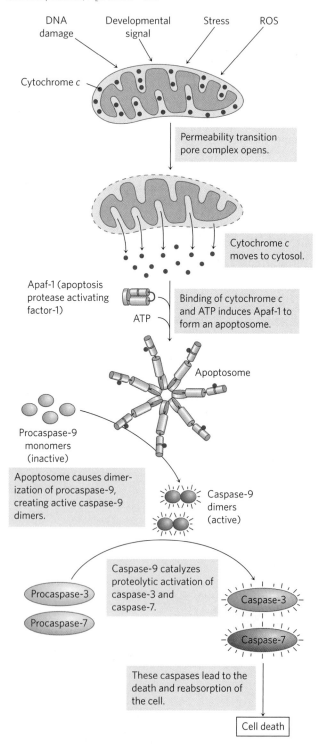

FIGURE 19–39 Role of cytochrome c in apoptosis. Cytochrome c is a small, soluble, mitochondrial protein, located in the intermembrane space, that carries electrons between Complex III and Complex IV during respiration. In a completely separate role, as outlined here, it acts as a trigger for apoptosis by stimulating the activation of a family of proteases called caspases.

side of *Asp* residues, thus the name "caspases." Activated caspase-9 initiates a cascade of proteolytic activations, with one caspase activating a second, and it in turn activating a third, and so forth (see Fig. 12–52). (This role of cytochrome *c* in apoptosis is a clear case of "moonlighting," in that one protein plays two very different roles in the cell; see Box 16–1.)

SUMMARY 19.4 Mitochondria in Thermogenesis, Steroid Synthesis, and Apoptosis

▶ In the brown adipose tissue of newborns, electron transfer is uncoupled from ATP synthesis and the energy of fuel oxidation is dissipated as heat.

▶ Hydroxylation reaction steps in the synthesis of steroid hormones in steroidogenic tissues (adrenal gland, gonads, liver, and kidney) take place in specialized mitochondria.

▶ Mitochondrial cytochrome *c*, released into the cytosol, participates in activation of caspase-9, one of the proteases involved in apoptosis.

19.5 Mitochondrial Genes: Their Origin and the Effects of Mutations

Mitochondria contain their own genome, a circular, double-stranded DNA (mtDNA) molecule. Each of the hundreds or thousands of mitochondria in a typical cell has about five copies of this genome. The human mitochondrial chromosome **(Fig. 19–40)** contains 37 genes

(16,569 bp), including 13 that encode subunits of proteins of the respiratory chain (Table 19–6); the remaining genes code for rRNA and tRNA molecules essential to the protein-synthesizing machinery of mitochondria. The great majority of mitochondrial proteins—about 1,100 different types—are encoded by nuclear genes, synthesized on cytoplasmic ribosomes, then imported into and assembled in the mitochondria (Chapter 27).

Mitochondria Evolved from Endosymbiotic Bacteria

The existence of mitochondrial DNA, ribosomes, and tRNAs supports the hypothesis of the endosymbiotic origin of mitochondria (see Fig. 1–38), which holds that the first organisms capable of aerobic metabolism, including respiration-linked ATP production, were bacteria. Primitive eukaryotes that lived anaerobically (by fermentation) acquired the ability to carry out oxidative phosphorylation when they established a symbiotic relationship with bacteria living in their cytosol. After much evolution and the movement of many bacterial genes into the nucleus of the "host" eukaryote, the endosymbiotic bacteria eventually became mitochondria.

This hypothesis presumes that early free-living bacteria had the enzymatic machinery for oxidative phosphorylation and predicts that their modern bacterial descendants must have respiratory chains closely similar

FIGURE 19–40 Mitochondrial genes and mutations. (a) Map of human mitochondrial DNA, showing the genes that encode proteins of Complex I, the NADH dehydrogenase (*ND1* to *ND6*); the cytochrome *b* of Complex III (*Cyt b*); the subunits of cytochrome oxidase (Complex IV) (*COI* to *COIII*); and two subunits of ATP synthase (*ATPase6* and *ATPase8*). The colors of the genes correspond to those of the complexes shown in Figure 19–7. Also included here are the genes for ribosomal RNAs (*rRNA*) and for some mitochondrion-specific transfer RNAs; tRNA specificity is indicated by the one-letter codes for amino acids. Arrows indicate the positions of mutations that cause Leber's hereditary optic neuropathy (LHON) and myoclonic epilepsy and ragged-red fiber disease (MERRF). Numbers in parentheses indicate the position of the altered nucleotides (nucleotide 1 is at the top of the circle and numbering proceeds counterclockwise). **(b)** Electron micrograph of an abnormal mitochondrion from the muscle of an individual with MERRF, showing the paracrystalline protein inclusions sometimes present in the mutant mitochondria.

Complex I
Complex III
Complex IV
ATP synthase
Transfer RNA
Ribosomal RNA
Control region of DNA

(a)

(b)

TABLE 19–6 Respiratory Proteins Encoded by Mitochondrial Genes in Humans

Complex	Number of subunits	Number of subunits encoded by mitochondrial DNA
I NADH dehydrogenase	43	7
II Succinate dehydrogenase	4	0
III Ubiquinone:cytochrome c oxidoreductase	11	1
IV Cytochrome oxidase	13	3
V ATP synthase	8	2

to those of modern eukaryotes. They do. Aerobic bacteria carry out NAD-linked electron transfer from substrates to O_2, coupled to the phosphorylation of cytosolic ADP. The dehydrogenases are located in the bacterial cytosol and the respiratory chain in the plasma membrane. The electron carriers translocate protons outward across the plasma membrane as electrons are transferred to O_2. Bacteria such as *Escherichia coli* have F_oF_1 complexes in their plasma membranes; the F_1 portion protrudes into the cytosol and catalyzes ATP synthesis from ADP and P_i as protons flow back into the cell through the proton channel of F_o.

The respiration-linked extrusion of protons across the bacterial plasma membrane also provides the driving force for other processes. Certain bacterial transport systems bring about uptake of extracellular nutrients (lactose, for example) against a concentration gradient, in symport with protons (see Fig. 11–41). And the rotary motion of bacterial flagella is provided by "proton turbines," molecular rotary motors driven not by ATP but directly by the transmembrane electrochemical potential generated by respiration-linked proton pumping **(Fig. 19–41)**. It seems likely that the chemiosmotic

mechanism evolved early, before the emergence of eukaryotes.

Mutations in Mitochondrial DNA Accumulate throughout the Life of the Organism

The respiratory chain is the major producer of reactive oxygen species in cells, so mitochondrial contents, including the mitochondrial genome, suffer the greatest exposure to, and damage by, ROS. Moreover, the mitochondrial DNA replication system is less effective than the nuclear system at correcting mistakes made during replication and at repairing DNA damage. As a consequence of these two factors, defects in mtDNA accumulate over time. One theory of aging is that this gradual accumulation of defects with increasing age is the primary cause of many of the "symptoms" of aging, which include, for example, progressive weakening of skeletal and heart muscle.

A unique feature of mitochondrial inheritance is the variation among individual cells, and between one individual organism and another, in the effects of a mtDNA mutation. A typical cell has hundreds or thousands of mitochondria, each with multiple copies of its own genome **(Fig. 19–42)**. Animals inherit essentially all of their mitochondria from the female parent. Eggs are large and contain 10^5 or 10^6 mitochondria, but sperm are

FIGURE 19–41 Rotation of bacterial flagella by proton-motive force. The shaft and rings at the base of the flagellum make up a rotary motor that has been called a "proton turbine." Protons ejected by electron transfer flow back into the cell through the turbine, causing rotation of the shaft of the flagellum. This motion differs fundamentally from the motion of muscle and of eukaryotic flagella and cilia, for which ATP hydrolysis is the energy source.

FIGURE 19–42 Single cells contain many mitochondria. A typical animal cell has hundreds or thousands of mitochondria, some fraction of which may contain genomes with mutations that affect mitochondrial function. This ovine (sheep) kidney epithelial cell was cultured in the laboratory, fixed, and then stained with fluorescent probes that show mitochondria as gold, microfilaments of actin as red, and nuclei as green when observed in the fluorescence microscope.

much smaller and contain far fewer mitochondria—perhaps 100 to 1,000. Furthermore, there is an active mechanism for targeting sperm-derived mitochondria for degradation in the fertilized egg. Just after fertilization, maternal phagosomes migrate to the site of sperm entry, engulf sperm mitochondria, and degrade them.

Suppose that, in a female organism, damage to one mitochondrial genome occurs in a germ cell from which oocytes develop, such that the germ cell contains mainly mitochondria with wild-type genes but one mitochondrion with a mutant gene. During the course of oocyte maturation, as this germ cell and its descendants repeatedly divide, the defective mitochondrion replicates and its progeny, all defective, are randomly distributed to daughter cells. Eventually, the mature egg cells contain different proportions of the defective mitochondria. When an egg cell is fertilized and undergoes the many divisions of embryonic development, the resulting somatic cells differ in their proportion of mutant mitochondria **(Fig. 19–43a)**. This **heteroplasmy** (in contrast to **homoplasmy**, in which every mitochondrial genome in every cell is the same) results in mutant phenotypes of varying degrees of severity. Cells (and tissues) containing mostly wild-type mitochondria have the wild-type phenotype; they are essentially normal. Other heteroplasmic cells will have intermediate phenotypes, some almost normal, others (with a high proportion of mutant mitochondria) abnormal (Fig. 19–41b). If the abnormal phenotype is associated with a disease (see below), individuals with the same mtDNA mutation may have disease symptoms of differing severity—depending on the number and distribution of affected mitochondria.

Some Mutations in Mitochondrial Genomes Cause Disease

A growing number of human diseases have been attributed to mutations in mitochondrial genes that reduce the cell's capacity to produce ATP. Some tissues and cell types—neurons, myocytes of both skeletal and cardiac muscle, and β cells of the pancreas—are less able than others to tolerate lowered ATP production and are therefore more affected by mutations in mitochondrial proteins.

A group of genetic diseases known as the **mitochondrial encephalomyopathies** affect primarily the brain and skeletal muscle. These diseases are invariably inherited from the mother, because, as noted above, a developing embryo derives all its mitochondria from the egg. The rare disease **Leber's hereditary optic neuropathy (LHON)** affects the central nervous system, including the optic nerves, causing bilateral loss of vision in early adulthood. A single base change in the mitochondrial gene *ND4* (Fig. 19–40a) changes an Arg residue to a His residue in a polypeptide of Complex I, and the result is mitochondria partially defective in electron transfer from NADH to ubiquinone. Although these mitochondria can produce some ATP by electron transfer from succinate, they apparently cannot supply sufficient ATP to support the very active metabolism of neurons. One result is damage to the

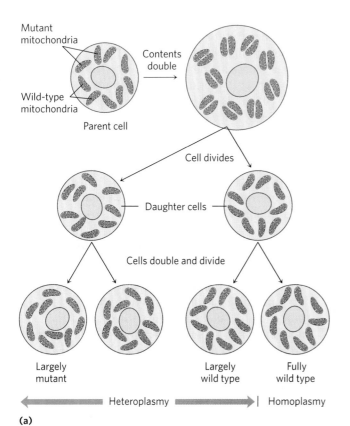

Largely mutant

Largely wild type

Fully wild type

◄━━━━ Heteroplasmy ━━━━► | Homoplasmy

(a)

(b)

FIGURE 19–43 Heteroplasmy in mitochondrial genomes. (a) When a mature egg cell is fertilized, all of the mitochondria in the resulting diploid cell (zygote) are maternal; none come from the sperm. If some fraction of the maternal mitochondria have a mutant gene, the random distribution of mitochondria during subsequent cell divisions yields some daughter cells with mostly mutant mitochondria, some with mostly wild-type mitochondria, and some in between; thus the daughter cells show varying degrees of heteroplasmy. **(b)** Different degrees of heteroplasmy produce different cellular phenotypes. This section of human muscle tissue is from an individual with defective cytochrome oxidase. The cells have been stained to make wild-type cells blue and cells with mutant cytochrome oxidase brown. As the micrograph shows, different cells in the same tissue are affected to different degrees by the mitochondrial mutation.

optic nerve, leading to blindness. A single base change in the mitochondrial gene for cytochrome b, a component of Complex III, also produces LHON, demonstrating that the pathology results from a general reduction of mitochondrial function, not specifically from a defect in electron transfer through Complex I.

A mutation (in *ATP6*) that affects the proton pore in ATP synthase leads to low rates of ATP synthesis while leaving the respiratory chain intact. The oxidative stress due to the continued supply of electrons from NADH increases the production of ROS, and the damage to mitochondria caused by ROS sets up a vicious cycle. Half of individuals with this mutant gene die within days or months of birth.

Myoclonic epilepsy with ragged-red fiber disease (MERRF) is caused by a mutation in the mitochondrial gene that encodes a tRNA specific for lysine (tRNA^{Lys}). This disease, characterized by uncontrollable muscular jerking, apparently results from defective production of several of the proteins that require mitochondrial tRNAs for their synthesis. Skeletal muscle fibers of individuals with MERRF have abnormally shaped mitochondria that sometimes contain paracrystalline structures (Fig. 19–40b). Other mutations in mitochondrial genes are believed to be responsible for the progressive muscular weakness that characterizes mitochondrial myopathy and for enlargement and deterioration of the heart muscle in hypertrophic cardiomyopathy. According to one hypothesis on the progressive changes that accompany aging, the accumulation of mutations in mtDNA during a lifetime of exposure to DNA-damaging agents such as $\cdot O_2^-$ results in mitochondria that cannot supply sufficient ATP for normal cellular function. Mitochondrial disease can also result from mutations in any of the ~1,100 nuclear genes that encode mitochondrial proteins. ∎

Diabetes Can Result from Defects in the Mitochondria of Pancreatic β Cells

The mechanism that regulates the release of insulin from pancreatic β cells hinges on the ATP concentration in those cells. When blood glucose is high, β cells take up glucose and oxidize it by glycolysis and the citric acid cycle, raising [ATP] above a threshold level **(Fig. 19–44)**. When [ATP] exceeds this threshold, an ATP-gated K$^+$ channel in the plasma membrane closes, depolarizing the membrane and triggering insulin release (see Fig. 23–27). Pancreatic β cells with defects in oxidative phosphorylation cannot increase [ATP] above this threshold, and the resulting failure of insulin release effectively produces diabetes. For example, defects in the gene for glucokinase, the hexokinase IV isozyme present in β cells, lead to a rare form of diabetes, MODY2 (see Box 15–3); low glucokinase activity prevents the generation of above-threshold [ATP], blocking insulin secretion. Mutations in the mitochondrial tRNA^{Lys} or tRNA^{Leu} genes also

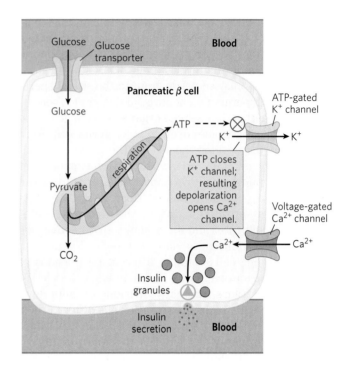

FIGURE 19–44 Defective oxidative phosphorylation in pancreatic β cells blocks insulin secretion. In the normal situation depicted here, when blood glucose rises, the production of ATP in β cells increases. ATP, by blocking K$^+$ channels, depolarizes the plasma membrane and thus opens the voltage-gated Ca^{2+} channels. The resulting influx of Ca^{2+} triggers exocytosis of insulin-containing secretory vesicles, releasing insulin. When oxidative phosphorylation in β cells is defective, [ATP] is never sufficient to trigger this process, and insulin is not released.

compromise mitochondrial ATP production, and type 2 diabetes mellitus is common among individuals with these defects (although these cases make up a very small fraction of all cases of diabetes).

When nicotinamide nucleotide transhydrogenase, which is part of the mitochondrial defense against ROS (see Fig. 19–18), is genetically defective, the accumulation of ROS damages mitochondria, slowing ATP production and blocking insulin release by β cells (Fig. 19–44). Damage caused by ROS, including damage to mtDNA, may also underlie other human diseases; there is some evidence for its involvement in Alzheimer, Parkinson, and Huntington diseases and in heart failure, as well as in aging. ∎

SUMMARY 19.5 Mitochondrial Genes: Their Origin and the Effects of Mutations

▶ A small proportion of human mitochondrial proteins, 13 in all, are encoded by the mitochondrial genome and synthesized in mitochondria. About 1,100 mitochondrial proteins are encoded in nuclear genes and imported into mitochondria after their synthesis.

▶ Mitochondria arose from aerobic bacteria that entered into an endosymbiotic relationship with ancestral eukaryotes.

► Mutations in the mitochondrial genome accumulate over the life of the organism. Mutations in the genes that encode components of the respiratory chain, ATP synthase, and the ROS-scavenging system, and even in tRNA genes, can cause a variety of human diseases, which often most severely affect muscle, heart, pancreatic β cells, and brain.

PHOTOSYNTHESIS: HARVESTING LIGHT ENERGY

We now turn to another reaction sequence in which the flow of electrons is coupled to the synthesis of ATP: light-driven phosphorylation. The capture of solar energy by photosynthetic organisms and its conversion to the chemical energy of reduced organic compounds is the ultimate source of nearly all biological energy on Earth. Photosynthetic and heterotrophic organisms live in a balanced steady state in the biosphere **(Fig. 19–45)**. Photosynthetic organisms trap solar energy and form ATP and NADPH, which they use as energy sources to make carbohydrates and other organic compounds from CO_2 and H_2O; simultaneously, they release O_2 into the atmosphere. Aerobic heterotrophs (humans, for example, as well as plants during dark periods) use the O_2 so formed to degrade the energy-rich organic products of photosynthesis to CO_2 and H_2O, generating ATP. The CO_2 returns to the atmosphere, to be used again by photosynthetic organisms. Solar energy thus provides the driving force for the continuous cycling of CO_2 and O_2 through the biosphere and provides the reduced substrates—fuels, such as glucose—on which nonphotosynthetic organisms depend.

Photosynthesis occurs in a variety of bacteria and in unicellular eukaryotes (algae) as well as in plants. Although the process in these organisms differs in detail, the underlying mechanisms are remarkably similar, and much of our understanding of photosynthesis in vascular plants is derived from studies of simpler organisms. The overall equation for photosynthesis in plants describes an oxidation-reduction reaction in which H_2O donates electrons (as hydrogen) for the reduction of CO_2 to carbohydrate (CH_2O):

$$CO_2 + H_2O \xrightarrow{\text{light}} O_2 + (CH_2O)$$

19.6 General Features of Photophosphorylation

Unlike NADH (the major electron donor in oxidative phosphorylation), H_2O is a poor donor of electrons; its standard reduction potential is 0.816 V, compared with -0.320 V for NADH. Photophosphorylation differs from oxidative phosphorylation in requiring the input of energy in the form of light to *create* a good electron donor and a good electron acceptor (see Fig. 19–1). In photophosphorylation, electrons flow through a series of membrane-bound carriers including cytochromes, quinones, and iron-sulfur proteins, while protons are pumped across a membrane to create an electrochemical potential. Electron transfer and proton pumping are catalyzed by membrane complexes homologous in structure and function to Complex III of mitochondria. The electrochemical potential they produce is the driving force for ATP synthesis from ADP and P_i, catalyzed by a membrane-bound ATP synthase complex closely similar to that of mitochondria and bacteria.

Photosynthesis in plants encompasses two processes: the **light-dependent reactions**, or **light reactions**, which occur only when plants are illuminated, and the **carbon-assimilation reactions** (or **carbon-fixation reactions**), sometimes misleadingly called the dark reactions, which are driven by products of the light reactions **(Fig. 19–46)**. In the light reactions, chlorophyll and other pigments of photosynthetic cells absorb light energy and conserve it as ATP and NADPH; simultaneously, O_2 is evolved. In the carbon-assimilation reactions, ATP and NADPH are used to reduce CO_2 to form triose phosphates, starch, and sucrose, and other products derived from them. In this chapter we are concerned only with the light-dependent reactions that lead to the synthesis of ATP and NADPH. The reduction of CO_2 is described in Chapter 20.

Photosynthesis in Plants Takes Place in Chloroplasts

In photosynthetic eukaryotic cells, both the light-dependent and the carbon-assimilation reactions take place in the chloroplasts **(Fig. 19–47)**, intracellular organelles that are variable in shape and generally a few micrometers in diameter. Like mitochondria, they are surrounded by two membranes, an outer membrane that is permeable to small molecules and ions, and an inner membrane that encloses the internal compartment.

FIGURE 19–45 Solar energy as the ultimate source of all biological energy. Photosynthetic organisms use the energy of sunlight to manufacture glucose and other organic products, which heterotrophic cells use as energy and carbon sources.

FIGURE 19–46 The light reactions of photosynthesis generate energy-rich NADPH and ATP at the expense of solar energy. NADPH and ATP are used in the carbon-assimilation reactions, which occur in light or darkness, to reduce CO_2 to form trioses and more complex compounds (such as glucose) derived from trioses.

This compartment contains many flattened, membrane-surrounded vesicles or sacs, the **thylakoids**, usually arranged in stacks called **grana** (Fig. 19–47b). Embedded in the thylakoid membranes (commonly called **lamellae**) are the photosynthetic pigments and the enzyme complexes that carry out the light reactions and ATP synthesis. The **stroma** (the aqueous phase enclosed by the inner membrane) contains

(a)

(b)

FIGURE 19–47 Chloroplast. (a) Schematic diagram. **(b)** Electron micrograph at high magnification showing grana, stacks of thylakoid membranes.

most of the enzymes required for the carbon-assimilation reactions.

Light Drives Electron Flow in Chloroplasts

In 1937 Robert Hill found that when leaf extracts containing chloroplasts were illuminated, they (1) evolved O_2 and (2) reduced a nonbiological electron acceptor added to the medium, according to the **Hill reaction**:

$$2H_2O + 2A \xrightarrow{\text{light}} 2AH_2 + O_2$$

where A is the artificial electron acceptor, or **Hill reagent**. One Hill reagent, the dye 2,6-dichlorophenol-indophenol, is blue when oxidized (A) and colorless when reduced (AH_2), making the reaction easy to follow.

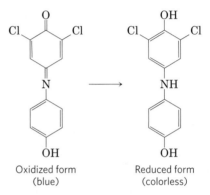

Oxidized form (blue) → Reduced form (colorless)

Dichlorophenolindophenol

When a leaf extract supplemented with the dye was illuminated, the blue dye became colorless and O_2 was evolved. In the dark, neither O_2 evolution nor dye reduction took place. This was the first evidence that absorbed light energy causes electrons to flow from H_2O to an electron acceptor. Moreover, Hill found that CO_2 was neither required nor reduced to a stable form under these conditions; O_2 evolution could be dissociated from CO_2 reduction. Several years later Severo Ochoa showed that $NADP^+$ is the biological electron acceptor in chloroplasts, according to the equation

$$2H_2O + 2NADP^+ \xrightarrow{\text{light}} 2NADPH + 2H^+ + O_2$$

To understand this photochemical process, we must first consider the more general topic of the effects of light absorption on molecular structure.

SUMMARY 19.6 General Features of Photophosphorylation

▶ Photosynthesis takes place in the chloroplasts of algae and plants, structures enclosed in double membranes and filled with stacked membranous discs (thylakoid membranes) containing the photosynthetic machinery.

▶ The light reactions of photosynthesis are those directly dependent on the absorption of light; the resulting photochemistry takes electrons from H_2O

and drives them through a series of membrane-bound carriers, producing NADPH and ATP.

▶ The carbon-assimilation reactions of photosynthesis reduce CO_2 with electrons from NADPH and energy from ATP, forming trioses, hexoses, and a wide variety of carbohydrates derived from them.

19.7 Light Absorption

Visible light is electromagnetic radiation of wavelengths 400 to 700 nm, a small part of the electromagnetic spectrum **(Fig. 19–48)**, ranging from violet to red. The energy of a single **photon** (a quantum of light) is greater at the violet end of the spectrum than at the red end; shorter wavelength (and higher frequency) corresponds to higher energy. The energy, E, in a single photon of visible light is given by the Planck equation:

$$E = h\nu = hc/\lambda$$

where h is Planck's constant (6.626×10^{-34} J·s), ν is the frequency of the light in cycles/s, c is the speed of light (3.00×10^8 m/s), and λ is the wavelength in meters. The energy of a photon of visible light ranges from 150 kJ/einstein for red light to ~300 kJ/einstein for violet light.

WORKED EXAMPLE 19–3 Energy of a Photon

The light used by vascular plants for photosynthesis has a wavelength of about 700 nm. Calculate the energy in a "mole" of photons (an einstein) of light of this wavelength, and compare this with the energy needed to synthesize a mole of ATP.

Solution: The energy in a single photon is given by the Planck equation. At a wavelength of 700×10^{-9} m, the energy of a photon is

$$
\begin{aligned}
E &= hc/\lambda \\
&= \frac{[(6.626 \times 10^{-34}\ \text{J·s})(3.00 \times 10^8\ \text{m/s})]}{(7.00 \times 10^{-7}\ \text{m})} \\
&= 2.84 \times 10^{-19}\ \text{J}
\end{aligned}
$$

An einstein of light is Avogadro's number (6.022×10^{23}) of photons; thus the energy of one einstein of photons at 700 nm is given by

$$
\begin{aligned}
(2.84 \times 10^{-19}\ &\text{J/photon})(6.022 \times 10^{23}\ \text{photons/einstein}) \\
&= 17.1 \times 10^4\ \text{J/einstein} \\
&= 171\ \text{kJ/einstein}.
\end{aligned}
$$

So, a "mole" of photons of red light has about five times the energy needed to produce a mole of ATP from ADP and P_i (30.5 kJ/mol).

When a photon is absorbed, an electron in the absorbing molecule (chromophore) is lifted to a higher energy level. This is an all-or-nothing event: to be absorbed, the photon must contain a quantity of energy (a **quantum**) that exactly matches the energy of the electronic transition. A molecule that has absorbed a photon is in an **excited state**, which is generally unstable. An electron lifted into a higher-energy orbital usually returns rapidly to its lower-energy orbital; the excited molecule decays to the stable **ground state**, giving up the absorbed quantum as light or heat or using it to do chemical work. Light emission accompanying decay of excited molecules (called **fluorescence**) is always at a longer wavelength (lower energy) than that of the absorbed light (see Box 12–3). An alternative mode of decay important in photosynthesis involves direct transfer of excitation energy from an excited molecule to a neighboring molecule. Just as the photon is a quantum of light energy, so the **exciton** is a quantum of energy passed from an excited molecule to another molecule in a process called **exciton transfer**.

Chlorophylls Absorb Light Energy for Photosynthesis

The most important light-absorbing pigments in the thylakoid membranes are the **chlorophylls**, green pigments with polycyclic, planar structures resembling the protoporphyrin of hemoglobin (see Fig. 5–1), except that Mg^{2+}, not Fe^{2+}, occupies the central position **(Fig. 19–49)**. The four inward-oriented nitrogen atoms of chlorophyll are coordinated with the Mg^{2+}. All chlorophylls have a long **phytol** side chain, esterified to a

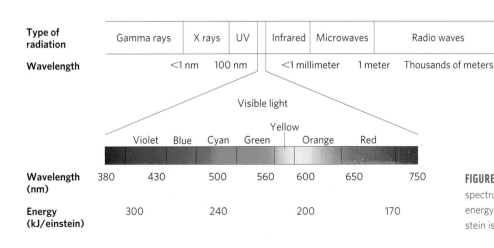

Type of radiation	Gamma rays	X rays	UV		Infrared	Microwaves	Radio waves

| Wavelength | | <1 nm | 100 nm | | <1 millimeter | 1 meter | Thousands of meters |

Visible light

	Violet	Blue	Cyan	Green	Yellow	Orange	Red

| Wavelength (nm) | 380 | 430 | | 500 | 560 | 600 | 650 | 750 |

| Energy (kJ/einstein) | | 300 | | 240 | | 200 | 170 | |

FIGURE 19–48 Electromagnetic radiation. The spectrum of electromagnetic radiation, and the energy of photons in the visible range. One einstein is 6.022×10^{23} photons.

FIGURE 19–49 Primary and secondary photopigments. (a) Chlorophylls *a* and *b* and bacteriochlorophyll are the primary gatherers of light energy. **(b)** Phycoerythrobilin and phycocyanobilin (phycobilins) are the antenna pigments in cyanobacteria and red algae. **(c)** *β*-Carotene (a carotenoid) and **(d)** lutein (a xanthophyll) are accessory pigments in plants. The conjugated systems (alternating single and double bonds) in these molecules largely account for the absorption of visible light.

carboxyl-group substituent in ring IV, and chlorophylls also have a fifth five-membered ring not present in heme.

The heterocyclic five-ring system that surrounds the Mg^{2+} has an extended polyene structure, with alternating single and double bonds. Such polyenes characteristically show strong absorption in the visible region of the spectrum **(Fig. 19–50)**; the chlorophylls have unusually high molar extinction coefficients (see Box 3–1) and are therefore particularly well-suited for absorbing visible light during photosynthesis.

Chloroplasts always contain both chlorophyll *a* and chlorophyll *b* (Fig. 19–49a). Although both are green, their absorption spectra are sufficiently different (Fig. 19–50) that they complement each other's range of light absorption in the visible region. Most plants contain about twice as much chlorophyll *a* as chlorophyll *b*. The pigments in

FIGURE 19–50 Absorption of visible light by photopigments. Plants are green because their pigments absorb light from the red and blue regions of the spectrum, leaving primarily green light to be reflected. Compare the absorption spectra of the pigments with the spectrum of sunlight reaching the earth's surface; the combination of chlorophylls (*a* and *b*) and accessory pigments enables plants to harvest most of the energy available in sunlight. The relative amounts of chlorophylls and accessory pigments are characteristic of a particular plant species. Variation in the proportions of these pigments is responsible for the range of colors of photosynthetic organisms, from the deep blue-green of spruce needles, to the greener green of maple leaves, to the red, brown, or purple color of some species of multicellular algae and the leaves of some foliage plants favored by gardeners.

algae and photosynthetic bacteria include chlorophylls that differ only slightly from the plant pigments.

Chlorophyll is always associated with specific binding proteins, forming **light-harvesting complexes (LHCs)** in which chlorophyll molecules are fixed in relation to each other, to other protein complexes, and to the membrane. One light-harvesting complex (LHCII; **Fig. 19–51**) contains seven molecules of chlorophyll *a*, five of chlorophyll *b*, and two of the accessory pigment lutein (see below).

Cyanobacteria and red algae employ **phycobilins** such as phycoerythrobilin and phycocyanobilin (Fig. 19–49b) as their light-harvesting pigments. These open-chain tetrapyrroles have the extended polyene system found in chlorophylls, but not their cyclic structure or central Mg^{2+}. Phycobilins are covalently linked to specific binding proteins, forming **phycobiliproteins**, which associate in highly ordered complexes called phycobilisomes **(Fig. 19–52)** that constitute the primary light-harvesting structures in these microorganisms.

Accessory Pigments Extend the Range of Light Absorption

In addition to chlorophylls, thylakoid membranes contain secondary light-absorbing pigments, or **accessory pigments**, called carotenoids. **Carotenoids** may be yellow, red, or purple. The most important are **β-carotene**, which is a red-orange isoprenoid, and the yellow carotenoid **lutein** (Fig. 19–49c, d). The carotenoid pigments absorb light at wavelengths not absorbed by the chlorophylls (Fig. 19–50) and thus are supplementary light receptors.

Experimental determination of the effectiveness of light of different colors in promoting photosynthesis

FIGURE 19–51 A light-harvesting complex, LHCII. (PDB ID 2BHW) The functional unit is an LHC trimer, with 36 chlorophyll and 6 lutein molecules. Shown here is a monomer, viewed in the plane of the membrane, with its three transmembrane α-helical segments, seven chlorophyll a molecules (light green), five chlorophyll b molecules (dark green), and two molecules of the accessory pigment lutein (yellow), which form an internal cross-brace.

FIGURE 19–52 A phycobilisome. In these highly structured assemblies found in cyanobacteria and red algae, phycobilin pigments bound to specific proteins form complexes called phycoerythrin (PE), phycocyanin (PC), and allophycocyanin (AP). The energy of photons absorbed by PE or PC is conveyed through AP (a phycocyanobilin-binding protein) to chlorophyll *a* of the reaction center by exciton transfer, a process discussed in the text.

yields an **action spectrum (Fig. 19–53)**, often useful in identifying the pigment primarily responsible for a biological effect of light. By capturing light in a region of the spectrum not used by other organisms, a photosynthetic organism can claim a unique ecological niche. For example, the phycobilins in red algae and cyanobacteria absorb light in the range 520 to 630 nm (Fig. 19–50), allowing them to occupy niches where light of lower or higher wavelength has been filtered out by the pigments of other organisms living in the water above them, or by the water itself.

Chlorophyll Funnels the Absorbed Energy to Reaction Centers by Exciton Transfer

The light-absorbing pigments of thylakoid or bacterial membranes are arranged in functional arrays called **photosystems**. In spinach chloroplasts, for example, each photosystem contains about 200 chlorophyll and 50 carotenoid molecules. All the pigment molecules in a photosystem can absorb photons, but only a few chlorophyll molecules associated with the **photochemical reaction center** are specialized to transduce light into chemical energy. The other pigment molecules in a photosystem are called **light-harvesting** or **antenna molecules**. They absorb light energy and transmit it rapidly and efficiently to the reaction center **(Fig. 19–54)**.

The chlorophyll molecules in light-harvesting complexes have light-absorption properties that are subtly different from those of free chlorophyll. When isolated chlorophyll molecules in vitro are excited by light, the

FIGURE 19–53 Two ways to determine the action spectrum for photosynthesis. (a) Results of a classic experiment performed by T. W. Englemann in 1882 to determine the wavelength of light that is most effective in supporting photosynthesis. Englemann placed cells of a filamentous photosynthetic alga on a microscope slide and illuminated them with light from a prism, so that one part of the filament received mainly blue light, another part yellow, another red. To determine which algal cells carried out photosynthesis most actively, Englemann also placed on the microscope slide bacteria known to migrate toward regions of high O_2 concentration. After a period of illumination, the distribution of bacteria showed highest O_2 levels (produced by photosynthesis) in the regions illuminated with violet and red light.

(b) Results of a similar experiment that used modern techniques (an oxygen electrode) for the measurement of O_2 production. An action spectrum (as shown here) describes the relative rate of photosynthesis for illumination with a constant number of photons of different wavelengths. An action spectrum is useful because, by comparison with absorption spectra (such as those in Fig. 19–50), it suggests which pigments can channel energy into photosynthesis.

absorbed energy is quickly released as fluorescence and heat, but when chlorophyll in intact leaves is excited by visible light **(Fig. 19–55,** step **❶)**, very little fluorescence is observed. Instead, the excited antenna chlorophyll transfers energy directly to a neighboring chlorophyll molecule, which becomes excited as the first molecule returns to its ground state (step **❷**). This transfer of energy, exciton transfer, extends to a third, fourth, or subsequent neighbor, until one of a special pair of

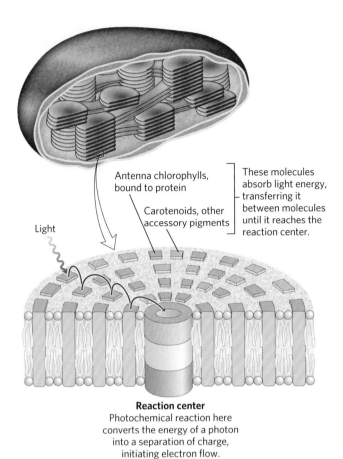

These molecules absorb light energy, transferring it between molecules until it reaches the reaction center.

Antenna chlorophylls, bound to protein

Carotenoids, other accessory pigments

Light

Reaction center
Photochemical reaction here converts the energy of a photon into a separation of charge, initiating electron flow.

FIGURE 19–54 Organization of photosystems in the thylakoid membrane. Photosystems are tightly packed in the thylakoid membrane, with several hundred antenna chlorophylls and accessory pigments surrounding a photoreaction center. Absorption of a photon by any of the antenna chlorophylls leads to excitation of the reaction center by exciton transfer (red arrow). Also embedded in the thylakoid membrane are the cytochrome b_6f complex and ATP synthase (see Fig. 19-61).

chlorophyll a molecules at the photochemical reaction center is excited (step ❸). In this excited chlorophyll molecule, an electron is promoted to a higher-energy orbital. This electron then passes to a nearby electron acceptor that is part of the electron-transfer chain, leaving the reaction-center chlorophyll with a missing electron (an "electron hole," denoted by + in Fig. 19–55) (step ❹). The electron acceptor acquires a negative charge in this transaction. The electron lost by the reaction-center chlorophyll is replaced by an electron from a neighboring electron-donor molecule (step ❺), which thereby becomes positively charged. In this way, *excitation by light causes electric charge separation and initiates an oxidation-reduction chain.*

FIGURE 19–55 Exciton and electron transfer. This generalized scheme shows conversion of the energy of an absorbed photon into separation of charges at the reaction center. The steps are further described in the text. Note that step ❶ may repeat between successive antenna molecules until the exciton reaches a reaction-center chlorophyll. The asterisk (٭) represents the excited state of a molecule.

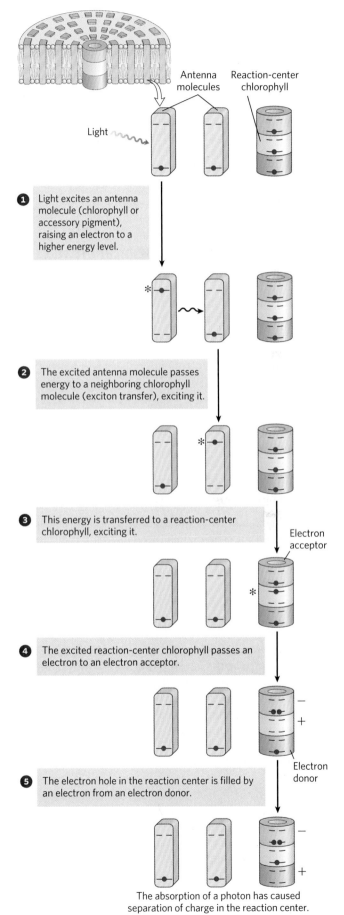

Antenna molecules   Reaction-center chlorophyll

Light

❶ Light excites an antenna molecule (chlorophyll or accessory pigment), raising an electron to a higher energy level.

❷ The excited antenna molecule passes energy to a neighboring chlorophyll molecule (exciton transfer), exciting it.

❸ This energy is transferred to a reaction-center chlorophyll, exciting it.

Electron acceptor

❹ The excited reaction-center chlorophyll passes an electron to an electron acceptor.

Electron donor

❺ The electron hole in the reaction center is filled by an electron from an electron donor.

The absorption of a photon has caused separation of charge in the reaction center.

▶ A photon of visible light possesses enough energy to bring about photochemical reactions, which in photosynthetic organisms lead eventually to ATP synthesis.

▶ In the light reactions of plants, absorption of a photon excites chlorophyll molecules and other (accessory) pigments, which funnel the energy into reaction centers in the thylakoid membranes. In the reaction centers, photoexcitation results in a charge separation that produces a strong electron donor (reducing agent) and a strong electron acceptor.

19.8 The Central Photochemical Event: Light-Driven Electron Flow

Light-driven electron transfer in plant chloroplasts during photosynthesis is accomplished by multienzyme systems in the thylakoid membrane. Our current picture of photosynthetic mechanisms is a composite, drawn from studies of plant chloroplasts and a variety of bacteria and algae. Determination of the molecular structures of bacterial photosynthetic complexes (by x-ray crystallography) has given us a much improved understanding of the molecular events in photosynthesis in general.

Bacteria Have One of Two Types of Single Photochemical Reaction Center

One major insight from studies of photosynthetic bacteria came in 1952 when Louis Duysens found that illumination of the photosynthetic membranes of the purple bacterium *Rhodospirillum rubrum* with a pulse of light of a specific wavelength (870 nm) caused a temporary decrease in the absorption of light at that wavelength; a pigment was "bleached" by 870 nm light. Later studies by Bessel Kok and Horst Witt showed similar bleaching of plant chloroplast pigments by light of 680 and 700 nm. Furthermore, addition of the (nonbiological) electron acceptor $[Fe(CN)_6]^{3-}$ (ferricyanide) caused bleaching at these wavelengths *without illumination*. These findings indicated that bleaching of the pigments was due to the loss of an electron from a photochemical reaction center. The pigments were named for the wavelength of maximum bleaching: P870, P680, and P700.

Photosynthetic bacteria have relatively simple phototransduction machinery, with one of two general types of reaction center. One type (found in purple bacteria) passes electrons through **pheophytin** (chlorophyll lacking the central Mg^{2+} ion) to a quinone. The other (in green sulfur bacteria) passes electrons through a quinone to an iron-sulfur center. Cyanobacteria and plants have two photosystems (PSI, PSII), one of each type, acting in tandem. Biochemical and biophysical

studies have revealed many of the molecular details of reaction centers of bacteria, which therefore serve as prototypes for the more complex phototransduction systems of plants.

The Pheophytin-Quinone Reaction Center (Type II Reaction Center) The photosynthetic machinery in purple bacteria consists of three basic modules (**Fig. 19–56a**): a single reaction center (P870), a cytochrome bc_1 electron-transfer complex similar to Complex III of the mitochondrial electron-transfer chain, and an ATP synthase, also similar to that of mitochondria. Illumination drives electrons through pheophytin and a quinone to the cytochrome bc_1 complex; after passing through the complex, electrons flow through cytochrome c_2 back to the reaction center, restoring its preillumination state. This light-driven cyclic flow of electrons provides the energy for proton pumping by the cytochrome bc_1 complex. Powered by the resulting proton gradient, ATP synthase produces ATP, exactly as in mitochondria.

The three-dimensional structures of the reaction centers of purple bacteria (*Rhodopseudomonas viridis* and *Rhodobacter sphaeroides*), deduced from x-ray crystallography, shed light on how phototransduction takes place in a pheophytin-quinone reaction center. The *R. viridis* reaction center (**Fig. 19–57a**) is a large protein complex containing four polypeptide subunits and 13 cofactors: two pairs of bacterial chlorophylls, a pair of pheophytins, two quinones, a nonheme iron, and four hemes in the associated *c*-type cytochrome.

The extremely rapid sequence of electron transfers shown in Figure 19–57b has been deduced from physical studies of the bacterial pheophytin-quinone centers, using brief flashes of light to trigger phototransduction and a variety of spectroscopic techniques to follow the flow of electrons through several carriers. A pair of bacteriochlorophylls—the "special pair," designated $(Chl)_2$—is the site of the initial photochemistry in the bacterial reaction center. Energy from a photon absorbed by one of the many antenna chlorophyll molecules surrounding the reaction center reaches $(Chl)_2$ by exciton transfer. When these two chlorophyll molecules—so close that their bonding orbitals overlap—absorb an exciton, the redox potential of $(Chl)_2$ is shifted, by an amount equivalent to the energy of the photon, converting the special pair to a very strong electron donor. The $(Chl)_2$ donates an electron that passes through a neighboring chlorophyll monomer to pheophytin (Pheo). This produces two radicals, one positively charged (the special pair of chlorophylls) and one negatively charged (the pheophytin):

$$(Chl)_2 + 1 \text{ exciton} \longrightarrow (Chl)_2^* \qquad \text{(excitation)}$$

$$(Chl)_2^* + Pheo \longrightarrow$$
$${}^\bullet(Chl)_2^+ + {}^\bullet Pheo^- \quad \text{(charge separation)}$$

The pheophytin radical now passes its electron to a tightly bound molecule of quinone (Q_A), converting it to

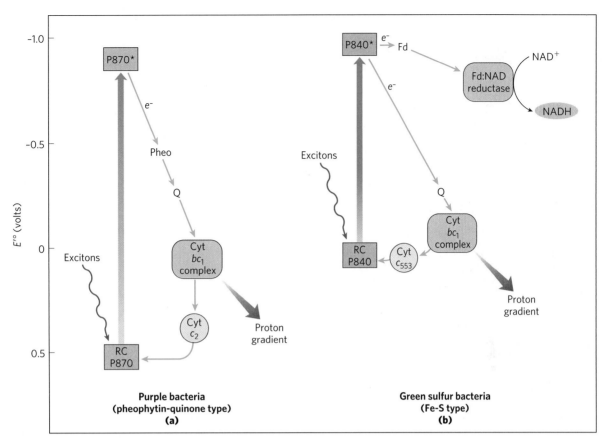

FIGURE 19-56 Functional modules of photosynthetic machinery in purple bacteria and green sulfur bacteria. (a) In purple bacteria, light energy drives electrons from the reaction-center P870 through pheophytin (Pheo), a quinone (Q), and the cytochrome bc_1 complex, then through cytochrome c_2 and thus back to the reaction center. Electron flow through the cytochrome bc_1 complex causes proton pumping, creating an electro-chemical potential that powers ATP synthesis. **(b)** Green sulfur bacteria have two routes for electrons driven by excitation of P840: a cyclic route through a quinone to the cytochrome bc_1 complex and back to the reaction center via cytochrome c, and a noncyclic route from the reaction center through the iron-sulfur protein ferredoxin (Fd), then to NAD^+ in a reaction catalyzed by ferredoxin:NAD reductase.

a semiquinone radical, which immediately donates its extra electron to a second, loosely bound quinone (Q_B). Two such electron transfers convert Q_B to its fully reduced form, Q_BH_2, which is free to diffuse in the membrane bilayer, away from the reaction center:

$$2\ {}^{\bullet}Pheo^- + 2H^+ + Q_B \longrightarrow 2\ Pheo + Q_BH_2$$
$$\text{(quinone reduction)}$$

The hydroquinone (Q_BH_2), carrying in its chemical bonds some of the energy of the photons that originally excited P870, enters the pool of reduced quinone (QH_2) dissolved in the membrane and moves through the lipid phase of the bilayer to the cytochrome bc_1 complex.

Like the homologous Complex III in mitochondria, the cytochrome bc_1 complex of purple bacteria carries electrons from a quinol donor (QH_2) to an electron acceptor, using the energy of electron transfer to pump protons across the membrane, producing a proton-motive force. The path of electron flow through this complex is believed to be very similar to that through mitochondrial Complex III, involving a Q cycle (Fig. 19–12) in which protons are consumed on one side of

the membrane and released on the other. The ultimate electron acceptor in purple bacteria is the electron-depleted form of P870, ${}^{\bullet}(Chl)_2^+$ (Fig. 19–56a). Electrons move from the cytochrome bc_1 complex to P870 via a soluble c-type cytochrome, cytochrome c_2. The electron-transfer process completes the cycle, returning the reaction center to its unbleached state, ready to absorb another exciton from antenna chlorophyll.

A remarkable feature of this system is that all the chemistry occurs in the *solid state*, with reacting species held close together in the right orientation for reaction. The result is a very fast and efficient series of reactions.

The Fe-S Reaction Center (Type I Reaction Center) Photosynthesis in green sulfur bacteria involves the same three modules as in purple bacteria, but the process differs in several respects and involves additional enzymatic reactions (Fig. 19–56b). Excitation causes an electron to move from the reaction center to the cytochrome bc_1 complex via a quinone carrier. Electron transfer through this complex powers proton transport and creates the proton-motive force used for ATP synthesis, just as in

P side

N side

(a)

Light

Hemes of
c-type
cytochrome

4 (270 ns) ——— Bacteriochlorophyll (2)
((Chl)$_2$, the special pair)

1 (3 ps) ——— Bacteriochlorophyll (2)
(accessory pigments)

——— Bacteriopheophytin (2)

2 (200 ps)

Fe

Q$_B$
(quinone)

Q$_A$
(quinone)

3

(b)   (6μs)

FIGURE 19–57 Photoreaction center of the purple bacterium *Rhodopseudomonas viridis*. (PDB ID 1PRC) **(a)** The system has four components: three subunits, H, M, and L (brown, blue, and gray, respectively), with a total of 11 transmembrane helical segments, and a fourth protein, cytochrome *c* (yellow), associated with the complex at the membrane surface. Subunits L and M are paired transmembrane proteins that together form a cylindrical structure with roughly bilateral symmetry about its long axis. Shown as space-filling models (and in (b) as ball-and-stick structures) are the prosthetic groups that participate in the photochemical events. Bound to the L and M chains are two pairs of bacteriochlorophyll molecules (green); one of the pairs—the "special pair," (Chl)$_2$—is the site of the first photochemical changes after light absorption. Also incorporated in the system are a pair of pheophytin *a* (Pheo *a*) molecules (blue); two

quinones, menaquinone (Q$_A$) and ubiquinone (Q$_B$) (orange and yellow), also arranged with bilateral symmetry; and a single nonheme Fe (red) located approximately on the axis of symmetry between the quinones. Shown at the top of the figure are four heme groups (red) associated with the *c*-type cytochrome of the reaction center

(b) Sequence of events following excitation of the special pair of bacteriochlorophylls (all components colored as in (a)), with the time scale of the electron transfers in parentheses. **1** The excited special pair passes an electron to pheophytin, **2** from which the electron moves rapidly to the tightly bound menaquinone, Q$_A$. **3** This quinone passes electrons much more slowly to the diffusible ubiquinone, Q$_B$, through the nonheme Fe. Meanwhile, **4** the "electron hole" in the special pair is filled by an electron from a heme of cytochrome *c*.

purple bacteria and in mitochondria. However, in contrast to the cyclic flow of electrons in purple bacteria, some electrons flow from the reaction center to an iron-sulfur protein, **ferredoxin**, which then passes electrons via ferredoxin:NAD reductase to NAD$^+$, producing NADH. The electrons taken from the reaction center to reduce NAD$^+$ are replaced by the oxidation of H$_2$S to elemental S, then to SO$_4^{2-}$, in the reaction that defines the green sulfur bacteria. This oxidation of H$_2$S by bacteria is chemically analogous to the oxidation of H$_2$O by oxygenic plants.

Kinetic and Thermodynamic Factors Prevent the Dissipation of Energy by Internal Conversion

The complex construction of reaction centers is the product of evolutionary selection for efficiency in the photosynthetic process. The excited state (Chl)$_2^*$ could in principle decay to its ground state by internal conversion,

a very rapid process (10 picoseconds; 1 ps = 10^{-12} s) in which the energy of the absorbed photon is converted to heat (molecular motion). Reaction centers are constructed to prevent the inefficiency that would result from internal conversion. The proteins of the reaction center hold the bacteriochlorophylls, bacteriopheophytins, and quinones in a fixed orientation relative to each other, allowing the photochemical reactions to take place in a virtually solid state. This accounts for the high efficiency and rapidity of the reactions; nothing is left to chance collision or random diffusion. Exciton transfer from antenna chlorophyll to the special pair of the reaction center is accomplished in less than 100 ps with >90% efficiency. Within 3 ps of the excitation of P870, pheophytin has received an electron and become a negatively charged radical; less than 200 ps later, the electron has reached the quinone Q$_B$ (Fig. 19–57b). The electron-transfer reactions not only are fast but are thermodynamically "downhill"; the excited special pair (Chl)$_2^*$ is a very good electron donor ($E'^\circ = -1\,\text{V}$) and each successive electron transfer is to an acceptor of substantially less negative E'°. The standard free-energy change for the process is therefore negative and large; recall from Chapter 13 that $\Delta G'^\circ = -n\,\Delta E'^\circ$; here, $\Delta E'^\circ$ is the difference between the standard reduction potentials of the two half-reactions

$$\text{(1) } (\text{Chl})_2^* \longrightarrow {}^\cdot(\text{Chl})_2^+ + e^- \quad E'^\circ = -1.0\,\text{V}$$
$$\text{(2) } Q + 2H^+ + 2e^- \longrightarrow QH_2 \quad E'^\circ = -0.045\,\text{V}$$

Thus

$$\Delta E'^\circ = -0.045\,\text{V} - (-1.0\,\text{V}) \approx 0.95\,\text{V}$$

and

$$\Delta G'^\circ = -2(96.5\,\text{kJ/V}\cdot\text{mol})(0.95\,\text{V}) = -180\,\text{kJ/mol}$$

The combination of fast kinetics and favorable thermodynamics makes the process virtually irreversible and highly efficient. The overall energy yield (the percentage of the photon's energy conserved in QH_2) is >30%, with the remainder of the energy dissipated as heat and entropy.

In Plants, Two Reaction Centers Act in Tandem

The photosynthetic apparatus of modern cyanobacteria, algae, and vascular plants is more complex than the one-center bacterial systems, and it seems to have evolved through the combination of two simpler bacterial photocenters. The thylakoid membranes of chloroplasts have two different kinds of photosystems, each with its own type of photochemical reaction center and set of antenna molecules. The two systems have distinct and complementary functions **(Fig. 19–58)**. **Photosystem II (PSII)** is a pheophytin-quinone type of system (like the single photosystem of purple bacteria) containing roughly equal amounts of chlorophylls a and b. Excitation of its reaction-center P680 drives electrons through the cytochrome b_6f complex with concomitant movement of protons across the thylakoid

FIGURE 19–58 Integration of photosystems I and II in chloroplasts. This "Z scheme" shows the pathway of electron transfer from H_2O (lower left) to $NADP^+$ (far right) in noncyclic photosynthesis. The position on the vertical scale of each electron carrier reflects its standard reduction potential. To raise the energy of electrons derived from H_2O to the energy level required to reduce $NADP^+$ to NADPH, each electron must be "lifted" twice (heavy arrows) by photons absorbed in PSII and PSI. One photon is required per electron in each photosystem. After excitation, the high-energy electrons flow "downhill" through the carrier chains shown. Protons move across the thylakoid membrane during the water-splitting reaction and during electron transfer through the cytochrome b_6f complex, producing the proton gradient that is essential to ATP formation. An alternative path of electrons is cyclic electron transfer, in which electrons move from ferredoxin back to the cytochrome b_6f complex, instead of reducing $NADP^+$ to NADPH. The cyclic pathway produces more ATP and less NADPH than the noncyclic.

membrane. **Photosystem I (PSI)** is structurally and functionally related to the type I reaction center of green sulfur bacteria. It has a reaction center designated P700 and a high ratio of chlorophyll a to chlorophyll b. Excited P700 passes electrons to the Fe-S protein ferredoxin, then to $NADP^+$, producing NADPH. The thylakoid membranes of a single spinach chloroplast have many hundreds of each kind of photosystem.

These two reaction centers in plants act in tandem to catalyze the light-driven movement of electrons from H_2O to $NADP^+$ (Fig. 19–58). Electrons are carried between the two photosystems by the soluble protein **plastocyanin**, a one-electron carrier functionally similar to cytochrome c of mitochondria. To replace the electrons that move from PSII through PSI to $NADP^+$, cyanobacteria and plants oxidize H_2O (as green sulfur bacteria oxidize H_2S), producing O_2 (Fig. 19–58, bottom left). This process is called **oxygenic photosynthesis** to distinguish it from the anoxygenic photosynthesis of purple and green sulfur bacteria. All O_2-evolving photosynthetic cells—those of plants, algae, and cyanobacteria—contain both PSI and PSII; organisms with only one photosystem do not evolve O_2. The diagram in Figure 19–58, often called the **Z scheme** because of its overall form, outlines the pathway of electron flow between the two photosystems and the energy relationships in the light reactions. The Z scheme thus describes the complete route by which electrons flow from H_2O to $NADP^+$, according to the equation

$$2H_2O + 2NADP^+ + 8 \text{ photons} \longrightarrow$$
$$O_2 + 2NADPH + 2H^+$$

For every two photons absorbed (one by each photosystem), one electron is transferred from H_2O to $NADP^+$. To form one molecule of O_2, which requires transfer of four electrons from two H_2O to two $NADP^+$, a total of eight photons must be absorbed, four by each photosystem.

The mechanistic details of the photochemical reactions in PSII and PSI are essentially similar to those of the two bacterial photosystems, with several important additions. In PSII, two very similar proteins, D1 and D2, form an almost symmetric dimer, to which all the electron-carrying cofactors are bound **(Fig. 19–59)**. Excitation of P680 in PSII produces P680*, an excellent electron donor that, within picoseconds, transfers an electron to pheophytin, giving it a negative charge ($\cdot Pheo^-$). With the loss of its electron, P680* is transformed into a radical cation, $P680^+$. $\cdot Pheo^-$ very rapidly passes its extra electron to a protein-bound **plastoquinone, PQ_A** (or Q_A), which in turn passes its electron to another, more loosely bound plastoquinone, PQ_B (or Q_B). When PQ_B has acquired two electrons in two such transfers from PQ_A and two protons from the solvent water, it is in its fully reduced quinol form, PQ_BH_2. The overall reaction initiated by light in PSII is

$$4 P680 + 4H^+ + 2 PQ_B + 4 \text{ photons} \longrightarrow$$
$$4 P680^+ + 2 PQ_BH_2 \quad (19\text{–}12)$$

FIGURE 19–59 Photosystem II of the cyanobacterium *Synechococcus elongates.* The monomeric form of the complex shown here has two major transmembrane proteins, D1 and D2, each with its set of cofactors. Although the two subunits are nearly symmetric, electron flow occurs through only one of the two branches of cofactors, that on the right (on D1). The arrows show the path of electron flow from the Mn_4Ca ion cluster of the water-splitting enzyme to the quinone PQ_B. The photochemical events occur in the sequence indicated by the step numbers. Notice the close similarity between the positions of cofactors here and the positions in the bacterial photoreaction center shown in Figure 19-57. The role of the Tyr residues and the detailed structure of the Mn_4Ca cluster are discussed later (see Fig. 19-64b).

Eventually, the electrons in PQ_BH_2 pass through the cytochrome b_6f complex (Fig. 19–58). The electron initially removed from P680 is replaced with an electron obtained from the oxidation of water, as described below. The binding site for plastoquinone is the point of action of many commercial herbicides that kill plants by blocking electron transfer through the cytochrome b_6f complex and preventing photosynthetic ATP production.

The photochemical events that follow excitation of PSI at the reaction-center P700 are formally similar to those in PSII. The excited reaction-center P700* loses an electron to an acceptor, A_0 (believed to be a special form of chlorophyll, functionally homologous to the pheophytin of PSII), creating A_0^- and $P700^+$ (Fig. 19–58, right side); again, excitation results in charge separation at the photochemical reaction center. $P700^+$ is a strong oxidizing agent, which quickly acquires an electron from plastocyanin, a soluble Cu-containing electron-transfer protein. A_0^- is an exceptionally strong reducing agent that passes its electron through a chain of carriers that leads to $NADP^+$. First, **phylloquinone** (A_1) accepts an electron and passes it to an iron-sulfur protein (through three Fe-S centers in PSI). From here, the electron moves to **ferredoxin** (Fd), another iron-sulfur protein loosely associated with the thylakoid membrane. Spinach ferredoxin (M_r 10,700) contains a 2Fe-2S center (Fig. 19–5) that undergoes one-electron oxidation and reduction reactions. The fourth electron carrier in the chain is the flavoprotein

ferredoxin:NADP$^+$ oxidoreductase, which transfers electrons from reduced ferredoxin (Fd$_{red}$) to NADP$^+$:

$$2Fd_{red} + 2H^+ + NADP^+ \longrightarrow 2Fd_{ox} + NADPH + H^+$$

This enzyme is homologous to the ferredoxin:NAD reductase of green sulfur bacteria (Fig. 19–56b).

Antenna Chlorophylls Are Tightly Integrated with Electron Carriers

The electron-carrying cofactors of PSI and the light-harvesting complexes are part of a supramolecular complex **(Fig. 19–60a)**, the structure of which has been

FIGURE 19–60 The supramolecular complex of PSI and its associated antenna chlorophylls. (a) Schematic drawing of the essential proteins and cofactors in a single unit of PSI. A large number of antenna chlorophylls surround the reaction center and convey to it (red arrows) the energy of absorbed photons. The result is excitation of the pair of chlorophyll molecules that constitute P700, greatly decreasing its reduction potential; P700 then passes an electron through two nearby chlorophylls to phylloquinone (Q$_K$; also called A$_1$). Reduced phylloquinone is reoxidized as it passes two electrons, one at a time (blue arrows), to an Fe-S protein (F$_X$) near the N side of the membrane. From F$_X$, electrons move through two more Fe-S centers (F$_A$ and F$_B$) to the Fe-S protein ferredoxin in the stroma. Ferredoxin then

donates electrons to NADP$^+$ (not shown), reducing it to NADPH, one of the forms in which the energy of photons is trapped in chloroplasts.

(b) The trimeric structure (derived from PDB ID 1JBO), viewed from the thylakoid lumen perpendicular to the membrane, showing all protein subunits (gray) and cofactors. **(c)** A monomer of PSI with all the proteins omitted, revealing the antenna and reaction-center chlorophylls (green with dark green Mg^{2+} ions in the center), carotenoids (yellow), and Fe-S centers of the reaction center (space-filling red and orange structures). The proteins in the complex hold the components rigidly in orientations that maximize efficient exciton transfers between excited antenna molecules and the reaction center.

solved crystallographically. The protein consists of three identical complexes, each composed of 11 different proteins (Fig. 19–60b). In this remarkable structure the many antenna chlorophyll and carotenoid molecules are precisely arrayed around the reaction center (Fig. 19–60c). The reaction center's electron-carrying cofactors are therefore tightly integrated with antenna chlorophylls. This arrangement allows very rapid and efficient exciton transfer from antenna chlorophylls to the reaction center. In contrast to the single path of electrons in PSII, the electron flow initiated by absorption of a photon is believed to occur through both branches of carriers in PSI.

The Cytochrome b_6f Complex Links Photosystems II and I

Electrons temporarily stored in plastoquinol as a result of the excitation of P680 in PSII are carried to P700 of PSI via the cytochrome b_6f complex and the soluble protein plastocyanin (Fig. 19–58, center). Like Complex III of mitochondria, the cytochrome b_6f complex **(Fig. 19–61)** contains a b-type cytochrome with two heme groups (designated b_H and b_L), a Rieske iron-sulfur protein (M_r 20,000), and cytochrome f (named for the Latin *frons*, "leaf"). Electrons flow through the cytochrome

FIGURE 19–61 Electron and proton flow through the cytochrome b_6f complex. (a) The crystal structure of the complex (PDB ID 1FV5) reveals the positions of the cofactors involved in electron transfers. In addition to the hemes of cytochrome b (heme b_H and b_L; also called heme b_N and b_P, respectively, because of their proximity to the N and P sides of the bilayer) and cytochrome f (heme f), there is a fourth (heme x) near heme b_H; there is also a β-carotene of unknown function. Two sites bind plastoquinone: the PQH$_2$ site near the P side of the bilayer, and the PQ site near the N side. The Fe-S center of the Rieske protein lies just outside the bilayer on the P side, and the heme f site is on a protein domain that extends well into the thylakoid lumen. The electron path is shown for one of the monomers, but both sets of carriers in the dimer carry electrons to plastocyanin (PDB ID 2Q5B). **(b)** The complex is a homodimer arranged to create a cavern connecting the PQH$_2$ and PQ sites (compare this with the structure of mitochondrial Complex III in Fig. 19–11). This cavern allows plastoquinone to move between the sites of its oxidation and reduction.

(c) Plastoquinol (PQH$_2$) formed in PSII is oxidized by the cytochrome b_6f complex in a series of steps like those of the Q cycle in the cytochrome bc_1 complex (Complex III) of mitochondria (see Fig. 19-12). One electron from PQH$_2$ passes to the Fe-S center of the Rieske protein, the other to heme b_L of cytochrome b_6. The net effect is passage of electrons from PQH$_2$ to the soluble protein plastocyanin, which carries them to PSI.

b_6f complex from PQ_BH_2 to cytochrome f, then to plastocyanin, and finally to P700, thereby reducing it.

Like Complex III of mitochondria, cytochrome b_6f conveys electrons from a reduced quinone—a mobile, lipid-soluble carrier of two electrons (Q in mitochondria, PQ_B in chloroplasts)—to a water-soluble protein that carries one electron (cytochrome c in mitochondria, plastocyanin in chloroplasts). As in mitochondria, the function of this complex involves a Q cycle (Fig. 19–12) in which electrons pass, one at a time, from PQ_BH_2 to cytochrome b_6. This cycle results in the pumping of protons across the membrane; in chloroplasts, the direction of proton movement is from the stromal compartment to the thylakoid lumen, up to four protons moving for each pair of electrons. The result is production of a proton gradient across the thylakoid membrane as electrons pass from PSII to PSI. Because the volume of the flattened thylakoid lumen is small, the influx of a small number of protons has a relatively large effect on lumenal pH. The measured difference in pH between the stroma (pH 8) and the thylakoid lumen (pH 5) represents a 1,000-fold difference in proton concentration—a powerful driving force for ATP synthesis.

Cyclic Electron Flow between PSI and the Cytochrome b_6f Complex Increases the Production of ATP Relative to NADPH

Electron flow from PSII through the cytochrome b_6f complex, then through PSI to $NADP^+$, is sometimes called **noncyclic electron flow**, to distinguish it from **cyclic electron flow**, which occurs to varying degrees depending primarily on the light conditions. The noncyclic path produces a proton gradient, which is used to drive ATP synthesis, and NADPH, which is used in reductive biosynthetic processes. Cyclic electron flow involves only PSI, not PSII (Fig. 19–58). Electrons passing from P700 to ferredoxin do not continue to $NADP^+$, but move back through the cytochrome b_6f complex to plastocyanin. (This electron path parallels that in green sulfur bacteria, shown in Fig. 19–56b.) Plastocyanin then donates electrons to P700, which transfers them to ferredoxin. In this way, electrons are repeatedly recycled through the cytochrome b_6f complex and the reaction center of PSI, each electron propelled around the cycle by the energy of one photon. Cyclic electron flow is not accompanied by net formation of NADPH or evolution of O_2. However, it *is* accompanied by proton pumping by the cytochrome b_6f complex and by phosphorylation of ADP to ATP, referred to as **cyclic photophosphorylation**. The overall equation for cyclic electron flow and photophosphorylation is simply

$$ADP + P_i \xrightarrow{\text{light}} ATP + H_2O$$

By regulating the partitioning of electrons between $NADP^+$ reduction and cyclic photophosphorylation, a plant adjusts the ratio of ATP to NADPH produced in the light-dependent reactions to match its needs for these products in the carbon-assimilation reactions and other biosynthetic processes. As we shall see in Chapter 20, the carbon-assimilation reactions require ATP and NADPH in the ratio 3:2.

This regulation of electron-transfer pathways is part of a short-term adaptation to changes in light color (wavelength) and quantity (intensity), as further described below.

State Transitions Change the Distribution of LHCII between the Two Photosystems

The energy required to excite PSI (P700) is less (light of longer wavelength, lower energy) than that needed to excite PSII (P680). If PSI and PSII were physically contiguous, excitons originating in the antenna system of PSII would migrate to the reaction center of PSI, leaving PSII chronically underexcited and interfering with the operation of the two-center system. This imbalance in the supply of excitons is prevented by separation of the two photosystems in the thylakoid membrane **(Fig. 19–62)**. PSII is located almost exclusively in the tightly appressed membrane stacks of thylakoid grana; its associated light-harvesting complex (LHCII) mediates the tight association of adjacent membranes in the grana. PSI and the ATP synthase complex are located almost exclusively in the nonappressed thylakoid membranes (the stromal lamellae), where they have access to the contents of the stroma, including ADP and $NADP^+$. The cytochrome b_6f complex is present primarily in the grana.

The association of LHCII with PSI and PSII depends on light intensity and wavelength, which can change in the short term, leading to **state transitions** in the chloroplast. In state 1, a critical Ser residue in LCHII is not phosphorylated, and LHCII associates with PSII. Under conditions of intense or blue light, which favor absorption by PSII, that photosystem reduces plastoquinone to plastoquinol (PQH_2) faster than PSI can oxidize it. The resulting accumulation of PQH_2 activates a protein kinase that triggers the transition to state 2 by phosphorylating a Thr residue on LHCII **(Fig. 19–63)**. Phosphorylation weakens the interaction of LHCII with PSII, and some LHCII dissociates and moves to the stromal lamellae; here it captures photons (excitons) for PSI, speeding the oxidation of PQH_2 and reversing the imbalance between electron flow in PSI and PSII. In less intense light (in the shade, with more red light), PSI oxidizes PQH_2 faster than PSII can make it, and the resulting increase in [PQ] triggers dephosphorylation of LHCII, reversing the effect of phosphorylation.

The state transition in LCHII localization is mutually regulated with the transition from cyclic to noncyclic photophosphorylation, described above; the path of electrons is primarily noncyclic in state 1 and primarily cyclic in state 2.

(a)

(b)

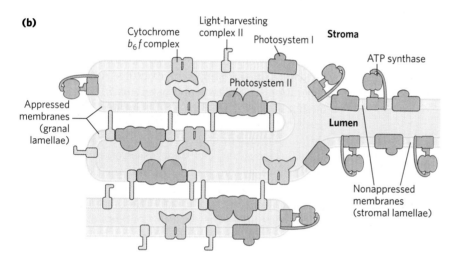

FIGURE 19–62 Localization of PSI and PSII in thylakoid membranes. **(a)** The structures of the complexes and soluble proteins of the photosynthetic apparatus of a vascular plant or alga, drawn to the same scale. The structures are photosystem II (PDB ID 2AXT), cytochrome b_{6f}, complex (PDB ID 2E74), plastocyanin (PDB ID 1AG6), photosystem I (PDB ID 1QZV), ferredoxin (PDB ID 1A70), and ferredoxin : NADP reductase (PDB ID 1QG0). The ATP synthase is the composite shown in Figure 19-25c.

(b) Light-harvesting complex LHCII and ATP synthase are located both in appressed regions of the thylakoid membrane (granal lamellae, in which several membranes are in contact) and in nonappressed regions (stromal lamellae), and have ready access to ADP and $NADP^+$ in the stroma. PSII is present almost exclusively in the appressed regions, and PSI almost exclusively in nonappressed regions, exposed to the stroma. LHCII is the "adhesive" that holds appressed lamellae together (see Fig. 19–63).

FIGURE 19–63 Balancing of electron flow in PSI and PSII by state transition. A hydrophobic domain of LHCII in one thylakoid lamella inserts into the neighboring lamella and closely appresses the two membranes (state 1). Accumulation of plastoquinol (not shown) stimulates a protein kinase that phosphorylates a Thr residue in the hydrophobic domain of LHCII, which reduces its affinity for the neighboring thylakoid membrane and converts appressed regions to nonappressed regions (state 2). A specific protein phosphatase reverses this regulatory phosphorylation when the $[PQ]/[PQH_2]$ ratio increases.

Water Is Split by the Oxygen-Evolving Complex

The ultimate source of the electrons passed to NADPH in plant (oxygenic) photosynthesis is water. Having given up an electron to pheophytin, $P680^+$ (of PSII)

must acquire an electron to return to its ground state in preparation for capture of another photon. In principle, the required electron might come from any number of organic or inorganic compounds. Photosynthetic bacteria use a variety of electron donors for this purpose—

(a)

FIGURE 19–64 Water-splitting activity of the oxygen-evolving complex. (a) Shown here is the process that produces a four-electron oxidizing agent—a multinuclear center with four Mn ions, one Ca ion, and four oxygen atoms—in the water-splitting complex of PSII. The sequential absorption of four photons (excitons), each absorption causing the loss of one electron from the Mn_4Ca cluster, produces an oxidizing agent that can remove four electrons from two molecules of water, producing O_2. The electrons lost from the Mn_4Ca cluster pass one at a time to an oxidized Tyr residue in a PSII protein, then to $P680^+$. **(b)** The metallic center of the water-splitting complex, obtained by x-ray crystallography (PDB ID 3ARC). Tyr^{161}, known to participate in the oxidation of water, is seen hydrogen-bonded to a network of water molecules, including several directly in contact with the Mn_4Ca cluster. This is the site of one of the most important reactions in the biosphere!

acetate, succinate, malate, or sulfide—depending on what is available in a particular ecological niche. About 3 billion years ago, evolution of primitive photosynthetic bacteria (the progenitors of the modern cyanobacteria) produced a photosystem capable of taking electrons from a donor that is always available—water. Two water molecules are split, yielding four electrons, four protons, and molecular oxygen:

$$2H_2O \longrightarrow 4H^+ + 4e^- + O_2$$

A single photon of visible light does not have enough energy to break the bonds in water; four photons are required in this photolytic cleavage reaction.

The four electrons abstracted from water do not pass directly to $P680^+$, which can accept only one electron at a time. Instead, a remarkable molecular device, the **oxygen-evolving complex** (also called the **water-splitting complex**), passes four electrons *one at a time* to $P680^+$ **(Fig. 19–64)**. The immediate electron donor to $P680^+$ is a Tyr residue (often designated Z or Tyr_Z) in subunit D1 of the PSII reaction center. The Tyr residue loses both a proton and an electron, generating the electrically neutral Tyr free radical, •Tyr:

$$4\,P680^+ + 4\,Tyr \longrightarrow 4\,P680 + 4\,{}^{\bullet}Tyr \quad (19\text{–}13)$$

The Tyr radical regains its missing electron and proton by oxidizing a cluster of four manganese ions and one calcium ion in the water-splitting complex. With each single-electron transfer, the Mn_4Ca cluster becomes more oxidized; four single-electron transfers, each corresponding to the absorption of one photon, produce a charge of 4+ on the Mn_4Ca cluster (Fig. 19–64):

$$4\,{}^{\bullet}Tyr + [Mn_4Ca]^0 \longrightarrow 4\,Tyr + [Mn_4Ca]^{4+} \quad (19\text{–}14)$$

In this state, the Mn_4Ca cluster can take four electrons from a pair of water molecules, releasing $4H^+$ and O_2:

$$[Mn_4Ca]^{4+} + 2H_2O \longrightarrow$$
$$[Mn_4Ca]^0 + 4H^+ + O_2 \quad (19\text{–}15)$$

Because the four protons produced in this reaction are released into the thylakoid lumen, the oxygen-evolving complex acts as a proton pump, driven by electron transfer. The sum of Equations 19–12 through 19–15 is

$$2H_2O + 2PQ_B + 4\ photons \longrightarrow$$
$$O_2 + 2PQ_BH_2 \quad (19\text{–}16)$$

The detailed structure of the oxygen-evolving cluster has been obtained by high-resolution x-ray crystallography. The metal cluster takes the shape of a chair (Fig. 19–64b). The seat and legs of the chair are made up of three Mn ions, one Ca ion, and four O atoms; the fourth Mn and another O form the back of the chair. Four water molecules are also seen in the crystal structure, two associated with one of the Mn ions, the other two with the Ca ion. It is possible that one or more of these water molecules is the one that undergoes oxidation to produce O_2. This metal cluster is associated with a peripheral membrane protein (M_r 33,000) on the lumenal side of the thylakoid membrane that presumably stabilizes the cluster. The Tyr residue designated Z, through which electrons move between water and the PSII reaction center, is part of a network of hydrogen-bonded water molecules that includes the four associated with the Mn_4Ca cluster. The detailed mechanism of water oxidation by the Mn_4Ca cluster is not known but is under intense investigation. The reaction is central to life on Earth and may involve novel bioinorganic chemistry. Determination of the structure of the polymetallic center has inspired several reasonable and testable hypotheses. Stay tuned.

SUMMARY 19.8 The Central Photochemical Event: Light-Driven Electron Flow

▶ Bacteria have a single reaction center; in purple bacteria, it is of the pheophytin-quinone type, and in green sulfur bacteria, the Fe-S type.

▶ Structural studies of the reaction center of a purple bacterium have provided information about light-driven electron flow from an excited special pair of chlorophyll molecules, through pheophytin, to quinones. Electrons then pass from quinones through the cytochrome bc_1 complex, and back to the photoreaction center.

▶ An alternative path, in green sulfur bacteria, sends electrons from reduced quinones to NAD^+.

▶ Cyanobacteria and plants have two different photoreaction centers, arranged in tandem.

▶ Plant photosystem I passes electrons from its excited reaction center, P700, through a series of carriers to ferredoxin, which then reduces $NADP^+$ to NADPH.

▶ The reaction center of plant photosystem II, P680, passes electrons to plastoquinone, and the electrons lost from P680 are replaced by electrons from H_2O (electron donors other than H_2O are used in other organisms).

▶ Flow of electrons through the photosystems produces NADPH and ATP. Cyclic electron flow produces ATP only and allows variability in the proportions of NADPH and ATP formed.

▶ The localization of PSI and PSII between the granal and stromal lamellae can change and is indirectly controlled by light intensity, optimizing the distribution of excitons between PSI and PSII for efficient energy capture.

▶ The light-driven splitting of H_2O is catalyzed by a Mn- and Ca-containing protein complex; O_2 is produced. The reduced plastoquinone carries electrons to the cytochrome b_6f complex; from here they pass to plastocyanin, and then to P700 to replace those lost during its photoexcitation.

▶ Electron flow through the cytochrome b_6f complex drives protons across the plasma membrane, creating a proton-motive force that provides the energy for ATP synthesis by an ATP synthase.

19.9 ATP Synthesis by Photophosphorylation

The combined activities of the two plant photosystems move electrons from water to $NADP^+$, conserving some of the energy of absorbed light as NADPH (Fig. 19–58). Simultaneously, protons are pumped across the thyla-

Daniel Arnon, 1910–1994

koid membrane and energy is conserved as an electrochemical potential. We turn now to the process by which this proton gradient drives the synthesis of ATP, the other energy-conserving product of the light-dependent reactions.

In 1954 Daniel Arnon and his colleagues discovered that ATP is generated from ADP and P_i during photosynthetic electron transfer in illuminated spinach chloroplasts. Support for these findings came from the work of Albert Frenkel, who detected light-dependent ATP production in pigment-containing membranous structures called **chromatophores**, derived from photosynthetic bacteria. Investigators concluded that some of the light energy captured by the photosynthetic systems of these organisms is transformed into the phosphate bond energy of ATP. This process is called **photophosphorylation**, to distinguish it from oxidative phosphorylation in respiring mitochondria.

A Proton Gradient Couples Electron Flow and Phosphorylation

Several properties of photosynthetic electron transfer and photophosphorylation in chloroplasts indicate that a proton gradient plays the same role as in mitochondrial oxidative phosphorylation. (1) The reaction centers, electron carriers, and ATP-forming enzymes are located in a proton-impermeable membrane—the thylakoid membrane—which must be intact to support photophosphorylation. (2) Photophosphorylation can be uncoupled from electron flow by reagents that promote the passage of protons through the thylakoid membrane. (3) Photophosphorylation can be blocked by venturicidin and similar agents that inhibit the formation of ATP from ADP and P_i by the mitochondrial ATP synthase (Table 19–4). (4) ATP synthesis is catalyzed by F_oF_1 complexes, located on the outer surface of the thylakoid membranes, that are very similar in structure and function to the F_oF_1 complexes of mitochondria.

Electron-transferring molecules in the chain of carriers connecting PSII and PSI are oriented asymmetrically in the thylakoid membrane, so photoinduced electron flow results in the net movement of protons across the membrane, from the stromal side to the thylakoid lumen **(Fig. 19–65)**. In 1966 André Jagendorf showed that a pH gradient across the thylakoid membrane (alkaline outside) could furnish the driving force to generate ATP. His early observations provided some of the most important experimental evidence in support of Mitchell's chemiosmotic hypothesis.

FIGURE 19–65 Proton and electron circuits during photophosphorylation. Electrons (blue arrows) move from H_2O through PSII, through the intermediate chain of carriers, through PSI, and finally to $NADP^+$. Protons (red arrows) are pumped into the thylakoid lumen by the flow of electrons through the carriers linking PSII and PSI, and reenter the stroma through proton channels formed by the F_o (designated CF_o) of ATP synthase. The F_1 subunit (CF_1) catalyzes synthesis of ATP.

The Approximate Stoichiometry of Photophosphorylation Has Been Established

As electrons move from water to $NADP^+$ in plant chloroplasts, about 12 H^+ move from the stroma to the thylakoid lumen per four electrons passed (that is, per O_2 formed). Four of these protons are moved by the oxygen-evolving complex, and up to eight by the cytochrome b_6f complex. The measurable result is a 1,000-fold difference in proton concentration across the thylakoid membrane ($\Delta pH = 3$). Recall that the energy stored in a proton gradient (the electrochemical potential) has two components: a proton concentration difference (ΔpH) and an electrical potential ($\Delta\psi$) due to charge separation. In chloroplasts, ΔpH is the dominant component; counter-ion movement apparently dissipates most of the electrical potential. In illuminated chloroplasts, the energy stored in the proton gradient per mole of protons is

$$\Delta G = 2.3RT\,\Delta pH + Z\mathcal{F}\,\Delta\psi = -17 \text{ kJ/mol}$$

so the movement of 12 mol of protons across the thylakoid membrane represents conservation of about 200 kJ of energy—enough energy to drive the synthesis of several moles of ATP ($\Delta G'^\circ = 30.5$ kJ/mol). Experimental measurements yield values of about 3 ATP per O_2 produced.

At least eight photons must be absorbed to drive four electrons from H_2O to NADPH (one photon per electron at each reaction center). The energy in eight photons of visible light is more than enough for the synthesis of three molecules of ATP.

ATP synthesis is not the only energy-conserving reaction of photosynthesis in plants; the NADPH formed in the final electron transfer is also energetically rich. The overall equation for noncyclic photophosphorylation (a term explained below) is

$$2H_2O + 8 \text{ photons} + 2NADP^+ + {\sim}3ADP^+ + {\sim}3P_i \longrightarrow$$
$$O_2 + {\sim}3ATP + 2NADPH \quad (19\text{–}17)$$

The ATP Synthase of Chloroplasts Is Like That of Mitochondria

The enzyme responsible for ATP synthesis in chloroplasts is a large complex with two functional components, CF_o and CF_1 (C denoting its location in chloroplasts). CF_o is a transmembrane proton pore composed of several integral membrane proteins and is homologous to mitochondrial F_o. CF_1 is a peripheral membrane protein complex very similar in subunit composition, structure, and function to mitochondrial F_1.

Electron microscopy of sectioned chloroplasts shows ATP synthase complexes as knoblike projections on the *outside* (stromal, or N) surface of thylakoid membranes; these complexes correspond to the ATP synthase complexes seen to project on the *inside*

André Jagendorf

Jagendorf incubated chloroplasts, in the dark, in a pH 4 buffer; the buffer slowly penetrated into the inner compartment of the thylakoids, lowering their internal pH. He added ADP and P_i to the dark suspension of chloroplasts and then suddenly raised the pH of the outer medium to 8, momentarily creating a large pH gradient across the membrane. As protons moved out of the thylakoids into the medium, ATP was generated from ADP and P_i. Because the formation of ATP occurred in the dark (with no input of energy from light), this experiment showed that a pH gradient across the membrane is a high-energy state that, as in mitochondrial oxidative phosphorylation, can mediate the transduction of energy from electron transfer into the chemical energy of ATP.

(matrix, or N) surface of the inner mitochondrial membrane. Thus the relationship between the orientation of the ATP synthase and the direction of proton pumping is the same in chloroplasts and mitochondria. In both cases, the F_1 portion of ATP synthase is located on the more alkaline (N) side of the membrane through which protons flow down their concentration gradient; the direction of proton flow relative to F_1 is the same in both cases: P to N **(Fig. 19–66)**.

The mechanism of chloroplast ATP synthase is also believed to be essentially identical to that of its mitochondrial analog; ADP and P_i readily condense to form ATP on the enzyme surface, and the release of this enzyme-bound ATP requires a proton-motive force. Rotational catalysis sequentially engages each of the three β subunits of the ATP synthase in ATP synthesis, ATP release, and ADP + P_i binding (Figs 19–26, 19–27).

The chloroplast ATP synthase of spinach, with 14 c subunits in its F_o complex, is predicted to have a lower ratio of ATP formed to electrons transferred, compared with the bovine, yeast, or *E. coli* F_o complexes, with 8, 10, and 10 c subunits, respectively (Fig. 19–29).

SUMMARY 19.9 ATP Synthesis by Photophosphorylation

▶ In plants, both the water-splitting reaction and electron flow through the cytochrome $b_6 f$ complex are accompanied by proton pumping across the thylakoid membrane. The proton-motive force thus created drives ATP synthesis by a $CF_o CF_1$ complex similar to the mitochondrial $F_o F_1$ complex.

▶ The catalytic mechanism of $CF_o CF_1$ is very similar to that of the ATP synthases of mitochondria and

bacteria. Physical rotation driven by the proton gradient is accompanied by ATP synthesis at sites that cycle through three conformations, one with high affinity for ATP, one with high affinity for ADP + P_i and one with low affinity for both nucleotides.

19.10 The Evolution of Oxygenic Photosynthesis

The appearance of oxygenic photosynthesis on Earth about 2.5 billion years ago was a crucial event in the evolution of the biosphere. Until then, the earth had been essentially devoid of molecular oxygen and lacked the ozone layer that protects living organisms from solar UV radiation. Oxygenic photosynthesis made available a nearly limitless supply of reducing agent (H_2O) to drive the production of organic compounds by reductive biosynthetic reactions. And mechanisms evolved that allowed organisms to use O_2 as a terminal electron acceptor in highly energetic electron transfers from organic substrates, employing the energy of oxidation to support their metabolism. The complex photosynthetic apparatus of a modern vascular plant is the culmination of a series of evolutionary events, the most recent of which was the acquisition by eukaryotic cells of a cyanobacterial endosymbiont.

Chloroplasts Evolved from Ancient Photosynthetic Bacteria

Chloroplasts in modern organisms resemble mitochondria in several properties, and are believed to have originated by the same mechanism that gave rise to mitochondria: endosymbiosis. Like mitochondria, chloroplasts contain their own DNA and protein-synthesizing machinery. Some of the polypeptides of chloroplast proteins are encoded by chloroplast genes and synthesized in the chloroplast; others are encoded by nuclear genes, synthesized outside the chloroplast, and imported (Chapter 27). When plant cells grow and divide, chloroplasts give rise to new chloroplasts by division, during which their DNA is replicated and divided between daughter chloroplasts. The machinery and mechanism for light capture, electron flow, and ATP synthesis in modern cyanobacteria are similar in many respects to those in plant chloroplasts. These observations led to the now widely accepted hypothesis that the evolutionary progenitors of modern plant cells were primitive eukaryotes that engulfed photosynthetic cyanobacteria and established stable endosymbiotic relationships with them (see Fig. 1–38).

At least half of the photosynthetic activity on Earth now occurs in microorganisms—algae, other photosynthetic eukaryotes, and photosynthetic bacteria. Cyanobacteria have PSII and PSI in tandem, and the PSII has

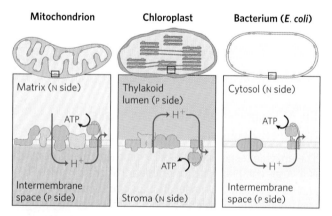

| Mitochondrion | Chloroplast | Bacterium (*E. coli*) |

FIGURE 19–66 Orientation of the ATP synthase is fixed relative to the proton gradient. Superficially, the direction of proton pumping may seem to be different in chloroplasts relative to mitochondria and bacteria. In mitochondria and bacteria, protons are pumped *out of* the organelle or cell, and F_1 is on the *inside* of the membrane; in plant thylakoids, protons are pumped *into* the flattened discs within chloroplasts, and CF_1 is on the *outside* of the disc membrane. However, exactly the same mechanism of energy conversion (from proton gradient to ATP) occurs in all three cases. ATP is synthesized in the matrix of mitochondria, the stroma of chloroplasts, and the cytosol of bacteria.

an associated water-splitting activity resembling that of plants. However, the other groups of photosynthetic bacteria have single reaction centers and do not split H_2O or produce O_2. Many are obligate anaerobes and cannot tolerate O_2; they must use some compound other than H_2O as electron donor. Some photosynthetic bacteria use inorganic compounds as electron (and hydrogen) donors. For example, green sulfur bacteria use hydrogen sulfide:

$$2H_2S + CO_2 \xrightarrow{\text{light}} (CH_2O) + H_2O + 2S$$

These bacteria, instead of producing molecular O_2, form elemental sulfur as the oxidation product of H_2S. (They further oxidize the S to SO_4^{2-}.) Other photosynthetic bacteria use organic compounds such as lactate as electron donors:

$$2 \, \text{Lactate} + CO_2 \xrightarrow{\text{light}} (CH_2O) + H_2O + 2 \, \text{pyruvate}$$

The fundamental similarity of photosynthesis in plants and bacteria, despite the differences in the electron donors they employ, becomes more obvious when the equation of photosynthesis is written in the more general form

$$2H_2D + CO_2 \xrightarrow{\text{light}} (CH_2O) + H_2O + 2D$$

in which H_2D is an electron (and hydrogen) donor and D is its oxidized form. H_2D may be water, hydrogen sulfide, lactate, or some other organic compound, depending on the species. Most likely, the bacteria that first developed photosynthetic ability used H_2S as their electron source.

The ancient relatives of modern cyanobacteria probably arose by the combination of genetic material from two types of photosynthetic bacteria, with systems of the type seen in modern purple bacteria (with a PSII-like electron path) and green sulfur bacteria (with an electron path resembling that in PSI). The bacterium with two independent photosystems may have used one in one set of conditions, the other in different conditions. Over time, a mechanism to connect the two photosystems for simultaneous use evolved, and the PSII-like system acquired the water-splitting capacity found in modern cyanobacteria.

Modern cyanobacteria can synthesize ATP by oxidative phosphorylation or by photophosphorylation, although they have neither mitochondria nor chloroplasts. The enzymatic machinery for both processes is in a highly convoluted plasma membrane (Fig. 19–67). Three protein components function in both processes, giving evidence that the processes have a common evolutionary origin (Fig. 19–68). First, the proton-pumping cytochrome b_6f complex carries electrons from plastoquinone to cytochrome c_6 in photosynthesis, and also carries electrons from ubiquinone to cytochrome c_6 in oxidative phosphorylation—the role played by cytochrome bc_1 in mitochondria. Second, cytochrome

FIGURE 19–67 The photosynthetic membranes of a cyanobacterium. In these thin sections of a cyanobacterium, viewed with a transmission electron microscope, the multiple layers of the internal membranes are seen to fill half of the total volume of the cell. The extensive membrane system serves the same role as the thylakoids of vascular plants, providing a large surface area containing all of the photosynthetic machinery. (Bar = 100 nm.)

c_6, homologous to mitochondrial cytochrome c, carries electrons from Complex III to Complex IV in cyanobacteria; it can also carry electrons from the cytochrome b_6f complex to PSI—a role performed in plants by plastocyanin. We therefore see the functional homology between the cyanobacterial cytochrome b_6f complex and the mitochondrial cytochrome bc_1 complex, and between cyanobacterial cytochrome c_6 and plant plastocyanin. The third conserved component is the ATP synthase, which functions in oxidative phosphorylation and photophosphorylation in cyanobacteria, and in the mitochondria and chloroplasts of photosynthetic eukaryotes. The structure and remarkable mechanism of this enzyme have been strongly conserved throughout evolution.

In *Halobacterium*, a Single Protein Absorbs Light and Pumps Protons to Drive ATP Synthesis

In some modern archaea, a quite different mechanism for converting the energy of light into an electrochemical gradient has evolved. The halophilic ("salt-loving") archaeon *Halobacterium salinarum* is descended from ancient evolutionary progenitors. This archaeon (commonly referred to as a halobacterium) lives only in brine ponds and salt lakes (Great Salt Lake and the Dead Sea, for example), where the

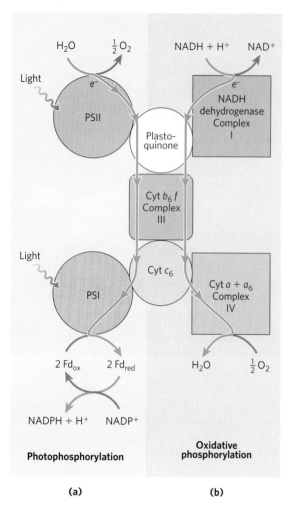

(a)   **(b)**

FIGURE 19–68 Dual roles of cytochrome b_6f and cytochrome c_6 in cyanobacteria reflect evolutionary origins. Cyanobacteria use cytochrome b_6f, cytochrome c_6, and plastoquinone for both oxidative phosphorylation and photophosphorylation. **(a)** In photophosphorylation, electrons flow (top to bottom) from water to $NADP^+$. **(b)** In oxidative phosphorylation, electrons flow from NADH to O_2. Both processes are accompanied by proton movement across the membrane, accomplished by a Q cycle.

high salt concentration—which can exceed 4 M—results from water loss by evaporation; indeed, halobacteria cannot live in NaCl concentrations lower than 3 M. These organisms are aerobes and normally use O_2 to oxidize organic fuel molecules. However, the solubility of O_2 is so low in brine ponds that sometimes oxidative metabolism must be supplemented by sunlight as an alternative source of energy.

The plasma membrane of *H. salinarum* contains patches of the light-absorbing pigment **bacteriorhodopsin**, which contains retinal (the aldehyde derivative of vitamin A; see Fig. 10–21) as a light-harvesting prosthetic group. When the cells are illuminated, all-*trans*-retinal bound to the bacteriorhodopsin absorbs a photon and undergoes photoisomerization to 13-*cis*-retinal, forcing a conformational change in the protein. The restoration of

all-*trans*-retinal is accompanied by the outward movement of protons through the plasma membrane. Bacteriorhodopsin, with only 247 amino acid residues, is the simplest light-driven proton pump known. The difference in the three-dimensional structure of bacteriorhodopsin in the dark and after illumination **(Fig. 19–69a)** suggests a pathway by which a concerted series of proton "hops" could effectively move a proton across the membrane. The chromophore retinal is bound through a Schiff-base linkage to the ε-amino group of a Lys residue. In the dark, the nitrogen of this Schiff base is protonated; in the light, photoisomerization of retinal lowers the pK_a of this group and it releases its proton to a nearby Asp residue, triggering a series of proton hops that ultimately result in the release of a proton at the outer surface of the membrane (Fig. 19–69b).

The electrochemical potential across the membrane drives protons back into the cell through a membrane ATP synthase complex very similar to that of mitochondria and chloroplasts. Thus, when O_2 is limited, halobacteria can use light to supplement the ATP synthesized by oxidative phosphorylation. Halobacteria do not evolve O_2, nor do they carry out photoreduction of $NADP^+$; their phototransducing machinery is therefore much simpler than that of cyanobacteria or plants. Nevertheless, its proton-pumping mechanism may prove to be prototypical for the many other, more complex, ion pumps.

🖱 **Bacteriorhodopsin**

SUMMARY 19.10 The Evolution of Oxygenic Photosynthesis

▶ Modern cyanobacteria are derived from an ancient organism that acquired two photosystems, one of the type now found in purple bacteria, the other of the type found in green sulfur bacteria.

▶ Many photosynthetic microorganisms obtain electrons for photosynthesis not from water but from donors such as H_2S.

▶ Cyanobacteria with the tandem photosystems and a water-splitting activity that released oxygen into the atmosphere appeared on Earth about 2.5 billion years ago.

▶ Chloroplasts, like mitochondria, evolved from bacteria living endosymbiotically in early eukaryotic cells. The ATP synthases of bacteria, cyanobacteria, mitochondria, and chloroplasts share a common evolutionary precursor and a common enzymatic mechanism.

▶ An entirely different mechanism for converting light energy to a proton gradient has evolved in the modern archaea, in which the light-harvesting pigment is retinal.

(a)

FIGURE 19–69 **A different mechanism for light-driven proton pumping evolved independently in a halophilic archaeon. (a)** Bacteriorhodopsin (M_r 26,000) of *Halobacterium halobium* has seven membrane-spanning α helices (PDB ID 1C8R). The chromophore all-*trans*-retinal (purple) is covalently attached via a Schiff base to the ε-amino group of a Lys residue deep in the membrane interior. Running through the protein are a series of Asp and Glu residues and a series of closely associated water molecules that together provide the transmembrane path for protons (red arrows). Steps ❶ through ❺ indicate proton movements, described below.

(b) In the dark (left panel), the Schiff base is protonated. Illumination (right panel) photoisomerizes the retinal, forcing subtle conformational changes in the protein that alter the distance between the Schiff base and its neighboring amino acid residues. Interaction with these neighbors (Leu[93] and Val[49]) lowers the pK_a of the protonated Schiff base, and the base gives up its proton to a nearby carboxyl group on Asp[85] (step ❶ in (a)). This initiates a series of concerted proton hops between water molecules (see Fig. 2-14) in the interior of the protein, which ends with ❷ the release of a proton that was shared by Glu[194] and Glu[204] near the extracellular surface. (Tyr[83] forms a hydrogen bond with Glu[194] that facilitates this proton release.) ❸ The Schiff base reacquires a proton from Asp[96], which ❹ takes up a proton from the cytosol. ❺ Finally, Asp[85] gives up its proton, leading to reprotonation of the Glu[204]-Glu[194] pair. The system is now ready for another round of proton pumping.

Dark

Light

(b)

Key Terms

Terms in bold are defined in the glossary.

Further Reading

History and General Background

Arnon, D.I. (1984) The discovery of photosynthetic
phosphorylation. *Trends Biochem. Sci.* **9**, 258–262.

Harold, F.M. (1986) *The Vital Force: A Study in Bioenergetics*,
W. H. Freeman and Company, New York.

   A very readable synthesis of the principles of bioenergetics and
their application to energy transductions.

Heldt, H.-W. & Piechulla, B. (2010) *Plant Biochemistry*, 4th edn,
Elsevier, New York.

   A textbook of plant biochemistry with excellent discussions of
photophosphorylation.

Lane, N. (2005) *Power, Sex, Suicide: Mitochondria and the
Meaning of Life*, Oxford University Press, Oxford.

   An entry-level description of the roles of mitochondria in energy
conservation and in apoptosis.

Mitchell, P. (1979) Keilin's respiratory chain concept and its
chemiosmotic consequences. *Science* **206**, 1148–1159.

   Mitchell's Nobel lecture, outlining the evolution of the
chemiosmotic hypothesis.

Scheffler, I.E. (2008) *Mitochondria*, 2nd edn, John Wiley & Sons,
Hoboken, NJ.

   An excellent survey of mitochondrial structure and function.

Slater, E.C. (1987) The mechanism of the conservation of energy of
biological oxidations. *Eur. J. Biochem.* **166**, 489–504.

   A clear and critical account of the evolution of the chemiosmotic
model.

OXIDATIVE PHOSPHORYLATION

Electron-Transfer Reactions in Mitochondria

Adam-Vizi, V. & Chinopoulos, C. (2006) Bioenergetics and the
formation of mitochondrial reactive oxygen species. *Trends
Pharmacol.* **27**, 639–645.

Cramer, W.A., Hasan, S.S., & Yamashita, E. (2011) The Q cycle
of cytochrome *bc* complexes: a structure perspective. *Biochim.
Biophys. Acta* **1807**, 788–802.

   Advanced discussion of the Q cycle in oxidative phosphorylation
and photophosphorylation.

**Dudkina, N.V., Kouřil, R., Peters, K., Braun, H.-P., & Boekema,
E.J.** (2010) Structure and function of mitochondrial
supercomplexes. *Biochim. Biophys. Acta* **1797**, 664–670.

Efremov, R.S., Baradaran, R., & Sazanov, L.A. (2010) The
architecture of respiratory complex I. *Nature* **465**, 441–447.

Hamanaka, R.B. & Chandel, N.S. (2010) Mitochondrial reactive
oxygen species regulate cellular signaling and dictate biological
outcomes. *Trends Biochem. Sci.* **35**, 505–513.

Millar, A.H., Whelan, J., Soole, K.L., & Day, D.A. (2011)
Organization and regulation of mitochondrial respiration in plants.
Annu. Rev. Plant Biol. **62**, 79–104.

**Sun, F., Huo, X., Zhai, Y., Wang, A., Xu, J., Su, D., Bartlam, M.,
& Rao, Z.** (2005) Crystal structure of mitochondrial respiratory
protein Complex II. *Cell* **121**, 1043–1057.

**Tsukihara, T., Aoyama, H., Yamashita, E., Tomizaki, T.,
Yamaguchi, H., Shinzawa-Itoh, K., Nakashima, R., Yaono, R., &
Yoshikawa, S.** (1996) The whole structure of the 13-subunit
oxidized cytochrome *c* oxidase at 2.8 Å. *Science* **272**, 1136–1144.

   The solution by x-ray crystallography of the structure of this
huge membrane protein.

**Xia, D., Yu, C.-A., Kim, H., Xia, J.-Z., Kachurin, A.M., Zhang, L.,
Yu, L., & Deisenhofer, J.** (1997) Crystal structure of the cytochrome
bc₁ complex from bovine heart mitochondria. *Science* **277**, 60–66.

   Report revealing the crystallographic structure of Complex III.

Yoshikawa, S., Muramoto, K., & Shinzawa-Itoh, K. (2011)
Proton-pumping mechanism of cytochrome *c* oxidase. *Annu. Rev.
Biophys.* **40**, 205–223.

ATP Synthesis

Abrahams, J.P., Leslie, A.G.W., Lutter, R., & Walker, J.E.
(1994) The structure of F$_1$-ATPase from bovine heart mitochondria
determined at 2.8 Å resolution. *Nature* **370**, 621–628.

Boyer, P.D. (1997) The ATP synthase—a splendid molecular
machine. *Annu. Rev. Biochem.* **66**, 717–749.

   An account of the historical development and current state of
the binding-change model, written by its principal architect.

Hinkle, P.C., Kumar, M.A., Resetar, A., & Harris, D.L. (1991)
Mechanistic stoichiometry of mitochondrial oxidative
phosphorylation. *Biochemistry* **30**, 3576–3582.

   A careful analysis of experimental results and theoretical
considerations on the question of nonintegral P/O ratios.

Okuno, D., Iino, R., & Noji, H. (2011) Rotation and structure of
F$_o$F$_1$-ATP synthase. *J. Biochem.* **149**, 655–664.

Silverstein, T. (2005) The mitochondrial phosphate-to-oxygen ratio
is not an integer. *Biochem. Mol. Biol. Educ.* **33**, 416–417.

von Ballmoos, C., Wiedenmann, A., & Dimroth, P. (2009)
Essentials for ATP synthesis by F$_1$F$_o$ ATP synthase. *Annu. Rev.
Biochem.* **78**, 649–672.

**Watt, I.N., Montgomery, M.G., Runswick, M.J., Leslie, A.G.W.,
& Walker, J.E.** (2010) Bioenergetic cost of making an adenosine

triphosphate molecule in animal mitochondria. *Proc. Natl. Acad. Sci. USA* **107**, 16,823–16,827.

Regulation of Oxidative Phosphorylation

Taylor, C.T. (2008) Mitochondria and cellular oxygen sensing in the HIF pathway. *Biochem. J.* **409**, 19–26.

Mitochondria in Thermogenesis, Steroid Synthesis, and Apoptosis

Azzu, V. & Brand, M.D. (2010) The on-off switches of the mitochondrial uncoupling proteins. *Trends Biochem. Sci.* **35**, 298–307.

Intermediate-level review of how cold, overfeeding, and starvation affect the expression of thermogenin genes.

Wang, C. & Youle, R.J. (2009) The role of mitochondria in apoptosis. *Annu. Rev. Genet.* **43**, 95–118.

Mitochondrial Genes: Their Origin and Effects of Mutations

Abou-Sleiman, P.M., Muqit, M.M.K., & Wood, N.W. (2006) Expanding insights of mitochondrial dysfunction in Parkinson's disease. *Nat. Rev. Neurosci.* **7**, 207–219.

Becker, T., Böttinger, L., & Pfanner, N. (2012) Mitochondrial protein import: from transport pathways to an integrated network. *Trends Biochem. Sci.* **37**, 85–91.

Intermediate-level review of how proteins encoded in the nucleus get into mitochondria.

Chen, Z.J. & Butow, R.A. (2005) The organization and inheritance of the mitochondrial genome. *Nat. Rev. Genet.* **6**, 815–825.

Intermediate-level review.

Seyfried, T.N. & Shelton, L.M. (2010) Cancer as a metabolic disease. *Nutr. Metab.* **7**, 7–29.

Impaired energy metabolism is typical of a wide variety of cancers; a review of the role of mitochondria.

PHOTOSYNTHESIS

Light-Driven Electron Flow

Barber, J. & Anderson, J.M. (eds). (2002) Photosystem II: Molecular Structure and Function. Proceedings of a Meeting, 13–14 March 2002. *Philos. Trans. R. Soc. (Biol. Sci.)* **357** (1426).

A collection of 16 papers on photosystem II.

Biochim. Biophys. Acta Bioenerg. (2007) **1767** (6).

This journal issue contains 10 reviews on the structure and function of photosystems.

Busch, A. & Hipler, M. (2011) The structure and function of eukaryotic photosystem I. *Biochim. Biophys. Acta* **1807**, 864–877.

Huber, R. (1990) A structural basis of light energy and electron transfer in biology. *Eur. J. Biochem.* **187**, 283–305.

Huber's Nobel lecture, describing the physics and chemistry of phototransductions; an exceptionally clear and well-illustrated discussion, based on crystallographic studies of reaction centers.

Johnson, G.N. (2011) Physiology of PSI cyclic electron transport in higher plants. *Biochim. Biophys. Acta* **1807**, 906–911.

Review of the paths and roles of cyclic photophosphorylation.

Kramer, D.M., Avenson, T.J., & Edwards, G.E. (2007) Dynamic flexibility in the light reactions of photosynthesis governed by both electron and proton transfer reactions. *Trends Plant Sci.* **9**, 349–357.

Intermediate-level review of regulation of state transitions.

Rochaix, J.-D. (2011) Regulation of photosynthetic electron transport. *Biochim. Biophys. Acta* **1807**, 878–886.

Intermediate-level review of plant's responses to changes in quality and quantity of light.

Umena, Y., Kawakami, K., Shen, J.-R., & Kamiya, N. (2011) Crystal structure of oxygen-evolving photosystem II at a resolution of 1.9 Å. *Nature* **473**, 55–61.

ATP Synthesis by Photophosphorylation

Jagendorf, A.T. (1967) Acid-base transitions and phosphorylation by chloroplasts. *Fed. Proc.* **26**, 1361–1369.

Classic experiment establishing the ability of a proton gradient to drive ATP synthesis in the dark.

The Evolution of Oxygenic Photosynthesis

Hohmann-Marriott, M.F. & Blankenship, R.E. (2011) Evolution of photosynthesis. *Annu. Rev. Plant Biol.* **62**, 515–548.

Nelson, N. (2011) Photosystems and global effects of oxygenic photosynthesis. *Biochim. Biophys. Acta* **1807**, 856–863.

Intermediate-level review of the evolutionary forces that shaped oxygenic photosynthesis.

Problems

1. Oxidation-Reduction Reactions The NADH dehydrogenase complex of the mitochondrial respiratory chain promotes the following series of oxidation-reduction reactions, in which Fe^{3+} and Fe^{2+} represent the iron in iron-sulfur centers, Q is ubiquinone, QH_2 is ubiquinol, and E is the enzyme:

(1) $\quad NADH + H^+ + E\text{-}FMN \longrightarrow NAD^+ + E\text{-}FMNH_2$

(2) $\quad E\text{-}FMNH_2 + 2Fe^{3+} \longrightarrow E\text{-}FMN + 2Fe^{2+} + 2H^+$

(3) $\quad 2Fe^{2+} + 2H^+ + Q \longrightarrow 2Fe^{3+} + QH_2$

Sum: $NADH + H^+ + Q \longrightarrow NAD^+ + QH_2$

For each of the three reactions catalyzed by the NADH dehydrogenase complex, identify (a) the electron donor, (b) the electron acceptor, (c) the conjugate redox pair, (d) the reducing agent, and (e) the oxidizing agent.

2. All Parts of Ubiquinone Have a Function In electron transfer, only the quinone portion of ubiquinone undergoes oxidation-reduction; the isoprenoid side chain remains unchanged. What is the function of this chain?

3. Use of FAD Rather Than NAD^+ in Succinate Oxidation All the dehydrogenases of glycolysis and the citric acid cycle use NAD^+ (E'° for $NAD^+/NADH$ is -0.32 V) as electron acceptor except succinate dehydrogenase, which uses covalently bound FAD (E'° for $FAD/FADH_2$ in this enzyme is 0.050 V). Suggest why FAD is a more appropriate electron acceptor than NAD^+ in the dehydrogenation of succinate, based on the E'° values of fumarate/succinate ($E'^\circ = 0.031$ V), $NAD^+/NADH$, and the succinate dehydrogenase $FAD/FADH_2$.

4. Degree of Reduction of Electron Carriers in the Respiratory Chain The degree of reduction of each carrier in the respiratory chain is determined by conditions in the mitochondrion. For example, when NADH and O_2 are abundant, the steady-state degree of reduction of the carriers decreases as electrons pass from the substrate to O_2. When electron transfer is blocked, the carriers before the block become more reduced and those beyond the block become more oxidized (see Fig. 19–6). For each of the conditions below, predict the state of oxidation of ubiquinone and cytochromes b, c_1, c, and $a + a_3$.

(a) Abundant NADH and O_2, but cyanide added

(b) Abundant NADH, but O_2 exhausted

(c) Abundant O_2, but NADH exhausted

(d) Abundant NADH and O_2

5. Effect of Rotenone and Antimycin A on Electron Transfer Rotenone, a toxic natural product from plants, strongly inhibits NADH dehydrogenase of insect and fish mitochondria. Antimycin A, a toxic antibiotic, strongly inhibits the oxidation of ubiquinol.

(a) Explain why rotenone ingestion is lethal to some insect and fish species.

(b) Explain why antimycin A is a poison.

(c) Given that rotenone and antimycin A are equally effective in blocking their respective sites in the electron-transfer chain, which would be a more potent poison? Explain.

6. Uncouplers of Oxidative Phosphorylation In normal mitochondria the rate of electron transfer is tightly coupled to the demand for ATP. When the rate of use of ATP is relatively low, the rate of electron transfer is low; when demand for ATP increases, electron-transfer rate increases. Under these conditions of tight coupling, the number of ATP molecules produced per atom of oxygen consumed when NADH is the electron donor—the P/O ratio—is about 2.5.

(a) Predict the effect of a relatively low and a relatively high concentration of uncoupling agent on the rate of electron transfer and the P/O ratio.

(b) Ingestion of uncouplers causes profuse sweating and an increase in body temperature. Explain this phenomenon in molecular terms. What happens to the P/O ratio in the presence of uncouplers?

(c) The uncoupler 2,4-dinitrophenol was once prescribed as a weight-reducing drug. How could this agent, in principle, serve as a weight-reducing aid? Uncoupling agents are no longer prescribed, because some deaths occurred following their use. Why might the ingestion of uncouplers lead to death?

7. Effects of Valinomycin on Oxidative Phosphorylation When the antibiotic valinomycin is added to actively respiring mitochondria, several things happen: the yield of ATP decreases, the rate of O_2 consumption increases, heat is released, and the pH gradient across the inner mitochondrial membrane increases. Does valinomycin act as an uncoupler or as an inhibitor of oxidative phosphorylation? Explain the experimental observations in terms of the antibiotic's ability to transfer K^+ ions across the inner mitochondrial membrane.

8. Mode of Action of Dicyclohexylcarbodiimide (DCCD) When DCCD is added to a suspension of tightly coupled, actively respiring mitochondria, the rate of electron transfer (measured by O_2 consumption) and the rate of ATP production dramatically decrease. If a solution of 2,4-dinitrophenol is now added to the preparation, O_2 consumption returns to normal but ATP production remains inhibited.

(a) What process in electron transfer or oxidative phosphorylation is affected by DCCD?

(b) Why does DCCD affect the O_2 consumption of mitochondria? Explain the effect of 2,4-dinitrophenol on the inhibited mitochondrial preparation.

(c) Which of the following inhibitors does DCCD most resemble in its action: antimycin A, rotenone, or oligomycin?

9. Compartmentalization of Citric Acid Cycle Components Isocitrate dehydrogenase is found only in the mitochondrion, but malate dehydrogenase is found in both the cytosol and mitochondrion. What is the role of cytosolic malate dehydrogenase?

10. The Malate–α-Ketoglutarate Transport System The transport system that conveys malate and α-ketoglutarate across the inner mitochondrial membrane (see Fig. 19–31) is inhibited by n-butylmalonate. Suppose n-butylmalonate is added to an aerobic suspension of kidney cells using glucose exclusively as fuel. Predict the effect of this inhibitor on (a) glycolysis, (b) oxygen consumption, (c) lactate formation, and (d) ATP synthesis.

11. Cellular ADP Concentration Controls ATP Formation Although both ADP and P_i are required for the synthesis of ATP, the rate of synthesis depends mainly on the concentration of ADP, not P_i. Why?

12. Time Scales of Regulatory Events in Mitochondria Compare the likely time scales for the adjustments in respiratory rate caused by (a) increased [ADP] and (b) reduced pO_2. What accounts for the difference?

13. The Pasteur Effect When O_2 is added to an anaerobic suspension of cells consuming glucose at a high rate, the rate of glucose consumption declines greatly as the O_2 is used up, and accumulation of lactate ceases. This effect, first observed by Louis Pasteur in the 1860s, is characteristic of most cells capable of both aerobic and anaerobic glucose catabolism.

(a) Why does the accumulation of lactate cease after O_2 is added?

(b) Why does the presence of O_2 decrease the rate of glucose consumption?

(c) How does the onset of O_2 consumption slow down the rate of glucose consumption? Explain in terms of specific enzymes.

14. Respiration-Deficient Yeast Mutants and Ethanol Production Respiration-deficient yeast mutants (p^-; "petites") can be produced from wild-type parents by treatment with mutagenic agents. The mutants lack cytochrome oxidase, a deficit that markedly affects their metabolic behavior. One striking effect is that fermentation is not suppressed by O_2—that is, the mutants lack the Pasteur effect (see Problem 13). Some companies are very interested in using these mutants to ferment wood chips to ethanol for energy use. Explain the advantages of using these mutants rather than wild-type yeast for large-scale ethanol production. Why does the absence of cytochrome oxidase eliminate the Pasteur effect?

15. Advantages of Supercomplexes for Electron Transfer There is growing evidence that mitochondrial Complexes I, II, III, and IV are part of a larger supercomplex. What might be the advantage of having all four complexes within a supercomplex?

16. How Many Protons in a Mitochondrion? Electron transfer translocates protons from the mitochondrial matrix to the external medium, establishing a pH gradient across the inner membrane (outside more acidic than inside). The tendency of protons to diffuse back into the matrix is the driving force for ATP synthesis by ATP synthase. During oxidative phosphorylation by a suspension of mitochondria in a medium of pH 7.4, the pH of the matrix has been measured as 7.7.

(a) Calculate $[H^+]$ in the external medium and in the matrix under these conditions.

(b) What is the outside-to-inside ratio of $[H^+]$? Comment on the energy inherent in this concentration difference. (Hint: See Eqn 11–4, p. 410.)

(c) Calculate the number of protons in a respiring liver mitochondrion, assuming its inner matrix compartment is a sphere of diameter 1.5 μm.

(d) From these data, is the pH gradient alone sufficient to generate ATP?

(e) If not, suggest how the necessary energy for synthesis of ATP arises.

17. Rate of ATP Turnover in Rat Heart Muscle Rat heart muscle operating aerobically fills more than 90% of its ATP needs by oxidative phosphorylation. Each gram of tissue consumes O_2 at the rate of 10.0 μmol/min, with glucose as the fuel source.

(a) Calculate the rate at which the heart muscle consumes glucose and produces ATP.

(b) For a steady-state concentration of ATP of 5.0 μmol/g of heart muscle tissue, calculate the time required (in seconds) to completely turn over the cellular pool of ATP. What does this result indicate about the need for tight regulation of ATP production? (Note: Concentrations are expressed as micromoles per gram of muscle tissue because the tissue is mostly water.)

18. Rate of ATP Breakdown in Insect Flight Muscle ATP production in the flight muscle of the fly *Lucilia sericata* results almost exclusively from oxidative phosphorylation. During flight, 187 mL of O_2/h·g of body weight is needed to maintain an ATP concentration of 7.0 μmol/g of flight muscle. Assuming that flight muscle makes up 20% of the weight of the fly, calculate the rate at which the flight-muscle ATP pool turns over. How long would the reservoir of ATP last in the absence of oxidative phosphorylation? Assume that reducing equivalents are transferred by the glycerol 3-phosphate shuttle and that O_2 is at 25 °C and 101.3 kPa (1 atm).

19. Mitochondrial Disease and Cancer Mutations in the genes that encode certain mitochondrial proteins are associated with a high incidence of some types of cancer. How might defective mitochondria lead to cancer?

20. Variable Severity of a Mitochondrial Disease Individuals with a disease caused by a specific defect in the mitochondrial genome may have symptoms ranging from mild to severe. Explain why.

21. Transmembrane Movement of Reducing Equivalents Under aerobic conditions, extramitochondrial NADH must be oxidized by the mitochondrial electron-transfer chain. Consider a preparation of rat hepatocytes containing mitochondria and all the cytosolic enzymes. If $[4\text{-}^3H]NADH$ is introduced, radioactivity soon appears in the mitochondrial matrix. However, if $[7\text{-}^{14}C]NADH$ is introduced, no radioactivity appears in the matrix. What do these observations reveal about the oxidation of extramitochondrial NADH by the electron-transfer chain?

[4-^{3}H]NADH   [7-^{14}C]NADH

22. High Blood Alanine Level Associated with Defects in Oxidative Phosphorylation Most individuals with genetic defects in oxidative phosphorylation are found to have relatively high concentrations of alanine in their blood. Explain this in biochemical terms.

23. NAD Pools and Dehydrogenase Activities Although both pyruvate dehydrogenase and glyceraldehyde 3-phosphate dehydrogenase use NAD^+ as their electron acceptor, the two enzymes do not compete for the same cellular NAD pool. Why?

24. Diabetes as a Consequence of Mitochondrial Defects Glucokinase is essential in the metabolism of glucose in pancreatic β cells. Humans with two defective copies of the glucokinase gene exhibit a severe, neonatal diabetes, whereas those with only one defective copy of the gene have a much milder form of the disease (mature onset diabetes of the young, MODY2). Explain this difference in terms of the biology of the β cell.

25. Effects of Mutations in Mitochondrial Complex II Single nucleotide changes in the gene for succinate dehydrogenase (Complex II) are associated with midgut carcinoid tumors. Suggest a mechanism to explain this observation.

26. Photochemical Efficiency of Light at Different Wavelengths The rate of photosynthesis, measured by O_2 production, is higher when a green plant is illuminated with light of wavelength 680 nm than with light of 700 nm. However, illumination by a combination of light of 680 nm and 700 nm gives a higher rate of photosynthesis than light of either wavelength alone. Explain.

27. Balance Sheet for Photosynthesis In 1804 Theodore de Saussure observed that the total weight of oxygen and dry organic matter produced by plants is greater than the weight of carbon dioxide consumed during photosynthesis. Where does the extra weight come from?

28. Role of H_2S in Some Photosynthetic Bacteria Illuminated purple sulfur bacteria carry out photosynthesis in the presence of H_2O and $^{14}CO_2$, but only if H_2S is added and O_2 is absent. During the course of photosynthesis, measured by formation of

[^{14}C]carbohydrate, H$_2$S is converted to elemental sulfur, but no O$_2$ is evolved. What is the role of the conversion of H$_2$S to sulfur? Why is no O$_2$ evolved?

29. Boosting the Reducing Power of Photosystem I by Light Absorption When photosystem I absorbs red light at 700 nm, the standard reduction potential of P700 changes from 0.40 V to about -1.2 V. What fraction of the absorbed light is trapped in the form of reducing power?

30. Electron Flow through Photosystems I and II Predict how an inhibitor of electron passage through pheophytin would affect electron flow through (a) photosystem II and (b) photosystem I. Explain your reasoning.

31. Limited ATP Synthesis in the Dark In a laboratory experiment, spinach chloroplasts are illuminated in the absence of ADP and P$_i$, then the light is turned off and ADP and P$_i$ are added. ATP is synthesized for a short time in the dark. Explain this finding.

32. Mode of Action of the Herbicide DCMU When chloroplasts are treated with 3-(3,4-dichlorophenyl)-1,1-dimethylurea (DCMU, or diuron), a potent herbicide, O$_2$ evolution and photophosphorylation cease. Oxygen evolution, but not photophosphorylation, can be restored by addition of an external electron acceptor, or Hill reagent. How does DCMU act as a weed killer? Suggest a location for the inhibitory action of this herbicide in the scheme shown in Figure 19–58. Explain.

33. Effect of Venturicidin on Oxygen Evolution Venturicidin is a powerful inhibitor of the chloroplast ATP synthase, interacting with the CF$_o$ part of the enzyme and blocking proton passage through the CF$_o$CF$_1$ complex. How would venturicidin affect oxygen evolution in a suspension of well-illuminated chloroplasts? Would your answer change if the experiment were done in the presence of an uncoupling reagent such as 2,4-dinitrophenol (DNP)? Explain.

34. Bioenergetics of Photophosphorylation The steady-state concentrations of ATP, ADP, and P$_i$ in isolated spinach chloroplasts under full illumination at pH 7.0 are 120.0, 6.0, and 700.0 μM, respectively.

(a) What is the free-energy requirement for the synthesis of 1 mol of ATP under these conditions?

(b) The energy for ATP synthesis is furnished by light-induced electron transfer in the chloroplasts. What is the minimum voltage drop necessary (during transfer of a pair of electrons) to synthesize ATP under these conditions? (You may need to refer to Eqn 13–7, p. 531.)

35. Light Energy for a Redox Reaction Suppose you have isolated a new photosynthetic microorganism that oxidizes H$_2$S and passes the electrons to NAD$^+$. What wavelength of light would provide enough energy for H$_2$S to reduce NAD$^+$ under standard conditions? Assume 100% efficiency in the photochemical event, and use E'° of -243 mV for H$_2$S and -320 mV for NAD$^+$. See Figure 19–48 for the energy equivalents of wavelengths of light.

36. Equilibrium Constant for Water-Splitting Reactions The coenzyme NADP$^+$ is the terminal electron acceptor in chloroplasts, according to the reaction

$$2H_2O + 2NADP^+ \longrightarrow 2NADPH + 2H^+ + O_2$$

Use the information in Table 19–2 to calculate the equilibrium constant for this reaction at 25 °C. (The relationship between K'_{eq} and $\Delta G'^\circ$ is discussed on p. 508.) How can the chloroplast overcome this unfavorable equilibrium?

37. Energetics of Phototransduction During photosynthesis, eight photons must be absorbed (four by each photosystem) for every O$_2$ molecule produced:

$$2H_2O + 2NADP^+ + 8 \text{ photons} \longrightarrow 2NADPH + 2H^+ + O_2$$

Assuming that these photons have a wavelength of 700 nm (red) and that the light absorption and use of light energy are 100% efficient, calculate the free-energy change for the process.

38. Electron Transfer to a Hill Reagent Isolated spinach chloroplasts evolve O$_2$ when illuminated in the presence of potassium ferricyanide (a Hill reagent), according to the equation

$$2H_2O + 4Fe^{3+} \longrightarrow O_2 + 4H^+ + 4Fe^{2+}$$

where Fe^{3+} represents ferricyanide and Fe^{2+}, ferrocyanide. Is NADPH produced in this process? Explain.

39. How Often Does a Chlorophyll Molecule Absorb a Photon? The amount of chlorophyll a (M_r 892) in a spinach leaf is about 20 μg/cm^2 of leaf surface. In noonday sunlight (average energy reaching the leaf is 5.4 J/cm^2 · min), the leaf absorbs about 50% of the radiation. How often does a single chlorophyll molecule absorb a photon? Given that the average lifetime of an excited chlorophyll molecule in vivo is 1 ns, what fraction of the chlorophyll molecules are excited at any one time?

40. Effect of Monochromatic Light on Electron Flow The extent to which an electron carrier is oxidized or reduced during photosynthetic electron transfer can sometimes be observed directly with a spectrophotometer. When chloroplasts are illuminated with 700 nm light, cytochrome f, plastocyanin, and plastoquinone are oxidized. When chloroplasts are illuminated with 680 nm light, however, these electron carriers are reduced. Explain.

41. Function of Cyclic Photophosphorylation When the [NADPH]/[NADP$^+$] ratio in chloroplasts is high, photophosphorylation is predominantly cyclic (see Fig. 19–58). Is O$_2$ evolved during cyclic photophosphorylation? Is NADPH produced? Explain. What is the main function of cyclic photophosphorylation?

Data Analysis Problem

42. Photophosphorylation: Discovery, Rejection, and Rediscovery In the 1930s and 1940s, researchers were beginning to make progress toward understanding the mechanism of photosynthesis. At the time, the role of "energy-rich phosphate bonds" (today, "ATP") in glycolysis and cellular respiration was just becoming known. There were many theories about the mechanism of photosynthesis, especially about the role of light. This problem focuses on what was then called the "primary photochemical process"—that is, on what it is, exactly, that the energy from captured light produces in the photosynthetic cell. Interestingly, one important part of the modern model of photosynthesis was proposed early on, only to be rejected, ignored for several years, then finally revived and accepted.

In 1944, Emerson, Stauffer, and Umbreit proposed that "the function of light energy in photosynthesis is the formation of 'energy-rich' phosphate bonds" (p. 107). In their model (hereafter, the "Emerson model"), the free energy necessary to drive both CO_2 fixation *and* reduction came from these "energy-rich phosphate bonds" (i.e., ATP), produced as a result of light absorption by a chlorophyll-containing protein.

This model was explicitly rejected by Rabinowitch (1945). After summarizing Emerson and coauthors' findings, Rabinowitch stated: "Until more positive evidence is provided, we are inclined to consider as more convincing a general argument against this hypothesis, which can be derived from energy considerations. Photosynthesis is eminently a problem of energy *accumulation*. What good can be served, then, by converting light quanta (even those of red light, which amount to about 43 kcal per Einstein) into 'phosphate quanta' of only 10 kcal per mole? This appears to be a start in the wrong direction—toward *dissipation* rather than toward accumulation of energy" (Vol. I, p. 228). This argument, along with other evidence, led to the abandonment of the Emerson model until the 1950s, when it was found to be correct—albeit in a modified form.

For each piece of information from Emerson and coauthors' article presented in (a) through (d), answer the following three questions:

1. How does this information support the Emerson model, in which light energy is used directly by chlorophyll *to make ATP*, and the ATP then provides the energy to drive CO_2 fixation and reduction?

2. How would Rabinowitch explain this information, based on his model (and most other models of the day), in which light energy is used directly by chlorophyll *to make reducing compounds*? Rabinowitch wrote: "Theoretically, there is no reason why *all* electronic energy contained in molecules excited by the absorption of light should not be available for oxidation-reduction" (Vol. I, p. 152). In this model, the reducing compounds are then used to fix and reduce CO_2, and the energy for these reactions comes from the large amounts of free energy released by the reduction reactions.

3. How is this information explained by our modern understanding of photosynthesis?

(a) Chlorophyll contains a Mg^{2+} ion, which is known to be an essential cofactor for many enzymes that catalyze phosphorylation and dephosphorylation reactions.

(b) A crude "chlorophyll protein" isolated from photosynthetic cells showed phosphorylating activity.

(c) The phosphorylating activity of the "chlorophyll protein" was inhibited by light.

(d) The levels of several different phosphorylated compounds in photosynthetic cells changed dramatically in response to light exposure. (Emerson and coworkers were not able to identify the specific compounds involved.)

As it turned out, the Emerson and Rabinowitch models were both partly correct and partly incorrect.

(e) Explain how the two models relate to our current model of photosynthesis.

In his rejection of the Emerson model, Rabinowitch went on to say: "The difficulty of the phosphate storage theory appears most clearly when one considers the fact that, in weak light, eight or ten quanta of light are sufficient to reduce one molecule of carbon dioxide. If each quantum should produce one molecule of high-energy phosphate, the accumulated energy would be only 80–100 kcal per Einstein—while photosynthesis requires *at least* 112 kcal per mole, and probably more, because of losses in irreversible partial reactions" (Vol. 1, p. 228).

(f) How does Rabinowitch's value of 8 to 10 photons per molecule of CO_2 reduced compare with the value accepted today? You need to consult Chapter 20 for some of the information required here.

(g) How would you rebut Rabinowitch's argument, based on our current knowledge about photosynthesis?

References

Emerson, R.L., Stauffer, J.F., & Umbreit, W.W. (1944) Relationships between phosphorylation and photosynthesis in *Chlorella. Am. J. Botany* **31**, 107–120.

Rabinowitch, E.I. (1945) *Photosynthesis and Related Processes*, Interscience Publishers, New York.

20

Carbohydrate Biosynthesis in Plants and Bacteria

W e have now reached a turning point in our study of cellular metabolism. Thus far in Part II we have described how the major metabolic fuels—carbohydrates, fatty acids, and amino acids—are degraded through converging *catabolic* pathways to enter the citric acid cycle and yield their electrons to the respiratory chain, and how this exergonic flow of electrons to oxygen is coupled to the endergonic synthesis of ATP. We now turn to *anabolic* pathways, which use chemical energy in the form of ATP and NADH or NADPH to synthesize cellular components from simple precursor molecules. Anabolic pathways are generally reductive rather than oxidative. Catabolism and anabolism proceed simultaneously in a dynamic steady state, so the energy-yielding degradation of cellular components is counterbalanced by biosynthetic processes, which create and maintain the intricate orderliness of living cells.

Plants must be especially versatile in their handling of carbohydrates, for several reasons. First, plants are autotrophs, able to convert inorganic carbon (as CO$_2$) into organic compounds. Second, biosynthesis occurs primarily in plastids, membrane-bounded organelles unique to photosynthetic organisms, and the movement of intermediates between cellular compartments is an important aspect of metabolism. Third, plants are not motile: they cannot move to find better supplies of water, sunlight, or nutrients. They must have sufficient metabolic flexibility to allow them to adapt to changing conditions in the place where they are rooted. Finally, plants have thick cell walls made of carbohydrate polymers, which must be assembled outside the plasma membrane and which constitute a significant proportion of the cell's carbohydrate.

The chapter begins with a description of the process by which CO$_2$ is assimilated into trioses and hexoses, then considers photorespiration, an important side reaction during CO$_2$ fixation, and the ways in which certain plants avoid this side reaction. We then look at how the biosynthesis of sucrose (for sugar transport) and starch (for energy storage) is accomplished by mechanisms analogous to those employed by animal cells to make glycogen. The next topic is the synthesis of the cellulose of plant cell walls and the peptidoglycan of bacterial cell walls, illustrating the problems of energy-dependent biosynthesis outside the plasma membrane. Finally, we discuss how the various pathways that share pools of common intermediates are segregated within organelles yet integrated with one another.

20.1 Photosynthetic Carbohydrate Synthesis

The synthesis of carbohydrates in animal cells always employs precursors having at least three carbons, all of which are less oxidized than the carbon in CO$_2$. Plants and photosynthetic microorganisms, by contrast, can synthesize carbohydrates from CO$_2$ and water, reducing CO$_2$ at the expense of the energy and reducing power furnished by the ATP and NADPH that are generated by the light-dependent reactions of photosynthesis **(Fig. 20–1)**. Plants (and other autotrophs) can use CO$_2$ as the sole source of the carbon atoms required for the biosynthesis of cellulose and starch, lipids and proteins, and the many other organic components of plant cells. By contrast, heterotrophs cannot bring

Melvin Calvin, 1911–1997

FIGURE 20–1 Assimilation of CO_2 into biomass in plants. The light-driven synthesis of ATP and NADPH, described in Chapter 19, provides energy and reducing power for the fixation of CO_2 into trioses, from which all the carbon-containing compounds of the plant cell are synthesized. The processes shown with red arrows are the focus of this chapter.

about the net reduction of CO_2 to achieve a net synthesis of glucose.

Green plants contain in their chloroplasts unique enzymatic machinery that catalyzes the conversion of CO_2 to simple (reduced) organic compounds, a process called **CO_2 assimilation**. This process has also been called **CO_2 fixation** or **carbon fixation**, but we reserve these terms for the specific reaction in which CO_2 is incorporated (fixed) into a three-carbon organic compound, the triose phosphate 3-phosphoglycerate. This simple product of photosynthesis is the precursor of more complex biomolecules, including sugars, polysaccharides, and the metabolites derived from them, all of which are synthesized by metabolic pathways similar to those of animal tissues. Carbon dioxide is assimilated via a cyclic pathway, its key intermediates constantly regenerated. The pathway was elucidated in the early 1950s by Melvin Calvin, Andrew Benson, and James A. Bassham, and is often called the **Calvin cycle** or, more descriptively, the **photosynthetic carbon reduction cycle**.

Carbohydrate metabolism is more complex in plant cells than in animal cells or in nonphotosynthetic microorganisms. In addition to the universal pathways of glycolysis and gluconeogenesis, plants have the unique reaction sequences for reduction of CO_2 to triose phosphates and the associated reductive pentose phosphate pathway—all of which must be coordinately regulated to ensure proper allocation of carbon to energy production and synthesis of starch and sucrose. Key enzymes are regulated, as we shall see, by (1) reduction of disulfide bonds by electrons flowing from photosystem I and (2) changes in pH and Mg^{2+} concentration that result from illumination. When we look at other aspects of plant carbohydrate metabolism, we also find enzymes that are modulated by (3) conventional allosteric regulation by one or more metabolic intermediates and (4) covalent modification (phosphorylation).

Plastids Are Organelles Unique to Plant Cells and Algae

Most of the biosynthetic activities in plants (including CO_2 assimilation) occur in **plastids**, a family of self-reproducing organelles bounded by a double membrane and containing a small genome that encodes some of their proteins. Most proteins destined for plastids are encoded in nuclear genes, which are transcribed and translated like other nuclear genes; then the proteins are imported into plastids. Plastids reproduce by binary fission, replicating their genome (a single circular DNA molecule) and using their own enzymes and ribosomes to synthesize the proteins encoded by that genome. **Chloroplasts** (see Fig. 19–47) are the sites of CO_2 assimilation. The enzymes for this process are contained in the stroma, the soluble phase bounded by the inner chloroplast membrane. **Amyloplasts** are colorless plastids (that is, they lack chlorophyll and other pigments found in chloroplasts). They have no internal membranes analogous to the photosynthetic membranes (thylakoids) of chloroplasts, and in plant tissues rich in starch these plastids are packed with starch granules **(Fig. 20–2)**. Chloroplasts can be converted to

FIGURE 20–2 Amyloplasts filled with starch (dark granules) are stained with iodine in this section of *Ranunculus* root cells. Starch granules in various tissues range from 1 to 100 μm in diameter.

proplastids by the loss of their internal membranes and chlorophyll, and proplastids are interconvertible with amyloplasts **(Fig. 20–3)**. In turn, both amyloplasts and proplastids can develop into chloroplasts. The relative abundance of the plastid types depends on the type of plant tissue and on the intensity of illumination. Cells of green leaves are rich in chloroplasts, whereas amyloplasts dominate in nonphotosynthetic tissues that store starch in large quantities, such as potato tubers.

The inner membranes of all types of plastids are impermeable to polar and charged molecules. Traffic across these membranes is mediated by sets of specific transporters.

Carbon Dioxide Assimilation Occurs in Three Stages

The first stage in the assimilation of CO_2 into biomolecules **(Fig. 20–4)** is the **carbon-fixation reaction**: condensation of CO_2 with a five-carbon acceptor, **ribulose 1,5-bisphosphate**, to form two molecules of **3-phosphoglycerate**. In the second stage, the 3-phosphoglycerate is reduced to triose phosphates. Overall, three molecules of CO_2 are fixed to three molecules of ribulose 1,5-bisphosphate to form six molecules of glyceraldehyde 3-phosphate (18 carbons) in equilibrium with dihydroxyacetone phosphate. In the third stage, five of the six molecules of triose phosphate (15 carbons) are used to regenerate three molecules of ribulose 1,5-bisphosphate (15 carbons), the starting material. The sixth molecule of triose phosphate, the net product of photosynthesis, can be used to make hexoses for fuel and building materials, sucrose for transport to nonphotosynthetic tissues, or starch for storage. Thus the overall process is cyclical, with the continuous conversion of CO_2 to triose and hexose phosphates. Fructose 6-phosphate is a key intermediate in stage 3 of CO_2 assimilation; it stands at a branch point, leading either to regeneration of ribulose 1,5-bisphosphate or to synthesis of starch. The pathway from hexose phosphate to pentose bisphosphate involves many of the same reactions used in animal cells for the conversion of pentose phosphates to hexose phosphates during the nonoxidative phase of the **pentose phosphate pathway** (see Fig. 14–23). In the photosynthetic assimilation of CO_2, essentially the same set of reactions operates in the other direction, converting hexose phosphates to pentose phosphates. This **reductive pentose phosphate cycle** uses the same enzymes as the oxidative pathway, and several more enzymes that make the reductive cycle irreversible. All 13 enzymes of the pathway are in the chloroplast stroma.

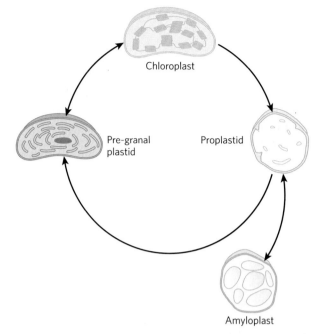

FIGURE 20–3 Plastids: their origins and interconversions. All types of plastids are bounded by a double membrane, and some (notably the mature chloroplast) have extensive internal membranes. The internal membranes can be lost (when a mature chloroplast becomes a proplastid) and resynthesized (as a proplastid gives rise to a pre-granal plastid and then a mature chloroplast). Proplastids in nonphotosynthetic tissues (such as root) give rise to amyloplasts, which contain large quantities of starch. All plant cells have plastids, and these organelles are the sites, not only of photosynthesis, but of other processes, including the synthesis of essential amino acids, thiamine, pyridoxal phosphate, flavins, and vitamins A, C, E, and K.

Stage 1: Fixation of CO_2 into 3-Phosphoglycerate An important clue to the nature of the CO_2-assimilation mechanisms in photosynthetic organisms came in the late 1940s. Calvin and his associates illuminated a suspension of green algae in the presence of radioactive carbon dioxide ($^{14}CO_2$) for

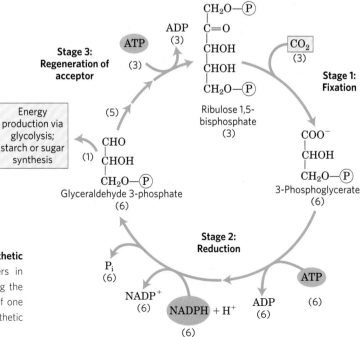

FIGURE 20–4 The three stages of CO_2 assimilation in photosynthetic organisms. Stoichiometries of three key intermediates (numbers in parentheses) reveal the fate of carbon atoms entering and leaving the cycle. As shown here, three CO_2 are fixed for the net synthesis of one molecule of glyceraldehyde 3-phosphate. This cycle is the photosynthetic carbon reduction cycle, or the Calvin cycle.

just a few seconds, then quickly killed the cells, extracted their contents, and with the help of chromatographic methods searched for the metabolites in which the labeled carbon first appeared. The first compound that became labeled was **3-phosphoglycerate**, with the ^{14}C predominantly located in the carboxyl carbon atom. These experiments strongly suggested that 3-phosphoglycerate is an early intermediate in photosynthesis. The many plants in which this three-carbon compound is the first intermediate are called **C₃ plants**, in contrast to the C₄ plants described below.

The enzyme that catalyzes incorporation of CO_2 into an organic form is **ribulose 1,5-bisphosphate carboxylase/oxygenase**, a name mercifully shortened to **rubisco**. As a carboxylase, rubisco catalyzes the covalent attachment of CO_2 to the five-carbon sugar ribulose 1,5-bisphosphate and cleavage of the unstable six-carbon intermediate to form two molecules of 3-phosphoglycerate, one of which bears the carbon introduced as CO_2 in its carboxyl group (Fig. 20–4). The enzyme's oxygenase activity is discussed in Section 20.2.

There are two distinct forms of rubisco. Form I is found in vascular plants, algae, and cyanobacteria; form II is confined to certain photosynthetic bacteria. Plant rubisco, the crucial enzyme in the production of biomass from CO_2, has a complex form I structure **(Fig. 20–5a)**, with eight identical large subunits (M_r 53,000; encoded in the chloroplast genome, or plastome), each containing a catalytic site, and eight identical small subunits (M_r 14,000; encoded in the nuclear genome) of uncertain function. The form II rubisco of photosynthetic bacteria is simpler in structure, having two subunits that in many respects resemble the large subunits of the plant enzyme (Fig. 20–5b). This similarity is consistent with the endosymbiont hypothesis for the origin of chloroplasts (p. 36). The plant enzyme has an exceptionally low turnover number; only three molecules of CO_2 are fixed per second per molecule of rubisco at 25 °C. To achieve high rates of CO_2 fixation, plants therefore need large amounts of this enzyme. In fact, rubisco makes up almost 50% of soluble protein in chloroplasts and is probably one of the most abundant enzymes in the biosphere.

Central to the proposed mechanism for plant rubisco is a carbamoylated Lys side chain with a bound Mg^{2+} ion. The Mg^{2+} ion brings together and orients the reactants at

(a)   Top view   Side view

Ribulose 1,5-bisphosphate

Mg^{2+}

Carbamoyl-Lys residue

(b)

FIGURE 20–5 Structure of ribulose 1,5-bisphosphate carboxylase (rubisco). (a) Top and side views of a ribbon model of form I rubisco from spinach (PDB ID 8RUC). The enzyme has eight large subunits (blue) and eight small ones (gray), tightly packed into a structure of M_r greater than 500,000. Rubisco is present at a concentration of about 250 mg/mL in the chloroplast stroma, corresponding to an extraordinarily high concentration of active sites (~4 mM). A transition-state analog, 2-carboxyarabinitol bisphosphate, is shown in yellow bound to each of the eight substrate-binding sites. Mg^{2+} is shown in green. **(b)** Ribbon model of form II rubisco from the bacterium *Rhodospirillum rubrum* (PDB ID 9RUB). The subunits are in gray and blue.

FIGURE 20–6 Central role of Mg^{2+} in the catalytic mechanism of rubisco. (Derived from PDB ID 1RXO) Mg^{2+} is coordinated in a roughly octahedral complex with six oxygen atoms: one oxygen in the carbamate on Lys^{201}; two in the carboxyl groups of Glu^{204} and Asp^{203}; two at C-2 and C-3 of the substrate, ribulose 1,5-bisphosphate; and one in the other substrate, CO_2. A water molecule occupies the CO_2-binding site in this crystal structure. In this figure a CO_2 molecule is modeled in its place. (Residue numbers refer to the spinach enzyme.)

the active site **(Fig. 20–6)** and polarizes the CO_2, opening it to nucleophilic attack by the five-carbon enediolate reaction intermediate formed on the enzyme **(Fig. 20–7)**. The resulting six-carbon intermediate breaks down to yield two molecules of 3-phosphoglycerate.

MECHANISM FIGURE 20–7 First stage of CO_2 assimilation: rubisco's carboxylase activity. The CO_2-fixation reaction is catalyzed by ribulose 1,5-bisphosphate carboxylase/oxygenase (rubisco). The overall reaction accomplishes the combination of one CO_2 and one ribulose 1,5-bisphosphate to form two molecules of 3-phosphoglycerate, one of which contains the carbon atom from CO_2 (red). Additional proton transfers (not shown), involving Lys^{201}, Lys^{175}, and His^{294}, occur in several of these steps.
🔵 Rubisco Mechanism; Rubisco Tutorial

As the catalyst for the first step of photosynthetic CO_2 assimilation, rubisco is a prime target for regulation. The enzyme is inactive until carbamoylated on the ε amino group of Lys[201] (**Fig. 20–8**). Ribulose 1,5-bisphosphate inhibits carbamoylation by binding tightly to the active site and locking the enzyme in the "closed"

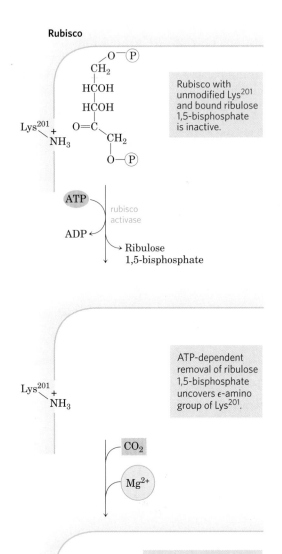

Rubisco

Rubisco with unmodified Lys[201] and bound ribulose 1,5-bisphosphate is inactive.

ATP-dependent removal of ribulose 1,5-bisphosphate uncovers ϵ-amino group of Lys[201].

ϵ-Amino group of Lys[201] is carbamoylated by CO_2; Mg^{2+} binds to carbamoyl-Lys, activating rubisco.

FIGURE 20–8 Role of rubisco activase in the carbamoylation of Lys[201] of rubisco. When the substrate ribulose 1,5-bisphosphate is bound to the active site, Lys[201] is not accessible. Rubisco activase couples ATP hydrolysis to expulsion of the bound sugar bisphosphate, exposing Lys[201]; this Lys residue can now be carbamoylated with CO_2 in a reaction that is apparently not enzyme-mediated. Mg^{2+} is attracted to and binds to the negatively charged carbamoyl-Lys, and the enzyme is thus activated.

conformation, in which Lys[201] is inaccessible. **Rubisco activase** overcomes the inhibition by promoting ATP-dependent release of the ribulose 1,5-bisphosphate, exposing the Lys amino group to nonenzymatic carbamoylation by CO_2; this is followed by Mg^{2+} binding, which activates the rubisco. Rubisco activase in some species is activated by light through a redox mechanism similar to that shown in Figure 20.19.

Another regulatory mechanism involves the "nocturnal inhibitor" 2-carboxyarabinitol 1-phosphate, a naturally occurring transition-state analog (p. 210) with a structure similar to that of the β-keto acid intermediate of the rubisco reaction (Fig. 20–7). This compound, synthesized in the dark in some plants, is a potent inhibitor of carbamoylated rubisco. It is either broken down when light returns or is expelled by rubisco activase, activating the rubisco.

2-Carboxyarabinitol 1-phosphate

Stage 2: Conversion of 3-Phosphoglycerate to Glyceraldehyde 3-Phosphate

The 3-phosphoglycerate formed in stage 1 is converted to glyceraldehyde 3-phosphate in two steps that are essentially the reversal of the corresponding steps in glycolysis, with one exception: the nucleotide cofactor for the reduction of 1,3-bisphosphoglycerate is NADPH rather than NADH (**Fig. 20–9**). The chloroplast stroma contains all the glycolytic enzymes except phosphoglycerate mutase. The stromal and cytosolic enzymes are isozymes; both sets of enzymes catalyze the same reactions, but they are the products of different genes.

In the first step of stage 2, the stromal **3-phosphoglycerate kinase** catalyzes the transfer of a phosphoryl group from ATP to 3-phosphoglycerate, yielding 1,3-bisphosphoglycerate. Next, NADPH donates electrons in a reduction catalyzed by the chloroplast-specific isozyme of **glyceraldehyde 3-phosphate dehydrogenase**, producing glyceraldehyde 3-phosphate and P_i. The high concentrations of NADPH and ATP in the chloroplast stroma allow this thermodynamically unfavorable pair of reactions to proceed in the direction of 1,3-bisphosphoglycerate. Triose phosphate isomerase then interconverts glyceraldehyde 3-phosphate and dihydroxyacetone phosphate. Most of the triose phosphate thus produced is used to regenerate ribulose 1,5-bisphosphate; the rest is either converted to starch in the chloroplast and stored for later use or immediately exported to the cytosol and converted to sucrose for transport to growing regions of the plant. In

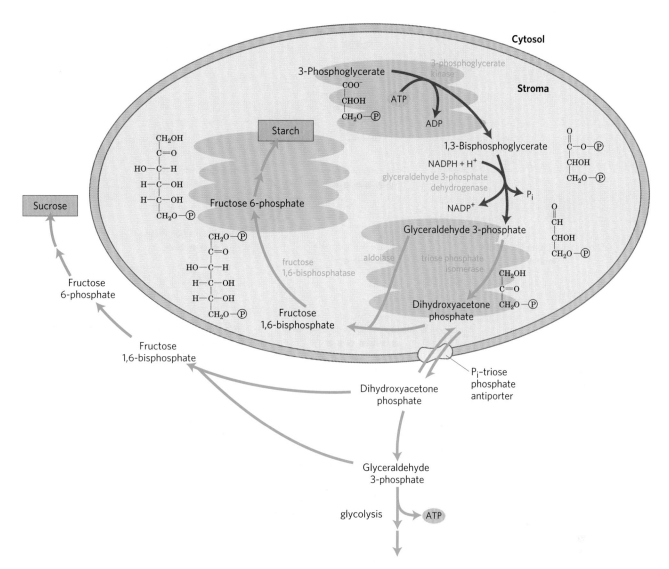

FIGURE 20–9 Second stage of CO₂ assimilation. 3-Phosphoglycerate is converted to glyceraldehyde 3-phosphate (red arrows). Also shown are the alternative fates of the fixed carbon of glyceraldehyde 3-phosphate (blue arrows). Most of the glyceraldehyde 3-phosphate is recycled to ribulose 1,5-bisphosphate as shown in Figure 20-10. A small fraction of the "extra" glyceraldehyde 3-phosphate may be used immediately as a source of energy, but most is converted to sucrose for transport or is stored in the chloroplast as starch. In the latter case, glyceraldehyde 3-phosphate condenses with dihydroxyacetone phosphate in the stroma to form fructose 1,6-bisphosphate, a precursor of starch. In other situations the glyceraldehyde 3-phosphate is converted to dihydroxyacetone phosphate, which leaves the chloroplast via a specific transporter (see Fig. 20-15) and, in the cytosol, can be degraded glycolytically to provide energy or used to form fructose 6-phosphate and hence sucrose.

developing leaves, a significant portion of the triose phosphate may be degraded by glycolysis to provide energy.

Stage 3: Regeneration of Ribulose 1,5-Bisphosphate from Triose Phosphates

The first reaction in the assimilation of CO₂ into triose phosphates consumes ribulose 1,5-bisphosphate and, for continuous flow of CO₂ into carbohydrate, ribulose 1,5-bisphosphate must be constantly regenerated. This is accomplished in a series of reactions (**Fig. 20–10**) that, together with stages 1 and 2, constitute the cyclic pathway shown in Figure 20–4. The product of the first assimilation reaction (3-phosphoglycerate) thus undergoes transformations that regenerate ribulose 1,5-bisphosphate. The intermediates

in this pathway include three-, four-, five-, six-, and seven-carbon sugars. In the following discussion, all step numbers refer to Figure 20–10.

Steps ❶ and ❹ are catalyzed by the same enzyme, **aldolase**. It first catalyzes the reversible condensation of glyceraldehyde 3-phosphate with dihydroxyacetone phosphate, yielding fructose 1,6-bisphosphate (step ❶); this is cleaved to fructose 6-phosphate and Pᵢ by fructose 1,6-bisphosphatase (FBPase-1) in step ❷. The reaction is strongly exergonic and essentially irreversible. Step ❸ is catalyzed by **transketolase**, which contains thiamine pyrophosphate (TPP) as its prosthetic group (see Fig. 14–15a) and requires Mg²⁺. Transketolase catalyzes the reversible transfer of a 2-carbon ketol group

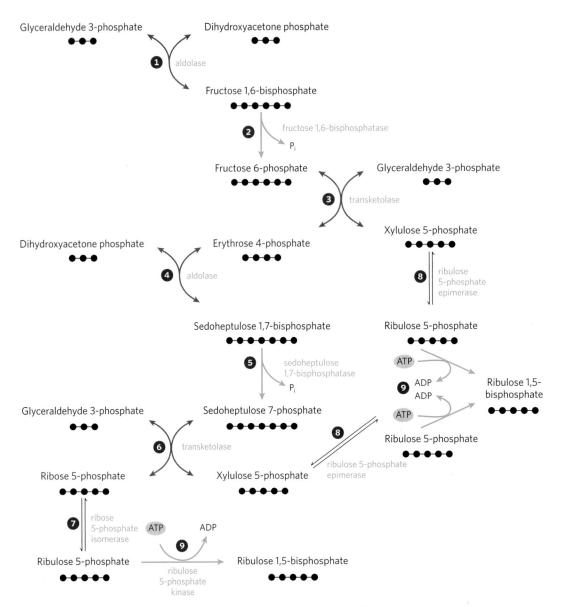

FIGURE 20–10 Third stage of CO_2 assimilation. This schematic diagram shows the interconversions of triose phosphates and pentose phosphates. Black dots represent the number of carbons in each compound. The starting materials are glyceraldehyde 3-phosphate and dihydroxyacetone phosphate. Reactions catalyzed by aldolase (❶ and ❹) and transketolase (❸ and ❻) produce pentose phosphates that are converted to ribulose 1,5-bisphosphate—ribose 5-phosphate by ribose 5-phosphate isomerase (❼) and xylulose 5-phosphate by ribulose 5-phosphate epimerase (❽). In step ❾, ribulose 5-phosphate is phosphorylated, regenerating ribulose 1,5-bisphosphate. The steps with blue arrows are exergonic and make the whole process irreversible: steps ❷ fructose 1,6-bisphosphatase, ❺ sedoheptulose bisphosphatase, and ❾ ribulose 5-phosphate kinase.

(CH_2OH—CO—) from a ketose phosphate donor, fructose 6-phosphate, to an aldose phosphate acceptor, glyceraldehyde 3-phosphate **(Fig. 20–11a, b)**, forming the pentose xylulose 5-phosphate and the tetrose erythrose 4-phosphate. In step ❹, aldolase acts again, combining erythrose 4-phosphate with dihydroxyacetone phosphate to form the seven-carbon **sedoheptulose 1,7-bisphosphate**. An enzyme unique to plastids, sedoheptulose 1,7-bisphosphatase, converts the bisphosphate to sedoheptulose 7-phosphate (step ❺); this is the second irreversible reaction in the pathway. Transketolase now acts again, converting sedoheptulose 7-phosphate and glyceraldehyde 3-phosphate to two pentose phosphates in step ❻ (Fig. 20–11c). **Figure 20–12** shows how a two-carbon fragment is temporarily carried on the transketolase cofactor TPP and condensed with the three carbons of glyceraldehyde 3-phosphate in step ❻.

The pentose phosphates formed in the transketolase reactions—ribose 5-phosphate and xylulose 5-phosphate—are converted to **ribulose 5-phosphate** (steps ❼ and ❽), which in the final step ❾ of the cycle is phosphorylated to ribulose 1,5-bisphosphate by ribulose 5-phosphate kinase **(Fig. 20–13)**. This is the third very exergonic reaction of the pathway, as the phosphate anhydride bond in ATP is swapped for a phosphate ester in ribulose 1,5-bisphosphate.

FIGURE 20–11 Transketolase-catalyzed reactions of the Calvin cycle. (a) General reaction catalyzed by transketolase: the transfer of a two-carbon group, carried temporarily on enzyme-bound TPP, from a ketose donor to an aldose acceptor. **(b)** Conversion of a hexose and a triose to a four-carbon and a five-carbon sugar (step ❸ of Fig. 20-10). **(c)** Conversion of seven-carbon and three-carbon sugars to two pentoses (step ❻ of Fig. 20-10).

FIGURE 20–12 TPP as a cofactor for transketolase. Transketolase transfers a two-carbon group from sedoheptulose 7-phosphate to glyceraldehyde 3-phosphate, producing two pentose phosphates (step ❻ in Fig. 20-10). Thiamine pyrophosphate serves as a temporary carrier of the two-carbon unit and as an electron sink (see Fig. 14-15) to facilitate the reactions.

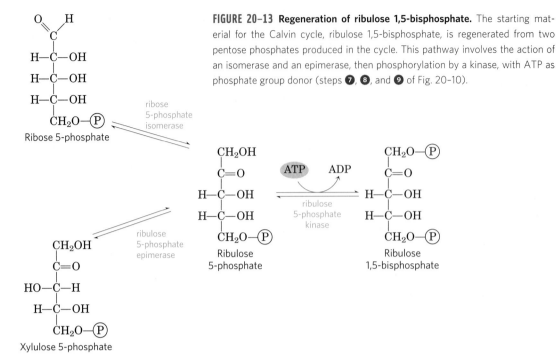

FIGURE 20–13 Regeneration of ribulose 1,5-bisphosphate. The starting material for the Calvin cycle, ribulose 1,5-bisphosphate, is regenerated from two pentose phosphates produced in the cycle. This pathway involves the action of an isomerase and an epimerase, then phosphorylation by a kinase, with ATP as phosphate group donor (steps **7**, **8**, and **9** of Fig. 20–10).

Synthesis of Each Triose Phosphate from CO_2 Requires Six NADPH and Nine ATP

The net result of three turns of the Calvin cycle is the conversion of three molecules of CO_2 and one molecule of phosphate to a molecule of triose phosphate. The stoichiometry of the overall path from CO_2 to triose phosphate, with regeneration of ribulose 1,5-bisphosphate, is shown in **Figure 20–14**. Three molecules of ribulose 1,5-bisphosphate (a total of 15 carbons) condense with three CO_2 (3 carbons) to form six molecules of 3-phosphoglycerate (18 carbons). These six molecules of 3-phosphoglycerate are reduced to six molecules of glyceraldehyde 3-phosphate (which is in equilibrium with dihydroxyacetone phosphate), with the expenditure of six ATP (in the synthesis of 1,3-bisphosphoglycerate) and six NADPH (in the reduction of 1,3-bisphosphoglycerate to glyceraldehyde 3-phosphate). The isozyme of glyceraldehyde 3-phosphate dehydrogenase present in chloroplasts can use NADPH

FIGURE 20–14 Stoichiometry of CO_2 assimilation in the Calvin cycle. For every three CO_2 molecules fixed, one molecule of triose phosphate (glyceraldehyde 3-phosphate) is produced and nine ATP and six NADPH are consumed.

as its electron carrier and normally functions in the direction of 1,3-bisphosphoglycerate reduction. The cytosolic isozyme uses NAD, as does the glycolytic enzyme of animals and other eukaryotes, and in the dark this isozyme acts in glycolysis to oxidize glyceraldehyde 3-phosphate. Both glyceraldehyde 3-phosphate dehydrogenase isozymes, like all enzymes, catalyze the reaction in both directions.

One molecule of glyceraldehyde 3-phosphate is the net product of the carbon-assimilation pathway. The other five triose phosphate molecules (15 carbons) are rearranged in steps ❶ to ❾ of Figure 20–10 to form three molecules of ribulose 1,5-bisphosphate (15 carbons). The last step in this conversion requires one ATP per ribulose 1,5-bisphosphate, or a total of three ATP. Thus, in summary, for every molecule of triose phosphate produced by photosynthetic CO_2 assimilation, six NADPH and nine ATP are required.

NADPH and ATP are produced in the light-dependent reactions of photosynthesis in about the same ratio (2:3) as they are consumed in the Calvin cycle. Nine ATP molecules are converted to ADP and phosphate in the generation of a molecule of triose phosphate; eight of the phosphates are released as P_i and combined with eight ADP to regenerate ATP. The ninth phosphate is incorporated into the triose phosphate itself. To convert the ninth ADP to ATP, a molecule of P_i must be imported from the cytosol, as we shall see.

In the dark, the production of ATP and NADPH by photophosphorylation, and the incorporation of CO_2 into triose phosphate (by the so-called dark reactions), cease. The "dark reactions" of photosynthesis were so named to distinguish them from the *primary* light-driven reactions of electron transfer to $NADP^+$ and synthesis of ATP, described in Chapter 19. They do not, in fact, occur at significant rates in the dark and are thus more appropriately called the **carbon-assimilation reactions**. Later in this section we describe the regulatory mechanisms that turn on carbon assimilation in the light and turn it off in the dark.

The chloroplast stroma contains all the enzymes necessary to convert the triose phosphates produced by CO_2 assimilation (glyceraldehyde 3-phosphate and dihydroxyacetone phosphate) to starch, which is temporarily stored in the chloroplast as insoluble granules. Aldolase condenses the trioses to fructose 1,6-bisphosphate; fructose 1,6-bisphosphatase produces fructose 6-phosphate; phosphohexose isomerase yields glucose 6-phosphate; and phosphoglucomutase produces glucose 1-phosphate, the starting material for starch synthesis (see Section 20.3).

All the reactions of the Calvin cycle except those catalyzed by rubisco, sedoheptulose 1,7-bisphosphatase, and ribulose 5-phosphate kinase also take place in animal tissues. Lacking these three enzymes, animals cannot carry out net conversion of CO_2 to glucose.

A Transport System Exports Triose Phosphates from the Chloroplast and Imports Phosphate

The inner chloroplast membrane is impermeable to most phosphorylated compounds, including fructose 6-phosphate, glucose 6-phosphate, and fructose 1,6-bisphosphate. It does, however, have a specific antiporter that catalyzes the one-for-one exchange of P_i with a triose phosphate, either dihydroxyacetone phosphate or 3-phosphoglycerate (**Fig. 20–15**; see also Fig. 20–9). This antiporter simultaneously moves P_i into the chloroplast, where it is used in photophosphorylation, and moves triose phosphate into the cytosol, where it can be used to synthesize sucrose, the form in which the fixed carbon is transported to distant plant tissues.

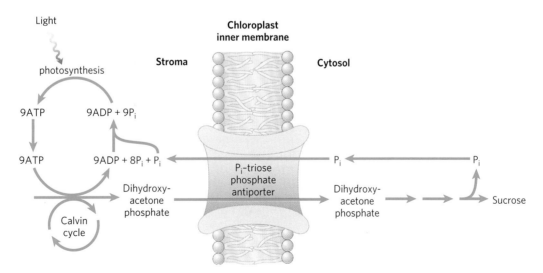

FIGURE 20–15 The P_i-triose phosphate antiport system of the inner chloroplast membrane. This transporter facilitates the exchange of cytosolic P_i for stromal dihydroxyacetone phosphate. The products of photosynthetic carbon assimilation are thus moved into the cytosol, where they serve as a starting point for sucrose biosynthesis, and P_i required for photophosphorylation is moved into the stroma. This same antiporter can transport 3-phosphoglycerate and acts in the shuttle for exporting ATP and reducing equivalents (see Fig. 20-16).

Sucrose synthesis in the cytosol and starch synthesis in the chloroplast are the major pathways by which the excess triose phosphate from photosynthesis is "harvested." Sucrose synthesis (described below) releases four P_i molecules from the four triose phosphates required to make sucrose. For every molecule of triose phosphate removed from the chloroplast, one P_i is transported into the chloroplast, providing the ninth P_i mentioned above, to be used in regenerating ATP. If this exchange were blocked, triose phosphate synthesis would quickly deplete the available P_i in the chloroplast, slowing ATP synthesis and suppressing assimilation of CO_2 into starch.

The P_i–triose phosphate antiport system serves one additional function. ATP and reducing power are needed in the cytosol for a variety of synthetic and energy-requiring reactions. These requirements are met to an as yet undetermined degree by mitochondria, but a second potential source of energy is the ATP and NADPH generated in the chloroplast stroma during the light reactions. However, neither ATP nor NADPH can cross the chloroplast membrane. The P_i–triose phosphate antiport system has the indirect effect of moving ATP equivalents and reducing equivalents from the chloroplast to the cytosol (**Fig. 20–16**). Dihydroxyacetone phosphate formed in the stroma is transported to the cytosol, where it is converted by glycolytic enzymes to 3-phosphoglycerate, generating ATP and NADH. 3-Phosphoglycerate reenters the chloroplast, completing the cycle.

Four Enzymes of the Calvin Cycle Are Indirectly Activated by Light

The reductive assimilation of CO_2 requires a lot of ATP and NADPH, and their stromal concentrations increase when chloroplasts are illuminated (**Fig. 20–17**). The light-induced transport of protons across the thylakoid membrane (Chapter 19) also increases the stromal pH from about 7 to about 8, and it is accompanied by a flow of Mg^{2+} from the thylakoid compartment into the stroma, raising the $[Mg^{2+}]$ from 1 to 3 mM to 3 to 6 mM. Several

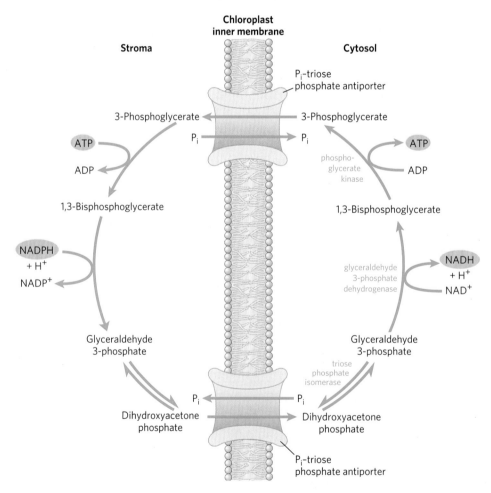

FIGURE 20–16 Role of the P_i–triose phosphate antiporter in the transport of ATP and reducing equivalents. Dihydroxyacetone phosphate leaves the chloroplast and is converted to glyceraldehyde 3-phosphate in the cytosol. The cytosolic glyceraldehyde 3-phosphate dehydrogenase and phosphoglycerate kinase reactions then produce NADH, ATP, and 3-phosphoglycerate. The latter reenters the chloroplast and is reduced to dihydroxyacetone phosphate, completing a cycle that effectively moves ATP and reducing equivalents (NAD(P)H) from chloroplast to cytosol.

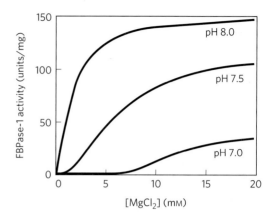

FIGURE 20–18 Activation of chloroplast fructose 1,6-bisphosphatase. Reduced fructose 1,6-bisphosphatase (FBPase-1) is activated by light and by the combination of high pH and high $[Mg^{2+}]$ in the stroma, both of which are results of illumination.

FIGURE 20–17 Source of ATP and NADPH. ATP and NADPH produced by the light reactions are essential substrates for the reduction of CO_2. The photosynthetic reactions that produce ATP and NADPH are accompanied by movement of protons (red) from the stroma into the thylakoid, creating alkaline conditions in the stroma. Magnesium ions pass from the thylakoid into the stroma, increasing the stromal $[Mg^{2+}]$.

stromal enzymes have evolved to take advantage of these light-induced conditions, which signal the availability of ATP and NADPH: the enzymes are more active in an alkaline environment and at high $[Mg^{2+}]$. For example, activation of rubisco by formation of the carbamoyllysine is faster at alkaline pH, and high stromal

$[Mg^{2+}]$ favors formation of the enzyme's active Mg^{2+} complex. Fructose 1,6-bisphosphatase requires Mg^{2+} and is very dependent on pH **(Fig. 20–18)**; its activity increases more than 100-fold when pH and $[Mg^{2+}]$ rise during chloroplast illumination.

Four Calvin cycle enzymes are subject to a special type of regulation by light. Ribulose 5-phosphate kinase, fructose 1,6-bisphosphatase, sedoheptulose 1,7-bisphosphatase, and glyceraldehyde 3-phosphate dehydrogenase are activated by light-driven reduction of disulfide bonds between two Cys residues critical to their catalytic activities. When these Cys residues are disulfide-bonded (oxidized), the enzymes are inactive; this is the normal situation in the dark. With illumination, electrons flow from photosystem I to ferredoxin (see Fig. 19–58), which passes electrons to a small, soluble, disulfide-containing protein called **thioredoxin (Fig. 20–19)**, in a reaction catalyzed by **ferredoxin-thioredoxin reductase**. Reduced thioredoxin donates electrons for the reduction

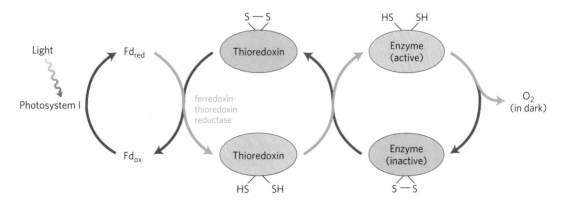

FIGURE 20–19 Light activation of several enzymes of the Calvin cycle. The light activation is mediated by thioredoxin, a small, disulfide-containing protein. In the light, thioredoxin is reduced by electrons moving from photosystem I through ferredoxin (Fd) (blue arrows), then thioredoxin reduces critical disulfide bonds in each of the enzymes sedoheptulose

1,7-bisphosphatase, fructose 1,6-bisphosphatase, ribulose 5-phosphate kinase, and glyceraldehyde 3-phosphate dehydrogenase, activating these enzymes. In the dark, the —SH groups undergo reoxidation to disulfides, inactivating the enzymes.

of the disulfide bonds of the light-activated enzymes, and these reductive cleavage reactions are accompanied by conformational changes that increase enzyme activities. At nightfall, the Cys residues in the four enzymes are reoxidized to their disulfide forms, the enzymes are inactivated, and ATP is not expended in CO_2 assimilation. Instead, starch synthesized and stored during the daytime is degraded to fuel glycolysis at night.

Glucose 6-phosphate dehydrogenase, the first enzyme in the *oxidative* pentose phosphate pathway, is also regulated by this light-driven reduction mechanism, but in the opposite sense. During the day, when photosynthesis produces plenty of NADPH, this enzyme is not needed for NADPH production. Reduction of a critical disulfide bond by electrons from ferredoxin *inactivates* the enzyme.

SUMMARY 20.1 Photosynthetic Carbohydrate Synthesis

▶ Photosynthesis in vascular plants takes place in chloroplasts. In the CO_2-assimilating reactions (the Calvin cycle), ATP and NADPH are used to reduce CO_2 to triose phosphates. These reactions occur in three stages: the fixation reaction itself, catalyzed by rubisco; reduction of the resulting 3-phosphoglycerate to glyceraldehyde 3-phosphate; and regeneration of ribulose 1,5-bisphosphate from triose phosphates.

▶ Rubisco condenses CO_2 with ribulose 1,5-bisphosphate, forming an unstable hexose bisphosphate that splits into two molecules of 3-phosphoglycerate. Rubisco is activated by covalent modification (carbamoylation of Lys[201]) catalyzed by rubisco activase and is inhibited by a natural transition-state analog, the concentration of which rises in the dark and falls during daylight.

▶ Stromal isozymes of the glycolytic enzymes catalyze reduction of 3-phosphoglycerate to glyceraldehyde 3-phosphate; each molecule reduced requires one ATP and one NADPH.

▶ Stromal enzymes, including transketolase and aldolase, rearrange the carbon skeletons of triose phosphates, generating intermediates of three, four, five, six, and seven carbons and eventually yielding pentose phosphates. The pentose phosphates are converted to ribulose 5-phosphate, then phosphorylated to ribulose 1,5-bisphosphate to complete the Calvin cycle.

▶ The cost of fixing three CO_2 into one triose phosphate is nine ATP and six NADPH, which are provided by the light-dependent reactions of photosynthesis.

▶ An antiporter in the inner chloroplast membrane exchanges P_i in the cytosol for 3-phosphoglycerate or dihydroxyacetone phosphate produced by CO_2

assimilation in the stroma. Oxidation of dihydroxyacetone phosphate in the cytosol generates ATP and NADH, thus moving ATP and reducing equivalents from the chloroplast to the cytosol.

▶ Four enzymes of the Calvin cycle are activated indirectly by light and are inactive in the dark, so that hexose synthesis does not compete with glycolysis—which is required to provide energy in the dark.

20.2 Photorespiration and the C₄ and CAM Pathways

As we have seen, photosynthetic cells produce O_2 (by the splitting of H_2O) during the light-driven reactions (Chapter 19) and use CO_2 during the light-independent processes (described above), so the net gaseous change during photosynthesis is the uptake of CO_2 and release of O_2:

$$CO_2 + H_2O \longrightarrow O_2 + (CH_2O)$$

In the dark, plants also carry out **mitochondrial respiration**, the oxidation of substrates to CO_2 and the conversion of O_2 to H_2O. And there is another process in plants that, like mitochondrial respiration, consumes O_2 and produces CO_2 and, like photosynthesis, is driven by light. This process, **photorespiration**, is a costly side reaction of photosynthesis, a result of the lack of specificity of the enzyme rubisco. In this section we describe this side reaction and the strategies plants use to minimize its metabolic consequences.

Photorespiration Results from Rubisco's Oxygenase Activity

Rubisco is not absolutely specific for CO_2 as a substrate. Molecular oxygen (O_2) competes with CO_2 at the active site, and about once in every three or four turnovers, rubisco catalyzes the condensation of O_2 with ribulose 1,5-bisphosphate to form 3-phosphoglycerate and **2-phosphoglycolate** (Fig. 20–20), a metabolically useless product. This is the oxygenase activity referred to in the full name of the enzyme: ribulose 1,5-bisphosphate carboxylase/oxygenase. The reaction with O_2 results in no fixation of carbon and seems to be a net liability to the cell; salvaging the carbons from 2-phosphoglycolate (by the pathway outlined below) consumes significant amounts of cellular energy and releases some previously fixed CO_2.

Given that the reaction with oxygen is deleterious to the organism, why did the evolution of rubisco produce an active site unable to discriminate well between CO_2 and O_2? Perhaps much of this evolution occurred before the time, about 2.5 billion years ago, when production of O_2 by photosynthetic organisms started to raise the oxygen content of the atmosphere. Before

CH$_2$O—Ⓟ
|
C=O
|
H—C—OH   Ribulose
| 1,5-bisphosphate
H—C—OH
|
CH$_2$O—Ⓟ

⇕

CH$_2$O—Ⓟ
|
C—OH
‖
C—OH   Enediol form
|
H—C—OH
|
CH$_2$O—Ⓟ

↓ O$_2$

CH$_2$O—Ⓟ
|
H—O—O—C—OH
|
C=O   Enzyme-bound
| intermediate
H—C—OH
|
CH$_2$O—Ⓟ

↓ OH$^-$

↘ H$_2$O

CH$_2$O—Ⓟ   ⁻O   O
| \ //
C   +   C
/ \\ |
⁻O   O   H—C—OH
|
CH$_2$O—Ⓟ

2-Phosphoglycolate   3-Phosphoglycerate

FIGURE 20–20 Oxygenase activity of rubisco. Rubisco can incorporate O$_2$ rather than CO$_2$ into ribulose 1,5-bisphosphate. The unstable intermediate thus formed splits into 2-phosphoglycolate (recycled as described in Fig. 20-21) and 3-phosphoglycerate, which can reenter the Calvin cycle.

that time, there was no selective pressure for rubisco to discriminate between CO$_2$ and O$_2$. The K_m for CO$_2$ is about 9 μM, and that for O$_2$ is about 350 μM. The modern atmosphere contains about 20% O$_2$ and only 0.04% CO$_2$, so an aqueous solution in equilibrium with air at room temperature contains about 250 μM O$_2$ and 11 μM CO$_2$—concentrations that allow significant O$_2$ "fixation" by rubisco and thus a significant waste of energy. The temperature dependence of the solubilities of O$_2$ and CO$_2$ is such that at higher temperatures, the ratio of O$_2$ to CO$_2$ in solution increases. In addition, the affinity of rubisco for CO$_2$ decreases with increasing temperature, exacerbating its tendency to catalyze the wasteful oxygenase reaction. And as CO$_2$ is consumed in the assimilation reactions, the ratio of O$_2$ to CO$_2$ in the air spaces of a leaf increases, further favoring the oxygenase reaction.

The Salvage of Phosphoglycolate Is Costly

The **glycolate pathway** converts two molecules of 2-phosphoglycolate to a molecule of serine (three carbons) and a molecule of CO$_2$ (**Fig. 20–21**). In the

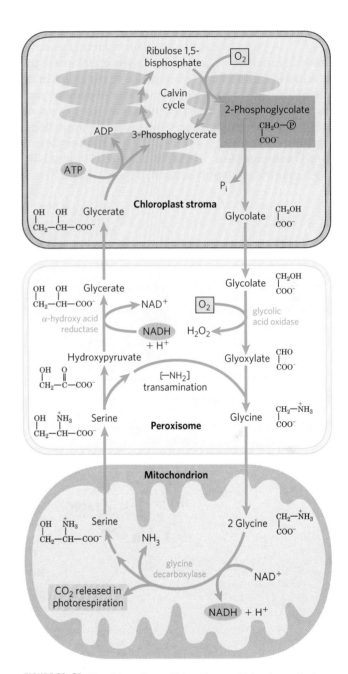

FIGURE 20–21 Glycolate pathway. This pathway, which salvages 2-phosphoglycolate (shaded light red) by its conversion to serine and eventually 3-phosphoglycerate, involves three cellular compartments. Glycolate formed by dephosphorylation of 2-phosphoglycolate in chloroplasts is oxidized to glyoxylate in peroxisomes and then transaminated to glycine. In mitochondria, two glycine molecules condense to form serine and the CO$_2$ released during photorespiration (shaded green). This reaction is catalyzed by glycine decarboxylase, an enzyme present at very high levels in the mitochondria of C$_3$ plants (see text). The serine is converted to hydroxypyruvate and then to glycerate in peroxisomes; glycerate reenters the chloroplasts to be phosphorylated, rejoining the Calvin cycle. Oxygen (shaded blue) is consumed at two steps during photorespiration.

chloroplast, a phosphatase converts 2-phosphoglycolate to glycolate, which is exported to the peroxisome. There, glycolate is oxidized by molecular oxygen, and the resulting aldehyde (glyoxylate) undergoes transamination to glycine. The hydrogen peroxide formed as a side product of glycolate oxidation is rendered harmless by peroxidases in the peroxisome. Glycine passes from the peroxisome to the mitochondrial matrix, where it undergoes oxidative decarboxylation by the glycine decarboxylase complex, an enzyme similar in structure and mechanism to two mitochondrial complexes we have already encountered: the pyruvate dehydrogenase complex and the α-ketoglutarate dehydrogenase complex (Chapter 16). The **glycine decarboxylase complex** oxidizes glycine to CO_2 and NH_3, with the concomitant reduction of NAD^+ to NADH and

transfer of the remaining carbon from glycine to the cofactor tetrahydrofolate **(Fig. 20–22)**. The one-carbon unit carried on tetrahydrofolate is then transferred to a second glycine by serine hydroxymethyltransferase, producing serine. The net reaction catalyzed by the glycine decarboxylase complex and serine hydroxymethyltransferase is

$$2 \text{ Glycine} + NAD^+ + H_2O \longrightarrow$$
$$\text{serine} + CO_2 + NH_3 + NADH + H^+$$

The serine is converted to hydroxypyruvate, to glycerate, and finally to 3-phosphoglycerate, which is used to regenerate ribulose 1,5-bisphosphate, completing the long, expensive cycle (Fig. 20–21).

In bright sunlight, the flux through the glycolate salvage pathway can be very high, producing about

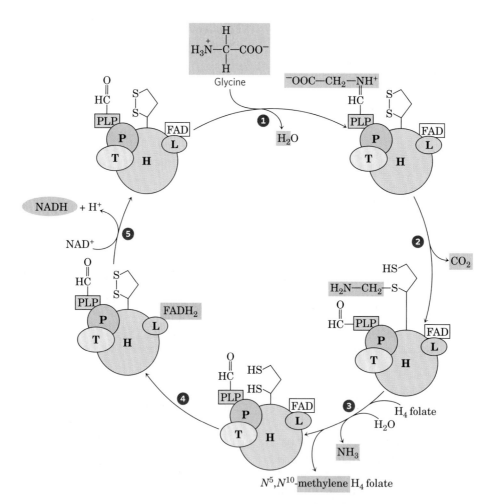

N^5,N^{10}-methylene H_4 folate

FIGURE 20–22 The glycine decarboxylase system. Glycine decarboxylase in plant mitochondria is a complex of four types of subunits, with the stoichiometry $P_4H_{27}T_9L_2$. Protein H has a covalently attached lipoic acid residue that can undergo reversible oxidation. Step ❶ is formation of a Schiff base between pyridoxal phosphate (PLP) and glycine, catalyzed by protein P (named for its bound PLP). In step ❷, protein P catalyzes oxidative decarboxylation of glycine, releasing CO_2; the remaining methylamine group is attached to one of the —SH groups of reduced lipoic acid. ❸ Protein T (which uses tetrahydrofolate (H_4 folate) as cofactor) now releases NH_3 from the methylamine moiety and transfers

the remaining one-carbon fragment to tetrahydrofolate, producing N^5,N^{10}-methylenetetrahydrofolate. ❹ Protein L oxidizes the two —SH groups of lipoic acid to a disulfide, passing electrons through FAD to NAD^+ ❺, thus completing the cycle. The N^5,N^{10}-methylenetetrahydrofolate formed in this process is used by serine hydroxymethyltransferase to convert a molecule of glycine to serine, regenerating the tetrahydrofolate that is essential for the reaction catalyzed by protein T. The L subunit of glycine decarboxylase is identical to the dihydrolipoyl dehydrogenase (E_3) of pyruvate dehydrogenase and α-ketoglutarate dehydrogenase (see Fig. 16–6).

five times more CO_2 than is typically produced by all the oxidations of the citric acid cycle. To generate this large flux, mitochondria contain prodigious amounts of the glycine decarboxylase complex: the four proteins of the complex make up *half* of all the protein in the mitochondrial matrix in the leaves of pea and spinach plants! In nonphotosynthetic parts of a plant, such as potato tubers, mitochondria have very low concentrations of the glycine decarboxylase complex.

The combined activity of the rubisco oxygenase and the glycolate salvage pathway consumes O_2 and produces CO_2—hence the name **photorespiration**. This pathway is perhaps better called the **oxidative photosynthetic carbon cycle** or **C₂ cycle**, names that do not invite comparison with respiration in mitochondria. Unlike mitochondrial respiration, "photorespiration" does not conserve energy and may actually inhibit net biomass formation as much as 50%. This inefficiency has led to evolutionary adaptations in the carbon-assimilation processes, particularly in plants that have evolved in warm climates. The apparent inefficiency of rubisco, and its effect in limiting biomass production, has inspired efforts to genetically engineer a "better" rubisco, but this goal is not, as yet, within reach (Box 20–1).

In C₄ Plants, CO₂ Fixation and Rubisco Activity Are Spatially Separated

In many plants that grow in the tropics (and in temperate-zone crop plants native to the tropics, such as maize, sugarcane, and sorghum) a mechanism has evolved to circumvent the problem of wasteful photorespiration. The step in which CO_2 is fixed into a three-carbon product, 3-phosphoglycerate, is preceded by several steps, one of which is temporary fixation of CO_2 into a four-carbon compound. Plants that use this process are referred to as **C₄ plants**, and the assimilation process as **C₄ metabolism** or the **C₄ pathway**. Plants that use the carbon-assimilation method we have described thus far, in which the *first step* is reaction of CO_2 with ribulose 1,5-bisphosphate to form 3-phosphoglycerate, are called **C₃ plants**.

The C₄ plants, which typically grow at high light intensity and high temperatures, have several important characteristics: high photosynthetic rates, high growth rates, low photorespiration rates, low rates of water loss, and a specialized leaf structure. Photosynthesis in the leaves of C₄ plants involves two cell types: mesophyll and bundle-sheath cells **(Fig. 20–23a)**. There are three

(a)

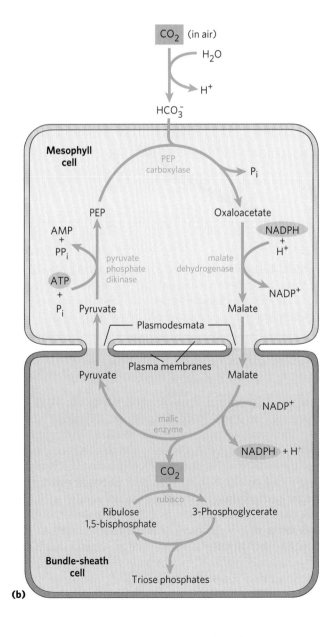

(b)

FIGURE 20–23 Carbon assimilation in C₄ plants. The C₄ pathway, involving mesophyll cells and bundle-sheath cells, predominates in plants of tropical origin. **(a)** Electron micrograph showing chloroplasts of adjacent mesophyll and bundle-sheath cells. The bundle-sheath cell contains starch granules. Plasmodesmata connecting the two cells are visible. **(b)** The C₄ pathway of CO_2 assimilation, which occurs through a four-carbon intermediate.

BOX 20–1   Will Genetic Engineering of Photosynthetic Organisms Increase Their Efficiency?

Three pressing world problems have prompted serious attention to the possibility of engineering plants to be more efficient in converting sunlight into biomass: the "greenhouse" effect of increasing atmospheric CO_2 on global climate change, the dwindling supply of oil for the generation of energy, and the need for more and better food for the world's growing population.

The concentration of CO_2 in the earth's atmosphere has risen steadily over the past 50 years (Fig. 1), the combined effect of the use of fossil fuels for energy and the clearing and burning of tropical forests to allow use of the land for agriculture. As atmospheric CO_2 increases, the atmosphere absorbs more heat radiated from the earth's surface and reradiates more toward the surface of the planet (and in all other directions). Retention of heat raises the temperature

at the surface of the earth; this is the greenhouse effect. One way to limit the increase in atmospheric CO_2 would be to engineer plants or microorganisms with a greater capacity for sequestering CO_2.

The estimated amount of total carbon in all terrestrial systems (atmosphere, soil, biomass) is about 3,200 gigatons (GT)—3,200 billion metric tons. The atmosphere contains another 760 GT of CO_2.

The flux of carbon through these terrestrial reservoirs (Fig. 2) is largely due to the photosynthetic activities of plants and the degradative activities of microorganisms. Plants fix some 123 GT of carbon annually, then immediately release about half of that to the atmosphere as they respire. Much of the remainder is gradually released to the atmosphere by microbial action on dead plant materials, but biomass is sequestered in woody plants and trees for decades or centuries. Anthropogenic carbon flux, the amount of CO_2 released into the atmosphere by human activities, is 9 GT per year—small compared with total biomass, but enough to tip the balance of nature toward increased CO_2 in the atmosphere. It is estimated that the forests of North America sequester 0.7 GT of carbon annually, which represents about a tenth of the annual *global* production of CO_2 from fossil fuels. Clearly, preservation of forests and reforestation are effective ways to limit the flow of CO_2 back into the atmosphere.

A second approach to limiting the increase of atmospheric CO_2, while also addressing the need to replace dwindling fossil fuels, is to use renewable biomass as a source of ethanol to replace fossil fuels in internal combustion engines. This reduces the *unidirectional* movement of carbon from fossil fuels into the atmospheric pool of CO_2, replacing it with the *cyclic* flow of CO_2 from ethanol to CO_2 and back to biomass. When maize, wheat, or switchgrass is fermented to ethanol for fuel, every increase in biomass

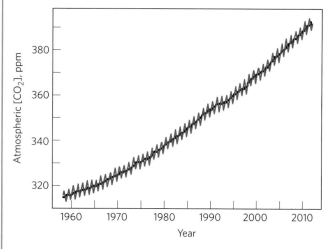

FIGURE 1 The concentration of CO_2 in the atmosphere measured at the Mauna Loa Observatory in Hawaii. Data from the National Oceanic and Atmospheric Administration and the Scripps Institution of Oceanography CO_2 Program.

variants of C_4 metabolism, worked out in the 1960s by Marshall Hatch and Rodger Slack (Fig. 20–23b).

In plants of tropical origin, the first intermediate into which $^{14}CO_2$ is fixed is oxaloacetate, a four-carbon compound. This reaction, which occurs in the cytosol of leaf mesophyll cells, is catalyzed by **phosphoenolpyruvate carboxylase**, for which the substrate is HCO_3^-, not CO_2. The oxaloacetate thus formed is either reduced to malate at the expense of NADPH (as shown in Fig. 20–23b) or converted to aspartate by transamination:

Oxaloacetate + α-amino acid $\longrightarrow$

L-aspartate + α-keto acid

The malate or aspartate formed in the mesophyll cells then passes into neighboring bundle-sheath cells through plasmodesmata, protein-lined channels that connect two plant cells and provide a path for movement of metabolites and even small proteins between cells. In the bundle-sheath cells, malate is oxidized and decarboxylated to yield pyruvate and CO_2 by the action of **malic enzyme**, reducing $NADP^+$. In plants that use aspartate as the CO_2 carrier, aspartate arriving in bundle-sheath cells is transaminated to form oxaloacetate and reduced to malate, then the CO_2 is released by malic enzyme or PEP carboxykinase. As labeling experiments show, the free CO_2 released in the bundle-sheath cells is

FIGURE 2 The terrestrial carbon cycle. Carbon stocks (boxes) are shown as gigatons (GT), and fluxes (arrows) are shown in GT per year. Animal biomass is negligible here—less than 0.5 GT.

production brought about by more efficient photosynthesis should result in a corresponding decrease in the use of fossil fuels.

Finally, engineering of food crops to yield more food per acre of land, or per hour of work, could improve human nutrition worldwide.

In principle, these goals might be accomplished by developing a rubisco that didn't also catalyze the wasteful reaction with O_2, or by increasing the turnover number for rubisco, or by increasing the level of rubisco or other enzymes in the pathway for carbon fixation. Rubisco, we have said, is an unusually inefficient enzyme, with a turnover number of $3\ \text{s}^{-1}$ at 25°C; most enzymes have turnover numbers orders of magnitude larger. It also catalyzes the wasteful reaction with oxygen, which further reduces its efficiency in fixing CO_2 and producing biomass. If

rubisco could be genetically engineered to turn over faster or to be more selective for CO_2 relative to O_2, would the effect be greater photosynthetic production of biomass and thus greater sequestration of CO_2, production of more nonfossil fuel, and improved nutrition?

We noted in Chapter 15 that the traditional view of metabolic pathways held that one step in any pathway was the slowest and therefore the limiting factor in material flow through the pathway. However, heroic efforts to engineer cells or organisms have often given discouraging results; organisms engineered to produce more of the "limiting" enzyme in a pathway often show little or no change in the flux through that pathway. The Calvin cycle is an instructive case in point. Increasing the amount of rubisco in plant cells through genetic engineering has little or no effect on the rate of CO_2 conversion into carbohydrate. Similarly, changes in the levels of enzymes known to be regulated by light and therefore suspected of playing key roles in the regulation of the pathway (fructose 1,6-bisphosphatase, 3-phosphoglycerate kinase, and glyceraldehyde 3-phosphate dehydrogenase) also lead to little or no significant improvement in photosynthetic rate. However, changed levels of sedoheptulose 1,7-bisphosphatase, not thought to be a regulatory enzyme, have a significant impact on photosynthesis. Metabolic control analysis (Section 15.2) suggests that this result is not unexpected; in the living organism, pathways can be limited by more than one step, since every change in one step results in compensating changes in other steps. Careful determination of the flux control coefficient (see Box 15–1) helps to pinpoint which enzymes in a pathway to target for genetic engineering. Clearly, genetic engineers and metabolic control analysts will need to work together on problems like this one!

the same CO_2 molecule originally fixed into oxaloacetate in the mesophyll cells. This CO_2 is now fixed again, this time by rubisco, in exactly the same reaction that occurs in C_3 plants: incorporation of CO_2 into C-1 of 3-phosphoglycerate.

The pyruvate formed by decarboxylation of malate in bundle-sheath cells is transferred back to the mesophyll cells, where it is converted to PEP by an unusual enzymatic reaction catalyzed by **pyruvate phosphate dikinase** (Fig. 20–23b). This enzyme is called a dikinase because two different molecules are simultaneously phosphorylated by one molecule of ATP: pyruvate to PEP, and phosphate to pyrophosphate. The pyrophosphate is

subsequently hydrolyzed to phosphate, so two high-energy phosphate groups of ATP are used in regenerating PEP. The PEP is now ready to receive another molecule of CO_2 in the mesophyll cell.

The PEP carboxylase of mesophyll cells has a high affinity for HCO_3^- (which is favored relative to CO_2 in aqueous solution) and can fix CO_2 more efficiently than can rubisco. Unlike rubisco, it does not use O_2 as an alternative substrate, so there is no competition between CO_2 and O_2. The PEP carboxylase reaction, then, serves to fix and concentrate CO_2 in the form of malate. Release of CO_2 from malate in the bundle-sheath cells yields a sufficiently high local concentration of CO_2 for rubisco to function near its maximal rate, and for suppression of the enzyme's oxygenase activity.

Once CO_2 is fixed into 3-phosphoglycerate in the bundle-sheath cells, the other reactions of the Calvin cycle take place exactly as described earlier. Thus in C_4 plants, mesophyll cells carry out CO_2 assimilation by the C_4 pathway and bundle-sheath cells synthesize starch and sucrose by the C_3 pathway.

Three enzymes of the C_4 pathway are regulated by light, becoming more active in daylight. Malate dehydrogenase is activated by the thioredoxin-dependent reduction mechanism shown in Figure 20–19; PEP carboxylase is activated by phosphorylation of a Ser residue; and pyruvate phosphate dikinase is activated by dephosphorylation. In the latter two cases, the details of how light effects phosphorylation or dephosphorylation are not known.

The pathway of CO_2 assimilation has a greater energy cost in C_4 plants than in C_3 plants. For each molecule of CO_2 assimilated in the C_4 pathway, a molecule of PEP must be regenerated at the expense of two phosphoanhydride bonds in ATP. Thus C_4 plants need five ATP molecules to assimilate one molecule of CO_2, whereas C_3 plants need only three (nine per triose phosphate). As the temperature increases (and the affinity of rubisco for CO_2 decreases, as noted above), a point is reached (at about 28 to 30 °C) at which the gain in efficiency from the elimination of photorespiration more than compensates for this energetic cost. C_4 plants (crabgrass, for example) outgrow most C_3 plants during the summer, as any experienced gardener can attest.

In CAM Plants, CO_2 Capture and Rubisco Action Are Temporally Separated

Succulent plants such as cactus and pineapple, which are native to very hot, very dry environments, have another variation on photosynthetic CO_2 fixation, which reduces loss of water vapor through the pores (stomata) by which CO_2 and O_2 must enter leaf tissue. Instead of separating the initial trapping of CO_2 and its fixation by rubisco across space (as do the C_4 plants), they separate these two events over time. At night, when the air is cooler and moister, the stomata open to allow entry of CO_2, which is then fixed into oxaloacetate by PEP carboxylase. The oxaloacetate is reduced to malate and stored in the vacuoles, to protect cytosolic and plastid enzymes from the low pH produced by malic acid dissociation. During the day the stomata close, preventing the water loss that would result from high daytime temperatures, and the CO_2 trapped overnight in malate is released as CO_2 by the NADP-linked malic enzyme. This CO_2 is now assimilated by the action of rubisco and the Calvin cycle enzymes. Because this method of CO_2 fixation was first discovered in stonecrops, perennial flowering plants of the family *Crassulaceae*, it is called

crassulacean acid metabolism, and the plants are called **CAM plants**.

SUMMARY 20.2 Photorespiration and the C_4 and CAM Pathways

▶ When rubisco uses O_2 rather than CO_2 as substrate, the 2-phosphoglycolate so formed is disposed of in an oxygen-dependent pathway. The result is increased consumption of O_2—photorespiration or, more accurately, the oxidative photosynthetic carbon cycle or C_2 cycle. The 2-phosphoglycolate is converted to glyoxylate, to glycine, and then to serine in a pathway that involves enzymes in the chloroplast stroma, the peroxisome, and the mitochondrion.

▶ In C_4 plants, the carbon-assimilation pathway minimizes photorespiration: CO_2 is first fixed in mesophyll cells into a four-carbon compound, which passes into bundle-sheath cells and releases CO_2 in high concentrations. The released CO_2 is fixed by rubisco, and the remaining reactions of the Calvin cycle occur as in C_3 plants.

▶ In CAM plants, CO_2 is fixed into malate in the dark and stored in vacuoles until daylight, when the stomata are closed (minimizing water loss) and malate serves as a source of CO_2 for rubisco.

20.3 Biosynthesis of Starch and Sucrose

During active photosynthesis in bright light, a plant leaf produces more carbohydrate (as triose phosphates) than it needs for generating energy or synthesizing precursors. The excess is converted to sucrose and transported to other parts of the plant, to be used as fuel or stored. In most plants, starch is the main storage form, but in a few plants, such as sugar beet and sugarcane, sucrose is the primary storage form. The synthesis of sucrose and starch occurs in different cellular compartments (cytosol and plastids, respectively), and these processes are coordinated by a variety of regulatory mechanisms that respond to changes in light level and photosynthetic rate. The synthesis of sucrose and starch is important to the plant but also to humans: starch provides more than 80% of human dietary calories worldwide.

ADP-Glucose Is the Substrate for Starch Synthesis in Plant Plastids and for Glycogen Synthesis in Bacteria

Starch, like glycogen, is a high molecular weight polymer of D-glucose in $(\alpha 1 \rightarrow 4)$ linkage. It is synthesized in chloroplasts for temporary storage as one of the stable end products of photosynthesis, and for long-term storage it is synthesized in the amyloplasts of the nonphotosynthetic parts of plants—seeds, roots, and tubers (underground stems).

The mechanism of glucose activation in starch synthesis is similar to that in glycogen synthesis. An activated **nucleotide sugar**, in this case **ADP-glucose**, is formed by condensation of glucose 1-phosphate with ATP in a reaction made essentially irreversible by the presence in plastids of inorganic pyrophosphatase (Fig. 15–31). **Starch synthase** then transfers glucose residues from ADP-glucose to preexisting starch molecules. The monomeric units are almost certainly added to the nonreducing end of the growing polymer, as they are in glycogen synthesis (see Fig. 15–32).

The amylose of starch is unbranched, but amylopectin has numerous ($\alpha 1 \rightarrow 6$)-linked branches (see Fig. 7–13). Chloroplasts contain a branching enzyme, similar to glycogen-branching enzyme (see Fig. 15–33), that introduces the ($\alpha 1 \rightarrow 6$) branches of amylopectin. Taking into account the hydrolysis by inorganic pyrophosphatase of the PP_i produced during ADP-glucose synthesis, the overall reaction for starch formation from glucose 1-phosphate is

$$\text{Starch}_n + \text{glucose 1-phosphate} + \text{ATP} \longrightarrow$$
$$\text{starch}_{n+1} + \text{ADP} + 2P_i$$
$$\Delta G'^{\circ} = -50 \text{ kJ/mol}$$

Starch synthesis is regulated at the level of ADP-glucose formation, as discussed below.

Many types of bacteria store carbohydrate in the form of glycogen (essentially highly branched starch), which they synthesize in a reaction analogous to that catalyzed by glycogen synthase in animals. Bacteria, like plant plastids, use ADP-glucose as the activated form of glucose, whereas animal cells use UDP-glucose. Again, the similarity between plastid and bacterial metabolism is consistent with the endosymbiont hypothesis for the origin of organelles (p. 36).

UDP-Glucose Is the Substrate for Sucrose Synthesis in the Cytosol of Leaf Cells

Most of the triose phosphate generated by CO_2 fixation in plants is converted to sucrose **(Fig. 20–24)** or starch. In the course of evolution, sucrose may have been selected as the transport form of carbon because of its unusual linkage between the anomeric C-1 of glucose and the anomeric C-2 of fructose. This bond is not hydrolyzed by amylases or other common carbohydrate-cleaving enzymes, and the unavailability of the anomeric carbons prevents sucrose from reacting nonenzymatically (as does glucose) with amino acids and proteins.

Sucrose is synthesized in the cytosol, beginning with dihydroxyacetone phosphate and glyceraldehyde 3-phosphate exported from the chloroplast. After condensation of two triose phosphates to form fructose 1,6-bisphosphate (catalyzed by aldolase), hydrolysis by fructose 1,6-bisphosphatase yields fructose 6-phosphate. **Sucrose 6-phosphate synthase** then catalyzes

FIGURE 20–24 Sucrose synthesis. Sucrose is synthesized from UDP-glucose and fructose 6-phosphate, which are synthesized from triose phosphates in the plant cell cytosol by pathways shown in Figures 15–31 and 20-9. The sucrose 6-phosphate synthase of most plant species is allosterically regulated by glucose 6-phosphate and P_i.

the reaction of fructose 6-phosphate with **UDP-glucose** to form **sucrose 6-phosphate** (Fig. 20–24). Finally, **sucrose 6-phosphate phosphatase** removes the phosphate group, making sucrose available for export to other tissues. The reaction catalyzed by sucrose 6-phosphate synthase is a low-energy process ($\Delta G'^{\circ} = -5.7$ kJ/mol), but the hydrolysis of sucrose 6-phosphate to sucrose is sufficiently exergonic ($\Delta G'^{\circ} = -16.5$ kJ/mol) to make the overall synthesis of sucrose essentially irreversible. Sucrose synthesis is regulated and closely coordinated with starch synthesis, as we shall see.

One remarkable difference between the cells of plants and animals is the absence in the plant cell cytosol of the enzyme inorganic pyrophosphatase, which catalyzes the reaction

$$PP_i + H_2O \longrightarrow 2P_i \qquad \Delta G'^{\circ} = -19.2 \text{ kJ/mol}$$

For many biosynthetic reactions that liberate PP_i, pyrophosphatase activity makes the process more favorable energetically, tending to make these reactions irreversible.

In plants, this enzyme is present in plastids but absent from the cytosol. As a result, the cytosol of leaf cells contains a substantial concentration of PP$_i$—enough (~0.3 mM) to make reactions such as that catalyzed by UDP-glucose pyrophosphorylase (see Fig. 15–31) readily reversible. Recall from Chapter 14 (p. 550) that the cytosolic isozyme of phosphofructokinase in plants uses PP$_i$, not ATP, as the phosphoryl donor.

Conversion of Triose Phosphates to Sucrose and Starch Is Tightly Regulated

Triose phosphates produced by the Calvin cycle in bright sunlight, as we have noted, may be stored temporarily in the chloroplast as starch, or converted to sucrose and exported to nonphotosynthetic parts of the plant, or both. The balance between the two processes is tightly regulated, and both must be coordinated with the rate of carbon fixation. Five-sixths of the triose phosphate formed in the Calvin cycle must be recycled to ribulose 1,5-bisphosphate (Fig. 20–14); if more than one-sixth of the triose phosphate is drawn out of the cycle to make sucrose and starch, the cycle will slow or stop. However, *insufficient* conversion of triose phosphate to starch or sucrose would tie up phosphate, leaving a chloroplast deficient in P$_i$, which is also essential for operation of the Calvin cycle.

The flow of triose phosphates into sucrose is regulated by the activity of fructose 1,6-bisphosphatase (FBPase-1) and the enzyme that effectively reverses its action, PP$_i$-dependent phosphofructokinase (PP-PFK-1; p. 550). These enzymes are therefore critical points for determining the fate of triose phosphates produced by photosynthesis. Both enzymes are regulated by **fructose 2,6-bisphosphate (F26BP)**, which inhibits FBPase-1 and stimulates PP-PFK-1. In vascular plants, the concentration of F26BP varies inversely with the rate of photosynthesis **(Fig. 20–25)**. Phosphofructokinase-2, responsible for F26BP synthesis, is inhibited by dihydroxyacetone phosphate or 3-phosphoglycerate and stimulated by fructose 6-phosphate and P$_i$. During active photosynthesis, dihydroxyacetone phosphate is produced and P$_i$ is consumed, resulting in inhibition of PFK-2 and lowered concentrations of F26BP. This favors greater flux of triose phosphate into fructose 6-phosphate formation and sucrose synthesis. With this regulatory system, sucrose synthesis occurs when the level of triose phosphate produced by the Calvin cycle exceeds that needed to maintain the operation of the cycle.

Sucrose synthesis is also regulated at the level of sucrose 6-phosphate synthase, which is allosterically activated by glucose 6-phosphate and inhibited by P$_i$. This enzyme is further regulated by phosphorylation and dephosphorylation; a protein kinase phosphorylates the enzyme on a specific Ser residue, making it less active, and a phosphatase reverses this inactivation by

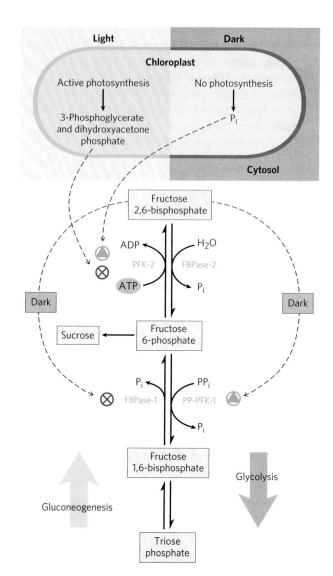

FIGURE 20–25 Fructose 2,6-bisphosphate as regulator of sucrose synthesis. The concentration of the allosteric regulator fructose 2,6-bisphosphate in plant cells is regulated by the products of photosynthetic carbon assimilation and by P$_i$. Dihydroxyacetone phosphate and 3-phosphoglycerate produced by CO_2 assimilation inhibit phosphofructokinase-2 (PFK-2), the enzyme that synthesizes the regulator; P$_i$ stimulates PFK-2. The concentration of the regulator is therefore inversely proportional to the rate of photosynthesis. In the dark, the concentration of fructose 2,6-bisphosphate increases and stimulates the glycolytic enzyme PP$_i$-dependent phosphofructokinase-1 (PP-PFK-1), while inhibiting the gluconeogenic enzyme fructose 1,6-bisphosphatase (FBPase-1). When photosynthesis is active (in the light), the concentration of the regulator drops and the synthesis of fructose 6-phosphate and sucrose is favored.

removing the phosphate **(Fig. 20–26)**. Inhibition of the kinase by glucose 6-phosphate, and of the phosphatase by P$_i$, enhances the effects of these two compounds on sucrose synthesis. When hexose phosphates are abundant, sucrose 6-phosphate synthase is activated by glucose 6-phosphate; when P$_i$ is elevated (as when photosynthesis is slow), sucrose synthesis is slowed.

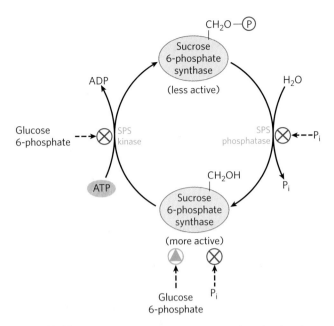

FIGURE 20–26 Regulation of sucrose phosphate synthase by phosphorylation. A protein kinase (SPS kinase) specific for sucrose phosphate synthase (SPS) phosphorylates a Ser residue in SPS, inactivating it; a specific phosphatase (SPS phosphatase) reverses this inhibition. The kinase is inhibited allosterically by glucose 6-phosphate, which also activates SPS allosterically. The phosphatase is inhibited by P_i, which also inhibits SPS directly. Thus when the concentration of glucose 6-phosphate is high as a result of active photosynthesis, SPS is activated and produces sucrose phosphate. A high P_i concentration, which occurs when photosynthetic conversion of ADP to ATP is slow, inhibits sucrose phosphate synthesis.

During active photosynthesis, triose phosphates are converted to fructose 6-phosphate, which is rapidly equilibrated with glucose 6-phosphate by phosphohexose isomerase. Because the equilibrium lies far toward glucose 6-phosphate, as soon as fructose 6-phosphate accumulates, the level of glucose 6-phosphate rises and sucrose synthesis is stimulated.

The key regulatory enzyme in starch synthesis is **ADP-glucose pyrophosphorylase (Fig. 20–27)**; it is activated by 3-phosphoglycerate (which accumulates during active photosynthesis) and inhibited by P_i (which accumulates when light-driven condensation of ADP and P_i slows). When sucrose synthesis slows, 3-phosphoglycerate formed by CO_2 fixation accumulates, activating this enzyme and stimulating the synthesis of starch.

SUMMARY 20.3 Biosynthesis of Starch and Sucrose

▶ Starch synthase in chloroplasts and amyloplasts catalyzes the addition of single glucose residues, donated by ADP-glucose, probably to the nonreducing end. Branches in amylopectin are introduced by a second enzyme.

▶ Sucrose is synthesized in the cytosol in two steps from UDP-glucose and fructose 1-phosphate.

FIGURE 20–27 Regulation of ADP-glucose phosphorylase by 3-phosphoglycerate and P_i. This enzyme, which produces the precursor for starch synthesis, is rate-limiting in starch production. The enzyme is stimulated allosterically by 3-phosphoglycerate (3-PGA) and inhibited by P_i; in effect, the ratio [3-PGA]/[P_i], which rises with increasing rates of photosynthesis, controls starch synthesis at this step.

▶ The partitioning of triose phosphates between sucrose synthesis and starch synthesis is regulated by fructose 2,6-bisphosphate (F26BP), an allosteric effector of the enzymes that determine the level of fructose 6-phosphate. F26BP concentration varies inversely with the rate of photosynthesis, and F26BP inhibits the synthesis of fructose 6-phosphate, the precursor to sucrose.

20.4 Synthesis of Cell Wall Polysaccharides: Plant Cellulose and Bacterial Peptidoglycan

Cellulose is a major constituent of plant cell walls, providing strength and rigidity and preventing the swelling of the cell and rupture of the plasma membrane that might result when osmotic conditions favor water entry into the cell. Each year, worldwide, plants synthesize more than 10^{11} metric tons of cellulose, making this simple polymer one of the most abundant compounds in the biosphere. The structure of cellulose is simple: linear polymers of thousands of ($\beta 1 \rightarrow 4$)-linked D-glucose units, assembled into bundles of about 36 chains, which aggregate side by side to form a microfibril **(Fig. 20–28)**.

The biosynthesis of cellulose is less well understood than that of glycogen or starch. As a major component of the plant cell wall, cellulose must be synthesized from intracellular precursors but deposited and assembled outside the plasma membrane. The enzymatic machinery for initiation, elongation, and export of cellulose chains is more complicated than that needed to synthesize starch or glycogen (which are not exported). Bacteria face a similar set of problems when they synthesize the complex polysaccharides that make up their cell walls, and they may employ some of the same mechanisms to solve these problems.

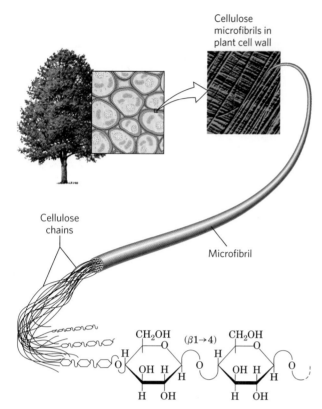

FIGURE 20–28 **Cellulose structure.** The plant cell wall is made up in part of cellulose molecules arranged side by side to form paracrystalline arrays—cellulose microfibrils. Many microfibrils combine to form a cellulose fiber, seen in the scanning electron microscope as a structure 5 to 12 nm in diameter, laid down on the cell surface in several layers distinguishable by the different orientations of their fibers.

Cellulose Is Synthesized by Supramolecular Structures in the Plasma Membrane

The complex enzymatic machinery that assembles cellulose chains spans the plasma membrane, with one part positioned to bind the substrate, UDP-glucose, in the cytosol and another part extending to the outside, responsible for elongating and crystallizing cellulose molecules in the extracellular space. Freeze-fracture electron microscopy shows these **terminal complexes**, also called **rosettes**, to be composed of six large particles arranged in a regular hexagon with a diameter of about 30 nm **(Fig. 20–29)**. Several proteins, including the catalytic subunit of **cellulose synthase**, make up the terminal complex. Much of the recent progress in understanding cellulose synthesis stems from genetic and molecular genetic studies of the plant *Arabidopsis thaliana*, which is especially amenable to genetic dissection and whose genome has been sequenced. The gene family that encodes this cellulose-synthesizing activity has been cloned and found to encode proteins with eight transmembrane segments and a central domain on the cytosolic side of the plasma membrane that includes sequences expected in a glycosyltransferase (Fig. 20–29).

In one working model of cellulose synthesis, cellulose chains are initiated by the formation of a lipid-linked intermediate unlike anything involved in starch or glycogen synthesis. As shown in step ❶ of Figure 20–29, glucose is transferred from UDP-glucose to a membrane lipid, probably the plant sterol sitosterol, on the inner face of the plasma membrane. Here, intracellular cellulose synthase adds several more glucose residues to the first one, in (β1→4) linkage, forming a short oligosac-

FIGURE 20–29 **One proposed model for synthesis of cellulose in a vascular plant.** This schematic is derived from a combination of genetic and biochemical studies of *Arabidopsis thaliana* and of other vascular plants.

charide chain attached to the sitosterol (sitosterol dextrin). Next, the whole sitosterol dextrin flips across to the outer face of the plasma membrane, where it now associates with another form of cellulose synthase.

The UDP-glucose used for cellulose synthesis (step ❷) is generated from sucrose produced during photosynthesis, by the reaction catalyzed by sucrose synthase (named for the reverse reaction):

$$\text{Sucrose} + \text{UDP} \longrightarrow \text{UDP-glucose} + \text{fructose}$$

Cellulose synthase spans the plasma membrane and uses cytosolic UDP-glucose as the precursor for extracellular cellulose synthesis. A membrane-bound form of sucrose synthase forms a complex with cellulose synthase, feeding UDP-glucose from sucrose directly into cell wall synthesis.

In step ❸ a second form of cellulose synthase extends the polymer to 500 to 15,000 glucose units, extruding it onto the outer surface of the cell. The action of the enzyme is processive: one enzyme molecule adds many glucose units before releasing the growing cellulose chain. The direction of chain growth (whether addition occurs at the reducing end or at the nonreducing end) has not been definitively established.

Each of the six globules of the rosette consists of multiple protein subunits that together synthesize six cellulose chains. The large enzyme complex that catalyzes this process (step ❹) actually moves along the plasma membrane, following the course of microtubules in the cortex, the cytoplasmic layer just under the membrane. Because these microtubules lie perpendicular to the axis of the plant's growth, the cellulose microfibrils are laid down across the axis of growth. The motion of the cellulose synthase complexes is believed to be driven by energy released in the polymerization reaction, not by a molecular motor such as kinesin.

The finished cellulose is in the form of crystalline microfibrils (Fig. 20–28), each consisting of 36 separate cellulose chains lying side by side, all with the same (parallel) orientation of nonreducing and reducing ends. It seems likely that the 36 separate polymers synthesized at one rosette arrive together on the outer surface of the cell, already aligned and (step ❺) ready to crystallize as a microfibril of the cell wall. When the 36 polymers reach some critical length, their synthesis is terminated by an unknown mechanism; crystallization into a microfibril follows.

FIGURE 20–30 Peptidoglycan structure. This is the peptidoglycan of the cell wall of *Staphylococcus aureus*, a gram-positive bacterium. Peptides (strings of colored spheres) covalently link *N*-acetylmuramic acid residues in neighboring polysaccharide chains. Note the mixture of L and D amino acids in the peptides. Gram-positive bacteria such as *S. aureus* have a pentaglycine chain in the cross-link. Gram-negative bacteria, such as *E. coli*, lack the pentaglycine; instead, the terminal D-Ala residue of one tetrapeptide is attached directly to a neighboring tetrapeptide through either L-Lys or a lysine-like amino acid, diaminopimelic acid. The peptide bond of glutamate is unusual here; it involves the carboxyl group of the glutamate side chain.

In the activated precursor of cellulose (UDP-glucose), the glucose is α-linked to the nucleotide, but in the product (cellulose), glucose residues are ($\beta 1 \rightarrow 4$)-linked, so there is an inversion of configuration at the anomeric carbon (C-1) as the glycosidic bond forms. Glycosyltransferases that invert configuration are generally assumed to use a single-displacement mechanism, with nucleophilic attack by the acceptor species at the anomeric carbon of the donor sugar (UDP-glucose).

Certain bacteria (*Acetobacter*, *Agrobacteria*, *Rhizobia*, and *Sarcina*) and many simple eukaryotes also carry out cellulose synthesis, apparently by a mechanism similar to that in plants. If the bacteria use a membrane lipid to initiate new chains, it cannot be a sterol—bacteria do not contain sterols.

Lipid-Linked Oligosaccharides Are Precursors for Bacterial Cell Wall Synthesis

Like plants, many bacteria have thick, rigid extracellular walls that protect them from osmotic lysis. The **peptidoglycan** that gives bacterial envelopes their strength and rigidity is an alternating linear copolymer of *N*-acetylglucosamine (GlcNAc) and *N*-acetylmuramic acid (Mur2Ac), linked by ($\beta 1 \rightarrow 4$) glycosidic bonds and cross-linked by short peptides attached to the Mur2Ac **(Fig. 20–30)**. During assembly of the polysaccharide

Staphylococcus aureus

N-Acetylglucosamine (GlcNAc)

N-Acetylmuramic acid (Mur2Ac)

($\beta 1 \rightarrow 4$)

Reducing end

L-Ala
D-Glu
L-Lys
D-Ala

Pentaglycine cross-link

backbone of this complex macromolecule, both GlcNAc and Mur2Ac are activated by attachment of a uridine nucleotide at their anomeric carbons. First, GlcNAc 1-phosphate condenses with UTP to form UDP-GlcNAc (**Fig. 20–31**, step ❶), which reacts with phosphoenolpyruvate to form UDP-Mur2Ac (step ❷); five amino acids are then added (step ❸). The Mur2Ac-pentapeptide moiety is transferred from the uridine nucleotide to the membrane lipid dolichol, a long-chain isoprenoid alcohol (see Fig. 10–22f) (step ❹), and a GlcNAc residue is donated by UDP-GlcNAc (step ❺). In many bacteria, five glycines are added in peptide linkage to the amino group of the Lys residue of the pentapeptide (step ❻). Finally, this disaccharide decapeptide is added to the nonreducing end of an existing peptidoglycan molecule (step ❼). A transpeptidation reaction cross-links adjacent polysaccharide chains (step ❽), contributing to a huge, strong, macromolecular wall around the bacterial cell. Many of the most effective antibiotics in use today

act by inhibiting reactions in the synthesis of the peptidoglycan.

Many other oligosaccharides and polysaccharides are synthesized by similar routes in which sugars are activated for subsequent reactions by attachment to nucleotides. In the glycosylation of proteins, for example (see Fig. 27–39), the precursors of the carbohydrate moieties include sugar nucleotides and lipid-linked oligosaccharides.

SUMMARY 20.4 Synthesis of Cell Wall Polysaccharides: Plant Cellulose and Bacterial Peptidoglycan

▶ Cellulose synthesis takes place in terminal complexes (rosettes) in the plasma membrane. Each cellulose chain begins as a sitosterol dextrin formed inside the cell. It then flips to the outside, where the oligosaccharide portion is transferred to cellulose synthase in the rosette and is then

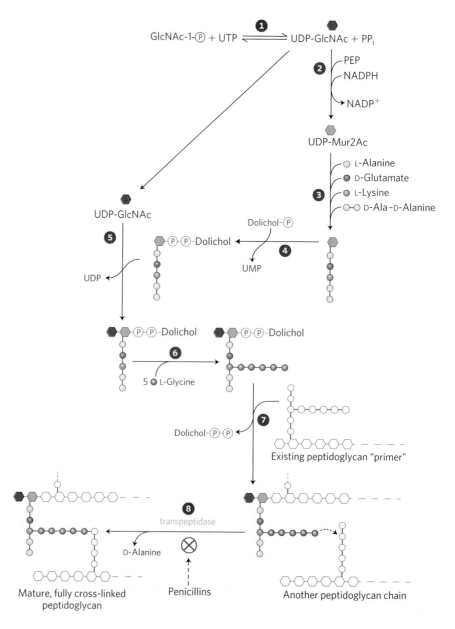

FIGURE 20–31 Synthesis of bacterial peptidoglycan. In the early steps of this pathway (❶ through ❹), *N*-acetylglucosamine (GlcNAc) and *N*-acetylmuramic acid (Mur2Ac) are activated by attachment of a uridine nucleotide (UDP) to their anomeric carbons and, in the case of Mur2Ac, of a long-chain isoprenoid alcohol (dolichol) through a phosphodiester bond. These activating groups participate in the formation of glycosidic linkages; they serve as excellent leaving groups. After steps ❺ and ❻, assembly of a disaccharide with a peptide side chain (10 amino acid residues), ❼ this precursor is transferred to the nonreducing end of an existing peptidoglycan chain, which serves as a primer for the polymerization reaction. Finally, ❽ in a transpeptidation reaction between the peptide side chains on two different peptidoglycan molecules, a Gly residue at the end of one chain displaces a terminal D-Ala in the other chain, forming a cross-link. This transpeptidation reaction is inhibited by the penicillins, which kill bacteria by weakening their cell walls (see Fig. 6-30).

extended. Each rosette produces 36 separate cellulose chains simultaneously and in parallel. The chains crystallize into one of the microfibrils that form the cell wall.

▶ Synthesis of the bacterial cell wall peptidoglycan also involves lipid-linked oligosaccharides formed inside the cell and flipped to the outside for assembly.

20.5 Integration of Carbohydrate Metabolism in the Plant Cell

Carbohydrate metabolism in a typical plant cell is more complex in several ways than that in a typical animal cell. The plant cell carries out the same processes that generate energy in animal cells (glycolysis, citric acid cycle, and oxidative phosphorylation); it can generate hexoses from three- or four-carbon compounds by gluconeogenesis; it can oxidize hexose phosphates to pentose phosphates with the generation of NADPH (the oxidative pentose phosphate pathway); and it can produce a polymer of (α1→4)-linked glucose (starch) and degrade it to generate hexoses. But besides these carbohydrate transformations that it shares with animal cells, the photosynthetic plant cell can fix CO_2 into organic compounds (the rubisco reaction); use the products of fixation to generate trioses, hexoses, and pentoses (the Calvin cycle); and convert acetyl-CoA generated from fatty acid breakdown to four-carbon compounds (the glyoxylate cycle) and the four-carbon compounds to hexoses (gluconeogenesis). These processes, unique to the plant cell, are segregated in several compartments not found in animal cells: the glyoxylate cycle in glyoxysomes, the Calvin cycle in chloroplasts, starch synthesis in amyloplasts, and organic acid storage in vacuoles. The integration of events among these various compartments requires specific transporters in the membranes of each organelle, to move products from one organelle to another or into the cytosol.

Gluconeogenesis Converts Fats and Proteins to Glucose in Germinating Seeds

Many plants store lipids and proteins in their seeds, to be used as sources of energy and as biosynthetic precursors during germination, before photosynthetic mechanisms have developed. Active gluconeogenesis in germinating seeds provides glucose for the synthesis of sucrose, polysaccharides, and many metabolites derived from hexoses. In plant seedlings, sucrose provides much of the chemical energy needed for initial growth.

We noted earlier (Chapter 14) that animal cells can carry out gluconeogenesis from three- and four-carbon precursors, but not from the two acetyl carbons of acetyl-CoA. Because the pyruvate dehydrogenase reaction is effectively irreversible (Section 16.1), animal cells have no way to convert acetyl-CoA to pyruvate or oxaloace-

tate. Unlike animals, plants and some microorganisms *can* convert acetyl-CoA derived from fatty acid oxidation to glucose (**Fig. 20–32**). Some of the enzymes essential

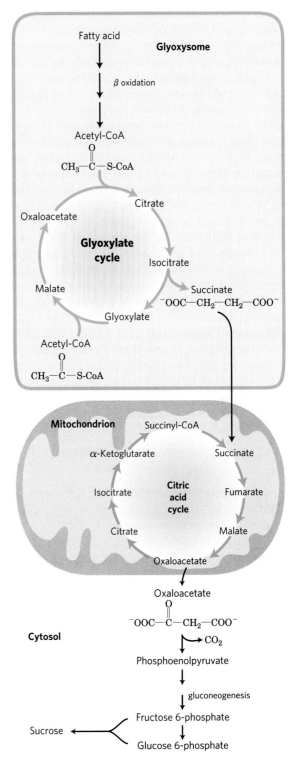

FIGURE 20–32 Conversion of stored fatty acids to sucrose in germinating seeds. This pathway begins in glyoxysomes. Succinate is produced and exported to mitochondria, where it is converted to oxaloacetate by enzymes of the citric acid cycle. Oxaloacetate enters the cytosol and serves as the starting material for gluconeogenesis and for the synthesis of sucrose, the transport form of carbon in plants.

to this conversion are sequestered in glyoxysomes, where glyoxysome-specific isozymes of β oxidation break down fatty acids to acetyl-CoA (see Fig. 16–24). The physical separation of the glyoxylate cycle and β-oxidation enzymes from the mitochondrial citric acid cycle enzymes prevents further oxidation of acetyl-CoA to CO_2. Instead, the acetyl-CoA is converted to succinate in the glyoxylate cycle (see Fig. 16–22). The succinate passes into the mitochondrial matrix, where it is converted by citric acid cycle enzymes to oxaloacetate, which moves into the cytosol. Cytosolic oxaloacetate is converted by gluconeogenesis to fructose 6-phosphate, the precursor of sucrose. Thus the integration of reaction sequences in three subcellular compartments is required for the production of fructose 6-phosphate or sucrose from stored lipids. Because only three of the four carbons in each molecule of oxaloacetate are converted to hexose in the cytosol, about 75% of the carbon in the fatty acids stored as seed lipids is converted to carbohydrate by the combined pathways of Figure 20–32. The other 25% is lost as CO_2 in the conversion of oxaloacetate to phosphoenolpyruvate. Hydrolysis of storage triacylglycerols also produces glycerol 3-phosphate, which can enter the gluconeogenic pathway after its oxidation to dihydroxyacetone phosphate (Fig. 20–33).

Glucogenic amino acids (see Table 14–4) derived from the breakdown of stored seed proteins also yield precursors for gluconeogenesis, following transamination and oxidation to succinyl-CoA, pyruvate, oxaloacetate, fumarate, and α-ketoglutarate (Chapter 18)—all good starting materials for gluconeogenesis.

Pools of Common Intermediates Link Pathways in Different Organelles

Although we have described metabolic transformations in plant cells in terms of individual pathways, these pathways interconnect so completely that we should instead consider pools of metabolic intermediates shared among these pathways and connected by readily reversible reactions (Fig. 20–34). One such **metabolite pool** includes the hexose phosphates glucose 1-phosphate, glucose 6-phosphate, and fructose 6-phosphate; a second includes the 5-phosphates of the pentoses ribose, ribulose, and xylulose; a third includes the triose phosphates dihydroxyacetone phosphate and glyceraldehyde 3-phosphate. Metabolite fluxes through these pools change in magnitude and direction in response to changes in the circumstances of the plant, and they vary with tissue type. Transporters in the membranes of each organelle move specific compounds in and out, and the regulation of these transporters presumably influences the degree to which the pools mix.

During daylight hours, triose phosphates produced in leaf tissue by the Calvin cycle move out of the chloroplast and into the cytosolic hexose phosphate pool, where they are converted to sucrose for transport to nonphotosynthetic tissues. In these tissues, sucrose is

FIGURE 20–33 Conversion of the glycerol moiety of triacylglycerols to sucrose in germinating seeds. The glycerol of triacylglycerols is oxidized to dihydroxyacetone phosphate, which enters the gluconeogenic pathway at the triose phosphate isomerase reaction.

converted to starch for storage or is used as an energy source via glycolysis. In growing plants, hexose phosphates are also withdrawn from the pool for the synthesis of cell walls. At night, starch is metabolized by glycolysis to provide energy, essentially as in nonphotosynthetic organisms, and NADPH and ribose 5-phosphate are obtained through the oxidative pentose phosphate pathway.

SUMMARY 20.5 Integration of Carbohydrate Metabolism in the Plant Cell

▶ Plants can synthesize sugars from acetyl-CoA, the product of fatty acid breakdown, by the combined actions of the glyoxylate cycle and gluconeogenesis.

▶ The individual pathways of carbohydrate metabolism in plants overlap extensively; they

FIGURE 20–34 Pools of hexose phosphates, pentose phosphates, and triose phosphates. The compounds in each pool are readily interconvertible by reactions that have small standard free-energy changes. When one component of the pool is temporarily depleted, a new equilibrium is quickly established to replenish it. Movement of the sugar phosphates between intracellular compartments is limited; specific transporters must be present in an organelle membrane.

share pools of common intermediates, including hexose phosphates, pentose phosphates, and triose phosphates. Transporters in the membranes of chloroplasts, mitochondria, and amyloplasts mediate the movement of sugar phosphates between organelles. The direction of metabolite flow through the pools changes from day to night.

Further Reading

General References

Blankenship, R.E. (2002) *Molecular Mechanisms of Photosynthesis,* Blackwell Science, Oxford.

Very readable, well-illustrated, intermediate-level treatment of all aspects of photosynthesis, including the carbon metabolism covered in this chapter and the light-driven reactions described in Chapter 19.

Buchanan, B.B., Gruissem, W., & Jones, R.L. (eds). (2002) *Biochemistry and Molecular Biology of Plants,* American Society of Plant Physiology, Rockville, MD.

This wonderful, authoritative book covers all aspects of plant biochemistry and molecular biology. The following chapters cover carbohydrate structure and synthesis in greater depth: **Carpita, N. & McCann, M.,** Chapter 2, The cell wall (pp. 52–109); **Malkin, R. & Niyogi, K.,** Chapter 12, Photosynthesis (pp. 568–629); **Dennis, D.T. & Blakeley, S.D.,** Chapter 13, Carbohydrate Metabolism (pp. 630–675); **Siedow, J.N. & Day, D.A.,** Chapter 14, Respiration and Photorespiration (pp. 676–729).

Heldt, H.-W. & Piechulla, B. (2010) *Plant Biochemistry,* 4th edn., Academic Press, New York.

An *excellent* textbook of plant biochemistry.

Morton, O. (2007) *Eating the Sun: How Plants Power the Planet,* Harper, New York.

An engaging account of the discoveries in photosynthesis in the context of the history of science.

Photosynthetic Carbohydrate Synthesis

Andersson, I. & Backlund, A. (2008) Structure and function of rubisco. *Plant Physiol. Biochem.* **46**, 275–291.
Intermediate-level review.

Bassham, J.A. (2003) Mapping the carbon reduction cycle: a personal retrospective. *Photosynth. Res.* **76**, 35–52.

Benson, A.A. (2002) Following the path of carbon in photosynthesis: a personal story—history of photosynthesis. *Photosynth. Res.* **73**, 31–49.

Calvin, M. (1989) Forty years of photosynthesis and related activities. *Photosynth. Res.* **21**, 3–16.

Cleland, W.W., Andrews, T.J., Gutteridge, S., Hartman, F.C., & Lorimer, G.H. (1998) Mechanism of rubisco—the carbamate as general base. *Chem. Rev.* **98**, 549–561.
Review with a special focus on the carbamate at the active site.

Dietz, K.J., Link, G., Pistorius, E.K., & Scheibe, R. (2002) Redox regulation in oxygenic photosynthesis. *Prog. Botany* **63**, 207–245.

Hartman, F.C. & Harpel, M.R. (1994) Structure, function, regulation and assembly of D-ribulose-1,5-bisphosphate carboxylase/oxygenase. *Annu. Rev. Biochem.* **63**, 197–234.

Horecker, B.L. (2002) The pentose phosphate pathway. *J. Biol. Chem.* **277**, 47,965–47,971.

Jansson, C., Wullschleger, S.D., Kalluri, U.S., & Tuskan, G.A. (2010) Phytosequestration: carbon biosequestration by plants and the prospects of genetic engineering. *Bioscience* **60**, 685–696.
Global fluxes of carbon and the contribution of biomass to total terrestrial carbon.

Portis, A.R., Jr. (2003) Rubisco activase: rubisco's catalytic chaperone. *Photosynth. Res.* **75**, 11–27.
Structure, regulation, mechanism, and importance of rubisco activase.

Portis, A.R., Jr. & Parry, M.A.J. (2007) Discoveries in rubisco (ribulose 1,5-bisphosphate carboxylase/oxygenase): a historical perspective. *Photosynth. Res.* **94**, 121–143.

Raines, C.A. (2003) The Calvin cycle revisited. *Photosynth. Res.* **75**, 1–10.
Metabolic control analysis applied to the Calvin cycle.

Sage, R.F., Way, D.A., & Kubien, D.S. (2008) Rubisco, rubisco activase, and global climate change. *J. Exper. Bot.* **59**, 1581–1595.

Smith, A.M., Denyer, K., & Martin, C. (1997) The synthesis of the starch granule. *Annu. Rev. Plant Physiol. Plant Mol. Biol.* **48**, 67–87.
Review of the role of ADP-glucose pyrophosphorylase in the synthesis of amylose and amylopectin in starch granules.

Spreitzer, R.J. & Salvucci, M.E. (2002) Rubisco: structure, regulatory interactions, and possibilities for a better enzyme. *Annu. Rev. Plant Biol.* **53**, 449–475.
Advanced review on rubisco and rubisco activase.

Whitney, S.M., Houtz, R.L., & Alonso, H. (2011) Advancing our understanding and capacity to engineer nature's CO_2-sequestering enzyme, rubisco. *Plant Physiol.* **155**, 27–35.

Photorespiration and the C₄ and CAM Pathways

Ainsworth, E.A. & Long, S.P. (2005) What have we learned from 15 years of free-air CO_2 enrichment (FACE)? A meta-analytic review of the response of photosynthesis, canopy properties, and plant production to rising CO_2. *New Phytol.* **165**, 351–372.

Bauwe, H., Hagemann, M., & Fernie, A.R. (2010) Photorespiration: players, partners and origin. *Trends Plant Sci.* **15**, 330–336.

Black, C.C. & Osmond, C.B. (2003) Crassulacean acid metabolism and photosynthesis: working the night shift. *Photosynth. Res.* **76**, 329–341.

Douce, R., Bourguignon, J., Neuburger, M., & Rébeillé, F. (2001) The glycine decarboxylase system: a fascinating complex. *Trends Plant Sci.* **6**, 167–176.
Intermediate-level description of the structure and the reaction mechanism of the enzyme.

Hatch, M.D. (1987) C₄ photosynthesis: a unique blend of modified biochemistry, anatomy and ultrastructure. *Biochim. Biophys. Acta* **895**, 81–106.
Intermediate-level review by one of the discoverers of the C₄ pathway.

Hatch, M.D. & Slack, S.R. (1966) Photosynthesis by sugar-cane leaves: a new carboxylation reaction and the pathway of sugar formation. *Biochem. J.* **101**, 103–111.
The classic description of the path named for these authors.

Langdale, J.A. (2011) C4 cycles: past, present, and future research on C4 photosynthesis. *Plant Cell* **23**, 3879–3892.
Intermediate-level review

Tolbert, N.E. (1997) The C₂ oxidative photosynthetic carbon cycle. *Annu. Rev. Plant Physiol. Plant Mol. Biol.* **48**, 1–25.
A fascinating personal account of the development of our understanding of photorespiration, by one who was central in this development.

Biosynthesis of Starch and Sucrose

Doblin, M.S., Kurek, I., Javob-Wilk, D., & Delmer, D.P. (2002) Cellulose biosynthesis in plants: from genes to rosettes. *Plant Cell Physiol.* **43**, 1407–1420.

Huber, S.C. & Huber, J.L. (1996) Role and regulation of sucrose-phosphate synthase in higher plants. *Annu. Rev. Plant Physiol. Plant Mol. Biol.* **47**, 431–444.
Short review of factors that regulate this critical enzyme.

Keeling, P.L & Myers, A.M. (2010) Biochemistry and genetics of starch synthesis. *Annu. Rev. Food Sci. Technol.* **1**, 271–303.
Advanced review

Kotting, O., Kossmann, J., Zeeman, S.C., & Lloyd, J.R. (2010) Regulation of starch metabolism: the age of enlightenment? *Curr. Opin. Plant Biol.* **13**, 321–329.

Leloir, L.F. (1971) Two decades of research on the biosynthesis of saccharides. *Science* **172**, 1299–1303.
Leloir's Nobel address, including a discussion of the role of sugar nucleotides in metabolism.

Synthesis of Cellulose and Peptidoglycan

Endler, A. & Persson, S. (2011) Cellulose synthases and synthesis in *Arabidopsis*. *Mol. Plant* **4**, 199–211.

Joshi, C.P. & Mansfield, S.D. (2007) The cellulose paradox—simple molecule, complex biosynthesis. *Curr. Opin. Plant Biol.* **10**, 220–226.

Liepman, A.H., Andrews, T.J., Gutteridge, S., Hartman, F.C., & Lorimer, G.H. (2010) *Arabidopsis*—a powerful model system for plant cell wall research. *Plant J.* **61**, 1107–1121.

Scheible, W.-R. & Pauly, M. (2004) Glycosyltransferases and cell wall biosynthesis: novel players and insights. *Curr. Opin. Plant Biol.* **7**, 285–296.

Vollmer, W. & Seligman, S.J. (2009) Architecture of peptidoglycan: more data and more models. *Trends Microbiol.* **18**, 59–66.

Problems

1. Segregation of Metabolism in Organelles What are the advantages to the plant cell of having different organelles to carry out different reaction sequences that share intermediates?

2. Phases of Photosynthesis When a suspension of green algae is illuminated in the absence of CO_2 and then incubated with $^{14}CO_2$ in the dark, $^{14}CO_2$ is converted to $[^{14}C]$glucose for a brief time. What is the significance of this observation with regard to the CO_2-assimilation process, and how is it related to the light reactions of photosynthesis? Why does the conversion of $^{14}CO_2$ to $[^{14}C]$glucose stop after a brief time?

3. Identification of Key Intermediates in CO_2 Assimilation Calvin and his colleagues used the unicellular green alga *Chlorella* to study the carbon-assimilation reactions of photosynthesis. They incubated $^{14}CO_2$ with illuminated suspensions of algae and followed the time course of appearance of ^{14}C in two compounds, X and Y, under two sets of conditions. Suggest the identities of X and Y, based on your understanding of the Calvin cycle.

(a) Illuminated *Chlorella* were grown with unlabeled CO_2, then the light was turned off and $^{14}CO_2$ was added (vertical dashed line in the graph below). Under these conditions, X was the first compound to become labeled with ^{14}C; Y was unlabeled.

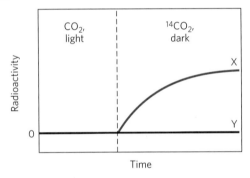

(b) Illuminated *Chlorella* cells were grown with $^{14}CO_2$. Illumination was continued until all the $^{14}CO_2$ had disappeared (vertical dashed line in the graph below). Under these conditions, X became labeled quickly but lost its radioactivity with time, whereas Y became more radioactive with time.

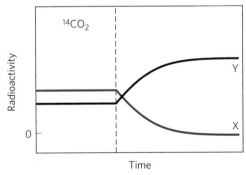

4. Regulation of the Calvin Cycle Iodoacetate reacts irreversibly with the free —SH groups of Cys residues in proteins.

Iodoacetate   Inactive enzyme

Predict which Calvin cycle enzyme(s) would be inhibited by iodoacetate, and explain why.

5. Thioredoxin in Regulation of Calvin Cycle Enzymes Motohashi and colleagues used thioredoxin as a hook to fish out from plant extracts the proteins that are activated by thioredoxin. To do this, they prepared a mutant thioredoxin in which one of the reactive Cys residues was replaced with a Ser. Explain why this modification was necessary for their experiments. Source: **Motohashi, K., Kondoh, A., Stumpp, M.T., & Hisabori, T.** (2001) Comprehensive survey of proteins targeted by chloroplast thioredoxin. *Proc. Natl. Acad. Sci. USA* **98**, 11,224–11,229.

6. Comparison of the Reductive and Oxidative Pentose Phosphate Pathways The *reductive* pentose phosphate pathway generates a number of intermediates identical to those of the *oxidative* pentose phosphate pathway (Chapter 14). What role does each pathway play in cells where it is active?

7. Photorespiration and Mitochondrial Respiration Compare the oxidative photosynthetic carbon cycle (C_2 cycle), also called *photorespiration*, with the *mitochondrial respiration* that drives ATP synthesis. Why are both processes referred to as respiration? Where in the cell do they occur, and under what circumstances? What is the path of electron flow in each?

8. Rubisco and the Composition of the Atmosphere N. E. Tolbert has argued that the dual specificity of rubisco for CO_2 and O_2 is not simply a leftover from evolution in a low-oxygen environment. He suggests that the relative activities of the carboxylase and oxygenase activities of rubisco actually have set, and now maintain, the ratio of CO_2 to O_2 in the earth's atmosphere. Discuss the pros and cons of this hypothesis, in molecular terms and in global terms. How does the existence of C_4 organisms bear on the hypothesis? Source: **Tolbert, N.E.** (1994) The role of photosynthesis and photorespiration in regulating atmospheric CO_2 and O_2. In *Regulation of Atmospheric CO_2 and O_2 by Photosynthetic Carbon Metabolism* (Tolbert, N.E., & Preiss, J., eds), pp. 8–33, Oxford University Press, New York.

9. Role of Sedoheptulose 1,7-Bisphosphatase What effect on the cell and the organism might result from a defect in sedoheptulose 1,7-bisphosphatase in (a) a human hepatocyte and (b) the leaf cell of a green plant?

10. Pathway of CO_2 Assimilation in Maize If a maize (corn) plant is illuminated in the presence of $^{14}CO_2$, after about 1 second, more than 90% of all the radioactivity incorporated in the leaves is found at C-4 of malate, aspartate, and oxaloacetate. Only after 60 seconds does ^{14}C appear at C-1 of 3-phosphoglycerate. Explain.

11. Identifying CAM Plants Given some $^{14}CO_2$ and all the tools typically present in a biochemistry research lab, how would you design a simple experiment to determine whether a plant is a typical C_4 plant or a CAM plant?

12. Chemistry of Malic Enzyme: Variation on a Theme Malic enzyme, found in the bundle-sheath cells of C_4 plants, carries out a reaction that has a counterpart in the citric acid cycle. What is the analogous reaction? Explain your choice.

13. The Cost of Storing Glucose as Starch Write the sequence of steps and the net reaction required to calculate the cost, in ATP molecules, of converting a molecule of cytosolic glucose 6-phosphate to starch and back to glucose 6-phosphate. What fraction of the maximum number of ATP molecules available from complete catabolism of glucose 6-phosphate to CO_2 and H_2O does this cost represent?

14. Inorganic Pyrophosphatase The enzyme inorganic pyrophosphatase contributes to making many biosynthetic reactions that generate inorganic pyrophosphate essentially irreversible in cells. By keeping the concentration of PP_i very low, the enzyme "pulls" these reactions in the direction of PP_i formation. The synthesis of ADP-glucose in chloroplasts is one reaction that is pulled in the forward direction by this mechanism. However, the synthesis of UDP-glucose in the plant cytosol, which produces PP_i, is readily reversible in vivo. How do you reconcile these two facts?

15. Regulation of Starch and Sucrose Synthesis Sucrose synthesis occurs in the cytosol and starch synthesis in the chloroplast stroma, yet the two processes are intricately balanced. What factors shift the reactions in favor of (a) starch synthesis and (b) sucrose synthesis?

16. Regulation of Sucrose Synthesis In the regulation of sucrose synthesis from the triose phosphates produced during photosynthesis, 3-phosphoglycerate and P_i play critical roles (see Fig. 20–25). Explain why the concentrations of these two regulators reflect the rate of photosynthesis.

17. Sucrose and Dental Caries The most prevalent infection in humans worldwide is dental caries, which stems from the colonization and destruction of tooth enamel by a variety of acidifying microorganisms. These organisms synthesize and live within a water-insoluble network of dextrans, called dental plaque, composed of $(\alpha 1 \rightarrow 6)$-linked polymers of glucose with many $(\alpha 1 \rightarrow 3)$ branch points. Polymerization of dextran requires dietary sucrose, and the reaction is catalyzed by a bacterial enzyme, dextran-sucrose glucosyltransferase.

(a) Write the overall reaction for dextran polymerization.

(b) In addition to providing a substrate for the formation of dental plaque, how does dietary sucrose also provide oral bacteria with an abundant source of metabolic energy?

18. Differences between C_3 and C_4 Plants The plant genus *Atriplex* includes some C_3 and some C_4 species. From the data in the following plots (species 1, black curve; species

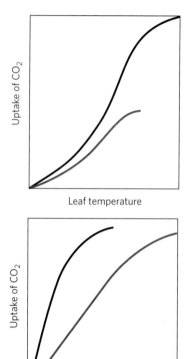

2, red curve), identify which is a C_3 plant and which is a C_4 plant. Justify your answer in molecular terms that account for the data in all three plots.

19. C_4 Pathway in a Single Cell In typical C_4 plants, the initial capture of CO_2 occurs in one cell type, and the Calvin cycle reactions occur in another (see Fig. 20–23). Voznesenskaya and colleagues[‡] have described a plant, *Bienertia cycloptera*—which grows in salty depressions of semidesert in Central Asia—that shows the biochemical properties of a C_4 plant but unlike typical C_4 plants does not segregate the reactions of CO_2 fixation into two cell types. PEP carboxylase and rubisco are present in the same cell. However, the cells have two types of chloroplasts, which are localized differently. One type, relatively poor in grana (thylakoids), is confined to the periphery; the more typical chloroplasts are clustered in the center of the cell, separated from the peripheral chloroplasts by large vacuoles. Thin cytosolic bridges pass through the vacuoles, connecting the peripheral and central cytosol. A micrograph of a *B. cycloptera* cell, with arrows pointing to peripheral chloroplasts, is shown below.

In this plant, where would you expect to find (a) PEP carboxylase, (b) rubisco, and (c) starch granules? Explain your answers with a model for CO_2 fixation in these C_4 cells. Source: **Voznesenskaya, E.V., Fraceschi, V.R., Kiirats, O., Artyusheva, E.G., Freitag, H., & Edwards, G.E.** (2002) Proof of C_4 photosynthesis without Kranz anatomy in *Bienertia cycloptera* (Chenopodiaceae). *Plant J.* **31,** 649–662.

Data Analysis Problem

20. Rubisco of Bacterial Endosymbionts of Hydrothermal Vent Animals Undersea hydrothermal vents support remarkable ecosystems. At these extreme depths there is no light to support photosynthesis, yet thriving vent communities are found. Much of their primary productivity occurs through chemosynthesis carried out by bacterial symbionts that live in specialized organs (trophosomes) of certain vent animals.

Chemosynthesis in these bacteria involves a process that is virtually identical to photosynthesis. Carbon dioxide is fixed by rubisco and reduced to glucose, and the necessary ATP and NADPH are produced by electron-transfer processes similar to those of the light-dependent reactions of photosynthesis. The key difference is that in chemosynthesis, the energy driving electron transfer comes from a highly exergonic chemical reaction rather than from light. Different chemosynthetic bacteria use different reactions for this purpose. The bacteria found in hydrothermal vent animals typically use the oxidation of H_2S (abundant in the vent water) by O_2, producing elemental sulfur. These bacteria also use the conversion of H_2S to sulfur as a source of electrons for chemosynthetic CO_2 reduction.

(a) What is the overall reaction for chemosynthesis in these bacteria? You do not need to write a balanced equation; just give the starting materials and products.

(b) Ultimately, these endosymbiotic bacteria obtain their energy from sunlight. Explain how this occurs.

Robinson and colleagues (2003) explored the properties of rubisco from the bacterial endosymbiont of the giant tube worm *Riftia pachyptila*. Rubisco, from any source, catalyzes the reaction of either CO_2 (Fig. 20–7) or O_2 (Fig. 20–20) with ribulose 1,5-bisphosphate. In general, rubisco reacts more readily with CO_2 than O_2. The degree of selectivity (Ω) can be expressed in the equation

$$\frac{V_{carboxylation}}{V_{oxygenation}} = \Omega \frac{[CO_2]}{[O_2]}$$

where V is the reaction velocity.

Robinson and coworkers measured the Ω value for the rubisco of the bacterial endosymbionts. They purified rubisco from tube-worm trophosomes, reacted it with mixtures of different ratios of O_2 and CO_2 in the presence of [1-^{3}H]ribulose 1,5-bisphosphate, and measured the ratio of [^{3}H]phosphoglycerate to [^{3}H]phosphoglycolate.

(c) The measured ratio of [^{3}H]phosphoglycerate to [^{3}H] phosphoglycolate is equal to the ratio $V_{carboxylation}/V_{oxygenation}$. Explain why.

(d) Why would [5-^{3}H]ribulose 1,5-bisphosphate not be a suitable substrate for this assay?

The Ω for the endosymbiont rubisco had a value of 8.6 ± 0.9.

(e) The atmospheric (molar) concentration of O_2 is 20% and that of CO_2 is about 380 parts per million. If the endosymbiont were to carry out chemosynthesis under these atmospheric conditions, what would be the value of $V_{carboxylation}/V_{oxygenation}$?

(f) Based on your answer to (e), would you expect Ω for the rubisco of a terrestrial plant to be higher than, equal to, or lower than 8.6? Explain your reasoning.

Two stable isotopes of carbon are commonly found in the environment: the more abundant ^{12}C and the rare ^{13}C. All rubisco enzymes catalyze the fixation of $^{12}CO_2$ faster than that of $^{13}CO_2$. As a result, the carbon in glucose is slightly enriched in ^{12}C compared with the isotopic composition of CO_2 in the environment. Several factors are involved in this "preferential" use of $^{12}CO_2$, but one factor is the fundamental physics of gases. The temperature of a gas is related to the kinetic energy of its molecules. Kinetic energy is given by $\frac{1}{2}mv^2$, where m is molecular mass and v is velocity. Thus, at the same temperature (same kinetic energy), the molecules of a lighter gas will be moving faster than those of a heavier gas.

(g) How could this contribute to rubisco's "preference" for $^{12}CO_2$ over $^{13}CO_2$? Some of the first convincing evidence that the tube-worm hosts were obtaining their fixed carbon from the endosymbionts was that the ^{13}C/^{12}C ratio in the animals was much closer to that of the bacteria than that of nonvent marine animals.

(h) Why is this more convincing evidence for a symbiotic relationship than earlier studies that simply showed the presence of rubisco in the bacteria found in trophosomes?

Reference

Robinson, J.J., Scott, K.M., Swanson, S.T., O'Leary, M.H., Horken, K., Tabita, F.R., & Cavanaugh, C.M. (2003) Kinetic isotope effect and characterization of form II RubisCO from the chemoautotrophic endosymbionts of the hydrothermal vent tubeworm *Riftia pachyptila. Limnol. Oceanogr.* **48,** 48–54.

Lipid Biosynthesis

L ipids play a variety of cellular roles, some only recently recognized. They are the principal form of stored energy in most organisms and major constituents of cellular membranes. Specialized lipids serve as pigments (retinal, carotene), cofactors (vitamin K), detergents (bile salts), transporters (dolichols), hormones (vitamin D derivatives, sex hormones), extracellular and intracellular messengers (eicosanoids, phosphatidylinositol derivatives), and anchors for membrane proteins (covalently attached fatty acids, prenyl groups, and phosphatidylinositol). The ability to synthesize a variety of lipids is essential to all organisms. This chapter describes the biosynthetic pathways for some of the most common cellular lipids, illustrating the strategies employed in assembling these water-insoluble products from water-soluble precursors such as acetate. Like other biosynthetic pathways, these reaction sequences are endergonic and reductive. They use ATP as a source of metabolic energy and a reduced electron carrier (usually NADPH) as a reductant.

We first describe the biosynthesis of fatty acids, the primary components of both triacylglycerols and phospholipids, then examine the assembly of fatty acids into triacylglycerols and the simpler membrane phospholipids. Finally, we consider the synthesis of cholesterol, a component of some membranes and the precursor of steroids such as the bile acids, sex hormones, and adrenocortical hormones.

21.1 Biosynthesis of Fatty Acids and Eicosanoids

After the discovery that fatty acid oxidation takes place by the oxidative removal of successive two-carbon (acetyl-CoA) units (see Fig. 17–8), biochemists thought the biosynthesis of fatty acids might proceed by a simple reversal of the same enzymatic steps. However, as they were to find out, fatty acid biosynthesis and breakdown occur by different pathways, are catalyzed by different sets of enzymes, and take place in different parts of the cell. Moreover, biosynthesis requires the participation of a three-carbon intermediate, **malonyl-CoA**, that is not involved in fatty acid breakdown.

Malonyl-CoA

We focus first on the pathway of fatty acid synthesis, then turn our attention to regulation of the pathway and to the biosynthesis of longer-chain fatty acids, unsaturated fatty acids, and their eicosanoid derivatives.

Malonyl-CoA Is Formed from Acetyl-CoA and Bicarbonate

The formation of malonyl-CoA from acetyl-CoA is an irreversible process, catalyzed by **acetyl-CoA carboxylase**. The bacterial enzyme has three separate polypeptide subunits **(Fig. 21–1)**; in animal cells, all three activities are part of a single multifunctional polypeptide. Plant cells contain both types of acetyl-CoA carboxylase. In all cases, the enzyme contains a biotin prosthetic group covalently bound in amide linkage to the ε-amino group of a Lys residue in one of the three polypeptides or domains of the enzyme molecule. The two-step reaction catalyzed by this enzyme is very similar to other biotin-dependent carboxylation reactions, such as those catalyzed by pyruvate carboxylase (see Fig. 16–17) and propionyl-CoA carboxylase (see Fig. 17–12). A carboxyl group, derived from bicarbonate (HCO_3^-), is first transferred to biotin in an ATP-dependent reaction. The biotinyl group serves as a temporary carrier of CO_2, transferring it to acetyl-CoA in the second step to yield malonyl-CoA.

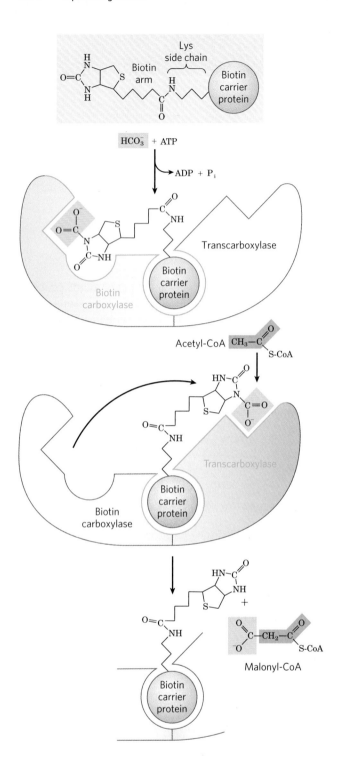

FIGURE 21–1 The acetyl-CoA carboxylase reaction. Acetyl-CoA carboxylase has three functional regions: biotin carrier protein (gray); biotin carboxylase, which activates CO_2 by attaching it to a nitrogen in the biotin ring in an ATP-dependent reaction (see Fig. 16–17); and transcarboxylase, which transfers activated CO_2 (shaded green) from biotin to acetyl-CoA, producing malonyl-CoA. The long, flexible biotin arm carries the activated CO_2 from the biotin carboxylase region to the transcarboxylase active site. The active enzyme in each step is shaded in blue.

Both the electron-carrying cofactor and the activating groups in the reductive anabolic sequence differ from those in the oxidative catabolic process. Recall that in β oxidation, NAD^+ and FAD serve as electron acceptors and the activating group is the thiol (—SH) group of coenzyme A (see Fig. 17–8). By contrast, the reducing agent in the synthetic sequence is NADPH and the activating groups are two different enzyme-bound —SH groups, as described in the following section.

There are two major variants of fatty acid synthase: fatty acid synthase I (FAS I), found in vertebrates and fungi, and fatty acid synthase II (FAS II), found in plants and bacteria. The FAS I found in vertebrates consists of a single multifunctional polypeptide chain (M_r 240,000). The mammalian FAS I is the prototype. Seven active sites for different reactions lie in separate domains **(Fig. 21–3a)**. The mammalian polypeptide functions as a homodimer (M_r 480,000). The subunits appear to function independently. When all the active sites in one subunit are inactivated by mutation, fatty acid synthesis is only modestly reduced. A somewhat different FAS I is found in yeast and other fungi, and is made up of two multifunctional polypeptides that form a complex with an architecture distinct from the vertebrate systems (Fig. 21–3b). Three of the seven required active sites are found on the α subunit and four on the β subunit.

With FAS I systems, fatty acid synthesis leads to a single product, and no intermediates are released. When the chain length reaches 16 carbons, that product (palmitate, 16:0; see Table 10–1) leaves the cycle. Carbons C-16 and C-15 of the palmitate are derived from the methyl and carboxyl carbon atoms, respectively, of an acetyl-CoA used directly to prime the system at the outset **(Fig. 21–4)**; the rest of the carbon atoms in the chain are derived from acetyl-CoA via malonyl-CoA.

FAS II, in plants and bacteria, is a dissociated system; each step in the synthesis is catalyzed by a separate and freely diffusible enzyme. Intermediates are also diffusible and may be diverted into other pathways (such as lipoic acid synthesis). Unlike FAS I, FAS II generates a variety of products, including saturated fatty acids of several lengths, as well as unsaturated, branched, and hydroxy fatty acids. An FAS II system is also found in vertebrate mitochondria. The discussion to follow will focus on the mammalian FAS I.

Fatty Acid Synthesis Proceeds in a Repeating Reaction Sequence

In all organisms, the long carbon chains of fatty acids are assembled in a repeating four-step sequence **(Fig. 21–2)**, catalyzed by a system collectively referred to as **fatty acid synthase**. A saturated acyl group produced by each four-step series of reactions becomes the substrate for subsequent condensation with an activated malonyl group. With each passage through the cycle, the fatty acyl chain is extended by two carbons.

The Mammalian Fatty Acid Synthase Has Multiple Active Sites

The multiple domains of mammalian FAS I function as distinct but linked enzymes. The active site for each

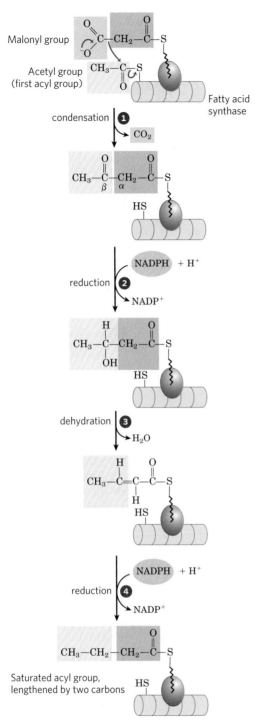

FIGURE 21–2 Addition of two carbons to a growing fatty acyl chain: a four-step sequence. Each malonyl group and acetyl (or longer acyl) group is activated by a thioester that links it to fatty acid synthase, a multienzyme system described later in the text. ❶ Condensation of an activated acyl group (an acetyl group from acetyl-CoA is the first acyl group) and two carbons derived from malonyl-CoA, with elimination of CO_2 from the malonyl group, extends the acyl chain by two carbons. The mechanism of the first step of this reaction is given to illustrate the role of decarboxylation in facilitating condensation. The β-keto product of this condensation is then reduced in three more steps nearly identical to the reactions of β oxidation, but in the reverse sequence: ❷ the β-keto group is reduced to an alcohol, ❸ elimination of H_2O creates a double bond, and ❹ the double bond is reduced to form the corresponding saturated fatty acyl group.

FIGURE 21–3 The structure of fatty acid synthase type I systems. Shown here are low-resolution structures of (a) the mammalian (porcine; dimer, derived from PDB ID 2CF2) and (b) fungal (derived from PDB IDs 2UV9, 2UVA, 2UVB, and 2UVC) enzyme systems. **(a)** All of the active sites in the mammalian system are located in different domains within a single large polypeptide chain. The different enzymatic activites are: β-ketoacyl-ACP synthase (KS), malonyl/acetyl-CoA–ACP transferase (MAT), β-hydroxyacyl-ACP dehydratase (DH), enoyl-ACP reductase (ER), and β-ketoacyl-ACP reductase (KR). ACP is the acyl carrier protein. The linear arrangement of the domains in the polypeptide is shown in the lower panel. The seventh domain (TE) is a thioesterase that releases the palmitate product from ACP when the synthesis is completed. The ACP and TE domains are disordered in the crystal and are therefore not shown in the structure. **(b)** In the structure of the FAS I from the fungus *Thermomyces lanuginosus*, the same active sites are divided between two multifunctional polypeptide chains that function together. Six copies of each polypeptide are found in the heterododecameric complex. A wheel of six α subunits, which include ACP as well as the KS and KR active sites, is found at the center of the complex. In the wheel, three subunits are found on one face, three on the other. On either side of the wheel are domes formed by trimers of the β subunits (containing the ER and DH active sites, as well as two domains with active sites analogous to MAT in the mammalian enzyme). The domains of one of each type of subunit are colored according to the active site colors of the mammalian enzyme in (a).

FIGURE 21–4 The overall process of palmitate synthesis. The fatty acyl chain grows by two-carbon units donated by activated malonate, with loss of CO_2 at each step. The initial acetyl group is shaded yellow, C-1 and C-2 of malonate are shaded light red, and the carbon released as CO_2 is shaded green. After each two-carbon addition, reductions convert the growing chain to a saturated fatty acid of four, then six, then eight carbons, and so on. The final product is palmitate (16:0).

enzyme is found in a separate domain within the larger polypeptide. Throughout the process of fatty acid synthesis, the intermediates remain covalently attached as thioesters to one of two thiol groups. One point of attachment is the —SH group of a Cys residue in one of the synthase domains (β-ketoacyl-ACP synthase; KS); the other is the —SH group of acyl carrier protein, a separate domain of the same polypeptide. Hydrolysis of thioesters is highly exergonic, and the energy released helps to make two different steps (❶ and ❺ in Fig. 21–6) in fatty acid synthesis (condensation) thermodynamically favorable.

Acyl carrier protein (ACP) is the shuttle that holds the system together. The *Escherichia coli* ACP is a small protein (M_r 8,860) containing the prosthetic group **4'-phosphopantetheine** (**Fig. 21–5**; compare this with the panthothenic acid and β-mercaptoethylamine moiety of coenzyme A in Fig. 8–38). The 4'-phosphopantetheine prosthetic group of *E. coli* ACP is believed to serve as a flexible arm, tethering the growing fatty acyl chain to the surface of the fatty acid synthase complex while carrying the reaction intermediates from one enzyme active site to the next. The ACP of mammals has a similar function and the same prosthetic group; as we have seen, however, it is embedded as a domain in a much larger multifunctional polypeptide.

FIGURE 21–5 Acyl carrier protein (ACP). The prosthetic group is 4'-phosphopantetheine, which is covalently attached to the hydroxyl group of a Ser residue in ACP. Phosphopantetheine contains the B vitamin pantothenic acid, also found in the coenzyme A molecule. Its —SH group is the site of entry of malonyl groups during fatty acid synthesis.

Fatty Acid Synthase Receives the Acetyl and Malonyl Groups

Before the condensation reactions that build up the fatty acid chain can begin, the two thiol groups on the enzyme complex must be charged with the correct acyl groups (**Fig. 21–6**, top). First, the acetyl group of acetyl-CoA is

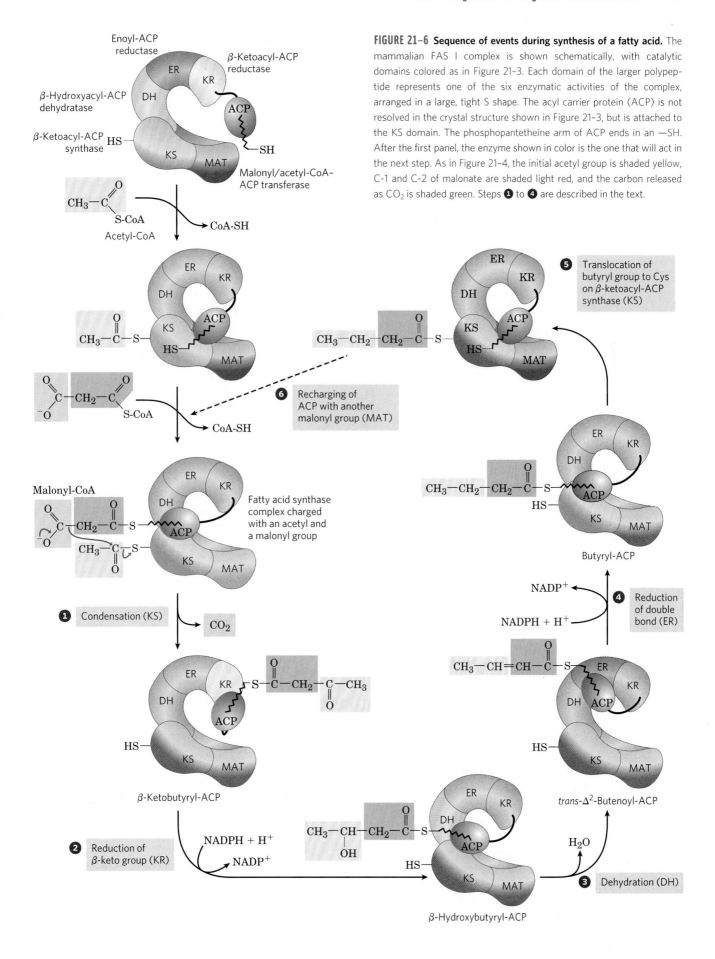

FIGURE 21-6 Sequence of events during synthesis of a fatty acid. The mammalian FAS I complex is shown schematically, with catalytic domains colored as in Figure 21–3. Each domain of the larger polypeptide represents one of the six enzymatic activities of the complex, arranged in a large, tight S shape. The acyl carrier protein (ACP) is not resolved in the crystal structure shown in Figure 21–3, but is attached to the KS domain. The phosphopantetheine arm of ACP ends in an —SH. After the first panel, the enzyme shown in color is the one that will act in the next step. As in Figure 21–4, the initial acetyl group is shaded yellow, C-1 and C-2 of malonate are shaded light red, and the carbon released as CO_2 is shaded green. Steps ❶ to ❹ are described in the text.

transferred to ACP in a reaction catalyzed by the **malonyl/acetyl-CoA–ACP transferase** (MAT in Fig. 21–6) domain of the multifunctional polypeptide. The acetyl group is then transferred to the Cys —SH group of the **β-ketoacyl-ACP synthase** (KS). The second reaction, transfer of the malonyl group from malonyl-CoA to the —SH group of ACP, is also catalyzed by malonyl/acetyl-CoA–ACP transferase. In the charged synthase complex, the acetyl and malonyl groups are activated for the chain-lengthening process. The first four steps of this process are now considered in some detail; all step numbers refer to Figure 21–6.

Step ❶ Condensation The first reaction in the formation of a fatty acid chain is a formal Claisen condensation involving the activated acetyl and malonyl groups to form **acetoacetyl-ACP**, an acetoacetyl group bound to ACP through the phosphopantetheine —SH group; simultaneously, a molecule of CO_2 is produced. In this reaction, catalyzed by β-ketoacyl-ACP synthase, the acetyl group is transferred from the Cys —SH group of the enzyme to the malonyl group on the —SH of ACP, becoming the methyl-terminal two-carbon unit of the new acetoacetyl group.

The carbon atom of the CO_2 formed in this reaction is the same carbon originally introduced into malonyl CoA from HCO_3^- by the acetyl-CoA carboxylase reaction (Fig. 21–1). Thus CO_2 is only transiently in covalent linkage during fatty acid biosynthesis; it is removed as each two-carbon unit is added.

Why do cells go to the trouble of adding CO_2 to make a malonyl group from an acetyl group, only to lose the CO_2 during the formation of acetoacetate? The use of activated malonyl groups rather than acetyl groups is what makes the condensation reactions thermodynamically favorable. The methylene carbon (C-2) of the malonyl group, sandwiched between carbonyl and carboxyl carbons, forms a good nucleophile. In the condensation step (step ❶), decarboxylation of the malonyl group facilitates the nucleophilic attack of the methylene carbon on the thioester linking the acetyl group to β-ketoacyl-ACP synthase, displacing the enzyme's —SH group. (This is a classic Claisen ester condensation; see Fig. 13–4.) Coupling the condensation to the decarboxylation of the malonyl group renders the overall process highly exergonic. A similar carboxylation-decarboxylation sequence facilitates the formation of phosphoenolpyruvate from pyruvate in gluconeogenesis (see Fig. 14–18).

By using activated malonyl groups in the synthesis of fatty acids and activated acetate in their degradation, the cell makes both processes energetically favorable, although one is effectively the reversal of the other. The extra energy required to make fatty acid synthesis favorable is provided by the ATP used to synthesize malonyl-CoA from acetyl-CoA and HCO_3^- (Fig. 21–1).

Step ❷ Reduction of the Carbonyl Group The acetoacetyl-ACP formed in the condensation step now undergoes reduction of the carbonyl group at C-3 to form D-β-hydroxybutyryl-ACP. This reaction is catalyzed by **β-ketoacyl-ACP reductase** (KR) and the electron donor is NADPH. Notice that the D-β-hydroxybutyryl group does not have the same stereoisomeric form as the L-β-hydroxyacyl intermediate in fatty acid oxidation (see Fig. 17–8).

Step ❸ Dehydration The elements of water are now removed from C-2 and C-3 of D-β-hydroxybutyryl-ACP to yield a double bond in the product, ***trans*-Δ²-butenoyl-ACP**. The enzyme that catalyzes this dehydration is **β-hydroxyacyl-ACP dehydratase** (DH).

Step ❹ Reduction of the Double Bond Finally, the double bond of *trans*-Δ²-butenoyl-ACP is reduced (saturated) to form **butyryl-ACP** by the action of **enoyl-ACP reductase** (ER); again, NADPH is the electron donor.

The Fatty Acid Synthase Reactions Are Repeated to Form Palmitate

Production of the four-carbon, saturated fatty acyl–ACP marks completion of one pass through the fatty acid synthase complex. In step ❺, the butyryl group is transferred from the phosphopantetheine —SH group of ACP to the Cys —SH group of β-ketoacyl-ACP synthase, which initially bore the acetyl group (Fig. 21–6). To start the next cycle of four reactions that lengthens the chain by two more carbons (step ❻), another malonyl group is linked to the now unoccupied phosphopantetheine —SH group of ACP **(Fig. 21–7)**. Condensation occurs as the butyryl group, acting like the acetyl group in the first cycle, is linked to two carbons of the malonyl-ACP group with concurrent loss of CO_2. The product of this condensation is a six-carbon acyl group, covalently bound to the phosphopantetheine —SH group. Its β-keto group is reduced in the next three steps of the synthase cycle to yield the saturated acyl group, exactly as in the first round of reactions—in this case forming the six-carbon product.

Seven cycles of condensation and reduction produce the 16-carbon saturated palmitoyl group, still bound to ACP. For reasons not well understood, chain elongation by the synthase complex generally stops at this point and free palmitate is released from the ACP by a hydrolytic activity (thioesterase; TE) in the multifunctional protein.

We can consider the overall reaction for the synthesis of palmitate from acetyl-CoA in two parts. First, the formation of seven malonyl-CoA molecules:

$$7 \text{ Acetyl-CoA} + 7CO_2 + 7ATP \longrightarrow$$
$$7 \text{ malonyl-CoA} + 7ADP + 7P_i \quad (21\text{–}1)$$

then seven cycles of condensation and reduction:

$$\text{Acetyl-CoA} + 7 \text{ malonyl-CoA} + 14NADPH + 14H^+ \longrightarrow$$
$$\text{palmitate} + 7CO_2 + 8 \text{ CoA} + 14NADP^+ + 6H_2O$$
$$(21\text{–}2)$$

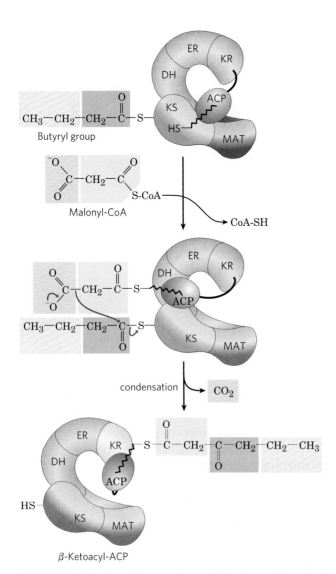

FIGURE 21–7 Beginning of the second round of the fatty acid synthesis cycle. The butyryl group is on the Cys —SH group. The incoming malonyl group is first attached to the phosphopantetheine —SH group. Then, in the condensation step, the entire butyryl group on the Cys —SH is exchanged for the carboxyl group of the malonyl residue, which is lost as CO_2 (green). This step is analogous to step ❶ in Figure 21–6. The product, a six-carbon β-ketoacyl group, now contains four carbons derived from malonyl-CoA and two derived from the acetyl-CoA that started the reaction. The β-ketoacyl group then undergoes steps ❷ through ❹, as in Figure 21-6.

Note that only six net water molecules are produced, because one is used to hydrolyze the thioester linking the palmitate product to the enzyme. The overall process (the sum of Eqns 21–1 and 21–2) is

$$8 \text{ Acetyl-CoA} + 7 \text{ ATP} + 14\text{NADPH} + 14\text{H}^+ \longrightarrow$$
$$\text{palmitate} + 8 \text{ CoA} + 7\text{ADP} + 7\text{P}_i + 14\text{NADP}^+ + 6\text{H}_2\text{O}$$
$$(21\text{–}3)$$

The biosynthesis of fatty acids such as palmitate thus requires acetyl-CoA and the input of chemical energy in two forms: the group transfer potential of ATP and the reducing power of NADPH. The ATP is required to attach CO_2 to acetyl-CoA to make malonyl-CoA; the NADPH molecules are required to reduce the β-keto group and the double bond.

In nonphotosynthetic eukaryotes there is an additional cost to fatty acid synthesis, because acetyl-CoA is generated in the mitochondria and must be transported to the cytosol. As we will see, this extra step consumes two ATP per molecule of acetyl-CoA transported, increasing the energetic cost of fatty acid synthesis to three ATP per two-carbon unit.

Fatty Acid Synthesis Occurs in the Cytosol of Many Organisms but in the Chloroplasts of Plants

In most higher eukaryotes the fatty acid synthase complex is found exclusively in the cytosol **(Fig. 21–8)**, as are the biosynthetic enzymes for nucleotides, amino acids, and glucose. This location segregates synthetic processes from degradative reactions, many of which take place in the mitochondrial matrix. There is a corresponding segregation of the electron-carrying cofactors used in anabolism (generally a reductive process) and those used in catabolism (generally oxidative).

Usually, NADPH is the electron carrier for anabolic reactions, and NAD^+ serves in catabolic reactions. In hepatocytes, the $[\text{NADPH}]/[\text{NADP}^+]$ ratio is very high (about 75) in the cytosol, furnishing a strongly reducing environment for the reductive synthesis of fatty acids and other biomolecules. The cytosolic $[\text{NADH}]/[\text{NAD}^+]$ ratio is much smaller (only about 8×10^{-4}), so the NAD^+-dependent oxidative catabolism of glucose can take place in the same compartment, and at the same time, as fatty acid synthesis. The $[\text{NADH}]/[\text{NAD}^+]$ ratio in the mitochondrion is much higher than in the cytosol, because of the flow of electrons to NAD^+ from the oxidation of fatty acids, amino acids, pyruvate, and acetyl-CoA. This high mitochondrial $[\text{NADH}]/[\text{NAD}^+]$ ratio favors the reduction of oxygen via the respiratory chain.

In hepatocytes and adipocytes, cytosolic NADPH is largely generated by the pentose phosphate pathway (see Fig. 14–22) and by **malic enzyme (Fig. 21–9a)**. The NADP-linked malic enzyme that operates in the carbon-assimilation pathway of C_4 plants (see Fig. 20–23) is unrelated in function. The pyruvate produced in the reaction shown in Figure 21–9a reenters the mitochondrion. In hepatocytes and in the mammary gland of lactating animals, the NADPH required for fatty acid biosynthesis is supplied primarily by the pentose phosphate pathway (Fig. 21–9b).

In the photosynthetic cells of plants, fatty acid synthesis occurs not in the cytosol but in the chloroplast stroma (Fig. 21–8). This makes sense, given that NADPH is produced in chloroplasts by the light-dependent reactions of photosynthesis:

$$\text{H}_2\text{O} + \text{NADP}^+ \xrightarrow{\text{light}} \tfrac{1}{2}\text{O}_2 + \text{NADPH} + \text{H}^+$$

Animal cells, yeast cells

Plant cells

Mitochondria
- No fatty acid oxidation

- Fatty acid oxidation
- Acetyl-CoA production
- Ketone body synthesis
- Fatty acid elongation

Endoplasmic reticulum
- Phospholipid synthesis
- Sterol synthesis (late stages)
- Fatty acid elongation
- Fatty acid desaturation

Cytosol
- NADPH production
  (pentose phosphate pathway; malic enzyme)
- [NADPH]/[NADP$^+$] high
- Isoprenoid and sterol synthesis (early stages)
- Fatty acid synthesis

Chloroplasts
- NADPH, ATP production
- [NADPH]/[NADP$^+$] high
- Fatty acid synthesis

Peroxisomes
- Fatty acid oxidation
  ($\longrightarrow$ H$_2$O$_2$)
- Catalase, peroxidase:
  H$_2$O$_2$ $\longrightarrow$ H$_2$O

FIGURE 21–8 Subcellular localization of lipid metabolism. Yeast and vertebrate cells differ from higher plant cells in the compartmentation of lipid metabolism. Fatty acid synthesis takes place in the compartment in which NADPH is available for reductive synthesis (i.e., where the [NADPH]/[NADP$^+$] ratio is high); this is the cytosol in animals and yeast, and the chloroplast in plants. Processes in red type are covered in this chapter.

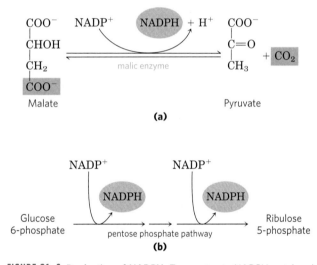

(a)

Malate → Pyruvate + CO$_2$ (via malic enzyme, NADP$^+$ → NADPH + H$^+$)

(b)

Glucose 6-phosphate → (pentose phosphate pathway) → Ribulose 5-phosphate (NADP$^+$ → NADPH, two steps)

FIGURE 21–9 Production of NADPH. Two routes to NADPH, catalyzed by **(a)** malic enzyme and **(b)** the pentose phosphate pathway.

Acetate Is Shuttled out of Mitochondria as Citrate

In nonphotosynthetic eukaryotes, nearly all the acetyl-CoA used in fatty acid synthesis is formed in mitochondria from pyruvate oxidation and from the catabolism of the carbon skeletons of amino acids. Acetyl-CoA arising from the oxidation of fatty acids is not a significant source of acetyl-CoA for fatty acid biosynthesis in animals, because the two pathways are reciprocally regulated, as described below.

The mitochondrial inner membrane is impermeable to acetyl-CoA, so an indirect shuttle transfers acetyl group equivalents across the inner membrane **(Fig. 21–10)**. Intramitochondrial acetyl-CoA first reacts with oxaloacetate to form citrate, in the citric acid cycle reaction catalyzed by **citrate synthase** (see Fig. 16–7). Citrate then passes through the inner membrane on the **citrate transporter**. In the cytosol, citrate cleavage by **citrate lyase** regenerates acetyl-CoA and oxaloacetate in an ATP-dependent reaction. Oxaloacetate cannot return to the mitochondrial matrix directly, as there is no oxaloacetate transporter. Instead, cytosolic malate dehydrogenase reduces the oxaloacetate to malate, which can return to the mitochondrial matrix on the malate–α-ketoglutarate transporter in exchange for citrate. In the matrix, malate is reoxidized to oxaloacetate to complete the shuttle. However, most of the malate produced in the cytosol is used to generate cytosolic NADPH through the activity of malic enzyme (Fig. 21–9a). The pyruvate produced is transported to the mitochondria by the pyruvate transporter (Fig. 21–10), and converted back into oxaloacetate by pyruvate carboxylase in the matrix. The resulting cycle results in the consumption of two ATP (by citrate lyase and pyruvate carboxylase) for every molecule of acetyl-CoA delivered to fatty acid synthesis. After citrate cleavage to generate acetyl-CoA, conversion of the four remaining carbons to pyruvate and CO$_2$ via malic enzyme generates about half the NADPH required for fatty acid synthesis. The pentose phosphate pathway contributes the rest of the needed NADPH.

Fatty Acid Biosynthesis Is Tightly Regulated

When a cell or organism has more than enough metabolic fuel to meet its energy needs, the excess is generally converted to fatty acids and stored as lipids such as triacylglycerols. The reaction catalyzed by acetyl-CoA

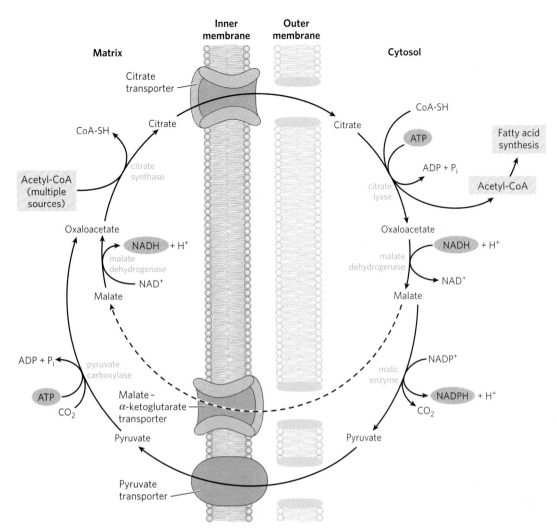

FIGURE 21–10 Shuttle for transfer of acetyl groups from mitochondria to the cytosol. The outer mitochondrial membrane is freely permeable to all these compounds. Pyruvate derived from amino acid catabolism in the mitochondrial matrix, or from glucose by glycolysis in the cytosol, is converted to acetyl-CoA in the matrix. Acetyl groups pass out of the mitochondrion as citrate; in the cytosol they are delivered as acetyl-CoA for fatty acid synthesis. Oxaloacetate is reduced to malate, which can return to the mitochondrial matrix and is converted to oxaloacetate. The major fate for cytosolic malate is oxidation by malic enzyme to generate cytosolic NADPH; the pyruvate produced returns to the mitochondrial matrix.

carboxylase is the rate-limiting step in the biosynthesis of fatty acids, and this enzyme is an important site of regulation. In vertebrates, palmitoyl-CoA, the principal product of fatty acid synthesis, is a feedback inhibitor of the enzyme; citrate is an allosteric activator **(Fig. 21–11a)**, increasing V_{max}. Citrate plays a central role in diverting cellular metabolism from the consumption (oxidation) of metabolic fuel to the storage of fuel as fatty acids. When the concentrations of mitochondrial acetyl-CoA and ATP increase, citrate is transported out of mitochondria; it then becomes both the precursor of cytosolic acetyl-CoA and an allosteric signal for the activation of acetyl-CoA carboxylase. At the same time, citrate inhibits the activity of phosphofructokinase-1 (see Fig. 15–16), reducing the flow of carbon through glycolysis.

Acetyl-CoA carboxylase is also regulated by covalent modification. Phosphorylation, triggered by the hormones glucagon and epinephrine, inactivates the enzyme and reduces its sensitivity to activation by citrate, thereby slowing fatty acid synthesis. In its active (dephosphorylated) form, acetyl-CoA carboxylase polymerizes into long filaments (Fig. 21–11b); phosphorylation is accompanied by dissociation into monomeric subunits and loss of activity.

The acetyl-CoA carboxylase of plants and bacteria is not regulated by citrate or by a phosphorylation-dephosphorylation cycle. The plant enzyme is activated by an increase in stromal pH and $[Mg^{2+}]$, which occurs on illumination of the plant (see Fig. 20–17). Bacteria do not use triacylglycerols as energy stores. In *E. coli,* the primary role of fatty acid synthesis is to provide precursors for membrane lipids; the regulation of this process is complex, involving guanine nucleotides (such as ppGpp) that coordinate cell growth with membrane formation (see Figs 8–39, 28–22).

(a)   **(b)**

FIGURE 21–11 Regulation of fatty acid synthesis. (a) In the cells of vertebrates, both allosteric regulation and hormone-dependent covalent modification influence the flow of precursors into malonyl-CoA. In plants, acetyl-CoA carboxylase is activated by the changes in [Mg^{2+}] and pH that accompany illumination (not shown here). **(b)** Filaments of acetyl-CoA carboxylase from chicken hepatocytes (the active, dephosphorylated form) as seen with the electron microscope.

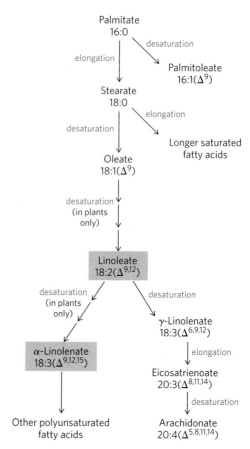

FIGURE 21–12 Routes of synthesis of other fatty acids. Palmitate is the precursor of stearate and longer-chain saturated fatty acids, as well as the monounsaturated acids palmitoleate and oleate. Mammals cannot convert oleate to linoleate or α-linolenate (shaded light red), which are therefore required in the diet as essential fatty acids. Conversion of linoleate to other polyunsaturated fatty acids and eicosanoids is outlined. Unsaturated fatty acids are symbolized by indicating the number of carbons and the number and position of the double bonds, as in Table 10–1.

In addition to the moment-by-moment regulation of enzymatic activity, these pathways are regulated at the level of gene expression. For example, when animals ingest an excess of certain polyunsaturated fatty acids, the expression of genes encoding a wide range of lipogenic enzymes in the liver is suppressed. This gene regulation is mediated by a family of nuclear receptor proteins called PPARs, which are described in more detail in Chapter 23 (see Fig. 23–42).

If fatty acid synthesis and β oxidation were to proceed simultaneously, the two processes would constitute a futile cycle, wasting energy. We noted earlier (see Fig. 17–13) that β oxidation is blocked by malonyl-CoA, which inhibits carnitine acyltransferase I. Thus during fatty acid synthesis, the production of the first intermediate, malonyl-CoA, shuts down β oxidation at the level of a transport system in the mitochondrial inner membrane. This control mechanism illustrates another advantage of segregating synthetic and degradative pathways in different cellular compartments.

Long-Chain Saturated Fatty Acids Are Synthesized from Palmitate

Palmitate, the principal product of the fatty acid synthase system in animal cells, is the precursor of other long-chain fatty acids **(Fig. 21–12)**. It may be length-ened to form stearate (18:0) or even longer saturated fatty acids by further additions of acetyl groups, through the action of **fatty acid elongation systems** present in the smooth endoplasmic reticulum and in mitochondria. The more active elongation system of the ER extends the 16-carbon chain of palmitoyl-CoA by two carbons, forming stearoyl-CoA. Although different enzyme systems are involved, and coenzyme A rather than ACP is the acyl carrier in the reaction, the mechanism of elongation in the ER is otherwise identical to that in palmitate synthesis: donation of two carbons by malonyl-CoA, followed by reduction, dehydration, and reduction to the saturated 18-carbon product, stearoyl-CoA.

Desaturation of Fatty Acids Requires a Mixed-Function Oxidase

Palmitate and stearate serve as precursors of the two most common monounsaturated fatty acids of animal tissues: palmitoleate, 16:1(Δ^9), and oleate, 18:1(Δ^9); both of these fatty acids have a single cis double bond

$$O_2 + 2H^+ +$$

$$CH_3-(CH_2)_n-CH_2-CH_2-(CH_2)_m-C\bigg\langle{\!\!\!}^O_{\text{S-CoA}}$$

Saturated
fatty acyl–CoA

$$2H_2O +$$

$$CH_3-(CH_2)_n-CH=CH-(CH_2)_m-C\bigg\langle{\!\!\!}^O_{\text{S-CoA}}$$

Monounsaturated
fatty acyl–CoA

fatty acyl–
CoA desaturase

2 Cyt b_5 (Fe^{2+}) → Cyt b_5 reductase (FAD) → **NADPH** + H$^+$

2 Cyt b_5 (Fe^{3+}) → Cyt b_5 reductase (FADH$_2$) → NADP$^+$

FIGURE 21–13 Electron transfer in the desaturation of fatty acids in vertebrates. Blue arrows show the path of electrons as two substrates— a fatty acyl–CoA and NADPH—undergo oxidation by molecular oxygen. These reactions take place on the lumenal face of the smooth ER. A similar pathway, but with different electron carriers, occurs in plants.

between C-9 and C-10 (see Table 10–1). The double bond is introduced into the fatty acid chain by an oxidative reaction catalyzed by **fatty acyl–CoA desaturase (Fig. 21–13)**, a **mixed-function oxidase** (Box 21–1). Two different substrates, the fatty acid and NADPH, simultaneously undergo two-electron oxidations. The path of electron flow includes a cytochrome (cytochrome b_5) and a flavoprotein (cytochrome b_5 reductase), both of which, like fatty acyl–CoA desaturase, are in the smooth ER. In plants, oleate ($18:1(\Delta^9)$) is produced by a **stearoyl-ACP desaturase (SCD)** that uses reduced ferredoxin as the electron donor in the chloroplast stroma.

The SCD of animals (mice) has an important role in the development of obesity and the insulin resistance that often accompanies obesity and precedes development of type 2 diabetes mellitus. Mice have four isozymes, SCD1 through SCD4, of which SCD1 is the best understood. Its synthesis is induced by dietary saturated fatty acids, and also by the action of SREBP and LXR, two protein regulators of lipid metabolism that activate transcription of lipid-synthesizing enzymes (described in Section 21.4). Mice with mutant forms of SCD1 are resistant to diet-induced obesity, and do not develop diabetes under conditions that cause both obesity and diabetes in mice with normal SCD1.

Mammalian hepatocytes can readily introduce double bonds at the Δ^9 position of fatty acids but cannot introduce additional double bonds between C-10 and the methyl-terminal end. Thus mammals cannot synthesize linoleate, $18:2(\Delta^{9,12})$, or α-linolenate, $18:3(\Delta^{9,12,15})$. Plants, however, can synthesize both; the desaturases that introduce double bonds at the Δ^{12} and Δ^{15} positions are located in the ER and the chloroplast. The ER enzymes act not on free fatty acids but on a phospholipid, phosphatidylcholine, that contains at least one oleate linked to the glycerol **(Fig. 21–14)**. Both plants and bacteria must synthesize polyunsaturated fatty acids to ensure membrane fluidity at reduced temperatures.

FIGURE 21–14 Action of plant desaturases. Desaturases in plants oxidize phosphatidylcholine-bound oleate to polyunsaturated fatty acids. Some of the products are released from the phosphatidylcholine by hydrolysis.

In this chapter we encounter several enzymes that carry out oxidation-reduction reactions in which molecular oxygen is a participant. The stearoyl-CoA desaturase (SCD) that introduces a double bond into a fatty acyl chain (see Fig. 21–13) is one such enzyme.

The nomenclature for enzymes that catalyze reactions of this general type is often confusing to students. **Oxidase** is the general name for enzymes that catalyze oxidations in which molecular oxygen is the electron acceptor but oxygen atoms do not appear in the oxidized product. The enzyme that creates a double bond in fatty acyl–CoA during the oxidation of fatty acids in peroxisomes (see Fig. 17–14) is an oxidase of this type; a second example is the cytochrome oxidase of the mitochondrial electron-transfer chain (see Fig. 19–14). In the first case, the transfer of two electrons to H_2O produces hydrogen peroxide, H_2O_2; in the second, two electrons reduce $\frac{1}{2}O_2$ to H_2O. Many, but not all, oxidases are flavoproteins.

Oxygenases catalyze oxidative reactions in which oxygen atoms *are* directly incorporated into the product molecule, forming a new hydroxyl or carboxyl group, for example. **Dioxygenases** catalyze reactions in which both oxygen atoms of O_2 are incorporated into the organic product. An example of a dioxygenase is tryptophan 2,3-dioxygenase, which catalyzes the opening of the five-membered ring of tryptophan in the catabolism of this amino acid. When this reaction takes place in the presence of $^{18}O_2$, the isotopic oxygen atoms are found in the two carbonyl groups of the product (shown in red):

Tryptophan

O_2 → tryptophan 2,3-dioxygenase

N-Formylkynurenine

Monooxygenases, more common and more complex in their action, catalyze reactions in which only one of the two oxygen atoms of O_2 is incorporated into the organic product, the other being reduced to H_2O. Monooxygenases require two substrates to serve as

reductants of the two oxygen atoms of O_2. The main substrate accepts one of the two oxygen atoms, and a cosubstrate furnishes hydrogen atoms to reduce the other oxygen atom to H_2O. The general reaction equation for monooxygenases is

$$AH + BH_2 + O{-}O \longrightarrow A{-}OH + B + H_2O$$

where AH is the main substrate and BH_2 the cosubstrate. Because most monooxygenases catalyze reactions in which the main substrate becomes hydroxylated, they are also called **hydroxylases**. They are also sometimes called **mixed-function oxidases** or **mixed-function oxygenases**, to indicate that they oxidize two different substrates simultaneously. (Note here the use of "oxidase," which violates the general definition given above).

There are different classes of monooxygenases, depending on the nature of the cosubstrate. Some use reduced flavin nucleotides ($FMNH_2$ or $FADH_2$), others use NADH or NADPH, and still others use α-ketoglutarate as the cosubstrate. The enzyme that hydroxylates the phenyl ring of phenylalanine to form tyrosine is a monooxygenase for which tetrahydrobiopterin serves as cosubstrate (see Fig. 18–23). (This is the enzyme that is defective in the human genetic disease phenylketonuria.)

The most numerous and most complex monooxygenation reactions are those employing a type of heme protein called **cytochrome P-450**. Like mitochondrial cytochrome oxidase, enzymes containing a cytochrome P-450 domain can react with O_2 and bind carbon monoxide, but they can be differentiated from cytochrome oxidase because the carbon monoxide complex of their reduced form absorbs light strongly at 450 nm—thus the name P-450.

Cytochrome P-450 enzymes catalyze hydroxylation reactions in which an organic substrate, RH, is hydroxylated to R—OH, incorporating one oxygen atom of O_2; the other oxygen atom is reduced to H_2O by reducing equivalents that are furnished by NADH or NADPH but are usually passed to cytochrome P-450 by an iron-sulfur protein. Figure 1 shows a simplified outline of the action of cytochrome P-450.

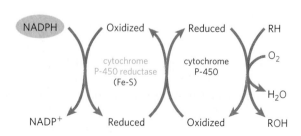

FIGURE 1

There is a large family of P-450–containing proteins of two general types: those highly specific for a single substrate (like typical enzymes) and those with more promiscuous binding sites that accept a variety of substrates, generally similar in being hydrophobic. In the adrenal cortex, for example, a specific cytochrome P-450 participates in the hydroxylation of steroids to yield the adrenocortical hormones (see Fig. 21–49). There are dozens of P-450 enzymes that act on very specific substrates in the biosynthetic pathways to steroid hormones and eicosanoids. Cytochrome P-450 enzymes with broader specificity are important in the hydroxylation of many different drugs, such as barbiturates and other xenobiotics (substances foreign to the organism), particularly if they are hydrophobic and relatively insoluble. The environmental carcinogen benzo[a]pyrene (found in cigarette smoke) undergoes cytochrome P-450–dependent hydroxylation during detoxification. Hydroxylation of xenobiotics, sometimes combined with the attachment of a polar compound such as glucuronic acid to the hydroxyl group, makes them more soluble in water and allows their excretion in the urine. Hydroxylation (and glucuronidation) inactivates most drugs, and the rate at which it occurs can determine how long a given dose of a drug remains in the blood at therapeutic levels. Humans differ in their levels of drug-metabolizing enzymes, both because of their genetics and because past exposure to substrates can induce the synthesis of higher levels of P-450 enzymes in a given individual. Both ethanol and barbiturate drugs share a P-450 enzyme. Long-term heavy drinking induces synthesis of that P-450 enzyme. Then, because the barbiturate is inactivated and cleared faster, larger doses of the drug are required to get the same therapeutic effect. If an individual takes this larger-than-usual dose of barbiturate and then also drinks alcohol, competition between the alcohol and the barbiturate for the limited amount of enzyme means that both alcohol and barbiturate are cleared more slowly. The resulting high levels of these two central nervous system depressants can be lethal. Similar complications arise when an individual takes two drugs that happen to be inactivated by the same P-450 enzyme; each drug increases the effective dose of the other by slowing its inactivation. It is therefore essential for physicians and pharmacists to know about all of a patient's prescribed and over-the-counter drugs and supplements (as well as a history of heavy drinking, or smoking, or exposure to environmental toxins).

Reactions described in this chapter that are catalyzed by mixed-function oxidases are those involved in fatty acyl–CoA desaturation (Fig. 21–13), leukotriene synthesis (Fig. 21–16), plasmalogen synthesis (Fig. 21–30), conversion of squalene to cholesterol (Fig. 21–37), and steroid hormone synthesis (Fig. 21–49).

Because they are necessary precursors for the synthesis of other products, linoleate and α-linolenate are **essential fatty acids** for mammals; they must be obtained from dietary plant material. Once ingested, linoleate may be converted to certain other polyunsaturated acids, particularly γ-linolenate, eicosatrienoate, and arachidonate (eicosatetraenoate), all of which can be made only from linoleate (Fig. 21–12). Arachidonate, $20:4(\Delta^{5,8,11,14})$, is an essential precursor of regulatory lipids, the eicosanoids. The 20-carbon fatty acids are synthesized from linoleate (and α-linolenate) by fatty acid elongation reactions analogous to those described on page 842.

Eicosanoids Are Formed from 20-Carbon Polyunsaturated Fatty Acids

Eicosanoids are a family of very potent biological signaling molecules that act as short-range messengers, affecting tissues near the cells that produce them. In response to hormonal or other stimuli, phospholipase A_2, present in most types of mammalian cells, attacks membrane phospholipids, releasing arachidonate from the middle carbon of glycerol. Enzymes of the smooth ER then convert arachidonate to **prostaglandins**, beginning with the formation of prostaglandin H_2 (PGH$_2$), the immediate precursor of many other prostaglandins and of thromboxanes **(Fig. 21–15a)**. The two reactions that lead to PGH$_2$ are catalyzed by a bifunctional enzyme, **cyclooxygenase (COX)**, also called **prostaglandin H_2 synthase**. In the first of two steps, the cyclooxygenase activity introduces molecular oxygen to convert arachidonate to PGG$_2$. The second step, catalyzed by the peroxidase activity of COX, converts PGG$_2$ to PGH$_2$.

Mammals have two isozymes of prostaglandin H_2 synthase, COX-1 and COX-2. These have different functions but closely similar amino acid sequences (60% to 65% sequence identity) and similar reaction mechanisms at both of their catalytic centers. COX-1 is responsible for the synthesis of the prostaglandins that regulate the secretion of gastric mucin, and COX-2 for the prostaglandins that mediate inflammation, pain, and fever.

Pain can be relieved by inhibiting COX-2. The first drug widely marketed for this purpose was aspirin (acetylsalicylate; Fig. 21–15b). The name aspirin (from *a* for acetyl and *spir* for *Spirsaüre*, the German

FIGURE 21–15 The "cyclic" pathway from arachidonate to prostaglandins and thromboxanes. (a) After arachidonate is released from phospholipids by the action of phospholipase A_2, the cyclooxygenase and peroxidase activities of COX (also called prostaglandin H_2 synthase) catalyze the production of PGH_2, the precursor of other prostaglandins and thromboxanes. **(b)** Aspirin inhibits the first reaction by acetylating an essential Ser residue on the enzyme. Ibuprofen and naproxen inhibit the same step, probably by mimicking the structure of the substrate or an intermediate in the reaction. **(c)** COX-2–specific cyclooxygenase inhibitors developed as pain relievers (see text). Vioxx was withdrawn from the market in 2004, and Bextra in 2005, because of their side effects on the cardiovascular system.

word for the salicylates prepared from the plant *Spiraea ulmaria*) appeared in 1899 when the drug was introduced by the Bayer company. Aspirin irreversibly inactivates the cyclooxygenase activity of both COX isozymes, by acetylating a Ser residue and blocking each enzyme's active site. The synthesis of prostaglandins and thromboxanes is thereby inhibited. Ibuprofen, another widely used *nonsteroidal antiinflammatory drug* (NSAID; Fig. 21–15b), inhibits the same pair of enzymes. However, the inhibition of COX-1 can result in undesired side effects including stomach irritation and more serious conditions. In the 1990s, following discovery of the crystal structures of COX-1 and COX-2, NSAID compounds that had a greater specificity for COX-2 were developed as advanced therapies for severe pain. Three of these drugs were approved for use worldwide: rofecoxib (Vioxx), valdecoxib (Bextra), and celecoxib (Celebrex) (Fig. 21–15c). Launched in the late

1990s, the new compounds were initially a success for the pharmaceutical firms that produced them. However, enthusiasm turned to concern as field reports and clinical studies connected the drugs with an increased risk of heart attack and stroke. The reasons for the problems are still not clear, but some researchers speculated that the COX-2 inhibitors were altering the fine balance maintained between the hormone prostacyclin, which dilates blood vessels, prevents blood clotting, and is reduced by COX-2 inhibitors, and the thromboxanes, produced on the pathway involving COX-1, that help to form blood clots. Vioxx was withdrawn from the market in 2004, and Bextra was withdrawn soon after. As of early 2012, Celebrex is still on the market but is being used with increased caution.

Thromboxane synthase, present in blood platelets (thrombocytes), converts PGH_2 to thromboxane A_2, from which other **thromboxanes** are derived (Fig. 21–15a). Thromboxanes induce constriction of blood vessels and platelet aggregation, early steps in blood clotting. Low doses of aspirin, taken regularly, reduce the probability of heart attacks and strokes by reducing thromboxane production. ■

Thromboxanes, like prostaglandins, contain a ring of five or six atoms; the pathway from arachidonate to these two classes of compounds is sometimes called the "cyclic" pathway, to distinguish it from the "linear" pathway that leads from arachidonate to the **leukotrienes**, which are linear compounds **(Fig. 21–16)**. Leukotriene synthesis begins with the action of several lipoxygenases that catalyze the incorporation of molecular oxygen into arachidonate. These enzymes, found in leukocytes and in heart, brain, lung, and spleen, are mixed-function oxidases of the cytochrome P-450 family (Box 21–1). The various leukotrienes differ in the position of the peroxide group introduced by the lipoxygenases. This linear pathway from arachidonate, unlike the cyclic pathway, is not inhibited by aspirin or other NSAIDs.

Plants also derive important signaling molecules from fatty acids. As in animals, a key step in the initiation of signaling involves activation of a specific phospholi-pase. In plants, the fatty acid substrate that is released is α-linolenate. A lipoxygenase then catalyzes the first step in a pathway that converts linolenate to jasmonate, a substance known to have signaling roles in insect defense, resistance to fungal pathogens, and pollen maturation. Jasmonate (see Fig. 12–33) also affects seed germination, root growth, and fruit and seed development.

SUMMARY 21.1 Biosynthesis of Fatty Acids and Eicosanoids

▶ Long-chain saturated fatty acids are synthesized from acetyl-CoA by a cytosolic system of six enzymatic activities plus acyl carrier protein (ACP). There are two types of fatty acid synthase. FAS I, found in vertebrates and fungi, consists of multifunctional polypeptides. FAS II is a dissociated system found in bacteria and plants. Both contain two types of —SH groups (one furnished by the phosphopantetheine of ACP, the other by a Cys residue of β-ketoacyl-ACP synthase) that function as carriers of the fatty acyl intermediates.

▶ Malonyl-ACP, formed from acetyl-CoA (shuttled out of mitochondria) and CO_2, condenses with an acetyl bound to the Cys —SH to yield acetoacetyl-ACP, with release of CO_2. This is followed by reduction to the D-β-hydroxy derivative, dehydration to the *trans*-Δ^2-unsaturated acyl-ACP, and reduction to butyryl-ACP. NADPH is the electron donor for both reductions. Fatty acid synthesis is regulated at the level of malonyl-CoA formation.

▶ Six more molecules of malonyl-ACP react successively at the carboxyl end of the growing fatty acid chain to form palmitoyl-ACP—the end product of the fatty acid synthase reaction. Free palmitate is released by hydrolysis.

▶ Palmitate may be elongated to the 18-carbon stearate. Palmitate and stearate can be desaturated

FIGURE 21–16 **The "linear" pathway from arachidonate to leukotrienes.**

to yield palmitoleate and oleate, respectively, by the action of mixed-function oxidases.

▶ Mammals cannot make linoleate and must obtain it from plant sources; they convert exogenous linoleate to arachidonate, the parent compound of eicosanoids (prostaglandins, thromboxanes, and leukotrienes), a family of very potent signaling molecules. The synthesis of prostaglandins and thromboxanes is inhibited by NSAIDs that act on the cyclooxygenase activity of prostaglandin H_2 synthase.

21.2 Biosynthesis of Triacylglycerols

Most of the fatty acids synthesized or ingested by an organism have one of two fates: incorporation into triacylglycerols for the storage of metabolic energy or incorporation into the phospholipid components of membranes. The partitioning between these alternative fates depends on the organism's current needs. During rapid growth, synthesis of new membranes requires the production of membrane phospholipids; when an organism has a plentiful food supply but is not actively growing, it shunts most of its fatty acids into storage fats. Both pathways begin at the same point: the formation of fatty acyl esters of glycerol. In this section we examine the route to triacylglycerols and its regulation, and the production of glycerol 3-phosphate in the process of glyceroneogenesis.

Triacylglycerols and Glycerophospholipids Are Synthesized from the Same Precursors

Animals can synthesize and store large quantities of triacylglycerols, to be used later as fuel (see Box 17–1). Humans can store only a few hundred grams of glycogen in liver and muscle, barely enough to supply the body's energy needs for 12 hours. In contrast, the total amount of stored triacylglycerol in a 70 kg man of average build is about 15 kg, enough to support basal energy needs for as long as 12 weeks (see Table 23–5). Triacylglycerols have the highest energy content of all stored nutrients—more than 38 kJ/g. Whenever carbohydrate is ingested in excess of the organism's capacity to store glycogen, the excess is converted to triacylglycerols and stored in adipose tissue. Plants also manufacture triacylglycerols as an energy-rich fuel, mainly stored in fruits, nuts, and seeds.

In animal tissues, triacylglycerols and glycerophospholipids such as phosphatidylethanolamine share two precursors (fatty acyl–CoA and L-glycerol 3-phosphate) and several biosynthetic steps. The vast majority of the glycerol 3-phosphate is derived from the glycolytic intermediate dihydroxyacetone phosphate (DHAP) by the action of the cytosolic NAD-linked **glycerol 3-phosphate dehydrogenase**; in liver and kidney, a small amount of glycerol 3-phosphate is also formed from glycerol by the action of **glycerol kinase (Fig. 21–17).** The other precursors of triacylglycerols are fatty acyl–CoAs, formed from fatty acids by **acyl-CoA synthetases,** the same

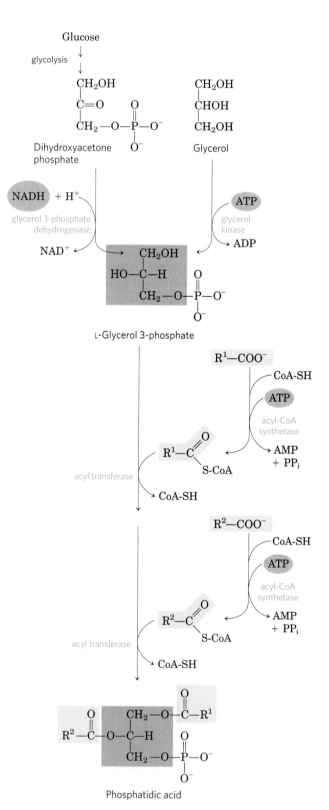

FIGURE 21–17 Biosynthesis of phosphatidic acid. A fatty acyl group is activated by formation of the fatty acyl–CoA, then transferred to ester linkage with L-glycerol 3-phosphate, formed in either of the two ways shown. Phosphatidic acid is shown here with the correct stereochemistry at C-2 of the glycerol molecule. (The intermediate product with only one esterified fatty acyl group is lysophosphatidic acid.) To conserve space in subsequent figures (and in Fig. 21–14), both fatty acyl groups of glycerophospholipids, and all three acyl groups of triacylglycerols, are shown projecting to the right.

enzymes responsible for the activation of fatty acids for β oxidation (see Fig. 17–5).

The first stage in the biosynthesis of triacylglycerols is the acylation of the two free hydroxyl groups of L-glycerol 3-phosphate by two molecules of fatty acyl–CoA to yield **diacylglycerol 3-phosphate**, more commonly called **phosphatidic acid** or phosphatidate (Fig. 21–17). Phosphatidic acid is present in only trace amounts in cells but is a central intermediate in lipid biosynthesis; it can be converted either to a triacylglycerol or to a glycerophospholipid. In the pathway to triacylglycerols, phosphatidic acid is hydrolyzed by **phosphatidic acid phosphatase** (also called lipin) to form a 1,2-diacylglycerol **(Fig. 21–18)**. Diacylglycerols are then converted to triacylglycerols by transesterification with a third fatty acyl–CoA.

Triacylglycerol Biosynthesis in Animals Is Regulated by Hormones

In humans, the amount of body fat stays relatively constant over long periods, although there may be minor short-term changes as caloric intake fluctuates. Carbohydrate, fat, or protein ingested in excess of energy needs is stored in the form of triacylglycerols that can be drawn upon for energy, enabling the body to withstand periods of fasting.

Biosynthesis and degradation of triacylglycerols are regulated such that the favored path depends on the metabolic resources and requirements of the moment. The rate of triacylglycerol biosynthesis is profoundly altered by the action of several hormones. Insulin, for example, promotes the conversion of carbohydrate to triacylglycerols **(Fig. 21–19)**. People with severe diabetes mellitus, due to failure of insulin secretion or action, not only are unable to use glucose properly but also fail to synthesize fatty acids from carbohydrates or amino acids. If the diabetes is untreated, these individuals have increased rates of fat oxidation and ketone body formation (Chapter 17) and therefore lose weight. ∎

An additional factor in the balance between biosynthesis and degradation of triacylglycerols is that approximately 75% of all fatty acids released by lipolysis are reesterified to form triacylglycerols rather than used for fuel. This ratio persists even under starvation conditions, when energy metabolism is shunted from the use of carbohydrate to the oxidation of fatty acids. Some of this fatty acid recycling takes place in adipose tissue, with the reesterification occurring before release into the bloodstream; some takes place via a systemic cycle in

FIGURE 21–18 Phosphatidic acid in lipid biosynthesis. Phosphatidic acid is the precursor of both triacylglycerols and glycerophospholipids. The mechanisms for head-group attachment in phospholipid synthesis are described later in this section.

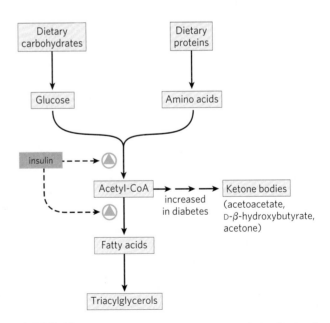

FIGURE 21–19 Regulation of triacylglycerol synthesis by insulin. Insulin stimulates conversion of dietary carbohydrates and proteins to fat. Individuals with diabetes mellitus either lack insulin or are insensitive to it. This results in diminished fatty acid synthesis, and the acetyl-CoA arising from catabolism of carbohydrates and proteins is shunted instead to ketone body production. People in severe ketosis smell of acetone, so the condition is sometimes mistaken for drunkenness (p. 959).

which free fatty acids are transported to liver, recycled to triacylglycerol, exported again to the blood (transport of lipids in the blood is discussed in Section 21.4), and taken up again by adipose tissue after release from triacylglycerol by extracellular lipoprotein lipase (**Fig. 21–20**; see also Fig. 17–1). Flux through this **triacylglycerol cycle** between adipose tissue and liver may be quite low when other fuels are available and the release of fatty acids from adipose tissue is limited, but as noted above, the proportion of released fatty acids that are reesterified remains roughly constant at 75% under all metabolic conditions. The level of free fatty acids in the blood thus reflects both the rate of release of fatty acids and the balance between the synthesis and breakdown of triacylglycerols in adipose tissue and liver.

When the mobilization of fatty acids is required to meet energy needs, release from adipose tissue is stimulated by the hormones glucagon and epinephrine (see Figs 17–3, 17–13). Simultaneously, these hormonal signals decrease the rate of glycolysis and increase the rate of gluconeogenesis in the liver (providing glucose for the brain, as further elaborated in Chapter 23). The released fatty acid is taken up by a number of tissues, including muscle, where it is oxidized to provide energy. Much of the fatty acid taken up by liver is not oxidized but is recycled to triacylglycerol and returned to adipose tissue.

The function of the apparently futile triacylglycerol cycle (futile cycles are discussed in Chapter 15) is not well understood. However, as we learn more about how the triacylglycerol cycle is sustained via metabolism in two separate organs and is coordinately regulated, some possibilities emerge. For example, the excess capacity in the triacylglycerol cycle (the fatty acid that is eventually reconverted to triacylglycerol rather than oxidized as fuel) could represent an energy reserve in the bloodstream during fasting, one that would be more rapidly mobilized in a "fight or flight" emergency than would stored triacylglycerol.

The constant recycling of triacylglycerols in adipose tissue even during starvation raises a second question: what is the source of the glycerol 3-phosphate required for this process? As noted above, glycolysis is suppressed in these conditions by the action of glucagon and epinephrine, so little DHAP is available, and glycerol released during lipolysis cannot be converted directly to glycerol 3-phosphate in adipose tissue, because these cells lack glycerol kinase (Fig. 21–17). So, how is sufficient glycerol 3-phosphate produced? The answer lies in a pathway discovered more than three decades ago and given little attention until recently, a pathway intimately linked to the triacylglycerol cycle and, in a larger sense, to the balance between fatty acid and carbohydrate metabolism.

Adipose Tissue Generates Glycerol 3-Phosphate by Glyceroneogenesis

Glyceroneogenesis is a shortened version of gluconeogenesis, from pyruvate to DHAP (see Fig. 14–17), followed by conversion of the DHAP to glycerol 3-phosphate by cytosolic NAD-linked glycerol 3-phosphate dehydrogenase (**Fig. 21–21**). Glycerol 3-phosphate is subsequently used in triacylglycerol synthesis. Glyceroneogenesis was discovered in the 1960s by Lea Reshef, Richard Hanson, and John Ballard, and simultaneously by Eleazar Shafrir and his coworkers, who were intrigued by the presence of two gluconeogenic enzymes, pyruvate carboxylase and phosphoenolpyruvate (PEP) carboxykinase, in adipose tissue, where glucose is not synthesized. After a long period of inattention, interest in this pathway has been renewed by the demonstration of a link between glyceroneogenesis and type 2 diabetes, as we shall see.

Glyceroneogenesis has multiple roles. In adipose tissue, glyceroneogenesis coupled with reesterification of free fatty acids controls the rate of fatty acid release to the blood. In brown adipose tissue, the same pathway may control the rate at which free fatty acids are delivered to mitochondria for use in thermogenesis (see Fig. 19–36). And in fasting humans, glyceroneogenesis in the liver alone supports the synthesis of enough glycerol 3-phosphate to account for up to 65% of fatty acids reesterified to triacylglycerol.

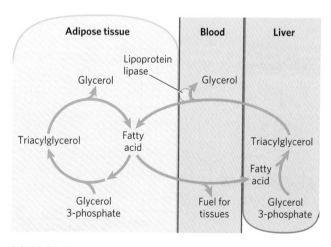

FIGURE 21–20 The triacylglycerol cycle. In mammals, triacylglycerol molecules are broken down and resynthesized in a triacylglycerol cycle during starvation. Some of the fatty acids released by lipolysis of triacylglycerol in adipose tissue pass into the bloodstream, and the remainder are used for resynthesis of triacylglycerol. Some of the fatty acids released into the blood are used for energy (in muscle, for example), and some are taken up by the liver and used in triacylglycerol synthesis. The triacylglycerol formed in the liver is transported in the blood back to adipose tissue, where the fatty acid is released by extracellular lipoprotein lipase, taken up by adipocytes, and reesterified into triacylglycerol.

Pyruvate

↓ pyruvate carboxylase

Oxaloacetate

↓ PEP carboxykinase

Phosphoenolpyruvate

↓ multistep

Dihydroxyacetone phosphate

↓ glycerol 3-phosphate
dehydrogenase

CH₂OH
|
CHOH   O
| ∥
CH₂—O—P—O⁻
         |
         O⁻

Glycerol 3-phosphate

↓

Triacylglycerol synthesis

FIGURE 21–21 Glyceroneogenesis. The pathway is essentially an abbreviated version of gluconeogenesis, from pyruvate to dihydroxyacetone phosphate (DHAP), followed by conversion of DHAP to glycerol 3-phosphate, which is used for the synthesis of triacylglycerol.

Flux through the triacylglycerol cycle between liver and adipose tissue is controlled to a large degree by the activity of PEP carboxykinase, which limits the rate of both gluconeogenesis and glyceroneogenesis. Glucocorticoid hormones such as cortisol (a biological steroid derived from cholesterol; see Fig. 21–48) and dexamethasone (a synthetic glucocorticoid) regulate the levels of PEP carboxykinase reciprocally in the liver and adipose tissue. Acting through the glucocorticoid receptor, these steroid hormones increase the expression of the gene encoding PEP carboxykinase in the liver, thus increasing gluconeogenesis and glyceroneogenesis **(Fig. 21–22).**

Cortisol   Dexamethasone

Stimulation of glyceroneogenesis leads to an increase in the synthesis of triacylglycerol molecules in the liver and their release into the blood. At the same time, glucocorticoids suppress the expression of the gene encoding PEP carboxykinase in adipose tissue. This results in a decrease in glyceroneogenesis in adipose tissue; recycling of fatty acids declines as a result, and more free fatty acids are released into the blood.

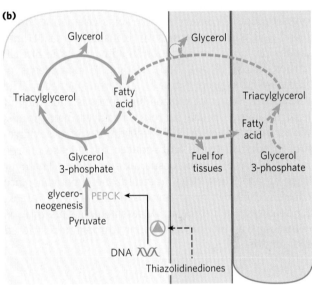

FIGURE 21–22 Regulation of glyceroneogenesis. (a) Glucocorticoid hormones stimulate glyceroneogenesis and gluconeogenesis in the liver, while suppressing glyceroneogenesis in the adipose tissue (by reciprocal regulation of the gene expressing PEP carboxykinase (PEPCK) in the two tissues); this increases the flux through the triacylglycerol cycle. The glycerol freed by the breakdown of triacylglycerol in adipose tissue is released to the blood and transported to the liver, where it is primarily converted to glucose, although some is converted to glycerol 3-phosphate by glycerol kinase. **(b)** A class of drugs called thiazolidinediones are now used to treat type 2 diabetes. In this disease, high levels of free fatty acids in the blood interfere with glucose utilization in muscle and promote insulin resistance. Thiazolidinediones activate a nuclear receptor called peroxisome proliferator-activated receptor γ (PPARγ), which induces the activity of PEP carboxylase. Therapeutically, thiazolidinediones increase the rate of glyceroneogenesis, thus increasing the resynthesis of triacylglycerol in adipose tissue and reducing the amount of free fatty acid in the blood.

Thus glyceroneogenesis is regulated reciprocally in the liver and adipose tissue, affecting lipid metabolism in opposite ways: a lower rate of glyceroneogenesis in adipose tissue leads to more fatty acid release (rather than recycling), whereas a higher rate in the liver leads to more synthesis and export of triacylglycerols. The net result is an increase in flux through the triacylglycerol cycle. When the glucocorticoids are no longer present, flux through the cycle declines as the expression of PEP carboxykinase increases in adipose tissue and decreases in the liver.

Thiazolidinediones Treat Type 2 Diabetes by Increasing Glyceroneogenesis

The recent attention to glyceroneogenesis has arisen in part from the connection between this pathway and diabetes. High levels of free fatty acids in the blood interfere with glucose utilization in muscle and promote the insulin resistance that leads to type 2 diabetes. A new class of drugs called **thiazolidinediones** reduce the levels of fatty acids circulating in the blood and increase sensitivity to insulin. Thiazolidinediones promote the induction in adipose tissue of PEP carboxykinase (Fig. 21–22), leading to increased synthesis of the precursors of glyceroneogenesis. The therapeutic effect of thiazolidinediones is thus due, at least in part, to the increase in glyceroneogenesis, which in turn increases the resynthesis of triacylglycerol in adipose tissue and reduces the release of free fatty acid from adipose tissue into the blood. The benefits of one such drug, rosiglitazone (Avandia), are countered in part by an increased risk of heart attack, for reasons not yet clear. Assessment of this drug is continuing, and it is available only through a restricted distribution system.

Rosiglitazone (Avandia)

Pioglitazone (Actos)

Thiazolidinediones

SUMMARY 21.2 Biosynthesis of Triacylglycerols

▶ Triacylglycerols are formed by reaction of two molecules of fatty acyl–CoA with glycerol 3-phosphate to form phosphatidic acid; this product is dephosphorylated to a diacylglycerol, then acylated by a third molecule of fatty acyl–CoA to yield a triacylglycerol.

▶ The synthesis and degradation of triacylglycerol are hormonally regulated.

▶ Mobilization and recycling of triacylglycerol molecules results in a triacylglycerol cycle. Triacylglycerols are resynthesized from free fatty acids and glycerol 3-phosphate even during starvation. The dihydroxyacetone phosphate precursor of glycerol 3-phosphate is derived from pyruvate via glyceroneogenesis.

21.3 Biosynthesis of Membrane Phospholipids

In Chapter 10 we introduced two major classes of membrane phospholipids: glycerophospholipids and sphingolipids. Many different phospholipid species can be constructed by combining various fatty acids and polar head groups with the glycerol or sphingosine backbone (see Figs 10–9, 10–13). All the biosynthetic pathways follow a few basic patterns. In general, the assembly of phospholipids from simple precursors requires (1) synthesis of the backbone molecule (glycerol or sphingosine); (2) attachment of fatty acid(s) to the backbone through an ester or amide linkage; (3) addition of a hydrophilic head group to the backbone through a phosphodiester linkage; and, in some cases, (4) alteration or exchange of the head group to yield the final phospholipid product.

In eukaryotic cells, phospholipid synthesis occurs primarily on the surfaces of the smooth endoplasmic reticulum and the mitochondrial inner membrane. Some newly formed phospholipids remain at the site of synthesis, but most are destined for other cellular locations. The process by which water-insoluble phospholipids move from the site of synthesis to the point of their eventual function is not fully understood, but we will discuss some mechanisms that have emerged in recent years.

Cells Have Two Strategies for Attaching Phospholipid Head Groups

The first steps of glycerophospholipid synthesis are shared with the pathway to triacylglycerols (Fig. 21–17): two fatty acyl groups are esterified to C-1 and C-2 of L-glycerol 3-phosphate to form phosphatidic acid. Commonly but not invariably, the fatty acid at C-1 is saturated and that at C-2 is unsaturated. A second route to phosphatidic acid is the phosphorylation of a diacylglycerol by a specific kinase.

The polar head group of glycerophospholipids is attached through a phosphodiester bond, in which each of two alcohol hydroxyls (one on the polar head group and one on C-3 of glycerol) forms an ester with phosphoric acid **(Fig. 21–23)**. In the biosynthetic process, one of the hydroxyls is first activated by attachment of

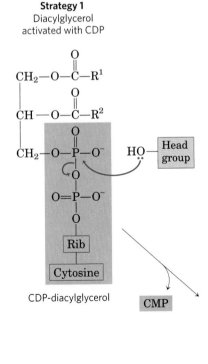

FIGURE 21–23 Head-group attachment. The phospholipid head group is attached to a diacylglycerol by a phosphodiester bond, formed when phosphoric acid condenses with two alcohols, eliminating two molecules of H₂O.

a nucleotide, cytidine diphosphate (CDP). Cytidine monophosphate (CMP) is then displaced in a nucleophilic attack by the other hydroxyl **(Fig. 21–24)**. The

CDP is attached either to the diacylglycerol, forming the activated phosphatidic acid **CDP-diacylglycerol** (strategy 1), or to the hydroxyl of the head group (strategy 2). Eukaryotic cells employ both strategies, whereas bacteria use only the first. The central importance of cytidine nucleotides in lipid biosynthesis was discovered by Eugene P. Kennedy in the early 1960s, and this pathway is commonly referred to as the Kennedy pathway.

Phospholipid Synthesis in *E. coli* Employs CDP-Diacylglycerol

The first strategy for head-group attachment is illustrated by the synthesis of phosphatidylserine, phosphatidylethanolamine, and phosphatidylglycerol in *E. coli*. The diacylglycerol is activated by condensation of phosphatidic acid with cytidine triphosphate (CTP) to form CDP-diacylglycerol, with the elimination of pyrophosphate **(Fig. 21–25)**. Displacement of CMP through nucleophilic attack by the hydroxyl group of serine or by the C-1 hydroxyl of glycerol 3-phosphate yields **phosphatidylserine** or phosphatidylglycerol 3-phosphate, respectively. The latter is processed further by cleavage of the phosphate monoester (with release of P$_i$) to yield **phosphatidylglycerol**.

Phosphatidylserine and phosphatidylglycerol can serve as precursors of other membrane lipids in bacteria (Fig. 21–25). Decarboxylation of the serine moiety in

Eugene P. Kennedy, 1919–2011

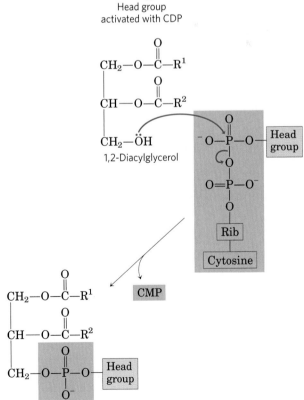

FIGURE 21–24 Two general strategies for forming the phosphodiester bond of phospholipids. In both cases, CDP supplies the phosphate group of the phosphodiester bond.

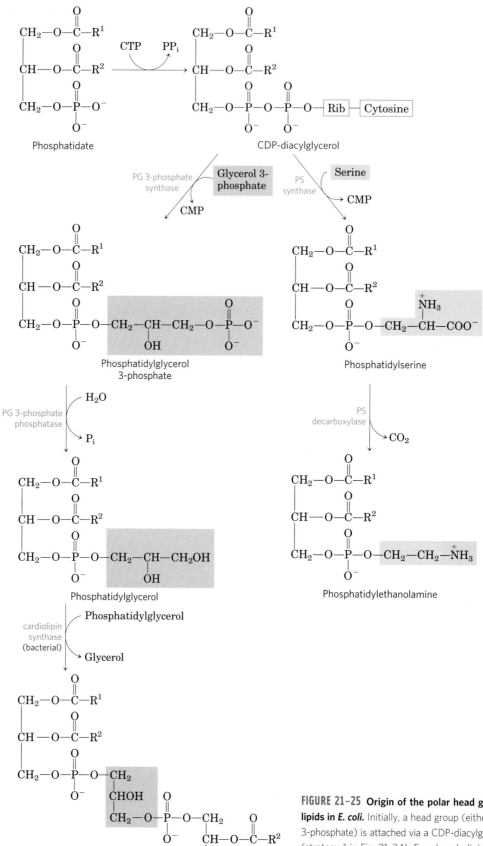

FIGURE 21–25 Origin of the polar head groups of phospholipids in *E. coli.* Initially, a head group (either serine or glycerol 3-phosphate) is attached via a CDP-diacylglycerol intermediate (strategy 1 in Fig. 21–24). For phospholipids other than phosphatidylserine, the head group is further modified, as shown here. In the enzyme names, PG represents phosphatidylglycerol; PS, phosphatidylserine.

FIGURE 21–26 Synthesis of cardiolipin and phosphatidylinositol in eukaryotes. These glycerophospholipids are synthesized using strategy 1 in Figure 21–24. Phosphatidylglycerol is synthesized as in bacteria (see Fig. 21–25). PI represents phosphatidylinositol.

phosphatidylserine, catalyzed by phosphatidylserine decarboxylase, yields **phosphatidylethanolamine**. In *E. coli*, condensation of two molecules of phosphatidylglycerol, with elimination of one glycerol, yields **cardiolipin**, in which two diacylglycerols are joined through a common head group.

Eukaryotes Synthesize Anionic Phospholipids from CDP-Diacylglycerol

In eukaryotes, phosphatidylglycerol, cardiolipin, and the phosphatidylinositols (all anionic phospholipids; see Fig. 10–9) are synthesized by the same strategy used for phospholipid synthesis in bacteria. Phosphatidylglycerol is made exactly as in bacteria. Cardiolipin synthesis in eukaryotes differs slightly: phosphatidylglycerol condenses with CDP-diacylglycerol **(Fig. 21–26)**, not another molecule of phosphatidylglycerol as in *E. coli* (Fig. 21–25).

Phosphatidylinositol is synthesized by condensation of CDP-diacylglycerol with inositol (Fig. 21–26). Specific **phosphatidylinositol kinases** then convert phosphatidylinositol to its phosphorylated derivatives. Phosphatidylinositol and its phosphorylated products in

the plasma membrane play a central role in signal transduction in eukaryotes (see Figs 12–10, 12–16).

Eukaryotic Pathways to Phosphatidylserine, Phosphatidylethanolamine, and Phosphatidylcholine Are Interrelated

Yeast, like bacteria, can produce phosphatidylserine by condensation of CDP-diacylglycerol and serine, and can synthesize phosphatidylethanolamine from phosphatidylserine in the reaction catalyzed by phosphatidylserine decarboxylase **(Fig. 21–27)**. Phosphatidylethanolamine may be converted to **phosphatidylcholine** (lecithin) by the addition of three methyl groups to its amino group; *S*-adenosylmethionine is the methyl group donor (see Fig. 18–18) for all three methylation reactions. The pathways to phosphatidylcholine and phosphatidylethanolamine in yeast are summarized in **Figure 21–28**. These paths are the major sources of phosphatidylethanolamine and phosphatidylcholine in all eukaryotic cells.

In mammals, phosphatidylserine is not synthesized from CDP-diacylglycerol; instead, it is derived from phosphatidylethanolamine or phosphatidylcholine via one of two head-group exchange reactions carried out

FIGURE 21–27 The major path from phosphatidylserine to phosphatidylethanolamine and phosphatidylcholine in all eukaryotes. AdoMet is *S*-adenosylmethionine; adoHcy, *S*-adenosylhomocysteine.

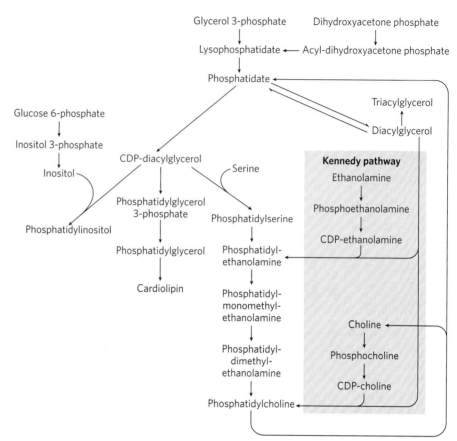

FIGURE 21–28 Summary of the pathways for synthesis of major phospholipids and triacylglycerides in a eukaryote (yeast). Phosphatidic acid is formed by transacylation of L-glycerol 3-phosphate with two fatty acyl groups donated from fatty acyl-CoA. The enzyme phosphatidic acid phosphatase (lipin) converts phosphatidic acid to diacylglycerol, which in the Kennedy pathway condenses with a CDP-activated head group (ethanolamine or choline) to form phosphatidylethanolamine or phosphatidylcholine. Alternatively, phosphatidic acid can be activated with a CDP moiety, which is displaced by condensation with a head group alcohol—inositol, glycerol 3-phosphate, or serine, forming phosphatidylinositol, phosphatidylglycerol, or phosphatidylserine. Decarboxylation of phosphatidylserine yields phosphatidylethanolamine, and methylation of phosphatidylethanolamine produces phosphatidylcholine. Not shown here are the head-group exchange reactions (see Fig. 21–29a) that interconvert phosphatidylethanolamine, phosphatidylserine, and phosphatidylcholine in mammals. Lysophosphatidic acid is phosphatidic acid missing one of the two fatty acyl groups.

in the endoplasmic reticulum **(Fig. 21–29a)**. Synthesis of phosphatidylethanolamine and phosphatidylcholine in mammals occurs by strategy 2 of Figure 21–24: phosphorylation and activation of the head group, followed by condensation with diacylglycerol. For example, choline is reused ("salvaged") by being phosphorylated then converted to CDP-choline by condensation with CTP. A diacylglycerol displaces CMP from CDP-choline, producing phosphatidylcholine (Fig. 21–29b). An analogous salvage pathway converts ethanolamine obtained in the diet to phosphatidylethanolamine. In the liver, phosphatidylcholine is also produced by methylation of phosphatidylethanolamine (with *S*-adenosylmethionine, as described above), but in all other tissues phosphatidylcholine is produced only by condensation of diacylglycerol and CDP-choline.

Plasmalogen Synthesis Requires Formation of an Ether-Linked Fatty Alcohol

The biosynthetic pathway to ether lipids, including **plasmalogens** and the **platelet-activating factor** (see Fig. 10–10), involves the displacement of an esterified fatty acyl group by a long-chain alcohol to form the ether linkage **(Fig. 21–30)**. Head-group attachment follows, by mechanisms essentially like those used in formation of the common ester-linked phospholipids. Finally, the characteristic double bond of plasmalogens (shaded blue in Fig. 21–30) is introduced by the action of a mixed-function oxidase similar to that responsible

(a)

$$CH_2-O-\overset{\overset{\displaystyle O}{\|}}{C}-R^1$$
$$CH-O-\overset{\overset{\displaystyle O}{\|}}{C}-R^2$$
$$CH_2-O-\overset{\underset{\displaystyle O^-}{|}}{P}-O-CH_2-CH_2-\overset{+}{N}H_3$$

Phosphatidylethanolamine

$$CH_2-O-\overset{\overset{\displaystyle O}{\|}}{C}-R^1$$
$$CH-O-\overset{\overset{\displaystyle O}{\|}}{C}-R^2$$
$$CH_2-O-\overset{\underset{\displaystyle O^-}{|}}{P}-O-CH_2-CH_2-\overset{+}{N}(CH_3)_3$$

Phosphatidylcholine

Serine ⇄ (PSS1, PSS2) Ethanolamine

Serine ⇄ (PSS1) Choline

$$CH_2-O-\overset{\overset{\displaystyle O}{\|}}{C}-R^1$$
$$CH-O-\overset{\overset{\displaystyle O}{\|}}{C}-R^2$$
$$CH_2-O-\overset{\underset{\displaystyle O^-}{|}}{P}-O-CH_2-\overset{\overset{\displaystyle +}{N}H_3}{CH}-COO^-$$

Phosphatidylserine

FIGURE 21–29 Pathways for phosphatidylserine and phosphatidylcholine synthesis in mammals. (a) Phosphatidylserine is synthesized by Ca^{2+}-dependent head-group exchange reactions promoted by phosphatidylserine synthase 1 (PSS1) or phosphatidylserine synthase 2 (PSS2). PSS1 can use either phosphatidylethanolamine or phosphatidylcholine as a substrate. **(b)** The same strategy shown here for phosphatidylcholine synthesis (strategy 2 in Fig. 21–24) is also used for salvaging ethanolamine in phosphatidylethanolamine synthesis.

(b)

$HO-CH_2-CH_2-\overset{+}{N}(CH_3)_3$ Choline

choline kinase: ATP → ADP

$^-O-\overset{\underset{\displaystyle O}{\|}}{\underset{}{P}}-O-CH_2-CH_2-\overset{+}{N}(CH_3)_3$ Phosphocholine

CTP-choline cytidylyl transferase: CTP → PPi

CDP-choline

CDP-choline-diacylglycerol phosphocholine transferase: Diacylglycerol → CMP

$$CH_2-O-\overset{\overset{\displaystyle O}{\|}}{C}-R^1$$
$$CH-O-\overset{\overset{\displaystyle O}{\|}}{C}-R^2$$
$$CH_2-O-\overset{\underset{\displaystyle O^-}{|}}{P}-O-CH_2-CH_2-\overset{+}{N}(CH_3)_3$$

Phosphatidylcholine

for desaturation of fatty acids (Fig. 21–13). The peroxisome is the primary site of plasmalogen synthesis.

Sphingolipid and Glycerophospholipid Synthesis Share Precursors and Some Mechanisms

The biosynthesis of sphingolipids takes place in four stages: (1) synthesis of the 18-carbon amine **sphinganine** from palmitoyl-CoA and serine; (2) attachment of a fatty acid in amide linkage to yield **N-acylsphinganine**; (3) desaturation of the sphinganine moiety to form **N-acylsphingosine** (ceramide); and (4) attachment of a head group to produce a sphingolipid such as a **cerebroside** or **sphingomyelin (Fig. 21–31)**. The first few steps of this pathway occur in the endoplasmic reticulum, while the attachment of head groups in stage 4 occurs in the Golgi complex. The pathway shares several features with the pathways leading to glycerophospholipids: NADPH provides reducing power, and fatty acids enter as their activated CoA derivatives. In cerebroside formation, sugars enter as their activated nucleotide derivatives. Head-group attachment in sphingolipid synthesis has several novel aspects. Phosphatidylcholine, rather than CDP-choline, serves as the donor of phosphocholine in the synthesis of sphingomyelin.

In glycolipids—the cerebrosides and **gangliosides** (see Fig. 10–13)—the head-group sugar is attached directly to the C-1 hydroxyl of sphingosine in glycosidic linkage rather than through a phosphodiester bond. The sugar donor is a UDP-sugar (UDP-glucose or UDP-galactose).

Polar Lipids Are Targeted to Specific Cellular Membranes

After synthesis on the smooth ER, the polar lipids, including the glycerophospholipids, sphingolipids, and glycolipids, are inserted into specific cellular membranes in specific proportions, by mechanisms not yet understood. Membrane lipids are insoluble in water, so they cannot simply diffuse from their point of synthesis (the ER) to their point of insertion. Instead, they are transported from the ER to the Golgi complex, where additional synthesis can take place. They are then delivered in membrane vesicles that bud from the Golgi complex then move to and fuse with the target membrane. Sphingolipid transfer proteins carry ceramide from the ER to the Golgi complex, where sphingomyelin synthesis occurs. Cytosolic proteins also bind phospholipids and sterols and transport them between cellular membranes. These mechanisms contribute to the

FIGURE 21–30 Synthesis of ether lipids and plasmalogens. The newly formed ether linkage is shaded light red. The intermediate 1-alkyl-2-acyl-glycerol 3-phosphate is the ether analog of phosphatidic acid. Mechanisms for attaching head groups to ether lipids are essentially the same as for their ester-linked analogs. The characteristic double bond of plasmalogens (shaded blue) is introduced in a final step by a mixed-function oxidase system similar to that shown in Figure 21–13.

establishment of the characteristic lipid compositions of organelle membranes (see Fig. 11–2).

SUMMARY 21.3 Biosynthesis of Membrane Phospholipids

▶ Diacylglycerols are the principal precursors of glycerophospholipids.

▶ In bacteria, phosphatidylserine is formed by the condensation of serine with CDP-diacylglycerol; decarboxylation of phosphatidylserine produces phosphatidylethanolamine. Phosphatidylglycerol is formed by condensation of CDP-diacylglycerol with glycerol 3-phosphate, followed by removal of the phosphate in monoester linkage.

▶ Yeast pathways for the synthesis of phosphatidylserine, phosphatidylethanolamine, and phosphatidylglycerol are similar to those in bacteria; phosphatidylcholine is formed by methylation of phosphatidylethanolamine.

▶ Mammalian cells have some pathways similar to those in bacteria, but somewhat different routes for synthesizing phosphatidylcholine and phosphatidylethanolamine. The head-group alcohol (choline or ethanolamine) is activated as the CDP derivative, then condensed with diacylglycerol.

FIGURE 21-31 Biosynthesis of sphingolipids. Condensation of palmitoyl-CoA and serine (forming β-ketosphinganine) followed by reduction with NADPH yields sphinganine, which is then acylated to N-acylsphinganine (a ceramide). In animals, a double bond (shaded light red) is created by a mixed-function oxidase before the final addition of a head group: phosphatidylcholine, to form sphingomyelin, or glucose, to form a cerebroside.

Phosphatidylserine is derived only from phosphatidylethanolamine.

▶ Synthesis of plasmalogens involves formation of their characteristic double bond by a mixed-function oxidase. The head groups of sphingolipids are attached by unique mechanisms.

▶ Phospholipids travel to their intracellular destinations via transport vesicles or specific proteins.

21.4 Cholesterol, Steroids, and Isoprenoids: Biosynthesis, Regulation, and Transport

Cholesterol is doubtless the most publicized lipid, notorious because of the strong correlation between high levels of cholesterol in the blood and the incidence of human cardiovascular diseases. Less well advertised is cholesterol's crucial role as a component of cellular membranes and as a precursor of steroid hormones and bile acids. Cholesterol is an essential molecule in many animals, including humans, but is not required in the mammalian diet—all cells can synthesize it from simple precursors.

The structure of this 27-carbon compound suggests a complex biosynthetic pathway, but all of its carbon atoms are provided by a single precursor—acetate. The **isoprene** units that are the essential intermediates in the pathway from acetate to cholesterol are also precursors to many other natural lipids, and the mechanisms by which isoprene units are polymerized are similar in all these pathways.

$$CH_2{=}C{-}CH{=}CH_2$$
$$\overset{\displaystyle CH_3}{|}$$
Isoprene

We begin with an account of the main steps in the biosynthesis of cholesterol from acetate, and then discuss the transport of cholesterol in the blood, its uptake by cells, the normal regulation of cholesterol synthesis, and its regulation in those with defects in cholesterol uptake or transport. We next consider other cellular components derived from cholesterol, such as bile acids and steroid hormones. Finally, an outline of the biosynthetic pathways to some of the many compounds derived from isoprene units, which share early steps with the pathway to cholesterol, illustrates the extraordinary versatility of isoprenoid condensations in biosynthesis.

Cholesterol Is Made from Acetyl-CoA in Four Stages

Cholesterol, like long-chain fatty acids, is made from acetyl-CoA. But the assembly plan of cholesterol is quite different from that of long-chain fatty acids. In early experiments, animals were fed acetate labeled with ^{14}C in either the methyl carbon or the carboxyl carbon. The pattern of labeling in the cholesterol isolated from the two groups of animals (**Fig. 21–32**) provided the blueprint for working out the enzymatic steps in cholesterol biosynthesis.

Synthesis takes place in four stages, as shown in **Figure 21–33**: ❶ condensation of three acetate units to form a six-carbon intermediate, mevalonate; ❷ conversion of mevalonate to activated isoprene units; ❸ polymerization of six 5-carbon isoprene units to form the 30-carbon linear squalene; and ❹ cyclization of squalene to form the four rings of the steroid nucleus, with a further series of changes (oxidations, removal or migration of methyl groups) to produce cholesterol.

Stage ❶ Synthesis of Mevalonate from Acetate The first stage in cholesterol biosynthesis leads to the intermediate **mevalonate (Fig. 21–34)**. Two molecules of acetyl-CoA condense to form acetoacetyl-CoA, which condenses with a third molecule of acetyl-CoA to yield the six-carbon compound **β-hydroxy-β-methylglutaryl-CoA (HMG-CoA)**. These first two reactions are catalyzed by **acetyl-CoA acetyl transferase** and **HMG-CoA synthase**, respectively. The cytosolic HMG-CoA synthase in this pathway is distinct from the mitochondrial isozyme that catalyzes HMG-CoA synthesis in ketone body formation (see Fig. 17–19).

FIGURE 21–33 Summary of cholesterol biosynthesis. The four stages are discussed in the text. Isoprene units in squalene are set off by red dashed lines.

The third reaction is the committed step: reduction of HMG-CoA to mevalonate, for which each of two molecules of NADPH donates two electrons. **HMG-CoA reductase**, an integral membrane protein of the smooth ER, is the major point of regulation on the pathway to cholesterol, as we shall see.

Stage ❷ Conversion of Mevalonate to Two Activated Isoprenes In the next stage of cholesterol synthesis, three phosphate groups are transferred from three ATP molecules to mevalonate (**Fig. 21–35**). The phosphate attached to the C-3 hydroxyl group of mevalonate in the intermediate 3-phospho-5-pyrophosphomevalonate is a good leaving group; in the next step, both this phosphate and the nearby carboxyl group leave, producing a double bond in the five-carbon product, **Δ³-isopentenyl pyrophosphate**. This is the first of the two activated isoprenes central to cholesterol formation. Isomerization

FIGURE 21–32 Origin of the carbon atoms of cholesterol. This can be deduced from tracer experiments with acetate labeled in the methyl carbon (black) or the carboxyl carbon (red). The individual rings in the fused-ring system are designated A through D.

FIGURE 21–34 Formation of mevalonate from acetyl-CoA. The origin of C-1 and C-2 of mevalonate from acetyl-CoA is shaded light red.

of Δ^3-isopentenyl pyrophosphate yields the second activated isoprene, **dimethylallyl pyrophosphate**. Synthesis of isopentenyl pyrophosphate in the cytoplasm of plant cells follows the pathway described here. However, plant chloroplasts and many bacteria use a mevalonate-independent pathway. This alternative pathway does not occur in animals, so it is an attractive target for the development of new antibiotics.

Stage ❸ Condensation of Six Activated Isoprene Units to Form Squalene Isopentenyl pyrophosphate and dimethylallyl pyrophosphate now undergo a head-to-tail condensation, in which one pyrophosphate group is displaced and a 10-carbon chain, **geranyl pyrophosphate**, is formed **(Fig. 21–36)**. (The "head" is the end to which pyrophosphate is joined.) Geranyl pyrophosphate undergoes another head-to-tail condensation with isopentenyl pyrophosphate, yielding the 15-carbon intermediate **farnesyl pyrophosphate**. Finally, two molecules of farnesyl pyro-

FIGURE 21–35 Conversion of mevalonate to activated isoprene units. Six of these activated units combine to form squalene (see Fig. 21–36). The leaving groups of 3-phospho-5-pyrophosphomevalonate are shaded light red. The bracketed intermediate is hypothetical.

phosphate join head to head, with the elimination of both pyrophosphate groups, to form **squalene**.

The common names of these intermediates derive from the sources from which they were first isolated.

FIGURE 21-36 Formation of squalene. This 30-carbon structure arises through successive condensations of activated isoprene (five-carbon) units.

Geraniol, a component of rose oil, has the aroma of geraniums, and farnesol is an aromatic compound found in the flowers of the Farnese acacia tree. Many natural scents of plant origin are synthesized from isoprene units. Squalene, first isolated from the liver of sharks (genus *Squalus*), has 30 carbons, 24 in the main chain and 6 in the form of methyl group branches.

Stage ④ Conversion of Squalene to the Four-Ring Steroid Nucleus

When the squalene molecule is represented as in **Figure 21–37**, the relationship of its linear structure to the cyclic structure of the sterols becomes apparent. All sterols have the four fused rings that form the steroid nucleus, and all are alcohols, with a hydroxyl group at C-3—thus the name "sterol." The action of **squalene monooxygenase** adds one oxygen atom from O_2 to the

end of the squalene chain, forming an epoxide. This enzyme is another mixed-function oxidase (Box 21–1); NADPH reduces the other oxygen atom of O_2 to H_2O. The double bonds of the product, **squalene 2,3-epoxide**, are positioned so that a remarkable concerted reaction can convert the linear squalene epoxide to a cyclic structure. In animal cells, this cyclization results in the formation of **lanosterol**, which contains the four rings characteristic of the steroid nucleus. Lanosterol is finally converted to cholesterol in a series of about 20 reactions that include the migration of some methyl groups and the removal of others. Elucidation of this extraordinary biosynthetic pathway, one of the most complex known, was accomplished by Konrad Bloch, Feodor Lynen, John Cornforth, and George Popják in the late 1950s.

Konrad Bloch, 1912–2000

Feodor Lynen, 1911–1979

John Cornforth

George Popják, 1914–1998

Cholesterol is the sterol characteristic of animal cells; plants, fungi, and protists make other, closely related sterols instead. They use the same synthetic pathway as far as squalene 2,3-epoxide, at which point the pathways diverge slightly, yielding other sterols, such as stigmasterol in many plants and ergosterol in fungi (Fig. 21–37).

Squalene

NADPH + H$^+$

squalene monooxygenase

O$_2$
H$_2$O
NADP$^+$

Squalene 2,3-epoxide

multistep (plants)

cyclase (animals)

multistep (fungi)

HO— Stigmasterol   C$_2$H$_5$

HO—

HO— Ergosterol

cyclase

HO— Lanosterol

multistep

HO— Cholesterol

FIGURE 21–37 Ring closure converts linear squalene to the condensed steroid nucleus. The first step in this sequence is catalyzed by a mixed-function oxidase (a monooxygenase), for which the cosubstrate is NADPH. The product is an epoxide, which in the next step is cyclized to the steroid nucleus. The final product of these reactions in animal cells is cholesterol; in other organisms, slightly different sterols are produced, as shown.

Energetic Cost of Squalene Synthesis

What is the energetic cost of the synthesis of squalene from acetyl-CoA, in number of ATP per molecule of squalene synthesized?

Solution: In the pathway from acetyl-CoA to squalene, ATP is consumed only in the steps that convert mevalonate to the activated isoprene precursors of squalene. Three ATP molecules are used to create each of the six activated isoprenes required to construct squalene, for a total cost of 18 ATP molecules.

Cholesterol Has Several Fates

Most of the cholesterol synthesis in vertebrates takes place in the liver. A small fraction of the cholesterol made there is incorporated into the membranes of hepatocytes, but most of it is exported in one of three forms: bile acids, biliary cholesterol, or cholesteryl esters **(Fig. 21–38)**. Small quantities of oxysterols such as 25-hydroxycholesterol are formed in the liver, and act as regulators of cholesterol synthesis (see below). In other tissues, cholesterol is converted into steroid hormones (in the adrenal cortex and gonads, for example; see Fig. 10–19) or vitamin D hormone (i.e., in the liver and kidney; see Fig. 10–20). Such hormones are extremely potent biological signals acting through nuclear receptor proteins.

One of the three forms of cholesterol exported from the liver is bile, a fluid stored in the gallbladder and excreted into the small intestine to aid in the digestion of fat-containing meals. Its principal components are **bile acids** and their salts—both relatively hydrophilic cholesterol derivatives synthesized in the liver that serve as emulsifiers in the intestine, converting large particles of fat into tiny micelles and thereby greatly increasing the surface at which digestive lipases can act (see Fig. 17–1). Bile also contains much smaller amounts of cholesterol.

Cholesteryl esters are formed in the liver through the action of **acyl-CoA–cholesterol acyl transferase (ACAT)**. This enzyme catalyzes the transfer of a fatty acid from coenzyme A to the hydroxyl group of cholesterol (Fig. 21–38), converting the cholesterol to a more hydrophobic form and preventing it from entering membranes. Cholesteryl esters are transported in secreted lipoprotein particles to other tissues that use cholesterol, or they are stored in the liver in lipid droplets.

Cholesterol and Other Lipids Are Carried on Plasma Lipoproteins

Cholesterol and cholesteryl esters, like triacylglycerols and phospholipids, are essentially insoluble in water, yet must be moved from the tissue of origin to the tissues in which they will be stored or consumed. To facilitate their transport, they are carried in the blood plasma as **plasma lipoproteins**, macromolecular complexes of specific carrier proteins, called **apolipoproteins**, and various combinations of phospholipids, cholesterol, cholesteryl esters, and triacylglycerols.

Apolipoproteins ("apo" designates the protein in its lipid-free form) combine with lipids to form several

FIGURE 21–38 Metabolic fates of cholesterol. Modifications of the cholesterol structure are shown in red. Esterification converts cholesterol to an even more hydrophobic form for storage and transport; each of the other modifications yields a less hydrophobic product.

(a)

Chylomicrons (×60,000) VLDL (×180,000)

LDL (×180,000) HDL (×180,000)

(b)

FIGURE 21–39 Lipoproteins. (a) Structure of a low-density lipoprotein (LDL). Apolipoprotein B-100 (apoB-100) is one of the largest single polypeptide chains known, with 4,636 amino acid residues (M_r 512,000). One particle of LDL contains a core with about 1,500 molecules of cholesteryl esters, surrounded by a shell composed of about 500 more molecules of cholesterol, 800 molecules of phospholipids, and one molecule of apoB-100. **(b)** Four classes of lipoproteins, visualized in the electron microscope after negative staining. Clockwise from top left: chylomicrons, 50 to 200 nm in diameter; VLDL, 28 to 70 nm; HDL, 8 to 11 nm; and LDL, 20 to 25 nm. The particle sizes given are those measured for these samples; particle sizes vary considerably in different preparations. For properties of lipoproteins, see Table 21-1.

classes of lipoprotein particles, spherical complexes with hydrophobic lipids in the core and hydrophilic amino acid side chains at the surface **(Fig. 21–39a)**. Different combinations of lipids and proteins produce particles of different densities, ranging from chylomicrons to high-density lipoproteins. These particles can be separated by ultracentrifugation (Table 21–1) and visualized by electron microscopy (Fig. 21–39b).

Each class of lipoprotein has a specific function, determined by its point of synthesis, lipid composition, and apolipoprotein content. At least ten distinct apolipoproteins are found in the lipoproteins of human plasma (Table 21–2), distinguishable by their size, their reactions with specific antibodies, and their characteristic distribution in the lipoprotein classes. These protein components act as signals, targeting lipoproteins to specific tissues or activating enzymes that act on the lipoproteins. They have also been implicated in disease; Box 21–2 describes a link between apoE and Alzheimer disease. **Figure 21–40** provides an overview of the formation and transport of the lipoproteins in mammals. The numbered steps in the following discussion refer to this figure.

Chylomicrons, discussed in Chapter 17 in connection with the movement of dietary triacylglycerols from the intestine to other tissues, are the largest of the lipoproteins and the least dense, containing a high proportion of triacylglycerols (see Fig. 17–2). ❶ Chylomicrons

TABLE 21–1 Major Classes of Human Plasma Lipoproteins: Some Properties

Lipoprotein	Density (g/mL)	Composition (wt %)				
		Protein	Phospholipids	Free cholesterol	Cholesteryl esters	Triacylglycerols
Chylomicrons	<1.006	2	9	1	3	85
VLDL	0.95–1.006	10	18	7	12	50
LDL	1.006–1.063	23	20	8	37	10
HDL	1.063–1.210	55	24	2	15	4

Source: Modified from Kritchevsky, D. (1986) Atherosclerosis and nutrition. *Nutr. Int.* 2, 290–297.

TABLE 21–2 Apolipoproteins of the Human Plasma Lipoproteins

Apolipoprotein	Polypeptide molecular weight	Lipoprotein association	Function (if known)
ApoA-I	28,100	HDL	Activates LCAT; interacts with ABC transporter
ApoA-II	17,400	HDL	Inhibits LCAT
ApoA-IV	44,500	Chylomicrons, HDL	Activates LCAT; cholesterol transport/clearance
ApoB-48	242,000	Chylomicrons	Cholesterol transport/clearance
ApoB-100	512,000	VLDL, LDL	Binds to LDL receptor
ApoC-I	7,000	VLDL, HDL	
ApoC-II	9,000	Chylomicrons, VLDL, HDL	Activates lipoprotein lipase
ApoC-III	9,000	Chylomicrons, VLDL, HDL	Inhibits lipoprotein lipase
ApoD	32,500	HDL	
ApoE	34,200	Chylomicrons, VLDL, HDL	Triggers clearance of VLDL and chylomicron remnants

Source: Modified from Vance, D.E. & Vance, J.E. (eds) (2008) *Biochemistry of Lipids and Membranes,* 5th edn, Elsevier Science Publishing.

are synthesized from dietary fats in the ER of enterocytes, epithelial cells that line the small intestine. The chylomicrons then move through the lymphatic system and enter the bloodstream via the left subclavian vein. The apolipoproteins of chylomicrons include apoB-48 (unique to this class of lipoproteins), apoE, and apoC-II (Table 21–2). ❷ ApoC-II activates lipoprotein lipase in the capillaries of adipose, heart, skeletal muscle, and lactating mammary tissues, allowing the release of free fatty acids (FFA) to these tissues. Chylomicrons thus carry dietary fatty acids to tissues where they will be consumed or stored as fuel. ❸ The remnants of chylomicrons (depleted of most of their triacylglycerols but still containing cholesterol, apoE, and apoB-48) move through the bloodstream to the liver. Receptors in the liver bind to the apoE in the chylomicron remnants and mediate their uptake by endocytosis. ❹ In the liver, the remnants release their cholesterol and are degraded in lysosomes. This pathway from dietary cholesterol to the liver is the **exogenous pathway** (blue arrows in Fig. 21–40).

When the diet contains more fatty acids and cholesterol than are needed immediately as fuel or precursors to other molecules, they are ❺ converted to triacylglycerols or cholesteryl esters in the liver and packaged with specific apolipoproteins into **very-low-density lipoprotein (VLDL)**. Excess carbohydrate in the diet can also be converted to triacylglycerols in the liver and exported as VLDL. In addition to triacylglycerols and cholesteryl esters, VLDL contains apoB-100, apoC-I,

BOX 21–2 🩺 MEDICINE ApoE Alleles Predict Incidence of Alzheimer Disease

In the human population there are three common variants, or alleles, of the gene encoding apolipoprotein E. The most common, accounting for about 78% of human apoE alleles, is *APOE3*; alleles *APOE4* and *APOE2* account for 15% and 7%, respectively. The *APOE4* allele is particularly common in humans with Alzheimer disease, and the link is highly predictive. Individuals who inherit *APOE4* have an increased risk of late-onset Alzheimer disease. Those who are homozygous for *APOE4* have a 16-fold increased risk of developing the disease; for those who do, the mean age of onset is just under 70 years. For people who inherit two copies of *APOE3*, by contrast, the mean age of onset of Alzheimer disease exceeds 90 years.

The molecular basis for the association between apoE-4 and Alzheimer disease is not yet known. It is also not clear how apoE-4 might affect the growth of the amyloid fibers that appear to be the primary causative agents of the disease (see Fig. 4–32). Speculation has focused on a possible role for apoE in stabilizing the cytoskeletal structure of neurons. The apoE-2 and apoE-3 proteins bind to a number of proteins associated with neuronal microtubules, whereas apoE-4 does not. This may accelerate the death of neurons. Whatever the mechanism proves to be, these observations promise to expand our understanding of the biological functions of apolipoproteins.

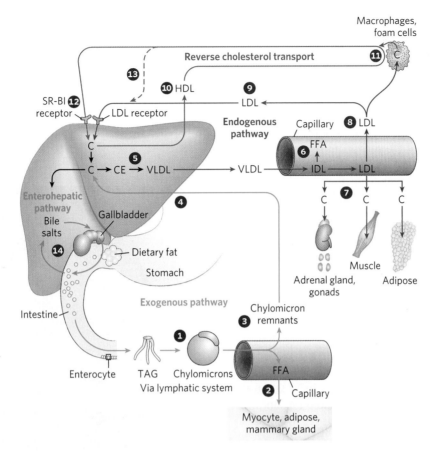

FIGURE 21–40 Lipoproteins and lipid transport. Lipids are transported in the bloodstream as lipoproteins, which exist as several variants that have different functions, different protein and lipid compositions (see Tables 21–1, 21–2), and thus different densities. Numbered steps are described in the text. In the exogenous pathway (blue arrows), dietary lipids are packaged into chylomicrons; much of their triacylglycerol content is released by lipoprotein lipase to adipose and muscle tissues during transport through capillaries. Chylomicron remnants (containing largely protein and cholesterol) are taken up by the liver. Bile salts produced in the liver aid in dispersing dietary fats, and are then reabsorbed in the enterohepatic pathway (green arrows). In the endogenous pathway (red arrows), lipids synthesized or packaged in the liver are delivered to peripheral tissues by VLDL. Extraction of lipid from VLDL (along with loss of some apolipoproteins) gradually converts some of it to LDL, which delivers cholesterol to extrahepatic tissues or returns to the liver. Excess cholesterol in extrahepatic tissues is transported back to the liver as HDL in reverse cholesterol transport (purple arrows). C represents cholesterol; CE represents cholesteryl ester.

apoC-II, apoC-III, and apoE (Table 21–2). VLDL is transported in the blood from the liver to muscle and adipose tissue. ❻ In the capillaries of these tissues, apoC-II activates lipoprotein lipase, which catalyzes the release of free fatty acids from triacylglycerols in the VLDL. Adipocytes take up these fatty acids, reconvert them to triacylglycerols, and store the products in intracellular lipid droplets; myocytes, in contrast, primarily oxidize the fatty acids to supply energy. When the insulin level is high (after a meal), VLDL serves primarily to convey lipids from the diet to adipose tissue for storage there. In the fasting state between meals, the fatty acids used to produce VLDL in the liver originate primarily from the adipose tissue, and the principal VLDL target is myocytes of the heart and skeletal muscle.

The loss of triacylglycerol converts some VLDL to VLDL remnants (also called intermediate-density lipoprotein, IDL). Further removal of triacylglycerol from IDL (remnants) produces **low-density lipoprotein (LDL)**. Rich in cholesterol and cholesteryl esters, and containing apoB-100 as their major apolipoprotein, ❼ LDL carries cholesterol to extrahepatic tissues such as muscle, adrenal glands, and adipose tissue. These tissues have plasma membrane LDL receptors that recognize apoB-100 and mediate the uptake of cholesterol and cholesteryl esters. ❽ LDL also delivers cholesterol to macrophages, sometimes converting them into foam cells (see Fig. 21–46). ❾ LDL not taken up by peripheral tissues and cells returns to the liver and is taken up via **LDL receptors** in the hepatocyte plasma membrane. Cholesterol that enters hepatocytes by this path may be incorporated into membranes, converted to bile acids, or reesterified by ACAT (Fig. 21–38) for storage within cytosolic lipid droplets. This pathway, from VLDL formation in the liver to LDL return to the liver, is the **endogenous pathway** of cholesterol metabolism and transport (red arrows in Fig. 21–40). Accumulation of excess intracellular cholesterol is prevented by reducing the rate of cholesterol synthesis when sufficient cholesterol is available from LDL in the blood. Regulatory mechanisms to accomplish this are described below. We will return to Figure 21–40 and other pathways of lipoprotein transport after a discussion of LDL uptake by cells.

Cholesteryl Esters Enter Cells by Receptor-Mediated Endocytosis

Each LDL particle in the bloodstream contains apoB-100, which is recognized by LDL receptors present in the plasma membranes of cells that need to take up cholesterol. **Figure 21–41** shows such a cell, in which ❶ LDL receptors are synthesized in the Golgi complex and transported to the plasma membrane, where they are available to bind apoB-100. ❷ The binding of LDL to an LDL receptor initiates endocytosis, which ❸ conveys the LDL and its receptor into the cell within an endosome. ❹ The receptor-containing portions of the endosome membrane bud off and are returned to the cell surface, to function again in LDL uptake. ❺ The endosome fuses with a lysosome, which ❻ contains enzymes that hydrolyze the cholesteryl esters, releasing cholesterol and fatty acids into the cytosol. The apoB-100 protein is also degraded to amino acids that are released to the cytosol. ApoB-100 is also present in VLDL, but its receptor-binding domain is not available for binding to the LDL receptor; conversion of VLDL to LDL exposes the receptor-binding domain of apoB-100.

This pathway for the transport of cholesterol in blood and its **receptor-mediated endocytosis** by target tissues was elucidated by Michael Brown and Joseph Goldstein. They discovered that individuals with the genetic disease **familial hypercholesterolemia (FH)** have mutations in the LDL receptor that prevent the normal uptake of LDL by liver and peripheral tissues. The result of defective LDL uptake is very high blood levels of LDL (and of the cholesterol it carries). Individuals with FH have a greatly increased probability of developing atherosclerosis, a disease of

Michael Brown and Joseph Goldstein

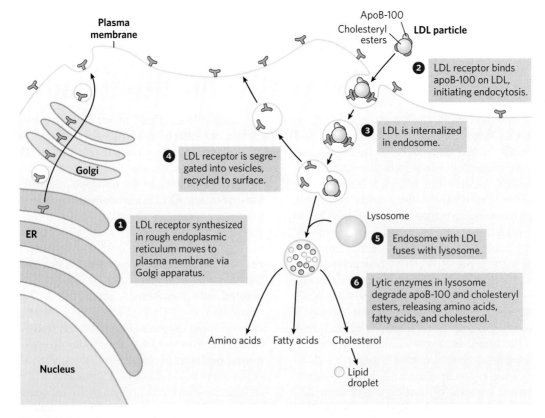

FIGURE 21–41 Uptake of cholesterol by receptor-mediated endocytosis.

the cardiovascular system in which blood vessels are occluded by cholesterol-rich plaques (see Fig. 21–46).

Niemann-Pick type-C (NPC) disease is an inherited defect in lipid storage, in which cholesterol is not transported out of the lysosomes and instead accumulates in lysosomes of liver, brain, and lung, bringing about early death. NPC is the result of a mutation in either of two genes (*NPC1*, *NPC2*) essential to moving cholesterol out of the lysosome and into the cytosol, where it can be further metabolized. *NPC1* encodes a transmembrane lysosomal protein, and *NPC2* encodes a soluble protein. These proteins act in tandem to transfer cholesterol out of the lysosome and into the cytosol for further processing or metabolism. ■

HDL Carries Out Reverse Cholesterol Transport

A fourth major lipoprotein in mammals, **high-density lipoprotein (HDL)**, originates in the liver and small intestine as small, protein-rich particles that contain relatively little cholesterol and no cholesteryl esters (❿, Fig. 21–40). HDLs contain primarily apoA-I and other apolipoproteins (Table 21–2). They also contain the enzyme **lecithin-cholesterol acyl transferase (LCAT)**, which catalyzes the formation of cholesteryl esters from lecithin (phosphatidylcholine) and cholesterol **(Fig. 21–42)**. LCAT on the surface of nascent (newly forming) HDL particles converts the cholesterol and phosphatidylcholine of chylomicron and VLDL remnants encountered in the bloodstream to cholesteryl esters, which begin to form a core, transforming the disk-shaped nascent HDL to a mature, spherical HDL particle. ⓫ Nascent HDL can also pick up cholesterol from cholesterol-rich extrahepatic cells (including macrophages and foam cells formed from them; see below). ⓬ Mature HDL then returns to the liver, where the cholesterol is unloaded via the scavenger receptor SR-BI. ⓭ Some of the cholesteryl esters in HDL can also be transferred to LDL by the cholesteryl ester transfer protein. The HDL circuit is **reverse cholesterol transport** (purple arrows in Fig. 21–40). Much of this cholesterol is converted to bile salts in the liver and stored in the gallbladder. When a meal is ingested, bile salts are excreted into the intestine, where they disperse macroscopic pieces of fat into microscopic micelles that can be attacked by lipases. Bile salts are reabsorbed by the liver ⓮ and recirculate through the gallbladder in this **enterohepatic circulation** (green arrows in Fig. 21–40).

The mechanism of sterol unloading via the receptor SR-BI in liver and other tissues does not involve endocytosis, the mechanism used for LDL uptake. Instead, when HDL binds to SR-BI receptors in the plasma membranes of hepatocytes or steroidogenic tissues such as the adrenal gland, these receptors mediate partial and selective transfer of cholesterol and other lipids in HDL into the cell. Depleted HDL then dissociates to recirculate in the bloodstream and extract more lipids from

FIGURE 21–42 Reaction catalyzed by lecithin-cholesterol acyl transferase (LCAT). This enzyme is present on the surface of HDL and is stimulated by the HDL component apoA-I. Cholesteryl esters accumulate within nascent HDLs, converting them to mature HDLs.

remnants of chylomicrons and VLDL, and from cells overloaded with cholesterol, as described below.

Cholesterol Synthesis and Transport Are Regulated at Several Levels

Cholesterol synthesis is a complex and energy-expensive process. Excess cholesterol cannot be catabolized for use as fuel, and must therefore be excreted. Therefore, it is clearly advantageous to an organism to regulate the biosynthesis of cholesterol to complement dietary intake. In mammals, cholesterol production is regulated by intracellular cholesterol concentration, by the supply of ATP, and by the hormones glucagon and insulin. The committed step in the pathway to cholesterol (and a major site of regulation) is the conversion of HMG-CoA to mevalonate (Fig. 21–34), the reaction catalyzed by HMG-CoA reductase.

Short-term regulation of the *activity* of existing HMG-CoA reductase is accomplished by reversible covalent alteration—phosphorylation by the AMP-dependent protein kinase (AMPK), which senses high AMP concentration (indicating low ATP concentration). Thus, when ATP levels drop, the synthesis of cholesterol slows, and catabolic pathways for the generation of ATP are stimulated **(Fig. 21–43)**. Hormones that mediate global regulation of lipid and carbohydrate metabolism also act on HMG-CoA reductase; glucagon stimulates its phosphorylation (inactivation), and insulin promotes dephosphorylation, activating the enzyme and favoring cholesterol synthesis. These covalent regulatory mechanisms are probably not as important, quantitatively, as the mechanisms that affect the synthesis and degradation of the enzyme.

In the longer term, the *number of molecules* of HMG-CoA reductase is increased or decreased in response to the cellular concentrations of cholesterol. Regulation of HMG-CoA reductase synthesis by cholesterol is mediated by an elegant system of transcriptional regulation of the HMG-CoA gene **(Fig. 21–44)**. This gene, along with more than 20 other genes encoding enzymes that mediate the uptake and synthesis of cholesterol and unsaturated fatty acids, is controlled by a small family of proteins called **sterol regulatory element-binding proteins (SREBPs)**. When newly synthesized, these proteins are embedded in the ER. Only the soluble regulatory domain fragment of an SREBP functions as a transcriptional activator, using mechanisms discussed in Chapter 28. When cholesterol and oxysterol levels are high, SREBPs are held in the ER in a complex with another protein called **SREBP**

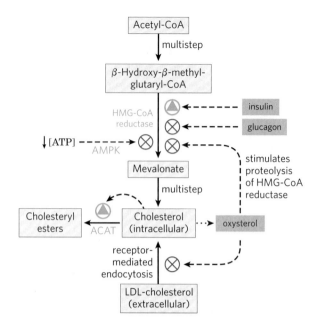

FIGURE 21–43 Regulation of cholesterol formation balances synthesis with dietary uptake and energy state. Insulin promotes dephosphorylation (activation) of HMG-CoA reductase; glucagon promotes its phosphorylation (inactivation); and the AMP-dependent protein kinase AMPK, when activated by low [ATP] relative to [AMP], phosphorylates and inactivates it. Oxysterol metabolites of cholesterol (for example, 24(S)-hydroxycholesterol) stimulate proteolysis of HMG-CoA reductase.

cleavage-activating protein (SCAP), which in turn is anchored in the ER membrane by its interaction with a third membrane protein, **Insig (*ins*ulin-*i*nduced *g*ene protein)** (Fig. 21–44a). SCAP and Insig act as sterol sensors. When sterol levels are high, the

(a) High [sterol] in ER
SCAP/SREBP retained in ER, bound to Insig

(b) Low [sterol] in ER
Regulatory domain of SREBP released by proteolysis

(c) Increased cholesterol synthesis in ER

FIGURE 21–44 Regulation of cholesterol synthesis by SREBP. Sterol regulatory element-binding proteins (SREBPs, shown in green) are embedded in the ER when first synthesized, in a complex with the protein SREBP cleavage-activating protein (SCAP, red), which is in turn bound to Insig (blue). (*N* and *C* represent the amino and carboxyl termini of the proteins.) **(a)** When bound to SCAP and Insig, SREBPs are inactive. **(b)** When sterol levels decline, sterol-binding sites on Insig and SCAP are unoccupied, the complex migrates to the Golgi complex, and SREBP is cleaved to produce a regulatory domain, which **(c)** acts in the nucleus to increase the transcription of sterol-regulated genes. Insig is targeted for degradation by the attachment of several ubiquitin molecules.

Insig-SCAP-SREBP complex is retained in the ER membrane. When the level of sterols in the cell declines (Fig. 21–44b), the SCAP-SREBP complex is escorted by secretory proteins to the Golgi complex. There, two proteolytic cleavages of SREBP release a regulatory fragment, which enters the nucleus and activates transcription of its target genes, including HMG-CoA reductase, the LDL receptor protein, and a number of other proteins needed for lipid synthesis. When sterol levels increase sufficiently, the proteolytic release of SREBP amino-terminal domains is again blocked, and proteasome degradation of the existing active domains results in a rapid shutdown of the gene targets.

In the long term, the level of HMG-CoA reductase is also regulated by proteolytic degradation of the enzyme itself. High levels of cellular cholesterol are sensed by Insig, which triggers attachment of ubiquitin molecules to HMG-CoA reductase, leading to its degradation by proteasomes (see Fig. 27–48).

Liver X receptor (LXR) is a nuclear transcription factor activated by oxysterol ligands (reflecting high cholesterol levels), which integrates the metabolism of fatty acids, sterols, and glucose. LXRα is expressed primarily in liver, adipose tissue, and macrophages; LXRβ is present in all tissues. When bound to an oxysterol ligand, LXRs form heterodimers with a second type of nuclear receptor, the **retinoid X receptors (RXR)**, and the LXR-RXR dimer activates transcription from a set of genes **(Fig. 21–45)** including those for acetyl-CoA carboxylase (the first enzyme in fatty acid synthe-

sis); fatty acid synthase; the cytochrome P-450 enzyme CYP7A1, required for sterol conversion to bile acid; apoproteins involved in cholesterol transport (apoC-I, apoC-II, apoD, and apoE); the ABC transporters ABCA1 and ABCG1, involved in reverse cholesterol transport (see below); GLUT4, the insulin-stimulated glucose transporter of muscle and adipose tissue; and SREBP1C. The transcriptional regulators LXR and SREBP therefore work together to achieve and maintain cholesterol homeostasis; SREBPs are activated by low levels of cellular cholesterol, and LXRs are activated by high cholesterol levels.

Finally, two other regulatory mechanisms influence cellular cholesterol level: (1) high intracellular concentrations of cholesterol activate ACAT, which increases esterification of cholesterol for storage, and (2) high cellular cholesterol levels diminish (via SREBP) transcription of the gene that encodes the LDL receptor, reducing production of the receptor and thus the uptake of cholesterol from the blood.

Dysregulation of Cholesterol Metabolism Can Lead to Cardiovascular Disease

When the sum of cholesterol synthesized and cholesterol obtained in the diet exceeds the amount required for the synthesis of membranes, bile salts, and steroids, pathological accumulations of cholesterol (plaques) can obstruct blood vessels, a condition called **atherosclerosis**. Heart failure due to occluded coronary arteries is a leading cause of death in industrialized societies. Atherosclerosis is linked to high levels of cholesterol in the blood, and particularly to high levels of LDL-cholesterol ("bad cholesterol"); there is a *negative* correlation between HDL ("good cholesterol") levels and arterial disease. Plaque formation in blood vessels is initiated when LDL containing partially oxidized fatty acyl groups adheres to and accumulates in the extracellular matrix of epithelial cells lining arteries **(Fig. 21–46)**. Immune cells (monocytes) are attracted to regions with such LDL accumulations, and they differentiate into macrophages, which take up the oxidized LDL and the cholesterol they contain. Macrophages cannot limit their uptake of sterols, and with increasing accumulation of cholesteryl esters and free cholesterol, the macrophages become **foam cells** (they appear foamy in the microscope). As excess free cholesterol accumulates in foam cells and their membranes, they undergo apoptosis. Over long periods of time, arteries become progressively occluded as plaques consisting of extracellular matrix material, scar tissue formed from smooth muscle tissue, and foam cell remnants gradually become larger. Occasionally a plaque breaks loose from the site of its formation and is carried through the blood to a narrowed region of an artery in the brain or the heart, causing a stroke or a heart attack.

In familial hypercholesterolemia, blood levels of cholesterol are extremely high and severe atherosclerosis

FIGURE 21–45 Action of RXR-LXR dimer on expression of genes for lipid and glucose metabolism. When their ligands are absent, RXR and LXR associate with a corepressor protein, preventing transcription of the genes associated with the LXR element (LXRE). When their respective ligands are present (9-*cis*-retinoic acid for RXR, cholesterol or oxysterols for LXR), the dimer dissociates from the corepressor, then associates with a coactivator protein. This complex binds to the LXR element and turns on the expression of the associated genes. Regulation of gene expression is a topic discussed in more detail in Chapter 28.

FIGURE 21–46 Role of foam cells in formation of atherosclerotic plaques.

BOX 21–3 ⚕ MEDICINE The Lipid Hypothesis and the Development of Statins

Coronary heart disease is the leading cause of death in developed countries. The coronary arteries that bring blood to the heart become narrowed due to the formation of fatty deposits called atherosclerotic plaques (containing cholesterol, fibrous proteins, calcium deposits, blood platelets, and cell debris). Developing the link between artery occlusion (atherosclerosis) and blood cholesterol levels was a project of the twentieth century, triggering a dispute that was resolved only with the development of effective cholesterol-lowering drugs. The Framingham Heart Study is a longitudinal study, begun in 1948 and continuing today, aimed at identifying factors correlated with cardiovascular disease. About 5,000 participants from the city of Framingham, Massachusetts, underwent periodic physical examinations and lifestyle interviews. By 2002, participants of the third generation were included in the study. This monumental study led to the identification of risk factors for cardiovascular disease, including smoking, obesity, physical inactivity, diabetes, high blood pressure, and high blood cholesterol.

In 1913, N. N. Anitschkov, an experimental pathologist in St. Petersburg, Russia, published a study showing that rabbits fed a diet rich in cholesterol developed lesions very similar to the atherosclerotic plaques seen in aging humans. Anitschkov continued his work over the next few decades, publishing it in prominent western journals. Nevertheless, the work was not accepted as a model for human atherosclerosis, due to a prevailing view that the disease was a simple consequence of aging and could not be prevented. The link between serum cholesterol and atherosclerosis (the lipid hypothesis) was gradually strengthened, until some researchers in the 1960s openly suggested that therapeutic intervention might be helpful. However, controversy persisted until the results of a large study of cholesterol lowering, sponsored by the United States National Institutes of Health, was published in 1984: the Coronary Primary Prevention Trial. This study conclusively showed a statistically significant decrease in heart attacks and strokes as a result of decreasing cholesterol level. The study made use of a bile acid–binding resin, cholestyramine, to control cholesterol. The results triggered a search for more effective therapeutic interventions. Some controversy persisted until the development of the statins in the late 1980s and 1990s.

Dr. Akira Endo, working at the Sankyo company in Tokyo, discovered the first statin and reported the work in 1976. Endo had been interested in cholesterol metabolism for some time, and speculated in 1971 that the fungi being screened at that time for new antibiotics might also contain an inhibitor of cholesterol synthesis. Over a period of several years, he screened more than 6,000 fungal cultures until a positive result was found. The compound that resulted was named compactin (Fig. 1). The compound eventually proved effective in

$R_1 = H$	$R_2 = H$	Compactin
$R_1 = CH_3$	$R_2 = CH_3$	Simvastatin (Zocor)
$R_1 = H$	$R_2 = OH$	Pravastatin (Pravachol)
$R_1 = H$	$R_2 = CH_3$	Lovastatin (Mevacor)

FIGURE 1 Statins as inhibitors of HMG-CoA reductase. A comparison of the structures of mevalonate and four pharmaceutical compounds (statins) that inhibit HMG-CoA reductase.

develops in childhood. These individuals have a defective LDL receptor and lack receptor-mediated uptake of cholesterol carried by LDL. Consequently, cholesterol is not cleared from the blood; it accumulates in foam cells and contributes to the formation of atherosclerotic plaques. Endogenous cholesterol synthesis continues despite the excessive cholesterol in the blood, because extracellular cholesterol cannot enter cells to regulate intracellular synthesis (Fig. 21–44). A class of drugs called **statins**, some isolated from natural sources and some synthesized by the pharmaceutical industry, is used to treat patients with familial hypercholesterolemia and other conditions involving elevated serum cholesterol. The statins resemble mevalonate (Box 21–3) and are competitive inhibitors of HMG-CoA reductase. ■

Reverse Cholesterol Transport by HDL Counters Plaque Formation and Atherosclerosis

HDL plays a critical role in the reverse cholesterol transport pathway **(Fig. 21–47)**, reducing the potential damage from foam cell buildup. Depleted HDL (low in cholesterol) picks up cholesterol stored in extrahepatic tissues (including foam cells at nascent plaques) and carries it to the liver. Two ATP-binding cassette (ABC) transporters are involved in cholesterol exit from cells. In this process, apoA-I interacts with an ABC transporter (ABCA1) in a cholesterol-rich cell. ABCA-1 transports a load of cholesterol from inside the cell to the outer surface of the plasma membrane, where lipid-free or lipid-poor apoA-I picks it up and transports it to the liver. Another ABC transporter (ABCG1) interacts

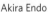

Akira Endo   Alfred Alberts   P. Roy Vagelos

Statins inhibit HMG-CoA reductase in part by mimicking the structure of mevalonate (Fig. 1), and thus inhibit cholesterol synthesis. Lovastatin treatment lowers serum cholesterol by as much as 30% in individuals with hypercholesterolemia resulting from one defective copy of the gene for the LDL receptor. When combined with an edible resin that binds bile acids and prevents their reabsorption from the intestine, the drug is even more effective.

reducing cholesterol levels in dogs and monkeys, and the work came to the attention of Michael Brown and Joseph Goldstein at the University of Texas–Southwestern medical school. Brown and Goldstein began to work with Endo, and confirmed his results. Some dramatic results in the first limited clinical trials convinced several pharmaceutical firms to join the hunt for statins. A team at Merck led by Alfred Alberts and P. Roy Vagelos began screening fungal cultures and found a positive result after screening just 18 cultures. The new statin was eventually called lovastatin (Fig. 1). A rumor that compactin, at very high doses, was carcinogenic in dogs almost sidelined the race to develop statins in 1980, but the benefits to patients with familial hypercholesterolemia were already evident. After much consultation with experts around the world and the U.S. Food and Drug Administration, Merck proceeded carefully to develop lovastatin. Extensive testing over the next two decades revealed no carcinogenic effects from lovastatin or from the newer generations of statins that have appeared since.

Statins are now the most widely used drugs for lowering serum cholesterol levels. Side effects are always a concern with drugs, but statins represent a case where many of the side effects are positive. These drugs can improve blood flow, enhance the stability of atherosclerotic plaques (so they don't rupture and obstruct blood flow), reduce platelet aggregation, and reduce vascular inflammation. Some of these effects occur before cholesterol levels drop in patients taking statins for the first time, and may be related to a secondary inhibition of isoprenoid synthesis by statins. Not all the effects of statins are positive. A few patients, usually among those taking statins in combination with other cholesterol-lowering drugs, experience muscle pain or weakness that can become severe and even debilitating. A fairly long list of other side effects has been documented in patients; most are rare. However, for the vast majority of patients, the statin-mediated decrease in the risks associated with coronary heart disease can be dramatic. As with all medications, statins should be used only in consultation with a physician.

FIGURE 21–47 **Reverse cholesterol transport.** ApoA-I and HDLs pick up excess cholesterol from peripheral cells, with the participation of ABCA1 and ABCG1 transporters, and return it to the liver. In individuals with genetically defective ABCA1, the failure of reverse cholesterol transport leads to severe and early cardiovascular diseases: Tangier disease and familial HDL deficiency disease.

with mature HDL, facilitating the movement of cholesterol out of the cell and into the HDL. This efflux process is particularly critical when it involves reverse cholesterol transport away from foam cells at the sites of plaques that form in blood vessels in individuals with cardiovascular disease.

In **familial HDL deficiency**, HDL levels are very low, and in **Tangier disease** they are almost undetectable (Fig. 21–47). Both genetic disorders are the result of mutations in the ABCA1 protein. ApoA-I in cholesterol-depleted HDL cannot take up cholesterol from cells that lack ABCA1 protein, and apoA-I and cholesterol-poor HDL are rapidly removed from the blood and destroyed. Both familial HDL deficiency and Tangier disease are very rare (worldwide, fewer than 100 families with Tangier disease are known), but the existence of these diseases establishes a role for ABCA1 and ABCG1 proteins in the regulation of plasma HDL levels. ∎

Steroid Hormones Are Formed by Side-Chain Cleavage and Oxidation of Cholesterol

Humans derive all their steroid hormones from cholesterol (**Fig. 21–48**). Two classes of steroid hormones are synthesized in the cortex of the adrenal gland: **mineralocorticoids**, which control the reabsorption of inorganic ions (Na^+, Cl^-, and HCO_3^-) by the kidney, and **glucocorticoids**, which help regulate gluconeogenesis and reduce the inflammatory response. Sex hormones are produced in male and female gonads and the placenta. They include **progesterone**, which regulates the female reproductive cycle, and **androgens** (such as testosterone) and **estrogens** (such as estradiol), which influence the development of secondary sexual characteristics in males and females, respectively. Steroid hormones are effective at very low concentrations and are therefore synthesized in relatively small quantities.

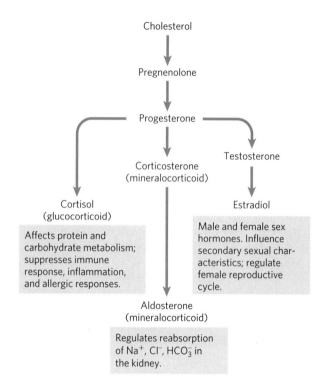

FIGURE 21–48 **Some steroid hormones derived from cholesterol.** The structures of some of these compounds are shown in Figure 10-19.

In comparison with the bile salts, their production consumes relatively little cholesterol.

Progesterone

Synthesis of steroid hormones requires removal of some or all of the carbons in the "side chain" on C-17 of the D ring of cholesterol. Side-chain removal takes place in the mitochondria of steroidogenic tissues. Removal involves the hydroxylation of two adjacent carbons in the side chain (C-20 and C-22) followed by cleavage of the bond between them (**Fig. 21–49**). Formation of the various hormones also involves the introduction of oxygen atoms. All the hydroxylation and oxygenation reactions in steroid biosynthesis are catalyzed by mixed-function oxidases (Box 21–1) that use NADPH, O_2, and mitochondrial cytochrome P-450.

Intermediates in Cholesterol Biosynthesis Have Many Alternative Fates

In addition to its role as an intermediate in cholesterol biosynthesis, isopentenyl pyrophosphate is the activated precursor of a huge array of biomolecules with diverse biological roles (**Fig. 21–50**). They include vitamins A, E, and K; plant pigments such as carotene and the phytol

Cholesterol

mixed-function oxidase

cyt P-450 adrenodoxin (Fe-S)

adrenodoxin reductase (flavoprotein)

$2O_2$

$2H_2O$

2 NADPH
+
$2H^+$

$2NADP^+$

20,22-Dihydroxycholesterol

NADPH $+ H^+ + O_2$

desmolase

$NADP^+ + H_2O$

Isocaproaldehyde

Pregnenolone

FIGURE 21–49 Side-chain cleavage in the synthesis of steroid hormones. Cytochrome P-450 acts as electron carrier in this mixed-function oxidase system that oxidizes adjacent carbons. The process also requires the electron-transferring proteins adrenodoxin and adrenodoxin reductase. This system for cleaving side chains is found in mitochondria of the adrenal cortex, where active steroid production occurs. Pregnenolone is the precursor of all other steroid hormones (see Fig. 21–48).

chain of chlorophyll; natural rubber; many essential oils (such as the fragrant principles of lemon oil, eucalyptus, and musk); insect juvenile hormone, which controls metamorphosis; dolichols, which serve as lipid-soluble carriers in complex polysaccharide synthesis; and ubiquinone and plastoquinone, electron carriers in mitochondria and chloroplasts. Collectively, these molecules are called isoprenoids. More than 20,000 different isoprenoid molecules have been discovered in nature, and hundreds of new ones are reported each year.

Prenylation (covalent attachment of an isoprenoid; see Fig. 27–35) is a common mechanism by which pro-

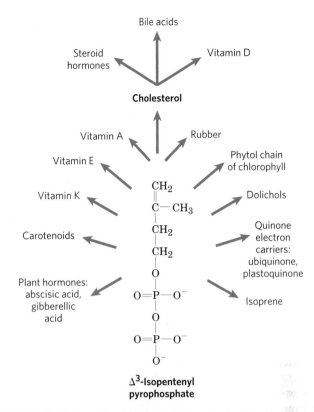

FIGURE 21–50 Overview of isoprenoid biosynthesis. The structures of most of the end products shown here are given in Chapter 10.

teins are anchored to the inner surface of cellular membranes in mammals (see Fig. 11–15). In some of these proteins the attached lipid is the 15-carbon farnesyl group; others have the 20-carbon geranylgeranyl group. Different enzymes attach the two types of lipids. It is possible that prenylation reactions target proteins to different membranes, depending on which lipid is attached. Protein prenylation is another important role for the isoprene derivatives of the pathway to cholesterol.

SUMMARY 21.4 Cholesterol, Steroids, and Isoprenoids: Biosynthesis, Regulation and Transport

▶ Cholesterol is formed from acetyl-CoA in a complex series of reactions, through the intermediates β-hydroxy-β-methylglutaryl-CoA, mevalonate, and two activated isoprenes, dimethylallyl pyrophosphate and isopentenyl pyrophosphate. Condensation of isoprene units produces the noncyclic squalene, which is cyclized to yield the steroid ring system and side chain.

▶ Cholesterol and cholesteryl esters are carried in the blood as plasma lipoproteins. VLDL carries cholesterol, cholesteryl esters, and triacylglycerols from the liver to other tissues, where the triacylglycerols are degraded by lipoprotein lipase, converting VLDL to LDL. The LDL, rich in cholesterol and its esters, is taken up by receptor-mediated endocytosis, in which the apolipoprotein B-100 of LDL is recognized by receptors in the plasma membrane.

▶ Cholesterol synthesis and transport are under complex regulation by hormones, cellular cholesterol content, and energy level (AMP concentration). HMG-CoA reductase is regulated allosterically and by covalent modification. Furthermore, both its synthesis and degradation rates are controlled by a complex of three proteins: Insig, SCAP, and SREBP, which sense cholesterol levels and trigger increased synthesis or degradation of HMG-CoA reductase. The number of LDL receptors per cell is also regulated by cholesterol content.

▶ Dietary conditions or genetic defects in cholesterol metabolism may lead to atherosclerosis and heart disease. In reverse cholesterol transport, HDL removes cholesterol from peripheral tissues, carrying it to the liver. By reducing the cholesterol content of foam cells, HDL protects against atherosclerosis.

▶ The steroid hormones (glucocorticoids, mineralocorticoids, and sex hormones) are produced from cholesterol by alteration of the side chain and introduction of oxygen atoms into the steroid ring system. In addition to cholesterol, a wide variety of isoprenoid compounds are derived from mevalonate through condensations of isopentenyl pyrophosphate and dimethylallyl pyrophosphate.

▶ Prenylation of certain proteins targets them for association with cellular membranes and is essential for their biological activity.

Key Terms

Terms in bold are defined in the glossary.

Further Reading

The general references in Chapters 10 and 17 are also useful.

General

Vance, D.E. & Vance, J.E. (eds) (2008) *Biochemistry of Lipids, Lipoproteins, and Membranes*, 5th edn, New Comprehensive Biochemistry, Elsevier Science Publishing Co., Inc., New York.
   Excellent reviews of lipid structure, biosynthesis, and function.

Biosynthesis of Fatty Acids and Eicosanoids

Chan, D.I. & Vogel, H.J. (2010) Current understanding of fatty acid biosynthesis and the acyl carrier protein. *Biochem. J.* **430**, 1–19.
   Intermediate-level review

Maier, T., Jenni, S., & Ban, N. (2006) Architecture of mammalian fatty acid synthase at 4.5 Å resolution. *Science* **311**, 1258–1262.
   The large multiprotein complexes that synthesize fatty acids in fungi have interesting and very different architectures compared with those in mammals.

Munday, M.R. (2002) Regulation of mammalian acetyl-CoA carboxylase. *Biochem. Soc. Trans.* **30**, 1059–1064.

Reshef, L., Olswang, Y., Cassuto, H., Blum, B., Croniger, C.M., Kalhan, S.C., Tilghman, S.M., & Hanson, R.M. (2003) Glyceroneogenesis and the triglyceride/fatty acid cycle. *J. Biol. Chem.* **278**, 30,413–30,416.

Sampath, H. & Ntambi, J.M. (2011) The role of stearoyl-CoA desaturase in obesity, insulin resistance, and inflammation. *Ann. N.Y. Acad. Sci.* **1243**, 47–53.
   Brief review of roles of this enzyme in lipogenesis in several physiological processes.

Smith, W.L., Urade, Y., & Jakovsson, P.-J. (2011) Enzymes of the cyclooxygenase pathways of prostanoid biosynthesis. *Chem. Rev.* **111**, 5821–5865.
   Advanced review.

Warner, T.D. & Mitchell, J.A. (2004) Cyclooxygenases: new forms, new inhibitors, and lessons from the clinic. *FASEB J.* **18**, 790–804.

White, S.W., Zheng, J., Zhang, Y.-M., & Rock, C.O. (2005) The structural biology of type II fatty acid biosynthesis. *Annu. Rev. Biochem.* **74**, 791–831.

Biosynthesis of Triacylglycerols and Membrane Phospholipids

Carman, G.M. & Han, G.-S. (2011) Regulation of phospholipid synthesis in the yeast *Saccharomyces cerevisiae*. *Annu. Rev. Biochem.* **89**, 859–883.

Advanced review.

Coleman, R.S. & Mashek, D.G. (2011) Mammalian triacylglycerol metabolism: synthesis, lipolysis, and signaling. *Chem. Rev.* **111**, 6359–6386.

Advanced review.

Dowhan, W. (1997) Molecular basis for membrane phospholipid diversity: why are there so many lipids? *Annu. Rev. Biochem.* **66**, 199–232.

Gibellini, F. & Smith, T.K. (2010) The Kennedy pathway—*de novo* synthesis of phosphatidylethanolamine and phosphatidylcholine. *IUBMB Life* **62**, 414–428.

Kennedy, E.P. (1962) The metabolism and function of complex lipids. *Harvey Lect.* **57**, 143–171.

A classic description of the role of cytidine nucleotides in phospholipid synthesis.

Raetz, C.R.H. & Dowhan, W. (1990) Biosynthesis and function of phospholipids in *Escherichia coli. J. Biol. Chem.* **265**, 1235–1238.

A brief review of bacterial biosynthesis of phospholipids and lipopolysaccharides.

Zechner, R., Zimmermann, R., Eichmann, T.O., Kohlwein, S.D., Haemmerle, G., Lass, A., & Madeo, R. (2012) Fat signals—lipases and lipolysis in lipid metabolism and signaling. *Cell Metab.* **15**, 279–291.

Biosynthesis of Cholesterol, Steroids, and Isoprenoids

Bloch, K. (1965) The biological synthesis of cholesterol. *Science* **150**, 19–28.

The author's Nobel address; a classic description of cholesterol synthesis in animals.

Boutte, Y. & Grebe, M. (2009) Cellular processes relying on sterol function in plants. *Curr. Opin. Plant Biol.* **12**, 705–713.

Review of synthesis and role of sterols in plants.

Brown, M.S. & Goldstein, J.L. (2009) Cholesterol feedback: from Schoenheimer's bottle to Scap's MELADL. *J. Lipid Res.* **50**, S15–S27.

Short historical summary of the roles of SCAP and SREBP in regulation of cholesterol synthesis.

Calandra, S., Tarugi, P., Speedy, H.E., Dean, A.F., Bertolini, S., & Shoulders, C.C. (2011) Mechanisms and genetic determinants regulating sterol absorption, circulating LDL levels, and sterol elimination: implications for classification and disease risk. *J. Lipid Res.* **52**, 1885–1926.

Extensive review of the role of LDL in human disease.

Calkin, A.C. & Tontonoz, P. (2012) Transcriptional integration of metabolism by the nuclear sterol-activated receptors LXR and FXR. *Nat. Rev. Mol. Cell Biol.* **13**, 213–224.

The role of these nuclear receptors in cholesterol metabolism and its integration into overall metabolism.

Chang, T.Y., Chang, C.C.Y., & Cheng, D. (1997) Acyl–coenzyme A: cholesterol acyltransferase. *Annu. Rev. Biochem.* **66**, 613–638.

Choi, S.H. & Ginsberg, H.N. (2011) Increased very low density lipoprotein (VLDL) secretion, hepatic steatosis, and insulin resistance. *Trends Endocrinol. Metab.* **22**, 353–363.

Review of insulin action of lipoprotein synthesis and circulation.

Getz, G.S. & Reardon, C.A. (2011) ABC transporters and the thickening cholesterol plot. *Curr. Opin. Lipidol.* **22**, 72–73.

Goldstein, J.L. & Brown, M.S. (2008) From fatty steak to fatty liver: 33 years of joint publications in the JCI. *J. Clin. Invest.* **118**, 1220–1222.

Short, excellent recap of seminal work on cholesterol and disease.

Goldstein, J.L. & Brown, M.S. (1990) Regulation of the mevalonate pathway. *Nature* **343**, 425–430.

Description of the allosteric and covalent regulation of the enzymes of the mevalonate pathway; includes a short discussion of the prenylation of Ras and other proteins.

Hanson, J.R. (2010) Classics of cholesterol biosynthesis. *Biochem. J.*, doi:10.1042/BJ20091543.

Review of the classic work of Popják and Cornforth on cholesterol synthesis.

Hegele, R.A. (2012) Plasma lipoproteins: genetic influences and clinical implications. *Nat. Rev. Genet.* **10**, 109–120.

Review of genetic basis for variations in plasma lipids and lipoproteins.

Jeon, T.-I. & Osborne, T.F. (2012) SREBPs: metabolic integrators in physiology and metabolism. *Trends Endocrinol. Metab.* **23**, 65–72.

Intermediate-level review of general roles of SREBPs, including cholesterol homeostasis.

Leduc, V., Jasmin-Belanger, S., & Poirier, J. (2010) APOE and cholesterol homeostasis in Alzheimer's disease. *Trends Mol. Med.* **16**, 469–477.

Maxfield, F.R. & Tabas, I. (2005) Role of cholesterol and lipid organization in disease. *Nature* **438**, 612–621.

Maxfield, F.R. & van Meer, G. (2010) Cholesterol, the central lipid of mammalian cells. *Curr. Opin. Cell Biol.* **22**, 422–429.

Miller, W.L. & Auchus, R.J. (2011) The molecular biology, biochemistry, and physiology of human steroidogenesis and its disorders. *Endocrine Rev.* **32**, 81–151.

Extensive, advanced review.

Miziorko, H.M. (2011) Enzymes of the mevalonate pathway of isoprenoid biosynthesis. *Arch. Biochem. Biophys.* **505**, 131–143.

Nes, D. (2011) Biosynthesis of cholesterol and other sterols. *Chem. Rev.* **111**, 6423–6451.

Olson, R.E. (1998) Discovery of the lipoproteins, their role in fat transport and their significance as risk factors. *J. Nutr.* **128** (2 Suppl.), 439S–443S.

Brief, clear, historical background on studies of lipoprotein function.

Raghow, R., Yellaturu, C., Deng, X., Park, E.A., & Elam, M.B. (2008) SREBPs: the crossroads of physiological and pathological lipid homeostasis. *Trends Endocrinol. Metab.* **19**, 65–73.

Russell, D.W. (2003) The enzymes, regulation, and genetics of bile acid synthesis. *Annu. Rev. Biochem.* **72**, 137–174.

Steinberg, D. (2006) An interpretive history of the cholesterol controversy, part V: the discovery of the statins and the end of the controversy. *J. Lipid Res.* **47**, 1339–1351.

The last in a five-part series detailing the history of the lipid controversy that led to the development of statins.

Tall, A.R., Yvan-Charvet, L., Terasaka, N., Pagler, T., & Wang, N. (2008) HDL, ABC transporters, and cholesterol efflux: implications for the treatment of atherosclerosis. *Cell Metab.* **7**, 365–375.

Reverse cholesterol transport: the role of ABC transporters.

Young, S.G. & Fielding, C.J. (1999) The ABCs of cholesterol efflux. *Nat. Genet.* **22**, 316–318.

A brief review of three papers in this journal issue that establish mutations in ABC1 as the cause of Tangier disease and familial HDL deficiency.

Shao, Y., Van Berkel, T.J.C., & Van Eck, M. (2010) Relative roles of various efflux pathways in net cholesterol efflux from macrophage foam cells in atherosclerotic lesions. *Curr. Opin. Lipidol.* **21**, 441–453.

Problems

1. Pathway of Carbon in Fatty Acid Synthesis Using your knowledge of fatty acid biosynthesis, provide an explanation for the following experimental observations:

(a) Addition of uniformly labeled [^{14}C]acetyl-CoA to a soluble liver fraction yields palmitate uniformly labeled with ^{14}C.

(b) However, addition of a *trace* of uniformly labeled [^{14}C]acetyl-CoA in the presence of an excess of unlabeled malonyl-CoA to a soluble liver fraction yields palmitate labeled with ^{14}C only in C-15 and C-16.

2. Synthesis of Fatty Acids from Glucose After a person has ingested large amounts of sucrose, the glucose and fructose that exceed caloric requirements are transformed to fatty acids for triacylglycerol synthesis. This fatty acid synthesis consumes acetyl-CoA, ATP, and NADPH. How are these substances produced from glucose?

3. Net Equation of Fatty Acid Synthesis Write the net equation for the biosynthesis of palmitate in rat liver, starting from mitochondrial acetyl-CoA and cytosolic NADPH, ATP, and CO_2.

4. Pathway of Hydrogen in Fatty Acid Synthesis Consider a preparation that contains all the enzymes and cofactors necessary for fatty acid biosynthesis from added acetyl-CoA and malonyl-CoA.

(a) If [2-^{2}H]acetyl-CoA (labeled with deuterium, the heavy isotope of hydrogen)

$$^2H-\overset{\overset{\displaystyle ^2H}{|}}{\underset{\underset{\displaystyle ^2H}{|}}{C}}-\overset{\overset{\displaystyle O}{\|}}{C}{\diagdown}_{S\text{-}CoA}$$

and an excess of unlabeled malonyl-CoA are added as substrates, how many deuterium atoms are incorporated into every molecule of palmitate? What are their locations? Explain.

(b) If unlabeled acetyl-CoA and [2-^{2}H]malonyl-CoA

$$^-OOC-\overset{\overset{\displaystyle ^2H}{|}}{\underset{\underset{\displaystyle ^2H}{|}}{C}}-\overset{\overset{\displaystyle O}{\|}}{C}{\diagdown}_{S\text{-}CoA}$$

are added as substrates, how many deuterium atoms are incorporated into every molecule of palmitate? What are their locations? Explain.

5. Energetics of β-Ketoacyl-ACP Synthase In the condensation reaction catalyzed by β-ketoacyl-ACP synthase (see Fig. 21–6), a four-carbon unit is synthesized by the combination of a two-carbon unit and a three-carbon unit, with the release of CO_2. What is the thermodynamic advantage of this process over one that simply combines two two-carbon units?

6. Modulation of Acetyl-CoA Carboxylase Acetyl-CoA carboxylase is the principal regulation point in the biosynthesis of fatty acids. Some of the properties of the enzyme are described below.

(a) Addition of citrate or isocitrate raises the V_{max} of the enzyme as much as 10-fold.

(b) The enzyme exists in two interconvertible forms that differ markedly in their activities:

Protomer (inactive) $\rightleftharpoons$ filamentous polymer (active)

Citrate and isocitrate bind preferentially to the filamentous form, and palmitoyl-CoA binds preferentially to the protomer.

Explain how these properties are consistent with the regulatory role of acetyl-CoA carboxylase in the biosynthesis of fatty acids.

7. Shuttling of Acetyl Groups across the Mitochondrial Inner Membrane The acetyl group of acetyl-CoA, produced by the oxidative decarboxylation of pyruvate in the mitochondrion, is transferred to the cytosol by the acetyl group shuttle outlined in Figure 21–10.

(a) Write the overall equation for the transfer of one acetyl group from the mitochondrion to the cytosol.

(b) What is the cost of this process in ATPs per acetyl group?

(c) In Chapter 17 we encountered an acyl group shuttle in the transfer of fatty acyl–CoA from the cytosol to the mitochondrion in preparation for β oxidation (see Fig. 17–6). One result of that shuttle was separation of the mitochondrial and cytosolic pools of CoA. Does the acetyl group shuttle also accomplish this? Explain.

8. Oxygen Requirement for Desaturases The biosynthesis of palmitoleate (see Fig. 21–12) a common unsaturated fatty acid with a cis double bond in the Δ^9 position, uses palmitate as a precursor. Can palmitoleate synthesis be carried out under strictly anaerobic conditions? Explain.

9. Energy Cost of Triacylglycerol Synthesis Use a net equation for the biosynthesis of tripalmitoylglycerol (tripalmitin) from glycerol and palmitate to show how many ATPs are required per molecule of tripalmitin formed.

10. Turnover of Triacylglycerols in Adipose Tissue When [^{14}C]glucose is added to the balanced diet of adult rats, there is no increase in the total amount of stored triacylglycerols, but the triacylglycerols become labeled with ^{14}C. Explain.

11. Energy Cost of Phosphatidylcholine Synthesis Write the sequence of steps and the net reaction for the biosynthesis of phosphatidylcholine by the salvage pathway from oleate, palmitate, dihydroxyacetone phosphate, and choline. Starting from these precursors, what is the cost (in number of ATPs) of the synthesis of phosphatidylcholine by the salvage pathway?

12. Salvage Pathway for Synthesis of Phosphatidylcholine A young rat maintained on a diet deficient in methionine fails to thrive unless choline is included in the diet. Explain.

13. Synthesis of Isopentenyl Pyrophosphate If 2[^{14}C] acetyl-CoA is added to a rat liver homogenate that is synthesizing cholesterol, where will the ^{14}C label appear in Δ^3-isopentenyl pyrophosphate, the activated form of an isoprene unit?

14. Activated Donors in Lipid Synthesis In the biosynthesis of complex lipids, components are assembled by transfer of the appropriate group from an activated donor. For example, the activated donor of acetyl groups is acetyl-CoA. For each of the following groups, give the form of the activated donor: (a) phosphate; (b) D-glucosyl; (c) phosphoethanolamine; (d) D-galactosyl; (e) fatty acyl; (f) methyl; (g) the two-carbon group in fatty acid biosynthesis; (h) Δ^3-isopentenyl.

15. Importance of Fats in the Diet When young rats are placed on a totally fat-free diet, they grow poorly, develop a scaly dermatitis, lose hair, and soon die—symptoms that can be prevented if linoleate or plant material is included in the diet. What makes linoleate an essential fatty acid? Why can plant material be substituted?

16. Regulation of Cholesterol Biosynthesis Cholesterol in humans can be obtained from the diet or synthesized de novo. An adult human on a low-cholesterol diet typically synthesizes 600 mg of cholesterol per day in the liver. If the amount of cholesterol in the diet is large, de novo synthesis of cholesterol is drastically reduced. How is this regulation brought about?

17. Lowering Serum Cholesterol Levels with Statins Patients treated with a statin drug generally exhibit a dramatic lowering of serum cholesterol. However, the amount of the enzyme HMG-CoA reductase present in cells can increase substantially. Suggest an explanation for this effect.

18. Roles of Thiol Esters in Cholesterol Biosynthesis Draw a mechanism for each of the three reactions shown in Figure 21–34, detailing the pathway for the synthesis of mevalonate from acetyl-CoA.

19. Potential Side Effects of Treatment with Statins Although clinical trials have not yet been carried out to document benefits or side effects, some physicians have suggested that patients being treated with statins also take a supplement of coenzyme Q. Suggest a rationale for this recommendation.

Data Analysis Problem

20. Engineering *E. coli* to Produce Large Quantities of an Isoprenoid There is a huge variety of naturally occurring isoprenoids, some of which are medically or commercially important and produced industrially. The production methods include in vitro enzymatic synthesis, which is an expensive and low-yield process. In 1999, Wang, Oh, and Liao reported their experiments to engineer the easily grown bacterium *E. coli* to produce large amounts of astaxanthin, a commercially important isoprenoid.

Astaxanthin is a red-orange carotenoid pigment (an antioxidant) produced by marine algae. Marine animals such as shrimp, lobster, and some fish that feed on the algae get their orange color from the ingested astaxanthin. Astaxanthin is composed of eight isoprene units; its molecular formula is $C_{40}H_{52}O_4$:

Astaxanthin

(a) Circle the eight isoprene units in the astaxanthin molecule. Hint: Use the projecting methyl groups as a guide.

Astaxanthin is synthesized by the pathway shown on the next page, starting with Δ^3-isopentenyl pyrophosphate (IPP).

Steps ❶ and ❷ are shown in Figure 21–36, and the reaction catalyzed by IPP isomerase is shown in Figure 21–35.

(b) In step ❹ of the pathway, two molecules of geranylgeranyl pyrophosphate are linked to form phytoene. Is this a head-to-head or a head-to-tail joining? (See Figure 21–36 for details.)

(c) Briefly describe the chemical transformation in step ❺.

(d) The synthesis of cholesterol (Fig. 21–37) includes a cyclization (ring closure) that involves a net oxidation by O_2. Does the cyclization in step ❻ of the astaxanthin synthetic pathway require a net oxidation of the substrate (lycopene)? Explain your reasoning.

E. coli does not make large quantities of many isoprenoids, and does not synthesize astaxanthin. It is known to synthesize small amounts of IPP, DMAPP, geranyl pyrophosphate, farnesyl pyrophosphate, and geranylgeranyl pyrophosphate. Wang and colleagues cloned several of the *E. coli* genes that encode enzymes needed for astaxanthin synthesis in plasmids that allow their overexpression. These genes included *idi*, which encodes IPP isomerase, and *ispA*, which encodes a prenyl transferase that catalyzes steps ❶ and ❷.

To engineer an *E. coli* capable of the complete astaxanthin pathway, Wang and colleagues cloned several genes from other bacteria into plasmids that would allow their overexpression in *E. coli*. These genes included *crtE* from *Erwinia uredovora*, which encodes an enzyme that catalyzes step ❸; and *crtB*, *crtI*, *crtY*, *crtZ*, and *crtW* from *Agrobacterium aurantiacum*, which encode enzymes for steps ❹, ❺, ❻, ❼, and ❽, respectively.

The investigators also cloned the gene *gps* from *Archaeoglobus fulgidus*, overexpressed this gene in *E. coli*, and extracted the gene product. When this extract was reacted with [^{14}C]IPP and DMAPP, or geranyl pyrophosphate, or farnesyl pyrophosphate, only ^{14}C-labeled geranylgeranyl pyrophosphate was produced in all cases.

(e) Based on these data, which step(s) in the pathway are catalyzed by the enzyme encoded by *gps*? Explain your reasoning.

Wang and coworkers then constructed several *E. coli* strains overexpressing different genes and measured the orange color of the colonies (wild-type *E. coli* colonies are off-white) and the amount of astaxanthin produced. Their results are shown below.

Strain	Gene(s) overexpressed	Orange color	Astaxanthin yield (μg/g dry weight)
1	*crtBIZYW*	−	ND
2	*crtBIZYW, ispA*	−	ND
3	*crtBIZYW, idi*	−	ND
4	*crtBIZYW, idi, ispA*	−	ND
5	*crtBIZYW, crtE*	+	32.8
6	*crtBIZYW, crtE, ispA*	+	35.3
7	*crtBIZYW, crtE, idi*	++	234.1
8	*crtBIZYW, crtE, idi, ispA*	+++	390.3
9	*crtBIZYW, gps*	+	35.6
10	*crtBIZYW, gps, idi*	+++	1,418.8

Note: ND, not determined.

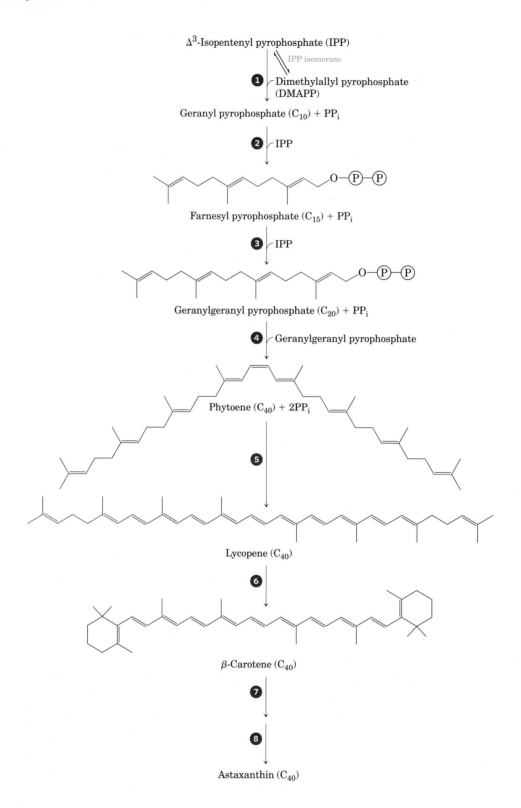

Δ^3-Isopentenyl pyrophosphate (IPP)

IPP isomerase

1 Dimethylallyl pyrophosphate (DMAPP)

Geranyl pyrophosphate (C_{10}) + PP_i

2 IPP

Farnesyl pyrophosphate (C_{15}) + PP_i

3 IPP

Geranylgeranyl pyrophosphate (C_{20}) + PP_i

4 Geranylgeranyl pyrophosphate

Phytoene (C_{40}) + $2PP_i$

5

Lycopene (C_{40})

6

β-Carotene (C_{40})

7

8

Astaxanthin (C_{40})

(f) Comparing the results for strains 1 through 4 with those for strains 5 through 8, what can you conclude about the expression level of an enzyme capable of catalyzing step **3** of the astaxanthin synthetic pathway in wild-type *E. coli*? Explain your reasoning.

(g) Based on the data above, which enzyme is rate-limiting in this pathway, IPP isomerase or the enzyme encoded by *idi*? Explain your reasoning.

(h) Would you expect a strain overexpressing *crtBIZYW*, *gps*, and *crtE* to produce low (+), medium (++), or high (+++) levels of astaxanthin, as measured by its orange color? Explain your reasoning.

Reference

Wang, C.-W., Oh, M.-K., & Liao, J.C. (1999) Engineered isoprenoid pathway enhances astaxanthin production in *Escherichia coli. Biotechnol. Bioeng.* **62**, 235–241.

Biosynthesis of Amino Acids, Nucleotides, and Related Molecules

Nitrogen ranks behind only carbon, hydrogen, and oxygen in its contribution to the mass of living systems. Most of this nitrogen is bound up in amino acids and nucleotides. In this chapter we address all aspects of the metabolism of these nitrogen-containing compounds except amino acid catabolism, which is covered in Chapter 18.

Discussing the biosynthetic pathways for amino acids and nucleotides together is a sound approach, not only because both classes of molecules contain nitrogen (which arises from common biological sources) but because the two sets of pathways are extensively intertwined, with several key intermediates in common. Certain amino acids or parts of amino acids are incorporated into the structure of purines and pyrimidines, and in one case part of a purine ring is incorporated into an amino acid (histidine). The two sets of pathways also share much common chemistry, in particular a preponderance of reactions involving the transfer of nitrogen or one-carbon groups.

The pathways described here can be intimidating to the beginning biochemistry student. Their complexity arises not so much from the chemistry itself, which in many cases is well understood, but from the sheer number of steps and variety of intermediates. These pathways are best approached by maintaining a focus on metabolic principles we have already discussed, on key intermediates and precursors, and on common classes of reactions. Even a cursory look at the chemistry can be rewarding, for some of the most unusual chemical transformations in biological systems occur in these pathways; for instance, we find prominent examples of the rare biological use of

the metals molybdenum, selenium, and vanadium. The effort also offers a practical dividend, especially for students of human or veterinary medicine. Many genetic diseases of humans and animals have been traced to an absence of one or more enzymes of amino acid and nucleotide metabolism, and many pharmaceuticals in common use to combat infectious diseases are inhibitors of enzymes in these pathways—as are a number of the most important agents in cancer chemotherapy.

Regulation is crucial in the biosynthesis of the nitrogen-containing compounds. Because each amino acid and each nucleotide is required in relatively small amounts, the metabolic flow through most of these pathways is not nearly as great as the biosynthetic flow leading to carbohydrate or fat in animal tissues. Because the different amino acids and nucleotides must be made in the correct ratios and at the right time for protein and nucleic acid synthesis, their biosynthetic pathways must be accurately regulated and coordinated with each other. And because amino acids and nucleotides are charged molecules, their levels must be regulated to maintain electrochemical balance in the cell. As discussed in earlier chapters, pathways can be controlled by changes in either the activity or the amounts of specific enzymes. The pathways we encounter in this chapter provide some of the best-understood examples of the regulation of enzyme activity. Control of the *amounts* of different enzymes in a cell (that is, of their synthesis and degradation) is a topic covered in Chapter 28.

22.1 Overview of Nitrogen Metabolism

The biosynthetic pathways leading to amino acids and nucleotides share a requirement for nitrogen. Because soluble, biologically useful nitrogen compounds are generally scarce in natural environments, most organisms maintain strict economy in their use of ammonia, amino acids, and nucleotides. Indeed, as we shall see, free amino acids, purines, and pyrimidines formed during metabolic turnover of proteins and nucleic acids are often salvaged and reused. We first examine the pathways by which nitrogen from the environment is introduced into biological systems.

The Nitrogen Cycle Maintains a Pool of Biologically Available Nitrogen

Although Earth's atmosphere is four-fifths molecular nitrogen (N_2), relatively few species can convert this atmospheric nitrogen into forms useful to living organisms. In the biosphere, the metabolic processes of different species function interdependently to salvage and reuse biologically available nitrogen in a vast **nitrogen cycle (Fig. 22–1)**. The first step in the cycle is **fixation** (reduction) of atmospheric nitrogen by nitrogen-fixing bacteria to yield ammonia (NH_3 or NH_4^+). Although ammonia can be used by most living organisms, soil bacteria that derive their energy by oxidizing ammonia to nitrite (NO_2^-) and ultimately nitrate (NO_3^-) are so abundant and active that nearly all ammonia reaching the soil is oxidized to nitrate. This process is known as **nitrification**. Plants and many bacteria can take up and readily reduce nitrate and nitrite to ammonia through the action of nitrate and nitrite reductases. This ammonia is incorporated into amino acids by plants. Animals then use plants as a source of amino acids, both nonessential and essential, to build their proteins. When organisms die, microbial degradation of their proteins returns ammonia to the soil, where nitrifying bacteria again convert it to nitrite and nitrate. A balance is maintained between fixed nitrogen and atmospheric nitrogen by bacteria that reduce nitrate to N_2 under anaerobic conditions, a process called **denitrification** (Fig. 22–1). These soil bacteria use NO_3^- rather than O_2 as the ultimate electron acceptor in a series of reactions that (like oxidative phosphorylation) generates a transmembrane proton gradient, which is used to synthesize ATP.

The nitrogen cycle is short-circuited by a group of bacteria that promote anaerobic ammonia oxidation, or **anammox** (Fig. 22–1), a process that converts ammonia and nitrite to N_2. As much as 50% to 70% of the NH_3-to-N_2 conversion in the biosphere may occur through this pathway, undetected until the 1980s. The obligate anaerobes that promote anammox are fascinating in their own right and are providing some useful solutions to waste-treatment problems (Box 22–1).

Now let's examine the processes that generate the ammonia that is incorporated into microorganisms, plants, and the animals that eat them.

More than 90% of the NH_4^+ generated by vascular plants, algae, and microorganisms comes from nitrate assimilation, a two-step process. First NO_3^- is reduced to NO_2^- by **nitrate reductase**, then the NO_2^- is reduced to NH_4^+ in a six-electron transfer catalyzed by **nitrite reductase (Fig. 22–2)**. Both reactions involve chains of electron carriers and cofactors we have not yet encountered. Nitrate reductase is a large, soluble protein (M_r 220,000). Within the enzyme, a pair of electrons, donated by NADH, flows through —SH groups of cysteine, FAD, and a cytochrome (cyt b_{557}), then to a novel cofactor containing molybdenum, before reducing the substrate NO_3^- to NO_2^-.

The nitrite reductase of plants is located in the chloroplasts and receives its electrons from ferredoxin (which is reduced in the light-dependent reactions of photosynthesis; see Section 19.8). Six electrons, donated one at a time by ferredoxin, pass through a 4S-4Fe center in the enzyme, then through a novel heme-like molecule (siroheme) before reducing NO_2^- to NH_4^+ (Fig. 22–2). In nonphotosynthetic microbes, NADPH provides the electrons for this reaction.

Nitrogen Is Fixed by Enzymes of the Nitrogenase Complex

Only certain bacteria and archaea can fix atmospheric N_2. These organisms, called diazotrophs, include the cyanobacteria of soils and fresh and salt waters, methanogenic archaea (strict anaerobes that obtain energy and carbon by converting H_2 and CO_2 to methane), other kinds of free-living soil bacteria such as *Azotobacter* species, and the nitrogen-fixing bacteria that live as **symbionts** in the root nodules of leguminous plants. The first important product of nitrogen fixation is ammonia,

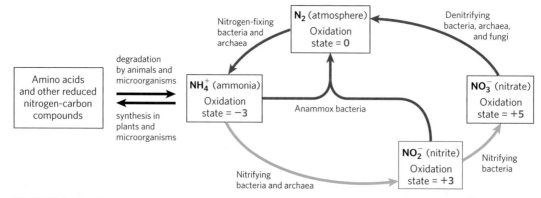

FIGURE 22–1 The nitrogen cycle. The total amount of nitrogen fixed annually in the biosphere exceeds 10^{11} kg. Reactions with red arrows occur largely or entirely in anaerobic environments.

FIGURE 22–2 Nitrate assimilation by nitrate reductase and nitrite reductase. (a) Nitrate reductases of plants and bacteria catalyze the two-electron reduction of NO_3^- to NO_2^-, in which a novel Mo-containing cofactor plays a central role. NADH is the electron donor. **(b)** Nitrite reductase converts the product of nitrate reductase into NH_4^+ in a six-electron, eight-proton transfer process in which the metallic center in siroheme carries electrons, and the carboxyl groups of siroheme may donate protons. The initial source of electrons is reduced ferredoxin.

which can be used by all organisms either directly or after its conversion to other soluble compounds such as nitrites, nitrates, or amino acids.

The reduction of nitrogen to ammonia is an exergonic reaction:

$$N_2 + 3H_2 \longrightarrow 2NH_3 \qquad \Delta G'^{\circ} = -33.5 \text{ kJ/mol}$$

The $N\equiv N$ triple bond, however, is very stable, with a bond energy of 930 kJ/mol. Nitrogen fixation therefore has an extremely high activation energy, and atmospheric nitrogen is almost chemically inert under normal conditions. Ammonia is produced industrially by the Haber process (named for its inventor, Fritz Haber), which requires temperatures of 400 to 500 °C and nitrogen and hydrogen at pressures of tens of thousands of kilopascals (several hundred atmospheres) to provide the necessary activation energy. Biological nitrogen fixation, however, must occur at biological temperatures and at 0.8 atm of nitrogen, and the high activation barrier is overcome by other means. This is accomplished, at least in part, by the binding and hydrolysis of ATP. The overall reaction can be written

$$N_2 + 10H^+ + 8e^- + 16ATP \longrightarrow$$
$$2NH_4^+ + 16ADP + 16P_i + H_2$$

Biological nitrogen fixation is carried out by a highly conserved complex of proteins called the **nitrogenase complex**; its central components are **dinitrogenase reductase** and **dinitrogenase (Fig. 22–3a)**. Dinitrogenase reductase (M_r 60,000) is a dimer of two identical subunits. It contains a single 4Fe-4S redox center (see Fig. 19–5), bound between the subunits, and can be oxidized and reduced by one electron. It also has two

binding sites for ATP/ADP (one site on each subunit). Dinitrogenase (M_r 240,000), an $\alpha_2\beta_2$ tetramer, has two Fe-containing cofactors that transfer electrons (Fig. 22–3b). One, the **P cluster**, has a pair of 4Fe-4S centers; these share a sulfur atom, making an 8Fe-7S center. The second cofactor in dinitrogenase, the **FeMo cofactor**, is a novel structure composed of 7 Fe atoms, 9 inorganic S atoms, a Cys side chain, and a single carbon atom in the center of the FeS cluster. Also part of the cofactor is a molybdenum atom, with ligands that include three inorganic S atoms, a His side chain, and two oxygen atoms from a molecule of homocitrate that is an intrinsic part of the FeMo cofactor. There is also a form of nitrogenase that contains vanadium rather than molybdenum, and some bacterial species can produce both types. The vanadium-containing enzyme may be the primary nitrogen-fixing system under some conditions. The vanadium nitrogenase of *Azotobacter vinelandii* has the remarkable capacity to catalyze the reduction of carbon monoxide (CO) to ethylene (C_2H_4), ethane, and propane.

Nitrogen fixation is carried out by a highly reduced form of dinitrogenase and requires eight electrons: six for the reduction of N_2 and two to produce one molecule of H_2. Production of H_2 is an obligate part of the reaction mechanism, but its biological role in the process is not understood.

Dinitrogenase is reduced by the transfer of electrons from dinitrogenase reductase **(Fig. 22–4)**. The dinitrogenase tetramer has two binding sites for the reductase. The required eight electrons are transferred from reductase to dinitrogenase one at a time: a reduced reductase molecule binds to the dinitrogenase and transfers a single electron, then the oxidized reductase

BOX 22–1   Unusual Lifestyles of the Obscure but Abundant

Air-breathers that we are, we can easily overlook the bacteria and archaea that thrive in anaerobic environments. Although rarely featured in introductory biochemistry textbooks, these organisms constitute much of the biomass of this planet, and their contributions to the balance of carbon and nitrogen in the biosphere are essential to all forms of life.

As detailed in earlier chapters, the energy used to maintain living systems relies on the generation of proton gradients across membranes. Electrons derived from a reduced substrate are made available to electron carriers in membranes and pass through a series of electron transfers to a final electron acceptor. As a byproduct of this process, protons are released on one side of the membrane, generating the transmembrane proton gradient. The proton gradient is used to synthesize ATP or to drive other energy-requiring processes. For all eukaryotes, the reduced substrate is generally a carbohydrate (glucose or pyruvate) or a fatty acid and the electron acceptor is oxygen.

Many bacteria and archaea are much more versatile. In anaerobic environments such as marine and freshwater sediments, the variety of life strategies is extraordinary. Almost any available redox pair can be an energy source for some specialized organism or group of organisms. For example, a large number of lithotrophic bacteria (a lithotroph is a chemotroph that uses inorganic energy sources; see Fig. 1–5) have a hydrogenase that uses molecular hydrogen to reduce NAD^+:

$$H_2 + NAD^+ \xrightarrow{\text{hydrogenase}} NADH + H^+$$

The NADH is a source of electrons for a variety of membrane-bound electron acceptors, generating the proton gradient needed for ATP synthesis. Other lithotrophs oxidize sulfur compounds (H_2S, elemental sulfur, or thiosulfate) or ferrous iron. A widespread group of archaea called methanogens, all strict anaerobes, extract energy from the reduction of CO_2 to methane. And this is just a small sampling of what anaerobic

organisms do for a living. Their metabolic pathways are replete with interesting reactions and highly specialized cofactors unknown in our own world of obligate aerobic metabolism. Study of these organisms can yield practical dividends. It can also provide clues about the origins of life on an early Earth, in an atmosphere that lacked molecular oxygen.

The nitrogen cycle depends on a wide range of specialized bacteria. There are two groups of nitrifying bacteria: those that oxidize ammonia to nitrites and those that oxidize the resulting nitrites to nitrates (see Fig. 22–1). Nitrate is second only to O_2 as a biological electron acceptor, and a great many bacteria and archaea can catalyze the denitrification of nitrates to nitrogen, which the nitrogen-fixing bacteria then convert back into ammonia. Ammonia is a major pollutant in sewage and in farm animal waste, and is a byproduct of fertilizer manufacture and oil refining. Waste-treatment plants have generally made use of communities of nitrifying and denitrifying bacteria to convert ammonia waste to atmospheric nitrogen. The process is expensive, requiring inputs of organic carbon and oxygen.

In the 1960s and 1970s, a few articles appeared in the research literature suggesting that ammonia could be oxidized to nitrogen anaerobically, using nitrite as an electron acceptor; this process was called anammox. The reports received little notice until bacteria promoting anammox were discovered in a waste-treatment system in Delft, the Netherlands, in the mid-1980s. A team of Dutch microbiologists led by Gijs Kuenen and Mike Jetten began to study these bacteria, which were soon identified as belonging to an unusual bacterial phylum, Planctomycetes. Some surprises were to follow.

The biochemistry underlying the anammox process was slowly unraveled (Fig. 1). Hydrazine (N_2H_4),

FIGURE 1 The anammox reactions. Ammonia and hydroxylamine are converted to hydrazine and H_2O by hydrazine hydrolase, and the hydrazine is oxidized by hydrazine-oxidizing enzyme, generating N_2 and protons. The protons generate a proton gradient for ATP synthesis. On the anammoxosome exterior, protons are used by the nitrite-reducing enzyme, producing hydroxylamine and completing the cycle. All of the anammox enzymes are embedded in the anammoxosome membrane.

(a)

(b)

FIGURE 2 **(a)** Ladderane lipids of the anammoxosome membrane. The mechanism for synthesis of the unstable fused cyclobutane ring structures is unknown. **(b)** Ladderanes can stack to form a very dense, impermeable, hydrophobic membrane structure, allowing sequestration of the hydrazine produced in the anammox reactions.

a highly reactive molecule used as a rocket fuel, was an unexpected intermediate. As a small molecule, hydrazine is both highly toxic and difficult to contain. It readily diffuses across typical phospholipid membranes. The anammox bacteria solve this problem by sequestering hydrazine in a specialized organelle, dubbed the **anammoxosome**. The membrane of this organelle is composed of lipids known as **ladderanes** (Fig. 2), never before encountered in biology. The fused cyclobutane rings of ladderanes stack tightly to form a very dense barrier, greatly slowing the release of hydrazine. Cyclobutane rings are strained and difficult to synthesize; the bacterial mechanisms for synthesizing these lipids are not yet known.

The anammoxosome was a surprising finding. Bacterial cells generally do not have compartments, and the lack of a membrane-enclosed nucleus is often cited as the primary distinction between eukaryotes and bacteria. One type of organelle in a bacterium was interesting enough, but planctomycetes also have a nucleus: their chromosomal DNA is contained within a membrane (Fig. 3). Discovery of this subcellular organization has prompted further research to trace the origin of the planctomycetes and the evolution of eukaryotic nuclei. Planctomycetes are an ancient bacterial line with multiple genera, three of which are known to carry out the anammox reactions. Further study of this group may ultimately bring us closer to a key goal of evolutionary biology: a description of the organism affectionately referred to as LUCA—the last universal common ancestor of all life on our planet.

For now, the anammox bacteria offer a major advance in waste treatment, reducing the cost of ammonia removal by as much as 90% (the conventional denitrification steps are eliminated completely, and the aeration costs associated with nitrification are lower) and reducing the release of polluting byproducts. Clearly, a greater familiarity with the bacterial underpinnings of the biosphere can pay big dividends as we deal with the environmental challenges of the twenty-first century.

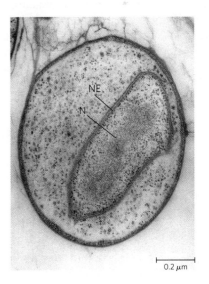

0.2 μm

FIGURE 3 Transmission electron micrograph of a cross section through *Gemmata obscuriglobus*, showing the DNA in a nucleus (N) with enclosing nuclear envelope (NE). Bacteria of the *Gemmata* genus (phylum Planctomycetes) do not promote the anammox reactions.

(a)

Dinitrogenase
(α subunit)

Dinitrogenase
(β subunit)

Dinitrogenase
reductase

(b)

FIGURE 22–3 Enzymes and cofactors of the nitrogenase complex. (PDB ID 1FP6 and PDB ID 1M1N) **(a)** The holoenzyme consists of two identical dinitrogenase reductase molecules (green), each with a 4Fe-4S redox center and binding sites for two ATP, and two identical dinitrogenase heterodimers (purple and blue), each with a P cluster (Fe-S center) and an FeMo cofactor. In this structure, ADP is bound in the ATP site, to make the crystal more stable. **(b)** The electron-transfer cofactors. A P cluster is shown here in its reduced (top) and oxidized (middle) forms. The FeMo cofactor (bottom) has a Mo atom with three S ligands, a His ligand, and two oxygen ligands from a molecule of homocitrate. In some organisms, the Mo atom is replaced with a vanadium atom. (Fe is shown in orange, S in yellow.)

dissociates from dinitrogenase, in a repeating cycle. Each turn of the cycle requires the hydrolysis of two ATP molecules by the dimeric reductase. The immediate source of electrons to reduce dinitrogenase reductase varies, with reduced **ferredoxin** (see Section 19.8), reduced flavodoxin, and perhaps other sources playing a role. In at least one species, the ultimate source of electrons to reduce ferredoxin is pyruvate (Fig. 22–4).

The role of ATP in this process is somewhat unusual. Recall that ATP can contribute not only chemical energy, through the hydrolysis of one or more of its phosphoanhydride bonds, but also binding energy (p. 195), through noncovalent interactions that lower the activation energy. In the reaction carried out by dinitrogenase reductase, both ATP binding and ATP hydrolysis bring about protein conformational changes that help overcome the high activation energy of nitrogen fixation. The binding of two ATP molecules to the reductase shifts the reduction potential ($E'°$) of this protein from -300 to -420 mV, an enhancement of its reducing power that is

FIGURE 22–4 Electron path in nitrogen fixation by the nitrogenase complex. Electrons are transferred from pyruvate to dinitrogenase via ferredoxin (or flavodoxin) and dinitrogenase reductase. Dinitrogenase reductase reduces dinitrogenase one electron at a time, with at least six electrons required to fix one molecule of N_2. Two additional electrons are used to reduce $2H^+$ to H_2 in a process that obligatorily accompanies nitrogen fixation in anaerobes, making a total of eight electrons required per N_2 molecule. The subunit structures and metal cofactors of the dinitrogenase reductase and dinitrogenase proteins are described in the text and in Figure 22–3.

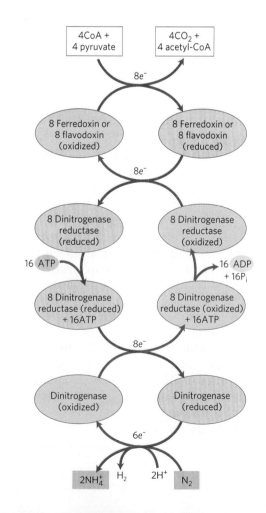

required to transfer electrons through dinitrogenase to N_2; the standard reduction potential for the half-reaction $N_2 + 6H^+ + 6e^- \longrightarrow 2NH_3$ is -0.34 V. The ATP molecules are then hydrolyzed just before the actual transfer of one electron to dinitrogenase.

ATP binding and hydrolysis change the conformation of nitrogenase reductase in two regions, which are structurally homologous with switch 1 and switch 2 regions of the GTP-binding proteins involved in biological signaling (see Box 12–2). ATP binding produces a conformational change that brings the 4Fe-4S center of the reductase closer to the P cluster of dinitrogenase (from 18 Å to 14 Å away), which facilitates electron transfer between the reductase and dinitrogenase. The details of electron transfer from the P cluster to the FeMo cofactor, and the means by which eight electrons are accumulated by nitrogenase, are not known, nor are the intermediates in the reaction known with certainty; two reasonable hypotheses are being tested, both involving the Mo atom as a central player **(Fig. 22–5)**.

The nitrogenase complex is remarkably unstable in the presence of oxygen. The reductase is inactivated in air, with a half-life of 30 seconds; dinitrogenase has a half-life of only 10 minutes in air. Free-living bacteria that fix nitrogen cope with this problem in a variety of ways. Some live only anaerobically or repress nitrogenase synthesis when oxygen is present. Some aerobic species, such as *Azotobacter vinelandii*, partially uncouple electron transfer from ATP synthesis so that oxygen is burned off as rapidly as it enters the cell (see Box 19–1). When fixing nitrogen, cultures of these bacteria actually increase in temperature as a result of their efforts to rid themselves of oxygen.

The symbiotic relationship between leguminous plants and the nitrogen-fixing bacteria in their root nodules **(Fig. 22–6)** takes care of both the energy requirements and the oxygen lability of the nitrogenase complex. The energy required for nitrogen fixation was probably the evolutionary driving force for this plant-bacteria association. The bacteria in root nodules have access to a large

FIGURE 22–5 Two reasonable hypotheses for the intermediates involved in N$_2$ reduction. In both scenarios, the FeMo cofactor (abbreviated as **M** here) plays a central role, binding directly to one of the nitrogen atoms of N$_2$ and remaining bound throughout the sequence of reduction steps.

reservoir of energy in the form of abundant carbohydrate and citric acid cycle intermediates made available by the plant. This may allow the bacteria to fix hundreds of times more nitrogen than do their free-living cousins under

FIGURE 22–6 Nitrogen-fixing nodules. (a) Root nodules of the common pea, *Pisum saliva*, a legume. The flower of this plant is shown in the inset. **(b)** Artificially colorized electron micrograph of a thin section through a pea root nodule. Symbiotic nitrogen-fixing bacteria, or bacteroids (red), live inside the nodule cell, surrounded by the peribacteroid membrane (blue). Bacteroids produce the nitrogenase complex that converts atmospheric nitrogen (N_2) to ammonium (NH_4^+); without the bacteroids, the plant is unable to utilize N_2. The infected root cell provides some factors essential for nitrogen fixation, including leghemoglobin; this heme protein has a very high binding affinity for oxygen, which strongly inhibits nitrogenase. (The cell nucleus is shown in yellow/green. Not visible in this micrograph are other organelles of the infected root cell that are normally found in plant cells.)

(a)   **(b)**   2 µm

conditions generally encountered in soils. To solve the oxygen-toxicity problem, the bacteria in root nodules are bathed in a solution of the oxygen-binding heme protein **leghemoglobin**, produced by the plant (although the heme may be contributed by the bacteria). Leghemoglobin binds all available oxygen so that it cannot interfere with nitrogen fixation, and efficiently delivers the oxygen to the bacterial electron-transfer system. The benefit to the plant, of course, is a ready supply of reduced nitrogen. In fact, the bacterial symbionts typically produce far more NH_3 than is needed by their symbiotic partner; the excess is released into the soil. The efficiency of the symbiosis between plants and bacteria is evident in the enrichment of soil nitrogen brought about by leguminous plants. This enrichment of NH_3 in the soil is the basis of crop rotation methods, in which plantings of nonleguminous plants (such as maize) that extract fixed nitrogen from the soil are alternated every few years with plantings of legumes such as alfalfa, peas, or clover.

Nitrogen fixation is energetically costly: 16 ATP and 8 electron pairs yield only 2 NH_3. It is therefore not surprising that the process is tightly regulated, so that NH_3 is produced only when needed. High [ADP], an indicator of low [ATP], is a strong inhibitor of nitrogenase. NH_4^+ represses the expression of the ~20 nitrogen fixation (*nif*) genes, effectively shutting down the pathway. Covalent alteration of nitrogenase is also used in some diazotrophs to control nitrogen fixation in response to the availability of NH_4^+ in the surroundings. Transfer of an ADP-ribosyl group from NADH to a specific Arg residue in the nitrogenase reductase shuts down N_2 fixation in *Rhodospirillum*, for example. This is the same covalent modification that we saw in the case of G protein inhibition by the toxins of cholera and pertussis (see Box 12–2).

Nitrogen fixation is the subject of intense study because of its immense practical importance. Industrial production of ammonia for use in fertilizers requires a large and expensive input of energy, and this has spurred a drive to develop recombinant or transgenic organisms that can fix nitrogen. In principle, recombinant DNA techniques (Chapter 9) might be used to transfer the DNA that encodes the enzymes of nitrogen fixation into non-nitrogen-fixing bacteria and plants. However, those genes alone will not suffice. About 20 genes are essential to nitrogenase activity in bacteria, many of them needed for the synthesis, assembly, and insertion of the cofactors. There is also the problem of protecting the enzyme in its new setting from destruction by oxygen. In all, there are formidable challenges in engineering new nitrogen-fixing plants. Success in these efforts will depend on overcoming the problem of oxygen toxicity in any cell that produces nitrogenase.

Ammonia Is Incorporated into Biomolecules through Glutamate and Glutamine

Reduced nitrogen in the form of NH_4^+ is assimilated into amino acids and then into other nitrogen-containing biomolecules. Two amino acids, **glutamate** and **glutamine**,

provide the critical entry point. Recall that these same two amino acids play central roles in the catabolism of ammonia and amino groups in amino acid oxidation (Chapter 18). Glutamate is the source of amino groups for most other amino acids, through transamination reactions (the reverse of the reaction shown in Fig. 18–4). The amide nitrogen of glutamine is a source of amino groups in a wide range of biosynthetic processes. In most types of cells, and in extracellular fluids in higher organisms, one or both of these amino acids are present at higher concentrations—sometimes an order of magnitude or more higher—than other amino acids. An *Escherichia coli* cell requires so much glutamate that this amino acid is one of the primary solutes in the cytosol. Its concentration is regulated not only in response to the cell's nitrogen requirements but also to maintain an osmotic balance between the cytosol and the external medium.

The biosynthetic pathways to glutamate and glutamine are simple, and all or some of the steps occur in most organisms. The most important pathway for the assimilation of NH_4^+ into glutamate requires two reactions. First, **glutamine synthetase** catalyzes the reaction of glutamate and NH_4^+ to yield glutamine. This reaction takes place in two steps, with enzyme-bound γ-glutamyl phosphate as an intermediate (see Fig. 18–8):

(1) Glutamate + ATP $\longrightarrow$
$$\text{γ-glutamyl phosphate + ADP}$$

(2) γ-Glutamyl phosphate + NH_4^+ $\longrightarrow$
$$\text{glutamine} + P_i + H^+$$

Sum: Glutamate + NH_4^+ + ATP $\longrightarrow$
$$\text{glutamine} + \text{ADP} + P_i + H^+ \quad (22\text{–}1)$$

Glutamine synthetase is found in all organisms. In addition to its importance for NH_4^+ assimilation in bacteria, it has a central role in amino acid metabolism in mammals, converting free NH_4^+, which is toxic, to glutamine for transport in the blood (Chapter 18).

In bacteria and plants, glutamate is produced from glutamine in a reaction catalyzed by **glutamate synthase**. (An alternative name for this enzyme, glutamate:oxoglutarate aminotransferase, yields the acronym GOGAT, by which the enzyme also is known.) α-Ketoglutarate, an intermediate of the citric acid cycle, undergoes reductive amination with glutamine as nitrogen donor:

α-Ketoglutarate + glutamine + NADPH + H^+ $\longrightarrow$
$$\text{2 glutamate} + \text{NADP}^+ \quad (22\text{–}2)$$

The net reaction of glutamine synthetase and glutamate synthase (Eqns 22–1 and 22–2) is

α-Ketoglutarate + NH_4^+ + NADPH + ATP $\longrightarrow$
$$\text{L-glutamate} + \text{NADP}^+ + \text{ADP} + P_i$$

Glutamate synthase is not present in animals, which instead maintain high levels of glutamate by processes such as the transamination of α-ketoglutarate during amino acid catabolism.

Glutamate can also be formed in yet another, albeit minor, pathway: the reaction of α-ketoglutarate and NH_4^+ to form glutamate in one step. This is catalyzed by

L-glutamate dehydrogenase, an enzyme present in all organisms. Reducing power is furnished by NADPH:

$$\alpha\text{-Ketoglutarate} + NH_4^+ + NADPH \longrightarrow$$
$$\text{L-glutamate} + NADP^+ + H_2O$$

We encountered this reaction in the catabolism of amino acids (see Fig. 18–7). In eukaryotic cells, L-glutamate dehydrogenase is located in the mitochondrial matrix. The reaction equilibrium favors the reactants, and the K_m for NH_4^+ (~1 mM) is so high that the reaction probably makes only a modest contribution to NH_4^+ assimilation into amino acids and other metabolites. (Recall that the glutamate dehydrogenase reaction, in reverse (see Fig. 18–10), is one source of NH_4^+ destined for the urea cycle.) Concentrations of NH_4^+ high enough for the glutamate dehydrogenase reaction to make a significant contribution to glutamate levels generally occur only when NH_3 is added to the soil or when organisms are grown in a laboratory in the presence of high NH_3 concentrations. In general, soil bacteria and plants rely on the two-enzyme pathway outlined above (Eqns. 22–1, 22–2).

Glutamine Synthetase Is a Primary Regulatory Point in Nitrogen Metabolism

The activity of glutamine synthetase is regulated in virtually all organisms—as expected, given its central metabolic role as an entry point for reduced nitrogen. In enteric bacteria such as *E. coli*, the regulation is unusually complex. Type I enzyme (from bacteria) has 12 identical subunits of M_r 50,000 (**Fig. 22–7**) and is regulated both allosterically and by covalent modification. (Type II enzyme, from eukaryotes and some bacteria, has 10 identical subunits.) Alanine, glycine, and at least

FIGURE 22–7 Subunit structure of bacterial type I glutamine synthetase. (PDB ID 2GLS) This view shows 6 of 12 the identical subunits; a second layer of 6 subunits lies directly beneath the six shown. Each of the 12 subunits has an active site, where ATP and glutamate are bound in orientations that favor transfer of a phosphoryl group from ATP to the side-chain carboxyl of glutamate. In this crystal structure, ADP occupies the ATP site.

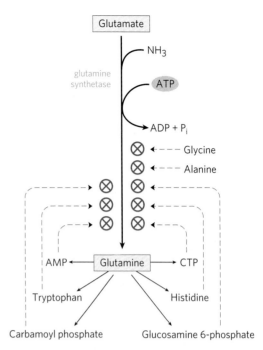

FIGURE 22–8 Allosteric regulation of glutamine synthetase. The enzyme undergoes cumulative regulation by six end products of glutamine metabolism. Alanine and glycine probably serve as indicators of the general status of amino acid metabolism in the cell.

six end products of glutamine metabolism are allosteric inhibitors of the enzyme (**Fig. 22–8**). Each inhibitor alone produces only partial inhibition, but the effects of multiple inhibitors are more than additive, and all eight together virtually shut down the enzyme. This is an example of cumulative feedback inhibition. This control mechanism provides a constant adjustment of glutamine levels to match immediate metabolic requirements.

Superimposed on the allosteric regulation is inhibition by adenylylation of (addition of AMP to) Tyr[397], located near the enzyme's active site (**Fig. 22–9**). This covalent modification increases sensitivity to the allosteric inhibitors, and activity decreases as more subunits are adenylylated. Both adenylylation and deadenylylation are promoted by **adenylyltransferase** (AT in Fig. 22–9), part of a complex enzymatic cascade that responds to levels of glutamine, α-ketoglutarate, ATP, and P_i. The activity of adenylyltransferase is modulated by binding to a regulatory protein called P_{II}, and the activity of P_{II}, in turn, is regulated by covalent modification (uridylylation), again at a Tyr residue. The adenylyltransferase complex with uridylylated P_{II} (P_{II}-UMP) stimulates deadenylylation, whereas the same complex with deuridylylated P_{II} stimulates adenylylation of glutamine synthetase. Both uridylylation and deuridylylation of P_{II} are brought about by a single enzyme, **uridylyltransferase**. Uridylylation is inhibited by binding of glutamine and P_i to uridylyltransferase and is stimulated by binding of α-ketoglutarate and ATP to P_{II}.

The regulation does not stop there. The uridylylated P_{II} also mediates the activation of transcription of the gene encoding glutamine synthetase, thus increasing the cellular concentration of the enzyme; the deuridylylated

(a)

(b)

FIGURE 22–9 **Second level of regulation of glutamine synthetase: covalent modifications. (a)** An adenylylated Tyr residue. **(b)** Cascade leading to adenylylation (inactivation) of glutamine synthetase. AT represents adenylyltransferase; UT, uridylyltransferase. P_{II} is a regulatory protein, itself regulated by uridylylation. Details of this cascade are discussed in the text.

P_{II} brings about a decrease in transcription of the same gene. This mechanism involves an interaction of P_{II} with additional proteins involved in gene regulation, of a type described in Chapter 28. The net result of this elaborate system of controls is a decrease in glutamine synthetase activity when glutamine levels are high, and an increase in activity when glutamine levels are low and α-ketoglutarate and ATP (substrates for the synthetase reaction) are available. The multiple layers of regulation permit a sensitive response in which glutamine synthesis is tailored to cellular needs.

Several Classes of Reactions Play Special Roles in the Biosynthesis of Amino Acids and Nucleotides

The pathways described in this chapter include a variety of interesting chemical rearrangements. Several of these recur and deserve special note before we progress to the pathways themselves. These are (1) transamination reactions and other rearrangements promoted by enzymes containing pyridoxal phosphate; (2) transfer of one-carbon groups, with either tetrahydrofolate (usually at the —CHO and —CH_2OH oxidation levels) or S-adenosylmethionine (at the —CH_3 oxidation level) as cofactor; and (3) transfer of amino groups derived from the amide nitrogen of glutamine. Pyridoxal phosphate (PLP), tetrahydrofolate (H_4 folate), and S-adenosylmethionine (adoMet) are described in some detail in Chapter 18 (see

Figs 18–6, 18–17, and 18–18). Here we focus on amino group transfer involving the amide nitrogen of glutamine.

More than a dozen known biosynthetic reactions use glutamine as the major physiological source of amino groups, and most of these occur in the pathways outlined in this chapter. As a class, the enzymes catalyzing these reactions are called **glutamine amidotransferases**. All have two structural domains: one binding glutamine, the other binding the second substrate, which serves as amino group acceptor **(Fig. 22–10)**. A conserved Cys

Glutamine amidotransferase

The γ-amido nitrogen of glutamine (red) is released as NH_3 in a reaction that probably involves a covalent glutamyl-enzyme intermediate. The NH_3 travels via a channel to the second active site.

Glutamyl-enzyme intermediate

NH_3 reacts with any of several acceptors.

MECHANISM FIGURE 22–10 **Proposed mechanism for glutamine amidotransferases.** Each enzyme has two domains. The glutamine-binding domain contains structural elements conserved among many of these enzymes, including a Cys residue required for activity. The NH_3-acceptor (second-substrate) domain varies. Two types of amino acceptors are shown. X represents an activating group, typically a phosphoryl group derived from ATP, that facilitates displacement of a hydroxyl group from R—OH by NH_3.

residue in the glutamine-binding domain is believed to act as a nucleophile, cleaving the amide bond of glutamine and forming a covalent glutamyl-enzyme intermediate. The NH_3 produced in this reaction is not released, but instead is transferred through an "ammonia channel" to a second active site, where it reacts with the second substrate to form the aminated product. The covalent intermediate is hydrolyzed to the free enzyme and glutamate. If the second substrate must be activated, the usual method is the use of ATP to generate an acyl phosphate intermediate (R—OX in Fig. 22–10, with X as a phosphoryl group). The enzyme glutaminase acts in a similar fashion but uses H_2O as the second substrate, yielding NH_4^+ and glutamate (see Fig. 18–8).

SUMMARY 22.1 Overview of Nitrogen Metabolism

▶ The molecular nitrogen that makes up 80% of Earth's atmosphere is unavailable to most living organisms until it is reduced. This fixation of atmospheric N_2 takes place in certain free-living bacteria and in symbiotic bacteria in the root nodules of leguminous plants.

▶ In soil bacteria and vascular plants, the sequential action of nitrate reductase and nitrite reductase converts NO_3^- to NH_3, which can be assimilated into nitrogen-containing compounds.

▶ The nitrogen cycle entails formation of ammonia by bacterial fixation of N_2, nitrification of ammonia to nitrate by soil organisms, conversion of nitrate to ammonia by higher plants, synthesis of amino acids from ammonia by all organisms, and conversion of nitrate to N_2 by denitrifying soil bacteria. The anammox bacteria anaerobically oxidize ammonia to nitrogen, using nitrite as an electron acceptor.

▶ Fixation of N_2 as NH_3 is carried out by the nitrogenase complex, in a reaction that requires large investments of ATP and of reducing power. The nitrogenase complex is highly labile in the presence of O_2, and is subject to regulation by the supply of NH_3.

▶ In living systems, reduced nitrogen is incorporated first into amino acids and then into a variety of other biomolecules, including nucleotides. The key entry point is the amino acid glutamate. Glutamate and glutamine are the nitrogen donors in a wide range of biosynthetic reactions. Glutamine synthetase, which catalyzes the formation of glutamine from glutamate, is a main regulatory enzyme of nitrogen metabolism.

▶ The amino acid and nucleotide biosynthetic pathways make repeated use of the biological cofactors pyridoxal phosphate, tetrahydrofolate, and S-adenosylmethionine. Pyridoxal phosphate is required for transamination reactions involving glutamate and for other amino acid transformations. One-carbon transfers require S-adenosylmethionine and tetrahydrofolate. Glutamine amidotransferases catalyze reactions that incorporate nitrogen derived from glutamine.

22.2 Biosynthesis of Amino Acids

All amino acids are derived from intermediates in glycolysis, the citric acid cycle, or the pentose phosphate pathway **(Fig. 22–11)**. Nitrogen enters these pathways by way of glutamate and glutamine. Some pathways are simple, others are not. Ten of the amino acids are just one or several steps removed from the common metabolite from which they are derived. The biosynthetic pathways for others, such as the aromatic amino acids, are more complex.

Organisms vary greatly in their ability to synthesize the 20 common amino acids. Whereas most bacteria and plants can synthesize all 20, mammals can synthesize only about half of them—generally those

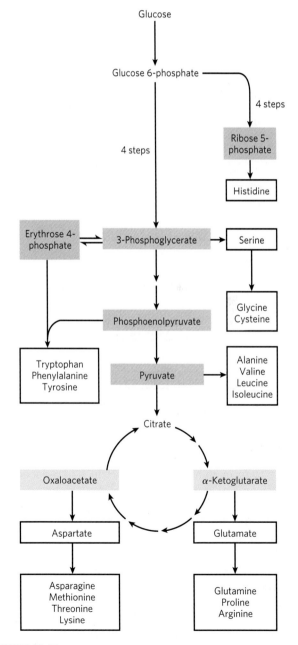

FIGURE 22–11 Overview of amino acid biosynthesis. The carbon skeleton precursors derive from three sources: glycolysis (light red), the citric acid cycle (blue), and the pentose phosphate pathway (purple).

with simple pathways. These are the **nonessential amino acids**, not needed in the diet (see Table 18–1). The remainder, the **essential amino acids**, must be obtained from food. Unless otherwise indicated, the pathways for the 20 common amino acids presented below are those operative in bacteria.

A useful way to organize these biosynthetic pathways is to group them into six families corresponding to their metabolic precursors (Table 22–1), and we use this approach to structure the detailed descriptions that follow. In addition to these six precursors, there is a notable intermediate in several pathways of amino acid and nucleotide synthesis: **5-phosphoribosyl-1-pyrophosphate (PRPP)**:

PRPP is synthesized from ribose 5-phosphate derived from the pentose phosphate pathway (see Fig. 14–22), in a reaction catalyzed by **ribose phosphate pyrophosphokinase**:

Ribose 5-phosphate + ATP $\longrightarrow$
    5-phosphoribosyl-1-pyrophosphate + AMP

This enzyme is allosterically regulated by many of the biomolecules for which PRPP is a precursor.

α-Ketoglutarate Gives Rise to Glutamate, Glutamine, Proline, and Arginine

```
         α-Ketoglutarate
                |
                ↓
            Glutamate
          /     |     \
         ↓      ↓      ↓
   Glutamine  Proline  Arginine
```

We have already described the biosynthesis of **glutamate** and **glutamine**. **Proline** is a cyclized derivative of glutamate **(Fig. 22–12)**. In the first step of proline synthesis, ATP reacts with the γ-carboxyl group of glutamate to form an acyl phosphate, which is reduced by NADPH or NADH to glutamate γ-semialdehyde. This intermediate undergoes rapid spontaneous cyclization and is then reduced further to yield proline.

Arginine is synthesized from glutamate via ornithine and the urea cycle in animals (Chapter 18). In principle, ornithine could also be synthesized from glutamate γ-semialdehyde by transamination, but the spontaneous cyclization of the semialdehyde in the proline pathway precludes a sufficient supply of this intermediate for ornithine synthesis. Bacteria have a de novo biosynthetic pathway for ornithine (and thus arginine) that parallels some steps of the proline pathway but includes two additional steps

TABLE 22–1	Amino Acid Biosynthetic Families, Grouped by Metabolic Precursor

α-Ketoglutarate	**Pyruvate**
Glutamate	Alanine
Glutamine	Valine*
Proline	Leucine*
Arginine	Isoleucine*
3-Phosphoglycerate	**Phosphoenolpyruvate**
Serine	**and erythrose**
Glycine	**4-phosphate**
Cysteine	Tryptophan*
Oxaloacetate	Phenylalanine*
Aspartate	Tyrosine†
Asparagine	**Ribose 5-phosphate**
Methionine*	Histidine*
Threonine*	
Lysine*	

*Essential amino acids in mammals.
†Derived from phenylalanine in mammals.

that avoid the problem of the spontaneous cyclization of glutamate γ-semialdehyde (Fig. 22–12). In the first step, the α-amino group of glutamate is blocked by an acetylation requiring acetyl-CoA; then, after the transamination step, the acetyl group is removed to yield ornithine.

The pathways to proline and arginine are somewhat different in mammals. Proline can be synthesized by the pathway shown in Figure 22–12, but it is also formed from arginine obtained from dietary or tissue protein. Arginase, a urea cycle enzyme, converts arginine to ornithine and urea (see Figs 18–10, 18–26). The ornithine is converted to glutamate γ-semialdehyde by the enzyme **ornithine δ-aminotransferase (Fig. 22–13)**. The semialdehyde cyclizes to Δ^1-pyrroline-5-carboxylate, which is then converted to proline (Fig. 22–12). The pathway for arginine synthesis shown in Figure 22–12 is absent in mammals. When arginine from dietary intake or protein turnover is insufficient for protein synthesis, the ornithine δ-aminotransferase reaction operates in the direction of ornithine formation. Ornithine is then converted to citrulline and arginine in the urea cycle.

Serine, Glycine, and Cysteine Are Derived from 3-Phosphoglycerate

FIGURE 22–12 Biosynthesis of proline and arginine from glutamate in bacteria. All five carbon atoms of proline arise from glutamate. In many organisms, glutamate dehydrogenase is unusual in that it uses *either* NADH or NADPH as a cofactor. The same may be true of other enzymes in these pathways. The γ-semialdehyde in the proline pathway undergoes a rapid, reversible cyclization to Δ^1-pyrroline-5-carboxylate (P5C), with the equilibrium favoring P5C formation. Cyclization is averted in the ornithine/arginine pathway by acetylation of the α-amino group of glutamate in the first step and removal of the acetyl group after the transamination. Although some bacteria lack arginase and thus the complete urea cycle, they can synthesize arginine from ornithine in steps that parallel the mammalian urea cycle, with citrulline and argininosuccinate as intermediates (see Fig. 18–10).

Here, and in subsequent figures in this chapter, the reaction arrows indicate the linear path to the final products, without considering the reversibility of individual steps. For example, the step of the pathway leading to arginine that is catalyzed by *N*-acetylglutamate dehydrogenase is chemically similar to the glyceraldehyde 3-phosphate dehydrogenase reaction in glycolysis (see Fig. 14–8), and is readily reversible.

FIGURE 22–13 Ornithine δ-aminotransferase reaction: a step in the mammalian pathway to proline. This enzyme is found in the mitochondrial matrix of most tissues. Although the equilibrium favors P5C formation, the reverse reaction is the only mammalian pathway for synthesis of ornithine (and thus arginine) when arginine levels are insufficient for protein synthesis.

Ornithine — Glutamate γ-semialdehyde — Δ^1-Pyrroline-5-carboxylate (P5C)

The major pathway for the formation of **serine** is the same in all organisms **(Fig. 22–14)**. In the first step, the hydroxyl group of 3-phosphoglycerate is oxidized by a dehydrogenase (using NAD^+) to yield 3-phosphohydroxypyruvate. Transamination from glutamate yields 3-phosphoserine, which is hydrolyzed to free serine by phosphoserine phosphatase.

Serine (three carbons) is the precursor of **glycine** (two carbons) through removal of a carbon atom by **serine hydroxymethyltransferase** (Fig. 22–14). Tetrahydrofolate accepts the β carbon (C-3) of serine, which forms a methylene bridge between N-5 and N-10 to yield N^5,N^{10}-methylenetetrahydrofolate (see Fig. 18–17). The overall reaction, which is reversible, also requires pyridoxal phosphate. In the liver of vertebrates, glycine can be made by another route: the reverse of the reaction shown in Figure 18–20c, catalyzed by **glycine synthase** (also called **glycine cleavage enzyme**):

$$CO_2 + NH_4^+ + N^5,N^{10}\text{-methylenetetrahydrofolate} + NADH + H^+ \longrightarrow glycine + tetrahydrofolate + NAD^+$$

Plants and bacteria produce the reduced sulfur required for the synthesis of **cysteine** (and methionine, described later) from environmental sulfates; the pathway is shown on the right side of **Figure 22–15**. Sulfate is activated in two steps to produce 3′-phosphoadenosine 5′-phosphosulfate (PAPS), which undergoes an eight-electron reduction to sulfide. The sulfide is then used in the formation of cysteine from serine in a two-step pathway. Mammals synthesize cysteine from two amino acids: methionine furnishes the sulfur atom, and serine furnishes the carbon skeleton. Methionine is first converted to *S*-adenosylmethionine (see Fig. 18–18), which can lose its methyl group to any of a number of acceptors to form *S*-adenosylhomocysteine (adoHcy). This demethylated product is hydrolyzed to free homocysteine, which undergoes a reaction with serine, catalyzed by **cystathionine β-synthase**, to yield cystathionine **(Fig. 22–16)**. Finally, **cystathionine γ-lyase**, a PLP-requiring enzyme, catalyzes removal of ammonia and cleavage of cystathionine to yield free cysteine.

FIGURE 22–14 Biosynthesis of serine from 3-phosphoglycerate and of glycine from serine in all organisms. Glycine is also made from CO_2 and NH_4^+ by the action of glycine synthase, with N^5,N^{10}-methylenetetrahydrofolate as methyl group donor (see text).

FIGURE 22–15 **Biosynthesis of cysteine from serine in bacteria and plants.** The origin of reduced sulfur is shown in the pathway on the right.

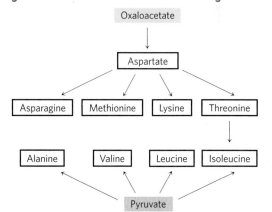

FIGURE 22–16 **Biosynthesis of cysteine from homocysteine and serine in mammals.** The homocysteine is formed from methionine, as described in the text.

Three Nonessential and Six Essential Amino Acids Are Synthesized from Oxaloacetate and Pyruvate

Alanine and **aspartate** are synthesized from pyruvate and oxaloacetate, respectively, by transamination from glutamate. **Asparagine** is synthesized by amidation of aspartate, with glutamine donating the NH_4^+. These are nonessential amino acids, and their simple biosynthetic pathways occur in all organisms.

For reasons incompletely understood, the malignant lymphocytes present in childhood acute lymphoblastic leukemia (ALL) require serum asparagine for growth. The chemotherapy for ALL is administered together with an L-asparaginase derived from bacteria,

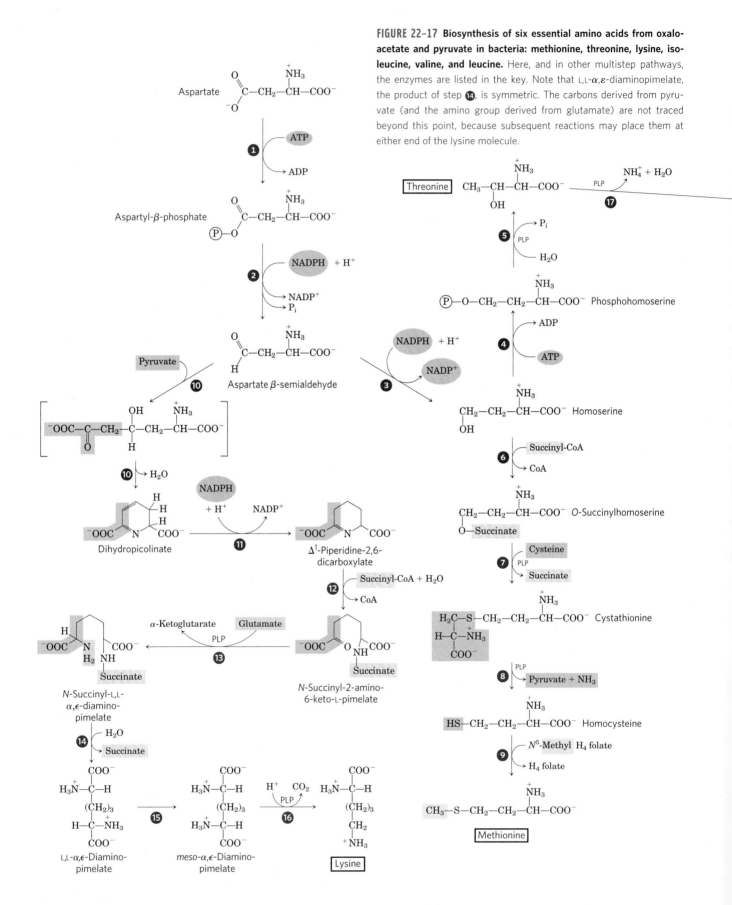

FIGURE 22–17 Biosynthesis of six essential amino acids from oxalo-acetate and pyruvate in bacteria: methionine, threonine, lysine, iso-leucine, valine, and leucine. Here, and in other multistep pathways, the enzymes are listed in the key. Note that L,L-α,ε-diaminopimelate, the product of step **14**, is symmetric. The carbons derived from pyruvate (and the amino group derived from glutamate) are not traced beyond this point, because subsequent reactions may place them at either end of the lysine molecule.

with the enzyme functioning to reduce serum asparagine. The combined treatment results in a greater than 95% remission rate in cases of childhood ALL (L-asparaginase treatment alone produces remission in 40% to 60% of cases). However, the asparaginase treatment has some deleterious side effects, and about 10% of patients who achieve remission eventually suffer relapse, with tumors resistant to drug therapy. Researchers are now

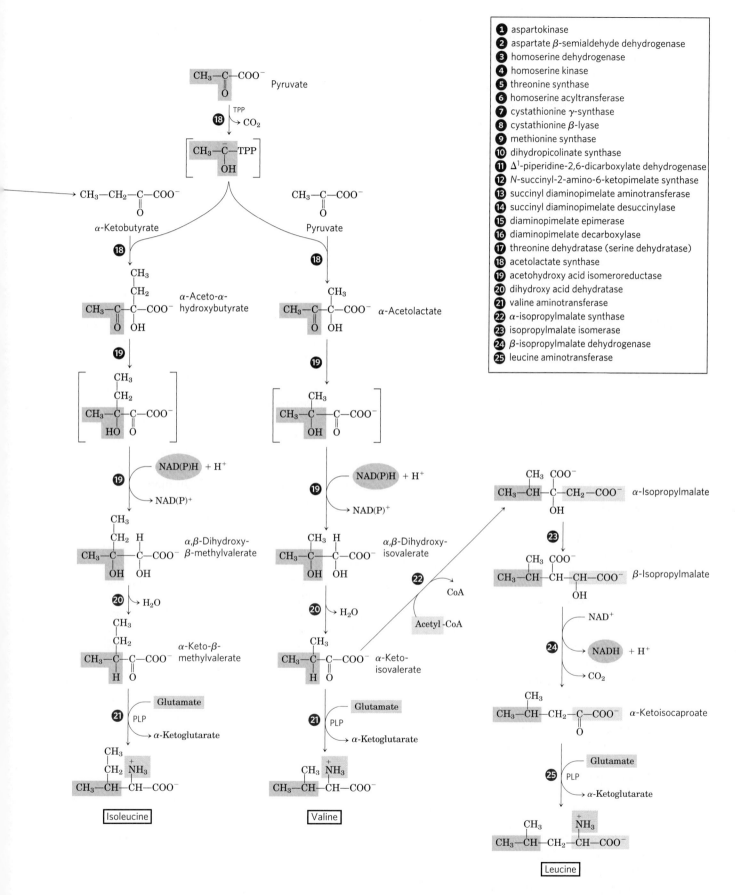

developing inhibitors of human asparagine synthetase to augment these therapies for childhood ALL. ∎

Methionine, threonine, lysine, isoleucine, valine, and leucine are essential amino acids; humans cannot synthesize them. Their biosynthetic pathways are complex and interconnected (**Fig. 22–17**). In some cases, the pathways in bacteria, fungi, and plants differ significantly. Figure 22–17 shows the bacterial pathways.

Aspartate gives rise to **methionine, threonine,** and **lysine**. Branch points occur at aspartate β-semial-dehyde, an intermediate in all three pathways, and at homoserine, a precursor of threonine and methionine. Threonine, in turn, is one of the precursors of isoleu-cine. The **valine** and **isoleucine** pathways share four enzymes (Fig. 22–17, steps ⑱ to ㉑). Pyruvate gives rise to valine and isoleucine in pathways that begin with condensation of two carbons of pyruvate (in the form of hydroxyethyl thiamine pyrophosphate; see Fig. 14–15) with another molecule of pyruvate (the valine path) or with α-ketobutyrate (the isoleucine path). The α-ketobutyrate is derived from threonine in a reaction that requires pyridoxal phosphate (Fig. 22–17, step ⑰). An intermediate in the valine pathway, α-ketoisovalerate, is the starting point for a four-step branch pathway lead-ing to **leucine** (steps ㉒ to ㉕).

Chorismate Is a Key Intermediate in the Synthesis of Tryptophan, Phenylalanine, and Tyrosine

Aromatic rings are not readily available in the environ-ment, even though the benzene ring is very stable. The branched pathway to tryptophan, phenylalanine, and tyrosine, occurring in bacteria, fungi, and plants, is the main biological route of aromatic ring formation. It pro-ceeds through ring closure of an aliphatic precursor followed by stepwise addition of double bonds. The first four steps produce shikimate, a seven-carbon molecule derived from erythrose 4-phosphate and phosphoenol-pyruvate **(Fig. 22–18)**. Shikimate is converted to cho-rismate in three steps that include the addition of three more carbons from another molecule of phosphoenol-pyruvate. Chorismate is the first branch point of the pathway, with one branch leading to tryptophan, the other to phenylalanine and tyrosine.

In the **tryptophan** branch **(Fig. 22–19)**, choris-mate is converted to anthranilate in a reaction in which glutamine donates the nitrogen that will become part of the indole ring. Anthranilate then condenses with PRPP. The indole ring of tryptophan is derived from the ring carbons and amino group of anthranilate plus two carbons derived from PRPP. The final reaction in the sequence is catalyzed by **tryptophan synthase**. This enzyme has an $\alpha_2\beta_2$ subunit structure and can be dis-sociated into two α subunits and a β_2 unit that catalyze different parts of the overall reaction:

$$\text{Indole-3-glycerol phosphate} \xrightarrow[\alpha \text{ subunit}]{}$$
$$\text{indole} + \text{glyceraldehyde 3-phosphate}$$
$$\text{Indole} + \text{serine} \xrightarrow[\beta_2 \text{ subunit}]{} \text{tryptophan} + \text{H}_2\text{O}$$

The second part of the reaction requires pyridoxal phosphate **(Fig. 22–20)**. Indole formed in the first part is not released by the enzyme, but instead moves through a channel from the α-subunit active site to one of the β-subunit active sites, where it condenses with a Schiff base intermediate derived from serine and PLP. Intermediate channeling of this type may be a feature of the entire pathway from chorismate to tryptophan. Enzyme active sites catalyzing different steps (some-times not sequential steps) of the pathway to trypto-phan are found on single polypeptides in some species of fungi and bacteria, but are separate proteins in other species. In addition, the activity of some of these enzymes requires a noncovalent association with other enzymes of the pathway. These observations suggest that all the pathway enzymes are components of a large, multienzyme complex in both bacteria and eukaryotes. Such complexes are generally not preserved intact when the enzymes are isolated using traditional bio-chemical methods, but evidence for the existence of multienzyme complexes is accumulating for this and other metabolic pathways (see Section 16.3).

In plants and bacteria, **phenylalanine** and **tyro-sine** are synthesized from chorismate in pathways much less complex than the tryptophan pathway. The common intermediate is prephenate **(Fig. 22–21)**. The final step in both cases is transamination with glutamate.

Animals can produce tyrosine directly from phenyl-alanine through hydroxylation at C-4 of the phenyl group by **phenylalanine hydroxylase**; this enzyme also participates in the degradation of phenylalanine (see Figs 18–23, 18–24). Tyrosine is considered a con-ditionally essential amino acid, or as nonessential inso-far as it can be synthesized from the essential amino acid phenylalanine.

Histidine Biosynthesis Uses Precursors of Purine Biosynthesis

The pathway to **histidine** in all plants and bacteria dif-fers in several respects from other amino acid biosyn-thetic pathways. Histidine is derived from three precur-sors **(Fig. 22–22)**: PRPP contributes five carbons, the purine ring of ATP contributes a nitrogen and a carbon, and glutamine supplies the second ring nitrogen. The key steps are condensation of ATP and PRPP, in which

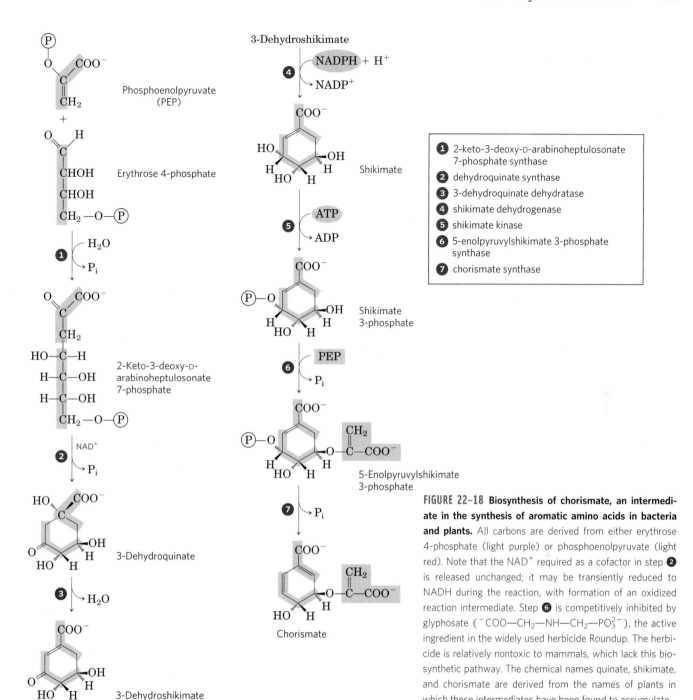

FIGURE 22–18 **Biosynthesis of chorismate, an intermediate in the synthesis of aromatic amino acids in bacteria and plants.** All carbons are derived from either erythrose 4-phosphate (light purple) or phosphoenolpyruvate (light red). Note that the NAD^+ required as a cofactor in step ❷ is released unchanged; it may be transiently reduced to NADH during the reaction, with formation of an oxidized reaction intermediate. Step ❻ is competitively inhibited by glyphosate ($^-COO—CH_2—NH—CH_2—PO_3^{2-}$), the active ingredient in the widely used herbicide Roundup. The herbicide is relatively nontoxic to mammals, which lack this biosynthetic pathway. The chemical names quinate, shikimate, and chorismate are derived from the names of plants in which these intermediates have been found to accumulate.

N-1 of the purine ring is linked to the activated C-1 of the ribose of PRPP (step ❶ in Fig. 22–22); purine ring opening that ultimately leaves N-1 and C-2 of adenine linked to the ribose (step ❸); and formation of the imidazole ring, a reaction in which glutamine donates a nitrogen (step ❺). The use of ATP as a metabolite rather than a high-energy cofactor is unusual—but not wasteful, because it dovetails with the purine biosynthetic pathway. The remnant of ATP that is released after the transfer of N-1 and C-2 is 5-aminoimidazole-4-carboxamide ribonucleotide (AICAR), an intermediate of purine biosynthesis (see Fig. 22–35) that is rapidly recycled to ATP.

Amino Acid Biosynthesis Is under Allosteric Regulation

As detailed in Chapter 15, the control of flux through a metabolic pathway often reflects the activity of multiple enzymes in that pathway. In the case of amino acid synthesis, regulation takes place in part through feedback inhibition of the first reaction by the end product of the pathway. This first reaction is often catalyzed by an allosteric enzyme that plays an important role in the overall control of flux through that pathway. As an example, **Figure 22–23** shows the allosteric regulation of isoleucine synthesis from threonine (detailed in Fig. 22–17).

The end product, isoleucine, is an allosteric inhibitor of the first reaction in the sequence. In bacteria, such allosteric modulation of amino acid synthesis contributes to the minute-to-minute adjustment of pathway activity to cellular needs.

Allosteric regulation of an individual enzyme can be considerably more complex. An example is the remarkable set of allosteric controls exerted on glutamine synthetase of *E. coli* (Fig. 22–8). Six products derived from glutamine serve as negative feedback modulators of the enzyme, and the overall effects of these and other modulators are more than additive. Such regulation is called **concerted inhibition**.

Additional mechanisms contribute to the regulation of the amino acid biosynthetic pathways. Because the 20 common amino acids must be made in the correct proportions for protein synthesis, cells have developed ways not only of controlling the rate of synthesis of individual amino acids but also of coordinating their formation. Such coordination is especially well developed in fast-growing bacterial cells. **Figure 22–24** shows how *E. coli* cells coordinate the synthesis of lysine, methionine, threonine, and isoleucine, all made from aspartate. Several important types of inhibition patterns are evident. The step from aspartate to aspartyl-β-phosphate is catalyzed by three isozymes, each independently controlled by different modulators. This **enzyme multiplicity** prevents one biosynthetic end product from shutting down key steps in a pathway when other products of the same pathway are required. The steps from aspartate β-semialdehyde to homoserine and from threonine to α-ketobutyrate (detailed in Fig. 22–17) are also catalyzed by dual, independently controlled isozymes. One isozyme for the conversion of aspartate to aspartyl-β-phosphate is allosterically inhibited by two different modulators, lysine and isoleucine, whose action is more than additive—another example of concerted inhibition. The sequence from aspartate to isoleucine undergoes multiple, overlapping negative feedback inhibitions; for example, isoleucine inhibits the conversion of threonine to α-ketobutyrate (as described above), and threonine inhibits its own formation at three points: from homoserine, from aspartate β-semialdehyde, and from aspartate (steps ❹, ❸, and ❶ in Fig. 22–17). This overall regulatory mechanism is called **sequential feedback inhibition**.

Similar patterns are evident in the pathways leading to the aromatic amino acids. The first step of the

FIGURE 22–19 **Biosynthesis of tryptophan from chorismate in bacteria and plants.** In *E. coli*, enzymes catalyzing steps ❶ and ❷ are subunits of a single complex.

❶ anthranilate synthase
❷ anthranilate phosphoribosyltransferase
❸ *N*-(5'-phosphoribosyl)-anthranilate isomerase
❹ indole-3-glycerol phosphate synthase
❺ tryptophan synthase

MECHANISM FIGURE 22–20 Tryptophan synthase reaction. (a) This enzyme catalyzes a multistep reaction with several types of chemical rearrangements. The PLP-facilitated transformations occur at the β carbon (C-3) of the amino acid, as opposed to the α-carbon reactions described in

Figure 18–6. The β carbon of serine is attached to the indole ring system. **(b)** (PDB ID 1KFJ) Indole generated on the α subunit (white) moves through a tunnel to the β subunit (blue), where it condenses with the serine moiety. 🖱 Tryptophan Synthase Mechanism

FIGURE 22–21 Biosynthesis of phenylalanine and tyrosine from chorismate in bacteria and plants. Conversion of chorismate to prephenate is a rare biological example of a Claisen rearrangement.

early pathway to the common intermediate chorismate is catalyzed by the enzyme 2-keto-3-deoxy-D-arabino-heptulosonate 7-phosphate (DAHP) synthase (❶ in Fig. 22–18). Most microorganisms and plants have three DAHP synthase isozymes. One is allosterically inhibited (feedback inhibition) by phenylalanine, another by tyrosine, and the third by tryptophan. This scheme helps the overall pathway to respond to cellular requirements for one or more of the aromatic amino acids. Additional regulation takes place after the pathway branches at chorismate. For example, the enzymes catalyzing the first two steps of the tryptophan branch are subject to allosteric inhibition by tryptophan.

SUMMARY 22.2 Biosynthesis of Amino Acids

▶ Plants and bacteria synthesize all 20 common amino acids. Mammals can synthesize about half;

the others are required in the diet (essential amino acids).

▶ Among the nonessential amino acids, glutamate is formed by reductive amination of α-ketoglutarate and serves as the precursor of glutamine, proline, and arginine. Alanine and aspartate (and thus asparagine) are formed from pyruvate and oxaloacetate, respectively, by transamination. The carbon chain of serine is derived from 3-phosphoglycerate. Serine is a precursor of glycine; the β-carbon atom of serine is transferred to tetrahydrofolate. In microorganisms, cysteine is produced from serine and from sulfide produced by the reduction of environmental sulfate. Mammals produce cysteine from methionine and serine by a series of reactions requiring S-adenosylmethionine and cystathionine.

▶ Among the essential amino acids, the aromatic amino acids (phenylalanine, tyrosine, and tryptophan) form by a pathway in which chorismate occupies a key branch point. Phosphoribosyl pyrophosphate is a precursor of tryptophan and histidine. The pathway to histidine is interconnected with the purine synthetic pathway. Tyrosine can also be formed by hydroxylation of phenylalanine (and thus is considered conditionally essential). The pathways for the other essential amino acids are complex.

▶ The amino acid biosynthetic pathways are subject to allosteric end-product inhibition; the regulatory enzyme is usually the first in the sequence. Regulation of the various synthetic pathways is coordinated.

22.3 Molecules Derived from Amino Acids

In addition to their role as the building blocks of proteins, amino acids are precursors of many specialized biomolecules, including hormones, coenzymes, nucleotides, alkaloids, cell wall polymers, porphyrins, antibiotics, pigments, and neurotransmitters. We describe here the pathways to a number of these amino acid derivatives.

Glycine Is a Precursor of Porphyrins

The biosynthesis of **porphyrins**, for which glycine is a major precursor, is our first example because of the central importance of the porphyrin nucleus in heme proteins such as hemoglobin and the cytochromes. The porphyrins are constructed from four molecules of the monopyrrole derivative **porphobilinogen**, which itself is derived from two molecules of δ-aminolevulinate. There are two major pathways to δ-aminolevulinate. In higher eukaryotes **(Fig. 22–25a)**, glycine reacts with succinyl-CoA in the first step to yield α-amino-β-ketoadipate,

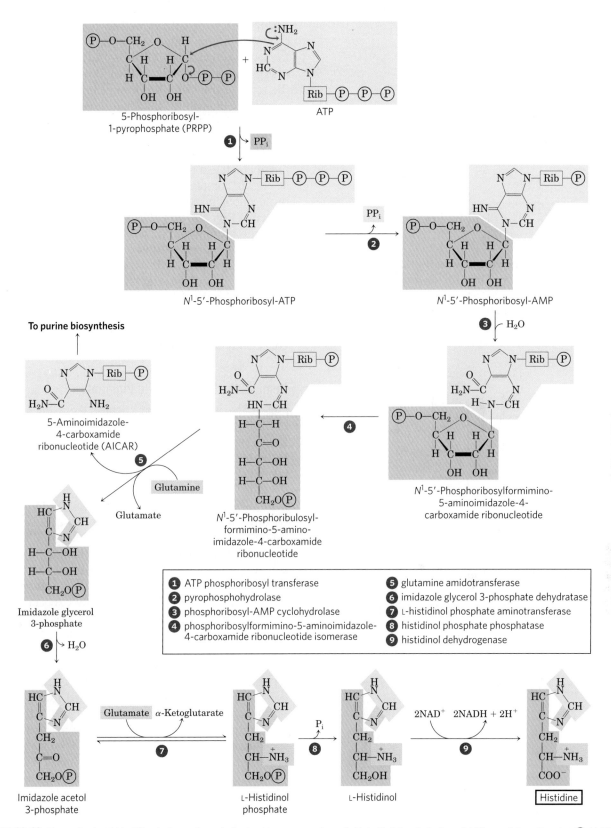

FIGURE 22–22 Biosynthesis of histidine in bacteria and plants. Atoms derived from PRPP and ATP are shaded light red and blue, respectively. Two of the histidine nitrogens are derived from glutamine and glutamate (green). Note that the derivative of ATP remaining after step ❺ (AICAR) is an intermediate in purine biosynthesis (see Fig. 22–35, step ❾), so ATP is rapidly regenerated.

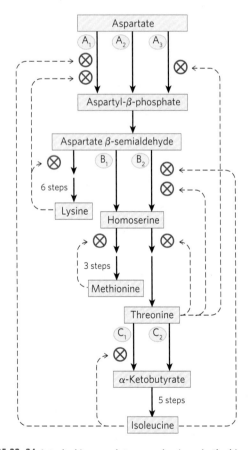

FIGURE 22–23 Allosteric regulation of isoleucine biosynthesis. The first reaction in the pathway from threonine to isoleucine is inhibited by the end product, isoleucine. This was one of the first examples of allosteric feedback inhibition to be discovered. The steps from α-ketobutyrate to isoleucine correspond to steps ⑱ through ㉑ in Figure 22-17 (five steps, because ⑲ is a two-step reaction).

FIGURE 22–24 Interlocking regulatory mechanisms in the biosynthesis of several amino acids derived from aspartate in *E. coli*. Three enzymes (A, B, C) have either two or three isozyme forms, indicated by numerical subscripts. In each case, one isozyme (A_2, B_1, and C_2) has no allosteric regulation; these isozymes are regulated by changes in the amount of enzyme synthesized (Chapter 28). Synthesis of isozymes A_2 and B_1 is repressed when methionine levels are high, and synthesis of isozyme C_2 is repressed when isoleucine levels are high. Enzyme A is aspartokinase; B, homoserine dehydrogenase; C, threonine dehydratase.

which is then decarboxylated to δ-aminolevulinate. In plants, algae, and most bacteria, δ-aminolevulinate is formed from glutamate (Fig. 22–25b). The glutamate is first esterified to glutamyl-tRNA^{Glu} (see Chapter 27 on the topic of transfer RNAs); reduction by NADPH converts the glutamate to glutamate 1-semialdehyde, which is cleaved from the tRNA. An aminotransferase converts the glutamate 1-semialdehyde to δ-aminolevulinate.

In all organisms, two molecules of δ-aminolevulinate condense to form porphobilinogen and, through a series of complex enzymatic reactions, four molecules of porphobilinogen come together to form **protoporphyrin (Fig. 22–26)**. The iron atom is incorporated after the protoporphyrin has been assembled, in a step catalyzed by ferrochelatase. Porphyrin biosynthesis is regulated in higher eukaryotes by heme, which serves as a feedback inhibitor of early steps in the synthetic pathway. Genetic defects in the biosynthesis of porphyrins can lead to the accumulation of pathway intermediates, causing a variety of human diseases known collectively as **porphyrias** (Box 22–2).

Heme Is the Source of Bile Pigments

The iron-porphyrin (heme) group of hemoglobin, released from dying erythrocytes in the spleen, is degraded to yield free Fe^{2+} and, ultimately, **bilirubin**. This pathway is arresting for its capacity to inject color into human biochemistry.

The first step in the two-step pathway, catalyzed by heme oxygenase, converts heme to biliverdin, a linear (open) tetrapyrrole derivative **(Fig. 22–27)**. The other products of the reaction are free Fe^{2+} and CO. The Fe^{2+} is quickly bound by ferritin. Carbon monoxide is a poison that binds to hemoglobin (see Box 5–1), and the production of CO by heme oxygenase ensures that, even in the absence of environmental exposure, about 1% of an individual's heme is complexed with CO.

Biliverdin is converted to bilirubin in the second step, catalyzed by biliverdin reductase. You can monitor this reaction colorimetrically in a familiar in situ experiment. When you are bruised, the black and/or purple color results from hemoglobin released from damaged erythrocytes. Over time, the color changes to the green of biliverdin, and then to the yellow of bilirubin. Bilirubin is largely insoluble, and it travels in the bloodstream as a complex with serum albumin. In the liver, bilirubin is transformed to the bile pigment bilirubin diglucuronide. This product is sufficiently water-soluble to be secreted with other components of bile into the small intestine, where microbial enzymes convert it to several products, predominantly urobilinogen. Some urobilinogen is reabsorbed into the blood and transported to the kidney, where it is converted to urobilin, the compound that gives urine its yellow color (Fig. 22–27). Urobilinogen remaining in the intestine is converted (in another microbe-dependent reaction) to stercobilin (Fig. 22–27), which imparts the red-brown color to feces.

(a) In most animals, including mammals, δ-aminolevulinate is synthesized from glycine and succinyl-CoA. The atoms furnished by glycine are shown in red. **(b)** In bacteria and plants, the precursor of δ-aminolevulinate is glutamate.

FIGURE 22–25 Biosynthesis of δ-aminolevulinate. (a) In most animals, including mammals, δ-aminolevulinate is synthesized from glycine and succinyl-CoA. The atoms furnished by glycine are shown in red. **(b)** In bacteria and plants, the precursor of δ-aminolevulinate is glutamate.

❶ porphobilinogen synthase
❷ uroporphyrinogen synthase
❸ uroporphyrinogen III cosynthase
❹ uroporphyrinogen decarboxylase
❺ coproporphyrinogen oxidase
❻ protoporphyrinogen oxidase
❼ ferrochelatase

FIGURE 22–26 Biosynthesis of heme from δ-aminolevulinate. Ac represents acetyl (—CH₂COO⁻); Pr, propionyl (—CH₂CH₂COO⁻).

BOX 22–2 ⚕ MEDICINE On Kings and Vampires

Porphyrias are a group of genetic diseases that result from defects in enzymes of the biosynthetic pathway from glycine to porphyrins; specific porphyrin precursors accumulate in erythrocytes, body fluids, and the liver. The most common form is acute intermittent porphyria. Most individuals inheriting this condition are heterozygotes and are usually asymptomatic, because the single copy of the normal gene provides a sufficient level of enzyme function. However, certain nutritional or environmental factors (as yet poorly understood) can cause a buildup of δ-aminolevulinate and porphobilinogen, leading to attacks of acute abdominal pain and neurological dysfunction. King George III, British monarch during the American Revolution, suffered several episodes of apparent madness that tarnished the record of this otherwise accomplished man. The symptoms of his condition suggest that George III suffered from acute intermittent porphyria.

One of the rarer porphyrias results in an accumulation of uroporphyrinogen I, an abnormal isomer of a protoporphyrin precursor. This compound stains the urine red, causes the teeth to fluoresce strongly in ultraviolet light, and makes the skin abnormally sensitive to sunlight. Many individuals with this porphyria are anemic because insufficient heme is synthesized.

This genetic condition may have given rise to the vampire myths of folk legend.

The symptoms of most porphyrias are now readily controlled with dietary changes or the administration of heme or heme derivatives.

Impaired liver function or blocked bile secretion causes bilirubin to leak from the liver into the blood, resulting in a yellowing of the skin and eyeballs, a condition called jaundice. In cases of jaundice, determination of the concentration of bilirubin in the blood may be useful in the diagnosis of underlying liver disease. Newborn infants sometimes develop jaundice because they have not yet produced enough glucuronyl bilirubin transferase to process their bilirubin. A traditional treatment to reduce excess bilirubin, exposure to a fluorescent lamp, causes a photochemical conversion of bilirubin to compounds that are more soluble and easily excreted.

These pathways of heme breakdown play significant roles in protecting cells from oxidative damage and in regulating certain cellular functions. The CO produced by heme oxygenase is toxic at high concentrations, but at the very low concentrations generated during heme degradation it seems to have some regulatory and/or signaling functions. It acts as a vasodilator, much the same as (but less potent than) nitric oxide (discussed below). Low levels of CO also have some regulatory effects on neurotransmission. Bilirubin is the most abundant antioxidant in mammalian tissues and is responsible for most of the antioxidant activity in serum. Its protective effects seem to be especially important in the developing brain of newborn infants.

The cell toxicity associated with jaundice may be due to bilirubin levels in excess of the serum albumin needed to solubilize it.

Given these varied roles of heme degradation products, the degradative pathway is subject to regulation, mainly at the first step. Humans have at least three isozymes of heme oxygenase (HO). HO-1 is highly regulated; the expression of its gene is induced by a wide range of stress conditions (shear stress, angiogenesis (uncontrolled development of blood vessels), hypoxia, hyperoxia, heat shock, exposure to ultraviolet light, hydrogen peroxide, and many other metabolic insults). HO-2 is found mainly in brain and testes, where it is continuously expressed. The third isozyme, HO-3, is not yet well characterized. ∎

Amino Acids Are Precursors of Creatine and Glutathione

Phosphocreatine, derived from **creatine**, is an important energy buffer in skeletal muscle (see Box 23–2). Creatine is synthesized from glycine and arginine (**Fig. 22–28**); methionine, in the form of S-adenosylmethionine, acts as methyl group donor.

Glutathione (GSH), present in plants, animals, and some bacteria, often at high levels, can be thought

FIGURE 22–27 Bilirubin and its breakdown products. M represents methyl; V, vinyl; Pr, propionyl; E, ethyl. For ease of comparison, these structures are shown in linear form, rather than in their correct stereochemical conformations.

D-Amino Acids Are Found Primarily in Bacteria

of as a redox buffer. It is derived from glutamate, cysteine, and glycine **(Fig. 22–29)**. The γ-carboxyl group of glutamate is activated by ATP to form an acyl phosphate intermediate, which is then attacked by the α-amino group of cysteine. A second condensation reaction follows, with the α-carboxyl group of cysteine activated to an acyl phosphate to permit reaction with glycine. The oxidized form of glutathione (GSSG), produced in the course of its redox activities, contains two glutathione molecules linked by a disulfide bond.

Glutathione probably helps maintain the sulfhydryl groups of proteins in the reduced state and the iron of heme in the ferrous (Fe^{2+}) state, and it serves as a reducing agent for glutaredoxin in deoxyribonucleotide synthesis (see Fig. 22–41). Its redox function is also used to remove toxic peroxides formed in the normal course of growth and metabolism under aerobic conditions:

$$2\ GSH + R{-}O{-}O{-}H \longrightarrow GSSG + H_2O + R{-}OH$$

This reaction is catalyzed by **glutathione peroxidase**, a remarkable enzyme in that it contains a covalently bound selenium (Se) atom in the form of selenocysteine (see Fig. 3–8a), which is essential for its activity.

Although D-amino acids do not generally occur in proteins, they do serve some special functions in the structure of bacterial cell walls and peptide antibiotics. Bacterial peptidoglycans (see Fig. 20–30) contain both D-alanine and D-glutamate. D-Amino acids arise directly from the L isomers by the action of amino acid racemases, which have pyridoxal phosphate as cofactor (see Fig. 18–6). Amino acid racemization is uniquely important to bacterial metabolism, and enzymes such as alanine racemase are prime targets for pharmaceutical agents. One such agent, **L-fluoroalanine**, is being tested as an antibacterial drug. Another, **cycloserine**, is used to treat tuberculosis. Because these inhibitors also affect some PLP-requiring human enzymes, however, they have potentially undesirable side effects. ∎

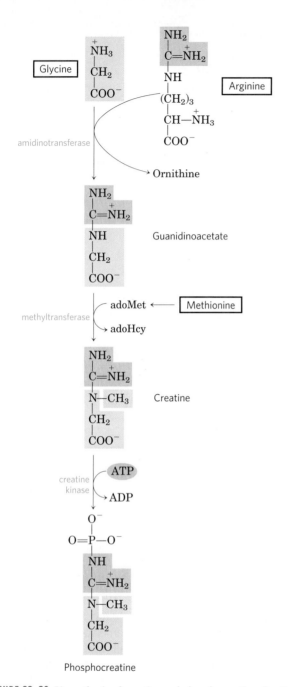

FIGURE 22–28 Biosynthesis of creatine and phosphocreatine. Creatine is made from three amino acids: glycine, arginine, and methionine. This pathway shows the versatility of amino acids as precursors of other nitrogenous biomolecules.

FIGURE 22–29 Glutathione metabolism. (a) Biosynthesis of glutathione. **(b)** Oxidized form of glutathione.

Aromatic Amino Acids Are Precursors of Many Plant Substances

Phenylalanine, tyrosine, and tryptophan are converted to a variety of important compounds in plants. The rigid polymer **lignin**, derived from phenylalanine and tyrosine, is second only to cellulose in abundance in plant tissues. The structure of the lignin polymer is complex and not well understood. Tryptophan is also the precursor of the plant growth hormone indole-3-acetate, or **auxin (Fig. 22–30a)**, which is important

in the regulation of a wide range of biological processes in plants.

Phenylalanine and tyrosine also give rise to many commercially significant natural products, including the tannins that inhibit oxidation in wines; alkaloids such as morphine, which have potent physiological effects; and the flavoring of cinnamon oil (Fig. 22–30b), nutmeg, cloves, vanilla, cayenne pepper, and other products.

Biological Amines Are Products of Amino Acid Decarboxylation

Many important neurotransmitters are primary or secondary amines, derived from amino acids in simple pathways. In addition, some polyamines that

FIGURE 22–30 Biosynthesis of two plant substances from amino acids. (a) Indole-3-acetate (auxin) and (b) cinnamate (cinnamon flavor).

form complexes with DNA are derived from the amino acid ornithine, a component of the urea cycle. A common denominator of many of these pathways is amino acid decarboxylation, another PLP-requiring reaction (see Fig. 18–6).

The synthesis of some neurotransmitters is illustrated in **Figure 22–31**. Tyrosine gives rise to a family of catecholamines that includes **dopamine, norepinephrine**, and **epinephrine**. Levels of catecholamines are correlated with, among other things, changes in blood pressure. The neurological disorder Parkinson disease is associated with an underproduction of dopamine, and it has traditionally been treated by administering L-dopa. Overproduction of dopamine in the brain may be linked to psychological disorders such as schizophrenia.

Glutamate decarboxylation gives rise to **γ-aminobutyrate (GABA)**, an inhibitory neurotransmitter. Its underproduction is associated with epileptic seizures. GABA analogs are used in the treatment of epilepsy and hypertension. Levels of GABA can also be increased by administering inhibitors of the GABA-degrading enzyme GABA aminotransferase. Another important neurotransmitter, **serotonin**, is derived from tryptophan in a two-step pathway.

Histidine undergoes decarboxylation to **histamine**, a powerful vasodilator in animal tissues. Histamine is released in large amounts as part of the allergic response, and it also stimulates acid secretion in the stomach. A growing array of pharmaceutical agents are being designed to interfere with either the synthesis or the action of histamine. A prominent example is the histamine receptor antagonist **cimetidine** (Tagamet), a structural analog of histamine:

It promotes the healing of duodenal ulcers by inhibiting secretion of gastric acid.

Polyamines such as **spermine** and **spermidine**, involved in DNA packaging, are derived from methionine and ornithine by the pathway shown in **Figure 22–32**. The first step is decarboxylation of ornithine, a precursor of arginine (Fig. 22–12). **Ornithine decarboxylase**, a PLP-requiring enzyme, is the target of several powerful inhibitors used as pharmaceutical agents (see Box 6–3). ■

Arginine Is the Precursor for Biological Synthesis of Nitric Oxide

A surprise finding in the mid-1980s was the role of nitric oxide (NO)—previously known mainly as a component of smog—as an important biological messenger. This simple gaseous substance diffuses readily through membranes, although its high reactivity limits its range of diffusion to about a 1 mm radius from the site of synthesis. In humans NO plays a role in a range of physiological processes, including neurotransmission, blood clotting, and the control of blood pressure. Its mode of action is described in Chapter 12 (see Section 12.4).

Nitric oxide is synthesized from arginine in an NADPH-dependent reaction catalyzed by nitric oxide synthase **(Fig. 22–33)**, a dimeric enzyme structurally related to NADPH cytochrome P-450 reductase (see Box 21–1). The reaction is a five-electron oxidation. Each subunit of the enzyme contains one bound molecule of each of four different cofactors: FMN, FAD, tetrahydrobiopterin, and Fe^{3+} heme. NO is an unstable molecule and cannot be stored. Its synthesis is stimulated by interaction of nitric oxide synthase with Ca^{2+}-calmodulin (see Fig. 12–11).

SUMMARY 22.3 Molecules Derived from Amino Acids

▶ Many important biomolecules are derived from amino acids. Glycine is a precursor of porphyrins. Degradation of iron-porphyrin (heme) generates bilirubin, which is converted to bile pigments, with several physiological functions.

FIGURE 22–31 Biosynthesis of some neurotransmitters from amino acids. The key step is the same in each case: a PLP-dependent decarboxylation (shaded light red).

▶ Glycine and arginine give rise to creatine and phosphocreatine, an energy buffer. Glutathione, formed from three amino acids, is an important cellular reducing agent.

▶ Bacteria synthesize D-amino acids from L-amino acids in racemization reactions requiring pyridoxal phosphate. D-Amino acids are commonly found in certain bacterial walls and certain antibiotics.

▶ The aromatic amino acids give rise to many plant substances. The PLP-dependent decarboxylation of some amino acids yields important biological amines, including neurotransmitters.

▶ Arginine is the precursor of nitric oxide, a biological messenger.

22.4 Biosynthesis and Degradation of Nucleotides

As discussed in Chapter 8, nucleotides have a variety of important functions in all cells. They are the precursors of DNA and RNA. They are essential carriers of chemical energy—a role primarily of ATP and to some extent GTP. They are components of the cofactors NAD, FAD, *S*-adenosylmethionine, and coenzyme A, as well as of activated biosynthetic intermediates such as UDP-glucose and CDP-diacylglycerol. Some, such as cAMP and cGMP, are also cellular second messengers.

Two types of pathways lead to nucleotides: the **de novo pathways** and the **salvage pathways**. De novo

FIGURE 22–32 Biosynthesis of spermidine and spermine. The PLP-dependent decarboxylation steps are shaded light red. In these reactions, S-adenosylmethionine (in its decarboxylated form) acts as a source of propylamino groups (shaded blue).

synthesis of nucleotides begins with their metabolic precursors: amino acids, ribose 5-phosphate, CO_2, and NH_3. Salvage pathways recycle the free bases and nucleosides released from nucleic acid breakdown. Both types of pathways are important in cellular metabolism and both are discussed in this section.

The de novo pathways for purine and pyrimidine biosynthesis seem to be nearly identical in all living organisms. Notably, the free bases guanine, adenine, thymine, cytidine, and uracil are *not* intermediates in these pathways; that is, the bases are not synthesized and then attached to ribose, as might be expected. The purine ring structure is built up one or a few atoms at a time, attached to ribose throughout the process. The

pyrimidine ring is synthesized as **orotate**, attached to ribose phosphate, and then converted to the common pyrimidine nucleotides required in nucleic acid synthesis. Although the free bases are not intermediates in the de novo pathways, they are intermediates in some of the salvage pathways.

Several important precursors are shared by the de novo pathways for synthesis of pyrimidines and purines. Phosphoribosyl pyrophosphate (PRPP) is important in both, and in these pathways the structure of ribose is retained in the product nucleotide, in contrast to its fate in the tryptophan and histidine biosynthetic pathways discussed earlier. An amino acid is an important precursor in each type of pathway: glycine for purines and

FIGURE 22–33 Biosynthesis of nitric oxide. Both steps are catalyzed by nitric oxide synthase. The nitrogen of the NO is derived from the guanidinium group of arginine.

aspartate for pyrimidines. Glutamine again is the most important source of amino groups—in five different steps in the de novo pathways. Aspartate is also used as the source of an amino group in the purine pathways, in two steps.

Two other features deserve mention. First, there is evidence, especially in the de novo purine pathway, that the enzymes are present as large, multienzyme complexes in the cell, a recurring theme in our discussion of metabolism. Second, the cellular pools of nucleotides (other than ATP) are quite small, perhaps 1% or less of the amounts required to synthesize the cell's DNA. Therefore, cells must continue to synthesize nucleotides during nucleic acid synthesis, and in some cases nucleotide synthesis may limit the rates of DNA replication and transcription. Because of the importance of these processes in dividing cells, agents that inhibit nucleotide synthesis have become particularly important in medicine.

We examine here the biosynthetic pathways of purine and pyrimidine nucleotides and their regulation, the formation of the deoxynucleotides, and the degradation of purines and pyrimidines to uric acid and urea. We end with a discussion of chemotherapeutic agents that affect nucleotide synthesis.

De Novo Purine Nucleotide Synthesis Begins with PRPP

John M. Buchanan, 1917–2007

The two parent purine nucleotides of nucleic acids are adenosine 5'-monophosphate (AMP; adenylate) and guanosine 5'-monophosphate (GMP; guanylate), containing the purine bases adenine and guanine. **Figure 22–34** shows the origin of the carbon and nitrogen atoms of the purine ring system, as determined by John M. Buchanan using isotopic tracer experiments in birds. The detailed pathway of purine biosynthesis was worked out primarily by Buchanan and G. Robert Greenberg in the 1950s.

In the first committed step of the pathway, an amino group donated by glutamine is attached at C-1 of PRPP **(Fig. 22–35)**. The resulting **5-phosphoribosyl-amine** is highly unstable, with a half-life of 30 seconds at pH 7.5. The purine ring is subsequently built up on this structure. The pathway described here is identical in all organisms, with the exception of one step that differs in higher eukaryotes, as noted below.

The second step is the addition of three atoms from glycine (Fig. 22–35, step ❷). An ATP is consumed to activate the glycine carboxyl group (in the form of an acyl phosphate) for this condensation reaction. The added glycine amino group is then formylated by N^{10}-

FIGURE 22–34 Origin of the ring atoms of purines. This information was obtained from isotopic experiments with ^{14}C- or ^{15}N-labeled precursors. Formate is supplied in the form of N^{10}-formyltetrahydrofolate.

formyltetrahydrofolate ❸, and a nitrogen is contributed by glutamine (step ❹), before dehydration and ring closure yield the five-membered imidazole ring of the purine nucleus, as 5-aminoimidazole ribonucleotide (AIR; step ❺).

At this point, three of the six atoms needed for the second ring in the purine structure are in place. To complete the process, a carboxyl group is first added (step ❻). This carboxylation is unusual in that it does not require biotin, but instead uses the bicarbonate generally present in aqueous solutions. A rearrangement transfers the carboxylate from the exocyclic amino group to position 4 of the imidazole ring (step ❼). Steps ❻ and ❼ are found only in bacteria and fungi. In higher eukaryotes, including humans, the 5-aminoimidazole ribonucleotide product of step ❺ is carboxylated directly to carboxyaminoimidazole ribonucleotide in one step instead of two (step ❻ₐ). The enzyme catalyzing this reaction is AIR carboxylase.

Aspartate now donates its amino group in two steps (❽ and ❾): formation of an amide bond, followed by elimination of the carbon skeleton of aspartate (as fumarate). (Recall that aspartate plays an analogous role in two steps of the urea cycle; see Fig. 18–10.) The final carbon is contributed by N^{10}-formyltetrahydrofolate (step ❿), and a second ring closure takes place to yield the second fused ring of the purine nucleus (step ⓫). The first intermediate with a complete purine ring is **inosinate (IMP)**.

As in the tryptophan and histidine biosynthetic pathways, the enzymes of IMP synthesis seem to be organized as large, multienzyme complexes in the cell. Once again, evidence comes from the existence of single polypeptides with several functions, some catalyzing nonsequential steps in the pathway. In eukaryotic cells ranging from yeast to fruit flies to chickens, steps ❶, ❸, and ❺ in Figure 22–35 are catalyzed by a multifunctional protein. An additional multifunctional protein catalyzes steps ❿ and ⓫. In humans, a multifunctional enzyme combines the activities of AIR carboxylase and SAICAR synthetase (steps ❻ₐ and ❽). In bacteria,

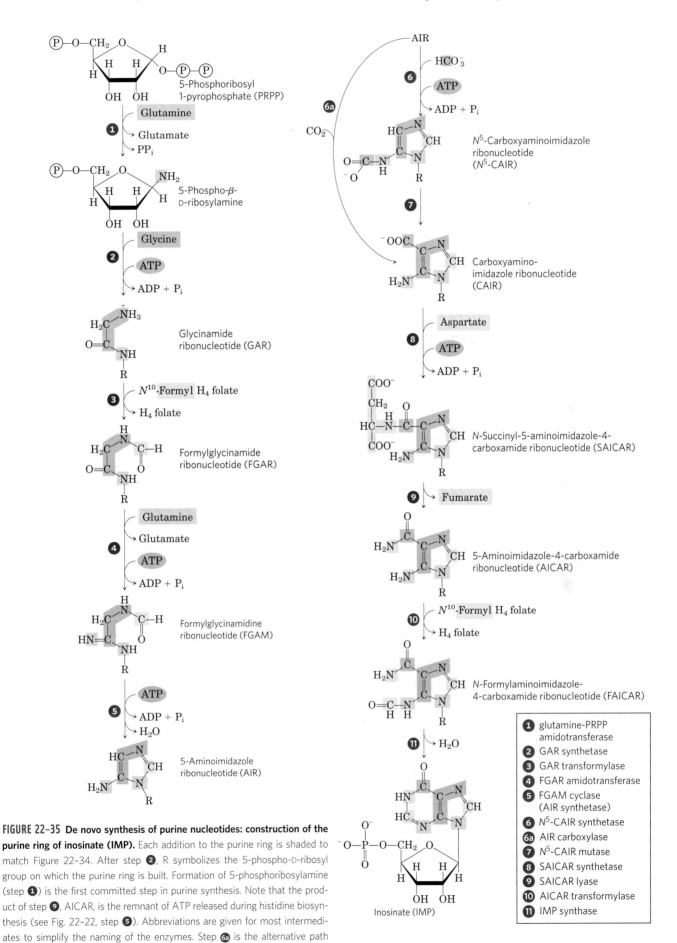

FIGURE 22–35 De novo synthesis of purine nucleotides: construction of the purine ring of inosinate (IMP). Each addition to the purine ring is shaded to match Figure 22–34. After step ❷, R symbolizes the 5-phospho-D-ribosyl group on which the purine ring is built. Formation of 5-phosphoribosylamine (step ❶) is the first committed step in purine synthesis. Note that the product of step ❾, AICAR, is the remnant of ATP released during histidine biosynthesis (see Fig. 22–22, step ❺). Abbreviations are given for most intermediates to simplify the naming of the enzymes. Step ❻ₐ is the alternative path from AIR to CAIR occurring in higher eukaryotes.

❶ glutamine-PRPP amidotransferase
❷ GAR synthetase
❸ GAR transformylase
❹ FGAR amidotransferase
❺ FGAM cyclase (AIR synthetase)
❻ N^5-CAIR synthetase
❻ₐ AIR carboxylase
❼ N^5-CAIR mutase
❽ SAICAR synthetase
❾ SAICAR lyase
❿ AICAR transformylase
⓫ IMP synthase

FIGURE 22–36 Biosynthesis of AMP and GMP from IMP.

these activities are found on separate proteins, but the proteins may form a large noncovalent complex. The channeling of reaction intermediates from one enzyme to the next permitted by these complexes is probably especially important for unstable intermediates such as 5-phosphoribosylamine.

Conversion of inosinate to adenylate requires the insertion of an amino group derived from aspartate **(Fig. 22–36)**; this takes place in two reactions similar to those used to introduce N-1 of the purine ring (Fig. 22–35, steps **8** and **9**). A crucial difference is that GTP rather than ATP is the source of the high-energy phosphate in synthesizing adenylosuccinate. Guanylate is formed by the NAD$^+$-requiring oxidation of inosinate at C-2, followed by addition of an amino group derived from glutamine. ATP is cleaved to AMP and PP$_i$ in the final step (Fig. 22–36).

Purine Nucleotide Biosynthesis Is Regulated by Feedback Inhibition

Three major feedback mechanisms cooperate in regulating the overall rate of de novo purine nucleotide synthesis and the relative rates of formation of the two end products, adenylate and guanylate **(Fig. 22–37)**. The first mechanism is exerted on the first reaction that is unique to purine synthesis: transfer of an amino group to PRPP to form 5-phosphoribosylamine. This reaction is catalyzed by the allosteric enzyme glutamine-PRPP amidotransferase, which is inhibited by the end products IMP, AMP, and GMP. AMP and GMP act synergistically in this concerted inhibition. Thus, whenever either AMP or GMP accumulates to excess, the first step in its biosynthesis from PRPP is partially inhibited.

In the second control mechanism, exerted at a later stage, an excess of GMP in the cell inhibits formation of xanthylate from inosinate by IMP dehydrogenase, with-

out affecting the formation of AMP. Conversely, an accumulation of adenylate inhibits formation of adenylosuccinate by adenylosuccinate synthetase, without affecting the biosynthesis of GMP. When both products are present in sufficient quantities, IMP builds up, and it inhibits an earlier step in the pathway; this regulatory strategy is called **sequential feedback inhibition**. In the third mechanism, GTP is required in the conversion

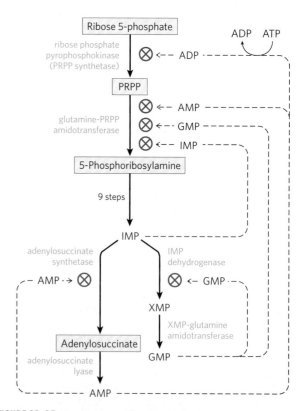

FIGURE 22–37 Regulatory mechanisms in the biosynthesis of adenine and guanine nucleotides in *E. coli*. Regulation of these pathways differs in other organisms.

of IMP to AMP, whereas ATP is required for conversion of IMP to GMP (Fig. 22–36), a reciprocal arrangement that tends to balance the synthesis of the two ribonucleotides.

The final control mechanism is the inhibition of PRPP synthesis by the allosteric regulation of ribose phosphate pyrophosphokinase. This enzyme is inhibited by ADP and GDP, in addition to metabolites from other pathways for which PRPP is a starting point.

Pyrimidine Nucleotides Are Made from Aspartate, PRPP, and Carbamoyl Phosphate

The common pyrimidine ribonucleotides are cytidine 5′-monophosphate (CMP; cytidylate) and uridine 5′-monophosphate (UMP; uridylate), which contain the pyrimidines cytosine and uracil. De novo pyrimidine nucleotide biosynthesis **(Fig. 22–38)** proceeds in a somewhat different manner from purine nucleotide synthesis; the six-membered pyrimidine ring is made first and then attached to ribose 5-phosphate. Required in this process is carbamoyl phosphate, also an intermediate in the urea cycle. However, as we noted in Chapter 18, in animals the carbamoyl phosphate required in urea synthesis is made in mitochondria by carbamoyl phosphate synthetase I, whereas the carbamoyl phosphate required in pyrimidine biosynthesis is made in the cytosol by a different form of the enzyme, **carbamoyl phosphate synthetase II**. In bacteria, a single enzyme supplies carbamoyl phosphate for the synthesis of arginine and pyrimidines. The bacterial enzyme has three separate active sites, spaced along a channel nearly 100 Å long **(Fig. 22–39)**. Bacterial carbamoyl phosphate synthetase provides a vivid illustration of the channeling of unstable reaction intermediates between active sites.

Carbamoyl phosphate reacts with aspartate to yield N-carbamoylaspartate in the first committed step of pyrimidine biosynthesis (Fig. 22–38). This reaction is catalyzed by **aspartate transcarbamoylase**. In bacteria, this step is highly regulated, and bacterial aspartate transcarbamoylase is one of the most thoroughly studied allosteric enzymes (see below). By removal of water from N-carbamoylaspartate, a reaction catalyzed by **dihydroorotase**, the pyrimidine ring is closed to form L-dihydroorotate. This compound is oxidized to the pyrimidine derivative orotate, a reaction in which NAD⁺ is the ultimate electron acceptor. In eukaryotes, the first three enzymes in this pathway—carbamoyl phosphate synthetase II, aspartate transcarbamoylase, and dihydroorotase—are part of a single trifunctional protein, known by the acronym CAD, contains three identical polypeptide chains (each of M_r 230,000), each with active sites for all three reactions. This suggests that large, multienzyme complexes may be the rule in this pathway.

Once orotate is formed, the ribose 5-phosphate side chain, provided once again by PRPP, is attached to yield orotidylate (Fig. 22–38). Orotidylate is then

FIGURE 22–38 De novo synthesis of pyrimidine nucleotides: biosynthesis of UTP and CTP via orotidylate. The pyrimidine is constructed from carbamoyl phosphate and aspartate. The ribose 5-phosphate is then added to the completed pyrimidine ring by orotate phosphoribosyltransferase. The first step in this pathway (not shown here; see Fig. 18–11a) is the synthesis of carbamoyl phosphate from CO_2 and NH_4^+. In eukaryotes, the first step is catalyzed by carbamoyl phosphate synthetase II.

FIGURE 22–39 Channeling of intermediates in bacterial carbamoyl phosphate synthetase. (Derived from PDB ID 1M6V) The reaction catalyzed by this enzyme (and its mitochondrial counterpart) is illustrated in Figure 18–11a. The small and large subunits are shown in tan and blue, respectively; the tunnel between active sites (almost 100 Å long) is shown as white. In this reaction, a glutamine molecule binds to the small subunit, donating its amido nitrogen as NH_4^+ in a glutamine amidotransferase–type reaction. The NH_4^+ enters the tunnel, which takes it to a second active site, where it combines with bicarbonate in a reaction requiring ATP. The carbamate then reenters the tunnel to reach the third active site, where it is phosphorylated by ATP to carbamoyl phosphate. To solve this structure, the enzyme was crystallized with ornithine bound to the glutamine-binding site and ADP bound to the ATP-binding sites.

decarboxylated to uridylate, which is phosphorylated to UTP. CTP is formed from UTP by the action of **cytidylate synthetase**, by way of an acyl phosphate intermediate (consuming one ATP). The nitrogen donor is normally glutamine, although the cytidylate synthetases in many species can use NH_4^+ directly.

Pyrimidine Nucleotide Biosynthesis Is Regulated by Feedback Inhibition

Regulation of the rate of pyrimidine nucleotide synthesis in bacteria occurs in large part through aspartate transcarbamoylase (ATCase), which catalyzes the first reaction in the sequence and is inhibited by CTP, the end product of the sequence (Fig. 22–38). The bacterial ATCase molecule consists of six catalytic subunits and six regulatory subunits (see Fig. 6–33). The catalytic subunits bind the substrate molecules, and the allosteric subunits bind the allosteric inhibitor, CTP. The entire ATCase molecule, as well as its subunits, exists in two conformations, active and inactive. When CTP is not bound to the regulatory subunits, the enzyme is maximally active. As CTP accumulates and binds to the regulatory subunits, they undergo a change in conformation. This change is transmitted to the catalytic subunits, which then also shift to an inactive conformation.

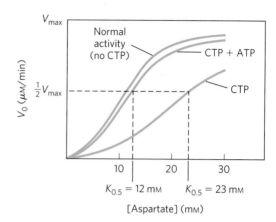

FIGURE 22–40 Allosteric regulation of aspartate transcarbamoylase by CTP and ATP. Addition of 0.8 mM CTP, the allosteric inhibitor of ATCase, increases the $K_{0.5}$ for aspartate (lower curve) thereby reducing the rate of conversion of aspartate to N-carbamoylaspartate. ATP at 0.6 mM fully reverses this inhibition by CTP (middle curve).

ATP prevents the changes induced by CTP. **Figure 22–40** shows the effects of the allosteric regulators on the activity of ATCase.

Nucleoside Monophosphates Are Converted to Nucleoside Triphosphates

Nucleotides to be used in biosynthesis are generally converted to nucleoside triphosphates. The conversion pathways are common to all cells. Phosphorylation of AMP to ADP is promoted by **adenylate kinase**, in the reaction

$$ATP + AMP \rightleftharpoons 2\ ADP$$

The ADP so formed is phosphorylated to ATP by the glycolytic enzymes or through oxidative phosphorylation.

ATP also brings about the formation of other nucleoside diphosphates by the action of a class of enzymes called **nucleoside monophosphate kinases**. These enzymes, which are generally specific for a particular base but nonspecific for the sugar (ribose or deoxyribose), catalyze the reaction

$$ATP + NMP \rightleftharpoons ADP + NDP$$

The efficient cellular systems for rephosphorylating ADP to ATP tend to pull this reaction in the direction of products.

Nucleoside diphosphates are converted to triphosphates by the action of a ubiquitous enzyme, **nucleoside diphosphate kinase**, which catalyzes the reaction

$$NTP_D + NDP_A \rightleftharpoons NDP_D + NTP_A$$

This enzyme is notable in that it is not specific for the base (purines or pyrimidines) or the sugar (ribose or deoxyribose). This nonspecificity applies to both phosphate acceptor (A) and donor (D), although the donor (NTP_D) is almost invariably ATP because it is present in higher concentration than other nucleoside triphosphates under aerobic conditions.

Ribonucleotides Are the Precursors of Deoxyribonucleotides

Deoxyribonucleotides, the building blocks of DNA, are derived from the corresponding ribonucleotides by direct reduction at the 2'-carbon atom of the D-ribose to form the 2'-deoxy derivative. For example, adenosine diphosphate (ADP) is reduced to 2'-deoxyadenosine diphosphate (dADP), and GDP is reduced to dGDP. This reaction is somewhat unusual in that the reduction occurs at a nonactivated carbon; no closely analogous chemical reactions are known. The reaction is catalyzed by **ribonucleotide reductase**, best characterized in *E. coli*, in which its substrates are ribonucleoside diphosphates.

The reduction of the D-ribose portion of a ribonucleoside diphosphate to 2'-deoxy-D-ribose requires a pair of hydrogen atoms, which are ultimately donated by NADPH via an intermediate hydrogen-carrying protein, **thioredoxin**. This ubiquitous protein serves a similar redox function in photosynthesis (see Fig. 20–19) and other processes. Thioredoxin has pairs of —SH groups that carry hydrogen atoms from NADPH to the ribonucleoside diphosphate. Its oxidized (disulfide) form is reduced by NADPH in a reaction catalyzed by **thioredoxin reductase (Fig. 22–41)**, and reduced thioredoxin is then used by ribonucleotide reductase to reduce the nucleoside diphosphates (NDPs) to deoxyribonucleoside diphosphates (dNDPs). A second source of reducing equivalents for ribonucleotide reductase is glutathione (GSH). Glutathione serves as the reductant for a protein closely related to thioredoxin, **glutaredoxin**, which then transfers the reducing power to ribonucleotide reductase (Fig. 22–41).

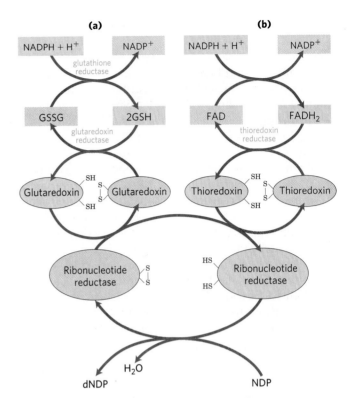

FIGURE 22–41 Reduction of ribonucleotides to deoxyribonucleotides by ribonucleotide reductase. Electrons are transmitted (red arrows) to the enzyme from NADPH via **(a)** glutaredoxin or **(b)** thioredoxin. The sulfide groups in glutaredoxin reductase are contributed by two molecules of bound glutathione (GSH; GSSG indicates oxidized glutathione). Note that thioredoxin reductase is a flavoenzyme, with FAD as prosthetic group.

Ribonucleotide reductase is notable in that its reaction mechanism provides the best-characterized example of the involvement of free radicals in biochemical transformations, once thought to be rare in biological systems. The enzyme in *E. coli* and most eukaryotes is an $\alpha_2\beta_2$ dimer, with catalytic subunits α_2 and radical-generation subunits β_2 **(Fig. 22–42)**. Each catalytic subunit contains two kinds of regulatory sites,

FIGURE 22–42 Ribonucleotide reductase. (a) A schematic diagram of the subunit structures. The catalytic subunit (α; also called R1) contains the two regulatory sites described in Fig. 22–44 and two Cys residues central to the reaction mechanism. The radical-generation subunits β (also called R2) contain the critical Tyr[122] residue and the binuclear iron center. **(b)** Structures of α_2 and β_2 give the likely structure of $\alpha_2\beta_2$ (derived from PDB ID 3UUS). **(c)** The likely path of radical formation from the initial Tyr[122] in a β subunit to the active-site Cys[439], which is used in the mechanism shown in Figure 22–43. Several aromatic amino acid residues participate in long-range radical transfer from the point of its formation at Tyr[122] to the active site, where the nucleotide substrate is bound.

as described below. The two active sites of the enzyme are formed at the interface between the catalytic (α_2) and radical-generation (β_2) subunits. At each active site, an α subunit contributes two sulfhydryl groups required for activity and the β_2 subunits contribute a stable tyrosyl radical. The β_2 subunits also have a binuclear iron (Fe^{3+}) cofactor that helps generate and stabilize the Tyr^{122} radical (Fig. 22–42). The tyrosyl radical is too far from the active site to interact directly with the site, but several aromatic residues form a long-range radical transfer pathway to the active site (Fig. 22–42c). A likely mechanism for the ribonucleotide reductase reaction is illustrated in **Figure 22–43**. In *E. coli*, the sources of the required reducing equivalents for this reaction are thioredoxin and glutaredoxin, as noted above.

Three classes of ribonucleotide reductase have been reported. Their mechanisms (where known) generally conform to the scheme in Figure 22–43, but they differ in the identity of the group supplying the active-site radical and in the cofactors used to generate it. The *E. coli*

enzyme (class I) requires oxygen to regenerate the tyrosyl radical if it is quenched, so this enzyme functions only in an aerobic environment. Class II enzymes, found in other microorganisms, have 5'-deoxyadenosylcobalamin (see Box 17–2) rather than a binuclear iron center. Class III enzymes have evolved to function in an anaerobic environment. *E. coli* contains a separate class III ribonucleotide reductase when grown anaerobically; this enzyme contains an iron-sulfur cluster (structurally distinct from the binuclear iron center of the class I enzyme) and requires NADPH and *S*-adenosylmethionine for activity. It uses nucleoside triphosphates rather than nucleoside diphosphates as substrates. The evolution of different classes of ribonucleotide reductase for production of DNA precursors in different environments reflects the importance of this reaction in nucleotide metabolism.

Regulation of *E. coli* ribonucleotide reductase is unusual in that not only its *activity* but its *substrate specificity* is regulated by the binding of effector molecules. Each α subunit has two types of regulatory

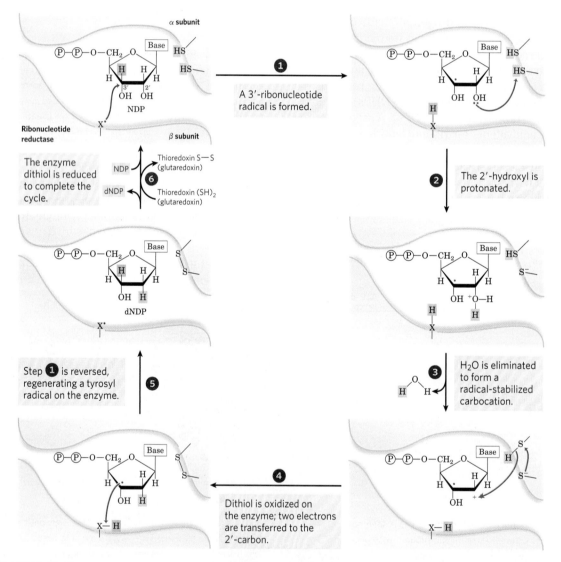

MECHANISM FIGURE 22–43 Proposed mechanism for ribonucleotide reductase. In the enzyme of *E. coli* and most eukaryotes, the active thiol groups are on the α subunit; the active-site radical ($-X^{\cdot}$) is on the β subunit and in *E. coli* is probably a thiyl radical of Cys^{439} (see Fig. 22-42).

FIGURE 22–44 Regulation of ribonucleotide reductase by deoxynucleoside triphosphates. The overall activity of the enzyme is affected by binding at the primary regulatory site (left). The substrate specificity of the enzyme is affected by the nature of the effector molecule bound at the second type of regulatory site, the substrate-specificity site (right). The diagram indicates inhibition or stimulation of enzyme activity with the four different substrates. The pathway from dUDP to dTTP is described later (see Figs. 22–46, 22–47).

site (Fig. 22–42). One type affects overall enzyme activity and binds either ATP, which activates the enzyme, or dATP, which inactivates it. The second type alters substrate specificity in response to the effector molecule—ATP, dATP, dTTP, or dGTP—that is bound there **(Fig. 22–44)**. When ATP or dATP is bound, reduction of UDP and CDP is favored. When dTTP or dGTP is bound, reduction of GDP or ADP, respectively, is stimulated. The scheme is designed to provide a balanced pool of precursors for DNA synthesis. ATP is also a general activator for biosynthesis and ribonucleotide reduction. The presence of dATP in small amounts increases the reduction of pyrimidine nucleotides. An oversupply of the pyrimidine

dNTPs is signaled by high levels of dTTP, which shifts the specificity to favor reduction of GDP. High levels of dGTP, in turn, shift the specificity to ADP reduction, and high levels of dATP shut the enzyme down. These effectors are thought to induce several distinct enzyme conformations with altered specificities.

These regulatory effects are accompanied by, and presumably mediated by, large structural rearrangements in the enzyme. When the active form of the *E. coli* enzyme ($\alpha_2\beta_2$) is inhibited by the addition of the allosteric inhibitor dATP, a ringlike $\alpha_4\beta_4$ structure forms, with alternating α_2 and β_2 subunits **(Fig. 22–45)**. In this altered structure, the radical-forming path from β to α is disrupted and the residues in the path are

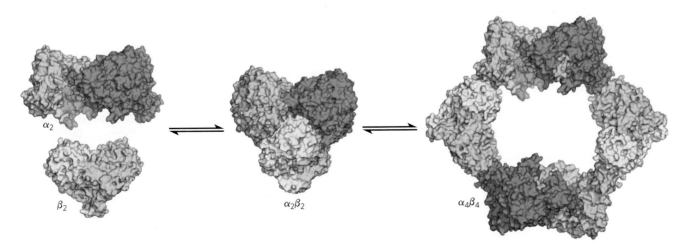

FIGURE 22–45 Oligomerization of ribonucleotide reductase induced by the allosteric inhibitor dATP. In high concentrations of dATP (50 μM), ring-shaped $\alpha_4\beta_4$ structures form (PDB ID 3UUS). In this conformation, the residues in the radical-forming path are exposed to the solvent, blocking the radical reaction and inhibiting the enzyme. The oligomerization is reversed at lower dATP concentrations.

exposed to solvent, effectively preventing radical transfer and thus inhibiting the reaction. The formation of ringlike $\alpha_4\beta_4$ structures is reversed when dATP levels are reduced. The yeast ribonucleotide reductase also undergoes oligomerization in the presence of dATP, forming a hexameric ring structure, $\alpha_6\beta_6$.

Thymidylate Is Derived from dCDP and dUMP

DNA contains thymine rather than uracil, and the de novo pathway to thymine involves only deoxyribonucleotides. The immediate precursor of thymidylate (dTMP) is dUMP. In bacteria, the pathway to dUMP begins with formation of dUTP, either by deamination of dCTP or by phosphorylation of dUDP **(Fig. 22–46)**. The dUTP is converted to dUMP by a dUTPase. The latter reaction must be efficient to keep dUTP pools low and prevent incorporation of uridylate into DNA.

Conversion of dUMP to dTMP is catalyzed by **thymidylate synthase**. A one-carbon unit at the hydroxymethyl (—CH_2OH) oxidation level (see Fig. 18–17) is transferred from N^5,N^{10}-methylenetetrahydrofolate to dUMP, then reduced to a methyl group **(Fig. 22–47)**. The reduction occurs at the expense of oxidation of tetrahydrofolate to dihydrofolate, which is unusual in tetrahydrofolate-requiring reactions. (The mechanism of this reaction is shown in Fig. 22–53.) The dihydrofolate is reduced to tetrahydrofolate by **dihydrofolate reductase**—a regeneration that is essential for the many processes that require tetrahydrofolate. In plants and at least one protist, thymidylate synthase and dihydrofolate reductase reside on a single bifunctional protein.

About 10% of the human population (and up to 50% of people in impoverished communities) suffers from folic acid deficiency. When the deficiency is severe, the symptoms can include heart disease, cancer, and some types of brain dysfunction. At least some of these symptoms arise from a reduction of thymidylate synthesis, leading to an abnormal incorporation of uracil into DNA. Uracil is recognized by DNA repair pathways (described in Chapter 25) and is cleaved from the DNA. The presence of high levels of uracil in DNA

FIGURE 22–47 Conversion of dUMP to dTMP by thymidylate synthase and dihydrofolate reductase. Serine hydroxymethyltransferase is required for regeneration of the N^5,N^{10}-methylene form of tetrahydrofolate. In the synthesis of dTMP, all three hydrogens of the added methyl group are derived from N^5,N^{10}-methylenetetrahydrofolate (light red and gray).

leads to strand breaks that can greatly affect the function and regulation of nuclear DNA, ultimately causing the observed effects on the heart and brain, as well as increased mutagenesis that leads to cancer. ∎

Degradation of Purines and Pyrimidines Produces Uric Acid and Urea, Respectively

Purine nucleotides are degraded by a pathway in which they lose their phosphate through the action of **5′-nucleotidase (Fig. 22–48)**. Adenylate yields adenosine, which is deaminated to inosine by **adenosine deaminase**, and inosine is hydrolyzed to hypoxanthine (its purine base) and D-ribose. Hypoxanthine is

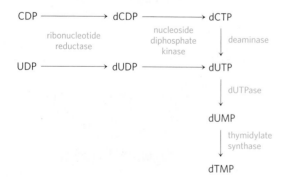

FIGURE 22–46 Biosynthesis of thymidylate (dTMP). The pathways are shown beginning with the reaction catalyzed by ribonucleotide reductase. Figure 22–47 gives details of the thymidylate synthase reaction.

FIGURE 22–48 Catabolism of purine nucleotides. Note that primates excrete much more nitrogen as urea via the urea cycle (Chapter 18) than as uric acid from purine degradation. Similarly, fish excrete much more nitrogen as NH_4^+ than as urea produced by the pathway shown here.

oxidized successively to xanthine and then uric acid by **xanthine oxidase**, a flavoenzyme with an atom of molybdenum and four iron-sulfur centers in its prosthetic group. Molecular oxygen is the electron acceptor in this complex reaction.

GMP catabolism also yields uric acid as end product. GMP is first hydrolyzed to guanosine, which is then cleaved to free guanine. Guanine undergoes hydrolytic removal of its amino group to yield xanthine, which is converted to uric acid by xanthine oxidase (Fig. 22–48).

Uric acid is the excreted end product of purine catabolism in primates, birds, and some other animals. A healthy adult human excretes uric acid at a rate of about 0.6 g/24 h; the excreted product arises in part from ingested purines and in part from turnover of the purine nucleotides of nucleic acids. In most mammals and many other vertebrates, uric acid is further degraded to **allantoin** by the action of **urate oxidase**. In other organisms the pathway is further extended, as shown in Figure 22–48.

The pathways for degradation of pyrimidines generally lead to NH_4^+ production and thus to urea synthesis. Thymine, for example, is degraded to methylmalonylsemialdehyde **(Fig. 22–49)**, an intermediate of valine catabolism. It is further degraded through propionyl-CoA and methylmalonyl-CoA to succinyl-CoA (see Fig. 18–27).

Genetic aberrations in human purine metabolism have been found, some with serious consequences. For example, **adenosine deaminase (ADA) deficiency** leads to severe immunodeficiency disease in

FIGURE 22–49 Catabolism of a pyrimidine. Shown here is the pathway for thymine. The methylmalonylsemialdehyde is further degraded to succinyl-CoA.

which T lymphocytes and B lymphocytes do not develop properly. Lack of ADA leads to a 100-fold increase in the cellular concentration of dATP, a strong inhibitor of ribonucleotide reductase (Fig. 22–44). High levels of dATP produce a general deficiency of other dNTPs in T lymphocytes. The basis for B-lymphocyte toxicity is less clear. Individuals with ADA deficiency lack an effective immune system and do not survive unless isolated in a sterile "bubble" environment. ADA deficiency was one of the first targets of human gene therapy trials (in 1990), which yielded mixed results. In more recent trials, some ADA-deficient individuals regained normal immune function after being given a functional gene for ADA. ■

Purine and Pyrimidine Bases Are Recycled by Salvage Pathways

Free purine and pyrimidine bases are constantly released in cells during the metabolic degradation of nucleotides. Free purines are in large part salvaged and reused to make nucleotides, in a pathway much simpler than the de novo synthesis of purine nucleotides described earlier. One of the primary salvage pathways consists of a single reaction catalyzed by **adenosine phosphoribosyltransferase**, in which free adenine reacts with PRPP to yield the corresponding adenine nucleotide:

$$\text{Adenine} + \text{PRPP} \longrightarrow \text{AMP} + \text{PP}_i$$

Free guanine and hypoxanthine (the deamination product of adenine; Fig. 22–48) are salvaged in the same way by **hypoxanthine-guanine phosphoribosyltransferase**. A similar salvage pathway exists for pyrimidine bases in microorganisms, and possibly in mammals.

A genetic lack of hypoxanthine-guanine phosphoribosyltransferase activity, seen almost exclusively in male children, results in a bizarre set of symptoms called **Lesch-Nyhan syndrome**. Children with this genetic disorder, which becomes manifest by the age of 2 years, are sometimes poorly coordinated and have intellectual deficits. In addition, they are extremely hostile and show compulsive self-destructive tendencies: they mutilate themselves by biting off their fingers, toes, and lips.

The devastating effects of Lesch-Nyhan syndrome illustrate the importance of the salvage pathways. Hypoxanthine and guanine arise constantly from the breakdown of nucleic acids. In the absence of hypoxanthine-guanine phosphoribosyltransferase, PRPP levels rise and purines are overproduced by the de novo pathway, resulting in high levels of uric acid production and goutlike damage to tissue (see below). The brain is especially dependent on the salvage pathways, and this may account for the central nervous system damage in children with Lesch-Nyhan syndrome. This syndrome is another potential target for gene therapy. ■

Excess Uric Acid Causes Gout

Long thought (erroneously) to be due to "high living," gout is a disease of the joints caused by an elevated concentration of uric acid in the blood and tissues. The joints become inflamed, painful, and arthritic, owing to the abnormal deposition of sodium urate crystals. The kidneys are also affected, as excess uric acid is deposited in the kidney tubules. Gout occurs predominantly in males. Its precise cause is not known, but it often involves an underexcretion of urate. A genetic deficiency of one or another enzyme of purine metabolism may also be a factor in some cases.

Gout is effectively treated by a combination of nutritional and drug therapies. Foods especially rich in nucleotides and nucleic acids, such as liver or glandular products, are withheld from the diet. Major alleviation

FIGURE 22–50 Allopurinol, an inhibitor of xanthine oxidase. Hypoxanthine is the normal substrate of xanthine oxidase. Only a slight alteration in the structure of hypoxanthine (shaded light red) yields the medically effective enzyme inhibitor allopurinol. At the active site, allopurinol is converted to oxypurinol, a strong competitive inhibitor that remains tightly bound to the reduced form of the enzyme.

of the symptoms is provided by the drug **allopurinol** (Fig. 22–50), which inhibits xanthine oxidase, the enzyme that catalyzes the conversion of purines to uric acid. Allopurinol is a substrate of xanthine oxidase, which converts allopurinol to oxypurinol (alloxanthine). Oxypurinol inactivates the reduced form of the enzyme by remaining tightly bound in its active site. When xanthine oxidase is inhibited, the excreted products of purine metabolism are xanthine and hypoxanthine, which are more water-soluble than uric acid and less likely to form crystalline deposits. Allopurinol was developed by Gertrude Elion and George Hitchings, who also developed acyclovir, used in treating people with genital and oral herpes infections, and other purine analogs used in cancer chemotherapy. ■

Many Chemotherapeutic Agents Target Enzymes in the Nucleotide Biosynthetic Pathways

The growth of cancer cells is not controlled in the same way as cell growth in most normal tissues. Cancer cells have greater requirements for nucleotides as precursors of DNA and RNA, and consequently are generally more sensitive than normal cells to inhibitors of nucleotide biosynthesis. A growing array of important chemotherapeutic agents—for cancer and other diseases—act by inhibiting one or more enzymes in these pathways. We describe here several well-studied examples that illustrate productive approaches to treatment and help us understand how these enzymes work.

The first set of agents includes compounds that inhibit glutamine amidotransferases. Recall that glutamine is a nitrogen donor in at least half a dozen separate reactions in nucleotide biosynthesis. The binding sites for glutamine and the mechanism by which NH_4^+ is extracted are quite similar in many of these enzymes. Most are strongly inhibited by glutamine analogs such as **azaserine** and **acivicin** (Fig. 22–51). Azaserine, characterized by John Buchanan in the 1950s, was one of the first examples of a mechanism-based enzyme inactivator (suicide inactivator; p. 210 and Box 6–3). Acivicin shows promise as a cancer chemotherapeutic agent.

Other useful targets for pharmaceutical agents are thymidylate synthase and dihydrofolate reductase, enzymes that provide the only cellular pathway for thymine synthesis (Fig. 22–52). One inhibitor that acts on thymidylate synthase, **fluorouracil**, is an important chemotherapeutic agent. Fluorouracil itself is not the enzyme inhibitor. In the cell, salvage pathways convert it to the deoxynucleoside monophosphate FdUMP, which binds to and inactivates the enzyme. Inhibition by FdUMP (Fig. 22–53) is a classic example of mechanism-based enzyme inactivation. Another prominent chemotherapeutic agent, **methotrexate**, is an inhibitor of dihydrofolate reductase.

Gertrude Elion, 1918–1999, and George Hitchings, 1905–1998

FIGURE 22–51 Azaserine and acivicin, inhibitors of glutamine amidotransferases. These analogs of glutamine interfere in several amino acid and nucleotide biosynthetic pathways.

(a)

(b)

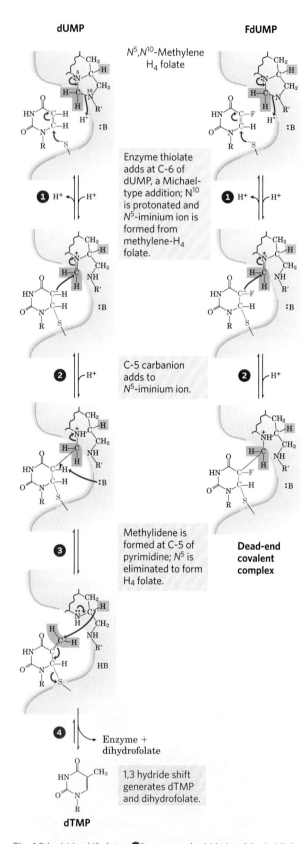

FIGURE 22-52 **Thymidylate synthesis and folate metabolism as targets of chemotherapy. (a)** During thymidylate synthesis, N^5,N^{10}-methylenetetrahydrofolate is converted to 7,8-dihydrofolate; the N^5,N^{10}-methylenetetrahydrofolate is regenerated in two steps (see Fig. 22-47). This cycle is a major target of several chemotherapeutic agents. **(b)** Fluorouracil and methotrexate are important chemotherapeutic agents. In cells, fluorouracil is converted to FdUMP, which inhibits thymidylate synthase. Methotrexate, a structural analog of tetrahydrofolate, inhibits dihydrofolate reductase; the shaded amino and methyl groups replace a carbonyl oxygen and a proton, respectively, in folate (see Fig. 22-47). Another important folate analog, aminopterin, is identical to methotrexate except that it lacks the shaded methyl group. Trimethoprim, a tight-binding inhibitor of bacterial dihydrofolate reductase, was developed as an antibiotic.

MECHANISM FIGURE 22-53 Conversion of dUMP to dTMP and its inhibition by FdUMP. The left side is the normal reaction mechanism of thymidylate synthase. The nucleophilic sulfhydryl group contributed by the enzyme in ❶ and the ring atoms of dUMP taking part in the reaction are shown in red; **:B** denotes an amino acid side chain that acts as a base to abstract a proton after step ❸. The hydrogens derived from the methyl group of N^5,N^{10}-methylenetetrahydrofolate are shaded in gray. The 1,3 hydride shift (step ❹), moves a hydride ion (shaded light red) from C-6 of H₄ folate to the methyl group of thymidine, resulting in the oxidation of tetrahydrofolate to dihydrofolate. This hydride shift is blocked when FdUMP is the substrate (right). Steps ❶ and ❷ proceed normally, but result in a stable complex—consisting of FdUMP linked covalently to the enzyme and to tetrahydrofolate—that inactivates the enzyme. 🖱 **Thymidylate Synthase Mechanism**

This folate analog acts as a competitive inhibitor; the enzyme binds methotrexate with about 100 times higher affinity than dihydrofolate. **Aminopterin** is a related compound that acts similarly.

The medical potential of inhibitors of nucleotide biosynthesis is not limited to cancer treatment. All fast-growing cells (including bacteria and protists) are potential targets. **Trimethoprim**, an antibiotic developed by Hitchings and Elion, binds to bacterial dihydrofolate reductase nearly 100,000 times better than to the mammalian enzyme. It is used to treat certain urinary and middle-ear bacterial infections. Parasitic protists, such as the trypanosomes that cause African sleeping sickness (African trypanosomiasis), lack pathways for de novo nucleotide biosynthesis and are particularly sensitive to agents that interfere with their scavenging of nucleotides from the surrounding environment using salvage pathways. Allopurinol (Fig. 22–50) and several similar purine analogs have shown promise for the treatment of African trypanosomiasis and related afflictions. See Box 6–3 for another approach to combating African trypanosomiasis, made possible by advances in our understanding of metabolism and enzyme mechanisms. ■

SUMMARY 22.4 Biosynthesis and Degradation of Nucleotides

▶ The purine ring system is built up step-by-step beginning with 5-phosphoribosylamine. The amino acids glutamine, glycine, and aspartate furnish all the nitrogen atoms of purines. Two ring-closure steps form the purine nucleus.

▶ Pyrimidines are synthesized from carbamoyl phosphate and aspartate, and ribose 5-phosphate is then attached to yield the pyrimidine ribonucleotides.

▶ Nucleoside monophosphates are converted to their triphosphates by enzymatic phosphorylation reactions. Ribonucleotides are converted to deoxyribonucleotides by ribonucleotide reductase, an enzyme with novel mechanistic and regulatory characteristics. The thymine nucleotides are derived from dCDP and dUMP.

▶ Uric acid and urea are the end products of purine and pyrimidine degradation.

▶ Free purines can be salvaged and rebuilt into nucleotides. Genetic deficiencies in certain salvage enzymes cause serious disorders such as Lesch-Nyhan syndrome and ADA deficiency.

▶ Accumulation of uric acid crystals in the joints, possibly caused by another genetic deficiency, results in gout.

▶ Enzymes of the nucleotide biosynthetic pathways are targets for an array of chemotherapeutic agents used to treat cancer and other diseases.

Key Terms

Terms in bold are defined in the glossary.

nitrogen cycle 882
nitrogen fixation 882
anammox 882
symbionts 882
nitrogenase complex 883
P cluster 883
FeMo cofactor 883
leghemoglobin 888
glutamine synthetase 888
glutamate synthase 888
glutamine
    amidotransferases 890
5-phosphoribosyl-1-
    pyrophosphate
    (PRPP) 892
tryptophan synthase 898
porphyrin 902
porphyria 904
bilirubin 904
phosphocreatine 906
creatine 906
glutathione (GSH) 906
auxin 908
dopamine 909
norepinephrine 909
epinephrine 909
γ-aminobutyrate
    (GABA) 909
serotonin 909
histamine 909
cimetidine 909
spermine 909

spermidine 909
ornithine
    decarboxylase 909
de novo pathway 910
salvage pathway 910
inosinate (IMP) 912
carbamoyl phosphate
    synthetase II 915
aspartate
    transcarbamoylase 915
**nucleoside
    monophosphate
    kinase** 916
**nucleoside diphosphate
    kinase** 916
ribonucleotide
    reductase 917
thioredoxin 917
thymidylate synthase 920
dihydrofolate
    reductase 920
adenosine deaminase
    (ADA) deficiency 921
Lesch-Nyhan
    syndrome 922
allopurinol 923
azaserine 923
acivicin 923
fluorouracil 923
methotrexate 923
aminopterin 925

Further Reading

Nitrogen Fixation

Arp, D.J. & Stein, L.Y. (2003) Metabolism of inorganic N compounds by ammonia-oxidizing bacteria. *Crit. Rev. Biochem. Mol. Biol.* **38**, 491–495.

Burris, R.H. (1995) Breaking the N–N bond. *Annu. Rev. Plant Physiol. Plant Mol. Biol.* **46**, 1–19.

Eisenberg, D.S., Harindarpal, S., Gill, G., Pfluegl, M.U., & Rothstein, H. (2000) Structure-function relationships of glutamine synthetases. *Biochim. Biophys. Acta* **1477**, 122–145.

Fuerst, J.A. (2005) Intracellular compartmentation in planctomycetes. *Annu. Rev. Microbiol.* **59**, 299–328.
    Advanced discussion of the structure and biochemistry of anammox.

Igarishi, R.Y. & Seefeldt, L.C. (2003) Nitrogen fixation: the mechanism of the Mo-dependent nitrogenase. *Crit. Rev. Biochem. Mol. Biol.* **38**, 351–384.

Lin, J.T. & Stewart, V. (1998) Nitrate assimilation in bacteria. *Adv. Microbiol. Physiol.* **39**, 1–30.

Patriarca, E.J., Tate, R., & Iaccarino, M. (2002) Key role of bacterial NH$_4^+$ metabolism in rhizobium-plant symbiosis. *Microbiol. Mol. Biol. Rev.* **66**, 203–222.

A good overview of ammonia assimilation in bacterial systems and its regulation.

Prell, J. & Poole, P. (2006) Metabolic changes of rhizobia in legume nodules. *Trends Microbiol.* **14**, 161–168.

A good summary of the intricate symbiotic relationship between rhizobial bacteria and their hosts.

Reese, D.C. & Howard, J.B. (2000) Nitrogenase: standing at the crossroads. *Curr. Opin. Chem. Biol.* **4**, 559–566.

Schwarz, G., Mendel, R.R., & Ribbe, M.W. (2009) Molybdenum cofactors, enzymes and pathways. *Nature* **460**, 839–847.

Seefeldt, L.C., Hoffman, B.M., & Dean, D.R. (2009) Mechanism of Mo-dependent nitrogenase. *Annu. Rev. Biochem.* **78**, 701–722.

Sinha, S.C. & Smith, J.L. (2001) The PRT protein family. *Curr. Opin. Struct. Biol.* **11**, 733–739.

Description of a protein family that includes many amidotransferases, with channels for the movement of NH_3 from one active site to another.

Amino Acid Biosynthesis

Frey, P.A. & Hegeman, A.D. (2007) *Enzymatic Reaction Mechanisms*, Oxford University Press, New York.

An updated summary of reaction mechanisms, including one-carbon metabolism and pyridoxal phosphate enzymes.

Neidhardt, F.C. (ed.). (1996) Escherichia coli *and* Salmonella: *Cellular and Molecular Biology*, 2nd edn, ASM Press, Washington, DC.

Volume 1 of this two-volume set has 13 chapters devoted to detailed descriptions of amino acid and nucleotide biosynthesis in bacteria. The Web-based version at www.ecosal.org is updated regularly. A valuable resource.

Pan P., Woehl, E., & Dunn, M.F. (1997) Protein architecture, dynamics and allostery in tryptophan synthase channeling. *Trends Biochem. Sci.* **22**, 22–27.

Richards, N.G.J. & Kilberg, M.S. (2006) Asparagine synthetase chemotherapy. *Annu. Rev. Biochem.* **75**, 629–654.

Stadtman, E.R. (2001) The story of glutamine synthetase regulation. *J. Biol. Chem.* **276**, 44,357–44,364.

Compounds Derived from Amino Acids

Ajioka R.S., Phillips, J.D., & Kushner, J.P. (2006) Biosynthesis of heme in mammals. *Biochim. Biophys. Acta Mol. Cell Res.* **1763**, 723–736.

Bredt, D.S. & Snyder, S.H. (1994) Nitric oxide: a physiologic messenger molecule. *Annu. Rev. Biochem.* **63**, 175–195.

Meister, A. & Anderson, M.E. (1983) Glutathione. *Annu. Rev. Biochem.* **52**, 711–760.

Morse, D. & Choi, A.M.K. (2002) Heme oxygenase-1—the "emerging molecule" has arrived. *Am. J. Respir. Cell Mol. Biol.* **27**, 8–16.

Rondon, M.R., Trzebiatowski, J.R., & Escalante-Semerena, J.C. (1997) Biochemistry and molecular genetics of cobalamin biosynthesis. *Prog. Nucleic Acid Res. Mol. Biol.* **56**, 347–384.

Stadtman, T.C. (1996) Selenocysteine. *Annu. Rev. Biochem.* **65**, 83–100.

Nucleotide Biosynthesis

Aiuti, A., Cattaneo, R., Galimberti, S., Benninghoff, U., Cassani, B., Callegaro, L., Scaramuzza, S., Andolfi, G., Mirolo, M., Brigida, I., et al. (2009) Gene therapy for immunodeficiency due to adenosine deaminase deficiency. *N. Engl. J. Med.* **360**, 447–458.

Ando, N., Brignole, E.J., Zimanyi, C.M., Funk, M.A., Yokoyama, K., Asturias, F.J., Stubbe, J., & Drennan, C.L. (2011) Structural interconversions modulate activity of *Escherichia coli*, ribonucleotide reductase. *Proc. Natl. Acad. Sci. USA* **108**, 21,046–21,051.

Carreras, C.W. & Santi, D.V. (1995) The catalytic mechanism and structure of thymidylate synthase. *Annu. Rev. Biochem.* **64**, 721–762.

Cotruvo, J.A., Jr. & Stubbe, J. (2011) Class I ribonucleotide reductases: metallocofactor assembly and repair in vitro and in vivo. *Annu. Rev. Biochem.* **80**, 733–767.

Fairman, J.W., Wijerathna, S.R., Ahmad, M.F., Xu, H., Nakano, R., Jha, S., Prendergast, J., Welin, M., Flodin, S., Roos, A., et al. (2011) Structural basis for allosteric regulation of human ribonucleotide reductase by nucleotide-induced oligomerization. *Nat. Struct. Mol. Biol.* **18**, 316–322.

Herrick, J. & Sclavi, B. (2007) Ribonucleotide reductase and the regulation of DNA replication: an old story and an ancient heritage. *Mol. Microbiol.* **63**, 22–34.

Holmgren, A. (1989) Thioredoxin and glutaredoxin systems. *J. Biol. Chem.* **264**, 13,963–13,966.

Kappock, T.J., Ealick, S.E., & Stubbe, J. (2000) Modular evolution of the purine biosynthetic pathway. *Curr. Opin. Chem. Biol.* **4**, 567–572.

Kornberg, A. & Baker, T.A. (1991) *DNA Replication*, 2nd edn, W. H. Freeman and Company, New York.

This text includes a good summary of nucleotide biosynthesis.

Licht, S., Gerfen, G.J., & Stubbe, J. (1996) Thiyl radicals in ribonucleotide reductases. *Science* **271**, 477–481.

Nordlund, P. & Reichard, P. (2006) Ribonucleotide reductases. *Annu. Rev. Biochem.* **75**, 681–706.

Schachman, H.K. (2000) Still looking for the ivory tower. *Annu. Rev. Biochem.* **69**, 1–29.

A lively description of research on aspartate transcarbamoylase, accompanied by delightful tales of science and politics.

Weeks, A., Lund, L., & Raushel, F.M. (2006) Tunneling of intermediates in enzyme-catalyzed reactions. *Curr. Opin. Chem. Biol.* **10**, 465–472.

Genetic Diseases

Löffler, M., Fairbanks, L.D., Zameitat, E., Marinaki, A.M., & Simmonds, H.A. (2005) Pyrimidine pathways in health and disease. *Trends Mol. Med.* **11**, 430–437.

Valle, D., Beaudet, A.L., Vogelstein, B., Kinzler, K.W., Antonarakis, S.E., & Ballabio, A. (eds). *Scriver's Online Metabolic and Molecular Bases of Inherited Disease*, www.ommbid.com. Published January 2006. Updated March 28, 2011.

This classic medical encyclopedia, last published in print in 2001 as a four-volume set, is now maintained online. It has good chapters on disorders of amino acid, porphyrin, and heme metabolism. See also the chapters on inborn errors of purine and pyrimidine metabolism.

Problems

1. ATP Consumption by Root Nodules in Legumes Bacteria residing in the root nodules of the pea plant consume more than 20% of the ATP produced by the plant. Suggest why these bacteria consume so much ATP.

2. Glutamate Dehydrogenase and Protein Synthesis The bacterium *Methylophilus methylotrophus* can synthesize protein from methanol and ammonia. Recombinant DNA techniques have improved the yield of protein by introducing into *M. methylotrophus* the glutamate dehydrogenase gene from *E. coli*. Why does this genetic manipulation increase the protein yield?

3. PLP Reaction Mechanisms Pyridoxal phosphate can help catalyze transformations one or two carbons removed from the α carbon of an amino acid. The enzyme threonine synthase (see Fig. 22–17) promotes the PLP-dependent conversion of phosphohomoserine to threonine. Suggest a mechanism for this reaction.

4. Transformation of Aspartate to Asparagine There are two routes for transforming aspartate to asparagine at the expense of ATP. Many bacteria have an asparagine synthetase that uses ammonium ion as the nitrogen donor. Mammals have an asparagine synthetase that uses glutamine as the nitrogen donor. Given that the latter requires an extra ATP (for the synthesis of glutamine), why do mammals use this route?

5. Equation for the Synthesis of Aspartate from Glucose Write the net equation for the synthesis of aspartate (a nonessential amino acid) from glucose, carbon dioxide, and ammonia.

6. Asparagine Synthetase Inhibitors in Leukemia Therapy Mammalian asparagine synthetase is a glutamine-dependent amidotransferase. Efforts to identify an effective inhibitor of human asparagine synthetase for use in chemotherapy for patients with leukemia has focused not on the amino-terminal glutaminase domain but on the carboxyl-terminal synthetase active site. Explain why the glutaminase domain is not a promising target for a useful drug.

7. Phenylalanine Hydroxylase Deficiency and Diet Tyrosine is normally a nonessential amino acid, but individuals with a genetic defect in phenylalanine hydroxylase require tyrosine in their diet for normal growth. Explain.

8. Cofactors for One-Carbon Transfer Reactions Most one-carbon transfers are promoted by one of three cofactors: biotin, tetrahydrofolate, or S-adenosylmethionine (Chapter 18). S-Adenosylmethionine is generally used as a methyl group donor; the transfer potential of the methyl group in N^5-methyltetrahydrofolate is insufficient for most biosynthetic reactions. However, one example of the use of N^5-methyltetrahydrofolate in methyl group transfer is in methionine formation by the methionine synthase reaction (step ❾ of Fig. 22–17); methionine is the immediate precursor of S-adenosylmethionine (see Fig. 18–18). Explain how the methyl group of S-adenosylmethionine can be derived from N^5-methyltetrahydrofolate, even though the transfer potential of the methyl group in N^5-methyltetrahydrofolate is one-thousandth of that in S-adenosylmethionine.

9. Concerted Regulation in Amino Acid Biosynthesis The glutamine synthetase of *E. coli* is independently modulated by various products of glutamine metabolism (see Fig. 22–8). In this concerted inhibition, the extent of enzyme inhibition is greater than the sum of the separate inhibitions caused by each product. For *E. coli* grown in a medium rich in histidine, what would be the advantage of concerted inhibition?

10. Relationship between Folic Acid Deficiency and Anemia Folic acid deficiency, believed to be the most common vitamin deficiency, causes a type of anemia in which hemoglobin synthesis is impaired and erythrocytes do not mature properly. What is the metabolic relationship between hemoglobin synthesis and folic acid deficiency?

11. Nucleotide Biosynthesis in Amino Acid Auxotrophic Bacteria Wild-type *E. coli* cells can synthesize all 20 common amino acids, but some mutants, called amino acid auxotrophs, are unable to synthesize a specific amino acid and require its addition to the culture medium for optimal growth. Besides their role in protein synthesis, some amino acids are also precursors for other nitrogenous cell products. Consider the three amino acid auxotrophs that are unable to synthesize glycine, glutamine, and aspartate, respectively. For each mutant, what nitrogenous products other than proteins would the cell fail to synthesize?

12. Inhibitors of Nucleotide Biosynthesis Suggest mechanisms for the inhibition of (a) alanine racemase by L-fluoroalanine and (b) glutamine amidotransferases by azaserine.

13. Mode of Action of Sulfa Drugs Some bacteria require p-aminobenzoate in the culture medium for normal growth, and their growth is severely inhibited by the addition of sulfanilamide, one of the earliest sulfa drugs. Moreover, in the presence of this drug, 5-aminoimidazole-4-carboxamide ribonucleotide (AICAR; see Fig. 22–35) accumulates in the culture medium. These effects are reversed by addition of excess p-aminobenzoate.

p-Aminobenzoate   Sulfanilamide

(a) What is the role of p-aminobenzoate in these bacteria? (Hint: See Fig. 18–16.)

(b) Why does AICAR accumulate in the presence of sulfanilamide?

(c) Why are the inhibition and accumulation reversed by addition of excess p-aminobenzoate?

14. Pathway of Carbon in Pyrimidine Biosynthesis Predict the locations of ^{14}C in orotate isolated from cells grown on a small amount of uniformly labeled [^{14}C]succinate. Justify your prediction.

15. Nucleotides as Poor Sources of Energy Under starvation conditions, organisms can use proteins and amino acids as sources of energy. Deamination of amino acids produces carbon skeletons that can enter the glycolytic pathway and the citric acid cycle to produce energy in the form of ATP. Nucleotides, on the other hand, are not similarly degraded for use as energy-yielding fuels. What observations about cellular physiology support this statement? What aspect of the structure of nucleotides makes them a relatively poor source of energy?

16. Treatment of Gout Allopurinol (see Fig. 22–50), an inhibitor of xanthine oxidase, is used to treat chronic gout. Explain the biochemical basis for this treatment. Patients treated with allopurinol sometimes develop xanthine stones in the kidneys, although the incidence of kidney damage is much lower than in untreated gout. Explain this observation in the light of the following solubilities in urine: uric acid, 0.15 g/L; xanthine, 0.05 g/L; and hypoxanthine, 1.4 g/L.

17. Inhibition of Nucleotide Synthesis by Azaserine
The diazo compound O-(2-diazoacetyl)-L-serine, known also as azaserine (see Fig. 22–51), is a powerful inhibitor of glutamine amidotransferases. If growing cells are treated with azaserine, what intermediates of nucleotide biosynthesis will accumulate? Explain.

Data Analysis Problem

18. Use of Modern Molecular Techniques to Determine the Synthetic Pathway of a Novel Amino Acid Most of the biosynthetic pathways described in this chapter were determined before the development of recombinant DNA technology and genomics, so the techniques were quite different from those that researchers would use today. Here we explore an example of the use of modern molecular techniques to investigate the pathway of synthesis of a novel amino acid, (2S)-4-amino-2-hydroxybutyrate (AHBA). The techniques mentioned here are described in various places in the book; this problem is designed to show how they can be integrated in a comprehensive study.

AHBA is a γ-amino acid that is a component of some aminoglycoside antibiotics, including the antibiotic butirosin. Antibiotics modified by the addition of an AHBA residue are often more resistant to inactivation by bacterial antibiotic-resistance enzymes. As a result, understanding how AHBA is synthesized and added to antibiotics is useful in the design of pharmaceuticals.

In an article published in 2005, Li and coworkers describe how they determined the synthetic pathway of AHBA from glutamate.

Glutamate   AHBA

(a) Briefly describe the chemical transformations needed to convert glutamate to AHBA. At this point, don't be concerned about the *order* of the reactions.

Li and colleagues began by cloning the butirosin biosynthetic gene cluster from the bacterium *Bacillus circulans*, which makes large quantities of butirosin. They identified five genes that are essential for the pathway: *btrI*, *btrJ*, *btrK*, *btrO*, and *btrV*. They cloned these genes into *E. coli* plasmids that allow overexpression of the genes, producing proteins with "histidine tags" fused to their amino termini to facilitate purification (see Section 9.1).

The predicted amino acid sequence of the BtrI protein showed strong homology to known acyl carrier proteins (see Fig. 21–5). Using mass spectrometry, Li and colleagues found a molecular mass of 11,812 for the purified BtrI protein (including the His tag). When the purified BtrI was incubated with coenzyme A and an enzyme known to attach CoA to other acyl carrier proteins, the majority molecular species had an M_r of 12,153.

(b) How would you use these data to argue that BtrI can function as an acyl carrier protein with a CoA prosthetic group?

Using standard terminology, Li and coauthors called the form of the protein lacking CoA apo-BtrI and the form with CoA (linked as in Fig. 21–5) holo-BtrI. When holo-BtrI was incubated with glutamine, ATP, and purified BtrJ protein, the holo-BtrI species of M_r 12,153 was replaced with a species of M_r 12,281, corresponding to the thioester of glutamate and holo-BtrI. Based on these data, the authors proposed the following structure for the M_r 12,281 species (γ-glutamyl-S-BtrI):

γ-Glutamyl-S-BtrI

(c) What other structure(s) is (are) consistent with the data above?

(d) Li and coauthors argued that the structure shown here (γ-glutamyl-S-BtrI) is likely to be correct because the α-carboxyl group must be removed at some point in the synthetic process. Explain the chemical basis of this argument. (Hint: See Fig. 18–6, reaction C.)

The BtrK protein showed significant homology to PLP-dependent amino acid decarboxylases, and BtrK isolated from *E. coli* was found to contain tightly bound PLP. When γ-glutamyl-S-BtrI was incubated with purified BtrK, a molecular species of M_r 12,240 was produced.

(e) What is the most likely structure of this species?

(f) Interestingly, when the investigators incubated glutamate and ATP with purified BtrI, BtrJ, and BtrK, they found a molecular species of M_r 12,370. What is the most likely structure of this species? Hint: Remember that BtrJ can use ATP to γ-glutamylate nucleophilic groups.

Li and colleagues found that BtrO is homologous to monooxygenase enzymes (see Box 21–1) that hydroxylate alkanes, using FMN as a cofactor, and BtrV is homologous to an NAD(P)H oxidoreductase. Two other genes in the cluster, *btrG* and *btrH*, probably encode enzymes that remove the γ-glutamyl group and attach AHBA to the target antibiotic molecule.

(g) Based on these data, propose a plausible pathway for the synthesis of AHBA and its addition to the target antibiotic. Include the enzymes that catalyze each step and any other substrates or cofactors needed (ATP, NAD, etc.).

Reference

Li, Y., Llewellyn, N.M., Giri, R., Huang, F., & Spencer, J.B. (2005) Biosynthesis of the unique amino acid side chain of butirosin: possible protective-group chemistry in an acyl carrier protein–mediated pathway. *Chem. Biol.* **12**, 665–675.

Hormonal Regulation and Integration of Mammalian Metabolism

I n Chapters 13 through 22 we have discussed metabolism at the level of the individual cell, emphasizing central pathways common to almost all cells—bacterial, archaeal, and eukaryotic. We have seen how metabolic processes within cells are regulated at the level of individual enzyme reactions by substrate availability, by allosteric mechanisms, and by phosphorylation or other covalent modifications of enzymes.

To appreciate fully the significance of individual metabolic pathways and their regulation, we must view these pathways in the context of the whole organism. An essential characteristic of multicellular organisms is cell differentiation and division of labor. The specialized functions of the tissues and organs of complex organisms such as humans impose characteristic fuel requirements and patterns of metabolism. Hormonal signals integrate and coordinate the metabolic activities of different tissues and optimize the allocation of fuels and precursors to each organ.

In this chapter we focus on mammals, looking at the specialized metabolism of several major organs and tissues and the integration of metabolism in the whole organism. We begin by examining the broad range of

hormones and hormonal mechanisms, then turn to the tissue-specific functions regulated by these mechanisms. We discuss the distribution of nutrients to various organs—emphasizing the central role played by the liver—and the metabolic cooperation among these organs. To illustrate the integrative role of hormones, we describe the interplay of insulin, glucagon, and epinephrine in coordinating fuel metabolism in muscle, liver, and adipose tissue. The metabolic disturbances in diabetes further illustrate the importance of hormonal regulation of metabolism. We discuss the long-term hormonal regulation of body mass and, finally, the role of obesity in development of the metabolic syndrome and diabetes.

23.1 Hormones: Diverse Structures for Diverse Functions

Hormones are small molecules or proteins that are produced in one tissue, released into the circulation, and carried to other tissues, where they act through receptors to bring about changes in cellular activities. They serve to coordinate the metabolic activities of several tissues or organs. Virtually every process in a complex organism is regulated by one or more hormones: maintenance of blood pressure, blood volume, and electrolyte balance; embryogenesis; sexual differentiation, development, and reproduction; hunger, eating behavior, digestion, and fuel allocation—to name but a few. We examine here the methods for detecting and measuring hormones and their interaction with receptors, and we consider a representative selection of hormone types.

The coordination of metabolism in mammals is achieved by the **neuroendocrine system**. Individual cells in one tissue sense a change in the organism's circumstances and respond by secreting a chemical messenger that passes to another cell in the same or different tissue, where the messenger binds to a receptor molecule and triggers a change in this second cell. These chemical messengers may relay information over very short or very long distances. In neuronal signaling **(Fig. 23–1a)**, the chemical messenger is a neurotransmitter (acetylcholine, for example) and may travel only a fraction of a micrometer, across the synaptic cleft to the next neuron in a network. In hormonal signaling, the messengers—hormones—are carried in the bloodstream to neighboring cells or to distant organs and tissues; they may travel a meter or more before encountering their target cell (Fig. 23–1b). Except for this anatomic difference, these two chemical signaling mechanisms are remarkably similar, and the same molecule can sometimes act as both neurotransmitter and hormone. Epinephrine and norepinephrine, for example, serve as neurotransmitters at certain synapses of the brain and neuromuscular junctions of smooth muscle and as hormones that regulate fuel metabolism in liver and muscle. The following discussion of cellular signaling emphasizes hormone action, drawing on discussions of fuel metabolism in earlier chapters, but most of the fundamental mechanisms described here also occur in neurotransmitter action.

The Detection and Purification of Hormones Requires a Bioassay

How is a hormone detected and isolated? First, researchers find that a physiological process in one tissue depends on a signal that originates in another tissue. Insulin, for example, was first recognized as a substance that is produced in the pancreas and affects the concentration of glucose in blood and urine (Box 23–1). Once a physiological effect of the putative hormone is discovered, a quantitative bioassay for the hormone can be developed. In the case of insulin, the assay consisted of injecting extracts of pancreas (a crude source of insulin) into experimental animals deficient in insulin, then quantifying the resulting changes in glucose concentration in blood and urine. To isolate a hormone, the biochemist fractionates extracts containing the putative hormone, with the same techniques used to purify other biomolecules (solvent fractionation, chromatography, and electrophoresis), and then assays each fraction for hormone activity. Once the chemical has been purified, its composition and structure can be determined.

This protocol for hormone characterization is deceptively simple. Hormones are extremely potent and are produced in very small amounts. Obtaining sufficient hormone to allow its chemical characterization often involves biochemical isolations on a heroic scale. When Andrew Schally and Roger Guillemin independently purified and characterized thyrotropin-releasing hormone (TRH) from the hypothalamus, Schally's group processed about 20 tons of hypothalamus from nearly two million sheep, and Guillemin's group extracted the hypothalamus from about a million pigs! TRH proved to be a simple derivative of the tripeptide Glu–His–Pro **(Fig. 23–2)**. Once the structure of the hormone was known, it could be chemically synthesized in large quantities for use in physiological and biochemical studies. For their work on hypothalamic hormones, Schally and Guillemin shared the Nobel Prize in Physiology or Medicine in 1977, along with Rosalyn Yalow, who (with Solomon A. Berson) developed the extraordinarily sensitive **radioimmunoassay (RIA)** for peptide hormones and used it to study hormone action. RIA revolutionized hormone research by making possible the rapid, quantitative, and specific measurement of hormones in minute amounts.

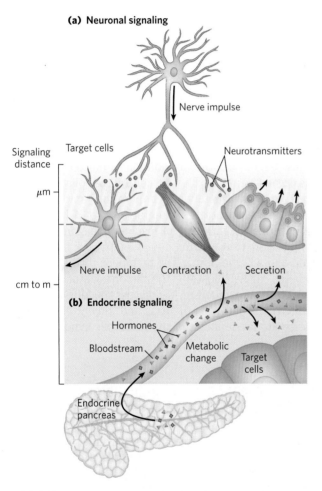

(a) Neuronal signaling

Nerve impulse

Signaling distance

Target cells

Neurotransmitters

μm

Nerve impulse Contraction Secretion

cm to m

(b) Endocrine signaling

Hormones

Bloodstream

Metabolic change

Target cells

Endocrine pancreas

FIGURE 23–1 Signaling by the neuroendocrine system. (a) In neuronal signaling, electrical signals (nerve impulses) originate in the cell body of a neuron and travel very rapidly over long distances to the axon tip, where neurotransmitters are released and diffuse to the target cell. The target cell (another neuron, a myocyte, or a secretory cell) is only a fraction of a micrometer or a few micrometers away from the site of neurotransmitter release. **(b)** In endocrine signaling, hormones (such as insulin produced in pancreatic β cells) are secreted into the bloodstream, which carries them throughout the body to target tissues that may be a meter or more away from the secreting cell. Both neurotransmitters and hormones interact with specific receptors on or in their target cells, triggering responses.

BOX 23–1 MEDICINE How Is a Hormone Discovered? The Arduous Path to Purified Insulin

Millions of people with type 1 diabetes mellitus inject themselves daily with pure insulin to compensate for the lack of production of this critical hormone by their own pancreatic β cells. Insulin injection is not a cure for diabetes, but it allows people who otherwise would have died young to lead long and productive lives. The discovery of insulin, which began with an accidental observation, illustrates the combination of serendipity and careful experimentation that led to the discovery of many of the hormones.

In 1889, Oskar Minkowski, a young assistant at the Medical College of Strasbourg, and Josef von Mering, at the Hoppe-Seyler Institute in Strasbourg, had a friendly disagreement about whether the pancreas, known to contain lipases, was important in fat digestion in dogs. To resolve the issue, they began an experiment on the digestion of fats. They surgically removed the pancreas from a dog, but before their experiment got any farther, Minkowski noticed that the dog was now producing far more urine than normal (a common symptom of untreated diabetes). Also, the dog's urine had glucose levels far above normal (another symptom of diabetes). These findings suggested that lack of some pancreatic product caused diabetes.

Minkowski tried unsuccessfully to prepare an extract of dog pancreas that would reverse the effect of removing the pancreas—that is, would lower the urinary or blood glucose levels. We now know that insulin is a protein and that the pancreas is very rich in proteases (trypsin and chymotrypsin), normally released directly into the small intestine to aid in digestion. These proteases doubtless degraded the insulin in the pancreatic extracts in Minkowski's experiments.

Despite considerable effort, no significant progress was made in the isolation or characterization of the "antidiabetic factor" until the summer of 1921, when Frederick G. Banting, a young scientist working in the laboratory of J. J. R. MacLeod at the University of Toronto, and a student assistant, Charles Best, took up the problem. By that time, several lines of evidence pointed to a group of specialized cells in the pancreas (the islets of Langerhans; see Fig. 23–26) as the source of the antidiabetic factor, which came to be called insulin (from Latin *insula*, "island").

Taking precautions to prevent proteolysis, Banting and Best (later aided by biochemist J. B. Collip) succeeded in December 1921 in preparing a purified pancreatic extract that cured the symptoms of experimental diabetes in dogs. On January 25, 1922 (just one month later!), their insulin preparation was injected into Leonard Thompson, a 14-year-old boy severely ill with diabetes mellitus. Within days, the levels of ketone bodies and glucose in Thompson's urine dropped dramatically; the extract saved his life and the lives of a number of other seriously ill children who also received these early preparations (Fig. 1). In 1923, Banting and MacLeod won the Nobel Prize for their isolation of insulin. Banting immediately announced that he would share his prize with Best; MacLeod shared his with Collip.

By 1923, pharmaceutical companies were supplying thousands of patients throughout the world with insulin extracted from porcine pancreas. With the development of genetic engineering techniques in the 1980s (Chapter 9), it became possible to produce unlimited quantities of human insulin by inserting the cloned human gene for insulin into a microorganism, which was then cultured on an industrial scale. Some patients with diabetes are now fitted with implanted insulin pumps, which release adjustable amounts of insulin on demand to meet changing needs at meal times and during exercise. There is a reasonable prospect that, in the future, transplantation of pancreatic tissue will provide diabetic patients with a source of insulin that responds as well as the normal pancreas, releasing insulin into the bloodstream only when blood glucose rises.

FIGURE 1 A child with type 1 diabetes before (left) and after (right) three months of treatment with an early preparation of insulin.

Pyroglutamate   Histidine   Prolylamide

pyroGlu-His-Pro-NH$_2$

FIGURE 23–2 The structure of thyrotropin-releasing hormone (TRH).
Purified (by heroic efforts) from extracts of hypothalamus, TRH proved to be a derivative of the tripeptide Glu–His–Pro. The side-chain carboxyl group of the amino-terminal Glu forms an amide (red bond) with the residue's α-amino group, creating pyroglutamate, and the carboxyl group of the carboxyl-terminal Pro is converted to an amide (red —NH$_2$). Such modifications are common among the small peptide hormones. In a typical protein of M_r ~50,000, the charges on the amino- and carboxyl-terminal groups contribute relatively little to the overall charge on the molecule, but in a tripeptide these two charges dominate the properties of the molecule. Formation of the amide derivatives removes these charges.

Hormone-specific antibodies are the key to the radioimmunoassay and its modern equivalent, the enzyme-linked immunosorbent assay (ELISA; see Fig. 5–26b). Purified hormone, injected into rabbits or mice, elicits antibodies that bind to the hormone with very high affinity and specificity. These antibodies may be purified and either radioisotopically labeled (for RIA) or conjugated with an enzyme that produces a colored product (for ELISA). The tagged antibodies are then allowed to interact with extracts containing the hormone. The fraction of antibody bound by the hormone in the extract is quantified by radiation detection or photometry. Because of the high affinity of the antibody for the hormone, such assays can be made sensitive to picograms of hormone in a sample.

Hormones Act through Specific High-Affinity Cellular Receptors

As we saw in Chapter 12, all hormones act through highly specific receptors in hormone-sensitive target cells, to which the hormones bind with high affinity. Each cell type has its own combination of hormone receptors, which define the range of its hormone responsiveness. Moreover, two cell types with the same type of receptor may have different intracellular targets of hormone action and thus may respond differently to the same hormone. The specificity of hormone action results from structural complementarity between the hormone and its receptor; this interaction is extremely selective, so even structurally similar hormones can have different effects if they preferentially bind to different receptors. The high affinity of the interaction allows cells to respond to very low concentrations of hormone. In the design of drugs intended to intervene in hormonal regulation, we need to know the

relative specificity and affinity of the drug and the natural hormone. Recall that hormone-receptor interactions can be quantified by **Scatchard analysis** (see Box 12–1), which, under favorable conditions, yields a quantitative measure of affinity (the dissociation constant for the complex) and the number of hormone-binding sites in a preparation of receptor.

The intracellular consequences of ligand-receptor interaction are of at least five general types: (1) a second messenger (such as cAMP, cGMP, or inositol trisphosphate) generated inside the cell acts as an allosteric regulator of one or more enzymes, (2) a receptor tyrosine kinase is activated by the extracellular hormone, (3) a change in membrane potential results from the opening or closing of a hormone-gated ion channel, (4) an adhesion receptor on the cell surface conveys information from the extracellular matrix to the cytoskeleton, or (5) a steroid or steroidlike molecule causes a change in the level of expression (transcription of DNA into mRNA) of one or more genes, mediated by a nuclear hormone receptor protein (see Fig. 12–2).

Water-soluble peptide and amine hormones (insulin and epinephrine, for example) act extracellularly by binding to cell surface receptors that span the plasma membrane **(Fig. 23–3)**. When the hormone binds to its extracellular domain, the receptor undergoes a conformational change analogous to that produced in an allosteric enzyme by binding of an effector molecule. The

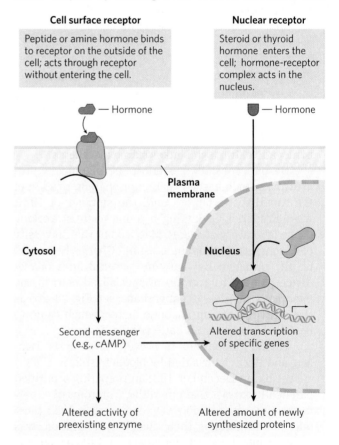

FIGURE 23–3 Two general mechanisms of hormone action. The peptide and amine hormones are faster acting than steroid and thyroid hormones.

conformational change triggers the downstream effects of the hormone.

A single hormone molecule, in forming a hormone-receptor complex, activates a catalyst that produces many molecules of second messenger, so the receptor serves not only as a signal transducer but also as a signal amplifier. The signal may be further amplified by a signaling cascade, a series of steps in which a catalyst activates a catalyst, resulting in very large amplifications of the original signal. A cascade of this type occurs in the regulation of glycogen synthesis and breakdown by epinephrine (see Fig. 12–7). Epinephrine activates (through its receptor) adenylyl cyclase, which produces many molecules of cAMP for each molecule of receptor-bound hormone. Cyclic AMP in turn activates cAMP-dependent protein kinase (protein kinase A), which activates glycogen phosphorylase b kinase, which activates glycogen phosphorylase b. The result is signal amplification: one epinephrine molecule causes the production of many thousands or millions of molecules of glucose 1-phosphate from glycogen.

Water-insoluble hormones (steroid, retinoid, and thyroid hormones) readily pass through the plasma membrane of their target cells to reach their receptor proteins in the nucleus (Fig. 23–3). With this class of hormones, the hormone-receptor complex itself carries the message: it interacts with DNA to alter the expression of specific genes, changing the enzyme complement of the cell and thereby changing cellular metabolism (see Fig. 12–30).

Hormones that act through plasma membrane receptors generally trigger very rapid physiological or biochemical responses. Just seconds after the adrenal medulla secretes epinephrine into the bloodstream, skeletal muscle responds by accelerating the breakdown of glycogen. By contrast, the thyroid hormones and the sex (steroid) hormones promote maximal responses in their target tissues only after hours or even days. These differences in response time correspond to different modes of action. In general, the fast-acting hormones lead to a change in the activity of one or more preexisting enzymes in the cell, by allosteric mechanisms or covalent modification. The slower-acting hormones generally alter gene expression, resulting in the synthesis of more (upregulation) or less (downregulation) of the regulated protein(s).

Hormones Are Chemically Diverse

Mammals have several classes of hormones, distinguishable by their chemical structures and their modes of action (Table 23–1). Peptide, catecholamine, and eicosanoid hormones act from outside the target cell via cell surface receptors. Steroid, vitamin D, retinoid, and thyroid hormones enter the cell and act through nuclear receptors. Nitric oxide (a gas) also enters the cell but activates a cytosolic enzyme, guanylyl cyclase (see Fig. 12–20).

Hormones can also be classified by the way they get from their point of release to their target tissue. **Endocrine** (from the Greek *endon*, "within," and *krinein*, "to release") hormones are released into the blood and carried to target cells throughout the body (insulin and glucagon are examples). **Paracrine** hormones are released into the extracellular space and diffuse to neighboring target cells (the eicosanoid hormones are of this type). **Autocrine** hormones affect the same cell that releases them, binding to receptors on the cell surface.

Mammals are hardly unique in possessing hormonal signaling systems. Insects and nematode worms have highly developed systems for hormonal regulation with fundamental mechanisms similar to those in mammals. Plants, too, use hormonal signals to coordinate the activities of their tissues (Chapter 12). The study of hormone action is not as advanced in plants as in animals, but we do know that some mechanisms are shared.

To illustrate the structural diversity and range of action of mammalian hormones, we consider representative examples of each major class listed in Table 23–1.

TABLE 23–1 Classes of Hormones

Type	Example	Synthetic path	Mode of action
Peptide	Insulin, glucagon	Proteolytic processing of prohormone	Plasma membrane receptors; second messengers
Catecholamine	Epinephrine	From tyrosine	
Eicosanoid	PGE$_1$	From arachidonate (20:4 fatty acid)	
Steroid	Testosterone	From cholesterol	Nuclear receptors; transcriptional regulation
Vitamin D	1α,25-Dihydroxyvitamin D$_3$	From cholesterol	
Retinoid	Retinoic acid	From vitamin A	
Thyroid	Triiodothyronine (T$_3$)	From Tyr in thyroglobulin	
Nitric oxide	Nitric oxide	From arginine + O$_2$	Cytosolic receptor (guanylyl cyclase) and second messenger (cGMP)

Peptide Hormones Peptide hormones may have from 3 to 200 or more amino acid residues. They include the pancreatic hormones insulin, glucagon, and somatostatin; the parathyroid hormone calcitonin; and all the hormones of the hypothalamus and pituitary (described below). These hormones are synthesized on ribosomes in the form of longer precursor proteins (prohormones), then packaged into secretory vesicles and proteolytically cleaved to form the active peptides. **Insulin** is a small protein (M_r 5,800) with two polypeptide chains, A and B, joined by two disulfide bonds. It is synthesized in the pancreas as an inactive single-chain precursor, preproinsulin **(Fig. 23–4)**, with an amino-terminal "signal sequence" that directs its passage into secretory vesicles. (Signal sequences are discussed in Chapter 27; see Fig. 27–38.) Proteolytic removal of the signal sequence and formation of three disulfide bonds produces proinsulin, which is stored in secretory granules (membrane vesicles filled with protein synthesized in the endoplasmic reticulum) in pancreatic β cells. When blood glucose is elevated sufficiently to trigger insulin secretion,

FIGURE 23–5 Proteolytic processing of the pro-opiomelanocortin (POMC) precursor. The initial gene product of the POMC gene is a long polypeptide that undergoes cleavage by a series of specific proteases to produce ACTH, β- and γ-lipotropin, α-, β-, and γ-MSH (melanocyte-stimulating hormone, or melanocortin), CLIP (corticotropin-like intermediary peptide), β-endorphin, and Met-enkephalin. The points of cleavage are paired basic residues, Arg–Lys, Lys–Arg, or Lys–Lys.

proinsulin is converted to active insulin by specific proteases, which cleave two peptide bonds to form the mature insulin molecule and C peptide, which are released into the blood by exocytosis.

In some cases, prohormone proteins, rather than yielding a single peptide hormone, produce several active hormones. Pro-opiomelanocortin (POMC) is a spectacular example of multiple hormones encoded by a single gene. The POMC gene encodes a large polypeptide that is progressively carved up into at least nine biologically active peptides **(Fig. 23–5)**. In many peptide hormones the terminal residues are modified, as in TRH (Fig. 23–2).

The concentration of peptide hormones in secretory granules is so high that the vesicle contents are virtually crystalline; when the contents are released by exocytosis, a large amount of hormone is released suddenly. The capillaries that serve peptide-producing endocrine glands are fenestrated (punctuated with tiny holes or "windows"), so the hormone molecules readily enter the bloodstream for transport to target cells elsewhere. As noted earlier, all peptide hormones act by binding to receptors in the plasma membrane. They cause the generation of a second messenger in the cytosol, which changes the activity of an intracellular enzyme, thereby altering the cell's metabolism.

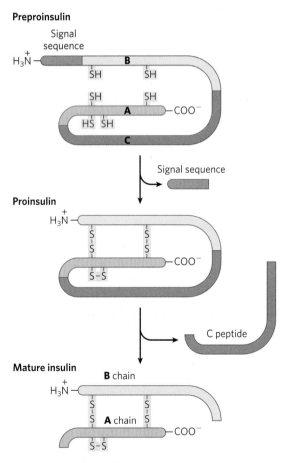

FIGURE 23–4 Insulin. Mature insulin is formed from its larger precursor preproinsulin by proteolytic processing. Removal of a 23 amino acid segment (the signal sequence) at the amino terminus of preproinsulin and formation of three disulfide bonds produces proinsulin. Further proteolytic cuts remove the C peptide from proinsulin to produce mature insulin, composed of A and B chains. The amino acid sequence of bovine insulin is shown in Figure 3–24.

Catecholamine Hormones The water-soluble compounds **epinephrine (adrenaline)** and **norepinephrine (noradrenaline)** are **catecholamines**, named for the structurally related compound catechol. They are synthesized from tyrosine (see Fig. 22–31).

Tyrosine $\longrightarrow$ L-Dopa $\longrightarrow$ Dopamine $\longrightarrow$
Norepinephrine $\longrightarrow$ Epinephrine

Catecholamines produced in the brain and in other neural tissues function as neurotransmitters, but epinephrine and norepinephrine are also hormones, synthesized and secreted by the adrenal glands. Like the peptide hormones, catecholamines are highly concentrated in secretory vesicles and released by exocytosis, and they act through surface receptors to generate intracellular second messengers. They mediate a wide variety of physiological responses to acute stress (see Table 23–6).

Eicosanoid Hormones The eicosanoid hormones (prostaglandins, thromboxanes, and leukotrienes) are derived from the 20-carbon polyunsaturated fatty acid arachidonate.

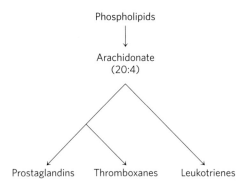

Unlike the hormones described above, they are not synthesized in advance and stored; they are produced, when needed, from arachidonate enzymatically released from membrane phospholipids by phospholipase A_2. The enzymes of the pathway leading to prostaglandins and thromboxanes are very widely distributed in mammalian tissues; most cells can produce these hormone signals, and cells of many tissues can respond to them through specific plasma membrane receptors. The eicosanoid hormones are paracrine hormones, secreted into the interstitial fluid (not primarily into the blood) and acting on nearby cells.

Prostaglandins promote the contraction of smooth muscle, including that of the intestine and uterus (and can therefore be used medically to induce labor). They also mediate pain and inflammation in all tissues. Many antiinflammatory drugs act by inhibiting steps in the prostaglandin synthetic pathway (see Fig. 21–15). Thromboxanes regulate platelet function and therefore blood clotting (see Fig. 6–39). Leukotrienes LTC_4 and LTD_4 act through plasma membrane receptors to stimulate contraction of smooth muscle in the intestine, pulmonary airways, and trachea. They are mediators of anaphylaxis, an immune overresponse that can include airway constriction, altered heartbeat, shock, and sometimes death. ■

Steroid Hormones The steroid hormones (adrenocortical hormones and sex hormones) are synthesized from cholesterol in several endocrine tissues.

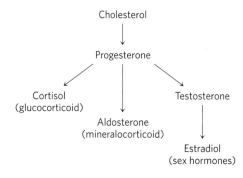

They travel to their target cells through the bloodstream, bound to carrier proteins. More than 50 corticosteroid hormones are produced in the adrenal cortex by reactions that remove the side chain from the D ring of cholesterol and introduce oxygen to form keto and hydroxyl groups. Many of these reactions involve cytochrome P-450 enzymes (see Box 21–1). The corticosteroids are of two general types, defined by their actions. Glucocorticoids (such as cortisol) primarily affect the metabolism of carbohydrates; mineralocorticoids (such as aldosterone) regulate the concentrations of electrolytes (K^+, Na^+, Ca^{2+}, Cl^-) in the blood. Androgens (such as testosterone) and estrogens (such as estradiol; see Fig. 10–19) are synthesized in the testes and ovaries. They affect sexual development, sexual behavior, and a variety of other reproductive and nonreproductive functions. Their synthesis also involves cytochrome P-450 enzymes that cleave the side chain of cholesterol and introduce oxygen atoms.

All steroid hormones act through nuclear receptors to change the level of expression of specific genes (see Fig. 12–30). They can also have more rapid effects, mediated by receptors in the plasma membrane.

Vitamin D Hormone Calcitriol ($1\alpha,25$-dihydroxycalcitriol) is produced from vitamin D by enzyme-catalyzed hydroxylation in the liver and kidneys (see Fig. 10–20a). Vitamin D is obtained in the diet or by photolysis of 7-dehydrocholesterol in skin exposed to sunlight.

7-Dehydrocholesterol
$\downarrow$ UV light
Vitamin D_3
(cholecalciferol)
$\downarrow$
25-Hydroxycholecalciferol
$\downarrow$
$1\alpha,25$-Dihydroxycalcitriol

Calcitriol works in concert with parathyroid hormone in Ca^{2+} homeostasis, regulating $[Ca^{2+}]$ in the blood and the balance between Ca^{2+} deposition and Ca^{2+} mobilization from bone. Acting through nuclear receptors, calcitriol activates the synthesis of an intestinal Ca^{2+}-binding protein essential for uptake of dietary Ca^{2+}. Inadequate dietary vitamin D or defects in the biosynthesis of calcitriol result in serious diseases such as rickets, in which bones are weak and malformed (see Fig. 10–20b). ■

Retinoid Hormones Retinoids are potent hormones that regulate the growth, survival, and differentiation of cells via nuclear retinoid receptors. The prohormone retinol is synthesized from β-carotene, primarily in liver (see Fig. 10–21), and many tissues convert retinol to the hormone retinoic acid (RA).

$$\beta\text{-Carotene}$$
$$\downarrow$$
$$\begin{array}{c}\text{Vitamin A}_1\\ \text{(retinol)}\end{array}$$
$$\downarrow$$
$$\text{Retinoic acid}$$

 All tissues are retinoid targets, as all cell types have at least one form of nuclear retinoid receptor. In adults, the most significant targets include cornea, skin, epithelia of the lungs and trachea, and the immune system, in all of which there is continuing replacement of cells. RA regulates the synthesis of proteins essential for growth or differentiation. Excessive vitamin A (the precursor to retinoid hormones) can cause birth defects, and pregnant women are advised not to use the retinoid creams that have been developed for treatment of severe acne. ■

Thyroid Hormones The thyroid hormones T$_4$ (thyroxine) and T$_3$ (triiodothyronine) are synthesized from the precursor protein thyroglobulin (M_r 660,000). Up to 20 Tyr residues in thyroglobulin are enzymatically iodinated in the thyroid gland, then two iodotyrosine residues condense to form the precursor to thyroxine. When needed, thyroxine is released by proteolysis. Condensation of monoiodotyrosine with diiodothyronine produces T$_3$, which is also an active hormone released by proteolysis.

$$\text{Thyroglobulin}-\text{Tyr}$$
$$\downarrow$$
$$\begin{array}{c}\text{Thyroglobulin}-\text{Tyr}-\text{I}\\ \text{(iodinated Tyr residues)}\end{array}$$
$$\downarrow\text{proteolysis}$$
$$\begin{array}{c}\text{Thyroxine (T}_4\text{),}\\ \text{triiodothyronine (T}_3\text{)}\end{array}$$

The thyroid hormones act through nuclear receptors to stimulate energy-yielding metabolism, especially in liver and muscle, by increasing the expression of genes encoding key catabolic enzymes.

Nitric Oxide (NO·) Nitric oxide is a relatively stable free radical synthesized from molecular oxygen and the guanidinium nitrogen of arginine (see Fig. 22–33) in a reaction catalyzed by **NO synthase**.

$$\text{Arginine} + 1\tfrac{1}{2}\text{NADPH} + 2\text{O}_2 \longrightarrow$$
$$\text{NO}^{\bullet} + \text{citrulline} + 2\text{H}_2\text{O} + 1\tfrac{1}{2}\text{NADP}^+$$

This enzyme is found in many tissues and cell types: neurons, macrophages, hepatocytes, myocytes of smooth muscle, endothelial cells of the blood vessels, and epithelial cells of the kidney. NO acts near its point of release, entering the target cell and activating the cytosolic enzyme guanylyl cyclase, which catalyzes the formation of the second messenger cGMP (see Fig. 12–20). A cGMP-dependent protein kinase mediates the effects of NO by phosphorylating key proteins and altering their activities. For example, phosphorylation of contractile proteins in the smooth muscle surrounding blood vessels relaxes the muscle, thereby lowering blood pressure.

Hormone Release Is Regulated by a Hierarchy of Neuronal and Hormonal Signals

The changing levels of specific hormones regulate specific cellular processes, but what regulates the level of each hormone? The brief answer is that the central nervous system receives input from many internal and external sensors—signals about danger, hunger, dietary intake, blood composition and pressure, for example—and orchestrates the production of appropriate hormonal signals by the endocrine tissues. For a more complete answer, we must look at the hormone-producing systems of the human body and some of their functional interrelationships.

Figure 23–6 shows the anatomic location of the major endocrine glands in humans, and **Figure 23–7** represents the "chain of command" in the hormonal signaling hierarchy. The **hypothalamus**, a small region of

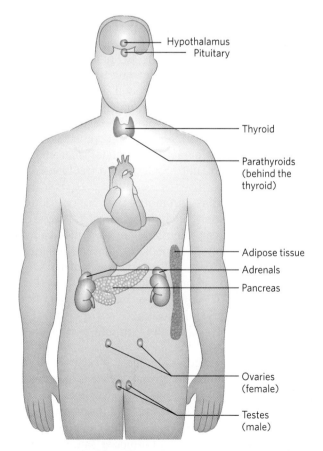

FIGURE 23–6 The major endocrine glands. The glands are shaded pink.

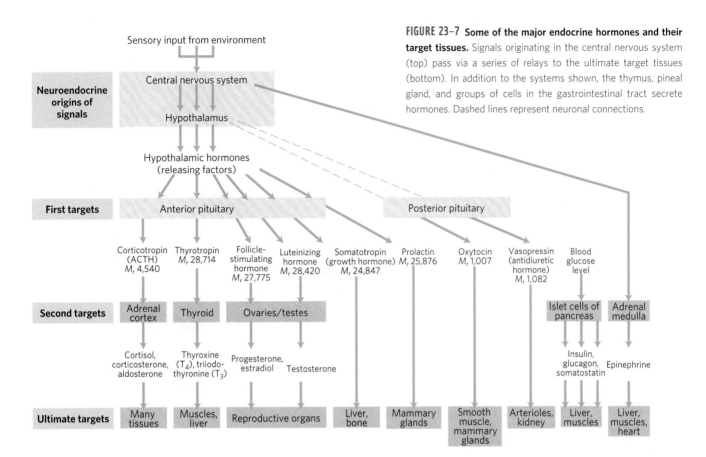

FIGURE 23–7 Some of the major endocrine hormones and their target tissues. Signals originating in the central nervous system (top) pass via a series of relays to the ultimate target tissues (bottom). In addition to the systems shown, the thymus, pineal gland, and groups of cells in the gastrointestinal tract secrete hormones. Dashed lines represent neuronal connections.

the brain **(Fig. 23–8)**, is the coordination center of the endocrine system; it receives and integrates messages from the central nervous system. In response to these messages, the hypothalamus produces regulatory hormones (releasing factors) that pass directly to the nearby pituitary gland through special blood vessels and neurons that connect the two glands (Fig. 23–8b). The pituitary gland has two functionally distinct parts. The **posterior pituitary** contains the axonal endings of many neurons that originate in the hypothalamus. These neurons produce the short peptide hormones oxytocin and vasopressin **(Fig. 23–9)**, which move down the axon to the nerve endings in the pituitary, where they are stored in secretory granules to await the signal for their release.

The **anterior pituitary** responds to hypothalamic hormones carried in the blood, producing **tropic hormones**, or **tropins** (from the Greek *tropos*, "turn"). These relatively long polypeptides activate the next rank of endocrine glands (Fig. 23–7), which includes the adrenal cortex, thyroid gland, ovaries, and testes. These glands in turn secrete their specific hormones, which are carried in the bloodstream to the target tissues. For example, corticotropin-releasing hormone secreted from the hypothalamus stimulates the anterior pituitary to release ACTH, which travels through the blood to the zona fasciculata of the adrenal cortex and triggers the release of cortisol. Cortisol, the ultimate hormone in this cascade, acts through its receptor in many types of target cells to alter their metabolism. In hepatocytes, one effect of cortisol is to increase the rate of gluconeogenesis.

Hormonal cascades such as those responsible for the release of cortisol and epinephrine result in large amplifications of the initial signal and allow exquisite fine-tuning of the output of the ultimate hormone **(Fig. 23–10)**. At each level in the cascade, a small signal elicits a larger response. For example, the initial electrical signal to the hypothalamus results in the release of a few *nanograms* of corticotropin-releasing hormone, which elicits the release of a few *micrograms* of corticotropin. Corticotropin acts on the adrenal cortex to cause the release of *milligrams* of cortisol, for an overall amplification of at least a millionfold.

At each level of a hormonal cascade, feedback inhibition of earlier steps in the cascade is possible; an unnecessarily elevated level of the ultimate hormone or of an intermediate hormone inhibits the release of earlier hormones in the cascade. These feedback mechanisms accomplish the same end as those that limit the output of a biosynthetic pathway (compare Fig. 23–10 with Fig. 22–37): a product is synthesized (or released) only until the necessary concentration is reached.

SUMMARY 23.1 Hormones: Diverse Structures for Diverse Functions

▶ Hormones are chemical messengers secreted by certain tissues into the blood or interstitial fluid, serving to regulate the activity of other cells or tissues.

(a)

(b)

FIGURE 23–8 Neuroendocrine origins of hormone signals. (a) Location of the hypothalamus and pituitary gland. **(b)** Details of the hypothalamus-pituitary system. Signals from connecting neurons stimulate the hypothalamus to secrete releasing factors into a blood vessel that carries the hormones directly to a capillary network in the anterior pituitary. In response to each hypothalamic releasing factor, the anterior pituitary releases the appropriate hormone into the general circulation. Posterior pituitary hormones are synthesized in neurons arising in the hypothalamus, transported along axons to nerve endings in the posterior pituitary, and stored there until released into the blood in response to a neuronal signal.

▶ Radioimmunoassay and ELISA are two very sensitive techniques for detecting and quantifying hormones.

▶ Peptide, catecholamine, and eicosanoid hormones bind to specific receptors in the plasma membrane of target cells, altering the level of an intracellular second messenger, without actually entering the cell.

▶ Steroid, vitamin D, retinoid, and thyroid hormones enter target cells and alter gene expression by interacting with specific nuclear receptors.

▶ Some hormones are synthesized as prohormones and activated by enzymatic cleavage. In some cases, such as insulin, a single hormone is produced by proteolytic cleavages; in others, such as POMC, several distinct hormones are produced by cleavage of a single prohormone.

▶ Hormones are regulated by a hierarchy of interactions between the brain and endocrine glands: nerve impulses stimulate the hypothalamus to send specific hormones to the pituitary gland, thus stimulating (or inhibiting) the release of tropic

FIGURE 23–9 Two hormones of the posterior pituitary gland. The carboxyl-terminal residue of both peptides is glycinamide, $-NH-CH_2-CONH_2$ (as noted in Fig. 23-2, amidation of the carboxyl terminus is common in short peptide hormones). These two hormones, identical in all but two residues (shaded light red), have very different biological effects. Oxytocin acts on the smooth muscle of the uterus and mammary gland, causing uterine contractions during labor and promoting milk release during lactation. Vasopressin (also called antidiuretic hormone) increases water reabsorption in the kidney and promotes the constriction of blood vessels, thereby increasing blood pressure.

Human oxytocin

Human vasopressin (antidiuretic hormone)

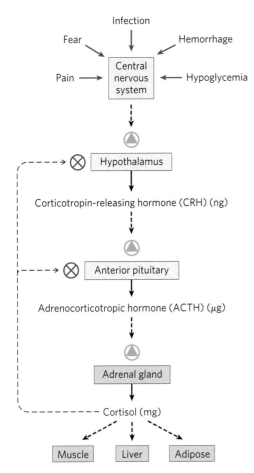

FIGURE 23–10 Cascade of hormone release following central nervous system input to the hypothalamus. Solid black arrows indicate hormone production and release; broken black arrows indicate the action of hormones on target tissues. In each endocrine tissue along the pathway, a stimulus from the level above is received, amplified, and transduced into the release of the next hormone in the cascade. The cascade is sensitive to regulation at several levels through feedback inhibition (thin, dashed arrows) by the ultimate hormone (in this case, cortisol). The product therefore regulates its own production, as in feedback inhibition of biosynthetic pathways within a single cell.

hormones. The anterior pituitary hormones in turn stimulate other endocrine glands (thyroid, adrenals, pancreas) to secrete their characteristic hormones, which in turn stimulate specific target tissues.

▶ Hormonal cascades, in which catalysts activate catalysts, amplify the initial stimulus by several orders of magnitude, often in a very short time (seconds).

23.2 Tissue-Specific Metabolism: The Division of Labor

Each tissue of the human body has a specialized function, reflected in its anatomy and metabolic activity **(Fig. 23–11)**. Skeletal muscle allows directed motion; adipose tissue stores and distributes energy in the form of fats, which serve as fuel throughout the body as well as thermal insulation; in the brain, cells pump ions across their plasma membranes to produce electrical signals. The liver plays a central processing and distributing role in metabolism and furnishes all other organs and tissues with an appropriate mix of nutrients via the bloodstream. The functional centrality of the liver is indicated by the common reference to all other tissues and organs as "extrahepatic." We therefore begin our discussion of the division of metabolic labor by considering the transformations of carbohydrates, amino acids, and fats in the mammalian liver. This is followed by brief descriptions of the primary metabolic functions of adipose tissue, muscle, brain, and the medium that interconnects all others: the blood.

The Liver Processes and Distributes Nutrients

During digestion in mammals, the three main classes of nutrients (carbohydrates, proteins, and fats) undergo enzymatic hydrolysis into their simple constituents. This breakdown is necessary because the epithelial cells lining the intestinal lumen absorb only relatively small molecules. Many of the fatty acids and monoacylglycerols released by digestion of fats in the intestine are reassembled within these epithelial cells into triacylglycerols (TAGs).

After being absorbed, most sugars and amino acids and some reconstituted TAGs pass from intestinal epithelial cells into blood capillaries and travel in the bloodstream to the liver; the remaining TAGs enter adipose tissue via the lymphatic system. The portal vein (Fig. 23–11) is a direct route from the digestive organs to the liver, and liver therefore has first access to ingested nutrients. The liver has two main cell types. Kupffer cells are phagocytes, important in immune function. **Hepatocytes**, of primary interest here, transform dietary nutrients into the fuels and precursors required by other tissues and export them via the blood. The kinds and amounts of nutrients supplied to the liver vary with several factors, including the diet and the time between meals. The demand of extrahepatic tissues for fuels and precursors varies among organs and with the level of activity and overall nutritional state of the individual.

To meet these changing circumstances, the liver has remarkable metabolic flexibility. For example, when the diet is rich in protein, hepatocytes supply themselves with high levels of enzymes for amino acid catabolism and gluconeogenesis. Within hours after a shift to a high-carbohydrate diet, the levels of these enzymes begin to drop and the hepatocytes increase their synthesis of enzymes essential to carbohydrate metabolism and fat synthesis. Liver enzymes turn over (are synthesized and degraded) at 5 to 10 times the rate of enzyme turnover in other tissues, such as muscle. Extrahepatic tissues also can adjust their metabolism to prevailing conditions, but none is as adaptable as the liver, and none is so central to the organism's overall metabolism. What follows is

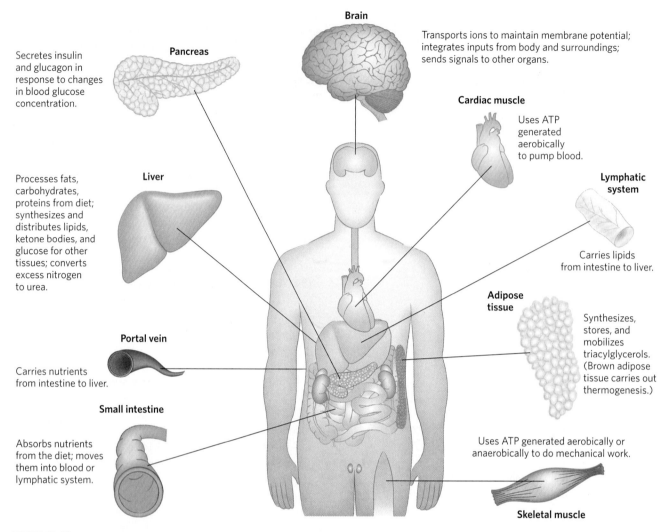

FIGURE 23–11 Specialized metabolic functions of mammalian tissues.

a survey of the possible fates of sugars, amino acids, and lipids that enter the liver from the bloodstream. To help you recall the metabolic transformations discussed here, Table 23–2 shows the major pathways and processes and indicates by figure number where each pathway is presented in detail. Here, we provide summaries of the pathways, referring to the numbered pathways and reactions in Figures 23–12 to 23–14.

Sugars The glucose transporter of hepatocytes (GLUT2) allows rapid, passive diffusion of glucose so that the concentration of glucose in a hepatocyte is essentially the same as that in the blood. Glucose entering hepatocytes is phosphorylated by hexokinase IV (glucokinase) to yield glucose 6-phosphate. Glucokinase has a much higher K_m for glucose (10 mM) than do the hexokinase isozymes in other cells (p. 603) and, unlike these other isozymes, it is not inhibited by its product, glucose 6-phosphate. The presence of glucokinase allows hepatocytes to continue phosphorylating glucose when the glucose concentration rises well above levels that would overwhelm other hexokinases. The high K_m of glucokinase also ensures that the phosphorylation of glucose in

hepatocytes is minimal when the glucose concentration is low, preventing the liver from consuming glucose as fuel via glycolysis. This spares glucose for other tissues. Fructose, galactose, and mannose, all absorbed from the small intestine, are also converted to glucose 6-phosphate by enzymatic pathways examined in Chapter 14. Glucose 6-phosphate is at the crossroads of carbohydrate metabolism in the liver. It may take any of several major metabolic routes **(Fig. 23–12)**, depending on the current metabolic needs of the organism. By the action of various allosterically regulated enzymes, and through hormonal regulation of enzyme synthesis and activity, the liver directs the flow of glucose into one or more of these pathways.

❶ Glucose 6-phosphate is dephosphorylated by glucose 6-phosphatase to yield free glucose (see Fig. 15–30), which is exported to replenish blood glucose. Export is the predominant pathway when glucose 6-phosphate is in limited supply, because the blood glucose concentration must be kept sufficiently high (4 mM) to provide adequate energy for the brain and other tissues. ❷ Glucose 6-phosphate not immediately needed to form blood glucose is converted to

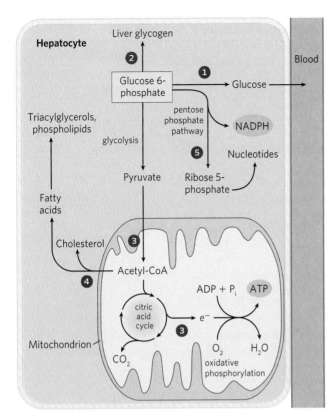

FIGURE 23–12 Metabolic pathways for glucose 6-phosphate in the liver. Here and in Figures 23–13 and 23–14, anabolic pathways are generally shown leading upward, catabolic pathways leading downward, and distribution to other organs horizontally. The numbered processes in each figure are described in the text.

liver glycogen or has one of several other fates. Following glycolysis and the pyruvate dehydrogenase reaction, ❸ the acetyl-CoA so formed can be oxidized for ATP production by the citric acid cycle, with ensuing electron transfer and oxidative phosphorylation yielding ATP. (Normally, however, fatty acids are the preferred fuel for ATP production in hepatocytes.) ❹ Acetyl-CoA can also serve as the precursor of fatty acids, which are incorporated into TAGs and phospholipids, and of cholesterol. Much of the lipid synthesized in the liver is transported to other tissues by blood lipoproteins. ❺ Alternatively, glucose 6-phosphate can enter the pentose phosphate pathway, yielding both reducing power (NADPH), needed for the biosynthesis of fatty acids and cholesterol, and D-ribose 5-phosphate, a precursor for nucleotide biosynthesis. NADPH is also an essential cofactor in the detoxification and elimination of many drugs and other xenobiotics metabolized in the liver.

Amino Acids Amino acids that enter the liver follow several important metabolic routes **(Fig. 23–13)**. ❶ They are precursors for protein synthesis, a process discussed in Chapter 27. The liver constantly renews its own proteins, which have a relatively high turnover rate (average half-life of hours to days), and is also the site

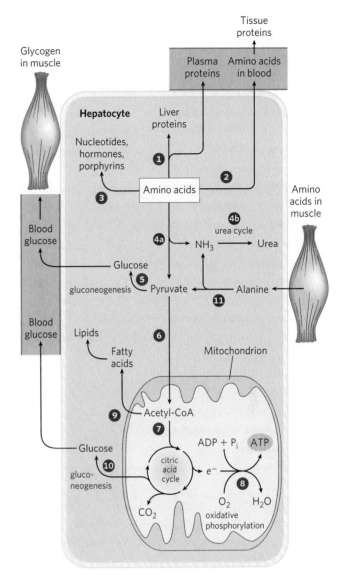

FIGURE 23–13 Metabolism of amino acids in the liver.

of biosynthesis of most plasma proteins. ❷ Alternatively, amino acids pass in the bloodstream to other organs to be used in the synthesis of tissue proteins. ❸ Other amino acids are precursors in the biosynthesis of nucleotides, hormones, and other nitrogenous compounds in the liver and other tissues.

❹ₐ Amino acids not needed as biosynthetic precursors are transaminated or deaminated and degraded to yield pyruvate and citric acid cycle intermediates, with various fates; ❹ᵦ the ammonia released is converted to the excretory product urea. ❺ Pyruvate can be converted to glucose and glycogen via gluconeogenesis, or ❻ it can be converted to acetyl-CoA, which has several possible fates: ❼ oxidation via the citric acid cycle and ❽ oxidative phosphorylation to produce ATP or ❾ conversion to lipids for storage. ❿ Citric acid cycle intermediates can be siphoned off into glucose synthesis by gluconeogenesis.

The liver also metabolizes amino acids that arrive intermittently from other tissues. The blood is adequately supplied with glucose just after the digestion

TABLE 23–2 Pathways of Carbohydrate, Amino Acid, and Fat Metabolism Illustrated in Earlier Chapters

Pathway	Figure reference(s)
Citric acid cycle: acetyl-CoA $\longrightarrow$ 2CO$_2$	16–7
Oxidative phosphorylation: ATP synthesis	19–19
Carbohydrate catabolism	
Glycogenolysis: glycogen $\longrightarrow$ glucose 1-phosphate $\longrightarrow$ blood glucose	15–27; 15–28
Hexose entry into glycolysis: fructose, mannose, galactose $\longrightarrow$ glucose 6-phosphate	14–11
Glycolysis: glucose $\longrightarrow$ pyruvate	14–2
Pyruvate dehydrogenase reaction: pyruvate $\longrightarrow$ acetyl-CoA	16–2
Lactic acid fermentation: glucose $\longrightarrow$ lactate + 2ATP	14–4
Pentose phosphate pathway: glucose 6-phosphate $\longrightarrow$ pentose phosphates + NADPH	14–22
Carbohydrate anabolism	
Gluconeogenesis: citric acid cycle intermediates $\longrightarrow$ glucose	14–17
Glucose-alanine cycle: glucose $\longrightarrow$ pyruvate $\longrightarrow$ alanine $\longrightarrow$ glucose	18–9
Glycogen synthesis: glucose 6-phosphate $\longrightarrow$ glucose 1-phosphate $\longrightarrow$ glycogen	15–32
Amino acid and nucleotide metabolism	
Amino acid degradation: amino acids $\longrightarrow$ acetyl-CoA, citric acid cycle intermediates	18–15
Amino acid synthesis	22–11
Urea cycle: NH$_3$ $\longrightarrow$ urea	18–10
Glucose-alanine cycle: alanine $\longrightarrow$ glucose	18–9
Nucleotide synthesis: amino acids $\longrightarrow$ purines, pyrimidines	22–35; 22–38
Hormone and neurotransmitter synthesis	22–31
Fat catabolism	
β Oxidation of fatty acids: fatty acids $\longrightarrow$ acetyl-CoA	17–8
Oxidation of ketone bodies: β-hydroxybutyrate $\longrightarrow$ acetyl-CoA $\longrightarrow$ CO$_2$ via citric acid cycle	17–20
Fat anabolism	
Fatty acid synthesis: acetyl-CoA $\longrightarrow$ fatty acids	21–6
Triacylglycerol synthesis: acetyl-CoA $\longrightarrow$ fatty acids $\longrightarrow$ triacylglycerol	21–18; 21–19
Ketone body formation: acetyl-CoA $\longrightarrow$ acetoacetate, β-hydroxybutyrate	17–19
Cholesterol and cholesteryl ester synthesis: acetyl-CoA $\longrightarrow$ cholesterol $\longrightarrow$ cholesteryl esters	21–33 to 21–37
Phospholipid synthesis: fatty acids $\longrightarrow$ phospholipids	21–17; 21–23 to 21–28

and absorption of dietary carbohydrate or, between meals, by the conversion of liver glycogen to blood glucose. During the interval between meals, especially if prolonged, some muscle protein is degraded to amino acids. These amino acids donate their amino groups (by transamination) to pyruvate, the product of glycolysis, to yield alanine, which ⑪ is transported to the liver and deaminated. Hepatocytes convert the resulting pyruvate to blood glucose (via gluconeogenesis ❺), and the ammonia to urea for excretion ❹b. One benefit of this glucose-alanine cycle (see Fig. 18–9) is the smoothing out of fluctuations in blood glucose between meals. The amino acid deficit incurred in muscles is made up after the next meal by incoming dietary amino acids.

Lipids The fatty acid components of lipids entering hepatocytes also have several different fates **(Fig. 23–14)**. ❶ Some are converted to liver lipids. ❷ Under most circumstances, fatty acids are the primary oxidative fuel in the liver. Free fatty acids may be activated and oxidized to yield acetyl-CoA and NADH. ❸ The acetyl-CoA is further oxidized via the citric acid cycle, and ❹ oxidations in the cycle drive the synthesis of ATP by oxidative phosphorylation. ❺ Excess acetyl-CoA, not required by the liver, is converted to acetoacetate and β-hydroxybutyrate; these ketone bodies circulate in the blood to other tissues to be used as fuel for the citric acid cycle. Ketone bodies, unlike fatty acids, can pass the blood-brain barrier, providing the brain with a source of acetyl-CoA for energy-yielding oxidation. Ketone bodies can supply a significant fraction of the energy in some extrahepatic tissues—up to one-third in the heart and as much as 60% to 70% in the brain during prolonged fasting. ❻ Some of the acetyl-CoA derived from fatty acids (and from glucose) is used for the biosynthesis of cholesterol, which is required for membrane synthesis. Cholesterol is also the precursor of all steroid hormones and of the bile salts, which are essential for the digestion and absorption of lipids.

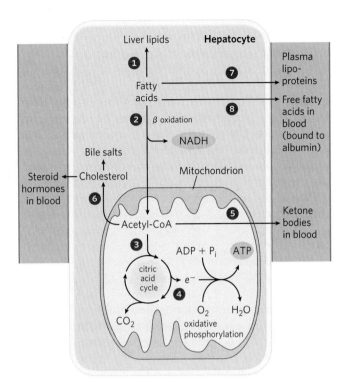

FIGURE 23–14 Metabolism of fatty acids in the liver.

(a) White adipocyte   **(b)** Brown adipocyte

(c)   **(d)**

FIGURE 23–15 Adipocytes of white and brown adipose tissue. Schematic views of typical mouse adipocytes from **(a)** white adipose tissue (WAT) and **(b)** brown adipose tissue (BAT). White adipocytes are larger and contain a single huge lipid droplet, which squeezes the mitochondria and nucleus against the plasma membrane. In brown adipocytes, mitochondria are much more prominent, the nucleus is near the center of the cell, and multiple small fat droplets are present. Below are shown scanning electron micrographs of adipocytes in **(c)** WAT and **(d)** BAT. In fat tissues, capillaries and collagen fibers form a supporting network around spherical adipocytes.

The other two metabolic fates of lipids involve specialized mechanisms for the transport of insoluble lipids in blood. ❼ Fatty acids are converted to the phospholipids and TAGs of plasma lipoproteins, which carry lipids to adipose tissue for storage. ❽ Some free fatty acids are bound to serum albumin and carried to the heart and skeletal muscles, which take up and oxidize free fatty acids as a major fuel. Serum albumin is the most abundant plasma protein; one molecule can carry up to 10 molecules of free fatty acid.

The liver thus serves as the body's distribution center, exporting nutrients in the correct proportions to other organs, smoothing out fluctuations in metabolism caused by intermittent food intake, and processing excess amino groups into urea and other products to be disposed of by the kidneys. Certain nutrients are stored in the liver, including iron ions and vitamin A. The liver also detoxifies foreign organic compounds, such as drugs, food additives, preservatives, and other possibly harmful agents with no food value. Detoxification often involves the cytochrome P-450–dependent hydroxylation of relatively insoluble organic compounds, making them sufficiently soluble for further breakdown and excretion (see Box 21–1).

Adipose Tissues Store and Supply Fatty Acids

There are two distinct types of adipose tissue, white and brown **(Fig. 23–15)**, with quite distinct roles, and we focus first on the more abundant of the two. **White adipose tissue (WAT)** is amorphous and widely distributed in the body: under the skin, around the deep blood vessels, and in the abdominal cavity. The **adipocytes** of

WAT are large (diameter 30 to 70 μm), spherical cells, completely filled with a single large lipid (TAG) droplet that constitutes about 65% of the cell mass and squeezes the mitochondria and nucleus into a thin layer against the plasma membrane (Fig. 23–15a, c). In humans, WAT typically makes up about 15% of the mass of a healthy young adult. The adipocytes are metabolically very active, responding quickly to hormonal stimuli in a metabolic interplay with the liver, skeletal muscles, and heart.

Like other cell types, adipocytes have an active glycolytic metabolism, oxidize pyruvate and fatty acids via the citric acid cycle, and carry out oxidative phosphorylation. During periods of high carbohydrate intake, adipose tissue can convert glucose (via pyruvate and acetyl-CoA) to fatty acids, convert the fatty acids to TAGs, and store the TAGs as large fat globules—although in humans, much of the fatty acid synthesis occurs in hepatocytes. Adipocytes store TAGs arriving from the liver (carried in the blood as VLDL; see Fig. 21–40) and from the intestinal tract (carried in chylomicrons), particularly after meals rich in fat.

When the demand for fuel rises (between meals, for example), lipases in adipocytes hydrolyze stored TAGs to release free fatty acids, which can travel in the bloodstream to skeletal muscle and the heart. The release of fatty acids from adipocytes is greatly accelerated by epinephrine, which stimulates the cAMP-dependent phosphorylation of perilipin and thus gives lipases specific for tri-, di-, and monoglycerides access to TAGs in the lipid droplet (see Fig. 17–3). Hormone-sensitive lipase is also stimulated by phosphorylation, but this is not the main cause of increased lipolysis. Insulin counterbalances this effect of epinephrine, decreasing the activity of the lipase.

The breakdown and synthesis of TAGs in adipose tissue constitute a substrate cycle; up to 70% of the fatty acids released by the three lipases are reesterified in adipocytes, re-forming TAGs. Recall from Chapter 15 that such substrate cycles allow fine regulation of the rate and direction of flow of intermediates through a bidirectional pathway. In adipose tissue, glycerol liberated by adipocyte lipases cannot be reused in the synthesis of TAGs, because adipocytes lack glycerol kinase. Instead, the glycerol phosphate required for TAG synthesis is made from pyruvate by glyceroneogenesis, involving the cytosolic PEP carboxykinase (see Fig. 21–22).

In addition to its central function as a fuel depot, adipose tissue plays an important role as an endocrine organ, producing and releasing hormones that signal the state of energy reserves and coordinate metabolism of fats and carbohydrates throughout the body. We return to this function later in the chapter when we discuss the hormonal regulation of body mass.

Brown Adipose Tissue Is Thermogenic

In small vertebrates and hibernating animals, a significant proportion of the adipose tissue is **brown adipose tissue (BAT)**, distinguished from WAT by its smaller (diameter 20 to 40 μm), differently shaped (polygonal, not round) adipocytes (Fig. 23–15b, d). Like white adipocytes, brown adipocytes store TAGs, but in several smaller lipid droplets per cell rather than as a single central droplet. BAT cells have more mitochondria and a richer supply of capillaries and innervation than WAT cells, and it is the cytochromes of mitochondria and the hemoglobin in capillaries that give BAT its characteristic brown color. A unique feature of brown adipocytes is their strong expression of the gene *UCP1*, which encodes **thermogenin**, the mitochondrial uncoupling protein (see Fig. 19–36). Thermogenin activity is responsible for one of the principal functions of BAT: **thermogenesis**.

In brown adipocytes, fatty acids stored in lipid droplets are released, enter mitochondria, and undergo complete conversion to CO_2 via β oxidation and the citric acid cycle. The reduced $FADH_2$ and NADH so generated pass their electrons through the respiratory chain to molecular oxygen. In WAT, protons pumped out of the mitochondria during electron transfer reenter the matrix through ATP synthase, with the energy of electron transfer conserved in ATP synthesis. In BAT, thermogenin provides an alternative route for protons to reenter the matrix that bypasses ATP synthase; the energy of the proton gradient is thus dissipated as heat,

which can maintain the body (especially the nervous system and viscera) at its optimal temperature when the ambient temperature is relatively low.

In the human fetus, differentiation of fibroblast preadipocytes into BAT begins at the twentieth week of gestation, and at the time of birth BAT represents 1% to 5% of total body mass. The brown fat deposits are located where the heat generated by thermogenesis can ensure that vital tissues—blood vessels to the head, major abdominal blood vessels, and the viscera, including the pancreas, adrenal glands, and kidneys—are not chilled as the newborn enters a world of lower ambient temperature **(Fig. 23–16)**.

At birth, WAT development begins and BAT begins to disappear. Young adult humans have much-diminished deposits of BAT, from 3% of all adipose tissue in males to 7% in females, making up less than 0.1% of body mass. However, adults apparently have BAT that can be activated by cold exposure and suppressed by increasing the core body temperature (Fig. 23–16b). Adipocytes of BAT produce heat by oxidation of their own fatty acids, but they take up and oxidize both fatty acids and glucose from the blood at rates out of proportion to their mass. In fact, the detection of BAT by PET scanning depends on the adipocytes' relatively high rate of *glucose* uptake and metabolism (Fig. 23–16b). Humans with pheochromocytoma (tumors of the adrenal gland) overproduce epinephrine and norepinephrine, and one effect is differentiation of preadipocytes into discrete regions of BAT, localized roughly as in newborns. In the adaptation to warm or cold surroundings, and in the normal differentiation of WAT and BAT, the nuclear transcription factor PPARγ (described later in the chapter) plays a central role.

Muscles Use ATP for Mechanical Work

Metabolism in the cells of skeletal muscle—**myocytes**—is specialized to generate ATP as the immediate source of energy for contraction. Moreover, skeletal muscle is adapted to do its mechanical work in an intermittent fashion, on demand. Sometimes skeletal muscles must work at their maximum capacity for a short time, as in a 100 m sprint; at other times more prolonged work is required, as in running a marathon or in extended physical labor.

There are two general classes of muscle tissue, which differ in physiological role and fuel utilization. **Slow-twitch muscle**, also called red muscle, provides relatively low tension but is highly resistant to fatigue. It produces ATP by the relatively slow but steady process of oxidative phosphorylation. Red muscle is very rich in mitochondria and is served by very dense networks of blood vessels, which bring the oxygen essential to ATP production. **Fast-twitch muscle**, or white muscle, has fewer mitochondria than red muscle and is less well supplied with blood vessels, but it can develop greater tension and do so faster. White muscle is quicker to fatigue because when active, it uses ATP faster than it can

(a)

(b)

FIGURE 23–16 Brown adipose tissue in infants and adults. (a) At birth, human infants have brown fat distributed as shown here, to protect the spine, major blood vessels, and the internal organs. **(b)** PET scans of a 45-year-old woman injected with ^{18}F-deoxyglucose (to detect tissue that rapidly metabolizes glucose; see Fig. 23–23) reveal tumors of the left lung and lymph node, the right adrenal gland, and a lumbar vertebra (left). The heart and bladder were also intensely labeled as expected, but in addition there was striking metabolic activity in the regions that normally have brown fat in infants. When the same patient was warmed for 48 hours before the PET scan (right), these typical brown fat areas were not active, indicating that this adult had brown fat deposits, which were metabolically active only when the core body temperature was relatively low.

replace it. There is a genetic component to the proportion of red and white muscle in any individual; with training, the endurance of fast-twitch muscle can be improved.

Skeletal muscle can use free fatty acids, ketone bodies, or glucose as fuel, depending on the degree of muscular activity **(Fig. 23–17)**. In resting muscle, the primary fuels are free fatty acids from adipose tissue and ketone bodies from the liver. These are oxidized and degraded to yield acetyl-CoA, which enters the citric acid cycle for oxidation to CO_2. The ensuing transfer of electrons to O_2 provides the energy for ATP synthesis by oxidative phosphorylation. Moderately active muscle uses blood glucose in addition to fatty acids and ketone bodies. The glucose is phosphorylated, then degraded by glycolysis to pyruvate, which is converted to acetyl-CoA and oxidized via the citric acid cycle and oxidative phosphorylation.

In maximally active fast-twitch muscles, the demand for ATP is so great that the blood flow cannot provide O_2 and fuels fast enough to supply sufficient ATP by

FIGURE 23–17 Energy sources for muscle contraction. Different fuels are used for ATP synthesis during bursts of heavy activity and during light activity or rest. Phosphocreatine can rapidly supply ATP.

BOX 23–2 Creatine and Creatine Kinase: Invaluable Diagnostic Aids and the Muscle Builder's Friends

In animal tissues that have a high and fluctuating need for ATP, primarily skeletal muscle, cardiac muscle, and brain, there are several isozymes of creatine kinase. A cytosolic isozyme is present in regions of high ATP use (myofibrils and sarcoplasmic reticulum, for example). By converting ADP produced during periods of high ATP use back to ATP, this isozyme prevents the accumulation of ADP to concentrations that could inhibit ATP-using enzymes by mass action. Another isozyme of creatine kinase is located in regions where the inner and outer membranes of mitochondria come into contact. This mitochondrial isozyme (mCK) probably serves to shuttle ATP equivalents produced in the mitochondria to cytosolic sites of ATP use (Fig. 1). The species that diffuses from the mitochondrion to ATP-consuming activities in the cytosol is therefore creatine phosphate, not ATP. The mCK isozyme colocalizes with the adenine nucleotide transporter (in the inner mitochondrial membrane) and porin (in the outer mitochondrial membrane), suggesting that these three components may function together to transport ATP formed in the mitochondria into the cytosol.

In knockout mice lacking the mitochondrial isozyme, myocytes compensate by producing more mitochondria, closely associated with myofibrils and sarcoplasmic reticulum, allowing quick diffusion of mitochondrial ATP to those sites of ATP use. Nevertheless, these mice have reduced capacity for running, indicating a defect in some aspect of energy-supplying metabolism.

Creatine and phosphocreatine spontaneously break down to form creatinine (Fig. 2). To maintain high creatine levels, these losses have to be replaced, either by dietary creatine, obtained primarily from meats (muscle) and dairy products, or by the de novo synthesis of creatine from glycine, arginine, and methionine (Fig. 22–28), which occurs primarily in liver and kidney. De novo synthesis of creatine is a major consumer of these amino acids, particularly in

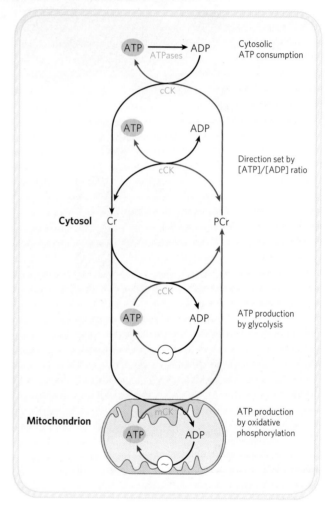

FIGURE 1 Mitochondrial creatine kinase (mCK) transfers a phosphoryl group from ATP to creatine (Cr) to form creatine phosphate (PCr), which diffuses to sites of ATP use, where cytosolic creatine kinase (cCK) then passes the phosphoryl group back into ATP. Cytosolic CK can also use ATP produced by glycolysis to synthesize PCr. During periods of little ATP demand, the pools of ATP and PCr are replenished in preparation for the next period of intense demand for ATP. In resting muscle, the concentration of PCr is 3 to 5 times that of ATP, buffering the cell against rapid depletion of ATP during short bursts of ATP demand.

aerobic respiration alone. Under these conditions, stored muscle glycogen is broken down to lactate by fermentation (p. 548). Each glucose unit degraded yields three ATP, because phosphorolysis of glycogen produces glucose 6-phosphate (via glucose 1-phosphate), sparing the ATP normally consumed in the hexokinase reaction. Lactic acid fermentation thus responds more quickly than oxidative phosphorylation to an increased need for ATP, supplementing basal ATP production by aerobic oxidation of other fuels via the citric acid cycle and respiratory chain. The use of blood glucose and muscle glycogen as fuels for muscular activity is greatly enhanced by the secretion of epinephrine, which stimulates both the release of glucose from liver glycogen and the breakdown of glycogen in muscle tissue. (Epinephrine mediates the so-called fight-or-flight response, discussed more fully below.)

The relatively small amount of glycogen (about 1% of the total weight of skeletal muscle) limits the amount of glycolytic energy available during all-out exertion. Moreover, the accumulation of lactate and consequent decrease in pH in maximally active muscles reduces their efficiency. Skeletal muscle, however, contains another source of ATP, phosphocreatine (10 to 30 mM),

FIGURE 2 Spontaneous (nonenzymatic) formation of creatinine from phosphocreatine or creatine consumes a few percent of total creatine per day, which must be replaced by biosynthesis or from the diet.

FIGURE 3 Body builders often take supplemental creatine to supply creatine phosphate in new muscle tissue.

vegans, for whom this is the only source of creatine; plants do not contain creatine. Muscle tissue has a specific system to take up creatine, exported by liver or kidney, from the blood into cells, against a substantial concentration gradient. Efficient uptake of dietary creatine requires continuous exercise; without exercise, creatine supplementation is of little value.

Heart muscle contains a unique isozyme of creatine kinase (the MB isozyme), which is not normally found in the blood but appears there when released from heart muscle damaged by a heart attack. The blood level of MB begins to rise within 2 hours of the heart attack, typically peaks 12 to 36 hours after the heart attack, and returns to normal levels in 3 to 5 days. Measurement of the MB isozyme in blood therefore confirms the diagnosis of a heart attack and indicates approximately when it occurred.

Body builders who are adding muscle mass have a greater need for creatine and commonly take creatine

supplements of up to 20 g per day for a few days, followed by lower maintenance doses. The combination of exercise and creatine supplementation increases muscle mass (Fig. 3) and improves performance in high-intensity, short-duration work. Children with inborn errors in the enzymes of creatine synthesis or uptake suffer severe intellectual disability and seizures. These individuals have much-reduced levels of brain creatine as measured by NMR (see Fig. 23–18). Creatine supplementation raises their brain creatine and creatine phosphate concentrations and brings about partial improvement of the symptoms.

In the healthy kidney, creatinine from creatine breakdown is efficiently cleared from the blood into the urine. When renal function is defective, creatinine levels in the blood rise above the normal range of 0.8 to 1.4 mg/dL. Elevated blood creatinine is associated with renal failure in diabetes and other conditions in which renal function is temporarily or permanently compromised. Renal clearance of creatinine varies slightly with age, race, and gender, so correcting the calculation for those factors yields a more sensitive measure of the extent of renal function, the **glomerular filtration rate (GFR)**.

which can rapidly regenerate ATP from ADP by the creatine kinase reaction:

During periods of active contraction and glycolysis, this reaction proceeds predominantly in the direction of ATP synthesis; during recovery from exertion, the same enzyme resynthesizes phosphocreatine from creatine and ATP. Given the relatively high levels of ATP and phosphocreatine in muscle, these compounds can be detected in intact muscle, in real time, by NMR spectroscopy **(Fig. 23–18)**. Creatine serves to shuttle ATP equivalents from the mitochondrion to sites of ATP consumption and can be the limiting factor in the development of new muscle tissue (Box 23–2).

FIGURE 23–18 Phosphocreatine buffers ATP concentration during exercise. A "stack plot" of magnetic resonance spectra (of ^{31}P) showing inorganic phosphate (P_i), phosphocreatine (PCr), and ATP (each of its three phosphates giving a signal). The series of plots represents the passage of time, from a period of rest to one of exercise, and then of recovery. Note that the ATP signal hardly changes during exercise, kept high by continued respiration and by the reservoir of phosphocreatine, which diminishes during exercise. During recovery, when ATP production by catabolism is greater than ATP utilization by the (now resting) muscle, the phosphocreatine reservoir is refilled.

After a period of intense muscular activity, the individual continues breathing heavily for some time, using much of the extra O_2 for oxidative phosphorylation in the liver. The ATP produced is used for gluconeogenesis (in the liver) from lactate that has been carried in the blood from the muscles. The glucose thus formed returns to the muscles to replenish their glycogen, completing the Cori cycle (**Fig. 23–19**; see also Box 15–4).

FIGURE 23–19 Metabolic cooperation between skeletal muscle and the liver: the Cori cycle. Extremely active muscles use glycogen as their energy source, generating lactate via glycolysis. During recovery, some of this lactate is transported to the liver and converted to glucose via gluconeogenesis. This glucose is released to the blood and returned to the muscles to replenish their glycogen stores. The overall pathway (glucose → lactate → glucose) constitutes the Cori cycle.

Actively contracting skeletal muscle generates heat as a byproduct of imperfect coupling of the chemical energy of ATP with the mechanical work of contraction. This heat production can be put to good use when ambient temperature is low: skeletal muscle carries out **shivering thermogenesis**, rapidly repeated muscle contraction that produces heat but little motion, helping to maintain the body at its preferred temperature of 37 °C.

Heart muscle differs from skeletal muscle in that it is continuously active in a regular rhythm of contraction and relaxation and it has a completely aerobic metabolism at all times. Mitochondria are much more abundant in heart muscle than in skeletal muscle, making up almost half the volume of the cells (**Fig. 23–20**). The heart uses mainly free fatty acids, but also some glucose and ketone bodies taken up from the blood, as sources of energy; these fuels are oxidized via the citric acid cycle and oxidative phosphorylation to generate ATP. Like skeletal muscle, heart muscle does not store lipids or glycogen in large amounts. It does have small amounts of reserve energy in the form of phosphocreatine, enough for a few seconds of contraction. Because the heart is normally aerobic and obtains its energy from oxidative phosphorylation, the failure of O_2 to reach part of the heart muscle when the blood vessels are blocked by lipid deposits (atherosclerosis) or blood clots (coronary thrombosis) can cause that region of the heart muscle to die. This is what happens in myocardial infarction, more commonly known as a heart attack. ■

The Brain Uses Energy for Transmission of Electrical Impulses

The metabolism of the brain is remarkable in several respects. The neurons of the adult mammalian brain

FIGURE 23–20 Electron micrograph of heart muscle. In the profuse mitochondria of heart tissue, pyruvate (from glucose), fatty acids, and ketone bodies are oxidized to drive ATP synthesis. This steady aerobic metabolism allows the human heart to pump blood at a rate of nearly 6 L/min, or about 350 L/h—or 200×10^6 L of blood over 70 years.

normally use only glucose as fuel **(Fig. 23–21)**. (Astrocytes, the other major cell type in the brain, can oxidize fatty acids.) The brain has a very active respiratory metabolism **(Fig. 23–22)**; it uses O_2 at a fairly constant rate, accounting for almost 20% of the total O_2 consumed by the body at rest. Because the brain contains very little glycogen, it is constantly dependent on incoming glucose in the blood. Should blood glucose fall significantly below a critical level for even a short time, severe and sometimes irreversible changes in brain function may result.

Although the neurons of the brain cannot directly use free fatty acids or lipids from the blood as fuels, they can, when necessary, use β-hydroxybutyrate (a ketone body), formed from fatty acids in the liver. The capacity of the brain to oxidize β-hydroxybutyrate via acetyl-CoA becomes important during prolonged fasting or starvation, after liver glycogen has been depleted, because it allows the brain to use body fat as an energy source. This spares muscle proteins—until they become the brain's ultimate source of glucose (via gluconeogenesis in the liver) during severe starvation.

Neurons oxidize glucose by glycolysis and the citric acid cycle, and the flow of electrons from these oxidations through the respiratory chain provides almost all the ATP used by these cells. Energy is required to create and maintain an electrical potential across the neuronal plasma membrane. The membrane contains an electrogenic ATP-driven antiporter, the Na^+K^+ ATPase, which simultaneously pumps 2 K^+ ions into and 3 Na^+ ions out of the neuron (see Fig. 11–38). The resulting transmembrane potential changes transiently as an electrical signal (action potential) sweeps from one end of a neuron to the other (see Fig. 12–26). Action poten-

(a)

(b)

12.00 ▮▮▮▮▮ 2.00
mg/100 g/min

FIGURE 23–22 Glucose metabolism in the brain. The technique of positron emission tomography (PET) scanning shows metabolic activity in specific regions of the brain. PET scans allow visualization of isotopically labeled glucose in precisely localized regions of the brain of a living person, in real time. A positron-emitting glucose analog (2-[^{18}F]-fluoro-2-deoxy-D-glucose) is injected into the bloodstream; a few seconds later, a PET scan shows how much of the glucose has been taken up by each region of the brain—a measure of metabolic activity. Shown here are PET scans of front-to-back cross sections of the brain at three levels, from the top (at the left) downward (to the right). The scans compare glucose metabolism when the experimental subject **(a)** is rested and **(b)** has been deprived of sleep for 48 hours.

tials are the chief mechanism of information transfer in the nervous system, so depletion of ATP in neurons would have disastrous effects on all activities coordinated by neuronal signaling.

Blood Carries Oxygen, Metabolites, and Hormones

Blood mediates the metabolic interactions among all tissues. It transports nutrients from the small intestine to the liver and from the liver and adipose tissue to other organs; it also transports waste products from the extrahepatic tissues to the liver for processing and to the kidneys for excretion. Oxygen moves in the bloodstream from the lungs to the tissues, and CO_2 generated by tissue respiration returns via the bloodstream to the lungs for exhalation. Blood also carries hormonal signals from one tissue to another. In its role as signal carrier, the circulatory system resembles the nervous system: both regulate and integrate the activities of different organs.

The average adult human has 5 to 6 L of blood. Almost half of this volume is occupied by three types of blood cells **(Fig. 23–23)**: **erythrocytes** (red cells), filled with hemoglobin and specialized for carrying O_2 and CO_2; much smaller numbers of **leukocytes**

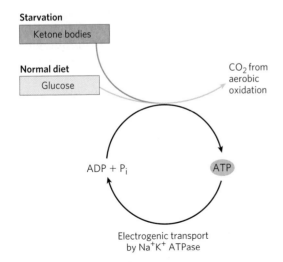

FIGURE 23–21 The fuels that supply ATP in the brain. The energy source used by the brain varies with nutritional state. The ketone body used during starvation is β-hydroxybutyrate. Electrogenic transport by the Na^+K^+ ATPase maintains the transmembrane potential essential to information transfer among neurons.

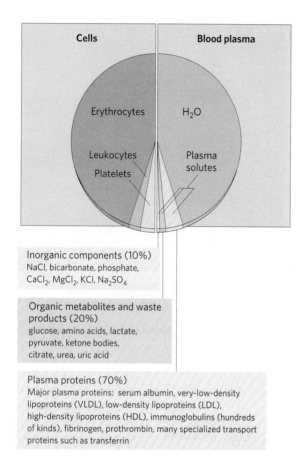

Inorganic components (10%)
NaCl, bicarbonate, phosphate, CaCl$_2$, MgCl$_2$, KCl, Na$_2$SO$_4$

Organic metabolites and waste products (20%)
glucose, amino acids, lactate, pyruvate, ketone bodies, citrate, urea, uric acid

Plasma proteins (70%)
Major plasma proteins: serum albumin, very-low-density lipoproteins (VLDL), low-density lipoproteins (LDL), high-density lipoproteins (HDL), immunoglobulins (hundreds of kinds), fibrinogen, prothrombin, many specialized transport proteins such as transferrin

FIGURE 23–23 The composition of blood (by weight). Whole blood can be separated into blood plasma and cells by centrifugation. About 10% of blood plasma is solutes, of which about 10% consists of inorganic salts, 20% small organic molecules, and 70% plasma proteins. The major dissolved components are listed. Blood contains many other substances, often in trace amounts. These include other metabolites, enzymes, hormones, vitamins, trace elements, and bile pigments. Measurements of the concentrations of components in blood plasma are important in the diagnosis and treatment of many diseases.

(white cells) of several types (including **lymphocytes**, also found in lymphatic tissue), which are central to the immune system to defend against infections; and **platelets**, which help to mediate blood clotting. The liquid portion is the **blood plasma**, which is 90% water and 10% solutes. Dissolved or suspended in the plasma is a large variety of proteins, lipoproteins, nutrients, metabolites, waste products, inorganic ions, and hormones. More than 70% of the plasma solids are **plasma proteins**, primarily immunoglobulins (circulating antibodies), serum albumin, apolipoproteins involved in the transport of lipids, transferrin (for iron transport), and blood-clotting proteins such as fibrinogen and prothrombin.

The ions and low molecular weight solutes in blood plasma are not fixed components but are in constant flux between blood and various tissues. Dietary uptake of the inorganic ions that are the predominant electrolytes of blood and cytosol (Na$^+$, K$^+$, and Ca^{2+}) is, in general, counterbalanced by their excretion in the urine. For many blood components, something near a dynamic steady state is achieved: the concentration of the component changes little, although a continuous flux occurs between the digestive tract, blood, and urine. The plasma levels of Na$^+$, K$^+$, and Ca^{2+} remain close to 140, 5, and 2.5 mM, respectively, with little change in response to dietary intake. Any significant departure from these values can result in serious illness or death. The kidneys play an especially important role in maintaining ion balance by selectively filtering waste products and excess ions out of the blood while preventing the loss of essential nutrients and ions.

The human erythrocyte loses its nucleus and mitochondria during differentiation. It therefore relies on glycolysis alone for its supply of ATP. The lactate produced by glycolysis returns to the liver, where gluconeogenesis converts it to glucose, to be stored as glycogen or recirculated to the peripheral tissues. The erythrocyte has constant access to glucose in the bloodstream.

The concentration of glucose in plasma is subject to tight regulation. We have noted the constant requirement of the brain for glucose and the role of the liver in maintaining blood glucose in the normal range, 60 to 90 mg/100 mL of whole blood (~4.5 mM). (Because erythrocytes make up a significant fraction of blood volume, their removal by centrifugation leaves a supernatant fluid, the plasma, containing the "blood glucose" in a smaller volume. To convert blood glucose to plasma glucose concentration, multiply the blood glucose level by 1.14.) When blood glucose in a human drops to 40 mg/100 mL (the hypoglycemic condition), the person experiences discomfort and mental confusion **(Fig. 23–24)**; further reductions lead to coma, convulsions, and, in extreme hypoglycemia, death. Maintaining the normal concentration of glucose in blood is therefore a very high priority of the organism, and a variety of regulatory mechanisms have evolved to achieve that end. Among the most important regulators of blood glucose are the hormones insulin, glucagon, and epinephrine, as discussed in the next section. ■

SUMMARY 23.2 Tissue-Specific Metabolism: The Division of Labor

▶ In mammals there is a division of metabolic labor among specialized tissues and organs. The liver is the central distributing and processing organ for nutrients. Sugars and amino acids produced in digestion cross the intestinal epithelium and enter the blood, which carries them to the liver. Some triacylglycerols derived from ingested lipids also make their way to the liver, where the constituent fatty acids are used in a variety of processes.

▶ Glucose 6-phosphate is the key intermediate in carbohydrate metabolism. It may be polymerized into glycogen, dephosphorylated to blood glucose, or converted to fatty acids via acetyl-CoA. It may

FIGURE 23–24 Physiological effects of low blood glucose in humans. Blood glucose levels of 40 mg/100 mL and below constitute severe hypoglycemia.

undergo oxidation by glycolysis, the citric acid cycle, and respiratory chain to yield ATP, or enter the pentose phosphate pathway to yield pentoses and NADPH.

▶ Amino acids are used to synthesize liver and plasma proteins, or their carbon skeletons are converted to glucose and glycogen by gluconeogenesis; the ammonia formed by deamination is converted to urea.

▶ The liver converts fatty acids to triacylglycerols, phospholipids, or cholesterol and its esters for transport as plasma lipoproteins to adipose tissue for storage. Fatty acids can also be oxidized to yield ATP or to form ketone bodies, which are circulated to other tissues.

▶ White adipose tissue stores large reserves of triacylglycerols and releases them into the blood in response to epinephrine or glucagon. Brown adipose tissue is specialized for thermogenesis, the result of fatty acid oxidation in uncoupled mitochondria.

▶ Skeletal muscle is specialized to produce and use ATP for mechanical work. During low to moderate muscular activity, oxidation of fatty acids and glucose is the primary source of ATP. During strenuous muscular activity, glycogen is the ultimate fuel, supplying ATP through lactic acid fermentation. During recovery, the lactate is reconverted (through gluconeogenesis) to glycogen and glucose in the liver for use in replenishing muscle glycogen supplies.

Phosphocreatine is an immediate source of ATP during active contraction.

▶ Heart muscle obtains nearly all its ATP from oxidative phosphorylation, with fatty acids as the primary fuel.

▶ The neurons of the brain use only glucose and β-hydroxybutyrate as fuels, the latter being important during fasting or starvation. The brain uses most of its ATP for the active transport of Na^+ and K^+ to maintain the electrical potential across the neuronal membrane.

▶ The blood transfers nutrients, waste products, and hormonal signals among tissues and organs. It is made up of cells (erythrocytes, leukocytes, and platelets) and electrolyte-rich water (plasma) containing many proteins.

23.3 Hormonal Regulation of Fuel Metabolism

The minute-by-minute adjustments that keep the blood glucose level near 4.5 mM involve the combined actions of insulin, glucagon, epinephrine, and cortisol on metabolic processes in many body tissues but especially in liver, muscle, and adipose tissue. Insulin signals these tissues that blood glucose is higher than necessary; as a result, cells take up excess glucose from the blood and convert it to glycogen and triacylglycerols for storage. Glucagon signals that blood glucose is too low, and tissues respond by producing glucose through glycogen breakdown and (in the liver) gluconeogenesis and by oxidizing fats to reduce the use of glucose. Epinephrine is released into the blood to prepare the muscles, lungs, and heart for a burst of activity. Cortisol mediates the body's response to longer-term stresses. We discuss these hormonal regulations in the context of three normal metabolic states—well fed, fasted, and starving—and look at the metabolic consequences of diabetes mellitus, a disorder that results from derangements in the signaling pathways that control glucose metabolism.

Insulin Counters High Blood Glucose

Acting through plasma membrane receptors (see Figs 12–15, 12–16), insulin stimulates glucose uptake by muscle and adipose tissue (Table 23–3), where the glucose is converted to glucose 6-phosphate. In the liver, insulin also activates glycogen synthase and inactivates glycogen phosphorylase, so that much of the glucose 6-phosphate is channeled into glycogen.

Insulin also stimulates the storage of excess fuel as fat in adipose tissue (**Fig. 23–25**). In the liver, insulin activates both the oxidation of glucose 6-phosphate to pyruvate via glycolysis and the oxidation of pyruvate to acetyl-CoA. The excess of acetyl-CoA not needed for energy production is used for fatty acid synthesis, and the fatty acids are exported from the liver as the TAGs

TABLE 23–3 Effects of Insulin on Blood Glucose: Uptake of Glucose by Cells and Storage as Triacylglycerols and Glycogen

Metabolic effect	Target enzyme
↑ Glucose uptake (muscle, adipose)	↑ Glucose transporter (GLUT4)
↑ Glucose uptake (liver)	↑ Glucokinase (increased expression)
↑ Glycogen synthesis (liver, muscle)	↑ Glycogen synthase
↓ Glycogen breakdown (liver, muscle)	↓ Glycogen phosphorylase
↑ Glycolysis, acetyl-CoA production (liver, muscle)	↑ PFK-1 (by ↑ PFK-2) ↑ Pyruvate dehydrogenase complex
↑ Fatty acid synthesis (liver)	↑ Acetyl-CoA carboxylase
↑ Triacylglycerol synthesis (adipose tissue)	↑ Lipoprotein lipase

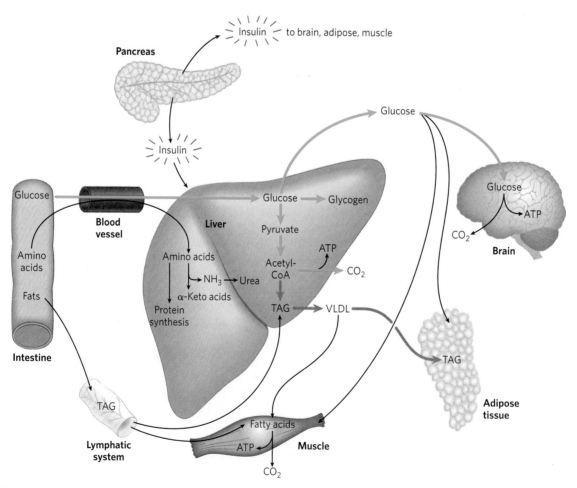

FIGURE 23–25 The well-fed state: the lipogenic liver. Immediately after a calorie-rich meal, glucose, fatty acids, and amino acids enter the liver. Blue arrows follow the path of glucose; orange arrows follow the path of lipids. Insulin released in response to the high blood glucose concentration stimulates glucose uptake by the tissues. Some glucose is exported to the brain for its energy needs and some to adipose and muscle tissue. In the liver, excess glucose is oxidized to acetyl-CoA, which is used to synthesize fatty acids for export as triacylglycerols in VLDLs to adipose and muscle tissue. The NADPH necessary for lipid synthesis is obtained by oxidation of glucose in the pentose phosphate pathway. Excess amino acids are converted to pyruvate and acetyl-CoA, which are also used for lipid synthesis. Dietary fats move via the lymphatic system, as chylomicrons, from the intestine to the liver, muscle, and adipose tissues.

of plasma lipoproteins (VLDL; see Fig. 21–40) to adipose tissue. Insulin stimulates the synthesis of TAGs in adipocytes, from fatty acids released from the VLDL triacylglycerols. These fatty acids are ultimately derived from the excess glucose taken up from blood by the liver. In summary, the effect of insulin is to favor the conversion of excess blood glucose to two storage forms: glycogen (in the liver and muscle) and triacylglycerols (in adipose tissue).

Besides acting directly on muscle and liver to change their metabolism of carbohydrates and fats, insulin can act in the brain to signal these tissues indirectly, as described later.

Pancreatic β Cells Secrete Insulin in Response to Changes in Blood Glucose

When glucose enters the bloodstream from the intestine after a carbohydrate-rich meal, the resulting increase in blood glucose causes increased secretion of insulin (and decreased secretion of glucagon) by the pancreas. Insulin release is largely regulated by the level of glucose in the blood supplying the pancreas. The peptide hormones insulin, glucagon, and somatostatin are produced by clusters of specialized pancreatic cells, the islets of Langerhans **(Fig. 23–26)**. Each cell type of the islets produces a single hormone: α cells produce glucagon; β cells, insulin; and δ cells, somatostatin.

As shown in **Figure 23–27**, when blood glucose rises, ❶ GLUT2 transporters carry glucose into the

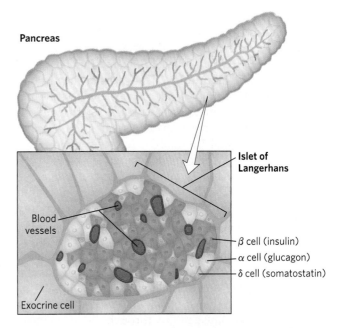

FIGURE 23–26 The endocrine system of the pancreas. The pancreas contains both exocrine cells (see Fig. 18–3b), which secrete digestive enzymes in the form of zymogens, and clusters of endocrine cells, the islets of Langerhans. The islets contain α, β, and δ cells (also known as A, B, and D cells, respectively), each cell type secreting a specific peptide hormone.

FIGURE 23–27 Glucose regulation of insulin secretion by pancreatic β cells. When the blood glucose level is high, active metabolism of glucose in the β cell raises intracellular [ATP], closing K$^+$ channels in the plasma membrane and thus depolarizing the membrane. In response to this membrane-depolarization triggered by high [ATP], voltage-gated Ca^{2+} channels open, allowing Ca^{2+} to flow into the cell. (Ca^{2+} is also released from the endoplasmic reticulum, in response to the initial elevation of [Ca^{2+}] in the cytosol.) Cytosolic [Ca^{2+}] is now high enough to trigger insulin release by exocytosis. The numbered processes are discussed in the text.

β cells, where it is immediately converted to glucose 6-phosphate by hexokinase IV (glucokinase) and enters glycolysis. With the higher rate of glucose catabolism, ❷ [ATP] increases, causing the closing of **ATP-gated K⁺ channels** in the plasma membrane. ❸ Reduced efflux of K⁺ depolarizes the membrane. (Recall from Section 12.6 that exit of K⁺ through an open K⁺ channel hyperpolarizes the membrane; closing the K⁺ channel therefore effectively depolarizes the membrane.) Membrane depolarization opens voltage-gated Ca^{2+} channels, and ❹ the resulting increase in cytosolic $[Ca^{2+}]$ triggers ❺ the release of insulin by exocytosis. The brain integrates inputs on energy supply and demand, and signals from the parasympathetic and sympathetic nervous system also affect (stimulate and inhibit, respectively) insulin release. A simple feedback loop limits hormone release: insulin lowers blood glucose by stimulating glucose uptake by the tissues; the reduced blood glucose is detected by the β cell as a diminished flux through the hexokinase reaction; this slows or stops the release of insulin. This feedback regulation holds blood glucose concentration nearly constant despite large fluctuations in dietary intake.

The activity of ATP-gated K⁺ channels is central to the regulation of insulin secretion by β cells. The channels are octamers of four identical Kir6.2 subunits and four identical SUR1 subunits and are constructed along the same lines as the K⁺ channels of bacteria and those of other eukaryotic cells (see Figs 11–47 and 11–48). The four Kir6.2 subunits form a cone around the K⁺ channel and function as the selectivity filter and ATP-gating mechanism **(Fig. 23–28)**. When [ATP] rises (indicating increased blood glucose), the K⁺ channels close, depolarizing the plasma membrane and triggering insulin release as shown in Figure 23–27.

The **sulfonylurea drugs**, oral medications used in the treatment of type 2 diabetes mellitus, bind to the SUR1 (sulfonylurea *r*eceptor) subunits of the K⁺ channels, closing the channels and stimulating insulin release. The first generation of these drugs (tolbutamide, for example) was developed in the 1950s. The second-generation drugs, including glyburide (Micronase), glipizide (Glucotrol), and glimepiride (Amaryl), are more potent and have fewer side effects. (The sulfonylurea moiety is screened light red in the following structures.) The sulfonylureas are sometime used in combination with injected insulin but often suffice alone for controlling type 2 diabetes.

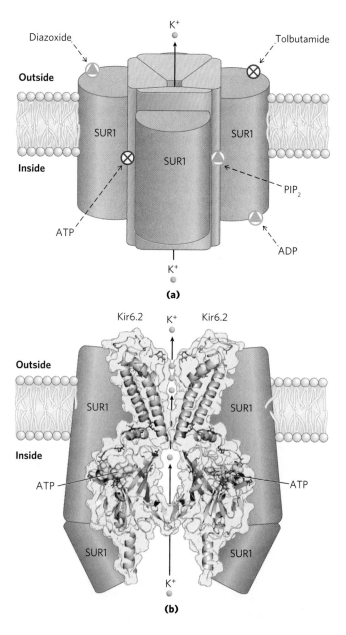

FIGURE 23–28 ATP-gated K⁺ channels in β cells. (a) The ATP-gated channel, viewed in the plane of the membrane. The channel is formed by four identical Kir6.2 subunits, and outside those are four SUR1 (sulfonylurea receptor) subunits. The SUR1 subunits have binding sites for ADP and the drug diazoxide, which favor the open channel, and tolbutamide, a sulfonylurea drug that favors the closed channel. The Kir6.2 subunits constitute the actual channel, and they contain, on the cytosolic side, binding sites for ATP and phosphatidylinositol 4,5-bisphosphate (PIP₂), which favor the closed and the open channel, respectively. **(b)** The structure of the Kir6.2 portion of the channel, viewed in the plane of the membrane. For clarity, only two transmembrane domains and two cytosolic domains are shown. Three K⁺ ions (green) are shown in the region of the selectivity filter. Mutation in certain amino acid residues (shown in red) leads to neonatal diabetes; mutation in others (shown in blue) leads to hyperinsulinism of infancy. This structure (coordinates courtesy of Frances Ashford and her colleagues at the University of Oxford) was obtained not directly by crystallography but by mapping the known Kir6.2 sequence onto the crystal structures of a bacterial Kir channel (KirBac1.1; PDB ID 1P7B) and the amino and carboxyl domains of another Kir protein, Kir3.1 (PDB ID 1U4E). Compare this structure with the gated K⁺ channel in Figure 11–48.

Glyburide

Glipizide

Mutations in the ATP-gated K^+ channels of β cells are, fortunately, rare. Mutations in Kir6.2 that result in constantly *open* K^+ channels (red residues in Fig. 23–28b) lead to neonatal diabetes mellitus, with severe hyperglycemia that requires insulin therapy. Other mutations in Kir6.2 or SUR1 (blue residues in Fig. 23–28b) produce permanently *closed* K^+ channels and continuous release of insulin. If untreated, individuals with these mutations develop congenital hyperinsulinemia (hyperinsulinism of infancy); excessive insulin causes severe hypoglycemia (low blood glucose) leading to irreversible brain damage. One effective treatment is surgical removal of part of the pancreas to reduce insulin production. ■

Glucagon Counters Low Blood Glucose

Several hours after the intake of dietary carbohydrate, blood glucose levels fall slightly because of the ongoing oxidation of glucose by the brain and other tissues. Lowered blood glucose triggers secretion of **glucagon** and decreases insulin release **(Fig. 23–29)**.

Glucagon causes an *increase* in blood glucose concentration in several ways (Table 23–4). Like epinephrine, it stimulates the net breakdown of liver glycogen by activating glycogen phosphorylase and inactivating glycogen synthase; both effects are the result of phosphorylation of the regulated enzymes, triggered by cAMP. Glucagon inhibits glucose breakdown by glycolysis in the liver and stimulates glucose synthesis by gluconeogenesis. Both effects result from lowering the concentration of fructose 2,6-bisphosphate, an allosteric inhibitor of the gluconeogenic enzyme fructose 1,6-bisphosphatase (FBPase-1) and an activator of the glycolytic enzyme phosphofructokinase-1. Recall that [fructose 2,6-bisphosphate] is ultimately controlled by a cAMP-dependent protein phosphorylation reaction (see Fig. 15–19) Glucagon also inhibits the glycolytic enzyme pyruvate kinase (by promoting its cAMP-dependent phosphorylation), thus blocking the conversion of phosphoenolpyruvate to pyruvate and preventing oxidation of pyruvate via the citric acid cycle. The resulting

FIGURE 23–29 The fasting state: the glucogenic liver. After some hours without a meal, the liver becomes the principal source of glucose for the brain. Liver glycogen is broken down, and the glucose 1-phosphate produced is converted to glucose 6-phosphate, then to free glucose, which is released into the bloodstream. Amino acids from the degradation of proteins in liver and muscle, and glycerol from the breakdown of TAGs in adipose tissue, are used for gluconeogenesis. The liver uses fatty acids as its principal fuel, and excess acetyl-CoA is converted to ketone bodies for export to other tissues; the brain is especially dependent on this fuel when glucose is in short supply (see Fig. 23–21). (Blue arrows follow the path of glucose; orange arrows, the path of lipids; and purple arrows, the path of amino acids.)

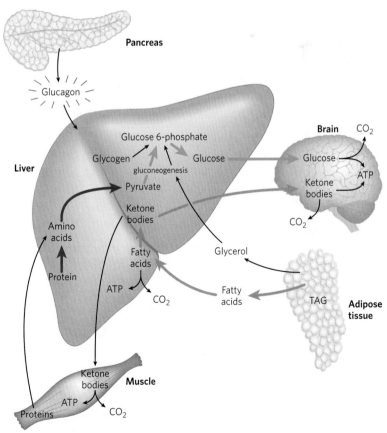

TABLE 23–4 Effects of Glucagon on Blood Glucose: Production and Release of Glucose by the Liver

Metabolic effect	Effect on glucose metabolism	Target enzyme
↑ Glycogen breakdown (liver)	Glycogen ⟶ glucose	↑ Glycogen phosphorylase
↓ Glycogen synthesis (liver)	Less glucose stored as glycogen	↓ Glycogen synthase
↓ Glycolysis (liver)	Less glucose used as fuel in liver	↓ PFK-1
↑ Gluconeogenesis (liver)	Amino acids Glycerol ⎬ ⟶ glucose Oxaloacetate	↑ FBPase-2 ↓ Pyruvate kinase ↑ PEP carboxykinase
↑ Fatty acid mobilization (adipose tissue)	Less glucose used as fuel by liver, muscle	↑ Hormone-sensitive lipase ↑ PKA (perilipin–Ⓟ)
↑ Ketogenesis	Provides alternative to glucose as energy source for brain	↓ Acetyl-CoA carboxylase

accumulation of phosphoenolpyruvate favors gluconeogenesis. This effect is augmented by glucagon's stimulation of the synthesis of the gluconeogenic enzyme PEP carboxykinase. By stimulating glycogen breakdown, preventing glycolysis, and promoting gluconeogenesis in hepatocytes, glucagon enables the liver to export glucose, restoring blood glucose to its normal level.

Although its primary target is the liver, glucagon (like epinephrine) also affects adipose tissue, activating TAG breakdown by causing cAMP-dependent phosphorylation of perilipin and hormone-sensitive lipase. The activated lipase liberates free fatty acids, which are exported to the liver and other tissues as fuel, sparing glucose for the brain. The net effect of glucagon is therefore to stimulate glucose synthesis and release by the liver and to mobilize fatty acids from adipose tissue, to be used instead of glucose by tissues other than the brain. All these effects of glucagon are mediated by cAMP-dependent protein phosphorylation.

During Fasting and Starvation, Metabolism Shifts to Provide Fuel for the Brain

The fuel reserves of a healthy adult human are of three types: glycogen stored in the liver and, in smaller quantities, in muscles; large quantities of triacylglycerols in adipose tissues; and tissue proteins, which can be degraded when necessary to provide fuel (Table 23–5).

TABLE 23–5 Available Metabolic Fuels in a Normal-Weight, 70 kg Man and in an Obese, 140 kg Man at the Beginning of a Fast

Type of fuel	Weight (kg)	Caloric equivalent (thousands of kcal (kJ))	Estimated survival (months)*
Normal-weight, 70 kg man			
Triacylglycerols (adipose tissue)	15	140 (590)	
Proteins (mainly muscle)	6	24 (100)	
Glycogen (muscle, liver)	0.23	0.90 (3.8)	
Circulating fuels (glucose, fatty acids, triacylglycerols, etc.)	0.023	0.10 (0.42)	
Total		165 (690)	3
Obese, 140 kg man			
Triacylglycerols (adipose tissue)	80	750 (3,100)	
Proteins (mainly muscle)	8	32 (130)	
Glycogen (muscle, liver)	0.23	0.92 (3.8)	
Circulating fuels	0.025	0.11 (0.46)	
Total		783 (3,200)	14

*Survival time is calculated on the assumption of a basal energy expenditure of 1,800 kcal/day.

Two hours after a meal, the blood glucose level is diminished slightly, and tissues receive glucose released from liver glycogen. There is little or no synthesis of triacylglycerols. By four hours after a meal, blood glucose has fallen further, insulin secretion has slowed, and glucagon secretion has increased. These hormonal signals mobilize triacylglycerols from adipose tissue, which now become the primary fuel for muscle and liver. **Figure 23–30** shows the responses to prolonged fasting. ❶ To provide glucose for the brain, the liver degrades certain proteins—those most expendable in an organism not ingesting food. Their nonessential amino acids are transaminated or deaminated (Chapter 18), and

❷ the extra amino groups are converted to urea, which is exported via the bloodstream to the kidneys and excreted in the urine.

Also in the liver, and to some extent in the kidneys, the carbon skeletons of glucogenic amino acids are converted to pyruvate or intermediates of the citric acid cycle. ❸ These intermediates (as well as the glycerol derived from TAGs in adipose tissue) provide the starting materials for gluconeogenesis in the liver, ❹ yielding glucose for export to the brain. ❺ Fatty acids released from adipose tissue are oxidized to acetyl-CoA in the liver, but as oxaloacetate is depleted by the use of citric acid cycle intermediates for gluconeogenesis, ❻ entry of

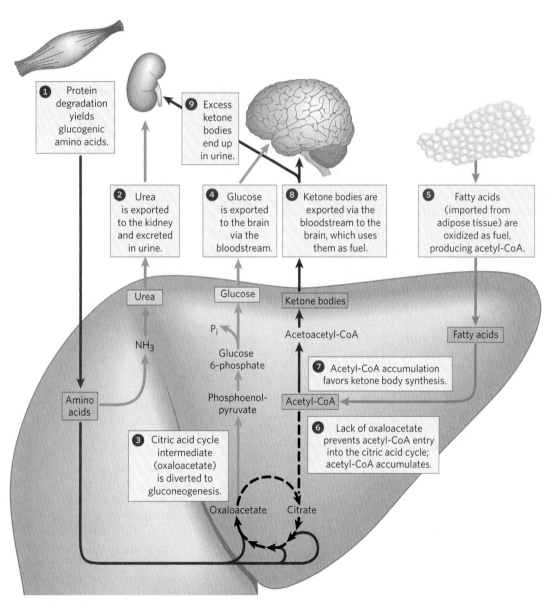

FIGURE 23–30 Fuel metabolism in the liver during prolonged fasting or in uncontrolled diabetes mellitus. The numbered steps are described in the text. After depletion of stored carbohydrates (glycogen), gluconeogenesis in the liver becomes the main source of glucose for the brain (blue arrows). NH₃ from amino acid deamination is converted into urea and excreted (green arrows). Glucogenic amino acids from protein breakdown (purple arrows) provide substrates for gluconeogenesis), and glu-cose is exported to the brain. Fatty acids from adipose tissue are imported into the liver and oxidized to acetyl-CoA (orange arrows), and acetyl-CoA is the starting material for ketone body formation in the liver and export to brain to serve as the principal energy source (red arrows). When the concentration of ketone bodies in the blood exceeds the ability of the kidneys to reabsorb ketones, these compounds begin to appear in the urine.

acetyl-CoA into the cycle is inhibited and acetyl-CoA accumulates. ❼ This favors the formation of acetoacetyl-CoA and ketone bodies. After a few days of fasting, the levels of ketone bodies in the blood rise (**Fig. 23–31**) as they are exported from the liver to the heart, skeletal muscle, and brain, which use these fuels instead of glucose (Fig. 23–30, ❽).

Acetyl-CoA is a critical regulator of the fate of pyruvate: it allosterically inhibits pyruvate dehydrogenase and stimulates pyruvate carboxylase. In these ways, acetyl-CoA prevents its own further production from pyruvate while stimulating the conversion of pyruvate to oxaloacetate, the first step in gluconeogenesis.

Triacylglycerols stored in the adipose tissue of a normal-weight adult could provide enough fuel to maintain a basal rate of metabolism for about three months; a very obese adult has enough stored fuel to endure a fast of more than a year (Table 23–5). When fat reserves are gone, the degradation of *essential* proteins begins; this leads to loss of heart and liver function and, in prolonged starvation, to death. Stored fat can provide adequate energy (calories) during a fast or rigid diet, but vitamins and minerals must be provided, and sufficient dietary glucogenic amino acids are needed to replace those being used for gluconeogenesis. Rations for those on a weight-reduction diet are commonly fortified with vitamins, minerals, and amino acids or proteins.

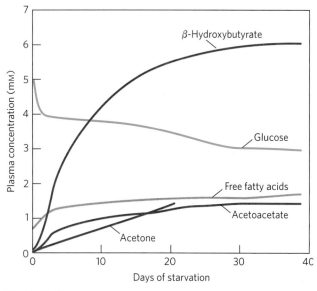

FIGURE 23–31 Plasma concentrations of fatty acids, glucose, and ketone bodies during six weeks of starvation. Despite the hormonal mechanisms for maintaining the level of glucose in blood, glucose begins to diminish within 2 days of fasting (blue). The level of ketone bodies, almost unmeasurable before the fast, rises dramatically after 2 to 4 days of fasting (red), with β-hydroxybutyrate as the major contributor. These water-soluble ketones, acetoacetate and β-hydroxybutyrate, supplement glucose as an energy source for the brain during a long fast. Acetone, a minor ketone body, is not metabolized but is eliminated in the breath. A much smaller rise in blood fatty acids (orange) also occurs, but this rise does not contribute to energy metabolism in the brain, as fatty acids do not cross the blood-brain barrier.

Epinephrine Signals Impending Activity

When an animal is confronted with a stressful situation that requires increased activity—fighting or fleeing, in the extreme case—neuronal signals from the brain trigger the release of epinephrine and norepinephrine from the adrenal medulla. Both hormones dilate the respiratory passages to facilitate the uptake of O_2, increase the rate and strength of the heartbeat, and raise the blood pressure, thereby promoting the flow of O_2 and fuels to the tissues (Table 23–6). This is the "fight or flight" response.

Epinephrine acts primarily on muscle, adipose, and liver tissues. It activates glycogen phosphorylase and inactivates glycogen synthase by cAMP-dependent phosphorylation of the enzymes, thus stimulating the conversion of *liver* glycogen to blood glucose, the fuel for anaerobic muscular work. Epinephrine also promotes the anaerobic breakdown of *muscle* glycogen by lactic acid fermentation, stimulating glycolytic ATP formation. The stimulation of glycolysis is accomplished by raising the concentration of fructose 2,6-bisphosphate, a potent allosteric activator of the key glycolytic enzyme phosphofructokinase-1. Epinephrine also stimulates fat mobilization in adipose tissue, activating (by cAMP-dependent phosphorylation) hormone-sensitive lipase and moving aside the perilipin covering the lipid droplet surface (see Fig. 17–3). Finally, epinephrine stimulates glucagon secretion and inhibits insulin secretion, reinforcing its effect of mobilizing fuels and inhibiting fuel storage.

Cortisol Signals Stress, Including Low Blood Glucose

A variety of stressors (anxiety, fear, pain, hemorrhage, infection, low blood glucose, starvation) stimulate release of the corticosteroid hormone **cortisol** from the adrenal cortex. Cortisol acts on muscle, liver, and adipose tissue to supply the organism with fuel to withstand the stress. Cortisol is a relatively slow-acting hormone that alters metabolism by changing the kinds and amounts of certain enzymes synthesized in its target cells rather than by regulating the activity of existing enzyme molecules.

In adipose tissue, cortisol leads to an increase in the release of fatty acids from stored TAGs. The exported fatty acids serve as fuel for other tissues, and the glycerol is used for gluconeogenesis in the liver. Cortisol stimulates the breakdown of nonessential muscle proteins and the export of amino acids to the liver, where they serve as precursors for gluconeogenesis. In the liver, cortisol promotes gluconeogenesis by stimulating synthesis of the key enzyme PEP carboxykinase; glucagon has the same effect, whereas insulin has the opposite effect. Glucose produced in this way is stored in the liver as glycogen or exported immediately to tissues that need glucose for fuel. The net effect of these metabolic changes is to restore blood glucose to its normal level and to increase glycogen stores, ready to support the fight-or-flight response commonly associated with stress. The effects of

TABLE 23-6	Physiological and Metabolic Effects of Epinephrine: Preparation for Action	
Immediate effect	**Overall effect**	
Physiological		
↑ Heart rate		
↑ Blood pressure	Increase delivery of O_2 to tissues (muscle)	
↑ Dilation of respiratory passages		
Metabolic		
↑ Glycogen breakdown (muscle, liver)		
↓ Glycogen synthesis (muscle, liver)	Increase production of glucose for fuel	
↑ Gluconeogenesis (liver)		
↑ Glycolysis (muscle)	Increases ATP production in muscle	
↑ Fatty acid mobilization (adipose tissue)	Increases availability of fatty acids as fuel	
↑ Glucagon secretion		
↓ Insulin secretion	Reinforce metabolic effects of epinephrine	

cortisol therefore counterbalance those of insulin. During extended periods of stress, the continued release of cortisol loses its positive adaptive value and begins to cause damage to muscle and bone and to impair endocrine and immune function.

Diabetes Mellitus Arises from Defects in Insulin Production or Action

Diabetes mellitus is a relatively common disease: nearly 6% of the U.S. population shows some degree of abnormality in glucose metabolism that is indicative of diabetes or a tendency toward the condition. There are two major clinical classes of diabetes mellitus: **type 1 diabetes**, sometimes referred to as insulin-dependent diabetes mellitus (IDDM), and **type 2 diabetes**, or non–insulin-dependent diabetes mellitus (NIDDM), also called insulin-resistant diabetes.

Type 1 diabetes begins early in life, and symptoms quickly become severe. This disease responds to insulin injection, because the metabolic defect stems from an autoimmune destruction of pancreatic β cells and a consequent inability to produce sufficient insulin. Type 1 diabetes requires both insulin therapy and careful, lifelong control of the balance between dietary intake and insulin dose. Characteristic symptoms of type 1 (and type 2) diabetes are excessive thirst and frequent urination (polyuria), leading to the intake of large volumes of water (polydipsia) ("diabetes mellitus" means "excessive excretion of sweet urine"). These symptoms are due to the excretion of large amounts of glucose in the urine, a condition known as **glucosuria**.

Type 2 diabetes is slow to develop (typically in older, obese individuals), and the symptoms are milder and often go unrecognized at first. This is really a group of diseases in which the regulatory activity of insulin is disordered: insulin is produced, but some feature of the insulin-response system is defective. Individuals with this disorder are insulin-resistant. The connection between type 2 diabetes and obesity (discussed below) is an active and exciting area of research.

Individuals with either type of diabetes are unable to take up glucose efficiently from the blood; recall that insulin triggers the movement of GLUT4 glucose transporters to the plasma membrane in muscle and adipose tissue (see Fig. 12–16). With glucose unavailable, fatty acids become the principal fuel, which leads to another characteristic metabolic change in diabetes: excessive but incomplete oxidation of fatty acids in the liver. The acetyl-CoA produced by β oxidation cannot be completely oxidized by the citric acid cycle, because the high $[NADH]/[NAD^+]$ ratio produced by β oxidation inhibits the cycle (recall that three steps of the cycle convert NAD^+ to NADH). Accumulation of acetyl-CoA leads to overproduction of the ketone bodies, acetoacetate and β-hydroxybutyrate, which cannot be used by extrahepatic tissues as fast as they are made in the liver. In addition to β-hydroxybutyrate and acetoacetate, the blood of individuals with diabetes also contains acetone, which results from the spontaneous decarboxylation of acetoacetate:

$$\underset{\text{Acetoacetate}}{H_3C-\overset{\overset{\displaystyle O}{\|}}{C}-CH_2-COO^- } + H_2O \longrightarrow \underset{\text{Acetone}}{H_3C-\overset{\overset{\displaystyle O}{\|}}{C}-CH_3} + HCO_3^-$$

Acetone is volatile and is exhaled, and in uncontrolled diabetes the breath has a characteristic odor sometimes mistaken for ethanol. A diabetic individual who is experiencing mental confusion due to high blood glucose is occasionally misdiagnosed as intoxicated, an error that can be fatal. The overproduction of ketone bodies, called **ketosis**, results in greatly increased concentrations of ketone bodies in the blood (ketonemia) and urine (ketonuria).

The ketone bodies are carboxylic acids, which ionize, releasing protons. In uncontrolled diabetes this acid production can overwhelm the capacity of the blood's bicarbonate buffering system and produce a lowering of blood pH called **acidosis** or, in combination with ketosis, **ketoacidosis**, a potentially life-threatening condition.

Biochemical measurements on blood and urine samples are essential in the diagnosis and treatment of diabetes. A sensitive diagnostic criterion is provided by the **glucose-tolerance test**. The individual fasts overnight, then drinks a test dose of 100 g of glucose dissolved in a glass of water. The blood glucose concentration is measured before the test dose and at 30 min intervals for several hours thereafter. A healthy individual assimilates the glucose readily, the blood glucose rising to no more than about 9 or 10 mM; little or no glucose appears in the urine. In diabetes, individuals assimilate the test dose of glucose poorly; their blood glucose level rises dramatically and returns to the fasting level very slowly. Because the blood glucose levels exceed the kidney threshold (about 10 mM), glucose also appears in the urine. ■

SUMMARY 23.3 Hormonal Regulation of Fuel Metabolism

▶ The concentration of glucose in blood is hormonally regulated. Fluctuations in blood glucose (normally 60 to 90 mg/100 mL, or about 4.5 mM) due to dietary intake or vigorous exercise are counterbalanced by a variety of hormonally triggered changes in metabolism in several organs.

▶ High blood glucose elicits the release of insulin, which speeds the uptake of glucose by tissues and favors the storage of fuels as glycogen and triacylglycerols while inhibiting fatty acid mobilization in adipose tissue.

▶ Low blood glucose triggers release of glucagon, which stimulates glucose release from liver glycogen and shifts fuel metabolism in liver and muscle to fatty acid oxidation, sparing glucose for use by the brain. In prolonged fasting, triacylglycerols become the principal fuel; the liver converts the fatty acids to ketone bodies for export to other tissues, including the brain.

▶ Epinephrine prepares the body for increased activity by mobilizing glucose from glycogen and other precursors, releasing it into the blood.

▶ Cortisol, released in response to a variety of stressors (including low blood glucose), stimulates gluconeogenesis from amino acids and glycerol in the liver, thus raising blood glucose and counterbalancing the effects of insulin.

▶ In diabetes, insulin is either not produced or not recognized by the tissues, and the uptake of blood glucose is compromised. When blood glucose levels are high, glucose is excreted. Tissues then depend on fatty acids for fuel (producing ketone bodies) and degrade cellular proteins to provide glucogenic amino acids for glucose synthesis. Uncontrolled diabetes is characterized by high glucose levels in the blood and urine and the production and excretion of ketone bodies.

23.4 Obesity and the Regulation of Body Mass

In the U.S. population, 30% of adults are obese and another 35% are overweight, as defined in terms of body mass index (BMI), calculated as (weight in kg)/(height in m)^2. A BMI below 25 is considered normal; an individual with a BMI of 25 to 30 is overweight; a BMI greater than 30 indicates **obesity**. Obesity is life-threatening. It significantly increases the chances of developing type 2 diabetes as well as heart attack, stroke, and cancers of the colon, breast, prostate, and endometrium. Consequently, there is great interest in understanding how body mass and the storage of fats in adipose tissue are regulated. ■

To a first approximation, obesity is the result of taking in more calories in the diet than are expended by the body's fuel-consuming activities. The body can deal with an excess of dietary calories in three ways: (1) convert excess fuel to fat and store it in adipose tissue, (2) burn excess fuel by extra exercise, and (3) "waste" fuel by diverting it to heat production (thermogenesis) by uncoupled mitochondria. In mammals, a complex set of hormonal and neuronal signals acts to keep fuel intake and energy expenditure in balance so as to hold the amount of adipose tissue at a suitable level. Dealing effectively with obesity requires understanding these various checks and balances under normal conditions and how these homeostatic mechanisms can fail.

Adipose Tissue Has Important Endocrine Functions

One early hypothesis to explain body-mass homeostasis, the "adiposity negative-feedback" model, postulated a mechanism that inhibits eating behavior and increases energy expenditure whenever body weight exceeds a certain value (the set point); the inhibition is relieved when body weight drops below the set point (**Fig. 23–32**). This model predicts that a feedback signal originating in adipose tissue influences the brain centers that control eating behavior and activity (metabolic and motor activity). The first such factor, leptin, was discovered in 1994, and subsequent research revealed that adipose tissue is an important endocrine organ that produces peptide hormones, known as **adipokines**. Adipokines may act locally (autocrine and paracrine action) or systemically (endocrine action), carrying information about the adequacy of the energy reserves (TAGs) stored in adipose tissue to other tissues and to the brain. Normally, adipokines produce changes in fuel metabolism and feeding behavior that reestablish adequate fuel reserves and maintain body mass. When adipokines are over- or underproduced, the resulting dysregulation may result in life-threatening disease.

FIGURE 23-32 Set-point model for maintaining constant mass. When the mass of adipose tissue increases (dashed outline), released leptin inhibits feeding and fat synthesis and stimulates oxidation of fatty acids. When the mass of adipose tissue decreases (solid outline), lowered leptin production favors greater food intake and less fatty acid oxidation.

Leptin (Greek *leptos*, "thin") is an adipokine (167 amino acids) that, on reaching the brain, acts on receptors in the hypothalamus to curtail appetite. Leptin was first identified as the product of a gene designated *OB* (obese) in laboratory mice. Mice with two defective copies of this gene (*ob/ob* genotype; lowercase letters signify a mutant form of the gene) show the behavior and physiology of animals in a constant state of starvation: their plasma cortisol levels are elevated; they are unable to stay warm, exhibit unrestrained appetite, grow abnormally, and do not reproduce. As a consequence of the unrestrained appetite, they become severely obese, weighing as much as three times more than normal mice **(Fig. 23–33)**. They also have metabolic disturbances very similar to those seen in diabetes, and they are insulin-resistant. When leptin is injected into *ob/ob* mice, they eat less, lose weight, and increase their locomotor activity and thermogenesis.

FIGURE 23-33 Obesity caused by defective leptin production. Both of these mice, which are the same age, have defects in the *OB* gene. The mouse on the right was injected daily with purified leptin and weighs 35 g. The mouse on the left got no leptin and consequently ate more food and was less active; it weighs 67 g.

A second mouse gene, designated *DB* (diabetic), also has a role in appetite regulation. Mice with two defective copies (*db/db*) are obese and diabetic. The *DB* gene encodes the **leptin receptor**. When the receptor is defective, the signaling function of leptin is lost.

The leptin receptor is expressed primarily in regions of the brain known to regulate feeding behavior—neurons of the arcuate nucleus of the hypothalamus **(Fig. 23–34a)**. Leptin carries the message that fat reserves are sufficient, and it promotes a reduction in

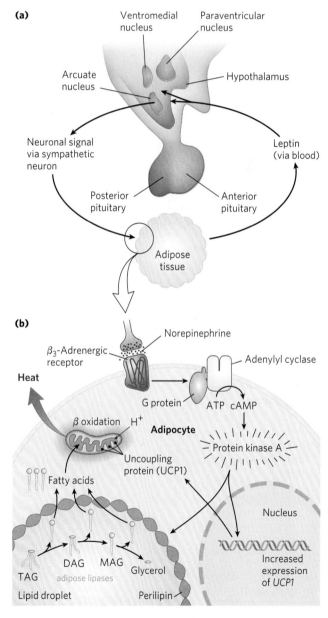

FIGURE 23-34 Hypothalamic regulation of food intake and energy expenditure. (a) Role of the hypothalamus in its interaction with adipose tissue. The hypothalamus receives input (leptin) from adipose tissue and responds with neuronal signals to adipocytes. **(b)** Effects of neuronal stimulation of the adipocyte include activation of protein kinase A, which triggers mobilization of fatty acids from TAG and their uncoupled oxidation in mitochondria, generating heat but not ATP.

fuel intake and increased expenditure of energy. Leptin-receptor interaction in the hypothalamus alters the release of neuronal signals to the region of the brain that affects appetite. Leptin also stimulates the sympathetic nervous system, increasing blood pressure, heart rate, and thermogenesis by uncoupling the mitochondria of brown adipocytes (Fig. 23–34b). Recall that thermogenin, or UCP1, forms a channel in the inner mitochondrial membrane that allows protons to reenter the mitochondrial matrix without passing through the ATP synthase complex. This permits continual oxidation of fuel (fatty acids in a brown adipocyte) without ATP synthesis, dissipating energy as heat and consuming dietary calories or stored fats in potentially very large amounts.

Leptin Stimulates Production of Anorexigenic Peptide Hormones

Two types of neurons in the arcuate nucleus control fuel intake and metabolism **(Fig. 23–35)**. The **orexigenic** (appetite-stimulating) neurons stimulate eating by producing and releasing **neuropeptide Y (NPY)**, which causes the next neuron in the circuit to send the signal to the brain: Eat! The blood level of NPY rises during starvation and is elevated in both *ob/ob* and *db/db* mice. The high NPY concentration presumably contributes to the obesity of these mice, who eat voraciously.

The **anorexigenic** (appetite-suppressing) neurons in the arcuate nucleus produce **α-melanocyte–stimulating hormone (α-MSH**; also known as melanocortin), formed from its polypeptide precursor pro-opiomelanocortin (POMC; Fig. 23–5). Release of α-MSH causes the next neuron in the circuit to send the signal to the brain: Stop eating!

The amount of leptin released by adipose tissue depends on both the number and the size of adipocytes. When weight loss decreases the mass of lipid tissue, leptin levels in the blood decrease, the production of NPY increases, and the processes in adipose tissue shown in Figure 23–34 are reversed. Uncoupling is diminished, slowing thermogenesis and saving fuel, and fat mobilization slows in response to reduced signaling by cAMP. Consumption of more food combined with more efficient utilization of fuel results in replenishment of the fat reserve in adipose tissue, bringing the system back into balance.

Leptin may also be essential to the normal development of hypothalamic neuronal circuits. In mice, the outgrowth of nerve fibers from the arcuate nucleus during early brain development is slower in the absence of leptin, affecting both the orexigenic and (to a lesser extent) anorexigenic outputs of the hypothalamus. It is possible that the leptin levels *during development* of these circuits determine the details of the hardwiring of this regulatory system.

Leptin Triggers a Signaling Cascade That Regulates Gene Expression

The leptin signal is transduced by a mechanism also used by receptors for interferon and growth factors, the JAK-STAT system **(Fig. 23–36**; see Fig. 12–18). The leptin receptor, which has a single transmembrane segment, dimerizes when leptin binds to the extracellular domains of two monomers. Both monomers are phosphorylated on a Tyr residue of their intracellular domain by a **Janus kinase (JAK)**. The Ⓟ–Tyr residues become docking

FIGURE 23–35 Hormones that control eating. In the arcuate nucleus of the hypothalamus, two sets of neurosecretory cells receive hormonal input and relay neuronal signals to the cells of muscle, adipose tissue, and liver. Leptin and insulin are released from adipose tissue and pancreas, respectively, in proportion to the mass of body fat. The two hormones act on anorexigenic neurosecretory cells to trigger release of α-MSH (melanocyte-stimulating hormone); α-MSH carries the signal to second-order neurons in the hypothalamus, which puts out the signals to eat less and metabolize more fuel. Leptin and insulin also act on orexigenic neurosecretory cells to inhibit the release of NPY, reducing the "eat more" signal sent to the tissues. As described later in the text, the gastric hormone ghrelin *stimulates* appetite by activating the NPY-expressing cells; PYY$_{3-36}$, released from the colon, *inhibits* these neurons and thereby decreases appetite. Each of the two types of neurosecretory cells inhibits hormone production by the other, so any stimulus that activates orexigenic cells inactivates anorexigenic cells, and vice versa. This strengthens the effect of stimulatory inputs.

FIGURE 23–36 The JAK-STAT mechanism of leptin signal transduction in the hypothalamus. Leptin binding induces dimerization of the leptin receptor, followed by phosphorylation of specific Tyr residues in the receptor's cytosolic domain, catalyzed by Janus kinase (JAK). STATs bound to the phosphorylated leptin receptor are now phosphorylated on Tyr residues by a separate activity of JAK. The STATs dimerize, binding each other's ℗-Tyr residues, and enter the nucleus. Here, they bind specific regulatory regions in the DNA and alter the expression of certain genes. The products of these genes ultimately influence the organism's feeding behavior and energy expenditure.

sites for three proteins that are *s*ignal *t*ransducers and *a*ctivators of *t*ranscription (**STATs** 3, 5, and 6; sometimes called fat-STATS). The docked STATs are then phosphorylated on Tyr residues by the same JAK. After phosphorylation, the STATs dimerize, then move to the nucleus, where they bind to specific DNA sequences and stimulate the expression of target genes, including the gene for POMC, from which α-MSH is produced.

The increased catabolism and thermogenesis triggered by leptin are due in part to increased synthesis of the mitochondrial uncoupling protein thermogenin (product of the *UCP1* gene) in brown adipocytes. Leptin stimulates the synthesis of thermogenin by altering synaptic transmissions from neurons in the arcuate nucleus to adipose and other tissues via the sympathetic nervous system. The resulting increased release of norepinephrine in these tissues acts through β₃-adrenergic receptors to stimulate transcription of the *UCP1* gene. The resulting uncoupling of electron transfer from oxidative phosphorylation consumes fat and is thermogenic (Fig. 23–34).

Might human obesity be the result of insufficient leptin production, and therefore treatable by the injection of leptin? Blood levels of leptin are in fact usually much *higher* in obese animals (including humans) than

in animals of normal body mass (except, of course, in *ob/ob* mutants, which cannot make leptin). Some downstream element in the leptin response system must be defective in obese individuals, and the elevation in leptin is the result of an (unsuccessful) attempt to overcome the leptin resistance. In those very rare humans with extreme obesity who have a defective leptin gene (*OB*), leptin injection does result in dramatic weight loss. In the vast majority of obese individuals, however, the *OB* gene is intact. In clinical trials, the injection of leptin did not have the weight-reducing effect observed in obese *ob/ob* mice. Clearly, most cases of human obesity involve one or more factors in addition to leptin.

The Leptin System May Have Evolved to Regulate the Starvation Response

The leptin system probably evolved to adjust an animal's activity and metabolism during periods of fasting and starvation, not as a means to restrict weight gain. The *reduction* in leptin level triggered by nutritional deficiency reverses the thermogenic processes illustrated in Figure 23–34, allowing fuel conservation. In the hypothalamus, decreased leptin signal also triggers decreased production of thyroid hormone (slowing basal metabolism), decreased production of sex hormones (preventing reproduction), and increased production of glucocorticoids (mobilizing the body's fuel-generating resources). By minimizing energy expenditure and maximizing the use of endogenous reserves of energy, these leptin-mediated responses may allow an animal to survive periods of severe nutritional deprivation. In liver and muscle, leptin stimulates **AMP-activated protein kinase (AMPK)** and through its action inhibits fatty acid synthesis and activates fatty acid oxidation, favoring energy-generating processes.

Insulin Acts in the Arcuate Nucleus to Regulate Eating and Energy Conservation

Insulin secretion may reflect both the size of fat reserves (adiposity) and the current energy balance (blood glucose level). In addition to its endocrine actions on various tissues, insulin acts in the central nervous system (in the hypothalamus) to inhibit eating (Fig. 23–35). Insulin receptors in the orexigenic neurons of the arcuate nucleus *inhibit* the release of NPY, and insulin receptors in the anorexigenic neurons *stimulate* α-MSH production, thereby decreasing fuel intake and increasing thermogenesis. By mechanisms discussed in Section 23.3, insulin also signals muscle, liver, and adipose tissues to increase the conversion of glucose to acetyl-CoA, providing the starting material for fat synthesis.

The actions of leptin and insulin are not wholly independent. Leptin makes the cells of liver and muscle more sensitive to insulin. One hypothesis to explain this effect suggests cross talk between the protein tyrosine kinases activated by leptin and those activated by insulin

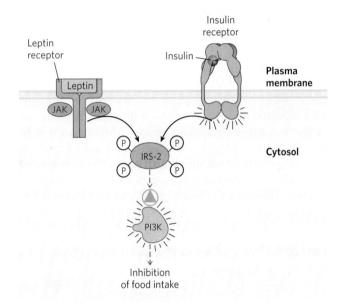

FIGURE 23–37 A possible mechanism for cross talk between receptors for insulin and leptin. The insulin receptor has intrinsic Tyr kinase activity (see Fig. 12-15), and the leptin receptor, when occupied by its ligand, is phosphorylated by a soluble Tyr kinase (JAK). One possible explanation for the observed interaction between leptin and insulin is that both may phosphorylate the same substrate—in the case shown here, insulin receptor substrate-2 (IRS-2). When phosphorylated, IRS-2 activates PI3K, which has downstream consequences that include inhibition of food intake. IRS-2 serves here as an integrator of the input from two receptors.

(Fig. 23–37); common second messengers in the two signaling pathways allow leptin to trigger some of the same downstream events that are triggered by insulin, through insulin receptor substrate-2 (IRS-2) and phosphatidylinositol 3-kinase (PI3K) (see Fig. 12–16).

Adiponectin Acts through AMPK to Increase Insulin Sensitivity

Adiponectin is a peptide hormone (224 amino acids) produced almost exclusively in adipose tissue, an adipokine that sensitizes other organs to the effects of insulin, protects against atherosclerosis, and inhibits inflammatory responses (monocyte adhesion, macrophage transformation, and the proliferation and migration of the cells of vascular smooth muscle). Adiponectin circulates in the blood and powerfully affects the metabolism of fatty acids and carbohydrates in liver and muscle. It increases the uptake of fatty acids from the blood by myocytes and the rate at which fatty acids undergo β oxidation in muscle. It also blocks fatty acid synthesis and gluconeogenesis in hepatocytes, and stimulates glucose uptake and catabolism in muscle and liver.

These effects of adiponectin are indirect and not fully understood, but AMPK clearly mediates many of them. Acting through its plasma membrane receptor, adiponectin triggers phosphorylation and activation of AMPK. Recall that AMPK is activated by factors that

signal the need to shift metabolism away from biosynthesis and toward energy generation **(Fig. 23–38)**. When activated, AMPK phosphorylates target proteins critical to the metabolism of lipids and carbohydrates, with profound effects on the metabolism of the whole animal **(Fig. 23–39)**. Adiponectin receptors are also present in the brain; the hormone activates AMPK in the hypothalamus, stimulating food intake and reducing energy expenditure.

One enzyme regulated by AMPK in the liver and in white adipose tissue is acetyl-CoA carboxylase, which produces malonyl-CoA, the first intermediate committed to fatty acid synthesis. Malonyl-CoA is a powerful inhibitor of the enzyme carnitine acyltransferase I, which starts the process of β oxidation by transporting fatty acids into the mitochondrion (see Fig. 17–6). By phosphorylating and inactivating acetyl-CoA carboxylase, AMPK inhibits fatty acid synthesis while relieving the inhibition (by malonyl-CoA) of β oxidation (see Fig. 17–13). Cholesterol synthesis is also inhibited by AMPK, which phosphorylates and inactivates HMG-CoA reductase, an enzyme in the path to cholesterol (see Fig. 21–34). Similarly, AMPK inhibits fatty acid synthase and acyl transferase, effectively blocking the synthesis of triacylglycerols. In short, lipid synthesis is inhibited and lipid use as fuel is stimulated.

Mice with defective adiponectin genes are less sensitive to insulin than those with normal adiponectin, and they show poor glucose tolerance: ingestion of dietary carbohydrate causes a long-lasting rise in blood glucose. These metabolic defects resemble those of humans with type 2 diabetes, who are similarly **insulin-insensitive** and clear glucose from the blood only slowly. Indeed, individuals with obesity or type 2 diabetes have lower blood adiponectin levels than nondiabetic controls. Moreover, drugs used in the treatment of type 2 diabetes—the thiazolidinediones, such as rosiglitazone (Avandia) and pioglitazone (Actos) (p. 852)—increase the expression of adiponectin mRNA

FIGURE 23–38 The role of AMP-activated protein kinase (AMPK) in regulating ATP metabolism. ADP produced in synthetic reactions is converted to AMP by adenylate kinase. AMP activates AMPK, which regulates ATP-consuming and ATP-generating pathways by phosphorylating key enzymes (see Fig. 23–39).

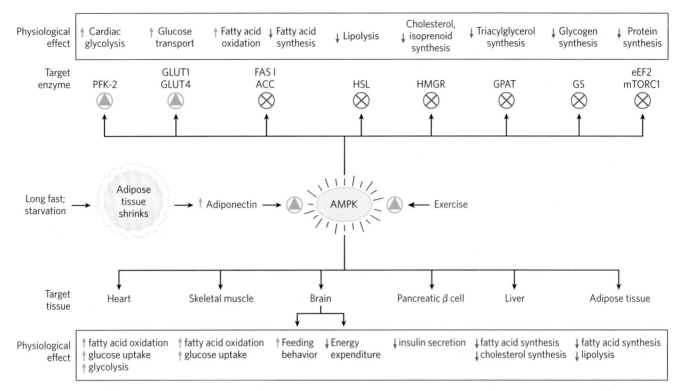

FIGURE 23–39 Formation of adiponectin and its actions through AMPK. Extended fasting or starvation results in decreased reserves of triacylglycerols in adipose tissue, which triggers adiponectin production and release from adipocytes. The rise in plasma adiponectin acts through its plasma membrane receptors in various cell types and organs to inhibit energy-consuming processes and stimulate energy-producing processes. Adiponectin acts through its receptors in the brain to stimulate feeding behavior and inhibit energy-consuming physical activity, and to inhibit thermogenesis in brown fat. Adiponectin exerts its metabolic effects by activating AMPK, which regulates (by phosphorylation) specific enzymes in key metabolic processes (see

Fig. 15–8). PFK-2, phosphofructokinase-2; GLUT1 and GLUT4, glucose transporters; FAS I, fatty acid synthase I; ACC, acetyl-CoA carboxylase; HSL, hormone-sensitive lipase; HMGR, HMG-CoA reductase; GPAT, an acyl transferase; GS, glycogen synthase; eEF2, eukaryotic elongation factor 2 (required for protein synthesis; see Chapter 27); mTOR1, mammalian target of rapamycin (a protein kinase that regulates protein synthesis on the basis of nutrient availability; see Fig. 23–40). Thiazolidinedione drugs activate the transcription factor PPARγ (see Figs 23–41 and 23–42), which then turns on adiponectin synthesis, indirectly activating AMPK. Exercise, through conversion of ATP to ADP and AMP, also stimulates AMPK.

in adipose tissue and increase blood adiponectin levels in experimental animals; they also activate AMPK (Fig. 23–39). (In 2011 the FDA concluded that Avandia use was associated with an increased risk of heart attacks, and it therefore limited the use of the drug to those diabetes patients for whom other available treatments were ineffective.) It seems likely that adiponectin will prove to be an important link between type 2 diabetes and its most important predisposing factor, obesity. ∎

mTORC1 Activity Coordinates Cell Growth with the Supply of Nutrients and Energy

The activity of a second novel protein kinase helps to mediate cell proliferation and increased cell size in response to growth factors and energy and nutrient availability. The highly conserved Ser/Thr kinase **mTORC1** is activated by abundant energy and nutrient supplies (such as high concentrations of branched-chain amino acids). When activated, mTORC1 phos-

phorylates several transcription factors, leading to increased expression of genes encoding the enzymes of lipid synthesis and mitochondrial proliferation and increased ribosome biogenesis **(Fig. 23–40)**. Fasting results in inactivation of mTORC1, leading to increased breakdown of protein and glycogen in liver and muscle and mobilization of triglycerides in adipose tissue. Chronic activation of mTORC1 by overeating results in excess deposition of triglycerides in adipose tissue and also in liver and muscle. This abnormal lipid accumulation in liver and muscle may contribute to insulin insensitivity and type 2 diabetes, as described in Section 23.5. Mutations that produce constantly activated mTORC1 are also commonly associated with human cancers.

Diet Regulates the Expression of Genes Central to Maintaining Body Mass

Proteins in a family of ligand-activated transcription factors, the **peroxisome proliferator-activated receptors (PPARs)**, respond to changes in dietary lipid by

FIGURE 23–40 mTORC1 stimulates cell growth and proliferation when adequate nutrition is available. mTORC1 is a Ser/Thr protein kinase activated by growth factors and metabolites that signal a state of adequate nutrition. By phosphorylating key target proteins, mTORC1 activates energy (ATP and NADPH) production for biosynthesis and stimulates the synthesis of proteins and lipids, allowing cell growth and proliferation. (PPP is pentose phosphate pathway; FA is fatty acids.)

altering the expression of genes involved in fat and carbohydrate metabolism. These transcription factors were first recognized for their roles in peroxisome synthesis—thus their name. Their normal ligands are fatty acids or fatty acid derivatives, but they can also bind synthetic agonists and can be activated in the laboratory by genetic manipulation. PPARα, PPARδ, and PPARγ are members of this nuclear receptor superfamily. They act in the nucleus by forming heterodimers with another nuclear receptor, RXR (retinoid X receptor), binding to regulatory regions of DNA near the genes under their control and changing the rate of transcription of those genes (**Fig. 23–41**).

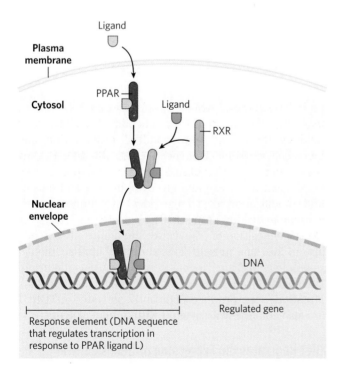

FIGURE 23–41 Mode of action of PPARs. PPARs are transcription factors that, when bound to their cognate ligand (L), form heterodimers with the nuclear receptor RXR. The dimer binds specific regions of DNA known as response elements, stimulating transcription of genes in those regions.

PPARγ, expressed primarily in liver and adipose tissue (both brown and white), is involved in turning on genes necessary to the differentiation of fibroblasts into adipocytes and genes that encode proteins required for lipid synthesis and storage in adipocytes (**Fig. 23–42**). PPARγ is activated by the thiazolidinedione drugs that are used to treat type 2 diabetes (discussed below).

PPARα is expressed in liver, kidney, heart, skeletal muscle, and brown adipose tissue. The ligands that activate this transcription factor include eicosanoids, free fatty acids, and the class of drugs called fibrates, such as fenofibrate (TriCor) and ciprofibrate (Modalim), which are used to treat coronary heart disease by raising HDL and lowering blood triacylglycerols. In hepatocytes, PPARα turns on the genes necessary for the uptake and β oxidation of fatty acids and formation of ketone bodies during fasting.

PPARδ and **PPARβ** are key regulators of fat oxidation, which act by sensing changes in dietary lipid. PPARδ acts in liver and muscle, stimulating the transcription of at least nine genes encoding proteins for β oxidation and for energy dissipation through uncoupling of mitochondria. Normal mice overfed on high-fat diets accumulate massive amounts of white fat, and fat droplets accumulate in the liver. But when the same overfeeding experiment is carried out with mice that have a genetically altered, always active PPARδ, this fat accumulation is prevented. In mice with a nonfunctioning leptin receptor (*db/db*), activated PPARδ prevents the development of obesity that would otherwise occur. By stimulating fatty acid breakdown in uncoupled mitochondria, PPARδ causes fat depletion, weight loss, and thermogenesis. Seen in this light, thermogenesis is both a means of keeping warm and a defense against obesity. Clearly, PPARδ is a potential target for drugs to treat obesity.

Short-Term Eating Behavior Is Influenced by Ghrelin and PYY$_{3-36}$

Ghrelin is a peptide hormone (28 amino acids) produced in cells lining the stomach. It was originally recognized as the stimulus for the release of growth hormone (*ghre* is the Proto-Indo-European root of "grow"), then subsequently shown to be a powerful appetite stimulant

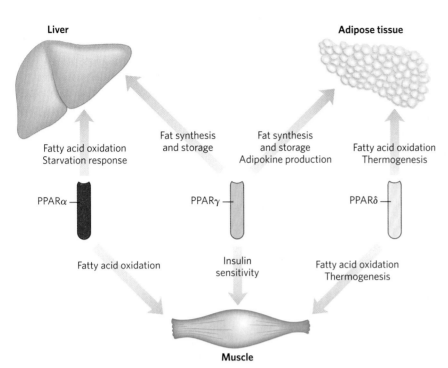

FIGURE 23–42 Metabolic integration by PPARs. The three PPAR isoforms regulate lipid and glucose homeostasis through their coordinated effects on gene expression in liver, muscle, and adipose tissue. PPARα and PPARδ (and its closely related isoform PPARβ) regulate lipid utilization; PPARγ regulates lipid storage and the insulin sensitivity of various tissues.

that works on a shorter time scale (between meals) than leptin and insulin. Ghrelin receptors are located in the pituitary gland (presumably mediating growth hormone release) and in the hypothalamus (affecting appetite), as well as in heart muscle and adipose tissue. Ghrelin acts through a G protein–coupled receptor to generate the second messenger IP_3, which mediates ghrelin action. The concentration of ghrelin in the blood varies strikingly between meals, peaking just before a meal and dropping sharply just after the meal (**Fig. 23–43**). Injection of ghrelin into humans produces immediate sensations of intense hunger. Individuals with Prader-Willi syndrome, whose blood levels of ghrelin are exceptionally high, have an uncontrollable appetite, leading to extreme obesity that often results in death before the age of 30.

PYY$_{3-36}$ is a peptide hormone (2 amino-terminal Y residues plus 34 amino acids) secreted by endocrine cells in the lining of the small intestine and colon in response to food entering from the stomach. The level of PYY$_{3-36}$ in the blood rises after a meal and remains high for some hours. It is carried in the blood to the arcuate nucleus, where it acts on orexigenic neurons, inhibiting NPY release and reducing hunger (Fig. 23–35). Humans injected with PYY$_{3-36}$ feel little hunger and eat less than normal amounts for about 12 hours.

FIGURE 23–43 Variations in blood concentrations of glucose, ghrelin, and insulin relative to meal times. (a) Plasma levels of ghrelin rise sharply just *before* the normal time for meals (7 AM breakfast, 12 noon lunch, 5:30 PM dinner) and drop precipitously just after meals, paralleling the subjective feelings of hunger. **(b)** Blood glucose rises sharply *after* a meal, **(c)** followed immediately by a rise in insulin levels in response to the increase in blood glucose concentration.

Microbial Symbionts in the Gut Influence Energy Metabolism and Adipogenesis

Lean and obese individuals harbor different combinations of microbial symbionts in the gut. Investigation of this observation led to the discovery that certain microorganisms of the gut release fermentation products, namely the short-chain fatty acids acetate, propionate, butyrate, and lactate, which enter the bloodstream and trigger metabolic changes in adipose tissue. Propionate, for example, drives the expansion of WAT by acting on G protein–coupled receptors (GPR43 and GPR41) in the plasma membranes of several cell types, including adipocytes. These receptors trigger differentiation of precursor cells into adipocytes and inhibit lipolysis in existing adipocytes, leading to an increase in WAT mass—obesity.

These findings suggest possible approaches to the prevention of obesity that involve altering the makeup of the microbial community in the gut, either by direct addition of microbial species (probiotics) that disfavor adipogenesis or by adding to the diet nutrients (prebiotics) that favor the dominance of the probiotic microbes. For example, certain polymers of fructose (fructans) that are indigestible to animals favor such a microbial community in mice and lead to decreased fat storage in WAT and liver (and to the decrease in insulin sensitivity associated with obesity and lipid deposition in the liver; see below).

Endocrine cells present in the lining of the GI tract secrete products—the anorexigenic peptides PYY_{3-36} and glucagon-like peptide-1 (GLP-1) and the orexigenic peptide ghrelin—that modulate food intake and energy expenditure in the animal. Interaction with specific microbes in the gut, or with their fermentation products, may be the trigger for release of these peptides. The investigations of effects of diet on microbial symbionts of the gut, and of the effects of microbial products on energy metabolism and adipogenesis, may lead to a deeper understanding of the effects of diet and microbial symbionts on the development of obesity and of its pathological sequels, the metabolic syndrome and type 2 diabetes.

This interlocking system of neuroendocrine controls of food intake and metabolism presumably evolved to protect against starvation and to eliminate counterproductive accumulation of fat (extreme obesity). The difficulty most people face in trying to lose weight testifies to the remarkable effectiveness of these controls.

SUMMARY 23.4 Obesity and the Regulation of Body Mass

▶ Obesity is increasingly common in the United States and other developed countries and predisposes people toward several life-threatening conditions, including cardiovascular disease and type 2 diabetes.

▶ Adipose tissue produces leptin, a hormone that regulates feeding behavior and energy expenditure so as to maintain adequate reserves of fat. Leptin production and release increase with the number and size of adipocytes.

▶ Leptin acts on receptors in the arcuate nucleus of the hypothalamus, causing the release of anorexigenic (appetite-suppressing) peptides, including α-MSH, that act in the brain to inhibit eating. Leptin also stimulates sympathetic nervous system action on adipocytes, leading to uncoupling of mitochondrial oxidative phosphorylation, with consequent thermogenesis.

▶ The signal-transduction mechanism for leptin involves phosphorylation of the JAK-STAT system. On phosphorylation by JAK, STATs can bind to regulatory regions in nuclear DNA and alter the expression of genes for proteins that set the level of metabolic activity and determine feeding behavior. Insulin acts on receptors in the arcuate nucleus, with results similar to those caused by leptin.

▶ The hormone adiponectin stimulates fatty acid uptake and oxidation and inhibits fatty acid synthesis. It also sensitizes muscle and liver to insulin. Adiponectin's actions are mediated by AMPK, which is also activated by low [AMP] and exercise.

▶ Ghrelin, a hormone produced in the stomach, acts on orexigenic (appetite-stimulating) neurons in the arcuate nucleus to produce hunger before a meal. PYY_{3-36}, a peptide hormone of the intestine, acts at the same site to lessen hunger after a meal.

▶ The specific types of microbial symbionts in the gut can influence adipogenesis, the increase in the mass of body fat.

23.5 Obesity, the Metabolic Syndrome, and Type 2 Diabetes

In the industrialized world, where the food supply is more than adequate, there is a growing epidemic of obesity and the type 2 diabetes associated with it. As many as 300 million people in the world now have diabetes, and reasonable projections predict a dramatic rise in cases over the next decade, following the world epidemic of obesity. The pathology of diabetes includes cardiovascular disease, renal failure, blindness, poor healing in the extremities that requires amputations, and neuropathy. In 2011, the global mortality from diabetes was estimated at 4 million, a number that is sure to rise in coming years. Clearly, it is essential to understand type 2 diabetes and its relationship to obesity and to find countermeasures that prevent or reverse the damage done by this disease. ■

In Type 2 Diabetes the Tissues Become Insensitive to Insulin

The hallmark of type 2 diabetes is the development of insulin resistance: a state in which more insulin is needed to bring about the biological effects produced by a lower amount of insulin in the normal, healthy state. In the early stages of the disease, pancreatic β cells secrete enough insulin to overcome the lower insulin sensitivity of muscle and liver. But the β cells eventually fail, and the lack of insulin becomes apparent in the body's inability to regulate blood glucose. The intermediate stage, preceding type 2 diabetes mellitus, is sometimes called the **metabolic syndrome**, or **syndrome X**. This is typified by obesity, especially in the abdomen; hypertension (high blood pressure); abnormal blood lipids (high TAG and LDL, low HDL); slightly high blood glucose; and a reduced ability to clear glucose in the glucose-tolerance test. Individuals with metabolic syndrome often also show changes in blood proteins, changes that are associated with abnormal clotting (high fibrinogen concentration) or inflammation (high concentrations of the C-reactive peptide, which typically increases with an inflammatory response). About 27% of the adult population in the United States has these symptoms of metabolic syndrome.

What predisposes individuals with metabolic syndrome to develop type 2 diabetes? According to the "lipid toxicity" hypothesis **(Fig. 23–44)**, the action of PPARγ on adipocytes normally keeps the cells ready to synthesize and store triacylglycerols—the adipocytes are insulin-sensitive and produce leptin, which leads to their continued intracellular deposition of TAG. However, excess caloric intake in obese individuals causes adipocytes to become filled with TAG, leaving adipose tissue unable to meet any further demand for TAG storage. Lipid-filled adipose tissue releases protein factors that attract macrophages, which infiltrate the tissue and may come to represent as much as 50% of the adipose tissue by mass. Macrophages trigger the inflammatory response, which impairs TAG deposition in adipocytes and favors release of free fatty acids into the blood. These excess fatty acids enter liver and muscle cells, where they are converted to TAGs that accumulate as lipid droplets. This ectopic (Greek *ektopos*, "out of place") deposition of TAGs leads to insulin insensitivity in liver and muscle, the hallmark of type 2 diabetes.

Moreover, according to this hypothesis, excess stored fatty acids and TAGs are toxic to liver and muscle. Some individuals are less well equipped genetically to handle this burden of ectopic lipids and are more susceptible to the cellular damage that leads to the development of type 2 diabetes. Insulin resistance probably involves impairment of several of the mechanisms by which

FIGURE 23-44 Overloading adipocytes with triacylglycerols triggers inflammation in fat tissue and ectopic lipid deposition and insulin resistance in muscle. In an individual of healthy body mass, triacylglycerol (TAG) uptake in the diet equals TAG oxidation for energy. In overweight individuals, excess caloric intake results in enlarged adipocytes, engorged with TAG and unable to store more. Enlarged adipocytes secrete MCP-1 (monocyte chemotaxis protein-1), attracting macrophages. Macrophages infiltrate adipose tissue and produce TNF α (tumor necrosis factor α), which triggers lipid breakdown and fatty acid release into the blood. The fatty acids released from adipose tissue enter muscle cells, where they accumulate in small lipid droplets. This ectopic storage of lipids somehow causes insulin resistance, perhaps by triggering lipid-activated protein kinases that inactivate some element in the insulin-signaling pathway. GLUT4 glucose transporters leave the muscle cell surface, preventing glucose entry into muscle; the myocyte has now become insulin-resistant. It cannot use blood glucose for its fuel, so fatty acids are mobilized from adipose tissue and become the primary fuel. The increased influx of fatty acids into muscle leads to further deposition of ectopic lipids. In some individuals, insulin resistance develops into type 2 diabetes. Other individuals are genetically less susceptible to the deleterious effects of ectopic lipid storage or are genetically better equipped to manage this storage and do not develop diabetes.

insulin acts on metabolism, which include changes in protein levels and changes in the activities of signaling enzymes and transcription factors. For example, both adiponectin synthesis in adipocytes and adiponectin level in the blood decrease with obesity and increase with weight loss.

Several of the drugs that are effective in improving insulin sensitivity in type 2 diabetes are known to act on specific proteins in signaling pathways, and their actions are consistent with the lipotoxicity model. Thiazolidinediones bind to PPARγ, turning on a set of adipocyte-specific genes and promoting the differentiation of preadipocytes to small adipocytes, thereby increasing the body's capacity for absorbing fatty acids from the diet and storing them as TAGs.

There are clearly genetic factors that predispose toward type 2 diabetes. Although 80% of individuals with type 2 diabetes are obese, most obese individuals do not develop type 2 diabetes. Given the complexity of the regulatory mechanisms we have discussed in this chapter, it is not surprising that the genetics of diabetes is complex, involving interactions between variant genes and environmental factors, including diet and lifestyle. At least 10 genetic loci have been reliably linked to type 2 diabetes; variation in any of these "diabetogenes" alone would cause a relatively small increase in the likelihood of developing type 2 diabetes. For example, people with a variant PPARγ in which an Ala residue replaces Pro at position 12 are at a slightly, though significantly, increased risk of developing type 2 diabetes. ■

Type 2 Diabetes Is Managed with Diet, Exercise, and Medication

Studies show that three factors improve the health of individuals with type 2 diabetes: dietary restriction, regular exercise, and drugs that increase insulin sensitivity or insulin production. Dietary restriction (and accompanying weight loss) reduces the overall burden of handling fatty acids. The lipid composition of the diet influences, through PPARs and other transcription factors, the expression of genes that encode proteins involved in fatty acid oxidation and in energy expenditure via thermogenesis. Exercise activates AMPK, as does adiponectin; AMPK shifts metabolism toward fat oxidation and inhibits fat synthesis.

Several classes of drugs are used in the management of type 2 diabetes, some of which we have discussed earlier in the chapter (Table 23–7). Sulfonylureas act on the ATP-gated K^+ channels in β cells to stimulate insulin release. Biguanides such as metformin (Glucophage) activate AMPK, mimicking the effects of adiponectin. Thiazolidinediones act through PPARγ to increase the concentration of adiponectin in plasma and to stimulate adipocyte differentiation, thereby increasing the capacity for TAG storage. Inhibitors of dipeptide protease IV (DPP IV) prevent the proteolytic degradation of GLP-1, a peptide hormone produced in the gut that stimulates pancreatic insulin secretion. Inhibition of the peptidase prolongs the action of GLP-1, effectively increasing insulin secretion.

TABLE 23–7   Treatments for Type 2 Diabetes Mellitus

Intervention/treatment	Direct target	Effect of treatment
Weight loss	Adipose tissue; reduces TAG content	Reduces lipid burden; increases capacity for lipid storage in adipose tissue; restores insulin sensitivity
Exercise	AMPK, activated by increasing [AMP]/[ATP]	Aids weight loss; see Fig. 23–39
Sulfonylureas: glipizide (Glucotrol), glyburide (several brands), glimepiride (Amaryl)	Pancreatic β cells; K^+ channels blocked	Stimulates insulin secretion by pancreas; see Fig. 23–27
Biguanides: metformin (Glucophage)	AMPK, activated	Increases glucose uptake by muscle; decreases glucose production in liver
Thiazoladinediones: troglitazone (Rezulin),* rosiglitazone (Avandia),† pioglitazone (Actos)	PPARγ	Stimulates expression of genes, potentiating the action of insulin in liver, muscle, adipose tissue; increases glucose uptake; decreases glucose synthesis in liver
GLP-1 modulators: exenatide (Byetta), sitagliptin (Januvia)	Glucagon-like peptide-1, dipeptide protease IV	Enhances insulin secretion by pancreas

*Voluntarily withdrawn because of side effects.

†Prescriptions limited to patients not helped by other treatment, because of possible increased risk of cardiovascular disease.

The combination of weight loss and exercise is the preferred way to *prevent* development of the metabolic syndrome and type 2 diabetes. Recent findings concerning BAT in adults suggest an interesting possibility for aiding weight loss and reducing the amount of TAG that must be stored. The protein PRDM16 is expressed strongly, and perhaps uniquely, in brown adipose tissue. When overexpressed in the adipose tissue of mice, PRDM16 induces the differentiation of preadipocytes in white adipose tissue into *brown* adipocytes, with high levels of thermogenin and strikingly uncoupled respiration. Such cells could, in principle, consume fatty acids above the amount needed for ATP production, converting the energy of oxidation to heat. Given the widespread and increasing occurrence of type 2 diabetes, the possibility of stimulating fat oxidation by activating BAT certainly deserves exploration. ■

SUMMARY 23.5 Obesity, the Metabolic Syndrome, and Type 2 Diabetes

▶ The metabolic syndrome, which includes obesity, hypertension, elevated blood lipids, and insulin resistance, is often the prelude to type 2 diabetes.

▶ The insulin resistance that characterizes type 2 diabetes may be a consequence of abnormal lipid storage in muscle and liver, in response to a lipid intake that cannot be accommodated by adipose tissue.

▶ Expression of the enzymes of lipid synthesis is under tight and complex regulation. PPARs are transcription factors that determine the rate of synthesis of many enzymes involved in lipid metabolism and adipocyte differentiation.

▶ Effective treatments for type 2 diabetes include exercise, appropriate diet, and drugs that increase insulin sensitivity or insulin production.

Key Terms

Terms in bold are defined in the glossary.

Further Reading

General Background and History

Blumenthal, S. (2009) The insulin immunoassay after 50 years: a reassessment. *Perspect. Biol. Med.* **52**, 343–354.
Fascinating historical account of development of the radioimmunoassay by Yalow and Berson and its use in studies of insulin and diabetes.

Chew, S.L. & Leslie, D. (2006) *Clinical Endocrinology and Diabetes: An Illustrated Colour Text*, Churchill Livingstone, Edinburgh.

Schmidt, T.J. (2010) Biochemistry of hormones. In *Textbook of Biochemistry with Clinical Correlations*, 7th edn (Devlin, T.M., ed.), pp. 883–938, John Wiley & Sons, Inc., New York.

Tissue-Specific Metabolism

Brosnan, J.T. & and Brosnan, M.E. (2007) Creatine: endogenous metabolite, dietary, and therapeutic supplement. *Annu. Rev. Nutr.* **27**, 241–261.

Cannon, B. & Nedergaard, K. (2004) Brown adipose tissue: function and physiological significance. *Physiol. Rev.* **84**, 277–359.

Elia, M. (1995) General integration and regulation of metabolism at the organ level. *Proc. Nutr. Soc.* **54**, 213–234.

Enerbäck, S. (2010) Human brown adipose tissue (minireview). *Cell Metab.* **11**, 248–252.

Randle, P.J. (1995) Metabolic fuel selection: general integration at the whole-body level. *Proc. Nutr. Soc.* **54**, 317–327.

Ravussin, E. & Galgani, J.E. (2011) The implication of brown adipose tissue for humans. *Annu. Rev. Nutr.* **31**, 33–37.

Hormonal Regulation of Fuel Metabolism

Cahill, G.F., Jr. (2006) Fuel metabolism in starvation. *Annu. Rev. Nutr.* **26**, 1–22.

Desvergne, B., Michalik, L., & Wahli, W. (2006) Transcriptional regulation of metabolism. *Physiol. Rev.* **86**, 465–514.

Feige, J.N. & Auwerx, J. (2007) Transcriptional coregulators in the control of energy homeostasis. *Trends Cell Biol.* **17**, 292–301.

Hardie, D.G. & Sakamoto, K. (2006) AMPK: a key sensor of fuel and energy status in skeletal muscle. *Physiology* **21**, 48–60.

Kadowaki, T., Yamauchi, T., Kubota, N., Hara, K., Ueki, K., & Tobe, K. (2006) Adiponectin and adiponectin receptors in insulin resistance, diabetes, and the metabolic syndrome. *J. Clin. Invest.* **116**, 1784–1792.

Kola, B., Boscaro, M., Rutter, G.A., Grossman, A.B., & Korbonits, M. (2006) Expanding role of AMPK in endocrinology. *Trends Endocrinol. Metab.* **17**, 205–215.

Kraemer, D.K., Al-Khalili, L., Guigas, B., Leng, Y., Garcia-Roves, P.M., & Krook, A. (2007) Role of AMP kinase and PPARδ in the regulation of lipid and glucose metabolism in human skeletal muscle. *J. Biol. Chem.* **282**, 19,313–19,320.

Lin, H.V. & Accill, D. (2011) Hormonal regulation of hepatic glucose production in health and disease. *Cell Metab.* **14**, 9–19.

Advanced review of the mechanisms that regulate production of glucose by the liver.

Maccarone, M., Gasperi, V., Catani, M.V., Diep, T.A., Dainese, E., Hansen, H.S., & Avigliano, L. (2010) The endocannabinoid system and its relevance for nutrition. *Annu. Rev. Nutr.* **30**, 423–440.

Nguyen, A.D., Herzog, H., & Sainsbury, A. (2011) Neuropeptide Y and peptide YY: important regulators of energy metabolism. *Curr. Opin. Endocrinol. Diabetes. Obes.* **18**, 56–60.

Intermediate-level survey of the effects of NPY and PYY in the brain on energy metabolism in the mouse.

Nichols, C.G. (2006) K$_{ATP}$ channels as molecular sensors of cellular metabolism. *Nature* **440**, 470–476.

Control of Body Mass

Ahima, R.S. (2005) Central actions of adipocyte hormones. *Trends Endocrinol. Metab.* **16**, 307–313.

A decrease in leptin is an important signal for the switch between fed and fasted states.

Ahima, R.S. (2006) Ghrelin—a new player in glucose homeostasis? *Cell Metab.* **3**, 301–307.

Azzu, V. & Brand, M.D. (2010) The on-off switches of the mitochondrial uncoupling proteins. *Trends Biochem. Sci.* **35**, 298–307.

Intermediate-level review of how cold, overfeeding, and starvation affect the expression of thermogenin genes.

Biddinger, S.B. & Kahn, C.R. (2006) From mice to men: insights into the insulin resistance syndromes. *Annu. Rev. Physiol.* **68**, 123–158.

Review of the structure, function, and role of uncoupling proteins.

Delzenne, N.M. & Cani, P.D. (2011) Interaction between obesity and the gut microbiota: relevance in nutrition. *Annu. Rev. Nutr.* **31**, 15–31.

Friedman, J.M. (2002) The function of leptin in nutrition, weight, and physiology. *Nutr. Rev.* **60**, S1–S14.

Intermediate-level review of all aspects of leptin biology.

Howell, J.J. & Manning, B.D. (2011) mTOR couples cellular nutrient sensing to metabolic homeostasis. *Trends Endocrinol. Metab.* **22**, 94–102.

Intermediate-level review of how mTORC1 mediates the metabolic responses to fluctuations in nutrient availability.

Jequier, E. & Tappy, L. (1999) Regulation of body weight in humans. *Physiol. Rev.* **79**, 451–480.

Detailed review of the role of leptin in body-weight regulation, the control of food intake, and the roles of white and brown adipose tissues in energy expenditure.

Klok, M.D., Jakobsdottir, S., & Drent, M.L. (2007) Role of leptin and ghrelin in the regulation of food intake and body weight. *Obes. Rev.* **8**, 21–34.

Marx, J. (2003) Cellular warriors at the battle of the bulge. *Science* **299**, 846–849.

Short review of biochemistry of weight control and introduction to several papers in the same issue of *Science* on obesity in humans.

Morton, G.J., Cummings, D.E., Baskin, D.G., Barsh, G.S., & Schwartz, M.W. (2006) Central nervous system control of food intake and body weight. *Nature* **443**, 289–295.

Zoncu, R., Efeyan, A., & Sabatini, D.M. (2011) mTOR: from growth signal integration to cancer, diabetes and ageing. *Nat. Rev. Mol. Cell Biol.* **12**, 21–35.

Obesity, the Metabolic Syndrome, and Type 2 Diabetes

Attie, A.D. & Scherer, P.E. (2009) Adipocyte metabolism and obesity. *J. Lipid Res.* **50**, S395–S399.

Barish, G.D., Narkar, V.A., & Evans, R.M. (2006) PPARδ: a dagger in the heart of the metabolic syndrome. *J. Clin. Invest.* **116**, 590–597.

Clee, S.M. & Attie, A.D. (2007) The genetic landscape of type 2 diabetes in mice. *Endocrinol. Rev.* **28**, 48–83.

Cypess, A.M. & Kahn, C.R. (2010) Brown fat as a therapy for obesity and diabetes. *Curr. Opin. Endocrinol. Diabetes. Obes.* **17**, 143–149.

Intermediate-level discussion of the possibility of stimulating fat oxidation by BAT as a treatment for obesity.

Guilherme, A., Virbasius, J.V., Puri, V., & Czech, M.P. (2008) Adipocyte dysfunctions linking obesity to insulin resistance and type 2 diabetes. *Nat. Rev. Mol. Cell Biol.* **9**, 367–377.

Harper, M.E., Green, K., & Brand, M.D. (2008) The efficiency of cellular energy transduction and its implications for obesity. *Annu. Rev. Nutr.* **28**, 13–33.

Kasuga, M. (2006) Insulin resistance and pancreatic β cell failure. *J. Clin. Invest.* **116**, 1756–1760.

Introduction to a series of reviews about type 2 diabetes in this issue of the journal.

Klein, J., Perwitz, N., Kraus, D., & Fasshauer, M. (2006) Adipose tissue as source and target for novel therapies. *Trends Endocrinol. Metab.* **17**, 26–32.

Intermediate-level review.

Lusis, A.J., Attie, A.D., & Reue, K. (2008) Metabolic syndrome: from epidemiology to systems biology. *Nat. Rev. Genet.* **9**, 819–830.

McAllister, E.J., Dhurandhar, N.V., Keith, S.W., Aronne, L.J., Barger, J., Baskin, M., Benca, R.M., Biggio, J., Boggiano, M.M., Eisenmann, J.C., et al. (2009) Ten putative contributors to the obesity epidemic. *Crit. Rev. Food Sci. Nutr.* **49**, 868–913.

Advanced and lengthy review of factors responsible for the international epidemic of obesity.

Muoio, D.M. & Newgard, C.B. (2008) Molecular and metabolic mechanisms of insulin resistance and β-cell failure in diabetes. *Nat. Rev. Mol. Cell Biol.* **9**, 193–205.

Savage, D.B., Petersen, K.F., & Shulman, G.I. (2007) Disordered lipid metabolism and the pathogenesis of insulin resistance. *Physiol. Rev.* **87**, 507–520.

Semple, R.K., Chatterjee, V.K.K., & O'Rahilly, S. (2006) PPARγ and human metabolic disease. *J. Clin. Invest.* **116**, 581–589.

Sharma, A.M. & Staels, B. (2007) Peroxisome proliferator-activated receptor γ and adipose tissue—understanding obesity-related changes in regulation of lipid and glucose metabolism. *J. Clin. Endocrinol. Metab.* **92**, 386–395.

Verdin, E., Hirschey, M.D., Finley, L.W., & Haigis, M.C. (2010) Sirtuin regulation of mitochondria: energy production, apoptosis, and signaling. *Trends Biochem. Sci.* **35**, 669–675.

Vidal-Puig, A., & Unger, R.H. (2010). *Biochim. Biophys. Acta*, **180**, 207–208.

This entire issue of the journal is about lipotoxicity and the metabolic syndrome.

Virtue, S. & Vidal-Puig, A. (2008) It's not how fat you are, it's what you do with it that counts. *PLoS Biol.*, **6**, e237. (doi:10.1371/journal.pbio.0060237)

Voight, B.F., Scott, L.J., Steinthorsdottir, V., Morris, A.P., Dina C., Welch, R.P., Zeggini, E., Huth, C., Aulchencko, Y.S., Thorleifsson, G., et al. (2010) Twelve type 2 diabetes susceptibility loci identified through large-scale association analysis. *Nat. Genet.* **42**, 579–589.

Results of a large survey (47,000 individuals) to identify genes that predispose humans toward type 2 diabetes.

Whittle, A.J., Lopez, M., & Vidal-Puig, A. (2011) Using brown adipose tissue to treat obesity—the central issue. *Trends Mol. Med.* **8**, 405–411.

Intermediate-level review of the possibility of treating obesity by stimulating brown fat activity.

Xu, H., Barnes, G.T., Yang, Q., Tan, G., Yang, D., Chou, C.J., Sole, J., Nichols, A., Ross, J.S., Tartaglia, L.A., & Chen, H. (2003) Chronic inflammation in fat plays a crucial role in the development of obesity-related insulin resistance. *J. Clin. Invest.* **112**, 1821–1830.

Problems

1. Peptide Hormone Activity Explain how two peptide hormones as structurally similar as oxytocin and vasopressin can have such different effects (see Fig. 23–9).

2. ATP and Phosphocreatine as Sources of Energy for Muscle During muscle contraction, the concentration of phosphocreatine in skeletal muscle drops while the concentration of ATP remains fairly constant. However, in a classic experiment, Robert Davies found that if he first treated muscle with 1-fluoro-2,4-dinitrobenzene (p. 98), the concentration of ATP declined rapidly while the concentration of phosphocreatine remained unchanged during a series of contractions. Suggest an explanation.

3. Metabolism of Glutamate in the Brain Brain tissue takes up glutamate from the blood, transforms it into glutamine, and releases the glutamine into the blood. What is accomplished by this metabolic conversion? How does it take place? The amount of glutamine produced in the brain can actually exceed the amount of glutamate entering from the blood. How does this extra glutamine arise? (Hint: You may want to review amino acid catabolism in Chapter 18; recall that NH_4^+ is very toxic to the brain.)

4. Proteins as Fuel during Fasting When muscle proteins are catabolized in skeletal muscle during a fast, what are the fates of the amino acids?

5. Absence of Glycerol Kinase in Adipose Tissue Glycerol 3-phosphate is required for the biosynthesis of triacylglycerols. Adipocytes, specialized for the synthesis and degradation of triacylglycerols, cannot use glycerol directly because they lack glycerol kinase, which catalyzes the reaction

$$\text{Glycerol} + \text{ATP} \longrightarrow \text{glycerol 3-phosphate} + \text{ADP}$$

How does adipose tissue obtain the glycerol 3-phosphate necessary for triacylglycerol synthesis?

6. Oxygen Consumption during Exercise A sedentary adult consumes about 0.05 L of O_2 in 10 seconds. A sprinter, running a 100 m race, consumes about 1 L of O_2 in 10 seconds. After finishing the race, the sprinter continues to breathe at an elevated (but declining) rate for some minutes, consuming an extra 4 L of O_2 above the amount consumed by the sedentary individual.

(a) Why does the need for O_2 increase dramatically during the sprint?

(b) Why does the demand for O_2 remain high after the sprint is completed?

7. Thiamine Deficiency and Brain Function Individuals with thiamine deficiency show some characteristic neurological signs and symptoms, including loss of reflexes, anxiety, and mental confusion. Why might thiamine deficiency be manifested by changes in brain function?

8. Potency of Hormones Under normal conditions, the human adrenal medulla secretes epinephrine ($C_9H_{13}NO_3$) at a rate sufficient to maintain a concentration of 10^{-10} M in circulating blood. To appreciate what that concentration means, calculate the diameter of a round swimming pool, with a water depth of 2.0 m, that would be needed to dissolve 1.0 g (about 1 teaspoon) of epinephrine to a concentration equal to that in blood.

9. Regulation of Hormone Levels in the Blood The half-life of most hormones in the blood is relatively short. For example, when radioactively labeled insulin is injected into an animal, half of the labeled hormone disappears from the blood within 30 min.

(a) What is the importance of the relatively rapid inactivation of circulating hormones?

(b) In view of this rapid inactivation, how is the level of circulating hormone kept constant under normal conditions?

(c) In what ways can the organism make rapid changes in the level of a circulating hormone?

10. Water-Soluble versus Lipid-Soluble Hormones On the basis of their physical properties, hormones fall into one of two categories: those that are very soluble in water but relatively insoluble in lipids (e.g., epinephrine) and those that are relatively insoluble in water but highly soluble in lipids (e.g., steroid hormones). In their role as regulators of cellular activity, most water-soluble hormones do not enter their target cells. The lipid-soluble hormones, by contrast, do enter their target cells and ultimately act in the nucleus. What is the correlation between solubility, the location of receptors, and the mode of action of these two classes of hormones?

11. Metabolic Differences between Muscle and Liver in a "Fight-or-Flight" Situation When an animal confronts a "fight-or-flight" situation, the release of epinephrine promotes glycogen breakdown in the liver, heart, and skeletal muscle. The end product of glycogen breakdown in the liver is glucose; the end product in skeletal muscle is pyruvate.

(a) What is the reason for the different products of glycogen breakdown in the two tissues?

(b) What is the advantage to an animal that must fight or flee of these specific glycogen breakdown routes?

12. Excessive Amounts of Insulin Secretion: Hyperinsulinism Certain malignant tumors of the pancreas cause excessive production of insulin by the β cells. Affected individuals exhibit shaking and trembling, weakness and fatigue, sweating, and hunger.

(a) What is the effect of hyperinsulinism on the metabolism of carbohydrates, amino acids, and lipids by the liver?

(b) What are the causes of the observed symptoms? Suggest why this condition, if prolonged, leads to brain damage.

13. Thermogenesis Caused by Thyroid Hormones Thyroid hormones are intimately involved in regulating the basal metabolic rate. Liver tissue of animals given excess thyroxine shows an increased rate of O_2 consumption and increased heat output (thermogenesis), but the ATP concentration in the tissue is normal. Different explanations have been offered for the thermogenic effect of thyroxine. One is that excess thyroxine causes uncoupling of oxidative phosphorylation in mitochondria. How could such an effect account for the observations? Another explanation suggests that the thermogenesis is due to an increased rate of ATP utilization by the thyroxine-stimulated tissue. Is this a reasonable explanation? Why?

14. Function of Prohormones What are the possible advantages of synthesizing hormones as prohormones?

15. Sources of Glucose during Starvation The typical human adult uses about 160 g of glucose per day, 120 g of which is used by the brain. The available reserve of glucose (~20 g of circulating glucose and ~190 g of glycogen) is adequate for about one day. After the reserve has been depleted during starvation, how would the body obtain more glucose?

16. Parabiotic *ob/ob* Mice By careful surgery, researchers can connect the circulatory systems of two mice so that the same blood circulates through both animals. In these **parabiotic** mice, products released into the blood by one animal reach the other animal via the shared circulation. Both animals are free to eat independently. If a mutant *ob/ob* mouse (both copies of the *OB* gene are defective) and a normal *OB/OB* mouse (two good copies of the *OB* gene) were made parabiotic, what would happen to the weight of each mouse?

17. Calculation of Body Mass Index A portly biochemistry professor weighs 260 lb (118 kg) and is 5 feet 8 inches (173 cm) tall. What is his body mass index? How much weight would he have to lose to bring his body mass index down to 25 (normal)?

18. Insulin Secretion Predict the effects on insulin secretion by pancreatic β cells of exposure to the potassium ionophore valinomycin (Fig. 11–44). Explain your prediction.

19. Effects of a Deleted Insulin Receptor A strain of mice specifically lacking the insulin receptor of liver is found to have mild fasting hyperglycemia (blood glucose = 132 mg/dL, vs. 101 mg/dL in controls) and a more striking hyperglycemia in the fed state (glucose = 363 mg/dL, vs. 135 mg/dL in controls). The mice have higher than normal levels of glucose 6-phosphatase in the liver and elevated levels of insulin in the blood. Explain these observations.

20. Decisions on Drug Safety The drug Avandia (rosiglitazone) is effective in lowering blood glucose in patients with type 2 diabetes but also carries an increased risk of heart attack. If it were your responsibility to decide whether this drug should remain on the market (labeled with suitable warnings of its side effects) or should be withdrawn, what factors would you weigh in making your decision?

21. Type 2 Diabetes Medication The drugs acarbose (Precose) and miglitol (Glyset), used in the treatment of type 2 diabetes mellitus, inhibit α-glucosidases in the brush border of the small intestine. These enzymes degrade oligosaccharides derived from glycogen or starch to monosaccharides. Suggest a possible mechanism for the salutary effect of these drugs for individuals with diabetes. What side effects, if any, would you expect from these drugs. Why? (Hint: Review lactose intolerance, pp. 561–562).

Data Analysis Problem

22. Cloning the Pancreatic β-Cell Sulfonylurea Receptor Glyburide, a member of the sulfonylurea family of drugs shown on p. 955, is used to treat type 2 diabetes. It binds to and closes the ATP-gated K^+ channel shown in Figures 23–27 and 23–28.

(a) Given the mechanism shown in Figure 23–27, would treatment with glyburide result in increased or decreased insulin secretion by pancreatic β cells? Explain your reasoning.

(b) How does treatment with glyburide help reduce the symptoms of type 2 diabetes?

(c) Would you expect glyburide to be useful for treating type 1 diabetes? Why or why not?

Aguilar-Bryan and coauthors (1995) cloned the gene for the sulfonylurea receptor (SUR) portion of the ATP-gated K^+ channel from hamsters. The research team went to great lengths to ensure that the gene they cloned was in fact the SUR-encoding gene. Here we explore how it is possible for researchers to demonstrate that they have actually cloned the gene of interest rather than another gene.

The first step was to obtain pure SUR protein. As was already known, drugs such as glyburide bind SUR with very high affinity ($K_d < 10$ nM), and SUR has a molecular weight of 140 to 170 kDa. Aguilar-Bryan and coworkers made use of the high-affinity glyburide binding to tag the SUR protein with a radioactive label that would serve as a marker to purify the protein from a cell extract. First, they made a radiolabeled derivative of glyburide, using radioactive iodine (^{125}I):

$[^{125}I]$5-Iodo-2-hydroxyglyburide

(d) In preliminary studies, the ^{125}I-labeled glyburide derivative (hereafter, $[^{125}I]$glyburide) was shown to have the same K_d and binding characteristics as unaltered glyburide. Why was it necessary to demonstrate this (what alternative possibilities did it rule out)?

Even though $[^{125}I]$glyburide bound to SUR with high affinity, a significant amount of the labeled drug would probably dissociate from the SUR protein during purification. To prevent this, [^{125}I]glyburide had to be covalently cross-linked to SUR. There are many methods for covalent cross-linking; Aguilar-Bryan and coworkers used UV light. When aromatic molecules are exposed to short-wave UV, they enter an excited state and readily form covalent bonds with nearby molecules. By cross-linking the radiolabeled glyburide to the SUR protein, the researchers could simply track the ^{125}I radioactivity to follow SUR through the purification procedure.

Aguilar-Bryan and colleagues treated hamster HIT cells (which express SUR) with [^{125}I]glyburide and UV light, purified the ^{125}I-labeled 140 kDa protein, and sequenced its amino-terminal 25 amino acid segment; they found the sequence PLAFCGTENHSAAYRVDQGVLNNGC. The investigators then generated antibodies that bound to two short peptides in this sequence, one that bound to PLAFCGTE and the other to HSAAYRVDQGV, and showed that these antibodies bound the purified ^{125}I-labeled 140 kDa protein.

(e) Why was it necessary to include this antibody-binding step?

Next, the researchers designed PCR primers based on the sequences above and cloned a gene from a hamster cDNA library that encoded a protein that included these sequences (see Chapter 9 on biotechnology methods). The cloned putative *SUR* cDNA hybridized to an mRNA of the appropriate length that was present in cells known to contain SUR. The putative *SUR* cDNA did not hybridize to any mRNA fraction of the mRNAs isolated from hepatocytes, which do not express SUR.

(f) Why was it necessary to include this putative *SUR* cDNA–mRNA hybridization step?

Finally, the cloned gene was inserted into and expressed in COS cells, which do not normally express the *SUR* gene. The investigators mixed these cells with [^{125}I]glyburide with or without a large excess of unlabeled glyburide, exposed the cells to UV light, and measured the radioactivity of the 140 kDa protein produced. Their results are shown in the table.

Experiment	Cell type	Added putative *SUR* cDNA?	Added excess unlabeled glyburide?	^{125}I label in 140 kDa protein
1	HIT	No	No	+ + +
2	HIT	No	Yes	−
3	COS	No	No	−
4	COS	Yes	No	+ + +
5	COS	Yes	Yes	−

(g) Why was no ^{125}I-labeled 140 kDa protein found in experiment 2?

(h) How would you use the information in the table to argue that the cDNA encoded SUR?

(i) What other information would you want to collect to be more confident that you had cloned the *SUR* gene?

Reference

Aguilar-Bryan, L., Nichols, C.G., Wechsler, S.W., Clement, J.P. IV, Boyd, A.E. III, González, G., Herrera-Sosa, H., Nguy, K., Bryan, J., & Nelson, D.A. (1995) Cloning of the β cell high-affinity sulfonylurea receptor: a regulator of insulin secretion. *Science* **268**, 423–426.

INFORMATION PATHWAYS

The third and final part of this book explores the biochemical mechanisms underlying the apparently contradictory requirements for both genetic continuity and the evolution of living organisms. What is the molecular nature of genetic material? How is genetic information transmitted from one generation to the next with high fidelity? How do the rare changes in genetic material that are the raw material of evolution arise? How is genetic information ultimately expressed in the amino acid sequences of the astonishing variety of protein molecules in a living cell?

Today's understanding of information pathways has arisen from the convergence of genetics, physics, and chemistry in modern biochemistry. This was epitomized by the discovery of the double-helical structure of DNA, postulated by James Watson and Francis Crick in 1953 (see Fig. 8–13). Genetic theory contributed the concept of coding by genes. Physics permitted the determination of molecular structure by x-ray diffraction analysis. Chemistry revealed the composition of DNA. The profound impact of the Watson-Crick hypothesis arose from its ability to account for a wide range of observations derived from studies in these diverse disciplines.

This revolutionized our understanding of the structure of DNA and inevitably stimulated questions about its function. The double-helical structure itself clearly suggested how DNA might be copied so that the information it contains can be transmitted from one generation to the next. Clarification of how the information in DNA is converted into functional proteins came with the discovery of messenger RNA and transfer RNA and with the deciphering of the genetic code.

These and other major advances gave rise to the central dogma of molecular biology, comprising the three major processes in the cellular utilization of genetic information. The first is replication, the copying of parental DNA to form daughter DNA molecules with identical nucleotide sequences. The second is transcription, the process by which parts of the genetic message encoded in DNA are copied precisely into RNA. The third is translation, whereby the genetic message encoded in messenger RNA is translated on the ribosomes into a polypeptide with a particular sequence of amino acids.

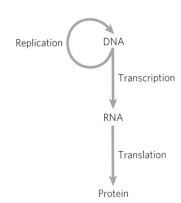

The central dogma of molecular biology, showing the general pathways of information flow via replication, transcription, and translation. The term "dogma" is a misnomer and is retained for historical reasons only. Introduced by Francis Crick at a time when little evidence supported these ideas, the dogma has become a well-established principle.

Part III explores these and related processes. In Chapter 24 we examine the structure, topology, and packaging of chromosomes and genes. The processes underlying the central dogma are elaborated in Chapters 25 through 27. Finally, we turn to regulation, examining how the expression of genetic information is controlled (Chapter 28).

A major theme running through these chapters is the added complexity inherent in the biosynthesis of macromolecules that contain information. Assembling nucleic acids and proteins with particular sequences of nucleotides and amino acids represents nothing less than preserving the faithful expression of the template upon which life itself is based. We might expect the formation of phosphodiester bonds in DNA or peptide bonds in proteins to be a trivial feat for cells, given the arsenal of enzymatic and chemical tools described in Part II. However, the framework of patterns and rules established in our examination of metabolic pathways thus far must be enlarged considerably to take into account molecular information. Bonds must be formed between *particular* subunits in informational biopolymers, avoiding either the occurrence or the persistence of sequence errors. This has an enormous impact on the thermodynamics, chemistry, and enzymology of the biosynthetic processes. Formation of a peptide bond requires an energy input of only about 21 kJ/mol of bonds and can be catalyzed by relatively simple enzymes. But to synthesize a bond between two specific amino acids at a particular point in a polypeptide, the cell invests about 125 kJ/mol while making use of more than 200 enzymes, RNA molecules, and specialized proteins. The chemistry involved in peptide bond formation does not change because of this requirement, but additional processes are layered over the basic reaction to ensure that the peptide bond is formed between particular amino acids. Biological information is expensive.

The dynamic interaction between nucleic acids and proteins is another central theme of Part III. Regulatory and catalytic RNA molecules are gradually taking a more prominent place in our understanding of these pathways (discussed in Chapters 26 and 27). However, most of the processes that make up the pathways of cellular information flow are catalyzed and regulated by proteins. An understanding of these enzymes and other proteins can have practical as well as intellectual rewards, because they form the basis of recombinant DNA technology (introduced in Chapter 9).

Evolution again constitutes an overarching theme. Many of the processes outlined in Part III can be traced back billions of years, and a few can be traced to LUCA, the last universal common ancestor. The ribosome, most of the translational apparatus, and some parts of the transcriptional machinery are shared by every living organism on this planet. Genetic information is a kind of molecular clock that can help define ancestral relationships among species. Shared information pathways connect humans to every other species now living on Earth, and to all species that came before. Exploration of these pathways is allowing scientists to slowly open the curtain on the first act—the events that may have heralded the beginning of life on Earth.

Genes and Chromosomes

The size of DNA molecules presents an interesting biological puzzle. Given that these molecules are generally much longer than the cells or viral particles that contain them **(Fig. 24–1)**, how do they fit into their cellular or viral packages? To address this question, we shift our focus from the secondary structure of DNA, considered in Chapter 8, to the extraordinary degree of organization required for the tertiary packaging of DNA into **chromosomes**—the repositories of genetic information. The chapter begins with an examination of the elements that make up viral and cellular chromosomes, and then considers chromosomal size and organization. We then discuss DNA topology, describing the coiling and supercoiling of DNA molecules. Finally, we consider the protein-DNA interactions that organize chromosomes into compact structures.

24.1 Chromosomal Elements

Cellular DNA contains genes and intergenic regions, both of which may serve functions vital to the cell. The more complex genomes, such as those of eukaryotic cells, demand increased levels of chromosomal organization, and this is reflected in the chromosomes' structural features. We begin by considering the different types of DNA sequences and structural elements within chromosomes.

Genes Are Segments of DNA That Code for Polypeptide Chains and RNAs

Our understanding of genes has evolved tremendously over the last century. Classically, a gene was defined as a portion of a chromosome that determines or affects a single character or **phenotype** (visible property), such as eye color. George Beadle and Edward Tatum proposed a molecular definition of a gene in 1940. After exposing spores of the fungus *Neurospora crassa* to x rays and other agents now known to damage DNA and cause alterations in DNA sequence **(mutations)**, they detected mutant fungal strains that lacked one or another specific enzyme, sometimes resulting in the failure of an entire metabolic pathway. Beadle and Tatum concluded that a gene is a segment of genetic material that determines, or codes for, one enzyme: the

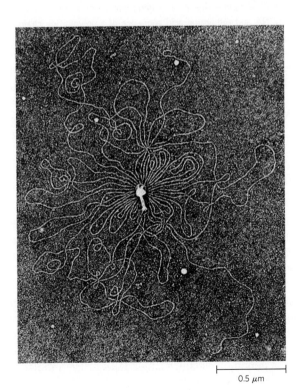

0.5 μm

FIGURE 24–1 Bacteriophage T2 protein coat surrounded by its single, linear molecule of DNA. The DNA was released by lysing the bacteriophage particle in distilled water and allowing the DNA to spread on the water surface. An undamaged T2 bacteriophage particle consists of a head structure that tapers to a tail by which the bacteriophage attaches itself to the outer surface of a bacterial cell. All the DNA shown in this electron micrograph is normally packaged inside the phage head.

one gene–one enzyme hypothesis. Later this concept was broadened to **one gene–one polypeptide**, because many genes code for a protein that is not an enzyme or for one polypeptide of a multisubunit protein.

George W. Beadle, 1903–1989

Edward L. Tatum, 1909–1975

The modern biochemical definition of a gene is even more precise. A **gene** is all the DNA that encodes the primary sequence of some final gene product, which can be either a polypeptide or an RNA with a structural or catalytic function. DNA also contains other segments or sequences that have a purely regulatory function. **Regulatory sequences** provide signals that may denote the beginning or the end of genes, or influence the transcription of genes, or function as initiation points for replication or recombination (Chapter 28). Some genes can be expressed in different ways to generate multiple gene products from a single segment of DNA. The special transcriptional and translational mechanisms that allow this are described in Chapters 26 through 28.

We can estimate directly the minimum overall size of genes that encode proteins. As described in detail in Chapter 27, each amino acid of a polypeptide chain is coded for by a sequence of three consecutive nucleotides in a single strand of DNA **(Fig. 24–2)**, with these "codons" arranged in a sequence that corresponds to the sequence of amino acids in the polypeptide that the gene encodes. A polypeptide chain of 350 amino acid residues (an average-size chain) corresponds to 1,050 bp of DNA. Many genes in eukaryotes and a few in bacteria and archaea are interrupted by noncoding DNA segments and are therefore considerably longer than this simple calculation would suggest.

How many genes are in a single chromosome? The *Escherichia coli* chromosome, one of the bacterial genomes that have been completely sequenced, is a circular DNA molecule (in the sense of an endless loop rather than a perfect circle) with 4,639,675 bp. These base pairs encode about 4,300 genes for proteins and another 157 genes for structural or catalytic RNA molecules. Among eukaryotes, the approximately 3.1 billion base pairs of the human genome include approximately 25,000 genes on the 24 different chromosomes.

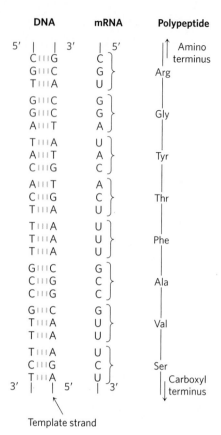

FIGURE 24–2 Colinearity of the coding nucleotide sequences of DNA and mRNA and the amino acid sequence of a polypeptide chain. The triplets of nucleotide units in DNA determine the amino acids in a protein through the intermediary mRNA. One of the DNA strands serves as a template for synthesis of mRNA, which has nucleotide triplets (codons) complementary to those of the DNA. In some bacterial and many eukaryotic genes, coding sequences are interrupted at intervals by regions of noncoding sequences (called introns).

DNA Molecules Are Much Longer Than the Cellular or Viral Packages That Contain Them

Chromosomal DNAs are often many orders of magnitude longer than the cells or viruses in which they are located (Fig. 24–1; Table 24–1). This is true of every class of organism or viral parasite.

Viruses Viruses are not free-living organisms; rather, they are infectious parasites that use the resources of a host cell to carry out many of the processes they require to propagate. Many viral particles consist of no more than a genome (usually a single RNA or DNA molecule) surrounded by a protein coat.

Almost all plant viruses and some bacterial and animal viruses have RNA genomes. These genomes tend to be particularly small. For example, the genomes of mammalian retroviruses such as HIV are about 9,000 nucleotides long, and the genome of the bacteriophage Qβ has 4,220 nucleotides. Both types of virus have single-stranded RNA genomes.

TABLE 24–1	The Sizes of DNA and Viral Particles for Some Bacterial Viruses (Bacteriophages)		
Virus	Size of viral DNA (bp)	Length of viral DNA (nm)	Long dimension of viral particle (nm)
ϕX174	5,386	1,939	25
T7	39,936	14,377	78
λ (lambda)	48,502	17,460	190
T4	168,889	60,800	210

Note: Data on size of DNA are for the replicative form (double-stranded). The contour length is calculated assuming that each base pair occupies a length of 3.4 Å (see Fig. 8–13).

The genomes of DNA viruses vary greatly in size (Table 24–1). Many viral DNAs are circular for at least part of their life cycle. During viral replication within a host cell, specific types of viral DNA called **replicative forms** may appear; for example, many linear DNAs become circular and all single-stranded DNAs become double-stranded. A typical medium-size DNA virus is bacteriophage λ (lambda), which infects *E. coli*. In its replicative form inside cells, λ DNA is a circular double helix. This double-stranded DNA contains 48,502 bp and has a contour length of 17.5 μm. Bacteriophage ϕX174 is a much smaller DNA virus; the DNA in the viral particle is a single-stranded circle, and the double-stranded replicative form contains 5,386 bp. Although viral genomes are small, the contour lengths of their DNAs are typically hundreds of times longer than the long dimensions of the viral particles that contain them (Table 24–1).

Bacteria A single *E. coli* cell contains almost 100 times as much DNA as a bacteriophage λ particle. The chromosome of an *E. coli* cell is a single double-stranded circular DNA molecule. Its 4,639,675 bp have a contour length of about 1.7 mm, some 850 times the length of the *E. coli* cell **(Fig. 24–3)**. In addition to the very large, circular DNA chromosome in their nucleoid, many bacteria contain one or more small circular DNA molecules that are free in the cytosol. These extrachromosomal elements are called **plasmids** **(Fig. 24–4**; see also p. 317). Most plasmids are only a few thousand base pairs long, but some contain more than 10,000 bp. They carry genetic information and undergo replication to yield daughter plasmids, which pass into the daughter cells at cell division. Plasmids have been found in yeast and other fungi as well as in bacteria.

In many cases plasmids confer no obvious advantage on their host, and their sole function seems to be self-propagation. However, some plasmids carry genes that are useful to the host bacterium. For example, some plasmid genes make a host bacterium resistant to antibacterial agents. Plasmids carrying the gene for the enzyme β-lactamase confer resistance to β-lactam antibiotics such as penicillin, ampicillin, and amoxicillin (see Fig. 6–31). These and similar plasmids may pass from an antibiotic-resistant cell to an antibiotic-sensitive cell of the same or another bacterial species, making the recipient cell antibiotic resistant. The extensive use of antibiotics in some human populations has served as a strong selective force, encouraging the spread of antibiotic resistance–coding plasmids (as well as transposable elements, described below, that harbor similar genes) in disease-causing bacteria. Physicians are becoming increasingly reluctant to prescribe antibiotics unless a clear clinical need is confirmed. For similar reasons, the widespread use of antibiotics in animal feeds is being curbed.

Eukaryotes A yeast cell, one of the simplest eukaryotes, has 2.6 times more DNA in its genome than an *E. coli* cell (Table 24–2). Cells of *Drosophila*, the fruit fly used in classical genetic studies, contain more than 35 times as much DNA as *E. coli* cells, and human cells have almost 700 times as much. The cells of many plants and amphibians contain even more. The genetic material of eukaryotic cells is apportioned into chromosomes, the diploid ($2n$) number depending on the species (Table 24–2). A human somatic cell, for example, has 46 chromosomes **(Fig. 24–5)**. Each chromosome of a eukaryotic cell, such as that shown in Figure 24–5a, contains a single, very large, duplex DNA molecule. The DNA molecules in the 24 different types of human chromosomes (22 matching pairs plus the X and Y sex chromosomes) vary in length over a 25-fold range. Each type of chromosome in eukaryotes carries a characteristic set of genes.

The DNA molecules of one human genome (22 chromosomes plus X and Y or two X chromosomes), placed end to end, would extend for about a meter. Most human cells are diploid, and each cell contains a total of 2 m of DNA. An adult human body contains approximately 10^{14} cells and thus a total DNA length of 2×10^{11} km. Compare this with the circumference of the earth (4×10^4 km) or the distance between the earth and the sun (1.5×10^8 km)—a dramatic illustration of the extraordinary degree of DNA compaction in our cells.

FIGURE 24–3 The length of the *E. coli* chromosome (1.7 mm) depicted in linear form relative to the length of a typical *E. coli* cell (2 μm).

E. coli

E. coli
DNA

FIGURE 24–4 DNA from a lysed *E. coli* cell. In this electron micrograph several small, circular plasmid DNAs are indicated by white arrows. The black spots and white specks are artifacts of the preparation.

TABLE 24–2   DNA, Gene, and Chromosome Content in Some Genomes

	Total DNA (bp)	Number of chromosomes*	Approximate number of genes
Escherichia coli K12 (bacterium)	4,639,675	1	4,435
Saccharomyces cerevisiae (yeast)	12,080,000	16[†]	5,860
Caenorhabditis elegans (nematode)	90,269,800	12[‡]	23,000
Arabidopsis thaliana (plant)	119,186,200	10	33,000
Drosophila melanogaster (fruit fly)	120,367,260	18	20,000
Oryza sativa (rice)	480,000,000	24	57,000
Mus musculus (mouse)	2,634,266,500	40	27,000
Homo sapiens (human)	3,070,128,600	46	29,000

Note: This information is constantly being refined. For the most current information, consult the websites for the individual genome projects.

*The diploid chromosome number is given for all eukaryotes except yeast.

[†]Haploid chromosome number. Wild yeast strains generally have eight (octoploid) or more sets of these chromosomes.

[‡]Number for females, with two X chromosomes. Males have an X but no Y, thus 11 chromosomes in all.

(b)

(a)

FIGURE 24–5 Eukaryotic chromosomes. (a) A pair of linked and condensed sister chromatids of a human chromosome. Eukaryotic chromosomes are in this state after replication at metaphase during mitosis. **(b)** A complete set of chromosomes from a leukocyte from one of the authors. There are 46 chromosomes in every normal human somatic cell.

Eukaryotic cells also have organelles, mitochondria **(Fig. 24–6)** and chloroplasts, that contain DNA. Mitochondrial DNA (mtDNA) molecules are much smaller than the nuclear chromosomes. In animal cells, mtDNA contains fewer than 20,000 bp (16,569 bp

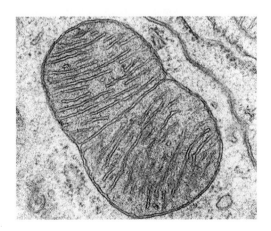

FIGURE 24–6 A dividing mitochondrion. Some mitochondrial proteins and RNAs are encoded by one of the copies of the mitochondrial DNA (none of which are visible here). The DNA (mtDNA) is replicated each time the mitochondrion divides, before cell division.

in human mtDNA) and is a circular duplex. Each mitochondrion typically has 2 to 10 copies of this mtDNA molecule, and the number can rise to hundreds in certain cells of an embryo that is undergoing cell differentiation. In a few organisms (trypanosomes, for example) each mitochondrion contains thousands of copies of mtDNA, organized into a complex and interlinked matrix known as a kinetoplast. Plant cell mtDNA ranges in size from 200,000 to 2,500,000 bp. Chloroplast DNA (cpDNA) also exists as circular duplexes and ranges in size from 120,000 to 160,000 bp. The evolutionary origin of mitochondrial and chloroplast DNAs has been the subject of much speculation. A widely accepted view is that they are vestiges of the chromosomes of ancient bacteria that gained access to the cytoplasm of host cells and became the precursors of these organelles (see Fig. 1–38). Mitochondrial DNA codes for the mitochondrial tRNAs and rRNAs and for a few mitochondrial proteins. More than 95% of mitochondrial proteins are encoded by nuclear DNA. Mitochondria and chloroplasts divide when the cell divides. Their DNA is replicated before and during division, and the daughter DNA molecules pass into the daughter organelles.

FIGURE 24–7 Introns in two eukaryotic genes. The gene for ovalbumin has seven introns (A to G), splitting the coding sequences into eight exons (L, and 1 to 7). The gene for the β subunit of hemoglobin has two introns and three exons, including one intron that alone contains more than half the base pairs of the gene.

Eukaryotic Genes and Chromosomes Are Very Complex

Many bacterial species have only one chromosome per cell and, in nearly all cases, each chromosome contains only one copy of each gene. A very few genes, such as those for rRNAs, are repeated several times. Genes and regulatory sequences account for almost all the DNA in bacteria. Moreover, almost every gene is precisely colinear with the amino acid sequence (or RNA sequence) for which it codes (Fig. 24–2).

The organization of genes in eukaryotic DNA is structurally and functionally much more complex. The study of eukaryotic chromosome structure, and more recently the sequencing of entire eukaryotic genomes, has yielded many surprises. Many, if not most, eukaryotic genes have a distinctive and puzzling structural feature: their nucleotide sequences contain one or more intervening segments of DNA that do not code for the amino acid sequence of the polypeptide product. These nontranslated inserts interrupt the otherwise colinear relationship between the nucleotide sequence of the gene and the amino acid sequence of the polypeptide it encodes. Such nontranslated DNA segments in genes are called **intervening sequences** or **introns**, and the coding segments are called **exons**. Few bacterial genes contain introns. In higher eukaryotes, the typical gene has much more intron sequence than sequences devoted to exons. For example, in the gene coding for the single polypeptide chain of ovalbumin, an avian egg protein **(Fig. 24–7)**, the introns are much longer than the exons; altogether, seven introns make up 85% of the gene's DNA. The gene for the muscle protein titin is the intron champion, with 178 introns. Genes for histones seem to have no introns. In most cases the function of introns is not clear. In total, only about 1.5% of human DNA is "coding" or exon DNA, carrying information for protein products. However, when the much larger introns are included in the count, as much as 30% of the human genome consists of genes. A great deal of work remains to be done to understand the other genomic sequences. Much of the DNA that is not within genes is made up of repeated sequences of several kinds. These include transposable elements (transposons), molecular parasites that account for nearly half of the DNA in the human genome (see Fig. 9–29 and Chapters 25 and 26).

Approximately 3% of the human genome consists of **highly repetitive** sequences, also referred to as **simple-sequence DNA** or **simple sequence repeats (SSR)**. These short sequences, generally less than 10 bp long, are sometimes repeated millions of times per cell. The simple-sequence DNA has also been called **satellite DNA**, so named because its unusual base composition often causes it to migrate as "satellite" bands (separated from the rest of the DNA) when fragmented cellular DNA samples are centrifuged in a cesium chloride density gradient. Studies suggest that simple-sequence DNA does not encode proteins or RNAs. Unlike the transposable elements, the highly repetitive DNA can have identifiable functional importance in human cellular metabolism, because much of it is associated with two defining features of eukaryotic chromosomes: centromeres and telomeres.

The **centromere (Fig. 24–8)** is a sequence of DNA that functions during cell division as an attachment point for proteins that link the chromosome to the mitotic spindle. This attachment is essential for the equal and orderly distribution of chromosome sets to daughter cells. The centromeres of *Saccharomyces cerevisiae* have been isolated and studied. The sequences essential to centromere function are about 130 bp long and are very rich in A=T pairs. The centromeric sequences of higher eukaryotes are much longer and, unlike those of yeast, generally contain simple-sequence DNA, which consists of thousands of tandem copies of one or a few short sequences of 5 to 10 bp, in the same orientation. The precise role of simple-sequence DNA in centromere function is not yet understood.

Telomeres (Greek *telos,* "end") are sequences at the ends of eukaryotic chromosomes that help stabilize the chromosome. Telomeres end with multiple repeated sequences of the form

$$(5')(T_xG_y)_n$$
$$(3')(A_xC_y)_n$$

FIGURE 24–8 Important structural elements of a yeast chromosome.

TABLE 24–3   Telomere Sequences

Organism	Telomere repeat sequence
Homo sapiens (human)	$(TTAGGG)_n$
Tetrahymena thermophila (ciliated protozoan)	$(TTGGGG)_n$
Saccharomyces cerevisiae (yeast)	$((TG)_{1-3}(TG)_{2-3})_n$
Arabidopsis thaliana (plant)	$(TTTAGGG)_n$

where x and y are generally between 1 and 4 (Table 24–3). The number of telomere repeats, n, is in the range of 20 to 100 for most single-celled eukaryotes and is generally more than 1,500 in mammals. The ends of a linear DNA molecule cannot be routinely replicated by the cellular replication machinery (which may be one reason why bacterial DNA molecules are circular). Repeated telomeric sequences are added to eukaryotic chromosome ends primarily by the enzyme telomerase (see Fig. 26–38).

Artificial chromosomes (Chapter 9) have been constructed as a means of better understanding the functional significance of many structural features of eukaryotic chromosomes. A reasonably stable artificial linear chromosome requires only three components: a centromere, a telomere at each end, and sequences that allow the initiation of DNA replication. Yeast artificial chromosomes (YACs; see Fig. 9–6) have been developed as a research tool in biotechnology. Similarly, human artificial chromosomes (HACs) are being developed for the treatment of genetic diseases. These may eventually provide a new path to the intracellular replacement of missing or defective gene products or somatic gene therapy.

SUMMARY 24.1 Chromosomal Elements

▶ Genes are segments of a chromosome that contain the information for a functional polypeptide or RNA molecule. In addition to genes, chromosomes contain a variety of regulatory sequences involved in replication, transcription, and other processes.

▶ Genomic DNA and RNA molecules are generally orders of magnitude longer than the viral particles or cells that contain them.

▶ Many genes in eukaryotic cells (but few in bacteria and archaea) are interrupted by noncoding sequences, or introns. The coding segments separated by introns are called exons.

▶ Only about 1.5% of human genomic DNA encodes proteins. Even when introns are included, less than one-third of human genomic DNA consists of genes. Much of the remainder consists of repeated sequences of various types. Nucleic acid parasites known as transposons account for about half of the human genome.

▶ Eukaryotic chromosomes have two important special-function repetitive DNA sequences: centromeres, which are attachment points for the mitotic spindle, and telomeres, located at the ends of chromosomes.

24.2 DNA Supercoiling

Cellular DNA, as we have seen, is extremely compacted, implying a high degree of structural organization. The folding mechanism not only must pack the DNA but also must permit access to the information in the DNA. Before considering how this is accomplished in processes such as replication and transcription, we need to examine an important property of DNA structure known as **supercoiling**.

"Supercoiling" means the coiling of a coil. An old-fashioned telephone cord, for example, is typically a coiled wire. The path taken by the wire between the base of the phone and the receiver often includes one or more supercoils **(Fig. 24–9)**. DNA is coiled in the form of a double helix, with both strands of the DNA coiling around an axis. The further coiling of that axis upon itself **(Fig. 24–10)** produces DNA supercoiling. As detailed below, DNA supercoiling is generally a manifestation of structural strain. When there is no net bending of the DNA axis upon itself, the DNA is said to be in a **relaxed** state.

FIGURE 24–9 Supercoils. A typical phone cord is coiled like a DNA helix, and the coiled cord can itself coil in a supercoil. The illustration is especially appropriate because an examination of phone cords helped lead Jerome Vinograd and his colleagues to the insight that many properties of small circular DNAs can be explained by supercoiling. They first detected DNA supercoiling—in small circular viral DNAs—in 1965.

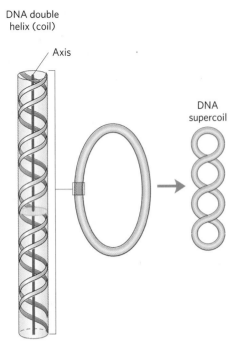

FIGURE 24–10 Supercoiling of DNA. When the axis of the DNA double helix is coiled on itself, it forms a new helix (superhelix). The DNA superhelix is usually called a supercoil.

We might have predicted that DNA compaction involved some form of supercoiling. Perhaps less predictable is that replication and transcription of DNA also affect and are affected by supercoiling. Both processes require a separation of DNA strands—a process complicated by the helical interwinding of the strands (as demonstrated in **Fig. 24–11**).

That a DNA molecule would bend on itself and become supercoiled in tightly packaged cellular DNA would seem logical, then, and perhaps even trivial, were it not for one additional fact: many circular DNA molecules remain highly supercoiled even after they are extracted and purified, freed from protein and other cellular components. This indicates that supercoiling is an intrinsic property of DNA tertiary structure. It occurs in all cellular DNAs and is highly regulated by each cell.

Several measurable properties of supercoiling have been established, and the study of supercoiling has provided many insights into DNA structure and function. This work has drawn heavily on concepts derived from a branch of mathematics called **topology**, the study of the properties of an object that do not change under continuous deformations. For DNA, continuous deformations include conformational changes due to thermal motion or an interaction with proteins or other molecules; discontinuous deformations involve DNA strand breakage. For circular DNA molecules, a topological property is one that is unaffected by deformations of the DNA strands as long as no breaks are introduced. Topological properties are changed only by breakage and rejoining of the backbone of one or both DNA strands.

We now examine the fundamental properties and physical basis of supercoiling.

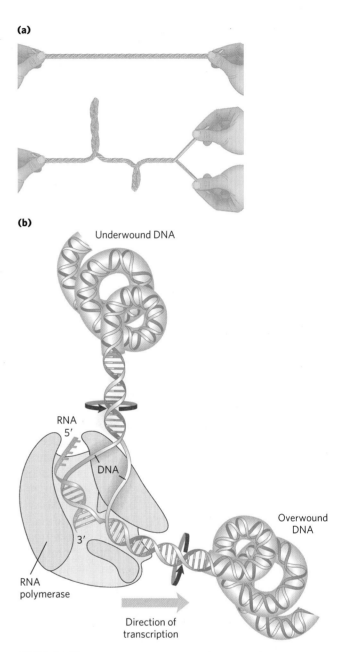

FIGURE 24–11 The effects of replication and transcription on DNA supercoiling. Because DNA is a double-helical structure, strand separation leads to added stress and supercoiling if the DNA is constrained (not free to rotate) ahead of the strand separation. **(a)** The general effect can be illustrated by twisting two strands of a rubber band about each other to form a double helix. If one end is constrained, separating the two strands at the other end will lead to twisting. **(b)** In a DNA molecule, the progress of a DNA polymerase or RNA polymerase (as shown here) along the DNA involves separation of the strands. As a result, the DNA becomes overwound ahead of the enzyme (upstream) and underwound behind it (downstream). Red arrows indicate the direction of winding.

Most Cellular DNA Is Underwound

To understand supercoiling, we must first focus on the properties of small circular DNAs such as plasmids and small viral DNAs. When these DNAs have no breaks in either strand, they are referred to as **closed-circular DNAs**. If the DNA of a closed-circular molecule conforms

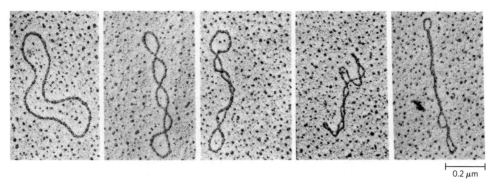

0.2 μm

FIGURE 24–12 Relaxed and supercoiled plasmid DNAs. The molecule in the leftmost electron micrograph is relaxed; the degree of supercoiling increases from left to right.

closely to the B-form structure (the Watson-Crick structure; see Fig. 8–13), with one turn of the double helix per 10.5 bp, the DNA is relaxed rather than supercoiled **(Fig. 24–12).** Supercoiling results when DNA is subject to some form of structural strain. Purified closed-circular DNA is rarely relaxed, regardless of its biological origin. Furthermore, DNAs derived from a given cellular source have a characteristic degree of supercoiling. DNA structure is therefore strained in a manner that is regulated by the cell to induce the supercoiling.

In almost every instance, the strain is a result of **underwinding** of the DNA double helix in the closed circle. In other words, the DNA has *fewer* helical turns than would be expected for the B-form structure. The effects of underwinding are summarized in **Figure 24–13**. An 84 bp segment of a circular DNA in the relaxed state would contain eight double-helical turns, or one for every 10.5 bp. If one of these turns were removed, there would be (84 bp)/7 = 12.0 bp per turn, rather than the 10.5 found in B-DNA (Fig. 24–13b). This is a deviation from the most stable DNA form, and the molecule is thermodynamically strained as a result. Generally, much of this strain would be accommodated by coiling the axis of the DNA on itself to form a supercoil (Fig. 24–13c; some of the strain in this 84 bp segment would simply become dispersed in the untwisted structure of the larger DNA molecule). In principle, the strain could also be accommodated by separating the two DNA strands over a distance of about 10 bp (Fig. 24–13d). In isolated closed-circular DNA, strain introduced by underwinding is generally accommodated by supercoiling rather than strand separation, because coiling the axis of the DNA usually requires less energy than breaking the hydrogen bonds that stabilize paired bases. Note, however, that the underwinding of DNA in vivo makes separation of the DNA strands easier, facilitating access to the information they contain.

Every cell actively underwinds its DNA with the aid of enzymatic processes (described below), and the resulting strained state represents a form of stored energy. Cells maintain DNA in an underwound state to facilitate its compaction by coiling. The underwinding of DNA is also important to enzymes of DNA metabolism

that must bring about strand separation as part of their function.

The underwound state can be maintained only if the DNA is a closed circle or if it is bound and stabilized by proteins so that the strands are not free to rotate about each other. If there is a break in one strand of an isolated, protein-free circular DNA, free rotation at that point will cause the underwound DNA to revert spontaneously to the relaxed state. In a closed-circular DNA molecule, however, the number of helical turns cannot be changed without at least transiently breaking one of

(a) Relaxed (8 turns)

(b) Strained (7 turns)

(c) Supercoil

(d) Strand separation

FIGURE 24–13 Effects of DNA underwinding. (a) A segment of DNA in a closed-circular molecule, 84 bp long, in its relaxed form with eight helical turns. **(b)** Removal of one turn induces structural strain. **(c)** The strain is generally accommodated by formation of a supercoil. **(d)** DNA underwinding also makes the separation of strands somewhat easier. In principle, each turn of underwinding should facilitate strand separation over about 10 bp, as shown. However, the hydrogen-bonded base pairs would generally preclude strand separation over such a short distance, and the effect becomes important only for longer DNAs and higher levels of DNA underwinding.

(a) $Lk = 1$

(b) $Lk = 6$

FIGURE 24–14 Linking number, Lk. Here, as usual, each blue ribbon represents one strand of a double-stranded DNA molecule. For the molecule in **(a)**, $Lk = 1$. For the molecule in **(b)**, $Lk = 6$. One of the strands in (b) is kept untwisted for illustrative purposes, to define the border of an imaginary surface (shaded blue). The number of times the twisting strand penetrates this surface provides a rigorous definition of linking number.

the DNA strands. The number of helical turns in a DNA molecule therefore provides a precise description of supercoiling.

DNA Underwinding Is Defined by Topological Linking Number

The field of topology provides some ideas that are useful to the discussion of DNA supercoiling, particularly the concept of **linking number**. Linking number is a topological property of double-stranded DNA, because it does not vary when the DNA is bent or deformed, as long as both DNA strands remain intact. Linking number (Lk) is illustrated in **Figure 24–14**.

Let's begin by visualizing the separation of the two strands of a double-stranded circular DNA. If the two strands are linked as shown in Figure 24–14a, they are effectively joined by what can be described as a topological bond. Even if all hydrogen bonds and base-stacking interactions were abolished such that the strands were not in physical contact, this topological bond would still link the two strands. Visualize one of the circular strands as the boundary of a surface (such as a soap film spanning the space framed by a circular wire before you blow a soap bubble). The linking number can be defined as the number of times the second strand pierces this surface. For the molecule in Figure 24–14a, $Lk = 1$; for that in Figure 24–14b, $Lk = 6$. The linking number for a closed-circular DNA is always an integer. By convention, if the links between two DNA strands are arranged so that the strands are interwound in a right-handed helix, the linking number is defined as positive ($+$); for strands interwound in a left-handed helix, the linking number is negative ($-$). Negative linking numbers are, for all practical purposes, not encountered in DNA.

Nick

strand break

(b) Lk undefined

$\Delta Lk = -2$

(a) $Lk = 200 = Lk_0$

(c) $Lk = 198$

FIGURE 24–15 Linking number applied to closed-circular DNA molecules. A 2,100 bp circular DNA is shown in three forms: **(a)** relaxed, $Lk = 200$; **(b)** relaxed with a nick (break) in one strand, Lk undefined; and **(c)** underwound by two turns, $Lk = 198$. The underwound molecule generally exists as a supercoiled molecule, but underwinding also facilitates the separation of DNA strands.

We can now extend these ideas to a closed-circular DNA with 2,100 bp **(Fig. 24–15a)**. When the molecule is relaxed, the linking number is simply the number of base pairs divided by the number of base pairs per turn, which is close to 10.5; so in this case, $Lk = 200$. For a circular DNA molecule to have a topological property such as linking number, neither strand may contain a break. If there is a break in either strand, the strands can, in principle, be unraveled and separated completely. In this case, no topological bond exists and Lk is undefined (Fig. 24–15b).

We can now describe DNA underwinding in terms of changes in the linking number. The linking number in relaxed DNA, Lk_0, is used as a reference. For the molecule shown in Figure 24–15a, $Lk_0 = 200$; if two turns are removed from this molecule, $Lk = 198$. The change can be described by the equation

$$\Delta Lk = Lk - Lk_0 \qquad (24\text{–}1)$$
$$= 198 - 200 = -2$$

It is often convenient to express the change in linking number in terms of a quantity that is independent of the length of the DNA molecule. This quantity, called the **specific linking difference** or **superhelical density** (σ), is a measure of the number of turns removed relative to the number present in relaxed DNA:

$$\sigma = \frac{\Delta Lk}{Lk_0} \qquad (24\text{–}2)$$

In the example in Figure 24–15c, $\sigma = -0.01$, which means that 1% (2 of 200) of the helical turns present in the DNA (in its B form) have been removed. The degree

FIGURE 24-16 Negative and positive supercoils. For the relaxed DNA molecule of Figure 24-15a, underwinding or overwinding by two helical turns (Lk = 198 or 202) will produce negative or positive supercoiling, respectively. Note that the DNA axis twists in opposite directions in the two cases.

of underwinding in cellular DNAs generally falls in the range of 5% to 7%; that is, σ = −0.05 to −0.07. The negative sign indicates that the change in linking number is due to underwinding of the DNA. The supercoiling induced by underwinding is therefore defined as negative supercoiling. Conversely, under some conditions DNA can be overwound, resulting in positive supercoiling. Note that the twisting path taken by the axis of the DNA helix when the DNA is underwound (negative supercoiling) is the mirror image of that taken when the DNA is overwound (positive supercoiling) **(Fig. 24-16)**. Supercoiling is not a random process; the path of the supercoiling is largely prescribed by the torsional strain imparted to the DNA by decreasing or increasing the linking number relative to B-DNA.

Linking number can be changed by ±1 by breaking one DNA strand, rotating one of the ends 360° about the unbroken strand, and rejoining the broken ends. This change has no effect on the number of base pairs or the number of atoms in the circular DNA molecule. Two forms of a circular DNA that differ only in a topological property such as linking number are referred to as **topoisomers**.

WORKED EXAMPLE 24-1   Calculation of Superhelical Density

What is the superhelical density (σ) of a closed-circular DNA with a length of 4,200 bp and a linking number (Lk) of 374? What is the superhelical density of the same DNA when Lk = 412? Are these molecules negatively or positively supercoiled?

Solution: First, calculate Lk_0 by dividing the length of the closed-circular DNA (in bp) by 10.5 bp/turn: (4,200 bp)/(10.5 bp/turn) = 400. We can now calculate ΔLk from Equation 24-1: $\Delta Lk = Lk - Lk_0 = 374 - 400 = -26$. Substituting the values for ΔLk and Lk_0 into Equation

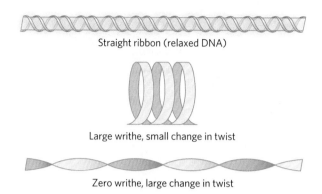

Straight ribbon (relaxed DNA)

Large writhe, small change in twist

Zero writhe, large change in twist

FIGURE 24-17 Ribbon model for illustrating twist and writhe. The tan ribbon represents the axis of a relaxed DNA molecule. Strain introduced by twisting the ribbon (underwinding the DNA) can be manifested as writhe or twist. Topological changes in linking number are usually accompanied by geometric changes in both writhe and twist.

24-2: $\sigma = \Delta Lk/Lk_0 = -26/400 = -0.065$. Since the superhelical density is negative, this DNA molecule is negatively supercoiled.

When the same DNA molecule has an Lk of 412, $\Delta Lk = 412 - 400 = 12$ and $\sigma = 12/400 = 0.03$. The superhelical density is positive, and the molecule is positively supercoiled.

Linking number can be broken down into two structural components, **twist (Tw)** and **writhe (Wr)** **(Fig. 24-17)**. These are more difficult to describe than linking number, but writhe may be thought of as a measure of the coiling of the helix axis, and twist as determining the local twisting or spatial relationship of neighboring base pairs. When the linking number changes, some of the resulting strain is usually compensated for by writhe (supercoiling) and some by changes in twist, giving rise to the equation

$$Lk = Tw + Wr$$

Tw and Wr need not be integers. Twist and writhe are geometric rather than topological properties, because they may be changed by deformation of a closed-circular DNA molecule.

In addition to causing supercoiling and making strand separation somewhat easier, the underwinding of DNA facilitates structural changes in the molecule. These are of less physiological importance but help illustrate the effects of underwinding. Recall that a cruciform (see Fig. 8-19) generally contains a few unpaired bases; DNA underwinding helps to maintain the required strand separation **(Fig. 24-18)**. Underwinding of a right-handed DNA helix also facilitates the formation of short stretches of left-handed Z-DNA in regions where the base sequence is consistent with the Z form (see Chapter 8).

Topoisomerases Catalyze Changes in the Linking Number of DNA

DNA supercoiling is a precisely regulated process that influences many aspects of DNA metabolism. Every cell

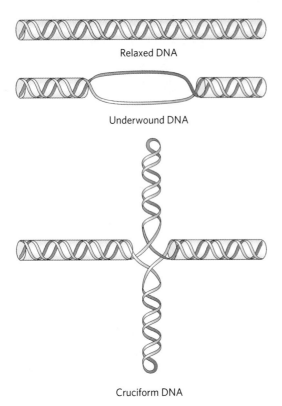

FIGURE 24–18 Promotion of cruciform structures by DNA underwinding. In principle, cruciforms can form at palindromic sequences (see Fig. 8-19), but they seldom occur in relaxed DNA because the linear DNA accommodates more paired bases than does the cruciform structure. Underwinding of the DNA facilitates the partial strand separation needed to promote cruciform formation at appropriate sequences.

has enzymes with the sole function of underwinding and/or relaxing DNA. The enzymes that increase or decrease the extent of DNA underwinding are **topoisomerases**; the property of DNA that they change is the linking number. These enzymes play an especially important role in processes such as replication and DNA packaging. There are two classes of topoisomerases. **Type I topoisomerases** act by transiently breaking one of the two DNA strands, passing the unbroken strand through the break and rejoining the broken ends; they change Lk in increments of 1. **Type II topoisomerases** break both DNA strands and change Lk in increments of 2.

The effects of these enzymes can be demonstrated with agarose gel electrophoresis **(Fig. 24–19)**. A population of identical plasmid DNAs with the same linking number migrates as a discrete band during electrophoresis. Topoisomers with Lk values differing by as little as 1 can be separated by this method, so changes in linking number induced by topoisomerases are readily detected.

E. coli has at least four individual topoisomerases (I through IV). Those of type I (topoisomerases I and III) generally relax DNA by removing negative supercoils (increasing Lk). The way in which bacterial type I topoisomerases change linking number is illustrated in **Figure 24–20**. A bacterial type II enzyme, called either topoisomerase II or DNA gyrase, can introduce negative

FIGURE 24–19 Visualization of topoisomers. In this experiment, all DNA molecules have the same number of base pairs but exhibit some range in the degree of supercoiling. Because supercoiled DNA molecules are more compact than relaxed molecules, they migrate more rapidly during gel electrophoresis. The gels shown here separate topoisomers (moving from top to bottom) over a limited range of superhelical density. In lane 1, highly supercoiled DNA migrates in a single band, even though different topoisomers are probably present. Lanes 2 and 3 illustrate the effect of treating the supercoiled DNA with a type I topoisomerase; the DNA in lane 3 was treated for a longer time than that in lane 2. As the superhelical density of the DNA is reduced to the point where it corresponds to the range in which the gel can resolve individual topoisomers, distinct bands appear. Individual bands in the region indicated by the bracket next to lane 3 each contain DNA circles with the same linking number; the linking number changes by 1 from one band to the next.

supercoils (decrease Lk). It uses the energy of ATP to accomplish this. To alter DNA linking number, type II topoisomerases cleave both strands of a DNA molecule and pass another duplex through the break. The degree of supercoiling of bacterial DNA is maintained by regulation of the net activity of topoisomerases I and II.

Eukaryotic cells also have type I and type II topoisomerases. The type I enzymes are topoisomerases I and III; the single type II enzyme has two isoforms in vertebrates, called IIα and IIβ. Most of the type II enzymes, including a DNA gyrase in archaea, are related and define a family called type IIA. Archaea also have an unusual enzyme, topoisomerase VI, which alone defines the type IIB family. The eukaryotic type II topoisomerases cannot underwind DNA (introduce negative supercoils), but they can relax both positive and negative supercoils **(Fig. 24–21)**.

As we will show in the next few chapters, topoisomerases play a critical role in every aspect of DNA metabolism. As a consequence, they are important drug targets for the treatment of bacterial infections and cancer (Box 24–1).

DNA Compaction Requires a Special Form of Supercoiling

Supercoiled DNA molecules are uniform in a number of respects. The supercoils are right-handed in a negatively supercoiled DNA molecule (Fig. 24–16), and they

① Active-site Tyr attacks a phosphodiester bond in one DNA strand, cleaving it and creating a covalent 5′-phosphotyrosyl protein-DNA linkage.

④ Enzyme in closed conformation; liberated 3′-OH attacks the 5′-phosphotyrosyl protein-DNA linkage to religate the cleaved DNA strand.

MECHANISM FIGURE 24–20   The type I topoisomerase reaction. Bacterial topoisomerase I increases *Lk* by breaking one DNA strand, passing the unbroken strand through the break, then resealing the break.

Nucleophilic attack by the active-site Tyr residue breaks one DNA strand. The ends are ligated by a second nucleophilic attack. At each step, one high-energy bond replaces another.

FIGURE 24–21   Alteration of the linking number by a eukaryotic type IIα topoisomerase. The general mechanism features the passage of one intact duplex DNA segment through a transient double-strand break in another segment. The DNA segment enters and leaves the topoisomerase through gated cavities above and below the bound DNA, which are called the N gate and the C gate. Two ATPs are bound and hydrolyzed during this cycle. The enzyme structure and use of ATP are specific to this reaction.

BOX 24–1 🏥 MEDICINE | Curing Disease by Inhibiting Topoisomerases

The topological state of cellular DNA is intimately connected with its function. Without topoisomerases, cells cannot replicate or package their DNA, or express their genes—and they die. Inhibitors of topoisomerases have therefore become important pharmaceutical agents, targeted at infectious agents and malignant cells.

Two classes of bacterial topoisomerase inhibitors have been developed as antibiotics. The coumarins, including novobiocin and coumermycin A1, are natural products derived from *Streptomyces* species. They inhibit the ATP binding of the bacterial type II topoisomerases, DNA gyrase and topoisomerase IV. These antibiotics are not often used to treat infections in humans, but research continues to identify clinically effective variants.

The quinolone antibiotics, also inhibitors of bacterial DNA gyrase and topoisomerase IV, first appeared in 1962 with the introduction of nalidixic acid. This compound had limited effectiveness and is no longer used clinically in the United States, but the continued development of this class of drugs eventually led to the introduction of the fluoroquinolones, exemplified by ciprofloxacin (Cipro). The quinolones act by blocking the last step of the topoisomerase reaction, the resealing of the DNA strand breaks. Ciprofloxacin is a wide-spectrum antibiotic. It is one of the few antibiot-

ics reliably effective in treating anthrax infections and is considered a valuable agent in protection against possible bioterrorism. Quinolones are selective for the bacterial topoisomerases, inhibiting the eukaryotic enzymes only at concentrations several orders of magnitude greater than the therapeutic doses.

Some of the most important chemotherapeutic agents used in cancer treatment are inhibitors of human topoisomerases. Topoisomerases are generally

Irinotecan

Nalidixic acid

Ciprofloxacin

Topotecan

tend to be extended and narrow rather than compacted, often with multiple branches **(Fig. 24–22)**. At the superhelical densities normally encountered in cells, the length of the supercoil axis, including branches, is about 40% of the length of the DNA. This type of super-

coiling is referred to as **plectonemic** (from the Greek *plektos*, "twisted," and *nema*, "thread"). This term can be applied to any structure with strands intertwined in some simple and regular way, and it is a good description of the general structure of supercoiled DNA in solution.

Branch points

Supercoil axis

(a)   (b)   (c)

FIGURE 24–22 Plectonemic supercoiling. (a) Electron micrograph of plectonemically supercoiled plasmid DNA and **(b)** an interpretation of the observed structure. The purple lines show the axis of the supercoil; note the branching of the supercoil. **(c)** An idealized representation of this structure.

present at elevated levels in tumor cells, and agents targeted to these enzymes are much more toxic to the tumors than to most other tissue types. Inhibitors of both type I and type II topoisomerases have been developed as anticancer drugs.

Camptothecin, isolated from a Chinese ornamental tree and first tested clinically in the 1970s, is an inhibitor of eukaryotic type I topoisomerases. Clinical trials indicated limited effectiveness, despite its early promise in preclinical work on mice. However, two effective derivatives, irinotecan (Campto) and topotecan (Hycamtin)—used to treat colorectal cancer and ovarian cancer, respectively—were developed in the 1990s. Additional derivatives are likely to be approved for clinical use in the coming years. All of these drugs act by trapping the topoisomerase-DNA complex in which the DNA is cleaved, inhibiting religation.

The human type II topoisomerases are targeted by a variety of antitumor drugs, including doxorubicin (Adriamycin), etoposide (Etopophos), and ellipticine. Doxorubicin, effective against several kinds of human tumors, is an anthracycline in clinical use. Most of these drugs stabilize the covalent topoisomerase-DNA (cleaved) complex.

All of these anticancer agents generally increase the levels of DNA damage in the targeted, rapidly growing tumor cells. However, noncancerous tissues can also be affected, leading to a more general toxicity and unpleasant side effects that must be managed during therapy. As cancer therapies become more effective and survival statistics for cancer patients improve, the independent appearance of new tumors is becoming a greater problem. In the continuing search for new cancer therapies, the topoisomerases are likely to remain prominent targets for research.

Doxorubicin

Etoposide

Ellipticine

Plectonemic supercoiling, the form observed in isolated DNAs in the laboratory, does not produce sufficient compaction to package DNA in the cell. A second form of supercoiling, **solenoidal (Fig. 24–23)**, can be adopted by an underwound DNA. Instead of the extended right-handed supercoils characteristic of the plectonemic form, solenoidal supercoiling involves tight left-handed turns, similar to the shape taken up by a garden hose neatly wrapped on a reel. Although their structures are dramatically different, plectonemic and solenoidal supercoiling are two forms of negative supercoiling that can be taken up by the *same* segment of underwound DNA. The two forms are readily interconvertible. Although the plectonemic form is more stable in solution, the solenoidal form can be stabilized by protein binding as it is in eukaryotic chromosomes. It provides a much greater degree of compaction (Fig. 24–23). Solenoidal supercoiling is the mechanism by which underwinding contributes to DNA compaction.

SUMMARY 24.2 DNA Supercoiling

▶ Most cellular DNAs are supercoiled. Underwinding decreases the total number of helical turns in the DNA relative to the relaxed, B form. To maintain an underwound state, DNA must be either a closed circle or bound to protein. Underwinding is quantified by a topological parameter called linking number, Lk.

▶ Underwinding is measured in terms of specific linking difference, σ (also called superhelical density), which is $(Lk - Lk_0)/Lk_0$. For cellular DNAs, σ is typically -0.05 to -0.07, which means that approximately 5% to 7% of the helical turns in the DNA have been removed. DNA underwinding facilitates strand separation by enzymes of DNA metabolism.

▶ DNAs that differ only in linking number are called topoisomers. Enzymes that underwind

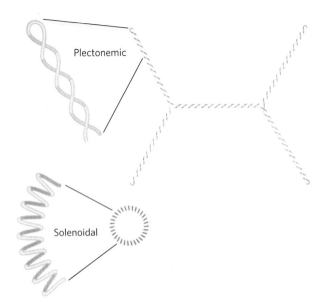

FIGURE 24–23 Plectonemic and solenoidal supercoiling of the same DNA molecule, drawn to scale. Plectonemic supercoiling takes the form of extended right-handed coils. Solenoidal negative supercoiling takes the form of tight left-handed turns about an imaginary tubelike structure. The two forms are readily interconverted, although the solenoidal form is generally not observed unless certain proteins are bound to the DNA. Solenoidal supercoiling provides a much greater degree of compaction.

and/or relax DNA, the topoisomerases, catalyze changes in linking number. The two classes of topoisomerases, type I and type II, change Lk in increments of 1 or 2, respectively, per catalytic event.

24.3 The Structure of Chromosomes

The term "chromosome" is used to refer to a nucleic acid molecule that is the repository of genetic information in a virus, a bacterium, a eukaryotic cell, or an organelle. It also refers to the densely colored bodies seen in the nuclei of dye-stained eukaryotic cells, as visualized using a light microscope.

Chromatin Consists of DNA and Proteins

The eukaryotic cell cycle (see Fig. 12–44) produces remarkable changes in the structure of chromosomes **(Fig. 24–24)**. In nondividing eukaryotic cells (in G0) and those in interphase (G1, S, and G2), the chromosomal material, **chromatin**, is amorphous and seems to be randomly dispersed in certain parts of the nucleus. In the S phase of interphase the DNA in this amorphous state replicates, each chromosome producing two sister chromosomes (called sister chromatids) that remain associated with each other after replication is complete. The chromosomes become much more condensed

during prophase of mitosis, taking the form of a species-specific number of well-defined pairs of sister chromatids (Fig. 24–5).

Chromatin consists of fibers containing protein and DNA in approximately equal proportions (by mass), along with a small amount of RNA. The DNA in the chromatin is very tightly associated with proteins called **histones**, which package and order the DNA into structural units called **nucleosomes (Fig. 24–25)**. Also found in chromatin are many nonhistone proteins, some of which help maintain chromosome structure

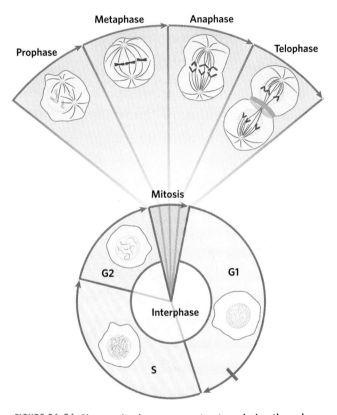

FIGURE 24–24 Changes in chromosome structure during the eukaryotic cell cycle. The relative lengths of the phases shown here are for convenience only. The duration of each phase varies with cell type and growth conditions (for single-celled organisms) or metabolic state (for multicellular organisms); mitosis is typically the shortest. Cellular DNA is uncondensed throughout interphase, as shown in the cartoons of the nucleus in the diagram. The interphase period can be divided (see Fig. 12–44) into the G1 (gap) phase; the S (synthesis) phase, when the DNA is replicated; and the G2 phase, throughout which the replicated chromosomes (chromatids) cohere to each other. Mitosis can be divided into four stages. The DNA undergoes condensation in prophase. During metaphase, the condensed chromosomes line up in pairs along the plane halfway between the spindle poles. The two chromosomes of each pair are linked to different spindle poles via microtubules that extend between the spindle and the centromere. The sister chromatids separate at anaphase, each drawn toward the spindle pole to which it is connected. The process is completed in telophase. After cell division is complete, the chromosomes decondense and the cycle begins anew.

(a) Histone core of nucleosome

Linker DNA of nucleosome

(b)

50 nm

FIGURE 24–25 Nucleosomes. (a) Regularly spaced nucleosomes consist of core histone proteins bound to DNA. **(b)** In this electron micrograph, the DNA-wrapped histone octamer structures are clearly visible.

are nearly identical in amino acid sequence in all eukaryotes, as are the H4 histones, suggesting strict conservation of their functions. For example, only 2 of 102 amino acid residues differ between the H4 histone molecules of peas and cows, and only 8 differ between the H4 histones of humans and yeast. Histones H1, H2A, and H2B show less sequence similarity among eukaryotic species.

Each type of histone is subject to enzymatic modification by methylation, acetylation, ADP-ribosylation, phosphorylation, glycosylation, sumoylation, or ubiquitination. Such modifications affect the net electric charge, shape, and other properties of histones, as well as the structural and functional properties of the chromatin, and they play a role in the regulation of transcription.

In addition, eukaryotes generally have several variant forms of certain histones, most notably histones H2A and H3, described in more detail below. The variant forms, along with their modifications, have specialized roles in DNA metabolism.

Nucleosomes Are the Fundamental Organizational Units of Chromatin

The eukaryotic chromosome depicted in Figure 24–5 represents the compaction of a DNA molecule about 10^5 μm long into a cell nucleus that is typically 5 to 10 μm in diameter. This compaction is achieved by means of several levels of highly organized folding. Subjection of chromosomes to treatments that partially unfold them reveals a structure in which the DNA is bound tightly to beads of protein, often regularly spaced. The beads in this "beads-on-a-string" arrangement are complexes of histones and DNA. The bead plus the connecting DNA that leads to the next bead form the nucleosome, the fundamental unit of organization on which the higher-order packing of chromatin is built **(Fig. 24–26)**. The bead of each nucleosome contains eight histone molecules: two copies each of H2A, H2B, H3, and H4. The spacing of the nucleosome beads provides a repeating unit typically of about 200 bp, of which 146 bp are bound tightly around the eight-part histone core and the remainder serve as linker DNA between nucleosome beads. Histone H1 binds to the linker DNA. Brief treatment of chromatin with enzymes that digest DNA causes the linker DNA to degrade preferentially, releasing histone particles containing 146 bp of bound DNA that have been protected from digestion. Researchers have crystallized nucleosome cores obtained in this way, and x-ray diffraction analysis reveals a particle made up of the eight histone molecules with the DNA wrapped around

and others that regulate the expression of specific genes (Chapter 28). Beginning with nucleosomes, eukaryotic chromosomal DNA is packaged into a succession of higher-order structures that ultimately yield the compact chromosome seen with the light microscope. We now turn to a description of this structure in eukaryotes and compare it with the packaging of DNA in bacterial cells.

Histones Are Small, Basic Proteins

Found in the chromatin of all eukaryotic cells, histones have molecular weights between 11,000 and 21,000 and are very rich in the basic amino acids arginine and lysine (together these make up about one-fourth of the amino acid residues). All eukaryotic cells have five major classes of histones, differing in molecular weight and amino acid composition (Table 24–4). The H3 histones

TABLE 24–4 Types and Properties of the Common Histones

Histone	Molecular weight	Number of amino acid residues	Content of basic amino acids (% of total)	
			Lys	Arg
H1*	21,130	223	29.5	11.3
H2A*	13,960	129	10.9	19.3
H2B*	13,774	125	16.0	16.4
H3	15,273	135	19.6	13.3
H4	11,236	102	10.8	13.7

*The sizes of these histones vary somewhat from species to species. The numbers given here are for bovine histones.

(a)

(b)

(c)

(d)

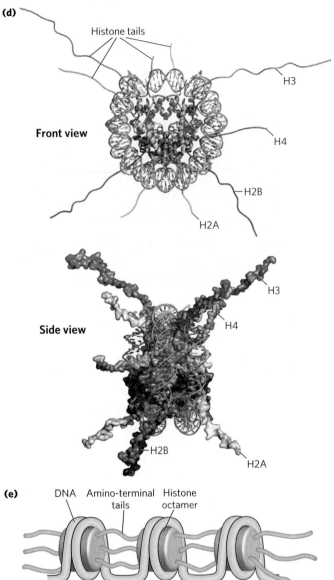

Front view

Side view

(e)

FIGURE 24–26 **DNA wrapped around a histone core. (a)** The simplified structure of a nucleosome octamer (left) with DNA wrapped around the histone core (right). **(b)** A ribbon representation of the nucleosome from the African frog *Xenopus laevis*. Different colors represent the different histones, matching the colors in (a). **(c)** Surface representation of the nucleosome. The view in (c) is rotated relative to the view in (b) to match the orientation shown above in (a). A 146 bp segment of DNA in the form of a left-handed solenoidal supercoil wraps around the histone complex 1.67 times. **(d)** Two views of histone amino-terminal tails protruding from between the two DNA duplexes that supercoil around the nucleosome. Some tails pass between the supercoils through holes formed by alignment of the minor grooves of adjacent helices. The H3 and H2B tails emerge between the two coils of DNA wrapped around the histone; the H4 and H2A tails emerge between adjacent histone subunits. **(e)** The amino-terminal tails of one nucleosome extrude from the particle and interact with adjacent nucleosomes, helping to define higher-order DNA packaging.

it in the form of a left-handed solenoidal supercoil (Fig. 24–26). Extending out from the nucleosome core are the amino-terminal tails of the histones, which are intrinsically disordered (Fig. 24–26d). Most of the histone modifications occur in these tails. The tails, in turn, play a key role in forming contacts between nucleosomes in the chromatin (Fig. 24–26e).

A close inspection of this structure reveals why eukaryotic DNA is underwound even though eukaryotic cells lack enzymes that underwind DNA. Recall that the solenoidal wrapping of DNA in nucleosomes is but one form of supercoiling that can be taken up by underwound (negatively supercoiled) DNA. The tight wrapping of DNA around the histone core requires the removal of about one helical turn in the DNA. When the protein core of a nucleosome binds in vitro to a relaxed, closed-circular DNA, the binding introduces a negative supercoil. Because this binding process does

not break the DNA or change the linking number, the formation of a negative solenoidal supercoil must be accompanied by a compensatory positive supercoil in the unbound region of the DNA (**Fig. 24–27**). As mentioned earlier, eukaryotic topoisomerases can relax positive supercoils. Relaxing the unbound positive supercoil leaves the negative supercoil fixed (through its binding to the nucleosome histone core) and results in an overall decrease in linking number. Indeed, topoisomerases have proved necessary for assembling chromatin from purified histones and closed-circular DNA in vitro.

Another factor that affects the binding of DNA to histones in nucleosome cores is the sequence of the bound DNA. Histone cores do not bind at random positions on the DNA; rather, some locations are more likely to be bound than others. This positioning is not fully understood but in some cases seems to depend on a

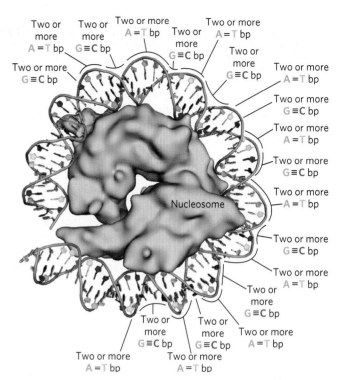

FIGURE 24–28 The effect of DNA sequence on nucleosome binding. (PDB ID 1AOI) Runs of two or more A=T base pairs facilitate the bending of DNA, while runs of two or more G≡C base pairs have the opposite effect. When spaced at about 10 bp intervals, consecutive A=T base pairs help bend DNA into a circle. When consecutive G≡C base pairs are spaced 10 bp apart, and offset by 5 bp from runs of A=T base pairs, DNA binding to the chromosome is facilitated.

FIGURE 24–27 Chromatin assembly. (a) Relaxed, closed-circular DNA. **(b)** Binding of a histone core to form a nucleosome induces one negative supercoil. In the absence of any strand breaks, a positive supercoil must form elsewhere in the DNA ($\Delta Lk = 0$). **(c)** Relaxation of this positive supercoil by a topoisomerase leaves one net negative supercoil ($\Delta Lk = -1$).

local abundance of A=T base pairs in the DNA helix where it is in contact with the histones **(Fig. 24–28)**. A cluster of two or three A=T base pairs facilitates the compression of the minor groove that is needed for the DNA to wrap tightly around the nucleosome's histone core. Nucleosomes bind particularly well to sequences where AA or AT or TT dinucleotides are staggered at 10 bp intervals, an arrangement that can account for up to 50% of the positions of bound histones in vivo.

Other proteins are required for the positioning of some nucleosome cores on DNA. In several organisms, certain proteins bind to a specific DNA sequence and facilitate the formation of a nucleosome core nearby. Nucleosomes are deposited on the DNA during replication, or following other processes that require a transient displacement of nucleosomes. Nucleosomes seem to be deposited in a stepwise manner. A tetramer of two H3 and two H4 histones binds first, followed by H2A-H2B dimers. The incorporation of nucleosomes into chromosomes after chromosomal replication is mediated by a complex of histone chaperones that include proteins known as chromatin assembly factor 1 (CAF1), RTT106 (*r*egulation of *T*y1 *t*ransposition), and anti-silencing factor 1 (ASF1).

These bind to acetylated variants of histones H3 and H4. The mechanism of nucleosome deposition is not understood in detail, although parts of this complex are known to directly interact with parts of the replication machinery. Some of the same histone chaperones, or different ones, may help assemble nucleosomes after DNA repair, transcription, or other processes. Histone exchange factors permit the substitution of histone variants for core histones in some contexts. Proper placement of these variant histones is important. Studies have shown that mice lacking one of these variant histones die as early embryos (Box 24–2). Precise positioning of nucleosome cores also plays a role in the expression of some eukaryotic genes (Chapter 28).

Nucleosomes Are Packed into Successively Higher-Order Structures

Wrapping of DNA around a nucleosome core compacts the DNA length about sevenfold. The overall compaction in a chromosome, however, is greater than 10,000-fold—ample evidence for even higher orders of structural organization. In chromosomes isolated by very gentle methods, nucleosome cores seem to be organized into a

BOX 24–2 | METHODS | Epigenetics, Nucleosome Structure, and Histone Variants

Information that is passed from one generation to the next— to daughter cells at cell division or from parent to offspring—but is not encoded in DNA sequences is referred to as **epigenetic** information. Much of it is in the form of covalent modification of histones and/or the placement of histone variants in chromosomes.

The chromatin regions where active gene expression (transcription) is occurring tend to be partially decondensed and are called **euchromatin**. In these regions, histones H3 and H2A are often replaced by the histone variants H3.3 and H2AZ, respectively (Fig. 1). The complexes that deposit nucleosomes containing histone variants on the DNA are similar to those that deposit nucleosomes with the more common histones. Nucleosomes containing histone H3.3 are deposited by a complex in which chromatin assembly factor 1 (CAF1) is replaced by the protein HIRA (the name is derived from a class of proteins called HIR, for *h*istone *r*epressor). Both CAF1 and HIRA can be considered histone chaperones, helping to ensure the proper assembly and placement of nucleosomes. Histone H3.3 differs in sequence from H3 by only four amino acid residues, but these residues all play key roles in histone deposition.

Like histone H3.3, H2AZ is associated with a distinct nucleosome deposition complex, and it is generally associated with chromatin regions involved in active transcription. Incorporation of H2AZ stabilizes the nucleosome octamer, but impedes some cooperative interactions between nucleosomes that are needed to compact the chromosome. This leads to a more open chromosome structure that facilitates the expression of genes in the region where H2AZ is located. The gene encoding H2AZ is essential in mammals. In fruit flies, loss of H2AZ prevents development beyond the larval stages.

Another H2A variant is H2AX, which is associated with DNA repair and genetic recombination. In mice, the absence of H2AX results in genome instability and male infertility. Modest amounts of H2AX seem to be scattered throughout the genome. When a double-strand break occurs, nearby molecules of H2AX become phosphorylated at Ser[139] in the carboxyl-terminal region. If this phosphorylation is blocked experimentally, formation of the protein complexes necessary for DNA repair is inhibited.

The H3 histone variant known as CENPA is associated with the repeated DNA sequences in centromeres. The chromatin in the centromere region contains the histone chaperones CAF1 and HIRA, and both proteins could be involved in the deposition of nucleosomes containing CENPA. Elimination of the gene for CENPA is lethal in mice.

The function and positioning of the histone variants can be studied by an application of technologies used in genomics. One useful technology is chromatin immunoprecipitation, or chromatin IP (ChIP). Nucleosomes containing a particular histone variant are precipitated by an antibody that binds specifically to this variant. These nucleosomes can be studied in isolation from their DNA, but more commonly the DNA associated with them is included in the study to determine where the nucleosomes of interest bind. The DNA can be labeled and used to probe a microarray (see Fig. 9–23), yielding a map of genomic sequences to which

FIGURE 1 Several variants of histones H3, H2A, and H2B are known. Shown here are the core histones and a few of the known variants. Sites of Lys/Arg residue methylation and Ser phosphorylation are indicated. HFD denotes the histone-fold domain, a structural domain common to all core histones.

structure called the **30 nm fiber (Fig. 24–29)**. This packing includes one molecule of histone H1 per nucleosome core. Two current models for the organization of histones and DNA in 30 nm fibers are presented in Figure 24–29. Organization into 30 nm fibers does not extend over the entire chromosome but is punctuated by regions bound by sequence-specific (nonhistone) DNA-binding proteins. The 30 nm structure also seems

to depend on the transcriptional activity of the particular region of DNA. Regions in which genes are being transcribed are apparently in a less-ordered state that contains little, if any, histone H1.

The 30 nm fiber—a second level of chromatin organization—provides an approximately 100-fold compaction of the DNA. The higher levels of folding are not yet understood, but certain regions of DNA those particular nucleosomes bind. Because microarrays are often referred to as chips, this technique is called a ChIP-chip experiment (Fig. 2).

The histone variants, along with the many covalent modifications that histones undergo, help define and isolate the functions of chromatin. They mark the chromatin, facilitating or suppressing specific functions such as chromosome segregation, transcription, and DNA repair. The histone modifications do not disappear at cell division or during meiosis, and thus they become part of the information transmitted from one generation to the next in all eukaryotic organisms.

FIGURE 2 A ChIP-chip experiment is designed to reveal the genomic DNA sequences to which a particular histone variant binds. **(a)** A histone variant with an epitope tag (a protein or chemical structure recognized by an antibody; see Chaptes 5 and 9) is introduced into a particular cell type, where it is incorporated into nucleosomes. (In some cases, an epitope tag is unnecessary because antibodies may be available that bind directly to the histone modification of interest.) Chromatin is isolated from the cells and digested briefly with micrococcal nuclease (MNase). The DNA bound by nucleosomes is protected from digestion, but the linker DNA is cleaved, releasing segments of DNA bound to one or two nucleosomes. An antibody that binds to the epitope tag is added, and the nucleosomes containing the epitope-tagged histone variant are selectively precipitated. The DNA in these nucleosomes is extracted from the precipitate, labeled, and used to probe a microarray representing all or selected parts of the genomic sequences of that particular cell type. **(b)** In this example, the binding of histone H3.3 is characterized in a short segment of chromosome 2L from *Drosophila melanogaster*. Numbers at the top correspond to numbered nucleotide positions in this chromosome arm. Each spot in the microarray represents 100 bp of genomic sequence, so the data here represent more than 1700 separate spots in the microarray. At each spot, the signal from the labeled DNA that was precipitated with antibody to histone H3.3 is presented as a ratio of that signal relative to the control signal generated when total genomic DNA is isolated witout immunoprecipitation, sheared, labeled with a different color of lable, and used to probe the same microarray. Signals above the horizontal line indicate genomic positions where histone H3.3 binding is enriched relative to the control. Signals below the line are regions where histone H3.3 is relatively absent. Annotated (known) genes in this segment of the genome are shown in the bottom panel (thickened bars). Bars above the line are genes transcribed 5′ to 3′ left to right, and boxes below the line are transcribed right to left. Red bars are genes where RNA polymerase II is also abundant, indicating active transcription. The histone H3.3 binding is concentrated in and near these genes undergoing active transcription.

seem to associate with a chromosomal scaffold **(Fig. 24–30)**. The scaffold-associated regions are separated by loops of DNA with perhaps 20 to 100 kbp. The DNA in a loop may contain a set of related genes. The scaffold itself may contain several proteins, notably topoisomerase II and SMC proteins, described below. The presence of topoisomerase II further emphasizes the relationship between DNA underwinding and chromatin structure.

Topoisomerase II is so important to the maintenance of chromatin structure that inhibitors of this enzyme can kill rapidly dividing cells. Several drugs used in cancer chemotherapy are topoisomerase II inhibitors that allow the enzyme to promote strand breakage but not the resealing of the breaks (see Box 24–1).

Evidence exists for additional layers of organization in eukaryotic chromosomes, each dramatically enhancing

(a) 30 nm
(b) Solenoid model
(one-start helix)
(c) Zigzag model
(two-start helix)

30 nm

30 nm

FIGURE 24–29 The 30 nm fiber, a higher-order organization of nucleo-somes. The compact fiber is formed by the tight packing of nucleosomes. **(a)** The 30 nm fiber, as seen by electron microscopy. There are two proposed models for the structure that are consistent with available data: **(b)** the solenoid model featuring one helical array of nucleosomes and **(c)** the zigzag model featuring two helical arrays of nucleosomes wrapped about each other. The black line is intended only to trace the proposed general path of the organized structure.

the degree of compaction. One model for achieving this compaction is illustrated in **Figure 24–31**. Higher-order chromatin structure probably varies from chromosome to chromosome, from one region to the next in a single chromosome, and from moment to moment in the life of a cell. No single model can adequately describe these structures. Nevertheless, the principle is clear: DNA compaction in eukaryotic chromosomes is likely to involve coils upon coils upon coils . . . 🖱 **Three-Dimensional Packaging of Nuclear Chromosomes**

Condensed Chromosome Structures Are Maintained by SMC Proteins

A third major class of chromatin proteins, in addition to the histones and topoisomerases, is the **SMC proteins** (structural maintenance of chromosomes). The primary structure of SMC proteins consists of five distinct domains **(Fig. 24–32a)**. The amino- and carboxyl-terminal globular domains, N and C, each of which contains part of an ATP-hydrolytic site, are connected by two regions of α-helical coiled-coil motifs (see Fig. 4–11) that are joined by a hinge domain. The proteins are generally dimeric, forming a V-shaped complex that is thought to be tied together through the protein's hinge domains (Fig. 24–32b). One N and one C domain come together to form a complete ATP-hydrolytic site at each free end of the V.

Proteins in the SMC family are found in all types of organisms, from bacteria to humans. Eukaryotes have two major types, cohesins and condensins, both of which are bound by regulatory and accessory proteins (Fig. 24–32c). The **cohesins** play a substantial role in linking together sister chromatids immediately after replication and keeping them together as the chromosomes condense to metaphase. This linkage is essential if chromosomes are to segregate properly at cell division. The cohesins, along with a third protein, kleisin, are thought to form a ring around the replicated chromosomes that

(a)

30 nm fibers

(c)

DNA loops

(b)

Chromosomal scaffold

DNA

FIGURE 24–30 Loops of DNA attached to a chromosomal scaffold. (a) A swollen chromosome, produced in a buffer of low ionic strength, as seen in the electron microscope. Notice the appearance of 30 nm fibers (chromatin loops) at the margins. **(b)** Extraction of the histones leaves a proteinaceous chromosomal scaffold surrounded by naked DNA. **(c)** The DNA appears to be organized in loops attached at their base to the scaffold in the upper left corner. Scale bar = 1 μm. The three images are at different magnifications.

FIGURE 24–31 Compaction of DNA in a eukaryotic chromosome. This model shows the levels of organization that could provide the observed degree of DNA compaction in the chromosomes of eukaryotes. First the DNA is wrapped around histone octamers, then H1 stimulates formation of the 30 nm fiber. Further levels of organization are not well understood but seem to involve further coiling and loops in the form of rosettes, which also coil into thicker structures. Overall, progressive levels of organization take the form of coils upon coils upon coils. It should be noted that in cells, the higher-order structures (above the 30 nm fiber) are unlikely to be as uniform as depicted here.

FIGURE 24–32 Structure of SMC proteins. (a) SMC proteins have five domains. **(b)** Each SMC polypeptide is folded so that the two coiled-coil domains wrap around each other and the N and C domains come together to form a complete ATP-binding site. Two polypeptides are linked at the hinge region to form the dimeric V-shaped SMC molecule. **(c)** Bacterial SMC proteins form a homodimer. The six different eukaryotic SMC proteins form heterodimers. Cohesins are made up of SMC1-SMC3 pairs, and condensins consist of SMC2-SMC4 pairs. The SMC5-SMC6 pair is involved in DNA repair. **(d)** Electron micrographs of SMC dimers from the bacterium *Bacillus subtilis*.

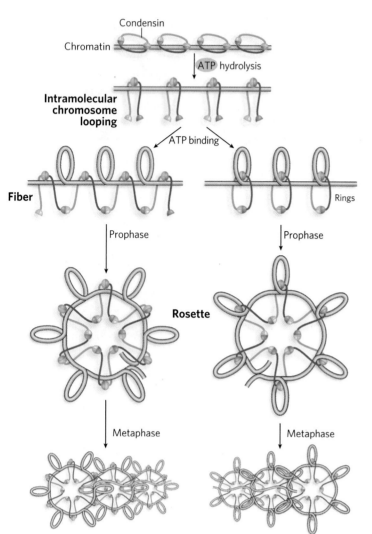

FIGURE 24–33 The possible role of condensins in chromatin condensation. Initially, the DNA is bound at the hinge region of the SMC protein, in the interior of what can become an intramolecular SMC ring. ATP binding leads to head-to-head association, forming supercoiled loops in the bound DNA. Subsequent rearrangement of the head-to-head interactions to form rosettes condenses the DNA. Condensins may organize the looping of the chromosome segments in a number of ways. Two current models are shown.

ties them together until separation is required at cell division. The ring may expand and contract in response to ATP hydrolysis. The **condensins** are essential to the condensation of chromosomes as cells enter mitosis. In the laboratory, condensins bind to DNA in a manner that creates positive supercoils; that is, condensin binding causes the DNA to become overwound, in contrast to the underwinding induced by the binding of nucleosomes. One model for the role of condensins in chromatin compaction is presented in **Figure 24–33**. The cohesins and condensins are essential in orchestrating the many changes in chromosome structure during the eukaryotic cell cycle **(Fig. 24–34)**.

Bacterial DNA Is Also Highly Organized

We now turn briefly to the structure of bacterial chromosomes. Bacterial DNA is compacted in a structure called the **nucleoid**, which can occupy a significant fraction of the cell volume **(Fig. 24–35)**. The DNA seems to be attached at one or more points to the inner surface of the plasma membrane. Much less is known about the structure of the nucleoid than of eukaryotic chromatin, but a complex organization is slowly being revealed. In *E. coli*, a scaffoldlike structure

FIGURE 24–34 The roles of cohesins and condensins in the eukaryotic cell cycle. Cohesins are loaded onto the chromosomes during G1 (see Fig. 24–24), tying the sister chromatids together during replication. At the onset of mitosis, condensins bind and maintain the chromatids in a condensed state. During anaphase, the enzyme separase removes the cohesin links. Once the chromatids separate, condensins begin to unload and the daughter chromosomes return to the uncondensed state.

FIGURE 24–35 *E. coli* nucleoids. The DNA of these cells is stained with a dye that fluoresces when exposed to UV light. The light area defines the nucleoid. Note that some cells have replicated their DNA but have not yet undergone cell division and hence have multiple nucleoids.

FIGURE 24–36 Looped domains of the *E. coli* chromosome. Each domain is about 10,000 bp in length. The domains are not static, but move along the DNA as replication proceeds. Barriers at the boundaries of the domains, of unknown composition, prevent the relaxation of DNA beyond the boundaries of the domain where a strand break occurs. The putative boundary complexes are shown as gray-shaded ovoids. The arrows denote movement of DNA through the boundary complexes.

seems to organize the *circular* chromosome into a series of about 500 looped domains, each encompassing 10,000 bp on average **(Fig. 24–36)**, as described above for chromatin. The domains are topologically constrained; for example, if the DNA is cleaved in one domain, only the DNA within that domain will be relaxed. The domains do not have fixed end points. Instead, the boundaries are most likely in constant motion along the DNA, coordinated with DNA replication. Bacterial DNA does not seem to have any structure comparable to the local organization provided by nucleosomes in eukaryotes. Histonelike proteins are abundant in *E. coli*—the best-characterized example is a two-subunit protein called HU (M_r 19,000)—but these proteins bind and dissociate within minutes, and no regular, stable DNA-histone structure has been found. The dynamic structural changes in the bacterial chromosome may reflect a requirement for more ready access to its genetic information. The bacterial cell division cycle can be as short as 15 min, whereas a typical eukaryotic cell may not divide for hours or even months. In addition, a much greater fraction of bacterial DNA is used to encode RNA and/or protein products. Higher rates of cellular metabolism in bacteria mean that a much higher proportion of the DNA is being transcribed or replicated at a given time than in most eukaryotic cells.

With this overview of the complexity of DNA structure, we are now ready to turn, in the next chapter, to a discussion of DNA metabolism.

SUMMARY 24.3 The Structure of Chromosomes

▶ The fundamental unit of organization in the chromatin of eukaryotic cells is the nucleosome, which consists of histones and a 200 bp segment of DNA. A core protein particle containing eight histones (two copies each of histones H2A, H2B, H3, and H4) is encircled by a segment of DNA (about 146 bp) in the form of a left-handed solenoidal supercoil.

▶ Nucleosomes are organized into 30 nm fibers, and the fibers are extensively folded to provide the 10,000-fold compaction required to fit a typical eukaryotic chromosome into a cell nucleus. The higher-order folding involves attachment to a chromosomal scaffold that contains histone H1, topoisomerase II, and SMC proteins. The SMC proteins, principally cohesins and condensins, play important roles in keeping the chromosomes organized during each stage of the cell cycle.

▶ Bacterial chromosomes are extensively compacted into the nucleoid, but the chromosome seems to be much more dynamic and irregular in structure than eukaryotic chromatin, reflecting the shorter cell cycle and very active metabolism of a bacterial cell.

Key Terms

Terms in bold are defined in the glossary.

chromosome 979	**centromere** 984
phenotype 979	**telomere** 984
mutation 979	**supercoil** 985
gene 980	**relaxed DNA** 985
regulatory	**topology** 986
sequence 980	underwinding 987
plasmid 981	**linking number** 988
intron 984	specific linking
exon 984	difference 988
simple-sequence	**superhelical density**
DNA 984	(σ) 988
satellite	**topoisomers** 989
DNA 984	twist 989

Further Reading

General

Cox, M.M., Doudna, J., & O'Donnell, M. (2012) *Molecular Biology: Principles and Practice*, W. H. Freeman & Company, New York.

Cozzarelli, N.R. & Wang, J.C. (eds). (1990) *DNA Topology and Its Biological Effects*, Cold Spring Harbor Laboratory Press, Cold Spring Harbor, NY.

Kornberg, A. & Baker, T.A. (1991) *DNA Replication*, 2nd edn, W. H. Freeman & Company, New York.

A good place to start for further information on the structure and function of DNA.

Genes and Chromosomes

Campbell, A., Lichten, M., & Schupbach, G. (2010) Telomeric strategies: means to an end. *Annu. Rev. Genet.* **44**, 243–269.

Levin, H.L. & Moran, J.V. (2011) Dynamic interactions between transposons and their hosts. *Nat. Rev. Genet.* **12**, 615–627.

McEachern, M.J., Krauskopf, A., & Blackburn, E.H. (2000) Telomeres and their control. *Annu. Rev. Genet.* **34**, 331–358.

Roy, S.W. & Gilbert, W. (2006) The evolution of spliceosomal introns: patterns, puzzles, and progress. *Nat. Rev. Genet.* **7**, 211–221.

Verdaasdonk, J.S. & Bloom, K. (2011) Centromeres: unique chromatin structures that drive chromosome segregation. *Nat. Rev. Mol. Cell Biol.* **12**, 320–332.

Supercoiling and Topoisomerases

Boles, T.C., White, J.H., & Cozzarelli, N.R. (1990) Structure of plectonemically supercoiled DNA. *J. Mol. Biol.* **213**, 931–951.

A study that defines several fundamental features of supercoiled DNA.

Garcia H.G., Grayson, P., Han, L., Inamdar, M., Kondev, J., Nelson, P.C., Phillips, R., Widom, W., & Wiggins, P.A. (2007) Biological consequences of tightly bent DNA: the other life of a macromolecular celebrity. *Biopolymers* **85**, 115–130.

A nice description of the physics of bent DNA.

Kohanski, M.A., Dwyer, D.J., & Collins, J.J. (2010) How antibiotics kill bacteria: from targets to networks. *Nat. Rev. Microbiol.* **8**, 423–435.

Lebowitz, J. (1990) Through the looking glass: the discovery of supercoiled DNA. *Trends Biochem. Sci.* **15**, 202–207.

A short and interesting historical note.

Pommier, Y. (2006) Topoisomerase I inhibitors: camptothecins and beyond. *Nat. Rev. Cancer* **6**, 789–802.

Vos, S.M., Tretter, E.M., Schmidt, B.H., & Berger, J.M. (2011) All tangled up: how cells direct, manage, and exploit topoisomerase function. *Nat. Rev. Mol. Cell Biol.* **12**, 827–841.

Chromatin and Nucleosomes

Campos, E.I. & Reinberg, D. (2009) Histones: annotating chromatin. *Annu. Rev. Genet.* **43**, 559–599.

Carter, S.D. & Sjogren, C. (2012) The SMC complexes, DNA and chromosome topology: right or knot? *Crit. Rev. Biochem. Mol. Biol.* **47**, 1–16.

Dillon, S.C. & Dorman, C.J. (2010) Bacterial nucleoid-associated proteins: nucleoid structure and gene expression. *Nat. Rev. Microbiol.* **8**, 185–195.

Kornberg, R.D. (1974) Chromatin structure: a repeating unit of histones and DNA. *Science* **184**, 868–871.

A classic paper that introduced the subunit model for chromatin.

Losada, A. & Hirano, T. (2005) Dynamic molecular linkers of the genome: the first decade of SMC proteins. *Genes Dev.* **19**, 1269–1287.

Luijsterburg, M.S., White, M.F., van Driel, R., & Remus, T.D. (2008) The major architects of chromatin: architectural proteins in bacteria, archaea, and eukaryotes. *Crit. Rev. Biochem. Mol. Biol.* **43**, 393–418.

Margueron, R. & Reinberg, D. (2010) Chromatin structure and the inheritance of epigenetic information. *Nat. Rev. Genet.* **11**, 285–296.

Rando, O.J. (2007) Chromatin structure in the genomics era. *Trends Genet.* **23**, 67–73.

A description of the imaginative methods being employed to study nucleosome modification patterns, nucleosome positioning, and other aspects of chromosome structure on a genomic scale.

Segal, E., Fondufe-Mittendorf, Y., Chen, L., Thåström, A., Field, Y., Moore, I.K., Wang, J.Z., & Widom, J. (2006) A genomic code for nucleosome positioning. *Nature* **442**, 772–778.

Problems

1. Packaging of DNA in a Virus Bacteriophage T2 has a DNA of molecular weight 120×10^6 contained in a head about 210 nm long. Calculate the length of the DNA (assume the molecular weight of a nucleotide pair is 650) and compare it with the length of the T2 head.

2. The DNA of Phage M13 The base composition of phage M13 DNA is A, 23%; T, 36%; G, 21%; C, 20%. What does this tell you about the DNA of phage M13?

3. The *Mycoplasma* Genome The complete genome of the simplest bacterium known, *Mycoplasma genitalium*, is a circular DNA molecule with 580,070 bp. Calculate the molecular weight and contour length (when relaxed) of this molecule. What is Lk_0 for the *Mycoplasma* chromosome? If $\sigma = -0.06$, what is Lk?

4. Size of Eukaryotic Genes An enzyme isolated from rat liver has 192 amino acid residues and is coded for by a gene with 1,440 bp. Explain the relationship between the number of amino acid residues in the enzyme and the number of nucleotide pairs in its gene.

5. Linking Number A closed-circular DNA molecule in its relaxed form has an Lk of 500. Approximately how many base pairs are in this DNA? How is the linking number altered (increases, decreases, doesn't change, becomes undefined) when (a) a protein complex binds to form a nucleosome, (b) one DNA strand is broken, (c) DNA gyrase and ATP are added to the DNA solution, or (d) the double helix is denatured by heat?

6. DNA Topology In the presence of a eukaryotic condensin and a type II topoisomerase, the Lk of a relaxed closed-circular DNA molecule does not change. However, the DNA becomes highly knotted.

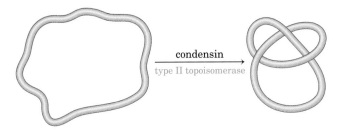

Formation of the knots requires breakage of the DNA, passage of a segment of DNA through the break, and religation by the topoisomerase. Given that every reaction of the topoisomerase would be expected to result in a change in linking number, how can Lk remain the same?

7. Superhelical Density Bacteriophage λ infects *E. coli* by integrating its DNA into the bacterial chromosome. The success of this recombination depends on the topology of the *E. coli* DNA. When the superhelical density (σ) of the *E. coli* DNA is greater than −0.045, the probability of integration is <20%; when σ is less than −0.06, the probability is −70%. Plasmid DNA isolated from an *E. coli* culture is found to have a length of 13,800 bp and an Lk of 1,222. Calculate σ for this DNA and predict the likelihood that bacteriophage λ will be able to infect this culture.

8. Altering Linking Number (a) What is the Lk of a 5,000 bp circular duplex DNA molecule with a nick in one strand? (b) What is the Lk of the molecule in (a) when the nick is sealed (relaxed)? (c) How would the Lk of the molecule in (b) be affected by the action of a single molecule of *E. coli* topoisomerase I? (d) What is the Lk of the molecule in (b) after eight enzymatic turnovers by a single molecule of DNA gyrase in the presence of ATP? (e) What is the Lk of the molecule in (d) after four enzymatic turnovers by a single molecule of bacterial type I topoisomerase? (f) What is the Lk of the molecule in (d) after binding of one nucleosome?

9. Chromatin Early evidence that helped researchers define nucleosome structure is illustrated by the agarose gel below, in which the thick bands represent DNA. It was generated by briefly treating chromatin with an enzyme that degrades DNA, then removing all protein and subjecting the purified DNA to electrophoresis. Numbers at the side of the gel denote the position to which a linear DNA of the indicated size would migrate. What does this gel tell you about chromatin structure? Why are the DNA bands thick and spread out rather than sharply defined?

1,000 bp —
800 bp —
600 bp —
400 bp —
200 bp —

10. DNA Structure Explain how the underwinding of a B-DNA helix might facilitate or stabilize the formation of Z-DNA.

11. Maintaining DNA Structure (a) Describe two structural features required for a DNA molecule to maintain a negatively supercoiled state. (b) List three structural changes that become more favorable when a DNA molecule is negatively supercoiled. (c) What enzyme, with the aid of ATP, can generate negative superhelicity in DNA? (d) Describe the physical mechanism by which this enzyme acts.

12. Yeast Artificial Chromosomes (YACs) YACs are used to clone large pieces of DNA in yeast cells. What three types of DNA sequence are required to ensure proper replication and propagation of a YAC in a yeast cell?

13. Nucleoid Structure in Bacteria In bacteria, the transcription of a subset of genes is affected by DNA topology, with expression increasing or (more often) decreasing when the DNA is relaxed. When a bacterial chromosome is cleaved at a specific site by a restriction enzyme (one that cuts at a long, and thus rare, sequence), only nearby genes (within 10,000 bp) exhibit either an increase or decrease in expression. The transcription of genes elsewhere in the chromosome is unaffected. Explain. (Hint: See Fig. 24–36.)

14. DNA Topology When DNA is subjected to electrophoresis in an agarose gel, shorter molecules migrate faster than longer ones. Closed-circular DNAs of the same size but with different linking numbers also can be separated on an agarose gel: topoisomers that are more supercoiled, and thus more condensed, migrate faster through the gel. In the gel shown below, purified plasmid DNA has migrated from top to bottom. There are two bands, with the faster band much more prominent.

(a) What are the DNA species in the two bands? (b) If topoisomerase I is added to a solution of this DNA, what will happen to the upper and lower bands after electrophoresis? (c) If DNA ligase is added to the DNA, will the appearance of the bands change? Explain your answer. (d) If DNA gyrase plus ATP is added to the DNA after the addition of DNA ligase, how will the band pattern change?

15. DNA Topoisomers When DNA is subjected to electrophoresis in an agarose gel, shorter molecules migrate faster than longer ones. Closed-circular DNAs of the same size but different linking number also can be separated on an agarose gel: topoisomers that are more supercoiled, and thus more condensed, migrate faster through the gel—from top to bottom in the gels shown on the right. A dye, chloroquine, was added to these gels. Chloroquine intercalates between base pairs and stabilizes a more underwound DNA structure. When the dye binds to a relaxed closed-circular DNA, the DNA is underwound where the dye binds, and unbound regions take on positive supercoils to compensate. In the experiment shown here, topoisomerases were used to make preparations of the same DNA circle with different superhelical densities (σ). Completely relaxed DNA migrated to the position labeled N (for *n*icked), and highly supercoiled DNA (above the limit where individual topoisomers can be distinguished) to the position labeled X.

0.5 μg/mL chloroquine

Average σ

Native 0 −0.009 −0.023 −0.037 −0.049 −0.066 −0.080 −0.115

N→

X→

Runs as +ΔLk Runs as −ΔLk

Gel A

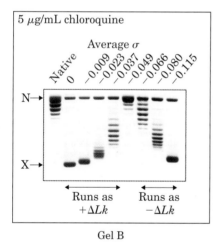

5 μg/mL chloroquine

Average σ

Native 0 −0.009 −0.023 −0.037 −0.049 −0.066 −0.080 −0.115

N→

X→

Runs as +ΔLk Runs as −ΔLk

Gel B

(a) In gel A, why does the $\sigma = 0$ lane (i.e., DNA prepared so that $\sigma = 0$, on average) have multiple bands?

(b) In gel B, is the DNA from the $\sigma = 0$ preparation positively or negatively supercoiled in the presence of the intercalating dye?

(c) In both gels, the $\sigma = -0.115$ lane has two bands, one a highly supercoiled DNA and one relaxed. Propose a reason for the presence of relaxed DNA in these lanes (and others).

(d) The native DNA (leftmost lane in each gel) is the same DNA circle isolated from bacterial cells and untreated. What is the approximate superhelical density of this native DNA?

16. Nucleosomes The human genome contains just over 3.1 billion base pairs. Assuming it is covered with nucleosomes that are spaced as described in this chapter, how many molecules of histone H2A are present in one somatic human cell?

(Do not consider reductions in H2A due to its replacement in some regions by H2A variants.) How would the number change after DNA replication but before cell division?

Data Analysis Problem

17. Defining the Functional Elements of Yeast Chromosomes Figure 24–8 shows the major structural elements of a chromosome of baker's yeast (*Saccharomyces cerevisiae*). Heiter, Mann, Snyder, and Davis (1985) determined the properties of some of these elements. They based their study on the finding that in yeast cells, plasmids (which have genes and an origin of replication) act differently from chromosomes (which have these elements plus centromeres and telomeres) during mitosis. The plasmids are not manipulated by the mitotic apparatus and segregate randomly between daughter cells. Without a selectable marker to force the host cells to retain them (see Fig. 9–4), these plasmids are rapidly lost. In contrast, chromosomes, even without a selectable marker, are manipulated by the mitotic apparatus and are lost at a very low rate (about 10^{-5} per cell division).

Heiter and colleagues set out to determine the important components of yeast chromosomes by constructing plasmids with various parts of chromosomes and observing whether these "synthetic chromosomes" segregated properly during mitosis. To measure the rates of different types of failed chromosome segregation, the researchers needed a rapid assay to determine the number of copies of synthetic chromosomes present in different cells. This assay took advantage of the fact that wild-type yeast colonies are white whereas certain adenine-requiring (ade$^-$) mutants yield red colonies on nutrient media. Specifically, *ade2$^-$* cells lack functional AIR carboxylase (the enzyme of step **6a** in Figure 22–35) and accumulate AIR (5-aminoimidazole ribonucleotide) in their cytoplasm. This excess AIR is converted to a conspicuous red pigment. The other part of the assay involved the gene *SUP11*, which encodes an ochre suppressor (a type of nonsense suppressor; see Box 27–4) that suppresses the phenotype of some *ade2$^-$* mutants.

Heiter and coworkers started with a diploid strain of yeast homozygous for *ade2$^-$*; these cells are red. When the mutant cells contain one copy of *SUP11*, the metabolic defect is partly suppressed and the cells are pink. When the cells contain two or more copies of *SUP11*, the defect is completely suppressed and the cells are white.

The researchers inserted one copy of *SUP11* into synthetic chromosomes containing various elements thought to be important in chromosome function, and then observed how well these chromosomes were passed from one generation to the next. These pink cells were plated on nonselective media, and the behavior of the synthetic chromosomes was observed. Specifically, Heiter and coworkers looked for colonies in which the synthetic chromosomes segregated improperly at the first division after plating, giving rise to a colony that is half one genotype and half the other. Because yeast cells are nonmotile, this will be a sectored colony, with one half one color and the other half another color.

(a) One way for the mitotic process to fail is *nondisjunction*: the chromosome replicates but the sister chromatids fail to separate, so both copies of the chromosome end up in the same daughter cell. Explain how nondisjunction of the synthetic chromosome would give rise to a colony that is half red and half white.

(b) Another way for the mitotic process to fail is *chromosome loss*: the chromosome does not enter the daughter nucleus or is not replicated. Explain how loss of the synthetic chromosome would give rise to a colony that is half red and half pink.

By counting the frequency of the different colony types, Heiter and colleagues could estimate the frequency of these aberrant mitotic events with different types of synthetic chromosome. First, they explored the requirement for centromeric sequences by constructing synthetic chromosomes with different-sized DNA fragments containing a known centromere. Their results are shown below.

Synthetic chromosome	Size of centromere-containing fragment (kbp)	Chromosome loss (%)	Nondisjunction (%)
1	none	—	>50
2	0.63	1.6	1.1
3	1.6	1.9	0.4
4	3.0	1.7	0.35
5	6.0	1.6	0.35

(c) Based on these data, what can you conclude about the size of the centromere required for normal mitotic segregation? Explain your reasoning.

(d) Interestingly, all the synthetic chromosomes created in these experiments were circular and lacked telomeres. Explain how they could be replicated more-or-less properly.

Heiter and colleagues next constructed a series of linear synthetic chromosomes that included the functional centromeric sequence and telomeres, and measured the total mitotic error frequency (% loss + % nondisjunction) as a function of size:

Synthetic chromosome	Size (kbp)	Total error frequency (%)
6	15	11.0
7	55	1.5
8	95	0.44
9	137	0.14

(e) Based on these data, what can you conclude about the chromosome size required for normal mitotic segregation? Explain your reasoning.

(f) Normal yeast chromosomes are linear, range from 250 kbp to 2,000 kbp in length, and have a mitotic error rate of about 10^{-5} per cell division. Extrapolating the results from (e), do the centromeric and telomeric sequences used in these experiments explain the mitotic stability of normal yeast chromosomes, or must other elements be involved? Explain your reasoning. (Hint: A plot of log (error rate) vs. length will be helpful.)

Reference

Heiter, P., Mann, C., Snyder, M., & Davis, R.W. (1985) Mitotic stability of yeast chromosomes: a colony color assay that measures nondisjunction and chromosome loss. *Cell* **40**, 381–392.

DNA Metabolism

As the repository of genetic information, DNA occupies a unique and central place among biological macromolecules. The nucleotide sequences of DNA encode the primary structures of all cellular RNAs and proteins and, through enzymes, indirectly affect the synthesis of all other cellular constituents. This passage of information from DNA to RNA and protein guides the size, shape, and functioning of every living thing.

DNA is a marvelous device for the stable storage of genetic information. The phrase "stable storage," however, conveys a static and misleading picture. It fails to capture the complexity of processes by which genetic information is preserved in an uncorrupted state and then transmitted from one generation of cells to the next. DNA metabolism comprises both the process that gives rise to faithful copies of DNA molecules (replication) and the processes that affect the inherent structure of the information (repair and recombination). Together, these activities are the focus of this chapter.

The metabolism of DNA is shaped by the requirement for an exquisite degree of accuracy. The chemistry of joining one nucleotide to the next in DNA replication is elegant and simple, almost deceptively so. However, as is the case with all information-containing polymers, forming a covalent link between two monomeric units is just a small part of the biochemical process. As we shall see, complexity arises in the form of enzymatic devices to ensure that the *correct* nucleotide is added and that genetic information is transmitted intact. Uncorrected errors that arise during DNA synthesis can have dire consequences, not only because they can permanently affect or eliminate the function of a gene but also because the change is inheritable.

The enzymes that synthesize DNA may copy DNA molecules that contain millions of bases. They do so with extraordinary fidelity and speed, even though the DNA substrate is highly compacted and bound with other proteins. Formation of phosphodiester bonds to link nucleotides in the backbone of a growing DNA strand is therefore only one part of an elaborate process that requires myriad proteins and enzymes.

Maintaining the integrity of genetic information lies at the heart of DNA repair. As detailed in Chapter 8, DNA is susceptible to many types of damaging reactions. Such reactions are infrequent but significant nevertheless, because of the very low biological tolerance for changes in DNA sequence. DNA is the only macromolecule for which repair systems exist; the number, diversity, and complexity of DNA repair mechanisms reflect the wide range of insults that can harm DNA.

Cells can rearrange their genetic information by processes collectively called recombination—seemingly undermining the principle that the stability and integrity of genetic information are paramount. However, most DNA rearrangements in fact play constructive roles in maintaining genomic integrity, contributing in special ways to DNA replication, DNA repair, and chromosome segregation.

Special emphasis is given in this chapter to the *enzymes* of DNA metabolism. They merit careful study not only because of their intrinsic biological importance and interest but also for their increasing importance in medicine and for their everyday use as reagents in a wide range of modern biochemical technologies. Many of the seminal discoveries in DNA metabolism have been made with *Escherichia coli*, so its well-understood enzymes are generally used to illustrate the ground rules. A quick look at some relevant genes on the *E. coli* genetic map **(Fig. 25–1)** provides just a hint of the complexity of the enzymatic systems involved in DNA metabolism.

Before taking a closer look at replication, we must make a short digression into the use of abbreviations in naming bacterial genes and proteins—you will encounter many of these in this and later chapters. Similar conventions exist for naming eukaryotic genes, although the exact form of the abbreviations may vary with the species and no single convention applies to all eukaryotic systems.

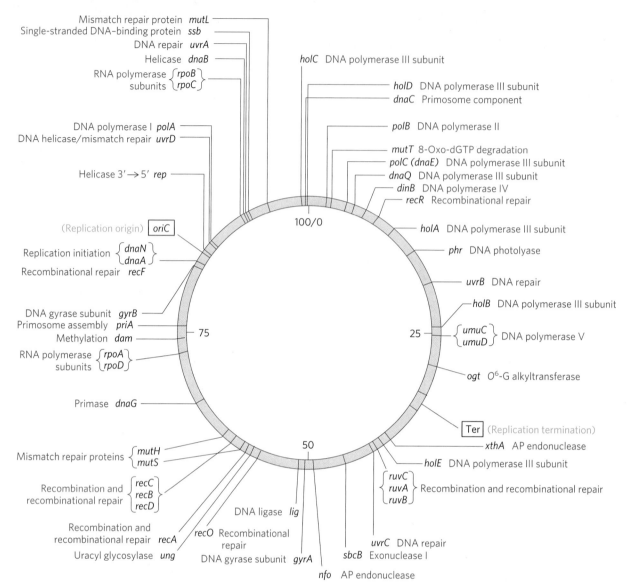

FIGURE 25–1 Map of the E. coli chromosome. The map shows the relative positions of genes encoding many of the proteins important in DNA metabolism. The number of genes known to be involved provides a hint of the complexity of these processes. The numbers 0 to 100 inside the circular chromosome denote a genetic measurement called minutes. Each minute corresponds to ~40,000 bp along the DNA molecule. The three-letter names of genes and other elements generally reflect some aspect of their function. These include *mut*, *mutagenesis*; *dna*, DNA replication; *pol*, DNA *polymerase*; *rpo*, RNA *polymerase*; *uvr*, UV resistance; *rec*, *recombination*; *dam*, DNA *adenine methylation*; *lig*, DNA *ligase*; Ter, *termination* of replication; and *ori*, *origin* of replication (*oriC* in *E. coli*, as shown here).

KEY CONVENTION: Bacterial genes generally are named using three lowercase, italicized letters that often reflect a gene's apparent function. For example, the *dna*, *uvr*, and *rec* genes affect *DNA* replication, *resistance* to the damaging effects of *UV* radiation, and *recombination*, respectively. Where several genes affect the same process, the letters *A*, *B*, *C*, and so forth, are added—as in *dnaA*, *dnaB*, *dnaQ*, for example—usually reflecting their order of discovery rather than their order in a reaction sequence. ■

The use of abbreviations in naming proteins is less straightforward. During genetic investigations, the protein product of each gene is usually isolated and characterized. Many bacterial genes have been identified and named before the roles of their protein products are understood in detail. Sometimes the gene product is found to be a previously isolated protein, and some renaming occurs. Often, however, the product turns out to be an as yet unknown protein, with an activity not easily described by a simple enzyme name.

KEY CONVENTION: Bacterial proteins often retain the name of their genes. When referring to the protein product of an *E. coli* gene, roman type is used and the first letter is capitalized: for example, the *dnaA* and *recA* gene products are the DnaA and RecA proteins, respectively. ■

25.1 DNA Replication

Long before the structure of DNA was known, scientists wondered at the ability of organisms to create faithful copies of themselves and, later, at the ability of cells to produce many identical copies of large, complex macromolecules. Speculation about these problems centered around the concept of a **template**, a structure that would allow molecules to be lined up in a specific order and joined to create a macromolecule with a unique sequence and function. The 1940s brought the revelation that DNA was the genetic molecule, but not until James Watson and Francis Crick deduced its structure did the way in which DNA could act as a template for the replication and transmission of genetic information become clear: *one strand is the complement of the other.* The strict base-pairing rules mean that each strand provides the template for a new strand with a predictable and complementary sequence (see Figs 8–14, 8–15).

🖱 Nucleotides: Building Blocks of Nucleic Acids

The fundamental properties of the DNA replication process and the mechanisms used by the enzymes that catalyze it have proved to be essentially identical in all species. This mechanistic unity is a major theme as we proceed from general properties of the replication process, to *E. coli* replication enzymes, and, finally, to replication in eukaryotes.

DNA Replication Follows a Set of Fundamental Rules

Early research on bacterial DNA replication and its enzymes helped to establish several basic properties that have proven applicable to DNA synthesis in every organism.

DNA Replication Is Semiconservative Each DNA strand serves as a template for the synthesis of a new strand, producing two new DNA molecules, each with one new strand and one old strand. This is **semiconservative replication**.

Watson and Crick proposed the hypothesis of semiconservative replication soon after publication of their 1953 paper on the structure of DNA, and the hypothesis was proved by ingeniously designed experiments carried out by Matthew Meselson and Franklin Stahl in 1957. Meselson and Stahl grew *E. coli* cells for many generations in a medium in which the sole nitrogen source (NH_4Cl) contained ^{15}N, the "heavy" isotope of nitrogen, instead of the normal, more abundant "light" isotope, ^{14}N. The DNA isolated from these cells had a density about 1% greater than that of normal [^{14}N]DNA **(Fig. 25–2a)**. Although this is only a small difference, a mixture of heavy [^{15}N]DNA and light [^{14}N]DNA can be separated by centrifugation to equilibrium in a cesium chloride density gradient.

The *E. coli* cells grown in the ^{15}N medium were transferred to a fresh medium containing only the ^{14}N

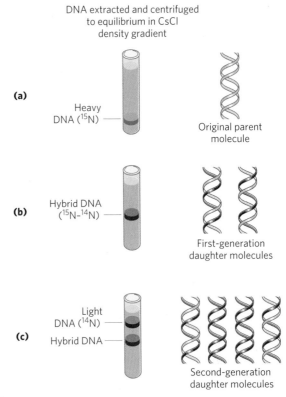

FIGURE 25–2 The Meselson-Stahl experiment. (a) Cells were grown for many generations in a medium containing only heavy nitrogen, ^{15}N, so that all the nitrogen in their DNA was ^{15}N, as shown by a single band (blue) when centrifuged in a CsCl density gradient. **(b)** Once the cells had been transferred to a medium containing only light nitrogen, ^{14}N, cellular DNA isolated after one generation equilibrated at a higher position in the density gradient (purple band). **(c)** A second cycle of replication yielded a hybrid DNA band (purple) and another band (red), containing only [^{14}N]DNA, confirming semiconservative replication.

isotope, where they were allowed to grow until the cell population had just doubled. The DNA isolated from these first-generation cells formed a *single* band in the CsCl gradient at a position indicating that the double-helical DNA molecules of the daughter cells were hybrids containing one new ^{14}N strand and one parent ^{15}N strand (Fig. 25–2b).

This result argued against conservative replication, an alternative hypothesis in which one progeny DNA molecule would consist of two newly synthesized DNA strands and the other would contain the two parent strands; this would not yield hybrid DNA molecules in the Meselson-Stahl experiment. The semiconservative replication hypothesis was further supported in the next step of the experiment (Fig. 25–2c). Cells were again allowed to double in number in the ^{14}N medium. The isolated DNA product of this second cycle of replication exhibited *two* bands in the density gradient, one with a density equal to that of light DNA and the other with the density of the hybrid DNA observed after the first cell doubling.

Replication Begins at an Origin and Usually Proceeds Bidirectionally Following the confirmation of a semiconservative mechanism of replication, a host of questions arose. Are the parent DNA strands completely unwound before each is replicated? Does replication begin at random places or at a unique point? After initiation at any point in the DNA, does replication proceed in one direction or both?

An early indication that replication is a highly coordinated process in which the parent strands are simultaneously unwound and replicated was provided by John Cairns, using autoradiography. He made *E. coli* DNA radioactive by growing cells in a medium containing thymidine labeled with tritium (^{3}H). When the DNA was carefully isolated, spread, and overlaid with a photographic emulsion for several weeks, the radioactive thymidine residues generated "tracks" of silver grains in the emulsion, producing an image of the DNA molecule. These tracks revealed that the intact chromosome of *E. coli* is a single huge circle, 1.7 mm long. Radioactive DNA isolated from cells during replication showed an extra loop (**Fig. 25–3**). Cairns concluded that the loop resulted from the formation of two radioactive daughter strands, each complementary to a parent strand. One or both ends of the loop are dynamic points, termed **replication forks**, where parent DNA is being unwound and the separated strands quickly replicated. Cairns's results demonstrated that both DNA strands are replicated simultaneously, and variations on his experiment indicated that replication of bacterial chromosomes is bidirectional: both ends of the loop have active replication forks.

The determination of whether the replication loops originate at a unique point in the DNA required landmarks along the DNA molecule. These were provided by a technique called **denaturation mapping**, developed by Ross Inman and colleagues. Using the 48,502 bp chromosome of bacteriophage λ, Inman showed that DNA could be selectively denatured at sequences unusually rich in A≡T base pairs, generating a reproducible pattern of single-strand bubbles (see Fig. 8–28). Isolated DNA containing replication loops can be partially denatured in the same way. This allows the position and progress of the replication forks to be measured and mapped, using the denatured regions as points of reference. The technique revealed that in this system the replication loops always initiate at a unique point, which was termed an **origin**. It also confirmed the earlier observation that replication is usually bidirectional. For circular DNA molecules, the two replication forks meet at a point on the side of the circle opposite to the origin. Specific origins of replication have since been identified and characterized in bacteria and lower eukaryotes.

DNA Synthesis Proceeds in a 5′→3′ Direction and Is Semidiscontinuous A new strand of DNA is always synthesized in the 5′→3′ direction, with the free 3′ OH as the point at which the DNA is elongated (the 5′ and 3′ ends of a

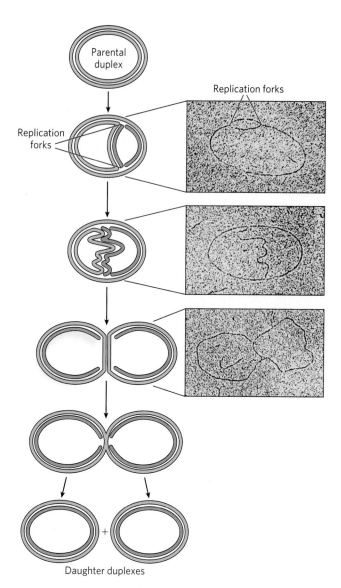

FIGURE 25–3 Visualization of DNA replication. Stages in the replication of circular DNA molecules have been visualized by electron microscopy. Replication of a circular chromosome produces a structure resembling the Greek letter theta, θ, as both strands are replicated simultaneously (new strands shown in light red). The electron micrographs show images of plasmid DNA being replicated from a single replication origin.

DNA strand are defined in Fig. 8–7). Because the two DNA strands are antiparallel, the strand serving as the template is read from its 3′ end toward its 5′ end.

If synthesis always proceeds in the 5′→3′ direction, how can both strands be synthesized simultaneously? If both strands were synthesized *continuously* while the replication fork moved, one strand would have to undergo 3′→5′ synthesis. This problem was resolved by Reiji Okazaki and colleagues in the 1960s. Okazaki found that one of the new DNA strands is synthesized in short pieces, now called **Okazaki fragments**. This work ultimately led to the conclusion that one strand is synthesized continuously and the other discontinuously (**Fig. 25–4**). The continuous strand, or **leading strand**, is the one in which 5′→3′ synthesis proceeds in the *same* direction as

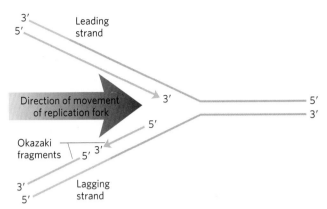

FIGURE 25–4 Defining DNA strands at the replication fork. A new DNA strand (light red) is always synthesized in the 5′→3′ direction. The template is read in the opposite direction, 3′→5′. The leading strand is continuously synthesized in the direction taken by the replication fork. The other strand, the lagging strand, is synthesized discontinuously in short pieces (Okazaki fragments) in a direction opposite to that in which the replication fork moves. The Okazaki fragments are spliced together by DNA ligase. In bacteria, Okazaki fragments are ~1,000 to 2,000 nucleotides long. In eukaryotic cells, they are 150 to 200 nucleotides long.

replication fork movement. The discontinuous strand, or **lagging strand**, is the one in which 5′→3′ synthesis proceeds in the direction *opposite* to the direction of fork movement. Okazaki fragments range in length from a few hundred to a few thousand nucleotides, depending on the cell type. As we shall see later, leading and lagging strand syntheses are tightly coordinated.

DNA Is Degraded by Nucleases

To explain the enzymology of DNA replication, we first introduce the enzymes that degrade DNA rather than synthesize it. These enzymes are known as **nucleases**, or **DNases** if they are specific for DNA rather than RNA. Every cell contains several different nucleases, belonging to two broad classes: exonucleases and endonucleases. **Exonucleases** degrade nucleic acids from one end of the molecule. Many operate in only the 5′→3′ or the 3′→5′ direction, removing nucleotides only from the 5′ or the 3′ end, respectively, of one strand of a double-stranded nucleic acid or of a single-stranded DNA. **Endonucleases** can begin to degrade at specific internal sites in a nucleic acid strand or molecule, reducing it to smaller and smaller fragments. A few exonucleases and endonucleases degrade only single-stranded DNA. There are a few important classes of endonucleases that cleave only at specific nucleotide sequences (such as the restriction endonucleases that are so important in biotechnology; see Chapter 9, Fig. 9–2). You will encounter many types of nucleases in this and subsequent chapters.

DNA Is Synthesized by DNA Polymerases

The search for an enzyme that could synthesize DNA began in 1955. Work by Arthur Kornberg and colleagues

Arthur Kornberg, 1918–2007

led to the purification and characterization of a DNA polymerase from *E. coli* cells, a single-polypeptide enzyme now called **DNA polymerase I** (M_r 103,000; encoded by the *polA* gene). Much later, investigators found that *E. coli* contains at least four other distinct DNA polymerases, described below.

Detailed studies of DNA polymerase I revealed features of the DNA synthetic process that are now known to be common to all DNA polymerases. The fundamental reaction is a phosphoryl group transfer. The nucleophile is the 3′-hydroxyl group of the nucleotide at the 3′ end of the growing strand. Nucleophilic attack occurs at the α phosphorus of the incoming deoxynucleoside 5′-triphosphate **(Fig. 25–5a)**. Inorganic pyrophosphate is released in the reaction. The general reaction is

$$(\text{dNMP})_n + \text{dNTP} \longrightarrow (\text{dNMP})_{n+1} + \text{PP}_i \quad (25\text{--}1)$$
$$\underset{\text{DNA}}{} \qquad \underset{\substack{\text{Lengthened} \\ \text{DNA}}}{}$$

where dNMP and dNTP are deoxynucleoside 5′-monophosphate and 5′-triphosphate, respectively. Catalysis by virtually all DNA polymerases prominently involves two Mg^{2+} ions at the active site (Fig. 25–5a). One of these helps to deprotonate the 3′-hydroxyl group, rendering it a more effective nucleophile. The other binds to the incoming dNTP and facilitates departure of the pyrophosphate.

The reaction seems to proceed with only a minimal change in free energy, given that one phosphodiester bond is formed at the expense of a somewhat less stable phosphate anhydride. However, noncovalent base-stacking and base-pairing interactions provide additional stabilization to the lengthened DNA product relative to the free nucleotide. Also, the formation of products is facilitated in the cell by the 19 kJ/mol generated in the subsequent hydrolysis of the pyrophosphate product by the enzyme pyrophosphatase (p. 524).

Early work on DNA polymerase I led to the definition of two central requirements for DNA polymerization (Fig. 25–5). First, all DNA polymerases require a template. The polymerization reaction is guided by a template DNA strand according to the base-pairing rules predicted by Watson and Crick: where a guanine is present in the template, a cytosine deoxynucleotide is added to the new strand, and so on. This was a particularly important discovery, not only because it provided a chemical basis for accurate semiconservative DNA replication but also because it represented the first example of the use of a template to guide a biosynthetic reaction.

(a) DNA synthesis

MECHANISM FIGURE 25–5 Elongation of a DNA chain. (a) The catalytic mechanism for addition of a new nucleotide by DNA polymerase involves two Mg^{2+} ions, coordinated to the phosphate groups of the incoming nucleotide triphosphate, the 3'-hydroxyl group that will act as a nucleophile, and three Asp residues, two of which are highly conserved in all DNA polymerases. The Mg^{2+} ion depicted at the top facilitates attack of the 3'-hydroxyl group of the primer on the α phosphate of the nucleotide triphosphate; the other Mg^{2+} ion facilitates displacement of the pyrophosphate. Both ions stabilize the structure of the pentacovalent transition state. RNA polymerases use a similar mechanism (see Fig. 26–1a). **(b)** DNA polymerase I activity also requires a single unpaired strand to act as template and a primer strand to provide the free hydroxyl group at the 3' end, to which the new nucleotide unit is added. Each incoming nucleotide is selected in part by base-pairing to the appropriate nucleotide in the template strand. The reaction product has a new free 3' hydroxyl, allowing the addition of another nucleotide. The newly formed base pair migrates to make the active site available to the next pair to be formed. **(c)** The core of most DNA polymerases is shaped like a human hand that wraps around the active site. The structure shown is the DNA polymerase I of *Thermus aquaticus*, bound to DNA (PDB ID 4KTQ). **(d)** A cartoon interpretation of the polymerase structure shows the insertion and postinsertion parts of the active site. The insertion site is where the nucleotide addition occurs, and the postinsertion site is where the newly formed base pair migrates after it appears. 🖱 Nucleotide Polymerization by DNA Polymerase

Second, the polymerases require a **primer**. A primer is a strand segment (complementary to the template) with a free 3'-hydroxyl group to which a nucleotide can be added; the free 3' end of the primer is called the **primer terminus**. In other words, part of the new strand must already be in place: all DNA polymerases can only add nucleotides to a preexisting strand. Many primers are oligonucleotides of RNA rather than DNA, and specialized enzymes synthesize primers when and where they are required.

A DNA polymerase active site has two parts (Fig. 25–5b). The incoming nucleotide is initially positioned in the **insertion site**. Once the phosphodiester bond is formed, the polymerase slides forward on the DNA and the new base pair is positioned in the **postinsertion site**. These elements are located in a pocket that resembles the palm of a hand (Fig. 25–5c).

After adding a nucleotide to a growing DNA strand, a DNA polymerase either dissociates or moves along the template and adds another nucleotide. Dissociation and reassociation of the polymerase can limit the overall polymerization rate—the process is generally faster when a polymerase adds more nucleotides without dissociating from the template. The average number of nucleotides added before a polymerase dissociates defines its **processivity**. DNA polymerases vary greatly in processivity; some add just a few nucleotides before dissociating, others add many thousands. 🖱 **Nucleotide Polymerization by DNA Polymerase**

Replication Is Very Accurate

Replication proceeds with an extraordinary degree of fidelity. In *E. coli*, a mistake is made only once for every 10^9 to 10^{10} nucleotides added. For the *E. coli* chromosome of $\sim 4.6 \times 10^6$ bp, this means that an error occurs only once per 1,000 to 10,000 replications. During polymerization, discrimination between correct and incorrect nucleotides relies not just on the hydrogen bonds that specify the correct pairing between complementary bases but also on the common geometry of the standard A=T and G≡C base pairs **(Fig. 25–6)**. The active site of DNA polymerase I accommodates only base pairs with this geometry. An incorrect nucleotide may be able to hydrogen-bond with a base in the template, but it generally will not fit into the active site. Incorrect bases can be rejected before the phosphodiester bond is formed.

The accuracy of the polymerization reaction itself, however, is insufficient to account for the high degree of fidelity in replication. Careful measurements in vitro have shown that DNA polymerases insert one incorrect nucleotide for every 10^4 to 10^5 correct ones. These mistakes sometimes occur because a base is briefly in an unusual tautomeric form (see Fig. 8–9), allowing it to hydrogen-bond with an incorrect partner. In vivo, the error rate is reduced by additional enzymatic mechanisms.

One mechanism intrinsic to virtually all DNA polymerases is a separate 3′→5′ exonuclease activity that double-checks each nucleotide after it is added. This nuclease activity permits the enzyme to remove a newly added nucleotide and is highly specific for mismatched base pairs **(Fig. 25–7)**. If the polymerase has added the wrong nucleotide, translocation of the enzyme to the position where the next nucleotide is to be added is inhibited. This kinetic pause provides the opportunity for a correction. The 3′→5′ exonuclease activity removes the mispaired nucleotide, and the polymerase begins again. This activity, known as **proofreading**, is not simply the reverse of the polymerization reaction (Eqn 25–1), because pyrophosphate is not involved. The polymerizing and proofreading activities of a DNA polymerase can be measured separately. Proofreading improves the inherent accuracy of the polymerization reaction 10^2- to 10^3-fold. In the monomeric DNA polymerase I, the polymerizing and proofreading activities have separate active sites within the same polypeptide.

When base selection and proofreading are combined, DNA polymerase leaves behind one net error for every 10^6 to 10^8 bases added. Yet the measured accuracy of replication in *E. coli* is higher still. The additional accuracy is provided by a separate enzyme system that repairs the mismatched base pairs remaining after replication. We describe this mismatch repair, along with other DNA repair processes, in Section 25.2.

(a) Correct base pairs

Active site shape

(b) Incorrect base pairs

FIGURE 25–6 Contribution of base-pair geometry to the fidelity of DNA replication. (a) The standard A=T and G≡C base pairs have very similar geometries, and an active site sized to fit one will generally accommodate the other. **(b)** The geometry of incorrectly paired bases can exclude them from the active site, as occurs on DNA polymerase.

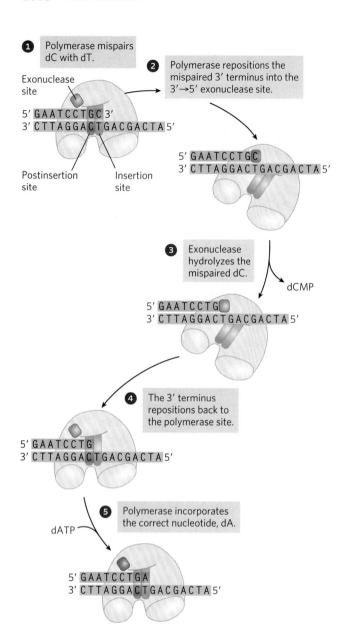

① Polymerase mispairs dC with dT.

Exonuclease site

5′ G A A T C C T G C 3′
3′ C T T A G G A C T G A C G A C T A 5′

Postinsertion site   Insertion site

② Polymerase repositions the mispaired 3′ terminus into the 3′→5′ exonuclease site.

5′ G A A T C C T G C
3′ C T T A G G A C T G A C G A C T A 5′

③ Exonuclease hydrolyzes the mispaired dC.

dCMP

5′ G A A T C C T G
3′ C T T A G G A C T G A C G A C T A 5′

④ The 3′ terminus repositions back to the polymerase site.

5′ G A A T C C T G
3′ C T T A G G A C T G A C G A C T A 5′

⑤ Polymerase incorporates the correct nucleotide, dA.

dATP

5′ G A A T C C T G A
3′ C T T A G G A C T G A C G A C T A 5′

FIGURE 25–7 An example of error correction by the 3′→5′ exonuclease activity of DNA polymerase I. Structural analysis has located the exonuclease activity behind the polymerase activity as the enzyme is oriented in its movement along the DNA. A mismatched base (here, a C–T mismatch) impedes translocation of DNA polymerase I to the next site. The DNA bound to the enzyme slides backward into the exonuclease site, and the enzyme corrects the mistake with its 3′→5′ exonuclease activity. The enzyme then resumes its polymerase activity in the 5′→3′ direction.

E. coli Has at Least Five DNA Polymerases

More than 90% of the DNA polymerase activity observed in *E. coli* extracts can be accounted for by DNA polymerase I. Soon after the isolation of this enzyme in 1955, however, evidence began to accumulate that it is not suited for replication of the large *E. coli* chromosome. First, the rate at which it adds nucleotides (600 nucleotides/min) is too slow (by a factor of 100 or more) to account for the rates at which the replication fork moves in the bacterial cell. Second, DNA polymerase I has a relatively low processivity. Third, genetic studies have demonstrated that many genes, and therefore many proteins, are involved in replication: DNA polymerase I clearly does not act alone. Fourth, and most important, in 1969 John Cairns isolated a bacterial strain with an altered gene for DNA polymerase I that produced an inactive enzyme. Although this strain was abnormally sensitive to agents that damaged DNA, it was nevertheless viable!

A search for other DNA polymerases led to the discovery of *E. coli* **DNA polymerase II** and **DNA polymerase III** in the early 1970s. DNA polymerase II is an enzyme involved in one type of DNA repair (Section 25.3). DNA polymerase III is the principal replication enzyme in *E. coli*. The properties of these three DNA polymerases are compared in Table 25–1. DNA polymerases IV and V, identified in 1999, are involved in an unusual form of DNA repair (Section 25.2).

TABLE 25–1	Comparison of Three DNA Polymerases of *E. coli*		
	DNA polymerase		
	I	II	III
Structural gene*	*polA*	*polB*	*polC (dnaE)*
Subunits (number of different types)	1	7	≥10
M_r	103,000	88,000[†]	791,500
3′→5′ Exonuclease (proofreading)	Yes	Yes	Yes
5′→3′ Exonuclease	Yes	No	No
Polymerization rate (nucleotides/s)	10–20	40	250–1,000
Processivity (nucleotides added before polymerase dissociates)	3–200	1,500	≥500,000

*For enzymes with more than one subunit, the gene listed here encodes the subunit with polymerization activity. Note that *dnaE* is an earlier designation for the gene now referred to as *polC*.

[†]Polymerization subunit only. DNA polymerase II shares several subunits with DNA polymerase III, including the β, γ, δ, δ′, χ, and Ψ subunits (see Table 25–2).

DNA polymerase I, then, is not the primary enzyme of replication; instead it performs a host of cleanup functions during replication, recombination, and repair. The polymerase's special functions are enhanced by its $5'\rightarrow3'$ exonuclease activity. This activity, distinct from the $3'\rightarrow5'$ proofreading exonuclease (Fig. 25–7), is located in a structural domain that can be separated from the enzyme by mild protease treatment. When the $5'\rightarrow3'$ exonuclease domain is removed, the remaining fragment (M_r 68,000), the **large fragment** or **Klenow fragment**, retains the polymerization and proofreading activities. The $5'\rightarrow3'$ exonuclease activity of intact DNA polymerase I can replace a segment of DNA (or RNA) paired to the template strand, in a process known as nick translation **(Fig. 25–8)**. Most other DNA polymerases lack a $5'\rightarrow3'$ exonuclease activity.

DNA polymerase III is much more complex than DNA polymerase I (Table 25–2). Its polymerization and proofreading activities reside in its α and ε (epsilon) subunits, respectively. The θ subunit associates with α and ε to form a core polymerase, which can polymerize DNA but with limited processivity. Two core polymerases can be linked by another set of subunits, a clamp-loading complex, or γ complex, consisting of five subunits of four different types, $\tau_2\gamma\delta\delta'$. The core polymerases are linked through the τ (tau) subunits.

Two additional subunits, χ (chi) and ψ (psi), are bound to the clamp-loading complex. The entire assembly of 13 protein subunits (nine different types) is called DNA polymerase III* **(Fig. 25–9a)**.

DNA polymerase III* can polymerize DNA, but with a much lower processivity than one would expect for the organized replication of an entire chromosome. The necessary increase in processivity is provided by the addition of the β subunits, four of which complete the DNA polymerase III holoenzyme. The β subunits associate in pairs to form donut-shaped structures that encircle the DNA and act like clamps (Fig. 25–9b). Each dimer associates with a core subassembly of polymerase III* (one dimeric clamp per core subassembly) and slides along the DNA as replication proceeds. The β sliding clamp prevents the dissociation of DNA polymerase III from DNA, dramatically increasing processivity—to greater than 500,000 (Table 25–1).

DNA Replication Requires Many Enzymes and Protein Factors

Replication in *E. coli* requires not just a single DNA polymerase but 20 or more different enzymes and proteins, each performing a specific task. The entire complex has been termed the **DNA replicase system** or **replisome**. The enzymatic complexity of replication reflects the constraints imposed by the structure of DNA and by the requirements for accuracy. The main classes of replication enzymes are considered here in terms of the problems they overcome.

Access to the DNA strands that are to act as templates requires separation of the two parent strands. This is generally accomplished by **helicases**, enzymes that move along the DNA and separate the strands, using chemical energy from ATP. Strand separation creates topological stress in the helical DNA structure (see Fig. 24–11), which is relieved by the action of **topoisomerases**. The separated strands are stabilized by **DNA-binding proteins**. As noted earlier, before DNA polymerases can begin synthesizing DNA, primers must be present on the template—generally, short

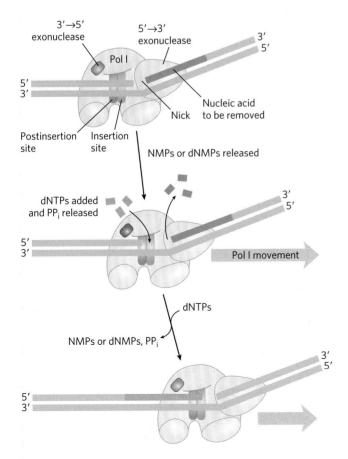

FIGURE 25–8 Nick translation. The bacterial DNA polymerase I has three domains, catalyzing its DNA polymerase, $5'\rightarrow3'$ exonuclease, and $3'\rightarrow5'$ exonuclease activities. The $5'\rightarrow3'$ exonuclease domain is in front of the enzyme as it moves along the DNA and is not shown in Figure 25–5. By degrading the DNA strand ahead of the enzyme and synthesizing a new strand behind, DNA polymerase I can promote a reaction called nick translation, where a break or nick in the DNA is effectively moved along with the enzyme. This process has a role in DNA repair and in the removal of RNA primers during replication (both described later). The strand of nucleic acid to be removed (either DNA or RNA) is shown in purple, the replacement strand in red. DNA synthesis begins at a nick (a broken phosphodiester bond, leaving a free 3' hydroxyl and a free 5' phosphate). A nick remains where DNA polymerase I eventually dissociates, and the nick is later sealed by another enzyme.

TABLE 25–2 Subunits of DNA Polymerase III of *E. coli*

Subunit	Number of subunits per holoenzyme	M_r of subunit	Gene	Function of subunit	
α	2	129,900	*polC (dnaE)*	Polymerization activity	⎫
ε	2	27,500	*dnaQ (mutD)*	$3' \rightarrow 5'$ Proofreading exonuclease	⎬ Core polymerase
θ	2	8,600	*holE*	Stabilization of ε subunit	⎭
τ	2	71,100	*dnaX*	Stable template binding; core enzyme dimerization	⎫
γ	1	47,500	*dnaX**	Clamp loader	⎪ Clamp-loading (γ) complex
δ	1	38,700	*holA*	Clamp opener	⎬ that loads β subunits on
δ'	1	36,900	*holB*	Clamp loader	⎪ lagging strand at each
χ	1	16,600	*holC*	Interaction with SSB	⎪ Okazaki fragment
ψ	1	15,200	*holD*	Interaction with γ and χ	⎭
β	4	40,600	*dnaN*	DNA clamp required for optimal processivity	

*The γ subunit is encoded by a portion of the gene for the τ subunit, such that the amino-terminal 66% of the τ subunit has the same amino acid sequence as the γ subunit. The γ subunit is generated by a translational frameshifting mechanism (p. 1111) that leads to premature translational termination.

segments of RNA synthesized by enzymes known as **primases**. Ultimately, the RNA primers are removed and replaced by DNA; in *E. coli*, this is one of the many functions of DNA polymerase I. After an RNA primer is removed and the gap is filled in with DNA, a nick remains in the DNA backbone in the form of a broken phosphodiester bond. These nicks are sealed by **DNA ligases**. All these processes require coordination and regulation, an interplay best characterized in the *E. coli* system.

(a)

β sliding clamp

Pol III core (α, ε, θ) Pol III core (α, ε, θ)

τ τ

γ τ

τ δ

δ'

β sliding clamp

Clamp loader

(b)

FIGURE 25–9 DNA polymerase III. (a) Architecture of bacterial DNA polymerase III (Pol III). Two core domains, composed of subunits α, ϵ, and θ, are linked by a five-subunit clamp-loading complex (also known as the γ complex) with the composition $\tau_2\gamma\delta\delta'$. The core subunits and clamp-loader complex constitute DNA polymerase III*. The γ and τ subunits are encoded by the same gene. The γ subunit is a shortened version of the τ subunit; τ thus contains a domain identical to γ, along with an additional segment that interacts with the core polymerase. The other two subunits of DNA polymerase III*, χ and ψ (not shown), also bind to the clamp-loading complex. Two β clamps interact with the two-core subassembly, each clamp a dimer of the β subunit. The complex interacts with the DnaB helicase (described later in the text) through the τ subunits. **(b)** Two β subunits of *E. coli* polymerase III form a circular clamp that surrounds the DNA. The clamp slides along the DNA molecule, increasing the processivity of the polymerase III holoenzyme to greater than 500,000 nucleotides by preventing its dissociation from the DNA. The two β subunits are shown in two shades of purple as ribbon structures (left) and surface images (right), surrounding the DNA (derived from PDB ID 2POL).

FIGURE 25–10 Arrangement of sequences in the *E. coli* replication origin, *oriC*. Consensus sequences (p. 104) for key repeated elements are shown. N represents any of the four nucleotides. The horizontal arrows indicate the orientations of the nucleotide sequences (left-to-right arrow denotes sequence in top strand; right-to-left, bottom strand). FIS and IHF are binding sites for proteins described in the text. R sites are bound by DnaA. I sites are additional DnaA-binding sites (with different sequences), bound by DnaA only when the protein is complexed with ATP.

Replication of the *E. coli* Chromosome Proceeds in Stages

The synthesis of a DNA molecule can be divided into three stages: initiation, elongation, and termination, distinguished both by the reactions taking place and by the enzymes required. As you will find here and in the next two chapters, synthesis of the major information-containing biological polymers—DNAs, RNAs, and proteins—can be understood in terms of these same three stages, with the stages of each pathway having unique characteristics. The events described below reflect information derived primarily from in vitro experiments using purified *E. coli* proteins, although the principles are highly conserved in all replication systems.

Initiation The *E. coli* replication origin, *oriC*, consists of 245 bp and contains DNA sequence elements that are highly conserved among bacterial replication origins. The general arrangement of the conserved sequences is illustrated in **Figure 25–10**. Two types of sequences are of special interest: five repeats of a 9 bp sequence (R sites) that serve as binding sites for the key initiator protein DnaA, and a region rich in A=T base pairs called the **DNA unwinding element (DUE)**. There are three additional DnaA-binding sites (I sites), and binding sites for the proteins IHF (integration host factor) and FIS (factor for inversion stimulation). These two proteins were discovered as required components of certain recombination reactions described later in this chapter, and their names reflect those roles. Another DNA-binding protein, HU (a histonelike bacterial protein originally dubbed factor U), also participates but does not have a specific binding site.

At least 10 different enzymes or proteins (summarized in Table 25–3) participate in the initiation phase of replication. They open the DNA helix at the origin and establish a prepriming complex for subsequent reactions. The crucial component in the initiation process is the DnaA protein, a member of the **AAA+ ATPase** protein family (ATPases associated with diverse cellular activities). Many AAA+ ATPases, including DnaA,

TABLE 25–3	Proteins Required to Initiate Replication at the *E. coli* Origin		
Protein	M_r	Number of subunits	Function
DnaA protein	52,000	1	Recognizes ori sequence; opens duplex at specific sites in origin
DnaB protein (helicase)	300,000	6*	Unwinds DNA
DnaC protein	174,000	6*	Required for DnaB binding at origin
HU	19,000	2	Histonelike protein; DNA-binding protein; stimulates initiation
FIS	22,500	2*	DNA-binding protein; stimulates initiation
IHF	22,000	2	DNA-binding protein; stimulates initiation
Primase (DnaG protein)	60,000	1	Synthesizes RNA primers
Single-stranded DNA-binding protein (SSB)	75,600	4*	Binds single-stranded DNA
DNA gyrase (DNA topoisomerase II)	400,000	4	Relieves torsional strain generated by DNA unwinding
Dam methylase	32,000	1	Methylates (5′)GATC sequences at *oriC*

*Subunits in these cases are identical.

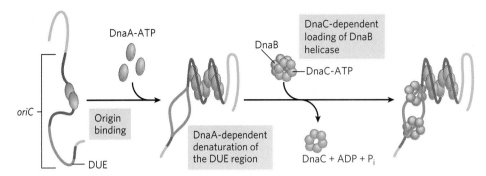

FIGURE 25–11 Model for initiation of replication at the *E. coli* origin, *oriC*. Eight DnaA protein molecules, each with a bound ATP, bind at the R and I sites in the origin (see Fig. 25–10). The DNA is wrapped around this complex, which forms a right-handed helical structure. The A≡T-rich DUE region is denatured as a result of the strain imparted by the adjacent DnaA binding. Formation of the helical DnaA complex is facilitated by the proteins HU, IHF, and FIS, which are not shown here because their detailed structural roles have not yet been defined. Hexamers of the DnaB protein bind to each strand, with the aid of DnaC protein. The DnaB helicase activity further unwinds the DNA in preparation for priming and DNA synthesis.

form oligomers and hydrolyze ATP relatively slowly. This ATP hydrolysis acts as a switch mediating interconversion of the protein between two states. In the case of DnaA, the ATP-bound form is active and the ADP-bound form is inactive.

Eight DnaA protein molecules, all in the ATP-bound state, assemble to form a helical complex encompassing the R and I sites in *oriC* **(Fig. 25–11)**. DnaA has a higher affinity for the R sites than I sites, and binds R sites equally well in its ATP- or ADP-bound form. The I sites, which bind only the ATP-bound DnaA, allow discrimination between the active and inactive forms of DnaA. The tight right-handed wrapping of the DNA around this complex introduces an effective positive supercoil (see Chapter 24). The associated strain in the nearby DNA leads to denaturation in the A≡T-rich DUE region. The complex formed at the replication origin also includes several DNA-binding proteins—HU, IHF, and FIS—that facilitate DNA bending.

The DnaC protein, another AAA+ ATPase, then loads the DnaB protein onto the separated DNA strands in the denatured region. A hexamer of DnaC, each subunit bound to ATP, forms a tight complex with the hexameric, ring-shaped DnaB helicase. This DnaC-DnaB interaction opens the DnaB ring, the process being aided by a further interaction between DnaB and DnaA. Two of the ring-shaped DnaB hexamers are loaded in the DUE, one onto each DNA strand. The ATP bound to DnaC is hydrolyzed, releasing the DnaC and leaving the DnaB bound to the DNA.

Loading of the DnaB helicase is the key step in replication initiation. As a replicative helicase, DnaB migrates along the single-stranded DNA in the 5′→3′ direction, unwinding the DNA as it travels. The DnaB helicases loaded onto the two DNA strands thus travel in opposite directions, creating two potential replication forks. All other proteins at the replication fork are linked directly or indirectly to DnaB. The DNA polymerase III

holoenzyme is linked through the τ subunits; additional DnaB interactions are described below. As replication begins and the DNA strands are separated at the fork, many molecules of single-stranded DNA–binding protein (SSB) bind to and stabilize the separated strands, and DNA gyrase (DNA topoisomerase II) relieves the topological stress induced ahead of the fork by the unwinding reaction.

Initiation is the only phase of DNA replication that is known to be regulated, and it is regulated such that replication occurs only once in each cell cycle. The mechanism of regulation is not yet entirely understood, but genetic and biochemical studies have provided insights into several separate regulatory mechanisms.

Once DNA polymerase III has been loaded onto the DNA, along with the β subunits (signaling completion of the initiation phase), the protein Hda binds to the β subunits and interacts with DnaA to stimulate hydrolysis of its bound ATP. Hda is yet another AAA+ ATPase closely related to DnaA (its name is derived from *h*omologous to *DnaA*). This ATP hydrolysis leads to disassembly of the DnaA complex at the origin. Slow release of ADP by DnaA and rebinding of ATP cycles the protein between its inactive (with bound ADP) and active (with bound ATP) forms on a time scale of 20 to 40 minutes.

The timing of replication initiation is affected by DNA methylation and interactions with the bacterial plasma membrane. The *oriC* DNA is methylated by the Dam methylase (Table 25–3), which methylates the N^6 position of adenine within the palindromic sequence (5′)GATC. (Dam is not a biochemical expletive; it stands for *D*NA *a*denine *m*ethylation.) The *oriC* region of *E. coli* is highly enriched in GATC sequences—it has 11 of them in its 245 bp, whereas the average frequency of GATC in the *E. coli* chromosome as a whole is 1 in 256 bp.

Immediately after replication, the DNA is hemimethylated: the parent strands have methylated *oriC* sequences but the newly synthesized strands do not. The hemimethylated *oriC* sequences are now sequestered by interaction with the plasma membrane (the mechanism is unknown) and by binding of the protein SeqA. After a time, *oriC* is released from the plasma membrane, SeqA dissociates, and the DNA must be fully methylated by Dam methylase before it can again bind DnaA and initiate a new round of replication.

Elongation The elongation phase of replication includes two distinct but related operations: leading strand synthesis and lagging strand synthesis. Several enzymes at the replication fork are important to the synthesis of both strands. Parent DNA is first unwound by DNA helicases, and the resulting topological stress is relieved by topoisomerases. Each separated strand is then stabilized by SSB. From this point, synthesis of leading and lagging strands is sharply different.

Leading strand synthesis, the more straightforward of the two, begins with the synthesis by primase (DnaG protein) of a short (10 to 60 nucleotide) RNA primer at the replication origin. DnaG interacts with DnaB helicase to carry out this reaction, and the primer is synthesized in the direction opposite to that in which the DnaB helicase is moving. In effect, the DnaB helicase moves along the strand that becomes the lagging strand in DNA synthesis; however, the first

primer laid down in the first DnaG-DnaB interaction serves to prime leading strand DNA synthesis in the opposite direction. Deoxyribonucleotides are added to this primer by a DNA polymerase III complex linked to the DnaB helicase tethered to the opposite DNA strand. Leading strand synthesis then proceeds continuously, keeping pace with the unwinding of DNA at the replication fork.

Lagging strand synthesis, as we have noted, is accomplished in short Okazaki fragments (**Fig. 25–12a**). First, an RNA primer is synthesized by primase and, as in leading strand synthesis, DNA polymerase III binds to the RNA primer and adds deoxyribonucleotides (Fig. 25–12b). On this level, the synthesis of each Okazaki fragment seems straightforward, but the reality is quite complex. The complexity lies in the *coordination* of leading and lagging strand synthesis. Both strands are produced by a *single* asymmetric DNA polymerase III dimer; this is accomplished by looping the DNA of the lagging strand as shown in **Figure 25–13**, bringing together the two points of polymerization.

The synthesis of Okazaki fragments on the lagging strand entails some elegant enzymatic choreography. DnaB helicase and DnaG primase constitute a functional unit within the replication complex, the **primosome**. DNA polymerase III uses one set of its core

FIGURE 25–12 Synthesis of Okazaki fragments. **(a)** At intervals, primase synthesizes an RNA primer for a new Okazaki fragment. Note that if we consider the two template strands as lying side by side, lagging strand synthesis formally proceeds in the opposite direction from fork movement. **(b)** Each primer is extended by DNA polymerase III. **(c)** DNA synthesis continues until the fragment extends as far as the primer of the previously added Okazaki fragment. A new primer is synthesized near the replication fork to begin the process again.

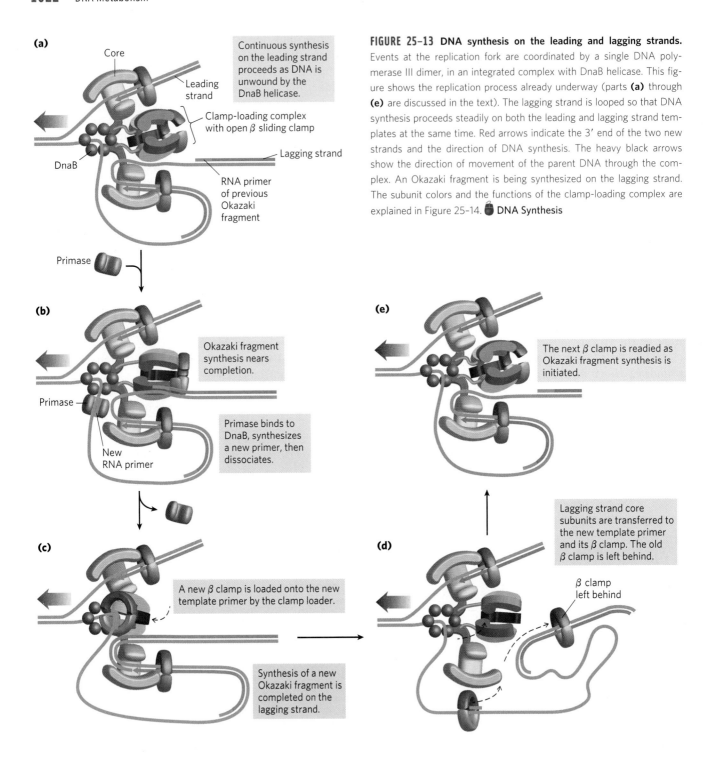

(a)

Core

Leading strand

Continuous synthesis on the leading strand proceeds as DNA is unwound by the DnaB helicase.

Clamp-loading complex with open β sliding clamp

DnaB

Lagging strand

RNA primer of previous Okazaki fragment

Primase

(b)

Okazaki fragment synthesis nears completion.

Primase

Primase binds to DnaB, synthesizes a new primer, then dissociates.

New RNA primer

(c)

A new β clamp is loaded onto the new template primer by the clamp loader.

Synthesis of a new Okazaki fragment is completed on the lagging strand.

(e)

The next β clamp is readied as Okazaki fragment synthesis is initiated.

Lagging strand core subunits are transferred to the new template primer and its β clamp. The old β clamp is left behind.

(d)

β clamp left behind

FIGURE 25–13 DNA synthesis on the leading and lagging strands. Events at the replication fork are coordinated by a single DNA polymerase III dimer, in an integrated complex with DnaB helicase. This figure shows the replication process already underway (parts **(a)** through **(e)** are discussed in the text). The lagging strand is looped so that DNA synthesis proceeds steadily on both the leading and lagging strand templates at the same time. Red arrows indicate the 3' end of the two new strands and the direction of DNA synthesis. The heavy black arrows show the direction of movement of the parent DNA through the complex. An Okazaki fragment is being synthesized on the lagging strand. The subunit colors and the functions of the clamp-loading complex are explained in Figure 25–14. ● DNA Synthesis

subunits (the core polymerase) to synthesize the leading strand continuously, while the other set of core subunits cycles from one Okazaki fragment to the next on the looped lagging strand. DnaB helicase, bound in front of DNA polymerase III, unwinds the DNA at the replication fork (Fig. 25–13a) as it travels along the lagging strand template in the 5'→3' direction. DnaG primase occasionally associates with DnaB helicase and synthesizes a short RNA primer (Fig. 25–13b). A new β sliding clamp is then positioned at the primer by the clamp-loading complex of DNA

polymerase III (Fig. 25–13c). When synthesis of an Okazaki fragment has been completed, replication halts, and the core subunits of DNA polymerase III dissociate from their β sliding clamp (and from the completed Okazaki fragment) and associate with the new clamp (Fig. 25–13d, e). This initiates synthesis of a new Okazaki fragment. As noted earlier, the entire complex responsible for coordinated DNA synthesis at a replication fork is known as the replisome. The proteins acting at the replication fork are summarized in Table 25–4.

TABLE 25-4 Proteins of the *E. coli* Replisome

Protein	M_r	Number of subunits	Function
SSB	75,600	4	Binding to single-stranded DNA
DnaB protein (helicase)	300,000	6	DNA unwinding; primosome constituent
Primase (DnaG protein)	60,000	1	RNA primer synthesis; primosome constituent
DNA polymerase III	791,500	17	New strand elongation
DNA polymerase I	103,000	1	Filling of gaps; excision of primers
DNA ligase	74,000	1	Ligation
DNA gyrase (DNA topoisomerase II)	400,000	4	Supercoiling

Source: Modified from Kornberg, A. (1982) *Supplement to DNA Replication*, Table S11-2, W. H. Freeman and Company, New York.

FIGURE 25-14 The DNA polymerase III clamp loader. The five subunits of the clamp-loading complex are the γ, δ, and δ' subunits and the amino-terminal domain of each τ subunit (see Fig. 25-9). The complex binds to three molecules of ATP and to a dimeric β clamp. This binding forces the β clamp open at one of its two subunit interfaces. Hydrolysis of the bound ATP allows the β clamp to close again around the DNA.

The clamp-loading complex of DNA polymerase III, consisting of parts of the two τ subunits along with the γ, δ, and δ' subunits, is also an AAA+ ATPase. This complex binds to ATP and to the new β sliding clamp. The binding imparts strain on the dimeric clamp, opening up the ring at one subunit interface **(Fig. 25-14)**. The newly primed lagging strand is slipped into the ring through the resulting break. The clamp loader then hydrolyzes ATP, releasing the β sliding clamp and allowing it to close around the DNA.

The replisome promotes rapid DNA synthesis, adding ~1,000 nucleotides/s to each strand (leading and lagging). Once an Okazaki fragment has been completed, its RNA primer is removed and replaced with DNA by DNA polymerase I, and the remaining nick is sealed by DNA ligase **(Fig. 25-15)**.

DNA ligase catalyzes the formation of a phosphodiester bond between a 3′ hydroxyl at the end of one DNA strand and a 5′ phosphate at the end of another strand. The phosphate must be activated by adenylylation. DNA ligases isolated from viruses and eukaryotes use ATP for this purpose. DNA ligases from bacteria are unusual in that many use NAD^+—a cofactor that usually functions in hydride transfer reactions (see Fig. 13-24)—as the source of the AMP activating group **(Fig. 25-16)**. DNA ligase is another enzyme of DNA metabolism that has become an important reagent in recombinant DNA experiments (see Fig. 9-1).

Termination Eventually, the two replication forks of the circular *E. coli* chromosome meet at a terminus region

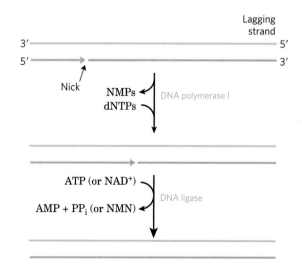

FIGURE 25-15 Final steps in the synthesis of lagging strand segments. RNA primers in the lagging strand are removed by the 5′→3′ exonuclease activity of DNA polymerase I and are replaced with DNA by the same enzyme. The remaining nick is sealed by DNA ligase. The role of ATP or NAD^+ is shown in Figure 25-16.

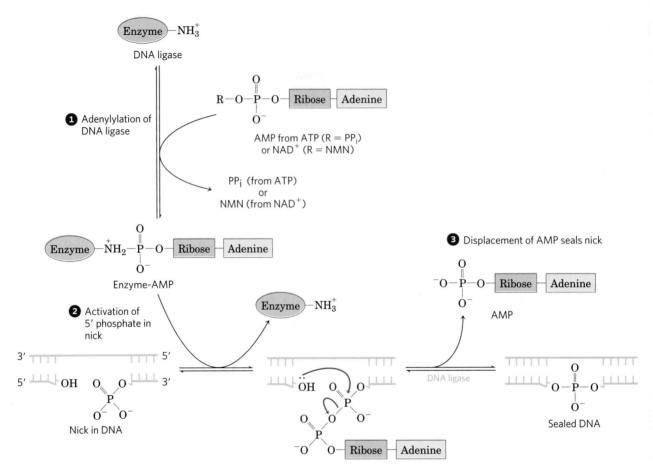

FIGURE 25–16 Mechanism of the DNA ligase reaction. In each of the three steps, one phosphodiester bond is formed at the expense of another. Steps ❶ and ❷ lead to activation of the 5′ phosphate in the nick. An AMP group is transferred first to a Lys residue on the enzyme and then to the 5′ phosphate in the nick. In step ❸, the 3′-hydroxyl group attacks this phosphate and displaces AMP, producing a phospho-diester bond to seal the nick. In the *E. coli* DNA ligase reaction, AMP is derived from NAD⁺. The DNA ligases isolated from some viral and eukaryotic sources use ATP rather than NAD⁺ and they release pyro-phosphate rather than nicotinamide mononucleotide (NMN) in step ❶.

containing multiple copies of a 20 bp sequence called Ter **(Fig. 25–17)**. The Ter sequences are arranged on the chromosome to create a trap that a replication fork

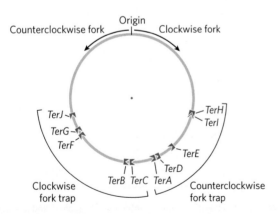

FIGURE 25–17 Termination of chromosome replication in *E. coli*. The Ter sequences (*TerA* through *TerJ*) are positioned on the chromosome in two clusters with opposite orientations.

can enter but cannot leave. The Ter sequences function as binding sites for the protein Tus (terminus utilization substance). The Tus-Ter complex can arrest a replica-tion fork from only one direction. Only one Tus-Ter complex functions per replication cycle—the complex first encountered by either replication fork. Given that opposing replication forks generally halt when they col-lide, Ter sequences would not seem to be essential, but they may prevent overreplication by one fork in the event that the other is delayed or halted by an encoun-ter with DNA damage or some other obstacle.

So, when either replication fork encounters a func-tional Tus-Ter complex, it halts; the other fork halts when it meets the first (arrested) fork. The final few hundred base pairs of DNA between these large protein complexes are then replicated (by an as yet unknown mechanism), completing two topologically interlinked (catenated) circular chromosomes **(Fig. 25–18)**. DNA circles linked in this way are known as **catenanes**. Separation of the catenated circles in *E. coli* requires

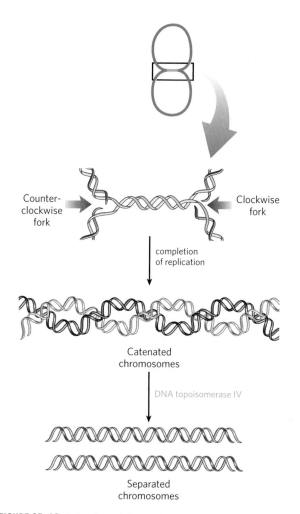

FIGURE 25-18 Role of topoisomerases in replication termination. Replication of the DNA separating opposing replication forks leaves the completed chromosomes joined as catenanes, or topologically interlinked circles. The circles are not covalently linked, but because they are interwound and each is covalently closed, they cannot be separated—except by the action of topoisomerases. In *E. coli*, a type II topoisomerase known as DNA topoisomerase IV plays the primary role in the separation of catenated chromosomes, transiently breaking both DNA strands of one chromosome and allowing the other chromosome to pass through the break.

topoisomerase IV (a type II topoisomerase). The separated chromosomes then segregate into daughter cells at cell division. The terminal phase of replication of other circular chromosomes, including many of the DNA viruses that infect eukaryotic cells, is similar.

Replication in Eukaryotic Cells Is Similar but More Complex

The DNA molecules in eukaryotic cells are considerably larger than those in bacteria and are organized into complex nucleoprotein structures (chromatin; p. 994). The essential features of DNA replication are the same in eukaryotes and bacteria, and many of the protein complexes are functionally and structurally conserved. However, eukaryotic replication is regulated and coordinated with the cell cycle, introducing some additional complexities.

Origins of replication have a well-characterized structure in some lower eukaryotes, but they are much less defined in higher eukaryotes. In vertebrates, a variety of A=T-rich sequences may be used for replication initiation, and the sites may vary from one cell division to the next. Yeast (*Saccharomyces cerevisiae*) has defined replication origins called autonomously replicating sequences (ARS), or **replicators**. Yeast replicators span ~150 bp and contain several essential, conserved sequences. About 400 replicators are distributed among the 16 chromosomes of the haploid yeast genome.

Regulation ensures that all cellular DNA is replicated once per cell cycle. Much of this regulation involves proteins called cyclins and the cyclin-dependent kinases (CDKs) with which they form complexes (p. 484). The cyclins are rapidly destroyed by ubiquitin-dependent proteolysis at the end of the M phase (mitosis), and the absence of cyclins allows the establishment of **pre-replicative complexes (pre-RCs)** on replication initiation sites. In rapidly growing cells, the pre-RC forms at the end of M phase. In slow-growing cells, it does not form until the end of G1. Formation of the pre-RC renders the cell competent for replication, an event sometimes called **licensing**.

As in bacteria, the key event in the initiation of replication in all eukaryotes is the loading of the replicative helicase, a heterohexameric complex of **minichromosome maintenance (MCM) proteins** (MCM2 to MCM7). The ring-shaped MCM2–7 helicase, functioning much like the bacterial DnaB helicase, is loaded onto the DNA by another six-protein complex called **ORC (origin recognition complex) (Fig. 25–19)**. ORC has five AAA+ ATPase domains among its subunits and is functionally analogous to the bacterial DnaA. Two other proteins, CDC6 (*c*ell *d*ivision *c*ycle) and CDT1 (*CDC10-d*ependent *t*ranscript *1*), are also required to load the MCM2–7 complex, and the yeast CDC6 is another AAA+ ATPase.

Commitment to replication requires the synthesis and activity of S-phase cyclin-CDK complexes (such as the cyclin E–CDK2 complex; see Fig. 12–46) and CDC7-DBF4. Both types of complexes help to activate replication by binding to and phosphorylating several proteins in the pre-RC. Other cyclins and CDKs function to inhibit the formation of more pre-RC complexes once replication has been initiated. For example, CDK2 binds to cyclin A as cyclin E levels decline during S phase, inhibiting CDK2 and preventing the licensing of additional pre-RC complexes.

The rate of movement of the replication fork in eukaryotes (~50 nucleotides/s) is only one-twentieth that observed in *E. coli*. At this rate, replication of an

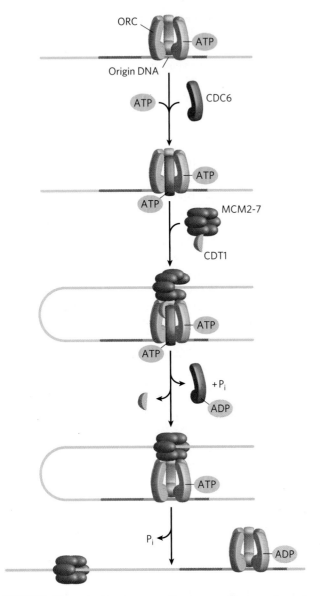

FIGURE 25–19 Assembly of a pre-replicative complex at a eukaryotic replication origin. The initiation site (origin) is bound by ORC, CDC6, and CDT1. These proteins, many of them AAA+ ATPases, promote loading of the replicative helicase, MCM2–7, in a reaction that is analogous to the loading of the bacterial DnaB helicase by DnaC protein. Loading of the MCM helicase complex onto the DNA forms the pre-replicative complex, or pre-RC, and is the key step in the initiation of replication.

average human chromosome proceeding from a single origin would take more than 500 hours. Replication of human chromosomes in fact proceeds bidirectionally from many origins, spaced 30 to 300 kbp apart. Eukaryotic chromosomes are almost always much larger than bacterial chromosomes, so multiple origins are probably a universal feature of eukaryotic cells.

Like bacteria, eukaryotes have several types of DNA polymerases. Some have been linked to particular

functions, such as the replication of mitochondrial DNA. The replication of nuclear chromosomes involves DNA polymerase α, in association with DNA polymerase δ. **DNA polymerase α** is typically a multisubunit enzyme with similar structure and properties in all eukaryotic cells. One subunit has a primase activity, and the largest subunit (M_r ~180,000) contains the polymerization activity. However, this polymerase has no proofreading $3'\rightarrow5'$ exonuclease activity, making it unsuitable for high-fidelity DNA replication. DNA polymerase α is believed to function only in the synthesis of short primers (either RNA or DNA) for Okazaki fragments on the lagging strand. These primers are then extended by the multi-subunit **DNA polymerase δ**. This enzyme is associated with and stimulated by proliferating cell nuclear antigen (PCNA; M_r 29,000), a protein found in large amounts in the nuclei of proliferating cells. The three-dimensional structure of PCNA is remarkably similar to that of the β subunit of *E. coli* DNA polymerase III (Fig. 25–9b), although primary sequence homology is not evident. PCNA has a function analogous to that of the β subunit, forming a circular clamp that greatly enhances the processivity of the polymerase. DNA polymerase δ has a $3'\rightarrow5'$ proofreading exonuclease activity and seems to carry out both leading and lagging strand synthesis in a complex comparable to the dimeric bacterial DNA polymerase III.

Yet another polymerase, **DNA polymerase ε**, replaces DNA polymerase δ in some situations, such as in DNA repair. DNA polymerase ε may also function at the replication fork, perhaps playing a role analogous to that of the bacterial DNA polymerase I, removing the primers of Okazaki fragments on the lagging strand.

Two other protein complexes also function in eukaryotic DNA replication. RPA (replication protein A) is a eukaryotic single-stranded DNA–binding protein, equivalent in function to the *E. coli* SSB protein. RFC (replication factor C) is a clamp loader for PCNA and facilitates the assembly of active replication complexes. The subunits of the RFC complex have significant sequence similarity to the subunits of the bacterial clamp-loading (γ) complex.

The termination of replication on linear eukaryotic chromosomes involves the synthesis of special structures called **telomeres** at the ends of each chromosome, as discussed in the next chapter.

Viral DNA Polymerases Provide Targets for Antiviral Therapy

Many DNA viruses encode their own DNA polymerases, and some of these have become targets for pharmaceuticals. For example, the DNA polymerase of the herpes simplex virus is inhibited by acyclovir, a compound developed by Gertrude Elion and

George Hitchings (p. 923). Acyclovir consists of guanine attached to an incomplete ribose ring.

Acyclovir

It is phosphorylated by a virally encoded thymidine kinase; acyclovir binds to this viral enzyme with an affinity 200-fold greater than its binding to the cellular thymidine kinase. This ensures that phosphorylation occurs mainly in virus-infected cells. Cellular kinases convert the resulting acyclo-GMP to acyclo-GTP, which is both an inhibitor and a substrate of DNA polymerases; acyclo-GTP competitively inhibits the herpes DNA polymerase more strongly than cellular DNA polymerases. Because it lacks a 3′ hydroxyl, acyclo-GTP also acts as a chain terminator when incorporated into DNA. Thus viral replication is inhibited at several steps. ∎

SUMMARY 25.1 DNA Replication

▶ Replication of DNA occurs with very high fidelity and at a designated time in the cell cycle. Replication is semiconservative, each strand acting as template for a new daughter strand. It is carried out in three identifiable phases: initiation, elongation, and termination. The process starts at a single origin in bacteria and usually proceeds bidirectionally.

▶ DNA is synthesized in the 5′→3′ direction by DNA polymerases. At the replication fork, the leading strand is synthesized continuously in the same direction as replication fork movement; the lagging strand is synthesized discontinuously as Okazaki fragments, which are subsequently ligated.

▶ The fidelity of DNA replication is maintained by (1) base selection by the polymerase, (2) a 3′→5′ proofreading exonuclease activity that is part of most DNA polymerases, and (3) specific repair systems for mismatches left behind after replication.

▶ Most cells have several DNA polymerases. In *E. coli*, DNA polymerase III is the primary replication enzyme. DNA polymerase I is responsible for special functions during replication, recombination, and repair.

▶ The separate initiation, elongation, and termination phases of DNA replication involve an array of enzymes and protein factors, many belonging to the AAA+ ATPase family.

▶ The major replicative DNA polymerase in eukaryotes is DNA polymerase δ. DNA polymerase α functions to synthesize primers. DNA polymerase ε functions in DNA repair.

25.2 DNA Repair

Most cells have only one or two sets of genomic DNA. Damaged proteins and RNA molecules can be quickly replaced by using information encoded in the DNA, but DNA molecules themselves are irreplaceable. Maintaining the integrity of the information in DNA is a cellular imperative, supported by an elaborate set of DNA repair systems. DNA can become damaged by a variety of processes, some spontaneous, others catalyzed by environmental agents (Chapter 8). Replication itself can very occasionally damage the information content in DNA when errors introduce mismatched base pairs (such as G paired with T).

The chemistry of DNA damage is diverse and complex. The cellular response to this damage includes a wide range of enzymatic systems that catalyze some of the most interesting chemical transformations in DNA metabolism. We first examine the effects of alterations in DNA sequence and then consider specific repair systems.

Mutations Are Linked to Cancer

The best way to illustrate the importance of DNA repair is to consider the effects of *unrepaired* DNA damage (a lesion). The most serious outcome is a change in the base sequence of the DNA, which, if replicated and transmitted to future generations of cells, becomes permanent. A permanent change in the nucleotide sequence of DNA is called a **mutation**. Mutations can involve the replacement of one base pair with another (substitution mutation) or the addition or deletion of one or more base pairs (insertion or deletion mutations). If the mutation affects nonessential DNA or if it has a negligible effect on the function of a gene, it is known as a **silent mutation**. Rarely, a mutation confers some biological advantage. Most nonsilent mutations, however, are neutral or deleterious.

In mammals there is a strong correlation between the accumulation of mutations and cancer. A simple test developed by Bruce Ames measures the potential of a given chemical compound to promote certain easily detected mutations in a specialized bacterial strain **(Fig. 25–20)**. Few of the chemicals that we encounter in daily life score as mutagens in this test. However, of the compounds known to be carcinogenic from extensive animal trials, more than 90% are also found to be mutagenic in the Ames test. Because of this strong correlation between mutagenesis and carcinogenesis, the Ames test for bacterial mutagens is widely used as a

(a)   **(b)**

(c)   **(d)**

FIGURE 25–20 Ames test for carcinogens, based on their mutagenicity.
A strain of *Salmonella typhimurium* having a mutation that inactivates an
enzyme of the histidine biosynthetic pathway is plated on a histidine-
free medium. Few cells grow. **(a)** The few small colonies of *S. typhimurium*
that do grow on a histidine-free medium carry spontaneous back-
mutations that permit the histidine biosynthetic pathway to operate.
Three identical nutrient plates **(b)**, **(c)**, and **(d)** have been inoculated
with an equal number of cells. Each plate then receives a disk of filter
paper containing progressively lower concentrations of a mutagen. The
mutagen greatly increases the rate of back-mutation and hence the number of colonies. The clear areas around the filter paper indicate where the
concentration of mutagen is so high that it is lethal to the cells. As the
mutagen diffuses away from the filter paper, it is diluted to sublethal
concentrations that promote back-mutation. Mutagens can be compared
on the basis of their effect on mutation rate. Because many compounds
undergo a variety of chemical transformations after entering cells, compounds
are sometimes tested for mutagenicity after first incubating them with a
liver extract. Some substances have been found to be mutagenic only
after this treatment.

rapid and inexpensive screen for potential human
carcinogens.

The genomic DNA in a typical mammalian cell accumulates many thousands of lesions during a 24-hour
period. However, as a result of DNA repair, fewer than
1 in 1,000 become a mutation. DNA is a relatively stable
molecule, but in the absence of repair systems, the
cumulative effect of many infrequent but damaging
reactions would make life impossible. ∎

All Cells Have Multiple DNA Repair Systems

The number and diversity of repair systems reflect both
the importance of DNA repair to cell survival and the
diverse sources of DNA damage (Table 25–5). Some
common types of lesions, such as pyrimidine dimers
(see Fig. 8–31), can be repaired by several distinct

TABLE 25–5 Types of DNA Repair Systems in *E. coli*

Enzymes/proteins	Type of damage
Mismatch repair	
Dam methylase	
MutH, MutL, MutS proteins	
DNA helicase II	
SSB	
DNA polymerase III	Mismatches
Exonuclease I	
Exonuclease VII	
RecJ nuclease	
Exonuclease X	
DNA ligase	
Base-excision repair	
DNA glycosylases	Abnormal bases (uracil, hypoxanthine,
AP endonucleases	xanthine); alkylated
DNA polymerase I	bases; in some
DNA ligase	other organisms, pyrimidine dimers
Nucleotide-excision repair	
ABC excinuclease	DNA lesions that cause
DNA polymerase I	large structural
DNA ligase	changes (e.g., pyrimidine dimers)
Direct repair	
DNA photolyases	Pyrimidine dimers
O^6-Methylguanine-DNA methyltransferase	O^6-Methylguanine
AlkB protein	1-Methylguanine, 3-methylcytosine

---

systems. Many DNA repair processes also seem to be
extraordinarily inefficient energetically—an exception
to the pattern observed in the vast majority of metabolic pathways, where every ATP is generally accounted
for and used optimally. When the integrity of the genetic information is at stake, the amount of chemical
energy invested in a repair process seems almost
irrelevant.

DNA repair is possible largely because the DNA
molecule consists of two complementary strands. DNA
damage in one strand can be removed and accurately
replaced by using the undamaged complementary
strand as a template. We consider here the principal
types of repair systems, beginning with those that repair
the rare nucleotide mismatches that are left behind by
replication.

Mismatch Repair Correction of the rare mismatches left
after replication in *E. coli* improves the overall fidelity

of replication by an additional factor of 10^2 to 10^3. The mismatches are nearly always corrected to reflect the information in the old (template) strand, so the repair system must somehow discriminate between the template and the newly synthesized strand. The cell accomplishes this by tagging the template DNA with methyl groups to distinguish it from newly synthesized strands. The mismatch repair system of *E. coli* includes at least 12 protein components (Table 25–5) that function either in strand discrimination or in the repair process itself.

The strand discrimination mechanism has not been worked out for most bacteria or eukaryotes, but is well understood for *E. coli* and some closely related bacterial species. In these bacteria, strand discrimination is based on the action of Dam methylase, which, as you will recall, methylates DNA at the N^6 position of all adenines within (5′)GATC sequences. Immediately after passage of the replication fork, there is a short period (a few seconds or minutes) during which the template strand is methylated but the newly synthesized strand is not **(Fig. 25–21)**. The transient unmethylated state of GATC sequences in the newly synthesized strand permits the new strand to be distinguished from the template strand. Replication mismatches in the vicinity of a hemimethylated GATC sequence are then repaired according to the information in the methylated parent (template) strand. Tests in vitro show that if both strands are methylated at a GATC sequence, few mismatches are repaired; if neither strand is methylated, repair occurs but does not favor either strand. The cell's methyl-directed mismatch repair system efficiently repairs mismatches up to 1,000 bp from a hemimethylated GATC sequence.

How is the mismatch correction process directed by relatively distant GATC sequences? A mechanism is illustrated in **Figure 25–22**. MutL protein forms a complex with MutS protein, and the complex binds to all mismatched base pairs (except C–C). MutH protein binds to MutL and to GATC sequences encountered by the MutL-MutS complex. DNA on both sides of the mismatch is threaded through the MutL-MutS complex, creating a DNA loop; simultaneous movement of both legs of the loop through the complex is equivalent to the complex moving in both directions at once along the DNA. MutH has a site-specific endonuclease activity that is inactive until the complex encounters a hemimethylated GATC sequence. At this site, MutH

catalyzes cleavage of the unmethylated strand on the 5′ side of the G in GATC, which marks the strand for repair. Further steps in the pathway depend on where

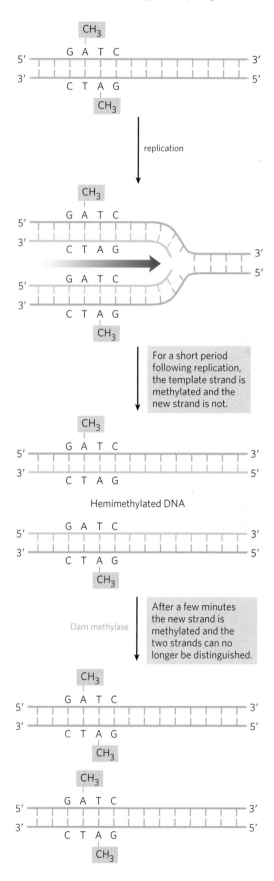

FIGURE 25–21 Methylation and mismatch repair. Methylation of DNA strands can serve to distinguish parent (template) strands from newly synthesized strands in *E. coli* DNA, a function that is critical to mismatch repair (see Fig. 25–22). The methylation occurs at the N^6 of adenines in (5′)GATC sequences. This sequence is a palindrome (see Fig. 8-18), present in opposite orientations on the two strands.

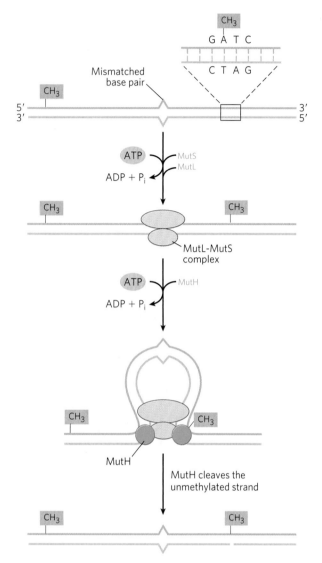

FIGURE 25–22 A model for the early steps of methyl-directed mismatch repair. The proteins involved in this process in *E. coli* have been purified (see Table 25–5). Recognition of the sequence (5′)GATC and of the mismatch are specialized functions of the MutH and MutS proteins, respectively. The MutL protein forms a complex with MutS at the mismatch. DNA is threaded through this complex such that the complex moves simultaneously in both directions along the DNA until it encounters a MutH protein bound at a hemimethylated GATC sequence. MutH cleaves the unmethylated strand on the 5′ side of the G in this sequence. A complex consisting of DNA helicase II and one of several exonucleases then degrades the unmethylated DNA strand from that point toward the mismatch (see Fig. 25–23).

the mismatch is located relative to this cleavage site **(Fig. 25–23)**.

When the mismatch is on the 5′ side of the cleavage site (Fig. 25–23, right side), the unmethylated strand is unwound and degraded in the 3′→5′ direction from the cleavage site through the mismatch, and this segment is replaced with new DNA. This process requires the combined action of DNA helicase II (also called UvrD helicase), SSB, exonuclease I or exonuclease X (both of which degrade strands of DNA in the 3′→5′ direction),

DNA polymerase III, and DNA ligase. The pathway for repair of mismatches on the 3′ side of the cleavage site is similar (Fig. 25–23, left), except that the exonuclease is either exonuclease VII (which degrades single-stranded DNA in the 5′→3′ or 3′→5′ direction) or RecJ nuclease (which degrades single-stranded DNA in the 5′→3′ direction).

Mismatch repair is a particularly expensive process for *E. coli* in terms of energy expended. The mismatch may be 1,000 bp or more from the GATC sequence. The degradation and replacement of a strand segment of this length require an enormous investment in activated deoxynucleotide precursors to repair a *single* mismatched base. This again underscores the importance to the cell of genomic integrity.

All eukaryotic cells have several proteins structurally and functionally analogous to the bacterial MutS and MutL (but not MutH) proteins. Alterations in human genes encoding proteins of this type produce some of the most common inherited cancer-susceptibility syndromes (see Box 25–1, p. 1037), further demonstrating the value to the organism of DNA repair systems. The main MutS homologs in most eukaryotes, from yeast to humans, are MSH2 (*MutS h*omolog), MSH3, and MSH6. Heterodimers of MSH2 and MSH6 generally bind to single base-pair mismatches, and bind less well to slightly longer mispaired loops. In many organisms the longer mismatches (2 to 6 bp) may be bound instead by a heterodimer of MSH2 and MSH3, or are bound by both types of heterodimers in tandem. Homologs of MutL, predominantly a heterodimer of MLH1 (*MutL h*omolog) and PMS1 (*post-m*eiotic *seg*regation), bind to and stabilize the MSH complexes. Many details of the subsequent events in eukaryotic mismatch repair remain to be worked out. In particular, we do not know the mechanism by which newly synthesized DNA strands are identified, although research has revealed that this strand identification does not involve GATC sequences.

Base-Excision Repair Every cell has a class of enzymes called **DNA glycosylases** that recognize particularly common DNA lesions (such as the products of cytosine and adenine deamination; see Fig. 8–30a) and remove the affected base by cleaving the *N*-glycosyl bond. This cleavage creates an apurinic or apyrimidinic site in the DNA, commonly referred to as an **AP site** or **abasic site**. Each DNA glycosylase is generally specific for one type of lesion.

Uracil DNA glycosylases, for example, found in most cells, specifically remove from DNA the uracil that results from spontaneous deamination of cytosine. Mutant cells that lack this enzyme have a high rate of G≡C to A=T mutations. This glycosylase does not remove uracil residues from RNA or thymine residues from DNA. The capacity to distinguish thymine from uracil, the product of cytosine deamination—necessary for the selective repair of the latter—may be one reason

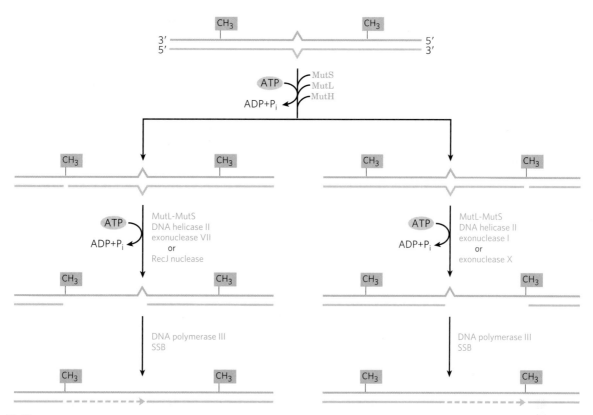

FIGURE 25–23 Completing methyl-directed mismatch repair. The combined action of DNA helicase II, SSB, and one of four different exonucleases removes a segment of the new strand between the MutH cleavage site and a point just beyond the mismatch. The exonuclease that is used depends on the location of the cleavage site relative to the mismatch, as shown by the alternative pathways here. The resulting gap is filled in (dashed line) by DNA polymerase III, and the nick is sealed by DNA ligase (not shown).

why DNA evolved to contain thymine instead of uracil (p. 299).

Most bacteria have just one type of uracil DNA glycosylase, whereas humans have at least four types, with different specificities—an indicator of the importance of uracil removal from DNA. The most abundant human uracil glycosylase, UNG, is associated with the human replisome, where it eliminates the occasional U residue inserted in place of a T during replication. The deamination of C residues is 100-fold faster in single-stranded DNA than in double-stranded DNA, and humans have the enzyme hSMUG1, which removes any U residues that occur in single-stranded DNA during replication or transcription. Two other human DNA glycosylases, TDG and MBD4, remove either U or T residues paired with G, generated by deamination of cytosine or 5-methylcytosine, respectively.

Other DNA glycosylases recognize and remove a variety of damaged bases, including formamidopyrimidine and 8-hydroxyguanine (both arising from purine oxidation), hypoxanthine (arising from adenine deamination), and alkylated bases such as 3-methyladenine and 7-methylguanine. Glycosylases that recognize other lesions, including pyrimidine dimers, have also been identified in some classes of organisms. Remember that AP sites also arise from the slow, spontaneous hydrolysis of the *N*-glycosyl bonds in DNA (see Fig. 8–30b).

Once an AP site has been formed by a DNA glycosylase, another type of enzyme must repair it. The repair is *not* made by simply inserting a new base and re-forming the *N*-glycosyl bond. Instead, the deoxyribose 5′-phosphate left behind is removed and replaced with a new nucleotide. This process begins with one of the **AP endonucleases**, enzymes that cut the DNA strand containing the AP site. The position of the incision relative to the AP site (5′ or 3′ to the site) varies with the type of AP endonuclease. A segment of DNA including the AP site is then removed, DNA polymerase I replaces the DNA, and DNA ligase seals the remaining nick **(Fig. 25–24)**. In eukaryotes, nucleotide replacement is carried out by specialized polymerases, as described below.

Nucleotide-Excision Repair DNA lesions that cause large distortions in the helical structure of DNA generally are repaired by the nucleotide-excision system, a repair pathway critical to the survival of all free-living organisms. In nucleotide-excision repair **(Fig. 25–25)**, a multisubunit enzyme (excinuclease) hydrolyzes two phosphodiester bonds, one on either side of the distortion caused by the lesion. In *E. coli* and other bacteria, the enzyme system hydrolyzes the fifth phosphodiester bond on the 3′ side and the eighth phosphodiester bond on the 5′ side to generate a fragment of 12 to 13 nucleotides (depending

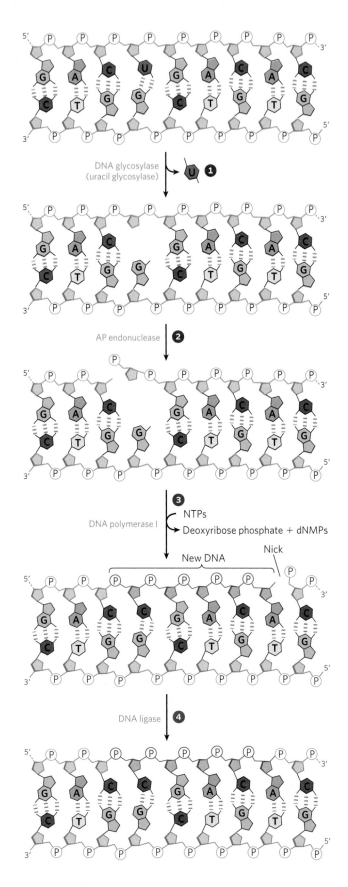

DNA glycosylase (uracil glycosylase) → U ❶

AP endonuclease ❷

DNA polymerase I ❸

NTPs

Deoxyribose phosphate + dNMPs

New DNA — Nick

DNA ligase ❹

FIGURE 25–24 DNA repair by the base-excision repair pathway. ❶ A DNA glycosylase recognizes a damaged base (in this case, a uracil) and cleaves between the base and deoxyribose in the backbone. ❷ An AP endonuclease cleaves the phosphodiester backbone near the AP site. ❸ DNA polymerase I initiates repair synthesis from the free 3' hydroxyl at the nick, removing (with its 5'→3' exonuclease activity) and replacing a portion of the damaged strand. ❹ The nick remaining after DNA polymerase I has dissociated is sealed by DNA ligase.

producing a fragment of 27 to 29 nucleotides. Following the dual incision, the excised oligonucleotides are released from the duplex and the resulting gap is filled—by DNA polymerase I in *E. coli* and DNA polymerase ε in humans. DNA ligase seals the nick.

In *E. coli*, the key enzymatic complex is the ABC excinuclease, which has three subunits, UvrA (M_r 104,000), UvrB (M_r 78,000), and UvrC (M_r 68,000). The term "excinuclease" is used to describe the unique capacity of this enzyme complex to catalyze two specific endonucleolytic cleavages, distinguishing this activity from that of standard endonucleases. A complex of the UvrA and UvrB proteins (A_2B) scans the DNA and binds to the site of a lesion. The UvrA dimer then dissociates, leaving a tight UvrB-DNA complex. UvrC protein then binds to UvrB, and UvrB makes an incision at the fifth phosphodiester bond on the 3' side of the lesion. This is followed by a UvrC-mediated incision at the eighth phosphodiester bond on the 5' side. The resulting 12 to 13 nucleotide fragment is removed by UvrD helicase. The short gap thus created is filled in by DNA polymerase I and DNA ligase. This pathway (Fig. 25–25, left) is a primary repair route for many types of lesions, including cyclobutane pyrimidine dimers, 6-4 photoproducts (see Fig. 8–31), and several other types of base adducts including benzo[a]pyrene-guanine, which is formed in DNA by exposure to cigarette smoke. The nucleolytic activity of the ABC excinuclease is novel in the sense that two cuts are made in the DNA.

The mechanism of eukaryotic excinucleases is quite similar to that of the bacterial enzyme, although 16 polypeptides with no similarity to the *E. coli* excinuclease subunits are required for the dual incision. As described in Chapter 26, some of the nucleotide-excision repair and base-excision repair in eukaryotes is closely tied to transcription. Genetic deficiencies in nucleotide-excision repair in humans give rise to a variety of serious diseases (see Box 25–1).

Direct Repair Several types of damage are repaired without removing a base or nucleotide. The best-characterized example is direct photoreactivation of cyclobutane pyrimidine dimers, a reaction promoted by **DNA photolyases**. Pyrimidine dimers result from a UV-induced reaction, and photolyases use energy derived from absorbed light to reverse the damage **(Fig. 25–26)**. Photolyases generally contain two cofactors that serve as light-absorbing agents, or chromophores.

on whether the lesion involves one or two bases). In humans and other eukaryotes, the enzyme system hydrolyzes the sixth phosphodiester bond on the 3' side and the twenty-second phosphodiester bond on the 5' side,

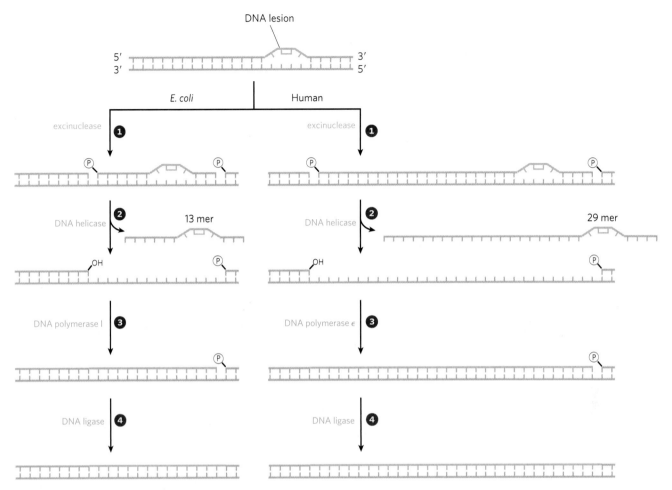

FIGURE 25–25 Nucleotide-excision repair in *E. coli* and humans.
The general pathway of nucleotide-excision repair is similar in all organisms. ❶ An excinuclease binds to DNA at the site of a bulky lesion and cleaves the damaged DNA strand on either side of the lesion. ❷ The DNA segment—of 13 nucleotides (13 mer) or 29 nucleotides (29 mer)—is removed with the aid of a helicase. ❸ The gap is filled in by DNA polymerase, and ❹ the remaining nick is sealed with DNA ligase.

One of the chromophores is always FADH$_2$. In *E. coli* and yeast, the other chromophore is a folate. The reaction mechanism entails the generation of free radicals. DNA photolyases are not present in the cells of placental mammals (which include humans).

Additional examples can be seen in the repair of nucleotides with alkylation damage. The modified nucleotide O^6-methylguanine forms in the presence of alkylating agents and is a common and highly mutagenic lesion (p. 302). It tends to pair with thymine rather than cytosine during replication, and therefore causes G≡C to A=T mutations **(Fig. 25–27)**. Direct repair of O^6-methylguanine is carried out by O^6-methylguanine-DNA methyltransferase, a protein that catalyzes transfer of the methyl group of O^6-methylguanine to one of its own Cys residues. This methyltransferase is not strictly an enzyme, because a single methyl transfer event permanently methylates the protein, making it inactive in this pathway. The consumption of an entire protein molecule to correct a single damaged base is another vivid illustration of the priority given to maintaining integrity of cellular DNA.

O^6-Methylguanine nucleotide

Guanine nucleotide

A very different but equally direct mechanism is used to repair 1-methyladenine and 3-methylcytosine. The amino groups of A and C residues are sometimes methylated when the DNA is single-stranded, and the methylation directly affects proper base pairing. In *E. coli*, oxidative demethylation of these alkylated nucleotides is mediated by the AlkB protein, a member of the α-ketoglutarate-Fe^{2+}–dependent dioxygenase superfamily **(Fig. 25–28)**. (See Box 4–3 for a description of another member of this enzyme family.)

MECHANISM FIGURE 25–26 Repair of pyrimidine dimers with photolyase. Energy derived from absorbed light is used to reverse the photoreaction that caused the lesion. The two chromophores in *E. coli* photolyase (M_r 54,000), N^5,N^{10}-methenyltetrahydrofolylpolyglutamate (MTHFpolyGlu) and FADH⁻, perform complementary functions. MTHF-polyGlu functions as a photoantenna to absorb blue-light photons. The excitation energy passes to FADH⁻, and the excited flavin (*FADH⁻) donates an electron to the pyrimidine dimer, leading to the rearrangement as shown.

The Interaction of Replication Forks with DNA Damage Can Lead to Error-Prone Translesion DNA Synthesis

The repair pathways considered to this point generally work only for lesions in double-stranded DNA, the undamaged strand providing the correct genetic information to restore the damaged strand to its original state. However, in certain types of lesions, such as double-strand breaks, double-strand cross-links, or lesions in a single-stranded DNA, the complementary strand is itself damaged or is absent. Double-strand breaks and lesions in single-stranded DNA most often arise when a replication fork encounters an unrepaired DNA lesion **(Fig. 25–29)**. Such lesions and DNA cross-links can also result from ionizing radiation and oxidative reactions.

At a stalled bacterial replication fork, there are two avenues for repair. In the absence of a second strand, the information required for accurate repair must come from a separate, homologous chromosome. The repair system thus involves homologous genetic recombination. This recombinational DNA repair is considered in detail in Section 25.3. Under some conditions, a second repair pathway, **error-prone translesion DNA**

(a)

Guanine

Cytosine

↓ methylation and replication

O^6-Methylguanine

Thymine

FIGURE 25–27 Example of how DNA damage results in mutations. (a) The methylation product O^6-methylguanine pairs with thymine rather than cytosine. **(b)** If not repaired, this leads to a G≡C to A=T mutation after replication.

synthesis (often abbreviated TLS), becomes available. When this pathway is active, DNA repair becomes significantly less accurate and a high mutation rate can result. In bacteria, error-prone translesion DNA synthesis is part of a cellular stress response to extensive DNA damage known, appropriately enough, as the **SOS response**. Some SOS proteins, such as the UvrA and UvrB proteins already described, are normally present in the cell but are induced to higher levels as part of the SOS response (Table 25–6). Additional SOS proteins participate in the pathway for error-prone repair; these include the UmuC and UmuD proteins ("Umu" from *unmu*table; lack of the *umu* gene function eliminates error-prone repair). The UmuD protein is cleaved

(b)

FIGURE 25–28 Direct repair of alkylated bases by AlkB. The AlkB protein is an α-ketoglutarate-Fe^{2+}-dependent hydroxylase (see Box 4–3). It catalyzes the oxidative demethylation of 1-methyladenine and 3-methylcytosine residues.

1-Methyladenine → Adenine

3-Methylcytosine → Cytosine

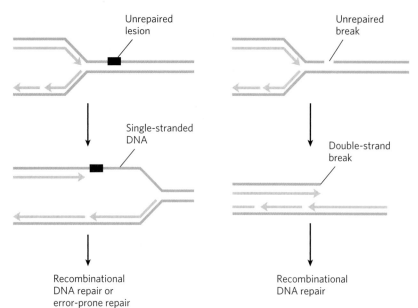

FIGURE 25–29 DNA damage and its effect on DNA replication. If the replication fork encounters an unrepaired lesion or strand break, replication generally halts and the fork may collapse. A lesion is left behind in an unreplicated, single-stranded segment of the DNA (left); a strand break becomes a double-strand break (right). In each case the damage to one strand cannot be repaired by mechanisms described earlier in this chapter, because the complementary strand required to direct accurate repair is damaged or absent. There are two possible avenues for repair: recombinational DNA repair (one pathway is described in Fig. 25–30) or, when lesions are unusually numerous, error-prone repair. The latter mechanism involves a novel DNA polymerase (DNA polymerase V, encoded by the *umuC* and *umuD* genes) that can replicate, albeit inaccurately, over many types of lesions. The repair mechanism is "error-prone" because mutations often result.

in an SOS-regulated process to a shorter form called UmuD′, which forms a complex with UmuC and a protein called RecA (described in Section 25.3) to create a specialized DNA polymerase, DNA polymerase V (UmuD′$_2$UmuCRecA), that can replicate past many of the DNA lesions that would normally block replication. Proper base pairing is often impossible at the site of such a lesion, so this translesion replication is error-prone.

Given the emphasis on the importance of genomic integrity throughout this chapter, the existence of a system that increases the rate of mutation may seem incongruous. However, we can think of this system as a desperation strategy. The *umuC* and *umuD* genes are fully

TABLE 25–6 Genes Induced as Part of the SOS response in *E. coli*

Gene name	Protein encoded and/or role in DNA repair
Genes of known function	
polB (dinA)	Encodes polymerization subunit of DNA polymerase II, required for replication restart in recombinational DNA repair
uvrA *uvrB*	Encode ABC excinuclease subunits UvrA and UvrB
umuC *umuD*	Encode core and polymerization subunits of DNA polymerase V
sulA	Encodes protein that inhibits cell division, possibly to allow time for DNA repair
recA	Encodes RecA protein, required for error-prone repair and recombinational repair
dinB	Encodes DNA polymerase IV
ssb	Encodes single-stranded DNA–binding protein (SSB)
himA	Encodes subunit of integration host factor (IHF), involved in site-specific recombination, replication, transposition, regulation of gene expression
Genes involved in DNA metabolism, but role in DNA repair unknown	
uvrD	Encodes DNA helicase II (DNA-unwinding protein)
recN	Required for recombinational repair
Genes of unknown function	
dinD	
dinF	

Note: Some of these genes and their functions are further discussed in Chapter 28.

induced only late in the SOS response, and they are not activated for translesion synthesis initiated by UmuD cleavage unless the levels of DNA damage are particularly high and all replication forks are blocked. The mutations resulting from DNA polymerase V–mediated replication kill some cells and create deleterious mutations in others, but this is the biological price a species pays to overcome an otherwise insurmountable barrier to replication, as it permits at least a few mutant daughter cells to survive.

Yet another DNA polymerase, DNA polymerase IV, is also induced during the SOS response. Replication by DNA polymerase IV, a product of the *dinB* gene, is also highly error-prone. The bacterial DNA polymerases IV and V are part of a family of TLS polymerases found in all organisms. These enzymes lack a proofreading exonuclease and possess a more open active site that accommodates damaged template nucleotides. The fidelity of base selection during replication can be reduced by a factor of 10^2, lowering overall replication fidelity to one error in ~1,000 nucleotides.

Mammals have many low-fidelity DNA polymerases of the TLS polymerase family. However, the presence of these enzymes does not necessarily translate into an unacceptable mutational burden, because most of these enzymes also have specialized functions in DNA repair. DNA polymerase η (eta), for example,

found in all eukaryotes, promotes translesion synthesis primarily across cyclobutane T–T dimers. Few mutations result in this case, because the enzyme preferentially inserts two A residues across from the linked T residues. Several other low-fidelity polymerases, including DNA polymerases β, ι (iota), and λ, have specialized roles in eukaryotic base-excision repair. Each of these enzymes has a 5′-deoxyribose phosphate lyase activity in addition to its polymerase activity. After base removal by a glycosylase and backbone cleavage by an AP endonuclease, these polymerases remove the abasic site (a 5′-deoxyribose phosphate) and fill in the very short gap. The frequency of mutation due to DNA polymerase η activity is minimized by the very short lengths (often one nucleotide) of DNA synthesized.

What emerges from research into cellular DNA repair systems is a picture of a DNA metabolism that maintains genomic integrity with multiple and often redundant systems. In the human genome, more than 130 genes encode proteins dedicated to the repair of DNA. In many cases, the loss of function of one of these proteins results in genomic instability and an increased occurrence of oncogenesis (Box 25–1). These repair systems are often integrated with the DNA replication systems and are complemented by recombination systems, which we turn to next.

BOX 25–1   MEDICINE   DNA Repair and Cancer

Human cancers develop when genes that regulate normal cell division (oncogenes and tumor suppressor genes; Chapter 12) fail to function, are activated at the wrong time, or are altered. As a consequence, cells may grow out of control and form a tumor. The genes controlling cell division can be damaged by spontaneous mutation or overridden by the invasion of a tumor virus (Chapter 26). Not surprisingly, alterations in DNA repair genes that result in an increased rate of mutation can greatly increase an individual's susceptibility to cancer. Defects in the genes encoding the proteins involved in nucleotide-excision repair, mismatch repair, recombinational repair, and error-prone translesion DNA synthesis have all been linked to human cancers. Clearly, DNA repair can be a matter of life and death.

Nucleotide-excision repair requires a larger number of proteins in humans than in bacteria, although the overall pathways are very similar. Genetic defects that inactivate nucleotide-excision repair have been associated with several genetic diseases, the best-studied of which is xeroderma pigmentosum (XP). Because nucleotide-excision repair is the sole repair pathway for pyrimidine dimers in humans, people with XP are extremely sensitive to light and readily develop sunlight-induced skin cancers. Most people

with XP also have neurological abnormalities, presumably because of their inability to repair certain lesions caused by the high rate of oxidative metabolism in neurons. Defects in the genes encoding any of at least seven different protein components of the nucleotide-excision repair system can result in XP, giving rise to seven different genetic groups denoted XPA to XPG. Several of these proteins (notably those defective in XPB, XPD, and XPG) also play roles in transcription-coupled base-excision repair of oxidative lesions, described in Chapter 26.

Most microorganisms have redundant pathways for the repair of cyclobutane pyrimidine dimers—making use of DNA photolyases and sometimes base-excision repair as alternatives to nucleotide-excision repair—but humans and other placental mammals do not. This lack of a backup for nucleotide-excision repair for removing pyrimidine dimers has led to speculation that early mammalian evolution involved small, furry, nocturnal animals with little need to repair UV damage. However, mammals do have a pathway for the translesion bypass of cyclobutane pyrimidine dimers, which involves DNA polymerase η. This enzyme preferentially inserts two A residues opposite a T–T pyrimidine dimer, minimizing mutations. People with a genetic condition in which DNA polymerase η

(continued on next page)

function is missing exhibit an XP-like illness known as XP-variant or XP-V. Clinical manifestations of XP-V are similar to those of the classic XP diseases, although mutation levels are higher in XP-V when cells are exposed to UV light. Apparently, the nucleotide-excision repair system works in concert with DNA polymerase η in normal human cells, repairing and/or bypassing pyrimidine dimers as needed to keep cell growth and DNA replication going. Exposure to UV light introduces a heavy load of pyrimidine dimers, and some must be bypassed by translesion synthesis to keep replication on track. When one system is missing, it is partly compensated for by the other. A loss of DNA polymerase η activity leads to stalled replication forks and bypass of UV lesions by different, more mutagenic, translesion synthesis (TLS) polymerases. As when other DNA repair systems are absent, the resulting increase in mutations often leads to cancer.

One of the most common inherited cancer-susceptibility syndromes is hereditary nonpolyposis colon cancer (HNPCC). This syndrome has been traced to defects in mismatch repair. Human and other eukaryotic cells have several proteins analogous to the bacterial MutL and MutS proteins (see Fig. 25–22). Defects in at least five different mismatch repair genes can give rise to HNPCC. The most prevalent are defects in the *hMLH1* (human MutL homolog 1) and *hMSH2* (human MutS homolog 2) genes. In individuals with HNPCC, cancer generally develops at an early age, with colon cancers being most common.

Most human breast cancer occurs in women with no known predisposition. However, about 10% of cases are associated with inherited defects in two genes, *BRCA1* and *BRCA2*. Human BRCA1 and BRCA2 are large proteins (1,834 and 3,418 amino acid residues, respectively) that interact with a wide range of other proteins involved in transcription, chromosome maintenance, DNA repair, and control of the cell cycle. BRCA2 has been implicated in the recombinational DNA repair of double-strand breaks. However, the precise molecular function of BRCA1 and BRCA2 in these various cellular processes is not yet clear. Women with defects in either the *BRCA1* or *BRCA2* gene have a greater than 80% chance of developing breast cancer.

SUMMARY 25.2 DNA Repair

▸ Cells have many systems for DNA repair. Mismatch repair in *E. coli* is directed by transient nonmethylation of (5′)GATC sequences on the newly synthesized strand.

▸ Base-excision repair systems recognize and repair damage caused by environmental agents (such as radiation and alkylating agents) and spontaneous reactions of nucleotides. Some repair systems recognize and excise only damaged or incorrect bases, leaving an AP (abasic) site in the DNA. This is repaired by excision and replacement of the DNA segment containing the AP site.

▸ Nucleotide-excision repair systems recognize and remove a variety of bulky lesions and pyrimidine dimers. They excise a segment of the DNA strand including the lesion, leaving a gap that is filled in by DNA polymerase and ligase activities.

▸ Some DNA damage is repaired by direct reversal of the reaction causing the damage: pyrimidine dimers are directly converted to monomeric pyrimidines by a photolyase, and the methyl group of O^6-methylguanine is removed by a methyltransferase.

▸ In bacteria, error-prone translesion DNA synthesis, involving TLS DNA polymerases, occurs in response to very heavy DNA damage. In eukaryotes, similar polymerases have specialized roles in DNA repair that minimize the introduction of mutations.

25.3 DNA Recombination

The rearrangement of genetic information within and among DNA molecules encompasses a variety of processes, collectively placed under the heading of genetic recombination. The practical applications of DNA rearrangements in altering the genomes of increasing numbers of organisms are now being explored (Chapter 9).

Genetic recombination events fall into at least three general classes. **Homologous genetic recombination** (also called general recombination) involves genetic exchanges between any two DNA molecules (or segments of the same molecule) that share an extended region of nearly identical sequence. The actual sequence of bases is irrelevant, as long as it is similar in the two DNAs. In **site-specific recombination** the exchanges occur only at a *particular* DNA sequence. **DNA transposition** is distinct from both other classes in that it usually involves a short segment of DNA with the remarkable capacity to move from one

Barbara McClintock, 1902–1992

location in a chromosome to another. These "jumping genes" were first observed in maize in the 1940s by Barbara McClintock. There is in addition a wide range of unusual genetic rearrangements for which no mechanism or purpose has yet been proposed. Here we focus on the three general classes.

The functions of genetic recombination systems are as varied as their mechanisms. They include roles in specialized DNA repair systems, specialized activities in DNA replication, regulation of expression of certain genes, facilitation of proper chromosome segregation during eukaryotic cell division, maintenance of genetic diversity, and implementation of programmed genetic rearrangements during embryonic development. In most cases, genetic recombination is closely integrated with other processes in DNA metabolism, and this becomes a theme of our discussion.

Bacterial Homologous Recombination Is a DNA Repair Function

In bacteria, homologous genetic recombination is primarily a DNA repair process and in this context (as noted in Section 25.2) is referred to as **recombinational DNA repair**. It is usually directed at the reconstruction of replication forks that have stalled or collapsed at the site of DNA damage. Homologous genetic recombination can also occur during conjugation (mating), when chromosomal DNA is transferred from one bacterial cell (donor) to another (recipient). Recombination during conjugation, although rare in wild bacterial populations, contributes to genetic diversity.

An example of what happens when a replication fork encounters DNA damage is shown in **Figure 25–30**. A common feature of the DNA repair pathways

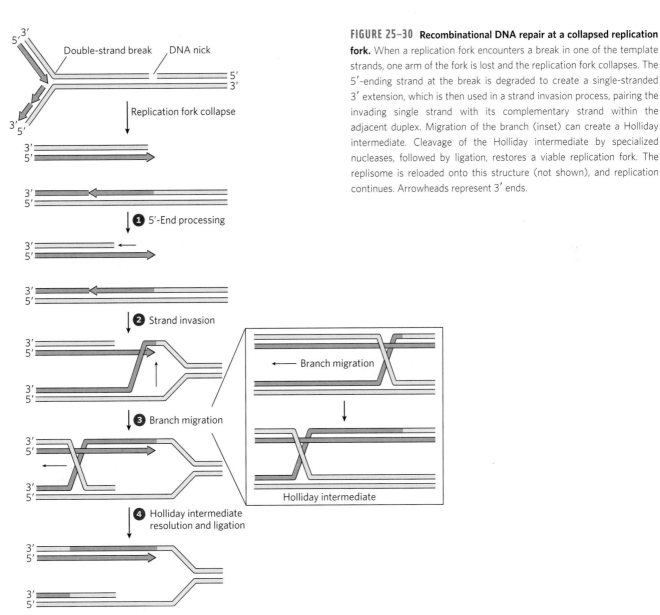

FIGURE 25–30 Recombinational DNA repair at a collapsed replication fork. When a replication fork encounters a break in one of the template strands, one arm of the fork is lost and the replication fork collapses. The 5'-ending strand at the break is degraded to create a single-stranded 3' extension, which is then used in a strand invasion process, pairing the invading single strand with its complementary strand within the adjacent duplex. Migration of the branch (inset) can create a Holliday intermediate. Cleavage of the Holliday intermediate by specialized nucleases, followed by ligation, restores a viable replication fork. The replisome is reloaded onto this structure (not shown), and replication continues. Arrowheads represent 3' ends.

illustrated in Figures 25–22 to 25–25 is that they introduce a transient break into one of the DNA strands. If a replication fork encounters a damaged site under repair near a break in one of the template strands, one arm of the replication fork becomes disconnected by a double-strand break and the fork collapses. The end of that break is processed by degrading the 5'-ending strand. The resulting 3' single-strand extension is bound by a recombinase that uses it to promote strand invasion: the 3' end invades the intact duplex DNA connected to the other arm of the fork and pairs with its complementary sequence. This creates a branched DNA structure (a point where three DNA segments come together). The DNA branch can be moved in a process called **branch migration** to create an X-like crossover structure known as a **Holliday intermediate**, named after researcher Robin Holliday, who first postulated its existence. The Holliday intermediate is then resolved by a special class of nuclease. The overall process reconstructs the replication fork.

In *E. coli*, the DNA end-processing is promoted by the RecBCD nuclease/helicase. The RecBCD enzyme binds to linear DNA at a free (broken) end and moves inward along the double helix, unwinding and degrading the DNA in a reaction coupled to ATP hydrolysis **(Fig. 25–31)**. The RecB and RecD subunits are helicase motors, with RecB moving 3'→5' along one strand and RecD moving 5'→3' along the other strand. The activity of the enzyme is altered when it interacts with a sequence referred to as **chi**, (5')GCTGGTGG, which binds tightly to a site on the RecC subunit. From that point, degradation of the strand with a 3' terminus is greatly reduced, but degradation of the 5'-terminal strand is increased. This process creates a single-stranded DNA with a 3' end, which is used during subsequent steps in recombination. The 1,009 chi sequences scattered throughout the *E. coli* genome enhance the frequency of recombination about 5- to 10-fold within 1,000 bp of the chi site. The enhancement declines as the distance from the chi site increases. Sequences that enhance recombination frequency have also been identified in several other organisms.

The bacterial recombinase is the RecA protein. RecA is unusual among the proteins of DNA metabolism

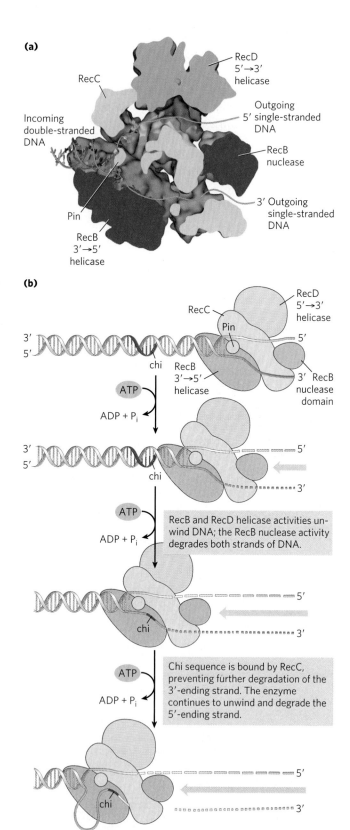

FIGURE 25–31 The RecBCD helicase/nuclease. (a) A cutaway view of the RecBCD enzyme structure as it is bound to DNA (PDB ID 1W36). The subunits are shown in different colors, with the DNA entering the structure at left and the unwound DNA strands (not part of the solved structure) are modeled as exiting to the right. A bulbous protein structure called a pin, part of the RecC subunit, facilitates the separation of strands. **(b)** Activities of the RecBCD enzyme at a DNA end. The RecB and RecD subunits are helicases, molecular motors that propel the complex along the DNA, a process that requires ATP; RecB degrades both strands as the complex travels, cleaving the 3'-ending strand more often than the 5'-ending strand. When a chi site is encountered on the 3'-ending strand, the RecC subunit binds to it, halting the advance of this strand through the complex. The 5'-ending strand continues to be degraded as the 3'-ending strand is looped out, eventually creating a 3' single-stranded extension. RecA protein (not shown) is finally loaded onto the processed DNA by the RecBCD enzyme.

in that its active form is an ordered, helical filament of up to several thousand subunits that assemble cooperatively on DNA **(Fig. 25–32)**. This filament usually forms on single-stranded DNA, such as that produced by the RecBCD enzyme. Its formation is not as straightforward as shown in Figure 25–32, since the single-stranded DNA–binding protein (SSB) is normally present and specifically impedes the binding of the first few subunits to DNA (filament nucleation). The RecBCD enzyme acts directly as a RecA loader, facilitating the nucleation of a RecA filament on single-stranded DNA that is coated with SSB. The filaments assemble and disassemble in a $5'\rightarrow3'$ direction. Many other bacterial proteins regulate the formation and disassembly of RecA filaments. RecA protein promotes the central steps of homologous recombination, including the DNA strand invasion step of Figure 25–31 and a number of other strand exchange reactions in vitro.

After strand invasion has occurred, branch migration is promoted by a complex called RuvAB **(Fig. 25–33a)**. Once a Holliday intermediate has been created, it can be cleaved by a specialized nuclease called RuvC (Fig. 25–33b), and nicks are sealed with DNA ligase. A viable replication fork structure is thus reconstructed, as outlined in Figure 25–31.

After the recombination steps are completed, the replication fork reassembles in a process called **origin-independent restart of replication.** Four proteins (PriA, PriB, PriC, and DnaT) act with DnaC to load the DnaB helicase onto the reconstructed replication fork. The DnaG primase then synthesizes an RNA primer, and DNA polymerase reassembles on DnaB to restart DNA synthesis. Originally discovered as a component required for the replication of ϕX174 DNA in vitro, a complex of PriA, PriB, PriC, and DnaT, along with DnaB, DnaC, and DnaG, is now termed the **replication restart primosome.** Restart of the replication fork also requires DNA polymerase II, in a role not yet defined; this polymerase II activity gives way to DNA polymerase III activity for the extensive replication generally required to complete the chromosome. In this way, the process of recombination is tightly intertwined with replication. One process of DNA metabolism supports the other.

Eukaryotic Homologous Recombination Is Required for Proper Chromosome Segregation during Meiosis

In eukaryotes, homologous genetic recombination can have several roles in replication and cell division, including the repair of stalled replication forks. Recombination

FIGURE 25–32 RecA protein filaments. RecA and other recombinases in this class function as filaments of nucleoprotein. **(a)** Filament formation proceeds in discrete nucleation and extension steps. Nucleation is the addition of the first few RecA subunits. Extension occurs by adding RecA subunits so that the filament grows in the $5'\rightarrow3'$ direction. When disassembly occurs, subunits are subtracted from the trailing end. **(b)** Colorized electron micrograph of a RecA filament bound to DNA. **(c)** Segment of a RecA filament with four helical turns (24 RecA subunits; derived from PDB ID 3CMX). Notice the bound double-stranded DNA in the center. The core domain of RecA is structurally related to the domains in helicases.

(a)

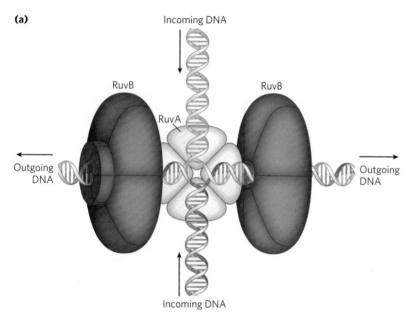

Incoming DNA

RuvB

RuvA

RuvB

Outgoing DNA

Outgoing DNA

Incoming DNA

FIGURE 25–33 Catalysis of DNA branch migration and Holliday intermediate resolution by the RuvA, RuvB, and RuvC proteins. (a) The RuvA protein binds directly to a Holliday intermediate where the four DNA arms come together. The hexameric RuvB protein is a DNA translocase. Two hexamers bind to opposing arms of the Holliday intermediate and propel the DNA outward in a reaction coupled to ATP hydrolysis. The branch thus moves. **(b)** RuvC is a specialized nuclease that binds to the RuvAB complex and cleaves the Holliday intermediate on opposing sides of the junction (red arrows) so that two contiguous DNA arms remain in each product.

(b)

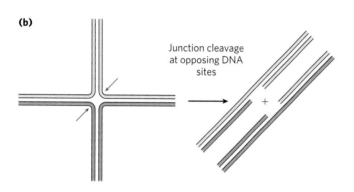

Junction cleavage at opposing DNA sites

+

occurs with the highest frequency during **meiosis**, the process by which diploid germ-line cells with two sets of chromosomes divide to produce haploid gametes (sperm cells or ova) in animals (haploid spores in plants)—each gamete having only one member of each chromosome pair **(Fig. 25–34)**. Meiosis begins with replication of the DNA in the germ-line cell so that each DNA molecule is present in four copies. Each set of four homologous chromosomes (tetrad) exists as two pairs of sister chromatids, and the sister chromatids remain associated at their centromeres. The cell then goes through two rounds of cell division without an intervening round of DNA replication. In the first cell division, the two pairs of sister chromatids are segregated into daughter cells. In the second cell division, the two chromosomes in each sister chromatid pair are segregated into new daughter cells. In each division, the chromosomes to be segregated are drawn into the daughter cells by spindle fibers attached to opposite poles of the dividing cell. The two successive divisions reduce the DNA content to the haploid level in each gamete. Proper chromosome segregation into daughter cells requires that physical links exist between the homologous chromosomes to be segregated. As the

spindle fibers attach to the centromeres of chromosomes and start to pull, the links between homologous chromosomes create tension. This tension, sensed by cellular mechanisms not yet understood, signals that this pair of chromosomes or sister chromatids is properly aligned for segregation. Once the tension is sensed, the links are gradually dissolved and segregation proceeds. If improper spindle fiber attachment occurs (e.g., if the centromeres of a chromosome pair are attached to the same cellular pole), a cellular kinase senses the lack of tension and activates a system that removes the spindle attachments, allowing the cell to try again.

During the second meiotic division, the centromeric attachments between the sister chromatids, augmented by cohesins deposited during replication (see Fig. 24–34), provide the needed physical links to guide segregation. However, during the first meiotic cell division, the two pairs of sister chromatids to be segregated are not related by a recent replication event and are not linked by cohesins or any other physical association. Instead, the homologous pairs of sister chromatids are aligned and new links are created by recombination, a process involving the breakage and rejoining of DNA

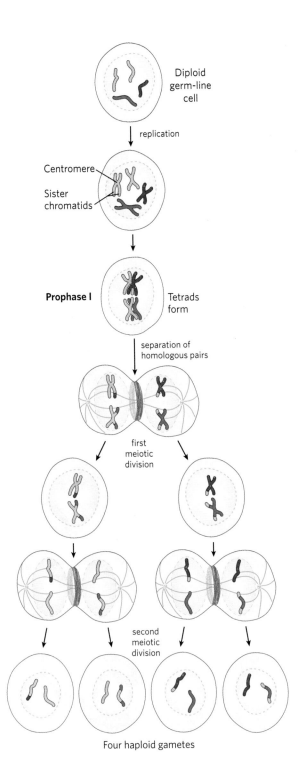

Diploid germ-line cell

replication

Centromere

Sister chromatids

Prophase I

Tetrads form

separation of homologous pairs

first meiotic division

second meiotic division

Four haploid gametes

FIGURE 25–34 Meiosis in animal germ-line cells. The chromosomes of a hypothetical diploid germ-line cell (four chromosomes; two homologous pairs) replicate and are held together at their centromeres. Each replicated double-stranded DNA molecule is called a chromatid (sister chromatid). In prophase I, just before the first meiotic division, the two homologous sets of chromatids align to form tetrads, held together by covalent links at homologous junctions (chiasmata). Crossovers occur within the chiasmata (see Fig. 25–35). These transient associations between homologs ensure that the two tethered chromosomes segregate properly in the next step, when attached spindle fibers pull them toward opposite poles of the dividing cell in the first meiotic division. The products of this division are two daughter cells each with two pairs of different sister chromatids. The pairs now line up across the equator of the cell in preparation for separation of the chromatids (now called chromosomes). The second meiotic division produces four haploid daughter cells that can serve as gametes. Each has two chromosomes, half the number of the diploid germ-line cell. The chromosomes have re-sorted and recombined.

Crossing over is not an entirely random process, and "hot spots" have been identified on many eukaryotic chromosomes. However, the assumption that crossing over can occur with equal probability at almost any point along the length of two homologous chromosomes remains a reasonable approximation in many cases, and it is this assumption that permits the genetic mapping of genes on a particular chromosome. The frequency of homologous recombination in any region separating two points on a chromosome is roughly proportional to the distance between the points, and this allows determination of the relative positions of and distances between different genes. The independent assortment of unlinked genes on different chromosomes **(Fig. 25–36)** makes another major contribution to the genetic diversity of the gametes. These genetic realities guide many of the modern applications of genomics, such as defining haplotypes (see Fig. 9–30) or searching for disease genes in the human genome (see Fig. 9–34).

Homologous recombination thus serves at least three identifiable functions in eukaryotes: (1) it contributes to the repair of several types of DNA damage; (2) it provides, in eukaryotic cells, a transient physical link between chromatids that promotes the orderly segregation of chromosomes at the first meiotic cell division; and (3) it enhances genetic diversity in a population.

Recombination during Meiosis Is Initiated with Double-Strand Breaks

A likely pathway for homologous recombination during meiosis is outlined in Figure 25–35a. The model has four key features. First, homologous chromosomes are aligned. Second, a double-strand break in a DNA molecule is created and then the exposed ends are processed by an exonuclease, leaving a single-strand extension with a free 3'-hydroxyl group at the broken end (step ❶). Third, the exposed 3' ends invade the intact duplex DNA of the homolog, and this is followed by

(Fig. 25–35). This exchange, also referred to as crossing over, can be observed with the light microscope. Crossing over links the two pairs of sister chromatids together at points called chiasmata (singular, chiasma). Also during crossovers, genetic material is exchanged between the pairs of sister chromatids. These exchanges also increase genetic diversity in the resulting gametes. The importance of meiotic recombination to proper chromosomal segregation is well illustrated by the physiological and societal consequences of their failure (Box 25–2).

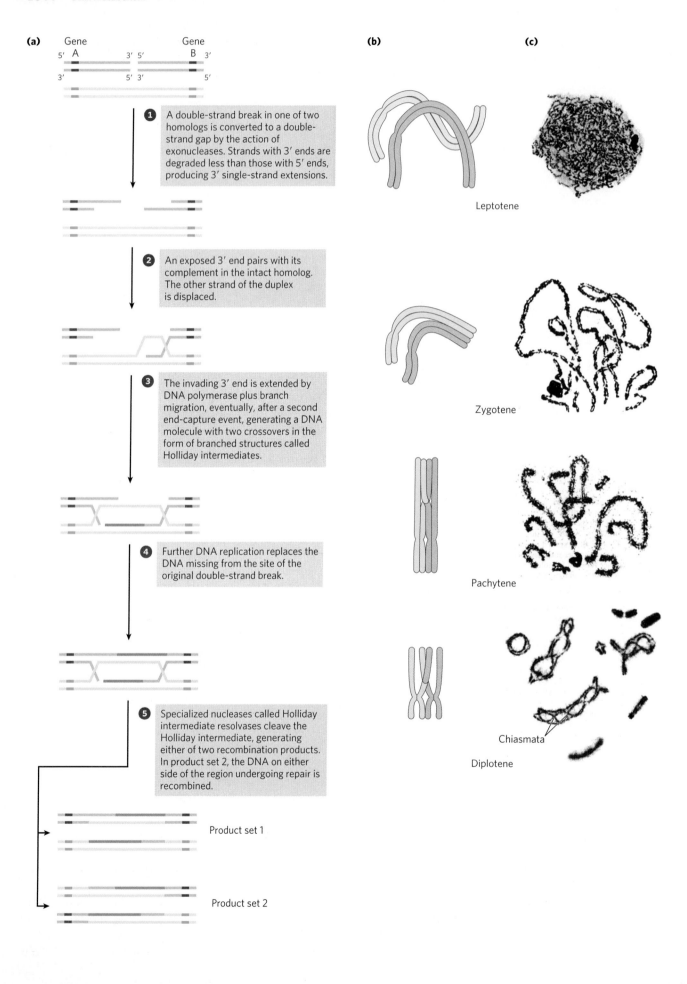

(a)

Gene A   Gene B

1. A double-strand break in one of two homologs is converted to a double-strand gap by the action of exonucleases. Strands with 3′ ends are degraded less than those with 5′ ends, producing 3′ single-strand extensions.

2. An exposed 3′ end pairs with its complement in the intact homolog. The other strand of the duplex is displaced.

3. The invading 3′ end is extended by DNA polymerase plus branch migration, eventually, after a second end-capture event, generating a DNA molecule with two crossovers in the form of branched structures called Holliday intermediates.

4. Further DNA replication replaces the DNA missing from the site of the original double-strand break.

5. Specialized nucleases called Holliday intermediate resolvases cleave the Holliday intermediate, generating either of two recombination products. In product set 2, the DNA on either side of the region undergoing repair is recombined.

Product set 1

Product set 2

(b)

(c)

Leptotene

Zygotene

Pachytene

Chiasmata

Diplotene

FIGURE 25–35 Recombination during prophase I in meiosis. (a) Model of double-strand break repair for homologous genetic recombination. The two homologous chromosomes (one shown in red, the other blue) involved in this recombination event have identical or very nearly identical sequences. Each of the two genes shown has different alleles on the two chromosomes. The steps are described in the text. **(b)** Crossing over occurs during prophase of meiosis I. The several stages of prophase I are aligned with the recombination processes that are occurring in part (a). Double-strand breaks are introduced and processed in the leptotene stage. The strand invasion and completion of crossover occur later. As homologous sequences in the two pairs of sister chromatids are aligned in the zygotene stage, synaptonemal complexes form and strand invasion occurs. The homologous chromosomes are tightly aligned by the pachytene stage. **(c)** Homologous chromosomes of a grasshopper are viewed at successive stages of meiotic prophase I. The chiasmata become visible in the diplotene stage.

BOX 25–2 ⚕ MEDICINE Why Proper Chromosomal Segregation Matters

When chromosomal alignment and recombination are not correct and complete in meiosis I, segregation of chromosomes can go awry. One result may be aneuploidy, a condition in which a cell has the wrong number of chromosomes. The haploid products of meiosis (gametes or spores) may have no copies or two copies of a chromosome. When a gamete with two copies of a chromosome joins with a gamete with one copy of a chromosome during fertilization, cells in the resulting embryo have three copies of that chromosome (they are trisomic).

In *S. cerevisiae*, aneuploidy resulting from errors in meiosis occurs at a rate of about 1 in 10,000 meiotic events. In fruit flies, the rate is about 1 in a few thousand. Rates of aneuploidy in mammals are considerably higher. In mice, the rate is 1 in 100, and it is even higher in other mammals. The rate of aneuploidy in fertilized human eggs has been estimated as 10% to 30%; most of these aneuploid cells are monosomies (they have a single copy of a chromosome) or trisomies. This is almost certainly an underestimate. Most trisomies are lethal and many result in miscarriage long before the pregnancy is detected. Aneuploidy is the leading cause of pregnancy loss. The few trisomic fetuses that survive to birth generally have three copies of chromosome 13, 18, or 21 (trisomy 21 is Down syndrome). Abnormal complements of the sex chromosomes are also found in the human population. Almost all monosomies are fatal in the early stages of fetal development. The societal consequences of aneuploidy in humans are considerable. Aneuploidy is the leading genetic cause of developmental and mental disabilities. At the heart of these high rates is a feature of meiosis in female mammals that has special significance for the human species.

In a human male, germ-line cells begin to undergo meiosis at puberty, and each meiotic event requires a relatively short time. In contrast, meiosis in the germ-line cells of human females is a highly protracted process. The production of an egg begins before a female is born—with the onset of meiosis in the fetus, at 12 to 13 weeks of gestation. Meiosis is initiated in all the developing fetal germ-line cells over a period of a few weeks. The cells proceed through much of meiosis I. Chromosomes line up and generate crossovers, continuing just beyond the pachytene phase (see Fig. 25–35)—and then the process stops. The chromosomes enter an arrested phase called the dictyate stage, with the crossovers in place, a kind of suspended animation where they remain as the female matures—typically from 13 to 50 years. It is not until sexual maturity that individual germ-line cells continue through the two meiotic cell divisions to produce egg cells.

Between the onset of the dictyate stage and the final completion of meiosis, something may happen that disrupts or damages the crossovers linking homologous chromosomes in the germ-line cells. As a woman ages, the rate of trisomy in the egg cells she produces increases, dramatically so as she approaches menopause (Fig. 1). There are many hypotheses on why this occurs, and several different factors may play a role. However, most of the hypotheses are centered on recombination crossovers in meiosis I and their stability over the protracted dictyate stage.

It is not yet clear what medical steps could be taken to reduce the incidence of aneuploidy in females of child-bearing age. What is revealed is the inherent importance of recombination and crossover generation in human meiosis.

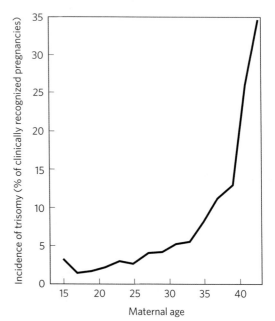

FIGURE 1 The increasing incidence of human trisomy with increasing age of the mother.

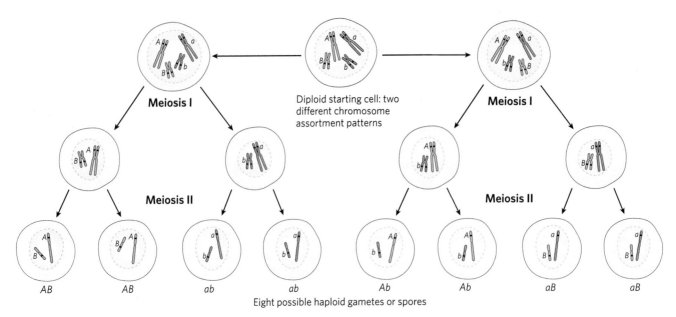

Diploid starting cell: two
different chromosome
assortment patterns

Meiosis I

Meiosis II

AB   AB   ab   ab   Ab   Ab   aB   aB

Eight possible haploid gametes or spores

FIGURE 25–36 The contribution of independent assortment to genetic diversity. In this example, the chromosomes have already been replicated to create two pairs of sister chromatids for two chromosomes. Blue and red distinguish the sister chromatids of each pair. One gene on each chromosome is highlighted, with different alleles (*A* or *a*, *B* or *b*) in the homologs. Independent assortment can lead to gametes with any combination of the alleles present on the two different chromosomes. Crossing over (not shown here; see Fig. 25-34) would also contribute to genetic diversity in a typical meiotic sequence.

branch migration and/or replication to create a pair of Holliday intermediates steps ❷ to ❹). Fourth, cleavage of the two crossovers creates either of two pairs of complete recombinant products (step ❺). Note the similarity of these steps to the bacterial recombinational repair processes outlined in Figure 25–30.

In this **double-strand break repair model** for recombination, the 3′ ends are used to initiate the genetic exchange. Once paired with the complementary strand on the intact homolog, a region of hybrid DNA is created that contains complementary strands from two different parent DNAs (the product of step ❷ in Fig. 25–35a). Each of the 3′ ends can then act as a primer for DNA replication. Meiotic homologous recombination can vary in many details from one species to another, but most of the steps outlined above are generally present in some form. There are two ways to cleave, or "resolve," the Holliday intermediate with a RuvC-like nuclease so that the two products carry genes in the same linear order as in the substrates—the original, un-recombined chromosomes (step ❺). If cleaved one way, the DNA flanking the region containing the hybrid DNA is not recombined; if cleaved the other way, the flanking DNA is recombined. Both outcomes are observed in vivo.

The homologous recombination illustrated in Figure 25–35 is a very elaborate process that is essential to accurate chromosome segregation. Its molecular consequences for the generation of genetic diversity are subtle. To understand how this process contributes to diversity, we should keep in mind that the two homologous chromosomes that undergo recombination are not necessarily *identical*. The linear array of genes may be the same, but the base sequences in some of the genes may differ slightly (in different alleles). In a human, for example, one chromosome may contain the allele for hemoglobin A (normal hemoglobin) while the other contains the allele for hemoglobin S (the sickle-cell mutation). The difference may consist of no more than one base pair among millions. Homologous recombination does not change the linear array of genes, but it can determine which alleles become linked on a single chromosome and are thereby passed to the next generation together. The independent assortment of different chromosomes (Fig. 25–36) determines which gene alleles from different chromosomes are inherited together.

Site-Specific Recombination Results in Precise DNA Rearrangements

Homologous genetic recombination, the type we have discussed so far, can involve any two homologous sequences. The second general type of recombination, site-specific recombination, is a very different type of process: recombination is limited to specific sequences. Recombination reactions of this type occur in virtually every cell, filling specialized roles that vary greatly from one species to another. Examples include regulation of the expression of certain genes and promotion of programmed DNA rearrangements in embryonic development or in the replication cycles of some viral and plasmid DNAs. Each site-specific recombination system consists of an enzyme called a recombinase and a short

(a)

Holliday intermediates

(b)

FIGURE 25–37 A site-specific recombination reaction. (a) The reaction shown here is for a common class of site-specific recombinases called integrase-class recombinases (named after bacteriophage λ integrase, the first recombinase characterized). These enzymes use Tyr residues as nucleophiles at the active site. The reaction is carried out within a tetramer of identical subunits. Recombinase subunits bind to a specific sequence, the recombination site. ❶ One strand in each DNA is cleaved at particular points in the sequence. The nucleophile is the —OH group of an active-site Tyr residue, and the product is a covalent phosphotyrosine link between protein and DNA ❷. After isomerization ❸, the cleaved strands join to new partners, producing a Holliday intermediate. Steps ❹ and ❺ complete the reaction by a process similar to the first two steps. The original sequence of the recombination site is regenerated after recombining the DNA flanking the site. These steps occur within a complex of multiple recombinase subunits that sometimes includes other proteins not shown here. **(b)** Surface contour model of a four-subunit integrase-class recombinase called the Cre recombinase, bound to a Holliday intermediate (shown with light blue and dark blue helix strands). The protein has been rendered transparent so that the bound DNA is visible (derived from PDB ID 3CRX). Another group of recombinases, called the resolvase/invertase family, use a Ser residue as nucleophile at the active site.

(20 to 200 bp), unique DNA sequence where the recombinase acts (the recombination site). One or more auxiliary proteins may regulate the timing or outcome of the reaction.

There are two general classes of site-specific recombination systems, which rely on either Tyr or Ser residues in the active site. In vitro studies of many site-specific recombination systems in the tyrosine class have elucidated some general principles, including the fundamental reaction pathway **(Fig. 25–37a)**. Several of these enzymes have been crystallized, revealing structural details of the reaction. A separate recombinase recognizes and binds to each of two recombination sites on two different DNA molecules or within the same DNA. One DNA strand in each site is cleaved at a specific point within the site, and the recombinase becomes covalently linked to the DNA at the cleavage site through a phosphotyrosine bond (step ❶). The transient protein-DNA linkage preserves the phosphodiester bond that is lost in cleaving the DNA, so high-energy cofactors such as ATP are unnecessary in subsequent steps. The cleaved DNA strands are rejoined to new partners to form a Holliday intermediate, with new phosphodiester bonds created at the expense of the protein-DNA linkage (step ❷). An isomerization then occurs (step ❸), and the process is repeated at a second point within each of the two recombination sites (steps ❹ and ❺). In the systems that employ an active-site Ser residue, both strands of each recombination site are cut concurrently and rejoined to new partners without the Holliday intermediate. In both types of system, the exchange is always reciprocal and precise, regenerating the recombination sites when the reaction is complete. We can view a recombinase as a site-specific endonuclease and ligase in one package.

The sequences of the recombination sites recognized by site-specific recombinases are partially asymmetric (nonpalindromic), and the two recombining sites align in the same orientation during the recombinase reaction. The outcome depends on the location and orientation of the recombination sites **(Fig. 25–38)**. If the two sites are on the same DNA molecule, the reaction either inverts or deletes the

Inversion

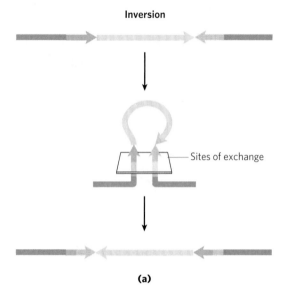

Sites of exchange

(a)

Deletion and insertion

insertion ‖ deletion

+

(b)

FIGURE 25–38 Effects of site-specific recombination. The outcome of site-specific recombination depends on the location and orientation of the recombination sites (red and green) in a double-stranded DNA molecule. Orientation here (shown by arrowheads) refers to the order of nucleotides in the recombination site, not the 5'→3' direction. **(a)** Recombination sites with opposite orientation in the same DNA molecule. The result is an inversion. **(b)** Recombination sites with the same orientation, either on one DNA molecule, producing a deletion, or on two DNA molecules, producing an insertion.

intervening DNA, determined by whether the recombination sites have the opposite or the same orientation, respectively. If the sites are on different DNAs, the recombination is intermolecular; if one or both DNAs are circular, the result is an insertion. Some recombinase systems are highly specific for one of these reaction types and act only on sites with particular orientations.

Complete chromosomal replication can require site-specific recombination. Recombinational DNA repair of a circular bacterial chromosome, while essential, sometimes generates deleterious byproducts. The resolution of a Holliday intermediate at a replication fork by a nuclease such as RuvC, followed by completion of replication, can give rise to one of two products: the usual two monomeric chromosomes or a contiguous dimeric chromosome **(Fig. 25–39)**. In the latter case, the covalently linked chromosomes cannot be segregated to daughter cells at cell division and the dividing cells become "stuck." A specialized site-specific recombination system in *E. coli*, the XerCD system, converts the dimeric chromosomes to monomeric chromosomes so that cell division can proceed. The reaction is a site-specific deletion (Fig. 25–38b). This

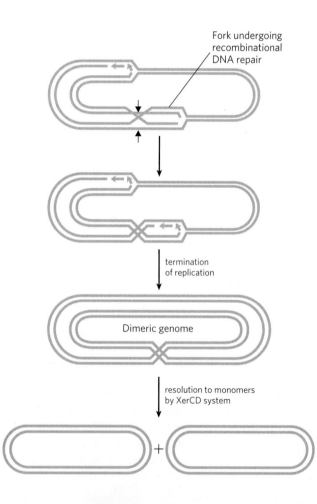

Fork undergoing recombinational DNA repair

termination of replication

Dimeric genome

resolution to monomers by XerCD system

+

FIGURE 25–39 DNA deletion to undo a deleterious effect of recombinational DNA repair. The resolution of a Holliday intermediate during recombinational DNA repair (if cut at the points indicated by red arrows) can generate a contiguous dimeric chromosome. A specialized site-specific recombinase in *E. coli*, XerCD, converts the dimer to monomers, allowing chromosome segregation and cell division to proceed.

is another example of the close coordination between DNA recombination processes and other aspects of DNA metabolism.

Transposable Genetic Elements Move from One Location to Another

We now consider the third general type of recombination system: recombination that allows the movement of transposable elements, or **transposons**. These segments of DNA, found in virtually all cells, move, or "jump," from one place on a chromosome (the donor site) to another on the same or a different chromosome (the target site). DNA sequence homology is not usually required for this movement, called **transposition**; the new location is determined more or less randomly. Insertion of a transposon in an essential gene could kill the cell, so transposition is tightly regulated and usually very infrequent. Transposons are perhaps the simplest of molecular parasites, adapted to replicate passively within the chromosomes of host cells. In some cases they carry genes that are useful to the host cell, and thus exist in a kind of symbiosis with the host.

Bacteria have two classes of transposons. **Insertion sequences** (simple transposons) contain only the sequences required for transposition and the genes for the proteins (transposases) that promote the process. **Complex transposons** contain one or more genes in addition to those needed for transposition. These extra genes might, for example, confer resistance to antibiotics and thus enhance the survival chances of the host cell. The spread of antibiotic-resistance elements among disease-causing bacterial populations that is rendering some antibiotics ineffectual (p. 981) is mediated in part by transposition. ∎

Bacterial transposons vary in structure, but most have short repeated sequences at each end that serve as binding sites for the transposase. When transposition occurs, a short sequence at the target site (5 to 10 bp) is duplicated to form an additional short repeated sequence that flanks each end of the inserted transposon **(Fig. 25–40)**. These duplicated segments result from the cutting mechanism used to insert a transposon into the DNA at a new location.

There are two general pathways for transposition in bacteria. In direct (or simple) transposition **(Fig. 25–41**, left), cuts on each side of the transposon excise it, and the transposon moves to a new location. This leaves a double-strand break in the donor DNA that must be repaired. At the target site, a staggered cut is made (as in Fig. 25–40), the transposon is inserted into the break, and DNA replication fills in the gaps to duplicate the target site sequence. In replicative transposition (Fig. 25–41, right), the entire transposon is replicated, leaving a copy behind at the donor location. A **cointegrate** is an intermediate in this process, consisting of the donor region covalently linked to DNA at the

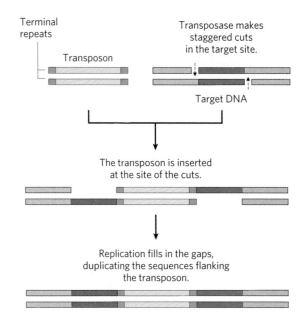

FIGURE 25–40 Duplication of the DNA sequence at a target site when a transposon is inserted. The sequences that are duplicated following transposon insertion are shown in red. These sequences are generally only a few base pairs long, so their size relative to that of a typical transposon is greatly exaggerated in this drawing.

target site. Two complete copies of the transposon are present in the cointegrate, both having the same relative orientation in the DNA. In some well-characterized transposons, the cointegrate intermediate is converted to products by site-specific recombination, in which specialized recombinases promote the required deletion reaction.

Eukaryotes also have transposons, structurally similar to bacterial transposons, and some use similar transposition mechanisms. In other cases, however, the mechanism of transposition seems to involve an RNA intermediate. Evolution of these transposons is intertwined with the evolution of certain classes of RNA viruses. Both are described in the next chapter.

Immunoglobulin Genes Assemble by Recombination

Some DNA rearrangements are a programmed part of development in eukaryotic organisms. An important example is the generation of complete immunoglobulin genes from separate gene segments in vertebrate genomes. A human (like other mammals) is capable of producing *millions* of different immunoglobulins (antibodies) with distinct binding specificities, even though the human genome contains only ~29,000 genes. Recombination allows an organism to produce an extraordinary diversity of antibodies from a limited DNA-coding capacity. Studies of the recombination mechanism reveal a close relationship to DNA transposition and suggest that this system for generating antibody diversity may have evolved from an ancient cellular invasion of transposons.

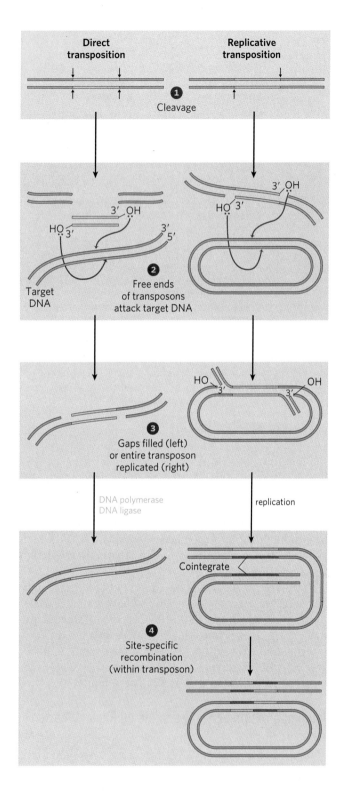

Direct transposition / **Replicative transposition**

❶ Cleavage

Target DNA

Free ends of transposons attack target DNA ❷

❸ **Gaps filled (left) or entire transposon replicated (right)**

DNA polymerase
DNA ligase

replication

Cointegrate

❹ **Site-specific recombination (within transposon)**

FIGURE 25–41 Two general pathways for transposition: direct (simple) and replicative. ❶ The DNA is first cleaved on each side of the transposon, at the sites indicated by arrows. ❷ The liberated 3'-hydroxyl groups at the ends of the transposon act as nucleophiles in a direct attack on phosphodiester bonds in the target DNA. The target phosphodiester bonds are staggered (not directly across from each other) in the two DNA strands. ❸ The transposon is now linked to the target DNA. In direct transposition (left), replication fills in gaps at each end to complete the process. In replicative transposition (right), the entire transposon is replicated to create a cointegrate intermediate. ❹ The cointegrate is often resolved later, with the aid of a separate site-specific recombination system. The cleaved host DNA left behind after direct transposition is either repaired by DNA end-joining or degraded (not shown). The latter outcome can be lethal to an organism.

lambda, which differ somewhat in the sequences of their constant regions. For all three types of polypeptide chain (heavy chain, and kappa and lambda light chains), diversity in the variable regions is generated by a similar mechanism. The genes for these polypeptides are divided into segments, and the genome contains clusters with multiple versions of each segment. The joining of one version of each gene segment creates a complete gene.

Figure 25–42 depicts the organization of the DNA encoding the kappa light chains of human IgG and shows how a mature kappa light chain is generated. In undifferentiated cells, the coding information for this polypeptide chain is separated into three segments. The V (variable) segment encodes the first 95 amino acid residues of the variable region, the J (joining) segment encodes the remaining 12 residues of the variable region, and the C segment encodes the constant region. The genome contains ~300 different V segments, 4 different J segments, and 1 C segment.

As a stem cell in the bone marrow differentiates to form a mature B lymphocyte, one V segment and one J segment are brought together by a specialized recombination system (Fig. 25–42). During this programmed DNA deletion, the intervening DNA is discarded. There are about $300 \times 4 = 1,200$ possible V–J combinations. The recombination process is not as precise as the site-specific recombination described earlier, so additional variation occurs in the sequence at the V–J junction. This increases the overall variation by a factor of at least 2.5, so the cells can generate about $2.5 \times 1,200 = 3,000$ different V–J combinations. The final joining of the V–J combination to the C region is accomplished by an RNA-splicing reaction after transcription, a process described in Chapter 26.

The recombination mechanism for joining the V and J segments is illustrated in **Figure 25–43**. Just beyond each V segment and just before each J segment lie recombination signal sequences (RSS). These are bound by proteins called RAG1 and RAG2 (products of the *r*ecombination *a*ctivating *g*ene). The RAG proteins catalyze the formation of a double-strand break between

We can use the human genes that encode proteins of the immunoglobulin G (IgG) class to illustrate how antibody diversity is generated. Immunoglobulins consist of two heavy and two light polypeptide chains (see Fig. 5–21). Each chain has two regions, a variable region, with a sequence that differs greatly from one immunoglobulin to another, and a region that is virtually constant within a class of immunoglobulins. There are also two distinct families of light chains, kappa and

FIGURE 25–42 Recombination of the V and J gene segments of the human IgG kappa light chain. This process is designed to generate antibody diversity. At the top is shown the arrangement of IgG-coding sequences in a stem cell of the bone marrow. Recombination deletes the DNA between a particular V segment and a J segment. After transcription, the transcript is processed by RNA splicing, as described in Chapter 26; translation produces the light-chain poly-peptide. The light chain can combine with any of 5,000 possible heavy chains to produce an antibody molecule.

the signal sequences and the V (or J) segments to be joined. The V and J segments are then joined with the aid of a second complex of proteins.

The genes for the heavy chains and the lambda light chains form by similar processes. Heavy chains have more gene segments than light chains, with more than 5,000 possible combinations. Because any heavy chain can combine with any light chain to generate an immunoglobulin, each human has at least $3,000 \times 5,000 = 1.5 \times 10^7$ possible IgGs. And additional diversity is generated by high mutation rates (of unknown mechanism) in the V sequences during B-lymphocyte differentiation. Each mature B lymphocyte produces only one type of antibody, but the range of antibodies produced by the B lymphocytes of an individual organism is clearly enormous.

Did the immune system evolve in part from ancient transposons? The mechanism for generation of the double-strand breaks by RAG1 and RAG2 does mirror several reaction steps in transposition (Fig. 25–43). In addition, the deleted DNA, with its terminal RSS, has a sequence structure found in most transposons. In the test tube, RAG1 and RAG2 can associate with this

deleted DNA and insert it, transposonlike, into other DNA molecules (probably a rare reaction in B lympho-cytes). Although we cannot know for certain, the prop-erties of the immunoglobulin gene rearrangement system suggest an intriguing origin in which the dis-tinction between host and parasite has become blurred by evolution.

SUMMARY 25.3 DNA Recombination

▶ DNA sequences are rearranged in recombination reactions, usually in processes tightly coordinated with DNA replication or repair.

▶ Homologous genetic recombination can take place between any two DNA molecules that share sequence homology. In bacteria, recombination serves mainly as a DNA repair process, focused on reactivating stalled or collapsed replication forks or on the general repair of double-strand breaks. In eukaryotes, recombination is essential to ensure accurate chromosomal segregation during the first meiotic cell division. It also helps to create genetic diversity in the resulting gametes.

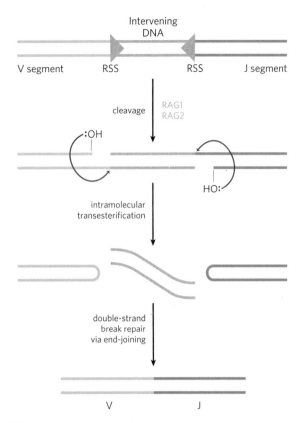

FIGURE 25–43 Mechanism of immunoglobulin gene rearrangement. The RAG1 and RAG2 proteins bind to the recombination signal sequences (RSS) and cleave one DNA strand between the RSS and the V (or J) segments to be joined. The liberated 3′ hydroxyl then acts as a nucleophile, attacking a phosphodiester bond in the other strand to create a double-strand break. The resulting hairpin bends on the V and J segments are cleaved, and the ends are covalently linked by a complex of proteins specialized for end-joining repair of double-strand breaks. The steps in the generation of the double-strand break catalyzed by RAG1 and RAG2 are chemically related to steps in transposition reactions.

▶ Site-specific recombination occurs only at specific target sequences, and this process can also involve a Holliday intermediate. Recombinases cleave the DNA at specific points and ligate the strands to new partners. This type of recombination is found in virtually all cells, and its many functions include DNA integration and regulation of gene expression.

▶ In virtually all cells, transposons use recombination to move within or between chromosomes. In vertebrates, a programmed recombination reaction related to transposition joins immunoglobulin gene segments to form immunoglobulin genes during B-lymphocyte differentiation.

Key Terms

Terms in bold are defined in the glossary.

Further Reading

General

Cox, M.M., Doudna, J.A., & O'Donnell, M.E. (2012) *Molecular Biology: Principles and Practice*, W. H. Freeman and Company, New York, NY.

  An excellent introduction to the science and how it is done.

Friedberg, E.C., Walker, G.C., Siede, W., Wood, R.D., Schultz, R.A., & Ellenberger, T. (2006) *DNA Repair and Mutagenesis*, 2nd edn, American Society for Microbiology, Washington, DC.

  A thorough treatment of DNA metabolism and a good place to start exploring this field.

DNA Replication

Bloom, L.B. (2006) Dynamics of loading the *Escherichia coli* DNA polymerase processivity clamp. *Crit. Rev. Biochem. Mol. Biol.* **41**, 179–208.

Heller, R.C. & Marians, K.J. (2006) Replisome assembly and the direct restart of stalled replication forks. *Nat. Rev. Mol. Cell Biol.* **7**, 932–943.

  Mechanisms for the restart of replication forks before the repair of DNA damage.

Hübscher, U., Maga, G., & Spadari, S. (2002) Eukaryotic DNA polymerases. *Annu. Rev. Biochem.* **71**, 133–163.

  Good summary of the properties and roles of the more than one dozen known eukaryotic DNA polymerases.

Indiani, C. & O'Donnell, M. (2006) The replication clamp-loading machine at work in the three domains of life. *Nat. Rev. Mol. Cell Biol.* **7**, 751–761.

Kaguni, J. (2011) Replication initiation at the *Escherichia coli* chromosomal origin. *Curr. Opin. Chem. Biol.* **15**, 606–613.

Kool, E.T. (2002) Active site tightness and substrate fit in DNA replication. *Annu. Rev. Biochem.* **71**, 191–219.

Excellent summary of the molecular basis of replication fidelity by a DNA polymerase—base-pair geometry as well as hydrogen bonding.

Kunkel, T.A. & Burgers, P.M. (2008) Dividing the workload at a eukaryotic replication fork. *Trends Cell Biol.* **18**, 521–527.

Masai, H., Matsumoto, S., You, Z.Y., Yoshizawa-Sugata, N., & Oda, M. (2010) Eukaryotic chromosome DNA replication: where, when, and how? *Annu. Rev. Biochem.* **79**, 89–130.

O'Donnell, M. (2006) Replisome architecture and dynamics in *Escherichia coli*. *J. Biol. Chem.* **281**, 10,653–10,656.

An excellent summary of what goes on at a replication fork.

Stillman, B. (2005) Origin recognition and the chromosome cycle. *FEBS Lett.* **579**, 877–884.

Good summary of the initiation of eukaryotic DNA replication.

DNA Repair

Erzberger, J.P. & Berger, J.M. (2006) Evolutionary relationships and structural mechanisms of AAA+ proteins. *Annu. Rev. Biophys. Biomol. Struct.* **35**, 93–114.

Lynch, M. (2010) Evolution of the mutation rate. *Trends Genet.* **26**, 345–352.

Kunkel, T.A. & Erie, D.A. (2005) DNA mismatch repair. *Annu. Rev. Biochem.* **74**, 681–710.

Schlacher, K. & Goodman, M.J. (2007) Lessons from 50 years of SOS DNA-damage-induced mutagenesis *Nat. Rev. Mol. Cell Biol.* **8**, 587–594.

Sutton, M.D., Smith, B.T., Godoy, V.G., & Walker, G.C. (2000) The SOS response: recent insights into umuDC-dependent mutagenesis and DNA damage tolerance. *Annu. Rev. Genet.* **34**, 479–497.

Wilson, D.M. III & Bohr, V.A. (2007) The mechanics of base excision repair, and its relation to aging and disease. *DNA Repair* **6**, 544–559.

DNA Recombination

Cox, M.M. (2001) Historical overview: searching for replication help in all of the rec places. *Proc. Natl. Acad. Sci. USA* **98**, 8173–8180.

A review of how recombination was shown to be a replication-fork repair process.

Cox, M.M. (2007) Regulation of bacterial RecA protein function. *Crit. Rev. Biochem. Mol. Biol.* **42**, 41–63.

Grindley, N.D.F., Whiteson, K.L., & Rice, P.A. (2006) Mechanisms of site-specific recombination. *Annu. Rev. Biochem.* **75**, 567–605.

Haniford, D.B. (2006) Transpososome dynamics and regulation in Tn10 transposition. *Crit. Rev. Biochem. Mol. Biol.* **41**, 407–424.

A detailed look at one well-studied bacterial transposon.

Heyer, W.-D., Ehmsen, K.T., & Liu, J. (2010) Regulation of homologous recombination in eukaryotes. *Annu. Rev. Genet.* **44**, 113–139.

Levin, H.L. & Moran, J.V. (2011) Dynamic interactions between transposable elements and their hosts. *Nat. Rev. Genet.* **12**, 615–627.

Lusetti, S.L. & Cox, M.M. (2002) The bacterial RecA protein and the recombinational DNA repair of stalled replication forks. *Annu. Rev. Biochem.* **71**, 71–100.

Mimitou, E.P. & Symington, L.S. (2009) Nucleases and helicases take center stage in homologous recombination. *Trends Biochem. Sci.* **34**, 264–272.

Montano, S.P. & Rice, P.A. (2011) Moving DNA around: DNA transposition and retroviral integration. *Curr. Opin. Struct. Biol.* **21**, 370–378.

San Filippo, J., Sung, P., & Klein, H. (2008) Mechanism of eukaryotic homologous recombination. *Annu. Rev. Biochem.* **77**, 229–257.

Singleton, M.R., Dillingham, M.S., Gaudier, M., Kowalczykowski, S.C., & Wigley, D.B. (2004) Crystal structure of RecBCD enzyme reveals a machine for processing DNA breaks. *Nature* **432**, 187–193.

Problems

1. Conclusions from the Meselson-Stahl Experiment The Meselson-Stahl experiment (see Fig. 25–2) proved that DNA undergoes semiconservative replication in *E. coli*. In the "dispersive" model of DNA replication, the parent DNA strands are cleaved into pieces of random size, then joined with pieces of newly replicated DNA to yield daughter duplexes. Explain how the results of Meselson and Stahl's experiment ruled out such a model.

2. Heavy Isotope Analysis of DNA Replication A culture of *E. coli* growing in a medium containing $^{15}NH_4Cl$ is switched to a medium containing $^{14}NH_4Cl$ for three generations (an eightfold increase in population). What is the molar ratio of hybrid DNA (^{15}N–^{14}N) to light DNA (^{14}N–^{14}N) at this point?

3. Replication of the *E. coli* Chromosome The *E. coli* chromosome contains 4,639,221 bp.

(a) How many turns of the double helix must be unwound during replication of the *E. coli* chromosome?

(b) From the data in this chapter, how long would it take to replicate the *E. coli* chromosome at 37 °C if two replication forks proceeded from the origin? Assume replication occurs at a rate of 1,000 bp/s. Under some conditions *E. coli* cells can divide every 20 min. How might this be possible?

(c) In the replication of the *E. coli* chromosome, about how many Okazaki fragments would be formed? What factors guarantee that the numerous Okazaki fragments are assembled in the correct order in the new DNA?

4. Base Composition of DNAs Made from Single-Stranded Templates Predict the base composition of the total DNA synthesized by DNA polymerase on templates provided by an equimolar mixture of the two complementary strands of bacteriophage ϕX174 DNA (a circular DNA molecule). The base composition of one strand is A, 24.7%; G, 24.1%; C, 18.5%; and T, 32.7%. What assumption is necessary to answer this problem?

5. DNA Replication Kornberg and his colleagues incubated soluble extracts of *E. coli* with a mixture of dATP, dTTP, dGTP, and dCTP, all labeled with ^{32}P in the α-phosphate group. After a time, the incubation mixture was treated with trichloroacetic acid, which precipitates the DNA but not the nucleotide precursors. The precipitate was collected, and the extent of precursor incorporation into DNA was determined from the amount of radioactivity present in the precipitate.

(a) If any one of the four nucleotide precursors were omitted from the incubation mixture, would radioactivity be found in the precipitate? Explain.

(b) Would ^{32}P be incorporated into the DNA if only dTTP were labeled? Explain.

(c) Would radioactivity be found in the precipitate if ^{32}P labeled the β or γ phosphate rather than the α phosphate of the deoxyribonucleotides? Explain.

6. The Chemistry of DNA Replication All DNA polymerases synthesize new DNA strands in the $5'{\rightarrow}3'$ direction. In some respects, replication of the antiparallel strands of duplex DNA would be simpler if there were also a second type of polymerase, one that synthesized DNA in the $3'{\rightarrow}5'$ direction. The two types of polymerase could, in principle, coordinate DNA synthesis without the complicated mechanics required for lagging strand replication. However, no such $3'{\rightarrow}5'$-synthesizing enzyme has been found. Suggest two possible mechanisms for $3'{\rightarrow}5'$ DNA synthesis. Pyrophosphate should be one product of both proposed reactions. Could one or both mechanisms be supported in a cell? Why or why not? (Hint: You may suggest the use of DNA precursors not actually present in extant cells.)

7. Activities of DNA Polymerases You are characterizing a new DNA polymerase. When the enzyme is incubated with [^{32}P]-labeled DNA and no dNTPs, you observe the release of [^{32}P]dNMPs. This release is prevented by adding unlabeled dNTPs. Explain the reactions that most likely underlie these observations. What would you expect to observe if you added pyrophosphate instead of dNTPs?

8. Leading and Lagging Strands Prepare a table that lists the names and compares the functions of the precursors, enzymes, and other proteins needed to make the leading strand versus the lagging strand during DNA replication in *E. coli*.

9. Function of DNA Ligase Some *E. coli* mutants contain defective DNA ligase. When these mutants are exposed to ^{3}H-labeled thymine and the DNA produced is sedimented on an alkaline sucrose density gradient, two radioactive bands appear. One corresponds to a high molecular weight fraction, the other to a low molecular weight fraction. Explain.

10. Fidelity of Replication of DNA What factors promote the fidelity of replication during synthesis of the leading strand of DNA? Would you expect the lagging strand to be made with the same fidelity? Give reasons for your answers.

11. Importance of DNA Topoisomerases in DNA Replication DNA unwinding, such as that occurring in replication, affects the superhelical density of DNA. In the absence of topoisomerases, the DNA would become overwound ahead of a replication fork as the DNA is unwound behind it. A bacterial replication fork will stall when the superhelical density (σ) of the DNA ahead of the fork reaches +0.14 (see Chapter 24).

Bidirectional replication is initiated at the origin of a 6,000 bp plasmid in vitro, in the absence of topoisomerases. The plasmid initially has a σ of –0.06. How many base pairs will be unwound and replicated by each replication fork before the forks stall? Assume that both forks travel at the same rate and that each includes all components necessary for elongation except topoisomerase.

12. The Ames Test In a nutrient medium that lacks histidine, a thin layer of agar containing $\sim 10^9$ *Salmonella typhimurium* histidine auxotrophs (mutant cells that require histidine to survive) produces ~ 13 colonies over a two-day incubation period at 37°C (see Fig. 25–20). How do these colonies arise in the absence of histidine? The experiment is repeated in the presence of 0.4 μg of 2-aminoanthracene. The number of colonies produced over two days exceeds 10,000. What does this indicate about 2-aminoanthracene? What can you surmise about its carcinogenicity?

13. DNA Repair Mechanisms Vertebrate and plant cells often methylate cytosine in DNA to form 5-methylcytosine (see Fig. 8–5a). In these same cells, a specialized repair system recognizes G–T mismatches and repairs them to G≡C base pairs. How might this repair system be advantageous to the cell? (Explain in terms of the presence of 5-methylcytosine in the DNA.)

14. DNA Repair in People with Xeroderma Pigmentosum The condition known as xeroderma pigmentosum (XP) arises from mutations in at least seven different human genes (see Box 25–1). The deficiencies are generally in genes encoding enzymes involved in some part of the pathway for human nucleotide-excision repair. The various types of XP are denoted A through G (XPA, XPB, etc.), with a few additional variants lumped under the label XP-V.

Cultures of fibroblasts from healthy individuals and from patients with XPG are irradiated with ultraviolet light. The DNA is isolated and denatured, and the resulting single-stranded DNA is characterized by analytical ultracentrifugation.

(a) Samples from the normal fibroblasts show a significant reduction in the average molecular weight of the single-stranded DNA after irradiation, but samples from the XPG fibroblasts show no such reduction. Why might this be?

(b) If you assume that a nucleotide-excision repair system is operative in fibroblasts, which step might be defective in the cells from the patients with XPG? Explain.

15. Holliday Intermediates How does the formation of Holliday intermediates in homologous genetic recombination differ from their formation in site-specific recombination?

16. Cleavage of Holliday Intermediates A Holliday intermediate is formed between two homologous chromosomes, at a point between genes *A* and *B*, as shown below. The chromosomes have different alleles of the two genes (*A* and *a*, *B* and *b*). Where would the Holliday intermediate have to be cleaved (points X and/or Y) to generate a chromosome that would convey (a) an *Ab* genotype or (b) an *ab* genotype?

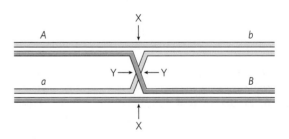

17. A Connection between Replication and Site-Specific Recombination Most wild strains of *Saccharomyces cerevisiae* have multiple copies of the circular plasmid 2μ (named for its contour length of about 2 μm), which has $\sim 6{,}300$ bp of DNA. For its replication the plasmid uses the host replication system, under the same strict control as the host cell chromosomes, replicating only once per cell cycle. Replication of the plasmid is bidirectional, with both replication forks initiating at a single, well-defined origin. However, one replication cycle of a 2μ plasmid can result in more than two copies of the plasmid, allowing amplification of the plasmid copy number (number of plasmid copies per cell) whenever plasmid segregation at cell division leaves one daughter cell with fewer than the normal complement of plasmid copies. Amplification requires a site-specific recombination system encoded by the plasmid, which serves to invert one part of the plasmid relative to the other. Explain how a site-specific inversion event could result in amplification of the plasmid copy number. (Hint: Consider the situation when replication forks have duplicated one recombination site but not the other.)

Data Analysis Problem

18. Mutagenesis in _Escherichia coli_ Many mutagenic compounds act by alkylating the bases in DNA. The alkylating agent R7000 (7-methoxy-2-nitronaphtho[2,1-_b_]furan) is an extremely potent mutagen.

R7000

In vivo, R7000 is activated by the enzyme nitroreductase, and this more reactive form covalently attaches to DNA—primarily, but not exclusively, to G≡C base pairs.

In a 1996 study, Quillardet, Touati, and Hofnung explored the mechanisms by which R7000 causes mutations in _E. coli_. They compared the genotoxic activity of R7000 in two strains of _E. coli_: the wild-type (Uvr$^+$) and mutants lacking _uvrA_ activity (Uvr$^-$; see Table 25–6). They first measured rates of mutagenesis. Rifampicin is an inhibitor of RNA polymerase (see Chapter 26). In its presence, cells will not grow unless certain mutations occur in the gene encoding RNA polymerase; the appearance of rifampicin-resistant colonies thus provides a useful measure of mutagenesis rates.

The effects of different concentrations of R7000 were determined, with the results shown in the graph below.

(a) Why are some mutants produced even when no R7000 is present?

Quillardet and colleagues also measured the survival rate of bacteria treated with different concentrations of R7000 with the following results.

(b) Explain how treatment with R7000 is lethal to cells.

(c) Explain the differences in the mutagenesis curves and in the survival curves for the two types of bacteria, Uvr$^+$ and Uvr$^-$, as shown in the graphs.

The researchers then went on to measure the amount of R7000 covalently attached to the DNA in Uvr$^+$ and Uvr$^-$ _E. coli_. They incubated bacteria with [^{3}H]R7000 for 10 or 70 minutes, extracted the DNA, and measured its ^{3}H content in counts per minute (cpm) per μg of DNA.

	^{3}H in DNA (cpm/μg)	
Time (min)	**Uvr$^+$**	**Uvr$^-$**
10	76	159
70	69	228

(d) Explain why the amount of ^{3}H drops over time in the Uvr$^+$ strain and rises over time in the Uvr$^-$ strain.

Quillardet and colleagues then examined the particular DNA sequence changes caused by R7000 in the Uvr$^+$ and Uvr$^-$ bacteria. For this, they used six different strains of _E. coli_, each with a different point mutation in the _lacZ_ gene, which encodes β-galactosidase (this enzyme catalyzes the same reaction as lactase; see Fig. 14–11). Cells with any of these mutations have a nonfunctional β-galactosidase and are unable to metabolize lactose (i.e., a Lac$^-$ phenotype). Each type of point mutation required a specific reverse mutation to restore _lacZ_ gene function and Lac$^+$ phenotype. By plating cells on a medium containing lactose as the sole carbon source, it was possible to select for these reverse-mutated, Lac$^+$ cells. And by counting the number of Lac$^+$ cells following mutagenesis of a particular strain, the researchers could measure the frequency of each type of mutation.

First, they looked at the mutation spectrum in Uvr$^-$ cells. The following table shows the results for the six strains, CC101 through CC106 (with the point mutation required to produce Lac$^+$ cells indicated in parentheses).

Number of Lac⁺ cells (average ± SD)

R7000 (μg/mL)	CC101 (A=T to C≡G)	CC102 (G≡C to A=T)	CC103 (G≡C to C≡G)	CC104 (G≡C To T=A)	CC105 (A=T to T=A)	CC106 (A=T to G≡C)
0	6 ± 3	11 ± 9	2 ± 1	5 ± 3	2 ± 1	1 ± 1
0.075	24 ± 19	34 ± 3	8 ± 4	82 ± 23	40 ± 14	4 ± 2
0.15	24 ± 4	26 ± 2	9 ± 5	180 ± 71	130 ± 50	3 ± 2

(e) Which types of mutation show significant increases above the background rate due to treatment with R7000? Provide a plausible explanation for why some have higher frequencies than others.

(f) Can all of the mutations you listed in (e) be explained as resulting from covalent attachment of R7000 to a G≡C base pair? Explain your reasoning.

(g) Figure 25–27b shows how methylation of guanine residues can lead to a G≡C to A=T mutation. Using a similar pathway, show how a G–R7000 adduct could lead to the G≡C to A=T or T=A mutations shown above. Which base pairs with the G–R7000 adduct?

The results for the Uvr⁺ bacteria are shown in the table below.

Number of Lac⁺ cells (average ± SD)

R7000 (μg/mL)	CC101 (A=T to C≡G)	CC102 (G≡C to A=T)	CC103 (G≡C to C≡G)	CC104 (G≡C to T=A)	CC105 (A=T to T=A)	CC106 (A=T to G≡C)
0	2 ± 2	10 ± 9	3 ± 3	4 ± 2	6 ± 1	0.5 ± 1
1	7 ± 6	21 ± 9	8 ± 3	23 ± 15	13 ± 1	1 ± 1
5	4 ± 3	15 ± 7	22 ± 2	68 ± 25	67 ± 14	1 ± 1

(h) Do these results show that all mutation types are repaired with equal fidelity? Provide a plausible explanation for your answer.

Reference

Quillardet, P., Touati, E., & Hofnung, M. (1996) Influence of the *uvr*-dependent nucleotide excision repair on DNA adducts formation and mutagenic spectrum of a potent genotoxic agent: 7-methoxy-2-nitronaphtho[2,1-*b*]furan (R7000). *Mutat. Res.* **358**, 113–122.

RNA Metabolism

Expression of the information in a gene generally involves production of an RNA molecule transcribed from a DNA template. Strands of RNA and DNA may seem quite similar at first glance, differing only in that RNA has a hydroxyl group at the 2′ position of the aldopentose, and uracil instead of thymine. However, unlike DNA, most RNAs carry out their functions as single strands, strands that fold back on themselves and have the potential for much greater structural diversity than DNA (Chapter 8). RNA is thus suited to a variety of cellular functions.

RNA is the only macromolecule known to have a role both in the storage and transmission of information and in catalysis, which has led to much speculation about its possible role as an essential chemical intermediate in the development of life on this planet. The discovery of catalytic RNAs, or ribozymes, has changed the very definition of an enzyme, extending it beyond the domain of proteins. Proteins nevertheless remain essential to RNA and its functions. In the modern cell, all nucleic acids, including RNAs, are complexed with proteins. Some of these complexes are quite elaborate, and RNA can assume both structural and catalytic roles within complicated biochemical machines.

All RNA molecules except the RNA genomes of certain viruses are derived from information permanently stored in DNA. During **transcription**, an enzyme system converts the genetic information in a segment of double-stranded DNA into an RNA strand with a base sequence complementary to one of the DNA strands. Three major kinds of RNA are produced. **Messenger RNAs (mRNAs)** encode the amino acid sequence of one or more polypeptides specified by a gene or set of genes. **Transfer RNAs (tRNAs)** read the information encoded in the mRNA and transfer the appropriate amino acid to a growing polypeptide chain during protein synthesis. **Ribosomal RNAs (rRNAs)** are constituents of ribosomes, the intricate cellular machines that synthesize proteins. Many additional, specialized RNAs have regulatory or catalytic functions or are precursors to the three main classes of RNA. These special-function RNAs are no longer thought of as minor species in the catalog of cellular RNAs. In vertebrates, RNAs that do not fit into one of the classical categories (mRNA, tRNA, rRNA) seem to vastly outnumber those that do.

During replication the entire chromosome is usually copied, but transcription is more selective. Only particular genes or groups of genes are transcribed at any one time, and some portions of the DNA genome are never transcribed. The cell restricts the expression of genetic information to the formation of gene products needed at any particular moment. Specific regulatory sequences mark the beginning and end of the DNA segments to be transcribed and designate which strand in duplex DNA is to be used as the template. The transcript itself may interact with other RNA molecules as part of the overall regulatory program. The regulation of transcription is described in detail in Chapter 28.

The sum of all the RNA molecules produced in a cell under a given set of conditions is called the cellular **transcriptome**. Given the relatively small fraction of the human genome devoted to protein-encoding genes, we might expect that only a small part of the human genome is transcribed. This is not the case. Modern microarray analysis of transcription patterns has revealed that much of the genome of humans and other mammals is transcribed into RNA. The products are predominantly not mRNAs, tRNAs, or rRNAs, but rather special-function RNAs, a host of which are being discovered. Many of these seem to be involved in regulation of gene expression; however, the rapid pace of discovery has forced the realization that we do not yet know what many of these RNAs do.

In this chapter we examine the synthesis of RNA on a DNA template and the postsynthetic processing and turnover of RNA molecules. In doing so, we encounter many of the specialized functions of RNA, including catalytic functions. Interestingly, the substrates for RNA enzymes are often other RNA molecules. We also describe systems in which RNA is the template and DNA the product, rather than vice versa. The information pathways thus come full circle, and reveal that template-dependent nucleic acid synthesis has standard rules, regardless of the nature of template or product (RNA or DNA). This examination of the biological interconversion of DNA and RNA as information carriers leads to a discussion of the evolutionary origin of biological information.

26.1 DNA-Dependent Synthesis of RNA

Our discussion of RNA synthesis begins with a comparison between transcription and DNA replication (Chapter 25). Transcription resembles replication in its fundamental chemical mechanism, its polarity (direction of synthesis), and its use of a template. And like replication, transcription has initiation, elongation, and termination phases—though in the literature on transcription, initiation is further divided into discrete phases of DNA binding and initiation of RNA synthesis. Transcription differs from replication in that it does not require a primer and, generally, involves only limited segments of a DNA molecule. Additionally, within transcribed segments, only one DNA strand serves as a template for a particular RNA molecule.

RNA Is Synthesized by RNA Polymerases

The discovery of DNA polymerase and its dependence on a DNA template spurred a search for an enzyme that synthesizes RNA complementary to a DNA strand. By 1960, four research groups had independently detected an enzyme in cellular extracts that could form an RNA polymer from ribonucleoside 5'-triphosphates. Subsequent work on the purified *Escherichia coli* RNA polymerase helped to define the fundamental properties of transcription **(Fig. 26–1)**. **DNA-dependent RNA polymerase** requires, in addition to a DNA template, all four ribonucleoside 5'-triphosphates (ATP, GTP, UTP, and CTP) as precursors of the nucleotide units of RNA, as well as Mg^{2+}. The protein also binds one Zn^{2+}. The chemistry and mechanism of RNA synthesis closely

MECHANISM FIGURE 26–1 Transcription by RNA polymerase in *E. coli*. For synthesis of an RNA strand complementary to one of two DNA strands in a double helix, the DNA is transiently unwound. **(a)** Catalytic mechanism of RNA synthesis by RNA polymerase. Note that this is essentially the same mechanism used by DNA polymerases (see Fig. 25–5a). The reaction involves two Mg^{2+} ions, coordinated to the phosphate groups of the incoming nucleoside triphosphates (NTPs) and to three Asp residues, which are highly conserved in the RNA polymerases of all species. One Mg^{2+} ion facilitates attack by the 3'-hydroxyl group on the α phosphate of the NTP; the other Mg^{2+} ion facilitates displacement of the pyrophosphate, and both metal ions stabilize the pentacovalent transition state.

(b) About 17 bp of DNA are unwound at any given time. RNA polymerase and the transcription bubble move from left to right along the DNA as shown, facilitating RNA synthesis. The DNA is unwound ahead and rewound behind as RNA is transcribed. As the DNA is rewound, the RNA-DNA hybrid is displaced and the RNA strand is extruded.

(c) Movement of an RNA polymerase along DNA tends to create positive supercoils (overwound DNA) ahead of the transcription bubble and negative supercoils (underwound DNA) behind it. The RNA polymerase is in close contact with the DNA ahead of the transcription bubble as well as with the separated DNA strands and the RNA within and immediately behind the bubble. A channel in the protein funnels new NTPs to the polymerase active site. The polymerase footprint encompasses about 35 bp of DNA during elongation.

(5′) C G C T A T A G C G T T T (3′)   DNA nontemplate (coding) strand

(3′) G C G A T A T C G C A A A (5′)   DNA template strand

(5′) C G C U A U A G C G U U U (3′)   RNA transcript

FIGURE 26-2 Template and nontemplate (coding) DNA strands.
The two complementary strands of DNA are defined by their function
in transcription. The RNA transcript is synthesized on the template
strand and is identical in sequence (with U in place of T) to the nontemplate strand, or coding strand.

resemble those used by DNA polymerases (see Fig.
25–5a). RNA polymerase elongates an RNA strand by
adding ribonucleotide units to the 3′-hydroxyl end,
building RNA in the 5′→3′ direction. The 3′-hydroxyl
group acts as a nucleophile, attacking the α phosphate of
the incoming ribonucleoside triphosphate (Fig. 26–1a)
and releasing pyrophosphate. The overall reaction is

$$(NMP)_n + NTP \longrightarrow (NMP)_{n+1} + PP_i$$
$$\text{RNA} \qquad\qquad \text{Lengthened RNA}$$

RNA polymerase requires DNA for activity and is most
active when bound to a double-stranded DNA. As noted
above, only one of the two DNA strands serves as a template. The template DNA strand is copied in the 3′→5′
direction (antiparallel to the new RNA strand), just as in
DNA replication. Each nucleotide in the newly formed
RNA is selected by Watson-Crick base-pairing interactions: U residues are inserted in the RNA to pair with A
residues in the DNA template, G residues are inserted to
pair with C residues, and so on. Base-pair geometry (see
Fig. 25–6) may also play a role in base selection.

Unlike DNA polymerase, RNA polymerase does not
require a primer to initiate synthesis. Initiation occurs
when RNA polymerase binds at specific DNA sequences
called promoters (described below). The 5′-triphosphate
group of the first residue in a nascent (newly formed)
RNA molecule is not cleaved to release PP_i, but instead
remains intact throughout the transcription process.
During the elongation phase of transcription, the growing
end of the new RNA strand base-pairs temporarily with
the DNA template to form a short hybrid RNA-DNA
double helix, estimated to be 8 bp long (Fig. 26–1b). The
RNA in this hybrid duplex "peels off" shortly after its
formation, and the DNA duplex re-forms.

To enable RNA polymerase to synthesize an RNA
strand complementary to one of the DNA strands, the
DNA duplex must unwind over a short distance, forming
a transcription "bubble." During transcription, the *E. coli*
RNA polymerase generally keeps about 17 bp unwound.
The 8 bp RNA-DNA hybrid occurs in this unwound
region. Elongation of a transcript by *E. coli* RNA polymerase proceeds at a rate of 50 to 90 nucleotides/s.
Because DNA is a helix, movement of a transcription
bubble requires considerable strand rotation of the
nucleic acid molecules. DNA strand rotation is restricted
in most DNAs by DNA-binding proteins and other structural barriers. As a result, a moving RNA polymerase
generates waves of positive supercoils ahead of the transcription bubble and negative supercoils behind (Fig.
26–1c). This has been observed both in vitro and in vivo
(in bacteria). In the cell, the topological problems caused
by transcription are relieved through the action of topoisomerases (Chapter 24).

KEY CONVENTION: The two complementary DNA strands
have different roles in transcription. The strand that
serves as template for RNA synthesis is called the **template strand**. The DNA strand complementary to the
template, the **nontemplate strand**, or **coding strand**,
is identical in base sequence to the RNA transcribed
from the gene, with U in the RNA in place of T in the
DNA (**Fig. 26–2**). The coding strand for a particular
gene may be located in either strand of a given chromosome (as shown in **Fig. 26–3** for a virus). By convention, the regulatory sequences that control transcription
(described later in this chapter) are designated by the
sequences in the coding strand. ■

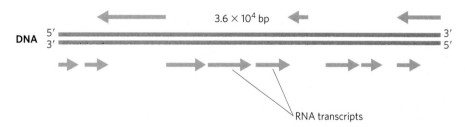

**FIGURE 26-3 Organization of coding information in the adenovirus ge-
nome.** The genetic information of the adenovirus genome (a conveniently simple example) is encoded by a double-stranded DNA molecule of
36,000 bp, both strands of which encode proteins. The information for
most proteins is encoded by (that is, identical to) the top strand—by convention, the strand oriented 5′ to 3′ from left to right. The bottom strand
acts as template for these transcripts. However, a few proteins are en-

coded by the bottom strand, which is transcribed in the opposite direction
(and uses the top strand as template). Synthesis of mRNAs in adenovirus
is actually much more complex than shown here. Many of the mRNAs
derived using the upper strand as template are initially synthesized as a
single, long transcript (25,000 nucleotides), which is then extensively
processed to produce the separate mRNAs. Adenovirus causes upper
respiratory tract infections in some vertebrates.

FIGURE 26-4 Structure of the RNA polymerase holoenzyme of the bacterium *Thermus aquaticus*. (Derived from PDB ID 1HQM) The overall structure of this enzyme is very similar to that of the *E. coli* RNA polymerase; no DNA or RNA is shown here. The several subunits of the bacterial RNA polymerase give the enzyme the shape of a crab claw. The pincers are formed from the large β and β' subunits. The subunits are shown in the same colors in the schematic and the ribbon structure.

The DNA-dependent RNA polymerase of *E. coli* is a large, complex enzyme with five core subunits ($\alpha_2\beta\beta'\omega$; M_r 390,000) and a sixth subunit, one of a group designated σ, with variants designated by size (molecular weight). The σ subunit binds transiently to the core and directs the enzyme to specific binding sites on the DNA (described below). These six subunits constitute the RNA polymerase holoenzyme **(Fig. 26–4)**. The RNA polymerase holoenzyme of *E. coli* thus exists in several forms, depending on the type of σ subunit. The most common subunit is σ^{70} (M_r 70,000), and the upcoming discussion focuses on the corresponding RNA polymerase holoenzyme.

RNA polymerases lack a separate proofreading $3'{\rightarrow}5'$ exonuclease active site (such as that of many DNA polymerases), and the error rate for transcription is higher than that for chromosomal DNA replication— approximately one error for every 10^4 to 10^5 ribonucleotides incorporated into RNA. Because many copies of an RNA are generally produced from a single gene and all RNAs are eventually degraded and replaced, a mistake in an RNA molecule is of less consequence to the cell than a mistake in the permanent information stored in DNA. Many RNA polymerases, including bacterial RNA polymerase and the eukaryotic RNA polymerase II (discussed below), do pause when a mispaired base is added during transcription, and they can remove mismatched nucleotides from the 3' end of a transcript by direct reversal of the polymerase reaction. But we do not yet know whether this activity is a true proofreading function and to what extent it may contribute to the fidelity of transcription.

RNA Synthesis Begins at Promoters

Initiation of RNA synthesis at random points in a DNA molecule would be an extraordinarily wasteful process. Instead, an RNA polymerase binds to specific sequences in the DNA called **promoters**, which direct the transcription of adjacent segments of DNA (genes). The sequences where RNA polymerases bind can be quite variable, and much research has focused on identifying the particular sequences that are critical to promoter function.

In *E. coli*, RNA polymerase binding occurs within a region stretching from about 70 bp before the transcription start site to about 30 bp beyond it. By convention, the DNA base pairs that correspond to the beginning of an RNA molecule are given positive numbers, and those preceding the RNA start site are given negative numbers. The promoter region thus extends between positions -70 and $+30$. Analyses and comparisons of the most common class of bacterial promoters (those recognized by an RNA polymerase holoenzyme containing σ^{70}) have revealed similarities in two short sequences centered about positions -10 and -35 **(Fig. 26–5)**. These sequences are important interaction sites for the σ^{70} subunit. Although the sequences are not identical for all bacterial promoters in this class, certain nucleotides that are particularly common at each position form a **consensus sequence** (recall the *E. coli* oriC consensus sequence; see Fig. 25–10). The consensus sequence at the -10 region is (5')TATAAT(3'); the consensus sequence at the -35 region is (5')TTGACA(3'). A third AT-rich recognition element, called the UP (upstream promoter) element, occurs between positions -40 and -60 in the promoters of certain highly expressed genes. The UP element is bound by the α subunit of RNA polymerase. The efficiency with which an RNA polymerase containing σ^{70} binds to a promoter and initiates transcription is determined in large measure by these sequences, the spacing between them, and their distance from the transcription start site.

Many independent lines of evidence attest to the functional importance of the sequences in the -35 and -10 regions. Mutations that affect the function of a given promoter often involve a base pair in these regions. Variations in the consensus sequence also affect the efficiency of RNA polymerase binding and transcription initiation. A change in only one base pair can decrease the rate of binding by several orders of magnitude. The promoter sequence thus establishes a basal level of expression that can vary greatly from one *E. coli* gene to the next. A method that provides information about the interaction between RNA polymerase and promoters is illustrated in Box 26–1.

The pathway of transcription initiation and the fate of the σ subunit are becoming much better defined **(Fig. 26–6)**. The pathway consists of two major parts, binding and initiation, each with multiple steps. First, the polymerase, directed by its bound σ factor, binds to the promoter. A closed complex (in which the bound DNA is intact) and an open complex (in which the bound DNA is intact and partially unwound near the -10 sequence) form in succession. Second, transcription

FIGURE 26–5 Typical *E. coli* promoters recognized by an RNA polymerase holoenzyme containing σ70. Sequences of the nontemplate strand are shown, read in the 5′→3′ direction, as is the convention for representations of this kind. The sequences differ from one promoter to the next, but comparisons of many promoters reveal similarities, particularly in the −10 and −35 regions. The sequence element UP, not present in all *E. coli* promoters, is shown in the P1 promoter for the highly expressed rRNA gene *rrnB*. UP elements, generally occurring in the region between −40 and −60, strongly stimulate transcription at the promoters that contain them. The UP element in the *rrnB* P1 promoter encompasses the region between −38 and −59. The consensus sequence for *E. coli* promoters recognized by σ70 is shown second from the top. Spacer regions contain slightly variable numbers of nucleotides (N). Only the first nucleotide coding the RNA transcript (at position +1) is shown.

is initiated within the complex, leading to a conformational change that converts the complex to the elongation form, followed by movement of the transcription complex away from the promoter (promoter clearance). Any of these steps can be affected by the specific makeup of the promoter sequences. The σ subunit dissociates stochastically (at random) as the polymerase enters the elongation phase of transcription. The protein NusA (M_r 54,430) binds to the elongating RNA polymerase, competitively with the σ subunit. Once transcription is complete, NusA dissociates from the enzyme, the RNA polymerase dissociates from the DNA, and a σ factor (σ^{70} or another) can again bind to the enzyme to initiate transcription.

E. coli has other classes of promoters, bound by RNA polymerase holoenzymes with different σ subunits (Table 26–1). An example is the promoters of the heat shock genes. The products of this set of genes are made at higher levels when the cell has received an insult, such as a sudden increase in temperature. RNA polymerase binds to the promoters of these genes only when σ^{70} is replaced with the σ^{32} (M_r 32,000) subunit, which is specific for the heat shock promoters (see Fig. 28–3). By using different σ subunits, the cell can coordinate the expression of sets of genes, permitting major changes in cell physiology. Which sets of genes are expressed is determined by the availability of the various σ subunits, which is determined by several factors: regulated rates of synthesis and degradation, postsynthetic modifications that switch individual σ subunits between active and inactive forms, and a specialized class of anti-σ proteins, each type binding to and sequestering a particular σ subunit (rendering it unavailable for transcription initiation).

Transcription Is Regulated at Several Levels

Requirements for any gene product vary with cellular conditions or developmental stage, and transcription of each gene is carefully regulated to form gene products only in the proportions needed. Regulation can occur at any step in transcription, including elongation and termination. However, much of the regulation is directed at the polymerase binding and transcription initiation steps outlined in Figure 26–6. Differences in promoter sequences are just one of several levels of control.

The binding of proteins to sequences both near to and distant from the promoter can also affect levels of gene expression. Protein binding can *activate* transcription by facilitating either RNA polymerase binding or steps further along in the initiation process, or it can *repress* transcription by blocking the activity of the polymerase. In *E. coli*, one protein that activates transcription is the **cAMP receptor protein (CRP)**, which increases the transcription of genes coding for enzymes that metabolize sugars other than glucose when cells are grown in the absence of glucose. **Repressors** are proteins that block the synthesis of RNA at specific genes. In the case of the Lac repressor (Chapter 28), transcription of the genes for the enzymes of lactose metabolism is blocked when lactose is unavailable.

BOX 26–1 | METHODS | RNA Polymerase Leaves Its Footprint on a Promoter

Footprinting, a technique derived from principles used in DNA sequencing, identifies the DNA sequences bound by a particular protein. Researchers isolate a DNA fragment thought to contain sequences recognized by a DNA-binding protein and radiolabel one end of one strand (Fig. 1). They then use chemical or enzymatic reagents to introduce random breaks in the DNA fragment (averaging about one per molecule).

Separation of the labeled cleavage products (broken fragments of various lengths) by high-resolution electrophoresis produces a ladder of radioactive bands. In a separate tube, the cleavage procedure is repeated on copies of the same DNA fragment in the presence of the DNA-binding protein. The researchers then subject the two sets of cleavage products to electrophoresis and compare them side by side. A gap ("footprint") in the series of radioactive bands derived from the DNA-protein sample, attributable to protection of the DNA by the bound protein, identifies the sequences that the protein binds.

The precise location of the protein-binding site can be determined by directly sequencing (see Fig. 8–34) copies of the same DNA fragment and including the sequencing lanes (not shown here) on the same gel with the footprint. Figure 2 shows footprinting results for the binding of RNA polymerase to a DNA fragment containing a promoter. The polymerase covers 60 to 80 bp; protection by the bound enzyme includes the −10 and −35 regions.

Solution of identical DNA fragments radioactively labeled ✱ at one end of one strand.

DNase I

(+)   (−)

Treat with DNase under conditions in which each strand is cut once (on average). No cuts are made in the area where RNA polymerase has bound.

Site of DNase cut

Isolate labeled DNA fragments and denature. Only labeled strands are detected in next step.

Separate fragments by polyacrylamide gel electrophoresis and visualize radiolabeled bands on x-ray film.

Uncut DNA fragment

DNA migration

Missing bands indicate where RNA polymerase was bound to DNA.

FIGURE 1 Footprint analysis of the RNA polymerase–binding site on a DNA fragment. Separate experiments are carried out in the presence (+) and absence (−) of the polymerase.

Nontemplate strand

− + C

+1
−10
−20
−30
−40
−50

Regions bound by RNA polymerase

FIGURE 2 Footprinting results of RNA polymerase binding to the *lac* promoter (see Fig. 26-5). In this experiment, the 5′ end of the nontemplate strand was radioactively labeled. Lane C is a control in which the labeled DNA fragments were cleaved with a chemical reagent that produces a more uniform banding pattern.

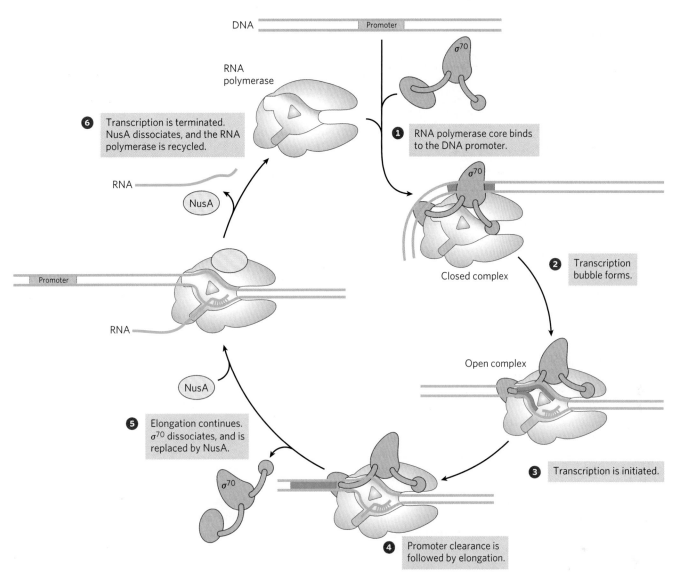

FIGURE 26–6 Transcription initiation and elongation by *E. coli* RNA polymerase. Initiation of transcription requires several steps generally divided into two phases, binding and initiation. In the binding phase, the initial interaction of the RNA polymerase with the promoter leads to formation of a closed complex, in which the promoter DNA is stably bound but not unwound. A 12 to 15 bp region of DNA—from within the −10 region to position +2 or +3—is then unwound to form an open complex. Additional intermediates (not shown) have been detected in the pathways leading to the closed and open complexes, along with several changes in protein conformation. The initiation phase encompasses transcription initiation and promoter clearance (steps ❶ through ❹ here). Once elongation commences, the σ subunit is released and it is replaced by the protein NusA. The polymerase leaves the promoter and becomes committed to elongation of the RNA (step ❺). When transcription is complete, the RNA is released, the NusA protein dissociates, and the RNA polymerase dissociates from the DNA (step ❻). Another σ subunit binds to the RNA polymerase and the process begins again.

Transcription is the first step in the complicated and energy-intensive pathway of protein synthesis, so much of the regulation of protein levels in both bacterial and eukaryotic cells is directed at transcription, particularly its early stages. In Chapter 28 we describe many mechanisms by which this regulation is accomplished.

Specific Sequences Signal Termination of RNA Synthesis

RNA synthesis is processive; that is, the RNA polymerase will introduce a large number of nucleotides into a growing RNA molecule before dissociating (p. 1014). This is necessary because, if an RNA polymerase released an RNA transcript prematurely, it could not resume synthesis of the same RNA but instead would have to start again. However, an encounter with certain DNA sequences results in a pause in RNA synthesis, and at some of these sequences transcription is terminated. Our focus here is again on the well-studied systems in bacteria. *E. coli* has at least two classes of termination signals: one class relies on a protein factor called ρ (rho) and the other is ρ-independent.

TABLE 26–1 The Seven σ Subunits of *Escherichia coli*

σ subunit	K_d (nM)	Molecules/cell*	Holoenzyme ratio (%)*	Function
σ^{70}	0.26	700	78	Housekeeping
σ^{54}	0.30	110	8	Modulation of cellular nitrogen levels
σ^{38}	4.26	<1	0	Stationary phase genes
σ^{32}	1.24	<10	0	Heat shock genes
σ^{28}	0.74	370	14	Flagella and chemotaxis genes
σ^{24}	2.43	<10	0	Extracytoplasmic functions; some heat shock functions
σ^{18}	1.73	<1	0	Extracytoplasmic functions, including ferric citrate transport

Source: Adapted from Maeda, H., Fujita, N., & Ishihama, A. (2000) *Nucleic Acids Res.* 28, 3500.

Note: σ factors are widely distributed in bacteria; the number varies from a single σ factor in *Mycoplasma genitalium* to 63 distinct σ factors in *Streptomyces coelicolor*.

*Approximate number of each σ subunit per cell and the fraction of RNA polymerase holoenzyme complexed with each σ subunit during exponential growth. The numbers change as growth conditions change. The fraction of RNA polymerase complexed with each σ subunit reflects both the amount of the particular subunit and its affinity for the enzyme.

Most ρ-independent terminators have two distinguishing features. The first is a region that produces an RNA transcript with self-complementary sequences, permitting the formation of a hairpin structure (see Fig. 8–19a) centered 15 to 20 nucleotides before the projected end of the RNA strand. The second feature is a highly conserved string of three A residues in the template strand that are transcribed into U residues near the 3′ end of the hairpin. When a polymerase arrives at a termination site with this structure, it pauses **(Fig. 26–7a)**. Formation of the hairpin structure in the RNA disrupts several A=U base pairs in the RNA-DNA hybrid segment and may disrupt important interactions between RNA and the RNA polymerase, facilitating dissociation of the transcript.

The ρ-dependent terminators lack the sequence of repeated A residues in the template strand but usually include a CA-rich sequence called a *rut* (*r*ho *u*tilization) element. The ρ protein associates with the RNA at specific binding sites and migrates in the 5′→3′ direction until it reaches the transcription complex that is paused at a termination site (Fig. 26–7b). Here it contributes to release of the RNA transcript. The ρ protein has an ATP-dependent RNA-DNA helicase activity that promotes translocation of the protein along the RNA, and ATP is hydrolyzed by ρ protein during the termination process. The detailed mechanism by which the protein promotes the release of the RNA transcript is not known.

Eukaryotic Cells Have Three Kinds of Nuclear RNA Polymerases

The transcriptional machinery in the nucleus of a eukaryotic cell is much more complex than that in bacteria. Eukaryotes have three RNA polymerases, designated I, II, and III, which are distinct complexes but have certain subunits in common. Each polymerase has a specific function and is recruited to a specific promoter sequence.

RNA polymerase I (Pol I) is responsible for the synthesis of only one type of RNA, a transcript called preribosomal RNA (or pre-rRNA), which contains the precursor for the 18S, 5.8S, and 28S rRNAs (see Fig. 26–24). Pol I promoters differ greatly in sequence from one species to another. The principal function of RNA polymerase II (Pol II) is synthesis of mRNAs and some specialized RNAs. This enzyme can recognize thousands of promoters that vary greatly in sequence. Some Pol II promoters have a few sequence features in common, including a TATA box (eukaryotic consensus sequence TATA(A/T)A(A/T)(A/G)) near base pair −30 and an Inr sequence (initiator) near the RNA start site at +1 **(Fig. 26–8)**. However, such promoters are in the minority, and Pol II operates at many promoters that lack these features.

RNA polymerase III (Pol III) makes tRNAs, the 5S rRNA, and some other small specialized RNAs. The promoters recognized by Pol III are well characterized. Interestingly, some of the sequences required for the regulated initiation of transcription by Pol III are located within the gene itself, whereas others are in more conventional locations upstream of the RNA start site (Chapter 28).

RNA Polymerase II Requires Many Other Protein Factors for Its Activity

RNA polymerase II is central to eukaryotic gene expression and has been studied extensively. Although this polymerase is strikingly more complex than its bacterial counterpart, the complexity masks a remarkable conservation of structure, function, and mechanism. Pol II isolated from yeast is a huge enzyme with 12 subunits. The largest subunit (RBP1) exhibits a high degree of homology to the β′ subunit of bacterial RNA polymerase.

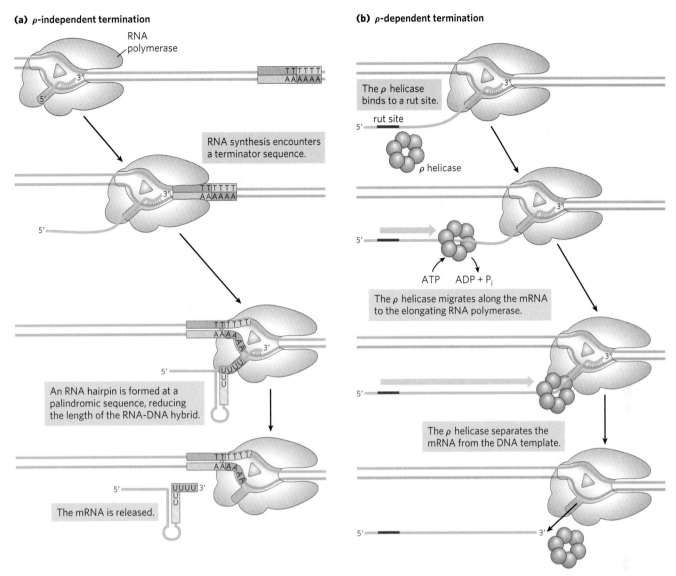

(a) ρ-independent termination

RNA polymerase

RNA synthesis encounters a terminator sequence.

An RNA hairpin is formed at a palindromic sequence, reducing the length of the RNA-DNA hybrid.

The mRNA is released.

(b) ρ-dependent termination

The ρ helicase binds to a rut site.

rut site

ρ helicase

ATP   ADP + P$_i$

The ρ helicase migrates along the mRNA to the elongating RNA polymerase.

The ρ helicase separates the mRNA from the DNA template.

FIGURE 26–7 Termination of transcription in *E. coli*. (a) ρ-Independent termination. RNA polymerase pauses at a variety of DNA sequences, some of which are terminators. One of two outcomes is then possible: the polymerase bypasses the site and continues on its way, or the complex undergoes a conformational change (isomerization). In the latter case, intramolecular pairing of complementary sequences in the newly formed RNA transcript may form a hairpin that disrupts the RNA-DNA hybrid or the interactions between RNA and polymerase, or both, resulting in isomerization. An A=U hybrid region at the 3' end of the new transcript is relatively unstable, and the RNA dissociates from the complex completely, leading to termination. This is the usual outcome at terminators. At other pause sites, the complex may escape after the isomerization step to continue RNA synthesis. **(b)** ρ-Dependent termination. RNAs that include a rut site (purple) recruit the ρ helicase. The ρ helicase migrates along the mRNA in the 5'→3' direction and separates it from the polymerase.

FIGURE 26–8 Some common sequences in promoters recognized by eukaryotic RNA polymerase II. The TATA box is the major assembly point for the proteins of the preinitiation complexes of Pol II. The DNA is unwound at the initiator sequence (Inr), and the transcription start site is usually within or very near this sequence. In the Inr consensus sequence shown here, N represents any nucleotide; Y, a pyrimidine nucleotide. Many additional sequences serve as binding sites for a wide variety of proteins that affect the activity of Pol II. These sequences are important in regulating Pol II promoters and differ greatly in type and number, and in general the eukaryotic promoter is much more complex than suggested here (see Fig. 15–25). Many of the sequences are located within a few hundred base pairs of the TATA box on the 5' side; others may be thousands of base pairs away. The sequence elements summarized here are more variable among the Pol II promoters of eukaryotes than among the *E. coli* promoters (see Fig. 26–5). The majority of Pol II promoters lack a TATA box or a consensus Inr element or both. Additional sequences around the TATA box and downstream (to the right as drawn) of Inr may be recognized by one or more transcription factors.

TABLE 26–2 Proteins Required for Initiation of Transcription at the RNA Polymerase II (Pol II) Promoters of Eukaryotes

Transcription protein	Number of subunits	Subunit(s) M_r	Function(s)
Initiation			
Pol II	12	10,000–220,000	Catalyzes RNA synthesis
TBP (TATA-binding protein)	1	38,000	Specifically recognizes the TATA box
TFIIA	3	12,000, 19,000, 35,000	Stabilizes binding of TFIIB and TBP to the promoter
TFIIB	1	35,000	Binds to TBP; recruits Pol II–TFIIF complex
TFIIE	2	34,000, 57,000	Recruits TFIIH; has ATPase and helicase activities
TFIIF	2	30,000, 74,000	Binds tightly to Pol II; binds to TFIIB and prevents binding of Pol II to nonspecific DNA sequences
TFIIH	12	35,000–89,000	Unwinds DNA at promoter (helicase activity); phosphorylates Pol II (within the CTD); recruits nucleotide-excision repair proteins
Elongation*			
ELL[†]	1	80,000	
pTEFb	2	43,000, 124,000	Phosphorylates Pol II (within the CTD)
SII (TFIIS)	1	38,000	
Elongin (SIII)	3	15,000, 18,000, 110,000	

*The function of all elongation factors is to suppress the pausing or arrest of transcription by the Pol II–TFIIF complex.

[†]Name derived from *e*leven-nineteen *l*ysine-rich *l*eukemia. The gene for ELL is the site of chromosomal recombination events frequently associated with acute myeloid leukemia.

Another subunit (RBP2) is structurally similar to the bacterial β subunit, and two others (RBP3 and RBP11) show some structural homology to the two bacterial α subunits. Pol II must function with genomes that are more complex and with DNA molecules more elaborately packaged than in bacteria. The need for protein-protein contacts with the numerous other protein factors required to navigate this labyrinth accounts in large measure for the added complexity of the eukaryotic polymerase.

The largest subunit of Pol II (RBP1) also has an unusual feature, a long carboxyl-terminal tail consisting of many repeats of a consensus heptad amino acid sequence –YSPTSPS–. There are 27 repeats in the yeast enzyme (18 exactly matching the consensus) and 52 (21 exact) in the mouse and human enzymes. This carboxyl-terminal domain (CTD) is separated from the main body of the enzyme by an inherently unstructured linker sequence. The CTD has many important roles in Pol II function, as outlined below.

RNA polymerase II requires an array of other proteins, called **transcription factors**, in order to form the active transcription complex. The **general transcription factors** required at every Pol II promoter (factors usually designated TFII with an additional identifier) are highly conserved in all eukaryotes (Table 26–2). The process of transcription by Pol II can be described in terms of several phases—assembly, initiation, elongation, termination—each associated with characteristic proteins (**Fig. 26–9**). The step-by-step pathway described below leads to active transcription in vitro. In the cell, many of the proteins may be present in larger, preassembled complexes, simplifying the pathways for assembly on promoters. As you read about this process, consult Figure 26–9 and Table 26–2 to help keep track of the many participants.

FIGURE 26–9 Transcription at RNA polymerase II promoters. (a) The sequential assembly of TBP (often with TFIIA), TFIIB, TFIIF plus Pol II, TFIIE, and TFIIH results in a closed complex. Within the complex, the DNA is unwound at the Inr region by the helicase activity of TFIIH and perhaps of TFIIE, creating an open complex. The carboxyl-terminal domain of the largest Pol II subunit is phosphorylated by TFIIH, and the polymerase then escapes the promoter and begins transcription. Elongation is accompanied by the release of many transcription factors and is also enhanced by elongation factors (see Table 26–2). After termination, Pol II is released, dephosphorylated, and recycled. **(b)** Human TBP bound to DNA. The DNA is bent in this complex, opening the minor groove to allow specific hydrogen-bonding between protein and DNA (PDB ID 1TGH). **(c)** A cutaway view of transcription elongation promoted by the Pol II core enzyme.

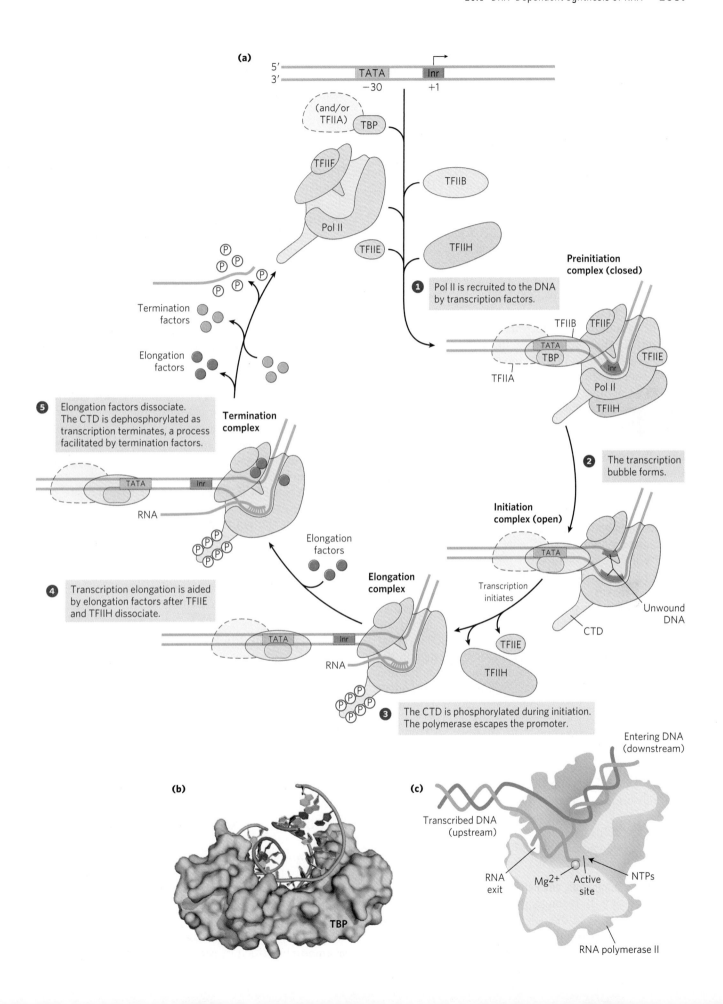

Assembly of RNA Polymerase and Transcription Factors at a Promoter The formation of a closed complex begins when the TATA-binding protein (TBP) binds to the TATA box (Figs 26–9a, step ❶, and 26–9b). At promoters lacking a TATA box, TBP arrives as part of a multisubunit complex called TFIID (not shown in Fig. 26–9). The sequence elements that direct the binding of TFIID at TATA-less promoters are poorly understood. TBP is bound in turn by the transcription factor TFIIB, which also binds to DNA on either side of TBP. TFIIA binds, and along with TFIIB helps to stabilize the TBP-DNA complex. TFIIB provides an important link to DNA polymerase II, and the TFIIB-TBP complex is next bound by another complex consisting of TFIIF and Pol II. TFIIF helps target Pol II to its promoters, both by interacting with TFIIB and by reducing the binding of the polymerase to nonspecific sites on the DNA. Finally, TFIIE and TFIIH bind to create the closed complex. TFIIH has multiple subunits and includes a DNA helicase activity that promotes the unwinding of DNA near the RNA start site (a process requiring the hydrolysis of ATP), thereby creating an open complex (Fig. 26–9a, step ❷). Counting all the subunits of the various essential factors (excluding TFIIA and some subunits of TFIID), this minimal active assembly has more than 30 polypeptides. Structural studies by Roger Kornberg and his collaborators have provided a more detailed look at the core structure of RNA polymerase II during elongation (Fig. 26–9c).

RNA Strand Initiation and Promoter Clearance TFIIH has an additional function during the initiation phase. A kinase activity in one of its subunits phosphorylates Pol II at many places in the CTD (Fig. 26–9a). Several other protein kinases, including CDK9 (cyclin-dependent kinase 9), which is part of the complex pTEFb (*positive transcription elongation factor b*), also phosphorylate the CTD, primarily on the Ser residues of the CTD repeat sequence. This causes a conformational change in the overall complex, initiating transcription. Phosphorylation of the CTD is also important during the subsequent elongation phase, with the phosphorylation state of the CTD changing as transcription proceeds. The changes affect the interactions between the transcription complex and other proteins and enzymes, such that different sets of proteins are bound at initiation than at later stages. Some of these proteins are involved in processing the transcript (as described below).

During synthesis of the initial 60 to 70 nucleotides of RNA, first TFIIE and then TFIIH is released, and Pol II enters the elongation phase of transcription.

Elongation, Termination, and Release TFIIF remains associated with Pol II throughout elongation. During this stage, the activity of the polymerase is greatly enhanced by proteins called elongation factors (Table 26–2). The elongation factors, some bound to the phosphorylated CTD, suppress pausing during transcription and also coordinate interactions between protein complexes involved in the posttranscriptional processing of mRNAs. Once the RNA transcript is completed, transcription is terminated. Pol II is dephosphorylated and recycled, ready to initiate another transcript (Fig. 26–9a, steps ❸ to ❺).

Regulation of RNA Polymerase II Activity Regulation of transcription at Pol II promoters is quite elaborate. It involves the interaction of a wide variety of other proteins with the preinitiation complex. Some of these regulatory proteins interact with transcription factors, others with Pol II itself. The regulation of transcription is described in more detail in Chapter 28.

Diverse Functions of TFIIH In eukaryotes, the repair of damaged DNA (see Table 25–5) is more efficient within genes that are actively being transcribed than for other damaged DNA, and the template strand is repaired somewhat more efficiently than the nontemplate strand. These remarkable observations are explained by the alternative roles of the TFIIH subunits. Not only does TFIIH participate in formation of the closed complex during assembly of a transcription complex (as described above), but some of its subunits are also essential components of the separate nucleotide-excision repair complex (see Fig. 25–25).

When Pol II transcription halts at the site of a DNA lesion, TFIIH can interact with the lesion and recruit the entire nucleotide-excision repair complex. Genetic loss of certain TFIIH subunits can produce human diseases. Some examples are xeroderma pigmentosum (see Box 25–1) and Cockayne syndrome, which is characterized by arrested growth, photosensitivity, and neurological disorders. ■

DNA-Dependent RNA Polymerase Undergoes Selective Inhibition

The elongation of RNA strands by RNA polymerase in both bacteria and eukaryotes is inhibited by the antibiotic **actinomycin D** (Fig. 26–10). The planar portion of this molecule inserts (intercalates) into the double-helical DNA between successive G≡C base pairs, deforming the DNA. This prevents movement of the polymerase along the template. Because actinomycin D inhibits RNA elongation in intact cells as well as in cell extracts, it is used to identify cell processes that depend on RNA synthesis. **Acridine** inhibits RNA synthesis in a similar fashion.

Rifampicin inhibits bacterial RNA synthesis by binding to the β subunit of bacterial RNA polymerases, preventing the promoter clearance step of transcription (Fig. 26–6). It is sometimes used as an antibiotic.

The mushroom *Amanita phalloides* has a very effective defense mechanism against predators. It produces **α-amanitin**, which disrupts mRNA formation in animal

FIGURE 26–10 Actinomycin D and acridine, inhibitors of DNA transcription. (a) The shaded portion of actinomycin D is planar and intercalates between two successive G≡C base pairs in duplex DNA. The two cyclic peptide moieties of actinomycin D bind to the minor groove of the double helix. Sarcosine (Sar) is *N*-methylglycine; meVal is methylvaline. Acridine also acts by intercalation in DNA. **(b)** A complex of actinomycin D with DNA (PDB ID 1DSC). The DNA backbone is shown in blue, the bases are white, the intercalated part of actinomycin (shaded in (a)) is orange, and the remainder of the actinomycin is red. The DNA is bent as a result of the actinomycin binding.

Actinomycin D

Acridine

(a)

(b)

cells by blocking Pol II and, at higher concentrations, Pol III. Neither Pol I nor bacterial RNA polymerase is sensitive to α-amanitin—nor is the RNA polymerase II of *A. phalloides* itself!

SUMMARY 26.1 DNA-Dependent Synthesis of RNA

▶ Transcription is catalyzed by DNA-dependent RNA polymerases, which use ribonucleoside 5′-triphosphates to synthesize RNA complementary to the template strand of duplex DNA. Transcription occurs in several phases: binding of RNA polymerase to a DNA site called a promoter, initiation of transcript synthesis, elongation, and termination.

▶ Bacterial RNA polymerase requires a special subunit to recognize the promoter. As the first committed step in transcription, binding of RNA polymerase to the promoter and initiation of transcription are closely regulated. Transcription stops at sequences called terminators.

▶ Eukaryotic cells have three types of RNA polymerases. Binding of RNA polymerase II to its promoters requires an array of proteins called transcription factors. Elongation factors participate in the elongation phase of transcription. The largest subunit of Pol II has a long carboxyl-terminal domain, which is phosphorylated during the initiation and elongation phases.

26.2 RNA Processing

Many of the RNA molecules in bacteria and virtually all RNA molecules in eukaryotes are processed to some degree after synthesis. Some of the most interesting molecular events in RNA metabolism occur during this postsynthetic processing. Intriguingly, several of the enzymes that catalyze these reactions consist of RNA rather than protein. The discovery of these catalytic RNAs, or **ribozymes**, has brought a revolution in thinking about RNA function and about the origin of life.

A newly synthesized RNA molecule is called a **primary transcript**. Perhaps the most extensive processing of primary transcripts occurs in eukaryotic mRNAs and in tRNAs of both bacteria and eukaryotes. Special-function RNAs are also processed.

The primary transcript for a eukaryotic mRNA typically contains sequences encompassing one gene, although the sequences encoding the polypeptide may not be contiguous. Noncoding tracts that break up the coding region of the transcript are called introns, and the coding segments are called exons (see the discussion of introns and exons in DNA in Chapter 24). In a process called **RNA splicing**, the introns are removed from the primary transcript and the exons are joined to form a continuous sequence that specifies a functional polypeptide. Eukaryotic mRNAs are also modified at each end. A modified residue called a 5′ cap is added at the 5′ end. The 3′ end is cleaved, and 80 to 250 A residues are added to create a poly(A) "tail." The sometimes elaborate protein complexes that carry out each of these three mRNA-processing reactions do not operate independently. They seem to be organized in association with each other and with the phosphorylated CTD of Pol II; each complex affects the function of the others. Proteins involved in mRNA transport to the cytoplasm are also associated with the mRNA in the nucleus, and the processing of the transcript is coupled to its transport. In effect, a eukaryotic mRNA, as it is synthesized, is ensconced in an elaborate complex involving dozens of proteins. The composition of the complex changes as the primary transcript is processed, transported to the cytoplasm, and delivered to the ribosome for translation. The associated proteins modulate all aspects of the function and fate of the mRNA. These processes are outlined in **Figure 26–11** and described in more detail below.

The primary transcripts of bacterial and eukaryotic tRNAs are processed by the removal of sequences from each end (cleavage) and in a few cases by the removal of introns (splicing). Many bases and sugars in tRNAs are also modified; mature tRNAs are replete with unusual bases not found in other nucleic acids (see Fig. 26–22). Many of the special-function RNAs also undergo elaborate processing, often involving the removal of segments from one or both ends.

The ultimate fate of any RNA is its complete and regulated degradation. The rate of turnover of RNAs plays a critical role in determining their steady-state levels and the rate at which cells can shut down expression of a gene whose product is no longer needed. During the development of multicellular organisms, for example, certain proteins must be expressed at one stage only, and the mRNA encoding such a protein must be made and destroyed at the appropriate times.

Eukaryotic mRNAs Are Capped at the 5′ End

Most eukaryotic mRNAs have a **5′ cap**, a residue of 7-methylguanosine linked to the 5′-terminal residue of

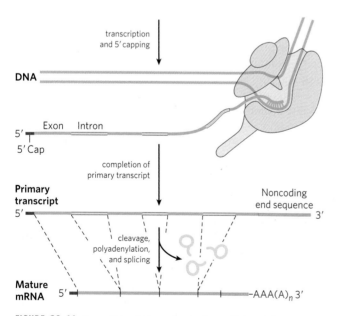

FIGURE 26–11 Formation of the primary transcript and its processing during maturation of mRNA in a eukaryotic cell. The 5′ cap (red) is added before synthesis of the primary transcript is complete. A noncoding end sequence (intron) following the last exon is shown in orange. Splicing can occur either before or after the cleavage and polyadenylation steps. All the processes shown here take place in the nucleus.

the mRNA through an unusual 5′,5′-triphosphate linkage **(Fig. 26–12)**. The 5′ cap helps protect mRNA from ribonucleases. It also binds to a specific cap-binding complex of proteins and participates in binding of the mRNA to the ribosome to initiate translation (Chapter 27).

The 5′ cap is formed by condensation of a molecule of GTP with the triphosphate at the 5′ end of the transcript. The guanine is subsequently methylated at N-7, and additional methyl groups are often added at the 2′ hydroxyls of the first and second nucleotides adjacent to the cap (Fig. 26–12a). The methyl groups are derived from *S*-adenosylmethionine. All these reactions occur very early in transcription, after the first 20 to 30 nucleotides of the transcript have been added. All three of the capping enzymes, and through them the 5′ end of the transcript itself, are associated with the RNA polymerase II CTD until the cap is synthesized. The capped 5′ end is then released from the capping enzymes and bound by the cap-binding complex (Fig. 26–12c).

Both Introns and Exons Are Transcribed from DNA into RNA

In bacteria, a polypeptide chain is generally encoded by a DNA sequence that is colinear with the amino acid sequence, continuing along the DNA template without interruption until the information needed to specify the polypeptide is complete. However, the notion that *all* genes are continuous was disproved in 1977 when Phillip Sharp and Richard Roberts independently discovered that many genes for polypeptides in eukaryotes are interrupted by noncoding sequences (introns).

The vast majority of genes in vertebrates contain introns; among the few exceptions are those that encode histones. The occurrence of introns in other eukaryotes varies. Many genes in the yeast *Saccharomyces cerevisiae* lack introns, although in some other yeast species introns are more common. Introns are also found in a few bacterial and archaeal genes. Introns in DNA are transcribed along with the rest of the gene by RNA polymerases. The introns in the primary RNA transcript are then spliced, and the exons are joined to form a mature, functional RNA. In eukaryotic mRNAs, most exons are less than 1,000 nucleotides long, with many in the 100 to 200 nucleotide size range, encoding stretches of 30 to 60 amino acids within a longer polypeptide. Introns vary in size from 50 to 20,000 nucleotides. Genes of higher eukaryotes, including humans, typically have much more DNA devoted to introns than to exons. Many genes have introns; some genes have dozens of them.

RNA Catalyzes the Splicing of Introns

There are four classes of introns. The first two, the group I and group II introns, differ in the details of their splicing mechanisms but share one surprising characteristic: they are *self-splicing*—no protein enzymes are involved. Group I introns are found in some nuclear,

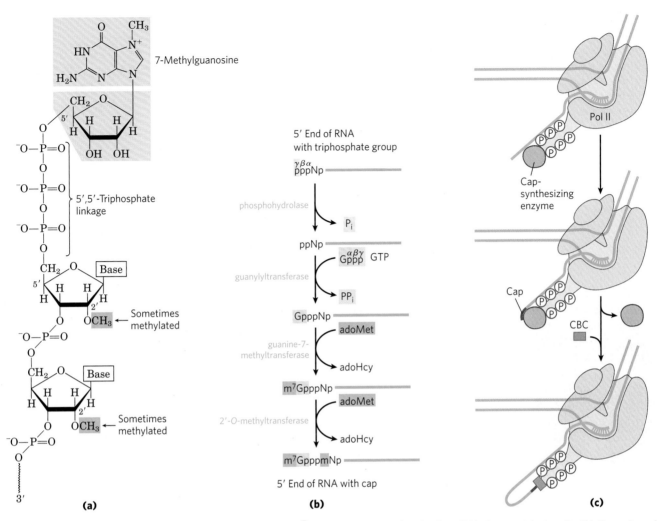

FIGURE 26–12 The 5′ cap of mRNA. (a) 7-Methylguanosine (m^7G) is joined to the 5′ end of almost all eukaryotic mRNAs in an unusual 5′,5′-triphosphate linkage. Methyl groups (light red) are often found at the 2′ position of the first and second nucleotides. RNAs in yeast cells lack the 2′-methyl groups. The 2′-methyl group on the second nucleotide is generally found only in RNAs from vertebrate cells. **(b)** Generation of the 5′ cap involves four to five separate steps (adoHcy is *S*-adenosylhomocysteine). **(c)** Synthesis of the cap is carried out by enzymes tethered to the CTD of Pol II. The cap remains tethered to the CTD through an association with the cap-binding complex (CBC).

mitochondrial, and chloroplast genes that code for rRNAs, mRNAs, and tRNAs. Group II introns are generally found in the primary transcripts of mitochondrial or chloroplast mRNAs in fungi, algae, and plants. Group I and group II introns are also found among the rare examples of introns in bacteria. Neither class requires a high-energy cofactor (such as ATP) for splicing. The splicing mechanisms in both groups involve two transesterification reaction steps **(Fig. 26–13)**, in which a ribose 2′- or 3′-hydroxyl group makes a nucleophilic attack on a phosphorus and a new phosphodiester bond is formed at the expense of the old, maintaining the balance of energy. These reactions are very similar to the DNA breaking and rejoining reactions promoted by topoisomerases (see Fig. 24–20) and site-specific recombinases (see Fig. 25–37).

The group I splicing reaction requires a guanine nucleoside or nucleotide cofactor, but the cofactor is not used as a source of energy; instead, the 3′-hydroxyl group of guanosine is used as a nucleophile in the first step of the splicing pathway. The guanosine 3′-hydroxyl group forms a normal 3′,5′-phosphodiester bond with the 5′ end of the intron **(Fig. 26–14)**. The 3′ hydroxyl of the exon that is displaced in this step then acts as a nucleophile in a similar reaction at the 3′ end of the intron. The result is precise excision of the intron and ligation of the exons.

In group II introns the reaction pattern is similar except for the nucleophile in the first step, which in this case is the 2′-hydroxyl group of an A residue *within* the intron **(Fig. 26–15)**. A branched lariat structure is formed as an intermediate.

Self-splicing of introns was first revealed in 1982 in studies of the splicing mechanism of the group I rRNA intron from the ciliated protozoan *Tetrahymena thermophila*, conducted by Thomas Cech and colleagues.

FIGURE 26–13 Transesterification reaction. Shown here is the first step in the two-step splicing of group I introns. In this example, the 3' OH of a guanosine molecule acts as nucleophile, attacking the phosphodiester linkage between U and A residues at an exon-intron junction of an mRNA molecule (see Fig. 26–14).

Thomas Cech

These workers transcribed isolated *Tetrahymena* DNA (including the intron) in vitro using purified bacterial RNA polymerase. The resulting RNA spliced itself accurately without any protein enzymes from *Tetrahymena*. The discovery that RNAs could have catalytic functions was a milestone in our understanding of biological systems.

Most introns are *not* self-splicing, and these types are not designated with a group number. The third and largest class of introns includes those found in nuclear mRNA primary transcripts. These are called **spliceosomal introns**, because their removal occurs within and is catalyzed by a large protein complex called a **spliceosome**. Within the spliceosome, the introns undergo splicing by the same lariat-forming mechanism as the group II introns. The spliceosome is made up of specialized RNA-protein complexes, small *nuclear ribonucleoproteins* (snRNPs, often pronounced *snurps*). Each snRNP contains one of a class of eukaryotic RNAs, 100 to 200

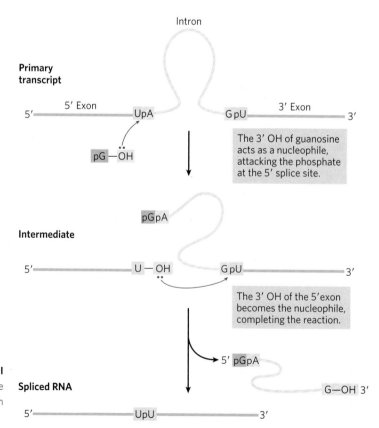

FIGURE 26–14 Splicing mechanism of group I introns. The nucleophile in the first step may be guanosine, GMP, GDP, or GTP. The spliced intron is eventually degraded.

Intron

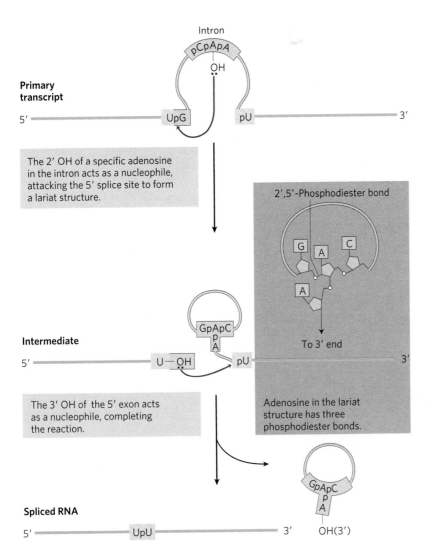

FIGURE 26-15 Splicing mechanism of group II introns. The chemistry is similar to that of group I intron splicing, except for the identity of the nucleophile in the first step and formation of a lariatlike intermediate, in which one branch is a 2′,5′-phosphodiester bond.

Primary transcript

The 2′ OH of a specific adenosine in the intron acts as a nucleophile, attacking the 5′ splice site to form a lariat structure.

2′,5′-Phosphodiester bond

Intermediate

To 3′ end

The 3′ OH of the 5′ exon acts as a nucleophile, completing the reaction.

Adenosine in the lariat structure has three phosphodiester bonds.

Spliced RNA

OH(3′)

nucleotides long, known as **small nuclear RNAs (snRNAs)**. Five snRNAs (U1, U2, U4, U5, and U6) involved in splicing reactions are generally found in abundance in eukaryotic nuclei. The RNAs and proteins in snRNPs are highly conserved in eukaryotes from yeasts to humans. 🖱 mRNA Splicing

Spliceosomal introns generally have the dinucleotide sequence GU at the 5′ end and AG at the 3′ end, and these sequences mark the sites where splicing occurs. The U1 snRNA contains a sequence complementary to sequences near the 5′ splice site of nuclear mRNA introns **(Fig. 26–16a)**, and the U1 snRNP binds to this region in the primary transcript. Addition of the U2, U4, U5, and U6 snRNPs leads to formation of the spliceosome (Fig. 26–16b). The snRNPs together contribute five RNAs and about 50 proteins to the core spliceosome, a supramolecular assembly nearly as complex as the ribosome (described in Chapter 27). Perhaps 50 additional proteins are associated with the spliceosome at different stages in the splicing process, with some of these proteins having multiple functions: in splicing, mRNA transport to the cytoplasm, translation, and eventual mRNA degradation. ATP is required for assembly of the spliceo-

some, but the RNA cleavage-ligation reactions do not seem to require ATP. Some mRNA introns are spliced by a less common type of spliceosome, in which the U1 and U2 snRNPs are replaced by the U11 and U12 snRNPs. Whereas U1- and U2-containing spliceosomes remove introns with (5′)GU and AG(3′) terminal sequences, as shown in Figure 26–16, the U11- and U12-containing spliceosomes remove a rare class of introns that have (5′)AU and AC(3′) terminal sequences to mark the intronic splice sites. The spliceosomes used in nuclear RNA splicing may have evolved from more ancient group II introns, with the snRNPs replacing the catalytic domains of their self-splicing ancestors.

Some components of the splicing apparatus are tethered to the CTD of RNA polymerase II, indicating that splicing, like other RNA processing reactions, is tightly coordinated with transcription (Fig. 26–16c). As the first splice junction is synthesized, it is bound by a tethered spliceosome. The second splice junction is then captured by this complex as it passes, facilitating the juxtaposition of the intron ends and the subsequent splicing process. After splicing, the intron remains in the nucleus and is eventually degraded.

FIGURE 26–16 Splicing mechanism in mRNA primary transcripts. (a) RNA pairing interactions in the formation of spliceosome complexes. The U1 snRNA has a sequence near its 5′ end that is complementary to the splice site at the 5′ end of the intron. Base pairing of U1 to this region of the primary transcript helps define the 5′ splice site during spliceosome assembly (ψ is pseudouridine; see Fig. 26–22). U2 is paired to the intron at a position encompassing the A residue (shaded light red) that becomes the nucleophile during the splicing reaction. Base pairing of U2 snRNA causes a bulge that displaces and helps to activate the adenylate, the 2′ OH of which will form the lariat structure through a 2′,5′-phosphodiester bond.

(b) Assembly of spliceosomes. The U1 and U2 snRNPs bind, then the remaining snRNPs (the U4-U6 complex and U5) bind to form an inactive spliceosome. Internal rearrangements convert this species to an active spliceosome in which U1 and U4 have been expelled and U6 is paired with both the 5′ splice site and U2. This is followed by the catalytic steps, which parallel those of the splicing of group II introns (see Fig. 26–15).

(c) Coordination of splicing and transcription brings the two splice sites together. See the text for details. The spliceosome is much larger than indicated here.

The fourth class of introns, found in certain tRNAs, is distinguished from the group I and II introns in that the splicing reaction requires ATP and an endonuclease. The splicing endonuclease cleaves the phosphodiester bonds at both ends of the intron, and the two exons are joined by a mechanism similar to the DNA ligase reaction (see Fig. 25–16).

Although spliceosomal introns seem to be limited to eukaryotes, the other intron classes are not. Genes with group I and II introns have now been found in both bacteria and bacterial viruses. Bacteriophage T4, for example, has several protein-encoding genes with group I introns. Introns may be more common in archaea than in bacteria.

Eukaryotic mRNAs Have a Distinctive 3′ End Structure

At their 3′ end, most eukaryotic mRNAs have a string of 80 to 250 A residues, making up the **poly(A) tail**. This tail serves as a binding site for one or more specific proteins. The poly(A) tail and its associated proteins probably help protect mRNA from enzymatic destruction. Many bacterial mRNAs also acquire poly(A) tails, but these tails stimulate decay of mRNA rather than protecting it from degradation.

The poly(A) tail is added in a multistep process. The transcript is extended beyond the site where the poly(A) tail is to be added, then is cleaved at the poly(A) addition site by an endonuclease component of a large enzyme complex, again associated with the CTD of RNA polymerase II **(Fig. 26–17)**. The mRNA site where cleavage occurs is marked by two sequence elements: the highly conserved sequence (5′)AAUAAA(3′), 10 to 30 nucleotides on the 5′ side (upstream) of the cleavage site, and a less well-defined sequence rich in G and U residues, 20 to 40 nucleotides downstream of the cleavage site. Cleavage generates the free 3′-hydroxyl group that defines the end of the mRNA, to which A residues are immediately added by **polyadenylate polymerase**, which catalyzes the reaction

$$\text{RNA} + n\text{ATP} \longrightarrow \text{RNA} - (\text{AMP})_n + n\text{PP}_i$$

where $n = 80$ to 250. This enzyme does not require a template but does require the cleaved mRNA as a primer.

The overall processing of a typical eukaryotic mRNA is summarized in **Figure 26–18**. In some cases the polypeptide-coding region of the mRNA is also modified by RNA "editing" (see Section 27.1 for details). This editing includes processes that add or delete bases in the coding regions of primary transcripts or that change the sequence (by, for example, enzymatic deamination of a C residue to create a U residue). A particularly dramatic example occurs in trypanosomes, which are parasitic protozoa: large regions of an mRNA are synthesized without any uridylate, and the U residues are inserted later by RNA editing.

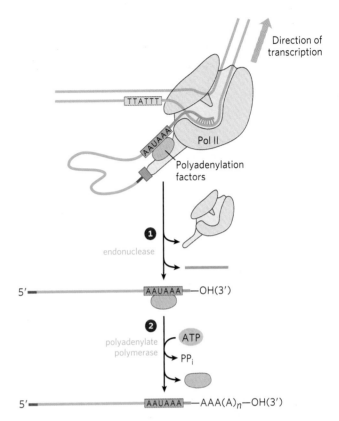

FIGURE 26–17 Addition of the poly(A) tail to the primary RNA transcript of eukaryotes. Pol II synthesizes RNA beyond the segment of the transcript containing the cleavage signal sequences, including the highly conserved upstream sequence (5′)AAUAAA. This cleavage signal sequence is bound by an enzyme complex that includes an endonuclease, a polyadenylate polymerase, and several other multisubunit proteins involved in sequence recognition, stimulation of cleavage, and regulation of the length of the poly(A) tail, all of which are tethered to the CTD. ❶ The RNA is cleaved by the endonuclease at a point 10 to 30 nucleotides 3′ to (downstream of) the sequence AAUAAA. ❷ The polyadenylate polymerase synthesizes a poly(A) tail 80 to 250 nucleotides long, beginning at the cleavage site.

A Gene Can Give Rise to Multiple Products by Differential RNA Processing

One of the paradoxes of modern genomics is that the apparent complexity of organisms does not correlate with the number of protein-encoding genes or even with the amount of genomic DNA. However, the traditional focus on protein-encoding genes ignores the complexity of an organism's transcriptome. As the functions of RNA are better appreciated, new genomic complexities become apparent.

Some eukaryotic mRNA transcripts produce only one mature mRNA and one corresponding polypeptide, but others can be processed in more than one way to produce *different* mRNAs and thus different polypeptides. The primary transcript contains molecular signals for all the alternative processing pathways, and the pathway favored in a given cell is determined by processing factors, RNA-binding proteins that promote one particular path.

FIGURE 26–18 Overview of the processing of a eukaryotic mRNA. The ovalbumin gene, shown here, has introns A to G and exons 1 to 7 and L (L encodes a signal peptide sequence that targets the protein for export from the cell; see Fig. 27–38). About three-quarters of the RNA is removed during processing. Pol II extends the primary transcript well beyond the cleavage and polyadenylation site ("extra RNA") before terminating transcription. Termination signals for Pol II have not yet been defined.

Complex transcripts can have either more than one site for cleavage and polyadenylation or alternative splicing patterns, or both. If there are two or more sites for cleavage and polyadenylation, use of the one closest to the 5′ end will remove more of the primary transcript sequence **(Fig. 26–19a)**. This mechanism, called poly(A) site choice, generates diversity in the variable domains of immunoglobulin heavy chains (see Fig. 25–42). Alternative splicing patterns (Fig. 26–19b) produce, from a common primary transcript, three different forms of the myosin heavy chain at different stages of fruit fly development. *Both* mechanisms come into play when a single

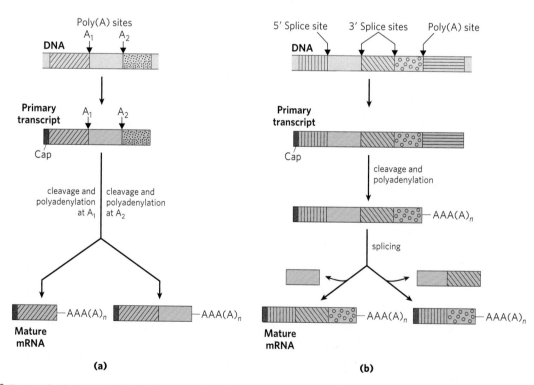

(a)   **(b)**

FIGURE 26–19 Two mechanisms for the alternative processing of complex transcripts in eukaryotes. (a) Alternative cleavage and polyadenylation patterns. Two poly(A) sites, A_1 and A_2, are shown. **(b)** Alternative splicing patterns. Two different 3′ splice sites are shown. In both mechanisms, different mature mRNAs are produced from the same primary transcript.

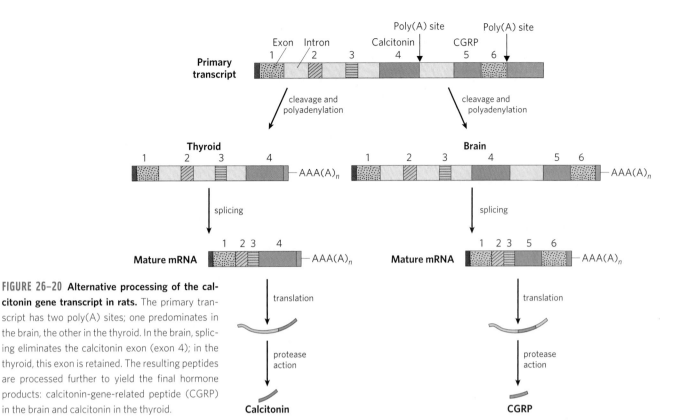

FIGURE 26–20 Alternative processing of the calcitonin gene transcript in rats. The primary transcript has two poly(A) sites; one predominates in the brain, the other in the thyroid. In the brain, splicing eliminates the calcitonin exon (exon 4); in the thyroid, this exon is retained. The resulting peptides are processed further to yield the final hormone products: calcitonin-gene-related peptide (CGRP) in the brain and calcitonin in the thyroid.

RNA transcript is processed differently to produce two different hormones: the calcium-regulating hormone calcitonin in rat thyroid and calcitonin-gene-related peptide (CGRP) in rat brain **(Fig. 26–20)**. There are many additional patterns of alternative splicing **(Fig. 26–21)**. Many, perhaps most, of the genes in mammalian genomes are subject to alternative splicing, substantially increasing the number of proteins encoded by the genes. The same processes play a much smaller role in lower eukaryotes, with only a few genes subject to alternative splicing in yeast.

Ribosomal RNAs and tRNAs Also Undergo Processing

Posttranscriptional processing is not limited to mRNA. Ribosomal RNAs of bacterial, archaeal, and eukaryotic cells are made from longer precursors called **preribosomal RNAs**, or pre-rRNAs. Transfer RNAs are similarly derived from longer precursors. These RNAs may also contain a variety of modified nucleosides; some examples are shown in **Figure 26–22**.

Ribosomal RNAs In bacteria, 16S, 23S, and 5S rRNAs (and some tRNAs, although most tRNAs are encoded elsewhere) arise from a single 30S RNA precursor of

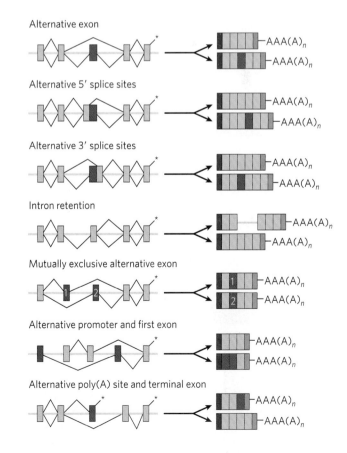

FIGURE 26–21 Summary of splicing patterns. Exons are shown in shades of green, and introns/untranslated regions as yellow lines. Positions where polyadenosine is to be added are marked with asterisks. Exons joined in a particular splicing scheme are linked with black lines. The alternative linkage patterns above and below the transcript lead to the top and bottom spliced mRNA products, respectively. In the products, red and orange boxes represent the 5′ cap and 3′ untranslated regions, respectively.

FIGURE 26–22 Some modified bases of rRNAs and tRNAs, produced in posttranscriptional reactions. The standard symbols are shown in parentheses. Note the unusual ribose attachment point in pseudouridine. This is just a small sampling of the 96 modified nucleosides known to occur in different RNA species, with 81 different types known in tRNAs and 30 observed to date in rRNAs. A complete listing of these modified bases can be found in the RNA modification database (http://rna-mdb.cas.albany.edu/RNAmods/).

about 6,500 nucleotides. RNA at both ends of the 30S precursor and segments between the rRNAs are removed during processing **(Fig. 26–23)**. The 16S and 23S rRNAs contain modified nucleosides. In *E. coli*, the 11 modifications in the 16S rRNA include a pseudouridine and 10 nucleosides methylated on the base or the 2′-hydroxyl group or both. The 23S rRNA has 10 pseudouridines, 1 dihydrouridine, and 12 methylated nucleosides. In bacteria, each modification is generally catalyzed by a distinct enzyme. Methylation reactions use *S*-adenosylmethionine as cofactor. No cofactor is required for pseudouridine formation.

The genome of *E. coli* encodes seven pre-rRNA molecules. All of these genes have essentially identical rRNA-coding regions, but they differ in the segments between these regions. The segment between the 16S and 23S rRNA genes generally encodes one or two tRNAs, with different tRNAs produced from different pre-rRNA transcripts. Coding sequences for tRNAs are also found on the 3′ side of the 5S rRNA in some precursor transcripts.

The situation in eukaryotes is more complicated. A 45S pre-rRNA transcript is synthesized by RNA polymerase I and processed in the nucleolus to form the

FIGURE 26–23 Processing of pre-rRNA transcripts in bacteria. ❶ Before cleavage, the 30S RNA precursor is methylated at specific bases (red tick marks), and some uridine residues are converted to pseudouridine (blue tick marks) or dihydrouridine (black tick mark) residues. The methylation reactions are of multiple types, some occurring on bases and some on 2′-hydroxyl groups. ❷ Cleavage liberates precursors of rRNAs and tRNA(s). Cleavage at the points labeled 1, 2, and 3 is carried out by the enzymes RNase III, RNase P, and RNase E, respectively. As discussed later in the text, RNase P is a ribozyme. ❸ The final 16S, 23S, and 5S rRNA products result from the action of a variety of specific nucleases. The seven copies of the gene for pre-rRNA in the *E. coli* chromosome differ in the number, location, and identity of tRNAs included in the primary transcript. Some copies of the gene have additional tRNA gene segments between the 16S and 23S rRNA segments and at the far 3′ end of the primary transcript.

18S, 28S, and 5.8S rRNAs characteristic of eukaryotic ribosomes **(Fig. 26–24)**. As in bacteria, the processing includes cleavage reactions mediated by endo- or exoribonucleases and nucleoside modification reactions. Some pre-rRNAs also include introns that must be spliced. The entire process is initiated in the nucleolus, in large complexes that assemble on the rRNA precursor as it is synthesized by Pol I. There is a tight coupling between rRNA transcription, rRNA maturation, and

FIGURE 26–24 Processing of pre-rRNA transcripts in vertebrates. During transcription, the 45S primary transcript is incorporated into a nucleolar 90S preribosomal complex, in which rRNA processing and ribosome assembly are tightly coupled. ❶ The 45S precursor is methylated at more than 100 of its 14,000 nucleotides, either on the bases or on the 2′-OH groups (red ticks), some uridines are converted to pseudouridine (blue ticks), and a few other modifications occur (green ticks are dihydrouridine). ❷ and ❸ A series of enzymatic cleavages of the 45S precursor produces the 18S, 5.8S, and 28S rRNAs, and the ribosomal subunits gradually take shape with the assembling ribosomal proteins. The cleavage reactions and all of the modifications require small nucleolar RNAs (snoRNAs) found in protein complexes (snoRNPs) in the nucleolus that are reminiscent of spliceosomes. The 5S rRNA is produced separately.

ribosome assembly in the nucleolus. Each complex includes the ribonucleases that cleave the rRNA precursor, the enzymes that modify particular bases, large numbers of **small nucleolar RNAs**, or **snoRNAs**, that guide nucleoside modification and some cleavage reactions, and ribosomal proteins. In yeast, the entire process involves the pre-rRNA, more than 170 nonribosomal proteins, snoRNAs for each nucleoside modification (about 70 in all, since some snoRNAs guide two types of modification), and the 78 ribosomal proteins. Humans have an even greater number of modified nucleosides, about 200, and a greater number of associated snoRNAs. The composition of the complexes may change as the ribosomes are assembled, and many of the intermediate complexes may rival the ribosome itself, and the snRNPs, in complexity. The 5S rRNA of most eukaryotes is made as a completely separate transcript by a different polymerase (Pol III).

The most common nucleoside modifications in eukaryotic rRNAs are, again, conversion of uridine to pseudouridine and adoMet-dependent nucleoside methylation (often at 2′-hydroxyl groups). These reactions rely on snoRNA-protein complexes, or **snoRNPs**, each consisting of a snoRNA and four or five proteins, which include the enzyme that carries out the modification. There are two classes of snoRNPs, both defined by key conserved sequence elements referred to as lettered boxes. The box H/ACA snoRNPs are involved in pseudouridylylation, and box C/D snoRNPs function in 2′-*O*-methylations. Unlike the situation in bacteria, the same enzyme may participate in modifications at many sites, guided by the snoRNAs.

The snoRNAs are 60 to 300 nucleotides long. Many are encoded within the introns of other genes and cotranscribed with those genes. Each snoRNA includes a 10 to 21 nucleotide sequence that is perfectly complementary to some site on an rRNA. The conserved sequence elements in the remainder of the snoRNA fold into structures that are bound by the snoRNP proteins **(Fig. 26–25)**.

Transfer RNAs Most cells have 40 to 50 distinct tRNAs, and eukaryotic cells have multiple copies of many of the tRNA genes. Transfer RNAs are derived from longer RNA precursors by enzymatic removal of nucleotides from the 5′ and 3′ ends **(Fig. 26–26)**. In eukaryotes, introns are present in a few tRNA transcripts and must be excised. Where two or more different tRNAs are contained in a single primary transcript, they are separated by enzymatic cleavage. The endonuclease RNase P, found in all organisms, removes RNA at the 5′ end of tRNAs. This enzyme contains both protein and RNA. The RNA component is essential for activity, and in bacterial cells it can carry out its processing function with precision even without the protein component. RNase P is therefore another example of a catalytic RNA, as described in more detail below. The 3′ end of tRNAs is processed by one or more nucleases, including the exonuclease RNase D.

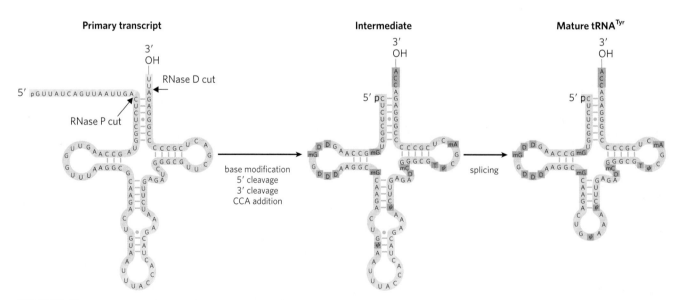

FIGURE 26–25 The function of snoRNAs in guiding rRNA modification. (a) RNA pairing with box C/D snoRNAs to guide methylation reactions. The methylation sites in the target rRNA (dark green) are in the regions paired with the C/D snoRNA. The highly conserved C and D (and C' and D') box sequences are binding sites for proteins that make up the larger snoRNP. **(b)** RNA pairing with box H/ACA snoRNAs to guide pseudouridylylations. The pseudouridine conversion sites in the target rRNA (green segments) are again in the regions paired with the snoRNA, and the conserved H/ACA box sequences are protein-binding sites.

Transfer RNA precursors may undergo further posttranscriptional processing. The 3'-terminal trinucleotide CCA(3') to which an amino acid is attached during protein synthesis (Chapter 27) is absent from some bacterial and all eukaryotic tRNA precursors and is added during processing (Fig. 26–26). This addition is carried out by tRNA nucleotidyltransferase, an unusual enzyme that binds the three ribonucleoside triphosphate precursors in separate active sites and catalyzes formation of the phosphodiester bonds to produce the

CCA(3') sequence. The creation of this defined sequence of nucleotides is therefore not dependent on a DNA or RNA template—the template is the binding site of the enzyme.

The final type of tRNA processing is the modification of some bases by methylation, deamination, or reduction (Fig. 26–22). In the case of pseudouridine, the base (uracil) is removed and reattached to the sugar through C-5. Some of these modified bases occur at characteristic positions in all tRNAs (Fig. 26–26).

FIGURE 26–26 Processing of tRNAs in bacteria and eukaryotes. The yeast tRNA^Tyr (the tRNA specific for tyrosine binding; see Chapter 27) is used to illustrate the important steps. The nucleotide sequences shown in yellow are removed from the primary transcript. The ends are processed first, the 5' end before the 3' end. CCA is then added to the 3' end, a necessary step in processing eukaryotic tRNAs and those bacterial tRNAs that lack this sequence in the primary transcript. While the ends are being processed, specific bases in the rest of the transcript are modified (see Fig. 26–22). For the eukaryotic tRNA shown here, the final step is splicing of the 14 nucleotide intron. Introns are found in some eukaryotic tRNAs but not in bacterial tRNAs.

Special-Function RNAs Undergo Several Types of Processing

The number of known classes of special-function RNAs is expanding rapidly, as is the variety of functions known to be associated with them. Many of these RNAs undergo processing.

The snRNAs and snoRNAs not only facilitate RNA processing reactions but are themselves synthesized as larger precursors and then processed. Many snoRNAs are encoded within the introns of other genes. As the introns are spliced from the pre-mRNA, the snoRNP proteins bind to the snoRNA sequences and ribonucleases remove the extra RNA at the 5′ and 3′ ends. The snRNAs destined for spliceosomes are synthesized as pre-snRNAs by RNA polymerase II, and ribonucleases remove the extra RNA at each end. Particular nucleosides in snRNAs are also subject to 11 types of modification, with 2′-O-methylation and conversion of uridine to pseudouridine predominating.

MicroRNAs (miRNAs) are a special class of RNAs involved in gene regulation. They are noncoding RNAs, about 22 nucleotides long, complementary in sequence to particular regions of mRNAs. They regulate mRNA function by cleaving the mRNA or suppressing its translation. The miRNAs are found in multicellular eukaryotes ranging from worms and fruit flies to plants and mammals. Up to 1% of the human genome may encode miRNAs, and miRNAs may target up to one-third of human mRNAs. Their function in gene regulation is described in Chapter 28.

The miRNAs are synthesized from much larger precursors, in several steps **(Fig. 26–27)**. The primary transcripts for miRNAs (pri-miRNAs) vary greatly in size; some are encoded in the introns of other genes and are coexpressed with these host genes. Their roles in gene regulation also are detailed in Chapter 28.

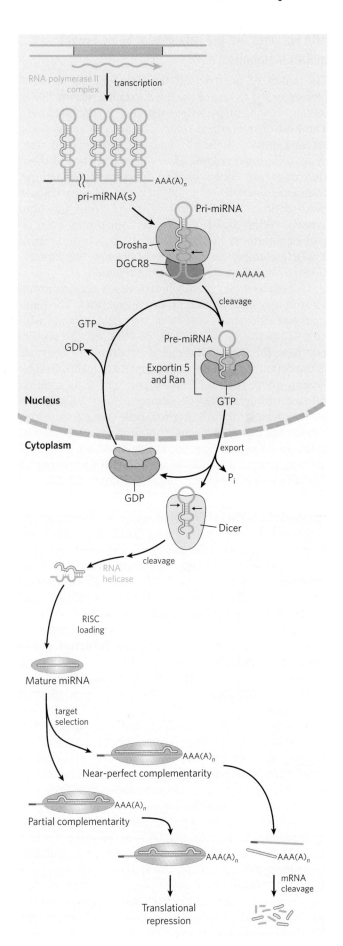

FIGURE 26–27 Synthesis and processing of miRNAs. The primary transcript of miRNAs is larger RNA of variable length, a pri-miRNA. Much of its processing is mediated by two endoribonucleases in the RNase III family, Drosha and Dicer. First, in the nucleus, the pri-miRNA is reduced to a 70 to 80 nucleotide precursor miRNA (pre-miRNA) by a protein complex including Drosha and another protein, DGCR8. The pre-miRNA is then exported to the cytoplasm in a complex with a protein called exportin-5 and the Ran GTPase (see Fig. 27–42). In the cytoplasm, Ran hydrolyzes the GTP, then exportin-5 protein and the pre-miRNA are released. The Ran-GDP and exportin-5 proteins are transported back into the nucleus. The pre-miRNA is acted on by Dicer to produce the nearly mature miRNA paired with a short RNA complement. The complement is removed by an RNA helicase, and the mature miRNA is incorporated into protein complexes, such as the RNA-induced silencing complex (RISC), which then bind a target mRNA. If the complementarity between miRNA and its target is nearly perfect, the target mRNA is cleaved. If the complementarity is only partial, the complex blocks translation of the target mRNA.

RNA Enzymes Are the Catalysts of Some Events in RNA Metabolism

The study of posttranscriptional processing of RNA molecules led to one of the most exciting discoveries in modern biochemistry—the existence of RNA enzymes. The best-characterized ribozymes are the self-splicing group I introns, RNase P, and the hammerhead ribozyme (discussed below). Most of the activities of these ribozymes are based on two fundamental reactions: transesterification (Fig. 26–13) and phosphodiester bond hydrolysis (cleavage). The substrate for ribozymes is often an RNA molecule, and it may even be part of the ribozyme itself. When its substrate is RNA, the RNA catalyst can make use of base-pairing interactions to align the substrate for the reaction.

Ribozymes vary greatly in size. A self-splicing group I intron may have more than 400 nucleotides. The hammerhead ribozyme consists of two RNA strands with only 41 nucleotides in all **(Fig. 26–28)**. As with protein enzymes, the three-dimensional structure of ribozymes is important for function. Ribozymes are inactivated by heating above their melting temperature or by addition of denaturing agents or complementary oligonucleotides, which disrupt normal base-pairing patterns. Ribozymes can also be inactivated if essential nucleotides are changed. The secondary structure of a self-splicing group I intron from the 26S rRNA precursor of *Tetrahymena* is shown in detail in **Figure 26–29**.

Enzymatic Properties of Group I Introns Self-splicing group I introns share several properties with enzymes besides accelerating the reaction rate, including their kinetic behavior and their specificity. Binding of the guanosine cofactor (Fig. 26–13) to the *Tetrahymena* group I rRNA intron is saturable ($K_m \approx 30\ \mu M$) and can be competitively inhibited by 3′-deoxyguanosine. The intron is very precise in its excision reaction, largely due to a segment called the **internal guide sequence** that can base-pair with exon sequences near the 5′ splice site (Fig. 26–29). This pairing promotes the alignment of specific bonds to be cleaved and rejoined.

Because the intron itself is chemically altered during the splicing reaction—its ends are cleaved—it may seem to lack one key enzymatic property: the ability to catalyze multiple reactions. Closer inspection has shown that after excision, the 414 nucleotide intron from *Tetrahymena* rRNA can, in vitro, act as a true enzyme (but in vivo it is quickly degraded). A series of intramolecular cyclization and cleavage reactions in the excised intron leads to the loss of 19 nucleotides from its 5′ end. The remaining 395 nucleotide, linear RNA—referred to as L-19 IVS (*intervening sequence*)—promotes nucleotidyl transfer reactions in which some oligonucleotides are lengthened at the expense of others **(Fig. 26–30)**. The best substrates are oligonucleotides, such as a synthetic $(C)_5$ oligomer, that can base-pair with the same

(a)

(b)

FIGURE 26–28 Hammerhead ribozyme. Certain viruslike elements, or virusoids, have small RNA genomes and usually require another virus to assist in their replication or packaging or both. Some virusoid RNAs include small segments that promote site-specific RNA cleavage reactions associated with replication. These segments are called hammerhead ribozymes, because their secondary structures are shaped like the head of a hammer. Hammerhead ribozymes have been defined and studied separately from the much larger viral RNAs. **(a)** The minimal sequences required for catalysis by the ribozyme. The boxed nucleotides are highly conserved and are required for catalytic function. The arrow indicates the site of self-cleavage. **(b)** Three-dimensional structure (PDB ID 1MME; see Fig. 8–25b for a space-filling view). The strands are colored as in (a). The hammerhead ribozyme is a metalloenzyme; Mg^{2+} ions are required for activity in vivo. The phosphodiester bond at the site of self-cleavage is indicated by an arrow. 🖱 Hammerhead Ribozyme

guanylate-rich internal guide sequence that held the 5′ exon in place for self-splicing.

The enzymatic activity of the L-19 IVS ribozyme results from a cycle of transesterification reactions mechanistically similar to self-splicing. Each ribozyme molecule can process about 100 substrate molecules per hour and is not altered in the reaction; therefore the intron acts as a catalyst. It follows Michaelis-Menten kinetics, is specific for RNA oligonucleotide substrates, and can be competitively inhibited. The k_{cat}/K_m (specificity constant) is $10^3\ M^{-1}s^{-1}$, lower than that of many enzymes, but the ribozyme accelerates hydrolysis by a factor of 10^{10} relative

FIGURE 26–29 Secondary structure of the self-splicing rRNA intron of Tetrahymena. (a) Intron sequences are shaded yellow and light red, exon sequences green. Each thin, light red line represents a bond between neighboring nucleotides in a continuous sequence (a device necessitated by showing this complex molecule in two dimensions). Short blue lines represent normal base pairing; blue dots indicate G–U base pairs. All nucleotides are shown. The catalytic core of the self-splicing activity is shaded in gray. Some base-paired regions are labeled (P1, P3, P2.1, P5a, and so forth) according to an established convention for this RNA molecule. The P1 region, which contains the internal guide sequence (purple), is the location of the 5′ splice site (black arrow). Part of the internal guide sequence pairs with the end of the 3′ exon, bringing the 5′ and 3′ splice sites (black arrows) into close proximity. **(b)** (PDB ID 1GID) The three-dimensional structure of the P4–P6 domain, shown in yellow in (a). **(c)** (PDB ID 1U6B) The three-dimensional structure of most of the remainder of the intron, shown in light red in (a).

to the uncatalyzed reaction. It makes use of substrate orientation, covalent catalysis, and metal-ion catalysis—strategies used by protein enzymes.

Characteristics of Other Ribozymes *E. coli* RNase P has both an RNA component (the M1 RNA, with 377 nucleotides) and a protein component (M_r 17,500). In 1983 Sidney Altman and Norman Pace and their coworkers discovered that under some conditions, the M1 RNA alone is capable of catalysis, cleaving tRNA precursors at the correct position. The protein component apparently serves to stabilize the RNA or facilitate its function in vivo. The RNase P ribozyme recognizes the three-dimensional shape of its pre-tRNA substrate, along with the CCA sequence, and thus can cleave the 5′ leaders from diverse tRNAs (Fig. 26–26).

The known catalytic repertoire of ribozymes continues to expand. Some virusoids, small RNAs associated with plant RNA viruses, include a structure that promotes a self-cleavage reaction; the hammerhead ribozyme illustrated in Figure 26–28 is in this class, catalyzing the hydrolysis of an internal phosphodiester bond. The splicing reaction that occurs in a spliceosome seems to rely on a catalytic center formed by the U2, U5, and U6 snRNAs (Fig. 26–16). And perhaps most important, an RNA component of ribosomes catalyzes the synthesis of proteins (Chapter 27).

Exploring catalytic RNAs has provided new insights into catalytic function in general and has important implications for our understanding of the origin and evolution of life on this planet, a topic discussed in Section 26.3.

FIGURE 26–30 In vitro catalytic activity of L-19 IVS. (a) L-19 IVS is generated by the autocatalytic removal of 19 nucleotides from the 5′ end of the spliced *Tetrahymena* intron. The cleavage site is indicated by the arrow in the internal guide sequence (boxed). The G residue (shaded light red) added in the first step of the splicing reaction (see Fig. 26–14) is part of the removed sequence. A portion of the internal guide sequence remains at the 5′ end of L-19 IVS. **(b)** L-19 IVS lengthens some RNA oligonucleotides at the expense of others in a cycle of transesterification reactions (steps ❶ through ❹). The 3′ OH of the G residue at the 3′ end of L-19 IVS plays a key role in this cycle (note that this is *not* the G residue added in the splicing reaction). (C)₅ is one of the ribozyme's better substrates because it can base-pair with the guide sequence remaining in the intron. Although this catalytic activity is probably irrelevant to the cell, it has important implications for current hypotheses on evolution, discussed at the end of this chapter.

Cellular mRNAs Are Degraded at Different Rates

The expression of genes is regulated at many levels. A crucial factor governing a gene's expression is the cellular concentration of its associated mRNA. The concentration of any molecule depends on two factors: its rate of synthesis and its rate of degradation. When synthesis and degradation of an mRNA are balanced, the concentration of the mRNA remains in a steady state. A change in either rate will lead to net accumulation or depletion of the mRNA. Degradative pathways ensure that mRNAs do not build up in the cell and direct the synthesis of unnecessary proteins.

The rates of degradation vary greatly for mRNAs from different eukaryotic genes. For a gene product that is needed only briefly, the half-life of its mRNA may be only minutes or even seconds. Gene products needed constantly by the cell may have mRNAs that are stable over many cell generations. The average half-life of the mRNAs of a vertebrate cell is about 3 hours, with the pool of each type of mRNA turning over about 10 times per cell generation. The half-life of bacterial mRNAs is much shorter—only about 1.5 min—perhaps because of regulatory requirements.

Messenger RNA is degraded by ribonucleases present in all cells. In *E. coli*, the process begins with one or several cuts by an endoribonuclease, followed by 3′→5′ degradation by exoribonucleases. In lower eukaryotes, the major pathway involves first shortening the poly(A) tail, then decapping the 5′ end and degrading the mRNA in the 5′→3′ direction. A 3′→5′ degradative pathway also exists and may be the major path in higher eukaryotes. All eukaryotes have a complex of up to 10 conserved 3′→5′ exoribonucleases, called the **exosome**, which is involved in the processing of the 3′ end of rRNAs, tRNAs, and some special-function RNAs (including snRNAs and snoRNAs), as well as the degradation of mRNAs.

A hairpin structure in bacterial mRNAs with a ρ-independent terminator (Fig. 26–7) confers stability against degradation. Similar hairpin structures can make some parts of a primary transcript more stable, leading to nonuniform degradation of transcripts. In eukaryotic cells, both the 3′ poly(A) tail and the 5′ cap are important to the stability of many mRNAs. 🖱 **Life Cycle of an mRNA**

Polynucleotide Phosphorylase Makes Random RNA-like Polymers

In 1955, Marianne Grunberg-Manago and Severo Ochoa discovered the bacterial enzyme **polynucleotide phosphorylase**, which in vitro catalyzes the reaction

$$(NMP)_n + NDP \rightleftharpoons \underset{\substack{\text{Lengthened}\\\text{polynucleotide}}}{(NMP)_{n+1}} + P_i$$

Polynucleotide phosphorylase was the first nucleic acid–synthesizing enzyme discovered (Arthur Kornberg's discovery of DNA polymerase followed soon thereafter). The reaction catalyzed by polynucleotide phosphorylase differs fundamentally from the polymerase activities discussed so far in that it is not template-dependent. The enzyme uses the 5′-diphosphates of ribonucleosides as substrates and cannot act on the homologous 5′-triphosphates or on deoxyribonucleoside 5′-diphosphates. The RNA polymer formed by polynucleotide phosphorylase contains the usual 3′,5′-phosphodiester linkages, which can be hydrolyzed by ribonuclease. The reaction is readily reversible and can be pushed in the direction of breakdown of the polyribonucleotide by increasing the phosphate concentration. The probable function of this enzyme in the cell is the degradation of mRNAs to nucleoside diphosphates.

Marianne Grunberg-Manago   Severo Ochoa, 1905–1993

Because the polynucleotide phosphorylase reaction does not use a template, the polymer it forms does not have a specific base sequence. The reaction proceeds equally well with any or all of the four nucleoside diphosphates, and the base composition of the resulting polymer reflects nothing more than the relative concentrations of the 5′-diphosphate substrates in the medium.

Polynucleotide phosphorylase can be used in the laboratory to prepare RNA polymers with many different base sequences and frequencies. Synthetic RNA polymers of this sort were critical for deducing the genetic code for the amino acids (Chapter 27).

SUMMARY 26.2 RNA Processing

▶ Eukaryotic mRNAs are modified by addition of a 7-methylguanosine residue at the 5′ end and by cleavage and polyadenylation at the 3′ end to form a long poly(A) tail.

▶ Many primary mRNA transcripts contain introns (noncoding regions), which are removed by splicing. Excision of the group I introns found in some rRNAs requires a guanosine cofactor. Some group I and group II introns are capable of self-splicing; no protein enzymes are required. Nuclear mRNA precursors have a third (the largest) class of introns, which are spliced with the aid of RNA-protein complexes called snRNPs, assembled into spliceosomes. A fourth class of introns, found in some tRNAs, consists of the only introns known to be spliced by protein enzymes.

▶ The function of many eukaryotic mRNAs is regulated by complementary microRNAs (miRNAs). The miRNAs are themselves derived from larger precursors through a series of processing reactions.

▶ Ribosomal RNAs and transfer RNAs are derived from longer precursor RNAs, trimmed by nucleases. Some bases are modified enzymatically during the maturation process. Some nucleoside modifications are guided by snoRNAs, within protein complexes called snoRNPs.

▶ The self-splicing introns and the RNA component of RNase P (which cleaves the 5′ end of tRNA precursors) are two examples of ribozymes. These biological catalysts have the properties of true enzymes. They generally promote hydrolytic cleavage and transesterification, using RNA as substrate. Combinations of these reactions can be promoted by the excised group I intron of *Tetrahymena* rRNA, resulting in a type of RNA polymerization reaction.

▶ Polynucleotide phosphorylase reversibly forms RNA-like polymers from ribonucleoside 5′-diphosphates, adding or removing ribonucleotides at the 3′-hydroxyl end of the polymer. The enzyme degrades RNA in vivo.

26.3 RNA-Dependent Synthesis of RNA and DNA

In our discussion of DNA and RNA synthesis up to this point, the role of the template strand has been reserved for DNA. However, some enzymes use an RNA template for nucleic acid synthesis. With the very important exception of viruses with an RNA genome, these enzymes play only a modest role in information pathways. RNA viruses are the source of most RNA-dependent polymerases characterized so far.

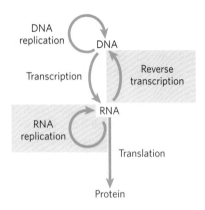

FIGURE 26–31 Extension of the central dogma to include RNA-dependent synthesis of RNA and DNA.

The existence of RNA replication requires an elaboration of the central dogma (**Fig. 26–31**; contrast this with the diagram on p. 977). The enzymes involved in RNA replication have profound implications for investigations into the nature of self-replicating molecules that may have existed in prebiotic times.

Reverse Transcriptase Produces DNA from Viral RNA

Certain RNA viruses that infect animal cells carry within the viral particle an RNA-dependent DNA polymerase called **reverse transcriptase**. On infection, the single-stranded RNA viral genome (~10,000 nucleotides) and the enzyme enter the host cell. The reverse transcriptase first catalyzes the synthesis of a DNA strand complementary to the viral RNA (**Fig. 26–32**), then degrades the RNA strand of the viral RNA-DNA hybrid and replaces it with DNA. The resulting duplex DNA often becomes incorporated into the genome of the eukaryotic host cell. These integrated (and dormant)

FIGURE 26–32 Retroviral infection of a mammalian cell and integration of the retrovirus into the host chromosome. Viral particles entering the host cell carry viral reverse transcriptase and a cellular tRNA (picked up from a former host cell) already base-paired to the viral RNA. The purple segments represent the long terminal repeats on the viral RNA. The tRNA facilitates immediate conversion of viral RNA to double-stranded DNA by the action of reverse transcriptase, as described in the text. Once converted to double-stranded DNA, the DNA enters the nucleus and is integrated into the host genome. The integration is catalyzed by a virally encoded integrase. Integration of viral DNA into host DNA is mechanis-

tically similar to the insertion of transposons in bacterial chromosomes (see Fig. 25–41). For example, a few base pairs of host DNA become duplicated at the site of integration, forming short repeats of 4 to 6 bp at each end of the inserted retroviral DNA (not shown). On transcription and translation of the integrated viral DNA, new viruses are formed and released by cell lysis (right). In the viruses, the viral RNA is enclosed by capsid proteins called Gag and outer envelope proteins called Env. Additional viral proteins (reverse transcriptase, integrase, and a viral protease needed for posttranslational processing of viral proteins) are packaged within the virus particle with the RNA.

viral genes can be activated and transcribed, and the gene products—viral proteins and the viral RNA genome itself—packaged as new viruses. The RNA viruses that contain reverse transcriptases are known as **retroviruses** (*retro* is the Latin prefix for "backward").

The existence of reverse transcriptases in RNA viruses was predicted by Howard Temin in 1962, and the enzymes were ultimately detected by Temin and, independently, by David Baltimore in 1970. Their discovery aroused much attention as dogma-shaking proof that genetic information can flow "backward" from RNA to DNA.

Howard Temin, 1934–1994

David Baltimore

Retroviruses typically have three genes: *gag* (a name derived from the historical designation *g*roup *a*ssociated anti*g*en), *pol*, and *env* (**Fig. 26–33**). The transcript that contains *gag* and *pol* is translated into a long "polyprotein," a single large polypeptide that is cleaved into six proteins with distinct functions. The proteins derived from the *gag* gene make up the interior core of the viral particle. The *pol* gene encodes the protease that cleaves the long polypeptide, an integrase that inserts the viral DNA into the host chromosomes, and reverse transcriptase. Many reverse transcriptases have two subunits, α and β. The *pol* gene specifies the β subunit (M_r 90,000), and the α subunit (M_r 65,000) is simply a proteolytic fragment of the β subunit. The *env* gene encodes the proteins of the viral envelope. At each end of the linear RNA genome are long terminal repeat (LTR) sequences of a few hundred nucleotides. Transcribed into the duplex DNA, these sequences facilitate integration of the viral chromosome into the host DNA and contain promoters for viral gene expression.

Reverse transcriptases catalyze three different reactions: (1) RNA-dependent DNA synthesis, (2) RNA degradation, and (3) DNA-dependent DNA synthesis. Like many DNA and RNA polymerases, reverse transcriptases contain Zn^{2+}. Each transcriptase is most active with the RNA of its own virus, but each can be used experimentally to make DNA complementary to a variety of RNAs. The DNA and RNA synthesis and RNA degradation activities use separate active sites on the protein. For DNA synthesis to begin, the reverse transcriptase requires a primer, a cellular tRNA obtained during an earlier infection and carried in the viral particle. This

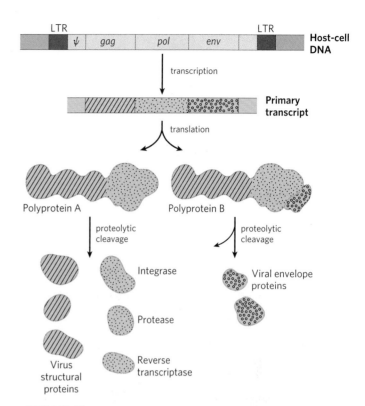

FIGURE 26–33 Structure and gene products of an integrated retroviral genome. The long terminal repeats (LTRs) have sequences needed for the regulation and initiation of transcription. The sequence denoted ψ is required for packaging of retroviral RNAs into mature viral particles. Transcription of the retroviral DNA produces a primary transcript encompassing the *gag*, *pol*, and *env* genes. Translation (Chapter 27) produces a polyprotein, a single long polypeptide derived from the *gag* and *pol* genes, which is cleaved into six distinct proteins. Splicing of the primary transcript yields an mRNA derived largely from the *env* gene, which is also translated into a polyprotein, then cleaved to generate viral envelope proteins.

tRNA is base-paired at its 3' end with a complementary sequence in the viral RNA. The new DNA strand is synthesized in the 5'→3' direction, as in all RNA and DNA polymerase reactions. Reverse transcriptases, like RNA polymerases, do not have 3'→5' proofreading exonucleases. They generally have error rates of about 1 per 20,000 nucleotides added. An error rate this high is extremely unusual in DNA replication and seems to be a feature of most enzymes that replicate the genomes of RNA viruses. A consequence is a higher mutation rate and faster rate of viral evolution, which is a factor in the frequent appearance of new strains of disease-causing retroviruses.

Reverse transcriptases have become important reagents in the study of DNA-RNA relationships and in DNA cloning techniques. They make possible the synthesis of DNA complementary to an mRNA template, and synthetic DNA prepared in this manner, called **complementary DNA (cDNA)**, can be used to clone cellular genes (see Fig. 9–14).

FIGURE 26–34 Rous sarcoma virus genome. The *src* gene encodes a tyrosine kinase, one of a class of enzymes that function in systems affecting cell division, cell-cell interactions, and intercellular communication (Chapter 12). The same gene is found in chicken DNA (the usual host for this virus) and in the genomes of many other eukaryotes, including humans. When associated with the Rous sarcoma virus, this oncogene is often expressed at abnormally high levels, contributing to unregulated cell division and cancer.

FIGURE 26–35 The genome of HIV, the virus that causes AIDS. In addition to the typical retroviral genes, HIV contains several small genes with a variety of functions (not identified here and not all known). Some of these genes overlap. Alternative splicing mechanisms produce many different proteins from this small (9.7×10^3 nucleotides) genome.

Some Retroviruses Cause Cancer and AIDS

Retroviruses have featured prominently in recent advances in the molecular understanding of cancer. Most retroviruses do not kill their host cells but remain integrated in the cellular DNA, replicating when the cell divides. Some retroviruses, classified as RNA tumor viruses, contain an oncogene that can cause the cell to grow abnormally. The first retrovirus of this type to be studied was the Rous sarcoma virus (also called avian sarcoma virus; **Fig. 26–34**), named for F. Peyton Rous, who studied chicken tumors now known to be caused by this virus. Since the initial discovery of oncogenes by Harold Varmus and Michael Bishop, many dozens of such genes have been found in retroviruses.

The human immunodeficiency virus (HIV), which causes acquired immune deficiency syndrome (AIDS), is a retrovirus. Identified in 1983, HIV has an RNA genome with standard retroviral genes along with several other unusual genes **(Fig. 26–35)**. Unlike many other retroviruses, HIV kills many of the cells it infects (principally T lymphocytes) rather than causing tumor formation. This gradually leads to suppression of the immune system in the host organism. The reverse transcriptase of HIV is even more error-prone than other known reverse transcriptases—10 times more so—resulting in high mutation rates in this virus. One or more errors are generally made every time the viral genome is replicated, so any two viral RNA molecules are likely to differ.

Many modern vaccines for viral infections consist of one or more coat proteins of the virus, produced by methods described in Chapter 9. These proteins are not infectious on their own but stimulate the immune system to recognize and resist subsequent viral invasions (Chapter 5). Because of the high error rate of the HIV reverse transcriptase, the *env* gene in this virus (along with the rest of the genome) undergoes very rapid mutation, complicating the development of an effective vaccine. However, repeated cycles of cell invasion and replication are needed to propagate an HIV infection, so inhibition of viral enzymes offers the most effective therapy currently available. The HIV protease is targeted by a class of drugs called protease inhibitors (see Fig. 6–24). Reverse transcriptase is the target of some additional drugs widely used to treat HIV-infected individuals (Box 26–2). ■

Many Transposons, Retroviruses, and Introns May Have a Common Evolutionary Origin

Some well-characterized eukaryotic DNA transposons from sources as diverse as yeast and fruit flies have a structure very similar to that of retroviruses; these are sometimes called retrotransposons **(Fig. 26–36)**. Retrotransposons encode an enzyme homologous to the retroviral reverse transcriptase, and their coding regions are flanked by LTR sequences. They transpose from one position to another in the cellular genome by means of an RNA intermediate, using reverse transcriptase to

FIGURE 26–36 Eukaryotic transposons. The Ty element of the yeast *Saccharomyces* and the copia element of the fruit fly *Drosophila* serve as examples of eukaryotic retrotransposons, which often have a structure similar to retroviruses but lack the *env* gene. The δ sequences of the Ty element are functionally equivalent to retroviral LTRs. In the copia element, *INT* and *RT* are homologous to the integrase and reverse transcriptase segments, respectively, of the *pol* gene.

BOX 26–2 ⚕ MEDICINE Fighting AIDS with Inhibitors of HIV Reverse Transcriptase

Research into the chemistry of template-dependent nucleic acid biosynthesis, combined with modern techniques of molecular biology, has elucidated the life cycle and structure of the human immunodeficiency virus, the retrovirus that causes AIDS. A few years after the isolation of HIV, this research resulted in the development of drugs capable of prolonging the lives of people infected by HIV.

The first drug to be approved for clinical use was AZT, a structural analog of deoxythymidine. AZT was first synthesized in 1964 by Jerome P. Horwitz. It failed as an anticancer drug (the purpose for which it was made), but in 1985 it was found to be a useful treatment for AIDS. AZT is taken up by T lymphocytes, immune system cells that are particularly vulnerable to

HIV infection, and converted to AZT triphosphate. (AZT triphosphate taken directly would be ineffective because it cannot cross the plasma membrane.) HIV's reverse transcriptase has a higher affinity for AZT triphosphate than for dTTP, and binding of AZT triphosphate to this enzyme competitively inhibits dTTP binding. When AZT is added to the 3′ end of the growing DNA strand, lack of a 3′ hydroxyl means that the DNA strand is terminated prematurely and viral DNA synthesis grinds to a halt.

AZT triphosphate is not as toxic to the T lymphocytes themselves because *cellular* DNA polymerases have a lower affinity for this compound than for dTTP. At concentrations of 1 to 5 μM, AZT affects HIV reverse transcription but not most cellular DNA replication. Unfortunately, AZT seems to be toxic to the bone marrow cells that are the progenitors of erythrocytes, and many individuals taking AZT develop anemia. AZT can increase the survival time of people with advanced AIDS by about a year, and it delays the onset of AIDS in those who are still in the early stages of HIV infection. Some other AIDS drugs, such as dideoxyinosine (DDI), have a similar mechanism of action. Newer drugs target and inactivate the HIV protease. Because of the high error rate of HIV reverse transcriptase and the resulting rapid evolution of HIV, the most effective treatments of HIV infection use a combination of drugs directed at both the protease and the reverse transcriptase.

3′-Azido-2′,3′-dideoxy-thymidine (AZT)

2′,3′-Dideoxyinosine (DDI)

make a DNA copy of the RNA, followed by integration of the DNA at a new site. Most transposons in eukaryotes use this mechanism for transposition, distinguishing them from bacterial transposons, which move as DNA directly from one chromosomal location to another (see Fig. 25–41).

Retrotransposons lack an *env* gene and so cannot form viral particles. They can be thought of as defective viruses, trapped in cells. Comparisons between retroviruses and eukaryotic transposons suggest that reverse transcriptase is an ancient enzyme that predates the evolution of multicellular organisms.

Interestingly, many group I and group II introns are also mobile genetic elements. In addition to their self-splicing activities, they encode DNA endonucleases that promote their movement. During genetic exchanges between cells of the same species, or when DNA is introduced into a cell by parasites or by other means, these endonucleases promote insertion of the intron into an identical site in another DNA copy of a homologous gene that does not contain the intron, in a process termed **homing (Fig. 26–37)**. Whereas group I intron homing is DNA-based, group II intron homing occurs

through an RNA intermediate. The endonucleases of the group II introns have associated reverse transcriptase activity. The proteins can form complexes with the intron RNAs themselves, after the introns are spliced from the primary transcripts. Because the homing process involves insertion of the RNA intron into DNA and reverse transcription of the intron, the movement of these introns has been called retrohoming. Over time, every copy of a particular gene in a population may acquire the intron. Much more rarely, the intron may insert itself into a new location in an unrelated gene. If this event does not kill the host cell, it can lead to the evolution and distribution of an intron in a new location. The structures and mechanisms used by mobile introns support the idea that at least some introns originated as molecular parasites whose evolutionary past can be traced to retroviruses and transposons.

Telomerase Is a Specialized Reverse Transcriptase

Telomeres, the structures at the ends of linear eukaryotic chromosomes (see Fig. 24–8), generally consist of many tandem copies of a short oligonucleotide sequence.

FIGURE 26–37 Introns that move: homing and retrohoming. Certain introns include a gene (shown in red) for enzymes that promote homing (certain group I introns) or retrohoming (certain group II introns). **(a)** The gene in the spliced intron is bound by a ribosome and translated. Group I homing introns specify a site-specific endonuclease, called a homing endonuclease. Group II retrohoming introns specify a protein with both endonuclease and reverse transcriptase activities (not shown here).

(b) Homing. Allele *a* of a gene *X* containing a group I homing intron is present in a cell containing allele *b* of the same gene, which lacks the intron. The homing endonuclease produced by *a* cleaves *b* at the position corresponding to the intron in *a*, and double-strand break repair (recombination with allele *a*; see Fig. 25–35) then creates a new copy of the intron in *b*. **(c)** Retrohoming. Allele *a* of gene *Y* contains a retrohoming group II intron; allele *b* lacks the intron. The spliced intron inserts itself into the coding strand of *b* in a reaction that is the reverse of the splicing that excised the intron from the primary transcript (see Fig. 26–15), except that here the insertion is into DNA rather than RNA. The noncoding DNA strand of *b* is then cleaved by the intron-encoded endonuclease/reverse transcriptase. This same enzyme uses the inserted RNA as a template to synthesize a complementary DNA strand. The RNA is then degraded by cellular ribonucleases and replaced with DNA.

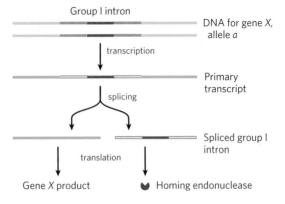

(a) Production of homing endonuclease

(b) Homing

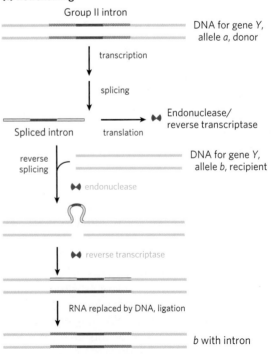

(c) Retrohoming

This sequence usually has the form T_xG_y in one strand and C_yA_x in the complementary strand, where x and y are typically in the range of 1 to 4 (p. 984). Telomeres vary in length from a few dozen base pairs in some ciliated protozoans to tens of thousands of base pairs in mammals. The TG strand is longer than its complement, leaving a region of single-stranded DNA of up to a few hundred nucleotides at the 3′ end.

The ends of a linear chromosome are not readily replicated by cellular DNA polymerases. DNA replication requires a template and primer, and beyond the end of a linear DNA molecule no template is available for the pairing of an RNA primer. Without a special mechanism for replicating the ends, chromosomes would be shortened somewhat in each cell generation. The enzyme **telomerase**, discovered by Carol Greider and Elizabeth Blackburn, solves this problem by adding telomeres to chromosome ends.

Carol Greider

Elizabeth Blackburn

The discovery and purification of this enzyme provided insight into a reaction mechanism that is remarkable and unprecedented. Telomerase, like some other enzymes described in this chapter, contains both RNA and protein components. The RNA component is about 150 nucleotides long and contains about 1.5 copies of the appropriate C_yA_x telomere repeat. This region of the RNA acts as a template for synthesis of the T_xG_y strand of the telomere. Telomerase thereby acts as a cellular reverse transcriptase that provides the active site for RNA-dependent DNA synthesis. Unlike retroviral reverse transcriptases, telomerase copies only a small segment of RNA that it carries within itself. Telomere synthesis requires the 3′ end of a chromosome as primer and proceeds in the usual 5′→3′ direction. Having synthesized one copy of the repeat, the enzyme repositions to resume extension of the telomere **(Fig. 26–38a)**.

After extension of the T_xG_y strand by telomerase, the complementary C_yA_x strand is synthesized by cellular DNA polymerases, starting with an RNA primer (see Fig. 25–12). The single-stranded region is protected by specific binding proteins in many lower eukaryotes, especially those species with telomeres of less than a few hundred base pairs. In higher eukaryotes (including mammals) with telomeres many thousands of base pairs long, the single-stranded end is sequestered in a specialized structure called a **T loop** (Fig. 26–38b). The single-stranded end is folded back and paired with its complement in the double-stranded portion of the telomere. The formation of a T loop involves invasion of the 3′ end of the telomere's single strand into the duplex

FIGURE 26–38 Telomere synthesis and structure. (a) The internal template RNA of telomerase binds to and base-pairs with the TG primer (T_xG_y) of DNA. Telomerase adds more T and G residues to the TG primer, then repositions the internal template RNA to allow the addition of more T and G residues that generate the TG strand of the telomere. The complementary strand is synthesized by cellular DNA polymerases after priming by an RNA primase. **(b)** Proposed structure of T loops in telomeres. The single-stranded tail synthesized by telomerase is folded back and paired with its complement in the duplex portion of the telomere. The telomere is bound by several telomere-binding proteins, including TRF1 and TRF2 (*telomere repeat binding factors*). **(c)** Electron micrograph of a T loop at the end of a chromosome isolated from a mouse hepatocyte. The bar at the bottom of the micrograph represents a length of 5,000 bp.

DNA, perhaps by a mechanism similar to the initiation of homologous genetic recombination (see Fig. 25–35). In mammals, the looped DNA is bound by two proteins, TRF1 and TRF2, with the latter protein involved in formation of the T loop. T loops protect the 3′ ends of chromosomes, making them inaccessible to nucleases and the enzymes that repair double-strand breaks.

In protozoans (such as *Tetrahymena*), loss of telomerase activity results in a gradual shortening of telomeres with each cell division, ultimately leading to the death of the cell line. A similar link between telomere length and cell senescence (cessation of cell division) has been observed in humans. In germ-line cells, which contain telomerase activity, telomere lengths are maintained; in somatic cells, which lack telomerase, they are not. There is a linear, inverse relationship between the length of telomeres in cultured fibroblasts and the age of the individual from whom the fibroblasts were taken: telomeres in human somatic cells gradually shorten as an individual ages. If the telomerase reverse transcriptase is introduced into human somatic cells in vitro, telomerase activity is restored and the cellular life span increases markedly.

Is the gradual shortening of telomeres a key to the aging process? Is our natural life span determined by the length of the telomeres we are born with? Further research in this area should yield some fascinating insights.

Some Viral RNAs Are Replicated by RNA-Dependent RNA Polymerase

Some *E. coli* bacteriophages, including f2, MS2, R17, and Qβ, as well as some eukaryotic viruses (including influenza and Sindbis viruses, the latter associated with a form of encephalitis) have RNA genomes. The single-stranded RNA chromosomes of these viruses, which also function as mRNAs for the synthesis of viral proteins, are replicated in the host cell by an **RNA-dependent RNA polymerase (RNA replicase)**. All RNA viruses—with the exception of retroviruses—must encode a protein with RNA-dependent RNA polymerase activity, because the host cells do not possess this enzyme.

The RNA replicase of most RNA bacteriophages has a molecular weight of ~210,000 and consists of four subunits. One subunit (M_r 65,000) is the product of the replicase gene encoded by the viral RNA and has the active site for replication. The other three subunits are host proteins normally involved in host-cell protein synthesis: the *E. coli* elongation factors Tu (M_r 45,000) and Ts (M_r 34,000) (which ferry amino acyl–tRNAs to the ribosomes) and the protein S1 (an integral part of the 30S ribosomal subunit). These three host proteins may help the RNA replicase locate and bind to the 3′ ends of the viral RNAs.

The RNA replicase isolated from Qβ-infected *E. coli* cells catalyzes the formation of an RNA complementary to the viral RNA, in a reaction equivalent to that catalyzed by DNA-dependent RNA polymerases.

New RNA strand synthesis proceeds in the 5′→3′ direction by a chemical mechanism identical to that used in all other nucleic acid synthetic reactions that require a template. RNA replicase requires RNA as its template and will not function with DNA. It lacks a separate proofreading endonuclease activity and has an error rate similar to that of RNA polymerase. Unlike the DNA and RNA polymerases, RNA replicases are specific for the RNA of their own virus; the RNAs of the host cell are generally not replicated. This explains how RNA viruses are preferentially replicated in the host cell, which contains many other types of RNA.

RNA Synthesis Offers Important Clues to Biochemical Evolution

The extraordinary complexity and order that distinguish living from inanimate systems are key manifestations of fundamental life processes. Maintaining the living state requires that *selected* chemical transformations occur very rapidly—especially those that use environmental energy sources and synthesize elaborate or specialized cellular macromolecules. Life depends on powerful and selective catalysts—enzymes—and on informational systems capable of both securely storing the blueprint for these enzymes and accurately reproducing the blueprint for generation after generation. Chromosomes encode the blueprint not for the cell but for the enzymes that construct and maintain the cell. The parallel demands for information and catalysis present a classic conundrum: what came first, the information needed to specify structure or the enzymes needed to maintain and transmit the information?

Carl Woese

Francis Crick, 1916–2004

The unveiling of the structural and functional complexity of RNA led Carl Woese, Francis Crick, and Leslie Orgel to propose in the 1960s that this macromolecule might serve as both information carrier and catalyst. The discovery of catalytic RNAs took this proposal from conjecture to hypothesis

Leslie Orgel, 1927–2007

and has led to widespread speculation that an "RNA world" might have been important in the transition from prebiotic chemistry to life (see Fig. 1–36). The parent of all life on this planet, in the sense that it could reproduce itself across the generations from the origin of life to the present, might have been a self-replicating RNA or a polymer with equivalent chemical characteristics.

How might a self-replicating polymer come to be? How might it maintain itself in an environment where the precursors for polymer synthesis are scarce? How could evolution progress from such a polymer to the modern DNA-protein world? These difficult questions can be addressed by careful experimentation, providing clues about how life on Earth began and evolved.

The probable origin of purine and pyrimidine bases is suggested by experiments designed to test hypotheses about prebiotic chemistry (pp. 33–34). Beginning with simple molecules thought to be present in the early atmosphere (CH_4, NH_3, H_2O, H_2), electrical discharges such as lightning generate, first, more reactive molecules such as HCN and aldehydes, then an array of amino acids and organic acids (see Fig. 1–34). When molecules such as HCN become abundant, purine and pyrimidine bases are synthesized in detectable amounts. Remarkably, a concentrated solution of ammonium cyanide, refluxed for a few days, generates adenine in yields of up to 0.5% **(Fig. 26–39)**. Adenine may well have been the first and most abundant nucleotide constituent to appear on Earth. Intriguingly, most enzyme cofactors contain adenosine as part of their structure, although it plays no direct role in the cofactor function (see Fig. 8–38). This may suggest an evolutionary relationship. Based on the simple synthesis of adenine from cyanide, adenine may simply have been abundant and available.

The RNA world hypothesis requires a nucleotide polymer to reproduce itself. Can a ribozyme bring about its own synthesis in a template-directed manner? Researchers are getting closer to finding such a ribozyme or ribozyme system. For example, Gerald Joyce and colleagues, in 2009, reported on the first set of two ribozymes that could cross-catalyze each other's formation **(Fig. 26–40)**. One ribozyme, E, catalyzes the joining of two oligonucleotides (A′ and B′) to create a second and complementary ribozyme called E′. E′ could then catalyze the joining of two other oligonucleotides (A and B) to form another molecule of E. In this system, the formation of E and E′ was templated, and the amounts grew exponentially as long as substrates were available and proteins were absent. The system evolved so that more efficient enzymes appeared in the population.

A more general RNA-polymerase-like ribozyme was described in 2011 by Philipp Holliger and colleagues. Indeed, the pace of discovery in this field is accelerating.

A self-replicating polymer would quickly use up available supplies of precursors provided by the relatively slow processes of prebiotic chemistry. Thus, from an early stage in evolution, metabolic pathways would be required to generate precursors efficiently, with the synthesis of precursors presumably catalyzed by ribozymes. The extant ribozymes found in nature have a limited repertoire of catalytic functions, and of the ribozymes that may once have existed, no trace is left. To explore the RNA world hypothesis more deeply, we need to know whether RNA has the potential to catalyze the many different reactions needed in a primitive system of metabolic pathways.

The search for RNAs with new catalytic functions has been aided by the development of a method that rapidly searches pools of random polymers of RNA and extracts those with particular activities: **SELEX** is nothing less than accelerated evolution in a test tube (Box 26–3). It has been used to generate RNA molecules that bind to amino acids, organic dyes, nucleotides, cyanocobalamin, and other molecules. Researchers have isolated ribozymes that catalyze ester and amide bond formation, S_N2 reactions, metallation of (addition of metal ions to) porphyrins, and carbon–carbon bond formation. The evolution of enzymatic cofactors with nucleotide "handles" that facilitate their binding to ribozymes might have further expanded the repertoire of chemical processes available to primitive metabolic systems.

As we shall see in the next chapter, some natural RNA molecules catalyze the formation of peptide bonds, offering an idea of how the RNA world might have been transformed by the greater catalytic potential of proteins. The synthesis of proteins would have been a major event in the evolution of the RNA world but would also have hastened its demise. The information-carrying role of RNA may have passed to DNA because DNA is chemically more stable. RNA replicase and reverse transcriptase may be modern versions of enzymes that once played important roles in making the transition to the modern DNA-based system.

The RNA world hypothesis is now supported by at least five lines of evidence. First, RNA catalysts exist. RNA thus clearly has the capacity to both store information and catalyze reactions. Second, the wide range of reactions for which RNA catalysts have been developed is sufficient to form the basis of a primordial metabolic system. Third, there are numerous extant RNAs that are likely vestiges of an RNA world, including ribozymes, retroviruses, RNA viruses, and retrotransposons. Fourth, an RNA catalyst is responsible for the synthesis of proteins (Chapter 27). Finally, RNA catalysts with a capacity for self-replication are coming to light in the laboratory. An active field of investigation— prebiotic chemistry (not described here)—is revealing

HCN
(NH_4CN) $\xrightarrow{\text{reflux}}$

FIGURE 26–39 Possible prebiotic synthesis of adenine from ammonium cyanide. Adenine is derived from five molecules of cyanide, denoted by shading.

(a)

(b)

FIGURE 26–40 Self-sustained replication of an RNA enzyme. This system has many of the properties of a living system. The RNA molecules incorporate information and catalytic function, and the reactions produce an exponential increase in the product RNAs. When variants of the RNA substrates are introduced, the system undergoes natural selection such that the best replicators come to dominate the population. **(a)** The reaction scheme is outlined. Oligoribonucleotides A and B anneal to ribozyme E' and are ligated catalytically to form ribozyme E. The joining of oligori-

bonucleotides A' and B' is similarly catalyzed by ribozyme E. The levels of E and E' grow exponentially, with a doubling time of about one hour at 42 °C, as long as there is a supply of the precursors A, B, A', and B'. **(b)** The ligation reaction involves attack of the 3' OH of one oligoribonucleotide on the α-phosphate of the 5'-triphosphate of the other oligoribonucleotide. Pyrophosphate is released. Base pairing of the substrates with the ribozyme plays a key role in aligning the substrates for the reaction.

plausible pathways for the appearance of nucleotide precursors in the prebiotic soup.

Molecular parasites may also have originated in an RNA world. With the appearance of the first inefficient self-replicators, transposition could have been a potentially important alternative to replication as a strategy for successful reproduction and survival. Early parasitic RNAs would simply hop into a self-replicating molecule via catalyzed transesterification and then passively undergo replication. Natural selection would have driven transposition to become site-specific, targeting sequences that did not interfere with the catalytic activities of the host RNA. Replicators and RNA transposons could have existed in a primitive symbiotic relationship, each contributing to the evolution of the other. Modern introns, retroviruses, and transposons may all be vestiges of a "piggyback" strategy pursued by early parasitic RNAs.

These elements continue to make major contributions to the evolution of their hosts.

Although the RNA world remains a hypothesis, with many gaps yet to be explained, experimental evidence supports a growing list of its key elements. Further experimentation should increase our understanding. Important clues to the puzzle will be found in the workings of fundamental chemistry, in living cells, and perhaps on other planets. Meanwhile, the extant RNA universe continues to expand (Box 26–4).

SUMMARY 26.3 RNA-Dependent Synthesis of RNA and DNA

▶ RNA-dependent DNA polymerases, also called reverse transcriptases, were first discovered in retroviruses, which must convert their RNA

BOX 26-3 | METHODS | The SELEX Method for Generating RNA Polymers with New Functions

SELEX (*systematic evolution of ligands by exponential enrichment*) is used to generate **aptamers**, oligonucleotides selected to tightly bind a specific molecular target. The process is generally automated to allow rapid identification of one or more aptamers with the desired binding specificity.

Figure 1 illustrates how SELEX is used to select an RNA species that binds tightly to ATP. In step ❶, a random mixture of RNA polymers is subjected to "unnatural selection" by passing it through a resin to which ATP is attached. The practical limit for the complexity of an RNA mixture in SELEX is about 10^{15} different sequences, which allows for the complete randomization of 25 nucleotides ($4^{25} = 10^{15}$). For longer RNAs, the RNA pool used to initiate the search does not include all possible sequences. ❷ RNA polymers that pass through the column are discarded; ❸ those that bind to ATP are washed from the column with salt solution and collected. ❹ The collected RNA polymers are amplified by reverse transcriptase to make many DNA complements to the selected RNAs; then an RNA polymerase makes many RNA complements of the resulting DNA molecules. ❺ This new pool of RNA is subjected to the same selection procedure, and the cycle is repeated a dozen or more times. At the end, only a few aptamers—in this case, RNA sequences with considerable affinity for ATP—remain.

Critical sequence features of an RNA aptamer that binds ATP are shown in Figure 2; molecules with this general structure bind ATP (and other adenosine nucleotides) with $K_d < 50$ μM. Figure 3 presents the three-dimensional structure of a 36 nucleotide RNA aptamer (shown as a complex with AMP) generated by SELEX. This RNA has the backbone structure shown in Figure 2.

In addition to its use in exploring the potential functionality of RNA, SELEX has an important practical side in identifying short RNAs with pharmaceutical uses. Finding an aptamer that binds specifically to every potential therapeutic target may be impossible, but the capacity of SELEX to rapidly select and amplify a specific oligonucleotide sequence from a highly complex pool of sequences makes this a promising approach for the generation of new therapies. For example, one could select an RNA that binds tightly to a receptor protein prominent in the plasma membrane of cells in a particular cancerous tumor. Blocking the activity of the receptor, or targeting a toxin to the tumor cells by attaching it to the aptamer, would kill the cells. SELEX also has been used to select DNA aptamers that detect anthrax spores. Many other promising applications are under development. ■

FIGURE 2 RNA aptamer that binds ATP. The shaded nucleotides are those required for the binding activity.

FIGURE 3 (Derived from PDB ID 1RAW) RNA aptamer bound to AMP. The bases of the conserved nucleotides (forming the binding pocket) are white; the bound AMP is red.

10^{15} random RNA sequences

repeat

ATP coupled to resin

RNA sequences enriched for ATP-binding function

amplify

RNA sequences that bind ATP

RNA sequences that do not bind ATP (discard)

FIGURE 1 The SELEX procedure.

BOX 26–4   An Expanding RNA Universe Filled with TUF RNAs

Current estimates for the number of genes in the human genome, and in many other genomes, are mentioned in multiple places throughout this text. The estimates presume that scientists know a gene when they see it, based on our current understanding of DNA, RNA, and proteins. Is the presumption correct?

As noted in Chapter 9, less than 2% of the human genome seems to encode proteins. Even when introns are factored in, one might expect only a tiny fraction of the genome to be transcribed into RNA, mostly mRNA to encode those proteins. The remainder of the genome has sometimes been referred to as junk DNA. The "junk" moniker simply reflects our ignorance, which is slowly giving way to the realization that most of the genome is fully functional.

In an effort to better map the boundaries of the human transcriptome, researchers have invented new tools to determine with higher accuracy which genomic sequences are transcribed into RNA. The answers are surprising. Much more of our genome is transcribed into RNA than anyone supposed. Much of this RNA seems not to encode proteins. Much of it lacks some of the structures (for example, the 3′ poly(A) tail) that characterize mRNA. So what is this RNA doing?

Most of the methods for looking into this matter fall into two broad categories: cDNA cloning and microarrays. The creation of a cDNA library to study the genes transcribed in a particular eukaryotic genome is described in Chapter 9 (see Fig. 9–14). However, the classical methods for generating cDNA often lead to the cloning of only part of the sequence of a given transcript. Because reverse transcriptase may stall at regions of secondary structure in mRNA, or may simply dissociate, often 20% or less of the clones in a cDNA library are full-length DNAs. This makes it difficult to use the library to map transcription start sites (TSSs) and to study the part of a gene that encodes the amino-terminal sequence of a protein. One of the many approaches developed to overcome this problem is illustrated in Figure 1. Such refinements in technology have resulted in the creation of cDNA libraries in which more than 95% of the clones are full-length, providing an enriched source of information about cellular RNAs. However, cDNAs are generally created from RNA transcripts that have poly(A) tails. The use of microarrays, coupled to methods of cDNA preparation that do not rely on poly(A) tails (Fig. 2), has revealed that much of the RNA in eukaryotic cells lacks the common end structures.

A complete picture has not yet emerged, but some conclusions are already clear. If one excludes the repetitive sequences (transposons, for example) that can make up half of a mammalian genome, at least

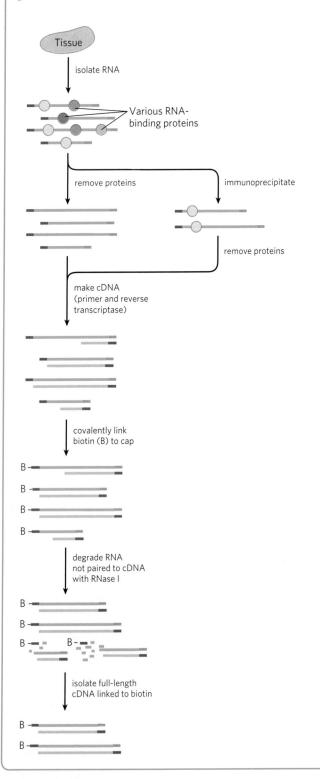

FIGURE 1 A strategy for cloning full-length cDNAs. A pool of mRNAs is isolated from a tissue sample. In some cases, mRNAs bound by a particular protein may be targeted by immunoprecipitating the protein and isolating the associated mRNAs. Biotin (B) is covalently attached to the 5′ ends of the mRNAs, making use of unique features of the 5′ cap. A poly(dT) primer (purple) is used to prime reverse transcription of the mRNAs. RNase I degrades RNA that is not part of a DNA-RNA hybrid and thus destroys the incomplete cDNA-RNA pairs. The full-length cDNA-RNA hybrids are collected with streptavidin beads (which bind biotin), converted to duplex DNA, and cloned.

40%—and perhaps the vast majority—of the remaining genomic DNA is transcribed into RNA. There seem to be more RNAs lacking poly(A) tails than RNAs with them. Much of this RNA is not transported to the cytoplasm but remains exclusively in the nucleus. Many segments of the genome are transcribed on both strands; one transcript is the complement of the other, a relationship referred to as antisense. Many of the antisense RNAs may be involved in regulation of the RNAs with which they pair. Many RNAs are produced in only one or a few tissues, and new transcripts are discovered every time a new tissue source is analyzed. Thus, the complete transcriptome has not yet been defined for any organism. Most important, many of the novel RNAs are transcribed from genomic segments, such as those illustrated in Figure 9–15, that share synteny in more than one organism. This evolutionary conservation strongly suggests that these RNAs have an important function.

Some of the novel RNAs are snoRNAs, snRNAs, or miRNAs, types of RNA recognized only in the past two decades. New TSS sequences are being discovered. New classes of RNA molecules are being defined. New patterns of alternative splicing are being elucidated. And all of these findings are challenging our definitions of a gene. In the mouse and human genomes, even the familiar protein-encoding mRNA transcripts may be much more numerous than initially

thought, and the number of known protein-encoding genes may soon increase. However, the function of most of the newly discovered transcripts is unknown, and they are simply called TUFs (transcripts of unknown function). This new RNA universe is a frontier that promises further insights into the workings of eukaryotic cells and perhaps a new glimpse of our origins in the RNA world of the distant past.

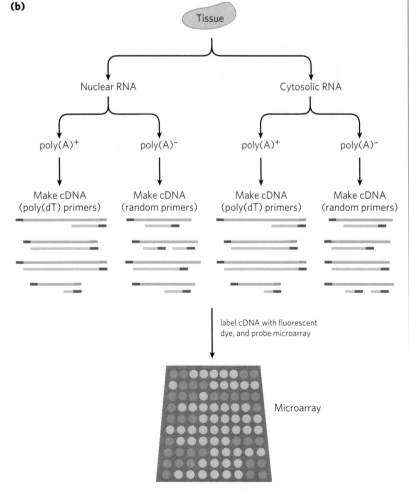

FIGURE 2 Defining the transcriptome with microarrays. **(a)** Tiled microarrays are synthesized, representing the nonrepetitive parts of the genome. In a tiled array, the successive oligonucleotides in the individual spots overlap in sequence, so that each nucleotide (such as the T shown in blue) is represented multiple times. **(b)** A tissue sample is fractionated to separate nuclear and cytoplasmic samples, and RNA is isolated from each. The RNA containing poly(A) tails is separated from RNA lacking the tails (by passing the RNA over a column with bound poly(dT)). RNA with a poly(A) tail is converted to cDNA, using methods described in Figure 1. RNA lacking a poly(A) tail is converted to cDNA by using primers with randomized sequences. The resulting DNA fragments do not correspond in length precisely to the RNAs from which they are derived, but the overall DNA pool includes most of the sequences present in the original RNAs. The cDNA samples are then labeled and used to probe the microarrays. The signals from the microarrays define the sequences that were transcribed into RNA.

genomes into double-stranded DNA as part of their life cycle. These enzymes transcribe the viral RNA into DNA, a process that can be used experimentally to form complementary DNA.

▶ Many eukaryotic transposons are related to retroviruses, and their mechanism of transposition includes an RNA intermediate.

▶ Telomerase, the enzyme that synthesizes the telomere ends of linear chromosomes, is a specialized reverse transcriptase that contains an internal RNA template.

▶ RNA-dependent RNA polymerases, such as the replicases of RNA bacteriophages, are template-specific for the viral RNA.

▶ The existence of catalytic RNAs and pathways for the interconversion of RNA and DNA has led to speculation that an important stage in evolution was the appearance of an RNA (or an equivalent polymer) that could catalyze its own replication. The biochemical potential of RNAs can be explored by SELEX, a method for rapidly selecting RNA sequences with particular binding or catalytic properties.

Key Terms

Terms in bold are defined in the glossary.

transcription 1057
messenger RNA (mRNA) 1057
transfer RNA (tRNA) 1057
ribosomal RNA (rRNA) 1057
transcriptome 1057
DNA-dependent RNA polymerase 1058
template strand 1059
nontemplate strand 1059
coding strand 1059
promoter 1060
consensus sequence 1060
footprinting 1062
cAMP receptor protein (CRP) 1063
repressor 1063
transcription factors 1066
ribozymes 1069
primary transcript 1069
RNA splicing 1069
5′ cap 1070
spliceosome 1072
small nuclear RNA (snRNA) 1073

poly(A) tail 1075
polyadenylate polymerase 1075
small nucleolar RNA (snoRNA) 1079
snoRNP 1079
microRNA (miRNA) 1081
internal guide sequence 1082
exosome 1084
polynucleotide phosphorylase 1085
reverse transcriptase 1086
retrovirus 1087
complementary DNA (cDNA) 1087
homing 1089
telomerase 1090
T loop 1091
RNA-dependent RNA polymerase (RNA replicase) 1092
SELEX 1093
aptamer 1095

Further Reading

General

Cox, M.M., Doudna, J.A., & O'Donnell, M. (2012) *Molecular Biology: Principles and Practice,* W. H. Freeman and Company, New York.

Jacob, F. & Monod, J. (1961) Genetic regulatory mechanisms in the synthesis of proteins. *J. Mol. Biol.* **3**, 318–356.
  A classic article that introduced many important ideas.

DNA-Directed RNA Synthesis

Gruber, T.M. & Gross, C.A. (2003) Multiple sigma subunits and the partitioning of bacterial transcription space. *Annu. Rev. Microbiol.* **57**, 441–466.

Herbert, K.M., Greenleaf, W.J., & Block, S.M. (2008) Single-molecule studies of RNA polymerase: motoring along. *Annu. Rev. Biochem.* **77**, 149–176.

Kuehner, J.N., Pearson, E.L., & Moore, C. (2011) Unravelling the means to an end: RNA polymerase II transcription termination. *Nat. Rev. Mol. Cell Biol.* **12**, 283–294.

Laine, J.P. & Egly, J.M. (2006) When transcription and repair meet: a complex system. *Trends Genet.* **22**, 430–436.

Mooney, R.A., Darst, S.A., & Landick, R. (2005) Sigma and RNA polymerase: an on-again, off-again relationship? *Mol. Cell* **20**, 335–345.

Nudler, E. (2009) RNA polymerase active center: the molecular engine of transcription. *Annu. Rev. Biochem.* **78**, 335–361.

Svejstrup, J.Q., Conaway, R.C., & Conaway, J.W. (2006) RNA polymerase II: a "Nobel" enzyme demystified. *Mol. Cell* **24**, 637–642.

Thomas, M.C. & Chiang, C.M. (2006) The general transcription machinery and general cofactors. *Crit. Rev. Biochem. Mol. Biol.* **41**, 105–178.

Zhou, Q., Li, T., & Price, D.H. (2012) RNA polymerase II elongation control. *Annu. Rev. Biochem.* **81**, 119–143.

RNA Processing

Berezikov, E. (2011) Evolution of microRNA diversity and regulation in animals. *Nat. Rev. Genet.* **12**, 846–860.

Buchan, J.R. & Parker, R. (2007) The two faces of miRNA. *Science* **318**, 1877–1878.

Butcher, S.E. & Brow, D.A. (2005) Towards understanding the catalytic core structure of the spliceosome. *Biochem. Soc. Trans.* **33**, 447–449.

Gogarten, J.P. & Hilario, E. (2006) Inteins, introns, and homing endonucleases: recent revelations about the life cycle of parasitic genetic elements. *BMC Evol. Biol.* **6**, 94.

Hoskins, A.A., Gelles, J., & Moore, M.J. (2011) New insights into the spliceosome by single molecule fluorescence microscopy. *Curr. Opin. Chem. Biol.* **15**, 864–870.

Huang, Y.Q. & Steitz, J.A. (2005) SRprises along a messenger's journey. *Mol. Cell* **17**, 613–615.

Kaberdin, V.R. & Blasi, U. (2006) Translation initiation and the fate of bacterial mRNAs. *FEMS Microbiol. Rev.* **30**, 967–979.

Kalsotra, A. & Cooper, T.A. (2011) Functional consequences of developmentally regulated alternative splicing. *Nat. Rev. Genet.* **12**, 715–729.

Reiter, N.J., Chan, C.W., & Mondragon, A. (2011) Emerging structural themes in large RNA molecules. *Curr. Opin. Struct. Biol.* **21**, 319–326.

Rodriguez-Trelles, F., Tarrio, R., & Ayala, F.J. (2006) Origins and evolution of spliceosomal introns. *Annu. Rev. Genet.* **40**, 47–76.

Schneider, D.A., Michel, A., Sikes, M.L., Vu, L., Dodd, J.A., Salgia, S., Osheim, Y.S., Beyer, A.L., & Nomura, M. (2007) Transcription elongation by RNA polymerase I is linked to efficient rRNA processing and ribosome assembly. *Mol. Cell* **26**, 217–229.

RNA-Directed RNA or DNA Synthesis

Blackburn, E.H., Greider, C.W., & Szostak, J.W. (2006) Telomeres and telomerase: the path from maize, *Tetrahymena* and yeast to human cancer and aging. *Nat. Med.* **12**, 1133–1138.

Boeke, J.D. & Devine, S.E. (1998) Yeast retrotransposons: finding a nice, quiet neighborhood. *Cell* **93**, 1087–1089.

Cordaux, R. & Batzer, M.A. (2009) The impact of retrotransposons on human genome evolution. *Nat. Rev. Genet.* **10**, 691–703.

Frankel, A.D. & Young, J.A.T. (1998) HIV-1: fifteen proteins and an RNA. *Annu. Rev. Biochem.* **67**, 1–25.

Mason, M., Schuller, A., & Skordalakes, E. (2011) Telomerase structure function. *Curr Opin. Struct. Biol.* **21**, 92–100.

O'Sullivan, R.J. & Karlseder, J. (2010) Telomeres: protecting chromosomes against genome instability. *Nat. Rev. Mol. Cell Biol.* **11**, 171–181.

Temin, H.M. (1976) The DNA provirus hypothesis: the establishment and implications of RNA-directed DNA synthesis. *Science* **192**, 1075–1080.

Discussion of the original proposal for reverse transcription in retroviruses.

Ribozymes and Evolution

Carninci, P. (2007) Constructing the landscape of the mammalian genome. *J. Exp. Biol.* **210**, 1497–1506.

A good summary of the work showing that mammalian transcriptomes are much more extensive than previously thought.

Doudna, J.A. & Lorsch, J.R. (2005) Ribozyme catalysis: not different, just worse. *Nat. Struct. Mol. Biol.* **12**, 395–402.

Green, R. & Doudna, J.A. (2006) RNAs regulate biology. *ACS Chem. Biol.* **1**, 335–338.

Huttenhofer, A. & Vogel, J. (2006) Experimental approaches to identify non-coding RNAs. *Nucleic Acids Res.* **34**, 635–646.

Joyce, G.F. (2002) The antiquity of RNA-based evolution. *Nature* **418**, 214–221.

Kazantsev, A.V. & Pace, N.R. (2006) Bacterial RNase P: a new view of an ancient enzyme. *Nat. Rev. Microbiol.* **4**, 729–740.

Lassila, J.K., Zalatan, J.G., & Herschlag, D. (2011) Biological phosphoryl-transfer reactions: understanding mechanism and catalysis. *Annu. Rev. Biochem.* **80**, 669–702.

Lincoln, T.A. & Joyce, G.F. (2009) Self-sustained replication of an RNA enzyme. *Science* **323**, 1229–1232.

Müller, U.F. (2006) Recreating an RNA world. *Cell. Mol. Life Sci.* **63**, 1278–1293.

Willingham, A.T. & Gingeras, T.R. (2006) TUF love for "junk" DNA. *Cell* **125**, 1215–1220.

Wilson, T.J. & Lilley, D.M. (2009) The evolution of ribozyme chemistry. *Science* **323**, 1436–1438.

Wochner, A., Attwater, J., Coulson, A., & Holliger, P. (2011) Ribozyme-catalyzed transcription of an active ribozyme. *Science* **332**, 209–212.

Yarus, M. (2002) Primordial genetics: phenotype of the ribocyte. *Annu. Rev. Genet.* **36**, 125–151.

Detailed speculations about what an RNA-based life form might have been like and a good summary of the research behind them.

Problems

1. **RNA Polymerase** (a) How long would it take for the *E. coli* RNA polymerase to synthesize the primary transcript for the *E. coli* genes encoding the enzymes for lactose metabolism (the 5,300 bp *lac* operon, considered in Chapter 28)? (b) How far along the DNA would the transcription "bubble" formed by RNA polymerase move in 10 seconds?

2. **Error Correction by RNA Polymerases** DNA polymerases are capable of editing and error correction, whereas the capacity for error correction in RNA polymerases seems to be quite limited. Given that a single base error in either replication or transcription can lead to an error in protein synthesis, suggest a possible biological explanation for this difference.

3. **RNA Posttranscriptional Processing** Predict the likely effects of a mutation in the sequence (5′)AAUAAA in a eukaryotic mRNA transcript.

4. **Coding versus Template Strands** The RNA genome of phage Qβ is the nontemplate strand, or coding strand, and when introduced into the cell, it functions as an mRNA. Suppose the RNA replicase of phage Qβ synthesized primarily template-strand RNA and uniquely incorporated this, rather than nontemplate strands, into the viral particles. What would be the fate of the template strands when they entered a new cell? What enzyme would have to be included in the viral particles for successful invasion of a host cell?

5. **Transcription** The gene encoding the *E. coli* enzyme β-galactosidase begins with the sequence ATGACCATGATTACG. What is the sequence of the RNA transcript specified by this part of the gene?

6. **The Chemistry of Nucleic Acid Biosynthesis** Describe three properties common to the reactions catalyzed by DNA polymerase, RNA polymerase, reverse transcriptase, and RNA replicase. How is the enzyme polynucleotide phosphorylase similar to and different from these four enzymes?

7. **RNA Splicing** What is the minimum number of transesterification reactions needed to splice an intron from an mRNA transcript? Explain.

8. **RNA Processing** If the splicing of mRNA in a vertebrate cell is blocked, the rRNA modification reactions are also blocked. Suggest a reason for this.

9. **RNA Genomes** The RNA viruses have relatively small genomes. For example, the single-stranded RNAs of retroviruses have about 10,000 nucleotides and the Qβ RNA is only 4,220 nucleotides long. Given the properties of reverse transcriptase and RNA replicase described in this chapter, can you suggest a reason for the small size of these viral genomes?

10. **Screening RNAs by SELEX** The practical limit for the number of different RNA sequences that can be screened in a SELEX experiment is 10^{15}. (a) Suppose you are working with oligonucleotides 32 nucleotides long. How many sequences exist in a randomized pool containing every sequence possible?

(b) What percentage of these can be screened in a SELEX experiment? (c) Suppose you wish to select an RNA molecule that catalyzes the hydrolysis of a particular ester. From what you know about catalysis, propose a SELEX strategy that might allow you to select the appropriate catalyst.

11. Slow Death The death cap mushroom, *Amanita phalloides*, contains several dangerous substances, including the lethal α-amanitin. This toxin blocks RNA elongation in consumers of the mushroom by binding to eukaryotic RNA polymerase II with very high affinity; it is deadly in concentrations as low as 10^{-8} M. The initial reaction to ingestion of the mushroom is gastrointestinal distress (caused by some of the other toxins). These symptoms disappear, but about 48 hours later, the mushroom-eater dies, usually from liver dysfunction. Speculate on why it takes this long for α-amanitin to kill.

12. Detection of Rifampicin-Resistant Strains of Tuberculosis Rifampicin is an important antibiotic used to treat tuberculosis and other mycobacterial diseases. Some strains of *Mycobacterium tuberculosis*, the causative agent of tuberculosis, are resistant to rifampicin. These strains become resistant through mutations that alter the *rpoB* gene, which encodes the β subunit of the RNA polymerase. Rifampicin cannot bind to the mutant RNA polymerase and so is unable to block the initiation of transcription. DNA sequences from a large number of rifampicin-resistant *M. tuberculosis* strains have been found to have mutations in a specific 69 bp region of *rpoB*. One well-characterized rifampicin-resistant strain has a single base pair alteration in *rpoB* that results in a His residue being replaced by an Asp residue in the β subunit.

(a) Based on your knowledge of protein chemistry, suggest a technique that would allow detection of the rifampicin-resistant strain containing this particular mutant protein.

(b) Based on your knowledge of nucleic acid chemistry, suggest a technique to identify the mutant form of *rpoB*.

Using the Web

13. The Ribonuclease Gene Human pancreatic ribonuclease has 128 amino acid residues.

(a) What is the minimum number of nucleotide pairs required to code for this protein?

(b) The mRNA expressed in human pancreatic cells was copied with reverse transcriptase to create a "library" of human DNA. The sequence of the mRNA coding for human pancreatic ribonuclease was determined by sequencing the complementary DNA (cDNA) from this library that included an open reading frame for the protein. Use the Entrez database system (www.ncbi.nlm.nih.gov/Entrez) to find the published sequence of this mRNA (search the CoreNucleotide

database for accession number D26129). What is the length of this mRNA?

(c) How can you account for the discrepancy between the size you calculated in (a) and the actual length of the mRNA?

Data Analysis Problem

14. A Case of RNA Editing The AMPA (α-amino-3-hydroxy-5-methyl-4-isoxazolepropionic acid) receptor is an important component of the human nervous system. It is present in several forms, in different neurons, and some of this variety results from posttranscriptional modification. This problem explores research on the mechanism of this RNA editing.

An initial report by Sommer and coauthors (1991) looked at the sequence encoding a key Arg residue in the AMPA receptor. The sequence of the cDNA (see Fig. 9–14) for the AMPA receptor showed a CGG (Arg; see Fig. 27–7) codon for this amino acid. Surprisingly, the genomic DNA showed a CAG (Gln) codon at this position.

(a) Explain how this result is consistent with posttranscriptional modification of the AMPA receptor mRNA.

Rueter and colleagues (1995) explored this mechanism in detail. They first developed an assay to differentiate between edited and unedited transcripts, based on the Sanger method of DNA sequencing (see Fig. 8–33). They modified the technique to determine whether the base in question was an A (as in CAG) or not. They designed two DNA primers based on the genomic DNA sequence of this region of the AMPA gene. These primers, and the genomic DNA sequence of the nontemplate strand for the relevant region of the AMPA receptor gene, are shown at the bottom of the page; the A residue that is edited is in red.

To detect whether this A was present or had been edited to another base, Rueter and coworkers used the following procedure:

1. Prepared cDNA complementary to the mRNA, using primer 1, reverse transcriptase, dATP, dGTP, dCTP, and dTTP.
2. Removed the mRNA.
3. Annealed ^{32}P-labeled primer 2 to the cDNA and reacted this with DNA polymerase, dGTP, dCTP, dTTP, and ddATP (dideoxy ATP; see Fig. 8–33).
4. Denatured the resulting duplexes and separated them with polyacrylamide gel electrophoresis (see Fig. 3–18).
5. Detected the ^{32}P-labeled DNA species with autoradiography.

They found that edited mRNA produced a 22 nucleotide [^{32}P] DNA, whereas unedited mRNA produced a 19 nucleotide [^{32}P] DNA.

(b) Using the sequences below, explain how the edited and unedited mRNAs resulted in these different products.

(5′) ...GTCTCTGGTTTTCCTTGGGTGCCTTTATGCAGCAAGGATGCGATATTTCGCCAAG...
Primer 1:   CGTTCCTACGCTATAAAGCGGTTC (5′)
Primer 2:   (5′)CCTTGGGTGCCTTTA

Using the same procedure, to measure the fraction of transcripts edited under different conditions, the researchers found that extracts of cultured epithelial cells (a common cell line called HeLa) could edit the mRNA at a high level. To determine the nature of the editing machinery, they pretreated an active HeLa cell extract as described in the table and measured its ability to edit AMPA mRNA. Proteinase K degrades only proteins; micrococcal nuclease, only DNA.

Sample	Pretreatment	% mRNA edited
1	None	18
2	Proteinase K	5
3	Heat to 65 °C	3
4	Heat to 85 °C	3
5	Micrococcal nuclease	17

(c) Use these data to argue that the editing machinery consists of protein. What is a key weakness in this argument?

To determine the exact nature of the edited base, Rueter and colleagues used the following procedure:

1. Produced mRNA, using $[\alpha\text{-}^{32}P]$ATP in the reaction mixture.
2. Edited the labeled mRNA by incubating with HeLa extract.
3. Hydrolyzed the edited mRNA to single nucleotide monophosphates with nuclease P1.
4. Separated the nucleotide monophosphates with thin-layer chromatography (TLC; see Fig. 10–25b).
5. Identified the resulting ^{32}P-labeled nucleotide monophosphates with autoradiography.

In unedited mRNA, they found only $[^{32}P]$AMP; in edited mRNA, they found mostly $[^{32}P]$AMP with some $[^{32}P]$IMP (inosine monophosphate; see Fig. 22–36).

(d) Why was it necessary to use $[\alpha\text{-}^{32}P]$ATP rather than $[\beta\text{-}^{32}P]$ATP or $[\gamma\text{-}^{32}P]$ATP in this experiment?

(e) Why was it necessary to use $[\alpha\text{-}^{32}P]$ATP rather than $[\alpha\text{-}^{32}P]$GTP, $[\alpha\text{-}^{32}P]$CTP, or $[\alpha\text{-}^{32}P]$UTP?

(f) How does the result exclude the possibility that the entire A nucleotide (sugar, base, and phosphate) was removed and replaced by an I nucleotide during the editing process?

The researchers next edited mRNA that was labeled with $[2,8\text{-}^{3}H]$ATP and repeated the above procedure. The only ^{3}H-labeled mononucleotides produced were AMP and IMP.

(g) How does this result exclude removal of the A base (leaving the sugar-phosphate backbone intact) followed by replacement with an I base as a mechanism of editing? What, then, is the most likely mechanism of editing in this case?

(h) How does changing an A to an I residue in the mRNA explain the Gln to Arg change in protein sequence in the two forms of AMPA receptor protein? (Hint: See Fig. 27–8.)

References

Rueter, S.M., Burns, C.M., Coode, S.A., Mookherjee, P., & Emesont, R.B. (1995) Glutamate receptor RNA editing in vitro by enzymatic conversion of adenosine to inosine. *Science* **267**, 1491–1494.

Sommer, B., Köhler, M., Sprengel, R., & Seeburg, P.H. (1991) RNA editing in brain controls a determinant of ion flow in glutamate-gated channels. *Cell* **67**, 11–19.

Protein Metabolism

Proteins are the end products of most information pathways. A typical cell requires thousands of different proteins at any given moment. These must be synthesized in response to the cell's current needs, transported (targeted) to their appropriate cellular locations, and degraded when no longer needed. Many of the fundamental components and mechanisms utilized by the protein biosynthetic machinery are remarkably well conserved in all life-forms from bacteria to higher eukaryotes, indicating that they were present in the last universal common ancestor (LUCA) of all extant organisms.

An understanding of protein synthesis, the most complex biosynthetic process, has been one of the greatest challenges in biochemistry. Eukaryotic protein synthesis involves more than 70 different ribosomal proteins; 20 or more enzymes to activate the amino acid precursors; a dozen or more auxiliary enzymes and other protein factors for the initiation, elongation, and termination of polypeptides; perhaps 100 additional enzymes for the final processing of different proteins; and 40 or more kinds of transfer and ribosomal RNAs. Overall, almost 300 different macromolecules cooperate to synthesize polypeptides. Many of these macromolecules are organized into the complex three-dimensional structure of the ribosome.

To appreciate the central importance of protein synthesis, consider the cellular resources devoted to this process. Protein synthesis can account for up to 90% of the chemical energy used by a cell for all biosynthetic reactions. Every bacterial, archaeal, and eukaryotic cell contains from several to thousands of copies of many different proteins and RNAs. The 15,000 ribosomes, 100,000 molecules of protein synthesis–related protein factors and enzymes, and 200,000 tRNA molecules in a typical bacterial cell can account for more than 35% of the cell's dry weight.

Despite the great complexity of protein synthesis, proteins are made at exceedingly high rates. A polypeptide of 100 residues is synthesized in an *Escherichia coli* cell (at 37 °C) in about 5 seconds. Synthesis of the thousands of different proteins in a cell is tightly regulated, so that just enough copies are made to match the current metabolic circumstances. To maintain the appropriate mix and concentration of proteins, the targeting and degradative processes must keep pace with synthesis. Research is gradually uncovering the finely coordinated cellular choreography that guides each protein to its proper cellular location and selectively degrades it when it is no longer required.

Harry Noller

The study of protein synthesis offers another important reward: a look at a world of RNA catalysts that may have existed before the dawn of life "as we know it." Elucidation of the three-dimensional structures of ribosomes, beginning in the year 2000, has given us an increasingly detailed look at the mechanics of protein synthesis. It has also confirmed a hypothesis first put forward by Harry Noller two decades earlier: proteins are synthesized by a gigantic RNA enzyme!

27.1 The Genetic Code

Three major advances set the stage for our present knowledge of protein biosynthesis. First, in the early 1950s, Paul Zamecnik and his colleagues designed a set of experiments to investigate where in the cell proteins are synthesized. They injected radioactive amino acids into rats and, at different time intervals after the injection, removed the liver, homogenized it, fractionated the homogenate by centrifugation, and examined the subcellular fractions for the presence of radioactive protein. When hours or days were allowed to elapse after injection of the labeled amino acids, *all* the subcellular

Paul Zamecnik, 1912–2009

fractions contained labeled proteins. However, when only minutes had elapsed, labeled protein appeared only in a fraction containing small ribonucleoprotein particles. These particles, visible in animal tissues by electron microscopy, were therefore identified as the site of protein synthesis from amino acids and later were named ribosomes (**Fig. 27–1**).

The second key advance was made by Mahlon Hoagland and Zamecnik when they found that amino acids were "activated" when incubated with ATP and the cytosolic fraction of liver cells. The amino acids became attached to a heat-stable soluble RNA of the type that had been discovered and characterized by Robert Holley and later called transfer RNA (tRNA), to form **aminoacyl-tRNAs**. The enzymes that catalyze this process are the **aminoacyl-tRNA synthetases**.

The third advance resulted from Francis Crick's reasoning on how the genetic information encoded in the 4-letter language of nucleic acids could be translated into the 20-letter language of proteins. A small nucleic acid (perhaps RNA) could serve the role of an adaptor, one part of the adaptor molecule binding a specific amino acid and another part recognizing the nucleotide sequence encoding that amino acid in an mRNA (**Fig. 27–2**). This idea was soon verified. The tRNA adaptor "translates" the nucleotide sequence of an mRNA into the amino acid sequence of a polypeptide. The overall process of mRNA-guided protein synthesis is often referred to simply as **translation**.

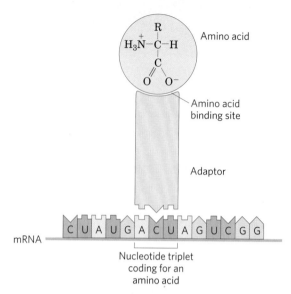

FIGURE 27–2 Crick's adaptor hypothesis. Today we know that the amino acid is covalently bound at the 3' end of a tRNA molecule and that a specific nucleotide triplet elsewhere in the tRNA interacts with a particular triplet codon in mRNA through hydrogen bonding of complementary bases.

These three developments soon led to recognition of the major stages of protein synthesis and ultimately to the elucidation of the genetic code that specifies each amino acid.

The Genetic Code Was Cracked Using Artificial mRNA Templates

By the 1960s it was apparent that at least three nucleotide residues of DNA are necessary to encode each amino acid. The four code letters of DNA (A, T, G, and C) in groups of two can yield only $4^2 = 16$ different combinations, insufficient to encode 20 amino acids. Groups of three, however, yield $4^3 = 64$ different combinations.

Several key properties of the genetic code were established in early genetic studies (**Figs 27–3, 27–4**).

FIGURE 27–1 Ribosomes and endoplasmic reticulum. Electron micrograph and schematic drawing of a portion of a pancreatic cell, showing ribosomes attached to the outer (cytosolic) face of the endoplasmic reticulum (ER). The ribosomes are the numerous small dots bordering the parallel layers of membranes.

FIGURE 27–3 Overlapping versus nonoverlapping genetic codes. In a nonoverlapping code, codons (numbered consecutively) do not share nucleotides. In an overlapping code, some nucleotides in the mRNA are shared by different codons. In a triplet code with maximum overlap, many nucleotides, such as the third nucleotide from the left (A), are shared by three codons. Note that in an overlapping code, the triplet sequence of the first codon limits the possible sequences for the second codon. A nonoverlapping code provides much more flexibility in the triplet sequence of neighboring codons and therefore in the possible amino acid sequences designated by the code. The genetic code used in all living systems is now known to be nonoverlapping.

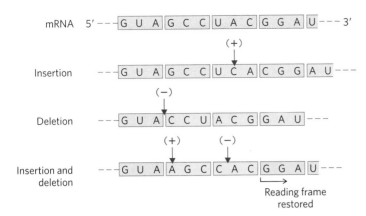

mRNA 5′ ---G U A G C C U A C G G A U--- 3′

(+)

Insertion ---G U A G C C U C A C G G A U---

(−)

Deletion ---G U A C C U A C G G A U---

(+) (−)

Insertion and deletion ---G U A A G C C A C G G A U---

Reading frame restored

FIGURE 27–4 The triplet, nonoverlapping code. Evidence for the general nature of the genetic code came from many types of experiments, including genetic experiments on the effects of deletion and insertion mutations. Inserting or deleting one base pair (shown here in the mRNA transcript) alters the sequence of triplets in a nonoverlapping code; all amino acids coded by the mRNA following the change are affected. Combining insertion and deletion mutations affects some amino acids but can eventually restore the correct amino acid sequence. Adding or subtracting three nucleotides (not shown) leaves the remaining triplets intact, providing evidence that a codon has three, rather than four or five, nucleotides. The triplet codons shaded in gray are those transcribed from the original gene; codons shaded in blue are new codons resulting from the insertion or deletion mutations.

A **codon** is a triplet of nucleotides that codes for a specific amino acid. Translation occurs in such a way that these nucleotide triplets are read in a successive, nonoverlapping fashion. A specific first codon in the sequence establishes the **reading frame**, in which a new codon begins every three nucleotide residues. There is no punctuation between codons for successive amino acid residues. The amino acid sequence of a protein is defined by a linear sequence of contiguous triplets. In principle, any given single-stranded DNA or mRNA sequence has three possible reading frames. Each reading frame gives a different sequence of codons **(Fig. 27–5)**, but only one is likely to encode a given protein. A key question remained: what were the three-letter code words for each amino acid?

Marshall Nirenberg

In 1961 Marshall Nirenberg and Heinrich Matthaei reported the first breakthrough. They incubated synthetic polyuridylate, poly(U), with an *E. coli* extract, GTP, ATP, and a mixture of the 20 amino acids in 20 different tubes, each tube containing a different radioactively labeled amino acid. Because poly(U) mRNA is made up of many successive UUU triplets, it should promote the synthesis of a polypeptide containing only the amino acid encoded by the triplet UUU. A radioactive polypeptide was indeed formed in only one of the 20 tubes, the one containing radioactive phenylalanine. Nirenberg and Matthaei therefore concluded that the triplet codon UUU encodes phenylalanine. The same approach soon revealed that polycytidylate, poly(C),

encodes a polypeptide containing only proline (polyproline), and polyadenylate, poly(A), encodes polylysine. Polyguanylate did not generate any polypeptide in this experiment because it spontaneously forms tetraplexes (see Fig. 8–20d) that cannot be bound by ribosomes.

The synthetic polynucleotides used in such experiments were prepared with polynucleotide phosphorylase (p. 1085), which catalyzes the formation of RNA polymers starting from ADP, UDP, CDP, and GDP. This enzyme, discovered by Severo Ochoa, requires no template and makes polymers with a base composition that directly reflects the relative concentrations of the nucleoside 5′-diphosphate precursors in the medium. If polynucleotide phosphorylase is presented with UDP only, it makes only poly(U). If it is presented with a mixture of five parts ADP and one part CDP, it makes a polymer in which about five-sixths of the residues are adenylate and one-sixth are cytidylate. This random polymer is likely to have many triplets of the sequence AAA, smaller numbers of AAC, ACA, and CAA triplets, relatively few ACC, CCA, and CAC triplets, and very few CCC triplets (Table 27–1). Using a variety of artificial mRNAs made by polynucleotide phosphorylase from different starting mixtures of ADP, GDP, UDP, and CDP, the Nirenberg and Ochoa groups soon identified the base compositions of the triplets coding for almost all the amino acids. Although these experiments revealed the base composition of the coding triplets, they usually could not reveal the sequence of the bases.

KEY CONVENTION: Much of the following discussion deals with tRNAs. The amino acid specified by a tRNA is indicated by a superscript, such as tRNA^{Ala}, and the aminoacylated tRNA by a hyphenated name: alanyl-tRNA^{Ala} or Ala-tRNA^{Ala}. ∎

Reading frame 1 5′ ---U U C U C G G A C C U G G A G A U U C A C A G U--- 3′

Reading frame 2 ---U U C U C G G A C C U G G A G A U U C A C A G U---

Reading frame 3 ---U U C U C G G A C C U G G A G A U U C A C A G U---

FIGURE 27–5 Reading frames in the genetic code. In a triplet, nonoverlapping code, all mRNAs have three potential reading frames, shaded here in different colors. The triplets, and hence the amino acids specified, are different in each reading frame.

TABLE 27–1   Incorporation of Amino Acids into Polypeptides in Response to Random Polymers of RNA

Amino acid	Observed frequency of incorporation (Lys = 100)	Tentative assignment for nucleotide composition of corresponding codon*	Expected frequency of incorporation based on assignment (Lys = 100)
Asparagine	24	A_2C	20
Glutamine	24	A_2C	20
Histidine	6	AC_2	4
Lysine	100	AAA	100
Proline	7	AC_2, CCC	4.8
Threonine	26	A_2C, AC_2	24

Note: Presented here is a summary of data from one of the early experiments designed to elucidate the genetic code. A synthetic RNA containing only A and C residues in a 5:1 ratio directed polypeptide synthesis, and both the identity and the quantity of incorporated amino acids were determined. Based on the relative abundance of A and C residues in the synthetic RNA, and assigning the codon AAA (the most likely codon) a frequency of 100, there should be three different codons of composition A_2C, each at a relative frequency of 20; three of composition AC_2, each at a relative frequency of 4.0; and CCC at a relative frequency of 0.8. The CCC assignment was based on information derived from prior studies with poly(C). Where two tentative codon assignments are made, both are proposed to code for the same amino acid.

*These designations of nucleotide composition contain no information on nucleotide sequence (except, of course, AAA and CCC).

In 1964 Nirenberg and Philip Leder achieved another experimental breakthrough. Isolated *E. coli* ribosomes would bind a specific aminoacyl-tRNA in the presence of the corresponding synthetic polynucleotide messenger. For example, ribosomes incubated with poly(U) and phenylalanyl-tRNA^{Phe} (Phe-tRNA^{Phe}) bind both RNAs, but if the ribosomes are incubated with poly(U) and some other aminoacyl-tRNA, the aminoacyl-tRNA is not bound, because it does not recognize the UUU triplets in poly(U) (Table 27–2). Even trinucleotides could promote specific binding of appropriate tRNAs, so these experiments could be carried out with chemically synthesized small oligonucleotides. With this technique researchers determined which aminoacyl-tRNA bound to 54 of the 64 possible triplet codons. For some codons, either no aminoacyl-tRNA or more than one would bind. Another method was needed to complete and confirm the entire genetic code.

H. Gobind Khorana, 1922–2011

At about this time, a complementary approach was provided by H. Gobind Khorana, who developed chemical methods to synthesize polyribonucleotides with defined, repeating sequences of two to four bases. The polypeptides produced by these mRNAs had one or a few amino acids in repeating patterns. These patterns, when combined with information from the random polymers used by Nirenberg and colleagues, permitted unambiguous codon assignments. The copolymer $(AC)_n$, for example, has alternating ACA and CAC codons: ACACACACACACACA. The polypeptide synthesized on this messenger contained equal amounts of threonine and histidine. Given that a histidine codon has one A and two Cs (Table 27–1), CAC must code for histidine and ACA for threonine.

Consolidation of the results from many experiments permitted the assignment of 61 of the 64 possible codons. The other three were identified as termination codons, in part because they disrupted amino acid coding patterns when they occurred in a synthetic RNA polymer (**Fig. 27–6**). Meanings for all the triplet codons (tabulated in **Fig. 27–7**) were established by 1966 and have been verified in many different ways.

The cracking of the genetic code is regarded as one of the most important scientific discoveries of the twentieth century.

Codons are the key to the translation of genetic information, directing the synthesis of specific proteins. The reading frame is set when translation of an mRNA

TABLE 27–2   Trinucleotides That Induce Specific Binding of Aminoacyl-tRNAs to Ribosomes

Trinucleotide	Relative increase in ^{14}C-labeled aminoacyl-tRNA bound to ribosome*		
	Phe-tRNA^{Phe}	Lys-tRNA^{Lys}	Pro-tRNA^{Pro}
UUU	4.6	0	0
AAA	0	7.7	0
CCC	0	0	3.1

Source: Modified from Nirenberg, M. & Leder, P. (1964) RNA code words and protein synthesis. *Science* 145, 1399.

*Each number represents the factor by which the amount of bound ^{14}C increased when the indicated trinucleotide was present, relative to a control with no trinucleotide.

FIGURE 27-6 **Effect of a termination codon in a repeating tetranucleotide.** Termination codons (light red) are encountered every fourth codon in three different reading frames (shown in different colors). Dipeptides or tripeptides are synthesized, depending on where the ribosome initially binds.

First letter of codon (5' end)

Second letter of codon

	U		C		A		G	
U	UUU	Phe	UCU	Ser	UAU	Tyr	UGU	Cys
	UUC	Phe	UCC	Ser	UAC	Tyr	UGC	Cys
	UUA	Leu	UCA	Ser	UAA	Stop	UGA	Stop
	UUG	Leu	UCG	Ser	UAG	Stop	UGG	Trp
C	CUU	Leu	CCU	Pro	CAU	His	CGU	Arg
	CUC	Leu	CCC	Pro	CAC	His	CGC	Arg
	CUA	Leu	CCA	Pro	CAA	Gln	CGA	Arg
	CUG	Leu	CCG	Pro	CAG	Gln	CGG	Arg
A	AUU	Ile	ACU	Thr	AAU	Asn	AGU	Ser
	AUC	Ile	ACC	Thr	AAC	Asn	AGC	Ser
	AUA	Ile	ACA	Thr	AAA	Lys	AGA	Arg
	AUG	Met	ACG	Thr	AAG	Lys	AGG	Arg
G	GUU	Val	GCU	Ala	GAU	Asp	GGU	Gly
	GUC	Val	GCC	Ala	GAC	Asp	GGC	Gly
	GUA	Val	GCA	Ala	GAA	Glu	GGA	Gly
	GUG	Val	GCG	Ala	GAG	Glu	GGG	Gly

FIGURE 27-7 **"Dictionary" of amino acid code words in mRNAs.** The codons are written in the 5'→3' direction. The third base of each codon (in bold type) plays a lesser role in specifying an amino acid than the first two. The three termination codons are shaded in light red, the initiation codon AUG in green. All the amino acids except methionine and tryptophan have more than one codon. In most cases, codons that specify the same amino acid differ only at the third base.

molecule begins and it is maintained as the synthetic machinery reads sequentially from one triplet to the next. If the initial reading frame is off by one or two bases, or if translation somehow skips a nucleotide in the mRNA, all the subsequent codons will be out of register; the result is usually a "missense" protein with a garbled amino acid sequence.

Several codons serve special functions (Fig. 27–7). The **initiation codon** AUG is the most common signal for the beginning of a polypeptide in all cells, in addition to coding for Met residues in internal positions of polypeptides. The **termination codons** (UAA, UAG, and UGA), also called stop codons or nonsense codons, normally signal the end of polypeptide synthesis and do not code for any known amino acids. Some deviations from these rules are discussed in Box 27–1.

As described in Section 27.2, initiation of protein synthesis in the cell is an elaborate process that relies on

initiation codons and other signals in the mRNA. In retrospect, the experiments of Nirenberg, Khorana, and others to identify codon function should not have worked in the absence of initiation codons. Serendipitously, experimental conditions caused the normal initiation requirements for protein synthesis to be relaxed. Diligence combined with chance to produce a breakthrough—a common occurrence in the history of biochemistry.

In a random sequence of nucleotides, 1 in every 20 codons in each reading frame is, on average, a termination codon. In general, a reading frame without a termination codon among 50 or more codons is referred to as an **open reading frame (ORF)**. Long open reading frames usually correspond to genes that encode proteins. In the analysis of sequence databases, sophisticated programs are used to search for open reading frames in order to find genes among the often huge background of nongenic DNA. An uninterrupted gene coding for a typical protein with a molecular weight of 60,000 would require an open reading frame with 500 or more codons.

A striking feature of the genetic code is that an amino acid may be specified by more than one codon, so the code is described as **degenerate**. This does *not* suggest that the code is flawed: although an amino acid may have two or more codons, each codon specifies only one amino acid. The degeneracy of the code is not uniform. Whereas methionine and tryptophan have single codons, for example, three amino acids (Arg, Leu, Ser) have six codons, five amino acids have four, isoleucine has three, and nine amino acids have two (Table 27–3).

TABLE 27-3 Degeneracy of the Genetic Code

Amino acid	Number of codons	Amino acid	Number of codons
Met	1	Tyr	2
Trp	1	Ile	3
Asn	2	Ala	4
Asp	2	Gly	4
Cys	2	Pro	4
Gln	2	Thr	4
Glu	2	Val	4
His	2	Arg	6
Lys	2	Leu	6
Phe	2	Ser	6

BOX 27–1   Exceptions That Prove the Rule: Natural Variations in the Genetic Code

In biochemistry, as in other disciplines, exceptions to general rules can be problematic for instructors and frustrating for students. At the same time, though, they teach us that life is complex and inspire us to search for more surprises. Understanding the exceptions can even reinforce the original rule in surprising ways.

One would expect little room for variation in the genetic code. Even a single amino acid substitution can have profoundly deleterious effects on the structure of a protein. Nevertheless, variations in the code do occur in some organisms, and they are both interesting and instructive. The types of variation and their rarity provide powerful evidence for a common evolutionary origin of all living things.

To alter the code, changes must occur in the gene(s) encoding one or more tRNAs, with the obvious target for alteration being the anticodon. Such a change would lead to the systematic insertion of an amino acid at a codon that, according to the standard code (see Fig. 27–7), does not specify that amino acid. The genetic code, in effect, is defined by two elements: (1) the anticodons on tRNAs (which determine where an amino acid is placed in a growing polypeptide) and (2) the specificity of the enzymes—the aminoacyl-tRNA synthetases—that charge the tRNAs, which determines the identity of the amino acid attached to a given tRNA.

Most sudden changes in the code would have catastrophic effects on cellular proteins, so code alterations are more likely to persist where relatively few proteins would be affected—such as in small genomes encoding only a few proteins. The biological consequences of a code change could also be limited by restricting changes to the three termination codons, which do not generally occur *within* genes (see Box 27–4 for exceptions to *this* rule). This pattern is in fact observed.

Of the very few variations in the genetic code that we know of, most occur in mitochondrial DNA (mtDNA), which encodes only 10 to 20 proteins. Mitochondria have their own tRNAs, so their code variations do not affect the much larger cellular genome. The most common changes in mitochondria involve termination codons. These changes affect termination in the products of only a subset of genes, and sometimes the effects are minor because the genes have multiple (redundant) termination codons.

Vertebrate mtDNAs have genes that encode 13 proteins, 2 rRNAs, and 22 tRNAs (see Fig. 19–40a). The small number of codon reassignments, along with an unusual set of wobble rules (p. 1110), makes the 22 tRNAs sufficient to decode the protein genes, as opposed to the 32 tRNAs required for the standard code. In mitochondria, these changes can be viewed as a kind of genomic streamlining, as a smaller genome confers a replication advantage on the organelle. Four codon families (in which the amino acid is determined entirely by the first two nucleotides) are decoded by a single tRNA with a U residue in the first (or wobble) position in the anticodon. Either the U pairs somehow with any of the four possible bases in the third position of the codon or a "two out of three" mechanism is used—that is, no base pairing is needed at the third position. Other tRNAs recognize codons with either A or G in the third position, and yet others recognize U or C, so that virtually all the tRNAs recognize either two or four codons.

In the standard code, only two amino acids are specified by single codons: methionine and tryptophan (see Table 27–3). If all mitochondrial tRNAs recognize two codons, we would expect additional Met and Trp codons in mitochondria. And we find that the single most common code variation is the normal termination codon UGA specifying tryptophan. The tRNA^Trp recognizes and inserts a Trp residue at either UGA or the normal Trp codon, UGG. The second most common variation is conversion of AUA from an Ile codon to a Met codon; the normal Met codon is AUG, and a single tRNA recognizes both codons. The known coding variations in mitochondria are summarized in Table 1.

Turning to the much rarer changes in the codes for cellular (as distinct from mitochondrial) genomes, we find that the only known variation in a bacterium is again the use of UGA to encode Trp residues, occurring in the simplest free-living cell, *Mycoplasma capricolum*. Among eukaryotes, rare extramitochondrial coding changes occur in a few species of ciliated protists, in which both termination codons UAA and UAG can specify glutamine. There are also rare but interesting cases where stop codons have been adapted to encode amino acids that are not among the standard 20, as detailed in Box 27–3.

Changes in the code need not be absolute; a codon might not always encode the same amino acid. For example, in many bacteria—including *E. coli*—GUG (Val) is sometimes used as an initiation codon that specifies Met. This occurs only for those genes in

---

The genetic code is nearly universal. With the intriguing exception of a few minor variations in mitochondria, some bacteria, and some single-celled eukaryotes (Box 27–1), amino acid codons are identical in all species examined so far. Human beings, *E. coli*, tobacco plants, amphibians, and viruses share the same genetic code. Thus it would appear that all life-forms have a common evolutionary ancestor, whose genetic code has been preserved throughout biological evolution. Even the variations reinforce this theme.

Wobble Allows Some tRNAs to Recognize More than One Codon

When several different codons specify one amino acid, the difference between them usually lies at the third

TABLE 1 Known Variant Codon Assignments in Mitochondria

	Codons*				
	UGA	AUA	AGA AGG	CUN	CGG
Normal code assignment	Stop	Ile	Arg	Leu	Arg
Animals					
Vertebrates	Trp	Met	Stop	+	+
Drosophila	Trp	Met	Ser	+	+
Yeasts					
Saccharomyces cerevisiae	Trp	Met	+	Thr	+
Torulopsis glabrata	Trp	Met	+	Thr	?
Schizosaccharomyces pombe	Trp	+	+	+	+
Filamentous fungi	Trp	+	+	+	+
Trypanosomes	Trp	+	+	+	+
Higher plants	+	+	+	+	Trp
Chlamydomonas reinhardtii	?	+	+	+	?

*N indicates any nucleotide; +, codon has the same meaning as in the normal code; ?, codon not observed in this mitochondrial genome.

which the GUG is properly located relative to particular mRNA sequences that affect the initiation of translation (as discussed in Section 27.2).

The most surprising alteration in the genetic code occurs in some fungal species of the genus *Candida*, as originally discovered for *Candida albicans*. *C. albicans* is an organism of high genomic complexity, yet its genetic code has undergone a dramatic change: the CUG codon, which normally encodes Leu, encodes Ser instead. The natural selection pressure for this change is completely unknown. Furthermore, the chemical nature of Ser and Leu are quite different. However, even this change can be understood based on the properties of a universal code. When several codons encode the same amino acid and utilize multiple tRNAs, not all of the codons are used with equal frequency. In a phenomenon called **codon bias**, some codons for a particular amino acid are used more frequently (sometimes much more frequently) than others. The tRNAs for the frequently used codons are often present at much higher concentrations than the tRNAs required for the rarely used codons. Code degeneracy leads to the presence of six codons for Leu. In bacteria, CUG is used often to encode Leu. However, in fungi in the genera that are very closely related to *Candida* but do not

have the coding change, the CUG codon is used quite rarely to encode Leu and is often entirely absent in highly expressed proteins. A change in the coding sense of CUG would thus have a much smaller effect on fungal cell metabolism than might be anticipated if all codons were used equally. The coding change may have occurred by a gradual loss of CUG codons in genes and the tRNA that recognizes CUG as a Leu codon, followed by a capture event—a mutation in the anticodon of a tRNA^Ser that allowed it to recognize CUG. Alternatively, there may have been an intermediate stage in which CUG was recognized as encoding both Leu and Ser, perhaps with contextual signals in the mRNAs that helped one tRNA or another to recognize specific CUG codons (see Box 27–3). Phylogenetic analysis indicates that the reassignment of CUG as a Ser codon occurred in *Candida* ancestors about 150 to 170 million years ago.

These variations tell us that the code is not quite as universal as once believed, but that its flexibility is severely constrained. The variations are obviously derivatives of the normal code, and no example of a completely different code has been found. The limited scope of code variants strengthens the principle that all life on this planet evolved on the basis of a single (slightly flexible) genetic code.

base position (at the 3′ end). For example, alanine is coded by the triplets GCU, GCC, GCA, and GCG. The codons for most amino acids can be symbolized by XY_G^A or XY_C^U. The first two letters of each codon are the primary determinants of specificity, a feature that has some interesting consequences.

Transfer RNAs base-pair with mRNA codons at a three-base sequence on the tRNA called the **antico-**

don. The first base of the codon in mRNA (read in the 5′→3′ direction) pairs with the third base of the anticodon (**Fig. 27–8a**). If the anticodon triplet of a tRNA recognized only one codon triplet through Watson-Crick base pairing at all three positions, cells would have a different tRNA for each amino acid codon. This is not the case, however, because the anticodons in some tRNAs include the nucleotide inosinate (designated I),

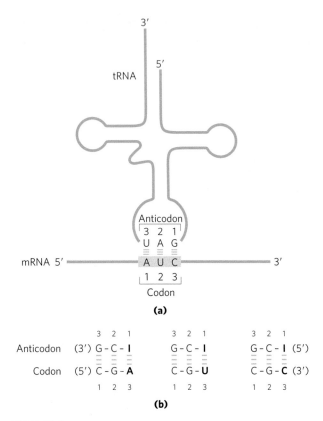

FIGURE 27–8 Pairing relationship of codon and anticodon. (a) Alignment of the two RNAs is antiparallel. The tRNA is shown in the traditional cloverleaf configuration. **(b)** Three different codon pairing relationships are possible when the tRNA anticodon contains inosinate.

which contains the uncommon base hypoxanthine (see Fig. 8–5b). Inosinate can form hydrogen bonds with three different nucleotides (U, C, and A; Fig. 27–8b), although these pairings are much weaker than the hydrogen bonds of Watson-Crick base pairs G≡C and A=U. In yeast, one tRNA^{Arg} has the anticodon (5′)ICG, which recognizes three arginine codons: (5′)CGA, (5′)CGU, and (5′)CGC. The first two bases are identical (CG) and form strong Watson-Crick base pairs with the corresponding bases of the anticodon, but the third base (A, U, or C) forms rather weak hydrogen bonds with the I residue at the first position of the anticodon.

Examination of these and other codon-anticodon pairings led Crick to conclude that the third base of most codons pairs rather loosely with the corresponding base of its anticodon; to use his picturesque word, the third base of such codons (and the first base of their corresponding anticodons) "wobbles." Crick proposed a set of four relationships called the **wobble hypothesis**:

1. The first two bases of an mRNA codon always form strong Watson-Crick base pairs with the corresponding bases of the tRNA anticodon and confer most of the coding specificity.

2. The first base of the anticodon (reading in the 5′→3′ direction; this pairs with the third base of

the codon) determines the number of codons recognized by the tRNA. When the first base of the anticodon is C or A, base pairing is specific and only one codon is recognized by that tRNA. When the first base is U or G, binding is less specific and two different codons may be read. When inosine (I) is the first (wobble) nucleotide of an anticodon, three different codons can be recognized—the maximum number for any tRNA. These relationships are summarized in Table 27–4.

3. When an amino acid is specified by several different codons, the codons that differ in either of the first two bases require different tRNAs.

4. A minimum of 32 tRNAs are required to translate all 61 codons (31 to encode the amino acids and 1 for initiation).

The wobble (or third) base of the codon contributes to specificity, but, because it pairs only loosely with its corresponding base in the anticodon, it permits rapid dissociation of the tRNA from its codon during protein synthesis. If all three bases of a codon engaged in strong Watson-Crick pairing with the three bases of the anticodon, tRNAs would dissociate too slowly and this would limit the rate of protein synthesis. Codon-anticodon interactions balance the requirements for accuracy and speed.

The Genetic Code Is Mutation-Resistant

The genetic code plays an interesting role in safeguarding the genomic integrity of every living organism. Evolution did not produce a code in which codon assignments

---

TABLE 27–4 How the Wobble Base of the Anticodon Determines the Number of Codons a tRNA Can Recognize

1. One codon recognized:

Anticodon	(3′) X−Y− **C** (5′)	(3′) X−Y− **A** (5′)
Codon	(5′) X′−Y′−**G** (3′)	(5′) X′−Y′−**U** (3′)

2. Two codons recognized:

Anticodon	(3′) X−Y− **U** (5′)	(3′) X−Y− **G** (5′)
Codon	(5′) X′−Y′−^A_G (3′)	(5′) X′−Y′−^C_U (3′)

3. Three codons recognized:

Anticodon	(3′) X−Y− **I** (5′)
Codon	(5′) X′−Y′−$^A_U{}_C$ (3′)

Note: X and Y denote bases complementary to and capable of strong Watson-Crick base pairing with X′ and Y′, respectively. Wobble bases—in the 3′ position of codons and 5′ position of anticodons—are shaded in white.

appeared at random. Instead, the code is strikingly resistant to the deleterious effects of the most common kinds of mutations—**missense mutations**, in which a single new base pair replaces another. In the third, or wobble, position of the codon, single base substitutions produce a change in the encoded amino acid only about 25% of the time. Most such changes are thus **silent mutations,** where the nucleotide is different but the encoded amino acid remains the same.

Due to the types of spontaneous DNA damage that affect genomes (see Chapter 8), the most frequent missense mutation is a **transition mutation**, where a purine is replaced by a purine or a pyrimidine by a pyrimidine (for example, G≡C changed to A=T). All three codon positions have evolved so that there is some resistance to transition mutations. A mutation in the first position of the codon will usually result in an amino acid coding change, but the change often involves an amino acid with similar chemical properties. This is especially true for the hydrophobic amino acids that dominate the first column of the code shown in Figure 27–7. Consider the Val codon GUU. A change to AUU would substitute Ile for Val. A change to CUU would replace Val with Leu. The resulting changes in the structure and/or function of the protein encoded by that gene would often (but not always) be small.

Computational studies have shown that genetic codes delineated at random are almost always less resistant to mutation than the existing code. The results indicate that the code underwent considerable streamlining prior to the appearance of LUCA, the ancestral cell.

The genetic code tells us how protein sequence information is stored in nucleic acids and provides some clues about how that information is translated into protein.

Translational Frameshifting and RNA Editing Affect How the Code Is Read

Once the reading frame has been set during protein synthesis, codons are translated without overlap or punctuation until the ribosomal complex encounters a termination codon. The other two possible reading frames usually contain no useful genetic information, but a few genes are structured so that ribosomes "hiccup" at a certain point in the translation of their mRNAs, changing the reading frame from that point on. This appears to be a mechanism either to allow two or more related but distinct proteins to be produced from a single transcript or to regulate the synthesis of a protein.

One of the best-documented examples of **translational frameshifting** occurs during translation of the mRNA for the overlapping *gag* and *pol* genes of the Rous sarcoma virus (see Fig. 26–34). The reading frame for *pol* is offset to the left by one base pair (–1 reading frame) relative to the reading frame for *gag* **(Fig. 27–9)**.

The product of the *pol* gene (reverse transcriptase) is translated as a larger polyprotein, on the same mRNA that is used for the *gag* protein alone (see Fig. 26–33). The polyprotein, or *gag-pol* protein, is then trimmed to the mature reverse transcriptase by proteolytic digestion. Production of the polyprotein requires a translational frameshift in the overlap region to allow the ribosome to bypass the UAG termination codon at the end of the *gag* gene (shaded light red in Fig. 27–9).

Frameshifts occur during about 5% of translations of this mRNA, and the *gag-pol* polyprotein (and ultimately reverse transcriptase) is synthesized at about one-twentieth the frequency of the *gag* protein, a level that suffices for efficient reproduction of the virus. In some retroviruses, another translational frameshift allows translation of an even larger polyprotein that includes the product of the *env* gene fused to the *gag* and *pol* gene products (see Fig. 26–33). A similar mechanism produces both the τ and γ subunits of *E. coli* DNA polymerase III from a single *dnaX* gene transcript (see Table 25–2).

Some mRNAs are edited before translation. **RNA editing** can involve the addition, deletion, or alteration of nucleotides in the RNA in a manner that affects the meaning of the transcript when it is translated. Addition or deletion of nucleotides has been most commonly observed in RNAs originating from the mitochondrial and chloroplast genomes of eukaryotes. The reactions require a special class of RNA molecules encoded by these same organelles, with sequences complementary to the edited mRNAs. These guide RNAs (gRNAs; **Fig. 27–10**) act as templates for the editing process.

The initial transcripts of the genes that encode cytochrome oxidase subunit II in some protist mitochondria provide an example of editing by insertion. These transcripts do not correspond precisely to the sequence needed at the carboxyl terminus of the protein product. A posttranscriptional editing process inserts four U residues that shift the translational reading

FIGURE 27–9 Translational frameshifting in a retroviral transcript. The *gag-pol* overlap region in Rous sarcoma virus RNA is shown.

DNA
coding 5'---|A A A|G T A|G A G|A A C|C T G|G T A|--- 3'
strand
---   Lys — Val   Glu   Asn   Leu   Val

Edited   ---|A A A|G U A|G A U|U G U|A U A|C C U|G G U|---
mRNA
---   Lys — Val — Asp — Cys — Ile — Pro — Gly

(a)

mRNA   5'--- A A A G U A G A U U G U A U A C C U G G U---3'

Guide RNA   U U A U A U C U A A U A U A U G G A U A U

    3'

(b)

FIGURE 27–10 RNA editing of the transcript of the cytochrome oxidase subunit II gene from *Trypanosoma brucei* mitochondria. **(a)** Insertion of four U residues (red) produces a revised reading frame. **(b)** A special class of guide RNAs, complementary to the edited product, act as templates for the editing process. Note the presence of two G≡U base pairs, signified by a blue dot to indicate non-Watson-Crick pairing.

frame of the transcript. Figure 27–10 shows the added U residues in the small part of the transcript that is affected by editing. Note that the base pairing between the initial transcript and the guide RNA involves a number of G≡U base pairs (blue dots), which are common in RNA molecules.

RNA editing by alteration of nucleotides most commonly involves the enzymatic deamination of adenosine or cytidine residues, forming inosine or uridine, respectively **(Fig. 27–11)**, although other base changes have been described. Inosine is interpreted as a G residue during translation. The adenosine deamination reactions are carried out by *a*denosine *d*eaminases that *a*ct on *R*NA **(ADARs)**. The cytidine deaminations are carried out by the *apoB* mRNA *e*diting *c*atalytic *p*eptide (**APOBEC**) family of enzymes, which includes the related *a*ctivation-*i*nduced *d*eaminase (**AID**) enzymes. Both groups of deaminase enzymes have a homologous zinc-coordinating catalytic domain.

A well-studied example of RNA editing by deamination occurs in the gene for the apolipoprotein B component of low-density lipoprotein in vertebrates. One form of apolipoprotein B, apoB-100 (M_r 513,000), is synthesized in the liver; a second form, apoB-48 (M_r 250,000), is synthesized in the intestine. Both are encoded by an mRNA produced from the gene for apoB-100. An APO-BEC cytidine deaminase found only in the intestine binds to the mRNA at the codon for amino acid residue 2,153 (CAA = Gln) and converts the C to a U to create the termination codon UAA. The apoB-48 produced in

FIGURE 27–11 Deamination reactions that result in RNA editing. (a) The conversion of adenosine nucleotides to inosine nucleotides is catalyzed by ADAR enzymes. **(b)** Cytidine-to-uridine conversions are catalyzed by the APOBEC family of enzymes.

FIGURE 27–12 RNA editing of the transcript of the gene for the apoB-100 component of LDL. Deamination, which occurs only in the intestine, converts a specific cytidine to uridine, changing a Gln codon to a stop codon and producing a truncated protein.

the intestine from this modified mRNA is simply an abbreviated form (corresponding to the amino-terminal half) of apoB-100 **(Fig. 27–12)**. This reaction permits tissue-specific synthesis of two different proteins from one gene.

The ADAR-promoted A-to-I editing is particularly common in transcripts derived from the genes of primates. Perhaps 90% or more of the editing occurs in Alu elements, a subset of the eukaryotic transposons called short interspersed elements (SINEs), that are particularly common in mammalian genomes. There are over a million of the 300 bp Alu elements in human DNA, making up about 10% of the genome. These are concentrated near protein-encoding genes, often appearing in introns and untranslated regions at the 3′ and 5′ ends of transcripts. When it is first synthesized (prior to processing), the *average* human mRNA includes 10 to 20 Alu elements. The ADAR enzymes bind to and promote A-to-I editing only in duplex regions of RNA. The abundant Alu elements offer many opportunities for intramolecular base pairing within the transcripts, providing the duplex targets required by the ADARs. Some of the editing affects the coding sequences of genes. Defects in ADAR function have been associated with a variety of human neurological conditions, including amyotrophic lateral sclerosis (ALS), epilepsy, and major depression.

The genomes of all vertebrates are replete with SINEs, but many different types of SINES are present in most of these organisms. The Alu elements predominate only in the primates. Careful screening of genes and transcripts indicates that A-to-I editing is 30 to 40 times more prevalent in humans than in mice, largely due to the presence of many Alu elements. Large-scale A-to-I editing and an increased level of alternative splicing (see Fig. 26–21) are two features that set primate genomes apart from those of other mammals. It is not yet clear whether these reactions are incidental or whether they played key roles in the evolution of primates and, ultimately, humans.

SUMMARY 27.1 The Genetic Code

▶ The particular amino acid sequence of a protein is constructed through the translation of information encoded in mRNA. This process is carried out by ribosomes.

▶ Amino acids are specified by mRNA codons consisting of nucleotide triplets. Translation requires adaptor molecules, the tRNAs, that recognize codons and insert amino acids into their appropriate sequential positions in the polypeptide.

▶ The base sequences of the codons were deduced from experiments using synthetic mRNAs of known composition and sequence.

▶ The codon AUG signals initiation of translation. The triplets UAA, UAG, and UGA are signals for termination.

▶ The genetic code is degenerate: it has multiple codons for almost every amino acid.

▶ The standard genetic code is universal in all species, with some minor deviations in mitochondria and a few single-celled organisms. The deviations occur in a pattern that reinforces the concept of a universal code.

▶ The third position in each codon is much less specific than the first and second and is said to wobble.

▶ The genetic code is resistant to the effects of missense mutations.

▶ Translational frameshifting and RNA editing affect how the genetic code is read during translation.

27.2 Protein Synthesis

As we have seen for DNA and RNA (Chapters 25 and 26), the synthesis of polymeric biomolecules can be considered in terms of initiation, elongation, and termination stages. These fundamental processes are typically bracketed by two additional stages: activation of precursors before synthesis and postsynthetic processing of the completed polymer. Protein synthesis follows the same pattern. The activation of amino acids before their incorporation into polypeptides and the posttranslational processing of the completed polypeptide play particularly important roles in ensuring both the fidelity of synthesis and the proper function of the protein product. The process is outlined in **Figure 27–13**. The cellular components involved in the five stages of protein synthesis in *E. coli* and other bacteria are listed in Table 27–5; the requirements in eukaryotic cells are quite similar, although the components are in some cases more numerous. An initial overview of the stages of protein synthesis provides a useful outline for the discussion that follows.

1 **Activation of amino acids:** the tRNA is aminoacylated

2 **Initiation:** the mRNA and the aminoacylated tRNA bind to the small ribsomal subunit. The large subunit then binds as well.

Large subunit

Aminoacyl-tRNA

Small subunit

5′   3′

Start   Stop

mRNA

4 **Termination:** translation stops when a stop codon is encountered. The mRNA and protein dissociate and the ribosomal subunits are recycled.

5 **Protein folding**

3 **Elongation:** successive cycles of aminoacyl-tRNA binding and peptide bond formation occur until ribosome reaches a stop codon.

FIGURE 27–13 **An overview of the five stages of protein synthesis.** **1** The tRNAs are aminoacylated. **2** Translation initiation occurs when an mRNA and an aminoacylated tRNA are bound to the ribosome. **3** In elongation, the ribosome moves along the mRNA, matching tRNAs to each codon and catalyzing peptide bond formation. **4** Translation is terminated at a stop codon, and the ribosomal subunits are released and recycled for another round of protein synthesis. **5** Following synthesis, the protein must fold into its active conformation and ribosome components are recycled.

Protein Biosynthesis Takes Place in Five Stages

Stage 1: Activation of Amino Acids For the synthesis of a polypeptide with a defined sequence, two fundamental chemical requirements must be met: (1) the carboxyl group of each amino acid must be activated to facilitate formation of a peptide bond, and (2) a link must be established between each new amino acid and the information in the mRNA that encodes it. Both these requirements are met by attaching the amino acid to a tRNA in the first stage of protein synthesis. Attaching the right amino acid to the right tRNA is critical. This reaction takes place in the cytosol, not on the ribosome. Each of the 20 amino acids is covalently attached to a specific tRNA at the expense of ATP energy, using Mg^{2+}-dependent activating enzymes known as aminoacyl-tRNA synthetases. When attached to their amino acid (aminoacylated) the tRNAs are said to be "charged."

Stage 2: Initiation The mRNA bearing the code for the polypeptide to be synthesized binds to the smaller of two ribosomal subunits and to the initiating aminoacyl-tRNA. The large ribosomal subunit then binds to form an initiation complex. The initiating aminoacyl-tRNA base-pairs with the mRNA codon AUG that signals the beginning of the polypeptide. This process, which requires GTP, is promoted by cytosolic proteins called initiation factors.

Stage 3: Elongation The nascent polypeptide is lengthened by covalent attachment of successive amino acid units, each carried to the ribosome and correctly positioned by its tRNA, which base-pairs to its corresponding codon in the mRNA. Elongation requires cytosolic proteins known as elongation factors. The binding of each incoming aminoacyl-tRNA and the movement of the ribosome along the mRNA are facilitated by the hydrolysis of GTP as each residue is added to the growing polypeptide.

Stage 4: Termination and Ribosome Recycling Completion of the polypeptide chain is signaled by a termination codon in the mRNA. The new polypeptide is released from the ribosome, aided by proteins called release factors, and the ribosome is recycled for another round of synthesis.

Stage 5: Folding and Posttranslational Processing In order to achieve its biologically active form, the new polypeptide must fold into its proper three-dimensional conformation.

TABLE 27–5 Components Required for the Five Major Stages of Protein Synthesis in *E. coli*

Stage	Essential components
1. Activation of amino acids	20 amino acids 20 aminoacyl-tRNA synthetases 32 or more tRNAs ATP Mg^{2+}
2. Initiation	mRNA *N*-Formylmethionyl-tRNA^{fMet} Initiation codon in mRNA (AUG) 30S ribosomal subunit 50S ribosomal subunit Initiation factors (IF-1, IF-2, IF-3) GTP Mg^{2+}
3. Elongation	Functional 70S ribosome (initiation complex) Aminoacyl-tRNAs specified by codons Elongation factors (EF-Tu, EF-Ts, EF-G) GTP Mg^{2+}
4. Termination and ribosome recycling	Termination codon in mRNA Release factors (RF-1, RF-2, RF-3, RRF) EF-G IF-3
5. Folding and posttranslational processing	Chaperones and folding enzymes (PPI, PDI); specific enzymes, cofactors, and other components for removal of initiating residues and signal sequences, additional proteolytic processing, modification of terminal residues, and attachment of acetyl, phosphoryl, methyl, carboxyl, carbohydrate, or prosthetic groups

Before or after folding, the new polypeptide may undergo enzymatic processing, including removal of one or more amino acids (usually from the amino terminus); addition of acetyl, phosphoryl, methyl, carboxyl, or other groups to certain amino acid residues; proteolytic cleavage; and/or attachment of oligosaccharides or prosthetic groups.

Before looking at these five stages in detail, we must examine two key components in protein biosynthesis: the ribosome and tRNAs.

The Ribosome Is a Complex Supramolecular Machine

Each *E. coli* cell contains 15,000 or more ribosomes, which comprise nearly a quarter of the dry weight of the cell. Bacterial ribosomes contain about 65% rRNA and 35% protein; they have a diameter of about 18 nm and are composed of two unequal subunits with sedimentation coefficients of 30S and 50S and a combined sedimentation coefficient of 70S. Both subunits contain dozens of ribosomal proteins and at least one large rRNA (Table 27–6).

Following Zamecnik's discovery that ribosomes are the complexes responsible for protein synthesis, and following elucidation of the genetic code, the study of ribosomes accelerated. In the late 1960s Masayasu Nomura and colleagues demonstrated that both ribosomal subunits can be broken down into their RNA and protein components, then reconstituted in vitro. Under appropriate experimental conditions, the RNA and protein spontaneously reassemble to form 30S or 50S subunits nearly identical in structure and activity to native subunits. This breakthrough fueled decades of research into the function and structure of ribosomal RNAs and proteins. At the same time, increasingly sophisticated structural methods revealed more and more details about ribosome structure.

Masayasu Nomura, 1927–2011

TABLE 27-6   RNA and Protein Components of the *E. coli* Ribosome

Subunit	Number of different proteins	Total number of proteins	Protein designations	Number and type of rRNAs
30S	21	21	S1–S21	1 (16S rRNA)
50S	33	36	L1–L36*	2 (5S and 23S rRNAs)

*The L1 to L36 protein designations do not correspond to 36 different proteins. The protein originally designated L7 is in fact a modified form of L12, and L8 is a complex of three other proteins. Also, L26 proved to be the same protein as S20 (and not part of the 50S subunit). This gives 33 different proteins in the large subunit. There are four copies of the L7/L12 protein, with the three extra copies bringing the total protein count to 36.

The dawn of a new millennium brought with it the elucidation of the first high-resolution structures of bacterial ribosomal subunits by Thomas Steitz, Ada Yonath, Venki Ramakrishnan, Harry Noller, and others. This work yielded a wealth of surprises **(Fig. 27–14a)**. First, a traditional focus on the protein components of ribosomes was shifted. The ribosomal subunits are huge RNA molecules. In the 50S subunit, the 5S and 23S rRNAs form the structural core. The proteins are secondary elements in the complex, decorating the surface. Second and most important, there is no protein within 18 Å of the active site for peptide bond formation. The high-resolution structure thus confirms what Harry Noller had predicted much earlier: the ribosome is a ribozyme. In addition to the insight that the detailed structures of the ribosome and its subunits provide into the mechanism of protein synthesis (as elaborated below), they have stimulated a new look at the evolution of life (Box 27–2). The ribosomes of eukaryotic cells have also yielded to structural analysis (Fig. 27–14b).

The bacterial ribosome is complex, with a combined molecular weight of ~2.7 million. The two irregularly shaped ribosomal subunits fit together to form a cleft through which the mRNA passes as the ribosome moves along it during translation (Fig. 27–14a). The 57 proteins in bacterial ribosomes vary enormously in size and structure. Molecular weights range from

Venkatraman Ramakrishnan   Thomas A. Steitz   Ada E. Yonath

FIGURE 27–14 The structure of ribosomes. Our understanding of ribosome structure has been greatly enhanced by multiple high-resolution images of the ribosomes from bacteria and yeast. **(a)** The bacterial ribosome (derived from PDB ID 2OW8 and PDB ID 1VSA). The 50S and 30S subunits come together to form the 70S ribosome. A cleft between them is where protein synthesis occurs. **(b)** The yeast ribosome has a similar structure with somewhat increased complexity (derived from PDB ID 3O58 and PDB ID 3O2Z).

BOX 27–2 From an RNA World to a Protein World

Extant ribozymes generally promote one of two types of reactions: hydrolytic cleavage of phosphodiester bonds or phosphoryl transfers (Chapter 26). In both cases, the substrates of the reactions are also RNA molecules. The ribosomal RNAs provide an important expansion of the catalytic range of known ribozymes. Coupled to the laboratory exploration of potential RNA catalytic function (see Box 26–3), the idea of an RNA world as a precursor to current life-forms becomes increasingly attractive.

A viable RNA world would require an RNA capable of self-replication, a primitive metabolism to generate the needed ribonucleotide precursors, and a cell boundary to aid in concentrating the precursors and sequestering them from the environment. The requirements for catalysis of reactions involving a growing range of metabolites and macromolecules could have led to larger and more complex RNA catalysts. The many negatively charged phosphoryl groups in the RNA backbone limit the stability of very large RNA molecules. In an RNA world, divalent cations or other positively charged groups could be incorporated into the structures to augment stability.

Certain peptides could stabilize large RNA molecules. For example, many ribosomal proteins in modern eukaryotic cells have long extensions, lacking a regular secondary structure, that snake into the rRNAs and help stabilize them (Fig. 1). Ribozyme-catalyzed synthesis of peptides could thus initially have evolved as part of a general solution to the structural maintenance of large RNA molecules. The synthesis of peptides may have helped stabilize large ribozymes, but this advance also marked the beginning of the end for the RNA world. Once peptide synthesis was possible, the greater catalytic potential of proteins would have set in motion an irreversible transition to a protein-dominated metabolic system.

Most enzymatic processes, then, were eventually surrendered to the proteins—but not all. In every organism, the critical task of synthesizing the proteins

remains, even now, a ribozyme-catalyzed process. There appears to be only one good arrangement (or just a very few) of nucleotide residues in a ribozyme active site that can catalyze peptide synthesis. The rRNA residues that seem to be involved in the peptidyl transferase activity of ribosomes are highly conserved in the large-subunit rRNAs of all species. Using in vitro evolution (SELEX; see Box 26–3), investigators have isolated artificial ribozymes that promote peptide synthesis. Intriguingly, most of them include the ribonucleotide octet (5′)AUAACAGG(3′), a highly conserved sequence found at the peptidyl transferase active site in the ribosomes of all cells. There may be just one optimal solution to the overall chemical problem of ribozyme-catalyzed synthesis of proteins of defined sequence. Evolution found this solution once, and no life-form has notably improved on it.

FIGURE 1 The 50S subunit of a bacterial ribosome (PDB ID 1Q7Y). The protein backbones are shown as blue wormlike structures; the rRNA components are transparent. The unstructured extensions of many of the ribosomal proteins snake into the rRNA structures, helping to stabilize them. Bound puromycin, in red, marks the rRNA active site for peptidyl transferase.

---

about 6,000 to 75,000. Most of the proteins have globular domains arranged on the ribosome surface. Some also have snakelike extensions that protrude into the rRNA core of the ribosome, stabilizing its structure. The functions of some of these proteins have not yet been elucidated in detail, although a structural role seems evident for many of them.

The sequences of the rRNAs of many organisms are now known. Each of the three single-stranded rRNAs of *E. coli* has a specific three-dimensional conformation featuring extensive intrachain base pairing. The folding patterns of the rRNAs are highly conserved in all organisms, particularly the regions implicated in key func-

tions **(Fig. 27–15)**. The predicted secondary structure of the rRNAs has largely been confirmed by structural analysis but fails to convey the extensive network of tertiary interactions apparent in the complete structure.

The ribosomes of eukaryotic cells (other than mitochondrial and chloroplast ribosomes) are larger and more complex than bacterial ribosomes **(Fig. 27–16**; compare Fig. 27–14b)**, with a diameter of about 23 nm and a sedimentation coefficient of about 80S. They also have two subunits, which vary in size among species but on average are 60S and 40S. Altogether, eukaryotic ribosomes contain more than 80 different proteins. The ribosomes of mitochondria and chloroplasts are somewhat smaller

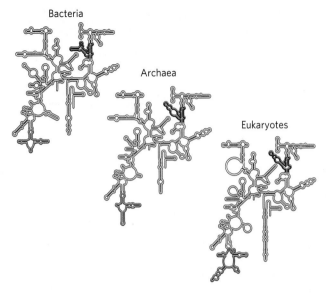

Bacteria

Archaea

Eukaryotes

FIGURE 27–15 Conservation of secondary structure in the small subunit rRNAs from the three domains of life. The red, yellow, and purple indicate areas where the structures of the rRNAs from bacteria, archaea, and eukaryotes have diverged. Conserved regions are shown in green.

Bacterial ribosome 70S M_r 2.7×10^6	Eukaryotic ribosome 80S M_r 4.2×10^6
50S	60S
M_r 1.8×10^6	M_r 2.8×10^6
5S rRNA (120 nucleotides) 23S rRNA (3,200 nucleotides) 36 proteins	5S rRNA (120 nucleotides) 28S rRNA (4,700 nucleotides) 5.8S rRNA (160 nucleotides) 47 proteins
30S	40S
M_r 0.9×10^6	M_r 1.4×10^6
16S rRNA (1,540 nucleotides) 21 proteins	18S rRNA (1,900 nucleotides) 33 proteins

FIGURE 27–16 Summary of the composition and mass of ribosomes in bacteria and eukaryotes. Ribosomal subunits are identified by their S (Svedberg unit) values, sedimentation coefficients that refer to their rate of sedimentation in a centrifuge. The S values are not additive when subunits are combined, because S values are approximately proportional to the ⅔ power of molecular weight and are also slightly affected by shape.

and simpler than bacterial ribosomes. Nevertheless, ribosomal structure and function are strikingly similar in all organisms and organelles.

Transfer RNAs Have Characteristic Structural Features

To understand how tRNAs can serve as adaptors in translating the language of nucleic acids into the language of proteins, we must first examine their structure in more detail. Transfer RNAs are relatively small and consist of a single strand of RNA folded into a precise three-dimensional structure (see Fig. 8–25a). The tRNAs in bacteria and in the cytosol of eukaryotes have between 73 and 93 nucleotide residues, corresponding to molecular weights of 24,000 to 31,000. Mitochondria and chloroplasts contain distinctive, somewhat smaller tRNAs. Cells have at least one kind of tRNA for each amino acid; at least 32 tRNAs are required to recognize all the amino acid codons (some recognize more than one codon), but some cells use more than 32.

Yeast alanine tRNA (tRNA^{Ala}) was the first nucleic acid to be completely sequenced, by Robert Holley in 1965. It contains 76 nucleotide residues, 10 of which have modified bases. Comparisons of tRNAs from various species have revealed many common structural features **(Fig. 27–17)**. Eight or more of the nucleotide residues have modified bases and sugars, many of which are methylated derivatives of the principal bases. Most tRNAs have a guanylate (pG) residue at the 5' end, and all have the tri-

Robert W. Holley, 1922–1993

nucleotide sequence CCA(3') at the 3' end. When drawn in two dimensions, the hydrogen-bonding pattern of all tRNAs forms a cloverleaf structure with four arms; the longer tRNAs have a short fifth arm, or extra arm. In three dimensions, a tRNA has the form of a twisted L **(Fig. 27–18)**.

Two of the arms of a tRNA are critical for its adaptor function. The **amino acid arm** can carry a specific amino acid esterified by its carboxyl group to the 2'- or 3'-hydroxyl group of the A residue at the 3' end of the tRNA. The **anticodon arm** contains the anticodon. The other major arms are the **D arm**, which contains the unusual nucleotide dihydrouridine (D), and the **TψC arm**, which contains ribothymidine (T), not usually present in RNAs, and pseudouridine (ψ), which has an unusual carbon–carbon bond between the base and ribose (see Fig. 26–22). The D and TψC arms contribute important interactions for the overall folding of tRNA molecules, and the TψC arm interacts with the large-subunit rRNA.

Having looked at the structures of ribosomes and tRNAs, we now consider in detail the five stages of protein synthesis.

FIGURE 27–17 **General cloverleaf secondary structure of tRNAs.** The large dots on the backbone represent nucleotide residues; the blue lines represent base pairs. Characteristic and/or invariant residues common to all tRNAs are shaded in light red. Transfer RNAs vary in length from 73 to 93 nucleotides. Extra nucleotides occur in the extra arm or in the D arm. At the end of the anticodon arm is the anticodon loop, which always contains seven unpaired nucleotides. The D arm contains two or three D (5,6-dihydrouridine) residues, depending on the tRNA. In some tRNAs, the D arm has only three hydrogen-bonded base pairs. Symbols are: Pu, purine nucleotide; Py, pyrimidine nucleotide; ψ, pseudouridylate; G*, either guanylate or 2′-O-methylguanylate.

Stage 1: Aminoacyl-tRNA Synthetases Attach the Correct Amino Acids to Their tRNAs

During the first stage of protein synthesis, taking place in the cytosol, aminoacyl-tRNA synthetases esterify the 20 amino acids to their corresponding tRNAs. Each enzyme is specific for one amino acid and one or more corresponding tRNAs. Most organisms have one aminoacyl-tRNA synthetase for each amino acid. For amino acids with two or more corresponding tRNAs, the same enzyme usually aminoacylates all of them.

The structures of all the aminoacyl-tRNA synthetases of *E. coli* have been determined. Researchers have divided them into two classes (Table 27–7) based on substantial differences in primary and tertiary structure and in reaction mechanism **(Fig. 27–19)**; these two classes are the same in all organisms. There is no

TABLE 27–7 The Two Classes of Aminoacyl-tRNA Synthetases

Class I		Class II	
Arg	Leu	Ala	Lys
Cys	Met	Asn	Phe
Gln	Trp	Asp	Pro
Glu	Tyr	Gly	Ser
Ile	Val	His	Thr

Note: Here, Arg represents arginyl-tRNA synthetase, and so forth. The classification applies to all organisms for which tRNA synthetases have been analyzed and is based on protein structural distinctions and on the mechanistic distinction outlined in Figure 27–19.

(a)

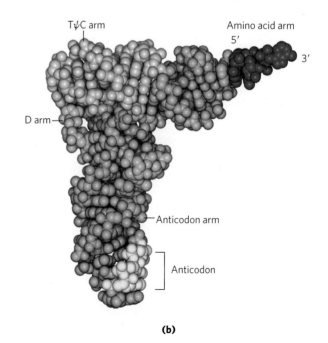

(b)

FIGURE 27–18 **Three-dimensional structure of yeast tRNA^Phe deduced from x-ray diffraction analysis.** The shape resembles a twisted L. **(a)** Schematic diagram with the various arms identified in Figure 27–17 shaded in different colors. **(b)** A space-filling model, with the same color coding (PDB ID 4TRA). The CCA sequence at the 3′ end (purple) is the attachment point for the amino acid.

Amino acid

MECHANISM FIGURE 27–19 Aminoacylation of tRNA by aminoacyl-tRNA synthetases. Step ❶ is formation of an aminoacyl adenylate, which remains bound to the active site. In the second step the aminoacyl group is transferred to the tRNA. The mechanism of this step is somewhat different for the two classes of aminoacyl-tRNA synthetases (see Table 27–7). For class I enzymes, ❷ⓐ the aminoacyl group is transferred initially to the 2'-hydroxyl group of the 3'-terminal A residue, then ❸ⓐ to the 3'-hydroxyl group by a transesterification reaction. For class II enzymes, ❷ⓑ the aminoacyl group is transferred directly to the 3'-hydroxyl group of the terminal adenylate.

Carboxyl of amino acid attacks α phosphorus of ATP, forming 5'-aminoacyl adenylate.

❶ ⟶ PP$_i$

5'-Aminoacyl adenylate (aminoacyl-AMP)

class I aminoacyl-tRNA synthetases

class II aminoacyl-tRNA synthetases

3' end of tRNA

Aminoacyl-AMP

tRNA

AMP ← ❷ⓐ Aminoacyl group is transferred to 2' —OH of the 3'-terminal A residue of tRNA, releasing AMP.

AMP ← ❷ⓑ Aminoacyl group is transferred directly to the 3' —OH of the 3'-terminal A residue of tRNA, generating the aminoacyl-tRNA product.

transesterification

❸ⓐ

Transesterification moves aminoacyl group to 3' —OH of the same tRNA residue, generating the aminoacyl-tRNA product.

Aminoacyl-tRNA

evidence that the two classes share a common ancestor, and the biological, chemical, or evolutionary reasons for two enzyme classes for essentially identical processes remain obscure.

The reaction catalyzed by an aminoacyl-tRNA synthetase is

$$\text{Amino acid} + \text{tRNA} + \text{ATP} \xrightarrow{\text{Mg}^{2+}}$$
$$\text{aminoacyl-tRNA} + \text{AMP} + \text{PP}_i$$

This reaction occurs in two steps in the enzyme's active site. In step ❶ (Fig. 27–19) an enzyme-bound intermediate, aminoacyl adenylate (aminoacyl-AMP), is formed. In the second step the aminoacyl group is transferred from enzyme-bound aminoacyl-AMP to its corresponding specific tRNA. The course of this second step depends on the class to which the enzyme belongs, as shown by pathways ❷ⓐ and ❷ⓑ in Figure 27–19. The resulting ester linkage between the amino acid and the tRNA **(Fig. 27–20)** has a highly negative standard free energy of hydrolysis ($\Delta G'^{\circ} = -29$ kJ/mol). The pyrophosphate formed in the activation reaction undergoes hydrolysis to phosphate by inorganic pyrophosphatase. Thus *two* high-energy phosphate bonds are ultimately expended for each amino acid molecule activated, rendering the overall reaction for amino acid activation essentially irreversible:

$$\text{Amino acid} + \text{tRNA} + \text{ATP} \xrightarrow{\text{Mg}^{2+}}$$
$$\text{aminoacyl-tRNA} + \text{AMP} + 2\text{P}_i$$
$$\Delta G'^{\circ} \approx -29 \text{ kJ/mol}$$

Proofreading by Aminoacyl-tRNA Synthetases The aminoacylation of tRNA accomplishes two ends: (1) it activates an amino acid for peptide bond formation and (2) it

ensures appropriate placement of the amino acid in a growing polypeptide. The identity of the amino acid attached to a tRNA is not checked on the ribosome, so attachment of the correct amino acid to the tRNA is essential to the fidelity of protein synthesis.

As you will recall from Chapter 6, enzyme specificity is limited by the binding energy available from enzyme-substrate interactions. Discrimination between two similar amino acid substrates has been studied in detail in the case of Ile-tRNA synthetase, which distinguishes between valine and isoleucine, amino acids that differ by only a single methylene group ($-\text{CH}_2-$):

Valine   Isoleucine

Ile-tRNA synthetase favors activation of isoleucine (to form Ile-AMP) over valine by a factor of 200—as we would expect, given the amount by which a methylene group (in Ile) could enhance substrate binding. Yet valine is erroneously incorporated into proteins in positions normally occupied by an Ile residue at a frequency of only about 1 in 3,000. How is this greater than 10-fold increase in accuracy brought about? Ile-tRNA synthetase, like some other aminoacyl-tRNA synthetases, has a proofreading function.

Recall a general principle from the discussion of proofreading by DNA polymerases (see Fig. 25–7): if available binding interactions do not provide sufficient discrimination between two substrates, the necessary specificity can be achieved by substrate-specific binding in *two successive* steps. The effect of forcing the system through two successive filters is multiplicative. In the case of Ile-tRNA synthetase, the first filter is the initial binding of the amino acid to the enzyme and its activation to aminoacyl-AMP. The second is the binding of any *incorrect* aminoacyl-AMP products to a separate active site on the enzyme; a substrate that binds in this second active site is hydrolyzed. The R group of valine is slightly smaller than that of isoleucine, so Val-AMP fits the hydrolytic (proofreading) site of the Ile-tRNA synthetase but Ile-AMP does not. Thus Val-AMP is hydrolyzed to valine and AMP in the proofreading active site, and tRNA bound to the synthetase does not become aminoacylated to the wrong amino acid.

In addition to proofreading after formation of the aminoacyl-AMP intermediate, most aminoacyl-tRNA synthetases can hydrolyze the ester linkage between amino acids and tRNAs in the aminoacyl-tRNAs. This hydrolysis is greatly accelerated for incorrectly charged tRNAs, providing yet a third filter to enhance the fidelity of the

FIGURE 27–20 General structure of aminoacyl-tRNAs. The aminoacyl group is esterified to the 3′ position of the terminal A residue. The ester linkage that both activates the amino acid and joins it to the tRNA is shaded light red.

overall process. The few aminoacyl-tRNA synthetases that activate amino acids with no close structural relatives (Cys-tRNA synthetase, for example) demonstrate little or no proofreading activity; in these cases, the active site for aminoacylation can sufficiently discriminate between the proper substrate and any incorrect amino acid.

The overall error rate of protein synthesis (~1 mistake per 10^4 amino acids incorporated) is not nearly as low as that of DNA replication. Because flaws in a protein are eliminated when the protein is degraded and are not passed on to future generations, they have less biological significance. The degree of fidelity in protein synthesis is sufficient to ensure that most proteins contain no mistakes and that the large amount of energy required to synthesize a protein is rarely wasted. One defective protein molecule is usually unimportant when many correct copies of the same protein are present.

Interaction between an Aminoacyl-tRNA Synthetase and a tRNA: A "Second Genetic Code" An individual aminoacyl-tRNA synthetase must be specific not only for a single amino acid but for certain tRNAs as well. Discriminating among dozens of tRNAs is just as important for the overall fidelity of protein biosynthesis as is distinguishing among amino acids. The interaction between aminoacyl-tRNA synthetases and tRNAs has been referred to

as the "second genetic code," reflecting its critical role in maintaining the accuracy of protein synthesis. The "coding" rules appear to be more complex than those in the "first" code.

Figure 27–21 summarizes what we know about the nucleotides involved in recognition by some aminoacyl-tRNA synthetases. Some nucleotides are conserved in all tRNAs and therefore cannot be used for discrimination. By observing changes in nucleotides that alter substrate specificity, researchers have identified nucleotide positions that are involved in discrimination by the aminoacyl-tRNA synthetases. These nucleotide positions seem to be concentrated in the amino acid arm and the anticodon arm, including the nucleotides of the anticodon itself, but are also located in other parts of the tRNA molecule. Determination of the crystal structures of aminoacyl-tRNA synthetases complexed with their cognate tRNAs and ATP has added a great deal to our understanding of these interactions **(Fig. 27–22)**.

Ten or more specific nucleotides may be involved in recognition of a tRNA by its specific aminoacyl-tRNA synthetase. But in a few cases the recognition mechanism is quite simple. Across a range of organisms from bacteria to humans, the primary determinant of tRNA recognition by the Ala-tRNA synthetases is a single G≡U base pair in the

(a)

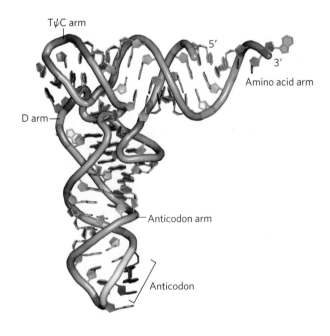

(b)

FIGURE 27–21 Nucleotide positions in tRNAs that are recognized by aminoacyl-tRNA synthetases. (a) Some positions (purple dots) are the same in all tRNAs and therefore cannot be used to discriminate one from another. Other positions are known recognition points for one (orange) or more (blue) aminoacyl-tRNA synthetases. Structural features other

than sequence are important for recognition by some of the synthetases. **(b)** (PDB ID 1EHZ) The same structural features are shown in three dimensions, with the orange and blue residues again representing positions recognized by one or more aminoacyl-tRNA synthetases, respectively.

(a)

(b)

FIGURE 27–22 Aminoacyl-tRNA synthetases. Both synthetases are complexed with their cognate tRNAs (green). Bound ATP (red) pinpoints the active site near the end of the aminoacyl arm. **(a)** Gln-tRNA synthetase from *E. coli*, a typical monomeric class I synthetase (PDB ID 1QRT). **(b)** Asp-tRNA synthetase from yeast, a typical dimeric class II synthetase (PDB ID 1ASZ).

amino acid arm of tRNA^{Ala} **(Fig. 27–23a)**. A short synthetic RNA with as few as 7 bp arranged in a simple hairpin minihelix is efficiently aminoacylated by the Ala-tRNA synthetase, as long as the RNA contains the critical G=U (Fig. 27–23b). This relatively simple alanine system may be an evolutionary relic of a period when RNA oligonucleotides, ancestors to tRNA, were aminoacylated in a primitive system for protein synthesis.

The interaction of aminoacyl-tRNA synthetases and their cognate tRNAs is critical to accurate reading of the genetic code. Any expansion of the code to include new amino acids would necessarily require a new aminoacyl-tRNA synthetase:tRNA pair. A limited expansion of the genetic code has been observed in nature; a more extensive expansion has been accomplished in the laboratory (Box 27–3).

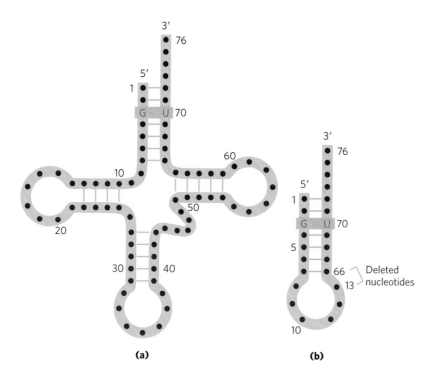

(a)

(b)

FIGURE 27–23 Structural elements of tRNA^{Ala} that are required for recognition by Ala-tRNA synthetase. **(a)** The tRNA^{Ala} structural elements recognized by the Ala-tRNA synthetase are unusually simple. A single G=U base pair (light red) is the only element needed for specific binding and aminoacylation. **(b)** A short synthetic RNA minihelix, with the critical G=U base pair but lacking most of the remaining tRNA structure. This is aminoacylated specifically with alanine almost as efficiently as the complete tRNA^{Ala}.

BOX 27–3   Natural and Unnatural Expansion of the Genetic Code

As we have seen, the 20 amino acids commonly found in proteins offer only limited chemical functionality. Living systems generally overcome these limitations by using enzymatic cofactors or by modifying particular amino acids after they have been incorporated into proteins. In principle, expansion of the genetic code to introduce new amino acids into proteins offers another route to new functionality, but it is a very difficult route to exploit. Such a change might just as easily result in the inactivation of thousands of cellular proteins.

Expanding the genetic code to include a new amino acid requires several cellular changes. A new aminoacyl-tRNA synthetase must generally be present, along with a cognate tRNA. Both of these components must be highly specific, interacting only with each other and the new amino acid. Significant concentrations of the new amino acid must be present in the cell, which may entail the evolution of new metabolic pathways. As outlined in Box 27–1, the anticodon on the tRNA would most likely pair with a codon that normally specifies termination. Making all of this work in a living cell seems unlikely, but it has happened both in nature and in the laboratory.

There are actually 22 rather than 20 amino acids specified by the known genetic code. The two extra ones are selenocysteine and pyrrolysine, each found in only very few proteins but both offering a glimpse into the intricacies of code evolution.

Selenocysteine

Pyrrolysine

A few proteins in all cells (such as formate dehydrogenase in bacteria and glutathione peroxidase in mammals) require selenocysteine for their activity. In *E. coli* selenocysteine is introduced into the enzyme formate dehydrogenase during translation, in response to an in-frame UGA codon. A special type of Ser-tRNA, present at lower levels than other Ser-tRNAs, recognizes UGA and no other codons. This tRNA is charged with serine by the normal serine aminoacyl-tRNA

synthetase, and the serine is enzymatically converted to selenocysteine by a separate enzyme before its use at the ribosome. The charged tRNA does not recognize just any UGA codon; some contextual signal in the mRNA, still to be identified, ensures that this tRNA recognizes only the few UGA codons, within certain genes, that specify selenocysteine. In effect, UGA doubles as a codon for both termination and (very occasionally) selenocysteine. This particular code expansion has a dedicated tRNA, as described above, but it lacks a dedicated cognate aminoacyl-tRNA synthetase. The process works for selenocysteine, but one might consider it an intermediate step in the evolution of a complete new codon definition.

Pyrrolysine is found in a group of anaerobic archaea called methanogens (see Box 22–1). These organisms produce methane as a required part of their metabolism, and the Methanosarcinaceae group can use methylamines as substrates for methanogenesis. Producing methane from monomethylamine requires the enzyme monomethylamine methyltransferase. The gene encoding this enzyme has an in-frame UAG termination codon. The structure of the methyltransferase was elucidated in 2002, revealing the presence of the novel amino acid pyrrolysine at the position specified by the UAG codon. Subsequent experiments demonstrated that—unlike selenocysteine—pyrrolysine was attached directly to a dedicated tRNA by a cognate pyrrolysyl-tRNA synthetase. These cells produce pyrrolysine via a metabolic pathway that remains to be elucidated. The overall system has all the hallmarks of an established codon assignment, although it only works for UAG codons in this particular gene. As in the case of selenocysteine, there are probably contextual signals that direct this tRNA to the correct UAG codon.

Can scientists match this evolutionary feat? Modification of proteins with various functional groups can provide important insights into the activity and/or structure of the proteins. However, protein modification is often quite laborious. For example, an investigator who wishes to attach a new group to a particular Cys residue will have to somehow block the other Cys residues that may be present on the same protein. If one could instead adapt the genetic code to enable a cell to insert a modified amino acid at a particular location in a protein, the process could be rendered much more convenient. Peter Schultz and coworkers have done just that.

To develop a new codon assignment, one again needs a new aminoacyl-tRNA synthetase and a novel cognate tRNA, both adapted to work only with a particular new amino acid. Efforts to create such an "unnatural" code expansion initially focused on *E. coli*. The codon UAG was chosen as the best target for

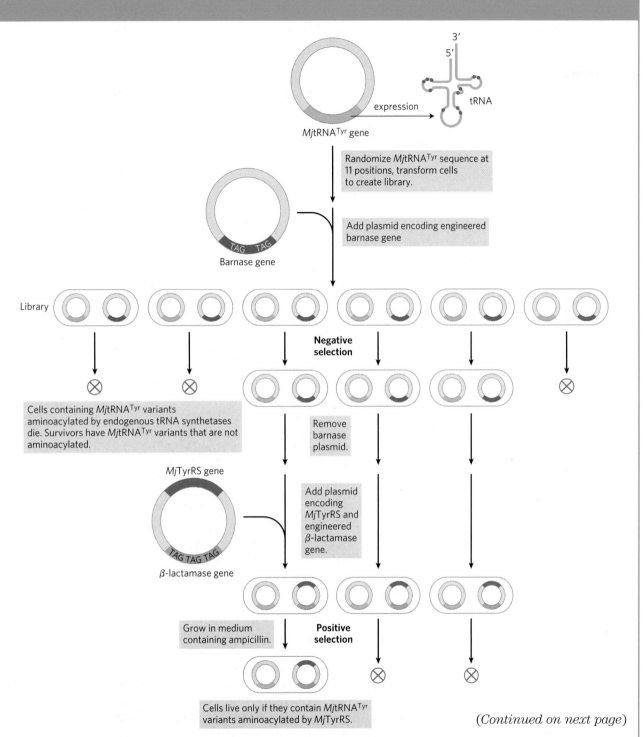

FIGURE 1 Selecting *Mj*tRNA^Tyr^ variants that function only with the tyrosyl-tRNA synthetase *Mj*TyrRS. The sequence of the gene encoding *Mj*tRNA^Tyr^, on a plasmid, is randomized at 11 positions that do not interact with *Mj*TyrRS (red dots). The mutagenized plasmids are introduced into *E. coli* cells to create a library of millions of *Mj*tRNA^Tyr^ variants, represented by the six cells shown here. The toxic barnase gene, engineered to include the sequence TAG so that its transcript includes UAG codons, is provided on a separate plasmid, providing a negative selection. If this gene is expressed, the cells die. It can only be expressed if the *Mj*tRNA^Tyr^ variant expressed by that particular cell is

aminoacylated by endogenous (*E. coli*) aminoacyl-tRNA synthetases, inserting an amino acid instead of stopping translation. Another gene, encoding β-lactamase, and also engineered with TAG sequences to produce UAG stop codons, is provided on yet another plasmid that also expresses the gene encoding *Mj*TyrRS. This serves as a means of positive selection for the remaining *Mj*tRNA^Tyr^ variants. Those variants that are aminoacylated by *Mj*TyrRS allow expression of the β-lactamase gene, which allows cells to grow on ampicillin. Multiple rounds of negative and positive selection yield the best *Mj*tRNA^Tyr^ variants that are aminoacylated uniquely by *Mj*TyrRS and used efficiently in translation.

(*Continued on next page*)

BOX 27–3   Natural and Unnatural Expansion of the Genetic Code *(Continued)*

encoding a new amino acid. UAG is the least used of the three termination codons, and strains with tRNAs selected to recognize UAG (see Box 27–4) do not exhibit growth defects. To create the new tRNA and tRNA synthetase, the genes for a tyrosyl-tRNA and its cognate tyrosyl-tRNA synthetase were taken from the archaeon *Methanococcus jannaschii* (*Mj*tRNA^{Tyr} and *Mj*TyrRS). *Mj*TyrRS does not bind to the anticodon loop of *Mj*tRNA^{Tyr}, allowing the anticodon loop to be modified to CUA (complementary to UAG) without affecting the interaction. Because the archaeal and bacterial systems are orthologous, the modified archaeal components could be transferred to *E. coli* cells without disrupting the intrinsic translation system of the cells.

First, the gene encoding *Mj*tRNA^{Tyr} had to be modified to generate an ideal product tRNA—one that was not recognized by any aminoacyl-tRNA synthetases endogenous to *E. coli*, but was aminoacylated by *Mj*TyrRS. Finding such a variant could be accomplished via a series of negative and positive selection cycles designed to efficiently sift through variants of the tRNA gene (Fig. 1). Parts of the *Mj*tRNA^{Tyr} sequence were randomized, allowing creation of a library of cells that each expressed a different version of the tRNA. A gene encoding barnase (a ribonuclease toxic to *E. coli*) was engineered so that its mRNA transcript contained several UAG codons, and this gene was also introduced into the cells on a plasmid. If the *Mj*tRNA^{Tyr} variant expressed in a particular cell in the library was aminoacylated by an endogenous tRNA synthetase, it would express the barnase gene and that cell would die (a negative selection). Surviving cells would contain tRNA variants that were not aminoacylated by endogenous tRNA synthetases, but could potentially be aminoacylated by *Mj*TyrRS. A positive selection (Fig. 1) was then set up by engineer-

ing the β-lactamase gene (which confers resistance to the antibiotic ampicillin) so that its transcript contained several UAG codons and introducing this gene into the cells along with the gene encoding *Mj*TyrRS. Those *Mj*tRNA^{Tyr} variants that could be aminoacylated by *Mj*TyrRS allowed growth on ampicillin only when *Mj*TyrRS was also expressed in the cell. Several rounds of this negative and positive selection scheme identified a new *Mj*tRNA^{Tyr} variant that was not affected by endogenous enzymes, was aminoacylated by *Mj*TyrRS, and functioned well in translation.

Second, the *Mj*TyrRS had to be modified to recognize the new amino acid. The gene encoding *Mj*TyrRS was now mutagenized to create a large library of variants. Variants that would aminoacylate the new *Mj*tRNA^{Tyr} variant with endogenous amino acids were eliminated using the barnase gene selection. A second positive selection (similar to the ampicillin selection above) was carried out so that cells would survive only if the *Mj*tRNA^{Tyr} variant was aminoacylated only in the presence of the unnatural amino acid. Several rounds of negative and positive selection generated a cognate tRNA synthetase–tRNA pair that recognized only the unnatural amino acid. These components were then renamed to reflect the unnatural amino acid used in the selection.

Using this approach, many *E. coli* strains have been constructed, each capable of incorporating one particular unnatural amino acid into a protein in response to a UAG codon. The same approach has been used to artificially expand the genetic code of yeast and even mammalian cells. Over 30 different amino acids (Fig. 2) can be introduced site-specifically and efficiently into cloned proteins in this way. The result is an increasingly useful and flexible tool kit with which to advance the study of protein structure and function.

FIGURE 2 A sampling of unnatural amino acids that have been added to the genetic code. These unnatural amino acids add uniquely reactive chemical groups such as **(a)** a ketone, **(b)** an azide, **(c)** a photo-crosslinker (a functional group designed to form a covalent bond with a nearby group when activated by light), **(d)** a highly fluorescent amino acid, **(e)** an amino acid with a heavy atom (Br) for use in crystallography, and **(f)** a long-chain cysteine analog that can form extended disulfide bonds.

Stage 2: A Specific Amino Acid Initiates Protein Synthesis

Protein synthesis begins at the amino-terminal end and proceeds by the stepwise addition of amino acids to the carboxyl-terminal end of the growing polypeptide, as determined by Howard Dintzis in 1961 **(Fig. 27–24)**. The AUG initiation codon thus specifies an *amino-terminal* methionine residue. Although methionine has only one codon, (5′)AUG, all organisms have two tRNAs for methionine. One is used exclusively when (5′)AUG is the initiation codon for protein synthesis. The other is used to code for a Met residue in an internal position in a polypeptide.

The distinction between an initiating (5′)AUG and an internal one is straightforward. In bacteria, the two types of tRNA specific for methionine are designated $tRNA^{Met}$ and $tRNA^{fMet}$. The amino acid incorporated in response to the (5′)AUG initiation codon is *N*-formylmethionine (fMet). It arrives at the ribosome as *N*-formylmethionyl-$tRNA^{fMet}$ (fMet-$tRNA^{fMet}$), which is formed in two successive reactions. First, methionine is attached to $tRNA^{fMet}$ by the Met-tRNA synthetase (which in *E. coli* aminoacylates both $tRNA^{fMet}$ and $tRNA^{Met}$):

$$\text{Methionine} + tRNAf^{Met} + ATP \longrightarrow$$
$$\text{Met-}tRNA^{fMet} + AMP + PP_i$$

Next, a transformylase transfers a formyl group from N^{10}-formyltetrahydrofolate to the amino group of the Met residue:

$$N^{10}\text{-Formyltetrahydrofolate} + \text{Met-}tRNA^{fMet} \longrightarrow$$
$$\text{tetrahydrofolate} + \text{fMet-}tRNA^{fMet}$$

The transformylase is more selective than the Met-tRNA synthetase; it is specific for Met residues attached to $tRNA^{fMet}$, presumably recognizing some unique structural feature of that tRNA. By contrast, Met-$tRNA^{Met}$ inserts methionine in interior positions in polypeptides.

N-Formylmethionine

Addition of the *N*-formyl group to the amino group of methionine by the transformylase prevents fMet from entering interior positions in a polypeptide while also allowing fMet-$tRNA^{fMet}$ to be bound at a specific ribosomal initiation site that accepts neither Met-$tRNA^{Met}$ nor any other aminoacyl-tRNA.

In eukaryotic cells, all polypeptides synthesized by cytosolic ribosomes begin with a Met residue (rather than fMet), but, again, the cell uses a specialized initiating tRNA that is distinct from the $tRNA^{Met}$ used at (5′)AUG codons at interior positions in the mRNA. Polypeptides synthesized by mitochondrial and chloroplast ribosomes, however, begin with *N*-formylmethionine. This strongly supports the view that mitochondria and chloroplasts originated from bacterial ancestors that were symbiotically incorporated into precursor eukaryotic cells at an early stage of evolution (see Fig. 1–38).

How can the single (5′)AUG codon determine whether a starting *N*-formylmethionine (or methionine, in eukaryotes) or an interior Met residue is ultimately inserted? The details of the initiation process provide the answer.

The Three Steps of Initiation The **initiation** of polypeptide synthesis in bacteria requires (1) the 30S ribosomal subunit, (2) the mRNA coding for the polypeptide to be made, (3) the initiating fMet-$tRNA^{fMet}$, (4) a set of three proteins called initiation factors (IF-1, IF-2, and IF-3), (5) GTP, (6) the 50S ribosomal subunit, and (7) Mg^{2+}. Formation of the initiation complex takes place in three steps **(Fig. 27–25)**.

In step ❶ the 30S ribosomal subunit binds two initiation factors, IF-1 and IF-3. Factor IF-3 prevents the 30S and 50S subunits from combining prematurely. The mRNA then binds to the 30S subunit. The initiating (5′)AUG is guided to its correct position by the **Shine-Dalgarno sequence** (named for Australian researchers John Shine and Lynn Dalgarno, who identified it) in the

Direction of chain growth

Amino terminus   Carboxyl terminus

4 min

7 min

16 min

60 min

1   146

Residue number

FIGURE 27–24 Proof that polypeptides grow by addition of amino acid residues to the carboxyl end: the Dintzis experiment. Reticulocytes (immature erythrocytes) actively synthesizing hemoglobin were incubated with radioactive leucine (selected because it occurs frequently in both the α- and β-globin chains). Samples of completed α chains were isolated from the reticulocytes at various times afterward, and the distribution of radioactivity was determined. The dark red zones show the portions of completed α-globin chains containing radioactive Leu residues. At 4 min, only a few residues at the carboxyl end of α-globin were labeled, because the only *complete* globin chains with incorporated label after 4 min were those that had nearly completed synthesis at the time the label was added. With longer incubation times, successively longer segments of the polypeptide contained labeled residues, always in a block at the carboxyl end of the chain. The unlabeled end of the polypeptide (the amino terminus) was thus defined as the initiating end, which means that polypeptides grow by successive addition of amino acids to the carboxyl end.

FIGURE 27–25 Formation of the initiation complex in bacteria. The complex forms in three steps (described in the text) at the expense of the hydrolysis of GTP to GDP and P$_i$. IF-1, IF-2, and IF-3 are initiation factors. E designates the exit site, P the peptidyl site, and A the aminoacyl site. Here the anticodon of the tRNA is oriented 3′ to 5′, left to right, as in Figure 27–8 but opposite to the orientation in Figures 27–21 and 27–23.

mRNA. This consensus sequence is an initiation signal of four to nine purine residues, 8 to 13 bp to the 5′ side of the initiation codon (**Fig. 27–26a**). The sequence base-pairs with a complementary pyrimidine-rich sequence near the 3′ end of the 16S rRNA of the 30S ribosomal subunit (Fig. 27–26b). This mRNA-rRNA interaction positions the initiating (5′)AUG sequence of the mRNA in the precise position on the 30S subunit where it is required for initiation of translation. The particular (5′)AUG where fMet-tRNA^{fMet} is to be bound is distinguished from other methionine codons by its proximity to the Shine-Dalgarno sequence in the mRNA.

Bacterial ribosomes have three sites that bind tRNAs, the **aminoacyl (A) site**, the **peptidyl (P) site**, and the **exit (E) site**. The A and P sites bind to aminoacyl-tRNAs, whereas the E site binds only to uncharged tRNAs that have completed their task on the ribosome. Both the 30S and the 50S subunits contribute to the characteristics of the A and P sites, whereas the E site is largely confined to the 50S subunit. The initiating (5′)AUG is positioned at the P site, the only site to which fMet-tRNA^{fMet} can bind (Fig. 27–25). The fMet-tRNA^{fMet} is the only aminoacyl-tRNA that binds first to the P site; during the subsequent elongation stage, all other incoming aminoacyl-tRNAs (including the Met-tRNA^{Met} that binds to interior AUG codons) bind first to the A site and only subsequently to the P and E sites. The E site is the site from which the "uncharged" tRNAs leave during elongation. Factor IF-1 binds at the A site and prevents tRNA binding at this site during initiation.

In step ❷ of the initiation process (Fig. 27–25), the complex consisting of the 30S ribosomal subunit, IF-3, and mRNA is joined by both GTP-bound IF-2 and the initiating fMet-tRNA^{fMet}. The anticodon of this tRNA now pairs correctly with the mRNA's initiation codon.

In step ❸ this large complex combines with the 50S ribosomal subunit; simultaneously, the GTP bound to IF-2 is hydrolyzed to GDP and P$_i$, which are released from the complex. All three initiation factors depart from the ribosome at this point.

Completion of the steps in Figure 27–25 produces a functional 70S ribosome called the **initiation complex**, containing the mRNA and the initiating fMet-tRNA^{fMet}. The correct binding of the fMet-tRNA^{fMet} to the P site in the complete 70S initiation complex is assured by at least three points of recognition and attachment: the codon-anticodon interaction involving the initiation AUG fixed in the P site, interaction between the Shine-Dalgarno sequence in the mRNA and the 16S rRNA, and binding interactions between the ribosomal P site and the fMet-tRNA^{fMet}. The initiation complex is now ready for elongation.

Initiation in Eukaryotic Cells Translation is generally similar in eukaryotic and bacterial cells; most of the significant differences are in the number of components and in

E. coli trpA (5')A G C A C G ⟦G A G G G G⟧ A A A U C U G ⟦A U G⟧ G A A C G C U A C (3')

E. coli araB U U U G G A U ⟦G G A G⟧ U G A A A C G ⟦A U G⟧ G C G A U U G C A

E. coli lacI C A A U U C A G ⟦G G U G G⟧ U G A A U G ⟦U G⟧ A A A C C A G U A

ϕX174 phage A protein A A U C U U ⟦G G A G G⟧ C U U U U U U ⟦A U G⟧ G U U C G U U C U

λ phage cro A U G U A C ⟦U A A G G A G G U⟧ U G U A ⟦U G⟧ G A A C A A C G C

Shine-Dalgarno sequence; pairs with 16S rRNA

Initiation codon; pairs with fMet-tRNA^fMet

(a)

3'
OH | G
| | A
Bacterial mRNA with consensus Shine-Dalgarno sequence   3' end of 16S rRNA   A | U
U | C
U C C U C C A

(5')G A U U C C U ⟦A G G A G G U⟧ U U G A C C U ⟦A U G C G A G C U U U U A G U⟧ (3')

(b)

FIGURE 27–26 Messenger RNA sequences that serve as signals for initiation of protein synthesis in bacteria. (a) Alignment of the initiating AUG (shaded in green) at its correct location on the 30S ribosomal subunit depends in part on upstream Shine-Dalgarno sequences (light red). Portions of the mRNA transcripts of five bacterial genes are shown. Note the unusual example of the *E. coli* LacI protein, which initiates with a GUG (Val) codon (see Box 27–1). In *E. coli*, AUG is the start codon in approximately 91% of the genes, with GUG (7%) and UUG (2%) assuming this role more rarely. **(b)** The Shine-Dalgarno sequence of the mRNA pairs with a sequence near the 3' end of the 16S rRNA.

mechanistic details. The initiation process in eukaryotes is outlined in **Figure 27–27**. Eukaryotic mRNAs are bound to the ribosome as a complex with a number of specific binding proteins. Eukaryotic cells have at least 12 initiation factors. Initiation factors eIF1A and eIF3 are the functional homologs of the bacterial IF-1 and IF-3, binding to the 40S subunit in step ❶ and blocking tRNA binding to the A site and premature joining of the large and small ribosomal subunits, respectively. The factor eIF1 binds to the E site. The charged initiator tRNA is bound by the initiation factor eIF2, which also has bound GTP. In step ❷ this ternary complex binds to the 40S ribosomal subunit, along with two other proteins involved in later steps, eIF5 (not shown in Fig. 27–27) and eIF5B. This creates a 43S preinitiation complex. The mRNA binds to the eIF4F complex, which, in step ❸, mediates its association with the 43S preinitiation complex. The eIF4F complex is made up of eIF4E (binding to the 5' cap), eIF4A (an ATPase and RNA helicase), and eIF4G (a linker protein). The eIF4G protein binds to eIF3 and eIF4E to provide the first link between the 43S preinitiation complex and the mRNA. The eIF4G also binds to the poly(A) binding protein (PABP) at the 3' end of the mRNA, circularizing the mRNA (**Fig. 27–28**) and facilitating the translational regulation of gene expression, as described in Chapter 28.

The addition of the mRNA and its associated factors creates a 48S complex. This complex scans the bound mRNA, starting at the 5' cap, until an AUG codon is encountered. The scanning process (step ❹ in Fig. 27–27)

may be facilitated by the RNA helicase of eIF4A and another bound factor (eIF4B, not shown in Fig. 27–27) whose precise molecular activity is not understood.

Once the initiating AUG site is encountered, the 60S ribosomal subunit associates with the complex in step ❺, accompanied by the release of many of the initiation factors. This requires the activity of eIF5 and eIF5B. The eIF5 protein promotes the GTPase activity of eIF2, producing an eIF2-GDP complex with reduced affinity for the initiator tRNA. The eIF5B protein is homologous to the bacterial IF-2. It hydrolyzes its bound GTP and triggers dissociation of eIF2-GDP and other initiation factors, followed closely by the association of the 60S subunit. This completes formation of the initiation complex.

The roles of the various bacterial and eukaryotic initiation factors in the overall process are summarized in Table 27–8. The mechanism by which these proteins act is an important area of investigation.

Stage 3: Peptide Bonds Are Formed in the Elongation Stage

The third stage of protein synthesis is **elongation**. Again, we begin with bacterial cells. Elongation requires (1) the initiation complex described above, (2) aminoacyl-tRNAs, (3) a set of three soluble cytosolic proteins called **elongation factors** (EF-Tu, EF-Ts, and EF-G in bacteria), and (4) GTP. Cells use three steps to add

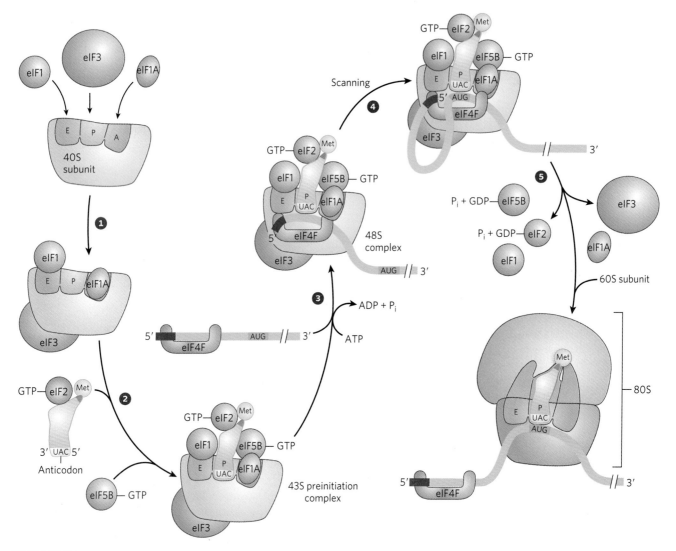

FIGURE 27–27 Initiation of protein synthesis in eukaryotes. The five steps are described in the text. Eukaryotic initiation factors mediate the association of first the charged initiator tRNA to form a 43S complex and then the mRNA (with the 5′ cap shown in red) to form a 48S complex. The final initiation complex is formed as the 60S subunit associates, coupled with the release of most of the initiation factors.

FIGURE 27–28 Circularization of the mRNA in the eukaryotic initiation complex. The 3′ and 5′ ends of eukaryotic mRNAs are linked by the eIF4F complex of proteins. The eIF4E subunit binds to the 5′ cap, and the eIF4G protein binds to the poly(A) binding protein (PABP) at the 3′ end of the mRNA. The eIF4G protein also binds to eIF3, linking the circularized mRNA to the 40S subunit of the ribosome.

each amino acid residue, and the steps are repeated as many times as there are residues to be added.

Elongation Step 1: Binding of an Incoming Aminoacyl-tRNA In the first step of the elongation cycle **(Fig. 27–29)**, the appropriate incoming aminoacyl-tRNA binds to a complex of GTP-bound EF-Tu. The resulting aminoacyl-tRNA–EF-Tu–GTP complex binds to the A site of the 70S initiation complex. The GTP is hydrolyzed and an EF-Tu–GDP complex is released from the 70S ribosome. The EF-Tu–GTP complex is regenerated in a process involving EF-Ts and GTP.

Elongation Step 2: Peptide Bond Formation A peptide bond is now formed between the two amino acids bound by their tRNAs to the A and P sites on the ribosome. This occurs by the transfer of the initiating *N*-formylmethionyl group from its tRNA to the amino group of the second

TABLE 27–8 Protein Factors Required for Initiation of Translation in Bacterial and Eukaryotic Cells

Factor	Function
Bacterial	
IF-1	Prevents premature binding of tRNAs to A site
IF-2	Facilitates binding of fMet-tRNA^{fMet} to 30S ribosomal subunit
IF-3	Binds to 30S subunit; prevents premature association of 50S subunit; enhances specificity of P site for fMet-tRNA^{fMet}
Eukaryotic	
eIF1	Binds to the E site of the 40S subunit; facilitates interaction between eIF2-tRNA-GTP ternary complex and the 40S subunit
eIF1A	Homolog of bacterial IF-1; prevents premature binding of tRNAs to A site
eIF2	GTPase; facilitates binding of initiating Met-tRNA^{Met} to 40S ribosomal subunit
eIF2B*, eIF3	First factors to bind 40S subunit; facilitate subsequent steps
eIF4F	Complex consisting of eIF4E, eIF4A, and eIF4G
eIF4A	RNA helicase activity; removes secondary structure in the mRNA to permit binding to 40S subunit; part of the eIF4F complex
eIF4B	Binds to mRNA; facilitates scanning of mRNA to locate the first AUG
eIF4E	Binds to the 5′ cap of mRNA; part of the eIF4F complex
eIF4G	Binds to eIF4E and to poly(A) binding protein (PABP); part of the eIF4F complex
eIF5*	Promotes dissociation of several other initiation factors from 40S subunit as a prelude to association of 60S subunit to form 80S initiation complex
eIF5b	GTPase homologous to bacterial IF-2; promotes dissociation of initiation factors prior to final ribosome assembly

*Not shown in Figure 27–27.

FIGURE 27–29 First elongation step in bacteria: binding of the second aminoacyl-tRNA. The second aminoacyl-tRNA (AA$_2$) enters the A site of the ribosome bound to GTP-bound EF-Tu (shown here as Tu). Binding of the second aminoacyl-tRNA to the A site is accompanied by hydrolysis of the GTP to GDP and P$_i$ and release of the EF-Tu–GDP complex from the ribosome. The bound GDP is released when the EF-Tu–GDP complex binds to EF-Ts, and EF-Ts is subsequently released when another molecule of GTP binds to EF-Tu. This recycles EF-Tu and makes it available to repeat the cycle. Accommodation involves a change in the second tRNA conformation that pulls its aminoacyl end into the peptidyl transferase site.

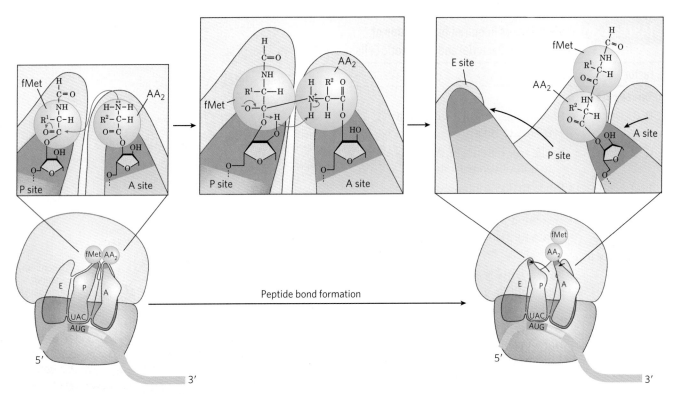

FIGURE 27–30 Second elongation step in bacteria: formation of the first peptide bond. The peptidyl transferase catalyzing this reaction is the 23S rRNA ribozyme. The *N*-formylmethionyl group is transferred to the amino group of the second aminoacyl-tRNA in the A site, forming a dipeptidyl-tRNA. At this stage, both tRNAs bound to the ribosome shift position in the 50S subunit to take up a hybrid binding state. The uncharged tRNA shifts so that its 3′ and 5′ ends are in the E site. Similarly, the 3′ and 5′ ends of the peptidyl tRNA shift to the P site. The anticodons remain in the P and A sites. Note the involvement of the 2′-hydroxyl group of the 3′-terminal adenosine as a general acid-base catalyst in this reaction.

amino acid, now in the A site **(Fig. 27–30)**. The α-amino group of the amino acid in the A site acts as a nucleophile, displacing the tRNA in the P site to form the peptide bond. This reaction produces a dipeptidyl-tRNA in the A site, and the now "uncharged" (deacylated) tRNA^{fMet} remains bound to the P site. The tRNAs then shift to a hybrid binding state, with elements of each spanning two different sites on the ribosome, as shown in Figure 27–30.

The enzymatic activity that catalyzes peptide bond formation has historically been referred to as **peptidyl transferase** and was widely assumed to be intrinsic to one or more of the proteins in the large ribosomal subunit. We now know that this reaction is catalyzed by the 23S rRNA, adding to the known catalytic repertoire of ribozymes. This discovery has interesting implications for the evolution of life (see Box 27–2).

Elongation Step 3: Translocation In the final step of the elongation cycle, **translocation**, the ribosome moves one codon toward the 3′ end of the mRNA **(Fig. 27–31a)**. This movement shifts the anticodon of the dipeptidyl-tRNA, which is still attached to the second codon of the mRNA, from the A site to the P site, and shifts the deacylated tRNA from the P site to the E site, from where the tRNA is released into the cytosol. The third codon of the mRNA now lies in the A site and the

second codon in the P site. Movement of the ribosome along the mRNA requires EF-G (also known as translocase) and the energy provided by hydrolysis of another molecule of GTP. A change in the three-dimensional conformation of the entire ribosome results in its movement along the mRNA. Because the structure of EF-G mimics the structure of the EF-Tu–tRNA complex (Fig. 27–31b), EF-G can bind the A site and presumably displace the peptidyl-tRNA.

After translocation, the ribosome, with its attached dipeptidyl-tRNA and mRNA, is ready for the next elongation cycle and attachment of a third amino acid residue. This process occurs in the same way as addition of the second residue (as shown in Figs 27–29, 27–30, and 27–31). For each amino acid residue correctly added to the growing polypeptide, two GTPs are hydrolyzed to GDP and P$_i$ as the ribosome moves from codon to codon along the mRNA toward the 3′ end.

The polypeptide remains attached to the tRNA of the most recent amino acid to be inserted. This association maintains the functional connection between the information in the mRNA and its decoded polypeptide output. At the same time, the ester linkage between this tRNA and the carboxyl terminus of the growing polypeptide activates the terminal carboxyl group for nucleophilic attack by the incoming amino acid to form a new peptide bond (Fig. 27–30). As the existing ester linkage between

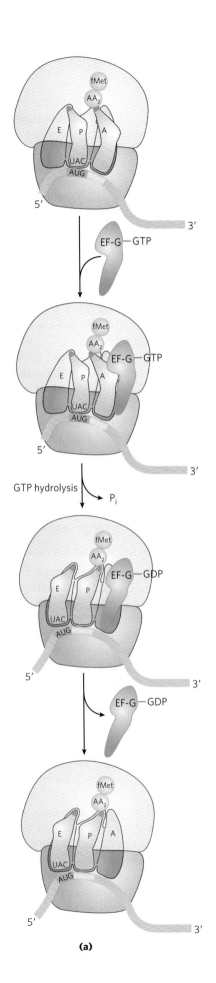

(a)

the polypeptide and tRNA is broken during peptide bond formation, the linkage between the polypeptide and the information in the mRNA persists, because each newly added amino acid is still attached to its tRNA.

The elongation cycle in eukaryotes is quite similar to that in bacteria. Three eukaryotic elongation factors (eEF1α, eEF1βγ, and eEF2) have functions analogous to those of the bacterial elongation factors (EF-Tu, EF-Ts, and EF-G, respectively). Eukaryotic ribosomes do not have an E site; uncharged tRNAs are expelled directly from the P site.

Proofreading on the Ribosome The GTPase activity of EF-Tu during the first step of elongation in bacterial cells (Fig. 27–29) makes an important contribution to the rate and fidelity of the overall biosynthetic process. Both the EF-Tu–GTP and EF-Tu–GDP complexes exist for a few milliseconds before they dissociate. These two intervals provide opportunities for the codon-anticodon interactions to be proofread. Incorrect aminoacyl-tRNAs normally dissociate from the A site during one of these periods. If the GTP analog guanosine 5′-*O*-(3-thiotriphosphate) (GTPγS) is used in place of GTP, hydrolysis is

FIGURE 27–31 Third elongation step in bacteria: translocation. (a) The ribosome moves one codon toward the 3′ end of the mRNA, using energy provided by hydrolysis of GTP bound to EF-G (translocase). The dipeptidyl-tRNA is now entirely in the P site, leaving the A site open for the incoming (third) aminoacyl-tRNA. The uncharged tRNA later dissociates from the E site, and the elongation cycle begins again. **(b)** The structure of EF-G mimics the structure of EF-Tu complexed with tRNA. Shown here are (left) EF-Tu complexed with tRNA (PDB ID 1B23) and (right) EF-G complexed with GDP (PDB ID 1DAR). The carboxyl-terminal part of EF-G mimics the structure of the anticodon loop of tRNA in both shape and charge distribution.

(b)

slowed, improving the fidelity (by increasing the proof-reading intervals) but reducing the rate of protein synthesis.

Guanosine 5′-O-(3-thiotriphosphate)
(GTPγS)

The process of protein synthesis (including the characteristics of codon-anticodon pairing already described) has clearly been optimized through evolution to balance the requirements for speed and fidelity. Improved fidelity might diminish speed, whereas increases in speed would probably compromise fidelity. And, recall that the proofreading mechanism on the ribosome establishes only that the proper codon-anticodon pairing has taken place, not that the correct amino acid is attached to the tRNA. If a tRNA is successfully aminoacylated with the wrong amino acid (as can be done experimentally), this incorrect amino acid is efficiently incorporated into a protein in response to whatever codon is normally recognized by the tRNA.

Stage 4: Termination of Polypeptide Synthesis Requires a Special Signal

Elongation continues until the ribosome adds the last amino acid coded by the mRNA. **Termination**, the fourth stage of polypeptide synthesis, is signaled by the presence of one of three termination codons in the mRNA (UAA, UAG, UGA), immediately following the final coded amino acid. Mutations in a tRNA anticodon that allow an amino acid to be inserted at a termination codon are generally deleterious to the cell (Box 27–4). In bacteria, once a termination codon occupies the ribosomal A site, three **termination factors**, or **release factors**—the proteins RF-1, RF-2, and RF-3—contribute to (1) hydrolysis of the terminal peptidyl-tRNA bond, (2) release of the free polypeptide and the last tRNA, now uncharged, from the P site, and (3) dissociation of the 70S ribosome into its 30S and 50S subunits, ready to start a new cycle of polypeptide synthesis **(Fig. 27–32)**. RF-1 recognizes the termination codons UAG and UAA, and RF-2 recognizes UGA and UAA. Either RF-1 or RF-2 (depending on which codon is present) binds at a termination codon and induces peptidyl transferase to transfer the growing polypeptide to a water molecule rather than to another amino acid. The release factors have domains thought to mimic the structure of tRNA, as shown for the elongation factor EF-G in Figure 27–31b. The specific function of

---

BOX 27–4   Induced Variation in the Genetic Code: Nonsense Suppression

When a mutation produces a termination codon in the interior of a gene, translation is prematurely halted and the incomplete polypeptide is usually inactive. These are called nonsense mutations. The gene can be restored to normal function if a second mutation either (1) converts the misplaced termination codon to a codon specifying an amino acid or (2) suppresses the effects of the termination codon. Such restorative mutations are called **nonsense suppressors**; they generally involve mutations in tRNA genes to produce altered (suppressor) tRNAs that can recognize the termination codon and insert an amino acid at that position. Most known suppressor tRNAs have single base substitutions in their anticodons.

Suppressor tRNAs constitute an experimentally induced variation in the genetic code to allow the reading of what are usually termination codons, much like the naturally occurring code variations described in Box 27–1. Nonsense suppression does not completely disrupt normal information transfer in a cell, because the cell usually has several copies of each tRNA gene; some of these duplicate genes are weakly expressed and account for only a minor part of the cellular pool of a particular tRNA. Suppressor mutations usually involve a "minor" tRNA, leaving the major tRNA to read its codon normally.

For example, *E. coli* has three identical genes for tRNA^{Tyr}, each producing a tRNA with the anticodon (5′)GUA. One of these genes is expressed at relatively high levels and thus its product represents the major tRNA^{Tyr} species; the other two genes are transcribed in only small amounts. A change in the anticodon of the tRNA product of one of these duplicate tRNA^{Tyr} genes, from (5′)GUA to (5′)CUA, produces a minor tRNA^{Tyr} species that will insert tyrosine at UAG stop codons. This insertion of tyrosine at UAG is carried out inefficiently, but it can produce enough full-length protein from a gene with a nonsense mutation to allow the cell to survive. The major tRNA^{Tyr} continues to translate the genetic code normally for the majority of proteins.

The mutation that leads to creation of a suppressor tRNA does not always occur in the anticodon. The suppression of UGA nonsense codons generally involves the tRNA^{Trp} that normally recognizes UGG. The alteration that allows it to read UGA (and insert Trp residues at these positions) is a G to A change at position 24 (in an arm of the tRNA somewhat removed from the anticodon); this tRNA can now recognize *both* UGG and UGA. A similar change is found in tRNAs involved in the most common naturally occurring variation in the genetic code (UGA = Trp; see Box 27–1).

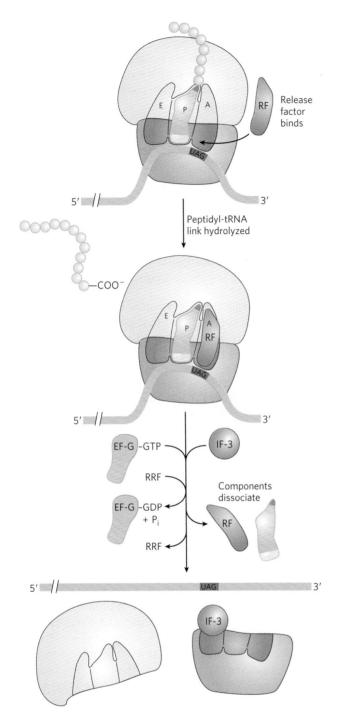

FIGURE 27–32 Termination of protein synthesis in bacteria. Termination occurs in response to a termination codon in the A site. First, a release factor, RF (RF-1 or RF-2, depending on which termination codon is present), binds to the A site. This leads to hydrolysis of the ester linkage between the nascent polypeptide and the tRNA in the P site and release of the completed polypeptide. Finally, the mRNA, deacylated tRNA, and release factor leave the ribosome, which dissociates into its 30S and 50S subunits, aided by ribosome recycling factor (RRF), IF-3, and energy provided by EF-G-mediated GTP hydrolysis. The 30S subunit complex with IF-3 is ready to begin another cycle of translation (see Fig. 27–25).

RF-3 has not been firmly established, although it is thought to release the ribosomal subunit. In eukaryotes, a single release factor, eRF, recognizes all three termination codons.

Ribosome recycling leads to dissociation of the translation components. The release factors dissociate from the posttermination complex (with an uncharged tRNA in the P site) and are replaced by EF-G and a protein called ribosome recycling factor (RRF; M_r 20,300). Hydrolysis of GTP by EF-G leads to dissociation of the 50S subunit from the 30S–tRNA–mRNA complex. EF-G and RRF are replaced by IF-3, which promotes the dissociation of the tRNA. The mRNA is then released. The complex of IF-3 and the 30S subunit is then ready to initiate another round of protein synthesis (Fig. 27–25).

Energy Cost of Fidelity in Protein Synthesis Synthesis of a protein true to the information specified in its mRNA requires energy. Formation of each aminoacyl-tRNA uses two high-energy phosphate groups. An additional ATP is consumed each time an incorrectly activated amino acid is hydrolyzed by the deacylation activity of an aminoacyl-tRNA synthetase as part of its proofreading activity. A GTP is cleaved to GDP and P_i during the first elongation step, and another during the translocation step. Thus, on average, the energy derived from the hydrolysis of more than four NTPs to NDPs is required for the formation of each peptide bond of a polypeptide.

This represents an exceedingly large thermodynamic "push" in the direction of synthesis: at least 4×30.5 kJ/mol $= 122$ kJ/mol of phosphodiester bond energy to generate a peptide bond, which has a standard free energy of hydrolysis of only about -21 kJ/mol. The net free-energy change during peptide bond synthesis is thus -101 kJ/mol. Proteins are information-containing polymers. The biochemical goal is not simply the formation of a peptide bond but the formation of a peptide bond between two *specified* amino acids. Each of the high-energy phosphate compounds expended in this process plays a critical role in maintaining proper alignment between each new codon in the mRNA and its associated amino acid at the growing end of the polypeptide. This energy permits very high fidelity in the biological translation of the genetic message of mRNA into the amino acid sequence of proteins.

Rapid Translation of a Single Message by Polysomes Large clusters of 10 to 100 ribosomes that are very active in protein synthesis can be isolated from both eukaryotic and bacterial cells. Electron micrographs show a fiber between adjacent ribosomes in the cluster, which is called a **polysome (Fig. 27–33a).** The connecting strand is a single molecule of mRNA that is being translated simultaneously by many closely spaced ribosomes, allowing the highly efficient use of the mRNA.

In bacteria, transcription and translation are tightly coupled. Messenger RNAs are synthesized and translated in the same $5' \rightarrow 3'$ direction. Ribosomes begin translating the 5' end of the mRNA before transcription is complete (Fig. 27–33b). The situation is quite different in eukaryotic cells, where newly transcribed mRNAs must leave the nucleus before they can be translated.

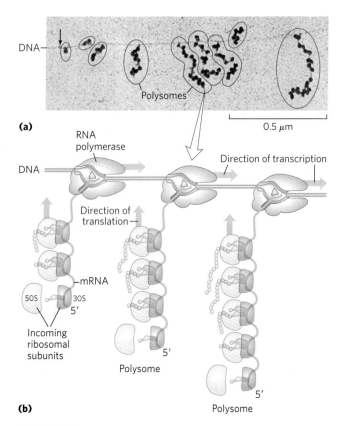

(a)

0.5 μm

(b)

FIGURE 27–33 Coupling of transcription and translation in bacteria.
(a) Electron micrograph of polysomes forming during the transcription of a segment of DNA from *E. coli.* Each mRNA is being translated by many ribosomes simultaneously. The nascent polypeptide chains forming on the ribosomes are difficult to see under the spreading conditions used to prepare the samples shown in these electron micrographs. The arrow marks the approximate beginning of the gene that is being transcribed. **(b)** Each mRNA is translated by ribosomes while it is still being transcribed from DNA by RNA polymerase. This is possible because the mRNA in bacteria does not have to be transported from a nucleus to the cytoplasm before encountering ribosomes. In this schematic diagram the ribosomes are depicted as smaller than the RNA polymerase. In reality the ribosomes (M_r 2.7×10^6) are an order of magnitude larger than the RNA polymerase (M_r 3.9×10^5).

Bacterial mRNAs generally exist for just a few minutes (p. 1084) before they are degraded by nucleases. In order to maintain high rates of protein synthesis, the mRNA for a given protein or set of proteins must be made continuously and translated with maximum efficiency. The short lifetime of mRNAs in bacteria allows a rapid cessation of synthesis when the protein is no longer needed.

Stage 5: Newly Synthesized Polypeptide Chains Undergo Folding and Processing

In the final stage of protein synthesis, the nascent polypeptide chain is folded and processed into its biologically active form. During or after its synthesis, the polypeptide progressively assumes its native conformation, with the formation of appropriate hydrogen bonds and van der Waals, ionic, and hydrophobic interactions. In this way the linear, or one-dimensional, genetic message in the mRNA is converted into the three-dimensional structure of the protein. Some newly made proteins, bacterial, archaeal, and eukaryotic, do not attain their final biologically active conformation until they have been altered by one or more processing reactions called **posttranslational modifications**.

Amino-Terminal and Carboxyl-Terminal Modifications The first residue inserted in all polypeptides is *N*-formylmethionine (in bacteria) or methionine (in eukaryotes). However, the formyl group, the amino-terminal Met residue, and often additional amino-terminal (and, in some cases, carboxyl-terminal) residues may be removed enzymatically in formation of the final functional protein. In as many as 50% of eukaryotic proteins, the amino group of the amino-terminal residue is *N*-acetylated after translation. Carboxyl-terminal residues are also sometimes modified.

Loss of Signal Sequences As we shall see in Section 27.3, the 15 to 30 residues at the amino-terminal end of some proteins play a role in directing the protein to its ultimate destination in the cell. Such **signal sequences** are eventually removed by specific peptidases.

Modification of Individual Amino Acids The hydroxyl groups of certain Ser, Thr, and Tyr residues of some proteins are enzymatically phosphorylated by ATP **(Fig. 27–34a)**; the phosphate groups add negative charges to these polypeptides. The functional significance of this modification varies from one protein to the next. For example, the milk protein casein has many phosphoserine groups that bind Ca^{2+}. Calcium, phosphate, and amino acids are all valuable to suckling young, so casein efficiently provides three essential nutrients. And as we have seen in numerous instances, phosphorylation-dephosphorylation cycles regulate the activity of many enzymes and regulatory proteins.

Extra carboxyl groups may be added to Glu residues of some proteins. For example, the blood-clotting protein prothrombin contains a number of γ-carboxyglutamate residues (Fig. 27–34b) in its amino-terminal region, introduced by an enzyme that requires vitamin K. These carboxyl groups bind Ca^{2+}, which is required to initiate the clotting mechanism.

Monomethyl- and dimethyllysine residues (Fig. 27–34c) occur in some muscle proteins and in cytochrome *c*. The calmodulin of most species contains one trimethyllysine residue at a specific position. In other proteins, the carboxyl groups of some Glu residues undergo methylation, removing their negative charge.

Attachment of Carbohydrate Side Chains The carbohydrate side chains of glycoproteins are attached covalently during or after synthesis of the polypeptide. In some glycoproteins, the carbohydrate side chain is attached enzymatically to Asn residues (*N*-linked oligosaccharides), in others to Ser or Thr residues (*O*-linked oligosaccharides) (see Fig. 7–30). Many proteins that function extracellularly, as well as the lubricating proteoglycans that coat mucous

FIGURE 27–35 Farnesylation of a Cys residue. The thioether linkage is shown in red. The Ras protein is the product of the *ras* oncogene.

(a)

(b)

(c)

FIGURE 27–34 Some modified amino acid residues. (a) Phosphorylated amino acids. **(b)** A carboxylated amino acid. **(c)** Some methylated amino acids.

membranes, contain oligosaccharide side chains (see Fig. 7–28).

Addition of Isoprenyl Groups A number of eukaryotic proteins are modified by the addition of groups derived from isoprene (isoprenyl groups). A thioether bond is formed between the isoprenyl group and a Cys residue of the protein (see Fig. 11–15). The isoprenyl groups are derived from pyrophosphorylated intermediates of the cholesterol biosynthetic pathway (see Fig. 21–35), such as farnesyl pyrophosphate **(Fig. 27–35)**. Proteins modified in this way include the Ras proteins (small G proteins), which are products of the *ras* oncogenes and proto-oncogenes, and the trimeric G proteins (both discussed in Chapter 12), as well as lamins, proteins found in the nuclear matrix. The isoprenyl group helps to anchor the protein in a membrane. The transforming (carcinogenic) activity of the *ras* oncogene is lost when isoprenylation of the Ras protein is blocked, a finding that has stimulated interest in identifying inhibitors of this posttranslational modification pathway for use in cancer chemotherapy.

Addition of Prosthetic Groups Many proteins require for their activity covalently bound prosthetic groups. Two examples are the biotin molecule of acetyl-CoA carboxylase and the heme group of hemoglobin or cytochrome *c*.

Proteolytic Processing Many proteins are initially synthesized as large, inactive precursor polypeptides that are proteolytically trimmed to form their smaller, active forms. Examples include proinsulin, some viral proteins, and proteases such as chymotrypsinogen and trypsinogen (see Fig. 6–38).

Formation of Disulfide Cross-Links After folding into their native conformations, some proteins form intrachain or interchain disulfide bridges between Cys residues. In eukaryotes, disulfide bonds are common in proteins to be exported from cells. The cross-links formed in this way help to protect the native conformation of the protein molecule from denaturation in the extracellular environment, which can differ greatly from intracellular conditions and is generally oxidizing.

Protein Synthesis Is Inhibited by Many Antibiotics and Toxins

Protein synthesis is a central function in cellular physiology and is the primary target of many naturally occurring antibiotics and toxins. Except as noted, these antibiotics inhibit protein synthesis in bacteria. The differences between bacterial and eukaryotic protein synthesis, though in some cases subtle, are sufficient that most of the compounds discussed below are relatively harmless to eukaryotic cells. Natural selection has favored the evolution of compounds that exploit minor differences in order to affect bacterial systems selectively, such that these biochemical weapons are synthesized by some microorganisms and are extremely toxic to others. Because nearly every step in protein synthesis can be specifically inhibited by one antibiotic or another, antibiotics have become valuable tools in the study of protein biosynthesis.

Puromycin, made by the mold *Streptomyces albo-niger*, is one of the best-understood inhibitory antibiotics. Its structure is very similar to the 3′ end of an aminoacyl-tRNA, enabling it to bind to the ribosomal A site and participate in peptide bond formation, producing peptidylpuromycin **(Fig. 27–36)**. However, because puromycin resembles only the 3′ end of the tRNA, it does not engage in translocation and dissociates from the ribosome shortly after it is linked to the carboxyl terminus of the peptide. This prematurely terminates polypeptide synthesis.

Tetracyclines inhibit protein synthesis in bacteria by blocking the A site on the ribosome, preventing the binding of aminoacyl-tRNAs. **Chloramphenicol** inhibits protein synthesis by bacterial (and mitochondrial and chloroplast) ribosomes by blocking peptidyl transfer; it does not affect cytosolic protein synthesis in eukaryotes. Conversely, **cycloheximide** blocks the peptidyl transferase of 80S eukaryotic ribosomes but not that of 70S bacterial (and mitochondrial and chloroplast) ribosomes. **Streptomycin**, a basic trisaccharide, causes misreading of the genetic code (in bacteria) at relatively low concentrations and inhibits initiation at higher concentrations.

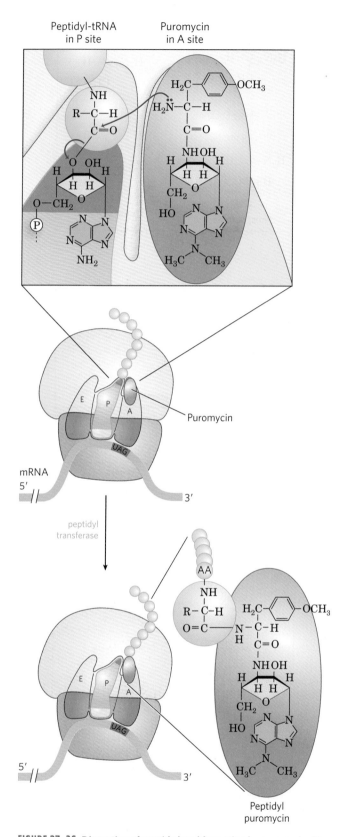

FIGURE 27–36 Disruption of peptide bond formation by puromycin. The antibiotic puromycin resembles the aminoacyl end of a charged tRNA, and it can bind to the ribosomal A site and participate in peptide bond formation. The product of this reaction, peptidyl puromycin, is not translocated to the P site. Instead, it dissociates from the ribosome, causing premature chain termination.

Cycloheximide

Streptomycin

Several other inhibitors of protein synthesis are notable because of their toxicity to humans and other mammals. **Diphtheria toxin** (M_r 58,330) catalyzes the ADP-ribosylation of a diphthamide (a modified histidine) residue of eukaryotic elongation factor eEF2, thereby inactivating it. **Ricin** (M_r 29,895), an extremely toxic protein of the castor bean, inactivates the 60S subunit of eukaryotic ribosomes by depurinating a specific adenosine in 23S rRNA. Ricin was used in the infamous 1978 murder of BBC journalist and Bulgarian dissident Georgi Markov, presumably by the Bulgarian secret police. Using a syringe hidden at the end of an umbrella, a member of the secret police injected Markov in the leg with a ricin-infused pellet. He died 4 days later.

SUMMARY 27.2 Protein Synthesis

▶ Protein synthesis occurs on the ribosomes, which consist of protein and rRNA. Bacteria have 70S ribosomes, with a large (50S) and a small (30S) subunit. Eukaryotic ribosomes are significantly larger (80S) and contain more proteins.

▶ Transfer RNAs have 73 to 93 nucleotide residues, some of which have modified bases. Each tRNA has an amino acid arm with the terminal sequence CCA(3′) to which an amino acid is esterified, an anticodon arm, a TψC arm, and a D arm; some tRNAs have a fifth arm. The anticodon is responsible for the specificity of interaction between the aminoacyl-tRNA and the complementary mRNA codon.

▶ The growth of polypeptides on ribosomes begins with the amino-terminal amino acid and proceeds by successive additions of new residues to the carboxyl-terminal end.

▶ Protein synthesis occurs in five stages.

1. Amino acids are activated by specific aminoacyl-tRNA synthetases in the cytosol. These enzymes catalyze the formation of aminoacyl-tRNAs, with simultaneous cleavage of ATP to AMP and PP$_i$. The fidelity of protein synthesis depends on the accuracy of this reaction, and some of these enzymes carry out proofreading steps at separate active sites.

2. In bacteria, the initiating aminoacyl-tRNA in all proteins is *N*-formylmethionyl-tRNA^{fMet}. Initiation of protein synthesis involves formation of a complex between the 30S ribosomal subunit, mRNA, GTP, fMet-tRNA^{fMet}, three initiation factors, and the 50S subunit; GTP is hydrolyzed to GDP and P$_i$.

3. In the elongation steps, GTP and elongation factors are required for binding the incoming aminoacyl-tRNA to the A site on the ribosome. In the first peptidyl transfer reaction, the fMet residue is transferred to the amino group of the incoming aminoacyl-tRNA. Movement of the ribosome along the mRNA then translocates the dipeptidyl-tRNA from the A site to the P site, a process requiring hydrolysis of GTP. Deacylated tRNAs dissociate from the ribosomal E site.

4. After many such elongation cycles, synthesis of the polypeptide is terminated with the aid of release factors. At least four high-energy phosphate equivalents (from ATP and GTP) are required to generate each peptide bond, an energy investment required to guarantee fidelity of translation.

5. Polypeptides fold into their active, three-dimensional forms. Many proteins are further processed by posttranslational modification reactions.

▶ Many well-studied antibiotics and toxins inhibit some aspect of protein synthesis.

27.3 Protein Targeting and Degradation

The eukaryotic cell is made up of many structures, compartments, and organelles, each with specific functions that require distinct sets of proteins and enzymes. These proteins (with the exception of those produced in mitochondria and plastids) are synthesized on ribosomes in the cytosol, so how are they directed to their final cellular destinations?

We are now beginning to understand this complex and fascinating process. Proteins destined for secretion, integration in the plasma membrane, or inclusion in lysosomes generally share the first few steps of a pathway that begins in the endoplasmic reticulum. Proteins destined for mitochondria, chloroplasts, or the nucleus use three separate mechanisms. And proteins destined for the cytosol simply remain where they are synthesized.

The most important element in many of these targeting pathways is a short sequence of amino acids called a **signal sequence**, whose function was first postulated by Günter Blobel and colleagues in 1970. The signal sequence directs a protein to its appropriate location in the cell and, for many proteins, is removed during transport or after the protein has reached its final destination. In proteins slated for transport into mitochondria, chloroplasts, or the ER, the signal sequence is at the amino terminus of a newly synthesized polypeptide. In many cases, the targeting capacity of particular signal sequences has been confirmed by fusing the signal sequence from one protein to a second protein and showing that the signal directs the second protein to the location where the first protein is normally found. The selective degradation of proteins no longer needed by the cell also relies largely on a set of molecular signals embedded in each protein's structure.

Günter Blobel

In this concluding section we examine protein targeting and degradation, emphasizing the underlying signals and molecular regulation that are so crucial to cellular metabolism. Except where noted, the focus is now on eukaryotic cells.

Posttranslational Modification of Many Eukaryotic Proteins Begins in the Endoplasmic Reticulum

Perhaps the best-characterized targeting system begins in the ER. Most lysosomal, membrane, or secreted proteins have an amino-terminal signal sequence (**Fig. 27–37**) that marks them for translocation into the lumen of the ER; hundreds of such signal sequences have been determined. The carboxyl terminus of the signal sequence is defined by a cleavage site, where protease action removes the sequence after the protein is imported into the ER. Signal sequences vary in length from 13 to 36 amino acid residues, but all have the following features: (1) about 10 to 15 hydrophobic amino acid residues; (2) one or more positively charged residues, usually near the amino terminus, preceding the hydrophobic sequence; and (3) a short sequence at the carboxyl terminus (near the cleavage site) that is relatively polar, typically having amino acid residues with short side chains (especially Ala) at the positions closest to the cleavage site.

As originally demonstrated by George Palade, proteins with these signal sequences are synthesized on ribosomes attached to the ER. The signal sequence itself helps to direct the ribosome to the ER, as illustrated by steps ❶ through ❽ in **Figure 27–38**. ❶ The targeting pathway begins with initiation of protein synthesis on free ribosomes. ❷ The sig-

George Palade

nal sequence appears early in the synthetic process, because it is at the amino terminus, which as we have seen is synthesized first. ❸ As it emerges from the ribosome, the signal sequence—and the ribosome itself—is bound by the large **signal recognition particle (SRP)**; SRP then binds GTP and halts elongation of the polypeptide when it is about 70 amino acids long and the signal sequence has completely emerged from the ribosome. ❹ The GTP-bound SRP now directs the ribosome (still bound to the mRNA) and the incomplete polypeptide to GTP-bound SRP receptors in the cytosolic face of the ER; the nascent polypeptide is delivered to a **peptide translocation complex** in the ER, which may interact directly with the ribosome. ❺ SRP dissociates from the ribosome, accompanied by hydrolysis of GTP in both SRP and the SRP receptor. ❻ Elongation of the polypeptide now resumes, with the ATP-driven translocation complex feeding the growing polypeptide into the ER lumen until the complete protein has been synthesized. ❼ The signal sequence is removed by a signal peptidase within the ER lumen; ❽ the ribosome dissociates and ❾ is recycled.

		cleavage site
Human influenza virus A	Met Lys Ala Lys Leu Leu Val Leu Leu Tyr Ala Phe Val Ala Gly	↓ Asp Gln --
Human preproinsulin	Met Ala Leu Trp Met Arg Leu Leu Pro Leu Leu Ala Leu Leu Ala Leu Trp Gly Pro Asp Pro Ala Ala Ala	↓ Phe Val --
Bovine growth hormone	Met Met Ala Ala Gly Pro Arg Thr Ser Leu Leu Leu Ala Phe Ala Leu Leu Cys Leu Pro Trp Thr Gln Val Val Gly	↓ Ala Phe --
Bee promellitin	Met Lys Phe Leu Val Asn Val Ala Leu Val Phe Met Val Val Tyr Ile Ser Tyr Ile Tyr Ala	↓ Ala Pro --
Drosophila glue protein	Met Lys Leu Leu Val Val Ala Val Ile Ala Cys Met Leu Ile Gly Phe Ala Asp Pro Ala Ser Gly	↓ Cys Lys --

FIGURE 27–37 Amino-terminal signal sequences of some eukaryotic proteins that direct their translocation into the ER. The hydrophobic core (yellow) is preceded by one or more basic residues (blue). Note the polar and short-side-chain residues immediately preceding (to the left of, as shown here) the cleavage sites (indicated by red arrows).

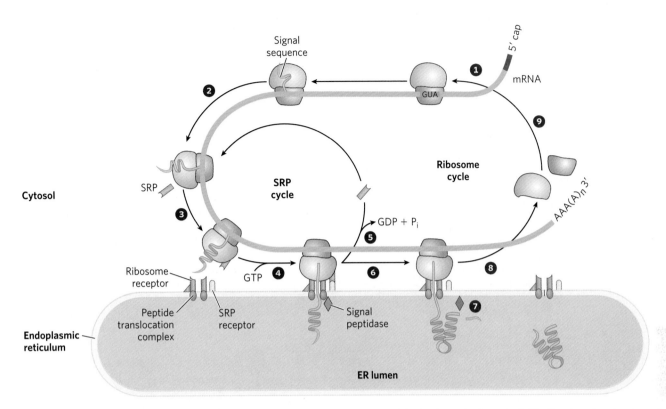

FIGURE 27–38 Directing eukaryotic proteins with the appropriate signals to the endoplasmic reticulum. This process involves the SRP cycle and translocation and cleavage of the nascent polypeptide. The steps are described in the text. SRP is a rod-shaped complex containing a 300 nucleotide RNA (7SL-RNA) and six different proteins (combined M_r 325,000). One protein subunit of SRP binds directly to the signal sequence, inhibiting elongation by sterically blocking the entry of aminoacyl-tRNAs and inhibiting peptidyl transferase. Another protein subunit binds and hydrolyzes GTP. The SRP receptor is a heterodimer of α (M_r 69,000) and β (M_r 30,000) subunits, both of which bind and hydrolyze multiple GTP molecules during this process.

Glycosylation Plays a Key Role in Protein Targeting

In the ER lumen, newly synthesized proteins are further modified in several ways. Following the removal of signal sequences, polypeptides are folded, disulfide bonds formed, and many proteins glycosylated to form glycoproteins. In many glycoproteins the linkage to their oligosaccharides is through Asn residues. These *N*-linked oligosaccharides are diverse (Chapter 7), but the pathways by which they form have a common first step. A 14 residue core oligosaccharide is built up in a stepwise fashion, then transferred from a dolichol phosphate donor molecule to certain Asn residues in the protein **(Fig. 27–39)**. The transferase is on the lumenal face of the ER and thus cannot catalyze glycosylation of cytosolic proteins. After transfer, the core oligosaccharide is trimmed and elaborated in different ways on different proteins, but all *N*-linked oligosaccharides retain a pentasaccharide core derived from the original 14 residue oligosaccharide. Several antibiotics act by interfering with one or more steps in this process and have aided in elucidating the steps of protein glycosylation. The best characterized is **tunicamycin**, which mimics the structure of UDP-*N*-acetylglucosamine and blocks the first step of the process (Fig. 27–39, step ❶). A few proteins are *O*-glycosylated in the ER, but most *O*-glycosylation occurs in the Golgi complex or in the cytosol (for proteins that do not enter the ER).

FIGURE 27–39 Synthesis of the core oligosaccharide of glycoproteins. The core oligosaccharide is built up by the successive addition of monosaccharide units. ❶, ❷ The first steps occur on the cytosolic face of the ER. ❸ Translocation moves the incomplete oligosaccharide across the membrane (mechanism not shown), and ❹ completion of the core oligosaccharide occurs within the lumen of the ER. The precursors that contribute additional mannose and glucose residues to the growing oligosaccharide in the lumen are dolichol phosphate derivatives. In the first step in the construction of the N-linked oligosaccharide moiety of a glycoprotein, ❺, the core oligosaccharide is transferred from dolichol phosphate to an Asn residue of the protein within the ER lumen. The core oligosaccharide is then further modified in the ER and the Golgi complex in pathways that differ for different proteins. The five sugar residues shown surrounded by a beige screen (after step ❼) are retained in the final structure of all N-linked oligosaccharides. ❽ The released dolichol pyrophosphate is again translocated so that the pyrophosphate is on the cytosolic face of the ER, then ❾ a phosphate is hydrolytically removed to regenerate dolichol phosphate.

Suitably modified proteins can now be moved to a variety of intracellular destinations. Proteins travel from the ER to the Golgi complex in transport vesicles **(Fig. 27–40)**. In the Golgi complex, oligosaccharides are O-linked to some proteins, and N-linked oligosaccharides are further modified. By mechanisms not yet fully understood, the Golgi complex also sorts proteins and sends them to their final destinations. The processes that segregate proteins targeted for secretion from those targeted for the plasma membrane or lysosomes must distinguish among these proteins on the basis of structural features other than signal sequences, which were removed in the ER lumen.

This sorting process is best understood in the case of hydrolases destined for transport to lysosomes. On arrival of a hydrolase (a glycoprotein) in the Golgi complex, an as yet undetermined feature (sometimes called a signal patch) of the three-dimensional structure of hydrolase is recognized by a phosphotransferase, which phosphorylates certain mannose residues in oligosaccharide **(Fig. 27–41)**. The presence of one or more mannose 6-phosphate residues in its N-linked oligosaccharide is the structural signal that targets a protein to lysosomes. A receptor protein in the membrane of the Golgi complex recognizes the mannose 6-phosphate signal and binds the hydrolase so marked. Vesicles containing these receptor-hydrolase complexes bud from the trans side of the Golgi complex and make their way to sorting vesicles. Here, the receptor-hydrolase complex dissociates in a process facilitated by the lower pH in the vesicle and by phosphatase-catalyzed removal of phosphate groups from the mannose 6-phosphate residues. The receptor is then recycled to the Golgi complex, and vesicles containing the hydrolases bud from the sorting vesicles and move to the lysosomes. In cells treated with tunicamycin (Fig. 27–39, step ❶), hydrolases that should be targeted to lysosomes are instead secreted, confirming that the N-linked oligosaccharide plays a key role in targeting these enzymes to lysosomes.

The pathways that target proteins to mitochondria and chloroplasts also rely on amino-terminal signal sequences. Although mitochondria and chloroplasts contain DNA, most of their proteins are encoded by

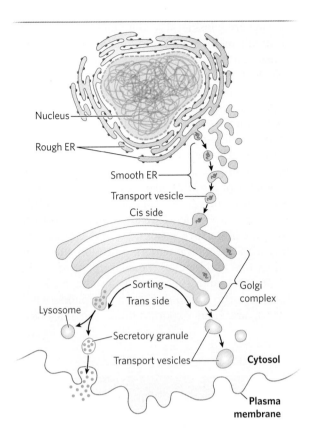

FIGURE 27–40 Pathway taken by proteins destined for lysosomes, the plasma membrane, or secretion. Proteins are moved from the ER to the cis side of the Golgi complex in transport vesicles. Sorting occurs primarily in the trans side of the Golgi complex.

nuclear DNA and must be targeted to the appropriate organelle. Unlike other targeting pathways, however, the mitochondrial and chloroplast pathways begin only *after* a precursor protein has been completely synthesized and released from the ribosome. Precursor proteins destined for mitochondria or chloroplasts are bound by cytosolic chaperone proteins and delivered to receptors on the exterior surface of the target organelle. Specialized translocation mechanisms then transport the protein to its final destination in the organelle, after which the signal sequence is removed.

Signal Sequences for Nuclear Transport Are Not Cleaved

Molecular communication between the nucleus and the cytosol requires the movement of macromolecules through nuclear pores. RNA molecules synthesized in the nucleus are exported to the cytosol. Ribosomal proteins synthesized on cytosolic ribosomes are imported

FIGURE 27–41 Phosphorylation of mannose residues on lysosome-targeted enzymes. *N*-Acetylglucosamine phosphotransferase recognizes some as yet unidentified structural feature of hydrolases destined for lysosomes.

into the nucleus and assembled into 60S and 40S ribosomal subunits in the nucleolus; completed subunits are then exported back to the cytosol. A variety of nuclear proteins (RNA and DNA polymerases, histones, topoisomerases, proteins that regulate gene expression, and so forth) are synthesized in the cytosol and imported into the nucleus. This traffic is modulated by a complex system of molecular signals and transport proteins that is gradually being elucidated.

In most multicellular eukaryotes, the nuclear envelope breaks down at each cell division, and once division is completed and the nuclear envelope reestablished, the dispersed nuclear proteins must be reimported. To allow this repeated nuclear importation, the signal sequence that targets a protein to the nucleus—the **nuclear localization sequence (NLS)**—is not removed after the protein arrives at its destination. An NLS, unlike other signal sequences, may be located almost anywhere along the primary sequence of the protein.

NLSs can vary considerably, but many consist of four to eight amino acid residues and include several consecutive basic (Arg or Lys) residues.

Nuclear importation is mediated by a number of proteins that cycle between the cytosol and the nucleus **(Fig. 27–42)**, including importin α and β and a small GTPase known as Ran (*Ras*-related *n*uclear protein). A heterodimer of importin α and β functions as a soluble receptor for proteins targeted to the nucleus, with the α subunit binding NLS-bearing proteins in the cytosol. The complex of the NLS-bearing protein and the importin docks at a nuclear pore and is translocated through the pore by an energy-dependent mechanism. In the nucleus, the importin β is bound by Ran GTPase, releasing importin β from the imported protein. Importin β is bound by Ran and by CAS (*c*ellular *a*poptosis *s*usceptibility protein) and separated from the NLS-bearing protein. Importin α and β, in their complexes with Ran and CAS, are then exported from the nucleus. Ran

(a)

(b)

0.2 μm

FIGURE 27–42 Targeting of nuclear proteins. (a) ❶ A protein with an appropriate nuclear localization signal (NLS) is bound by a complex of importin α and β. ❷ The resulting complex binds to a nuclear pore and translocates. ❸ Inside the nucleus, dissociation of importin β is promoted by the binding of Ran-GTP. ❹ Importin α binds to Ran-GTP and CAS (cellular apoptosis susceptibility protein), releasing the nuclear protein. ❺ Importin α and β and CAS are transported out of the nucleus and recycled. They are released in the cytosol when Ran hydrolyzes its bound GTP. ❻ Ran-GDP is bound by NTF2, and transported back into the nucleus. ❼ RanGEF promotes the exchange of GDP for GTP in the nucleus, and Ran-GTP is ready to process another NLS-bearing protein-importin complex. **(b)** Scanning electron micrograph of the surface of the nuclear envelope, showing numerous nuclear pores. The nuclear pore complex is one of the largest molecular aggregates in the cell ($M_r \sim 5 \times 10^7$). It is made up of multiple copies of over 30 different proteins.

Inner membrane proteins

Phage fd, major coat protein
Met Lys Lys Ser Leu Val Leu Lys Ala Ser Val Ala Val Ala Thr Leu Val Pro Met Leu Ser Phe Ala↓Ala Glu --

Phage fd, minor coat protein
Met Lys Lys Leu Leu Phe Ala Ile Pro Leu Val Val Pro Phe Tyr Ser His Ser↓Ala Glu --

Periplasmic proteins

Alkaline phosphatase
Met Lys Gln Ser Thr Ile Ala Leu Ala Leu Leu Pro Leu Leu Phe Thr Pro Val Thr Lys Ala↓Arg Thr --

Leucine-specific binding protein
Met Lys Ala Asn Ala Lys Thr Ile Ile Ala Gly Met Ile Ala Leu Ala Ile Ser His Thr Ala Met Ala↓Asp Asp --

β-Lactamase of pBR322
Met Ser Ile Gln His Phe Arg Val Ala Leu Ile Pro Phe Phe Ala Ala Phe Cys Leu Pro Val Phe Ala↓His Pro --

Outer membrane proteins

Lipoprotein
Met Lys Ala Thr Lys Leu Val Leu Gly Ala Val Ile Leu Gly Ser Thr Leu Leu Ala Gly↓Cys Ser --

LamB
Leu Arg Lys Leu Pro Leu Ala Val Ala Val Ala Ala Gly Val Met Ser Ala Gln Ala Met Ala↓Val Asp --

OmpA Met Met Ile Thr Met Lys Lys Thr Ala Ile Ala Ile Ala Val Ala Leu Ala Gly Phe Ala Thr Val Ala Gln Ala↓Ala Pro --

cleavage site

FIGURE 27–43 Signal sequences that target proteins to different locations in bacteria. Basic amino acids (blue) near the amino terminus and hydrophobic core amino acids (yellow) are highlighted. The cleavage sites marking the ends of the signal sequences are indicated by red arrows. Note that the inner bacterial cell membrane (see Fig. 1–6) is where phage fd coat proteins and DNA are assembled into phage particles. OmpA is outer membrane protein A; LamB is a cell surface receptor protein for λ phage.

hydrolyzes GTP in the cytosol to release the importins, which are then free to begin another importation cycle. Ran itself is also cycled back into the nucleus by the binding of Ran-GDP to nuclear transport factor 2 (NTF2). Inside the nucleus, the GDP bound to Ran is replaced with GTP through the action of Ran guanosine nucleotide–exchange factor (RanGEF; see Box 12–2).

Bacteria Also Use Signal Sequences for Protein Targeting

Bacteria can target proteins to their inner or outer membranes, to the periplasmic space between these membranes, or to the extracellular medium. They use signal sequences at the amino terminus of the proteins (Fig. 27–43), much like those on eukaryotic proteins targeted to the ER, mitochondria, and chloroplasts.

Most proteins exported from *E. coli* make use of the pathway shown in **Figure 27–44**. Following translation, a protein to be exported may fold only slowly, the amino-terminal signal sequence impeding the folding. The soluble chaperone protein SecB binds to the protein's signal sequence or other features of its incompletely folded structure. The bound protein is then delivered to SecA, a protein associated with the inner surface of the plasma membrane. SecA acts as both a receptor and a translocating ATPase. Released from SecB and bound to SecA, the protein is delivered to a translocation complex in the membrane, made up of SecY, E, and G, and is translocated stepwise through the membrane at the SecYEG complex

Cytosol

Periplasmic space

FIGURE 27–44 Model for protein export in bacteria. ❶ A newly translated polypeptide binds to the cytosolic chaperone protein SecB, which **❷** delivers it to SecA, a protein associated with the translocation complex (SecYEG) in the bacterial cell membrane. **❸** SecB is released, and SecA inserts itself into the membrane, forcing about 20 amino acid residues of the protein to be exported through the translocation complex. **❹** Hydrolysis of an ATP by SecA provides the energy for a conformational change that causes SecA to withdraw from the membrane, releasing the polypeptide. **❺** SecA binds another ATP, and the next stretch of 20 amino acid residues is pushed across the membrane through the translocation complex. Steps **❹** and **❺** are repeated until **❻** the entire protein has passed through and is released to the periplasm. The electrochemical potential across the membrane (denoted by + and −) also provides some of the driving force required for protein translocation.

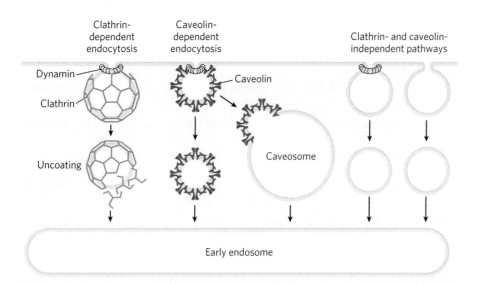

Clathrin-dependent endocytosis Caveolin-dependent endocytosis Clathrin- and caveolin-independent pathways

Dynamin Caveolin

Clathrin

Uncoating Caveosome

Early endosome

FIGURE 27–45 Summary of endocytosis pathways in eukaryotic cells. Pathways dependent on clathrin or caveolin make use of the GTPase dynamin to pinch vesicles from the plasma membrane. Some pathways do not use clathrin or caveolin; some of these make use of dynamin and some do not.

in lengths of about 20 amino acid residues. Each step is facilitated by the hydrolysis of ATP, catalyzed by SecA.

An exported protein is thus pushed through the membrane by a SecA protein located on the cytoplasmic surface, not pulled through the membrane by a protein on the periplasmic surface. This difference may simply reflect the need for the translocating ATPase to be where the ATP is. The transmembrane electrochemical potential can also provide energy for translocation of the protein, by an as yet unknown mechanism.

Although most exported bacterial proteins use this pathway, some follow an alternative pathway that uses signal recognition and receptor proteins homologous to components of the eukaryotic SRP and SRP receptor (Fig. 27–38).

Cells Import Proteins by Receptor-Mediated Endocytosis

Some proteins are imported into eukaryotic cells from the surrounding medium; examples include low-density lipoprotein (LDL), the iron-carrying protein transferrin, peptide hormones, and circulating proteins destined for degradation. There are several importation pathways **(Fig. 27–45)**. In one path, proteins bind to receptors in invaginations of the membrane called **coated pits**, which concentrate endocytic receptors in preference to other cell-surface proteins. The pits are coated on their cytosolic side with a lattice of the protein **clathrin**, which forms closed polyhedral structures **(Fig. 27–46)**. The clathrin lattice grows as more receptors are occupied by target proteins. Eventually, a complete membrane-bounded endocytic vesicle is pinched off the plasma membrane with the aid of the large GTPase **dynamin**, and enters the cytoplasm. The clathrin is quickly removed by uncoating enzymes, and the vesicle fuses with an endosome. ATPase activity in the endosomal membranes reduces the pH therein, facilitating dissociation of recep-

Heavy chain Light chain

~80 nm

(a) (b)

(c) 0.1 μm

FIGURE 27–46 Clathrin. (a) Three light (L) chains (M_r 35,000) and three heavy (H) chains (M_r 180,000) of the $(HL)_3$ clathrin unit, organized as a three-legged structure called a triskelion. **(b)** Triskelions tend to assemble into polyhedral lattices. **(c)** Electron micrograph of a coated pit on the cytosolic face of the plasma membrane of a fibroblast.

tors from their target proteins. In a related pathway, caveolin causes invagination of patches of membrane containing lipid rafts associated with certain types of receptors (see Fig. 11–22). These endocytic vesicles then fuse with caveolin-containing internal structures, caveosomes, where the internalized molecules are sorted and redirected to other parts of the cell and the caveolins are prepared for recycling to the membrane surface. There are also clathrin- and caveolin-independent pathways; some make use of dynamin and others do not.

The imported proteins and receptors then go their separate ways, their fates varying with the cell and protein type. Transferrin and its receptor are eventually recycled. Some hormones, growth factors, and immune complexes, after eliciting the appropriate cellular response, are degraded along with their receptors. LDL is degraded after the associated cholesterol has been delivered to its destination, but the LDL receptor is recycled (see Fig. 21–41).

Receptor-mediated endocytosis is exploited by some toxins and viruses to gain entry to cells. Influenza virus, diphtheria toxin, and cholera toxin all enter cells in this way.

Protein Degradation Is Mediated by Specialized Systems in All Cells

Protein degradation prevents the buildup of abnormal or unwanted proteins and permits the recycling of amino acids. The half-lives of eukaryotic proteins vary from 30 seconds to many days. Most proteins turn over rapidly relative to the lifetime of a cell, although a few (such as hemoglobin) can last for the life of the cell (about 110 days for an erythrocyte). Rapidly degraded proteins include those that are defective because of incorrectly inserted amino acids or because of damage accumulated during normal functioning. And enzymes that act at key regulatory points in metabolic pathways often turn over rapidly.

Defective proteins and those with characteristically short half-lives are generally degraded in both bacterial and eukaryotic cells by selective ATP-dependent cytosolic systems. A second system in vertebrates, operating in lysosomes, recycles the amino acids of membrane proteins, extracellular proteins, and proteins with characteristically long half-lives.

In *E. coli*, many proteins are degraded by an ATP-dependent protease called Lon (the name refers to the "long form" of proteins, observed only when this protease is absent). The protease is activated in the presence of defective proteins or those slated for rapid turnover; two ATP molecules are hydrolyzed for every peptide bond cleaved. The precise role of this ATP hydrolysis is not yet clear. Once a protein has been reduced to small inactive peptides, other ATP-independent proteases complete the degradation process.

The ATP-dependent pathway in eukaryotic cells is quite different, involving the protein **ubiquitin**, which, as its name suggests, occurs throughout the eukaryotic kingdoms. One of the most highly conserved proteins known, ubiquitin (76 amino acid residues) is essentially identical in organisms as different as yeasts and humans. Ubiquitin is covalently linked to proteins slated for destruction via an ATP-dependent pathway involving three separate types of enzymes, called E1 activating enzymes, E2 conjugating enzymes, and E3 ligases **(Fig. 27–47)**.

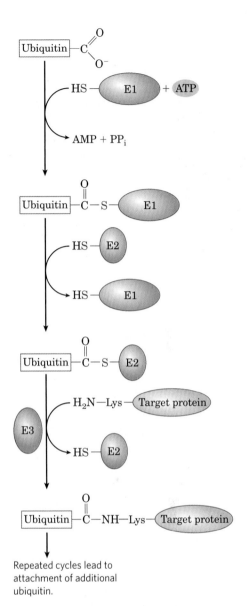

FIGURE 27–47 Three-step pathway by which ubiquitin is attached to a protein. Two different enzyme-ubiquitin intermediates are involved. The free carboxyl group of ubiquitin's carboxyl-terminal Gly residue first becomes linked to an E1-class activating enzyme via a thioester. The ubiquitin is then transferred to an E2 conjugating enzyme. An E3 ligase ultimately catalyzes the transfer of the ubiquitin from E2 to the target, linking ubiquitin through an amide (isopeptide) bond to an ε-amino group of a Lys residue of the target protein. Additional cycles produce polyubiquitin, a covalent polymer of ubiquitin subunits that targets the attached protein for destruction in eukaryotes. Multiple pathways of this sort, with different protein targets, are present in most eukaryotic cells.

Ubiquitinated proteins are degraded by a large complex known as the **26S proteasome** (M_r 2.5×10^6) **(Fig. 27–48)**. The eukaryotic proteasome consists of two copies each of at least 32 different subunits, most of which are highly conserved from yeasts to humans. The proteasome contains two main types of subcomplexes, a barrel-like core particle and regulatory particles on either end of the barrel. The 20S core particle consists of four rings; the outer rings are formed from seven α

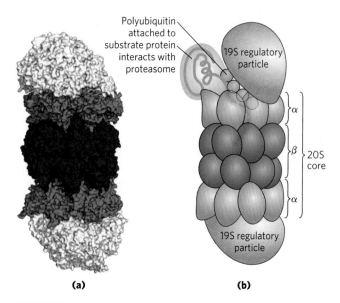

FIGURE 27–48 Three-dimensional structure of the eukaryotic protea-some. The 26S proteasome is highly conserved in all eukaryotes. The two subassemblies are the 20S core particle and the 19S regulatory particle, or cap. **(a)** (PDB ID 3L5Q) The core particle consists of four rings arranged to form a barrel-like structure. Each of the inner rings has seven different β subunits (dark brown), three of which have protease activities. The outer rings each have seven different α subunits (light brown). A regulatory particle forms a cap on each end of the core particle (gray). **(b)** The regu-latory particle binds ubiquitinated proteins, unfolds them, and translo-cates them into the core particle, where they are degraded to peptides of 3 to 25 amino acid residues.

subunits, and the inner rings from seven β subunits. Three of the seven subunits in each β ring have prote-ase activities, each with different substrate specificities. The stacked rings of the core particle form the barrel-like structure within which target proteins are degraded. The 19S regulatory particle on each end of the core particle contains approximately 18 subunits, including some that recognize and bind to ubiquitinated proteins. Six of the subunits are AAA+ ATPases (see Chapter 25) that probably function in unfolding the ubiquitinated proteins and translocating the unfolded polypeptide into the core particle for degradation. The 19S particle also deubiquitinates the proteins as they are degraded in the proteasome. Most cells have additional regulatory complexes that can replace the 19S particle. These alternative regulators do not hydrolyze ATP and do not bind to ubiquitin, but they are important for the degra-dation of particular cellular proteins. The 26S protea-some can be effectively "accessorized," with regulatory complexes changing with changing cellular conditions.

Although we do not yet understand all the signals that trigger ubiquitination, one simple signal has been found. For many proteins, the identity of the first resi-due that remains after removal of the amino-terminal Met residue, and any other posttranslational proteolytic processing of the amino-terminal end, has a profound

influence on half-life (Table 27–9). These amino-termi-nal signals have been conserved over billions of years of evolution and are the same in bacterial protein degrada-tion systems and in the human ubiquitination pathway. More complex signals, such as the destruction box dis-cussed in Chapter 12 (see Fig. 12–47), are also being identified.

Ubiquitin-dependent proteolysis is as important for the regulation of cellular processes as for the elimina-tion of defective proteins. Many proteins required at only one stage of the eukaryotic cell cycle are rapidly degraded by the ubiquitin-dependent pathway after completing their function. Ubiquitin-dependent destruc-tion of cyclin is critical to cell-cycle regulation (see Fig. 12–47). The E1, E2, and E3 components of the ubiqui-tination pathway (Fig. 27–47) are in fact large families of proteins. Different E1, E2, and E3 enzymes exhibit different specificities for target proteins and thus regu-late different cellular processes. Some of these enzymes are highly localized in certain cellular compartments, reflecting a specialized function.

Not surprisingly, defects in the ubiquitination pathway have been implicated in a wide range of disease states. An inability to degrade certain proteins that activate cell division (the products of oncogenes) can lead to tumor formation, whereas a too-rapid degra-dation of proteins that act as tumor suppressors can have the same effect. The ineffective or overly rapid degradation of cellular proteins also appears to play a role in a range of other conditions: renal diseases, asthma, neurodegenerative disorders such as Alzheimer and Parkinson diseases (associated with the formation of characteristic proteinaceous structures in neurons), cystic fibrosis (caused in some cases by a too-rapid deg-radation of a chloride ion channel, with resultant loss of function; see Box 11–2), Liddle syndrome (in which a

TABLE 27–9 Relationship between Protein Half-Life and Amino-Terminal Amino Acid Residue

Amino-terminal residue	Half-life*
Stabilizing	
Ala, Gly, Met, Ser, Thr, Val	>20 h
Destabilizing	
Gln, Ile	~30 min
Glu, Tyr	~10 min
Pro	~7 min
Asp, Leu, Lys, Phe	~3 min
Arg	~2 min

Source: Modified from Bachmair, A., Finley, D., & Varshavsky, A. (1986) In vivo half-life of a protein is a function of its amino-terminal residue. *Science* 234, 179–186.

*Half-lives were measured in yeast for the β-galactosidase protein modified so that in each experiment it had a different amino-terminal residue. Half-lives may vary for different proteins and in different organisms, but this general pattern appears to hold for all organisms.

sodium channel in the kidney is not degraded, leading to excessive Na$^+$ absorption and early-onset hypertension)—and many other disorders. Drugs designed to inhibit proteasome function are being developed as potential treatments for some of these conditions. In a changing metabolic environment, protein degradation is as important to a cell's survival as is protein synthesis, and much remains to be learned about these interesting pathways. ■

SUMMARY 27.3 Protein Targeting and Degradation

▶ After synthesis, many proteins are directed to particular locations in the cell. One targeting mechanism involves a peptide signal sequence, generally found at the amino terminus of a newly synthesized protein.

▶ In eukaryotic cells, one class of signal sequences is recognized by the signal recognition particle (SRP), which binds the signal sequence as soon as it appears on the ribosome and transfers the entire ribosome and incomplete polypeptide to the ER. Polypeptides with these signal sequences are moved into the ER lumen as they are synthesized; once in the lumen they may be modified and moved to the Golgi complex, then sorted and sent to lysosomes, the plasma membrane, or transport vesicles.

▶ Proteins targeted to mitochondria and chloroplasts in eukaryotic cells, and those destined for export in bacteria, also make use of an amino-terminal signal sequence.

▶ Proteins targeted to the nucleus have an internal signal sequence that, unlike other signal sequences, is not cleaved once the protein is successfully targeted.

▶ Some eukaryotic cells import proteins by receptor-mediated endocytosis.

▶ All cells eventually degrade proteins, using specialized proteolytic systems. Defective proteins and those slated for rapid turnover are generally degraded by an ATP-dependent system. In eukaryotic cells, the proteins are first tagged by linkage to ubiquitin, a highly conserved protein. Ubiquitin-dependent proteolysis is carried out by proteasomes, also highly conserved, and is critical to the regulation of many cellular processes.

Key Terms

Terms in bold are defined in the glossary.

aminoacyl-tRNA 1104
aminoacyl-tRNA synthetases 1104
translation 1104
codon 1105
reading frame 1105
initiation codon 1107
termination codons 1107
open reading frame (ORF) 1107
anticodon 1109
wobble 1110
translational frameshifting 1111
RNA editing 1111
initiation 1127
Shine-Dalgarno sequence 1127
aminoacyl (A) site 1128
peptidyl (P) site 1128
exit (E) site 1128
initiation complex 1128
elongation 1129
elongation factors 1129
peptidyl transferase 1132
translocation 1132
termination 1134
release factors 1134
nonsense suppressor 1134
polysome 1135
posttranslational modification 1136
puromycin 1138
tetracycline 1138
chloramphenicol 1138
cycloheximide 1138
streptomycin 1138
diphtheria toxin 1139
ricin 1139
signal sequence 1140
signal recognition particle (SRP) 1140
peptide translocation complex 1140
tunicamycin 1141
nuclear localization sequence (NLS) 1144
coated pits 1146
clathrin 1146
dynamin 1146
ubiquitin 1147
proteasome 1147

Further Reading

Genetic Code

Ambrogelly, A., Palioura, S., & Söll, D. (2007) Natural expansion of the genetic code. *Nat. Chem. Biol.* **3**, 29–35.

Blanc, V. & Davidson, N.O. (2003) C-to-U RNA editing: mechanisms leading to genetic diversity. *J. Biol. Chem.* **278**, 1395–1398.

Crick, F.H.C. (1966) The genetic code: III. *Sci. Am.* **215** (October), 55–62.
An insightful overview of the genetic code at a time when the code words had just been worked out.

Farajollahi, S. & Maas, S. (2010) Molecular diversity through RNA editing: a balancing act. *Trends Genet.* **26**, 221–230.

Hohn, M.J., Park, H.S., O'Donoghue, P., Schnitzbauer, M., & Söll, D. (2006) Emergence of the universal genetic code imprinted in an RNA record. *Proc. Natl. Acad. Sci. USA* **103**, 18,095–18,100.

Levanon, K., Eisenberg E., Rechavi G., & Levanon, E.Y. (2005) Letter from the editor: adenosine-to-inosine RNA editing in Alu repeats in the human genome. *EMBO Rep.* **6**, 831–835.

Liu, M. & Schatz, D.G. (2009) Balancing AID and DNA repair during somatic hypermutation. *Trends Immunol.* **30**, 173–181.

Lobanov, A.V., Turanov, A.A., Hatfield, D.L., & Gladyshev, V.N. (2010) Dual functions of codons in the genetic code. *Crit. Rev. Biochem. Mol. Biol.* **45**, 257–265.

Neeman, Y., Dahary, D., & Nishikura, K. (2006) Editor meets silencer: crosstalk between RNA editing and RNA interference. *Nat. Rev. Mol. Cell Biol.* **7**, 919–931.

Nirenberg, M. (2004) Historical review: deciphering the genetic code—a personal account. *Trends Biochem. Sci.* **29**, 46–54.

Schimmel, P. & Beebe, K. (2004) Molecular biology—genetic code seizes pyrrolysine. *Nature* **431**, 257–258.

Vetsigian, K., Woese, C., & Goldenfeld, N. (2006) Collective evolution and the genetic code. *Proc. Natl. Acad. Sci. USA* **103**, 10,696–10,701.

Yanofsky, C. (2007) Establishing the triplet nature of the genetic code. *Cell* **128**, 815–818.

Yarus, M., Caporaso, J.G., & Knight, R. (2005) Origins of the genetic code: the escaped triplet theory. *Annu. Rev. Biochem.* **74**, 179–198.

Protein Synthesis

Ban, N., Nissen, P., Hansen, J., Moore, P.B., & Steitz, T.A. (2000) The complete atomic structure of the large ribosomal subunit at 2.4 Å resolution. *Science* **289**, 905–920.

    The first high-resolution structure of a major ribosomal subunit.

Ben-Shem, A., de Loubresse, N.G., Melnikov, S., Jenner, L., Yusupova, G., & Yusupov, M. (2011) The structure of the eukaryotic ribosome at 3.0 Å resolution. *Science* **334**, 1524–1529.

Chapeville, F., Lipmann, F., von Ehrenstein, G., Weisblum, B., Ray, W.J., Jr., & Benzer, S. (1962) On the role of soluble ribonucleic acid in coding for amino acids. *Proc. Natl. Acad. Sci. USA* **48**, 1086–1092.

    Classic experiments providing proof for Crick's adaptor hypothesis and showing that amino acids are not checked after they are linked to tRNAs.

Decatur, W.A. & Fournier, M.J. (2002) rRNA modifications and ribosome function. *Trends Biochem. Sci.* **27**, 344–351.

Dintzis, H.M. (1961) Assembly of the peptide chains of hemoglobin. *Proc. Natl. Acad. Sci. USA* **47**, 247–261.

    A classic experiment establishing that proteins are assembled beginning at the amino terminus.

Dunkle, J.A. & Cate, J.H.D. (2010) Ribosome structure and dynamics during translocation and termination. *Annu. Rev. Biophys.* **39**, 227–244.

Gray, N.K. & Wickens, M. (1998) Control of translation initiation in animals. *Annu. Rev. Cell Dev. Biol.* **14**, 399–458.

Ibba, M. & Söll, D. (2000) Aminoacyl-tRNA synthesis. *Annu. Rev. Biochem.* **69**, 617–650.

Korostelev, A. & Noller, H.F. (2007) The ribosome in focus: new structures bring new insights. *Trends Biochem. Sci.* **32**, 434–441.

Korostelev, A., Trakhanov, S., Laurberg, M., & Noller, H.F. (2006) Crystal structure of a 70S ribosome-tRNA complex reveals functional interactions and rearrangements. *Cell* **126**, 1065–1077.

Liu, C.C. & Schultz, P.G. (2010) Adding new chemistries to the genetic code. *Annu. Rev. Biochem.* **79**, 413–444.

Poehlsgaard, J. & Douthwaite, S. (2005) The bacterial ribosome as a target for antibiotics. *Nat. Rev. Microbiol.* **3**, 870–881.

Rodnina, M.V. & Wintermeyer, W. (2001) Fidelity of aminoacyl-tRNA selection on the ribosome: kinetic and structural mechanisms. *Annu. Rev. Biochem.* **70**, 415–435.

Woese, C.R., Olsen, G.J., Ibba, M., & Söll, D. (2000) Aminoacyl-tRNA synthetases, the genetic code, and the evolutionary process. *Microbiol. Mol. Biol. Rev.* **64**, 202–236.

Protein Targeting and Degradation

Bedford, L., Lowe, J., Dick, L.R., Mayer, R.J., & Brownell, J.E. (2011) Ubiquitin-like protein conjugation and the ubiquitin-proteasome system as drug targets. *Nat. Rev. Drug Discov.* **10**, 29–46.

DeMartino, G.N. & Gillette, T.G. (2007) Proteasomes: machines for all reasons. *Cell* **129**, 659–662.

Ferguson, S.M. & De Camilli, P. (2012) Dynamin, a membrane-remodelling GTPase. *Nat. Rev. Mol. Cell Biol.* **13**, 75–88.

Hartmann-Petersen, R., Seeger, M., & Gordon C. (2003) Transferring substrates to the 26S proteasome. *Trends Biochem. Sci.* **28**, 26–31.

Komander, D. & Rape, M. (2012) The ubiquitin code. *Annu. Rev. Biochem.* **81**, 203–229.

Liu, C.W., Li, X.H., Thompson, D., Wooding, K., Chang, T., Tang, Z., Yu, H., Thomas, P.J., & DeMartino, G.N. (2006) ATP binding and ATP hydrolysis play distinct roles in the function of 26S proteasome. *Mol. Cell* **24**, 39–50.

Luzio, J.P., Pryor, P.R., & Bright, N.A. (2007) Lysosomes: fusion and function. *Nat. Rev. Mol. Cell Biol.* **8**, 622–632.

Mayor, S. & Pagano, R.E. (2007) Pathways of clathrin-independent endocytosis. *Nat. Rev. Mol. Cell Biol.* **8**, 603–612.

McMahon, H.T. & Boucrot, E. (2011) Molecular mechanism and physiological functions of clathrin-mediated endocytosis. *Nat. Rev. Mol. Cell Biol.* **12**, 517–533.

Pickart, C.M. & Cohen, R.E. (2004) Proteasomes and their kin: proteases in the machine age. *Nat. Rev. Mol. Cell Biol.* **5**, 177–187.

Royle, S.J. (2006) The cellular functions of clathrin. *Cell. Mol. Life Sci.* **63**, 1823–1832.

Schekman, R. (2007) How sterols regulate protein sorting and traffic. *Proc. Natl. Acad. Sci. USA* **104**, 6496–6497.

Stewart, M. (2007) Molecular mechanism of the nuclear protein import cycle. *Nat. Rev. Mol. Cell Biol.* **8**, 195–208.

Strambio-De-Castillia, C., Niepel, M., & Rout, M.P. (2010) The nuclear pore complex: bridging nuclear transport and gene regulation. *Nat. Rev. Mol. Cell Biol.* **11**, 490–501.

Tasaki, T., Sriram, S.M., Park, K.S., & Kwon, Y.T. (2012) The N-end rule pathway. *Annu. Rev. Biochem.* **81**, 261–289.

Problems

1. Messenger RNA Translation Predict the amino acid sequences of peptides formed by ribosomes in response to the following mRNA sequences, assuming that the reading frame begins with the first three bases in each sequence.

    (a) GGUCAGUCGCUCCUGAUU

    (b) UUGGAUGCGCCAUAAUUUGCU

    (c) CAUGAUGCCUGUUGCUAC

    (d) AUGGACGAA

2. How Many Different mRNA Sequences Can Specify One Amino Acid Sequence? Write all the possible mRNA sequences that can code for the simple tripeptide segment Leu–Met–Tyr. Your answer will give you some idea about the number of possible mRNAs that can code for one polypeptide.

3. Can the Base Sequence of an mRNA Be Predicted from the Amino Acid Sequence of Its Polypeptide Product? A given sequence of bases in an mRNA will code for one and only one sequence of amino acids in a polypeptide, if the reading frame is specified. From a given sequence of amino acid residues in a protein such as cytochrome *c*, can we predict the base sequence of the unique mRNA that coded it? Give reasons for your answer.

4. Coding of a Polypeptide by Duplex DNA The template strand of a segment of double-helical DNA contains the sequence

    (5′)CTTAACACCCCTGACTTCGCGCCGTCG(3′)

    (a) What is the base sequence of the mRNA that can be transcribed from this strand?

    (b) What amino acid sequence could be coded by the mRNA in (a), starting from the 5′ end?

(c) If the complementary (nontemplate) strand of this DNA were transcribed and translated, would the resulting amino acid sequence be the same as in (b)? Explain the biological significance of your answer.

5. Methionine Has Only One Codon Methionine is one of two amino acids with only one codon. How does the single codon for methionine specify both the initiating residue and interior Met residues of polypeptides synthesized by *E. coli*?

6. Synthetic mRNAs The genetic code was elucidated with polyribonucleotides synthesized either enzymatically or chemically in the laboratory. Given what we now know about the genetic code, how would you make a polyribonucleotide that could serve as an mRNA coding predominantly for many Phe residues and a small number of Leu and Ser residues? What other amino acid(s) would be coded for by this polyribonucleotide, but in smaller amounts?

7. Energy Cost of Protein Biosynthesis Determine the minimum energy cost, in terms of ATP equivalents expended, required for the biosynthesis of the β-globin chain of hemoglobin (146 residues), starting from a pool including all necessary amino acids, ATP, and GTP. Compare your answer with the direct energy cost of the biosynthesis of a linear glycogen chain of 146 glucose residues in $(\alpha1\rightarrow4)$ linkage, starting from a pool including glucose, UTP, and ATP (Chapter 15). From your data, what is the *extra* energy cost of making a protein, in which all the residues are ordered in a specific sequence, compared with the cost of making a polysaccharide containing the same number of residues but lacking the informational content of the protein?

In addition to the direct energy cost for the synthesis of a protein, there are indirect energy costs—those required for the cell to make the necessary enzymes for protein synthesis. Compare the magnitude of the indirect costs to a eukaryotic cell of the biosynthesis of linear $(\alpha1\rightarrow4)$ glycogen chains and the biosynthesis of polypeptides, in terms of the enzymatic machinery involved.

8. Predicting Anticodons from Codons Most amino acids have more than one codon and attach to more than one tRNA, each with a different anticodon. Write all possible anticodons for the four codons of glycine: (5′)GGU, GGC, GGA, and GGG.

(a) From your answer, which of the positions in the anticodons are primary determinants of their codon specificity in the case of glycine?

(b) Which of these anticodon-codon pairings has/have a wobbly base pair?

(c) In which of the anticodon-codon pairings do all three positions exhibit strong Watson-Crick hydrogen bonding?

9. Effect of Single-Base Changes on Amino Acid Sequence Much important confirmatory evidence on the genetic code has come from assessing changes in the amino acid sequence of mutant proteins after a single base has been changed in the gene that encodes the protein. Which of the following amino acid replacements would be consistent with the genetic code if the replacements were caused by a single base change? Which cannot be the result of a single-base mutation? Why?

(a) Phe→Leu (e) Ile→Leu
(b) Lys→Ala (f) His→Glu
(c) Ala→Thr (g) Pro→Ser
(d) Phe→Lys

10. Resistance of the Genetic Code to Mutation The following RNA sequence represents the beginning of an open reading frame. What changes (if any) can occur at each position without generating a change in the encoded amino acid residue?

(5′)AUGAUAUUGCUAUCUUGGACU

11. Basis of the Sickle-Cell Mutation Sickle-cell hemoglobin has a Val residue at position 6 of the β-globin chain instead of the Glu residue found in normal hemoglobin A. Can you predict what change took place in the DNA codon for glutamate to account for replacement of the Glu residue by Val?

12. Proofreading by Aminoacyl-tRNA Synthetases The isoleucyl-tRNA synthetase has a proofreading function that ensures the fidelity of the aminoacylation reaction, but the histidyl-tRNA synthetase lacks such a proofreading function. Explain.

13. Importance of the "Second Genetic Code" Some aminoacyl-tRNA synthetases do not recognize and bind the anticodon of their cognate tRNAs but instead use other structural features of the tRNAs to impart binding specificity. The tRNAs for alanine apparently fall into this category.

(a) What features of tRNA^Ala are recognized by Ala-tRNA synthetase?

(b) Describe the consequences of a C→G mutation in the third position of the anticodon of tRNA^Ala.

(c) What other kinds of mutations might have similar effects?

(d) Mutations of these types are never found in natural populations of organisms. Why? (Hint: Consider what might happen both to individual proteins and to the organism as a whole.)

14. The Role of Translation Factors A researcher isolates mutant variants of the bacterial translation factors IF-2, EF-Tu, and EF-G. In each case, the mutation allows proper folding of the protein and the binding of GTP but does not allow GTP hydrolysis. At what stage would translation be blocked by each mutant protein?

15. Maintaining the Fidelity of Protein Synthesis The chemical mechanisms used to avoid errors in protein synthesis are different from those used during DNA replication. DNA polymerases use a 3′→5′ exonuclease proofreading activity to remove mispaired nucleotides incorrectly inserted into a growing DNA strand. There is no analogous proofreading function on ribosomes and, in fact, the identity of an amino acid attached to an incoming tRNA and added to the growing polypeptide is never checked. A proofreading step that hydrolyzed the previously formed peptide bond after an incorrect amino acid had been inserted into a growing polypeptide (analogous to the proofreading step of DNA polymerases) would be impractical. Why? (Hint: Consider how the link between the growing polypeptide and the mRNA is maintained during elongation; see Figs 27–29 and 27–30.)

16. Predicting the Cellular Location of a Protein The gene for a eukaryotic polypeptide 300 amino acid residues long is altered so that a signal sequence recognized by SRP occurs at the polypeptide's amino terminus and a nuclear localization signal (NLS) occurs internally, beginning at residue 150. Where is the protein likely to be found in the cell?

17. Requirements for Protein Translocation across a Membrane The secreted bacterial protein OmpA has a precursor, ProOmpA, which has the amino-terminal signal sequence required for secretion. If purified ProOmpA is denatured with 8 M urea and the urea is then removed (such as by running the protein solution rapidly through a gel filtration column), the protein can be translocated across isolated bacterial inner membranes in vitro. However, translocation becomes impossible if ProOmpA is first allowed to incubate for a few hours in the absence of urea. Furthermore, the capacity for translocation is maintained for an extended period if ProOmpA is first incubated in the presence of another bacterial protein called trigger factor. Describe the probable function of this factor.

18. Protein-Coding Capacity of a Viral DNA The 5,386 bp genome of bacteriophage ϕX174 includes genes for 10 proteins, designated A to K, with sizes given in the table below. How much DNA would be required to encode these 10 proteins? How can you reconcile the size of the ϕX174 genome with its protein-coding capacity?

Protein	Number of amino acid residues	Protein	Number of amino acid residues
A	455	F	427
B	120	G	175
C	86	H	328
D	152	J	38
E	91	K	56

Data Analysis Problem

19. Designing Proteins by Using Randomly Generated Genes Studies of the amino acid sequence and corresponding three-dimensional structure of wild-type or mutant proteins have led to significant insights into the principles that govern protein folding. An important test of this understanding would be to *design* a protein based on these principles and see whether it folds as expected.

Kamtekar and colleagues (1993) used aspects of the genetic code to generate random protein sequences with defined patterns of hydrophilic and hydrophobic residues. Their clever approach combined knowledge about protein structure, amino acid properties, and the genetic code to explore the factors that influence protein structure.

They set out to generate a set of proteins with the simple four-helix bundle structure shown at the end of the paragraph, with α helices (shown as cylinders) connected by segments of random coil (light red). Each α helix is amphipathic—the

R groups on one side of the helix are exclusively hydrophobic (yellow) and those on the other side are exclusively hydrophilic (blue). A protein consisting of four of these helices separated by short segments of random coil would be expected to fold so that the hydrophilic sides of the helices face the solvent.

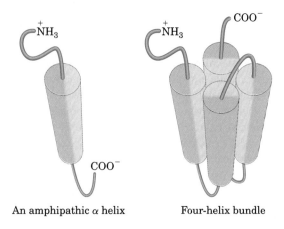

An amphipathic α helix   Four-helix bundle

(a) What forces or interactions hold the four α helices together in this bundled structure?

Figure 4–4a shows a segment of α helix consisting of 10 amino acid residues. With the gray central rod as a divider, four of the R groups (purple spheres) extend from the left side of the helix and six extend from the right.

(b) Number the R groups in Figure 4–4a, from top (amino terminus; 1) to bottom (carboxyl terminus; 10). Which R groups extend from the left side and which from the right?

(c) Suppose you wanted to design this 10 amino acid segment to be an amphipathic helix, with the left side hydrophilic and the right side hydrophobic. Give a sequence of 10 amino acids that could potentially fold into such a structure. There are many possible correct answers here.

(d) Give one possible double-stranded DNA sequence that could encode the amino acid sequence you chose for (c). (It is an internal portion of a protein, so you do not need to include start or stop codons.)

Rather than designing proteins with specific sequences, Kamtekar and colleagues designed proteins with partially random sequences, with hydrophilic and hydrophobic amino acid residues placed in a controlled pattern. They did this by taking advantage of some interesting features of the genetic code to construct a library of synthetic DNA molecules with partially random sequences arranged in a particular pattern.

To design a DNA sequence that would encode random hydrophobic amino acid sequences, the researchers began with the degenerate codon NTN, where N can be A, G, C, or T. They filled each N position by including an equimolar mixture of A, G, C, and T in the DNA synthesis reaction to generate a mixture of DNA molecules with different nucleotides at that position (see Fig. 8–35). Similarly, to encode random polar amino acid sequences, they began with the degenerate codon NAN and used an equimolar mixture of A, G, and C (but in this case, no T) to fill the N positions.

(e) Which amino acids can be encoded by the NTN triplet? Are all amino acids in this set hydrophobic? Does the set include *all* the hydrophobic amino acids?

(f) Which amino acids can be encoded by the NAN triplet? Are all of these polar? Does the set include *all* the polar amino acids?

(g) In creating the NAN codons, why was it necessary to leave T out of the reaction mixture?

Kamtekar and coworkers cloned this library of random DNA sequences into plasmids, selected 48 that produced the correct patterning of hydrophilic and hydrophobic amino acids, and expressed these in *E. coli*. The next challenge was to determine whether the proteins folded as expected. It would be very time-consuming to express each protein, crystallize it, and determine its complete three-dimensional structure. Instead, the investigators used the *E. coli* protein-processing machinery to screen out sequences that led to highly defective proteins. In this initial screening, they kept only those clones that resulted in a band of protein with the expected molecular weight on SDS polyacrylamide gel electrophoresis (see Fig. 3–18).

(h) Why would a grossly misfolded protein fail to produce a band of the expected molecular weight on electrophoresis?

Several proteins passed this initial test, and further exploration showed that they had the expected four-helix structure.

(i) Why didn't all of the random-sequence proteins that passed the initial screening test produce four-helix structures?

Reference

Kamtekar, S., Schiffer, J.M., Xiong, H., Babik, J.M., & Hecht, M.H. (1993) Protein design by binary patterning of polar and nonpolar amino acids. *Science* **262**, 1680–1685.

Regulation of Gene Expression

O f the 4,000 or so genes in the typical bacterial genome, or the perhaps 25,000 genes in the human genome, only a fraction are expressed in a cell at any given time. Some gene products are present in very large amounts: the elongation factors required for protein synthesis, for example, are among the most abundant proteins in bacteria, and ribulose 1,5-bisphosphate carboxylase/oxygenase (rubisco) of plants and photosynthetic bacteria is, as far as we know, the most abundant enzyme in the biosphere. Other gene products occur in much smaller amounts; for instance, a cell may contain only a few molecules of the enzymes that repair rare DNA lesions. Requirements for some gene products change over time. The need for enzymes in certain metabolic pathways may wax and wane as food sources change or are depleted. During development of a multicellular organism, some proteins that influence cellular differentiation are present for just a brief time in only a few cells. Specialization of cellular function can dramatically affect the need for various gene products; an example is the uniquely high concentration of a single protein—hemoglobin—in erythrocytes. Given the high cost of protein synthesis, regulation of gene expression is essential to making optimal use of available energy.

The cellular concentration of a protein is determined by a delicate balance of at least seven processes, each having several potential points of regulation:

1. Synthesis of the primary RNA transcript (transcription)
2. Posttranscriptional modification of mRNA
3. Messenger RNA degradation
4. Protein synthesis (translation)
5. Posttranslational modification of proteins
6. Protein targeting and transport
7. Protein degradation

These processes are summarized in **Figure 28–1**. We have examined several of these mechanisms in previous chapters. Posttranscriptional modification of mRNA, by processes such as alternative splicing patterns (see Fig. 26–21) or RNA editing (see Figs 27–10, 27–12), can affect which proteins are produced from an mRNA transcript and in what amounts. A variety of nucleotide sequences in an mRNA can affect the rate of its degradation (p. 1084). Many factors affect the rate at which an mRNA is translated into a protein, as well as the posttranslational modification, targeting, and eventual degradation of that protein (Chapter 27).

Of the regulatory processes illustrated in Figure 28–1, those operating at the level of transcription initiation are particularly well-documented. These processes are a major focus of this chapter, although other mechanisms are also considered. Researchers continue to discover complex and sometimes surprising regulatory mechanisms, leading to an increasing appreciation of the importance of posttranscriptional and translational regulation, especially in eukaryotes. For many genes, the regulatory processes are elaborate and redundant and can involve a considerable investment of chemical energy.

Control of transcription initiation permits the synchronized regulation of multiple genes encoding products with interdependent activities. For example, when their DNA is heavily damaged, bacterial cells require a coordinated increase in the levels of the many DNA repair enzymes. And perhaps the most sophisticated form of coordination occurs in the complex regulatory circuits that guide the development of multicellular eukaryotes, which can involve many types of regulatory mechanisms.

We begin by examining the interactions between proteins and DNA that are the key to transcriptional regulation. We next discuss the specific proteins that influence the expression of specific genes, first in bacterial

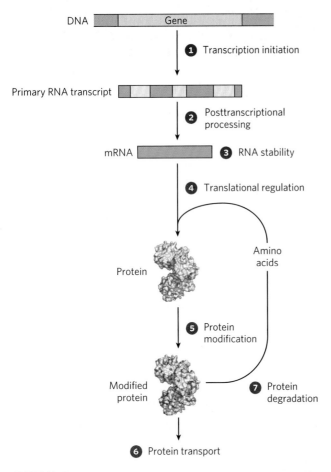

DNA

Gene

1 Transcription initiation

Primary RNA transcript

2 Posttranscriptional processing

mRNA   **3** RNA stability

4 Translational regulation

Amino acids

Protein

5 Protein modification

Modified protein   **7** Protein degradation

6 Protein transport

FIGURE 28–1 Seven processes that affect the steady-state concentration of a protein. Each process has several potential points of regulation.

and then in eukaryotic cells. Information about post-transcriptional and translational regulation is included in the discussion, where relevant, to provide a more complete overview of the rich complexity of regulatory mechanisms.

28.1 Principles of Gene Regulation

Genes for products that are required at all times, such as those for the enzymes of central metabolic pathways, are expressed at a more or less constant level in virtually every cell of a species or organism. Such genes are often referred to as **housekeeping genes**. Unvarying expression of a gene is called **constitutive gene expression**.

For other gene products, cellular levels rise and fall in response to molecular signals; this is **regulated gene expression**. Gene products that increase in concentration under particular molecular circumstances are referred to

as **inducible**; the process of increasing their expression is **induction**. The expression of many of the genes encoding DNA repair enzymes, for example, is induced by a system of regulatory proteins that responds to high levels of DNA damage. Conversely, gene products that decrease in concentration in response to a molecular signal are referred to as **repressible**, and the process is called **repression**. For example, in bacteria, ample supplies of tryptophan lead to repression of the genes for the enzymes that catalyze tryptophan biosynthesis.

Transcription is mediated and regulated by protein-DNA interactions, especially those involving the protein components of RNA polymerase (Chapter 26). We first consider how the activity of RNA polymerase is regulated, and proceed to a general description of the proteins participating in this regulation. We then examine the molecular basis for the recognition of specific DNA sequences by DNA-binding proteins.

RNA Polymerase Binds to DNA at Promoters

RNA polymerases bind to DNA and initiate transcription at promoters (see Fig. 26–5), sites generally found near points at which RNA synthesis begins on the DNA template. The regulation of transcription initiation often entails changes in how RNA polymerase interacts with a promoter.

The nucleotide sequences of promoters vary considerably, affecting the binding affinity of RNA polymerases and thus the frequency of transcription initiation. Some *Escherichia coli* genes are transcribed once per second, others less than once per cell generation. Much of this variation is due to differences in promoter sequence. In the absence of regulatory proteins, differences in promoter sequence may affect the frequency of transcription initiation by a factor of 1,000 or more. Most *E. coli* promoters have a sequence close to a consensus **(Fig. 28–2)**. Mutations that result in a shift away from the consensus sequence usually decrease promoter function; conversely, mutations toward consensus usually enhance promoter function.

KEY CONVENTION: By convention, DNA sequences are shown as they exist in the nontemplate strand, with the 5′ terminus on the left. Nucleotides are numbered from the transcription start site, with positive numbers to the right (in the direction of transcription) and negative numbers to the left. N indicates any nucleotide. ∎

Although housekeeping genes are expressed constitutively, the cellular concentrations of the proteins they

RNA start site

mRNA

−35 region   −10 region

DNA 5′ | UP element | TTGACA | N₁₇ | TATAAT | N₅₋₉ |

FIGURE 28–2 Consensus sequence for many *E. coli* promoters. Most base substitutions in the −10 and −35 regions have a negative effect on promoter function. Some promoters also include the UP (upstream promoter) element (see Fig. 26-5).

encode vary widely. For these genes, the RNA polymerase–promoter interaction strongly influences the rate of transcription initiation; differences in promoter sequence allow the cell to synthesize the appropriate level of each housekeeping gene product.

The basal rate of transcription initiation at the promoters of nonhousekeeping genes is also determined by the promoter sequence, but expression of these genes is further modulated by regulatory proteins. Many of these proteins work by enhancing or interfering with the interaction between RNA polymerase and the promoter.

The sequences of eukaryotic promoters are more variable than their bacterial counterparts (see Fig. 26–8). The three eukaryotic RNA polymerases usually require an array of general transcription factors in order to bind to a promoter. Yet, as with bacterial gene expression, the basal level of transcription is determined by the effect of promoter sequences on the function of RNA polymerase and its associated transcription factors.

Transcription Initiation Is Regulated by Proteins That Bind to or Near Promoters

At least three types of proteins regulate transcription initiation by RNA polymerase: **specificity factors** alter the specificity of RNA polymerase for a given promoter or set of promoters, **repressors** impede access of RNA polymerase to the promoter, and **activators** enhance the RNA polymerase–promoter interaction.

We introduced bacterial specificity factors in Chapter 26 (see Fig. 26–5), although we did not refer to them by that name. The σ subunit of the *E. coli* RNA polymerase holoenzyme is a specificity factor that mediates promoter recognition and binding. Most *E. coli* promoters are recognized by a single σ subunit (M_r 70,000), σ^{70}. Under some conditions, some of the σ^{70} subunits are replaced by one of six other specificity factors. One notable case arises when the bacteria are subjected to heat stress, leading to the replacement of σ^{70} by σ^{32} (M_r 32,000). When bound to σ^{32}, RNA polymerase is directed to a specialized set of promoters with a different consensus sequence **(Fig. 28–3)**. These promoters control the expression of a set of genes that encode proteins, including some protein chaperones (p. 146), that are part of a stress-induced system called the heat shock response. Thus, through changes in the binding affinity of the polymerase that direct it to different

promoters, a set of genes involved in related processes is coordinately regulated. In eukaryotic cells, some of the general transcription factors, in particular the TATA-binding protein (TBP; see Fig. 26–9), may be considered specificity factors.

Repressors bind to specific sites on the DNA. In bacterial cells, such binding sites, called **operators**, are generally near a promoter. RNA polymerase binding, or its movement along the DNA after binding, is blocked when the repressor is present. Regulation by means of a repressor protein that blocks transcription is referred to as **negative regulation**. Repressor binding to DNA is regulated by a molecular signal (or **effector**), usually a small molecule or a protein, that binds to the repressor and causes a conformational change. The interaction between repressor and signal molecule either increases or decreases transcription. In some cases, the conformational change results in dissociation of a DNA-bound repressor from the operator **(Fig. 28–4a)**. Transcription initiation can then proceed unhindered. In other cases, interaction between an inactive repressor and the signal molecule causes the repressor to bind to the operator (Fig. 28–4b). In eukaryotic cells, the binding site for a repressor may be some distance from the promoter. Binding of these repressors to their binding sites has the same effect as in bacterial cells: inhibiting the assembly or activity of a transcription complex at the promoter.

Activators provide a molecular counterpoint to repressors; they bind to DNA and *enhance* the activity of RNA polymerase at a promoter; this is **positive regulation**. In bacteria, activator-binding sites are often adjacent to promoters that are bound weakly or not at all by RNA polymerase alone, such that little transcription occurs in the absence of the activator. Many eukaryotic activators bind to DNA sites, called enhancers, that are quite distant from the promoter. These activators affect the rate of transcription at a promoter that may be located thousands of base pairs away. Some activators are usually bound to DNA, enhancing transcription until dissociation of the activator is triggered by the binding of a signal molecule (Fig. 28–4c). In other cases the activator binds to DNA only after interaction with a signal molecule (Fig. 28–4d). Signal molecules can therefore increase or decrease transcription, depending on how they affect the activator. Positive regulation is particularly common in eukaryotes, as we shall see.

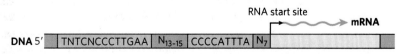

FIGURE 28–3 Consensus sequence for promoters that regulate expression of the *E. coli* heat shock genes. This system responds to temperature increases as well as some other environmental stresses, resulting in the induction of a set of proteins. Binding of RNA polymerase to heat shock promoters is mediated by a specialized σ subunit of the polymerase, σ^{32}, which replaces σ^{70} in the RNA polymerase initiation complex.

(a) Negative regulation
Molecular signal causes dissociation of repressor from DNA, inducing transcription.

(b) Negative regulation
Molecular signal causes binding of repressor to DNA, inhibiting transcription.

(c) Positive regulation
Molecular signal causes dissociation of activator from DNA, inhibiting transcription.

(d) Positive regulation
Molecular signal causes binding of activator to DNA, inducing transcription.

FIGURE 28–4 Common patterns of regulation of transcription initiation. Two types of negative regulation are illustrated. **(a)** Repressor binds to the operator in the absence of the molecular signal; the external signal causes dissociation of the repressor to permit transcription. **(b)** Repressor binds in the presence of the signal; the repressor dissociates and transcription ensues when the signal is removed. Positive regulation is mediated by gene activators. Again, two types are shown. **(c)** Activator binds in the absence of the molecular signal and transcription proceeds; when the signal is added, the activator dissociates and transcription is inhibited. **(d)** Activator binds in the presence of the signal; it dissociates only when the signal is removed. Note that "positive" and "negative" regulation refer to the type of regulatory protein involved: the bound protein either facilitates or inhibits transcription. In either case, addition of the molecular signal may increase or decrease transcription, depending on its effect on the regulatory protein.

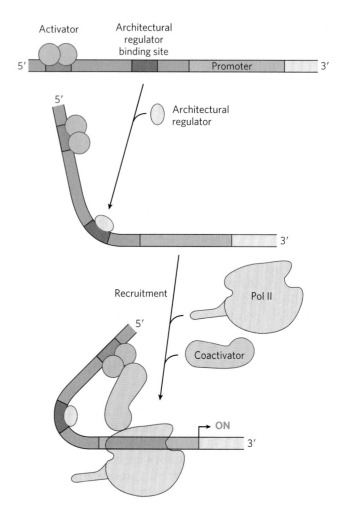

FIGURE 28–5 The interaction between activators/repressors and RNA polymerase in eukaryotes. Eukaryotic activators and repressors often bind sites thousands of base pairs from the promoters they regulate. DNA looping, often facilitated by architectural regulators, brings the sites together. The interaction between activators and RNA polymerase is often mediated by coactivators, as shown. Repression is sometimes mediated by repressors (described later) that bind to activators, thereby preventing the activating interaction with RNA polymerase.

In eukaryotes, the distance between the binding sites of activators or repressors and promoters is bridged by looping out the DNA in between **(Fig. 28–5)**. The looping is facilitated in some cases by proteins called **architectural regulators** that bind to intervening sites and facilitate the looping of the DNA. Most of the eukaryotic systems involve protein activators. The actual interaction between the activators and the RNA polymerase at the promoter is often mediated by intermediary proteins called coactivators. In some instances, protein repressors may take the place of coactivators, binding to the activators and preventing the activating interaction.

Many Bacterial Genes Are Clustered and Regulated in Operons

Bacteria have a simple general mechanism for coordinating the regulation of genes encoding products that participate in a set of related processes: these genes are clustered on the chromosome and are transcribed together. Many bacterial mRNAs are polycistronic—multiple genes on a single transcript—and the single promoter that initiates transcription of the cluster is the

FIGURE 28–6 Representative bacterial operon. Genes A, B, and C are transcribed on one polycistronic mRNA. Typical regulatory sequences include binding sites for proteins that either activate or repress transcription from the promoter.

site of regulation for expression of all the genes in the cluster. The gene cluster and promoter, plus additional sequences that function together in regulation, are called an **operon (Fig. 28–6).** Operons that include two to six genes transcribed as a unit are common; some operons contain 20 or more genes.

Many of the principles of bacterial gene expression were first defined by studies of lactose metabolism in *E. coli*, which can use lactose as its sole carbon source. In 1960, François Jacob and Jacques Monod published a short paper in the *Proceedings* of the French Academy of Sciences that described how two adjacent genes involved in lactose metabolism were coordinately regulated by a genetic element located at one end of the gene cluster. The genes were those for β-galactosidase, which cleaves lactose to galactose and glucose, and for galactoside permease (lactose permease, p. 416), which transports lactose into the cell **(Fig. 28–7).** The terms "operon" and "operator" were first introduced in this paper. With the operon model, gene regulation could, for the first time, be considered in molecular terms.

François Jacob

Jacques Monod, 1910–1976

The *lac* Operon Is Subject to Negative Regulation

The lactose (*lac*) operon **(Fig. 28–8a)** includes the genes for β-galactosidase (*Z*), galactoside permease (*Y*), and thiogalactoside transacetylase (*A*). The last of these enzymes seems to modify toxic galactosides to facilitate their removal from the cell. Each of the three genes is preceded by a ribosome-binding site (not shown in Fig. 28–8) that independently directs the translation of that gene (Chapter 27). Regulation of the *lac* operon by the *lac* repressor protein (Lac) follows the pattern outlined in Figure 28–4a.

The study of *lac* operon mutants has revealed some details of the workings of the operon's regulatory system. In the absence of lactose, the *lac* operon genes are repressed. Mutations in the operator or in another gene, the *I* gene, result in constitutive synthesis of the gene products. When the *I* gene is defective, repression can be restored by introducing a functional *I* gene into the cell on another DNA molecule, demonstrating that the *I* gene encodes a diffusible molecule that causes gene repression. This molecule proved to be a protein, now called the Lac repressor, a tetramer of identical monomers. The operator to which it binds most tightly (O_1) abuts the transcription start site (Fig. 28–8a). The *I* gene is transcribed from its own promoter (P_I) independent of the *lac* operon genes. The *lac* operon has two secondary binding sites for the Lac repressor. One (O_2) is centered near position +410, within the gene encoding β-galactosidase (*Z*); the other (O_3) is near position −90, within the *I* gene. To repress the operon, the Lac repressor seems to bind to both the main operator and one of the two secondary sites, with the intervening

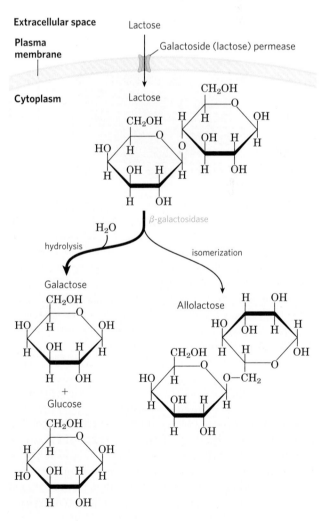

FIGURE 28–7 Lactose metabolism in *E. coli*. Uptake and metabolism of lactose require the activities of galactoside (lactose) permease and β-galactosidase. Conversion of lactose to allolactose by transglycosylation is a minor reaction also catalyzed by β-galactosidase.

(a)

(b)

(c)

FIGURE 28–8 The *lac* operon. (a) The *lac* operon. The *lacI* gene encodes the Lac repressor. The *lac Z, Y,* and *A* genes encode β-galactosidase, galactoside permease, and thiogalactoside transacetylase, respectively. P is the promoter for the *lac* genes, and P_I is the promoter for the *I* gene. O_1 is the main operator for the *lac* operon; O_2 and O_3 are secondary operator sites of lesser affinity for the Lac repressor. The inverted repeat to which the Lac repressor binds in O_1 is shown in the inset. **(b)** The Lac repressor binds to the main operator and O_2 or O_3, apparently forming a loop in the DNA. **(c)** (PDB ID 2PE5) Lac repressor (shades of red) is shown bound to short, discontinuous segments of DNA (blue and orange).

DNA looped out (Fig. 28–8b, c). Either binding arrangement blocks transcription initiation.

Despite this elaborate binding complex, repression is not absolute. Binding of the Lac repressor reduces the rate of transcription initiation by a factor of 10^3. If the O_2 and O_3 sites are eliminated by deletion or mutation, the binding of repressor to O_1 alone reduces transcription by a factor of about 10^2. Even in the repressed state, each cell has a few molecules of β-galactosidase and galactoside permease, presumably synthesized on the rare occasions when the repressor transiently dissociates from the operators. This basal level of transcription is essential to operon regulation.

When cells are provided with lactose, the *lac* operon is induced. An inducer (signal) molecule binds to a specific site on the Lac repressor, causing a conformational change that results in dissociation of the repressor from the operator. The inducer in the *lac* operon system is not lactose itself but allolactose, an isomer of lactose (Fig. 28–7). After entry into the *E. coli* cell (via the few existing molecules of lactose permease), lactose is converted to allolactose by one of the few existing β-galactosidase molecules. Release of the operator by Lac repressor, triggered as the repressor binds to allolactose, allows expression of the *lac* operon genes and leads to a 10^3-fold increase in the concentration of β-galactosidase.

Several β-galactosides structurally related to allolactose are inducers of the *lac* operon but are not substrates for β-galactosidase; others are substrates but not inducers. One particularly effective and nonmetabolizable inducer of the *lac* operon that is often used experimentally is isopropylthiogalactoside (IPTG).

Isopropyl-β-D-thiogalactoside
(IPTG)

An inducer that cannot be metabolized allows researchers to explore the physiological function of lactose as a carbon source for growth, separate from its function in the regulation of gene expression.

In addition to the multitude of operons now known in bacteria, a few polycistronic operons have been found in the cells of lower eukaryotes. In the cells of higher eukaryotes, however, almost all protein-encoding genes are transcribed separately.

The mechanisms by which operons are regulated can vary significantly from the simple model presented in Figure 28–8. Even the *lac* operon is more complex than indicated here, with an activator also contributing to the overall scheme, as we shall see in Section 28.2. Before any further discussion of the layers of regulation of gene expression, however, we examine the critical molecular interactions between DNA-binding proteins (such as repressors and activators) and the DNA sequences to which they bind.

Regulatory Proteins Have Discrete DNA-Binding Domains

Regulatory proteins generally bind to specific DNA sequences. Their affinity for these target sequences is roughly 10^4 to 10^6 times higher than their affinity for any other DNA sequence. Most regulatory proteins have discrete DNA-binding domains containing substructures

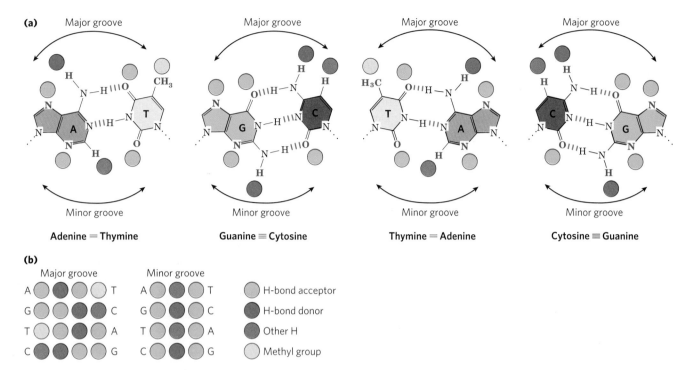

FIGURE 28–9 Groups in DNA available for protein binding. (a) Shown here are functional groups on all four base pairs that are displayed in the major and minor grooves of DNA (see Fig. 8–13). Hydrogen-bond acceptor and donor atoms are marked by blue and red disks, respectively. Other hydrogen atoms are marked with purple disks, and methyl groups with yellow disks. **(b)** Recognition patterns for each base pair, from left to right, are summarized at bottom. The much greater variation in the patterns for the major groove gives rise to a much greater discriminatory power in the major groove relative to the minor groove.

that interact closely and specifically with the DNA. These binding domains usually include one or more of a relatively small group of recognizable and characteristic structural motifs.

To bind specifically to DNA sequences, regulatory proteins must recognize surface features on the DNA. Most of the chemical groups that differ among the four bases and thus permit discrimination between base pairs are hydrogen-bond donor and acceptor groups exposed in the major groove of DNA **(Fig. 28–9)**, and most of the protein-DNA contacts that impart specificity are hydrogen bonds. A notable exception is the nonpolar surface near C-5 of pyrimidines, where thymine is readily distinguished from cytosine by its protruding methyl group. Protein-DNA contacts are also possible in the minor groove of the DNA, but the hydrogen-bonding patterns here generally do not allow ready discrimination between base pairs.

Within regulatory proteins, the amino acid side chains most often hydrogen-bonding to bases in the DNA are those of Asn, Gln, Glu, Lys, and Arg residues. Is there a simple recognition code in which a particular amino acid always pairs with a particular base? The two hydrogen bonds that can form between Gln or Asn and the N^6 and N-7 positions of adenine cannot form with any other base. And an Arg residue can form two hydrogen bonds with N-7 and O^6 of guanine **(Fig. 28–10)**. Examination of the structure of many DNA-binding proteins, however, has shown that a protein can recognize

each base pair in more than one way, leading to the conclusion that there is no simple amino acid–base code. For some proteins, the Gln-adenine interaction can specify A=T base pairs, but in others a van der Waals pocket for the methyl group of thymine can recognize A=T base pairs. Researchers cannot yet examine the structure of a DNA-binding protein and infer the DNA sequence to which it binds.

FIGURE 28–10 Specific amino acid residue–base pair interactions. The two examples shown have been observed in DNA-protein binding.

To interact with bases in the major groove of DNA, a protein requires a relatively small substructure that can stably protrude from the protein surface. The DNA-binding domains of regulatory proteins tend to be small (60 to 90 amino acid residues), and the structural motifs within these domains that are actually in contact with the DNA are smaller still. Many small proteins are unstable because of their limited capacity to form layers of structure to bury hydrophobic groups (p. 116). The DNA-binding motifs provide either a very compact stable structure or a way of allowing a segment of protein to protrude from the protein surface.

The DNA-binding sites for regulatory proteins are often inverted repeats of a short DNA sequence (a palindrome) at which multiple (usually two) subunits of a regulatory protein bind cooperatively. The Lac repressor is unusual in that it functions as a tetramer, with two dimers tethered together at the end distant from the DNA-binding sites (Fig. 28–8b). An *E. coli* cell usually contains about 20 tetramers of the Lac repressor. Each of the tethered dimers separately binds to a palindromic operator sequence, in contact with 17 bp of a 22 bp region in the *lac* operon. And each of the tethered dimers can independently bind to an operator sequence, with one generally binding to O_1 and the other to O_2 or O_3 (as in Fig. 28–8b). The symmetry of the O_1 operator sequence corresponds to the twofold axis of symmetry of two paired Lac repressor subunits. The tetrameric Lac repressor binds to its operator sequences in vivo with an estimated dissociation constant of about 10^{-10} M. The repressor discriminates between the operators and other sequences by a factor of about 10^6, so binding to these few base pairs among the 4.6 million or so of the *E. coli* chromosome is highly specific.

Several DNA-binding motifs have been described, but here we focus on two that play prominent roles in the binding of DNA by regulatory proteins: the **helix-turn-helix** and the **zinc finger**. We also consider a type of DNA-binding domain—the homeodomain—found in some eukaryotic proteins.

Helix-Turn-Helix This DNA-binding motif is crucial to the interaction of many bacterial regulatory proteins with DNA, and similar motifs occur in some eukaryotic regulatory proteins. The helix-turn-helix motif comprises about 20 amino acids in two short α-helical segments, each seven to nine amino acid residues long, separated by a β turn **(Fig. 28–11)**. This structure generally is not stable by itself; it is simply the reactive portion of a somewhat larger DNA-binding domain. One of the two α-helical segments is called the recognition helix because it usually contains many of the amino acids that interact with the DNA in a sequence-specific way. This α helix is stacked on other segments of the protein structure so that it protrudes from the protein surface. When bound to DNA, the recognition helix is positioned

FIGURE 28–11 Helix-turn-helix. (PDB ID 2PE5) **(a)** DNA-binding domain of the Lac repressor bound to DNA (blue and orange). The helix-turn-helix motif is shown in dark blue and purple; the DNA recognition helix is purple. **(b)** Entire Lac repressor. The DNA-binding domains are light blue, and the α helices involved in tetramer formation are green. The remainder of the protein (shades of red) has the binding sites for allolactose. The allolactose-binding domains are linked to the DNA-binding domains through linker helices (yellow).

in or nearly in the major groove. The Lac repressor has this DNA-binding motif (Fig. 28–11).

Zinc Finger In a zinc finger, about 30 amino acid residues form an elongated loop held together at the base by a single Zn^{2+} ion, which is coordinated to four of the residues (four Cys, or two Cys and two His). The zinc does not itself interact with DNA; rather, the coordination of zinc with the amino acid residues stabilizes this small structural motif. Several hydrophobic side chains in the core of the structure also lend stability. **Figure 28–12** shows the interaction between DNA and three zinc fingers of a single polypeptide from the mouse regulatory protein Zif268.

Many eukaryotic DNA-binding proteins contain zinc fingers. The interaction of a single zinc finger with DNA is typically weak, and many DNA-binding proteins, like Zif268, have multiple zinc fingers that substantially enhance binding by interacting simultaneously with the DNA. One DNA-binding protein of the frog *Xenopus* has 37 zinc fingers. There are few known examples of the zinc finger motif in bacterial proteins.

FIGURE 28–12 Zinc fingers. (PDB ID 1ZAA) Three zinc fingers (shades of red) of the regulatory protein Zif268, complexed with DNA (blue). Each Zn^{2+} coordinates with two His and two Cys residues.

The precise manner in which proteins with zinc fingers bind to DNA differs from one protein to the next. Some zinc fingers contain the amino acid residues that are important in sequence discrimination, whereas others seem to bind DNA nonspecifically (the amino acids required for specificity are located elsewhere in the protein). Zinc fingers can also function as RNA-binding motifs—for example, in certain proteins that bind eukaryotic mRNAs and act as translational repressors. We discuss this role later (Section 28.3).

Homeodomain Another type of DNA-binding domain has been identified in some proteins that function as transcriptional regulators, especially during eukaryotic development. This domain of 60 amino acids—called the **homeodomain** because it was discovered in homeotic genes (genes that regulate the development of body patterns)—is highly conserved and has now been identified in proteins from a wide variety of organisms, including humans **(Fig. 28–13)**. The DNA-binding segment of the domain is related to the helix-turn-helix motif. The DNA sequence that encodes this domain is known as the **homeobox**.

Regulatory Proteins Also Have Protein-Protein Interaction Domains

Regulatory proteins contain domains not only for DNA binding but also for protein-protein interactions—with RNA polymerase, other regulatory proteins, or other subunits of the same regulatory protein. Examples include many eukaryotic transcription factors that function as gene activators, which often bind as dimers to the DNA, through DNA-binding domains that contain zinc fingers. Some structural domains are devoted to the interactions required for dimer formation, which is generally a prerequisite for DNA binding. Like DNA-binding motifs, the structural motifs that mediate protein-protein interactions tend to fall within one of a few common categories. Two important examples

FIGURE 28–13 Homeodomains. (PDB ID 1FJL) Shown here are two homeodomains bound to DNA. In each homeodomain, one of the α helices (purple), layered on two others (dark blue and gray), can be seen protruding into the major groove. This is only a small part of a larger regulatory protein from a class called Pax, active in the regulation of development in fruit flies (see Section 28.3).

are the **leucine zipper** and the **basic helix-loop-helix**. Structural motifs such as these are the basis for classifying some regulatory proteins into structural families.

Leucine Zipper This motif is an amphipathic α helix with a series of hydrophobic amino acid residues concentrated on one side **(Fig. 28–14)**, with the hydrophobic surface forming the area of contact between the two polypeptides of a dimer. A striking feature of these α helices is the occurrence of Leu residues at every seventh position, forming a straight line along the hydrophobic surface. Although researchers initially thought the Leu residues interdigitated (hence the name "zipper"), we now know that they line up side by side as the interacting α helices coil around each other (forming a coiled coil; Fig. 28–14b). Regulatory proteins with leucine zippers often have a separate DNA-binding domain with a high concentration of basic (Lys or Arg) residues that can interact with the negatively charged phosphates of the DNA backbone. Leucine zippers have been found in many eukaryotic and a few bacterial proteins.

Basic Helix-Loop-Helix Another common structural motif occurs in some eukaryotic regulatory proteins implicated in the control of gene expression during the development of multicellular organisms. These proteins share a conserved region of about 50 amino acid residues important in both DNA binding and protein dimerization. This region can form two short amphipathic α helices linked by a loop of variable length, the helix-loop-helix (distinct

(a)

Source	Regulatory protein	Amino acid sequence		
		DNA-binding region	Connector (6 Amino acids)	Leucine zipper
Mammal	C/EBP	DKNSNEYRVRRERNNIAVRKSRDKAKQRNVE	TQQKVLE	LTSDNDRLRKRVEQLSRELDTLRG-
	Jun	SQERIKAERKRMRNRIAASKCRKRKLERIARL	EEKVKTL	KAQNSELASTANMLTEQVAQLKQ-
	Fos	EERRIRRIRRERNKMAAAKCRNRRRELTDTL	QAETDQL	EDKKSALQTEIANLLKEKEKLEF-
Yeast	GCN4	PESSDPAALKRARNTEAARRSRARKLQRMKQL	EDKVEEL	LSKNYHLENEVARLKKLVGER

Consensus sequence: - - - - - - - - R R - R - N - - - - - - R - R - R R - - - - - - - - L - - - - - - L - - - - - L - - - - - L - - - - - L

Invariant Asn

(b)

Invariant Asn

Zipper region

DNA-binding region

FIGURE 28–14 Leucine zippers. (a) Comparison of amino acid sequences of several leucine zipper proteins. Note the Leu (L) residues (red) at every seventh position in the zipper region, and the number of Lys (K) and Arg (R) residues in the DNA-binding region (yellow). **(b)** (PDB ID 1YSA) Leucine zipper from the yeast activator protein GCN4. Only the "zippered" α helices (gray), derived from different subunits of the dimeric protein, are shown. The two helices wrap around each other in a gently coiled coil. The interacting Leu side chains and the conserved residues in the DNA-binding region are colored to correspond to the sequence in (a).

from the helix-turn-helix motif associated with DNA binding). The helix-loop-helix motifs of two polypeptides interact to form dimers **(Fig. 28–15)**. In these proteins, DNA binding is mediated by an adjacent short amino acid sequence rich in basic residues, similar to the separate DNA-binding region in proteins containing leucine zippers.

Recognition helix

Loop

Dimer-forming helix

FIGURE 28–15 Helix-loop-helix. (PDB ID 1HLO) The human transcription factor Max, bound to its DNA target site. The protein is dimeric; one subunit is colored. The recognition helix (pink) is linked via the loop to the dimer-forming helix (light blue), which merges with the carboxyl-terminal end of the subunit. Interaction of the carboxyl-terminal helices of the two subunits describes a coiled coil very similar to that of a leucine zipper (see Fig. 28-14b), but with only one pair of interacting Leu residues (red side chains at right) in this example. The overall structure is sometimes called a helix-loop-helix/leucine zipper motif.

Protein-Protein Interactions in Eukaryotic Regulatory Proteins In eukaryotes most genes are regulated by activators, and most genes are monocistronic. If a different activator were required for each gene, the number of activators (and genes encoding them) would need to be equivalent to the number of regulated genes. However, in yeast about 300 transcription factors (many of them activators) are responsible for the regulation of many thousands of yeast genes. Many of the transcription factors regulate the induction of multiple genes, but most genes are subject to regulation by multiple transcription factors. Appropriate regulation of different genes is accomplished by utilizing different combinations of a limited repertoire of transcription factors at each gene, a phenomenon referred to as **combinatorial control**.

Combinatorial control is accomplished in part by mixing and matching the variants within a regulatory protein family to form a series of different active protein dimers. Several families of eukaryotic transcription factors have been defined based on close structural similarities. Within each family, dimers can sometimes form between two identical proteins (a homodimer) or between two different members of the family (a heterodimer). A hypothetical family of four different leucine-zipper proteins could thus form up to 10 different dimeric species. In many cases, the different combinations have distinct regulatory and functional properties and function in the regulation of different genes. As we shall see, multiple regulatory proteins of this kind function in the regulation of most eukaryotic genes, contributing further to combinatorial control.

In addition to having structural domains devoted to DNA binding and dimerization that direct a particular protein dimer to a particular gene, many regulatory proteins have domains that interact with RNA polymerase, with unrelated regulatory proteins, or with both. At least three types of additional domains for protein-protein interaction have been characterized (primarily in eukaryotes): glutamine-rich, proline-rich, and acidic domains, the names reflecting the amino acid residues that are especially abundant.

Protein-DNA binding interactions are the basis of the intricate regulatory circuits fundamental to gene function. We now turn to a closer examination of these gene regulatory schemes, first in bacterial, then in eukaryotic systems.

SUMMARY 28.1 Principles of Gene Regulation

▶ The expression of genes is regulated by processes that affect the rates at which gene products are synthesized and degraded. Much of this regulation occurs at the level of transcription initiation, mediated by regulatory proteins that either repress transcription (negative regulation) or activate transcription (positive regulation) at specific promoters.

▶ In bacteria, genes that encode products with interdependent functions are often clustered in an operon, a single transcriptional unit. Transcription of the genes is generally blocked by binding of a specific repressor protein at a DNA site called an operator. Dissociation of the repressor from the operator is mediated by a specific small molecule, an inducer. These principles were first elucidated in studies of the lactose (*lac*) operon. The Lac repressor dissociates from the *lac* operator when the repressor binds to its inducer, allolactose.

▶ Regulatory proteins are DNA-binding proteins that recognize specific DNA sequences; most have distinct DNA-binding domains. Within these domains, common structural motifs that bind DNA are the helix-turn-helix, zinc finger, and homeodomain.

▶ Regulatory proteins also contain domains for protein-protein interactions, including the leucine zipper and helix-loop-helix, which are involved in dimerization, and other motifs involved in activation of transcription. Mixing and matching of protein family variants in dimeric transcription factors provides for more efficient and responsive regulation through combinatorial control.

28.2 Regulation of Gene Expression in Bacteria

As in many other areas of biochemical investigation, the study of the regulation of gene expression advanced earlier and faster in bacteria than in other experimental organisms. The examples of bacterial gene regulation presented here are chosen from among scores of well-studied systems, partly for their historical significance, but primarily because they provide a good overview of the range of regulatory mechanisms in bacteria. Many of the principles of bacterial gene regulation are also relevant to understanding gene expression in eukaryotic cells.

We begin by examining the lactose and tryptophan operons; each system has regulatory proteins, but the overall mechanisms of regulation are very different. This is followed by a short discussion of the SOS response in *E. coli*, illustrating how genes scattered throughout the genome can be coordinately regulated. We then describe two bacterial systems of quite different types, illustrating the diversity of gene regulatory mechanisms: regulation of ribosomal protein synthesis at the level of translation, with many of the regulatory proteins binding to RNA (rather than DNA), and regulation of the process of "phase variation" in *Salmonella*, which results from genetic recombination. Finally, we examine some additional examples of posttranscriptional regulation in which the RNA modulates its own function.

The *lac* Operon Undergoes Positive Regulation

The operator-repressor-inducer interactions described earlier for the *lac* operon (Fig. 28–8) provide an intuitively satisfying model for an on/off switch in the regulation of gene expression. In truth, operon regulation is rarely so simple. A bacterium's environment is too complex for its genes to be controlled by one signal. Other factors besides lactose affect the expression of the *lac* genes, such as the availability of glucose. Glucose, metabolized directly by glycolysis, is the preferred energy source in *E. coli*. Other sugars can serve as the main or sole nutrient, but extra enzymatic steps are required to prepare them for entry into glycolysis, necessitating the synthesis of additional enzymes. Clearly, expressing the genes for proteins that metabolize sugars such as lactose or arabinose is wasteful when glucose is abundant.

What happens to the expression of the *lac* operon when both glucose and lactose are present? A regulatory mechanism known as **catabolite repression** restricts expression of the genes required for catabolism of lactose, arabinose, and other sugars in the presence of glucose, even when these secondary sugars are also present. The effect of glucose is mediated by cAMP, as a coactivator, and an activator protein known as **cAMP receptor protein**, or **CRP** (the protein is sometimes

FIGURE 28–16 CRP homodimer with bound cAMP. (PDB ID 1RUN) Note the bending of the DNA around the protein. The region that interacts with RNA polymerase is indicated.

called CAP, for *catabolite gene activator protein*). CRP is a homodimer (subunit M_r 22,000) with binding sites for DNA and cAMP. Binding is mediated by a helix-turn-helix motif in the protein's DNA-binding domain **(Fig. 28–16)**. When glucose is absent, CRP-cAMP binds to a site near the *lac* promoter **(Fig. 28–17)** and stimulates RNA transcription 50-fold. CRP-cAMP is therefore a positive regulatory element responsive to glucose levels, whereas the Lac repressor is a negative regulatory element responsive to lactose. The two act in concert. CRP-cAMP has little effect on the *lac* operon when the Lac repressor is blocking transcription, and dissociation of the repressor from the *lac* operator has little effect on transcription of the *lac* operon unless CRP-cAMP is present to facilitate transcription; when CRP is not bound, the wild-type *lac* promoter is a relatively weak promoter (Fig. 28–17a, c). The open complex of RNA polymerase and the promoter (see Fig. 26–6) does not form readily unless CRP-cAMP is present. CRP interacts directly with RNA polymerase (at the region shown in Fig. 28–16) through the polymerase's α subunit.

The effect of glucose on CRP is mediated by the cAMP interaction (Fig. 28–17). CRP binds to DNA most avidly when cAMP concentrations are high. In the presence of glucose, the synthesis of cAMP is inhibited and efflux of cAMP from the cell is stimulated. As [cAMP] declines, CRP binding to DNA declines, thereby decreasing the expression of the *lac* operon. Strong induction of the *lac* operon therefore requires both lactose (to inactivate the *lac* repressor) and a lowered concentration of glucose (to trigger an increase in [cAMP] and increased binding of cAMP to CRP).

CRP and cAMP are involved in the coordinated regulation of many operons, primarily those that encode enzymes for the metabolism of secondary sugars such as lactose and arabinose. A network of operons with a

(a) Glucose high, cAMP low, lactose absent

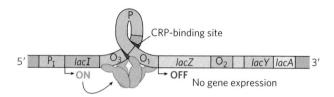

(b) Glucose low, cAMP high, lactose absent

(c) Glucose high, cAMP low, lactose present

(d) Glucose low, cAMP high, lactose present

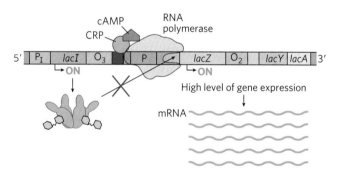

FIGURE 28–17 Positive regulation of the *lac* operon by CRP. The binding site for CRP-cAMP is near the promoter. The combined effects of glucose and lactose availability on *lac* operon expression are shown. When lactose is absent, the repressor binds to the operator and prevents transcription of the *lac* genes. It does not matter whether glucose is **(a)** present or **(b)** absent. **(c)** If lactose is present, the repressor dissociates from the operator. However, if glucose is also available, low cAMP levels prevent CRP-cAMP formation and DNA binding. RNA polymerase may occasionally bind and initiate transcription, resulting in a very low level of *lac* gene transcription. **(d)** When lactose is present and glucose levels are low, cAMP levels rise. The CRP-cAMP complex forms and facilitates robust binding of RNA polymerase to the *lac* promoter and high levels of transcription.

common regulator is called a **regulon**. This arrangement, which allows for coordinated shifts in cellular functions that can require the action of hundreds of genes, is a major theme in the regulated expression of dispersed networks of genes in eukaryotes. Other bacterial regulons include the heat shock gene system that

FIGURE 28–18 The *trp* operon. This operon is regulated by two mechanisms: when tryptophan levels are high, (1) the repressor (upper left) binds to its operator and (2) transcription of *trp* mRNA is attenuated (see Fig. 28–19). The biosynthesis of tryptophan by the enzymes encoded in the *trp* operon is diagrammed at the bottom (see also Fig. 22–19).

responds to changes in temperature (p. 1061) and the genes induced in *E. coli* as part of the SOS response to DNA damage, described later.

Many Genes for Amino Acid Biosynthetic Enzymes Are Regulated by Transcription Attenuation

The 20 common amino acids are required in large amounts for protein synthesis, and *E. coli* can synthesize all of them. The genes for the enzymes needed to synthesize a given amino acid are generally clustered in an operon and are expressed whenever existing supplies of that amino acid are inadequate for cellular requirements. When the amino acid is abundant, the biosynthetic enzymes are not needed and the operon is repressed.

The *E. coli* tryptophan (*trp*) operon **(Fig. 28–18)** includes five genes for the enzymes required to convert chorismate to tryptophan. Note that two of the enzymes catalyze more than one step in the pathway. The mRNA from the *trp* operon has a half-life of only about 3 min, allowing the cell to respond rapidly to changing needs for this amino acid. The Trp repressor is a homodimer. When tryptophan is abundant, it binds to the Trp repressor, causing a conformational change that permits the repressor to bind to the *trp* operator and inhibit expression of the *trp* operon. The *trp* operator site overlaps the promoter, so binding of the repressor blocks binding of RNA polymerase.

Once again, this simple on/off circuit mediated by a repressor is not the entire regulatory story. Different cellular concentrations of tryptophan can vary the rate of synthesis of the biosynthetic enzymes over a 700-fold range. Once repression is lifted and transcription begins, the rate of transcription is fine-tuned to cellular tryptophan requirements by a second regulatory process, called **transcription attenuation**, in which transcription is initiated normally but is abruptly halted *before* the operon genes are transcribed. The frequency with which transcription is attenuated is regulated by the availability of tryptophan and relies on the very close coupling of transcription and translation in bacteria.

The *trp* operon attenuation mechanism uses signals encoded in four sequences within a 162 nucleotide **leader** region at the 5′ end of the mRNA, preceding the initiation codon of the first gene **(Fig. 28–19a)**. Within the leader lies a region known as the **attenuator**, made up of sequences 3 and 4. These sequences base-pair to form a G≡C-rich stem-and-loop structure closely followed by a series of U residues. The attenuator structure acts as a transcription terminator (Fig. 28–19b). Sequence 2 is an alternative complement for sequence 3 (Fig. 28–19c). If sequences 2 and 3 base-pair, the attenuator structure cannot form and transcription continues into the *trp* biosynthetic genes; the loop formed by the pairing of sequences 2 and 3 does not obstruct transcription.

Site of transcription attenuation

End of leader region (*trpL*)

(a)

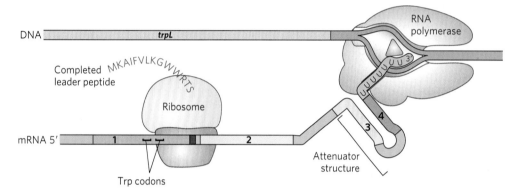

When tryptophan levels are high, the ribosome quickly translates sequence 1 (open reading frame encoding leader peptide) and blocks sequence 2 before sequence 3 is transcribed. Continued transcription leads to attenuation at the terminator-like attenuator structure formed by sequences 3 and 4.

When tryptophan levels are low, the ribosome pauses at the Trp codons in sequence 1. Formation of the paired structure between sequences 2 and 3 prevents attenuation, because sequence 3 is no longer available to form the attenuator structure with sequence 4. The 2:3 structure, unlike the 3:4 attenuator, does not prevent transcription.

(b)

3:4 Pair
(attenuator)

2:3 Pair

(c)

FIGURE 28–19 Transcriptional attenuation in the *trp* operon. Transcription is initiated at the beginning of the 162 nucleotide mRNA leader encoded by a DNA region called *trpL* (see Fig. 28-18). A regulatory mechanism determines whether transcription is attenuated at the end of the leader or continues into the structural genes. **(a)** The *trp* mRNA leader (*trpL*). The attenuation mechanism in the *trp* operon involves sequences 1 to 4 (highlighted). **(b)** Sequence 1 encodes a small peptide, the leader peptide, containing two Trp residues (W); it is translated immediately after transcription begins. Sequences 2 and 3 are complementary, as are sequences

3 and 4. The attenuator structure forms by the pairing of sequences 3 and 4 (top). Its structure and function are similar to those of a transcription terminator (see Fig. 26-7a). Pairing of sequences 2 and 3 (bottom) prevents the attenuator structure from forming. Note that the leader peptide has no other cellular function. Translation of its open reading frame has a purely regulatory role that determines which complementary sequences (2 and 3 or 3 and 4) are paired. **(c)** Base-pairing schemes for the complementary regions of the *trp* mRNA leader.

Regulatory sequence 1 is crucial for a tryptophan-sensitive mechanism that determines whether sequence 3 pairs with sequence 2 (allowing transcription to continue) or with sequence 4 (attenuating transcription). Formation of the attenuator stem-and-loop structure depends on events that occur during *translation* of regulatory sequence 1, which encodes a leader peptide (so called because it is encoded by the leader region of the mRNA) of 14 amino acids, two of which are Trp residues. The leader peptide has no other known cellular function; its synthesis is simply an operon regulatory device. This peptide is translated immediately after it is transcribed, by a ribosome that follows closely behind RNA polymerase as transcription proceeds.

When tryptophan concentrations are high, concentrations of charged tryptophan tRNA (Trp-tRNA^{Trp}) are also high. This allows translation to proceed rapidly past the two Trp codons of sequence 1 and into sequence 2, before sequence 3 is synthesized by RNA polymerase. In this situation, sequence 2 is covered by the ribosome and unavailable for pairing to sequence 3 when sequence 3 is synthesized; the attenuator structure (sequences 3 and 4) forms and transcription halts (Fig. 28–19b, top). When tryptophan concentrations are low, however, the ribosome stalls at the two Trp codons in sequence 1 because charged tRNA^{Trp} is less available. Sequence 2 remains free while sequence 3 is synthesized, allowing these two sequences to base-pair and permitting transcription to proceed (Fig. 28–19b, bottom). In this way, the proportion of transcripts that are attenuated declines as tryptophan concentration declines.

Many other amino acid biosynthetic operons use a similar attenuation strategy to fine-tune biosynthetic enzymes to meet the prevailing cellular requirements. The 15 amino acid leader peptide produced by the *phe* operon contains seven Phe residues. The *leu* operon leader peptide has four contiguous Leu residues. The leader peptide for the *his* operon contains seven contiguous His residues. In fact, in the *his* operon and a number of others, attenuation is sufficiently sensitive to be the *only* regulatory mechanism.

Induction of the SOS Response Requires Destruction of Repressor Proteins

Extensive DNA damage in the bacterial chromosome triggers the induction of many distantly located genes. This response, called the SOS response (p. 1035), provides another good example of coordinated gene regulation. Many of the induced genes are involved in DNA repair (see Table 25–6). The key regulatory proteins are the RecA protein and the LexA repressor.

The LexA repressor (M_r 22,700) inhibits transcription of all the SOS genes **(Fig. 28–20)**, and induction of the SOS response requires removal of LexA. This is not a simple dissociation from DNA in response to binding of a small molecule, as in the regulation of the *lac* operon described above. Instead, the LexA repressor is inactivated when it catalyzes its own cleavage at a specific Ala–Gly peptide bond, producing two roughly equal protein fragments. At physiological pH, this autocleavage reaction requires the RecA protein. RecA is not a

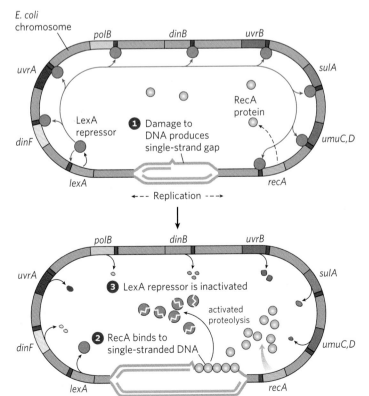

FIGURE 28–20 SOS response in *E. coli*. See Table 25–6 for the functions of many of these proteins. The LexA protein is the repressor in this system, which has an operator site (red) near each gene. Because the *recA* gene is not entirely repressed by the LexA repressor, the normal cell contains about 1,000 RecA monomers. ❶ When DNA is extensively damaged (such as by UV light), DNA replication is halted and the number of single-strand gaps in the DNA increases. ❷ RecA protein binds to this damaged, single-stranded DNA, activating the protein's coprotease activity. ❸ While bound to DNA, the RecA protein facilitates cleavage and inactivation of the LexA repressor. When the repressor is inactivated, the SOS genes, including *recA*, are induced; RecA levels increase 50- to 100-fold.

protease in the classical sense, but its interaction with LexA facilitates the repressor's self-cleavage reaction. This function of RecA is sometimes called a co-protease activity.

The RecA protein provides the functional link between the biological signal (DNA damage) and induction of the SOS genes. Heavy DNA damage leads to numerous single-strand gaps in the DNA, and only RecA that is bound to single-stranded DNA can facilitate cleavage of the LexA repressor (Fig. 28–20, bottom). Binding of RecA at the gaps eventually activates its co-protease activity, leading to cleavage of the LexA repressor and SOS induction.

During induction of the SOS response in a severely damaged cell, RecA also cleaves and thus inactivates the repressors that otherwise allow propagation of certain viruses in a dormant lysogenic state within the bacterial host. This provides a remarkable illustration of evolutionary adaptation. These repressors, like LexA, also undergo self-cleavage at a specific Ala–Gly peptide bond, so induction of the SOS response permits replication of the virus and lysis of the cell, releasing new viral particles. Thus the bacteriophage can make a hasty exit from a compromised bacterial host cell.

Synthesis of Ribosomal Proteins Is Coordinated with rRNA Synthesis

In bacteria, an increased cellular demand for protein synthesis is met by increasing the number of ribosomes rather than altering the activity of individual ribosomes. In general, the number of ribosomes increases as the cellular growth rate increases. At high growth rates, ribosomes make up approximately 45% of the cell's dry weight. The proportion of cellular resources devoted to making ribosomes is so large, and the function of ribosomes so important, that cells must coordinate the synthesis of the ribosomal components: the ribosomal proteins (r-proteins) and RNAs (rRNAs). This regulation is distinct from the mechanisms described so far, because it occurs largely at the level of *translation*.

The 52 genes that encode the r-proteins occur in at least 20 operons, each with 1 to 11 genes. Some of these operons also contain the genes for the subunits of DNA primase (see Fig. 25–12), RNA polymerase (see Fig. 26–4), and protein synthesis elongation factors (see Fig. 27–29)—revealing the close coupling of replication, transcription, and protein synthesis during cell growth.

The r-protein operons are regulated primarily through a translational feedback mechanism. One r-protein encoded by each operon also functions as a **translational repressor**, which binds to the mRNA transcribed from that operon and blocks translation of all the genes the messenger encodes (**Fig. 28–21**). In general, the r-protein that plays the role of repressor also binds directly to an rRNA. Each translational repressor r-protein binds with higher affinity to the appropriate rRNA than to its mRNA, so the mRNA is bound and

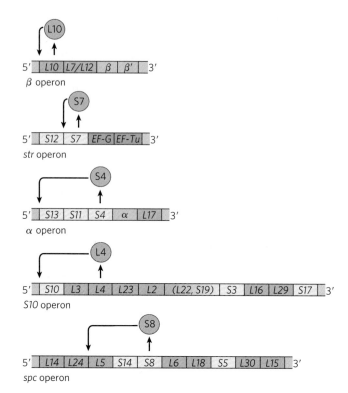

FIGURE 28–21 Translational feedback in some ribosomal protein operons. The r-proteins that act as translational repressors are shaded light red. Each translational repressor blocks the translation of all genes in that operon by binding to the indicated site on the mRNA. Genes that encode subunits of RNA polymerase are shown in purple; genes that encode elongation factors are blue. The r-proteins of the large (50S) ribosomal subunit are designated L1 to L34; those of the small (30S) subunit, S1 to S21.

translation repressed only when the level of the r-protein exceeds that of the rRNA. This ensures that translation of the mRNAs encoding r-proteins is repressed only when synthesis of these r-proteins exceeds that needed to make functional ribosomes. In this way, the rate of r-protein synthesis is kept in balance with rRNA availability.

The mRNA-binding site for the translational repressor is near the translational start site of one of the genes in the operon, usually the first gene (Fig. 28–21). In other operons this would affect only that one gene, because in bacterial polycistronic mRNAs most genes have independent translation signals. In the r-protein operons, however, the translation of one gene depends on the translation of all the others. The mechanism of this translational coupling is not yet understood in detail. However, in some cases the translation of multiple genes seems to be blocked by folding of the mRNA into an elaborate three-dimensional structure that is stabilized both by internal base-pairing (as in Fig. 8–24) and by binding of the translational repressor protein. When the translational repressor is absent, ribosome binding and translation of one or more of the genes disrupts the folded structure of the mRNA and allows all the genes to be translated.

Because the synthesis of r-proteins is coordinated with the available rRNA, the regulation of ribosome production reflects the regulation of rRNA synthesis. In *E. coli*, rRNA synthesis from the seven rRNA operons responds to cellular growth rate and to changes in the availability of crucial nutrients, particularly amino acids. The regulation coordinated with amino acid concentrations is known as the **stringent response (Fig. 28–22)**. When amino acid concentrations are low, rRNA synthesis is halted. Amino acid starvation leads to the binding of uncharged tRNAs to the ribosomal A site; this triggers a sequence of events that begins with the binding of an enzyme called **stringent factor** (RelA protein) to the ribosome. When bound to the ribosome, stringent factor catalyzes formation of the unusual nucleotide guanosine tetraphosphate (ppGpp; see Fig. 8–39); it adds pyrophosphate to the 3′ position of GTP, in the reaction

$$GTP + ATP \rightarrow pppGpp + AMP$$

then a phosphohydrolase cleaves off one phosphate to form ppGpp. The abrupt rise in ppGpp level in response to amino acid starvation results in a great reduction in rRNA synthesis, mediated at least in part by the binding of ppGpp to RNA polymerase.

The nucleotide ppGpp, along with cAMP, belongs to a class of modified nucleotides that act as cellular second messengers (p. 308). In *E. coli*, these two nucleotides serve as starvation signals; they cause large changes in cellular metabolism by increasing or decreasing the transcription of hundreds of genes. In eukaryotic cells, similar nucleotide second messengers also have multiple regulatory functions. The coordination of cellular metabolism with cell growth is highly complex, and further regulatory mechanisms undoubtedly remain to be discovered.

The Function of Some mRNAs Is Regulated by Small RNAs in Cis or in Trans

As described throughout this chapter, proteins play an important and well-documented role in regulating gene expression. But RNA also has a crucial role—one that is becoming better recognized as more examples of regulatory RNAs are discovered. Once an mRNA is synthesized, its functions can be controlled by RNA-binding proteins, as seen for the r-protein operons just described, or by an RNA. A separate RNA molecule may bind to the mRNA "in trans" and affect its activity. Alternatively, a portion of the mRNA itself may regulate its own function. When part of a molecule affects the function of another part of the same molecule, it is said to act "in cis."

A well-characterized example of RNA regulation in trans is seen in the regulation of the mRNA of the gene *rpoS* (*R*NA *po*lymerase *s*igma factor), which encodes σ^S, one of the seven *E. coli* sigma factors (see Table 26–1). The cell uses this specificity factor in certain stress situations, such as when it enters the stationary phase (a state of no growth, necessitated by lack of nutrients) and σ^S is needed to transcribe large numbers of stress response genes. The σ^S mRNA is present at low levels under most conditions but is not translated, because a large hairpin structure upstream of the coding region inhibits ribosome binding **(Fig. 28–23)**. Under certain stress conditions, one or both of two small special-function RNAs, DsrA (*d*ownstream *r*egion *A*) and RprA (*R*pos *r*egulator RNA *A*), are induced. Both can pair with one strand of the hairpin in the σ^S mRNA, disrupting the hairpin and thus allowing translation of *rpoS*. Another small RNA, OxyS (*ox*idative stress gene *S*), is induced under conditions of oxidative stress and inhibits the translation of *rpoS*, probably by pairing with and blocking the ribosome-binding site on the mRNA. OxyS is expressed as part of a system that responds to a different type of stress (oxidative damage) than does *rpoS*, and its task is to prevent expression of unneeded repair pathways. DsrA, RprA, and OxyS are all relatively small bacterial RNA molecules (less than 300 nucleotides), designated sRNAs (*s* for *s*mall; there are of course other "small" RNAs with other designations in eukaryotes). All require for their function a protein called Hfq, an RNA chaperone that facilitates RNA-RNA pairing. The known bacterial genes regulated in this way are few in number, just a few dozen in a typical bacterial species. However, these examples provide

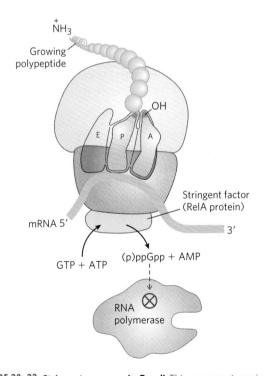

FIGURE 28–22 Stringent response in *E. coli*. This response to amino acid starvation is triggered by binding of an uncharged tRNA in the ribosomal A site. A protein called stringent factor binds to the ribosome and catalyzes the synthesis of pppGpp, which is converted by a phosphohydrolase to ppGpp. The signal ppGpp reduces transcription of some genes and increases that of others, in part by binding to the *β* subunit of RNA polymerase and altering the enzyme's promoter specificity. Synthesis of rRNA is reduced when ppGpp levels increase.

(a)

(b)

FIGURE 28–23 Regulation of bacterial mRNA function in trans by sRNAs. Several sRNAs (small RNAs)—DsrA, RprA, and OxyS—are involved in regulation of the *rpoS* gene. All require the protein Hfq, an RNA chaperone that facilitates RNA-RNA pairing. Hfq has a toroid structure, with a pore in the center. **(a)** DsrA promotes translation by pairing with one strand of a stem-loop structure that otherwise blocks the ribosome-binding site. RprA (not shown) acts in a similar way. **(b)** OxyS blocks translation by pairing with the ribosome-binding site.

FIGURE 28–24 Regulation of bacterial mRNA function in cis by riboswitches. The known modes of action are illustrated by several different riboswitches based on a widespread natural aptamer that binds thiamine pyrophosphate. TPP binding to the aptamer leads to a conformational change that produces the varied results illustrated in parts **(a)**, **(b)**, and **(c)** in the different systems in which the aptamer is utilized.

good model systems for understanding patterns present in the more complex and numerous examples of RNA-mediated regulation in eukaryotes.

Regulation in cis involves a class of RNA structures known as **riboswitches**. As described in Box 26–3, aptamers are RNA molecules, generated in vitro, that are capable of specific binding to a particular ligand. As one might expect, such ligand-binding RNA domains are also present in nature—in riboswitches—in a significant number of bacterial mRNAs (and even in some eukaryotic mRNAs). These natural aptamers are structured domains found in untranslated regions at the 5′ ends of certain bacterial mRNAs. Binding of an mRNA's riboswitch to its appropriate ligand results in a conformational change in the mRNA, and transcription is inhibited by stabilization of a premature transcription termination structure, or translation is inhibited (in cis) by occlusion of the ribosome-binding site **(Fig. 28–24)**. In most cases, the riboswitch acts in a kind of feedback loop. Most genes regulated in this way are involved in the synthesis or transport of the ligand that is bound by the riboswitch; thus, when the ligand is present in high concentrations, the riboswitch inhibits expression of the genes needed to replenish this ligand.

Each riboswitch binds only one ligand. Distinct riboswitches have been detected that respond to more than a dozen different ligands, including thiamine pyrophosphate (TPP, vitamin B$_1$), cobalamin (vitamin B$_{12}$), flavin mononucleotide, lysine, S-adenosylmethionine (adoMet), purines, N-acetylglucosamine 6-phosphate, and glycine. It is likely that many more remain to be discovered. The riboswitch that responds to TPP seems to be the most widespread; it is found in many bacteria, fungi, and some plants. The bacterial TPP riboswitch inhibits translation in some species and induces premature transcription termination in others (Fig. 28–24). The eukaryotic TPP riboswitch is found in the introns of certain genes and modulates the alternative splicing of those genes (see Fig. 26–21). It is not yet clear how common riboswitches are. However, estimates suggest that more than 4% of the genes of *Bacillus subtilis* are regulated by riboswitches.

As riboswitches become better understood, researchers are finding medical applications. For example, most of the riboswitches described to date, including the one that responds to adoMet, have been found only in bacteria. A drug that bound to and activated the adoMet riboswitch would shut down the genes encoding the enzymes that synthesize and transport adoMet, effectively starving the bacterial cells of this essential cofactor. Drugs of this type are being sought for use as a new class of antibiotics. ■

The pace of discovery of functional RNAs shows no signs of abatement and continues to enrich the hypothesis that RNA played a special role in the evolution of life (Chapter 26). The sRNAs and riboswitches, like ribozymes and ribosomes, may be vestiges of an RNA world obscured by time but persisting as a rich array of biological devices still functioning in the extant biosphere. The laboratory selection of aptamers and ribozymes with novel ligand-binding and enzymatic functions (see Box 26–3) tells us that the RNA-based activities necessary for a viable RNA world are possible. Discovery of many of the same RNA functions in living organisms tells us that key components for RNA-based metabolism do exist. For example, the natural aptamers of riboswitches may be derived from RNAs that, billions of years ago, bound to cofactors needed to promote the enzymatic processes required for metabolism in the RNA world.

Some Genes Are Regulated by Genetic Recombination

We turn now to another mode of bacterial gene regulation, at the level of DNA rearrangement—recombination. *Salmonella typhimurium*, which inhabits the mammalian intestine, moves by rotating the flagella on its cell surface **(Fig. 28–25)**. The many copies of the protein

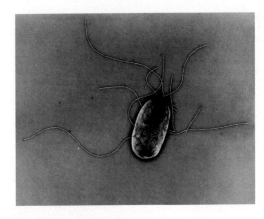

FIGURE 28–25 *Salmonella typhimurium,* with flagella evident.

flagellin (M_r 53,000) that make up the flagella are prominent targets of mammalian immune systems. But *Salmonella* cells have a mechanism that evades the immune response: they switch between two distinct flagellin proteins (FljB and FliC) roughly once every 1,000 generations, using a process called **phase variation**.

The switch is accomplished by periodic inversion of a segment of DNA containing the promoter for a flagellin gene. The inversion is a site-specific recombination reaction (see Fig. 25–37) mediated by the Hin recombinase at specific 14 bp sequences (*hix* sequences) at either end of the DNA segment. When the DNA segment is in one orientation, the gene for FljB flagellin and the gene encoding a repressor (FljA) are expressed **(Fig. 28–26a)**; the repressor shuts down expression of the gene for FliC flagellin. When the DNA segment is inverted (Fig. 28–26b), the *fljA* and *fljB* genes are no longer transcribed, and the *fliC* gene is induced as the repressor becomes depleted. The Hin recombinase, encoded

FIGURE 28–26 Regulation of flagellin genes in *Salmonella*: phase variation. The products of genes *fliC* and *fljB* are different flagellins. The *hin* gene encodes the recombinase that catalyzes inversion of the DNA segment containing the *fljB* promoter and the *hin* gene. The recombination sites (inverted repeats) are called *hix* (yellow). **(a)** In one orientation, *fljB* is expressed along with a repressor protein (product of the *fljA* gene) that represses transcription of the *fliC* gene. **(b)** In the opposite orientation, only the *fliC* gene is expressed; the *fljA* and *fljB* genes cannot be transcribed. The interconversion between these two states, known as phase variation, also requires two other nonspecific DNA-binding proteins (not shown), HU and FIS.

TABLE 28–1 Examples of Gene Regulation by Recombination

System	Recombinase/ recombination site	Type of recombination	Function
Phase variation (*Salmonella*)	Hin/*hix*	Site-specific	Alternative expression of two flagellin genes allows evasion of host immune response.
Host range (bacteriophage μ)	Gin/*gix*	Site-specific	Alternative expression of two sets of tail fiber genes affects host range.
Mating-type switch (yeast)	HO endonuclease, RAD52 protein, other proteins/*MAT*	Nonreciprocal gene conversion*	Alternative expression of two mating types of yeast, a and α, creates cells of different mating types that can mate and undergo meiosis.
Antigenic variation (trypanosomes)[†]	Varies	Nonreciprocal gene conversion*	Successive expression of different genes encoding the variable surface glycoproteins (VSGs) allows evasion of host immune response.

*In nonreciprocal gene conversion (a class of recombination events not discussed in Chapter 25), genetic information is moved from one part of the genome (where it is silent) to another (where it is expressed). The reaction is similar to replicative transposition (see Fig. 25–41).

[†]Trypanosomes cause African sleeping sickness and other diseases (see Box 6–3). The outer surface of a trypanosome is made up of multiple copies of a single VSG, the major surface antigen. A cell can change surface antigens to more than 100 different forms, precluding an effective defense by the host immune system.

by the *hin* gene in the DNA segment that undergoes inversion, is expressed when the DNA segment is in either orientation, so the cell can always switch from one state to the other.

This type of regulatory mechanism has the advantage of being absolute: gene expression is impossible when the gene is physically separated from its promoter (note the position of the *fljB* promoter in Fig. 28–26b). An absolute on/off switch may be important in this system (even though it affects only one of the two flagellin genes) because a flagellum with just one copy of the wrong flagellin might be vulnerable to host antibodies against that protein. The *Salmonella* system is by no means unique. Similar regulatory systems occur in some other bacteria and in some bacteriophages, and recombination systems with similar functions have been found in eukaryotes (Table 28–1). Gene regulation by DNA rearrangements that move genes and/or promoters is particularly common in pathogens that benefit by changing their host range or by changing their surface proteins, thereby staying ahead of host immune systems.

SUMMARY 28.2 Regulation of Gene Expression in Bacteria

▶ In addition to repression by the Lac repressor, the *E. coli lac* operon undergoes positive regulation by the cAMP receptor protein (CRP). When [glucose] is low, [cAMP] is high and CRP-cAMP binds to a specific site on the DNA, stimulating transcription of the *lac* operon and production of lactose-metabolizing enzymes. The presence of glucose depresses [cAMP], decreasing expression of *lac* and other genes involved in metabolism of secondary sugars. A group of coordinately regulated operons is referred to as a regulon.

▶ Operons that produce the enzymes of amino acid synthesis have a regulatory circuit called attenuation, which uses a transcription termination site (the attenuator) in the mRNA. Formation of the attenuator is modulated by a mechanism that couples transcription and translation while responding to small changes in amino acid concentration.

▶ In the SOS system, multiple unlinked genes repressed by a single repressor are induced simultaneously when DNA damage triggers RecA protein–facilitated autocatalytic proteolysis of the repressor.

▶ In the synthesis of ribosomal proteins, one protein in each r-protein operon acts as a translational repressor. The mRNA is bound by the repressor, and translation is blocked only when the r-protein is present in excess of available rRNA.

▶ Posttranscriptional regulation of some mRNAs is mediated by sRNAs that act in trans or by riboswitches, part of the mRNA structure itself, that act in cis.

▶ Some genes are regulated by genetic recombination processes that move promoters relative to the genes being regulated. Regulation can also take place at the level of translation.

28.3 Regulation of Gene Expression in Eukaryotes

Initiation of transcription is a crucial regulation point for gene expression in all organisms. Although eukaryotes and bacteria use some of the same regulatory mechanisms, the regulation of transcription in the two systems is fundamentally different.

We can define a transcriptional ground state as the inherent activity of promoters and transcriptional machinery in vivo in the absence of regulatory sequences. In bacteria, RNA polymerase generally has access to every promoter and can bind and initiate transcription at some level of efficiency in the absence of activators or repressors; the transcriptional ground state is therefore nonrestrictive. In eukaryotes, however, strong promoters are generally inactive in vivo in the absence of regulatory proteins; that is, the transcriptional ground state is restrictive. This fundamental difference gives rise to at least four important features that distinguish the regulation of gene expression in eukaryotes from that in bacteria.

First, access to eukaryotic promoters is restricted by the structure of chromatin, and activation of transcription is associated with many changes in chromatin structure in the transcribed region. Second, although eukaryotic cells have both positive and negative regulatory mechanisms, positive mechanisms predominate in all systems characterized so far. Thus, given that the transcriptional ground state is restrictive, virtually every eukaryotic gene requires activation in order to be transcribed. Third, eukaryotic cells have larger, more complex multimeric regulatory proteins than do bacteria. Finally, transcription in the eukaryotic nucleus is separated from translation in the cytoplasm in both space and time.

The complexity of regulatory circuits in eukaryotic cells is extraordinary, as the following discussion shows. We conclude the section with an illustrated description of one of the most elaborate circuits: the regulatory cascade that controls development in fruit flies.

Transcriptionally Active Chromatin Is Structurally Distinct from Inactive Chromatin

The effects of chromosome structure on gene regulation in eukaryotes have no clear parallel in bacteria. In the eukaryotic cell cycle, interphase chromosomes appear, at first viewing, to be dispersed and amorphous (see Fig. 24–24). Nevertheless, several forms of chromatin can be found along these chromosomes. About 10% of the chromatin in a typical eukaryotic cell is in a more condensed form than the rest of the chromatin. This form, **heterochromatin**, is transcriptionally inactive. Heterochromatin is generally associated with particular chromosome structures—the centromeres, for example. The remaining, less condensed chromatin is called **euchromatin**.

Transcription of a eukaryotic gene is strongly repressed when its DNA is condensed within heterochromatin. Some, but not all, of the euchromatin is transcriptionally active. Transcriptionally active chromosomal regions are distinguished from heterochromatin in at least three ways: the positioning of nucleosomes, the presence of histone variants, and the covalent modification of nucleosomes. These transcription-associated structural changes in chromatin are collectively called **chromatin remodeling**. The remodeling involves enzymes that promote these changes (Table 28–2).

Five known families of enzyme complexes actively reposition or displace nucleosomes, hydrolyzing ATP in the process. Three of these are particularly important in transcriptional activation (Table 28–2; see the table footnote for an explanation of the abbreviated names of the enzyme complexes described here). **SWI/SNF**, found in all eukaryotic cells, contains at least six core polypeptides that together remodel chromatin so that nucleosomes become more irregularly spaced. They also stimulate transcription factor binding. The complex includes a component called a bromodomain near the carboxyl terminus of the active ATPase subunit, which interacts with acetylated histone tails. SWI/SNF is not required for the transcription of every gene. **NURF**, a member of the ISW1 family, remodels chromatin in ways that complement and overlap the activity of SWI/SNF. These two enzyme complexes are crucial in preparing a region of chromatin for active transcription.

The third important protein family has a somewhat different role. Transcriptionally active chromatin tends to be deficient in histone H1, which binds to the linker DNA between nucleosome particles. These regions of chromatin are also enriched in the histone variants H3.3 and H2AZ (see Box 24–2). Alterations in histone content are again mediated by specialized enzymes and protein complexes. H2AZ deposition involves members of this third family of ATP-dependent remodeling enzymes, called **SWR1**.

The covalent modification of histones is altered dramatically within transcriptionally active chromatin. The core histones of nucleosome particles (H2A, H2B, H3, H4; see Fig. 24–26) are modified by methylation of Lys or Arg residues, phosphorylation of Ser or Thr residues, acetylation (see below), ubiquitination (see Fig. 27–47), or sumoylation. Each of the core histones has two distinct structural domains. A central domain is involved in histone-histone interaction and the wrapping of DNA around the nucleosome. A second, lysine-rich amino-terminal domain is generally positioned near the exterior of the assembled nucleosome particle; the covalent modifications occur at specific residues concentrated in this amino-terminal domain. The patterns of modification have led some researchers to propose the existence of a histone code, in which modification patterns are recognized by enzymes that alter the structure of chromatin. Indeed, some of the modifications are essential for interactions with proteins that play key roles in transcription.

The acetylation and methylation of histones figure prominently in the processes that activate chromatin

TABLE 28-2 Some Enzyme Complexes Catalyzing Chromatin Structural Changes Associated with Transcription

Enzyme complex*	Oligomeric structure (number of polypeptides)	Source	Activities
Histone modification			
GCN5-ADA2-ADA3	3	Yeast	GCN5 has type A HAT activity
SAGA/PCAF	>20	Eukaryotes	Includes GCN5-ADA2-ADA3; acetylates residues in H3 and H2B
NuA4	At least 12	Eukaryotes	EsaI component has HAT activity; acetylates H4, H2A, and H2AZ
Histone movement/replacement enzymes that require ATP			
SWI/SNF	$\geq$6; total M_r 2 $\times 10^6$	Eukaryotes	Nucleosome remodeling; transcriptional activation
ISWI family	Varies	Eukaryotes	Nucleosome remodeling; transcriptional repression; transcriptional activation in some cases (NURF)
SWR1 family	~12	Eukaryotes	H2AZ deposition
Histone chaperones that do not require ATP			
HIRA	1	Eukaryotes	Deposition of H3.3 during transcription

*The abbreviations for eukaryotic genes and proteins are often more confusing or obscure than those used for bacteria. The complex of GCN5 (general control nonderepressible) and ADA (alteration/deficiency activation) proteins was discovered during investigation of the regulation of nitrogen metabolism genes in yeast. These proteins can be part of the larger SAGA complex (*SPF, ADA2,3, GCN5, acetyltransferase*) in yeasts. The equivalent of SAGA in humans is PCAF (*p300/CBP-associated factor*). NuA4 is *nucleosome acetyltransferase of H4*; ESA1 is *essential SAS2-related acetyltransferase*. SWI (*switching*) was discovered as a protein required for expression of certain genes involved in mating-type switching in yeast, and SNF (*sucrose nonfermenting*) as a factor for expression of the yeast gene for sucrase. Subsequent studies revealed multiple SWI and SNF proteins that acted in a complex. The SWI/SNF complex has a role in the expression of a wide range of genes and has been found in many eukaryotes, including humans. ISWI is *imitation SWI*; NURF, *nuclear remodeling factor*; SWR1, *Swi2/Snf2-related ATPase 1*; and HIRA, *histone regulator A*.

for transcription. During transcription, histone H3 is methylated (by specific histone methylases) at Lys^4 in nucleosomes near the 5' end of the coding region and at Lys^{36} throughout the coding region. These methylations facilitate the binding of **histone acetyltransferases (HATs)**, enzymes that acetylate particular Lys residues. Cytosolic (type B) HATs acetylate newly synthesized histones before the histones are imported into the nucleus. The subsequent assembly of the histones into chromatin after replication is facilitated by histone chaperones: CAF1 for H3 and H4 (see Box 24–2), and NAP1 for H2A and H2B.

Where chromatin is being activated for transcription, the nucleosomal histones are further acetylated by nuclear (type A) HATs. The acetylation of multiple Lys residues in the amino-terminal domains of histones H3 and H4 can reduce the affinity of the entire nucleosome for DNA. Acetylation of particular Lys residues is critical for the interaction of nucleosomes with other proteins. When transcription of a gene is no longer required, the extent of acetylation of nucleosomes in that vicinity is reduced by the activity of **histone deacetylases (HDACs)**, as part of a general gene-silencing process that restores the chromatin to a transcriptionally inactive

state. In addition to the removal of certain acetyl groups, new covalent modification of histones marks chromatin as transcriptionally inactive. For example, Lys^9 of histone H3 is often methylated in heterochromatin.

The net effect of chromatin remodeling in the context of transcription is to make a segment of the chromosome more accessible and to "label" (chemically modify) it so as to facilitate the binding and activity of transcription factors that regulate expression of the gene or genes in that region.

Most Eukaryotic Promoters Are Positively Regulated

As already noted, eukaryotic RNA polymerases have little or no intrinsic affinity for their promoters; initiation of transcription is almost always dependent on the action of multiple activator proteins. One important reason for the apparent predominance of positive regulation seems obvious: the storage of DNA within chromatin effectively renders most promoters inaccessible, so genes are silent in the absence of other regulation. The structure of chromatin affects access to some promoters more than others, but repressors that bind to DNA so as to preclude access of RNA polymerase (negative regulation) would often be simply redundant. Other factors must be at play in the use of positive regulation, and speculation generally centers around two: the large size of eukaryotic genomes and the greater efficiency of positive regulation.

First, nonspecific DNA binding of regulatory proteins becomes a more important problem in the much larger genomes of higher eukaryotes. And the chance that a single specific binding sequence will occur randomly at an inappropriate site also increases with genome size. Combinatorial control thus becomes important in a large genome **(Fig. 28–27)**. Specificity for transcriptional activation can be improved if each of several positive-regulatory proteins must bind specific DNA sequences in order to activate a gene. The average number of regulatory sites for a gene in a multicellular organism is six, and genes that are regulated by a dozen such sites are common. The requirement for binding of several positive-regulatory proteins to specific DNA sequences vastly reduces the probability of the random occurrence of a functional juxtaposition of all the necessary binding sites. In addition, the actual number of regulatory proteins that must be encoded by a genome to regulate all of its genes can be reduced (Fig. 28–27). Thus, a new regulator is not needed for every gene, although regulation is complex enough in higher eukaryotes that regulator proteins may represent 5% to 10% of

all protein-encoding genes. In principle, a similar combinatorial strategy could be used by multiple negative-regulatory elements, but this brings us to the second reason for the use of positive regulation: it is simply more efficient. If the ~25,000 genes in the human genome were negatively regulated, each cell would have to synthesize, at all times, all of the different repressors in concentrations sufficient to permit specific binding to each "unwanted" gene. In positive regulation, most of the genes are usually inactive (that is, RNA polymerases do not bind to the promoters) and the cell synthesizes only the activator proteins needed to promote transcription of the subset of genes required in the cell at that time. These arguments notwithstanding, there are examples of negative regulation in eukaryotes, from yeasts to humans, as we shall see.

DNA-Binding Activators and Coactivators Facilitate Assembly of the General Transcription Factors

To continue our exploration of the regulation of gene expression in eukaryotes, we return to the interactions between promoters and RNA polymerase II (Pol II), the enzyme responsible for the synthesis of eukaryotic mRNAs. Although many (but not all) Pol II promoters include the TATA box and Inr (initiator) sequences, with their standard spacing (see Fig. 26–8), they vary

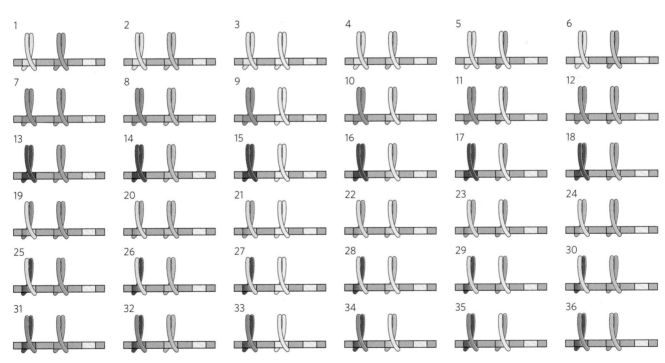

FIGURE 28–27 The advantages of combinatorial control. Combinatorial control allows specific regulation of many genes using a limited repertoire of regulatory proteins. Consider the possibilities inherent in regulation by two different families of leucine zipper proteins (red and green). If each regulatory gene family has three members (shown here as dark, medium, and light shades, each binding to a different DNA sequence) that can freely form either homo- or heterodimers, there are six possible dimeric

species in each family, each of which would recognize a different bipartite regulatory DNA sequence. If a gene had a regulatory site for each protein family, 36 different regulatory combinations would be possible, using just the six proteins from these two families. With six or more sites used in the regulation of a typical eukaryotic gene, the number of possible variants is much greater than this example suggests.

greatly in both the number and the location of additional sequences required for the regulation of transcription.

The additional regulatory sequences, generally bound by transcription activators, are usually called **enhancers** in higher eukaryotes and **upstream activator sequences (UASs)** in yeast. A typical enhancer may be found hundreds or even thousands of base pairs upstream from the transcription start site, or may even be downstream, within the gene itself. When bound by the appropriate regulatory proteins, an enhancer increases transcription at nearby promoters regardless of its orientation in the DNA. The UASs of yeast function in a similar way, although generally they must be positioned upstream and within a few hundred base pairs of the transcription start site.

Successful binding of active RNA polymerase II holoenzyme at one of its promoters usually requires the combined action of proteins of five types that have now been described: (1) **transcription activators**, which bind to enhancers or UASs and facilitate transcription; (2) **architectural regulators** that facilitate DNA looping; (3) **chromatin modification and remodeling proteins**, described above; (4) **coactivators**; and (5) **basal transcription factors** (see Fig. 26–9, Table 26–2), required at most Pol II promoters **(Fig. 28–28)**. The coactivators act indirectly—not by binding to the DNA—and are required for essential communication between the activators and the complex composed of Pol II and the basal (or general) transcription factors. Furthermore, a variety of repressor proteins can interfere with communication between the RNA polymerase and the activators, resulting in repression of transcription (Fig. 28–28b). Here we focus on the protein complexes

shown in Figure 28–28 and on how they interact to activate transcription.

Transcription Activators The requirements for activators vary greatly from one promoter to another. A few activators are known to facilitate transcription at hundreds of promoters, whereas others are specific for a few promoters. Many activators are sensitive to the binding of signal molecules, providing the capacity to activate or deactivate transcription in response to a changing cellular environment. Some enhancers bound by activators are quite distant from the promoter's TATA box. Multiple

FIGURE 28–28 Eukaryotic promoters and regulatory proteins. RNA polymerase II and its associated basal (general) transcription factors form a pre-initiation complex at the TATA box and Inr site of the cognate promoters, a process facilitated by transcription activators, acting through coactivators (Mediator, TFIID, or both). **(a)** A composite promoter with typical sequence elements and protein complexes found in both yeast and higher eukaryotes. The carboxyl-terminal domain (CTD) of Pol II (see Fig. 26–9) is an important point of interaction with Mediator and other protein complexes. Histone modification enzymes (not shown) catalyze methylation and acetylation; remodeling enzymes alter the content and placement of nucleosomes. The transcription activators have distinct DNA-binding domains and activation domains. Green arrows indicate common modes of interaction often required for the activation of transcription, as discussed in the text. The HMG proteins are a common type of architectural regulator (see Fig. 28-5), facilitating the looping of the DNA required to bring together system components bound at distant binding sites. **(b)** Eukaryotic transcriptional repressors function by a range of mechanisms. Some bind directly to DNA, displacing a protein complex required for activation (not shown); many others interact with various parts of the transcription or activation protein complexes to prevent activation. Possible points of interaction are indicated with red arrows. **(c)** The structure of an HMG protein complex with DNA shows how HMG proteins facilitate DNA looping. The binding is relatively nonspecific, although DNA sequence preferences have been identified for many HMG proteins. Shown is the HMG domain of the protein HMG-D of *Drosophila*, bound to DNA (PDB ID 1QRV).

enhancers (often six or more) are bound by a similar number of activators for a typical gene, providing combinatorial control and response to multiple signals.

Architectural Regulators How do the activators function at a distance? The answer in most cases seems to be that, as indicated earlier, the intervening DNA is looped so that the various protein complexes can interact directly. The looping is promoted by architectural regulators that are abundant in chromatin and bind to DNA with limited specificity. Most prominently, the **high mobility group (HMG)** proteins (Fig. 28–28c; "high mobility" refers to their electrophoretic mobility in polyacrylamide gels) play an important structural role in chromatin remodeling and transcriptional activation.

Coactivator Protein Complexes Most transcription requires the presence of additional protein complexes. Some major regulatory protein complexes that interact with Pol II have been defined both genetically and biochemically. These coactivator complexes act as intermediaries between the transcription activators and the Pol II complex.

A major eukaryotic coactivator, consisting of 20 to 30 or more polypeptides in a protein complex, is called **Mediator** (Fig. 28–28). Many of the 20 core polypeptides are highly conserved from fungi to humans. An additional complex of four subunits can interact with Mediator and inhibit transcription initiation. Mediator binds tightly to the carboxyl-terminal domain (CTD) of the largest subunit of Pol II. The Mediator complex is required for both basal and regulated transcription at many promoters used by Pol II, and it also stimulates phosphorylation of the CTD by TFIIH (a basal transcription factor). Transcription activators interact with one or more components of the Mediator complex, with the precise interaction sites differing from one activator to another. Coactivator complexes function at or near the promoter's TATA box.

Additional coactivators, functioning with one or a few genes, have also been described. Some of these operate in conjunction with Mediator, and some may act in systems that do not employ Mediator.

TATA-Binding Protein The first component to bind in the assembly of a **preinitiation complex (PIC)** at the TATA box of a typical Pol II promoter is the **TATA-binding protein (TBP)**. TBP is often, but not always, delivered as part of a larger complex (~15 subunits) called TFIID. The complete complex also includes the basal transcription factors TFIIB, TFIIE, TFIIF, TFIIH; Pol II; and perhaps TFIIA. This minimal PIC, however, is often insufficient for the initiation of transcription and generally does not form at all if the promoter is obscured

within chromatin. Positive regulation, leading to transcription, is imposed by the activators and coactivators.

Choreography of Transcriptional Activation We can now begin to piece together the sequence of transcriptional activation events at a typical Pol II promoter **(Fig. 28–29)**.

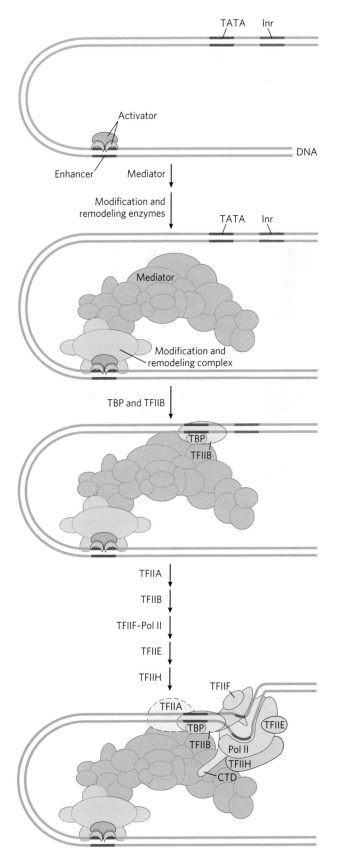

FIGURE 28–29 The components of transcriptional activation. Activators bind the DNA first. The activators recruit the histone modification/nucleosome remodeling complexes and a coactivator such as Mediator. Mediator facilitates the binding of TBP (or TFIID) and TFIIB, and the other basal transcription factors and Pol II then bind. Phosphorylation of the carboxyl-terminal domain (CTD) of Pol II leads to transcription initiation (not shown).

The exact order of binding of some components may vary, but the model in Figure 28–29 illustrates the principles of activation as well as one common path. Many transcription activators have significant affinity for their binding sites even when the sites are within condensed chromatin. The binding of activators is often the event that triggers subsequent activation of the promoter. Binding of one activator may enable the binding of others, gradually displacing some nucleosomes.

Crucial remodeling of the chromatin then takes place in stages, facilitated by interactions between activators and HATs or enzyme complexes such as SWI/SNF (or both). In this way, a bound activator can draw in other components necessary for further chromatin remodeling to permit transcription of specific genes. The bound activators interact with the large Mediator complex. Mediator, in turn, provides an assembly surface for the binding of first TBP (or TFIID), then TFIIB, and then other components of the PIC, including RNA polymerase II. Mediator stabilizes the binding of Pol II and its associated transcription factors and greatly facilitates formation of the PIC. Complexity in these regulatory circuits is the rule rather than the exception, with multiple DNA-bound activators promoting transcription.

The script can change from one promoter to another. For example, many promoters have a different set of recognition sequences and may not have a TATA box, and in multicellular eukaryotes the subunit composition of factors such as TFIID can vary from one tissue to another. However, most promoters seem to require a precisely ordered assembly of components to initiate transcription. The assembly process is not always fast. At some genes it may take minutes; at certain genes of higher eukaryotes the process can take days.

Reversible Transcriptional Activation Although rarer, some eukaryotic regulatory proteins that bind to Pol II promoters or that interact with transcriptional activators can act as repressors, inhibiting the formation of active PICs (Fig. 28–29). Some activators can adopt different conformations, enabling them to serve as transcription activators or as repressors. For example, some steroid hormone receptors (described later) function in the nucleus as transcription activators, stimulating the transcription of certain genes when a particular steroid hormone signal is present. When the hormone is absent, the receptor proteins revert to a repressor conformation, *preventing* the formation of PICs. In some cases this repression involves interaction with histone deacetylases and other proteins that help restore the surrounding chromatin to its transcriptionally inactive state. Mediator, when it includes the inhibitory subunits, may also block transcription initiation. This may be a regulatory mechanism to ensure ordered assembly of the PIC (by delaying transcriptional activation until all required factors are present), or it may be a mechanism that helps deactivate promoters when transcription is no longer required.

The Genes of Galactose Metabolism in Yeast Are Subject to Both Positive and Negative Regulation

Some of the general principles described above can be illustrated by one well-studied eukaryotic regulatory circuit (**Fig. 28–30**). The enzymes required for the importation and metabolism of galactose in yeast are

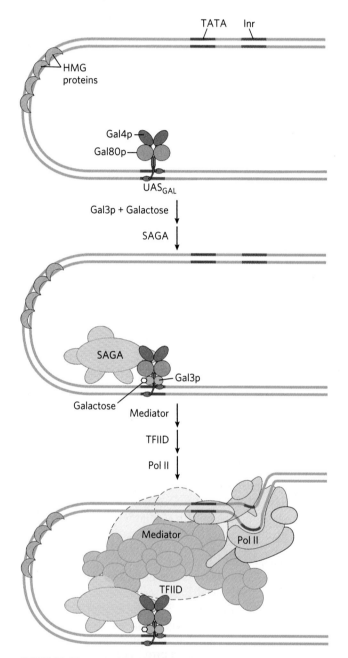

FIGURE 28–30 Regulation of transcription of *GAL* genes in yeast. Galactose imported into the yeast cell is converted to galactose 6-phosphate by a pathway involving six enzymes, whose genes are scattered over three chromosomes (see Table 28–3). Transcription of these genes is regulated by the combined actions of the proteins Gal4p, Gal80p, and Gal3p, with Gal4p playing the central role of transcription activator. The Gal4p-Gal80p complex is inactive. Binding of galactose to Gal3p leads to interaction of Gal3p with the Gal80p-Gal4p complex and activates Gal4p. The Gal4p subsequently recruits SAGA, Mediator, and TFIID to the galactose promoters, leading to DNA polymerase II recruitment and initiation of transcription.

| TABLE 28–3 | Genes of Galactose Metabolism in Yeast |

Gene	Protein function	Chromosomal location	Protein size (number of residues)	Relative protein expression in different carbon sources		
				Glucose	Glycerol	Galactose
Regulated genes						
GAL1	Galactokinase	II	528	–	–	+++
GAL2	Galactose permease	XII	574	–	–	+++
PGM2	Phosphoglucomutase	XIII	569	+	+	++
GAL7	Galactose 1-phosphate uridylyltransferase	II	365	–	–	+++
GAL10	UDP-glucose 4-epimerase	II	699	–	–	+++
MEL1	α-Galactosidase	II	453	–	+	++
Regulatory genes						
GAL3	Inducer	IV	520	–	+	++
GAL4	Transcriptional activator	XVI	881	+/–	+	+
GAL80	Transcriptional inhibitor	XIII	435	+	+	++

Source: Adapted from Reece, R. & Platt, A. (1997) Signaling activation and repression of RNA polymerase II transcription in yeast. *Bioessays* 19, 1001–1010.

encoded by genes scattered over several chromosomes (Table 28–3). Each of the *GAL* genes is transcribed separately, and yeast cells have no operons like those in bacteria. However, all the *GAL* genes have similar promoters and are regulated coordinately by a common set of proteins. The promoters for the *GAL* genes consist of the TATA box and Inr sequences, as well as an upstream activator sequence (UAS$_G$) recognized by the transcription activator Gal4 protein (Gal4p). Regulation of gene expression by galactose entails an interplay between Gal4p and two other proteins, Gal80p and Gal3p (Fig. 28–30). Gal80p forms a complex with Gal4p, preventing Gal4p from functioning as an activator of the *GAL* promoters. When galactose is present, it binds Gal3p, which then interacts with the Gal80p-Gal4p complex, allowing Gal4p to function as an activator at the various *GAL* promoters. As the various galactose genes are induced and their products build up, Gal3p may be replaced with Gal1p (a galactokinase needed for galactose metabolism that also functions in regulation) for sustained activation of the regulatory circuit.

Other protein complexes also have a role in activating transcription of the *GAL* genes. These include the SAGA complex for histone acetylation and chromatin remodeling, the SWI/SNF complex for chromatin remodeling, and the Mediator complex. The Gal4 protein is responsible for the recruitment of these additional factors needed for transcriptional activation. SAGA may be the first and primary recruitment target for Gal4p.

Glucose is the preferred carbon source for yeast, as it is for bacteria. When glucose is present, most of the *GAL* genes are repressed—whether galactose is present or not. The *GAL* regulatory system described above is effectively overridden by a complex catabolite repression system that includes several proteins (not depicted in Fig. 28–30).

Transcription Activators Have a Modular Structure

Transcription activators typically have a distinct structural domain for specific DNA binding and one or more additional domains for transcriptional activation or for interaction with other regulatory proteins. Interaction of two regulatory proteins is often mediated by domains containing leucine zippers (Fig. 28–14) or helix-loop-helix motifs (Fig. 28–15). We consider here three distinct types of structural domains used in activation by transcription activators (**Fig. 28–31a**): Gal4p, Sp1, and CTF1.

Gal4p contains a zinc finger–like structure in its DNA-binding domain, near the amino terminus; this domain has six Cys residues that coordinate two Zn^{2+}. The protein functions as a homodimer (with dimerization mediated by interactions between two coiled coils) and binds to UAS$_G$, a palindromic DNA sequence about 17 bp long. Gal4p has a separate activation domain with many acidic amino acid residues. Experiments that substitute a variety of different peptide sequences for the **acidic activation domain** of Gal4p suggest that the acidic nature of this domain is critical to its function, although its precise amino acid sequence can vary considerably.

Sp1 (M_r 80,000) is a transcription activator for a large number of genes in higher eukaryotes. Its DNA-binding site, the GC box (consensus sequence GGGCGG), is usually quite near the TATA box. The DNA-binding domain of the Sp1 protein is near its carboxyl terminus

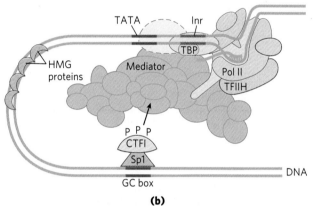

FIGURE 28–31 Transcription activators. (a) Typical activators such as CTF1, Gal4p, and Sp1 have a DNA-binding domain and an activation domain. The nature of the activation domain is indicated by symbols: – – –, acidic; Q Q Q, glutamine-rich; P P P, proline-rich. These proteins generally activate transcription by interacting with coactivator complexes such as Mediator. Note that the binding sites illustrated here are not generally found together near a single gene. **(b)** A chimeric protein containing the DNA-binding domain of Sp1 and the activation domain of CTF1 activates transcription if a GC box is present.

and contains three zinc fingers. Two other domains in Sp1 function in activation and are notable in that 25% of their amino acid residues are Gln. A wide variety of other activator proteins also have these **glutamine-rich domains**.

CTF1 (*CCAAT-binding transcription factor 1*) belongs to a family of transcription activators that bind a sequence called the CCAAT site (its consensus sequence is $TGGN_6GCCAA$, where N is any nucleotide). The DNA-binding domain of CTF1 contains many basic amino acid residues, and the binding region is probably arranged as an α helix. This protein has neither a helix-turn-helix nor a zinc finger motif; its DNA-binding mechanism is not yet clear. CTF1 has a **proline-rich activation domain**, with Pro accounting for more than 20% of the amino acid residues.

The discrete activation and DNA-binding domains of regulatory proteins often act completely independently, as has been demonstrated in "domain-swapping" experiments. Genetic engineering techniques (Chapter 9) can join the proline-rich activation domain of CTF1 to

the DNA-binding domain of Sp1 to create a protein that, like intact Sp1, binds to GC boxes on the DNA and activates transcription at a nearby promoter (as in Fig. 28–31b). The DNA-binding domain of Gal4p has similarly been replaced experimentally with the DNA-binding domain of the *E. coli* LexA repressor (of the SOS response; Fig. 28–20). This chimeric protein neither binds at UAS_G nor activates the yeast *GAL* genes (as would intact Gal4p) unless the UAS_G sequence in the DNA is replaced by the LexA recognition site.

Eukaryotic Gene Expression Can Be Regulated by Intercellular and Intracellular Signals

The effects of steroid hormones (and of thyroid and retinoid hormones, which have a similar mode of action) provide additional well-studied examples of the modulation of eukaryotic regulatory proteins by direct interaction with molecular signals (see Fig. 12–30). Unlike other types of hormones, steroid hormones do not have to bind to plasma membrane receptors. Instead, they can interact with intracellular receptors that are themselves transcription activators. Steroid hormones too hydrophobic to dissolve readily in the blood (estrogen, progesterone, and cortisol, for example) travel on specific carrier proteins from their point of release to their target tissues. In the target tissue, the hormone passes through the plasma membrane by simple diffusion. Once inside the cell, the hormones interact with one of two types of steroid-binding nuclear receptor **(Fig. 28–32)**. In both cases, the hormone-receptor complex acts by binding to highly specific DNA sequences called **hormone response elements (HREs)**, thereby altering gene expression. Acting at these sites, the receptors act as transcription activators, recruiting coactivators and DNA polymerase II (plus its associated transcription factors) to trigger transcription of the gene.

The DNA sequences (HREs) to which hormone-receptor complexes bind are similar in length and arrangement, but differ in sequence, for the various steroid hormones. Each receptor has a consensus HRE sequence (Table 28–4) to which the hormone-receptor complex binds well, with each consensus consisting of two six-nucleotide sequences, either contiguous or separated by three nucleotides, in tandem or in a palindromic arrangement. The hormone receptors have a highly conserved DNA-binding domain with two zinc fingers **(Fig. 28–33)**. The hormone-receptor complex binds to the DNA as a dimer, with the zinc finger domains of each monomer recognizing one of the six-nucleotide sequences. The ability of a given hormone to act through the hormone-receptor complex to alter the expression of a specific gene depends on the exact sequence of the HRE, its position relative to the gene, and the number of HREs associated with the gene.

The ligand-binding region of the receptor protein—always at the carboxyl terminus—is quite specific to the particular receptor. In the ligand-binding region, the

FIGURE 28–32 Mechanisms of steroid hormone receptor function. There are two types of steroid-binding nuclear receptors. **(a)** Monomeric type I receptors (NR) are found in the cytoplasm, in a complex with a heat shock protein (Hsp70). Receptors for estrogen, progesterone, androgens, and glucocorticoids are of this type. When the steroid hormone binds, the Hsp70 dissociates and the receptor dimerizes, exposing a nuclear localization signal. The dimeric receptor, with hormone bound, migrates to the nucleus, where it binds a hormone response element (HRE) and acts as a transcription activator. **(b)** Type II receptors, by contrast, are always in the nucleus, bound to an HRE in the DNA and to a corepressor that renders them inactive. The thyroid hormone receptor (TR) is of this type. The hormone migrates through the cytoplasm and diffuses across the nuclear membrane. In the nucleus it binds to a heterodimer consisting of the thyroid hormone receptor and the retinoid X receptor (RXR). A conformation change leads to dissociation of the corepressor, and the receptor then functions as a transcription activator.

TABLE 28–4	Hormone Response Elements (HREs) Bound by Steroid-Type Hormone Receptors
Receptor	**Consensus sequence bound***
Androgen	GG(A/T)ACAN₂TGTTCT
Glucocorticoid	GGTACAN₃TGTTCT
Retinoic acid (some)	AGGTCAN₅AGGTCA
Vitamin D	AGGTCAN₃AGGTCA
Thyroid hormone	AGGTCAN₃AGGTCA
RX†	AGGTCANAGGTCANAG GTCANAGGTCA

*N represents any nucleotide.

†Forms a dimer with the retinoic acid receptor or vitamin D receptor.

glucocorticoid receptor is only 30% similar to the estrogen receptor and 17% similar to the thyroid hormone receptor. The size of the ligand-binding region varies dramatically; in the vitamin D receptor it has only 25 amino acid residues, whereas in the mineralocorticoid receptor it has 603 residues. Mutations that change one amino acid in these regions can result in loss of responsiveness to a specific hormone. Some humans unable to respond to cortisol, testosterone, vitamin D, or thyroxine have mutations of this type.

Some hormone receptors, including the human progesterone receptor, activate transcription with the aid of an unusual coactivator—**steroid receptor RNA activator (SRA)**, an ~700 nucleotide RNA. SRA acts

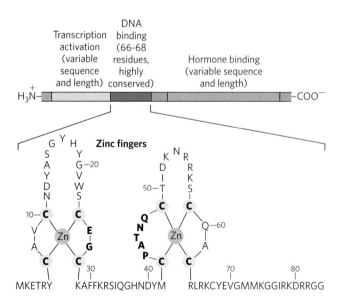

FIGURE 28–33 Typical steroid hormone receptors. These receptor proteins have a binding site for the hormone, a DNA-binding domain, and a region that activates transcription of the regulated gene. The highly conserved DNA-binding domain has two zinc fingers. The sequence shown here is that for the estrogen receptor, but the residues in bold type are common to all steroid hormone receptors.

as part of a ribonucleoprotein complex, but it is the RNA component that is required for transcription coactivation. The detailed set of interactions between SRA and other components of the regulatory systems for these genes remains to be worked out.

Regulation Can Result from Phosphorylation of Nuclear Transcription Factors

We noted in Chapter 12 that the effects of insulin on gene expression are mediated by a series of steps leading ultimately to the activation of a protein kinase in the nucleus that phosphorylates specific DNA-binding proteins and thereby alters their ability to act as transcription factors (see Fig. 12–15). This general mechanism mediates the effects of many nonsteroid hormones. For example, the β-adrenergic pathway that leads to elevated levels of cytosolic cAMP, which acts as a second messenger in eukaryotes as well as in bacteria (see Figs 12–4, 28–17), also affects the transcription of a set of genes, each of which is located near a specific DNA sequence called a **cAMP response element (CRE)**. The catalytic subunit of protein kinase A, released when cAMP levels rise (see Fig. 12–6), enters the nucleus and phosphorylates a nuclear protein, the **CRE-binding protein (CREB)**. When phosphorylated, CREB binds to CREs near certain genes and acts as a transcription factor, turning on the expression of these genes.

Many Eukaryotic mRNAs Are Subject to Translational Repression

Regulation at the level of translation assumes a much more prominent role in eukaryotes than in bacteria and is observed in a range of cellular situations. In contrast to the tight coupling of transcription and translation in bacteria, the transcripts generated in a eukaryotic nucleus must be processed and transported to the cytoplasm before translation. This can impose a significant delay on the appearance of a protein. When a rapid increase in protein production is needed, a translationally repressed mRNA already in the cytoplasm can be activated for translation without delay. Translational regulation may play an especially important role in regulating certain very long eukaryotic genes (a few are measured in the millions of base pairs), for which transcription and mRNA processing can require many hours. Some genes are regulated at both the transcriptional and translational stages, with the latter playing a role in the fine-tuning of cellular protein levels. In some anucleate cells, such as reticulocytes (immature erythrocytes), transcriptional control is entirely unavailable and translational control of stored mRNAs becomes essential. As described below, translational controls can also have spatial significance during development, when the regulated translation of prepositioned mRNAs creates a local gradient of the protein product.

FIGURE 28–34 Translational regulation of eukaryotic mRNA. One of the most important mechanisms for translational regulation in eukaryotes involves the binding of translational repressors (RNA-binding proteins) to specific sites in the 3′ untranslated region (3′UTR) of the mRNA. These proteins interact with eukaryotic initiation factors or with the ribosome to prevent or slow translation.

Eukaryotes have at least four main mechanisms of translational regulation.

1. Translation initiation factors are subject to phosphorylation by protein kinases. The phosphorylated forms are often less active and cause a general depression of translation in the cell.

2. Some proteins bind directly to mRNA and act as translational repressors, many of them binding at specific sites in the 3′ untranslated region (3′UTR). So positioned, these proteins interact with other translation initiation factors bound to the mRNA or with the 40S ribosomal subunit to prevent translation initiation **(Fig. 28–34)**.

3. Binding proteins, present in eukaryotes from yeast to mammals, disrupt the interaction between eIF4E and eIF4G (see Fig. 27–28). The mammalian versions are known as 4E-BPs (eIF4E binding proteins). When cell growth is slow, these proteins limit translation by binding to the site on eIF4E that normally interacts with eIF4G. When cell growth resumes or increases in response to growth factors or other stimuli, the binding proteins are inactivated by protein kinase–dependent phosphorylation.

4. RNA-mediated regulation of gene expression, considered later, often occurs at the level of translational repression.

The variety of translational regulation mechanisms provides flexibility, allowing focused repression of a few mRNAs or global regulation of all cellular translation.

Translational regulation has been particularly well studied in reticulocytes. One such mechanism in these cells involves eIF2, the initiation factor that binds to the initiator tRNA and conveys it to the ribosome; when

Met-tRNA has bound to the P site, the factor eIF2B binds to eIF2, recycling it with the aid of GTP binding and hydrolysis. The maturation of reticulocytes includes destruction of the cell nucleus, leaving behind a plasma membrane packed with hemoglobin. Messenger RNAs deposited in the cytoplasm before the loss of the nucleus allow for the replacement of hemoglobin. When reticulocytes become deficient in iron or heme, the translation of globin mRNAs is repressed. A protein kinase called **HCR (hemin-controlled repressor)** is activated, catalyzing the phosphorylation of eIF2. When phosphorylated, eIF2 forms a stable complex with eIF2B that sequesters the eIF2, making it unavailable for participation in translation. In this way, the reticulocyte coordinates the synthesis of globin with the availability of heme.

Many additional examples of translational regulation have been found in studies of the development of multicellular organisms, as discussed in more detail below.

Posttranscriptional Gene Silencing Is Mediated by RNA Interference

Craig Mello

Andrew Fire

In higher eukaryotes, including nematodes, fruit flies, plants, and mammals, a class of small RNAs called **microRNAs (miRNAs)** mediates the silencing of many genes. In a phenomenon first described and explained by Craig Mello and Andrew Fire, the RNAs function by interacting with mRNAs, often in the 3'UTR, resulting in either mRNA degradation or translation inhibition. In either case, the mRNA, and thus the gene that produces it, is silenced. This form of gene regulation controls developmental timing in at least some organisms. It is also used as a mechanism to protect against invading RNA viruses (particularly important in plants, which lack an immune system) and to control the activity of transposons. In addition, small RNA molecules may play a critical (but still undefined) role in the formation of heterochromatin.

Many miRNAs are present only transiently during development, and these are sometimes referred to as **small temporal RNAs (stRNAs)**. Thousands of different miRNAs have been identified in higher eukaryotes, and they may affect the regulation of a third of mammalian genes. They are transcribed as precursor RNAs about 70 nucleotides long, with internally complementary sequences that form hairpinlike structures. Details of the pathway for processing of miRNAs were described in Chapter 26 (see Fig. 26–27). The precursors are cleaved by endonucleases such as Dro-

sha and Dicer to form short duplexes about 20 to 25 nucleotides long. One strand of the processed miRNA is transferred to the target mRNA (or to a viral or transposon RNA), leading to inhibition of translation or degradation of the RNA **(Fig. 28–35)**. Some miRNAs bind to and affect a single mRNA and thus affect expression of only one gene. Others interact with multiple mRNAs and thus form the mechanistic core of regulons that coordinate the expression of multiple genes.

This gene regulation mechanism has an interesting and very useful practical side. If an investigator introduces into an organism a duplex RNA molecule corresponding in sequence to virtually any mRNA, Dicer cleaves the duplex into short segments, called **small interfering RNAs (siRNAs)**. These bind to the mRNA and silence it (Fig. 28–35b). The process is known as **RNA interference (RNAi)**. In plants, virtually any gene can be effectively shut down in this way. Nematodes readily ingest entire, functional RNAs, and simply introducing the duplex RNA into the worm's diet produces very effective suppression of the target gene. The technique has rapidly become an important tool in the ongoing efforts to study gene function, because it can disrupt gene function without creating a mutant organism. The procedure can be applied to humans as well. Laboratory-produced siRNAs have been used to block HIV and poliovirus infections in cultured human cells

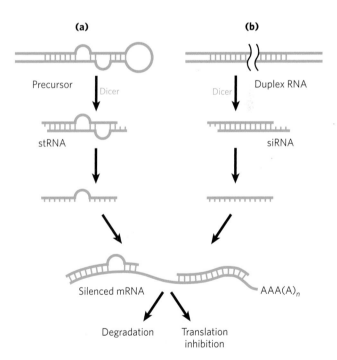

FIGURE 28–35 Gene silencing by RNA interference. (a) Small temporal RNAs (stRNAs, a class of miRNAs) are generated by Dicer-mediated cleavage of longer precursors that fold to create duplex regions. The stRNAs then bind to mRNAs, leading to degradation of mRNA or inhibition of translation. **(b)** Double-stranded RNAs can be constructed and introduced into a cell. Dicer processes the duplex RNAs into small interfering RNAs (siRNAs), which interact with the target mRNA. Again, either the mRNA is degraded or translation is inhibited.

for a week or so at a time. Rapid progress in research into RNA interference makes this a field to watch for future medical advances.

RNA-Mediated Regulation of Gene Expression Takes Many Forms in Eukaryotes

The special-function RNAs in eukaryotes include miRNAs, described above; snRNAs, involved in RNA splicing (see Fig. 26–16); and snoRNAs, involved in rRNA modification (see Fig. 26–25). All RNAs that do not encode proteins, including rRNAs and tRNAs, come under the general designation of **ncRNAs** (noncoding RNAs). Mammalian genomes seem to encode more ncRNAs than coding mRNAs (see Box 26–4). Not surprisingly, additional functional classes of ncRNAs are still being discovered.

Many of the newly found examples interact with proteins rather than with RNAs and affect the function of the bound proteins. The SRA that functions as a coactivator of steroid hormone–responsive genes is one example: it affects the activation of transcription. The heat shock response in human cells provides another example. Heat shock factor 1 (HSF1) is an activator protein that, in nonstressed cells, exists as a monomer bound by the chaperone Hsp90. Under stress conditions, HSF1 is released from Hsp90 and trimerizes. The HSF1 trimer binds to DNA and activates transcription of genes whose products are required to deal with the stress. An ncRNA called HSR1 (heat shock RNA 1; ~600 nucleotides) stimulates the HSF1 trimerization and DNA binding. HSR1 does not act alone; it functions in a complex with the translation elongation factor eEF1A.

Additional RNAs affect transcription in a variety of ways. Besides its role in splicing (see Fig. 26–16), the snRNA U1 directly binds to the transcription factor TFIIH. Its function in this context is not yet clear, but it may regulate TFIIH or affect the coupling between transcription and splicing, or both. A 331 nucleotide ncRNA called 7SK, abundant in mammals, binds to the Pol II transcription elongation factor pTEFb (see Table 26–2) and represses transcript elongation. Another ncRNA,

B2 (~178 nucleotides), binds directly to Pol II during heat shock and represses transcription. The B2-bound Pol II assembles into stable PICs, but transcription is blocked. The B2 RNA thus halts the transcription of many genes during heat shock, and the mechanism that allows HSF1-responsive genes to be expressed in the presence of B2 remains to be worked out.

The recognized roles of ncRNAs in gene expression and in many other cellular processes are rapidly expanding. At the same time, the study of the biochemistry of gene regulation is becoming much less protein-centric.

Development Is Controlled by Cascades of Regulatory Proteins

For sheer complexity and intricacy of coordination, the patterns of gene regulation that bring about development of a zygote into a multicellular animal or plant have no peer. Development requires transitions in morphology and protein composition that depend on tightly coordinated changes in expression of the genome. More genes are expressed during early development than in any other part of the life cycle. For example, in the sea urchin, an oocyte has about 18,500 *different* mRNAs, compared with about 6,000 different mRNAs in the cells of a typical differentiated tissue. The mRNAs in the oocyte give rise to a cascade of events that regulate the expression of many genes across both space and time.

Several organisms have emerged as important model systems for the study of development, because they are easy to maintain in a laboratory and have relatively short generation times. These include nematodes, fruit flies, zebra fish, mice, and the plant *Arabidopsis.* This discussion focuses on the development of fruit flies. Our understanding of the molecular events during development of *Drosophila melanogaster* is particularly well advanced and can be used to illustrate patterns and principles of general significance.

The life cycle of the fruit fly includes complete metamorphosis during its progression from an embryo to an adult **(Fig. 28–36)**. Among the most important characteristics of the embryo are its **polarity** (the

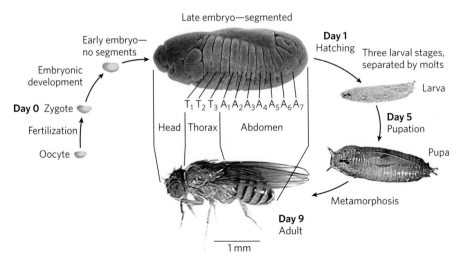

FIGURE 28–36 Life cycle of the fruit fly ***Drosophila melanogaster.*** *Drosophila* undergoes a complete metamorphosis, which means that the adult insect is radically different in form from its immature stages, a transformation that requires extensive alterations during development. By the late embryonic stage, segments have formed, each containing specialized structures from which the various appendages and other features of the adult fly will develop.

FIGURE 28-37 Early development in *Drosophila*. During development of the egg, maternal mRNAs (including the *bicoid* and *nanos* gene transcripts, discussed in the text) and proteins are deposited in the developing oocyte (unfertilized egg cell) by nurse cells and follicle cells. After fertilization, the two nuclei of the fertilized egg divide in synchrony within the common cytoplasm (syncytium), then migrate to the periphery. Membrane invaginations surround the nuclei to create a monolayer of cells at the periphery; this is the cellular blastoderm stage. During the early nuclear divisions, several nuclei at the far posterior become pole cells, which later become the germ-line cells.

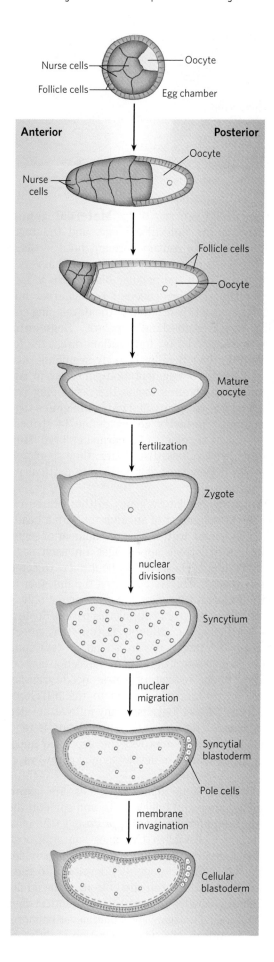

anterior and posterior parts of the animal are readily distinguished, as are its dorsal and ventral parts) and its **metamerism** (the embryo body is made up of serially repeating segments, each with characteristic features). During development, these segments become organized into a head, thorax, and abdomen. Each segment of the adult thorax has a different set of appendages. Development of this complex pattern is under genetic control, and a variety of pattern-regulating genes have been discovered that dramatically affect the organization of the body.

The *Drosophila* egg, along with 15 nurse cells, is surrounded by a layer of follicle cells **(Fig. 28-37)**. As the egg cell forms (before fertilization), mRNAs and proteins originating in the nurse and follicle cells are deposited in the egg cell, where some play a critical role in development. Once a fertilized egg is laid, its nucleus divides and the nuclear descendants continue to divide in synchrony every 6 to 10 min. Plasma membranes are not formed around the nuclei, which are distributed within the egg cytoplasm (forming a syncytium). Between the eighth and eleventh rounds of nuclear division, the nuclei migrate to the outer layer of the egg, forming a monolayer of nuclei surrounding the common yolk-rich cytoplasm; this is the syncytial blastoderm. After a few additional divisions, membrane invaginations surround the nuclei to create a layer of cells that form the cellular blastoderm. At this stage, the mitotic cycles in the various cells lose their synchrony. The developmental fate of the cells is determined by the mRNAs and proteins originally deposited in the egg by the nurse and follicle cells.

Proteins that, through changes in local concentration or activity, cause the surrounding tissue to take up a particular shape or structure are sometimes referred to as **morphogens**; they are the products of pattern-

Christiane Nüsslein-Volhard

regulating genes. As defined by Christiane Nüsslein-Volhard, Edward B. Lewis, and Eric F. Wieschaus, three major classes of pattern-regulating genes—maternal, segmentation, and homeotic genes—function in successive stages of development to specify the basic features of

Edward B. Lewis, 1918–2004

Eric F. Wieschaus

the *Drosophila* embryo body. **Maternal genes** are expressed in the unfertilized egg, and the resulting **maternal mRNAs** remain dormant until fertilization. These provide most of the proteins needed in very early development, until the cellular blastoderm is formed. Some of the proteins encoded by maternal mRNAs direct the spatial organization of the developing embryo at early stages, establishing its polarity. **Segmentation genes**, transcribed after fertilization, direct the formation of the proper number of body segments. At least three subclasses of segmentation genes act at successive stages: **gap genes** divide the developing embryo into several broad regions, and **pair-rule genes** together with **segment polarity genes** define 14 stripes that become the 14 segments of a normal embryo. **Homeotic genes** are expressed still later; they specify which organs and appendages will develop in particular body segments.

The many regulatory genes in these three classes direct the development of an adult fly, with a head, thorax, and abdomen, with the proper number of segments, and with the correct appendages on each segment. Although embryogenesis takes about a day to complete, all these genes are activated during the first four hours. Some mRNAs and proteins are present for only a few minutes at specific points during this period. Some of the genes code for transcription factors that affect the expression of other genes in a kind of developmental cascade. Regulation at the level of translation also occurs, and many of the regulatory genes encode translational repressors, most of which bind to the 3′UTR of the mRNA (Fig. 28–34). Because many mRNAs are deposited in the egg long before their translation is required, translational repression provides an especially important avenue for regulation in developmental pathways.

Maternal Genes Some maternal genes are expressed within the nurse and follicle cells, and some in the egg itself. In the unfertilized *Drosophila* egg, the maternal gene products establish two axes—anterior-posterior and dorsal-ventral—and thus define which regions of the radially symmetric egg will develop into the head and abdomen and the top and bottom of the adult fly.

If all cells divided to produce two identical daughter cells, multicellular organisms would never be more than a ball of identical cells. The generation of different cell fates requires programmed asymmetric cell divisions. A key event in very early development is establishment of mRNA and protein gradients along the body axes. Some maternal mRNAs have protein products that diffuse through the cytoplasm to create an asymmetric distribution in the egg. Different cells in the cellular blastoderm therefore inherit different amounts of these proteins, setting the cells on different developmental paths. The products of the maternal mRNAs include transcription activators or repressors as well as translational repressors, all regulating the expression of other pattern-regulating genes. The resulting specific patterns and sequences of gene expression therefore differ between cell lineages, ultimately orchestrating the development of each adult structure.

The anterior-posterior axis in *Drosophila* is defined at least in part by the products of the *bicoid* and *nanos* genes. The *bicoid* gene product is a major anterior morphogen, and the *nanos* gene product is a major posterior morphogen. The mRNA from the *bicoid* gene is synthesized by nurse cells and deposited in the unfertilized egg near its anterior pole. Nüsslein-Volhard found that this mRNA is translated soon after fertilization, and the Bicoid protein diffuses through the cell to create, by the seventh nuclear division, a concentration gradient radiating out from the anterior pole (**Fig. 28–38a**). The Bicoid protein is a transcription factor that activates the expression of several segmentation genes; the protein contains a homeodomain (p. 1163). Bicoid is also a translational repressor that inactivates certain mRNAs. The amounts of Bicoid protein in various parts of the embryo affect the subsequent expression of other genes in a threshold-dependent manner. Genes are transcriptionally activated or translationally repressed only where the Bicoid protein concentration exceeds the threshold. Changes in the shape of the Bicoid concentration gradient have dramatic effects on the body pattern. Lack of Bicoid protein results in development of an embryo with two abdomens but neither head nor thorax (Fig. 28–38b); however, embryos without Bicoid will develop normally if an adequate amount of *bicoid* mRNA is injected into the egg at the appropriate end. The *nanos* gene has an analogous role, but its mRNA is deposited at the posterior end of the egg and the anterior-posterior protein gradient peaks at the posterior pole. The Nanos protein is a translational repressor.

A broader look at the effects of maternal genes reveals the outline of a developmental circuit. In addition to the *bicoid* and *nanos* mRNAs, which are deposited in the egg asymmetrically, several other maternal mRNAs are deposited uniformly throughout the egg cytoplasm. Three of these mRNAs encode the Pumilio, Hunchback, and Caudal proteins, all affected by *nanos* and *bicoid* (**Fig. 28–39**). Caudal and Pumilio are involved in development of the posterior end of the fly. Caudal is a transcription activator with a homeodomain; Pumilio is a translational repressor. Hunchback protein plays an important role in development of the anterior end and is also a transcriptional regulator of a variety of

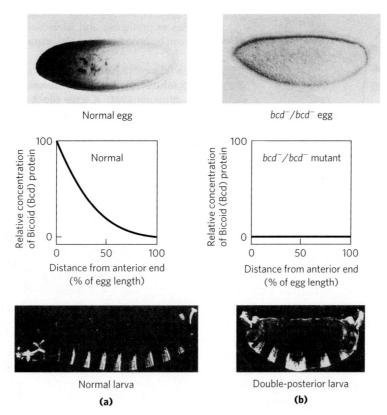

Normal egg

bcd⁻/bcd⁻ egg

Normal larva

Double-posterior larva

(a)   **(b)**

FIGURE 28–38 Distribution of a maternal gene product in a *Drosophila* egg. (a) Micrograph of an immunologically stained egg, showing distribution of the *bicoid* (*bcd*) gene product. The graph measures stain intensity. This distribution is essential for normal development of the anterior structures.

(b) If the *bcd* gene is not expressed by the mother (*bcd⁻/bcd⁻* mutant) and thus no *bicoid* mRNA is deposited in the egg, the resulting embryo has two posteriors (and soon dies).

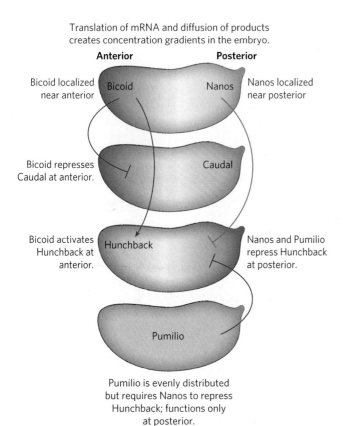

Translation of mRNA and diffusion of products creates concentration gradients in the embryo.

Anterior   **Posterior**

Bicoid localized near anterior   Bicoid   Nanos   Nanos localized near posterior

Bicoid represses Caudal at anterior.   Caudal

Bicoid activates Hunchback at anterior.   Hunchback   Nanos and Pumilio repress Hunchback at posterior.

Pumilio

Pumilio is evenly distributed but requires Nanos to repress Hunchback; functions only at posterior.

genes, in some cases a positive regulator, in other cases negative. Bicoid suppresses translation of *caudal* at the anterior end and also acts as a transcription activator of *hunchback* in the cellular blastoderm. Because *hunchback* is expressed both from maternal mRNAs and from genes in the developing egg, it is considered both a maternal and a segmentation gene. The result of the activities of Bicoid is an increased concentration of Hunchback at the anterior end of the egg. The Nanos and Pumilio proteins act as translational repressors of *hunchback*, suppressing synthesis of its protein near the posterior end of the egg. Pumilio does not function in the absence of the Nanos protein, and the gradient of Nanos expression confines the activity of both proteins to the posterior region. Translational repression of the

FIGURE 28–39 Regulatory circuits of the anterior-posterior axis in a *Drosophila* egg. The *bicoid* and *nanos* mRNAs are localized near the anterior and posterior poles, respectively. The *caudal*, *hunchback*, and *pumilio* mRNAs are distributed evenly throughout the egg cytoplasm. The gradients of Bicoid (Bcd) and Nanos proteins affect the expression of the *caudal* and *hunchback* mRNAs as shown, leading to accumulation of Hunchback protein in the anterior and Caudal protein in the posterior of the egg. Because Pumilio protein requires Nanos protein for its activity as a translational repressor of *hunchback*, it functions only at the posterior end.

hunchback gene leads to degradation of *hunchback* mRNA near the posterior end. However, lack of Bicoid in the posterior leads to expression of *caudal.* In this way, the Hunchback and Caudal proteins become asymmetrically distributed in the egg.

Segmentation Genes Gap genes, pair-rule genes, and segment polarity genes, three subclasses of segmentation genes in *Drosophila*, are activated at successive stages of embryonic development. Expression of the gap genes is generally regulated by the products of one or more maternal genes. At least some of the gap genes encode transcription factors that affect the expression of other segmentation or (later) homeotic genes.

One well-characterized segmentation gene is *fushi tarazu* (*ftz*), of the pair-rule subclass. When *ftz* is deleted, the embryo develops 7 segments instead of the normal 14, each segment twice the normal width. The Fushi-tarazu protein (Ftz) is a transcription activator with a homeodomain. The mRNAs and proteins derived from the normal *ftz* gene accumulate in a striking pattern of seven stripes that encircle the posterior two-thirds of the embryo **(Fig. 28–40)**. The stripes demarcate the positions of segments that develop later; these segments are eliminated if *ftz* function is lost. The Ftz protein and a few similar regulatory proteins directly or indirectly regulate the expression of vast numbers of genes in the continuing developmental cascade.

Homeotic Genes A set of 8 to 11 homeotic genes directs the formation of particular structures at specific locations in the body plan. These genes are now more commonly referred to as **Hox genes**, the term derived from "homeobox," the conserved gene sequence that

encodes the homeodomain and is present in all of these genes. Despite the name, these are not the only development-related proteins to include a homeodomain (for example, the *bicoid* gene product described above has a homeodomain), and "Hox" is more a functional than a structural classification. The *Hox* genes are organized in genomic clusters. *Drosophila* has one such cluster and mammals have four **(Fig. 28–41)**. The genes in these clusters are remarkably similar from nematodes to humans. In *Drosophila*, each of the *Hox* genes is

(a)

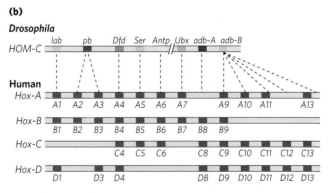

(b)

Drosophila

FIGURE 28–41 The *Hox* gene clusters and their effects on development. **(a)** Each *Hox* gene in the fruit fly is responsible for the development of structures in a defined part of the body and is expressed in defined regions of the embryo, as shown here with color coding. **(b)** *Drosophila* has one *Hox* gene cluster; the human genome has four. Many of these genes are highly conserved in multicellular animals. Evolutionary relationships, as indicated by sequence alignments, between genes in the *Drosophila Hox* gene cluster and those in the mammalian *Hox* gene clusters are shown by dashed lines. Similar relationships among the four sets of mammalian *Hox* genes are indicated by vertical alignment.

(a) Side view

(b) Cross-sectional dorsal view

ftz gene bands

ftz gene bands

100 μm

FIGURE 28–40 Distribution of the *fushi tarazu* (*ftz*) gene product in early *Drosophila* embryos. (a) Following a gene-specific staining procedure, the gene product can be detected in seven bands around the circumference of the embryo. These bands **(b)** appear as dark spots (generated by a radioactive label) in a cross-sectional autoradiograph and demarcate the anterior margins of the segments that will appear in the late embryo.

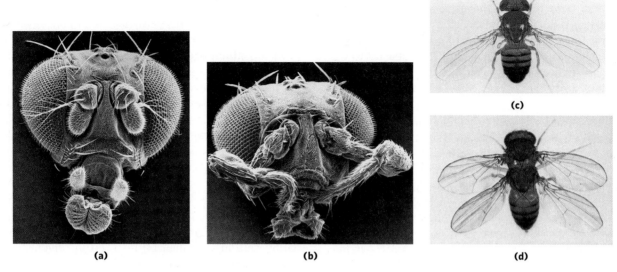

(a)   **(b)**   **(d)**

FIGURE 28–42 Effects of mutations in *Hox* genes in *Drosophila*. (a) Normal head. **(b)** Homeotic mutant (*antennapedia*) in which antennae are replaced by legs. **(c)** Normal body structure. **(d)** Homeotic mutant (*bithorax*) in which a segment has developed incorrectly to produce an extra set of wings.

expressed in a particular segment of the embryo and controls the development of the corresponding part of the mature fly. The terminology used to describe *Hox* genes can be confusing. They have historical names in the fruit fly (for example, *ultrabithorax*), whereas in mammals they are designated by two competing systems based on lettered (A, B, C, D) or numbered (1, 2, 3, 4) clusters.

Loss of *Hox* genes in fruit flies by mutation or deletion causes the appearance of a normal appendage or body structure at an inappropriate body position. An important example is the *ultrabithorax* (*ubx*) gene. When Ubx function is lost, the first abdominal segment develops incorrectly, having the structure of the third thoracic segment. Other known homeotic mutations cause the formation of an extra set of wings, or two legs at the position in the head where the antennae are normally found **(Fig. 28–42).** The *Hox* genes often span long regions of DNA. The *ubx* gene, for example, is 77,000 bp long. More than 73,000 bp of this gene are in introns, one of which is more than 50,000 bp long. Transcription of the *ubx* gene takes nearly an hour. The delay this imposes on *ubx* gene expression is believed to be a timing mechanism involved in the temporal regulation of subsequent steps in development. Many *Hox* genes are further regulated by miRNAs encoded by intergenic regions of the *Hox* gene clusters. All of the *Hox* gene products are themselves transcription factors that regulate the expression of an array of downstream genes. Identification of these downstream targets is ongoing.

Many of the principles of development outlined above apply to other eukaryotes, from nematodes to humans. Some of the regulatory proteins are conserved.

For example, the products of the homeobox-containing genes *HOXA7* in mouse and *antennapedia* in fruit fly differ in only one amino acid residue. Of course, although the molecular regulatory mechanisms may be similar, many of the ultimate developmental events are not conserved (humans do not have wings or antennae). The different outcomes are brought about by differences in the downstream target genes controlled by the *Hox* genes. The discovery of structural determinants with identifiable molecular functions is the first step in understanding the molecular events underlying development. As more genes and their protein products are discovered, the biochemical side of this vast puzzle will be elucidated in increasingly rich detail.

Stem Cells Have Developmental Potential That Can Be Controlled

If we can understand development, and the mechanisms of gene regulation behind it, we can control it. An adult human has many different types of tissues. Many of the cells are terminally differentiated and no longer divide. If an organ malfunctions due to disease or a limb is lost in an accident, the tissues are not readily replaced. Most cells, because of the regulatory processes that are in place, or even the loss of some or all genomic DNA, are not easily reprogrammed. Medical science has made organ transplants possible, but organ donors are a limited resource and organ rejection remains a major medical problem. If humans could regenerate their own organs or limbs or nervous tissue, rejection would no longer be an issue. Real cures for kidney failure or neurodegenerative disorders could become reality.

The key to tissue regeneration lies in **stem cells**—cells that have retained the capacity to differentiate into various tissues. In humans, after an egg is fertilized, the first few cell divisions create a ball of **totipotent** cells (the morula), cells that have the capacity to differentiate individually into any tissue or even into a complete organism **(Fig. 28–43)**. Continued cell

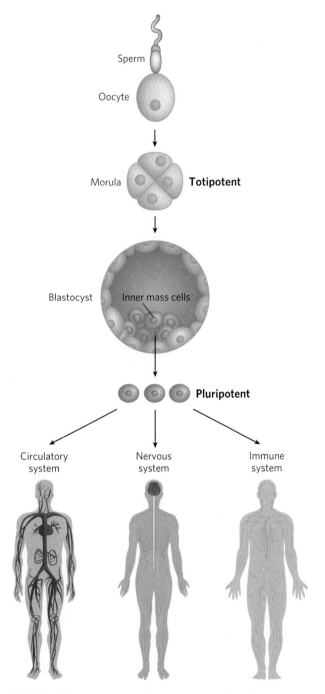

FIGURE 28–43 Totipotent and pluripotent stem cells. Cells of the morula stage are totipotent and have the capacity to differentiate into a complete organism. The source of pluripotent embryonic stem cells is the inner mass cells of the blastocyst. Pluripotent cells give rise to many tissue types but cannot form complete organisms.

division produces a hollow ball, a blastocyst. The outer cells of the blastocyst eventually form the placenta. The inner layers form the germ layers of the developing fetus—the ectoderm, mesoderm, and endoderm. These cells are **pluripotent**: they can give rise to cells of all three germ layers and can differentiate into many types of tissues. However, they cannot differentiate into a complete organism. Some of these cells are **unipotent**: they can develop into only one type of cell and/or tissue. It is the pluripotent cells of the blastocyst, the **embryonic stem cells**, that are currently used in embryonic stem cell research.

Stem cells have two functions: to replenish themselves and, at the same time, provide cells that can differentiate. These tasks are accomplished in multiple ways **(Fig. 28–44a)**. All or parts of the stem cell population can, in principle, be involved in replenishment, differentiation, or both.

Other types of stem cells can potentially be used for medical benefit. In the adult organism, **adult stem cells**, as products of additional differentiation, have a more limited potential for further development than do embryonic stem cells. For example, the hematopoietic stem cells of bone marrow can give rise to many types of blood cells and also to cells with the capacity to regenerate bone. They are referred to as **multipotent**. However, these cells cannot differentiate into a liver or kidney or neuron. Adult stem cells are often said to have a **niche**, a microenvironment that promotes stem cell maintenance while allowing differentiation of some daughter cells as replacements for cells in the tissue they serve (Fig. 28–44b). Hematopoietic stem cells in the bone marrow occupy a niche in which signaling from neighboring cells and other cues maintain the stem cell lineage. At the same time, some daughter cells differentiate to provide needed blood cells. Understanding the niche in which stem cells operate, and the signals the niche provides, is essential in efforts to harness the potential of stem cells for tissue regeneration.

All stem cells have problems with respect to human medical applications. Adult stem cells have a limited capacity to regenerate tissues, are generally present in small numbers, and are hard to isolate from an adult human. Embryonic stem cells have much greater differentiation potential and can be cultured to generate large numbers of cells. However, their use is accompanied by ethical concerns related to the necessary destruction of human embryos. Identifying a source of plentiful and medically useful stem cells that does not raise concerns remains a major goal of medical research.

Our ability to culture stem cells (i.e., maintain them in an undifferentiated state), and to manipulate them to grow and differentiate into particular tissues, is very much a function of our understanding of developmental biology. The identification and culturing of pluripotent

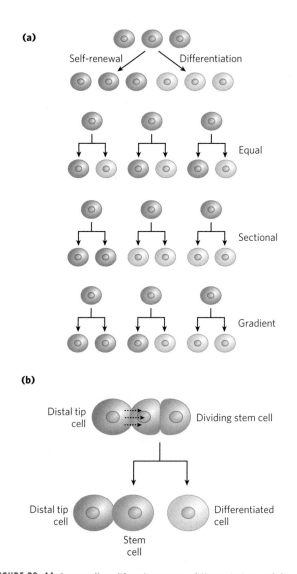

(a)

Self-renewal Differentiation

Equal

Sectional

Gradient

(b)

Distal tip cell — Dividing stem cell

Distal tip cell — Stem cell — Differentiated cell

FIGURE 28–44 Stem cell proliferation versus differentiation and development. Stem cells must strike a balance between self-renewal and differentiation. **(a)** Some possible cell division patterns that allow for replenishment of stem cells and production of some differentiated cells. Each cell may produce one stem cell and one differentiated cell, or two differentiated cells, or two stem cells in defined parts of the tissue or culture. Or a gradient of growth conditions can be established, with cell fates differing from one end of the gradient to the other. **(b)** Establishing a developmental niche through stem cell contact with a cell or group of cells. Molecular signals provided by the niche cells (in this case, for plants, a distal tip cell) help orient the mitotic spindle for stem cell division and ensure that one daughter cell retains stem cell properties.

James Thomson

stem cells from human blastocysts was reported by James Thomson and colleagues in 1998. This advance led to the long-term availability of established cell lines for research.

Thus far, mouse and human embryonic stem cells have been used for most research. Although both types

of stem cells are pluripotent, they require very different culture conditions, optimized to allow cell division indefinitely without differentiation. Mouse embryonic stem cells are grown on a layer of gelatin and require the presence of leukemia inhibitory factor (LIF). Human embryonic stem cells are grown on a feeder layer of mouse embryonic fibroblasts and require basic fibroblast growth factor (bFGF or FGF2). The use of a feeder cell layer implies that the mouse cells are providing a diffusible product or some surface signal, not yet known, that is needed by human stem cells to either promote cell division or prevent differentiation.

A significant advance, reported in 2007, centers on success in reversing differentiation. In effect, skin cells—first from mice, then from humans—have been reprogrammed to take on the characteristics of pluripotent stem cells. The reprogramming involves manipulations to get the cells to express at least four transcription factors (Oct4, Sox2, Nanog, and Lin28), all of which are known to help maintain the stem cell–like state. Gradual improvements in this technology may make the harvesting of embryonic stem cells unnecessary and provide a source of stem cells that is genetically matched to a prospective patient.

Our discussion of developmental regulation and stem cells brings us full circle, back to a biochemical beginning—both figuratively and literally. Evolution appropriately provides the first and last words of text in this book. If evolution is to generate the kind of changes in an organism that we associate with a different species, it is the developmental program that must be affected. Developmental and evolutionary processes are closely allied, each informing the other (Box 28–1). The continuing study of biochemistry has everything to do with enriching the future of humanity and understanding our origins.

SUMMARY 28.3 Regulation of Gene Expression in Eukaryotes

► In eukaryotes, positive regulation is more common than negative regulation, and transcription is accompanied by large changes in chromatin structure.

► Promoters for Pol II typically have a TATA box and Inr sequence, as well as multiple binding sites for transcription activators. The latter sites, sometimes located hundreds or thousands of base pairs away from the TATA box, are called upstream activator sequences in yeast and enhancers in higher eukaryotes.

► Large complexes of proteins are generally required to regulate transcriptional activity. The effects of transcription activators on Pol II are mediated by coactivator protein complexes such

BOX 28–1   METHODS   Of Fins, Wings, Beaks, and Things

South America has several species of seed-eating finches, commonly called grassquits. About 3 million years ago, a small group of grassquits, of a single species, took flight from the continent's Pacific coast. Perhaps driven by a storm, they lost sight of land and traveled nearly 1,000 km. Small birds such as these might easily have perished on such a journey, but the smallest of chances brought this group to a newly formed volcanic island in an archipelago later to be known as the Galápagos. It was a virgin landscape with untapped plant and insect food sources, and the newly arrived finches survived. Over the years, new islands formed and were colonized by new plants and insects—and by the finches. The birds exploited the new resources on the islands, and groups of birds gradually specialized and diverged into new species. By the time Charles Darwin stepped onto the islands in 1835, there were many different finch species to be found on the various islands of the archipelago, feeding on seeds, fruits, insects, pollen, or even blood.

The diversity of living creatures was a source of wonder for humans long before scientists sought to understand its origins. The extraordinary insight handed down to us by Darwin, inspired in part by his encounter with the Galápagos finches, provided a broad explanation for the existence of organisms with a vast array of appearances and characteristics. It also gave rise to many questions about the mechanisms underlying evolution. Answers to those questions have started to appear, first through the study of genomes and nucleic acid metabolism in the last half of the twentieth century, and more recently through an emerging field nicknamed evo-devo—a blend of evolutionary and developmental biology.

In its modern synthesis, the theory of evolution has two main elements: mutations in a population generate genetic diversity; natural selection then acts on this diversity to favor individuals with more useful genomic tools and to disfavor others. Mutations occur at significant rates in every individual's genome, in every cell (see Section 8.3 and Fig. 25–20). Advantageous mutations in single-celled organisms or in the germ line of multicellular organisms can be inherited, and they are more likely to be inherited (that is, are passed on to greater numbers of offspring) if they confer an advantage. It is a straightforward scheme. But many have wondered whether it is enough to explain, say, the many different beak shapes in the Galápagos finches or the diversity of size and shape among mammals. Until recent decades, there were several widely held assumptions about the evolutionary process: that many mutations and new genes would be needed to bring about a new physical structure, that more-complex organisms would have larger genomes, and that very different species would have few genes in common. All of these assumptions were wrong.

Modern genomics has revealed that the human genome contains fewer genes than expected—not many more than the fruit fly genome and fewer than some amphibian genomes. The genomes of every mammal, from mouse to human, are surprisingly similar in the number, types, and chromosomal arrangement of genes. Meanwhile, evo-devo is telling us how complex and very different creatures can evolve within these genomic realities.

The kinds of mutant organisms shown in Figure 28–42 were studied by the English biologist William Bateson in the late nineteenth century. Bateson used his observations to challenge the Darwinian notion that evolutionary change would have to be gradual. Recent studies of the genes that control organismal development have put an exclamation point on Bateson's ideas. Subtle changes in regulatory patterns during development, reflecting just one or a few mutations, can result in startling physical changes and fuel surprisingly rapid evolution.

The Galápagos finches provide a wonderful example of the link between evolution and development. There are at least 14 (some specialists list 15) species of Galápagos finches, distinguished in large measure by their beak structure. The ground finches, for example, have broad, heavy beaks adapted to crushing large, hard seeds. The cactus finches have longer, slender beaks ideal for probing cactus fruits

---

as Mediator. The modular structures of the activators have distinct activation and DNA-binding domains. Other protein complexes, including histone acetyltransferases and ATP-dependent complexes such as SWI/SNF and NURF, reversibly remodel and modify chromatin structure.

▶ Hormones affect the regulation of gene expression in one of two ways. Steroid hormones interact directly with intracellular receptors that are DNA-binding regulatory proteins; binding of the hormone has either positive or negative effects on the transcription of genes targeted by the hormone. Nonsteroid hormones bind to cell surface receptors, triggering a signaling pathway that can lead to phosphorylation of a regulatory protein, affecting its activity.

and flowers (Fig. 1). Clifford Tabin and colleagues carefully surveyed a set of genes expressed during avian craniofacial development. They identified a single gene, *Bmp4*, whose expression level correlated with formation of the more robust beaks of the ground finches. More robust beaks were also formed in chicken embryos when high levels of Bmp4 were artificially expressed in the appropriate tissues, confirming the importance of *Bmp4*. In a similar study, the formation of long, slender beaks was linked to the expression of calmodulin (see Fig. 12–11) in particular tissues at appropriate developmental stages. Thus, major changes in the shape and function of the beak can be brought about by subtle changes in the expression of just two genes involved in developmental regulation. Very few mutations are required, and the needed mutations affect regulation. New genes are *not* required.

The system of regulatory genes that guides development is remarkably conserved among all vertebrates. Elevated expression of *Bmp4* in the right tissue at the right time leads to more robust jaw parts in zebrafish. The same gene plays a key role in tooth development in mammals. The development of eyes is triggered by the expression of a single gene, *Pax6*, in fruit flies and in mammals. The mouse *Pax6* gene will trigger the development of fruit fly eyes in the fruit fly, and the fruit fly *Pax6* gene will trigger the development of mouse eyes in the mouse. In each organism, these genes are part of the much larger regulatory cascade that ultimately creates the correct structures in the correct locations in each organism. The cascade is ancient; for example, the *Hox* genes (described in the text) have been part of the developmental program of multicellular eukaryotes for more than 500 million years. Subtle changes in the cascade can have large effects on development, and thus on the ultimate appearance, of the organism. These same subtle changes can fuel remarkably rapid evolution. For example, the 400 to 500 described species of cichlids (spiny-finned fish) in Lake Malawi and Lake Victoria on the African continent are all derived from one or a few populations that colonized each lake in the past 100,000 to 200,000 years. The Galápagos finches simply followed a path of evolution and change that living creatures have been traveling for billions of years.

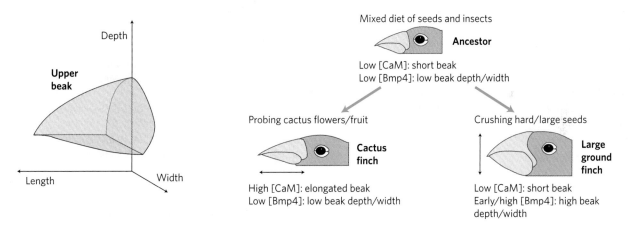

FIGURE 1 Evolution of new beak structures to exploit new food sources. In the Galápagos finches, the different beak structures of the cactus finch and the large ground finch, which feed on different, specialized food sources, were produced to a large extent by a few mutations that altered the timing and level of expression of just two genes: those encoding calmodulin (CaM) and Bmp4.

- ▶ RNA-mediated regulation plays an important role in eukaryotic gene expression, with the range of known mechanisms expanding.

- ▶ Development of a multicellular organism presents the most complex regulatory challenge. The fate of cells in the early embryo is determined by establishment of anterior-posterior and dorsal-ventral gradients of proteins that act as transcription activators or translational repressors, regulating the genes required for the development of structures appropriate to a particular part of the organism. Sets of regulatory genes operate in temporal and spatial succession, transforming given areas of an egg cell into predictable structures in the adult organism.

- ▶ The differentiation of stem cells into functional tissues can be controlled by extracellular signals and conditions.

Key Terms

Terms in bold are defined in the glossary

Further Reading

General

Carroll, S.B. (2005) *Endless Forms Most Beautiful: The New Science of Evo Devo and the Making of the Animal Kingdom*, W. W. Norton & Company, New York.

A fascinating look at how developmental biology informs evolutionary biology.

Neidhardt, F.C. (ed.). Escherichia coli *and* Salmonella typhimurium: *Cellular and Molecular Biology* (Curtiss, R., Ingraham, J.L., Lin, E.C.C., Magasanik, B., Low, K.B., Reznikoff, W.S., Riley, M., Schaechter, M., & Umbarger, H.E., vol. eds), American Society for Microbiology, Washington, DC.

A good source for information on bacterial gene regulation and many other topics. Last published in a print edition in 1996, this is now an important and continuously updated online resource at http://ecosal.org.

Regulation of Gene Expression in Bacteria

Babitske, P., Baker, C.S., & Romeo, T. (2009) Regulation of translation initiation by RNA binding proteins. *Annu. Rev. Microbiol.* **63**, 27–44.

Jacob, F. & Monod, J. (1961) Genetic regulatory mechanisms in the synthesis of proteins. *J. Mol. Biol.* **3**, 318–356.

The operon model and the concept of messenger RNA, first proposed in the *Proceedings* of the French Academy of Sciences in 1960, are presented in this historic paper.

Osterberg, S., del Peso-Santos, T., & Shingler, V. (2011) Regulation of alternative sigma factor use. *Annu. Rev. Microbiol.* **65**, 37–55.

Roth, A. & Breaker, R.R. (2009) The structural and functional diversity of metabolite-binding riboswitches. *Annu. Rev. Biochem.* **78**, 305–334.

Regulation of Gene Expression in Eukaryotes

Berezikov, E. (2011) Evolution of microRNA diversity and regulation in animals. *Nat. Rev. Genet.* **12**, 846–860.

Conaway, R.C. & Conaway, J.W. (2011) Function and regulation of the Mediator complex. *Curr. Opin. Genet. Dev.* **21**, 225–230.

Fabian, M.R., Sonenberg, N., & Filipowicz, W. (2010) Regulation of mRNA translation and stability by microRNAs. *Annu. Rev. Biochem.* **79**, 351–379.

Groppo, R. & Richter, J.D. (2009) Translational control from head to tail. *Curr. Opin. Cell Biol.* **21**, 444–451.

Keene, J.D. (2007) RNA regulons: coordination of post-transcriptional events. *Nat. Rev. Genet.* **8**, 533–543.

Krol, J., Loedige, I., & Filipowicz, W. (2010) The widespread regulation of microRNA biogenesis, function and decay. *Nat. Rev. Genet.* **11**, 597–610.

Pasquinelli, A.E. (2012) MicroRNAs and their targets: recognition, regulation and an emerging reciprocal relationship. *Nat. Rev. Genet.* **13**, 271–282.

Prud'homme, B., Gompel, N., & Carroll, S.B. (2007) Emerging principles of regulatory evolution. *Proc. Natl. Acad. Sci. USA* **104**, 8605–8612.

Rinn, J.L. & Chang, H.Y. (2012) Genome regulation by long noncoding RNAs. *Annu. Rev. Biochem.* **81**, 145–166.

Shahbazian, M.D. & Grunstein, M. (2007) Functions of site-specific histone acetylation and deacetylation. *Annu. Rev. Biochem.* **76**, 75–100.

Struhl, K. (1999) Fundamentally different logic of gene regulation in eukaryotes and prokaryotes. *Cell* **98**, 1–4.

Talbert, P.B. & Henikoff, S. (2010) Histone variants—ancient wrap artists of the epigenome. *Nat. Rev. Mol. Cell Biol.* **11**, 264–275.

Werner, F. & Grohmann, D. (2011) Evolution of multisubunit RNA polymerases in the three domains of life. *Nat. Rev. Microbiol.* **9**, 85–98.

Zhou, Q., Li, T., & Price, D.H. (2012) RNA polymerase II elongation control. *Annu. Rev. Biochem.* **81**, 119–143.

Problems

1. Effect of mRNA and Protein Stability on Regulation
E. coli cells are growing in a medium with glucose as the sole carbon source. Tryptophan is suddenly added. The cells continue to grow, and divide every 30 min. Describe (qualitatively) how the amount of tryptophan synthase activity in the cells changes with time under the following conditions:

(a) The *trp* mRNA is stable (degraded slowly over many hours).

(b) The *trp* mRNA is degraded rapidly, but tryptophan synthase is stable.

(c) The *trp* mRNA and tryptophan synthase are both degraded rapidly.

2. The Lactose Operon A researcher engineers a *lac* operon on a plasmid but inactivates all parts of the *lac* operator (*lacO*) and the *lac* promoter, replacing them with the binding site for the LexA repressor (which acts in the SOS response) and a promoter regulated by LexA. The plasmid is introduced into *E. coli* cells that have a *lac* operon with an inactive *lacZ* gene. Under what conditions will these transformed cells produce β-galactosidase?

3. Negative Regulation Describe the probable effects on gene expression in the *lac* operon of a mutation in (a) the *lac* operator that deletes most of O_1, (b) the *lacI* gene that inactivates the repressor, and (c) the promoter that alters the region around position −10.

4. Specific DNA Binding by Regulatory Proteins A typical bacterial repressor protein discriminates between its specific DNA-binding site (operator) and nonspecific DNA by a factor of 10^4 to 10^6. About 10 molecules of repressor per cell are sufficient to ensure a high level of repression. Assume that a very similar repressor existed in a human cell, with a similar specificity for its binding site. How many copies of the repressor would be required to elicit a level of repression similar to that in the bacterial cell? (Hint: The *E. coli* genome contains about 4.6 million bp; the human haploid genome has about 3.2 billion bp.)

5. Repressor Concentration in *E. coli* The dissociation constant for a particular repressor-operator complex is very low, about 10^{-13} M. An *E. coli* cell (volume 2×10^{-12} mL) contains 10 copies of the repressor. Calculate the cellular concentration of the repressor protein. How does this value compare with the dissociation constant of the repressor-operator complex? What is the significance of this answer?

6. Catabolite Repression *E. coli* cells are growing in a medium containing lactose but no glucose. Indicate whether each of the following changes or conditions would increase, decrease, or not change the expression of the *lac* operon. It may be helpful to draw a model depicting what is happening in each situation.

(a) Addition of a high concentration of glucose

(b) A mutation that prevents dissociation of the Lac repressor from the operator

(c) A mutation that completely inactivates β-galactosidase

(d) A mutation that completely inactivates galactoside permease

(e) A mutation that prevents binding of CRP to its binding site near the *lac* promoter

7. Transcription Attenuation How would transcription of the *E. coli trp* operon be affected by the following manipulations of the leader region of the *trp* mRNA?

(a) Increasing the distance (number of bases) between the leader peptide gene and sequence 2

(b) Increasing the distance between sequences 2 and 3

(c) Removing sequence 4

(d) Changing the two Trp codons in the leader peptide gene to His codons

(e) Eliminating the ribosome-binding site for the gene that encodes the leader peptide

(f) Changing several nucleotides in sequence 3 so that it can base-pair with sequence 4 but not with sequence 2

8. Repressors and Repression How would the SOS response in *E. coli* be affected by a mutation in the *lexA* gene that prevented autocatalytic cleavage of the LexA protein?

9. Regulation by Recombination In the phase variation system of *Salmonella*, what would happen to the cell if the Hin recombinase became more active and promoted recombination (DNA inversion) several times in each cell generation?

10. Initiation of Transcription in Eukaryotes A new RNA polymerase activity is discovered in crude extracts of cells derived from an exotic fungus. The RNA polymerase initiates transcription only from a single, highly specialized promoter. As the polymerase is purified its activity declines, and the purified enzyme is completely inactive unless crude extract is added to the reaction mixture. Suggest an explanation for these observations.

11. Functional Domains in Regulatory Proteins A biochemist replaces the DNA-binding domain of the yeast Gal4 protein with the DNA-binding domain from the Lac repressor and finds that the engineered protein no longer regulates transcription of the *GAL* genes in yeast. Draw a diagram of the different functional domains you would expect to find in the Gal4 protein and in the engineered protein. Why does the engineered protein no longer regulate transcription of the *GAL* genes? What might be done to the DNA-binding site recognized by this chimeric protein to make it functional in activating transcription of *GAL* genes?

12. Nucleosome Modification during Transcriptional Activation To prepare genomic regions for transcription, certain histones in the resident nucleosomes are acetylated and methylated at specific locations. Once transcription is no longer needed, these modifications need to be reversed. In mammals, the methylation of Arg residues in histones is reversed by peptidylarginine deiminases (PADIs). The reaction promoted by these enzymes does not yield unmethylated arginine. Instead, it produces citrulline residues in the histone. What is the other product of the reaction? Suggest a mechanism for this reaction.

13. Inheritance Mechanisms in Development A *Drosophila* egg that is bcd^-/bcd^- may develop normally, but the adult fruit fly will not be able to produce viable offspring. Explain.

Data Analysis Problem

14. Engineering a Genetic Toggle Switch in *Escherichia coli* Gene regulation is often described as an "on or off" phenomenon—a gene is either fully expressed or not expressed at all. In fact, repression and activation of a gene involve ligand-binding reactions, so genes can show intermediate levels of expression when intermediate levels of regulatory molecules are present. For example, for the *E. coli lac* operon, consider the binding equilibrium of the Lac repressor, operator

DNA, and inducer (see Fig. 28–8). Although this is a complex, cooperative process, it can be approximately modeled by the following reaction (R is repressor; IPTG is the inducer isopropyl-β-D-thiogalactoside):

$$K_d = 10^{-4}\,\text{M}$$
$$R + IPTG \rightleftharpoons R \cdot IPTG$$

Free repressor, R, binds to the operator and prevents transcription of the *lac* operon; the R · IPTG complex does not bind to the operator and thus transcription of the *lac* operon can proceed.

(a) Using Equation 5–8, we can calculate the relative expression level of the proteins of the *lac* operon as a function of [IPTG]. Use this calculation to determine over what range of [IPTG] the expression level would vary from 10% to 90%.

(b) Describe qualitatively the level of *lac* operon proteins present in an *E. coli* cell before, during, and after induction with IPTG. You need not give the amounts at exact times—just indicate the general trends.

Gardner, Cantor, and Collins (2000) set out to make a "genetic toggle switch"—a gene-regulatory system with two key characteristics of a light switch. (A) *It has only two states:* it is either fully on or fully off; it is not a dimmer switch. In biochemical terms, the target gene or gene system (operon) is either fully expressed or not expressed at all; it cannot be expressed at an intermediate level. (B) *Both states are stable:* although you must use a finger to flip the light switch from one state to the other, once you have flipped it and removed your finger, the switch stays in that state. In biochemical terms, exposure to an inducer or some other signal changes the expression state of the gene or operon, and it remains in that state once the signal is removed.

(c) Explain how the *lac* operon lacks both characteristics A and B.

To make their "toggle switch," Gardner and coworkers constructed a plasmid from the following components:

OP_{lac} The operator-promoter region of the *E. coli lac* operon

OP_λ The operator-promoter region of λ phage

lacI   The gene encoding the *lac* repressor protein, LacI. In the absence of IPTG, this protein strongly represses OP_{lac}; in the presence of IPTG, it allows full expression from OP_{lac}.

rep^{ts} The gene encoding a temperature-sensitive mutant λ repressor protein, rep^{ts}. At 37 °C this protein strongly represses OP_λ; at 42 °C it allows full expression from OP_λ.

GFP   The gene for green fluorescent protein (GFP), a highly fluorescent reporter protein (see Fig. 9–16)

T   Transcription terminator

The investigators arranged these components (see figure below) so that the two promoters were reciprocally repressed: OP_{lac} controlled expression of rep^{ts}, and OP_λ controlled expression of *lacI*. The state of this system was reported by the expression level of *GFP*, which was also under the control of OP_{lac}.

(d) The constructed system has two states: GFP-on (high level of expression) and GFP-off (low level of expression). For each state, describe which proteins are present and which promoters are being expressed.

(e) Treatment with IPTG would be expected to toggle the system from one state to the other. From which state to which? Explain your reasoning.

(f) Treatment with heat (42 °C) would be expected to toggle the system from one state to the other. From which state to which? Explain your reasoning.

(g) Why would this plasmid be expected to have characteristics A and B as described above?

To confirm that their construct did indeed exhibit these characteristics, Gardner and colleagues first showed that, once switched, the GFP expression level (high or low) was stable for long periods of time (characteristic B). Next, they measured GFP level at different concentrations of the inducer IPTG, with the following results.

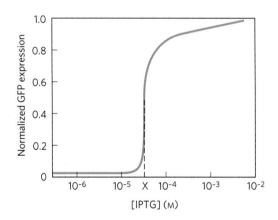

They noticed that the average GFP expression level was intermediate at concentration X of IPTG. However, when they measured the GFP expression level *in individual cells* at [IPTG] = X, they found either a high level or a low level of GFP—no cells showed an intermediate level.

(h) Explain how this finding demonstrates that the system has characteristic A. What is happening to cause the bimodal distribution of expression levels at [IPTG] = X?

Reference

Gardner, T.S., Cantor, C.R., & Collins, J.J. (2000) Construction of a genetic toggle switch in *Escherichia coli*. *Nature* **403**, 339–342.

Abbreviated Solutions to Problems

Fuller solutions to all chapter problems are published in the *Absolute Ultimate Study Guide to Accompany Principles of Biochemistry*. For all numerical problems, answers are expressed with the correct number of significant figures.

Chapter 1

1. **(a)** Diameter of magnified cell = 500 mm **(b)** 2.7×10^{12} actin molecules **(c)** 36,000 mitochondria **(d)** 3.9×10^{10} glucose molecules **(e)** 50 glucose molecules per hexokinase molecule

2. **(a)** 1×10^{-12} g = 1 pg **(b)** 10% **(c)** 5%

3. **(a)** 1.6 mm; 800 times longer than the cell; DNA must be tightly coiled. **(b)** 4,000 proteins

4. **(a)** Metabolic rate is limited by diffusion, which is limited by surface area. **(b)** 12 μm^{-1} for the bacterium; 0.04 μm^{-1} for the amoeba; surface-to-volume ratio 300 times higher in the bacterium.

5. 2×10^6 s (about 23 days)

6. The vitamin molecules from the two sources are identical; the body cannot distinguish the source; only associated impurities might vary with the source.

7.

8.

The two enantiomers have different interactions with a chiral biological "receptor" (a protein).

9. **(a)** Only the amino acids have amino groups; separation could be based on the charge or binding affinity of these groups. Fatty acids are less soluble in water than amino acids, and the two types of molecules also differ in size and shape—either of these property differences could be the basis for separation. **(b)** Glucose is a smaller molecule than a nucleotide; separation could be based on size. The nitrogenous base and/or the phosphate group also endows nucleotides with characteristics (solubility, charge) that could be used for separation from glucose.

10. It is improbable that silicon could serve as the central organizing element for life, especially in an O_2-containing atmosphere such as that of Earth. Long chains of silicon atoms are not readily synthesized; the polymeric macromolecules necessary for more complex functions would not readily form. Oxygen disrupts bonds between silicon atoms, and silicon–oxygen bonds are extremely stable and difficult to break, preventing the breaking and making of bonds that is essential to life processes.

11. Only one enantiomer of the drug was physiologically active. Dexedrine consisted of the single enantiomer; Benzedrine consisted of a racemic mixture.

12. **(a)** 3 Phosphoric acid groups; α-D-ribose; guanine **(b)** Choline; phosphoric acid; glycerol; oleic acid; palmitic acid **(c)** Tyrosine; 2 glycines; phenylalanine; methionine

13. **(a)** CH_2O; $C_3H_6O_3$

(b)

(c) X contains a chiral center; eliminates all but **6** and **8**. **(d)** X contains an acidic functional group; eliminates **8**; structure **6** is consistent with all data. **(e)** Structure **6**; we cannot distinguish between the two possible enantiomers.

14. The compound shown is (R)-propranolol; the carbon bearing the hydroxyl group is the chiral carbon. (S)-Propranolol has the structure:

15. The compound shown is (S,S)-methylphenidate. (R,R)-methylphenidate has the structure:

The chiral carbons are indicated with asterisks.

16. (a) A more negative $\Delta G°$ corresponds to a larger K_{eq} for the binding reaction, so the equilibrium is shifted more toward products and tighter binding—and thus greater sweetness and higher MRS. **(b)** Animal-based sweetness assays are time-consuming. A computer program to predict sweetness, even if not always completely accurate, would allow chemists to design effective sweeteners much faster. Candidate molecules could then be tested in the conventional assay. **(c)** The range 0.25 to 0.4 nm corresponds to about 1.5 to 2.5 single-bond lengths. The figure below can be used to construct an approximate ruler; any atoms in the light red rectangle are between 0.25 and 0.4 nm from the origin of the ruler.

There are many possible AH-B groups in the molecules; a few are shown here.

Deoxysucrose

Sucrose

D-Tryptophan   Saccharin   Aspartame

6-Chloro-D-tryptophan   Alitame

Neotame

Tetrabromosucrose   Sucronic acid

(d) First, each molecule has multiple AH-B groups, so it is difficult to know which is the important one. Second, because the AH-B motif is very simple, many nonsweet molecules will have this group. **(e)** Sucrose and deoxysucrose. Deoxysucrose lacks one of the AH-B groups present in sucrose and has a slightly lower MRS than sucrose—as is expected if the AH-B groups are important for sweetness. **(f)** There are many such examples; here are a few: (1) D-Tryptophan and 6-chloro-D-tryptophan have the same AH-B group but very different MRS values. (2) Aspartame and neotame have the same AH-B groups but very different MRS values. (3) Neotame has two AH-B groups and alitame has three, yet neotame is more than five times sweeter than alitame. (4) Bromine is less electronegative than oxygen and thus is expected to weaken an AH-B group, yet tetrabromosucrose is much sweeter than sucrose. **(g)** Given enough "tweaking" of parameters, any model can be made to fit a defined dataset. Because the objective was to create a model to predict $\Delta G°$ for molecules not tested in vivo, the researchers needed to show that the model worked well for molecules it had not been trained on. The degree of inaccuracy with the test set could give researchers an idea of how the model would behave for novel molecules. **(h)** MRS is related to K_{eq}, which is related exponentially to $\Delta G°$, so adding a constant amount to $\Delta G°$ multiplies the MRS by a constant amount. Based on the values given with the structures, a change in $\Delta G°$ of 1.3 kcal/mol corresponds to a 10-fold change in MRS.

Chapter 2

1. Ethanol is polar; ethane is not. The ethanol —OH group can hydrogen-bond with water.
2. (a) 4.76 **(b)** 9.19 **(c)** 4.0 **(d)** 4.82
3. (a) 1.51×10^{-4} M **(b)** 3.02×10^{-7} M **(c)** 7.76×10^{-12} M
4. 1.1
5. (a) HCl $\rightleftharpoons$ H$^+$ + Cl$^-$ **(b)** 3.3 **(c)** NaOH $\rightleftharpoons$ Na$^+$ + OH$^-$ **(d)** 9.8
6. 1.1
7. 1.7×10^{-9} mol of acetylcholine
8. 0.1 M HCl
9. (a) greater **(b)** higher **(c)** lower
10. 3.3 mL
11. (a) RCOO$^-$ **(b)** RNH$_2$ **(c)** H$_2$PO$_4^-$ **(d)** HCO$_3^-$
12. (a) 5.06 **(b)** 4.28 **(c)** 5.46 **(d)** 4.76 **(e)** 3.76
13. (a) 0.1 M HCl **(b)** 0.1 M NaOH **(c)** 0.1 M NaOH
14. (d) Bicarbonate, a weak base, titrates —OH to —O$^-$, making the compound more polar and more water-soluble.
15. Stomach; the neutral form of aspirin present at the lower pH is less polar and passes through the membrane more easily.
16. 9
17. 7.4
18. (a) pH 8.6 to 10.6 **(b)** 4/5 **(c)** 10 mL **(d)** pH = pK_a − 2
19. 8.9
20. 2.4
21. 6.9
22. 1.4
23. NaH$_2$PO$_4$ · H$_2$O, 5.8 g/L; Na$_2$HPO$_4$, 8.2 g/L
24. [A$^-$]/[HA] = 0.10

25. Mix 150 mL of 0.10 M sodium acetate and 850 mL of 0.10 M acetic acid.

26. Acetic acid; its pK_a is closest to the desired pH.

27. (a) 4.6 **(b)** 0.1 pH unit **(c)** 4 pH units

28. 4.3

29. 0.13 M acetate and 0.07 M acetic acid

30. 1.7

31. 7

32. (a)

Fully protonated	Fully deprotonated

(b) fully protonated **(c)** zwitterion **(d)** zwitterion **(e)** fully deprotonated

33. (a) Blood pH is controlled by the carbon dioxide–bicarbonate buffer system, $CO_2 + H_2O \rightleftharpoons H^+ + HCO_3^-$. During *hypoventilation*, $[CO_2]$ increases in the lungs and arterial blood, driving the equilibrium to the right, raising $[H^+]$ and lowering pH. **(b)** During *hyperventilation*, $[CO_2]$ decreases in the lungs and arterial blood, reducing $[H^+]$ and increasing pH above the normal 7.4 value. **(c)** Lactate is a moderately strong acid, completely dissociating under physiological conditions and thus lowering the pH of blood and muscle tissue. Hyperventilation removes H^+, raising the pH of blood and tissues in anticipation of the acid buildup.

34. 7.4

35. Dissolving more CO_2 in the blood increases $[H^+]$ in blood and extracellular fluids, lowering pH: $CO_2(d) + H_2O \rightleftharpoons H_2CO_3 \rightleftharpoons H^+ + HCO_3^-$

36. (a) Use the substance in its surfactant form to emulsify the spilled oil, collect the emulsified oil, then switch to the nonsurfactant form. The oil and water will separate and the oil can be collected for further use. **(b)** The equilibrium lies strongly to the right. The stronger acid (lower pK_a), H_2CO_3, donates a proton to the conjugate base of the weaker acid (higher pK_a), amidine. **(c)** The strength of a surfactant depends on the hydrophilicity of its head groups: the more hydrophilic, the more powerful the surfactant. The amidinium form of s-surf is much more hydrophilic than the amidine form, so it is a more powerful surfactant. **(d)** *Point A:* amidinium; the CO_2 has had plenty of time to react with the amidine to produce the amidinium form. *Point B:* amidine; Ar has removed CO_2 from the solution, leaving the amidine form. **(e)** The conductivity rises as uncharged amidine reacts with CO_2 to produce the charged amidinium form. **(f)** The conductivity falls as Ar removes CO_2, shifting the equilibrium to the uncharged amidine form. **(g)** Treat s-surf with CO_2 to produce the surfactant amidinium form and use this to emulsify the spill. Treat the emulsion with Ar to remove the CO_2 and produce the nonsurfactant amidine form. The oil will separate from the water and can be recovered.

Chapter 3

1. L; determine the absolute configuration at the α carbon and compare it with D- and L-glyceraldehyde.

2. (a) I **(b)** II **(c)** IV **(d)** II **(e)** IV **(f)** II and IV **(g)** III **(h)** III **(i)** V **(j)** III **(k)** V **(l)** II **(m)** III **(n)** V **(o)** I, III, and V

3. (a) pI > pK_a of the α-carboxyl group and pI < pK_a of the α-amino group, so both groups are charged (ionized). **(b)** 1 in 2.19×10^7. The pI of alanine is 6.01. From Table 3–1 and the Henderson-Hasselbalch equation, 1/4,680 carboxyl groups and 1/4,680 amino groups are uncharged. The fraction of alanine molecules with both groups uncharged is the product of these fractions.

4. (a)–(c)

pH	Structure	Net charge	Migrates toward
1	1	+2	Cathode
4	2	+1	Cathode
8	3	0	Does not migrate
12	4	−1	Anode

5. (a) Asp **(b)** Met **(c)** Glu **(d)** Gly **(e)** Ser

6. (a) 2 **(b)** 4 **(c)**

7. (a) Structure at pH 7:

(b) Electrostatic interaction between the carboxylate anion and the protonated amino group of the alanine zwitterion favorably affects the ionization of the carboxyl group. This favorable electrostatic interaction decreases as the length of the poly(Ala) increases, resulting in an increase in pK_1. **(c)** Ionization of the protonated amino group destroys the favorable electrostatic interaction noted in (b). With increasing distance between the charged groups, removal of the proton from the amino group in poly(Ala) becomes easier and thus pK_2 is lower. The intramolecular effects of the amide (peptide bond) linkages keep pK_a values lower than they would be for an alkyl-substituted amine.

8. 75,000

9. (a) 32,000. The elements of water are lost when a peptide bond forms, so the molecular weight of a Trp residue is not the same as the molecular weight of free tryptophan. **(b)** 2

10. The protein has four subunits, with molecular masses of 160, 90, 90, and 60 kDa. The two 90 kDa subunits (possibly identical) are linked by one or more disulfide bonds.

11. **(a)** at pH 3, +2; at pH 8, 0; at pH 11, −1 **(b)** pI = 7.8

12. pI ≈ 1; carboxylate groups; Asp and Glu

13. Lys, His, Arg; negatively charged phosphate groups in DNA interact with positively charged side groups in histones.

14. **(a)** (Glu)$_{20}$ **(b)** (Lys–Ala)$_3$ **(c)** (Asn–Ser–His)$_5$ **(d)** (Asn–Ser–His)$_5$

15. **(a)** Specific activity after step 1 is 200 units/mg; step 2, 600 units/mg; step 3, 250 units/mg; step 4, 4,000 units/mg; step 5, 15,000 units/mg; step 6, 15,000 units/mg **(b)** Step 4 **(c)** Step 3 **(d)** Yes. Specific activity did not increase in step 6; SDS polyacrylamide gel electrophoresis

16. **(a)** [NaCl] = 0.5 mM **(b)** [NaCl] = 0.05 mM.

17. C elutes first, B second, A last.

18. Tyr–Gly–Gly–Phe–Leu

19.
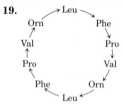

The arrows correspond to the orientation of the peptide bonds, —CO → NH—.

20. 88%, 97%. The percentage (x) of correct amino acid residues released in cycle n is x_n/x. All residues released in the first cycle are correct, even though the efficiency of cleavage is not perfect.

21. **(a)** Y (1), F (7), and R (9) **(b)** Positions 4 and 9; K (Lys) is more common at 4, R (Arg) is invariant at 9 **(c)** Positions 5 and 10; E (Glu) is more common at both positions **(d)** Position 2; S (Ser)

22. **(a)** peptide 2 **(b)** peptide 1 **(c)** peptide 2 **(d)** peptide 3

23. **(a)** Any linear polypeptide chain has only two kinds of free amino groups: a single α-amino group at the amino terminus, and an ε-amino group on each Lys residue present. These amino groups react with FDNB to form a DNP–amino acid derivative. Insulin gave two different α-amino-DNP derivatives, suggesting that it has two amino termini and thus two polypeptide chains—one with an amino-terminal Gly and the other with an amino-terminal Phe. Because the DNP-lysine product is ε-DNP-lysine, the Lys is not at an amino terminus. **(b)** Yes. The A chain has amino-terminal Gly; the B chain has amino-terminal Phe; and (nonterminal) residue 29 in the B chain is Lys. **(c)** Phe–Val–Asp–Glu–. Peptide B1 shows that the amino-terminal residue is Phe. Peptide B2 also includes Val, but since no DNP-Val is formed, Val is not at the amino terminus; it must be on the carboxyl side of Phe. Thus the sequence of B2 is DNP-Phe-Val. Similarly, the sequence of B3 must be DNP-Phe-Val–Asp, and the sequence of the A chain must begin Phe–Val–Asp–Glu–. **(d)** No. The known amino-terminal sequence of the A chain is Phe–Val–Asn–Gln–. The Asn and Gln appear in Sanger's analysis as Asp and Glu because the vigorous hydrolysis in step 7 hydrolyzed the amide bonds in Asn and Gln (as well as the peptide bonds), forming Asp and Glu. Sanger et al. could not distinguish Asp from Asn or Glu from Gln at this stage in their analysis. **(e)** The sequence exactly matches that in Fig. 3–24. Each peptide in the table gives specific information about which Asx residues are Asn or Asp and which Glx residues are Glu or Gln.

Ac1: residues 20–21. This is the only Cys–Asx sequence in the A chain; there is ~1 amido group in this peptide, so it must be Cys–Asn:

N–Gly–Ile–Val–Glx–Glx–Cys–Cys–Ala–Ser–Val–
  1   5   10

Cys–Ser–Leu–Tyr–Glx–Leu–Glx–Asx–Tyr–Cys–**Asn**–C
        15   20

Ap15: residues 14–15–16. This is the only Tyr–Glx–Leu in the A chain; there is ~1 amido group, so the peptide must be Tyr–Gln–Leu:

N–Gly–Ile–Val–Glx–Glx–Cys–Cys–Ala–Ser–Val–
  1   5   10

Cys–Ser–Leu–Tyr–**Gln**–Leu–Glx–Asx–Tyr–Cys–Asn–C
        15   20

Ap14: residues 14–15–16–17. It has ~1 amido group, and we already know that residue 15 is Gln, so residue 17 must be Glu:

N–Gly–Ile–Val–Glx–Glx–Cys–Cys–Ala–Ser–Val–
  1   5   10

Cys–Ser–Leu–Tyr–Gln–Leu–**Glu**–Asx–Tyr–Cys–Asn–C
        15   20

Ap3: residues 18–19–20–21. It has ~2 amido groups, and we know that residue 21 is Asn, so residue 18 must be Asn:

N–Gly–Ile–Val–Glx–Glx–Cys–Cys–Ala–Ser–Val–
  1   5   10

Cys–Ser–Leu–Tyr–Gln–Leu–Glu–**Asn**–Tyr–Cys–Asn–C
        15   20

Ap1: residues 17–18–19–20–21, which is consistent with residues 18 and 21 being Asn.

Ap5pa1: residues 1–2–3–4. It has ~0 amido group, so residue 4 must be Glu:

N–Gly–Ile–Val–**Glu**–Glx–Cys–Cys–Ala–Ser–Val–
  1   5   10

Cys–Ser–Leu–Tyr–Gln–Leu–Glu–Asn–Tyr–Cys–Asn–C
        15   20

Ap5: residues 1 through 13. It has ~1 amido group, and we know that residue 4 is Glu, so residue 5 must be Gln:

N–Gly–Ile–Val–Glu–**Gln**–Cys–Cys–Ala–Ser–Val–
  1   5   10

Cys–Ser–Leu–Tyr–Gln–Leu–Glu–Asn–Tyr–Cys–Asn–C
        15   20

Chapter 4

1. **(a)** Shorter bonds have a higher bond order (are multiple rather than single) and are stronger. The peptide C—N bond is stronger than a single bond and is midway between a single and a double bond in character. **(b)** Rotation about the peptide bond is difficult at physiological temperatures because of its partial double-bond character.

2. **(a)** The principal structural units in the wool fiber polypeptide (α-keratin) are successive turns of the α helix, at 5.4 Å intervals; coiled coils produce the 5.2 Å spacing. Steaming and stretching the fiber yields an extended polypeptide chain with the β conformation, with a distance between adjacent R groups of about 7.0 Å. As the polypeptide reassumes an α-helical structure, the fiber shortens. **(b)** Processed wool shrinks when polypeptide chains are converted from an extended β conformation to the native α-helical conformation in the presence of moist heat. The structure of silk—β sheets, with their small, closely packed amino acid side chains—is more stable than that of wool.

3. ~42 peptide bonds per second

4. At pH > 6, the carboxyl groups of poly(Glu) are deprotonated; repulsion among negatively charged carboxylate groups leads to unfolding. Similarly, at pH 7, the amino groups of poly(Lys) are protonated; repulsion among these positively charged groups also leads to unfolding.

5. **(a)** Disulfide bonds are covalent bonds, which are much stronger than the noncovalent interactions that stabilize most proteins. They cross-link protein chains, increasing their stiffness, mechanical strength, and hardness. **(b)** Cystine residues (disulfide bonds) prevent the complete unfolding of the protein.

6. $\phi =$ (f) and $\psi =$ (e).

7. (a) Bends are most likely at residues 7 and 19; Pro residues in the cis configuration accommodate turns well. **(b)** The Cys residues at positions 13 and 24 can form disulfide bonds. **(c)** External surface: polar and charged residues (Asp, Gln, Lys); interior: nonpolar and aliphatic residues (Ala, Ile); Thr, though polar, has a hydropathy index near zero and thus can be found either on the external surface or in the interior of the protein.

8. 30 amino acid residues; 0.87

9. Myoglobin is all three. The folded structure, the "globin fold," is a motif found in all globins. The polypeptide folds into a single domain, which for this protein represents the entire three-dimensional structure.

10. Protein (a), a β barrel, is described by Ramachandran plot (c), which shows most of the allowable conformations in the upper left quadrant where the bond angles characteristic of the β conformation are concentrated; (b), a series of α helices, is described by plot (d), where most of the allowable conformations are in the lower left quadrant.

11. The bacterial enzyme is a collagenase; it destroys the connective tissue barrier of the host, allowing the bacterium to invade the tissues. Bacteria do not contain collagen.

12. (a) The number of moles of DNP-valine formed per mole of protein equals the number of amino termini and thus the number of polypeptide chains. **(b)** 4 **(c)** Different chains would probably run as discrete bands on an SDS polyacrylamide gel.

13. (a); it has more amino acid residues that favor α-helical structure (see Table 4–1).

14. (a) Aromatic residues seem to play an important role in stabilizing amyloid fibrils. Thus, molecules with aromatic substituents may inhibit amyloid formation by interfering with the stacking or association of the aromatic side chains. **(b)** Amyloid is formed in the pancreas in association with type 2 diabetes, as it is in the brain in Alzheimer disease. Although the amyloid fibrils in the two diseases involve different proteins, the fundamental structure of the amyloid is similar and similarly stabilized in both, and so they are potential targets for similar drugs designed to disrupt this structure.

15. (a) NFκB transcription factor, also called RelA transforming factor. **(b)** No. You will obtain similar results, but with additional related proteins listed. **(c)** The protein has two subunits. There are multiple variants of the subunits, with the best characterized being 50, 52, or 65 kDa. These pair with each other to form a variety of homodimers and heterodimers. The structures of a number of different variants can be found in the PDB. **(d)** The NFκB transcription factor is a dimeric protein that binds specific DNA sequences, enhancing transcription of nearby genes. One such gene is the immunoglobulin κ light chain, from which the transcription factor gets its name.

16. (a) Aba is a suitable replacement because Aba and Cys have side chains of approximately the same size and are similarly hydrophobic. However, Aba cannot form disulfide bonds, so it will not be a suitable replacement if these are required. **(b)** There are many important differences between the synthesized protein and HIV protease produced by a human cell, any of which could result in an inactive synthetic enzyme: (1) Although Aba and Cys have similar size and hydrophobicity, Aba may not be similar enough for the protein to fold properly. (2) HIV protease may require disulfide bonds for proper functioning. (3) Many proteins synthesized by ribosomes fold as they are produced; the protein in this study folded only after the chain was complete. (4) Proteins synthesized by ribosomes may interact with the ribosomes as they fold; this is not possible for the protein in the study. (5) Cytosol is a more complex solution than the buffer used in the study; some proteins may require specific, unknown proteins for proper folding. (6) Proteins synthesized in cells often require chaperones for proper folding; these are not present in the study buffer. (7) In cells, HIV protease is synthesized as part of a larger chain that is then proteolytically processed; the protein in the study was synthesized as a single molecule. **(c)** Because the enzyme *is* functional with Aba substituted for Cys, disulfide bonds do not play an important role in the structure of HIV protease. **(d)** *Model 1:* It would fold like the L-protease. *Argument for:* The covalent structure is the same (except for chirality), so it should fold like the L-protease. *Argument against:* Chirality is not a trivial detail; three-dimensional shape is a key feature of biological molecules. The synthetic enzyme will not fold like the L-protease. *Model 2:* It would fold to the mirror image of the L-protease. *For:* Because the individual components are mirror images of those in the biological protein, it will fold in the mirror-image shape. *Against:* The interactions involved in protein folding are very complex, so the synthetic protein will most likely fold in another form. *Model 3:* It would fold to something else. *For:* The interactions involved in protein folding are very complex, so the synthetic protein will most likely fold in another form. *Against:* Because the individual components are mirror images of those in the biological protein, it will fold in the mirror-image shape. **(e)** *Model 1*. The enzyme is active, but with the enantiomeric form of the biological substrate, and it is inhibited by the enantiomeric form of the biological inhibitor. This is consistent with the D-protease being the mirror image of the L-protease. **(f)** Evans blue is achiral; it binds to both forms of the enzyme. **(g)** No. Because proteases contain only L-amino acids and recognize only L-peptides, chymotrypsin would not digest the D-protease. **(h)** Not necessarily. Depending on the individual enzyme, any of the problems listed in (b) could result in an inactive enzyme.

Chapter 5

1. Protein B has a higher affinity for ligand X; it will be half-saturated at a much lower concentration of X than will protein A. Protein A has $K_a = 10^6 \ \text{M}^{-1}$; protein B has $K_a = 10^9 \ \text{M}^{-1}$.

2. (a), (b), and **(c)** all have $n_H < 1.0$. Apparent negative cooperativity in ligand binding can be caused by the presence of two or more types of ligand-binding sites with different affinities for the ligand on the same or different proteins in the same solution. Apparent negative cooperativity is also commonly observed in heterogeneous protein preparations. There are few well-documented cases of true negative cooperativity.

3. (a) decreases **(b)** increases **(c)** decreases **(d)** increases

4. $k_d = 8.9 \times 10^{-5} \ \text{s}^{-1}$.

5. (a) 0.5 nM (shortcut: the K_d is equivalent to the ligand concentration where $\theta = 0.5$). **(b)** Protein 2 has the highest affinity, as it has the lowest K_d.

6. The cooperative behavior of hemoglobin arises from subunit interactions.

7. (a) The observation that hemoglobin A (HbA; maternal) is about 60% saturated when the pO_2 is 4 kPa, whereas hemoglobin F (HbF; fetal) is more than 90% saturated under the same physiological conditions, indicates that HbF has a higher O_2 affinity than HbA. **(b)** The higher O_2 affinity of HbF ensures that oxygen will flow from maternal blood to fetal blood in the placenta. Fetal blood approaches full saturation where the O_2 affinity of HbA is low. **(c)** The observation that the O_2-saturation curve of HbA undergoes a larger shift on BPG binding than that of HbF suggests that HbA binds BPG more tightly than does HbF. Differential binding of BPG to the two hemoglobins may determine the difference in their O_2 affinities.

8. (a) Hb Memphis **(b)** HbS, Hb Milwaukee, Hb Providence, possibly Hb Cowtown **(c)** Hb Providence

9. More tightly. An inability to form tetramers would limit the cooperativity of these variants, and the binding curve would become more hyperbolic. Also, the BPG-binding site would be disrupted. Oxygen binding would probably be tighter, because the default state in the absence of bound BPG is the tight-binding R state.

10. **(a)** 1×10^{-8} M **(b)** 5×10^{-8} M **(c)** 8×10^{-8} M **(d)** 2×10^{-7} M. Note that a rearrangement of Eqn 5–8 gives $[L] = \theta K_d/(1 - \theta)$.

11. The epitope is likely to be a structure that is buried when G-actin polymerizes to F-actin.

12. Many pathogens, including HIV, have mechanisms by which they can repeatedly alter the surface proteins to which immune system components initially bind. Thus the host organism regularly faces new antigens and requires time to mount an immune response to each one. As the immune system responds to one variant, new variants are created.

13. Binding of ATP to myosin triggers dissociation of myosin from the actin thin filament. In the absence of ATP, actin and myosin bind tightly to each other.

14.

   (a)   **(b)**   **(c)**   **(d)**

15. **(a)** Chain L is the light chain and chain H is the heavy chain of the Fab fragment of this antibody molecule. Chain Y is lysozyme. **(b)** β structures are predominant in the variable and constant regions of the fragment. **(c)** Fab heavy-chain fragment, 218 amino acid residues; light-chain fragment, 214; lysozyme, 129. Less than 15% of the lysozyme molecule is in contact with the Fab fragment. **(d)** In the H chain, residues that seem to be in contact with lysozyme include Gly^{31}, Tyr^{32}, Arg^{99}, Asp^{100}, and Tyr^{101}. In the L chain the residues that seem to be in contact with lysozyme include Tyr^{32}, Tyr^{49}, Tyr^{50}, and Trp^{92}. In lysozyme, residues Asn^{19}, Gly^{22}, Tyr^{23}, Ser^{24}, Lys^{116}, Gly^{117}, Thr^{118}, Asp^{119}, Gln^{121}, and Arg^{125} seem to be situated at the antigen-antibody interface. Not all these residues are adjacent in the primary structure. Folding of the polypeptide chain into higher levels of structure brings the nonconsecutive residues together to form the antigen-binding site.

16. **(a)** The protein with a K_d of 5 μM will have the highest affinity for ligand L. When the K_d is 10 μM, doubling [L] from 0.2 to 0.4 μM (values well below the K_d) will nearly double θ (the actual increase factor is 1.96). This is a property of the hyperbolic curve; at low ligand concentrations, θ is an almost linear function of [L]. On the other hand, doubling [L] from 40 to 80 μM (well above the K_d, where the binding curve is approaching its asymptotic limit) will increase θ by a factor of only 1.1. The increase factors are identical for the curves generated from Eqn 5–11. **(b)** $\theta = 0.998$. **(c)** A variety of answers will be obtained, depending on the values entered for the different parameters.

17. **(a)**

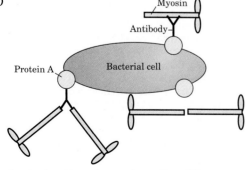

Myosin
Antibody
Protein A
Bacterial cell

The drawing is not to scale; any given cell would have many more myosin molecules on its surface. **(b)** ATP is needed to provide the chemical energy to drive the motion (see Chapter 13). **(c)** An antibody that bound to the myosin tail, the actin-binding site,

would block actin binding and prevent movement. An antibody that bound to actin would also prevent actin-myosin interaction and thus movement. **(d)** There are two possible explanations: (1) Trypsin cleaves only at Lys and Arg residues (see Table 3–6) so would not cleave at many sites in the protein. (2) Not all Arg or Lys residues are equally accessible to trypsin; the most-exposed sites would be cleaved first. **(e)** The S1 model. The hinge model predicts that bead-antibody-HMM complexes (with the hinge) would move, but bead-antibody-SHMM complexes (no hinge) would not. The S1 model predicts that because both complexes include S1, both would move. The finding that the beads move with SHMM (no hinge) is consistent only with the S1 model. **(f)** With fewer myosin molecules bound, the beads could temporarily fall off the actin as a myosin let go of it. The beads would then move more slowly, as time is required for a second myosin to bind. At higher myosin density, as one myosin lets go, another quickly binds, leading to faster motion. **(g)** Above a certain density, what limits the rate of movement is the intrinsic speed with which myosin molecules move the beads. The myosin molecules are moving at a maximum rate and adding more will not increase speed. **(h)** Because the force is produced in the S1 head, damaging the S1 head would probably inactivate the resulting molecule, and SHMM would be incapable of producing movement. **(i)** The S1 head must be held together by noncovalent interactions that are strong enough to retain the active shape of the molecule.

Chapter 6

1. The activity of the enzyme that converts sugar to starch is destroyed by heat denaturation.

2. 2.4×10^{-6} M

3. 9.5×10^8 years

4. The enzyme-substrate complex is more stable than the enzyme alone.

5. **(a)** 190 Å **(b)** Three-dimensional folding of the enzyme brings the amino acid residues into proximity.

6. The reaction rate can be measured by following the decrease in absorption by NADH (at 340 nm) as the reaction proceeds. Determine the K_m value; using substrate concentrations well above the K_m, measure initial rate (rate of NADH disappearance with time, measured spectrophotometrically) at several known enzyme concentrations, and make a plot of initial rate versus concentration of enzyme. The plot should be linear, with a slope that provides a measure of LDH concentration.

7. (b), (e), (g)

8. **(a)** 1.7×10^{-3} M **(b)** 0.33; 0.67; 0.91 **(c)** The upper curve corresponds to enzyme B ([X] > K_m for this enzyme); the lower curve, enzyme A.

9. **(a)** 400 s^{-1} **(b)** 10 μM **(c)** $\alpha = 2$, $\alpha' = 3$ **(d)** Mixed inhibitor

10. **(a)** 24 nM **(b)** 4 μM (V_0 is exactly half V_{max}, so [A] = K_m) **(c)** 40 μM (V_0 is exactly half V_{max}, so [A] = 10 times K_m in the presence of inhibitor) **(d)** No. $k_{cat}/K_m = (0.33\ s^{-1})/(4 \times 10^{-6}$ M) $= 8.25 \times 10^4\ M^{-1}\ s^{-1}$, well below the diffusion-controlled limit.

11. $V_{max} \approx 140\ \mu$M/min; $K_m \approx 1 \times 10^{-5}$ M

12. **(a)** $V_{max} = 51.5$ mM/min; $K_m = 0.59$ mM **(b)** Competitive inhibition

13. $K_m = 2.2$ mM; $V_{max} = 0.50\ \mu$mol/min

14. Curve A

15. $k_{cat} = 2.0 \times 10^7\ min^{-1}$

16. The basic assumptions of the Michaelis-Menten equation still hold. The reaction is at steady state, and the rate is determined by $V_0 = k_2$ [ES]. The equations needed to solve for [ES] are

$$[E_t] = [E] + [ES] + [EI] \quad \text{and} \quad [EI] = \frac{[E][I]}{K_I}$$

[E] can be obtained by rearranging Eqn 6–19. The rest follows the pattern of the Michaelis-Menten equation derivation in the text.

17. Minimum $M_r = 29,000$

18. Activity of the prostate enzyme equals total phosphatase activity in a blood sample minus phosphatase activity in the presence of enough tartrate to completely inhibit the prostate enzyme.

19. The inhibition is mixed. Because K_m seems not to change appreciably, this could be the special case of mixed inhibition called noncompetitive.

20. The [S] at which $V_0 = V_{max}/2\alpha'$ is obtained when all terms except V_{max} on the right side of Eqn 6–30—that is, $[S]/(\alpha K_m + \alpha'[S])$—equal $\frac{1}{2}\alpha'$. Begin with $[S]/(\alpha K_m + \alpha'[S]) = \frac{1}{2}\alpha'$ and solve for [S].

21. The optimum activity occurs when Glu^{35} is protonated and Asp^{52} is unprotonated.

22. **(a)** Increase factor = 1.96; $V_0 = 50\ \mu M\ s^{-1}$; increase factor = 1.048.
(b) When $\alpha = 2.0$, the curve is shifted to the right as the K_m is increased by a factor of 2. When $\alpha' = 3.0$, the asymptote of the curve (the V_{max}) declines by a factor of 3. When $\alpha = 2.0$ and $\alpha' = 3.0$, the curve briefly rises above the curve where both α and $\alpha' = 1.0$, due to a decline in K_m. However, the asymptote is lower, because V_{max} declines by a factor of 3. **(c)** When $\alpha = 2.0$, the x intercept moves to the right. When $\alpha = 2.0$ and $\alpha' = 3.0$, the x intercept moves to the left.

23. **(a)** In the wild-type enzyme, the substrate is held in place by a hydrogen bond and an ion-dipole interaction between the charged side chain of Arg^{109} and the polar carbonyl of pyruvate. During catalysis, the charged Arg^{109} side chain also stabilizes the polarized carbonyl transition state. In the mutant, the binding is reduced to just a hydrogen bond, substrate binding is weaker, and ionic stabilization of the transition state is lost, reducing catalytic activity. **(b)** Because Lys and Arg are roughly the same size and have a similar positive charge, they probably have very similar properties. Furthermore, because pyruvate binds to Arg^{171} by (presumably) an ionic interaction, an Arg to Lys mutation would probably have little effect on substrate binding. **(c)** The "forked" arrangement aligns two positively charged groups of Arg residues with the negatively charged oxygens of pyruvate and facilitates two combined hydrogen-bond and ion-dipole interactions. When Lys is present, only one such combined hydrogen-bond and ion-dipole interaction is possible, thus reducing the strength of the interaction. The positioning of the substrate is less precise. **(d)** Ile^{250} interacts hydrophobically with the ring of NADH. This type of interaction is not possible with the hydrophilic side chain of Gln. **(e)** The structure is shown below. **(f)** The mutant enzyme rejects pyruvate because pyruvate's hydrophobic methyl group will not interact with the highly hydrophilic guanidinium group of Arg^{102}. The mutant binds oxaloacetate because of the strong ionic interaction between the Arg^{102} side chain and the carboxyl of oxaloacetate. **(g)** The protein must be flexible enough to accommodate the added bulk of the side chain and the larger substrate.

Chapter 7

1. With reduction of the carbonyl oxygen to a hydroxyl group, the chemistry at C-1 and C-3 is the same; the glycerol molecule is not chiral.

2. Epimers differ by the configuration about only *one* carbon.
(a) D-altrose (C-2), D-glucose (C-3), D-gulose (C-4)
(b) D-idose (C-2), D-galactose (C-3), D-allose (C-4)
(c) D-arabinose (C-2), D-xylose (C-3)

3. Osazone formation destroys the configuration around C-2 of aldoses, so aldoses differing only at the C-2 configuration give the same derivative, with the same melting point.

4. To convert α-D-glucose to β-D-glucose, the bond between C-1 and the hydroxyl on C-5 (as in Fig. 7–6). To convert D-glucose to D-mannose, either the —H or the —OH on C-2. Conversion between chair conformations does not require bond breakage; this is the critical distinction between configuration and conformation.

5. No; glucose and galactose differ at C-4.

6. **(a)** Both are polymers of D-glucose, but they differ in the glycosidic linkage: ($\beta1\rightarrow4$) for cellulose, ($\alpha1\rightarrow4$) for glycogen. **(b)** Both are hexoses, but glucose is an aldohexose, fructose a ketohexose. **(c)** Both are disaccharides, but maltose has two ($\alpha1\rightarrow4$)-linked D-glucose units; sucrose has ($\alpha1\leftrightarrow2\beta$)-linked D-glucose and D-fructose.

7.

8. A hemiacetal is formed when an aldose or ketose condenses with an alcohol; a glycoside is formed when a hemiacetal condenses with an alcohol (see Fig. 7–5).

9. Fructose cyclizes to either the pyranose or the furanose structure. Increasing the temperature shifts the equilibrium in the direction of the furanose, the less sweet form.

10. The rate of mutarotation is sufficiently high that, as the enzyme consumes β-D-glucose, more α-D-glucose is converted to the β form and, eventually, all the glucose is oxidized. Glucose oxidase is specific for glucose and does not detect other reducing sugars (such as galactose) that react with Fehling's reagent.

11. **(a)** Measure the change in optical rotation with time. **(b)** The optical rotation of the mixture is negative (inverted) relative to that of the sucrose solution. **(c)** $-2.0°$

12. Prepare a slurry of sucrose and water for the core; add a small amount of sucrase (invertase); immediately coat with chocolate.

13. Sucrose has no free anomeric carbon to undergo mutarotation.

14.

Yes; yes

15. N-Acetyl-β-D-glucosamine is a reducing sugar; its C-1 can be oxidized (see p. 252). D-Gluconate is not a reducing sugar; its C-1 is already at the oxidation state of a carboxylic acid. GlcN($\alpha1\leftrightarrow1\alpha$)Glc is not a reducing sugar; the anomeric carbons of both monosaccharides are involved in the glycosidic bond.

16. Humans lack cellulase in the gut and cannot break down cellulose.

17. Native cellulose consists of glucose units linked by ($\beta1\rightarrow4$) glycosidic bonds, which force the polymer chain into an extended conformation. Parallel series of these extended chains form intermolecular hydrogen bonds, aggregating into long, tough, insoluble fibers. Glycogen consists of glucose units linked by ($\alpha1\rightarrow4$) glycosidic bonds, which cause bends in the chain and prevent formation of long fibers. In addition, glycogen is highly branched and, because many of its hydroxyl groups are exposed to water, is highly hydrated and disperses in water.

 Cellulose is a structural material in plants, consistent with its side-by-side aggregation into insoluble fibers. Glycogen is a storage fuel in animals. Highly hydrated glycogen granules with their many nonreducing ends are rapidly hydrolyzed by glycogen phosphorylase to release glucose 1-phosphate.

18. Cellulose is several times longer; it assumes an extended conformation, whereas amylose has a helical structure.

19. 6,000 residues/s

20. 11 s

21. The ball-and-stick model of the disaccharide in Fig. 7–18b shows no steric interactions, but a space-filling model, showing atoms with their real relative sizes, would show several strong steric hindrances in the $-170°$, $-170°$ conformer that are not present in the $30°$, $-40°$ conformer.

22. The negative charges on chondroitin sulfate repel each other and force the molecule into an extended conformation. The polar molecule attracts many water molecules, increasing the molecular volume. In the dehydrated solid, each negative charge is counterbalanced by a positive ion, and the molecule condenses.

23. Positively charged amino acid residues would bind the highly negatively charged groups on heparin. In fact, Lys residues of antithrombin III interact with heparin.

24. 8 possible sequences, 144 possible linkages, and 64 stereochemical possibilities, for a total of 73,728 permutations!

25.

Electrophoresis

Oligosaccharide chains:
- 0
- 1
- 2
- 3

⊕

26. Oligosaccharides; their subunits can be combined in more ways than the amino acid subunits of oligopeptides. Each hydroxyl group can participate in glycosidic bonds, and the configuration of each glycosidic bond can be either α or β. The polymer can be linear or branched.

27. **(a)** Branch-point residues yield 2,3-di-*O*-methylglucose; the unbranched residues yield 2,3,6-tri-*O*-methylglucose. **(b)** 3.75%

28. Chains of ($1\rightarrow6$)-linked D-glucose residues with occasional ($1\rightarrow3$)-linked branches, with about one branch every 20 residues

29. **(a)** The tests involve trying to dissolve only part of the sample in a variety of solvents, then analyzing both dissolved and undissolved materials to see whether their compositions differ. **(b)** For a pure substance, all molecules are the same and any dissolved fraction will have the same composition as any undissolved fraction. An impure substance is a mixture of more than one compound. When treated with a particular solvent, more of one component may dissolve, leaving more of the other component(s) behind. As a result, the dissolved and undissolved fractions have different compositions. **(c)** A quantitative assay allows researchers to be sure that none of the activity has been lost through degradation. When determining the structure of a molecule, it is important that the sample under analysis consist only of intact (undegraded) molecules. If the sample is contaminated with degraded material,

this will give confusing and perhaps uninterpretable structural results. A qualitative assay would detect the presence of activity, even if it had become significantly degraded. **(d)** Results 1 and 2. Result 1 is consistent with the known structure, because type B antigen has three molecules of galactose; types A and O each have only two. Result 2 is also consistent, because type A has two amino sugars (*N*-acetylgalactosamine and *N*-acetylglucosamine); types B and O have only one (*N*-acetylglucosamine). Result 3 is *not* consistent with the known structure: for type A, the glucosamine: galactosamine ratio is 1:1; for type B, it is 1:0. **(e)** The samples were probably impure and/or partly degraded. The first two results were correct possibly because the method was only roughly quantitative and thus not as sensitive to inaccuracies in measurement. The third result is more quantitative and thus more likely to differ from predicted values because of impure or degraded samples. **(f)** An exoglycosidase. If it were an endoglycosidase, one of the products of its action on O antigen would include galactose, *N*-acetylglucosamine, or *N*-acetylgalactosamine, and at least one of those sugars would be able to inhibit the degradation. Given that the enzyme is not inhibited by any of these sugars, it must be an exoglycosidase, removing only the terminal sugar from the chain. The terminal sugar of O antigen is fucose, so fucose is the only sugar that could inhibit the degradation of O antigen. **(g)** The exoglycosidase removes *N*-acetylgalactosamine from A antigen and galactose from B antigen. Because fucose is not a product of either reaction, it will not prevent removal of these sugars, and the resulting substances will no longer be active as A or B antigen. However, the products should be active as O antigen, because degradation stops at fucose. **(h)** All the results are consistent with Fig. 10–15. (1) D-Fucose and L-galactose, which would protect against degradation, are not present in any of the antigens. (2) The terminal sugar of A antigen is *N*-acetylgalactosamine, and this sugar alone protects this antigen from degradation. (3) The terminal sugar of B antigen is galactose, which is the only sugar capable of protecting this antigen.

Chapter 8

1. N-3 and N-7

2. (5′)GCGCAATATTTTGAGAAATATTGCGC(3′); it contains a palindrome. The individual strands can form hairpin structures; the two strands can form a cruciform.

3. 9.4×10^{-4} g

4. **(a)** 40° **(b)** 0°

5. The RNA helix is in the A conformation; the DNA helix is generally in the B conformation.

6. In eukaryotic DNA, about 5% of C residues are methylated. 5-Methylcytosine can spontaneously deaminate to form thymine; the resulting G–T pair is one of the most common mismatches in eukaryotic cells.

7. Higher

8. Without the base, the ribose ring can be opened to generate the noncyclic aldehyde form. This, and the loss of base-stacking interactions, could contribute significant flexibility to the DNA backbone.

9. CGCGCGTGCGCGCGCG

10. Base stacking in nucleic acids tends to reduce the absorption of UV light. Denaturation involves loss of base stacking, and UV absorption increases.

11. 0.35 mg/mL

12.

Deoxyribose   Guanine   Phosphate

Solubilities: phosphate > deoxyribose > guanine. The highly polar phosphate groups and sugar moieties are on the outside of the double helix, exposed to water; the hydrophobic bases are in the interior of the helix.

13. If dCTP is omitted, when the first G residue is encountered in the template, ddCTP will be added, and polymerization will halt. Only one band will be seen in the sequencing gel.

14.

15. (5′)P—GCGCCAUUGC(3′)—OH
(5′)P—GCGCCAUUG(3′)—OH
(5′)P—GCGCCAUU(3′)—OH
(5′)P—GCGCCAU(3′)—OH
(5′)P—GCGCCA(3′)—OH
(5′)P—GCGCC(3′)—OH
(5′)P—GCGC(3′)—OH
(5′)P—GCG(3′)—OH
(5′)P—GC(3′)—OH

and the nucleoside 5′-phosphates

16. (a) Water is a participant in most biological reactions, including those that cause mutations. The low water content in endospores reduces the activity of mutation-causing enzymes and slows the rate of nonenzymatic depurination reactions, which are hydrolysis reactions. **(b)** UV light induces formation of cyclobutane pyrimidine dimers. Because *B. subtilis* is a soil organism, spores can be lofted to the top of the soil or into the air, where they may be subject to prolonged UV exposure.

17. DMT is a blocking group that prevents reaction of the incoming base with itself.

18. (a) Right-handed. The base at one 5′ end is adenine; at the other 5′ end, cytidine. **(b)** Left-handed **(c)** If you cannot see the structures in stereo, see additional tips in the expanded solutions manual, or use a search engine to find tips online.

19. (a) It would not be easy! The data for different samples from the same organism show significant variation, and the recovery is never 100%. The numbers for C and T show much more consistency than those for A and G, so for C and T it is much easier to make the case that samples from the same organism have the same composition. But even with the less consistent values for A and G, (1) the range of values for different tissues does overlap substantially; (2) the difference between different preparations of the same tissue is about the same as the difference between samples from different tissues; and (3) in samples for which recovery is high, the numbers are more consistent. **(b)** This technique would not be sensitive enough to detect a difference between normal and cancerous cells. Cancer is caused by mutations, but these changes in DNA—a few base pairs out of several billion—would be too small to detect with these techniques. **(c)** The ratios of A:G and T:C vary widely among different species. For example, in the bacterium *Serratia marcescens*, both ratios are 0.4, meaning that the DNA contains mostly G and C. In *Haemophilus influenzae*, by contrast, the ratios are 1.74 and 1.54, meaning that the DNA is mostly A and T. **(d)** Conclusion 4 has three requirements. A = T: The table shows an A:T ratio very close to 1 in all cases. Certainly, the variation in this ratio is substantially less than the variation in

the A:G and T:C ratios. G = C: Again, the G:C ratio is very close to 1, and the other ratios vary widely. (A + G) = (T + C): This is the purine:pyrimidine ratio, which also is very close to 1. **(e)** The different "core" fractions represent different regions of the wheat germ DNA. If the DNA were a monotonous repeating sequence, the base composition of all regions would be the same. Because different core regions have different sequences, the DNA sequence must be more complex.

Chapter 9

1. (a) (5′) – – – G(3′)   and (5′)AATTC – – – (3′)
(3′) – – – CTTAA(5′)   (3′)G – – – (5′)

(b) (5′) – – – GAATT(3′)   and   (5′)AATTC – – – (3′)
(3′) – – – CTTAA(5′)   (3′)TTAAG – – – (5′)

(c) (5′) – – – GAATTAATTC – – – (3′)
(3′) – – – CTTAATTAAG – – – (5′)

(d) (5′) – – – G(3′)   and   (5′)C – – – (3′)
(3′) – – – C(5′)   (3′)G – – – (5′)

(e) (5′) – – – GAATTC – – – (3′)
(3′) – – – CTTAAG – – – (5′)

(f) (5′) – – – CAG(3′)   and   (5′)CTG – – – (3′)
(3′) – – – GTC(5′)   (3′)GAC – – – (5′)

(g) (5′) – – – CAGAATTC – – – (3′)
(3′) – – – GTCTTAAG – – – (5′)

(h) Method 1: Cut the DNA with *Eco*RI as in (a). At this point, one could treat the DNA as in (b) or (d), then ligate a synthetic DNA fragment with the *Bam*HI recognition sequence between the two resulting blunt ends. Method 2 (more efficient): Synthesize a DNA fragment with the structure

(5′)AATTGGATCC(3′)

(3′)CCTAGGTTAA(5′)

This would ligate efficiently to the sticky ends generated by *Eco*RI cleavage, would introduce a *Bam*HI site, but would not regenerate the *Eco*RI site. **(i)** The four fragments (with N = any nucleotide), in order of discussion in the problem, are

(5′)AATTCNNNNCTGCA(3′)
(3′)GNNNNG(5′)
(5′)AATTCNNNNGTGCA(3′)
(3′)GNNNNC(5′)
(5′)AATTGNNNNCTGCA(3′)
(3′)CNNNNG(5′)
(5′)AATTGNNNNGTGCA(3′)
(3′)CNNNNC(5′)

2. λ phage DNA can be packaged into infectious phage particles only if it is between 40,000 and 53,000 bp in length. Since bacteriophage vectors generally include about 30,000 bp (in two pieces), they will not be packaged into phage particles unless they contain a sufficient length of inserted DNA (10,000 to 23,000 bp).

3. (a) Plasmids in which the original pBR322 was regenerated without insertion of a foreign DNA fragment; these would retain resistance to ampicillin. Also, two or more molecules of pBR322 might be ligated together with or without insertion of foreign DNA. **(b)** The clones in lanes 1 and 2 each have one DNA fragment inserted in different orientations. The clone in lane 3 has two DNA fragments, ligated such that the *Eco*RI proximal ends are joined.

4.   (5′)GAAAGTCCGCGTTATAGGCATG(3′)
(3′)ACGTCTTTCAGGCGCAATATCCGTACTTAA(5′)

5. Your test would require DNA primers, a heat-stable DNA polymerase, deoxynucleoside triphosphates, and a PCR machine (thermal cycler). The primers would be designed to amplify a DNA segment encompassing the CAG repeat. The DNA strand shown is the coding strand, oriented 5′→3′ left to right. The primer targeted to DNA to the left of the repeat would be identical to any 25-nucleotide sequence shown in the region to the left of the CAG repeat. The primer on the right side must be *complementary* and *antiparallel* to a 25-nucleotide sequence to the right of the CAG repeat. Using the primers, DNA including the CAG repeat would be amplified by PCR, and its size would be determined by comparison to size markers after electrophoresis. The length of the DNA would reflect the length of the CAG repeat, providing a simple test for the disease.

6. Design PCR primers that are complementary to the DNA in the deleted segment but that would direct DNA synthesis away from each other. No PCR product will be generated unless the ends of the deleted segment are joined to create a circle.

7. The plant expressing firefly luciferase must take up luciferin, the substrate of luciferase, before it can "glow" (albeit weakly). The plant expressing green fluorescent protein glows without requiring any other compound.

8. Primer 1: CCTCGAGTCAATCGATGCTG
Primer 2: CGCGCACATCAGACGAACCA
Recall that all DNA sequences are always written in the 5′ to 3′ direction, left to right; that the two strands of a DNA molecule are antiparallel; and that both PCR primers must target the end sequences so that their 3′ ends are oriented toward the segment to be amplified. In a lab, writing a sequence in the wrong orientation on an order form when ordering a synthetic oligonucleotide primer can be a very expensive mistake.

9.

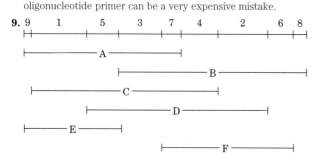

10. The production of labeled antibodies is difficult and expensive. The labeling of every antibody to every protein target would be impractical. By labeling one antibody preparation for binding to all antibodies of a particular class, the same labeled antibody preparation can be used in many different immunofluorescence experiments.

11. Express the protein in yeast strain 1 as a fusion protein with one of the domains of Gal4p—say, the DNA-binding domain. Using yeast strain 2, make a library in which essentially every protein of the fungus is expressed as a fusion protein with the interaction domain of Gal4p. Mate strain 1 with the strain 2 library, and look for colonies that are colored due to expression of the reporter gene. These colonies will generally arise from mated cells containing a fusion protein that interacts with your target protein.

12. Cover spot 4, add solution containing activated T, irradiate, wash.

  1. A–T   2. G–T   3. A–T   4. G–C

Cover spots 2 and 4, add solution containing activated G, irradiate, wash.

  1. A–T–G   2. G–T   3. A–T–G   4. G–C

Cover spot 3, add solution containing activated C, irradiate, wash.

  1. A–T–G–C   2. G–T–C   3. A–T–G   4. G–C–C

Cover spots 1, 3, and 4, add solution containing activated C, irradiate, wash.

  1. A–T–G–C   2. G–T–C–C   3. A–T–G   4. G–C–C

Cover spots 1 and 2, add solution containing activated G, irradiate, wash.

  1. A–T–G–C   2. G–T–C–C   3. A–T–G–C   4. G–C–C–C

13. The primers can be used to probe libraries containing long genomic clones to identify contig ends that lie close to each other. If the contigs flanking the gap are close enough, the primers can be used in PCR to directly amplify the intervening DNA separating the contigs, which can then be cloned and sequenced.

14. ATSAAG**W**DEWEGGK**V**LIHL**DG**KLQNRGALLELDIGAV

15. The same disease condition can be caused by defects in two or more genes, which are on different chromosomes.

16. **(a)** DNA solutions are highly viscous because the very long molecules are tangled in solution. Shorter molecules tend to tangle less and form a less viscous solution, so decreased viscosity corresponds to shortening of the polymers—as caused by nuclease activity. **(b)** An endonuclease. An exonuclease removes single nucleotides from the 5′ or 3′ end and would produce TCA-soluble ^{32}P-labeled nucleotides. An endonuclease cuts DNA into oligonucleotide fragments and produces little or no TCA-soluble ^{32}P-labeled material. **(c)** The 5′ end. If the phosphate were left on the 3′ end, the kinase would incorporate significant ^{32}P as it added phosphate to the 5′ end; treatment with the phosphatase would have no effect on this. In this case, samples A and B would incorporate significant amounts of ^{32}P. When the phosphate is left on the 5′ end, the kinase does not incorporate any ^{32}P: it cannot add a phosphate if one is already present. Treatment with the phosphatase removes 5′ phosphate, and the kinase then incorporates significant amounts of ^{32}P. Sample A will have little or no ^{32}P, and B will show substantial ^{32}P incorporation—as was observed.
(d) Random breaks would produce a distribution of fragments of random size. The production of specific fragments indicates that the enzyme is site-specific. **(e)** Cleavage at the site of recognition. This produces a specific sequence at the 5′ end of the fragments. If cleavage occurred near but not within the recognition site, the sequence at the 5′ end of the fragments would be random. **(f)** The results are consistent with two recognition sequences, as shown below, cleaved where shown by the arrows,

$$\downarrow$$
$$(5') \text{ - - - GTT AAC - - - } (3')$$
$$(3') \text{ - - - CAA TTG - - - } (5')$$
$$\uparrow$$

which gives the (5′)pApApC and (3′)TpTp fragments, and

$$\downarrow$$
$$(5') \text{ - - - GTC GAC - - - } (3')$$
$$(3') \text{ - - - CAG CTG - - - } (5')$$
$$\uparrow$$

which gives the (5′)pGpApC and (3′)CpTp fragments

Chapter 10

1. The term "lipid" does not specify a particular chemical structure. Compounds are categorized as lipids based on their greater solubility in organic solvents than in water.

2. **(a)** The number of cis double bonds. Each cis double bond causes a bend in the hydrocarbon chain, lowering the melting temperature. **(b)** Six different triacylglycerols can be constructed, in order of increasing melting points:

OOO < OOP = OPO < PPO = POP < PPP

where O = oleic and P = palmitic acid. The greater the content of saturated fatty acid, the higher is the melting point.
(c) Branched-chain fatty acids increase the fluidity of membranes because they decrease the extent of membrane lipid packing.

3. Lecithin, an amphipathic compound, is an emulsifying agent, facilitating the solubilization of butter.

4.

Squalene

5. Spearmint is (*R*)-carvone; caraway is (*S*)-carvone.

6.

(*R*)-2-Aminopropanoic acid   (*S*)-2-Aminopropanoic acid

(*R*)-2-Hydroxypropanoic acid   (*S*)-2-Hydroxypropanoic acid

7. *Hydrophobic units:* (a) 2 fatty acids; (b), (c), and (d) 1 fatty acid and the hydrocarbon chain of sphingosine; (e) steroid nucleus and acyl side chain. *Hydrophilic units:* (a) phosphoethanolamine; (b) phosphocholine; (c) D-galactose; (d) several sugar molecules; (e) alcohol group (OH)

8.

9. It reduces double bonds, which increases the melting point of lipids containing the fatty acids.

10. The triacylglycerols of animal fats (grease) are hydrolyzed by NaOH (saponified) to form soaps, which are much more soluble in water than are triacylglycerols.

11. It could only be a sphingolipid (sphingomyelin).

12.

Phosphatidylserine

13. Long, saturated acyl chains, nearly solid at air temperature, form a hydrophobic layer in which a polar compound such as H_2O cannot dissolve or diffuse.

14. (**a**) The free —OH group on C-2 and the phosphocholine head group on C-3 are hydrophilic; the fatty acid on C-1 of lysolecithin is hydrophobic. (**b**) Certain steroids such as prednisone inhibit the action of phospholipase A_2, inhibiting the release of arachidonic acid from C-2. Arachidonic acid is converted to a variety of eicosanoids, some of which cause inflammation and pain. (**c**) Phospholipase A_2 releases arachidonic acid, a precursor of other eicosanoids with vital protective functions in the body; it also breaks down dietary glycerophospholipids.

15. The part of the membrane lipid that determines blood type is the oligosaccharide in the head group of the membrane sphingolipids (see Fig. 10–15). This same oligosaccharide is attached to certain membrane glycoproteins, which also serve as points of recognition by the antibodies that distinguish blood groups.

16. Diacylglycerol is hydrophobic and remains in the membrane. Inositol 1,4,5-trisphosphate is highly polar, very soluble in water, and more readily diffusible in the cytosol. Both are second messengers.

17. Water-soluble vitamins are more rapidly excreted in the urine and are not stored effectively. Fat-soluble vitamins have very low solubility in water and are stored in body lipids.

18. (**a**) Glycerol and the sodium salts of palmitic and stearic acids. (**b**) D-Glycerol 3-phosphocholine and the sodium salts of palmitic and oleic acids.

19. Solubilities in water: monoacylglycerol > diacylglycerol > triacylglycerol.

20. First eluted to last eluted: cholesteryl palmitate and triacylglycerol; cholesterol and *n*-tetradecanol; phosphatidylcholine and phosphatidylethanolamine; sphingomyelin; phosphatidylserine and palmitate.

21. (**a**) Subject acid hydrolysates of each compound to chromatography (GLC or silica gel TLC) and compare the result with known standards. *Sphingomyelin hydrolysate:* sphingosine, fatty acids, phosphocholine, choline, and phosphate; *cerebroside hydrolysate:* sphingosine, fatty acids, sugars, but no phosphate. (**b**) Strong alkaline hydrolysis of sphingomyelin yields sphingosine; phosphatidylcholine yields glycerol. Detect hydrolysate components on thin-layer chromatograms by comparing with standards or by their differential reaction with FDNB (only sphingosine reacts to form a colored product). Treatment with phospholipase A_1 or A_2 releases free fatty acids from phosphatidylcholine, but not from sphingomyelin.

22. Phosphatidylethanolamine and phosphatidylserine.

23. (**a**) GM1 and globoside. Both glucose and galactose are hexoses, so "hexose" in the molar ratio refers to glucose + galactose. The ratios for the four gangliosides are: GM1, 1:3:1:1; GM2, 1:2:1:1; GM3, 1:2:0:1; globoside, 1:3:1:0. (**b**) Yes. The ratio matches GM2, the ganglioside expected to build up in Tay-Sachs disease (see Box 10–1, Fig. 1). (**c**) This analysis is similar to that used by Sanger to determine the amino acid sequence of insulin. The analysis of each fragment reveals only its *composition*, not its *sequence*, but because each fragment is formed by sequential removal of one sugar, we can draw conclusions about sequence. The structure of the normal asialoganglioside is ceramide–glucose–galactose–galactosamine–galactose, consistent with Box 10–1 (excluding Neu5Ac, removed before hydrolysis). (**d**) The Tay-Sachs asialoganglioside is ceramide–glucose–galactose–galactosamine, consistent with Box 10–1. (**e**) The structure of the normal asialoganglioside, GM1, is: *ceramide–glucose* (2 —OH involved in glycosidic links; 1 —OH involved in ring structure; 3 —OH (2,3,6) free for methylation)–*galactose* (2 —OH in links; 1 —OH in ring; 3 —OH (2,4,6) free for methylation)–*galactosamine* (2 —OH in links; 1 —OH in ring; 1 —NH₂ instead of an —OH; 2 —OH (4,6) free for methylation)–*galactose* (1 —OH in link; 1 —OH in ring; 4 —OH (2,3,4,6) free for methylation). (**f**) Two key pieces of information are missing: What are the linkages between the sugars? Where is Neu5Ac attached?

Chapter 11

1. The area per molecule would be calculated from the known amount (number of molecules) of lipid used and the area occupied by a monolayer when it begins to resist compression (when the required force increases dramatically, as shown in the plot of force vs. area).

2. The data support a bilayer of lipid in the dog erythrocytes: a single cell, with surface area 98 μm^2, has a lipid monolayer area of 200 μm^2. In the case of sheep and human erythrocytes, the data suggest a monolayer, not a bilayer. In fact, significant experimental errors occurred in these early experiments; recent, more accurate measurements support a bilayer in all cases.

3. 63 SDS molecules per micelle

4. (a) Lipids that form bilayers are amphipathic molecules: they contain a hydrophilic and a hydrophobic region. To minimize the hydrophobic area exposed to the water surface, these lipids form two-dimensional sheets, with the hydrophilic regions exposed to water and the hydrophobic regions buried in the interior of the sheet. Furthermore, to avoid exposing the hydrophobic edges of the sheet to water, lipid bilayers close on themselves. **(b)** These sheets form the closed membrane surfaces that envelop cells and compartments within cells (organelles).

5. 2 nm. Two palmitates placed end to end span about 4 nm, approximately the thickness of a typical bilayer.

6. Decrease. Movement of individual lipids in bilayers occurs much faster at 37 °C, when the lipids are in the "fluid" phase, than at 10 °C, when they are in the "solid" phase.

7. 35 kJ/mol, neglecting the effects of transmembrane electrical potential; 0.60 mol.

8. 13 kJ/mol.

9. Most of the O_2 consumed by a tissue is for oxidative phosphorylation, the source of most of the ATP. Therefore, about two-thirds of the ATP synthesized by the kidney is used for pumping K^+ and Na^+.

10. No. The symporter may carry more than one equivalent of Na^+ for each mole of glucose transported.

11. Salt extraction indicates a peripheral location, and inaccessibility to protease in intact cells indicates an internal location. X seems to be a peripheral protein on the cytosolic face of the membrane.

12. The hydrophobic interactions among membrane lipids are noncovalent and reversible, allowing membranes to spontaneously reseal.

13. The temperature of body tissues at the extremities is lower than that of tissues closer to the center of the body. If lipid is to remain fluid at this lower temperature, it must contain a higher proportion of unsaturated fatty acids; unsaturated fatty acids lower the melting point of lipid mixtures.

14. The energetic cost of moving the highly polar, sometimes charged, head group through the hydrophobic interior of the bilayer is prohibitive.

15. At pH 7, tryptophan bears a positive and a negative charge, but indole is uncharged. The movement of the less polar indole through the hydrophobic core of the bilayer is energetically more favorable.

16. 3×10^{-2} s

17. Treat a suspension of cells with unlabeled NEM in the presence of excess lactose, remove the lactose, then add radiolabeled NEM. Use SDS-PAGE to determine the M_r of the radiolabeled band (the transporter).

18. Construct a hydropathy plot; hydrophobic regions of 20 or more residues suggest transmembrane segments. Determine whether the protein in intact erythrocytes reacts with a membrane-impermeant reagent specific for primary amines; if so, the transporter is of type I.

19. The leucine transporter is specific for the L isomer, but the binding site can accommodate either L-leucine or L-valine. Reduction of V_{max} in the absence of Na^+ indicates that leucine (or valine) is transported by symport with Na^+.

20. V_{max} reduced; K_t unaffected.

21. ~1%; estimated by calculating the surface area of the cell and of 10,000 transporter molecules (using the dimensions of hemoglobin (5.5 nm diameter, p. 163) as a model globular protein).

22.

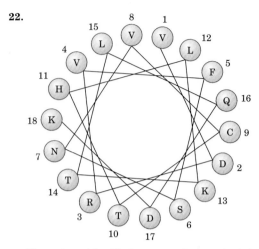

The amino acids with the greatest hydropathy index (V, L, F, and C) are clustered on one side of the helix. This amphipathic helix is likely to dip into the lipid bilayer along its hydrophobic surface while exposing the other surface to the aqueous phase. Alternatively, a group of helices may cluster with their polar surfaces in contact with one another and their hydrophobic surfaces facing the lipid bilayer.

23. ~22. To estimate the fraction of membrane surface covered by phospholipids, you would need to know (or estimate) the average cross-sectional area of a phospholipid molecule in a bilayer (e.g., from an experiment such as that diagrammed in problem 1 in this chapter) and the average cross-sectional area of a 50 kDa protein.

24. (a) The rise-per-residue for an α helix (Chapter 4) is about 1.5 Å = 0.15 nm. To span a 4 nm bilayer, an α helix must contain about 27 residues; thus for seven spans, about 190 residues are required. A protein of M_r 64,000 has about 580 residues. **(b)** A hydropathy plot is used to locate transmembrane regions. **(c)** Because about half of this portion of the epinephrine receptor consists of charged residues, it probably represents an intracellular loop that connects two adjacent membrane-spanning regions of the protein. **(d)** Because this helix is composed mostly of hydrophobic residues, this portion of the receptor is probably one of the membrane-spanning regions of the protein.

25. (a) *Model A:* supported. The two dark lines are either the protein layers or the phospholipid heads, and the clear space is either the bilayer or the hydrophobic core, respectively. *Model B:* not supported. This model requires a more-or-less uniformly stained band surrounding the cell. *Model C:* supported, with one reservation. The two dark lines are the phospholipid heads; the clear zone is the tails. This assumes that the membrane proteins are not visible, because they do not stain with osmium or do not happen to be in the sections viewed. **(b)** *Model A:* supported. A "naked" bilayer (4.5 nm) + two layers of protein (2 nm) sums to 6.5 nm, which is within the observed range of thickness. *Model B:* neither. This model makes no predictions about membrane thickness. *Model C:* unclear. The result is hard to reconcile with this model, which predicts a membrane as thick as, or slightly thicker than (due to the projecting ends of embedded proteins), a "naked" bilayer. The model is supported only if the smallest values for membrane thickness are correct or if a substantial amount of protein projects from the bilayer. **(c)** *Model A:* unclear. The result is hard to reconcile with this model. If the proteins are bound to the membrane by ionic interactions, the model predicts that the proteins contain a high proportion of charged amino acids, in contrast to what was observed. Also, because the protein layer must be very thin (see (b)), there would not be much room for a hydrophobic protein core, so hydrophobic residues would be exposed to the solvent. *Model B:* supported. The proteins have a mixture of hydrophobic residues (interacting with lipids) and charged residues (interacting with water). *Model C:* supported. The proteins have a mixture of hydrophobic residues (anchoring in the membrane) and charged residues (interacting with water). **(d)** *Model A:* unclear. The result is hard to reconcile with this model, which predicts a ratio of exactly 2.0; this would be hard to achieve under physiologically relevant pressures. *Model B:* neither. This model makes no predictions about amount of lipid in the membrane. *Model C:* supported. Some membrane surface area is taken up with proteins, so the ratio would be less than 2.0, as was observed under more physiologically relevant conditions. **(e)** *Model A:* unclear. The model predicts proteins in extended conformations rather than globular, so supported only if one assumes that proteins layered on the surfaces include helical segments. *Model B:* supported. The model predicts mostly globular proteins (containing some helical segments). *Model C:* supported. The model predicts mostly globular proteins. **(f)** *Model A:* unclear. The phosphorylamine head groups are protected by the protein layer, but only if the proteins completely cover the surface will the phospholipids be completely protected from phospholipase. *Model B:* supported. Most head groups are accessible to phospholipase. *Model C:* supported. All head groups are accessible to phospholipase. **(g)** *Model A:* not supported. Proteins are entirely accessible to trypsin digestion and virtually all will undergo multiple cleavage, with no protected hydrophobic segments. *Model B:* not supported. Virtually all proteins are in the bilayer and inaccessible to trypsin. *Model C:* supported. Segments of protein that penetrate or span the bilayer are protected from trypsin; those exposed at the surfaces will be cleaved. The trypsin-resistant portions have a high proportion of hydrophobic residues.

Chapter 12

1. X is cAMP; its production is stimulated by epinephrine.
(a) Centrifugation sediments adenylyl cyclase (which catalyzes cAMP formation) in the particulate fraction. **(b)** Added cAMP stimulates glycogen phosphorylase. **(c)** cAMP is heat stable; it can be prepared by treating ATP with barium hydroxide.

2. Unlike cAMP, dibutyryl cAMP passes readily through the plasma membrane.

3. **(a)** It increases [cAMP]. **(b)** cAMP regulates Na^+ permeability.
(c) Replace lost body fluids and electrolytes.

4. **(a)** The mutation makes R unable to bind and inhibit C, so C is constantly active. **(b)** The mutation prevents cAMP binding to R, leaving C inhibited by bound R.

5. Albuterol raises [cAMP], leading to relaxation and dilation of the bronchi and bronchioles. Because β-adrenergic receptors control many other processes, this drug would have undesirable side effects. To minimize them, find an agonist specific for the subtype of β-adrenergic receptors found in the bronchial smooth muscle.

6. Hormone degradation; hydrolysis of GTP bound to a G protein; degradation, metabolism, or sequestration of second messenger; receptor desensitization; removal of receptor from the cell surface.

7. Fuse CFP to β-arrestin and YFP to the cytoplasmic domain of the β-adrenergic receptor, or vice versa. In either case, illuminate at 433 nm and observe at both 476 and 527 nm. If the interaction occurs, emitted light intensity will decrease at 476 nm and increase at 527 nm on addition of epinephrine to cells expressing the fusion proteins. If the interaction does not occur, the wavelength of emitted light will remain at 476 nm. Some reasons why this might fail: The fusion proteins (1) are inactive or otherwise unable to interact, (2) are not translocated to their normal subcellular location, or (3) are not stable to proteolytic breakdown.

8. Vasopressin acts by elevating cytosolic $[Ca^{2+}]$ to 10^{-6} M, activating protein kinase C. EGTA injection blocks vasopressin action but should not affect the response to glucagon, which uses cAMP, *not* Ca^{2+}, as second messenger.

9. Amplification results as one molecule of a catalyst activates many molecules of another catalyst, in an amplification cascade involving, in order, insulin receptor, IRS-1, Raf, MEK, ERK; ERK activates a transcription factor, which stimulates mRNA production.

10. A mutation in *ras* that inactivates the Ras GTPase activity creates a protein that, once activated by the binding of GTP, continues to give, through Raf, the insulin-response signal.

11. *Shared properties of Ras and G_s:* Both bind either GDP or GTP; both are activated by GTP; both, when active, activate a downstream enzyme; both have intrinsic GTPase activity that shuts them off after a short period of activation. *Differences between Ras and G_s:* Ras is a small, monomeric protein; G_s is heterotrimeric. *Functional difference between G_s and G_i:* G_s activates adenylyl cyclase, G_i inhibits it.

12. *Kinase (factor in parentheses):* PKA (cAMP); PKG (cGMP); PKC (Ca^{2+}, DAG); Ca^{2+}/CaM kinase (Ca^{2+}, CaM); cyclin-dependent kinase (cyclin); protein Tyr kinase (ligand for the receptor, such as insulin); MAPK (Raf); Raf (Ras); glycogen phosphorylase kinase (PKA).

13. G_s remains in its activated form when the nonhydrolyzable analog is bound. The analog therefore prolongs the effect of epinephrine on the injected cell.

14. **(a)** Use the α-bungarotoxin–bound beads for affinity purification (see Fig. 3–17c) of AChR. Extract proteins from the electric organs and pass the mixture through the chromatography column; the AChR binds selectively to the beads. Elute the AChR with a solute that weakens its interaction with α-bungarotoxin. **(b)** Use binding of $[^{125}I]\alpha$-bungarotoxin as a *quantitative assay* for AChR during purification by various techniques. At each step, assay AChR by measuring $[^{125}I]\alpha$-bungarotoxin binding to the proteins in the sample. Optimize purification for the highest specific activity of AChR (counts/min of bound $[^{125}I]\alpha$-bungarotoxin per mg of protein) in the final material.

15. **(a)** No. If V_m were set by permeability to (primarily) K^+, the Nernst equation would predict a V_m of -90 mV, not the observed -95 mV, so some other conductance must contribute to V_m. **(b)** Chloride ion is probably the determinant of V_m; the predicted E_{Cl^-} is -94 mV.

16. **(a)** V_m of the oocyte membrane changes from -60 mV to -10 mV—that is, the membrane is depolarized. **(b)** The effect of KCl depends on influx of Ca^{2+} from the extracellular medium.

17. Hyperpolarization results in the closing of voltage-dependent Ca^{2+} channels in the presynaptic region of the rod cell. The resulting decrease in $[Ca^{2+}]_{in}$ diminishes release of an inhibitory neurotransmitter that suppresses activity in the next neuron of the visual circuit. When this inhibition is removed in response to a light stimulus, the circuit becomes active and visual centers in the brain are excited.

18. **(a)** This would prevent influx of Na^+ and Ca^{2+} into the cells in response to light; the cone cells would fail to signal the brain that light had been received. Because rod cells are unaffected, the individuals would be able to see but would not have color vision. **(b)** This would prevent efflux of K^+, which would lead to depolarization of the β-cell membrane and constitutive release of insulin into the blood. **(c)** ATP is responsible for closing this channel, so the channels will remain open, preventing depolarization of the β-cell membrane and release of insulin.

19. Individuals with Oguchi disease might have a defect in rhodopsin kinase or in arrestin.

20. Rod cells would no longer show any change in membrane potential in response to light. This experiment has been done. Illumination did activate PDE, but the enzyme could not significantly reduce the 8-Br-cGMP level, which remained well above that needed to keep the gated ion channels open. Thus, light had no impact on membrane potential.

21. (a) On exposure to heat, TRPV1 channels open, causing an influx of Na^+ and Ca^{2+} into the sensory neuron. This depolarizes the neuron, triggering an action potential. When the action potential reaches the axon terminus, neurotransmitter is released, signaling the nervous system that heat has been sensed. **(b)** Capsaicin mimics the effects of heat by opening TRPV1 at low temperature, leading to the false sensation of heat. The extremely low EC_{50} indicates that even very small amounts of capsaicin will have dramatic sensory effects. **(c)** At low levels, menthol should open the TRPM8 channel, leading to a sensation of cool; at high levels, both TRPM8 and TRPV3 will open, leading to a mixed sensation of cool and heat, such as you may have experienced with very strong peppermints.

22. (a) These mutations might lead to permanent activation of the PGE_2 receptor, leading to unregulated cell division and tumor formation. **(b)** The viral gene might encode a constitutively active form of the receptor, causing a constant signal for cell division and thus tumor formation. **(c)** E1A protein might bind to pRb and prevent E2F from binding, so E2F is constantly active and cells divide uncontrollably. **(d)** Lung cells do not normally respond to PGE_2 because they do not express the PGE_2 receptor; mutations resulting in a constitutively active PGE_2 receptor do not affect lung cells.

23. A normal tumor suppressor gene encodes a protein that restrains cell division. A mutant form of the protein fails to suppress cell division, but if either of the two alleles encodes normal protein, normal function will continue. A normal oncogene encodes a regulator protein that triggers cell division, but only when an appropriate signal (growth factor) is present. The mutant version of the oncogene product constantly sends the signal to divide, whether or not growth factors are present.

24. In a child who develops multiple tumors in both eyes, every retinal cell had a defective copy of the *Rb* gene at birth. Early in the child's life, several cells independently underwent a second mutation that damaged the one good *Rb* allele, producing a tumor. A child who develops a single tumor had, at birth, two good copies of the *Rb* gene in every cell; mutation in both *Rb* alleles in one cell (extremely rare) caused a single tumor.

25. Two cells expressing the same surface receptor may have different complements of target proteins for protein phosphorylation.

26. (a) The cell-based model, which predicts different receptors present on different cells. **(b)** This experiment addresses the issue of the independence of different taste sensations. Even though the receptors for sweet and/or umami are missing, the animals' other taste sensations are normal; thus, pleasant and unpleasant taste sensations are independent. **(c)** Yes. Loss of either T1R1 or T1R3 subunits abolishes umami taste sensation. **(d)** Both models. With either model, removing one receptor would abolish that taste sensation. **(e)** Yes. Loss of either the T1R2 or T1R3 subunits almost completely abolishes the sweet taste sensation; complete elimination of sweet taste requires deletion of both subunits. **(f)** At very high sucrose concentrations, T1R2 and, to a lesser extent, T1R3 receptors, as homodimers, can detect sweet taste. **(g)** The results are consistent with either model of taste encoding, but do strengthen the researchers' conclusions. Ligand binding can be completely separated from taste sensation. If the ligand for the receptor in "sweet-tasting cells" binds a molecule, mice prefer that molecule as a sweet compound.

Chapter 13

1. Consider the developing chick as the system; the nutrients, egg shell, and outside world are the surroundings. Transformation of the single cell into a chick drastically reduces the entropy of the system. Initially, the parts of the egg outside the embryo

(the surroundings) contain complex fuel molecules (a low-entropy condition). During incubation, some of these complex molecules are converted to large numbers of CO_2 and H_2O molecules (high entropy). This increase in the entropy of the surroundings is larger than the decrease in entropy of the chick (the system).

2. (a) -4.8 kJ/mol **(b)** 7.56 kJ/mol **(c)** -13.7 kJ/mol

3. (a) 262 **(b)** 608 **(c)** 0.30

4. $K'_{eq} = 21$; $\Delta G'^{\circ} = -7.6$ kJ/mol

5. -31 kJ/mol

6. (a) -1.68 kJ/mol **(b)** -4.4 kJ/mol **(c)** At a given temperature, the value of $\Delta G'^{\circ}$ for any reaction is fixed and is defined for standard conditions (here, both fructose 6-phosphate and glucose 6-phosphate at 1 M). In contrast, ΔG is a variable that can be calculated for any set of reactant and product concentrations.

7. $K'_{eq} \approx 1$; $\Delta G'^{\circ} \approx 0$

8. Less. The overall equation for ATP hydrolysis can be approximated as

$$ATP^{4-} + H_2O \rightarrow ADP^{3-} + HPO_4^{2-} + H^+$$

(This is only an approximation, because the ionized species shown here are the major, but not the only, forms present.) Under standard conditions (i.e., $[ATP] = [ADP] = [P_i] = 1$ M), the concentration of water is 55 M and does not change during the reaction. Because H^+ ions are produced in the reaction, at a higher $[H^+]$ (pH 5.0) the equilibrium would be shifted to the left and less free energy would be released.

9. 10

10.

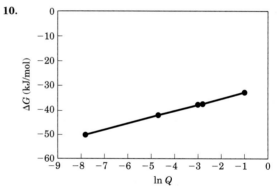

ΔG for ATP hydrolysis is lower when [ATP]/[ADP] is low ($<<1$) than when [ATP]/[ADP] is high. The energy available to the cell from a given [ATP] is lower when the [ATP]/[ADP] ratio falls and greater when it rises.

11. (a) 3.85×10^{-3} M^{-1}; [glucose 6-phosphate] $= 8.9 \times 10^{-8}$ M; no. **(b)** 14 M; because the maximum solubility of glucose is less than 1 M, this is not a reasonable step. **(c)** 837 ($\Delta G'^{\circ} = -16.7$ kJ/mol); [glucose] $= 1.2 \times 10^{-7}$ M; yes. **(d)** No. This would require such high $[P_i]$ that the phosphate salts of divalent cations would precipitate. **(e)** By directly transferring the phosphoryl group from ATP to glucose, the phosphoryl group transfer potential ("tendency" or "pressure") of ATP is utilized without generating high concentrations of intermediates. The essential part of this transfer is, of course, the enzymatic catalysis.

12. (a) -12.5 kJ/mol **(b)** -14.6 kJ/mol

13. (a) 3×10^{-4} **(b)** 68.7 **(c)** 7.4×10^4

14. -13 kJ/mol

15. 46.7 kJ/mol

16. Isomerization moves the carbonyl group from C-1 to C-2, setting up a carbon–carbon bond cleavage between C-3 and C-4. Without isomerization, bond cleavage would occur between C-2 and C-3, generating one two-carbon and one four-carbon compound.

17. The mechanism is the same as that of the alcohol dehydrogenase reaction (see Fig. 14–14).

18. The first step is the reverse of an aldol condensation (see the aldolase mechanism, Fig. 14–6); the second step is an aldol condensation (see Fig. 13–4).

19. **(a)** 46 kJ/mol **(b)** 46 kg; 68% **(c)** ATP is synthesized as it is needed, then broken down to ADP and P_i; its concentration is maintained in a steady state.

20. The ATP system is in a dynamic steady state; [ATP] remains constant because the rate of ATP consumption equals its rate of synthesis. ATP consumption involves release of the terminal (γ) phosphoryl group; synthesis of ATP from ADP involves replacement of this phosphoryl group. Hence the terminal phosphate undergoes rapid turnover. In contrast, the central (β) phosphate undergoes only relatively slow turnover.

21. **(a)** 1.7 kJ/mol **(b)** Inorganic pyrophosphatase catalyzes the hydrolysis of pyrophosphate and drives the net reaction toward the synthesis of acetyl-CoA.

22. 36 kJ/mol

23. **(a)** NAD^+/NADH **(b)** Pyruvate/lactate **(c)** Lactate formation **(d)** −26.1 kJ/mol **(e)** 3.63×10^4

24. **(a)** 1.14 V **(b)** −220 kJ/mol **(c)** ~4

25. **(a)** −0.35 V **(b)** −0.320 V **(c)** −0.29 V

26. In order of increasing tendency: (a), (d), (b), (c)

27. (c) and (d)

28. **(a)** The lowest-energy, highest-entropy state occurs when the dye concentration is the same in both cells. If a "fish trap" gap junction allowed unidirectional transport, more of the dye would end up in the oligodendrocyte and less in the astrocyte. This would be a higher-energy, lower-entropy state than the starting state, violating the second law of thermodynamics. The model proposed by Robinson et al. requires an impossible spontaneous decrease in entropy. In terms of energy, the model entails a spontaneous change from a lower-energy to a higher-energy state without an energy input—again, thermodynamically impossible. **(b)** Molecules, unlike fish, do not exhibit *directed behavior*; they move randomly by Brownian motion. Diffusion results in *net* movement of molecules from a region of higher concentration to a region of lower concentration simply because it is more likely that a molecule on the high-concentration side will enter the connecting channel. Look at this as a pathway with a rate-limiting step: the narrow end of the channel. The narrower end limits the rate at which molecules pass through because random motion of the molecules is less likely to move them through the smaller cross section. The wide end of the channel does *not* act like a funnel for molecules, although it may for fish, because molecules are not "crowded" by the sides of the narrowing funnel as fish would be. The narrow end limits the rate of movement equally in both directions. When the concentrations on both sides are equal, the rates of movement in both directions are equal and there will be no change in concentration. **(c)** Fish exhibit *nonrandom behavior*, adjusting their actions in response to the environment. Fish that enter the large opening of the channel tend to move forward because fish have behavior that tends to make them prefer forward movement, and they experience "crowding" as they move through the narrowing channel. It is easy for fish to enter the large opening, but they don't move out of the trap as readily because they are less likely to enter the small opening. **(d)** There are many possible explanations, some of which were proposed by the letter-writers who criticized the article. Here are two: (1) *The dye could bind to a molecule in the oligodendrocyte.* Binding effectively removes the dye from the bulk solvent, so it doesn't "count" as a solute for thermodynamic considerations yet remains visible in the fluorescence microscope. (2) *The dye could be sequestered in a subcellular organelle of the oligodendrocyte,* either actively pumped at the expense of ATP or drawn in by its attraction to other molecules in that organelle.

Chapter 14

1. Net equation: Glucose + 2ATP → 2 glyceraldehyde 3-phosphate + 2ADP + $2H^+$; $\Delta G'^\circ = 2.1$ kJ/mol

2. Net equation: 2 Glyceraldehyde 3-phosphate + 4ADP + $2P_i$ → 2 lactate + $2NAD^+$; $\Delta G'^\circ = -114$ kJ/mol

3. GLUT2 (and GLUT1) is found in liver and is always present in the plasma membrane of hepatocytes. GLUT3 is always present in the plasma membrane of certain brain cells. GLUT4 is normally sequestered in vesicles in cells of muscle and adipose tissue and enters the plasma membrane only in response to insulin. Thus, liver and brain can take up glucose from blood regardless of insulin level, but muscle and adipose tissue take up glucose only when insulin levels are elevated in response to high blood glucose.

4. $CH_3CHO + NADH + H^+ \rightleftharpoons CH_3CH_2OH + NAD^+$; $K'_{eq} = 1.45 \times 10^4$

5. −8.6 kJ/mol

6. **(a)** $^{14}CH_3CH_2OH$ **(b)** [3-^{14}C]glucose or [4-^{14}C]glucose

7. Fermentation releases energy, some conserved in the form of ATP but much of it dissipated as heat. Unless the fermenter contents are cooled, the temperature would become high enough to kill the microorganisms.

8. Soybeans and wheat contain starch, a polymer of glucose. The microorganisms break down starch to glucose, glucose to pyruvate via glycolysis, and—because the process is carried out in the absence of O_2 (i.e., it is a fermentation)—pyruvate to lactic acid and ethanol. If O_2 were present, pyruvate would be oxidized to acetyl-CoA, then to CO_2 and H_2O. Some of the acetyl-CoA, however, would also be hydrolyzed to acetic acid (vinegar) in the presence of oxygen.

9. C-1. This experiment demonstrates the reversibility of the aldolase reaction. The C-1 of glyceraldehyde 3-phosphate is equivalent to C-4 of fructose 1,6-bisphosphate (see Fig. 14–7). The starting glyceraldehyde 3-phosphate must have been labeled at C-1. The C-3 of dihydroxyacetone phosphate becomes labeled through the triose phosphate isomerase reaction, thus giving rise to fructose 1,6-bisphosphate labeled at C-3.

10. No. There would be no anaerobic production of ATP; aerobic ATP production would be diminished only slightly.

11. No. Lactate dehydrogenase is required to recycle NAD^+ from the NADH formed during the oxidation of glyceraldehyde 3-phosphate.

12. The transformation of glucose to lactate occurs when myocytes are low in oxygen, and it provides a means of generating ATP under O_2-deficient conditions. Because lactate can be oxidized to pyruvate, glucose is not wasted; pyruvate is oxidized by aerobic reactions when O_2 becomes plentiful. This metabolic flexibility gives the organism a greater capacity to adapt to its environment.

13. It rapidly removes the 1,3-bisphosphoglycerate in a favorable subsequent step, catalyzed by phosphoglycerate kinase.

14. **(a)** 3-Phosphoglycerate is the product. **(b)** In the presence of arsenate there is no net ATP synthesis under anaerobic conditions.

15. **(a)** Ethanol fermentation requires 2 mol of P_i per mole of glucose. **(b)** Ethanol is the reduced product formed during reoxidation of NADH to NAD^+, and CO_2 is the byproduct of the conversion of pyruvate to ethanol. Yes; pyruvate must be converted to ethanol, to produce a continuous supply of NAD^+ for the oxidation of glyceraldehyde 3-phosphate. Fructose 1,6-bisphosphate accumulates; it is formed as an intermediate in glycolysis. **(c)** Arsenate replaces P_i in the glyceraldehyde 3-phosphate dehydrogenase reaction to yield an acyl arsenate, which spontaneously hydrolyzes. This

prevents formation of ATP, but 3-phosphoglycerate continues through the pathway.

16. Dietary niacin is used to synthesize NAD^+. Oxidations carried out by NAD^+ are part of cyclic processes, with NAD^+ as electron carrier (reducing agent); one molecule of NAD^+ can oxidize many thousands of molecules of glucose, and thus the dietary requirement for the precursor vitamin (niacin) is relatively small.

17. Dihydroxyacetone phosphate + NADH + $H^+ \rightarrow$ glycerol 3-phosphate + NAD^+ (catalyzed by a dehydrogenase)

18. *Galactokinase deficiency:* galactose (less toxic); *UDP-glucose: galactose 1-phosphate uridylyl deficiency:* galactose 1-phosphate (more toxic).

19. The proteins are degraded to amino acids and used for gluconeogenesis.

20. (a) In the pyruvate carboxylase reaction, $^{14}CO_2$ is added to pyruvate, but PEP carboxykinase removes the *same* CO_2 in the next step. Thus, ^{14}C is not (initially) incorporated into glucose.

(b)

21. 4 ATP equivalents per glucose molecule

22. Gluconeogenesis would be highly endergonic, and it would be impossible to separately regulate gluconeogenesis and glycolysis.

23. The cell "spends" 1 ATP and 1 GTP in converting pyruvate to PEP.

24. (a), (b), (d) are glucogenic; (c) (e) are not.

25. Consumption of alcohol forces competition for NAD^+ between ethanol metabolism and gluconeogenesis. The problem is compounded by strenuous exercise and lack of food, because at these times the level of blood glucose is already low.

26. (a) The rapid increase in glycolysis; the rise in pyruvate and NADH results in a rise in lactate. **(b)** Lactate is transformed to

glucose via pyruvate; this is a slower process, because formation of pyruvate is limited by NAD^+ availability, the LDH equilibrium is in favor of lactate, and conversion of pyruvate to glucose is energy-requiring. **(c)** The equilibrium for the LDH reaction is in favor of lactate formation.

27. Lactate is transformed to glucose in the liver by gluconeogenesis (see Figs 14–16, 14–17). A defect in FBPase-1 would prevent entry of lactate into the gluconeogenic pathway in hepatocytes, causing lactate to accumulate in the blood.

28. Succinate is transformed to oxaloacetate, which passes into the cytosol and is converted to PEP by PEP carboxykinase. Two moles of PEP are then required to produce a mole of glucose by the route outlined in Fig. 14–17.

29. If the catabolic and anabolic pathways of glucose metabolism are operating simultaneously, futile cycling of ATP occurs, with extra O_2 consumption.

30. At the very least, accumulation of ribose 5-phosphate would tend to force this reaction in the reverse direction by mass action (see Eqn 13–4). It might also affect other metabolic reactions that involve ribose 5-phosphate as a substrate or product—such as the pathways of nucleotide synthesis.

31. (a) Ethanol tolerance is likely to involve many more genes, and thus the engineering would be a much more involved project. **(b)** L-Arabinose isomerase (the *araA* enzyme) converts an aldose to a ketose by moving the carbonyl of a nonphosphorylated sugar from C-1 to C-2. No analogous enzyme is discussed in this chapter; all the enzymes described here act on phosphorylated sugars. An enzyme that carries out a similar transformation with phosphorylated sugars is phosphohexose isomerase. L-Ribulokinase (*araB*) phosphorylates a sugar at C-5 by transferring the γ phosphate from ATP. Many such reactions are described in this chapter, including the hexokinase reaction. L-Ribulose 5-phosphate epimerase (*araD*) switches the —H and —OH groups on a chiral carbon of a sugar. No analogous reaction is described in the chapter, but it is described in Chapter 20 (see Fig. 20–13). **(c)** The three *ara* enzymes would convert arabinose to xylulose 5-phosphate by the following pathway: Arabinose $\xrightarrow{\text{L-arabinose isomerase}}$ L-ribulose $\xrightarrow{\text{L-ribulokinase}}$ L-ribulose 5-phosphate $\xrightarrow{\text{epimerase}}$ xylulose 5-phosphate. **(d)** The arabinose is converted to xylulose 5-phosphate as in (c), which enters the pathway in Fig. 14–23; the glucose 6-phosphate product is then fermented to ethanol and CO_2. **(e)** 6 molecules of arabinose + 6 molecules of ATP are converted to 6 molecules of xylulose 5-phosphate, which feed into the pathway in Fig. 14–23 to yield 5 molecules of glucose 6-phosphate, each of which is fermented to yield 3 ATP (they enter as glucose 6-phosphate, not glucose)—15 ATP in all. Overall, you would expect a yield of 15 ATP − 6 ATP = 9 ATP from the 6 arabinose molecules. The other products are 10 molecules of ethanol and 10 molecules of CO_2. **(f)** Given the lower ATP yield, for an amount of growth (i.e., of available ATP) equivalent to growth without the added genes, the engineered *Z. mobilis* must ferment more arabinose, and thus it produces more ethanol. **(g)** One way to allow the use of xylose would be to add the genes for two enzymes: an analog of the *araD* enzyme that converts xylose to ribose by switching the —H and —OH on C-3, and an analog of the *araB* enzyme that phosphorylates ribose at C-5. The resulting ribose 5-phosphate would feed into the existing pathway.

Chapter 15

1. (a) 0.0293 **(b)** 308 **(c)** No. Q is much lower than K'_{eq}, indicating that the PFK-1 reaction is far from equilibrium in cells; this reaction is slower than the subsequent reactions in glycolysis. Flux through the glycolytic pathway is largely determined by the activity of PFK-1.

2. (a) 1.4×10^{-9} M **(b)** The physiological concentration (0.023 mM) is 16,000 times the equilibrium concentration; this reaction does not reach equilibrium in the cell. Many reactions in the cell are not at equilibrium.

3. In the absence of O_2, the ATP needs are met by anaerobic glucose metabolism (fermentation to lactate). Because aerobic oxidation of glucose produces far more ATP than does fermentation, less glucose is needed to produce the same amount of ATP.

4. (a) There are two binding sites for ATP: a catalytic site and a regulatory site. Binding of ATP to a regulatory site inhibits PFK-1, by reducing V_{max} or increasing K_m for ATP at the catalytic site. **(b)** Glycolytic flux is reduced when ATP is plentiful. **(c)** The graph indicates that increased [ADP] suppresses the inhibition by ATP. Because the adenine nucleotide pool is fairly constant, consumption of ATP leads to an increase in [ADP]. The data show that the activity of PFK-1 may be regulated by the [ATP]/[ADP] ratio.

5. The phosphate group of glucose 6-phosphate is completely ionized at pH 7, giving the molecule an overall negative charge. Because membranes are generally impermeable to electrically charged molecules, glucose 6-phosphate cannot pass from the bloodstream into cells and hence cannot enter the glycolytic pathway and generate ATP. (This is why glucose, once phosphorylated, cannot escape from the cell.)

6. (a) *In muscle:* Glycogen breakdown supplies energy (ATP) via glycolysis. Glycogen phosphorylase catalyzes the conversion of stored glycogen to glucose 1-phosphate, which is converted to glucose 6-phosphate, an intermediate in glycolysis. During strenuous activity, skeletal muscle requires large quantities of glucose 6-phosphate. *In the liver:* Glycogen breakdown maintains a steady level of blood glucose between meals (glucose 6-phosphate is converted to free glucose). **(b)** In actively working muscle, ATP flux requirements are very high and glucose 1-phosphate must be produced rapidly, requiring a high V_{max}.

7. (a) $[P_i]/[\text{glucose 1-phosphate}] = 3.3/1$ **(b), (c)** The value of this ratio in the cell (>100:1) indicates that [glucose 1-phosphate] is far below the equilibrium value. The rate at which glucose 1-phosphate is removed (through entry into glycolysis) is greater than its rate of production (by the glycogen phosphorylase reaction), so metabolite flow is from glycogen to glucose 1-phosphate. The glycogen phosphorylase reaction is probably the regulatory step in glycogen breakdown.

8. (a) increases **(b)** decreases **(c)** increases

9. *Resting:* [ATP] high; [AMP] low; [acetyl-CoA] and [citrate] intermediate. *Running:* [ATP] intermediate; [AMP] high; [acetyl-CoA] and [citrate] low. Glucose flux through glycolysis increases during the anaerobic sprint because (1) the ATP inhibition of glycogen phosphorylase and PFK-1 is partially relieved, (2) AMP stimulates both enzymes, and (3) lower citrate and acetyl-CoA levels relieve their inhibitory effects on PFK-1 and pyruvate kinase, respectively.

10. The migrating bird relies on the highly efficient aerobic oxidation of fats, rather than the anaerobic metabolism of glucose used by a sprinting rabbit. The bird reserves its muscle glycogen for short bursts of energy during emergencies.

11. *Case A:* (f), (3); *Case B:* (c), (3); *Case C:* (h), (4); *Case D:* (d), (6)

12. (a) (1) Adipose: fatty acid synthesis slower. (2) Muscle: glycolysis, fatty acid synthesis, and glycogen synthesis slower. (3) Liver: glycolysis faster; gluconeogenesis, glycogen synthesis, and fatty acid synthesis slower; pentose phosphate pathway unchanged. **(b)** (1) Adipose and (3) liver: fatty acid synthesis slower because lack of insulin results in inactive acetyl-CoA carboxylase, the first enzyme of fatty acid synthesis. Glycogen synthesis inhibited by cAMP-dependent phosphorylation (thus activation) of glycogen synthase. (2) Muscle: glycolysis slower because GLUT4 is inactive, so glucose uptake is inhibited. (3) Liver: glycolysis slower because the bifunctional PFK-2/FBPase-2 is converted to the form with active FBPase-2,

decreasing [fructose 2,6-bisphosphate], which allosterically stimulates phosphofructokinase and inhibits FBPase-1; this also accounts for the stimulation of gluconeogenesis.

13. (a) elevated **(b)** elevated **(c)** elevated

14. (a) PKA cannot be activated in response to glucagon or epinephrine, and glycogen phosphorylase is not activated. **(b)** PP1 remains active, allowing it to dephosphorylate glycogen synthase (activating it) and glycogen phosphorylase (inhibiting it). **(c)** Phosphorylase remains phosphorylated (active), increasing the breakdown of glycogen. **(d)** Gluconeogenesis cannot be stimulated when blood glucose is low, leading to dangerously low blood glucose during periods of fasting.

15. The drop in blood glucose triggers release of glucagon by the pancreas. In the liver, glucagon activates glycogen phosphorylase by stimulating its cAMP-dependent phosphorylation and stimulates gluconeogenesis by lowering [fructose 2,6-bisphosphate], thus stimulating FBPase-1.

16. (a) Reduced capacity to mobilize glycogen; lowered blood glucose between meals **(b)** Reduced capacity to lower blood glucose after a carbohydrate meal; elevated blood glucose **(c)** Reduced fructose 2,6-bisphosphate (F26BP) in liver, stimulating glycolysis and inhibiting gluconeogenesis **(d)** Reduced F26BP, stimulating gluconeogenesis and inhibiting glycolysis **(e)** Increased uptake of fatty acids and glucose; increased oxidation of both **(f)** Increased conversion of pyruvate to acetyl-CoA; increased fatty acid synthesis.

17. (a) Given that each particle contains about 55,000 glucose residues, the equivalent free glucose concentration would be $55,000 \times 0.01\ \mu\text{M} = 550$ mM, or 0.55 M. This would present a serious osmotic challenge for the cell! (Body fluids have a substantially lower osmolarity.) **(b)** The lower the number of branches, the lower the number of free ends available for glycogen phosphorylase activity, and the slower the rate of glucose release. With no branches, there would be just one site for phosphorylase to act. **(c)** The outer tier of the particle would be too crowded with glucose residues for the enzyme to gain access to cleave bonds and release glucose. **(d)** The number of chains doubles in each succeeding tier: tier 1 has one chain (2^0), tier 2 has two (2^1), tier 3 has four (2^2), and so on. Thus, for t tiers, the number of chains in the outermost tier, C_A, is 2^{t-1}. **(e)** The total number of chains is $2^0 + 2^1 + 2^2 + \ldots$ $2^{t-1} = 2^t - 1$. Each chain contains g_c glucose molecules, so the total number of glucose molecules, G_T, is $g_c(2t - 1)$. **(f)** Glycogen phosphorylase can release all but four of the glucose residues in a chain of length g_c. Therefore, from each chain in the outer tier it can release $(g_c - 4)$ glucose molecules. Given that there are 2^{t-1} chains in the outer tier, the number of glucose molecules the enzyme can release, G_{PT}, is $(g_c - 4)(2^{t-1})$. **(g)** The volume of a sphere is $\frac{4}{3}\pi r^3$. In this case, r is the thickness of one tier times the number of tiers, or $(0.12g_c + 0.35)t$ nm. Thus $V_s = \frac{4}{3}\pi t^3 (0.12g_c + 0.35)^3$ nm^3. **(h)** You can show algebraically that the value of g_c that maximizes f is independent of t. Choosing $t = 3$:

g_c	C_A	G_T	G_{PT}	V_s	f
5	4	35	4	11	5.8
6	4	42	8	19	9.7
7	4	49	12	24	12
8	4	56	16	28	14
9	4	63	20	32	15
10	4	70	24	34	16
11	4	77	28	36	16
12	4	84	32	38	17
13	4	91	36	40	17
14	4	98	40	41	17
15	4	100	44	42	16
16	4	110	48	43	16

The optimum value of g_c (i.e., at maximum f) is 13. In nature, g_c varies from 12 to 14, which corresponds to f values very close to the optimum. If you choose another value for t, the numbers will differ but the optimal g_c will still be 13.

Chapter 16

1. (a)

❶ *Citrate synthase:*
Acetyl-CoA + oxaloacetate + H_2O → citrate + CoA

❷ *Aconitase:*
Citrate → isocitrate

❸ *Isocitrate dehydrogenase:*
Isocitrate + NAD^+ → α-ketoglutarate + CO_2 + NADH

❹ *α-Ketoglutarate dehydrogenase:*
α-Ketoglutarate + NAD^+ + CoA → succinyl-CoA + CO_2 + NADH

❺ *Succinyl-CoA synthetase:*
Succinyl-CoA + P_i + GDP → succinate + CoA + GTP

❻ *Succinate dehydrogenase:*
Succinate + FAD → fumarate + $FADH_2$

❼ *Fumarase:*
Fumarate + H_2O → malate

❽ *Malate dehydrogenase:*
Malate + NAD^+ → oxaloacetate + NADH + H^+

(b), (c) **❶** CoA, condensation; **❷** none, isomerization; **❸** NAD^+, oxidative decarboxylation; **❹** NAD^+, CoA, and thiamine pyrophosphate, oxidative decarboxylation; **❺** CoA, substrate-level phosphorylation; **❻** FAD, oxidation; **❼** none, hydration; **❽** NAD^+, oxidation

(d) Acetyl-CoA + $3NAD^+$ + FAD + GDP + P_i + $2H_2O$ →
$2CO_2$ + CoA + 3NADH + $FADH_2$ + GTP + $2H^+$

2. Glucose + 4ADP + $4P_i$ + $10NAD^+$ + 2FAD →
4ATP + 10NADH + $2FADH_2$ + $6CO_2$

3. (a) Oxidation; methanol → formaldehyde + [H—H]

(b) Oxidation; formaldehyde → formate + [H—H]

(c) Reduction; CO_2 + [H—H] → formate + H^+

(d) Reduction; glycerate + H^+ + [H—H] →
glyceraldehyde + H_2O

(e) Oxidation; glycerol → dihydroxyacetone + [H—H]

(f) Oxidation; $2H_2O$ + toluene → benzoate + H^+ + 3[H—H]

(g) Oxidation; succinate → fumarate + [H—H]

(h) Oxidation; pyruvate + H_2O → acetate + CO_2 + [H—H]

4. From the structural formulas, we see that the carbon-bound H/C ratio of hexanoic acid (11/6) is higher than that of glucose (7/6). Hexanoic acid is more reduced and yields more energy on complete combustion to CO_2 and H_2O.

5. (a) Oxidized; ethanol + NAD^+ → acetaldehyde + NADH + H^+

(b) Reduced; 1,3-bisphosphoglycerate + NADH + H^+ →
glyceraldehyde 3-phosphate + NAD^+ + HPO_4^{2-}

(c) Unchanged; pyruvate + H^+ → acetaldehyde + CO_2

(d) Oxidized; pyruvate + NAD^+ →
acetate + CO_2 + NADH + H^+

(e) Reduced; oxaloacetate + NADH + H^+ → malate + NAD^+

(f) Unchanged; acetoacetate + H^+ → acetone + CO_2

6. *TPP:* thiazolium ring adds to α carbon of pyruvate, then stabilizes the resulting carbanion by acting as an electron sink. *Lipoic acid:* oxidizes pyruvate to level of acetate (acetyl-CoA), and activates acetate as a thioester. *CoA-SH:* activates acetate as thioester. *FAD:* oxidizes lipoic acid. *NAD^+:* oxidizes FAD.

7. Lack of TPP inhibits pyruvate dehydrogenase; pyruvate accumulates.

8. Oxidative decarboxylation; NAD^+ or $NADP^+$; α-ketoglutarate dehydrogenase

9. Oxygen consumption is a measure of the activity of the first two stages of cellular respiration: glycolysis and the citric acid cycle. The addition of oxaloacetate or malate stimulates the citric acid cycle and thus stimulates respiration. The added oxaloacetate or malate serves a catalytic role, because it is regenerated in the latter part of the citric acid cycle.

10. (a) 5.6×10^{-6} **(b)** 1.1×10^{-8} M **(c)** 28 molecules

11. ADP (or GDP), P_i, CoA-SH, TPP, NAD^+; *not* lipoic acid, which is covalently attached to the isolated enzymes that use it

12. The flavin nucleotides, FMN and FAD, would not be synthesized. Because FAD is required in the citric acid cycle, flavin deficiency would strongly inhibit the cycle.

13. Oxaloacetate might be withdrawn for aspartate synthesis or for gluconeogenesis. Oxaloacetate is replenished by the anaplerotic reactions catalyzed by PEP carboxykinase, PEP carboxylase, malic enzyme, or pyruvate carboxylase (see Fig. 16–16).

14. The terminal phosphoryl group in GTP can be transferred to ADP in a reaction catalyzed by nucleoside diphosphate kinase, with an equilibrium constant of 1.0:
$$GTP + ADP \rightarrow GDP + ATP$$

15. (a) $^-OOC—CH_2—CH_2—COO^-$ (succinate) **(b)** Malonate is a competitive inhibitor of succinate dehydrogenase. **(c)** A block in the citric acid cycle stops NADH formation, which stops electron transfer, which stops respiration. **(d)** A large excess of succinate (substrate) overcomes the competitive inhibition.

16. (a) Add uniformly labeled [^{14}C]glucose and check for the release of $^{14}CO_2$. **(b)** Equally distributed in C-2 and C-3 of oxaloacetate; an infinite number

17. Oxaloacetate equilibrates with succinate, in which C-1 and C-4 are equivalent. Oxaloacetate derived from succinate is labeled at C-1 and C-4, and the PEP derived from it has label at C-1, which gives rise to C-3 and C-4 of glucose.

18. (a) C-1 **(b)** C-3 **(c)** C-3 **(d)** C-2 (methyl group) **(e)** C-4 **(f)** C-4 **(g)** equally distributed in C-2 and C-3

19. Thiamine is required for the synthesis of TPP, a prosthetic group in the pyruvate dehydrogenase and α-ketoglutarate dehydrogenase complexes. A thiamine deficiency reduces the activity of these enzyme complexes and causes the observed accumulation of precursors.

20. No. For every two carbons that enter as acetate, two leave the cycle as CO_2; thus there is no net synthesis of oxaloacetate. Net synthesis of oxaloacetate occurs by the carboxylation of pyruvate, an anaplerotic reaction.

21. Yes, the citric acid cycle would be inhibited. Oxaloacetate is present at relatively low concentrations in mitochondria, and removing it for gluconeogenesis would tend to shift the equilibrium for the citrate synthase reaction toward oxaloacetate.

22. (a) Inhibition of aconitase **(b)** Fluorocitrate; competes with citrate; by a large excess of citrate **(c)** Citrate and fluorocitrate are inhibitors of PFK-1. **(d)** All catabolic processes necessary for ATP production are shut down.

23. *Glycolysis:*
Glucose + $2P_i$ + 2ADP + $2NAD^+$ →
2 pyruvate + 2ATP + 2NADH + $2H^+$ + $2H_2O$

Pyruvate carboxylase reaction:
2 Pyruvate + $2CO_2$ + 2ATP + $2H_2O$ →
2 oxaloacetate + 2ADP + $2P_i$ + $4H^+$

Malate dehydrogenase reaction:
2 Oxaloacetate + 2NADH + $2H^+$ → 2 L-malate + $2NAD^+$

This recycles nicotinamide coenzymes under anaerobic conditions. The overall reaction is
$$Glucose + 2CO_2 \rightarrow 2\ L\text{-malate} + 4H^+$$

This produces four H^+ per glucose, increasing the acidity and thus the tartness of the wine.

24. Net reaction: 2 Pyruvate + ATP + 2NAD$^+$ + H$_2$O →
α-ketoglutarate + CO$_2$ + ADP + P$_i$ + 2NADH + 3H$^+$

25. The cycle participates in catabolic and anabolic processes. For example, it generates ATP by substrate oxidation, but also provides precursors for amino acid synthesis (see Fig. 16–16).

26. (a) decreases (b) increases (c) decreases

27. (a) Citrate is produced through the action of citrate synthase on oxaloacetate and acetyl-CoA. Citrate synthase can be used for net synthesis of citrate when (1) there is a continuous influx of new oxaloacetate and acetyl-CoA and (2) isocitrate synthesis is restricted, as in a medium low in Fe^{3+}. Aconitase requires Fe^{3+}, so an Fe^{3+}-restricted medium restricts the synthesis of aconitase.

(b) Sucrose + H$_2$O → glucose + fructose
Glucose + 2P$_i$ + 2ADP + 2NAD$^+$ →
2 pyruvate + 2ATP + 2NADH + 2H$^+$ + 2H$_2$O
Fructose + 2P$_i$ + 2ADP + 2NAD$^+$ →
2 pyruvate + 2ATP + 2NADH + 2H$^+$ + 2H$_2$O
2 Pyruvate + 2NAD$^+$ + 2CoA →
2 acetyl-CoA + 2NADH + 2H$^+$ + 2CO$_2$
2 Pyruvate + 2CO$_2$ + 2ATP + 2H$_2$O →
2 oxaloacetate + 2ADP + 2P$_i$ + 4H$^+$
2 Acetyl-CoA + 2 oxaloacetate + 2H$_2$O → 2 citrate + 2CoA

The overall reaction is

Sucrose + H$_2$O + 2P$_i$ + 2ADP + 6NAD$^+$ →
2 citrate + 2ATP + 6NADH + 10H$^+$

(c) The overall reaction consumes NAD$^+$. Because the cellular pool of this oxidized coenzyme is limited, it must be recycled by the electron-transfer chain with consumption of O$_2$. Consequently, the overall conversion of sucrose to citric acid is an aerobic process and requires molecular oxygen.

28. Succinyl-CoA is an intermediate of the citric acid cycle; its accumulation signals reduced flux through the cycle, calling for reduced entry of acetyl-CoA into the cycle. Citrate synthase, by regulating the primary oxidative pathway of the cell, regulates the supply of NADH and thus the flow of electrons from NADH to O$_2$.

29. Fatty acid catabolism increases [acetyl-CoA], which stimulates pyruvate carboxylase. The resulting increase in [oxaloacetate] stimulates acetyl-CoA consumption by the citric acid cycle, and [citrate] rises, inhibiting glycolysis at the level of PFK-1. In addition, increased [acetyl-CoA] inhibits the pyruvate dehydrogenase complex, slowing the utilization of pyruvate from glycolysis.

30. Oxygen is needed to recycle NAD$^+$ from the NADH produced by the oxidative reactions of the citric acid cycle. Reoxidation of NADH occurs during mitochondrial oxidative phosphorylation.

31. Increased [NADH]/[NAD$^+$] inhibits the citric acid cycle by mass action at the three NAD$^+$-reducing steps; high [NADH] shifts equilibrium toward NAD$^+$.

32. Toward citrate; ΔG for the citrate synthase reaction under these conditions is about −8 kJ/mol.

33. Steps ❹ and ❺ are essential in the reoxidation of the enzyme's reduced lipoamide cofactor.

34. The citric acid cycle is so central to metabolism that a serious defect in any cycle enzyme would probably be lethal to the embryo.

35. The first enzyme in each path is under reciprocal allosteric regulation. Inhibition of one path shunts isocitrate into the other path.

36. (a) The only reaction in muscle tissue that consumes significant amounts of oxygen is cellular respiration, so O$_2$ consumption is a good proxy for respiration. (b) Freshly prepared muscle tissue contains some residual glucose; O$_2$ consumption is due to oxidation of this glucose. (c) Yes. Because the amount of O$_2$ consumed increased when citrate or 1-phosphoglycerol was added, both can serve as substrate

for cellular respiration in this system. (d) *Experiment I:* Citrate is causing much more O$_2$ consumption than would be expected from its complete oxidation. Each molecule of Citrate seems to be acting as though it were more than one molecule. The only possible explanation is that each molecule of citrate functions more than once in the reaction—which is how a catalyst operates. *Experiment II:* The key is to calculate the excess O$_2$ consumed by each sample compared with the control (sample 1).

Sample	Substrate(s) added	μL O$_2$ absorbed	Excess μL O$_2$ consumed
1	No extra	342	0
2	0.3 mL 0.2 M 1-phosphoglycerol	757	415
3	0.15 mL 0.02 M citrate	431	89
4	0.3 mL 0.2 M 1-phosphoglycerol + 0.15 mL 0.02 M citrate	1,385	1,043

If both citrate and 1-phosphoglycerol were simply substrates for the reaction, you would expect the excess O$_2$ consumption by sample 4 to be the sum of the individual excess consumptions by samples 2 and 3 (415 μL + 89 μL = 504 μL). However, the excess consumption when both substrates are present is roughly twice this amount (1,043 μL). Thus citrate increases the ability of the tissue to metabolize 1-phosphoglycerol. This behavior is typical of a catalyst. Both experiments (I and II) are required to make this case convincing. Based on experiment I only, citrate is somehow accelerating the reaction, but it is not clear whether it acts by helping substrate metabolism or by some other mechanism. Based on experiment II only, it is not clear which molecule is the catalyst, citrate or 1-phosphoglycerol. Together, the experiments show that citrate is acting as a "catalyst" for the oxidation of 1-phosphoglycerol. (e) Given that the pathway can consume citrate (see sample 3), if citrate is to act as a catalyst it must be regenerated. If the set of reactions first consumes then regenerates citrate, it must be a circular rather than a linear pathway. (f) When the pathway is blocked at α-ketoglutarate dehydrogenase, citrate is converted to α-ketoglutarate but the pathway goes no further. Oxygen is consumed by reoxidation of the NADH produced by isocitrate dehydrogenase.

(g)

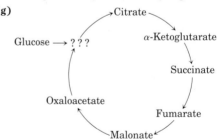

This differs from Fig. 16–7 in that it does not include *cis*-aconitate and isocitrate (between citrate and α-ketoglutarate), or succinyl-CoA, or acetyl-CoA. (h) Establishing a quantitative conversion was essential to rule out a branched or other, more complex pathway.

Chapter 17

1. The fatty acid portion; the carbons in fatty acids are more reduced than those in glycerol.

2. (a) 4.0×10^5 kJ (9.6×10^4 kcal) (b) 48 days (c) 0.48 lb/day

3. The first step in fatty acid oxidation is analogous to the conversion of succinate to fumarate; the second step, to the conversion of fumarate to malate; the third step, to the conversion of malate to oxaloacetate.

4. 8 cycles; the last releases 2 acetyl-CoA.

5. (a) R—COO$^-$ + ATP → acyl-AMP + PP$_i$

Acyl-AMP + CoA → acyl-CoA + AMP

(b) Irreversible hydrolysis of PP$_i$ to 2P$_i$ by cellular inorganic pyrophosphatase

6. *cis*-Δ^3-Dodecanoyl-CoA; it is converted to *cis*-Δ^2-dodecanoyl-CoA, then β-hydroxydodecanoyl-CoA.

7. 4 acetyl-CoA and 1 propionyl-CoA

8. Yes. Some of the tritium is removed from palmitate during the dehydrogenation reactions of β oxidation. The removed tritium appears as tritiated water.

9. Fatty acyl groups condensed with CoA in the cytosol are first transferred to carnitine, releasing CoA, then transported into the mitochondrion, where they are again condensed with CoA. The cytosolic and mitochondrial pools of CoA are thus kept separate, and no radioactive CoA from the cytosolic pool enters the mitochondrion.

10. (a) In the pigeon, β oxidation predominates; in the pheasant, anaerobic glycolysis of glycogen predominates. **(b)** Pigeon muscle would consume more O$_2$. **(c)** Fat contains more energy per gram than glycogen does. In addition, the anaerobic breakdown of glycogen is limited by the tissue's tolerance to lactate buildup. Thus the pigeon, operating on the oxidative catabolism of fats, is the long-distance flyer. **(d)** These enzymes are the regulatory enzymes of their respective pathways and thus limit ATP production rates.

11. Malonyl-CoA would no longer inhibit fatty acid entry into the mitochondrion and β oxidation, so there might be a futile cycle of simultaneous fatty acid synthesis in the cytosol and fatty acid breakdown in mitochondria.

12. (a) The carnitine-mediated entry of fatty acids into mitochondria is the rate-limiting step in fatty acid oxidation. Carnitine deficiency slows fatty acid oxidation; added carnitine increases the rate. **(b)** All increase the metabolic need for fatty acid oxidation. **(c)** Carnitine deficiency might result from a deficiency of lysine, its precursor, or from a defect in one of the enzymes in the biosynthesis of carnitine.

13. Oxidation of fats releases metabolic water; 1.4 L of water per kg of tripalmitoylglycerol (ignores the small contribution of glycerol to the mass).

14. The bacteria can be used to completely oxidize hydrocarbons to CO$_2$ and H$_2$O. However, contact between hydrocarbons and bacterial enzymes may be difficult to achieve. Bacterial nutrients such as nitrogen and phosphorus may be limiting and inhibit growth.

15. (a) M_r 136; phenylacetic acid **(b)** Even

16. Because the mitochondrial pool of CoA is small, CoA must be recycled from acetyl-CoA via the formation of ketone bodies. This allows the operation of the β-oxidation pathway, necessary for energy production.

17. (a) Glucose yields pyruvate via glycolysis, and pyruvate is the main source of oxaloacetate. Without glucose in the diet, [oxaloacetate] drops and the citric acid cycle slows. **(b)** Odd-numbered; propionate conversion to succinyl-CoA provides intermediates for the citric acid cycle and four-carbon precursors for gluconeogenesis.

18. For the odd-chain heptanoic acid, β oxidation produces propionyl-CoA, which can be converted in several steps to oxaloacetate, a starting material for gluconeogenesis. The even-chain fatty acid cannot support gluconeogenesis, because it is entirely oxidized to acetyl-CoA.

19. β Oxidation of ω-fluorooleate forms fluoroacetyl-CoA, which enters the citric acid cycle and produces fluorocitrate, a powerful inhibitor of aconitase. Inhibition of aconitase shuts down the citric acid cycle. Without reducing equivalents from the citric acid cycle, oxidative phosphorylation (ATP synthesis) is fatally slowed.

20. Ser to Ala: blocks β oxidation in mitochondria. Ser to Asp: blocks fatty acid synthesis, stimulates β oxidation.

21. Response to glucagon or epinephrine would be prolonged, giving a greater mobilization of fatty acids in adipocytes.

22. Enz-FAD, having a more positive standard reduction potential, is a better electron acceptor than NAD$^+$, and the reaction is driven in the direction of fatty acyl–CoA oxidation. This more favorable equilibrium is obtained at the cost of 1 ATP; only 1.5 ATP are produced per FADH$_2$ oxidized in the respiratory chain (vs. 2.5 per NADH).

23. 9 turns; arachidic acid, a 20-carbon saturated fatty acid, yields 10 molecules of acetyl-CoA, the last two formed in the ninth turn.

24. See Fig. 17–12. [3-^{14}C]Succinyl-CoA is formed, which gives rise to oxaloacetate labeled at C-2 and C-3.

25. Phytanic acid → pristanic acid → propionyl-CoA → → → succinyl-CoA → succinate → fumarate → malate. All malate carbons would be labeled, but C-1 and C-4 would have only half as much label as C-2 and C-3.

26. ATP hydrolysis in the energy-requiring reactions of a cell takes up water in the reaction ATP + H$_2$O → ADP + P$_i$; thus, in the steady state, there is no *net* production of H$_2$O.

27. Methylmalonyl-CoA mutase requires the cobalt-containing cofactor formed from vitamin B$_{12}$.

28. Mass lost per day is about 0.66 kg, or about 140 kg in 7 months. Ketosis could be avoided by degradation of nonessential body proteins to supply amino acid skeletons for gluconeogenesis.

29. (a) Fatty acids are converted to their CoA derivatives by enzymes in the cytoplasm; the acyl-CoAs are then imported into mitochondria for oxidation. Given that the researchers were using isolated mitochondria, they had to use CoA derivatives. **(b)** Stearoyl-CoA was rapidly converted to 9 acetyl-CoA by the β-oxidation pathway. All intermediates reacted rapidly and none were detectable at significant levels. **(c)** Two rounds. Each round removes two carbon atoms, thus two rounds convert an 18-carbon to a 14-carbon fatty acid and 2 acetyl-CoA. **(d)** The K_m is higher for the trans isomer than for the cis, so a higher concentration of trans isomer is required for the same rate of breakdown. Roughly speaking, the trans isomer binds less well than the cis, probably because differences in shape, even though not at the target site for the enzyme, affect substrate binding to the enzyme. **(e)** The substrate for LCAD/VLCAD builds up differently, depending on the particular substrate; this is expected for the rate-limiting step in a pathway. **(f)** The kinetic parameters show that the trans isomer is a poorer substrate than the cis for LCAD, but there is little difference for VLCAD. Because it is a poorer substrate, the trans isomer accumulates to higher levels than the cis. **(g)** One possible pathway is shown below (indicating "inside" and "outside" mitochondria).

Elaidoyl-CoA $\xrightarrow{\text{carnitine acyltransferase I}}$ elaidoyl-carnitine $\xrightarrow{\text{transport}}$
(outside)   (outside)

elaidoyl-carnitine $\xrightarrow{\text{carnitine acyltransferase II}}$ elaidoyl-CoA $\xrightarrow{\text{2 rounds of }\beta\text{ oxidation}}$
(inside)   (inside)

5-*trans*-tetradecenoyl-CoA $\xrightarrow{\text{thioesterase}}$ 5-*trans*-tetradecanoic acid $\xrightarrow{\text{diffusion}}$
(inside)   (inside)

5-*trans*-tetradecanoic acid
(outside)

(h) It is correct insofar as trans fats are broken down less efficiently than cis fats, and thus trans fats may "leak" out of mitochondria. It is incorrect to say that trans fats are not broken down by cells; they are broken down, but at a slower rate than cis fats.

Chapter 18

1.

(a) $^-OOC-CH_2-\overset{\overset{\textstyle O}{\|}}{C}-COO^-$ Oxaloacetate

(b) $^-OOC-CH_2-CH_2-\overset{\overset{\textstyle O}{\|}}{C}-COO^-$ α-Ketoglutarate

(c) $CH_3-\overset{\overset{\textstyle O}{\|}}{C}-COO^-$ Pyruvate

(d) (phenyl ring)$-CH_2-\overset{\overset{\textstyle O}{\|}}{C}-COO^-$ Phenylpyruvate

2. This is a coupled-reaction assay. The product of the slow transamination (pyruvate) is rapidly consumed in the subsequent "indicator reaction" catalyzed by lactate dehydrogenase, which consumes NADH. Thus the rate of disappearance of NADH is a measure of the rate of the aminotransferase reaction. The indicator reaction is monitored by observing the decrease in absorption of NADH at 340 nm with a spectrophotometer.

3. Alanine and glutamine play special roles in the transport of amino groups from muscle and from other nonhepatic tissues, respectively, to the liver.

4. No. The nitrogen in alanine can be transferred to oxaloacetate via transamination, to form aspartate.

5. 15 mol of ATP per mole of lactate; 13 mol of ATP per mole of alanine, when nitrogen removal is included

6. **(a)** Fasting resulted in low blood glucose; subsequent administration of the experimental diet led to rapid catabolism of glucogenic amino acids. **(b)** Oxidative deamination caused the rise in NH_3 levels; the absence of arginine (an intermediate in the urea cycle) prevented conversion of NH_3 to urea; arginine is not synthesized in sufficient quantities in the cat to meet the needs imposed by the stress of the experiment. This suggests that arginine is an essential amino acid in the cat's diet. **(c)** Ornithine is converted to arginine by the urea cycle.

7. H_2O + glutamate + $NAD^+ \rightarrow$
    α-ketoglutarate + NH_4^+ + NADH + H^+
NH_4^+ + 2ATP + H_2O + $CO_2 \rightarrow$
    carbamoyl phosphate + 2ADP + P_i + $3H^+$
Carbamoyl phosphate + ornithine → citrulline + P_i + H^+
Citrulline + aspartate + ATP →
    argininosuccinate + AMP + PP_i + H^+
Argininosuccinate → arginine + fumarate
Fumarate + H_2O → malate
Malate + NAD^+ → oxaloacetate + NADH + H^+
Oxaloacetate + glutamate → aspartate + α-ketoglutarate
Arginine + H_2O → urea + ornithine

---

2 Glutamate + CO_2 + $4H_2O$ + $2NAD^+$ + 3ATP →
    2 α-ketoglutarate + 2NADH + $7H^+$ + urea + 2ADP +
        AMP + PP_i + $2P_i$ (1)

Additional reactions that need to be considered:
AMP + ATP → 2ADP   (2)
O_2 + $8H^+$ + 2NADH + 6ADP + $6P_i$ →
        $2NAD^+$ + 6ATP + $8H_2O$ (3)
H_2O + PP_i → $2P_i$ + H^+ (4)

Summing equations (1) through (4),
2 Glutamate + CO_2 + O_2 + 2ADP + $2P_i$ →
        2 α-ketoglutarate + urea + $3H_2O$ + 2ATP

8. The second amino group introduced into urea is transferred from aspartate, which is generated during the transamination of glutamate to oxaloacetate, a reaction catalyzed by aspartate aminotransferase. Approximately one-half of all the amino groups excreted as urea must pass through the aspartate aminotransferase reaction, making this the most highly active aminotransferase.

9. **(a)** A person on a diet consisting only of protein must use amino acids as the principal source of metabolic fuel. Because the catabolism of amino acids requires the removal of nitrogen as urea, the process consumes abnormally large quantities of water to dilute and excrete the urea in the urine. Furthermore, electrolytes in the "liquid protein" must be diluted with water and excreted. If the daily water loss through the kidney is not balanced by a sufficient water intake, a net loss of body water results. **(b)** When considering the nutritional benefits of protein, one must keep in mind the total amount of amino acids needed for protein synthesis and the distribution of amino acids in the dietary protein. Gelatin contains a nutritionally unbalanced distribution of amino acids. As large amounts of gelatin are ingested and the excess amino acids are catabolized, the capacity of the urea cycle may be exceeded, leading to ammonia toxicity. This is further complicated by the dehydration that may result from excretion of large quantities of urea. A combination of these two factors could produce coma and death.

10. Lysine and leucine

11. **(a)** Phenylalanine hydroxylase; a low-phenylalanine diet **(b)** The normal route of phenylalanine metabolism via hydroxylation to tyrosine is blocked, and phenylalanine accumulates. **(c)** Phenylalanine is transformed to phenylpyruvate by transamination, and then to phenyllactate by reduction. The transamination reaction has an equilibrium constant of 1.0, and phenylpyruvate is formed in significant amounts when phenylalanine accumulates. **(d)** Because of the deficiency in production of tyrosine, which is a precursor of melanin, the pigment normally present in hair.

12. Catabolism of the carbon skeletons of valine, methionine, and isoleucine is impaired because a functional methylmalonyl-CoA mutase (a coenzyme B_{12} enzyme) is absent. The physiological effects of loss of this enzyme are described in Table 18–2 and Box 18–2.

13. The vegan diet lacks vitamin B_{12}, leading to the increase in homocysteine and methylmalonate (reflecting the deficiencies in methionine synthase and methylmalonic acid mutase, respectively) in individuals on the diet for several years. Dairy products provide some vitamin B_{12} in the lactovegetarian diet.

14. The genetic forms of pernicious anemia generally arise as a result of defects in the pathway that mediates absorption of dietary vitamin B_{12} (see Box 17–2). Because dietary supplements are not absorbed in the intestine, these conditions are treated by injecting supplementary B_{12} directly into the bloodstream.

15. The mechanism is identical to that for serine dehydratase (see Fig. 18–20a) except that the extra methyl group of threonine is retained, yielding α-ketobutyrate instead of pyruvate.

16. **(a)** $^{15}NH_2-CO-^{15}NH_2$

(b) $^-OO^{14}C-CH_2-CH_2-^{14}COO^-$

(c) $R-NH-\overset{\overset{\textstyle ^{15}NH}{\|}}{C}-^{15}NH_2$

(d) $R-NH-\overset{\overset{\textstyle O}{\|}}{C}-^{15}NH_2$

(e) No label

(f) $^-OO^{14}C-\overset{\overset{\textstyle ^{15}NH_2}{|}}{\underset{\underset{\textstyle H}{|}}{C}}-CH_2-^{14}COO^-$

17. (a) Isoleucine ⟶ II ⟶ IV ⟶ I ⟶ V ⟶ III ⟶ (steps numbered ① ② ③ ④ ⑤ ⑥)
acetyl-CoA + propionyl-CoA **(b)** Step ① transamination, no analogous reaction, PLP; ② oxidative decarboxylation, analogous to the pyruvate dehydrogenase reaction, NAD⁺, TPP, lipoate, FAD; ③ oxidation, analogous to the succinate dehydrogenase reaction, FAD; ④ hydration, analogous to the fumarase reaction, no cofactor; ⑤ oxidation, analogous to the malate dehydrogenase reaction, NAD⁺; ⑥ thiolysis (reverse aldol condensation), analogous to the thiolase reaction, CoA.

18. A likely mechanism is:

[Chemical mechanism: Serine + PLP, showing reaction with Enz-Lys, through PLP-aldimine intermediates, formaldehyde elimination, and formation of Glycine-PLP product. The formaldehyde (HCHO) produced in the second step reacts rapidly with tetrahydrofolate at the enzyme active site to produce N^5,N^{10}-methylenetetrahydrofolate (see Fig. 18–17).]

19. (a) Transamination; no analogies; PLP. **(b)** Oxidative decarboxylation; analogous to oxidative decarboxylation of pyruvate to acetyl-CoA prior to entry into the citric acid cycle, and of α-ketoglutarate to succinyl-CoA in the citric acid cycle; NAD⁺, FAD, lipoate, and TPP. **(c)** Dehydrogenation (oxidation); analogous to dehydrogenation of succinate to fumarate in the citric acid cycle, and of fatty acyl–CoA to enoyl-CoA in β oxidation; FAD. **(d)** Carboxylation; no analogies in citric acid cycle or β oxidation; ATP and biotin. **(e)** Hydration; analogous to hydration of fumarate to malate in the citric acid cycle, and of enoyl-CoA to 3-hydroxyacyl-CoA in β oxidation; no cofactors. **(f)** Reverse aldol reaction; analogous to reverse of citrate synthase reaction in the citric acid cycle; no cofactors

20. (a) Leucine; valine; isoleucine **(b)** Cysteine (derived from cystine). If cysteine were decarboxylated as shown in Fig. 18–6, it would yield $H_3N–CH_2–CH_2–SH$, which could be oxidized to taurine. **(c)** The January 1957 blood shows significantly elevated levels of isoleucine, leucine, methionine, and valine; the January 1957 urine, significantly elevated isoleucine, leucine, taurine, and valine. **(d)** All patients had high levels of isoleucine, leucine, and valine in both blood and urine, suggesting a defect in the breakdown of these amino acids. Given that the urine also contained high levels of the keto forms of these three amino acids, the block in the pathway must occur after deamination but before dehydrogenation (as shown in Fig. 18–28). **(e)** The model does not explain the high levels of methionine in blood and taurine in urine. The high taurine levels may be due to the death of brain cells during the end stage of the disease. However, the reason for high levels of methionine in blood are unclear; the pathway of methionine degradation is not linked with the degradation of branched-chain amino acids. Increased methionine could be a secondary effect of buildup of the other amino acids. It is important to keep in mind that the January 1957 samples were from an individual who was dying, so comparing blood and urine results with those of a healthy individual may not be appropriate. **(f)** The following information is needed (and was eventually obtained by other workers): (1) The dehydrogenase activity is significantly reduced or missing in individuals with maple syrup urine disease. (2) The disease is inherited as a single-gene defect. (3) The defect occurs in a gene encoding all or part of the dehydrogenase. (4) The genetic defect leads to production of inactive enzyme.

Chapter 19

1. *Reaction (1):* (a), (d) NADH; (b), (e) E-FMN; (c) NAD⁺/NADH and E-FMN/FMNH₂
Reaction (2): (a), (d) E-FMNH₂; (b), (e) Fe^{3+}; (c) E-FMN/FMNH₂ and Fe^{3+}/Fe^{2+}
Reaction (3): (a), (d) Fe^{2+}; (b), (e) Q; (c) Fe^{3+}/Fe^{2+} and Q/QH₂

2. The side chain makes ubiquinone soluble in lipids and allows it to diffuse in the semifluid membrane.

3. From the difference in standard reduction potential ($\Delta E'^\circ$) for each pair of half-reactions, one can calculate $\Delta G'^\circ$. The oxidation of succinate by FAD is favored by the negative standard free-energy change ($\Delta G'^\circ = -3.7$ kJ/mol). Oxidation by NAD⁺ would require a large, positive, standard free-energy change ($\Delta G'^\circ = 68$ kJ/mol).

4. (a) All carriers reduced; CN⁻ blocks the reduction of O_2 catalyzed by cytochrome oxidase. **(b)** All carriers reduced; in the absence of O_2, the reduced carriers are not reoxidized. **(c)** All carriers oxidized. **(d)** Early carriers more reduced; later carriers more oxidized.

5. (a) Inhibition of NADH dehydrogenase by rotenone decreases the rate of electron flow through the respiratory chain, which in turn decreases the rate of ATP production. If this reduced rate is unable to meet the organism's ATP requirements, the organism dies. **(b)** Antimycin A strongly inhibits the oxidation of Q in the respiratory chain, reducing the rate of electron transfer and leading to the consequences described in (a). **(c)** Because antimycin A blocks *all* electron flow to oxygen, it is a more potent poison than rotenone, which blocks electron flow from NADH but not from FADH₂.

6. (a) The rate of electron transfer necessary to meet the ATP demand increases, and thus the P/O ratio decreases. **(b)** High concentrations of uncoupler produce P/O ratios near zero. The P/O ratio decreases, and more fuel must be oxidized to generate the same amount of ATP. The extra heat released by this oxidation raises the body temperature. **(c)** Increased activity of the respiratory chain in the presence of an uncoupler requires the degradation of additional fuel. By oxidizing more fuel (including fat reserves) to produce the same amount of ATP, the body loses weight. When the P/O ratio approaches zero, the lack of ATP results in death.

7. Valinomycin acts as an uncoupler. It combines with K^+ to form a complex that passes through the inner mitochondrial membrane, dissipating the membrane potential. ATP synthesis decreases, which causes the rate of electron transfer to increase. This results in an increase in the H^+ gradient, O_2 consumption, and amount of heat released.

8. **(a)** The formation of ATP is inhibited. **(b)** The formation of ATP is tightly coupled to electron transfer: 2,4-dinitrophenol is an uncoupler of oxidative phosphorylation. **(c)** Oligomycin

9. Cytosolic malate dehydrogenase plays a key role in the transport of reducing equivalents across the inner mitochondrial membrane via the malate-aspartate shuttle.

10. **(a)** Glycolysis becomes anaerobic. **(b)** Oxygen consumption ceases. **(c)** Lactate formation increases. **(d)** ATP synthesis decreases to 2 ATP/glucose.

11. The steady-state concentration of P_i in the cell is much higher than that of ADP. The P_i released by ATP hydrolysis changes total $[P_i]$ very little.

12. The response to (a) increased [ADP] is faster because the response to (b) reduced pO_2 requires protein synthesis.

13. **(a)** NADH is reoxidized via electron transfer instead of lactic acid fermentation. **(b)** Oxidative phosphorylation is more efficient. **(c)** The high mass-action ratio of the ATP system inhibits phosphofructokinase-1.

14. Fermentation to ethanol could be accomplished in the presence of O_2, which is an advantage because strict anaerobic conditions are difficult to maintain. The Pasteur effect is not observed, since the citric acid cycle and electron-transfer chain are inactive.

15. More-efficient electron transfer between complexes.

16. **(a)** External medium: 4.0×10^{-8} M; matrix: 2.0×10^{-8} M **(b)** $[H^+]$ gradient contributes 1.7 kJ/mol toward ATP synthesis **(c)** 21 **(d)** No **(e)** From the overall transmembrane potential

17. **(a)** $0.91\ \mu$mol/s · g **(b)** 5.5 s; to provide a constant level of ATP, regulation of ATP production must be tight and rapid.

18. 53 μmol/s · g. With a steady state [ATP] of 7.0 μmol/g, this is equivalent to 10 turnovers of the ATP pool per second; the reservoir would last about 0.13 s.

19. Reactive oxygen species react with macromolecules, including DNA. If a mitochondrial defect leads to increased production of ROS, the nuclear genes that encode proto-oncogenes can be damaged, producing oncogenes and leading to unregulated cell division and cancer (see Section 12.12).

20. Different extents of heteroplasmy for the defective gene produce different degrees of defective mitochondrial function.

21. The inner mitochondrial membrane is impermeable to NADH, but the reducing equivalents of NADH are transferred (shuttled) through the membrane indirectly: they are transferred to oxaloacetate in the cytosol, the resulting malate is transported into the matrix, and mitochondrial NAD^+ is reduced to NADH.

22. The citric acid cycle is stalled for lack of an acceptor of electrons from NADH. Pyruvate produced by glycolysis cannot enter the cycle as acetyl-CoA; accumulated pyruvate is transaminated to alanine and exported to the liver.

23. Pyruvate dehydrogenase is located in mitochondria, glyceraldehyde 3-phosphate dehydrogenase in the cytosol. The NAD pools are separated by the inner mitochondrial membrane.

24. Complete lack of glucokinase (two defective alleles) makes it impossible to carry out glycolysis at a sufficient rate to raise [ATP] to the threshold required for insulin secretion.

25. Defects in Complex II result in increased production of ROS, damage to DNA, and mutations that lead to unregulated cell division (cancer; see Section 12.12). It is not clear why the cancer tends to occur in the midgut.

26. For the maximum photosynthetic rate, PSI (which absorbs light of 700 nm) and PSII (which absorbs light of 680 nm) must be operating simultaneously.

27. The extra weight comes from the water consumed in the overall reaction.

28. Purple sulfur bacteria use H_2S as the hydrogen donor in photosynthesis. No O_2 is evolved, because the single photosystem lacks the manganese-containing water-splitting complex.

29. 0.44

30. **(a)** Stops **(b)** Slows; some electron flow continues by the cyclic pathway.

31. During illumination, a proton gradient is established. When ADP and P_i are added, ATP synthesis is driven by the gradient, which becomes exhausted in the absence of light.

32. DCMU blocks electron transfer between PSII and the first site of ATP production.

33. In the presence of venturicidin, proton movement through the CF_oCF_1 complex is blocked, and electron flow (oxygen evolution) continues only until the free-energy cost of pumping protons against the rising proton gradient equals the free energy available in a photon. DNP, by dissipating the proton gradient, restores electron flow and oxygen evolution.

34. **(a)** 56 kJ/mol **(b)** 0.29 V

35. From the difference in reduction potentials, one can calculate that $\Delta G'^\circ = 15$ kJ/mol for the redox reaction. Fig. 19–48 shows that the energy of photons in any region of the visible spectrum is more than sufficient to drive this endergonic reaction.

36. 1.35×10^{-77}; the reaction is highly unfavorable! In chloroplasts, the input of light energy overcomes this barrier.

37. -920 kJ/mol

38. No. The electrons from H_2O flow to the artificial electron acceptor Fe^{3+}, not to $NADP^+$.

39. About once every 0.1 s; 1 in 10^8 is excited.

40. Light of 700 nm excites PSI but not PSII; electrons flow from P700 to $NADP^+$, but no electrons flow from P680 to replace them. When light of 680 nm excites PSII, electrons tend to flow to PSI, but the electron carriers between the two photosystems quickly become completely reduced.

41. No. The excited electron from P700 returns to refill the electron "hole" created by illumination. PSII is not needed to supply electrons, and no O_2 is evolved from H_2O. NADPH is not formed, because the excited electron returns to P700.

42. **(a)** (1) The presence of Mg^{2+} supports the hypothesis that chlorophyll is directly involved in catalysis of the phosphorylation reaction: ADP + $P_i \rightarrow$ ATP. (2) Many enzymes (or other proteins) that contain Mg^{2+} are not phosphorylating enzymes, so the presence of Mg^{2+} in chlorophyll does not prove its role in phosphorylation reactions. (3) The presence of Mg^{2+} is essential to chlorophyll's photochemical properties: light absorption and electron transfer. **(b)** (1) Enzymes catalyze reversible reactions, so an isolated enzyme that can, under certain laboratory conditions, catalyze removal of a phosphoryl group could probably, under different conditions (such as in cells), catalyze addition of a phosphoryl group. So it is plausible that chlorophyll could be involved in the phosphorylation of ADP. (2) There are two possible explanations: the chlorophyll protein is a phosphatase only and does not catalyze ADP phosphorylation under cellular conditions, or the crude preparation contains a contaminating phosphatase activity that is unconnected to the photosynthetic reactions. (3) It is likely that the preparation was contaminated with a nonphotosynthetic

phosphatase activity. **(c)** (1) This light inhibition is what one would expect if the chlorophyll protein catalyzed the reaction ADP + P_i + light → ATP. Without light, the reverse reaction, a dephosphorylation, would be favored. In the presence of light, energy is provided and the equilibrium would shift to the right, reducing the phosphatase activity. (2) This inhibition must be an artifact of the isolation or assay methods. (3) It is unlikely that the crude preparation methods in use at the time preserved intact chloroplast membranes, so the inhibition must be an artifact. **(d)** (1) In the presence of light, ATP is synthesized and other phosphorylated intermediates are consumed. (2) In the presence of light, glucose is produced and is metabolized by cellular respiration to produce ATP, with changes in the levels of phosphorylated intermediates. (3) In the presence of light, ATP is produced and other phosphorylated intermediates are consumed. **(e)** Light energy is used to produce ATP (as in the Emerson model) *and* is used to produce reducing power (as in the Rabinowitch model). **(f)** The approximate stoichiometry for photophosphorylation (Chapter 19) is that 8 photons yield 2 NADPH and about 3 ATP. Two NADPH and 3 ATP are required to reduce 1 CO_2 (Chapter 20). Thus, at a minimum, 8 photons are required per CO_2 molecule reduced. This is in good agreement with Rabinowitch's value. **(g)** Because the energy of light is used to produce *both* ATP and NADPH, each photon absorbed contributes more than just 1 ATP for photosynthesis. The process of energy extraction from light is more efficient than Rabinowitch supposed, and plenty of energy is available for this process—even with red light.

Chapter 20

1. Within subcellular organelles, concentrations of specific enzymes and metabolites are elevated; separate pools of cofactors and intermediates are maintained; regulatory mechanisms affect only one set of enzymes and pools.

2. This observation suggests that ATP and NADPH are generated in the light and are essential for CO_2 fixation; conversion stops as the supply of ATP and NADPH becomes exhausted. Furthermore, some enzymes are switched off in the dark.

3. X is 3-phosphoglycerate; Y is ribulose 1,5-bisphosphate.

4. Ribulose 5-phosphate kinase, fructose 1,6-bisphosphatase, sedoheptulose 1,7-bisphosphatase, and glyceraldehyde 3-phosphate dehydrogenase; all are activated by reduction of a critical disulfide bond to a pair of sulfhydryls; iodoacetate reacts irreversibly with free sulfhydryls.

5. To carry out the disulfide exchange reaction that activates the Calvin cycle enzymes, thioredoxin needs both of its sulfhydryl groups.

6. Reductive pentose phosphate pathway regenerates ribulose 1,5-bisphosphate from triose phosphates produced during photosynthesis. Oxidative pentose phosphate pathway provides NADPH for reductive biosynthesis and pentose phosphates for nucleotide synthesis.

7. Both types of "respiration" occur in plants, consume O_2, and produce CO_2. (Mitochondrial respiration also occurs in animals.) Mitochondrial respiration occurs continuously; electrons derived from various fuels are passed through a chain of carriers in the inner mitochondrial membrane to O_2. Photorespiration occurs in chloroplasts, peroxisomes, and mitochondria. Photorespiration occurs during the daytime, when photosynthetic carbon fixation is occurring; mitochondrial respiration occurs primarily at night, or during cloudy days. The path of electron flow in photorespiration is shown in Fig. 20–21; that for mitochondrial respiration, in Fig. 19–19.

8. This hypothesis assumes directed evolution, or evolution with a purpose—ideas not generally accepted by evolutionary biologists. Other processes, such as burning fossil fuels and global deforestation, affect the global atmospheric composition.

C_4 plants, by fixing CO_2 under conditions when rubisco prefers O_2 as substrate, also contributes to setting atmospheric CO_2/O_2 ratios.

9. **(a)** Without production of NADPH by the pentose phosphate pathway, cells would be unable to synthesize lipids and other reduced products. **(b)** Without generation of ribulose 1,5-bisphosphate, the Calvin cycle is effectively blocked.

10. In maize, CO_2 is fixed by the C_4 pathway elucidated by Hatch and Slack, in which PEP is carboxylated rapidly to oxaloacetate (some of which undergoes transamination to aspartate) and reduced to malate. Only after subsequent decarboxylation does the CO_2 enter the Calvin cycle.

11. Measure the rate of fixation of ^{14}C carbon dioxide in the light (daytime) and the dark. Greater fixation in the dark identifies the CAM plant. One could also determine the titratable acidity; acids stored in the vacuole during the night can be quantified in this way.

12. Isocitrate dehydrogenase reaction

13. Storage consumes 1 mol of ATP per mole of glucose 6-phosphate; this represents 3.3% of the total ATP available from glucose 6-phosphate metabolism (i.e., the efficiency of storage is 96.7%).

14. $[PP_i]$ is high in the cytosol because the cytosol lacks inorganic pyrophosphatase.

15. **(a)** Low $[P_i]$ in the cytosol and high [triose phosphate] in the chloroplast **(b)** High [triose phosphate] in the cytosol

16. 3-Phosphoglycerate is the primary product of photosynthesis; $[P_i]$ rises when light-driven synthesis of ATP from ADP and P_i slows.

17. **(a)** Sucrose + (glucose)$_n$ → (glucose)$_{n+1}$ + fructose **(b)** Fructose generated in the synthesis of dextran is readily imported and metabolized by the bacteria.

18. Species 1 is C_4; species 2, C_3.

19. **(a)** In peripheral chloroplasts **(b)**, **(c)** In central sphere

20. **(a)** By analogy to the oxygenic photosynthesis carried out by plants ($H_2O + CO_2$ → glucose + O_2), the reaction would be $H_2S + O_2 + CO_2$ → glucose + H_2O + S. This is the sum of the reduction of CO_2 by H_2S ($H_2S + CO_2$ → glucose + S) and the energy input ($H_2S + O_2$ → S + H_2O). **(b)** The H_2S and CO_2 are produced chemically in deep-sea sediments, but the O_2, like the vast majority of O_2 on Earth, is produced by photosynthesis, which is driven by light energy. **(c)** In the assay used by Robinson et al., ^3H labels the C-1 of ribulose 1,5-bisphosphate, so reaction with CO_2 yields one molecule of $[^3H]$3-phosphoglycerate and one molecule of unlabeled 3-phosphoglycerate; reaction with O_2 produces one molecule of $[^3H]$2-phosphoglycolate and one molecule of unlabeled 3-phosphoglycerate. Thus the ratio of $[^3H]$3-phosphoglycerate to $[^3H]$2-phosphoglycolate equals the ratio of carboxylation to oxygenation. **(d)** If the ^3H labeled C-5, *both* oxygenation and carboxylation would yield $[^3H]$3-phosphoglycerate and it would be impossible to distinguish which reaction had produced the labeled product; the reaction could not be used to measure Ω.

(e) Substituting $\dfrac{[CO_2]}{[O_2]} = \dfrac{0.00038}{0.2} = 0.0019$ into

$$\frac{V_{carboxylation}}{V_{oxygenation}} = \Omega \frac{[CO_2]}{[O_2]} \text{ gives}$$

$$\frac{V_{carboxylation}}{V_{oxygenation}} = (8.6)(0.0019) = 0.016$$

Therefore, the rate of oxygenation would be roughly 60 times the rate of carboxylation! **(f)** If terrestrial plants had $\Omega = 8.6$, carboxylation would occur at a much lower rate than oxygenation. This would be extremely inefficient, so one would expect the rubisco of terrestrial plants to have an Ω substantially higher than 8.6. In fact, Ω values for land plants vary between 10 and 250. Even with these values, the expected rate of the oxygenation reaction is still very high. **(g)** The rubisco reaction occurs with CO_2 as a gas. At the same temperature, $^{13}CO_2$ molecules diffuse more slowly than the lighter $^{12}CO_2$ molecules, and thus $^{13}CO_2$ will enter the active site (and become incorporated into substrate) more slowly than $^{12}CO_2$. **(h)** For the relationship to be truly symbiotic, the tube worms must be getting a substantial amount of their carbon from the bacteria. The presence of rubisco in the endosymbionts simply shows that they are capable of chemosynthesis, not that they are supplying the host with a significant fraction of its carbon. On the other hand, showing that the ^{13}C:^{12}C ratio in the host is more similar to that in the endosymbiont than that in other marine animals strongly suggests that the tube worms are getting the majority of their carbon from the bacteria.

Chapter 21

1. **(a)** The 16 carbons of palmitate are derived from 8 acetyl groups of 8 acetyl-CoA molecules. The ^{14}C-labeled acetyl-CoA gives rise to malonyl-CoA labeled at C-1 and C-2. **(b)** The metabolic pool of malonyl-CoA, the source of all palmitate carbons except C-16 and C-15, does not become labeled with small amounts of ^{14}C-labeled acetyl-CoA. Hence, only [15,16-^{14}C] palmitate is formed.

2. Both glucose and fructose are degraded to pyruvate in glycolysis. Pyruvate is converted to acetyl-CoA by the pyruvate dehydrogenase complex. Some of this acetyl-CoA enters the citric acid cycle, which produces reducing equivalents (NADH and NADPH). Mitochondrial electron transfer to O_2 yields ATP.

3. 8 Acetyl-CoA + 15ATP + 14NADPH + 9H$_2$O $\rightarrow$
     palmitate + 8CoA + 15ADP + 15P$_i$ + 14NADP$^+$ + 2H$^+$

4. **(a)** 3 deuteriums per palmitate; all located on C-16; all other two-carbon units are derived from unlabeled malonyl-CoA. **(b)** 7 deuteriums per palmitate; located on all *even*-numbered carbons except C-16.

5. By using the three-carbon unit malonyl-CoA, the activated form of acetyl-CoA (recall that malonyl-CoA synthesis requires ATP), metabolism is driven in the direction of fatty acid synthesis by the exergonic release of CO_2.

6. The rate-limiting step in fatty acid synthesis is carboxylation of acetyl-CoA, catalyzed by acetyl-CoA carboxylase. High [citrate] and [isocitrate] indicate that conditions are favorable for fatty acid synthesis: an active citric acid cycle is providing a plentiful supply of ATP, reduced pyridine nucleotides, and acetyl-CoA. Citrate stimulates (increases the V_{max} of) acetyl-CoA carboxylase (a). Because citrate binds more tightly to the filamentous form of the enzyme (the active form), high [citrate] drives the protomer $\rightleftharpoons$ filament equilibrium in the direction of the active form (b). In contrast, palmitoyl-CoA (the end product of fatty acid synthesis) drives the equilibrium in the direction of the inactive (protomer) form. Hence, when the end product of fatty acid synthesis accumulates, the biosynthetic path slows.

7. **(a)** Acetyl-CoA$_{(mit)}$ + ATP + CoA$_{(cyt)}$ $\rightarrow$ acetyl-CoA$_{(cyt)}$ + ADP + P$_i$ + CoA$_{(mit)}$ **(b)** 1 ATP per acetyl group **(c)** Yes

8. The double bond in palmitoleate is introduced by an oxidation catalyzed by fatty acyl–CoA desaturase, a mixed-function oxidase that requires O_2 as a cosubstrate.

9. 3 Palmitate + glycerol + 7ATP + 4H$_2$O $\rightarrow$
     tripalmitin + 7ADP + 7P$_i$ + 7H$^+$

10. In adult rats, stored triacylglycerols are maintained at a steady-state level through a balance of the rates of degradation and biosynthesis. Hence, the triacylglycerols of adipose (fat) tissue are constantly turned over, which explains the incorporation of ^{14}C label from dietary glucose.

11. Net reaction:

     Dihydroxyacetone phosphate + NADH + palmitate + oleate + 3ATP + CTP + choline + 4H$_2$O $\rightarrow$
     phosphatidylcholine + NAD$^+$ + 2AMP + ADP + H$^+$ + CMP + 5P$_i$
     7ATP per molecule of phosphatidylcholine

12. Methionine deficiency reduces the level of adoMet, which is required for the de novo synthesis of phosphatidylcholine. The salvage pathway does not employ adoMet, but uses available choline. Thus phosphatidylcholine can be synthesized even when the diet is deficient in methionine, as long as choline is available.

13. ^{14}C label appears in three places in the activated isoprene:

$$^{14}CH_2 \diagdown \atop ^{14}CH_3 \diagup C - {}^{14}CH_2 - CH_2 -$$

14. **(a)** ATP **(b)** UDP-glucose **(c)** CDP-ethanolamine **(d)** UDP-galactose **(e)** fatty acyl–CoA **(f)** *S*-adenosylmethionine **(g)** malonyl-CoA **(h)** Δ^3-isopentenyl pyrophosphate

15. Linoleate is required in the synthesis of prostaglandins. Animals are unable to transform oleate to linoleate, so linoleate is an essential fatty acid. Plants can transform oleate to linoleate, and they provide animals with the required linoleate (see Fig. 21–12).

16. The rate-determining step in the biosynthesis of cholesterol is the synthesis of mevalonate, catalyzed by hydroxymethylglutaryl-CoA reductase. This enzyme is allosterically regulated by mevalonate and cholesterol derivatives. High intracellular [cholesterol] also reduces transcription of the gene encoding HMG-CoA reductase.

17. When cholesterol levels decline because of treatment with a statin, cells attempt to compensate by increasing expression of the gene encoding HMG-CoA reductase. The statins are good competitive inhibitors of HMG-CoA reductase activity nonetheless and reduce overall production of cholesterol.

18. Note: There are several plausible alternatives that a student might propose in the absence of a detailed knowledge of the literature on this enzyme. *Thiolase reaction:* Begins with nucleophilic attack of an active-site Cys residue on the first acetyl-CoA substrate, displacing —S-CoA and forming a covalent thioester link between Cys and the acetyl group. A base on the enzyme then extracts a proton from the methyl group of the second acetyl-CoA, leaving a carbanion that attacks the carbonyl carbon of the thioester formed in the first step. The sulfhydryl of the Cys residue is displaced, creating the product acetoacetyl-CoA. *HMG-CoA synthase reaction:* Begins in the same way, with a covalent thioester link formed between the enzyme's Cys residue and the acetyl group of acetyl-CoA, with displacement of the —S-CoA. The —S-CoA dissociates as CoA-SH, and acetoacetyl-CoA binds to the enzyme. A proton is abstracted from the methyl group of the enzyme-linked acetyl, forming a carbanion that attacks the ketone carbonyl of the acetoacetyl-CoA substrate. The carbonyl is converted to a hydroxyl ion in this reaction, and this is protonated to create —OH. The thioester link with the enzyme is then cleaved hydrolytically to generate the HMG-CoA product. *HMG-CoA reductase reaction:* Two successive hydride ions derived from NADPH first displace the —S-CoA, and then reduce the aldehyde to a hydroxyl group.

19. Statins inhibit HMG-CoA reductase, an enzyme in the pathway to the synthesis of activated isoprenes, which are precursors of cholesterol and a wide range of isoprenoids, including coenzyme Q (ubiquinone). Hence, statins might reduce the levels of coenzyme Q available for mitochondrial respiration. Ubiquinone is obtained in the diet as well as by direct biosynthesis, but it is not yet clear how much is required and how well dietary sources can substitute for reduced synthesis. Reductions in the levels of particular isoprenoids may account for some side effects of statins.

20. (a)

Astaxanthin

(b) Head-to-head. There are two ways to look at this. First, the "tail" of geranylgeranyl pyrophosphate has a branched dimethyl structure, as do both ends of phytoene. Second, no free —OH is formed by the release of PP_i, indicating that the two —O—℗—℗ "heads" are linked to form phytoene. **(c)** Four rounds of dehydrogenation convert four single bonds to double bonds. **(d)** No. A count of single and double bonds in the reaction below shows that one double bond is replaced by two single bonds—so there is no net oxidation or reduction:

Lycopene (C-40)

↓ bend ends around for cyclization

↓ cyclize

β-Carotene (C-40)

(e) Steps ❶ through ❸. The enzyme can convert IPP and DMAP to geranylgeranyl pyrophosphate, but catalyzes no further reactions in the pathway, as confirmed by results with the other substrates. **(f)** Strains 1 through 4 lack *crtE* and have much lower astaxanthin production than strains 5 through 8, all of which overexpress *crtE*. Thus, overexpression of *crtE* leads to a substantial increase in astaxanthin production. Wild-type *E. coli* has some step ❸ activity, but this conversion of farnesyl pyrophosphate to geranylgeranyl pyrophosphate is strongly rate-limiting. **(g)** IPP isomerase. Comparing strains 5 and 6 shows that adding *ispA*, which catalyzes steps ❶ and ❷, has little effect on astaxanthin production, so these steps are not rate-limiting. However, comparing strains 5 and 7 shows that adding *idi* substantially increases astaxanthin production, so IPP isomerase must be the rate-limiting step when *crtE* is overexpressed. **(h)** A low (+) level, comparable to that of strains 5, 6, and 9. Without overexpression of *idi*, production of astaxanthin is limited by low IPP isomerase activity and the resulting limited supply of IPP.

Chapter 22

1. In their symbiotic relationship with the plant, bacteria supply ammonium ion by reducing atmospheric nitrogen, which requires large quantities of ATP.

2. The transfer of nitrogen from NH_3 to carbon skeletons can be catalyzed by (1) glutamine synthetase and (2) glutamate dehydrogenase. The latter enzyme produces glutamate, the amino group donor in all transamination reactions, necessary to the formation of amino acids for protein synthesis.

3. A link between enzyme-bound PLP and the phosphohomoserine substrate is first formed, with rearrangement to generate the ketimine at the α carbon of the substrate. This activates the β carbon for proton abstraction, leading to displacement of the phosphate and formation of a double bond between the β and γ carbons. A rearrangement (beginning with proton abstraction at the pyridoxal carbon adjacent to the substrate amino nitrogen) moves the double bond between the α and β carbons, and converts the ketimine to the aldimine form of PLP. Attack of water at the β carbon is then facilitated by the linked pyridoxal, followed by hydrolysis of the imine link between PLP and the product, to generate threonine.

4. In the mammalian route, toxic ammonium ions are transformed to glutamine, reducing toxic effects on the brain.

5. Glucose + $2CO_2$ + $2NH_3$ → 2 aspartate + $2H^+$ + $2H_2O$

6. The amino-terminal glutaminase domain is quite similar in *all* glutamine amidotransferases. A drug that targeted this active site would probably inhibit many enzymes and thus be prone to producing many more side effects than a more specific inhibitor targeting the unique carboxyl-terminal synthetase active site.

7. If phenylalanine hydroxylase is defective, the biosynthetic route to tyrosine is blocked and tyrosine must be obtained from the diet.

8. In adoMet synthesis, triphosphate is released from ATP. Hydrolysis of the triphosphate renders the reaction thermodynamically more favorable.

9. If the inhibition of glutamine synthase were not concerted, saturating concentrations of histidine would shut down the enzyme and cut off production of glutamine, which the bacterium needs to synthesize other products.

10. Folic acid is a precursor of tetrahydrofolate (see Fig. 18–16), required in the biosynthesis of glycine (see Fig. 22–14), a precursor of porphyrins. A folic acid deficiency therefore impairs hemoglobin synthesis.

11. *For glycine auxotrophs:* adenine and guanine; *for glutamine auxotrophs:* adenine, guanine, and cytosine; *for aspartate auxotrophs:* adenine, guanine, cytosine, and uridine.

12. (a) See Fig. 18–6, step ❷, for the reaction mechanism of amino acid racemization. The F atom of fluoroalanine is an excellent leaving group. Fluoroalanine causes irreversible (covalent) inhibition of alanine racemase. One plausible mechanism is (Nuc denotes any nucleophilic amino acid side chain in the enzyme active site):

(b) Azaserine (see Fig. 22–51) is an analog of glutamine. The diazoacetyl group is highly reactive and forms covalent bonds with nucleophiles at the active site of a glutamine amidotransferase.

13. (a) As shown in Fig. 18–16, *p*-aminobenzoate is a component of N^5,N^{10}-methylenetetrahydrofolate, the cofactor involved in the transfer of one-carbon units. **(b)** In the presence of sulfanilamide, a structural analog of *p*-aminobenzoate, bacteria are unable to synthesize tetrahydrofolate, a cofactor necessary for converting AICAR to FAICAR; thus AICAR accumulates. **(c)** The competitive inhibition by sulfanilamide of the enzyme involved in tetrahydrofolate biosynthesis is overcome by the addition of excess substrate (*p*-aminobenzoate).

14. The ^{14}C-labeled orotate arises from the following pathway (the first three steps are part of the citric acid cycle):

$^-OO^{14}C-^{14}CH_2-^{14}CH_2-^{14}COO^-$
Succinate

$^-OO^{14}C$, H, $^{14}C=^{14}C$, H, $^{14}COO^-$
Fumarate

OH H
$^-OO^{14}C-^{14}C-^{14}C-^{14}COO^-$
H H
Malate

O H
$^-OO^{14}C-^{14}C-^{14}C-^{14}COO^-$
H
Oxaloacetate

transamination

$\overset{+}{N}H_3$
$^-OO^{14}C-^{14}C-^{14}CH_2-^{14}COO^-$
H
Aspartate

Orotate

---

15. Organisms do not store nucleotides to be used as fuel, and they do not completely degrade them, but rather hydrolyze them to release the bases, which can be recovered in salvage pathways. The low C:N ratio of nucleotides makes them poor sources of energy.

16. Treatment with allopurinol has two consequences. (1) It inhibits conversion of hypoxanthine to uric acid, causing accumulation of hypoxanthine, which is more soluble and more readily excreted; this alleviates the clinical problems associated with AMP degradation. (2) It inhibits conversion of guanine to uric acid, causing accumulation of xanthine, which is less soluble than uric acid; this is the source of xanthine stones. Because the amount of GMP degradation is low relative to AMP degradation, the kidney damage caused by xanthine stones is less than that caused by untreated gout.

17. 5-Phosphoribosyl-1-pyrophosphate; this is the first NH_3 acceptor in the purine biosynthetic pathway.

18. (a) The α-carboxyl group is removed and an —OH is added to the γ carbon. **(b)** BtrI has sequence homology with acyl carrier proteins. The molecular weight of BtrI increases when incubated under conditions in which CoA could be added to the protein. Adding CoA to a Ser residue would replace an —OH (formula weight (FW) 17) with a 4'-phosphopantetheine group (see Fig. 21–5). This group has the formula $C_{11}H_{21}N_2O_7PS$ (FW 356). Thus, $11,182 - 17 + 356 = 12,151$, which is very close to the observed M_r of 12,153. **(c)** The thioester could form with the α-carboxyl group. **(d)** In the most common reaction for removing the α-carboxyl group of an amino acid (see Fig. 18–6c), the carboxyl group must be free. Furthermore, it is difficult to imagine a decarboxylation reaction starting with a carboxyl group in its thioester form. **(e)** $12,240 - 12,281 = 41$, close to the M_r of CO_2 (44). Given that BtrK is probably a decarboxylase, its most likely structure is the decarboxylated form:

$H_3\overset{+}{N}$ ... O, S, BtrI

(f) $12,370 - 12,240 = 130$. Glutamic acid ($C_5H_9NO_4$; M_r 147), minus the —OH (FW 17) removed in the glutamylation reaction, leaves a glutamyl group of FW 130; thus, γ-glutamylating the molecule above would add 130 to its M_r. BtrJ is capable of γ-glutamylating other substrates, so it may γ-glutamylate the structure above. The most likely site for this is the free amino group, giving the following structure:

^-O, O, O, $\overset{+}{N}H_3$, N, H, O, S, BtrI

(g)

^-O, $\overset{+}{N}H_3$, O, O, O$^-$ → BtrJ, ATP, ADP → ^-O, $\overset{+}{N}H_3$, O, O, S, BtrI → CO_2, BtrK →

$H_3\overset{+}{N}$, O, S, BtrI → Glutamate, ATP, ADP, BtrJ →

The structures show: a reaction with O_2, H_2O, BtrO, FMNH$_2$, FMN, BtrV, NAD$^+$, NADH involving BtrI.

A second reaction: Antibiotic, BtrG + BtrH, Glutamate + BtrI, BtrI.

An Antibiotic structure with OH, NH_3^+.

Chapter 23

1. They are recognized by two different receptors, typically found in different cell types, and are coupled to different downstream effects.

2. Steady-state levels of ATP are maintained by phosphoryl group transfer to ADP from phosphocreatine. 1-Fluoro-2,4-dinitrobenzene inhibits creatine kinase.

3. Ammonia is very toxic to nervous tissue, especially the brain. Excess NH_3 is removed by transformation of glutamate to glutamine, which travels to the liver and is subsequently transformed to urea. The additional glutamine arises from the transformation of glucose to α-ketoglutarate, transamination of α-ketoglutarate to glutamate, and conversion of glutamate to glutamine.

4. Glucogenic amino acids are used to make glucose for the brain; others are oxidized in mitochondria via the citric acid cycle.

5. From glucose, by the following route: Glucose $\rightarrow$ dihydroxyacetone phosphate (in glycolysis); dihydroxyacetone phosphate + NADH + H$^+$ $\rightarrow$ glycerol 3-phosphate + NAD$^+$ (glycerol 3-phosphate dehydrogenase reaction)

6. (a) Increased muscular activity increases the demand for ATP, which is met by increased O_2 consumption. (b) After the sprint, lactate produced by anaerobic glycolysis is converted to glucose and glycogen, which requires ATP and therefore O_2.

7. Glucose is the primary fuel of the brain. TPP-dependent oxidative decarboxylation of pyruvate to acetyl-CoA is essential to complete glucose metabolism.

8. 190 m

9. (a) Inactivation provides a rapid means to change hormone concentrations. (b) Insulin level is maintained by equal rates of synthesis and degradation. (c) Changes in the rate of release from storage, rate of transport, and rate of conversion from prohormone to active hormone.

10. Water-soluble hormones bind to receptors on the outer surface of the cell, triggering the formation of a second messenger (e.g., cAMP) inside the cell. Lipid-soluble hormones can pass through the plasma membrane to act on target molecules or receptors directly.

11. (a) Heart and skeletal muscle lack glucose 6-phosphatase. Any glucose 6-phosphate produced enters the glycolytic pathway, and under O_2-deficient conditions is converted to lactate via pyruvate. (b) In a "fight or flight" situation, the concentration of glycolytic precursors must be high in preparation for muscular

activity. Phosphorylated intermediates cannot escape from the cell, because the membrane is not permeable to charged species, and glucose 6-phosphate is not exported on the glucose transporter. The liver, by contrast, must release the glucose necessary to maintain blood glucose level; glucose is formed from glucose 6-phosphate and enters the bloodstream.

12. (a) Excessive uptake and use of blood glucose by the liver, leading to hypoglycemia; shutdown of amino acid and fatty acid catabolism (b) Little circulating fuel is available for ATP requirements. Brain damage results because glucose is the main source of fuel for the brain.

13. Thyroxine acts as an uncoupler of oxidative phosphorylation. Uncouplers lower the P/O ratio, and the tissue must increase respiration to meet the normal ATP demands. Thermogenesis could also be due to the increased rate of ATP utilization by the thyroid-stimulated tissue, as increased ATP demands are met by increased oxidative phosphorylation and thus respiration.

14. Because prohormones are inactive, they can be stored in quantity in secretory granules. Rapid activation is achieved by enzymatic cleavage in response to an appropriate signal.

15. In animals, glucose can be synthesized from many precursors (see Fig. 14–16). In humans, the principal precursors are glycerol from triacylglycerols and glucogenic amino acids from protein.

16. The *ob/ob* mouse, which is initially obese, will lose weight. The *OB/OB* mouse will retain its normal body weight.

17. BMI = 39.3. For BMI of 25, weight must be 75 kg; must lose 43 kg = 95 lbs.

18. Reduced insulin secretion. Valinomycin has the same effect as opening the K$^+$ channel, allowing K$^+$ exit and consequent hyperpolarization.

19. The liver does not receive the insulin message and therefore continues to have high levels of glucose 6-phosphatase and gluconeogenesis, increasing blood glucose both during a fast and after a glucose-containing meal. The elevated blood glucose triggers insulin release from pancreatic β cells, hence the high level of insulin in the blood.

20. Some things to consider: What is the frequency of heart attack attributable to the drug? How does this frequency compare with the number of individuals spared the long-term consequences of type 2 diabetes? Are other, equally effective treatment options with fewer adverse effects available?

21. Without intestinal glucosidase activity, absorption of glucose from dietary glycogen and starch is reduced, blunting the usual rise in blood glucose after the meal. The undigested oligosaccharides are fermented by bacteria in the large intestine, and the gases released cause intestinal discomfort.

22. (a) Closing the ATP-gated K$^+$ channel would depolarize the membrane, leading to increased insulin release. (b) Type 2 diabetes results from decreased sensitivity to insulin, not a deficit of insulin production; increasing circulating insulin levels will reduce the symptoms associated with this disease. (c) Individuals with type 1 diabetes have deficient pancreatic β cells, so glyburide will have no beneficial effect. (d) Iodine, like chlorine (the atom it replaces in the labeled glyburide), is a halogen, but it is a larger atom and has slightly different chemical properties. It is possible that the iodinated glyburide would not bind to SUR. If it bound to another molecule instead, the experiment would result in cloning of the gene for this other, incorrect protein. (e) Although a protein has been "purified," the "purified" preparation might be a mixture of several proteins that co-purify under those experimental conditions. In this case, the amino acid sequence could be that of a protein that co-purifies with SUR. Using antibody binding to show that the peptide sequences are present in SUR excludes this possibility. (f) Although the cloned gene does encode the 25 amino acid sequence found in SUR, it could be a gene that, coincidentally, encodes the same sequence in another protein. In this case, this

other gene would most likely be expressed in different cells than the *SUR* gene. The mRNA hybridization results are consistent with the putative *SUR* cDNA actually encoding SUR. **(g)** The excess unlabeled glyburide competes with labeled glyburide for the binding site on SUR. As a result, there is significantly less binding of labeled glyburide, so little or no radioactivity is detected in the 140 kDa protein. **(h)** In the absence of excess unlabeled glyburide, labeled 140 kDa protein is found only in the presence of the putative *SUR* cDNA. Excess unlabeled glyburide competes with the labeled glyburide, and no ^{125}I-labeled 140 kDa protein is detected. This shows that the cDNA produces a glyburide-binding protein of the same molecular weight as SUR— strong evidence that the cloned gene encodes the SUR protein. **(i)** Several additional steps are possible, such as: (1) Express the putative *SUR* cDNA in CHO (Chinese hamster ovary) cells and show that the transformed cells have ATP-gated K$^+$ channel activity. (2) Show that HIT cells with mutations in the putative *SUR* gene lack ATP-gated K$^+$ channel activity. (3) Show that experimental animals or human patients with mutations in the putative *SUR* gene are unable to secrete insulin.

Chapter 24

1. 6.1×10^4 nm; 290 times longer than the T2 phage head

2. The number of A residues does not equal the number of T residues, nor does the number of G equal the number of C, so the DNA is not a base-paired double helix; the M13 DNA is single-stranded.

3. $M_r = 3.8 \times 10^8$; length = 200 μm; $Lk_0 = 55{,}200$; $Lk = 51{,}900$

4. The exons contain 3 bp/amino acid $\times$ 192 amino acids = 576 bp. The remaining 864 bp are in introns, possibly in a leader or signal sequence, and/or in other noncoding DNA.

5. 5,000 bp. **(a)** Doesn't change; Lk cannot change without breaking and re-forming the covalent backbone of the DNA. **(b)** Becomes undefined; a circular DNA with a break in one strand has, by definition, no Lk. **(c)** Decreases; in the presence of ATP, gyrase underwinds DNA. **(d)** Doesn't change; this assumes that neither of the DNA strands is broken in the heating process.

6. For Lk to remain unchanged, the topoisomerase must introduce the same number of positive and negative supercoils.

7. $\sigma = -0.067$; >70% probability

8. **(a)** Undefined; the strands of a nicked DNA could be separated and thus have no Lk. **(b)** 476 **(c)** 476; the DNA is already relaxed, so the topoisomerase does not cause a net change. **(d)** 460; gyrase plus ATP reduces the Lk in increments of 2. **(e)** 464; eukaryotic type I topoisomerases increase the Lk of underwound or negatively supercoiled DNA in increments of 1. **(f)** 460; nucleosome binding does not break any DNA strands and thus cannot change Lk.

9. A fundamental structural unit in chromatin repeats about every 200 bp; the DNA is accessible to the nuclease only at 200 bp intervals. The brief treatment was insufficient to cleave the DNA at every accessible point, so a ladder of DNA bands is created in which the DNA fragments are multiples of 200 bp. The thickness of the DNA bands suggests that the distance between cleavage sites varies somewhat. For instance, not all the fragments in the lowest band are exactly 200 bp long.

10. A right-handed helix has a positive Lk; a left-handed helix (such as Z-DNA) has a negative Lk. Decreasing the Lk of a closed circular B-DNA by underwinding it facilitates formation of regions of Z-DNA within certain sequences. (See Chapter 8, p. 291, for a description of sequences that permit the formation of Z-DNA.)

11. **(a)** Both strands must be covalently closed, and the molecule must be either circular or constrained at both ends. **(b)** Formation of cruciforms, left-handed Z-DNA, plectonemic or solenoidal supercoils, and unwinding of the DNA are favored. **(c)** *E. coli* DNA topoisomerase II or DNA gyrase. **(d)** It binds the DNA at a point where it crosses on itself, cleaves both strands of one of the crossing segments, passes the other segment through the break, then reseals the break. The result is a change in Lk of -2.

12. Centromere, telomeres, and an autonomous replicating sequence or replication origin

13. The bacterial nucleoid is organized into domains approximately 10,000 bp long. Cleavage by a restriction enzyme relaxes the DNA within a domain, but not outside the domain. Any gene in the cleaved domain for which expression is affected by DNA topology will be affected by the cleavage; genes outside the domain will not.

14. **(a)** The lower, faster-migrating band is negatively supercoiled plasmid DNA. The upper band is nicked, relaxed DNA. **(b)** DNA topoisomerase I relaxes the supercoiled DNA. The lower band will disappear, and all of the DNA will converge on the upper band. **(c)** DNA ligase produces little change in the pattern. Some minor additional bands may appear near the upper band, due to the trapping of topoisomers not quite perfectly relaxed by the ligation reaction. **(d)** The upper band will disappear, and all of the DNA will be in the lower band. The supercoiled DNA in the lower band may become even more supercoiled and migrate somewhat faster.

15. **(a)** When DNA ends are sealed to create a relaxed, closed circle, some DNA species are completely relaxed but others are trapped in slightly under- or overwound states. This gives rise to a distribution of topoisomers centered on the most relaxed species. **(b)** Positively supercoiled. **(c)** The DNA that is relaxed despite the addition of dye is DNA with one or both strands broken. DNA isolation procedures inevitably introduce small numbers of strand breaks in some of the closed-circular molecules. **(d)** Approximately -0.05. This is determined by simply comparing native DNA with samples of known σ. In both gels, the native DNA migrates most closely with the sample of $\sigma = -0.049$.

16. 62 million (the genome refers to the haploid genetic content of the cell; the cell is actually diploid, so the number of nucleosomes is doubled). The number comes from 3.1 billion base pairs divided by 200 base pairs per nucleosome (giving 15.5 million nucleosomes), multiplied by two copies of H2A per nucleosome and again multiplied by 2 to account for the diploid state of the cell. The 62 million would double upon replication.

17. **(a)** In nondisjunction, one daughter cell and all of its descendants get two copies of the synthetic chromosome and are white; the other daughter cell and all of its descendants get no copies of the synthetic chromosome and are red. This gives rise to a half-white, half-red colony. **(b)** In chromosome loss, one daughter cell and all of its descendants get one copy of the synthetic chromosome and are pink; the other daughter and all its descendants get no copies of the synthetic chromosome and are red. This gives rise to a half-pink, half-red colony. **(c)** The minimum functional centromere must be smaller than 0.63 kbp, since all fragments of this size or larger confer relative mitotic stability. **(d)** Telomeres are required to fully replicate only linear DNA; a circular molecule can replicate without them. **(e)** The larger the chromosome, the more faithfully it is segregated. The data show neither a minimum size below which the synthetic chromosome is completely unstable, nor a maximum size above which stability no longer changes.

(f)

As shown in the graph, even if the synthetic chromosomes were as long as the normal yeast chromosomes, they would not be as stable. This suggests that other, as yet undiscovered, elements are required for stability.

Chapter 25

1. In random, dispersive replication, in the second generation, all the DNAs would have the same density and would appear as a single band, not the two bands observed in the Meselson-Stahl experiment.

2. In this extension of the Meselson-Stahl experiment, after three generations the molar ratio of $^{15}N–^{14}N$ DNA to $^{14}N–^{14}N$ DNA is $2/6 = 0.33$.

3. (a) 4.42×10^5 turns; **(b)** 40 min. In cells dividing every 20 min, a replicative cycle is initiated every 20 min, each cycle beginning before the prior one is complete. **(c)** 2,000 to 5,000 Okazaki fragments. The fragments are 1,000 to 2,000 nucleotides long and are firmly bound to the template strand by base pairing. Each fragment is quickly joined to the lagging strand, thus preserving the correct order of the fragments.

4. A 28.7%; G 21.3%; C 21.3%; T 28.7%. The DNA strand made from the template strand: A 32.7%; G 18.5%; C 24.1%; T 24.7%; the DNA strand made from the complementary template strand: A 24.7%; G 24.1%; C 18.5%; T 32.7%. It is assumed that the two template strands are replicated completely.

5. (a) No. Incorporation of ^{32}P into DNA results from the synthesis of new DNA, which requires the presence of *all four* nucleotide precursors. **(b)** Yes. Although all four nucleotide precursors must be present for DNA synthesis, only one of them has to be radioactive in order for radioactivity to appear in the new DNA. **(c)** No. Radioactivity is incorporated only if the ^{32}P label is in the α phosphate; DNA polymerase cleaves off pyrophosphate—i.e., the β- and γ-phosphate groups.

6. *Mechanism 1:* 3′-OH of an incoming dNTP attacks the α phosphate of the triphosphate at the 5′ end of the growing DNA strand, displacing pyrophosphate. This mechanism uses normal dNTPs, and the growing end of the DNA always has a triphosphate on the 5′ end.

Mechanism 2: This uses a new type of precursor, nucleotide 3′-triphosphates. The growing end of the DNA strand has a 5′-OH, which attacks the α phosphate of an incoming deoxynucleotide

3′-triphosphate, displacing pyrophosphate. Note that this mechanism would require the evolution of new metabolic pathways to supply the needed deoxynucleotide 3′-triphosphates.

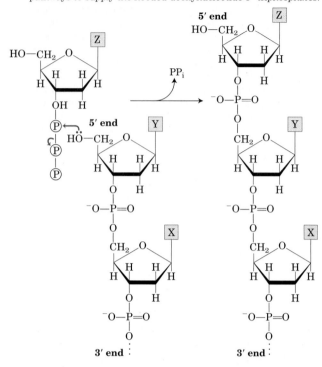

7. The DNA polymerase contains a 3′→5′ exonuclease activity that degrades DNA to produce [^{32}P]dNMPs. The activity is not a 5′→3′ exonuclease, because the addition of unlabeled dNTPs inhibits the production of [^{32}P]dNMPs (polymerization activity would suppress a proofreading exonuclease but not an exonuclease operating downstream of the polymerase). Addition of pyrophosphate would generate [^{32}P]dNTPs through reversal of the polymerase reaction.

8. *Leading strand:* Precursors: dATP, dGTP, dCTP, dTTP (also needs a template DNA strand and DNA primer); enzymes and other proteins: DNA gyrase, helicase, single-stranded DNA–binding protein, DNA polymerase III, topoisomerases, and pyrophosphatase. *Lagging strand:* Precursors: ATP, GTP, CTP, UTP, dATP, dGTP, dCTP, dTTP (also needs an RNA primer); enzymes and other proteins: DNA gyrase, helicase, single-stranded DNA–binding protein, primase, DNA polymerase III, DNA polymerase I, DNA ligase, topoisomerases, and pyrophosphatase. NAD$^+$ is also required as a cofactor for DNA ligase.

9. Mutants with defective DNA ligase produce a DNA duplex in which one of the strands remains in pieces (as Okazaki fragments). When this duplex is denatured, sedimentation results in one fraction containing the intact single strand (the high molecular weight band) and one fraction containing the unspliced fragments (the low molecular weight band).

10. Watson-Crick base pairing between template and leading strand; proofreading and removal of wrongly inserted nucleotides by the 3′-exonuclease activity of DNA polymerase III. Yes—perhaps. Because the factors ensuring fidelity of replication are operative in both the leading and the lagging strands, the lagging strand would probably be made with the same fidelity. However, the greater number of distinct chemical operations involved in making the lagging strand might provide a greater opportunity for errors to arise.

11. ~1,200 bp (600 in each direction)

12. A small fraction (13 of 10^9 cells) of the histidine-requiring mutants spontaneously undergo back-mutation and regain their capacity to synthesize histidine. 2-Aminoanthracene increases the rate of back-mutations about 1,800-fold and is therefore mutagenic. Since most carcinogens are mutagenic, 2-aminoanthracene is probably carcinogenic.

13. Spontaneous deamination of 5-methylcytosine (see Fig. 8–30) produces thymine, and thus a G–T mismatched pair. These are among the most common mismatches in the DNA of eukaryotes. The specialized repair system restores the G≡C pair.

14. **(a)** Ultraviolet irradiation produces pyrimidine dimers; in normal fibroblasts these are excised by cleavage of the damaged strand by a special excinuclease. Thus the denatured single-stranded DNA contains the many fragments caused by the cleavage, and the average molecular weight is lowered. These fragments of single-stranded DNA are absent from the XPG samples, as indicated by the unchanged average molecular weight. **(b)** The absence of fragments in the single-stranded DNA from the XPG cells after irradiation suggests the special excinuclease is defective or missing.

15. During homologous genetic recombination, a Holliday intermediate may be formed almost anywhere within the two paired, homologous chromosomes; the branch point of the intermediate can move extensively by branch migration. In site-specific recombination, the Holliday intermediate is formed between two specific sites, and branch migration is generally restricted by heterologous sequences on either side of the recombination sites.

16. **(a)** Points Y. **(b)** Points X.

17. Once replication has proceeded from the origin to a point where one recombination site has been replicated but the other has not, site-specific recombination not only inverts the DNA between the recombination sites but also changes the direction of one replication fork relative to the other. The forks will chase each other around the DNA circle, generating many tandem copies of the plasmid. The multimeric circle can be resolved to monomers by additional site-specific recombination events.

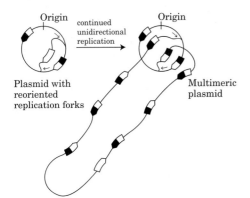

18. **(a)** Even in the absence of an added mutagen, background mutations occur due to radiation, cellular chemical reactions, and so forth. **(b)** If the DNA is sufficiently damaged, a substantial fraction of gene products are nonfunctional and the cell is nonviable. **(c)** Cells with reduced DNA repair capability are more sensitive to mutagens. Because they less readily repair lesions caused by R7000, uvr⁻ bacteria have an increased mutation rate and increased chance of lethal effects. **(d)** In the uvr⁺ strain, the excision-repair system removes DNA bases with attached [³H]R7000, decreasing the ³H in these cells over time.

In the uvr⁻ strain, the DNA is not repaired and the ³H level increases as [³H]R7000 continues to react with the DNA. **(e)** All mutations listed in the table except A=T to G≡C show significant increases over background. Each type of mutation results from a different type of interaction between R7000 and DNA. Because different types of interactions are not equally likely (due to differences in reactivity, steric constraints, etc.), the resulting mutations occur with different frequencies. **(f)** No. Only those that start with a G≡C base pair are explained by this model. Thus A=T to C≡G and A=T to T=A must be due to R7000 attaching to an A or a T. **(g)** R7000-G pairs with A. First, R7000 adds to G≡C to give R7000-G≡C. (Compare this with what happens with the CH₃-G in Fig. 25–27b.) If this is not repaired, one strand is replicated as R7000-G=A, which is repaired to T=A. The other strand is wild-type. If the replication produces R7000-G=T, a similar pathway leads to an A=T base pair. **(h)** No. Compare data in the two tables, and keep in mind that different mutations occur at different frequencies.

A=T to C≡G: moderate in both strains; but better repair in the uvr⁺ strain

G≡C to A=T: moderate in both; no real difference

G≡C to C≡G: higher in uvr⁺; certainly less repair!

G≡C to T=A: high in both; no real difference

A=T to T=A: high in both; no real difference

A=T to G≡C: low in both; no real difference

Certain adducts may be more readily recognized by the repair apparatus than others and thus are repaired more rapidly and result in fewer mutations.

Chapter 26

1. **(a)** 60 to 100 s; **(b)** 500 to 900 nucleotides

2. A single base error in DNA replication, if not corrected, would cause one of the two daughter cells, and all its progeny, to have a mutated chromosome. A single base error in RNA transcription would not affect the chromosome; it would lead to formation of some defective copies of one protein, but because mRNAs turn over rapidly, most copies of the protein would not be defective. The progeny of this cell would be normal.

3. Normal posttranscriptional processing at the 3′ end (cleavage and polyadenylation) would be inhibited or blocked.

4. Because the template-strand RNA does not encode the enzymes needed to initiate viral infection, it would probably be inert or simply degraded by cellular ribonucleases. Replication of the template-strand RNA and propagation of the virus could occur only if intact RNA replicase (RNA-dependent RNA polymerase) were introduced into the cell along with the template strand.

5. AUGACCAUGAUUACG

6. (1) Use of a template strand of nucleic acid; (2) synthesis in the 5′→3′ direction; (3) use of nucleoside triphosphate substrates, with formation of a phosphodiester bond and displacement of PP$_i$. Polynucleotide phosphorylase forms phosphodiester bonds but differs in all other listed properties.

7. Generally two: one to cleave the phosphodiester bond at one intron-exon junction, the other to link the resulting free exon end to the exon at the other end of the intron. If the nucleophile in the first step were water, this step would be a hydrolytic event and only one transesterification step would be required to complete the splicing process.

8. Many snoRNAs, required for rRNA modification reactions, are encoded in introns. If splicing does not occur, snoRNAs are not produced.

9. These enzymes lack a 3′→5′ proofreading exonuclease and have a high error rate; the likelihood of a replication error that would inactivate the virus is much less in a small genome than in a large one.

10. (a) $4^{32} = 1.8 \times 10^{19}$ **(b)** 0.006% **(c)** For the "unnatural selection" step, use a chromatographic resin with a bound molecule that is a transition-state analog of the ester hydrolysis reaction.

11. Though RNA synthesis is quickly halted by α-amanitin toxin, it takes several days for the critical mRNAs and the proteins in the liver to degrade, causing liver dysfunction and death.

12. (a) After lysis of the cells and partial purification of the contents, the protein extract could be subjected to isoelectric focusing. The β subunit could be detected by an antibody-based assay. The difference in amino acid residues between the normal β subunit and the mutated form (i.e., the different charges on the amino acids) would alter the electrophoretic mobility of the mutant protein in an isoelectric focusing gel, relative to the protein from a nonresistant strain. **(b)** Direct DNA sequencing (by the Sanger method).

13. (a) 384 nucleotide pairs **(b)** 1,620 nucleotide pairs **(c)** Most of the nucleotides are untranslated regions at the 3′ and 5′ ends of the mRNA. Also, most mRNAs code for a signal sequence (Chapter 27) in their protein products, which is eventually cleaved off to produce the mature and functional protein.

14. (a) cDNA is produced by reverse transcription of mRNA; thus, the mRNA sequence is probably CGG. Because the genomic DNA transcribed to make the mRNA has the sequence CAG, the primary transcript most likely has CAG, which is posttranscriptionally modified to CGG. **(b)** The unedited mRNA sequence is the same as that of the DNA (except for U replacing T). Unedited mRNA has the sequence (* indicates site of editing)

$$\overset{*}{}$$
(5′) --- GUCUCUGGUUUUCCUUGGGUGCCUUUAUGCAGCAAGGAUGCGAUAUUUCGCCAAG --- (3′)

In step 1, primer 1 anneals as shown:

(5′) --- GUCUCUGGUUUUCCUUGGGUGCCUUUAUGCAGCAAGGAUGCGAUAUUUCGCCAAG --- (3′)
                                          | | | | | | | | | | | | | | | | | | | | | | | | | | |
                                Primer 1: (3′)CGTTCCTACGCTATAAAGCGGTTC(5′)

cDNA (underlined) is synthesized from right to left:

(5′) --- GUCUCUGGUUUUCCUUGGGUGCCUUUAUGCAGCAAGGAUGCGAUAUUUCGCCAAG --- (3′)
        | | | | | | | | | | | | | | | | | | | | | | | | | | | | | | | | | | | | | | | | | | | | | | | | | | | | | | | |
(3′) --- CAGAGACCAAAAGGAACCCACGGAAATACGTCGTTCCTACGCTATAAAGCGGTTC(5′)

Then step 2 yields just the cDNA:

(3′) --- CAGAGACCAAAAGGAACCCACGGAAATACGTCGTTCCTACGCTATAAAGCGGTTC(5′)

In step 3, primer 2 anneals to the cDNA:

Primer 2: (5′)CCTTGGGTGCCTTTA(3′)
               | | | | | | | | | | | | | | |
(3′) --- CAGAGACCAAAAGGAACCCACGGAAATACGTCGTTCCTACGCTATAAAGCGGTTC (5′)

DNA polymerase adds nucleotides to the 3′ end of this primer. Moving from left to right, it inserts T, G, C, and A. However, because the A from ddATP lacks the 3′ —OH needed to attach the next nucleotide, the chain is not elongated past this point. This A is shown in *italic*; the new DNA is underlined:

Primer 2: (5′)CCTTGGGTGCCTTTA*TGCA*
               | | | | | | | | | | | | | | | | | | |
(3′) --- CAGAGACCAAAAGGAACCCACGGAAATACGTCGTTCCTACGCTATAAAGCGGTTC(5′)

This yields a 19 nucleotide fragment for the unedited transcript. In the edited transcript, the *A is changed to G; in the cDNA this corresponds to C. At the start of step 3:

Primer 2: (5′)CCTTGGGTGCCTTTA(3′)
               | | | | | | | | | | | | | | |
(3′) --- CAGAGACCAAAAGGAACCCACGGAAATACGCCGTTCCTACGCTATAAAGCGGTTC(5′)

In this case, DNA polymerase can elongate past the edited base and will stop at the next T in the cDNA. The dideoxy A is *italic*; the new DNA is underlined:

Primer 2: (5′)CCTTGGGTGCCTTTA*TGCGGCA*
               | | | | | | | | | | | | | | | | | | | | | |
(3′) --- CAGAGACCAAAAGGAACCCACGGAAATACGCCGTTCCTACGCTATAAAGCGGTTC(5′)

This gives the 22 nucleotide product. **(c)** Treatments (proteases, heat) known to disrupt protein function inhibit the editing activity, whereas treatments (nuclease) that do not affect proteins have little or no effect on editing. A key weakness of this argument is that the protein-disrupting treatments do not completely abolish editing. There could be some background editing or degradation of the mRNA even without the enzyme, or some of the enzyme might survive the treatments. **(d)** Only the α phosphate of NTPs is incorporated into polynucleotides. If the researchers had used the other types of [^{32}P]NTPs, none of the products would have been labeled. **(e)** Because only an A is being edited, only the fate of any A in the sequence is of interest. **(f)** Given that only ATP was labeled, if the entire nucleotide were removed, all radioactivity would have been removed from the mRNA, so only unmodified [^{32}P]AMP would be present on the chromatography plate. **(g)** If the base were removed and replaced, one would expect to see only [^{3}H]AMP. The presence of [^{3}H]IMP indicates that the A to I change occurs without removal of H at positions 2 and 8. The most likely mechanism is chemical modification of A to I by hydrolytic deamination (see Fig. 22–36). **(h)** CAG is changed to CIG. This codon is read as CGG.

Chapter 27

1. **(a)** Gly–Gln–Ser–Leu–Leu–Ile **(b)** Leu–Asp–Ala–Pro **(c)** His–Asp–Ala–Cys–Cys–Tyr **(d)** Met–Asp–Glu in eukaryotes; fMet–Asp–Glu in bacteria

2. UUAAUGUAU, UUGAUGUAU, CUUAUGUAU, CUCAUGUAU, CUAAUGUAU, CUGAUGUAU, UUAAUGUAC, UUGAUGUAC, CUUAUGUAC, CUCAUGUAC, CUAAUGUAC, CUGAUGUAC

3. No. Because nearly all the amino acids have more than one codon (e.g., Leu has six), any given polypeptide can be coded for by a number of different base sequences. However, some amino acids are encoded by only one codon and those with multiple codons often share the same nucleotide at two of the three positions, so *certain parts* of the mRNA sequence encoding a protein of known amino acid sequence can be predicted with high certainty.

4. **(a)** (5′)CGACGGCGCGAAGUCAGGGGUGUUAAG(3′)

   (b) Arg–Arg–Arg–Glu–Val–Arg–Gly–Val–Lys

   (c) No. The complementary antiparallel strands in double-helical DNA do not have the same base sequence in the 5′→3′ direction. RNA is transcribed from only one specific strand of duplex DNA. The RNA polymerase must therefore recognize and bind to the correct strand.

5. There are two tRNAs for methionine: tRNA^{fMet}, which is the initiating tRNA, and tRNA^{Met}, which can insert a Met residue in interior positions in a polypeptide. Only fMet-tRNA^{fMet} is recognized by the initiation factor IF-2 and is aligned with the initiating AUG positioned at the ribosomal P site in the initiation complex. AUG codons in the interior of the mRNA can bind and incorporate only Met-tRNA^{Met}.

6. Allow polynucleotide phosphorylase to act on a mixture of UDP and CDP in which UDP has, say, five times the concentration of CDP. The result would be a synthetic RNA polymer with many UUU triplets (coding for Phe), a smaller number of UUC (Phe), UCU (Ser), and CUU (Leu), a much smaller number of UCC (Ser), CUC (Leu), and CCU (Pro), and the smallest number of CCC (Pro).

7. A minimum of 583 ATP equivalents (based on 4 per amino acid residue added, except that there are only 145 translocation steps). Correction of each error requires 2 ATP equivalents. For glycogen synthesis, 292 ATP equivalents. The extra energy cost for β-globin synthesis reflects the cost of the information content of the protein. At least 20 activating enzymes, 70 ribosomal proteins, 4 rRNAs, 32 or more tRNAs, an mRNA, and 10 or more auxiliary enzymes must be made by the eukaryotic cell in order to synthesize a protein from amino acids. The synthesis of an (α1→4) chain of glycogen from glucose requires only 4 or 5 enzymes (Chapter 15).

8.

Glycine codons	Anticodons
(5′)GGU	(5′)ACC, GCC, ICC
(5′)GGC	(5′)GCC, ICC
(5′)GGA	(5′)UCC, ICC
(5′)GGG	(5′)CCC, UCC

(a) The 3′ and middle position **(b)** Pairings with anticodons (5′)GCC, ICC, and UCC **(c)** Pairings with anticodons (5′)ACC and CCC

9. (a), (c), (e), and (g) only; (b), (d), and (f) cannot be the result of single-base mutations: (b) and (f) would require substitutions of two bases, and (d) would require substitutions of all three bases.

10. (5′)AUGAUAUUGCUAUCUUGGACU

    Changes: CC AU U   C   C
              U   C A   A
                    G G   G

    14 of 63 possible one-base changes would result in no coding change.

11. The two DNA codons for Glu are GAA and GAG, and the four DNA codons for Val are GTT, GTC, GTA, and GTG. A single base change in GAA to form GTA or in GAG to form GTG could account for the Glu → Val replacement in sickle-cell hemoglobin. Much less likely are two-base changes, from GAA to GTG, GTT, or GTC; and from GAG to GTA, GTT, or GTC.

12. Isoleucine is similar in structure to several other amino acids, particularly valine. Distinguishing between valine and isoleucine in the aminoacylation process requires the second filter of a proofreading function. Histidine has a structure unlike that of any other amino acid, and this structure provides opportunities for binding specificity adequate to ensure accurate aminoacylation of the cognate tRNA.

13. **(a)** The Ala-tRNA synthetase recognizes the G^3–U^{70} base pair in the amino acid arm of tRNA^{Ala}. **(b)** The mutant tRNA^{Ala} would insert Ala residues at codons encoding Pro. **(c)** A mutation that might have similar effects is an alteration in tRNA^{Pro} that allowed it to be recognized and aminoacylated by Ala-tRNA synthetase. **(d)** Most of the proteins in the cell would be inactivated, so these would be lethal mutations and hence never observed. This represents a powerful selective pressure for maintaining the genetic code.

14. *IF-2:* The 70S ribosome would form, but initiation factors would not be released and elongation could not start. *EF-Tu:* The second aminoacyl-tRNA would bind to the ribosomal A site, but no peptide bond would form. *EF-G:* The first peptide bond would form, but the ribosome would not move along the mRNA to vacate the A site for binding of a new EF-Tu–tRNA.

15. The amino acid most recently added to a growing polypeptide chain is the only one covalently attached to a tRNA and thus is the only link between the polypeptide and the mRNA encoding it. A proofreading activity would sever this link, halting synthesis of the polypeptide and releasing it from the mRNA.

16. The protein would be directed into the ER, and from there the targeting would depend on additional signals. SRP binds the amino-terminal signal early in protein synthesis and directs the nascent polypeptide and ribosome to receptors in the ER. Because the protein is translocated into the lumen of the ER as it is synthesized, the NLS is never accessible to the proteins involved in nuclear targeting.

17. Trigger factor is a molecular chaperone that stabilizes an unfolded and translocation-competent conformation of ProOmpA.

18. DNA with a minimum of 5,784 bp; some of the coding sequences must be nested or overlapping.

19. **(a)** The helices associate through hydrophobic and van der Waals interactions. **(b)** R groups 3, 6, 7, and 10 extend to the left; 1, 2, 4, 5, 8, and 9 extend to the right. **(c)** One possible sequence is

    1 2 3 4 5 6 7 8 9 10
    N–Phe–Ile–Glu–Val–Met–Asn–Ser–Ala–Phe–Gln–*C*

    (d) One possible DNA sequence for the amino acid sequence in (c) is

    Nontemplate strand

    (5′)TTTATTGAAGTAATGAATAGTGCATTCC AG(3′)
        ||||||||||||||||||||||||||||||
    (3′)AAATAACTTCATTACTTATCACGTAAGGTC(5′)

    Template strand

(e) Phe, Leu, Ile, Met, and Val. All are hydrophobic, but the set does not include *all* the hydrophobic amino acids; Trp, Pro, Ala, and Gly are missing. **(f)** Tyr, His, Gln, Asn, Lys, Asp, and Glu. All of these are hydrophilic, although Tyr is less hydrophilic than the others. The set does not include *all* the hydrophilic amino acids; Ser, Thr, and Arg are missing. **(g)** Omitting T from the mixture excludes codons starting or ending with T—thus excluding Tyr, which is not very hydrophilic, and, more importantly, excluding the two possible stop codons (TAA and TAG). No other amino acids in the NAN set are excluded by omitting T. **(h)** Misfolded proteins are often degraded in the cell. Therefore, if a synthetic gene has produced a protein that forms a band on the SDS gel, it is likely that this protein is folded properly. **(i)** Protein folding depends on more than hydrophobic and van der Waals interactions. There are many reasons why a synthesized random-sequence protein might not fold into the four-helix structure. For example, hydrogen bonds between hydrophilic side chains could disrupt the structure. Also, not all sequences have an equal propensity to form an α helix.

Chapter 28

1. (a) Tryptophan synthase levels remain high in spite of the presence of tryptophan. **(b)** Levels again remain high. **(c)** Levels rapidly decrease, preventing wasteful synthesis of tryptophan.

2. The *E. coli* cells will produce β-galactosidase when they are subjected to high levels of a DNA-damaging agent, such as UV light. Under such conditions, RecA binds to single-stranded chromosomal DNA and facilitates the autocatalytic cleavage of the LexA repressor, releasing LexA from its binding site and allowing transcription of downstream genes.

3. (a) Constitutive, low-level expression of the operon; most mutations in the operator would make the repressor less likely to bind. **(b)** Either constitutive expression, as in (a), or constant repression, if the mutation destroyed the capability to bind to lactose and related compounds and hence the response to inducers. **(c)** Either increased or decreased expression of the operon (under conditions in which it is induced), depending on whether the mutation made the promoter more or less similar, respectively, to the consensus *E. coli* promoter.

4. 7,000 copies

5. 8×10^{-9} M, about 10^5 times greater than the dissociation constant. With 10 copies of active repressor in the cell, the operator site is always bound by the repressor molecule.

6. (a)–(e) Each condition decreases expression of *lac* operon genes.

7. (a) Less attenuation of transcription. The ribosome completing the translation of sequence 1 would no longer overlap and block sequence 2; sequence 2 would always be available to pair with sequence 3, preventing formation of the attenuator structure. **(b)** More attenuation of transcription. Sequence 2 would pair less efficiently with sequence 3; the attenuator structure would be formed more often, even when sequence 2 was not blocked by a ribosome. **(c)** No attenuation of transcription. The only regulation would be that afforded by the Trp repressor. **(d)** Attenuation loses its sensitivity to Trp tRNA. It might become sensitive to His tRNA. **(e)** Attenuation would rarely, if ever, occur. Sequences 2 and 3 always block formation of the attenuator. **(f)** Constant attenuation of transcription. Attenuator always forms, regardless of the availability of tryptophan.

8. Induction of the SOS response could not occur, making the cells more sensitive to high levels of DNA damage.

9. Each *Salmonella* cell would have flagella made up of both types of flagellar protein, and the cell would be vulnerable to antibodies generated in response to either protein.

10. A dissociable factor necessary for activity (e.g., a specificity factor similar to the σ subunit of the *E. coli* enzyme) may have been lost during purification of the polymerase.

11.

Gal4 protein

Gal4 DNA-binding domain	Gal4 activator domain

Engineered protein

Lac repressor DNA-binding domain	Gal4 activator domain

The engineered protein cannot bind to the Gal4 binding site in the *GAL* gene (UAS$_G$) because it lacks the Gal4 DNA-binding domain. Modify the Gal4p DNA binding site to give it the nucleotide sequence to which the Lac repressor normally binds (using methods described in Chapter 9).

12. Methylamine. The reaction proceeds with attack of water on the guanidinium carbon of the modified arginine.

13. The *bcd* mRNA needed for development is contributed to the egg by the mother. The egg develops normally even if its genotype is *bcd⁻/bcd⁻*, as long as the mother has one normal *bcd* gene and the *bcd⁻* allele is recessive. However, the adult *bcd⁻/bcd⁻* female will be sterile because she has no normal *bcd* mRNA to contribute to her own eggs.

14. (a) For 10% expression (90% repression), 10% of the repressor has bound inducer and 90% is free and available to bind the operator. The calculation uses Eqn 5–8 (p. 161), with $\theta = 0.1$ and $K_d = 10^{-4}$ M.

$$\theta = \frac{[\text{IPTG}]}{[\text{IPTG}] + K_d} = \frac{[\text{IPTG}]}{[\text{IPTG}] + 10^{-4} \text{ M}}$$

$$0.1 = \frac{[\text{IPTG}]}{[\text{IPTG}] + 10^{-4} \text{ M}} \quad \text{so } 0.9\,[\text{IPTG}] = 10^{-5} \text{ or } [\text{IPTG}] = 1.1 \times 10^{-5} \text{ M}$$

For 90% expression, 90% of the repressor has bound inducer, so $\theta = 0.9$. Entering the values for θ and K_d in Eqn 5–8 gives [IPTG] $= 9 \times 10^{-4}$ M. Thus, gene expression varies 10-fold over a roughly 10-fold [IPTG] range. **(b)** You would expect the protein levels to be low before induction, rise during induction, and then decay as synthesis stops and the proteins are degraded. **(c)** As shown in (a), the *lac* operon has more levels of expression than just on or off; thus it does not have characteristic A. As shown in (b), expression of the *lac* operon subsides once the inducer is removed; thus it lacks characteristic B. **(d)** *GFP-on:* rep^{ts} and GFP are expressed at high levels; rep^{ts} represses OP$_\lambda$, so no LacI protein is produced. *GFP-off:* LacI is expressed at a high level; LacI represses OP$_{lac}$, so rep^{ts} and GFP are not produced. **(e)** IPTG treatment switches the system from GFP-off to GFP-on. IPTG has an effect only when LacI is present, so affects only the GFP-off state. Adding IPTG relieves the repression of OP$_{lac}$, allowing high-level expression of rep^{ts}, which turns off expression of LacI, and high-level expression of GFP. **(f)** Heat treatment switches the system from GFP-on to GFP-off. Heat has an effect only when rep^{ts} is present, so affects only the GFP-on state. Heat inactivates rep^{ts} and relieves the repression of OP$_\lambda$, allowing high-level expression of LacI. LacI then acts at OP$_{lac}$ to repress synthesis of rep^{ts} and GFP. **(g)** *Characteristic A:* The system is not stable in the intermediate state. At some point, one repressor will act more strongly than the other due to chance fluctuations in expression; this shuts off expression of the other repressor and locks the system in one state. *Characteristic B:* Once one repressor is expressed, it prevents the synthesis of the other; thus the system remains in one state even after the switching stimulus has been removed. **(h)** At no time does any cell express an intermediate level of GFP—this is a confirmation of characteristic A. At the intermediate concentration (X) of inducer, some cells have switched to GFP-on while others have not yet made the switch and remain in the GFP-off state; none are in between. The bimodal distribution of expression levels at [IPTG] = X is caused by the mixed population of GFP-on and GFP-off cells.

Glossary

a

ABC transporters: Plasma membrane proteins with sequences that make up *ATP-binding cassettes*; serve to transport a large variety of substrates, including inorganic ions, lipids, and nonpolar drugs, out of the cell, using ATP as the energy source.

absolute configuration: The configuration of four different substituent groups around an asymmetric carbon atom, in relation to D- and L-glyceraldehyde.

absorption: Transport of the products of digestion from the intestinal tract into the blood.

acceptor control: Regulation of the rate of respiration by the availability of ADP as phosphate group acceptor.

accessory pigments: Visible light–absorbing pigments (carotenoids, xanthophylls, and phycobilins) in plants and photosynthetic bacteria that complement chlorophylls in trapping energy from sunlight.

acid dissociation constant: The dissociation constant (K_a) of an acid, describing its dissociation into its conjugate base and a proton.

acidosis: A metabolic condition in which the capacity of the body to buffer H^+ is diminished; usually accompanied by decreased blood pH.

actin: A protein that makes up the thin filaments of muscle; also an important component of the cytoskeleton of many eukaryotic cells.

action spectrum: A plot of the efficiency of light at promoting a light-dependent process such as photosynthesis as a function of wavelength.

activation energy ($\Delta G^{\ddagger}$): The amount of energy (in joules) required to convert all the molecules in 1 mol of a reacting substance from the ground state to the transition state.

activator: (1) A DNA-binding protein that positively regulates the expression of one or more genes; that is, transcription rates increase when an activator is bound to the DNA. (2) A positive modulator of an allosteric enzyme.

active site: The region of an enzyme surface that binds the substrate molecule and catalytically transforms it; also known as the catalytic site.

active transport: Energy-requiring transport of a solute across a membrane in the direction of increasing concentration.

activity: The true thermodynamic activity or potential of a substance, as distinct from its molar concentration.

acyl phosphate: Any molecule with the general chemical form $R-\overset{\overset{\textstyle O}{\|}}{C}-O-OPO_3^{2-}$.

adaptor proteins: Signaling proteins, generally without their own enzymatic activities, that have binding sites for two or more cellular components and serve to bring those components together.

adenosine 3′,5′-cyclic monophosphate: *See* cyclic AMP.

***S*-adenosylmethionine (adoMet):** An enzymatic cofactor involved in methyl group transfers.

adipocyte: An animal cell specialized for the storage of fats (triacylglycerols).

adipose tissue: Connective tissue specialized for the storage of large amounts of triacylglycerols. *See also* brown adipose tissue; white adipose tissue.

ADP (adenosine diphosphate): A ribonucleoside 5′-diphosphate serving as phosphate group acceptor in the cell energy cycle.

aerobe: An organism that lives in air and uses oxygen as the terminal electron acceptor in respiration.

aerobic: Requiring or occurring in the presence of oxygen.

agonist: A compound, typically a hormone or neurotransmitter, that elicits a physiological response when it binds to its specific receptor.

alcohol fermentation: *See* ethanol fermentation.

aldose: A simple sugar in which the carbonyl carbon atom is an aldehyde; that is, the carbonyl carbon is at one end of the carbon chain.

alkalosis: A metabolic condition in which the capacity of the body to buffer OH^- is diminished; usually accompanied by an increase in blood pH.

allosteric enzyme: A regulatory enzyme with catalytic activity modulated by the noncovalent binding of a specific metabolite at a site other than the active site.

allosteric protein: A protein (generally with multiple subunits) with multiple ligand-binding sites, such that ligand binding at one site affects ligand binding at another.

allosteric site: The specific site on the surface of an allosteric enzyme molecule to which the modulator or effector molecule is bound.

α helix: A helical conformation of a polypeptide chain, usually right-handed, with maximal intrachain hydrogen bonding; one of the most common secondary structures in proteins.

α oxidation: An alternative path for the oxidation of β-methyl fatty acids in peroxisomes.

Ames test: A simple bacterial test for carcinogenicity, based on the assumption that carcinogens are mutagens.

amino acid activation: ATP-dependent enzymatic esterification of the carboxyl group of an amino acid to the 3′-hydroxyl group of its corresponding tRNA.

amino acids: α-Amino–substituted carboxylic acids, the building blocks of proteins.

aminoacyl-tRNA: An aminoacyl ester of a tRNA.

aminoacyl-tRNA synthetases: Enzymes that catalyze synthesis of an aminoacyl-tRNA at the expense of ATP energy.

amino-terminal residue: The only amino acid residue in a polypeptide chain with a free α-amino group; defines the amino terminus of the polypeptide.

aminotransferases: Enzymes that catalyze the transfer of amino groups from α-amino to α-keto acids; also called transaminases.

ammonotelic: Excreting excess nitrogen in the form of ammonia.

AMP-activated protein kinase (AMPK): A protein kinase activated by 5′-adenosine monophosphate (AMP). AMPK action generally shifts metabolism away from biosynthesis toward energy production.

amphibolic pathway: A metabolic pathway used in both catabolism and anabolism.

amphipathic: Containing both polar and nonpolar domains.

amphitropic proteins: Proteins that associate reversibly with the membrane and thus can be found in the cytosol, in the membrane, or in both places.

ampholyte: A substance that can act as either a base or an acid.

amphoteric: Capable of donating and accepting protons, thus able to serve as an acid or a base.

AMPK: *See* AMP-activated protein kinase.

amyloidoses: A variety of progressive conditions characterized by abnormal deposits of misfolded proteins in one or more organs or tissues.

anabolism: The phase of intermediary metabolism concerned with the energy-requiring biosynthesis of cell components from smaller precursors.

anaerobe: An organism that lives without oxygen. Obligate anaerobes die when exposed to oxygen.

anaerobic: Occurring in the absence of air or oxygen.

analyte: A molecule to be analyzed by mass spectrometry.

anammox: Anaerobic oxidation of ammonia to N_2, using nitrite as electron acceptor; carried out by specialized chemolithotrophic bacteria.

anaplerotic reaction: An enzyme-catalyzed reaction that can replenish the supply of intermediates in the citric acid cycle.

angstrom (Å): A unit of length (10^{-8} cm) used to indicate molecular dimensions.

anhydride: The product of the condensation of two carboxyl or phosphate groups in which the elements of water are eliminated to form a compound with the general structure R—X—O—X—R, where X is either carbon or phosphorus.

anion-exchange resin: A polymeric resin with fixed cationic groups, used in the chromatographic separation of anions.

anomeric carbon: The carbon atom in a sugar at the new stereocenter formed when a sugar cyclizes to form a hemiacetal. This is the carbonyl carbon of aldehydes and ketones.

anomers: Two stereoisomers of a given sugar that differ only in the configuration about the carbonyl (anomeric) carbon atom.

antagonist: A compound that interferes with the physiological action of another substance (the agonist), usually at a hormone or neurotransmitter receptor.

antibiotic: One of many different organic compounds that are formed and secreted by various species of microorganisms and plants, are toxic to other species, and presumably have a defensive function.

antibody: A defense protein synthesized by the immune system of vertebrates. *See also* immunoglobulin.

anticodon: A specific sequence of three nucleotides in a tRNA, complementary to a codon for an amino acid in an mRNA.

antigen: A molecule capable of eliciting the synthesis of a specific antibody in vertebrates.

antiparallel: Describes two linear polymers that are opposite in polarity or orientation.

antiport: Cotransport of two solutes across a membrane in opposite directions.

apoenzyme: The protein portion of an enzyme, exclusive of any organic or inorganic cofactors or prosthetic groups that might be required for catalytic activity.

apolipoprotein: The protein component of a lipoprotein.

apoprotein: The protein portion of a protein, exclusive of any organic or inorganic cofactors or prosthetic groups that might be required for activity.

apoptosis: (app′-a-toe′-sis) Programmed cell death in which a cell brings about its own death and lysis, in response to a signal from outside or programmed in its genes, by systematically degrading its own macromolecules.

aptamer: Oligonucleotide that binds specifically to one molecular target, usually selected by an iterative cycle of affinity-based enrichment (SELEX).

aquaporin (AQP): A member of a family of integral membrane proteins that mediate the flow of water across membranes.

Archaea: One of the five kingdoms of living organisms; includes many species that thrive in extreme environments of high ionic strength, high temperature, or low pH.

asymmetric carbon atom: A carbon atom that is covalently bonded to four different

groups and thus may exist in two different tetrahedral configurations.

ATP (adenosine triphosphate): A ribonucleoside 5′-triphosphate functioning as a phosphate group donor in the cellular energy cycle; carries chemical energy between metabolic pathways by serving as a shared intermediate coupling endergonic and exergonic reactions.

ATPase: An enzyme that hydrolyzes ATP to yield ADP and phosphate, usually coupled to a process requiring energy.

ATP synthase: An enzyme complex that forms ATP from ADP and phosphate during oxidative phosphorylation in the inner mitochondrial membrane or the bacterial plasma membrane, and during photophosphorylation in chloroplasts.

attenuator: An RNA sequence involved in regulating the expression of certain genes; functions as a transcription terminator.

autophagy: Catabolic lysosomal degradation of cellular proteins and other components.

autophosphorylation: Strictly, the phosphorylation of an amino acid residue in a protein that is catalyzed by the same protein molecule; often extended to include phosphorylation of one subunit of a homodimer by the other subunit.

autotroph: An organism that can synthesize its own complex molecules from very simple carbon and nitrogen sources, such as carbon dioxide and ammonia.

auxin: A plant growth hormone.

auxotrophic mutant (auxotroph): A mutant organism defective in the synthesis of a particular biomolecule, which must therefore be supplied for the organism's growth.

Avogadro's number (N): The number of molecules in a gram molecular weight (a mole) of any compound (6.02×10^{23}).

b

Bacteria: One of the five kingdoms of living organisms; bacteria have a plasma membrane but no internal organelles or nucleus.

bacteriophage: A virus capable of replicating in a bacterial cell; also called phage.

baculovirus: Any of a group of double-stranded DNA viruses that infect invertebrates, particularly insects, and are widely used for purposes of protein expression in biotechnology.

basal metabolic rate: An animal's rate of oxygen consumption when at complete rest, long after a meal.

base pair: Two nucleotides in nucleic acid chains that are paired by hydrogen bonding of their bases; for example, A with T or U, and G with C.

BAT: *See* brown adipose tissue.

B cell: *See* B lymphocyte.

β conformation: An extended, zigzag arrangement of a polypeptide chain; a common secondary structure in proteins.

β oxidation: Oxidative degradation of fatty acids into acetyl-CoA by successive oxida-

tions at the β-carbon atom; as distinct from ω oxidation.

β turn: A type of protein secondary structure consisting of four amino acid residues arranged in a tight turn so that the polypeptide turns back on itself.

bilayer: A double layer of oriented amphipathic lipid molecules, forming the basic structure of biological membranes. The hydrocarbon tails face inward to form a continuous nonpolar phase.

bile acids: Polar derivatives of cholesterol, secreted by the liver into the intestine, that serve to emulsify dietary fats, facilitating lipase action on them.

bile salts: Amphipathic steroid derivatives with detergent properties, participating in digestion and absorption of lipids.

binding energy: The energy derived from noncovalent interactions between enzyme and substrate or receptor and ligand.

binding site: The crevice or pocket on a protein in which a ligand binds.

bioassay: A method for measuring the amount of a biologically active substance (such as a hormone) in a sample by quantifying the biological response to aliquots of that sample.

bioinformatics: The computerized analysis of biological data, using methods derived from statistics, linguistics, mathematics, chemistry, biochemistry, and physics. The data are often nucleic acid or protein sequence or structural data, but can also involve experimental data from many sources, patient statistics, and materials in the scientific literature. Bioinformatics research focuses on methods for data storage, retrieval, and analysis.

biosphere: All the living matter on or in the earth, the seas, and the atmosphere.

biotin: A vitamin; an enzymatic cofactor involved in carboxylation reactions.

B lymphocyte (B cell): One of a class of blood cells (lymphocytes), responsible for the production of circulating antibodies.

bond energy: The energy required to break a bond.

branch migration: Movement of the branch point in a branched DNA formed from two DNA molecules with identical sequences. *See also* Holliday intermediate.

brown adipose tissue (BAT): Thermogenic adipose tissue rich in mitochondria that contain the uncoupling protein thermogenin, which uncouples electron transfer through the respiratory chain from ATP synthesis. *Compare* white adipose tissue.

buffer: A system capable of resisting changes in pH, consisting of a conjugate acid-base pair in which the ratio of proton acceptor to proton donor is near unity.

c

calorie: The amount of heat required to raise the temperature of 1.0 g of water from 14.5 to 15.5 °C. One calorie (cal) equals 4.18 joules (J).

Calvin cycle: The cyclic pathway in plants that fixes carbon dioxide and produces triose phosphates.

CAM plants: Succulent plants of hot, dry climates, in which CO_2 is fixed into oxaloacetate in the dark, then fixed by rubisco in the light when stomata close to exclude O_2.

cAMP: *See* cyclic AMP.

cAMP receptor protein (CRP): In bacteria, a specific regulatory protein that controls initiation of transcription of the genes that produce the enzymes required for the cell to use some other nutrient when glucose is lacking; also called catabolite gene activator protein (CAP).

CAP: *See* cAMP receptor protein.

capsid: The protein coat of a virion or virus particle.

carbanion: A negatively charged carbon atom.

carbocation: A positively charged carbon atom; also called a carbonium ion.

carbohydrate: A polyhydroxy aldehyde or ketone, or substance that yields such a compound on hydrolysis. Many carbohydrates have the empirical formula $(CH_2O)_n$; some also contain nitrogen, phosphorus, or sulfur.

carbon-assimilation reactions: Reaction sequence in which atmospheric CO_2 is converted into organic compounds.

carbon-fixation reactions: The reactions, catalyzed by rubisco during photosynthesis or by other carboxylases, in which atmospheric CO_2 is initially incorporated (fixed) into an organic compound.

carbonium ion: *See* carbocation.

carboxyl-terminal residue: The only amino acid residue in a polypeptide chain with a free α-carboxyl group; defines the carboxyl terminus of the polypeptide.

cardiolipin: A membrane phospholipid in which two phosphatidic acid moieties share a single glycerol head group.

carnitine shuttle: Mechanism for moving fatty acids from the cytosol to the mitochondrial matrix as fatty esters of carnitine.

carotenoids: Lipid-soluble photosynthetic pigments made up of isoprene units.

cascade: *See* enzyme cascade.

catabolism: The phase of intermediary metabolism concerned with the energy-yielding degradation of nutrient molecules.

catabolite gene activator protein (CAP): *See* cAMP receptor protein.

catalytic site: *See* active site.

catecholamines: Hormones, such as epinephrine, that are amino derivatives of catechol.

catenane: Two or more circular polymeric molecules interlinked by one or more noncovalent topological links, resembling the links of a chain.

cation-exchange resin: An insoluble polymer with fixed negative charges, used in the chromatographic separation of cationic substances.

cDNA: *See* complementary DNA.

cDNA library: A collection of cloned DNA fragments derived entirely from the complement of mRNA being expressed in a particular organism or cell type under a defined set of conditions.

cellular differentiation: The process in which a precursor cell becomes specialized to carry out a particular function, by acquiring a new complement of proteins and RNA.

central dogma: The organizing principle of molecular biology: genetic information flows from DNA to RNA to protein.

centromere: A specialized site in a chromosome, serving as the attachment point for the mitotic or meiotic spindle.

cerebroside: Sphingolipid containing one sugar residue as a head group.

channeling: The direct transfer of a reaction product (common intermediate) from the active site of an enzyme to the active site of the enzyme catalyzing the next step in a pathway.

chaperone: Any of several classes of proteins or protein complexes that catalyze the accurate folding of proteins in all cells.

chaperonin: One of two major classes of chaperones found in virtually all organisms; a complex of proteins that functions in protein folding, either GroES/GroEL in bacteria or Hsp60 in eukaryotes.

chemiosmotic coupling: Coupling of ATP synthesis to electron transfer via a transmembrane difference in charge and pH.

chemiosmotic theory: The theory that energy derived from electron transfer reactions is temporarily stored as a transmembrane difference in charge and pH, which subsequently drives the formation of ATP in oxidative phosphorylation and photophosphorylation.

chemotaxis: A cell's sensing of and movement toward or away from a specific chemical agent.

chemotroph: An organism that obtains energy by metabolizing organic compounds derived from other organisms.

chiral center: An atom with substituents arranged so that the molecule is not superposable on its mirror image.

chiral compound: A compound that contains an asymmetric center (chiral atom or chiral center) and thus can occur in two nonsuperposable mirror-image forms (enantiomers).

chlorophylls: A family of green pigments that function as receptors of light energy in photosynthesis; magnesium-porphyrin complexes.

chloroplast: Chlorophyll-containing photosynthetic organelle in some eukaryotic cells.

chondroitin sulfate: One of a family of sulfated glycosaminoglycans, a major component of the extracellular matrix.

chromatin: A filamentous complex of DNA, histones, and other proteins, constituting the eukaryotic chromosome.

chromatography: A process in which complex mixtures of molecules are separated by many repeated partitionings between a flowing (mobile) phase and a stationary phase.

chromatophore: A compound or moiety (natural or synthetic) that absorbs visible or ultraviolet light.

chromosome: A single large DNA molecule and its associated proteins, containing many genes; stores and transmits genetic information.

chylomicron: A plasma lipoprotein consisting of a large droplet of triacylglycerols stabilized by a coat of protein and phospholipid; carries lipids from the intestine to the tissues.

circular dichroism spectroscopy: A method used to characterize the degree of folding in a protein, based on differences in the absorption of right-handed versus left-handed circularly polarized light.

cis and trans isomers: *See* geometric isomers.

cistron: A unit of DNA or RNA corresponding to one gene.

citric acid cycle: A cyclic pathway for the oxidation of acetyl residues to carbon dioxide, in which formation of citrate is the first step; also known as the Krebs cycle or tricarboxylic acid cycle.

clones: The descendants of a single cell.

cloning: The production of large numbers of identical DNA molecules, cells, or organisms from a single ancestral DNA molecule, cell, or organism.

closed system: A system that exchanges neither matter nor energy with the surroundings. *See also* system.

cobalamin: *See* coenzyme B_{12}.

coding strand: In DNA transcription, the DNA strand identical in base sequence to the RNA transcribed from it, with U in the RNA in place of T in the DNA; as distinct from the template strand. Also called the nontemplate strand.

codon: A sequence of three adjacent nucleotides in a nucleic acid that codes for a specific amino acid.

coenzyme: An organic cofactor required for the action of certain enzymes; often has a vitamin component.

coenzyme A: A pantothenic acid–containing coenzyme that serves as an acyl group carrier in certain enzymatic reactions.

coenzyme B_{12}: An enzymatic cofactor derived from the vitamin cobalamin, involved in certain types of carbon skeletal rearrangements.

cofactor: An inorganic ion or a coenzyme required for enzyme activity.

cognate: Describes two biomolecules that normally interact; for example, an enzyme and its usual substrate, or a receptor and its usual ligand.

cohesive ends: *See* sticky ends.

cointegrate: An intermediate in the migration of certain DNA transposons in which the donor DNA and target DNA are covalently attached.

colligative properties: Properties of a solution that depend on the number of solute particles per unit volume; for example, freezing-point depression.

combinatorial control: Use of combinations of a limited repertoire of regulatory proteins to provide gene-specific regulation of many individual genes.

competitive inhibition: A type of enzyme inhibition reversed by increasing the substrate concentration; a competitive inhibitor generally competes with the normal substrate or ligand for a protein's binding site.

complementary: Having a molecular surface with chemical groups arranged to interact specifically with chemical groups on another molecule.

complementary DNA (cDNA): A DNA complementary to a specific mRNA, used in DNA cloning; usually made by reverse transcriptase.

condensation: A reaction type in which two compounds are joined with the elimination of water.

configuration: The spatial arrangement of an organic molecule that is conferred by the presence of either (1) double bonds, about which there is no freedom of rotation, or (2) chiral centers, around which substituent groups are arranged in a specific sequence. Configurational isomers cannot be interconverted without breaking one or more covalent bonds.

conformation: A spatial arrangement of substituent groups that are free to assume different positions in space, without breaking any bonds, because of the freedom of bond rotation.

conjugate acid-base pair: A proton donor and its corresponding deprotonated species; for example, acetic acid (donor) and acetate (acceptor).

conjugated protein: A protein containing one or more prosthetic groups.

conjugate redox pair: An electron donor and its corresponding electron acceptor form; for example, Cu^+ (donor) and Cu^{2+} (acceptor), or NADH (donor) and NAD^+ (acceptor).

consensus sequence: A DNA or amino acid sequence consisting of the residues that most commonly occur at each position in a set of similar sequences.

conservative substitution: Replacement of an amino acid residue in a polypeptide by another residue with similar properties; for example, substitution of Glu by Asp.

constitutive enzymes: Enzymes required at all times by a cell and present at some constant level; for example, many enzymes of the central metabolic pathways. Sometimes called housekeeping enzymes.

contig: A series of overlapping clones or a continuous sequence defining an uninterrupted section of a chromosome.

contour length: The length of a helical polymeric molecule as measured along its helical axis.

cooperativity: The characteristic of an enzyme or other protein in which binding of the first molecule of a ligand changes the affinity for the second molecule. In positive cooperativity, the affinity for the second ligand molecule increases; in negative cooperativity, it decreases.

cotransport: The simultaneous transport, by a single transporter, of two solutes across a membrane. *See also* antiport; symport.

coupled reactions: Two chemical reactions that have a common intermediate and thus a means of energy transfer from one to the other.

covalent bond: A chemical bond that involves sharing of electron pairs.

C_4 plants: Plants (generally tropical) in which CO_2 is first fixed into a four-carbon compound (oxaloacetate or malate) before entering the Calvin cycle via rubisco.

cristae: Infoldings of the inner mitochondrial membrane.

CRP: *See* cAMP receptor protein.

cruciform: Secondary structure in double-stranded RNA or DNA in which the double helix is denatured at palindromic repeat sequences in each strand, and each separated strand is paired internally to form opposing hairpin structures. *See also* hairpin.

cyclic AMP (cAMP, adenosine 3′, 5′-cyclic monophosphate): A second messenger; its formation in a cell by adenylyl cyclase is stimulated by certain hormones or other molecular signals.

cyclic electron flow: In chloroplasts, the light-induced flow of electrons originating from and returning to photosystem I.

cyclic photophosphorylation: ATP synthesis driven by cyclic electron flow through photosystem I.

cyclin: One of a family of proteins that activate cyclin-dependent protein kinases and thereby regulate the cell cycle.

cytochrome P-450: A family of heme-containing enzymes, with a characteristic absorption band at 450 nm, that participate in biological hydroxylations with O_2.

cytochromes: Heme proteins serving as electron carriers in respiration, photosynthesis, and other oxidation-reduction reactions.

cytokine: One of a family of small secreted proteins (such as interleukins or interferons) that activate cell division or differentiation by binding to plasma membrane receptors in target cells.

cytokinesis: The final separation of daughter cells following mitosis.

cytoplasm: The portion of a cell's contents outside the nucleus but within the plasma membrane; includes organelles such as mitochondria.

cytoskeleton: The filamentous network that provides structure and organization to the cytoplasm; includes actin filaments, microtubules, and intermediate filaments.

cytosol: The continuous aqueous phase of the cytoplasm, with its dissolved solutes; excludes the organelles such as mitochondria.

d

dalton: Unit of atomic or molecular weight; 1 dalton (Da) is the weight of a hydrogen atom (1.66×10^{-24} g).

dark reactions: *See* carbon-assimilation reactions.

deamination: The enzymatic removal of amino groups from biomolecules such as amino acids or nucleotides.

degenerate code: A code in which a single element in one language is specified by more than one element in a second language.

dehydrogenases: Enzymes that catalyze the removal of pairs of hydrogen atoms from substrates.

deletion mutation: A mutation resulting from the deletion of one or more nucleotides from a gene or chromosome.

ΔG: *See* free-energy change.

$\Delta G^{\ddagger}$: *See* activation energy.

$\Delta G'^{\circ}$: *See* standard free-energy change.

denaturation: Partial or complete unfolding of the specific native conformation of a polypeptide chain, protein, or nucleic acid such that the function of the molecule is lost.

denatured protein: A protein that has lost enough of its native conformation by exposure to a destabilizing agent such as heat or detergent that its function is lost.

de novo pathway: Pathway for the synthesis of a biomolecule, such as a nucleotide, from simple precursors; as distinct from a salvage pathway.

deoxyribonucleic acid: *See* DNA.

deoxyribonucleotides: Nucleotides containing 2-deoxy-D-ribose as the pentose component.

desaturases: Enzymes that catalyze the introduction of double bonds into the hydrocarbon portion of fatty acids.

desensitization: Universal process by which sensory mechanisms cease to respond after prolonged exposure to the specific stimulus they detect.

desolvation: In aqueous solution, the release of bound water surrounding a solute.

diabetes mellitus: A group of metabolic diseases with symptoms that result from a deficiency in insulin production or utilization; characterized by a failure in glucose transport from the blood into cells at normal glucose concentrations.

dialysis: Removal of small molecules from a solution of a macromolecule by their diffusion through a semipermeable membrane into a suitably buffered solution.

differential centrifugation: Separation of cell organelles or other particles of different size by their different rates of sedimentation in a centrifugal field.

differentiation: Specialization of cell structure and function during growth and development.

diffusion: The net movement of molecules in the direction of lower concentration.

digestion: Enzymatic hydrolysis of major nutrients in the gastrointestinal system to yield their simpler components.

diploid: Having two sets of genetic information; describes a cell with two chromosomes of each type. *Compare* haploid.

disaccharide: A carbohydrate consisting of two covalently joined monosaccharide units.

dissociation constant: An equilibrium constant (K_d) for the dissociation of a complex of two or more biomolecules into its components;

for example, dissociation of a substrate from an enzyme.

disulfide bond: A covalent bond involving the oxidative linkage of two Cys residues, from the same or different polypeptide chains, forming a cystine residue.

DNA (deoxyribonucleic acid): A polynucleotide with a specific sequence of deoxyribonucleotide units covalently joined through $3',5'$-phosphodiester bonds; serves as the carrier of genetic information.

DNA chimera: DNA containing genetic information derived from two different species.

DNA chip: Informal term for a DNA microarray, referring to the small size of typical microarrays.

DNA cloning: *See* cloning.

DNA library: A collection of cloned DNA fragments.

DNA ligases: Enzymes that creates a phosphodiester bond between the $3'$ end of one DNA segment and the $5'$ end of another.

DNA looping: The interaction of proteins bound at distant sites on a DNA molecule so that the intervening DNA forms a loop.

DNA microarray: A collection of DNA sequences immobilized on a solid surface, with individual sequences laid out in patterned arrays that can be probed by hybridization.

DNA polymerase: An enzyme that catalyzes template-dependent synthesis of DNA from its deoxyribonucleoside $5'$-triphosphate precursors.

DNA supercoiling: The coiling of DNA upon itself, generally as a result of bending, underwinding, or overwinding of the DNA helix.

DNA transposition: *See* transposition.

domain: A distinct structural unit of a polypeptide; domains may have separate functions and may fold as independent, compact units.

double helix: The natural coiled conformation of two complementary, antiparallel DNA chains.

double-reciprocal plot: A plot of $1/V_0$ versus $1/[S]$, which allows a more accurate determination of V_{max} and K_m than a plot of V_0 versus [S]; also called the Lineweaver-Burk plot.

e

$E'°$: *See* standard reduction potential.

ECM: *See* extracellular matrix.

electrochemical gradient: The resultant of the gradients of concentration and of electric charge of an ion across a membrane; the driving force for oxidative phosphorylation and photophosphorylation.

electrochemical potential: The energy required to maintain a separation of charge and of concentration across a membrane.

electrogenic: Contributing to an electrical potential across a membrane.

electron acceptor: A substance that receives electrons in an oxidation-reduction reaction.

electron carrier: A protein, such as a flavoprotein or a cytochrome, that can reversibly gain and lose electrons; functions in the transfer of electrons from organic nutrients to oxygen or some other terminal acceptor.

electron donor: A substance that donates electrons in an oxidation-reduction reaction.

electron transfer: Movement of electrons from electron donor to electron acceptor; especially, from substrates to oxygen via the carriers of the respiratory (electron-transfer) chain.

electrophile: An electron-deficient group with a strong tendency to accept electrons from an electron-rich group (nucleophile).

electrophoresis: Movement of charged solutes in response to an electrical field; often used to separate mixtures of ions, proteins, or nucleic acids.

elongation factors: (1) Proteins that function in the elongation phase of eukaryotic transcription. (2) Specific proteins required in the elongation of polypeptide chains by ribosomes.

eluate: The effluent from a chromatographic column.

enantiomers: Stereoisomers that are nonsuperposable mirror images of each other.

endergonic reaction: A chemical reaction that consumes energy (that is, for which ΔG is positive).

endocrine: Pertaining to cellular secretions that enter the bloodstream and have their effects on distant tissues.

endocytosis: The uptake of extracellular material by its inclusion in a vesicle (endosome) formed by invagination of the plasma membrane.

endonucleases: Enzymes that hydrolyze the interior phosphodiester bonds of a nucleic acid—that is, act at bonds other than the terminal bonds.

endoplasmic reticulum: An extensive system of double membranes in the cytoplasm of eukaryotic cells; it encloses secretory channels and is often studded with ribosomes (rough endoplasmic reticulum).

endothermic reaction: A chemical reaction that takes up heat (that is, for which ΔH is positive).

end-product inhibition: *See* feedback inhibition.

enhancers: DNA sequences that facilitate the expression of a given gene; may be located a few hundred, or even thousand, base pairs away from the gene.

enthalpy (H): The heat content of a system.

enthalpy change (ΔH): For a reaction, approximately equal to the difference between the energy used to break bonds and the energy gained by the formation of new ones.

entropy (S): The extent of randomness or disorder in a system.

enzyme: A biomolecule, either protein or RNA, that catalyzes a specific chemical reaction. It does not affect the equilibrium of the catalyzed reaction; it enhances the rate of the reaction by providing a reaction path with a lower activation energy.

enzyme cascade: A series of reactions, often involved in regulatory events, in which one enzyme activates another (often by phosphorylation), which activates a third, and so on. The effect of a catalyst activating a catalyst is a large amplification of the signal that initiated the cascade.

epigenetic: Describes any inherited characteristic of a living organism that is acquired by means that do not involve the nucleotide sequence of the parental chromosomes; for example, covalent modifications of histones.

epimerases: Enzymes that catalyze the reversible interconversion of two epimers.

epimers: Two stereoisomers differing in configuration at one asymmetric center in a compound having two or more asymmetric centers.

epithelial cell: Any cell that forms part of the outer covering of an organism or organ.

epitope: An antigenic determinant; the particular chemical group or groups in a macromolecule (antigen) to which a given antibody binds.

epitope tag: A protein sequence or domain bound by some well-characterized antibody.

equilibrium: The state of a system in which no further net change is occurring; the free energy is at a minimum.

equilibrium constant (K_{eq}): A constant, characteristic for each chemical reaction, that relates the specific concentrations of all reactants and products at equilibrium at a given temperature and pressure.

erythrocyte: A cell containing large amounts of hemoglobin and specialized for oxygen transport; a red blood cell.

essential amino acids: Amino acids that cannot be synthesized by humans (and other vertebrates) and must be obtained from the diet.

essential fatty acids: The group of polyunsaturated fatty acids produced by plants, but not by humans; required in the human diet.

ethanol fermentation: The anaerobic conversion of glucose to ethanol via glycolysis; also called alcohol fermentation. *See also* fermentation.

euchromatin: The regions of interphase chromosomes that stain diffusely, as opposed to the more condensed, heavily staining, heterochromatin. These are often regions in which genes are being actively expressed.

eukaryote: A unicellular or multicellular organism with cells having a membrane-bounded nucleus, multiple chromosomes, and internal organelles.

excited state: An energy-rich state of an atom or molecule, produced by the absorption of light energy; as distinct from ground state.

exergonic reaction: A chemical reaction that proceeds with the release of free energy (that is, for which ΔG is negative).

exocytosis: The fusion of an intracellular vesicle with the plasma membrane, releasing the vesicle contents to the extracellular space.

exon: The segment of a eukaryotic gene that encodes a portion of the final product of the gene; a segment of RNA that remains after posttranscriptional processing and is transcribed into a protein or incorporated into the structure of an RNA. *See also* intron.

exonucleases: Enzymes that hydrolyze only those phosphodiester bonds that are in the terminal positions of a nucleic acid.

exothermic reaction: A chemical reaction that releases heat (that is, for which ΔH is negative).

expression vector: *See* vector.

extracellular matrix (ECM): An interwoven combination of glycosaminoglycans, proteoglycans, and proteins, just outside the plasma membrane, that provides cell anchorage, positional recognition, and traction during cell migration.

extrahepatic: Describes all tissues outside the liver; implies the centrality of the liver in metabolism.

f

facilitated diffusion: *See* passive transport.

FAD (flavin adenine dinucleotide): The coenzyme of some oxidation-reduction enzymes; contains riboflavin.

F$_1$ ATPase: The multiprotein subunit of ATP synthase that has the ATP-synthesizing catalytic sites. It interacts with the F$_o$ subunit of ATP synthase, coupling proton movement to ATP synthesis.

fatty acid: A long-chain aliphatic carboxylic acid found in natural fats and oils; also a component of membrane phospholipids and glycolipids.

feedback inhibition: Inhibition of an allosteric enzyme at the beginning of a metabolic sequence by the end product of the sequence; also known as end-product inhibition.

fermentation: Energy-yielding anaerobic breakdown of a nutrient molecule, such as glucose, without net oxidation; yields lactate, ethanol, or some other simple product.

fibrin: A protein factor that forms the cross-linked fibers in blood clots.

fibrinogen: The inactive precursor protein of fibrin.

fibroblast: A cell of the connective tissue that secretes connective tissue proteins such as collagen.

fibrous proteins: Insoluble proteins that serve a protective or structural role; contain polypeptide chains that generally share a common secondary structure.

first law of thermodynamics: The law stating that, in all processes, the total energy of the universe remains constant.

Fischer projection formulas: A method for representing molecules that shows the configuration of groups around chiral centers; also known as projection formulas.

5′ end: The end of a nucleic acid that lacks a nucleotide bound at the 5′ position of the terminal residue.

flagellum: A cell appendage used in propulsion. Bacterial flagella have a much simpler structure than eukaryotic flagella, which are similar to cilia.

flavin-linked dehydrogenases: Dehydrogenases requiring one of the riboflavin coenzymes, FMN or FAD.

flavin nucleotides: Nucleotide coenzymes (FMN and FAD) containing riboflavin.

flavoprotein: An enzyme containing a flavin nucleotide as a tightly bound prosthetic group.

flippases: Membrane proteins in the ABC transporter family that catalyze the movement of phospholipids from the extracellular leaflet to the cytosolic leaflet of a membrane bilayer.

floppases: Membrane proteins in the ABC transporter family that catalyze movement of phospholipids from the cytosolic leaflet to the extracellular leaflet of a membrane bilayer.

fluid mosaic model: A model describing biological membranes as a fluid lipid bilayer with embedded proteins; the bilayer exhibits both structural and functional asymmetry.

fluorescence: Emission of light by excited molecules as they revert to the ground state.

fluorescence recovery after photobleaching: *See* FRAP.

fluorescence resonance energy transfer: *See* FRET.

FMN (flavin mononucleotide): Riboflavin phosphate, a coenzyme of certain oxidation-reduction enzymes.

fold: *See* motif.

footprinting: A technique for identifying the nucleic acid sequence bound by a DNA- or RNA-binding protein.

fraction: A portion of a biological sample that has been subjected to a procedure designed to separate macromolecules based on a property such as solubility, net charge, molecular weight, or function.

fractionation: The process of separating the proteins or other components of a complex molecular mixture into fractions based on differences in properties such as solubility, net charge, molecular weight, or function.

frame shift: A mutation caused by insertion or deletion of one or more paired nucleotides, changing the reading frame of codons during protein synthesis; the polypeptide product has a garbled amino acid sequence beginning at the mutated codon.

FRAP (fluorescence recovery after photobleaching): A technique used to quantify the diffusion of membrane components (lipids or proteins) in the plane of the bilayer.

free energy (G): The component of the total energy of a system that can do work at constant temperature and pressure.

free energy of activation ($\Delta G^{\ddagger}$): *See* activation energy.

free-energy change (ΔG): The amount of free energy released (negative ΔG) or absorbed (positive ΔG) in a reaction at constant temperature and pressure.

free radical: *See* radical.

FRET (fluorescence resonance energy transfer): A technique for estimating the distance between two proteins or two domains of a protein by measuring the nonradiative transfer of energy between reporter chromophores when one is excited and the fluorescence emitted from the other is quantified.

functional group: The specific atom or group of atoms that confers a particular chemical property on a biomolecule.

fusion protein: (1) One of a family of proteins that facilitate membrane fusion. (2) The protein product of a gene created by the fusion of two distinct genes or portions of genes.

futile cycle: A cycle of enzyme-catalyzed reactions that results in release of thermal energy by the hydrolysis of ATP.

g

G$_i$: *See* inhibitory G protein.

G$_s$: *See* stimulatory G protein.

gametes: Reproductive cells with a haploid gene content; sperm or egg cells.

ganglioside: Sphingolipid containing a complex oligosaccharide as a head group; especially common in nervous tissue.

GEFs: *See* guanosine nucleotide–exchange factors.

gel filtration: *See* size-exclusion chromatography.

gene: A chromosomal segment that codes for a single functional polypeptide chain or RNA molecule.

gene expression: Transcription, and in the case of proteins, translation, to yield the product of a gene; a gene is expressed when its biological product is present and active.

gene fusion: The enzymatic attachment of one gene, or part of a gene, to another.

general acid-base catalysis: Catalysis involving proton transfer(s) to or from a molecule other than water.

genetic code: The set of triplet code words in DNA (or mRNA) coding for the amino acids of proteins.

genetic engineering: Any process by which genetic material, particularly DNA, is altered by a molecular biologist.

genetic map: A diagram showing the relative sequence and position of specific genes along a chromosome.

genome: All the genetic information encoded in a cell or virus.

genomic library: A DNA library containing DNA segments that represent all (or most) of the sequences in an organism's genome.

genomics: A science devoted broadly to the understanding of cellular and organism genomes.

genotype: The genetic constitution of an organism, as distinct from its physical characteristics, or phenotype.

geometric isomers: Isomers related by rotation about a double bond; also called cis and trans isomers.

germ-line cell: A type of animal cell that is formed early in embryogenesis and may multiply by mitosis or produce by meiosis cells that develop into gametes (egg or sperm cells).

GFP: *See* green fluorescent protein.

globular proteins: Soluble proteins with a globular (somewhat rounded) shape.

glucogenic: Capable of being converted into glucose or glycogen by the process of gluconeogenesis.

gluconeogenesis: The biosynthesis of a carbohydrate from simpler, noncarbohydrate precursors such as oxaloacetate or pyruvate.

GLUT: Designation for a family of membrane proteins that transport glucose.

glycan: A polymer of monosaccharide units joined by glycosidic bonds; polysaccharide.

glyceroneogenesis: The synthesis in adipocytes of glycerol 3-phosphate from pyruvate for use in triacylglycerol synthesis.

glycerophospholipid: An amphipathic lipid with a glycerol backbone; fatty acids are ester-linked to C-1 and C-2 of glycerol, and a polar alcohol is attached through a phosphodiester linkage to C-3.

glycoconjugate: A compound containing a carbohydrate component bound covalently to a protein or lipid, forming a glycoprotein or glycolipid.

glycogenesis: The process of converting glucose to glycogen.

glycogenin: The protein that both primes the synthesis of new glycogen chains and catalyzes the polymerization of the first few sugar residues of each chain before glycogen synthase continues the extension.

glycogenolysis: The enzymatic breakdown of stored (not dietary) glycogen.

glycolate pathway: The metabolic pathway in photosynthetic organisms that converts glycolate produced during photorespiration into 3-phosphoglycerate.

glycolipid: A lipid containing a carbohydrate group.

glycolysis: The catabolic pathway by which a molecule of glucose is broken down into two molecules of pyruvate.

glycome: The full complement of carbohydrates and carbohydrate-containing molecules of a cell or tissue under a particular set of conditions.

glycomics: The systematic characterization of the glycome.

glycoprotein: A protein containing a carbohydrate group.

glycosaminoglycan: A heteropolysaccharide of two alternating units: one is either *N*-acetylglucosamine or *N*-acetylgalactosamine; the other is a uronic acid (usually glucuronic acid). Formerly called a mucopolysaccharide.

glycosidic bonds: *See* O-glyosidic bonds.

glycosphingolipid: An amphipathic lipid with a sphingosine backbone to which are attached a long-chain fatty acid and a polar alcohol.

glyoxylate cycle: A variant of the citric acid cycle, for the net conversion of acetate into succinate and, eventually, new carbohydrate; present in bacteria and some plant cells.

glyoxysome: A specialized peroxisome containing the enzymes of the glyoxylate cycle; found in cells of germinating seeds.

Golgi complex: A complex membranous organelle of eukaryotic cells; functions in the post-translational modification of proteins and their secretion from the cell or incorporation into the plasma membrane or organellar membranes.

GPCRs: *See* G protein–coupled receptors.

GPI-anchored protein: A protein held to the outer monolayer of the plasma membrane by its covalent attachment through a short oligosaccharide chain to a phosphatidylinositol molecule in the membrane.

G protein–coupled receptor kinases (GRKs): A family of protein kinases that phosphorylate Ser and Thr residues near the carboxyl terminus of G protein–coupled receptors, initiating their internalization.

G protein–coupled receptors (GPCRs): A large family of membrane receptor proteins with seven transmembrane helical segments, often associating with G proteins to transduce an extracellular signal into a change in cellular metabolism; also called heptahelical receptors.

G proteins: A large family of GTP-binding proteins that act in intracellular signaling pathways and in membrane trafficking. Active when GTP is bound, they self-inactivate by converting GTP to GDP. Also called guanosine nucleotide–binding proteins.

gram molecular weight: For a compound, the weight in grams that is numerically equal to its molecular weight; the weight of one mole.

grana: Stacks of thylakoids, flattened membranous sacs or disks, in chloroplasts.

green fluorescent protein (GFP): A small protein from a marine organism that produces a bright fluorescence in the green region of the visible spectrum. Fusion proteins with GFP are commonly used to determine the subcellular location of the fused protein by fluorescence microscopy.

ground state: The normal, stable form of an atom or molecule; as distinct from the excited state.

group transfer potential: A measure of the ability of a compound to donate an activated group (such as a phosphate or acyl group); generally expressed as the standard free energy of hydrolysis.

growth factors: Proteins or other molecules that act from outside a cell to stimulate cell growth and division.

GTPase activator proteins (GAPs): Regulatory proteins that bind activated G proteins and stimulate their intrinsic GTPase activity, speeding their self-inactivation.

guanosine nucleotide–binding proteins: *See* G proteins.

guanosine nucleotide–exchange factors (GEFs): Regulatory proteins that bind to and activate G proteins by stimulating the exchange of bound GDP for GTP.

h

hairpin: Secondary structure in single-stranded RNA or DNA, in which complementary parts of a palindromic repeat fold back and are paired to form an antiparallel duplex helix that is closed at one end.

half-life: The time required for the disappearance or decay of one-half of a given component in a system.

haploid: Having a single set of genetic information; describes a cell with one chromosome of each type. *Compare* diploid.

haplotype: A combination of alleles of different genes located sufficiently close together on a chromosome that they tend to be inherited together.

hapten: A small molecule that, when linked to a larger molecule, elicits an immune response.

Haworth perspective formulas: A method for representing cyclic chemical structures so as to define the configuration of each substituent group; commonly used for representing sugars.

helicases: Enzymes that catalyze the separation of strands in a DNA molecule before replication.

heme: The iron-porphyrin prosthetic group of heme proteins.

heme protein: A protein containing a heme as a prosthetic group.

hemoglobin: A heme protein in erythrocytes; functions in oxygen transport.

Henderson-Hasselbalch equation: An equation relating the pH, the pK_a, and the ratio of the concentrations of proton-acceptor (A^-) and proton-donor (HA) species in a solution:

$$pH = pK_a + \log \frac{[A^-]}{[HA]}.$$

heparan sulfate: A sulfated polymer of alternating *N*-acetylglucosamine and a uronic acid, either glucuronic or iduronic acid; typically found in the extracellular matrix.

hepatocyte: The major cell type of liver tissue.

heptahelical receptors: *See* G protein–coupled receptors.

heteropolysaccharide: A polysaccharide containing more than one type of sugar.

heterotroph: An organism that requires complex nutrient molecules, such as glucose, as a source of energy and carbon.

heterotropic: Describes an allosteric modulator that is distinct from the normal ligand.

heterotropic enzyme: An allosteric enzyme requiring a modulator other than its substrate.

hexose: A simple sugar with a backbone containing six carbon atoms.

hexose monophosphate pathway: *See* pentose phosphate pathway.

high-performance liquid chromatography (HPLC): Chromatographic procedure, often conducted at relatively high pressures using automated equipment, which permits refined and highly reproducible profiles.

Hill coefficient: A measure of cooperative interaction between protein subunits.

Hill reaction: The evolution of oxygen and photoreduction of an artificial electron acceptor by a chloroplast preparation in the absence of carbon dioxide.

histones: The family of basic proteins that associate tightly with DNA in the chromosomes of all eukaryotic cells.

Holliday intermediate: An intermediate in genetic recombination in which two double-stranded DNA molecules are joined by a reciprocal crossover involving one strand of each molecule.

holoenzyme: A catalytically active enzyme, including all necessary subunits, prosthetic groups, and cofactors.

homeobox: A conserved DNA sequence of 180 base pairs that encodes a protein domain found in many proteins that play a regulatory role in development.

homeodomain: The protein domain encoded by the homeobox; a regulatory unit that determines the segmentation of a body plan.

homeostasis: The maintenance of a dynamic steady state by regulatory mechanisms that compensate for changes in external circumstances.

homeotic genes: Genes that regulate development of the pattern of segments in the *Drosophila* body plan; similar genes are found in most vertebrates.

homologs: Genes or proteins that possess a clear sequence and functional relationship to each other.

homologous genetic recombination: Recombination between two DNA molecules of similar sequence, occurring in all cells; occurs during meiosis and mitosis in eukaryotes.

homologous proteins: Proteins having similar sequences and functions in different species; for example, the hemoglobins.

homotropic: Describes an allosteric modulator that is identical to the normal ligand.

homotropic enzyme: An allosteric enzyme that uses its substrate as a modulator.

hormone: A chemical substance synthesized in small amounts by an endocrine tissue and carried in the blood to another tissue, where it acts as a messenger to regulate the function of the target tissue or organ.

hormone receptor: A protein in, or on the surface of, target cells that binds a specific hormone and initiates the cellular response.

hormone response element (HRE): A short (12 to 20 bp) DNA sequence that binds receptors for steroid, retinoid, thyroid, and vitamin D hormones, altering the expression of the contiguous genes. Each hormone has a consensus sequence preferred by the cognate receptor.

HPLC: *See* high-performance liquid chromatography.

HRE: *See* hormone response element.

hyaluronan: A high molecular weight, acidic polysaccharide typically composed of the alternating disaccharide GlcUA(β1→3)GlcNAc; a major component of the extracellular matrix, forming larger complexes (proteoglycans) with proteins and other acidic polysaccharides. Also called hyaluronic acid.

hydrogen bond: A weak electrostatic attraction between one electronegative atom (such as oxygen or nitrogen) and a hydrogen atom covalently linked to a second electronegative atom.

hydrolases: Enzymes (proteases, lipases, phosphatases, nucleases, for example) that catalyze hydrolysis reactions.

hydrolysis: Cleavage of a bond, such as an anhydride or peptide bond, by the addition of the elements of water, yielding two or more products.

hydronium ion: The hydrated hydrogen ion (H_3O^+).

hydropathy index: A scale that expresses the relative hydrophobic and hydrophilic tendencies of a chemical group.

hydrophilic: Polar or charged; describes molecules or groups that associate with (dissolve easily in) water.

hydrophobic: Nonpolar; describes molecules or groups that are insoluble in water.

hydrophobic interactions: The association of nonpolar groups or compounds with each other in aqueous systems, driven by the tendency of the surrounding water molecules to seek their most stable (disordered) state.

hyperchromic effect: The large increase in light absorption at 260 nm occurring as a double-helical DNA unwinds (melts).

hypoxia: The metabolic condition in which the supply of oxygen is severely limited.

i

immune response: The capacity of a vertebrate to generate antibodies to an antigen, a macromolecule foreign to the organism.

immunoblotting: A technique that employs antibodies to detect the presence of a protein in a biological sample after the proteins in the sample have been separated by gel electrophoresis, transferred to a membrane and immobilized; also called Western blotting.

immunoglobulin: An antibody protein generated against, and capable of binding specifically to, an antigen.

induced fit: A change in the conformation of an enzyme in response to substrate binding that renders the enzyme catalytically active; also used to denote changes in the conformation of any macromolecule in response to ligand binding such that the binding site of the macromolecule better conforms to the shape of the ligand.

inducer: A signal molecule that, when bound to a regulatory protein, produces an increase in the expression of a given gene.

induction: An increase in the expression of a gene in response to a change in the activity of a regulatory protein.

informational macromolecules: Biomolecules containing information in the form of specific sequences of different monomers; for example, many proteins, lipids, polysaccharides, and nucleic acids.

inhibitory G protein (G_i): A trimeric GTP-binding protein that, when activated by an associated plasma membrane receptor, inhibits a neighboring membrane enzyme such as adenylyl cyclase. Compare stimulatory G protein (G_s).

initiation codon: AUG (sometimes GUG or, even more rarely, UUG in bacteria and archaea); codes for the first amino acid in a polypeptide sequence: *N*-formylmethionine in bacteria; methionine in archaea and eukaryotes.

initiation complex: A complex of a ribosome with an mRNA and the initiating Met-tRNA^{Met} or fMet-tRNA^{fMet}, ready for the elongation steps.

inorganic pyrophosphatase: An enzyme that hydrolyzes a molecule of inorganic pyrophosphate to yield two molecules of (ortho) phosphate; also known as pyrophosphatase.

insertion mutation: A mutation caused by insertion of one or more extra bases, or a mutagen, between successive bases in DNA.

insertion sequence: Specific base sequences at either end of a transposable segment of DNA.

in situ: "In position"; that is, in its natural position or location.

integral proteins: Proteins firmly bound to a membrane by hydrophobic interactions; as distinct from peripheral proteins.

integrin: One of a large family of heterodimeric transmembrane proteins that mediate adhesion of cells to other cells or to the extracellular matrix.

intercalation: Insertion between stacked aromatic or planar rings; for example, the insertion of a planar molecule between two successive bases in a nucleic acid.

intermediary metabolism: In cells, the enzyme-catalyzed reactions that extract chemical energy from nutrient molecules and use it to synthesize and assemble cell components.

intrinsically disordered proteins: Proteins, or segments of proteins, that lack a definable three-dimensional structure in solution.

intron: A sequence of nucleotides in a gene that is transcribed but excised before the gene is translated; also called intervening sequence. *See also* exon.

in vitro: "In glass"; that is, in the test tube.

in vivo: "In life"; that is, in the living cell or organism.

ion channel: An integral protein that provides for the regulated transport of a specific ion, or ions, across a membrane.

ion-exchange chromatography: A process for separating complex mixtures of ionic compounds by many repeated partitionings between a flowing (mobile) phase and a stationary phase consisting of a polymeric resin that contains fixed charged groups.

ionizing radiation: A type of radiation, such as x rays, that causes loss of electrons from some organic molecules, thus making them more reactive.

ionophore: A compound that binds one or more metal ions and is capable of diffusing across a membrane, carrying the bound ion.

ion product of water (K_w): The product of the concentrations of H^+ and OH^- in pure water: $K_w = [H^+][OH^-] = 1 \times 10^{-14}$ at 25 °C.

iron-sulfur protein: One of a large family of electron-transfer proteins in which the electron carrier is one or more iron ions associated with two or more sulfur atoms of Cys residues or of inorganic sulfide.

isoelectric focusing: An electrophoretic method for separating macromolecules on the basis of isoelectric pH.

isoelectric pH (isoelectric point, pI): The pH at which a solute has no net electric charge and thus does not move in an electric field.

isoenzymes: *See* isozymes.

isomerases: Enzymes that catalyze the transformation of compounds into their positional isomers.

isomers: Any two molecules with the same molecular formula but a different arrangement of molecular groups.

isoprene: The hydrocarbon 2-methyl-1,3-butadiene, a recurring structural unit of terpenoids.

isoprenoid: Any of a large number of natural products synthesized by enzymatic polymerization of two or more isoprene units; also called terpenoid.

isozymes: Multiple forms of an enzyme that catalyze the same reaction but differ in amino acid sequence, substrate affinity, V_{max}, and/or regulatory properties; also called isoenzymes.

k

ketogenic: Yielding acetyl-CoA, a precursor for ketone body formation, as a breakdown product.

ketone bodies: Acetoacetate, D-β-hydroxybutyrate, and acetone; water-soluble fuels normally exported by the liver but overproduced during fasting or in untreated diabetes mellitus.

ketose: A simple monosaccharide in which the carbonyl group is a ketone.

ketosis: A condition in which the concentration of ketone bodies in the blood, tissues, and urine is abnormally high.

kinases: Enzymes that catalyze the phosphorylation of certain molecules by ATP.

kinetics: The study of reaction rates.

Krebs cycle: *See* citric acid cycle.

K_t ($K_{transport}$): A kinetic parameter for a membrane transporter analogous to the Michaelis constant, K_m, for an enzymatic reaction. The rate of substrate uptake is half-maximal when the substrate concentration equals the K_t.

l

lagging strand: The DNA strand that, during replication, must be synthesized in the direction opposite to that in which the replication fork moves.

law of mass action: The law stating that the rate of any given chemical reaction is proportional to the product of the activities (or concentrations) of the reactants.

leader: A short sequence near the amino terminus of a protein or the 5' end of an RNA that has a specialized targeting or regulatory function.

leading strand: The DNA strand that, during replication, is synthesized in the same direction as the replication fork moves.

leaky mutant: A mutant gene that gives rise to a product with a detectable level of biological activity.

leaving group: The departing or displaced molecular group in a unimolecular elimination or bimolecular substitution reaction.

lectin: A protein that binds a carbohydrate, commonly an oligosaccharide, with very high affinity and specificity, mediating cell-cell interactions.

lethal mutation: A mutation that inactivates a biological function essential to the life of the cell or organism.

leucine zipper: A protein structural motif involved in protein-protein interactions in many eukaryotic regulatory proteins; consists of two interacting α helices in which Leu residues in every seventh position are a prominent feature of the interacting surfaces.

leukocyte: White blood cell; involved in the immune response in mammals.

leukotriene: Any of a class of signaling lipids derived from arachidonate in the noncyclic pathway; modulate smooth muscle activity.

ligand: A small molecule that binds specifically to a larger one; for example, a hormone is the ligand for its specific protein receptor.

ligases: Enzymes that catalyze condensation reactions in which two atoms are joined using the energy of ATP or another energy-rich compound.

light-dependent reactions: The reactions of photosynthesis that require light and cannot occur in the dark; also known as light reactions.

Lineweaver-Burk equation: An algebraic transform of the Michaelis-Menten equation, allowing determination of V_{max} and K_m by extrapolation of [S] to infinity:

$$\frac{1}{V_0} = \frac{K_m}{V_{max}[S]} + \frac{1}{V_{max}}.$$

linking number: The number of times one closed circular DNA strand is wound about another; the number of topological links holding the circles together.

lipases: Enzymes that catalyze the hydrolysis of triacylglycerols.

lipid: A small water-insoluble biomolecule generally containing fatty acids, sterols, or isoprenoid compounds.

lipidome: The full complement of lipid-containing molecules in a cell, organ, or tissue under a particular set of conditions.

lipidomics: The systematic characterization of the lipidome.

lipoate (lipoic acid): A vitamin for some microorganisms; an intermediate carrier of hydrogen atoms and acyl groups in α-keto acid dehydrogenases.

lipoprotein: A lipid-protein aggregate that serves to carry water-insoluble lipids in the blood. The protein component alone is an apolipoprotein.

liposome: A small, spherical vesicle composed of a phospholipid bilayer, forming spontaneously when phospholipids are suspended in an aqueous buffer.

lyases: Enzymes that catalyze the removal of a group from a molecule to form a double bond, or the addition of a group to a double bond.

lymphocytes: A subclass of leukocytes involved in the immune response. *See also* B lymphocytes; T lymphocytes.

lysis: Destruction of a plasma membrane or (in bacteria) cell wall, releasing the cellular contents and killing the cell.

lysosome: A membrane-bounded organelle of eukaryotic cells; it contains many hydrolytic enzymes and serves as a degrading and recycling center for unneeded components.

m

macromolecule: A molecule having a molecular weight in the range of a few thousand to many millions.

mass-action ratio (Q): For the reaction $aA + bB \rightleftharpoons cC + dD$, the ratio $[C]^c[D]^d/[A]^a[B]^b$.

matrix: The space enclosed by the inner membrane of the mitochondrion.

mechanistic target of rapamycin complex 1: *See* mTORC1.

meiosis: A type of cell division in which diploid cells give rise to haploid cells destined to become gametes or spores.

membrane potential (V_m): The difference in electrical potential across a biological membrane, commonly measured by the insertion of a microelectrode. Typical membrane potentials vary from -25 mV (by convention, the negative sign indicates that the inside is negative relative to the outside) to greater than -100 mV across some plant vacuolar membranes.

membrane transport: Movement of a polar solute across a membrane via a specific membrane protein (a transporter).

messenger RNA (mRNA): A class of RNA molecules, each of which is complementary to one strand of DNA; carries the genetic message from the chromosome to the ribosomes.

metabolic control: The mechanisms by which the flux through a metabolic pathway is changed to reflect a cell's altered circumstances.

metabolic regulation: The mechanisms by which a cell resists changes in the concentration of individual metabolites that would otherwise occur when metabolic control mechanisms alter the flux through a pathway.

metabolic syndrome: A combination of medical conditions that together predispose to cardiovascular disease and type 2 diabetes. They include high blood pressure, high concentrations of LDL and triacylglycerol in the blood, slightly elevated fasting blood glucose concentration, and obesity.

metabolism: The entire set of enzyme-catalyzed transformations of organic molecules in living cells; the sum of anabolism and catabolism.

metabolite: A chemical intermediate in the enzyme-catalyzed reactions of metabolism.

metabolome: The complete set of small-molecule metabolites (metabolic intermediates, signals, secondary metabolites) present in a given cell or tissue under specific conditions.

metabolomics: The systematic characterization of the metabolome of a cell or tissue.

metabolon: A supramolecular assembly of sequential metabolic enzymes.

metalloprotein: A protein with a metal ion as its prosthetic group.

metamerism: Division of the body into segments, as in insects, for example.

micelle: An aggregate of amphipathic molecules in water, with the nonpolar portions in the interior and the polar portions at the exterior surface, exposed to water.

Michaelis constant (K_m): The substrate concentration at which an enzyme-catalyzed reaction proceeds at one-half its maximum velocity.

Michaelis-Menten equation: The equation describing the hyperbolic dependence of the initial reaction velocity, V_0, on substrate concentration, [S], in many enzyme-catalyzed reactions:

$$V_0 = \frac{V_{max}[S]}{K_m + [S]}.$$

Michaelis-Menten kinetics: A kinetic pattern in which the initial rate of an enzyme-catalyzed reaction exhibits a hyperbolic dependence on substrate concentration.

microRNA: A class of small RNA molecules (20 to 25 nucleotides after processing is complete) involved in gene silencing by inhibiting translation and/or promoting the degradation of particular mRNAs.

microsomes: Membranous vesicles formed by fragmentation of the endoplasmic reticulum of eukaryotic cells; recovered by differential centrifugation.

miRNA: *See* microRNA.

mismatch: A base pair in a nucleic acid that cannot form a normal Watson-Crick pair.

mismatch repair: An enzymatic system for repairing base mismatches in DNA.

mitochondrion: Membrane-bounded organelle of eukaryotic cells; contains the enzyme systems required for the citric acid cycle, fatty acid oxidation, electron transfer, and oxidative phosphorylation.

mitosis: In eukaryotic cells, the multistep process that results in the replication of chromosomes and cell division.

mixed-function oxygenases: Enzymes (a monooxygenase, for example) that catalyze reactions in which two reductants—one of which is generally NADPH, the other the

substrate—are oxidized. One oxygen atom is incorporated into the product, the other is reduced to H_2O. These enzymes often use cytochrome P-450 to carry electrons from NADPH to O_2.

mixed inhibition: The reversible inhibition pattern resulting when an inhibitor molecule can bind to either the free enzyme or the enzyme-substrate complex (not necessarily with the same affinity).

modulator: A metabolite that, when bound to the allosteric site of an enzyme, alters its kinetic characteristics.

molar solution: One mole of solute dissolved in water to give a total volume of 1,000 mL.

mole: One gram molecular weight of a compound. *See also* Avogadro's number.

monocistronic mRNA: An mRNA that can be translated into only one protein.

monoclonal antibodies: Antibodies produced by a cloned hybridoma cell, which therefore are identical and directed against the same epitope of the antigen. (Hybridoma cells are stable antibody-producing cell lines that grow well in tissue culture; created by fusing an antibody-producing B cell with a myeloma cell.)

monosaccharide: A carbohydrate consisting of a single sugar unit.

moonlighting enzymes: Enzymes that play two distinct roles, at least one of which is catalytic; the other may be catalytic, regulatory, or structural.

motif: Any distinct folding pattern for elements of secondary structure, observed in one or more proteins. A motif can be simple or complex, and can represent all or just a small part of a polypeptide chain. Also called a fold or supersecondary structure.

mRNA: *See* messenger RNA.

mTORC1 (mechanistic target of rapamycin complex 1): A multiprotein complex of mTOR (mechanistic target of rapamycin) and several regulatory subunits, which together have activity as a Ser/Thr protein kinase. Stimulated by nutrients and energy-sufficient conditions, it triggers cell growth and proliferation.

mucopolysaccharide: *See* glycosaminoglycan.

multidrug transporters: Plasma membrane transporters in the ABC transporter family that expel several commonly used antitumor drugs, thereby interfering with antitumor therapy.

multienzyme system: A group of related enzymes participating in a given metabolic pathway.

mutarotation: The change in specific rotation of a pyranose or furanose sugar or glycoside accompanying the equilibration of its α- and β-anomeric forms.

mutases: Enzymes that catalyze the transposition of functional groups.

mutation: An inheritable change in the nucleotide sequence of a chromosome.

myocyte: A muscle cell.

myofibril: A unit of thick and thin filaments of muscle fibers.

myosin: A contractile protein; the major component of the thick filaments of muscle and other actin-myosin systems.

n

NAD, NADP (nicotinamide adenine dinucleotide, nicotinamide adenine dinucleotide phosphate): Nicotinamide-containing coenzymes functioning as carriers of hydrogen atoms and electrons in some oxidation-reduction reactions.

Na^+K^+ ATPase: The electrogenic ATP-driven active transporter in the plasma membrane of most animal cells that pumps three Na^+ outward for every two K^+ moved inward.

native conformation: The biologically active conformation of a macromolecule.

ncRNA (noncoding RNA): Any RNA that does not encode instructions for a protein product.

negative cooperativity: A property of some multisubunit enzymes or proteins in which binding of a ligand or substrate to one subunit impairs binding to another subunit.

negative feedback: Regulation of a biochemical pathway in which a reaction product inhibits an earlier step in the pathway.

neuron: A cell of nervous tissue specialized for transmission of a nerve impulse.

neurotransmitter: A low molecular weight compound (usually containing nitrogen) secreted from the axon terminal of a neuron and bound by a specific receptor on the next neuron or on a myocyte; serves to transmit a nerve impulse.

nitrogenase complex: A system of enzymes capable of reducing atmospheric nitrogen to ammonia in the presence of ATP.

nitrogen cycle: The cycling of various forms of biologically available nitrogen through the plant, animal, and microbial worlds, and through the atmosphere and geosphere.

nitrogen fixation: Conversion of atmospheric nitrogen (N_2) into a reduced, biologically available form by nitrogen-fixing organisms.

NMR: *See* nuclear magnetic resonance spectroscopy.

noncoding RNA: *See* ncRNA.

noncyclic electron flow: The light-induced flow of electrons from water to $NADP^+$ in oxygen-evolving photosynthesis; involves both photosystems I and II.

nonessential amino acids: Amino acids that can be made by humans and other vertebrates from simpler precursors and are thus not required in the diet.

nonheme iron proteins: Proteins, usually acting in oxidation-reduction reactions, containing iron but no porphyrin groups.

nonpolar: Hydrophobic; describes molecules or groups that are poorly soluble in water.

nonsense codon: A codon that does not specify an amino acid, but signals the termination of a polypeptide chain.

nonsense mutation: A mutation that results in the premature termination of a polypeptide chain.

nonsense suppressor: A mutation, usually in the gene for a tRNA, that causes an amino acid to be inserted into a polypeptide in response to a termination codon.

nontemplate strand: *See* coding strand.

nuclear magnetic resonance (NMR) spectroscopy: A technique that utilizes certain quantum mechanical properties of atomic nuclei to study the structure and dynamics of the molecules of which they are a part.

nucleases: Enzymes that hydrolyze the internucleotide (phosphodiester) linkages of nucleic acids.

nucleic acids: Biologically occurring polynucleotides in which the nucleotide residues are linked in a specific sequence by phosphodiester bonds; DNA and RNA.

nucleoid: In bacteria, the nuclear zone that contains the chromosome but has no surrounding membrane.

nucleolus: In eukaryotic cells, a densely staining structure in the nucleus; involved in rRNA synthesis and ribosome formation.

nucleophile: An electron-rich group with a strong tendency to donate electrons to an electron-deficient nucleus (electrophile); the entering reactant in a bimolecular substitution reaction.

nucleoplasm: The portion of a eukaryotic cell's contents enclosed by the nuclear membrane.

nucleoside: A compound consisting of a purine or pyrimidine base covalently linked to a pentose.

nucleoside diphosphate kinase: An enzyme that catalyzes the transfer of the terminal phosphate of a nucleoside 5′-triphosphate to a nucleoside 5′-diphosphate.

nucleoside diphosphate sugar: A coenzymelike carrier of a sugar molecule, functioning in the enzymatic synthesis of polysaccharides and sugar derivatives.

nucleoside monophosphate kinase: An enzyme that catalyzes the transfer of the terminal phosphate of ATP to a nucleoside 5′-monophosphate.

nucleosome: In eukaryotes, structural unit for packaging chromatin; consists of a DNA strand wound around a histone core.

nucleotide: A nucleoside phosphorylated at one of its pentose hydroxyl groups.

nucleus: In eukaryotes, a membrane-bounded organelle that contains chromosomes.

o

O-glycosidic bonds: Bonds between a sugar and another molecule (typically an alcohol, purine, pyrimidine, or sugar) through an intervening oxygen.

oligomer: A short polymer, usually of amino acids, sugars, or nucleotides; the definition of "short" is somewhat arbitrary, but usually fewer than 50 subunits.

oligomeric protein: A multisubunit protein having two or more identical polypeptide chains.

oligonucleotide: A short polymer of nucleotides (usually fewer than 50).

oligopeptide: A few amino acids joined by peptide bonds.

oligosaccharide: Several monosaccharide groups joined by glycosidic bonds.

ω oxidation: An alternative mode of fatty acid oxidation in which the initial oxidation is at the carbon most distant from the carboxyl carbon; as distinct from β oxidation.

oncogene: A cancer-causing gene; any of several mutant genes that cause cells to exhibit rapid, uncontrolled proliferation. *See also* proto-oncogene.

open reading frame (ORF): A group of contiguous nonoverlapping nucleotide codons in a DNA or RNA molecule that does not include a termination codon.

open system: A system that exchanges matter and energy with its surroundings. *See also* system.

operator: A region of DNA that interacts with a repressor protein to control the expression of a gene or group of genes.

operon: A unit of genetic expression consisting of one or more related genes and the operator and promoter sequences that regulate their transcription.

opsin: The protein portion of the visual pigment, which becomes rhodopsin with the addition of the chromophore retinal.

optical activity: The capacity of a substance to rotate the plane of plane-polarized light.

optimum pH: The characteristic pH at which an enzyme has maximal catalytic activity.

orexigenic: Tending to increase appetite and food consumption.

ORF: *See* open reading frame.

organelles: Membrane-bounded structures found in eukaryotic cells; contain enzymes and other components required for specialized cell functions.

origin: The nucleotide sequence or site in DNA where DNA replication is initiated.

orthologs: Genes or proteins from different species that possess a clear sequence and functional relationship to each other.

osmosis: Bulk flow of water through a semipermeable membrane into another aqueous compartment containing solute at a higher concentration.

osmotic pressure: Pressure generated by the osmotic flow of water through a semipermeable membrane into an aqueous compartment containing solute at a higher concentration.

oxidases: Enzymes that catalyze oxidation reactions in which molecular oxygen serves as the electron acceptor, but neither of the oxygen atoms is incorporated into the product. *Compare* oxygenases.

oxidation: The loss of electrons from a compound.

oxidation-reduction reaction: A reaction in which electrons are transferred from a donor to an acceptor molecule; also called a redox reaction.

oxidative phosphorylation: The enzymatic phosphorylation of ADP to ATP coupled to electron transfer from a substrate to molecular oxygen.

oxidizing agent (oxidant): The acceptor of electrons in an oxidation-reduction reaction.

oxygenases: Enzymes that catalyze reactions in which oxygen atoms are directly incorporated into the product, forming a hydroxyl or carboxyl group. In reactions catalyzed by a monooxygenase, only one of the two O atoms is incorporated; the other is reduced to H_2O. In reactions catalyzed by a dioxygenase, both O atoms are incorporated into the product. *Compare* oxidases.

oxygenic photosynthesis: Light-driven ATP and NADPH synthesis in organisms that use water as the electron source, producing O_2.

p

palindrome: A segment of duplex DNA in which the base sequences of the two strands exhibit twofold rotational symmetry about an axis.

paradigm: In biochemistry, an experimental model or example.

paralogs: Genes or proteins present in the same species that possess a clear sequence and functional relationship to each other.

partition coefficient: A constant that expresses the ratio in which a given solute will be partitioned or distributed between two given immiscible liquids at equilibrium.

passive transport: Diffusion of a polar substance across a biological membrane through a protein transporter; also called facilitated diffusion.

pathogenic: Disease-causing.

PCR: *See* polymerase chain reaction.

PDB (Protein Data Bank): An international database (www.pdb.org) that archives the data describing the three-dimensional structure of nearly all macromolecules for which structures have been published.

pentose: A simple sugar with a backbone containing five carbon atoms.

pentose phosphate pathway: A pathway present in most organisms that serves to interconvert hexoses and pentoses and is a source of reducing equivalents (NADPH) and pentoses for biosynthetic processes; it begins with glucose 6-phosphate and includes 6-phosphogluconate as an intermediate. Also called the phosphogluconate pathway and the hexose monophosphate pathway.

peptidases: Enzymes that hydrolyze peptide bonds.

peptide: Two or more amino acids covalently joined by peptide bonds.

peptide bond: A substituted amide linkage between the α-amino group of one amino acid and the α-carboxyl group of another, with the elimination of the elements of water.

peptidoglycan: A major component of bacterial cell walls; generally consists of parallel heteropolysaccharides cross-linked by short peptides.

peptidyl transferase: The enzyme activity that synthesizes the peptide bonds of proteins; a ribozyme, part of the rRNA of the large ribosomal subunit.

peripheral proteins: Proteins loosely or reversibly bound to a membrane by hydrogen bonds or electrostatic forces; generally water soluble once released from the membrane. *Compare* integral proteins.

permeases: *See* transporters.

peroxisome: Membrane-bounded organelle of eukaryotic cells; contains peroxide-forming and peroxide-destroying enzymes.

peroxisome proliferator-activated receptor: *See* PPAR.

pH: The negative logarithm of the hydrogen ion concentration of an aqueous solution.

phage: *See* bacteriophage.

phenotype: The observable characteristics of an organism.

phosphatases: Enzymes that cleave phosphate esters by hydrolysis, the addition of the elements of water.

phosphodiester linkage: A chemical grouping that contains two alcohols esterified to one molecule of phosphoric acid, which thus serves as a bridge between them.

phosphogluconate pathway: *See* pentose phosphate pathway.

phospholipid: A lipid containing one or more phosphate groups.

phosphorolysis: Cleavage of a compound with phosphate as the attacking group; analogous to hydrolysis.

phosphorylases: Enzymes that catalyze phosphorolysis.

phosphorylation: Formation of a phosphate derivative of a biomolecule, usually by enzymatic transfer of a phosphoryl group from ATP.

phosphorylation potential (ΔG_p): The actual free-energy change of ATP hydrolysis under the nonstandard conditions prevailing in a cell.

photochemical reaction center: The part of a photosynthetic complex where the energy of an absorbed photon causes charge separation, initiating electron transfer.

photon: The ultimate unit (a quantum) of light energy.

photophosphorylation: The enzymatic formation of ATP from ADP coupled to the light-dependent transfer of electrons in photosynthetic cells.

photoreduction: The light-induced reduction of an electron acceptor in photosynthetic cells.

photorespiration: Oxygen consumption occurring in illuminated temperate-zone plants, largely due to oxidation of phosphoglycolate.

photosynthesis: The use of light energy to produce carbohydrates from carbon dioxide and a reducing agent such as water. *Compare* oxygenic photosynthesis.

photosynthetic phosphorylation: *See* photophosphorylation.

photosystem: In photosynthetic cells, a functional set of light-absorbing pigments and its reaction center, where the energy of an absorbed photon is transduced into a separation of electric charges.

phototroph: An organism that can use the energy of light to synthesize its own fuels from simple molecules such as carbon dioxide, oxygen, and water; as distinct from a chemotroph.

pI: *See* isoelectric pH.

pK_a: The negative logarithm of an equilibrium constant.

plasmalogen: A phospholipid with an alkenyl ether substituent on C-1 of glycerol.

plasma membrane: The exterior membrane surrounding the cytoplasm of a cell.

plasma proteins: The proteins present in blood plasma.

plasmid: An extrachromosomal, independently replicating, small circular DNA molecule; commonly employed in genetic engineering.

plastid: In plants, a self-replicating organelle; may differentiate into a chloroplast or amyloplast.

platelets: Small, enucleated cells that initiate blood clotting; they arise from bone marrow cells called megakaryocytes. Also known as thrombocytes.

pleated sheet: The side-by-side, hydrogen-bonded arrangement of polypeptide chains in the extended β conformation.

plectonemic: Describes a structure in a molecular polymer that has a net twisting of strands about each other in some simple and regular way.

PLP: *See* pyridoxal phosphate.

polar: Hydrophilic, or "water-loving"; describes molecules or groups that are soluble in water.

polarity: (1) In chemistry, the nonuniform distribution of electrons in a molecule; polar molecules are usually soluble in water. (2) In molecular biology, the distinction between the 5′ and 3′ ends of nucleic acids.

poly(A) tail: A length of adenosine residues added to the 3′ end of many mRNAs in eukaryotes (and sometimes in bacteria).

polycistronic mRNA: A contiguous mRNA with more than two genes that can be translated into proteins.

polyclonal antibodies: A heterogeneous pool of antibodies produced in an animal by different B lymphocytes in response to an antigen. Different antibodies in the pool recognize different parts of the antigen.

polylinker: A short, often synthetic, fragment of DNA containing recognition sequences for several restriction endonucleases.

polymerase chain reaction (PCR): A repetitive laboratory procedure that results in a geometric amplification of a specific DNA sequence.

polymorphic: Describes a protein for which amino acid sequence variants exist in a population of organisms, but the variations do not destroy the protein's function.

polynucleotide: A covalently linked sequence of nucleotides in which the 3′ hydroxyl of the pentose of one nucleotide residue is joined by a phosphodiester bond to the 5′ hydroxyl of the pentose of the next residue.

polypeptide: A long chain of amino acids linked by peptide bonds; the molecular weight is generally less than 10,000.

polyribosome: *See* polysome.

polysaccharide: A linear or branched polymer of monosaccharide units linked by glycosidic bonds.

polysome: A complex of an mRNA molecule and two or more ribosomes; also called polyribosome.

polyunsaturated fatty acid: *See* PUFA.

P/O ratio: The number of moles of ATP formed in oxidative phosphorylation per $\frac{1}{2}O_2$ reduced (thus, per pair of electrons passed to O_2). Experimental values used in this text are 2.5 for passage of electrons from NADH to O_2, and 1.5 for passage of electrons from FADH to O_2.

porphyria: Inherited condition resulting from the lack of one or more enzymes required to synthesize porphyrins.

porphyrin: Complex nitrogenous compound containing four substituted pyrroles covalently joined into a ring; often complexed with a central metal atom.

positive cooperativity: A property of some multisubunit enzymes or proteins in which binding of a ligand or substrate to one subunit facilitates binding to another subunit.

positive-inside rule: General observation that most plasma membrane proteins are oriented so that most of their positively charged residues (Lys and Arg) are on the cytosolic face.

posttranscriptional processing: The enzymatic processing of the primary RNA transcript to produce functional RNAs, including mRNAs, tRNAs, rRNAs, and many other classes of RNAs.

posttranslational modification: Enzymatic processing of a polypeptide chain after translation from its mRNA.

PPAR (peroxisome proliferator-activated receptor): A family of nuclear transcription factors, activated by lipidic ligands, that alter the expression of specific genes, including those encoding enzymes of lipid synthesis and breakdown.

primary structure: A description of the covalent backbone of a polymer (macromolecule), including the sequence of monomeric subunits and any interchain and intrachain covalent bonds.

primary transcript: The immediate RNA product of transcription before any posttranscriptional processing reactions.

primases: Enzymes that catalyze the formation of RNA oligonucleotides used as primers by DNA polymerases.

primer: A short oligomer (of sugars or nucleotides, for example) to which an enzyme adds additional monomeric subunits.

primer terminus: The end of a primer to which monomeric subunits are added.

priming: (1) In protein phosphorylation, the phosphorylation of an amino acid residue that becomes the binding site and point of reference for phosphorylation of other residues in the same protein. (2) In DNA replication, the synthesis of a short oligonucleotide to which DNA polymerases can add additional nucleotides.

primosome: An enzyme complex that synthesizes the primers required for lagging strand DNA synthesis.

processivity: For any enzyme that catalyzes the synthesis of a biological polymer, the property of adding multiple subunits to the polymer without dissociating from the substrate.

prochiral molecule: A symmetric molecule that can react asymmetrically with an enzyme having an asymmetric active site, generating a chiral product.

projection formulas: *See* Fischer projection formulas.

prokaryote: A term used historically to refer to any species in the kingdoms Bacteria and Archaea. The differences between bacteria (formerly referred to as "eubacteria") and archaea are sufficiently great that the inclusive term is of marginal usefulness. A tendency to use "prokaryote" when referring only to bacteria is common and misleading; "prokaryote" also implies an ancestral relationship to eukaryotes, which is incorrect. In this text, "prokaryote" and "prokaryotic" are not used.

promoter: A DNA sequence at which RNA polymerase may bind, leading to initiation of transcription.

proofreading: The correction of errors in the synthesis of an information-containing biopolymer by removing incorrect monomeric subunits after they have been covalently added to the growing polymer.

prostaglandin: One of a class of polyunsaturated, cyclic lipids derived from arachidonate that act as paracrine hormones.

prosthetic group: A metal ion or an organic compound (other than an amino acid) that is covalently bound to a protein and is essential to its activity.

proteases: Enzymes that catalyze the hydrolytic cleavage of peptide bonds in proteins.

proteasome: Supramolecular assembly of enzymatic complexes that function in the degradation of damaged or unneeded cellular proteins.

protein: A macromolecule composed of one or more polypeptide chains, each with a characteristic sequence of amino acids linked by peptide bonds.

Protein Data Bank: *See* PDB.

protein kinases: Enzymes that transfer the terminal phosphoryl group of ATP or another nucleoside triphosphate to a Ser, Thr, Tyr, Asp, or His side chain in a target protein, thereby regulating the activity or other properties of that protein.

protein phosphatases: Enzymes that hydrolyze a phosphate ester or anhydride bond on a protein, releasing inorganic phosphate, P_i.

protein targeting: The process by which newly synthesized proteins are sorted and transported to their proper locations in the cell.

proteoglycan: A hybrid macromolecule consisting of a heteropolysaccharide joined to a polypeptide; the polysaccharide is the major component.

proteome: The full complement of proteins expressed in a given cell, or the complete complement of proteins that can be expressed by a given genome.

proteomics: Broadly, the study of the protein complement of a cell or organism.

proteostasis: The maintenance of a cellular steady-state collection of proteins that are required for cell functions under a given set of conditions.

protomer: A general term describing any repeated unit of one or more stably associated protein subunits in a larger protein structure. If a protomer has multiple subunits, the subunits may be identical or different.

proton acceptor: An anionic compound capable of accepting a proton from a proton donor; that is, a base.

proton donor: The donor of a proton in an acid-base reaction; that is, an acid.

proton-motive force: The electrochemical potential inherent in a transmembrane gradient of H^+ concentration; used in oxidative phosphorylation and photophosphorylation to drive ATP synthesis.

proto-oncogene: A cellular gene, usually encoding a regulatory protein, that can be converted into an oncogene by mutation.

PUFA (polyunsaturated fatty acid): A fatty acid with more than one double bond, generally nonconjugated.

purine: A nitrogenous heterocyclic base found in nucleotides and nucleic acids; contains fused pyrimidine and imidazole rings.

puromycin: An antibiotic that inhibits polypeptide synthesis by being incorporated into a growing polypeptide chain, causing its premature termination.

pyridine nucleotide: A nucleotide coenzyme containing the pyridine derivative nicotinamide; NAD or NADP.

pyridoxal phosphate (PLP): A coenzyme containing the vitamin pyridoxine (vitamin B_6); functions in reactions involving amino group transfer.

pyrimidine: A nitrogenous heterocyclic base found in nucleotides and nucleic acids.

pyrimidine dimer: A covalently joined dimer of two adjacent pyrimidine residues in DNA, induced by absorption of UV light; most commonly derived from two adjacent thymines (a thymine dimer).

pyrophosphatase: *See* inorganic pyrophosphatase.

q

Q: *See* mass-action ratio.

quantitative PCR (qPCR): A PCR procedure that allows the determination of how much of the amplified template was in the original sample.

quantum: The ultimate unit of energy.

quaternary structure: The three-dimensional structure of a multisubunit protein, particularly the manner in which the subunits fit together.

r

racemic mixture (racemate): An equimolar mixture of the D and L stereoisomers of an optically active compound.

radical: An atom or group of atoms possessing an unpaired electron; also called a free radical.

radioactive isotope: An isotopic form of an element with an unstable nucleus that stabilizes itself by emitting ionizing radiation.

radioimmunoassay (RIA): A sensitive, quantitative method for detecting trace amounts of a biomolecule, based on its capacity to displace a radioactive form of the molecule from combination with its specific antibody.

Ras superfamily of G proteins: Small (M_r ~20,000), monomeric guanosine nucleotide–binding proteins that regulate signaling and membrane trafficking pathways. Inactive with GDP bound, they are activated by displacement of GDP by GTP, then inactivated by their intrinsic GTPase. Also called small G proteins.

rate constant: The proportionality constant that relates the velocity of a chemical reaction to the concentration(s) of the reactant(s).

rate-limiting step: (1) Generally, the step in an enzymatic reaction with the greatest activation energy or the transition state of highest free energy. (2) The slowest step in a metabolic pathway.

reaction intermediate: Any chemical species in a reaction pathway that has a finite chemical lifetime.

reactive oxygen species (ROS): Highly reactive products of the partial reduction of O_2, including hydrogen peroxide (H_2O_2), superoxide ($^{\bullet}O_2^-$), and hydroxyl free radical $^{\bullet}OH$, produced as minor byproducts during oxidative phosphorylation.

reading frame: A contiguous, nonoverlapping set of three-nucleotide codons in DNA or RNA.

receptor Tyr kinase (RTK): A large family of plasma membrane proteins with ligand-binding sites on the extracellular domain, a single transmembrane helix, and a cytoplasmic domain with protein Tyr kinase activity controlled by the extracellular ligand.

recombinant DNA: DNA formed by the joining of genes into new combinations.

recombination: Any enzymatic process by which the linear arrangement of nucleic acid sequences in a chromosome is altered by cleavage and rejoining.

recombinational DNA repair: Recombinational processes directed at the repair of DNA strand breaks or cross-links, especially at inactivated replication forks.

redox pair: An electron donor and its corresponding oxidized form; for example, NADH and NAD$^+$.

redox reaction: *See* oxidation-reduction reaction.

reducing agent (reductant): The electron donor in an oxidation-reduction reaction.

reducing end: The end of a polysaccharide having a terminal sugar with a free anomeric carbon; the terminal residue can act as a reducing sugar.

reducing equivalent: A general term for an electron or an electron equivalent in the form of a hydrogen atom or a hydride ion.

reducing sugar: A sugar in which the carbonyl (anomeric) carbon is not involved in a glycosidic bond and can therefore undergo oxidation.

reduction: The gain of electrons by a compound or ion.

regulator of G protein signaling (RGS): Protein structural domain that stimulates the GTPase activity of heterotrimeric G proteins.

regulatory cascade: A multistep regulatory pathway in which a signal leads to activation of a series of proteins in succession, with each protein in the succession catalytically activating the next, such that the original signal is amplified exponentially.

regulatory enzyme: An enzyme with a regulatory function, through its capacity to undergo a change in catalytic activity by allosteric mechanisms or by covalent modification.

regulatory gene: A gene that gives rise to a product involved in the regulation of the expression of another gene; for example, a gene encoding a repressor protein.

regulatory sequence: A DNA sequence involved in regulating the expression of a gene; for example, a promoter or operator.

regulon: A group of genes or operons that are coordinately regulated even though some, or all, may be spatially distant in the chromosome or genome.

relaxed DNA: Any DNA that exists in its most stable and unstrained structure, typically the B form under most cellular conditions.

release factors: Protein factors of the cytosol required for the release of a completed polypeptide chain from a ribosome; also known as termination factors.

renaturation: Refolding of an unfolded (denatured) globular protein so as to restore its native structure and function.

replication: Synthesis of daughter nucleic acid molecules identical to the parental nucleic acid.

replication fork: The Y-shaped structure generally found at the point where DNA is being synthesized.

replicative form: Any of the full-length structural forms of a viral chromosome that serve as distinct replication intermediates.

replisome: The multiprotein complex that promotes DNA synthesis at the replication fork.

repressible enzyme: In bacteria, an enzyme whose synthesis is inhibited when its reaction product is readily available to the cell.

repression: A decrease in the expression of a gene in response to a change in the activity of a regulatory protein.

repressor: The protein that binds to the regulatory sequence or operator for a gene, blocking its transcription.

residue: A single unit in a polymer; for example, an amino acid in a polypeptide chain. The term reflects the fact that sugars, nucleotides, and amino acids lose a few atoms (generally the elements of water) when incorporated in their respective polymers.

respiration: Any metabolic process that leads to the uptake of oxygen and the release of CO_2.

respiration-linked phosphorylation: ATP formation from ADP and P_i, driven by electron flow through a series of membrane-bound carriers, with a proton gradient as the direct source of energy driving rotational catalysis by ATP synthase.

respiratory chain: The electron-transfer chain; a sequence of electron-carrying proteins that transfers electrons from substrates to molecular oxygen in aerobic cells.

response element: A region of DNA, near (upstream from) a gene, that is bound by specific proteins that influence the rate of transcription of the gene.

restriction endonucleases: Site-specific endodeoxyribonucleases that cleave both strands of DNA at points in or near the specific site recognized by the enzyme; important tools in genetic engineering.

restriction fragment: A segment of double-stranded DNA produced by the action of a restriction endonuclease on a larger DNA.

retinal: A 20-carbon isoprene aldehyde derived from carotene, which serves as the light-sensitive component of the visual pigment rhodopsin. Illumination converts 11-*cis*-retinal to all-*trans*-retinal.

retrovirus: An RNA virus containing a reverse transcriptase.

reverse transcriptase: An RNA-directed DNA polymerase in retroviruses; capable of making DNA complementary to an RNA.

reversible inhibition: Inhibition by a molecule that binds reversibly to the enzyme, such that the enzyme activity returns when the inhibitor is no longer present.

R group: (1) Formally, an abbreviation denoting any alkyl group. (2) Occasionally, used in a more general sense to denote virtually any organic substituent (the R groups of amino acids, for example).

RGS: *See* regulator of G protein signaling

rhodopsin: The visual pigment, composed of the protein opsin and the chromophore retinal.

RIA: *See* radioimmunoassay.

ribonuclease: A nuclease that catalyzes the hydrolysis of certain internucleotide linkages of RNA.

ribonucleic acid: *See* RNA.

ribonucleotide: A nucleotide containing D-ribose as its pentose component.

ribosomal RNA (rRNA): A class of RNA molecules serving as components of ribosomes.

ribosome: A supramolecular complex of rRNAs and proteins, approximately 18 to 22 nm in diameter; the site of protein synthesis.

riboswitch: A structured segment of an mRNA that binds to a specific ligand and affects the translation or processing of the mRNA.

ribozymes: Ribonucleic acid molecules with catalytic activities; RNA enzymes.

ribulose 1,5-bisphosphate carboxylase/ oxygenase (rubisco): The enzyme that fixes inorganic CO_2 into organic form (3-phosphoglycerate) in those organisms (plants and some microorganisms) capable of CO_2 fixation.

Rieske iron-sulfur protein: A type of iron-sulfur protein in which two of the ligands to the central iron ion are His side chains; act in many electron-transfer sequences, including oxidative phosphorylation and photophosphorylation.

RNA (ribonucleic acid): A polyribonucleotide of a specific sequence linked by successive 3′,5′-phosphodiester bonds.

RNA editing: Posttranscriptional modification of an mRNA that alters the meaning of one or more codons during translation.

RNA polymerase: An enzyme that catalyzes the formation of RNA from ribonucleoside 5′-triphosphates, using a strand of DNA or RNA as a template.

RNA splicing: Removal of introns and joining of exons in a primary transcript.

ROS: *See* reactive oxygen species.

rRNA: *See* ribosomal RNA.

RTK: *See* receptor Tyr kinase.

rubisco: *See* ribulose 1,5-bisphosphate carboxylase/oxygenase.

S

salvage pathway: Pathway for synthesis of a biomolecule, such as a nucleotide, from intermediates in the degradative pathway for the biomolecule; a recycling pathway, as distinct from a de novo pathway.

sarcomere: A functional and structural unit of the muscle contractile system.

satellite DNA: Highly repeated, nontranslated segments of DNA in eukaryotic chromosomes; most often associated with the centromeric region. Its function is unknown. Also called simple-sequence DNA.

saturated fatty acid: A fatty acid containing a fully saturated alkyl chain.

scaffold proteins: Noncatalytic proteins that nucleate formation of multienzyme complexes by providing two or more specific binding sites for those proteins.

scramblases: Membrane proteins that catalyze the movement of phospholipids across the membrane bilayer, leading to uniform distribution of a lipid between the two membrane leaflets.

secondary metabolism: Pathways that lead to specialized products not found in every living cell.

secondary structure: The local spatial arrangement of the main-chain atoms in a segment of a polypeptide chain; also applied to polynucleotide structure.

second law of thermodynamics: The law stating that, in any chemical or physical process, the entropy of the universe tends to increase.

second messenger: An effector molecule synthesized in a cell in response to an external signal (first messenger) such as a hormone.

sedimentation coefficient: A physical constant specifying the rate of sedimentation of a particle in a centrifugal field under specified conditions.

selectins: A large family of membrane proteins, lectins that bind oligosaccharides on other cells tightly and specifically and serve to carry signals across the plasma membrane.

SELEX: A method for rapid experimental identification of nucleic acid sequences (usually RNA) that have particular catalytic or ligand-binding properties.

sequence polymorphisms: Any alterations in genomic sequence (base-pair changes, insertions, deletions, rearrangements) that help distinguish subsets of individuals in a population or distinguish one species from another.

serine proteases: One of four major classes of proteases, featuring a reaction mechanism in which an active-site Ser residue acts as a covalent catalyst.

Shine-Dalgarno sequence: A sequence in an mRNA that is required for binding bacterial ribosomes.

short tandem repeat (STR): A short (typically 3 to 6 bp) DNA sequence, repeated many times in tandem at a particular location in a chromosome.

SH2 domain: A protein domain that binds tightly to a phosphotyrosine residue in certain proteins such as the receptor Tyr kinases, initiating the formation of a multiprotein complex that acts in a signaling pathway.

shuttle vector: A recombinant DNA vector that can be replicated in two or more different host species. *See also* vector.

sickle-cell anemia: A human disease characterized by defective hemoglobin molecules in individuals homozygous for a mutant allele coding for the β chain of hemoglobin.

σ: (1) *See* superhelical density. (2) A subunit of the bacterial RNA polymerase that confers specifity for certain promoters; usually designated by a superscript indicating its size (for example, σ^{70} has a molecular weight of 70,000).

signal sequence: An amino acid sequence, often at the amino terminus, that signals the cellular fate or destination of a newly synthesized protein.

signal transduction: The process by which an extracellular signal (chemical, mechanical, or electrical) is amplified and converted to a cellular response.

silent mutation: A mutation in a gene that causes no detectable change in the biological characteristics of the gene product.

simple diffusion: The movement of solute molecules across a membrane to a region of lower concentration, unassisted by a protein transporter.

simple protein: A protein yielding only amino acids on hydrolysis.

simple sequence DNA: *See* satellite DNA.

single nucleotide polymorphism (SNP): A genomic base-pair change that helps distinguish one species from another or one subset of individuals in a population.

site-directed mutagenesis: A set of methods used to create specific alterations in the sequence of a gene.

site-specific recombination: A type of genetic recombination that occurs only at specific sequences.

size-exclusion chromatography: A procedure for the separation of molecules by size, based on the capacity of porous polymers to exclude solutes above a certain size; also called gel filtration.

small G proteins: *See* Ras superfamily of G proteins.

small nuclear RNA (snRNA): A class of short RNAs, typically 100 to 200 nucleotides long, found in the nucleus and involved in the splicing of eukaryotic mRNAs.

small nucleolar RNA (snoRNA): A class of short RNAs, generally 60 to 300 nucleotides long, that guide the modification of rRNAs in the nucleolus.

SNP: *See* single nucleotide polymorphism.

somatic cells: All body cells except the germ-line cells.

SOS response: In bacteria, a coordinated induction of a variety of genes in response to high levels of DNA damage.

Southern blot: A DNA hybridization procedure in which one or more specific DNA fragments are detected in a larger population by hybridization to a complementary, labeled nucleic acid probe.

specific acid-base catalysis: Acid or base catalysis involving the constituents of water (hydroxide or hydronium ions).

specific activity: The number of micromoles (μmol) of a substrate transformed by an enzyme preparation per minute per milligram of protein at 25 °C; a measure of enzyme purity.

specificity: The ability of an enzyme or receptor to discriminate among competing substrates or ligands.

specific rotation: The rotation, in degrees, of the plane of plane-polarized light (D-line of sodium) by an optically active compound at 25 °C, with a specified concentration and light path.

sphingolipid: An amphipathic lipid with a sphingosine backbone to which are attached a long-chain fatty acid and a polar alcohol.

spliceosome: A complex of RNAs and proteins involved in the splicing of mRNAs in eukaryotic cells.

splicing: *See* RNA splicing.

standard free-energy change ($\Delta G°$): The free-energy change for a reaction occurring under a set of standard conditions: temperature, 298 K; pressure, 1 atm or 101.3 kPa; and all solutes at 1 M concentration. $\Delta G'°$ denotes the standard free-energy change at pH 7.0 in 55.5 M water.

standard reduction potential ($E'°$): The electromotive force exhibited at an electrode by 1 M concentrations of a reducing agent and its oxidized form at 25 °C and pH 7.0; a measure of the relative tendency of the reducing agent to lose electrons.

statin: Any of a class of drugs used to reduce blood cholesterol in humans; act by inhibiting the enzyme HMG-CoA reductase, an early step in sterol synthesis.

steady state: A nonequilibrium state of a system through which matter is flowing and in which all components remain at a constant concentration.

stem cells: The common, self-regenerating cells in bone marrow that give rise to differentiated blood cells such as erythrocytes and lymphocytes.

stereoisomers: Compounds that have the same composition and the same order of atomic connections but different molecular arrangements.

sterol: A lipid containing the steroid nucleus.

sticky ends: Two DNA ends in the same DNA molecule, or in different molecules, with short overhanging single-stranded segments that are complementary to one another, facilitating ligation of the ends; also known as cohesive ends.

stimulatory G protein (G_s): A trimeric regulatory GTP-binding protein that, when activated by an associated plasma membrane receptor, stimulates a neighboring membrane enzyme such as adenylyl cyclase. Its effects oppose those of G_i.

stop codons: *See* termination codons.

STR: *See* short tandem repeat.

stroma: The space and aqueous solution enclosed within the inner membrane of a chloroplast, not including the contents in the thylakoid membranes.

structural gene: A gene coding for a protein or RNA molecule; as distinct from a regulatory gene.

substitution mutation: A mutation caused by the replacement of one base by another.

substrate: The specific compound acted upon by an enzyme.

substrate channeling: Movement of the chemical intermediates in a series of enzyme-catalyzed reactions from the active site of one enzyme to that of the next enzyme in the pathway, without leaving the surface of a protein complex that includes both enzymes.

substrate-level phosphorylation: Phosphorylation of ADP or some other nucleoside 5′-diphosphate coupled to the dehydrogenation of an organic substrate; independent of the electron-transfer chain.

suicide inactivator: A relatively inert molecule that is transformed by an enzyme, at its active site, into a reactive substance that irreversibly inactivates the enzyme.

sulfonylurea drugs: A group of oral medications used in the treatment of type 2 diabetes; act by closing K^+ channels in pancreatic β cells, stimulating insulin secretion.

supercoil: The twisting of a helical (coiled) molecule on itself; a coiled coil.

supercoiled DNA: DNA that twists upon itself because it is under- or overwound (and thereby strained) relative to B-form DNA.

superhelical density (σ): In a helical molecule such as DNA, the number of supercoils (superhelical turns) relative to the number of coils (turns) in the relaxed molecule.

supersecondary structure: *See* motif.

suppressor mutation: A mutation that totally or partially restores a function lost by a primary mutation; located at a site different from the site of the primary mutation.

Svedberg (S): A unit of measure of the rate at which a particle sediments in a centrifugal field.

symbionts: Two or more organisms that are mutually interdependent; usually living in physical association.

symport: Cotransport of solutes across a membrane in the same direction.

synteny: Conserved gene order along the chromosomes of different species.

synthases: Enzymes that catalyze condensation reactions in which no nucleoside triphosphate is required as an energy source.

synthetases: Enzymes that catalyze condensation reactions using ATP or another nucleoside triphosphate as an energy source.

system: An isolated collection of matter; all other matter in the universe apart from the system is called the surroundings.

systems biology: The study of complex biochemical systems, integrating the functions of several to all of the macromolecules in a cell (RNA, DNA, proteins).

t

tag: An extra segment of protein that is fused via modification of its gene to a protein of interest, usually for purposes of purification.

T cell: *See* T lymphocyte.

telomere: Specialized nucleic acid structure found at the ends of linear eukaryotic chromosomes.

template: A macromolecular mold or pattern for the synthesis of an informational macromolecule.

template strand: A strand of nucleic acid used by a polymerase as a template to synthesize a complementary strand, as distinct from the coding strand.

terminal transferase: An enzyme that catalyzes the addition of nucleotide residues of a single kind to the 3′ end of DNA chains.

termination codons: UAA, UAG, and UGA; in protein synthesis, these codons signal the termination of a polypeptide chain. Also known as stop codons.

termination factors: *See* release factors.

termination sequence: A DNA sequence, at the end of a transcriptional unit, that signals the end of transcription.

tertiary structure: The three-dimensional conformation of a polymer in its native folded state.

tetrahydrobiopterin: The reduced coenzyme form of biopterin.

tetrahydrofolate: The reduced, active coenzyme form of the vitamin folate.

thermogenesis: The biological generation of heat by muscle activity (shivering), uncoupled oxidative phosphorylation, or the operation of futile cycles.

thermogenin: A protein of the inner mitochondrial membrane in brown adipose tissue that allows transmembrane movement of protons, short-circuiting the normal use of protons to drive ATP synthesis and dissipating the energy of substrate oxidation as heat; also called uncoupling protein 1 (UCP1).

thiamine pyrophosphate (TPP): The active coenzyme form of vitamin B_1; involved in aldehyde transfer reactions.

thioester: An ester of a carboxylic acid with a thiol or mercaptan.

3′ end: The end of a nucleic acid that lacks a nucleotide bound at the 3′ position of the terminal residue.

thrombocytes: *See* platelets.

thromboxane: Any of a class of molecules derived from arachidonate and involved in platelet aggregation during blood clotting.

thylakoid: Closed cisterna, or disk, formed by the pigment-bearing internal membranes of chloroplasts.

thymine dimer: *See* pyrimidine dimer.

tissue culture: Method by which cells derived from multicellular organisms are grown in liquid media.

titration curve: A plot of pH versus the equivalents of base added during titration of an acid.

T lymphocyte (T cell): One of a class of blood cells (lymphocytes) of thymic origin, involved in cell-mediated immune reactions.

tocopherol: Any of several forms of vitamin E.

topoisomerases: Enzymes that introduce positive or negative supercoils in closed, circular duplex DNA.

topoisomers: Different forms of a covalently closed, circular DNA molecule that differ only in their linking number.

topology: The study of the properties of an object that do not change under continuous deformations such as twisting or bending.

TPP: *See* thiamine pyrophosphate.

trace element: A chemical element required by an organism in only trace amounts.

transaminases: *See* aminotransferases.

transamination: Enzymatic transfer of an amino group from an α-amino acid to an α-keto acid.

transcription: The enzymatic process whereby the genetic information contained in one strand of DNA is used to specify a complementary sequence of bases in an mRNA chain.

transcriptional control: The regulation of a protein's synthesis by regulation of the formation of its mRNA.

transcription factor: In eukaryotes, a protein that affects the regulation and transcription initiation of a gene by binding to a regulatory sequence near or within the gene and interacting with RNA polymerase and/or other transcription factors.

transcriptome: The entire complement of RNA transcripts present in a given cell or tissue under specific conditions.

transducin: The trimeric G protein activated when light is absorbed by visual rhodopsin; activated transducin activates cGMP phosphodiesterase.

transduction: (1) Generally, the conversion of energy or information from one form to another. (2) The transfer of genetic information from one cell to another by means of a viral vector.

transfer RNA (tRNA): A class of RNA molecules (M_r 25,000 to 30,000), each of which combines covalently with a specific amino acid as the first step in protein synthesis.

transformation: Introduction of an exogenous DNA into a cell, causing the cell to acquire a new phenotype.

transgenic: Describes an organism that has genes from another organism incorporated in its genome as a result of recombinant DNA procedures.

transition state: An activated form of a molecule in which the molecule has undergone a partial chemical reaction; the highest point on the reaction coordinate.

transition-state analog: A stable molecule that resembles the transition state of a particular reaction, and therefore binds the enzyme that catalyzes the reaction more tightly than does the substrate in the ES complex.

translation: The process in which the genetic information present in an mRNA molecule specifies the sequence of amino acids during protein synthesis.

translational control: The regulation of a protein's synthesis by regulation of the rate of its translation on the ribosome.

translational frameshifting: A programmed change in the reading frame during translation of an mRNA on a ribosome, occurring by any of several mechanisms.

translational repressor: A repressor that binds to an mRNA, blocking translation.

translocase: (1) An enzyme that catalyzes membrane transport. (2) An enzyme that causes a movement such as the movement of a ribosome along an mRNA.

transpiration: Passage of water from the roots of a plant to the atmosphere via the vascular system and the stomata of the leaves.

transporters: Proteins that span a membrane and transport specific nutrients, metabolites, ions, or proteins across the membrane; sometimes called permeases.

transposition: The movement of a gene or set of genes from one site in the genome to another.

transposon (transposable element): A segment of DNA that can move from one position in the genome to another.

triacylglycerol: An ester of glycerol with three molecules of fatty acid; also called a triglyceride or neutral fat.

tricarboxylic acid cycle: *See* citric acid cycle.

trimeric G proteins: Members of the G protein family with three subunits, which function in a variety of signaling pathways. Inactive with GDP bound, they are activated by associated receptors as bound GDP is displaced by GTP, then inactivated by their intrinsic GTPase activity.

triose: A simple sugar with a backbone containing three carbon atoms.

tRNA: *See* transfer RNA.

tropic hormone (tropin): A peptide hormone that stimulates a specific target gland to secrete its hormone; for example, thyrotropin produced by the pituitary stimulates secretion of thyroxine by the thyroid.

t-SNAREs: Protein receptors in a targeted membrane (typically the plasma membrane) that bind to v-SNAREs in the membrane of a secretory vesicle and mediate fusion of the vesicle and target membranes.

tumor suppressor gene: One of a class of genes that encode proteins that normally regulate the cell cycle by suppressing cell division. Mutation of one copy of the gene is usually without effect, but when both copies are defective, the cell is allowed to continue dividing without limitation, producing a tumor.

turnover number: The number of times an enzyme molecule transforms a substrate molecule per unit time, under conditions giving maximal activity at substrate concentrations that are saturating.

two-component signaling systems: Signal-transducing systems found in bacteria and plants, composed of a receptor His kinase that phosphorylates an internal His residue when occupied by its ligand. It then catalyzes phosphoryl transfer to a second component, the response regulator, which, when phosphorylated, alters the output of the signaling system.

u

ubiquitin: A small, highly conserved eukaryotic protein that targets an intracellular protein for degradation by proteasomes. Several ubiquitin molecules are covalently attached in tandem to a Lys residue in the target protein by a specific ubiquitinating enzyme.

ultraviolet (UV) radiation: Electromagnetic radiation in the region of 200 to 400 nm.

uncompetitive inhibition: The reversible inhibition pattern resulting when an inhibitor molecule can bind to the enzyme-substrate complex but not to the free enzyme.

uncoupling agent: A substance that uncouples phosphorylation of ADP from electron transfer; for example, 2,4-dinitrophenol.

uncoupling protein 1: *See* thermogenin.

uniport: A transport system that carries only one solute, as distinct from cotransport.

unsaturated fatty acid: A fatty acid containing one or more double bonds.

urea cycle: A cyclic metabolic pathway in vertebrate liver, synthesizing urea from amino groups and carbon dioxide.

ureotelic: Excreting excess nitrogen in the form of urea.

uricotelic: Excreting excess nitrogen in the form of urate (uric acid).

v

V_{max}**:** The maximum velocity of an enzymatic reaction when the binding site is saturated with substrate.

van der Waals interaction: Weak intermolecular forces between molecules as a result of each inducing polarization in the other.

vector: A DNA molecule known to replicate autonomously in a host cell, to which a segment of DNA may be spliced to allow its replication; for example, a plasmid or an artificial chromosome.

vectorial: Describes an enzymatic reaction or transport process in which the protein has a specific orientation in a biological membrane such that the substrate is moved from one side of the membrane to the other as it is converted into product.

vectorial metabolism: Metabolic transformations in which the location (not the chemical composition) of a substrate changes relative to the plasma membrane or a membrane between two cellular compartments. Transporters catalyze vectorial reactions, as do the proton pumps of oxidative phosphorylation and photophosphorylation.

vesicle: A small, spherical membrane-bounded particle with an internal aqueous compartment that contains components such as hormones or neurotransmitters to be moved within or out of a cell.

viral vector: A viral DNA altered so that it can act as a vector for recombinant DNA.

virion: A virus particle.

virus: A self-replicating, infectious, nucleic acid–protein complex that requires an intact host cell for its replication; its genome is either DNA or RNA.

vitamin: An organic substance required in small quantities in the diet of some species; generally functions as a component of a coenzyme.

v-SNAREs: Protein receptors in the membrane of a secretory vesicle (typically the plasma membrane) that bind to t-SNAREs in a targeted membrane (typically the plasma membrane) of a secretory vesicle and mediate fusion of the vesicle and target membranes.

w

Western blotting: *See* immunoblotting.

white adipose tissue (WAT): Nonthermogenic adipose tissue rich in triacylglycerols stored and mobilized in response to hormonal signals. Transfer of electrons in the mitochondrial respiratory chain is tightly coupled to ATP synthesis. *Compare* brown adipose tissue.

wild type: The normal (unmutated) genotype or phenotype.

wobble: The relatively loose base pairing between the base at the 3′ end of a codon and the complementary base at the 5′ end of the anticodon.

x

x-ray crystallography: The analysis of x-ray diffraction patterns of a crystalline compound, used to determine the molecule's three-dimensional structure.

z

zinc finger: A specialized protein motif involved in DNA recognition by some DNA-binding proteins; characterized by a single atom of zinc coordinated to four Cys residues or to two His and two Cys residues.

Z scheme: The path of electrons in oxygenic photosynthesis from water through photosystem II and the cytochrome b_6f complex to photosystem I and finally to NADPH. When the sequence of electron carriers is plotted against their reduction potentials, the path of electrons looks like a sideways Z.

zwitterion: A dipolar ion with spatially separated positive and negative charges.

zymogen: An inactive precursor of an enzyme; for example, pepsinogen, the precursor of pepsin.

Credits

Molecular Models

MOLECULAR GRAPHICS Unless indicated, all molecular graphics were produced by H. Adam Steinberg, artforscience.com, or Jean-Yves Sgro, Ph.D., University of Wisconsin–Madison, Biotechnology Center.

ATOMIC COORDINATES Unless indicated, all atomic coordinates were obtained from the Research Collaboratory for Structural Bioinformatics (RCSB) Protein Data Bank (PDB), www.pdb.org. See Berman, H.M., Westbrook, J., Feng, Z., Gilliland, G., Bhat, T.N., Weissig, H., Shindyalov, I.N., & Bourne, P.E. (2000) The Protein Data Bank. *Nucleic Acids Res.* 28, 235–242. The RCSB PDB is a member of the worldwide PDB (wwPDB), www.wwpdb.org. See Berman, H., Henrick, K., & Nakamura, H. (2003) Announcing the worldwide Protein Data Bank. *Nat. Struct. Biol.* 10, 980.

Some structures were generated using PyMOL, pymol.sourceforge.net; Sybyl 6.2, Tripos Inc., www.tripos.com; Visual Molecular Dynamics, www.ks.uiuc.edu/Research/vmd/; or RasMol, rasmol.org.

CHAPTER 1 Figure 1–1a Dennis Kunkel Microscopy, Inc./Visuals Unlimited; **Figure 1–1b** W. Perry Conway/Corbis; **Figure 1–1c** Dave Pape/Wikipedia; **Figure 1–2** The Bridgeman Art Library; **Figure 1–4** Adapted from Woese, C.R. (1987) Bacterial evolution. *Microbiol. Rev.* 51, 221, Fig. 4; **Figure 1–6a** David S. Goodsell; **Figure 1–6b,c,d** Adapted from Albers, S.-V. & Meyer, B.H. (2011) The archaeal cell envelope. *Nat. Rev. Microbiol.* 9, 414, Fig. 2; **Figure 1–8** Adapted from Alberts, B., Bray, D., Lewis, J., Raff, M., Roberts, K., & Watson, J.D. (1989) *Molecular Biology of the Cell*, 2nd edn, Garland Publishing, Inc., New York, pp. 165–166; **Figure 1–9a** Courtesy of Invitrogen; **Figure 1–9b** Dr. Alexey Khodjakov, Wadsworth Center, New York State Department of Health; **Figure 1–11** Adapted from Becker, W.M. & Deamer, D.W. (1991) *The World of the Cell*, 2nd edn, Fig. 2–15, Benjamin/Cummings Publishing Company, Menlo Park, CA; **Figure 1–12** © David S. Goodsell 1999; **Figure 1–17** Acetyl-CoA extracted from PDB ID 1DM3, Modis, Y. & Wierenga, R.K. (2000) Crystallographic analysis of the reaction pathway of *Zoogloea ramigera* biosynthetic thiolase. *J. Mol. Biol.* 297, 1171; **Figure 1–21** Adapted from Carroll, F. (1998) *Perspectives on Structure and Mechanism in Organic Chemistry*, Brooks/Cole Publishing Co., Pacific Grove, CA, p. 63; **Box 1–2** (Pasteur) The Granger Collection; **Figure 1–23** PDB ID 3B8A, Kuser, P., Cupri, F., Bleicher, L., & Polikarpov, I. (2008) Crystal structure of yeast hexokinase PI in complex with glucose: a classical "induced fit" example revised. *Proteins* 72, 731; **p. 23** (Gibbs) Historical Pictures Service/Stock Montage; **Figure 1–30a** Erich Lessing/Art Resource, New York; **Figure 1–30b** Dr. Gopal Murti–CNRI/Phototake New York; **p. 32** Theodosius Dobzhansky (1973) Nothing in biology makes sense except in the light of evolution. *The American Biology Teacher* (March) 35, 125–129; **Figure 1–34b** Bettmann/Corbis; **Figure 1–35** P. Rona/OAR/National Undersea Research Program (NURP); NOAA; **p. 36** (Margulis) Ben Barnhart/UMass Magazine.

CHAPTER 2 p. 54 Linus Pauling (1939) *The Nature of the Chemical Bond and the Structure of Molecules and Crystals: An Introduction to Modern Structural Chemistry*, Cornell University Press, Ithaca, NY; **Figure 2–9** PDB ID 1A3N, Tame, J.R.H. & Vallone, B. The structures of deoxy human haemoglobin and the mutant Hb Tyrα42His at 120 K; **Figure 2–10** Adapted from Nicolls, P. (2000) Introduction: the biology of the water molecule. *Cell. Mol. Life Sci.* 57, 987, Fig. 6a (redrawn from information in the PDB and a Kinemage file published by Martinez, S.E., Huang, D., Ponomarev, M., Cramer, W.A., & Smith, J.L. (1996) The heme redox center of chloroplast cytochrome *f* is linked to a buried five-water chain. *Protein Sci.* 5, 1081); **Figure 2–11** Adapted from Ball, P. (2008) Water as an active constituent in cell biology. *Chem. Rev.* 108, 94, Fig. 16; **Box 2–1** J. B. S. Haldane (1928) *Possible Worlds*, Harper and Brothers, New York and London, pp. 113–126.

CHAPTER 3 Figure 3–1a Runk/Schoenburger/Grant Heilman Photography; **Figure 3–1b** Bill Longcore/Photo Researchers; **Figure 3–1c** Jerry Cooke, Inc./Animals Animals; **p. 76** (Dayhoff) Courtesy of Ruth E. Dayhoff; **Figure 3–18b** Julia Cox, University of Wisconsin–Madison, Department of Biochemistry; **Figure 3–21** Courtesy of Axel Mogk, from Mogk, A., Tomoyasu, T., Goloubinoff, P., Rüdiger, S., Röder, D., Lanen, H., & Bukau, B. (1999) Identification of thermolabile *Escherichia coli* proteins: prevention and reversion of aggregation by DnaK and ClpB. *EMBO J.* 18, 6934, Fig. 7A; **Figure 3–23** PDB ID 1HGA, Liddington, R., Derewenda, Z., Dodson, E., Hubbard, R., & Dodson, G. (1992) High resolution crystal structures and comparisons of T-state deoxyhaemoglobin

and two liganded T-state haemoglobins: T(α-oxy) haemoglobin and T(met) haemoglobin. *J. Mol. Biol.* 228, 551; **p. 98** (Sanger) UPI/Corbis-Bettmann; **Figures 3–30** Adapted from Mann, M. & Wilm, M. (1995) Electrospray mass spectrometry for protein characterization. *Trends Biochem. Sci.* 20, 219; **Figure 3–31a** Adapted from Keough, T., Youngquist, R.S., & Lacey, M.P. (1999) A method for high-sensitivity peptide sequencing using postsource decay matrix-assisted laser desorption ionization mass spectrometry. *Proc. Natl. Acad. Sci. USA* 96, 7131, Fig. 3; **p. 103** (Merrifield) Corbis/Bettmann; **Box 3–2 Figure 1** Sequence data for **(a)** from document ID PDOC00017 and for **(b)** from document ID PDOC00018, www.expasy.org/prosite, Hulo, N., Bairoch, A., Bulliard, V., Cerutti, L., De Castro, E., Langendijk-Genevaux, P.S., Pagni, M., & Sigrist, C.J.A. (2006) The PROSITE database. *Nucleic Acids Res.* 34, D227; WebLogo from http://weblogo.berkeley.edu, Crooks, G.E., Hon, G., Chandonia, J.M., & Brenner, S.E. (2004) WebLogo: a sequence logo generator. *Genome Res.* 14, 1188; **Figures 3–33, 3–34, 3–35** Adapted from Gupta, R.S. (1998) Protein phylogenies and signature sequences: a reappraisal of evolutionary relationships among archaebacteria, eubacteria, and eukaryotes. *Microbiol. Mol. Biol. Rev.* 62, 1435, Figs 2, 7, 11, respectively; **Figure 3–36** Adapted from Delsuc, F., Brinkmann, H., & Philippe, H. (2005) Phylogenomics and the reconstruction of the tree of life. *Nat. Rev. Genet.* 6, 366; **p. 113, problem 21**, see citation for Box 3–2 Figure 1, document ID PDOC00270; **p. 113, problem 22**, see citation for Box 3–2 Figure 1, document ID PDOC00017.

CHAPTER 4 Figure 4–1 PDB ID 6GCH, Brady, K., Wei, A., Ringe, D., & Abeles, R.H. (1990) Structure of chymotrypsin-trifluoromethyl ketone inhibitor complexes: comparison of slowly and rapidly equilibrating inhibitors. *Biochemistry* 29, 7600; glycine coordinates from Sybyl; **p. 117** (Pauling) Corbis/Bettmann; (Corey) AP/Wide World Photos; **Figure 4–3** Adapted from Creighton, T.E. (1984) *Proteins*, p. 166. © 1984 by W. H. Freeman and Company. Reprinted by permission; **Figure 4–4b,c** PDB ID 4TNC, Satyshur, K.A., Rao, S.T., Pyzalska, D., Drendel, W., Greaser, M., & Sundaralingam, M. (1988) Refined structure of chicken skeletal muscle troponin C in the two-calcium state at 2-angstroms resolution. *J. Biol. Chem.* 263, 1628; **Figure 4–9a** See citation for Figure 4–3; **Figure 4–9b** Courtesy of Hazel Holden, University of Wisconsin–Madison, Department of Biochemistry and Enzyme Institute; **Figure 4–12** PDB ID 1CGD (modified), Bella, J., Brodsky, B., & Berman, H.M. (1995) Hydration structure of a collagen peptide. *Structure* 3, 893; **p. 128** Ethel Wedgwood (1906) *The Memoirs of the Lord of Joinville: A New English Version*, E. P. Dutton and Company, New York; (Lind) Courtesy of the Royal College of Physicians of Edinburgh; **Figure 4–13** Science Source/Photo Researchers; **Figure 4–14a** PDB ID 1SLK (model), Fossey, S.A., Nemethy, G., Gibson, K.D., & Scheraga, H.A. (1991) Conformational energy studies of β-sheets of model silk fibroin peptides: I. sheets of poly(Ala-Gly) chains. *Biopolymers* 31, 1529; **Figure 4–14b** Dr. Dennis Kunkel/Phototake NYC; **Figure 4–16** PDB ID 1MBO, Phillips, S.E.V. (1980) Structure and refinement of oxymyoglobin at 1.6 angstroms resolution. *J. Mol. Biol.* 142, 531; **Box 4–5 Figure 1a,b,c** George N. Phillips, Jr., University of Wisconsin–Madison, Department of Biochemistry; **Box 4–5 Figure 1d** PDB ID 2MBW, Brucker, E.A., Olson, J.S., Phillips, G.N., Jr., Dou, Y., & Ikeda-Saito, M. (1996) High resolution crystal structures of the deoxy-, oxy-, and aquomet-forms of cobalt myoglobin. *J. Biol. Chem.* 271, 25,419; **Box 4–5 Figures 2, 3a** Volkman, B.F., Alam, S.L., Satterlee, J.D., & Markley, J.L. (1998) Solution structure and backbone dynamics of component IV-glycera dibranchiata monomeric hemoglobin-CO. *Biochemistry* 37, 10,906; **Box 4–5 Figure 3b,c** Created by Brian Volkman, National Magnetic Resonance Facility at Madison, using MOLMOL; PDB ID 1VRF **(b)** and 1VRE **(c)**, see citation for Box 4–5 Figures 2, 3a; **Figure 4–18b** PDB ID 7AHL, Song, L., Hobaugh, M.R., Shustak, C., Cheley, S., Bayley, H., & Gouaux, J.E. (1996) Structure of staphylococcal α hemolysin, a heptameric transmembrane pore. *Science* 274, 1859; **Figure 4–19** PDB ID 4TNC, see citation for Figure 4–4b,c; **Figure 4–20c** PDB ID 1DNP, Park, H.W., Kim, S.T., Sancar, A., & Deisenhofer, J. (1995) Crystal structure of DNA photolyase from *Escherichia coli*. *Science* 268, 1866; **Figure 4–21** PDB ID 1PKN, Larsen, T.M., Laughlin, L.T., Holden, H.M., Rayment, I., & Reed, G.H. (1994) Structure of rabbit muscle pyruvate kinase complexed with Mn^{23}, K^3, and pyruvate. *Biochemistry* 33, 6301; **Figure 4–22 (all α)** PDB ID 1AO6, Sugio, S., Kashima, A., Mochizuki, S., Noda, M., & Kobayashi, K. (1999) Crystal structure of human serum albumin at 2.5 angstrom resolution. *Protein Eng.* 12, 439; PDB ID 1BCF, Frolow, F., Kalb (Gilboa), A.J., & Yariv, J. (1994) The structure of a unique, two-fold symmetric, haem-binding site. *Nat. Struct. Biol.*

1, 453; PDB ID 1GAI, Aleshin, A.E., Stoffer, B., Firsov, L.M., Svensson, B., & Honzatko, R.B. (1996) Crystallographic complexes of glucoamylase with maltooligosaccharide analogs: relationship of stereochemical distortions at the nonreducing end to the catalytic mechanism. *Biochemistry* 35, 8319; **(all β)** PDB ID 1LXA, Raetz, C.R.H. & Roderick, S.L. (1995) A left-handed parallel β-helix in the structure of UDP N-acetylglucosamine acyltransferase. *Science* 270, 997; PDB ID 1PEX, Gomis-Ruth, F.X., Gohlke, U., Betz, M., Knauper, V., Murphy, G., Lopez-Otin, C., & Bode, W. (1996) The helping hand of collagenase-3 (Mmp-13): 2.7 Å crystal structure of its C-terminal haemopexin-like domain. *J. Mol. Biol.* 264, 556; PDB ID 1CD8, Leahy, D.J., Axel, R., & Hendrickson, W.A. (1992) Crystal structure of a soluble form of the human T cell co-receptor Cd8 at 2.6 angstroms resolution. *Cell* 68, 1145; **(α/β)** PDB ID 1DEH, Davis, G.J., Stone, C.J., Bosron, W.F., & Hurley, T.D. (1996) X-ray structure of human $β_3β_3$ alcohol dehydrogenase: the contribution of ionic interactions to coenzyme binding. *J. Biol. Chem.* 271, 17,057; PDB ID 1DUB, Engel, C.K., Mathieu, M., Zeelen, J.P., Hiltunen, J.K., & Wierenga, R.K. (1996) Crystal structure of enoyl-coenzyme A (CoA) hydratase at 2.5 angstroms resolution: a spiral fold defines the CoA-binding pocket. *EMBO J.* 15, 5135; PDB ID 1PFK, Shirakihara, Y. & Evans, P.R. (1988) Crystal structure of the complex of phosphofructokinase from *Escherichia coli* with its reaction products. *J. Mol. Biol.* 204, 973; **(α + β)** PDB ID 2PIL, Forest, K.T., Dunham, S.A., Koomey, M., & Tainer, J.A. (1999) Crystallographic structure of phosphorylated pilin from *Neisseria*: phosphoserine sites modify type IV pilus surface chemistry and morphology. *Mol. Microbiol.* 31, 743; PDB ID 1SYN, Stout, T.J. & Stroud, R.M. (1996) The complex of the anti-cancer therapeutic, BW1843U89, with thymidylate synthase at 2.0 Å resolution: implications for a new mode of inhibition. *Structure* 4, 67; PDB ID 1EMA, Ormo, M., Cubitt, A.B., Kallio, K., Gross, L.A., Tsien, R.Y., & Remington, S.J. (1996) Crystal structure of the *Aequorea victoria* green fluorescent protein. *Science* 273, 1392; **p. 141** (Perutz and Kendrew) Corbis/Hulton Deutsch Collection; **Figure 4–23** PDB ID 2HHB, Fermi, G., Perutz, M.F., Shaanan, B., & Fourme, R. (1984) The crystal structure of human deoxyhaemoglobin at 1.74 angstroms resolution. *J. Mol. Biol.* 175, 159; **Figure 4–24** Adapted from Uversky, V.N. (2011) Intrinsically disordered proteins from A to Z. *Intl. J. Biochem. Cell Biol.* 43, 1090, Fig. 5; **Figure 4–24a** PDB ID 1XQH, Chuikov, S., Kurash, J.K., Wilson, J.R., Xiao, B., Justin, N., Ivanov, G.S., McKinney, K., Tempst, P., Prives, C., Gamblin, S.J., Barlev, N.A., & Reinberg, D. (2004) Regulation of p53 activity through lysine methylation. *Nature* 432, 353; **Figure 4–24c** PDB ID 1H26, Lowe, E.D., Tews, I., Cheng, K.Y., Brown, N.R., Gul, S., Noble, M.E.M., Gamblin, S., & Johnson, L.N. (2002) Specificity determinants of recruitment peptides bound to phospho-CDK2/cyclin A. *Biochemistry* 41, 15,625; PDB ID 1MA3, Avalos, J.L., Celic, I., Muhammad, S., Cosgrove, M.S., Boeke, J.D., & Wolberger, C. (2002) Structure of a Sir2 enzyme bound to an acetylated p53 peptide. *Mol. Cell* 10, 523; PDB ID 1JSP, Mujtaba, S., He, Y., Zeng, L., Yan, S., Plotnikova, O., Sachchidanand, S.R., Zeleznik-Le, N.J., Ronai, Z., & Zhou, M.M. (2004) Structural mechanism of the bromodomain of the coactivator CBP in p53 transcriptional activation. *Mol. Cell* 13, 251; PDB ID 1DT7, Rustandi, R.R., Baldisseri, D.M., & Weber, D.J. (2000) Structure of the negative regulatory domain of p53 bound to S100B(ββ). *Nat. Struct. Biol.* 7, 570; **Figure 4–25** Adapted from Hartl, F.U., Bracher, A., & Hayer-Hartl, M. (2011) Molecular chaperones in protein folding and proteostasis. *Nature* 475, 324, Fig. 6; **Figure 4–26a** Data from Sendak, R.A., Rothwarf, D.M., Wedemeyer, W.J., Houry, W.A., & Scheraga, H.A. (1996) Kinetic and thermodynamic studies of the folding/unfolding of a tryptophan-containing mutant of ribonuclease A. *Biochemistry* 35, 12,978; Nishii, I., Kataoka, M., & Goto, Y. (1995) Thermodynamic stability of the molten globule states of apomyoglobin. *J. Mol. Biol.* 250, 223; **Figure 4–26b** Data from Houry, W.A., Rothwarf, D.M., & Scheraga, H.A. (1996) Circular dichroism evidence for the presence of burst-phase intermediates on the conformational folding pathway of ribonuclease A. *Biochemistry* 35, 10,125; **Figures 4–28, 4–29** Adapted from Dill, K.A., Ozkan, S.B., Shell, M.S., & Weiki, T.R. (2008) The protein folding problem. *Annu. Rev. Biophys.* 37, 289, Figs 5, 9; **Figures 4–30, 4–31a** See citation for Figure 4–25, Figs 2, 3; **Figure 4–31b** PDB ID 1AON, Xu, Z., Horwich, A.L., & Sigler, P.B. (1997) The crystal structure of the asymmetric GroEL-GroES-(ADP)7 chaperonin complex. *Nature* 388, 741; **Figure 4–32a** Adapted from Selkoe, D.J. (2003) Folding proteins in fatal ways. *Nature* 426, 903, Fig. 1; **Figure 4–32b** PDB ID 1IYT, Crescenzi, O., Tomaselli, S., Guerrini, R., Salvadori, S., D'Ursi, A.M., Temussi, P.A., & Picone, D. (2002) Solution structure of the Alzheimer amyloid β-peptide (1–42) in an apolar microenvironment: similarity with a virus fusion domain. *Eur. J. Biochem.* 269, 5642; **Figure 4–32c** PDB ID 2BEG, Lührs, T., Ritter, C., Adrian, M., Riek-Loher, D., Bohrmann, B., Döbeli, H., Schubert, D., & Riek, R. (2005) 3D structure of Alzheimer's amyloid-β (1–42) fibrils. *Proc. Natl. Acad. Sci. USA* 102, 17,342; **Box 4–6 Figure 1** Stephen J. DeArmond; **Box 4–6 Figure 2** PDB ID 1QLX, Zahn, R., Lieu, A., Luhrs, T., Riek, R., Von Schroetter, C., Garcia, F.L., Billeter, M., Calzolai, L., Wider, G., & Wuthrich, K. (2000) NMR solution structure of the human prion protein. *Proc. Natl. Acad. Sci. USA* 97, 145, and models from Govaerts, C., Wille,

H., Prusiner, S.B., & Cohen, F.E. (2004) Evidence for assembly of prions with left-handed β-helices into trimers. *Proc. Natl. Acad. Sci. USA* 101, 8342.

CHAPTER 5 Figure 5–1c Heme extracted from PDB ID 1CCR, Ochi, H., Hata, Y., Tanaka, N., Kakudo, M., Sakurai, T., Aihara, S., & Morita, Y. (1983) Structure of rice ferricytochrome c at 2.0 angstroms resolution. *J. Mol. Biol.* 166, 407; **Figures 5–3, 5–5c, 5–6 (left)** PDB ID 1MBO, Phillips, S.E.V. (1980) Structure and refinement of oxymyoglobin at 1.6 angstroms resolution. *J. Mol. Biol.* 142, 531; **Figures 5–6 (right), 5–8, 5–9a, 5–10 (T state), 5–11 (T state)** PDB ID 1HGA, Liddington, R., Derewenda, Z., Dodson, E., Hubbard, R., & Dodson, G. (1992) High resolution crystal structures and comparisons of T state deoxyhaemoglobin and two liganded T-state haemoglobins: T(α-oxy) haemoglobin and T(met) haemoglobin. *J. Mol. Biol.* 228, 551; **Figures 5–10 (R state), 5–11 (R state)** PDB ID 1BBB, Silva, M.M., Rogers, P.H., & Arnone, A. (1992) A third quaternary structure of human hemoglobin a at 1.7-angstroms resolution. *J. Biol. Chem.* 267, 17,248; in Fig. 5–11, R-state was modified to represent O_2 instead of CO; **Box 5–1 Figure 1** Adapted from Coburn, R.F., Forster, R.E., & Kane, P.B. (1965) Considerations of the physiological variables that determine the blood carboxyhemoglobin concentration in man. *J. Clin. Invest.* 44, 1899; **Box 5–1 Figure 2** Adapted from Roughton, F.J.W. & Darling, R.C. (1944) The effect of carbon monoxide on the oxyhemoglobin dissociation curve. *Am. J. Physiol.* 141, 17; **Figure 5–18a** PDB ID 1B86, Richard, V., Dodson, G.G., & Mauguen, Y. (1993) Human deoxyhaemoglobin-2,3-disphosphoglycerate complex low-salt structure at 2.5 Å resolution. *J. Mol. Bio.* 233, 270; **Figure 5–18b** See citation for Figure 5–10 (R state); **Figure 5–19a** Andrew Syred/Science Photo Library/Custom Medical Stock Photo; **Figure 5–19b** Custom Medical Stock Photo; **Figure 5–21b** PDB ID 1IGT, Harris, L.J., Larson, S.B., Hasel, K.W., & McPherson, A. (1997) Refined structure of an intact IgG2A monoclonal antibody. *Biochemistry* 36, 1581; **Figure 5–25a** PDB ID 1GGC, Stanfield, R.L., Takimoto-Kamimura, M., Rini, J.M., Profy, A.T., & Wilson, I.A. (1993) Major antigen-induced domain rearrangements in an antibody. *Structure* 1, 83; **Figure 5–25b,c** PDB ID 1GGI, Rini, J.M., Stanfield, R.L., Stura, E.A., Salinas, P.A., Profy, A.T., & Wilson, I.A. (1993) Crystal structure of an human immunodeficiency virus type 1 neutralizing antibody, 50.1, in complex with its V3 loop peptide antigen. *Proc. Natl. Acad. Sci. USA* 90, 6325; **p. 178** (Kohler and Milstein) Corbis/UPI/Bettmann; **Figure 5–26b** State of Wisconsin Laboratory of Hygiene, Madison, WI; **Figure 5–26c** Son, M., Gundersen, R.E., & Nelson, D.L. (1993) A second member of the novel Ca^{23}-dependent protein kinase family from *Paramecium tetraurelia*: purification and characterization. *J. Biol. Chem.* 268, 5940; **Figure 5–27a** David Shotton, University of Oxford, Department of Zoology; **Figure 5–27c** Courtesy of Ivan Rayment, University of Wisconsin–Madison, Enzyme Institute and Department of Biochemistry (see also PDB ID 2MYS, Rayment, I., Rypniewski, W.R., Schmidt-Base, K., Smith, R., Tomchick, D.R., Benning, M.M., Winkelmann, D.A., Wesenberg, G., & Holden, H.M. (1993) Three-dimensional structure of myosin subfragment-1: a molecular motor. *Science* 261, 50); **Figure 5–28a** Eisaku Katayama, Institute of Medical Science, The University of Tokyo, Department of Fine Morphology; **Figure 5–28b** Roger Craig, University of Massachusetts Medical School; **Figure 5–28c** See citation for Figure 5–27c; **Figure 5–29b,c** James E. Dennis/Phototake NYC.

CHAPTER 6 p. 190 (Buchner) Science Photo Library/Photo Researchers; (Sumner) Courtesy of the Division of Rare and Manuscript Collections, Carl A. Kroch Library, Cornell University, Ithaca, NY; (Haldane) AP Photo/Jacob Harris; **Figure 6–1** PDB ID 7GCH, Brady, K., Wei, A., Ringe, D., & Abeles, R.H. (1990) Structure of chymotrypsin-trifluoromethyl ketone inhibitor complexes: comparison of slowly and rapidly equilibrating inhibitors. *Biochemistry* 29, 7600; **Figure 6–4** PDB ID 1RA2, Sawaya, M.R. & Kraut, J. (1997) Loop and subdomain movements in the mechanism of *Escherichia coli* dihydrofolate reductase: crystallographic evidence. *Biochemistry* 36, 586; **p. 201** (Michaelis) Rockefeller University Archive Center; (Menten) Courtesy of Dorothy C. Craig; **Box 6–3 Figure 1** John Mansfield, University of Wisconsin–Madison, Department of Veterinary Science; **Figure 6–19b,c,d** PDB ID 7GCH, see citation for Figure 6–1; **Figure 6–25a** PDB ID 2YHX, Anderson, C.M., Stenkamp, R.E., & Steitz, T.A. (1978) Sequencing a protein by x-ray crystallography: II. refinement of yeast hexokinase B coordinates and sequence at 2.1 angstroms. *J. Mol. Biol.* 123, 15; **Figure 6–25b** PDB ID 1HKG, Steitz, T.A., Shoham, M., & Bennett, W.S., Jr. (1981) Structural dynamics of yeast hexokinase during catalysis. *Philos. Trans. R. Soc. London Ser. B* 293, 43; glucose (GLC) coordinates transformed from PDB ID 1GLK, St. Charles, R., Harrison, R.W., Bell, G.I., Pilkis, S.J., & Weber, I.T. (1994) Molecular model of human beta-cell glucokinase built by analogy to the crystal structure of yeast hexokinase B. *Diabetes* 43, 784; **Figure 6–26b** PDB ID 1ONE, Larsen, T.M., Wedekind, J.E., Rayment, I., & Reed, G.H. (1996) A carboxylate oxygen of the substrate bridges the magnesium ions at the active site of enolase: structure of the yeast enzyme complexed with the equilibrium mixture of 2-phosphoglycerate and phosphoenolpyruvate at 1.8 Å resolution. *Biochemistry* 35, 4349; **Figure 6–27a** PDB ID 1LZE, Maenaka, K., Matsushima, M., Song, H., Sunada, F.,

Watanabe, K., & Kumagai, I. (1995) Dissection of protein-carbohydrate interactions in mutant hen egg-white lysozyme complexes and their hydrolytic activity. *J. Mol. Biol.* 247, 281; **Figure 6–28a,b** PDB ID 1H6M, Vocadlo, D.J., Davies, G.J., Laine, R., & Withers, S.G. (2001) Catalysis by hen egg-white lysozyme proceeds via a covalent intermediate. *Nature* 412, 835; **Figure 6–33a** PDB ID 1RAB, Kosman, R.P., Gouaux, J.E., & Lipscomb, W.N. (1993) Crystal structure of CTP-ligated T state aspartate transcarbamoylase at 2.5 Å resolution: implications for ATCase mutants and the mechanism of negative cooperativity. *Proteins* 15, 147; **Figure 6–33b** PDB ID 1F1B, Jin, L., Stec, B., & Kantrowitz, E.R. (2000) A cis-proline to alanine mutant of *E. coli* aspartate transcarbamoylase: kinetic studies and three-dimensional crystal structures. *Biochemistry* 39, 8058; **Figure 6–39a** CNRI/Photo Researchers.

CHAPTER 7 Box 7–2 Figure 1 David Cook/blueshiftstudios/Alamy; **Box 7–2 Figure 2** Adapted from Assadi-Porter, F.M., Maillet, E.L., Radek, J.T., Quijaada, J., Markley, J.L., & Max, M. (2010) Key amino acid residues involved in multi-point binding interactions between brazzein, a sweet protein, and the T1R2–T1R3 human sweet receptor. *J. Mol. Biol.* 398, 584, Fig. 1; **Box 7–2 Figure 3** Adapted from www.elmhurst.edu/~chm/vchembook/549receptor. html, copyright Charles E. Ophardt, Elmhurst College; **Figure 7–15** Richard Howey; **Figure 7–16b** Leroy Somon/Visuals Unlimited; **Figure 7–18** Courtesy of H.-J. Gabius and Herbert Kaltner, University of Munich, from a figure provided by C.-W. von der Lieth, Heidelberg; **Figure 7–19b** PDB ID 1C58, Gessler, K., Uson, I., Takaha, T., Krauss, N., Smith, S.M., Okada, S., Sheldrick, G.M., & Saenger, W. (1999) V-Amylose at atomic resolution: x-ray structure of a cycloamylose with 26 glucose residues (cyclomaltohexaicosaose). *Proc. Natl. Acad. Sci. USA* 96, 4246; **Figure 7–22** PDB ID 1HPN, Mulloy, B., Forster, M.J., Jones, C., & Davies, D.B. (1993) N.m.r. and molecular-modelling studies of the solution conformation of heparin. *Biochem. J.* 293, 849; **Figure 7–23** PDB ID 1E0O, Pellegrini, L., Burke, D.F., von Delft, F., Mulloy, B., & Blundell, T.L. (2000) Crystal structure of fibroblast growth factor receptor ectodomain bound to ligand and heparin. *Nature* 407, 1029; **Figure 7–26a** Adapted from Häker, U., Nybakken, K., & Perrimon, N. (2005) Heparan sulphate proteoglycans: the sweet side of development. *Nat. Rev. Mol. Cell Biol.* 6, 532; **Figures 7–26b, 7–27** Adapted from Turnbull, J., Powell, A., & Guimond, S. (2001) Heparan sulfate: decoding a dynamic multifunctional cell regulator. *Trends Cell Biol.* 11, 75; **Figure 7–28 inset** Courtesy of Laurel Ng. Reprinted with permission from Ng, L., Grodzinsky, A., Patwari, P., Sandy, J., Plaas, A.H.K., & Ortiz, C. (2003) Individual cartilage aggrecan macromolecules and their constituent glycosaminoglycans visualized via atomic force microscopy. *J. Struct. Biol.* 143, 242–257, Fig. 7a left; **Figure 7–33b** PDB ID 2BAT, Varghese, J.N, McKimm-Breschkin, J.L., Caldwell, J.B., Kortt, A.A., & Colman, P.M. (1992) The structure of the complex between influenza virus neuraminidase and sialic acid, the viral receptor. *Proteins* 14, 327; **Figure 7–33c** PDB ID 2HU4, Russell, R.J., Haire, L.F., Stevens, D.J., Collins, P.J., Lin, Y.P., Blackburn, G.M., Hay, A.J., Gamblin, S.J., & Skehel, J.J. (2006) The structure of H5N1 avian influenza neuraminidase suggests new opportunities for drug design. *Nature* 443, 45; **Figure 7–33d** PDB ID 3CL0, Collins, P.J., Haire, L.F., Lin, Y.P., Liu, J., Russell, R.J., Walker, P.A., Skehel, J.J., Martin, S.R., Hay, A.J., & Gamblin, S.J. (2008) Crystal structures of oseltamivir-resistant influenza virus neuraminidase mutants. *Nature* 453, 1258; **Figure 7–34** R.M. Genta and D.Y. Graham, Veterans Affairs Medical Center, Houston, TX; **Figure 7–35a,b** PDB ID 1M6P, Roberts, D.L., Weix, D.J., Dahms, N.M., & Kim, J.J. (1998) Molecular basis of lysosomal enzyme recognition: three-dimensional structure of the cation-dependent mannose 6-phosphate receptor. *Cell* 93, 639; **Figure 7–36** Adapted from a figure provided by Dr. C.-W. von der Lieth, Heidelberg, in Gabius, H.-J. (2000) Biological information transfer beyond the genetic code: the sugar code. *Naturwissenschaften* 87, 108, Fig. 6; **Figure 7–37** Adapted from Sharon, N. & Lis, H. (1993) Carbohydrates in cell recognition. *Sci. Am.* 268 (January), 82; **Figure 7–39** Courtesy of Anne Dell. Reprinted with permission from Comelli, E.M., Head, S.R., Gilmartin, T., Whisenant, T., Haslam, S.M., North, S.J., Wong, N.-K., Kudo, T., Narimatsu, H., Esko, J.D., Drickamer, K., Dell, A., & Paulson, J.C. (2006) A focused microarray approach to functional glycomics: transcriptional regulation of the glycome. *Glycobiology* 16, 117, Fig. 3; **Figure 7–40** Adapted from Seeberger, P.H. (2009) Chemical glycobiology: Why now? *Nat. Chem. Biol.* 5, 368, Fig. 2a.

CHAPTER 8 p. 287 (Watson and Crick) Corbis/UPI/Bettmann; **Figure 8–12** Science Source/Photo Researchers; **p. 288** (Franklin) Science Photo Library/Photo Researchers; (Wilkins) Corbis/UPI/Bettmann; **Figures 8–13b,c, 8–17** Coordinates generated by Sybyl; **Figure 8–20b** PDB ID 1BCE, Asensio, J.L., Brown, T., & Lane, A.N. (1998) Comparison of the solution structures of intramolecular DNA triple helices containing adjacent and non-adjacent CG.C3 triplets. *Nucleic Acids Res.* 26, 3677; **Figure 8–20d** PDB ID 244D, Laughlan, G., Murchie, A.I., Norman, D.G., Moore, M.H., Moody, P.C., Lilley, D.M., & Luisi, B. (1994) The high-resolution crystal structure of a parallel-stranded guanine tetraplex. *Science* 265, 520; **Figure 8–22** Coordinates generated by Sybyl;

Figure 8–23b Modified from PDB ID 1GID, Cate, J.H., Gooding, A.R., Podell, E., Zhou, K., Golden, B.L., Kundrot, C.E., Cech, T.R., & Doudna, J.A. (1996) Crystal structure of a group I ribozyme domain: principles of RNA packing. *Science* 273, 1678; **Figure 8–24** James, B., Olsen, G.J., Liu, J., & Pace, N.R. (1988) The secondary structure of ribonuclease P RNA, the catalytic element of a ribonucleoprotein enzyme. *Cell* 52, 19; **Figure 8–25a** PDB ID 1TRA, Westhof, E. & Sundaralingam, M. (1986) Restrained refinement of the monoclinic form of yeast phenylalanine transfer RNA: temperature factors and dynamics, coordinated waters, and base-pair propeller twist angles. *Biochemistry* 25, 4868; **Figure 8–25b** PDB ID 1MME, Scott, W.G., Finch, J.T., & Klug, A. (1995) The crystal structure of an all-RNA hammerhead ribozyme: a proposed mechanism for RNA catalytic cleavage. *Cell* 81, 991; **Figure 8–25c** PDB ID 1GRZ, Golden, B.L., Gooding, A.R., Podell, E.R., & Cech, T.R. (1998) A preorganized active site in the crystal structure of the *Tetrahymena* ribozyme. *Science* 282, 259; **Figure 8–27b** Adapted from Marmur, J. & Doty, P. (1962) Determination of the base composition of deoxyribonucleic acid from its thermal denaturation temperature. *J. Mol. Biol.* 5, 109; **Figure 8–28** Ross B. Inman, University of Wisconsin–Madison, Department of Molecular Biology; **Figure 8–31b** PDB ID 1TTD, McAteer, K., Jing, Y., Kao, J., Taylor, J.S., & Kennedy, M.A. (1998) Solution-state structure of a DNA dodecamer duplex containing a Cis-syn thymine cyclbutane dimer, the major UV photoproduct of DNA. *J. Mol. Biol.* 282, 1013; **Figure 8–33c** Lloyd Smith, University of Wisconsin–Madison, Department of Chemistry; **Figure 8–34** Data provided by Lloyd Smith, University of Wisconsin–Madison, Department of Chemistry.

CHAPTER 9 p. 313 (Berg) Courtesy of Stanford Visual Art Services; (Boyer) Courtesy of Genentech, Inc.; (Cohen) Courtesy of Stanford Visual Art Services; **Figure 9–4** Elizabeth A. Wood, University of Wisconsin–Madison, Department of Biochemistry; **Figure 9–8** Courtesy of Rachel Britt, University of Wisconsin–Madison, Department of Biochemistry; **Figure 9–9b** Courtesy of Arthur McIntosh, University of Missouri–Columbia, Department of Entomology; **Figure 9–10c** Elizabeth A. Wood, University of Wisconsin–Madison, Department of Biochemistry; **Box 9–1 Figure 1** Courtesy of Carol Bingham, Promega Corporation; **Figure 9–15** Adapted from Wolfsberg, T.G., McEntyre, J., & Schuler, G.D. (2001) Guide to the draft human genome. *Nature* 409, 824, Fig. 1; **Figure 9–16a** PDB ID 1GFL, Yang, F., Moss, L.G., & Phillips, G.N., Jr. (1996) The molecular structure of green fluorescent protein. *Nat. Biotechnol.* 14, 1246; **Figure 9–16b** Courtesy of Roger Tsien, University of California, San Diego, Department of Pharmacology and Department of Chemistry & Biochemistry; **Figure 9–16c (left)** Courtesy of Penelope J. Brockie and Andres V. Maricq, University of Utah, Department of Biology; **(right)** Courtesy of Joseph A. Pogliano, from Pogliano, J., Ho, T.Q., Zhong, Z., & Helinski, D.R. (2001) Multicopy plasmids are clustered and localized in *Escherichia coli*. *Proc. Natl. Acad. Sci. USA* 98, 4486, Fig. 2A; **Figure 9–17b** Fuss, J. & Linn, S. (2002) Human DNA polymerase ε colocalizes with proliferating cell nuclear antigen and DNA replication late, but not early, in S phase. *J. Biol. Chem.* 277, 8658; **Figure 9–18** Courtesy of Kevin Strange and Michael Christensen, Vanderbilt University Medical Center, Department of Pharmacology; **Figure 9–24** Courtesy of Patrick O. Brown, Stanford University School of Medicine, Department of Biochemistry; **p. 339** (Collins) Alex Wong/Newsmakers; (Venter) Mike Theiler/Reuters; **Box 9–2 Figure 1** Data from the National Human Genome Research Institute; **Figure 9–26c** Courtesy of Illumina, Inc.; **Figure 9–27** Courtesy of Guy Plunkett III, University of Wisconsin–Madison, Genome Center of Wisconsin; **Figure 9–29a** Adapted from Gregory, T.R. (2005) Synergy between sequence and size in large-scale genomics. *Nat. Rev. Genet.* 6, 699; **Figure 9–29b** Adapted from data obtained from the PANTHER Classification System website at www.pantherdb.org/; **Figure 9–30** Adapted from International HapMap Consortium, The International HapMap Project (2003) *Nature* 426, 789; **Figure 9–31a** Adapted from Chen, C., Opazo, J.C., Erez, O., Uddin, M., Santolaya-Forgas, J., Goodman, M., Grossman, L.I., Romero, R., & Wildman, D.E. (2008) The human progesterone receptor shows evidence of adaptive evolution associated with its ability to act as a transcription factor. *Mol. Phylogenet. Evol.* 47, 637; **Figure 9–33** Adapted from Marques-Bone, T., Ryder, O.A., & Eichler, E.E. (2009) Sequencing primate genomes: what have we learned? *Annu. Rev. Genomics Hum. Genet.* 10, 355; **Figure 9–34a,b** Adapted from Schellenberg, G.D., Bird, T.D., Wijsman, E.M., Orr, H.T., Anderson, L., Nemens, E., White, J.A., Bonnycastle, L., Weber, J.L., Alonso, M.E., et al. (1992) Genetic linkage evidence for a familial Alzheimer's disease locus on chromosome 14. *Science* 258, 668; **Figure 9–34c** Adapted from Sherrington, R., Rogaev, E.I., Liang, Y., Rogaeva, E.A., Levesque, G., Ikeda, M., Chi, H., Lin, C., Li, G., Holman, K., et al. (1995) Cloning of a gene bearing missense mutations in early-onset familial Alzheimer's disease. *Nature* 375, 754; **Figure 9–35** Adapted from Stix, G. (2008) Traces of a distant past. *Sci. Am.* 299 (July), 56; **Box 9–3 Figure 2** Adapted from Noonan, J.P., Coop, G., Kudaravalli, S., Smith, D., Krause, J. Alessi, J., Chen, F., Platt, D., Pääbo, S., Pritchard, J.K., & Rubin, E.M. (2006) Sequencing and analysis of Neanderthal genomic DNA. *Science* 314, 1113.

CHAPTER 10 Figures 10–2a,b, 10–3 Coordinates from Sybyl; **Figure 10–4a** Dr. Alvin Telser/Visuals Unlimited, Inc.; **Figure 10–4b** Courtesy of Howard Goodman, Harvard Medical School, Department of Genetics; **Figure 10–6b** iStockphoto/Thinkstock; **p. 367** (Thudichum) From Drabkin, D.L. (1958) *Thudichum: Chemist of the Brain*, University of Pennsylvania Press, credited to Thudichum, J.L.W. (1898) *Briefe über öffentliche Gesundheitspflege: ihre bisherigen Leistungen und heutigen Aufgaben*, F. Pietzcker, Tübingen; **Box 10–1 Figure 2** Herbert A. Fischler, Isaac Albert Research Institute of the Kingsbrook Jewish Medical Center; **p. 371** (Vane, Bergström, and Samuelsson) Ira Wyman/Sygma/Corbis; **Figure 10–20b** Courtesy of Media Center, University of Wisconsin–Madison, Department of Biochemistry; **p. 375** (Dam) AP Photo/John Rooney; (Doisy) AP/Wide World Photos; **Figure 10–23** (cardinal) Dr. Dan Suda; (goldfinch) Richard Day/VIREO; **Figure 10–26** Christie, W.W. (1996) Beginners' guide to mass spectrometry of fatty acids: 2. general purpose derivatives. *Lipid Technol.* 8, 64.

CHAPTER 11 Figure 11–1 Don W. Fawcett/Photo Researchers; **Figure 11–5** Data from Zachowski, A. (1993) Phospholipids in animal eukaryotic membranes: transverse asymmetry and movement. *Biochem. J.* 294, 1; **Figure 11–6** Adapted from van Meer, G. & de Kroon, A.I.P.M. (2011) Lipid map of the mammalian cell. *J. Cell Sci.* 124, 5; **Figure 11–8** Adapted from Marchesi, V.T., Furthmayr, H., & Tomita, M. (1976) The red cell membrane. *Annu. Rev. Biochem.* 45, 667; **Figure 11–10** PDB ID 2AT9, Mitsuoka, K., Hirai, T., Murata, K., Miyazawa, A., Kidera, A., Kimura, Y., & Fujiyoshi, Y. (1999) The structure of bacteriorhodopsin at 3.0 Å resolution based on electron crystallography: implication of the charge distribution. *J. Mol. Biol.* 286, 861; **Figure 11–11a** PDB ID 2B6O, Gonen, T., Cheng, Y., Sliz, P., Hiroaki, Y., Fujiyoshi, Y., Harrison, S.C., & Walz, T. (2005) Lipid-protein interactions in double-layered two-dimensional AQP0 crystals. *Nature* 438, 633; **Figure 11–11b** PDB ID 2BL2, Murata, T., Yamato, I., Kakinuma, Y., Leslie, A.G., & Walker, J.E. (2005) Structure of the rotor of the V-type Na$^+$-ATPase from *Enterococcus hirae*. *Science* 308, 654; **Figure 11–13** PDB ID 1BL8, Doyle, D.A., Morais Cabral, J., Pfuetzner, R.A., Kuo, A., Gulbis, J.M., Cohen, S.L., Chait, B.T., & MacKinnon, R. (1998) The structure of the potassium channel: molecular basis of K$^+$ conduction and selectivity. *Science* 280, 69; PDB ID 1AF6, Wang, Y.F., Dutzler, R., Rizkallah, P.J., Rosenbusch, J.P., & Schirmer, T. (1997) Channel specificity: structural basis for sugar discrimination and differential flux rates in maltoporin. *J. Mol. Biol.* 272, 56; PDB ID 1QD5, Snijder, H.J., Ubarretxena-Belandia, I., Blaauw, M., Kalk, K.H., Verheij, H.M., Egmond, M.R., Dekker, N., & Dijkstra, B.W. (1999) Structural evidence for dimerization-regulated activation of an integral membrane phospholipase. *Nature* 401, 717; PDB ID 1QJ9, Vogt, J. & Schulz, G.E. (1999) The structure of the outer membrane protein OmpX from *Escherichia coli* reveals mechanisms of virulence. *Structure* 7, 1301; PDB ID 1PHO, Cowan, S.W., Schirmer, T., Rummel, G., Steiert, M., Ghosh, R., Pauptit, R.A., Jansonius, J.N., & Rosenbusch, J.P. (1992) Crystal structures explain functional properties of two *E. coli* porins. *Nature* 358, 727; **Figure 11–14** PDB ID 1FEP, Buchanan, S.K., Smith, B.S., Venkatramani, L., Xia, D., Esser, L., Palnitkar, M., Chakraborty, R., van der Helm, D., & Deisenhofer, J. (1999) Crystal structure of the outer membrane active transporter FepA from *Escherichia coli*. *Nat. Struct. Biol.* 6, 56; PDB ID 1QD5, Snijder, H.J., Ubarretxena-Belandia, I., Blaauw, M., Kalk, K.H., Verheij, H.M., Egmond, M.R., Dekker, N., & Dijkstra, B.W. (1999) Structural evidence for dimerization-regulated activation of an integral membrane phospholipase. *Nature* 401, 717; PDB ID 1MAL, Schirmer, T., Keller, T.A., Wang, Y.F., & Rosenbusch, J.P. (1995) Structural basis for sugar translocation through maltoporin channels at 3.1 Å resolution. *Science* 267, 512; **Figure 11–16** Heller, H., Schaefer, M., & Schulten, K. (1993) Molecular dynamics simulation of a bilayer of 200 lipids, in the gel and in the liquid-crystal phases. *J. Phys. Chem.* 97, 8343; **Figure 11–19** Courtesy of Takahiro Fujiwara, Ken Ritchie, Hideji Murakoshi, Ken Jacobson, and Akihiro Kusumi; **Figure 11–21b** Adapted from a micrograph courtesy of J. M. Richardson, University of Cambridge, Department of Pharmacology; **Figure 11–22a** Courtesy of R. G. Parton. Reprinted with permission from Parton, R.G. & Simons, K. (2007) The multiple faces of caveolae. *Nat. Rev. Mol. Cell Biol.* 8, 185, Fig. 1a; **Figure 11–24a,b** Adapted from Qualmann, B., Koch, D., & Manfred Kessels, M. (2011) Let's go bananas: revisiting the endocytic BAR code. *EMBO J.* 30, 3501, Fig. 1; **Figure 11–24c** Adapted from Peter, B.J., Kent, H.M, Mills, I.G., Vallis, Y., Butler, P.J.G., Evans, P.R., & McMahon, H.T. (2004) BAR domains as sensors of membrane curvature: the amphiphysin BAR structure. *Science* 303, 495, Fig. 1A; **Figure 11–25** Adapted from Chen, Y.A. & Scheller, R.H. (2001) SNARE-mediated membrane fusion. *Nature* 2, 98; **Figure 11–29** Adapted from Gadsby, D.C. (2009) Ion channels versus ion pumps: the principal difference, in principle. *Nat. Rev. Mol. Cell Biol.* 10, 344, Fig. 1; **Figure 11–30a,c** Adapted from Mueckler, M. (1994) Facilitative glucose transporters. *Eur. J. Biochem.* 219, 713; **Box 11–1 Figure 1** Adapted from Lienhard, F.E., Slot, J.W., James, D.E., & Mueckler, M.M. (1992) How cells absorb glucose. *Sci. Am.* 266 (January), 86; **Figure 11–36a** Adapted from Bublitz, M., Poulsen, H., Preben Morth, J., & Nissen, P. (2010) In and out of the cation pumps: P-type

ATPase structure revisited. *Curr. Opin. Struct. Biol.* 20, 431, Fig. 1; **Figure 11–36b** PDB ID 1SU4, Toyoshima, C., Nakasako, M., Nomura, H., & Ogawa, H. (2000) Crystal structure of the calcium pump of sarcoplasmic reticulum at 2.6 angstrom resolution. *Nature* 405, 647; **Figure 11–36c** PDB ID 3KDP, Preben Morth, J., Pedersen, B.P., Toustrup-Jensen, M.S., Sorensen, T.L., Petersen, J., Andersen, J.P., Vilsen, B., & Nissen, P. (2007) Crystal structure of the sodium-potassium pump. *Nature* 450, 1043; PDB ID 3B8C, Pedersen, B.P., Buch-Pedersen, M.J., Preben Morth, J., Palmgren, M.G., & Nissen, P. (2007) Crystal structure of the plasma membrane proton pump. *Nature* 450, 1111; derived from PDB ID 3IXZ, Abe, K., Tani, K., Nishizawa, T., & Fujiyoshi, Y. (2009) Inter-subunit interaction of gastric H$^+$, K$^+$-ATPase prevents reverse reaction of the transport cycle. *EMBO J.* 28, 1637, modeled following PDB ID 3B8E, see citation for PDB ID 3KDP; **p. 411** (Skou) Courtesy of Information Office, University of Aarhus, Denmark; **Figure 11–37** Adapted from Kühlbrandt, W. (2004) Biology, structure and mechanism of P-type ATPases. *Nat. Rev. Mol. Cell Biol.* 5, 291; **Figure 11–40a** PDB ID 3G60, Aller, S.G., Yu, J., Ward, A., Weng, Y., Chittaboina, S., Zhuo, R., Harrell, P.M., Trinh, Y.T., Zhang, Q., Urbatsch, I.L., & Chang, G. (2009) Structure of P-glycoprotein reveals a molecular basis for poly-specific drug binding. *Science* 323, 1718; **Figure 11–40b** PDB ID 1L7V, Locher, K.P., Lee, A.T., & Rees, D.C. (2002) The *E. coli* BtuCD structure: a framework for ABC transporter architecture and mechanism. *Science* 296, 1091; **Figure 11–41c** Adapted from Rees, D.C., Johnson, E., & Lewinson, O. (2009) ABC transporters: the power to change. *Nat. Rev. Mol. Cell Biol.* 10, 218, Fig. 1; **Box 11–2 Figure 2** Tom Moninger, University of Iowa, Ames; **Figure 11–42a** PDB ID 1PV7, Abramson, J., Smirnova, I., Kasho, V., Verner, G., Kaback, H.R., & Iwata, S. (2003) Structure and mechanism of the lactose permease of *Escherichia coli*. *Science* 301, 610; **Figure 11–42b** PDB ID 2CFQ, Mirza, O., Guan, L., Verner, G., Iwata, S., & Kaback, H.R. (2006) Structural evidence for induced fit and a mechanism for sugar/H$^+$ symport in LacY. *EMBO J.* 25, 1177; **Figure 11–44** Coordinates prepared for The Virtual Museum of Minerals and Molecules, www.soils.wisc.edu/virtual_museum, by Phillip Barak, University of Wisconsin–Madison, Department of Soil Science, using data from Neupert-Laves, K. & Dobler, M. (1975) The crystal structure of a K$^+$ complex of valinomycin. *Helv. Chim. Acta* 58, 432; **p. 418** (Agre) Courtesy of the Royal Swedish Academy of Sciences; **Figure 11–45a** PDB ID 2B5F, Tornroth-Horsefield, S., Wang, Y., Hedfalk, K., Johanson, U., Karlsson, M., Tajkhorshid, E., Neutze, R., & Kjellbom, P. (2006) Structural mechanism of plant aquaporin gating. *Nature* 439, 688; **Figure 11–45b** Adapted from PDB ID 1J4N, Sui, H., Han, B.-G., Lee, J.K., Walian, P., & Jap, B.K. (2001) Structural basis of water-specific transport through the AQP1 water channel. *Nature* 414, 872; **p. 421** (Neher) Courtesy Boettcher-Gajewski/Max Planck Institut für Biophysikalische Chemie; (Sakmann) Courtesy Max Planck Institut für Neurobiologie; **Figure 11–46** Witzemann, V., Schwarz, H., Koenen, M., Berberich, C., Villarroel, A., Wernig, A., Brenner, H.R., & Sakmann, B. (1996) Acetylcholine receptor ε-subunit deletion causes muscle weakness and atrophy in juvenile and adult mice. *Proc. Natl. Acad. Sci. USA* 93, 13,286; **p. 422** (MacKinnon) Courtesy of the Royal Swedish Academy of Sciences; **Figure 11–47a,b** PDB ID 1BL8, Doyle, D.A., Cabral, J.M., Pfuetzner, R.A., Kuo, A., Gulbis, J.M., Cohen, S.L., Chait, B.T., & Mackinnon, R. (1998) The structure of the potassium channel: molecular basis of K$^+$ conduction and selectivity. *Science* 280, 69; **Figure 11–47c** Adapted from Yellen, G. (2002) The voltage-gated potassium channels and their relatives. *Nature* 419, 37, and PDB ID 1J95, Zhou, M., Morais-Cabral, J.H., Mann, S., & MacKinnon, R. (2001) Potassium channel receptor site for the inactivation gate and quaternary amine inhibitors. *Nature* 411, 657; **Figure 11–48a,b,d** PDB ID 2A79, Long, S.B., Campbell, E.B., & MacKinnon, R. (2005) Crystal structure of a mammalian voltage-dependent Shaker family K$^+$ channel. *Science* 309, 897; **Figure 11–48c** Adapted from Gandhi, C.S. & Isacoff, E.Y. (2005) Shedding light on membrane proteins. *Trends Neurosci.* 28, 476.

CHAPTER 12 Figure 12–6b PDB ID 1U7E, Kim, C., Xuong, N.-H., & Taylor, S.S. (2005) Crystal structure of a complex between the catalytic and regulatory (RIα) subunits of PKA. *Science* 307, 690; **Box 12–2 Figure 1** (Gilman) Office of News and Publications, The University of Texas Southwestern Medical Center at Dallas; (Rodbell) Courtesy of Andrew Rodbell; **Box 12–2 Figure 2** PDB ID 5P21, Pai, E.F., Krengel, U., Petsko, G.A., Goody, R.S., Kabsch, W., & Wittinghofer, A. (1990) Refined crystal structure of the triphosphate conformation of H-ras p21 at 1.35 Å resolution: implications for the mechanism of GTP hydrolysis. *EMBO J.* 9, 2351; **Box 12–2 Figure 3** Adapted from Vetter, I.R. & Wittinghofer, A. (2001) The guanine nucleotide–binding switch in three dimensions. *Science* 294, 1300, Fig. 3; **Box 12–3 Figure 1** Chris Parks/ImageQuest Marine; **Box 12–3 Figure 2** Derived from PDB ID 1GFL, Yang, F., Moss, L.G., & Phillips, G.N., Jr. (1996) The molecular structure of green fluorescent protein. *Nat. Biotechnol.* 14, 1246; **Figure 12–11a** PDB ID 1CLL, Chattopadhyaya, R., Meador, W.E., Means, A.R., & Quiocho, F.A. (1992) A calmodulin structure refined at 1.7 angstroms resolution. *J. Mol. Biol.* 228, 1177; **Figure 12–11b,c** PDB ID 1CDL, Meador, W.E., Means, A.R., & Quiocho, F.A. (1992) Target enzyme recognition by calmodulin: 2.4 angstroms structure of a calmodulin-peptide complex. *Science* 257, 1251; **Figure 12–12a** Courtesy of Michael D. Cahalan, University of California, Irvine, Department of Physiology and Biophysics; **Figure 12–12b** Rooney, T.A., Sass, E.J., & Thomas, A.P. (1989) Characterization of cytosolic calcium oscillations induced by phenylephrine and vasopressin in single fura-2-loaded hepatocytes. *J. Biol. Chem.* 264, 17,131; **Figure 12–13a** PDB ID 3SN6, Rasmussen, S.G.F., DeVree, B.T., Zou, Y., Kruse, A.C., Chung, K.Y., Kobilka, T.S., Thian, F.S., Chae, P.S., Pardon, E., Calinski, D., et al. (2011) Crystal structure of the β_2 adrenergic receptor–Gs protein complex. *Nature* 477, 549; **Figure 12–13b** PDB ID 4DKL, Manglik, A., Kruse, A.C., Kobilka, T.S., Thian, F.S., Mathiesen, J.M., Sunahara, R.K., Pardo, L., Weis, W.I., Kobilka, B.K., & Granier, S. (2012) Crystal structure of the μ-opioid receptor bound to a morphinan antagonist. *Nature* 485, 321; **Figure 12–13c** PDB ID 3RZE, Shimamura, T., Shiroishi, M., Weyand, S., Tsujimoto, H., Winter, G., Katritch, V., Abagyan, R., Cherezov, V., Liu, W., Han, G.W., et al. (2011) Structure of the human histamine H1 receptor complex with doxepin. *Nature* 475, 65; **Figure 12–13d** PDB ID 3EML, Jaakola, V.P., Griffith, M.T., Hanson, M.A., Cherezov, V., Chien, E.Y., Lane, J.R. Ijzerman, A.P., & Stevens, R.C. (2008) The 2.6 angstrom crystal structure of a human A_{2A} adenosine receptor bound to an antagonist. *Science* 322, 1211; PDB ID 2VT4, Warne, A., Serrano-Vega, M.J., Baker, J.G., Moukhametzianov, R., Edwards, P.C., Henderson, R., Leslie, A.G.W., Tage, C.G., & Schertler, G.F.X. (2008) Structure of the β_1-adrenergic G protein–coupled receptor. *Nature* 454, 486; PDB ID 2RH1, Cherezov, V., Rosenbaum, D.M., Hanson, M.A., Rasmussen, S.G., Thian, F.S., Choi, H.J., Kuhn, P., Weis, W.I., Kobilka, B.K., & Stevens, R.C. (2007) High-resolution crystal structure of an engineered human β_2-adrenergic G protein–coupled receptor. *Science* 318, 1258; PDB ID 2Z73, Murakami, M. & Kouama, T. (2008) Crystal structure of squid rhodopsin. *Nature* 453, 363; PDB ID 1U19, Okada, T., Sugihara, M., Bondar, A.N., Eistner, M., Entel, P., & Buss, V. (2004) The retinal conformation and its environment in rhodopsin in light of a new 2.2 Å crystal structure. *J. Mol. Biol.* 342, 571; **Figure 12–14b** (insulin receptor) Derived from PDB ID 2DTG, McKern, N.M., Lawrence, M.C., Streltsov, V.A., Lou, M.-Z., Adams, T.E., Lovrecz, G.O., Elleman, T.C., Richards, K.M., Bentley, J.D., Pilling, P.A., et al. (2006) Structure of the insulin receptor ectodomain reveals a folded-over conformation. *Nature* 443, 218; (insulin) PDB ID 2CEU, Whittingham, J.L., Zhang, Y., Zakova, L., Dodson, E.J., Turkenburg, J.P., & Dodson, G.G. (2006) I222 crystal form of despentapeptide (B26-B30) insulin provides new insights into the properties of monomeric insulin. *Acta Crystallogr. D Biol. Crystallogr.* 62, 505; **Figure 12–14c** PDB ID 1IRK, Hubbard, S.R., Wei, L., Ellis, L., & Hendrickson, W.A. (1994) Crystal structure of the tyrosine kinase domain of the human insulin receptor. *Nature* 372, 746; **Figure 12–14d** PDB ID 1IR3, Hubbard, S.R. (1997) Crystal structure of the activated insulin receptor tyrosine kinase in complex with peptide substrate and ATP analog. *EMBO J.* 16, 5572; **Figure 12–21** PDB ID 1SHC, Zhou, M.M., Ravichandran, K.S., Olejniczak, E.F., Petros, A.M., Meadows, R.P., Sattler, M., Harlan, J.E., Wade, W.S., Burakoff, S.J., & Fesik, S.W. (1995) Structure and ligand recognition of the phosphotyrosine binding domain of Shc. *Nature* 378, 584; **Figure 12–23** Adapted from Pawson, T., Gish, G.D., & Nash, P. (2001) SH2 domains, interaction modules and cellular wiring. *Trends Cell Biol.* 11, 504, Fig. 5; **Figure 12–24** Adapted from Good, M.C., Zalatan, J.G., & Lim, W.A. (2011) Scaffold proteins: hubs for controlling the flow of cellular information. *Science* 332, 680, Fig. 2E; **Figure 12–27a,c,d** Adapted from Taylor, R. (1994) Evolutions: the voltage-gated sodium channel. *J. NIH Res.* 6, 112; **Figure 12–27b** PDB ID 3RW0, Payandeh, J., Scheuer, T., Zheng, N., & Catterall, W.A. (2011) The crystal structure of a voltage-gated Na^+ channel. *Nature* 475, 353; **Figure 12–28a,b** Adapted from Changeux, J.P. (1993) Chemical signaling in the brain. *Sci. Am.* 269 (November), 58; **Figure 12–28c,e** PDB ID 1UV6, Celie, P.H.N., Van Rossum-Fikkert, S.E., Van Dijk, W.J., Brejc, K., Smit, A.B., & Sixma, T.K. (2004) Nicotine and carbamylcholine binding to nicotinic acetylcholine receptors as studied in AchBP crystal structures. *Neuron* 41, 907; **Figure 12–29** Adapted from Shattil, S.J., Kim, C., & Ginsberg, M.H. (2010) The final steps of integrin activation: the end game. *Nat. Rev. Mol. Cell Biol.* 11, 288, Box 2; **Figure 12–34** Adapted from Ouaked, F., Rozhon, W., Lecourieux, D., & Hirt, H. (2003) A MAPK pathway mediates ethylene signaling in plants. *EMBO J.* 22, 1282; **Figure 12–35** Adapted from Tichtinsky, G., Vanoosthuyse, V., Cock, J.M., & Gaude, T. (2003) Making inroads into plant receptor kinase signalling pathways. *Trends Plant Sci.* 8, 231, Fig. 1; **Figure 12–38** PDB ID 1BAC, Chou, K.-C., Carlacci, L., Maggiora, G.M., Parodi, L.A., & Schulz, M.W. (1992) An energy-based approach to packing the 7-helix bundle of bacteriorhodopsin. *Protein Sci.* 1, 810; **Figure 12–40** Adapted from Nathans, J. (1989) The genes for color vision. *Sci. Am.* 260 (February), 42; **Box 12–4 Figure 1** Courtesy of Professor J. D. Mollon, Cambridge University, Department of Experimental Psychology; **Figure 12–45a** PDB ID 1HCK, Schulze-Gahmen, U., De Bondt, H.L., & Kim, S.-H. (1996) High-resolution crystal structures of human cyclin-dependent kinase 2 with and without ATP: bound waters and natural ligand as guides for inhibitor design. *J. Med. Chem.* 39, 4540; **Figure 12–45b** PDB ID 1FIN, Jeffrey, P.D., Russo, A.A., Polyak, K., Gibbs, E., Hurwitz, J., Massague, J., & Pavletich, N.P. (1995) Mechanism of Cdk activation revealed by the structure of a cyclin

a–Cdk2 complex. *Nature* 376, 313; **Figure 12–45c** PDB ID 1JST, Russo, A.A., Jeffrey, P.D., & Pavletich, N.P. (1996) Structural basis of cyclin-dependent kinase activation by phosphorylation. *Nat. Struct. Biol.* 3, 696; **Figure 12–46** Data from Pines, J. (1999) Four-dimensional control of the cell cycle. *Nat. Cell Biol.* 1, E73; **Box 12–5 Figure 1** CNRI/Photo Researchers; **Box 12–5 Figure 2** PDB ID 1S9I, Ohren, J.F., Chen, H., Pavlovsky, A., Whitehead, C., Zhang, E., Kuffa, P., Yan, C., McConnell, P., Spessard, C., Banotai, C., et al. (2004) Structures of human MAP kinase kinase 1 (MEK1) and MEK2 describe novel noncompetitive kinase inhibition. *Nat. Struct. Mol. Biol.* 11, 1192; **Box 12–5 Figure 3a** PDB ID 1IEP, Nagar, B., Bornmann, W., Pellicena, P., Schindler, T., Veach, D.R., Miller, W.T., Clarkson, B., & Kuriyan, J. (2002) Crystal structures of the kinase domain of c-Abl in complex with the small molecule inhibitors PD173955 and imatinib (STI-571). *Cancer Res.* 62, 4236; **Box 12–5 Figure 3b** PDB ID 1M17, Stamos, J., Sliwkowski, M.X., & Eigenbrot, C. (2002) Structure of the epidermal growth factor receptor kinase domain alone and in complex with a 4-anilinoquinazoline inhibitor. *J. Biol. Chem.* 277, 46,265; **Box 12–5 Figure 3c** PDB ID 1S9I, see citation for Box 12–5 Figure 2; **Box 12–5 Figure 3d** PDB ID 2A4L, De Azevedo, W.F., Leclerc, S., Meijer, L., Havlicek, L., Strnad, M., & Kim, S.H. (1997) Inhibition of cyclin-dependent kinases by purine analogues: crystal structure of human cdk2 complexed with roscovitine. *Eur. J. Biochem.* 243, 518; **Figure 12–51** Adapted from Markowitz, S.D. & Bertagnolli, M.M. (2009) Molecular basis of colorectal cancer. *N. Engl. J. Med.* 361, 2449, Fig. 2.

CHAPTER 13 p. 505 (Lavoisier) INTERFOTO/Alamy; **p. 506** © Sidney Harris; **Figure 13–7** Adapted from Layer, G., Heinz, D.W., Hahn, D., & Schubert, W.-D. (2004) Structure and function of radical SAM enzymes. *Curr. Opin. Chem. Biol.* 8, 472, Fig. 4; **Box 13–1** (firefly) Cathy Keifer/Fotolia; **Figure 13–25** PDB ID 3LDH, White, J.L., Hackert, M.L., Buehner, M., Adams, M.J., Ford, G.C., Lentz, P.J., Jr., Smiley, I.E., Steindel, S.J., & Rossmann, M.G. (1976) A comparison of the structures of apo dogfish M4 lactate dehydrogenase and its ternary complexes. *J. Mol. Biol.* 102, 759; **p. 535** (Strong, Elvehjem) Courtesy of University of Wisconsin–Madison, Department of Biochemistry; (Woolley) Rockefeller Archive Center.

CHAPTER 14 p. 544 (von Euler-Chelpin) Austrian Archives/Corbis; (Embden) Courtesy of Institut für Biochemie I: Molekulare Bioenergetik, Universitätsklinikum Frankfurt, ZBC; (Meyerhof) Hulton-Seutsch Collection/ Corbis; **p. 548** (Harden) Hulton Archives/Getty Images; (Young) Courtesy of Medical History Museum, The University of Melbourne; **p. 555** (Warburg) Hulton Archive/ Getty Images; **Box 14–1 Figure 3** ISM/Phototake; **Box 14–2** Fritz Prenzel/Animals Animals; **Box 14–3 Figure 1** Charles O'Rear/Corbis.

CHAPTER 15 Figure 15–1 www.genome.ad.jp/kegg/pathway/map/map01100. html; **Figures 15–3, 15–4** Adapted from Bennett, B.D., Kimball E.H., Gao, M., Osterhout, R., Van Dien, S., & Rabinowitz, J.D. (2009) Absolute metabolite concentrations and implied enzyme active site occupancy in *Escherichia coli*. *Nat. Chem. Biol.* 5, 593, Figs 1 and 2; **p. 596** (Buchner) The Nobel Foundation; **Figure 15–9** Data from Torres, N.V., Mateo, F., Melendez-Hevia, E., & Kacser, H. (1986) Kinetics of metabolic pathways: a system in vitro to study the control of flux. *Biochem. J.* 234, 169; **Box 15–1 Figure 2** Fell, D. (1997) *Understanding the Control of Metabolism*, Portland, London, p. 103; **Figure 15–16a** PDB ID 1PFK, Shirakihara, Y. & Evans, P.R. (1988) Crystal structure of the complex of phosphofructokinase from *Escherichia coli* with its reaction products. *J. Mol. Biol.* 204, 973; **Figure 15–20a** PDB ID 2NPP, Xu, Y., Xing, Y., Chen, Y., Chao, Y., Lin, Z., Fan, E., Yu, J.W., Strack, S., Jeffrey, P.D., & Shi, Y. (2006) Structure of the protein phosphatase 2A holoenzyme. *Cell* 127, 1239; **Figure 15–25** Adapted from Chakravarty, K., Cassuto, H., Reshef, L., & Hanson, R.W. (2005) *Crit. Rev. Biochem. Mol. Biol.* 40, 133, Fig. 2; **Figure 15–26** BCC Microimaging, reproduced with permission; **p. 615** (Leloir) AP Photo/John Lindsay; **Box 15–4** (Coris) AP; **Figure 15–34** PDB ID 1LL2, Gibbons, B.J., Roach, P.J., & Hurley, T.D. (2002) Crystal structure of the autocatalytic initiator of glycogen biosynthesis, glycogenin. *J. Mol. Biol.* 319, 463; **p. 621** (Sutherland) Case Western Reserve University School of Medicine/ National Institute of Health.

CHAPTER 16 p. 633 (Krebs) Keystone Pictures USA/Alamy; **Figure 16–5a,b** Courtesy of Dr. Z. Hong Zhou, University of Texas–Houston Medical School, Department of Pathology and Laboratory Medicine; **Figure 16–8a** PDB ID 5CSC, Liao, D.-I., Karpusas, M., & Remington, S.J. (1991) Crystal structure of an open conformation of citrate synthase from chicken heart at 2.8-Å resolution. *Biochemistry* 30, 6031; **Figure 16–8b** PDB ID 5CTS, Karpusas, M., Branchaud, B., & Remington, S.J. (1990) Proposed mechanism for the condensation reaction of citrate synthase: 1.9-Å structure of the ternary complex with oxaloacetate and carboxymethyl coenzyme A. *Biochemistry* 29, 2213; **Figure 16–9** Adapted from Remington, J.S. (1992) Mechanisms of citrate synthase and related enzymes (triose phosphate isomerase and mandelate racemase). *Curr. Opin. Struct. Biol.* 2, 730; **Box 16–1 Figure 1** Adapted from Eisenstein, R.S. (2000) Iron regulatory proteins and the molecular control of mammalian iron metabolism. *Annu. Rev. Nutr.* 20, 637, Fig. 1; **Box 16–1 Figure 2a** PDB ID 2B3Y, Dupuy, J., Volbeda, A., Carpentier,

P., Darnault, C., Moulis, J.M., & Fontecilla-Camps, J.C. (2006) Crystal structure of human iron regulatory protein 1 as cytosolic aconitase. *Structure* 14, 129; **Box 16–1 Figure 2b** PDB ID 2IPY, Walden, W.E., Selezneva, A.I., Dupuy, J., Volbeda, A., Fontecilla-Camps, J.C., Theil, E.C., & Volz, K. (2006) Structure of dual function iron regulatory protein 1 complexed with ferritin IRE-RNA. *Science* 314, 1903; **Figure 16–13b** PDB ID 1SCU, Wolodko, W.T., Fraser, M.E., James, M.N.G., & Bridger, W.A. (1994) The crystal structure of succinyl-CoA synthetase from *Escherichia coli* at 2.5-Å resolution. *J. Biol. Chem.* 269, 10,883; **Figure 16–23** Richard N. Trelease, Arizona State University, Department of Botany.

CHAPTER 17 Box 17–1 Stouffer Productions/Animals Animals; **Box 17–2** (Hodgkin) The Nobel Foundation.

CHAPTER 18 Figure 18–5c,d,e PDB ID 1AJS, Rhee, S., Silva, M.M., Hyde, C.C., Rogers, P.H., Metzler, C.M., Metzler, D.E., & Arnone, A. (1997) Refinement and comparisons of the crystal structures of pig cytosolic aspartate aminotransferase and its complex with 2-methylaspartate. *J. Biol. Chem.* 272, 17, 293.

CHAPTER 19 p. 732 (Lehninger) Alan Mason Chesney Medical Archives of The Johns Hopkins Medical Institutions; **Figure 19–2b** Quest/Photo Researchers; **Figure 19–2c** Dr. Donald Fawcett/Visuals Unlimited, Inc.; **Figure 19–5d** PDB ID 1FRD, Jacobson, B.L., Chae, Y.K., Markley, J.L., Rayment, I., & Holden, H.M. (1993) Molecular structure of the oxidized, recombinant, heterocyst (2Fe-2S) ferredoxin from Anabaena 7120 determined to 1.7 ångstroms resolution. *Biochemistry* 32, 6788; **Figure 19–9** PDB ID 3M9S, Efremov, R.G., Baradaran, R., & Sazanov, L.A. (2010) The architecture of respiratory complex I. *Nature* 465, 441; **Figure 19–10** PDB ID 1ZOY, Sun, F., Huo, X., Zhai, Y., Wang, A., Xu, J., Su, D., Bartlam, M., & Rao, Z. (2005) Crystal structure of mitochondrial respiratory membrane protein Complex II. *Cell* 121, 1043; **Figure 19–11** PDB ID 1BGY, Iwata, S., Lee, J.W., Okada, K., Lee, J.K., Iwata, M., Rasmussen, B., Link, T.A., Ramaswamy, S., & Jap, B.K. (1998) Complete structure of the 11-subunit bovine mitochondrial cytochrome bc_1 complex. *Science* 281, 64; **Figure 19–13** PDB ID 1OCC, Tsukihara, T., Aoyama, H., Yamashita, E., Tomizaki, T., Yamaguchi, H., Shinzawa-Itoh, K., Nakashima, R., Yaono, R., & Yoshikawa, S. (1996) The whole structure of the 13-subunit oxidized cytochrome *c* oxidase at 2.8 Å. *Science* 272, 1136; **Figure 19–14** Adapted from Williams, R.J.P. (1995) *Nature* 378, 235, a correction to Williams, R.J.P. (1995) Purpose of proton pathways. *Nature* 376, 643; **Figure 19–15a,b** Courtesy of Egbert Boekema. Reprinted with permission from Heinemeyer, J., Braun, H.-P., Boekema, E.J., & Kouřil, R. (2007) A structural model of the cytochrome *c* reductase/oxidase supercomplex from yeast mitochondria. *J. Biol. Chem.* 282, 12,240, Figs 4A and 5A; **Figure 19–16** PDB ID 3M9S, 1ZOY, 1BGY, and 1OCC; see citations for Figures 19–9, 19–10, 19–11, and 19–13; PDB ID 1HRC, Bushnell, G.W., Louie, G.V., & Brayer, G.D. (1990) High-resolution three-dimensional structure of horse heart cytochrome *c*. *J. Mol. Biol.* 214, 585; **Box 19–1 Figure 1** D. Cavagnaro/Visuals Unlimited; **p. 748** (Mitchell) AP/Wide World Photos; **p. 749** (Lardy) Courtesy of Department of Biochemistry, University of Wisconsin–Madison; **p. 750** (Racker) Courtesy of E. Racker; **Figure 19–23b** PDB ID 1BMF, Abrahams, J.P., Leslie, A.G., Lutter, R., & Walker, J.E. (1994) Structure at 2.8 Å resolution of F₁-ATPase from bovine heart mitochondria. *Nature* 370, 621; **p. 752** (Walker) Courtesy of Professor John E. Walker; (Boyer) AP Photo/Lacy Atkins; **Figure 19–25b** PDB ID 1BMF, see citation for Figure 19–23b; PDB ID 1JNV, Hausrath, A.C., Capaldi, R.A., & Matthews, B.W. (2001) The conformation of the ε- and γ-subunits within the *Escherichia coli* F₁ ATPase. *J. Biol. Chem.* 276, 47,227; **Figure 19–25c** PDB ID 1BMF, see citation for Figure 19–23b; PDB ID 1JNV, see citation for Figure 19–23b; PDB ID 2A7U, Wilkens, S., Borchardt, D., Weber, J., & Senior, A.E. (2005) Structural characterization of the interaction of the δ and α subunits of the *Escherichia coli* F₁F₀-ATP synthase by NMR spectroscopy. *Biochemistry* 44, 11,786; PDB ID 2CLY, Kane Dickson, V., Silvester, J.A., Fearnley, I.M., Leslie, A.G.W., & Walker, J.E. (2006) On the structure of the stator of the mitochondrial ATP synthase. *EMBO J.* 25, 2911; PDB ID 1C17, Rastogi, V.K. & Girvin, M.E. (1999) Structural changes linked to proton translocation by subunit c of the ATP synthase. *Nature* 402, 263; PDB ID 1B9U, Dmitriev, O., Jones, P.C., Jiang, W., & Fillingame, R.H. (1999) Structure of the membrane domain of subunit b of the *Escherichia coli* F₀F₁ ATP synthase. *J. Biol. Chem.* 274, 15,598; PDB ID 1YCE, Meier, T., Polzer, P., Diederichs, K., Welte, W., & Dimroth, P. (2005) Structure of the rotor ring of F-type Na⁺-ATPase from *Ilyobacter tartaricus*. *Science* 308, 659; **Figure 19–25d** PDB ID 1C17 and PDB ID 1YCE, see citations for Figure 19–25c; **Figure 19–27 (left)** Adapted from Sambongi, Y., Iko, Y., Tanabe, M., Omote, H., Iwamoto-Kihara, A., Ueda, I., Yanagida, T., Wada, Y., & Futai, M. (1999) Mechanical rotation of the c subunit oligomer in ATP synthase (F₀F₁): direct observation. *Science* 286, 1722–1724; **Figure 19–27 (right)** Courtesy of Ryohei Yasuda and Kazuhiko Kinosita, from Yasuda, R., Noji, H., Kinosita, K., Jr., & Yoshida, M. (1998) F₁-ATPase is a highly efficient molecular motor that rotates with discrete 120° steps. *Cell* 93, 1117; **Figures 19–28, 19–29a,b,c,d, 19–33** PDB ID 1OHH, Cabezon, E., Montgomery, M.G., Leslie, A.G.W., & Walker, J.E.

(2003) The structure of bovine F1-ATPase in complex with its regulatory protein IF1. *Nat. Struct. Biol.* 10, 744; **Figure 19–34** Adapted from Harris, D.A. (1995) *Bioenergetics at a Glance,* Blackwell Science, London, p. 36; **Figure 19–37** Don W. Fawcett/Photo Researchers; **Figure 19–39** Adapted from Riedl, S.J. & Salvesen, G.S. (2007) The apoptosome: signaling platform of cell death. *Nat. Rev. Mol. Cell Biol.* 8, 409, Fig. 3; **Figure 19–40a** Morris, M.A. (1990) Mitomutations in neuro-ophthalmological diseases: a review. *J. Clin. Neuroophthalmol.* 10, 159; **Figure 19–40b** From Wallace, D., Zheng, X., Lott, M.T., Shoffner, J.M., Hodge, J.A., Kelley, R.I., Epstein, C.M., & Hopkins, L.C. (1988) Familial mitochondrial encephalomyopathy (MERRF): genetic, pathophysiological, and biochemical characterization of a mitochondrial DNA disease. *Cell* 55, 601; **Figure 19–42** Michael W. Davidson, Florida State University; **Figure 19–43b** Courtesy of Rob Taylor. Reprinted with permission from Taylor, R.W. & Turnbull, D.M. (2005) Mitochondrial DNA mutations in human disease. *Nat. Rev. Genet.* 6, 389, Fig. 2a; **Figure 19–47b** Biological Photo Service; **Figure 19–51** PDB ID 2BHW, Standfuss, J., Terwissscha van Scheltinga, A.C., Lamborghini, M., & Kuhlbrandt, W. (2005) Mechanisms of photoprotection and nonphotochemical quenching in pea light-harvesting complex at 2.5 Å resolution. *EMBO J.* 24, 919; **Figures 19–52, 19–56a,b, 19–62b, 19–63, 19–68** Adapted from Heldt, H.-W. (1997) *Plant Biochemistry and Molecular Biology,* Oxford University Press, Oxford, pp. 57, 62, 63, 100, 101, 133; **Figure 19–57** PDB ID 1PRC, Deisenhofer, J., Epp, O., Sinning, I., & Michel, H. (1995) Crystallographic refinement at 2.3 angstroms resolution and refined model of the photosynthetic reaction center from *Rhodopseudomonas viridis.* *J. Mol. Biol.* 246, 429; **Figure 19–59** Adapted from Rutherford, A.W. & Faller, P. (2001) The heart of photosynthesis in glorious 3D. *Trends Biochem. Sci.* 26, 341, Fig. 1; **Figure 19–60a** Adapted from Kuhlbrandt, W. (2001) Structural biology: chlorophylls galore. *Nature* 411, 896, Fig. 1; **Figure 19–60b,c** PDB ID 1JBO, Nield, J., Rizkallah, P.J., Barber, J., & Chayen, N.E. (2003) The 1.45Å three-dimensional structure of c-phycocyanin from the thermophilic cyanobacterium *Synechococcus elongatus.* *J. Struct. Biol.* 141, 149; **Figure 19–61a,b** PDB ID 1FV5, Kurisu, G., Zhang, H., Smith, J.L., & Cramer, W.A. (2003) Structure of the cytochrome b6f complex of oxygenic photosynthesis: tuning the cavity. *Science* 302, 1009; **Figure 19–62a** PDB ID 2AXT, Loll, B., Kern, J., Saenger, W., Zouni, A., & Biesiadka, J. (2005) Towards complete cofactor arrangement in the 3.0 Å resolution structure of photosystem II. *Nature* 438, 1040; PDB ID 2E74, Yamashita, E., Zhang, H., & Cramer, W.A. (2007) Structure of the cytochrome b6f complex: quinone analogue inhibitors as ligands of heme c_n. *J. Mol. Biol.* 370, 39; PDB ID 1A70, Binda, C., Coda, A., Aliverti, A., Zanetti, G., & Mattevi, A. (1998) Structure of the mutant E92K of [2Fe-2S] ferredoxin I from *Spinacia oleracea* at 1.7 Å resolution. *Acta Crystallogr. D Biol. Crystallogr.* 54, 1353; PDB ID 1AG6, Xue, Y., Okvist, M., Hansson, O., & Young, S. (1998) Crystal structure of spinach plastocyanin at 1.7 Å resolution. *Protein Sci.* 7, 2099; PDB ID 2O01, Amunts, A., Drory, O., & Nelson, N. (2007) The structure of a plant photosystem I supercomplex at 3.4 Å resolution. *Nature* 447, 58; PDB ID 1QG0, Deng, Z., Aliverti, A., Zanetti, G., Arakaki, A.K., Ottado, J., Orellano, E.G., Calcaterra, N.B., Ceccarelli, E.A., Carrillo, N., & Karplus, P.A. (1999) A productive NADP⁺ binding mode of ferredoxin-NADP⁺ reductase revealed by protein engineering and crystallographic studies. *Nat. Struct. Biol.* 6, 847; PDB ID 1QO1, see citation for Figure 19–25d; **Figure 19–64b** PDB ID 3ARC, Umena, Y., Kawakami, K., Shen, J.-R., & Kamiya, N. (2011) Crystal structure of oxygen-evolving photosystem II at a resolution of 1.9 Å. *Nature* 473, 55; **p. 786** (Arnon) University of California, Berkeley; (Jagendorf) Cornell University; **Figure 19–67** Miller, S.R., Augustine, S., Le Olson, T., Blankenship, R.E., Selker, J., & Wood, A.M. (2005) Discovery of a free-living chlorophyll *d*–producing cyanobacterium with a hybrid proteobacterial/cyanobacterial small-subunit rRNA gene. *Proc. Natl. Acad. Sci. USA* 102, 850, Fig. 2; **Figure 19–69a** PDB ID 1C8R, Luecke, H., Schobert, B., Richter, H.-T., Cartailler, J.-P., & Lanyi, J.K. (1999) Structural changes in bacteriorhodopsin during ion transport at 2 Å resolution. *Science* 286, 255; **Figure 19–69b** Adapted from Gennis, R.B. & Ebrey, T.G. (1999) Proton pump caught in the act. *Science* 286, 252.

CHAPTER 20 Figure 20–1 PhotoDisc; **p. 800** (Calvin) Ted Spiegel/Corbis; **Figure 20–2** Ken Wagner/Visuals Unlimited; **Figure 20–5a** PDB ID 8RUC, Andersson, I. (1996) Large structures at high resolution: the 1.6 Å crystal structure of spinach ribulose-1,5-bisphosphate carboxylase/oxygenase complexed with 2-carboxyarabinitol bisphosphate. *J. Mol. Biol.* 259, 160; **Figure 20–5b** PDB ID 9RUB, Lundqvist, T. & Schneider, G. (1991) Crystal structure of activated ribulose-1,5-bisphosphate carboxylase complexed with its substrate, ribulose-1,5-bisphosphate. *J. Biol. Chem.* 266, 12,604; **Figure 20–6** PDB ID 1RCX, Taylor, T.C. & Andersson, I. (1997) The structure of the complex between rubisco and its natural substrate ribulose 1,5-bisphosphate. *J. Mol. Biol.* 265, 432; **Figure 20–18** Halliwell, B. (1984) *Chloroplast Metabolism: The Structure and Function of Chloroplasts in Green Leaf Cells,* Clarendon Press, Oxford, p. 97; **Figure 20–23a** Ray Evert, University of Wisconsin–Madison, Department of Botany; **Box 20–1 Figure 2** Adapted from Jansson, C., Wullschleger, S.D, Kalluri, U.C., & Tuskan, G.A. (2010) Phytosequestration:

carbon biosequestration by plants and the prospects of genetic engineering. *BioScience* 60, 683, Fig. 1; **Figure 20–28** (cellulose) Ken Wagner/Visuals Unlimited; art adapted from Becker, W. M. & Deamer, D.W. (1991) *The World of the Cell*, 2nd edn, The Benjamin/Cummings Publishing Company, Inc., Redwood City, CA, p. 60, Fig. 3–20; **Figure 20–29 inset** Courtesy of Mark J. Grimson, Texas Tech University, and Candace H. Haigler, North Carolina State University; **p. 830 (problem 19)** Courtesy of Elena V. Voznesenskaya, Vincent R. Franceschi, Olavi Kiirats, Elena G. Artyusheva, Helmut Freitag, and Gerald E. Edwards.

CHAPTER 21 Figure 21–3a PDB ID 2CF2, Maier, T., Jenni, S., & Ban, N. (2006) Architecture of mammalian fatty acid synthase at 4.5 Å resolution. *Science* 311, 1258; **Figure 21–3b** PDB IDs 2UV9, 2UVA, 2UVB, and 2UVC, Jenni, S., Leibundgut, M., Boehringer, D., Frick, C., Mikolasek, B., & Ban, N. (2007) Structure of fungal fatty acid synthase and implications for iterative substrate shuttling. *Science* 316, 254; **Figure 21–11b** Daniel Lane, The Johns Hopkins University, School of Medicine; **p. 853** (Kennedy) Harvard Medical School; **Figure 21–28** Adapted from Carman, G.M. & Han, G.-S. (2011) Regulation of phospholipid synthesis in the yeast *Saccharomyces cerevisiae. Annu. Rev. Biochem.* 80, 859, Fig. 2; **p. 863** (Bloch, Lynen, Cornforth) AP/Wide World Photos; (Popják) *Arterioscler. Thromb. Vasc. Biol.* 19, 830–831, 1999, ©1999 Wolters Kluwer Health; **Figure 21–39a** ApoB-100 model from Johs, A., Hammel, M., Waldner, I., May, R.P., Laggner, P., & Prassl, R. (2006) Modular structure of solubilized human apolipoprotein B-100: low resolution model revealed by small angle neutron scattering. *J. Biol. Chem.* 281, 19,732; **Figure 21–39b** Courtesy of Robert L. Hamilton and the Arteriosclerosis Specialized Center of Research, University of California, San Francisco; **p. 868** (Brown and Goldstein) Courtesy of Michael Brown and Joseph Goldstein, University of Texas Southwestern Medical Center; **Figure 21–44** Adapted from Raghow, R., Yellaturu, C., Deng, X., Park, E.A., & Elam, M.B. (2008) SREBPs: the crossroads of physiological and pathological lipid homeostasis. *Trends Endocrinol. Metab.* 19, 65, Fig. 2; **Figure 21–45** Adapted from Calkin, A.C. & Tontonoz, P. (2012) Transcriptional integration of metabolism by the nuclear sterol-activated receptors LXR and FXR. *Nat. Rev. Mol. Cell Biol.* 13, 213, Fig. 1; **Figure 21–46** Adapted from Maxfield, F.R. & Tabas, I. (2005) Role of cholesterol and lipid organization in disease. *Nature* 438, 612, Fig. 3; **Box 21–3** (Endo) Courtesy of Akira Endo, Ph.D.; (Alberts) Courtesy of Alfred W. Alberts; (Vagelos) Courtesy of P. Roy Vagelos. **Figure 21–47** Adapted from Tall, A.R., Yvan-Charvet, L., Terasaka, N., Pagler, T., & Wang, N. (2008) HDL, ABC transporters, and cholesterol efflux: implications for the treatment of atherosclerosis. *Cell Metab.* 7, 365, Fig. 1.

CHAPTER 22 Box 22–1 Figures 1, 2 Adapted from van Niftrik, L.A., Fuerst, J.A., Damsté, J.S.S., Kuenen, J.G., Jetten, M.S.M., & Strous, M. (2004) The anammoxosome: an intracytoplasmic compartment in anammox bacteria. *FEMS Microbiol. Lett.* 233, 10, Figs 4 and 3; **Box 22–1 Figure 3** Courtesy of John Fuerst. Reprinted with permission from Lindsay, M.R., Webb, R.I., Strous., M., Jetten, M.S., Butler, M.K., Forde, R.J., & Fuerst, J.A. (2001) Cell compartmentalisation in planctomycetes: novel types of structural organisation for the bacterial cell. *Arch. Microbiol.* 175, 421, Fig. 6A; **Figure 22–3** PDB ID 1FP6, Jang, S.B., Seefeldt, L.C., & Peters, J.W. (2000) Insights into nucleotide signal transduction in nitrogenase: structure of an iron protein with MgADP bound. *Biochemistry* 39, 14,745; PDB ID 1M1N, Einsle, O., Tezcan, F.A., Andrade, S.L., Schmid, B., Yoshida, M., Howard, J.B., & Rees, D.C. (2002) *Science* 297, 1696; **Figure 22–5** Adapted from Seefeldt, L.C., Hoffman, B.M., & Dean, D.R. (2009) Mechanism of Mo-dependent nitrogenase. *Annu. Rev. Biochem.* 78, 701, Fig. 9; **Figure 22–6a (including inset)** Wally Eberhart/Visuals Unlimited; **Figure 22–6b** Jeremy Burgess/Photo Researchers; **Figure 22–7** PDB ID 2GLS, Yamashita, M.M., Almassy, R.J., Janson, C.A., Cascio, D., & Eisenberg, D. (1989) Refined atomic model of glutamine synthetase at 3.5 Å resolution. *J. Biol. Chem.* 264, 17,681; **Figure 22–20b** PDB ID 1KFJ, Kulik, V., Weyand, M., Seidel, R., Niks, D., Arac, D., Dunn, M.F., & Schlichting, I. (2002) On the role of αThr183 in the allosteric regulation and catalytic mechanism of tryptophan synthase. *J. Mol. Biol.* 324, 677; **p. 912** (Buchanan) Courtesy of Massachusetts Institute of Technology Museum Collection; **Figure 22–39** PDB ID 1M6V, Thoden, J.B., Huang, X., Raushel, F.M., & Holden, H.M. (2002) Carbamoyl-phosphate synthetase: creation of an escape route for ammonia. *J. Biol. Chem.* 277, 39,722; Jim Thoden and Hazel Holden, University of Wisconsin–Madison, Department of Biochemistry and Enzyme Institute, provided preliminary data for the channel path; **Figure 22–42a** Thelander, L. & Reichard, P. (1979) Reduction of ribonucleotides. *Annu. Rev. Biochem.* 48, 133; **Figures 22–42b,c, 22–45** PDB ID 3UUS, Ando, N., Brignole, E.J., Zimanyi, C.M., Funk, M.A., Yokoyama, K., Asturias, F.J., Stubbe, J., & Drennan, C.L. (2001) Structural interconversions modulate activity of *Escherichia coli* ribonucleotide reductase. *Proc. Natl. Acad. Sci. USA* 108, 21,046; **p. 923** (Elion and Hitchings) Courtesy of Kathy Bendo Hitchings.

CHAPTER 23 Box 23–1 Figure 1 Allen, F.N. & Sherrill, J.W. (1922) Clinical observations with insulin. 1. The use of insulin in diabetic treatment. *J. Metabol.*

Res. II, 804. Photo courtesy of Ebling Library, University of Wisconsin–Madison; **Figure 23–15c,d** Dr. Fred Hossler/Visuals Unlimited, Inc.; **Figure 23–16b** Christensen, C.R., Clark, P.B., & Morton, K.A. (2006) Reversal of hypermetabolic brown adipose tissue in F-18 FDG PET imaging. *Clin. Nucl. Med.* 31, 193, Fig. 2. © Wolters Kluwer Health; **Box 23–2 Figure 1** Adapted from Schlattner, U., Tokarska-Schlattner, M., & Wallimann, T. (2006) Mitochondrial creatine kinase in human health and disease. *Biochim. Biophys. Acta* 1762, 164, Fig. 1; **Box 23–2 Figure 3** Photodisc/Getty Images; **Figure 23–18** Reprinted with permission from Blei, M.L., Conley, K.E., & Kushmerick, M.J. (1993) Separate measures of ATP utilization and recovery in human skeletal muscle. *J. Physiol.* 465, 210, Fig. 4; **Figure 23–20** D. W. Fawcett/Photo Researchers; **Figure 23–22** Courtesy of M. L. Thomas, H. C. Sing, G. Belenky, Walter Reed Army Institute of Research, U.S. Army Medical Research Materiel Command, Division of Neuropsychiatry; **Figure 23–28b** Coordinates courtesy of Frances M. Ashcroft, Oxford University, used with permission of S. Haider and M. S. P. Sansom to re-create a model published in Antcliff, J.F., Haider, S., Proks. P., Sansom, M.S.P., & Ashcroft, F.M. (2005) Functional analysis of a structural model of the ATP-binding site of the K_{ATP} channel Kir6.2 subunit. *EMBO J.* 24, 229; **Figure 23–31** Adapted from Cahill, G.F., Jr. (2006) Fuel metabolism in starvation. *Annu. Rev. Nutr.* 26, 1, Fig. 2; **Figure 23–33** John Sholtis, The Rockefeller University, New York; **Figure 23–34** Adapted from Ezzell, C. (1995) Fat times for obesity research: tons of new information, but how does it all fit together? *J. NIH Res.* 7, 39; **Figure 23–36** Adapted from Auwerx, J. & Staels, B. (1998) Leptin. *Lancet* 351, 737; **Figure 23–39** Adapted from http://web.indstate.edu/thcme/mwking/ampk.html and Steinberg, G.R. & Kemp, B.E. (2007) Adiponectin: starving for attention. *Cell Metab.* 6, 4, Fig. 1; **Figure 23–40** Adapted from Yecies, J.L. & Manning, B.D. (2011) mTOR links oncogenic signaling to tumor cell metabolism. *J. Mol. Med.* 89, 221, Fig. 2; **Figure 23–41** Adapted from Evans, R.M., Barish, G.D., & Wang, Y.-X. (2004) PPARs and the complex journey to obesity. *Nat. Med.* 10, 355, Fig. 3; **Figure 23–43a,c** Adapted from Cummings, D.E., Purnell, J.Q., Frayo, R.S., Schmidova, K., Wisse, B.E., & Weigle, D.S. (2001) A preprandial rise in plasma ghrelin levels suggests a role in meal initiation in humans. *Diabetes* 50, 1714, Fig. 1; **Figure 23–43b** Adapted from Feher, M.D. & Bailey, C.J. (2004) Reclassifying insulins. *Br. J. Diabet. Vasc. Dis.* 4, 39; **Figure 23–44** Adapted from Guilherme, A., Virbasius, J.V., Puri, V., & Czech, M.P. (2008) Adipocyte dysfunctions linking obesity to insulin resistance and type 2 diabetes. *Mol. Cell Biol.* 9, 367, Fig. 1.

CHAPTER 24 Figure 24–1 From Kleinschmidt, A.K., Land, D., Jackerts, D., & Zahn, R.K. (1962) Darstellung und Längenmessungen des gesamten Desoxyribonucleinsäure-Inhalter von T2-Bakteriophagen. *Biochim. Biophys. Acta* 61, 857; **p. 980** (Beadle) Archive Photos; (Tatum) Corbis/UPI/Bettmann; **Figure 24–4** Huntington Potter and David Dressler, Harvard Medical School, Department of Neurobiology; **Figure 24–5a** G. F. Bahr/Biological Photo Service; **Figure 24–5b** Michael M. Cox; **Figure 24–6** D. W. Fawcett/Photo Researchers; **Figure 24–10** Adapted from Cozzarelli, N.R., Boles, T.C., & White, J.H. (1990) Primer on the topology and geometry of DNA supercoiling. In *DNA Topology and Its Biological Effects* (Cozzarelli, N.R. & Wang, J.C., eds.), Cold Spring Harbor Laboratory Press, Cold Spring Harbor, NY, pp. 139–184; **Figure 24–11a** Adapted from Saenger, W. (1984) *Principles of Nucleic Acid Structure*, Springer-Verlag, New York, p. 452; **Figure 24–12** Laurien Polder, from Kornberg, A. (1980) *DNA Replication*, W. H. Freeman & Company, New York, p. 29; **Figures 24–13, 24–14** See citation for Figure 24–10; **Figure 24–19** Keller, W. (1975) Characterization of purified DNA-relaxing enzyme from human tissue culture cells. *Proc. Natl. Acad. Sci. USA* 72, 2553; **Figures 24–20, 24–21** Adapted from Champoux, J.J. (2001) DNA topoisomerases: structure, function, and mechanism. *Annu. Rev. Biochem.* 70, 369, Figs 3, 11; **Figure 24–22a** James H. White, T. Christian Boles, and N. R. Cozzarelli, University of California, Berkeley, Department of Molecular and Cell Biology; **Figure 24–25b** Ada L. Olins and Donald E. Olins, Oak Ridge National Laboratory; **Figure 24–26** PDB ID 1AOI, Luger, K., Maeder, A.W., Richmond, R.K., Sargent, D.F., & Richmond, T.J. (1997) Crystal structure of the nucleosome core particle at 2.8 Å resolution. *Nature* 389, 251; **Box 24–2 Figure 1** Adapted from Sarma, K. & Reinberg, D. (2005) Histone variants meet their match. *Nat. Rev. Mol. Cell Biol.* 6, 140; **Box 24–2 Figure 2b** Data courtesy of Steve Henikoff. Reprinted with permission from Mito, Y., Henikoff, J.G., & Henikoff, S. (2005) Genome-scale profiling of histone H3.3 replacement patterns. *Nat. Genet.* 37, 1092; **Figure 24–28** See citation for Figure 24–26; **Figure 24–29a** Barbara Hamkalo, University of California, Irvine, Department of Molecular Biology and Biochemistry; **Figure 24–30a** G. F. Bahr/Biological Photo Service; **Figure 24–30b** D. W. Fawcett/Visuals Unlimited; **Figure 24–30c** Laemmli, U.K., Cheng, S.M., Adolph, K.W., Paulson, J.R., Brown, J.A., & Baumbach, W.R. (1978) Metaphase chromosome structure: the role of nonhistone proteins. *Cold Spring Harb. Symp. Quant. Biol.* 42, 351. © Cold Spring Harbor Laboratory Press.; **Figure 24–31** Photo from G. F. Bahr/Biological Photo Service; **Figures 24–32, 24–33** Adapted from Hirano, T. (2006) *Nat. Rev. Mol. Cell. Biol.* 7, 311, Figs 1, 6; **Figure 24–32d** Courtesy of Harold P. Erickson, Johns Hopkins University, Department of Cell Biology; **Figure 24–34**

CHAPTER 28 p. 1159 (Jacob, Monod) Corbis/Bettmann; **Figure 28–8c** PDB ID 2PE5, Daber, R., Stayrook, S., Rosenberg, A., & Lewis, M. (2007) Structural analysis of lac repressor bound to allosteric effectors. *J. Mol. Biol.* 370, 609; **Figure 28–9** Adapted from Huret, J.L. (2006) DNA: molecular structure. *Atlas Genet. Cytogenet. Oncol. Haematol.,* http://atlasgeneticsoncology.org/Educ/DNAEngID30001ES.html; **Figure 28–11** PDB ID 2PE5, see citation for Figure 28–8c; **Figure 28–12** PDB ID 1ZAA, Pavletich, N.P. & Pabo, C.O. (1991) Zinc finger-DNA recognition: crystal structure of a Zif268-DNA complex at 2.1 Å. *Science* 252, 809; **Figure 28–13** PDB ID 1FJL, Wilson, D.S., Guenther, B., Desplan, C., & Kuriyan, J. (1995) High resolution crystal structure of a paired (Pax) class cooperative homeodomain dimer on DNA. *Cell* 82, 709; **Figure 28–14a** McKnight, S.L. (1991) Molecular zippers in gene regulation. *Sci. Am.* 264 (April), 54–64; **Figure 28–14b** PDB ID 1YSA, Ellenberger, T.E., Brandl, C.J., Struhl, K., & Harrison, S.C. (1992) The GCN4 basic region leucine zipper binds DNA as a dimer of uninterrupted α helices: crystal structure of the protein-DNA complex. *Cell* 71, 1223; **Figure 28–15** PDB ID 1HLO, Brownlie, P., Ceska, T.A., Lamers, M., Romier, C., Theo, H., & Suck, D. (1997) The crystal structure of an intact human max-DNA complex: new insights into mechanisms of transcriptional control. *Structure* 5, 509; **Figure 28–16** PDB ID 1RUN, Parkinson, G., Gunasekera, A., Vojtechovsky, J., Zhang, X., Kunkel, T.A., Berman, H., & Ebright, R.H. (1996) Aromatic hydrogen bond in sequence-specific protein-DNA recognition. *Nat. Struct. Biol.* 3, 837; **Figure 28–19a** Watson, J.D., Hopkins, N.H., Roberts, J.W., Steitz, J.A., & Weiner, A.M. (1987) *Molecular Biology of the Gene,* 4th edn, The Benjamin/Cummings Publishing Company, Menlo Park, CA, p. 487; **Figure 28–21** Adapted from Nomura, M., Gourse, R., & Baughman, G. (1984) Regulation of the synthesis of ribosomes and ribosomal components. *Annu. Rev. Biochem.* 53, 75; **Figure 28–23** Adapted from Szymański, M. & Barciszewski, J. (2002) Beyond the proteome: non-coding regulatory RNAs. *Genome Biol.* 3, 6; **Figure 28–24** Adapted from Winkler, W.C. & Breaker, R.R. (2005) Regulation of bacterial gene expression by riboswitches. *Annu. Rev. Microbiol.* 59, 493; **Figure 28–25** Eye of Science; **Figure 28–28c** PDB ID 1QRV, Murphy IV, F.V., Sweet, R.M., & Churchill, M.E. (1999) The structure of a chromosomal high mobility group protein-DNA complex reveals sequence-neutral mechanisms important for non-sequence-specific DNA recognition. *EMBO J.* 18, 6610; **Figure 28–29** Adapted from D'Alessio, J.A., Wright, K.J., & Tjian, R. (2009) Shifting players and paradigms in cell-specific transcription. *Mol. Cell* 36, 924; **Figure 28–33** Schwabe, J.W.R. & Rhodes, D. (1991) Beyond zinc fingers: steroid hormone receptors have a novel structural motif for DNA recognition. *Trends Biochem. Sci.* 16, 291; **p. 1185** (Mello) Courtesy of Craig Mello; (Fire) Linda A. Cicero/Stanford News Service; **Figure 28–36** Courtesy of F. R. Turner, Department of Biology, University of Indiana, Bloomington (late embryo), and Prof. Dr. Christian Klambt, Westfälische Wilhelms-Universität Münster, Institut für Neuro- und Verhaltensbiologie (other photos); **p. 1187** (Nüsslein-Volhard) Courtesy of Christiane Nüsslein-Volhard/Micheline Pelletier; **p. 1188** (Lewis) CalTech Archives; (Wieschaus) Courtesy of Eric F. Wieschaus; **Figure 28–38** Wolfgang Driever and Christiane Nüsslein-Volhard, Max-Planck-Institut; **Figure 28–40a** Courtesy of Stephen J. Small, Department of Biology, New York University; **Figure 28–40b** Courtesy of Phillip Ingham, Imperial Cancer Research Fund, Oxford University; **Figure 28–41a** Photo from F. R. Turner, University of Indiana, Bloomington, Department of Biology; **Figure 28–42a,b** Photo from F. R. Turner, University of Indiana, Bloomington, Department of Biology; **Figure 28–42c,d** E. B. Lewis, California Institute of Technology, Division of Biology; **p. 1193** (Thomson) Courtesy of James Thomson; **Box 28–1 Figure 1** Adapted from Abzhanov, A., Kuo, W.P., Hartmann, C., Grant, B.R., Grant, P.R., & Tabin, C.J. (2006) The calmodulin pathway and evolution of elongated beak morphology in Darwin's finches. *Nature* 442, 565, Fig. 4.

Index

Key: b = boxed material; f = figures; s = structural
formulas; t = tables; **boldface** = boldfaced terms
in text

A

A bands, **181,** 181f
A (aminoacyl) binding site, ribosomal, **1128**
A-DNA, **291,** 291f
A kinase anchoring proteins (AKAPs), **447**
AAA+ ATPase, **1019**–1020
abasic site, in base-excision repair,
    1030–1031, 1032f
ABC excinuclease, 1032
ABC transporter, **413,** 414f
ABCA1 protein, 874
abiotic synthesis, 33–34, 33f
absolute configuration, **78**
absolute temperature, units of, 507t
absorbance (*A*), 80b
absorption
    of fat, in small intestine, 668–669
    of light, 771–776. *See also* light, absorption of
absorption spectra
    of cytochrome *c*, 736f
    of nucleotides, 286, 286f
    of opsins, 480, 481f
ACAT (acyl-CoA–cholesterol acyl transferase), **864**
acceptor control, **760**
acceptor control ratio, **760**
accessory pigments, 772f, **773**–774, 774f
acetaldehyde, 529s
acetals, 245–246, 245f
acetate, 521s
    activated. *See* acetyl-CoA
    in citric acid cycle, 637f
    in cholesterol synthesis, 859–860, 860f
    in fatty acid synthesis, 840, 841f
    oxidation of, 649–651
    as source of phosphoenolpyruvate, 656–657
    transport of, 840, 841f
acetic acid, 521s, 529s
    p*K*$_a$ of, 83–84, 84f
    titration curve for, 62, 62f
acetic acid–acetate buffer system, 64, 64f
acetoacetate, **686,** 959s
acetoacetate decarboxylase, **686**
acetoacetyl-ACP, 837f, **838**
acetoacetyl-CoA, 307
acetone, 529s, **686,** 959s
    in diabetic ketoacidosis, 959
N-acetyl-*β*-D-glucosamine, 249s
acetyl-CoA, 14s
    amino acid degradation to, 717–719, 718f, 719f
    AMPK and, 963
    β oxidation yielding, 673f
    in cholesterol synthesis, 860–864, 860f
    decarboxylation of pyruvate to, 636–637, 637f
    in fatty acid synthesis, 833–834, 834f, 835f,
        838–839, 951–953
    in glucose metabolism, 951–952, 952f,
        957f, 958
    hepatic metabolism of, 942
    hydrolysis of, 521, 521s, 521t
    oxidation of, in citric acid cycle, 675–677, 676t
    oxidation of pyruvate to, 634, 634f
    oxidation yielding, 674–675
    in plant gluconeogenesis, 825–826
    production of, 840
        in citric acid cycle, 633–638, 634f–637f, 650f
        by pyruvate dehydrogenase complex,
            654–655, 654f
acetyl-CoA acetyl transferase, **860**

acetyl-CoA–ACP transacetylase, 837f
acetyl-CoA carboxylase
    in bacteria, 840–841
    in fatty acid synthesis, **833,** 834f, 835f,
        840–841, 842f
    in plants, 841
acetyl-coenzyme A. *See* acetyl-CoA
acetyl groups
    in fatty acid synthesis, 840, 841f
    transport of, 840, 841f
acetylation, enzyme, 229
acetylcholine, 468s
acetylcholine receptor, **424**
    defective, 426t
    open/closed conformation of, 468, 469f
    in signaling, 467–468, 469f
    structure of, 467–468, 469f
    synaptic aggregation of, 398
acetylene, 529s
N-acetylgalactosamine, in glycosaminoglycans,
    260, 261f
N-acetylglucosamine (GlcNAc), 249f, 250
    in bacterial cell walls, 259–260
    in glycosaminoglycans, 260, 261f
    in peptidoglycan synthesis, 823–824, 824f
N-acetylglutamate, **708**
N-acetylglutamate synthase, **708**
N-acetylmuramic acid (Mur2Ac), 249f
    in peptidoglycan synthesis, 823–824, 824f
N-acetylneuraminic acid (Neu5Ac), 249s, 250,
    269–270, 270s, 366s
    in gangliosides, 366, 366s
acetylsalicylate, 845–846
acetylsalicylic acid, 846s
achiral molecules, 17f
acid(s). *See also specific acids, e.g.,* acetic acid
    amino acids as, 84–85
    as buffers, 63–69, 65f
    definition of, 61
    dissociation constant (*K*$_a$) of, 61f–63f,
        62–63
    fatty. *See* fatty acid(s)
    relative strength of. *See* p*K*$_a$
    strong, 61–62
    titration curve for, 62–63, 62f, 63f
        Henderson-Hasselbalch equation for,
            64–65, 84
    weak, 61–63
acid anhydride, standard free-energy
        changes of, 509t
acid-base catalysis
    general, **199,** 199f
    specific, **199**
acid-base pairs, conjugate, **61,** 61f
    as buffer systems, 63–69
acid-base titration, 62–63, 62f, 63f, 83–84, 83f
acid dissociation constant (*K*$_a$), 61f–63f, **62**–63
acidemia
    argininosuccinic, 717t
    methylmalonic, 717t, 724b–725b
acidic activation domain, **1181**
acidic sugars, 249s
acidosis, **61,** 67–68, 68b, **688**
    diabetic, 67, 688, **960**
acivicin, **923,** 952
aconitase, **641,** 642b–643b, 642f
cis-aconitate, **641,** 641s
    isocitrate formation via, 641, 641f
aconitate hydratase, **641**
ACP (acyl carrier protein), 837f
acquired immunodeficiency syndrome (AIDS),
    218–219, 1088, 1089b
acridine, **1068,** 1069f

actin, 8, 8f, **181**–183, 183f
    in ATP hydrolysis, 181, 182, 183, 183f
    in muscle contraction, 182–183, 183f
    structure of, 181–182
    in thin filaments, 182–183, 183f
actin-myosin complex
    ATP in, 525–526
    phosphorylation of, 487–488
actin-myosin interactions, 182–183, 183f
α-actinin, **181**
actinomycin D, **1068,** 1069f
action spectrum, **774,** 774f
activated acetate. *See* acetyl-CoA
activation barrier, 27, 27f
activation energy, **27,** 27f, **193**
    of enzymatic reactions, **193**
    in membrane transport, 403–404, 404f
    rate constant and, **194**
activators, **1157,** 1158f
active site, **158, 192,** 192f
active transport, **405, 409**
    ATP in, 525–526, 757
    primary, 405, **409**
    secondary, 405, **409**
    transporters in, **404**
activity, **95**
Actos (pioglitazone), 852, 852s, 964–965, 970t
acute lymphoblastic leukemia, 895–897
acute pancreatitis, 699
acyclovir, 1026–1027
acyl-carnitine/carnitine transporter, **671**
acyl carrier protein (ACP), **836,** 836f, 837f
acyl-CoA, fatty, **670**–671
    conversion of fatty acids into, 670–671, 671f
acyl-CoA acetyltransferase, **674**
acyl-CoA–cholesterol acyl transferase (ACAT), **864**
acyl-CoA dehydrogenase, **674, 740**
    medium-chain, **682**
acyl-CoA synthetases, **670, 849**–850
    in triacylglycerol synthesis, 848f, 849–850
acyl-enzyme intermediate, in chymotrypsin
        mechanism, 214–218, 215f, 216f, 217
acyl phosphate, **552**
N-acylsphinganine, **857,** 895f
N-acylsphingosine, **857,** 859f
adaptor hypothesis, 1104, 1104f
adaptor proteins, in signaling, **446,** 460–464
ADARs, **1112**–1113
adenine, 10s, **282,** 282t, 533s. *See also* purine
        bases
    deamination of, 299, 300f
    evolutionary significance of, 1093
adenine nucleotide(s)
    biosynthesis of, regulatory mechanisms in,
        914–915, 915f
    cellular, 518t
    in metabolic regulation, 594–595
adenine nucleotide translocase, **757**
adenosine, 283s
    anti form of, 290, 290f
    as enzyme cofactor, 306–308, 307f
    evolutionary significance of, 307–308
    methylation of, 302
    syn form of, 290, 290f
adenosine 2′,3′-cyclic monophosphate, 284s
adenosine 3′,5′-cyclic monophosphate. *See* cAMP
        (adenosine 3′,5′-cyclic monophosphate)
adenosine 2′-monophosphate, 284s
adenosine 3′-monophosphate, 284s
adenosine 5′-monophosphate, 284s. *See also* AMP
        (adenosine monophosphate)
adenosine deaminase, **920**
adenosine deaminase deficiency, **922**